ENCYCLOPEDIA OF NANOTECHNOLOGY
VOLUME 1

Encyclopedia of Nanotechnology
Volume 1

Elwood D. Carlson
Editor

Nova Science Publishers, Inc.
New York

For permission to use material from this book please contact us:
Telephone 631-231-7269; Fax 631-231-8175
Web Site: http://www.novapublishers.com

LIBRARY OF CONGRESS CATALOGING-IN-PUBLICATION DATA

Encyclopedia of nanotechnology / editor, Elwood D. Carlson.
p. cm.
ISBN 978-1-60692-079-4 (hardcover)
1. Nanotechnology. I. Carlson, Elwood D.
T174.7.E53 2009
620'.503--dc22

2009004646

Published by Nova Science Publishers, Inc. ✢ New York

CONTENTS

Volume 1

Preface xi

Expert Commentary Nanostructured Polymers: Developments and Trends, Highlights and Shadows 1
Maria Vittoria Russo and Ilaria Fratoddi

Research and Review Articles

Chapter 1 TiO2 Nanostructured Materials: Design, Characterization and Applications 7
Aicheng Chen, Xinsheng Peng and Peter Holt-Hindle

Chapter 2 Molecular Foundations for Onset Wear. A Review 37
Mario D'Acunto

Chapter 3 Application of Coordination Chemistry in the Fabrication of Inorganic Anostructures 65
Yi Xie, Yujie Xiong and Zhengquan Li

Chapter 4 1-D Zinc Sulfide Nanostructures: Synthesis and Characterization 85
Soumitra Kar and Subhadra Chaudhuri

Chapter 5 Advance and Application Prospect of Dendrimers 113
Bifeng Pan and Daxiang Cui

Chapter 6 Thermodynamic and Kinetic Theories of Nanowire Growth 201
G. W. Yang

Chapter 7 Novel Manufacturing and Processing Technologies of Nanoporous Silicon 237
Jia-Chuan Lin and Wei-Chih Tsai

Chapter 8 MOVPE Growth of High Density InAs/InGaAs/InP Quantum Dots for Mid-Infrared Emission 277
Tang Xiaohong and Yin Zongyou

Chapter 9	Self-Assembled Hierarchical Architectures by Liquid-Phase Pulsed Laser Ablation *Li Yang, Paul W. May and Lei Yin*	**307**
Chapter 10	Formation and Assembly of Hierarchical Nanoarchitectures by Bioinspired Techniques *Huilan Su, Di Zhang and Qun Dong*	**329**
Chapter 11	Preparation, Characterization and Thermal Properties of Nanofluids *Hai-tao Zhu, Shi-quan Liu, Lei Xu and Can-ying Zhang*	**393**
Chapter 12	III-Nitride Nanowires Research *Ho-Young Cha and Huaqiang Wu*	**427**
Chapter 13	Nanostructures Grown via Electrochemical Template Methods: Synthesis, Measurements and Applications *Stefan Mátéfi-Tempfli*	**459**
Chapter 14	Synthesis of Embedded Gold and AgAu and CuAu Alloy Nanoclusters in Soda-Lime Glass by Ion-Exchange: A Novel Route *S. Mohan, D. Manikandan, K. G. M. Nair and Sujin P. Jose*	**483**
Chapter 15	Investigation of P-Type ZnO Layers Obtained by RBQE Method *T. V. Butkhuzi, T. G. Khulordava, M. M. Sharvashidze, N. G. Gapishvili, N. G. Bukhsianidze, L. T. Trapaidze, E. E. Kekelidze, G. G. Natsvlishvili, L. G. Aptsiauri and L. N. Gapishvili*	**491**
Chapter 16	Low-Dimensional ZnO Nanostructures: Design, Structures and Properties *Guozhen Shen, Di Chen and Cheol Jin Lee*	**501**
Chapter 17	Structure and Opto-Electronic Behavior of Diamondoids, with Applications as MEMs and at the Nanoscale Level *G. Ali Mansoori, Thomas F. George, Guoping Zhang and Lahsen Assoufid*	**527**
Chapter 18	Three-Dimensional Controllable Submicron Gold Films: Sulfhydryl-Containing Molecular Assembly and Controllable Surface Plasmon Resonance and Electrocatalysis Properties *Jianqiang Hu, Juan Liu, Xianbo Lu, Wu Lu, Jianhua Zhou and Jinghong Li*	**547**
Chapter 19	Nanocrystalline Zinc Oxide Thin Films Prepared by Chemical Solution Deposition from a Zinc Naphthenate Precursor *Seung Hwangbo and Kyu-Seog Hwang*	**561**
Chapter 20	Fabrication of Core-Shell Structured Perovskite Nanoparticles with Heterogeneous Oxide Shells and Their Ceramics *Jian Quan Qi, Helen Lai Wah Chan, Hu Yong Tian, Wan Ping Chen and Long Tu Li*	**581**

Chapter 21 Wear Dependent Frictional Properties of Alkanethiols Self-Assembled Films **623**
Mario D'Acunto

Chapter 22 Preparation and Characterization of Nanoparticles and Nanocrystalline Bulks of Pure Rare-Earth Metals **633**
Xiaoyan Song

Chapter 23 Refluxing Polyol Synthesis of Transition Metal Chalcogenides Nanocrystallites: A General Process **665**
Guozhen Shen and Di Chen

Chapter 24 Molecular Dynamics Simulation for Mechanical Behavior of Nanocrystals **687**
Wei Yang, Xiaoyan Li, Xinling Ma and Zengcai Guo

Chapter 25 Carbon Nanotubes Based Electrochemical Sensors **749**
Chengguo Hu and Shengshui Hu

Volume 2

Chapter 26 Synthesis of Carbon Nanotubes Using Novel Alloy Hydride Catalysts by Thermal CVD and their Energy-Related Applications **809**
Sundara Ramaprabhu and Mancheri Manikoth Shaijumon

Chapter 27 Properties, Synthesis, Purification, and Integration of Carbon Nanotubes for the Electronic Device Applications **845**
Wanjun Park and Yo-sep Min

Chapter 28 Field Enhanced Thermal Conductivity in Heat Transfer Nanofluids with Magnetic Coating on Carbon Nanotubes **885**
Brian Wright , Dustin Thomas, Haiping Hong, Lori Groven, Jan Puszynski, Edward Duke, Xiangrong Ye , Sungho Jin, Pauline Smith, Walter Roy

Chapter 29 Synthesis and Investigation of Magnetic Properties of GD - Substituted Mn - Zn Ferrite Nanoparticles as a Potential Low Curie Temperature Agent for Magnetic Fluid Hyperthermia **895**
F. S. Bayburtskiy, Yu. V. Baldokhin, D. B. Korman, A. E. Chekanova and O. A. Shlyakhtin

Chapter 30 Lipid Nanotubes as Scaffolds to Create Structured Inorganic Nanomaterials **905**
Yong Zhou and Xiaqin Wang

Chapter 31 Multifunctional Reaction Platform Based on Carbon Nanotubes **921**
Wenwen Li, Yu Zhang and Chao Gao

Chapter 32 Super Carbon Nanotube: Its Concept, Geometric Conservation Laws and Mechanical Properties **941**
Yuli Chen, Yajun Yin, Bin Liu and Keh-Chih Hwang

Chapter 33	Improving Dielectric Properties of Polymer Nanocomposites through Nanoparticle Surface Modification *Dongling Ma, Linda S. Schadler, Richard W. Siegel, Nianqiang Wu, Diana F. Rogers,and Brian C. Benicewicz*	**959**
Chapter 34	Formation of Nanoparticles Under Laser Ablation of Solids in Liquids *G. A. Shafeev*	**991**
Chapter 35	Carbon Nanoparticles as Substrates for Cell Adhesion and Growth *Lucie Bacakova', Lubica Grausova, Marta Vandrovcova, Jiri Vacik', Aneta Frazcek, Stanislaw Blazewicz, Alexander Kromka', Bohuslav Rezek, Milan Vanecek, Milos Nesladek', Vaclav Svorcik, Vladimir Vorlicek and Milan Kopecek*	**1027**
Chapter 36	Organic-Shell Inorganic-Core Hybrid Nanoparticles with Advanced Functions Designed by Wet Process *Mami Yamada*	**1095**
Chapter 37	Highly Stabilized Gold Nanoparticles Synthesized and Modified by PEG-b-polyamine *Daisuke Miyamoto and Yukio Nagasaki*	**1129**
Chapter 38	Nanoparticles and Quantum Dots as Biomolecule Labels for Electrochemical Biosensing *Martin Pumera*	**1151**
Chapter 39	Photoluminescence Behavior of Water-Soluble Thiol-Capped CdTe Quantum Dots in Living Cells *Pei-Nan Wang and Ji-Yao Chen*	**1157**
Chapter 40	Hybrid Nanoparticle Based on Silica *Shaojun Guo, Erkang Wang*	**1185**
Chapter 41	Heat Transfer of Nanoparticle Suspensions (Nanofluids) *S. M. Sohel Murshed, Kai Choong Leong and Chun Yang*	**1215**
Chapter 42	Recent Developments in the Effective Thermal Conductivity of Nanoparticle Suspensions (Nanofluids) Research *Calvin H. Li and G. P. Peterson*	**1257**
Chapter 43	Enhancement and Temperature Variation in the Thermal Conductivity of Nanofluids *Pei Tillman, Miccal T. Matthews and James M. Hil*	**1291**
Chapter 44	Nanotecnology: Obtaining of Nanoparticles and Nanocomposites and their Use in Food and Drug Packaging *A. Edwin Moncada'*	**1321**
Chapter 45	The Applications of Nanoparticles in Electrochemistry *Ruhai Tian and Jinfang Zhi*	**1349**

Chapter 46 Polymeric Nanoparticles for Oral Delivery of Protein Drugs **1350**
Ji-Shan Quan, Hu- Lin Jiang, Jia-Hui Yu, Ding-Ding Guo, Rohidas Arote, Yun-Jai Choi, Chong-Su Cho

Chapter 47 Fabrication of Micro/Nano-Composite Structures and Their Photovoltaic Behaviors **1350**
Jin Zhai, Xiangliang Sheng, Yong Zhao and Liping Heng

Chapter 48 Charge Separation and Transport in Organic/Inorganic Nanocomposites and Its Applications in Photovoltaic Devices **1350**
Chun-Wei Chen, Yun-Yue Lin, Yu-Ting Lin, Tsung-Wei Zeng, Chia-Hao Chang and Wei-Fang Su

Chapter 49 Polymeric Nanocomposites, Stabilized Organic Derivatives of Five-Valent Phosphorus **1350**
A. Kh. Shaov, Kh. Kh. Gurdaliev and A. M. Kharaev

Chapter 50 Polymer/Silicate Nanocomposites Based on Organomodified Clays **1350**
A. K. Mikitaev, O. B. Lednev, A. Yu. Bedanokov and M. A. Mikitaev

Chapter 51 Fundamental Aspects of Filling of Nanocomposites with High-Elasticity Matrix: Fractal Models **1350**
Georgi V. Kozlov, Yurii G. Yanovskii and Gennadi E. Zaikov

Chapter 52 Structure Formation of Polymer Nanocomposites Based on Polypropylene **1350**
Ahmed Kh. Malamatov, Georgi V. Kozlov and Eugeni M. Antipov

Chapter 53 The Interrelation of Elasticity Modulus and Amorphous Chain's Tightness for Nanocomposites Based on the Polypropylene **1350**
Georgi V. Kozlov, Ahmed Kh. Malamatov, Eugeni M. Antipov and Abdulah K. Mikitaev

Chapter 54 Thermal Degradation and Combustion of Polypropylene Nanocomposite **1350**
S. M. Lomakin, I. L. Dubnikova, S. M. Berezina, G. E. Zaikov, R. Kozlowski, Gyeong-Man Kim and G. H. Michler

Chapter 55 Biomedical Application of Nanotechnology in Cancer Early Diagnosis **1350**
Min Song and Xuemei Wang

Chapter 56 Gold Nanorods: Synthesis, Optical Properties, Assembly and Bioapplications **1350**
Zhanfang Ma and Chungang Wang

Chapter 57 Nanotechnology Based Gas Sensors **1350**
John T.W. Yeow and Yun Wang

Index **1350**

PREFACE

Chapter 1 - Nanostructured materials have attracted tremendous attention in both fundamental studies and applications due to their unique physical and chemical properties. Of the various types of nanostructured materials, TiO_2 nanomaterials are extensively studied for their broad range of applications. A number of methods have been developed for preparing TiO_2 nanoparticles, nanorods, whiskers, nanowires, nanotubes and mesoporous TiO_2 films. In this chapter, the authors describe two novel methods recently developed in our laboratory to grow TiO_2 nanostructured materials directly on titanium (Ti) substrate. Dense and well-aligned TiO_2 nanorod arrays form on a Ti substrate when using organic molecules, such as ethanol, acetaldehyde, acetone and dibutyltin dilaurate, as the oxygen source in the oxidation of Ti at high temperatures. In contrast, polycrystalline grains are formed when pure oxygen and formic acid are used as the oxygen source; elongated micro-crystals are produced when water vapor is used as the oxygen source; whereas the use of argon with a low concentration of oxygen produces random nanofibers growing from the ledges of the TiO_2 grains. Measurements of the water contact angle of these samples show that the wetting properties of the TiO_2 films are dictated by their structure, varying from a hydrophilic surface (polycrystalline grains) to hydrophobic (random nanofibers), and towards a highly-hydrophobic surface (well-aligned nanorod arrays). Secondly, the authors have synthesized large-scale $Na_2Ti_6O_{13}$ nanoplates, nanowires, and continuous nanowire network films directly on Ti substrate hydrothermally. The morphology of the formed $Na_2Ti_6O_{13}$ nanostructures is easily tuned by varying the experimental parameters of temperature, reaction duration and NaOH concentrations. Their study shows that the synthesized $Na_2Ti_6O_{13}$ nanostructures are easily converted into $H_2Ti_3O_7$ nanostructures, a desirable precursor for the fabrication of various TiO_2-based nanomaterials such as anatase, a mixture of anatase and rutile, and rutile TiO_2 nanomaterials, with shape-preservation through an ion-exchange process. The approaches described in this chapter provide simple and novel methods to synthesize various large-scale TiO_2-based nanostructured materials. This tunable growth of TiO_2 nanostructures is desirable for the promising applications of TiO_2 nanomaterials in the development of photo-catalysts, optical devices, sensors and self-cleaning coatings.

Chapter 2- One definition of wear in a tribological sense is that it is the progressive loss of mass from the operating surface of a body as a result of relative motion at the surface. Generally, tribology may be defined as the study of friction wear and lubrication, but in comparison with friction forces or lubrication methods, very little work has been done on the extremely important area of wear and surface damage. For systems consisting of common materials (metals, polymers, ceramics), there are at least four main mechanisms by which wear and surface damage can occur between solids in relative motion: abrasive- and/or adhesive-, chemical- or corrosive-, fatigue-wear mechanisms. The occurrence of such different wear mechanisms involves the fundamental mechanics of molecular and supra-molecular scale. As a consequence, the understanding of surface properties on nanoscale level should be generated to have also a satisfactory knowledge of responses of materials and machine components on macrometer scale. Investigation of the fundamental characteristics of wear at the nanoscale is complicated by some factors and forces that act on such scale level have not yet been addressed in the tribology of macro-systems. Since these forces are sensitive to the environment and surface condition of the specimens, it is quite difficult to determine them accurately. Further, quantification of wear is not straightforward since the amount of wear is often too small to be detected by surface-sensitive instruments. Nanowear measurements have been object of rapid development of precise measuring tools following to the introduction of scanning probe microscopies, such as atomic force microscopy (AFM). The advent of AFM opened new opportunity to study wear mechanisms on nanoscale, making possible to simulate one surface sliding on other one and controlling many physical parameters during the contact dynamics. Further, AFM made possible to a have a direct connection between experimental data and theoretical models.

The chapter is devoted to give a panoramic lecture of some experimental results and models on the onset wear occurrence on nanoscale.

Chapter 3 - Nanostructures of metals and their oxides, chalcogenides have been of extensive interest for their widespread use in catalysis, photography, optics, electronics, optoelectronics, information storage, biological and chemical sensing, and surface-enhanced Raman scattering (SERS). By tailoring the size, shape and/or structure, one can, in principle, tune their intrinsic properties for a range of applications. To this end, coordination structures have been rationally applied as precursors to fabricate diversified nanostructures in solution. In this chapter, the authors discuss the synthetic strategy and related reaction mechanism through the illustration of some detailed cases developed by their group in the recent years. The simple process, mild synthetic conditions, clean reactions, excellent reproducibility and high product quality make their strategies significant. It is believed that introduction of coordination structures into the synthesis of metal, oxide and chalcogenide nanocrystals may provide a versatile tool to control their growth and assembly into well-defined nanostructures. The authors also expect that the related synthetic strategy is applicable to extend to chemical vapor transport or laser-assisted catalytic growth.

Chapter 4 - With the growing interest for miniaturization in electronic and optoelectronic devices, scientists are more and more interested in fabricating different materials in nano-scale dimensions with different morphologies with a view to use them as the basic building block for nano-scale devices. One-dimensional (1-D) nanostructures such as nanorods, nanowires, nanobelts or nanoribbons are of special importance because of their improved fundamental, physical and chemical properties, which make them potential candidate for nano-scale devices. Zinc sulfide (ZnS) is a unique luminescent material well known for its photoluminescence, electroluminescence, etc., which enable wide applications in the fields of displays, sensors and lasers. This article reviews the synthesis, structure and properties of 1-D ZnS nanostructures. ZnS 1-D nanostructures have been fabricated by using different physical (vapor-solid, vapor-liquid-solid etc.) and chemical (solvothermal etc.) routes. Optical properties such as optical absorption, photoluminescence and Raman spectra of the 1-D ZnS nanoforms are also discussed here. Wherever possible, the possible applications of these nanostructures are also included.

Chapter 5- Nanotechnology quite certainly will become an important field in the science and technology of the next decades, allowing one to manipulate matter at the nanometer scale and to incorporate nanostructure and nanoprocesses into working technological innovation. Dendrimers are branched, synthetic polymers with layered architectures that show a fast growing field covering many areas of chemistry. Dendrimers are considered as a modern and an elegant class of branched macromolecules in the nanoscale dimension. They resemble treelike molecular architectures since they are built from repetitive monomers with branching point units that are radially connected around a template core. The hydrodynamic volume is dictated by the core, monomer units, branching points, dendrons, generation numbers, and peripheral functions. Dendrimers are discrete macromolecules with a high degree of molecular uniformity and monodispersity. Those features make for an easy characterization and analysis of their tailor-made properties. They are often macromolecules with few defects, whose dense outward (and/or inward) functional groups generate specific properties and functions such as a polyvalent ligand, a molecular receptor, molecular confinement, and a molecular translocator (vector). Multivalency as well as adhesive, amplification, recognition, additive, and cooperative effects have been observed. The expression "dendritic effects" was coined to illustrate those specific chemical behaviors. Dendrimers that used to be employed as standard, innovative stabilizers, media, ligand and supports have appeared to provided remarkable results with extremely high metal loading. By regulating dendrimer synthesis, it is possible to precisely manipulate both their molecular weight and chemical composition, thereby allowing predictable tuning of their biocompatibility and pharmacokinetics. Advances in their understanding of the role of molecular weight and architecture on the in vivo behavior of dendrimers, together with recent progress in the design of biodegradable chemistries, has enabled the application of these branched polymers as anti-viral drugs, tissue repair scaffolds, targeted carriers of chemotherapeutics and electrochemical sensors. Various types of dendrimers differing in composition and structure are mentioned, together with their practical use spanning from catalysis, transport vehicles to synthetic vaccines. This paper discusses the research advances development of functional and multifunctional dendrimeric and hyperbranched polymers with the objective of being applied in catalysis, biosensing, imaging, drug and gene delivery, nanoprinting, self-assembly, energy transfer, etc.

Chapter 6 - The universal models of nucleation thermodynamics and growth kinetics have been established for nanowire growth upon metal-catalyst-assisted thermal chemical vapor transport (CCVT) on the basis of vapor-liquid-solid (VLS) mechanism, with respect to the size-effects induced by the curvature of critical nuclei and nanowire in the nucleation and growth. Based on the proposed thermodynamic and kinetic analyses on nanoscale, the thermodynamic and kinetic size limit of the nanowire growth was addressed. It was found that the critical size of the nanowire nucleation is suggested to be the thermodynamic size limit, i.e., the theoretical minimum size of the nanowire grown by CCVT both thermodynamically and kinetically, and the size of the catalyst droplet could limit the growing size, i.e., the theoretical possible or maximum size of the nanowire to a certain extent in kinetically. For the issue of catalyst nanoparticles on nanowires tip or substrate, a nucleation thermodynamic criterion and a diffusion kinetic criterion were deduced for the selection growth of catalyst nanoparticles on nanowires tip or substrate upon CCVT.

Importantly, theoretical predictions on the basis of the proposed models are in well agreement with the experimental data. Additionally, a growth kinetic approach was developed to theoretically elucidate the growth of one-dimensional nanostructures inside carbon nanotubes upon vapor deposition. Theoretical analysis showed that the growth rate of one-dimensional nanostructures would go much higher once nuclei formed inside carbon nanotubes, due to the effect of surface tension induced by the nano-sized curvature of carbon nanotubes. Theoretical results are in well agreement with the experimental data. The size-dependence of the shape evolution of nanowires such as SnO2 was pursued in theoretical. It was found that the thermodynamic driving force, the Gibbs free energy difference between two phases, is always responsible for the shape choice in the different growth stages such as nucleation, growth, and structural transition of 1D nanostructures, implying that the thermodynamic driving force is one of the physical origins causing the interesting morphological transformation in the growth of 1D nanostructures.

Chapter 7 - Porous silicon (PS) films consisting of many pores and pillars are widely used to yield efficient visible photoluminescence (PL) and electroluminescence (EL) at room temperature. Such light-emission behaviors are primarily attributed to electron confinement in the nanocrystals that constitute the PS film. Also, the PS film shows a unique electrical property of negative difference conductance (NDC) for the carrier mobility difference in different sizes of pores and pillars.

The techniques of PS formation have been developed by many different methods. Among them, electrochemical anodization is the most commonly used. The anodization is performed in a hydrogen fluoride (HF) solution with an anodic current on the sample. However, it is very difficult to control the manufacturing parameters precisely in the manufacturing processes. Especially in nano-scale, the anisotropic etching or selective etching cannot be easily achieved in the wet-etching method. However, it is well known that nano-scaled PS (NPS) shows light emission more readily in the visible range than macro-PS does. Therefore, precise control of the NPS formation is very important.

In this chapter, the authors introduce three novel manufacturing technologies. In the first method, the Hall-effect is applied in the electrochemical anodization. By Lorentz force, the carriers can be well swept to the allocated areas. In the second method, a forward biased pn-junction is used for bottom-hole assistance. The NPS films with nano-scaled pores and high-aspect-ratio pillars can be formed. In the third method, the authors proposed a simple and mask-free method for the fabrication of patterned PS using a localized electric field. The electric field is applied by patterned electrodes (anode and cathode) which are placed horizontally underneath the sample. No masking-layer or related photo-lithography processes are needed in this method. Strong visible photoluminescence emissions in PS can be obtained by these novel methods.

In addition, the surface roughness of NPS leads to a high contact resistance to the metal layer of the applications on EL, and hence a waste of an electrical power. To enhance the EL efficiency and stabilize performance, a good metal contact to the light-emitting layer (NPS layer) is required. In the authors' study, a supercritical fluid (SCF) technique is explored to overcome the processing problem on metal contact. The SCF is a liquid or gas material used in a state above the critical temperature and critical pressure where gases and liquids can coexist. The use of a CO2 SCF and silver (Ag) nanoparticles is shown to improve the electrical contact between the metal and NPS. Therefore, the contact resistance can be largely reduced, and the power consumption of EL on the PS structure can be greatly reduced by the method.

The morphology, porosity, and photoluminescence of the NPS prepared by the proposed method are investigated. This novel technology will have much benefit in the applications of nano-electronic and opto-electronic devices.

Chapter 8 - Emission in mid-infrared range, 2 – 5□m, has very important applications in variety of areas including range-finding, laser surgery, remote trace-gas sensing, secure free-space communications, etc. Owing to reduction of the dimensionality carriers' confinement, quantum dots (QDs) based optoelectronic devices, e.g. semiconductor lasers, show novel properties when compared with their counterpart quantum well based devices.

In this article, self-assembly InAs quantum dots grown on InGaAs/InP matrix for mid-infrared emission using metal-organic vapor phase epitaxy is introduced. Effects of the growth conditions on the dot density and size uniformity of the quantum dots have been investigated in detail. A two-step growth method for growing high density and uniform quantum dots has been introduced. Improvement in QDs density and size uniformity in the two-step growth has been discussed in details. By changing the indium content of the InxGa1-xAs matrix layer of the InAs/InxGa1-xAs/InP quantum dot structure, emission wavelength of the InAs QDs structure has been extended to the longest >2.35 μm measured at 77K. Full-width at half-maximum (HWHM) of the photoluminescence emission spectrum of the InAs quantum dot structure was measured as narrow as 25.5 meV.

Chapter 9 - Over the past decade, liquid phase pulsed laser ablation (LP-PLA) has been used as an effective method for preparation of many nanostructured materials. This technique involves the interaction of a laser with a solid target and surrounding liquid. The laser causes the target surface to vaporise in the form of an energetic ablation plume, containing species such as atoms, ions, and clusters, travelling with high kinetic energy. These can collide and react with molecules of the surrounding liquid phase, producing new materials, often on the nanometre scale. Due the intensity of the laser and the nanosecond timescales, high temperature, high pressure, and high density conditions within the reaction volume are produced which provide a 'brute force' method of synthesising novel materials that have hitherto been inaccessible using milder, more conventional techniques. In this chapter, the authors discuss more recent progress in the synthesis of hierarchical, complex, and ordered crystal architectures by this novel technique. It is believed that the ability of nanostructures to self-assemble in this manner provide potential building blocks for microscale or nanoscale devices. The authors show that for C3N4, a Group IV-V covalent compound, a range of self-assembled arrays of nanoparticles, nanorods, nanoleaf, and mesoflowers can be produced, and this also suggests that the route described here can be applicable to the synthesis of nanostructures of other similar materials. Therefore, the authors expect LP-PLA may play a key role in advancing complex nanomaterials into many applications in optics, electronics, biology, medicine and energy/chemical conversion.

Chapter 10 - The bioinspired technique has developed into an intersection of materials science, soft chemistry technique, and biotechnology. In this chapter, the bioinspired strategies integrating biotemplate, biomineralization, and biomimesis are adopted to fabricate functional nanomaterials with hierarchical architectures and interrelated unique properties.

Some natural biomaterials, such as eggshell membrane, silk fiber, etc., are introduced into different liquid processes to construct a series of functional hierarchical nanoarchitectures (containing oxides, chalcogenides, noble metals, and/or their composites). Under the direction of the biotemplates or biosubstrates, the nanocrystallites units of about 2-10 nm are formed, developed, and assembled into hierarchical architectures from the nanoscale to the macroscale. The nucleation, growth, and assembly of nanocrystallite units depend on the collaboration of special proteins structures and functional biomacromolecules of relevant biomaterials as well as the processing conditions. Moreover, the biomaterials usually provide both the physical substrates and the inductive substances to direct some complicated reactions. In addition, some amino acids of the biomaterials function as the capping agents to achieve the organized growth and assembly of small-sized nanocrystallites. As-synthesized hierarchical nanoarchitectures take on outstanding chemical or physical properties and have valuable applications in photocatalysis, gas sensing, ductile ceramics, and semiconductor technology.

This method and the relevant ideas provide a novel and versatile avenue to synthesize a new family of functional nanomaterials by integrating material science, chemistry, and biotechnology.

Chapter 11 - Nanofluids are fluids containing suspended solid nanoparticles. It is expected that the nanofluids will be the next generation of heat transfer fluids due to their unique thermal properties. The nanofluids may have potential applications in many important fields, such as micro-electronics, aerospace, transportation, and medicine, etc. In this chapter, a brief introduction of the background and concept of nanofluids is first presented, followed by reviewing the preparation, characterization and thermal properties of the nanofluids. In the preparation section, the stabilization mechanism of nanoparticles in liquids is described. Then several preparation methods are discussed. Emphasis is put on the stability of the nanofluids, which is influenced by the preparation methods and affects the properties of the nanofluids. Several typical characterization technologies of nanofluids are summarized in the characterization section. The thermal conductivity, convection, and boiling properties of nanofluids are depicted in the section of the thermal properties.

Chapter 12 - Semiconductor nanowires have great potential for highly integrated nanoscale systems, high efficiency sensors, and advanced optical and electronic applications due to their reduced dimensionality incorporated with inherent doping capability and defect free material quality. Due to their unique properties, intensive research efforts have been made on III-nitride nanowires. Research progress has been impressive, both n-type and p-type doped III-nitride nanowires have been fabricated, and heterostructure nanowires have also been successfully synthesized. This chapter reviews various methods of III-nitride nanowire synthesis, and outstanding research on optical and electronic device demonstration.

Chapter 13 - Nanoscaled materials often exhibit very different properties from their macroscopic and even micrometer sized counterparts. This occurs when the size of an object is reduced down to the atomic level such that one or more of its dimensions become smaller than some characteristic lengths, i.e. mean free paths, coherence lengths and crystallite sizes. In this case new properties show up where no more scattering on defects and statistical averaging of various fundamental phenomena are dominant. Below these characteristic lengths quantizations of different parameters appear. The possibility of tuning the properties of nanomaterials by changing their sizes is very interesting on both fundamental and applied aspects. Nanostructures are even more attractive because there is virtually an infinity of possibilities to combine different nanomaterials together, thus creating new and smart nano devices and structures.

One of the great challenges of today is to find reliable techniques for the fabrication of such nanomaterials and nanostructures. Methods based on template synthesis and on self organization are the most promising due to the easiness and low cost of fabrication. This chapter will focus on various electrochemical synthesis methods of nanowires and nanostructured nanowires using nanoporous host materials as templates. A short overview of the templates and their fabrication methods will be presented. Among them, the nanoporous aluminum oxide shows a self-ordering of the nanopores in a honeycomb structure. It stands high temperatures and is insoluble in organic solvents. A recent development of a supported nanoporous aluminum oxide on silicon will be described. It will be shown that this new approach solves the handling problems associated with the brittle nanoporous aluminum oxide thin layers and allows its integration in silicon processes for microelectronic manufacturing.

Combining electrochemical template synthesis with nanolithography and patterning techniques opens up new ways for the fabrication of nanowires-based complex structures and devices. Two approaches will be presented. The first is aiming at contacting and measuring a large amount of single nanowires and compositionally modulated nanowires while the second is showing a localized growth of bundles or single nanowires for contacting them in a high density matrix arrangement.

Several results of measurements and applications will illustrate the presented methods and their use. Finally, concluding remarks and prospects for the electrochemical template synthesis will close the chapter.

Chapter 14 - Gold nanoclusters were formed in the soda-lime glass matrix by ion-exchange technique. The ion-exchanged glasses were subjected to heat treatment for different temperatures in air and the effect of it on the optical absorption spectra was analyzed. Theoretical optical absorption was calculated using the Mie's and Maxwell Garnett's effective medium theories. The simulated spectra matched with the experimentally recorded spectra confirming the presence of gold clusters in the glass matrix. As a novel route, hetero system nanoclusters like AgAu and CuAu were formed by ion-exchange method. The optical absorption measurements showed signatures of an incomplete alloy formation between the respective metal species.

Chapter 15 - The authors have investigated p-type ZnO layers obtained at T=400°C by Radical Beam Quasiepitaxy (RBQE). Hall Effect measurements performed at T=77K and 300K show resistivity $\rho=5\times10^{-2}$ and 3.1×10^{-3} Ωcm accordingly. At T=77K the hole concentration and mobility was $5\times10^{18}cm^{-3}$ and 22 cm^2/Vs. At T=300K the hole concentration and mobility was $8\times10^{18}cm^{-3}$ and 250 cm^2/Vs. Following bands in PL spectrum of p-type ZnO at T=77 K, λ=369.1 nm, λ=374.5 nm, λ=383.5 nm, λ=392.5 nm and λ=401 nm, were observed and identified. It is shown that RBQE technology enables the obtainment of high monocrystallinity ZnO samples with the low concentration of uncontrollable impurities.

Chapter 16 - Zinc oxide (ZnO) based one-dimensional (1-D) nanoscale materials, as important functional oxide nanostructures, have received increasing attention over the past few years due to their potential applications in optoelectronic switches, high efficiency photonic devices, near-UV lasers, and for assembling complex three-dimensional nanoscale systems. Various kinds of ZnO nanostructures have been realized, such as nanodots, nanorods, nanowires, nanohelixes, seamless nanorings, mesoporous nanowires and so on.

This paper presents a brief review of their recent research achievements on low-dimensional ZnO nanostructures. By using a metal vapor deposition process, a group of unique nanoarchitectures of ZnO are synthesized, including several types of hierarchical ZnO nanostructures, vertically aligned nanowires, microtubes, nanonails, nanoribbons-nanowires junctions, etc. ZnO-based heterostructures are also briefly discussed in this paper. Structures of these nanostructures are discussed in detail using several techniques including scanning electron miscroscopy, high-resolution transmission electron microscopy and several structure models are presented. Finally, structural related photoluminescence properties are demonstrated. The synthesized ZnO nanostructures show excellent field emission properties. These novel nanoarchitectures could be unique for many applications in nanotechnology.

Chapter 17 - Diamondoids have been of great interest in recent years due to their role in nanotechnology, drug-delivery and medicine. Due to their six or more linking groups, they have found major applications as templates and as molecular building blocks in nanotechnology, polymers synthesis, drug delivery, drug targeting, DNA-directed assembly, DNA-amino acid nanostructure formation, and in host-guest chemistry. In this paper, the molecular nature of diamondoids, their molecular specificities, their intermolecular interactions and their opto-electronic properties are introduced. These will help in the understanding of the structure-property relations and self-assembly of diamondoids, which is essential for designing functional molecular gears for micro-electro-mechanical systems (MEMS) and sensitive bionanosensors, and for developing new nanodrugs, just to name a few.

Chapter 18 - Surfactant-free capped gold nanoparticles (SFGNPs) of about 60 nm in diameter were synthesized through using a seed-induced route and subsequently self-assembled on sulfhydryl-containing ITO glass slices. Two-layer or multi-layer submicron gold films could be obtained by repeatedly self-assembling the freshly prepared SFGNPs on the as-prepared one-layer gold nanofilm in the function of p-aminothiophenol (p-ATP) molecules, and the self-assembled thickness or layer number of which was tunable. The thickness control in the assemblies was demonstrated by the atomic force microscopy (AFM), scanning electron microscopy (SEM), and UV-visible absorption spectroscopy measurements. Moreover, UV-visible absorption spectroscopy, surface-enhanced Raman scattering (SERS) spectroscopy, and electrocatalysis property studies indicated that surface plasmon resonance (SPR) and electrocatalysis properties of the self-assembled films prepared thus were enhanced along with the increase in their layer-numbers. These controlled self-assembled films are not only helpful to study physicochemical properties of the uniform and controllable assemblies, but also are expected to explore applications in many fields such as chemical and biological sensors, transmission surface plasmon resonance transducer, SERS, biological detection, and electrochemistry.

Chapter 19 - Nanocrystalline ZnO thin films on silica glass substrates were prepared by spin coating-pyrolysis process from a zinc naphthenate precursor. As-deposited films were heat treated at 500°C ~ 1000°C. The ZnO thin films were analyzed by X-ray diffraction, field emission-scanning electron microscope, scanning probe microscope, and ultra violet-visible-near infrared spectrophotometer. Only (002) peak is observed for all samples on silica glass substrates. With the increase of heat-treatment temperature, the peak intensity of (002) plane increases. No evidently aggregated particles are present. From SPM analyses, three-dimensional grain growth, which is thought to be due to inhomogeneous substrate surface and *c*-axis oriented grain growth of the ZnO phase, is independent on heat treatment temperature.

The most homogeneous surface of the highly oriented ZnO film was observed at the heat-treatment temperature of 800°C. All the films exhibit a high transmittance (>80%) in the visible region except film heat-treated at 1000°C and show a sharp fundamental absorption edge at about 0.38 ~ 0.40 μm. The estimated energy band gap for all the films are within the range reported for ZnO films and single crystal. ZnO films, consisting of densely packed grains with smooth surface morphology, were obtained by heat treatment at 600 ~ 800°C, expected to be ideal for practical usage.

Chapter 20 - The perovskite nanoparticles such as BaTiO3, SrTiO3, BaZrO3 and their solid solutions can be prepared using a new method, DSS, near room temperature. Hydrogen interstitials are introduced into the lattice of perovskite nanoparticles synthesized by DSS method so that the nanoparticles take positive charges and adsorb the hydroxyl group with negative charges on the surface. Hydroxyl groups adsorbed on the surface of nanoparticles can be ionized freely, and the nanoparticles exhibit an alkalescent property. When this kind of perovskite nanopaticles disperses in a certain solution of inorganic salt such as Mg(NO3)2•6H2O, Zn(NO3)2•6H2O, Al(NO3)3•9H2O, Y(NO3)3•6H2O etc., the hydroxide related to these salts spontaneously deposit on the surface of the nanoparticles, as a result, a perovskite structure nanoscaled core materials are successfully self-wrapped by a non-ferroelectric hydroxide derived from a nitrate salt solution. After annealing at high temperature, core-shell structured perovskite nanoparticles with heterogeneous oxide shells can be synthesized. When the core-shell structured perovskite nanoparticles with heterogeneous oxide shells are composited with doped shell oxide materials, the pinning effect during sintering can be avoided, as a result, the sintering temperature are decreased and dense ceramics are obtained. The core-shell structures are analyzed by a conjunction of XRD, HRTEM and FE-SEM. The lattice match of shell and core in their ceramics of MZO wrapped BST nanoparticles probably follows the second match mode. The tunability and FOM values of the ceramics of MZO wrapped BST nanoparticles have been improved significantly compared with that of pure BST ceramics. The ceramics of MZO wrapped BST nanoparticles are promising materials for tunable microwave device applications and high energy density ceramic capacitors for their distinct dielectric loss improvement.

Chapter 21 - Self-assembled monolayers (SAMs) present a 2D regular lattice, and the lateral force measured with an atomic force microscopy (AFM) scanning probe tip displays the typical stick-slip behavior. In this paper, the authors study the possible connection between an AFM induced damaged SAM surface and irregular behavior of the sticking phases.

In fact, the repeated passages of an AFM probe tip on an SAM surface can produce wear stripping off radicals from the carbon chain backbones, and irregularities of the SAM lattice take place. As a consequence, the frictional properties are changed because new extra space is made available to the final alkanethiol chain groups and the stick and slip phases are not time invariant and hysteresis between the forward and backward scanning direction is altered. The paper discusses if and how such friction irregularities can monitor the progressive modifications of the SAM worn surface.

Chapter 22- Because of the extremely high chemical reactivity and hence the considerably rigorous requirements for the equipments to preserve a high purity of the product, the preparation and characterization of nanostructured pure rare-earth metals are still a big challenge in nanoscience and nanotechnology. In this chapter, followed a thermodynamic description of the energy state and the thermal stability of rare-earth nanoparticles and nanocrystalline bulks, the practical preparation and characterization of nanoscale rare-earths are depicted in detail. Different from a number of existing techniques that have been developed to synthesize nanocrystalline bulk materials, a home-designed configuration of the "oxygen-free" in-situ synthesis system, where the inert-gas condensation is combined with the SPS technology into an entirely closed and vacuumized system, is mainly introduced. With the technically designed route of sequentially performed processes, i.e. amorphization of nanoparticles, nucleation and growth of the short-range ordered "clusters" inside the nanoparticle, and the complete nanocrystallization, the ultrafine nanocrystalline bulk can be obtained with the grain size much smaller than the initial nanoparticle size. Microstructures and properties of some typical members of the rare-earth family are presented ordinally. The prominently changed thermal, physical and mechanical properties imply the promising applications of the nanocrystalline rare-earth metals and alloys in the fields of functional and structural materials. Finally, although the chapter is focused on rare-earth nanoparticles and nanocrystalline bulks, the proposed thermodynamic models and preparation techniques are potentially applicable to a large variety of nano metal materials, hence may establish a new way for studies on microstructures and performances of metallic nanomaterials, especially of those highly reactive in the air.

Chapter 23 - Transition metal chalcogenides, including sulfides, selenides, and tellurides are important optoelectronic, luminescent, and semiconducting materials. With regard to all of the applications of chalcogenides, highly crystalline particles with almost monodisperse size distribution and regular morphology are required. The synthesis of chalcogenides nanocrystallites under mild conditions is of great interest in both basic research and engineering applications. In this paper, the synthesis of chalcogenides nanocrystallites, such as nanoparticles, nanoplatelets, nanorods, nanowires, and their assembled patterns, via the simple, rapid and convenient refluxing polyol method was reviewed based on their works. Several techniques including powder X-ray diffraction (XRD), scanning electron microscopy (SEM), transmission electron microscope (TEM), X-ray photoelectron (XPS), Raman, photoluminescence (PL), etc. have been used to characterize the produced nanocrystallites. Possible growth mechanism of the refluxing polyol technique to transition metal chalcogenides nanocrystallites is discussed. Combining with other techniques, the refluxing polyol method could be extended to synthesize other important inorganic materials nanocrystallites. Given to the generality of this simple refluxing polyol method, coating and modification of the surfaces of transition metal chalcogenide nanocrystallites may also be fulfilled.

Chapter 24 - The endeavor toward the understanding of micro- and nano-scale phenomena in mechanics is powered by three technology drives. The first drive comes from the field of material sciences, accompany with the endless search for stronger, lighter, tougher and multi-functional materials. The second drive emerges from the rapid growth in information technology whose products involving microelectronic packaging, MEMS and NEMS technologies and nano-electronics. Nano-structured information materials can offer more information storage capacity. The third, but more recent technology drive is formed upon the emergence of nano-science and engineering, where an understanding for nano-scale failure and deformation is in urgent demands.

The characterization of the mechanical properties, such as the Young's modulus, the flow stress and the hardness, of materials in a nanometer scale becomes an emerging research land as the field of nanotechnology unfolds. Thin films have small dimensions whose properties might differ from those of a bulk material of the same compositions. Fundamental understanding of nanoscale friction and wear processes calls for the knowledge of contact nanomechanics. Instruments with atomic resolution, such as the atomic force microscopy, high resolution electron microscopy and nano-moire method, enable the investigation of surface and bulk material properties at a nanoscale.

In this brief review, current methodological developments in MD are summarized, with complementary computer simulations for mechanical behavior of nanocrystals. From aforementioned simulations, the formation and deformation processes of nanocrystals are depicted at atomic level. Their discussions will be focused on two aspects: (1) dislocation-assisted grain growth in nanocrystalline copper under fast stretching, and (2) the MPM/MD handshaking method applied to impacts of high energy nanocluster.

Chapter 25 - Carbon nanotubes (CNTs) have become the subject of intense researches in the last decades because of their unique properties and promising applications in any aspect of nanotechnology. As a consequence of their small dimensions and unique structural properties, CNTs display fascinating electronic and optical properties that are distinct from other carbonaceous materials and nanoparticles of other types. CNTs are currently being used in electronic and optoelectronic, biomedical, pharmaceutical, energy, catalytic, analytical and materials applications. Specifically, the properties of small dimensions, functional surfaces, good conductivity, excellent biocompatibility, modifiable sidewall and high reactivity make CNTs ideal candidates for constructing sensors with high performance. As an example, CNTs have been widely employed in constructing various electrochemical sensors.

As a new type of carbonaceous materials, carbon nanotubes (CNTs) possess some unique properties that are much different from either the conventional-scaled materials or other types of nanoparticles, such as the special and well-defined tubular structure of nanosizes, the excellent chemical stability, the modifiable surface and the strong electrocatalytic activity. These special properties foresee their promising applications in electroanalytical chemistry for constructing CNTs-based sensors. The fact is that the modification of electrodes with CNTs has been observed to apparently improve the responses of substrates from small H_2O_2 molecules to huge redox proteins. Most of current works mainly focused on the fabrication of randomly dispersed CNTs-based electrochemical sensors and few reports deal with the fabrications of well organized nanostructures of CNTs. As a tubular nanomaterials, the key advantages of CNTs are their small diameter and huge length to diameter ratio that allow them to be used as molecular wires for facilitating electron transfer between proteins and electrodes or to be organized in a low density manner for using as a single nanoelectrode or nanoelectrode arrays with ultra-sensitivity. Since the direct growth by CVD methods inevitably surfers from the need of rigorous conditions, development of simple physical or chemical immobilization methods that can control the orientation of CNTs on electrode surfaces has become a matter of great urgency.

Chapter 26 - This chapter presents the recent trends in the carbon nanotubes growth by chemical vapour deposition (CVD) technique. The use of CVD technique and novel alloy hydride catalysts for the synthesis of multiwalled carbon nanotubes (MWNTs) by the pyrolysis of acetylene is presented. The advantages of this novel approach to catalyst preparation using hydrogen decrepitation with reference to the increase in the catalytic reactivity and active sites for the formation of CNTs are highlightened. A growth process of CNTs with alloy hydride catalysts by vapour-liquid-solid (VLS) mechanism is proposed. The results of the different characterization of the as-grown and purified MWNTs are described. The dependence of the yield and the purity of the MWNTs synthesized on the alloy hydride catalysts are discussed. The various experimental techniques of the energy- related applications of MWNTs are presented.

Chapter 27 - Since the discovery of carbon nanotubes (CNTs), extensive studies on this new carbon allotrope have formed a new epoch in various fields of science and technology with their excellent physical and electrical properties, and chemical functionality. In particular, understanding of their electrical properties has aroused many research groups in academy and industry to explore the nano-scale electronics based on CNTs such as transistor, memory, interconnect, logic array, and so on. Although the CNT-based electronics is regarded as a strong potential candidate for the next generation electronic devices, many technological barriers are still remained.

First of all, pure nanotubes should be separated from a mixture of single-walled, multi-walled nanotubes, and carbonaceous impurities or should be grown without impurities for reliable device performance. Another critical barrier to integrate nanotubes in devices is uncertainty and uncontrollability in positioning nanotubes on a right region. In addition, transport mechanism in CNTs should be more understood to modulate band gap of them to fabricate various device elements.

In this article, the authors present recent efforts to overcome the barriers focusing on transistors based on single-walled carbon nanotubes (SWNTs). Experimental and theoretical studies to control the CNT electronic structure are covered for CNT-field effect transistors. The large scale fabrication of CNT electronic elements are also given for the selective growth of CNT that is extendable to device integration. For the purity issue of nanotubes, sono-chemical route to single-walled carbon nanotube and purification method of nanotubes integrated in transistors via sulfidation are demonstrated. For the positioning problems, direct photolithographic method using a mixture of catalytic precursors and conventional resists as a catalytic resist is introduced with other positioning methods of nanotubes.

Chapter 28 - In this chapter, the authors report that the thermal conductivity (TC) of heat transfer nanofluids containing Ni coated single wall carbon nanotube (SWNT) can be enhanced by applied magnetic field. A reasonable explanation for these interesting results is that Ni coated nanotubes form aligned chains under applied magnetic field, which improves thermal conductivity via increased contacts. On longer holding in magnetic field, the nanotubes gradually move and form large clumps of nanotubes, which eventually decreases the TC. When the authors reduce the magnetic field strength and maintain a smaller field right after TC reaches the maximum, the TC value can be kept longer compared to without magnetic field. The authors attribute gradual magnetic clumping to the gradual cause of the TC decrease in the magnetic field. The authors also found that the time to reach the maximum peak value of TC is increased as the applied magnetic field is reduced. Scanning Electron Microscopy (SEM) images show that the Ni coated nantubes are aligned well under the influence of a magnetic field. Transmission Electron Microscopy (TEM) images indicate that nickel remains attached onto the nanotubes after the magnetic field exposure.

Chapter 29 - Gd-substituted Mn–Zn ferrite nanoparticles of different compositions were synthesized by a chemical co-precipitation method. To study the reduction of the Curie temperature (T_C) for different samples, their magnetic properties in dependence to the composition and cationic distribution were investigated. An attempt to lower the T_C of superparamagnetic particles to the optimal temperature required in magnetic fluid hyperthermia (44–47 °C) was made.

Chapter 30 - Harnessing nature's amazing ability to form self-assembled structures for nanotechnology applications is an attractive alternative to conventional fabrication methods. Lipid nanotubes (LNTs) are especially intriguing assemblies, which represent a potentially powerful architecture generated through noncovalent self-assembly of amphiphilic molecules. They can be considered among the largest self-organized non-living structures yet observed. LNTs have hydrophilically internal and external membrane surfaces, and can provide the wide scope for chemical modifications, in sharp contrast to carbon nanotubes. These unique properties make themselves as ideal candidates for a variety of applications in chemistry, biochemistry, materials science and medicine. The lipid headgroups can function as templates for the nucleation, growth and deposition of inorganic substances on the external and internal surfaces of preformed LNTs, and the linear hollow cylinders and channels inside the nanotubes may also supply specific nanospaces for nanomaterials reaction and orrientation. This chapter assesses recent progress in application of LNTs as scaffolds for synthesis of structured inorganic nanomaterials, aiming to present the significant importance of the key technology of the 21st century, so-called "nanobiotechnology" by combination of biotechnology and nanoscicence for material fabrication.

Chapter 31 - Functionalization of carbon nanotubes (CNTs) opens a versatile avenue for the preparation of soluble CNTs and novel CNT-based hybrid nanomaterials and nanodevices. In this work, the authors present a multifunctional reaction platform based on CNTs, poly(glycidyl methacrylate)-grafted multiwalled carbon nanotubes (MWNT-g-PGMA). The reaction platform of MWNT-g-PGMA was achieved by surface-initiated atom transfer radical polymerization (ATRP). The polymerization kinetics was investigated. The epoxy groups of MWNT-g-PGMA can be converted into azido groups by reaction with NaN_3. The "click" reaction between azido groups and alkyne modified coumarin 343 affords luminescent CNTs. The epoxy groups can also be readily converted into hydroxyl and amino groups in the presence of hydrochloric acid and 1,2-ethylenediamine, respectively. Further reaction of the hydroxyl groups with succinic anhydride results in the formation of multicarboxylic acid polymer-functionalized CNTs. The chemical structure, morphology and the grafted quantities of epoxy, hydroxyl, amino and carboxylic acid groups functionalized MWNTs were determined by FTIR, ^{1}H NMR, SEM, TEM and TGA measurements. Magnetic CNTs and silver nanocrystals-decorated CNTs were fabricated through subsequent assembly of magnetic nanoparticles (e.g., Fe_3O_4) and in situ reduction of Ag^+ in aqueous solution using the amino and carboxylic groups functionalized CNTs as templates, respectively.

Chapter 32 - Based on the integral theorems for mean curvature and Gauss curvature, geometric conservation laws for spontaneous Y-branched carbon nanotubes are presented. From these laws the concept of "super carbon nanotube" is proposed. It may be proved that such a super structure might be of the lowest energy both locally and globally. To make clear the response of the super carbon nanotube under loading, the mechanical properties of the fundamental cell in super carbon nanotube are studied by atomic-scale finite element method.

Chapter 33 - In this study, the potential of using nanoparticle surface chemistry to tune the dielectric properties of polymer nanocomposites is explored. Composites with 5 wt% of titania (TiO_2) nanoparticles and low-density polyethylene (LDPE) were fabricated by melt mixing. Both polar (N-(2-aminoethyl)3-aminopropyl-trimethoxysilane (AEAPS)) and nonpolar (decyltrichlorosilane (DCS)) silane coupling agents were used to modify the nanoparticle surface to yield either hydrophilic or hydrophobic surfaces. Two coating methods were employed and the resulting surface structures were characterized in detail. It was found that the surface structure was controlled primarily by the experimental conditions of the coating process. In addition, a previously un-described surface reaction that resulted in the formation of $Si-CH_3$ bonds was proposed, which may be due to the reaction of AEAPS with oxygen vacancies on the TiO_2 nanoparticle surface. A large observed change in the chemical shift (over 40 ppm) in the ^{29}Si NMR and the –2.3 ppm chemical shift in ^{13}C NMR supported this conclusion. The method leading to a "dry" compact monolayer surface structure was chosen for preparing TiO_2 nanoparticles with either polar or non-polar surfaces for composite sample preparation and subsequent electrical studies. In both cases, surface modification had a positive effect on dielectric breakdown strength although the possible underlying mechanisms were different. Compared to the uncoated TiO_2 filled samples, AEAPS-TiO_2 and DCS-TiO_2 filled samples showed a 35% and 43% increase in the short-term Weibull breakdown strength, respectively. The space charge distribution patterns were also clearly different among the samples. The microstructure characteristics of the composite samples, associated with different TiO_2 surfaces, were studied and correlated to the corresponding dielectric properties.

Chapter 34 - The process of nanoparticle formation under laser ablation of solids in liquids is described. Critical parameters are discussed that govern the properties of nanoparticles ejected into the surrounding liquid. These parameters are laser wavelength, pulse duration interaction of individual nanoparticles with laser beam inside the liquid. A review of previous results is presented on the properties of nanoparticles of noble metals. Recent data on laser-assisted generation of other metals is given, including the formation of alloyed nanoparticles. Micro- and nanostructuring of the target upon its laser ablation in liquid environment is discussed. Examples are given of the influence of the surrounding liquid on the chemical composition of generated nanoparticles. The laser control over the size distribution of nanoparticles in liquids is demonstrated either by spatial profiling of laser beam intensity or proper tuning of laser wavelength into plasmon resonance of nanoparticles. Recent results are given on excitation of high energy levels of media under laser exposure to laser pulses of picosecond range of duration.

Chapter 35 - Carbon nanoparticles, such as fullerenes, nanotubes and nanodiamonds, have been considered as promising building blocks in nanotechnology for broadscale nanostuctured devices and materials, e.g., microchips, nanorobots and biosensors, carriers for controlled drug and gene delivery or tracers for novel imaging technologies. Soon after macroscopic production was established, fullerenes, as donors and acceptors of electrons, started to be targeted for use in photodynamic therapy against tumor and microbial cells and also for quenching oxygen radicals. It has been discovered that nanodiamonds can also act as antioxidant and anti-inflammatory agents.

The interaction of carbon nanoparticles with cells and tissues has been investigated mainly using suspensions of these particles in cell culture media or other fluids. Relatively little is known about the influence of layers consisting of carbon nanoparticles on cell-substrate adhesion. Carbon nanoparticles can be advantageously used for surface modification of various artificial materials developed for the construction of tissue replacements and other body implants. In the form of films deposited on the surface of implants, carbon nanoparticles can not only improve the mechanical and other physical properties of a body implant, but can also enhance the attractiveness of the implant for cell colonization. The latter effect is probably mediated by the surface nanostructure of the films, which, at least to a certain degree, mimics the architecture of physiological cell adhesion substrates, such as extracellular matrix molecules. On nanostructured substrates, the cell adhesion-mediating molecules are adsorbed in the appropriate spectrum, amount and spatial conformation that make specific sites on these molecules (e.g. amino acid sequences like RGD) accessible to cell adhesion receptors. In addition, the nanostructured surfaces have been reported to enhance the adsorption of vitronectin, preferred by osteoblasts for their adhesion. The electrical conductivity of carbon nanotube films has also had beneficial effects on the growth and maturation of osteoblasts. Thus, the nanoparticles deposited on the bone-anchoring parts of bone, joint or dental replacements can improve the integration of these devices with the surrounding bone tissue. In addition, carbon nanoparticles admixed into polymers, designed for the fabrication of three-dimensional scaffolds for bone tissue engineering, could decorate the walls of the pores in these materials, and thus promote the ingrowth of bone cells. The interaction of cells with substrates modified with carbon nanoparticles could be further intensified by functionalizing these particles with various chemical functional groups or biomolecules, including KRSR-containing adhesion oligopeptides, recognized preferentially by osteoblasts.

This interdisciplinary review, based mainly on their own results, deals with the interaction of bone-derived cells with glass, polymeric and silicon substrates modified with fullerenes C_{60}, nanotubes, nanocrystalline diamond as well as other carbon allotropes, such as amorphous hydrogenated carbon or pyrolytic graphite. In general, these modifications have resulted in improved adhesion, growth, viability and osteogenic differentiation of these cells. The use of carbon and/or metal-carbon composite nanoparticles for the construction of patterned surfaces for regionally-selective adhesion and directed growth of cells is also discussed in this work.

Chapter 36 - This chapter reviews the novel synthesis of inorganic nanoparticles stabilized by an organic shell layer with investigation into their specific characteristics. How to design organic-inorganic hybrid nanoparticles in chemical solution process effectively is indispensable to uncover their unprecedented natures due to the fusion of merits of both organic and inorganic substances. Expressly, the two synthetic parameters are a focus of this chapter: the physical construction of the inorganic core and the chemical constituents introduced. Concretely, the catalytic activities of nanoparticles are controllable by modifying the crystal faces and the surface area of their inorganic cores in connection with their shape and size. In another case, when functional units, as an organic shell, are attached to the surface of nanoparticles, external stimuli such as electric field or light are utilized to transform the characteristics of nanoparticles themselves. It is also interesting to bring metal coordination polymers into an inorganic core, leading to the first isolation of alkyl chain stabilized-metal coordination nano-polymers with ferromagnetism. The nanomaterials presented here are listed as Pt nano-cube, size-selected Au nanoparticles, metal hexacyanoferrate coordination nano-polymers, and metal nanoparticles functionalized by bifferocene, anthraquinone, porphyrin and triphenylene derivatives. The synthetic procedure and their remarkable characteristics with some application examples are demonstrated.

Chapter 37 - Colloidal gold nanoparticles (GNPs) with a size range of several to hundreds of nanometers show a bright-pinkish color due to plasmon resonance and have been widely utilized for preparation of stained glasses for several hundred years. Recently, much attention has been given to gold nanoparticles (GNPs) because of their unique characteristics, and

GNPs have been applied in many areas, such as biomedical materials science, optics and electronics.

In the 20th century, various methods for the preparation of gold colloids were reported and reviewed. Most of these methods were based on the reduction of tetrachloroauric acid ($HAuCl_4$). The most popular one for a long time has been that using citrate reduction of $HAuCl_4$ in water. It leads to GNPs of ca. 20 nm. To obtain GNPs of prechosen size between 16 and 147 nm via their controlled formation, a method was proposed where the ratio between the reducing and stabilizing agents was varied. GNPs thus prepared are dispersed in the solution by ionic repulsion of adsorbed ions such as citric acid on their surface. Under physiological condition, however, the ionically stabilized GNPs tend to aggregate because of the charge shielding. To improve the dispersion stability of ionically stabilized GNPs, much kind of polymer has been used as GNPs-stabilizer.

Monodispersed GNPs were readily prepared by the simple addition of a ligand installed PEG-*b*-PAMA in $HAuCl_4$ aqueous solution at room temperature without any additional reducing reagent. The ligand-installed PEG layer locating on the outer surface of the obtained GNPs substantially contributed to sterically stabilize them even under very high salt concentrations. Furthermore, biotin as the ligand molecule installed at the end of tethered PEG chain showed specific recognition of streptavidin under physiological salt concentration indicating the high utility of these ligand-installed PEG/PAMA-gold nanoparticles as colloidal biosensor systems.

The polyvalency coordination of tertiary amino groups on the gold surface was confirmed to lead to stable modification with PEG-*b*-PAMA, and not electrostatic interaction between the protonated PAMA and the gold surface. PEG-GNPs modified with PEG-*b*-PAMA having a short PAMA chain length showed excellent dispersion stability under physiological conditions, even when kept in 95% human serum for 4 days. It was confirmed that the increased PEG chain density resulting from PEG-*b*-PAMA with a short PAMA chain length is one of the factors which improve the dispersion stability of GNPs, indicating that three amino nitrogens are enough to stabilize the PEG brush on the gold colloid surface via the polyvalency effect. The obtained PEG-GNPs may be useful in a variety of biological applications, for example, as carriers of gene delivery, model carriers of drug delivery systems, and contrast agents for *in vivo* X-ray computed tomography.

Chapter 38 - The aim of this chapter is to discuss the advantages of use of nanoparticles and quantum dots as tags for electrochemical bioassays.

Nanoscale materials bring new opportunities for electrochemical biosensing. Quantum dots are very stable (comparing to enzyme labels), they offer high sensitivity (thousands of atoms can be released from one nanoparticle) and wide variety of nanoparticles brings possibility for multiplex detection of several biomolecules.

Chapter 39 - Water-soluble thiol-capped colloidal CdTe QDs were synthesized by the hydrothermal route. The PL wavelength was determined by the QD size due to the quantum confinement effect. The PL properties of the CdTe QDs were not changed upon aging in air for several months. When studying time-dependent PL spectroscopy of thiol-capped CdTe QDs in living cells, it was observed that the PL was first enhanced and then redshifted until it was fully quenched. This characteristic PL behavior was proved to be originated from the effect of photooxidation. Deoxygenation could effectively prevent the PL blueshift to occur. It was revealed that singlet oxygen is responsible for the PL blueshift. The singlet oxygen molecules were presumably formed from the oxygen molecules intercalated in the thiol layer at the QD core surface. The photostability of QDs in living cells was also studied. Photobleaching of QDs occurred in living cells. The higher irradiation dose and environment oxygen density and the lower QD concentration resulted in the higher photobleaching rate. It was proved that photobleaching resulted mainly from the photooxidation of the cellular QDs. The photostability of cellular QDs was well improved when the cells were deoxygenated by purging with nitrogen. However, the photostability of cellular QDs is better than that of the chlorophyll and green fluorescence protein (GFP) in living cells, and is much better than that of FITC when the local concentration of QDs is not too low.

Chapter 40 - Inorganic nanoparticles have received great interests due to their attractive electronic, optical, and thermal properties as well as catalytic properties and potential applications in the fields of physics, chemistry, biology, medicine, and material science and their different interdisciplinary fields. Great deals of synthetic methods such as aqueous- and organic-phase approaches have been reported to synthesize high-quality inorganic nanoparticles, but the stability of colloid nanoparticles obtained is still a great problem. In addition, most of these methods for preparing nanoparticles were based on the use of organic ligand, surfactant or polymeric templates. However, the addition of organic ligand, surfactant or polymeric templates may introduce contamination in the final product due to the formation of undesired byproduct in the reaction medium, which makes them unsuitable for further application, especially in biology. On the other hand, fluorohores such as dyes has been employed as optical probe by many groups for carrying out bioassay such as detecting target DNA, protein, etc. However, because one DNA or other probes can only be labeled with one or a few fluorophores, the fluorescence signal is too weak to be detected when the target concentration is low. Furthermore, most organic dyes suffer serious photobleaching, often resulting in irreproducible signals for ultratrace bioanalysis. Silica coating of inorganic nanomaterial or fluorohores appears as an attractive alternative for providing enhanced colloidal or photostable stability and suitable functional groups for bioconjugation. Hybrid nanoparticles based on silica has been pursued vigorously in the last few years and in this review chapter the authors provide a perspective on the present status of the subject. Described approaches include encapsulating inorganic nanomaterials using SiO_2 sphere, fluorophore-doped silica nanoparticles, constructing muti-functional hybrid materials using SiO_2 sphere as supporting material, and some important applications of hybrid nanoparticle based on silica. Finally, summary and outlook of these hybrid nanoparticles are also presented.

Chapter 41 - Nanofluids, a dispersion of nanoparticles in conventional heat transfer fluids, promises to offer enhanced heat transfer performance. Nanofluids are a new, innovative class of heat transfer fluids and represent a rapidly emerging field where nanoscale science and thermal engineering meet. In recent years, nanofluids have attracted great interest from researchers of multi-disciplines because of their superior thermal properties and potential applications in important fields such as microelectronics, microfluidics, transportation, and biomedical. Published research works have shown that nanofluids possess higher effective thermal conductivity and effective thermal diffusivity compared to their base fluids and the magnitudes of these properties increase remarkably with increasing nanoparticle volume fraction. Particle size and shape as well as fluid temperature are also found to influence the enhancement of thermal conductivity of the nanofluids. Despite numerous theoretical studies on model development for nanofluids, there is no widely accepted model available due to inconclusive heat transfer mechanisms of nanofluids. There are also many inconsistencies in reported experimental results and controversies in the proposed mechanisms for enhanced thermophysical properties of nanofluids. In this chapter, the current state-of-research in nanofluids including synthesis, potential applications, experimental and analytical studies on the effective thermal conductivity, thermal diffusivity, and viscosity of nanofluids are critically reviewed. Results from the authors' extensive theoretical and experimental studies on thermophysical and electrokinetic properties of nanofluids are also summarized.

Chapter 42 - Since the first report in 1995 that indicated that the addition of nanometer copper metal particles to traditional heat transfer liquids, such as water and Ethylene Glycol, could result in an enhanced effective thermal conductivity which was considerably higher than that predicted by traditional mean-field theory, tremendous experimental studies have been conducted. Most of these support this conclusion, but there is some variation in the degree of the level of the enhancement. These investigations include experiments on metal nanoparticle/liquid suspensions, metal oxide nanoparticle/liquid suspensions, liquid nanoparticle/liquid suspensions, and carbon nanotube/liquid suspensions, and cover a wide range of nanoparticle shapes and volume fractions. These investigations include variations in the size of the nanoparticle, the type of nanoparticle material, the type of base fluid, the effect of surfactants, the effect of pH value of the suspension, the temperature of the suspension, and the ultrasonic vibrating time. Moreover, a number of experimental techniques and methods have been adopted to measure the effective thermal conductivity of different types of suspensions, including steady-state cut-bar methods, transient hot-wire methods and other less well known methods.

The theoretical exploration of the fundamental mechanisms for enhanced effective thermal conductivity of nanoparticle suspensions has resulted in a number of theoretical models. There are basically two different mechanisms proposed for these variations, the Brownian motion effect and the agglomeration effect. In addition to these, several other alternatives, such as the effect of the adsorption layer of the fluid molecules on the surface of the nanoparticles and the thermophoresis of the nanoparticles inside the suspension have been proposed.

The Brownian motion mechanism in nanoparticle suspensions is a relatively new concept, which has grown in acceptance with the addition of additional experimental evidence. While the agglomeration mechanism is not new, it assumes that the agglomeration of the nanoparticles replaces the micron size and/or millimeter size particles in the mean-field theory. Both major mechanisms are supported by the comparisons between predictions of models and reported experimental results. Moreover, there is almost equal support for both from numerical simulation studies and experimental observations.

An extensive review of the available experimental reports and theoretical developments indicates that a majority of the experiments have confirmed the enhanced effective thermal conductivities of nanoparticle suspensions and indicated that these are greater than the values predicted from mean-field theory models. In addition, the mean-field theory can not explain why the effective thermal conductivity of nanoparticle suspensions could have this unusually high enhancement, or why it increases with increases in the bulk temperature, and with decreases in the nanoparticle size. Finally, the evolution process of newly developed Brownian motion models and agglomeration models is discussed in detail.

As a result, what is clear from the available research is that it is necessary to illuminate why effective thermal conductivities of nanoparticle suspensions are higher than the predictions of mean-field theory models. Extensive study on the contributions of the mechanisms of Brownian motion, agglomeration and correlation of both is critically needed. This review will help the development of a more robust understanding of the new phenomenon and the further application of effective thermal conductivity of nanoparticle suspensions in the future.

Chapter 43 - One of the most important findings for nanofluids is the discovery of a significant enhancement and a strong temperature effect in their thermal conductivity. Research confirms that the standard rules for mixtures is not capable of describing the nanoparticle's contribution to the thermal conductivity, and some researchers believe that the Brownian motion is the key mechanism affecting the variation of thermal conductivity with temperature. In this paper, the authors review existing experimental results and explore different approaches to estimating the effective thermal conductivity of nanofluids from recent theoretical models. Focus is primarily on an explanation of the temperature-dependence on the thermal conductivity of nanofluids. It is found that the mobility of nanoparticles is theoretically conjectured as having a key role in determining the temperature effect on the thermal conductivity of a nanofluid. The limitations of existing models and future research directions are discussed.

Chapter 44 - This work aims to review the immense influence that nanotechnology has having along the last years in all areas of knowledge, as will as, the method of obtaining of nanoparticles, trough of clays modification and Sol-gel synthesis and the used of the nanoparticles obtained in the formation of nanocomposites with polyolefin, also is discussed the interaction nanoparticle-polymer, and the used of a new compatibilizer agent in the properties of nanocomposites. Is presented one of its main applications and the main capital investment from the governments is the packaging industry of food and pharmaceutical products. This is due, among many other factors, to the possibility of using nano-sensors that when combined with the package communicate to the consumer/producer, as well an in control situations, the characteristics of package material such as: microbiological quality, storage conditions (humidity, temperature, light, etc), time-temperature treatment and shelf-life, among others. Finally is shows some trends for the application of nanoparticles.

Chapter 45 - techniques are promising in many aspects, such as electroanalysis, bio-analysis, power & energy, etc. At the same time, the introduction of Nano-sciences & techniques in electrochemistry also expands the unknown filed of the interdiscipline.

In this comment paper, the authors will discuss the applications of nanoparticles (NPs) in electrochemistry including the following threads 1) why scientists introduce the NPs into electrochemistry; in this part, the authors summarize the develop of Nano-sciences & techniques; mainly discuss some basic theories and concepts about NPs and electrochemistry and how scientists use NPs in electrochemistry; 2) progresses of NPs application in electrochemistry, in this part, the authors summarize the achieved marvelous achievements in this area; 3) challenge of the NPs application in electrochemistry, in this part, the authors summarize the challenges that electrochemists are encountering with and bring up their opinions and strategies.

Chapter 46 - Proteins and peptides as therapeutic agents have been used for clinical applications. While due to their short *in vivo* half life, easily undergoing hydrolysis, low permeability to biological membrane barriers and poor stability in gastrointestinal tract, parenteral administration such as subcutaneous, intravenous, or intramuscular injections is the common way for these drugs delivery. Alternative injection, the oral route presents many advantages such as avoidance injection pain and discomfort, elimination of possible infections caused by the use of needles. Therefore, many strategies to improve their oral delivery efficiency have been studied. One of the most potential strategies is usage of micro- and nanoparticles of functional polymers as protein drug carriers. This chapter summarizes the nanoparticles based on different polymers as an oral administration system for protein and peptide drugs delivery. It primarily describes the pH-sensitive nanoparticles. Depending on the fact that pH of the human gastrointestinal tract increases progressively from the stomach to the intestine, nanoparticles remain insoluble in the stomach and disintegrate at the higher pH. Secondly, mucoadhesive polymer-coated nanoparticles are described, which can intensify the contact between the polymer and the mucus surface either by non covalent bonds or covalent disulfide bonds. Thirdly, bacteria degradable polymer-coated nanoparticles for colon delivery are briefly reported. Fourthly, a pH-sensitive and mucoadhesive polymer based on nano- and microparticles is included. Finally, some success of oral protein drug *in vivo* deliveries are enumerates. All of these nanoparticles can be expected to improve the proteins absorption.

Chapter 47 - Photovoltaic behavior in both organic thin films and organic/inorganic hybrid systems is an important subject not only in basic research but also in application fields such as organic solar cells, dyes sensitized solar cells etc. In order to improve the light collection efficiency, tuning film morphology is an efficient way to improve light harvest, then conversion efficiency. In this paper, some fabrication methods for different photoanodes with controlled morphology and their photovoltaic behaviors will be reviewed. While using electro-hydrodynamics (EHD) technique to prepare lotus leaf-like photoanodes will be discussed in detail to illustrate the effect of micro-/nano- composite structure on the enhancement of photo-electric conversion behavior in dye-sensitized solar cells (DSSCs). Further more, there will some discussion on the morphology control for the organic thin film system to increase photocurrent generation.

Chapter 48 - All solution processed photovoltaic devices based on conjugated polymer/nanocrystal hybrid materials offer the potential toward fabricating low-cost, large area and mechanically flexible solar cells. In this article, the mechanisms of charge separation and transport in the solar cells based on polymer/nanocrystal (nanoparticles, nanorods) composites will be present. Carrier dynamics in the polymer/nanocrystal composites are investigated by temperature dependent and time-resolved photoluminescence (PL) spectroscopy and time of flight (TOF) technique. Rapid exciton dissociation at the polymer/nanocrystal interfaces leading to efficient charge separation, accompanying with a significant quench in the photoluminescence (PL) intensity and a decrease of PL lifetime with increasing nanocrystal concentrations. The separated carriers are transported through the device to the electrodes without recombining with oppositely charged carriers. The improvement in carrier transport can be further achieved by the inclusion of one-dimensional semiconductor nanorods which can offer direct pathways for electric conduction, supported by TOF mobility measurement. In addition, the efficiency of charge separation and transport in the polymer/nanocrystal composite can be further improved by surface treatment at polymer/nanocrystal interfaces.

Chapter 49 - On the basis of aromatic polyesters and the block-copolyesters, and also polyethylene of high density compositions c were prepared the contents of organic compounds of five-valent phosphorus with cyclohexyl a radical at atom of phosphorus. The sizes of molecules phosphor organic compounds make 8-10 Å3, that allows them to relate to so-called nanoparticles. It is established, that investigated phosphorus organic compounds appeared effective stabilizers thermal and mechanical properties of the investigated polymers, and also good antipyrens.

Chapter 50 - It should be known that a lot of studies devoted to the preparation of polymer nanocomposite materials have been investigated at resent years. The amount of such works increases intensively. The possibility of preparation such materials was shown for practically all kinds of polymerized and polycondensed polymer materials. Investigators demonstrate particular interest to the organomodified montmorillonyte as an element of nanotechnology and bearer of nanostructure with great differ between its length and thickness. In this case the organomodification is carried out by using of ionic surfactants. The using of nonionic surfactants for hydrofobization of clay's surface also found reflection in some works. Common steady understanding is formed about investigation technique and structure of polymer nanocomposite materials and how thermo-mechanical properties depend on its structure. The increasing in amount of such investigations shows that this perspective technology will find reflection in industrial application.

Chapter 51 - In the present paper are considered three fundamental factors controlling filling processes of nanocomposites with high-elasticity matrix. They are: filler particles aggregation, strain localization and change of structure type from Euclidean object up to fractal. The last postulate is the most physically fundamental, since it demonstrates noncorrectness of description of structure and properties of nanocomposites with high-elasticity matrix within the framework of classical theory of entropic high-elasticity.

Chapter 52 - It is shown, that the structure formation of nanocomposites based on the polypropylene takes in Euclidean space with dimension $d=3$. Within the framework of the offered for these material structural model is shown identity of local order regions, interfacial areas and crystalline β-phase. This model allows quantitative description of nanocomposites properties.

Chapter 53 - The interrelation of elasticity modulus and amorphous chain's tightness characterized by fractal dimension of chain part between its fixation points for nanocomposites based on the polypropylene is shown. This assumes the polymeric matrix structure change in comparison with initial polymer: the role of densely-packed regions for it is played by interphase areas. An offered fractal model allows estimation of elasticity modulus limiting values.

Chapter 54 - Polypropylene (PP) has wide acceptance for use in many application areas. However, low thermal resistance complicates its general practice. The new approach in thermal stabilization of PP is based on the synthesis of PP nanocomposites. This paper discusses new advances in the study of the thermo-oxidative degradation of PP nanocomposite. The observed results are interpreted by a proposed kinetic model, and the predominant role of the one-dimensional diffusion type reaction. According to the kinetic analysis, PP nanocomposites had superior thermal and fireproof behavior compared with neat PP. Evidently, the mechanism of nanocomposite flame retardancy is based on shielding role of high-performance carbonaceous-silicate char which insulates the underlying polymeric material and slows down the mass loss rate of decomposition products.

Chapter 55 - Nanomaterials have attracted much attention in the biomedical areas due to their unique advantages of application in the biomedical related fields. It is already known that nanoparticles could present a versatile nanoscale interface for respective biomolecular recognition, drug delivery, etc. The strategy of nanotechnology could result in an increased concentration of anticancer drugs in selected target tissues and a reduction of systemic side effects and toxicity. Furthermore, with the development of the advanced nanotechnology, nanocomposites functioned with different biomarkers or relevant functional groups could be used to adsorb different biomolecules for specifically targeting tumor cells, which may provide a new way for the cancer diagnosis and efficiently facilitate drug accumulation on target cancer cells and inhibit the multidrug resistance of the cancer chemotherapy.

In the meanwhile, nanotechnology is in a unique position to transform cancer diagnostics and produce a new generation of biosensors and medical imaging techniques with higher sensitivity and precision of recognition. Nanomaterial-based biomolecule delivery carriers show promising perspective in target biomarker and cancer therapy. Based on the recent development of nanotechnology, nanomaterials display promising and far-ranging prospect in the fields of biomedical engineering, especially in the diagnosis and treatment of cancers.

Chapter 56 - The considerable interest in designing anisotropic metal nanoparticles, especially gold nanorods, stems from their unique properties and potential applications in biosensing, catalysis, biomedical imaging, cancer diagnosis and therapy and drug delivery. This chapter describes recent advances in the synthesis, assembly and bioapplications of gold naorods. At the beginning, the authors outline briefly the most popular solution methods to synthesize gold nanorods, as well as their shape-dependent optical properties. In addition, gold nanorods as attractive building blocks to form hybrid nanomaterials such as the assembly of gold nanorods on inorganic materials or core-shell particles are elucidated. Finally, particular emphasis is placed on the rapid development of the intergration gold nanorods with biomolecules such as protein and DNA and the utilization of such bionanohybrid assemblies for bioanalytical applications.

Chapter 57 - This chapter focuses on novel gas/chemical sensing systems that are based on nanotechnology. Recently, there has been an extensive interest in nanomaterials such as carbon nanotubes, nanowires, nanofibers, and nanoparticles. Their application on gas sensing systems has also been widely studied. Due to their unique micro/nano porous structures and ultra high surface-to-volume ratio, nanostructured materials have a high potential to enhance conventional gas sensors' performance, including the sensitivity, the operation temperature and the recovery time. In this chapter, various nanomaterials-based gas/chemical sensors are presented, including their structure, fabrication, and operation principles.

In: Encyclopedia of Nanotechnology
Editor: Elwood D. Carlson ISBN: 978-1-60692-079-4

Expert Commentary

NANOSTRUCTURED POLYMERS: DEVELOPMENTS AND TRENDS, HIGHLIGHTS AND SHADOWS

Maria Vittoria Russo and Ilaria Fratoddi
Department of Chemistry, University of Rome"
Sapienza", P.le A.Moro 5, 00185 Rome, Italy

The world of nano-materials burst out quite recently as a novelty in the scientific community, although macromolecules at the nanoscale have been built up by chemists and biologists and quantum effects have been studied by physicists since decades. Nowadays, since the growth of nanomaterial phases and structures can be controlled, applications in nanotechnology related to their chemical and physical properties are almost ready to be achieved. Dendrimers, fullerenes, carbon nanotubes and graphenes, metal and ceramic nanoparticles are all available for the development of advanced technological instruments. Atomic force and scanning probe microscopes, atomic-layer-deposition devices and nanolithography tools can manipulate matter in the atomic or molecular regime. These nanotools are able to control matter at the nano or atomic regime.

The first nanodevices on the market were quantum dot fluorescent biodetectors. MEMS devices are used as accelerometers in automotive airbags. Many other promising applications are pending such as nanoelectric memory devices, nanosensors, catalysts and drug delivery systems. Components for nanodevices envisage the use of semiconducting organic molecules, polymers and composites.

Among the variety of nano-materials, polymers are suitable for multipurpose technologies because their chemical structure can be varied to achieve functionalities that are improved by the nanosize morphology. Nanostructured polymers are easily synthesized in bulk amounts, while classical colloids or other nano-objects need the aid of ancillary techniques for the achievement of their nano shape. Moreover, unforeseen properties may be revealed merely because the polymers are formed in this peculiar structure. Indeed, nanoscience applied to polymers leads to a vision of unity of knowledge, involving integration of chemistry, physics, engineering and a bottom-up mastery of matter. Through two different mastering approaches, i.e. bottom-up and top-down, in principle any type of polymer can be prepared in a nanostructured fashion and some representative examples, which however are not exhaustive, will be reported in this commentary.

Polyamides, polystyrene and polymethylmethacrylate (S.Gross *et al.*, Eur. Polym. Journal (2007) 43, 673) are common multipurpose synthetic polymers used in a wide range of photonic crystals, gel opals, medical carriers, chromatographic industrial settings and consumer products such as spacers for display and biomedical media.

There have been many reports on the methods of preparation of their nanoscale particles, mainly concerning optimization of the emulsion polymerization conditions (bottom-up approach). Variations of temperature, reaction time, solvent/co-solvent ratio, addition of surfactants, water content as the dispersion medium, nature of the initiator, ionic strength of the suspension, functionalization of the monomer with hydrophilic/hydrophobic groups, allow the control of the shape, of the monodispersion and of the wideness of ordered domains for optical purposes. In general, the shape of nanoparticles for these materials is a sphere due to the isotropic surface energy. Otherwise, polymer nanofibers (with submicron scale diameters) can be formed with the electrospinning process. As a curiosity, some peculiar morphologies, such as raspberry-like nanospheres can be tuned for polymer composites obtained from a charge compensation process (Gouliang Li *et al*, J. Colloid and Interface Sci., (2006) 297, 705) and for magnetic polystyrene microspheres (Zhizhong Xu *et al*, Mat.Chem.Phys., (2007) in press).

Polymers, unlike other materials, are not considered suitable materials for hydrogen storage; however, when a intrinsic microporosity can be achieved by tailoring the chemical structure (for example, made from fused-ring subunits) so that large areas of free space are made available, hydrogen physisorption occurs rapidly and reversibly in amounts that compete with those of activated carbon (McKeown *et al.*, Angew.Chem.Int.Ed., (2006) 45,1804). A sector apart is represented by carbon nanotubes (CNT), which can be considered polymeric carbon and show a wide range of remarkable properties, such as high charge-carrier mobility and high mechanical strength associated with semiconductivity. The scientific literature on SWCNT and MWCNT is extremely ample and these materials are now finding successful commercialization, despite their cost.

The versatile properties of π-conjugated polymers, such as solubility, semiconductivity and easy processability, have led to the development and improvement of nanoscale organic transistors applied to chemical and biological sensors and light emitting diodes (A.Dodabalapur, Mat.Today (2006) 9(4), 24). The research on the synthesis and properties of polymeric semiconductors has been carried out thoroughly since the last two decades, because these materials are very promising and multipurpose for the achievement of flexible electronics. The fundamental figure of merit of charge transport is carrier mobility which is a prerequisite for the technological application of conjugated polymers. Despite the considerable amount of excellent experimental and theoretical work produced so far, there is still not a clear knowledge on what limits the charge transport in thin films of conjugated polymers. A basic picture could allow a universal view of the correlation between morphology, microstructure and electrical properties. In this contest, some indications come from the studies concerning the formation of crystallites as large as a few nanometers of a polythiophene with a particular chemical structure, i.e. PBTTT, which shows the best performance in semiconducting properties for this class of materials just because of its peculiar lamellar morphology (M. Heeney *et al.*, J.Am.Chem.Soc., (2005) 127, 1078; I. McCulloch *et al.,* Nat.Mater., (2006) 5, 328).

As a further example, among conducting polymers, polyaniline is also a carefully investigated material due to its excellent optical and electronic properties. Recently much research has been conducted on the nanostructure of polyaniline (PANI) because it exhibits enhanced performance in applications (sensors, catalysis) where a high surface contact area is needed between the nanostructures and its environment. To this purpose, polyaniline nanostructured particles i.e. rods, wires, fibres, spheres, belts, with unusual physical and chemical properties have been produced by various strategies including template synthesis, interfacial polymerization, self-assembly and stepwise electrochemical deposition. Although various methods for the synthesis of nanostructures have been proposed, yet there is lack of full understanding of the rationale which drives the nanostructure formation and most of the reports are based on phenomenological experiments. An attempt to explain the reasons about forming polyaniline nanofibers was suggested by applying a shaping theory based on hydrogen bond formation between polymer chains and intercalated alcohol molecules (J.Kan *et al.*, Eur. Polym. Journal, (2007) 43, 395). However, this theory cannot attain to every type of polymers.

A relationship between the supramolecular interactions between the individual π-conjugated molecules/polymers and the microscopic morphology appearing upon aggregation has been proposed recently. Approaches to control the self-assembling of conjugated systems in the solid-state have been developed with a joint experimental–theoretical approach, combining scanning probe microscopy characterization of thin deposits and force-field molecular modeling of supramolecular aggregates and adsorbates. With this methodology, a strong correlation is established between the polymer molecular structure, the degree of supramolecular order and the solid-state photoluminescence properties (for light-emitting applications), or the charge transport properties (for the design of field-effect transistors). (Ph.Leclerc *et al.*, Mat.Sci.Engin. R, (2007) in press).

It is expected that the use of nanosized polymers will further enhance the performance of these electronic devices and represents the challenge of the future, providing a real breakthrough in this field of science and technology. However, more effort in research has to be done for the assessment of a general pathway allowing the achievement of a desired nanostructure for any kind of polymeric material.

Block copolymers are able to self-assemble in ordered structures at the molecular scale with no phase separation, to form domain dimensions of 5-50 nm. Due to this peculiar property, these materials can be used for the fabrication of templates which give rise to patterns of periodic arrays of spots; these patterns based on block copolymers in some cases are being commercialized for the development of quantum dots, magnetic storage and flash memory devices, semiconductor capacitors, nanowires, photonic crystals and nanopores, besides the premises for integrating these materials into a conventional lithographic process by a bottom-up mode (M.P. Stoykovich and P.F.Nealey, Mat. Today (2006) 9(9), 20; *ibidem* M.Li and C.K.Ober, page 30).

An up to date and controversial topic in medicine has to deal with the growth and differentiation of stem cells which are able to repair damaged tissues and organs. The expectations of people for the resolution of health problems by the use of these cells are high, although accompanied by some ethic concerns. In the field of tissue engineering , of primary importance is the choice of a proper scaffold which enables the cell attachment and growth, mimicking the *in vivo* environment. In this contest, ideal scaffolds can be biocompatible nanostructured polymers e.g. natural proteins, polysaccharides and synthetic polymers. Some examples will give a glance on this topic. Synthetic hydrogels like polyethylene glycol, are water swollen polymeric networks, with micro/nano pores particularly suited for tissue engineering, because their surface can be modified to match with protein sequences of specific cell types. Peptide building blocks can self-assemble into nanofibers through hydrogen bonds and π-electron interactions, so that the hydrophobicity of the scaffolds can be driven by the types of amino acid used (N.D.Evans, E.Gentleman, J.M.Polak, Mat.Today, (2006) 9 (12), 26; *ibidem* J.R.Jones, page 34).

The electrospinning technique can be successfully used to form a three-dimensional hyaluronic acid nanofibrous scaffold which mimics the architecture of a natural extracellular matrix. NIH 3T3 fibroblasts were seeded on these fibronectin-adsorbed nanofibrous scaffolds showing that upon attachment the fibroblasts exhibit a dendritic morphology. Thus, potential applications of nanofibrous scaffolds in cell encapsulation and tissue regeneration is suggested also in this case (Yuan Ji *et al.*, Biomaterials (2006) 27(20) 3782).

Scaffolds based on a synthetic polymer can be prepared by a different way. A colloidal template-assisted method allows to obtain two-dimensional nanopore arrays, consisting of hydrophilic SiO2-like holes within hydrophobic poly(hydroxymethylsiloxane) (PHMS) surfaces. The nanostructured films are able to induce a general increase of micrometer width adhered cells, e.g. bacteria Pseudomonas aeruginosa, with respect to the unmodified hydrophobic surfaces and to endorse a spot grown of biofilm-like aggregates (C.Satriano *et al.*, Materials Science and Engineering C, (2006) 26, 942).

In general, the presence of nanoscale features increases cell attachment and proliferation. However, protein attachment and cell response to three dimensional nano-materials at the interface needs deeper insight and is still a great challenge for scientists.

Natural nanostructured polymers are also excellent carriers of catalyst such as enzymes. For example, spherical chitosan nano-particles prepared by the ionization gelation method are used for the immobilization of neutral lipase. This hybrid material exhibits improved catalytic performance in comparison with the free neutral lipase, i.e. remarkably improved stability properties to various parameters, such as temperature, reuse, and storage time. (Zhen-Xing Tang *et al.*, Materials Letters (2007) 61,37).

Among the important topics attaining at nano size, particularly appealing for technological applications are polymer-based nanocomposites. These materials are in some way apart from the main field of nano-polymers, since the nanostructure which provides new mechanical and physical properties to the final material, is a feature of the fillers. A recent review article reports the most investigated classes of nanocomposites with special attention to structure-property relationship (S.C.Tjong, Mat.Sci.Engin.R, (2007) 53, 73).

The reported spots on the world of nanostructured polymers may give an idea of the expectations arose from this field of science. In the ideal framework of structurally well defined nano-polymers there is still quandary about the number and complexity of methods for achieving architectures hierarchically organised in terms of structure and functions. At the state of art, too many parameters may affect a desired morphology and it is difficult to predict which one is the clue.

It must be mentioned that there is some concern in the society on the risks for health connected to the use of nanoparticles. At the moment, no evidence of potential hazard due to the interaction of nanoparticles with cells and organic tissues has been detected and nanosized polymers are probably safe. However, this topic cannot be ignored and needs to be investigated and submitted to regulation.

Looking to the future, there is no doubt that this amazing size discovered for new and known polymers, born from the fruitful activities in this research field still in its infancy, will open a land of promising applications in many areas: optics, electronics, energy, environment, biology, medicine, functional smart coatings, fuel and solar cells, catalysts, sensors, etc, that could not be all cited in this commentary, but are open to the fantasy and the curiosity of scientists.

In: Encyclopedia of Nanotechnology
Editor: Elwood D. Carlson

ISBN: 978-1-60692-079-4

Chapter 1

TiO_2 NANOSTRUCTURED MATERIALS: DESIGN, CHARACTERIZATION AND APPLICATIONS

Aicheng Chen*, Xinsheng Peng and Peter Holt-Hindle
Department of Chemistry, Lakehead University, Thunder Bay, Ontario, Canada

ABSTRACT

Nanostructured materials have attracted tremendous attention in both fundamental studies and applications due to their unique physical and chemical properties. Of the various types of nanostructured materials, TiO_2 nanomaterials are extensively studied for their broad range of applications. A number of methods have been developed for preparing TiO_2 nanoparticles, nanorods, whiskers, nanowires, nanotubes and mesoporous TiO_2 films. In this chapter, we describe two novel methods recently developed in our laboratory to grow TiO_2 nanostructured materials directly on titanium (Ti) substrate. Dense and well-aligned TiO_2 nanorod arrays form on a Ti substrate when using organic molecules, such as ethanol, acetaldehyde, acetone and dibutyltin dilaurate, as the oxygen source in the oxidation of Ti at high temperatures. In contrast, polycrystalline grains are formed when pure oxygen and formic acid are used as the oxygen source; elongated micro-crystals are produced when water vapor is used as the oxygen source; whereas the use of argon with a low concentration of oxygen produces random nanofibers growing from the ledges of the TiO_2 grains. Measurements of the water contact angle of these samples show that the wetting properties of the TiO_2 films are dictated by their structure, varying from a hydrophilic surface (polycrystalline grains) to hydrophobic (random nanofibers), and towards a highly-hydrophobic surface (well-aligned nanorod arrays). Secondly, we have synthesized large-scale $Na_2Ti_6O_{13}$ nanoplates, nanowires, and continuous nanowire network films directly on Ti substrate hydrothermally. The morphology of the formed $Na_2Ti_6O_{13}$ nanostructures is easily tuned by varying the experimental parameters of temperature, reaction duration and NaOH concentrations. Our study shows that the synthesized $Na_2Ti_6O_{13}$ nanostructures are easily converted into $H_2Ti_3O_7$ nanostructures, a desirable precursor for the fabrication of various TiO_2-based nanomaterials such as anatase, a mixture of anatase and rutile, and rutile TiO_2 nanomaterials, with shape-preservation through an ion-exchange process. The approaches

* E-mail address: achen@lakeheadu.ca; Fax: (807) 346-7775; Tel: (807) 343-8318. Corresponding author.

described in this chapter provide simple and novel methods to synthesize various large-scale TiO_2-based nanostructured materials. This tunable growth of TiO_2 nanostructures is desirable for the promising applications of TiO_2 nanomaterials in the development of photo-catalysts, optical devices, sensors and self-cleaning coatings.

1. INTRODUCTION

Nanocrystalline materials have attracted tremendous attention in many disciplines [1-6]. Nanocrystals exhibit novel electronic, magnetic, optical, chemical, and mechanical properties, making them highly attractive for many important technological applications. Of the various types of nanocrystals, titanium oxide (TiO_2) and TiO_2-based materials are widely investigated for applications in photovoltaic cells, batteries, electrochemical sensors, optical emitters, photonic crystals, catalysis and photocatalysis, selective adsorption, ion exchange, ultraviolet blockers, smart surface coatings, and as functional filling materials in textiles, paints, paper and cosmetics [7-18]. TiO_2 exists mainly in three crystallographic forms: rutile (tetragonal, a = b = 4.584 Å, c = 2.952 Å), anatase (tetragonal, a = b = 3.782 Å, c = 9.502 Å), and brookite (rhombohedrical, a = 5.436 Å, b = 9.166 Å, c = 5.135 Å). However, only rutile and anatase play a role in the applications of TiO_2 and thus they have been extensively studied with surface science techniques. The surface science aspects of rutile and anatase TiO_2 have been comprehensively reviewed lately by Diebold [7]. The photoactivity of TiO_2 is one of its most attractive properties. Several review papers have discussed the technical and scientific aspects of TiO_2 photocatalysis [19-22]. Compared to anatase, rutile TiO_2 has attracted less attention in producing photocatalysts [23, 24] and dye-sensitized solar cells (DSCs) [25]. This is attributed to the fact that the rutile phase exhibits lower photo-electrochemical performance than the anatase phase, which is primarily due to the differences in their electronic structures. Most of the basic understanding of the electronic structure of TiO_2 surfaces is discussed by Hedrich and Cox [26]. On the other hand, rutile TiO_2 has some advantages over anatase, such as higher chemical stability, a higher refractive index and cheaper production cost [27]. Rutile TiO_2 is also preferred for optical applications. Rutile single crystals are used as polarizers in many optical devices such as isolators, modulators, switches and interferometers [28].

By far, the most actively pursued applied research on TiO_2 is its use for the photo-assisted degradation of organic molecules. TiO_2 photocatalysis provides a promising technology for the purification and treatment of both contaminated air and water [29]. TiO_2 catalysts have also been used for selective oxidation of olefins, alkyl-substituted benzenes, and alkanes at room temperature [30, 31]. TiO_2 photocatalysis has been applied in the treatment of air stripping of gas for the remediation of contaminated soil and solvents from contaminated ground water [32, 33]. When TiO_2 is exposed to UV light, electrons in the uppermost valence band jump to the conduction band, creating conduction band electrons and valence band holes. In most cases, the valence band holes and conduction band electrons simply recombine liberating heat or light, which is known as recombination. The efficiency of the photocatalytic process is measured as quantum yield, which is defined as the number of reaction events occurring per photon absorbed. In an ideal system, the quantum yield is proportional to the rate of the charge transfer processes and inversely proportional to the sum of the charge-transfer rate and the electron-hole recombination rate. Without recombination, the quantum yield could approach the ideal value of one for UV light, with a small enough wavelength to activate the catalyst [33]. Recombination is responsible for the low quantum yields. A positive charged TiO_2 coated electrode can suppress the rate of hole-electron recombination since the electrode can act as a sink for the photogenerated electrons. This concept has been used to enhance the efficiency of formic acid destruction under UV illumination, with a positive bias potential applied to the electrode [34]. Recent research suggests that this approach can achieve higher efficiency with TiO_2 photoanodes [35]. Although this application has achieved some degree of success, the true potential of the process has not been reached, probably because the resistance path for the electrons is relatively long and the surface area for reaction is low. An electron generated in one TiO_2 particle must hop or tunnel through boundaries from one TiO_2 conduction band to another until it finally enters an electrode which has a limited surface area. Depending on how far it has to travel to the electrode, an electron will still have an opportunity to interact with holes present in other TiO_2 particles and may recombine.

The partial oxidation of light alkanes and aromatics by molecular oxygen is a key process in the chemical industry [36, 37]. Unfortunately, partial oxidation reactions have always challenged industrial chemists because many oxidation reactions tend to proceed past the desired product to low- or no-value complete combustion products. Other selective oxidation reactions rely on the use of highly reactive chlorinated intermediates with the cost of producing chlorinated byproducts. Currently, UV-light-assisted oxidation shows promise for the oxidation of the cyclohexane and epidation of small olefins, whose selectivities differ from those obtained by other oxidation routes [38, 39]. This route offers a potential method for synthesis, provided that the expected product can be obtained within an acceptable quantum yield [40-42]. However, the reaction selectivity is still low due to the poor selectivity of the photocatalysts. An improvement in oxidation reaction activity and selectivity of a few percent could lead to a tremendous increase in efficiency, reduction of by-products, and improve the economics of production. Studying the relationship between reaction activity and the size of nanomaterials may lead to the improvement of their reaction selectivity.

The quantum yield for TiO_2-based photocatalysis is still very low. An initial approach to shifting the optical response of TiO_2 from UV to the visible spectral range by doping TiO_2 with transition-metal elements has been proposed [43-49]. However, metal doping has several drawbacks. The doped materials suffer from thermal instability and the metal centers act as electron traps, reducing the photocatalytic efficiency. Furthermore, the preparation of transition-metal-doped TiO_2 requires expensive ion-implantation facilities [50, 51]. Recent studies have shown that the desired narrow band gap of TiO_2 can be achieved by using anionic dopant species rather than metals ions [52-53]. Asahi et al. have demonstrated that TiO_2 films can be doped with nitrogen by sputtering methods and exhibit thereafter enhanced photoactivity in the visible spectral range [54-56]. The substitutional doping of nitrogen is very effective as its p states contribute to the band gap narrowing by mixing with O 2p states. Considerable effort has been made to dope TiO_2 thin films and powders with nitrogen by annealing TiO_2 at elevated temperatures under NH_3 flow. Nevertheless, this doping process only results in a small amount (2%) of nitrogen incorporation in micrometer-sized TiO_2 systems [54].

While notable advances have been made in these areas, three broad challenges still remain: (i) to increase the quantum yields, (ii) to improve the reaction selectively to produce desired products when used for chemical synthesis, and (iii) to shift the irradiation light from

UV to visible light as UV light accounts for only a small portion (8%) of solar energy compared to visible light (45%). To address these three challenges, a number of methods have been developed for preparing TiO_2 thin films, for instance, electron beam evaporation [57], low temperature oxidation of metal Ti thin film in a mixture of water and oxygen gas [58], chemical vapor deposition [59, 60], physical vapor deposition [61], dip coating [62], spin coating [63], and atmosphere pressure chemical vapor deposition [64]. The thin films are usually composed of TiO_2 nanoparticles. Recently, one-dimensional TiO_2-based nanostructures have begun to attract a great deal of attention because of their promising applications in many areas, such as highly efficient photocatalysis [65] and photovoltaic cells [66, 67]. Two main approaches have been reported for preparing titania nanorods and nanowires, one using a template synthesis [68-71] and the other employing a hydrothermal reaction from TiO_2 [72-78]. Tian et al. applied a hydrothermal method to synthesize TiO_2-based nanotubes with TiO_2 nanoparticles as seeds [79]. Akbar and co-workers developed a process for 'carving' oriented arrays of single crystal TiO_2 nanofibers from bulk TiO_2 crystals [80]. Nanostructured TiO_2 is also produced using the sol-gel method, where titanium alkoxide or halide ($TiCl_4$, TiF_4) is hydrolyzed in the presence of a template, such as nanospheric nanorods or anodic porous alumina [18, 68, 81-86]. The growing widespread utilization of these advanced nanomaterials demands novel methods for the fabrication of large-scale TiO_2-based nanostructures. The new technology should involve a minimal number of steps and be able to manufacture large amounts through continuous production. It is well known that the use of aqueous suspensions limits practical applications because of the difficulties in separating TiO_2 fine particles and recycling the photo-catalyst. Many techniques have been proposed for the immobilization of TiO_2 on solid substrates to overcome this problem [87, 88]. Generally speaking all the proposed methods involve a tedious procedure: first prepare TiO_2 nanostructures from solution, then re-disperse them into certain agents and finally precipitate, paste or spin coat them onto a substrate.

In this chapter, we summarize our recent work on the development of templateless and seedless methods to directly prepare large arrays of oriented TiO_2 nanomaterials on Ti substrates. We have synthesized well-aligned rutile TiO_2 nanorod arrays on Ti substrates by a simple gas-solid reaction process using small organic molecules as the oxygen source. Our experimental results show that tunable and oriented rutile nanostructures are formed by oxidizing metallic titanium at high temperatures using different small organic molecules as the oxygen source, and that the wetting properties strongly depend on the surface structure. The growth mechanism of the TiO_2 nanostructures is discussed. Secondly, we have synthesized $Na_2Ti_6O_{13}$ nanostructures using metal Ti as the Ti source in NaOH solution under hydrothermal conditions. This study demonstrates that the morphology of the products can be easily tuned by controlling the reaction parameters of temperature, reaction duration, and the NaOH concentration. Through the ion-exchanged process, these $Na_2Ti_6O_{13}$ nanostructures are converted into $H_2Ti_3O_7$ nanoplates, nanowire arrays and nanowire networks, while preserving the morphology of the nanostructures. The formed $H_2Ti_3O_7$ nanostructures are further annealed and reconstructed into anatase and rutile TiO_2 at 450 and 750 °C, respectively. The corresponding optical properties and photocatalytic activity of the synthesized TiO_2 nanostructures is presented. In addition, electrochemical methods are promising in fabricating one-dimensional TiO_2 nanotubes. The electrochemical anodization of titanium and the applications of TiO_2 nanotubes in Green Chemistry and electrochemical biosensoring is also briefly addressed.

2. EXPERIMENTAL SECTION

2.1. Thermal Oxidation Method

The thermal oxidation process used in the study is described in our previous reports [89-91]. Briefly, the oxide films were grown on Grade 1 titanium substrate. Small titanium plates of 30 x 15 x 0.8 mm^3 were first degreased using acetone, then washed with distilled water, etched in 30 wt% HCl at 80 °C for 10 minutes to remove the oxide layer on the surface, then completely washed with distilled water, and finally dried in a vacuum oven at 40 °C. The oxidation was carried out in a horizontal tube furnace equipped with a ceramic tube (diameter 35 mm, length 700mm). Prior to heating, the pretreated Ti plates were placed into the ceramic tube, and the system was purged using ultrapure argon (99.999%) at a flow rate of 200 sccm for 4 hours. The furnace was heated to a predetermined temperature at a rate of 30 °C/minute. Then a mixture of the ultrapure argon (200 sccm) and the selected oxygen source was introduced into the system for a pre-determined period of time. Finally the system was cooled to room temperature under a purge of the ultrapure argon. Oxygen, water vapour, formic acid, ethanol, acetaldehyde, acetone and dibutyltin dilaurate (DBTDL) were used as the oxygen sources. DBTDL, $(C_4H_9)_2Sn(OOC_{11}H_{23})_2$, is a commercially available tin organic compound that easily decomposes at high temperatures.

2.2. Hydrothermal Method

The hydrothermal method was used to prepare $Na_2Ti_6O_{13}$ nanostructures directly from Ti in NaOH solution. Commercial Ti (Grade 1) plates sized 3 cm (width) x 3 cm (length) x 0.8 mm (thickness) were used as the Ti source and the substrate. Sodium hydroxide solutions were prepared by dissolving NaOH in nanopure water (18.2 MΩ.cm). To investigate the influence of the NaOH concentration, six samples were prepared in the following steps: (i) after being washed, the Ti plates were put into Teflon autoclaves filled with 40 mL of x M sodium hydroxide aqueous solution, where x is 0.0, 0.01, 1.25, 2.5, 7.5 and 10.0; (ii) heated up to 150 °C at a rate of 10 °C per minute; (iii) held at 150 °C for 15 hours. All the heating rates in this study are the same. To investigate the effect of temperature, four samples were prepared under the same NaOH concentration (2.5 M) and the same length of heating time but at different temperatures: 125, 150, 175 and 200 °C. To determine the effect of varying time, five samples were synthesized under the same NaOH concentration (2.5 M) and at the same temperature (150 °C) but were kept for 0.5, 1.5, 15, 30 and 48 hours. After the autoclave cooled naturally to room temperature, all the samples were taken out and immersed into Nanopure water for three days, then washed several times in pure water and air dried. We then put the synthesized $Na_2Ti_6O_{13}$ samples into 1.0 M HNO_3 solution for three days. The resulting $H_2Ti_3O_7$ nanostructures, from an ion-exchanged process [92-94], were washed several times with Nanopure water and dried at room temperature. Some synthesized hydrogen titanate samples were transferred into a furnace and treated at 450, 600 and 750 °C for 5 hours in air and converted into anatase, a mixture of anatase and rutile, and rutile TiO_2 nanostrutures, respectively.

2.3. Characterization

The synthesized nanostructures were characterized by X-ray diffraction (XRD) (Philips PW 1050-37 10 Diffractometer with Cu Ka radiation), scanning electron microscopy (SEM) (JEOL JSM 5900LV), transmission electron microscopy (TEM) and high-resolution TEM (HRTEM) (JEOL 2010F TEM/STEM) using a 2048 x 2048 CCD camera equipped with a GIF imaging filter. The HRTEM was equipped with an energy dispersive X-ray spectrometer (EDS, Oxford Links ISIS). For SEM observation, the samples were cut into 10 mm x 10 mm squares and pasted onto an Al holder using carbon conducting paste; then a conducting gold film was sputtered on the surface. Specimens for the TEM and HRTEM investigation were scratched from the substrate and ultra-sonicated in ethanol, then a droplet of the suspension was placed on a holey copper grid covered with porous carbon film. A dye, Eosin Y, was employed as a probe to study the optical and photocatalytic properties of the $H_2Ti_3O_7$ nanowires and TiO_2-based nanostructures. The synthesized $H_2Ti_3O_7$ nanowires, anatase and rutile TiO_2 nanostructures were immersed in an ethanol solution containing Eosin Y (9 mM) for 30 minutes. The nanomaterials loaded with dye were irradiated using an UV-Xe lamp (365 nm and 5 mW/cm^2). After irradiation for selected periods of time, diffuse reflectance spectra were recorded to examine the changes of the absorption of Eosin Y using Cary5e UV-VIS.

2.4. Electrochemical Measurements

TiO_2 nanotubes were fabricated by anodizing Ti sheets in a dilute HF solution, and then we co-immobilized horseradish peroxidase (HRP) and thionine on the TiO_2 nanotube arrays by immersing the Ti/TiO_2 electrodes in a mixture of HRP and thionine solution for biosensoring [18]. Electrochemical measurements were performed with a Solartron 1287 potentiostat and a Solartron 1252B frequency response analyzer. A three-electrode system was used for all electrochemical experiments, employing one of the above electrodes, platinum wire and an Ag/AgCl electrode (3 M KCl) as the working, auxiliary and reference electrodes, respectively. All experiments were performed at room temperature (22±2 ^{o}C) with a 0.1 M phosphate buffer (PB) solution as the background electrolyte. All experimental solutions were deoxygenated by bubbling ultra-pure argon for 15 minutes and maintained under an argon atmosphere during the course of the experiments. Amperometric experiments were performed in a stirred cell with the successive addition of H_2O_2 into 8.0 ml supporting solution while the electrode potential was set at –450 mV.

3. RESULTS AND DISCUSSION

3.1. TiO_2 Nanostructures Prepared by Thermal Oxidation of Ti

3.1.1. The Effect of Oxygen Source

Figure 1 presents the general SEM images of the samples synthesized at 850 °C for 1.5 hours using different oxygen sources. Figures 1a and 1b show that polycrystalline grains were formed when pure oxygen gas and formic acid were used as the oxygen sources, respectively.

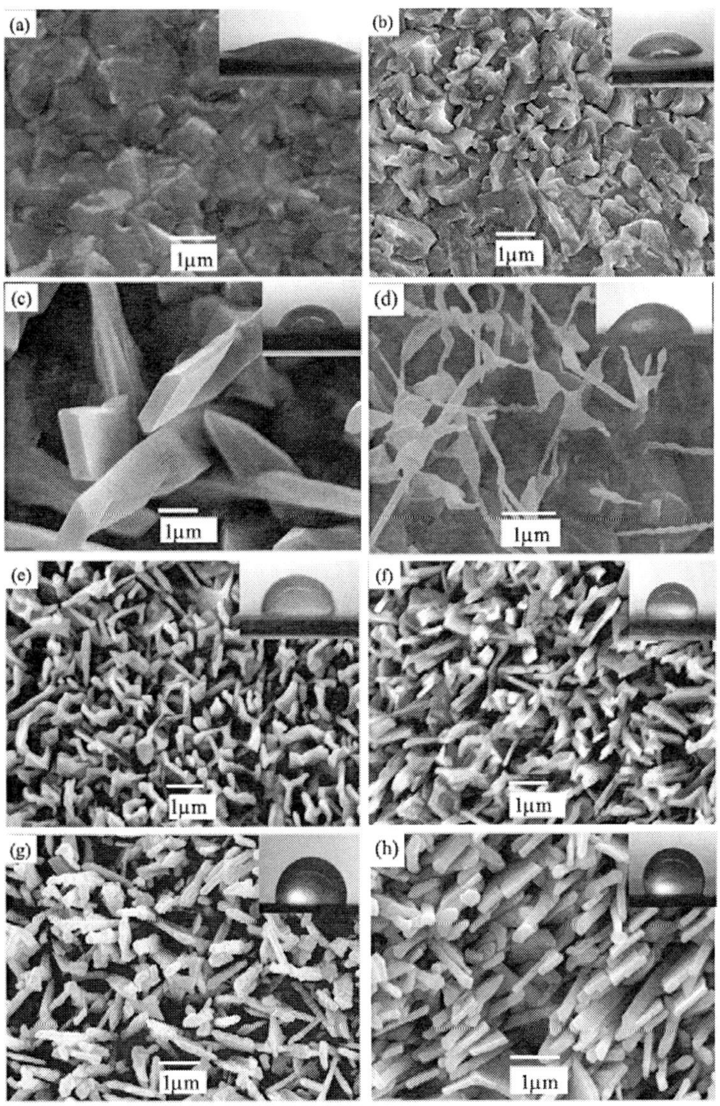

Figure 1. SEM images of the TiO2 samples prepared by oxidizing Ti using (a) O2, (b) HCOOH, (c) H2O, (d) a mixture gas of Ar (200 sccm) and O2 (1 sccm), (e) CH3CH2OH, (f) CH3CHO, (g) CH3COCH3, and (h) dibutyltin dilaurate as oxygen sources at 850 °C for 1.5 hours. The insets show the corresponding contact angle of water droplets on their surfaces. (Figure 1a, b, c, e, f and g reprinted with permission from [91]).

When water vapor was used as the oxygen source, as shown in Figure 1c, large and long micro-crystals with diameters of about one micrometer and length of about five micrometers were formed. Figure 1d shows that long and random nanofibers were formed when diluted oxygen (a mixture gas of Ar and O_2) was used as the oxygen source. In contrast to Figure 1a and 1b (polycrystalline films), 1c (elongated micro-crystalline fibers) and 1d (long nanofibers), oriented and aligned TiO_2 nanorod arrays were yielded as seen in Figure 1e, 1f, 1g and 1h, when ethanol, acetaldehyde, acetone and DBTDL were used as the oxygen sources, respectively. As shown in Figure 1e, the nanorods formed using ethanol as the oxygen source are plate-like with approximately 150 nm thickness, 500 nm width and 1 micrometer length. Figure 1f demonstrates that the nanorods formed in acetaldehyde are similar to those formed in ethanol, although slightly larger. Figure 1g shows that the nanorods formed in acetone have two typical morphologies: plate-like and column-like. As shown in Figure 1h, more and larger plate-like nanorod arrays with thickness of approximately 150 nm, width of several hundred nanometers and length of a few micrometers were formed when DBTDL was used as the oxygen source. All these SEM images demonstrate that the morphology of the formed titanium oxide nanomaterials is tunable by choosing different oxygen sources.

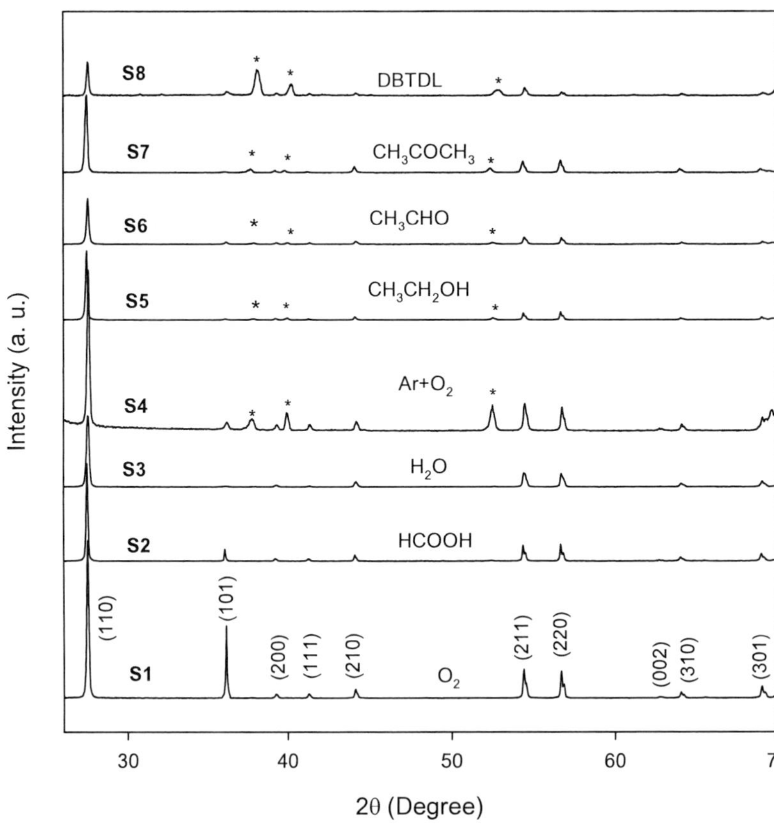

Figure 2. XRD patterns of the samples synthesized by oxidizing Ti using (S1) O_2, (S2) HCOOH, (S3) H_2O, (S4) a mixture gas of Ar (200 sccm) and O2 (1 sccm), (S5) CH_3CH_2OH, (S6) CH_3CHO, (S7) CH_3COCH_3, and (S8) dibutyltin dilaurate as oxygen sources at 850 °C for 1.5 h.

Figure 2 presents the corresponding XRD patterns of those samples shown in Figure 1. The diffraction peaks, (110), (101), (200), (111), (210), (211) and (220) can be indexed to those of the tetragonal rutile TiO_2 phase, with lattice constants a = 4.593 Å, c = 2.959 Å and a space group of P42/mnm (136), (JCPDS file No. 21-1276). The peaks marked with a star are derived from the Ti substrate. All of the XRD spectra show that the formed titanium oxide films have the same rutile phase. The changes of the intensity ratio from the (110) to (101) peaks shown in Figure 2 are consistent with the aforementioned SEM observations presented in Figure 1. The TiO_2 growth is performed from the random polycrystalline films S1, S2 and S4, to the oriented nanofibers S3, and to the well-aligned nanorod arrays S5, S6, S7, and S8.

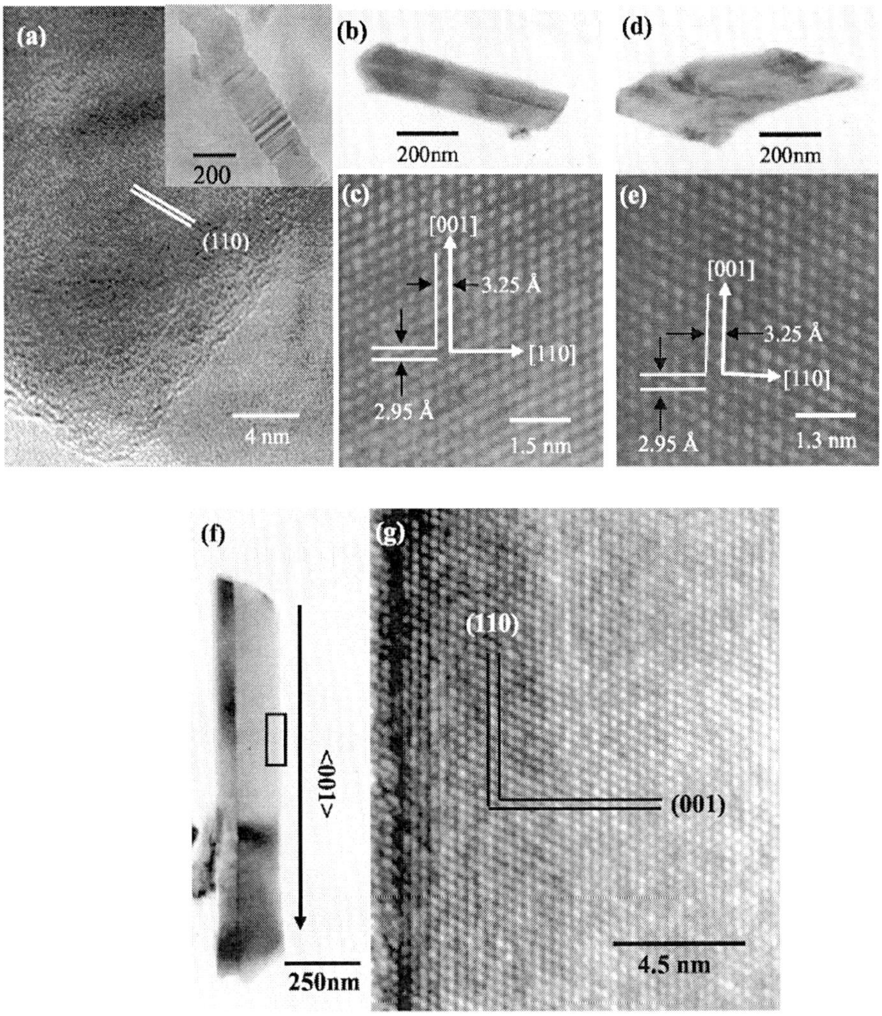

Figure 3. TEM and HRTEM images: (a) the chain-like nanofibers of S4 fabricated using diluted oxygen gas as the oxygen source; (b) and (c) a sample nanorod; (d) and (e) a nanoplate of S7 synthesized using acetone as the oxygen source; (f) and (g) an individual rod of S8 formed using dibutyltin dilaurate as the oxygen source at 850 °C for 1.5 hours. The inset to Figure 1a is a chain-like nanofiber of S4 fabricated using diluted oxygen gas as the oxygen source. (Figure 3b, 3c, 3d and 3e reprinted with permission from [90]).

In addition, the peaks indexed to the Ti substrate appear in the XRD spectra of the samples S4-S8; but they are invisible in the spectra of the samples S1-S3. This indicates that the thickness of the TiO_2 films formed on the Ti substrate is greater when oxygen gas, formic acid and water vapor were used as the oxygen source versus those formed using diluted oxygen, ethanol, acetaldehyde, acetone and DBTDL as the oxygen source.

The detailed crystal structure, morphology and composition of S4, S7 and S8 were characterized using TEM and HRTEM. Figure 3a presents HRTEM images of the chain-like nanofibers (as shown in the inset to Figure 3a) of S4 fabricated using the diluted oxygen gas as the oxygen source. This nanofiber is crystalline with an inter-planar spacing of 0.325 nm, in agreement with the *d* value of the (110) planes of the rutile TiO_2 crystal. Figure 3b and 3d present low magnification TEM images of S7 prepared using acetone as the oxygen source. The corresponding HRTEM images are shown in Figure 3c and 3e. The planes with the lattice space of 0.325 nm and 0.295 nm (shown in Figure 3c and 3e) are assigned to the (110) and (001) planes of rutile TiO_2 crystal, respectively. These results show that the TiO_2 nanorod fabricated with acetone as the oxygen source is single crystalline with growth along the <001> direction. Figure 3f presents a TEM image of an individual rod of S8 (using DBTDL as the oxygen source) with a rectangle cross-section similar to the SEM results (Figure 1h). Figure 3g, the corresponding HRTEM image recorded from the rectangle zone, reveals that the rod is single crystalline with inter-planar spacings of 0.325 nm (110) plane and 0.295 nm (00 1) plane of the rutile TiO_2. These results indicate that these nanorods also grow along the <001> direction.

3.1.2. Temperature Effect

We further studied the temperature effect on the formation of the TiO_2 nanostructures. The oxidation behaviour of Ti at different temperatures using acetone as the oxygen source was investigated. Figure 4 shows the representative SEM images of the samples synthesized at (a) 650 °C, (b) 700 °C, (c) 750 °C, (d) 800 °C and (e) 850 °C for 1.5 hours using acetone as the oxygen source. Some small nanodots were formed at 650 °C as seen in Figure 4a. Increasing the temperature results in a significant increase in the dimension and the coverage of the formed nanostructures. Sparse and short aggregates composed of several nanoplates were formed at 700 °C (Figure 4b); dense and long aggregates were produced at 750 °C (Figure 4c). Further increasing the temperature to 800 °C, continuous, large and plate-like nanorod arrays were formed on the Ti substrate (Figure 4d). At 850 °C, continuous, long, column-like and plate-like nanorods were formed as seen in Figure 4e. These SEM images show that the reaction temperature significantly affects the morphology of the formed titanium oxide nanostructures. Figure 4f presents the corresponding XRD spectra of the five samples (Figures 4a to 4e). All the peaks (110), (101), (200), (111), (210), (211), (220) are indexed to those of the tetragonal rutile TiO_2 phase (JCPDS file No. 2 1-1276). Thus, all these formed titanium oxide films have the same phase. The peaks marked with a star are derived from the Ti substrate. The significant increase of the intensity of the TiO_2 peaks and the obvious decrease of the intensity of the Ti peaks from 650 °C to 850 °C indicates that the yield amount/thickness of the titanium oxide films on the Ti substrate increases with the increase of the reaction temperature. These XRD results presented in Figure 4f are consistent with the SEM observations shown in Figures 4a to 4e.

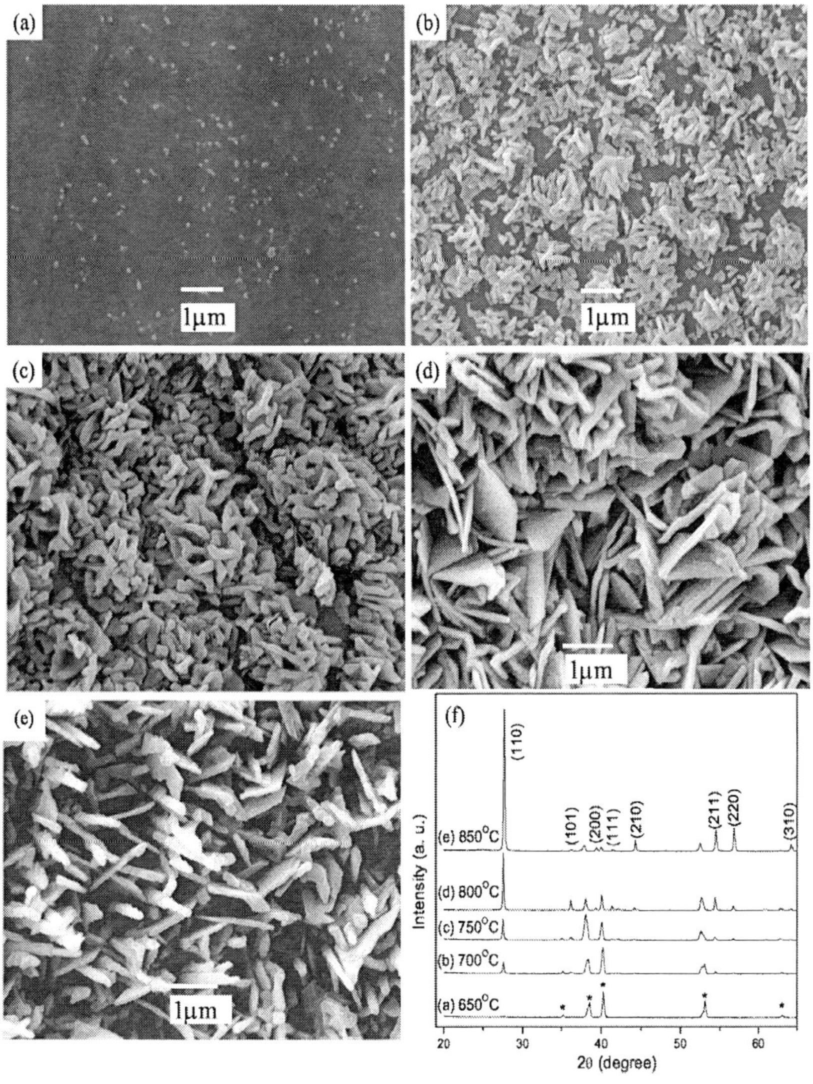

Figure 4. SEM images of Ti plates oxidized in acetone at (a) 650, (b) 700, (c) 750, (d) 800, (e) 850 °C for 1.5 hours. (f) The corresponding XRD results of the samples as shown in a to e.

Figure 5 presents the Raman spectra of the samples synthesized at (a) 650 °C, (b) 700 °C, (c) 750 °C, (d) 800 °C and (e) 850 °C for 1.5 hours using acetone as the oxygen source. The bands at 446.5 and 607.9 cm^{-1} are indexed to the Raman active modes E_g and A_{1g} of rutile TiO_2, respectively [95]. There is no observable TiO_2 Raman peak shown on the spectrum of Figure 5a, where the temperature is 650 °C (the lowest one among the five investigated temperatures) and only sparse TiO_2 nanodots were formed as seen in Figure 4a. By increasing the temperature from 700 °C (Figure 5b) to 850 °C (Figure 5e), the intensity of the Raman bands centered at 446.5 and 607.9 cm^{-1} dramatically increases, indicating that the yielded TiO_2 significantly increases with the increasing of the reaction temperature. This is consistent with the XRD studies shown in Figure 4f.

3.1.3. Growth Mechanism

Ti can easily react with oxygen to form a TiO_2 film. But for the growth of titanium oxide, the question is where the oxidation occurs, at the Ti and TiO_2 interface, on the TiO_2 surface or inside the TiO_2 film. If oxidation occurs at the Ti and TiO_2 interface, then the oxygen species needs to move to the interface through the formed TiO_2 film.

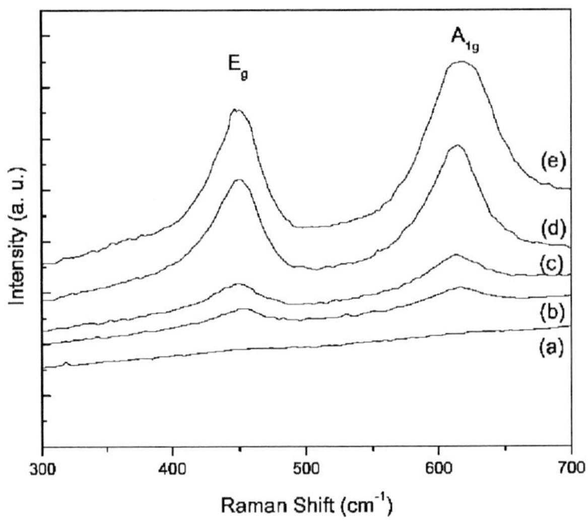

Figure 5. Raman Spectra of the samples prepared using acetone as the oxygen source at (a) 650, (b) 700, (c) 750, (d) 800, (e) 850°C for 1.5 hours. (Reprinted with permission from [91]).

On the other hand, if the oxidation occurs on the formed TiO_2 surface, it requires the Ti species to move to the oxide surface. The third alternative is that both the Ti and oxygen species are involved in the diffusion processes and encounter each other inside the oxide film. It is expected that the mechanism which dominates the growth process of TiO_2 depends on the applied temperature, the concentration of oxygen and the thickness of the TiO_2 film, and would affect the structure of the TiO_2 film. The oxidation of titanium in pure oxygen and air has been widely studied. The structure of the oxide layer (pin-shaped, lamellar and acicular crystals [96], pyramid crystals [97], nanoparticle thin films [98]) generally depends on the duration and temperature of the oxidation process. Several possible mechanisms for the oxidation of titanium on metallic surfaces have been proposed [96-98]. Badescu and Momirlan report that there is a predominant diffusion of oxygen through the grain boundaries of the oxide layer and the oxidation site of metal-oxide at temperatures in the range of 800-1000 °C; while at temperatures above 1000 °C, a considerable increase in the diffusion of Ti cations through the oxide layer is observed [96]. On the other hand, Czerwinski and Szpunar examined the surface topography of oxide layers on titanium using atomic force microscopy and X-ray diffraction and show that TiO_2 growth at 900 °C is controlled by the surface diffusion of Ti cations to the ledges located on the sidewalls of pyramidal grains [97]. Our experimental results presented show that the morphology of the synthesized TiO_2 nanostructures strongly depends on the oxygen source and reaction temperature, providing new insight into the growth mechanisms of TiO_2 nanostructures. Since the melting point of Ti and rutile TiO_2 is very high, at 1660 and 1840 °C, respectively, the vapour pressure of Ti and rutile TiO_2 at our experimental temperatures should be very low. Thus, the growth of rutile

TiO_2 nanorods is unlikely via either vapor-liquid-solid (VLS) [99] or vapour-solid (VS) mechanisms [100]. The VLS mechanism is also rejected due to the fact that no spherical particles were found at the tips of the nanorods; furthermore, the VS mechanism is also ruled out because all of the nanorods we observed grew directly from the Ti substrate and we did not find any traces of rutile TiO_2 anywhere else in the furnace tube. Titanium can easily react with oxygen to form a TiO_2 film. For the different oxygen sources, the initial formation of TiO_2 film is expected to be very similar (oxygen species reacting with Ti to form TiO_2). Then titanium oxide may grow at the oxide/gas interface, inside of the oxide film, and/or at the oxide/metal interface. It is known that grain boundaries are the most favorable paths for Ti and oxygen species diffusion [101]. It is expected that the structure of the formed titanium oxide depends on: (i) the diffusion rate of active oxygen species from the oxide/gas interface to the oxide/metal interface through the grain boundaries; (ii) the diffusion rate of the Ti species from the oxide/metal interface to the oxide/gas interface through the grain boundaries; (iii) the diffusion rate of the active oxygen species and Ti species on a specific crystal plane; and (iv) the amount of available active oxygen species.

Under the same temperature, the diffusion rate of the Ti species can be assumed to be the same. The oxygen source thus plays a key role in TiO_2 growth. Formic acid can be easily decomposed. When oxygen or formic acid is used as the oxygen source, a large amount of active oxygen species is available. The diffusion of the active oxygen species dominates the growth process and the oxidation occurs mainly at the oxide/metal interface, resulting in a thick polycrystalline TiO_2 film as shown in Figure 1a and 1b. In contrast, when ethanol, acetaldehyde, acetone or DBTDL is used as the oxygen source, the produced active oxygen species is expected to be much less; the diffusion of the Ti species to the oxide/gas interface via the network of the grain boundaries of the thin oxide layer predominates. Then the diffused Ti species react with the selectively adsorbed oxygen species on different planes of rutile TiO_2 [102]. This is supported by the formation of aligned nanorod arrays along the <001> direction on the substrate. The available active oxygen species is expected to be less at the bottom than that at the top of the nanorods; thus, smaller and longer rods are formed. The observation that TiO_2 growth is predominated by Ti diffusion is further supported by the effect of reaction time on the formed TiO_2 nanostructures [89]. When acetone was used as the oxygen source at 850 °C, short and oriented dots emerged with the reaction time of 10 minutes; dense and short nanorod arrays were formed with the reaction time of 30 minutes and long nanorod arrays were produced when the reaction time was increased to 60 minutes. Moreover, the dramatic difference of the formed TiO_2 nanostructures at different temperatures indicates that the diffusion rate of the Ti species is the determining factor for the growth of TiO_2 nanostructures. As shown in Figure 4, the different morphologies are due to the different diffusion rates of Ti atoms from the substrate to the oxide surfaces. At the low temperature of 650 °C very few nanodots are formed, because both the Ti diffusion rate and oxidation rate are very low. At 700 and 750 °C the surface oxidation mainly produces rutile nanoplates since the increased diffusion rate results in a faster growth rate. When the temperature is increased to 800 °C, the increased diffusion rate of Ti facilitates the TiO_2 nanoplate growth such that the aligned nanoplates uniformly cover the whole substrate surface. When the temperature is further increased to 850 °C, the TiO_2 nanorod arrays are formed; moreover, the Ti diffusion rate is high enough to move Ti species from the Ti substrate to the nanorod tips and to sustain the nanorod growth. On the other hand, when water and diluted oxygen gas is used as the oxygen source, TiO_2 polycrystalline and micro/nano-fibres are formed as shown Figure 1 c and 1 d, respectively; indicating that both active oxygen species diffusion and Ti species diffusion are heavily involved in the course of the TiO_2 growth. The formation of the fiber-like structures is due to the difference in diffusion rate of Ti atoms along the different crystal directions of TiO_2 [7].

3.1.4. Wetting Properties

The surface wettability conversion of TiO_2 has been widely investigated based on the change of the water contact angle before and after UV irradiation using single crystal [103], polycrystalline films [102], and nanocrystal films [57]. However, very little work has been done on how the surface micro-structures of TiO_2 films affect their wetting properties. The insets to Figure 1 show that the wetting abilities of these oxide films strongly depend on their morphology. The TiO_2 surfaces formed using oxygen gas, formic acid, water and diluted oxygen as the oxygen source are hydrophilic; and their water contact angles increase from 10° (Figure 1 a), to 30° (Figure 1b), to 40° (Figure 1 c), and to 60° (Figure 1 d). They are all smaller than the water contact angle of a smooth rutile TiO_2 surface, 70° [103]. In contrast, the TiO_2 films formed using acetaldehyde, ethanol, acetone and DBTDL as the oxygen source are hydrophobic as shown in the insets to Figure 1e, 1f, 1g and 1h, respectively. The water contact angles of the TiO_2 films produced using ethanol and acetone as the oxygen source are very similar (~120°) (Figure 1 e and Figure 1 g), which is slightly smaller than the TiO_2 film formed using DBTDL (~125°) as the oxygen source (Figure 1h), but slightly larger than the TiO_2 film formed using acetaldehyde as the oxygen source (~108°) (Figure 1f). It is known that roughness has a significant influence on the wetting behavior of a given surface [104-106]. The dramatic difference of the wetting properties of the formed TiO_2 films using different oxygen sources can be explained by two different models: Wenzel's Equation [107] and Cassie and Baxter's Law [108]. In Wenzel's Equation, the apparent water contact angle θ^* is given by:

$$\cos\theta^* = r\cos\theta \tag{1}$$

Where r is the roughness factor (the ratio of total surface area to the projected area in the horizontal plane), and θ is the contact angle measured on a flat surface. Wenzel's Equation assumes that liquid fills up the rough surface forming a fully wetted contact and that there is no air entrained in the pits of the rough surface below the liquid. As seen in Figure 1(a), the formed TiO_2 polycrystalline grains are big and dense, and the height of the hollows presented on the surface is very low. Thus, the air trapping ability of the formed TiO_2 surface, using pure oxygen gas as the oxygen source, is very low. The very small water contact angle (~10°) is consistent with Wenzel's Equation. When this θ^* is substituted into Wenzel's Equation with the θ value 70° for a smooth TiO_2 surface, the calculated r is 2.88, which is in agreement with the surface roughness r as estimated by our SEM observations. However, Wenzel's Equation is invalid for the samples shown in Figures 1e – 1h, where the surfaces are hydrophobic with the apparent angle θ^* over 100°. The high hydrophobicity of these surfaces originates from the air trapped in the interspace of the nanorod arrays similar to that reported for Gold bead arrays [105], Carbon nanotube arrays [109] and SnO_2 nanoflower arrays [110]. This is consistent with Cassie and Baxter's Law [108]:

$$\cos \theta^* = f(\cos\theta + 1) - 1 \tag{2}$$

Where f is the ratio of the liquid-solid contact area to the projected surface area. Cassie and Baxter's Law describes the apparent contact angle θ^* at a heterogeneous surface as being composed of two different materials. The surfaces shown in Figure 1e – 1h are composed of two phases: solid (TiO_2) and gas phase (trapped air). When the θ^* is substituted into Cassie and Baxter's Law with $\theta = 70°$ for a smooth TiO_2 surface, the calculated f is 0.51 for the surface shown in Figure 1e and 0.37 for the surfaces shown in Figures 1f and 1g, and 0.27 for the surface shown in Figure 1h. This is consistent with their SEM images. The slightly higher θ^* for the surfaces shown in Figures 1b, 1c and 1d compared to the surface shown in Figure 1a results from partial air trapping and is similar to the transition from Wenzel's model to Cassie and Baxter's air trapping model [109]. All these results show that the change of the surface structure from polycrystalline grains to well-aligned nanorod arrays switches the dominant wetting mode from Wenzel's to Cassie and Baxter's and results in the significant change of the wetting behavior from hydrophility to hydrophobicity.

3.2. TiO_2 Nanostructures Synthesized by Hydrothermal Method

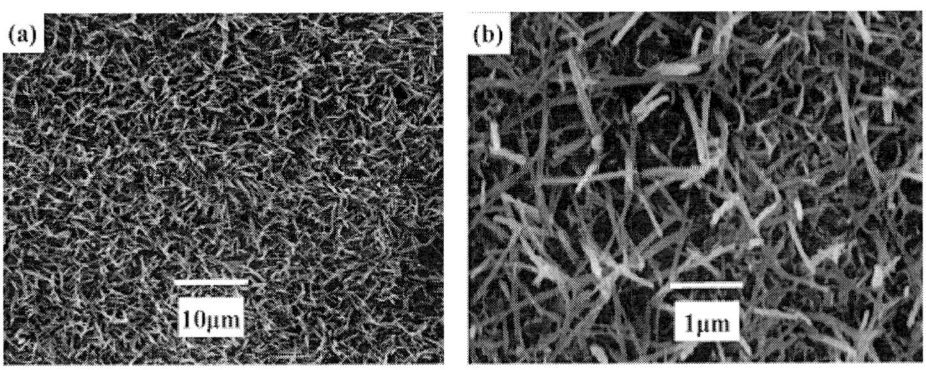

Figure 6. (a) low and (b) high magnification SEM images of the Ti surface after hydrothermal treatment in 2.5 M NaOH at 150 °C for 15 hours.

3.2.1. Preparation of Na2Ti6O13 Nanowire Arrays

Figure 6 shows the SEM images of the sample synthesized in a 2.5 M NaOH solution under hydrothermal conditions at 150 °C for 15 hours at (a) a low magnification and (b) a high magnification. Nanowire arrays are uniformly formed in a large-scale on the titanium substrate. These nanowires have diameters of tens of nanometers, and length of about 2 micrometers. The crystal phase of these nanowire arrays is characterized using XRD. A representative XRD pattern is shown in Figure 7. All the diffraction peaks can be indexed to $Na_2Ti_6O_{13}$ with monoclinic lattice (JCPDS 37- 951, a = 15.12, b = 3.738 and c = 9.16, $\beta = 99.3^o$), demonstrating that the as-synthesized nanowires have the structure of $Na_2Ti_6O_{13}$. The general reaction is expressed as Equation 3:

$$6\,Ti + 2\,NaOH + 11\,H_2O \rightarrow Na_2Ti_6O_{13} + 12\,H_2 \tag{3}$$

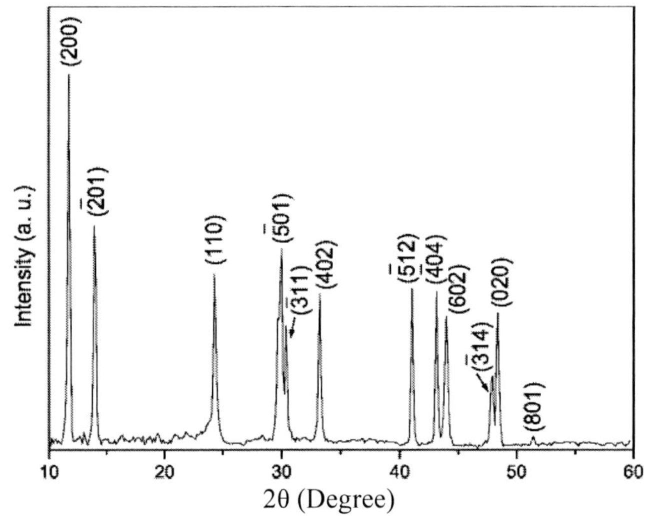

Figure 7. A representative XRD pattern of the as-synthesized $Na_2Ti_6O_{13}$ nanowire arrays prepared in a 2.5 M NaOH solution at 150 °C for 15 hours.

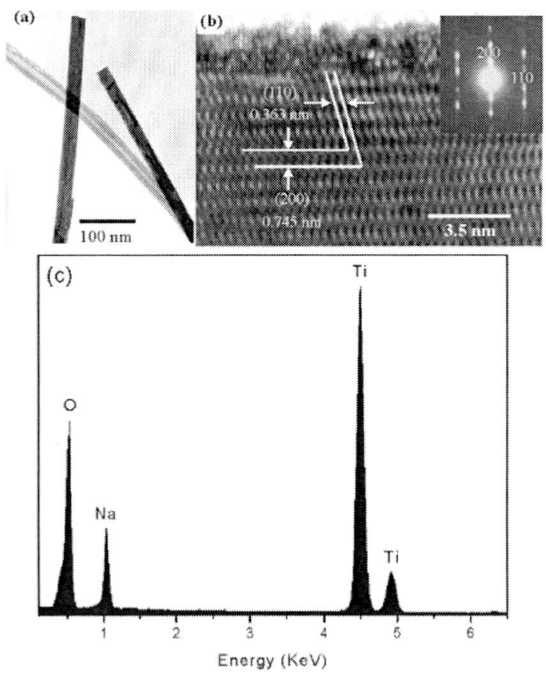

Figure 8. (a) TEM and (b) HRTEM and SAED (the inset) of synthesized $Na_2Ti_6O_{13}$ nanowires synthesized in 2.5 M NaOH at 150 °C for 15 hours. (c) The corresponding EDS of one of these $Na_2Ti_6O_{13}$ nanowires.

The crystal structure and the composition of these nanowires were further characterized using TEM, HRTEM, SAED and EDS. Figure 8a shows the TEM image of wire-like nanostructures with diameters of about 30 nm. Figure 8b, the HRTEM image recorded perpendicularly to the nanowire shows that the nanowires have a layered lattice structure with inter-planar spacing of 0.363 and 0.745 nm, in excellent agreement with the *d* values of the

(110) and (200) planes of the $Na_2Ti_6O_{13}$ crystal. The corresponding SAED pattern presented in the inset of Figure 8b shows the monoclinic *hk0* spots of $Na_2Ti_6O_{13}$. All these results confirm that the formed $Na_2Ti_6O_{13}$ nanowires are single crystals and grew along the <110> direction. Figure 8c presents the corresponding EDS spectrum, showing that the nanowire is composed of 9.61% Na, 29.12% Ti and 61.2 % O. This is very close to the chemical ratios of the elements in $Na_2Ti_6O_{13}$ and further supports our XRD results described above.

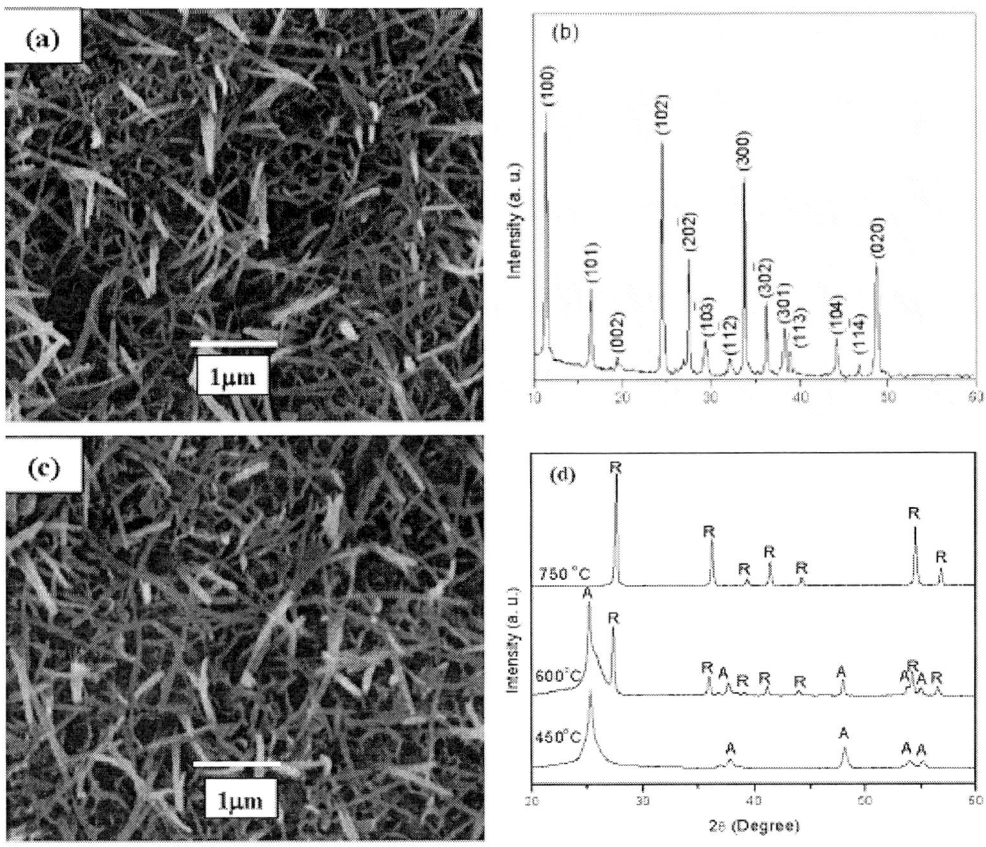

Figure 9. (a) SEM image and (b) XRD pattern of the $H_2Ti_3O_7$ nanowire arrays synthesized from $Na_2Ti_6O_{13}$ nanowires arrays through ion exchanging in 1 M HNO3 solution for 3 days. (c) SEM images of the sample synthesized by annealing $H_2Ti_3O_7$ nanowire arrays at 750 °C for 5 hours. (d) the XRD patterns recorded from the three samples synthesized by annealing $H_2Ti_3O_7$ nanowire arrays at 450 °C, 600 °C and 750 °C.

To protonate the $Na_2Ti_6O_{13}$ nanostuctured materials, we simply immersed the synthesized monoclinic $Na_2Ti_6O_{13}$ samples in 1 M HNO_3 solution for three days. Figure 9a shows a typical SEM image of the $Na_2Ti_6O_{13}$ sample after the acid treatment. The morphology of the nanowire arrays is maintained without any obvious change before and after the acid treatment (as compared to Figure 6b). The XRD pattern of this sample is shown in Figure 9b. All the peaks can be indexed to the monoclinic $H_2Ti_3O_7$ phase (JCPDS 3 6-654) with lattice constants a = 8.08, b = 3.725, c = 9.17 and β = 101.27. EDS analysis shows that there is no Na element present in the sample after the acid treatment and that the composition of the synthesized nanowires is composed of 30.25% Ti and 69.75% O, which is close to the ratio of the Ti and

O in $H_2Ti_3O_7$. The absence of a Na peak in the EDS spectra further demonstrates that the $Na_2Ti_6O_{13}$ is completely converted into $H_2Ti_3O_7$ during the ion-exchange process. These results indicate that the $Na_2Ti_6O_{13}$ nanowires are successfully converted into $H_2Ti_3O_7$ nanowires with morphology-preservation through an ion-change process, where the Na^+ ions in $Na_2Ti_6O_{13}$ are replaced by H^+ or H_3O^+. The general reaction is expressed as Equation 4:

$$Na_2Ti_6O_{13} + 2\ HNO_3 + H_2O \rightarrow 2\ H_2Ti_3O_7 + 2\ NaNO_3 \tag{4}$$

3.2.2 Synthesis of Anatase and Rutile Nanowire Arrays

Figure 9c shows a typical SEM image of the synthesized $H_2Ti_3O_7$ nanowires after being annealed for 5 hours at 750 °C. It is interesting to note that the morphology of the sample does not change; it is still very similar to the $Na_2Ti_6O_{13}$ and $H_2Ti_3O_7$ nanowires. The XRD patterns recorded from the three samples synthesized by annealing $H_2Ti_3O_7$ nanowire arrays at 450 °C, 600 °C and 750 °C are presented in Figure 9d. At 450 °C, all the peaks shown in the XRD spectrum can be indexed to anatase (JCPDS 2 1-1272), demonstrating that the $H_2Ti_3O_7$ nanowires are completely converted into anatase nanowires. At 750 °C, all the XRD peaks can be indexed to rutile (JCPDS 4-673), indicating that the synthesized $H_2Ti_3O_7$ nanowire arrays are completely converted into rutile nanowire arrays. In addition, the XRD spectrum shows that a mixture of anatase and rutile nanowires are formed from the $H_2Ti_3O_7$ nanowires after being annealed at 600 °C. The general reaction is expressed as Equation 5:

$$H_2Ti_3O_7 \rightarrow 3\ TiO_2 + H_2O \tag{5}$$

We did further characterization on the formed nanowires. Figure 10 presents the typical TEM, HRTEM and SAED results of the samples (a)-(b) as-synthesized $H_2Ti_3O_7$, after being further annealed at 450 °C (c and d) and 750 °C (e and f) for 5 hours. The size of the $H_2Ti_3O_7$ nanowires as seen in Figure 10a is similar to that of the $Na_2Ti_6O_{13}$ nanowires shown in Figure 8a, with diameters of about 30 nm. In contrast, HRTEM analysis (Figure 10b) shows that the microstructure of the $H_2Ti_3O_7$ nanowires seen in Figure 10a is different from that of the $Na_2Ti_6O_{13}$ nanowires observed in Figure 8a. The lattice structure is parallel to the long axis of the $H_2Ti_3O_7$ nanowire. The distances of 2.25 and 1.88 between the adjacent lattice fringes can be assigned to the interplaner distances of monoclinic $H_2Ti_3O_7$ (004) and (020), respectively. The inset of Figure 10b shows the corresponding SAED pattern recorded from the same $H_2Ti_3O_7$ nanowire, indicating that the $H_2Ti_3O_7$ nanowire is a single-crystalline. As seen in Figure 10c, the diameter of the anatase nanowires is similar to that of $H_2Ti_3O_7$ nanowires in the range of several tens of nanometers. The HRTEM image shown in Figure 10d and the SAED pattern (the inset of Figure 10d) confirm that these nanowires are single-crystalline anatase with the growth along the <101> direction. Figure 10e shows that the formed rutile nanomaterials have an analogous nanowire structure to that of the anatase samples. The lattice image in Figure 10f recorded from one of the nanowires in Figure 10e has lattice fringes with inter planner distances of 3.25 and 2.51 Å, which is in excellent accordance with the *d* values of the rutile (110) and (101) planes. This is further supported by the SAED pattern as seen in the inset of Figure 10f. In addition, EDS analysis shows that the sample annealed at 450 °C is almost identical to that annealed at 750 °C. A quantitative analysis shows that all of the annealed nanowires are composed of Ti and O, with a ratio very close to 1:2. All of these results indicate that, through this facile annealing treatment, different phase titanium oxide based nanomaterials are synthesized and that the $Na_2Ti_6O_{13}$ nanostructures directly grown on a Ti surface under hydrothermal conditions can serve as a precursor to synthesize various desirable TiO_2 nanomaterials for photocatalytic applications. We further studied the effect of NaOH concentration, temperature and time on the morphology of these products synthesized under hydrothermal conditions.

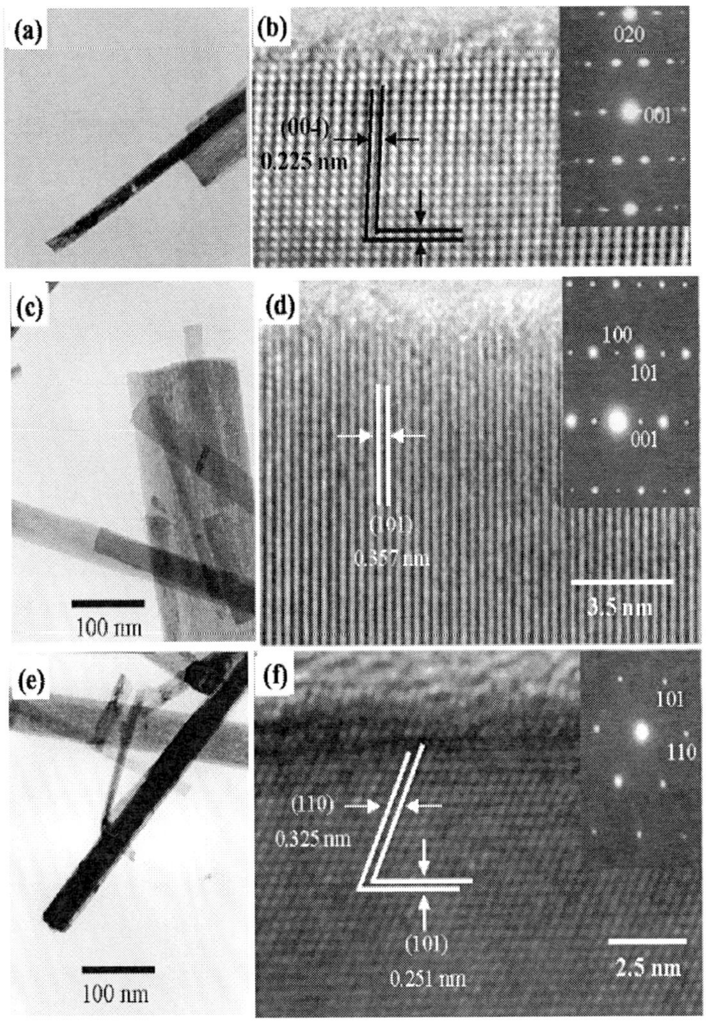

Figure 10. (a) TEM and (b) HRTEM and SAED (the inset) of $H_2Ti_3O_7$ nanowires; (c) TEM and (d) HRTEM and SAED (the inset) of the sample synthesized by annealing $H_2Ti_3O_7$ nanowires arrays at 450 °C for 5 hours; (e) TEM and (f) HRTEM and SAED (the inset) of the sample synthesized by annealing the $H_2Ti_3O_7$ nanowire arrays at 750 °C for 5 hours.

3.2.3. NaOH Concentration Effect

Figure 11 shows the SEM images of the surfaces of the Ti substrates after hydrothermal treatment at 150 °C for 15 hours in (a) 0.0, (b) 0.01, (c) 1.25, (d) 2.5, (e) 7.5, and (f) 10.0 M NaOH solutions. The Ti surface does not change as shown in Figure 11a in the absence of NaOH. At a low concentration of NaOH 0.01 M, particles and nanodots were formed on the Ti substrate as seen in Figure 11b. Increasing the NaOH concentration to 1.25 M, well-aligned dense and short nanorod arrays were formed (Figure 11c). Compared with the nanorod arrays formed in 1.25 M NaOH, longer and thinner nanowire arrays were produced in 2.5 M NaOH as shown in Figure 11d. Further increasing the NaOH concentration to 7.5 M, continuous nanowire network films were formed on the Ti surface as observed in Figure 11e. When a 10.0 M NaOH solution was used, the films formed on the Ti surface were composed of long and large nanowires as shown in Figure 11f. All of these results show that the morphology of the synthesized nanodots, nanorod arrays, nanowire arrays and continuous nanowire network films, are strongly affected by the concentration of NaOH solution.

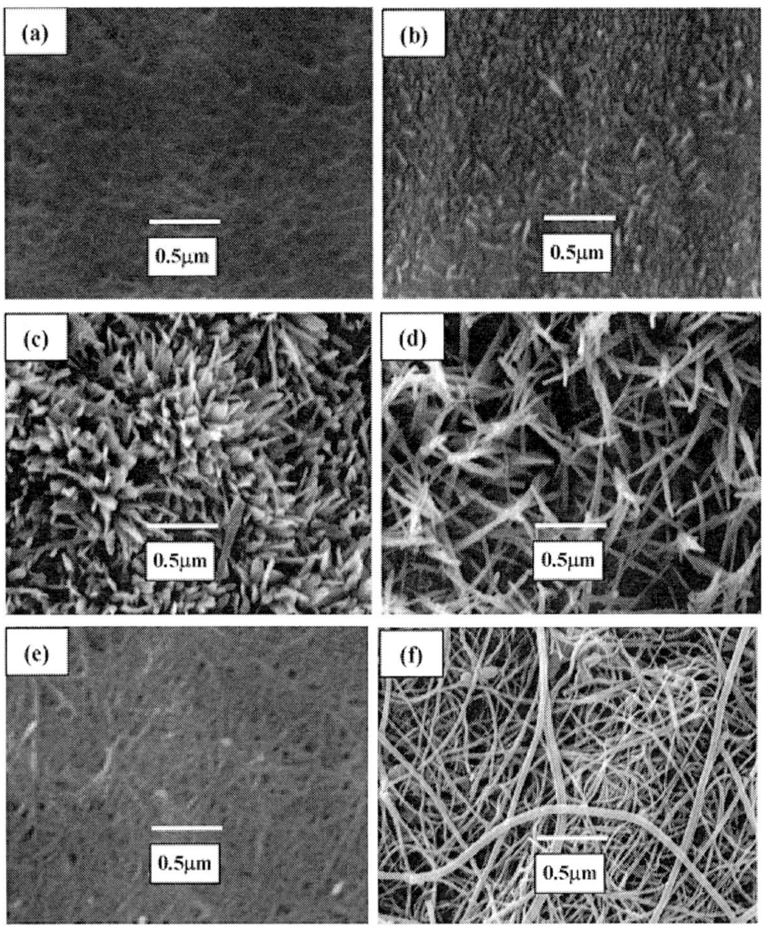

Figure 11. SEM images of the surface structures of the samples synthesized hydrothermally from Ti in (a) 0.0, (b) 0.01, (c) 1.25, (d) 2.5, (c) 7.5 and (f) 10.0 M NaOH solution at 150 °C for 15 hours.

3.2.4. Effect of Temperature and Reaction Time

Figure 12 shows the SEM images of the Ti substrates after the hydrothermal treatment in a 2.5 M NaOH solution at temperatures varying from 125 to 200 °C for 3 hours. At 125 °C, dense nanograss was formed as shown in Figure 12a and 12b. Increasing the temperature to 175 °C (Figures 12c and 12d), some flower-like structures were formed on top of the nanowire films. When the temperature was further increased to 200 °C, Figures 12e and 12f show that many nanospheres composed of irradiated short nanorods were grafted onto the surface of the continuous nanowire network films. These results indicate that the reaction temperature dramatically influences the structure of the hydrothermal products. We further investigated the influence of the reaction time in a 2.5 M NaOH solution at 150 °C. A mixture of nanoplates and a few nanowires were formed on the surface of the Ti substrate when the reaction time was 0.5 hours. Increasing the time to 1.5 hours, many nanowires accompanied with a few small nanoplates covered the Ti substrate. Large-scale nanowires were formed when the reaction time extended to 15 hours. When the time was further increased to 30 hours and 48 hours, continuous nanowire network films were formed. These results show that the reaction time is another critical parameter that affects the morphology of the formed nanomaterials.

3.3. Applications of Tio2 Nanomaterials

The photocatalytic and photoelectrochemical properties of semiconducting oxide materials provide a strong incentive for ongoing scientific research into Green Chemistry applications [17, 22, 111, 112]. TiO_2 belongs to one of the most extensively used and investigated systems and has the advantages of being inexpensive, chemically stable, and non-toxic. The photoactivity of TiO_2 is one of its technologically most attractive properties. Intense research was initiated by Fujishima and Honda's discovery that water can be photocatalytically split into hydrogen and oxygen [8]. The band gap of TiO_2 is relatively wide and its absorption properties are not well matched to the frequency spectrum of natural sun light. The functionalization of the TiO_2 surface with dye molecules has made TiO_2-based solar cells much more efficient for the conversion of solar energy into electric energy. Another special interest of the employment of TiO_2 is photo-degradation of organic pollutants. Nanostructured materials possess a large surface area and are promising in photocatalytic applications. Figure 13a shows the diffuse reflectance spectra of the nanowires, curve 1: anatase TiO_2 resulting from annealing $H_2Ti_3O_7$ nanowires at 450 °C, curve 2: a mixture of anatase and rutile TiO_2 synthesized by annealing $H_2Ti_3O_7$ nanowires at 600 °C, and curve 3: rutile TiO_2 fabricated by annealing $H_2Ti_3O_7$ nanowires at 750 °C. The reflection spectra show that the band gap of these nanowires changes in the following order:

$$\text{anatase} > \text{mixture of anatase and rutile} \approx \text{rutile}$$

These results show that different materials with different optical properties can be formed through this simple annealing treatment. The decomposition of Eosin Y, an organic dye, was employed as a probe to study the photocatalytical activity of these nanomaterials. To immobilize Eosin Y on the nanowire arrays, all of the samples were immersed in an ethanol solution containing Eosin Y (9 mM) for 30 minutes. Then, the samples loaded with dye were irradiated by an UV-Xe lamp (365 nm and 0.2 mW/cm^2). After irradiation with UV light for selected periods of time, diffuse reflectance spectra were recorded to examine the changes in

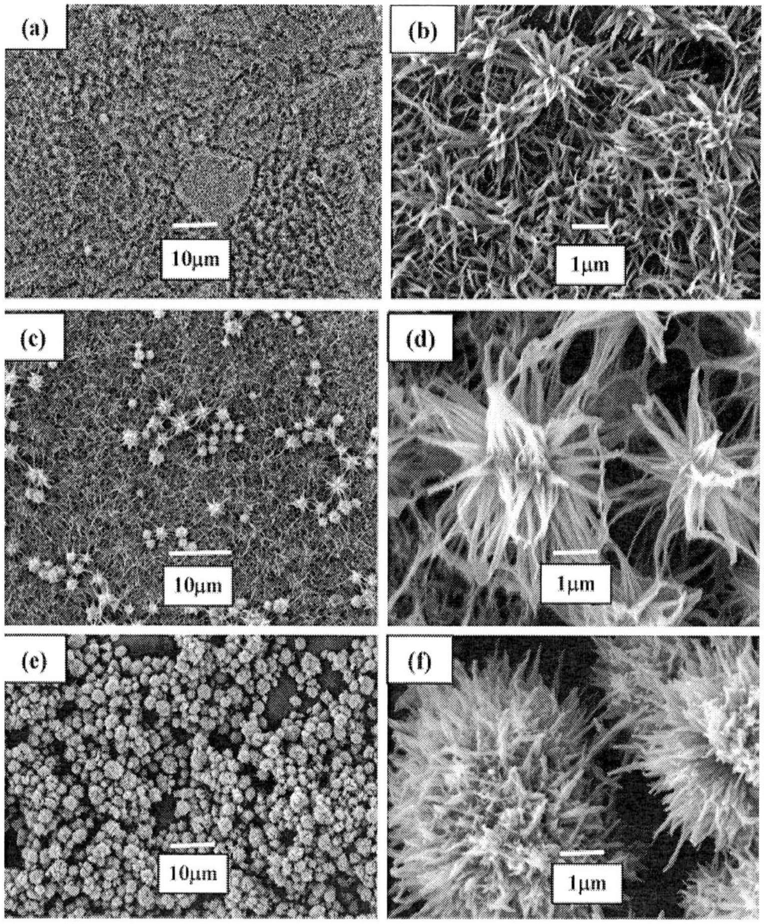

Figure 12. SEM images of the samples synthesized hydrothermally in 2.5 M NaOH solution at (a) – (b) 125 °C; (c) - (d) 175 °C; and (e) - (f) 200 °C for 3 hours.

absorption of Eosin Y. Figures 13b, 13c and 13d show the performances of the anatase nanowires, mixture of anatase and rutile nanowires, and rutile TiO_2 nanowiress for the photocatalytic decomposition of Eosin Y, respectively, after being irradiated for selected periods of time. Before the UV irradiation, as shown in curves 0 h, all the three samples show strong absorption in the wavelength ranging from 420 to 580 nm due to the immobilized Eosin Y. Upon increasing the irradiation time, the absorbance decreases monotonically, indicating that the Eosin Y adsorbed on the nanowires gradually decomposed; however, with different degrading rates. Comparison of all the diffuse reflectance spectra shows that the anatase TiO_2 nanomaterials possess the highest activity for Eosin Y photo-degradation (Figure 13b). After being annealed at higher temperatures, the photocatalytic activity of the nanostructured materials decreases from anatase (Figure 13b), to a mixture of anatase and rutile (Figure 13c), and to rutile TiO_2 nanowires (Figure 13d). This is in good accordance with current literature reports [112-114].

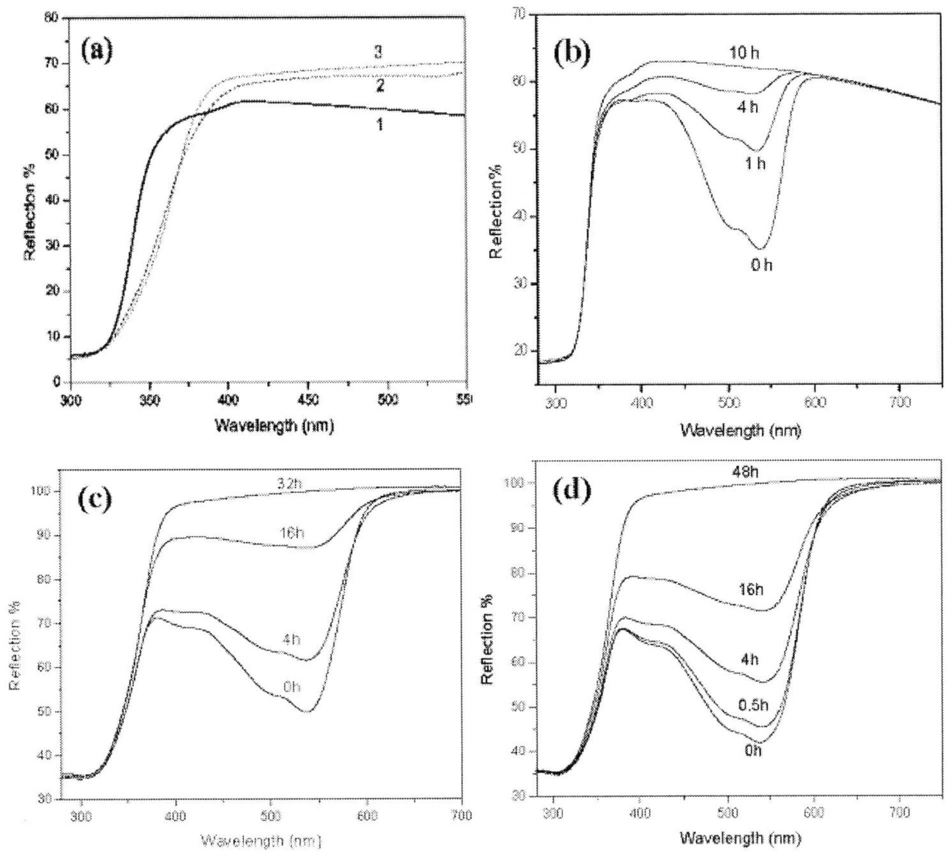

Figure 13. (a) Diffuse reflectance spectra of the samples: (1) anatase synthesized by annealing $H_2Ti_3O_7$ nanowires at 450 °C , (2) mixture of anatase and rutile fabricated by annealing $H_2Ti_3O_7$ at 600 °C, and (3) rutile synthesized by annealing $H_2Ti_3O_7$ at 750 °C; Diffuse reflectance spectra of the Eosin Y loaded samples: (b) anatase, (c) composite of anatase and rutile, and (d) rutile nanowires after irradiation with UV light (365 nm) for selected periods of time.

In addition, electrochemical anodization of titanium sheet has been widely investigated in synthesizing TiO_2 nanotubes [18, 115-121]. Solutions containing floride ions are used as the electrolytes since the dissolution of TiO_2 as soluble $[TiF_6]^{2-}$ is required for pore formation. For instance, Figure 14a shows the cross-sectional and top view of the ordered porous TiO_2 layers fabricated by Schmuki and co-workers using potentiostatic anodic etching of polycrystalline titanium [118]. The Ti sample was anodized at 20 V for 2 hours in 1M $(NH_4)_2SO_4$ + 0.5 wt% NH_4F after a potential sweep from open-circuit potential to 20 V at a scan rate of 0.1 V/s. The thickness of the porous layer is in the order of 2.5 μm with an average pore size of 100 nm. The morphology of the formed porous TiO_2 layer depends on the applied voltage and the electrolyte used. Nanotube arrays were synthesized by Bard et al. [121] by anodization of a Ti substrate in 1M $(NH_4)H_2PO_4$ + 0.5 wt% NH_4F at 15 V with the porous layer thickness in the order of 3 μm. The pore diameter was about 70 nm, and the wall thickness was 20 nm. In contrast, anodizing Ti substrate at 20 V in 1M $(NH_4)H_2PO_4$ + 0.5 wt% NH_4F only produced a porous TiO_2 film. Bard and co-workers further compared the photo-activity of the TiO_2 nanotube arrays, the porous TiO_2 film and one commercial P-25

TiO_2 photoanode under Xe lamp irradiation (~100 mW/cm^2) as shown in Figure 14b. The TiO_2 nanotube arrays produced much higher photocurrent than the nanoporous TiO_2 film and the 15-μm-thick P-25 TiO_2 film. This suggests that TiO_2 nanotube arrays can harvest solar light more effectively than the standard P-25 TiO2 photoanode.

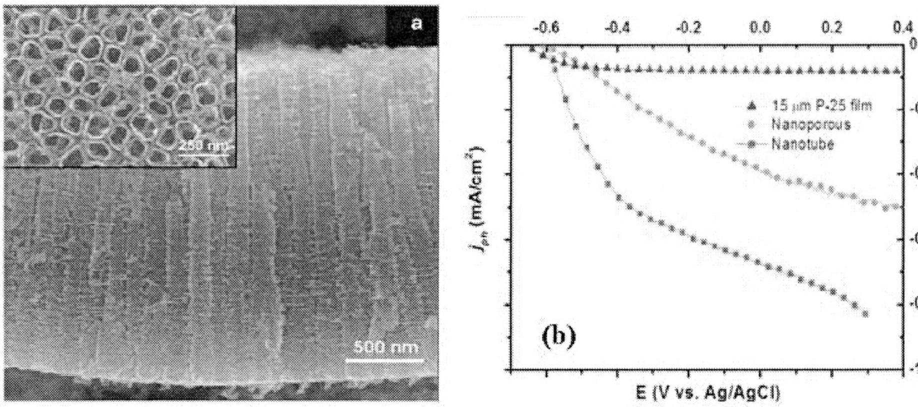

Figure 14. (a) Cross-sectional and top view of a 2.5μm thick self-organized porous TiO_2 layer (Ti samples were anodized at 20 V for 2 hours in 1M $(NH_4)_2SO_4$ + 0.5 wt% NH4F) (b) Variation of photocurrent density vs measured potential for three different TiO_2 photoanodes in 1 M KOH electrolyte. (Figure 14a reprinted with permission from [118]; Figure 14b reprinted with permission from [121]).

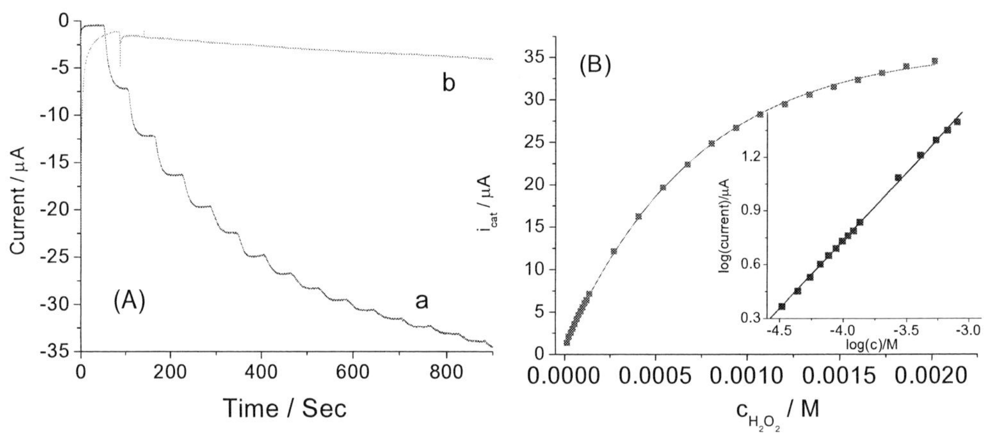

Figure 15. (A) Amperometric response of (a) Th/HRP/TiO_2 and (b) Th/TiO_2 at -0.45V upon successive additions of 0.134 mM H_2O_2 into 0.1 M pH 6.8 PB. (B) Plot of the reduction current versus the H_2O_2 concentration. (Inset) Linear calibration curve. (Reprinted with permission from [18]).

Due to non-toxicity, nanostructured TiO_2 materials show high biocompatibility and good retention of biological activity for protein binding [122-123]. Our recent studies have shown that TiO_2 nanotube arrays provide excellent matrices for the co-immobilization of thionine and horseradish peroxidase (HRP) [18]. Figure 15a presents the chronoamperometric response of the Th/HRP/TiO_2 (a) and Th/TiO_2 (b) electrode with successive additions of

0.134 mM H_2O_2 into 0.1 M pH 6.8 PB at the potential of –0.45 V. The sensor shows an increasing amperometric response to H_2O_2 from 1.1×10^{-5} to 2×10^{-3} M as shown in Figure 15b. The linear response range of the sensor to H_2O_2 concentration is from 1.1×10^{-5} to 1.1×10^{-3} M as seen in the inset of Figure 15b. The linear regression equation is log I (μA) = 3.679 + 0.736 log c (M), with a correlation coefficient of 0.9987. The detection limit is estimated as 1.2×10^{-6} M at 3σ based on the slope. These results show that the immobilized HRP on the TiO_2 nanotube arrays not only retains its bioactivity but also has high affinity for H_2O_2 reduction, resulting in the possibility of developing novel H_2O_2 sensors with very good stability.

4. Conclusion

In summary, simple and facile methods have been developed to directly synthesize TiO_2 nanostructures on titanium substrates by oxidizing Ti foil using small organic molecules as the oxygen source. The morphologies of the nanostructures can be controlled by varying the oxygen source, reaction temperature and time. Polycrystalline grains are formed when pure oxygen and formic acid are used as the oxygen source; elongated micro-crystals are produced when water vapor is used as the oxygen source; oriented and aligned TiO_2 nanorod arrays are synthesized when ethanol, acetaldehyde, acetone and DBTDL are used as the oxygen source. The diffusion of Ti atoms to the oxide/gas interface *via* the network of the grain boundaries of the thin oxide layer is the determining factor for the formation of well-aligned TiO_2 nanorod arrays. The nanorods all grow along the <001> direction through the surface diffusion of Ti atoms or ions to the tips of the nanostructures. The wetting properties of the formed TiO_2 nanostructured surfaces are dictated by their structure, varying from a hydrophilic surface to a strongly hydrophobic surface as the surface structure changes from polycrystalline grains to well-aligned nanorod arrays. The controlled formation of the TiO_2 nanorod arrays promises applications in the area of sensors and catalysts. It is anticipated that this method could be used to synthesize ordered nanorod arrrays of other oxides.

In addition, large-scale $Na_2Ti_6O_{13}$ nanoplates, nanowires, and continuous nanowire network films have been directly formed on Ti substrate hydrothermally for the first time. The morphology of the formed $Na_2Ti_6O_{13}$ nanostructures is easily tunable by varying the experimental parameters of temperature, reaction duration and the NaOH concentrations. This study has demonstrated that the synthesized $Na_2Ti_6O_{13}$ nanostructures are easily converted into $H_2Ti_3O_7$ nanostructures, a desirable precursor for fabrication of various TiO_2-based nanomaterials, with shape-preservation through an ion-exchange process. Anatase, a mixture of anatase and rutile, and rutile TiO_2 nanowires were formed when the $H_2Ti_3O_7$ nanowires were annealed at 450 °C, 600 °C and 750 °C, respectively. The formed TiO_2 anatase nanowires show much higher photocatalytic activity for degradation of Eosin Y, compared to the rutile TiO_2 nanowires. In this chapter, we have demonstrated that hydrothermal treatment of Ti substrate in NaOH solutions provides a simple and easy method to synthesize various large-scale TiO_2-based nanostructured materials, ready for practical applications such as the photocatalytic purification of wastewater. We have also illustrated that electrochemical anodization of titanium is an effective method in fabricating TiO_2 nanotubes. The formed TiO_2 nanotube arrays are very promisng in Green Chemistry applications and biosensoring.

Ackowledgements

This work was supported by a Discovery Grant from the Natural Sciences and Engineering Research Council of Canada (NSERC). A. Chen acknowledges the Ontario Premier's Research Excellence Award.

References

[1] Schimid, G. *Nanoparticle: from Theory to Application;* Wiley-VCH: Weinheim, Germany, 2004.
[2] Zhang, J. Z.; Wang, Z. L.; Liu, J.; Chen, S.; Liu, G.-Y. *Self-Assembled Nanostructures;* Kluwer Academic/Plenum Publishers: New York City, NY, 2003.
[3] Nirmal, M.; Brus, L. E. *Acc. Chem. Res.* 1999, 32, 407-414.
[4] Dumestre, F.; Chaudret, B.; Amiens, C.; Renaud, P.; Fejes, P. *Science* 2004, 303, 821-823.
[5] Sun, S.; Murray, C. B.; Weller, D.; Folks, L.; Moser, A. *Science* 2000, 287, 1989-1992.
[6] Peng, X. S.; Meng, G. W.; Wang, X. F.; Wang, Y. W.; Zhang, J.; Liu, X.; Zhang, L. D. *Chem. Mater.* 2002, 14, 4490 – 4493.
[7] Diebold, U. *Surface Science Reports* 2003, 48, 53-229.
[8] Fujishima, A.; Honda, K. *Nature* 1972, 238, 37-38.
[9] Schiavello, M. *Photoelectrochemistry, Photocatalysis, and Photoreactors: Fundamentals and Developments;* NATO ASI Series; D. Reidel Publishing Company: Dordrecht, The Netherlands, 1985.
[10] Ollis, D. S.; Al-Ekabi, H. *Photocatalytic Purification and Treatment of Water and Air;* Elsevier: Amsterdam, The Netherlands, 1993.
[11] Khan, S. U. M.; Akikusa, J. *J. Phys. Chem. B,* 1999, 103, 7184-7189.
[12] Licht, S.; Wang, B.; Mukerji, S.; Soga, T.; Umeno, M.; Tributsch, H. *J. Phys. Chem. B,* 2000, 104, 8920-8924.
[13] Wilcoxon, J. P. *Photocatalysis Using Semiconductor Nanoclusters,* In *Advanced Catalytic Materials;* Lednor, P. W.; Nagaki, D. A.; Thompson, L. T.; Materials Research Society Symposium Proceedings: Boston, MA, 1999.
[14] Serpone, N.; Pelizzetti, E. *Photocatalysis: Fundamentals and Applications;* Wiley: New York City, NY, 1989.
[15] Kozhukharov, V.; Vitanov, P.; Stefchev, P.; Kabasanova, E.; Kabasanov, K.; Machkova, M.; Blaskov, V.; Simeonov, D.; Tzaneva, G. *J. Environ. Prot. Eco.* 2001, 2, 107-115.
[16] Schiavello, M.; Dordrecht, H. *Photoelectrochemistry, Photocatalysis, and Photoreactors: Fundamentals and Developments;* Kluwer Academic: Boston, MA, 1985.
[17] Linsebigler, A. L.; Lu, G.; Yates, J. T. *Chem. Rev.* 1995, 95, 735-738.
[18] Liu, S.; Chen, A. *Langmuir* 2005, 21, 8409-8413.
[19] Mills, A.; Davies, H. R.; Worsley, D. *Chem. Soc. Rev.* 1993, 22, 417.
[20] Legrini, O.; Oliveros, E.; Braun, A. M. *Chem. Rev.* 1993, 93, 671.
[21] Heller, A. *Acc. Chem. Res.* 1995, 28, 503.

[22] Hoffman, M.; Martin, S.; Choi, W.; Bahnemann, D. *Chem. Rev.* 1995, 95, 69-96.
[23] Bonhote, P.; Gogniat, E.; Gräztel, M.; Ashrit, P. V. *Thin Solid Films,* 1999, 350, 269-275.
[24] Jia, J. G.; Ohno, T.; Matsumura, M. *Chem. Lett.* 2000, 908.
[25] Sun, J.; Gao, L.; Zhang, Q. *J. Am. Ceram. Soc.* 2003, 86, 1677-1681.
[26] Henrich, V. E.; Cox, P. A. *The Surface Science of Metal Oxides*; Cambridge University Press: Cambridge, 1996.
[27] Kim, K. J.; Benksten, K. D.; van de Lagemaat, J.; Frank, A. J. *J. Chem. Mater.* 2002, 14, 1042-1047.
[28] Sato, T.; Baba, K.; Hirozawa, T.; Kawakami, S. *IEEE J. Quantum Electron.* 1993, 29, 175-181.
[29] Hagfeldt, A.; Gratzel, M. *Chem. Rev.* 1995, 95, 49-68.
[30] Sahle-Demessie, E.; Gonzalez, M.; Wang, Z. M.; Biswas, P. *Ind. Eng. Chem. Res.* 1999, 38, 3276-3284.
[31] Mills, A.; Hunte, S. L. *J. Photochem. Photobiol.* 1997, *A*108, 1-35.
[32] Hoffmann, M. R.; Martin, S. T.; Choi, W.; Bahnemann, D. W. *Chem. Rev.* 1995, 95, 69-96.
[33] Linsebigler, A. L.; Lu, G.; Yates, J. T. Jr. *Chem. Rev.* 1995, 95, 735-758.
[34] Kim, D.; Anderson, M. *Environ. Sci. Technol.* 1994, 28, 479-483.
[35] Vinodgopal, K.; Kamat, P. *CHEMTECH* 1996, 26, 18-22.
[36] Wada, K.; Yoshida, K.; Watanabe, Y. *J. Chem. Soc., Faraday Trans.* 1995, 91, 1647-1654.
[37] Dartt, C. B.; Davis, M. E. *Ind. Eng. Chem. Res.* 1994, 33, 2887-2899.
[38] Maldotii, A.; Bartocci, C.; Amadelli, R.; Polo, E.; Battioni, P.; Mansuy, D. *J. Chem. Soc., Chem. Commun.* 1991, 20, 1487-1489.
[39] Groves, J. T.; Quinn, R. *J. Am. Chem. Soc.* 1985, 107, 5790-5794.
[40] Chen, Y. S.; Crittenden, J. C.; Perram, D. L.; Hand, D. W. *U.S. Provisional Application No.* 60/121,500 1998.
[41] Fox, M. A. *Acc. Chem. Res.* 1983, 16, 314-321.
[42] Ollis, D. F.; Hsiao, C. Y.; Budiman, L.; Lee, C. L. *J. Catal.* 1984, 88, 89-96.
[43] Shah, S. I.; Li, W.; Huang, C. P.; Jung, O.; Ni, C. *Proc. Natl. Acad Sci. U.S.A.* 2002, 99, 6482-6486.
[44] Xu, A.; Zhu, J.; Gao, Y.; Liu, H. *Chem. Res. Chin. Univ.* 2001, 17, 281.
[45] Wang, C.; Bahnemann, D. W.; Dohrmann, J. K. *Chem. Commun.* 2000, 16, 1539-1541.
[46] Wang, Y.; Hao, Y.; Cheng, H.; Ma, H.; Xu, B.; Li, W.; Cai, S. *J. Mater. Sci.* 1991, 34, 2773.
[47] Coloma, F.; Marquez, F.; Rochester, C. H.; Anderson, J. A. *Phys. Chem. Chem. Phys.* 2000, 2, 5320-5324.
[48] Altynnikov, A. A.; Zenkovets, G. A.; Anufrienko, V. F. *React. Kinet. Catal. Lett.* 1999, 67, 273-279.
[49] Umebayashi, T.; Yamaki, T.; Itoh, H.; Asai, K. *J. Phys. Chem. Solids* 2002, 63, 1909-1920.
[50] Wang, Y.; Cheng, H.; Hao, Y.; Ma, J.; Li, W.; Cai, S. *Thin Solid Films* 1999, 349, 120-125.
[51] Yamashita, H.; Honda, M.; Harada, M.; Ichihashi, Y.; Anpo, M.; Hirao, T.; Itoh, N.; Iwamoto, N. *J. Phys. Chem. B* 1998, 102, 10707-10711.

[52] Yu, J. C.; Yu, J. G.; Ho, W. K.; Jiang, Z. T.; Zhang, L. Z. *Chem. Mater.* 2002, 14, 3808-3816.
[53] Khan, S. U. M.; Al-Shahry, M.; Ingler, W. B. Jr. *Science* 2002, 297, 2243-2245.
[54] Asahi, R.; Morikawa, T.; Ohwaki, T.; Aoki, K.; Taga, Y. *Science* 2001, 293, 269-271.
[55] Irie, H.; Wanatabe, Y.; Hashimoto, K. *J. Phys. Chem. B* 2003, 107, 5483-5486.
[56] Burda, C.; Lou, Y. B.; Chen, X. B.; Samia, A. C. S.; Stout, T.; Gole, J. L. *Nano Lett.* 2003, 3, 1049-1051.
[57] Yang, T. S.; Shiu, C. B.; Wong, M. S. *Surf. Sci.* 2004, 548, 75-82.
[58] Lee, K. S.; Park, I. S. *Scripta Materialia* 2003, 48, 659-663.
[59] de Lucas, M. C.; Fabreguette, F.; Collin, S.; Bourgeois, S. *Int. J. Inorg. Mater.* 2000, 2, 255-259.
[60] Byun, D.; Jin, Y.; Kim, B.; Lee, J. L.; Park, D. *J. Hazard Mater.* 2000, 73, 199-206.
[61] Damiriv, D.; Bally, A. R.; Ballif, C.; Homes, P.; Schmid, P. E.; Sanjines, R.; Levy, F.; Parvulescu, V. I. *Appl. Catal.* 2000, 25, 83-92.
[62] Wang, X. P.; Yu, Y.; Hu, X. F.; Gao, L. *Thin Solid Films* 2000, 371, 148-152.
[63] Kato, K.; Tsuzuki, A.; Taoda, H.; Torrii, Y.; Kato, T.; Batsugn,Y. *J. Mater. Sci.* 1994, 29, 5911-5915.
[64] Hoyer, P. *Langmuir* 1996, 12, 1411-1413.
[65] Burnside, S. D.; Shklower, V.; Barbe, C.; Comte, P.; Arendse, F.; Brookes, K.; Grätzel, M. *Chem. Mater.* 1998, 10, 2419-2425.
[66] Yoo, S.; Akbar, S. A.; Sandhage, K. H. *Adv. Mater.* 2004, 16, 260-264.
[67] Liu, S. M.; Gan, L. M.; Lu, L. H.; Zhang, W. D.; Zheng, H. C. *Chem. Mater.* 2002, 14, 1391-1397.
[68] Uchida, S.; Chiba, R.; Tomiha, M.; Masaki, N.; Shirai, M. *Electrochemistry* 2002, 70, 418-420.
[69] Imai, H.; Takei, Y.; Shimizu, K.; Matsuda, M.; Hirashima, H. *J. Mater. Chem.* 1999, 9, 2971-2972.
[70] Chae, W. S.; Lee, S. W.; Kim, Y. R. *Chem. Mater.* 2005, 17, 3072.
[71] Liu, S. M.; Gan, L. M.; Liu, L. H.; Zhang, W. D.; Zeng, H. C. *Chem. Mater.* 2002, 14, 1391-1397.
[72] Kasuga, T.; Hiramatsu, M.; Hoson, A.; Sekino, T.; Niihara, K. *Langmuir* 1998, 14, 3 160-3163.
[73] Kasuga, T.; Hiramatsu, M.; Hoson, A.; Sekino, T.; Niihara, K. *Adv. Mater.* 1999, 11, 1307-1311.
[74] Zhang, Y. X.; Li, G. H.; Jin, Y.; Zhang, X. Y.; Zhang, J.; Zhang, L. D. *Chem. Phys. Lett.* 2004, 365, 300-304.
[75] Lei, Y.; Zhang, L. D.; Meng, G. W.; Li, G. H.; Zhang, X. Y.; Liang, C. H.; Chen, W.; Wang, S. X. *Appl. Phys. Lett.* 2001, 78, 1125-1127.
[76] Zhang, S. L.; Zhou, J. F.; Zhang, Z. J.; Du, Z. L.; Vorontsov, A. V.; Jin, Z. S. *Chin. Sci. Bull.* 2000, 45, 1533-1536.
[77] Seo, D. S.; Lee, J. K.; Kim, H. *J. Cryst. Growth* 2001, 229, 428-432.
[78] Zhang, Q. H.; Gao, L. A.; Sun, J.; Zheng, S. *Chem. Lett.* 2002, 2, 226-227.
[79] Tian, Z. R. R.; Voigt, J. A.; Liu, J.; McKenzie, B.; Xu, H. F. *J. Am. Chem. Soc.* 2003, 125, 12384-12385.
[80] Adachi, M.; Murata, Y.; Yoshikawa, S. *Chem. Lett.* 2000, 8, 942-943.

[81] Adachi, M.; Okada, I.; Ngamsinlapasathian, S.; Murata, Y.; Yoshikawa, S. *Electrochemistry* 2002, 70, 449-452.
[82] Hippe, C.; Wark, M.; Lork, E.; Schulz-Ekloff, G. *Microporous Mesoporous Mater.* 1999, 31, 235-239.
[83] Khitrov, G. *MRS Bull.* 2000, 25, 8, 3-4.
[84] Zhang, M.; Bando, Y.; Wada, K. *J. Mater. Sci. Lett.* 2001, 20, 167-170.
[85] Zhang, M.; Bando, Y.; Wada, K. *J. Mater. Res.* 2001, 16, 1408-1412.
[86] Li, D.; Xia, Y. N. *Nano Lett.* 2003, 3, 555-560.
[87] Corcoran, D. J. D.; Tunstall, D. P.; Irvine, J. T. S. *Solid State Ionics* 2000, 136-137.
[88] Pozo, R. L.; Baltanás, M. A.; Cassano, A. E. *Catal. Today* 1997, 39, 219-231.
[89] Peng, X.; Chen, A. *J. Mater. Chem.* 2004, 14, 2542-2548.
[90] Peng, X.; Chen, A. *Appl. Phys. A* 2005, 80, 473-476.
[91] Peng, X.; Wang, J.; Thomas, D. F.; Chen, A. *Nanotechnology* 2005, 16, 2389-2395.
[92] Fiest, T. P.; Davies, P. K. *J. Solid. State Chem.* 1992, 101, 275-295.
[93] Suzuki, Y.; Yoshikawa, S. *J. Mater. Res.* 2004, 19, 982-985.
[94] Armstrong, A. R.; Armstrong, G.; Canales, J.; Bruce. P. G. *Angew. Chem., Int. Ed.* 2004, 43, 2286-2288.
[95] Parker, J. C.; Segel, R. W. *Appl. Phys. Lett.* 1990, 57, 943-945.
[96] Badescu, V.; Mormirlan, M. *J. Crystal Growth* 1996, 169, 309-316.
[97] Czerwinski, F.; Szpunar, J. A. *Micron* 1998, 29, 201-206.
[98] Gracia, F.; Holgado, J. P.; Contreras, L.; Girardeau, T.; González-Elipe, A. R. *Thin Solid Films* 2003, 429, 84-90.
[99] Peng, X. S.; Zhang, L. D.; Meng, G. W.; Wang, X. F.; Wang, Y. W.; Wang, C. Z.; Wu, G. S. *J. Phys. Chem. B* 2002, 106, 11163-11167.
[100] Peng, X. S.; Meng, G. W.; Zhang, J.; Wang, X. F.; Wang, Y. W.; Wang, C. Z.; Zhang, L. D. *J. Mater. Chem.* 2002, 12, 1602-1605.
[101] Smeltzer, W. W. *Mater. Sci. Forum* 1988, 29, 151.
[102] Sakai, N.; Fujishima, A.; Watanabe, T.; Hashimoto, K. *J. Phys. Chem. B* 2003, 107, 1028-1035.
[103] Wang, R.; Sakai, N.; Fujishima, A.; Watanabe, T.; Hashimoto, K. *J. Phys. Chem. B* 1999, 103, 2188-2194.
[104] Zhang, X.; Shi, F.; Yu, X.; Liu, H.; Fu, Y.; Wang, Z. Q.; Jiang, L.; Li, X. Y. *J. Am. Chem. Soc.* 2004, 126, 3064-3065.
[105] Shiu, J. Y.; Kuo, C. W.; Chen, P. L.; Mou, C. Y. *Chem. Mater.* 2004, 16, 561-564.
[106] Yoshimisu, Z.; Nakajima, A.; Watanabe, T.; Hashimoto, K. *Langmuir* 2002, 18, 5818-5822.
[107] Wenzel, R. N. *Ind. Eng. Chem.* 1936, 28, 988-994.
[108] Cassie, A. B. D.; Baxter, S. *Trans. Faraday Soc.* 1944, 40, 546-551.
[109] Otten, A.; Herminghaus, S. *Langmuir* 2004, 20, 2405-2408.
[110] Chen, A.; Peng, X.; Koczkur, K.; Miller, B. *Chem. Commun.* 2004, 17, 1964-1965.
[111] Peng, X.; Chen, A. *Adv. Funct. Mater.* Adfm200500464 (*in press*).
[112] Serpore, N.; Pilezzeti, E. *Photocatalysis: Fundamentals and Applications;* Wiley-Interscience: New York City, NY, 1989.
[113] Hosono, E.; Fujihara, S.; Kakiuchi, K.; Imai, H. *J. Am. Chem. Soc.* 2004, 126, 7790-7791.

[114] Cozzoli, P. D.; Comparelli, R.; Fanizza, E.; Locia-Curri, M.; Agostiano, A.; Laub, D. *J. Am. Chem. Soc.* 2004, 126, 3868-3879.

[115] Diwarld, O.; Thompson, T. L.; Zubkov, T.; Goralski, E. G.; Walck, S. D.; Yates, J. T. Jr. *J. Phys. Chem. B* 2004, 108, 6004-6008.

[116] Gong, D.; Grimes, C. A.; Varghese, O. K; Hu, W. C.; Singh, R. S.; Chen, Z.; Dickey, E. C. *J. Mater. Res.* 2001, 16, 3331-3334.

[117] Wang, W. Z.; Varghese, O. K.; Paulose, M.; Grimes, C. A.; Wang, Q. L.; Dickey, E. C. *J. Mater. Res.* 2004, 19, 417-422.

[118] Balaur, E.; Macak, J. M.; Taveira, L.; Schmuki, P. *Electrochem. Commun.* 2005, 7, 1066-1017.

[119] Macak, J. M.; Tsuchiya, H.; Ghicov, A.; Schmuki, P. *Electrochem. Commun.* 2005, 7, 1133-1137.

[120] Beranek, R.; Tsuchiya, H.; Sugishima, T.; Macak, J. M.; Taveira, L.; Fujimoto, S.; Kisch, H.; Schmuki, P. *Appl. Phys. Lett.* 2005, 87, 243114.

[121] Park, J. H.; Sungwook, K.; Bard, A. J. *Nano Lett.* 2006, 6, 24-28.

[122] Topoglidis, E.; Cass, A. E. G.; Gilardi, G.; Sadeghi, S.; Beaumont, N.; Durrant, J. R. *Anal. Chem.* 1998, 70, 5111-5113.

[123] Zhang, J. K.; Cass, A. E. G. *Anal. Biochem.* 2001, 292, 307-310.

In: Encyclopedia of Nanotechnology
Editor: Elwood D. Carlson

ISBN: 978-1-60692-079-4

Chapter 2

MOLECULAR FOUNDATIONS FOR ONSET WEAR. A REVIEW

Mario D'Acunto[1]
Department of Chemical Engineering and Materials Science, University of Pisa, via Diotisalvi 2, 56100 Pisa, Italy
and
Department of Physics, Univeristy of Pisa, largo B. Pontecorvo 3, 56127 Pisa, Italy

ABSTRACT

One definition of wear in a tribological sense is that it is the progressive loss of mass from the operating surface of a body as a result of relative motion at the surface. Generally, tribology may be defined as the study of friction wear and lubrication, but in comparison with friction forces or lubrication methods, very little work has been done on the extremely important area of wear and surface damage. For systems consisting of common materials (metals, polymers, ceramics), there are at least four main mechanisms by which wear and surface damage can occur between solids in relative motion: abrasive- and/or adhesive-, chemical- or corrosive-, fatigue-wear mechanisms. The occurrence of such different wear mechanisms involves the fundamental mechanics of molecular and supra-molecular scale. As a consequence, the understanding of surface properties on nanoscale level should be generated to have also a satisfactory knowledge of responses of materials and machine components on macrometer scale. Investigation of the fundamental characteristics of wear at the nanoscale is complicated by some factors and forces that act on such scale level have not yet been addressed in the tribology of macro-systems. Since these forces are sensitive to the environment and surface condition of the specimens, it is quite difficult to determine them accurately. Further, quantification of wear is not straightforward since the amount of wear is often too small to be detected by surface-sensitive instruments. Nanowear measurements have been object of rapid development of precise measuring tools following to the introduction of scanning probe microscopies, such as atomic force microscopy (AFM). The advent of AFM opened new opportunity to study wear mechanisms on nanoscale, making possible to simulate one surface sliding on other one and controlling many physical parameters during the contact

[1] E-mail address: m.dacunto@ing.unipi.it

dynamics. Further, AFM made possible to a have a direct connection between experimental data and theoretical models.

The chapter is devoted to give a panoramic lecture of some experimental results and models on the onset wear occurrence on nanoscale.

1. INTRODUCTION

The atomistic mechanisms and dynamics of the interactions of two materials during relative motion, need to be understood in order to develop fundamental understanding of adhesion, friction, wear, indentation and lubrication processes. At most solid-solid interfaces of technologies relevance, contact occurs at many asperities. Consequently the importance of investigating single asperity contacts in studies of the fundamental mechanical and tribological properties of surfaces and interfaces on nanoscale has long been recognized. The recent emergence and proliferation of scanning probe microscopies (SPMs) and computational techniques for simulating tip-surface interactions and interfacial properties, has allowed systematic investigations of interfacial problems with high resolution as well as ways and means for modifying and manipulating nanoscale structures. These advances have led to the appearance of the new field of nanotribology, which pertains to experimental and theoretical investigations of interfacial processes occurring during adhesion, friction, wear, scratching, nanoindentation and thin-film lubrication at sliding surfaces on scales ranging from the atomic- and molecular- to the microscale [Bhushan et al., 1995; Bhushan, 2003]. The nanotribological studies are needed to develop fundamental understanding of interfacial phenomena on atomic, molecular and supramolecular scale and to study interfacial phenomena in nanostructures used in the area commonly defined nanotechnology. Friction and wear of lightly loaded nanocomposites are highly dependent on the surface interactions involving few atomic layers. Consequently, these structures are generally lubricated with molecularly thin films.

A definition of wear mechanisms can be as follows: the succession of events whereby atoms, products of chemical conversion, fragments, debris, are induced to leave the system and identified in a manner that embody or suggest solutions [Ludema, 2001]. These solutions may include choice of materials, choice of lubricants, choice of contact condition, choice of the manner of operation of the mechanical system. This definition includes also: a) the conditions imposed upon surfaces, such as contact stress, shear stress and number of cycles of these stress applications; b) the changing composition of substances in the sliding interfaces due to both sliding and these will likely influence the friction and shearing stresses imposed in sliding contact; c) the processes whereby particles (atoms, molecules) are separated from a surface; d) the circulation and path of those particles before they are expelled as debris. There are more than 24 common terms to describe wear mechanisms [Ludema 1996]. Generally, such wear mechanisms are listed in six categories of terms. Among the most important categories, we limit to discuss the most important, such as adhesion based wear mechanisms, fatigue derived wear mechanisms and abrasion.

Adhesive wear is the most difficult term to define. It may denote a particular type of material loss due to high local friction (which is often attributed to adhesion) and is a tempting term to use because high local friction produces tearing and fragmentation whereas lubricants diminish tearing. Some time, lubricated wear is taken to be the opposite of adhesive wear. An often overlooked aspect of adhesive wear is that if adhesion is effective then the particles torn loose by adhesion will re-adhere and not be expelled as debris.

Wear mechanisms deriving from cyclic stressing imply fatigue of materials such as fretting. Fretting is the oscillatory sliding motion of small slip amplitude between two nominally stationary contacting surfaces. It occurs when two quasi-static surfaces in contact are actually subjected to deformation and/or vibrations. Damage caused by fretting is one of the severe problems in engineering since it contributes substantially to shorten life of mechanical components. Analogous to fretting is the delamination mechanisms, which describes a type of wear debris developed by low cycle fatigue when surfaces are rubbed repeatedly by a small shaped slider.

Abrasive-wear mechanisms denote a wide class of wearing mechanisms occurring when a hard solid, particle, or asperity rubs against a softer surface. Examples of abrasive wear include the smoothing of wood by sandpaper, the action of emery paper in polishing metals, grinding processes, and harmful effects of hard particles (e.g. sand, dirt, carbide or silica inclusions, oxides and work-hardened wear debris) in lubricated systems. Abrasive wear is a form of microcutting, involving a hard particle or shape which indents, grooves, and then cuts material out of the surface.

On macroscale, wear has been the subject of much study and many general mechanisms are reasonably well understood [Hutchings, 1992]. On nanoscale, this is yet not so. The mechanisms are largely unknown and not readily determined. Nevertheless, the occurrence of wear processes at nanoscale has been demonstrated [Kim et al., 1991; Bhushan and Koinkar, 1994; Hu et al., 1995; Carpick and Salmeron, 1997; Schöfer and Santer, 1998; Gåhlin and Jacobson, 1998].

Let us consider the possible mechanisms of wear to be essentially two types: abrasive- and adhesive-wear mechanism. On macroscale, if the contribution to wear is essentially of adhesive nature, it is possible to quantify the volume of wear for plastic contacts as

$$V = \frac{k_{adh} L x}{H} \tag{1}$$

where k_{adh} is a non-dimensional wear coefficient dependent on the materials in contact, L is the load applied, H is the hardness coefficient and x is the sliding distance. Equation (1) is commonly referred as Archard's equation [Archard, 1953]. It suggests that the wear rate is independent of the apparent area, directly proportional to the applied load and the wear rate is constant with sliding distance (or time). These properties seem to be analogous to Amontons' laws of friction. Furthermore, equation (1) suggests that the probability of decohesion of a certain volume of material and formation of a wear particle increases with real area of contact, A_r ($=L/H$) and the sliding distance. An equation of the form similar to Archard's equation for adhesive wear is found to cover a wide range of abrasive situations

$$V = \frac{k_{abr} L x}{H} \tag{2}$$

In analogy to macroscopic scale case, the basic mechanisms for the occurrence of adhesive- or abrasive-wear mechanisms quantifying the wear volume rate correspondent to material loss during the scanning probe tip have been performed in micro/nanoscale [Chung and Kim, 2003].

Several techniques are used for fundamental studies of tribological features. Proximal probes- including scanning tunnelling microscopy (STM), AFM and the surface forces apparatus (SFA) that operate in ultrahigh vacuum (STM, AFM) and in controlled environments (AFM, SFA) – constitute the basic imaging and force measuring tools. These techniques have made possible the study of single-asperity contacts in a controlled fashion, eliminating complications encountered in more traditional methods, such as pin-on-disk, indenters, profilometers, etc., in which the contact is made between rough surfaces involving a large and uncontrolled number of asperities, figure 1. The chapter is structured as follows: in section 2, prominent wear mechanisms on nanoscale are discussed; in section 3, activation process describing onset wear on the nanoscale will be presented.

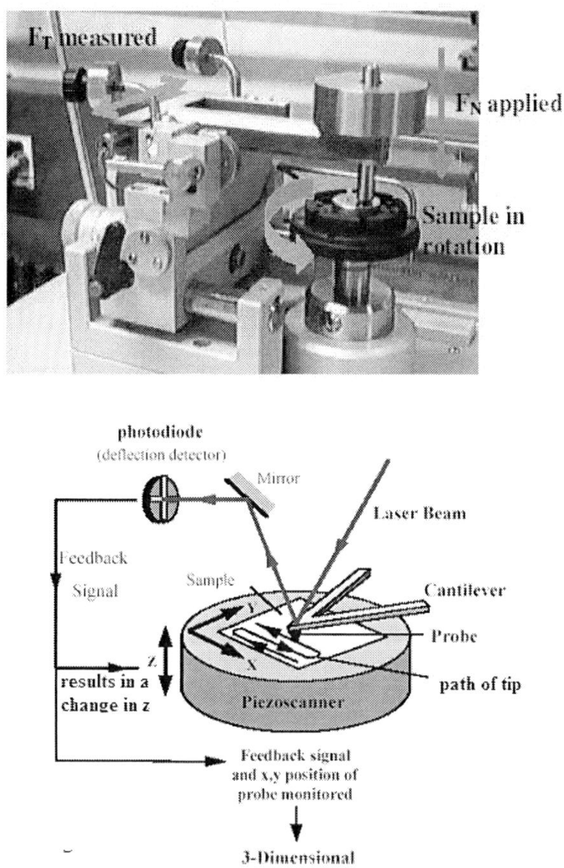

Figure 1. (Upper image) typical wear machine where a flat or a sphere shaped indenter is loaded on to the test sample with a precisely known weight. A pin is mounted on a stiff lever, designed as a frictionless force transducer. As the sample is either rotating or sliding, resulting frictional forces acting between the pin and the sample are measured by very small deflections of the lever. Wear coefficients for both the sample and material are calculated from the volume of the material lost during a specific friction run. (Reprinted from CSM instruments, Switzerland, with permission). Below, typical experimental set up for AFM configuration.

2. MOLECULAR AND SUPRAMOLECULAR FEATURES FOR WEAR

One main SPM involved method for the observation of the nanowear processes is given by the production of defects in sliding contacts making use of the probe tip. Generally, the probe tip is used to scan repeatedly the same area of the sample. AFM has been used to study such processes on mica both in controlled atmosphere [Kopta and Salmeron, 2000] and in aqueous environment [Helt and Batteas, 2005]. Muscovite mica, a layered alumino-silicate in the form of $KAl_2(Si_3AlO_{10})(OH)_2$ has been frequently employed as a model surface for the investigation of the atomic scale relationship between friction [Erlandsson et al., 1988; Carpick et al., 1996a], adhesion [Carpick et al., 1996b; Xu and Salmeron, 1998; Barthel and Roux, 2000] and wear [Miyake, 1994; Hu et al., 1995] by the AFM and SFA [Israelashvili, 1992] measurements due to the ability to readily generate large domains of atomically smooth surface. The layered sheet structure of mica has a ~1nm periodicity normal to the *(001)* crystal plane, which is defined by the repeating boundary layers of K^+, figure 2. The K^+ ions electrostatically bind the layers togheter and act to neutralize the net negative charge associated with the partial substitution of Si with Al (on average one out of every four) in the basal plane. The K^+ interlayer demarcates the preferred cleavage planes and defines the atomic boundaries along which shearing of the layers most readily occurs. Under aqueous conditions, the outmost K^+ layer is exchanged into solution and therefore has yet to be imaged due to its ready displacement. With the K^+, ions solvated, the outermost mica surface exposes a hexagonally arrayed pattern of oxygen atoms with a 0.52nm periodicity.

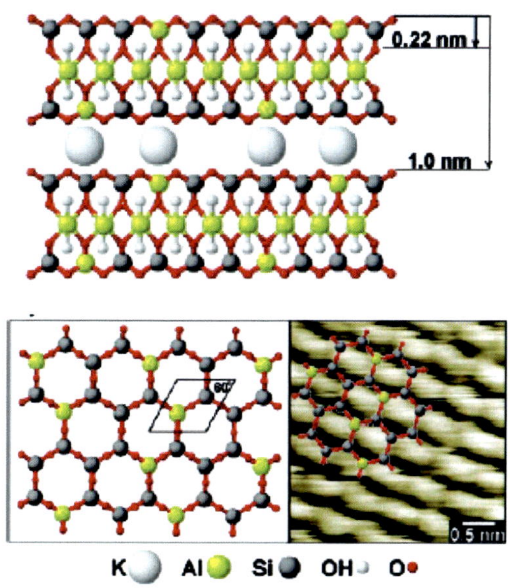

Figure 2. Cross-section depicting micas layered structure along the *c* axis with interlayer distances referenced to the SiO_3/AlO_3 tetrahedral basal plane (upper image). A graphic representation and an unfiltered AFM topographic image (3.4nm×3.4nm) of the mica (001) crystal plane showing the 0.52nm lattice periodicity (lower image). (Reprinted from [Helt and Batteas, 2005], with permission).

AFM studies of mica surfaces have shown that under sufficiently high loads the surface becomes damaged and ~0.2 or ~1nm deep holes are formed, depending on the imaging conditions. Wear scars have been produced by repeatedly scanning over the same region at a fixed load. Point defects are created as a result of the lowering of the activation energy for bond breaking by the applied stresses [Dickinson et al., 1997]. In the tip sample contact area, the number of defects created is given by [Kopta and Salmeron, 2000]

$$N_{def}(L) = t_{res} n_0 A(L) \nu e^{\{-[\varepsilon_0 - \delta\varepsilon(L)]/k_B T\}} \tag{3}$$

where t_{res} is the residence time of the tip, n_0 ($\sim 10^{20} m^{-2}$) the surface density of atoms, and $A(L)$ the contact area as a function of total load L, ν is the attempt frequency ($\sim 10^{13} s^{-1}$) to overcome the energy barrier ε_0 to break a *Si-O* bond, $\delta\varepsilon(L)$ is the decrease in activation energy due to the applied stress, which can be estimated by multiplying the force acting on an atom by the resulting bond stretch. The average force per atom is given by the pressure $P(L)$ in the contact area divided by n_0. The relative deformation (bond stretch) $\Delta l/l$, should be on the same order as the Hertzian indentation depth h, divided by the characteristic length of the affected area, which is on the order of the contact radius $r(L)$: $\Delta l/l \approx h/r(L)$, with $h \approx r^2(L)/2R$ for a tip radius $R >> h$. Thus the decrease in activation energy due to the applied stress $\delta\varepsilon(L)$ changes as $L^{2/3}$. Wear damage will occur when the defect density, $N_{def}/n_0 A(L)$, reaches a critical value η_{crit}, and gives rise to nucleation of a hole. Thus the number of consecutive scans necessary to cause a visible damage (0.2nm holes) on the muscovite mica surface is given by the equality of critical value with the product of number scans with the defect density. Such threshold value for number scans can be expressed by the following cumulative curve

$$N_{scans} = AL^{-2/3} e^{-BL^{2/3}} \tag{4}$$

where A and B are two constants.

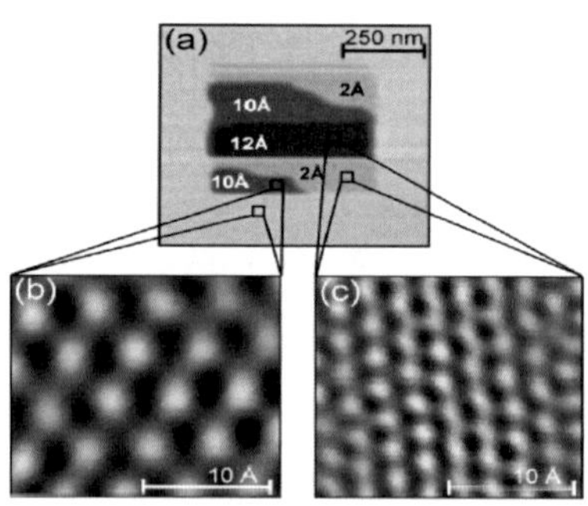

Figure 3. AFM images of a hole scratched into muscovite mica with an AFM tip. The images were recorded making use of a silicon cantilever with a nominal force constant of 2.1N/m and a nominal tip radius of 50nm was used. The hole was created after 8 consecutive scans with a load of 230nN, over an area of 500×500nm^2 at a scan speed of 60ms/line. (a) Image size is 0.75×0.75μm2. The hole obtained with these parameters shows 2, 10 and 12Å deep regions (denoted by gray levels). Marked with a square are regions where the atomic lattice resolution images, shown Fourier filtered in (b) and (c), were obtained. The periodicities in these images are 5. Å (b) and 3.0 Å (c). The latter is rotated by~30°. (Reprinted from [Kopta and Salmeron, 2000], with permission).

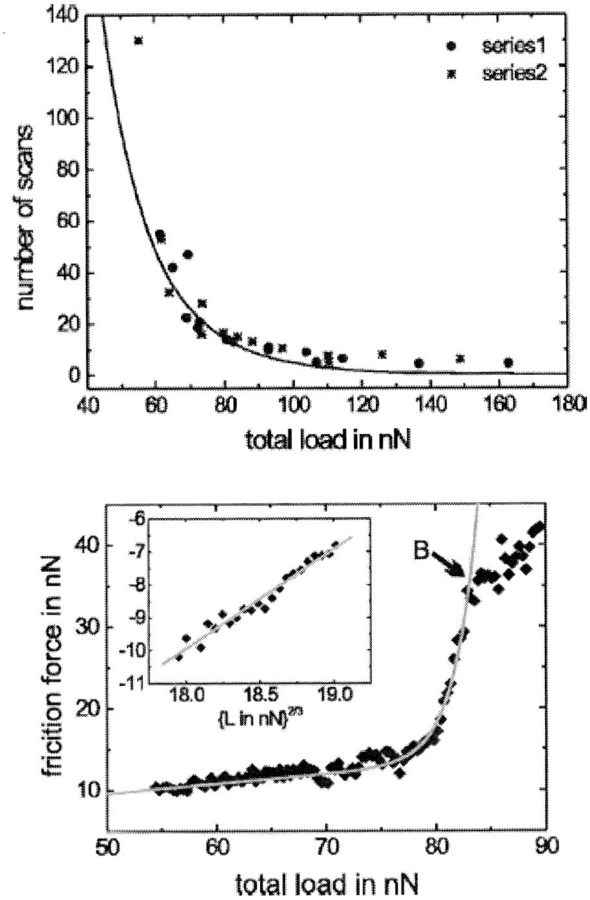

Figure 4. In the upper image, it is shown the number of consecutive scans necessary to cause visible damage to the mica surface (2Å holes). Data from two series of experiments measured with the same tip (force constant 0.4N/m) on two consecutive days are shown. In the lower image, friction vs load curve obtained during the creation of 2Å and 10Å deep holes. Rapid increase in friction is observed above 75nN load, corresponding to a large increase in the production of defects. When the density of defects reaches a critical value, the friction levels off and a wear scar detectable by AFM is produced. The inset shows the region, where the friction increases rapidly, in a semilog representation: $\ln\{[F(L) - cL^{2/3}]/L^{2/3}\}$ is plotted vs $L^{2/3}$ to allow for a simple linear fit of the coefficients B, eq.(4). (Adapted from [Kopta and Salmeron, 2000], with permission).

Helt and Batteas reported the wear of mica surfaces under aqueous environments of varying pH [Helt and Batteas, 2005]. Using the AFM, they have evaluated the simultaneous changes in surface topography, friction and adhesion at the mica surface under controlled pH conditions. By varying pH and applied loads, the incidence and mechanisms of the observed wear were found to be influenced dramatically. In fact, with the appropriate control of applied load, number of scans, and pH, they have been able, for the first time, to directly observe defect production, consisting in the breaking of terminal surface bonds, as detected in the form of a surface restructuring prior to the formation of 2Å pits or scars, generally, considered gross wear. A further result shown by Helt and Batteas is that the degree of wear can be completely controlled by modifications of imaging conditions and pH. Under low pH conditions (pH ≤3), the wear threshold pressure relating to the formation of 2Å pits, is equivalent to that of a simply humid environment and complete inhibition of wear is observed for tests conducted at the threshold pressure for higher pH studies.

At the highest pH conditions (pH≈8), the mica surface undergoes rapid wear. Even under low loads, wear scars ~2Å in depth form readily. The threshold pressure for wear at pH=8 was found to be nearly 1.2GPa. With additional scans (>10), wear scars of 10Å or more are rapidly formed, figure 5. Interestingly, the depth of the wear scars formed under these pH conditions is found to scale roughly linearly with the number of scans for wear scars of up to 60-70Å in depth.

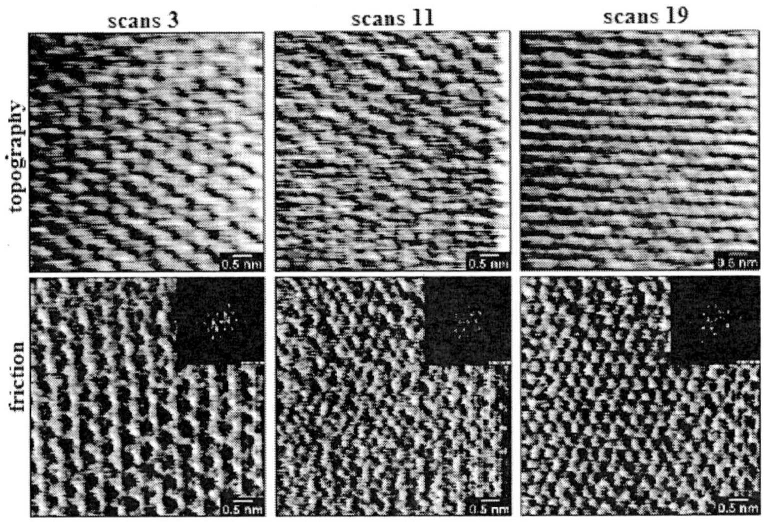

Figure 5. Topography and frictional images monitoring the tribochemical surface reconstruction of the native mica lattices from a 5.2Å (scans 3) to an ~3 Å (scans 19) periodicity under the local stress of an AFM tip. Images represent the 3rd, 11th and 19th scans during continuous scanning at a rate of 33Hz over a 7×7 nm^2 region at pH=5 and an applied load of 9nN (mean pressure of 1.25GPa). The 2D FFT insert of the friction micrographs illustrates the lattices transition where the intermediate state (11th scan) possesses a mixed 3Å and 5.2Å periodicity. (Reprinted from [Helt and Batteas, 2005], with permission).

Dickinson and co-worker performed a series of fundamental experiments on nanowear on single-crystal calcite in satured CaCO3 solutions [Park et al., 1996a; Park et al., 1996b; Dickinson et al., 1997], to quantify the shear stress required to detach single crystal salt particles from a soda lime glass substrate under controlled atmospheres of known humidity [Dickinson et al, 1999], and corrosive wear on a single-crystal brushite ($CaHPO_4{\cdot}2H_2O$) in aqueous solution [Scudiero et al., 1999]. Generally, in such experiments the tribological stimulus was provided by the scanning probe tip itself. In the case of single-crystal calcite, the dissolution and growth of calcite crystals in aqueous solution have been extensively studied due to their importance in mineral formation, global CO2 exchange, and interactions with environmental contaminants. Localized corrosive wear can readily be induced by the SFM tip on calcite surfaces imaged in aqueous solutions. Localised dissolution along pre-existing steps can also be produced by simply indenting the surface near the edge of an etch pit. Wear produced by drawning the probe tip back and forth in a repetitive, linear fashion across a monolayer step can be unambiguously interpreted in terms of volume activated, double kink nucleation in the strain field of the SFM tip. Experimental details can be found in [Dickinson et al, 1997]; here, we like to discuss the activation process.

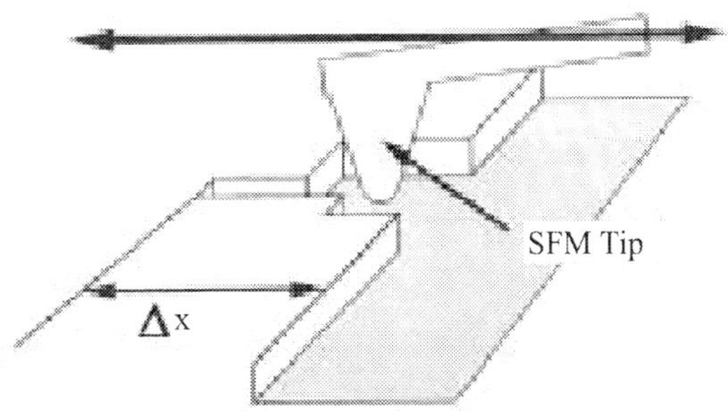

Figure 6. Schematical sketch of wear geometry of the experiment design from Dickinson and co-workers. The SFM tip is drawn back and forth in a linear fashion across the edge of pre-existing etch pit. (Reprinted from [Dickinson et al., 1997], with permission).

Wear tracks can be easily imaged in a geometry as presented in figure 6. The wear track growth rate ($\Delta x/\Delta t$) is a strong function of contact force. In figure 8, it is presented a plot of the wear track growth rate for wear tracks force.

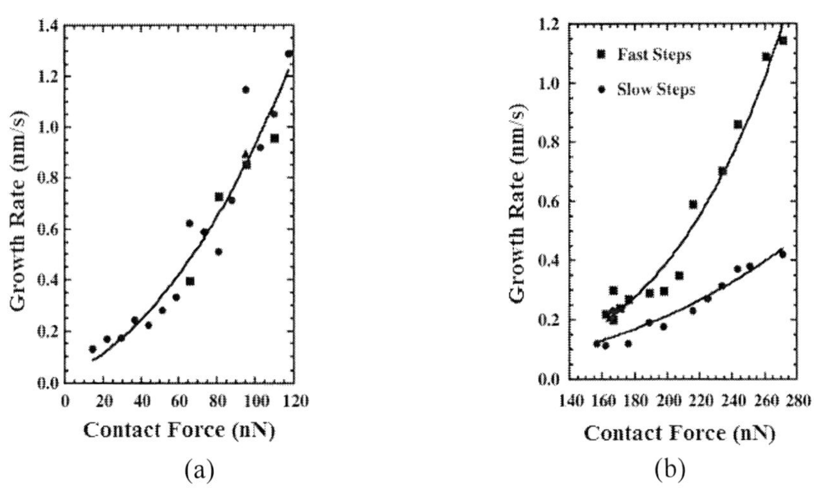

Figure 7. (a) Wear track growth rate versus contact force during flow of solution and repeated linear scanning. The data points represented by different symbols were made on different calcite samples with different cantilevers. The continue line is a least-squares fit of eq. (5) to the data. The measurements employed 500×500nm2 scans acquired at a tip velocity of 12μm/s. (b) Wear track growth rate for both fast and slow steps versus contact force under high degree of saturation. As in the case (a), the continue lines represent least-squares fits of eq. (5) to the data. (Reprinted from [Dickinson et al., 1997], with permission).

The magnitude of the stresses surrounding the SFM tip can be estimated assuming isotropic, elastic behaviour. Under such assumptions, the maximum tensile stress involves the radial components, σ_r, along the circle where the SFM tip contacts the substrate. Double-kink nucleation (the removal of ions from initially complete or intact steps) should be enhanced over a modest range of distances from the tip. The radial stress, σ_r, can be calculated by the Hertzian relation [Johnson, 1985]:

$$\sigma_r = \frac{(1-2v)}{\pi}\left(\frac{2PE^2}{9R^2(1-v^2)^2}\right)^{1/3} \tag{5}$$

where v is the Poisson's ratio and E is the reduced Young's modulus (approximately 58GPa), R is the tip radius and $P=F_N/$*real area of contact* is the normal pressure. With an applied normal force of 100nN yields a maximum radial stress of about 500MPa. The dependence of the growth rate of the wear track V, on contact force is readily modelled with an Arrhenius-type expression

$$V = V_0 \exp\left[-\left(\frac{E_{act} - \upsilon^* \sigma_r}{k_B T}\right)\right] \tag{6}$$

where V_0 is the appropriate pre-exponential, E_{act} is the zero stress activation energy for double-kink nucleation and υ^* is an activation volume. The best-fit curve of eq. (6) to the data of fig. 7a (unsatured solution), corresponds to $V_0\exp(-E_{act}/k_BT)\approx 5.8$pm/s and $\upsilon^* \approx 4.4\times10^{-29}m^3$. This activation volume is slightly larger than the average volume per ion in the $CaCO_3$ lattice ($3.3\times10^{-29}m^3$), making it reasonable to suppose that this activation volume corresponds to the displacement of one or perhaps two ions from a step site. A similar analysis of the wear track growth rates at slow steps under satured conditions (fig. 7b) yields $\upsilon^* \approx 4.4\times10^{-29}m^3$. This agreement is consistent with the expectation that the degree of saturation controls the lifetime of nucleated kinks and not the nucleation rate.

The description of length of the wear track in terms of the rate of mechanically enhanced double-kink nucleation was found particularly efficient to explain the corrosive wear of single-crystal brushite in aqueous solution. In the case of atomic-scale corrosion, an understanding of fundamental effects is particularly necessary, requiring accurate details of interactions of the water molecules with stressed surfaces. The measurements of E_{act}, υ^*, and the factors for the key dissolution processes (such as pit nucleation, double-kink nucleation) provide an experimental foundation for a complete predictive model for corrosive wear of crystalline materials.

Chemical factors have been shown to be relevant for the particles removal, as in the case of single-crystal NaCl salt [Dickinson et al., 1999]. The nominal shear stress required to remove such particles from a soda lime glass substrate with an SFM probe tip is a strong function of particle size and relative humidity. At 3% of relative humidity (RH), particles larger than about 500nm on an edge could not be removed by the probe tip at any accessible contact force. Increasing the RH to 30% dramatically reduces the stress required for particle removal and promotes the removal of much larger particles. Further increases in RH yield smaller decreases in the failure stress. The humidity dependence of the shear stress can be described in terms of the expected change in interfacial energy due to water absorption. The stress required for particle removal scales roughly with the inverse square root of the particle area. Fracture mechanics can help to suggest that this is associated with an initial flaw size that is roughly proportional to their area. An improved understanding for particle removal could be an important factor in a number of technologies including integrated circuit manufacture and optical component manufacture. The experiments of Dickinson and co-workers showed in special way how the combination of chemical and mechanical effects plays a key role for tribological features on molecular scale.

Recently, a series of nanowear tests where the SPM in UHV has been used to realize and detect atomic-scale abrasion on KBr(001) have been prformed [Gnecco et al, 2002; Socoliuc et al., 2003]. They showed how the debris extracted from the KBr(001) surface by a silicon tip is reorganized on the atomic scale, figure 8. The dependence of wear vs time was investigated scratching across a single line for a long time at fixed load and velocity (5120 scans at F_N=11.2nN and scanning velocity of 2.5μm/s). The mean lateral force $<F_L>$ increased continuously from an initial value as a function of the number of scratches N. The behavior of the lateral force is a clear demonstration of the abrasive character of the wear process. In fact, the ions seem to be removed and released by pairs or by small cluster of ion pairs, and no dramatic crack events were observed in thousands of scans, and the regular arrangement of the deposited material resembles an epitaxial growth of thin ionic films, where the ions are piled up pair by pair. The ion removal can also be recognized in the irregularity of the friction loops, where the number of displaced ion pairs is comparable to the number of slip events, as the ions were detached when the tip jumps from one stick position to the next one. Generally, the stick-slip appears more irregular compared to atomic stick-slip observed without wear.

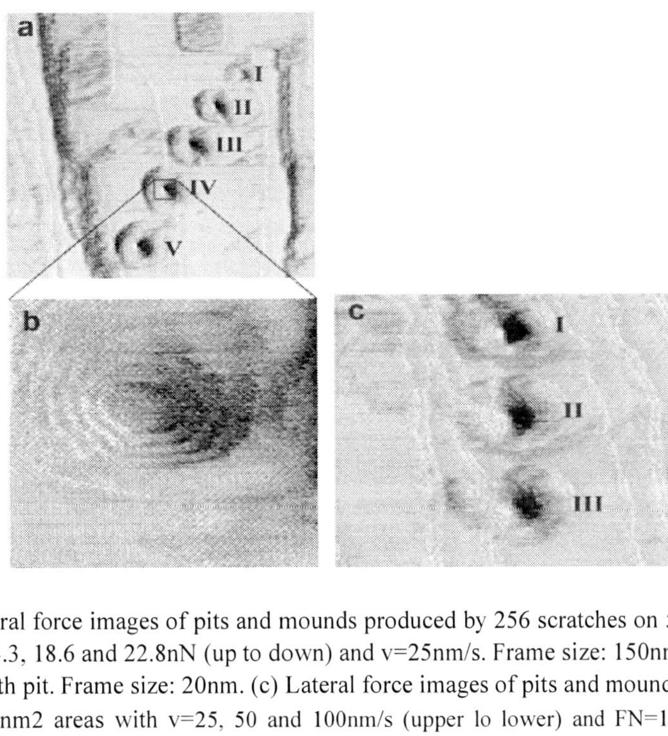

Figure 8. (a) Lateral force images of pits and mounds produced by 256 scratches on 5×5nm2 areas with FN =5.7, 10.0, 14.3, 18.6 and 22.8nN (up to down) and v=25nm/s. Frame size: 150nm. (b) Lateral force image of the fourth pit. Frame size: 20nm. (c) Lateral force images of pits and mounds produced by 256 scratches on 5×5nm2 areas with v=25, 50 and 100nm/s (upper lo lower) and FN=14.1nN. Frame size: 74nm.

The AFM scratch process was also used to produce nanoscale regular patterns on KBr(001) in UHV. In such tests, wear of single atomic layers has been observed when the AFM tip is repeatedly scanned across a line. The initially flat surface is rearranged in a quasiperiodic pattern of mounds and pits, figure 9. The distance between the pits was observed to be about 40nm when normal forces of a few nNs are applied, and it slowly increases with the load. If a square area is scanned, a pattern of regular ripples was formed. Ripple formation due to perturbations acting on a surface is commonly observed on macroscopic scale. Well-known examples are given by wind-blown sand dunes observed in the desert and in the submarine sands [Bagnold, 1941]. In the case of the continuous scratching of AFM probe tip, such features can be interpreted within an erosion-diffusion process induced by a periodic increase of the strain produced by the scanning tip.

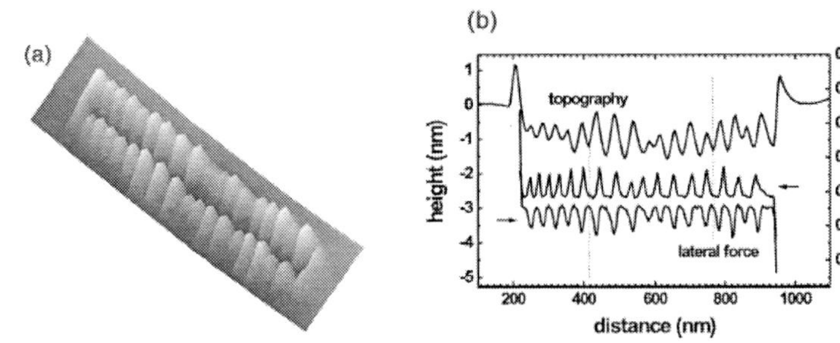

Figure 9. (a) Topography image of a groove formed on KBr(001) after 512 scratches along the (100) direction with a normal load of FN=26.6nN and a scanning velocity of 2.67μm/s. the groove is surrounded by a pattern of 34 mounds aligned along the two sides, and by two mounds piled up at its ends. The height difference between the top of the mounds and the bottom of the pits in the groove is about 2.5nm. The distance between two adjacent pits or mounds is about 40nm, as well as the distance between the two rods or mounds. (b) Cross section of the topography along the ground of the groove and lateral force acquired while scratching the groove. The lower and upper parts of the lateral force loop were obtained while scanning forwards and backwards, as indicated by the arrows. The dotted lines indicate positions in the pits where the tip is decelerated as derived from an increase of the lateral force in either direction. (Reprinted from [Socoliuc et al, 2003], with permission).

Figure 10. On left, topography map of a region scratched with the following loads (up to down): 1.7, 5.3, 8.8, 12.4, 15.9, 19.5, 23, 26.6, 30.1nN; frame size: 1.2μm. On right, development of lateral force with time while scratching a single line of 800nm length forwards and backwards. In horizontal direction the lateral force during forward scan is plotted, where darker colour means stronger lateral force; the vertical direction is the time axis. (Reprinted from [Socoliuc et al., 2003], with permission).

One common recognized explanation of the results showed in the various experiments mentioned in the section converge on the attribution of onset wear as an activated process, where the stripping off atomic or molecular debris is due essentially to the scanning probe tip induced stresses. The next section is devoted to examine thoroughly the activated process when applied to the onset wear problem.

3. ONSET WEAR AS AN ACTIVATED PROCESS

In the preview section, there have been shown some experimental results linking the onset wear to the activation rate for the breaking of bonds and consequent stripping off of atomic debris. The onset wear so described seems to be very close to activation for chemical reactions. Ever since Kramers seminal paper [Kramers, 1940], the fluctuational escape over a potential barrier has been a paradigm for a thermal activation process. Kramers understood chemical reactions, as the climbing of an energy barrier between two energy wells in a large configuration space, where the first well represents the state of the reactant molecules and the second well those of the product molecules. Activated escape underlies diffusion in crystals, protein folding, nucleation rate in phase transitions and provides a paradigm for activated chemical reactions. In a wear experiment, we can be mainly interested to quantify the amount of mass stripped off from a surface in relative motion with another generally harder surface. As a consequence of the wearing debris is formed, one possibility is to describe the change of spatial location for the defects produced during the damaging as described in figure 11 and figure 12.

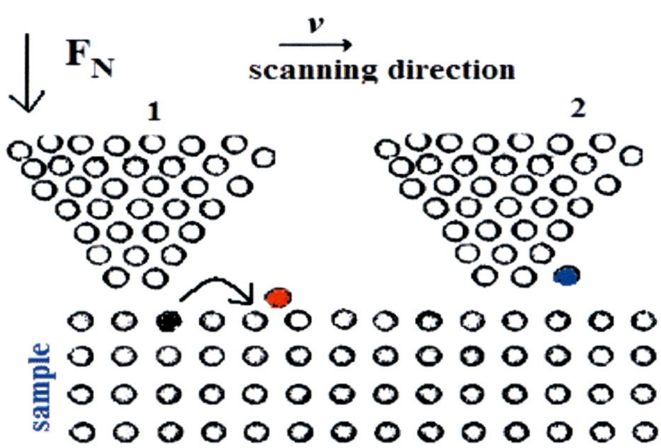

Figure 11. Schematic sketch on the production of debris (defects) due to abrasion wear-mechanism or adhesion wear-mechanism. The scanning tip in contact with the sample produces a damage on the regular surface moving the black atom, relocating in a new position (red atom), or adsorbed by the scanning probe tip (blue atom). The AFM-induced activated possible detachment of the black atom should be described by a conditional probability $P_1(x,t_1/x_0,t_0)$, where at time t the atomic debris is in the x position after the tip stress contact is made starting from x_0 position before tip contact, and the possibility of relocation on the native surface or adhesion to the stressing tip after the contact should be given by an other conditional probability, $P_2(y,t_2/x,t_1)$. The whole conditional probability is the convolution between these two probability functions.

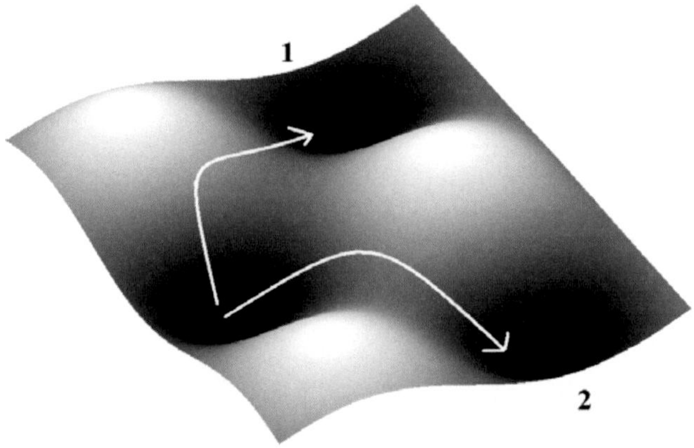

Figure 12. Schematic sketch to describe the well potential transition applied to describe the possible location of a debris stripped off from the native surface during sliding motion between two interacting surfaces, as in figure 11. For molecular scale contact, the debris has molecular dimensions, and the different final location of debris are connected to the specific wear-mechanisms. As a consequence, the final wells 1 and 2 represent different surfaces (in analogy to figure 11, re-positioning on sample surface or tip surface) and wear mechanisms.

In activated process, meanly we have a system of interest at thermal equilibrium connected to a driving force responsible for the activation of process from the system of interest. The mechanism responsible can be readily understood for adiabatically slow driving, where the driving frequency is small compared to the relaxation rate in the absence of fluctuations and the system of interest remains in quasi-equilibrium. For systems in thermal equilibrium, the fluctuation probabilities are given by the activation law, $W \propto \exp(-E/k_BT)$, where E is the activation energy of escape, that is tuned by driven force, and even where the small fluctuation of amplitude $|\delta E|$ is small compared to E, it may still substantially exceed k_BT, in which case W will be changed very strongly. In our picture, W represents the probability that a portion of mass (debris) can be detached by one surface when it is in relative motion respect to other one surface. The formation of wear debris is a random process and stochastic picture should be the most adequate tool to approach quantitative results. A stochastic process is the correspondence between a random experiment and a group of functions, as a consequence, the system can be predicted in terms of probability as for example the velocity of a Brownian particle [van Kampen, 1983; Risken, 1984].

Effectively, the connection between Brownian motion and debris can be made only a level of product of wear process, when a debris is a formed detached from the native surface and its mechanical features are subject to environmental fluctuations. Two systems in contact, such as the AFM tip and the scanned sample share a common surface. Each exerts an equal and opposite force on the other determined by the pressure it exerts on every elements of the surface of separation as required by the physics of an open systems [Bader, 2000]. Like general approach to open systems, we can consider a sample of interest in contact with external or environmental forces. In the tip-sample contact, the external forces are represented by the tip-induced stresses. Some general assumption can be made in order to simplify the tip-sample systems looking at the different time scale for the motion of the AFM tip and the atomic or molecular components of the sample investigated. The AFM has an effective cantilever mass around 10^{-11} Kg, many orders of magnitude larger than the atomic mass unit (~10^{-27} Kg). The resonant frequency of the cantilever is around 10^{5} Hz, while the atomic vibration frequency of the order of 10^{13} Hz. Practically, in the contact area we have two kinds of dynamics, the slow dynamics of the cantilever and the fast vibrations of atoms. At the time scale of picoseconds, we encounter thermal vibrations of atoms, while the cantilever is stationary. On contrary, at the time scale of microsecond, we observe the scanning vibrations of the cantilever in contact with a cluster of atoms. If the relaxation time of this atomic cluster is much smaller than the cantilever time scale, then we will always observe an equilibrium cluster of atoms that follows the slow motion of the cantilever. Therefore, the energy modes between the cantilever and the atomic substrate can be decoupled. In Molecular Dynamics simulations, integration of dynamic equations of motion for the cantilever and the atomic substrate can be performed separately. As the consequence, we can limit to study the possible escape process for the sample atoms considering the effect of the AFM tip in terms of induced stresses lowering the energy activation for the surface atoms and considering the tip rigid.

For example, let us consider a general situation where the fundamental wear-mechanisms are abrasive and adhesive wear mechanisms. The quantification of wear due to adhesive or abrasive forces has been calculated in terms of rate of transition between two subsequent well potential. The problem can be reduced to a bi-dimensional question. Along the z-axis, the vertical forces are defined by a van der Waals term leading to the adhesion mechanism. Shear forces along the x-axis, i.e., the direction of motion of the tip leading to the abrasive mechanism for the atomic debris. The occurrence of one or both of two mechanisms depends on the intensity of the forces along the two dimensions, respectively. Such situation can be resumed by the sketch in figure 12, where is shown the simultaneous double-well for the two wear processes. In both the cases, the wear volume rate could be calculated within the microscopic diffusive processes can be expressed by the formula [Bassani and D'Acunto, 2000; D'Acunto 2003; D'Acunto 2004]

$$\Psi = n_{tot} w \Theta \tag{7}$$

where n_{tot} is the total number of possible debris lying on the primitive sample surface, w is rate of transition associated to the activated process, and Θ is the mean debris volume (pair ionic atomic volume). In experiments performed on ionic crystal, the mean debris volume can be identified in ionic pairs, so the wear volume rate can be reduced in the calculation of wear rate for activated de-bonding under applied tip stresses. Equation (7) should be considered an equivalent of macroscopic wear volume equation (1) and (2) on nanoscale. In a simplified vision, wear contributions due to adhesive- and abrasive-wear mechanisms can be considered separated and summarised as $\Psi=\Psi_{abr}+\Psi_{adh}=(n_{abr}+n_{adh})\Theta$, where Ψ_{abr} and Ψ_{adh} are the abrasive- and adhesive-wear volumes, respectively, and n_{abr} n_{adh} are the number of atoms involved for the two distinct wear mechanisms. In reality, the overlap of the two wear processes can be easily supposed. The experimental results performed on ionic compound in ultrahigh vacuum (UHV) by Gnecco et al., (2002), show clearly that $n_{abr} >> n_{adh}$ to be valid, i.e., the abrasive-wear mechanism is stronger compared to the activation of adhesive-wear mechanism. The advantage of modelling the interface between two surfaces by a double-well potential is that the wear volume can be quantified essentially in terms of rate of transition between the well minima. Moreover, the rate and (as a consequence) the wear volume depend on the potential barrier height, and the potential barrier height would be connected to experimental and materials quantities such as hardness, elasticity, load applied, and time (or sliding velocity). Obviously, the wear volume will be a function of the energy barrier. Essentially, vertical forces lead to adhesion-wear mechanism (transition from surface sample to tip), and shear forces lead to abrasion-wear mechanism (transition from surface sample to surface sample). Independently by the wear mechanisms activated, the rate of transition in equation (7) is given anyway by the Kramers picture

$$w_{abr,adh} \approx e^{-\Delta E_{x,z}/k_B T} \tag{8}$$

where the index is referred to x-dimension for abrasion and z-dimension for adhesion. To find the force of interaction between an SFM tip and a flat sample, the interatomic potential has to be summed over the whole number of atoms involved in the contact between sample and tip. The accurate derivation of such rate of transition for a double well potential model were derived in [Bassani and D'Acunto, 2000; D'Acunto 2003; D'Acunto, 2004].

The probabilistic approach to the production of wear debris can be performed within the classical double-well potential with time fluctuating barrier. The barrier is fluctuating because during the contact between the tip and the sample we have two different time scales. Here, we limit to comment the wear rate as obtained as function of the frictional state in the initial well potential. Each crossing particle is defined by a frictional term, proportional to $\gamma\dot{x}$, where $\dot{x}$ is the velocity of the particle living in the initial well and γ is the damping factor. The physical nature of such frictional term can be explained in terms of irreversible transferring of energy to the environment, other particles or phonon states. Nevertheless, there is another irreversible process, which costs energy leading to a non-conservative lateral force during the contact between the probe tip and the sample. This supplement of friction is manifest during the transition between well potential. The total amount of lost energy for the sample can be quantified like

$$F = n_A w \gamma \dot{x} / L \tag{9}$$

where L is the spatial distance between the minima of the well potential, n_A described the atom cluster on the native surface, (initial well potential) and $n_A w$ is the debris formed during the contact with the tip. The frictional state of particles in the initial well modifies the wear rate. In this sense, we can define a correlation between molecular friction and adhesive wear. In condition of high internal viscosity, i.e., when the cluster velocity relaxes much more rapidly than the position, constant wear rate as

$$w(t) = \frac{\Omega^2}{2\pi\gamma} \exp\left(-\frac{\Delta U(t)}{k_B T} \right) \tag{10}$$

where Ω is instantaneous effective oscillation frequency and ΔU is the energy barrier between the two wells. As can be expected, the wear rate depends by an Arrhenius factor and it is interesting to note that the rate does not depend on the shape of the potential barrier, but only on the difference of the top barrier. The total number of particles passing the barrier in one second is equal to $w \cdot n_A$. This is the case of high viscosity. It is also possible to treat in a similar fashion the limiting case of low viscosity with result obtaining the expression [Risken, 1984]:

$$w(t) = \gamma \frac{\Delta U(t)}{k_B T} \exp\left(-\frac{\Delta U(t)}{k_B T} \right) \tag{11}$$

where now, the wear rate is directly proportional to the friction term γ. Equations (10) and (11) can be slightly modified if we consider additive fluctuating terms to potential barriers due to noise of other nature respect to thermal noise due to the complexity of dynamic interaction between tip and sample in a sliding process, where anomalous diffusion processes take place [D'Acunto, 2003]. The presence of such fluctuations in the potential barrier is important since it may be hypothesised that in certain catalytic reactions, the catalyst-environment interaction could give rise to effects which can be represented with an effective potential which contains a supply noise coupled with position variable [Stein, et al., 1990; San and Wortis, 2004].

As mentioned above, the combined effects of the sliding tip and interatomic forces can be summarized in terms of shearing forces and vertical forces. The first ones could be considered responsible for the detachment of the ionic atoms from its crystal site and subsequent dragging until a new equilibrium crystal site is reached. The new position is achieved when the equilibrium is approached, this new position is a step on the flat atomic surface, figure 11. The ionic crystal shows that the process involves pair by pair ionic atoms. Vertical forces are due to the interaction of the tip probe approaching the surface sample. This interaction is led by a van der Waals force and stimulates the ionic atoms to follow the tip adhering to it, figure 11. In terms of effective wear rendering, the adhesion causes a vacancy along the crystal surface. Following the equation (10), the abrasive- and adhesive-wear mechanisms can be quantified considering the rate of transition as given by Kramers picture for the ionic atoms in a two-dimensional well potential. The correspondent height of energy barriers determines the incidence of wear both for abrasion or adhesive case. In the case of shear forces, the energy barrier can be deduced by the surface potential U_0, and a reasonable energy barrier along the x-direction $\Delta E_x \approx U_0$.

At the minimum value of the barrier, the vertical force is given essentially by the energy binding of the attractive term of the Van der Waals force. The apex tip is considered a cylinder of radius $R \cong 10$nm, the potential barrier can be expressed like $\Delta E_z \approx A R^{1/2} r_0 / z^{3/2}$, where z is the distance of separation between the surface and the tip and A is the Hamaker constant [Israelachvili, 1992]. The distance of separation, z, is directly proportional to the applied load. After the minimum value of the barrier is reached, correspondent to the minimum position z_{min}, the first term of a Lennard-Jones potential is not negligible. Moreover, the temperature should increase the rate lowering the well potential. Although the effect of temperature could engrave in a different way on the shear forces compared to the vertical forces. Nevertheless, the role of thermal fluctuations should be investigated more appropriately. Let us consider a volume of 120nm×120nm×1.8nm, the density of atoms involved in the wear processes is given by the summation on the first three layers of the quantity $n_L \exp(-(i-1))$, where n_L is the number of atoms for layer, and i=1,2,3, denotes the layers. The exponential correction is introduced because the atoms lying on the inner layer are less involved in the possible wear mechanism compared to the first layer one. The lattice constant is considered to be a=0.6nm, any layer contains 200 atoms, and the total number of atoms could be involved in the wear process for the volume considered is nearly 6×10^4. The calculation for an applied load of F_N=0.65nN is performed at room temperature, T=298°K, the thermal energy is $k_BT=0.041\times10^{-19}$J, the lattice potential barrier is $U_0=0.432\times10^{-19}$J, and the Hamaker constant $A=10^{-19}$J. Inserting these values in the expression of the barriers ΔE_x and ΔE_z we obtain that $\Delta E_x/k_BT\cong19$, while $\Delta E_z/k_BT$ is nearly 15 times $\Delta E_x/k_BT$. Because the functional dependence of the rates by the ratio between the energy barrier and the thermal energy is $x\exp(-x)$, the second term correspondent to vertical transitions toward the tip is practically negligible. The rate of atoms passing a potential barrier and undergoing a wear mechanism is given dividing the wear volume by the atomic volume

$$V/\Theta = nw_{abr} \approx n\left(\frac{2\pi/T_0T_1}{\beta}\exp\left(-\frac{U_0}{k_BT}\right)\right) \tag{12}$$

For a load of F_N=0.65nN and a cycle of 1000 scans, equation (12) gives nearly 2400 atoms involved in the abrasion-wear mechanism, i.e., the 3.9% of the total number of atoms, while the number of atoms undergoing an adhesive-type process is practically irrelevant. Unfortunately, it is difficult to perform a direct comparison with experimental results obtained by Gnecco et al., 2002. In fact, their experimental wear tests were made at higher applied loads, as $F_N\approx5\div25$nN. Nevertheless, it is remarkable that this model describes in a direct manner that the incidence of abrasive-type mechanism for wear is the most prominent mechanism for wear on nanoscale for low applied loads.

At low loads, it can be interesting to know the effects of the tip on the atoms, which are not able to pass the energy barrier. These atoms do not form the material loss but can be subject to a lattice modification, like dislocations, and consequent formation of gradient stress.

3.1. Self-Assembled Monolayers as a Frame to Study Wear in Viscoelastic Materials

The use of viscoelastic materials such as polymeric materials constantly increases in the field of nanotechnology. These materials are softer than metallic and inorganic ones, and, because of that, they are easier to wear and deform. Therefore, the wear mechanisms occurring for viscoelastic materials are rather complex, and, generally, present more complications respect to metals or ceramics materials. As a consequence, such mechanisms should be analyzed in a simplified model. Self-Assembled Monolayers (SAMs) are very useful for the systematic modification of the physical, chemical and structural properties of a surface by varying chain length, tail group and composition. Many of these properties can be studied making use of the AFM, and the interaction between AFM probe tip and SAMs can be considered also an excellent reference to study fundamental properties of dissipation phenomena and onset wear for viscoelastic materials on the nanoscale.

In this subsection, we deal out a numerical study showing that the fundamental mechanism for the onset wear is a process of nucleation of domains starting by initial defects. A SAM surface repeatedly sheared by an AFM probe tip with enough applied loads shows the formation of progressive damages enucleating in domains. The AFM induced surface damages involve primarily the formation of radicals from the carbon chain backbones, but also the deformations of the chains resulting in changes of period lattice have to be taken in consideration. The nucleation of the wear domains starts generally by initial surface defects where the energy cohesion between chains is lower. Moreover, the presence of surface defects is consistent with changes in lateral force increasing the probability of the activation for the removal of carbon debris from the chain backbone. The quantification of progressive worn area can be performed making use of the Kolmogorov-Johnson-Mehl-Avrami (KJMA) theory for phase transition kinetic processes [Sear, 2004; Jun et al., 2005]. The advantage of the knowledge of general conditions for onset wear on SAM surfaces can help to know fundamental mechanisms for the tribological properties of viscoelastic materials, in solid lubrication applications and biopolymer mechanics.

When the AFM probe tip slides on a deformable viscoelastic surface, relaxation of inner polymer structure leads to inevitable contact damping, which makes surface forces non-conservative. The class of SAMs provides a useful tool for controlling both physical and chemical properties of a solid viscoelastic surface [Ulmann, 1991]. Since the pioneering paper of Nuzzo and Allara [Nuzzo and Allara, 1983], the fabrication of nanoscale structures using SAMs has attracted attention due to their scientific importance and potential applications [Haussling et al., 1991; Ulman, 1996; Schlesser et al., 1996; Tidwell et al., 1997; Lavrich et al., 1998]. SAM and Langmuir-Blodgett (LB) techniques are the most common methods for forming monomolecular films on solid substrates for use as model lubricants [Bliznyuk et al., 1998; Barrena et al., 1999; Bhushan and Liu, 2001]. In the case of LB films, the molecular interaction with the substrate is through weak van der Waals forces, and repeated shearing results in rapid wear of the molecules. On contrary, SAMs are ordered molecular assemblies formed by the chemical adsorption of an active surfactant on a solid surface and are therefore better candidates as lubricants. SAMs can be formed by immersion of an appropriate substrate into a solution of an active surfactant in an organic solvent. SAMs consist of three building groups: a head group that binds strongly to a substrate, a tail group that constitutes the outer surface of the film and a spacer chain (backbone chain) that connects the head and tail groups, figures 13-14. It is now well established that complete alkanethiol monolayers formed on Au(*111*) adopt a commensurate $(\sqrt{3}\times\sqrt{3})R30^\circ$ structure, with a nearest neighbor spacing of 0.497nm, which corresponds to a surface area of $0.2nm^2$ per chain [Hautman and Klein, 1989], figure 13.

The self-assembly mechanism of alkanethiol C_n, $[-SH(CH_2)_{n-1}CH_3]$, molecules on Au(*111*) has also been studied using several techniques including scanning force microscopy techniques [Delamarche et al., 1994; Poirier, 1997; Takano et al., 1999; Guo et al., 2002].AFM sliding friction experiments on C_n-alkanethiol SAMs showed a friction behavior dependent by chain-length [Lio et al., 1997]. The coefficient of friction decreased as the chain length was increased and the low friction of methylic group terminated SAMs is due largely to low surface energy coming from their hydrophobic tail groups and high load bearing capacity due to their compact structure minimizing van der Waals interaction energy.

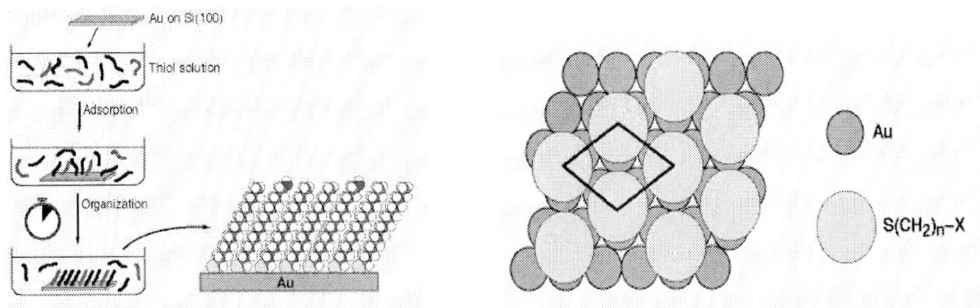

(a) (b)

Figure 13. (a) Schematical sketch for the preparation of SAMs. The substrate, Au on Si, is immersed into an ethanol solution of the desired thiol(s). Initial adsorption is fast (seconds); then an organization phase follows which should be allowed to continue for >15 h for best results. A schematic of a fully assembled SAM is shown on the right. (b) A schematic model of the $(\sqrt{3}\times\sqrt{3})R30°$ structure formed by alkanethiolate SAMs on Au(*111*).

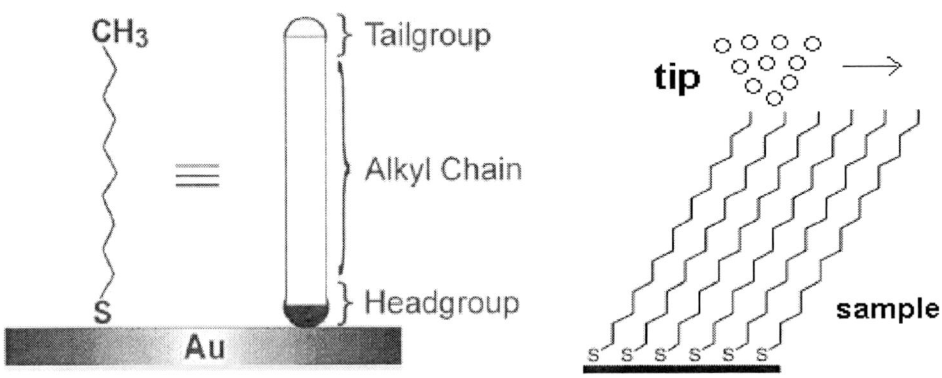

(a) (b)

Figure 14. (a) Single C_{15}-alkanethiol chain, and the three main components composing a chain: the headgroup, the alkyl chain or carbon backbone and the tailgroup, that in our case is a methyl group. (b) Schematic sketch of the AFM probe tip-SAM system. The contact with the probe tip involves several possible deformations for the alkanethiol chains. The close-packed arrangement of molecules should hamper the large deformations because these require space and are energetically expensive.

Onset wear tests can be simulated on a SAM surface taking as an example an experimental AFM test consisting in many scanning cycles on the same area and acquiring data of such progressive damaged area scan after scan. In the simulated approach, the percentage of the worn area can be studied as a function of several parameters, such as applied loads, lateral friction force and chain lengths, and curves are plotted indicating evolution of progressive damages. Consequently, the progressive damaging of the surface due to the continuous passage of the AFM tip can be monitored scan by scan. Because SAM lattice structure, this implies the possibility of a 2D quantification for the wear damage rather than classical macroscopic volume test quantification. In a wear test making use of an AFM, the regular lattice of SAM surface represents an excellent system that can display the time evolution of damages induced by the continuous passage of the probe tip. For this reason, we can quantify the progressive wear formation on the SAM surface observing how the primitive regular lattice area is progressively changed and reduced during the wear test. When the applied load is not enough low, deformations of the SAM structure or defects or formation of free radicals due to applied stress lead to changes in the 2D regular structure. The connection between dissipation mechanisms and wear is essentially given by the fact that initial surface defects correspond to the points where the lateral force and exchange of energy between tip and sample are higher.

It is well known that alkanethiol monolayers on Au(*111*) exhibit a distribution of pit-like defects [Endinger et al., 1993]. These pit-like defects should be a consequence of already existing defects of underlying Au surface. Other defects can be formed during alkanethiol monolayer assembly. They diminish locally the physical and chemical properties of alkanethiols chains, other than to change the height profiles. During the AFM induced compression, the molecular spring assembly can experience orientation and compression under normal load [Salmeron, 2001], so, the SAM chains experiment further deformations that can be considered in the class of defects. Generally, three types of AFM induced viscoelastic defects can be considered. The first type involves the creation of gauge distorsions at the free end groups. This results in a 120° rotation of the methyl group around the C-C backbone. The height shortening of the chain is expected by about 0.1nm, and the energy activation for such rotation is around 0.1×10^{-19}J. The second type of AFM induced defect is internal gauge deformations, which require displacement of the neighboring molecules. The third type is given by the rigid chain tilts. These two last types of defects are more difficult to produce, because their energy activation is considerable higher. Nevertheless, in the surrounding of structural initial defects such as pit-like defects, the alteration of the molecular packing density can lower their energy activation. Barrena *et al.* observed that the heights of self-assembled alkysilanes decrease in discrete amounts with normal load [Barrena et al., 1999]. This step-like behavior is due to the discrete molecular tilts, which are dictated by the geometrical requirements of the close packing of molecules. Only certain angles are allowed due to the zigzag arrangement of the carbon atoms.

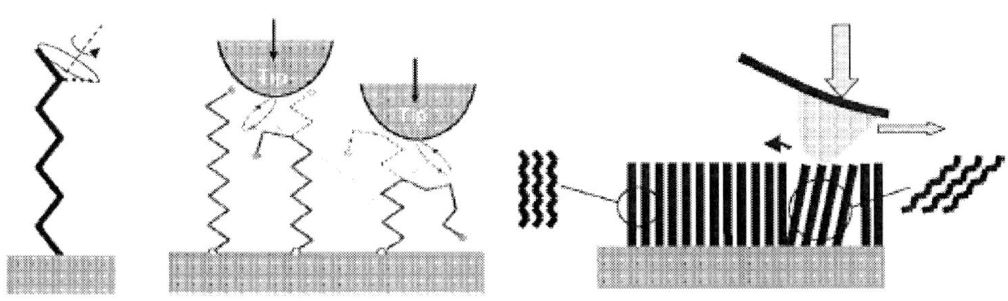

(a) (b) (c)

Figure 15. Schematic representation of three types of vicoelastic deformations of alkane molecules proposed by Salmeron, [Salmeron, 2001]. (a) Terminal gauche distorsions, (b) internal gauche distorsions, which require displacement of the neighboring molecules, and (c) rigid molecular tilts. The requirement of displacing neighboring molecules is less stringent in this case due to the concerted motion. (Reprinted from [Salmeron, 2001], with permission).

It is reasonable suppose that the energy exchanged between the probe tip and a SAM sample involves the initial defects points for nucleation activation for larger defects when the scanning probe tip slides more and more times. The idea is that at the edge of a defect the energy cohesion due to van der Waals interaction and responsible of chain packaging should be lowered by the extra-space available due to the presence of defects. The van der Waals energy cohesion ε_{vdW} is nearly 0.6-1.3×10^{-20}J, and the van der Waals radius of the methyl groups is approximately 0.2nm in a SAM packed structure [Salmeron, 2001]. The bound energy of C-C is nearly 1÷2×10^{-19}J, and S-C and S-Au binding energy fall in the range 2÷4×10^{-19}J [Lavrich et al., 1998].

Under opportune applied loads, the shearing stress induced by the passages of the probe tip should provoke the enlargement of the defects area provoking modifications of period lattice with subsequent changes of orientation of the chains and, over threshold loads, the breaking of covalent bonds with the formation of holes. These holes are due to the stripping off of radicals, such as CH_3-, macroradicals such as $(CH_2)_x$-CH_3-, or simply H-, here subscript x denotes the fraction stripped off from the original chain backbone. But, it is reasonable to suppose that the AFM induced de-bonding regards essentially C-C connections. Any carbon end group can be considered in a well potential and the passage of probe tip should make possible to overcome this potential barrier with the probability of transition enhanced by the stress applied, figure 12. Generally, the formation of active debris should involve the adhesive attachment of such SAM debris to the sliding probe tip changing the shape of the probe tip (adhesion tip wear) and as a consequence the real area of contact. Adhesion tip wear due to the attachment of progressive formed debris to the probe tip could be describe satisfactory by a double well potential, where any well potential corresponds to one surface in dynamic contact with other one, as described in the preview section on activated process [Bassani and D'Acunto, 2000; D'Acunto, 2003; D'Acunto, 2004]. Although the adhesion tip wear is a common occurring wear mechanism, in this review, we will not include its description, limiting ourselves to study general condition for progressive damaging of SAM surface without considering the dynamics of SAM debris when detached by the carbon backbone.

It is reasonable to suppose that in an AFM abrasive test consisting of a certain number of scans the density of defects should increase before gross wear will take place. This process is essentially a process of nucleation of defects forming holes increasing in time, where the time is the product of the residence time of the probe tip for the number of scanning cycles. Nucleation is an activated process [Dickinson et al. 1996; Dickinson et al. 1997; Dickinson et al., 1999; Scudiero et al., 1999], as such its rate has an exponential dependence on the free-energy barrier to nucleation defects. If at the *i-th* site of the SAM surface the free energy barrier is ΔU_i, and the frequency of attempts at unfolding is ν_i(~10^{13}Hz), then the rate of nucleation at the *i-th* site is

$$\Gamma_i \approx \nu_i \exp\{-\lfloor \varepsilon_{0,i} - \delta\varepsilon_i(F_N, F_l)\rfloor / k_B T\} \tag{13}$$

where $\Delta U_i = \varepsilon_{0,i} - \delta\varepsilon_i(F_N, F_l)$ is expressed by the mean value ε_0 of the C-C bond and the stress-induced deviation represented by the term $\delta\varepsilon(F_N, F_l)$ that is a function of the applied load F_N and lateral force F_l, k_B is the Boltzmann's constant, T is the temperature. The stress-induced deviation is a combined effect of compression and shear stresses and it lowers the energy barrier making possible the carbon stripping off from the chain backbone. The lateral stress is generally higher in correspondence of defects respect to regular lattices, as a consequence, the rate of nucleation must be calculated locally and cannot be considered homogeneous. The number of defects produced in the absence of stresses is rather small, under the tip stress the maximum possible rate of debris production, $\Delta N_i/\Delta t$ at the i-*th* point contact, should be given by the relation for thermal activated detachment [Weeks et al., 2002]:

$$\frac{\Delta N_i}{\Delta t} \cong \frac{R^2}{a^2}\Gamma_i \tag{14}$$

where $\Delta t=(A_r/A_{\text{image}})t_{\text{image}}$ is the residence time of the probe tip on the sample, A_r is the real area of contact, A_{image} is the area to be scanned and t_{image} is the image required for an image [Kopta and Salmeron, 2000], R is the radius of the real area of contact between the tip and the SAM, a is the van der Waals average radius of single methyl groups, and Γ_i is the rate of activation for the creation of AFM induced defects. Here, we assume that the continuum mechanics could give a good approximate value for the real area of contact and the depth of penetration even if the contact is discrete [Leng and Jiang, 2000]. For a paraboloid shape of the probe tip in non-adhesive contact with a flat surface one finds that the real area of contact and penetration depth are respectively [Johnson, 1985]:

$$A_r = \pi\left(\frac{3F_N\rho}{4E^*}\right)^{2/3} \qquad z_d = \frac{1}{2}\left(\frac{3F_N}{4\rho^{1/2}E^*}\right)^{2/3} \tag{15}$$

where $E^*=\left[\left(1-v_1^2\right)/E_1+\left(1-v_2^2\right)/E_2\right]^{-1}$ is the reduced modulus, $E_{1,2}$ are the Young moduli and $v_{1,2}$ the Poisson's ratios, F_N is the applied load and ρ (=1nm) is the tip radius. For typical moduli of Young and an applied load of 2nN, we obtain a value for the reduced modulus of nearly 29.5Gpa [Leng and Jiang, 2000], as a consequence the real area is approximately 0.4nm^2, a single C_n-alkanethiol head group formed by $-CH_3$ has an area of 0.2nm^2, this means that approximately 2 final head groups are simultaneously in contact with the probe tip and support the applied load. It should be noted that differently to friction behavior, SAM films show chain-length independence of elastic modulus [Leng and Jiang, 2000].

We suppose that the decrease of activation energy $\delta\varepsilon(F_N,F_l)$ in eq. (13) has two major contributions depending by the normal stress or shearing stress. The first contribution, due essentially to the normal stress, $\delta\varepsilon_1$ (divided by the C-C bond energy, $\varepsilon_{C\text{-}C}$) can be estimated by multiplying the force acting on the methyl group carbon atoms involved simultaneously in contact with the tip by the resulting bond stretch

$$\frac{\delta\varepsilon_1}{\varepsilon_{C-C}} \approx \frac{a^2}{R^2}\frac{z_d}{\left(A_r/\pi\right)^{1/2}} \tag{16}$$

where z_d is the penetration depth, and A_r is the real area of contact, R is the radius of the real area of contact between the tip and the SAM, and a is the van der Waals average radius of single methyl groups. The relation (16) can be justified by the assumption that the applied load induced bond stretch should be on the same order as the Hertzian penetration depth z_d, divided by the characteristic length of the contact area, which is on the order of the contact radius. The ratio between the areas of contact and the van der Waals region of single methyl group takes in account the number of methyl groups supporting the tip-SAM contact. The relation (16) shows that the contribution of applied load stress to decrease in activation energy changes as $F_N^{1/3}$. In the second term of the decrease of activation energy, the shearing stress plays an important role and should contribute for a lowering factor as:

$$\frac{\delta\varepsilon_2}{\varepsilon_{C-C}} \approx \frac{d_{C-C}}{(A_l/\pi)^{1/2}} \frac{z_d \times stickingforce}{\varepsilon_{vdW}} \tag{17}$$

where $d_{C\text{-}C}$=0.153nm is the carbon-carbon bond length along chain axis, A_l is the lateral real area calculated at the edge of penetration depth, ε_{vdW} is the van der Waals contribution to the energy of cohesion between chains, and sticking force is the maximum value of the lateral force in sticking phase during the typical stick-slip motion of the probe tip sliding on the SAM surface. The existence of a penetration depth implies that the carbon atoms set at the edge of the penetration depth should support entirely the shearing stress provoked by the coming forward probe tip along the sliding direction [Nie et al., 2005]. The cohesion energy ε_{vdW} between chains involves a resistance to the shearing inducing debris production on the edge of the penetration depth as the tip slides. An advantage of a probe scanning based wear test is that the state of a progressively worn surface can be monitored scan after scan. The

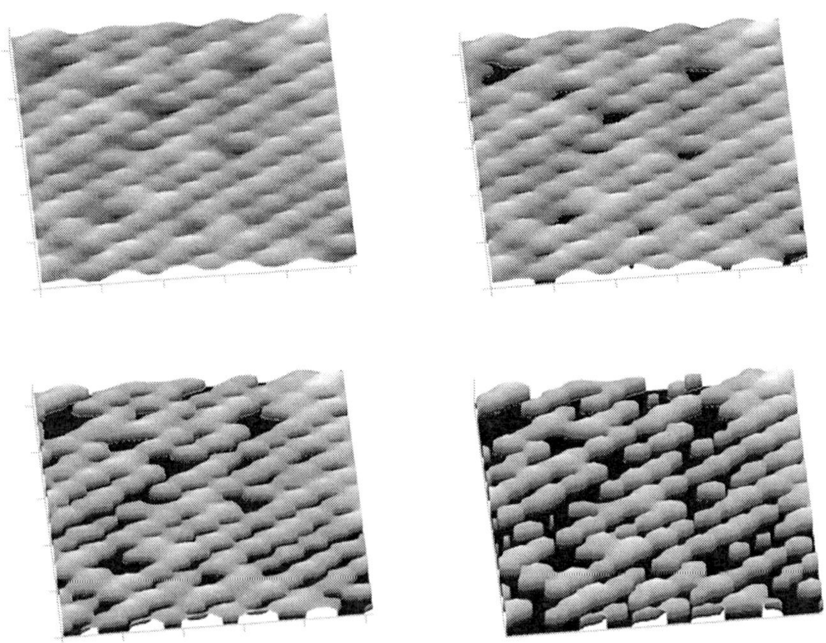

Figure 16. C_{18}-chain progressive damage of simulate methyl group end surface. The initial image (left upper) is based on an AFM experimental acquired data on a C_{18} sample, the typical commensurate c(4×2) uperlattice of (√3×√3)R30° arrangement is slightly altered by the applied load. The final image (right down) shows that nearly 20% of methyl groups have been stripped off by the AFM tip after 100 scans with an applied load of 5nN. The black portions represent the wear domains.

time evolution of such worn areas progressively damaged can be visualized as in a real AFM experiment. For example, figure 16 proves the progressive AFM induced damaging for a C_{18} surface for an applied load of 5nN.

4. Conclusion

The chapter has been devoted to give a review about experimental data and models on the onset wear occurring on nanoscale. Wear is the progressive loss of mass from the operating surface of a body as a result of relative motion at the surface. The occurrence of different wear mechanisms involves the fundamental mechanics of molecular and supra-molecular scale. As a consequence, a fundamental understanding of surface properties on nanoscale level should be generated to have a satisfactory knowledge of responses of materials and machine components also at macrometer scale. Investigation of the fundamental characteristics of wear at the micro-scale is complicated by some factors and forces that act on a nanoscale level have not yet been addressed in the tribology of macro-systems. Since these forces are sensitive to the environment and surface condition of the specimens, it is quite difficult to determine them accurately. Further, quantification of wear is not straightforward since the amount of wear is often too small to be detected by surface-sensitive instruments. Nano-wear measurements have been object of rapid development of precise measuring tools following to the introduction of atomic force microscopy (AFM). The advent of AFM opened new opportunity to study wear mechanisms on nanoscale, making possible to simulate one surface sliding on other one and controlling many physical parameters during the contact dynamics. Further, AFM made possible to a have a direct connection between experimental data and theoretical models.

Muscovite mica, ionic crystals, self-assembled monolayers represent ideal samples where to study the fundamental features for wear on nanoscale.

References

Archard, J.F.; 1953, *J. Appl. Phys.*, **24**, 981-8.

Bader, R.F.; 2000, *Physical Review* **B**, **61**, 7795-7802.

Bagnold, R.; *The Physics of Blown Sand and Desert Dunes*, Methuen, London, 1941.

Barrena, E.; Kopta, S.; Ogletree, D.F.; Charych, D.H.; Salmeron, M.; 1999, *Phys. Rev. Lett.*, **82**, 2880-3.

Barthel, E.; Roux, S.; 2000, *Langmuir*, **16**, 8134-38.

Bassani R.; and D'Acunto, M.; 2000, *Tribology International,* **33**, 434-443.

Bhushan, B.; Israelachvili, J.N.; Landman, U.; 1995, *Nature,* **374**, 607-16.

Bhushan, B.; and Koinkar V.N.; 1994, *J. Appl. Phys*. **75**, 5741-8.

Bhushan, B.; and Liu, H.; 2001, *Physical Review B*, **63**, 245412(1-11).

Bhushan, B.; *Springer Handbook of Nanotechnology*, Springer-Verlag, Berlin, 2003.

Bliznyuk, V.N.; Everson, M.P.; and Tsukruk, V.V.; 1998, *J. of Tribology,* **120**, 489-96.

Carpick, R.W., Agrait, N.; Ogletree, D.F.; Salmeron, M.; 1996a, *J. V. Sci. Technol.* **B 14**, 1289-95.

Carpick, R.W., Agrait, N.; Ogletree, D.F.; Salmeron, M.; 1996b, *Langmuir*, **12**, 3334-40.
Carpick, R.W.; and Salmeron, M.; 1997, *Chemical Reviews*, **97**, 1163-94.
Chung K.H.; Kim D.E.; 2003, *Tribology Letters*, **15**(2), 135-144.
D'Acunto, M.; 2004, *Nanotechnology*, **15**, 793-804.
Delamarche, E.; Michel, B.; Gerber, C.; Anselmetti, D.; Guentherodt, H.J.; Wolf, H.; Ringsdorf, G.E.; 1994, *Langmuir*, **10**, 2869-71.
Dickinson, J.T.; Park, N.S.; Kim M.W.; Langford, S.C.; 1997, *Tribology Letters* **3**, 69-80.
Dickinson, J.T.; Hariadi, R.F.; Scudiero, L.; Langford, S.C.; 1999, *Tribology Letters* **7**, 113-119.
Edinger, K.; Golzhauser, A.; Demota, K.; Woll C.; Grunze, M.; 1993, *Langmuir,* 9, 4-8.
Erlandsson, R.; Hadziioannou, G.; Mate, C.M.; McClelland, G.M.; Chiang, S.; 1988, *J. Chem. Phys.* **89**, 5190-93.
Gåhlin, R.; and Jacobson S.; 1998, *Wear,* **222**, 93-102.
Gnecco, E.; Bennewitz, R.; Meyer, E.; 2002, *Phys. Rev. Lett*., **88**, 215501(1-4).
Guo, Q.; Sun, X.; Palmer, R.E.; 2002, *Phys. Rev.* **B, 71**, 035406.
Haussling, L.; Ringsdorf, H.; Schmitt, F.J.; Knoll, W.; 1991, *Langmuir*, **7**, 1837-40.
Hautman J.; and Klein, M.L.; 1989, *J. Chem. Phys.,* **91**, 4994-5001.
Helt, J.M.; and Batteas, J.D.; 2005, *Langmuir,* **21**, 633-639.
Hu, J.; Xiao X.D.; Ogletree, D.F.; Salmeron, M.; 1995, *Surface Science*, **327**, 358-70.
Hutchings, I.M.; *Tribology: Friction and Wear of Engineering Materials,* CRC Press, 1992.
Israelashvili, J.N.; 1992, *Intermolecular and Surfaces Forces*, 2nd ed., Academic Press, NY.
Jiang, S.; 2002, *Molecular Physics*, **100**, 2261-75.
Johnson, K.L.; *Contact Mechanics*, Cambridge University Press, 1985.
Jun, S.; Zhang, H.; Bechhoefer, J.; 2005, *Phys. Rev.* **E, 71**, 011908(1-6).
Kim, Y.; Huang, W.N.; and Lieber, C.M.; 1991, *Applied Phys. Lett.,* **59**, 3404-6.
Kopta, S.; and Salmeron, M.; 2000, *J. Chemical Physics*, **113**, 8249-52.
Kramers H.A., 1940, *Physica* **7,** 284-304.
Lavrich, D.J.; Wetterer, S.M.; Bernasek, S.L.; Scoles, G.; 1998, *J. Phys. Chem.* **B, 102**, 3456-65.
Leng, Y.S.; and Jiang, S.; 2000, *J. Chem. Phys*., **113**, 8800-6.
Lio, A.; Charych, D.H.; Salmeron, M.; 1997, *J. Phys. Chem.* **B, 101**, 3800-5.
Ludema K.; *Friction, Wear and Lubrication, a Textbook in Tribology,* CRC Press, 1996.
Ludema K., in Fundamentals of Tribology and Bridging the Gap between the Macro- and Micro/Nanoscale, Bhushan ed., *Kluwer Academic Publishers*, 2001.
Miyake, S.; 1994, *Appl. Phys. Lett.* **65**, 980-2.
Nie, X.; Zhang, P.; Wiener A.M.; Cheng, Y.T.; 2005, *Nano Letters,* **5**, 1992-6.
Nuzzo, R.G.; and Allara, D.L.; 1983, *J. Am. Chem. Soc.,* **105**, 4481-3.
Park, N.-S.; Kim, M.-W.; Langford, S.C.; Dickinson, J.T.; 1996a, *Langmuir*, **12**, 4599-604.
Park, N.-S.; Kim, M.-W.; Langford, S.C.; Dickinson, J.T.; 1996b, *J. Appl. Phys.,* **80**, 2680-6.
Poirier, G.E.; 1997, Langmuir, **13**, 2019-26.
Risken, H.; *The Fokker-Planck Equation. Methods of Solution and Applications.* Springer-Verlag, Berlin Heidelberg, 1984.
Salmeron, M.; 2001, *Tribology Letters,* **10**, 69-79.
San, A.; and Wortis, M.; 2004, *Phys. Rev.* **E, 70**, 031102(1-11).
Sang, Y.; Dubè, M.; Grant, M.; 2001, *Phys. Rev. Lett.,* **87**, 174301(1-4).
Schessler, H.M.; Karpovich, D.S.; Blanchard, G.J.; 1996, *J. Am. Chem. Soc*., **118**, 9645-51.

Schöfer, J.; and Santner, E.; 1998, *Wear,* **222**, 74-83.

Scudiero, L.; Langford, S.C.; Dickinson, J.T.; 1999, *Tribology Letters* **6**, 41-55.

Sear, R.P.; 2004, *Phys. Rev*. **E, 70**, 021605(1-6).

Shevade A.V.; Zhou, J.; Zin M.T.; Jiang, S.; 2001, *Langmuir* **17**, 7566-7562.

Sokoliuc, A.; Bennewitz, R.; Gnecco, E.; Meyer, E.; 2004, *Phys. Rev. Lett.,* **92**, 134301(1-4).

Sokolov, I.Y.; 2002, *Tribology Letters,* **12**, 131-134.

Stein, D.L.; Doering, C.R.; Palmers, R.G.; van Hemmen, J.L.; McLaughlin, R.M.; 1990, *J. Phys. A, Math. Gen*. **23**, L203-L208.

Takano, H.; Kenseth, J.R.; Wong, S.; O'Brien, J.C.; Porter, M.D.; 1999, *Chem. Rev*. **99**, 2845-90.

Tidwell, H.C.D.; Ertel, S.I.; Ratner, B.D.; Tarasevinch, B.J.; Atre, S.; Allara, D.L.; 1997, *Langmuir*, **13**, 3404-13.

Ulman, A.; *Introduction to Ultrathin Organic Films: From Langmuir-Blodgett to Self-Assembly*, Academic Press, San Diego, CA, 1991.

Ulman, A.; 1996, *Chemical Review*, **96**, 1533-54.

Unertl, W.N.; 1999*, J. Vac.Sci. Technol*. **A, 17**, 1779-86.

van Kampen, N.G.; *Stochastic Processes in Physics and Chemistry*. North-Holland Publishing Company, 1983.

Wang, W.; Lee T.; Reed, M.A.; 2003, *Phys. Rev. B*, **68**, 035416(1-7).

Weeks, B.L.; Noy, A.; Miller, A.E.; De Yoreo, J.J.; 2002, *Phys. Rev. Lett.,* **88**, 255505(1-4).

Xu, L.; Salmeron, M.; 1998, *Langmuir*, **14**, 2187-90.

In: Encyclopedia of Nanotechnology
Editor: Elwood D. Carlson
ISBN: 978-1-60692-079-4

Chapter 3

APPLICATION OF COORDINATION CHEMISTRY IN THE FABRICATION OF INORGANIC ANOSTRUCTURES

Yi Xie[†], Yujie Xiong[‡] and Zhengquan Li[†]

[†] School of Chemical and Material Engineering, Southern Yangtze University, Wuxi, Jiangsu 214036, P. R. China

[‡] Department of Chemistry, University of Washington, Seattle, Washington 98195, USA

ABSTRACT

Nanostructures of metals and their oxides, chalcogenides have been of extensive interest for their widespread use in catalysis, photography, optics, electronics, optoelectronics, information storage, biological and chemical sensing, and surface-enhanced Raman scattering (SERS). By tailoring the size, shape and/or structure, one can, in principle, tune their intrinsic properties for a range of applications. To this end, coordination structures have been rationally applied as precursors to fabricate diversified nanostructures in solution. In this chapter, we discuss the synthetic strategy and related reaction mechanism through the illustration of some detailed cases developed by our group in the recent years. The simple process, mild synthetic conditions, clean reactions, excellent reproducibility and high product quality make our strategies significant. It is believed that introduction of coordination structures into the synthesis of metal, oxide and chalcogenide nanocrystals may provide a versatile tool to control their growth and assembly into well-defined nanostructures. We also expect that the related synthetic strategy is applicable to extend to chemical vapor transport or laser-assisted catalytic growth.

1. INTRODUCTION

Nanostructures of metals and their oxides, chalcogenides have been of extensive interest for their widespread use in catalysis, photography, optics, electronics, optoelectronics, information storage, biological and chemical sensing, and surface-enhanced Raman scattering

(SERS).[1-6] By tailoring the size, shape and/or structure, one can, in principle, tune their intrinsic properties for a range of applications. To this end, diverse strategies, including crystal-structure directed growth, template-directed synthesis (chemical or physical, using solid substrates, porous materials, self-assembled molecular structures or existing nanostructures, such as tubes, wires or rods), electrochemistry, vapor phase growth, capping reagents, laser–assisted catalysis growth, vapor-liquid-solid (VLS), solution-liquid-solid (SLS), and solvothermal methods, have been designed to fabricate diversified nanostructures.[7-14]

Coordination chemistry is an important and expansive branch of chemistry, focusing on transition metals. Transition metals differ from main group metals in several key properties. One of the most interesting aspects of transition metals is their ability to form coordination compounds. The formation of coordination compounds occurs between a metal atom/cation and a molecule with one or more unshared electron pairs that is so-called "ligand". According to the number of their donor atoms, ligands may be classified into two types: monodentate ligand donating a single electron pair to the metal or metal ion, such as NH_3, H_2O, CN^- and SCN^-, bidentate ligand donating two electron pairs, for example, ethylenediamine. Therefore, there may be many coordination forms via diversified connection ways between metal atoms/cations and ligands: simple complexes with single metal-center, linear coordination cluster compounds with multiple metal-centers bridged by bidentate ligands, metal-polymer coordination compounds, coordination polymers with one-, two-, or three-dimensional (1D, 2D or 3D) structures bridged by bidentate ligands. The coordination forms in these compounds are absolutely different, observed from not only their atom numbers (or molecular weight), but also their structural characteristics.

Coordination compounds have applied in the synthesis of nanomaterials long before. For instance, coordination polymers were used as precursors to be decomposed into thin film via a chemical vapor deposition (CVD) route.[15] In general, these techniques could not take advantage of the structural characteristics of coordination compounds, because the reaction condition was extremely rigorous. We believe that intrinsic structural characteristics in specific coordination compounds are advantageous for the oriented growth and simultaneous assembly of nanostructures as they are used as precursors in solution-based reactions, even if the diversified coordination compounds have absolutely different structures. On the other hand, the coordination between metal cations and ligands usually increases their solubility and slows down the hydrolysis in aqueous solution, so that the reactions can proceed more slowly to provide enough time for the crystals to grow, or the ligands can adsorb to specific surfaces to affect the morphology of the crystals. Both the two factors provide useful ways to achieve shape control. Interestingly, Chaudret group also used the tools of organometallic chemistry in preparing 1D nanostructures.[16] The bonding ways of organometallic compounds are very similar to those of coordination structures. The successful synthesis in their related works shows that our strategy is reasonable and general. In this chapter, we review the development of synthesizing inorganic nanostructures from coordination structures by our group. We will describe in detail, six typical strategies of our work on the synthesis of a variety of nanostructures. Each section highlights a unique approach for the fabrication of nanostructures from specific coordination structures: (*i*) simple complex; (*ii*) linear coordination cluster compound; (*iii*) metal-polymer coordination compound; (*iv*) 1D coordination polymer; (*v*) 2D coordination polymer; (*vi*) 3D coordination polymer.

2. Simple Complex (0D)

Simple complex consists of single metal atom/cation center and ligands, in which no extension structure could be found from any dimension. Ethylenediamine is a ligand composing simple complex that has been introduced to synthesize metal chalcogenide 1D nanostructures long before.17,18 For instance, Qian group prepared CdS, CdSe and CdTe nanorods in ethylenediamine by means of solvothermal reactions.17 The role of ethylenediamine was confirmed by further study of the temporal evolution of CdS nanocrystals in solvothermal reactions.18 It demonstrated that ethylenediamine molecules adsorbed on the surface of CdS played a critical role in the formation of nanorods.18 Infrared (IR) spectra of these ethylenediamine molecules show that they are not in chelating (cis) configuration but in trans configuration.18 Since the ethylenediamine molecules are not in chelating configuration, it is possible to introduce monodentate ligands as shape-controller for the synthesis of metal chalcogenide 1D nanostructures. Qian and Yu employed n-butylamine that has only one anchor atom as ligand and solvent, and successfully obtained CdS, CdSe, ZnSe and PbSe nanorods.19 It indicates that one anchor atom in a ligand is necessary and sufficient for the formation of nanorods, even though more anchor atoms may be present in a ligand. Reetz et al. also reported that a non-chelating coordination mode of α-hydroxycarboxylates on a metal surface is likely to be the morphology-determining factor in shape-selective preparation.20 Our group also developed a two-step reaction route to synthesize PbSe nanorods from Pb2+ and Se in the presence of ligand ethylenediamine, in which the ultrasonic radiation and γ-irradiation techniques were combined.21 The similar route was applicable to be extended to produce PbS rods by refluxing.22

As a special case, ammonia could act as structure-directing agent in the formation of NiS layer-rolled structures,23 in which the aqueous solution contains NH3, CS2 and Ni2+. The following reactions were involved in the fabrication process.

$$4NH_3 + CS_2 \rightarrow NH_4SCN + (NH_4)_2S$$

$$Ni^{2+} + (NH_4)_2S \rightarrow NiS + 2NH_4^+$$

It is considered that hydrogen-bonding interactions between $Ni(NH_3)_6^{2+}$ can organize complex ions in space and the linked ones adopt the "least destabilizing" mutual arrangement (Figure 1). Hence the growth of NiS particles is limited in-plane and the surface features of the particles offer the possibility to roll up into tubular structures. This route can also be applied to produce metal sulfide hollow spheres.[24]

Note that the above ligands have no rigid structures, thereby the coordination between metal cations and ligands mainly makes ligands adsorbed on the surface of crystals, affecting the crystal growth, which is so-called "shape-controller". Some other ligands with rigid structures, such as 1, 10-phenanthroline, can coordinate with transition metal cations in chelating configuration. The formed coordination structures have obvious steric effect when chemical reactions occur, due to the rigid characteristics of ligands. For example, it is known that 1,10-phenanthroline can coordinate with Ag^{2+} to form a type of stable complex, $[Ag(phen)_2]^{2+}$.[25] The complex is of d^9 structure, taking a approximate square-planar geometry and having steric inhibition on its *xy* plane. Therefore, if the Ag^{2+} ions in the complex could be reduced to atomic Ag, it would be preferential to orientedly grow along the vertical direction (*z* axis), due to their steric inhibition on the *xy* plane. It was reported that Ag^{+} ions could be oxidized to Ag^{2+} ions by oxygen under conditions of ultrasound.[26] Thus we designed an ultrasound-assisted oxidation-reduction route to prepare silver nanorods with 1,10-phenanthroline as ligand,[27] in which $[Ag(phen)_2]^{2+}$ formed at first and then was reduced to atomic Ag.

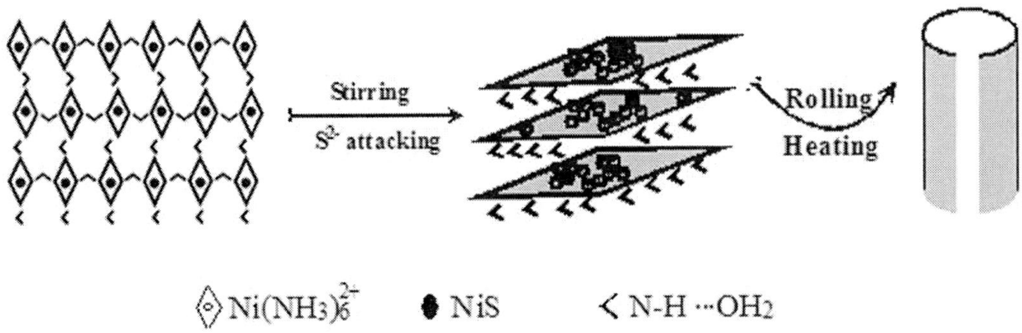

Figure 1. Illustration of the formation of layer-rolled NiS structures. Adapted in part from ref. 23, © 2001 Wiley-VCH. Used with permission.

The coordination situation of iron cations and 1,10-phenanthroline was more complicated. From the structural characteristics, $[Fe(phen)_3]^{2+}$ has a distorted octahedral structure, which can be gradually hydrated to $[Fe(H_2O)_2(phen)_2]^{2+}$, $[Fe(H_2O)_4phen]^{2+}$ and $[Fe(H_2O)_6]^{2+}$. The resulting $[Fe(H_2O)_6]^{2+}$ can be oxidized and hydrolyzed into β-FeOOH in air,[25] which ensures the completion of the hydration from $[Fe(phen)_3]^{2+}$ to $[Fe(H_2O)_6]^{2+}$. It is noteworthy that the ligand of phen can stabilize Fe(II), preventing $[Fe(phen)_3]^{2+}$ from oxidization by air.[25] With the dissociation of phen and the hydration of $[Fe(phen)_3]^{2+}$, the complexes of Fe(II) are more and more inclined to be oxidized by air. As a result, a little $[Fe(H_2O)_2(phen)_2]^{2+}$ can be oxidized to $[Fe(H_2O)_2(phen)_2]^{3+}$ that can also produce β-FeOOH finally. Thus, the first-degree hydration process will most probably form two products: $[Fe(H_2O)_2(phen)_2]^{2+}$ and $[Fe(H_2O)_2(phen)_2]^{3+}$, shown as the Complex (I) and (II) in Figure 2A, respectively. In Complex (I), the Fe^{2+} cation has a d^6 high spin structure and the two ligands of phen lie on an approximate square-plane.[28] Thus there exists planar steric-inhibition for the later oxidization-hydrolysis reactions, leading to the preferential oxidization-hydrolysis along its normal direction. It is advantageous for the growth of β-FeOOH nanowires along [001] axis via oxidization and hydrolysis processes. In Complex (II), the Fe^{2+} cations with d^5 structure has larger splitting energy than Fe^{3+} and the two ligands of phen tend to keep away form a plane,[28] hence the steric-inhibition along the normal direction of the results in the preferential hydrolysis on the plane and hence the formation of β-FeOOH flakes with (001) orientation. Obviously, the oxidization degree from Complex (I) to Complex (II) is relatively low and the first-hydration process is more inclined to produce (I). Thus, it should be more undemanding to produce the nanowires than the flakes, resulting in the production of a mass of nanowires and a small quantity of flakes. Since the axis growth plane of nanowires is consistent with the orientation plane of flakes, the nanowires are facile to grow on flakes directionally. Applied the above strategy, self-supported patterns of oriented alignment of β-FeOOH nanowires (Figure 2B and C) were fabricated from the complex of $[Fe(phen)_3]^{2+}$ at 60 °C.[29] The characterizations of electrochemical performance and UV-vis spectrum show that the product is expected to have potential applications in lithium batteries and semiconductor industry.

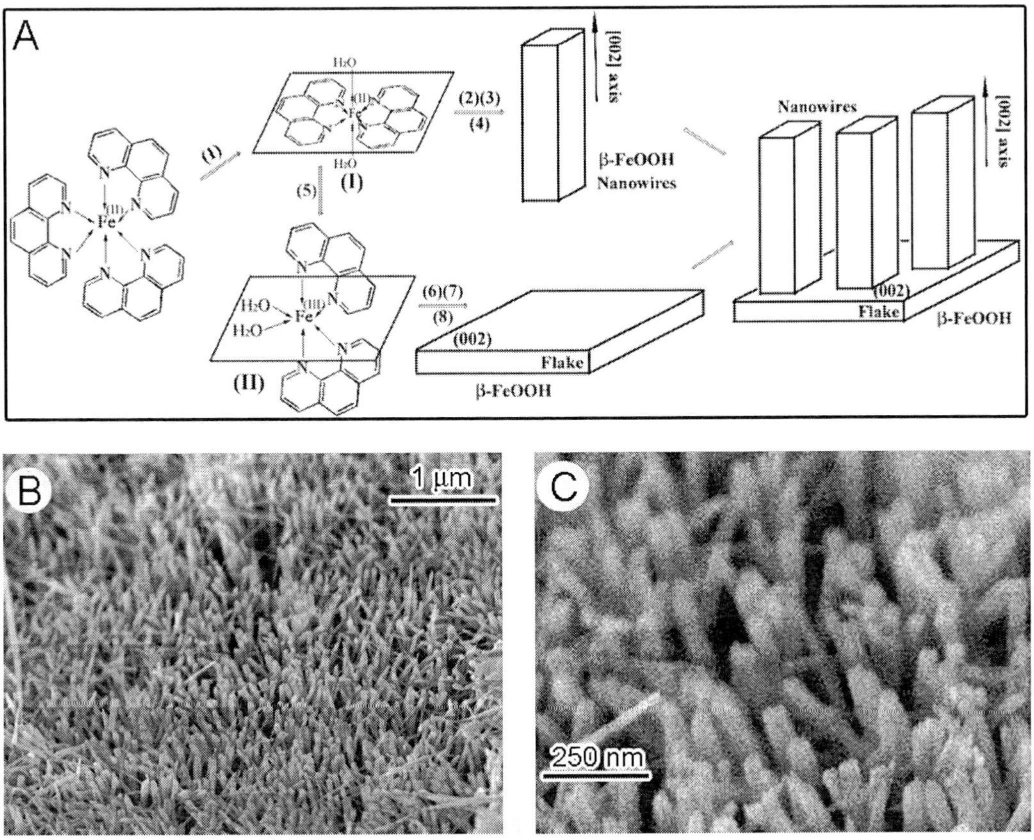

Figure 2. (A) Schematic formation illustration of the oriented alignment of nanowires growing on self-formed flakes. (B), (C) SEM images of the oriented alignment of β-FeOOH nanowires without any treatment. Adapted in part from ref. 29, © 2003 Wiley-VCH. Used with permission.

From the strategy, one can see that a kinetic control is needed to form both the Complex (I) and (II) at the same time, which results in the simultaneous formation of nanowires and flakes. Our experiments indicate that the products prepared at room temperature are all nanowires without flakes supporting. Through the additional reduction or reduction-oxidation process, α-Fe and γ-Fe_2O_3 nanowires were successfully synthesized.[30] It is interesting that there existed some folded ribbon-like nanostructures as well as nanowires in the obtained α-Fe and γ-Fe_2O_3 products, which indicates that this route might provide a new mild approach to metal and metal oxide nanoribbons. Magnetic behavior of as-prepared α-Fe and γ-Fe_2O_3 nanowires was investigated by the temperature dependence of magnetization, showing that the as-prepared α-Fe and γ-Fe_2O_3 nanowires perhaps have more applications in magnetic science and technology, due to their shape anisotropy. Through pyrolysis on β-FeOOH nanowires, α- Fe_2O_3 hollow nanowires could be obtained.[31] The products can catalyze the oxidation of almost 100% carbon monoxide at 320 °C, exhibiting excellent catalytic performances.

The simple coordination structures also have potential applications in the fabrication of micropatterns. In a typical case, novel branch-like CdS micropatterns were synthesized using thiosemicarbazide ($NH_2NHCSNH_2$) as both a sulfur source and a capping ligand in a methanol/water system.[32] The thiosemicarbazide/$CdCl_2$ ratios were found to effectively play crucial roles in the morphologies of CdS crystals. With the ratios decreasing from eight to one, the CdS crystals exhibited different morphologies, from spheres to spindle-like morphologies mediated by cauliflower- and branch-like micropatterns (Figure 3A and B). Control experiments using different sulfur sources showed that the NH_2 groups, which could form hydrogen bonds, were necessary for the formation of the novel branch-like patterns. There involved probably two key processes responsible for the formation of micropatterns, namely "oriented aggregation" and "diffusion-control growth".[33] Figure 3C illustrates the whole growth process of CdS micropatterns.

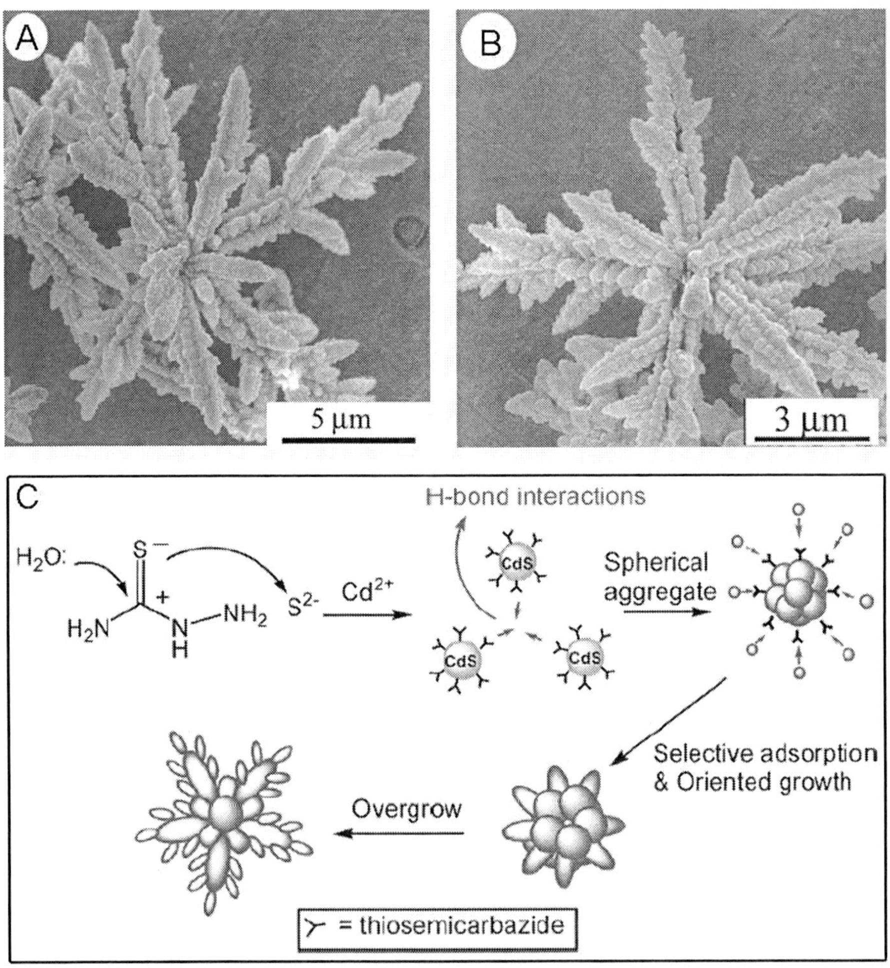

Figure 3. (A), (B) SEM images of branch-like CdS micropatterns. (C) Proposed growth process of the CdS micropatterns. Adapted in part from ref. 32, © 2004 American Chemical Society. Used with permission.

3. LINEAR COORDINATION CLUSTER COMPOUNDS (QUASI-1D)

Distinct from simple complex, linear coordination cluster compounds will directly provide orientation for the growth of 1D nanostructure due to their linear structures. This type of compounds is very familiar to us all, and usually applied in the detection of metal cations in the analytic chemistry. For example, the cluster precursor of cadmium diethyldithiocarbamate, $Cd_2(S_2CNEt_2)_4$, could be treated as an inorganic core $[Cd_2S_2]$ with four capping groups.[34] It was found that nucleophilic attack by ethylenediamine at the thione carbon will strip the capping groups,[35] shown as the following reaction.

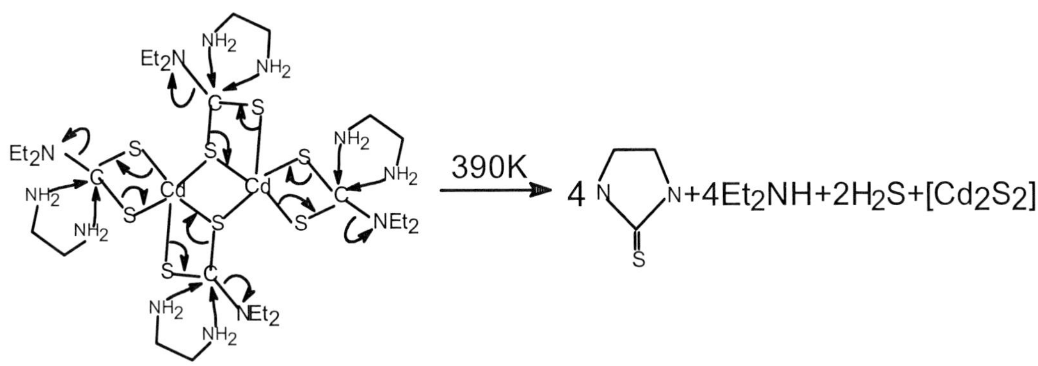

The resulting organic product has been characterized in organic chemistry.[36] The inorganic core $[Cd_2S_2]$ can be combined with each other to form chains after removing the capping groups. Hence under appropriate chemical environment, the $[Cd_2S_2]$ may preferentially grow along the chain direction and result in 1D structure. By using this strategy, quantum-confined CdS nanowires with diameters around 4 nm and lengths ranging from 150 to 250 nm were grown for the first time by our group.[35]

After the successful synthesis of 4 nm CdS nanowires, we did more research works on the application of linear coordination cluster compounds in the preparation of other 1D nanostructure. Unfortunately, no desirable result was obtained. For example, metal diethyldithiocarbamate (M=Pb, Hg, Cu, Mn, Ag) was used as precursors to synthesize metal sulfides, however, some nanoparticles rather than 1D nanostructures were obtained.[37] It should result from the fact that other metal sulfides is not as inclined to grow orientedly as cadmium sulfide. Another important reason is that the coordination cluster is too small to provide orientation for the growth of nanowires.

Thus, the small clusters with several linear-aligned metal cations can most probably used to synthesize 1D nanostructure, but the key problem is how to drive them to connect with each other to form longer chains. Some bridging anions, such as halogen ions, can connect the units to lengthen the linear chains containing metal cations. However, in disorder reaction system, the bridging effect is limited. The previous colloidal research shows that the rod-like micelle can offer a unique microenvironment for chemical reactions and reduce the systematic disorder to small area. It should be an excellent candidate for driving the small linear units' connecting action by bridging anions. Thus, the long chains bridged by bridging anions in rod-like micelles can provide orientation for the growth of nanowires via suitable reactions.

In a detailed case, Cu^{2+} cations can be combined with N and O atoms of dimethylglyoxime (dmgH) to form the trinuclear complex of $Cu_3(dmg)_2Cl_4$ whose 3 copper cations arrange on a line as a unit (Figure 4A).[25] The units composed of 3 copper cations on lines can be bridged by Cl anions, resulting in the formation of longer chains containing copper cations. It is thought that if the copper (II) cations on chains in the complex may be transformed to Cu_2O, the 1D array of copper cations in the precursor will provide orientation for the growth of Cu_2O nanowires. However, the bridging effect of Cl anions is too limited to connect copper cations of every unit to form enough long chains. To solve this problem, an efficient way is to build up rod-like micelles composed of surfactant to confine the linear units of $Cu_3(dmg)_2Cl_4$ in microenvironments and reduce the systematic disorder. Hence the $Cu_3(dmg)_2Cl_4$ units are driven to connect with each other by Cl anions to form $[Cu_3(dmg)_2Cl_2]_n^{2n+}$ (Figure 4B) whose remain Cl^- anions are dissociated near and remain the electric neutrality of reaction system. Note that the rod-like micelles are both hydrophobic and hydrophilic (W/O), and form when the concentration of surfactant is as ten times as critical micelle concentration.[38] As the complex-precursor forms in water phase, it will be enwrapped in rod-like micelles. As a typical result, the copper cations in complex can arrange on long chains of $[Cu_3(dmg)_2Cl_2]_n^{2n+}$ in the rod-like micelles and provide orientation for the growth of Cu_2O nanowires, when the copper (II) cations in the complex are reduced to Cu_2O by glucose.[39] Further studies indicate that the rod-like micelles cannot only drive linear units of $Cu_3(dmg)_2Cl_4$ to connect with each other, but also confine the diameter of nanowires. The characterization results of IR and laser light scattering (LLS) spectra revealed that the linear units of $Cu_3(dmg)_2Cl_4$ actually form longer chains in $[Cu_3(dmg)_2Cl_2]_n^{2n+}$ by the bridge of Cl anions, enwrapped in rod-like SDS micelles. In contrast, without building SDS micelles, only some Cu_2O short rods could be obtained. Thus, the formation of long chains from linear coordination cluster compounds is crucial to the growth of 1D nanostructure. The band gap of Cu_2O nanowires was determined to 2.34 eV by UV-vis absorption spectrum, showing its promising application for the reversible conversion between solar energy and electrical or chemical energy.

4. 1D Metal-Polymer Coordination Chain

An alternative to the precursors of linear coordination cluster compounds is metal-polymer coordination structure with long chains. In a typical case, poly(ethylene glycol) (PEG) as a non-ionic polymer has hydrophilic –O– and hydrophobic $-CH_2$-CH_2- on the long chains. According to Pearson's results,[40] the –O– of the ether group on the PEG has a strong basic property and possesses a pair of electrons. Its ability to be a donor of oxygen atoms enables it to couple with transition metal cations to form coordination structures. However, the long polymer chain has a high degree of flexibility because its C–O bond is easy to rotate, thereby the chains always incline to be entwined and curled. Therefore, building an appropriate microenvironment for the extension of PEG chains is technically necessary for the application of metal-polymer coordination structures.

The metal-PEG complexes generally form in aqueous media within a narrow range of solvent composition, pH, and ionic strength.[41] Butylacetate is a good solvent for PEG, favoring the full extension of the polymer chain, while cyclohexane is a poor solvent for

PEG, favoring the curl and twist of the polymer chain. Thus, Ag^+-PEG coordination chains can extend in the mixture solvent of butylacetate, cyclohexane and water via the curl and extension of PEG. By the strategy, a polymer-surfactant-assisted sandwich-reduction (PSASR) route has been reported to prepare silver nanowires, in which the sandwich structure of reducing agent $(\eta^5\text{-}C_5H_5)_2Fe$ results in the formation of atomic silver along certain axis in *xy* plane and the linear arrangement of silver cations along PEG chains in mixture solvents provide orientation for the 1D growth of silver nanowires (Figure 5).[42] The as-prepared silver nanowires exhibited surface plasma Raman (SPR) peak at ~410 nm. SPR is an important optical property of metal nanostructures, which could lead to applications in colorimetric sensing, nanoscale waveguiding, enhancement of electromagnetic fields, and light transmission. It is noteworthy that the extension of metal-polymer coordination chains is crucial to the successful synthesis of 1D nanostructures. The strategy is applicable to be extended to grow metal oxide 1D nanostructures. For instance, $\gamma\text{-}Fe_2O_3$ nanorods and ZnO nanowires (Figure 6A) were prepared via a similar process.[43,44]

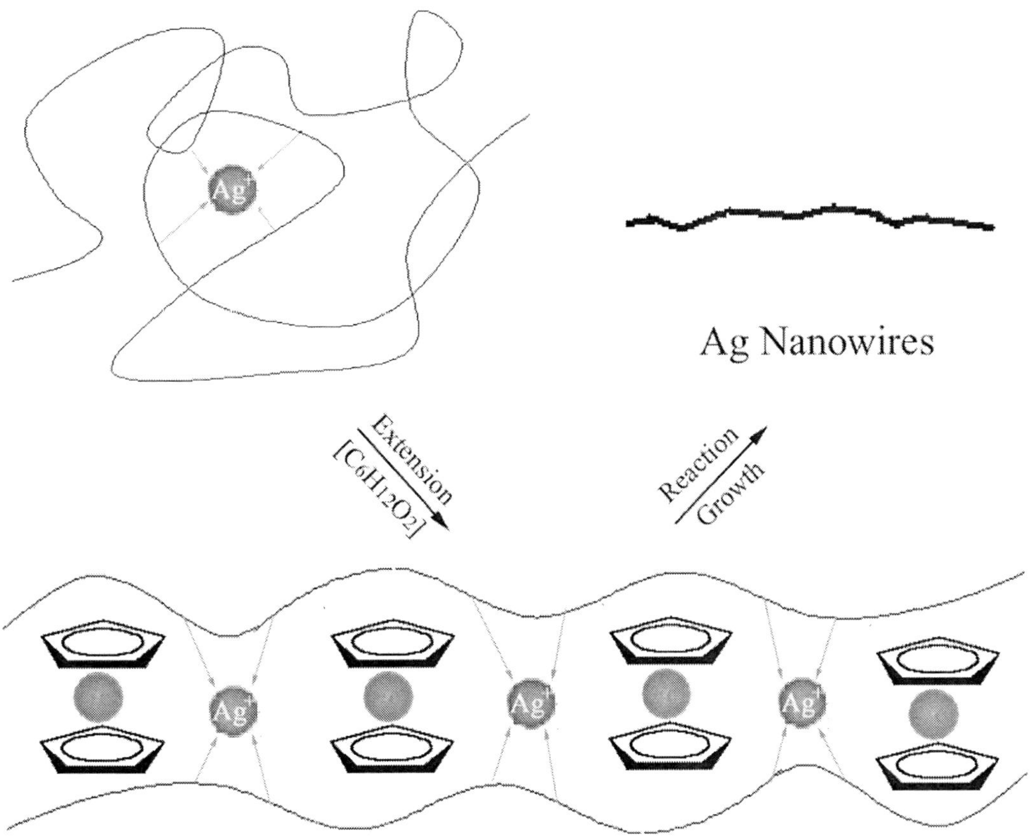

Figure 5. Schematic illustration of preparing silver nanowires from metal-polymer coordination chains. Adapted in part from ref. 42, © 2003 Wiley-VCH. Used with permission.

The polymer of PEG or polyvinylpyrrolidone (PVP) has also been applied to synthesize $\gamma\text{-}MnO_2$ and α-NiS hierarchical structures.[45,46] In our cases, urchin-like and sisal-like $\gamma\text{-}MnO_2$ structures (Figure 6B and C), and flower-like and tennis-ball α-NiS structures (Figure 6D) were obtained. The IR spectra of the precipitates composed of metal cations and polymers showed that coordination structures actually formed.[45,46] The formation of assemblies was found to result from the emergence of aggregated particle cores.[45] We found that the key parameter to the formation of these assembly structures was the amount of polymers and solvents.

5. 1D Coordination Polymer

The "1D coordination polymer" means complex with infinite structure in one dimension. For example, a novel supramolecule $[HgCl_2(ptz)]_2{\cdot}HgCl_2$ (ptz = phenothiazine) with uncoordinated inorganic salt $HgCl_2$ presented in a 1D chain was prepared by our group.[47] Single crystal X-ray analysis showed that the 1D chain structure is stabilized by hydrogen bonds between adjacent chains and the coordination mode of the ligand phenothiazine is unusual with large steric inhibition other than the chain directions. From its structure, the preorientation of 1D chain structure in the precursor may facilitate the formation of 1D nanostructure. Hence the submicrometer Hg_2Cl_2 rods could be obtained with $[HgCl_2(ptz)]_2{\cdot}HgCl_2$ as precursor by heating.[47]

Figure 6. SEM images of (A) ZnO nanowires, (B) γ-MnO_2 urchin-like hierarchical structures, (C) γ-MnO_2 sisal-like hierarchical structures, and (D) α-NiS tennis-ball hierarchical structures. Adapted in part from ref. 44 and 45, © 2003 American Chemical Society. Used with permission. Adapted in part from ref. 46, © 2003 Royal Society of Chemistry. Used with permission.

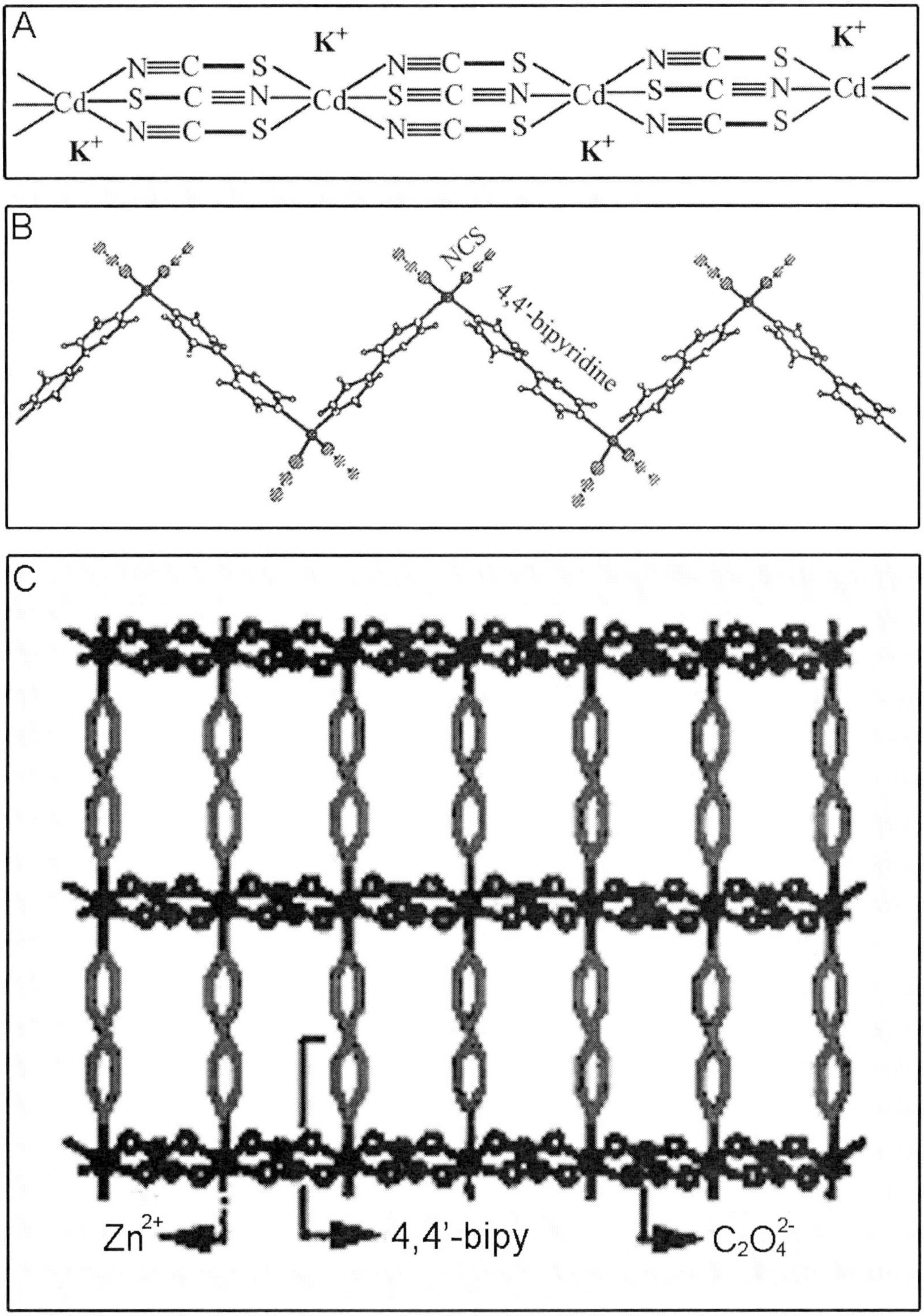

Figure 7. (A) Structural illustration of $KCd(SCN)_3$. Adapted in part from ref. 48, © 2003 Royal Society of Chemistry. Used with permission. (B) Structural illustration of $[Zn(4,4'\text{-bipy})(NCS)_2]_n$. Adapted in part from ref. 51, © 2003 Royal Society of Chemistry. Used with permission. (C) Structural illustration of $^{2}_{\infty}[Zn(ox)(4,4'\text{-bipy})]$.

Besides precursor, 1D coordination structure could also be applied as template for the growth of metal sulfide 1D nanostructures. For example, we synthesized CdS and HgS aligned nanowire bundles using 1D-chain-structural $KCd(SCN)_3$ (Figure 7A) as templates.[48] In the reaction process, with the addition of metal cations (Cd^{2+} or Hg^{2+}) and S^{2-}, immediate solution supersaturation of metal sulfide occurred due to its low solubility in the water system.

Most of the metal sulfide nanoparticles produced in the solution then was attached to surrounding negatively charged coordination polymer chains. At the same time, the anionic $Cd(NCS)_3^-$ chain adsorbs selectively to positively charged crystal faces, which directs the aggregation of the MS building blocks along one principal axis to produce the chain-like aggregates. As the $Cd(NCS)_3^-$ template is a 1D linear chain, these interactions are likely to be sufficient to offset the accompanying steric repulsion. The separating surfaces of adjacent particles may have been eliminated after repulsive interactions were overcome as the result of random, Brownian motion-driven particle collisions.[49] Jiggling of nanoparticles by Brownian motion may also allow adjacent particles to rotate to find the low-energy configuration represented by a coherent particle–particle interface. Rotation of particles within aggregates may also be driven by short-range interactions between adjacent surfaces.[50] Diffusion-controlled growth of these metal sulfide nanoparticles then occurs in the aggregates as the amorphous particles are in close proximity and the surface area reduces by particle fusion and structure rearrangement.[33] In aggregation-based growth, reduction in surface free energy is achieved by complete removal of pairs of surfaces.[49] Because the director $Cd(NCS)_3^-$ chains were arranged along one direction and the distances between them were almost equal in this system, which should play a definitive role in the formation of aligned nanowire bundles. The whole process of the formation of MS nanowires can be summarized as follows: primary nucleation → oriented aggregation → particle-particle diffusion → aligned-growth.

Note that the orientation of 1D coordination structure as precursor or template is crucial for the production of 1D nanostructures, therefore, the bonding along the 1D chains is required to be strong enough for reactions. Otherwise, during the reaction process, 1D chains may break down so that they cannot provide orientation for 1D crystal growth. For example, we ever tried to use $[Zn(4,4'\text{-bipy})(NCS)_2]_n$ with zigzag chain (Figure 7B) as precursor in the reaction with ammonia.[51] The NCS^- ligand has a stronger coordination with Zn^{2+} than 4,4'-bipyridine, as seen from their bond lengths [Zn–N (NCS) 1.908 Å, Zn–N (4,4'-bipy) 2.033 Å] in the coordination polymer.[52] Thus, 4,4'-bipyridine could be preferentially removed when attacked by OH^- and the infinite chains of the coordination polymer would gradually be broken. The broken chains with two -OH at either end were likely to aggregate through a dehydration process in the heated basic environment. As the removal of ligands (including 4,4'-bipyridine and –NCS) and the dehydration of –OH were repeated, produced ZnO nanoparticles tended to form hollow spheres, due to their lower surface tension. Therefore, as a typical result, ZnO hollow spheres were prepared from the precursor $[Zn(4,4'\text{-bipy})(NCS)_2]_n$.[51]

6. 2D Coordination Polymer

The coordination polymers with 2D networks have been directly employed as precursors for the growth of metal oxide 1D nanostructures by our group. In a typical case, we used the coordination polymer, ${}^{2}{}_{\infty}$[Zn(ox)(4,4′-bipy)], with a 2D coordination network (Figure 7C) to synthesize the radial assemblies of nanoneedles.[53] The experimental results showed that the 2D network of the coordination polymer was crucial to the formation of 1D nanostructures. The coordination polymer ${}^{2}{}_{\infty}$[Zn(ox)(4,4′-bipy)] had a 2D coordination network structure constructed by two bridging ligands, oxalate and 4,4′-bipy. All Zn^{2+} ions had an octahedral coordination environment with four oxygen atoms from two oxalate ions and two nitrogen atoms from two 4,4′-bipy molecules along two vertical directions. Zn^{2+} cations were strongly coordinated with both strong ligands that could be removed slowly, thus the newly formed ZnO colloids could not reach monodispersed situation quickly. Therefore, the produced ZnO colloids were inclined to be bound to each other. As ZnO crystal had the anisotropic growth habit, 1D ZnO nanostructures were likely to epitaxially grown from the surface of aggregated colloids during the succedent matured process, such as Ostwald ripenning. It was noteworthy that the role of framework in coordination polymer here was to assemble those produced ZnO crystals that were matured to assembly structures. There is no evidence to show that they acted as template.

7. 3D Coordination Polymer

3D coordination polymers have stable ordered infinite structures in three dimensions. In detail, the metal cations in some 3D coordination polymers bridged by rigid ligands (e. g. 4,4'-bpy) are not only ordered, but also have different connecting fashions. If selecting an appropriate coordination one as precursor, it is possible to make the polymer dissociated and form new metal-oxygen bonds under alkaline hydrothermal conditions. Thus, the formation of the as-desired metal oxide could be achieved by this route, which is significant as there exist many phases of metal oxides. For example, MnO_2 and related manganese (IV) oxide are known to exist in a wide variety of structural forms, such as α-, β-, γ-, δ-type and ramsdellite, etc.

Especially in the process of oxidization reaction, the different coordination abilities of ligands in the coordination polymer will probably lead to a faster oxidization along some direction, thus resulting in an oriented growth of metal oxide nanowires. Therefore, we design a hydrothermal route to synthesize metal oxide nanowires via an alkaline replacement reaction on coordination polymer-precursors. In a specific case, γ-MnO_2 nanowires (Figure 8A) was prepared from the reaction between coordination polymer $[Mn(SO_4)(4,4'\text{-}bpy)(H_2O)_2]_n$ and boiling NaOH solution, in which both the crystal phase and the shape of product are well controlled.[54] The tops of nanowires were found to almost align along one direction (Figure 8B). This ordered alignment of γ-MnO_2 nanowires probably leads to more applications in Lithium batteries.

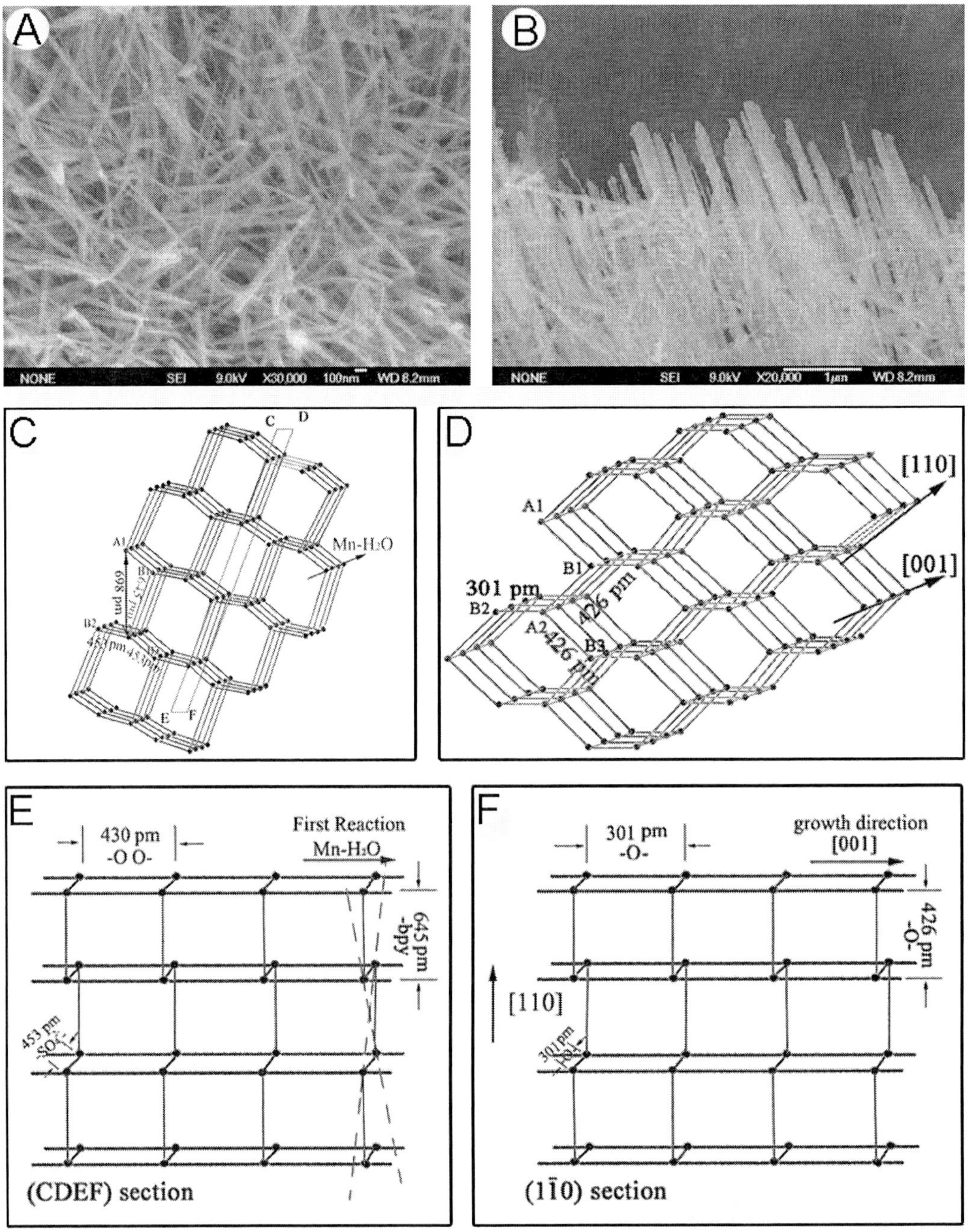

Figure 8. (A), (B) SEM images of γ-MnO_2 nanowires. (C) The Mn cations in $[Mn(SO_4)(4,4'\text{-}bpy)(H_2O)_2]_n$ having more chances to combine with each other and form stable manganese (IV) oxide, when the coordination polymer is dissociated in boiling NaOH solution. (D) The frameworks of Mn cations in γ-MnO_2 crystal lattices. (E) The (CDEF) section in the structure of $[Mn(SO_4)(4,4'\text{-}bpy)(H_2O)_2]_n$ (The dashlines correspond to two directions of –Mn-bpy-Mn-bpy- chains). (F) The (110) section of γ-MnO_2 crystal lattices. The thin lines only represent the distance of Mn cations. Adapted in part from ref. 54, © 2003 Wiley-VCH. Used with permission.

The supplementary experiments revealed that the crystal structure of coordination polymer is crucial to the formation of γ-MnO_2 nanowires. To reveal the reaction process more clearly, the real-time IR spectra were taken, showing that the Mn-water ligands, –Mn-SO_4^{2-}-Mn-SO_4^{2-} and -Mn-bpy-Mn-bpy were dissociated in turn while the new Mn-O of γ-MnO_2 formed, which agrees with the fact that the coordination polymers containing 4,4'-bpy are usually stable. These characterization results reveal that the dissociation of coordination polymer and the formation of γ-MnO_2 occurred at the same time, and the Mn-water ligands were dissociated at first. Therefore, the structural correlation between the polymer and the oxide is important for the formation of product.

The coordination polymer $[Mn(SO_4)(4,4'\text{-bpy})(H_2O)_2]_n$ has 3D infinite layered structure,[55] in which each layer is formed in the *ac* plane by connecting Mn centers via bridging bpy ligands to generate parallel –Mn-bpy-Mn-bpy chains, whereas between adjacent layers the –Mn-bpy-Mn-bpy chains have an angle of 45°. –Mn-bpy-Mn-bpy chains in different layers are connected by zigzag –Mn-SO_4^{2-}-Mn-SO_4^{2-} chains, leading to a 3D polymer framework. The Mn cations having more chances to form Mn-O-Mn through alkaline hydrothermal reaction could be described as distorted hexagonal units in layers (Figure 8C). γ-MnO_2 is considered to be a disordered intergrowth of the β-MnO_2 and ramsdellite structures, consisting of a random arrangement of single and double chains of MnO_6 octahedra.[56] Each Mn cation is connected with other three adjacent ones by oxygen anions to build up distorted Mn hexagonal units in layers, which are composed of Mn cation frameworks shown in Figure 8D. The layers (parallel to *ab* plane) are also connected by sharing oxygen anions. It is found that the structural figure of these Mn cations are similar to that in γ-MnO_2 crystal lattices. More careful studies on the structure of $[Mn(SO_4)(4,4'\text{-bpy})(H_2O)_2]_n$ show that the Mn distorted hexagonal units arrange in layers. These layers deviate from the *ac* plane and each of them is separated by oxygen donors from two water ligands. This characteristic can be clearly shown in the (CDEF) section of the structure of coordination polymer-precursor (Figure 8E). Comparing this section (Figure 8F) with (110) section of γ-MnO_2 crystal lattices, it is found that the two sections are similar in structure and the coordination direction of Mn-water ligands in coordination polymer-precursor can correspond to the [001] axis of γ-MnO_2 crystal lattices. Mn-water ligands perpendicular to the layers were firstly dissociated in boiling NaOH solution and could lead to the oriented growth of γ-MnO_2 crystal lattices to nanowires along [002] axis. The ordered alignment of γ-MnO_2 nanowires might also result from the connecting of –Mn-bpy-Mn-bpy, the last one that was dissociated.

8. Conclusion and Outlook

Coordination chemistry is an ancient subject that has been widely used in many fields. The application of coordination chemistry in the rational synthesis and simultaneous assembly of novel nanostructures provides a new approach to nanomaterials. Especially, it considers more information of chemistry that is so-called "soft-chemistry approach", exhibiting the significance of nano-chemistry. In this chapter, we review the development of synthesizing inorganic nanostructures from coordination structures, including simple complex, linear coordination cluster compound, metal-polymer coordination compound and

1D, 2D, 3D coordination polymers. In detail, simple complex works by steric-inhibition effect and hydrogen bonding; linear coordination cluster compound, metal-polymer coordination compound and 1D coordination polymers directly provide orientation for the growth of 1D nanostructures or act as template; the network of 2D coordination polymer is helpful for the assembly of produced nanostructures; the different coordination ability of ligands in 3D coordination polymer results in the unequal crystal growth rate in three dimensions. A variety of nanostructures have been fabricated via these techniques, which display excellent optical, catalytic, electrochemical, magnetic and SPR properties.

It is believed that these techniques can be readily extended to produce more complex nanostructures covering a wide range of materials by involving more reactions into the fabrication system, incorporating more desired materials into the synthetic scheme, or studying detailed mechanism. It is also expected that the related synthetic strategy is applicable to extend to chemical vapor transport or laser-assisted catalytic growth. As a matter of fact, Lieber group successfully synthesized CdS and ZnS nanowires with cadmium or zinc diethyldithiocarbamate ($Cd_2(S_2CNEt_2)_4$ or $Zn_2(S_2CNEt_2)_4$) as single-source precursors and gold nanoclusters as catalyst by VLS technique recently.[57] In general, by tailoring the size, shape and/or structure, one can, in principle, tune their intrinsic properties in a range of applications. We expect that further research on these techniques can provide more versatile ways to generate desirable nanostructures for potential applications.

ACKNOWLEDGEMENT

This work was supported in part by the National Natural Science Foundation of China.

REFERENCES

[1] Reviews: (a) Hu, J. T.; Odom, T. W.; Lieber, C. M. *Acc. Chem. Res.* 1999, 32, 435. (b) Templeton, A. C.; Wuelfing, W. P.; Murray, R. W. *Acc. Chem. Res.* 2000, 33, 27.

[2] Catalysis and optical: (a) Nicewarner-Peña, S. R. R.; Freeman, G.; Reiss, B. D.; He, L.; Peña, D. J.; Walton, I. D.; Cromer, R.; Keating, C. D.; Natan, M. J. *Science* 2001, 294, 137. (b) Huang, M. H.; Mao, S.; Feick, H.; Yan, H. Q.; Wu, Y. Y.; Kind, H.; Weber, E.; Russo, R.; Yang, P. D. *Science* 2001, 292, 1897. (c) Ahmadi, T. S.; Wang, Z. L.; Green, T. C.; Henglein, A.; El-Sayed, M. A. *Science* 1996, 272, 1924. (d) Peng, X.; Manna, L.; Yang, W.; Wickham, J.; Scher, E.; Kadavanich, A; Alivisatos, A. P. *Nature* 2000, 404, 59. (e) Lewis, L. N. *Chem. Rev.* 1993, 93, 2693.

[3] Electronics and optoelectronics: (a) El-Sayed, M. A. *Acc. Chem. Res.* 2001, 34, 257. (b) Chen, S.; Yang, Y. *J. Am. Chem. Soc.* 2002, 124, 5280. (c) Wu, Y.; Xiang, J.; Yang, C.; Lu, W.; Lieber, C. M. *Nature* 2004, 430, 61.

[4] Information storage: (a) Peyser, L. A.; Vinson, A. E.; Bartko, A. P.; Dickson, R. M. *Science* 2001, 291, 103. (b) Chiba, D.; Yamanouchi, M.; Matsukura, F.; Ohno, H. *Science* 2003, 301, 943.

[5] Sensing and clinical diagnostics: (a) Tkachenko, A. G.; Xie, H.; Coleman, D.; Glomm, W.; Ryan, J.; Anderson, M. F.; Franzen, S.; Feldheim, D. L. *J. Am. Chem. Soc.* 2003,

125, 4700. (b) Taton, T. A.; Mirkin, C. A.; Letsinger, R. L. *Science* 2000, 289, 1757. (c) Wang, Y.; Jiang, X.; Xia, Y. *J. Am. Chem. Soc.* 2003, 125, 16176. (d) Zhang, X.; Young, M. A.; Lyandres, O.; Van Duyne, R. P. *J. Am. Chem. Soc.* 2005, 127, 4484.

[6] SERS: (a) Cao, Y. C.; Jin, R.; Mirkin, C. A. *Science* 2002, 297, 1536. (b) Nie, S.; Emory, S. R. *Science* 1997, 275, 1102. (c) Tessier, P. M.; Velev, O. D.; Kalambur, A. T.; Rabolt, J. F.; Lenhoff, A. M.; Kaler, E. W. *J. Am. Chem. Soc.* 2000, 122, 9554. (d) Dick, L. A.; McFarland, A. D.; Haynes, C. L.; Van Duyne, R. P. *J. Phys. Chem. B* 2002, 106, 853.

[7] Crystal-structure directed growth: (a) Song, J. H.; Messer, B.; Wu, Y.; Kind, H.; Yang, P. *J. Am. Chem. Soc.* 2001, 123, 9714. (b) Gates, B.; Yin, Y.; Xia, Y. *J. Am. Chem. Soc.* 2000, 122, 12582.(c) Xiong, Y.; Xie, Y.; Du, G.; Su, H. *Inorg. Chem.* 2002, 41, 2953.

[8] Template-directed synthesis: (a) Jones, E. T. T.; Chyan, O. M.; Wrighten, M. S. *J. Am. Chem. Soc.* 1987, 109, 5526. (b) Dai, H.; Wong, E. W.; Lu, Y. Z.; Fan, S.; Lieber, C. M. *Nature* 1995, 375, 769. (c) Song, J. H.; Wu, Y.; Messer, B.; Kind, H.; Yang, P. *J. Am. Chem. Soc.* 2001, 123, 10397.

[9] Electrochemistry: (a) Zhou, Y.; Yu, S. H.; Cui, X. P.; Wang, C. Y.; Chen, Z. Y. *Chem. Mater.* 1999, 11, 545. (b) Zhu, J. J.; Liu, S. W.; Palchik, O.; Koltypin, Y.; Gedanken, A. *Langmuir* 2000, 16, 6396. (c) Xu, D.; Xu, Y.; Chen, D.; Guo, G.; Gui, L.; Tang, Y. *Adv. Mater.* 2000, 12, 520.

[10] Vapor phase growth: (a) Pan, Z. W.; Dai, Z. R.; Wang, Z. L. *Science* 2001, 291, 1947. (b) Jiang, X.; Herricks, T.; Xia, Y. *Nano Lett.* 2002, 2, 1333.

[11] Capping reagents: (a) Peng, Z. A.; Peng, X. G *J. Am. Chem. Soc.* 2001, 123, 1389. (b) Sun, Y.; Gates, B.; Mayers, B.; Xia, Y. *Nano Lett.* 2002, 2, 165.

[12] Laser–assisted catalysis growth and VLS: (a) Morales, A. M.; Lieber, C. M. *Science* 1998, 279, 208. (b) Gudiken, M. S.; Lieber, C. M. *J. Am. Chem. Soc.* 2000, 122, 8801.

[13] SLS: (a) Trentler, T. J.; Hickman, K. M.; Geol, S. C.; Viano, A. M.; Gibbons, P. C.; Buhro, W. E. *Science* 1995, 270, 1791. (b) Yu, H.; Buhro, W. E. *Adv. Mater.* 2003, 15, 416.

[14] Solvothermal method: (a) Wang, X.; Li, Y. *J. Am. Chem. Soc.* 2002, 124, 2880. (b) Xie, Y.; Yan, P.; Lu, J.; Wang, W.; Qian, Y. *Chem. Mater.* 1999, 11, 2619.

[15] Bae, B. J.; Park, J. E.; Kim, B.; Park, J. T. *J. Organomet. Chem.* 2000, 616, 128. (b) Grocholl, L.; Wang, J.; Gillan, E. G. *Chem. Mater.* 2001, 13, 4290.

[16] Soulantica, K.; Maisonnat, A.; Senocq, F.; Fromen, M. C.; Casanove, M. J.; Chaudret, B. *Angew. Chem. Int. Ed.* 2001, 40, 2983.

[17] Li, Y. D.; Liao, H. W.; Ding, Y.; Fan, Y.; Zhang, Y.; Qian, Y. T. *Inorg. Chem.* 1999, 38, 1382.

[18] Yang, J.; Zeng, J. H.; Yu, S. H.; Yang, L.; Zhou, G.; Qian, Y. T. *Chem. Mater.* 2000, 12, 3259.

[19] Yang, J.; Xue, C.; Yu, S. H.; Zeng, J. H.; Qian, Y. T. *Angew. Chem. Int. Ed.* 2002, 41, 4697.

[20] Bradley, J. S.; Tesche, B.; Busser, W.; Maase, M.; Reetz, M. T. *J. Am. Chem. Soc.* 2000, 122, 4631.

[21] Chen, M.; Xie, Y.; Lu, J. C.; Zhu, Y. J.; Qian, Y. T. *J. Mater. Chem.* 2001, 11, 518.

[22] Chen, M.; Xie, Y.; Yao, Z. Y.; Qian, Y. T.; Zhou, G. E. *Mater. Res. Bull.* 2002, 37, 247.

[23] Jiang, X. C.; Xie, Y.; Lu, J.; Zhu, L. Y.; He, W.; Qian, Y. T. *Adv. Mater.* 2001, 13, 1278.

[24] Jiang, X. C.; Xie, Y.; Lu, J.; Zhu, L. Y.; He, W.; Liu, X. M. *Can. J. Chem.* 2002, 80, 263.
[25] Cotton, F. A.; Wilkinson, G. *Advanced Inorganic Chemistry*, John Wilkey and Sons Inc., 1972.
[26] Lorimer, J. P.; Mason, T. J. *Chem. Soc. Rev.* 1987, 16, 275.
[27] Xiong, Y. J.; Xie, Y.; Du, G.; Liu, X. M.; Tian, X. B. *Chem. Lett.* 2002, 98.
[28] König, E. *Coordin. Chem. Rev.* 1968, 3, 471. (b) Madeja, K.; König, E. *J. Inorg. Nucl. Chem.* 1963, 25, 377.
[29] Xiong, Y. J.; Xie, Y.; Chen, S. W.; Li, Z. Q. *Chem.-Eur. J.* 2003, 9, 4991.
[30] Xiong, Y. J.; Xie, Y.; Li, Z. Q.; Zhang, R.; Yang, J. C.; Wu, Z. *New J. Chem.* 2003, 27, 588.
[31] Xiong, Y. J.; Li, Z. Q.; Li, X. X.; Hu, B.; Xie, Y. *Inorg. Chem.* 2004, 43, 6540.
[32] Zhang, X. J.; Zhao, Q. R.; Tian, Y. P.; Xie, Y. *Crys. Growth Des.* 2004, 4, 355.
[33] Manna, L.; Scher, E. C.; Alivisatos, A. P. *J. Am. Chem. Soc.* 2000, 122, 12700. (b) Penn, R. L.; Banfield, J. F. *Am. Mineral.* 1998, 83, 1077. (c) Alivisatos, A. P. *Science* 2000, 289, 736. (d) Chemseddine, A.; Morita, T. *Eur. J. Inorg. Chem.* 1999, 235. (e) Park, S. J.; Kim, S.; Lee, S.; Khim, Z. G.; Char, K.; Hyeon, T. *J. Am. Chem. Soc.* 2000, 122, 8581.
[34] Domenicano, A.; Torelli, L.; Vaciago, A.; Zambonelli, L. *J. Chem. Soc. A* 1968, 1351.
[35] Yang, P.; Xie, Y.; Qian, Y. T.; Liu, X. M. *Chem. Comm.* 1999, 1293.
[36] Ramadas, K.; Janarthanan, N. *J. Chem. Res. S* 1998, 228.
[37] Yan, P.; Xie, Y.; Qian, Y. T. *Chem. Lett.* 1999, 655.
[38] Bourrel, M.; Schechter, R. S. *Microemulsion and Related Systems: Formation, Solvency, and Physical Properties*, Surfactant Science Series Vol. 30, Marcel Dekker Inc., New York and Basel, 1988.
[39] Xiong, Y. J.; Li, Z. Q.; Zhang, R.; Xie, Y.; Yang, J.; Wu, C. Z. *J. Phys. Chem. B* 2003, 107, 3697.
[40] Pearson, P. G. *J. Chem. Educ.* 1968, 45, 581.
[41] Lowman, A. M.; Peppas, N. A. *Polymer* 2000, 41, 73.
[42] Xiong, Y. J.; Xie, Y.; Wu, C. Z.; Yang, J.; Li, Z.; Xu, F. *Adv. Mater.* 2003, 15, 405.
[43] Xu, F.; Zhang, X.; Xie, Y.; Tian, X. B.; Li, Y. Z. *J. Colloid Interface Sci.* 2003, 260, 160.
[44] Li, Z. Q.; Xiong, Y. J.; Xie, Y. *Inorg. Chem.* 2003, 42, 8105.
[45] Wu, C. Z.; Xie, Y.; Wang, D.; Yang, J.; Li, T. W. *J. Phys, Chem. B* 2003, 107, 13583.
[46] Xu, F.; Xie, Y.; Zhang, X.; Wu, C. Z.; Xi, W.; Hong, J.; Tian, X. B. *New. J. Chem.* 2003, 27, 1331.
[47] Zhang, X. J.; Xie, Y.; Yu, W. T.; Zhao, Q. R.; Jiang, M. H.; Tian, Y. P. *Inorg. Chem.* 2003, 42, 3734.
[48] Zhang, X. J.; Xie, Y.; Zhao, Q. R.; Tian, Y. P. *New J. Chem.* 2003, 27, 827.
[49] Banfield, J. F.; Welch, S. A.; Zhang, H.; Ebert, T. T.; Penn, R. L. *Science* 2000, 289, 751.
[50] Zhu, H. L.; Averback, R. S. *Philos. Mag.* 1996, 73, 27.
[51] Li, Z. Q.; Xie, Y.; Xiong, Y. J.; Zhang, R. *New J. Chem.* 2003, 27, 1518.
[52] Pan, L.; Zheng, N. W.; Wu, Y. G.; Jin, X. L.; Huang, X. Y. *J. Coord. Chem.* 1999, 47, 269.
[53] Li, Z. Q.; Xiong, Y. J.; Xie, Y. *unpublished work.*

[54] Xiong, Y. J.; Xie, Y.; Li, Z. Q.; Wu, C. Z. *Chem.-Eur. J.* 2003, 9, 1645.

[55] Hou, H.; Wei, Y.; Fan, Y.; Du, C.; Zhu, Y.; Song, Y.; Niu, Y.; Xin, X. *Inorg. Chim. Acta* 2001, 319, 212.

[56] Hunter, J. C. *J. Solid State Chem.* 1981, 39, 142. (b) Burns, R. G.; Burns, V. M. *Manganese Dioxide Symposium*, I. C. Sample Office, Cleveland, 1970, Vol. 1. (c) Qian, Y. *Introduction to Crystallography*, University of Science and Technology Press, Hefei, 1999 (In Chinese).

[57] Barrelet, C. J.; Wu, Y.; Bell, D. C.; Lieber, C. M. *J. Am. Chem. Soc.* 2003, 125, 11498.

In: Encyclopedia of Nanotechnology
Editor: Elwood D. Carlson

ISBN: 978-1-60692-079-4

Chapter 4

1-D ZINC SULFIDE NANOSTRUCTURES: SYNTHESIS AND CHARACTERIZATION

Soumitra Kar[1] and Subhadra Chaudhuri[1]
Department of Materials Science, Indian Association for the Cultivation of Science, Kolkata- 700 032, India

ABSTRACT

With the growing interest for miniaturization in electronic and optoelectronic devices, scientists are more and more interested in fabricating different materials in nano-scale dimensions with different morphologies with a view to use them as the basic building block for nano-scale devices. One-dimensional (1-D) nanostructures such as nanorods, nanowires, nanobelts or nanoribbons are of special importance because of their improved fundamental, physical and chemical properties, which make them potential candidate for nano-scale devices. Zinc sulfide (ZnS) is a unique luminescent material well known for its photoluminescence, electroluminescence, etc., which enable wide applications in the fields of displays, sensors and lasers. This article reviews the synthesis, structure and properties of 1-D ZnS nanostructures. ZnS 1-D nanostructures have been fabricated by using different physical (vapor-solid, vapor-liquid-solid etc.) and chemical (solvothermal etc.) routes. Optical properties such as optical absorption, photoluminescence and Raman spectra of the 1-D ZnS nanoforms are also discussed here. Wherever possible, the possible applications of these nanostructures are also included.

1. INTRODUCTION

Nanostructures have received tremendous interest among the researchers working in all fields including physics to bioscience, from basic science to technologists for their improved physical and chemical properties and applications superior to their bulk counterparts [1-4].

[1] E-mail address: mssc2@iacs.res.in (S. C.); kar_mitra@yahoo.co.in (S. K.) Corresponding author. Tel: +91-033-2473-4971, Fax: +91-033-2473-2805

Since the first discovery of carbon nanotubes, one-dimensional (1-D) semiconductor materials such as nanorods, nanowires, nanotubes and nanobelts/nanoribbons have attracted extensive interest because of their fundamental physical, chemical, optical, electrical and magnetic properties, which are distinct from their bulk counterparts, and also due to their potential applications in nano-scale devices [5-16]. It is well known that 1-D nanostructures can play an important role both as interconnect and functional units in fabricating electronic, optoelectronic, electrochemical and electromechanical devices with nanoscale dimension. 1-D nanostructures have been synthesized through a variety of synthesis techniques, such as, template-directed synthesis [17-19], Vapor-solid growth [20-22], vapor-liquid –solid (VLS) growth [23-26], solution-liquid-solid (SLS) growth [27], solvothermal process [28-31] etc. A variety of metal [32-35], semiconducting oxides [22, 36-40], nitrides [41-44] and sulfides [45-55] in different 1-D nanoforms were reported so far.

Out of these materials, zinc sulfide a II-VI semiconductor is one of the most studied material for its wide range of technologically important properties. Zinc Sulfide (ZnS) has a wide band gap of 3.72 eV for cubic phase [56] and 3.77 eV for the hexagonal wurtzite phase [57] at room temperature. It is a key material for ultraviolet light-emitting diodes, injection lasers [58], cathode ray tubes, flat panel displays [59] and IR windows [60]. ZnS is also important for photoluminescence [61], electroluminescence [62], triboluminescence [63], etc. Recently, optical wave confining and lasing has been demonstrated in ZnS nanoribbons [64]. Moreover, zinc sulfide is also widely used in photocatalysis [65-66]. In recent years, nanocrystalline ZnS attracted much attention because properties in nanoforms differ significantly from those of their bulk counterparts [67]. Therefore, much effort has been made to control the size, morphology and crystallinity of the ZnS nanocrystals with a view to tune their physical properties.

Thus keeping in mind the recent trend towards one-dimensional nanostructured materials and applicability of ZnS as an technologically important semiconductor, this article reviews the different aspects of 1-D ZnS nanoforms i.e. synthesis techniques for these nanoforms, their properties and possible applications.

2. ZnS 1-D Nanoforms

The research on the synthesis and characterization of ZnS 1-D nanoforms have progressed much and a variety of 1-D and quasi 1-D nanoforms of ZnS are reported so far. In the following section we will discuss different 1-D nanoforms of ZnS and their synthesis scheme in brief.

2.1. ZnS Nanorods

1-D structures with diameters ranging from several nanometers to few hundred nanometers without having any bending along their lengths are referred as the nanorods. ZnS nanorods were prepared by several routes [68-70, 52]. ZnS nanorods were prepared by ethylenediamine assisted hydrothermal process [68-69] at temperature ~ 200 °C. The fabricated nanorods were found to have diameters within 10-50 nm and lengths ~ few hundred nanometers. These nanorods were single crystalline and no staking faults were revealed from the HRTEM images. Figure 1a-b shows the TEM and HRTEM images of ZnS nanorods synthesized in the author's laboratory via solvothermal route [69]. Interestingly, these nanorods synthesized at ~ 200 °C possessed hexagonal wurtzite structure, which is the meta-stable polymorph of ZnS as wurtzite ZnS is generally stable beyond 1020 °C. ZnS nanorods with cubic zinc blende structure were also reported by chemical routes [70]. The 1-D nanoforms can show quantum confinement effect if their diameter lies well within the limit of 10 nm [71]. ZnS nanorods having quantum size effect have also been synthesized by several groups [72-73] through chemical approach. ZnS nanorods were often appeared as an intermediate product when ZnS nanowires were being prepared by VLS approach as reported in our earlier work [52]. These nanorods further grew in length giving rise to ZnS nanowires.

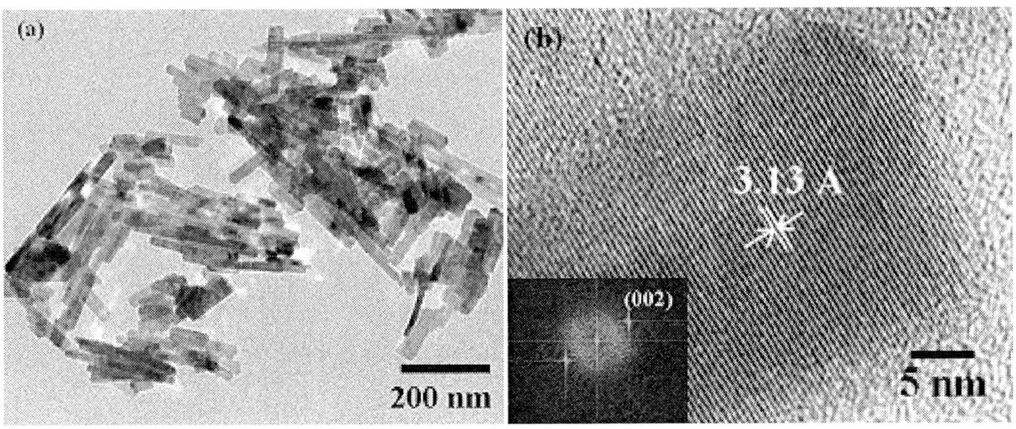

Figure 1. (a) TEM image of the ZnS nanorods synthesized by solvothermal process and (b) HRTEM image of one ZnS nanorod revealing its single crystalline nature.

2.2. ZnS Nanowires

Nanowires are basically elongated version of the nanorods having wire like bending throughout their lengths (Figure 2a-b). The aspect ratio (length/diameter) of the nanowires is generally very high compared to those of the nanorods. Cubic ZnS nanowires with zinc blende structure were prepared by liquid crystal template [74], and miecelle-template [75] based chemical routes. Meng et al. [76] also reported cubic ZnS nanowires by a VLS approach. Unlike the cubic nanowires produced by the chemical approach, the nanowires produced by the VLS approach were found to have better control over their morphology. The nanowires produced by this high temperature (Figure 2c) based approach were smooth and uniform throughout their length and possessed Au nanoparticles at their tips, which were utilized to catalyze the nucleation of the nanowires. Similar VLS approach was also utilized by other groups [52, 77-78] to fabricate ZnS nanowires with hexagonal wurtzite structure. The nanowires produced by these groups were found to have circular cross-section whereas Xiong et al. [79-80] have reported square and hexagonal cross-sectional nanowires by pulsed laser vaporization (PLV) method. Jiang et al. [81] have reported polytype modulated ZnS nanowires via intermittent laser ablation-catalytic growth. The polytype modulated ZnS nanowires are composed of closed packed planes of the wurtzite structure stacking and wurtzite variant half cells of 8H and 10H appearing along the entire wire lengths in turn. Basically the growth mechanism of these ZnS nanowires, produced by laser ablation methods, is VLS mechanism. The nanowires produced by all these groups were non-aligned lying allover the substrate in an irregular manner as revealed by the SEM image (Fig 2a). Freestanding aligned 1-D semiconductor nanostructures are now attracting special attention because of their promising application as building blocks for nano-scale devices and as model system to study the 1-D nano-scale effect [82-83]. Efforts are on to achieve 1-D ZnS aligned nanoforms with a few recent successes [84-85]. Ding et al. [84] and Z. L. Wang's [85] research group have reported synthesis of aligned ZnS nanowires by thermal evaporation of ZnS powder on to aniodic alumina template (AAA) and CdSe crystals respectively. Ding et al [84] have also showed room temperature lasing from the aligned ZnS nanowires.

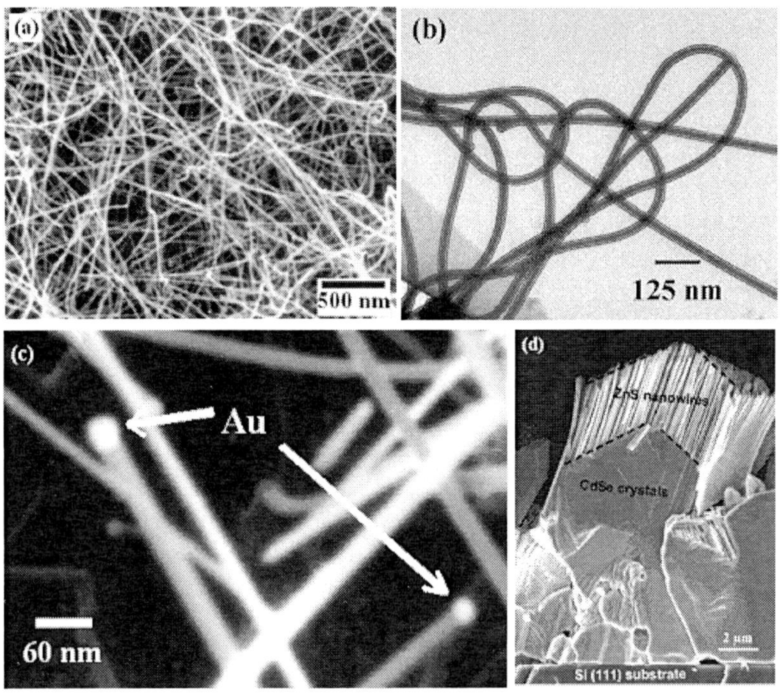

Figure 2. SEM (a) and TEM (b) images of ZnS nanowires. The presence of the Au nanoparticle at the tip of the ZnS nanowires could be seen in SEM image of the ZnS nanowire shown in (c). ZnS nanowire arrays grown on to a lattice matched CdSe crystals is shown in (d) [Figure d reprinted with permission from J. Am. Chem. Soc. 126 (2004) 14372].

2.3. ZnS Nanotubes

Nanotube was the first 1-D nanostructure, the discovery of which by S. Iijima [5] in 1991 triggered the research of 1-D systems. These 1-D systems are basically wire or rod like structures with hollow channel throughout their lengths. Though carbon nanotube is the most popular in this category, other inorganic nanotubes are also attracting interest to explore their physical, chemical properties and technological importance. Dloczik et al. [86] have synthesized ZnS nanotubes by chemical conversion of ZnO nanocolumns with H_2S at 400 °C temperature. ZnS nanotubes were also produced at near room temperature through chemical conversion of ZnS nanobelts in H_2S atmosphere [87]. The wall of these ZnS nanotubes were not single crystalline but, polycrystalline in nature and they had hexagonal wurtzite phase. Single crystalline wurtzite ZnS nanotubes with hexagonal cross-section (Figure 3) was reported by Yin et al [88]. They produced the ZnS nanotubes by thermal evaporation of ZnS powder at 1500-1700 °C in an H_2O-Ar atmosphere. These ZnS nanotubes were few micrometer long having diameter within 100-120 nm and wall thickness ~ 10 nm. The walls of these ZnS nanotubes were perfectly single crystalline with hexagonal crystal structure. These nanotubes showed strong green emission at room temperature.

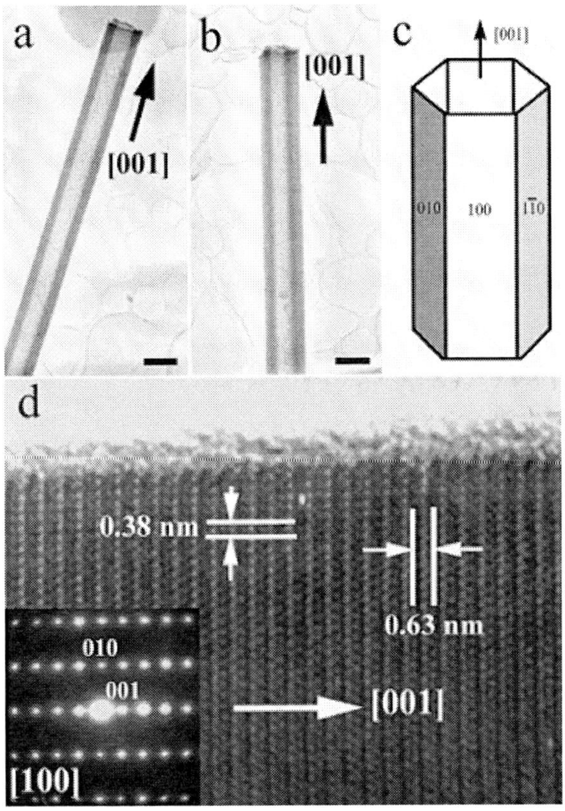

Figure 3. (a, b) TEM images of the ZnS nanotubes, (c) Schematic view of the hexagonal cross-sectional ZnS nanotubes and its growth habit; (d) HRTEM of the wall of the ZnS nanotube with the SAED pattern in its inset. [reprinted with permission from Adv. Mater. 17 (2005) 1972].

2.4. ZnS Nanobelts/Nanoribbons

Semiconductor nanoribbons/nanobelts have been reported as a new subgroup of 1-D nanostructures, which are distinctly different from the nanotubes and nanowires. These nanobelts/nanoribbons gained much interest after Z. L. Wang's group first reported it in 2001 [22]. The nanobelts have specific growth direction and the top, bottom and side surfaces have well-defined crystallographic facets. Whereas, for the nanowires, the side surfaces need not be well defined. For the use of these 1-D nanoforms as a building block for the fabrication of nano-scale devices such as lasers etc. nanoribbons are more suitable candidates than nanowires or nanorods because of their larger widths, which makes the nanobelts/nanoribbons structurally more controllable. ZnS nanobelts/ribbons have been fabricated by several groups [52, 54-55, 64, 89-98]. Almost all the groups have fabricated ZnS nanobelts/nanoribbons by thermal evaporation of ZnS powder at high temperature (900-1600°C) in an inert or reducing atmosphere. For the synthesis, catalytic and non-catalytic both types of approach have been applied. The catalyst-assisted approach is called vapor-liquid-solid (VLS) process whereas; non-catalytic thermal evaporation is termed as vapor-solid (VS) approach. Gong et al. [93] have used elementary Zn and S as the source materials to fabricate ZnS nanobelts by catalyst free thermal evaporation process. The basic advantage of using Zn and S powder rather than using ZnS powder is that ZnS sublimes rather slowly than those of Zn and S resulting in the faster growth of the ZnS nanobelts with Zn and S as the precursors. Recently Yao et al. [94] have fabricated ZnS nanobelts using a low temperature based solvothermal process. Fig 4a shows the SEM image of the ZnS nanobelts/nanoribbons with flat side surfaces fabricated by our group by thermal evaporation of ZnS powder at 1225 °C in flowing Ar atmosphere. Though the side facets of the nanobelts/ribbons are expected to be flat, but, saw-teeth like ZnS ribbons were also reported by Moore et al. [92]. They have suggested that the saw-teeth structure was a direct result from the difference in the surface chemical activities as determined by surface termination. The Zn-terminated surface was chemically active, while the S-terminated surface was relatively inactive, resulting in the growth of saw-teeth like structure. Figure 4b shows the TEM image of such saw-like ZnS nanobelts. These nanobelts were mainly composed of hexagonal wurtzite phase of ZnS but in some cases cubic structured zones could be visible from the HRTEM images. These cubic structured zones were found within two wurtzite regions separated by few staking faults [54-55, 91]. ZnS nanoribbons with a bicrystalline structure was also reported by few groups [54, 97-98]. TEM image of a ZnS bicrystal nanoribbon is displayed in Figure 4c along with the schematic representation of the bi-crystal nanoribbon in the inset. Figure 4d shows the HRTEM image of one representative ZnS bicrystal nanoribbon revealing two identical lattice planes on both sides of a twin boundary. The twin boundary was defined by two stacking faults (marked by drawing two arrows in the image). The SAED pattern recorded on a bicrystal nanoribbon is shown in the inset of Figure 4d revealing two sets of points for the two planes.

3. SYNTHESIS

The basic concept behind the 1-D nanostructures formation is the controlled and restricted crystal growth. Crystal growth means, the evolution of a solid from a vapor, liquid or solid phase resulting into nucleation and subsequently crystal growth. When the concentration of the elementary species or building blocks of a solid becomes sufficient, they assembled together to form a small cluster, which act as the nucleation center or seed. For the successful fabrication of 1-D nanoforms the building blocks needs to be supplied at controlled rate so that the resulting crystal have a homogeneous composition and uniform morphology. The basic need for the growth of the 1-D nanoforms is to prevent the growth of the seeds in two dimensions. This could be achieved by using various templates, use of appropriate capping agents to kinetically control the growth rates of various facets of the seed, use of supersaturation condensation to modify the growth habit of the seed, introducing a liquid-solid interface to reduce the symmetry of a seed etc. [99]. For the synthesis ZnS 1-D nanoforms various synthetic approaches have been adopted and are discussed below.

3.1. Vapor-Solid (VS) Method

The vapor-solid (VS) growth is nothing but an evaporation condensation process. This is one kind of spontaneous growth process, where the growth of 1-D nanoforms is driven by the decrease in Gibbs free energy, which arises from either recrystalization or decrease in supersaturation. The 1-D nanoforms grown by this process are generally single crystalline with very few imperfections. The formation of the 1-D nanostructure via VS process is due to the anisotropic growth. Several mechanisms may be responsible for the anisotropic growth such as, different growth rates at different facets in a crystal, presence of imperfection or defects in a specific crystal directions or preferential accumulations of impurities on specific facets. Sears [20] was the first to explain the growth of mercury whisker by anisotropic growth in 1955.

ZnS nanobelts/nanoribbons are the commonly reported structure grown by VS method. The growth mechanism of the ZnS nanoribbons can be well understood on the basis of the vapor-solid (VS) mechanism [20-22]. In the VS method the source materials vaporize to the molecular level with stoichiometric cation-anion molecules which condense on the substrate and the molecules arranged themselves in such a way that the proper local charge balance and the structural symmetry is maintained resulting in a nucleation center. With further intake of the molecules, the surfaces that have lower energy e.g. side surfaces starts to form and the low energy surfaces tend to be flat. As the molecules have higher mobility at the higher growth temperature, the newly arrived molecules tend to diffuse to the rough growth front resulting in the increase in the surface area, which leads to the formation of the nanoribbons. The rough surface at the growth front leads to the rapid intake of the incoming molecules resulting in the rapid formation of the nanoribbons. The newly arrived molecules can stick to the rough growth front or the side surfaces. But the smooth side surfaces and the high mobilities prevent them from remaining on the surface. The molecules randomly diffuse on the surface and finally accumulate at the lower energy growth sites.

Generally, a horizontal quartz tube furnace fitted with a gas flow system and a vacuum system is used for the synthesis purpose. A simple schematic view of the experimental set up is presented in Figure 5. Jiang et al. [55] have first reported wurtzite ZnS nanobelts by hydrogen assisted thermal evaporation of ZnS powder at 1100 °C for 3-4 h in horizontal quartz tube furnace. Analytical grade ZnS powder was placed in an alumina wafer kept in the central part of the quartz tube, which was then evacuated and Ar gas pre-mixed with 5 % H_2 gas was flown through it at rate of 50 cm^3 /min. ZnS nanobelts were collected after the deposition from the inner wall of the quartz tube from a region having temperature ~ 900 °C. The synthesized products were found to have ribbon like morphology and they possessed wurtzite crystal structure. But, the HRTEM study revealed that the nanoribbons have two kinds of structure zones, one wurtzite and another cubic zone. These two zones were found coexisting in some of the nanoribbons by changing stacking sequences. Zhu et al. [90] also produced wurtzite ZnS nanobelts by thermal evaporation of ZnS powder in presence of graphite and B-N-O powder in an N_2 / H_2O flowing atmosphere. Ma et al. [91] have also reported ZnS nanobelts by simple evaporation of ZnS powder at flowing atmosphere at 1050 °C. But they could not achieved morphological control as other structures like nanocombs, windmills etc were also formed simultaneously. Recently, we have [54] reported controlled growth of ZnS nanoribbons by simple thermal evaporation of ZnS powder in pure Ar flowing condition. Also, the HRTEM observations revealed no cubic structured zone in the ZnS nanoribbons synthesized by us. We ascribed this to the fact that the growth region of the ZnS nanoribbons in our case was situated in a region with temperature ~ 1050 °C which was above the phase transition temperature of ZnS from cubic to wurtzite (1020 °C). Whereas, the growth region of the nanobelts in the previous reports were below this phase transition temperature resulting into some cubic phase formation along with the dominant wurtzite phase.

We have also reported [54] the growth of ZnS bi-crystal nanoribbons, where two single crystal planes were formed on both sides of a twin boundary along the length of the ribbon. Identical growth mechanism was responsible for the formation of single crystal and bicrystal nanoribbons at the higher Ar flow rate. The sword like end was another indication of the vapor-solid growth mechanism. The higher Ar flow rate had carried the ZnS molecular vapor to the distant end with temperature ~ 500 °C. This sudden fall of temperature of the nucleation site relative to the source region (1225 °C) might have introduced the stacking faults, which favors the bicrystal growth. The twin structure observed in the bicrystal nanoribbons (Figure 4d) provided energetically favorable sites onto which the incoming molecules and clusters stack uni-directionally into nanoribbons.

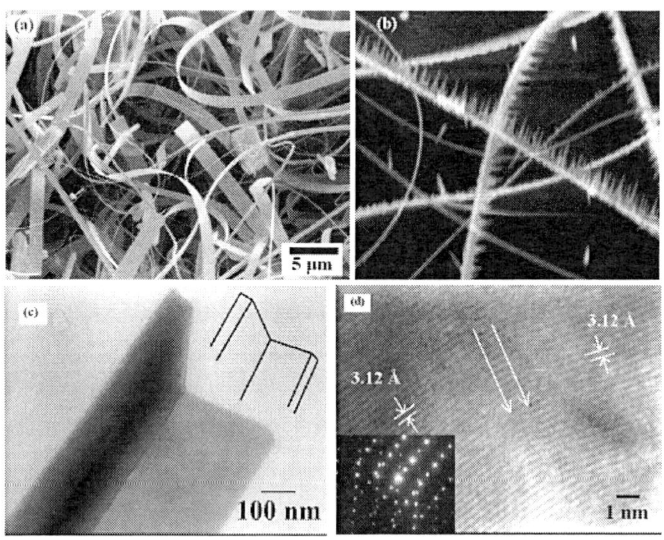

Figure 4. (a) SEM image of the smooth ZnS nanoribbons; (b) SEM image of the ZnS nanosaws; TEM (c) and HRTEM (d) image of ZnS bicrystal nanoribbons and the corresponding inset shows the schematic view of bicrystal ribbon and the SAED pattern showing two different set of points for the bicrystal ribbon respectively [Figure b reprinted with permission from Chem. Phys. Lett. 385 (2004) 8 and Fig c-d reprinted with permission from Chem. Phys. Lett. (2005) Doi:10.1016/ j.cplett.2005.08.021].

3.2. Vapor-Liquid-Solid (VLS) Method

For the VLS process similar experimental set up as that of VS process are used. In the VLS process, a second material commonly known as the catalyst is purposefully introduced to direct and confine the crystal growth onto a specific orientation and within a confined area. The catalyst forms a liquid droplet by itself or by alloying with the growth material during the growth, which acts as the trap of growth species. Enriched growth materials in the catalyst droplets subsequently precipitate upon super saturation at the surface of the substrate resulting in the 1-D growth. Wagner et al. [23-26] first proposed the VLS theory in 1964 to explain the experimental results and observations in the growth of Si nanowire. So in the VLS process, the presence of the catalyst is always necessary and a liquid droplet is always found at the tip of the nanowire.

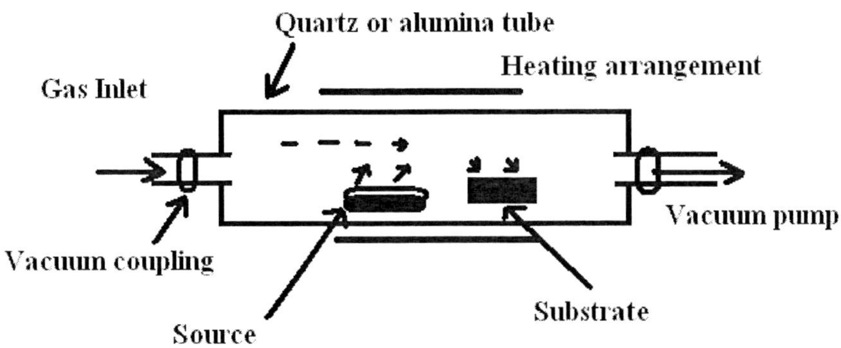

Figure 5. Schematic view of the deposition system for the VS and VLS growth of 1-D ZnS nanoforms.

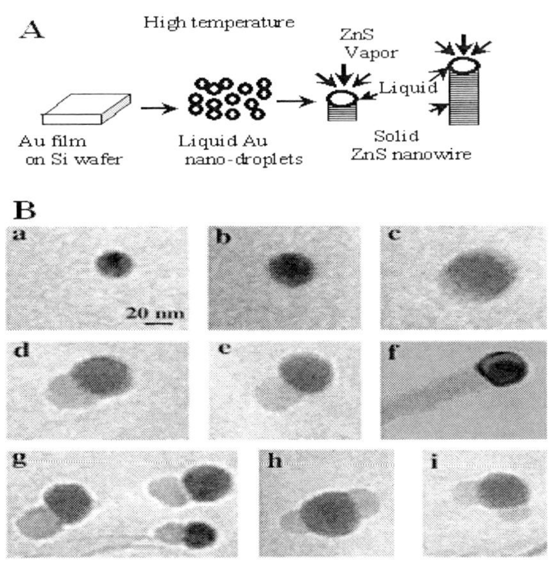

Figure 6. (A) Schematic view of the VLS process, (B) In situ TEM images recorded during the Ge nanowire growth revealing the role of the catalyst particle in the nanowire growth by VLS process [Figure B reprinted with permission from J. Am. Chem. Soc. 123 (2001) 3165].

Schematic representation of the VLS model is presented in Figure 6 on the basis of the in situ TEM observations of the Ge nanowire growth by VLS process [25].

For the growth of ZnS 1-D nanostructures generally gold (Au) was used as the catalyst and ZnS powder as the source material. A thin layer of Au film deposited on suitable substrates broke up at high temperature to form liquid Au nanodroplets on the surface. These liquid Au droplets at high temperature absorbed ZnS vapor and Au-ZnS nanocomposite alloy particles were formed. On super-saturation, pure ZnS nanostructures were originated from these composite particles. Depending on the experimental parameters such as, temperature and flow rate of the Ar gas, these liquid composite particles took different shapes and different morphologies of the ZnS nanostructures were originated from them. There are various reports on the fabrication of ZnS nanowire [52, 77-81] and nanobelt/nanoribbon [52, 89, 97] by VLS process. Generally higher temperature was needed for the fabrication of ZnS nanopbelts/nanoribbons than those of nanowires by the VLS process. We have reported [52] that ZnS nanowires were formed at 850 °C whereas; ZnS nanoribbons were achieved at 950 °C via VLS process. Figure 7 shows the SEM images of ZnS nanowires at 900 °C at various stage of their growth by VLS technique [78].

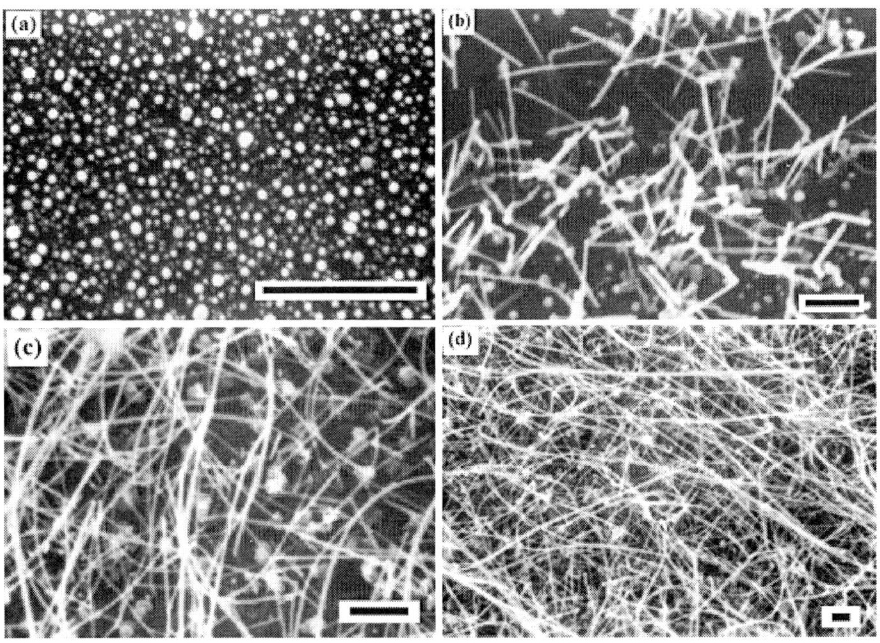

Figure 7. SEM images showing different stages of the ZnS nanowire growth after (a) 5 mins, (b) 15 mins, (c) 20 mins and (d) 30 mins. The scale bars in all these images represent 500 nm [reprinted with permission from Nanotechnology 16 (2005) 737].

We have also [52] shown that the initial shape of the ZnS nanostructures could determine the morphology of the final ZnS nanoforms and we postulated on the basis of experimental observation that in some favorable experimental condition the catalyst particle could be elongated to form cylindrical Au particles, which in turn induces the ZnS nanoribbon growth. In our recent work [53] we have proposed that the formation of ZnS nanobelts/nanoribbons having width greater than 500 nm could not be attributed to the VLS process alone. Instead, we propose that the ZnS nanoribbons with larger width were formed by a combined effect of

VLS and VS mechanism. Unlike other groups, instead of using ZnS powder as the source material, Barrelet et al. [100] have used a single source molecular precursor $Zn(S_2CNEt_2)_2$ for the fabrication of ZnS nanowire by the VLS process. Laser assisted catalytic growth process were also reported [79-80, 100-101] for the ZnS nanowire formation, where a target of mixed powder of ZnS and Au was ablated by a laser source. The nanowires obtained by this process were smooth and uniform throughout their length and they were perfectly single crystalline in nature. Both cubic [76] and hexagonal [52, 77-78] nanowire could be achieved by varying the experimental parameters like temperature of the source and the substrate, position of the substrate etc. ZnS nanoribbons/nanobelts were single crystalline, and possessed well-defined side as well as end surfaces. The lengths of the nanowires and nanoribbons could be of the order of few hundred micrometers.

3.3. Template Assisted Growth

Template based synthesis of nanostructures of different materials is a very general method used to fabricate nanorods, nanowires and nanotubes. Various templates with nanosized channels have been explored for the template-assisted growth of 1-D nanoforms. The most commonly used and commercially available templates are anodized alumina membrane, mesoporous silica, carbon nanotube etc. Liquid crystal template [74] and micelle templates [75] were used to fabricate ZnS nanowire at or near room temperature. Porous templates like aniodic alumina were generally used to fabricate nanowire arrays by electrochemical process, sol-gel deposition and other chemical routes. Ding et al. [84] have utilized Au particle filled aniodic alumina membrane as the substrate to synthesize aligned ZnS nanowire arrays by VLS approach at 1100 °C.

3.4. Chemical Conversion Method

1-D nanoforms can also be prepared by using a consumable template i.e. first nanowires or nanoribbons of one phase of the material is formed which are subsequently converted to another phase by chemical reactions. ZnS nanotubes were produced by chemical conversion of ZnO nanocolumns [86] and ZnO nanoribbons [87] with H_2S. Dloczik et al. [86] converted an existing ZnO column arrays by passing H_2S through it at 400 °C, when a thin layer of ZnS was formed on the ZnO core, which was then etched by using H_2SO_4 solution. Being more etch resistant, the ZnS shell remained intact giving rise to a tubular ZnS with ~ 15 nm wall thickness. They have also obtained ZnS nanotubes by passing S vapor at 450 °C. Wang et al. [87] have converted ZnO nanobelts to ZnS nanotubes by adding H_2S solution to an ethanol solution of ZnO nanobelts and subsequently kept it at room temperature for some time for complete conversion. The basic principle behind this kind of conversion can be expressed by the following equation.

$$ZnO + H_2S \rightarrow ZnS + H_2O$$

3.5. Solvothermal Process

Solvothermal synthesis utilizes a solvent under pressure and temperature above its critical point to increase the solubility of the solid reactants and speed up the reactions. In the typical procedure, the precursors and a solvent, possibly an amine capable of regulating or templating the crystal growth to generate 1-D nanoforms. The mixture is placed in an autoclave, which is placed at an elevated temperature and pressure to speed up the reaction to generate 1-D nanoforms. The major advantage of this approach is that most of the materials can be made soluble in a solvent by heating and pressurizing it close to the critical point. The products obtained by this approach are generally in the form of powder. Ethylenediamine (en) was found to promote one-dimensional growth of CdS by solvothermal process due to its chelating property [29-30]. But ZnS nanorods could not be synthesized with ethylenediamine following the same experimental conditions used for the CdS nanorods. Instead, flakes like two-dimensional complex $ZnS.(NH_2CH_2CH_2NH_2)_{0.5}$ were produced [31]. The differences in the coordination ability of ethylenediamine with Zn^{2+} and Cd^{2+} ions and some other kinetic factors related with the solvothermal growth might be the reason behind the difference [31]. However, by annealing this complex precursor in Ar atmosphere at high temperature or by treating the complex hydrothermally, a mixture of pure hexagonal ZnS nanorods and particles could be obtained. It was observed that by mixing water with the ethylenediamine (en), uniform hexagonal ZnS nanorods were produced at 200 °C [68-69]. The hexagonal phase of ZnS is stable at temperature higher than 1020 °C while the cubic phase is stable below this temperature. But the presence of the chelating solvent at high pressure might have brought down the free energy difference between the hexagonal and cubic phases of ZnS resulting in the formation of wurtzite ZnS at much lower temperature. Recently, ZnS nanoribbons were also fabricated by solvothermal process [94] in a mixed solvent of water and diethylenetriamine with a volume ratio 1:2.5.

4. Characterization

ZnS nanoforms were characterized by different techniques to explore their physical and chemical properties. Crystal structure and phase of the products are determined from the powder X-ray diffraction (XRD) studies. Chemical purity and stoichiometry of the ZnS nanoforms can be tested from the energy dispersive analysis of X-ray (EDAX) and X-ray photoelectron spectroscopy (XPS). Morphology of the ZnS nanoforms is determined from the scanning electron microscopy (SEM) and transmission electron microscopy (TEM). The HRTEM of the nanoforms helps to determine the crystal structure of the individual nanoforms. Various chemical characterizations are also performed on the ZnS nanoforms to determine the surface and interior atoms and compounds and their spatial distribution. Optical absorption spectroscopy, photoluminescence spectra (PL), Raman spectroscopy, infrared spectroscopy etc. are the most popular characterization techniques to explore the chemical properties of the ZnS nanoforms.

4.1. Phase and Crystal Structural Identification By XRD

X-ray diffraction XRD is a very important experimental technique that has been used to address all issues related to the crystal structure of samples, including lattice parameters, identification of unknown materials, orientations of single crystals, preferred orientations of polycrystals, defects, stress etc. Also, XRD is used to identify the specimen's crystalline phases. In case of ZnS nanostructures the phase of the as synthesized 1-D nanoforms is identified by XRD i.e. whether they are cubic or hexagonal. Figures 8a-b show the representative XRD patterns of the cubic and hexagonal ZnS. In most of the reports on ZnS 1-D nanoforms XRD was mainly used for the identification of their phases. We have observed some difference in the XRD pattern of the ZnS nanoribbons produced by VLS and VS methods. Figure 8c shows that the relative intensities of the different diffraction peaks obtained from the ZnS nanoribbons produced by VLS and VS approaches are not identical. This indicates that the growth rates of the different crystal planes of the wurtzite ZnS nanoribbons are very much dependent on the experimental condition i.e. the environment at which these nanoforms are formed.

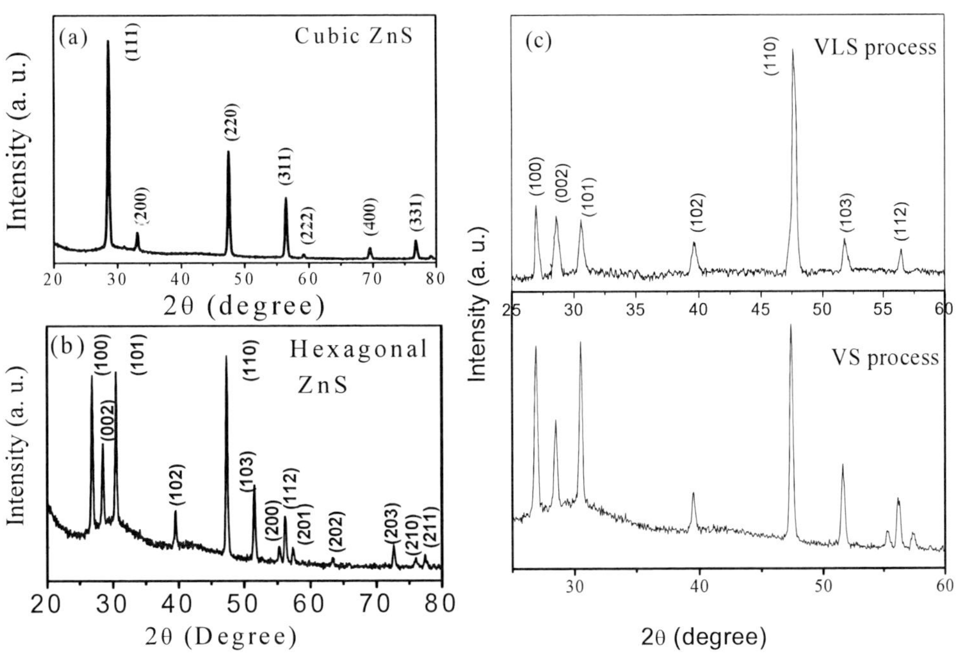

Figure 8. XRD pattern of the different polymorphs of ZnS: (a) cubic ZnS with sphalerite phase and (b) hexagonal ZnS with wurtzite phase and (c) XRD pattern of the wurtzite ZnS nanoribbons synthesized by VLS and VS techniques.

4.2. Chemical Purity and Stoichiometry Test by EDAX

Energy dispersive analysis of X-rays (EDAX) is one kind of electron spectroscopy, which relies on the unique energy levels of the emission of photons (X-rays) or electrons ejected from the atom under investigation. Elementary analysis by EDAX is very important for ZnS nanoforms as the presence of impurity or foreign elements like Mn, Cu etc can change the emission from the ZnS nanoforms drastically. Figure 9a shows the energy dispersive analysis of X-ray (EDAX) pattern recorded over a single ZnS nanoribbon revealing the presence of elementary Zn and S. The atomic ratio of the constituent atoms i.e. the stoichiometry of the sample can also be determined from the compositional analysis of the EDAX spectrum. The EDAX was also utilized to test the identity of the spherical particle obtained at the tip of the 1-D ZnS nanoforms produced by VLS approach. Figure 9b shows the EDAX spectra of the Mn-incorporated ZnS nanorods. Thus the role of the EDAX is not only to test the chemical composition of the ZnS nanoforms but also to investigate the reason behind the visible emission from the ZnS nanoforms. The main drawback of this technique is that it cannot trace the presence of elements having very low concentration.

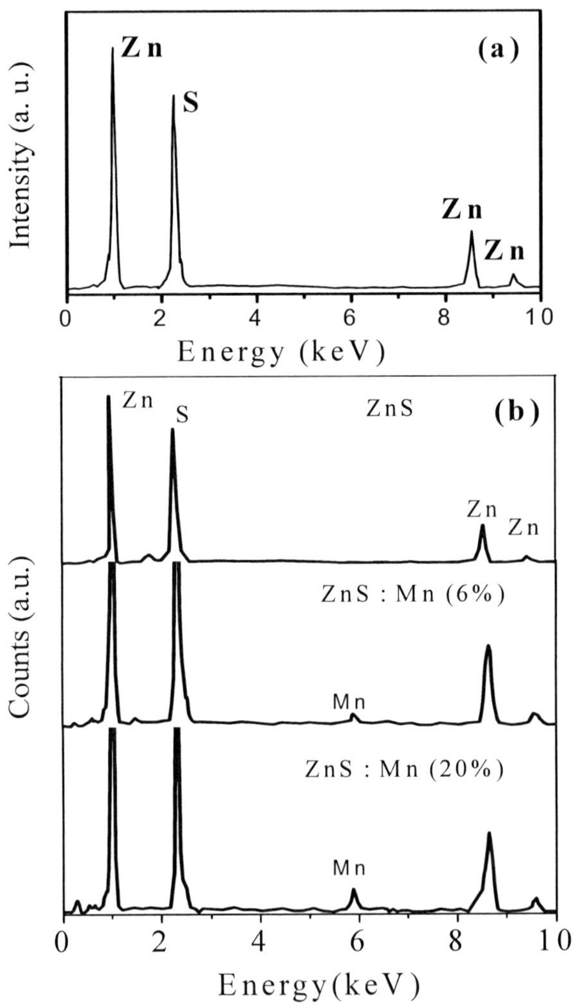

Figure 9. (a) EDAX spectrum of a representative ZnS 1-D nanoforms and (b) EDAX spectra of the $Mn_xZn_{1-x}S$ nanorods.

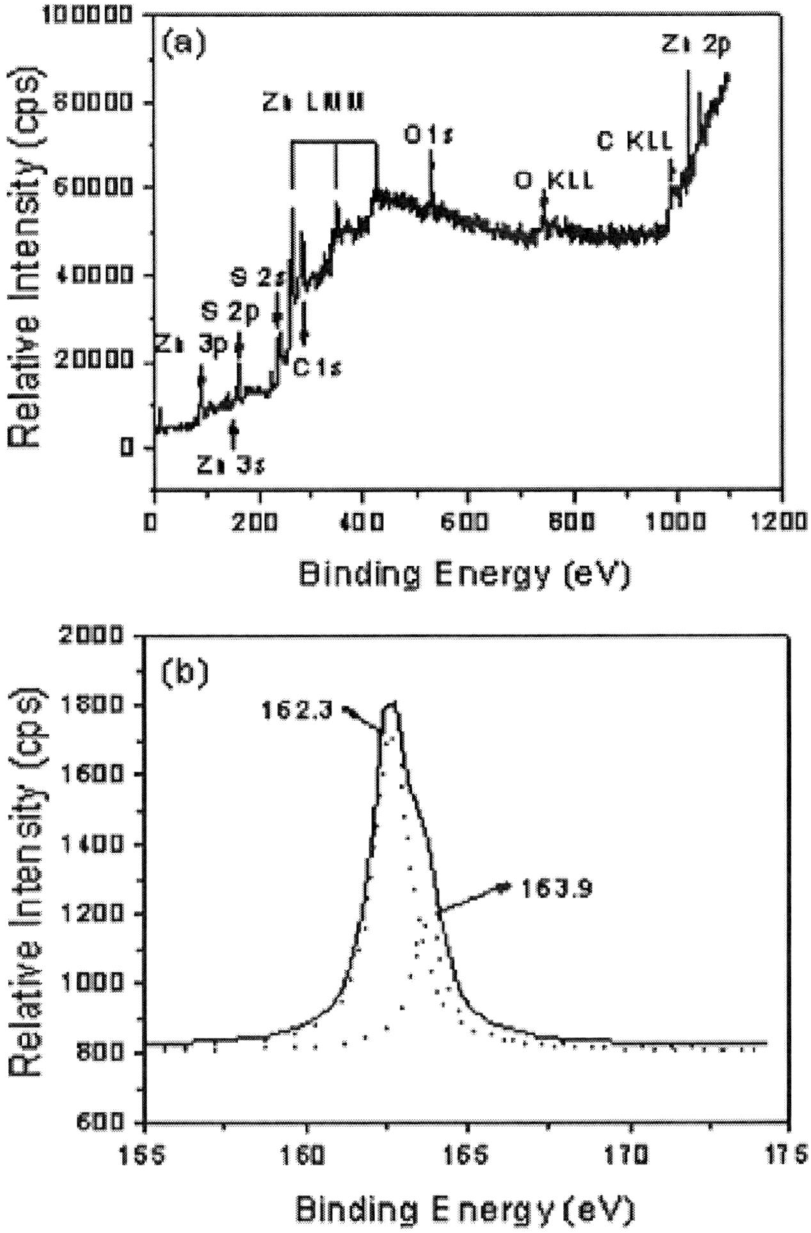

Figure 10. XPS spectra recorded over the ZnS nanobelts: (a) survey spectrum of ZnS nanobelt, and (b) S (2p) spectrum [reprinted with permission from Appl. Phys. Lett. 85 (2004) 3035].

4.3. XPS Study

Working principle of the XPS spectroscopy is somewhat similar to EDAX. The only difference is that in XPS, relatively low-energy X-rays are used to eject the electrons from the atom via the photoelectric effect. In XPS study the nature of bonding between the elements could also be distinguished. In addition to its ability to test the chemical composition, XPS pattern was successfully utilized to investigate the nature of chemical bonding in the ZnS 1-D nanoforms and their impact on the properties of the ZnS nanoforms. Ye et al. [96] have utilized the XPS to establish the origin of the green emission from their ZnS nanobelts. From the XPS spectrum (Figure 10) they have identified the presence of the elementary S along with the S^{2-} species in the ZnS nanobelts and they have attributed this elementary S as the reason behind the green emission from their sample.

4.4. Morphological Identification by SEM Study

Scanning electron microscopy (SEM) is one of the most widely used techniques to characterize nanomaterials. The resolution of the SEM approaches a few nanometers. The biggest advantage of the SEM is that one can investigate the morphology of the sample without disturbing the original nature of the products. Generally, ZnS 1-D nanoforms are synthesized over a flat surface or in the form of powder. In either case the as grown sample can be transferred to the SEM chamber without disturbing the original nature of the sample and this is most important to investigate the growth nature of the sample. From the SEM image along with the morphological identification of the products one can also deduce the size distribution of the nanoforms. Figures 2a,c-d, 4a and 7 show the SEM image of the different forms of ZnS 1-D nanoforms. The SEM images are widely utilized to distinguish between the nanorods, nanowires, nanobelts/nanoribbons and nanotubes. In addition to its unique ability to identify the morphology of the ZnS nanoforms, the SEM images can also be utilized to monitor the growth of ZnS nanoforms by taking images at different stage of their formation. We have utilized SEM images [78] to investigate the role of VLS process in the formation of ZnS nanowires by recording the images (Figure 7) after different deposition periods.

4.5. Morphology and Crystal Structure Identification by TEM and HRTEM Studies

For the TEM study the sample has to be (after it is scrapped from the substrate if the nanoforms grows over a substrate) dispersed ultrasonically in some volatile solvent and subsequently deposited on a thin layer of carbon coated Cu grid. Figure 11a shows the TEM image of a ZnS nanowire and nanoribbon.

The ripples like contrast observed in the TEM images were originated from the strains in the sample due to the bending of the structures. Selected area electron diffraction (SAED) pattern (inset of Figure 11a) were utilized to identify the crystal structureandthe growth direction of the ZnS 1-D nanoforms. The HRTEM images are more important as they can be utilized to identify the crystallographic structure of the individual nanofrom or even their selected area. Figure 11b and 11c show the HRTEM image of a cubic ZnS nanowire and wurtzite ZnS nanoribbon respectively. Figure 11b also reveals the presence of the catalyst particle at the tip of the ZnS nanowire, which was the essential evidence of the VLS growth mechanism. Also, the presence of defects in the samples could be identified from HRTEM images. From the HRTEM analysis of the ZnS 1-D nanoforms reported by different groups, it was observed that almost all the 1-D ZnS nanoforms were single crystalline in nature. Figure

11d shows the HRTEM image of a ZnS nanoribbon synthesized by us where cubic structured zones are visible within the hexagonal wurtzite zones separated by few staking faults. From the measured spacing of the lattice fringes and the position of the lattice image, growth direction of the 1-D nanoforms as well as their side facets could be identified accurately [102-103].

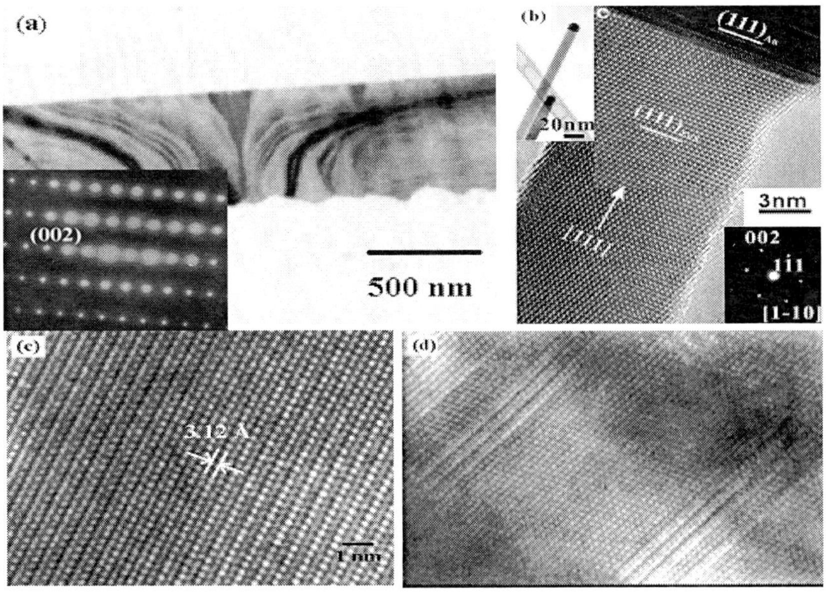

Figure 11. (a) TEM image of a ZNS nanoribbon showing the presence of ripple like contrast originated from the strains developed due to the bending of the ZnS nanoribbons, (b) HRTEM image of a cubic ZnS nanowire showing the presence of the Au catalyst at its tip indicating the role of VLS process, (c) HRTEM image of a ZnS nanoribbons showing the hexagonal crystal structure, and (d) HRTEM image showing the presence of cubic structured zone within wurtzite zone separated by few staking faults. [Figure b reprinted with permission from Chem. Phys. Lett. 382 (2003) 434].

4.6. Optical Absorption/Transmission Spectroscopy

For the optical absorption/transmission spectroscopic measurements ZnS 1-D nanoforms were ultrasonically dispersed in ethanol and subsequently the optical spectra were measured at room temperature [54, 69]. The optical absorption spectrum of the single crystalline ZnS nanoribbon sample is shown in Figure 12a. The absorption study revealed that the nanoribbons were transparent in the visible region. The absorption edge determined from the peak minimum of the first derivative of the absorption plot was at 330 nm (3.76 eV), which matches well with the standard bulk value of band-gap of the wurtzite ZnS. Also, the steep rise of the absorption edge is an indication of the defect free structure of the nanoribbons. If m represents the nature of optical transition in a semiconducting material, the absorption coefficient (α) may be expressed as [104].

$$\alpha = (A/h\nu)(h\nu - E_g)^m \quad (1)$$

Where, A is an optical constant, E_g is the band gap of the semiconductor corresponding to the particular value of m. The parameter m, representing the nature of optical transition may have certain specific values such as, 1/2, 2, 3/2 or 3 corresponding to the allowed direct, allowed indirect, forbidden direct and forbidden indirect transitions respectively.

From equation (1) we have:

$$d(\ln(\alpha h\nu)) / d(h\nu) = m / (h\nu - E_g) \quad (2)$$

So, the plot of $d(\ln(\alpha h\nu)) / d(h\nu)$ vs. $h\nu$ will show a discontinuity at $h\nu - E_g = 0$. The Figure 12b shows the above plot and the corresponding value of the E_g was 3.75 eV. Now, instead of pre-assuming the nature of optical transition, it may be determined from the $\ln(\alpha h\nu)$ vs. $\ln(h\nu - E_g)$ plot (inset of Figure 5b) with E_g = 3.76 eV. The slope of this plot gives m= 0.51 corresponding to the allowed direct transition.

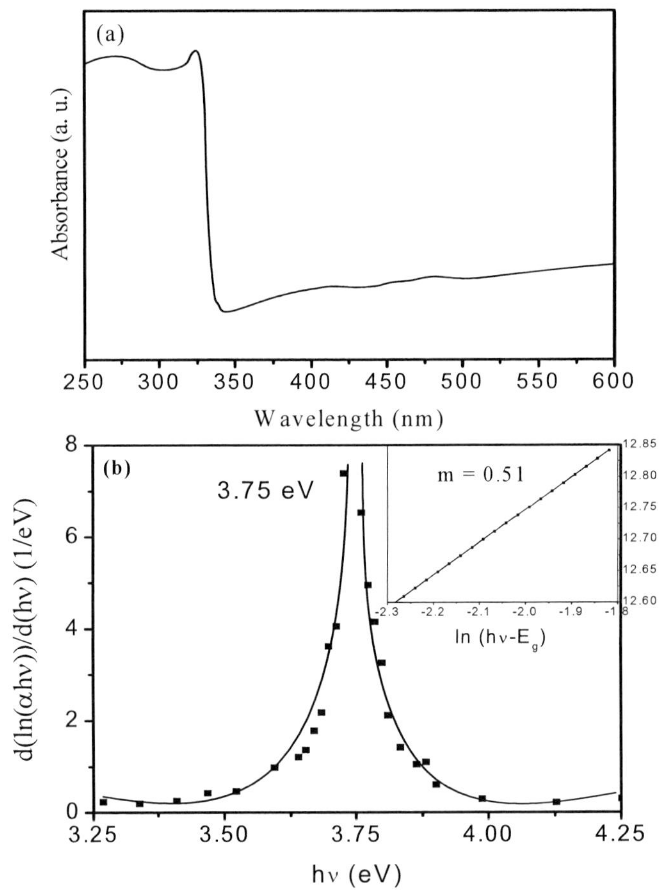

Figure 12. (a) The absorbance spectrum recorded at room temperature, (b) $d(\ln(\alpha h\nu)) / d(h\nu)$ vs. $h\nu$ plot with the discontinuity position assigned to the band gap value and inset shows the $\ln(\alpha h\nu)$ vs. $\ln(h\nu - E_g)$ plot for determining the nature of optical transition (m) [reprinted with permission from Chem. Phys. Lett. (2005) Doi:10.1016/j.cplett.2005.08.021].

4.7. Photoluminescence (PL) Study

In PL, a material gains energy by absorbing light at some wavelength by promoting an electron from a low to higher energy level. After a characteristic lifetime in the excited state, electrons will return to the ground state by radiative transition. ZnS is the most popular semiconductor well known for its luminescent properties. Luminescence in the wide range of UV to visible range is observed from the ZnS nanostructures. In this section we will discuss only the luminescence from the 1-D nanostructures. PL study of the ZnS 1-D nanoforms could be carried out either in solid state (for the samples deposited on a substrate) or by dispersing the ZnS nanoforms into a spectroscopic grade alcohol.

Two kinds of emissions are observed from a semiconductor nanoforms - excitonic and trapped luminescence [105-106]. The excitonic emission is sharp and is located near the absorption of the particles, while the trapped luminescence luminescence arising from different kinds of defects or impurity states is broad and stokes-shifted. Only the trapped luminescence is observed in pure ZnS nanoparticles [105-106] and in 1-D nanoforms, reported by most of the groups [52, 54, 77-78, 81, 87-88, 90, 93, 95-97]. Xiong et al. [79] have recently reported band edge emissions from the ZnS nanowires at room temperature. They have correlated the PL and optical absorption spectra (Figure 13) of the ZnS nanowires to explain the origin of the different emissions.

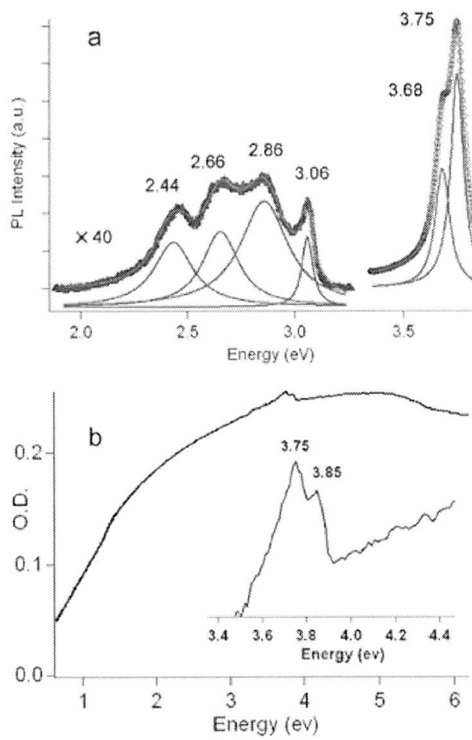

Figure 13. (a) PL spectrum of ZnS nanowires taken at room temperature excited by a 266 nm UV laser. Two strong emission bands near bandedge and four weak defect related luminescence bands were observed. (b) Optical density plot versus photon energy of ZnS nanowires. The inset shows the zoomed view of the vicinity of fundamental absorption edge [reprinted with permission from Nanoletters 4 (204) 1663].

They have observed a strong emission doublet at 3.68 eV and 3.75 eV along with few weaker defect related peaks at 2.44, 2.66, 2.86 and 3.06 eV. In case of wurtzite ZnS with direct bandgap ~ 3.75 eV, the top most valence band is split into three bands due to the crystal field and spin-orbit coupling. Exciton states associated with electron-hole pairs in each case of these valence bands involved are usually denoted as A, B, C excitons [107]. The spin-orbit coupling and crystal-field splitting energy given in energy ref [107] are 89 and 57 meV, respectively. The three features of 3.68, 3.75 and 3.85 eV observed in the PL and OA are assigned to the A, B, C excitons. This kind of bandedge Pl from ZnS was observed for very high quality single crystals [56]. Defect related PL peaks are reported over a wide region ranging from 400 – 600 nm. The blue emission peak at ~ 400 nm was attributed to the sulfur vacancy and interstitial lattice defects in the ZnS nanoforms [52, 78, 95]. The blue emission peaks within the region 440-460 nm are attributed to the trapped luminescence arising from the surface states [52,77-79]. However, various explanations were presented to reveal the origin of the green emission from the ZnS 1-D nanoforms. Few groups [77, 79] believed that the Au used as the catalyst in the VLS growth, might have present in the ZnS lattice as an impurity, which acts as the luminescence center for the green emission from the ZnS nanowires. There are several reports that the self activated defect centers [81], structural defects like point defects [97] or zinc vacancy related defect centers [54, 108] are the origin of the green emission from these 1-D ZnS nanoforms. Ye et al. [96] have reported that the presence of the elementary sulfur species in the ZnS nanobelts could also induces the green emission from the nanoforms. Also the presence of the foreign impurities like Cu^{2+}, Mn^{2+} or rear earth ions even in minute quantity could introduces green emission from the ZnS nanoforms [109-117].

4.8. Raman Spectroscopic Studies

Raman Spectroscopy is very sensitive to lengths, strengths and arrangement of chemical bonds in a material, but less sensitive to chemical composition. Wurtzite ZnS has the space group C_{6v}^4 ($C6_3mc$) [80] with two formula units per primitive cell and with all atoms occupying C_{3v} sites. Group theory predicts the following q=0 phonon symmetries: $\chi_{atom} = 1A_1 + 2B_1 + 1E_1 + 2E_2$, while the A_1 and E_1 symmetry phonons are both Raman- and IR active, the two E_2 pair of modes are only Raman-active, and two B_1 modes are optically silent [118]. There are reports on the Raman scattering study of bulk ZnS [119-121]. But only very recently, Xiong et al. [80] have reported Raman scattering from rectangular cross-sectional ZnS nanowires. They have reported strong first order longitudinal optical (LO), transverse optical (TO), and surface optical (SO) modes and second order combinations and overtones. Figure 14 shows the Raman spectra recorded over the rectangular cross-sectional ZnS nanowires at room temperature. In analogy with the bulk ZnS Raman peaks [118-121] the different Raman peaks were assigned to different phonon modes. The strong peak at 346 cm^{-1} was assigned to the A_1 and E_1 symmetry of LO modes. At lower frequency resolvable doublet at 269 cm^{-1} and 282 cm^{-1} were attributed to (A_1 and E_1) and E_2 TO modes respectively. The broad feature at 216 cm^{-1} was assigned to the LA mode. The continuum scattering within 350-450 cm^{-1} was assigned to the 2-phonon density of states (DOS). The Raman bands between 600-700 cm^{-1} were assigned to the combination and overtones. The peak at 335 cm^{-1} was assigned to the surface optical (SO) mode the line shape of which was sensitive to the real part of the dielectric function of the dielectric medium in contact with the surface.

5. Doped-ZnS 1-D Nanoforms

Zinc Sulfide, a II-VI group compound semiconductor is particularly suitable for use as host material for a large variety of dopants because of its wide band gap (3.75 eV).

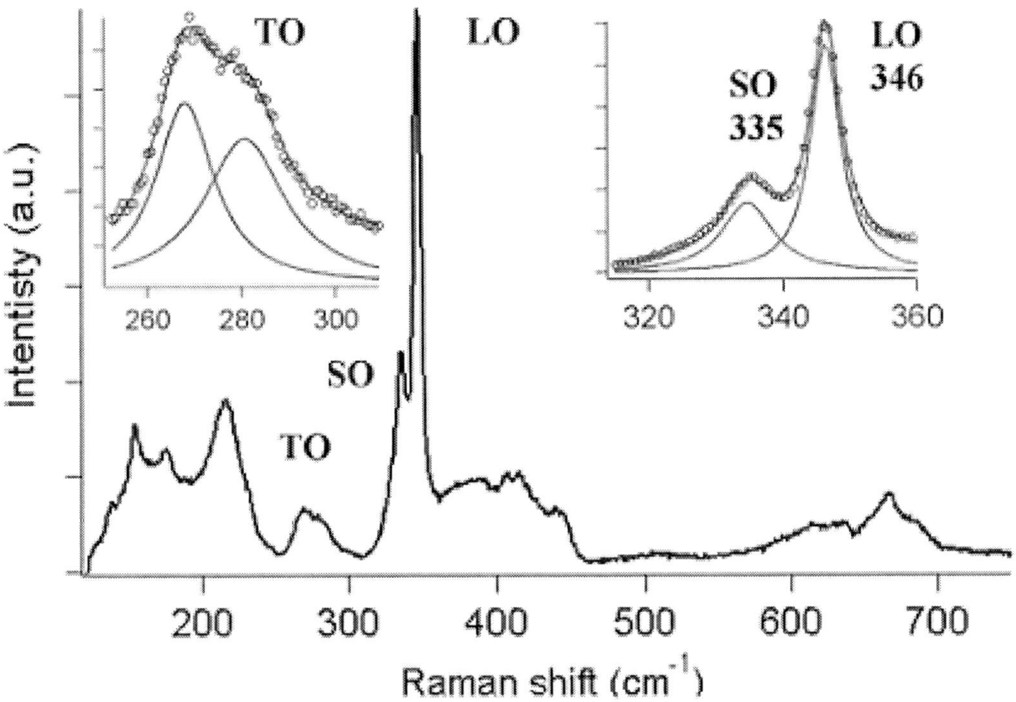

Figure 14. Raman spectrum of ZnS nanowires collected in air. Two insets show Lorentzian line shape analysis of LO (346 cm^{-1}) and SO (335 cm^{-1}) modes and TO doublet (269, 282 cm^{-1}) [reprinted with permission from Nanoletters 4 (2004) 1991].

In fact ZnS has found special importance in electroluminescent devices, lasers, flat panel displays when doped with divalent manganese ions [59, 61-62, 123-125]. Manganese usually occupies substitutionall sites in the Zn lattice as a divalent ion and the excitation and decay of this ion produces an orange luminescence at approximately 590 nm [109, 114]. This emission peak is generally associated with a transition between $^4T_1$ and $^6A_1$ energy levels. Other dopants such as Cu^{2+} [116, 126], Ni^{2+} [127], Cu-Al co-doped [128] and rear earth elements [129] were used to tailor the photoluminescence properties of the ZnS nanocrystals. 1-D ZnS doped with Mn^{2+} [130-133] and Cu^{2+} [131, 133] were also reported. All these processes are based on the thermal evaporation of the precursors at high temperature. Geng et al. [130] have fabricated Mn doped ZnS nanobelts by thermal evaporation of zinc acetylacetonate and manganese acetylacetonate in proper molar ratio in an H_2S atmosphere at 900 C. those doped ZnS nanobelts showed characteristic defect related emission at 450 nm along with the orange emission associated with the Mn^{2+} d-level transition from $^4T_1$ to $^6A_1$. Yuan et al [132] have fabricated Mn-doped ZnS nanowires by thermal evaporation of ZnS and $MnCl_2$ powders onto a Au coated Si substrate at high temperature in an Ar atmosphere. They have reported defect related emission at ~ 503 nm and Mn^{2+} related emissions at 570 nm. The defect related emissions was stronger than that due to ionic level transition at Mn^{2+}. Zhu et al. [131] have synthesized Mn and Cu doped ZnS nanobelt arrays by thermal evaporation of ZnS and MnO_2 or CuO powders in a mixed gas flow (N_2, CO and H_2) at high temperature. These nanobelts arrays showed mainly orange emission for the Mn-incorporated ZnS samles and green emission at 520 nm for the Cu-doped samples. Whereas, the un-doped ZnS samples sowed blue emissions at ~ 450 nm associated with the defect related emissions. Recently, we have synthesized Mn-incorporated ZnS nanorods by a simple solvothermal approach [134]. The sample without Mn corresponds to hexagonal ZnS. Mn^{+2} ion incorporation into the ZnS nanorods resulted into morphological as well as structural changes. The diameter of the ZnS nanorods increased gradually from 12-15 nm to ~ 100 nm for 0 to 20 % Mn incorporations. Mn incorporation introduces a phase transformation from wurtzite to cubic structure. Three different emissions in the blue (400 nm), green (500 nm) and orange (585 nm) regions were visible in the room temperature photoluminescence spectra (Figure 16a).

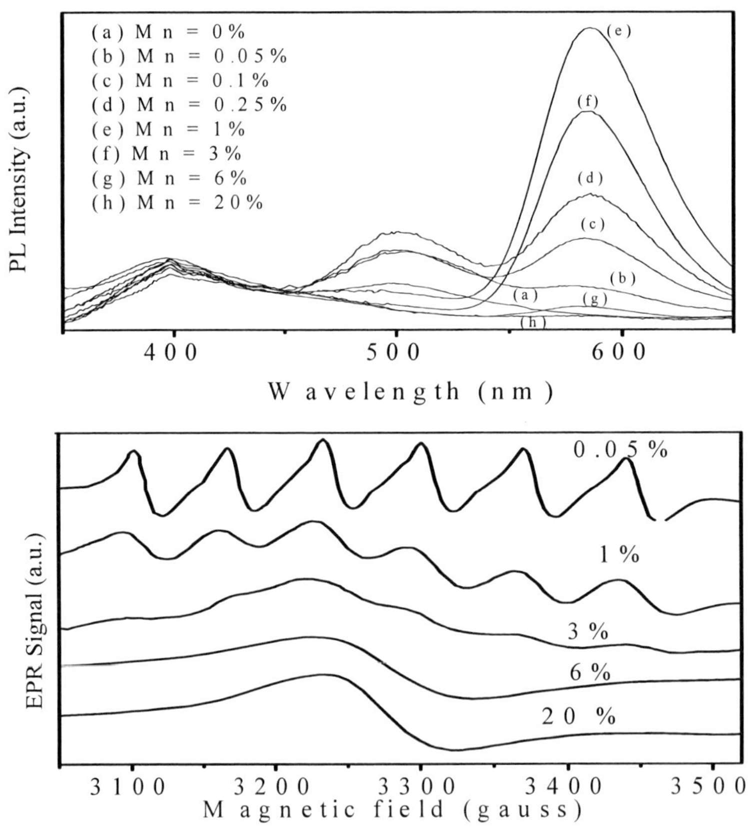

Figure 15. (a) Room-temperature of the $Mn_xZn_{1-x}S$ nanorods and (b) EPR spectra of the $Mn_xZn_{1-x}S$ nanorods [reprinted with permission from J. Phys. Chem. B (2005) DOI: 10.1021/jp053138i].

These blue, green and orange emissions were attributed to the surface defects, zinc vacancy related self activated defect sites and $^4T_1 - {}^6A_1$ transition of the Mn^{2+} ions respectively. The orange emission first increases with the Mn incorporation up to 1 mol % Mn and the decreased gradually for further addition of Mn. Quenching of the orange emission at higher Mn concentrations was attributed to Mn-Mn dipolar interaction. The isolated Mn^{2+} ion and Mn cluster was also detected from the EPR measurement. EPR (Figure 16b) results support the presence of magnetic dipole interaction for the Mn incorporated ZnS nanorods with higher Mn concentration. Also the basic advantage of our approach is that bulk amount of the nanorods can be prepared at low temperature without using any kind of template.

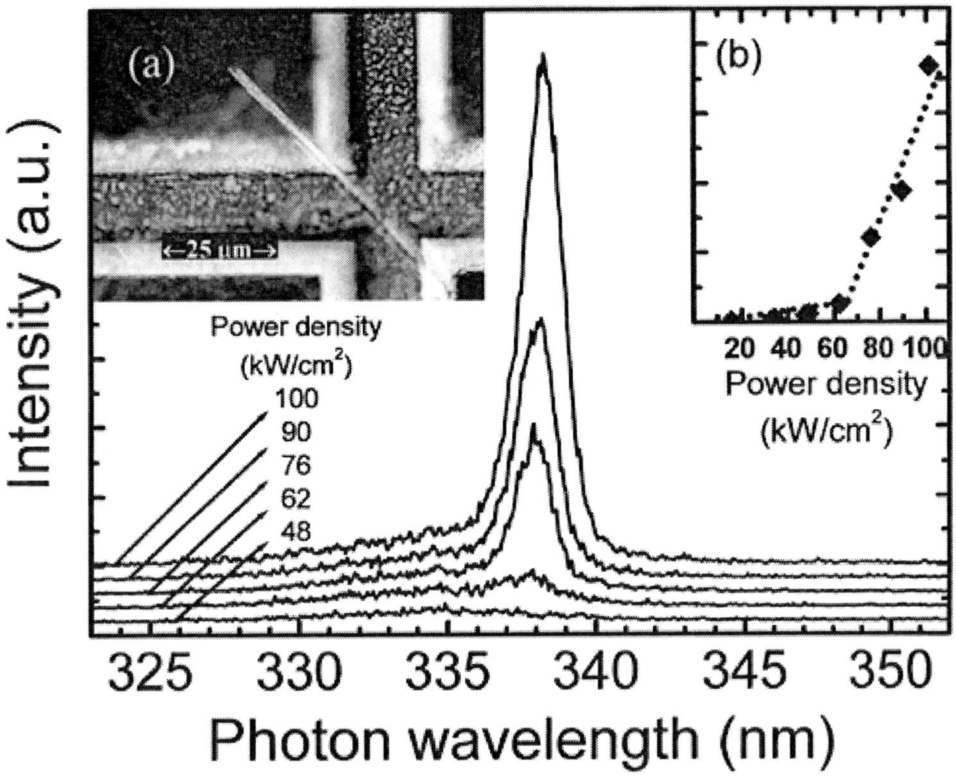

Figure 16. PL of a single ZnS nanoribbon excited by a 266 nm laser with different power densities. (a) The single ZnS nanoribbon dispersed on a TEM grid used for the measurements; the scale is 5 μm. (b) PL intensity versus input power density. Lasing starts at ~60 kW/cm^2 [reprinted with permission from Appl. Phys. Lett. 84 (2004) 1189].

6. Applications

Althouhg, ZnS nanocrystals have been utilized for various types of application [56-60], the 1-D ZnS nanostructures are of special importance for laser fabrication. This is because like ZnO, ZnS is also a high band gap (3.75eV) compound semiconductor with exciton binding energy 40 meV, which is much higher than GaN (25 meV) and ZnSe (22 meV). This information indicates that the excitons in ZnS are thermally stable at room temperature.

Nanowires and nanoribbons with flat ends are suitable candidate for laser fabrication. Lasing has been observed from the ZnS nanowire arrays [84] and single nanoribbons [64] at room temperature. Figure 15 shows the PL spectra recorded on a single ZnS nanoribbon showing the lasing from the ZnS nanoribbons. The measured threshold pumping power of the room temperature ZnS nanoribbon laser is 45 kW/cm^2, which is comparable to those of the ZnO based lasers. This low threshold and availability of good quality of ZnS nanoribbons could lead to their application in diodes and electrical pumping.

Another application of ZnS 1-D nanoforms is that the ZnS nanowires can be used as a removable template to fabricate other nanoforms. Single-crystalline Si tubular structure was formed by using ZnS nanowires as a template [122].

7. Conclusions

Different aspects of the ZnS 1-D nanoforms such as nanorods, nanowires, nanotubes, nanobelts or nanoribbons are discussed here. Various synthetic procedures ranging from low temperature based chemical approach to high temperature based thermal evaporation have been utilized to achieve different 1-D ZnS nanoforms. The ZnS 1-D nanoforms were found have different crystallographic structures. The synthesized ZnS nanoforms were single crystalline but these nanoforms could be defect free or could contain staking faults depending on the synthesis technique and conditions. These 1-D ZnS nanoforms are excellent visible light emitters and their emission could be tailored by introducing dopants like Mn^{2+} or Cu^{2+} ions in to the ZnS lattice. These emission properties of the ZnS nanoforms could be ideal for the fabrication of various optoelectronic devices like flat panel displays, LED etc. Room temperature UV lasing has been demonstrated from the ZnS nanowires and nanoribbons. So, the further investigation of the 1-D ZnS nanoforms could lead to the fabrication of technologically important nano-scale devices such as LEDs, lasers etc based on these ZnS nanoforms.

References

[1] P. Alvisatos, *Science* **271** (1996) 933.

[2] P.l Alivisatos, P. F. Barbara, A. W. Castleman, J. Chang, D. A. Dixon, M. L. Klein, G. L. McLendon, J. S. Miller, M. A. Ratner, P. J. Rossky, S. I. Stupp, and M. E. Thompson, *Adv. Mater.* **10** (1998) 1297.

[3] Handbook of Nanostructured Materials and Nanotechnology, Edited by H. S. Nalwa, *Academic Press*, New York, 2000.

[4] Nanostructures and Nanomaterials: Synthesis, Properties and Applications, G. Gao, *Imperial College Press*, London, 2004.

[5] S. Iijima, *Nature* 354 (1991) 56.

[6] W. Q. Han, S. S. Fan, Q. Q. Li, and Y. D. Hu, *Science* **277** (1997) 1287.

[7] S. S. Wang, E. Joselevich, A. T. Woolley, C. L. Cheung, and C. M. Lieber, *Nature* **394** (1998) 52.

[8] Y. Huang, X. Duan, Y. Cui, L. J. Lauhon, K. Kim, and C. M. Lieber, *Science* **294** (2001) 1313.
[9] J. Lee, T. J. Lee, S. C. Lyu, Y. Zhang, H. Ruh, and H. J. Lee, *Appl. Phys. Lett.* **81** (2002) 3648.
[10] S. Frank, P. Poncharal, Z. L. Wang, and W. A. Heer, *Science* **280** (1998) 1744.
[11] J. Hu, M. Quyang, P. Yang, and C. M. Lieber, *Nature* **399** (1999) 48.
[12] J. C. Johnson, K. P. Knutsen, H. Yan, M. Law, Y. Zhang, P. Yang, and R. J. Saykally, *Nanoletters* **4** (2004) 197.
[13] N. R. Rao, F. L. Deepak, G, Gundiah, and A. Govindaraj, *Prog. Sol. State Chem.* **31** (2003) 5.
[14] M. H. Huang, S. Mao, H. Feick, H. Yan, Y. Wu, H. Kind, E. Weber, R. Russo, and P. Yang, *Science* **292** (2001) 1897.
[15] Z. L. Wang, *Adv. Mater.* **12** (2000) 1295.
[16] J. Hu, T. W. Odom, and C. M. Lieber, *Acc. Chem. Res.* **32** (1999) 435.
[17] Xu, Y. Xu, D. Chen, G. Guo, L. Gui, and Y. Tang, *Adv. Mater.* **12** (2000) 520.
[18] H. Cao, Y. Xu, J. Hong, H. Liu, G. Yin, B. Li, C. Tie, and Z. Xu, *Adv. Mater.* **14** (2001) 1393.
[19] Y. Liang, C. Zhen, D. Zou, and D. Xu, *J. Am. Chem. Soc.* **126** (2004) 16338.
[20] G. W. Sears, *Acta Metal* **3** (1955) 361.
[21] P. Yang, and C. M. Lieber, *J. Mater. Res.* **12** (1997) 2981.
[22] Z. W. Pan, Z. R. Dai, and Z. L. Wang, *Science* **291** (2001) 1947.
[23] R. S. Wagner and W. C. Ellis, *Appl. Phys. Lett.* **4** (1964) 89.
[24] R. S. Wagner, W. C. Ellis, K. A. Jackson, and S.M. Arnold, *J. Appl. Phys.* **35** (1964) 2993.
[25] Y. Wu, and P. J. Yang, *J. Am. Chem. Soc.* **123** (2001) 3165.
[26] A.M. Morales, and C. M. Lieber, *Science* **279** (1998) 208.
[27] T. J. Trentler, K. M. Hickman, S. C. Goel, A. M. Viano, P. C. Gibbons, and W. E. Buhro, *Science* **270** (1995) 1791.
[28] J. R. Heath, and F. K. LeGoues, *Chem. Phys. Lett.* **208** (1993) 263.
[29] Y. D. Li, H. W. Liao, Y. Ding, Y. T. Qian, L. Yang, and G. E. Zhou, *Chem. Mater.* **10** (1998) 2301.
[30] Y. Li, H. Liao, Y. Ding, Y. Fan, Y. Zhang, and Y. Qian, *Inorg. Chem.* 38 (1999) 1382.
[31] Z. X. Deng, C. Wang, X. M. Sun, and Y. D. Li, *Inorg. Chem.* **41** (2002) 869.
[32] L. Qu, G. Shi, X. Wu, and B. Fan, *Adv. Mater.* **16** (2004) 1200.
[33] S. R. C. Vivekchand, G. Gundiah, A. Govindaraj, and C. N. R. Rao, *Adv. Mater.* **16** (2004) 1842.
[34] C. Cheng, R. K. Gonela, Q. Gu, and D. T. Haynie, *Nanoletters* **5** (2005) 175.
[35] Y. Wang, L. Zhang, G. Meng, C. Liang, G. Wang, and S. Sun, *Chem. Comm.* (2001) 2632.
[36] X. Wang, C. J. Summers, and Z. L. Wang, *Nanoletters* **3** (2003) 1315.
[37] Banerjee, S. H. Jo, and Z. F. Ren, *Adv. Mater.* **16** (2004) 2028.
[38] C. Liang, G. Meng, Y. Lei, F. Phillipp, and L. Zhang, *Adv. Mater.* **13** (2001) 1330.
[39] P. Guha, S. Kar, and S. Chaudhuria, *Appl. Phys. Lett.* **85** (2004) 3851.
[40] Y. C. Choi, W. S. Kim, Y. S. Park, S. M. Lee, D. J. Bae, Y. H. Lee, G. S. Park, W. B. Choi, N. S. Lee, and J. M. Kim, *Adv. Mater.* **12** (2000) 746.
[41] X. Chen, J. Xu, R. M. Wang, and D. Yu, *Adv. Mater.* **15** (2003) 419.

[42] Y. Huang, X. Duan, Y. Cui, and C. M. Lieber, *Nanoletters* **2** (2002) 101.
[43] Z. Zhong, F. Qian, D. Wang, and C. M. Lieber, *Nanoletters* **3** (2003) 343.
[44] Q. Wu, Z. Hu, X. Wang, Y. Lu, X. Chen, H. Xu, and Y. Chen, J. Am. *Chem. Soc.* **125** (2003) 10176.
[45] S. Kar, and S. Chaudhuri, *Chem. Phys. Lett.* **398** (2004) 22.
[46] M. Nath, A. Choudhury, A. Kundu, and C. N. R. Rao, *Adv. Mater.* **15** (2003) 2098.
[47] Y. Wang, G. Meng, L. Zhang, C. Liang, and J. Zhang, *Chem. Mater.* **14** (2002) 1773.
[48] C. Ye, G. Meng, Y. Wang, Z. Jiang, and L. Zhang, *J. Phys. Chem.* **B 106** (2002) 10338.
[49] L. Dong, J. Jiao, M. Coulter, and L. Love, *Chem. Phys. Lett.* **376** (2003) 653.
[50] X. Duan, C. Niu, V. Sahi, J. Chen, J. W. Parce, S. Empedocles and J. L. Goldman, *Nature* **425** (2003) 274.
[51] R. Agarwal, C. J. Barrelet, and C. M. Lieber, *Nanoletters* **5** (2005) 917.
[52] S. Kar, and S. Chaudhuri, *J. Phys. Chem.* **B 109** (2005) 3298.
[53] S. Kar, S. Biswas, and S. Chaudhuri, *communicated to Nanotechnology*.
[54] S. Kar, and S. Chaudhuri, *Chem. Phys. Lett. In press* (2005).
[55] Y. Jiang, X. M. Meng, J. Liu, Z. Y. Xie, C. S. Lee, and S. T. Lee, *Adv. Mater.* **15** (2003) 323.
[56] T. K.Tran, W. Park, W. Tong, M. M. Kyi, B. K. Wagner, and C. J. Summers, *J. Appl. Phys.* **81** (1997) 2803.
[57] H. C. Ong, and R. P. H. Chang, *Appl. Phys. Lett.* **79** (2001) 3612.
[58] T. Yamamoto, S. Kishimoto, and S. Iida, *Physica* **B 916** (2001) 308310.
[59] M. Bredol, and J. Merikhi, *J. Mater. Sci.* **33** (1998) 471.
[60] P. Calandra, M. Goffredi, and V. Turcoliver, *Colloids Surf.* **A 160** (1999) 9.
[61] C. Falcony, M. Garcia, A. Ortiz, and J. C. Alonso, *J Appl. Phys.* **72** (1992) 1525.
[62] W. Tang, and D. C. Cameron, *Thin Solid Films* **280** (1996) 221.
[63] C. N. Xu, T. Watanabe, M. Akiyama, and X. G. Zhang, *Appl. Phys. Lett.* **74** (1999) 1236.
[64] J. A. Zapien, Y. Jiang, X. M. Meng, W. Chen, F. C. K. Au, Y. Lifshitz, and S. T. Lee, *Appl. Phys. Lett.* **84** (2004) 1189.
[65] Y. Wada, H. Yin, T. Kitamura, and S. Yanagida, *Chem. Comm. (Cambridge)* **24** (1998) 2683.
[66] H. Fujiwara, H. Hosokawa, K. Murakoshi, Y. Wada, and S. Yanagida, *Langmuir* **14** (1998) 5154.
[67] N. A. Dhas, A. Zaban, and A. Gedanken, *Chem. Mater.* **11** (1999) 806.
[68] X. Chen, H. Xu, N. Xu, F. Zhao, W. Lin, G. Lin, Y. Fu, Z. Huang, H. Wang, and M. Wu, *Inorg. Chem.* **42** (2003) 3100.
[69] S. Biswas, S. Kar, and S. Chaudhuri, communicated to Chem. *Phys. Lett.* (2005).
[70] J. H. Yu,J. Joo, H. M. Park, S. Baik, Y. W. Kim, S. C. Kim, and T. Hyeon, J. *Am. Chem. Soc.* **127** (2005) 5662.
[71] X. Wang, Y. Ding, C. J. Summers, and Z. L. Wang, *J. Phys. Chem.* **B. 108** (2004) 8773.
[72] N. Pradhan, and S. Efrima, *J. Phys. Chem.* **B 108** (2004) 11964.
[73] Y. Li, X. Li, C. Yang, and Y. Li, *J. Phys. Chem.* **B 108** (2004) 16002.
[74] X. Jiang, Y. Xie, J. Lu, L. Zhu, W. He, and Y. Qian, *Chem. Mater.* **13** (2001) 1213.
[75] Q. Wu, N. Zheng, Y. Ding, and Y. Li, Inorg. *Chem. Comm.* **5** (2002) 671.
[76] X. M. Meng, J. Liu, Y. Jiang, W. W. Chen, C. S. Lee, I. Bello, and S. T. Lee, Chem. *Phys. Lett.* **382** (2003) 434.

[77] Y. Wang, L. Zhang, C. Liang, G. Wang, and X. Peng, *Chem. Phys. Lett.* **357** (2002) 314.
[78] S. Kar, S. Biswas, and S. Chaudhuri, *Nanotechnology* **16** (2005) 737.
[79] Q. Xiong, G. Chen, J. D. Acord, X. Liu, J. J. Zengel, H. R. Gutierrez, J. M. Redwing, L. C. L. Y. Voon, B. lassen, and P. C. Eklund, *Nanoletters* **4** (2004) 1663.
[80] Q. Xiong, J. Wang, O. Reese, L. C. L. Y. Voon, and P. C. Eklund, *Nanoletters* **4** (2004) 1991.
[81] Y. Jiang, X. M. Meng, J. Liu, Z. R. Ho, C. S. Lee, and S. T. Lee, *Adv. Mater.* **15** (2003) 1195.
[82] D. Appeli, *Nature* **419** (2002) 553.
[83] H. T. Ng, J. Han, T. Yamada, P. Nguyen Y. P. Chen, and M. Meyyappan *Nanoletters* 4 (2004) 1247.
[84] J. X. Ding, J. A. Zapien, W. W. Chen, Y. Lifshitz, S. T. Lee, and X. M. Meng, *Appl. Phys. Lett.* **85** (2004) 2361.
[85] D. F. Moore, Y. Ding, and Z. L. Wang, *J. Am. Chem. Soc.* **126** (2004) 14372.
[86] L. Dloczik, R. Engelhardt, K. Ernst, S. Fiechter, I. Sieber, and R. Ko¨nenkampa, *Appl. Phys. Lett.* **78** (2001) 3687.
[87] X. Wang, P. Gao, J. Li, C. J. Summers, and Z. L. Wang, *Adv. Mater.* **14** (2002) 1732.
[88] L. W. Yin, Y. Bando, J. H. Zhan, M. S. Li, and D. Golberg, *Adv. Mater.* **17** (2005) 1972.
[89] Q. Li, and C. Wang, *Appl. Phys. Lett.* **83** (2003) 359.
[90] Y. C. Zhu, Y. Bando, and D. F. Xue, *Appl. Phys. Lett.* **82** (2003) 1769.
[91] C. Ma, D. Moore, J. Li, and Z. L. Wang, *Adv. Mater.* **15** (2003) 228.
[92] D. Moore, C. Ronning, C. Ma, and Z. L. Wang, *Chem. Phys. Lett.* **385** (2004) 8.
[93] J. Gong, S. Yang, J. Duan, R. Zhang, and Y. Du, *Chem. Comm.* (2005) 351.
[94] W. T. Yao, S. H. Yu, L. Pan, J. Li, Q. S. Wu, L. Zhang, and J. Jiang, *Small* **1** (2005) 320.
[95] P. Hu, Y. Liu, L. Fu, L. Cao, and D. Zhu, *J. Phys. Chem.* **B 108** (2004) 936.
[96] C. Ye, X. Fang, G. Li, and L. Zhang, *Appl. Phys. Lett.* **85** (2004) 3035.
[97] C. Liang, Y. Shimizu, T. Sasaki, H. Umehara, and N. Koshizaki, *J. Phys. Chem.* **B 108** (2004) 9728.
[98] X. M. Meng, Y. Jiang, J. Liu, C. S. Lee, I. Bello, and S. T. Lee, *Appl. Phys. Lett.* **83** (2003) 2244.
[99] Y. Xia, P. Yang, Y. Sun, Y. Wu, B. Mayers, B. Gates, Y. Yin, F. Kim, and H. Yan, *Adv. Mater.* **15** (2003) 353.
[100] C. J. Barrelet, Y. Wu, D. C. Bell, and C. M. Lieber, *J. Am. Chem. Soc.* **125** (2003) 11498.
[101] X. Duan, and C. M. Lieber, *Adv. Mater.* **12** (2002) 298.
[102] D. C. Bell, Y. Wu, C. J. Barrelet, S. Gradecak, J. Xiang, B. P. timko, and C. M. Lieber, *Micros. Res. Tech.* **64** (2004) 373.
[103] Y. Ding, and Z. L. Wang, *J. Phys. Chem.* **B. 108** (2004) 12280.
[104] J. I. Pankove, *Optical Processes in Semiconductors*, Prentice Hall, New Jersey, 1971, p. 34-38.
[105] M. Nell, J. Marohn, and G. Mclendon, *J. Phys. Chem.* 109 (1990) 4359.
[106] L. Spanhel, M. Haase, H. Weller, and A. Henglein, *J. Am. Chem. Soc.* **109** (1987) 5649.

[107] H. Landolt; R. Bornstein, *Numerical data and fundamental relationship in science and technology, New Seriesed*; Springer-Verlag: Berlin, 1999.

[108] H. Samelson, and A. Lempicki, *Phys. Rev.* **125** (1962) 901.

[109] R. N. Bhargava, D. Gaillagher, X. Hong, and A. Nurmikko, *Phys. Rev. Lett.* **72** (1994) 416.

[110] Bol, and A. J. Meijerink, *J. Phys. Chem.* **B 105** (2001) 10197.

[111] Bol, and A. J. Meijerink, *J. Phys. Chem.* **B 105** (2001) 10203.

[112] Yu, and M. Senna, *Appl. Phys. Lett.* **66** (1995) 424.

[113] P. Yang, M. K. Lu, D. Xu, D. L. Yuan, and G. J. Zhou, *Chem. Phys. Lett.* **336** (2001) 76.

[114] S. Sapra, J. Nanda, A. Anand, S. V. Bhat, and D. D. Sarma *J. Nanosci Nanotechnol.* **3** (2003) 392.

[115] P. H. Borse, D. Srinivas, R. E. Shinde, S. K. Date, W. Vogel, and S. K. Kulkarni, *Phys. Rev.* **B 60** (1999) 8659.

[116] W. Que, Y. Zhou, Y. L. lam, Y. C. Chan, C. H. Kam, B. Liu, L. M. Gan, C. H. Chew, G. Q. Xu, S. J. Chua, S. J. Xu, and F. V. C. Mendis, *Appl. Phys. Lett.* **73** (1998) 2727.

[117] J. F. Suyver, S. F. Wuister, J. J. Kelly, and A. Meijerink, *Nanoletters* **1** (2001) 429.

[118] Arguello, D. L. Rousseau, and S. P. S. Porto, *Phys. Rev.* **181** (1969) 1351.

[119] J. Schneider, and R. D. Kirby, *Phys. Rev.* **B 6** (1972) 1290.

[120] Brafman, and S. S. Mitra, *Phys. Rev* **B 171** (1968) 931.

[121] R. L. Schmidt, K. Kunc, M. Cardona, and H. Bilz, *Phys. Rev.* **B 20** (1979) 3345.

[122] J. Hu, Y. Bando, Z. Liu, J. Zhan, D. Golberg, and T. Sekiguchi, *Angew. Chem. Int. Ed.* **43** (2004) 63.

[123] K. Sooklal, B. S. Cullum, S. M. Angel, and C. J. Murphy, *J. Phys. Chem.* **100 (**1996) 4551.

[124] F. J. Garlick, and Gibson, *J. Opt. Soc. Am.* **39** (1949) 935.

[125] P. McClean, and C. B. Thomas, *Semicond. Sci. Technol.* **7** (1992) 1394.

[126] Y. Zhang, and Y. Li, *J. Phys. Chem.* **B. 108** (2004) 17805.

[127] P. Yang, M. Lu, D. Xu, D. Yuan, J. Chang, G. Zhou, and M. Pan, *Appl. Phys.* **A 74** (2002) 257.

[128] Ishizumi, C. W. White, and Y. Kanemitsu. *Appl. Phys. Lett.* **84** (2004) 2397.

[129] W. Chen, A. G. Joly, J. O. Malm, and J. O. Bavin, *J. Appl. Phys.* **95** (2004) 667.

[130] Y. Geng, L. D. Zhang, G. Z. Wang, T. Xie, Y. G. Zhang, and G. W. Meng, *Appl. Phys. Lett.* **84** (2004) 2157.

[131] Y. C. Zhu, Y. Bando, D. F. Xue, and D. Golberg, *Adv. Mater.* **16** (2004) 831.

[132] J. Yuan, X. Q. Yan, Z. X. Zhang, D. F. Liu, Z. P. Zhou, L. Cao, J. X. Wang, Y. Gao, L. Song, L.F. Liu, X. W. Zhao, X. Y. Dou, W. Y. Zhou, and S. S. Xie, *J. cryst. Growth,* **271** (2004) 403.

[133] J. P. Ge, J. Wang, H. X. zhang, X. Wang, Q. Peng, and Y. D. Li, *Adv. Funct. Mater.* **15** (2005) 303.

[134] S. Biswas, S. Kar, and S. Chaudhuri, *J. Phys. Chem. B In press* (2005)

In: Encyclopedia of Nanotechnology
Editor: Elwood D. Carlson

ISBN: 978-1-60692-079-4

Chapter 5

ADVANCE AND APPLICATION PROSPECT OF DENDRIMERS

Bifeng Pan and Daxiang Cui*
Department of Bio-Nano Science and Technology,
National Dey Laboratory of Nano/Micro Fabrication Technology,
Key Laboratory for Thin Film and Microfabrication of Ministry of Education,
Institute of Micro and Nano Science and Technology,
Shanghai Jiao Tong University, 800 Dongchuan Road,
Shanghai 200030, China.

ABSTRACT

Nanotechnology quite certainly will become an important field in the science and technology of the next decades, allowing one to manipulate matter at the nanometer scale and to incorporate nanostructure and nanoprocesses into working technological innovation. Dendrimers are branched, synthetic polymers with layered architectures that show a fast growing field covering many areas of chemistry. Dendrimers are considered as a modern and an elegant class of branched macromolecules in the nanoscale dimension. They resemble treelike molecular architectures since they are built from repetitive monomers with branching point units that are radially connected around a template core. The hydrodynamic volume is dictated by the core, monomer units, branching points, dendrons, generation numbers, and peripheral functions. Dendrimers are discrete macromolecules with a high degree of molecular uniformity and monodispersity. Those features make for an easy characterization and analysis of their tailor-made properties. They are often macromolecules with few defects, whose dense outward (and/or inward) functional groups generate specific properties and functions such as a polyvalent ligand, a molecular receptor, molecular confinement, and a molecular translocator (vector). Multivalency as well as adhesive, amplification, recognition, additive, and cooperative effects have been observed. The expression "dendritic effects" was coined to illustrate those specific chemical behaviors. Dendrimers that used to be employed as standard, innovative stabilizers, media, ligand and supports have appeared to provided remarkable results with extremely high metal loading. By

* Tel and Fax: 0086-21-62933291 E-mail: dxcui@sjtu.edu.cn (D Cui) and panbifeng@tom.com (B Pan)

regulating dendrimer synthesis, it is possible to precisely manipulate both their molecular weight and chemical composition, thereby allowing predictable tuning of their biocompatibility and pharmacokinetics. Advances in our understanding of the role of molecular weight and architecture on the in vivo behavior of dendrimers, together with recent progress in the design of biodegradable chemistries, has enabled the application of these branched polymers as anti-viral drugs, tissue repair scaffolds, targeted carriers of chemotherapeutics and electrochemical sensors. Various types of dendrimers differing in composition and structure are mentioned, together with their practical use spanning from catalysis, transport vehicles to synthetic vaccines. This paper discusses the research advances development of functional and multifunctional dendrimeric and hyperbranched polymers with the objective of being applied in catalysis, biosensing, imaging, drug and gene delivery, nanoprinting, self-assembly, energy transfer, etc.

INTRODUCTION

During the past decade, there has been an explosive growth in studies of dendrimers, a new class of macromolecules that can be considered as monodisperse, precisely ordered polyfunctional polymers. Several applications of these relatively soluble compounds have been proposed. Their remarkable properties result from the presence of a large number of chain-end functionalities that provide highly controlled surfaces and interfaces, and from the presence of internal voids allowing, for example, encapsulation of guest molecules or molecular recognition at specific locations on the cascade superstructure.

1. SYNTHESIS ADVANCES OF DENDRIMER

A number of creative approaches to the synthesis of organic or inorganic "monodendrimers" have been reported. The reactivity of the core has been used to couple two or more dendrons leading, for example, to structures with two dissimilar halves. Coupling of cascade macromolecules through metal centers has also been achieved. Dendrons containing isophthalic acid units covalently attached to a rigid aromatic spacer have been shown to self-assemble through hydrogen bonding to form a hexameric dendritic aggregate whose stability varies in a generation- dependent manner. All these different routes to bis- or multidendrimers involve the association of dendrons by their core.

1.1. Control of Dendrimer Structure

Anne-Marie Caminade *et al.* [1] report two examples of the regioselective stepwise growth of six dendrimer units in the internal voids of a main dendrimer, illustrating that these cavities are accessible and that internal functional groups can be used for the synthesis of covalently linked multidendritic systems. The synthesis of the central dendrimer is based on the strategy using a divergent method. The formation of generation 1 involves three steps: (i) a condensation reaction between the flexible core, hexaaldehyde 1, and six equivalents of monomethylhydrazine; (ii) a "Mannich reaction" between the resulting product and six equivalents of the phosphane Ph2PCH2OH, giving an hexaaminophosphane; and (iii) a

"Staudinger reaction" between this latter compound and the azide N3P(S)(OC6H4CHO)2 to form the dendrimer 2 (generation 1) with six P5N–P(S) fragments in the framework..

Daniel Taton *et al.* [2] engineered polymer chains in star-like and dendrimerlike architectures using atom-transfer radical polymerization (ATRP) and anionic ring-opening polymerization (AROP) of ethylene oxide as synthetic tools. Daniel Taton *et al* first discussed the scope and limitations of ATRP when this method of controlled radical polymerization is applied to multifunctional initiators for the synthesis of star polymers by the core-first approach. The switch from ATRP to AROP of ethylene oxide and vice versa allows access to branched amphiphilic block copolymers exhibiting core–shell structures. Stress is also put on the methodologies for the selective branching of polymeric chain-ends, with a view to introducing ω-geminal functionalities from which further polymer branches can be grown. When linear polymer precursors are used, such a strategy leads to asymmetric and mikto-arm stars but the same can be applied to multi-arm stars so as to generate so-called 'dendrimer-like (co)polymers' that are dendrimers with true macromolecular generations.

George R. Newkome *et al* [3] presented an elongated structural design leading to more conical-shaped dendritic architectures by using a combination of 1→3, 1→(2+1), and 1→(2+1Me) C-branched monomers. Synthesis of the conifershaped macromolecule was achieved by reaction between isocyanate and amine in dry CH2Cl2. Aresultant extended focal adamantane-modified dendron was deprotected to generate the water-soluble product, which was subsequently complexed with β-cyclodextrin in D2O to create the desired tree-like product (Figure 1).

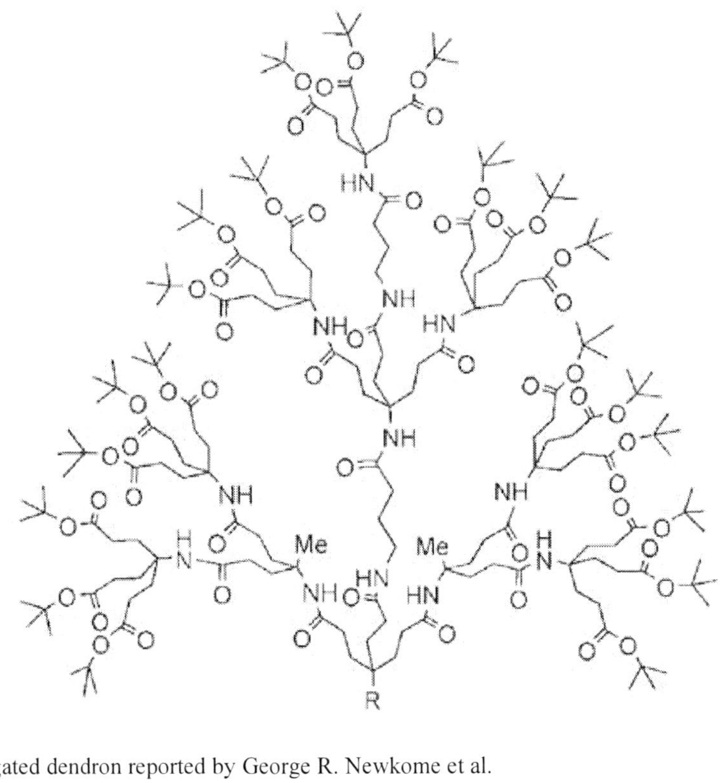

Figure 1. Elongated dendron reported by George R. Newkome et al.

Helmut Keul *et al.* [4] prepared monodisperse branched polyurethanes containing long alkyl chains by a new convergent synthesis. This synthesis comprises two steps, with hexamethylene diisocyanate uretdione as the starting molecule. The free isocyanate groups of this monomer are reacted with long-chain alkanols. These diurethane uretdiones are then reacted with amines or amino alcohols under ring opening and the formation of a biuret group. Branching points are not, as usual, part of the monomer but are formed during preparation.

A new reversible addition-fragmentation chain transfer (RAFT) agent, dendritic polyester with 16 dithiobenzoate terminal groups, was prepared by Chun-Yan Hong *et al* [5] for the RAFT polymerization of styrene (St) to produce star polystyrene (PSt) with a dendrimer core. It was found that this polymerization was of living characters, the molecular weight of the dendrimer-star polymers could be controlled and the polydispersities were narrow. The dendrimer-star block copolymers of St and methyl acrylate (MA) were also prepared by the successive RAFT polymerization using the dendrimer-star PSt as macro chain transfer agent (Figure 2).

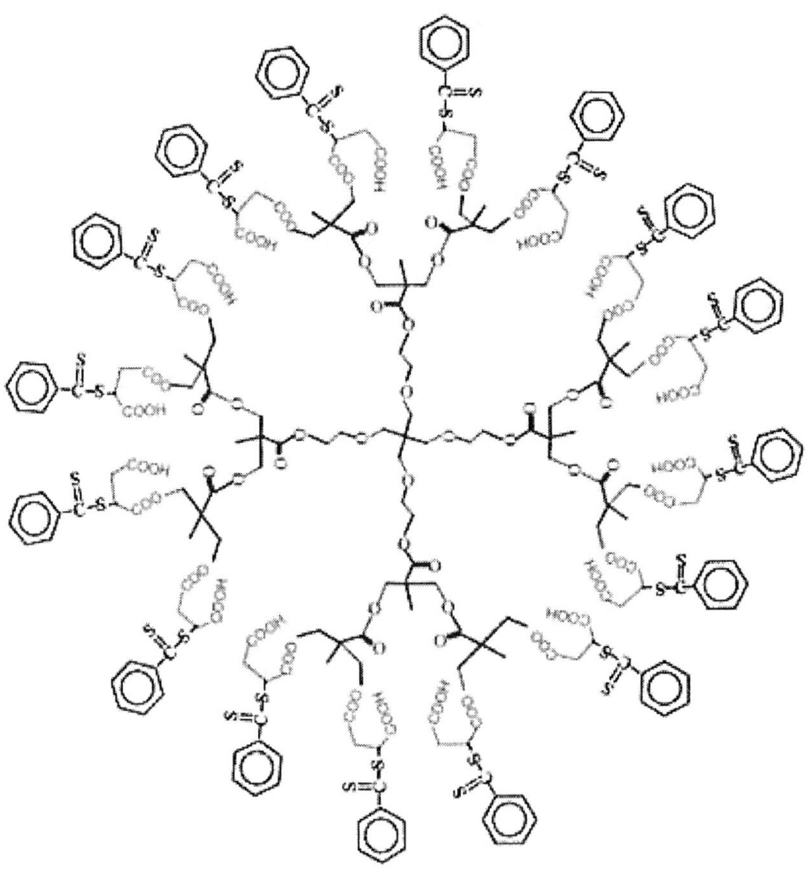

Figure 2. Dithiobenzoate-terminated dendrimer reported by Chun-Yan Hong et al.

Wenfang Shi *et al* [6] prepared a series of dendritic core-shell structural polymers with different shell densities based on dendritic polyester Boltorn H40 to form unimolecular micelles in chloroform. The encapsulation mechanism study using Congo red as a guest molecule by fluorescence and UV-vis methods showed that the interaction of Congo red with hydroxyl groups in the dendritic core-shell polymer led to encapsulation. Moreover, the results also indicated that the dendrimer with 43.4% hydroxyl groups endcapped by the long alkyl chains showed the best encapsulation capacity.

Fabrizia Negri *et al* [7] presented an atomistic molecular dynamics investigation of the structural time evolution of isolated polyphenylene dendrimers, carbon based dendrimers with a planar core formed by a 1,3,5 trisubstituted benzene ring. Fabrizia Negri *et al* presented a general classification of the conformations (core conformations) assumed by the three dendrimer branches with respect to the planar core.

1.2. Control of Surface Groups, Core, and Building Block

Chae Kyu Kim et al [8] described the synthesis and luminescence characterization of conjugated dendrimers with triazine peripheries and a distyrylanthracene core that are suitable for electroluminescence applications. The dendrimers consist of dendritic triazine wedges with high electron affinity, stilbene branches, and a distyrylanthracene core as an emitting moiety. The dendrimers have lowest unoccupied molecular orbital values of about -2.7 eV. Photoluminescence studies have indicated that a cascade energy transfer occurs from the triazine wedges to the stilbene bridges and finally to the distyrylanthracene core.

Cheol-Hee Ahn et al [9] reported the synthesis of dendrons and dendrimers with cationic amino groups at their periphery up to the third and second generation, respectively. The results obtained by 1H NMR spectroscopy and gel permeation chromatography analysis supported the formation of the targeted dendrons and dendrimers. The dendrons were grown via ester linkages, which endowed them with biodegradability in D2O at 37°C. The degradation rate depends upon the steric hindrance and reactivity caused by the bulkiness and compact structure of the dendrons. All of the synthesized dendrons were degraded within a month, while 60% of the ester groups in the sterically crowded dendrimers were degraded over the same time period.

Klaus Müllen et al [10] reported the systhesis first- and second-generation polyphenylene dendrimers starting from benzophenone as core. Variation of the size and the density of the dendrimer shell resulted in different isolation of the cores. To investigate the effect of shielding upon the reactivity of the core, chemical functionalizations as well as the alkali-metal reduction of the encapsulated benzophenone core were performed. The herein presented synthetic concept opens the way to spatially well-defined spherical nanoparticles bearing a single isolated function in the center.

Mackay B. Steffensen et al [11] reported a comprehensive and chronological account of dendrimers based on [1,3,5]-triazines. Mackay B. Steffensen et al discussed the synthetic strategies to install the triazine through cycloaddition, cyclotrimerization, and nucleophilic aromatic substitution of cyanuric chloride. Motivations and applications of these architectures are surveyed, including the preparation of supramolecular assemblies in the solution and solid states and their use in medicines, advanced materials, and separations when anchored to solid supports.

Well-defined, modular dendrimers enable processing techniques and electronic properties to be tuned independently by Weihong Zhu et al [12]. Moreover, the dendritic topology can isolate the core chromophore, thus reducing or eliminating strong intermolecular interactions.

Weihong Zhu et al presented the synthesis of three series of flexible, dendron-functionalized dendrimers as red-lightemitting materials by a convergent approach: (1) carbazole (CZ) or oxadiazole (OXZ) terminated imide-type dendrimers, (2) cascade energy-transferring imide-type dendrimers, and (3) CZ-terminated perylene bay-type dendrimers. They all consist of the luminescent core of perylenebis(dicarboximide)s with specific functional groups of CZ or OXZ at the periphery and are constructed from flexible Fréchet-type poly(aryl ether) dendrons (Figure 3).

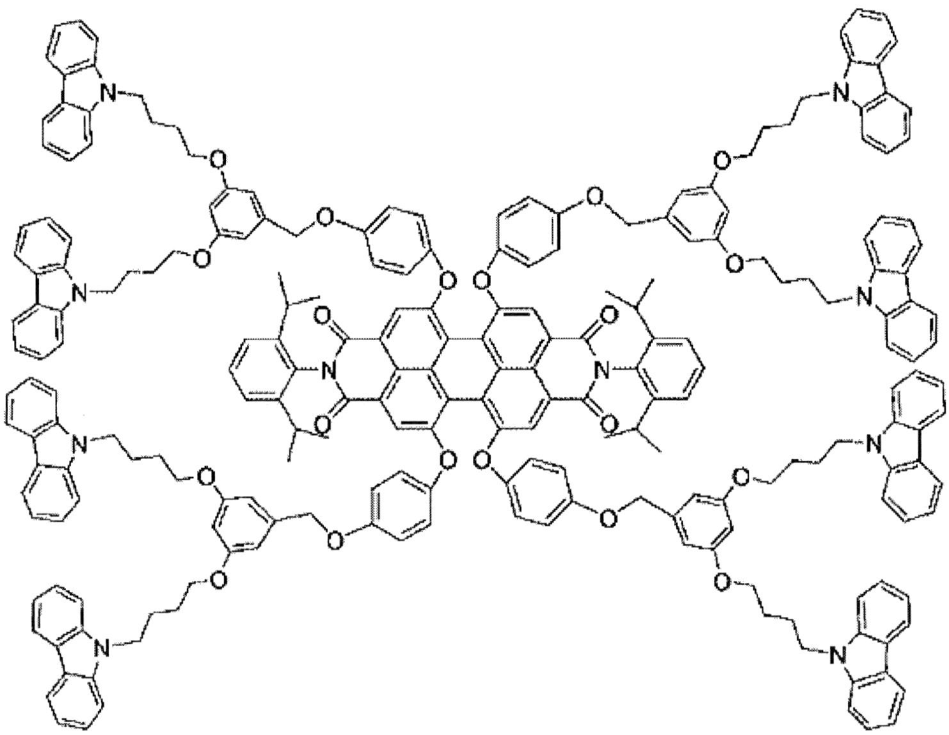

Figure 3. Carrier-transporting dendrimers with perylenebis(dicarboximide)s as a luminescent core.

Thermally labile dendrimers based on the reversible furan-maleimide Diels-Alder reaction are described by Dominic V. McGrath et al [13]. First through fourth generation benzyl aryl ether based dendrons that contained furan moieties at their focal point were allowed to react with bismaleimide central linker to provide the corresponding dendrimers.

As reported by Nicola Armaroli et al [14], multifullerene-terminated dendrimers Gn (n=1-5) were synthesized and structural, photophysical, and photoelectrochemical properties were studied for the fullerene dendrimers and their nanoclusters (Figure 4). The fullerene dendrimers formed clusters when toluene solutions of the fullerene dendrimers were injected into acetonitrile. Dynamic light scattering and atomic force and scanning electron microscopic measurements on these clusters revealed that the cluster size decreased with increasing the generation number of the dendrimers.

Cheol-Hee Ahn et al [15] reported the synthesis of aliphatic ester dendrimers based on 2,2-bis(hydroxymethyl)propionic acid (bis-MPA) growth units as a form of dendritic hybrids up to G4.0. Linear polystyrene supports in the dendritic hybrids were obtained through atom transfer radical polymerization at 130°C for 12 h using the CuCl/PMDETA catalyst/ligand system with 4-(chloromethyl)benzyl alcohol as the initiator.

Figure 4. Fullerene Dendrimer reported by Nicola Armaroli et al.

Two series of oligo (phenylene ethynylene)s (OPEs) with different dendrimer side groups have been designed and synthesized by Xiao-Ping Cao et al [16]. The molecules contain thiol groups at both ends to enable interconnection between nanoscale gapped metallic electrodes. The different dendrimer groups act as "shells", allowing tailoring to the nanoscopic environment surrounding the OPE "core". Meanwhile, the dendrimer shells also act as spacers for the precise control of the packing density and intermolecular interaction between the OPE cores.

Ye Tao et al [17] reported the synthesis and properties of a novel series of multi-triarylamine-substituted carbazole-based dendrimers G2-OTP(n)-G2(n=2-5) with an oligothiophene core. The length of the oligothiophene core varied from 2 to 5 thiophene units. It was found that both the absorption and fluorescence emission peaks red-shifted with increasing oligothiophene core length. As a result, the power conversion efficiencies of the as-fabricated solar cells using the composite of these dendrimers and PCBM as the photoactive layer increased with increasing oligothiophene core length.

As reported by Herbert Schumann et al [18], the catalytic hydrogermylation of Si(CHdCH2)4 with Ph3GeH yielded Si(CH2CH2GePh3)4. In contrast, Sn(CH=CH2)4 and

Sn(CH2CH=CH2)4 reacted with Ph3GeH under the same conditions, forming Ph3Ge(CH2)2GePh3 and Ph3Ge(CH2)3GePh3, respectively, whereas the reaction of Sn(CH2CH2CH=CH2)4 and Ph3GeH afforded the expected tin-based metallodendrimer Sn[(CH2)4- GePh3]4. The reactions of Si[(CH2)2SnBr3]4 with the Grignard reagents RMgBr (R: CH=CH2, CH2CH=CH2, (CH2)2CH=CH2)) produced Si[(CH2)2Sn(CH=CH2)3]4, Si[(CH2)2Sn(CH2CH=CH2)3]4, and Si[(CH2)2Sn(CH2CH2CH=CH2)3]4.

Stéphane Campidelli et al [19] described the synthesis and characterization of a series of single-wall carbon nanotubes (SWNTs) functionalized with polyamidoamine dendrimers. Importantly, the dendrimers are linked directly to the SWNT surface using a divergent methodology. This approach allows the number of functional groups on the nanotubes to be increased without provoking significant damage to the conjugated π-system of the SWNTs (Figure 5).

Very high molecular weight (M_n up to 700 000) and narrowly polydispersed (PDI=1.1-1.4) hyperbranched polyglycerols (HPG) were synthesized by ring-opening multibranching polymerization of glycidol using dioxane as an emulsifying agent as reported by Donald E. Brooks et al [20]. Broader molecular weight distributions with low molecular weight fractions were obtained when diglyme was used as the emulsifying agent. The results show that these polymers are very compact and have spherical conformations in water with no indications of aggregate formation. Despite their self-similar structures, depending on the type of solvent used to synthesize them (diglyme vs dioxane), topologically restricted configurations are produced that result in completely different entanglement dynamics.

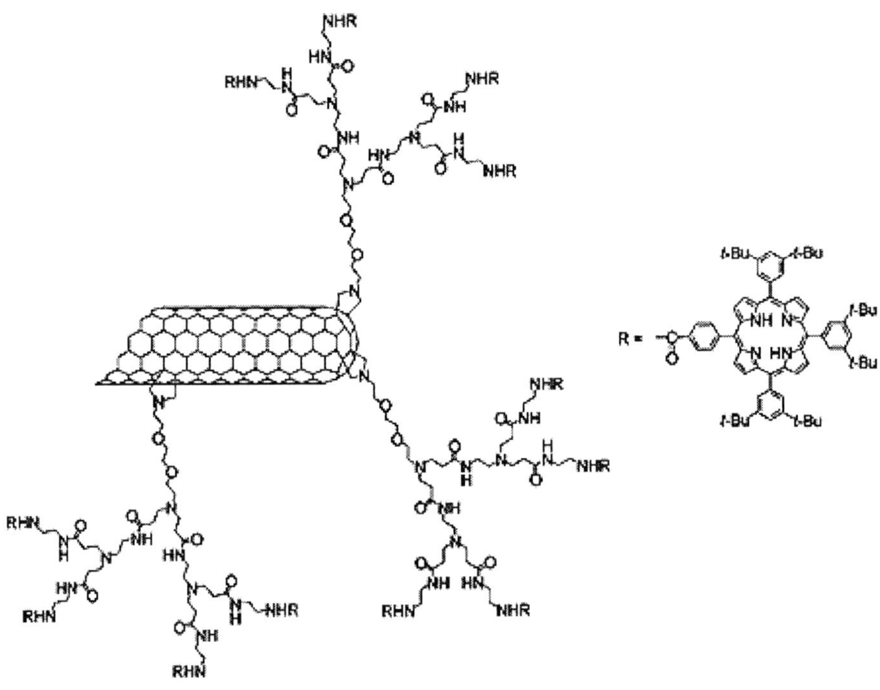

Figure 5. Dendrimer-functionalized single-wall carbon nanotubes reported by Stéphane Campidelli et al.

Two large π-conjugated dendrimers (G0 and G1) employing the stilbenoid moiety as the bridge unit have been developed by Yuguo Ma et al [21] through the Suzuki and the Horner-

Wadsworth-Emmons reactions. The molecular weight of G1 is 10 973 Da. Both G0 and G1 have good fluorescence quantum yields and exhibit similar absorption and emission behaviors in solutions and in thin films. They emit strong blue fluorescence in films under the irradiation of UV light.

As an application of a one-pot reaction involving Birch reduction and subsequent SN2 reaction in liquid ammonia, synthetic assembly of trisaccharidic moieties of globotriaosyl ceramide onto carbosilane dendrimers was accomplished using tris(3-bromopropyl)phenylsilane and tris[tris(3-bromopropyl)silylpropyl]phenylsilane as the core scaffolds as reported by Koji Matsuoka et al [22]. The common globotriaosyl derivative having benzylsulfide functionality at the terminal of the aglycon was efficiently prepared from D-galactose and D-lactose as starting materials. The glycosyl donor derived from galactose and the glycosyl acceptor derived from lactose were condensed in the presence of silver triflate as the best promoter to provide corresponding trisaccharide with newly formed α-1-4 linkages in 90% yield. Fully benzylated protection of the trisaccharide was deprotected under the Birch reduction condition followed by acetylation to give an acetate in which alkene was converted into benzyl sulfide by radical addition of α-toluenethiol in high yields. On the other hand, carbosilane dendrimers were prepared from appropriate chlorosilanes as starting materials by a combination of hydrosylation followed by alkenylation. The terminal C=C double bonds of the carbosilanes were converted into corresponding alcohols by means of the usual hydroboration reaction, and the alcohols underwent further chemical manipulation to give carbosilane dendrimers with peripheral bromine atoms.

Oleg Lukin et al [23] reported the synthesis of "designer" dendrimers and dendrons with sulfonimide units at every branching point (Figure 6). The synthesis is based on a series of (regio) selective functionalization reactions of amines and sulfonamides allowing precise control of the dendrimers' shape, the number of branches in each generation, and their peripheral decoration with functional groups. In principle, structurally different branches can be incorporated at any position within the dendrimer structure at will. Structurally perfect symmetrical and two-faced "Janus"-type dendrimers, as well as dendrimers and dendrons with intended interstices were synthesized on a preparative scale and fully characterized. Oligosulfonimide dendrons of various generations bearing an aryl bromide functional group at their focal points were attached to a p-phenylene core with the aid of Suzuki cross-coupling reactions resulting in dendrimers with photoactive terphenyl cores.

A novel family of diaminobutane-based poly(propyleneamine) ferrocenyl dendrimers with aza-crown ethers (Figure 7), DABdend-(Fc-A18C6)x (2-5: x=2, 4, 8, 16; Fc-A18C6) {[η5- C5H4CONH]Fe [η5-C5H4CONCH2 CH2(OCH2CH2)5]}), has been prepared by reaction of the noVel functionalized ferrocenyl derivative {[η5-C5H4COCl]Fe[η5-C5H4CONCH2CH2(OCH2- CH2)5]} as reported by Beatriz Alonso et al [24].

Mary J. Cloninger et al [25] functionalized dendrimers with mixtures of mannose, glucose, and galactose. Their association with concanavalin A was studied using precipitation and hemagglutination assays. With less idealized systems where the association was not optimized, mixtures of low- and high-affinity ligands caused smaller than the theoretically determined differences in binding activity, although linear binding trends were observed.

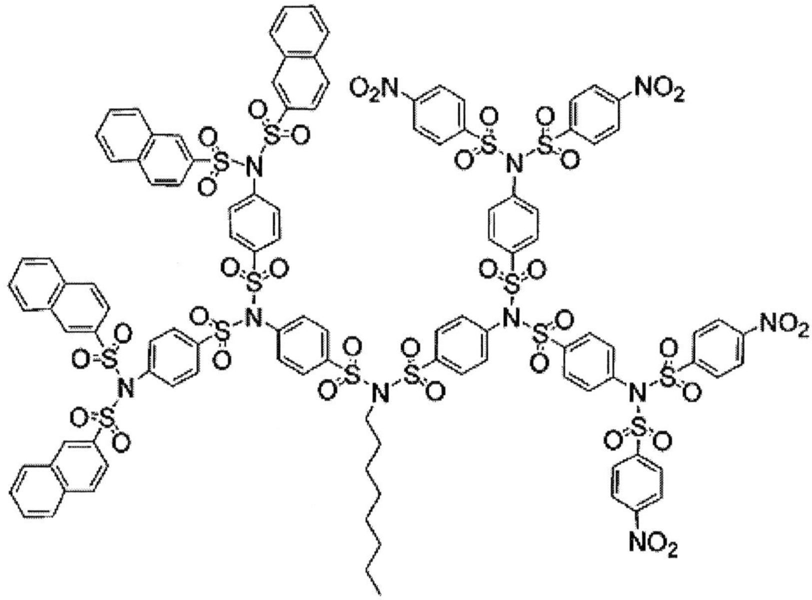

Figure 6. Branched Oligosulfonimides with Controllable Molecular Architectures.

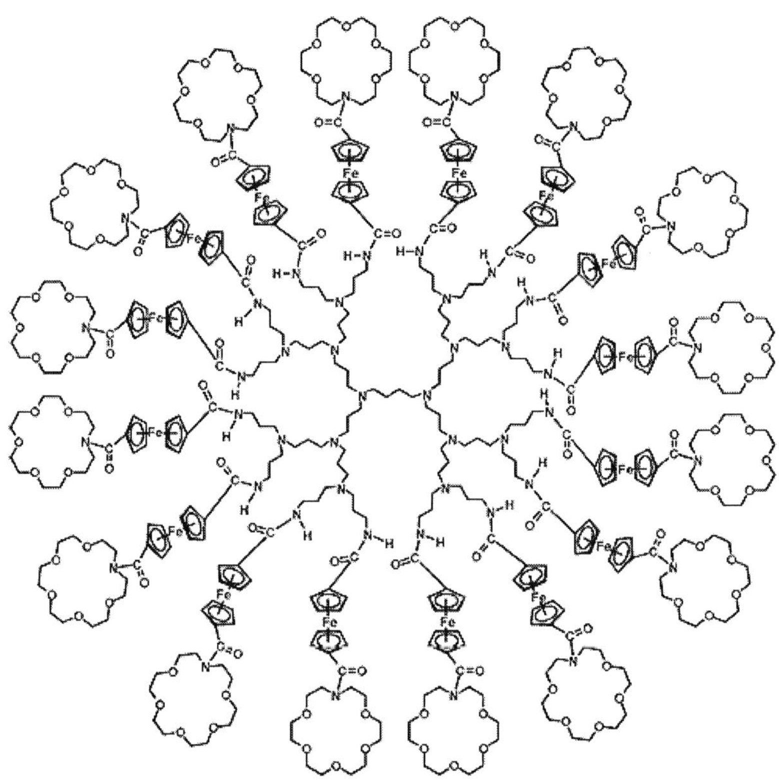

Figure 7. Aza-Crown Ethers Attached to Dendrimers through Amidoferrocenyl Units.

When systems were optimized so that high-affinity binding was achieved, then mixing low- and high-affinity ligands on the dendrimer's surface showed a predictable trend for lectin binding.

Jan H. van Maarseveen et al [26] described the convergent syntheses of three generations of carbosilane dendrimeric carbodiimides (Figure 8). The wedge-type building blocks were synthesized in a divergent way, starting from allyl chloride and a repetitive sequence of hydrosilylation with HSiCl3 and a Grignard reaction with allylmagnesium bromide. Hydrogenation of the terminal double bonds led to inert and stable wedges. Thc chloride substitutent at the focal point was transformed into several functional groups that eventually led to dendrimeric structures with a carbodiimide core.

Virgil Percec et al [27] reported the synthesis of methyl esters of 3-(4-hydroxyphenyl)propionic, 3-(3,4-dihydroxyphenyl)- propionic, 3-(3,5-dihydroxyphenyl)propionic, and 3-(3,4,5-trihydroxyphenyl)propionic acids and their use in a convergent iterative strategy to prepare up to four generations of three libraries, one of 3,4,5- and two of constitutional isomeric 3,4- and 3,5-substituted 3-phenylpropyl dendrons. Each library contains 3-[3,4,5- tris(dodecyl-1-oxy)phenyl]propyl-, 3-[3,4-bis(dodecyl-1-oxy)phenyl]propyl-, 3-{3,4-bis[3-(4-dodecyl-1-oxyphenyl) propyl-1-oxy]phenyl}propyl-, and 3-{3,4,5-tris[3-(4-dodecyl-1-oxyphenyl)propyl-1-oxy]phenyl}propyl ether first-generation dendrons on their periphery and -CO2CH3, -COOH, and -CH2OH groups at their apex. Regardless of their generation number and their periphery, internal, and apex structures, these dendrons self-assemble into supramolecular dendrimers that self-organize into all periodic and quasi-periodic assemblies encountered previously and in several unencountered with architecturally related benzyl etherbased supramolecular dendrimers. A variety of porous columnar lattices that were previously obtained only from dendritic dipeptides and hollow supramolecular spheres were also discovered from these building blocks.

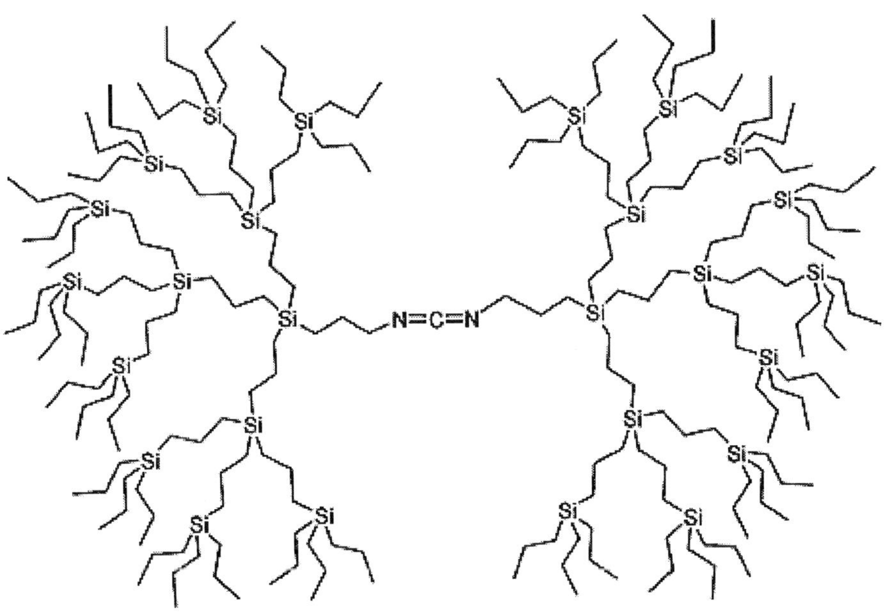

Figure 8. Dendrimeric Carbodiimide.

The more flexible and less compact 3-phenylpropyl ether repeat units are stable under acidic conditions, facilitate a simpler synthetic strategy, provide faster dynamics of self-assembly into higherorder supramolecular structures of larger dimensions, exhibit lower transition temperatures than the corresponding benzyl ether homologues, and demonstrate the generality of the self-assembly concept based on amphiphilic dendrons (Figure 9).

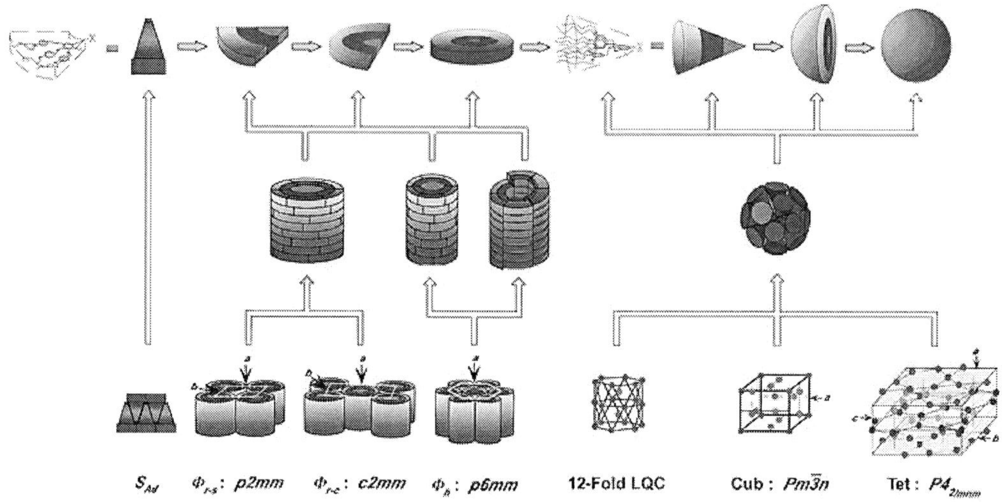

Figure 9. Retrostructural Analysis of 2D Smectic.

Vincenzo Balzani et al [28] have investigated the spectroscopic and electrochemical behavior of symmetric and unsymmetric first-, second-, and third-generation dendrimers comprising an electron-acceptor 4,4'- bipyridinium core (viologen type) and electron-donor 1,3-dimethyleneoxybenzene (Fréchet-type) dendrons. The quite strong fluorescence of the symmetrically and unsymmetrically disubstituted 1,3-dimethyleneoxybenzene units of the dendrons is completely quenched as a result of donor-acceptor interactions that are also evidenced by a low-energy tail in the absorption spectrum. In dichloromethane solution, the 4,4'- bipyridinium cores of the investigated dendrimers are hosted by a molecular tweezer comprising a naphthalene and four benzene components bridged by four methylene units. Host-guest formation causes the quenching of the tweezer fluorescence.

A new type of phenylazomethine dendrimer with a tetraphenylmethane core (Figure 10) was synthesized by a convergent method as reported by Kimihisa Yamamoto et al [29].

Anne-Marie Caminade et al [30] reported the synthesis of a new family of phosphorus dendrimers built from an octasubstituted metalfree phthalocyanine core up to generation 5. This core is used as a sensor and a probe for analyzing the properties of the internal structure and the influence of each structural part (core, branches, surface) upon the whole structure. UV-visible spectra show both a hyperchromic and bathochromic effect on the Q-bands with increasing generation, indicating that the chromophore is more isolated, and that the dendritic shell mimics a highly polar solvent. There is no evidence for aggregation, except for generation 0, showing again the isolation of the core (Figure 11).

Figure 10. Phenylazomethine Dendrimer with a Tetraphenylmethane Core.

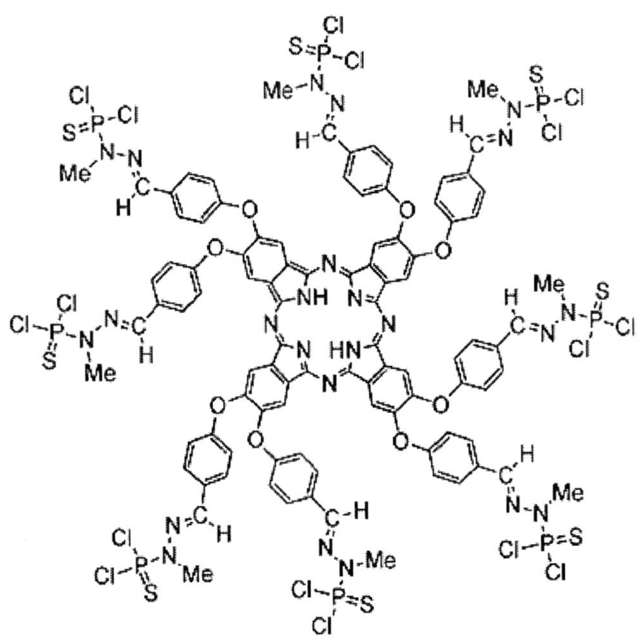

Figure 11. Octasubstituted metal-free phthalocyanine as core of phosphorus dendrimers.

Zinaida B. Shifrina et al [31] synthesized a new family of poly(phenylene-pyridyl) dendrimers (up to the fourth generation) with different periphery via a divergent route by using a cascade of Diels-Alder cycloadditions. Tetra(4-ethynylphen-1-yl)methane served as a core. A new A2B cyclopentadienone building block, namely, pyridine-containing cyclopentadienone with protected dienophile functions was prepared. By varying the building blocks at the last step of dendrimer construction, one could design molecules possessing either a phenyl- or a pyridyl-decorated outershell (Figure 12).

Natarajan Raju et al [32] described the synthesis of a new scaffold derived from iminodipropionic acid for the preparation of peptide dimers and tetramers.

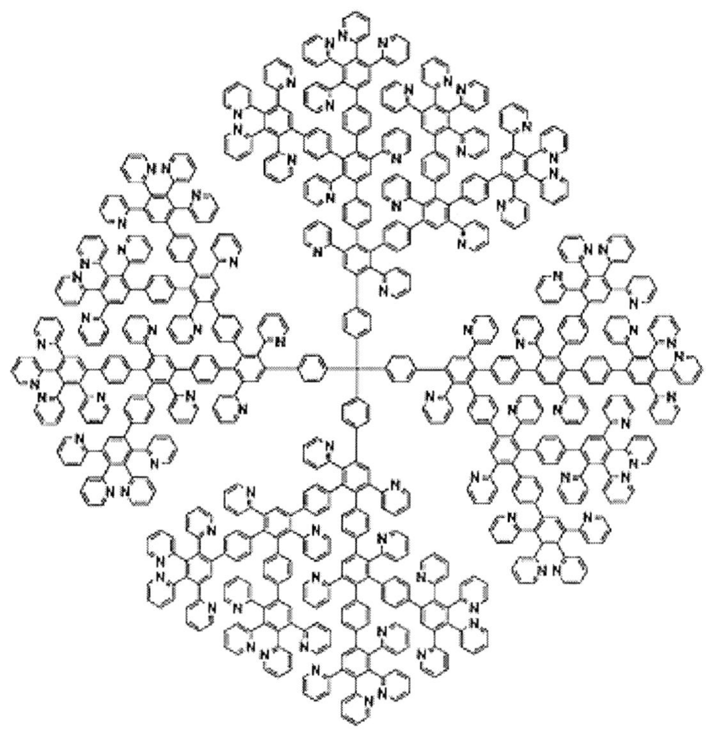

Figure 12. Poly(Phenylene-pyridyl) Dendrimers.

2. Characterization Advances of Dendrimer Physical-Chemical Properties

Johan Hofkens et al [33] used wide-field imaging of individual multichromophoric molecules and successive photobleaching to determine, accurately, the relative position of the chromophores in such systems. First, a polyphenylene dendrimer with well-defined geometry was used to establish the accuracy in localization that can be obtained by this methodology. For a signal-to-noise ratio of 20, interchromophoric distances could be measured with 4 nm accuracy. Next, the method was used to determine the end-to-end distribution of an end-capped polyfluorene polymer. From comparison between the experimental and simulated distributions, information on the conformation of the polymer could be deduced. It was found that the polymer has a nonlinear conformation. A conjugation length of sixmonomer units gave the best fit of the experimental data to the proposed model.

Shing Bor Chen et al [34] conducted brownian dynamics simulation to investigate the dynamics of dendrimer-polymer conjugates on the basis of a coarse-grained bead-spring model. The grafted polymers are found to slow the global dynamics of the dendrimer core. Because of the structural difference between the two moieties, the translational and rotational diffusivities of a conjugate show different scaling behaviors from those for a dendrimer, but in principle follow a Stokes-Einstein fashion, analogous to polymers of other types. When the grafted chains are long enough, they become brushlike, evidenced by the scaling behavior of the end-to-end distance as well as the chain-length dependence of the rotational relaxation time. For the range of chain length investigated, it is found that the brushlike grafted polymers still fall within the dilute regime. Thereby the interchain interactions are not strong enough to render a change in the scaling behavior of rotation relaxation in comparison with an isolated rodlike polymer.

Paul L. Burn et al [35] developed a simple convergent procedure for the formation of sterically encumbered phosphorescent dendrimers. The procedure is demonstrated with the preparation of a first-generation dendrimer composed of a fac-tris(2-phenylpyridyl)iridium(III) core and three dendrons (Figure 13). Each dendron is comprised of a branching phenyl unit with a further four phenyl groups attached. The lack of surface groups on the dendrons was found to reduce solubility and also reduced the level of control over the intermolecular interactions of the emissive and electroactive core in films. The 6-fold decrease in photoluminescence quantum yield in going from solution (69%) to the solid state (11%) showed that there were strong intermolecular interactions of the emissive cores in the solid state. Single-layer devices with the dendrimer blended with 4,4'-bis(N-carbazolyl)biphenyl showed an external quantum efficiency of 1.7% (5.4 cd/A) at 100 cd/m2 and 11.4 V, giving a power efficiency of 1.5 lm/W.

Ying Wang et al [36] reported nonlinear optical (NLO) and time-resolved fluorescence spectroscopic measurements of dendrimer-metal nanocomposites (DNCs). Third-order NLO effects were investigated by degenerate four-wave mixing (DFWM) measurements for DNCs and DNCs incorporated into thin polymeric films. The results show an enhancement of the third-order nonlinear susceptibility for the chromophore-functionalized dendrimer- metal nanocomposites. Investigations of the mechanism of the enhanced NLO effect suggested a strong contribution because of the metal's local field. These results show the potential usefulness of dendrimer-metal nanocomposites for nonlinear optical effects and biophotonic applications.

Natalia P. Yevlampieva et al [37] synthesized [60]fulleropyrrolidine-containing poly(benzyl ether) dendrimers via 1,3- dipolar cycloaddition of amino acid and aldehyde derivatives and [60]fullerene (Figure 14). Their dielectric and electrooptical (Kerr effect) solution properties were studied and compared with poly(propyleneimine) dendrimers, namely DAB dendr-(alkyl chain)n and DAB-dendr-(CN)n, and with known symmetrical and non-symmetrical poly(benzyl ether) dendrimers carrying benzyl groups ([Gx]-[C]-[Gx]) or benzyl and cyanobenzyl groups (CNn-[Gx]-[C]-[Gx+1]), respectively. It was shown that dendrimers display similar polar and electrooptical characteristics independent of their dendrimer generation. This is in contrast to the reference dendrimers for which a strong dependence of total dipole moment on dendrimer generation was observed.

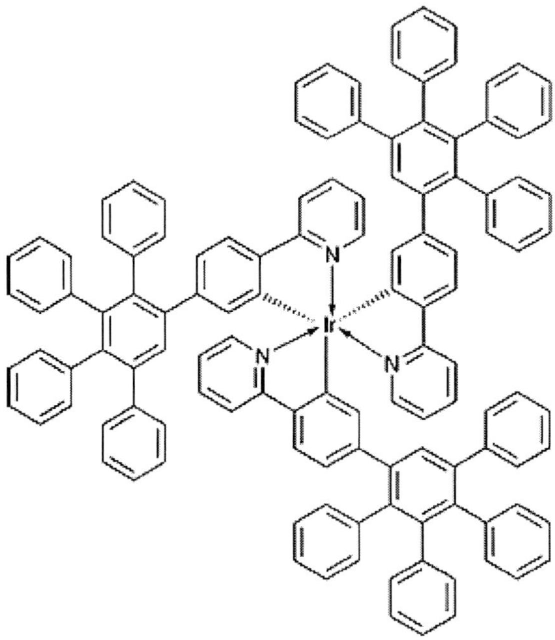

Figure 13. Phosphorescent dendrimer.

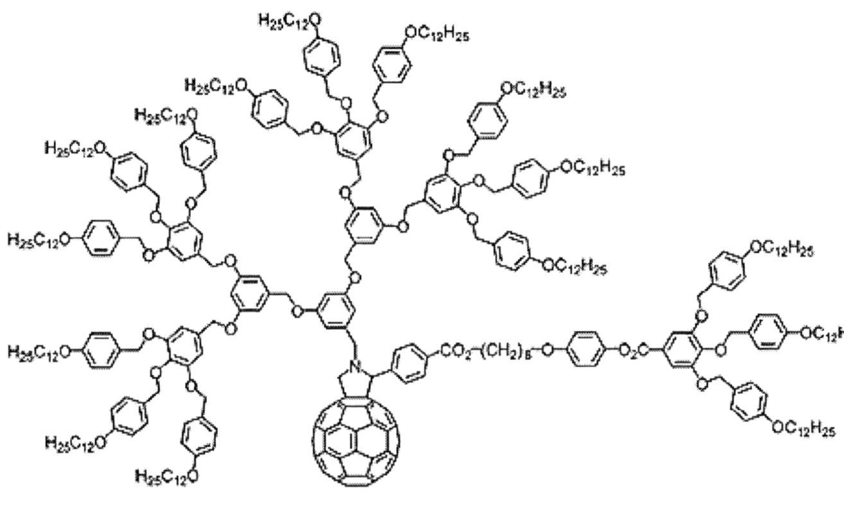

Figure 14. [60]Fullerene-containing poly(benzyl ether) dendrimers.

Robin L. McCarley et al [38] showed that generations 1-5 of poly(propyleneimine) dendrimers terminated with redox-labile, trimethyllocked quinone (TLQ) end groups released the TLQ units as lactone upon chemical or electrochemical redox action (Figure 15). Sodium dithionite was used as a chemical reducing agent to instantly convert the TLQ units to their deuteroquinone analogues (TLDQ), which were subsequently liberated from the dendrimers following zero-order kinetics in both DMSO-d_6-D2O (85:15, v/v, kobs =7 × 10–6 M s–1, t1/2 = 9 min) and THF-d_8-D2O (76:24, v/v, k*obs* = 4 × 10–6 M s–1, t1/2 = 15 min). Cyclic voltammetry of dendrimers in DMSO/0.1 M *n*-Bu4PF6 revealed that the reduction peak potential (Epc) values for the formation of TLQ2– were generally invariant (-1.85 V vs Fc/Fc+) with dendrimer generation.

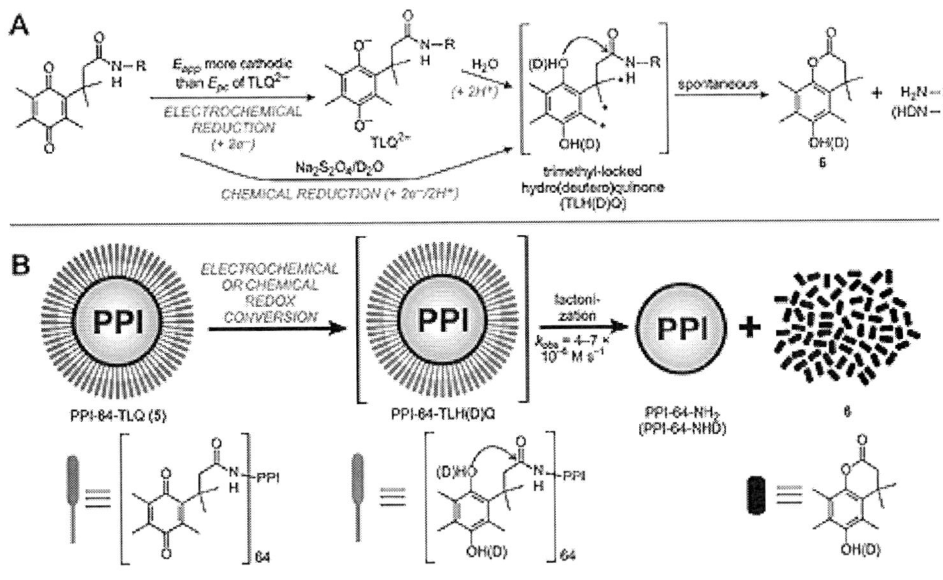

Figure 15. Chemically and electrochemically mediated release of dendrimer end groups.

Thus, bulk electrochemical reduction of 1-5 at an overpotential of -0.3 V vs the Epc of TLQ2 produced PPI dendrimers containing dianionic TLQ2– termini. Protonation of the electrochemically generated TLQ2– end-group units with H2O produced the transient and structurally labile hydroquinone (TLHQ) intermediates, which were liberated from the dendrimers as lactone 6 in a similar fashion as in the sodium dithionite-induced approach. In both the chemically and electrochemically mediated routes, 1H NMR spectroscopy verified that the redox-promoted disconnection of TLQ from the dendrimers was quantitative.

Jiri Horsky et al [39] reported the quantitative interpretation of MALDI-TOF mass spectra of imperfect carbosilane dendrimers. A general relationship is derived for the abundance of an imperfect dendrimer with a given number of missing constitutional repeating units in the two outmost layers. The relationship is used in the interpretation of the MALDI TOF mass spectrum of the second-generation carbosilane dendrimer prepared by the iterative divergent method. The model quantitatively describes the spectrum of the dendrimer and correctly predicts the MALDI TOF mass spectrum of its first-generation precursor. Thus, the use of well-resolved MALDI TOF mass spectra for assessing the purity of low-generation dendrimers with uniform end groups is substantiated for carbosilane dendrimers and to lesser extent for dendrimers in general.

Hak-Sung Kima et al [40] reported mass spectrometric analysis of affnity-captured proteins on a dendrimer-based immunosensing surface and the investigation of on-chip proteolytic digestion. The monolayer of fourth-generation poly(amidoamine) dendrimers was adopted to construct the immunoaffnity surface of an antibody layer. The antibody layer as a bait on the dendrimer monolayer was found to result in high binding capacity of antigenic proteins and a reliable detection. The affnity-captured protein at the immunosensing surface was subjected to direct on-chip tryptic digestion, and the resulting proteolytic peptides were analyzed by using matrix-assisted laser desorption/ionization time-of-flight mass spectrometry. The on-chip digestion method identiWed the affinity-captured bovine serum albumin (BSA), lysozyme, and ferritin at the level of around 100 fmol. The use of antibodies against the proteins involved in the metabolic pathway of L-threonine in Escherichia coli also led to discrimination of the respective target proteins from cell lysates.

Matthias Ballauff et al [41] reported the solution structure of stilbenoid dendrimers by a small-angle scattering study. The spatial structure of a stilbenoid dendrimer is investigated by small-angle X-ray scattering (SAXS) and small-angle neutron scattering (SANS) in dilute solution. All measurements are performed in toluene. The dendrimer consists of a stilbenoid scaffold with appended hexyloxy chains. SAXS is mainly sensitive to the dendrimer scaffold whereas SANS intensity, measured in fully deuterated toluene, derives from the solute molecules. The resulting SAXS and SANS intensities are analyzed by comparison with various models. It is found that the model of a circular disk gives the best description of the data. SAXS data demonstrate that the stilbenoid scaffold is flat as expected for benzene rings conjugated through vinylene units. SANS, on the other hand, shows that the hexyloxy chains are not confined in the plane defined by the core. The structure factor S(q), describing the interaction of the dendrimers at higher concentrations, is modeled quantitatively with the "polymer reference interaction site model" (PRISM) integral equation formalism for hard plates such as particles.

3. Advances in Nano Printing

Geoffrey A. Ozin et al [42] reported the synthesis of a new class of materials called as periodic mesoporous dendrisilicas. They are prepared from dendrimers with trialkoxysilyl groups at the outmost shell. Dendrimers of various cores and numbers of shells are used. The dendrisilica synthesis is based on acid- or base-catalyzed hydrolysis of the trialkoxysilyl groups and subsequent template-directed condensation of the dendrimers into an ordered template-dendrisilica nanocomposite. The template can be removed with organic solvents, producing a periodic mesoporous dendrisilica whose pore walls consist of interconnected dendrimer building blocks.

Byeongdu Lee et al [43] described the preparation of low- and ultralow-k (dielectric constant) nanoporous organosilicate dielectrics from blends of polymethylsilsesquioxane (PMSSQ) precursor with globular ethyl acrylate-terminated polypropylenimine dendrimers, which act as porogens (Figure 16). These dendrimers are found to mix well with the PMSSQ precursor and after their sacrificial thermal decompositions result in closed, spherical pores of <2.0 nm radius with a very narrow distribution even at high loading. This pore size and distribution are the smallest and the narrowest respectively ever achieved in porous spin-on dielectrics. The method therefore successfully delivers low- and ultralow-k PMSSQ dielectric films that should prove very useful in advanced integrated circuits.

Neal A. Rakow et al [44] described the dynamic imprinting of dendritic macromolecules with porphyrin templates to yield synthetic host molecules containing one binding site each (Figure 17). The process is based on our general strategy to prepare cored dendrimers, and involves covalent attachment of dendrons to a porphyrin core, cross-linking of the end-groups of the dendrons, and removal of the porphyrin template by hydrolysis.

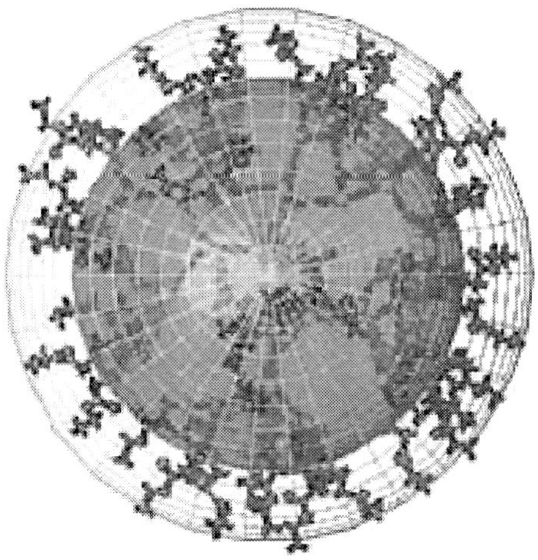

Figure 16. Globular ethyl acrylate-terminated polypropylenimine dendrimer.

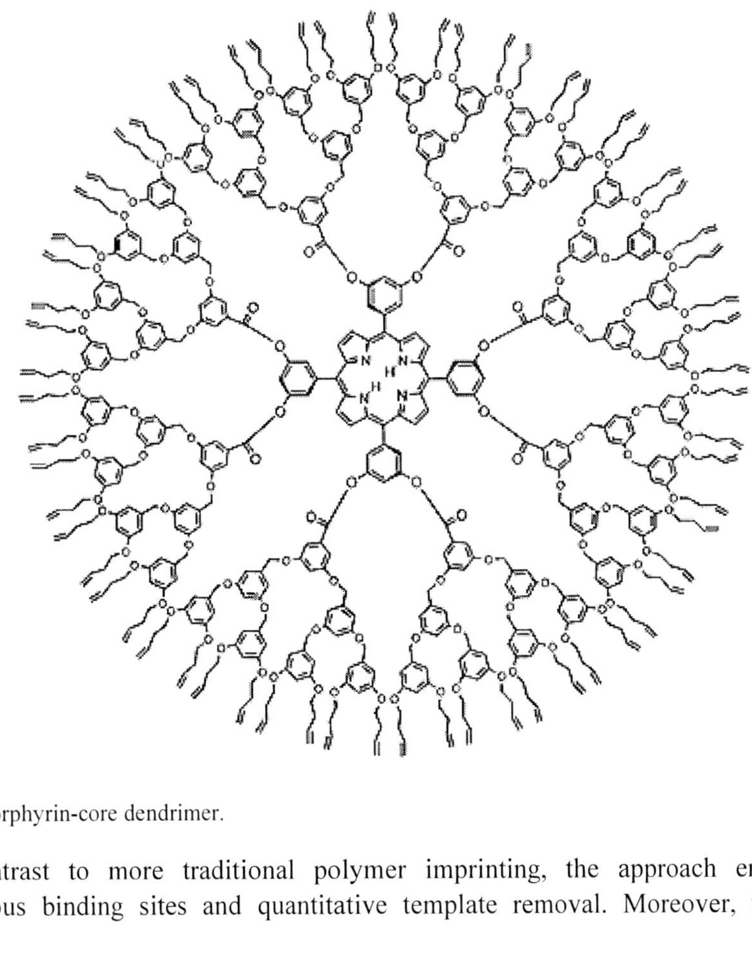

Figure 17. Porphyrin-core dendrimer.

In contrast to more traditional polymer imprinting, the approach ensures nearly homogeneous binding sites and quantitative template removal. Moreover, the hosts are soluble in common organic solvents and amenable to the incorporation of other functional groups, which should facilitate further development of this system for novel applications. When developing 'monomolecular imprinting' methodology, Neal A. Rakow et al selected dendrimers as macromolecular hosts because their molecular homogeneity and excellent solubility facilitate characterization. Ho-Cheol Kim et al [45] reported oriented nanoporous lamellar organosilicates templated from topologically unsymmetrical dendritic-linear block copolymers.

Jurriaan Huskens et al [46] reported heavyweight dendritic inks for positive microcontact printing. Poly(propylene imine) dendrimers with dialkyl sulfide end groups were prepared and developed as inks for positive microcontact printing ((+)μCP) on gold. Long (C10H21-S-C10H20-), medium (C3H7-S-C4H8-), and short (CH3- S-CH2-) dialkyl sulfide end groups were attached to second- and third-generation PPI dendrimers to create a family of dendritic sulfides.

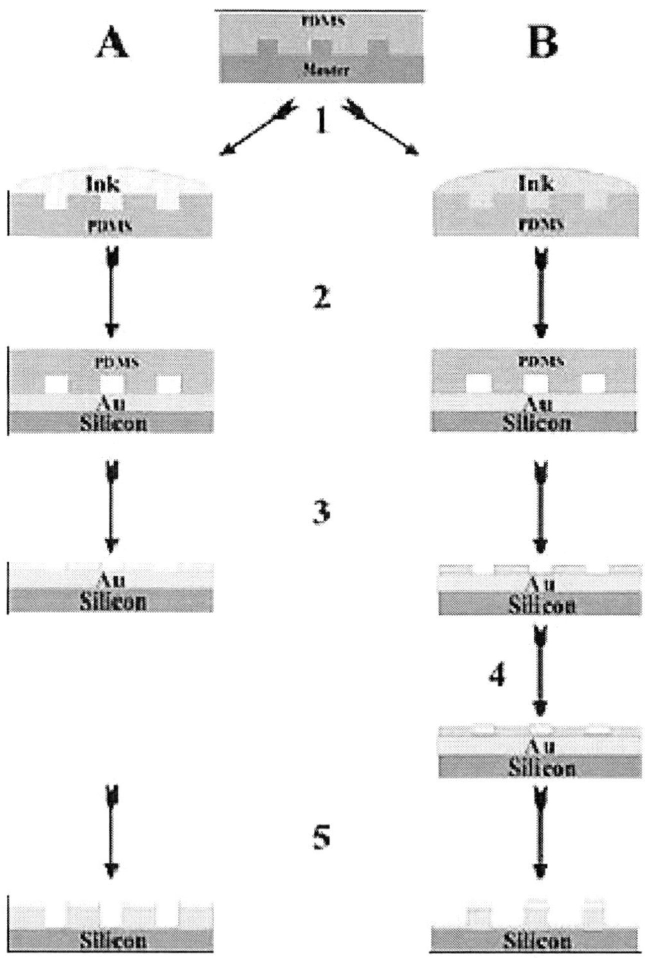

Figure 18. Heavyweight dendritic inks for positive microcontact printing.

The dendritic inks flatten upon adsorption and form monolayers on gold. (+)μCP was performed on gold using commercially available poly(dimethylsiloxane) as stamp material and *n*-octadecanethiol as etch resist. The gold beneath the dendrimers was selectively etched away with an acidic Fe(NO3)3/thiourea solution to give the positive copy of the original master pattern. The multivalent sulfide attachment and the relatively high molecular mass of these dendrimers ensured minimal lateral ink spreading and thus optimal feature reproducibility (Figure 18).

4. RESEARCH ADVANCES IN BIOLOGY AND MEDICINE

4.1. Imaging Contrast Reagent

Walter H. Backes et al [47] reported the evaluation of Gd(III) DTPA-terminated poly(propylene imine) dendrimers as contrast agents for MR imaging. Serial and dynamic CE-MRI were performed to monitor the distribution of MRI contrast agent in the heart, arteries, renal system, liver, spleen, bladder and tumor periphery. Dynamic CE-MRI showed timely renal clearance for all contrast agents, but a stronger and a prolonged blood signal enhancement for the higher generations of the dendritic contrast agent. Moreover, G0 and G1 showed a rapid tumor wash-in and wash-out, whereas G3 and G5 displayed a more gradual and prolonged tumor wash-in. Both G0 and dendritic contrast agents G1, G3 and G5 are well suited for non-tissue-specific MRI of submillimeter- sized blood vessels and evaluating tumor microcirculatory characteristics in mice. Higher generations of dendritic contrast agents display lower concentration detection limits, which suggests their future use for molecular imaging (Figure 19).

Hisataka Kobayashi et al [48] reported MR lymphangiography using dendrimer-based contrast agents to a comparison at 1.5T and 3.0T. Most macromolecular contrast agents (CAs) show lower r1 and higher r2 relaxivities at 3.0T than at 1.5T (Figure 20). MR lymphangiography in mice using a macromolecular G6 dendrimer-based CA was serially performed and compared at both 1.5T and 3.0T. The r1 and r2 relaxivities of the G6 CA were 25 and 78/s/mM at 1.5T and 17 and 82/s/mM at 3.0T, respectively.

Robert C. Brasch et al [49] designed, synthesized, and initially characterized a representative set of novel constructs for large-molecular radiographic/computed tomography (CT) contrast agents, intended for a primarily intravascular distribution. A new assembly of well-known and biocompatible components consists of paired, symmetrical dendritic polylysines initiated from both ends of a poly(ethylene glycol) (PEG) core, yielding an array of multiple free amino groups to which were conjugated highly soluble and stable triiodophthalamide ("triiodo") moieties. An array of six dendritic contrast agents was synthesized originally, using three different PEG cores (3, 6, 12 kDa) with t-Boc lysine-generated dendrimer "amplifiers" (from three to five generations) containing 16 to 64 amino groups for conjugation with reactive triiodo moieties. A clinically used, nonionic, small molecular CT contrast agent, iobitridol, was derivatized via a hydroxyl protection/deprotection strategy, introducing a new carboxyl group available for conjugation to the lysine amino groups of dendrimers.

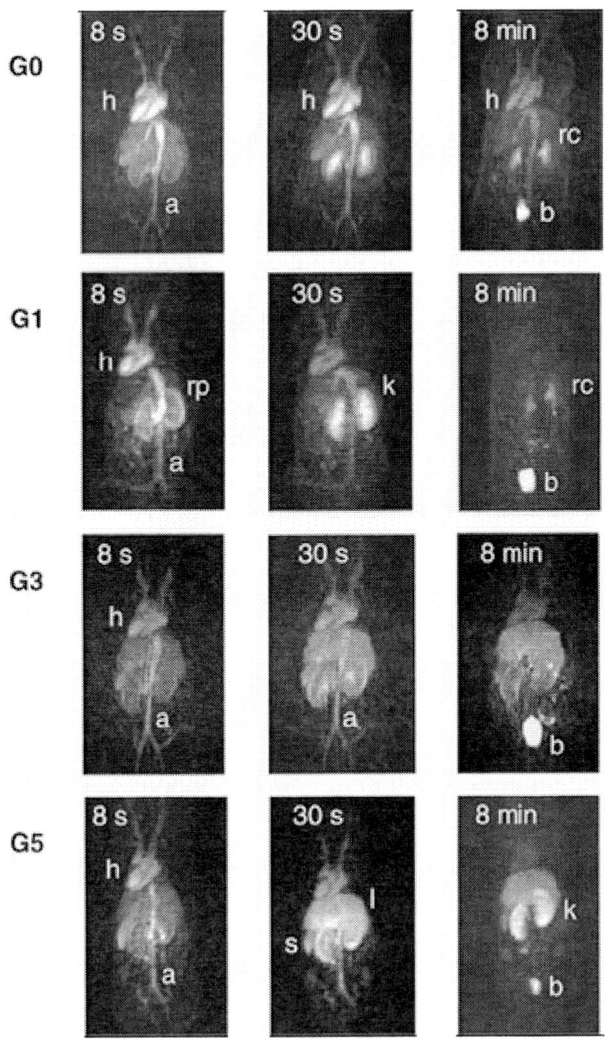

Figure 19. Digitally subtracted contrast-enhanced whole-body MR images.

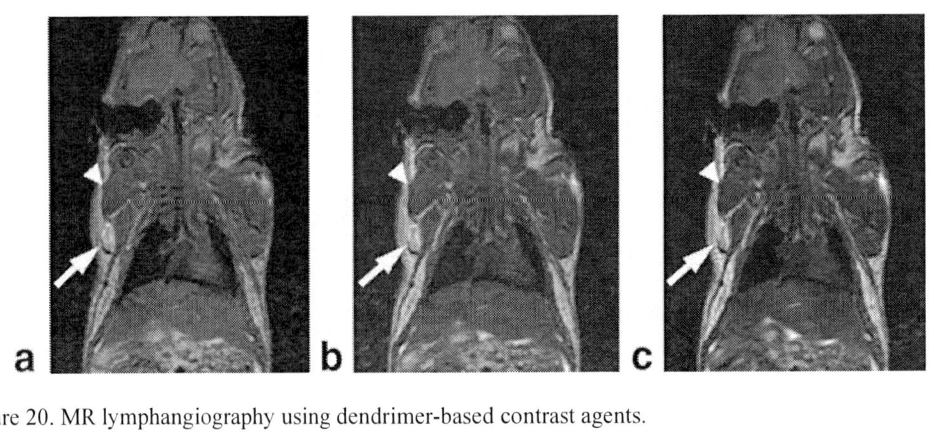

Figure 20. MR lymphangiography using dendrimer-based contrast agents.

The results showed the synthetic feasibility, desired basic characteristics, and potential utility for CT contrast enhancement achieved with a new type of iodinated, large-molecular PEG-core dendritic construct.

Peptide dendrimers have a variety of applications in biology such as the vehicles for drug and gene delivery, molecular inhibitors, protein mimics, and synthetic vaccines. The multiple antigenic peptide (MAP) system is a well-known example of a discrete, dendrimeric scaffold. Ching-Hsuan Tung et al [50] reported enzyme-targeted fluorescent imaging probes on a multiple antigenic peptide core. They explored a novel application of the MAP based scaffold by designing molecular probes that fluoresce only after enzymatic treatment. The probes, which were synthesized on solid support, incorporate a cathepsin S dipeptide substrate (Leu-Arg), and a poly(ethylene glycol) (PEG) chain in their dendritic arms. The fluorescence emission of the near-infrared fluorochromes attached to the N-termini of the dendritic arms was quenched. Mechanistic studies revealed formation of H-type dye aggregates within the tetravalent MAP system.

Vladimir S. Talanov et al [51] reported a dendrimer-based nanoprobe for dual modality magnetic resonance and fluorescence imaging. A novel PAMAM dendrimer-based nanoprobe for dual magnetic resonance and fluorescence imaging modalities was synthesized. Fluorescence studies revealed that Gd(III) complexation to the probe has no effect on the quantum yield; however, increases in the dye content resulted in partial quenching. The potential of the new nanoprobe as a dual modality imaging agent was demonstrated in vivo by the efficient visualization of sentinel lymph nodes in mice by both MRI and fluorescence imaging modalities (Figure 21).

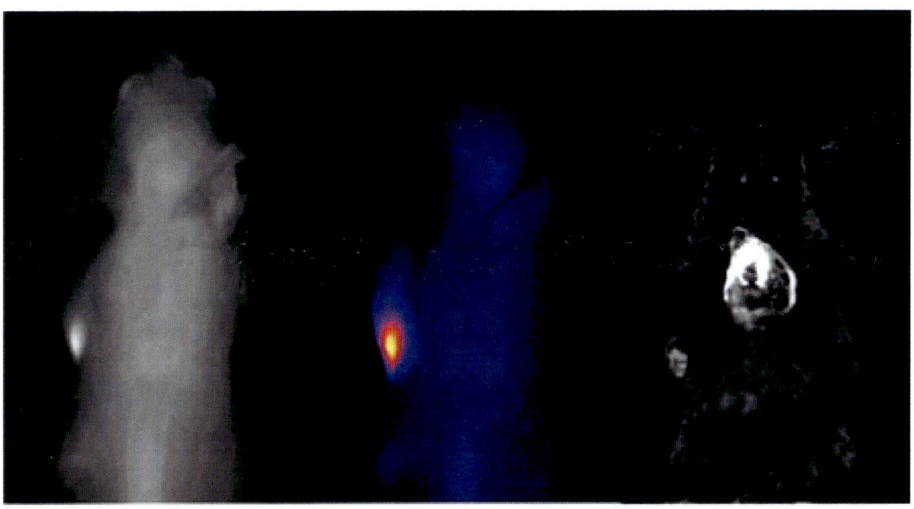

A B C

Figure 21. Dendrimer-based nanoprobe for dual modality magnetic resonance and fluorescence imaging.

4.2. Drug and Gene Delivery

The control of gene transfection in the body is a core issue in gene therapy. Photochemical internalization is a technology that allows light-induced delivery of DNA, drugs or other biological factors directly inside cells. Usually it requires that a photosensitizer be added to the drug-delivery system to photochemically destabilize the endosomal membrane. Nobuhiro Nishiyama et al [52] reported light-induced gene transfer from packaged DNA enveloped in a dendrimeric photosensitizer (Figure 22). A system for in vivo DNA delivery in which these two components are assembled into one structure was presented. This is a ternary complex composed of a core containing DNA packaged with cationic peptides and enveloped in the anionic dendrimer phthalocyanine, which provides the photosensitizing action. The ternary complex showed more than 100-fold photochemical enhancement of transgene expression in vitro with reduced photocytotoxicity. In an animal experiment, subconjuctival injection of the ternary complex followed by laser irradiation resulted in transgene expression only in the laser-irradiated site. Nobuhiro Nishiyama et al demonstrated a new biomedical application for dendrimers, and the first success in the photochemical-internalization-mediated gene delivery in vivo.

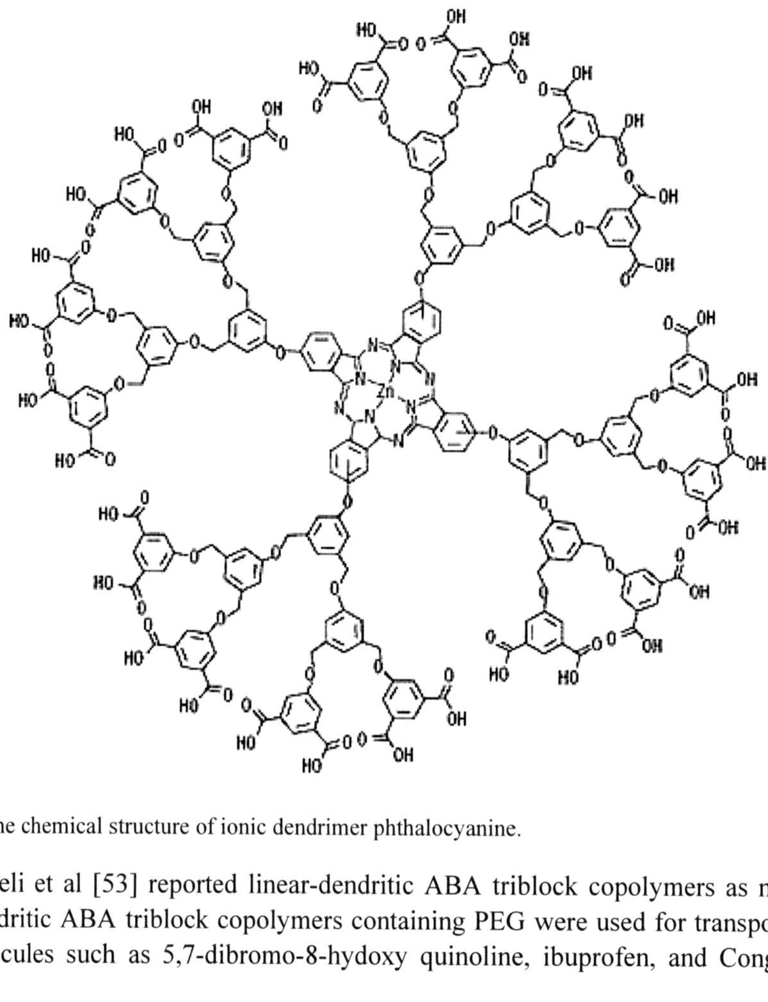

Figure 22. The chemical structure of ionic dendrimer phthalocyanine.

M. Adeli et al [53] reported linear-dendritic ABA triblock copolymers as nanocarriers. Linear-dendritic ABA triblock copolymers containing PEG were used for transport the small guest molecules such as 5,7-dibromo-8-hydoxy quinoline, ibuprofen, and Congo red. The linear-dendritic copolymers/ guest molecule complexes were stable at room temperature for about 10 months; during this time, guest molecules did not release from linear-dendritic copolymers/guest molecule complexes.

M. Adeli et al [54] reported multiarm star nanocarriers containing a poly(ethylene imine) core and polylactide arms. Star polymers (SPs) containing a hyperbranched poly(ethylene imine) (PEI; number-average molecular weight = 10,000) core and polylactide arms were synthesized via the ring-opening polymerization of lactide. PEI was used as a multifunctional macroinitiator for the ring-opening polymerization of lactide. Different lactide monomer/ amino-functional group (LA/NH_n; $n = 1$ or 2) ratios were used for preparing SPs with different molecular weights. SPs were able to encapsulate small guest molecules such as Rose Bengal; they also transported small, hydrophilic molecules from water to the organic phase. The transport capacity of all the nanocarriers depended on the LA/NH_n ratio used for synthesizing the SPs. Nanocarriers with a higher LA/NH_n ratio had higher transport capacities.

Robert Langer et al [55] reported a family of hierarchically self-assembling linear-dendritic hybrid polymers for highly efficient targeted gene delivery.

James R. Baker Jr et al [56] have demonstrated the applicability of poly(amidoamine) (PAMAM) dendrimers as a platform for the targeted delivery of chemotherapeutic drugs both in vitro and in vivo. They showed the synthesis, characterization, and in vitro targeting of an engineered PAMAM nanodevice in which folic acid (FA) is conjugated as the targeting molecule and a caspase-specific FRET-based agent (PhiPhiLux G1D2) is conjugated as the apoptosis-detecting agent. This conjugate specifically targets FA-receptor-positive, KB cells. In these cells, the apoptosis-inducing agent staurosporine caused a 5-fold increase in the cellular fluorescence. These results show, for the first time, the potential applicability of a targeted apoptosis-measuring nanodevice, which could be used for simultaneously monitoring the apoptotic potential of a delivered drug.

James R. Baker Jr et al [57] reported the synthesis and human growth factor receptor-2 (HER2) specific tumor targeting properties of a dendrimer conjugated to anti-HER2 mAb (monoclonal antibody) conjugate. The polyamidoamine (PAMAM) dendrimer generation five (G5) was labeled with alexaFluor 488 and conjugated to anti-HER2 mAb. Uniquely, the conjugate demonstrated cellular uptake and internalization in HER2-expressing cells as compared to free antibody. The time course of internalization and blocking experiments with free antibody suggest that the rapid and efficient cellular internalization of the dendrimer-antibody conjugate was achieved without alterations in specificity of targeting. Animal studies demonstrated that the conjugate targets HER2-expressing tumors.

Rangaramanujam M. Kannan et al [58] demonstrated the activity of dendrimer-methotrexate conjugates on methotrexate-sensitive and -resistant cell lines. They have investigated the effect of dendrimer end-functionality on the activity of polyamido amine (PAMAM) dendrimer-methotrexate (MTX) conjugates in MTX-sensitive and MTX-resistant human acute lymphoblastoid leukemia (CCRF-CEM) and Chinese hamster ovary (CHO) cell lines (Figure 23). Two amide-bonded PAMAM dendrimer-MTX conjugates were prepared using a dicyclohexylcarbodiimide (DCC) coupling reaction: one between a carboxylic acid-terminated G2.5 dendrimer and the amine groups of the MTX (conjugate A) and another between an amine-terminated G3 dendrimer and the carboxylic acid group of the MTX (conjugate B). The results suggested that conjugate A showed an increased drug activity compared to an equimolar amount of free MTX toward both sensitive and resistant cell lines, whereas conjugate B did not show significant activity on any of the cell lines. These findings demonstrated the feasibility of using dendrimers as drug delivery vehicles for achieving higher therapeutic effects in chemotherapy, especially in drug-resistant cells.

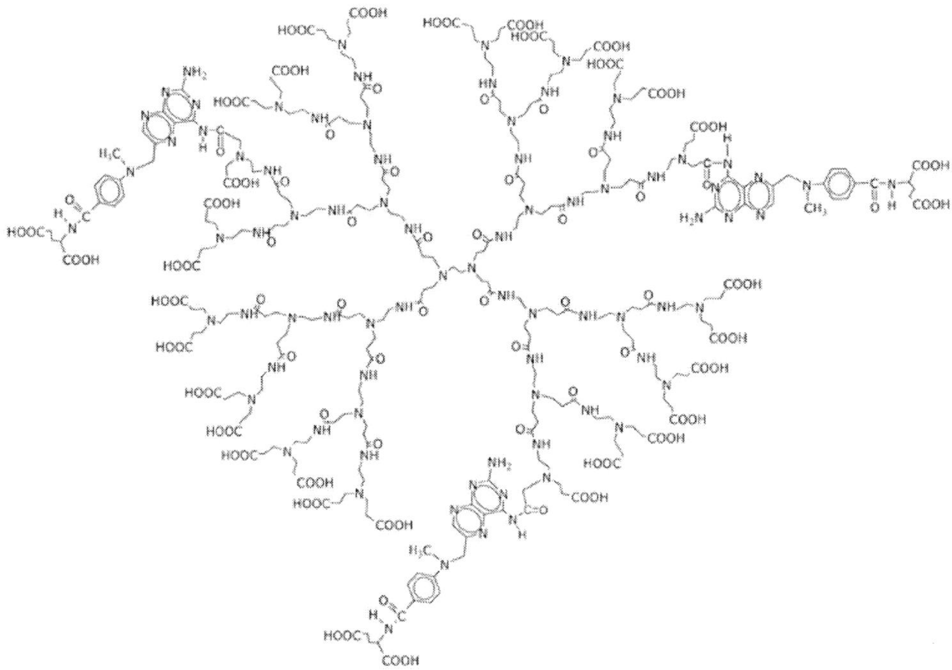

Figure 23. PAMAM-G2.5-COOH-MTX conjugate.

Patrice Hildgen et al [59] reported the synthesis and evaluation of novel dendrimers with a hydrophilic interior as nanocarriers for drug delivery. Novel polyester-co-polyether dendrimers consisting of a hydrophilic core were synthesized by a combination of convergent and divergent syntheses (Figure 24). The core was synthesized from biocompatible moieties, butanetetracarboxylic acid and aspartic acid, and the dendrons from PEO (poly(ethylene oxide)), dihydroxybenzoic acid or gallic acid, and PEG monomethacrylate. Dendrimers demonstrated good ability to encapsulate the guest molecule, with loading of 15.80 and 6.47% w/w for rhodamine and β-carotene, respectively. The release of the encapsulated compounds was found to be slow and sustained, suggesting that these dendrimers can serve as potential drug delivery vehicles.

István J. Majoros et al [60] have designed and synthesized poly(amidoamine) (PAMAM) dendrimer-based multifunctional cancer therapeutic conjugates. The primary amino groups on the surface of the generation 5 (G5) PAMAM dendrimer were neutralized through partial acetylation, providing enhanced solubility of the dendrimer (in conjugation of FITC (fluorescein isothiocyanate)) and preventing nonspecific targeting interactions (in vitro and in vivo) during delivery. The functional molecules fluorescein isothiocyanate (FITC, an imaging agent), folic acid (FA, targets overexpressed folate receptors on specific cancer cells), and paclitaxel (taxol, a chemotherapeutic drug) were conjugated to the remaining nonacetylated primary amino groups. These multifunctional dendrimer conjugates have been tested in vitro for targeted delivery of chemotherapeutic and imaging agents to specific cancer cells.

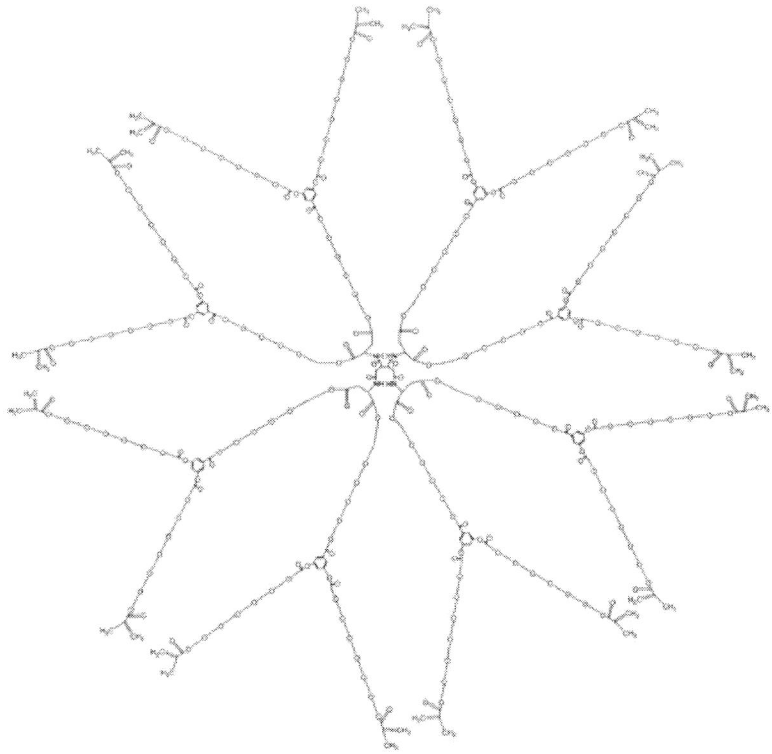

Figure 24. Novel dendrimers with a hydrophilic interior as nanocarriers for drug delivery.

Dendrimers, dendrons, and hyperbranched polymers are gaining popularity as novel drugs, imaging agents, and drug delivery systems. They present advantages of well-defined molecular weight, multivalent surfaces, and high drug carrying capacity. Moreover, it is emerging that such architectures can display unique endocytic properties. As poly(ethylene glycol) (PEG) is widely used for protein and drug conjugation, the aim of this study was for the first time to synthesize novel, branched PEG-based architectures, to define their cytotoxicity and, via preparation of Oregon green (OG) conjugates define the effect of structure on their cellular uptake. Ruth Duncan et al [61] reported novel monodisperse PEG-dendrons as new tools for targeted drug delivery. Five PEG-based dendrons were synthesized using monodisperse Fmoc-amino PEG propionic acid (Mw = 840) as a monomer, and cadaverine, tris(2-aminoethyl)amine or lysine as the branching moieties. These well-defined novel architectures have potential for further development as targetable drug delivery systems or tools for construction of structurally defined modified surfaces.

Jong-sang Park et al [62] synthesized arginine-conjugated polypropylenimine dendrimer G2 (DAB-8), PPI2-R for gene delivery systems. Rcsults demonstrate that arginine-conjugation of PPI2 is successful in developing a low toxic and highly transfection efficient gene delivery carrier.

Prakash V. Diwan et al [63] developed folate-PAMAM dendrimer conjugates for targeted delivery of anti-arthritic drugs and studied their pharmacokinetics and biodistribution in arthritic rats. The aim of this study was to synthesize folate-dendrimer conjugates as suitable vehicle for site specific delivery of anti-arthritic drug (indomethacin) to inflammatory regions and to determine its targeting efficiency, biodistribution in adjuvant induced arthritic rats.

This study demonstrated the superiority of active targeting over dendrimer mediated passive targeting and also for the first time, folate-mediated targeting of an anti-arthritic drug to the inflammatory tissues.

Mark M. Banaszak Holl et al [64] reported the binding avidity of a nanoparticle-based multivalent targeted drug delivery platform. Dendrimer-based anticancer nanotherapeutics containing ~5 folate molecules have shown in vitro and in vivo efficacy in cancer cell targeting. This study provides quantitative and systematic evidence for multivalent interactions between these nanodevices and folate-binding protein (FBP). Qualitative evidence is also provided for a multivalent targeting effect to KB cells using flow cytometry. These data support the hypothesis that multivalent enhancement of KD, not an enhanced rate of endocytosis, is the key factor resulting in the improved biological targeting by these drug delivery platforms.

Mitsuru Hashida et al [65] synthesized a sixth generation lysine dendrimer (KG6) and two PEGylated derivatives thereof and evaluated their biodistribution characteristics in both normal and tumor-bearing mice. The intact KG6 showed a rapid clearance from the blood stream and nonspecific accumulation in the liver and kidney. In contrast, the PEGylated derivatives showed a better retention in blood and low accumulativeness in organs dependent of the rate of PEGylation. In addition, PEGylated KG6 with a high modification rate was accumulated effectively in tumor tissue via the enhanced permeability and retention (EPR) effect. Moreover, Mitsuru Hashida et al clarified that multiple administrations did not affect the biodistribution characteristics of a second dose of PEGylated KG6. PEGylated lysine dendrimer would be a useful material for a clinically applicable tumor-targeting carrier.

Luca Costantino et al [66] described a general method for incorporating target moieties in a well-defined arrangement into the surface of biocompatible polyester poly(D,L-lactic-co-glycolic acid) (PLGA) materials using dendrons. In this way it is possible to obtain nanoparticles (NPs) with a high degree of surface coverage. This new strategy was successfully applied to the preparation of peptide- and β-D-glucose-covered NPs. The first application is based on the discovery of NPs made of conjugates between PLGA and short peptidic sequences able to cross the blood–brain barrier (BBB) after systemic administration. Luca Costantino et al used a branched structure (dendron) in order to prepare a derivative of PLGA able to form, by simple nanoprecipitation, NPs with a high degree of surface coverage, characteristic that could influence the uptake by the liver and spleen. The same strategy, namely the use of a branched spacer (a dendron or a G1 dendrimer) inserted between one end of the PLGA chain and a derivatizing molecule, was also successfully applied to obtain β-D-glucose-covered NPs.

Dendrimers are emerging as a relatively new class of polymeric biomaterials withapplications in drug delivery, and imaging. Achieving a high drug payload in dendrimers, and understanding the therapeutic effect of the dendrimer–drug conjugates are receiving increasing attention. Rangaramanujam M. Kannan et al [67] reported the preparation, cellular transport, and activity of polyamidoamine-based dendritic nanodevices with a high drug payload. A high drug payload nanodevice was obtained by covalent conjugation of ibuprofen to a polyamidoamine (PAMAM-G4-OH) dendrimer. Significant amounts of the conjugate entered the cells rapidly within 15 min. Suppression of prostaglandin was noted within 30 min for the dendrimer–drug conjugates versus 1 h for the free ibuprofen. The results suggest that dendrimers with high drug payload improve the drug's efficacy by enhanced cellular delivery, and may produce a rapid pharmacological response. These dendrimer–drug conjugates can potentially be further modified by attaching antibodies and ligands for targeted drug delivery.

H. Namazi et al [68] reported the dendrimers of citric acid and poly (ethylene glycol) as the new drug-delivery agents. Citric acid–polyethylene glycol–citric acid (CPEGC) triblock dendrimers as biocompatible compounds containing G1, G2 and G3 were applied as the drug-delivery systems. Some of the small size molecules and drugs are trapped with the above-synthesized dendrimers. The guest molecules, which are hydrophobic when trapped into the suitable sites of dendrimers, are becoming soluble in aqueous solution. The quantity of trapped molecules and drugs such as 5-amino salicylic acid (5-ASA), pyridine, mefenamic acid, and diclofenac was measured. The drug/dendrimer complexes remained in room temperature for about 10 months and after this long time they were stable and the drugs were not released.

James R. Baker, Jr. et al [69] reported the synthesis and functional evaluation of DNA-assembled polyamidoamine dendrimer clusters for cancer cell-specific targeting. They sought to produce dendrimers conjugated to different biofunctional moieties (fluorescein [FITC] and folic acid [FA]), and then link them together using complementary DNA oligonucleotides to produce clustered molecules that target cancer cells that overexpress the high-affinity folate receptor (Figure 25). Amine-terminated, generation 5 polyamidoamine (G5 PAMAM) dendrimers are first partially acetylated and then conjugated with FITC or FA, followed by the covalent attachment of complementary, 5'-phosphate-modified 34-base-long oligonucleotides. Hybridization of these oligonucleotide conjugates led to the self-assembly of the FITC- and FA-conjugated dendrimers. In vitro studies of the DNA-linked dendrimer clusters indicated specific binding to KB cells expressing the folate receptor. Confocal microscopy also showed the internalization of the dendrimer cluster. These results demonstrate the ability to design and produce supramolecular arrays of dendrimers using oligonucleotide bridges. This will also allow for further development of DNA-linked dendrimer clusters as imaging agents and therapeutics.

Saghir Akhtar et al [70] reported a novel anionic dendrimer for improved cellular delivery of antisense oligonucleotides. The optimal design of hybridisation-competent antisense oligonucleotides (ODNs) coupled with an efficient delivery system appear to be important prerequisites for the successful use of antisense reagents for gene silencing.

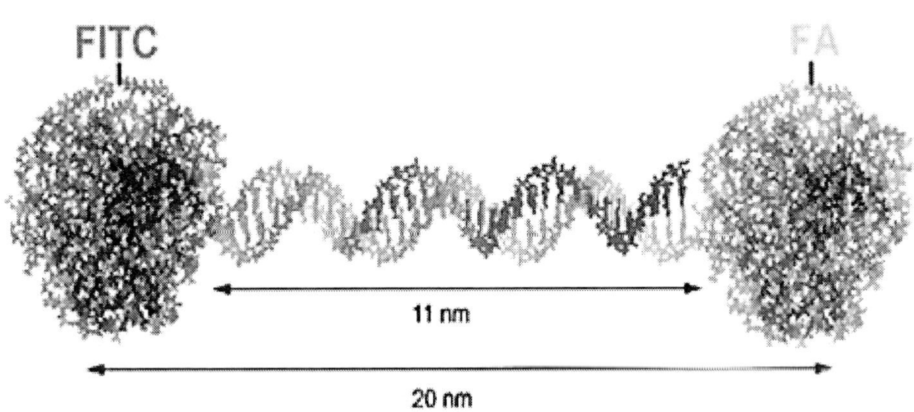

Figure 25. DNA-Assembled Polyamidoamine Dendrimer Clusters for Cancer Cell-Specific Targeting.

Saghir Akhtar et al selected an antisense ODN complementary to an accessible region of the epidermal growth factor receptor (EGFR) mRNA with the aid of an antisense oligonucleotide scanning array. This dendrimer conjugate was markedly more stable to serum nucleases compared to the free ODNs and the cellular uptake of ODN-dendrimer conjugates was up to 100-fold greater as compared to mannitol, a marker for fluid phase endocytosis, and up to 4-fold greater than naked ODN in cancer cells.

Eric E. Simanek et al [71] reported in vitro and in vivo evaluation of a melamine dendrimer as a vehicle for drug delivery. Cell-based and acute and subchronic in vivo toxicity profiles of a dendrimer based on melamine reveal that this class of molecules warrants additional study as vehicles for drug delivery (Figure 26).

Polymeric drug delivery systems are used not only to improve aqueous solubility of drug molecules but also to achieve desirable pharmacokinetics and an enhanced therapeutic index. New biodegradable polymers are needed to improve the biodistribution and targeting-ability of polymeric carriers. Chun Li et al [72] described the synthesis and characterization of branched poly(Lglutamic acid) (PG) containing multiple PG chains centered on a poly(amidoamine) (PAMAM) dendrimer or polyethyleneimine (PEI) cores. The branched PG polymers were obtained by ring-opening polymerization of benzyl ester of Lglutamic acid N-carboxyanhydride using PAMAM or PEI as the initiator. These polymers were degradable in the presence of the lysosomal enzyme cathepsin B, albeit more slowly than linear PG. Unlike conventional linear PG, each branched PG possessed multiple terminal amino groups.

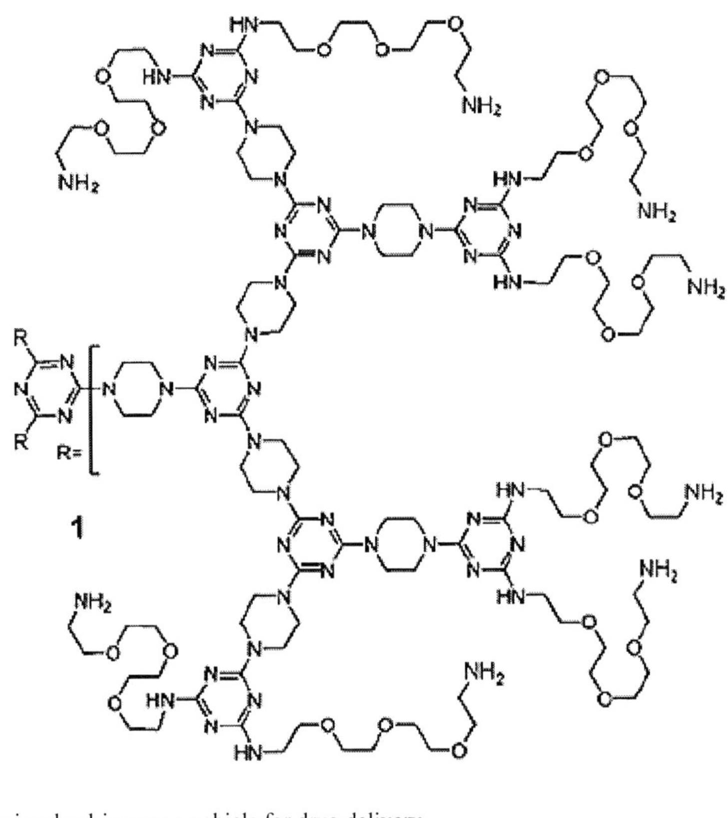

Figure 26. Melamine dendrimer as a vehicle for drug delivery.

This made it possible to attach multiple targeting moieties selectively to the termini of branched PG. Conjugation of monofunctional or heterodifunctional polyethylene glycol to the chain ends of branched PG was demonstrated in the presence of side chain carboxyl groups. Furthermore, folic acid, a model targeting moiety, and the nearinfrared dye indocyanine green, a model diagnostic agent, were successfully conjugated to the terminal amino groups and the side chain carboxyl groups of branched PG, respectively. The resulting conjugate had reduced nonspecific interaction and bound selectively to tumor cells expressing folate receptors (Figure 27).

To establish an effective nonviral gene transfer vector to hepatocytes, various oligo-carrier complexes were developed employing dendrimer (G4) and avidin–biotin systems (Av–bt), and their biodistribution were evaluated by Tsuneo Saga et al [73]. In-111-labeled-oligo, without any carriers, showed low uptake in normal organs other than the kidney. In contrast, 111In-oligo coupled with avidin through biotin (111In-oligo-bt-Av) showed very high accumulation in the liver. 111In-oligo complexed with G4 showed high uptake in the kidney and spleen, but its hepatic uptake was relatively low. When both G4 and Av–bt systems were employed, 111In-oligo/G4-bt-Av showed extremely high uptake in the lung. 111In-oligo-bt-Av, which exhibited the highest hepatic uptake in vivo, also showed high and rapid internalization into hepatocytes. The avidin–biotin system seems to have potential as a carrier of oligo-DNA to the liver.

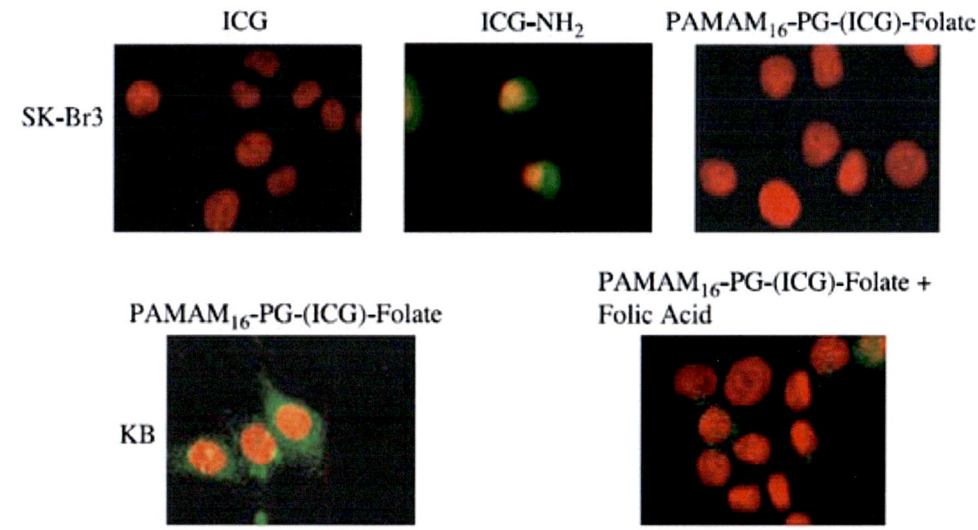

Figure 27. Branched poly(L-glutamic acid) as a biodegradable drug carrier.

4.3. Advances in Tissue Engineering

Sunil Shaunak et al [74] used anionic, polyamidoamine, generation 3.5 dendrimers to make novel water-soluble conjugates of D(+)-glucosamine and D(−)-glucosamine 6-sulfate with immuno-modulatory and antiangiogenic properties respectively (Figure 28).

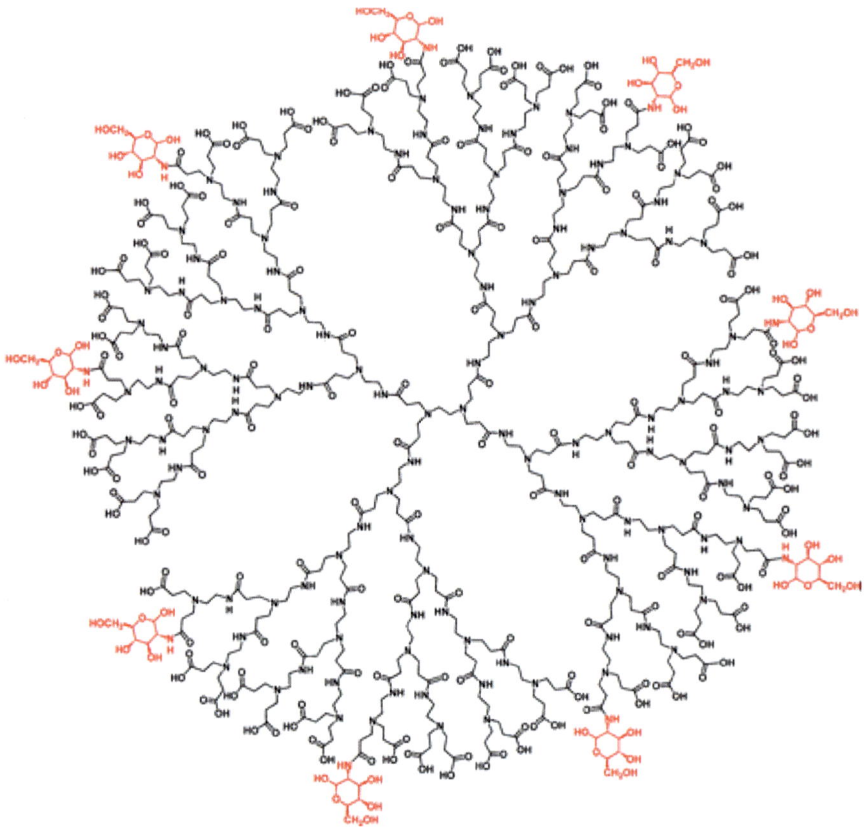

Figure 28. Polyvalent dendrimer glucosamine conjugates.

Dendrimer glucosamine inhibited Toll-like receptor 4–mediated lipopolysaccharide induced synthesis of pro-inflammatory chemokines (MIP-1α, MIP-1β, IL-8) and cytokines (TNF-α, IL-1β, IL-6) from human dendritic cells and macrophages but allowed upregulation of the costimulatory molecules CD25, CD80, CD83 and CD86. Sunil Shaunak et al concluded that synthetically engineered macromolecules such as the dendrimers described here can be tailored to have defined immuno-modulatory and antiangiogenic properties, and they can be used synergistically to prevent scar tissue formation.

Photo-crosslinkable dendritic macromolecules are attractive materials for the preparation of cartilage tissue engineering scaffolds that may be optimized for in situ formation of hydrated, mechanically stable, and wellintegrated hydrogel scaffolds supporting chondrocytes and chondrogenesis. Mark W. Grinstaff et al [75] reported biodendrimer-based hydrogel scaffolds for cartilage tissue repair. They designed and synthesized a novel hydrogel scaffold for cartilage repair, based on a multivalent and water-soluble tri-block copolymer consisting of a poly(ethylene glycol) core and methacrylated poly(glycerol succinic acid) dendrimer terminal blocks. The terminal methacrylates allow mild and biocompatible photo-crosslinking with a visible light, facilitating in vivo filling of irregularly shaped defects with the dendrimer-based scaffold. The multivalent dendrimer constituents allow high crosslink densities that inhibit swelling after crosslinking while simultaneously introducing biodegradation sites (Figure 29). The mechanical properties and water content of the hydrogel can easily be tuned by changing the biodendrimer concentration. In vitro chondrocyte encapsulation studies demonstrate significant synthesis of neocartilaginous material, containing proteoglycans and type II collagen.

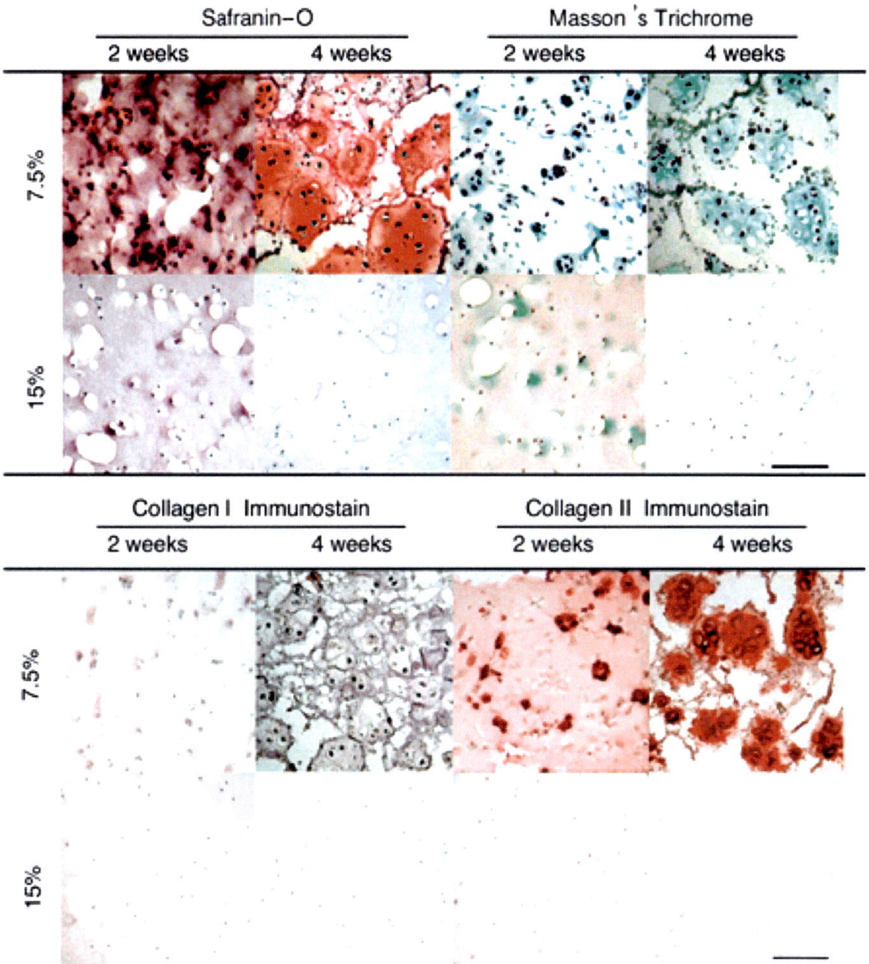

Figure 29. Biodendrimer-based hydrogel scaffolds for cartilage tissue repair.

4.4. Biological Evaluation of Dendrimer

The formation of particulate systems with well-defined sizes and shapes is of eminent interest in certain medical applications such as drug delivery, gene transfection, and imaging. The high level of control possible over the architectural design of dendrimers; their size, shape, branching length/density, and their surface functionality, clearly distinguishes these structures as unique and optimum carriers in those applications. The bioactive agents may be encapsulated into the interior of the dendrimers or chemically attached/physically adsorbed onto the dendrimer surface, with the option of tailoring the carrier to the specific needs of the active material and its therapeutic applications. In this regard, the high density of exo-presented surface groups allows attachment of targeting groups or functionality that may modify the solution behavior or toxicity of dendrimers. Quite remarkably, modified dendrimers have been shown to act as nano-drugs against tumors, bacteria, and viruses. Recent successes in simplifying and optimizing the synthesis of dendrimers such as the dlegoT and dclickT approaches, provide a large variety of structures while at the same time reducing the cost of their production. The reflections on biomedical applications of dendrimers given in this review clearly demonstrate the potential of this new fourth major class of polymer architecture and indeed substantiate the high hopes for the future of dendrimers [76].

The field of biomedical dendrimers is still in its infancy, but the explosion of interest in dendrimers and dendronised polymers as inherently active therapeutic agents, as vectors for targeted delivery of drugs, peptides and oligonucleotides, and as permeability enhancers able to promote oral and transdermal drug delivery makes it timely to review current knowledge regarding the toxicology of these dendrimer chemistries (currently under development for biomedical applications) [77]. Clinical experience with polymeric excipients, plasma expanders, and most recently the development of more dclassical polymerTderived therapeutics can be used to guide development of bsafeQ dendritic polymers. Moreover, in future it will only ever be possible to designate a dendrimer as bsafeQ when related to a specific application. The so far limited clinical experience using dendrimers make it impossible to designate any particular chemistry intrinsically bsafeQ or btoxicQ. Although there is widespread concern as to the safety of nano-sized particles, preclinical and clinical experience gained during the development of polymeric excipients, biomedical polymers and polymer therapeutics shows that judicious development of dendrimer chemistry for each specific application will ensure development of safe and important materials for biomedical and pharmaceutical use.

Dendrimers are branched, synthetic polymers with layered architectures that show promise in several biomedical applications. By regulating dendrimer synthesis, it is possible to precisely manipulate both their molecular weight and chemical composition, thereby allowing predictable tuning of their biocompatibility and pharmacokinetics. Advances in the understanding of the role of molecular weight and architecture on the in vivo behavior of dendrimers, together with recent progress in the design of biodegradable chemistries, has enabled the application of these branched polymers as anti-viral drugs, tissue repair scaffolds, targeted carriers of chemotherapeutics and optical oxygen sensors [78]. Before such products can reach the market, however, the field must not only address the cost of manufacture and quality control of pharmaceuticalgrade materials, but also assess the long-term human and environmental health consequences of dendrimer exposure in vivo.

Because of these well-defined characteristics, PAMAMs are finding utility in a variety of applications, many of which are biological in nature. Little is known, however, about the biological behavior of the PAMAMs, which is critical to their use in vivo. Jeanette C. Roberts et al [79] reported the preliminary biological evaluation of polyamidoamine (PAMAM) Starburst dendrimers. Generation 3, 5, and 7 PAMAMs were studied in V79 cells or in Swiss-Webster, mice for a number of biological properties, including (1) in vitro toxicity, (2) in vivo toxicity, (3) immunogenicity, and (4) biodistribution. Potential biological complications were observed only with G7 at the highest level tested. No evidence of immunogenicity was seen. The biodistribution properties of the Starburst dendrimers were rather unusual. G3 showed the highest accumulation in kidney tissue; G5 and G7 appeared to preferentially localize in the pancreas. In addition, G7 showed extremely high urinary excretion. In general, the dendrimers did not exhibit properties that would preclude their use in biological applications.

Muriel Blanzat et al [80] reported in vitro anti-HIV activity of phosphorus-containing dendrimers bearing Galβ1cer analogues. Two series of water-soluble dendritic catanionic assemblies, acting as multisite analogues of galactosylceramide (Galb1cer), have been prepared with the goal of blocking HIV infection prior to the entry of the virus into human cells. Trifunctional and hexafunctional cinnamic acid-terminated dendrimers have been synthesized from phosphorus-containing dendrimers bearing aldehyde end groups. A classical acid–base reaction performed in water between acid-terminated dendrimers and stoichiometric amounts of N-hexadecylamino-1-deoxylactitol provided the expected catanionic assemblies.

Koji Matsuoka et al [81] reported thiosialoside clusters using carbosilane dendrimer core scaffolds as a new class of influenza neuraminidase inhibitors. An efficient synthesis of a series of carbosilane dendrimers uniformly functionalized with α-thioglycoside of sialic acid was accomplished. The results of a preliminary study on biological responses against influenza virus sialidases using thiosialoside clusters showed that some of the glycodendrimers have inhibitory potencies against the sialidases.

Mitsuru Hashida et al [82] synthesized dendritic poly(L-lysine)s (DPKs), dendritic poly(L-ornithine)s (DPOs), which are constructed as novel amino acid dendrimers, and PEGylated KG6 (the sixth generation of DPKs), and evaluated the physicochemical properties and biodistribution characteristics of these dendrimers. After intravenous administration to mice, all tested dendrimers cleared rapidly from blood flow and mainly accumulated in the liver and kidney. The hepatic and renal accumulation changed in a generationdependent manner.

Costas Demetzos et al [83] reported the doxorubicin–PAMAM dendrimer complex attached to liposomes and the cytotoxic studies against human cancer cell lines. The low release rate of doxorubicin as well as the high incorporation efficiency of doxorubicin–PAMAM complex into liposomes are considered as beneficial factors concerning the activity of doxorubicin. Liposomal formulations composed of lipids and of a drug–dendrimer complex could be characterized as modulatory liposomal controlled release system, and could provide benefits to the delivery of drugs and modulate their release.

Natarajan Raju et al [84] reported the synthesis of a new scaffold derived from iminodipropionic acid for the preparation of peptide dimers and tetramers.

Toward the production of unimolecular nanocapsules with biocompatible surface, Kenji Kono et al [85] prepared a fourth generation polyamidoamine dendrimer having both a poly(ethylene glycol) (PEG) with the number average molecular weight of 2000 and a methacryloyl group at every chain end of the dendrimer through an L-lysine residue. Polymerization of the methacryloyl groups resulted in a significant reduction of the affinity of the dendrimer to a guest molecule, rose bengal, suggesting that the peripheral network formed by the linking of methacryloyl groups hides the dendrimer interior. In addition, when the methacryloyl groups of the dendrimer were polymerized in the presence of rose bengal, the guest molecule was found to be tightly associated with the dendrimer.

4.5. Pharmacokinetics of Dendrimers

Christopher J. H. Porter et al [86] reported the pharmacokinetics, biodistribution, and evidence for metabolism and bioresorption after intravenous administration to rats of cationic poly-L-lysine dendrimers. Cationic poly-L-lysine 3H-dendrimers with either 16 or 32 surface amine groups have been synthesized and their pharmacokinetics and biodistribution investigated after intravenous administration to rats. Plasma and whole blood pharmacokinetics were similar, precluding interaction with blood components as a causative factor in either the rapid removal or reappearance of radioactivity in plasma. For the L-lysine capped dendrimers, radiolabel was widely distributed throughout the major organs, with no apparent selectivity for organs of the reticuloendothelial system. In contrast, a greater proportion of the administered radiolabel was recovered in the organs of the reticuloendothelial system for the D-lysine capped system, as might be expected for a nondegrading circulating foreign colloid. This is the first data to demonstrate the biodegradation/bioresorption of poly-L-lysine dendrimers and has significant implications for the utility of these systems as drugs or drug delivery systems.

Single-triggered disassemble dendrimers were recently developed and introduced as a potential platform for a multi-prodrug. These unique structural dendrimers can release all of their tail units through a self-immolative chain fragmentation initiated by a single cleavage at the dendrimer's core. There are several examples for the bioactivation of first-generation self-immolative dendritic prodrugs. However, enzymatic activation failed for second-generation self-immolative dendrimers. The hydrophobic large molecular structure of the dendritic prodrugs results in aggregation under aqueous conditions and prevented the enzyme from reaching the triggering substrate. Doron Shabat et al [87] showed a simple solution for the enzymatic activation of second-generation self-immolative dendrimers. Poly(ethylene glycol) (PEG) was conjugated to the dendritic platform via click chemistry (Figure 30). The poly(ethylene glycol) tails significantly decreased the hydrophobic properties of the dendrimers and thereby prevented aggregate formation. Doron Shabat et al designed and synthesized a dendritic prodrug with four molecules of the anticancer agent camptothecin and a trigger that can be activated by penicillin-G-amidase. The PEG5000-conjugated, self-immolative dendritic prodrug was effectively activated by penicillin-G-amidase under physiological conditions and free camptothecin was released to the reaction media. Cell-growth inhibition assays demonstrated increased toxicity of the dendritic prodrug upon incubation with the enzyme.

Ijeoma F. Uchegbu et al [88] reported preferential liver gene expression with polypropylenimine dendrimers. Quaternisation of dendrimer proved to be critical in improving DNA binding, as evidenced by data from the ethidium bromide exclusion assay and dendrimer–DNA colloidal stability data. This improved colloidal stability had a major effect on vector tolerability, as dendrimer–DNA formulations were well tolerated on intravenous injection while a similar dendrimer–DNA dose was lethally toxic by the same route. Ijeoma F. Uchegbu et al concluded that the polypropylenimine dendrimers are promising gene delivery systems which may be used to target the liver and avoid the lung and also that molecular modifications conferring colloidal stability on gene delivery formulations have a profound effect on their tolerability on intravenous administration.

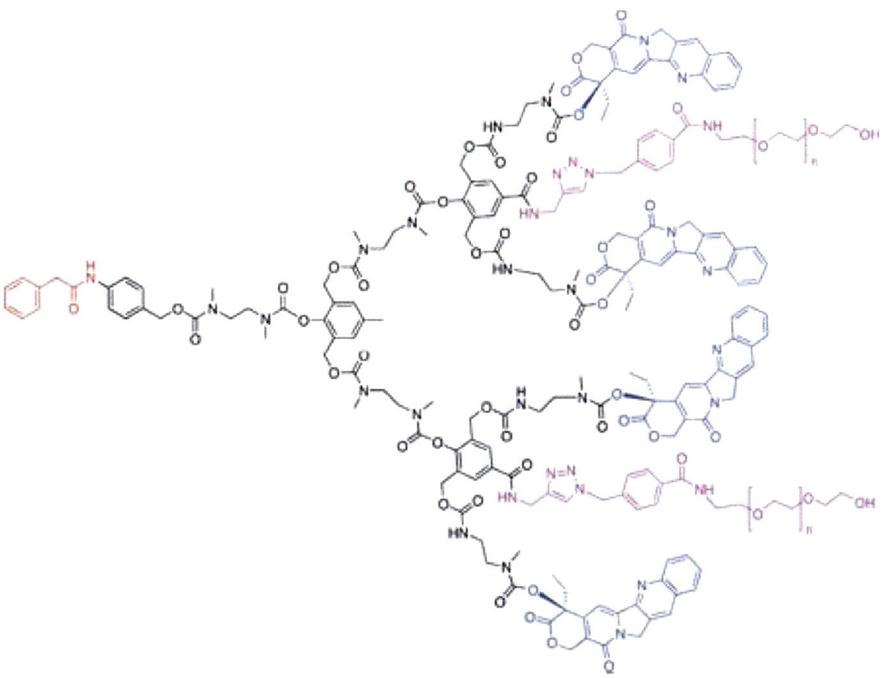

Figure 30. Conjugation of self-immolative dendrimers with poly(ethylene glycol) via click chemistry.

Antony D'Emanuele et al [89] reported the use of a dendrimer-propranolol prodrug to bypass efflux transporters and enhance oral bioavailability. The aim of the study was to determine the effects on the transport of propranolol across monolayers of the human colon adenocarcinoma cell line, Caco-2, of forming a prodrug by conjugating to generation 3 (G3) and lauroyl-G3 PAMAM dendrimers. The findings show that the A→B Papp of propranolol conjugates was reduced in the presence of the endocytosis inhibitor colchicines and was lower at 4 °C than at 37°C, suggesting that the enhancement mechanism involves endocytosis-mediated transepithelial transport. The A→B Papp of conjugated propranolol was not altered in the presence of the P-gp inhibitor cyclosporin A suggesting that conjugation of drug to dendrimer allows the bypassing of the efflux transporter. The results suggest that dendrimer-drug prodrugs may be used to increase drug solubility and bypass drug efflux transporters, therefore increasing drug bioavailability.

4.6. Peptides Dendrimers

The main stress is given to peptide dendrimers, namely, multiple antigenic peptides (MAPs). Their synthesis, physicochemical properties, biological activities, etc. have been described with many examples. MAPs can be used as diagnostics, mimetics, for complexation of different cations, as vaccines against parasites, bacteria, viruses, etc [90].

PEGylated polyamidoamine (PAMAM) dendrimers as drug carriers have been a topic of interest because of their biomedically favorable features, including minimal toxicity, reduced immunogenicity, and excellent solubility in aqueous and most organic solutions. A PEG shell on dendrimer surface may provide steric hindrance, known as stealth properties of PEG, to stabilize drug molecules to be delivered. Hu Yang et al [91] reported in vitro enzymatic stability of dendritic peptides. The effects of PEG and coupling sequence of drug, PEG, and dendrimer in modulating the stability of delivered drug molecules were evaluated. Nsuccinyl-Ala-Ala-Pro-Phe-p-nitroanilide was chosen as a model peptide. Dendritic peptides, that is, peptide-dendrimer, peptide-PAMAM-PEG, and peptide-PEG-dendrimer, were constructed based on Starburst G3.0 PAMAM dendrimer. The enzymatic stability of dendritic peptides was peptide-PAMAM-PEG > peptide- PAMAM > free peptide > peptide-PEG-PAMAM. The ratio of PEG/peptide could be reduced for increasing peptide loading while maintaining the delivered peptides' relatively high enzymatic stability.

Jean-Louis Reymond et al [92] reported peptide dendrimers as artificial enzymes, receptors, and drug-delivery agents. The dendritic architecture applied to peptides provides a practical entry into globular macromolecules resembling proteins (Figure 31).

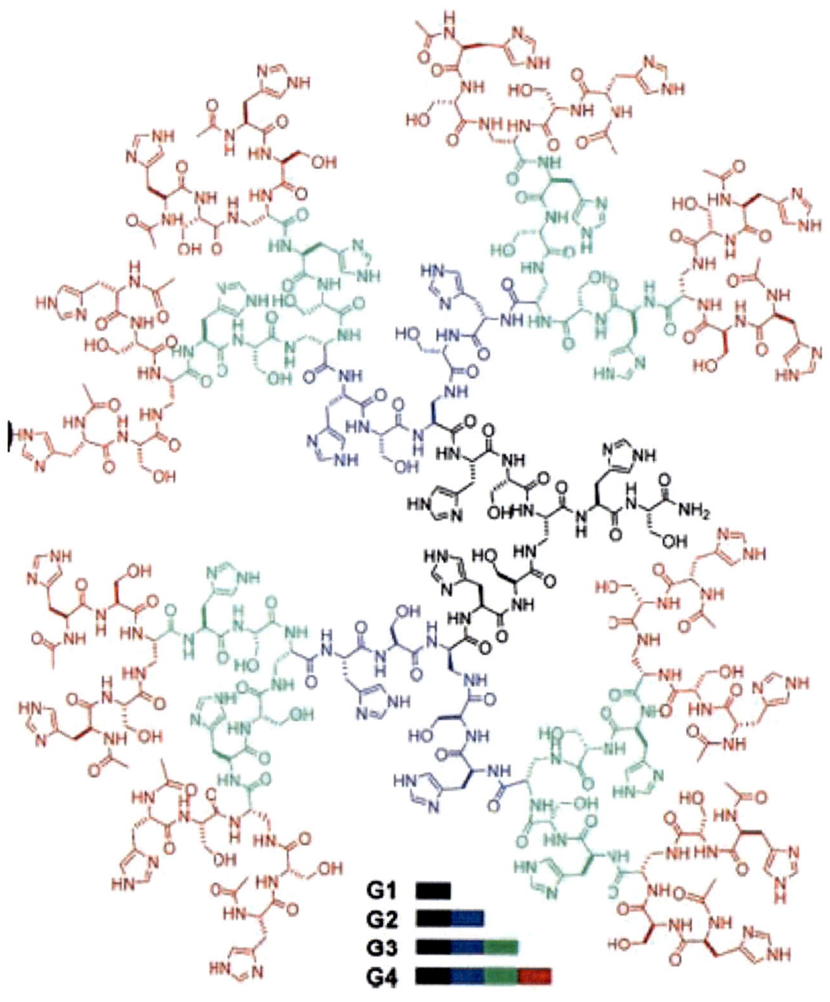

Figure 31. Peptide dendrimers G1-G4 with iterations of a (His-Ser)$_2$Dap dendron.

A modular design was chosen using a divergent synthesis on solid support alternating proteinogenic α-amino acids with branching diamino acids, producing peptide dendrimers with a molecular weight of 3-5 kDa. Initial studies focused on models for hydrolases and produced esterase peptide dendrimers featuring histidine as the key catalytic residue. Variations of amino acid composition and the branching diamino acid led to enantioselective catalysts. Rate accelerations of kcat/kuncat = 90 000 were obtained when the design was changed to monomeric peptide dendrimers alternating two amino acids with the branching unit. A combinatorial approach was developed allowing for the preparation of large libraries (>60 000 members), which were screened for B12 binding and catalytic activity. Glycopeptide dendrimers conjugated to colchicines selectively inhibited the proliferation of targeted cells, whereas colchicine alone displayed high toxicity.

M. Bryszewska et al [93] reported the biological properties of low molecular mass peptide dendrimers. A series of new, low molecular mass, lysine-based peptide dendrimers with varying distribution of cationic and aromatic groups in the structure were synthesized. They expressed antimicrobial activity against Gram-positive (*Staphylococcus aureus*) and Gram-negative (*Escherichia coli*) bacteria as well as against fungal pathogens (*Candida albicans*). Their cytotoxic, haematotoxic, and genotoxic effects were studied. It appears that degree of branching and steric distribution and types of hydrophobic (aromatic) groups and cationic centres are important components of dendrimeric structure and influence both antimicrobial potency and toxicity. Such 3D structure of our dendrimers mimics that of the natural antimicrobial peptides and can be achieved by application of dendrimer chemistry.

5. Advances of Metal Dendrimers

5.1. Nano Dendritic Catalysis

Interest in catalysis by metal nanoparticles (NPs)is increasing dramatically, as reflected by the large number of publications in the last five years. This field, "semi-heterogeneous catalysis", is at the frontier between homogeneous and heterogeneous catalysis, and progress has been made in the efficiency and selectivity of reactions and recovery and recyclability of the catalytic materials [94]. Usually NP catalysts are prepared from a metal salt, a reducing agent, and a stabilizer and are supported on an oxide, charcoal, or a zeolite. Besides the polymers and oxides that used to be employed as standard, innovative stabilizers, media, and supports have appeared, such as dendrimers, specific ligands, ionic liquids, surfactants, membranes, carbon nanotubes, and a variety of oxides. Ligand-free procedures have provided remarkable results with extremely low metal loading. The Review presents the recent developments and the use of NP catalysis in organic synthesis, for example, in hydrogenation and C-C coupling reactions, and the heterogeneous oxidation of CO on gold NPs.

Dirk E. De Vos et al [95] reported the synthesis and application of porphyrin-functionalized dendrimers for the application in recyclable photocatalysts in a nanofiltration membrane reactor. The convergent synthesis of a series of porphyrin-functionalized pyrimidine dendrimers has been accomplished by a procedure involving the nucleophilic aromatic substitution (NAS) as a key reaction step. The resulting dendritic porphyrin catalysts show high activity in the light-induced generation of singlet oxygen (1O2) from ground-state oxygen. These materials are synthetically useful photosensitizers for the oxidation of various olefinic compounds to the corresponding allylic hydroperoxides.

Oriol Rossell et al [96] reported carbosilane dendrimers peripherally functionalized with P-stereogenic monophosphines and the catalytic behavior of their allylpalladium complexes in the asymmetric hydrovinylation of styrene. The first dendrimers containing P-stereogenic monophosphines as peripheral groups were obtained by reacting the protected BH3-LiCH2PPhR (R = 2-biphenylyl, 9-phenanthryl) with zeroth- and first-generation chlorocarbosilane dendrimers, followed by treatment with morpholine and subsequent chromatographic purification (Figure 32). The reaction of the chiral dendrimers with the dinuclear allyl complex [Pd(μ-Cl)(η3-2-MeC3H4)]2 allowed us to graft PdCl- (η3-2-MeC3H4) units on their surface. The catalytic properties of the new palladodendrimers were tested in the asymmetric hydrovinylation of styrene. The results in terms of activity, selectivity, and enantiomeric excess depend strongly on the nature of the phosphine and the halide abstractor.

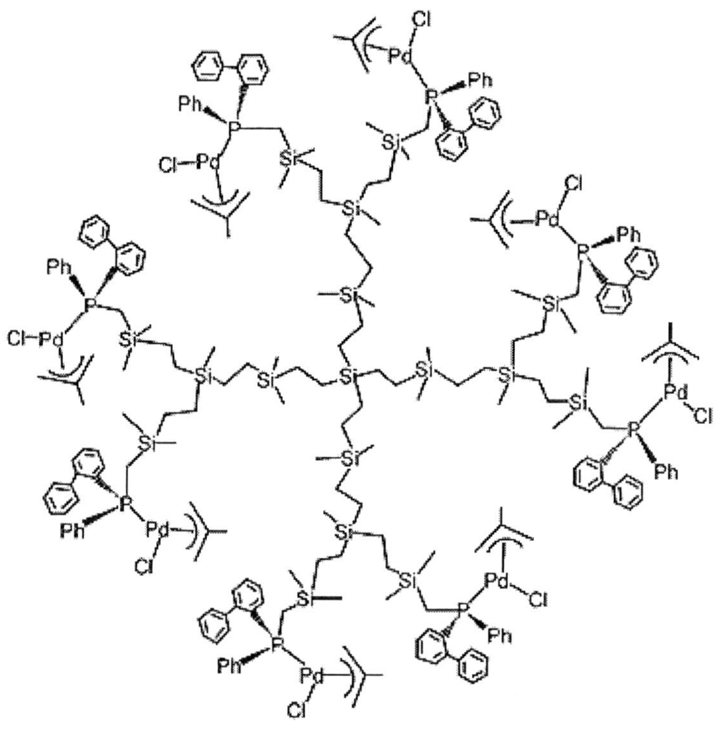

Figure 32. Carbosilane dendrimers peripherally functionalized with P-stereogenic monophosphines.

5.2. Metal Nanoparticles Preparation Using Dendrimer as Molecular Template

Dong-Mei Xu et al [97] reported the preparation of Ag nanoparticles in the presence of low generational poly(ester-amine) dendrimers. Stable Ag nanoparticles of 10–20 nm were prepared by reduction of AgNO3 with NaBH4 in water solution in the presence of low generational hydroxylterminated poly(ester-amine) dendrimer G1.0 (OH)16 and amino-terminated poly(ester-amine) dendrimer G1.5 (NH2)8 by optimizing preparation conditions. The results showed that the size of the Ag particles increased with Ag+/dendrimer molar ratio, and the size of Ag nanoparticles in Ag/G1.0 (OH)16 system was larger than that of Ag nanoparticles in Ag/G1.5 (NH2)8 system, while the polydispersities of two systems were similar.

Toyoko Imae et al [98] reported size-controlled in situ synthesis of metal nanoparticles on dendrimer-modified carbon nanotubes. Multiwalled carbon nanotubes (MWCNTs) were covalently modified with fourth-generation NH2-terminated poly(amido amine) (G4-PAMAM) dendrimers, named CNT/DEN. By simply tuning the metal ion-to-dendrimer terminal amine ratio, Au nanoparticles (NPs) with different sizes from 3 to 11 nm were in situ synthesized and deposited on the surface of CNT/DEN, as confirmed by a transmission electron microscopy (TEM) and an ultra violet-visible (UV-vis) absorption spectrometry. Furthermore, many metal NPs other than Au NPs, such as Ag, Cu, and Pt NPs, as well as binary NPs of Ag@Au, were synthesized in situ on CNT/DEN, yielding the water-dispersible and stable CNT/DEN/NPs (Figure 33).

M. P. Srinivasan et al [99] reported covalent molecular assembly in a supercritical medium and the formation of nanoparticles encapsulated in immobilized dendrimers. The work demonstrates the feasibility of forming metallic nanoparticles within the confines of dendrimerladen ultrathin films that are immobilized on a solid surface. The functional property of the poly amido amine dendrimer (PAMAM) as a sequestering agent for Cu nanoparticles was exploited by the introduction of copper acetylacetonate (Cu(acac)2) as a precursor. The nanoparticle precursors were introduced into the matrix using supercritical carbon dioxide and liquid tetrahydrofuran (THF) as processing media. The precursors were subsequently reduced to form the nanoparticles. Higher yield, a denser distribution of nanoparticles, and greater stability toward polar solvent attack were observed when the structures were assembled in supercritical CO2 than when THF was employed. TEM images revealed that the embedded nanoparticles are fairly tightly distributed in terms of size with an average diameter of 7 nm.

Richard M. Crooks et al [100] reported the synthesis, characterization, and magnetic properties of dendrimer-encapsulated nickel nanoparticles containing <150 atoms. Ni nanoparticles containing <150 atoms are prepared in an organic solvent using hydrophobic dendrimers as templates, which results in a high degree of size monodispersity. Magnetic (SQUID) studies indicate that the particles are ferromagnetic at 5 K with magnetic saturations that vary depending upon the particle size.

Hee-Tae Jung et al [101] prepared gold nanostructures from a taper-shaped supramolecular material (1-COOH) containing a carboxylic acid (-COOH) core, rigid biphenylene group, and three perfluorinated (-CF3) tails. The carboxylic core of the supramolecule serves as the site for reaction with the metal ions. Hee-Tae Jung et al showed that the self-assembling behavior of the material is strongly affected by substituting the carboxylic core of the supramolecule with a Na+ ion, followed by ion transfer of AuIII and subsequent reduction of AuIII with UV irradiation. This order and spacing are preserved during reduction of the metal salt with UV irradiation, although the intensities of the peaks increase. The size and shape of gold nanocrystals are controllable depending on UV irradiation time. These types of perfluorinated supramolecular templates have enhanced mesophase stability on account of the existence of strong intermolecular interactions, although many of these molecules are not traditional amphiphiles.

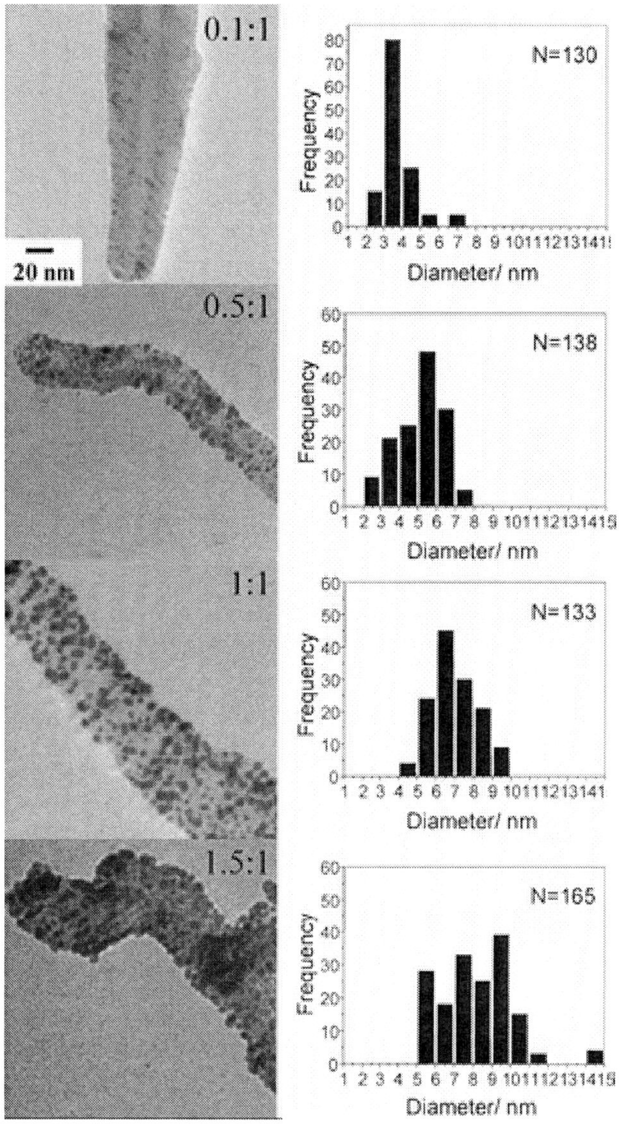

Figure 33. Size-controlled in situ synthesis of metal nanoparticles on dendrimer-modified carbon nanotubes.

Richard M. Crooks et al [102] reported the synthesis, characterization, and extraction of Au dendrimer-encapsulated nanoparticles (DENs) prepared in organic solvents. DENs composed of 31 and 55 Au atoms were prepared using organic solvents and poly(amidoamine) (PAMAM) dendrimer templates modified on their periphery with dodecyl groups. The spectral and microscopic properties of the resulting materials were identical to those prepared using water-soluble PAMAM dendrimers. It was possible to extract the organic-soluble DENs into water using the water-soluble thiols tiopronin and glutathione. The properties of the resulting monolayer-protected clusters were nearly identical to those of the precursor DENs. The synthetic methodology reported here provides a convenient method for preparing DENs of non noble metals such as Ni and Fe.

5.3. Complexation between Dendrimer and Metal Ions/Nanoclusters

Kimihisa Yamamoto et al [103] showed that tin ions, Sn2+, complex to the imine groups of a spherical polyphenylazomethine dendrimer in a stepwise fashion. This behaviour reflects a gradient in the electron density associated with the imine groups, with complexation in a more peripheral generation proceeding only after complexation in generations closer to the core has been completed. By attaching an electron-withdrawing group to the dendrimer core, Kimihisa Yamamoto et al are able to change the complexation pattern, so that the core imines are complexed last. By further extending this strategy, it should be possible to control the number and location of metal ions incorporated into dendrimer structures, which might find uses as tailored catalysts or building blocks for advanced materials (Figure 34).

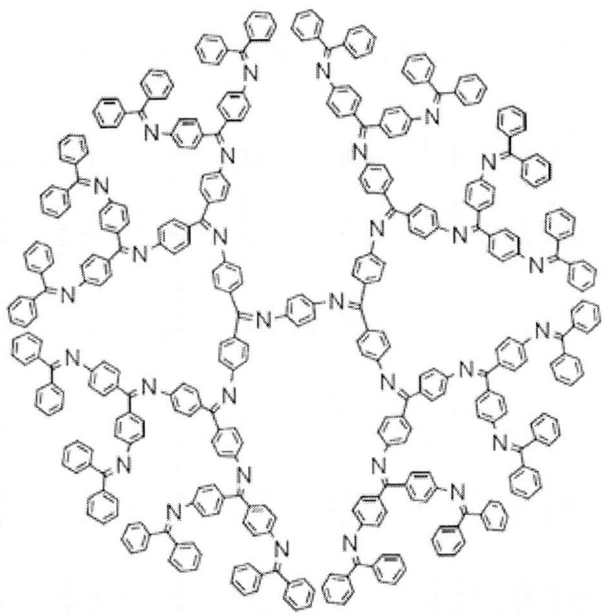

Figure 34. Chemical structures of the dendrimers and oligomers investigated by complexation with $SnCl_2$.

Jian Weng et al [104] synthesized a series of cyclens substituted with mixed stilbene and poly(ethylene glycol) dendritic arms. All dendrimers terminated with different peripheral groups had good solubility in common organic solvents, and dendrimers terminated with -CO2H groups (CO2H-dendrimers) were also soluble in alkaline solutions. The nickel coordination properties of these dendrimers were investigated in organic solvents. Dendrimers terminated with -CN groups (CN-dendrimers) and the second-generation CO2H-dendrimer [(CO2H)8L2] could produce pentacoordinated nickel complexes; the third-generation CO2H-dendrimer [(CO2H)16L3] could form tetra- and pentacoordinated nickel complexes, and the nickel complex of the fourth-generation CO2H-dendrimer [(CO2H)32L4] could not be obtained. The CN-dendrimers could not form hydrogen bonds on the surface, and the first- and secondgeneration CO2H-dendrimers could not form intramolecular hydrogen-bond networks.

Therefore, they had no selectivity for positive nickel ions and negative chloride ions. (CO2H)32L4 could not produce a nickel complex because it had a crammed backbone structure that could not penetrate nickel and chloride ions. Therefore, it was possible to control the ion access of cations and anions with the hydrogen-bond network of (CO2H)16L3.

Perla B. Balbuena et al [105] reported dendrimer-tetrachloroplatinate precursor interactions. Density functional theory is used to investigate the structure and energetics of the tetrachloroplatinate anion and its hydrolysis products at several degrees of hydration, as well as those of outer dendrimer pockets hosting such species.

Chunhui Huang et al [106] reported the synthesis and photopysical properties of polyaryl ether dendrimer with a 4-phenylacetyl-5-pyrazolone-based Terbium(III) complex as core. A series of novel dendritic β-diketone ligands, 1-phenyl-3-[G-*n*]-4-phenylacetyl-5-pyrazolone (*n* = 0-3, G stands for polyaryl ether), were synthesized by introducing Fréchet-type dendritic branches (Figure 35). These dendritic complexes, prepared from aqueous solution, exhibit high stability. Interestingly, the study of photophysical properties shows that the luminescence quantum yields of the dendritic Tb-complexes increase from 0.1 to 2.26% with an increase of the dendritic generation from 0 to 3. Importantly, an "energy-reservoir effect" was observed in the dendritic system using the method based on the resonance energy transfer from these complexes to rhodamine 6G. With the increase of the dendritic generation, the metal-centered luminescence quantum yield was almost the same, and the energy transfer (-transfer) from the ligand to Tb3+ increased.

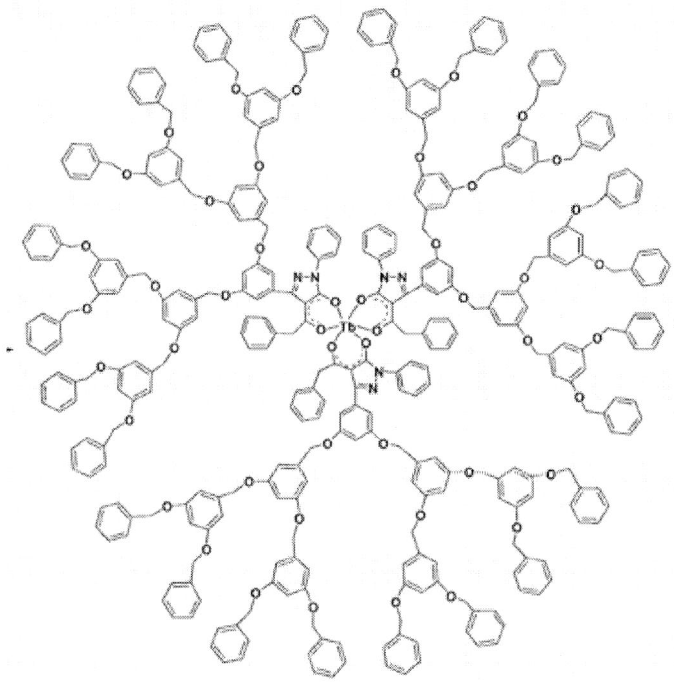

Figure 35. Polyaryl ether dendrimer with a 4-phenylacetyl-5-pyrazolone-based Terbium(III) complex as core.

Further measurements of the triplet state and oxygen quenching of these dendritic complexes verify that this enhancement of the energy transfer (-transfer) is attributed to both an "antenna effect" and a "shell effect".

Robert C. Hider et al [107] reported the iron binding dendrimers for the treatment of haemochromatosis. A range of iron binding dendrimers terminated with hexadentate ligands formed from hydroxypyridinone, hydroxypyranone, and catechol moieties have been synthesized in order to investigate their potential as clinically useful iron(III)-selective chelators capable of removing dietary iron from the gastrointestinal tract and preventing the development of iron overload typical of haemochromatosis and thalassaemia intermedia. Hydroxypyridinone-terminated dendrimers were found to possess a high affinity and selectivity for iron(III). A hydroxypyridinone-based dendrimer was demonstrated to be highly efficient at reducing the absorption of iron(III) in rat intestine. This family of dendrimers may find an application in the treatment of iron overload.

Fritz Vögtle et al [108] reported cyclam cored luminescent dendrimers as ligands for Co(II), Ni(II) and Cu(II) ions. Fritz Vögtle et al have investigated the photophysical properties of two dendrimers containing a cyclam core decorated with 4 naphthyl units (G0), 12 dimethoxybenzene and 16 naphthyl units (G2). These dendrimers show fluorescence bands that can be assigned to naphthyl localized excited states (λ_{max} = 337 nm), naphthyl-amine exciplexes (λ_{max} = 470 nm) and, for G2, naphthyl excimers (λ_{max} ca. 400 nm). Cyclam is a very good ligand for transition metal ions and we have investigated complex formation between these dendritic ligands and Ni(II), Co(II) and Cu(II), added as nitrate salts (Figure 36).

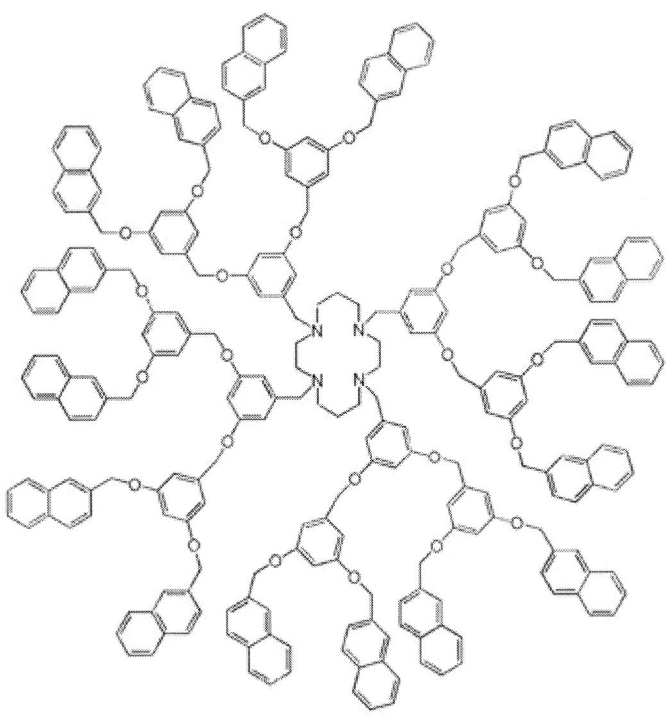

Figure 36. Cyclam cored luminescent dendrimers as ligands for Co(II), Ni(II) and Cu(II) ions.

This process can be monitored by the strong changes, both in shape and intensity, observed in the emission spectra of these dendrimers. Complexation with Cu(II) causes not only changes in the relative intensities of the fluorescence bands, but also the appearance of a new absorption band in the near UV spectral region. An analysis of the titration curves has allowed us to obtain clear evidence for the formation of not only 1:1 species, but also 1:2 metal to ligand species. G2 shows a clear preference, compared to G0, in forming complexes with a 1:2 metal-to-ligand stoichiometry, although it possesses very bulky dendrons appended to the cyclam central unit.

Wai-Yeung Wong et al [109] reported triphenylamine-dendronized pure red Iridium Phosphors with superior OLED efficiency/color purity trade-offs (Figure 37). Herbert Schumann et al [110] reported convergent synthesis and characterization of organotin dendrimers Sn{(CH2)nSn[(CH2)4SnPh3]3}4 (n=3,4). The reaction of the ω-haloalkyltin trihalides Br(CH2)3SnBr3 and Br(CH2)4SnBr3 with 3 equiv of but-3-enylmagnesium bromide yielded Br(CH2)3Sn(CH2CH2CH=CH2)3 and Br(CH2)4Sn(CH2CH2- CH=CH2)3. Both dendritic branches can be converted into their corresponding Grignard reagents, whose consequent treatment with 0.25 M amounts of SnCl4 resulted in the formation of the dendrimers Sn[(CH2)3Sn(CH2CH2CH=CH2)3]4 and Sn[(CH2)4Sn(CH2CH2CH=CH2)3]4, respectively.

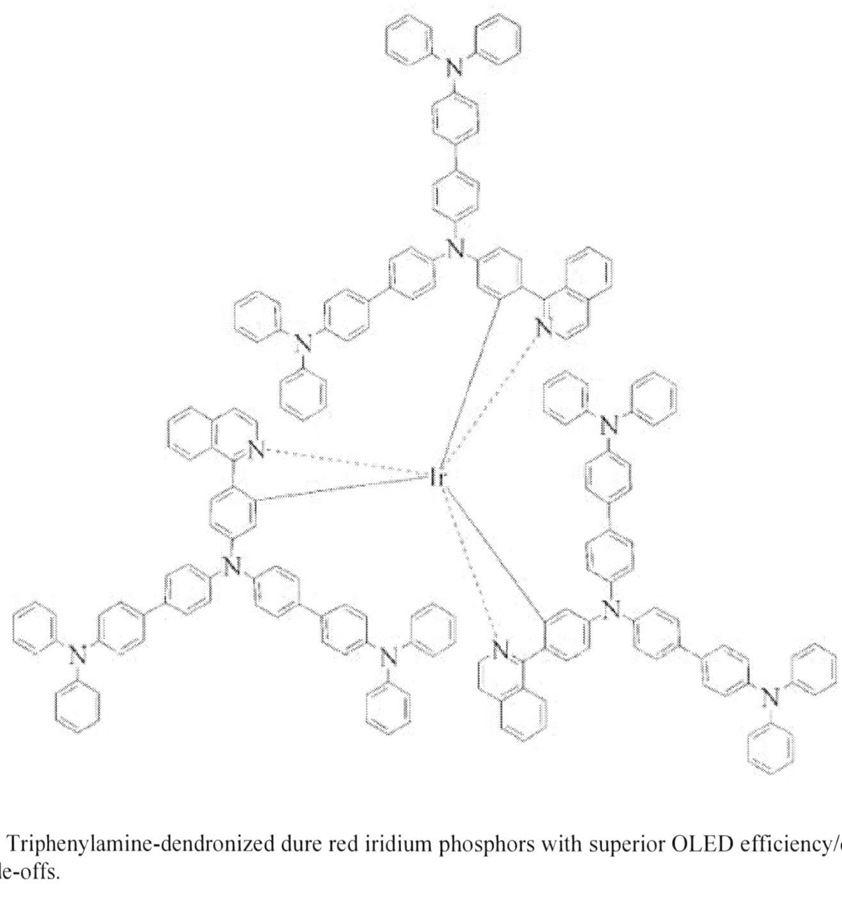

Figure 37. Triphenylamine-dendronized dure red iridium phosphors with superior OLED efficiency/color purity trade-offs.

Bradley D. Fahlman et al [111] reported facile synthesis of tin oxide nanoparticles stabilized by dendritic polymers. Kimihisa Yamamoto et al [112] reported the investigation of a molecular morphology effect on polyphenylazomethine dendrimers and their physical properties and metal-assembling processes. A series of novel dendritic polyphenylazomethines (DPA) with asymmetric morphologies was synthesized. Bis- and tris-substituted DPAs of the fourth generation also act as frameworks for stepwise assembly of a metal component (SnCl2), like the fully substituted symmetric DPA. The experimental results suggest a bent conformation of the dendrimer in which the core moiety is folded into the dendron structure. The structures of bis- and tris-substituted dendrimers which accurately follow the stepwise process are fixed. These observations provide a new insight into the finely controlled metal-assembly chemistry of dendritic macromolecules, and a rigid and fixed conformation is one of important factors for their unique properties.

Oriol Rossell et al [113] reported dendrimers containing two metallic layers, and the chloride migration from peripheral gold, palladium, or rhodium metals to internal ruthenium atoms (Figure 38). An efficient strategy for the synthesis of carbosilane dendrimers containing double metallic layers has been developed. The method consists of the attachment of the Ru(*p*-cymene)Cl2 unit to the branches of a phosphino-terminated dendrimer, followed by abstraction with AgOTf of the chloride ligands from the ruthenium and selective incorporation, through the nitrogen pyridine atom, of the bifunctional 4-pyPPh2 molecule.

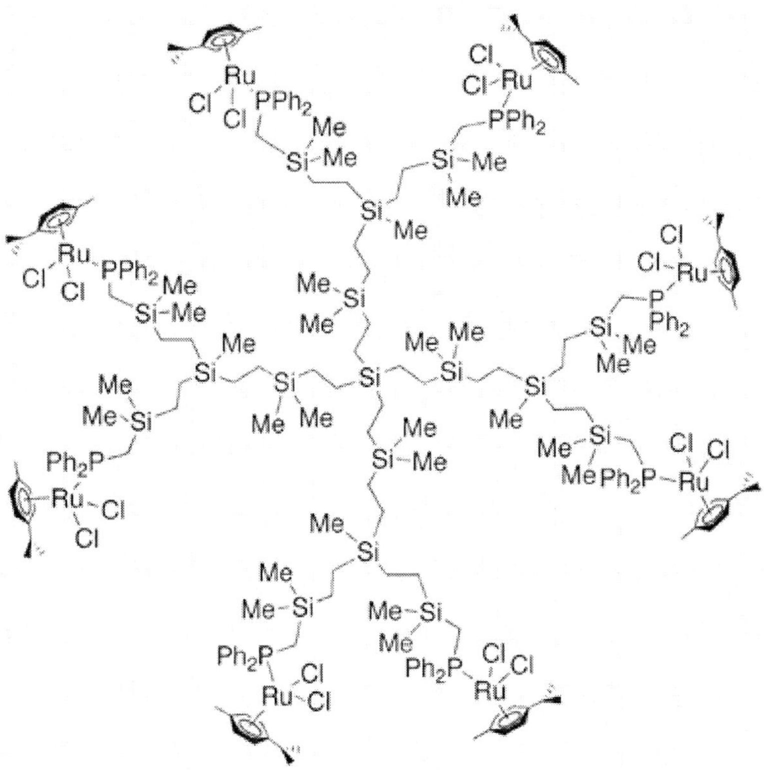

Figure 38. Dendrimers Containing Two Metallic Layers.

Finally, metalation of the resulting species with the complexes [(tht)AuCl], [PdCl(η3-2-MeC3H4)]2, and [RhCl(cod)]2 renders the targeted Ru/Au, Ru/Pd, and Ru/Rh dendrimers. An unexpected migration of the chloride ligand bonded to the peripheral metals to the internal ruthenium atoms was observed in terms of the results obtained from density functional calculations. In the case of the Ru/Au dendrimers, the different abilities for the ligand transfer process shown by the dendrimers in comparison with a mononuclear Ru/Au model compound are explained in terms of dendritic effects.

Novel Au-Pt bimetallic flower nanostructures fabricated on a polyamidoamine dendrimers-modified surface by electrodeposition are reported by Xiurong Yang et al [114]. These polyamidoamine dendrimers were stable, and they assisted the formation of Au-Pt bimetallic nanoflowers during the electrodeposition process. FE-SEM images showed that the bimetallic nanoflower included two parts: the "light" and the "pale" part. The two parts consisted of many small bimetallic nanoparticles, which was attributed to the progressive nucleation process. Moreover, the "light" part contained more bimetallic nanoparticles (Figure 39). The morphologies of bimetallic nanoflowers depended on the electrodeposition time and potential and the layer number of assembled dendrimers. The average size of nanoflowers increased with the increase in electrodeposition time.

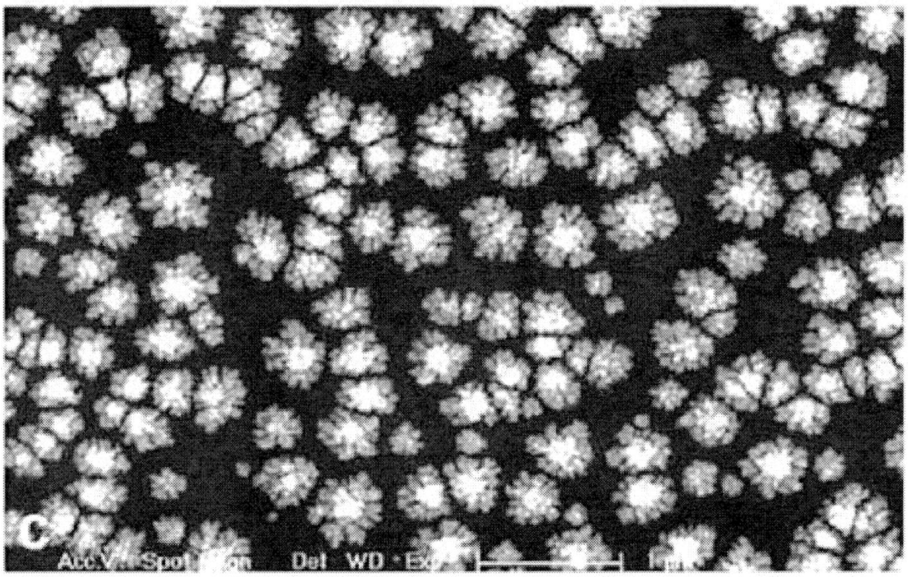

Figure 39. Polyamidoamine dendrimers-assisted electrodeposition of gold-platinum bimetallic nanoflowers.

Mark G. Humphrey et al [115] reported the electrochemical, spectroelectrochemical, and molecular quadratic and cubic nonlinear optical properties of alkynylruthenium dendrimers (Figure 40). Mo6X8i Nanocluster cores (X = Br, I): From inorganic solid state compounds to hybrids. The [(M6L8i)L6a] units (a = apical, i = inner) are the basic building blocks in the molybdenum, tungsten and rhenium octahedral cluster chemistry. Nano-sized metallic clusters are easily obtained by solid-state synthesis with transition elements associated with halogens or chalcogens (L). The intrinsic properties of M6 cluster units – one electron reversible redox process, magnetism and luminescence – depend on the nature of the metal and ligands. The solubilization of M6 solid state compounds provides [(M6L8i)L6a]n–

building blocks with individual properties that can be further used for the design of hybrid organic/inorganic materials [116].

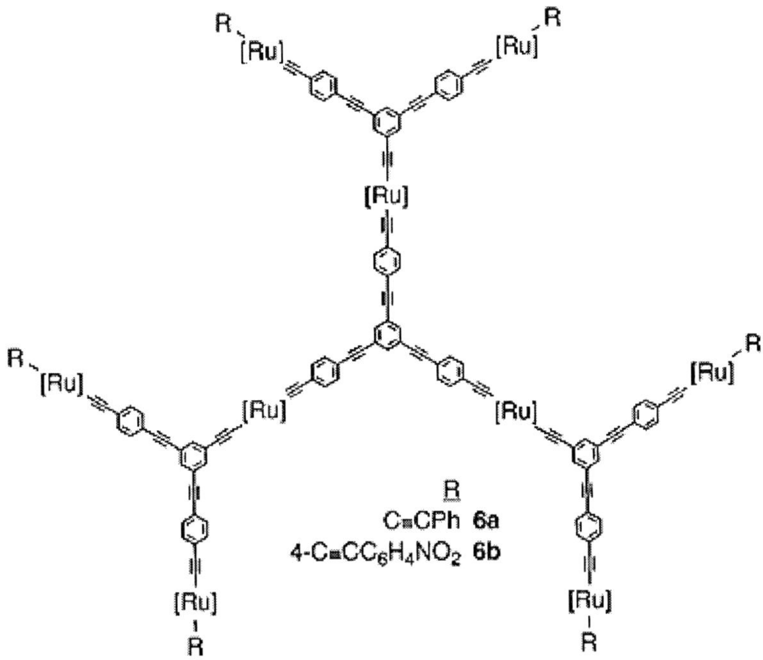

Figure 40. Alkynylruthenium dendrimers.

From Simple Monopyridine Clusters [Mo6Br13(Py-R)][*n*-Bu4N] and Hexapyridine Clusters [Mo6X8(Py-R)6][OSO2CF3]4 (X = Br or I) to Cluster-Cored Organometallic Stars, Dendrons, and Dendrimers were reported by Didier Astruc et al [117]. Hexasubstitution of apical triflate ligands in the octahedral clusters [M]2[Mo6X8(CF3SO3)6] (M = *n*-Bu4N or Cs, X = Br or I) and monosubstitution in [*n*-Bu4N]2[Mo6Br13(CF3SO3)] was carried out in tetrahydrofuran at 60 °C with simple pyridines and then extended to organometallic pyridines, yielding cluster-cored stars, and to dendronic polyallyland polyferrocenylpyridines, yielding cluster-cored polyallyl and polyferrocenyl dendrimers and dendrons (Figure 41).

Christiane Perrin et al [118] reported Mo6Br8-Cluster-cored organometallic stars and dendrimers. The octahedral molybdenum cluster [*n*-Bu4N]2[Mo6Br8(CF3SO3)6] undergoes substitution of all six terminal triflate ligands with the organometallic pyridine ligands [RuCp(PPh3)2(η1-C2-4-pyridinyl)] and 1-ferrocenyl-2-(4-pyridinyl) acetylene to give the new light and air sensitive hexa-functionalized Mo6 clusters, and with the dendronic phenolate ligand *p*-NaO–C6H4C{CH2CH2CH2Si(Me)2Fc}3 to give the air-sensitive Mo6-cluster-cored octadecylferrocenyl dendrimer that discloses a single CV wave in CH2Cl2 and recognizes the biologically important adenosyl triphosphate di-anion (ATP2–). The organometallic pyridines were also coordinated to Ag+ to give the new trinuclear AgRu2 and AgFc2 cationic complexes for comparison of the structures and electronic delocalization with those of the clusters.

Multiporphyrin dendrimers are among the most promising architectures to mimic the oxygenic lightharvesting complex because of their structural similarities and synthetic convenience. The overall geometries of dendrimers are determined by the core structure, the type of dendron, and the number of generations of interior repeating units. The rigid core and bulky volume of exterior porphyrin units in multiporphyrin dendrimers give rise to well-ordered three-dimensional structures. As the number of generations of interior repeating units increases, however, the overall structures of dendrimers become disordered and randomized due to the flexibility of the repeating units. Takuzo Aida et al [119] investigated the relationship between incoherent excitation energy migration processes and molecular structures in Zinc(II) Porphyrin dendrimers. To reveal the relationship between molecular structure and processes of excitation-energy migration in multiporphyrin dendrimers, Takuzo Aida et al calculated the molecular structure and measured the time-resolved transient absorption and fluorescence anisotropy decays for various hexaarylbenzeneanchored polyester zinc(II) porphyrin dendrimers along with three types of porphyrin dendrons as references.

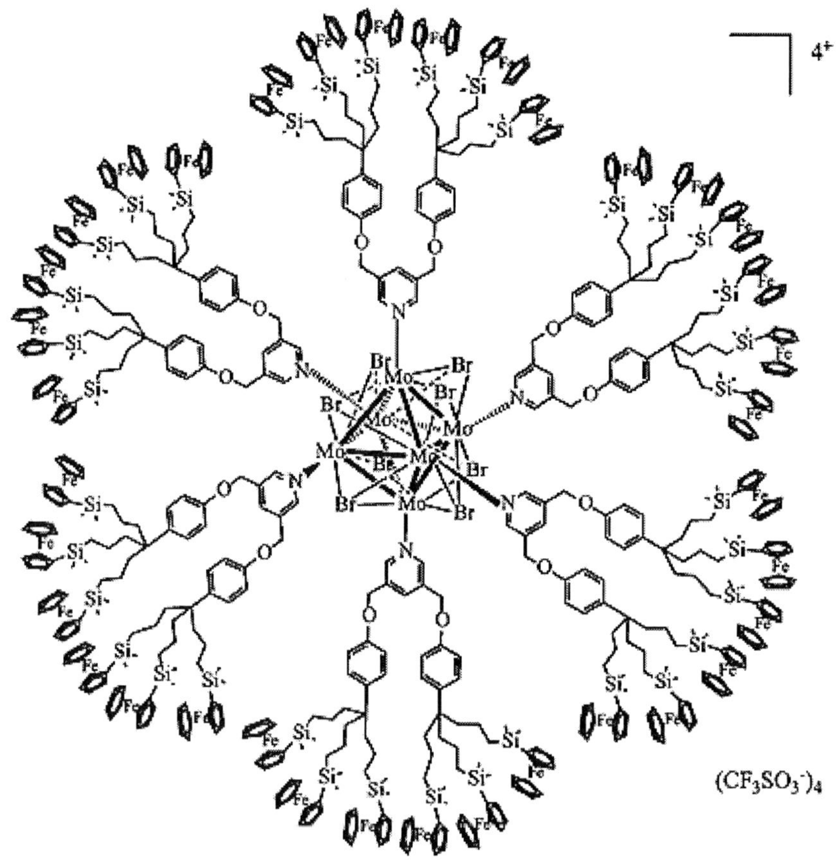

Figure 41. Mo-cluster-cored dendrimers.

Takuzo Aida et al found that the congested two-branched type dendrimers exhibit more efficient energy migration processes than one or three-branched type dendrimers because of multiple energy migration pathways, and the three-dimensional packing efficiency of dendrimers strongly depends on the type of dendrons (Figure 42).

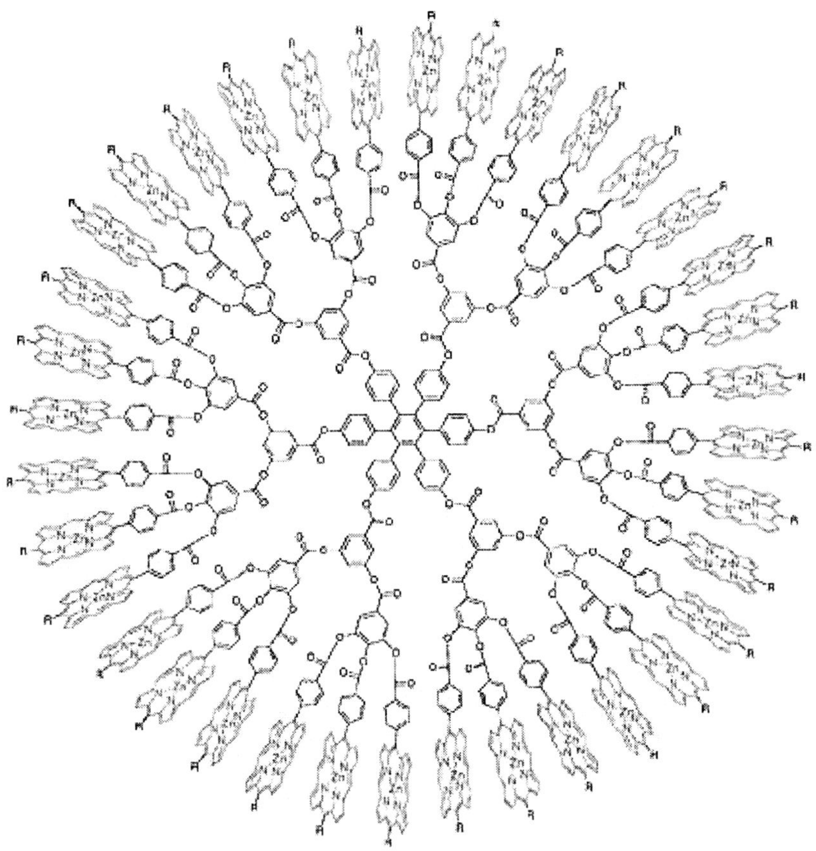

Figure 42. Zinc(II) porphyrin dendrimers.

6. SELF-ASSEMBLY OF DENDRIMERS

6.1. Biomolecular Self-Assembly and Recognition

Thomas Schrader et al [120] reported dendrimeric bisphosphonates for multivalent protein surface binding. A single weak-binding event is multiplied into an efficient receptor site for protein surfaces (<10–1 to >106M–1 in buffered aqueous solution) in a biomimetic fashion. This has hitherto been done with natural host/guest pairs, but not with artificial receptors. The organic reaction presented is one of very few that enable chemists to fuse multiple ionic building blocks covalently in highly polar solution; this one-pot reaction proceeds with virtually quantitative yield. According to this concept, other building blocks with aldehyde groups can likewise be multiplied into monodisperse functional dendrimers.

Mauri A. Kostiainen et al [121] reported multivalent dendrons for high-affinity adhesion of proteins to DNA. One of the keys for using deoxyribonucleic acid (DNA) as a nanomaterial relies on how the individual DNA chain can be aligned and how a multitude of DNA chains can be packed into ordered nanostructures. Hsin-Lung Chen et al [122] reported two-dimensional densely packed DNA nanostructure derived from DNA complexation with a low-generation poly(amidoamine). They presented a simple method for constructing a 2-D densely packed DNA nanostructure using the electrostatic complex of DNA with a poly(amidoamine) (PAMAM) dendrimer of generation two. Ordered DNA arrays are formed by drop-casting an aqueous solution containing positively overcharged complexes onto mica followed by a prolonged incubation. During the incubation, the complexes tend to adsorb onto the negatively charged mica surface through electrostatic attraction. The rodlike complexes organize to form ordered arrays to increase the surface density of the adsorbed complexes and hence the attractive free energy of adsorption. The densely packed nanostructure obtained here is distinguished from the previously reported spheroid or toroid structure derived from DNA complexations with the higher-generation dendrimers.

G. Floudas et al [123] reported the self-assembly and molecular dynamics of peptide-functionalized polyphenylene dendrimers. The self-assembly mechanism and the associated molecular dynamics are studied for a series of poly-L-lysine-functionalized polyphenylene dendrimer melts as a function of the core size (generation), functionality, and polypeptide length using X-rays, solid-state NMR, calorimetry, and dielectric spectroscopy. A striking dependence of the polyphenylene self-assembly on the poly-L-lysine length is shown. In addition, the type (α-helix/ β-sheet) of peptide secondary structure is controlled by the packing restrictions imposed by the polyphenylene core (Figure 43). G. Floudas et al showed that constrained poly-L-lysines can adopt different secondary structures from their linear analogues.

Cationic lipids and polymers are routinely used for cell transfection, and a variety of structure-activity relation data have been collected. Few studies, however, focus on the structural aspects of self-assembly as a crucial control parameter for gene delivery. Francois Diederich et al [124] reported the self-assembly, DNA complexation, and pH response of amphiphilic dendrimers for gene transfection. Francois Diederich et al presented here the observations collected for a set of cationic dendritic amphiphiles based on a stiff tolane core that are built from identical subunits but differ in the number and balance of their hydrophobic and cationic hydrophilic moieties. Francois Diederich et al found that the primary self-assembly of the amphiphile is of paramount importance and influences DNA binding, serum sensitivity, and pH response of the vector system.

Prabal K. Maiti et al [125] studed sequence-dependent complexation between oligonucleotides (single-strand DNA) and various generation ethylene diamine (EDA) cored poly amido amide (PAMAM) dendrimers through atomistic molecular dynamics simulations accompanied by free energy calculations and inherent structure determination. Simulations reveal formation of a stable complex and provide a detailed molecular level understanding of the structure and dynamics of such a complexation. The reaction free energy surface in the initial stage is found to be funnel-like, with a significant barrier arising in the late stage due to the occurrence of misfolded states of DNA. Complexation shows surprisingly strong sensitivity to the ssDNA sequence, which is found to arise from a competition between enthalpic versus entropic rigidity of ssDNA.

The formation of amyloid plaques is a key pathological event in neurodegenerative disorders, such as prion and Alzheimer's diseases. B. Klajnert et al [126] reported the pH Dependence of Molecular Interactions of Dendrimers with Amyloid Peptides. They have studied the effect of polypropyleneimine dendrimers on the formation of amyloid fibrils as a function of pH in order to gain further insight in the aggregation mechanism and its inhibition.

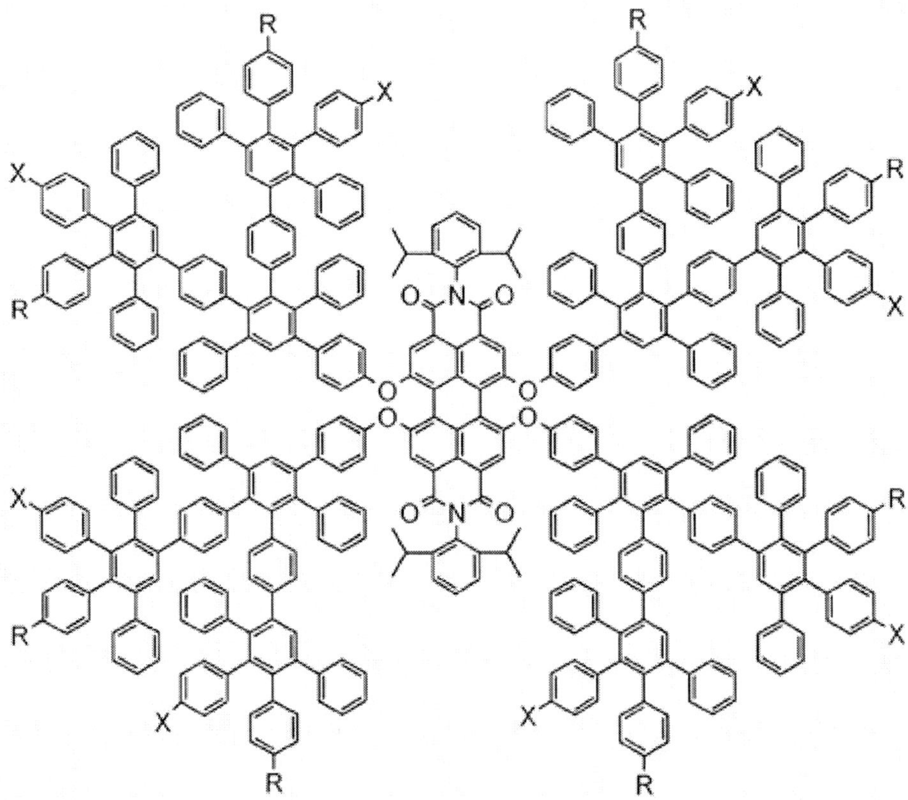

Figure 43. Polyphenylene scaffold of dendrimers.

The results showed that the level of protonation of His, Glu, and Asp residues is important for the final effect, especially at low dendrimer concentration when their inhibiting capacity depends on the pH. At the highest concentrations, dendrimers were very effective against fibril formations for both prion and Alzheimer's peptides.

The interactions between plasmid DNA and cationic polymers are of interest for their potential biological applications. Xiangliang Yang et al [127] reported studies on the interactions between DNA and PAMAM with fluorescent probe [Ru(phen)2dppz]2+. The interactions of DNA with polyamidoamine (PAMAM) dendrimer were studied by fluorescence spectroscopy and stopped-flow technique. A rapid and reproducible fluorescent assay method had been developed for assessing PAMAM and DNA interactions using [Ru(phen)2dppz]2+ as a probe. Xiangliang Yang et al further studied the kinetics of PAMAM binding to DNA and the reverse process of DNA dissociation from the complexes. The results indicated that DNA condensation was the rate-determining step during the complexation process, while DNA unfolding and expansion was the rate-determining step during the DNA dissociation process. And at N/P ratios above 2.0, DNA was fully condensed by PAMAM and dissociation was increasingly difficult. These results provided some useful instructions for self-assembling and disassembling of DNA as well as efficient gene delivery applications.

Hiroshi Tsukube et al [128] reported the proteo-dendrimers designed for complementary recognition of cytochrome c for preparing dendrimer architecture toward nanoscale protein complexation. "Proteo-dendrimers" in which polyanionic hepta(glutamic acids), fluorescent zinc porphyrinate cores, hydrophilic polyether surfaces, and nonpeptide hydrophobic dendrons are combined, were developed as a new series of synthetic receptors for protein recognition. They have polyanionic "patch" structures on their surfaces and undergo complementary electrostatic interactions with a positively charged cytochrome c patch, as observed in biological protein–protein recognition systems. These proteo-dendrimer receptors exhibited higher affinities with cytochrome c proteins in aqueous solutions than with biological cytochrome b5. Furthermore, they effectively blocked complexation of biological cytochrome b5 with cytochrome c, indicating that the proteo-dendrimers and cytochrome b5 similarly occupy the polycationic patch of cytochrome c.

Daiyo Terunuma et al [129] reported the syntheses and lectin-binding activities of lactotriaose-containing carbosilane dendrimers. Carbosilane dendrimers periphery-functionalized with lactotriaose (GlcNAcb1–3Galb1–4Glc) with valencies of three, four, six, and twelve were prepared for use in a lectin-binding assay. A lactotriaose derivative was prepared from D-glucosamine and D-lactose derivatives. The N-Troc-protected glucosamine glycosyl donor and 3'-O-unprotected lactose glycosyl acceptor were condensed in the presence of silver trifluoromethanesulfonate and methylsulfenyl bromide to provide corresponding trisaccharide with new β-1-3 linkages in 92% yield. The protection group of the trisaccharide was transformed into an acetyl group. The 4-pentenyl glycoside was prepared from the acetate via glycosyl bromide. The alkene was converted into acetyl sulfide by addition of thioacetic acid under radical conditions. The lactotriaose unit was linked with carbosilane dendrimers to afford acetyl-protected glycodendrimers. De-O-acetylation of the dendrimers was carried out in the presence of sodium methoxide and then aq NaOH to give the desired lactotriaose clusters using a carbosilane dendrimer backbone.

6.2. Supramoleclular Assembly

Dendrimers and dendrons are a unique class of molecular-scale building blocks for the construction of nanostructured materials. This makes these structures ideal for the construction of multiply crosslinked covalent assemblies, and also for the self-assembly of multiply bound non-covalent (supramolecular) assemblies [130].

Rigoberto Advincula et al [131] reported the conjugated polymer nanoparticles via intramolecular crosslinking of dendrimeric precursors.

Shiyong Liu et al [132] described the syntheses and solution behavior of model amphiphilic dendritic–linear diblock copolymers that self-assemble in aqueous solutions into micelles with thermoresponsive shells. The investigated materials are constructed of poly(benzyl ether) monodendrons of the second generation ([G-2]) or third generation ([G-3]) and linear poly(N-isopropylacrylamide) (PNIPAM). [G-2]-PNIPAM and [G-3]-PNIPAM dendritic–linear diblock copolymers have been prepared by reversible addition–fragmentation transfer (RAFT) polymerizations of N-isopropylacrylamide with a [G-2]- or [G-3]-based RAFT agent, respectively. The thermoresponsive collapse of the PNIPAM shell is a two-stage process; the first one occurs gradually in the temperature range of 20–29°C, which is much lower than the lower critical solution temperature of linear PNIPAM homopolymer, followed by the second process, in which the main collapse of PNIPAM chains takes place in the narrow temperature range of 29–31°C.

A new methodology for the preparation of well-defined core-shell nanoparticles was developed by Karen L. Wooley et al [133], based upon the employment of a multifunctional crosslinker to coincidently stabilize supramolecular polymer assemblies and imbed into the shell unique chemical functionalities. Amphiphilic diblock copolymers of poly(acrylic acid)80-*b*-poly(styrene)90 that had been assembled into micelles and partially functionalized throughout the corona with alkynyl groups were utilized as Click-readied nanoscaffolds for the formation of shell Click-crosslinked nanoparticles (SCCs). Divergently grown dendrimers of the zero, first, second, and third generations having increasing numbers of azide terminating groups ((N3)2-[G-0], (N3)4-[G-1], (N3)8- [G-2], and (N3)16-[G-3], respectively) were investigated as crosslinkers via Click reactions with the alkynyl groups to form covalent linkages throughout the block copolymer micelle corona, thus forming a crosslinked shell. The dendritic crosslinker is important as it also allows for the incorporation of excess functionality that can undergo complementary reactions.

Xin-Ru Jia et al [134] reported the self-assembly of a new class of amphiphilic poly(amidoamine) dendrimers and their electrochemical properties. Amphiphilic poly(amidoamine) (PAMAM) dendrimers consisting of a hydrophilic dendrimer core and hydrophobic aromatic dansyl or 1-(naphthalenyl)-2- phenyldiazene (NPD) shells have been synthesized. These amphiphilic dendrimers from the zero generation to the third generation self-assemble into vesicular aggregates in water. The self-assembly behavior of these dendrimers strongly depends on their generations. For the PAMAM–NPD aggregates, the photoisomerization process leads to a change in the aggregate size.

The solution behavior of spherical dendrimers as well as hybrid-linear dendritic diblock copolymers has been extensively studied, and the size, shape, and ability of these polymers to encapsulate small molecules have led to their comparison with traditional micelles. Paula T. Hammond et al [135] studied the solution behavior of linear-dendritic rod diblock copolymers in methanol. The diblock copolymers consist of a linear poly(ethylene oxide)-poly(ethylene imine) diblock copolymer backbone around which poly(amido amine) branches have been divergently synthesized from the poly(ethylene imine) block. The intrinsic viscosity of the amine and ester terminated polymers was found to follow two very different trends at low generation; however, at higher generations, they followed similar, yet slightly different, curves with the values for the amine terminated polymers only a little larger than those of the ester terminated polymers.

Daniel R. Blasini et al [136] studied the stepwise self-assembly of ordered supramolecular assemblies based on coordination chemistry. Dendrimer layers were deposited onto SiO2 wafers by alternately immersing the substrate into a solution of terpyridyl (tpy)-pendant poly(amido amine) (PAMAM) dendrimers (dend-n-tpy; n = 8, 16) dissolved in CH2Cl2, followed by the interfacial coordination reaction with cobalt (Co2+) from aqueous solution.

Virgil Percec et al [137] programmed the internal structure and stability of helical pores self-assembled from dendritic dipeptides via the protective groups of the peptide. The synthesis of dendritic dipeptides (4-3,4-3,5)12G2-CH2-X-L-Tyr-L-Ala-OMe with X = Boc, Moc, and Ac; their self-assembly in bulk and in solution; and the structural and retrostructural analysis of their supramolecular helical porous assemblies are reported. The dimensions, structure, internal order, thermal stability of the supramolecular helical pores, and conformations of the dendron and supramolecular dendrimer are programmed by the nature of the protective groups of the dipeptide. The ability of the protective groups to program the structure of the helical pore reveals the simplest design strategy that complements the more complex strategies based on the architecture of the dendron, the stereochemistry, and the structure of the dipeptide.

6.3. Liquid-Crystal Dendrimer

Carsten Tschierske et al [138] reported the first example of a third-generation liquid-crystalline carbosilane dendrimer with peripheral bent-core mesogenic units for understanding of "dark conglomerate phases".

Hee-Tae Jung et al [139] reported the large-area, highly aligned cylindrical perfluorinated supramolecular dendrimers using magnetic fields.

Milagros Pinol et al [140] reported the synthesis, characterization, and induction of stable anisotropy in liquid crystalline photo-addressable PPI dendrimers. A series of liquid crystalline azobenzene codendrimers with light-driven properties has been prepared by chemical modification of the third generation of an amine-terminated poly(propylene imine) (PPI) dendrimer. These photoactive dendrimers were obtained by grafting the photochromic 4-cyanoazobenzene and/or liquid crystalline 4-cyanobiphenyl units onto the periphery of PPI. The synthesized dendrimers include different percentages of each functional unit randomly distributed in the dendritic shell. Illumination with either blue (488 nm) or UV (350 nm) linearly polarized light at room temperature gave rise to a stable dichroism and moderate birefringence values when 80% (molar ratio) of the units are azobenzene. In addition, pre-irradiation of the films with unpolarized UV light gave rise to higher birefringence values.

6.4. Interfacial Behaviour of Dendrimer

Dendrimers are novel polymeric nanoarchitectures for solubility enhancement. Poor solubility and hydrophobicity of drugs/bioactives limit their possible applications in drug delivery and formulation development. Apart from conventional methods of solubility enhancement, there are some novel methods which can be used in solubilization. Dendrimers represent a novel type of polymeric material that has generated much interest in many diverse areas due to their unique structure and properties. Dendrimer-mediated solubility enhancement mainly depends on factors such as generation size, dendrimer concentration, pH, core, temperature, and terminal functionality [141]. Added advantage in solubilization can be achieved considering these factors. Available literature suggests that ionic interaction, hydrogen bonding, and hydrophobic interactions are the possible mechanisms by which a dendrimer exerts its solubilizing property.

A. I. Cooper et al [142] reported the extraction of a hydrophilic compound fromwater into liquid CO2 using dendritic surfactants.

Amphiphiles with a dendritic structure are attractive materials as they combine the features of dendrimers with the self-assembling properties and interfacial behavior of amphiphiles. Jean-Louis Gallani et al [143] have designed and synthesized three series of segmented amphiphilic block co-dendrimers (Janus-type) and studied their interfacial properties on the Langmuir trough. Various behaviors are observed with, as a rule, the lowest generation dendrimers behaving more or less like traditional amphiphiles while the larger molecules tend to exhibit more complicated isotherms, with a non-straighforward temperature dependence, one particular molecule seemingly forming supramolecular assemblies spontaneously (Figure 44).

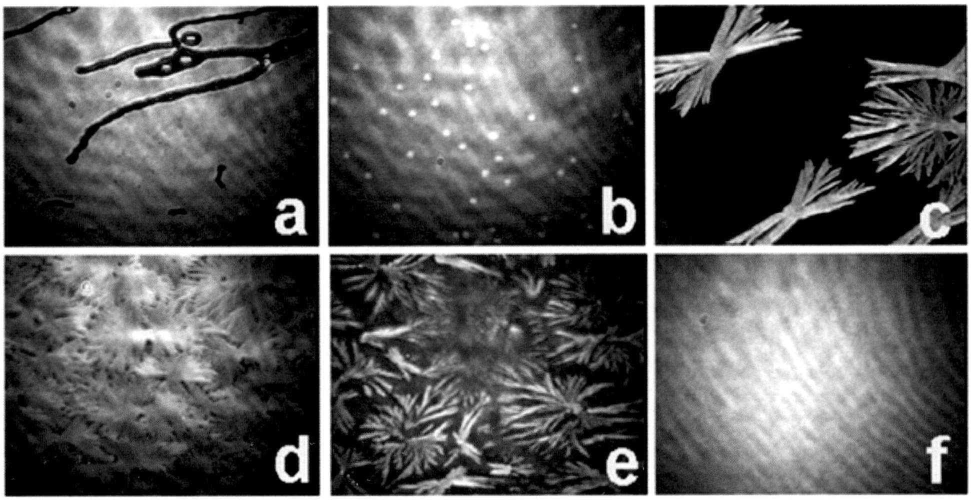

Figure 44. Interfacial behavior of a series of amphiphilic block co-dendrimers.

Dendritic folate rosettes as ion channels in lipid bilayers. Naomi Sakai et al [144] reported that π-stacked supramolecular rosettes can form synthetic ion channels. These esthetically pleasing supramolecules can self-assemble from heteroaromatic "sectors" with directional hydrogen-bond donor-acceptor arrays such as folate guanines, and so on. H-bond directed circular self-assembly of these sectors gives cyclic oligomers, which, in turn, may self-assemble on top of each other to give π-stacks. Templation of this process by cation binding to carbonyl lone pairs in the rosette centers accounts for the confirmed ionophoric properties of many π-stacked supramolecular rosettes in bulk liquid membranes.

6.5. Dendrimer Supramolecular Helix

Virgil Percec et al [145] reported the self-assembly of amphiphilic dendritic dipeptides into helical pores (Figure 45).

V. Percec et al [146] reported the self-organization of supramolecular helical dendrimers into complex electronic materials (Figure 46).

Virgil Percec et al [147] found that the internal structure of helical pores self-assembled from dendritic dipeptides is stereochemically programmed and allosterically regulated.

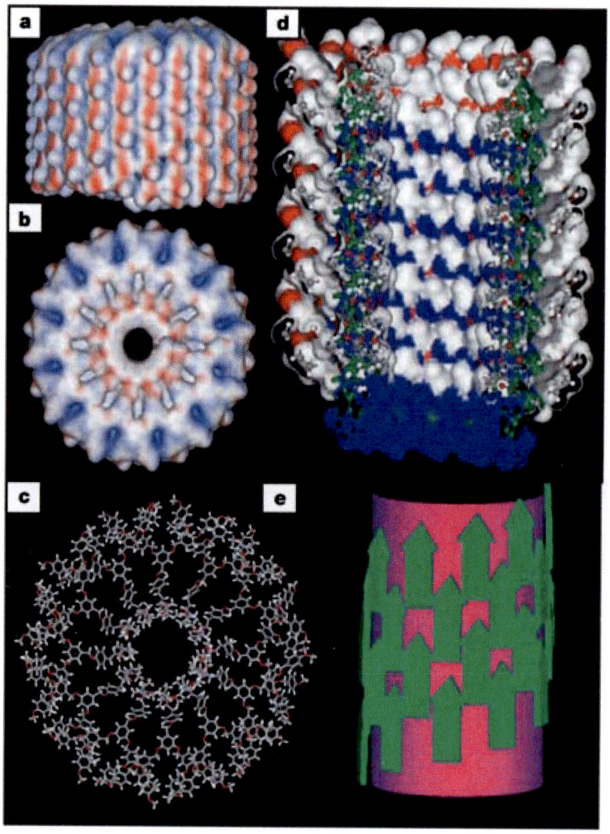

Figure 45. Self-assembly of amphiphilic dendritic dipeptides into helical pores.

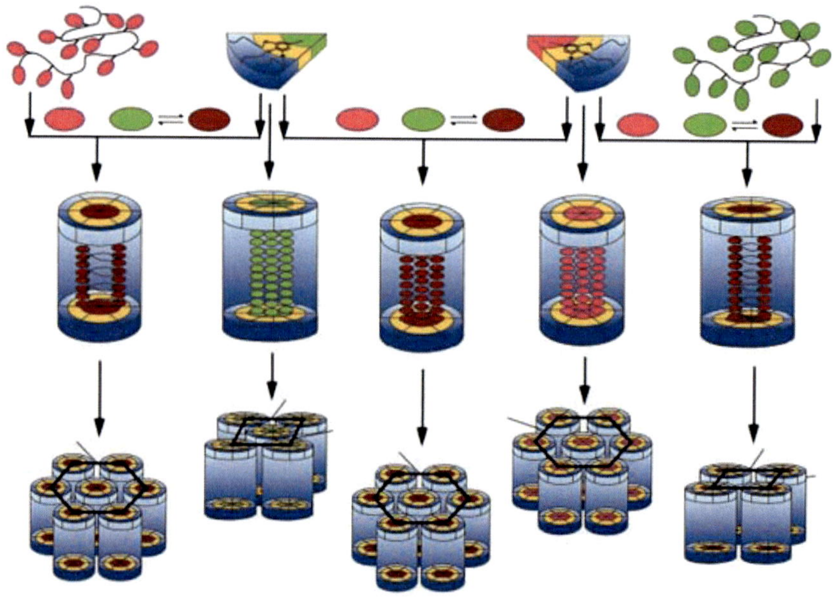

Figure 46. Self-organization of supramolecular helical dendrimers into complex electronic materials.

7. Charge and Energy Transfer

7.1. Fluorescence Properties

Kung-Hwa Wei et al [148] have been prepared highly luminescent dendron-substituted copolyfluorenes that incorporate surface-modified cadmium sulfide nanoparticles. A small percentage of these nanoparticles can be incorporated into the dendritic structures upon tailoring the interfaces between the ligands on the nanoparticles and the dendritic structures in the copolyfluorene. Both the photoluminescence and electroluminescence efficiencies of the polymer nanocomposites are dramatically enhanced—sometimes more than doubled—relative to the values of the pure polymer.

Donald A. Tomalia et al [149] have been synthesized Hydroxyl terminated, poly(alkyl ether) dendrons possessing single site phosphine functional groups at their focal points. These single site reactive dendrons are then used in ligand exchange reactions to produce stabilized surface capped CdSe/core CdS-shell quantum dots that exhibit high quantum yield, non-quenching photoluminescence (PL) properties. These enhanced features are in sharp contrast to quantum dots capped with analogous single site thiol functionalized dendrons that exhibit substantially quenched photoluminescence. These new dendron constructions begin with protected pentaerythritol moieties which are used as the initiator core as well as the branch cells. By protecting one of the four hydroxyl groups in the pentaerythritol moiety, the remaining three hydroxyl groups are utilized as divergent growth sites for the poly(ether) dendron construction. Desired generation levels (i.e., generations 1, 2) are synthesized, followed by deprotection of the focal point hydroxyl group to expose appropriate functional groups suitable for conversion to sulfhydryl, phosphine or other desired functional groups.

Toyoko Imae et al [150] studied the strong fluorescence emission from poly(amido amine) (PAMAM) dendrimers with different terminal groups or a poly(propylene imine) (PPI) dendrimer under different conditions by varying experimental parameters such as pH value, aging time, temperature, and concentration. The increase of fluorescence intensity was fast at low pH or high temperature but linear with respect to dendrimer concentration. It was reasonable that the formation of a fluorescence-emitting moiety had a close relation to protonated tertiary amine groups in PAMAM or PPI dendrimers. Furthermore, oxidation of the tertiary amines was confirmed to play an important role, which was evidently caused by oxygen in air. The results of fluorescence decay indicated that the deactivation of luminescence was raised with increasing temperature. Dendrimers emitted blue photoluminescence along fiber chain templates on a fluorescent microscope.

7.2. Energy Transfer

S. Thayumanavan et al [151] have designed and synthesized difunctionalized dendrimers containing two donors in the periphery and an acceptor at the core to serve as scaffolds for comparison with linear analogues to investigate the advantage of dendritic scaffolds for energy and charge transfer. Comparison of these dendrimers with the fully decorated dendrimers provides information on the advantage of chromophore density in energy/charge transfer from periphery to the core.

Jane Larsen et al [152] explored energy transfer processes within a series of Zn-porphyrin-appended dendrimers by means of excitation intensity dependent transient absorption measurements. Singlet-singlet annihilation on two distinct time scales of 18 ± 5 ps and 130 ± 10 ps in the dimer and the dendrimers is reported. The two distinct processes reflect the presence of two structural conformer distributions. Analysis of the singlet-singlet annihilation transient kinetics shows that sequential annihilation occurs within subunits up to four Zn-porphyrins in the dendrimers. The onset of the singlet-singlet annihilation process depending on the size of the molecule reveals a difference in the number of communicating Zn-porphyrins.

René M. Williams et al [153] described the synthesis and charge transfer properties of triarylamine-oligothiophene-perylenemonoimide dendrimers, TPA(T2-PMI)3 and TPA(T4-PMI)3. The fluorescence quantum yields indicate strong emission quenching by electron transfer [ΦTHF = 0.004 for TPA(T2-PMI)3, ΦTHF = 0.003 for TPA(T4-PMI)3, and ΦTHF = 0.8 for PMI]. Moreover, with the increase of the solvent polarity, the quantum yields decrease indicating that the A+•D–• (acceptor/donor) couple is more stabilized. The femtosecond transient absorption spectra show a very fast charge separation process (≈2 ps; kcs ≈ 5 × 1011 s–-1) and a charge recombination of more than 1 order of magnitude slower (≈50 ps; kcr ≈ 2 × 1010 s–1), as observed from the rise time and decay of the radical anion and radical cation absorption bands. The analysis of the transient absorption spectroscopy and of the energetics of the process using Marcus theory indicates that in the electron transfer process the thiophene unit is the first electron donor. The presence of the triarylamine and/or the proximity of the oligothiophenes does improve the donor capabilities of the oligothiophene unit slightly and enhances its conjugation as seen in the absorption spectra and the transients of the radial cations.

7.3. Photon Absorption

Lucia Flamigni et al [154] reported the photoinduced processes in a dendritic Zn porphyrin structure with a free-base porphyrin core. The photophysical characterization of a nonameric porphyrin assembly made up of a central free-base and eight peripheral zinc porphyrins connected by flexible nucleosidic linkers, and the identification of the photoinduced processes taking place in the array was reported (Figure 47). The kinetic parameters of singlet–singlet energy transfer are interpreted by assuming the existence of several nonequilibrated conformations in the array due to the flexibility of the linkages. A fast quenching of the donor beyond the time detection limit of the instrument is also hypothesized and assigned to strong interactions between the zinc chromophores in close contact which can give self-quenching phenomena. For the zinc porphyrin triplet excited state, a triplet–triplet annihilation is detected at moderate photon flux as a consequence of multi excitation of the same array. Drastic reduction of the photon flux allows us to determine the triplet energy transfer rate from zinc porphyrin to free-base porphyrin, which is the final recipient of the energy collected by the array. The decay of the triplet state of the core free-base porphyrin is prolonged with respect to the model, indicating a shielding effect of the peripheral groups.

Theodore Goodson III et al [155] reported the entangled photon absorption in an organic porphyrin dendrimer. Two-photon absorption spectroscopy is an intensity dependent nonlinear effect related to the excitation of virtual intermediate states. The classical two-photon absorption has an extremely low efficiency which is quantified by its cross-section (δ ~ 10–48 cm4 s at 800 nm).

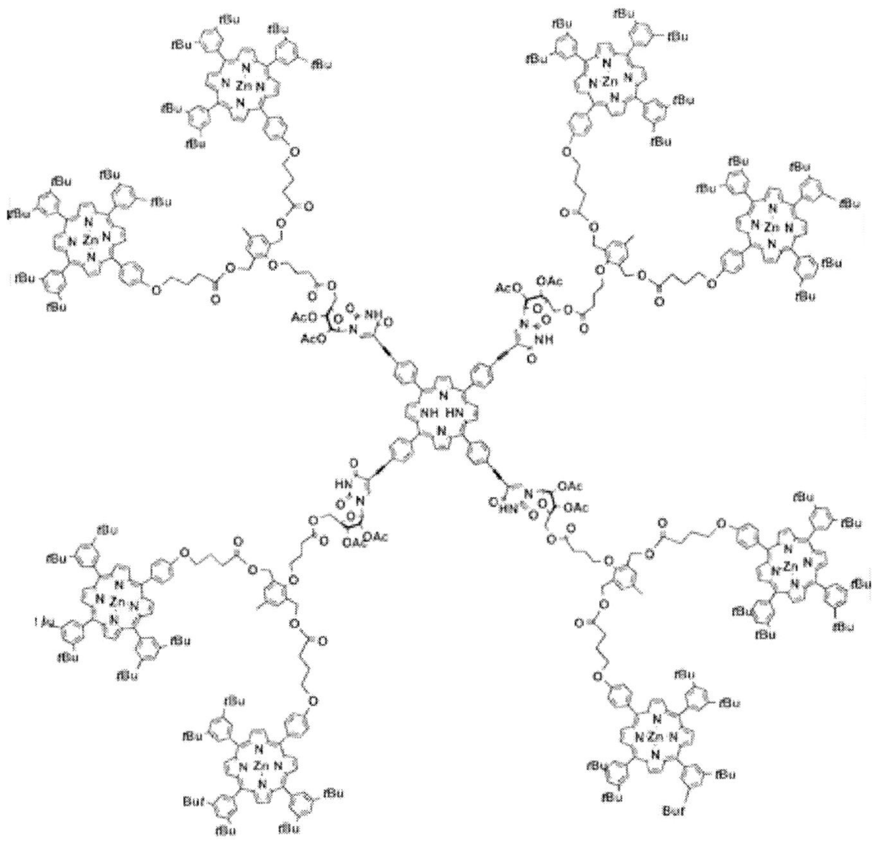

Figure 47. Dendritic Zn porphyrin structure with a free-base porphyrin core.

To overcome this limitation, Theodore Goodson III et al demonstrated a novel effect of the two-photon absorption method utilizing the high degree of quantum optical correlation between photon pairs created by the process of spontaneous parametric downconversion. A large entangled two-photon absorption cross-section (δ ~ 10–17 cm2 at 800 nm) was measured in an organic porphyrin dendrimer. This novel effect may open new avenues for ultrasensitive detection in chemical and biological systems. The use of entangled photons from a correlated source for the purpose of the two-photon effect is promising in this regard as one may obtain two-photon effects with very small numbers of photons.

Theodore Goodson, III et al [156] reported a strongly interacting new dendrimer system with an extended spectroscopic unit (coherent domain) beyond the trimer configuration. The report focused on the mechanism of enhancement of the two-photon absorption (TPA) response of this material as a function of the dendrimer generation number (Figure 48). This enhancement is strongly correlated to the size and geometry of the spectroscopic unit in a strongly interacting macromolecular system.

Uday Maitra et al [157] described a facile synthesis of bile acid-based dendritic structures with multiple naproxen groups at the periphery and a single anthracenyl moiety on the side chain. Absorption studies indicate that there are no groundstate interactions between the chromophores and the molar extinction coefficients increase linearly with increasing number of naproxen units. Upon excitation of the peripheral chromophores, there is efficient intramolecular energy transfer to the anthracenyl chromophore. These novel, non-conjugated dendrimers thus act as efficient molecular light harvesters.

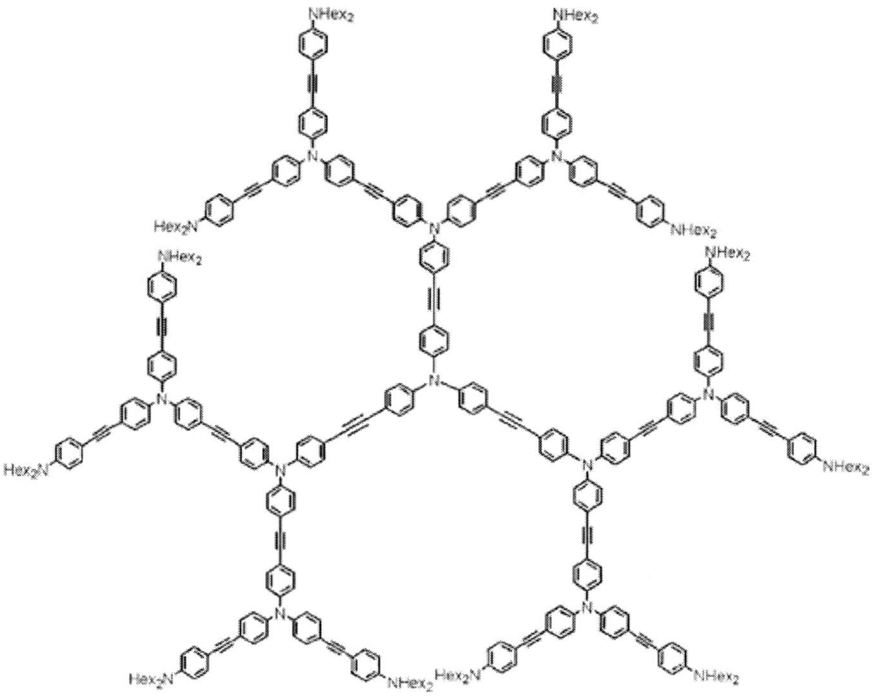

Figure 48. Organic conjugated dendrimers with enhanced two-photon absorption.

Utilization of two-photon (TP) excited fluorescence resonance energy transfer (FRET) within a lightharvesting dendrimer has proven to be a reliable method for the enhancement of the effective TP absorption efficiency of many FRET acceptor molecules. This light-harvesting approach has enabled TP photosensitization of singlet oxygen from a porphyrin in both aqueous and organic media using wavelengths more transmissive to human body tissue (750-1000 nm). This ability to utilize near-infrared irradiation to induce photochemical reactions is especially attractive for applications including in vivo photochemistry, oxygen sensing, and photodynamic cancer therapy. Jean M. J. Fréchet et al [158] reported light-harvesting chromophores with metalated porphyrin cores for tuned photosensitization of singlet oxygen via two-photon excited FRET. Jean M. J. Fréchet et al synthesized an array of novel photosensitizers by metalation of the porphyrin core with aluminum, silver, and zinc. Singlet oxygen photosensitization efficiency was found to be most efficient using aluminum followed by zinc and least efficient using silver.

7.4. Charge Transfer

Nicola Armaroli et al [159] reported the structure-dependent photoinduced electron transfer in fullerodendrimers with light-harvesting oligophenylenevinylene terminals. Oligophenylenevinylene (OPV)-terminated phenylenevinylene dendrons G1–G4 with one, two, four, and eight "side-arms", respectively, were prepared and attached to C60 by a 1,3-dipolar cycloaddition of azomethine ylides generated in situ from dendritic aldehydes and N-methylglycine. The relative electronic absorption of the OPV moiety increases progressively along the fullerodendrimer family C60G1–C60G4, reaching a 99:1 ratio for C60G4 (antenna effect). Regardless of solvent, all the fullerodendrimers exhibit ultrafast OPV→C60 singlet energy transfer. In CH2Cl2, the OPV→C60 electron transfer from the lowest fullerene singlet level (1C60*) is slightly exergonic (ΔGCS≈0.07 eV), but is observed, to an increasing extent, only in the largest systems C60G2–C60G4 with lower activation barriers for electron transfer (Figure 49). Monitoring of the lowest fullerene triplet state by sensitized singlet oxygen luminescence and transient absorption spectroscopy shows that this level is populated through intersystem crossing and is not involved in photoinduced electron transfer.

Clemens Krempner et al [160] reported hydroxy-substituted oligosilane dendrimers for controlling the electronic properties through hydrogen bonding. S. Thayumanavan et al [161] examined the photophysics of a series of molecules consisting of a benzthiadiazole core surrounded by a network of benzyl ether arms terminated by aminopyrene chromophores, which function as both energy and electron donors.

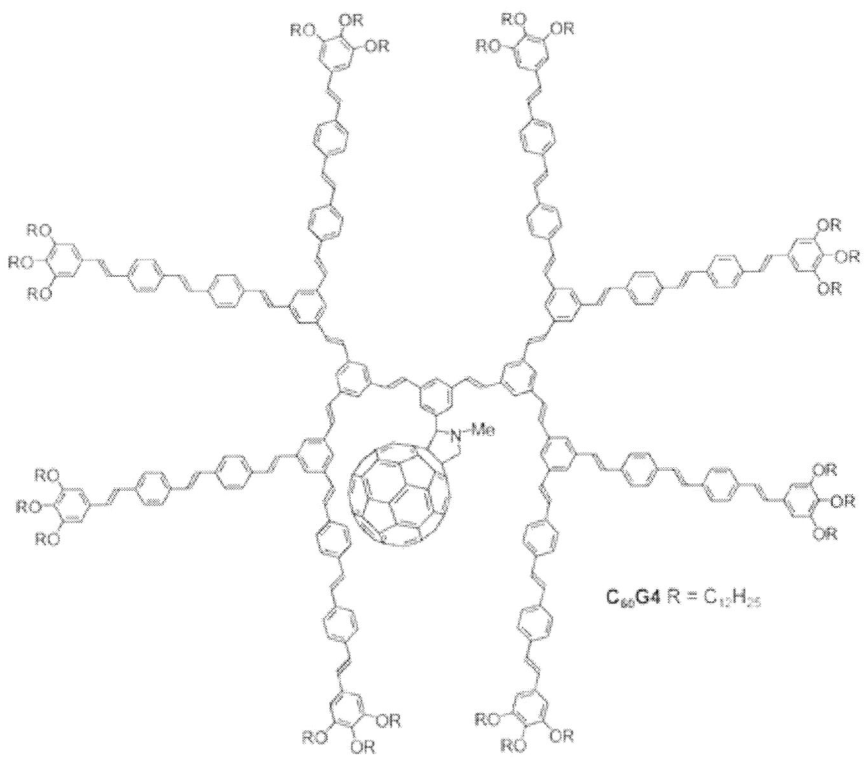

Figure 49. Fullerodendrimers with light-harvesting oligophenylenevinylene terminals.

Three classes of molecules are studied: dendrimers whose peripheries are fully decorated with aminopyrene donors (F), disubstituted dendrimers whose peripheries contain only two donors (D), and linear analogues in which a pair of benzyl ether arms link two donors to the central core (L). In all three types of molecules, the electronic energy transfer (EET) time scales as the square root of the generation number G, consistent with the flexible nature of the benzyl ether framework. Transient anisotropy measurements confirm that donor-donor energy hopping does not play a major role in determining the EET times. The charge transfer (CT) dynamics occur on the nanosecond time scale and lead to stretched exponential decays, probably due to conformational disorder. This work illustrated two important points about light-harvesting and charge-separation dendrimers. First, the use of a flexible dendrimer framework can lead to a more favorable scaling of the EET time (and thus the light-harvesting efficiency) with dendrimer size, relative to what would be expected for a fully extended dendrimer. Second, fully decorated dendrimers can compensate for the distancedependent slowdown in CT rate as G increases by providing additional pathways for the CT reaction to occur.

James K. Whitesell et al [162] reported intramolecular charge transfer in 1:1 Cu(II)/pyrenylcyclam dendrimer complexes. Two newly synthesized pyrenylcyclam dendrons exhibit a new emission band centered at 450 nm when coordinated with copper triflate. Observed fluorescence shifts induced by coordinative metalation indicate an unusual intramolecular charge transfer from a pyrenyl excited state to the coordinated metal ion that competes with pyrene excimer formation. This interaction likely proceeds by photoexcitation of a π-complex of the appended arene, followed by intramolecular charge transfer within the dendritic 1:1 cyclam/metal complex, effecting reduction of the bound Cu(II) metal ion. The appended dendritic groups not only decrease the equilibrium binding constant with Cu(II) but also participate in a new excited-state pathway as an alternative to energy-dissipative excimer formation.

Clifford P. Kubiak et al [163] reported charge gating and electronic delocalization over a denderimeric assembly of trinuclear Ruthenium clusters. A zeroth-order dendrimer was formed using a tridentate bridging ligand, 2,4,6-tri-4-pyridyl-S-triazine, and the redox-active trinuclear ruthenium cluster Ru3O(OAc)6(CO)(py)(H2O). The electronic properties of this dendrimer were probed using cyclic voltammetry. IR spectroelectrochemistry was performed at both low (−30°C) and room temperature. The IR spectroelectrochemical response at −30°C was straightforward, but at room temperature, the dendrimer exhibits an unusual and complex series of electronic behaviors, including intramolecular cluster-to-bridging-ligand charge transfer, gated electron transfer, and dynamic exchange on the IR time scale.

Wei-Shi Li et al [164] reported the construction of segregated arrays of multiple donor and acceptor units using a dendritic scaffold for remarkable dendrimer effects on photoinduced charge separation. Dendritic molecules appended with multiple zinc porphyrin units (DPm,m [number of zinc porphyrin units] = 6, 12, and 24) trap bipyridine compounds carrying multiple fullerene units (Py2Fn, n [number of C60 units] = 1-3), affording coordination complexes DPm⊃Py2Fn having a photoactive layer consisting of spatially segregated donor and acceptor arrays on their surface. Complexes DPm⊃Py2Fn are stable enough to be isolated by gel permeation chromatography. UHV-STM microscopy enables clear visualization of a petal-like structure of DP12⊃Py2F3. Photoexcitation of the zinc porphyrin units in DPm⊃Py2Fn results in a zinc porphyrin-to-fullerene electron transfer to generate a charge separation (Figure 50).

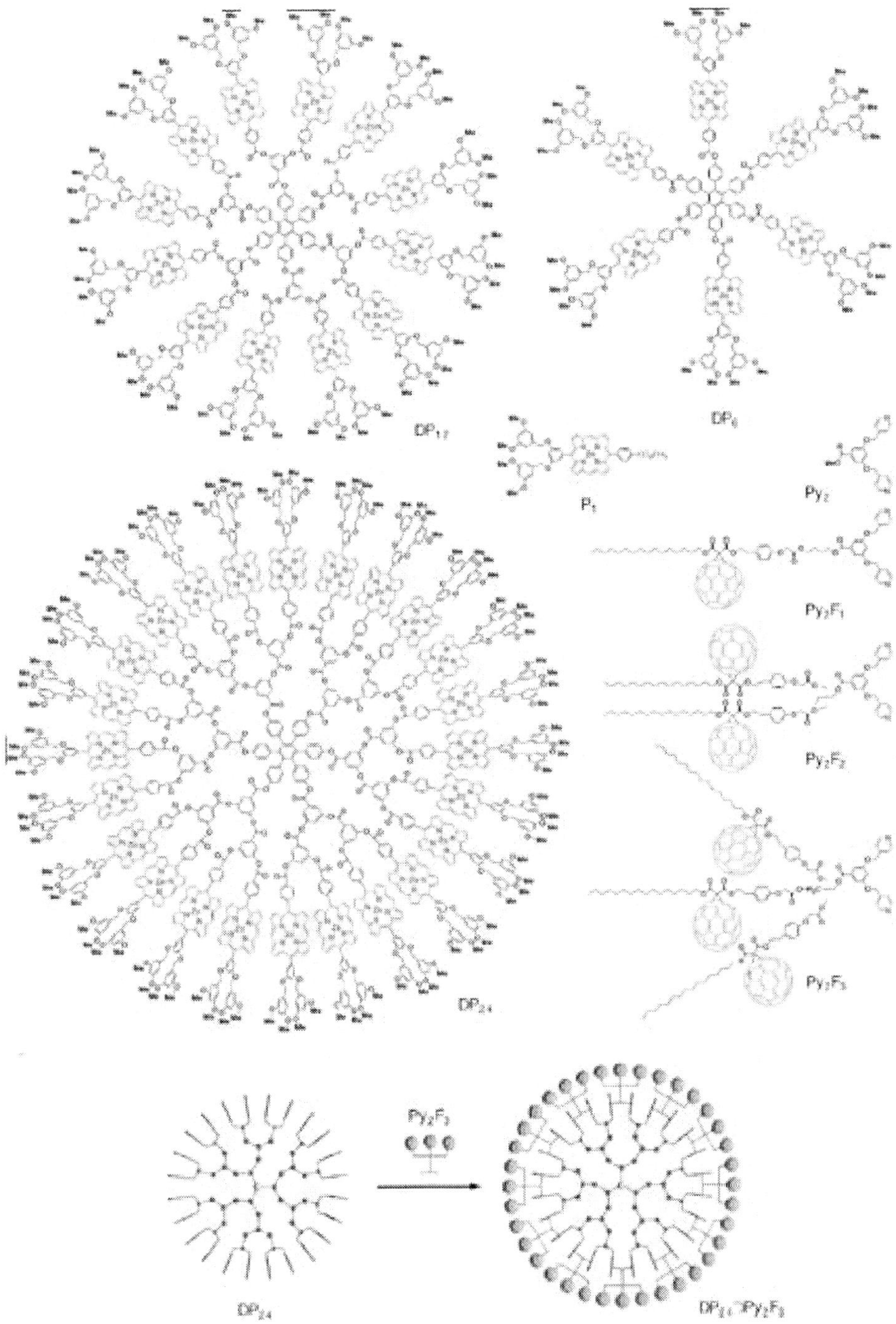

Figure 50. Construction of segregated arrays of multiple donor and acceptor units using a dendritic scaffold.

Chungkyun Kim et al [165] described the divergent synthesis of end-capped silole dendrimers Gn-2n+1Silole (n = 1-4) using 2,4,6,8-tetramethyl-2,4,6,8-tetravinylcyclotetrasiloxane [Me(CH2=CH)SiO]4 as a core molecule and allyl alcohol/ dichloromethylsilane as a building block. The reaction of the dichloride functionalities of the carbosiloxane dendrimers Gn-2n+2Cl (n = 1-4) with the 1,4-diphenylbutadiene- 1,4-dianions provided an effective route for the attachment of silole functionalities to a dendritic periphery. In addition, dendritic siloles GBn-2nSilole (n = 1-3) based on phenylethynylmethylsilane (PhC=C)2Me2Si as cores were synthesized. The building blocks of the dendrimers consisted of double bonds (-PhC=CHMeSi-) in the inner shell and silole groups on the outermost periphery. The generational limit of the dendrimer for the two-branching type core molecule was found to have eight silole groups for the third generation. Using this methodology, a series of silole-containing dendrimer systems showing green to greenish-blue fluorescence were synthesized.

7.5. Proton Harvest

π-conjugated dendrimers are molecular examples of tree-like structures known in physics as Bethe lattices. Electronic excitations in these systems can be spatially delocalized or localized depending on the branching topology. Without a priori knowledge of the localization pattern, understanding photoexcitation dynamics reflected in experimental optical spectra is difficult. ‘Supramolecular’-like quantum-chemical calculations quickly become intractable as the molecular size increases. Dong-Lin Jiang et al [166] reported photoisomerization in dendrimers by harvesting of low-energy photons.

Vladmir Y. Chernyak et al [167] reported exciton scattering and localization in branched dendrimeric structures. They developed a reduced excitonscattering (ES) model, which attributes excited states to standing waves in quasi-one-dimensional structures, assuming a quasiparticle picture of optical excitations. Direct quantum-chemical calculations of branched phenylacetylene chromophores are used to verify the model and to derive relevant parameters. Complex and non-trivial delocalization patterns of photoexcitations throughout the entire molecular tree can then be universally characterized and understood using the proposed ES method, completely bypassing ‘supramolecular’ calculations. This allows accurate modelling of excited-state dynamics in arbitrary branched structures.

8. Advances in Dendrimer Sensors

8.1. Electrochemistry Biosensors

Rigoberto Advincula et al [168] have achieved nanomolar detection and specific recognition of pinacolyl methylphosphonate (PMP), a hydrolysis product and an analog of a relatively persistent class of toxic nerve agents, In addition, picomolar sensitivity is observed with methylphosphonic acid (MPA), an end-hydrolysis product for all organophosphate-based nerve agents. This is achieved using a combined surface plasmon resonance (SPR) and potentiometry setup. A modified polyamidoamine (PAMAM) carbazole/Cu2+ dendrimer, which is electrochemically crosslinked on a self-assembled monolayer (SAM) modified Au substrate, is used as an active sensing element for trapping the nerve-agent analogs. The ultrathin films have been used to study the anchoring of nerve agents via non-covalent interactions. The carbazole to amine ratio is optimized to ensure free primary amines are available to interact with the analyte and the Cu2+ ions present in the system, which further enhances the selectivity. The carbazole group on the periphery serves a dual purpose: crosslinking the dendrimers to form a conjugated network film, and generating the potentiometric response. The adsorption kinetics are monitored by using an in situ SPR/potentiometric setup. This technique not only offers a real-time dual detection of highly toxic nerve-agent analogs, but also shows viability for future sensor-device applications.

Chongmok Lee et al [169] reported electrochemical deposition of end-capped triarylamine and carbazole dendrimers as the alternate technique for the manufacture of multilayer films (Figure 51).

Haesik Yang et al [170] reported a nanocatalyst-based assay for proteins based on DNA-free ultrasensitive electrochemical detection using catalytic reduction of p-nitrophenol by gold-nanoparticle labels (Figure 52). Lawrence D. Margerum et al [171] reported the effect of PAMAM dendrimer size and pH on the electrostatic binding of metal complexes using cyclic voltammetry. Cyclic voltammetry was employed to investigate the extent of electrostatic binding for metal complexes such as ferrocyanide, [Fe(CN)6]4−, to full generations of polyamidoamine (PAMAM) dendrimers (G1.0-G5.0) as a function of size and pH. Addition of dendrimer solutions to metal complexes caused a large net positive shift in *E*1/2 and reduced peak currents that varied with the size of the dendrimer. Binding ratios, *K*4−/*K*3−, calculated from the potential shifts changed from a ratio of five with small dendrimers (G1.0-G3.0) to 10 for G4.0 and G5.0, consistent with known structural changes in PAMAM dendrimers. Conditional binding constants for [Fe(CN)6]4−, calculated from the diffusion-limited oxidation currents, also increased from G3.0 (8.8 ×104 M−1) to G4.0 (31 × 104 M−1). At pH 5, the binding ratios increased dramatically at all generations, suggesting that the interior amine sites in the dendrimer are easily accessible to metal complexes and that dendrimers show some charge selectivity.

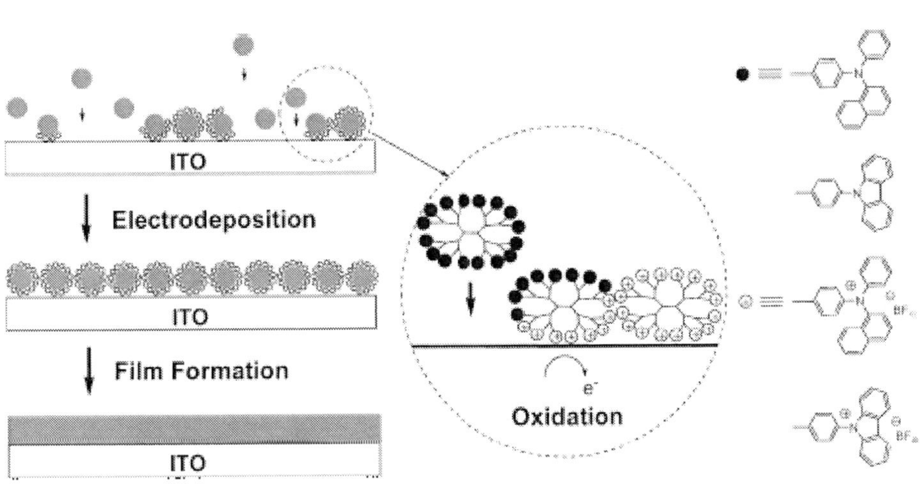

Figure 51. Electrochemical deposition of end-capped triarylamine and carbazole dendrimers.

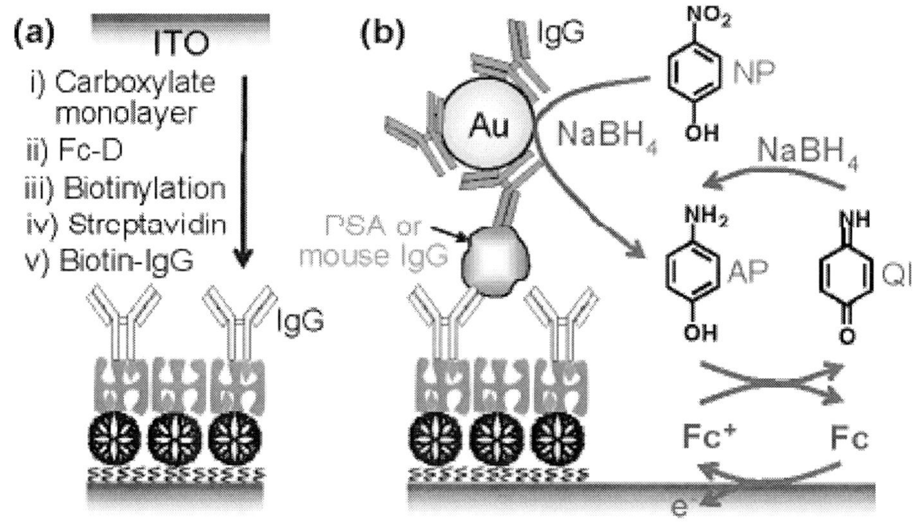

Figure 52. Preparation of an immunosensing layer for electrochemical detection of mouse IgG or PSA.

Francisco C. Nart et al [172] exploited the molecular engineering capability of the layer-by-layer (LbL) method to immobilize layers of gold nanoparticles on indium tin oxide (ITO) substrates, which exhibit enhanced charge transfer and may incorporate mediating redox substances. Polyamidoamine (PAMAM generation 4) dendrimers were used as template/stabilizers for Au nanoparticle growth, with PAMAM-Au nanoparticles serving as cationic polyelectrolytes to produce LbL films with poly(vinylsulfonic acid) (PVS) (Figure 53).

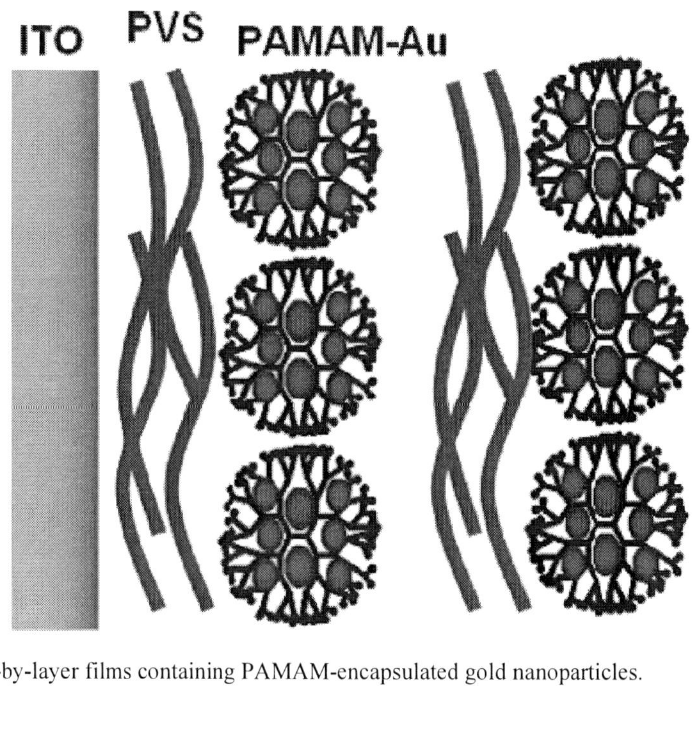

Figure 53. Layer-by-layer films containing PAMAM-encapsulated gold nanoparticles.

The cyclic voltammetry (CV) of ITO-PVS/PAMAM-Au electrodes in sulfuric acid presented a redox pair attributed to Au surface oxide formation. Electron hopping can be considered as the charge transport mechanism between the PVS/ PAMAM-Au layers within the LbL films. This charge transport was faster than that for nonmodified electrodes, shown by employing hexacyanoferrate(III) as the surface reaction marker.

Shaojun Dong et al [173] reported the assembly process of CuHCF/MPA multilayers on gold nanoparticles modified electrode and characterization by electrochemical SPR. Gold nanoparticles were deposited onto 2-mercaptoethylamine (MEA)-assembled planar gold thin film to construct gold nanoparticles modified electrode by virtue of a solution-based self-assembly strategy. Subsequently, 3-mercaptopropionic acid (MPA)-bridged copper hexacyanoferrate (CuHCF) multilayers were constructed on the as-prepared gold nanoparticles modified electrode. Compared with the multilayer system on a planar gold electrode, the different electrochemical and optical properties might result from higher curvature effect and extraordinary surface-to-volume ratio characteristic of gold nanoparticles and the nanoparticle- selective growth of CuHCF. A dendrimer-like assembly process was proposed to explain the experiment results. This new motif of multilayer on the gold nanoparticles modified electrode was different from that of on a planar gold electrode, indicating a potential application of EC-SPR technique in the study of nanocomposite materials.

Functionalization of multi-walled carbon nanotubes with poly(amidoamine) dendrimer for mediator-free glucose biosensor is of great interest. Yun-Long Zeng et al [174] have successfully prepared novel carbon nanotube (CNT) nano-composites by grafting with poly(amidoamine) (PAMAM) dendrimer. The novel functionalized matrix with plenty amino groups circumvents the troublesome solubility problem of CNTs in solvents, especially in water, greatly expanding the scope of the application of carbon nanotubes. The GOx and HRP immobilized CNT-PAMAM based on the functional CNTs was synthesized. The bi-enzymatic CNT-PAMAM nano-composites are highly dispersible in water and show very promising applications in the fabrication of mediator-free bi-enzymatic biosensors for sensitive detection of glucose. The cooperation of nano-composite between CNT and high dense GOx and HRP results in very high sensitivity to glucose with a current response of 2200 nA mM−1 and fast response (~1 s). The modified electrode exhibits a wide linear response range for glucose from 4.0 μM to 1.2 mM (R = 0.9971, N = 15), with a detection limit of 2.5 μM. The negative electrode potential of −0.34 V is favorable for glucose detection in real samples without interference caused by other biomolecules.

E. Lojou et al [175] used dendritic macromolecules of two adjacent (G3.5 and G4) generations to modify gold or carbon electrodes. Dendrimers have been shown to adsorb spontaneously on electrode materials as compressed macromolecular films. They are able to inhibit (G3.5) or promote (G4) electroactive anionic species such as Fe(CN)63−/4− used as a probe system. Mixed protein/dendrimer assemblies have been constructed with proteins differing in charge, nature of the prosthetic groups and sizes such as lysozyme, cytochrome c, polyhemic cytochrome c3 or glucose oxidase. Generally, the stability of adsorbed films seems to be limited to one dendrimer/protein bilayer. Owing to the satisfactory stability of composite cytochrome c3/G3.5 or glucose oxidase/G4 films, biosensing applications are described for metal bioremediation and glucose detection, respectively.

Christopher M.A. Brett et al [176] described a new approach to produce nanostructured electrocatalytic membranes using a combination of three methods. Using the layer-by-layer technique, PAMAM dendrimers with cobalt hexacyanoferrates-modified gold nanoparticles were alternated with poly(vinylsulfonic acid) layers on ITO (indium tin oxide) electrodes (Figure 54). This film was used as substrate for immobilization of glucose oxidase in the presence of bovine serum albumin and glutaraldehyde as cross-linker. The modified electrode was successfully applied as a biosensor for the amperometric measurement of glucose, using glucose oxidase enzyme, at 0.0 V vs. SCE.

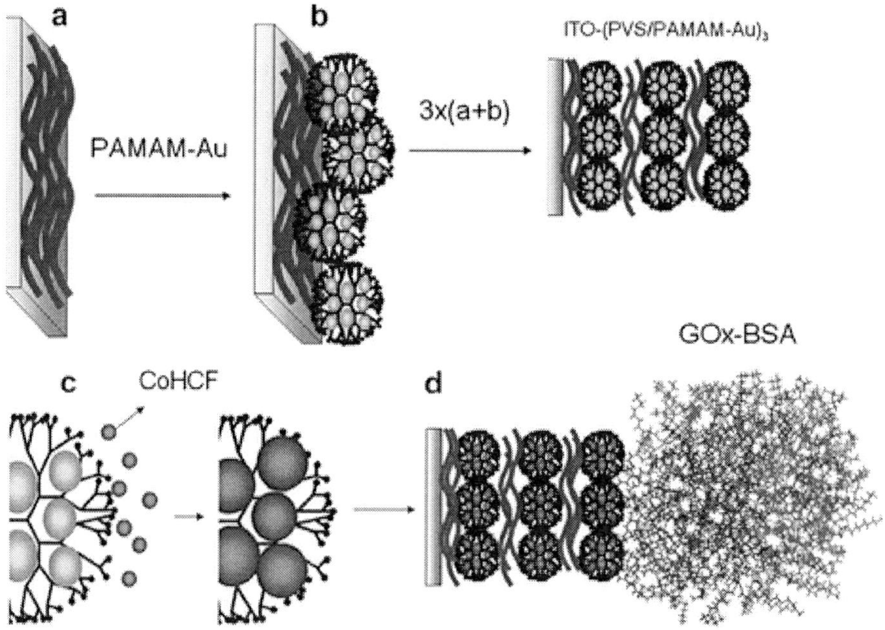

Figure 54. Fabrication of layer-by-layer films comprising PVS and PAMAM–Au.

M. Pilar García Armada et al [177] described the electrochemical characterization of polymethylferrocenyl dendrimers deposited onto a platinum electrode and their applications as hydrogen peroxide and glucose sensor. The redox dendrimers consist of flexible poly(propileneimine) dendrimer cores functionalised with octamethylferrocenyl units. Amperometric biosensors for glucose were prepared by immobilization of glucose oxidase onto these modified electrodes.

The detection of aromatic compounds from pesticides and industrial wastewater has become of great interest, since these compounds withstand chemical oxidation and biological degradation, accumulating in the environment. Valtencir Zucolotto et al [178] reported catechol biosensing using a nanostructured layer-by-layer film containing Cl-catechol 1,2-dioxygenase. A highly sensitive biosensor for detecting catechol was obtained with the immobilization of Cl-catechol 1,2-dioxygenase (CCD) in nanostructured films. CCD layers were alternated with poly(amidoamine) generation 4 (PAMAM G4) dendrimer using the electrostatic layer-by-layer (LbL) technique (Figure 55). Circular dichroism (CD) measurements indicated that the immobilized CCD preserved the same conformation as in solution. PAMAM/CCD 10-bilayer films were employed in detecting diluted catechol solutions using either an optical or electrical approach. Due to the mild immobilization conditions employed, especially regarding the pH and ionic strength of the dipping solutions, CCD remained active in the films for periods longer than 3 weeks.

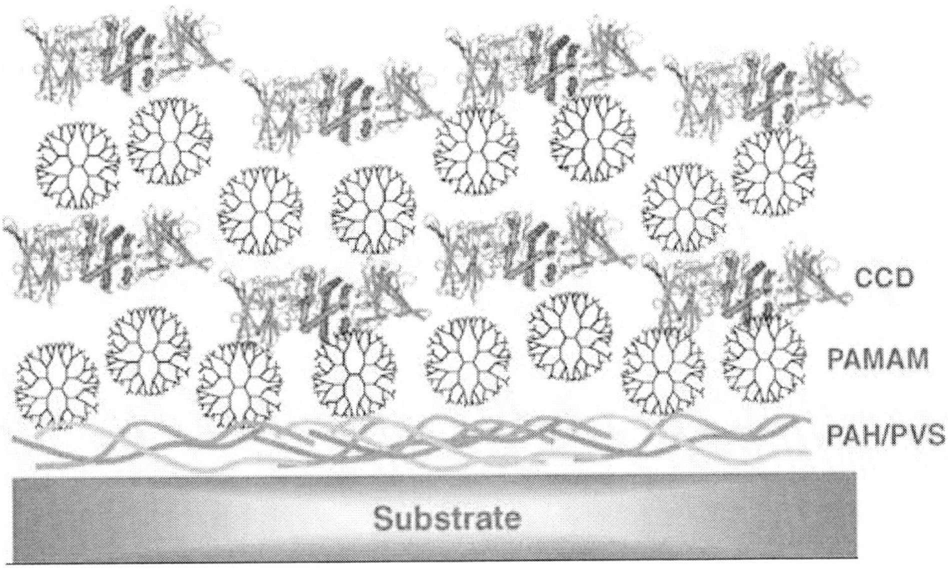

Figure 55. Catechol biosensing using a nanostructured layer-by-layer film containing Cl-catechol 1,2-dioxygenase.

Ru-Qin Yu et al [179] reported a hydrogen peroxide biosensor based on nano-Au/PAMAM dendrimer/cystamine modified gold electrode. Horseradish peroxidase (HRP) to a nano-Au monolayer, which was supported by PAMAM dendrimer/cystamine modified gold electrode. The high affinity of PAMAM dendrimer for nano-Au with its amine groups was used to realize the role of nano-Au as an intermediator to retain the high bioactivity of HRP. A hydrogen peroxide biosensor was fabricated based on the effective immobilization of the high bioactivity of HRP. In the presence of the hydroquinone mediator in the solution, the immobilized HRP exhibited excellent electrocatalytical response to the reduction of H2O2. The biosensor exhibited high sensitivity and good stability and reproducibility. Performance of the biosensor was evaluated with respect to possible interferences and the application to real sample analysis.

T. Hianik et al [180] reported an acetylcholinesterase sensors based on gold electrodes modified with dendrimer and polyaniline. Potentiometric and amperometric enzyme sensors based on modified gold electrodes have been developed and compared in pesticide determination. PAMAM dendrimer (generation G4) stabilized with 1-hexadecanethiol was used for the immobilization of acetylcholinesterase from electric eel and choline oxidase from Alcaligenes species in the assembly of amperometric sensor. Polyaniline-doped with camphorsulfonic acid was used to obtain potentiometric response. Trichlorfon, carbofuran and eserine suppress the biosensor response due to their inhibitory effect. The difference in the biosensor behavior and the high sensitivity of the dendrimer modified sensor to the inhibitors is due to the specific organization of protein layer at charged surface of the modifier macromolecules.

T. Hianik et al [181] studied the properties of mixed alkanethiol–dendrimer layers on a gold support and their application in biosensing. They showed that properties of glucose sensor can be modified using a different ratio of 1-hexadecanethiol (HDT) and poly(amidoamine) dendrimer of first generation (G1). The cyclic voltammetry in the presence of the redox couple, Fe(CN)63−/Fe(CN)64−, was used for estimating how effectively the layer blocks the redox probe's access to the electrode surface. A scanning electrochemical microscope (SECM) was used to image the resulting distribution of the organic compounds. T. Hianik et al found that with increasing content of dendrimers, the integrity of the layers was improved.

Naifei Hu et al [182] reported a heme protein films with polyamidoamine dendrimer for direct electrochemistry and electrocatalysis. Biocompatible nanosized polyamidoamine (PAMAM) dendrimer films provided a suitable microenvironment for heme proteins to transfer electron directly with underlying pyrolytic graphite (PG) electrodes. Hemoglobin (Hb), myoglobin (Mb), horseradish peroxidase (HRP), and catalase (Cat) incorporated in PAMAM films exhibited a pair of well-defined, quasi-reversible cyclic voltammetric peaks, respectively, characteristic of the protein heme Fe(III)/Fe(II) redox couples. Oxygen, hydrogen peroxide, and nitrite were catalytically reduced at the protein-PAMAM film electrodes, showing the potential applicability of the films as the new type of biosensors or bioreactors based on direct electrochemistry of the proteins.

Tibor Hianik et al [183] reported a QCM Immunosensor Based on Polyamidoamine Dendrimers. The poly(amidoamine)(PAMAM) dendrimers of fourth generation (G4) and 1-hexadecanethiol (HDT) has been used for formation of self assembled monolayers on a gold surface for preparation of quartz crystal microbalance (QCM) immunosensor for detection of human IgG. The anti-IgG antibodies were immobilized on a surface of G4-HDT layers or those covered by Protein A (Figure 56).

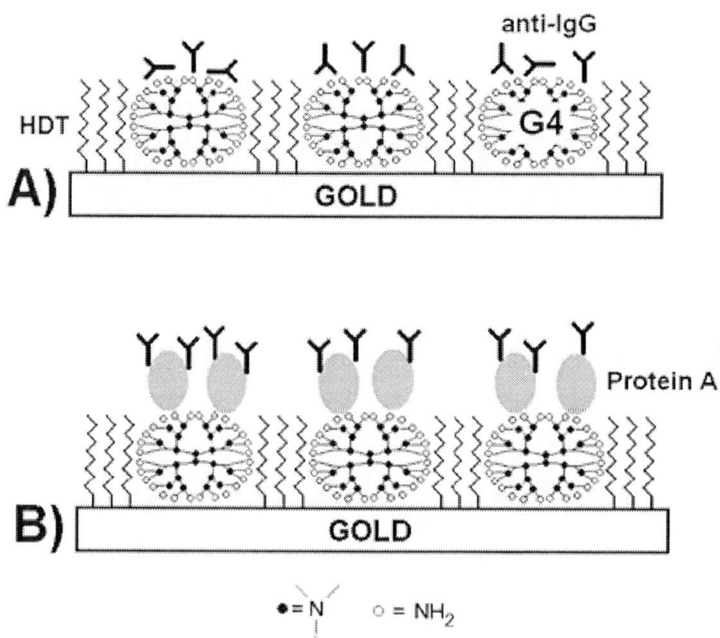

Figure 56. Quartz crystal microbalance immunosensor based on polyamidoamine dendrimers.

In later case the oriented immobilization of anti-IgG takes place due to preferred interaction of F_c portion of antibodies with the Protein A. The detection limit in a flow mode was 7 nM both for biosensor with and without Protein A. Increased concentration of IgG and formation of Ab-Ag complexes resulted in increase of motional resistance, that suggest increased contribution of viscoelasticity, as revealed from impedance analysis.

Mark H. Schoenfisch et al [184] reported the preparation and characterization of nitric oxide (NO)-releasing dendrimer conjugates. Generation 3 and 5 polypropylenimine dendrimers (DAB-Am-16 and DAB-Am-64) were modified at the exterior to impart different amine functionalities. The ability to store NO on a dendritic scaffold using N-diazeniumdiolate NO donors was examined via the reaction of primary amine, secondary amine, and amide functionalities with high pressures of NO (5 atm). The secondary amine dendrimer conjugates exhibited a high storage capacity for NO (up to 5.6 ímol NO/mg), greatly increasing the "payload" of released NO over existing macromolecular NO donors. The multivalent exterior of dendrimers allows for the future combination of NO donors and other functionalities on a single molecular scaffold, enabling diverse utility as NO storage/delivery systems.

8.2. Optical Biosensor

Andrei G. Kutateladze et al [185] reported photoinduced signal amplification through controlled externally sensitized fragmentation in masked sensitizers.

Atsushi Ikeda et al [186] used PAMAM dendrimers (G1-G4) as stationary phases to deposit porphyrinintercalated DNA onto ITO electrodes. The surface density of the external amino moieties of the dendrimers, i.e., the cationic charge density on the surface, increased with each generation, and correspondingly increased the surface coverage of DNA and porphyrin units through electrostatic interactions and intercalation. The resulting dendrimer-DNA-porphyrin membranes were sufficiently stable that they did not peel off the ITO electrodes. The resulting electrodes generated photocurrent waves in response to irradiation with visible light. The photocurrent increased exponentially upon increasing the generation of the dendrimer, i.e., a positive dendritic effect in these systems can be observed; actually, the photocurrent for the G4-DNA-porphyrin membrane (77.5 nA cm-2 at 440 nm) was ca. nine times larger than that for the G1-DNA-1 membrane (8.5 nA cm-2 at 440 nm), and the quantum yield for the former (1.0%) was five times larger than that for the latter (0.2%). Furthermore, the performance of the two-dimensional polymer G4 as a stationary phase was far superior to that of the one-dimensional polymer poly(allylamine hydrochloride) (PAH).

REFERENCES

[1] Christophe Galliot, Christophe Larre, Anne-Marie Caminade, Jean-Pierre Majoral. Regioselective Stepwise Growth of Dendrimer Units in the Internal Voids of a Main Dendrimer. *Science*, 1997, 277, 1981-1984

[2] Daniel Taton, Yves Gnanou, Rachid Matmour, Stephanie Angot, Sijian Hou, Raju Francis, Benedicte Lepoittevin, Delphine Moinard, and Jerome Babin. Controlled

polymerizations as tools for the design of star-like and dendrimer-like polymers. *Polym. Int.* 55:1138–1145 (2006)

[3] George R. Newkome, Kishore K. Kotta, and Charles N. Moorefield. Design, Synthesis, and Characterization of Conifer-Shaped Dendritic Architectures. *Chem. Eur. J.* 2006, 12, 3726 – 3734.

[4] Sandra Reemers, Ahmed Mourran, Helmut Keul, and Martin Moller. Novel Route to Dendritic Structures and Their Application for Surface Modification. *Journal of Polymer Science: Part A: Polymer Chemistry*, Vol. 44, 1372–1386 (2006).

[5] Chun-Yan Hong, Ye-Zi You, Jun Liu, Cai-Yuan Pan. Dendrimer-Star Polymer and Block Copolymer Prepared by Reversible Addition-Fragmentation Chain Transfer (RAFT) Polymerization with Dendritic Chain Transfer Agent. *Journal of Polymer Science: Part A: Polymer Chemistry,* Vol. 43, 6379–6393 (2005).

[6] Jianhua Zou, Yongbing Zhao, and Wenfang Shi. Encapsulation Mechanism of Molecular Nanocarriers Based on Unimolecular Micelle Forming Dendritic Core-Shell Structural Polymers. *J. Phys. Chem.* B 2006, 110, 2638-2642.

[7] Paola Carbone, Aurora Calabretta, Marco Di Stefano, Fabrizia Negri, and Klaus Mu1llen. Shape Persistence and Bistability of Planar Three-Fold Core Polyphenylene Dendrimers: A Molecular Dynamics Study. *J. Phys. Chem.* A 2006, 110, 2214-2224.

[8] Chae Kyu Kim, Eun Seok Song, Hyung Joo Kim, Chiyoung Park, Jai Kyeong Kim, Young Chul Kim, Jae Woong Yu, Chulhee Kim. Conjugated Dendrimers with Triazine Peripheries and a Distyrylanthracene Core. *Journal of Polymer Science: Part A: Polymer Chemistry,* Vol. 44, 5855–5862 (2006).

[9] Jin-Seong Lee, June Huh, Cheol-Hee Ahn, Minhyung Lee, Tae Gwan Park. Synthesis of Novel Biodegradable Cationic Dendrimers. Macromol. *Rapid Commun.* 2006, 27, 1608–1614.

[10] Stefan Bernhardt, Martin Baumgarten, and Klaus Müllen. Dendritic Encapsulation – "Postsynthetic" Functionalizations of a Single Benzophenone Shielded by Shape-Persistent Polyphenylene Dendrons. *Eur. J. Org. Chem.* 2006, 2523–2529.

[11] Mackay B. Steffensen, Emily Hollink, Frank Kuschel, Monika Bauer, Eric E. Simanek. Dendrimers Based on [1,3,5]-Triazines. *Journal of Polymer Science: Part A: Polymer Chemistry*, Vol. 44, 3411–3433 (2006).

[12] Jianfeng Pan, Weihong Zhu, Shangfeng Li, Jun Xu, and He Tian. Synthesis of Carrier-Transporting Dendrimers with Perylenebis(dicarboximide)s as a Luminescent Core. *Eur. J. Org. Chem.* 2006, 986–1001.

[13] Michael L. Szalai, Dominic V. McGrath, David R. Wheeler, Thomas Zifer, and James R. McElhanon. *Dendrimers Based on Thermally Reversible Furan-Maleimide Diels-Alder Adducts*. 2007; 40(4); 818-823.

[14] Kohei Hosomizu, Hiroshi Imahori, Uwe Hahn, Jean-Francüois Nierengarten, Andrea Listorti, Nicola Armaroli, Takashi Nemoto, and Seiji Isoda. Dendritic Effects on Structure and Photophysical and Photoelectrochemical Properties of Fullerene Dendrimers and Their Nanoclusters. *J. Phys. Chem.* C. 2007; 111(6); 2777-2786.

[15] Su-Hoon Yim, June Huh, Cheol-Hee Ahn, and Tae Gwan Park. Development of a Novel Synthetic Method for Aliphatic Ester Dendrimers. *Macromolecules* 2007, 40, 205-210.

[16] Zi-Fa Shi, Le-Jia Wang, Hong Wang, Xiao-Ping Cao, and Hao-Li Zhang. Synthesis of Oligo(phenylene ethynylene)s with Dendrimer "Shells" for Molecular Electronics. *Organic Letters*. 2007; 9(4); 595-598.

[17] Jianping Lu, Ping Fang Xia, Pik Kwan Lo, Ye Tao, and Man Shing Wong. Synthesis and Properties of Multi-Triarylamine-Substituted Carbazole-Based Dendrimers with an Oligothiophene Core for Potential Applications in Organic Solar Cells and Light-Emitting Diodes. *Chem. Mater*. 2006, 18, 6194-6203.

[18] Herbert Schumann, and Yilmaz Aksu. Synthesis and Characterization of Silicon-Based Group 14 Dendrimers Si{(CH2)2Sn[(CH2)4MPh3]3}4 (M = Ge, Sn). *Organometallics* 2007, 26, 397-402.

[19] Stephane Campidelli, Chloe Sooambar, Enrique Lozano Diz, Christian Ehli, Dirk M. Guldi, and Maurizio Prato. Dendrimer-Functionalized Single-Wall Carbon Nanotubes: Synthesis, Characterization, and Photoinduced Electron Transfer. *J. Am. Chem. Soc.* 2006, 128, 12544-12552.

[20] Rajesh Kumar Kainthan, Edward B. Muliawan, Savvas G. Hatzikiriakos, and Donald E. Brooks. Synthesis, Characterization, and Viscoelastic Properties of High Molecular Weight Hyperbranched Polyglycerols. *Macromolecules* 2006, 39, 7708-7717.

[21] Yang Jiang, Jie-Yu Wang, Yuguo Ma, Yu-Xin Cui, Qi-Feng Zhou, and Jian Pei. Large Rigid Blue-Emitting π-Conjugated Stilbenoid-Based Dendrimers: Synthesis and Properties. *Org. Lett.* 2004, 6, 3297-3300.

[22] Koji Matsuoka, Mikiko Terabatake, Atsushi Umino, Yasuaki Esumi, Ken Hatano, Daiyo Terunuma, and Hiroyoshi Kuzuhara. Carbosilane Dendrimers Bearing Globotriaoses: Syntheses of Globotrioasyl Derivative and Introduction into Carbosilane Dendrimers. *Biomacromolecules* 2006, 7, 2274-2283.

[23] Oleg Lukin, Volker Gramlich, Ramchandra Kandre, Igor Zhun, Thorsten Felder, Christoph A. Schalley, and Grigoriy Dolgonos. Designer Dendrimers: Branched Oligosulfonimides with Controllable Molecular Architectures. *J. Am. Chem. Soc.* 2006, 128, 8964-8974.

[24] Blanca Gonzalez, Beatriz Alonso, Jose Losada, M. Pilar Garcia-Armada, and Carmen M. Casado. Aza-Crown Ethers Attached to Dendrimers through Amidoferrocenyl Units. *Organometallics* 2006, 25, 3558-3561.

[25] Mark L. Wolfenden and Mary J. Cloninger. Carbohydrate-Functionalized Dendrimers To Investigate the Predictable Tunability of Multivalent Interactions. Bioconjugate Chem. 2006, 17, 958-966.

[26] Alessia Amore, Rieko van Heerbeek, Niek Zeep, Jan van Esch, Joost N. H. Reek, Henk Hiemstra, and Jan H. van Maarseveen. Carbosilane Dendrimeric Carbodiimides: Site Isolation as a Lactamization Tool. *J. Org. Chem.* 2006, 71, 1851-1860.

[27] Virgil Percec, Mihai Peterca, Monika J. Sienkowska, Marc A. Ilies, Emad Aqad, Jan Smidrkal, and Paul A. Heiney. Synthesis and Retrostructural Analysis of Libraries of AB3 and Constitutional Isomeric AB2 Phenylpropyl Ether-Based Supramolecular Dendrimers. *J. Am. Chem. Soc.* 2006, 128, 3324-3334.

[28] Vincenzo Balzani, Heinz Bandmann, Paola Ceroni, Carlo Giansante, Uwe Hahn, Frank-Gerrit Klarner, Ute Muller, Walter M. Muller, Carla Verhaelen, Veronica Vicinelli, and Fritz Vogtle. Host-Guest Complexes between an Aromatic Molecular Tweezer and Symmetric and Unsymmetric Dendrimers with a 4,4'-Bipyridinium Core. *J. Am. Chem. Soc.* 2006, 128, 637-648.

[29] Osamu Enoki, Hideaki Katoh, and Kimihisa Yamamoto. Synthesis and Properties of a Novel Phenylazomethine Dendrimer with a Tetraphenylmethane Core. *Organic Letters*. 2006, 8(4): 569-571.

[30] Julien Leclaire, Rodolphe Dagiral, Suzanne Fery-Forgues, Yannick Coppel, Bruno Donnadieu, Anne-Marie Caminade, and Jean-Pierre Majoral. Octasubstituted Metal-Free Phthalocyanine as Core of Phosphorus Dendrimers: A Probe for the Properties of the Internal Structure. *J. Am. Chem. Soc*. 2005, 127, 15762-15770.

[31] Zinaida B. Shifrina, Marina S. Rajadurai, Nina V. Firsova, Lyudmila M. Bronstein, Xinlei Huang, Alexander L. Rusanov, and Klaus Muellen. Poly(Phenylene-pyridyl) Dendrimers: Synthesis and Templating of Metal Nanoparticles. *Macromolecules* 2005, 38, 9920-9932.

[32] Natarajan Raju, Rama S. Ranganathan, Mike F. Tweedle and Rolf E. Swenson. A new dendrimer scaffold for preparing dimers or tetramers of biologically active molecules. *Tetrahedron Letters* 46 (2005) 1463–1465.

[33] Benoit Muls, Hiroshi Uji-i, Sergey Melnikov, Alain Moussa, Wendy Verheijen, Jean-Philippe Soumillion, Jacob Josemon, Klaus Mullen, and Johan Hofkens. Direct Measurement of the End-to-End Distance of Individual Polyfluorene Polymer Chains. *ChemPhysChem* 2005, 6, 2286 – 2294.

[34] Tong Zhou and Shing Bor Chen. A Simulation Study on Dynamics of Dendrimer-Polymer Conjugates. *Macromolecules* 2006, 39, 6686-6692.

[35] Neil Cumpstey, Raghu N. Bera, Paul L. Burn, and Ifor D. W. Samuel. Investigating the Effect of Steric Crowding in Phosphorescent Dendrimers. *Macromolecules* 2005, 38, 9564-9570.

[36] Ying Wang, Xinbing Xie, and Theodore Goodson, III. Enhanced Third-Order Nonlinear Optical Properties in Dendrimer-Metal Nanocomposites. *Nano Letters*. 2005, 5(12): 2379-2384.

[37] David Scanu, Natalia P. Yevlampieva, and Robert Deschenaux. Polar and Electrooptical Properties of [60]Fullerene-Containing Poly(benzyl ether) Dendrimers in Solution. *Macromolecules*. 2007; 40(4); 1133-1139.

[38] Winston Ong, and Robin L. McCarley. Chemically and Electrochemically Mediated Release of Dendrimer End Groups. *Macromolecules* 2006, 39, 7295-7301.

[39] Alena Krupkova, Jan Cermak, Zuzana Walterova, and Jiri Horsky. Quantitative Interpretation of MALDI-TOF Mass Spectra of Imperfect Carbosilane Dendrimers. *Analytical Chemistry*. 2007; 79(4); 1639-1645.

[40] Hak-Joon Seok, Mi-Young Hong, Young-Ja Kim, Min-Kyu Han, Dohoon Lee, Jung-Hwa Lee, Jong-Shin Yoo, Hak-Sung Kim. Mass spectrometric analysis of aYnity-captured proteins on a dendrimer-based immunosensing surface: investigation of on-chip proteolytic digestion. *Analytical Biochemistry* 337 (2005) 294–307.

[41] Sabine Rosenfeldt, Elena Karpuk, Matthias Lehmann, Herbert Meier, Peter Lindner, Ludger Harnau, and Matthias Ballauff. The Solution Structure of Stilbenoid Dendrimers: A Small-Angle Scattering Study. ChemPhysChem 2006, 7, 2097 – 2104.

[42] Kai Landskron and Geoffrey A. Ozin. Periodic Mesoporous Dendrisilicas. *Science*, 2004, 306: 1529-1532.

[43] Byeongdu Lee, Young-Hee Park, Yong-Taek Hwang, Weontae Oh, Jinhwan Yoon, and Moonhor Ree. Ultralow-k nanoporous organosilicate dielectric films imprinted with dendritic spheres. *Nature Mmaterials*, 2005, 4: 147-151.

[44] Steven C. Zimmerman, Michael S. Wendland, Neal A. Rakow, Ilya Zharov, Kenneth S. Suslick. Synthetic hosts by monomolecular imprinting inside dendrimers. *Nature*, 2002, 418: 399-403.

[45] Teddie Magbitang, Victor Y. Lee, Jennifer N. Cha, Hsiao-Lin Wang, W. Richard Chung, Robert D. Miller, Geraud Dubois, Willi Volksen, Ho-Cheol Kim, and James L. Hedrick. Oriented Nanoporous Lamellar Organosilicates Templated from Topologically Unsymmetrical Dendritic-Linear Block Copolymers. *Angew. Chem. Int. Ed.* 2005, 44, 7574 –7580.

[46] Andras Perl, Maria Peter, Bart Jan Ravoo, David N. Reinhoudt, and Jurriaan Huskens. Heavyweight Dendritic Inks for Positive Microcontact Printing. *Langmuir* 2006, 22, 7568-7573.

[47] Sander Langereis, Quido G. de Lussanet, Marcel H. P. van Genderen, E. W. Meijer, Regina G. H. Beets-Tan, Arjan W. Griffioen, Jos M. A. van Engelshoven, and Walter H. Backes. Evaluation of Gd(III)DTPA-terminated poly(propylene imine) dendrimers as contrast agents for MR imaging. *NMR Biomed.* 2006;19:133–141.

[48] Yukihiro Hama, Marcelino Bernardo, Celeste A.S. Regino, Yoshinori Koyama, Martin W. Brechbiel, Murali C. Krishna, Peter L. Choyke, and Hisataka Kobayashi. MR Lymphangiography Using Dendrimer-Based Contrast Agents: A Comparison at 1.5T and 3.0T. *Magnetic Resonance in Medicine* 57:431–436 (2007).

[49] Yanjun Fu, Danute E. Nitecki, David Maltby, Gerhard H. Simon, Kirill Berejnoi, Hans-Juergen Raatschen, Benjamin M. Yeh, David M. Shames, and Robert C. Brasch. Dendritic Iodinated Contrast Agents with PEG-Cores for CT Imaging: Synthesis and Preliminary Characterization. *Bioconjugate Chem.* 2006, 17, 1043-1056.

[50] Amit K. Galande, Scott A. Hilderbrand, Ralph Weissleder, and Ching-Hsuan Tung. Enzyme-Targeted Fluorescent Imaging Probes on a Multiple Antigenic Peptide Core. *J. Med. Chem.* 2006, 49, 4715-4720.

[51] Vladimir S. Talanov, Celeste A. S. Regino, Hisataka Kobayashi, Marcelino Bernardo, Peter L. Choyke, and Martin W. Brechbiel. Dendrimer-Based Nanoprobe for Dual Modality Magnetic Resonance and Fluorescence Imaging. *Nano Letters*. 2006, 6 (7): 1459-1463.

[52] Nobuhiro Nishiyama, Aya Iriyama, Woo-Dong Jang, Kanjiro Miyata, Keiji Itaka, Yuji Inoue, Hidenori Takahashi, Yasuo Yanagi, Yasuhiro Tamaki, Hiroyuki Koyama, and Kazunori Kataoka. Light-induced gene transfer from packaged DNA enveloped in a dendrimeric photosensitizer. *Nature Materials*. 2005, 4: 934-941.

[53] M. Adeli, Z. Zarnegar, A. Dadkhah, R. Hossieni, F. Salimi, A. Kanani. Linear-Dendritic ABA Triblock Copolymers as Nanocarriers. *Journal of Applied Polymer Science*, Vol. 104, 267–272 (2007).

[54] M. Adeli, R. Haag. Multiarm Star Nanocarriers Containing a Poly(ethylene imine) Core and Polylactide Arms. *Journal of Polymer Science: Part A: Polymer Chemistry*, Vol. 44, 5740–5749 (2006).

[55] Kris C. Wood, Steven R. Little, Robert Langer, and Paula T. Hammond. Family of Hierarchically Self-Assembling Linear-Dendritic Hybrid Polymers for Highly Efficient Targeted Gene Delivery. *Angew. Chem. Int. Ed.* 2005, 44, 6704 –6708.

[56] Andrzej Myc, Istvan J. Majoros, Thommey P. Thomas, and James R. Baker, Jr. Dendrimer-Based Targeted Delivery of an Apoptotic Sensor in Cancer Cells. *Biomacromolecules* 2007, 8, 13-18.

[57] Rameshwer Shukla, Thommey P. Thomas, Jennifer L. Peters, Ankur M. Desai, Jolanta Kukowska-Latallo, Anil K. Patri, Alina Kotlyar, and James R. Baker, Jr. HER2 Specific Tumor Targeting with Dendrimer Conjugated Anti-HER2 mAb. *Bioconjugate Chem.*, 2006, 17 (5): 1109-1115.

[58] Sezen Gurdag, Jayant Khandare, Sarah Stapels, Larry H. Matherly, and Rangaramanujam M. Kannan. Activity of Dendrimer-Methotrexate Conjugates on Methotrexate-Sensitive and -Resistant Cell Lines. *Bioconjugate Chem.* 2006, 17, 275-283.

[59] Renu Singh Dhanikula and Patrice Hildgen. Synthesis and Evaluation of Novel Dendrimers with a Hydrophilic Interior as Nanocarriers for Drug Delivery. *Bioconjugate Chem.* 2006, 17, 29-41.

[60] Istvan J. Majoros, Andrzej Myc, Thommey Thomas, Chandan B. Mehta, and James R. Baker, Jr. PAMAM Dendrimer-Based Multifunctional Conjugate for Cancer Therapy: Synthesis, Characterization, and Functionality. *Biomacromolecules* 2006, 7, 572-579.

[61] Manuela Berna, Daniele Dalzoppo, Gianfranco Pasut, Maria Manunta, Lorella Izzo, Arwyn T. Jones, Ruth Duncan, and Francesco M. Veronese. Novel Monodisperse PEG-Dendrons as New Tools for Targeted Drug Delivery: Synthesis, Characterization and Cellular Uptake. *Biomacromolecules* 2006, 7, 146-153.

[62] Tae-il Kim, Jung-un Baek, Cheng Zhe Bai, Jong-sang Park. Arginine-conjugated polypropylenimine dendrimer as a non-toxic and efficient gene delivery carrier. *Biomaterials* 28 (2007) 2061–2067.

[63] Durairaj Chandrasekar, Ramakrishna Sistla, Farhan J. Ahmad, Roop K. Khar, Prakash V. Diwan. The development of folate-PAMAM dendrimer conjugates for targeted delivery of anti-arthritic drugs and their pharmacokinetics and biodistribution in arthritic rats. *Biomaterials* 28 (2007) 504–512.

[64] Seungpyo Hong, Pascale R. Leroueil, Istvan J. Majoros, Bradford G. Orr, James R. Baker, Jr., and Mark M. Banaszak Holl. The Binding Avidity of a Nanoparticle-Based Multivalent Targeted Drug Delivery Platform. *Chemistry and Biology* 14, 107–115.

[65] Tatsuya Okuda, Shigeru Kawakami, Naomi Akimoto, Takuro Niidome, Fumiyoshi Yamashita, Mitsuru Hashida. PEGylated lysine dendrimers for tumor-selective targeting after intravenous injection in tumor-bearing mice. *Journal of Controlled Release* 116 (2006) 330–336.

[66] Luca Costantino, Francesca Gandolfi, Leila Bossy-Nobs, Giovanni Tosi, Robert Gurny, Francesco Rivasi, Maria Angela Vandelli, Flavio Forni. Nanoparticulate drug carriers based on hybrid poly(D,L-lactide-co-glycolide)-dendron structures. *Biomaterials* 27 (2006) 4635–4645.

[67] Parag Kolhe, Jayant Khandare, Omathanu Pillai, Sujatha Kannan, Mary Lieh-Lai, Rangaramanujam M. Kannan. Preparation, cellular transport, and activity of polyamidoamine-based dendritic nanodevices with a high drug payload. *Biomaterials* 27 (2006) 660–669.

[68] H. Namazi, M. Adeli. Dendrimers of citric acid and poly (ethylene glycol) as the new drug-delivery agents. *Biomaterials* 26 (2005) 1175–1183.

[69] Youngseon Choi, Thommey Thomas, Alina Kotlyar, Mohammad T. Islam, and James R. Baker, Jr. Synthesis and Functional Evaluation of DNA-Assembled Polyamidoamine Dendrimer Clusters for Cancer Cell-Specific Targeting. *Chemistry and Biology*, 12: 35–43.

[70] Majad Hussain, Mikhail S. Shchepinov, Muhammad Sohail, Ibrahim F. Benter, Andrew J. Hollins, Edwin M. Southern, Saghir Akhtar. A novel anionic dendrimer for improved cellular delivery of antisense oligonucleotides. *Journal of Controlled Release* 99 (2004) 139–155.

[71] Michael F. Neerman, Wen Zhang, Alan R. Parrish, Eric E. Simanek. In vitro and in vivo evaluation of a melamine dendrimer as a vehicle for drug delivery. *International Journal of Pharmaceutics* 281 (2004) 129–132.

[72] W. Tansey, S. Ke, X.-Y. Cao, Marites J. Pasuelo, Sidney Wallace, Chun Li. Synthesis and characterization of branched poly(L-glutamic acid) as a biodegradable drug carrier. *Journal of Controlled Release* 94 (2004) 39– 51.

[73] Marcelo Mamede, Tsuneo Saga, Takayoshi Ishimori, Tatsuya Higashi, Noriko Sato, Hisataka Kobayashi, Martin W. Brechbiel, Junji Konishi. Hepatocyte targeting of 111In-labeled oligo-DNA with avidin or avidin–dendrimer complex. *Journal of Controlled Release* 95 (2004) 133– 141.

[74] Sunil Shaunak, Sharyn Thomas, Elisabetta Gianasi, Antony Godwin, Emma Jones, Ian Teo, Kamiar Mireskandari, Philip Luthert, Ruth Duncan, Steve Patterson, Peng Khaw, and Steve Brocchini. Polyvalent dendrimer glucosamine conjugates prevent scar tissue formation. *Nature Biotechnology*, 2004, 22: 977-984.

[75] Serge H. M. Sontjens, Dana L. Nettles, Michael A. Carnahan, Lori A. Setton, and Mark W. Grinstaff. Biodendrimer-Based Hydrogel Scaffolds for Cartilage Tissue Repair. *Biomacromolecules* 2006, 7, 310-316.

[76] Sonke Svenson, Donald A. Tomalia. Dendrimers in biomedical applications—reflections on the field. *Advanced Drug Delivery Reviews* 57 (2005) 2106– 2129.

[77] Ruth Duncan, Lorella Izzo. Dendrimer biocompatibility and toxicity. *Advanced Drug Delivery Reviews* 57 (2005) 2215– 2237.

[78] Cameron C Lee, John A MacKay, Jean M J Fréchet, and Francis C Szoka. Designing dendrimers for biological applications. *Nature Biotechnology*. 2005, 12: 1517-1526.

[79] Jeanette C. Roberts, Mahesh K. Bhalgat, and Richard T. Zera. Preliminary biological evaluation of polyamidoamine (PAMAM) StarburstTM dendrimers. *Journal of Biomedical Materials Research*, 30, 53-65 (1996).

[80] Muriel Blanzat, Cedric-Olivier Turrin, Anne-Marie Aubertin, Christiane Couturier-Vidal, Anne-Marie Caminade, Jean-Pierre Majoral, Isabelle Rico-Lattes, and Armand Lattes. Dendritic Catanionic Assemblies: In vitro Anti-HIV Activity ofPhosphorus-Containing Dendrimers Bearing Galb1cer Analogues. *ChemBioChem* 2005, 6, 2207 – 2213.

[81] Jun-Ichi Sakamoto, Tetsuo Koyama, Daisei Miyamoto, Sangchai Yingsakmongkon, Kazuya I. P. J. Hidari, Wipawee Jampangern, Takashi Suzuki, Yasuo Suzuki, Yasuaki Esumi, Ken Hatano, Daiyo Terunuma and Koji Matsuoka. Thiosialoside clusters using carbosilane dendrimer core scaffolds as a new class of influenza neuraminidase inhibitors. *Bioorganic and Medicinal Chemistry Letters* 17 (2007) 717–721.

[82] Tatsuya Okuda, Shigeru Kawakami, Tadahiro Maeie, Takuro Niidome, Fumiyoshi Yamashita, Mitsuru Hashida. Biodistribution characteristics of amino acid dendrimers and their PEGylated derivatives after intravenous administration. *Journal of Controlled Release* 114 (2006) 69–77.

[83] Aristarchos Papagiannaros, Kostas Dimas, George Th Papaioannou, Costas Demetzos. Doxorubicin–PAMAM dendrimer complex attached to liposomes: Cytotoxic studies against human cancer cell line. *International Journal of Pharmaceutics* 302 (2005) 29–38.

[84] Natarajan Raju, Rama S. Ranganathan, Mike F. Tweedle and Rolf E. Swenson. A new dendrimer scaffold for preparing dimers or tetramers of biologically active molecules. *Tetrahedron Letters* 46 (2005) 1463–1465.

[85] Yasuhiro Haba, Atsushi Harada, Toru Takagishi, Kenji Kono. Synthesis of biocompatible dendrimers with a peripheral network formed by linking of polymerizable groups. *Polymer* 46 (2005) 1813–1820.

[86] Ben J. Boyd, Lisa M. Kaminskas, Peter Karellas, Guy Krippner, Romina Lessene, and Christopher J. H. Porter. Cationic Poly-L-lysine Dendrimers: Pharmacokinetics, Biodistribution, and Evidence for Metabolism and Bioresorption after Intravenous Administration to Rats. *Molecular Pharmaceutics* Vol. 3, NO. 5, 614-627.

[87] Anna Gopin, Sharon Ebner, Bernard Attali, and Doron Shabat. Enzymatic Activation of Second-Generation Dendritic Prodrugs: Conjugation of Self-Immolative Dendrimers with Poly(ethylene glycol) via Click Chemistry. *Bioconjugate Chem.* 2006, 17, 1432-1440.

[88] Andreas G. Schatzlein, Bernd H. Zinselmeyer, Adurrahim Elouzi, Christine Dufes, Ya Tsz A. Chim, Clive J. Roberts, Martyn C. Davies, Avril Munro, Alexander I. Grayb, Ijeom F. Uchegbu. Preferential liver gene expression with polypropylenimine dendrimers. *Journal of Controlled Release* 101 (2005) 247–258.

[89] Antony D'Emanuele, Rachaneekorn Jevprasesphant, Jeffrey Penny, David Attwood. The use of a dendrimer-propranolol prodrug to bypass efflux transporters and enhance oral bioavailability. *Journal of Controlled Release* 95 (2004) 447– 453.

[90] Petr Niederhafner, Jaroslav Sebestik and Jan Jezek. Peptide dendrimers. *J. Peptide Sci.* 11: 757–788 (2005).

[91] Hu Yang, Stephanie T. Lopina. In vitro enzymatic stability of dendritic peptides. *J. Biomed. Mater. Res.* 76A: 398–407, 2006.

[92] TAMIS DARBRE AND JEAN-LOUIS REYMOND. Peptide Dendrimers as Artificial Enzymes, Receptors, and Drug-Delivery Agents. *Acc. Chem. Res.* 2006, 39, 925-934.

[93] B. Klajnert, J. Janiszewska, Z. Urbanczyk-Lipkowska, M. Bryszewska, D. Shcharbin, M. Labieniec. Biological properties of low molecular mass peptide dendrimers. *International Journal of Pharmaceutics* 309 (2006) 208–217.

[94] Didier Astruc, Feng Lu, and Jaime Ruiz Aranzaes. Nanoparticles as Recyclable Catalysts: The Frontier between Homogeneous and Heterogeneous Catalysis. *Angew. Chem. Int.* Ed. 2005, 44, 7852 – 7872.

[95] Suhas A. Chavan, Wouter Maes, Lieven E. M. Gevers, Joos Wahlen, Ivo F. J. Vankelecom, Pierre A. Jacobs, Wim Dehaen, and Dirk E. De Vos. Porphyrin-Functionalized Dendrimers: Synthesis and Application as Recyclable Photocatalysts in a Nanofiltration Membrane Reactor. *Chem. Eur. J.* 2005, 11, 6754 – 6762.

[96] Lara-Isabel Rodriguez, Oriol Rossell, Miquel Seco, Arnald Grabulosa, Guillermo Muller, and Merce Rocamora. Carbosilane Dendrimers Peripherally Functionalized with P-Stereogenic Monophosphines. Catalytic Behavior of Their Allylpalladium

Complexes in the Asymmetric Hydrovinylation of Styrene. *Organometallics* 2006, 25, 1368-1376.

[97] Dong-Mei Xu, Ke-Da Zhang, Xiu-Lin Zhu. Preparation of Ag Nanoparticles in the Presence of Low Generational Poly(ester-amine) Dendrimers. *Journal of Applied Polymer Science,* Vol. 104, 422–426 (2007).

[98] Stephane Campidelli, Chloe Sooambar, Enrique Lozano Diz, Christian Ehli, Dirk M. Guldi, and Maurizio Prato. Dendrimer-Functionalized Single-Wall Carbon Nanotubes: Synthesis, Characterization, and Photoinduced Electron Transfer. *J. Am. Chem. Soc.* 2006, 128, 12544-12552.

[99] Sreenivasa Reddy Puniredd and M. P. Srinivasan. Covalent Molecular Assembly in a Supercritical Medium: Formation of Nanoparticles Encapsulated in Immobilized Dendrimers. *Ind. Eng. Chem. Res.* 2007, 46, 464-471.

[100] Marc R. Knecht, Joaquin C. Garcia-Martinez, and Richard M. Crooks. Synthesis, Characterization, and Magnetic Properties of Dendrimer-Encapsulated Nickel Nanoparticles Containing <150 Atoms. *Chem. Mater.* 2006, 18, 5039-5044.

[101] Seong A Kim, Dae-Hwan Jung, Yun-Ho Kim, Min Ah Kang, and Hee-Tae Jung. Core-Controlled Formation of Micellar Cubic Assemblies of Supramolecular Dendrimers: From Their Mechanism to Formation of Gold Nanostructures. *Macromolecules* 2006, 39, 6186-6190.

[102] Marc R. Knecht, Joaquin C. Garcia-Martinez, and Richard M. Crooks. Hydrophobic Dendrimers as Templates for Au Nanoparticles. *Langmuir* 2005, 21, 11981-11986.

[103] Kimihisa Yamamoto, Masayoshi Higuchi, Satoshi Shiki, Masanori Tsuruta and Hiroshi Chiba. Stepwiseradialcomplexationofimine groups in phenylazomethine dendrimers. *Nature* 2002, 415: 509-511.

[104] JIAN WENG, QIQING ZHANG. Syntheses and Nickel Coordination Properties of Cyclen Substituted with Mixed Stilbene/Poly(ethylene glycol) Dendritic Arms. *Journal of Polymer Science: Part A: Polymer Chemistry*, Vol. 43, 5414–5428 (2005).

[105] Francisco Tarazona-Vasquez and Perla B. Balbuena. Dendrimer-Tetrachloroplatinate Precursor Interactions. 1. Hydration of Pt(II) Species and PAMAM Outer Pockets. *J. Phys. Chem.* A 2007, 111, 932-944.

[106] Li Shen, Mei Shi, Fuyou Li, Dengqing Zhang, Xianghong Li, Enxian Shi, Tao Yi, Yukou Du, and Chunhui Huang. Polyaryl Ether Dendrimer with a 4-Phenylacetyl-5-pyrazolone-based Terbium(III) Complex as Core: Synthesis and Photopysical Properties. *Inorg. Chem.* 2006, 45, 6188-6197.

[107] Tao Zhou, Hendrik Neubert, Ding Yong Liu, Zu Dong Liu, Yong Min Ma, Xiao Le Kong, Wei Luo, Sykes Mark, and Robert C. Hider. Iron Binding Dendrimers: A Novel Approach for the Treatment of Haemochromatosis. *J. Med. Chem.* 2006, 49, 4171-4182.

[108] Giacomo Bergamini, Paola Ceroni, Mauro Maestri, Sang-Kyu Lee, Jeroen van Heyst, Fritz Vogtle. Cyclam cored luminescent dendrimers as ligands for Co(II), Ni(II) and Cu(II) ions. *Inorganica Chimica Acta* 360 (2007) 1043–1051.

[109] Guijiang Zhou, Wai-Yeung Wong, Bing Yao, Zhiyuan Xie, and Lixiang Wang. Triphenylamine-Dendronized Pure Red Iridium Phosphors with Superior OLED Efficiency/Color Purity Trade-Offs. *Angew. Chem. Int. Ed.* 2007, 46, 1149 –1151.

[110] Herbert Schumann, Yilmaz Aksu, and Birgit C. Wassermann. Convergent Synthesis and Characterization of Organotin Dendrimers Sn{(CH2)nSn[(CH2)4SnPh3]3}4 (n= 3, 4). *Organometallics* 2006, 25, 3428-3434.

[111] Venkateswarlu Juttukonda, Robert L. Paddock, Jeffery E. Raymond, Dan Denomme, Andrew E. Richardson, Laura E. Slusher, and Bradley D. Fahlman. Facile Synthesis of Tin Oxide Nanoparticles Stabilized by Dendritic Polymers. *J. Am. Chem. Soc.* 2006, 128, 420-421.

[112] Takane Imaoka, Reiko Tanaka, and Kimihisa Yamamoto. Investigation of a Molecular Morphology Effect on Polyphenylazomethine Dendrimers; Physical Properties and Metal-Assembling Processes. *Chem. Eur. J.* 2006, 12, 7328 – 7336.

[113] Inma Angurell, Oriol Rossell, Miquel Seco, and Eliseo Ruiz. Dendrimers Containing Two Metallic Layers. Chloride Migration from Peripheral Gold, Palladium, or Rhodium Metals to Internal Ruthenium Atoms. *Organometallics* 2005, 24, 6365-6373.

[114] Lei Qian and Xiurong Yang. Polyamidoamine Dendrimers-Assisted Electrodeposition of Gold-Platinum Bimetallic Nanoflowers. *J. Phys. Chem.* B 2006, 110, 16672-16678.

[115] Marie P. Cifuentes, Clem E. Powell, Joseph P. Morrall, Andrew M. McDonagh, Nigel T. Lucas, Mark G. Humphrey, Marek Samoc, Stephan Houbrechts, Inge Asselberghs, Koen Clays, Andre Persoons, and Takashi Isoshima. Electrochemical, Spectroelectrochemical, and Molecular Quadratic and Cubic Nonlinear Optical Properties of Alkynylruthenium Dendrimers. *J. Am. Chem. Soc.* 2006, 128, 10819-10832.

[116] Stephane Cordier, Kaplan Kirakci, Denise Mery, Christiane Perrin, Didier Astruc. Mo6Xi8 Nanocluster cores (X = Br, I): From inorganic solid state compounds to hybrids. *Inorganica Chimica Acta* 359 (2006) 1705–1709.

[117] Denise Mery, Lauriane Plault, Catia Ornelas, Jaime Ruiz, Sylvain Nlate, Didier Astruc, Jean-Claude Blais, Joao Rodrigues, Stephane Cordier, Kaplan Kirakci, and Christiane Perrin. From Simple Monopyridine Clusters [Mo6Br13(Py-R)][n-Bu4N] and Hexapyridine Clusters [Mo6X8(Py-R)6][OSO2CF3]4 (X = Br or I) to Cluster-Cored Organometallic Stars, Dendrons, and Dendrimers. *Inorg. Chem.* 2006, 45, 1156-1167.

[118] Denise Méry, Cátia Ornelas, Marie-Christine Daniel, Jaime Ruiz, João Rodrigues, Didier Astruc, Stéphane Cordier, Kaplan Kirakci, Christiane Perrin. Mo6Br8-Cluster-cored organometallic stars and dendrimers. *C. R. Chimie* 8 (2005) 1789–1797.

[119] Sung Cho, Wei-Shi Li, Min-Chul Yoon, Tae Kyu Ahn, Dong-Lin Jiang, Jiwon Kim, Takuzo Aida, and Dongho Kim. Relationship between Incoherent Excitation Energy Migration Processes and Molecular Structures in Zinc(II) Porphyrin Dendrimers. *Chem. Eur. J.* 2006, 12, 7576 – 7584.

[120] Markus Arendt, Wei Sun, Jens Thomann, Xiulan Xie, and Thomas Schrader. Dendrimeric Bisphosphonates for Multivalent Protein Surface Binding. *Chem. Asian J.* 2006, 1, 544 – 554.

[121] Mauri A. Kostiainen, Geza R. Szilvay, David K. Smith, Markus B. Linder, and Olli Ikkala. Multivalent Dendrons for High-Affinity Adhesion of Proteins to DNA. *Angew. Chem. Int. Ed.* 2006, 45, 3538 –3542.

[122] Chun-Jen Su, Yi-Chun Liu, Hsin-Lung Chen, Yen-Cheng Li, Hsien-Kuang Lin, Wen-Liang Liu, and Chain-Shu Hsu. Two-Dimensional Densely Packed DNA Nanostructure Derived from DNA Complexation with a Low-Generation Poly(amidoamine) Dendrimer. *Langmuir* 2007, 23, 975-978.

[123] M. Mondeshki, G. Mihov, R. Graf, H. W. Spiess, and K. Mullen. Self-Assembly and Molecular Dynamics of Peptide-Functionalized Polyphenylene Dendrimers. *Macromolecules* 2006, 39, 9605-9613.

[124] Marine Guillot-Nieckowski, Derk Joester, Meike Stohr, Myriam Losson, Marc Adrian, Bjoern Wagner, Manfred Kansy, Harry Heinzelmann, Raphael Pugin, Francois Diederich, and Jean-Louis Gallani. Self-assembly, DNA Complexation, and pH Response of Amphiphilic Dendrimers for Gene Transfection. *Langmuir* 2007, 23, 737-746.

[125] Prabal K. Maiti and Biman Bagchi. Structure and Dynamics of DNA-Dendrimer Complexation: Role of Counterions, Water, and Base Pair Sequence. *Nano Letters* 2006, 6 (11): 2478-2485.

[126] B. Klajnert, J. Cladera, and M. Bryszewska. Molecular Interactions of Dendrimers with Amyloid Peptides: pH Dependence. *Biomacromolecules* 2006, 7, 2186-2191.

[127] Linzhu Zhou, Lu Gan, Haibing Li, Xiangliang Yang. Studies on the interactions between DNA and PAMAM with fluorescent probe [Ru(phen)2dppz]2+. *Journal of Pharmaceutical and Biomedical Analysis* 43 (2007) 330–334.

[128] Dharam Paul, Hiroyuki Miyake, Satoshi Shinoda, and Hiroshi Tsukube. Proteo-Dendrimers Designed for Complementary Recognition of Cytochrome c: Dendrimer Architecture toward Nanoscale Protein Complexation. *Chem. Eur. J.* 2006, 12, 1328 – 1338.

[129] Akihiro Yamada, Ken Hatano, Tetsuo Koyama, Koji Matsuoka, Naonori Takahashi, Kazuya I. P. J. Hidari, Takashi Suzuki, Yasuo Suzukic, and Daiyo Terunuma. Lactotriaose-containing carbosilane dendrimers: Syntheses and lectin-binding activities. *Bioorganic and Medicinal Chemistry* 15 (2007) 1606–1614.

[130] David K. Smith. Dendritic Gels—Many Arms Make Light Work. *Adv. Mater.* 2006, 18, 2773–2778.

[131] Prasad Taranekar, Jin-Young Park, Derek Patton, Timothy Fulghum, Glenn J. Ramon, and Rigoberto Advincula. Conjugated Polymer Nanoparticles via Intramolecular Crosslinking of Dendrimeric Precursors. *Adv. Mater.* 2006, 18, 2461–2465.

[132] ZHISHEN GE, SHIZHONG LUO, SHIYONG LIU. Syntheses and Self-Assembly of Poly(benzyl ether)-b- Poly(N-isopropylacrylamide) Dendritic–Linear Diblock Copolymers. *J. Polym. Sci. Part A: Polym. Chem.* 44: 1357–1371, 2006.

[133] Maisie J. Joralemon, Rachel K. O'Reilly, Craig J. Hawker, and Karen L. Wooley. Shell Click-Crosslinked (SCC) Nanoparticles: A New Methodology for Synthesis and Orthogonal Functionalization. *J. Am. Chem. Soc.* 2005, 127, 16892-16899.

[134] BING-BING WANG, XIN ZHANG, XIN-RU JIA, ZI-CHEN LI, YAN JI, YEN WEI. Self-Assembly of a New Class of Amphiphilic Poly(amidoamine) Dendrimers and Their Electrochemical Properties. *J. Polym. Sci. Part A: Polym. Chem.* 43: 5512–5519, 2005.

[135] Catherine M. B. Santini, T. Alan Hatton, and Paula T. Hammond. Solution Behavior of Linear-Dendritic Rod Diblock Copolymers in Methanol. *Langmuir* 2006, 22, 7487-7498.

[136] Daniel R. Blasini, Samuel Flores-Torres, Detlef-M. Smilgies, and Hector D. Abruna. Stepwise Self-Assembly of Ordered Supramolecular Assemblies Based on Coordination Chemistry. *Langmuir* 2006, 22, 2082-2089.

[137] Virgil Percec, Andres E. Dulcey, Mihai Peterca, Monica Ilies, Monika J. Sienkowska, and Paul A. Heiney. Programming the Internal Structure and Stability of Helical Pores Self-Assembled from Dendritic Dipeptides via the Protective Groups of the Peptide. *J. Am. Chem. Soc.* 2005, 127, 17902-17909.

[138] Harald Hahn, Christina Keith, Heinrich Lang, Ramaiahgari Amaranatha Reddy, and Carsten Tschierske. First Example of a Third-Generation Liquid-Crystalline Carbosilane Dendrimer with Peripheral Bent-Core Mesogenic Units: Understanding of "Dark Conglomerate Phases". *Adv. Mater.* 2006, 18, 2629–2633.

[139] Dong Ki Yoon, Su Rim Lee, Yun Ho Kim, Sung-Min Choi, and Hee-Tae Jung. Large-Area, Highly Aligned Cylindrical Perfluorinated Supramolecular Dendrimers Using Magnetic Fields. *Adv. Mater.* 2006, 18, 509–513.

[140] Rafael Alcala, Raquel Gimenez, Luis Oriol, Milagros Pinol, Jose Luis Serrano, Belen Villacampa, and Ana I. Vinuales. Synthesis, Characterization, and Induction of Stable Anisotropy in Liquid Crystalline Photo-addressable PPI Dendrimers. *Chem. Mater.* 2007, 19, 235-246.

[141] Umesh Gupta, Hrushikesh Bharat Agashe, Abhay Asthana, and N. K. Jain. Dendrimers: Novel Polymeric Nanoarchitectures for Solubility Enhancement. *Biomacromolecules,* 2006, 7 (3): 649-658.

[142] I. Cooper, J. D. Londono, G. Wignall, J. B. McClain, E. T. Samulski, J. S. Lin, A. Dobrynin, M. Rubinstein, A. L. C. Burke, J. M. J. Frechet, and J. M. DeSimone. Extraction of a hydrophilic compound fromwater into liquidCO2 using dendritic surfactants. *Nature*, 1997, 389: 368-371.

[143] Izabela Bury, Bertrand Donnio, Jean-Louis Gallani, and Daniel Guillon. Interfacial Behavior of a Series of Amphiphilic Block Co-dendrimers. *Langmuir* 2007, 23, 619-625.

[144] Naomi Sakai, Yuko Kamikawa, Masayuki Nishii, Toru Matsuoka, Takashi Kato, and Stefan Matile. Dendritic Folate Rosettes as Ion Channels in Lipid Bilayers. *J. Am. Chem. Soc.* 2006, 128, 2218-2219.

[145] Virgil Percec, Andres E. Dulcey, Venkatachalapathy S. K. Balagurusamy, Yoshiko Miura, Jan Smidrkal, Mihai Peterca, Sami Nummelin, Ulrica Edlund, Steven D. Hudson, Paul A. Heiney, Hu Duan, Sergei N. Magonov, and Sergei A. Vinogradov. Self-assembly of amphiphilic dendritic dipeptides into helical pores. *Nature,* 2004, 430: 764-768.

[146] V. Percec, M. Glodde, T. K. Bera, Y. Miura, I. Shiyanovskaya, K. D. Singer, V. S. K. Balagurusamy, P. A. Heiney, I. Schnell, A. Rapp, H.-W. Spiess, S. D. Hudsonk, and H. Duank. Self-organization of supramolecular helical dendrimers into complex electronic materials. *Nature,* 2002, 419: 384-387.

[147] Virgil Percec, Andres E. Dulcey, Mihai Peterca, Monica Ilies, Janine Ladislaw, Brad M. Rosen, Ulrica Edlund, and Paul A. Heiney. The Internal Structure of Helical Pores Self-Assembled from Dendritic Dipeptides is Stereochemically Programmed and Allosterically Regulated. *Angew. Chem.* 2005, 117, 6674 –6679.

[148] Chia-Hung Chou, Hsu-Shen Wang, Kung-Hwa Wei, and Jung Y. Huang. Thiophenol-Modified CdS Nanoparticles Enhance the Luminescence of Benzoxyl Dendron-Substituted Polyfluorene Copolymers. *Adv. Funct. Mater.* 2006, 16, 909–916.

[149] Baohua Huang, Donald A. Tomalia. Poly(ether) dendrons possessing phosphine focal points for stabilization and reduced quenching of luminescent quantum dots. *Inorganica Chimica Acta* 359 (2006) 1961–1966.

[150] Dongjun Wang, Toyoko Imae, Masao Miki. Fluorescence emission from PAMAM and PPI dendrimers. Journal of Colloid and Interface, *Science* 306 (2007) 222–227.

[151] Arpornrat Nantalaksakul, Raghunath Reddy Dasari, Tai-Sang Ahn, Rabih Al-Kaysi, Christopher J. Bardeen, and S. Thayumanavan. Dendrimer Analogues of Linear Molecules to Evaluate Energy and Charge-Transfer Properties. *Organic Letters* 2006, 8(14): 2981-2984.

[152] Jane Larsen, Ben Bru1ggemann, Tomas Polivka, Villy Sundstro1m, Eva Akesson, Joseph Sly, and Maxwell J. Crossley. Energy Transfer within Zn-Porphyrin Dendrimers: Study of the Singlet-Singlet Annihilation Kinetics. *J. Phys. Chem.* A 2005, 109, 10654-10662.

[153] Andrea Petrella, Jens Cremer, Luisa De Cola, Peter Ba1uerle, and Rene M. Williams. Charge Transfer Processes in Conjugated Triarylamine-Oligothiophene-Perylenemonoimide Dendrimers. *J. Phys. Chem.* A 2005, 109, 11687-11695.

[154] Lucia Flamigni, Anna Maria Talarico, Barbara Ventura, Chloé Sooambar, and Nathalie Solladié. Photoinduced Processes in a Dendritic Zn Porphyrin Structure with a Free-Base Porphyrin Core. *Eur. J. Inorg. Chem.* 2006, 2155–2165.

[155] Dong-Ik Lee and Theodore Goodson III. Entangled Photon Absorption in an Organic Porphyrin Dendrimer. *J. Phys. Chem.* B, 2006, 110 (51): 25582-25585.

[156] Oleg Varnavski, Xingzhong Yan, Olivier Mongin, Mireille Blanchard-Desce, and Theodore Goodson, III. Strongly Interacting Organic Conjugated Dendrimers with Enhanced Two-Photon Absorption. *J. Phys. Chem.* C 2007, 111, 149-162.

[157] N. Vijayalakshmi and Uday Maitra. Multiple Chromophore Labeled Novel Bile Acid Dendrimers for Light Harvesting. *Macromolecules* 2006, 39, 7931-7940.

[158] Michael A. Oar, William R. Dichtel, Jason M. Serin, Jean M. J. Frechet, Joy E. Rogers, Jonathan E. Slagle, Paul A. Fleitz, Loon-Seng Tan, Tymish Y. Ohulchanskyy and Paras N. Prasad. Light-Harvesting Chromophores with Metalated Porphyrin Cores for Tuned Photosensitization of Singlet Oxygen via Two-Photon Excited FRET. *Chem. Mater.* 2006, 18, 3682-3692.

[159] Nicola Armaroli, Gianluca Accorsi, John N. Clifford, Jean-FranÅois Eckert, and Jean-Francois Nierengarten. Structure-Dependent Photoinduced Electron Transfer in Fullerodendrimers with Light-Harvesting Oligophenylenevinylene Terminals. *Chem. Asian J.* 2006, 1, 564 – 574.

[160] Ulrike Jager-Fiedler, Martin Kockerling, Ralf Ludwig, Alexander Wulf, and Clemens Krempner. Hydroxy-Substituted Oligosilane Dendrimers: Controlling the Electronic Properties through Hydrogen Bonding. *Angew. Chem. Int. Ed.* 2006, 45, 6755 –6759.

[161] Tai-Sang Ahn, Arpornrat Nantalaksakul, Raghunath R. Dasari, Rabih O. Al-Kaysi, Astrid M. Mu1ller, S. Thayumanavan, and Christopher J. Bardeen. Energy and Charge Transfer Dynamics in Fully Decorated Benzyl Ether Dendrimers and Their Disubstituted Analogues. *J. Phys. Chem.* B 2006, 110, 24331-24339.

[162] Tao Gu, James K. Whitesell, and Marye Anne Fox. Intramolecular Charge Transfer in 1:1 Cu(II)/Pyrenylcyclam Dendrimer Complexes. *J. Phys. Chem.* B 2006, 110, 25149-25152.

[163] Benjamin J. Lear and Clifford P. Kubiak. Charge Gating and Electronic Delocalization over a Denderimeric Assembly of Trinuclear Ruthenium Clusters. *Inorg. Chem.* 2006, 45, 7041-7043.

[164] Wei-Shi Li, Kil Suk Kim, Dong-Lin Jiang, Hiroyuki Tanaka, Tomoji Kawai, Jung Ho Kwon, Dongho Kim, and Takuzo Aida. Construction of Segregated Arrays of Multiple Donor and Acceptor Units Using a Dendritic Scaffold: Remarkable Dendrimer Effects on Photoinduced Charge Separation. *J. Am. Chem. Soc.* 2006, 128, 10527-10532.

[165] Ho-Jin Son, Won-Sik Han, Hyojeong Kim, Chungkyun Kim, Jaejung Ko, Chongmok Lee, and Sang Ook Kang. End-Capped Silole Dendrimers on a Carbosilane Periphery: Potential Electroluminescent Materials. *Organometallics* 2006, 25, 766-774.

[166] Dong-Lin Jiang, and Takuzo Aida. Photoisomerization in dendrimers by harvesting of low-energy photons. *Nature,* 1997, 388: 454-456.

[167] Chao Wu, Sergey V. Malinin, Sergei Tretiak, and Vladimir Y. Chernyak. Exciton scattering and localization in branched dendrimeric structures. *Nature Physics*. 2006, 2: 631-635.

[168] Prasad Taranekar, Akira Baba, Jin Young Park, Timothy M. Fulghum, and Rigoberto Advincula. Dendrimer Precursors for Nanomolar and Picomolar Real-Time Surface Plasmon Resonance/Potentiometric Chemical Nerve Agent Sensing Using Electrochemically Crosslinked Ultrathin Films. *Adv. Funct. Mater*. 2006, 16, 2000–2007.

[169] Ho-Jin Son, Won-Sik Han, Kuk Hwa Lee, Ha Jin Jung, Chongmok Lee, Jaejung Ko, and Sang Ook Kang. Electrochemical Deposition of End-Capped Triarylamine and Carbazole Dendrimers: Alternate Technique for the Manufacture of Multilayer Films. *Chem. Mater.*, Vol. 18, No. 25, 2006.

[170] Jagotamoy Das, Md. Abdul Aziz, and Haesik Yang. A Nanocatalyst-Based Assay for Proteins: DNA-Free Ultrasensitive Electrochemical Detection Using Catalytic Reduction of p-Nitrophenol by Gold-Nanoparticle Labels. *J. Am. Chem. Soc.* 2006, 128, 16022-16023.

[171] Agnieszka Kulczynska, Tony Frost, and Lawrence D. Margerum. Effect of PAMAM Dendrimer Size and pH on the Electrostatic Binding of Metal Complexes Using Cyclic Voltammetry. *Macromolecules* 2006, 39, 7372-7377.

[172] Frank N. Crespilho, Valtencir Zucolotto, Christopher M. A. Brett, Osvaldo N. Oliveira, Jr., and Francisco C. Nart. Enhanced Charge Transport and Incorporation of Redox Mediators in Layer-by-Layer Films Containing PAMAM-Encapsulated Gold Nanoparticles. *J. Phys. Chem.* B 2006, 110, 17478-17483.

[173] Fuan Wang, Jianlong Wang, Hongjun Chen, Shaojun Dong. Assembly process of CuHCF/MPA multilayers on gold nanoparticles modified electrode and characterization by electrochemical SPR. *Journal of Electroanalytical Chemistry* 600 (2007) 265–274.

[174] Yun-Long Zeng, Yu-Fang Huang, Jian-Hui Jiang, Xiao-Bing Zhang, Chun-Ran Tang, Guo-Li Shen, Ru-Qin Yu. Functionalization of multi-walled carbon nanotubes with poly(amidoamine) dendrimer for mediator-free glucose biosensor. *Electrochemistry Communications* 9 (2007) 185–190.

[175] E. Lojou, P. Bianco. Assemblies of dendrimers and proteins on carbon and gold electrodes. *Bioelectrochemistry* 69 (2006) 237–247.

[176] Frank N. Crespilho, M. Emilia Ghica, Monica Florescu, Francisco C. Nart, Osvaldo N. Oliveira Jr., Christopher M.A. Brett. A strategy for enzyme immobilization on layer-by-

layer dendrimer–gold nanoparticle electrocatalytic membrane incorporating redox mediator. *Electrochemistry Communications* 8 (2006) 1665–1670.

[177] M. Pilar García Armada, José Losada, Magdalena Zamora, Beatriz Alonso, Isabel Cuadrado, Carmen M. Casado. Electrocatalytical properties of polymethylferrocenyl dendrimers and their applications in biosensing. *Bioelectrochemistry* 69 (2006) 65–73.

[178] Valtencir Zucolotto, Andressa P.A. Pinto, Tathyana Tumolo, Marli L. Moraes, Maur´ıcio S. Baptista, Antonio Riul Jr., Ana Paula U. Araujo, Osvaldo N. Oliveira Jr. Catechol biosensing using a nanostructured layer-by-layer film containing Cl-catechol 1,2-dioxygenase. *Biosensors and Bioelectronics* 21 (2006) 1320–1326.

[179] Zhi-Min Liu, Yu Yang, Hua Wang, Yan-Li Liu, Guo-Li Shen, Ru-Qin Yu. A hydrogen peroxide biosensor based on nano-Au/PAMAM dendrimer/cystamine modified gold electrode. *Sensors and Actuators* B 106 (2005) 394–400.

[180] M. Snejdarkova, L. Svobodova, G. Evtugyn, H. Budnikov, A. Karyakin, D.P. Nikolelis, T. Hianik. Acetylcholinesterase sensors based on gold electrodes modified with dendrimer and polyaniline A comparative research. *Analytica Chimica Acta* 514 (2004) 79–88.

[181] L. Svobodova, M. Snejdarkova, K. Toth, R.E. Gyurcsanyi, T. Hianik. Properties of mixed alkanethiol–dendrimer layers and their applications in biosensing. *Bioelectrochemistry* 63 (2004) 285– 289.

[182] Li Shen, Naifei Hu. Heme protein films with polyamidoamine dendrimer: direct electrochemistry and electrocatalysis. *Biochimica et Biophysica Acta* 1608 (2004) 23–33.

[183] Lenka Svobodova, Maja Snejdarkova, Vladimira Polohova, Igor Grman, Peter Rybar, Tibor Hianik. QCM Immunosensor Based on Polyamidoamine Dendrimers. *Electroanalysis* 2006, 18 (19-20) 1943 – 1949.

[184] Nathan A. Stasko and Mark H. Schoenfisch. Dendrimers as a Scaffold for Nitric Oxide Release. *J. Am. Chem. Soc.* 2006, 128, 8265-8271.

[185] Rudresha Kottani, Janaki R. R. Majjigapu, Alexei Kurchan, Kavitha Majjigapu, Tiffany P. Gustafson, and Andrei G. Kutateladze. Photoinduced Signal Amplification through Controlled Externally Sensitized Fragmentation in Masked Sensitizers. *J. Am. Chem. Soc.* 2006, 128, 14794-14795.

[186] Shin Ogasawara, Atsushi Ikeda, and Jun-ichi Kikuchi. Positive Dendritic Effect in DNA/Porphyrin Composite Photocurrent Generators Containing Dendrimers as the Stationary Phase. *Chem. Mater.* 2006, 18, 5982-5987.

In: Encyclopedia of Nanotechnology
Editor: Elwood D. Carlson

ISBN: 978-1-60692-079-4

Chapter 6

THERMODYNAMIC AND KINETIC THEORIES OF NANOWIRE GROWTH

G. W. Yang

State Key Laboratory of Optoelectronic Materials and Technologies,
Institute of Optoelectronic and Functional Composite Materials,
School of Physics Science and Engineering, Zhongshan University,
Guangzhou 510275, China

ABSTRACT

The universal models of nucleation thermodynamics and growth kinetics have been established for nanowire growth upon metal-catalyst-assisted thermal chemical vapor transport (CCVT) on the basis of vapor-liquid-solid (VLS) mechanism, with respect to the size-effects induced by the curvature of critical nuclei and nanowire in the nucleation and growth. Based on the proposed thermodynamic and kinetic analyses on nanoscale, the thermodynamic and kinetic size limit of the nanowire growth was addressed. It was found that the critical size of the nanowire nucleation is suggested to be the thermodynamic size limit, i.e., the theoretical minimum size of the nanowire grown by CCVT both thermodynamically and kinetically, and the size of the catalyst droplet could limit the growing size, i.e., the theoretical possible or maximum size of the nanowire to a certain extent in kinetically. For the issue of catalyst nanoparticles on nanowires tip or substrate, a nucleation thermodynamic criterion and a diffusion kinetic criterion were deduced for the selection growth of catalyst nanoparticles on nanowires tip or substrate upon CCVT. Importantly, theoretical predictions on the basis of the proposed models are in well agreement with the experimental data. Additionally, a growth kinetic approach was developed to theoretically elucidate the growth of one-dimensional nanostructures inside carbon nanotubes upon vapor deposition. Theoretical analysis showed that the growth rate of one-dimensional nanostructures would go much higher once nuclei formed inside carbon nanotubes, due to the effect of surface tension induced by the nano-sized curvature of carbon nanotubes. Theoretical results are in well agreement with the experimental data. The size-dependence of the shape evolution of nanowires such as SnO2 was pursued in theoretical. It was found that the thermodynamic driving force, the Gibbs free energy difference between two phases, is always responsible for the shape choice in the different growth stages such as nucleation, growth, and structural transition of 1D nanostructures, implying that the thermodynamic driving force is one of the physical origins causing the interesting morphological transformation in the growth of 1D nanostructures.

1. INTRODUCTION

One-dimensional (1D) nanostructures such as wires, rods, belts, and tubes have become the focus of intensive research owing to their unique applications in mesoscopic physics and fabrication of nanoscaled devices. They not only provide a good system to study the electrical and thermal transport in one-dimensional confinement, but also are expected to play an important role in both interconnection and functional units in fabricating electronic, optoelectronic, and magnetic storage devices with nanoscale dimension [1]. Therefore, to attain various nanometer sized building blocks, a lot of self-assembly and synthesis processes have emerged in recent years [2]. Importantly, these assemblies and syntheses at the nanometer scale have revealed many unusual thermodynamic and kinetic behaviors of microphase growth [3-11], which plays an important role in the development of the classical thermodynamics [12]. A big challenge is how to flexibly control the self-assembly and synthesis processes of 1D nanostructure for these purposes of fundamental researches and potential applications. Basically, self-assembly processes require atomic or clustering interactions as the thermodynamic and kinetic driving forces to organize those atoms or clusters to form nanosized domain patterns [13]. Therefore, in order to control the growth of nanostructures, it is essential to clarify the thermodynamic and kinetic driving forces of the self-assembly process of nanostructures and develop new theoretical tools to address the nano-sized fabrications [14]. For this issue, in theoretical, we have developed a series of thermodynamic and kinetic tools to pursue the nucleation and growth of 1D nanostructures upon metal-catalyst-assisted thermal chemical vapor transport (CCVT) on the basis of vapor-liquid-solid (VLS) mechanism, with respect to the size-effects induced by the curvature of critical nuclei and nanowire in the nucleation and growth.

It is well known that CCVT has been one of leading methods to synthesize most semiconductor 1D nanostructures [2], and the process of VLS has been generally recognized to be the fundamental understanding of CCVT [15-16]. However, the VLS process is just an outline for the formation of 1D nanostructure upon CCVT, and many crucially physical and chemical issues involved in the nucleation and growth of nanowire have not been addressed in the present securable literatures, up to now. For instance, the nucleation position of nanowire on the surface of catalyst particles, the size limit of nanowire growth, the growth behavior of nanowire confined a nanotube, the size-dependence of nanowire growth rate, and the size-dependent shape evolution during nanowire growing. Importantly, these basic physical and chemical processes play key roles in controlling the fabrications of 1D nanostructure.

In this review chapter, we introduced many new theoretical tools to address the nucleation and growth of nanowire upon CCVT. The thermodynamic and kinetic theories of nanowire nucleation were described in section 2, in which thermodynamic and kinetic criteria are established to determine the selection nucleation of nanowire on the surface of catalyst. In section 3, we introduced the thermodynamic and kinetic theory of nanowire growth, which predicts the size limit of nanowire upon CCVT. Usually, the anomalous growth behavior would happen when a nanowire grows inside a carbon nantube. Therefore, a growth kinetic model was established to elucidate the formation of the nanowire confined nanotubes in section 4. The size-dependent growth behaviors of nanowire are theoretically pursued in section 5, in which a kinetic model of size-dependent growth rate of nanowire was introduced, and the thermodynamic and kinetic aspects of shape evolution of 1D nanostructure were addressed. Additionally, some remarked conclusions are summarized in section 6.

2. Thermodynamic and Kinetic Theory of Nanowire Nucleation

Based on the experimental data, it is not difficult to find that there mainly exist two general models of nanowire formations on substrates upon VLS process. One is that catalyst nanoparticles are on the bottom of nanowires, and another one that catalyst nanoparticles are on the tip of nanowires [1,17-21]. However, which physical and chemical origins result in two inverse growths of nanowires on catalyst nanoparticles? In other words, why would catalyst nanoparticles be on the tip of nanowires these times and on the bottom those times? To our knowledge, few studies have been reported to study the issue mentioned above [1,22]. We proposed the two universal theoretical criteria for the selection growth of nanowires upon CCVD in thermodynamical and kinetic, respectively, on the basis of the thermodynamical theory on nanoscale [10]. Importantly, taking Si nanowires as an example, the theoretical predictions are well consistent with experimental data. Noted that the thermodynamical theory on nanoscale (the capillary approximation) is valid when it is used to address the microphase growth, in which sizes of clusters are usually in the range of a couple of nanometers [8-10].

2.1. Thermodynamic Criteria of Selection Nucleation

Generally, the Gibbs free energy is an adaptable measure of the stability of a state in phase transition formation. Thermodynamically, the phase transformation is promoted by the difference of the free energy. Gibbs free energy of a phase can be expressed as a function of the pressure-temperature condition, and determined by a general coordinate or reactive coordinate [24]. Therefore, we will propose a thermodynamical criterion for the selection growth of nanowires by comparing the Gibbs free energy difference of the nanowire nucleation in two cases. The Gibbs free energy difference of a cluster can be expressed as

$$\Delta G = \Delta G_1 + \Delta G_2 \tag{2-1}$$

where the two terms on the right-hand side are the energy differences from surface and volume. According to the nucleation thermodynamics [24], the Gibbs energy difference aroused from the formation of spherical clusters in the gas phase can be expressed as follows [25]

$$\Delta G(r) = V \times \Delta g_v + A\sigma_{nv} \tag{2-2}$$

where σ_{nv}, V, and A are the interface free energy density of nucleus-vapor, volume, and area of spherical clusters, respectively, and Δg_v is the Gibbs free energy difference per unit volume, which can be expressed by $-RT/V_m \ln(P/P_e)$, where P, T, R, and V_m are the pressure, the temperature, the gas constant, and the mole volume of the nanowire, respectively. P_e is the vapor phase pressure in thermal equilibrium coexistence with the liquid of composition in a flat surface. Based on Tan's model [26] when the droplet is assumed to be an idea solution, Δg_v can be expressed as

$$\Delta g_v = -\frac{RT}{V_m}\ln\frac{p}{p_e} = -\frac{RT}{V_m}\ln\left(\frac{C}{C^{eq}}\right) \tag{2-3}$$

where C and Ceq are X concentrations X means nanowires materials_ on the solid and liquid lines of the eutectic catalyst-X phase diagram, respectively. Based on the Kelvin equation, taking the capillary effect of the nanosized curvature of the nanowire nuclei on the Gibbs free energy of the nucleation into account, Δg_v can be defined as

$$\Delta g_v = -\frac{1}{2}\left[\frac{RT}{V_m}\ln\left(\frac{C}{C^{eq}}\right) + \frac{2\sigma_{nv}}{r}\right] = \frac{1}{2}\Delta g_v^0 - \frac{\sigma_{nv}}{r} \tag{2-4}$$

where $\Delta g_v 0$ is defined as

$$\Delta g_v^0 = -\frac{RT}{V_m}\ln\left(\frac{C}{C^{eq}}\right) \tag{2-5}$$

The schematic illustration of the X nanowire growths by two models upon CCVT is shown in Figure 1. We will calculate the Gibbs free energies of the nanowire nucleation in two cases on the basis of the thermodynamic approach on nanoscale.

For the first case, i.e., the nanowire nucleation takes place on the surface of catalyst nanoparticles as shown in Figure 2 (a). When an atom cluster condensed with catalyst being on the substrate, the surface energy $\Delta G1_S$ can be calculated as follows

$$\Delta G_{1S} = 2\sigma_{nv}\pi r_1^2(1-\cos\theta_1) + \pi r_1^2 \sin^2\theta_1(\sigma_{cn} - \sigma_{cv}) \tag{2-6}$$

where $r1$ is the radius of the cluster, σ_{cn} and σ_{cv} are the interface energies of the catalyst-nucleus, the catalyst-vapor, respectively, and $\theta 1$ is the contact angle between the catalyst and nucleus. When the interface between nanowire and catalyst liquid droplet is assumed to be incoherent interface, the values of σ_{cn} can be denoted as

$$\sigma_{cn} = \frac{\sigma_{cv} + \sigma_{nv}}{2} \tag{2-7}$$

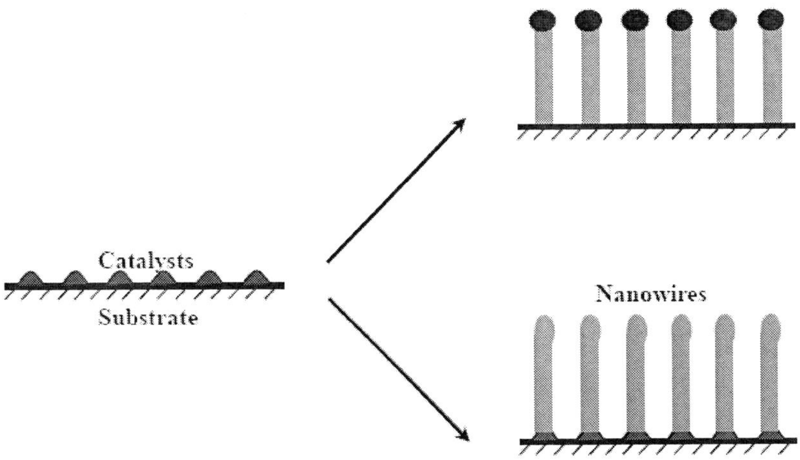

Figure 1. The selection growth of nanowires: catalyst nanoparticles on the tip and bottom of nanowires.

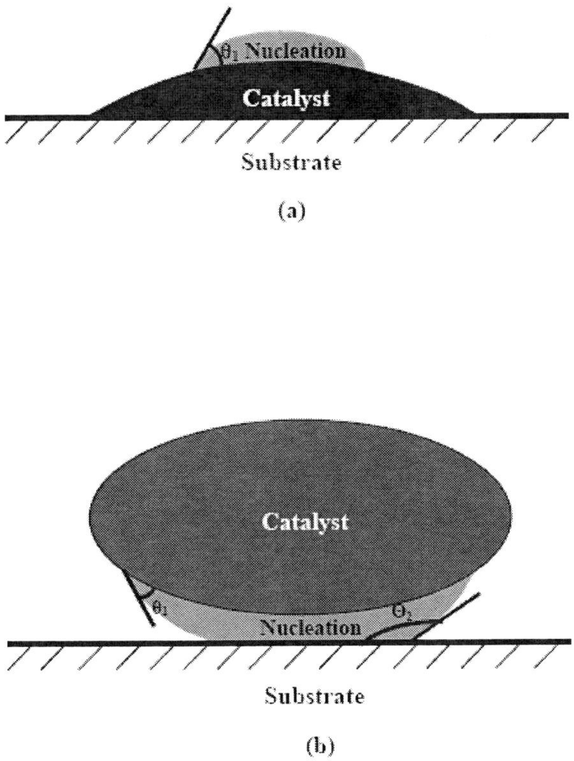

Figure 2. The schematic illustration of the nanowire nucleation by two models upon CCVD: nucleation on the surface of catalyst nanoparticles (a) and at the interface between catalyst nanoparticles and substrate (b).

The volume energy $\Delta G2_S$ can be given by

$$\Delta G_{2S} = \frac{4\pi r_1^3 \Delta g_v}{3} \frac{2 - 3\cos\theta_1 + \cos^3\theta_1}{4} \tag{2-8}$$

In this way, we obtain the Gibbs free energy can be expressed as

$$\Delta G_S = 2\sigma_{nv}\pi r_1^2(1-\cos\theta_1) + \pi r_1^2 \sin_2\theta_1 (\sigma_{cn} - \sigma_{cv}) + \frac{4\pi r_1^3 \Delta g_v}{3}\frac{2-3\cos\theta_1 + \cos\theta_1^3}{4}$$

(2-9)

According to the balance condition of forces, we have $\sigma_{cv} = \sigma_{nv}\cos\theta 1 + \sigma_{cn}$. Thus, the Gibbs free energy can be expressed as

$$\Delta G_S = 2\sigma_{nv}\pi r_1^2(1-\cos\theta_1) - \pi r_1^2 \sin^2\theta_1\, \sigma_{nv}\cos\theta_1 + \frac{4\pi r_1^3 \Delta g_v}{3}\frac{2-3\cos\theta_1 + \cos\theta_1^3}{4}$$

(2-10)

Based on the condition of the critical nucleus formation, $\partial\Delta G_s/\partial r1=0$, when the atomic cluster transforms into the critical nucleus, the radius of the critical nucleus can be calculated to be

$$r_1^* = -\frac{8}{3}\frac{\sigma_{nv}}{\Delta g_v^0} \qquad \textbf{(2-11)}$$

Accordingly, the formation energy $\Delta G_s^*(r1^*)$ of the critical nucleus can be given by

$$\Delta G_s^* = \left(\frac{256\pi\sigma_{nv}^3}{9(\Delta g_v^0)^2} + \frac{2048\pi\sigma_{nv}^3}{81(\Delta g_v^0)^3}\right)\left(\frac{2-3\cos\theta_1 + \cos^3\theta_1}{4}\right) \qquad \textbf{(2-12)}$$

For the second case, i.e., the nanowire nucleation takes place on the interface between the catalyst nanoparticle and the substrate as shown in Figure 2 (b). The surface energy $\Delta G1_T$ can be calculated as

$$\Delta G_{1T} = -2\sigma_{nv}\pi r_2^2(\cos\theta_1 + \cos\theta_2) - \pi r_2^2 \sin\theta_1^2 \cos\theta_1\sigma_{nv} - \pi r_2^2 \sin_2^2 \cos\theta_2\sigma_{nv}$$
$$= -\pi r_2^2\sigma_{nv} f_1(\theta)$$

(2-13)

where r_2 is the radius of the cluster, σ_{nv} is the interface energy of the nucleus-vapor, and θ1 and θ2 are the contact angles of catalyst-nucleus and nucleus-substrate, respectively. In addition, $f1(\theta)$ is defined as 2(cos θ1+cosθ2)+sin2θ1cosθ1+sin2θ2cosθ2. The volume Gibbs free energy ΔG_{2T} can be given by

$$\Delta G_{2T} = \frac{4\pi r_2^3 \Delta g_v}{3}\frac{\cos^3\theta_1 + \cos^3\theta_2 - 3\cos\theta_1 - 3\cos\theta_2}{4}$$

(2-14)

$$=\frac{4\pi r_2^3 \Delta g_v}{3} f_2(\theta)$$

where $f2(\theta)$ is defined as

$$f_2(\theta) = (\frac{\cos^3\theta_1 + \cos^3\theta_2 - 3\cos\theta_1 - 3\cos\theta_2}{4} \tag{2-15}$$

Summing the surface energy and the volume one, the whole Gibbs free energy difference can be obtained. Similarly, based on the condition of the critical nucleus formation, $\partial\Delta G_T/\partial r2=0$, when the atomic cluster transforms into the critical nucleus, the radius of the critical nucleus can be calculated to be

$$r_2^* = \frac{\sigma_{nv}}{2\Delta g_v^0}\left(\frac{2f_1(\theta)}{f_2(\theta)} + \frac{8}{3}\right) \tag{2-16}$$

Accordingly, the formation energy $\Delta G_T{}^*(r2^*)$ of the critical nucleus can be expressed as

$$\Delta G_T^* = -\pi\sigma_{nv}\left[\frac{\sigma_{nv}}{2\Delta g_v^0}\left(\frac{2f_1(\theta)}{f_2(\theta)} + \frac{8}{3}\right)\right]^2 f_1(\theta) + \frac{4\pi}{3}\left[\frac{\sigma_{nv}}{2\Delta g_v^0}\left(\frac{2f_1(\theta)}{f_2(\theta)} + \frac{8}{3}\right)\right]^3 \times\left(\frac{1}{2}\Delta g_v^0 - \frac{2\Delta g_v^0}{\frac{2f_1(\theta)}{f_2(\theta)} + \frac{8}{3}}\right) f_2(\theta) \tag{2-17}$$

From the deduction above, one can see that the critical energy only depends on these interfacial parameters of σ_{cv}, σ_{nv}, and $\theta 2$, and nanowire materials-catalyst phase diagram, without any adjustable parameters. Thus, we can obtain that the nanowire nucleation on the surface of catalyst nanoparticles (Figure 2 (a)) would be preferable when $\Delta G_S^* > \Delta G_T^*$. On the other hand, the nucleation on the interface between catalyst nanoparticles and substrates (Figure 2 (b)) would be preferable when $\Delta G_S^* < \Delta G_T^*$ from the point of view of thermodynamics. Therefore, we naturally deduce a thermodynamical criterion for the selection growth of nanowires upon CCVD as follows. Catalyst nanoparticles would be on the tip of nanowires when $\Delta G_S^* > \Delta G_T^*$, and on the bottom when $\Delta G_S^* < \Delta G_T^*$, thermodynamically. Additionally, from the classic thermodynamical nucleation [24], the value of $\theta 2$ shows the interaction between nuclei and substrates, and different interactions between nuclei and substrates would lead different values of $\theta 2$. For example, the value of $\theta 2$ should be different when substrates are crystalline and amorphous, respectively. Therefore, we could compare ΔG_S^* with ΔG_T^* based on the thermodynamic criterion above if the value of $\theta 2$ could be determined in experimental or theoretical.

2.2. Kinetic Criteria of Selection Nucleation

On the basis of the Louchev's model [27], we suggest the diffusion kinetic mechanism of nucleation and growth of nanowires upon CCVT. However, there exists the distinct difference between the diffusion kinetic mechanism in our case and the Louchev's model, as our model takes the size effect of nanostructures into account, which did not appear in Louchev's work. The schematic illustration of the *X* nanowire nucleation and growth by two models upon CCVT is shown in Figure 3. Two kinds of diffusion processes, surface diffusion and body diffusion, of depositing atoms on the surface of nanoparticles are important for the selection nucleation stage, defining whether the nanoparticle remains at the substrate, or on the tip of the growing *X* nanowires. For the nucleation of nanowires upon CCVT, the surface diffusion of adatoms on the surface of nanoparticles is express by the surface saturation time of t_S, meaning that the increase of the adatom content to the saturation concentration. Thus, the surface diffusion time of order is suggested [27]

$$t_s = \frac{C^{*2} D_b(r,T)}{Q^2} \tag{2-18}$$

where C^* is a concentration threshold triggering the precipitation of the growing atoms directly on the upper surface of nanoparticle. The $D_b(r,T)$ is the size-dependent body diffusion coefficient of atoms, and it can be defined as [28]

$$D_b(r,T) = D_0 \exp\left[\frac{-E(\infty)}{RT}\exp\left(\frac{-2S_{vib}(\infty)}{3R}\frac{1}{\frac{r}{r_0}-1}\right)\right] \tag{2-19}$$

where $D0$, $E(\infty)$, Svib(∞), r, and $r0$ denote a pre-exponential constant, diffusion activation energy of bulk material, the activation entropy of bulk material, the radius of the nanoparticle, and three times the atomic diameter, respectively. In addition, $Q=[Pc/(2mkBT)1/2$ is the atom flux from gas upon CCVT on a unit area in a unit time, in which m, Pc, k_B, and T are the atom mass, partial pressure of Si, Boltzmann constant, and deposition temperature, respectively. On the other hand, the body diffusion is expressed by the body diffusion time of t_d, meaning that the adatoms penetrate to the nanoparticle base from surface. According to the diffusion theory [29], the diffusion distance of atom l is expressed as $\sqrt{6D_b(r,T)t}$, where t is the diffusion time. Assuming the diameter of the nanoparticle being d, the diffusion distance on the average is $l' = \sqrt[3]{\frac{4}{3}\pi(d/2)^3}$. Therefore, we get the average diffusion time t_d as follows

$$t_d = \frac{1}{6D_b(r,T)}\sqrt[3]{\left[\frac{4\pi}{3}\left(\frac{d}{2}\right)^3\right]^2} \tag{2-20}$$

Definitely, when $t_d >> t_s$ as shown in Figure 3 (a), the nanoparticle surface saturates with atoms much faster than one penetrates to its base, suggesting that the nucleation of adatoms would take place on the upper surface of catalyst nanoparticles, and then provide a nanoscale template for the nanowire growth. In contrast, when $t_d << t_s$ as shown in Figure 3 (b), atoms would penetrate to the base from the upper surface of nanoparticles much faster than adatoms reach the saturation threshold on the upper surface, implying that growing atoms will precipitate at the bottom of nanoparticles, and then result in the nucleation at the interface between the nanoparticle and the substrate. Thus, nanoparticles are lifted in the later nuclei growing and maintained on the nanowire tip.

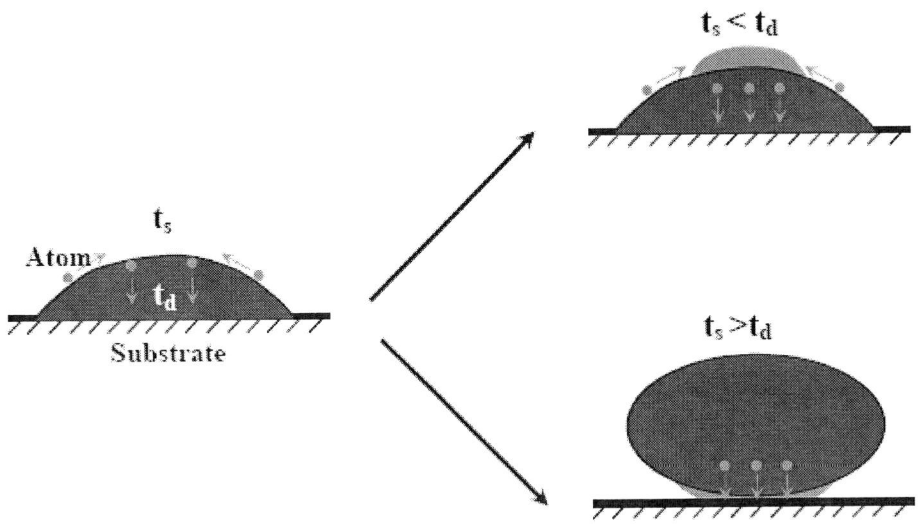

Figure 3. The schematic illustration of the nucleation kinetics of nanowires: the cluster forming on the surface of catalyst nanoparticles (a) when $t_d >> t_s$, and at the interface between catalyst nanoparticles and substrate (b) when $t_d << t_s$.

2.3. Comparison between Theory and Experiment

Note that CCVT on the basis of VLS mechanism is the steady-state process, implying that CCVT could be treated as a near-equilibrium process. In fact, the equilibrium thermodynamics has been validly employed to address the nucleation and growth upon CVD [9], especially, for the growth of 1D nanostructures upon CCVT [25,26]. Accordingly, it is convincing for our case to employ the equilibrium thermodynamic method to deal with the nanowire nucleation upon CCVT. Taking the Si nanowire growth upon CCVT as an example, we will compare the theoretical prediction with experimental data.

For the case of the Si nanowire nucleation on a Si substrate using Fe as catalyst, the typical growth conditions are T=1473 K and Pc=1.4 Torr [30]. The surface energy density of the silicon nanowire is σ_{nv}=1.610 J /m2 [31] and the liquid-catalyst surface energy density σ_{cv} is 0.885 J /m2 [32]. Note that the used Fe-Si phase diagram is from Ref. 33. Therefore, based on Eqs. (8) and (12), we can attain the relationship curves between θ2 and the critical radius of the two cases, as shown in Figure 4. Obviously, one can find that the critical radius $r2^*$ increases quickly with θ2 decreasing when θ2 is less than 80°, and it is close to $r1^*$ when θ2 is larger than 150°. In addition, we can obtain the dependence of θ2 and the critical energy on the basis of Eqs. (9) and (13), as shown in Figure 5. Similarly, we can see that the value of critical energy ΔG_T^* increases quickly with θ2 increasing when θ2 is more than 80°, and ΔG_S^* is much larger than ΔG_T^* when θ2 < 150°. When θ2 > 150°, ΔG_T^* is close to ΔG_S^*. Importantly, one can see that ΔG_T^* is always less than ΔG_S^* at given experiment parameters and θ2 values from 80° to 150° in Figure 5. However, comparing Figure 5 with Figure 6, we can see that the dependencies of the critical radius and critical energy of nuclei on the value of θ2 are inverse, i.e., the critical radius of nuclei decreases with increasing the value of θ2, and the critical energy of nuclei increases with increasing the value of θ2. Considering the physical process of nanowires nucleation shown as Figure 2, we propose that this anomalous behavior should be original from the interaction between nuclei and substrates. The large interface area between nuclei and substrates would reduce the formation energy of nuclei, while it would result in the critical radius of nuclei large.

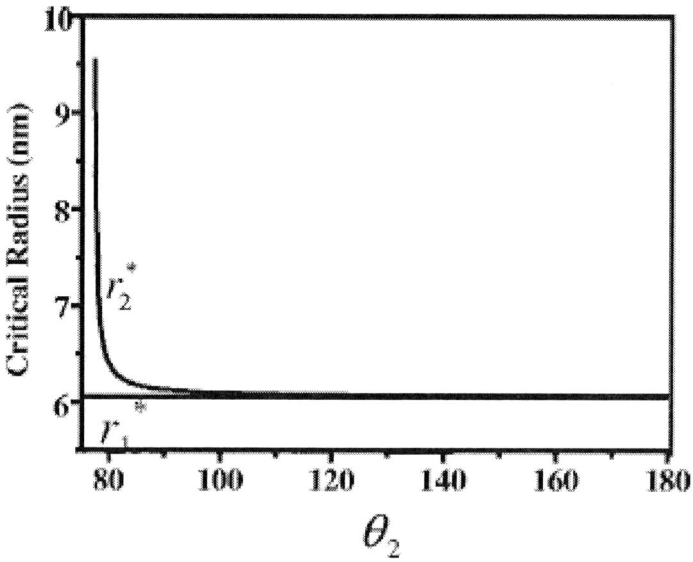

Figure 4. The dependence relationship between the contact angle θ2 and the critical nucleus $r1^*$, $r2^*$.

Therefore, these results indicate that the Gibbs free energy of the critical nuclei in the case of catalyst nanoparticles on the substrate is usually larger than that for catalyst nanoparticles on the tip of nanowires when θ2 is less than 180°. According to the thermodynamic criterion mentioned above, for the Si-Fe system, Fe catalyst nanoparticles would be on the tip of Si nanowires upon CCVT. More importantly, most of experimental cases of Si nanowires growing on Si substrates using Fe as catalysts by CCVT show that Fe nanoparticles are always on the tip of Si nanowires. [17,33–35].

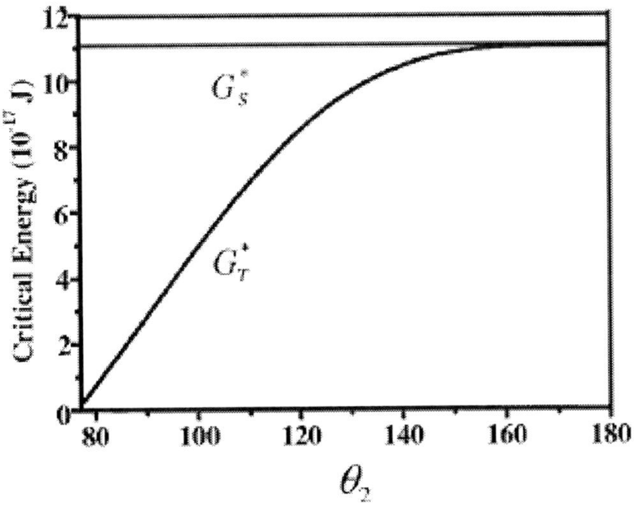

Figure 5. The relationship curves between the contact angle θ2 and the critical energy ΔG_S*, ΔG_T*.

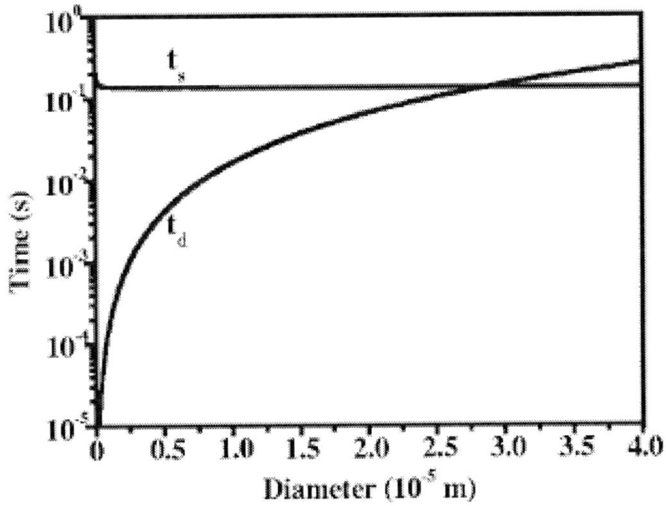

Figure 6. The comparison between the surface saturation time of t_s and the body diffusion time of t_d.

It is well known that the thermodynamical criterion just provides the probability for the selection growth of nanowires. On the other hand, the kinetic nucleation will play a key role in the achievement of the selection growth of nanowires upon CCVT, when the thermodynamic criterion operates. For instance, in the Si nanowire nucleation upon CCVT, when $t_d > t_s$, the Si atoms saturation on the catalyst surface is much faster than one penetrates to its base. Then, Si precipitates on the catalyst surface provides a nanoscaled template for the Si nanowire growth. In contrast, when $t_d < t_s$, the Si penetration to the base is much faster than the Si atoms saturation on the catalyst surface. Thus, Si precipitates at the bottom will lift the catalyst and maintain it on the nanowires tip later. In order to study the kinetics of Si nanowires growth, the diffusion time and saturation time are calculated at the given temperature and pressure in the Si-Fe system mentioned above on the basis of Eqs. (14–16).

The kinetics parameters are as follows: the value of *D*0, *E*(∞), *S*vib(∞), and *r*0 being 6.2×10−3 cm2/s [32], 8.4×104 J /mol [36], 34.32 J /mol K−1 [37] and 0.7449 nm [38], respectively, for the system of Si diffusion in Fe. Based on the Si-Fe phase diagram [39], Fe saturates at ≈50 at % of Si and the saturation concentration of Si is about 3.5×1022 cm−3 [40], we can calculate t_d and t_s under conditions of T=1473 K and P =1.4 Torr. Figure 6 show the dependence of t_d and t_s on the size of the Fe catalyst. From Figure 6, we can see that the diffusion time increases with the size increasing of Fe catalyst, while the saturation time hardly change with the diameter increasing of the nanoparticles. Very importantly, $t_d <$ t_s_(more two orders of magnitude) when the size of the Fe catalyst is less than 25 μm. According to the kinetic criterion above, this result displays that Si penetrates to the base much faster than Si on the catalyst surface reaches saturation threshold when the size of the nanoparticles is less than 25 μm. In other words, Si will precipitate at the bottom, lift catalysts, and later maintain catalyst droplet on nanowires tips when the size of the Fe catalyst is less than 25 μm.

Gösele *et al.* pointed out that the size of the Si nanowires is limited by the size of the metallic catalyst droplet, and the diameter of the Si nanowires is usually much smaller than that of the catalyst droplet [26]. Accordingly, the size of Si nanowires should be smaller than that of the catalyst droplet when catalyst droplets are on nanowires tips. Interestingly, up to date, almost all experimental data at given temperatures and pressures in the Si-Fe system show that the diameter of Si nanowires is less than 200 nm, and catalyst droplets are definitely located on the tip of the Si nanowires [17,33–35], which corresponds with the theory predictions above. On the other hand, from Figure 6, one can see that there exists a critical diameter of 30 μm for catalyst droplets, and the catalyst surface-saturation and Si penetration-base time is nearly the same order of magnitude when the size of the nanoparticle is larger than 30 μm. Namely, when the radius of droplet is larger than 30 μm, no solid conclusions can be drawn from the above model predictions, and maybe, the selection growth of nanowires would be determined by chance. Assuredly, in experiment, Wagner and Ellis have prepared Si whiskers with 200 μm-sized diameters, which always had the catalytic particle on top. In the experimental case of Wagner and Ellis, they selected Au as catalyst, and the deposition temperature is about 1223 K. One can see that their experimental parameters are very different from the parameters of our calculating cases. Thus, the theoretical prediction is not comparable with their experimental results. Naturally, we would have tried to compare their experimental case with our calculations by using their experimental parameters. However, we did not in this case due to the lack of the detailed experimental parameters, e.g., the partial pressure of SiCl4, and so on. In fact, since the diameters of Si nanowires reported in experiment are small, the catalyst nanoparticles would be lifted on the nanowires tip in most cases from the proposed thermodynamical and kinetic criterion.

Therefore, these comparisons show that the universal thermodynamical and kinetic criterion of the selection growth of nanowires forming upon CCVT seems to operate to determine catalyst nanoparticles on the tip or bottom of nanowires. Note that, in some Si nanowire synthesis by laser ablation [41,42], no catalyst nanoparticles were observed anywhere on nanowire tip or substrate even when metals were mixed in the target. Actually, this nucleation and growth of nanowires is different from the VLS mechanism. Moreover, Wang *et al.* [42] pointed out that metal was not necessary for Si nanowire synthesis by laser ablation. Instead, oxides such as SiO2 are the special and effective catalyst that largely enhanced the Si nanowire growth in these cases [41,42].

In summary, the general thermodynamical and kinetic approaches on nanoscale are introduced to elucidate the nanowires nucleation taking place in CCVT, with respect to the capillary effect of the nanosized curvature of the nanowire nuclei in the initial growing. A thermodynamic criterion and a diffusion kinetic criterion on the basis of the thermodynamic and kinetic analyses on nanoscale are deduced for the selection growth of catalyst nanoparticles on nanowires tip or substrate upon CCVT. Taking Si nanowires as the example, we compared the experimental data with the theoretical predictions and found that they are nicely consistence. We thus expected that the thermodynamic and kinetic criteria on the nanometer scale to be a general approach applicable to understand nanowires formation in the occasion of CCVT.

3. Thermodynamic and Kinetic Theory of Nanowire Growth

Up to date, most semiconductor nanowires have been synthesized using metal catalyst assistant thermal chemical vapor transport on the basis of so-called vapor-liquid-solid mechanisms [43-46]. Actually, VLS mechanisms for applicable to the growth of silicon crystal whiskers was initially proposed and demonstrated by Wagner and Ellis in 1964 [15-16], and has subsequently been used to address the growth of semiconductor nanowires by CCVT [18, 47-49]. Based on these studies above, for the nanowire formation upon CCVT, the VLS mechanism reveals that (i) reactants are supplied in the vapor phase and nanowires are formed by extracting of nanowires materials from liquid droplets of metal alloy, (ii) the nanowire growth results from the difference of sticking coefficients of liquid droplets and nanowires, namely, sticking coefficients of nanowires are orders of magnitude smaller than that of liquid droplets. As a result, the droplet and the solid nanowire can capture and reject nearly all the constituents of the growing material from the vapor phase, respectively, as shown in Figure 1(a). Generally, the intrinsic properties of 1D nanostructures are mainly determined by their size, shape, composition, and crystalline structure. In principle, one can control any one of these characters to fine-tune the properties of this nanostructure. It is therefore essential to pursue the basic physics and chemistry involved in the formation of nanowires. However, the size limit of the nanowire growth upon CCVT still remains much less understood in theory, even though there have been a lot of relevant experimental studies to try to control the nanowire growth [49-51]. For instance, in both theoretical and experimental, we haven't known how thin or thick nanowires could be synthesized using CCVT yet. Accordingly, we proposed the universal nucleation thermodynamics and growth kinetics for the nanowire growth upon CCVT based on the VLS mechanism, and then theoretically deduced the size limit of the nanowire growth. Importantly, taking the Si nanowire (SiNW) growth as an example, we found that the theoretical predictions are in excellent agreement with experimental cases.

3.1. Thermodynamic Size Limit of Nanowire Growth

In general, the minimal size of the nanowires is both dependent on nucleation thermodynamics and growth kinetics (energy stability theory), meanwhile, it is also dependent on the Rayleigh instability of nanowire itself (linearized stability theory) [31]. However, when the linearized system is symmetric, the prediction of the stability based upon the linearized stability theory and the energy stability theory are consistent [52-53]. In other words, linearized equations do not yield much information on the energy stability range. Namely, we could have an instability (usually called a subcritical instability) below the critical values of Rayleigh numbers determined by the linearized theory, which can be in the energy stable region. Thermodynamically, the phase transformation is promoted by the difference of Gibbs free energies. The Gibbs free energy of a phase can be expressed as a function of the pressure and temperature, and determined by a general coordinate or reactive coordinate. The schematic illustration of the X nanowire (XNW) nucleation upon CCVT is shown in Figure 7(b). The Gibbs free energy difference of a cluster can be expressed as [24]

$$\Delta G = (\sigma_{lx} - \sigma_{lv})S_1 + \sigma_{xv}S_2 + \Delta g_v V \tag{3-1}$$

where σ_{lx}, σ_{lv}, and σ_{xv} are the liquid-XNW, the liquid-vapor, and the XNW-vapor interface energy density, S_1 and S_2 are the corresponding interface areas (see Figure1b) expressed as: $S_1 = 2\pi Rh$ and $S_2 = 2\pi rH$, respectively. When the interface between nanowire and catalyst liquid droplet is assumed to be incoherent interface, the values of σ_{lx} can be denoted as: $\sigma_{lx} = \frac{\sigma_{lv} + \sigma_{xv}}{2}$. V is the volume of X clusters defined as: $V = \frac{1}{6}\pi H(H^2 + 3L^2) - \frac{1}{6}\pi h(h^2 + 3L^2)$. Note that the theoretical model in this case is formulated based on the assumptions of (i) the catalyst droplet and X cluster are perfectly spherical, and (ii) X nuclei are mutually non-interaction. From Figure 7(b) and considering $r<R_0$, h, H, and $L2$ can be defined respectively as:

$$h = R_0 - \sqrt{R_0^2 - r^2 + \frac{r^2(R_0\cos\theta - r)^2}{R_0^2 + r^2 - 2R_0 r\cos\theta}} \tag{3-2}$$

$$H = r\left(1 - \frac{R_0\cos\theta - r}{\sqrt{R_0^2 + r^2 - 2R_0 r\cos\theta}}\right) \tag{3-3}$$

$$L^2 = r^2\left(1 - \frac{(R_0\cos\theta - r)^2}{R_0^2 + r^2 - 2R_0 r\cos\theta}\right) \tag{3-4}$$

where R_0, r, θ denote the curvature radius of the liquid droplet, the curvature of the cluster, and the contact angle between cluster and liquid droplet, respectively. Meanwhile, one can obtain $\cos\theta = \frac{\sigma_{lv} - \sigma_{xv}}{2\sigma_{xv}}$, and Δg_v is the Gibbs free energy difference per unit volume from vapor to solid, which is expressed by $-RT/V_m \ln(P/P_e)$, where P, T, R, and V_m are the pressure, the temperature, the gas constant, and the mole volume of the XNW, respectively. P_e is the X vapor phase pressure in thermal equilibrium coexistence with the liquid of composition in a flat surface. Based on Tan's model [26], when the droplet is assumed to be an idea solution, Δg_v can be expressed as

$$\Delta g_v = -\frac{RT}{V_m}\ln\frac{p}{p_e} = -\frac{RT}{V_m}\ln\left(\frac{C}{C^{eq}}\right) \tag{3-5}$$

where C and C^{eq} are X concentrations on the solid and liquid lines of the eutectic catalyst-X phase diagram, respectively.

On the basis of the above deductions, the Gibbs free energy of an X cluster is expressed reasonably as:

$$\begin{aligned}
\Delta G = {} & \pi R(\sigma_{xv} - \sigma_{lv})\left(R - \sqrt{R^2 - r^2 + \frac{r^2(R\cos\theta - r)^2}{R^2 + r^2 - 2Rr\cos\theta}}\right) + \\
& 2\pi\sigma_{xv} r^2\left(1 - \frac{R\cos\theta - r}{\sqrt{R^2 + r^2 - 2Rr\cos\theta}}\right) \\
& + \frac{1}{6}\pi\left(\frac{RT}{V_m}\ln\left(\frac{C}{C^{eq}}\right)\right)\left(\left(R - \sqrt{R^2 - r^2 + \frac{r^2(R\cos\theta - r)^2}{R^2 + r^2 - 2Rr\cos\theta}}\right)^3\right. \\
& + 3r^2\left(R - \sqrt{R^2 - r^2 + \frac{r^2(R\cos\theta - r)^2}{R^2 + r^2 - 2Rr\cos\theta}}\right)\left(1 - \frac{(R\cos\theta - r)^2}{R^2 + r^2 - 2Rr\cos\theta}\right) - \\
& r^3\left(1 - \frac{R\cos\theta - r}{\sqrt{R^2 + r^2 - 2Rr\cos\theta}}\right)^3 - \\
& \left. 3r^3\left(1 - \frac{R\cos\theta - r}{\sqrt{R^2 + r^2 - 2Rr\cos\theta}}\right)\left(1 - \frac{(R\cos\theta - r)^2}{R^2 + r^2 - 2Rr\cos\theta}\right)\right)
\end{aligned} \tag{5-6}$$

From Eq. (6), one can see that the Gibbs free energy of a cluster of XNWs is related with two factors: one is the liquid droplet composition, and another one is the size of the liquid droplet. Therefore, when $\frac{\partial \Delta G(r)}{\partial r} = 0$, the critical size of the nuclei of XNWs can be attained

[10]. Distinctly, the critical size of the XNW nuclei is the thermodynamic size limit of the nanowire growth, the theoretical minimum size, from the view of point of thermodynamics.

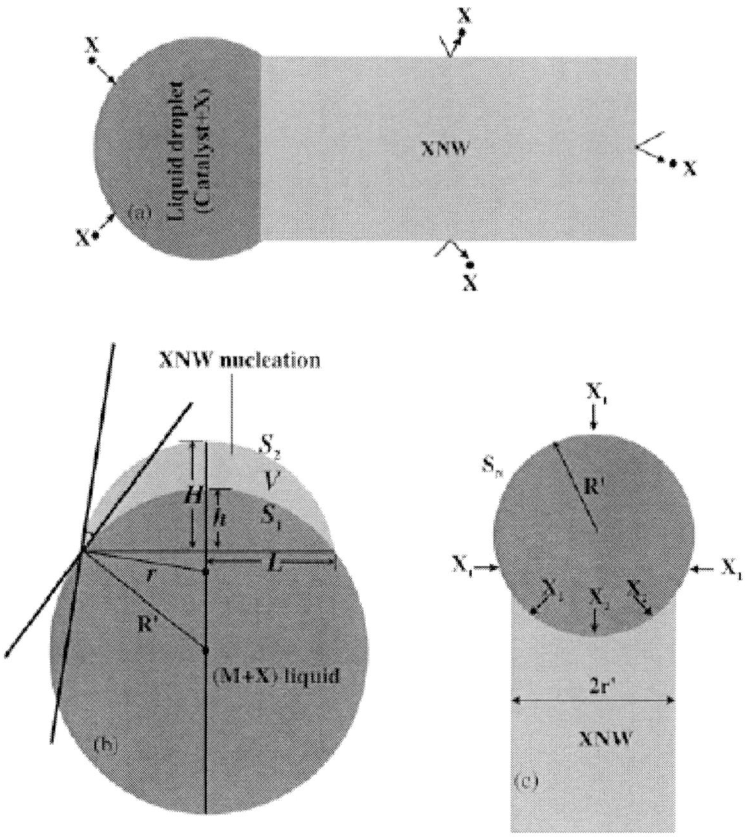

Figure 7. The schematic diagram showing the VLS mechanism, and XNW thermodynamic nucleation and kinetic growth processes. (a) The droplet and the solid nanowire capturing and rejecting nearly all the constituents of the growing material from the vapor phase, respectively, due to the sticking coefficient immense difference between them. (b) The thermodynamic nucleation case. (c) The kinetic growth case.

3.2. Kinetic Size Limit of Nanowire Growth

From the growth kinetics, when X atoms reach saturation in the liquid droplet at the given thermodynamic parameters, the number of X atoms of the droplet capturing (X_1) and extracting (X_2+ X_2^N) should be equal, as shown in Figure 7(c). Then, only X_2^N atoms contribute to the growth of XNWs in the extracting X atoms. The number of X_2^N atoms can be approximately equal to the net flux of impinging on the surface of the area whose value is equal to S_N. Reasonably, the number N of X_2^N atoms is expressed as

$$N = 2\pi R'\left(R' - \sqrt{R'^2 - r'^2}\right)\left(\theta - \theta_r\right) = \frac{10^6 \pi r'^2 S\rho N_m}{M} \quad \textbf{(5-7)}$$

where R', r', ρ, N_m, and M are the radius of droplet, the radius of XNW, the density of single crystal X, the Avogadoro constant, and the mole mass of single crystal X, respectively. θ is the number of X atoms from vapor impinging on a unit plane surface in an unit time. For the VLS progress of vapor growth or vapor deposition, an ideal gas should be a good approximation due to the low gas density. According to Maxwell's velocity distribution equation, at the given temperature T and pressure P, θ can be expressed as $\theta = \int_{-\infty}^{+\infty} du_x \int_{-\infty}^{+\infty} du_y \int_{0}^{+\infty} n u_z P(u) du_z = n\left(\frac{kT}{2\pi m}\right)^{1/2} = \frac{P}{\sqrt{2\pi m kT}}$ where $P(u)$ is the Maxwell's velocity distribution, n is the molecular number of the vapor phase per unit volume, and k is the Boltzmann constant, and m is the mass of X atom. θ_r is the number of desorbing flux from the surface of the droplet on an unit area in an unit time at a finite temperature T. Note that the desorbing flux (θ_r) is independent of the deposition flux from the vapor phase, but is a temperature function, and given by $\theta_r = \frac{P_r}{\sqrt{2\pi m kT}}$ where P_r is the saturation pressure in the liquid droplet surface. On the basis of the Gibbs-Thompson formula, we have

$$\ln\frac{P_r}{P_e} = \frac{2\sigma_{lv}\Omega}{r' kT} \qquad \textbf{(5-8)}$$

where Ω is the volume of the X atom. According to Tan's model [26] and Eq. (8), one can obtain

$$P_r = \frac{PC^{eq}}{C}\exp\left(\frac{2\sigma_{lv}\Omega}{R' kT}\right) \qquad \textbf{(5-9)}$$

In Eq. (7), S is the growth velocity of XNW, and can be expressed as [30]

$$S = h\nu \exp(-E_\alpha / RT)[1-\exp(-|\Delta g_m| / RT)] \qquad \textbf{(5-10)}$$

where h, v, E_α, R, and T are the lattice constant of the XNW in the growth direction, the thermal vibration frequency, the mole adsorption energy of adatoms attached of X atom at surface sites, gas constant, and the substrate temperature, respectively. Based on Eq. (5), the Gibbs free energy difference per mole, Δg_m, and it can be expressed as

$$\Delta g_m = -RT\ln\left(\frac{C}{C^{eq}}\right) \qquad \textbf{(5-11)}$$

According to Eq. (7) - (11), we can obtain the relationship between the size ($2R'$) of catalyst and the diameter ($2r'$) of XNW as follows:

$$2r' = 4R'\left(\frac{M\left(P-\frac{PC^{eq}}{C}\exp\left(\frac{2\sigma_{lv}\Omega}{R'kT}\right)\right)}{10^{6}\sqrt{2\pi mkT}\,\rho N_{m}h\nu\exp\left(-\frac{E_{\alpha}}{RT}\right)\left(1-\exp\left(-\frac{\left|-RT\ln\left(\frac{C}{C^{eq}}\right)\right|}{RT}\right)\right)} - \left(\frac{M\left(P-\frac{PC^{eq}}{C}\exp\left(\frac{2\sigma_{lv}\Omega}{R'kT}\right)\right)}{10^{6}\sqrt{2\pi mkT}\,\rho N_{m}h\nu\exp\left(-\frac{E_{\alpha}}{RT}\right)\left(1-\exp\left(-\frac{\left|-RT\ln\left(\frac{C}{C^{eq}}\right)\right|}{RT}\right)\right)}\right)^{2}\right)^{1/2} \quad (5\text{-}12)$$

Clearly, from Eq. (12), one can see that, the diameter of nanowires is related with the size of droplet. On other words, the growing size of nanowires, the theoretical possible or maximum size, depends on the droplet size from the view of point of kinetics.

3.3. Comparison between Theory and Experiment

In order to validate our model, we take the Si nanowire (SiNW) growth upon CCVT as an example to check its operation, as sufficient experiments or theoretical parameters of Si-Au system are presently available [51]. From Leiber [51] and other references, the thermodynamic parameters of our calculation by Eqs. (6) and (12) are as follows: the liquid droplet-gas surface energy density σ_{lv} is 1.12 J/m2 [32], the solid silicon-gas surface energy density σ_{Siv} is 1.610 J/m2 [31], the mole adsorption energy of adatoms attached of Si atom at surface sites E_{α} can be evaluated to be about 118 kJ/mol [58], the thermal vibration frequency of Si atom ν is about 2×1013 Hz [59], the lattice constant of the SNW in the growth direction (111) h is 0.312 nm, the mole volume V_m of single crystal Si is $1.2\times 10\text{-}5$ kg/m3, the mole mass M and density ρ of single crystal Si are 28.09 g/mol and 2.33 g/cm3, respectively. The experimental temperature T and pressure P are 713 K and 13.332 Pa, respectively.19 Note that the used Au-Si phase diagram is from Ref. [34]. On the basis of Eq. (6), we give the dependence of the radius of a cluster of Si (r) and the Gibbs free energy (ΔG) at the given temperature (713 K), pressure (13.332 Pa), and various Au-catalyst radius (R_0), as shown in Figure 8. We can see that the Gibbs free energy of Si clusters (ΔG) hardly changes with increasing of the Au-catalyst radius. In detail, the Gibbs free energy of Si clusters slightly little increases with the Au-catalyst radius decreasing. Namely, the nucleation barrier of nanowires does not actually depend on the size of metal catalyst. Based on the nucleation thermodynamics [10], the radius of Si clusters (r) corresponding to the peak value should be the critical radius (r') of the SiNW nucleation. Furthermore, from Eq. (6), the relationship curves between the Au-catalyst radius (R') and the critical radius of the SiNW nucleation (r') are given at the given temperature (713 K) and pressure (13.332 Pa) by numerical calculation, as shown in Figure 9. Definitely, the critical radius of the SiNW nucleation basically keeps unchanged with increasing the Au-catalyst radius when the catalyst radius more than 20 nm, and the critical radius decreases with increasing the catalyst radius when the catalyst radius less than 20 nm. However, the decreasing scope of the critical radius is very little when the catalyst radius is less than 20 nm, implying that the affect of the droplet size on the critical radius of the nanowire nucleation can be actually neglected. These results suggest that the critical radius of the nanowire nucleation seems a relatively independent value for a given system from thermodynamic viewpoint, e.g., Si-Au system. In other words, there is a certainly thermodynamic limit, a theoretical minimum value, for the nanowire growth size upon CCVT. For example, the thermodynamic limit is 1.4 nm in radius for the Si-Au system from Figure 9. Very interestingly, we haven't found any experiments that reported SiNWs less than 1.4 nm in radius was synthesized by CCVT yet, so far.

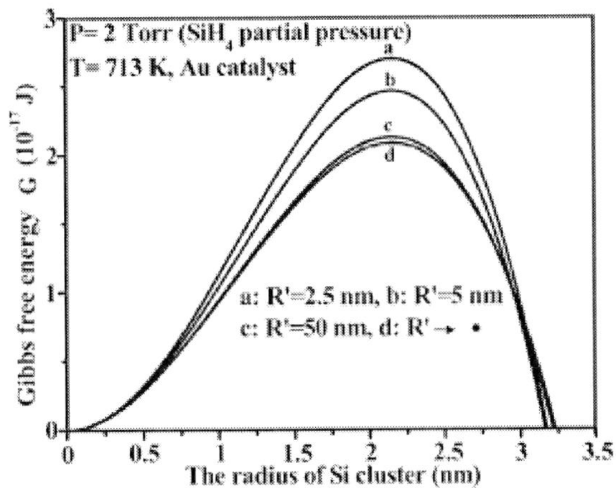

Figure 8. Dependence of the radius of the SiNW nucleation and the Gibbis free energy on the Si cluster size under conditions of various Au-catalyst sizes, temperature (713 K), and pressure (SiH_4 partial pressure of 2 Torr).

The dependence of the diameter of SiNWs on the Au-catalyst size at the given temperature (713 K) and pressure (13.332 Pa) can been attained from Eq. (12) as shown in Figure 10. One can see, the diameter of SiNWs increases with increasing the Au-catalyst diameter, and the SiNW diameter is always smaller than that of the Au-catalyst. Thus, these theoretical results are consistent with experimental cases [43,60]. Remarkably, the quantitative calculations are in excellent agreement with Lieber's experimental data.19 Similarly, Buhro *et. al.* [61] simulated the relationship curve between the diameter of catalyst and the size of GaAs nanowires on the basis of their empirically linear expression and experimental data. Interestingly, our theoretical predictions are consistent with Buhro's results. It is noticed that, in Figure 10, considering the oxidation effect after the SiNW growth, the data of Ref. [51] have been corrected on the basis of authors stated the thickness of oxidation layer. Accordingly, the size of the catalyst droplet can limit the possible growing size of the nanowire synthesized by CCVT to a certain extend from the kinetic viewpoint.

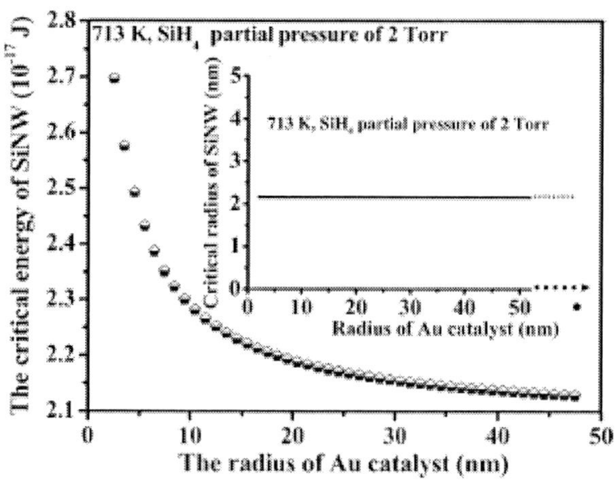

Figure 9. Relationship curve between the Au-catalyst radius and the critical energy of SiNW nucleation; the inset shows the dependence of the Au-catalyst radius and the critical nucleation of SiNW nucleation.

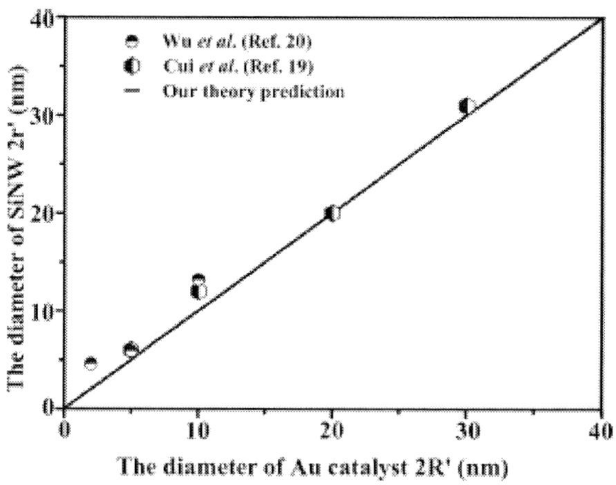

Figure 10. Experimental data of relationship between the size of the Au catalyst and the diameter of SiNW compared to the calculated curve by using eq 10 and Au-Si phase diagram, in which the Au-Si phase diagram from Ref [39] is used.

In summary, the critical size of the nanowire nucleation is suggested to be the thermodynamic size limit, the theoretical minimum size, of the nanowire grown by CCVT in both thermodynamic and kinetic, while the size of the catalyst droplet could limit the growing size, the theoretical possible or maximum size, of the nanowire to a certain extend in kinetics.

4. A Growth Kinetic Model of Nanowire Inside Nanotube

Carbon nanotubes (CNTs) regarded as a template have been demonstrated to be an effective route to fabricate one-dimensional nanostructures of metals and semiconductors [62-64]. Generally, fabrications of one-dimensional nanostructures by used CNTs regarded as template can be divided two cases. One is mechanic filled by capillarity of CNTs immersed liquid, and another one is vapor grown inside CNTs in gases by chemical and physical vapor depositions. In detail, the former is mainly employed to prepare metallic nanowires, and the latter is used to fabricate semiconductors nanowires [55-56,65-71]. We herein focus our attention on the last issue. Basically, it is important to control the growth of one-dimensional nanostructures inside CNTs both experimentally and theoretically for applications of one-dimensional nanostructures. However, compared with many studies of the experimental synthesis of nanowires by CNTs templates [71-72], few theoretical studies of nucleation thermodynamics and growth kinetics of one-dimensional nanostructures forming inside CNTs are found in the literature. To gain a better understanding of the formation of one-dimensional nanostructures inside CNTs in theory, we have proposed a nanothermodynamic nucleation theory to elucidate the nuclei formation of one-dimensional nanostructures inside CNTs [12]. Based on Wilson–Frenkel growth law [73] and our nanothermodynamic model [12], we developed a growth kinetic approach to clarify the growth of one-dimensional nanostructures forming inside CNTs upon vapor deposition in detail. Furthermore, the model is employed in the synthesis of Ga nanowires upon chemical vapor deposition by CNTs' template, and the theoretical results are in good agreement with experimental data.

4.1. Growth Kinetic Model

It is well known that CNTs have many potential applications as molecular sieves, membranes, and ''nanopipes'' for precise delivery of gases or liquid [74]. Some studies focused their attention on transport, adsorbed, and condensed phases of gases inside CNTs [75]. For instance, the transport rates of CH4 and H2 gases in CNTs were suggested to be exceptionally high [76]. Therefore, when one end or two ends of CNTs are open, one-dimensional nanostructures could be fabricated inside CNTs upon vapor deposition by nucleating and growing [75]. Based on Wilson–Frenkel growth law of crystals, here we present a simple kinetic model about phase equilibration and phase transition of nuclei to study growth of one-dimensional nanostructures inside CNTs in gases. This theory is formulated based on the assumption: nano-nuclei are perfectly spherical with no structural deformation of the internal structure from the bulk one. The schematic illustration of a nanowire grown inside a CNT upon vapor deposition is shown in Figure 1. When the reaction gases flow along the CNT, the clusters are condensed on the inner wall of the CNT by a series of surface reactions and diffusions. Sequentially, the nucleation of the clusters will occur by phase transition. Eventually, the nuclei form inside the CNT as shown in Figure 11. Naturally, the starting point in the case is the growth of the nucleus inside a CNT in gases. Generally, growth kinetics of crystals in gases follows the classical Wilson–Frenkel growth law [77].

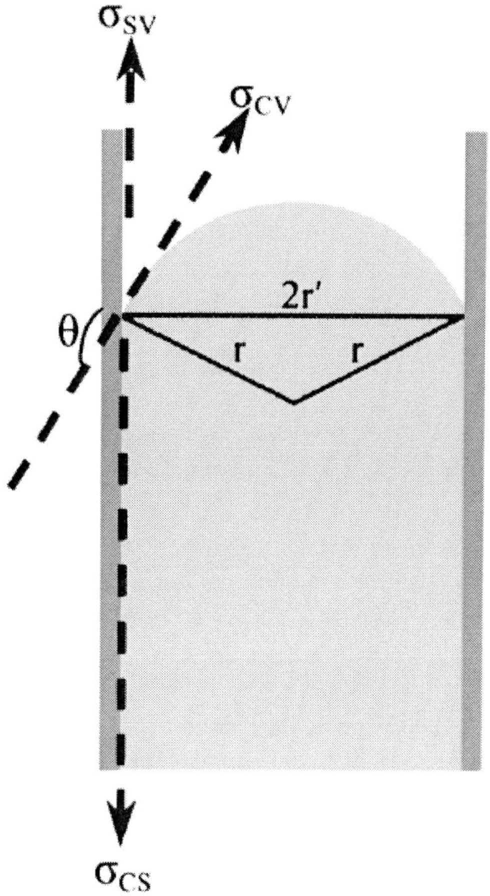

Figure 11. Schematic illustrations of one-dimensional nanostructure grown inside a carbon nanotube.

Accordingly, the growth velocity is driven by the difference in chemical potential between the metastable phase and equilibrium crystal. It is well known that the growth of crystalline nuclei is achieved by trapping their neighbor adatoms that are the mobile surface species and constantly diffuse from a surface site to another site [24]. Adatoms attached at surface sites need the adsorption energy E_a, while those adatoms transporting away from surface sites need to overcome an energy barrier E_b. Thus, the driving force Δg of phase transition from the metastable gas phase to equilibrium crystal is determinate by: $\Delta g = E_a - E_b$. If the adatom amount per unit area and time on the growth surface is N_0, the attachment atom amount N per unit area and time can be expressed by

$$N = N_0 v \exp(-E_a/kT)[1 - \exp(-|\Delta g|/kT)] \quad \textbf{(4-1)}$$

where v is the thermal vibration frequency, k is Boltzmann constant, T is the growth temperature, respectively. Therefore, when $N = N_0$, the crystalline nucleus will grow in the growing direction followed by the layer-by-layer mode. The thickness h of the lay should be equal to the lattice constant of the crystalline nucleus in the growth direction. Accordingly, the growth velocity R of the crystalline nucleus can be expressed by $R = (N/N_0)h$, as follows:

$$R = hv\exp(-E_a/kT)[1-\exp(-|\Delta g|/kT)] \quad \textbf{(4-2)}$$

Actually, when the actual vapor pressure is almost close to the saturation vapor pressure in vapor depositions, $\Delta g \ll kT$. According to the first-degree approximation of Eq. (2) by Thaler formula, we can obtain

$$R = \frac{hv}{kT}\exp(-E_a/kT)\cdot|\Delta g| \quad \textbf{(4-3)}$$

On the other hand, the driving force of phase transition can be expressed by $\Delta g = -kT \ln(P/P_0)$ where P and P_0 are super-saturation pressure and saturation pressure in vapor depositions, respectively. Based on the nanothermodynamic model [12], from the Laplace–Young equation, the sized-induced additional pressure ΔP of the nano-nucleus as shown in Figure 1 can be expressed by $\Delta P = 2\sigma_{cv}/r$ where σ_{cv} is the interface energy of crystalline nucleus vapor, r is the curvature radius of the nucleus, respectively. Taking account of the effect of nanosized-induced additional pressure on the Gibbs free energy of nucleus inside the CNT, Δg should be expressed by $\Delta g = -2kT\ln(P/P_0)+(2\sigma_{cv}/r)$. Furthermore, considering the Kelvin equation [12,36], the driving force Dg is finally expressed by

$$g_v = -\frac{RT}{V_m}\ln\frac{p}{p_e} - \frac{2\sigma_{cv}}{r'} + \frac{2\sigma_{cv}}{r} \quad \textbf{(4-4)}$$

where M and r are the mole mass and density, respectively. In this way, combining Eq. (3) with Eq. (4), we can describe the growth of the nucleus inside CNTs in Figure 1. Additionally, from Figure 1, one can see that $\cos\theta = (\sigma_{sv} - \sigma_{sc})/\sigma_{cv}$ where θ is the contact angle between the crystalline nucleus and the wall of the CNT, σ_{sv} and σ_{sc} are the interface energies of wall vapor and wall nucleus at a given temperature, respectively. Meanwhile, the curvature radius r of the nucleus relates to the column radius r' (inner radius of CNT) as defined as $r = -r'/\cos\theta$. Therefore, based on Eq.(3), Eq. (4), and the relationship of r and r' above, the growth velocity R of the crystalline nucleus inside CNTs is regarded as the function of the column diameter (inner radius of CNT) in the way.

4.2. Comparison between Theory and Experiment

For comparing our model with the corresponding experimental data, we select a typical growth of the onedimensional nanostructure inside CNTs, i.e., Ga nanowire grown CNTs [56], to check the model. According to the work of Gao and Bando [56] and other references, the thermodynamic parameters of our calculations by Eqs. (3) and (4) are as follows: the thermal vibration frequency v is 1013 (Hz) [27], the activation energy Ea is 2.3 (eV) [78], the temperature T and pressure P are 1073 (K) and 5×10-3 (Pa) [56,79], the saturation pressure P_0 is 4×10-6 (Pa) [36], the contact angle θ is 108° [56], the surface energy σ_{cv} is 0.7 (J/m2) [13], the crystalline spacing h along (001) is 0.766 (nm) [80]. Consequently, the growth rate of Ga nanowires inside CNTs upon chemical vapor deposition is calculated to be about 1.8×10-10 (m/s) based on our model. More importantly, the experimental value of the growth rate of Ga nanowires inside CNTs was in the range of 10-9–10-10 (m/s) [56]. Therefore, our calculations are in agreement with the corresponding experimental data.

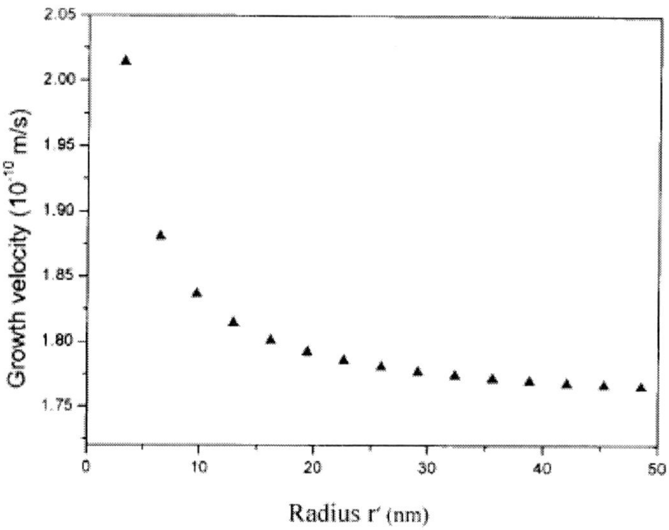

Figure 12. The growth velocity of the metallic Ga nanowire grown inside the CNT as a function of the radius of the CNT.

On the other hand, from Eq. (3), the dependence of the growth velocity R of the nucleus of metallic Ga inside the CNT on the inner radius r' of the CNT was shown in Figure 2, under the conditions of the given thermodynamic parameters such as interface tensions. Obviously, in the Ga nanowire case, the growth of one-dimensional nanostructures inside CNTs is much different from that of two- or three-dimensional structures on the flat of substrates from Figure 12. The diameter of CNTs could effectively influence the growth of one-dimensional nanostructures inside CNTs. Clearly, from Figure 12, we can see that the growth velocity of one-dimensional nanostructures increases with the size of the CNT decreasing at the given deposition conditions such as temperature T and pressure P, and other thermodynamic parameters such as surface and interface tensions. Definitely, when $r' < 10$ nm, the increase of the growth velocity goes much higher with the radius of CNTs continually decreasing. Then, the decreasing of the growth velocity becomes smooth with the radius of CNTs continually increasing, and the decreasing extent of the growth rate also becomes smooth, when $r' > 10$ nm. In fact, the growth is not correlative to the size of CNTs when the radius of CNTs becomes too large. In addition, we calculated the growth rates of GaN and Si3N4 nanowires grown inside CNTs upon vapor deposition. The calculations are consistent with experimental data [70] and the similar dependence of the growth velocity of the nucleus inside the CNT on the inner radius of CNTs is found. Accordingly, the model seems to apply for other nanotubes such as BN or WS2, then, there are the different thermodynamic parameters such as interface energies between nanowires and nanotubes in the different nanotube cases.

In summary, we propose the growth kinetics of one-dimensional nanostructures inside CNTs upon vapor deposition, with respect to the effect of nanosized-induced additional pressure on the growing kinetic parameters, based on the Wilson–Frenkel growth law. Interestingly, these theoretical results are very consistent with the corresponding experimental data. Consequently, we expect the growth kinetics to be a general approach to elucidate growth of one-dimensional nanostructures inside CNTs upon vapor deposition.

5. SIZE-DEPENDENT SHAPE EVOLUTION OF NANOWIRE

Well known, self-assembly processes require atomic or clustering interactions as the thermodynamic driving force to organize those atoms or clusters to form nano-sized domain morphologies [13]. For instance, Bartelt *et al.* recently reported the self-assembly of the regular arrays of two-dimensional (2D) vacancy islands via absorbate-driven dislocation reactions [81]. Therefore, it is essential to pursue the physical and chemical origin causing the nanostructure to control the growth. On the other hand, the surface shape of 1D nanostructures plays a particular important role in their application. For instance, nanowires as wraparound gate transistors or core-shell heterostructures require good control of the surface shape to achieve uniform cross section and minimize carrier scattering at rough interface [82]. Accordingly, a quantitative understanding of the surface shape formation of 1D nanostructures is needed for the application of interest.

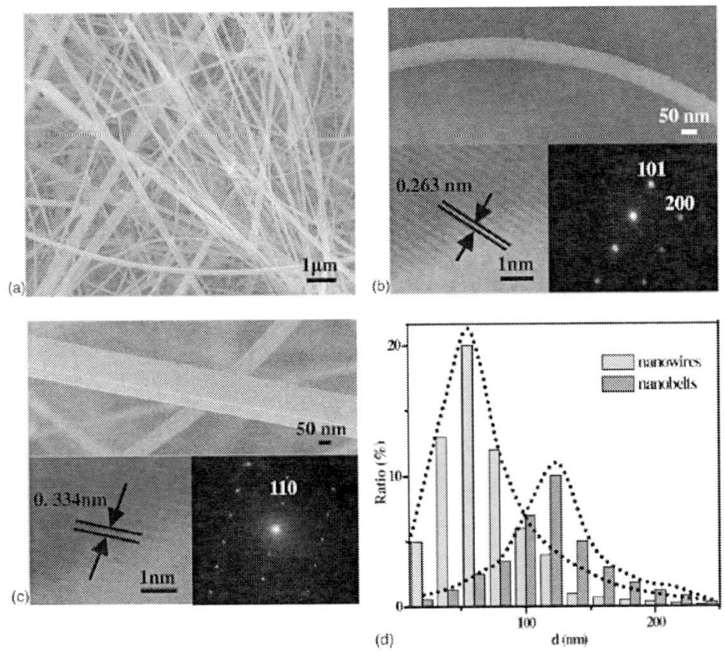

Figure 13. (a) Low-magnifying Field emission scanning electron microscopy (FESEM) image of SnO_2 nanowires and nanobelts. (b) High-magnifying FESEM image of a nanowire, in which a corresponding high-resolution transmission electron microscopy (HRTEM) image and selected area electronic diffraction (SAD) are shown below left right, respectively. (c) High-magnifying FESEM image of a nanobelt (the white line contrast on the nanobelt is another nanobelt whose wider plane of the cross-section inclines extremely towards left inside so as to seem thinner than the practical size), in which a corresponding HRTEM image and SAD are shown below left right, respectively. (d) The SEM statistics result of the radial size distribution of nanowires and nanobelts in the total prepared 1D nanostructures.

Recently, we experimentally observed the size-dependent morphological evolution of the 1D SnO2 nanostructure growth using metal catalysts assistant thermal chemical vapor transport and condensation (VLS process). In detail, we found that the shape formations of nanowires and nanobelts of SnO2 are size dependent, i.e., the wire is favorable when the size is less than 90 nm and the belt is favorable when the size is more than 90 nm (Figure 13). For this issue, we proposed the nucleation thermodynamics, the growth kinetics, and the shape transition thermodynamics to elucidate our findings on the basis of the thermodynamic analysis at the nanometer scale [8, 11, 12, 22, 83, 84]. Note that the size-dependent shape transformation from the initial stage of the nanowires nuclei to the nanobelts nuclei originates from the thermodynamic driving force of nucleation and growth of 1D nanostructures.

5.1. Nucleation Thermodynamic Consideration

Generally, the Gibbs free energy is an adaptable measure of the energy of a state in phase transformation among competing phases. Thermodynamically, the phase transformation is promoted by the difference of the Gibbs free energies [24]. Therefore, we perform the nucleation of SnO2 on the basis of the thermodynamic nucleation at the nanometer scale [22,84]. Considering the nanowire growth originates from the column-shape nuclei, and the formation of nuclei of the nanowire shown in Figure 1 (b) is a process of the reaction precursors extracting from the saturated Au catalyst [30], the Gibbs free energy difference of a column-shape nucleus is expressed

$$\Delta G_1 = -\Delta g_v \pi r_1^2 L_1 + \pi r_1^2 \left(\delta_1 + \delta_1^{''} \right) + 2\pi r_1 L_1 \delta_1 \tag{5-1}$$

where δ_1 and δ_1'' are the nucleus-vapor and the nucleus-liquid interface energies, and the r_1 and L_1 are the radius and the height of the nucleus. The $\Delta g_v = -RT/V_m ln(P/P_e)$ is the Gibbs free energy difference per unit volume, in which T, R and V_m are the temperature, the gas constant, and the mole volume, and . P and Pe are the partial pressure of the Sn and the Sn vapor phase pressure in thermal equilibrium coexistence with the liquid of composition in a flat surface. From Eq. (1), we deduce that the critical radius of r^*_1, the critical height of L^*_1 and the critical energy of ΔG^*_1 of nuclei are $2\delta_1/\Delta g_v$, $2(\delta_1+\delta_1'')/\Delta g_v$, and $4\pi\delta_1^2(\delta_1+\delta_1'')/\Delta g_v^2$.

Considering the nanobelt growth originates from the rectangle-shape nuclei. Note that, we use square-shape nuclei instead of rectangle-shape nuclei for simplifying the calculation in our case. Thus, The Gibbs free energy difference of a rectangle-shape nucleus is expressed

$$\Delta G_2 = -\Delta g_v r_2^2 L_2 + r_2^2 \left(\delta_2 + \delta_2^{''} \right) + 4 r_2 L_2 \delta_2 \tag{5-2}$$

where δ_2 and δ_2'' are the nucleus-vapor and nucleus-liquid interface energies, and the r_2 and L_2 are the side-length and the height of nucleus. Similarly, we have that the critical radius of r^*_2, the critical height of L^*_2 and the critical energy of ΔG^*_2 of nuclei are $4\delta_2/\Delta g_v$, $2(\delta_2+\delta_2'')/\Delta g_v$, and $16\delta_2^2(\delta_2+\delta_2'')/\Delta g_v^2$.

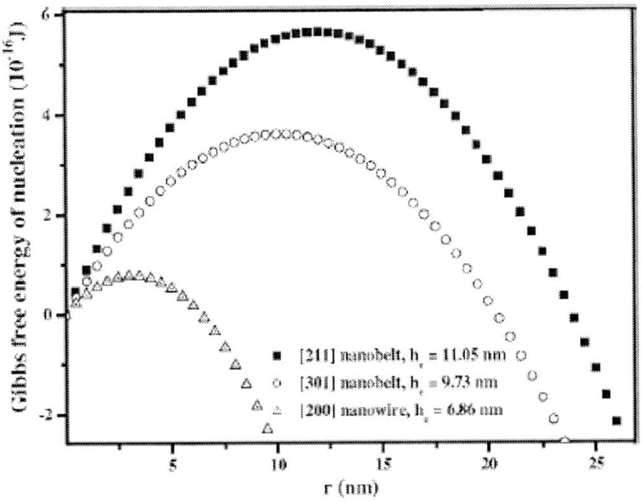

Figure 14. The nucleation energies of a column-shape nucleus of nanowires and a rectangle-shape nucleus of nanobelts growing along the (211) and the (301) directions, respectively.

Considering different growth directions of nanowires (the [200] direction) and nanobelts (the [211] or the [301] direction), Au catalyst, and the VLS nucleation mechanism [11,22], δ_1 and δ_2 are 1.14 and 2.14 (or 1.82) Jm-2 [85,86], δ_1'' and δ_2'' are 1.32 and 1.82 (or 1.66) Jm-2 [22,87], T and V_m are 1223 K, and 16.24 cm3/mole, and P and P_e are 3.29 and 1.047 Pa. Accordingly, r^*_1 and L^*_1 of nanowire nuclei are calculated to be 3.18 and 6.86 nm, r^*_2 and L^*_2 of nanobelt nuclei are calculated to be 11.94 (the [211] direction) or 10.18 nm (the [301] direction) and 11.05 (the [211] direction) or 9.73 nm (the [301] direction), and the size-dependence of the nanowire and nanobelt nucleation is shown in Figure 14. Clearly, the radial size of the critical nucleus of nanobelts is approximately two times larger than the diameter of the critical nucleus of nanowires. Moreover, the nucleation energy of nanowire nuclei is lower than that of nanobelt nuclei that means that the nucleation probabilities of nanowires is higher than that of nanobelts, which are well consistent with our observations that the ratios of wires and belts in the total synthesized nanostructures are 63 % and 37 %, respectively.

5.2. Growth Kinetic Consideration

Once the nucleation takes place, the increases in volume afforded by the continuous supply of the reaction precursors extracting from the saturated Au catalyst contributes to the increases in two feature sizes, radial size, and height, of nanowires and nanobelts on the basis of the VLS process [11,22,87]. The growth mechanism sketch map of the nanowire or the nanobelt is shown in Figure 15 (a). Thus, the radial growth rate of 1D nanostructures determines their final size. Well known, the surface free energy of nuclei and the concentration of the reaction precursors play key roles in the growth of nuclei [57], in which the surface free energy is consisted of the surface energy [88-89] and the strain energy [90,91]. Meanwhile, the supply of the reaction precursors comes from the axial direction reaction precursors extracting from the saturated Au catalyst. The supply mechanism lead to the growth of the radius and height of 1D nanostructures, i.e., (i) axial supply resulting in an augment in the height (P_{aa}), (ii) axial supply resulting in an augment in the radius (P_{ar}) as follows [90]

$$P_{ij} = \frac{\exp\left(-\frac{\Delta E_{ij}}{KT}\right)}{\sum_{i,j}\exp\left(-\frac{\Delta E_{ij}}{KT}\right)}, \Delta E_{ij} = \Delta E_{Sij} + \Delta E_{Tij}, i, j = a, r. \quad \textbf{(5-3)}$$

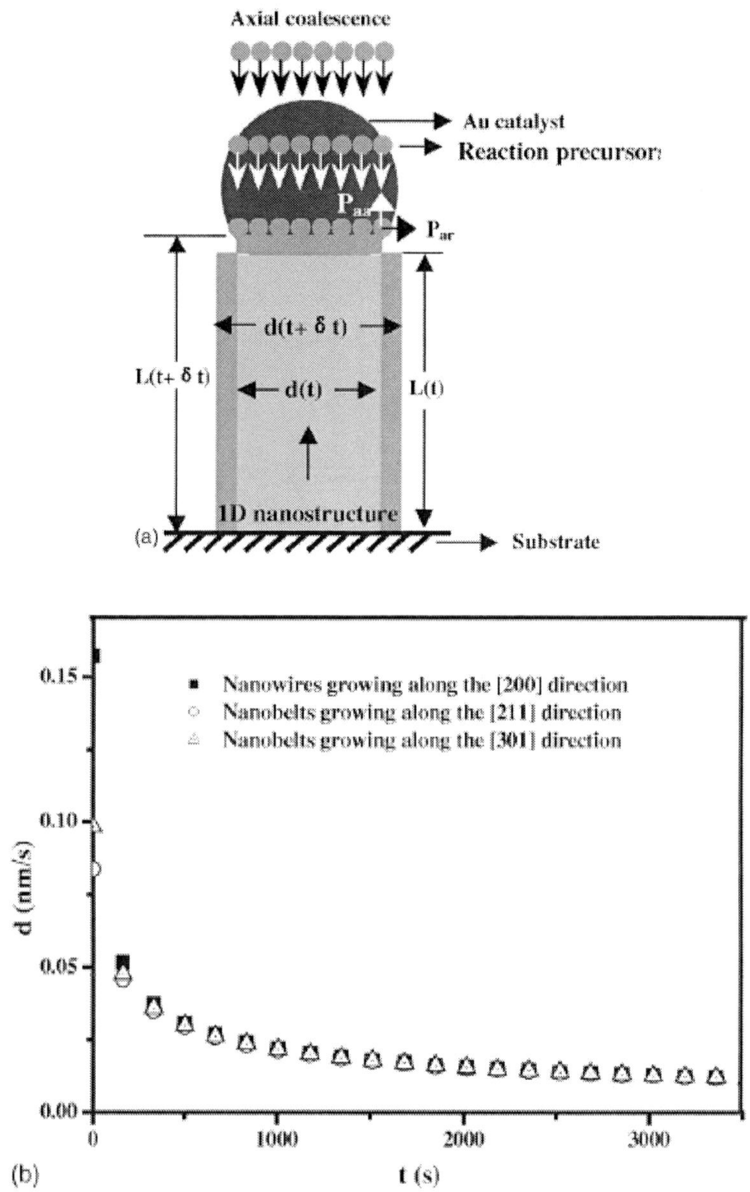

Figure 15. (a) The growth mechanism sketch map of the 1D nanostructure (the radial cross section of the nanowire or nanobelt is shown here). (b) The growth rates of nanowire and nanobelt.

K is Boltzman constant and T is the growth temperature. Thinking of the free energy ΔE_{ij} gaining in each case, we assume that the radial size of nanowires and nanobelts is large enough to make the surface energy ΔE_{Sij} much smaller than the strain energy ΔE_{Tij}. Thus, in each case, ΔE_{Tij} is written by [90]: $\Delta E_{Taa} = E_m(1-\sigma_m)\varepsilon_a^2, \Delta E_{Tar} = E_m\delta_m\varepsilon_a^2$, where E_m is the elastic modulus, δ_m is the Poisson's ratio of nanowires or nanobelts, and ε_a is the strain in the axial. Therefore, δt, the volume additivities of nanowires along the radial and axial are shown as follows [90]:

$$\frac{1}{4}L_1(t)d_1(t)\frac{\partial d_1(t)}{\partial t} = A\frac{1}{4}P_{ar}d_1(t)^2 \tag{5-4}$$

$$\frac{1}{4}d_1(t)^2\frac{\partial L_1(t)}{\partial t} = 2A\frac{1}{4}P_{aa}d_1(t)^2 \tag{5-5}$$

Similarly, in the case of nanobelts, we have

$$\frac{1}{4}L_2(t)d_2(t)\frac{\partial d_2(t)}{\partial t} = A\frac{1}{4}P_{ar}d_2(t)^2 \tag{5-6}$$

$$\frac{1}{4}d_2(t)^2\frac{\partial L_2(t)}{\partial t} = 2A\frac{1}{4}P_{aa}d_2(t)^2 \tag{5-7}$$

where $d_1(t)$ and $d_2(t)$ are the diameter (which is the function of the time t) of nanowires and the side-length (which is the function of the time t) of nanobelts, L1(t) and L2(t) are the height (which is the function of the time t) of the nanowires and the nanobelts, A is the constant that are relative with the growth. In the Eq. 4, the left section of the equation is the radial volume increase of the nanowire and the right section of the equation is the volume increase resulting from the axial supply precursors extracting from the Au catalyst, which leading to the augment of the radius of the nanowire. In the Eq. 5, the left section of the equation is the axial volume increase of the nanowire and the right section of the equation is the volume increase resulting from the axial supply precursors extracting from the Au catalyst, which leading to the augment of the height of the nanowire. Meanwhile, in the Eq. 6, the left section of the equation is the radial volume increase of the nanobelt and the right section of the equation is the volume increase resulting from the axial supply precursors extracting from the Au catalyst, which leading to the augment of the side-length of the nanobelt. In the Eq. 7, the left section of the equation is the axial volume increase of the nanobelt and the right section of the equation is the volume increase resulting from the axial supply precursors extracting from the Au catalyst, which leading to the augment of the height of the nanobelt.

Neglecting the difference in strain energies in every type of growth, we attain approximate conclusions as follows

$$d_1(t) \approx \left[2(t + 2r*_1^{\,2})\right]^{\frac{1}{2}},\; d_2(t) \approx \left[2(t + \frac{r*_2^{\,2}}{2})\right]^{\frac{1}{2}} \quad \textbf{(5-8)}$$

where t is the growth time of nanostructures after nucleating, and $r*_1$ and $r*_2$ are critical sizes of nanowires and nanobelts. According to Eq. (8), the growth rates of nanowires and nanobelts are shown in Figure 15(b). Distinctly, both growth rates of nanowire and nanobelt are basically the same during the growth of nanostructures. Therefore, the final size of nanobelts is approximately two times larger than that of nanowires, as $r*_2$ is double of $r*_1$. This theoretical result is in well agreement with our findings shown in Figure 13(d).

5.3. Shape Transition Thermodynamic Consideration

During the initial stage of the growth of nanowire nuclei, there are two probabilities of the shape transitions. One is from the wire nuclei to belt nuclei with an increase to the volume of nanowire nuclei, and another one is from the belt nuclei to wire nuclei with and increase to the volume of nanobelt nuclei. In order to clarify which shape transition is preferable in thermodynamical, we need to compare the Gibbs free energies of wire and belt nuclei. Therefore, the Gibbs free energy of the shape transition, i.e., the Gibbs free energy difference between wire nuclei and belt nuclei, is given as $\Delta G = V\Delta P + \delta\Delta S$ [92], where _P is a pressure difference between nanowire nuclei and nanobelt nuclei, which is small enough to ignore. $V = \pi d_1 2L_1/4 = d_2 2L_2$ is the volume of nanowire nuclei and nanobelt nuclei, in which d_1 and L_1 are the diameter and height of nanowire nuclei, and d_2 and L_2 are the side length and height of the nanobelt nuclei, and $L_2 = \pi d_1 2L_1/4d_2 2$. δ is the surface energy difference of the (200) plane of nanowire nuclei and the (211) plane of nanobelt nuclei. ΔS is the surface area difference between nanowires nuclei and nanobelt nuclei. Thus, ΔG can be expressed

$$\Delta G = \delta_2(4d_2L_2 + d_2^2) + \delta_2'' d_2^2 - \delta_1(\pi d_1 L_1 + \pi d_1^2/4) - \pi d_1^2 \delta_1''/4 \quad \textbf{(5-9)}$$

According to Eq. (9), when d_1 is equal to 150, 175 and 200 nm and L_1 = 12 nm, the relationships between ΔG and d_2 are shown in Figure 16(a). Clearly, three curves intersect at the A point (ΔG = 0 and d_2 = 43 nm), and $\Delta G < 0$ in these regions of AB, AC and AD. In detail, in these regions, $\Delta G < 0$ means the Gibbs free energy of nanowire nuclei is larger than that of nanobelt nuclei, suggesting that the shape transition from wire nuclei to belt nuclei is probable. In instance, the nanowire nuclei with diameters of 150, 175 and 200 nm could transform into the nanobelt nuclei with radial sizes of 60, 70, and 80 nm, respectively. Note that, $\Delta G > 0$, when d_1 < 120 nm, implying that the shape transition from wire nuclei to belt nuclei of nanowire nuclei with the diameter less than 120 nm is not thermodynamically expected. When the nanobelts grow along the [301] direction, according to the Eq. (9), the relationships between ΔG and d_2 are shown in Figure 16(b). Clearly, three curves intersect at the E point (ΔG = 0 and d_2 = 34 nm), and $\Delta G < 0$ in these regions of EF, EG and EH. The nanowire nuclei with diameters of 150, 175 and 200 nm could transform into the nanobelt nuclei with radial sizes of 60, 70, and 80 nm, respectively. Note that, $\Delta G > 0$, when d_1 < 90 nm, implying that the shape transition from the wire nuclei to the belt nuclei of nanowire nuclei with the diameter less than 90 nm is not thermodynamically expected. To sum up, the difference between the calculation results according to two different growth directions of the nanobelts is not so large that the conclusions are basically consistent and acceptable. Therefore, the larger nanowire nuclei sizes are, the easier the shape transition from wire nuclei to belt nuclei is, from the viewpoint of thermodynamic driving force. In fact, these theoretical predictions are not only consistent with experiments shown in Figure 13(d), but also are physical origins of the experimental observations that both the wire formation with small size and the belt formation with large size are favorable during the growth of nanostructures.

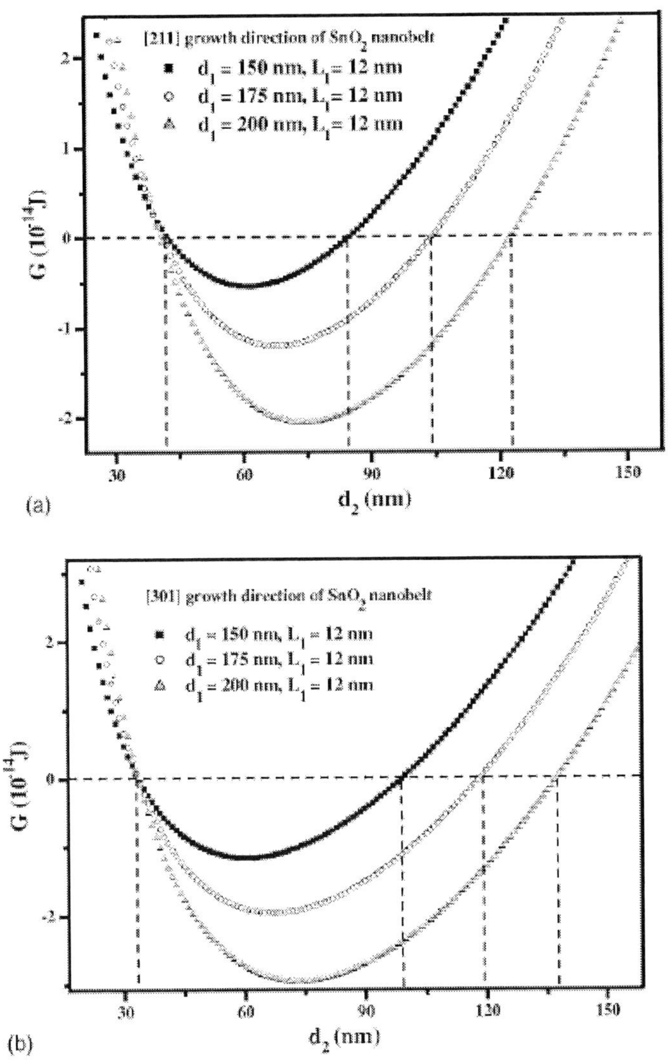

Figure 16. Dependence of the Gibbs free energy of the shape transition during the initial stage of the nuclei growth. (a) The shape transition from the wire nuclei to the belt nuclei that grows along the (211) direction on the radial size of the nanowires nuclei under conditions of d_1 =150, 175, and 200 nm, respectively, and L_1 = 12 nm. (b) The shape transition from the wire nuclei to the belt nuclei that growing along the (301) direction on the radial size of a nanowires nuclei under conditions of d_1 = 150, 175, and 200 nm, respectively, and L_1 = 12 nm.

In summary, we have systematically investigated the physical and chemical mechanisms of the shape evolution of 1D SnO2 nanostructures in both experimental and theoretical. Based on experiments and theories, we found that the thermodynamic driving force, the Gibbs free energy difference between two phases, is always responsible for the shape choice in the different growth stages such as nucleation, growth, and structural transition of 1D nanostructures, implying that the thermodynamic driving force is one of the physical origins causing the interesting morphological transformation in the growth of 1D nanostructures. Meanwhile, our studies suggested that the revealed physical mechanisms could be expected to generally apply to control the shape of 1D nanostructures during the growth for purpose of fundamental researches and potential applications.

6. Conclusion

1D nanostructures such as wires, rods, belts, and tubes have become the focus of intensive research owing to their unique applications in mesoscopic physics and fabrication of nanoscaled devices, as they not only provide a good system to study the electrical and thermal transport in 1D confinement, but also are expected to play an important role in both interconnection and functional units in fabricating electronic, optoelectronic, and magnetic storage devices with nanoscale dimension. Naturally, a great challenge for materials scientists is how to flexibly control the self-assembly and synthesis processes of 1D nanostructures for these purposes of fundamental researches and potential applications. Considering the self-assembly and synthesis processes of 1D nanostructures are relatively new material processing methods, it is important to develop new theoretical tools to investigate the physics and chemistry involved in these processes. For this issue, we have systemically established a series of new theoretical tools, i.e., thermodynamic and kinetic approaches, to address the nucleation, growth and phase transition of 1D nanostructures on the basis of the VLS process, which not only elucidate the physical and chemical mechanisms involved in the self-assembly and synthesis processes of 1D nanostructures, but also predict the growth behavior. Importantly, these theoretical tools provide a fundamental understanding to flexibly control the self-assembly and synthesis processes of 1D nanostructures. Therefore, we expected these new theoretical methods to be general approaches applicable to elucidate the formation of 1D nanostructures.

Acknowledgment

The National Science Foundation of China (90306006 and 50525206), the Ministry of Education (106126) and the Natural Science Foundation of Guangdong (01009737) supported this work.

REFERENCES

[1] Xia, Y.; Yang, P.; Sun, Y.; Wu, Y.; Mayers, B.; Gates, B.; Yin, Y.; Kim, F.; Yan, H. *Adv. Mater.* 2003, *15,* 353.
[2] Wang, Z. L. *J. Phys.: Condens. Matt.* 2004, *16*, R829.
[3] Chamberlin, R. V. *Nature* 2000, *408*, 337.
[4] Alivisato, P. *Nano Lett.* 2001, *1*, 109.
[5] Chamberlin, R. V. *Phys. Lett. A* 2003, *315*, 313.
[6] Abe, S.; Rajagopal, A. K. *Phys. Rev. Lett.* 2003, *91,* 120601.
[7] Wang, Z. L.; Kong, X.; Zou, J. M. *Phys. Rev. Lett.* 2003, *91,*185502.
[8] Zhang, C. Y.; Wang, C. X.; Yang, Y. H.; Yang, G. W. *J. Chem. Phys. B* 2004, *108*, 2589.
[9] Wang, C. X.; Yang, Y. H.; Liu, Q. X.; Yang, G. W. *J. Phys. Chem. B* 2004, *108*, 728.
[10] Wang, C. X.; Yang, Y. H.; Xu, N. S.; Yang, G. W. *J. Am. Chem. Soc.* 2004, *126*, 11303.
[11] Wang, C. X.; Chen, J.; Yang, G. W.; Xu, N. S. *Angew. Chem. Int. Edit.* 2005, *44*, 7414.
[12] Wang, C. X.; Yang, G. W. *Mater. Sci. and Eng. -R* 2005, *49*, 157.
[13] Pohl, K.; Bartelt, M. C.; Figuera, J. de la; Bartelt, N. C.; Hrbek, J.; Hwang, R. Q. *Nature* 1999, *397*, 238.
[14] Plass, R.; Last, J. A.; Bartelt, N. C.; Kellogg, G. L. *Nature* 2001, *412*, 875.
[15] Wagner, R. S.; Ellis, W. C. *Appl. Phys. Lett.* 1964, *4*, 89.
[16] Wagner, R. S.; Ellis, W. C.; Jackson, K. A.; Arnold, S. M. *J. Appl. Phys.* 1964, *35*, 2993.
[17] A. M. Morales and C. M. Lieber, *Science* 1998, *279*, 208.
[18] Duan, X. F.; Lieber, C. M. *Adv. Mater.* 2000, *12,* 298.
[19] Chen, C. C.; Yeh, C. C. *Adv. Mater.* 2000, *12,* 738.
[20] Wu, Y.; Yang, P. *Chem. Mater.* 2000, *12,* 605.
[21] Wu, X. C.; Song, W. H.; Wang, K. Y.; Hu, T.; Zhao, B.; Sun, Y. P.; Du, J. J. *Chem. Phys. Lett.* 2001, *336,* 53.
[22] Wang, C. X.; Wang, B.; Yang, Y. H.; Yang, G. W. *J. Phys. Chem. B* 2005, *109,* 9966.
[23] S. H. Tolbert and A. P. Alivisatos, *Annu. Rev. Phys. Chem.* 1995, *46*, 595.
[24] Yang, G. W.; Liu, B. X. *Phys. Rev. B* 2000, *61*, 4500.
[25] Kelton, K. F. *Solid State Phys.* 1991, *45,* 75.
[26] Tan, T. Y.; Li, N.; Gösele, U. *Appl. Phys. Lett.* 2003, *83,* 1199.
[27] Louchev, O. A.; Sato, Y.; Kanda, H. *Appl. Phys. Lett.* 2002, *80,* 2752.
[28] Jiang, Q.; Zhang, S. H.; Li, J. C. *Solid State Commun.* 2004, *130,* 581
[29] Kirkaldy, J. S. *Diffusion in the Condensed State,* The Universities Press, Belfast, 1987.
[30] Feng, S. Q.; Yu, P. D.; Zhang, H. Z.; Bai, Z. G.; Ding, Y. *J. Cryst. Growth* 2000, *209,* 513.
[31] Givargizov, E. I. *Highly Anisotropic Crystals,* Boston, 1986.
[32] Hardy, S. C. *J. Cryst. Growth* 1984, *69,* 456.
[33] Lew, K. K.; Pan, L.; Dicksy, E. C.; Redwing, J. M. *Adv. Mater.* 2003, *15,* 2073.
[34] Cui, Y.; Lauhon, L. J.; Gudiksen, M. S.; Wang, J. *Appl. Phys. Lett.* 2001, *78,* 2214.
[35] Lee, S. T.; Wang, N.; Lee, C. S. *Mater. Sci. Eng. A* 2000, *286,* 16.

[36] Lide, D. R. *CRC Handbook of Chemistry and Physics*, 71st ed. Chemical Rubber, Ohio, 1990–1991.

[37] *Sargent-Welch Scientific Company, Table of Periodic Properties of the lements,* Skokie, IL, 1980, pp. 1–2.

[38] King, H. W. in *Physical Metallurgy*, 3rd ed., edited by R. W. Cahn and P. Haasen, North-Holland Physics, Amsterdam, 1983, pp. 59–63.

[39] *Binary Alloy Phase Diagram*, edited by T. B. Massalsky, ASM, New York, 1986, p. 1108.

[40] The saturation concentration of Si (accounting for 50 at %) is defined by $C^{*}=NA/(M\mathrm{Si}/2\mathrm{Si})+(M\mathrm{Fe}/2\mathrm{Fe})$, where *NA*, *M*Si, *M*Fe, Si, and Fe, are the Avogadoro constant, mole mass of iron, mole mass of silicon, density of liquid silicon, and density of liquid iron, respectively.

[41] Wang, N.; Tang, Y. H.; Zhang, Y. F.; Yu, D. P.; Lee, C. S.; Bello, I.; Lee, S. T. *Chem. Phys. Lett.* 1998, *283,* 368.

[42] Wang, N.; Tang, Y. H.; Zhang, Lee, C. S.; Bello, I.; Lee, S. T. *Chem. Phys. Lett.* 1999, *299,* 237.

[43] Morales, A. M.; Lieber, C. M. *Science* 1998, *208*, 279.

[44] Sunkara, M. K.; Sharma, Miranda, S.; R.; Lian, G.; Dickey, E. C. *Appl. Phys. Lett.* 2001, *79*, 1546.

[45] Kamins, T. I.; Williams, R. S.; Basile, D. P.; Hesjedal, T.; Harris, J. S. *J. Appl. Phys.* 2001, *89*, 1008.

[46] Westwater, J.; Gosain, D. P.; Tomiya, S.; Usui, S.; Ruda, H. *J. Vac. Sci. Technol. B* 1997, *15*, 554.

[47] Hiruma, K.; Yazama, M.; Katsuyama, T.; Ogawa, K.; Haraguchi, K.; Koguchi, M.; Kakibayashi, H. *J. Appl. Phys.* 1995, *77*, 447.

[48] Duan, X.; Lieber, C. M. *J. Am. Chem. Soc.* 2000, *122*, 188.

[49] Ohlsson, B. J.; Björk, M. T.; Magnusson, M. H.; Deppert, K.; Samuelson, L.; Wallenberg, L. R. *Appl. Phys. Lett.* 2001, *79,* 3335.

[50] Gudiksen, M. S.; Lieber, C. M. *J. Am. Chem. Soc.* 2000, *122*, 8801.

[51] Cui, Y.; Lauhon, L. J.; Gudiksen, M. S.; Wang, J. F.; Lieber, C. M. *Appl. Phys. Lett.* 2001, *78*, 2214.

[52] Davis, S. H. *Proc. Roy. Soc. London. Ser. A* 1969, *310*, 341.

[53] Galdi, G. P.; Straughan, B. *Arch. Rational Mech. Anal.* 1989, *89*, 211.

[54] Hwang, N.M.; Hahn, J.H.; Poon, D.Y. *J. Cryst. Growth* 1996, *160*, 87.

[55] Gao, Y.H.; Bando, Y. *Nature* 2002, *415,* 599.

[56] Gao, Y.H.; Bando, Y. *Appl. Phys. Lett.* 2002, *81*, 3966.

[57] Liu, Q. X.; Wang, C. X.; Yang, Y. H.; Yang, G. W. *Appl. Phys. Lett.* 2004, *84*, 4568.

[58] Fissel, A.; Kaiser, U.; Ducke, E.; Schröter, B.; Richter, W. *J. Crystal Growth* 1995,*154*, 72.

[59] Ho, M. H.; Sun, Y. C. *Surf. Sci.* 2002, *516*, L540.

[60] Liu, Z. Q.; Zhou, W. Y.; Sun, L. F.; Tang, D. S.; Zou, X. P.; Li, Y. B.; Wang, C. Y.; Wang, G.; Xie, S. S. *Chem. Phys. Lett.* 2001, *341*, 523.

[61] Yu, H; Buhro, W. E. *Adv. Mater.* 2003, *15,* 416.

[62] Iijima, S. *Nature* 1991, *354,* 56.

[63] Ajayan, P. M.; Iijima, S. *Nature* 1993, *361*, 333.

[64] Dai, H.; Wong, E. W.; Lu, Y. Z.; Fan, S.; Lieber, C. *Nature*, 1995, *375*, 769.

[65] Tsang, S. C.; Chen, Y. K.; Harris, P. J.; Green, M. L. *Nature*, 1994, *372*, 159.
[66] Hsu, W. K.; Hare, J. P.; Terrones, M.; Harris, P. J.; Kroto, H. W.; Walton, D. R. *Nature* 1995, *377*, 687.
[67] Subramoney, S. *Adv. Mater*. 1998, *10*, 1157.
[68] Pradhan, B. K.; Kyotani, T.; Tomita, A. *Chem. Commun*. 1999, *67*, 595.
[69] Wong, E. W.; Maynor, B. W.; Burns, L. D.; Lieber, C. M. *Chem. Mater*. 1996, *8*, 2041.
[70] Han, W.; Fan, S.; Li, Q.; Hu, Y. *Science* 1997, *277*, 1287.
[71] Terrones, M.; Grobert, N.; Hsu, W. K.; Zhu, Y. Q.; Hu, W. B.; Terrones, H.; Hare, J. P.; Kroto, H. K.; Walton, D. R. *MRS Bull.* 1999, *24*, 42.
[72] See: A special issue on carbon nanotubes, *Acc. Chem. Res*. 2002, *35*, No. 12.
[73] Wilson, A. Philos. Mag. 1900, 50, 609; Frenkel, J. *Phys. Z.* 1932, *1*, 498.
[74] Wang, Q.; Challa, S. R.; Shol, D. S.; Johnson, J. K. *Phys. Rev. Lett.* 1999, *82*, 956.
[75] Calbi, M. M.; Cole, N. W.; Gatica, S. M.; Bojan, M. J.; Stan, G. *Rev. Mod. Phys.* 2001, *73*, 857.
[76] Skoulidas, A. I.; Ackerman, D. M.; Johnson, J. K.; Sholl, D. S. *Phys. Rev. Lett.* 2002, *89*, 185901.
[77] Halpin-Healy, T.; Zhang, Y. C. *Phys. Rep.* 1995, *254*, 215.
[78] Yang, H.; Labella, V. P.; Bullock, D. W.; Ding, Z.; Smathers, J. B.; Thibado, P. M. *J. Cryst. Growth* 1999, *201-202,* 88.
[79] Liu, Z.; Gao, Y.; Bando, Y. *Appl. Phys. Lett.* 2002, *81*, 4844.
[80] Mighell, A. D.; Santoro, A.; Donny, J. D. H. *International tables for x-ray crystallography*, 1969, Vol. I.
[81] Thurmer, K.; Carter, C. B.; Bartelt, N. C.; Hwang, R. Q. *Phys. Rev. Lett.* 2004, *92,* 106101.
[82] Ross, F. M.; Tersoff, J.; Reuter, M. C. *Phys. Rev. Lett.* 2005, *95,* 146104.
[83] Liu, Q. X.; Wang, C. X.; Yang, G. W. *Phys. Rev. B* 2005, *71*, 155422.
[84] Liu, Q. X.; Wang, C. X.; Xu, N. S.; Yang, G. W. *Phys. Rev. B* 2005, *72*, 085417.
[85] Oviedo, J.; Gillan, M. J. *Surf. Sci.* 2000, *463*, 93.
[86] Slater, B.; Catlow, C. R.; Gay, D. H.; Williams, D. E.; Dusastre, V. *J. Phys. Chem. B* 1999, *103*, 10644.
[87] Vitos, L.; Ruban, A. V.; Skriver, H. L.; Kollar, J. *Surf. Sci.* 1998, *411*, 186.
[88] Kwon, S. J.; Park, J. Q. *J. Chem. Phys.* 2005, *122*, 214714.
[89] Witt, F.; Vook, R. W. *J. Appl. Phys.* 1968, *39*, 2773.
[90] Kwon, S. J. *J. Phys. Chem. B* 2006, *110*, 3876.
[91] Pelleg, J.; Zerin, L. Z.; Lungo, S. *Thin Solid Films* 1991, *197*, 117.
[92] Liang, L. H.; Liu, F.; Shi, D. X.; Liu, W. M.; Xie, X. C.; Gao, H. J. *Phys. Rev. B* 2005, *72*, 035453.

In: Encyclopedia of Nanotechnology
Editor: Elwood D. Carlson

ISBN: 978-1-60692-079-4

Chapter 7

NOVEL MANUFACTURING AND PROCESSING TECHNOLOGIES OF NANOPOROUS SILICON

Jia-Chuan Lin and Wei-Chih Tsai
Department of Electronics Engineering, St. John's University,
Taipei 25135, Taiwan, R.O.C
Institute of Microelectronics, Department of Electrical Engineering,
National Cheng Kung University, Tainan 70101, Taiwan, R.O.C.

ABSTRACT

Porous silicon (PS) films consisting of many pores and pillars are widely used to yield efficient visible photoluminescence (PL) and electroluminescence (EL) at room temperature. Such light-emission behaviors are primarily attributed to electron confinement in the nanocrystals that constitute the PS film. Also, the PS film shows a unique electrical property of negative difference conductance (NDC) for the carrier mobility difference in different sizes of pores and pillars.

The techniques of PS formation have been developed by many different methods. Among them, electrochemical anodization is the most commonly used. The anodization is performed in a hydrogen fluoride (HF) solution with an anodic current on the sample. However, it is very difficult to control the manufacturing parameters precisely in the manufacturing processes. Especially in nano-scale, the anisotropic etching or selective etching cannot be easily achieved in the wet-etching method. However, it is well known that nano-scaled PS (NPS) shows light emission more readily in the visible range than macro-PS does. Therefore, precise control of the NPS formation is very important.

In this chapter, the authors introduce three novel manufacturing technologies. In the first method, the Hall-effect is applied in the electrochemical anodization. By Lorentz force, the carriers can be well swept to the allocated areas. In the second method, a forward biased pn-junction is used for bottom-hole assistance. The NPS films with nano-scaled pores and high-aspect-ratio pillars can be formed. In the third method, the authors proposed a simple and mask-free method for the fabrication of patterned PS using a localized electric field. The electric field is applied by patterned electrodes (anode and cathode) which are placed horizontally underneath the sample. No masking-layer or related photo-lithography processes are needed in this method. Strong visible photoluminescence emissions in PS can be obtained by these novel methods.

In addition, the surface roughness of NPS leads to a high contact resistance to the metal layer of the applications on EL, and hence a waste of an electrical power. To enhance the EL efficiency and stabilize performance, a good metal contact to the light-emitting layer (NPS layer) is required. In the authors' study, a supercritical fluid (SCF) technique is explored to overcome the processing problem on metal contact. The SCF is a liquid or gas material used in a state above the critical temperature and critical pressure where gases and liquids can coexist. The use of a CO2 SCF and silver (Ag) nanoparticles is shown to improve the electrical contact between the metal and NPS. Therefore, the contact resistance can be largely reduced, and the power consumption of EL on the PS structure can be greatly reduced by the method.

The morphology, porosity, and photoluminescence of the NPS prepared by the proposed method are investigated. This novel technology will have much benefit in the applications of nano-electronic and opto-electronic devices.

1. Introduction

1.1. Overview

Porous silicon (PS) consisting of many pores and silicon (Si) pillars is discovered in the 1950s [1]. As we know, PS structures are widely used to yield efficient visible photoluminescence (PL) at room temperature [1, 2]. Such light-emission behaviors are primarily attributed to electron confinement in the nano-crystals that constitute the porous structure [3-5]. The PS structure is prepared by electrochemical etching in a hydrogen fluoride (HF) solution. It is possible to control the size of the pores by adjusting the etching-process parameters including the formation conditions of PS strcutre. The PS structure with different scaled Si pillars/pores lead to different material properties. Besides, it is also known that the nano-scaled PS (NPS) layers are easily to show stronger emitting light in the visible range at room temperature.

In this chapter, some novel manufacturing and processing technologies of NPS would be discussed. Some new ideas for NPS manufacturing technologies are introduced. Among them, the Hall-effect is applied in the electrochemical anodization [6]. By Lorentz force, the carriers can be well swept to the allocated areas. In addition, a forward biased pn-junction is also used for bottom-hole assistance [7]. NPS films with nano-scaled pores and high-aspect-ratio pillars can be formed by this method. Moreover, a simple and mask-free method for the fabrication of patterned PS using a localized electric field is proposed [8]. The electric field is applied by patterned electrodes (anode and cathode) which are placed horizontally underneath the sample. No masking-layer or related photo-lithography processes are needed in this method. These ideas contain original information, theoretical and experimental in the field of nanoscaled electrochemistry. The principle, physical mechanism and experimental results by these methods are demonstrated.

In our experiments, the electrical properties of NPS samples on *current-voltage (I-V)* characteristics are measured by Kiethley 236 source-measure unit. The optical properties on photoluminescence (PL) feature are measured on the PL setup with a He-Cd laser (excitation wavelength of 325 nm) and a Hamamatsu R928 photomultiplier detector. All scanning electron microscope (SEM) imagines are performed by JEOL JSM-6335FNT.

1.2. Futrue Developments

The NPS is a promising material in the areas of optoelectronics and microelectronics due to their unique electrical and optical properties. The NPS shows room temperature visible luminescence characteristics which are not observed in bulk or single-crystalline Si. Such unique properties and low cost make the NPS become a good candidate for the realization of NPS-based optical devices. The electroluminescence (EL) of NPS is discovered in the 1990s [2]. Then, some optioelectronic devices based on the NPS materials such as light emitting diodes (LEDs), flat-panel displays, cathode-ray displays, nonvolatile light emission memories, photodetectors, solar cells, and sensors [9-13] are also being developed. Conventional bulk Si-based material which is an indirect bandgap has the disadvantages in a poor optical radiative efficiency and shows luminescence only outside visible range. However, NPS material overcomes those disadvantages. In addition, the surface area density of NPS is very large because of considerable quantities of porosities. It is about 600 times more than the surface area density of the normal polished Si surface [14]. It implies that the surface of NPS can lead to a better efficiency of solar cells [15].

Furthermore, some applications on the tunable active microcavity resonators, optical waveguides, laser mirrors, photonic crystals, silicon-on-insulator (SOI) structures, gas sensors, membranes, and molecular sieves also have been proposed for the furure developments [16-21].

2. Forming Mechanisms

The NPS forming mechanism is plotted in figure 2.1 schematicly. The illustrative equation of the overall process during PS formation can be expressed as below:

$$Si + 2HF + 2h^{+} \rightarrow SiF_2 + 2H^{+}$$
$$SiF_2 + 4HF \rightarrow H_2 + H_2SiF_6 \quad (\textit{for divalent anodic reaction})$$
$$Si + 2HF + 4h^{+} \rightarrow SiF_4 + 4H^{+}$$
$$SiF_4 + 2HF \rightarrow H_2SiF_6 \quad (\textit{for tetravalent anodic reaction})$$

The etching rate is determined by the hole (h^{+}) accumulation in the adjacent regions of the HF electrolyte and Si atoms. It implies that the etching rate and etching area can be controlled by the hole accumulation quantity and area. It is an important idea in this chapter's discussion.

As to the NPS formation mechanism, at least five mechanisms are proposed. (a) The Beale model (figure 2.2) [22-25] (b) The diffusion-limited model (figure 2.3) [26-29] (c) The quantum model (figure 2.4) [30-32] (d) The rate model [33] (e) The two-component resistor network model [34].

A brief summary for the first three common models are described below: Beale *et al.* pointed out that space charge regions exist between the PS layers and Si bulk. The electric field lines concentrate at surface irregularities, and focus current flow at the PS pore tips and locally enhance dissolution there.

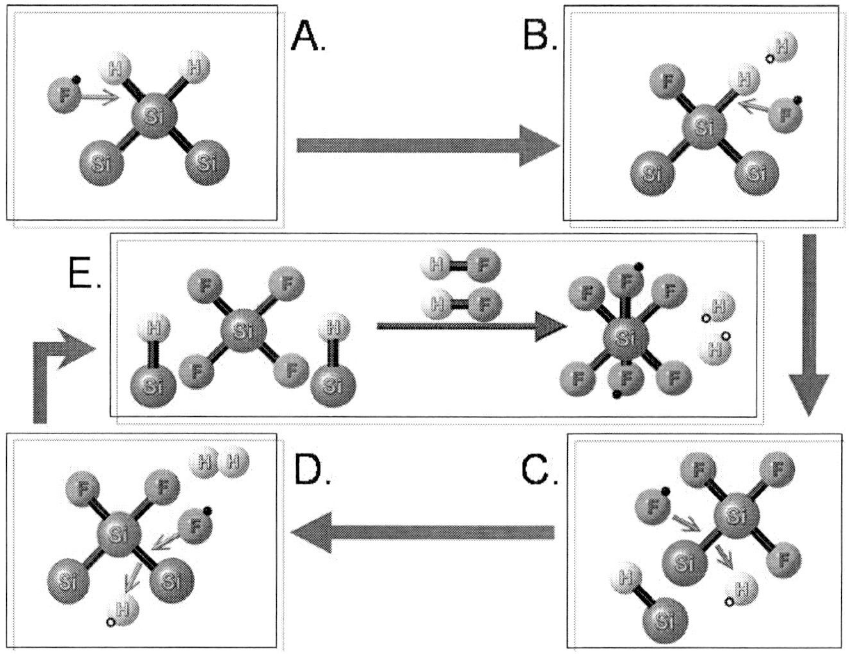

Figure 2.1. The schematic diagram of NPS forming mechanism.

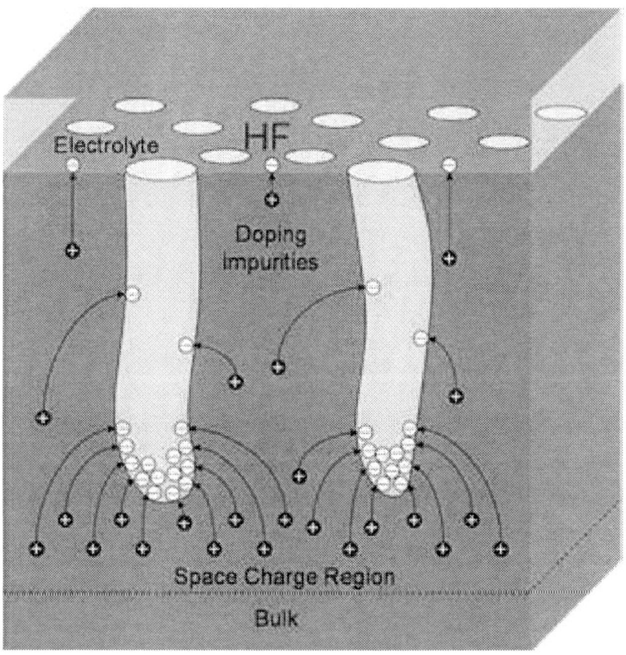

Figure 2.2. The Beale model.

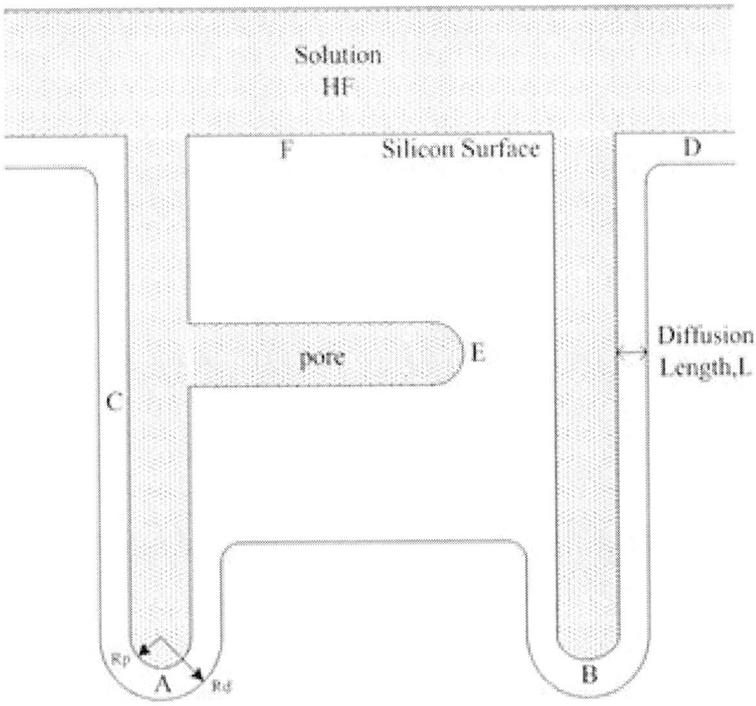

Figure 2.3. The diffusion-limited model.

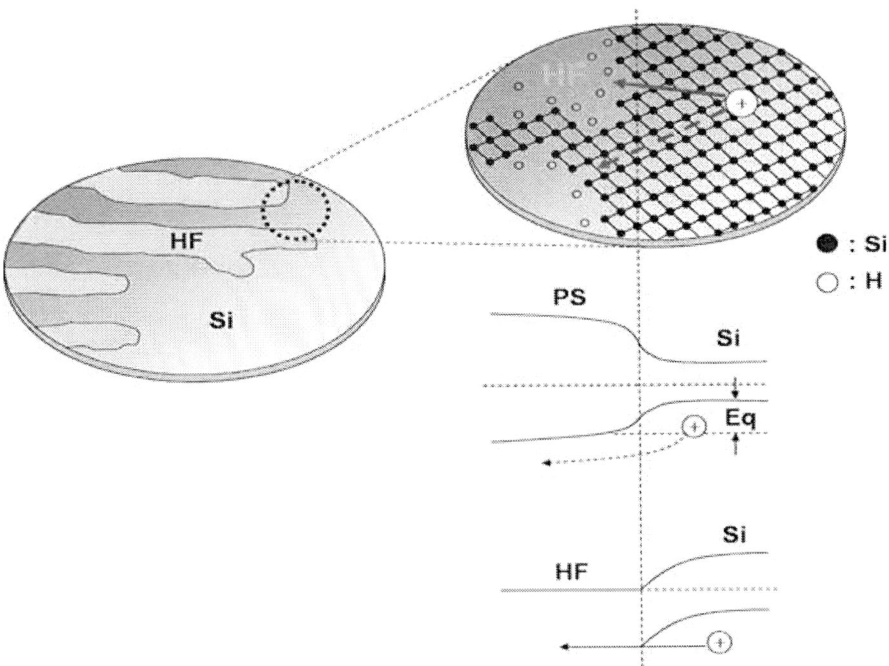

Figure 2.4. The quantum model.

Anodic current therefore anisotropicly directed to the pore tips. The diffusion-limited model proposed by Smith and Collins describes the formation of PS in terms of a random walk. The model is based on the the mechanism of diffusion-limited aggregation by Witten and Sander. It is the first computer simulated theoretical model. The holes diffuse to the Si surface and react with Si surface atoms. As to the quantum model proposed by Lehamann, it assumes the PS formation the possibility of a quantum confinement for charge carriers in the pore walls due to their small thickness.

3. Perpendicular-Electric-Field-Assisted Method

The NPS formation techniques have been developed by many methods, including electrochemical anodization [4, 35, 36], stain etching [37-39], spark-erosion [40] and vapor etching technique [41-42]. Among them, electrochemical anodization is the most commonly used.

The electrochemical etching is usually performed under an electrical field which is perpendicular to the sample surface. And the sample is dipped into a HF solution. The etching rates are controlled by adjusting the etching current densities and electrolyte compositions. It is well known that the etching current is governed by the hole generation at the Si surface to assist anodic oxidation during the electrochemical etching. It is possible to control the size of the pores and the material properties by adjusting process parameters including the formation conditions. The schematic diagram of the experimental setup for preparing NPS samples is given in figure 3.1. The main body of the HF electrolyte container is made of Teflon materials.

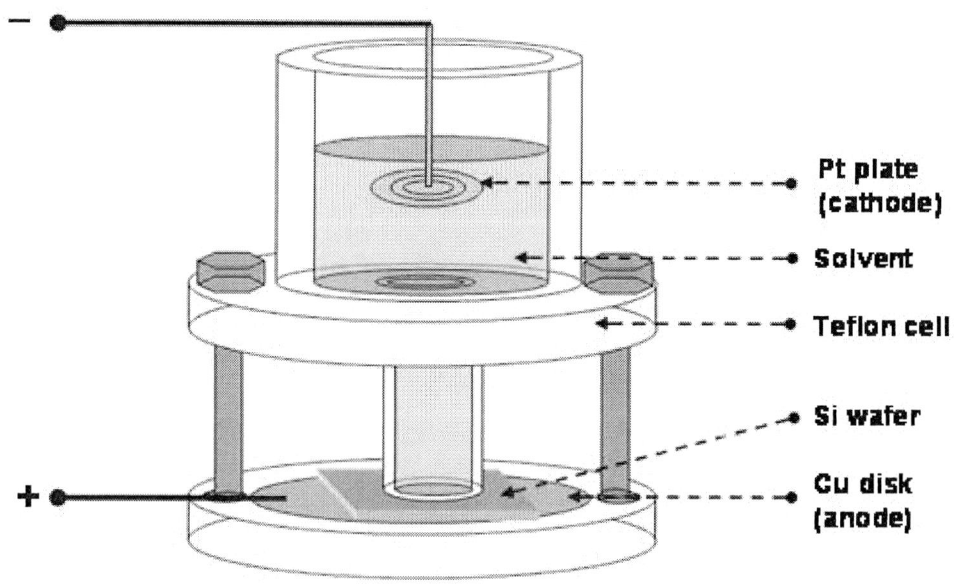

Figure 3.1. The schematic diagram of the experimental setup of perpendicular-electric-field-assisted method.

A mixture of $HF:C_2H_5OH=1:4$ is utilized as the etching solvent here. A copper (Cu) disk is served as anode, while cathode is made by platinum (Pt). It is noted that the apparatus is designed with a vertical-standing arrangement so as to easily remove the hydrogen bubbles from the NPS surface and to improve the uniformity of the NPS samples. In our study, the NPS samples are prepared on Si (100). Prior to the etching process, a sintered aluminum (Al) film is used to form an intimate backside ohmic contact to allow a homogeneous anodization current flow. The anodization etching current is supplied by an external constant current source. The schematic mechanism is plotted in figure 3.2. The typical top view and cross-section SEM images of the NPS sample are shown in figures 3.3.

For p-type Si, the holes are the majority charge carriers, and the p-type NPS (p-NPS) layers are easy to produce. Conversely, n-type NPS (n-NPS) is very difficult to form, as a result of the lack of holes. Previously, most of the research done on NPS has been made on hole-rich p-type Si. However, the control of n-NPS is necessary since an application based on n-type substrate is possible in light emitting diode (LED) technology and microelectronic technology. [43, 44] Therefore, some assistance methods are used for solving the problems and described as belows.

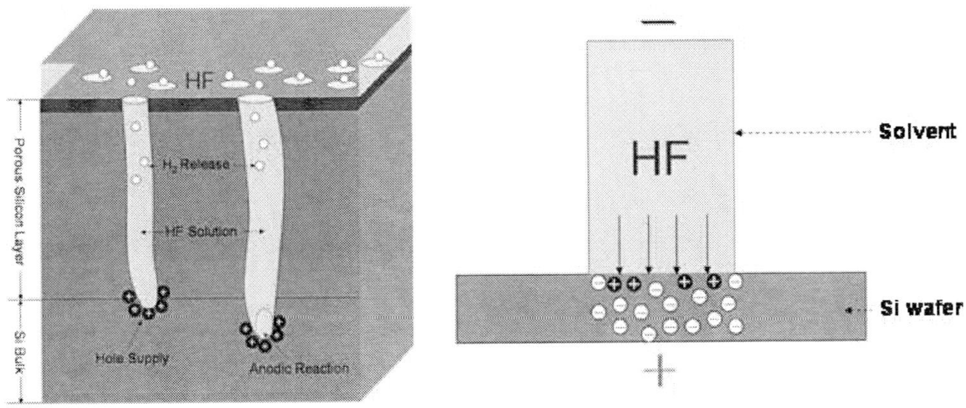

Figure 3.2. The schematic mechanism of perpendicular-electric-field-assisted method.

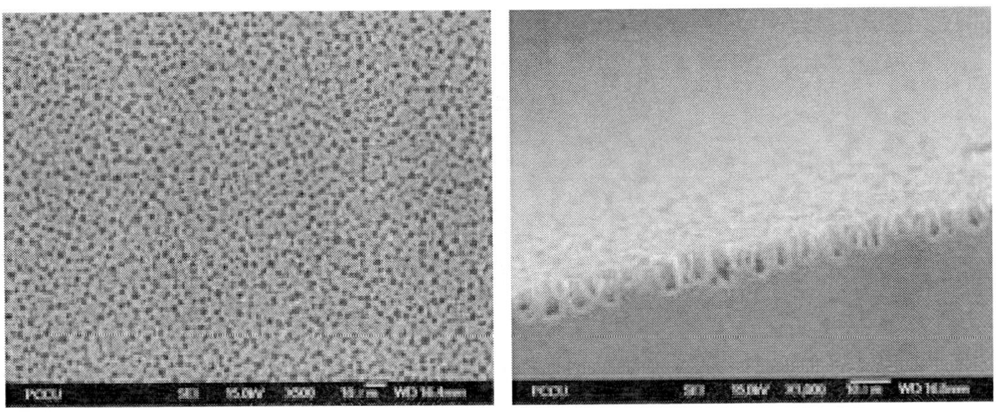

Figure 3.3. The SEM images of the NPS sample of perpendicular-electric-field-assisted method.

4. ILLUMINATION-ASSISTED METHOD

To get an n-NPS, illumination-assisted method is the most popular way to generate the holes required in the electrochemical etching process on the hole-poor n-type samples (as shown in figure 4.1.) [45, 46]. However, the actual photo-energy absorbing by atoms is very sensitive to the illumination-source intensity and electrolyte environment in which the conditions are not constant in the process. The top view and cross-section SEM images of the n-NPS sample are shown in figures 4.2. Since the illumination intensity is related to the distance, besides the absorption coefficient and the wavelength, only top-layer atoms are photo-excited and only the surface layers under illumination generate the electron-hole pairs. The actual photo-energy absorbing by Si atoms is very sensitive to the illumination-source intensity and electrolyte environment, and the conditions are not constant in the process. The etching rate will gradually decrease because the illumination is very difficult to reach the deeper layers. In other words, the only-illumination-assisted etching approach is depth-limited. The illumination related factors and limitation may affect the NPS morphology and formation. Therefore, cone-shape pores in NPS structures are often formed in this method [47]. Besides, it is also known that the n-PS obtained under illumination usually consists of a nano-PS layer and macro-PS layer. [48]. As we know, macro-PS is difficult to show room temperature light emitting in the visible range [4]. Therefore, in next sections, some ideas are proposed.

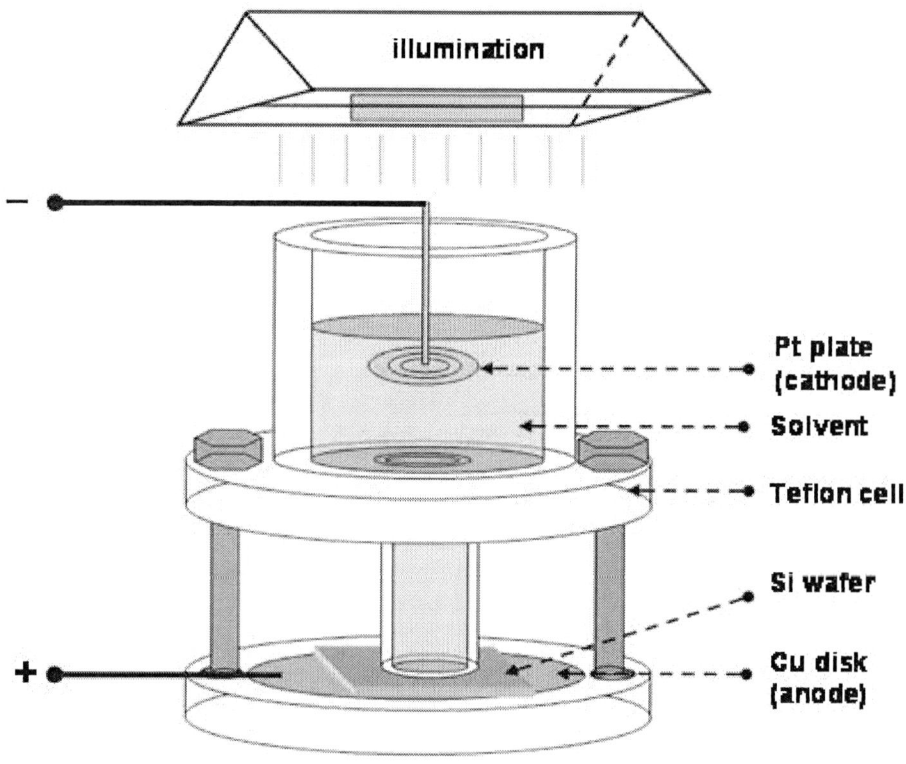

Figure 4.1. The schematic diagram of the experimental setup of illumination-assisted method.

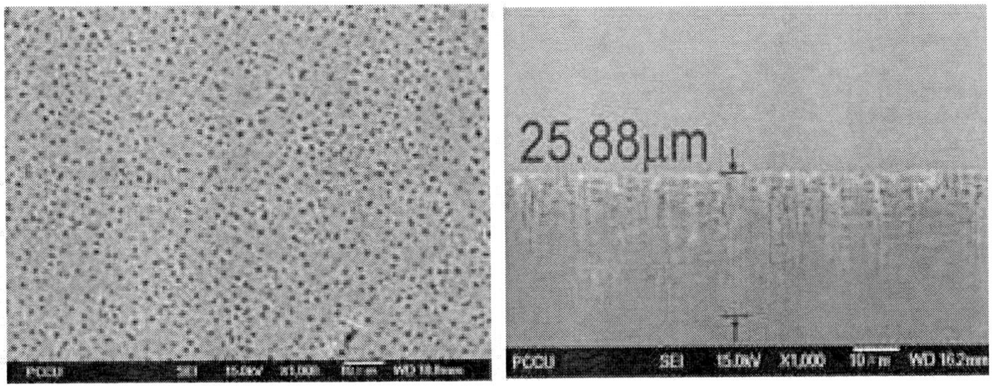

Figure 4.2. The SEM images of the NPS sample of illumination-assisted method.

5. Hall-Effect-Assisted Method

In this section, an alternative method based on Hall-effect is applied on n-NPS fabrication (as shown in figure 5.1). The majority carriers in n-type Si (electrons) are swept down by Lorentz force. Enough holes continuously appear on the surface layer to participate in chemical reaction during the etching process.

The schematic diagram of the experimental setup for preparing n-NPS is given in figure 5.1. A mixture of HF: C_2H_5OH=1:4 is also utilized as the etching solvent in the section 3. The NPS samples are prepared on n-type Si (100). The sample is exposed to the Hall-effect environment. An electrical field (a voltage V_x) is biased on the samples in x-direction and makes a current I_x. In y-direction, a magnetic field (B_y) is placed perpendicular to the current. Charge carriers flowing in the samples will experience a Lorentz force ($F_z = qE_z = qV_x \times B_y$), as indicated in the figure 5.1. The majority carriers (electrons) are swept down by $\vec{F}_z$. A Hall voltage (V_H) across the semiconductor is built in z-direction. By applying a sufficiently large bias-voltage and a magnetic field density, the upper layers of the semiconductor are depleted and then inverted from an n-type to a p-type semiconductor. Enough holes can reveal on the upper layers to participate in chemical reaction during the etching process. The schematic mechanism is plotted in figure 5.2. The etching rate is determined by the hole (h^+) accumulation in the adjacent regions of the HF electrolyte and Si atoms. It can be found that the hole accumulation increases with the magnetic field.

There are two special points in this method. First, the illumination sources are not necessary. Therefore, no illumination-limit has to be concerned in the formation of deep NPS layer. In the second, the hole supporting is owing to the inversion layer (from n-type to p-type) creating on the upper layers. So, no vertical metal electrodes are needed to be soaked in the electrolyte to serve as a cathode. By the way, the illumination-free and metal-electrode-free etching container can be well designed as a hermetic container with a strong cover for the safety of handling corrosive HF electrolyte. Also, the metal-electrode contamination can be prevented in HF electrolyte.

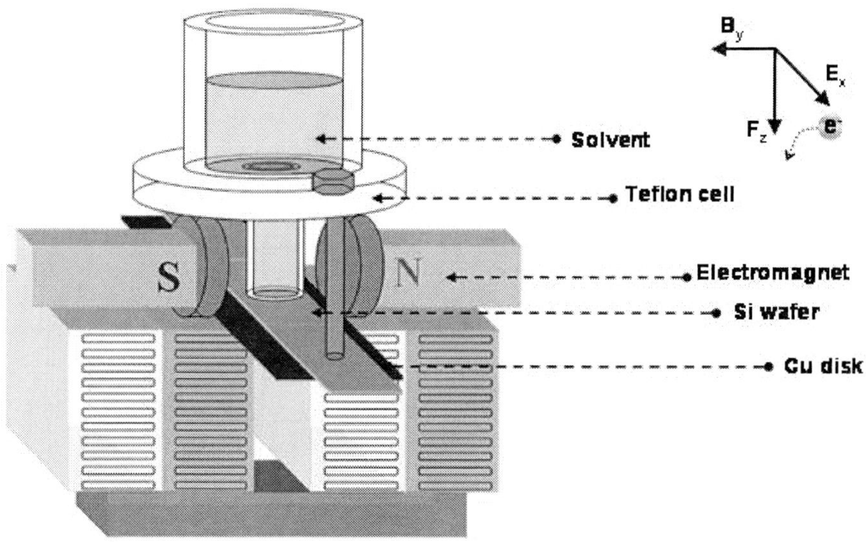

Figure 5.1. The schematic diagram of the experimental setup of Hall-effect-assisted method.

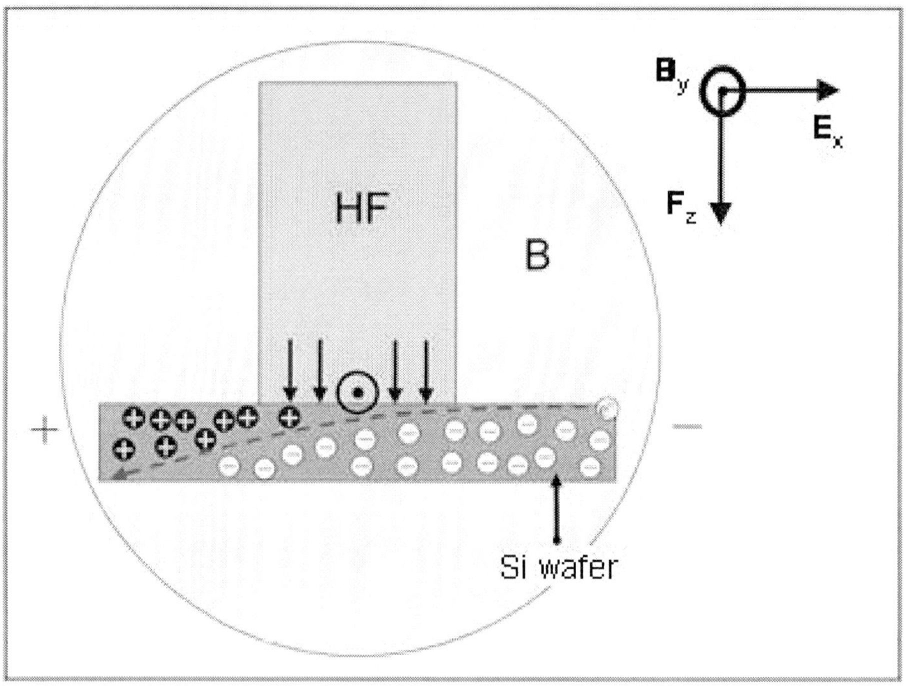

Figure 5.2. The schematic mechanism of Hall-effect-assisted Method.

Under a magnetic field (B_y) of about 20 milliTesla and a voltage (V_x) of about 60 Volts, NPS layers are successfully fabricated by the Hall-effect etching method without any illumination sources. The top view and cross-section SEM images of the NPS sample are shown in figure 5.3 and 5.4, respectively.

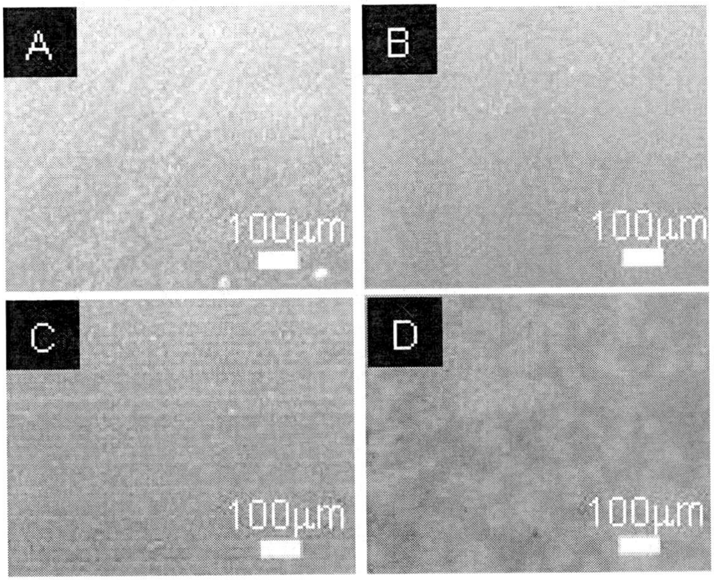

Figure 5.3. The top view SEM images of the NPS sample of Hall-effect-assisted method.

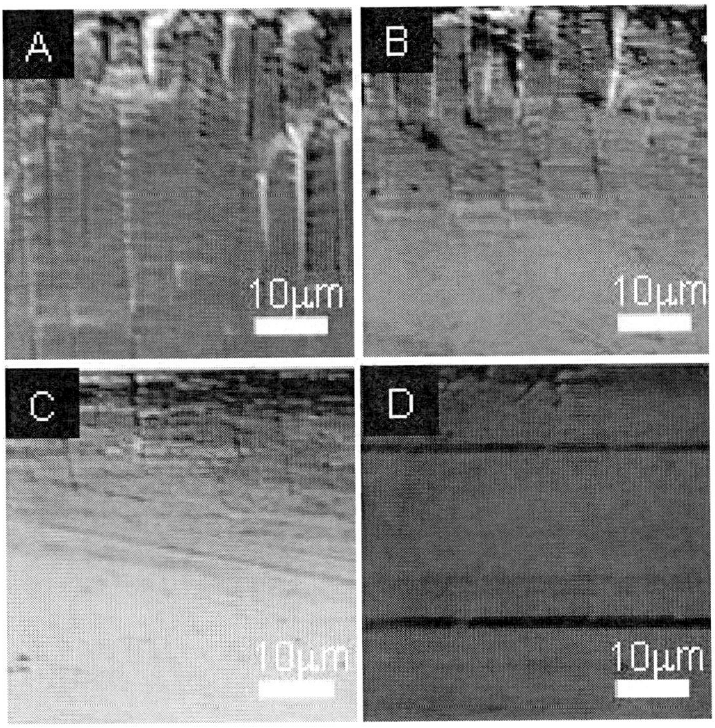

Figure 5.4. The cross section view SEM images of the NPS sample of Hall-effect-assisted method.

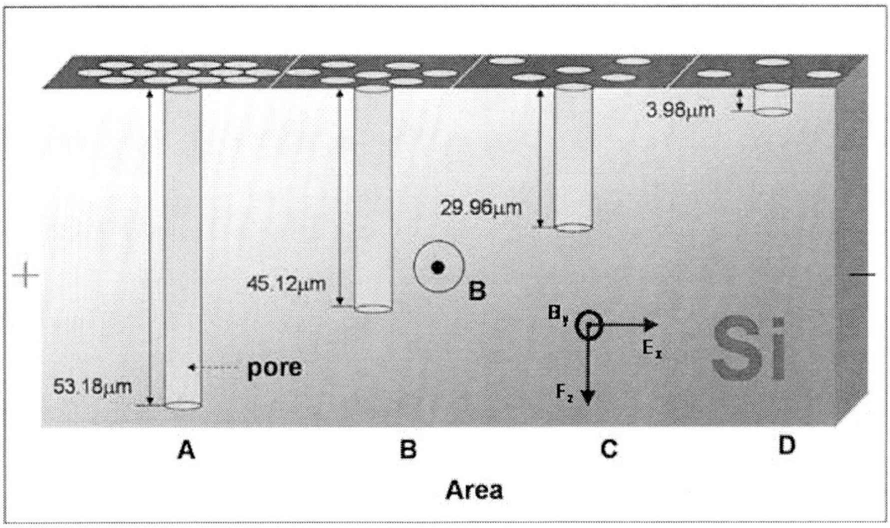

Figure 5.5. The schematic of the pore depths and distribution.

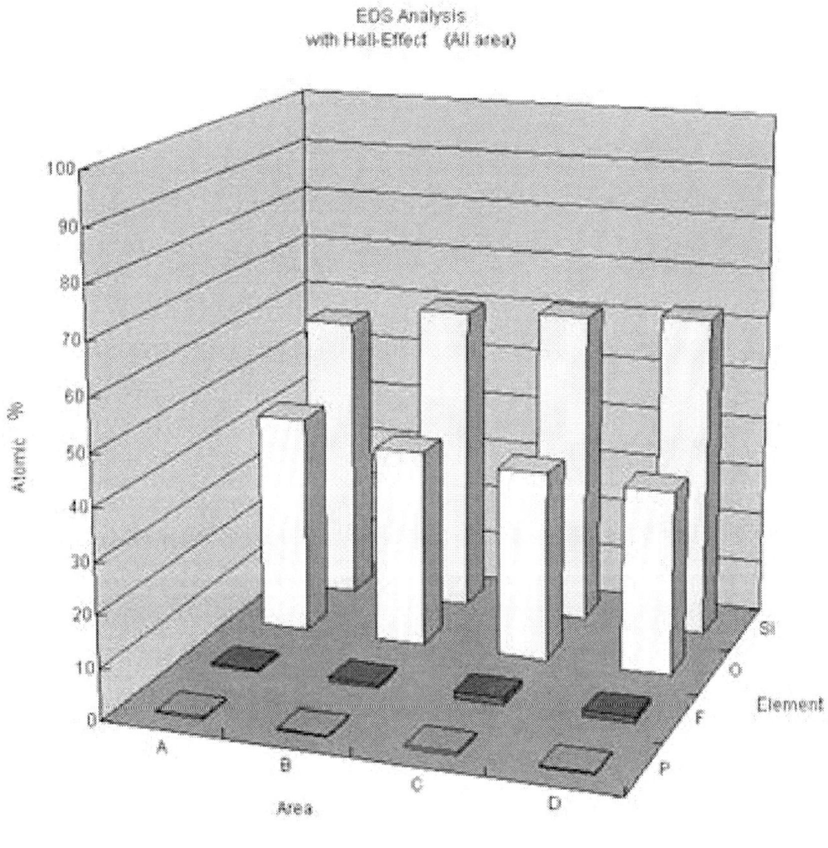

Figure 5.6. The EDS analysis of the NPS sample of Hall-effect-assisted method.

A larger voltage drop across the x-direction will cause a larger gradient-situation of etching extent of NPS. To easily show the gradient-situation for further investigation, a larger

V_x is adopted in this study. On the other hand, a small V_x is suggested for the film uniformity. Four regions (A, B, C and D) are defined on the sample for article describing, as shown in the figure 5.3 and 5.4. The pores are narrow and straight with a high aspect ratio because the hole-supplying mechanism is bottom-up. The mechanism is different from the only-illumination-assisted method in which it only relies on the illuminated surface atoms. The pore depths of the NPS layers are about 53 μm (region A), 45 μm (region B), 29 μm (region C) and 3 μm (region D), respectively. Figure 5.5 shows the schematic of the pore depths and distribution for the four regions. And figure 5.6 demonstrates the energy dispersive spectrometer (EDS) analysis. The experiment results are helpful to understand the variation of NPS morphology at different positions under Hall-effect. The variation of morphology in A~D regions implies that the areas closer to the anode can accumulate more holes to participate in etching reaction. Such plenty of holes following the bottom-up paths lead to high-aspect-ratio pores with a deeper and straighter shape. The hole concentration and originating source determine the pore shapes and NPS morphology.

In only-illumination-assisted n-NPS, after surface pores (of few μm) are formed, photo-energy is not completely received by the deeper layers. Therefore, the deeper layers are difficult to form the required NPS in the darkness. Only a thin part of NPS layer contributes in the PL. Meanwhile, a lateral etching keeps going on and cone-shape pores are produced. Thus, the n-NPS visible luminescence with a depth-limitation is usually less than p-NPS. In the Hall-effect-assisted method, the visible luminescence of the n-NPS is very similar to the hole-rich p-NPS. An excitation wavelength of 325 nm from a He-Cd laser is used for the PL measurements. Figure 5.7 are the PL spectra of A~D regions under the conditions of $B_y = 20$ milliTesla and $V_x = 60$ Volts. From regions A to D, the PL intensity, etching extent and the porosity decrease as shown in figure 5.6 and 5.7.

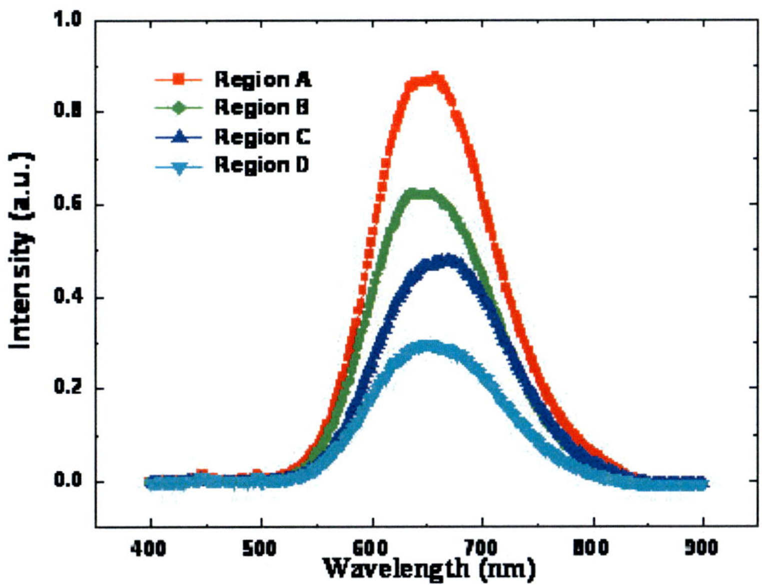

Figure 5.7. The PL spectra from A to D regions of n-NPS sample.

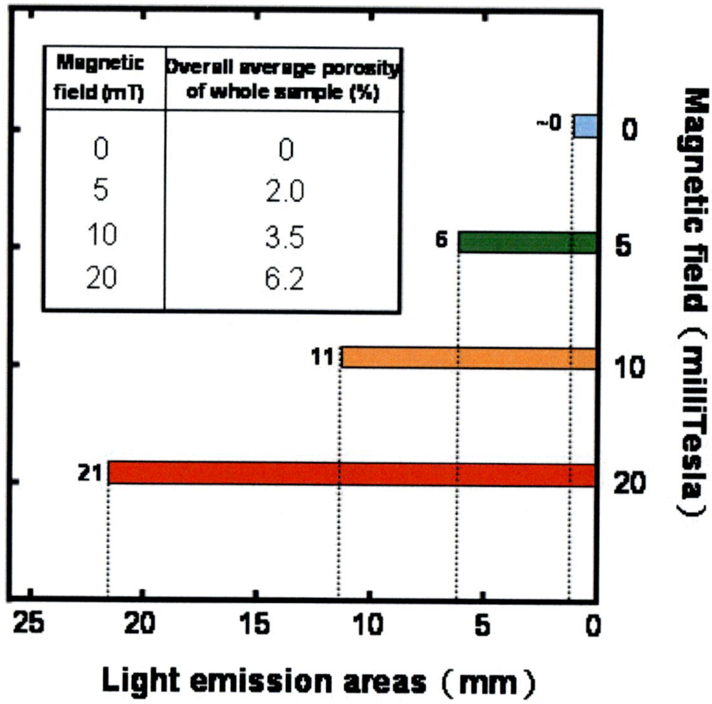

Figure 5.8. The influence of magnetic field (B_y) on the area size of light emission. The overall average porosities of the samples are also shown in the inset.

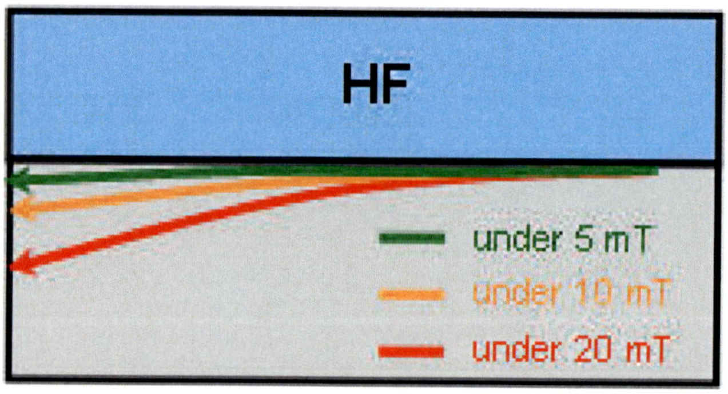

Figure 5.9. The schematic mechanism of the influences of magnetic field for Hall-effect-assisted method.

The high PL intensity in region A is attributed to the high hole-concentration, and hence the high porosity. Obviously, the PS sample produces a strong emission of orange-red luminance under ultraviolet (UV) illumination. The PL intensity is proportional to the hole concentration that is appearing on A~D regions gradually. This implies that the selective-etching and the gradient light-emission on the identical sample by this method is possible.

In the section, the Hall-effect-assisted method can provide an anisotropic and illumination-free technology for n-NPS manufacturing. A strong visible light is obtained on n-NPS at room temperature. Also, a gradient etching can be formed on the identical sample and that is controlled by a electric field and magnetic field.

6. Buried-P-Layer-Assisted Method

In this section, another method (named as buried-p-layer-assisted method) is proposed for manufacturing n-NPS layers. A mount of holes are supported by a forward biased pn-junction in which the buried p-layer plays the role as a hole-source underneath the n-layer. For the hole-assisting direction, it is bottom-up and straight-upward. Therefore, no depth-limitation should be taken into consideration in this approach. The schematic mechanism of the biased pn structures for bottom-hole-assistance is plotted in figure 6.1. The pillars and pores are narrower, straighter, deeper, and more condensed than those prepared by the conventional only-illumination-assisted approach. The proposed buried-p-layer-assisted method also can be superimposed on the illumination-assisted method in the electrochemical anodization (named as superimposition-assisted approach). The superimposition-assisted approach will be described in the section 7.

The schematic diagram of the experimental setup for preparing n-NPS layer is given in figure 6.2 (a). Figure 6.2 (b) shows the case of difficulty in electrochemical etching in n-Si sample without any hole-assistance (illumination-assistance and bottom-hole-assistance). When the illumination-assistance is applied, as shown in figure 6.2 (c), the electron-hole pairs are generated and accumulating only on the top surface layers in which some tiny lateral etching accompanies with the vertical etching. On the other hand, in the buried-p-layer-assisted method as figure 6.2 (d), an electrical field across the anode and cathode forward biases the np junction, and holes flow straight-upward from p- to n-region. Sufficient holes can reach to the pore tips to participate in chemical reaction during the etching process, even in the dark.

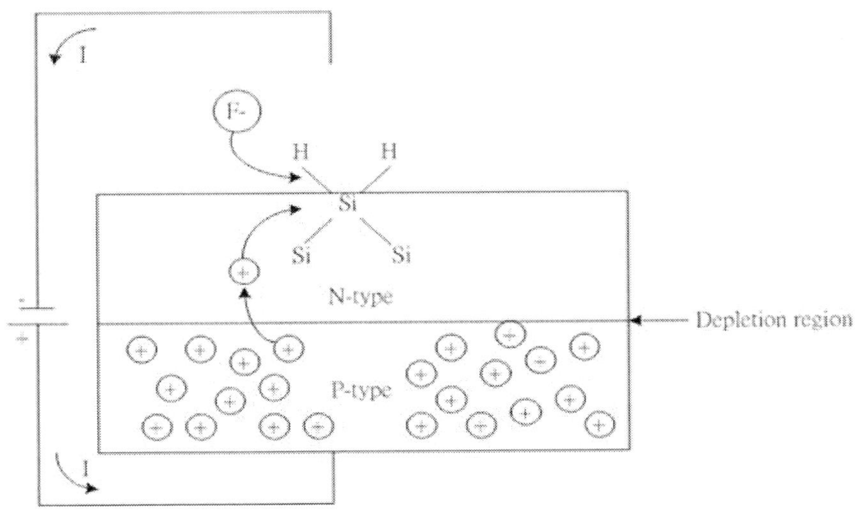

Figure 6.1. The schematic mechanism of the biased pn structures for buried-p-layer-assisted method.

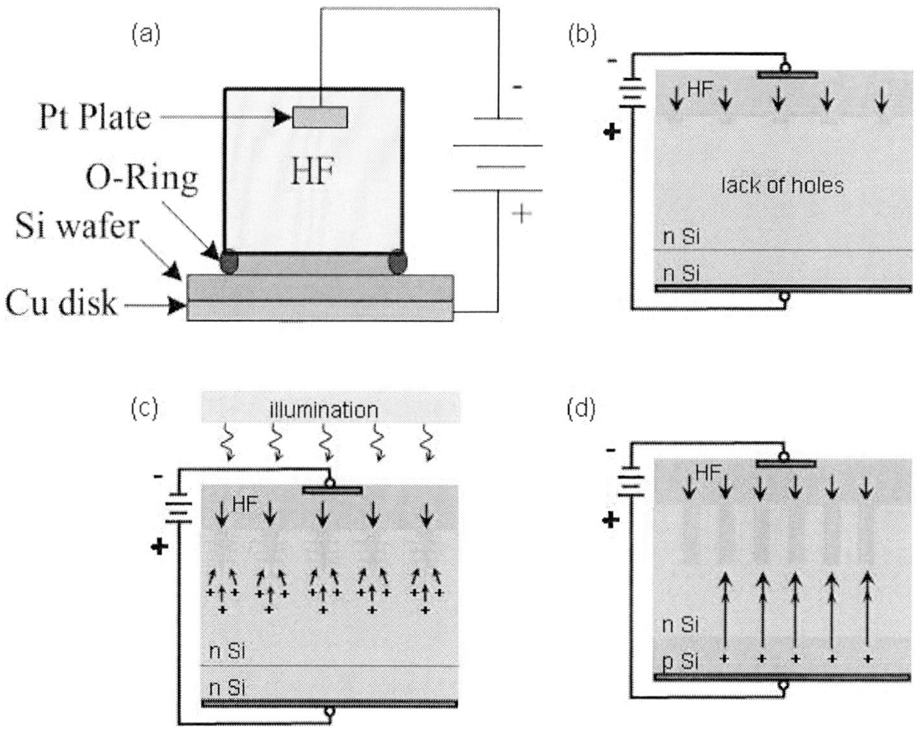

Figure 6.2. (a) The schematic diagram of experimental setup of buried-p-layer-assisted method, (b) the sketch of electrochemical etching in n-on-n Si without any hole-assistance, (c) the mechanism sketch with the illumination-assistance approach, and (d) the mechanism sketch with the bottom-hole-assisted approach.

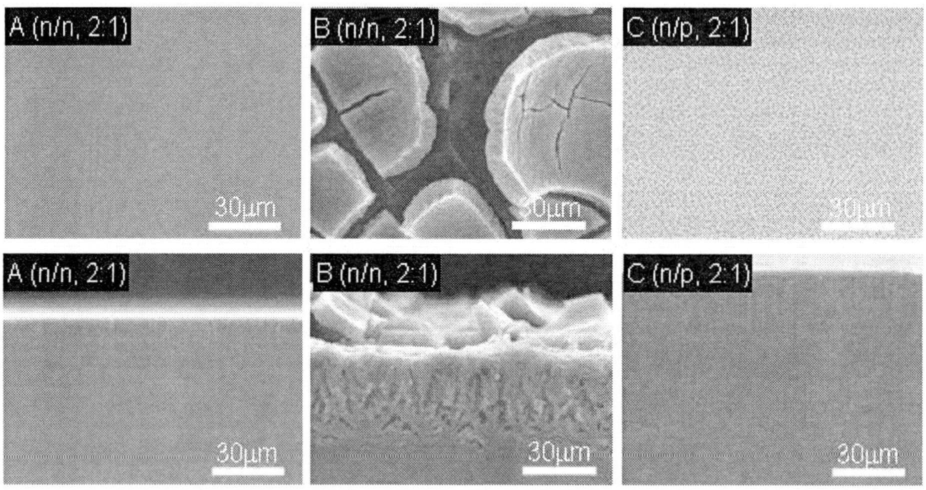

Figure 6.3. The SEM images of the NPS samples of buried-p-layer-assisted method.

The top view and cross section view of NPS films are successfully fabricated by the buried-p-layer-assisted method with or without illumination-assistance. The top view and cross-section SEM images of the NPS samples are shown in figure 6.3. Two NPS samples with n-epilayers on n-substrates (n-on-n) and one NPS samples with n-epilayers on p-substrates (n-on-p) are prepared, and named as samples A, B, and C, respectively.

In sample A (n-on-n), not any hole-assistance is applied on the sample, and then the electrochemical reaction can not successfully proceed. The NPS layer fails to be formed on sample A. For the conventional case based on the only-illumination-assistance, it is applied on sample B (n-on-n). In sample C (n-on-p), which is prepared by buried-p-layer-assisted method, the hole-supplying relies on the p-substrate without any illumination. Furthermore, etching solution ratio of C_2H_5OH: HF =2:1 is used in all samples.

In the sample B (with illumination-assistance), the macropores clearly appear on the surface. The surface pores are cone-shape (V-groove), especially even clearer in sample B. The electron-hole pairs only rely on the top-surface layers by illumination. After surface pores (of several ten um) are formed, photo-energy is not completely received by the deeper layers. Therefore the deeper layers are illumination-limited and the NPS pores are difficult to be formed. A lateral etching keeps on while the vertical etching is proceeding, hence crosswise-branched pores and tissue-like branch structures are formed.

Conversely, in sample C which is based on the buried-p-layer-assisted method, the hole-supplying path is bottom-up and no depth-limitation. The pore shapes are narrower, deeper and straighter. Also, high-aspect-ratio nano-scaled pores (aspect ratio = 300) can be obtained, and the pore structures are channel-like instead of tissue-like that always occurs in illumination-assisted approach. As shown in figure 6.3 (c), the pore depths of the PS layers can be up to 100 μm and the pore width can less than 300 nm. It is useful to serve as thick-layer light emitting applications.

An excitation wavelength of 325 nm from a He-Cd laser is used for the PL measurements. Figure 6.4 shows the PL spectra of samples A~C. Clearly, the buried-p-layer-assisted method can effectively improve the light-emitting density.

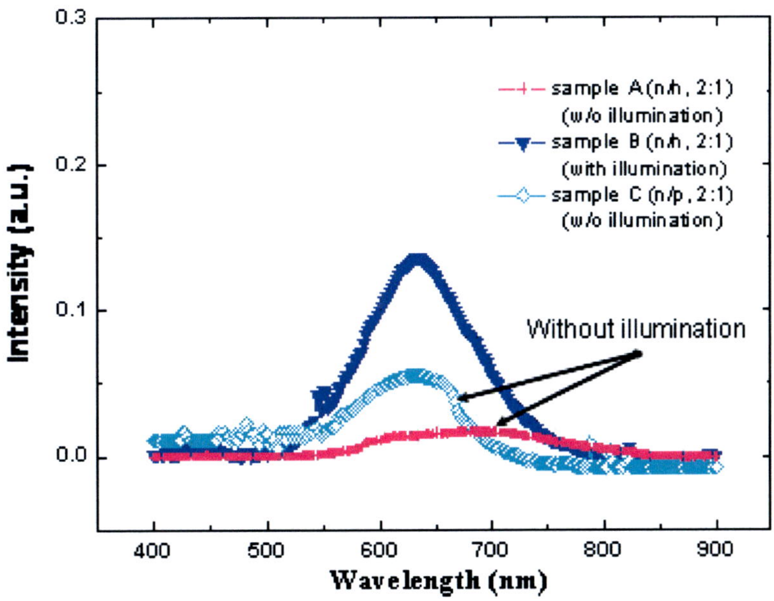

Figure 6.4. The PL spectra of the NPS of samples A~C of buried-p-layer-assisted method.

The measured PL that is usually attributed to quantum size effects is a combination result of light emissions from upper layer (macroporous) and lower layer (nanoporous). Thus no significant size dependence of the PL peak energy can be easily observed and concluded. Obviously, the PS samples produce a strong emission of orange-red luminance under UV illumination.

The buried-p-layer-assisted method based on a forward biased np-junction for manufacturing n-PS layer is proposed in this section. The buried-p-layer-assisted method can overcome the illumination-limitation and depth-limitation problems in only-illumination-assisted method. With the buried-p-layer-assistance, the anodization etching is almost anisotropic. This feature has great potential in fabricating processes.

In the section, clear visible PL emissions at room-temperature are obtained on the hole-poor n-type samples even in the dark. Strong visible photoluminescence emissions are demonstrated on the hole-poor n-type porous layer at about 650 nm.

7. Lateral-Electric-Field Method

In this section, a mask-free method is proposed for the fabrication of p-type NPS using a localized electric field. The electric field is applied by the patterned electrodes (anode and cathode) which are horizontally placed underneath the sample. No masking-layer and related photo-lithography process are needed in this method. Besides, no metal electrodes and hence no metal-pollution in electrolyte have to be concerned in the formation of NPS. The morphology of the NPS prepared by this method is investigated. Strong visible photoluminescence emissions in the selected areas of NPS are demonstrated on NPS at about 650 nm.

As we know in the previous section, the etching current is governed by the hole accumulation at the Si surface to assist anodic oxidation during the electrochemical etching. Most of the researches done on NPS have been made on the whole Si samples.

However, the patterning of NPS is necessary for the applications on microelectronic process, microelectromechanical system and optoelectronic integration. Some patterning method are done by masking layer (etch-resistant film) [49, 50]. As the patterned NPS is defined by masks, masking-design and related photo-lithography processes are required. Since no masking-layer is required, and hence no anodization-time limitation should be concerned in this method.

In the study, the schematic diagram of experimental setup for preparing NPS is given in figure 7.1. The samples are prepared on p-type Si (100). A mixture of $HF:C_2H_5OH=1:4$ is also utilized as the etching solvent here. The etching rate is determined by the hole (h^+) accumulation, so the hole accumulation strongly depends on the distribution and intensity of electric field in this method.

The patterned NPS is formed by a localized electric field during the manufacturing process as illustrated in figure 7.2. The hole distribution in Si is controlled by the electric field applied across the planar metal electrodes (anode and cathode) which are attached beneath the Si sample.

The geometry patterns of metal electrodes are designed on a reusable printed circuit board (PC board) as shown in figure 7.3.

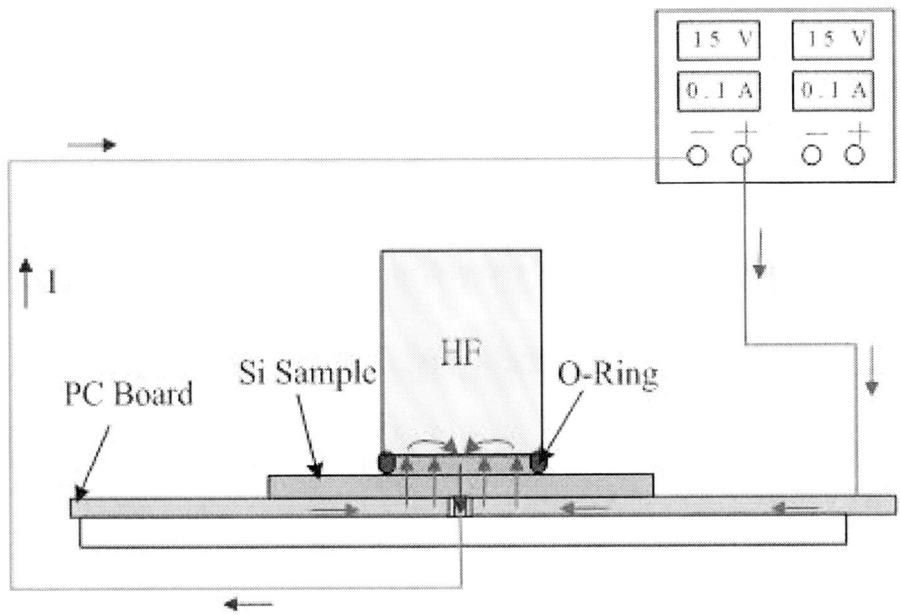

Figure 7.1. The schematic diagram of the experimental setup of lateral-electric-field method.

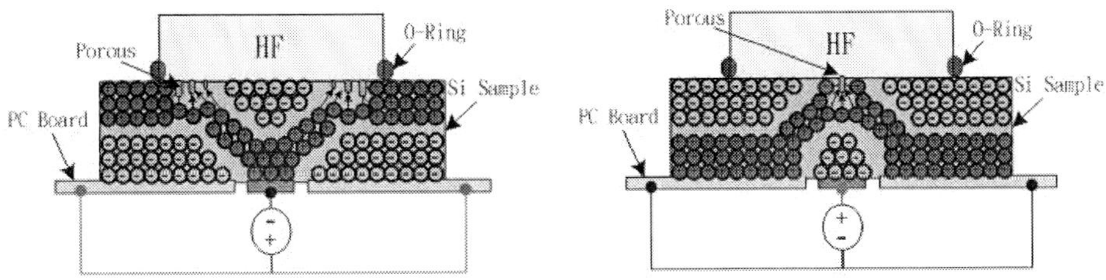

Figure 7.2. The schematic illustration of mechanism and the influence of a localized electric field.

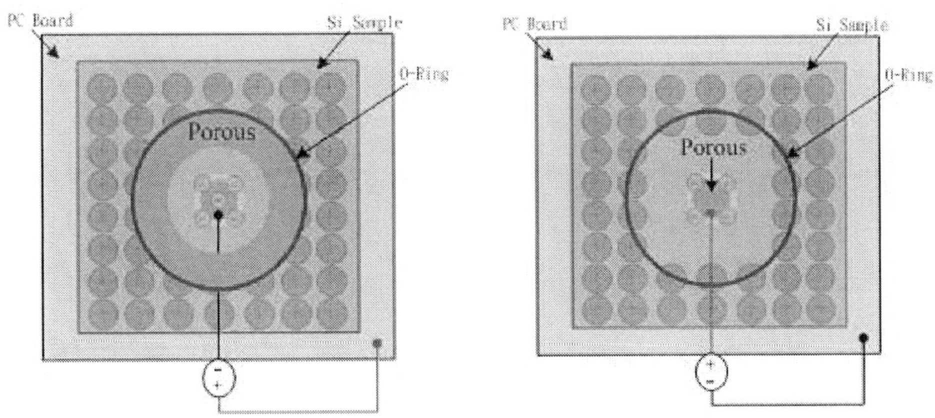

Figure 7.3. The geometry patterns of metal electrodes are designed on a reusable PC board.

For the simplification of fabrication procedure, the back-side surface of Si sample directly and closely touches on the PC board, and the procedure of back-side Ohmic contact can be omitted in our sample preparing. Though extra electric power may consume on the metal-semiconductor contact barrier without the Ohmic contact procedure, however, the experimental results do not change. In the cathode-related region (above cathode), holes are drifted downward the cathode-electrode and the top layers are depleted of holes. On the contrary, the holes in the anode-related region (above anode) drift upward the Si/HF interface layer to participate in chemical reaction during the etching process. Such a mechanism forms a patterned NPS. No else perpendicular electrodes are needed to be soaked in the electrolyte to serve as a cathode in this method.

Two samples (sample No.1 and sample No.2) are dealt with the etching processes under localized electric fields of 15 Volts for 30 minutes. On sample No.1, the inner-circle region is attached on the cathode electrode and negatively charged; on the contrary, the outer-circle region is attached on the anode electrode and positively charged. Meanwhile, the sample No.2 is biased in the same way but exchanging the electric polarity.

Patterned NPS layers are successfully formed on the two samples, and the naked-eye photos are shown in figure 7.4 (a) and (b), respectively. A strong emission of orange-red luminance appears in the selected areas (outer-circle region in sample No.1 and inner-circle region in sample No.2) under UV illumination without any mask.

The top view SEM images and the distribution of the four regions are shown in figure 7.5. For further studying and article describing, four regions (A, B, C and D) are defined on the sample No.1. A voltage drop across electric terminals would cause a non-uniform distribution of holes. To easy show the relationship between the etching degree and hole distribution for further investigation, a typical circle-shape light-emitting pattern is designed in this study. It is believed that other shape patterns (e.g. four-spot pattern in sample No.3) as shown in figure 7.4 (c) could be possible by proper designs of electric field distribution which are under our taken.

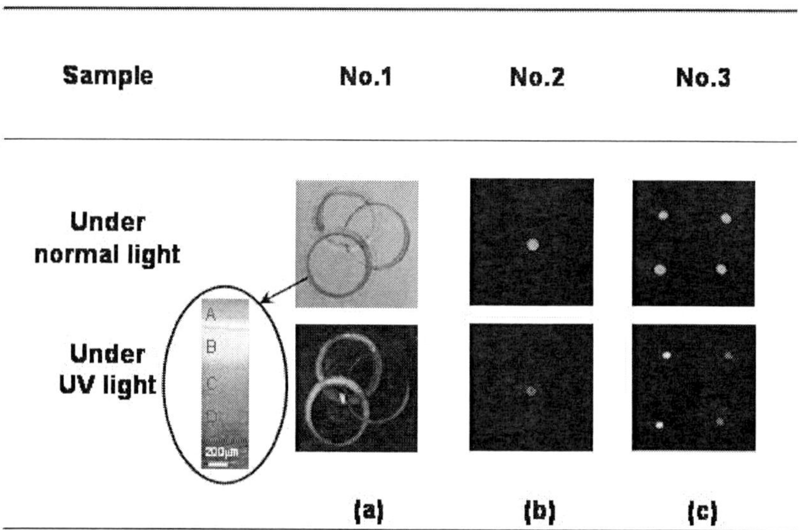

Figure 7.4. The naked-eye photos of the PS samples (a) sample No.1, (b) sample No.2 and (c) sample No.3 under normal light (top side figure) and UV light (bottom side figure) at room temperature.

Figure 7.6 shows the cross-section SEM images and the 3-D profile as well corresponding to regions A~D. In region A, no obvious NPS appears; on the contrary, the pores are drastically produced in region B, C, D for the sufficient hole-supplying. The pore depths of the NPS layers are about 0 μm (region A), 2.6 μm (region B), 8.4 μm (region C), 11.3 μm (region D), respectively. The experiment results are helpful to understand the variation of morphology under different etching conduction and extent.

In the method, the visible luminescence of the selected area is similar to the conventional method; however, the luminescence can not be found in non-selected areas. This indicates that the selective-etching and the gradient light-emission on the identical sample in one step is possible. With different geometry electrode, different NPS pattern can be produced.

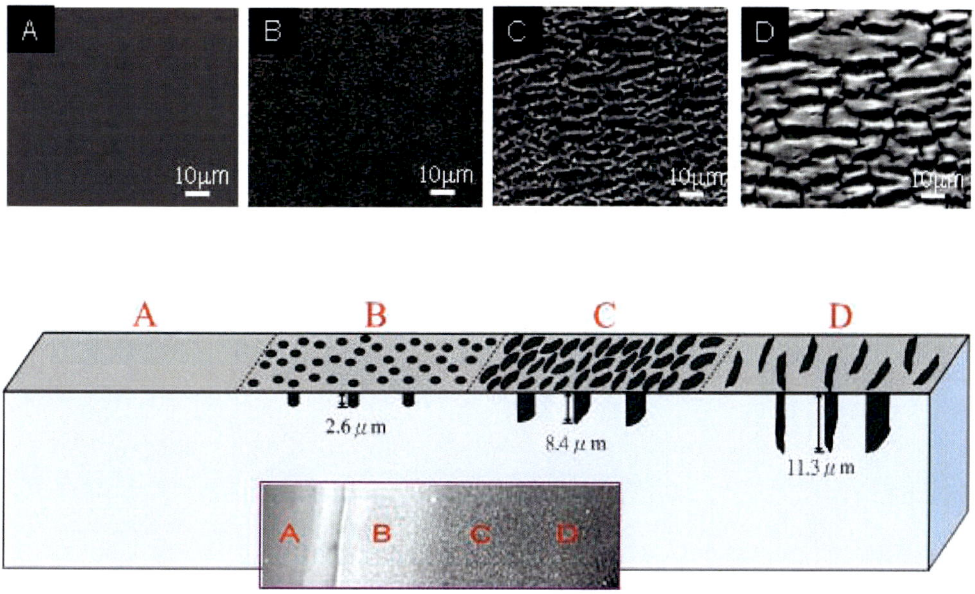

Figure 7.5. The top view SEM images and the distribution of the NPS sample of lateral-electric-field method.

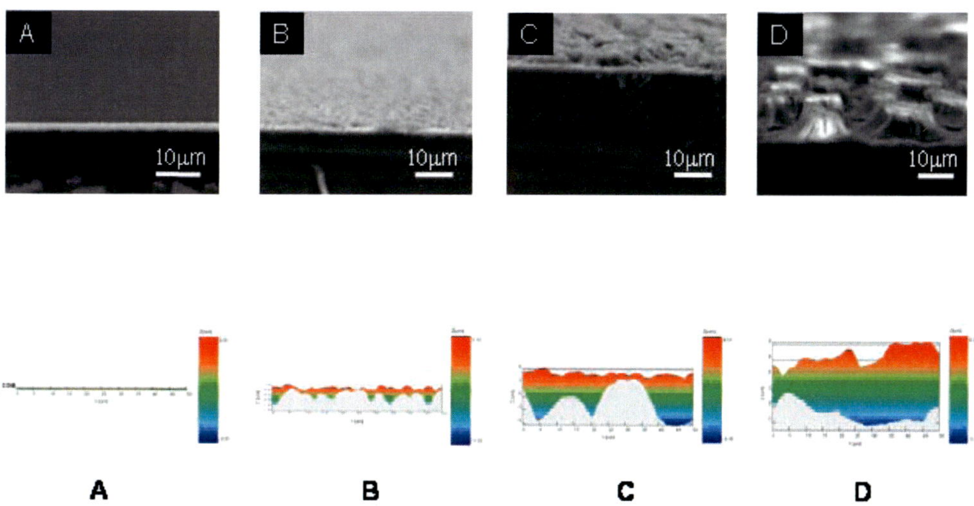

Figure 7.6. The cross-section SEM images and the 3-D profile of the NPS sample of lateral-electric-field method.

8. Superimposition-Assisted Method

As we mentioned above, perpendicular electric filed, lateral electric filed, magnetic field, illumination, buried p-layer are all important impact factors on the NPS formation. All factors are optional hole-source in the fabrication of NPS and can be super-imposed on the manufacturing processes. In this section, some examples are demonstrated for the superimposition-assisted method.

8.1. Illumination-Assisted Method with Buried-P-Layer-Assisted Method

The light-emission properties of n-NPS with an illumination-assistance and a buried-p-layer-assistance are explored in sections 4 and 6, respectively. In this section, illumination-assisted method and buried-p-layer-assisted method are superimposed on n-Si. The n-NPS film that has nano-scaled pores and high-aspect-ratio Si pillar arrays are stably fabricated. It is known that NPS is easier to show room temperature light emission in the visible range than macro-PS. The influences of anodic current density on the formation, morphology and properties of n-NPS are measured. The *I-V* characteristics and PL features of n-NPS films are investigated for different anodic-current cases. Since the anisotype-junction is forward biased during the anodization process, many holes can drift straight-upward from p-layer and participate in the electrochemical reaction. At room temperature, high peak-to-valley current ratios of about 117.3 can be obtained in negative difference conductance (NDC) region. In addition, strong visible light emissions are clearly observed under UV excitation.

The schematic diagram of the experimental setup for preparing n-NPS layers is given in figure 8.1. In figure 8.2, the left-side shows the case by the only-illumination-assisted method. The generation of top-photo-excited holes is almost isotropic and only occurs on the top layers. Therefore macro-pores with some tiny lateral branch etchings (tissue-like) are often observed in this way. And the pore has a depth limitation. On the other hand, as sketch in the right-side of figure 8.2, when the buried p-layer holes are superimposed by a forward biased pn-structure, some extra part of holes will drift straight-upward from p-layer to n-layer, and reach the pore tips to participate in electrochemical reaction during the anodization etching process. The supplying paths of buried p-layer holes are bottom-up and anisotropic, so the problems about depth limitation and lateral etching can be solved.

To analyze the impact of anodic-current densities on the etching process, three n-NPS formation cases which have different etching current densities are compared. They are specified as case-A (10 mA cm^{-2}), case-B (20 mA cm^{-2}) and case-C (40 mA cm^{-2}). An identical anodic time of 30 mins is used in all cases.

The top views and cross section views of SEM images on n-NPS samples are shown in figure 8.3 and 8.4, respectively. When the anodization current is 10 mA cm^{-2} (case-A), as shown in figures 8.3 (a) and 8.4 (a), the sample surface is shallowly etched, and some small pores with about 8 μm length are formed. In this case, the pores are formed on the top layer with illumination, so the width is wider than those in the deep layers of cases-B and C. In case-B, a clear macro-porous boundary reveals on the top view, as shown in figures 8.3 (b). Besides, in figures 8.4 (b), deep and straight pores and pillars appear in deep layers in which the anodization reaction relies on the holes from buried p-layer.

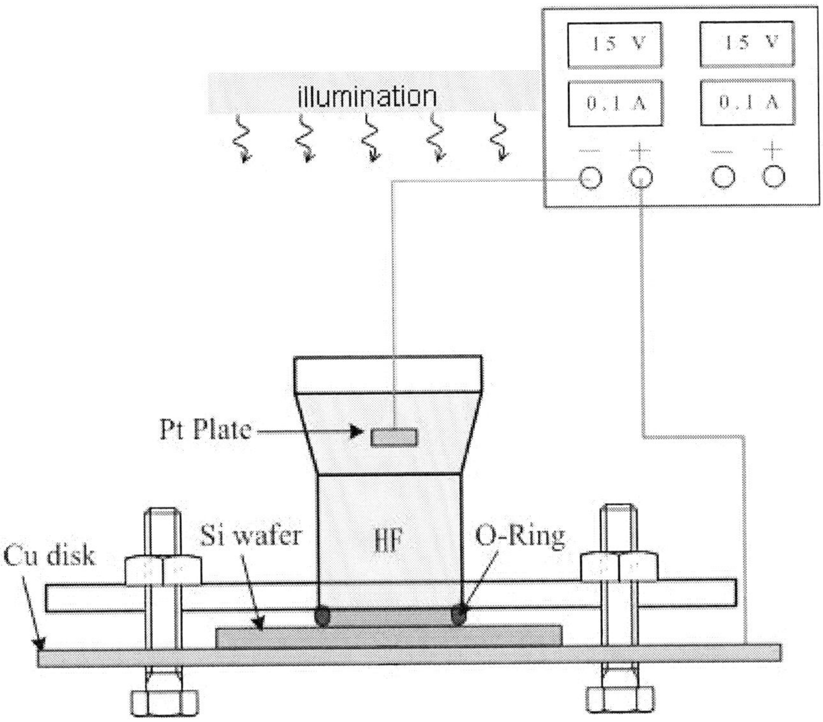

Figure 8.1. The schematic diagram of the experimental setup of illumination-assisted method with buried-p-layer-assisted method.

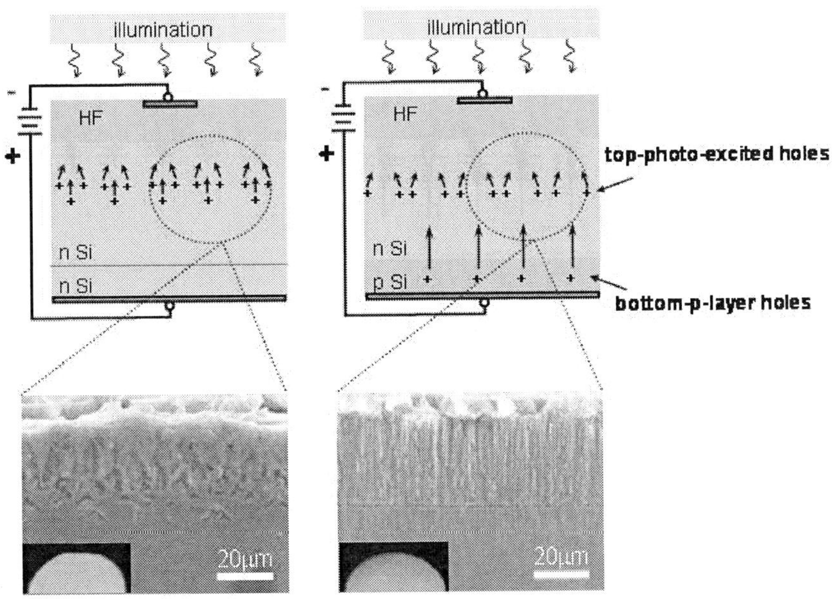

Figure 8.2. The illustrative mechanism of conventional method for n-NPS formation (left side figure) and the illustrative mechanism of n-NPS formation with the assistance of bottom-p-layer holes based on a pn structure (right side figure).

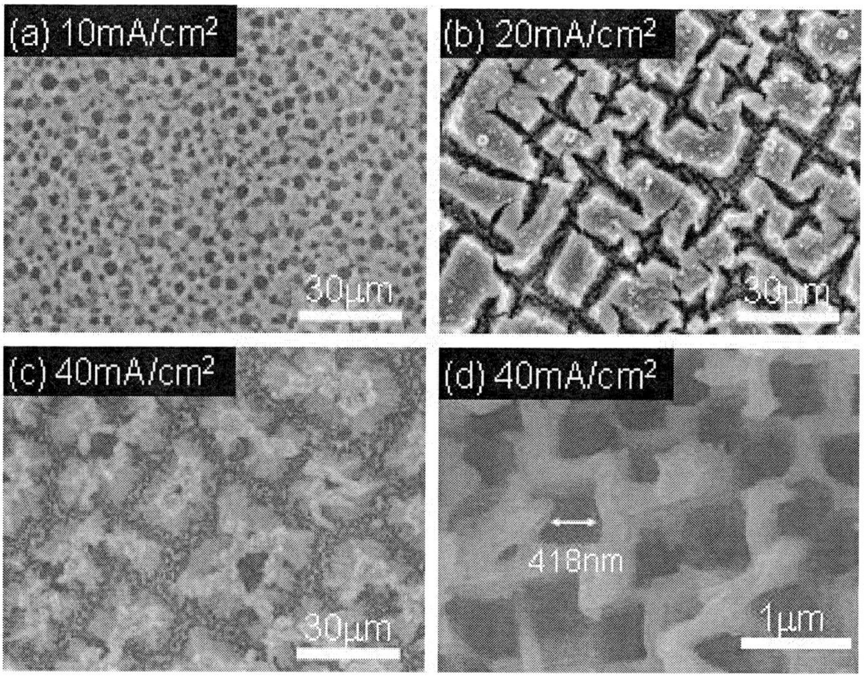

Figure 8.3. The top view SEM images of the NPS sample of illumination-assisted method with buried-p-layer-assisted method.

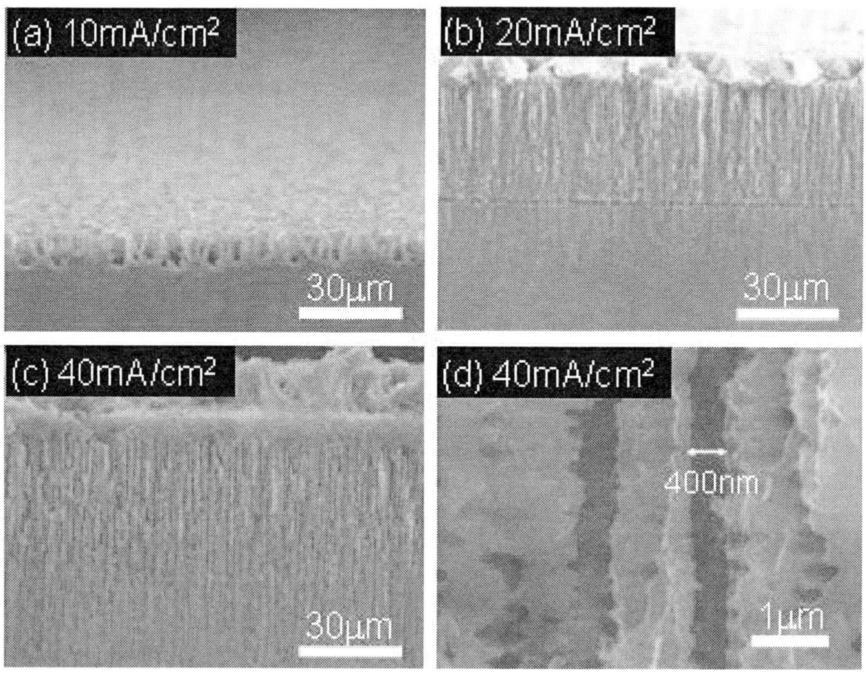

Figure 8.4. The cross section view SEM images of the NPS sample of illumination-assisted method with buried-p-layer-assisted method.

In case-C, with the highest anodic current density and drastic reaction, the vertical etchings on pore-tips proceed more rapidly. As shown in figures 8.3 (c) and 8.4 (c), the surface macro-pores expand larger than case-B. The SEM pictures of case-C at high magnification are shown in figures 8.3 (d) and 8.4 (d). It is clear that the samples formed with the superimposition of buried p-layer holes, the pore shapes are deep, straight and narrow to nano-scale, which profiles are very different from conventional method only with top-photo-excited holes. In this method, a high-aspect-ratio of 181 can be obtained. Besides, the n-NPS structures are channel-like instead of tissue-like that always occurs in conventional illumination method. The pore depths of the n-NPS layers can be up to 72 μm and the pore width can less than 400 nm. It is useful for the applications on deep-trench etching and thick-layer high-density light emission.

Figure 8.5 shows the *I-V* characteristic of cases A, B and C. All curves reveal the rectification properties due to the built-in potential in pn junction. With the smallest anodic current density in case-A, the NPS pores are the shallowest and hence a highest conductance than the other two cases is measured. It is well known that the resistivity of NPS layer is higher than bulk-Si layer [51]. In cases B and C, with higher anodic current densities than case-A, the PS layers are easily formed and that causes them to have higher resistances. Clear NDC phenomena are observed in case-B and case-C at room temperature. Their PVCRs are as high as 117.3 (case-B) and 9.5 (case-C), respectively. The NDC in NPS film is caused by the carrier transfer from high-mobility pillars to low-mobility pillars [52]. Since the widths of all pillars are not totally identical in NPS films, different (higher/lower) carrier mobilities appear in different (wider/narrower) pillars with corresponding lower/higher energy bandgaps. Among these different size pillars, the space confinement and surface scattering effect cause the carrier mobility in narrower pillars to be lower than wider pillars. Initially, when the applied voltage is turned on, the carriers prefer conducting by the wider pillars rather than narrower pillars. So the carriers are mainly concentrated in wider pillars during the first time stage. However, when the applied voltage (electric field) increases to a critical point, a drastic scattering occurs. In wider pillars, the carriers have sufficient energy to overcome the band discontinuity and to jump over the nanoscale pores into the narrower pillars. Many carriers are redistributed to all pillars (including narrower and wider ones) during the second time stage, and hence the average mobility as well as conductance reduces. The sudden switching from a higher conductance to a lower conductance produces the NDC phenomenon. The excellent NDC performance may have potential applications on microwave, multi-level switching, and memory devices [53-57].

However, NDC does not appear in case-A. Like real-space transfer diodes [24], the carriers in wider pillars must be accelerated and "heated" by the electric field so much that they are able to transfer to a "cooler" pillar. In case-A (the NPS prepared at 10 mA cm^{-2}), the porous etching depth is shallow, and the pillar length is not long enough for carrier acceleration. Therefore, no NDC phenomena are observed. However, in cases-B and case-C with higher anodic current densities than case-A, long Si pillars are easily formed by our method, and NDC phenomena are clearly observed at room temperature. In our study, the NDC can not be observed in every sample. In cases with pillars that are too short (shallow etching depths) or those with extremely uniform sizes of pillars, the NDC is difficult to be obtained.

Figure 8.6 shows the PL spectra of cases A, B and C. Such orange-red emissions are primarily attributed to the quantum confinement effect in the nano-scaled Si crystals that constitute the NPS film [2, 58, 59].

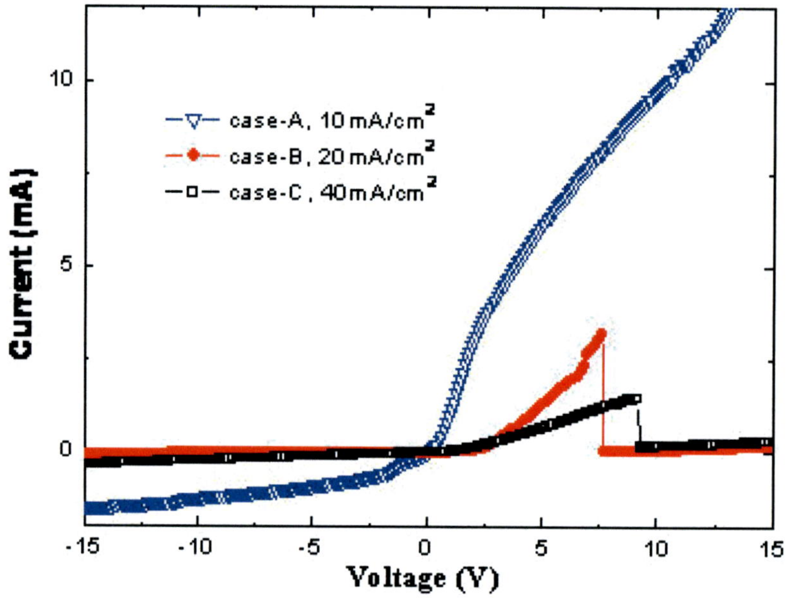

Figure 8.5. The I-V characteristics of cases A, B and C.

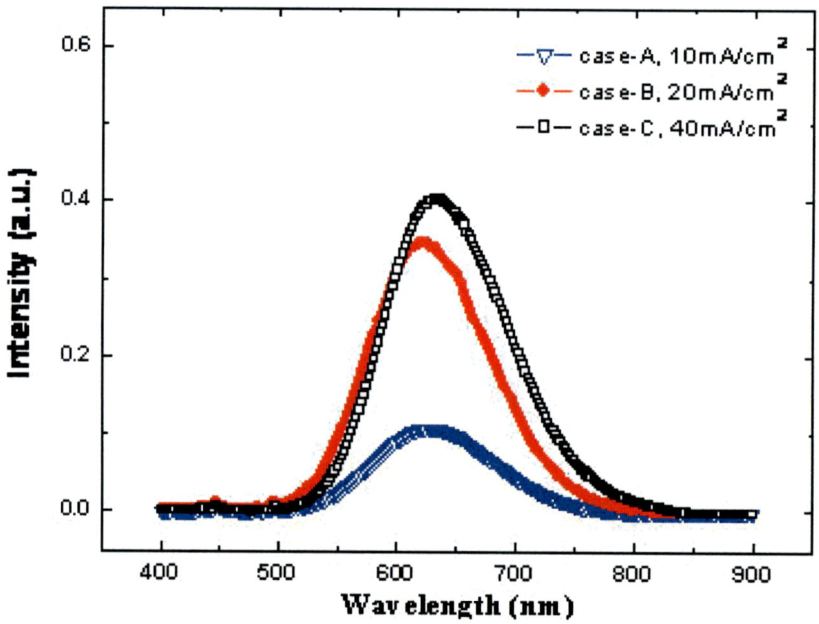

Figure 8.6. The PL spectra of cases A, B and C.

The quantum confinement effect for explaining the luminescence on NPS has ever been proposed by Canham [2, 60]. Radiative recombination in a confined nanoscale Si crystallite is proved to be more efficient than that in bulk Si. It is well known that the effective bandgap will increase when a Si structure is confined into a quantum scale. In this section, the energy bandgap is increased to about 1.9~2.0 eV by quantum confinement, and then the light with a wavelength of 650~620 nm emits. In addition, the case with higher current density that results in deep-size pores (pillars) produces a higher PL intensity. It also means that the assistance of bottom-p-layer holes can effectively improve the light emission intensity.

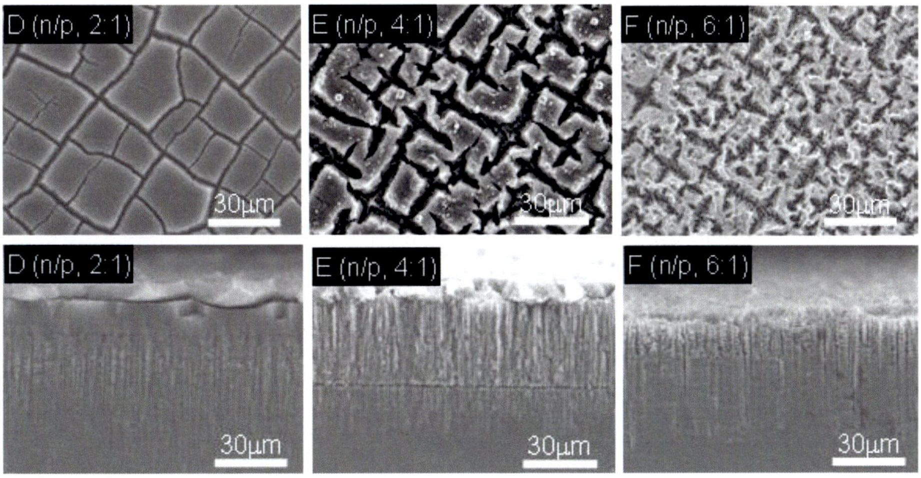

Figure 8.7. The top view SEM images of cases D, E and F.

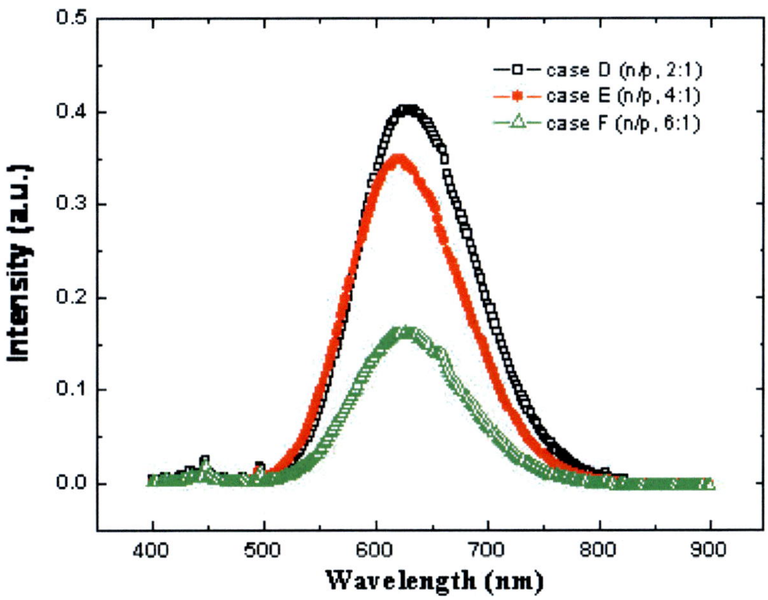

Figure 8.8. The PL spectra of cases D, E and F.

In addition, superimposition-assisted approach is applied on cases D, E and F. The top views and cross section views of SEM imagesof the samples are shown in figure 8.7. In cases D, E and F (n-on-p), the illumination-assistance and buried-p-layer-assisted are also simultaneously superimposed (superimposition-assisted approach). Furthermore, etching solution ratio of C_2H_5OH: HF =2:1 is used in cases D. In samples E and F, the ratios of 4:1 and 6:1 are used, respectively. To enhance the etching rate and etching depth, superimposition-assisted approach is useful to serve as thick-layer light emitting applications. The visible luminescence of the n-NPS can be drastically enhanced by the superimposition of buried-p-layer-assisted and illumination-assistance. This method can be used to improve the performance of photovoltaic devices that convert higher energy into visible wavelength light.Clear visible PL emissions at room-temperature are obtained. Strong visible photoluminescence emissions are demonstrated on the hole-poor n-type porous layer (as shown in figure 8.8).

8.2. Hall-Effect-Assisted Method with Perpendicular-Electrical-Field Method

The perpendicular-electric-filed-assisted method and Hall-effect-assisted method is desciribed in section 3 and 5, respectively. In this section, these two methods are superimposed to enhance the PL on n-NPS (without any illumination).

The schematic diagram of the experimental setup for preparing n-NPS is given in figure 8.9. A mixture of HF: C_2H_5OH=1:1 is utilized as the etching solvent here. No illumination equipment is needed in our process design. The NPS samples are prepared on n-type Si (100). A Si wafer with copper (Cu) contact is served as anode, while cathode is made by platinum (Pt). A sintered aluminum (Al) film is used to form an intimate backside ohmic contact to Si sample prior to the etching process. By adjusting the magnitude of the correspondent magnetic field and voltage, sharpened Si-tips can be fabricated.

Since $F_z = qV_x \times B_y$, different NPS structures reflect differences in the magnitude of the correspondent magnetic field and voltage. Electropolishing occurs for high magnetic field and voltage. At lower magnetic field and voltage, the dissolution process is changing and NPS layer is formed. By adjusting the magnitude of the correspondent magnetic field and voltage, sharpened Si-tips can be fabricated. Under a vertical anodization voltage of 30 Volts, Hall-effect assistance with a magnetic field (By) of 10 milliTesla and a voltage (Vx) of 30 Volts, which are the optimum conditions successfully support the etching process without any illumination sources. The top view and cross-section SEM images of the NPS sample are shown in figures figure 8.10 and 8.11, respectively. A larger voltage drop across the x-direction will cause a larger gradient-situation of etching extent of NPS. To clearly show the gradient-situation for further investigation, a larger Vx is adopted in this study. If a uniformed NPS film is needed, a small bias voltage (Vx), a small device area, or rotating Hall-effect apparatus are advised to solve this non-uniformity problem. Four regions (A, B, C and D) are defined on the sample for article describing, as shown in figure 8.12. The NPS films have high-aspect-ratio Si-tips because the hole-supplying mechanism is bottom-up.

All the Si-tips are vertical align to the substrate surface. The heights of Si-tips are about 78 μm (region A), 50 μm (region B), 31 μm (region C) and 16 μm (region D), respectively. Their diameters are 22 μm (region A), 18 μm (region B), 16 μm (region C) and 11 μm (region D), respectively. Figure 8.12 shows the schematic of the tip height and distribution for the four regions. And figure 8.13 shows the EDS analysis. The experiment results are helpful to understand the variation of morphology under different etching conduction and extent.

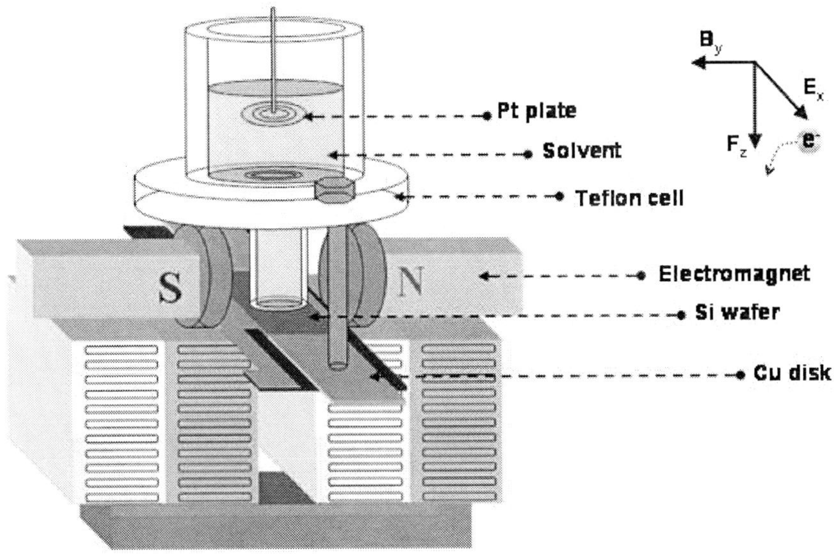

Figure 8.9. The schematic diagram of the experimental setup of Hall-effect-assisted method with buried-p-layer-assisted method.

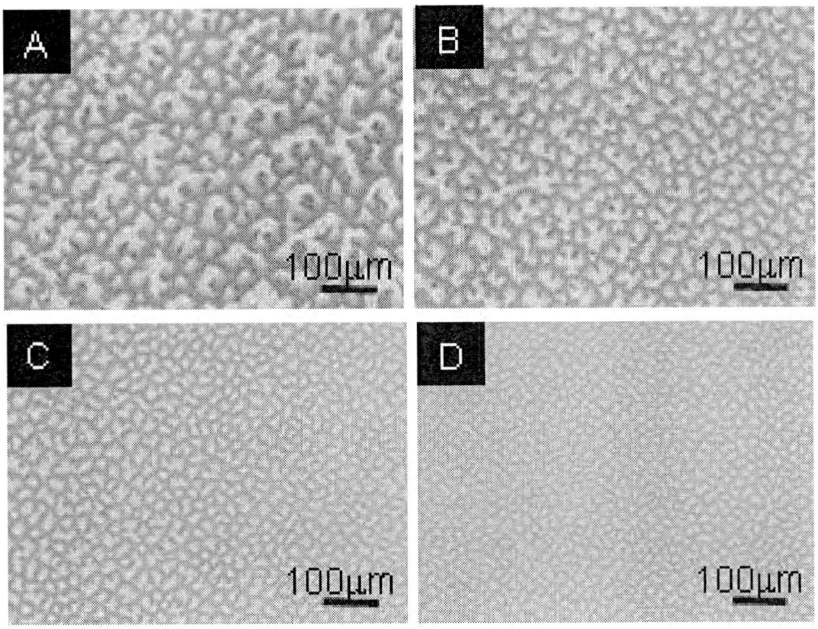

Figure 8.10. The top view SEM images of the NPS sample of Hall-effect-assisted method with buried-p-layer-assisted method.

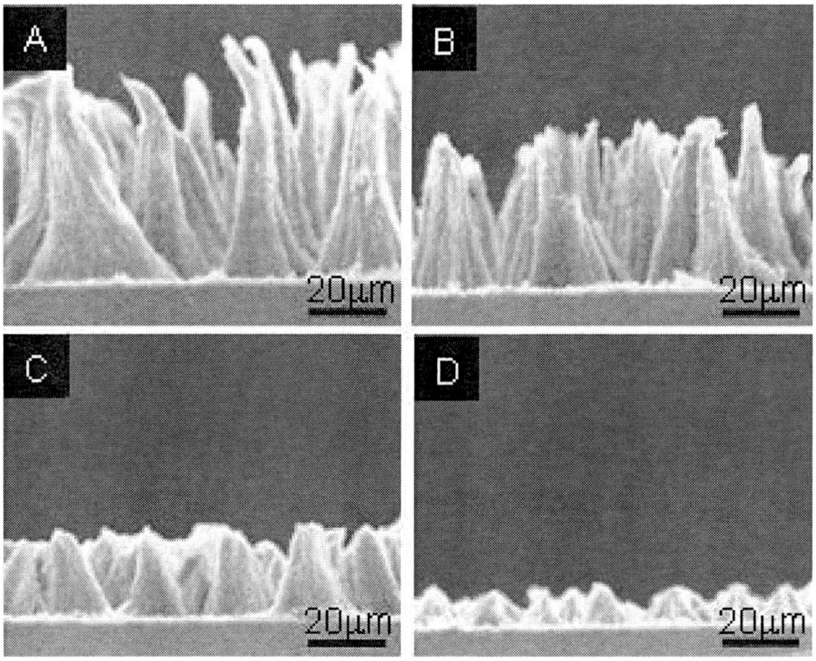

Figure 8.11. The cross section view SEM images of the NPS sample o of Hall-effect-assisted method with buried-p-layer-assisted method.

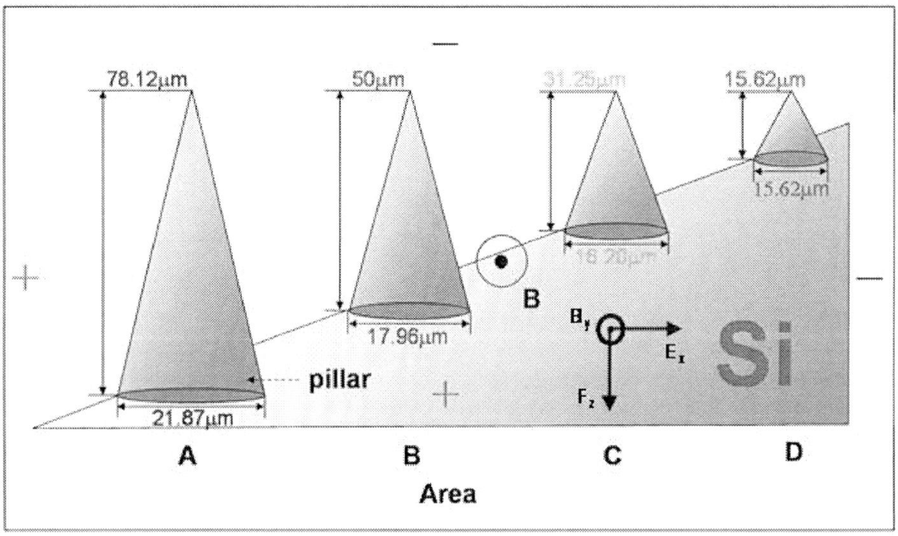

Figure 8.12. The schematic of the Si-tips and distribution.

Figure 8.14 shows the variation in time of anodization current under a constant voltage of 30 Volts. With the hole-effect-assistance on n-Si (case-1), the curve tendency of anodization current is very similar to that on p-Si (case-2) that is fairly different to conventional n-Si without any hole-assistance (case-3).

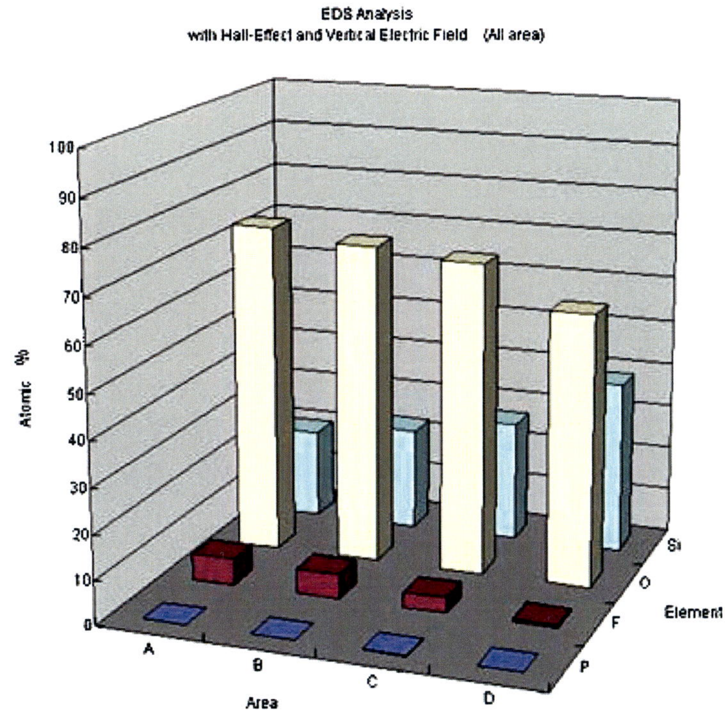

Figure 8.13. The EDS analysis of the NPS sample of Hall-effect-assisted method with buried-p-layer-assisted method.

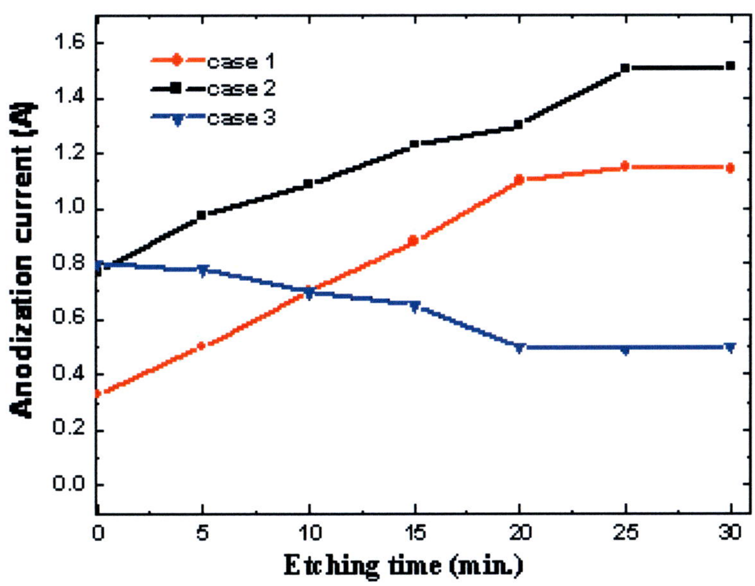

Figure 8.14. The variation in time of anodization current: the hole-effect-assistance on n-Si (case-1), conventional p-Si (case-2) and conventional n-Si without any hole-assistance (case-3).

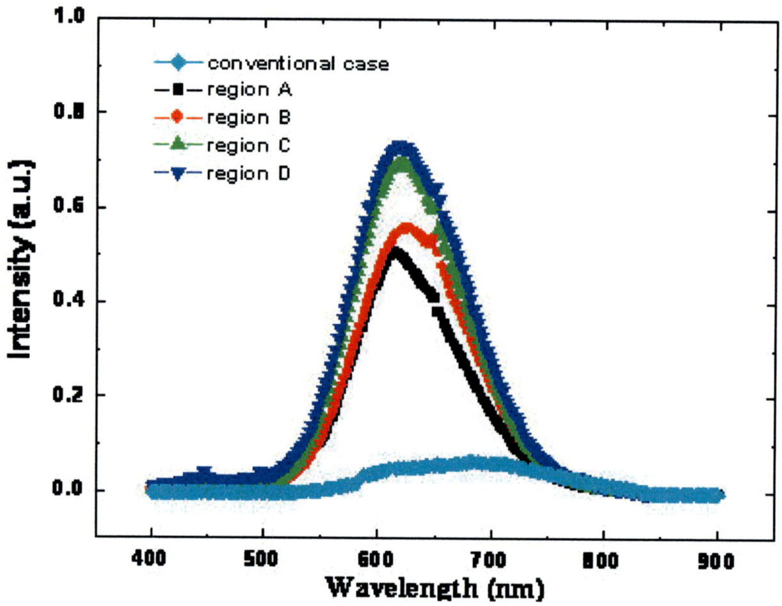

Figure 8.15. The PL spectra of A~D regions.

The increasing tendencies with time in case-1 and case-2 are resulted from the successful processing of chemical reaction which produces the hole drift current as well as the thinning down of bulk-Si layers which decrease the total resistances. The total resistance of the galvano-circuit can be express as follows:

$$R_{total} = R_{electrolyte} + R_{measurement-system} + R_{etched-layer} + R_{bulk-Si-layer}$$
$$\cong R_{etched-layer} + R_{bulk-Si-layer}$$

In case-1 and case-2, the etched layers that are formed with slim Si-tips let the electrolyte among the tips serve as the major conducting path. Therefore the dominant variation of total resistance (R_{total}) is the term of bulk-Si layer ($R_{bulk-Si-layer}$) and hence the current will increase with the decreasing of $R_{bulk-Si-layer}$. However, in case-3, the chemical reaction cannot keep going for the lack of holes. Slight etching forms some narrow pore and Si-pillars. The resistances of Si-pillars are larger than that of bulk-Si-layer ($R_{etched-layer} > R_{bulk-Si-layer}$), so the dominant variation of R_{total} is the changing from $R_{bulk-Si-layer}$ to $R_{etched-layer}$. Such situation is different from case-1 and case-2. As to the initial-state points (time=0) of current, a drop from 0.80 A (case-3) to 0.33 A (case-1) results from the electron depletion by Hall-effect on the surface layer in n-Si and hence the increasing of $R_{bulk-Si-layer}$.

In conventional illuminated n-NPS, after surface pores (of few um) are formed, photo-energy is not completely received by the deeper layers. Therefore, the deeper layers are difficult to form the required NPS in the darkness [27]. Only a thin part of NPS layer contributes in the PL. In our method, the visible luminescence of the n-NPS is similar to the hole-rich p-NPS.

An excitation wavelength of 325 nm from a He-Cd laser is used for the PL measurements. Figure 8.15 are the PL spectra of A~D regions under the Hall-effect assistance (By = 10 milliTesla and Vx = 30 Volts). The PL intensity is inversely proportional to the diameter of Si-tip that appears on A~D regions gradually. The highest PL intensity in region D can be attributed to its small size of tip-width. The results consist with the quantum confinement theory. In the quantum confinement model, the visible PL phenomenon is explained by the changes of the Si band structure when dimensions are reduced to the nano-scale size.

9. Supercritical-Fluid Method for Metal-Semiconductor Contact

The EL of NPS is discovered in the 1990s. In the applications on EL, the electrical consumption power is an important issue. However, the surface roughness of NPS leads to a high contact resistance to the metal layer and, hence, an electrical power waste. To enhance the EL efficiency and the stable performance, a good metal contact to the light-emitting layer (NPS layer) is required. Some efforts have been made to improve the metallic contact between the metal and NPS. Herino *et al.* [61] and Dhar *et al.* [62] use the electroless Ni plating technique to make electrical contacts with NPS. Gelloz *et al.* propose a low porosity superficial layer between NPS and the top contact to achieve low-resistance contacts [63, 64]. Metal films are always deposited on NPS layers by conventional physical vapor deposition (PVD) methods in most other experiments [65, 66]. However, the roughness surface of NPS layer consisting of nano scale Si pillars and pores is very difficult to attach with metal atoms by conventional metal deposition techniques. Conventional approaches always result in high contact resistances in metal-NPS contacts. Therefore, a supercritical fluid (SCF) technique is explored to overcome this problem. The results are now presented in this section.

In the section, supercritical carbon dioxide (CO2) is used as a solvent. Silver (Ag) nanoparticles are produced and successfully fill the NPS pores on the surface layer by CO2 SCF equipment which is made by the High Pressure Equipment Company (SFT-150 DLOW SHEET) (figure 9.1). The *I-V* characteristics of the samples are measured and analyzed at room temperature. The curve is also compared to that is formed by the conventional metal deposition approach (without SCF processes). Good metal contacts to the NPS layers are obtained by the new proposed approach. In addition, a NDC curve is clearly observed that is usually suppressed due to the high contact resistance between metal and NPS form by conventional deposition methods. The occurrence of NDC curve in NPS is also ever been observed by Kwei *et al.* In their presentation, they attribute the NDC phenomenon to the carrier mobility difference of thick and thin wires in the NPS layers [55].

The schematic diagram of SCF equipment setup is shown on the left side in figure 9.2. The CO2 pump can rapidly produce the high pressures which are required for SCF processing. The main part of the system is an interflow oven. The interflow oven includes both the sample holder and the SCF collector. As the chemical reaction is carried out, the valve of CO2 cylinder is closed and the decompression valve is opened quickly. Then, CO2 decompress in a transient time. Meanwhile, the Ag nanoparticles are produced and filled the pores in the NPS.

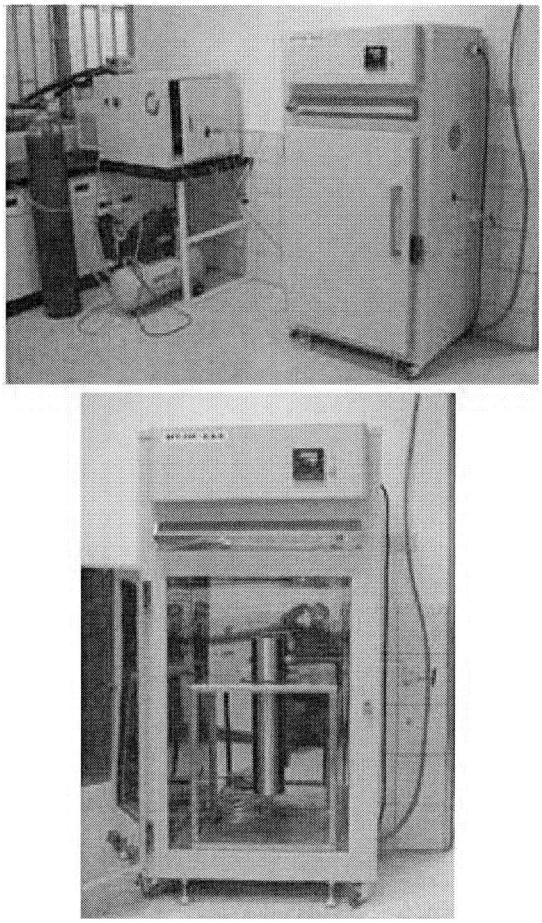

Figure 9.1. The CO_2 SCF equipment (SFT-150 DLOW SHEET).

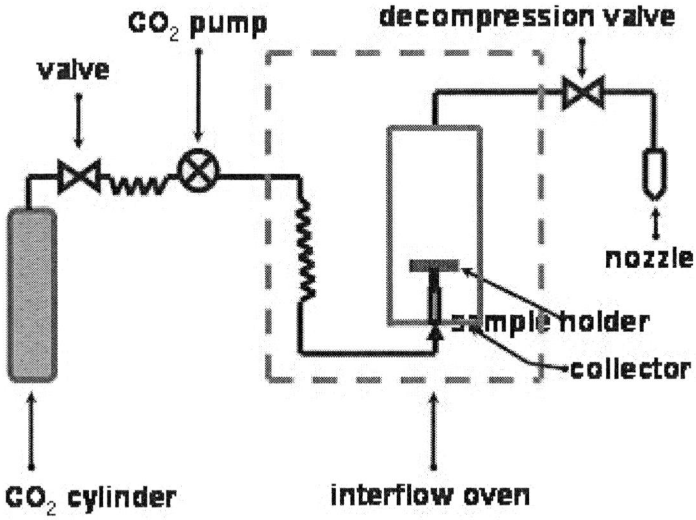

Figure 9.2. The schematic diagram of SCF equipment setup.

The SCF processing involves the chemical reduction of soluble compounds in supercritical CO2 to yield nano scaled Ag extraction. The CO2 solvent is the most commonly used in SCF systems because of its characteristics of nontoxic, nonflammable, low cost, low power consumption, room temperature operation, and simple production. In the study, silver nitrate (AgNO3) (solubility of 10%, 20% and 30%) are utilized as precursors of Ag nanoparticles extraction. Under a high pressure (200 bars) and a high temperature (100°C), Ag nanoparticles are extracted from AgNO3 by the chemistry reduction of CO2 solutions during the SCF process. The illustrative equation of the overall process during Ag nanoparticles extraction can be expressed as below:

$$8\ AgNO_3 + NaBH_4 + 7\ OH^- \rightarrow 8\ Ag + 8\ NO_3^- + 4\ H_2O + H_3BO_3 + Na^+$$

By contrast with the SCF process, when this reaction is carried out in an ultrasonic generator, many large-sized and randomly-distributed Ag particles are produced and accumulates on the top surface of NPS pillars.

The top view and cross-section scanning electron microscope (SEM) images of the NPS sample without SCF process are shown in figure 9.3 (a) and 9.3 (b), respectively. It shows that the pore diameters are about 800~2000 nm and the pore depths are about 10~20 μm. Narrow- and deep-trench NPS structures (with high aspect ratios of 1:25) are prepared and focused in our studies, whose surfaces are extremely rough, and it is always difficult to produce good electrical contacts between them and metal films by conventional methods.

Figure 9.4 (a)-(f) show the top view and cross-section SEM images of the NPS samples with SCF processes in which three kinds of AgNO3 concentration (10%, 20% and 30%) are used, respectively. Figure 9.4 (a) and (b) show the Ag nanoparticles with sub-100 nm fill the most part of the deep-trench NPS pores when using 10% AgNO3 solution.

The other cases with different solutions of AgNO3 (20% and 30%) are shown in figures (c)-(f). The results show that the sizes of these Ag nanoparticles are bigger than those in figures 9.4 (a) and 2 (b) when the concentration of AgNO3 increases. In figures 9.4 (e) and 2 (f), a mass of large-scaled Ag particles are extracted and congregated upon the NPS surface but cannot fill the trench due to the large size of Ag particles. Such a coating style is similar to a conventional deposition method.

It is concluded that the size and the morphology of metal particles can be controlled by changing the solution of reaction, and 10% AgNO3 is the best solution of Ag nanoparticles extraction in our study.

As illustrated in figure 9.4 (g)-(i), EDS mappings show the distribution analysis of Ag nanoparticles (shown in several white spot areas that are dependent on the resolution). The figures show the Ag nanoparticles (sub-100 nm in width) are widely distributed over the NPS surface homogenously when AgNO3 solution is 10%. A mass of large-scaled Ag particles are produced and assembled upon the surface of NPS layer when the concentration of AgNO3 increases. The above analyses can reach the same conclusions as figure 9.2 that the different AgNO3 solutions will lead to different grain sizes of Ag particles.

Table 1 demonstrates the EDS analysis of solids formed using solutions with different concentrations of AgNO3. The atomic percentage of Ag element decreases when the concentration of AgNO3 increases. When the concentration of the AgNO3 solution is higher than 10%, the filling of the NPS pores becomes more difficult.

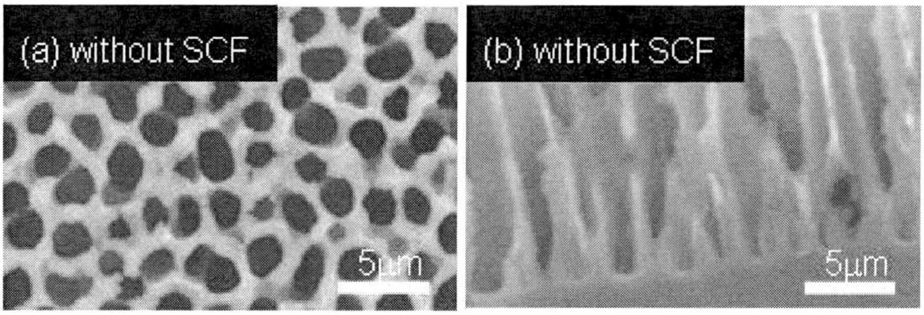

Figure 9.3. The SEM images of the NPS sample without SCF process.

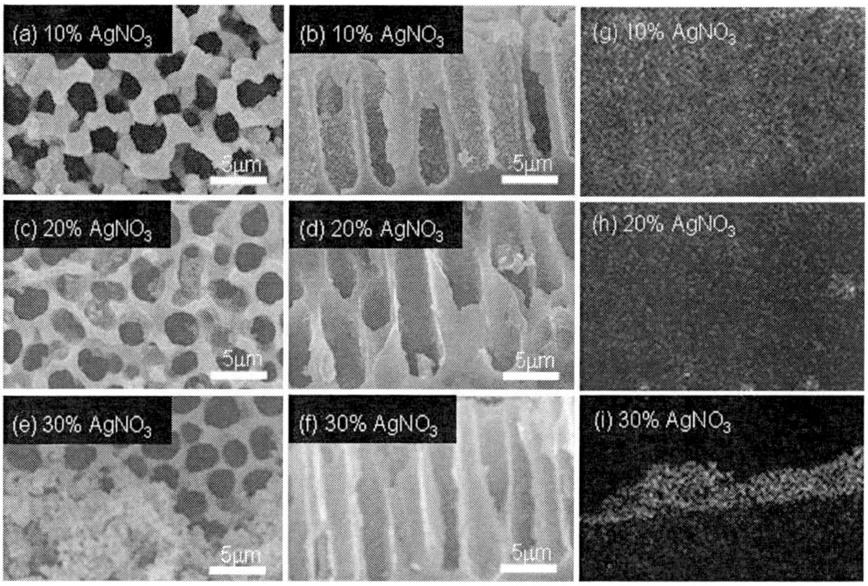

Figure 9.4. (a)-(f) The SEM images of the NPS sample with SCF process, and (g)-(i) EDS mappings of the NPS samples with different solutions of $AgNO_3$.

Table 9.1. The The EDS analysis of solids with different solutions of AgNO3

$AgNO_3$ (concentration)	10%	20%	30%
Element	Atomic(%)	Atomic(%)	Atomic(%)
O	52.49	26.41	57.35
Si	39.96	68.75	41.34
Ag	7.55	4.84	1.31
Total	100.00	100.00	100.00

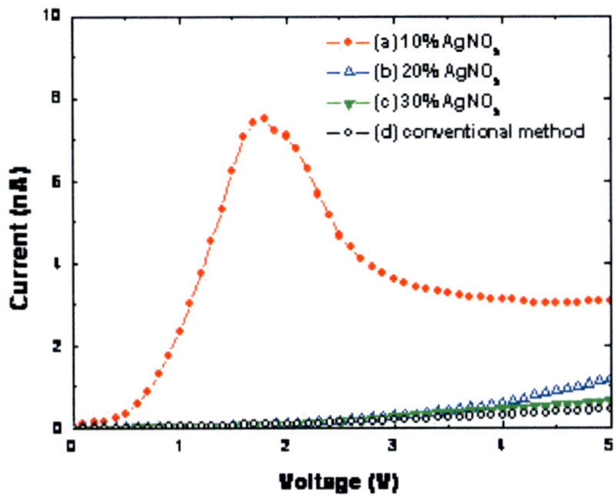

Figure 9.5. The *I-V* characteristics of NPS samples by SCF method.

The *I-V* measurement results of the NPS samples are shown in figure 9.5. The curves *a-c* show the *I-V* characteristics of the NPS samples that are treated with SCF processes. The curve *d* is the *I-V* characteristic of the NPS sample using conventional method for comparison in which the metal film is deposited on the NPS surface by PVD. It is clearly shown that the electrical contact is effectively improved with 10% AgNO3 solution (curve a). A nonlinear *I-V* characteristic with a peak current point at 1.8 V is clearly shown that is usually suppressed in poor metal contact. However, the SCF processes with 20% (curve b) and 30% (curve c) AgNO3 concentrations do not work in force because the Ag particles only accumulate on the top NPS pillars and cannot fill the NPS pores successfully. The results shown in curves *b* and *c* are very similar to curve *d* in which large-sized Ag particles are deposited and the metal film is formed on the top. Identically, good electrical contacts cannot be obtained in NPS pores and, hence, high contact resistances produce. The high contact resistance causes the suppression of NDC phenomenon. Apparently, the contact resistance is successfully reduced (about 250:1) due to the pore filling with Ag nanoparticles by SCF method (10% AgNO3). A good improvement of metal contact to the interfaces of NPS layers is achieved.

The method can be used to improve the metal contact to the NPS surface by the SCF technique. During the SCF process, Ag nanoparticles with sub-100 nm diameter are formed at 100oC under a pressure of 200 bars and fill the NPS pores. The size and the morphology of particles are controlled by changing the solutions. The best AgNO3 concentration of 10% is strongly advised in the study. In addition, the *I-V* characteristics are measured and discussed. It is demonstrated that the conventional metal deposition methods produce high contact resistances and, however, the new approach by SCF technique improves the performance substantially. Also, the NDC phenomenon of NPS structures can be clearly revealed resulting from the reduction of contact resistance. It is believed that the power consumption of EL on the NPS structure can be greatly reduced by this new approach. The study results will benefit the applications of electroluminescence, microelectronic and optoelectric devices.

10. Conclusions

In conclusion, several new ideas for NPS formations are introduced, and some related experimental results are presented. The conventional perpendicular-electric-field-assisted method for anodization process is described. Also, the conventional illumination-assisted method for n-NPS formation is mentioned. In addition, Hall-effect is applied to assist the manufacturing of n-NPS. The etching process is anisotropic and illumination-free. An uniform hole distribution and gradient etching are produced. Different etching strength in four regions on the same sample is explored. The electric field and magnetic field control the hole accumulation, and hence NPS formation and light-emission areas. A strong visible light is obtained on n-NPS at room temperature.

Also, a new buried-p-layer-assisted method based on a forward biased np-junction for manufacturing n-PS layer is studied. The method can overcome the illumination-limitation and depth-limitation problems in conventional only-illumination-assisted method. With the bottom-hole-assistance from the buried p-layer, the anodization etching is almost anisotropic. This feature has great potential in semiconductor manufacturing. The clear visible PL emissions at room-temperature are obtained on the hole-poor n-type samples, even in the dark. Strong visible PL emits at about 650 nm. Of course, the visible luminescence of the n-NPS can be drastically enhanced by the superimposition of buried-p-layer-assisted method and illumination-assisted method.

Remarkly, the buried-p-layer-assisted method can produce an n-NPS film with deep, straight and high-aspect-ratio Si pillar arrays. It is interesting that, at room temperature, both the properties of high PVCR in NDC and strong light emission in PL can exist on such n-NPS films together. The key factor for the n-NPS formation is the special design of anisotype pn structure that supports orderly holes from bottom p-layers.

Furthermore, with the superimposition of Hall-effect assistance and perpendicular-electric-field assistance, a film with the sharpened Si-tips can be obtained. A strong visible light is also obtained on the Si-tips at room temperature.

Alternatively, a new method for the fabrication of p-type patterned NPS only by electric field is proposed. The manufacturing process is anisotropic and illumination-free. No masking-layer and related photo-lithography processes are needed in this method.

On the other hand, the method is put forward to improve the metal contact to the NPS surface by the SCF technique. During the SCF process, Ag nanoparticles are formed at 100oC under a pressure of 200 bars and fill the NPS pores. The size and the morphology of particles are controlled by changing the solutions. In addition, by current-voltage characteristic measurements, it is proved that the SCF method improves the performance substantially. Also, the NDC phenomenon can be clearly revealed resulting from the reduction of contact resistance. It is believed that the power consumption of EL on the NPS structure can be greatly reduced.

In this chapter, these studies on NPS provide alternative methods of fabricating electronic and optoelectronic devices.

References

[1] L.T. Canham. (1990). *Appl. Phys. Lett.*, 57, 1046.

[2] P. McCord, S. L. Yau and A. J. Bard. (1992). *Science*, 257, 68.

[3] Bisi, Stefano Ossicini, and L. Pavesi. (2000). *Surf. Sci. Rep.*, 38, 1.

[4] G. Cullis, L. T. Canham and P. D. J. Calcott. (1997). *J. Appl. Phys.*, 82, 909.

[5] P. M. Fauchet. (1996). *J. Lumin.,* 70, 294.
[6] J. C. Lin, P. W. Lee and W. C. Tsai. (2006). *Appl. Phys. Lett.*, 89, 121119.
[7] J. C. Lin, W. L. Chen and W. C. Tsai. (2006). *Opt. Express*, 14, 9764.
[8] J. C. Lin, H. T. Hou and W. C. Tsai. (2007). *Microelectron. Eng.*, 84, 336.
[9] G. Smestad and H. Ries. (1992). *Solar Energy Materials Solar Cells*, 25, 51.
[10] G. Willeke, H. Nussbaumer, H. Bender and E. Bucher. (1992). *Solar Cells*, 25, 51.
[11] T. Taliercio, M. Dilhan, E. Massone, A. M. Gue, B. Fraisse and A. Foucaran. (1992). *Thin Solid Films*, 255, 310.
[12] L. Z. Yu and C. R. Wie. (1992). *Electron. Lett.*, 28, 911.
[13] Y. Watanabe, Y. Arita, T. Yokoyama and Y. Igarashi. (1975). *J. Electrochem. Soc.*, 122, 1351.
[14] H. J. Moon, K. H. Jung, H. P. Hyung, J. K. Joong, H. H. Sang and Y. C. Se. (1997). *Thin Solid Films*, 309, 490.
[15] H. Art. (2004) *Alternative Approaches for DMFC Design:Silicon-Based Systems 6th Annual SMALL FUEL CELLS.*
[16] S. Chan and P. M. Fauchet. (1999). *Appl. Phys. Lett.*, 75, 274.
[17] W. H. Zheng, P. Reece, B. Q. Sun, and M. Gal. (2004). *Appl. Phys. Lett.*, 84, 3519.
[18] V. Lehmann. (1993). *J. Electrochem. Soc.*, 140, 2836.
[19] J. D. Benjamin, J. M. Keen, A. G. Cullis, B. Innes, and N. G. Chew. (1986). *Appl. Phys. Lett.*, 49, 716.
[20] Z. Gaburro, P. Bettotti, M. Saiani, L. Pavesi, L. Pancheri, C. J. Oton, and N. Capuj. (2004). *Appl. Phys. Lett.*, 85, 555.
[21] Z. Gaburro, C. J. Oton, L. Pavesi, and L. Pancheri. (2004). *Appl. Phys. Lett.*, 84, 4388.
[22] C.Pickering, M.J. Beale, D.J. Robbins, P.J. Pearson, and R. Greef. (1984). *J. Phys. C: Solid State Phys.*, 17, 5535.
[23] M.J. Beale, J.D. Benjamin, M.J. Uren, N.G. Uren, N.G. Chew, and A.G. Cullis. (1985). *J. Cryst. Growth*, 73, 622.
[24] M.J. Beale, J.D. Benjamin, M.J. Uren, N.G. Uren, N.G. Chew, and A.G. Cullis. (1985). *Appl. Phys. Lett.*, 46, 86.
[25] M. Young, M. I. Beale and J.D. Benjamin. (1985). *Appl. Phys. Lett.*, 46, 1133.
[26] R.L. Smith, S.F. Chuang, and S.D. Collins. (1988). *J. Electron. Mater.*, 17, 533.
[27] R.L. Smith and S.D. Collins. (1989). *Phys. Rev. A*, 39, 5409.
[28] R.L. Smith and S.D. Collins. (1992). *Phys. J. Appl.Phys.*, 71, R1.
[29] T.A. Witten and L.M. Sander. (1983). *Phys. Rev. B*, 27, 5686.
[30] A.J. Read, R.J. Needs, K.J. Naish, L.T. Canham, P.D.J. Calcott, and A. Qteish. (1992). *Phys, Rev. Lett.*, 69, 1232.
[31] G. D. Sanders and Y. C. Chang. (1992). *Phys. Rev. B*, 45, 856.
[32] V. Lemann, U. Gösele. (1991). *Appl. Phys. Lett.*, 58, 856.
[33] V. P. Parkhutik, J. M. Albella, J. M. Martinez-Duart, J. M. Gomez-Rodriguez, A. M. Baro, and V.I. Shershulsky. (1993). *ibid.*, 2, 366.
[34] H. Yan and X. Hu. (1993). *J. Appl. Phys.*, 73, 4324.
[35] L. T. Canham, W. Y. Leong, T. I. Cox and L. Taylor. (1992). *Appl. Phys. Letts.*, 61, 2563.
[36] H. D. Fuchs, M. Stutzmann, M. S. Brandt, M. Rosenbauer, J. Weber, A. Breitschwerdt, P. Deak and M. Cardona. (1993). *Phys. Rev. B*, 48, 8172.

[37] Sarathy, S. Shih, K. H. Jung, C. Tsai, K. H. Li, D. L. Kwong and J. C. Campbell. (1992). *Appl. Phys. Lett.*, 60, 1532.
[38] Ksendzov, R. W. Fathauer, T. George, W. T. Pike and R. P. Vasquez. (1993). *Appl. Phys. Lett.*, 63, 200.
[39] S. Kalem and M. Rosenbauer. (1995). *Appl. Phys. Lett.*, 67, 2551.
[40] M. Saadoun, H. Ezzaouia, B. Bessais, M. F. Boujmil and R. Bennaceur. (1999). *Solar Energy Mater. Solar Cells*, 59, 377.
[41] S. Kalem and O. Yavuzcetin. (2000). *Opt. Express*, 6, 7.
[42] M. Saadoun, N. Mliki, H. Kaabi, K. Daoudi, B. Bessaïs, H. Ezzaouia and R. Bennaceur. (2002). *Thin Solid films*, 405, 29.
[43] N. Lalic and J. Linnros. (1996). *Thin Solid Films*, 276, 155.
[44] Richter, P. Steiner, F. Kozlowski and W. Lang. (1991). *IEEE EDL.*, 12, 691.
[45] M. J. Eddowes. (1990). *J. Electroanal. Chem.*, 280, 297.
[46] Stumper and L. M. Peter. (1991). *J. Electroanal. Chem.*, 309, 325.
[47] A. Yaron, S. Bastide, J. L. Maurice, C. Levy-Clement. (1993). *J. Lumim.*, 57, 67.
[48] E. Galun, A. Lagoubi, R. Tenne, C. Levy-Clement. (1993). *J. Lumim.* 57, 125.
[49] G. Nassiopoulos, S. Grigoropoulos, L. Canham, A. Halimaoui, I. Berbezier, E. Gogolides and D. Papadimitriou. (1995). *Thin Solid films*, 255, 329.
[50] Krüger, R. Arens-Fischer, M. Thönissen, H. Münder, M. G. Berger, H. Lüth, S. Hilbrich and W. Theiss. (1996). *Thin Solid films*, 276, 257.
[51] D. Mao, K. J. Kim, Y. S. Tsuo, and Arthur J. Frank. (1995). *J. Phys. Chem.*, 99, 3643.
[52] K. Lee, C. H. Chu, Y. C. Tseng, J. M. Shyr, and C. H. Kao. (2000). *IEEE EDL.*, 21, 587.
[53] Sibille, J. F. Palmier, H. Wang, J. C. Esnault and F. Mollot. (1990). *Appl. phys. Lett.* 56, 256.
[54] Reklaitis, A. Krotkus, A. Geizutis and M. Asche. (1999). *Semicond. Sci. Techno.*, 14, 341.
[55] S. J. Koester, K. Ismail, K. Y. Lee, and J. O. Chu. (1997). *IEEE EDL.*, 18, 432.
[56] Saitoh, H. Harata, and T. Hiramoto. (2004). *Appl. Phys. Lett.*, 85, 6233.
[57] H. Lee and Y. H. Jeong. (2004). *IEEE Trans. Nanotechnol.*, 3, 377.
[58] N. Saeta and A. C. Gallagher. (1995). *J. Appt. Phys.*, 77, 4639.
[59] G. D. Sanders and Y. C. Chang. (1992). *Phys. Rev. B*, 45, 9202.
[60] G. Cullis and L. T. Canham. (1991). *Nature*, 353, 335.
[61] R. Herino, P. Jan and G. Bomchil. (1980). *J. Electrochem. Soc.*, 127, 1339.
[62] S. Dhar and S. Chakrabarti. (1996) *Semicond. Sci. Technol.*, 11, 1231.
[63] Gelloza and N. Koshida. (1998). *Appl. Phys. Lett.,* 73, 2021.
[64] B. Gelloz and T. Nakagawa. (2000). *J. Appl. Phys.,* 88, 4319.
[65] Andsager, J. Hillard, J. M. Hetrick, L. H. AbuHassan, M. Plisch and M. H. Nayfeh. (1993). *J. Appl. Phys.,* 74, 4783.
[66] W. Lang, P. Steiner and F. Kozlowski. (1993). *J. Lumines,* 57, 341.

In: Encyclopedia of Nanotechnology
Editor: Elwood D. Carlson

ISBN: 978-1-60692-079-4

Chapter 8

MOVPE GROWTH OF HIGH DENSITY InAs/InGaAs/InP QUANTUM DOTS FOR MID-INFRARED EMISSION

Tang Xiaohong and Yin Zongyou
Photonics Research Center, School of Electrical and Electronic Engineering
Nanyang Technological University, Singapore 639798

ABSTRACT

Emission in mid-infrared range, 2 – 5μm, has very important applications in variety of areas including range-finding, laser surgery, remote trace-gas sensing, secure free-space communications, etc. Owing to reduction of the dimensionality carriers' confinement, quantum dots (QDs) based optoelectronic devices, e.g. semiconductor lasers, show novel properties when compared with their counterpart quantum well based devices.

In this article, self-assembly InAs quantum dots grown on InGaAs/InP matrix for mid-infrared emission using metal-organic vapor phase epitaxy is introduced. Effects of the growth conditions on the dot density and size uniformity of the quantum dots have been investigated in detail. A two-step growth method for growing high density and uniform quantum dots has been introduced. Improvement in QDs density and size uniformity in the two-step growth has been discussed in details. By changing the indium content of the InxGa1-xAs matrix layer of the InAs/InxGa1-xAs/InP quantum dot structure, emission wavelength of the InAs QDs structure has been extended to the longest >2.35 μm measured at 77K. Full-width at half-maximum (HWHM) of the photoluminescence emission spectrum of the InAs quantum dot structure was measured as narrow as 25.5 meV.

INTRODUCTION

Attempts to use InAs nanostructures based on InP substrates to extended the emission wavelength into mid-infrared (mid-IR) region of 2 –3 μm are attracting much attention, where mid-IR lasers have very important military, biomedical, environmental and industrial applications, including range finding, lidar atmospheric detection, laser surgery, molecular spectroscopy and remote sensing of atmospheric and planetary gases, etc. It has been shown that three-dimensional carrier confinement of quantum dot (QD) structures promise ultra-low threshold current and temperature independent semiconductor lasers with very high differential gain [1,2]. Also, using InP as the substrate, the device fabrication processes are mature and cheaper when compared with that of conventional antimonide substrate based quantum well (QW) mid-IR lasers. Stranski-Krastanov (S-K) self-assembled InAs quantum dots (QDs) and quantum-dash lasers have been demonstrated at various wavelengths from 1.60 to 2.04 μm [3–6]. To extend the emission wavelength of InAs/InP QDs longer than 2 μm, InGaAs has been inserted into the InAs/InP QD structure as the confinement layers of the InAs QDs structures [7-10].

However, InAs QDs self-assembled on an InGaAs/InP matrix generally have large size dispersion. Large size variation of the InAs QDs is due to the indium adatoms' migration [11,12] and the anisotropic surface diffusion [13,14] on top surface of the InGaAs/In matrix during the InAs QDs growth. In order to reduce the size dispersion of InAs QDs formed on an InGaAs/InP matrix, different techniques have been investigated [11,12,15]. But, the photoluminescence (PL) emission spectra from the InAs QD structures are still not narrow enough. Further improvement in the QD size uniformity and dot density is essential in order to make them suitable for mid-IR optoelectronic device applications.

In this research, InAs QDs formed on InP substrate by metal-organic vapor phase epitaxy (MOVPE) for mid-IR emission has been studied. Effects of the growth conditions on formation of the InAs QDs on InGaAs/InP matrix have been investigated in details. To achieve high dot density and narrow size dispersion, a two-step growth method has been developed and investigated. With the optimized growth conditions, the size uniformity and dot density of the InAs QDs formed by using the two-step growth has been improved by 50% and >80%, respectively, when compared to that of the InAs QDs grown on InP by using conventional S-K growth method with the similar MOVPE growth conditions. To extend emission wavelength of the InAs QD structure, indium graded content InxGa1-xAs layers are used as the matrix/barrier layers of the InAs QD structure. The indium content of the InxGa1-xAs matrix/barrier layers is linearly increased from x = 0.53 (lattice matched to InP) to x = 0.65 ~ 0.80. It has been received that, when compared to that of InAs QDs with the lattice matched In0.53Ga0.47As barriers, emission wavelength of the InAs QDs embedded in graded In0.53→(0.53+y)Ga0.47→(0.47-y)As barriers redshifts by 13.5 nm when the indium content gradation, y, of the In0.53→(0.53+y)Ga0.47→(0.47-y)As matrix/barrier layers is increased by every 0.01. The longest emission wavelength of the InAs/InxGa1-xAs/InP QD structure reaches > 2.35 μm when measured at 77K, which is the longest emission wavelength of InAs QD structure so far reported. The full width at half maximum (FWHM) of the PL spectrum of the InAs QD structure reaches 25.5 meV. These results would be promising for development of mid-IR QD optoelectronic devices on InP or GaAs substrates.

EXPERIMENT

All the samples were grown in a horizontal low-pressure MOVPE reactor (Aixtron, AIX200) with gas foil rotation of the susceptor. The metal-organic (MO) sources used in the growths were tertiarybutylarsine (TBA), tertiarybutylphosphine (TBP), trimethylgallium (TMGa) and trimethylindium (TMIn). To improve the safety of MOVPE growth process, high purity (99.999%) nitrogen gas further purified by a MonoTorr phase II Getter column with the dew point below –100oC was used as the carrier gas instead of using explosive hydrogen gas. The reactor pressure was set at 20 mbar for growing the InAs QDs, and it was set at 100 mbar for depositing InP buffer and InxGa1-xAs barrier layers. A 200 nm InP buffer layer was grown on InP (001) substrate with the temperature increased from 600 oC to 630 oC before growing the QD structure. Then, the reactor temperature was lowed down for growing the InAs QDs. As shown in Figure 1, the InAs QDs are grown on a 30 nm un-relaxed InxGa1-xAs/InP matrix. Atomic Force Microscopy (AFM) was used to take the surface morphology of the QDs. For photoluminescence measurement, the InAs QDs are covered by a symmetric InxGa1-xAs upper barrier and then capped with an InP cap layer.

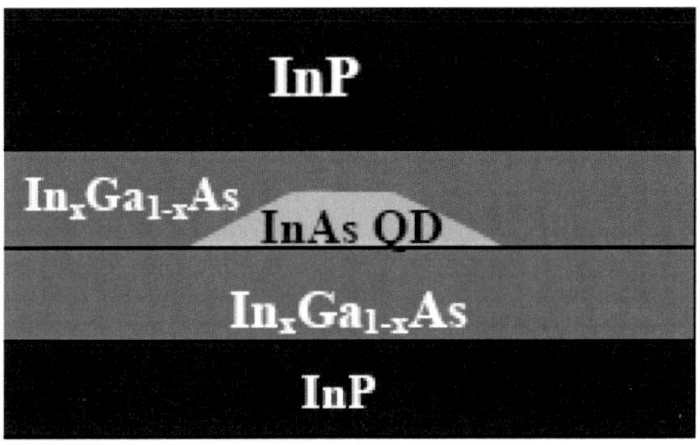

Figure 1. Schematic layer structure of the InAs/InGaAs/InP quantum dot structure.

A. S-K Self-Assembly Growth of InAs/ InxGa1-xAs/InP QDs

In S-K self-assembly growth of InAs QDs, the reactor pressure was set at 20 mbar and the V/III ratio was set at 10. The total gas flow rate in the reactor was Q_{total} = 3.1 slm. To grow the InA QDs, a very thin InAs layer, few monolayers (MLs), was grown as normal; and then the TMIn source was closed for the post-growth interruption before cooling down the system and the InAs QDs were formed self-assembly because of large lattice mismatch. Formation of the QDs is affected by changing the growth conditions including the growth time, post-growth interruption, substrate temperature and source flux, etc, which are discussed in details below.

1) Effect of the Growth Time

To study the effect of the growth time (or layer thickness) of InAs on the formation of InAs QDs, five samples, samples (a) to (e), were grown by changing the InAs layer growth time. The substrate temperature and InAs growth rate was kept at 500 oC and 0.15 ML/sec, respectively. All the other growth conditions were kept same for growing the five samples. Figure 2 (a) to (e) compares the surface morphology of the five samples with different InAs layer growth time. It is observed that when the growth time was 10 seconds or less, no QD was formed. When the InAs layer growth time was increased to 12 seconds, QDs started to be formed as shown in Figure 2(c), but with very low density, ~1×109cm-2. Average dot height, h_g, of the samples was calculated by using the following formula [17].

$$\log_{10} h_g = \frac{\Sigma(n \log_{10} h_n)}{N} \quad \mathbf{(1)}$$

where N is the total number of the dots counted on the sample, n is the number of the QDs with the same dot height of h_n. The average dot diameter was calculated using the similar equation. The calculated average dot height and diameter of sample (c) is 10.0 nm and 35.5 nm, respectively. Based on the dot size and number of the QDs counted, volume of the InAs QDs can be calculated by taking the QDs with pyramid shape. The total volume of InAs QDs grown of the sample was equivalent to an InAs nominal layer. The InAs norminal layer thickness of the InAs QDs is calculated by using the AFM scanned area to divide the calculated InAs QDs volume in the scanned area of the sample. We have taken that 1 ML = 0.3 nm in the calculations.

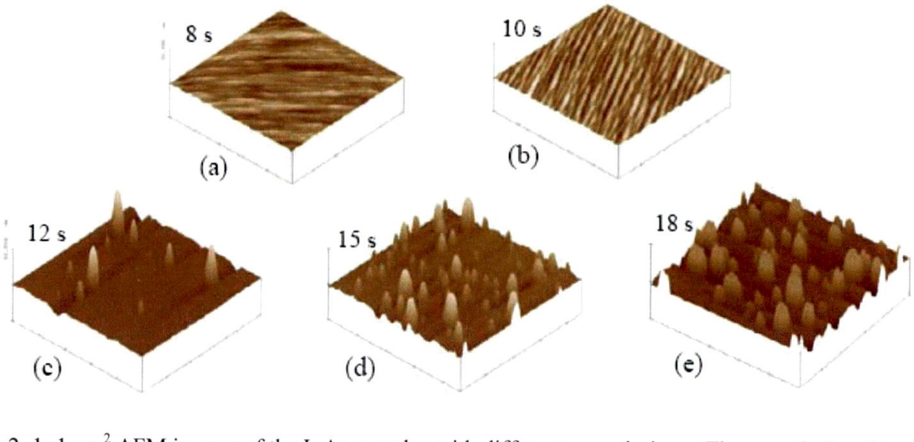

Figure 2. 1×1 μm^2 AFM images of the InAs samples with different growth times. The growth time for sample (a), (b), (c), (d) and (e) was 8 s (1.2 ML), 10 s (1.5 ML), 12 s (1.8 ML), 15 s (2.25 ML) and 18 s (2.7 ML), respectively.

The InAs wetting layer thickness of a sample was determined by deducting the QDs' nominal layer thickness from the total grown InAs layer thickness. Figure 3 shows the evolution of the grown InAs materials consumed for growing the wetting layer and the QDs as a function of the InAs growth time. It shows that if the total deposited InAs layer thickness is less than 1.5 ML, as in the cases of sample (a) and (b) in Figure 2, no QD is formed, only InAs wetting layer was grown. The QDs started to be formed when the total deposited InAs layer was thicker than 1.8 ML, as shown in samples (c), (d) and (e).

For sample (d), the total deposited InAs layer thickness is 2.25 ML, which is increased by 0.45 ML when compared with that of sample (c). But, the QDs' nominal layer thickness of sample (d) is 0.7 ML, which is 0.55 ML thicker than that of sample (c) whose QDs nominal layer thickness is only 0.15 ML. The wetting layer thickness of sample (c) and (d) is calculated of 1.65 ML and 1.55 ML, respectively, as shown in Figure 3.

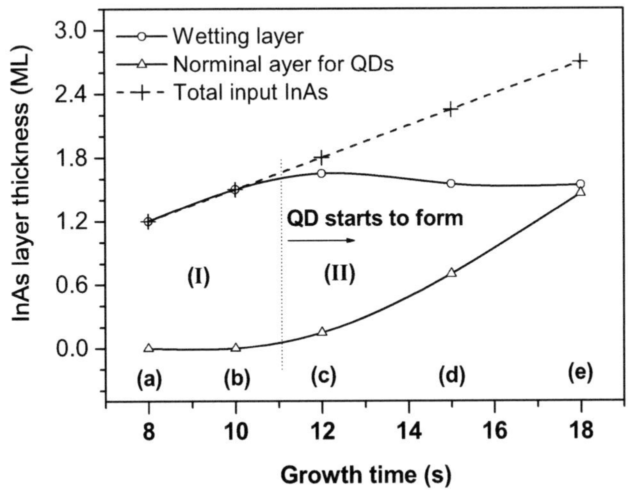

Figure 3. InAs material consumption for the wetting layer growth and the QD formation under different InAs deposition time. The two solid curves show the materials for growing wetting layer and QDs, and the dot line shows the total input InAs materials.

Reduction of the wetting layer thickness, or increase of the QDs nominal layer thickness from sample (c) to (d) shows that some InAs material was decomposed from the wetting layer and transferred to the formation of the QDs. Similar decomposition of wetting layer was observed in the InAs/GaAs QDs growth [17]. However, when the InAs growth time was increased to 18 seconds, the total grown InAs layer was very thick. The InAs islands no longer accommodated the strain, misfit dislocations were formed between the island-substrate interfaces. Incoherent islands were formed and grew into large clusters by consuming the InAs dots nearby, i.e. the ripening [18]. Therefore, the QDs of sample (e) are obviously larger than those of sample (d). An optimized growth time of 15 seconds has been received under these growth conditions.

2) Effect of the Post-Growth Interruption

To study the effects of the post-growth interruption time on the QDs formation, four samples, samples (f), (g), (h) and (i), were grown at 500 oC with the same grown InAs layer thickness of 2.25 ML. All the growth conditions of the four samples were the same except the post-growth interruption time. The post-growth interruption time for the four samples was changed from 0 to 20 seconds. For sample (f) with 0 second post-growth interruption, the TMIn source was closed at the same time when the heater of the reactor was switched off when the growth reached the target layer thickness. While for growing sample (g) to sample (i), after the TMIn source had been witched off to stop the InAs layer growth, the reactor was kept at the growth temperature for 5 seconds, 10 seconds and 20 seconds, respectively, before switched off the heater. Figure 4 shows the AFM measurement results of the four samples. The QDs' size distribution, dot density, dot size standard deviation and surface depth of the four samples are summarized in Table 1. The four samples have similar dot density when the post-growth interruption time was changed from 0 to 20 seconds, but the sample's dot size dispersion is sensitive to the post-growth interruption time.

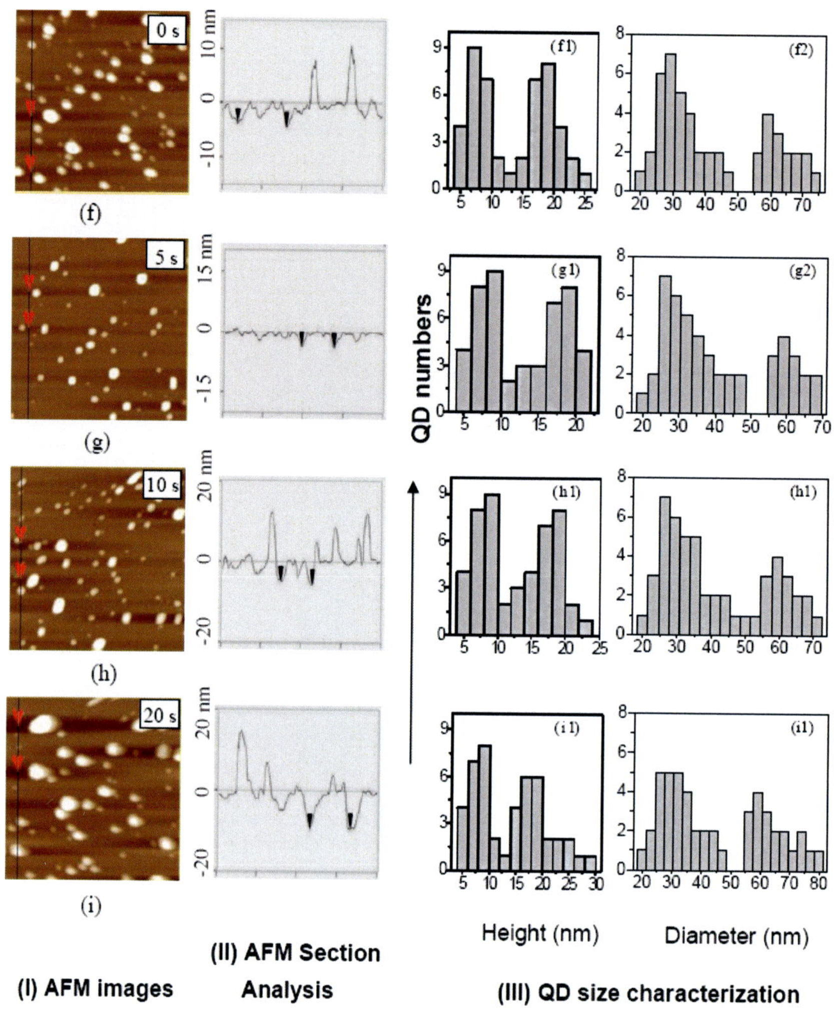

Figure 4. (I) Top-view 1×1 μm^2 AFM images, (II) AFM in-situ Section Analysis of the surface depth along a cross-sectional line across the image, and (III) The InAs QDs size characterization of samples (f), (g), (h) and (i).

Table 1. Comparison of QDs density, QDs' height & diameter standard deviations and wetting layer surface depth for the samples (f), (g), (h) and (i)

Compared parameters	Sample (f)	Sample (g)	Sample (h)	Sample (i)
QD density ($10^{10}/cm^2$)	0.63	0.61	0.62	0.6
QD's height standard deviation (nm)	6.2	5.2	6.0	7.8
QD's diameter standard deviation (nm)	15.2	14.3	15.1	16.9
Surface depth (nm)	3.7	3.5	4.8	8.7

With zero second post-growth interruption, standard deviation of the QDs' height of sample (f) is 6.2 nm. When the interruption time was 5 seconds, standard deviation of sample (g)'s QD height reduces to 5.2 nm. However, standard deviation of the QDs' height increases to 7.8 nm of sample (i) when the post-growth interruption time was increased to 20 seconds. The post-growth interruption time of 5 seconds gives sample (g) the minimum QDs' height standard-deviation, i.e. the narrowest QDs height distribution. Similar trend of the QDs' diameter standard deviation has been received with different post-growth interruption time. The results show that there exists a post-growth interruption time window for forming the uniform QDs. Out of this window, too long or too short of the post-growth interruption will cause large dot size distribution of the QDs formed. This can be explained by the surface kinetics of the QD epitaxy growth. On epitaxy surface, a 3-D nucleation center acts as a low energy site which attracts surrounding adatoms. Adatoms accumulate at the nucleation center to make it grows bigger [19]. For sample (f), the heater was turned off immediately when the InAs reached the target layer thickness, the surface adatoms had no time to diffuse across the surface to be efficiently absorbed by the 3-D nucleation center. Small islands or nuclei had no time to grow bigger. For sample (g), the reactor was maintained at the growth temperature for 5 seconds when the InAs growth was stopped, the small nuclei kept growing by absorbing the surface adatoms during the post-growth interruption. Since smaller dots have larger aspect ratio, they grow faster [20], while the bigger dots grow slower. So that, the QDs finally formed are more uniform in size. However, if the post-growth interruption time is too long, as for sample (i), when the surface adatoms are consumed out, some islands grow further into large clusters to relax the lattice mismatch strain between the QDs and the matrix layer underneath by consuming the material of the wetting layer. This process results in rough surface of the wetting layer underneath and a very broad QDs size distribution of sample (i). Dislocations will be generated in the large clusters formed. They act as non-radiative recombination centers and degrade the QDs' PL performance [21]. Using H2 as carrier gas in MOVPE growth of InAs QDs with similar growth conditions, high-quality and uniform QDs were formed with much longer post-growth interruption time, from 20 seconds to 30 seconds [22,.23]. The shorter post-growth interruption needed when the InAs QDs grown in N2 ambient may be due to the lower thermal conductivity and higher viscosity of nitrogen gas used in this study than that of hydrogen gas [24]. Because of the lower thermal conductivity and higher viscosity of nitrogen gas, the growth surface temperature cools down slower when the reactor heater is turned off and the unstable surface adatoms stay on the surface longer than adatoms in H2 ambient during the interruption before they are desorbed. Shorter interruption is needed for the atoms being finally absorbed by the nucleation center before the surface temperature becomes too low after switch off the heater.

In general, with current growth conditions, QDs formed with different post-growth interruption times in sample (f), (g), (h) and (i) are still with low dot density and broad size dispersion. In order to increase the dot density and improve the QDs size dispersion, the growth temperature, inlet source flux and growth mode are further investigated.

3) Effect of the Growth Temperature

To study the effects of growth temperature on the formation of InAs QDs, six samples, sample (j) to (o), were grown at different temperature in the range from 420 oC to 520 oC. The total InAs layer thickness was still target at 2.25 ML and the post-growth interruption time was set at 5 seconds. Figure 5 shows the top-view AFM images of the six samples grown at different temperatures.

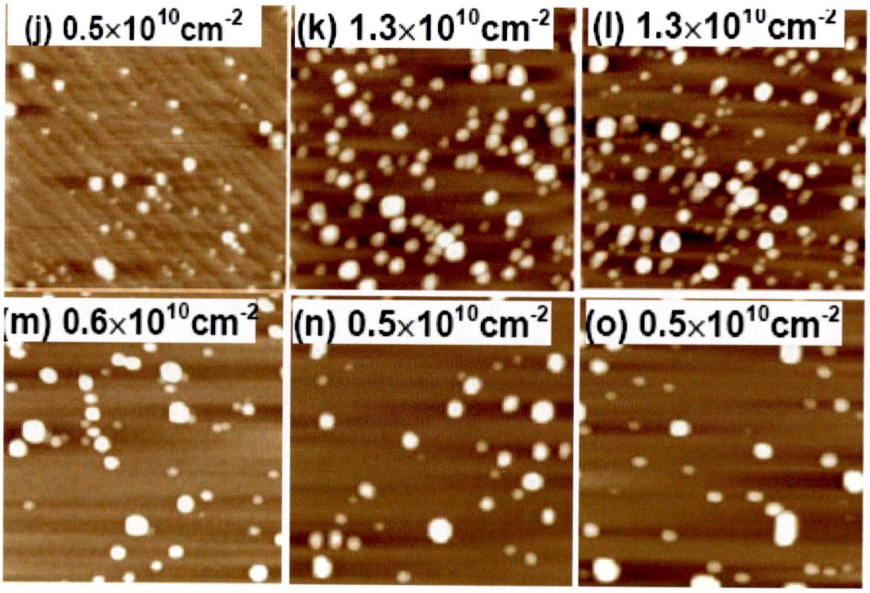

Figure 5. Top-view 1×1□m^2 AFM images of the InAs QD samples grown at different temperatures. Sample (j), (k), (l), (m), (n) and (o) was grown at 420°C, 440°C, 460°C, 480 °C, 500 °C and 520 °C, respectively. The measured dot density of each sample is inserted.

High QD density of around 1.3x1010 cm-2 is obtained when the samples were grown at the temperature range from 440 oC to 460 oC. Out of this temperature range, when the sample was grown at temperature either higher or lower by 20 oC, the samples' QD density reduces dramatically to around 0.5x1010 cm-2. When the growth temperature was above 480 oC, large InAs islands with low density were obtained. This is due to the longer diffusion length of the adatoms at higher growth temperature [25]. During the post-growth interruption, larger islands grew further by consuming small islands during the coarsening process [26]. This can be seen from the AFM images of samples (m), (n) and (o), where some large dots are surrounded with sparse distribution of very small dots. Low density QDs were formed at low growth temperature, ≤ 420 oC. This is because: (i), the critical thickness of the wetting layer for starting the QD formation increases when the growth temperature is lower due to the reduction of adatoms' diffusion length at lower temperature [26]; (ii), the diffusion coefficients of the source adatoms are smaller at lower the growth temperature, which reduces the growth efficiency [27]. Therefore, the materials available for growing the InAs QDs were reduced when the growth temperature was below 420 oC, which resulted in the low density of InAs QDs formed.

The distribution and standard deviation of the QDs' diameter and height versus the growth temperature are plotted in Figure 6. Figures 6(i) and (ii) show bimodal distributions of the QDs' diameter and height when the growth substrate temperature was increased. Sample (k), which was grown at 440 oC, has the smallest standard deviation of QDs' diameter and height, as shown in Figure 6(iii). The results show that more uniform QDs with high dot density have obtained when the sample was grown at 440 oC with the TMIn flux set at 30 sccm. But, the growth temperature of 440 oC is too low for growing the InP cap layer of the InAs QD structures. In the following investigations, the growth temperature was set at 500 oC.

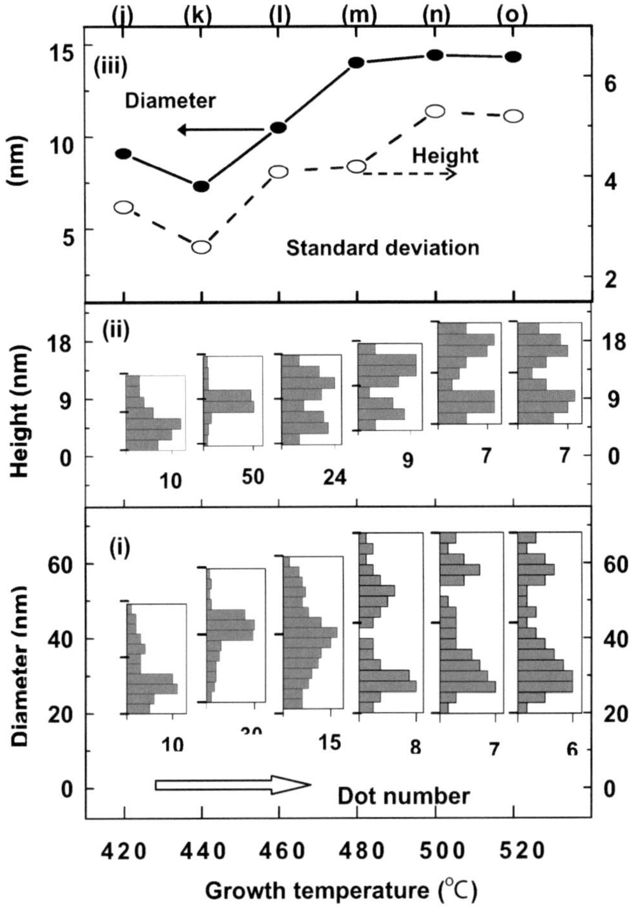

Figure 6. InAs QDs' size distribution and standard deviation of samples (j), (k), (l), (m), (n) and (o) which were grown at different temperatures: (i) Histograms of QDs' diameter distribution, (ii) histograms of QDs' height distribution and (iii) standard deviation of QDs' diameter and height as a function of deposition temperature.

4) Effect of the Source Flux

In MOVPE growths of QD structures, the molar ratio of the total input group-V sources versus the group-III sources, or V/III ratio, is generally larger than 1. The growth is with rich group-V source elements ambient. In this study, the V/III ratio was set at 10. The effects of input source flux on the InAs QDs formation were investigated by changing the input group-III (TMIn) and group V (TBA) source fluxes while the V/III ratio was kept same during the InAs QDs growth. Samples (p) and (q) were grown by setting the input TMIn flux at 30 sccm and 75 sccm, respectively. The substrate temperature was set at 500 oC and the post-growth interruption was kept 5 seconds for growing the two samples. To keep the same deposited InAs layer thickness, the InAs growth time was changed accordingly when the input TMIn source flux was changed. As shown in Table 2, higher density InAs QDs are formed in sample (q) than that of sample (p). This is because at temperature of 500 oC the surface reaction dominates the epitaxy growth, the growth rate is in mass flow control. The InAs growth rate is proportional to the inlet TMIn flux. When the TMIn flow flux was increased in growing sample (q), the growth rate was higher and more nuclei were formed. QD density of sample (q) is higher.

When the substrate temperature was lowered to 440 oC, as for growing sample (r) and (s), the epitaxy growth was in surface reaction limited growth. The InAs growth rate did not increase accordingly when the TMIn input flux was increased. The InAs layer deposited in sample (s) was too thin to form the QDs when the growth time was reduced to 6 seconds.

Table 2. Top-view 1×1□m2 AFM images of InAs QDs grown with different TMIn input flux & inflow time at 440 oC and 500 oC, respectively

		Growth methods	
		Slow growth	Fast growth
TMIn flux & inflow time		30 sccm &	75 sccm &
Growth temperature (°C)	500	(p) $0.5\times10^{10}cm^{-2}$ 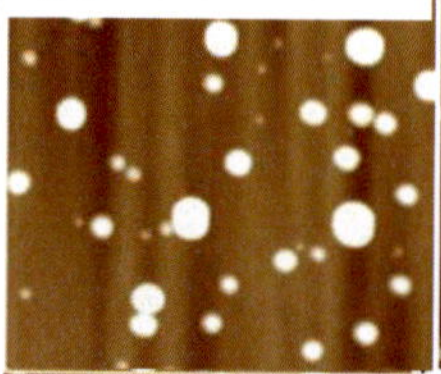	(q) $1.3\times10^{10}cm^{-2}$ 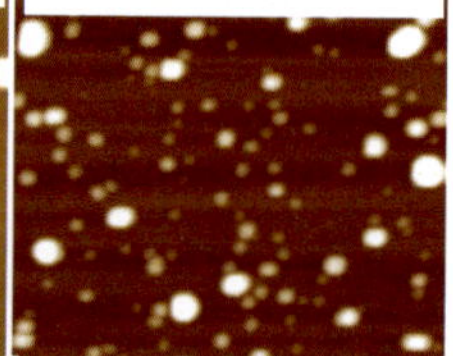
	440	(r) $1.3\times10^{10}cm^{-2}$ 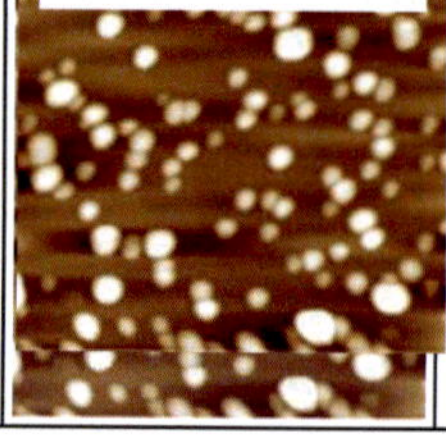	(s) $0cm^{-2}$ 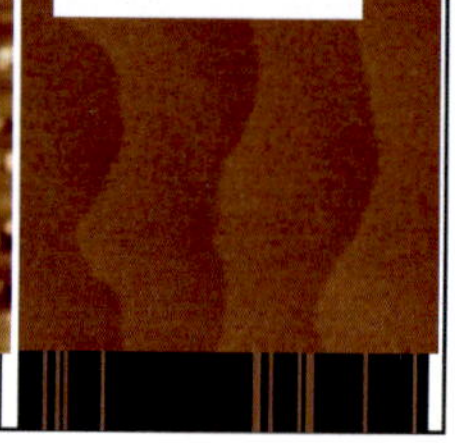

B. A New Two-Step Growth of QDs

To improve the dot density and uniformity, we proposed a two-step growth method for growing the InAs QDs [28]. In this QD growth method, InAs QD growth is divided into two steps. Figure 7 compares the inlet TMIn flux flow in conventional S-K growth (A) and in the two-step growth (B).

In step 1 growth of two-step growth QDs, a thin InAs layer is deposited with high TMIn source flux, and then the growth is paused for an internal interruption by closing the TMIn source. InAs nuclei are formed during the internal interruption. And then, the TMIn source is opened again for the step 2 growth. In the step 2 growth, the InAs layer is grown with lower growth rate. When the target layer thickness is reached, the TMIn source is closed for post-growth interruption and the InAs QDs are finally formed self-assembly. For comparison in this study, sample (t) and (u) was grown by the two-step growth method at 500 oC and 440 oC, respectively. The measured AFM images of the two samples are shown in Table 3. For comparison, AFM images of samples (q) and (s) which were grown by the conventional S-K self-assembly mentioned above are also shown in Table 3. Dot density of sample (u) is 0.2×1010 cm-2 which is higher than that of the sample (s) which was grown at the same temperature but by using the conventional S-K growth method. Increased the growth temperature to 500 oC, the QDs density of sample (t) which was grown by two-step growth

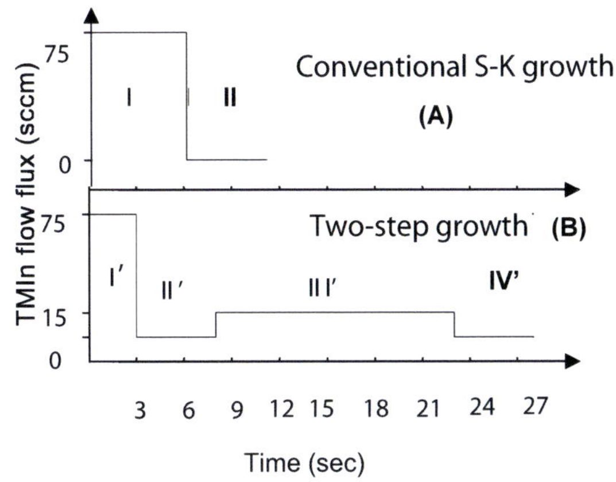

Figure 7. Schematic diagram of TMIn source flow settings during the InAs QDs growth. In conventional S-K growth (A), I: InAs deposition; II: post-growth interruption; in two-step growth (B): I′: step-1 InAs deposition; II′: internal interruption; III′: step-2 InAs growth; IV′: post-growth interruption.

Table 3. Top-view 1×1μm2 AFM images of InAs QDs grown with fast growth and two-step growth at 440 oC and 500 oC, respectively

		Growth methods	
		Fast growth	Two-step growth
TMIn flux & inflow time		75 sccm & 6 sec	75+15 sccm & 3+15 sec
Growth temperature (°C)	500	(q) $1.3\times10^{10}cm^{-2}$ 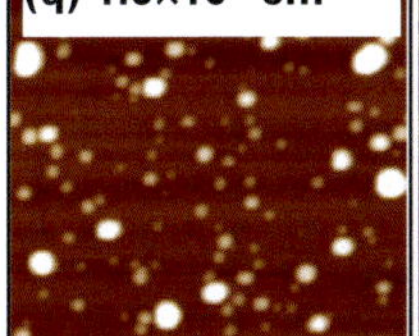	(t) $1.7\times10^{10}cm^{-2}$ 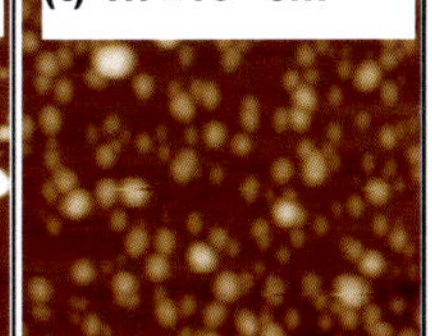
	440	(s) $0cm^{-2}$ 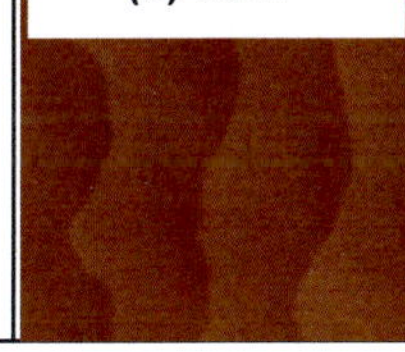	(u) $0.2\times10^{10}cm^{-2}$ 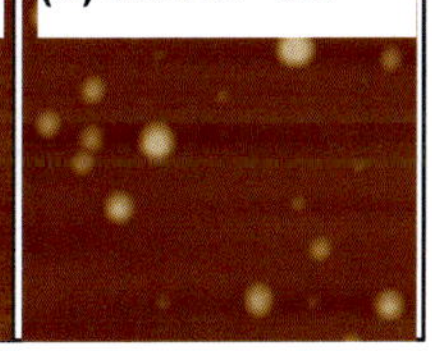

method is about 30% higher than that of sample (q) which was grown at same temperature but by using the conventional S-K mode. The QDs' diameter and height distributions of samples (q) and (t) are shown in Figure 8. The dot size dispersion of sample (t) is much narrower than that of sample (q). The standard deviation of the QDs' diameter and height of sample (q) is 12.9 nm and 7.6 nm, respectively, while that of sample (t) is 4.2 nm and 2.3 nm, respectively. These results show that the QDs' size uniformity grown by the two-step growth method has been improved.

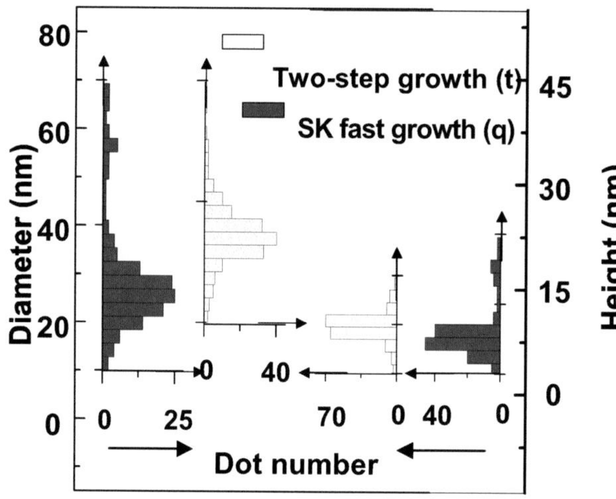

Figure 8. Histograms of the QDs' diameter (left) and height (right) of the samples grown by conventional S-K fast growth, sample (q), and by two-step growth, sample (t). The sampele were grown at 500 °C.

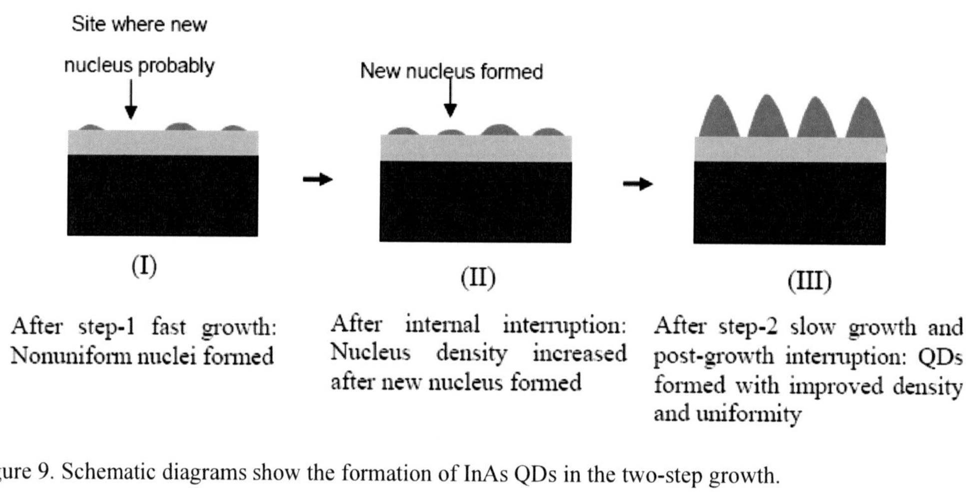

Figure 9. Schematic diagrams show the formation of InAs QDs in the two-step growth.

A schematic diagram of Figure 9 shows the QDs formation in the two-step growth: un-uniform nuclei are formed in the first step fast growth; during the internal interruption, these nuclei continue grow by absorbing the surface adatoms. At the same time, some of the adatoms self-nucleate between the formed dots, more QDs are formed in this way, which increases the QD density. All the formed dots and nuclei continue growing during the step-2 growth. In step-2 growth, the TMIn source is set at low flux so that the dots grow slowly into

QDs. At the same time, more new nuclei can be formed and grow also in step-2 growth and they all grow into QDs during the post-growth interruption. The higher edge energy barrier of the larger dots prevents the adatoms approaching the large dots and making them bigger[29,30]. As a result, the dot density is increased much by using the two-step growth and because the larger islands/dots grow much slower than the small islands/dots in setp 2 growth [30], the size uniformity of the QDs is also improved in two-step growth. To optimize the growth conditions of the two-step growth, effects of growth parameters of two-step growth on the QDs formations have been investigated.

1) Effect of the Internal Interruption Time

Different from conventional S-K self assembly growth of QDs, in two-step growth of QDs, an internal interruption is introduced between the two step growths. The internal interruption time is important in forming uniform QDs in the two-step growth. To study the effect of the internal interruption time on the QD formations in two-step growth, four samples, sample (v) to (y), were grown at 500 oC and the total InAs grown material was still kept at it's nominal layer thickness of 2.25 ML. The TMIn source flow flux and the growth time for each step of the MOVPE growths of the four samples are listed in Table 4. Figure 10 shows the 1×1 μm2 AFM surface images of the four InAs QDs samples grown by the tow-step growth under different internal interruption times. For comparison, a reference sample was grown using conventional S-K one-step growth under the similar growth conditions. Figure 10 shows that the dot density of sample (v), (w), (x) and (y) is 1.5×1010 cm-2, 1.7×1010cm-2, 1.7×1010cm-2 and 0.7×1010cm-2, respectively. It has been reported that in conventional S-K growth of QDs, higher QD density can be achieved if the InAs is grown with a faster growth rate [25,31].

Table 4. The TMIn source flow flux settings and the growth time for each growth step in MOCVD growth of the InAs QD samples

Compared samples	Two-step growth			
	Step 1	**Internal interruption**	Step 2	Post-growth interruption
(v)	**0.5**	**s**		
(w)	25 sccm	**s**	15 sccm	5 s
(x)	**5** & 3 s	**s**	& 15 s	
(y)		**60 s**		

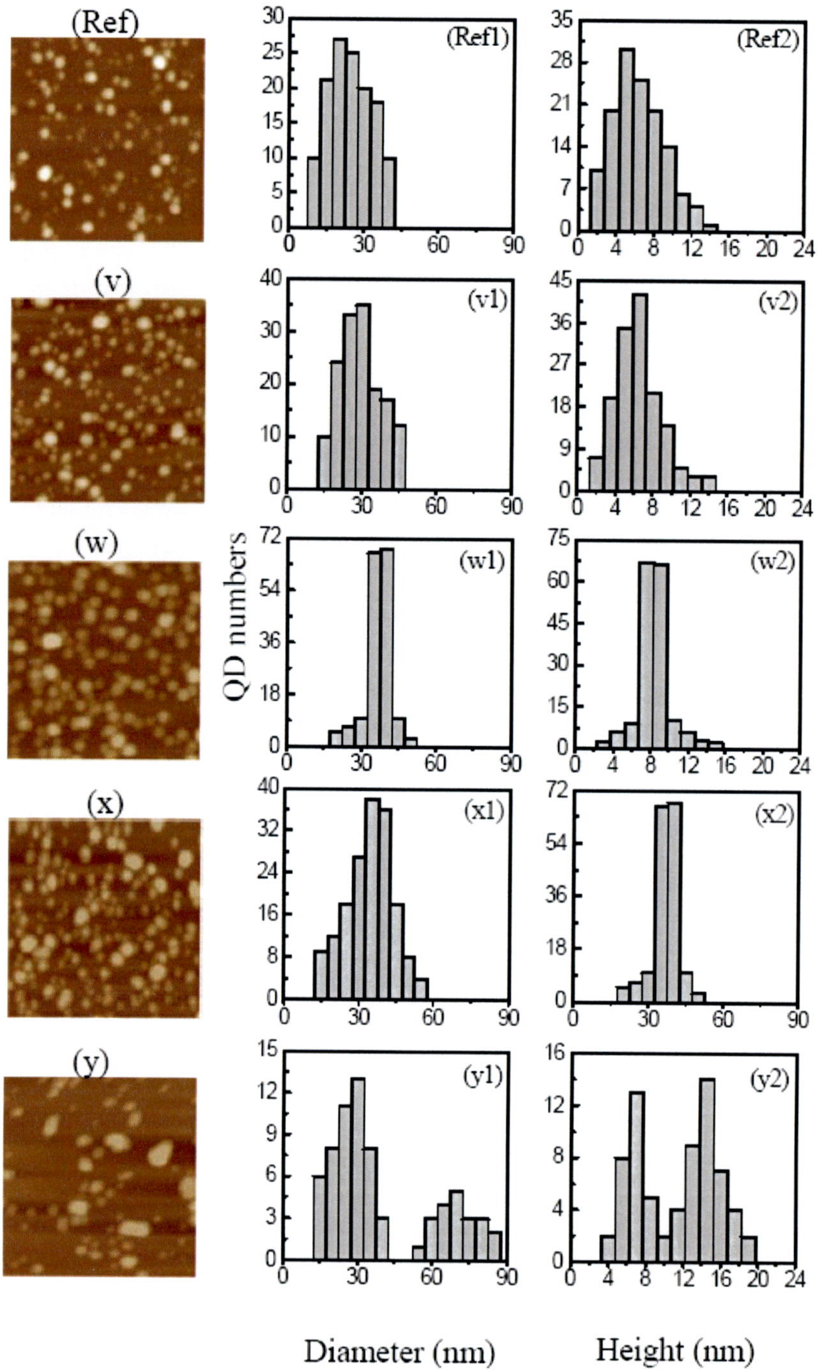

Figure 10. 1×1 μm2 AFM surface images of the Ref. InAs QDs sample which was grown by conventional S-K fast growth and the samples (v) to (y) which were grown by two-step growth with different internal interruption time (t_{int}). The dot diameter and height histograms associated to each sample are shown on right.

In this study, the Ref. sample was grown with the fast InAs growth rate and the dot density received is measured of 1.3×1010cm-2. However, for samples (v) and (x) ,which were grown by using the two-step growth method, they have even higher dot density than that of the Ref. sample, although the TMIn flow flux in step two growth of the two-step growth was only set at 15 sccm which was much smaller than that set from growing the Ref sample, where the TMIn was set at 75 sccm. As shown in the histograms in Figure 10, the average diameter of the InAs QDs grown by two-step growth is ~ 36 nm, while for the Ref sample, the average QDs diameter is ~ 25 nm. But, the QDs' diameter dispersion of sample (w) grown by using the two-step method with the internal interruption time *t*int = 2 seconds is much narrower than that of the QDs grown by the conventional S-K method. More uniform QDs are grown using the two-step growth. As shown in Figure 10, the internal interruption time, *t*int, strongly affects the size uniformity and dot density in two-step growth of QDs. When *t*int is too short, as for sample (v), there is not enough time to form the new nuclei during the internal interruption. In this case, the growth is similar to that of conventional S-K growth. If *t*int is too long, as in the case of sample (y), very large dots/islands or clusters are formed during the internal interruption by consuming the materials from the wetting layer or adjacent smaller islands/dots. These large dots/islands are ripened further by merging neighbor small dots/islands in the step 2 growth, resulting in very large dot diameter dispersion and very low dot density [32].

2) Effect of the Source Flux in Step-2 Growth

To investigate the effects of the InAs growth rate in step 2 growth on the formation of QDs in the two-step growth, four samples, samples (w)–(w‴), were grown with different TMIn source flow fluxes in step 2 growth of two-step growth of InAs QDs, while the other conditions were kept same. As shown in Table 5, the TMIn flux in step 2 growth of InAs was set at 15, 30, 56.3, 112.5 sccm and the InAs was grown for 15, 7.5, 4 and 2 seconds, respectively, for growing these four samples to keep the InAs thickness in step 2 growth the same. Figure 11 shows the 1×1μm2 AFM surface images of the four samples. When the TMIn flux in step 2 was increased from 15 sccm (sample (w)) to 112.5 sccm (sample (w‴)), the dot density increased by 47%, reaches 2.5x1010cm-2 of sample (w'''. The average dot diameter increases from 36 to 43 nm, but the average height of the QDs decreased from 8.4 nm for sample (w) to 5.6 nm for sample (w‴). The dot diameter dispersion was also observed to be increased when the growth rate in the step 2 growth was increased.

Table 5 The TMIn source flow flux settings and the growth time for each growth step in MOVPE growth of the InAs QD samples

Compared samples	Two-step growth			
	Step 1	Internal interruption	**Step 2**	Post-growth interruption
(w)	75 sccm & 3 s	2 s	**15 sccm & 15 s**	5 s
(w')			**30 sccm & 7.5 s**	
(w")			**56.3 sccm & 4 s**	
(w''')			**112.5 sccm & 2 s**	

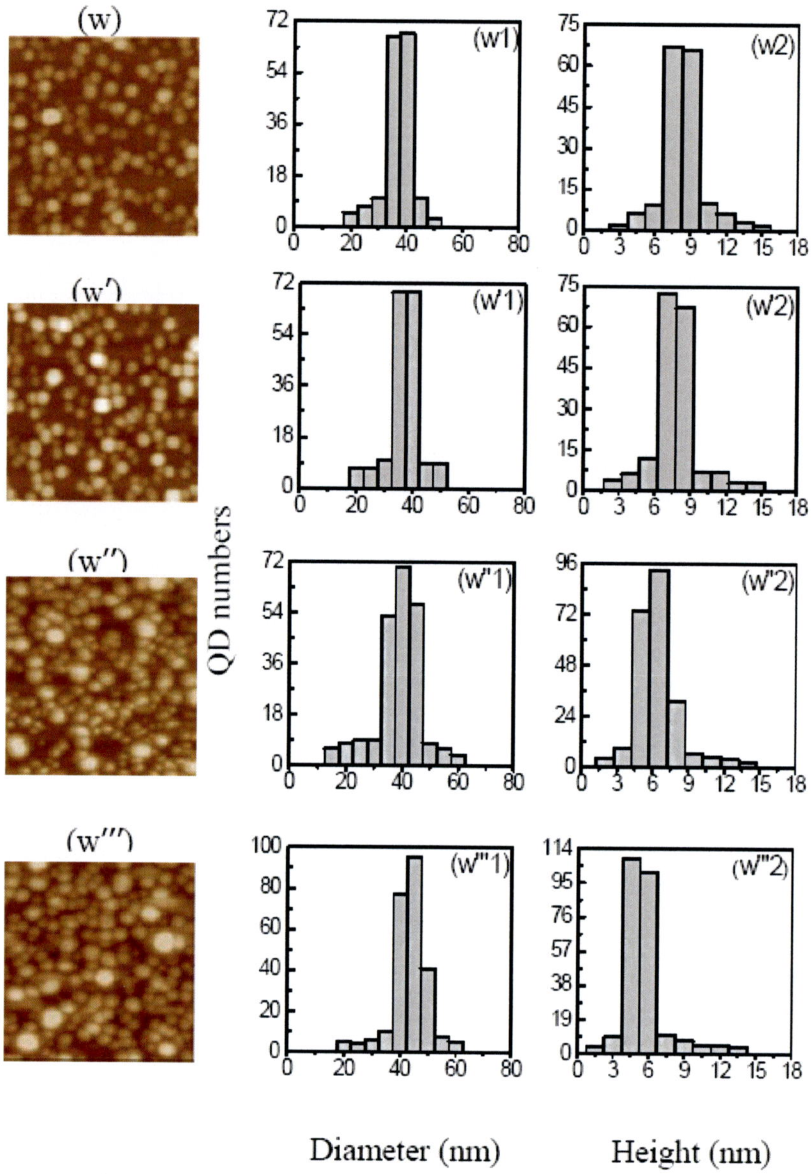

Figure 11. 1×1 μm2 AFM surface images of InAs QD samples grown by two-step growth with different TMIn source flux in step 2 growth. The total InAs layer thickness of all the sample are kept same. The dot diameter and height histograms associated to each image are also shown.

With the faster growth rate, the growth time for the step 2 growth of InAs was shorter; the dots formed in the step 2 growth do not have enough time to release all their strain energy by increasing the aspect ratio. Thus, QDs formed under such conditions are large in diameter but with low height [20].

So far the average diameter and height of the InAs QDs formed is around 40 nm and 7 nm, respectively, which is much larger than that of the InAs QDs grown by MOVPE using TBA and H2 carrier gas. It has been reported that the average dot diameter and height of InAs QDs grown by MOVPE in H2 ambient was around 20 nm and 5 nm, respectively, [23] for which the deposited InAs material thickness was of 2.6 MLs which is even thicker than the

InAs layer deposited of the samples in this study. The reason for larger dots formed in this study grown in N2 ambient than those formed in H2 ambient may be because of the lower thermal conductivity and larger viscosity of N2 as mentioned previously [22]. The adatoms stayed around the formed islands for longer time in N2 ambient, which increases the probability for them to be absorbed by the islands before they are desorbed or brought away to the exhaust. As a result, total efficiency of the material consumed in forming QDs in N2 ambient was increased and hence the larger QDs were formed.

3) Effects of Step-1 Nucleation in Two-Step Growth of QDs

In two-step growth of QDs, 3D island nucleation starts in step-1 growth and then followed by kinetic self-limiting growth of uniform QDs in step-2 atomic layer epitaxy (ALE) growth. Optimization of step-2 growth in the two-step growth of QDs has been investigated above. However, the first step nucleation has important impact in the morphology and crystal-quality of the QDs finally formed in the two-step growths.

In this section, effects of nucleation in the step-1 growth on the morphology and crystal quality of the InAs/InGaAs/InP QDs formed by the two-step growths have been investigated in detail. In the two-step growths, the optimized ALE growth conditions were used in the step 2 growth for all the samples. The InAs QDs are grown on a strained In0.53-0.72Ga0.47-0.28As/InP matrix. The detailed growth conditions for InAs QDs layer growths were listed in Table 6.

Table 6. Comparison of growth rate (samples (e)-(h)), growth temperature (samples (i)-(m)) and V/III flux ratio (samples (n)-(u)) for growing InAs QDs on In0.53-0.72Ga0.47-.28As/InP matrix. The other common growth conditions are also listed

Two-step growth / QDs samples	Step 1	Internal interruption	Step 2 (ALE)	Post-growth interruption
(e) – (h)	**0.42 1.68 ML/sec** (550 °C & V/III = 6.5)	2 sec	4-cycle monolayer growth:	5 sec
(i) - (m)	**Temp: 460 600 °C** (0.84 ML/sec & V/III = 6.5)			
(n) – (u)	**V/III ratio = 0 11** (0.84 ML/sec & 550 °C)			

(i) Effect of the Growth Rate

Four samples, sample (e), (f), (g) and (h) were grown by setting the growth rate in step-1 at 0.42, 0.84, 1.26 and 1.68 ML/sec, respectively. The other growth conditions for the four samples were kept the same. Top-view AFM images and the corresponding low temperature PL spectra of these samples are shown in Figure 12. The QDs' size standard deviation, dot density, PL peak intensity and linewidth are summarized in Figure 13.

The results show that when the growth rate is around 0.84 ML/sec in step-1 growth, sample (f) has the highest dot density, strongest PL emission intensity and narrowest PL spectrum, which indicates the narrowest dot size dispersion and high crystal-quality of the QDs of sample (f). The kinetic theory of stress driven QD growth shows that the 3D island surface density rises with increasing the growth rate, *V*, approximately as *V*3/2.[33].

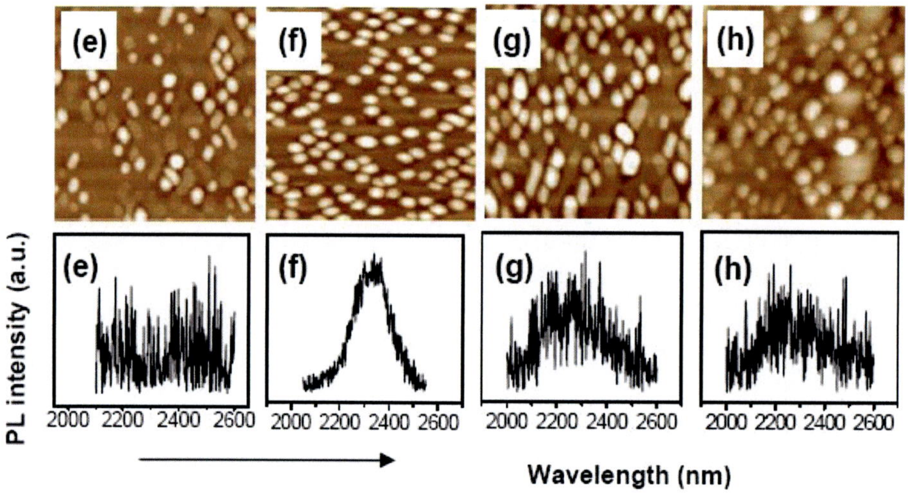

Figure 12. AFM (1×1μm2) top-view images and 20K-PL spectra for samples (e) - (h) with different growth rates in step-1 growth.

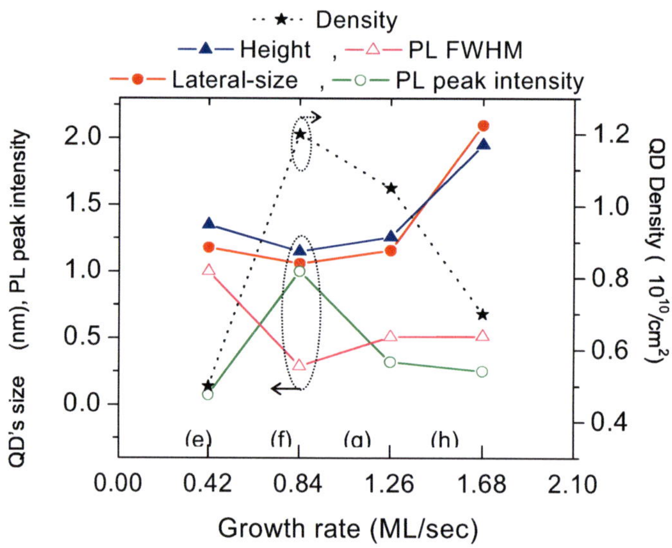

Figure 13. Comparison of the QDs density, dot size standard deviation (σ), PL peak intensity and linewidth of samples (e) – (h) grown with different growth rate.

In the two-step growth, QDs growth is divided into two steps. In step-1 growth, high-density nuclei are formed. Then the formed nuclei further develop into the QDs in step-2 ALE growth. As predicted in Ref. [33], the characteristic time of the nucleation during the QD formation is much shorter than the time for the nuclei being developed into the QDs. In two-step growth of QDs, if the growth rate, thus the nucleation rate, in step-1 growth is too fast, ≥ 1.26 ML/sec, a large amount of strain energy must be dissipated with the insufficient time for single-crystalline material formation. This causes the formation of non-crystalline defects in the QDs. These defects are with lower energy barrier and grow faster than the crystallographically-selective slow-growing single-crystals [22]. These defects will become the seeds for forming the incoherent islands or even the large clusters. As a result, the crystal quality, dot density and size uniformity of the QDs will be affected if the growth rate is too high in step-1 growth. This is shown from the measured results of samples (e), (f), (g) and (h) in Figure 12 and Figure 13. When the growth rate in step1 growth was increased properly from 0.42 ML/s to 0.84 ML/s, the dot density, size uniformity and PL emission of the sample has been improved much. But when the growth rate is increased to 1.26 ML/s and above with less than 2.7-second duration in step-1 growth, the dot density decreases due to the small QDs being consumed for the ripening growth of the incoherent islands growing into the large clusters. This obviously degrades the dot size-uniformity and PL emission intensity of the samples.

(ii) Effect of the Growth Temperature

To investigate the effects of the substrate temperature on the InAs/InGaAs/InP QDs formation, five samples, samples (i), (j), (k), (l) (= (f)) and (m), were grown at the temperature of 460, 500, 520, 550 and 600 oC, respectively, while the growth rate in step 1 was set at the optimized growth rate of 0.84 ML/s and all other growth conditions were kept the same for growing the five samples. Figure 14 shows the AFM images and the low temperature PL spectra of the five samples. The dot size standard deviation, dot density, PL peak intensity and PL spectrum linewidth of the samples are summarized in Figure 15. It has been observed: (I) when grown at low temperature, 460 oC, bi-modal sized QDs were formed in sample (i); when the temperature was increased to 550 oC, the dot size dispersion becomes narrower, but it becomes broader again when the temperature was increased to 600 oC; (II) The highest dot density of sample (k) was grown at 520 oC; (III) Sample (l) with the best PL performance was grown at 550 oC.

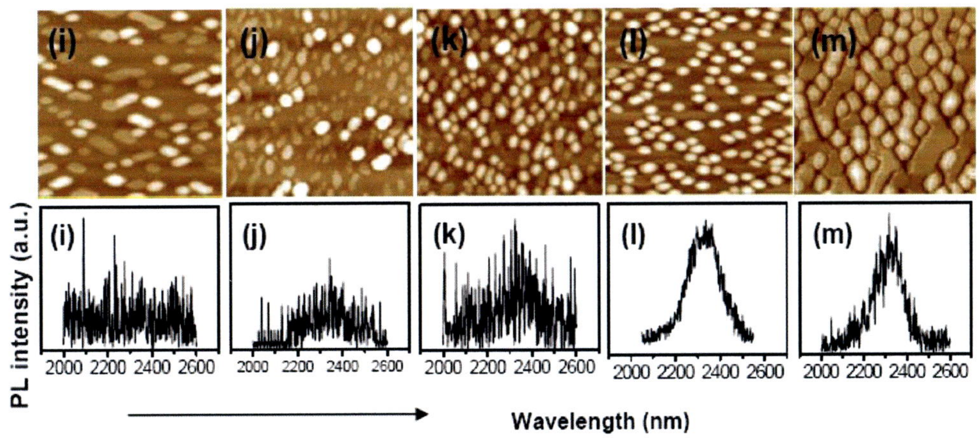

Figure 14. 1×1μm2 AFM top-view images and 20K-PL spectra for samples (i)–(m) grown at different temperatures.

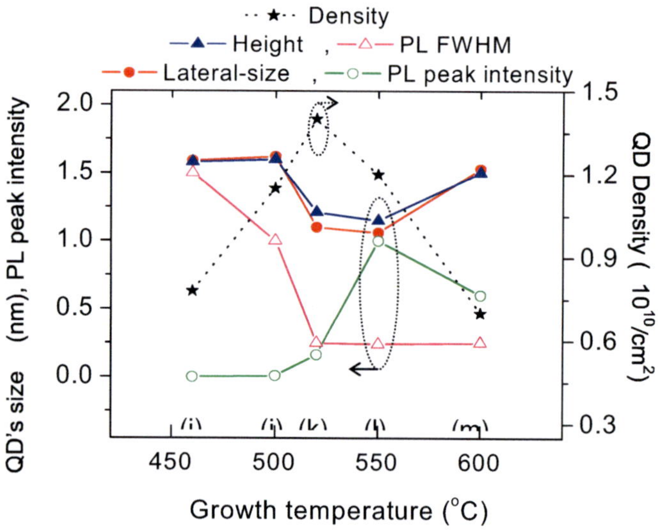

Figure 15. The samples' QD density, dot-size standard deviation (σ), PL peak intensity and linewidth changes with the growth tempertures.

At low growth temperature, ≤ 500 oC, the growth rate is low, there was not enough time to grow enough material for forming the QD nuclei in the step-1 growth, and only wetting layer was grown. The QDs were mainly grown in step-2 ALE growth. Therefore, the AFM image of the QDs of sample (i) which was grown at 460 oC looks similar to that of the QDs grown using normal ALE growth mode. When the temperature was increased to 520 oC, nuclei formed in step-1 were all developed into the QDs during step-2 growth. So, high density QDs of $1.4\times10^{10}/cm^2$ have been received in sample (k). However, the dot density of the samples decreases when the growth temperature was increased above 520 oC and the dot size becomes bigger, which is caused by the increased diffusion length of adatoms at higher temperature [34]. The better PL performance of sample (l) which was grown at 550 oC, rather than that of sample (k) with higher dot density, is because crystal quality of the formed QDs is better when grown at higher temperature. Material grown at higher temperature normally has better crystal quality due to the reduction of the inner grown defects, such as non-/poly-crystalline structures. Non-crystallines are generated at the QDs/matrix interface due to the large lattice mismatch strain between the InAs dots and the InGaAs/InP matrix. At higher growth temperature these non-crystalline structures are broken, and hence the crystal quality of the QDs has been improved. But, if the growth temperature is too high, ≥ 600 oC, as in the case of sample (m), the sample's dot-density is reduced because the larger size dots are formed at higher temperature. So that, sample (m)'s PL intensity is lower than that from sample (l). The results show that an optimized growth temperature of 550 oC which gives high dot density, uniform dot size and best PL performance of the QDs has been received under the current conditions.

(iii) Effect of the V/III Flux Ratio

Above investigations show that the dot density of the samples under optimized growth conditions received is around 1.2×1010/cm2. It can be calculated that the corresponding QDs' surface coverage of the sample is only ~23%. In order to enhance the sample's PL performance further, it is very important to grow the sample with higher dot density. It has shown that the dot density does not change much by optimizing the growth rate and substrate temperature during the growth. In this investigation, we have tried to increase the dot density by optimizing the inlet source flux of group-V source TBA flux in step-1 growth.

In this study, the TMIn group-III source flux and all the other growth conditions were kept unchanged for growing the samples. Only group-V source flux of TBA for step-1 growth was changed from 0 to 4.52×10-4 mol/min, with the V/III ratio correspondingly changed from 0 to 11. Samples (n), (o), (p), (q), (r), (s), (t) and (u) were grown by using the two-step growth with the input source flux V/III ratio in step-1 growth set at 0, 0.2, 0.5, 1.1, 2, 4, 6.5 and 11, respectively, by adjusting the TBA flux. Figure 16 shows the AFM top surface images of these samples and their low-temperature PL spectra. The measured dot density of the samples grown with different input V/III ratio in step 1 growth is summarized Figure 17.

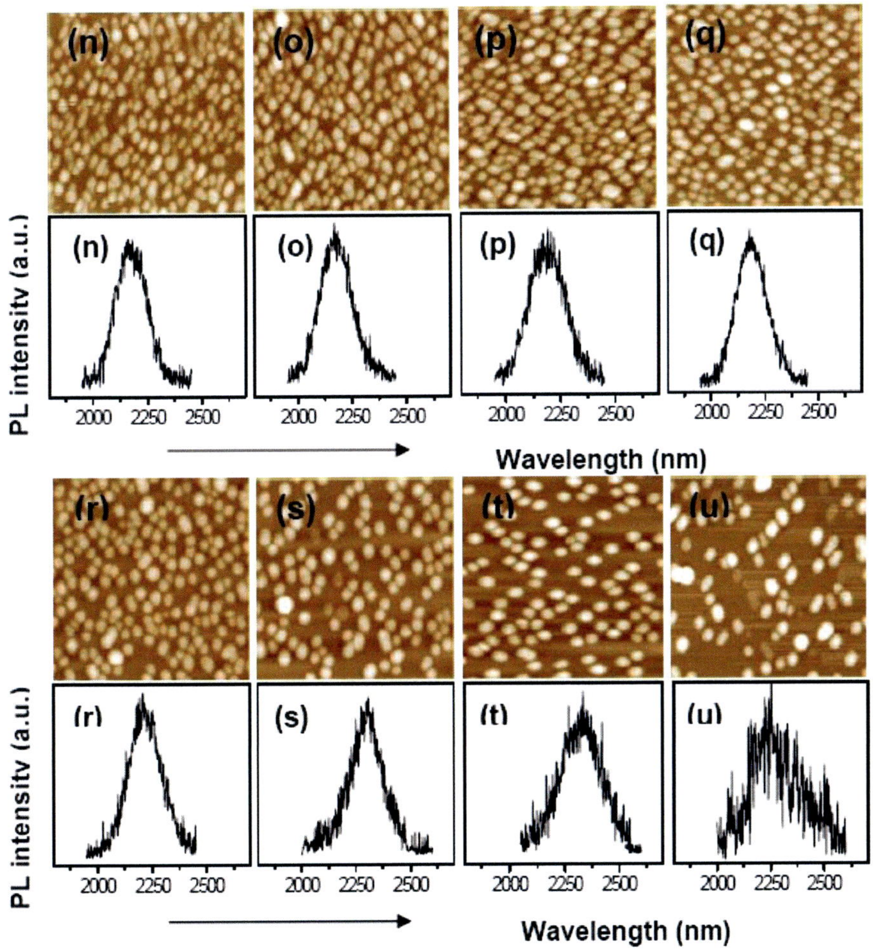

Figure 16. 1×1μm2 AFM top-view images and low temperature PL spectra of the two-step grown InAs QD samples (n), (o), (p), (q), (r), (s), (t) and (u) with different V/III source flux ratio in step-1 growth.

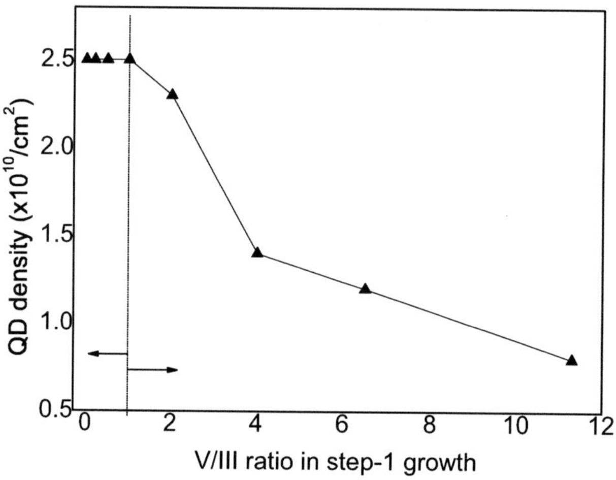

Figure 17. Dot density of the sample changes with the V/III ratio used in step-1 growth of samples (n), (o), (p), (q), (r), (s), (t) and (u).

Very high dot density, ~2.5×1010 cm-2, has been achieved with very low V/III ratio during the first step growth and it slightly changes when the V/III ratio changed from 0 to 1. For the V/III ratio is >1, the formed QD density drops sharply when the V/III ratio was increased. This observation is different from what has been reported where the dot density increased with the V/III source flow ratio [35, 36]. The drop of the dot density in the range of V/III >1 can be explained as the surface-diffusion length of indium adatom is increased when the group V atom concentration on the epitaxy growth surface is reduced by lowering the V/III ratio [37, 38]. With the longer surface diffucion length, the effective coverage of the indium adatoms on the growth surface is enlarged. The formed nuclei density is thus increased with the reduction of V/III ratio during the step-1 growth in two-step growth of InAs QDs, which results in higher dot density. Besides, it has also been shown that a thinner critical thickness is obtained when the group V source, TBAs, inlet flux is decreased in S-K growth of QDs [39]. Or in other wards, the low group V source flow can accelerate the onset of the 3D islands formation in S-K QDs growths, thus resulting in higher density nuclei/QDs with lower inlet V/III ratio.

When the input V/III ratio during the step 1 growth is very low, V/III<1, input indium adatoms are more than the arsenic adatoms on the reaction surface. The extra indium adatoms act as surface surfactant to enforce roughening the growth surface [40]. A roughened growth surface has much lower energy sites for 3D growth and at the same time reduces the critical thickness for the onset of 2D to 3D growth transformation [41]. As a result, the formed QDs' dot-density in the two-step growth becomes higher when the input V/III ratio in step-1 growth reduces to less than 1. It is calculated that the surface coverage of the QDs of sample (q) with setting the V/III input flow ratio in step-1 growth at V/III = 1.1 is ~ 60 %, which is very high. There are not many rooms to further increase the dot density of the sample by further reducing the input flow V/III ratio in step-1 growth. The dot-density of samples (n), (o) and (p) did not increase further when the V/III ratio in step-1 growth was reduced from 1 to zero.

Optical and crystal properties of the InAs QDs samples grown with different input V/III ratio in step-1 growths has been studied using PL measurement. The measured PL spectra of the samples are also shown in Figure 16. Figure 18 plots the PL peak position, peak emission intensity, integrated emission intensity and PL linewidth of the samples grown with different input V/III ratio in step-1 growth.

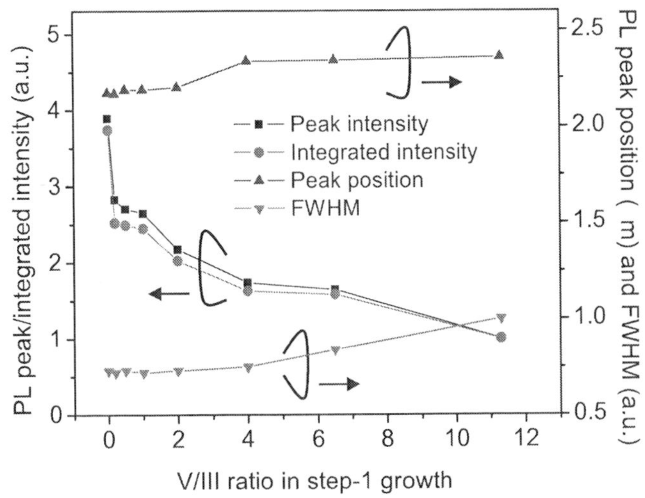

Figure 18. PL peak position, peak intensity, integrated emission intensity and FWHM of the QDs samples as a function of the input V/III ratio in step 1 growth.

It is observed that the PL peak position of the QDs samples slightly blueshifts when the input flow V/III ratio in step-1 growth was reduced in the twp-step growth of QDs. The blue shift of the PL emission is because the dot's height grown with lower input V/III ratio is lower. The peak emission intensity and the integrated emission intensity of the samples are also obersved increase with the input flow V/III ratio in step-1 growth decreasing. These results show that the crystal quality of the QDs grown is not degraded whem the input V/III ratio in step-1 growth was reduced in the two-step growth of InAs QDs. The large increase in the PL emission intensity of the sample grown with the input V/III ratio <1 in step-1 growth is mainly because of the much higher dot density of the sample received. The FWHM of the PL spectra of the samples grown with the input V/III ratio ≤ 2 are similar, which shows the uniformity of dot size and crystal quality of the samples are similar, even though the input V/III ratio was reduced. It is noticeable that for sample (n) with the input V/III ratio in step-1 growth was V/III = 0, the PL intensity is much higher than that of the other samples. The real mechanism of this sharp increasing in the PL emission intensity of the QDs grown with zero input V/III ratio in step-1 growth of the two-step growth InAs QDs needs to be further studied.

High density, uniform and high quality InAs QDs have been grown on the InGaAs/InP matrix with the optimized the growth conditions and by using a two-step growth method. In the following sessions, extending the emission wavelength of the InAs QDs into mid-infrared have been discussed.

C. Extend Emission Wavelength of the QDs Structures

To extend the emission wavelength of the InAs QD structures, InxGa1-xAs has been used as the InAs QDs' matrix layer and their upper barrier. To change the QD structure's emission wavelength, indium content of the InxGa1-xAs matrix layer and upper barrier layer of the QD structures was linearly increased from x = 0.53, which lattice matched to InP, to x = 0.65 ~ 0.80. Figure 19 shows schematically the InAs QD structure of the samples.

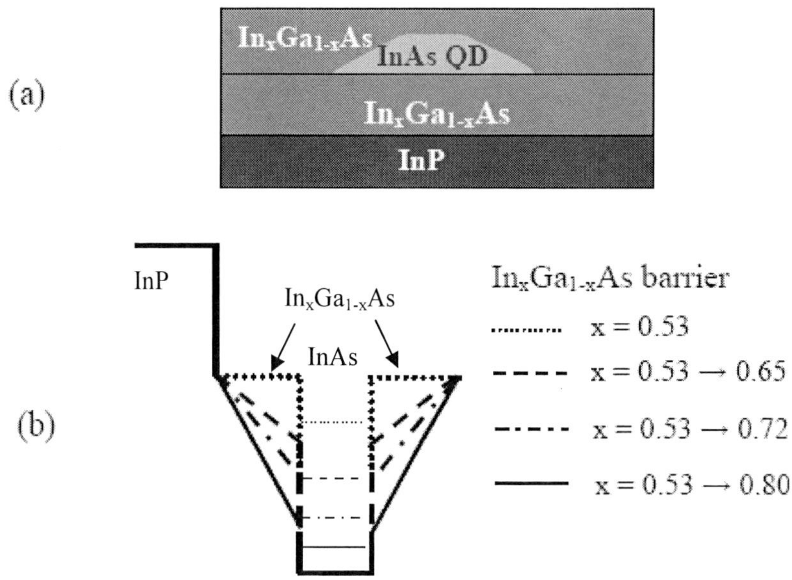

Figure 19. (a) Schematic structure of the InAs/In$_x$Ga$_{1-x}$As QD samples and (b) the conduction band diagrams of the structures with different In$_x$Ga$_{1-x}$As barriers.

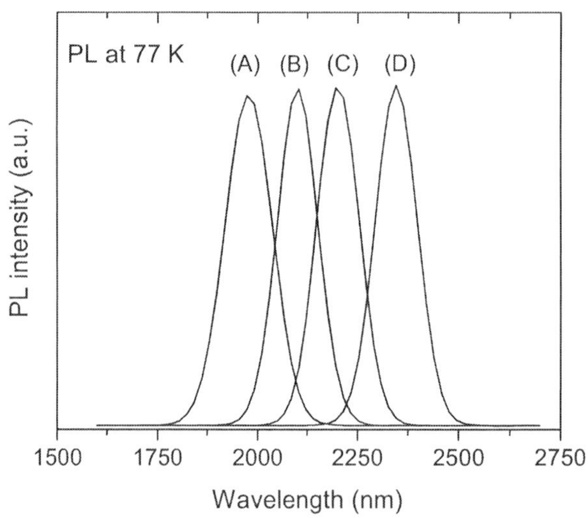

Figure 20. The normalized 77K photoluminescence spectra of the InAs QDs samples grown on different In$_x$Ga$_{1-x}$As/InP matrixes.

Figure 20 shows the normalized 77 K PL spectra of the InAs QDs sandwished between InxGa1-xAs matrixes and barriers with different In content. For sample (A), the InAs QDs are embedded in lattice matched In0.53Ga0.47As/InP matrix and barrier layer. Its peak emission wavelength is at 1.98 μm. To extend the emission wavelength, graded InxGa1-xAs layers are used as the matrix and barrier layer in samples (B), (C) and (D) with the indium content of the graded InxGa1-xAs layer increased linearly from 0.53 to 0.65, to 0.72 and to 0.8, respectively. As shown in Figure 21, the emission peak wavelength of the InAs QDs sandwished between In0.53→(0.53+y)Ga0.47→(0.47-y)As graded matrix and barriers red-shift from that of InAs QDs embedded in lattice matched In0.53Ga0.47As barriers of sample (A). Increasing the indium content gradation, y, of the In0.53→(0.53+y)Ga0.47→(0.47-y)As graded barriers of sample (B), (C) and (D) from 0.12 to 0.27, the InAs QD structure's emission wavelength redshifts by 120 nm, 220 nm and 370 nm, respectively. When the indium content gradation of the In0.53→(0.53+y)Ga0.47→(0.47-y)As graded barrier layer, y, increases by every 0.01, the peak emission wavelength of the InAs QD structure redshifts about 13.5 nm. When the indium gradation, y, increases to 0.27, emission wavelength of the In0.53→0.8Ga0.47→0.2As /InAs/ In0.53→0.8Ga0.47→0.2As/InP QD structure reaches 2.35 μm measured at 77 K, which is, to our best knowledge, the longest inter-band transition emission wavelength of InAs QD structures reported so far. FWHM of the emission spectra of the four QD samples are also plotted in Figure 21. The spectrum becomes narrower when the indium content gradation y of the In0.53→(0.53+y)Ga0.47→(0.47-y)As matrix/barrier increased. Reasons for the narrower emission spectrum of the In0.53→(0.53+y)Ga0.47→(0.47-y)As/InAs/In0.53→(0.53+y)Ga0.47→(0.47-y)As/InP QD structure with the higher indium content gradation of its matrix/barrier are because (i) less lattice mismatch or strain between InAs QDs and the upper InxGa1-xAs barrier layer and (ii) the larger QDs. With the higher indium content of the InxGa1-xAs upper barrier, lattice mismatch between the InxGa1-xAs barrier layer and the InAs QD layer is less, which suppresses the indium mixing or segregation between the InAs QDs and the $In_xGa_{1-x}As$ upper barrier layer [42]. The uniformity of the InAs QDs is not affected by covering the high indium gradation InxGa1-xAs upper barrier layer. The smaller lattic mismatch between the high indium content gradation In0.53→(0.53+y)Ga0.47→(0.47-y)As matrix and InAs makes larger InAs QDs being formed. Optical transition energies of larger QDs will be less affected by the dot size fluctuation [43, 44]. These result in narrower emission spectrum of the In0.53→(0.53+y)Ga0.47→(0.47-y)As/InAs/ In0.53→(0.53+y)Ga0.47→(0.47-y)As/InP QD structure with higher indium content gradation In0.53→(0.53+y)Ga0.47→(0.47-y)As matrix and upper barrier.

Emission wavelength of the InxGa1-xAs/InAs/ InxGa1-xAs/InP QD structure is mainly affected by:

(I) Height of the InAs QDs formed. Since the aspect ratio, which is defined as the QD height versus its diameter, of the InAs QDs formed in the four samples is very small, ~0.18, quantum confinement effect of the QDs is mainly determined by their QD height [45, 46]. Table 1 shows that the aspect ratio of the four samples A, B, C and D, is almost same, 0.167~0.171. The mean dot height of these four samples is also similar, 8.9 nm ~ 9.1 nm. Therefore, the measured red shift of PL peak of samples

(B), (C) and (D) versus that of sample (A) is not because of the sample's QD height variation.

(II) The second factor causes the sample's peak emission redshift is because its $In_xGa_{1-x}As$ barrier layers. Figure 19 (b) compares the conduction band profiles of the four samples under quantum well (QW) approximation [42, 46]. Quantum confinement of carriers in the InAs QDs structure with an $In_xGa_{1-x}As$ graded barriers is reduced, which lowers the transition energy of the QD structure. Thus, its emission wavelength redshifts. The lowest electron-hole transition energies of the four QD structures are calculated by using eight-band $k{\cdot}p$ calculation [47]. Figure 21 compares the calculated first electron-hole transition energies of sample (A), (B), (C), (D) and the measured PL emission peak positions of these four samples. The measured results agree with the calculated transition energies well. It has been noted that the calculated peak emission wavelength of the samples is systematically slightly longer than the experimental results, and the difference between the calculated transition energies and experimental results becomes smaller when the indium composition of the $In_xGa_{1-x}As$ barriers of the QD structure is increased. We believe that this systematically longer calculated emission wavelengths are because of the QW approximation used in the calculation. In the calculation, mean height of the sample's InAs QDs is used as the well width of the approximated QW of the sample. Since the QDs are generally in truncated pyramid shape, the effective height of the QDs is lower than their mean height. This causes the calculated transition energies smaller or the correcponding emission wavelength longer. Further more, we used the mean height of the un-capped InAs QDs of the sample in the calculation, while the actual QD height of the QD structure is expected to be higher because of the indium decomposition from the $In_xGa_{1-x}As$ barrier layer during the growth. This indium decomposition is depended on the $In_xGa_{1-x}As$ barrier layer's indium composition, which is described below. This explains the reduction of the difference between the calculated and measured peak positions of the sample when the indium content of the $In_xGa_{1-x}As$ barrier layer is increased as shown in Figure 21.

(III) Partially decomposition of the $In_xGa_{1-x}As$ barrier layer during the MOVPE growth of the $In_xGa_{1-x}As$ /InAs/$In_xGa_{1-x}As$/InP QD structure effectively increases the height of the InAs QDs [48, 49], which results in a redshift of the QDs' PL emission peaks. This indium decomposition from the $In_xGa_{1-x}As$ upper barrier layer depends on its indium composition. Higher the indium content of the $In_xGa_{1-x}As$ barrier layer, more indium decomposition during the QD structure growth, which makes the InAs QDs higher. This increasement of the QDs' height with the higher indium content of the $In_xGa_{1-x}As$ barrier layer effectively results in a larger the redshift of the QD structrue's emission wavelength.

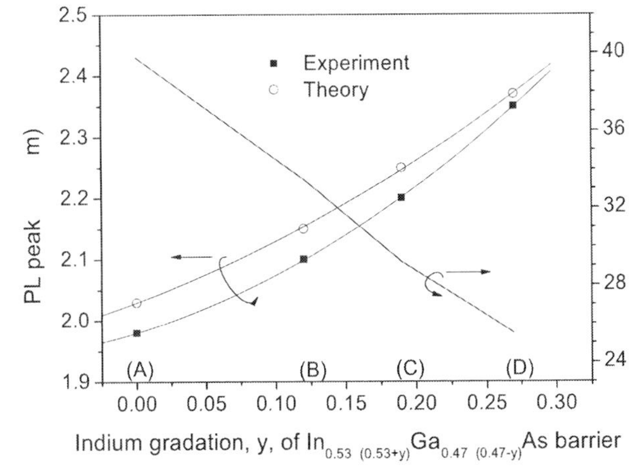

Figure 21. Experimental measured and the theoretical calculated PL peak positions and the measured FWHM of the PL spectra of the InAs QDs embedded in different $In_xGa_{1-x}As/InP$ matrixes.

Summary

In summary, MOVPE growths of mid-infrared emission InAs QDs on InxGa1-xAs /InP matrix have been investigated in detail. Effects of the MOVPE growth conditions on formation of the InAs QDs have been studied. A new two-step growth of QDs based on self-assembly growth mode has been developed and optimized. Under the optimized growth conditions, very high dot density (above 2.5×1010 cm-2) of InAs QDs emitting at >2.3μm wavelength has been received. The surface coverage of the mid-infrared emission InAs QDs reaches ~60% and the FWHM of the QDs' PL spectrum is only 25.5meV. By embedding the InAs QDs in graded In0.53→(0.53+y)Ga0.47→(0.47-y)As barriers, we have demonstrated the InAs QDs emitting in mid-IR range at the longest wavelength of 2.35 μm measured at 77 K. The InAs QDs structures' emission wavelength has been extended by increasing the indium content of the graded InxGa1-xAs barrier layers of the InAs QD structure. Compared with that of the InAs QDs embedded in lattice matched In0.53Ga0.47As barriers, the emission wavelength of the InAs QDs embedded in In0.53→(0.53+y)Ga0.47→(0.47-y)As graded barriers achieves much longer. Redshift of emission wavelength of the InAs QDs structures depends on the indium content gradation, y, of the In0.53→(0.53+y)Ga0.47→(0.47-y)As graded barrier layers. The results show guidance for further development of mid-IR QD lasers based on InP substrates.

Acknowledgement

This research was sponsored by the Science and Engineering Research Council (SERC) of Agency for Science, Technology & Research (A*STAR), Singapore (SERC Grant No. 052 101 0105) and the Academic Research Fund (Grant No. RGM 23/06) of Nanyang Technological University, Singapore.

REFERENCES

[1] Arakawa and S.Sakaki, *Appl. Phys. Lett.*, 40, (1982) 939.

[2] M. Asada, Y. Miyamoto and Y. Suematsu, *IEEE J. Quantum Elec.*, QE-22, (1986)1915.

[3] R. H. Wang, A. Stintz, P. M. Varangis, T. C. Newell, H. Li, K. J. Malloy, and L. F. Lester, *IEEE Photonics Technol. Lett.* 13, (2001) 767.

[4] R. Schwertberger, D. Gold, J. P. Reithmaier, and A. Forchel, *IEEE Photonics Technol. Lett.* 14, (2002) 735.

[5] Y. Qiu, D. Uhl, R. Chacon, and R. Q. Yang, *Appl. Phys. Lett.* 83, (2003)1074.

[6] Bradshaw J L, Yang R Q, Bruno J D, Pham J T, Wortman D E and Tober R L, *Appl. Phys. Lett.* 81 (2002) 397.

[7] Sun Z Z, Wu J, Liu F Q, Xu HZ, Chen YH, Ye X L, Jiang W H, Xu B and Wang Z G, *J. Appl. Phys.* 88 (2000) 533

[8] Zongyou Yin, Xiaohong Tang, Deny Sentosa and Jinghua Zhao, *Nanotechnology,* 17, (2006) 1646

[9] Qiu Y M and Uhl D, *Appl. Phys. Lett.* 84 (2004)1510.

[10] Landi S M, Pires M P, Tribuzy C V-B, Souza P L, Marega E Jr, Silva A G and Guimaraes P S S, *Phys. Status Solidi* 8 (2005) 317.

[11] Qiu Y M and Uhl D, *Appl. Phys. Lett.,* 84 (2004) 263.

[12] Qiu Y M and Uhl D, *J. Cryst. Growth,* 257 (2003) 225.

[13] Zhang Z H, Pickrell G W, Chang K L, Lin H C, Hsieh K C and Cheng K Y, *Appl. Phys. Lett.* 82 (2003) 4555.

[14] Li H X, Daniels-Race T and Hasan M A., *Appl. Phys. Lett.* 80 (2002) 1367.

[15] Z. H. Zhang, and K. Y. Cheng, *Appl. Phys. Lett.* 83,(2003) 3183.

[16] Y. Maeda, *Phys. Rev. B,* 51, (1995)1658.

[17] J. Johansson, and W. Seifert, *J. Cryst. Growth,* 234, (2002) 132.

[18] F. K. LeGoues, M. C. Reuter, J. Tersoff, M. Hammar, and R. M. Tromp, *Phys. Rev. Lett.*, 73, (1994) 300.

[19] T. T. Ngo, P. M. Petroff, H. Sakaki, and J. L. Merz, *Phys. Rev. B,* 53, (1996) 9618.

[20] F. M. Ross, J. Tersoff, and R. M. Tromp, *Phys. Rev. Lett.,* 80, (1998) 984.

[21] J. G. Cederberg, F. H. Kaats, and R. M. Biefeld, *J. Cryst. Growth,* 261 (2004) 197.

[22] R. L. Sellin, I. Kaiander, D. Ouyang, T. Kettler, U. W. Pohl, D. Bimberg, N. D. Zakharov, and P. Werner, *Appl. Phys. Lett.,* 82,(2003) 841.

[23] J. Tatebayashi, M. Nishioka, and Y. Arakawa, *Appl. Phys. Lett.,* 78,(2001) 3469.

[24] S. C. Saxena, and V. K. Saxena, *J. Phys. A: Gen. Phys.,* 3,(1970)309.

[25] A. A. El-Emawy, S. Birudavolu, P. S. Wong, Y.-B. Jiang, H. Xu, S. Huang, and D. L. Huffaker, *J. Appl. Phys.,* 93 (2003)3529.

[26] C. W. Snyder, J. F. Mansfield, and B. G. Orr, *Phys. Rev. B,* 46 (1992) 9551.

[27] C. C. Hsu, Y. F. Yang, H. J. Ou, E. S. Yang, and H. B. Lo, *Appl. Phys. Lett.*, 71 (1997)3248.

[28] Xiaohong Tang , Zongyou Yin, Jinghua Zhao and Sentosa Deny, *Nanotechnology,* 17,(2006) 295.

[29] M. Kastner, and G. Voigtlander, *Phys. Rev. Lett.*, 82,(1999) 2745.

[30] Y. Chen, and J. Washburn, *Phys. Rev. Lett.*, 77,(1996) 4046.

[31] A. A. El-Emawy, S. Birudavolu, S. Huang, H. Xu, and D. L. Huffakera. *J. Cryst. Growth*, 255, (2003) 213.
[32] A. V. Osipov, F. Schmitt, S. A. Kukushkin, and P. Hess, *Appl. Surf. Sci.*, 188,(2002) 156.
[33] V. G. Dubrovskii, G. E. Cirlin, and V. M. Ustinov, *Phys. Rev. B.* 68, (2003) 75409.
[34] Yin, Z. Y.; Tang, X. H.; Zhao, J. H.; Deny, S. *J. Appl. Phys.* 99 (2006)124306.
[35] M. C. Chen, H. H. Lin and C. W. Shie, *J. Appl. Phys.,* 83,(1998) 3061.
[36] R. Leon, C. Lobo, A. Clark, R. Bozek, A. Wysmolek, A. Kurpiewski and M. Kaminska, *J. Appl. Phys.,* 84,(1998) 248.
[37] Y. Horikoshi, M. Kawashima, and Hiroshi Yamaguchi, *Jpn. J. Appl. Phys.*, 25,(1986) 868.
[38] G. S. Solomon, J. A. Trezza and J. S. Harrris, Jr., *Appl. Phys. Lett.,* 66,(1995) 3161.
[39] R. Leon, C. Lobo, J. Zou, T. Romeo, and D. J. H. Cockayne, *Phys. Rev. Lett.*, 81, (1998)2486.
[40] C. S. Peng, Q. Huang, W. Q. Cheng, J. M. Zhou, Y. H. Zhang, T. T. Sheng and C. H. Tung, *Appl. Phys. Lett.,* 72,(1998) 2541.
[41] D. J. Eaglesham, F. C. Unterwald and D. C. Jacobson, *Phys. Rev. Lett.*, 70, (1993)966.
[42] K. Nishi, H. Saito, S. Sugou, and J. S. Lee, *Appl. Phys. Lett.*, 74, (1999) 1111.
[43] A E Zhukov, V M Ustinov, A R Kovsh, A Yu Egorov, N N Ledentsov, N A Maleev, A F Tsatsul'ikov, M V Maximov, Yu G Musikhin, N A Bert, P S Kop'v, D Bimberg, and Zh I Alferov, *Semicond. Sci. Technol.*, 14,(1999) 575.
[44] R. Heitz, I. Mukhametzhanov, J. Zeng, P. Chen, A. Madhukar, and D. Bimberg, *Superlattices Microstruct.*, 25,(1999) 97.
[45] F. Q. Liu, Z. G. Wang, J. Wu, B. Xu, Q. Gong, and J. B. Liang, *J. Appl. Phys.*, 85,(1999) 619.
[46] K. H. Schmidt, G. Medeiros-Ribeiro, J. Garcia, and P. M. Petroff, *Appl. Phys. Lett.*, 70,(1997) 727.
[47] Zongyou Yin, Xiaohong Tang, Wei Liu, Daohua Zhang, *J. of Appl. Phys.*, 100,(2006) 033109.
[48] M. V. Maximov, A. F. Tsatsul'nikov, B. V. Volovik, D. S. Sizov, Yu. M. Shernyakov, I. N. Kaiander, A. E. Zhukov, A. R. Kovsh, S. S. Mikhrin, V. M. Ustinov, Zh. I. Alferov, R. Heitz, V. A. Shchukin, N. N. Ledentsov, D. Bimberg, Yu. G. Musikhin, and W. Neumann, *Phys. Rev. B*, 62,(2000) 16 671.
[49] W. H. Chang, H. Y. Chen, H. S. Chang, W. Y. Chen, T. M. Hsu, T. P. Hsieh, J. I. Chyi, and N. T. Yeh, *Appl. Phys. Lett.*, 86,(2005) 131917-1-3.

In: Encyclopedia of Nanotechnology
Editor: Elwood D. Carlson

ISBN: 978-1-60692-079-4

Chapter 9

SELF-ASSEMBLED HIERARCHICAL ARCHITECTURES BY LIQUID-PHASE PULSED LASER ABLATION

Li Yang[1]*, Paul W. May[1] and Lei Yin[2]
[1]School of Chemistry, University of Bristol, Cantock's Close,
Bristol BS8 1TS, United Kingdom
[2]Department of Aerospace Engineering, University of Bristol,
Queen's Building, University Walk, Bristol BS8 1TR, United Kingdom

ABSTRACT

Over the past decade, liquid phase pulsed laser ablation (LP-PLA) has been used as an effective method for preparation of many nanostructured materials. This technique involves the interaction of a laser with a solid target and surrounding liquid. The laser causes the target surface to vaporise in the form of an energetic ablation plume, containing species such as atoms, ions, and clusters, travelling with high kinetic energy. These can collide and react with molecules of the surrounding liquid phase, producing new materials, often on the nanometre scale. Due the intensity of the laser and the nanosecond timescales, high temperature, high pressure, and high density conditions within the reaction volume are produced which provide a 'brute force' method of synthesising novel materials that have hitherto been inaccessible using milder, more conventional techniques. In this chapter, we discuss more recent progress in the synthesis of hierarchical, complex, and ordered crystal architectures by this novel technique. It is believed that the ability of nanostructures to self-assemble in this manner provide potential building blocks for microscale or nanoscale devices. We show that for C3N4, a Group IV-V covalent compound, a range of self-assembled arrays of nanoparticles, nanorods, nanoleaf, and mesoflowers can be produced, and this also suggests that the route described here can be applicable to the synthesis of nanostructures of other similar materials. Therefore, we expect LP-PLA may play a key role in advancing complex nanomaterials into many applications in optics, electronics, biology, medicine and energy/chemical conversion.

* E-mail: li.yang@bristol.ac.uk

1. INTRODUCTION

Although self assembly has occurred in nature for thousands of years (for instance, the formation of salt crystals and the intricate structure of snowflakes), the term 'self assembly' as used in chemical synthesis is relative new. Self assembly refers to the creation by physical or chemical reaction, of small building units, which may be nanoparticles (NPs), nanorods (NRs), *etc*, which then aggregate together in specific arrangements to give one-dimensional (1D), two-dimensional (2D) or even three-dimensional (3D) superstructures [1]. The aggregation may be spontaneous without human intervention, or may be as a result of changing local conditions, *e.g.* temperature, concentration, drying, *etc.* Current understanding extends only to the most rudimentary of these stages for its use in materials, biology, chemistry, and condensed matter science [2].

Due to the emergence of a new generation of high-technology materials, 2D and 3D self-assembled, aligned superstructure materials have begun to be widely investigated over the last decade [3, 4]. It is believed that the ability of nanostructures to self-assemble with controlled crystalline orientation, size, complexity and crystal morphology provide potential applications in data storage, functional devices, communications and technology. As a result, a rapidly growing field of science has emerged to understand and control these self-organized architectures, involving the dedicated efforts of chemists, physicists, material scientists and biologists. In general, physical methods (including chemical vapor deposition [5], vapour phase transport [6], and pulsed laser ablation in vacuum [7]) and chemical methods (including hydrothermal methods [8], soft-template [9] and use of various surfactants [10, 11]) have been developed to fabricate nano- to microscale materials with a range of morphologies, such as compact hexagonal networks [12], rings [13] , dandelion-shaped hollow structures [14], strips [15], tubes [16], and flower-like structures [17]. However, the former techniques require relatively high temperature, vacuum conditions, expensive equipment and sometimes complicated processes, which limit them to smaller scale fabrication. Also, conventional chemical methods usually involve the use of catalysts, surfactants and possibly complex chemical reactions, which often produce a significant amount of unwanted byproducts requiring further purification. Therefore, a technique which combines the merits of both physical and chemical methods, while giving high yield at low cost, would be desirable, if novel self assembled materials are to be produced on an industrial scale.

One such technique, liquid-phase pulsed laser ablation (LP-PLA), has been applied to the problem of producing self-organized nanomaterials only relatively recently. LP-PLA involves focusing a high power laser beam onto the surface of a solid target, which is submerged beneath a liquid. The interaction of the laser with the target causes the surface to vaporise in the form of an ablation plume, which contains species such as atoms, ions, and clusters, travelling with high kinetic energy. The species in the plume collide and react with molecules of the surrounding liquid, producing new compounds containing atoms from both the original target and the liquid. Due the intensity of the laser and the nanosecond timescales, the instantaneous temperatures and pressures within the reaction volume can be extreme (many thousands of K at tens of GPa) [18]. Such high temperature, high pressure, and high density conditions provide a 'brute force' method of synthesising novel materials that have hitherto been inaccessible using milder, more conventional techniques.

In this chapter, we first briefly review the use LP-PLA and its applications in synthesis of nanocrystals. Furthermore, we reported recent progress in the synthesis of hierarchical, complex, and ordered crystal architectures by this novel technique. We shall show that the chemical and physical properties of these superstructures are intrinsic to the self-assembly induced by the close vicinity of the NPs or NRs. Finally, a summary and expectation are discussed with regard to the application of LP-PLA as a synthesis route for highly desirable complex architectures.

2. Laser Ablation of Solids in Liquid Environments

Pulsed laser deposition was first developed in the 1960s, shortly after the invention of the pulsed ruby laser. Since then, laser ablation in a vacuum or dilute gas has been studied by many researchers. By using different target materials and background gases, and varying parameters such as the laser wavelength, fluence, and pulse duration, it is possible to produce a wide variety of thin films. These include high temperature superconductors [19], metals, semiconductors, oxides, and other ceramics [20], and diamond-like carbon [21]. The thin films have a variety of applications, for example semiconductor devices, electrodes, and wear-resistant coatings [22].

The introduction of pulsed laser ablation at the solid-liquid interface was first reported by Patil and co-workers in 1987, who used a pulsed laser to ablate a pure iron target in water to form iron oxides with metastable phases [23]. Following their work, Ogale [24] extended the potential of LP-PLA for the surface modification of metals, such as metallic oxidation, nitriding, and carbiding. This pioneering work opened new routes for materials processing based on the PLA of solids in various liquids. Since then, the LP-PLA method has been used to produce a wide range of novel materials, such as nanodiamond and related nanocrystals, metallic nanocrystals, nanocrystal alloys, and metal oxides. A summary of the nanocrystals synthesized by LP-PLA is listed in Table 1.

These studies clearly indicate that LP-PLA has become a successful material fabrication technique, allowing versatile design through choosing suitable solid targets and confining liquids. Compared to the conventional methods mentioned above, the technique of LP-PLA has many distinct advantages: (i) a chemically "simple and clean" synthesis, the final product is usually obtained without byproducts and no need for further purification; (ii) low cost of experimental setup and easy controllable parameters; (iii) the extreme confined conditions and induced high temperature, high pressure region favour the formation of unusual metastable phases. A recent review by Yang [25] gives a comprehensive understanding of the nucleation thermodynamics, the phase transition, and the growth kinetics of nanocrystals by laser ablation of liquids. Therefore, in this chapter we will not discuss the effects of liquid confinement, thermodynamic nucleation and growth mechanism in detail. Instead we shall focus upon the recent progress in the synthesis of hierarchical, complex, and ordered crystal architectures by LP-PLA.

Table 1. Nanomaterials synthesis via LP-PLA in various liquids. NP = nanoparticles, NR = nanrods

Product	Solid Target	Liquid Solution	Laser parameters
Ag NP	Ag	H_2O [26, 27]	Nd: YAG laser (355, 532, 1064 nm)
			Laser energy: 90-340 mJ/pulse
		C_2H_5OH [28]	Cu vapour laser (510.5 nm)
			Laser fluence: 1-20 J/cm^2
Au NP	Au	H_2O [29]	Nd: YAG laser (532, 266 nm)
			Laser fluence: 5-250 J/cm^2
		Alkane liquid [30]	Nd: YAG laser (532 nm)
			Laser fluence: 1-200 J/cm^2
Co_3O_4 NP	Co	H_2O [31]	Nd: YAG laser (355 nm)
			Laser energy: 30 mJ/pulse
Cubic-BN NP	Hexagonal-BN	Acetone [32]	Nd: YAG laser (532 nm)
			Power density: 10^{10} W/cm^2
Cubic-C_3N_4 NR	Graphite	Ammonia [33]	Nd: YAG laser (532 nm)

			Power density: 10^{10} W/cm^2
		Acetone [34]	Nd: YAG laser (532 nm)
			Power density: 10^{10} W/cm^2
Diamond NP	Graphite	H_2O [35]	Nd: YAG laser (532 nm)
			Laser energy: max 125 mJ/pulse
Hf_2S NP	HfS_3	Tert-butyl disulfide [36]	Nd: YAG laser (532 nm)
			Laser energy: 30-50 mJ/pulse
$Mg(OH)_2$ Tubular structures	Mg	H_2O +SDS [37, 38]	Nd: YAG laser (355 nm)
			Laser energy: 100 mJ/pulse
Pt/TiO_2 NP	Pt/TiO_2	H_2O [39]	Nd: YAG laser (355 nm)
			Laser power: 160 mJ/pulse
SnO_2 NP	Sn	H_2O +SDS [40]	Nd: YAG laser (355 nm)
			Laser energy: 100 mJ/pulse
TiO_2 NP	Ti	H_2O	Nd: YAG laser (355 nm)
		H_2O +SDS [41, 42]	Laser energy: max 150 mJ/pulse
ZnO NP core-shell	Zn	H_2O +SDS [43]	Nd: YAG laser (1064 nm)

			Laser energy: 70 mJ/pulse
ZnO NP	Zn	H_2O [44]	Nd: YAG laser (355 nm)
		H_2O +LDA [45]	Laser energy: max 3.2 J/cm^2
		H_2O +CTAB [46]	
$Zn(OH)_2$ layered nanocomposite	Zn	H_2O +SDS [47]	Nd: YAG laser (355 nm)
			Laser energy: 100 mJ/pulse

3. Highly Organized Growth of Carbon Nitride

To date, most nanomaterials made by LP-PLA have been either continuous film structures, zero-dimensional (0D) NPs, or 1D NRs. Since the 0D and 1D nanocrystals can serve as building blocks in forming 2D or 3D complex architectures with long term periodic structures, it should be expected that the LP-PLA approach would have great potential as a means to grow large arrays of hierarchical, complex, oriented and ordered superstructures. Indeed, our recent findings [48, 49] indicated that the instantaneous high temperature, high pressure and high density conditions that arise when a high-intensity focused laser beam impinges upon a graphite target confined by a thin layer of liquid ammonia can promote growth of crystalline carbon nitride nanoparticles.

Carbon nitride has been the subject of numerous publications since the prediction by Liu and Cohen [50] in 1989 that crystalline C3N4 should have superhard properties. However, successful synthesis of bulk amounts of this material still remains a challenge. Recently, we have developed two-step synthesis strategies for adding complexity to oriented nanocrystals, and shown the possibility of forming self-assembled nanostructures and microstructures of crystalline carbon nitride. This approach has involved initially using the LP-PLA process to prepare carbon nitride NP suspensions, and then sequential seeding these onto an appropriate substrate and carefully drying away the liquid. More complex nanostructures, such as observed in natural materials or biomineral 3D well-arranged architectures could also be generated [51, 52].

3.1. Nanocrystal Seed Preparation

The carbon nitride nanocrystal seed solution was prepared by LP-PLA, details of which are given elsewhere [53]. Briefly, a solid graphite target (Testbourne Ltd., 99.99%) was ablated at room temperature while submerged in a 5 ml solution of 25-35% ammonia solution (Fisher Scientific, used as received without further purification) inside a sealed stainless steel cell. In order to reduce the effect of target aging, the cell was rotated at 700 r.p.m. during ablation using a standard magnetic stirrer. A Q-switched Nd:YAG laser (532 nm, pulse duration 15 ns, frequency 10 Hz) was directed by a prism and then focused onto the graphite surface using a 25 mm-focal-length lens. The intense laser light passed through a quartz window in the top of the cell, then through ~5 mm layer of the liquid covering the graphite, to form a ~0.5 mm-diameter spot on the target surface. The ablation was typically carried out at laser fluences of 25-125mJ / pulse for reaction times $t = 0.5$-12 h. After ablation, a pale yellow colloidal suspension was obtained, which contained a mixture of unreacted graphite and ablation product, both in the form of NPs. The suspension was stable, with no precipitate being observed for months or even longer. The graphite sediments were filtered and removed as much as possible by boiling with 70% perchloric acid, before further characterization.

For analysis, a drop of the suspension was placed onto a transmission electron microscopy (TEM) copper grid. TEM revealed (Figure 1a) that pristine spherical particles appeared after only 10 min ablation (laser fluence at 75 mJ per pulse, 35% ammonia) with an average diameter 15.7 nm. The size distribution (Figure 1b) was obtained from statistical analysis of Figure1a by measuring the diameters of more than 900 particles, and fits well to a Gaussian curve. The area highlighted by the rectangle in Figure 1a indicates that some of the particles may be starting to line up. After 3 h of ablation, one-dimensional rod-like structures are formed of length ~200 nm and width 30-50 nm (see Figure1c). Since both ends of the flat structures taper to a point, they resemble the shape of a leaf, and so henceforth have been termed 'leaf-like' structures. The inset in Figure 1c shows the selected area electron diffraction (SAED) ring pattern (obtained without any tilting) from the leaf-like nanostructures, which indicates that these structures are polycrystalline. The four inner rings, with interlayer *d*-spacings of 2.788, 2.256, 1.610 and 1.348 Å, respectively, were indexed as (200), (101), (211) and (221) with reference to the calculated structure for β-C3N4 ($a = 6.4017$ Å, $c = 2.4041$ Å, space group: P63/m (176)) [54, 55], which are identical to that reported in the literature for the same β-C3N4 crystal [56].

When the reaction time was increased to 5 h (Figure 1d) the size of nanoleaf structures increased to ~100 ×~500 nm. The high magnification TEM image (Figure 1e) from a single nanoleaf shows that these structures are, themselves, composed of a large number of smaller NPs that have close-packed together in an ordered arrangement to form the leaf-like shapes. In contrast, Figure 1f shows that these leaf-like nanostructures are mostly made up of component NRs, while NPs are the building blocks for the NRs. These tiny NRs are well-aligned and all point in the same direction, perpendicular to the propagating *c*-axis. Although individual leaf-like structures within a local region seem to be randomly oriented, the overall NR aggregates still exhibit long-range ordering [57] (Figure 1f).

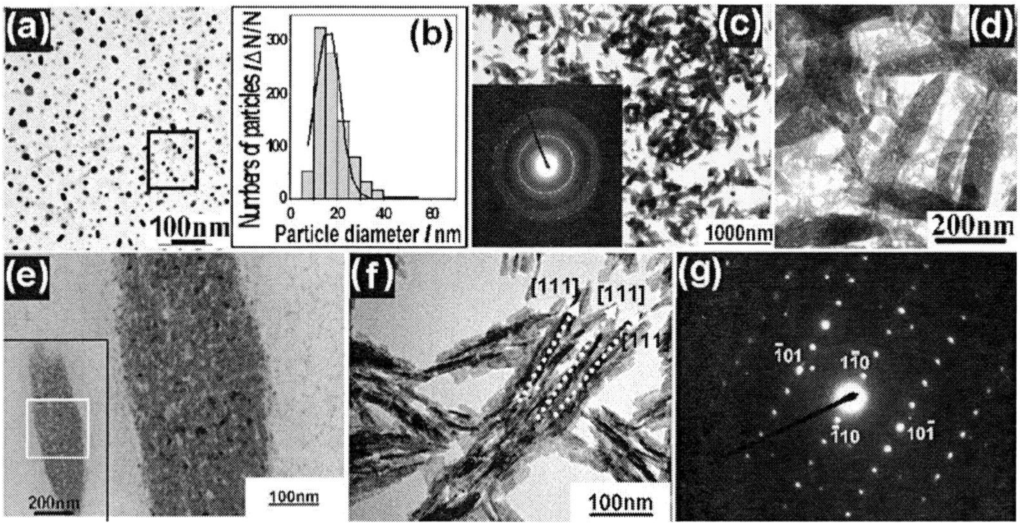

Figure 1. TEM images of C_3N_4 nanostructures synthesized with 5 ml 35% ammonia covering a graphite target, using 75 mJ / pulse laser fluence. (a) After 10 min ablation, spherical NPs appear, with an average size of 15.7 nm. Adjacent particles are aligned with one another along their main axes, highlighted by the rectangle. (b) Size distribution of the NPs, with a Gaussian fit to the data. (c) Overall appearance of the leaf-like nanostructures after 3 h ablation. Inset shows a typical SAED pattern taken from Figure1c. (d) Leaf-like structures of length ~500 nm produced following 5 h ablation time. (e) A single leaf-like unit formed from numerous small NPs (inset area framed with white box). (f) The NR aggregates exhibit a long-range order; arrows indicate the projecting direction along [111]. (g) A SAED pattern of a NR from the [111] zone-axis indicating the area is single crystal and consistent with crystalline β-carbon nitride. (Reproduced from [48] with the permission of American Chemical Society (ACS)).

A representative SAED pattern is given in Figure 1g, showing diffraction spots from the $(\bar{1}10)$, $(1\bar{1}0)$, $(\bar{1}01)$ and $(10\bar{1})$ planes of the [111] zone-axis, which are aligned parallel each other.

3.2. Chemical Bonding Environment in Carbon Nitride

X-ray photoelectron spectroscopy (XPS) measurements estimated that the overall elemental composition of the nanoleaf structures (laser fluence at 100 mJ per pulse, 35% ammonia) was 86.3% C, 11.4% N, and 2.3 % O after 5 h, whereas for 7 h ablation values were 90.5% C, 7.8% N and 1.7% O. One possibility for the presence of the O signal is that the graphite target reacted directly with the water (or OH radicals generated in the water) during the high temperature conditions in the ablation plume. However, since the O signal decreased with longer ablation time, this idea has been discounted. An alternative explanation is that prolonged exposure of the samples to the laboratory atmosphere led to possible contamination by absorbed or adsorbed water vapor. For pure C3N4, we would expect a composition of 42.8% C and 57.2% N, which shows that our samples are considerably carbon rich. Since we know from TEM observations that regions of the samples contain pure C3N4, we must conclude that our samples are a composite of crystalline C3N4 and an amorphous carbon background. This observation is consistent with the conclusions by other workers [58,

59], who found that it was difficult to generate highly ordered structures containing large concentrations of N incorporated into the C network.

The breadth and asymmetry of the C (1*s*) and N (1*s*) core-level XPS spectra shown in Figure 2 indicate the presence of different bonding states within the crystallites. The core-level lines were fitted to the Gaussian function [60]. The N 1*s* feature is difficult to assign unambiguously, since it is known that N can give a broad XPS peak around 399 eV in some materials. Alternatively, a better fit can be achieved if the feature is deconvoluted into three peaks (as shown in Figure 2b and d), located at ~398.5 eV, ~399.6 eV and ~400.2 eV, respectively. The peaks situated at ~398.5 eV can be assigned to *sp*3 C-N bonds, and one at 400.2 eV assigned to *sp*2 C=N bonds [61]. A peak centred at a binding energy ~399.6 eV has been assigned by some groups [62, 63] to N atoms bonded to *sp*-hybridized C atoms (*i.e.* -C≡N). These assignments are also supported by *ab initio* binding-energy calculations [64] Nevertheless, within experimental accuracy it is impossible to distinguish between these two alternatives. Also, it should be noted that C≡N bonds can produce 3 peaks in this region [63], 398.2 eV [65], 398.1 eV [66] and 400.1 eV [67]. However, these assignments have been discounted since they are outside of the experimental accuracy of the XPS spectrometer. The presence of *sp*3 C-N in the ablated nanostructures is, again, in agreement with a C3N4 structure.

Similarly, the C 1*s* core-level spectra obtained by XPS (Figure 2) was deconvoluted into three peaks located at ~284.6 eV, ~285.5 eV and ~288.5 eV, respectively. However, unlike nitrogen, verification of the assignment by independent C (1*s*) spectra was relatively difficult due to small shifts in binding energy and multiple bonding environments of the carbon. Carbon atoms may have zero, one, two, or three bonds with nitrogen atoms, which complicate the situation, so that there is a wide range of interpretation in the literature for carbon spectral feature identification.

The peaks at ~284.6 eV and ~285.5 eV can be identified unambiguously as *sp*2 and *sp*3 hybridized carbon [68, 69]. But the assignment for the peak at ~288.5 eV has not yet reached a consensus. In our case, it is believed this peak corresponds to N≡C *sp* bonding [67] supported by the Raman data [53]. Le Normand *et al* [70] have claimed that it is inappropriate to assign components appearing at higher energies to specific C-C or C-N bonding environments. After searching a comprehensive range of organic polymers, Beamson and Briggs [71] also found that no significant difference in C 1*s* position can be appreciated between *sp*2 and *sp*3-hybridized carbon bonded to nitrogen.

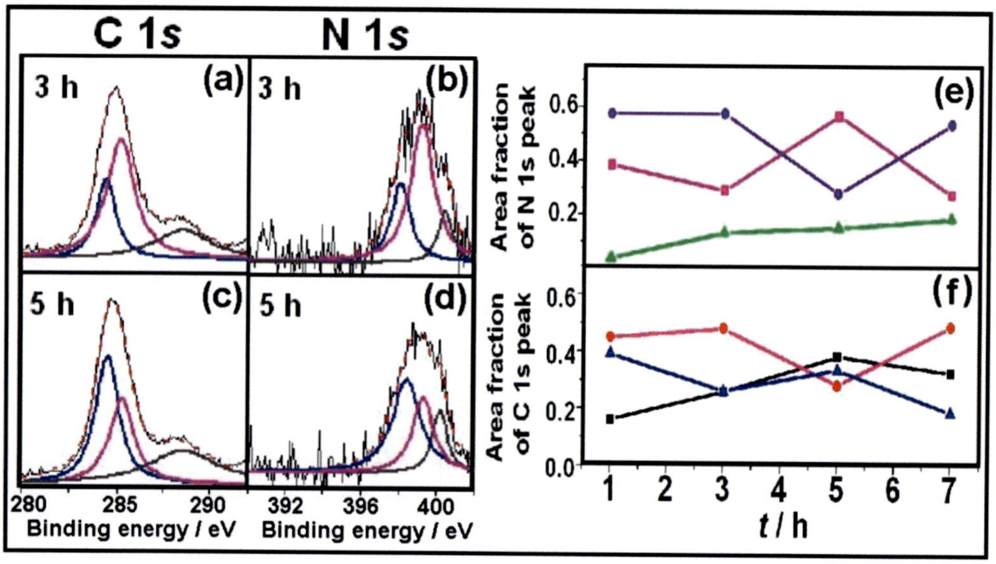

Figure 2. XPS spectra of C3N4 nanostructures (laser fluence at 100 mJ per pulse, 35% ammonia) for different ablation times *t* (a) C (1*s*) 3 h ; (b) N (1*s*) 3 h; (c) C (1*s*) 5 h; (d) N (1*s*) 3 h. Each spectrum has been deconvoluted by fitting to three Gaussian lineshape functions, with the total fitted line overlaid as a dashed line. Plots showing the time dependencies of the relative areas of the Gaussian functions used to fit (e) the N (1*s*) XPS peak (key: ■ 398.5 eV *sp*3 C-N bonds, ● 399.6 eV C≡N bonds, ▲ 400.2 eV *sp*2 C=N bonds), and (f) the C (1*s*) XPS peak (key: ■ 284.6 eV, ● 285.5 eV, ▲ 288.5 eV). (Reproduced from [53] with the permission of Institute of Physics (IOP)).

The relationship between ablation time and relative peak area fraction is shown in Figure 2e. The peak area corresponding to C=N (*sp*2) bonding indicates a marked increase with prolonged ablation time, whereas the trends for C≡N (*sp*) and C-N (*sp*3) are relatively complex. Nitrogen inclusion into the CN network changes the shape and position of the C 1*s* peak, as seen as in Figure 2f. However, only the signal attributed to C=C (*sp*2) bonding shows smooth variation, *i.e.* increasing with ablation time, while all other signals show a more complicated trend.

3.3. Control of Self-Assembled Structures

To understand the formation mechanism of nanocrystalline carbon nitride, experiments were performed for different laser energies with constant ablation time (3 h) and ammonia concentration (35%). Figure 3 summarises the different C3N4 nanostructures synthesized with varying laser fluences. For low laser fluence (25 mJ/pulse), the leaf-like structures had a length and width of ~300 and ~120 nm, respectively, with smooth surfaces, and were composed of smaller-sized NPs (Figure 3a). There were also plenty of voids situated between some of the NRs (Figure 3b). The sharp SAED pattern for the leaf-like structures (inset Figure 3b) was consistent with crystalline hexagonal β-phase carbon nitride [55, 72]. When the laser power was increased to 50 mJ / pulse, the aligned leaf-like structures increased to a length and width of ~600 and ~250 nm, respectively (Figures 3c-d), and now consisted of a substantial number of connected NRs, although there was no apparent change in size of the component NRs. The surface of the leaf-like structure was no longer smooth, but contained numerous protruding NRs. For still higher fluences (75 mJ / pulse), there was no further significant change in the size or shape of the leaf-like structures and NR building units (Figure 3e-f), but there was an increase the number formed. With further increases in fluence to 100 mJ per pulse, micron-scale carbon nitride spheres (Figure 3g) were observed, which were formed following aggregation of a large number of the leaf-like structures. 'Flower-like' spiked crystallites (Figure 3h) were also seen, where the leaf-like structures coalesced at a common centre with multi-fold symmetry [14]. In general, the trend is that the size of the overall nanocrystallites and their basic building blocks increases with increasing laser power.

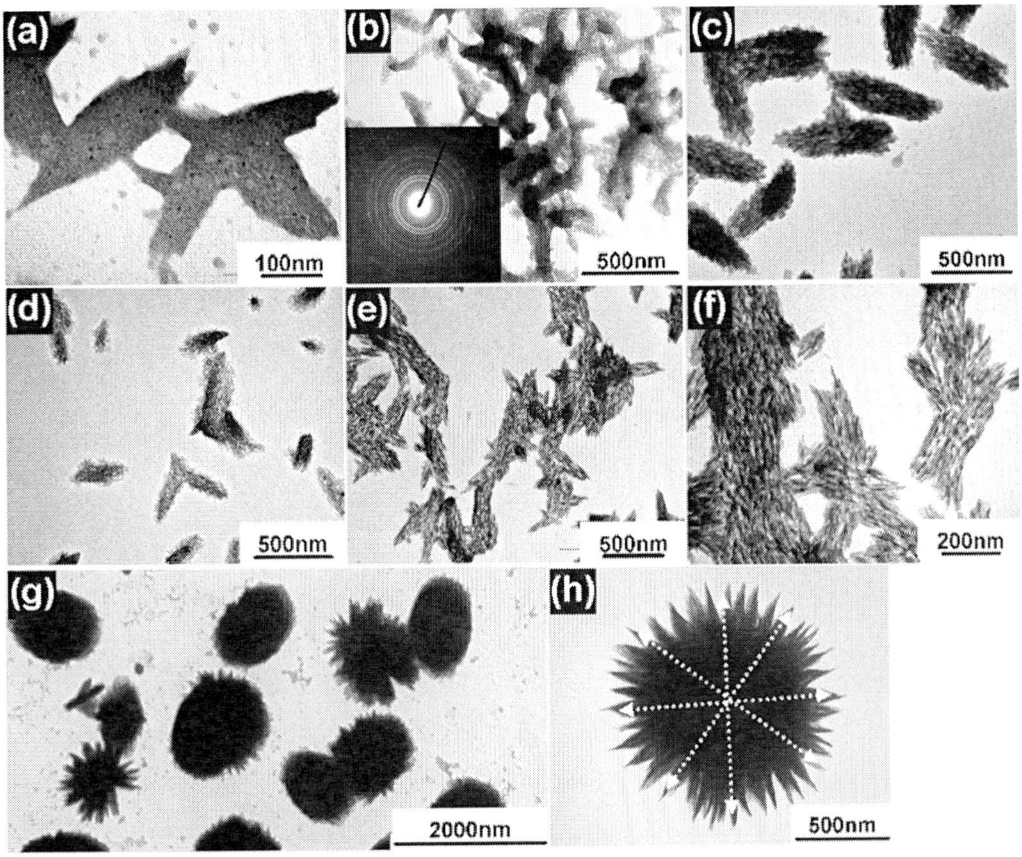

Figure 3. TEM images of C_3N_4 nanostructures synthesized for 3 h ablation time and 35% ammonia solution using different laser fluences. (a) and (b) 25 mJ/ pulse, leaf-like structures containing spherical NPs and short NRs. Inset in (b) is a sharp SAED pattern indicating the structures are crystalline. Note that there appear to be voids inside the leaf-like structures (the white areas in the grey area of image (b)). (c) and (d) 50 mJ/ pulse, denser leaf-like structures containing longer NRs which protrude from the surface. (e) and (f) 75 mJ/ pulse, leaf-like structures beginning to aggregate into larger clumps. (g) and (h) 100 mJ/ pulse, carbon nitride nanospheres and flower-like spiky crystallites. (Reproduced from [48] with the permission of ACS).

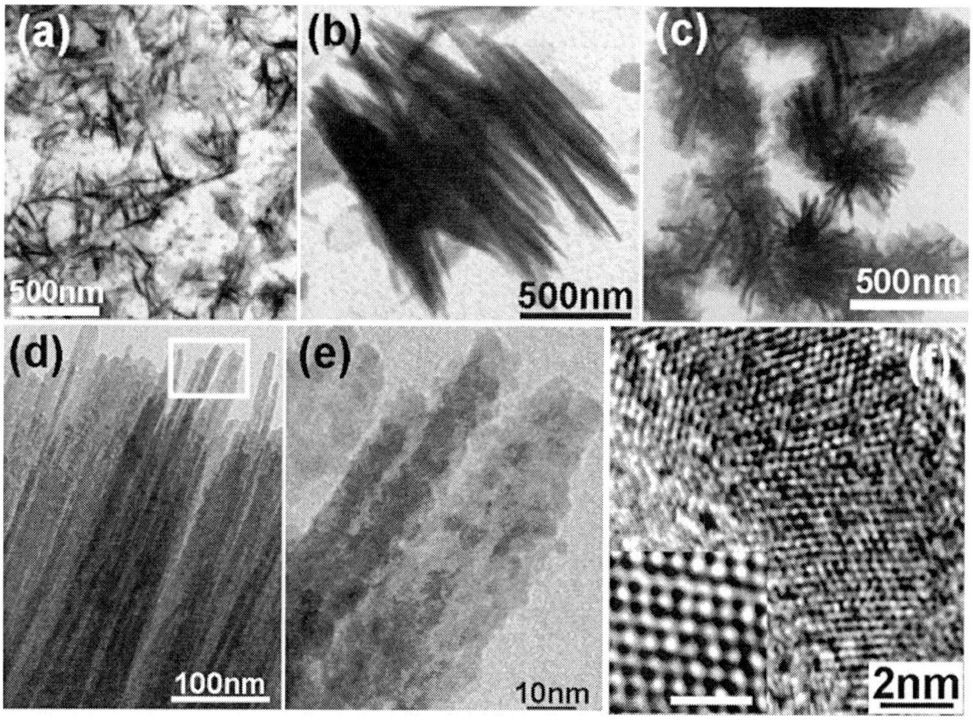

Figure 4. TEM images obtained by LP-PLA in 25% ammonia solution: (a) Isolated carbon nitride NRs (40 mJ/ pulse, $t = 1$ h) (b) Branched NRs (40 mJ/ pulse, $t = 3$ h) (c) Highly branched flowerlike architectures (100 mJ/ pulse, $t = 12$ h). (d) Rod-like structures showing straight, long and sharp tips. (e) Enlarged image of the region at the top of the NRs indicated by the open box in (d). (f) HRTEM image of a single NR, the inset shows the atomic arrangement, scale bar = 1 nm.

When 35% ammonia was replaced by 25% ammonia solution, carbon nitride was still formed on the copper grids; however, the morphology was different, as mentioned previously. For low laser power (40 mJ) and short ablation time (1 h), the product contained mostly a sparse collection of isolated NRs (Figure 4a). With longer ablation times (3 h), the NRs start to aggregate into branched structures (Figure 4b), serving as the starting points (or nucleation seeds) for the subsequent growth. However, with increasing laser power (100 mJ) and ablation time (12 h), the concentration of NRs increased, highly branched flower-like architectures completely made up of NRs were observed (Figure 4c). The NRs at the edge of the flower structures appeared to be protruding outward by ~10 nm. As shown in Figure 4d-4e, it should be pointed out that these NRs are themselves composed of a large number of smaller NPs that have packed together in an ordered arrangement to form the rod-like shapes. In particular, in the high-resolution (HR) TEM image in Figure 4f, taken from a single NR, the periodic lattices clearly show the atomic arrangement (Figure 4f inset) with very few defects, and reflect the relationship between the orientation of the NPs and the crystallography of the ordered NR array.

Based on above discussion, it was found that the size and morphology of the ablation products were a complicated function of laser power, ablation time, and ammonia concentration. Different experimental conditions can be used to optimize and control the nanocrystal nucleation and orientation independently.

4. Sequential Nucleation and Growth of Complex Nanostructures

The fabrication of well-defined 2D and 3D nanostructures is usually difficult because control of the nucleation and growth is still a challenge. Complex architectures are normally related to multistep syntheses, which begin with substrate surface preparation for heterogeneous nucleation of oriented nanocrystals. Two basic approaches have been used: one is deposition on chemically modified interfaces using organic self-assembled monolayers (SAMs) [73]. Another, more straightforward method, is to seed the substrate with NPs of the desired film material [74]. The fundamental advantage of seeded growth is enhanced control, obtained by separating the NP film nucleation and oriented rod growth into two steps.

Extending this concept, we performed a two-step synthesis to produce oriented carbon nitride nanostructures. Many more complex nanostructures than simple NPs and NRs described in the previous section can be produced in a controlled fashion by simply altering the drying time and drying method of the suspension of ablated product. In this methodology, the carbon nitride materials produced in the initial LP-PLA step were physically deposited onto a silicon p-(100) substrate. The liquid was then removed by drying. Four different drying processes were used: (1) dry naturally in air; (2) dry in a sealed tube; (3) dry in an oven or on a hotplate; (4) dry in a critical point dryer (CPD). CPD is a widely used drying technique in which the surface tension can be reduced to zero during evaporation [75]. Those procedures allowed the time taken to evaporate the liquid to be controlled.

4.1. General Structural Features

By drying the suspension under different conditions, four main classes of structure have been identified in the ablation product, categorised based upon their shape and size (Figure 5a). The first class of structure had the shape of thin plates with rounded edges. Since they were the components of the larger 'flower-like' structures (described later), they have been termed 'nano-petals'. The quantity of these nanopetals and their location with respect to the larger structures (see later) were dependent upon the deposition and drying conditions. As shown in Figure 5b-5c, this indicated that these nano-petals were 2D aligned crystallites of carbon nitride which preferentially aligned themselves perpendicularly to the surface of the Si substrate. The number and length of these nano-petals increased with increasing laser ablation time from 0.5-2 h for the same laser fluence. It is also possible that these nano-petals began to aggregate and self-assemble (Figure 5d). When the concentration of nano-petals in the suspension increased, they tried to minimize their interfacial energy upon subsequent drying of the liquid by preferential tilting with respect to each other. This produced the second class of nanostructure, which has a 'grass-like' shape and exhibits several different morphologies (see later). However, all are produced in large quantities and cover the whole substrate.

By carefully controlling the drying process, 'flower-like' spiked, crystalline superstructures were formed (Figure 5c-5f). This third class of structure – now fully 3-dimensional – with sizes 1-20 μm, were created when many nano-petal structures coalesced at a common centre with multi-fold symmetry. One possible explanation is that the presence of the solid substrate physically hinders growth in that direction; so many branches are tilted away from the substrate, towards the solution.

When the evaporation speed of the liquid was rapid (for example, drying in an oven or hotplate), a fourth class of structure was observed (Figure 5g-5h). Instead of 3D flowers, the carbon nitride now formed 2D 'star-like' or flattened flower-like structures. New dendrites emanated from the core and acted as nucleation centres, eventually allowing the structure to expand into 2D horizontal flowers (Figure 5h). It is suggested that the higher water evaporation rate increases the interparticle capillary forces [76]. As the continuous flux of particles fill up the spaces on the substrate, successive layers are expected to be formed rather than 3D complex shapes.

Although these four kinds of structures have different densities and morphologies, they all exhibit high surface-to-volume ratio and so might have potential uses in semiconductor devices, anticorrosion protective coatings and new applications.

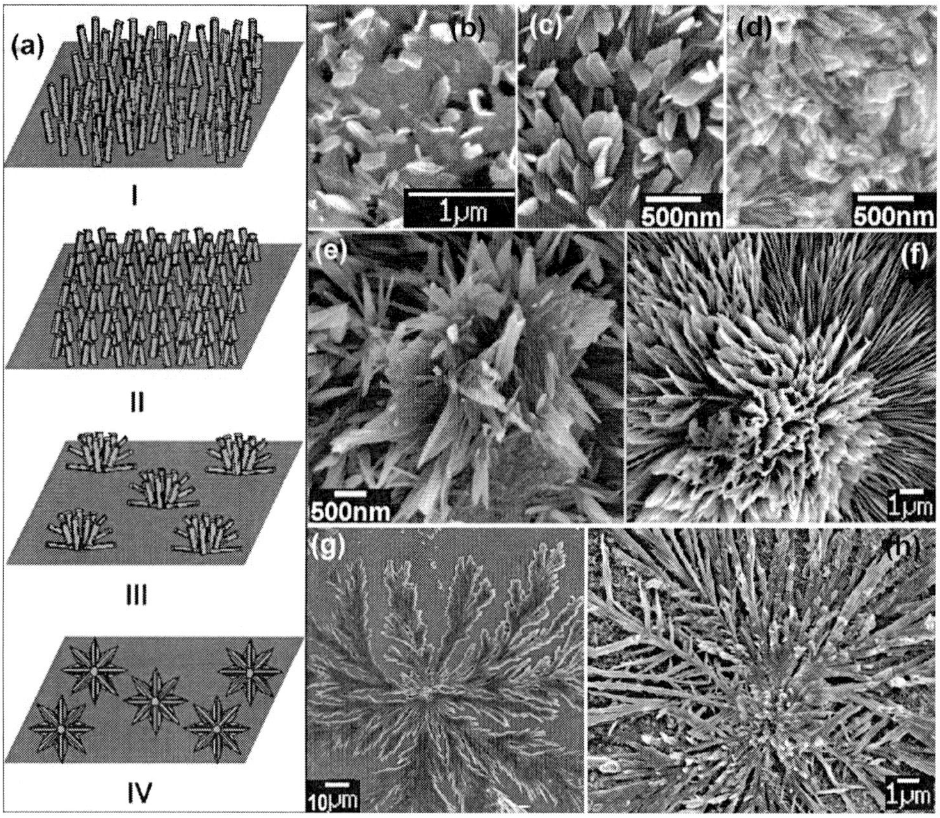

Figure 5. (a) Schematic illustration of the growth process leading to the observed four main classes (labelled as I-IV) of hierarchical structures, (b)-(d) SEM images of carbon nitride 'nano-petals' following ablation times of: (b) 0.5 h, (c) 2 h, and (d) 3 h. (e-f) Overall 'flower-like' structure following 5 h laser irradiation (synthesis conditions: laser power 125 mJ, 35% ammonia solution, drying in air). (g-h) 2D flattened flower, sample conditions identical to (e-f) except it was dried on a hotplate at 200°C (g) and an oven 80°C (h). (Reproduced from [51] with the permission of the Royal Society of Chemistry (RSC)).

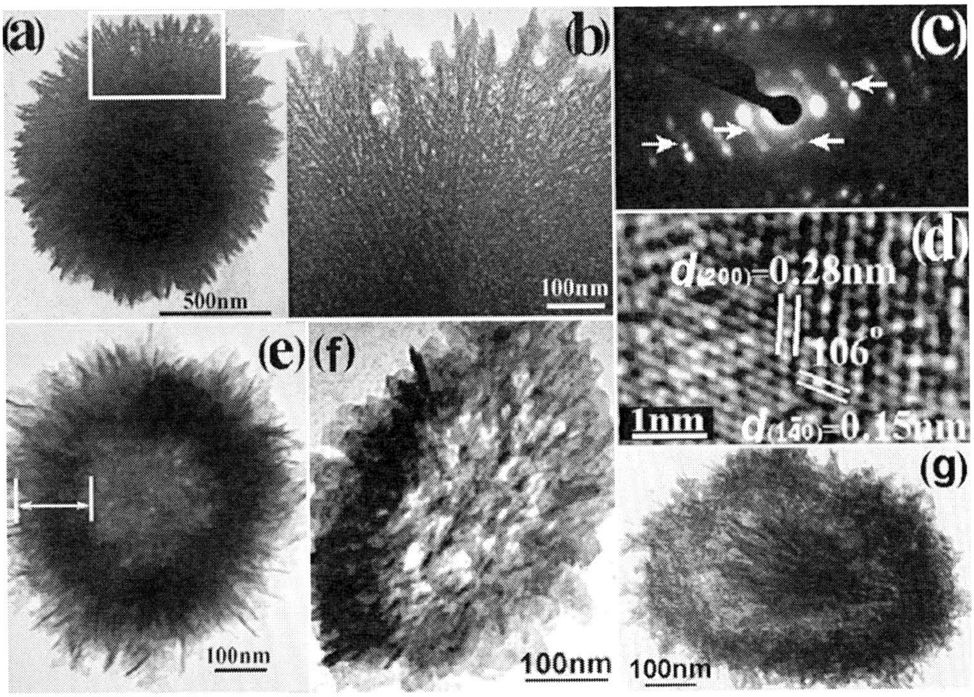

Figure 6. (a) TEM image of the flower-like structures produced by LP-PLA of a graphite target in 35% ammonia solution (laser fluence 125 mJ/pulse) for 5 h. (b) A higher magnification image of the framed region in (a), showing a high density of NPs surrounding the nano-petal framework. (c) [001] zone-axis MDA pattern from the tips of the nano-petals in (b), which corresponds closely to the calculated interlayer *d*-spacing of β-C3N4. Arrows point to different sets of [001] reflections (see text). (d) HR-TEM image recorded from the edge of the flower nanostructure that is oriented along [001]. (e) TEM image of a hollow flower formed after 8 h LP-PLA and prolonged drying. (f) Semi core-shell structure. (g) core-shell structure with a hollow tunnel. (Reproduced from [51] with the permission of the RSC).

Figure 6a shows a symmetric carbon nitride 'flower' together with its nano-petal building blocks (which look like flattened rods in the TEM). Figure 6b shows that these nano-petals appear fused together and 'interwoven' to form a lattice-like framework of the flower-like superstructure. The figure also shows the NPs that surround each nano-petal, and which fill in the holes within the framework to produce a dense, solid structure. Energy-dispersive X-ray (EDX) analysis confirms that carbon and nitrogen are present in all these structures, and micro diffraction analysis (MDA, Figure 6c) was also consistent with crystalline hexagonal β-phase carbon nitride oriented along the [001] zone-axis. Several [001] patterns in Figure 6c can be identified at the same time, indicating that the nano-petals consist of several domains, with different rotational orientations contributing to the diffraction pattern. The HR-TEM image in Figure 6d shows that the nano-petals at the very edge of the flower contain very few defects and are single crystalline, as was anticipated from the MDA pattern. Again, the lattice fringes ($d200$ = 0.28 nm, $d_{1\bar{4}0}$ = 0.15 nm) and their angles (106°) are in good agreement with the calculated values for hexagonal β-C3N4 [77].

The smaller NPs, which lie in and around the nano-petal-framework comprising the flowers, appeared to be mobile with respect to this framework, and diffused outward from the centre of the flower with longer drying times. The results of this diffusion can be seen in Figure 6e, where the solid carbon nitride flower (similar to that in Figure 6a) has become hollow. The NPs have diffused from the centre but remain loosely attached to the outside, making the outer shell of the flower appear fuzzy. The thickness of the fuzzy shell was ~ 140 nm and that of the hollow core was ~ 200 nm (shown as a lighter colour in the image). When the suspension was placed onto a hot-plate at 200°C for 0.5 h, the hollowing process was accelerated to form a semi-core-shell structure. The radial distribution of the NRs formed channels leading from the centre to the shell (Figure 6f). Another type of hollow structure was observed when the core-shell structures were essentially separated by a hollow tunnel, without linkage to the sphere (Figure 6g).

4.2. Arrangement of Components among the Architectures

The assembly process of hierarchical nanocrystals by repeated nucleation and growth is illustrated in Figure 7. By careful observation, it was found that the mesophase flower has a symmetrical structure (Figure 7a), consisting of many 2D nanopetals and radial growth from a central nucleus [78]. In addition, each nanopetal is constructed from many tiny nanocrystals or NR subunits (Figure 7b), as seen at the edge of the nanopetals (also see the later discussion). Surprisingly, just as for the nano-petals mentioned earlier in Figure 6b, these NRs also appeared to be 'interwoven' together into a similar lattice-like framework (Figure 7c). Interestingly, the NPs or NRs can self-assemble into a petal-like architecture along the *c*-axis direction. The tiny NPs attached side-by-side and fused into the wall of the nanopetal. This has the effect of making the nanopetal surfaces rough and curved, and can even make the two ends join together to form tubes (Figure 7d). Some nanopetals also exhibited structures inside the tubes, see the highlighted arrows in Figure 7d-f. The inset of Figure 7f shows a primary subunit with size of about 20 nm. It is worth pointing out that the length of the nanopetals is in the range of 1-20 μm, while the sizes of the subunit NPs or NRs have barely changed.

In principle, assembly is energetically favored because the formation of larger crystals can greatly reduce the interfacial energy of isolated NPs or NRs. Figure 7g summarizes a possible process for the carbon nitride self-assembly, mentioned above. By the aid of the solid substrate, hierarchically ordered nanocrystals assemble towards the surface of the substrate. New nanocrystals nucleate on the existing crystals and share the same edges. These structures are close-packed but still have spaces between them. It is still unclear why the NPs or NRs form such arrangements rather than random clumps. Further investigation of this mechanism is required.

The morphology of the self-assembled nanostructures is tuneable with the reaction conditions. By appling different building blocks, different final complex architectures on various scales can be achieved, as discussed in section 3.3. For lower ammonia concentration (25%), the carbon nitride formed 1D elongated nanoneedles or NRs. For low laser power (50 mJ) and short ablation time (2 h), the product contained mostly a sparse collection of these isolated NRs (Figure 8a). However, with increasing laser power (125 mJ) and ablation time (10 h), the concentration of NRs increased, and in some regions they began to coalesce into incomplete flowers (Figure 8b). The radial distribution of the NRs formed channels leading from the centre of the flower to the edge. These channels may form the conduits along which the NPs diffuse outwards. When a droplet of the suspension, identical to that shown in Figure 8a, was deposited on the substrate and dried for 12 h in a sealed tube, the carbon nitride produced flower-like structures constructed from pronounced tapered NRs (Figure 8c), instead of flowers made by the nanopetals (see Figure 7a). Figure 8d shows typical NR building blocks with a rod diameter of 10-20 nm.

Figure 7. (a) An individual flower exhibiting multi-fold symmetry. (Synthesis conditions: 35% ammonia solution, laser power 100 mJ, 3 h ablation and dried in a sealed tube) (b-f) Different arrangements of components that construct the final hierarchical superstructures. Arrows point to the inside surface of the tubular nanopetals. Inset of Figure 7f shows a single nanocrystal as building block, scale bar = 15 nm. (g) Schematic illustration of a suggested mechanism for the assembly of higher-order structures by nucleating new crystals through an edge-sharing configuration.

A new morphology was also observed when the carbon nitride was created using a laser power of 75 mJ for 3 h ablation and placed onto a silicon substrate, and then dried in a CPD. Figure 8e suggests that the product has a grass-like shape consisting of well-defined, protruding, short NRs. These structures can be assigned to the second class of morphology, mentioned previously. In contrast, if the substrate was dried in air (water evaporation time ~8 h), the individual NRs vertically assembled into bundles (see the highlighted black box region in Figure 8f) and formed stem-like structures. When the evaporation time of the liquid was increased to 24 h, the NRs developed a more complicated 3D structure, as shown in Figure 8g.

A possible explanation is that the longer duration for evaporation of the liquid allows sufficient time for the diffusion of all the NPs out of the flowers, leaving some liquid trapped in the gaps between the NR framework. With prolonged time, the wet NR framework has the opportunity to recrystallise or restructure into the shapes shown in Figure 8f-g.

Thus, by combining LP-PLA techniques with bottom-up self-assembly seeded growth, 3D complex well-arranged carbon nitride architectures can be created, with their structures and spatial organisation of the substrate controlled by deposition and drying conditions. The synthetic approach described above can also be extended to a range of other solid materials. For example, we have also demonstrated that a novel hierarchical and self-similar growth process can happen in zinc oxide materials by LP-PLA of a zinc plate submerged in aqueous SDS solution. The growth mechanism here also appears to involve an increase of the structural complexity from 0-dimensional nanoparticles to 1-dimensional nanorods, and then broadening of these into 2-dimensional 'nano-leaf' structures. Details can be seen in our previous report [79].

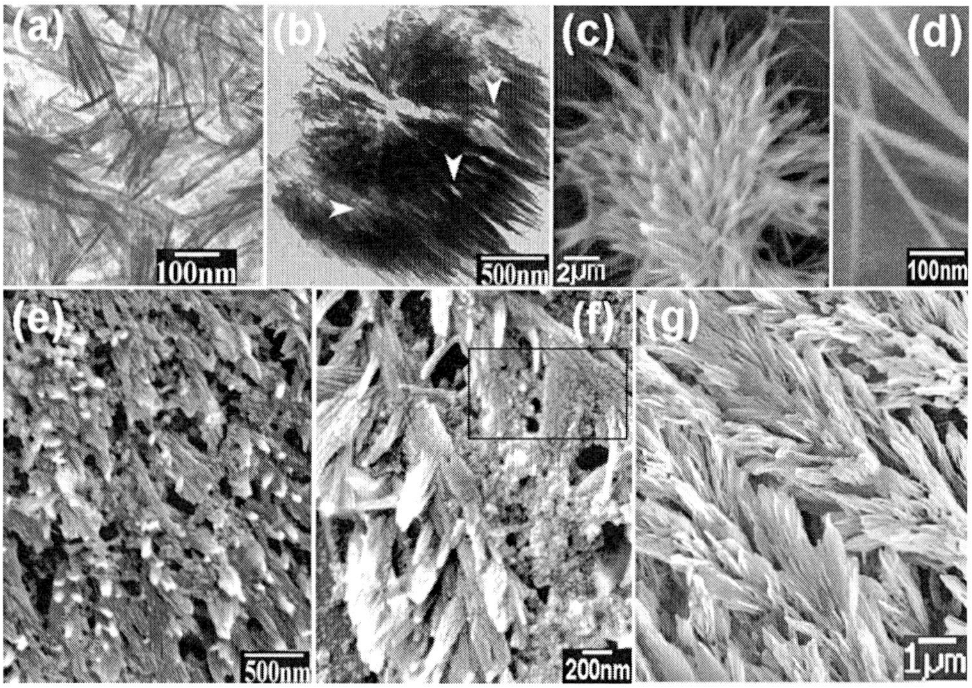

Figure 8. (a) TEM image of isolated carbon nitride NRs obtained by LP-PLA in 25% ammonia solution (laser fluence at 50 mJ per pulse, $t = 2$ h). (b) TEM image of an incomplete carbon nitride flower (25% ammonia solution, laser fluence at 125 mJ per pulse, $t = 10$ h) formed by the coalesced NRs. The arrows highlight the channels between the shell and the core. (c) After drying for 12 h (sample conditions identical to that in Figure 8a), numerous aligned NRs form the framework of the flower. (e) Grass-like structure constructed from many protruding NRs (25% ammonia solution, laser fluence at 75 mJ per pulse, $t = 3$ h, drying at CPD). (f) Stem-like structure consisting of vertical NR branches (region highlighted by black box), sample conditions identical to that in Figure 8e, but with drying in air. (g) After 24 h drying, the NRs have restructured to form 3D shapes. (Reproduced from [51] with the permission of the RSC).

5. CONCLUSION

With important discoveries continuing to be made in the field of nanomaterials, fulfilling the potential applications will require a greater control of material properties, along with a better understanding of the mechanisms for nanocrystal nucleation and growth. Extensive progress towards real-life applications have been made with ZnO, Cu2O, TiO2 and SiO2 nanostructures, because work on these materials is well established [74]. However, research on the understanding of the self-assembly processes of Group IV-V compounds, such as carbon nitride, is still in its infancy, and synthesis of complex nanostructures has only just begun. Our work has shown that it is possible to form self-assembled nanostructures and microstructures of crystalline carbon nitride. Through pulsed laser ablation of graphite in ammonia solutions, well-defined carbon nitride nanocrystal seeds were prepared. In the second step, 3D well-arranged architectures could be generated after these solutions were seeded onto a suitable substrate. The mechanism appears to involve an increase of the structural complexity from 0D NPs to 1D NRs, and then broadening of these into 2D nanopetals, which finally coalesce to 3D flowers. Factors such as the ammonia concentration, laser energy, the reaction time and the evaporation rate of the liquid are important in defining the structures observed. This unique process provides more insight into laser-induced chemical reactions in general, and of crystalline carbon nitride phases in particular, and may provide an alternative synthesis method for other composite nanostructures. A key advantage afforded by this aqueous multistage approach is the ability to systematically generate and control a diverse array of structures from a simple and inexpensive chemical route. With this improved understanding and concomitant seed control, LP-PLA may provide versatile and powerful industrial scale production processes for assembling complex architectures by specific design.

REFERENCES

[1] Whitesides, G. M.; Grzybowski, B. *Science* 2002, *295*, 2418-2421.

[2] Whitesides, G. M.; Boncheva, M. *Proc. Natl. Acad. Sci. USA* 2002, *99*, 4769-4774.

[3] Tang, Z. Y.; Kotov, N. A.; Giersig, M. *Science* 2002, *297*, 237-240.

[4] Liu, B.;Zeng, H.C. *J. Am. Chem. Soc.* 2005, *127*, 18262-18268.

[5] Kempa, K.; Kimball, B.; Rybczynski, J.; Huang, Z. P.; Wu, P. F.; Steeves, D.; Sennett, M.; Giersig, M.; Rao, D. V. G. L. N.; Carnahan, D. L.; Wang, D. Z.; Lao, J. Y.; Li, W. Z.; Ren, Z. F. *Nano Lett.* 2003, *3*, 13-18.

[6] Duan, X. F.; Lieber, C. M. *Adv. Mater.* 2000, *12*, 298-302.

[7] Sun, Y.; Fuge, G. M.; Fox, N. A.; Riley, D. J.; Ashfold, M. N. R. *Adv. Mater.* 2005, *17* 2477-2480.

[8] Yang, H. G.; Zeng, H. C. *J. Phys. Chem. B* 2004, *108*, 3492-3495.

[9] Mirkin, C. A.; Letsinger, R. L.; Mucic, R. C.; Storhoff, J. J. *Nature* 1996, *382*, 607-609.

[10] Sun, Y. G.; Gates, B.; Mayers, B.; Xia, Y. N. *Nano lett.* 2002, *2*, 165-168.

[11] Li, Y.D.; Li, X. L.; Deng, Z. X.; Zhou, B. C.; Fan, S. S.; Wang, J. W.; Sun, X. M *Angew. Chem. Int. Ed.* 2002, *41*, 333-335.

[12] Motte, L; Billoudet, F.; Pileni, M. P. *J. Phys. Chem.* 1995, *99*, 16425-16429.

[13] Maillard, M.; Motte, L.; Pileni, M. P. *Adv. Mater*. 2001, *13*, 200-204.
[14] Liu, B.;Zeng, H.C. *J. Am. Chem. Soc*. 2004, *126*, 16744-16746.
[15] Petit, C.; Legrand, J.; Russier, V.; Pileni, M. P. *J. Appl. Phys.* 2002, *91*, 1502-1508.
[16] Ngo, A. T.; Pileni, M. P. *Adv. Mater*. 2000, *12*, 276-279.
[17] Liu, J. P.; Huang, X. T.; Li, Y. Y.; Sulieman, K. M.; He, X.; Sun, F. L. *J. Mater. Chem.* 2006, *16*, 4427-4434.
[18] Wang, J. B.; Zhang, C. Y.; Zhong, X. L.;Yang, G. W. *Chem. Phys. Lett.* 2002, *361*, 86-90.
[19] Dijkkamp, D.; Venkatesan, T.; Wu, X. D.; Shaheen, S. A.; Jisrawi, N.; Minlee, Y. H.; Mclean, W. L.; Croft, M. *Appl. Phys. Lett.* 1987, *51*, 619-621.
[20] Radhakrishnan, G.; Adams, P. M. *Appl. Phys. A* 1999, *69*, S33-S38.
[21] Pappas, D. L; Saenger, K. L.; Bruley, J.; Krakow, W.; Cuomo, J. J.; Gu, T.; Collins, R. W. *J. Appl. Phys.*, 1992, *71*, 5675-5864.
[22] Loir, A. S.; Garrelie, F.; Donnet, C.; Belin, M.; Forest, B.; Rogemond, F.; Laporte, P. *Thin Solid Films*, 2004, *453*, 531-536.
[23] Patil, P. P.; Phase, D. M.; Kulkarni, S. A.; Ghaisas, S. V.; Kulkarni, S. K.; Kanetkar, S. M.; Ogale, S. B.; Bhide, V. G. *Phys. Rev. Lett.*, 1987, *58*, 238-241.
[24] Ogale, S. B. *Thin Solid Films* 1988, *163*, 215-227.
[25] Yang, G. W. *J. Prog. Mater. Sci.* 2007, *52*, 648-698.
[26] Pyatenko, A.; Shimokawa, K.; Yamaguchi, M.; Nishimura, O.; Suzuki, M. *Appl. Phys. A.* 2004, *79*, 803-806.
[27] Tsuji, T.; Iryo, K.; Watanabe, N.; Tsuji, M.; *Appl. Surf. Sci.* 2002, *202*, 80-85.
[28] Dolgaev, S. I.; Simakin, A.V.; Voronov, V. V.; Shafeev, G. A.; Bozon-Verduraz, F. *Appl. Surf. Sci.* 2002, *186*, 546-551.
[29] Tarasenko, N. V.; Butsen, A. V.; Nevar, E. A.; Savastenko, N. A. *Appl. Surf. Sci.* 2006, *252*, 4439-4444.
[30] Compagnini, G.; Scalisi, A. A.; Puglisi, O. *J. Appl. Phys.* 2003, *94*, 7874-7877.
[31] Tsuji, T.; Hamagami, T.; Kawamura, T.; Yamaki, J.; Tsuji, M. *Appl. Surf. Sci.* 2005, *243*, 214-219.
[32] Wang, J. B.; Zhang, C. Y.; Zhong, X. L.; Yang, G. W. *Chem. Phys. Lett.* 2003, *367*, 10-14.
[33] Yang, G. W.; Wang, J. B.; *Appl. Phys. A* 2000, *71*, 343-344.
[34] Wang, J. B.; Zhang, C. Y.; Zhong, X. L.; Yang, G. W. *Chem. Phys. Lett.* 2002, *361*, 86-90.
[35] Yang, L.; May, P. W.; Yin, L.; J. A.; Rosser, K. N. *Diamond Relat. Mater.* 2007, *16*, 725-729.
[36] Nath, M.; Rao, C. N. R.; Popvitz-biro, R.; Albu-Yaron, A.; Tenne, R. *Chem. Mater.* 2004, *16*, 2238-2243.
[37] SDS is sodium dodecyl sulfate.
[38] Liang, C. H.; Sasaki, T.; Shimizu, Y.; Koshizaki, N. *Chem. Phys. Lett.* 2004, *389*, 58-63.
[39] Sasaki, T.; Liang, C. H.; Nichols, W. T.; Shimizu, Y.; Koshizaki, N. *Appl. Phys. A* 2004, *79*, 1489-1492.
[40] Liang, C. H.; Shimizu, Y.; Sasaki, T.; Koshizaki, N. *J. Phys. Chem. B* 2003, *107*, 9220-9225.
[41] Liang, C. H.; Shimizu, Y.; Sasaki, T.; Koshizaki, N. *Appl. Phys. A* 2005, *80*, 819-822.

[42] Liang, C. H.; Shimizu, Y.; Sasaki, T.; Koshizaki, N. *J. Mater. Res.* 2004, *19*, 1551-1557.
[43] Zeng, H. B.; Cai, W. P.; Li, Y.; Hu, J. L.; Liu, P. S. *J. Phys. Chem. B* 2005, *109*, 18260-18266.
[44] Ishikawa, Y.; Shimizu, Y.; Sasaki, T.; Koshizaki, N. *J. Colloid Interf. Sci.* 2006, *300*, 612-615.
[45] LDA is lauryl dimethylaminoacetic acid.
[46] CTBA is cetyltrimethylammonium bromide.
[47] Usui, H.; Sasaki, T.; Koshizaki, N. *Appl. Phys. Lett.* 2005, *87*, 063105-3.
[48] Yang, L.; May, P. W.; Yin, L.; Brown, R.; Scott, T. B. *Chem. Mater.* 2006, *18*, 5058-5064.
[49] Yang, L.; May, P. W.; Yin, L.; J. A.; Rosser, K. N. *J. Nanopart. Res.* 2007, DOI 10.1007/s11051-006-9192-4.
[50] Liu, A. Y.; Cohen, M. L. *Science* 1989, *245*, 841-842.
[51] Yang, L.; May, P. W.; Huang, Y. Z.; Yin, L. *J. Mater. Chem.* 2007, *17*, 1255-1257.
[52] Yang, L.; May, P. W.; Yin, L.; Huang, Y. Z.; Smith, J. A.; Scott, T. B. *Nanotechnology* 2007, *18*, 335605 (1-5).
[53] Yang, L.; May, P. W.; Yin, L.; Scott, T. B.; Smith, J. A.; Rosser, K. N. *Nanotechnology* 2006, *17*, 5798-5804.
[54] Liu, A.Y.; Wentzcovitch, R. M. *Phys. Rev. B* 1994, *50*, 10362-10365.
[55] Wang, J. B.; Lei, J. L; Wang, R. H. *Phys. Rev. B* 1998, *58*, 11890-11895.
[56] Yin, L. W.; Bando, Y.; Li, M. S.; Liu, Y. X.; Qi, Y. X. *Adv. Mater.* 2003, *15*, 1840-1844.
[57] Zeng, H. C. *J. Mater. Chem*, 2006, *16*, 649-662.
[58] Sjöström, H.; Staftröm, S.; Boman, M.; Sundgren, J. E. *Phys. Rev. Lett.* 1995, *75* 1336-1339.
[59] Terrones, M.; Redlich, P.; Grobert, N.; Trasobares, S.; Hsu, W. K.; Terrones, H.; Zhu, Y. Q.; Hare, J. P.; Reeves, C. L.; Cheetham, A. K.; Rühle, M.; Kroto, H. W.; Walton, D. R. M. *Adv. Mater.* 1999, *11*, 655-658.
[60] Rodil, S. E.; Muhl, S. *Diamond Relat. Mater.* 2004, *13*, 1521-1531.
[61] Hellgren, N.; Johansson, M. P.; Broitman, E.; Hultman, L.; Sundgren, J. E. *Phys. Rev. B* 1999, *59*, 5162-5169.
[62] Muhl, S.; Méndez, J. M. *Diamond Relat. Mater.* 1999, *8*, 1809-1830.
[63] Hellgren, N.; Guo, J.; Såthe, C.; Agui, A.; Nordgren, J.; Luo, Y.; Ågren, H.; Sundgren, J. *Appl. Phys. Lett.* 2001, *79*, 4368-4350.
[64] Souto, S.; Pickholz, M.; Dos Santos, M. C.; Alvarez, F. *Phys. Rev. B* 1998, *57*, 2536-2540.
[65] Ripalda, J. M.; Montero, I.; Galan, L. *Diamond Relat. Mater.* 1998, *7*, 402-406.
[66] Ogata, K.; Chubaci, J. F. D.; Fujimoto, F. *J. Appl. Phys.* 1994, *76*, 3791-3796.
[67] Khabashesku, V. N.; Zimmerman, J. L.; Margrave, J. L. *Chem. Mater.* 2000, *12*, 3264-3270.
[68] Diaz, J.; Paolicelli, G.; Ferrer, S.; Comin, F. *Phys. Rev. B* 1996, *54*, 8064-8069.
[69] Tabbal, M.; Merel, P.; Moisa, S.; Chaker, M.; Ricard, A.; Moisan, M. *Appl. Phys. Lett.* 1996, *69*, 1698-1700.
[70] Le Normand, F.; Hommet, J.; Szörényi, T.; Fuchs, C.; Fogarassy, E. *Phys. Rev. B* 2001, *64*, 235416.

[71] Beamson, G.; Briggs, D.; *High Resolution XPS of Organic Polymers*; 1992, John Wiley and Sons, Chichester, pp.182.
[72] Guo, Y. J; Goddard, W.A. *Chem. Phys. Lett.*, 1995, *237*, 72-76.
[73] Niesen, T. P; De Guire, M. R. *J. Electroceram.* 2001, *6*, 169-207.
[74] Sounart, T. L.; Liu, J.; Voigt, J. A.; Hsu, J. W. P.; Spoerke, E. D.; Tian, Z.; Jiang, Y. B. *Adv. Funct. Mater.* 2006, *16*, 334-344.
[75] Lu, S. N; Chung, J.; Ruoff, R.S. *Nanotechnology* 2005, *16*, 1765-1770.
[76] Brinker, C. J.; Lu, Y. F.; Sellinger, A.; Fan, H. Y. *Adv. Mater.* 1999, *11*, 579-585.
[77] Liu, A. Y.; Cohen, M. L. *Phys. Rev. B* 1990, *41*, 10727-10734.
[78] Li, Z. Q.; Xiong, Y. J.; Xie, Y. *Nanotechnology* 2005, *16*, 2303-2308.
[79] Yang, L.; May, P. W.; Yin, L.; Scott, T. B. *Nanotechnology* 2007, *18*, 215602 (1-5).

In: Encyclopedia of Nanotechnology
Editor: Elwood D. Carlson
ISBN: 978-1-60692-079-4

Chapter 10

FORMATION AND ASSEMBLY OF HIERARCHICAL NANOARCHITECTURES BY BIOINSPIRED TECHNIQUES

Huilan Su, Di Zhang and Qun Dong
State Key Lab of Metal Matrix Composites,
Shanghai Jiaotong University, Shanghai 200030, China

ABSTRACT

The bioinspired technique has developed into an intersection of materials science, soft chemistry technique, and biotechnology. In this chapter, the bioinspired strategies integrating biotemplate, biomineralization, and biomimesis are adopted to fabricate functional nanomaterials with hierarchical architectures and interrelated unique properties.

Some natural biomaterials, such as eggshell membrane, silk fiber, etc., are introduced into different liquid processes to construct a series of functional hierarchical nano-architectures (containing oxides, chalcogenides, noble metals, and/or their composites). Under the direction of the biotemplates or biosubstrates, the nanocrystallites units of about 2-10 nm are formed, developed, and assembled into hierarchical architectures from the nanoscale to the macroscale. The nucleation, growth, and assembly of nanocrystallite units depend on the collaboration of special proteins structures and functional biomacromolecules of relevant biomaterials as well as the processing conditions. Moreover, the biomaterials usually provide both the physical substrates and the inductive substances to direct some complicated reactions. In addition, some amino acids of the biomaterials function as the capping agents to achieve the organized growth and assembly of small-sized nanocrystallites. As-synthesized hierarchical nanoarchitectures take on outstanding chemical or physical properties and have valuable applications in photocatalysis, gas sensing, ductile ceramics, and semiconductor technology.

This method and the relevant ideas provide a novel and versatile avenue to synthesize a new family of functional nanomaterials by integrating material science, chemistry, and biotechnology.

INTRODUCTION

Nanoscale functional materials have attracted much attention because of their potentially excellent and fascinating properties. In the progressive stage, the assembly of the nanoscale units into a hierarchical architecture would be required to obtain higher functionality and performance. Nevertheless, there are few feasible reports on the construction of hierarchical nanomaterials, especially practical and functional nanocomposites.

"Where nature finishes producing its own species, man begins, using natural things, with the help of this nature, and creates an infinite number of species". These words epitomizing the artist-scientist Leonardo da Vinci fully embody the competence of human beings to utilize natural substances for synthesizing new materials. Natural organisms or biological structures have evolved as a result of highly complex and elegant mechanism, the secrets gleaned from nature are a great inspiration for developing inorganic functional materials with hierarchical structures. Recently, mimicking vital functions in aqueous systems has been noteworthy as a promising nanofabrication technique for functional materials. Quite a few biological species have been utilized to construct inorganic materials, the list includes oyster shells, sea urchin spines, DNA, viruses, cuttlefish bone, nacre, and peptide nanotubes, diatoms and skeletal plates, butterfly wings, shell membrane, silk fiber, hair, wood, and other plants (leaves, paper, pollen grains), etc. These biogenic techniques enable the construction of hierarchical architectures including macro- and nanostructures by growing inorganic crystals with coexisting biospecies under ambient conditions.

Our bioinspired synthesis strategies integrating biotemplate, biomineralization, and biomimesis are based on the premise that amphiphilic and bioorganic macromolecules can be utilized for tailoring elaborate hierarchical biostructures through the self-assembly of nanoscale units to engender new functional properties.

1. SYNTHESIS OF HIERARCHICAL NANOSTRUCTURED METAL OXIDES WITH INTERWOVEN NETWORKS BY EGGSHELL MEMBRANE DIRECTED SOL–GEL TECHNIQUE

It has been discovered that in some biological systems such as diatoms and sponges, the formation of solid silica nanostructures with precisely controlled morphologies is the function of proteins and polysaccharides. These biomacromolecules can direct and bind the nucleation and the growth of inorganic particles. As a convenient and representative biomaterial, eggshell membrane (ESM) may be nominated to direct the fabrication of hierarchical nanostructured oxides.

The avian eggshell is composed of multilayered membranes and a calcified extracellular matrix, which are sequentially assembled during the motion of the egg along the oviduct. The examination of cross-fractured eggshell by FESEM, reveals that the calcified shell is composed of an inner mammillary layer (ML) and an outer palisade layer (PL) resting upon the shell membranes. The ESM is non-mineralized, collagen-based matrices interposed between the egg white and the mineralized shell, and consists of outer shell membrane (OSM), inner shell membrane (ISM), and limiting membrane (LM). The three layers are readily distinguished. Those deposited sparsely on the outer layer of the shell membranes are the mammillary knobs for penetrating into the tips of the mammillary cones.

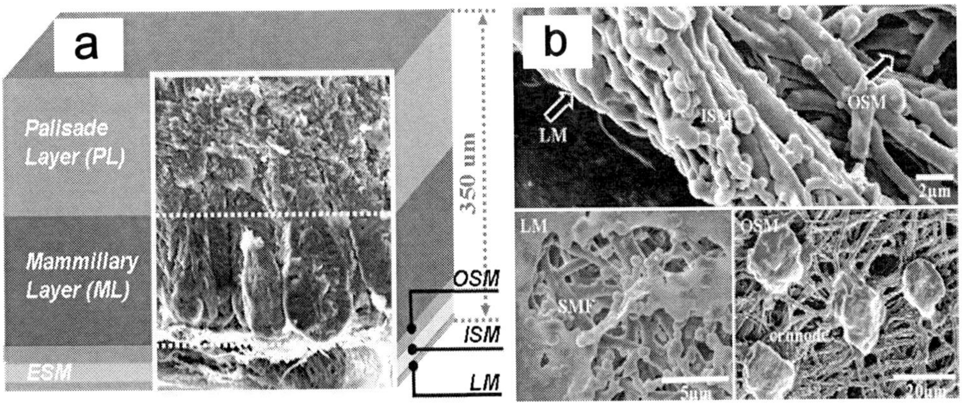

Figure 1. (a) FESEM images of eggshell membrane and the corresponding schematic illustration of eggshell structures. (b) FESEM images revealing the relative locations and the configurations of OSM, ISM, and LM.

In addition, the interwoven and coalescing nature of the shell membrane fibers (SMF) with an average width ranging from 500 nm to 2 μm is evident both in the OSM and the ISM. Furthermore, some of the broader fibers have a longitudinally striated appearance, where two or more narrower fibers join each other for part of their length. The amorphous LM shows an undulant appearance, which is due to a layer of spherical protein granules formed between the ISM and the LM. Within each layer of ESM, the fibers are randomly arranged.

Quoting from Cook et al. [1], "the synthetic replication of evidently useful biological structures by a simple casting process is expected to introduce some of the superb properties of biological structures into man-made materials." The most feasible method of attaining this goal at a high level seems to be the aqueous sol–gel technique. The colloid components can interact with some biomacromolecules and undergo surface-preferred gelation to form the coats on the biomaterials, even to veraciously replicate the biotemplate frameworks down to the nanoscale. Moreover, the sol–gel process offers several advantages over other methods, including simplicity of procedure, lower processing temperature, and better homogeneity. Commonly, metallic alkoxides are used as starting materials to prepare fine-scaled metallic oxides by the sol–gel processes. However, metallic alkoxides are both expensive and extremely sensitive to moisture, heat, and light, implying that the process is relatively difficult to control and not economical. Therefore, various metal salts are preferred as the precursors. Once biomaterials are introduced to the reformative sol–gel techniques, a series of intricate hydrolysis and condensation reactions happen in the sol–gel process. It seems to be a process that enable the easy preparation of hierarchical metal oxides, such as TiO2, SnO2, SiO2, ZnO, ZrO2, and so on. Therefore, biomimetic synthesis associated with the sol–gel technique may be an ideal approach for the design and fabrication of functional materials with predetermined physical and chemical properties. Our present study focuses on exploring the bottom up assembly of hierarchical metal oxides by a biogenic synthesis technique introducing the ESM as the template.

1.1. Synthesis of Hierarchical Mesoporous Titania with Tunable Anatase and Rutile

Mesoporous titania has outstanding chemical and physical properties, and widespread applications in photovoltaics, photocatalysis, ductile ceramics, pigmentation, optics, and so on, which depend more on the phases and crystallite sizes of building particles and relevant hierarchical structures. Therefore, the design of hierarchical mesoporous titania with tailored structures and morphologies can provide promising properties.

Herein, ESM is chosen as the biotemplate, in view of the special proteins patterns and functional residues of the ESM biomacromolecules, to synthesis hierarchical interwoven titania through a sol–gel approach followed by thermal decomposition. In a typical procedure, titanium tetrachloride ($TiCl_4$) is added slowly into ice water under stirring to obtain 0.04 M $TiCl_4$ colloid medium, and the pH values of the media are adjusted to 1, 2, and 3 by adding 1 M NaOH solution, respectively. The membrane is then immersed in the above colloid media for 13 h at room temperature, subsequently rinsed with distilled water and dried naturally to obtain gel-ESM hybrids. The hybrids are calcined at 400, 550, 700, and 800 °C (heating rate 35, 18, and 1 °C/ min), respectively, to form the resulting white sheets.

TGA results show that the ESM template start pyrolyzing at 200 °C and absolutely pyrolyzed by 550 °C. The gel-ESM hybrids containing titanic and titanyl precursors also decompose around 200 °C, and 18.3% (by mass) inorganic ingredients remain when the template ESM is completely removed at 550 °C. XRD results indicate that natural ESM and the hybrids are both totally amorphous. After calcination at 400 °C, TiO_2 crystallites come into being. However, the calcination temperature about 550 °C is appropriate for TiO_2 formation attended by the thorough removal of the biotemplate. The crystallization of TiO_2 can be well established at 550 °C. An average crystallite size of about 6 nm is estimated according to the line width analysis of the TiO_2 (101) diffraction peaks based on the Scherrer formula. Higher calcination temperature makes for the increase in the size of titania crystallites. When the calcination temperature is increased to 800 °C, some diffraction peaks of rutile TiO_2 increasing abruptly, indicates that the phase transformation corresponding to rutile TiO_2 mostly occurs at this temperature.

Representative FESEM images of titania prepared at pH 2 and 550 °C compared with the original ESM templates including OSM, ISM, and LM, suggest that the refined reprography of the hierarchy is fully achieved. The obtained OSM-based TiO2 are composed of interwoven fibers and sparsely crunodes depositing on the ceramic fibers with the average interval about 11 μm, which is smaller than the original spacing 31 μm due to the heat shrinkage. It is obvious that the refined meshworks of natural OSM are also duplicated besides the larger intersectional blocks. The interwoven and coalescing fibers of TiO2 replica range from 0.2 to 1.2 μm in diameter in comparison with original SMF approximate 0.5 to 3.0 μm in diameter. Although considerable shrinkage is engendered during the calcination, it can be lowed by repeatedly immersing in the sol with no changes in the configuration. A typical image of the cross section shows that the fibers complecting the networks are tubular titania. Moreover, the thickness of obtained TiO2 networks is decided by the thickness of primeval ISM. As ISM is composed of multilayers of interlaced fibers, the thickness of the replica TiO2 can be adjusted according to the actual thickness of ISM layers retained up to 20 μm. When using LM as the template, a lumpy titania layer composed of contiguously hemispherical caps is obtained. These special silkworm-like structures come into being and array according to the spherical protein granules distributing under LM. These hemispherical prominences have the diameter of about 900 nm, which is in agreement with the size of slightly smaller protein granules. Therefore, it can be concluded that a precision replication of ESM hierarchical structures from OSM to innermost LM could be achieved by nanocrystalline titania.

To investigate the detailed morphologies and the structures of ESM-morphic titania, representative FESEM images of the samples calcined at different temperature (heating rate 35 °C/min) show the interwoven and coalescing TiO_2 fibers obtained at 550 °C ranging from 0.2 to 1.2 μm in diameter in comparison with that of the original membrane fibers. Although there exists considerable shrinkage, the overall refined reprography of the interwoven hierarchy is retained.

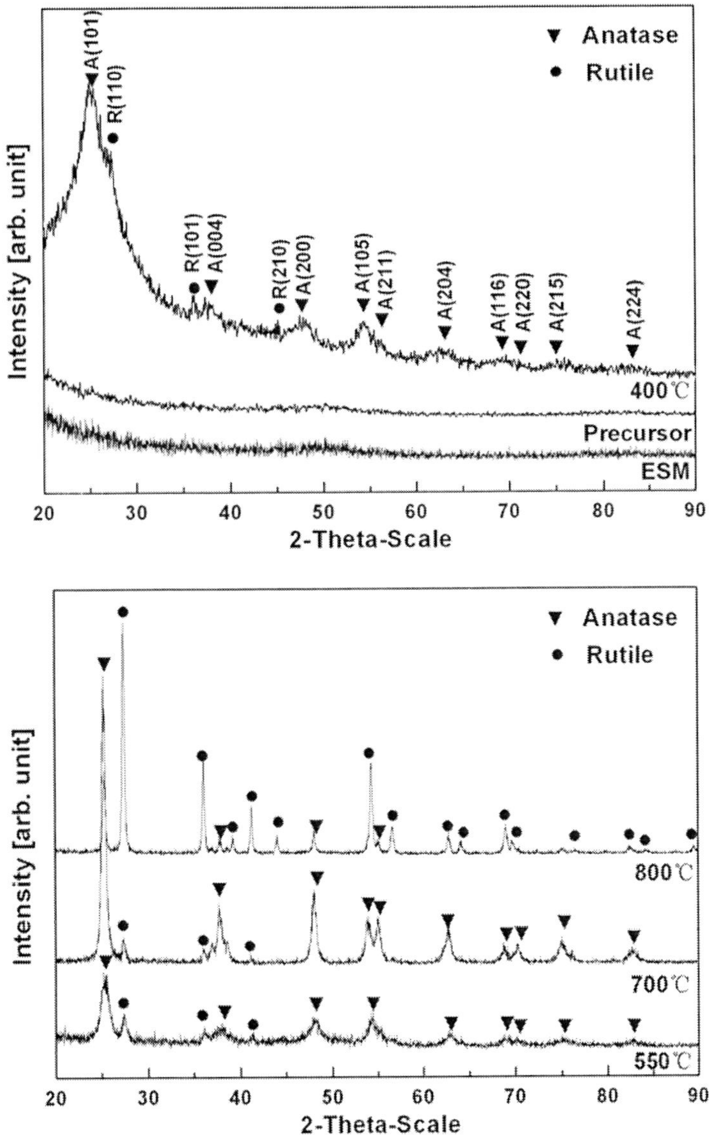

Figure 2. XRD patterns of ESM, gel-ESM precursor prepared at pH 2, and the sinters.

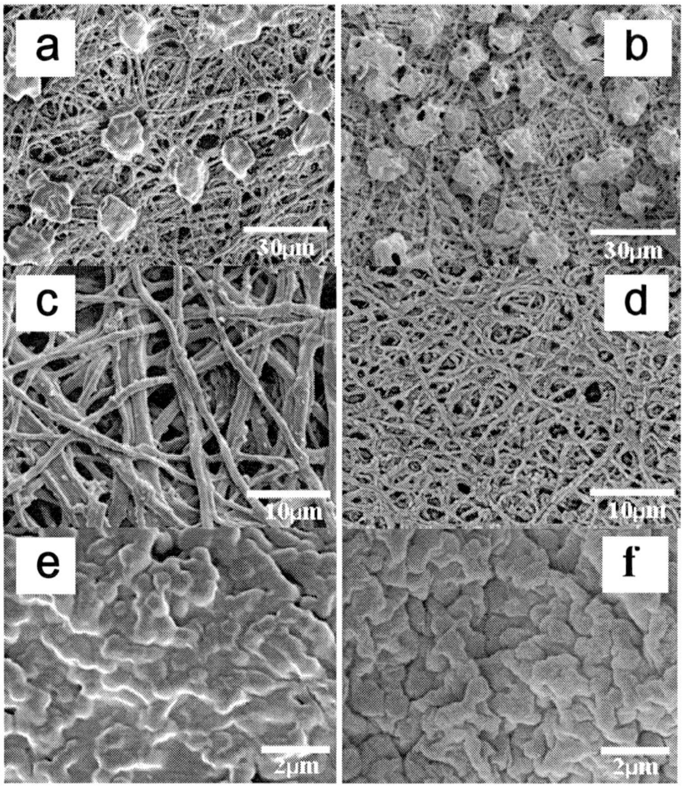

Figure 3. FESEM images showing the different layers of original ESM: (a) OSM, (b) ISM, and (c) LM; compared with the morphologies of as-prepared titania: (d) OSM-based, (e) ISM-based, and (f) LM-based TiO_2, respectively.

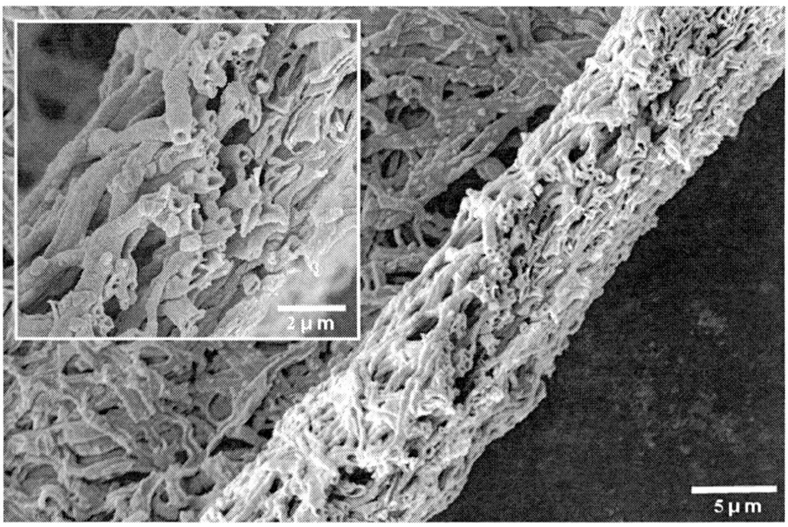

Figure 4. FESEM images of TiO_2 network by ESM using the sol-gel approach showing the overview of the cross section, and higher magnification of the cross section showing the broken hollow tubes.

When the calcination temperature is elevated to 700 °C, there are some coils due to higher temperature and presents interconnected networks with wavy figures. These characteristics are extended to an utmost as increasing calcination temperature to 800 °C. Obviously, titania networks have turned into unfolded mesh structures. With the higher magnification, it is observed that the fibers are vividly composed of short rods presenting porous figures. These fibers are inclined to stretch into belts, which are twice as large as ones at 550 °C in diameter. As a while, calcination temperature is crucial for the synthesis, which realizes the replication of ESM networks by titania below 700 °C while becoming laminated meshworks above 800 °C. At given calcination temperature, other parameters become more important. Heating rate of the whole calcination process may influence the morphologies and the crystal growth of target materials. Comparing with the higher heating speed 35 °C/ min, the heating rates about 18 and 1 °C/min are introduced to the corresponding process. With 18 °C/min at a medial level, ESM-morphic titania can also be formed, and some drapes along the fibers are found with careful observations.

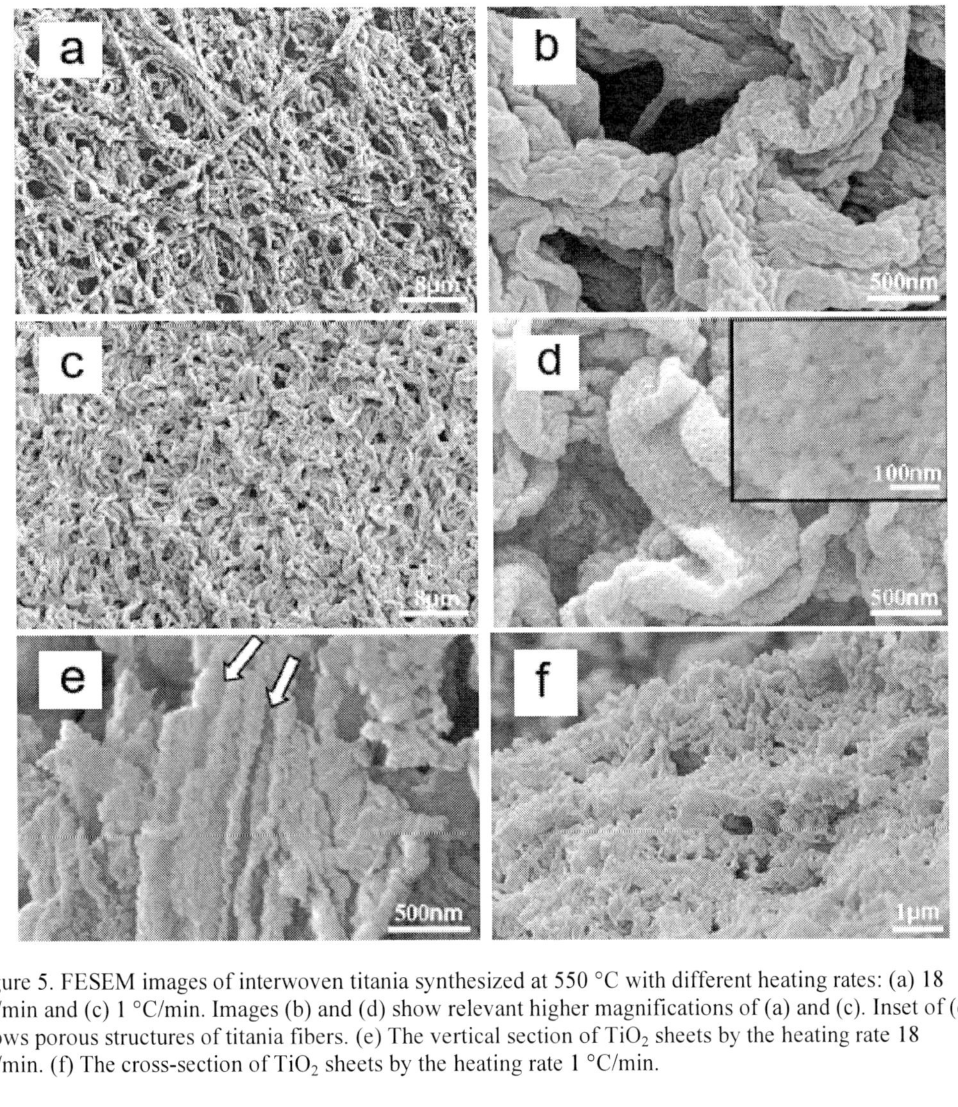

Figure 5. FESEM images of interwoven titania synthesized at 550 °C with different heating rates: (a) 18 °C/min and (c) 1 °C/min. Images (b) and (d) show relevant higher magnifications of (a) and (c). Inset of (d) shows porous structures of titania fibers. (e) The vertical section of TiO_2 sheets by the heating rate 18 °C/min. (f) The cross-section of TiO_2 sheets by the heating rate 1 °C/min.

Those subtle drapes may be formed during the sol–gel process, which contains the intricate interactions between the colloids and ESM biomacromolecules. When the heating rate reduced to 1 °C/min, the minutiae have vanished and more pores come into being. The low heating rate may result in particle agglomeration, with the crystallites grown at a larger scale to larger one, to blot out the minutiae and show porous structure. Moreover, interwoven fibers are arranged and fallen in loose folds. Probably based on the prominent dehydration of organic substances under the low heating rate, the framework of biotemplate ESM shrinks stepwisely before its completely pyrogenation. The TiO2 fibers constituting the porous networks might be tubular structures. The sample calcined at 550 °C with the heating rate 18 °C/min is consisted of channels about 200 nm in diameter, which are probably formed by the gelation and the crystallization of titania colloids along ESM fibers during the removal of the template. While the heating rate decreases to 1 °C/min, the sample calcined at 550 °C displays porous modality. Actually, these pores embody the channels as same as the above, and seems smaller as a result of overlay by the growth of building units with the lower heating rate.

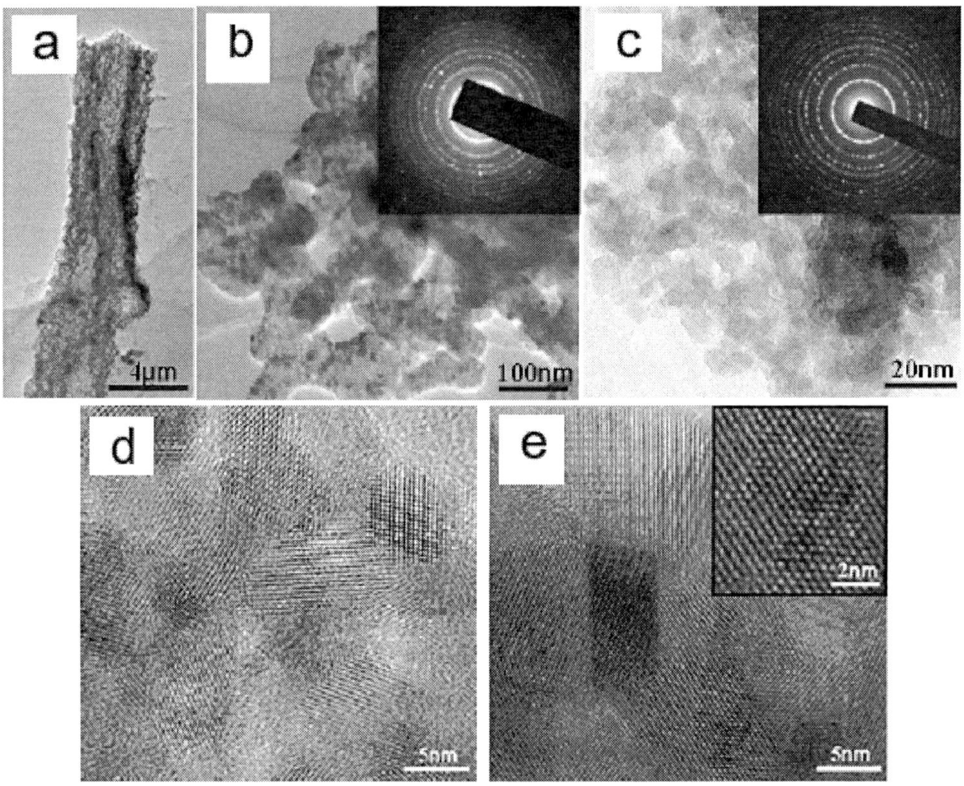

Figure 6. TEM images of (a) an individual TiO_2 fiber; the building particles of titania fibers synthesized at 550 °C with different heating rate: (b) 1 °C/min and (c) 35 °C/min. Insets in (b) and (c) show corresponding SAED patterns. HRTEM images of ESM-morphic TiO_2 obtained at 550 °C and pH 2 with the heating rate: (d) 35 °C /min; (e) 1 °C /min. Inset in (e) corresponds to anatase TiO_2.

The above analysis about the growth of the titania units is further proved by TEM results. The porous fibers synthesized at 550 °C with the heating rate 1 °C/ min are assembled by small particles. At higher magnification, these particles with an average size of about 50 nm, containing tiny nanocrystallites, are clearly seen. With the heating rate 35 °C/min, these particles show the serried and ordered assembly with the average size about 6 nm, which makes an allusion to the replication readily achieved by the small particles rather than large ones. The SAED patterns show that nanoscaled TiO_2 synthesized with different heating rate are polycrystalline on the basis of the uniform central and diffraction spots, and the diffraction rings are indexed to (101), (004), (200), (105), (204), and (116) diffractions of titania. HRTEM images show the difference of TiO_2 nanocrystallites prepared with distinct heating rate at the same calcination temperature 550 °C. The sizes of these nanocrystallites are close to 5–6 nm. However, there are more micropores of about 2 nm observed in the sample prepared at higher heating rate. A careful look also reveals that the present TiO_2 is characterized by ''nanostructured'' particles consisting of the nanocrystalline domains with structural defects such as twins.

The pH value of the impregnant has a significant effect on the phase of TiO_2 crystalline products. The high acidity is in favor of the formation of rutile phase. When the pH value of the precursor is about 1, the product is a mixture of rutile (majority) and anatase (minor). With increasing pH to 2, the product is formed by a remixture of rutile (minor) and anatase (majority). It is evident that the content of anatase phase in the products increases with the mount of the pH value. At pH = 4, the rutile phase disappears yet the anatase phase becomes dominant. As pH > 4, only anatase phase can be obtained. The concentration of $TiCl_4$ medium has direct affinity with the acidity of the impregnant. Decreasing the concentration of $TiCl_4$ medium favors the formation of anatase TiO_2. When the concentration is tuned from 0.04 to 0.03 M, the peak intensity of rutile titania is abruptly weakened. As the concentration is decreased to 0.02 M, further to 0.01 M, there are feeble diffraction peaks of rutile TiO_2. All the broader reflection peaks of different samples indicate that as-prepared TiO_2 nanocrystallites have small dimensions. The average crystallite size of about 5–8 nm is estimated according to the line width analysis of the TiO_2 (101) diffraction peaks based on the Scherrer formula.

FESEM image also indicates that the interwoven and coalescing TiO_2 fibers prepared at pH 1 and calcined at 550 °C range from 0.2 to 1.5 mm in diameter in comparison with the original membrane fibers. Although considerable shrinkage, there retains the overall refined reprography of the interwoven hierarchy. At pH 2, there are some drapes engendered on the surfaces of titania fibers while the replica of ESM hierarchy can also be obtained. However, when the pH value is elevated to 3, the morphology of obtained titania has a mutation, the titania fibers are distorted, and a sort of connected petalshaped structure comes into being. The two TiO_2 samples involving either 0.04 M or 0.01 M $TiCl_4$ impregnant display the analogical hierarchical interwoven conformations. However, there exists the evident distinction resulted from the effects of the impregnant concentration. The titania fibers synthesized at higher concentration are composed of lager particles while smaller titania nanocrystallites are obtained at lower concentration. The lumpy structures may form due to the agglomeration among TiO_2 particles at the higher concentration or higher acidity of the media precursor. A careful look at the surface of the biomorphic titania reveals that some mesopores are occurred, it is more clear for the sample involved in 0.01M $TiCl_4$ impregnant. So based on the alteration of the pH value and the concentration of $TiCl_4$ impregnant, different morphologies of TiO_2 samples can come forth.

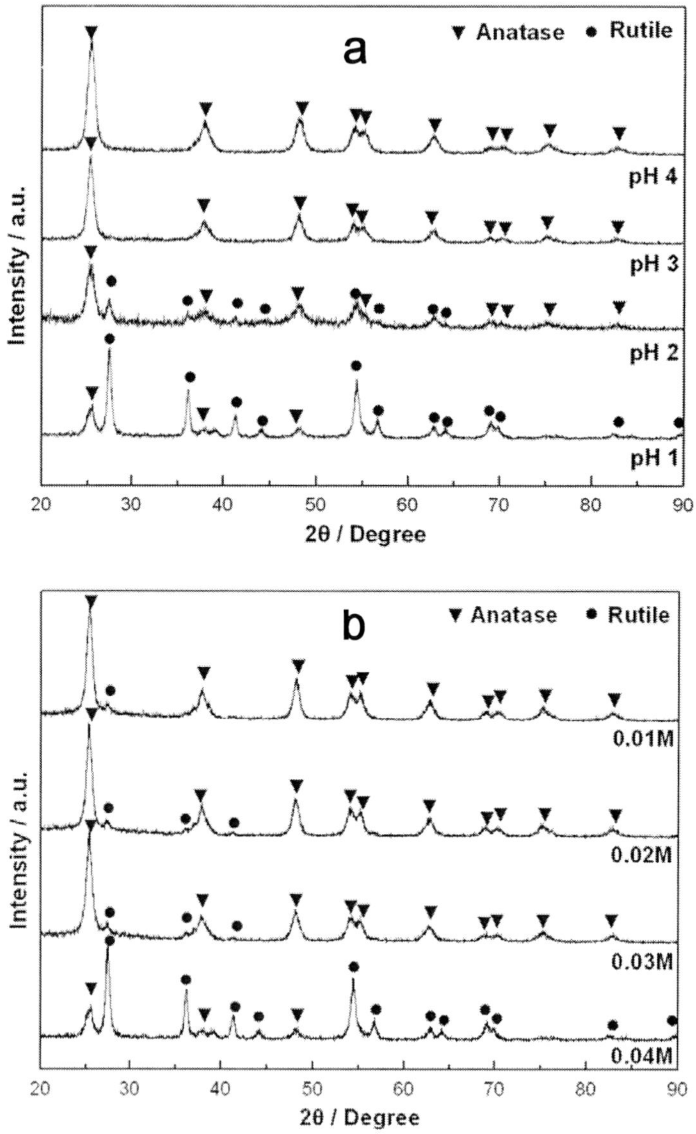

Figure 7. XRD patterns of Ti-ESM hybrids prepared under the different experimental conditions: (a) different pH value and (b) different concentration of the impregnant, and the sinter at 550 °C.

In balance, the synthesis conditions have momentous influences on the preparation of the interwoven titania. It observed that the higher acidity of the colloid impregnant is preferred to stabilize the ESM proteins, further to achieve the synthesis of interwoven titania. Dipping time is another concernful factor that may indirectly influence the morphology of resulting material. It has been affirmed that the proper dipping time can be tuned flexibly from 6 to 30 h. By decreasing the concentration of the colloid impregnant as well as shortening dipping time, an overall thinning of the wall and increased shrinkage are observed. The dilution of the nanoparticle dispersion leads to insufficient coating, thus the original structure is not maintained. Titania with ESM configuration is best obtained when the ESM immersed into pH 1 medium for 13 h, accompany a proper calcination treatment which is mainly responsible for the pyrogenation of the biotemplate and the crystallization of nanocrystalline TiO_2. The size of titania crystallites is mainly affected by the calcinations temperature, while the morphology is dominated not only by the calcination temperature, but also by the heating rate. However, the medium pH value and dipping time are crucial for the interfacial sol–gel process. The interactions between the colloids and ESM biomacromolecules present intricate and perplexing. It can be ascribed to a special sol–gel approach involving the biotemplate ESM, during which the biomacromolecules may direct the bottom-up assembly into hierarchical titania. To understand these investigations, a possible mechanism for this template-directed process is explored.

It is known that ESM are collagen-based matrices consisted of fibers of type I, V, and X collagen located in the core of each fiber. Each SMF is composed of a collagenous core, which contains smaller collateral channels, surrounded by a glycoprotein mantle. The core is predominantly composed of collagen and osteopontin proteins, while the mantle is rich in polyanions exhibiting a variety of keratan sulfate (KS) epitopes and Dermatan sulfate (DS). Lysozyme and clusterin are also present in ESM at high levels, and both have been detected within SMF, distributing discontinuously with no apparent partitioning between the core and the mantle. The cationic lysozyme can be bound to those anionic sites DS and particularly KS in the membrane. Elegant studies with clusterin have demonstrated that it exerts a powerful chaperone-like activity with lysozyme. In virtue of these special components and configurations, the inherent interactions of ESM and the colloids shall be daedal, and the correlations between the residues of SMF biomolecules and the colloid components are explored to understand the tamplate translation process. The FTIR spectrum of ISM shows the diagnostic peaks at 1641, 1532, and 1234 cm-1. These peaks are assigned to the amide I, amide II, and amide III bands of the proteins from natural ISM, respectively. The amide I peak, predominantly corresponding to the C=O stretch, shifts to 1632 cm-1 after ISM being immersed in the sol media. The red shift of the amide I peak indicates that the C=O bonds in the peptide chains are weakened because of the formation of new chelate bonds between C=O bonds and Ti ions involving in H-bonding with TiOH2+ or TiOH3+ surface groups. This indication implies that the carbonyl groups on the surface of glycoprotein biomolecules are the binding sites, and that the COO— symmetric stretching vibrations at 1446 shift to 1453 cm-1 of carboxylate anions complexed with surface Ti centers dominating in the spectra of the proteins with the sol precursor capped. The intensities of C—O—C and C—OH (at 1058 and 1400 cm-1, respectively) peaks are notably decreased, which indicates the interaction of hydroxy residues with cations and the transformation of glycoprotein configuration. It is evident not only the amide bands but also the characteristic COO— bands are not observable in the spectra of the sinters, thus confirming that the removal of the protein chains of the template. Above 2000 cm-1, both natural ISM and GMH exhibit the intense antisymmetric and symmetric —CH stretching vibrations (at 2926 and 2854 cm-1, respectively) of the —CH2— groups in the hydrocarbon moiety [2]. The shoulder at 2960 cm-1 can be associated with the asymmetric stretching of the terminal —CH3 group of the protein chains. In addition, the C—H stretching signals of the polypeptides at 3290 cm-1 are superimposed on an underlying surface —OH broad stretching band centered at 3400 cm-1. Below 900 cm-1, the characteristic vibrations of the Ti—O—Ti networks can be seen in the curves of GMH and the sinters according with the XRD results.

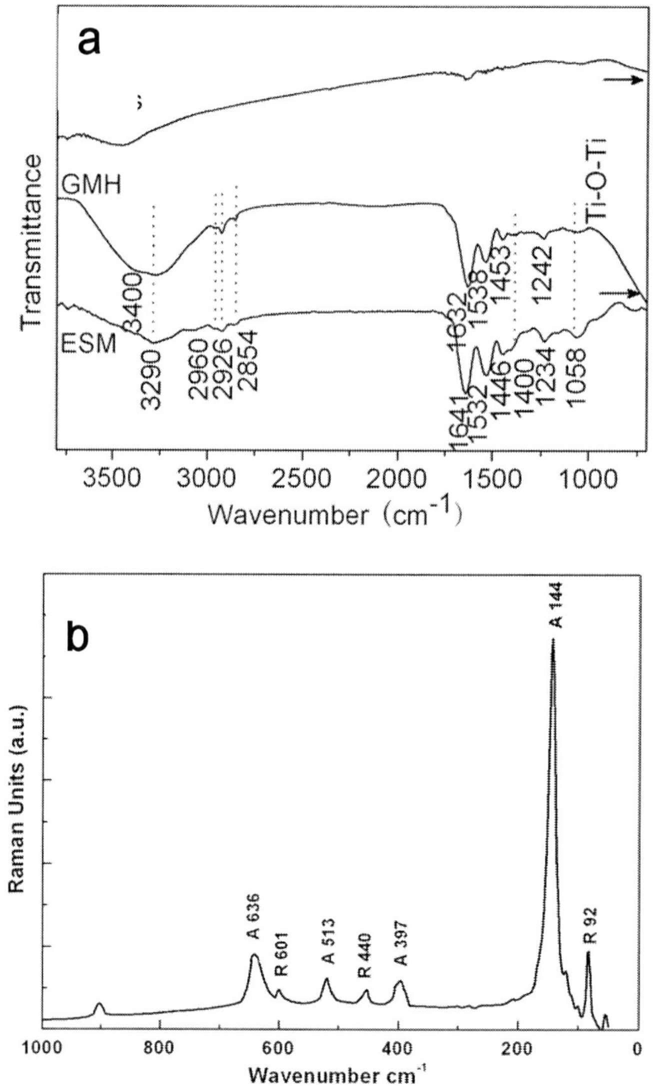

Figure 8. (a) FTIR spectra of original ISM, GMH prepared at pH 2, and the corresponding sinters at 550 °C, respectively. (b) Raman spectrum of ESM-templated titania prepared at pH 2 and 550 °C.

Herein, the interfacial sol-gel process is based on the adsorption of cationic matters from the sol media to the glycoprotein mantle of SMF containing abundant of carboxylic and hydroxy residues, subsequently a series of complicated reactions go along to drive the amorphous organogels for titanic and titanyl compositions, which further crystallize into titania nanocrystallites. The formation of nanocrystalline TiO2 can be denoted as: TiCl4 → TiOH3+ / TiOH2+ / TiO2+ → Ti (Ⅳ) oxo species → amorphous TiO2 → TiO2 nanocrystallites. The cationic TiOH3+ / TiOH2+ / TiO2+ in the sol system are prone to interact with carboxylic and hydroxy residues of glycoprotein, which has been proved by IR results. The interfacial sol-gel process would involve the strong affinity among some residues binding containing multiple hydrogen bonds, Van Der Waals interactions, and be enslaved to the ordering of surface polypeptide loops that bury a portion of the colloid particles in the protein interiors.

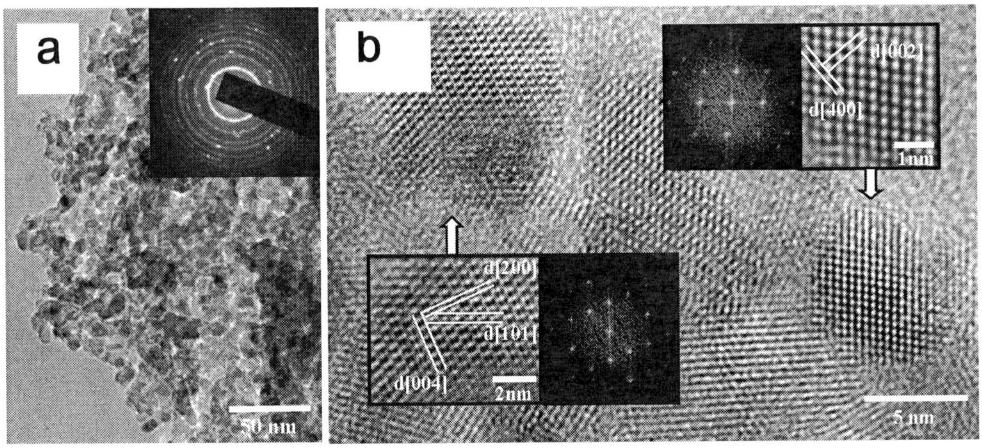

Figure 9. (a) TEM and (b) HRTEM Images of ISM-based TiO_2 obtained at 823K and pH 2. The inset in (a) showing the SAED pattern of as-prepared TiO_2; the insets in (b): FFT of the crystallites show verified the natase structure (left inset) and the rutile structure (right inset), which was established that the image was taken down the [100] zone axis.

The colloid particles can be bound to glycoprotein through the hydroxy groups of sugar chains linking to polypeptides, which are known to coordinate to titania gels [3]. Thereby the amorphous Ti-O networks can be derived from these reactions, and indeed an amorphous TiO2 film locating in the mantle is obtained with no change in the ESM configuration. The Ti (IV) oxo species is assumed to be an intermediate between TiO2+ and TiO2, consisting of partly dehydrated polymeric Ti (IV) hydroxide. Moreover, polyanions containing in the mantle tend to arrest the cationic H+. Therefore, in acidic media, the hydrolysis of Ti (IV) species is enhanced and the polycondensation of Ti–OH species is promoted, which further causes the homogeneous formation of a large number of seed nuclei, finally the crystallization of tiny titania nanocrystallites. Followed a calcination treatment, amorphous TiO2 further ripens into well-defined crystallites. Furthermore, high acidity is in favor of the formation of rutile TiO2 besides calcination temperature. Increasing calcination temperature usually accelerates phase transformation from thermodynamically metastable to stable phase and condenser phase. The heat treatment frequently brings on the agglomeration and large size. However, TiO2 in our work shows small size and good crystallization as well as a narrow size distribution. It probably ascribes to short-chained amino acides of glycoprotein, which functioned as the capping agents since the binding was strong enough for the control on the growth of titania nuclei. For comparison, the sample is synthesized through similar sol-gel approach and calcination treatment only without involving the template. Apparently, the above as-prepared TiO2 is distributed mussily with the average size of 28 nm, while TiO2 nanocrystallites with the average size of 5 nm in our work are arranged into orderly hierarchical structures, which further proves the significance of the template and supports the above analysis. HRTEM images show TiO2 nanocrystallites prepared at pH 2 and 550 °C are the admixture of rutile and anatase. Raman spectroscopy reveals that the four Raman bands at 636, 513, 397 and 144 cm-1 ranging between 100 and 800 cm-1 can be indexed as the anatase phase of TiO2. The Raman bands at 601, 440, and 92 cm-1 can also be indexed as the rutile phase of TiO2, which is in agreement with the XRD results.

Based on the above analysis, the interfacial sol-gel approach at different pH values are predominantly associated with the mantle proteins. The mantle of SMF containing lysozyme is prone to dissociate as the glucides in the sol or the solution other than the collagenous core, and lysozyme is a kind of protein with instability under alkaline medium. Therefore, the morphology of the mantle has the inherent affiliation with the pH value of the sol medium. The mantle can exist as a steady overcast at pH < 2.5 while the mantle protein seems flowing and reforming extremely at pH 3. Under that extreme circumstance, lysozyme is mostly activated, leading to the dissociation and disintegration of peptidoglycan-based mantle layer. Simultaneously, these proteins chains interact with the sol ingredients, and the clusterin exhibiting powerful chaperone-like activity with lysozyme can bind to and stabilize proteins aggregations. It prefers the irreversible precipitation by preferentially recognizing partially folded protein conformations that are responsible for this process, which conduces to moving and reforming of proteins molecules along the SMF intersections by the interfacial force. Thus, this special layer makes it possible to transform from interwoven to intermeshed structures. However, the interaction of strongly cationic lysozyme with collagenous matrix would be mediated by proteoglycans [4]. It is difficult for the colloid particles to interact with the core matrix under current concentration. Accordingly, hollow TiO_2 fibers form under higher acidity while the petal-shape titania is obtained with lower acidity after removing the templates.

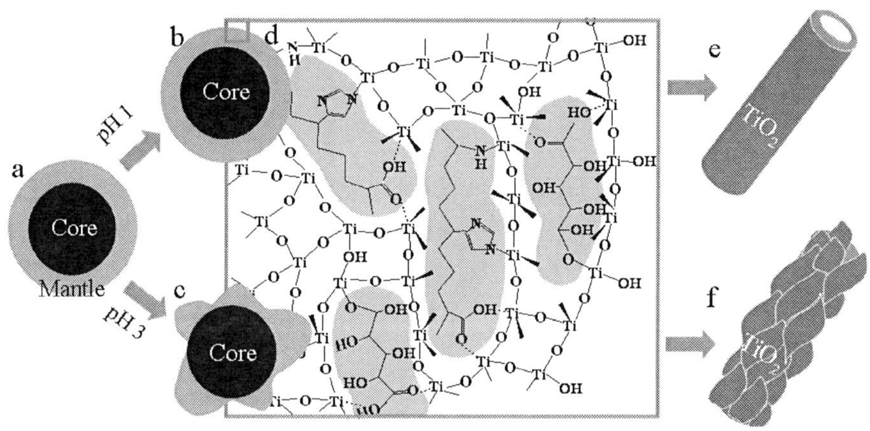

Figure 10. Schematic depiction of hierarchical TiO_2 synthesized through ESM directed interfacial sol-gel process: a) the cross-fractured SMF; b, c) the states of the mantle being immersed in the sol media at pH 1 and 3 for 13 h, respectively; d) the interaction between the sol ingredients and functional residues of ESM macromolecules; e, f) as-prepared tubular titania and petal-shaped titania by calcination treatment, respectively.

Based on N2 adsorption-desorption measurements, the isotherms and corresponding pore size distribution curves for interwoven TiO2 under different synthesis conditions are investigated in detail. Regardless of the increase of calcination temperature, the isotherms all exhibit IV-type with H3-type hysteresis loop in the Brunauer classification. As IV-type isotherm is caused by the capillary condensation of mesopores, and H3-type hysteresis loop is due to asymmetric channels and mesopores mentioned above in coincidence with the tubular characteristics pores during the synthesis process. The type of isotherms indicates an interconnected mesoporous system and very high pore connectivity according to the percolation theory, which goes well with the HRTEM results. The main pore sizes are about 5, 15, and 30 nm of the samples obtained at different calcinations temperature 550, 700 and 800 °C, while the corresponding pore volume decreases.

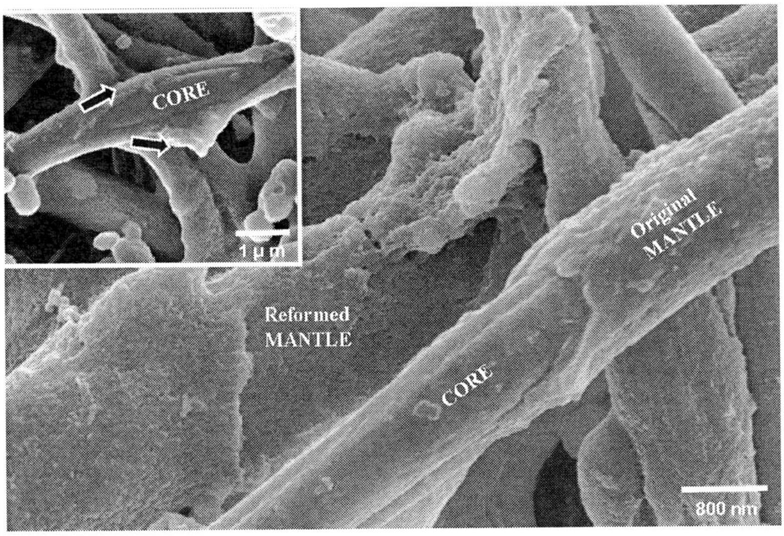

Figure 11. FESEM images with high magnification of SMF-based GMH in detail. The insetshows the mantle protein dissociating and disintegrating of the peptidoglycan-based mantle layer along the arrows direction.

Actually, pore size distribution is effected greatly by the size and morphology of building particles, which are directly dominated by the calcination temperature. Notably, the singular peak from IV-type curve of the sample at 550 °C is attributed to interwoven macropores compared with the other two mesoporous titania. Here, increasing temperature causes larger but less pores at the nanoscale.

By investigation on the isotherms and the pore size distribution curves for titania prepared at different heating rate, the trend variation of the isotherms is as same as that of the samples at different calcination temperature. By way of contrast, the size distribution is affected dominatingly by the calcination temperature rather than the heating rate. Pore sizes are less than 10 nm when reducing the heating rate below 18 °C/min. As the heating rate is further decreased to 1 °C/min, there is no evident difference in pore size distribution, due to the assembly of the target materials by the building particles with nearly same size.

The isotherms and corresponding pore size distribution curves for ESM-morphic titania at different media pH value show the presence of a pronounced hysteresis loop in each isotherm curve is associated with the filling and emptying of mesopores by capillary condensation. At relative low pressure (<0.5), the nitrogen adsorbing action of macropores is the same as the monolayer adsorption of mesopores. But above medium relative pressure (>0.5), there is no capillary condensation phenomenon in macropores. Hence, according to the types of isotherms, all these samples can be confirmed as a mixture of mesopores and macropores. The pores with larger size are formed whether medium pH value higher or lower than 2 is adapted to the system. It could be reasonably ascribed to the condensation of the hydrolytic products at low acidity and the contraction of frameworks at high acidity, which both bring on lager pores as demonstrated above. Herein the interwoven fibers are proved to be size-controllable porous, there exist several kinds of pores including micropore, mesopore and macropore. Prompted by the porous appearance, we suspect that as-prepared ESM-morphic hierachical titania must be useful in fabricating highly efficient catalyzer, photovoltaics, and so on. It makes the biotemplate-directed sol-gel synthesis to be a versatile and feasible route to the functional inorganic hierarchical materials.

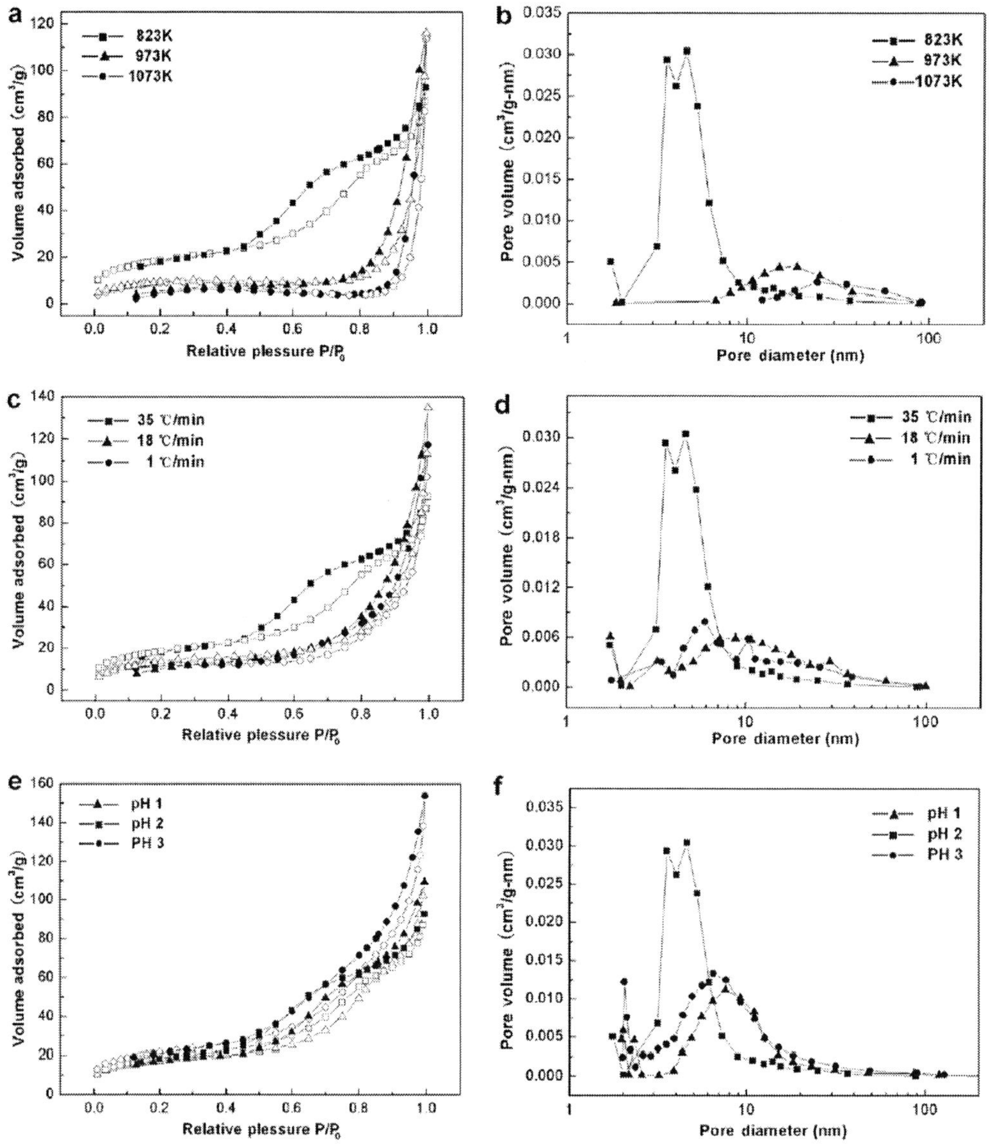

Figure 12. Nitrogen adsorption-desorption isotherms and corresponding pore size distribution plots of ESM-based TiO_2 prepared at (a, b) pH 2 and 35 °C/min, (c, d) pH 2 and 550 °C, and (e, f) 550 °C and 35 °C/min.

1.2. Fabrication and Gas-Sensing of ESM-Morphic SnO2

Having a large bandgap of 3.6 eV at 300 K, a high achievable carrier concentration of up to 6×1020 cm–3, and an expected exciton Bohr radius of about 2.4 nm, nanocrystalline tin oxide is of great technological and scientific interest as a key functional semiconductor that is used extensively for photovoltaic devices, transparent conductive electrodes, gas sensors, antistatic coatings, dye-sensitized solar cells, etc. In particular, as a gas sensor is one of its well-known applications. The involved mechanisms in the gas sensing processes are rather complex, which are related mainly with several steps such as gas adsorption, chemical reactions on the surface, and charge transport. All these processes mainly depend on the surface activity of gas-sensing materials, which is decided by the building unit and the assembly structure. It is usually claimed that nanometer-scaled materials as gas sensors have significant advantages because of their small grain size, which becomes important to enhance the adsorption capacity. Until now, few researches refer to the influence of surface activity derived from the structure and conformation on the gas sensing properties. Yet particular assembly of SnO2 nanocrystallites may provide promising gas-sensing performance. Attracted by bioinspiration, eggshell membrane in combination with a tin salt precursor is introduced to a sol-gel process, with the intent of synthesizing hierachical SnO2 semiconductors by suitable vehicles of special protein structures and functional residues of ESM biomacromolecules. In contrast to the metallic alkoxide precursors, this work employs a trilayer bio-scaffold in tin nitrate solution as the template for hierarchical nanostructured SnO2 with enhanced gas-sensing performance.

In a typical procedure, Sn powder is added stepwisely into ~2M HNO3 solution to obtain a clear solution incipiently, turn into a muddle subsequently, finally slight yellow Sn-colloid after being filtrated. The pH values of the colloid media are generally kept at 1 when used as the impregnants. For comparison, 1 M NaOH solution is added to modify the pH value of the media to 2 and 3. The prepared ESM immersed in the above-mentioned Sn-colloid systems for about 13 h, then being fished out, rinsed with distilled water, and dried, the soakage-treated ESM are subjected to calcination treatments in air at different temperatures to remove the organic ESM ingredients and achieve better crystallization of the inorganic sinters.

The calcination process was traced by conducting TGA and DTA surveys to follow the formation of hierarchical SnO_2. It is evident that the ESM starts pyrolyzing at 473 K and is absolutely decomposed by 823 K. Although GMH are also decomposed around 473 K, they can preserve up to 47.2% (by mass) inorganic matter at 823 K, at which temperature the organic ESM is completely removed. As seen from the differential curve, when GMH are heated from room temperature to 473 K, the weight loss of 9.8 wt-% is due to the removal of adsorbed water, and above 473 K they experience a two-step weight loss. Congruously, the DTA curve shows two peaks at 635 K and 731 K, which shall represent the decomposition of Sn–ligand hybrids and organic ESM components, respectively. After the calcination at 673 K, the amorphous hybrids are found to be crystallized, but the ESM is not thoroughly decomposed at this temperature. XRD results indicate the well-established crystallization process at 823 K. All the diffraction peaks can be indexed to tetragonal SnO_2 with a rutile structure (JCPDS 41–1445), which indicates that the specimens consist of pure single tetragonal SnO_2. The average crystallite sizes of about 2.8 nm at 673 K and 5.2 nm at 823 K, are estimated according to the line-width analysis of the diffraction peaks based on the Scherrer formula. The chemical states of tin and oxygen during the sol–gel process were investigated by XPS analysis. The corrected XPS spectra of the O1s emissions are different, where the peak with binding energy around 532.3 eV corresponds to the hydroxyl bonds of GMH, while the weaker peak around 531.1 eV corresponds to the Sn–O–Sn bonds of the sinters. Moreover, GMH appear as a spin-orbit doublet at 487.9 (3d5/2) and 496.4 eV (3d3/2) in the Sn3d emission spectra with an area ratio of 1.5:1. Thus, the chemical state of the element tin in as-prepared GMH corresponds to Sn^{IV}, but not to Sn^{II} or Sn^{III}. Representative FESEM images of as-prepared SnO_2 with various morphologies calcined at 823 K suggests that hierarchical SnO_2 accurately replicates the hierarchy of the biotemplate ESM from the OSM to the innermost LM, although the modest shrinkage. By repeating the immersing manipulation, heat constringency in size can be lowed to below 30%, and up to 47.2% of the mass of the sinters. Analysis with energy-dispersive X-ray spectroscopy also reveals that the atomic ratio Sn/O is close to 1:2, further confirming the stoichiometry of the replicas as SnO_2.

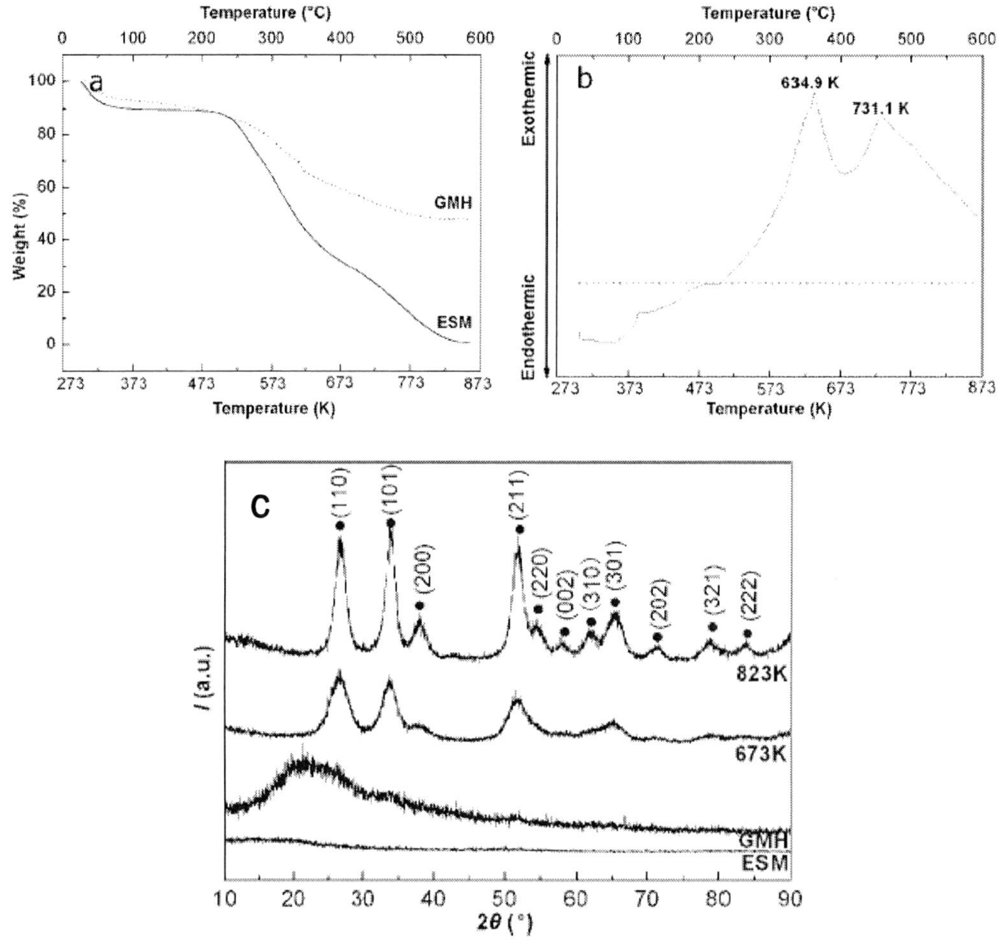

Figure 13. (a, b) TGA-DTA curves, (c) XRD patterns of the relevant samples.

FESEM images show the relative location of the hierarchical SnO_2 replicated from the three diverse membranes. It is clearly seen that the LM-based one has been sheltered by the upper layers composed of the ISM-based and OSM-based SnO_2 layers. However, the replica of the LM layer can be identified when the other side of this sheet is observed. The configuration of the intermeshed ISM consisting of interwoven fibers is intact, whereas the LM-based layer is curled up thoroughly. Moreover, the thickness of the obtained SnO_2 film can be tuned by removing the multilayers of the ISM. Thus, a SnO_2 film with a continuous thickness up to 20 μm can be achieved by varying the original thickness of remnant ISM layers. The tubular hierarchy can be observed clearly by the higher magnifications of the cross section and the vertical section of these SnO_2 interwoven fibrous films, respectively.

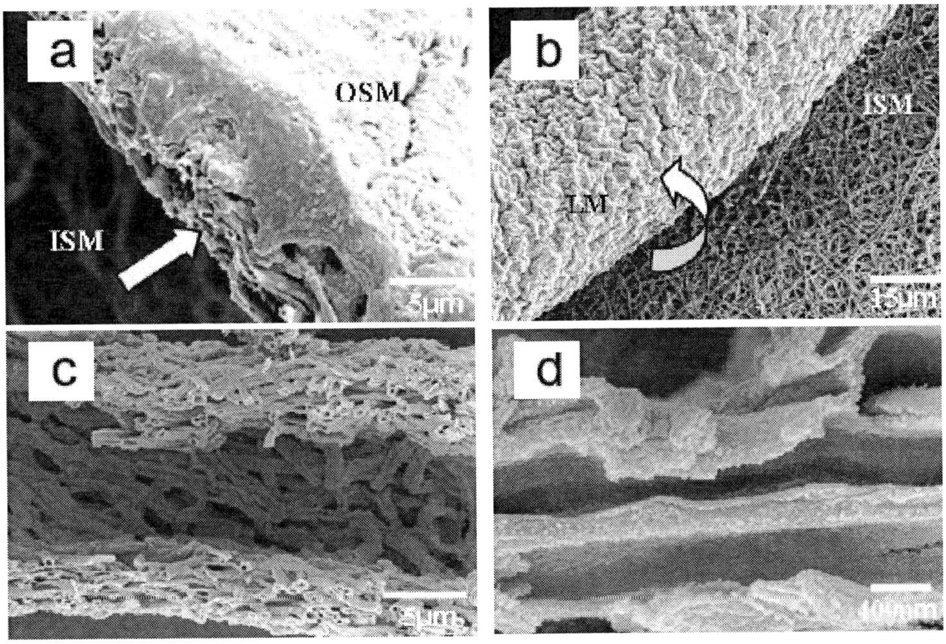

Figure 14. (a), (b) Typical FESEM images of as-prepared ISM-morphic SnO_2 revealing relative locations of the three distinct layers of the original ESM. (c), (d) FESEM images showing the cross-section and the vertical section of the ISM-based SnO_2, respectively.

The tubular structures have an inner-diameter range from 100 nm to 300 nm and an average wall thickness of less than 90 nm. These SnO2 tubes are rather uniform, and the hollow cores are glabrous and continuous throughout their length. Notably, these interwoven tubular structures are actually interconnected at junctions. A special kind of tube wrapping several smaller collateral cannulations is also observed, which likewise reproduces the original structures of the membrane fibers. To further investigate the replication process, the organic ISM is selected as the typical template to achieve the synthesis of hierarchical biomorphic SnO2 under different experimental conditions.

All the replicas at different calcination temperatures (823 to 1073 K) have similar microscopic features and preserve the interwoven tubular structures of the natural ISM, but their building blocks, SnO2 nanocrystallites, may grow with the increase in calcination temperature. The surface of SnO2 tubes synthesized below 973 K looks more glazed in virtue of the smaller particle units. Above 1073 K, the tubular structure is apparently composed of a network of interconnected SnO2 particles, and it may collapse down into zonary form at 1273 K. On the contrary, if the calcination temperature is lowered to 673 K, characteristic tubular structures appear partially, and some look puffy, which may be caused by the organic remnants. It is reasonable that some residues of organic ESM remain at 673 K even though SnO2 nanocrystallites have formed, which is in agreement with the TGA and XRD results.

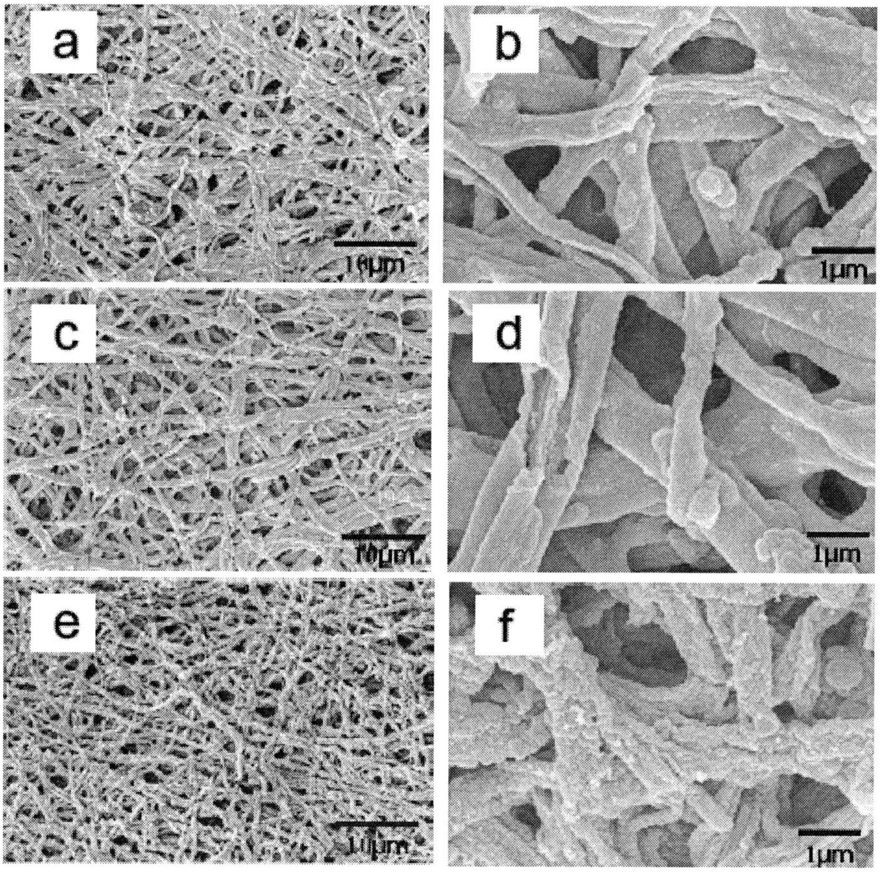

Figure 15. FESEM images of as-prepared ISM-morphic SnO_2 at different temperatures: (a) 823 K, (c) 973 K, and (e) 1073 K. Images (b), (d), and (f) show relevant higher magnifications of (a), (c), and (e), respectively.

All XRD diffraction peaks of the samples calcined at various temperatures between 823 and 1273 K, can be assigned to the tetragonal rutile structure of SnO2. The XRD peaks gradually become sharper with increasing temperature, indicating that the SnO2 nanocrystallites grow larger, and the crystallite sizes determined from the Scherrer equation are 5.2, 9.5, 10.3, and 20 nm, respectively. The TEM images further reveal the tubular interwoven character of ISM-based SnO2. The hollow tubular structure is clearly identified, and the wall thickness (50–60 nm) is found to be uniform along its entire length.

This indicates that at 1073 K each tube has been transformed from a dense structure with a smooth surface into a highly porous structure consisting of interconnected nanocrystallites of ca. 10 nm in size. In addition, the SAED patterns show that nanoscaled SnO2 at 1073 K has polycrystalline tetragonal rutile structure (cassiterite) due to the uniform central and diffraction spots, and the diffraction rings are indexed to the (110), (101), (200), (211), and (301) diffractions of rutile-phase SnO2. Although the sinters at 673 K contain some impurities from ESM residues, the characteristic crystalline structure of tetragonal SnO2 is observed in the electron diffraction image. This can be explained by the exothermic reactions occurring during the removal of the ESM: the oxidation of the organic biomaterial emits heat locally, causing an increase in the temperature around the adsorbed Sn colloids, thereby accelerating the gelation and bringing on nanocrystalline SnO2. These nanoparticles are ultrafine below 823 K while growing more quickly above 1073 K, and the average sizes of the nanoparticles are coincident with the crystallite sizes calculated from the XRD analyses. As the nanoparticles grow with the increase in calcination temperatures, the diffraction patterns gradually transform from ring (nanoscale) to spot (bulk scale). This indicates that accurately defined, size-controlled SnO2 nanoparticles in the range 2 to 20 nm are obtained to enable the construction of hierarchical nanomaterials at the corresponding calcinations temperatures. As observed the typical HRTEM images of SnO2 nanocrystallite units calcined at 823 K, the lattice image can be clearly seen as a result of phase contrast. The distances between the adjacent lattice fringes correspond to the interplanar distances of the tetragonal SnO2 (101) and (110), *d*101 = 0.264 nm and *d*110 = 0.335 nm, respectively. Considering that each domain having parallel lattice fringes is a single crystal of SnO2, the crystallite size is observed to be 5 nm. A careful look at the images also reveals that the present SnO2 is characterized by larger "nanostructured" particles consisting of nanocrystalline domains with structural defects such as twins.

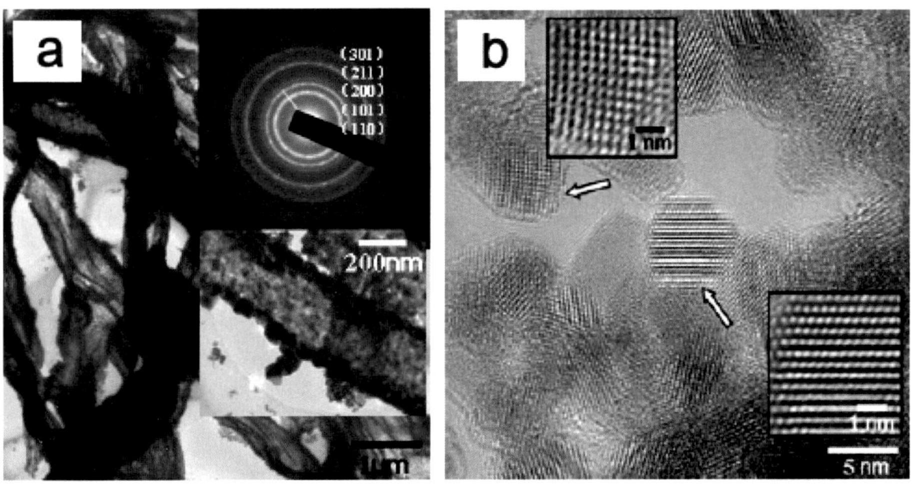

Figure 16. (a) TEM images of ISM-based SnO_2 synthesized at 1073 K and pH 1. Insets in (a) show an individual SnO_2 tube and corresponding SAED pattern. (b) HRTEM Image of ISM-based SnO_2 obtained at 823K and pH 1. Insets in (b) correspond to the tetragonal SnO_2 (101) plane (top left) and the (110) plane (bottom right), respectively.

To understand the multilevel interactions between the Sn colloids and the ESM biomacromolecules during the biomimetic sol–gel process, FTIR spectroscopic analysis was carried out on the original ESM, the GMH and the sinters. The spectrum of the original ISM shows the diagnostic peaks at 1641, 1532, and 1234 cm–1, which are assigned to the amide I, amide II, and amide III bands of the proteins from the natural ISM, respectively. After interaction with the sol precursor, the red shift to 1630 cm–1 of the peak in the spectrum of GMH may be partly attributed to the superposition of the bending mode of water with the broad band around 3450 cm–1 (not 3400 cm–1). The peak at 3430 to 3450 cm–1 is believed to be caused by adsorbed water. In general, the absorption band centered at 1630 cm–1, as well as the one at 3450 cm–1, is also ascribed to water adsorbed on the surface of the samples. Here, there may be no other corresponding peaks of amide I after immersion in the sol medium, if the peak at 1630 cm–1 (near 1620 cm–1) in the spectrum of GMH are only due to the bending mode of water. The spectrum shows an increase in intensity of the amide I band relative to that of the amide II band. It does confirm the presence of a water deformation mode. Below 700 cm–1, the intense bands appearing both in GMH and the sinters at lower frequencies around 600 cm–1 are typical of the Sn–O–Sn antisymmetric and symmetric vibrations. In addition, the bands appearing in the range between 850 and 1350 cm–1 are assigned to the bending mode of different types of surface hydroxy groups, which goes well with the previous research results. The surface of the GMH can be terminated with the –OH groups after the ESM is dipped in the Sn-sol medium. However, the calcinations treatment may bring on weak IR bands between 850 and 1350 cm–1, contributed by the decomposition of Sn–OH hydrate compounds into tin oxides.

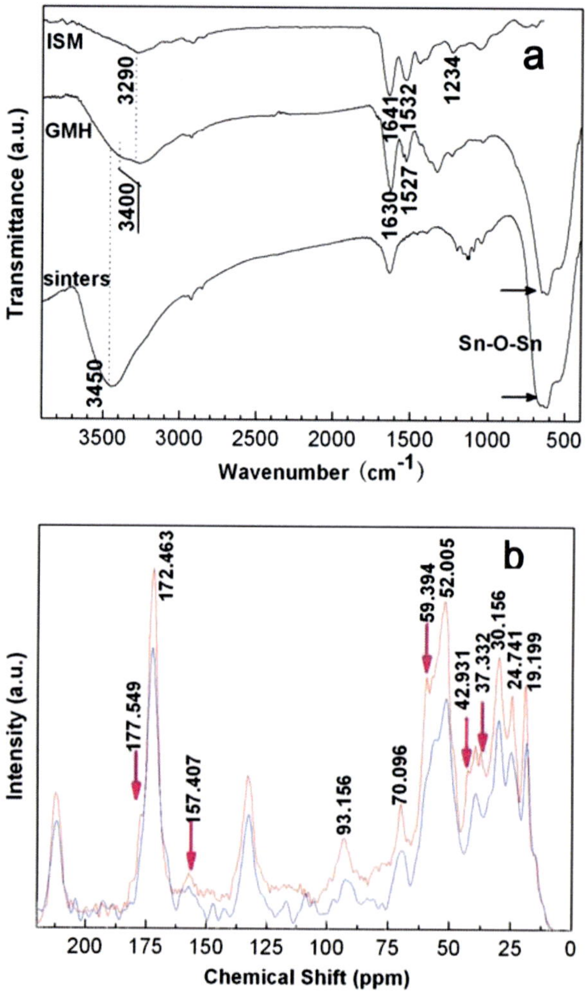

Figure 17. (a) FTIR spectra of original ISM, the corresponding GMH, and the sinters at 823 K. (b) MAS solid-state NMR spectra of original ESM (red line) and GMH (blue line).

Actually, the red shift of the amide I peak is probably overlapped by the above-mentioned water bending-mode peak. This red shift indicates that the C=O bonds in the peptide chains may be weakened because of the formation of new chelate bonds between C=O bonds and Sn colloids. This can be further corroborated by the NMR spectroscopic results. The chemical shift between 19 to 30 ppm is mainly due to the resonances of —CH3 and —(CH2)n— groups of Pro, Val, Arg and Metis. (a)—COOH 177.549; (b) —COH 59.394; (c) —CH2NH2 157.407, 42.931, 37.332. It is also detected that the peak corresponds to the Sn–OH vibration band at around 3400 cm–1, which is superimposed on the underlying C–H stretching signals of the polypeptides at 3290 cm–1 and close to the hydroxy bands of the water. Accordingly, the interactions between Sn colloids and ESM macromolecules are mostly ascribed to the coordination of carboxyl, hydroxy, and other groups.

Herein, the interfacial sol–gel process is presumably based on the adsorption of cationic groups from the colloid medium into the glycoprotein mantle of the ESM fiber containing an abundance of carboxylic, hydroxy, and amino residues; subsequently a series of complicated reactions proceed to give hierarchical SnO2 films composed of ultrafine nanoparticles. The formation of SnO2 can be denoted as follows:

$$\mathbf{Sn^{0} \rightarrow Sn^{IV}-ions\ (hydrate\ ions) \rightarrow Sn^{IV}-colloids \rightarrow Sn^{IV} oxo\ species\ (Sn-gel)}$$
$$\mathbf{\rightarrow amorphous\ SnO_2 \rightarrow nanocrystalline\ SnO_2}$$

The cationic SnIV species in the sol system are prone to interact with the residues of ESM glycoprotein proved by FTIR and NMR results. The formation of amorphous Sn-colloids/glycoprotein hybrids arises from strong affinity of some residues binding that involves multiple hydrogen bonds, Van Der Waals interactions, and the ordering of surface polypeptide loops. The Sn-colloids are bound on the biotemplate ESM by amino groups labeled glycoprotein through the -OH and -NH2 groups of sugar chains linked to polypeptides. Thereby the amorphous Sn-O networks can be derived from these reactions, and indeed an amorphous SnO2 film locating in the mantle is obtained without changing the original morphologies at room temperature. The Sn (IV) oxo species are assumed to be an intermediate, consisting of partly dehydrated polymeric Sn (IV) hydroxides.

Amorphous SnO2 ripens into well-defined nanocrystallites during the calcination treatment. Heat treatment frequently causes agglomeration and enlargement. However, SnO2 nanocrystallites in our work have small size and good crystallization as well as a narrow size distribution. This can probably be ascribed to some short-chained amino acids of glycoprotein, which function as capping agents, since the binding is strong enough for controlling the growth of tin dioxide nuclei. For comparison, a blank sample is synthesized through a similar sol–gel approach and calcination treatment at 823 K only without using the biomaterials. SnO2 nanoparticles in the blank sample are disorganized and have an average size of 21 nm, while the ESM-based SnO2 nanocrystallites with an average size of 5 nm are arranged into orderly hierarchical structures with higher surface area. This observation proves the significance of the biotemplate and supports the above analysis. It is also concluded that the ESM peptidoglycan functioning as the surfactant may play a key role in the aggregation and further crystallization of tin oxide nanoparticles. In addition, electrostatic forces assist the biomimetic sol–gel process. Because of the higher acidity of the system (with ample H+ ions), the tin oxide clusters in the colloid are likely to be positively charged at the surface by reference to the isoelectric point of cassiterite SnO2 (7.3) [5]. By electrostatic force, they will preferably attach onto the negatively charged DS– ions on the surface of the membrane. As surface coating is always dependent on surfactant concentration, it is difficult and complicated to determine the precise dosage.

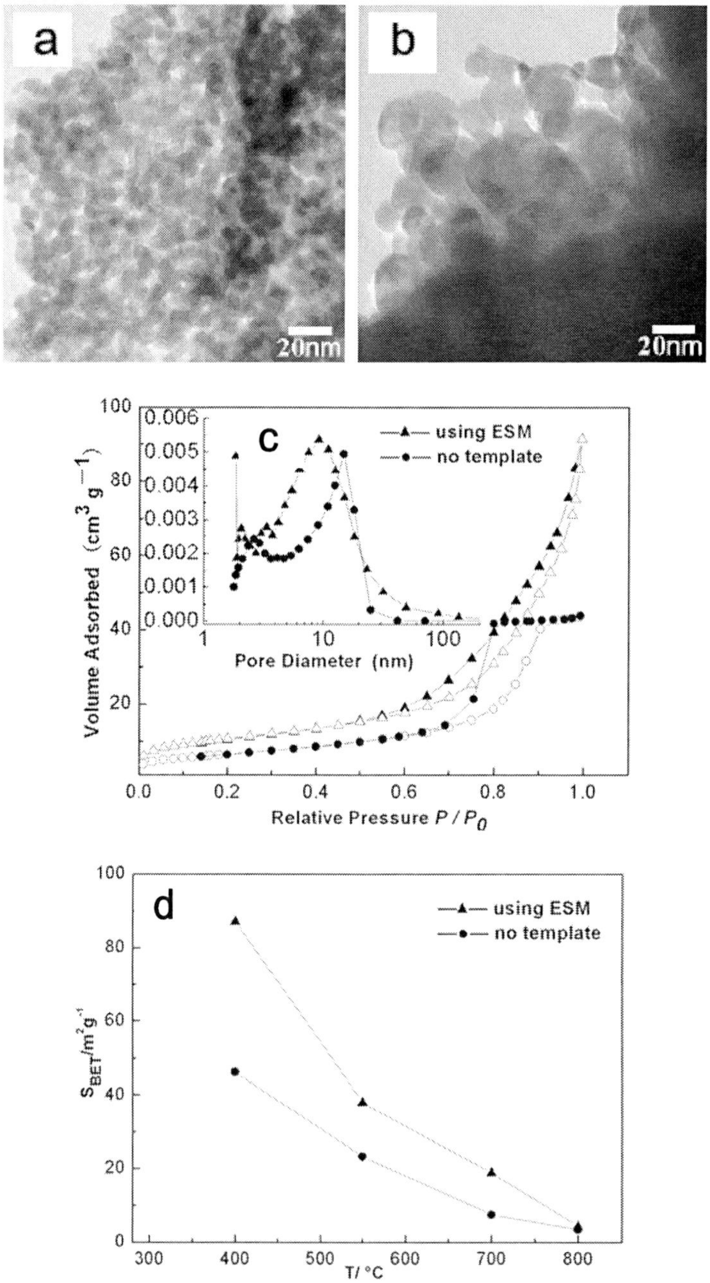

Figure 18. TEM images of (a) ISM-based SnO_2 nanocrystallites and the contrastive SnO_2 sample without introducing template through the same process at 823 K. (c) Nitrogen adsorption-desorption isotherms plots and corresponding pore size distribution plots of ESM-morphic SnO_2 and Blank sample, (d) Special surface area SBET vs calcination temperature.

However, in our work the "surfactant molecules" can be released systematically by the dissociation and reforming of ESM biomolecules. With the aid of "surfactant molecules", the construction of biostructure hierarchy can be achieved to synthesize subtle biomorphic functional nanomaterials, which is similar to biomoleculemicelle systems.

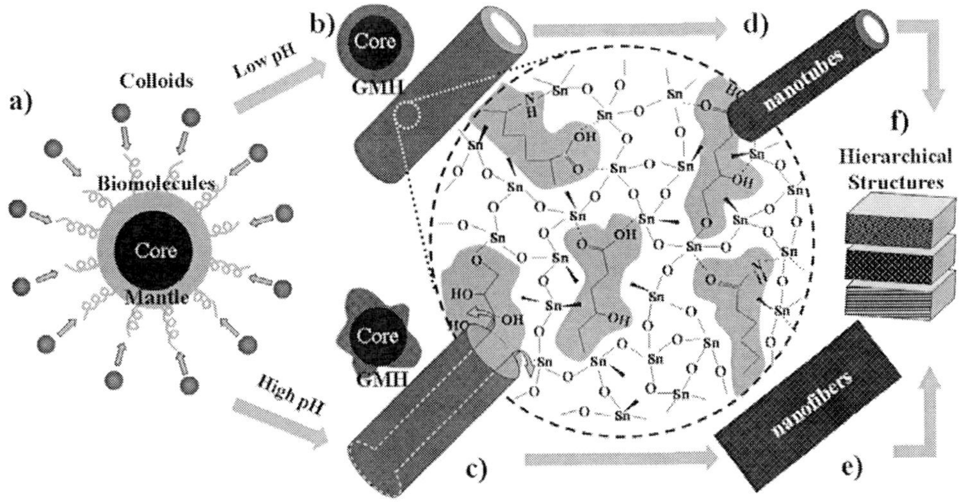

Figure 19. Schematic depiction of hierarchical SnO_2 synthesized through ESM directed interfacial sol-gel process: (a) the cross-fractured ESM fiber showing its biomolecules interacted with Sn-colloids; (b, c) the states of the mantle being immersed in the Sn-sol media at different pH; the inset in (b) illustrating the interaction between the sol ingredients and functional residues of ESM macromolecules; after the ESM removal, (d) tubular SnO_2 and (e) SnO_2 fibrous belts; (f) hierarchical SnO_2 achieved by nanotube or nanobelts arrays.

Notably, a completely morphologic conversion from SnO_2 tubes to SnO_2 belts occurred as a result of tuning the pH value of the sol media. As lysozyme is a kind of protein that is unstable in alkaline medium, it is clear that the mantle of ESM fiber containing lysozyme is prone to dissociation, its glucide contents not in the collagenous core end up in the colloid system. Therefore, the resultant morphologies of the SnO_2 products might have an inherent affiliation with the structures and constituents of the glycoprotein mantle. It is observed that the mantle can exist as a steady covering layer at pH < 2, while the distortions and dissociations of this protein layer occur as the pH value is elevated. The mantle protein seems to flow and reform evidently when pH > 3. Under these circumstances, lysozyme is most active, and it functions as a kind of enzyme leading to the dissociation and disintegration of the peptidoglycan-based mantle layer. These protein chains then interact with the Sn colloids. Simultaneously, the proteins formerly carrying tubular hybrids start to dissociate and stretch into planar fibers as a result of decreased acidity. Moreover, it is revealed that clusterin, exhibiting powerful chaperone-like action on lysozyme, binds to and stabilizes aggregating proteins, thus leading to irreversible precipitation by preferentially recognizing partially folded protein conformations. This process conduces to the stabilization of the final structures of SnO_2. The special layer thus makes it possible for SnO_2 fibers to transform from tubular to solid fibrous structures. However, the interaction of the strongly cationic lysozyme with the collagenous matrix may be mediated by proteoglycans. It is difficult for Sn colloids to interact with the core matrix under the current concentrations. Accordingly, hollow tubes form under higher acidity, while fibrous structures are obtained at lower acidity after the ESM is removed.

Prompted by their porous appearance and small-scale nanocrystalline units, as-prepared hierarchical ESM-based SnO_2 must be useful in fabricating semiconducting sensors. When the grain size is reduced to a scale comparable to the space-charge length (for SnO_2, this value is 6 nm), the sensitivity could be exponentially enhanced [6]. As an n-type semiconductor, SnO_2 has been extensively exploited in applications related to gas sensing. Most of these studies have focused on polycrystalline films made of small SnO_2 particles [7]. In contrast to some sensors made of individual single crystalline nanostructures or polycrystalline films, hierarchical nanostructured SnO_2 in our work, assembled by the small-scale, well-crystallized nanocrystallite units into tubular or zonary SnO_2 fibers with porous conformations and interconnected fibrous meshworks, would allow the sensors to be operated in the most sensitive, grain-controlled mode.

The sensor signal, *S*, is defined as the resistance ratio, *R*a/*R*g, where *R*a and *R*g stand for the electrical resistance in dry air and in the sample gas, respectively. The gas-sensing of as-prepared ESM-morphic SnO_2 films is measured continuously.

It is found that the dependence of the sensor signal (*S*) of the test gases on the synthesis temperature defining crystallite size (*D*). Here hierarchical SnO_2 film sensitivity to 50 ppm ethanol, 50 ppm H_2S, 500 ppm LPG, or 50 ppm gasoline 90#, is investigated at the working temperature 270 and 300 °C, respectively. At a lower working temperature, 270 °C, the sensitivity of as-prepared SnO_2 to ethanol is more prominent than that to H_2S, LPG and gasoline. However, when elevating the working temperature to 300 °C, the sensitivity of the sample SnO_2 calcined below 700 °C to LPG becomes remarkable among all these test gases. Regardless of the effects of the working temperature, the sensor signal to all other gases except H_2S shows downward trends with synthesis temperature. That is to say, the sensitivity to the test gas markedly increases with decreasing crystallite size of the sensor SnO_2 ranging from 26 to 3 nm.

There is the dependence of the sensitivity of hierarchical SnO_2 prepared at 550 °C to the test gases on the working temperature. It can clearly be seen that the sensor signal to LPG shows an upward trends, while that to ethanol shows a downward trends, but the sensor is insensitive to other gases at the working temperature. Combining the above analyses indicates that the target materials have good gassensing selectivity for LPG or ethanol by tuning the working temperature of the sensors.

As a powerful tool to reveal straightforwardly the surface characteristics of these nanocrystalline films, the evolution of Raman surface vibration modes is investigated to examine the surface reactivity with the target specimens. The relationships between the gas responses and the nanostructure surfaces derived from the hierarchy and small unit size characterized by the Raman band intensity are discussed, corroborating the significant influence of the assembly structures on the gas-sensing properties of SnO2 films.

The investigation on the evolution of Raman spectra versus the treatment temperature for the series, also confirms the rutile structure of the as-synthesized SnO2. The spectrum of well crystallized sample at 1000 °C, grain size 26 nm, is similar to that of microcrystalline SnO2. The three detectable Raman active modes have been identified: Eg (476 cm-1), A1g (629 cm-1), and B2g (772 cm-1), which usually appear in large single crystals or bulk polycrystalline SnO2 materials and in good agreement with previous reports.

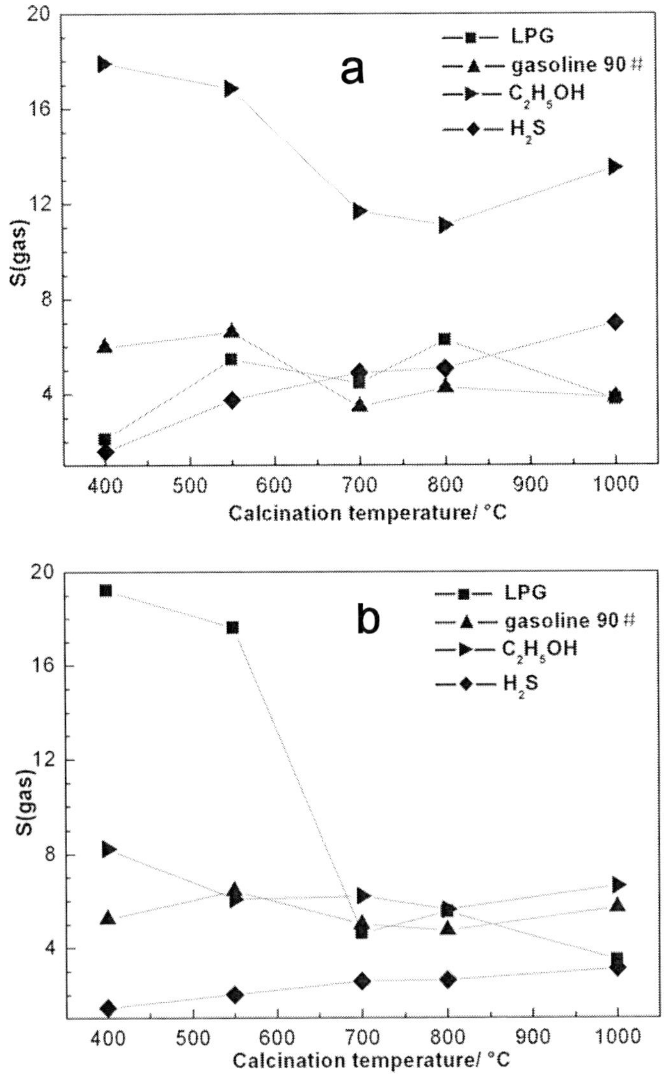

Figure 20. Dependence of the gas-sensitivity of SnO_2 films to 50 ppm ethanol, 50 ppm H_2S, 500 ppm LPG, and 50 ppm gasoline 90# on the synthesis temperature at the working temperature (a) 270 °C and (b) 300 °C.

As particle size decreased, the spectra of ESM700, ESM550, and ESM400 samples also exhibit the three bands and gradually descend around the A1g mode with a broad feature between 400 and 700 cm-1 attributed to surface modes. Additionally, three abnormal Raman lines are distinguished on the spectrum at 501.1, 541.9, and 693.9 cm-1, respectively, which have been detected before. Accordingly, these vibrational modes can be reasonably assigned to A2uTO (501.1), A2uLO (693.9) IR active, and B1u (541.9) Raman forbidden modes. The peaks are more sharpened with the smaller dimension of SnO2 particles. It is believed that the presence of IR modes and other forbidden Raman modes is attributed to the breaking down of the prevailing q0 = 0 selection rule as crystal size decreases to the nanoscale (limiting vibrations to the size of the crystal) or as the degree of disorder increases (reducing crystal symmetry). It is remarkable that high-frequency modes B2g, A2uLO, and A1g shifted to lower frequencies, whereas the low frequency modes, such as B1u, A2uTO, and Eg, shifted to higher wavenumbers accompanying the decrease in particle size. This is consistent with the findings reported by others, which can be explained using phonon dispersion curves.

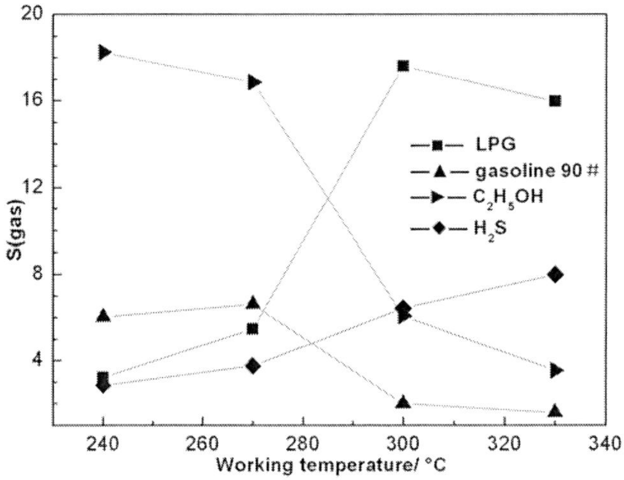

Figure 21. Dependence of the gas-sensitivity of SnO_2 films prepared at 550 °C to 50 ppm ethanol, 50 ppm H2S, 500 ppm LPG, and 50 ppm gasoline 90# on the working temperature.

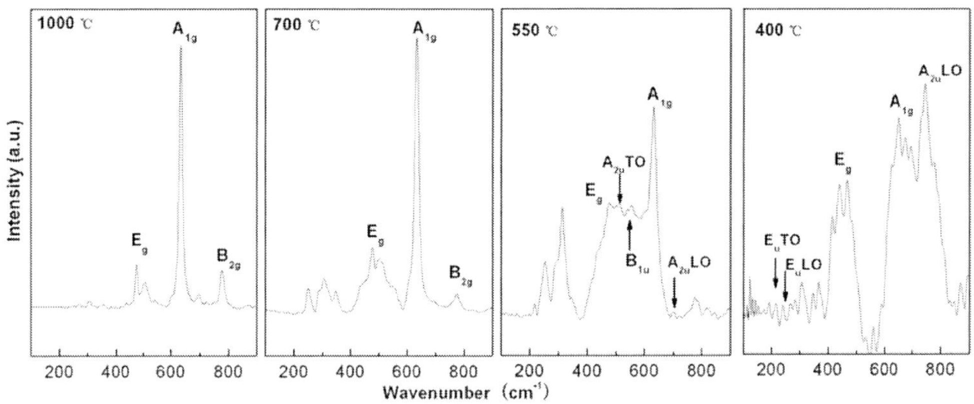

Figure 22. Raman spectrum evolution with the calcination temperature for ESM-templated SnO_2.

Furthermore, some new bands appear below the 400 cm-1 region and present a single broad feature composed of at least two components in the powders calcined at 400 °C. It reported that Raman bands located at 216.49 and 232.92 cm-1 in rutile SnO2 were attributed to EuTO and EuLO active modes. For SnO2 treated at higher temperatures, three bands are clearly observed at 245, 257, and 286 cm-1, together with the broad feature located at the region 310-350 cm-1, which has lost most of its intensity. These bands disappear for sample at 1000 °C. Other researchers associated the appearance of a 327 cm-1 centered band to the forced hydrolysis of Sn ions and the formation of SnO2 nanocluster, which might constitute a new kind of vibration mode. It is plausible to take into account that this band could be due to an active mode from the inactive Eu mode. Nevertheless, more work is still needed to clarify the origin of this band in very tiny SnO2 structures. It seems to strengthen the association of this band with the effects due to the nanostructure size against those due to the synthesis procedure. Those new bands, being related to the facet surface area of a crystal, arise from nanoscale SnO2 with small grain size and agree very well with the Matossi force constant model.

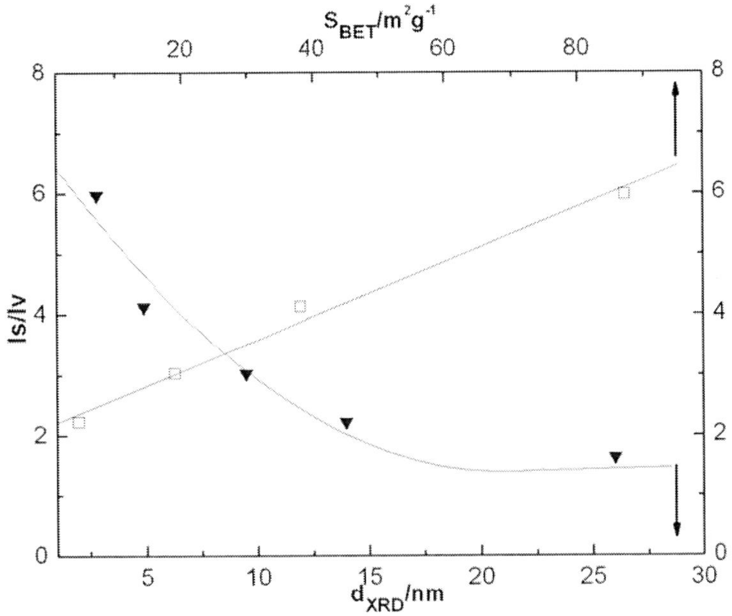

Figure 23. Raman relative intensity of surface modes I_S/I_V value vs SnO_2 crystallite size d_{XRD} and specific surface area S_{BET}.

The other peculiarity of Raman spectra of nanocrystalline SnO2 is the very broad feature between 400 and 700 cm-1. The origin of these bands has been connected to the low particle dimension of the materials. In a previous work, Diéguez et al. proposed that surface modes were due to the existence of a lattice softening taking place in the outer part of the grain, which is like a disordered lattice skin around each grain. Moreover, studies have shown that the relative intensity of surface bands depends strongly on the interactions with gases, which can modify the outer layers of the grain. So, to analyze these performances at the present cases, the relative intensity of surface modes have been taken as the ratio of the sum of their intensities *I*S to the intensity of A1g mode *I*V. These intensities are calculated by fitting A1g mode and the B2g mode when needed with Lorentzian-shaped peaks, and the surface modes with five large Gaussian peaks with Peakfit software. *I*V is taken as the area of the Lorentzian A1g peak, while *I*S is the sum of the five Gaussian peak areas. The variations of *I*S/*I*V value versus crystallite size *d*XRD and specific surface area *S*BET for SnO2 hierarchy are presented. There is a monotonic diminution of the surface modes as the grain size *d*XRD increases, which confirms a general trend. In addition, an excellent correlation between *I*S/*I*V and *S*BET suggests that surface modes are related to available surface sites to interact with target gas speciments. Therefore, it shall be corroborated from the behavior of the sensor response versus *I*S/*I*V.

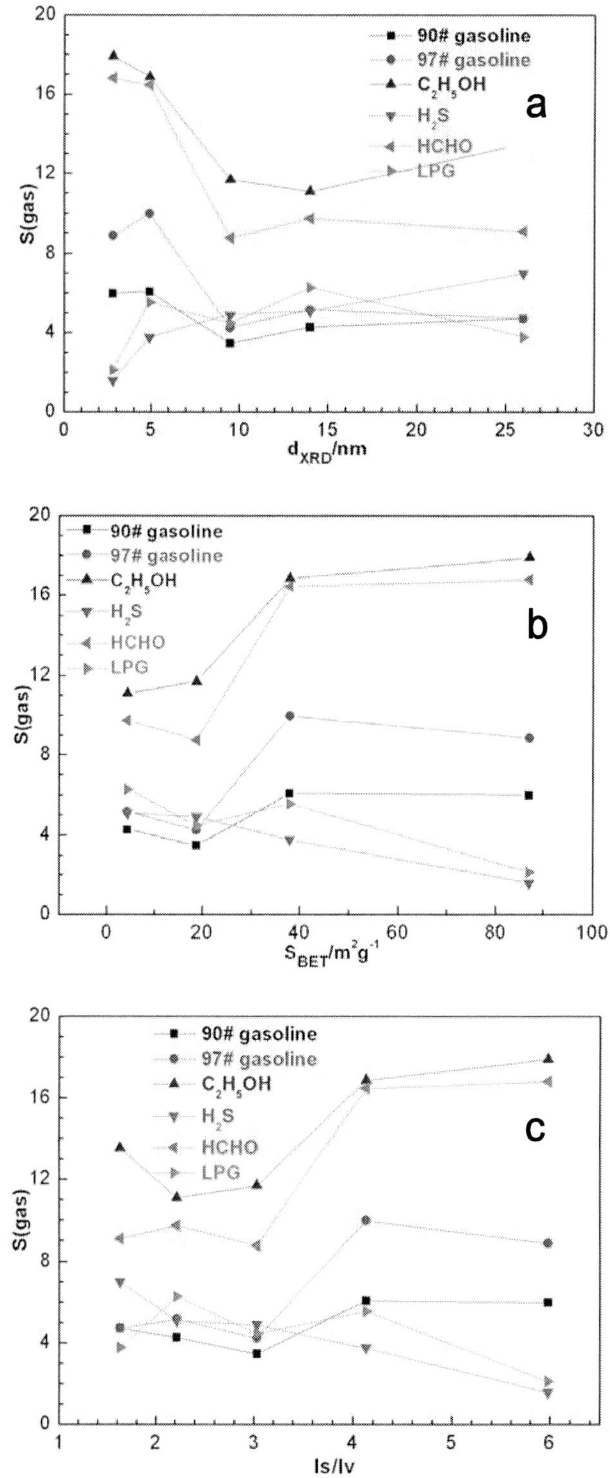

Figure 24. Sensing properties toward ethanol, formaldehyde, LPG, gasoline, and H_2S species S(gas) vs SnO_2 crystallite size d_{XRD} (a), specific surface area S_{BET} (b), and Raman parameter I_S/I_V (c).

With this aim, the values of sensor response of SnO2 hierarchy toward ethanol, formaldehyde, LPG, gasoline, and H2S species have been measured. The working temperatures are chosen at 270 °C for detection. Figure 19 presents the *S*(gas) value as a function of SnO2 crystallite size (a), specific surface area (b), and Raman parameter *I*S/*I*V (c) for the series. One can observe a general trend of the sensor response diminution as the SnO2 crystallite size increases. However, it is found that behavior of samples toward H2S molecules is different from that shown by the others. These series shows the inverse relationships between sensor response and microstructure parameters: *S* (H2S) increases as *d*XRD values rise too and *S*BET diminishes. The samples with *d*XRD 3-5 nm are characterized by an approximate sensor response value toward corresponding target gas. As for the larger crystallite size *d*XRD > 6 nm, the series display a lower *S* (gas) value. The dependency on *S*BET value appears to be more informative to characterize gas sensor properties of these tin oxide samples. All results can be approximatively presented as a linear function *S* (gas) – *S*BET taking even into account the accuracy of specific surface area and sensor response measurements. Additionally, an approximate linear correlation is also obtained plotting the sensor response against the Raman parameter *I*S/*I*V compared with the specific surface area. Herein, it confirms the usefulness of Raman spectroscopy for the characterization of nanocrystalline tin oxide interactions with gases as well as the specific surface area. Notably, the specific surface area is measured on powders with hierarchical structures demolished, while Raman spectra are taken on hierarchy and films. It corroborates that the special structures have some influence on valuable *S*BET of the materials, especially on the surface modes, further on gas-sensing response. Another important conclusion, which can be drawn from this strong correlation between *S*(gas) and *I*S/*I*V, is that the surface modes (*I*S) are straightforward connected with the outer layers of the grain, which, according to these results, depend on the hierarchical structures or introducing of the biotemplate ESM.

The examination of nanostructure influence on the sensor parameters of SnO_2 exhibits the opposite trends for detection of C_2H_5OH (or HCHO) and H_2S molecules. The *S* (C_2H_5OH) increases with the specific surface area growth, and vice versa, *S* (H_2S) increases as the specific surface area diminishes. These results can be explained from the mechanism analyzed in the literature from the point of view of the adsorption kinetics or from the temperature treatment influence.

The interaction of reducing or combustible gases with the surface chemisorbed oxygen, which exists in various forms, namely: O_2^-, O^-, O^{2-}, etc, can take place in different ways of physical or chemical reactions. Vancu et al. have suggested the following reaction to be favorable with the reducing gases: $R+O^- \Leftrightarrow RO + e^-$. It has also been pointed out that the gas-sensing phenomenon is intimately connected with the occurrence of surface catalyzed combustion. As we are aware from the catalytic chemistry, the surface acid/basic state could be advantageously utilized to favor some particular reactions. With the prior knowledge of the type of a target gas, the sensor surface could accordingly be modified in order to preferentially sense the reducing gas over the other interfering gases.

Based on the above theories, Gnanasekar et al.[8] have suggested that gasoline combustion on a basic surface where the following reactions would take place:

$$\text{Gasoline (or LPG)} + O_2 \rightarrow CO_2 + CO + H_2O$$

$$CO + O^-_{ads} \rightarrow CO_2 + e^-$$

Fu et al.[9] have suggested that aldehyde (CH_3CHO) was the intermediate product when ethanol was oxidated on the basic surface and the following reactions happened:

$$CH_3CH_2OH + O_2 \rightarrow CH_3CHO + H_2O$$

$$CH_3CHO + O^-_{ads} \rightarrow CO_2 + H_2O + e^-$$

H_2S and the chemisorbed oxygen O^- can take place the following reaction [10]:

$$H_2S + O^-_{ads} \rightarrow H_2O + S + e^-$$

Electrons released from the above reactions would annihilate the holes: $h\bullet + e^- \rightarrow 0$. Hence, the resistance of gas-sensing materials increased and then realized the detecting to different gases. For gasoline or LPG, the population of negatively charged oxygen adsorbates effectively controls the conductance of the target SnO_2. The chemical reactions occurring at the oxide surface have been assumed to involve CO adsorption followed by CO_2 desorption and replenishment of the resulting surface oxygen vacancies by the adsorption of molecular oxygen. For the oxide SnO_2, the electron concentration depends on the stoichiometry deviation determined by oxygen vacancies, which are predominant atomic defects. The electrical properties of nanocrystalline SnO_2 depend strongly on the surface states, produced by oxygen and other gas molecules chemisorbed at the grain boundaries that result in the space charge appearance and band modulation. Therefore, the variation of the chemisorbed molecule density is supposed to be the mainly responsible for the electrical response.

It is well known that ethanol can act as the reducing atmosphere and the reducing ability is stronger than formaldehyde (HCHO). So the sensitivity of hierarchical nanostructured SnO_2 in our work to ethanol was higher than that to formaldehyde. In the similar conditions, materials with higher effective surface area offer for ethanol molecules a higher probability of interaction with adsorbed oxygen and hence they are able to have a higher electrical conductivity change. Moreover, the excellent correlation with the Raman surface mode intensity also reveals the existence of a relationship between these modes and the density of the available oxygen adsorption sites.

Contrarily, the situation concerning H_2S sensing mechanisms is different. As concerned as the sensing behavior to H_2S, a certain amount of H_2S was oxidated to condensed H_2O and S at such working temperature to reduce actual amount of the working H_2S, as a result there shows lower sensitivity although H_2S is a type of acid gas and easy absorbed on the basic surface of the target material SnO_2. Generally, the density of real adsorbed gas is influenced by the concentration of available electrons to form and absorb, and by the amount of available adsorption sites at the grain surface that can be altered by the competition with the oxygen. For example, a nanograin surface that is highly covered by adsorbed oxygen has a high density of negative charge that can prevent or render more difficult any action for increasing the surface negative charge such as is required by the H_2S molecule adsorption. Therefore, the existence of an inverse correlation is quite plausible between the sensor responses for C_2H_5OH and H_2S. Experimental results seem to corroborate that for high I_S/I_V values the sensor response to H_2S is low and it increases when the surfaces modes diminish. It is proved that there is a better correlation between the sensor response for H_2S and the active surface than if it is considered against the grain size, d_{XRD}.

At working temperature 270 °C, the sensitivity of as-prepared SnO_2 to C_2H_5OH and HCHO is more prominent than that to H_2S, LPG and gasoline. Moreover, the sensitivity to the test gas markedly increases with the crystallite size of the sensor SnO_2 ranged from 26 nm to 3 nm depending on synthesis temperature. However, H_2S shows downward trends with synthesis temperature, which gives rise to surface modification revealed by the diminution of the surface disorder or by the faceting of the grains. It facilitates the availability of electrons and specific sites for H_2S adsorption. Thus, after annealing at high temperature 400-1000 °C, the sensor response to H_2S is improved.

As-prepared hierarchical nanostructured films consist of interwoven interconnected tubular SnO_2 assembled by nanocrystallites with the grain size of 3–26 nm and specific surface area up to 87 m^2 g^{-1}. The influence of the calcination temperature, crystallite size, and specific surface area on the intensity of Raman surface vibration modes is studied, and an excellent correlation is found between the Raman surface vibration mode intensity and the active surface. It has been considered to corroborate that the Raman surface vibration modes reveal the availability of active surface sites for interaction with the target gas more valuably than specific surface areas, based on the fact that the Raman spectra are taken from intrinsic sensors keeping special nanostructure, while the specific surface area is measured on powders forfeiting the hierarchy.

1.3. Fabrication of ZnO and Other Hierarchical Nanostructured Oxides

In our optimized procedure, the membrane pieces are immersed into 1 M Zn(NO3)2 ethanol solution kept for 30 h at room temperature. This immersing step is repeated three to five times. After being fished out and rinsed with distilled water, these hybrids are dried naturally and calcined at 550 °C for heat treatment with different time (1h, 2h, 3h, 4h, and 6h) to remove the template and achieve better crystallization of the sinters.

The calcination process is traced to suggest the formation of ZnO nanomaterials by conducting TGA and DTA surveys. It is explicit that the natural ESM start pyrolyzing at 200 °C and absolutely pyrolyzed by 550 °C, while the templated hybrids are also decomposed around 200 °C and remained 22 % (by mass) inorganic matters at 550 °C with template completely removed. The weight loss of the hybrids may result from the charing of organic ESM and the following pyrogenation of the charing products. In addition, DTA curve shows two peaks at 357 °C and 531 °C with an interval about 100 °C, which shall represent the decompositions of Zn(NO3)2 and ESM with different organic components, respectively.

The Zn-ESM hybrids are mainly amorphous matters containing zinc precursors and organic bioingredients. Based on the TGA results, the biotemplate ESM is thoroughly decomposed at 550 °C, there present no other peaks of the impurities in the corresponding XRD pattern. All XRD diffraction peaks of the sinters with various heating time can be indexed to hexagonal ZnO with a wurtzite structure (SG: *P*63*mc*, JCPDS 36-1451). With the increase in treating time up to 4h, the XRD peaks intensify sharply. It is evident that the diffraction peaks of the sample calcined at 550 °C for 1h are broadened, assigned to ZnO nanoseeds with very small dimensions, while the diffraction peaks of the sample calcined at 550 °C for 4h are narrow. The average crystallite sizes are estimated to be about 6 nm (treated for 1h) and about 40 nm (treated for 4h), srespectively. All three samples calcined at 550 °C for about 1h, 2h, and 4h, respectively, have similar microscopic features and preserve interwoven and interpenetrated structures of the natural ESM.

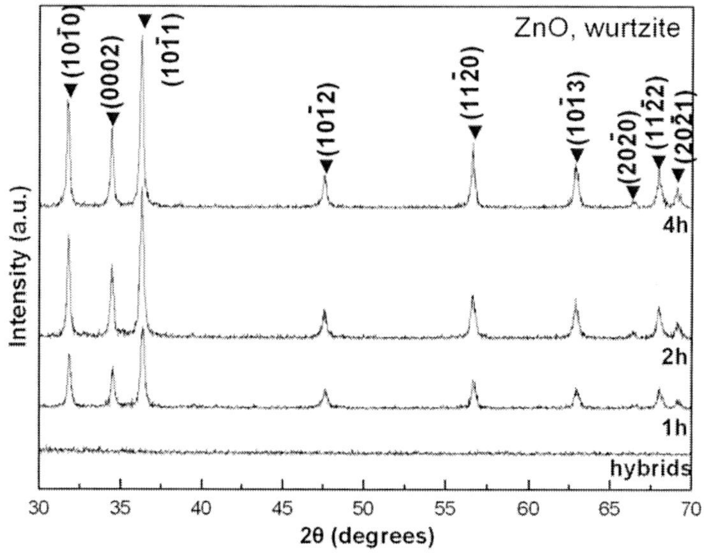

Figure 25. XRD patterns of the templated hybrids and the sinters at 550 °C for different time.

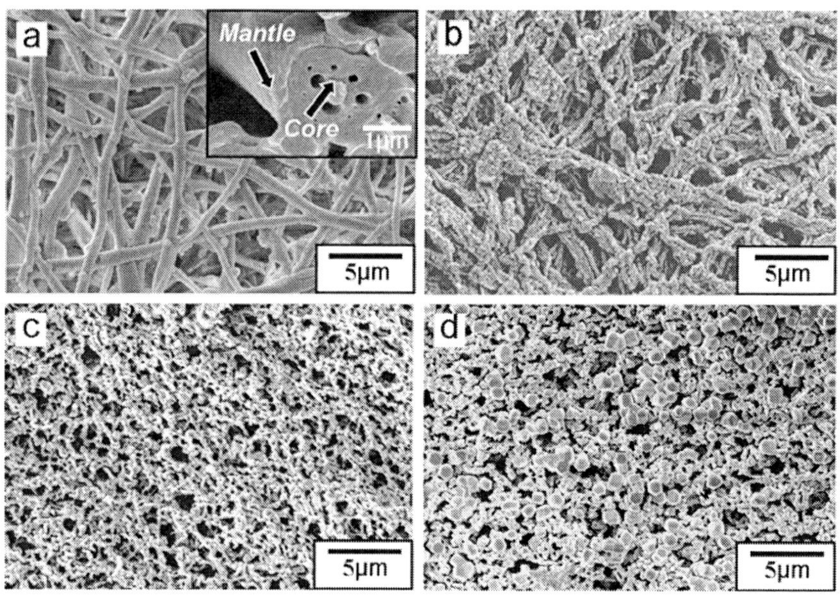

Figure 26. Typical FESEM images showing (a) the natural eggshell membrane and hierarchical interwoven ZnO calcined at 550 °C for different time: (b) 1h, (c) 2h, and (d) 4h. Inset in (a) reveals the conformations of one single membrane fiber.

A careful look at the FESEM images reveals the obvious morphologies evolution of the samples. The sample treated for 1h is characterized by repetitious networks consisting of the fibers ranging from 0.2 to 2 μm in diameter. The target materials are prone to exist as a film with the thickness of about 25 μm, and display hierarchical and subtle structures containing interconnected and penetrated meshworks at three dimensions. With the heat treating time prolonged to 2h, the ZnO fibers are slendered and exhibit porous structures. Notably, few particles with abmodality are appeared sticking on the networks. As heat treating time is further protracted to 4h, more ZnO polyhedron are *in situ* formed along the original ZnO fibers while the interwoven structures are weakened and nearly invisible. The density of the ZnO polyhedron is above 60% (volume).

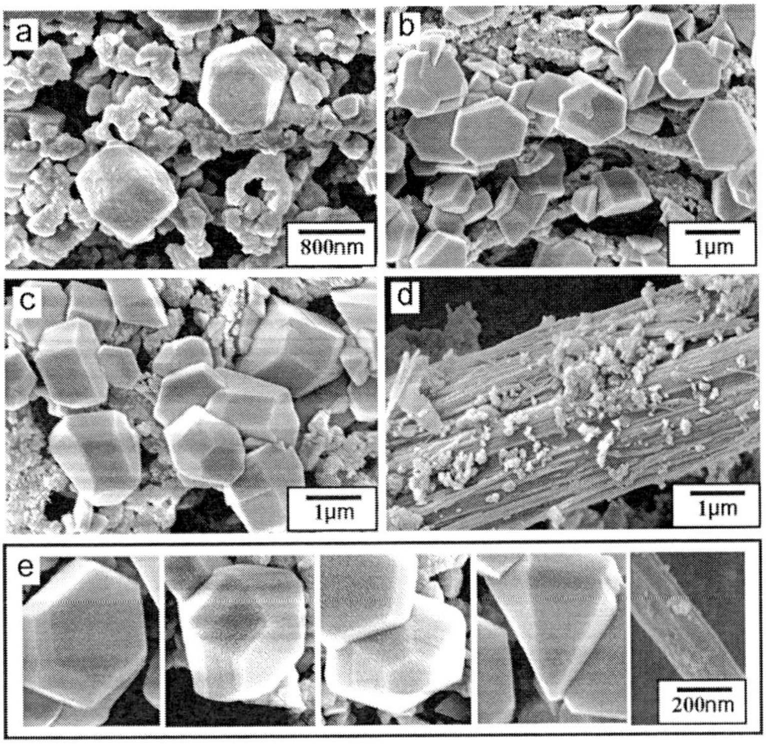

Figure 27. Typical FESEM images showing the extraordinary change in ESM-templated ZnO morphology by prolonging heat treating time: (a) 2h, (b) 3h, (c) 4h, and (d) 6h. Image (e) reveals the intuitionistic morphology evolution of one ZnO cell versus the heating time.

The higher magnification FESEM images show the extraordinary change in ESM-templated ZnO morphology by prolonging heat treating time. The ZnO crystallized along the interwoven matrix for about 2h consists of about 350 nm large polycrystalline agglomerates composed of about 20 nm large round-shaped nanoparticles. Notably, some ZnO crystals are formed like twinned platelets of hexagonal-shaped crystals with diameters of about 600 nm. It can be extended to a great amount of production as treated for about 3h. More hexagonal-shaped crystals with diameters of 400-500 nm come into being. Further treated for another hour, these hexagonal ZnO evolve into polyhedral crystals by the dominant growth with the whole thickness of about 1 μm. However, the use of a longer treating time of synthesis does not produce such regularly shaped ZnO crystals. For more than 6h heat treatment, a new kind of ZnO arrays are formed, which consists of nanorods with diameters of 100-200 nm and lengths of 10-20 μm.

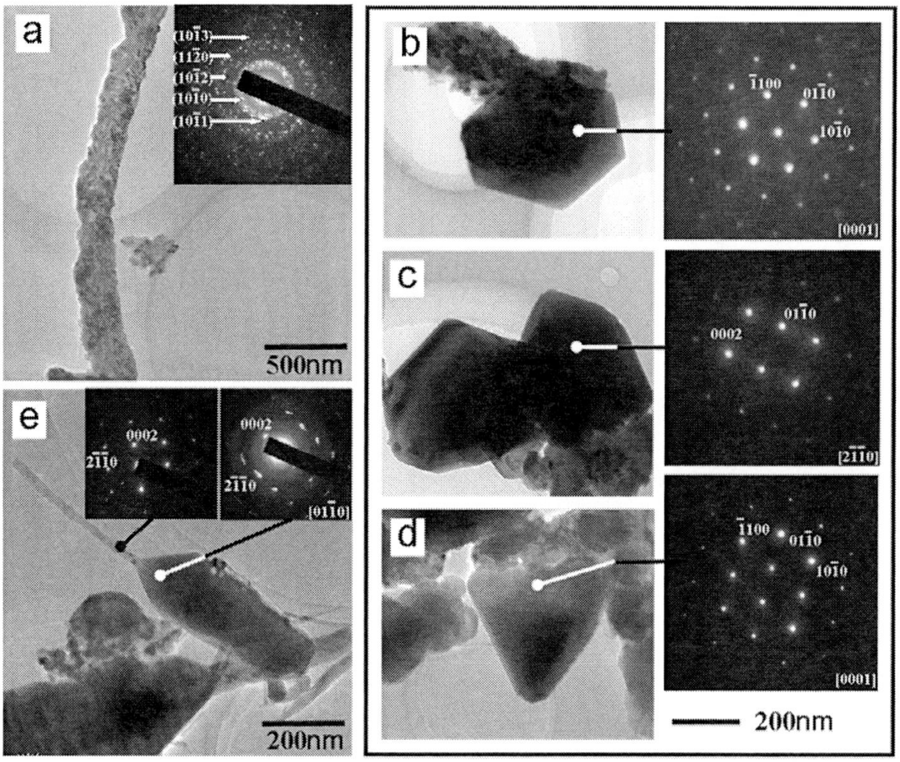

Figure 28. TEM images of the sinters at 550 °C for various heating time: (a) 1h, (b) 3h, (c,d) 4h, and (e) 6h. The insets show the corresponding SAED patterns.

Further structural characterization of the ZnO microcrystals is performed using TEM. One can that single ZnO fiber obtained at 550 °C for about 1h is assembled by nanocrystallites. The SAED patterns show that nanoscaled ZnO is polycrystalline nature, and the diffraction rings are indexed to (10-10), (10-11), (10-12), (11-20), and (10-13) diffraction of hexagonal ZnO, in agreement with the XRD results. The TEM images of the polyhedral ZnO microcrystals at different heat treated stage reveal the same size as determined from the FESEM images. The TEM image of hexagonal ZnO microcrystals as-treated for 3h shows disklike morphology with glazed surface and uniform size. The corresponding SAED pattern composed of bright spots in 6-fold symmetry indicates that the disklike ZnO microcrystal is a single crystal that is of hexagonal phase. The sample treated for 4h exhibits the TEM image of prism-like ZnO microcrystals, which actually grow along the [0001] orientation. The corresponding SAED pattern also indicates that the prism-like ZnO is of high crystallinity and hexagonal phase. There displays the TEM image of another kind of the ZnO crystals with hexagonal cone-shape viewed from the side of the hexagonal cone. The corresponding SAED pattern performed on the individual hexagonal cone-shaped crystal indicates that it is also single crystalline in nature and can be indexed as the hexagonal ZnO phase, which is analogous to the hexagonal disks treated for 3h. As continuously heat treated to 6h, morphology of ZnO crystals has changed. As treated for 6h, some ZnO nanorods are formed and grow along the [2-1-12] orientation basing on the former hexagonal cone structures. These nanorods incipiently have the diameters of 40-60 nm and the lengths of 500 nm, further grow up with the diameters of 200 nm and lengths of 10-20 μm.

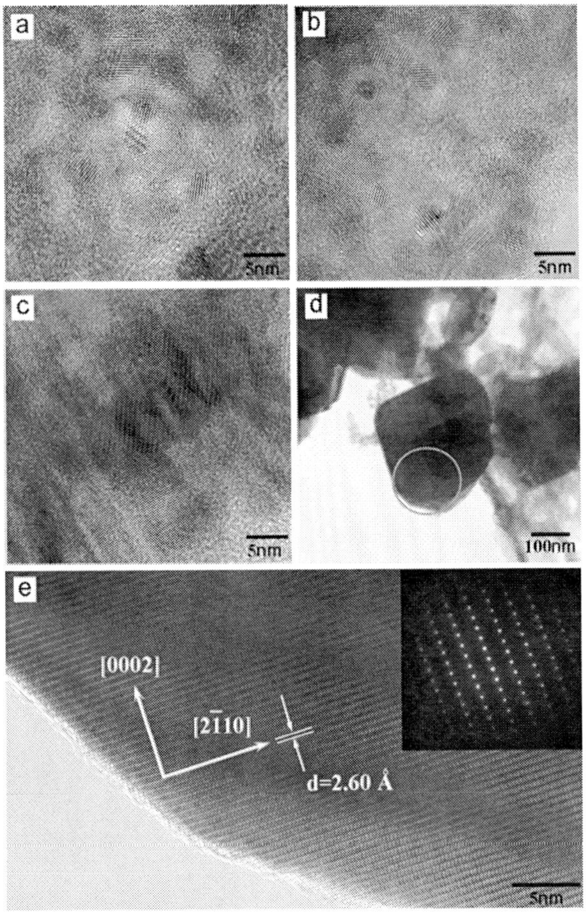

Figure 29. HRTEM images of the sinters at 550 °C and for various heating time: (a) 1h, (b) 2h, (c) 3h, and (d) 4h. Image (e) shows relative high magnification of circled area. The inset shows corresponding SAED patterns. The inset displays corresponding SAED pattern.

HRTEM image of the sinters at 550 °C for about 1h displays that the nanoseeds (ZnO nanocrystallites) are about 5-6.5 nm, and some nanocrystallites might be enwrapped by organic macromolecules of ESM remnants leading to the hazy lattice image. Herein the target material at 550 °C is assembled by such small nanocrystallites It probably ascribes to short-chained amino acides of ESM glycoprotein functioned as the capping agents to restrict the growth of zinc oxide nuclei. However, this limiting effect is weakened when the calcination is for longer time. As the heating time increased to 2h, the lattice image can be clearly seen due to the phase contrast. Considering that each domain having the parallel lattice fringes is a single crystal of ZnO, the average crystallite size is observed to be about 6.5 nm with better crystallization, which may result from the thorough pyrogenation of the organic matrix ESM. Notably, few grow up exceptionally. As-treated for 3h, obtained ZnO nanocrystallites have larger size up to 100 nm. TEM images reveal that the present ZnO crystals synthesized at 550 °C for about 4h is characterized by the polyhedral facets. The HRTEM and SAED images also indicate that the sample has larger crystal size and exhibits single-crystalline state. The (0002) planes are imaged of the hexagonal Wurtzite-type crystal with a lattice spacing of 2.60 Å.

The formation of the ZnO nanoseeds or the polyhedral ZnO structures is proposed to be a process comprised of solidification of biotemplated direction, surface oxidation, nucleation and growth. Herein, these interactions between dissolved zinc ions and specific surfaces affect not only the particle size and habit of nucleating crystals but also the stability of intermediate phases by dropping the activation energy required for the formation of specific crystal faces. Ascribing to the capping function of ESM glycoprotein short-chained amino acides, the growth of zinc oxide nanoseeds can be restricted. ZnO by itself tends to grow into rodlike structures like calcite, but ESM biomacromolecules specifically adsorb to the (0002) surface and force the crystal to grow into plates. The biomacromolecules of the bioorganic templates nucleate the mineral by controlling its crystallographic orientation and growth, by imitating the lattice of a two-dimensional face, or by the stereochemistry of the functional groups at the interface. Recent studies suggest that an epitaxial or geometrical match between macromolecules and crystalline faces does not occur. Instead, the charge density on the crystalline faces, and thus the electrostatic attraction, is alone sufficient for the heterogeneous nucleation. There has been much effort in trying to elucidate the mechanism of proteins and other biomolecules on biomineralization. As we know, in liquid medium, although the growth habit of ZnO microcrystals is mainly determined by its intrinsic structure, it is also affected by the external conditions such as temperature, pH value of solutions, and capping molecules. Among them, the capping molecules can tailor the surface energy according to capping the surface of nuclei and thus control the shapes of microcrystals. Many in vitro studies have demonstrated the importance of proteins and peptides in controlling the nucleation and/or growth of crystals.

Based on the directing and capping functions of ESM biomacromolecules for oxide nanocrystallites, natural ESM is chosen as the template to achieve 3D interwoven porous ceramic materials through similar bioinspired soakage processes with $Zn(NO_3)_2$, $Co(NO_3)_2$, and $PdCl_2$ solutions as the precursor impregnants, respectively. Thermogravimetric analysis of the impregnated ESM reveals that it begin to be pyrolyzed at 200 °C, completely decomposed at 500 °C, leaving behind 25.2%, 20.5%, and 36.8% (by mass) inorganic sinters of zinc, cobalt, and palladium oxides, respectively, along with the removal of organic ESM.

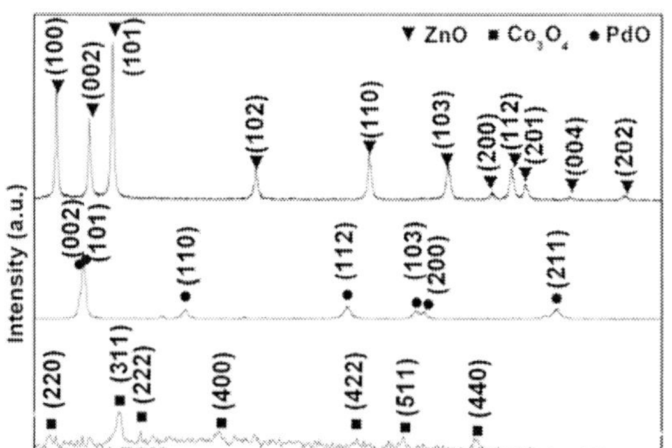

Figure 30. XRD patterns of the natural ESM template and the sinters at 550 °C.

All the XRD diffraction peaks of the calcined products at 550 °C can be assigned to the respective single phase of metal oxides without indication of other crystalline byproducts. All the peaks in the XRD pattern of zinc oxide can be attributed to the zincite structure of ZnO (JCPDS 36-1451), indicating well-established crystallization. For cobalt oxide, broadened peaks were found and are characteristic of smallsized crystallites with a cubic structure of Co_3O_4 (JCPDS 78-1970). The peaks for palladium oxide are fairly broadened and correspond to the tetragonal structure of PdO (JCPDS 75-0584). The average crystallite sizes of about 5.4 nm for ZnO, 9.5 nm for Co_3O_4, and 11 nm for PdO are estimated according to the line width analysis of the diffraction peaks based on the Scherrer formula, respectively. It indicates that by calcination at 550 °C, the ceramic products characterized as the target materials are composed of small-scaled nanocrystallites.

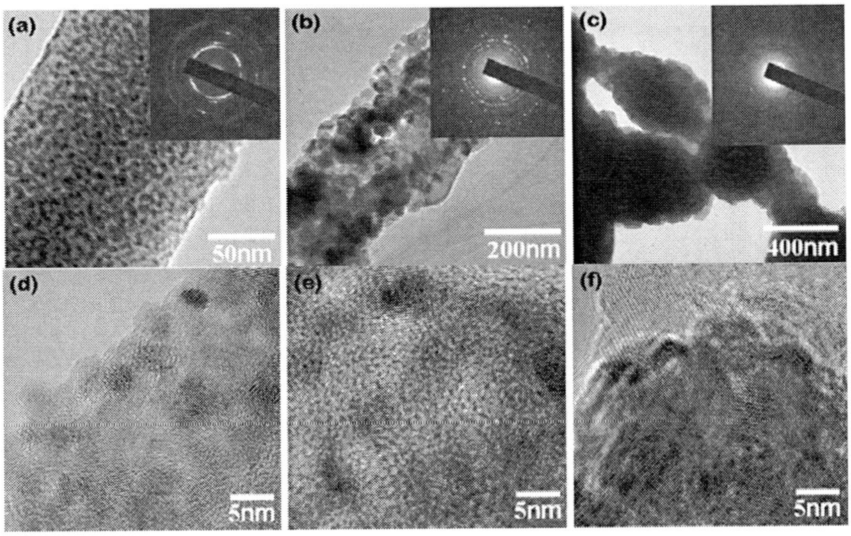

Figure 31. TEM overview images of (a) zinc oxide, (b) cobalt oxide, (c) palladium oxide fibers synthesized at 550 °C, the insets in (a,b,c) showing their respective SAED patterns. HRTEM images of above metal oxides are shown in (d), (e), and (f), respectively.

The overview TEM images of the series show that fibrous structures of about 200–400 nm in diameter are formed by the assembly of corresponding nanocrystallites. HRTEM images for the building unit nanocrystallites indicate that the lattice image can be clearly seen due to the phase contrast. Considering that each domain having the parallel lattice fringes is a single crystal, the average crystallite size is observed to be about 5, 9, and 11 nm for ZnO, Co_3O_4, and PdO, respectively. As expected from the XRD measurements, the zinc oxide nanocrystallites are particularly small, with the sizes ranging from 5 to 6 nm, and display more uniform morphology. It also reveals that the present ZnO hierarchy is characterized by the arrays of nanocrystallites with the structural defects such as micropores and mesopores. This indicates that accurately defined, size-controlled ZnO nanocrystallites of about 5 nm can be assembled to construct hierarchical and porous nanomaterials at the appropriate calcination temperature of 550 °C. However, cobalt oxide nanocrystallites exhibit a cubic shape with sizes of 9–10 nm and less pores compared with ZnO. For PdO, it is observed that the nanocrystallites of about 10 nm agglomerate compactly and the pores can hardly be found. In addition, the crystal structure of these nanoparticles is studied by selected area electron diffraction (SAED), in all cases, the SAED patterns coincide well with the XRD data.

The typical FESEM micrographs show that the size of coherent replicas depends on the original size of the natural ESM. Owing to the heat shrinkage of the biotemplated hybrids during the calcination process, the size of hierarchical replicas is up to about 70% of the original size of the natural ESM pieces. At low magnification, all the samples have similar microscopic features and preserve the interwoven structures of natural ESM fibers, but their detailed morphologies are quite different. At high magnification, fibrous ZnO and Co_3O_4 exhibited porous characteristics. On careful observation it is found that the pores in ZnO fibers are smaller than that for Co_3O_4, which may be ascribed to the lower agglomeration of ZnO building unit nanocrystallites, whereas the surfaces of PdO fibers appear more serried and glazed by virtue of the nonporous structures proved by TEM results.

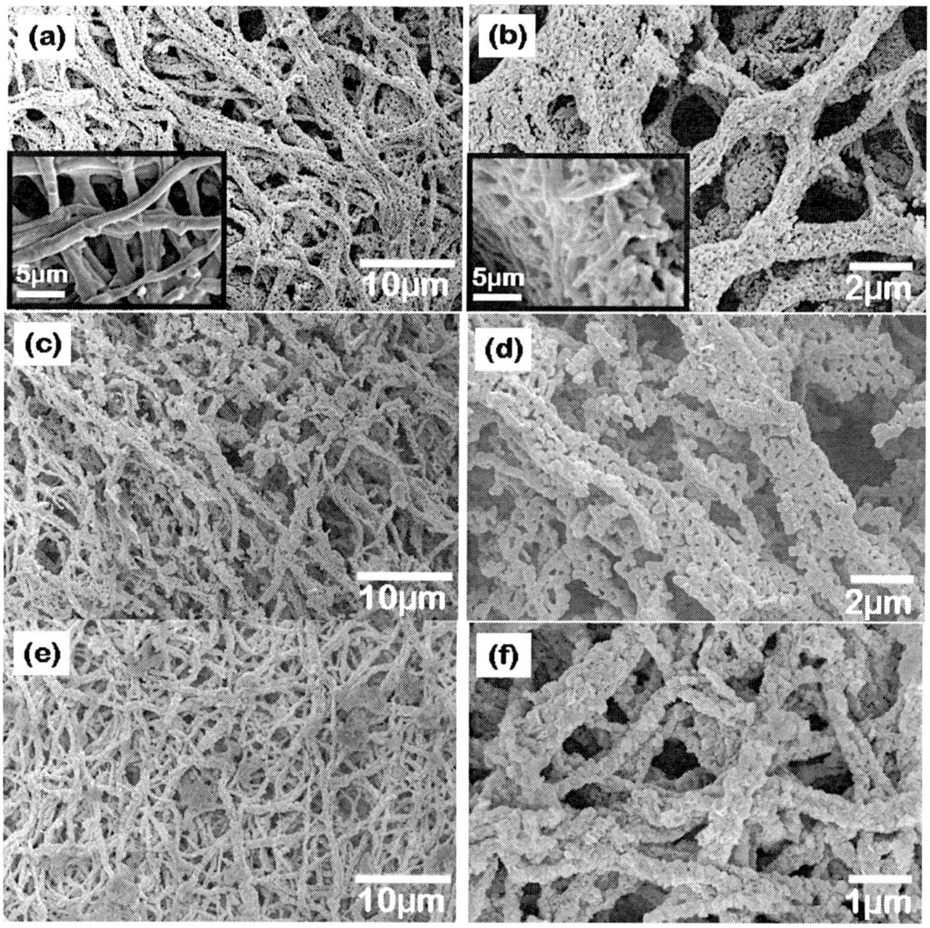

Figure 32. FESEM images showing (a) zinc oxide, (c) cobalt oxide, (e) palladium oxide with hierarchical interwoven structures prepared at 550 °C, respectively. Images (b), (d) and (f) show high magnifications of (a), (c) and (e), respectively. Insets in (a) and (b) displaying the feature of natural ESM and the cross-section of ESM-morphic ZnO, respectively.

From FTIR spectra of the samples calcined at 550 °C, it is found that the typical IR absorption peaks at about 460 cm^{-1}, 671 and 586 cm^{-1}, and 664 cm^{-1} and 603 cm^{-1} are characteristic vibrations of ZnO, Co_3O_4, and PdO, respectively, according to the above XRD results. For all the samples, the absorption bands centered at about 3440 and 1630 cm^{-1} are ascribed to adsorbed water on the surface of metal oxides. The bands appearing in the range between 850 and 1350 cm^{-1}, especially for Co_3O_4, are assigned to the bending mode of different types of surface hydroxyl groups [11], indicating that the surface of the nanoparticles is terminated with the —OH groups. Furthermore, an intense multiphonon scattering of ZnO nanoparticles is observed in the resonant Raman spectrum, where four major bands are observed to result mainly from the polar symmetry modes A_1(LO) and E_1(LO) at 582 cm^{-1} and their overtones. The first-order Raman mode is a superposition of the LO phonon mode in which Zn atoms and O atoms have the same vibration direction, respectively, to the neighbor lattices of the wurtzite ZnO [12]. The intense peak around 437 cm^{-1} is contributed to the E_2 (high) mode. Similarly, the Raman spectrum showing bands at 469, 512, 607, and 674 cm^{-1} with relative intensities corresponds to that of characteristic cubic Co_3O_4 [13], while the bands at 274, 418, 640, and 714 cm^{-1} correspond to that of characteristic tetragonal PdO. These results are consistent with the IR analyses.

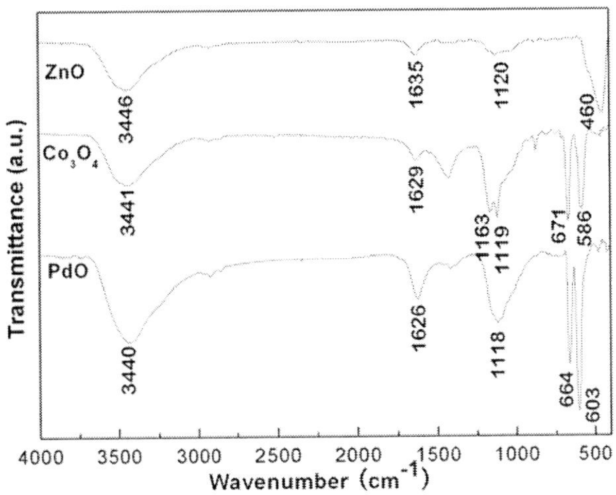

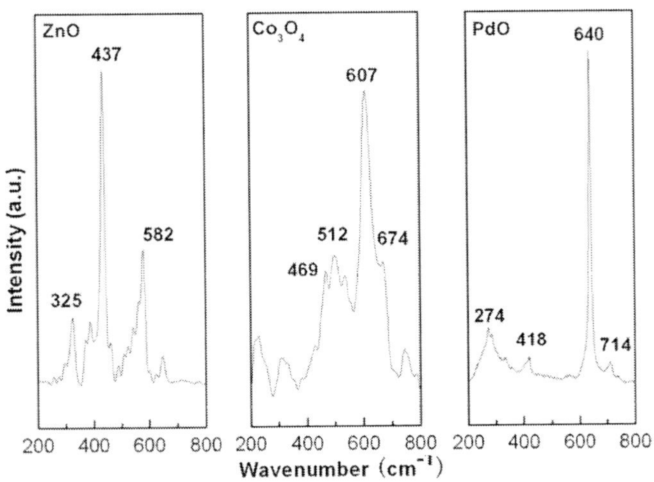

Figure 33. FTIR and Raman spectra of as-synthesized hierarchical metal oxides.

Typical nitrogen adsorption–desorption isotherms and corresponding pore-size distribution curves for ESM-templated metal oxides are examined. By the Brunauer classification, the isotherms change from IV-type for ZnO and Co_3O_4 to II-type for PdO where the hysteresis loop almost disappears completely. Different from IV-type curve based on mesoporous structures, the II-type isotherm is attributed to macroporous modes. At a low relative pressure (*o*0.5), the nitrogen-adsorbing action of macroporous materials is the same as the adsorption of mesoporous monolayer structures. But above the medium relative pressure (40.5), there is no capillary condensation phenomenon in macropores and then no hysteresis loop. Hence, according to the types of isotherms, three oxide samples can be confirmed as mesoporous ZnO and Co_3O_4, and macroporous PdO, individually. The average pore size increases from about 20–30 nm for ZnO, to 30 nm for Co_3O_4, to 80 nm for PdO, while the pore volume decreases. This means that the porous structures are quite different in nanometer scale depending on the different metal precursors and oxide materials. In addition, more than one pore diameter peaks in the pore-size distribution curves of ZnO and Co_3O_4 samples indicate several kinds of pore (micropore, mesopore, and macropore) characteristics. Actually, such mesopores and micropores have been observed in the above HRTEM images.

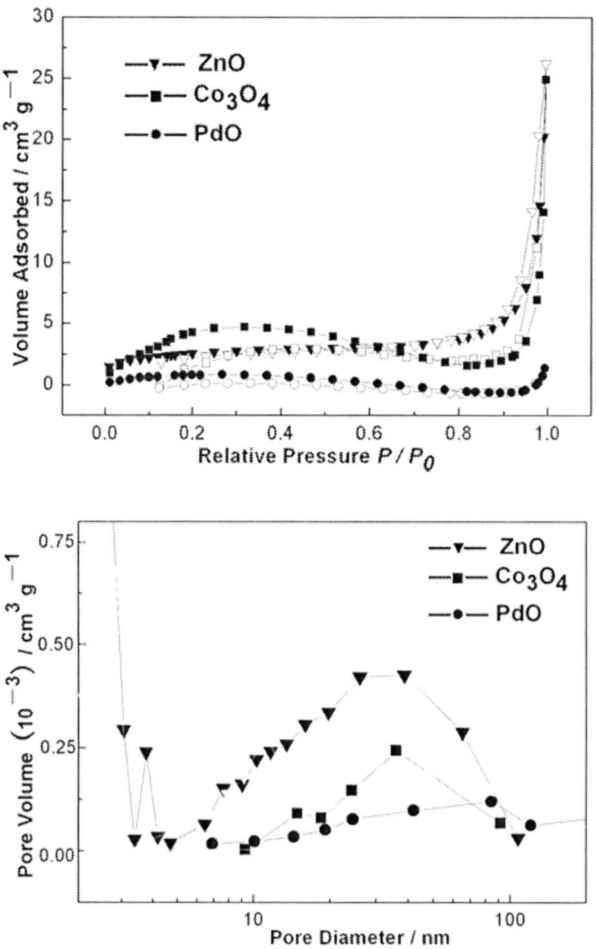

Figure 34. Nitrogen adsorption-desorption isotherms plots and corresponding pore size distribution plots of ESM-morphic metal oxides obtained at 550 °C.

The synthesis of biomorphic metal oxides can be separated into two steps: one contributes to the interplay of metal precursors and ESM constituents during the dipping pretreatment, and the other is the crystallization formation of metal oxides accompanied by the removal of the ESM template during the calcinations process. During the dipping treatment in the aqueous solution, due to the self-assembly behavior of the hydrophilic groups of glycoprotein, metal precursor ions are transported on the ESM surface; further, on calcination treatment at 550 °C, the final oxides can retain well the hierarchical structure of the biotemplate ESM when organic ESM is burned off. Moreover, it can probably be ascribed to short-chained amino acids of ESM glycoprotein, which functioned as capping agents to obtain well-distributed and small-scaled nanocrystallites as building blocks. However, there occur different reactions between corresponding metal precursor ions and ESM components; thus, final metal oxide unit nanocrystallites have divergent morphology and diameter. It has been found that the present ZnO or Co_3O_4 hierarchy is characterized by arrays of well-distributed and small-scaled nanocrystallites, and of micropores and mesopores. However, it is quite different for PdO that some Pd^{2+} ions could be reduced to Pd^0 by certain amino acids of ESM, which may further degrade the proteins by the catalysis. Thus, the conformations of ESM biomacromolecules may be altered and then the serried hybrids may be formed. The different porous conformation at the nanoscale can also be explained by the above mechanism. As a whole, ceramic oxides replicated well the hierarchical structure of the biotemplate ESM via a self-organized solution process, in which the biomaterial served both as a chemical revulsant and a physical substrate. Prompted by the special hierarchical buildup, we can speculate that as-prepared ESM-based metal oxides may be useful in fabricating optical materials (ZnO), magnetic storage devices (Co_3O_4), and catalysts or their carriers (PdO).

2. Formation and Assembly of Lead Chalcogenides on Eggshell Membrane Fibers

The synthesis control of metal chalcogenides and relevant nanocomposites has been the focus of intensive research due to their unique optical, electronic, and photovoltaic properties, which are chiefly determined by their shape, structure and distribution. Thereinto, nanostructured lead chalcogenides gain more and more research attention, due to their unique properties and valuable applications in highly efficient photovoltaic conversion devices, optical switching, biological labels, as well as electroluminescent devices and lasers.

A series of bioinspired methods involving some biospecies such as proteins, yeast, bacteria, virus, enzyme, eggshell membrane, rhodobacter sphaeroides, DNA, bean sprouts have succeeded in controlling small-scaled and well-dispersed chalcogenides nanocrystallites. In the progressive stage, the biogenic assembly of planned nanoarchitectures constitutes a main challenge to the "bottom-up" fabrication of nanocrystallites components for future nanodevices. Thus, research efforts are now invested in novel techniques for the deliberate organization and immobilization of metal chalcogenides nanoparticles on solid substrates to form hierarchical nanocomposites. However, it is still extremely difficult to find an appropriate biomaterial support on which the synthesis can undergo under ambient reaction conditions. In view of the special proteins patterns and functional residues of eggshell membrane (ESM) biomacromolecules, the convenient biological castoff can be utilized as an ideal biosubstrate for the preparation of metal chalcogenides and relevant nanocomposites.

Herein, a facile, moderate and green biosubstrate directed technique is designed to synthesize hybrid PbS/ESM nanocomposites at ambient conditions by immersing ESM into Pb(Ac)2-HAc and Na2S (or Na2SeSO3) solutions in turn. The functional residues such as amido and imido groups from glycoprotein mantle of ESM fibers, provide chelation and location functions to the formation and the assembly of nanocrystalline PbSe, and act as the surfactant and substrate to achieve fine nanocrystallites arraying well-distributed nanoclusters.

In a typical procedure for PbSe nanostructures, the substrate ESM is firstly immersed into the mixed solution of lead acetate and acetic acid at room temperature, taken out and rinsed with deionized water, subsequently the following two processes go along, respectively, namely that the membrane is dipped into fresh Na2SeSO3 solution (I); into ripen Na2SeSO3 solution laid still for 2 days (II). The final samples I and II as dark gray are obtained with correspondence to process I and II. XRD patterns indicate that the original ESM is totally amorphous, while all the reflection peaks of samples I and II can be indexed as cubic PbSe with the face-centered rock-salt structure (JCPDS No. 06-0354). The reflection peaks show markedly broadened, indicating that obtained products are small-sized nanocrystallites and sample I has smaller crystallite size than sample II.

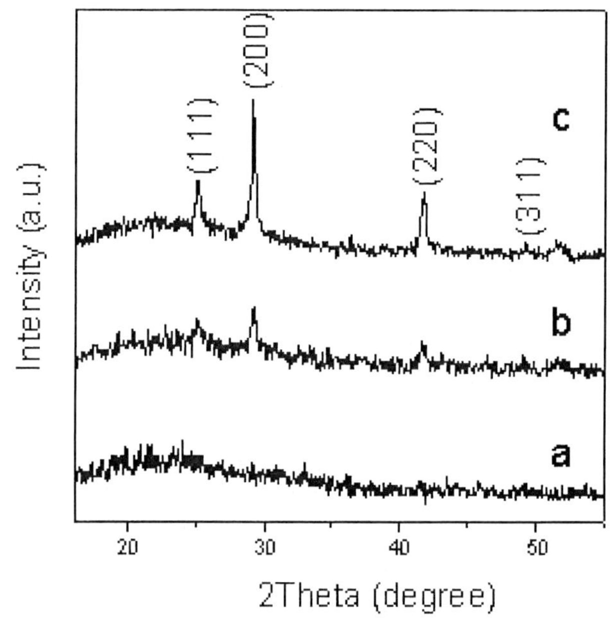

Figure 35. XRD patterns of the samples: (a) the original ESM; (b) sample I (PbSe/ESM); (c) sample II (PbSe/ESM), respectively.

The morphology and the configuration of the composite PbSe/ESM are investigated by SEM images. It can be clearly seen the ESM presents characteristic macroporous meshworks of diverse interwoven and coalescing fibers. PbSe nanocrystallites are stuck uniformly in ESM fibers while there is absent in the matrix among the membrane fibers. Fine PbSe nanocrystallites with the average size of 5 nm are assembled into 25–30 nm smaller nanoclusters distributing uniformly on the ESM support. And herein some smaller nanoclusters would like to arrange for bigger nanoclusters. While PbSe nanocrystallites are finally prone to ripen into nanocubes with lower thermodynamic energy. It is considered that these nanocrystallites can firstly be loosely assembled into smaller nanoclusters and subsequently compactly ripen into nanocubes with more glazed figures than the nanoclusters, as evidenced by the coexistence of PbSe nanoclusters and nanocubes when employing fresh Na2SeSO3 as the reactant with enough immersing time. All as-prepared nanocrystallites are well-dispersed and arrayed orderly on the surface of ESM fibers. The corresponding SAED pattern of PbSe nanocrystallites signifys the single-crystalline nature. The square spot array is indexed to (2 0 0) and (2 2 0) planes of the face-centered cubic (fcc) PbSe. The representative HRTEM images present well resolved lattice fringes revealing the high crystallinity, indexed to the (2 0 0) planes of the rock salt cubic phase with the interplanar spacing of 3.05 ° A, which are consistent with the XRD results of PbSe crystallites. On the basis of the TEM and HRTEM analyses, it can be considered that these nuclei bounded by (2 0 0) facets are assembled into nanoclusters and nanocubes on the ESM. In addition, as-prepared PbSe nanocrystallites with a characteristic feature of small size are displayed from the room temperature photoluminescence (PL) spectrum of sample I. PbSe nanocrystallites exhibit the emitting peak at 357 nm, with a 42 nm blue-shift in comparison with Guha's report, which reveals as-prepared PbSe nanocrystallites in Sample I have very small size and performed obvious 1Se~1Ph intraband transitions indicative of a short lifetime and obscuring the phonon bottleneck.

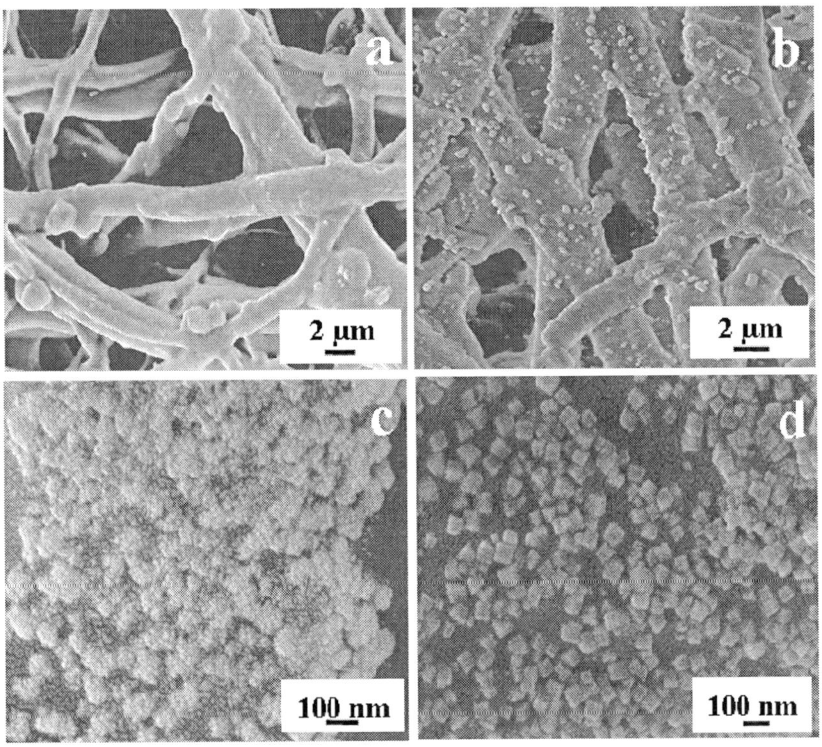

Figure 36. SEM images of (a) the overview of the surface of original ESM; (b) the overview of PbSe/ESM composites; (c) sample I and (d) sample II, respectively.

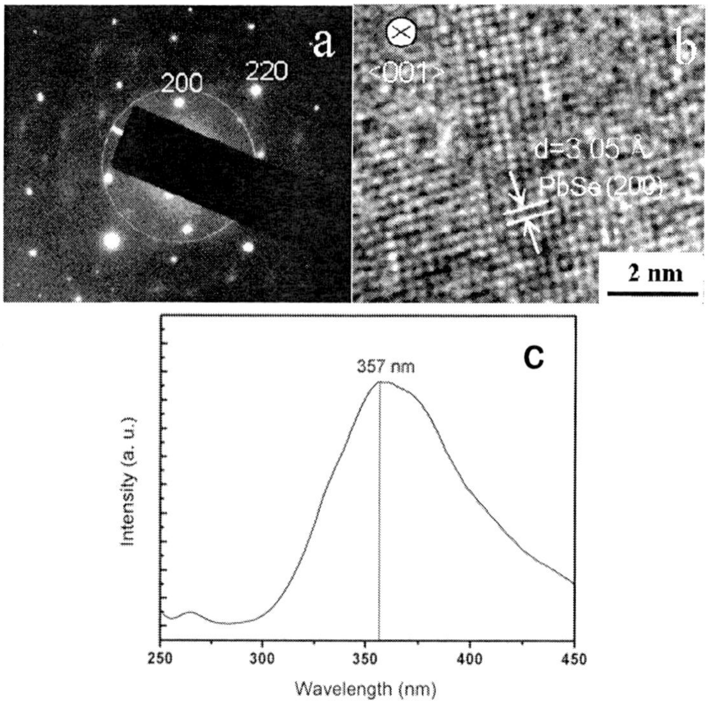

Figure 37. (a) Representative SAED pattern of PbSe nanocrystallites; (b) high-resolution TEM images showing PbSe nanocrystallites oriented attachment along (001) direction; (c) photoluminescence spectra of PbSe nanocrystallites, the excitation wavelength 250 nm.

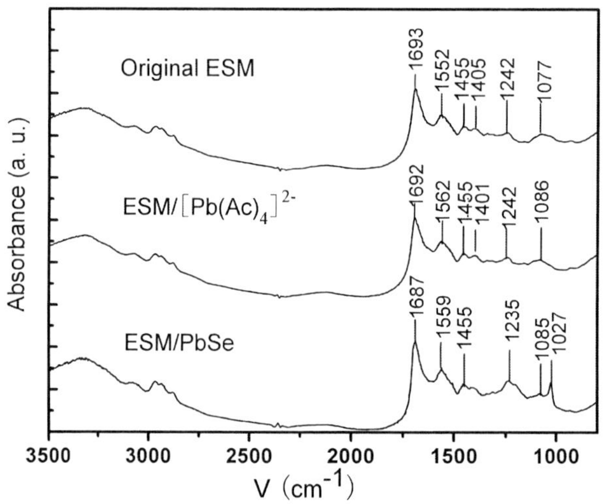

Figure 38. The typical FTIR spectra of (a) ESM; (b) [Pb(Ac)4]2-/ESM; (c) the composite ESM/PbSe, respectively.

FTIR spectra of [Pb(CH3COO)4]2^{-}/ESM and PbSe/ESM are also be probed into. The absorption bands at 1693, 1552, and 1242 cm^{-}1 are attributed to amide I, amide II, amide III bands of the glycoprotein mantle of original ESM, respectively. The peak appearing at 1455 cm^{-}1 is due to the *v*s(COO) stretching vibration, and the band around 1077 cm^{-}1 is assigned to the C–N stretching vibration in plane band. After the ESM being dipped into Pb-precursor solution, the shifts of amide II from 1552 to 1562 cm¯1 and νs(C–N) stretching vibration from 1077 to 1086 cm¯1 suggest that there occurred environmental changes around the imino groups and the C–N bond, which may be due to the condensation reaction of carboxyl groups from [Pb(CH3COO)4]2¯ with amine groups and imido residues from glycoprotein mantle of ESM [14]. Further soakage in the Na2SeSO3 solution, there clearly presents a new band at 1027 cm¯1 attended by the red shifts of amide I, amide II, amide III bands. These changes shall contribute to the formation of PbSe nanocrystallites on the ESM.

Herein, eggshell membrane as the biosubstrate, shall play crucial roles in the formation of PbSe nanocrystallites and the assembly into nanoclusters and nanocubes. The architecture of the ESM fiber can be depicted as the collagenous core which is mainly composed of collagen and osteopontin proteins, and the polyanionic glycoprotein mantle. When the ESM is dipped into Pb(Ac)2-HAc solution, the chelation prefers to occur on the surface of glycoprotein mantle of fibers while is hard to take place in the core.

Detailed amino acid composition (wt.%) of natural ESM fiber mantle is Asp 7.05, Thr 4.80, Ser 4.32, Glu 9.98, Pro 9.34, Gly 5.20, Ala 2.26, Cys 4.10, Val 5.30, Met 3.32, Ile 2.61, Leu 3.65, Tyr 1.87, Phe 1.35, His 2.97, Lys 2.98, Arg 5.93, Trp 1.80 [15]. It is well known that all amino acids are connected by peptide bonds. Especially, the lysine acid (2.98 wt.%), arginine acid (5.93 wt.%) and His (2.97 wt.%) containing two poorly alkaline amido groups, as functional moleculars, can beget the condensation reactions with the carboxyl residues from [Pb(CH3COO)4]2¯, thus the Pb-complexes in the solution may be caught on the surface of ESM fibers. Moreover, other acidic and neutral amino acids can also play some roles in attracting Pb-complexes but not evident. By sequential soakage treatments of the ESM into Na2SeSO3 solution, SeSO32¯ is gradually attracted towards lead sites on the ESM mantle and Se2¯ is prone to gently escape from SeSO32¯ under the electrostatic actions of lead cations and the inducement of imido residues and imino groups. Whereafter, in situ nucleus formation occurs and cubic PbSe comes into being at surface of ESM fiber mantle, since cubic PbSe has thermal stability which can effectively lower the system energy.

$$\left[Pb(CH_3COO)_4\right]^{2-} + 4R\text{-}NH_3^{+} \rightarrow \left[Pb(CH_3CONH\text{-}R)_4\right]^{2+} + 4H_2O \quad (1)$$

$$SeSO_3^{2-} + H_2O \rightarrow SO_4^{2-} + Se^{2-} + 2H^{+} \quad (2)$$

$$\left[Pb(CH_3CONH\text{-}R)_4\right]^{2+} + Se^{2-} \rightarrow (CH_3CONH\text{-}R)_4PbSe \quad (3)$$

$$\cdots(R = \text{alkyl residue})$$

ESM can also play an oriented role for the growth and the assembly of PbSe nanocrystallites. The shape of an fcc NC is determined by the ratio, *R*, of the growth rate in the <100> directions to that of the <111> directions [16]. During the growth stage, the facets of the crystallites tend to form on the low-index planes to minimize the surface energy, which is crucial for the growth rate of the crystallites. Herein, amido groups and imido residues contained in glycoprotein mantle may form weak binding ligands to the {100} faces while bind strongly to the {111} surfaces, thus compared with the {100} faces, the {111} faces can selectively have a low surface energy [17]. Accordingly, the growth rate in the {111} planes is significantly higher than that of the {100}, then {111} facets will vanish resulting in the formation of the nanocubes [18]. Moreover, the relative supersaturation of lead cations on glycoprotein mantle can in turn compensate for the negative electrostatic potential, which made seleniferous anions facile to contact with lead ions by getting over the spatial hurdle. As soon as seleniferous anions are captured on the ESM fiber mantle, the PbSe nucleus can come into being, which may also be an autocatalysis [19] for the growth of PbSe nuclei until well-distributed perfect nanocubes or uniform nanoclusters are achieved. In addition, the amino acid residues of the ESM macromolecules can be functioned as the surfactants to well-distributed and small-sized nanocrystallites. By capping roles, these fine nanocrystallites are feasible to evolve into nanoclusters and are finally self-assembled into thermodynamically stable perfect nanocubes in the absence of additional surfactants.

Moreover, the precursor Na2SeSO3 solution also plays an important role in the formation of PbSe nanoclusters and nanocubes. Upon the increase in the concentration of Na_2SeSO_3 solution and employing fresh Na2SeSO3, PbSe crystallites come into being at relatively fast speed corresponding to kinetic non-equilibrium of nucleus formation, thus large numbers of nuclei are formed, and the growth process is hard to control, it is less likely to form perfect crystal lattices and PbSe nanoclusters are obtained.

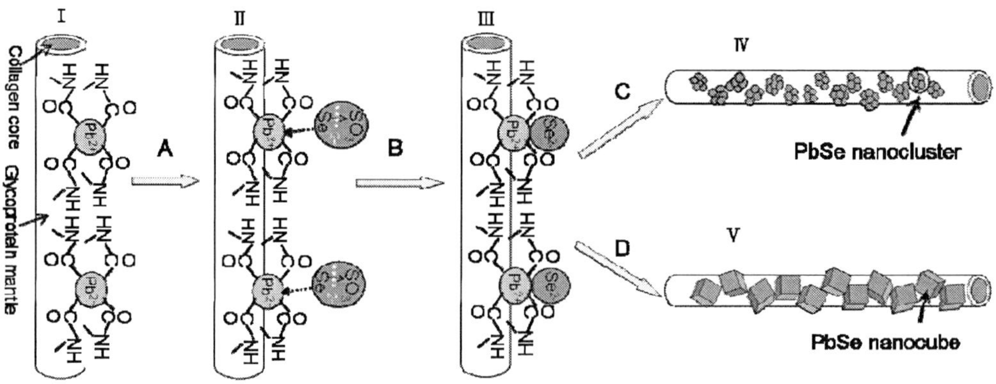

Figure 39. Formation illustration of lead selenide nanocubes and nanoclusters on ESM fiber. Models of (I) show the condensation reaction between the amido groups from glycoprotein mantle of a single ESM fiber and the carboxyls of $[Pb(CH_3COO)_4]^{2-}$ at the given locations, (II) $[(SeSO_3)^{2-}]$ ions in the solution approach to lead complexes on the mantle, based on the electrostatic actions, (III) in situ nucleation of PbSe nuclei, (IV) the formation of thermodynamic metastable PbSe nanoclusters, (V) tiny nanocrysatllites evolve into well-ordered nanocubes.

Since PbSe nanoclusters are thought to be thermodynamically metastable state, nanocubes are finally prepared, based on prolonging reaction time, using ripen precursor Na_2SeSO_3 and decreasing the concentration of Na_2SeSO_3 solution, which can meet all these conditions, such as the condense reactions between amidos and carboxyls, the electrostatic actions between lead cations and seleniferous anions, to the nucleation and the assembly of well-ordered PbSe nanocubes.

By similar procedure but Na2S solution as the corresponding impregnant, face-centered cubic PbS nanoparticles come into being on ESM fibers to obtain PbS/ESM hybrids. The formation scheme can be described as follows:

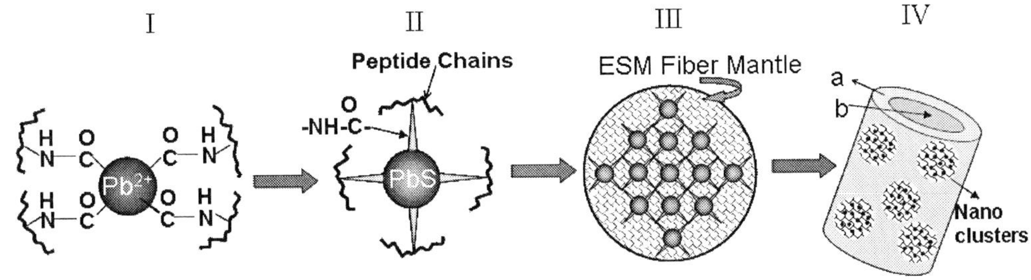

Figure 40. The illustration of the formation of well-distributed floriated lead sulfide nanoclusters on ESM fibers. Models of (I) the condensation reaction of the carboxyl of $[Pb(CH_3COO)_4]^{2-}$ with the amido groups from glycoprotein mantle of ESM fibers at the given locations; (II) in situ nucleation of PbS nuclei on the mantle by electrostatic actions; (III) the assembly of PbS nanocrystallites into the nanoclusters on the ESM fiber mantle; (IV) floriated PbS nanoclusters bound on the substrate ESM fibers to form the nanocomposites PbS/ESM. Single ESM fiber is structurally composed of glycoprotein mantle (marked as a) and collagen core (marked as b).

For further investigation on PbS nanoparticles, the Raman spectroscopy is thought to be effective and nondestructive. The strong peak at 155 cm^{-1} can be assigned to a combination of transverse acoustic and optic modes, TA+TO, over a large volume of the Brillouin zone. The strong band at 211 cm^{-1} can be indexed to fundamental longitudinal optical (LO) phonons near the Γ-point of the Brillouin zone [20]. It is noted that there exists a small shoulder at 204 cm^{-1} assigned to a surface phonon (SP) mode [21]. Based on Batonneau's reports [22], the bands observed at 155 and 204 cm^{-1} are presumably due to the luminescence of PbS semiconductor or caused by instrumental effects, and that the band at 450 cm^{-1} can be ascribed to the first overtone of the LO phonon (2LO) at the center (Γ) of the Brillouin zone, this overtone shall be induced by a Fröhlich-interaction and be Raman active. Meanwhile, the peak at 622 cm^{-1} can be labeled as the second overtone (3LO) [23]. Different to Krauss and Wise's report on the Raman spectrum of PbS quantum dots having three peaks at 215 (LO), 415 (2LO), and 630 cm^{-1} (3LO), respectively, the differences in the wavenumbers and the intensities to those of our Raman peaks can largely lie on the particle size, the excitation wavelength, and the operating temperature. In addition, it is demonstrated that the Raman frequency prefers to decrease with the increase in crystalline size and surface roughness. So it can be thought that in our work both PbS nanocrystallites and the nanoclusters have uniform small size and the nanoclusters are well-distributed on the ESM fibers. In addition, the observability of TO and SP modes proves that PbS nanocrystallites in the hybrids exhibit strong quantum size effects. Because of the Raman shifts of crystalline insulator or semiconductor, it is feasible to observe LO phonons while other modes such as TO and SP modes are usually not observable due to symmetry restrictions and their low intensities, respectively [24]. However, the SP mode by Raman scattering measurements can occur in Raman spectra of nanoscaled materials with quantum size effects [25]. Other peaks, such as located at 174, 511, 564 cm^{-1}, may attribute to some new bonds giving rise to the ESM intricate cooperative effects during the soakage treatment. Therefore, small-scaled PbS nanocrystallites on the ESM fibers are assembled into size-uniform nanoclusters distributing orderly on the ESM fibers.

Therefore, during the formation of lead chalcogenide nanocrystallites and nanoclusters on living ESM at room temperature, the ESM not only provides effectively revulsive sites to the formation of chalcogenides nuclei and the assembly into nanoclusters in virtue of the reciprocity of functional groups of ESM fiber mantle, precursors, but also acts as the surfactants bringing uniformly arranged and well-dispersed chalcogenides nanoclusters and nanocubes on the ESM. Other factors are also considered, such as reaction time, precursor concentration. The as-prepared chalcogenides nanoclusters behave good quantum size effect and dispersancy, and exhibit excellent biocompatibility when combined with the reactive biosubstrate ESM to be hybrid nanocomposites for applications in photoelectric, photonics, and electronics fields. This bio-inspired technique is a sort of green, moderate and effective strategy for the synthesis of other series of functional nanomaterials with pre-desired conformation and relevant valuable properties.

3. In Situ Synthesis of Noble Metal Nanoparticles on Biofibers at Room-Temperature

In a typical experiment for Pd nanoparticle, Ca. 15 mg ESM fiber is immerged into 200 ml 0.5 mM palladium chloride aqueous solution with the pH value tuned to about 8. The system is carried through in the dark for about 18 h. Then the as-immerged biofiber is fished out, subsequently rinsed with distilled water and directly dried with air flow to obtain final hybrids.

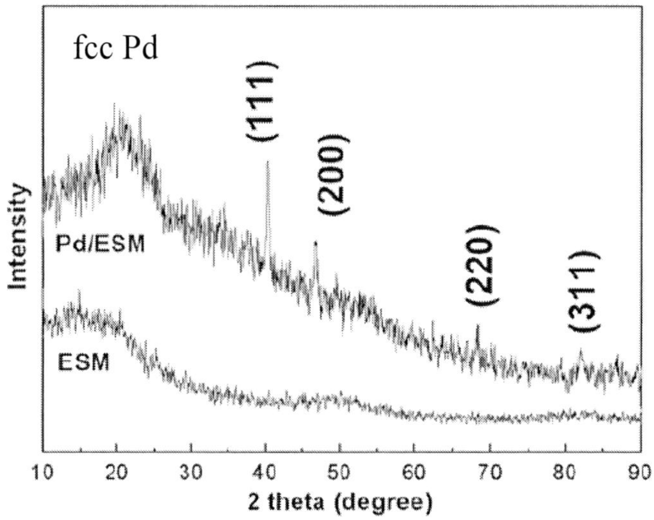

Figure 41. XRD patterns of natural eggshell membrane fibers (ESMF) and the composites Pd/ESMF.

One can be seen from the XRD patterns that the natural ESMF exhibits the diffractive peaks of crystalline domains around 20.64°, which are ascribed to the sequence and conformations of amino acids. As concerned as the hybrids samples being processed in dark for 18 h, it can be seen that the face-centered cubic (fcc) structure of palladium (JCPDS file No. 46-1043) comes into being, and the corresponding diffraction peaks can be indexed as (111), (200), (220), and (311) planes, respectively. In addition, the broadened reflection peaks indicate Pd nanocrystallites having small dimensions.

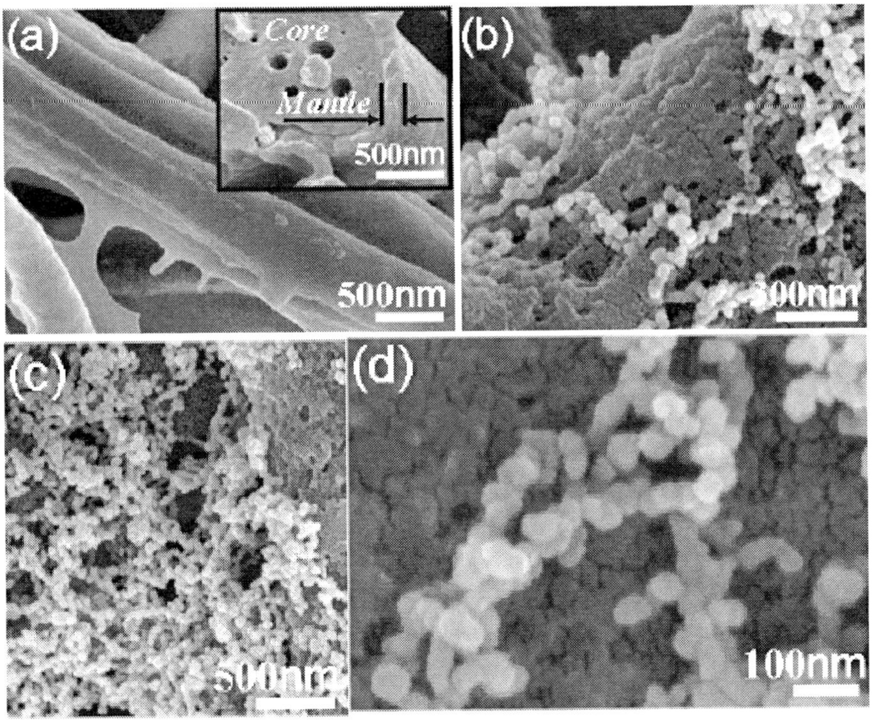

Figure 42. Typical FESEM images of (a) natural ESM fibers; as-prepared nanocomposites Pd/ESMF after reaction for (b) 5 h and (c) 18h; (d) high magnifications of image (c).

FESEM images reveal the morphologies of as-synthesized palladium nanoparticles on the biofibers with different reaction time about 5h and 18h. In the primeval stage the spherical Pd nanoparticles have formed and arrayed into the chained patterns, which is almost the same as the sample dealed with 18 h. One can see that there is no evident change in morphology by prolonging the reaction time besides a high yield of the target materials. A careful look at these fibers reveals that the previous glazed fibrous structures of the original ESMF have converted to the rugged forms, which may contribute to the nucleation and assembly of palladium nanocrystallites on ESMF. The high magnifications of the images, exhibit the pearl-necklace structures with the "pearls" about 40 nm in diameter and the "necklace" of 2-5 μm in the whole length. Figure 3 shows representative TEM and HRTEM pictures record from drop-coated films of the palladium synthesized by infusing ESM biofibers into $PdCl_2$ solutions for 18 h. At lower magnification, palladium nanoparticles show the pearl-necklace configuration with the "pearls" units of about 40 nm assembling into very open, quasilinear superstructures rather a dense, closely packed and chained form. The SAED pattern reveal that Pd particles are polycrystalline in nature and the rings could be indexed as the fcc structure of crystalline palladium, which are consistent with the XRD results. At higher magnification, it is striking that each "pearl" of the palladium necklaces virtually appear to be formed by the assembly of smaller nanoparticles containing a porous network, which can be described as "worm hole" variety. It is also observed that a second material coats the palladium nanoparticles.

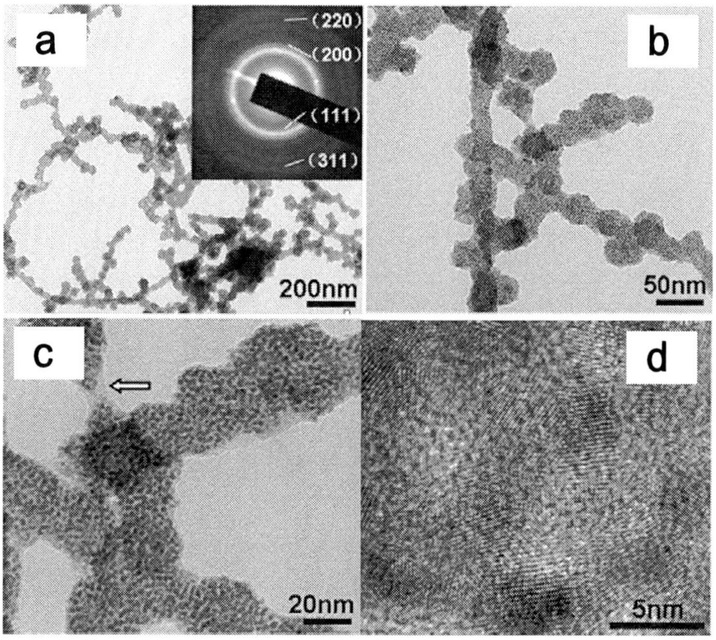

Figure 43. Representative TEM images of (a) necklace structured Pd nanoclusters; insets in (a) show the corresponding SAED patterns; (b, c) high magnifications of image (a); (d) HRTEM images of Pd nanoparticles.

It is believed that this coating material with an average thickness of 5 nm is made up of bioorganic components from ESMF glycoprotein and may form an amorphous shell around the palladium nanoparticles, giving rise to weak Bragg reflections. In addition, the lattice image can be clearly seen due to phase contrast. Considering that each domain having the parallel lattice fringes is a single crystal of Pd, the crystallite size is observed to be about 3 nm. It is noted that the Pd nanocrystallites are not in physical close contact but are monodispersed with a fairly uniform interparticle distance.

A recent review has highlighted on how intricate interplays during the crystallization and assembly of metal nanocrystallites under the directing of biomolecular template can give rise to cooperative transformation and reorganization of inorganic–organic hybrid building blocks to produce emergent nanostructures with complex form and hierarchy. An important point concerns the functionalization of inorganic nanoparticles by means of evolutionary optimized biological components. In this work, the aqueous synthesis of the hierarchical nanocomposites can be separated into two steps: one is the reduction execution of metal ions by reductive functional groups of the biofibers; and the other is the *in situ* crystallization and assembly of metallic nanoparticles under the direction of the biomacromolecules and arraying in special locations of the biofibers. As we known that shell membrane is a non-mineralized, collagen-based matrix consisted of type I, V, and X collagen located in the core of each fiber, which is surrounded by a soluble redox-active glycoprotein mantle. On the basis of the redox nature we expect that the glycoprotein mantle mainly containing amino and carboxyl groups can reduce Pd (II) to Pd (0) state. In above investigations, we have proved the assembly of hierarchical nanomaterials under the direction of the membrane macromolecules. Thus, we consider that the synthesis of hierarchical palladium nanostructures may rely on the reductive templating action of ESMF glycoprotein residues.

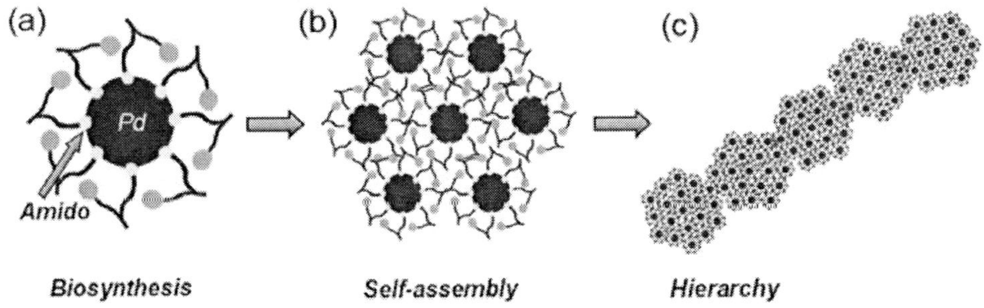

Figure 44. Schematic diagram of in situ synthesis of hierarchical nanocomposites involving redox-active biofibers at room temperature: (a) reductive mineralization of Pd nanocrystallites by amido residues of the eggshell membrane glycoprotein; (b) the assembly of Pd nanocrystallites into the nanoclusters under the direction of ESM biomacromolecules; (c) further assembly of Pd nanoclusters into Pd/ESMF nanocomposites.

The biomolecules are often coupled through noncovalent electrostatic interactions. At high pH values that are slightly above the isoelectric point of certain proteins, metal ions in aqueous solution are reduced by some biomolecule residues. However, as-formed palladium crystallites can be enwrapped by the soluble glycoprotein macromolecules in the hybrids composites. The palladium particles are overcastted by the cladding, which is maybe the organic glycoprotein remnants. It is also clearly displayed that the surface of formed hybrid composites exhibits rugged and rough pattern instead of the smooth appearance of the primeval ESMF.

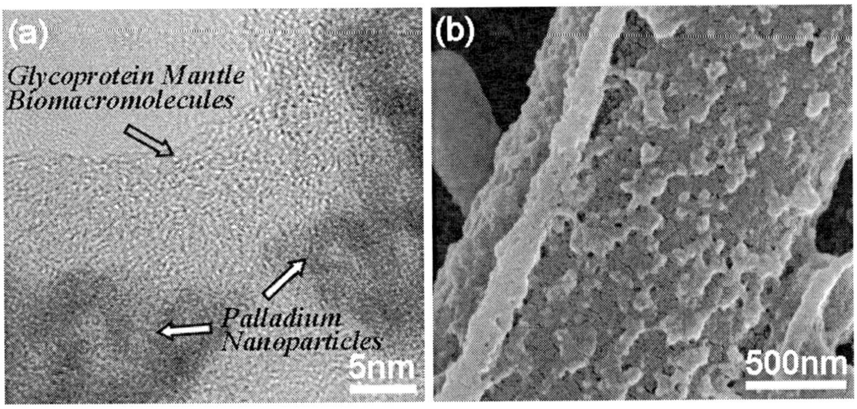

Figure 45. Representative (a) HRTEM and (b) FESEM images of Pd/ESMF hybrids prepared by infusing the eggshell membrane into PdC12 solution for 5h, respectively.

It confirms that ESMF glycoprotein have interacted with the inorganic matters during the biogenic process. Further, these formed palladium particles may be functioned as the catalyst to efficiently coupling the highly charged peptides under alkaline conditions. It can be deduced that small parts of glycoprotein macromolecules may disengage from the ESMF into the solution and coat Pd crystallites on the membrane fibers to assemble ordered inorganic/biomolecules hybrids. In that way, as-prepared noble metal nanoparticles can be stabilized on the biofibers over long periods of time. Since the nanoparticles and biomolecules typically meet at the same nanometer length scale, this interdisciplinary approach will contribute to the establishment of a novel field, descriptively termed nanobiotechnology.

4. In Situ Synthesis of Noble Metal Nanoparticles on Silk Fibroin Fibers

More attentions have been focused upon building noble metals nanostructures due to their unusual optical properties as well as their novel chemical and catalytic properties combining characteristic hierarchical structures. Actually, the utilization of biological organisms has emerged as a novel method for the synthesis of noble metal nanoparticles. In a break from tradition, it has hitherto relied more on the use of prokaryotes such as bacteria, viruses and so on, as well as proteins, recurring to the intracellular synthesis of sliver nanoparticles. The biosynthesis of silver nanoparticles has also been achieved using peptides with the ability to bind sliver ions and particles [26]. Moreover, it has been published that some living plants, such as Alfalfa sprouts [27], Geranium leaf [28], have the capability to take up sliver ions from solid media and produce sliver nanoparticles resulting in the possibility of an environmentally friendly method of recovering precious metals from mining wastes and a new alternative to produce nanoparticles. All the processes can be easily controlled under mild and eco-friendly conditions. Here silver nanoparticles are chosen as the candidate linked to the biotemplate, with a view that silver particles with nanoscale building blocks have received a great deal of attentions recently owing to their promising applications in modern science and technology, for example, electron microscopy (contrast agents), analysis (chemical and biological sensors), electronics (single-electron transistors, electrical connects), materials (dyes, conductive coatings), and even catalysis. Whereas sliver nanoparticles usually exist as the colloids in the liquid system, which restricts the applications in some domains, for example, as the solid catalyzer that the catalysis reckons on the sizes and morpologies of both sliver particles and solid supports, as well as their reciprocities. It is extremely difficult to find an appropriate solid support on which Ag(I) ions can be reduced to Ag(0) at ambient temperature [29]. However, it may be achieved by selecting the proper biological template as the substrate, on which silver nanoparticles can be deoxidized and *in situ* grow up.

A convenient biomaterial silk extracted from the silkworm *Bombyx mori* cocoons, what with redox-active silk fibroin fibers (SFF) containing manifold amino acids and chemical residues, among which tyrosine residue (Tyr) component having strong electron donating complex property, shall been given priority to prepare noble metal nanomaterials through bio-template techniques under mild conditions. Elicited for some bioinspired processes, we did not redissolve the silk into silk fibroin (SF) proteins, but directly utilized SFF as the reductive template to synthesize silver nanostructures. The redox-active SFF is obtained by removing the sericin from original silks, the gummy binding protein coating the silk fibroin filaments, with the degumming process that the cocoons are boiled in 0.05 mol·L-1 Na2CO3 solution. In a typical experiment, silver nitrate solution with volumes 20, 5, 1, and 0.2 mL of 0.01 M are injected into conical flasks and diluted to 100 mL, respectively. A certain amount of SFF ca. 8.5 mg is immersed into the solutions in the dark. The bio-reduction of sliver ions on SFF substrates is monitored by periodic sampling of aliquots (2 mL) of the aqueous component and measuring the UV-vis spectrum of the solution.

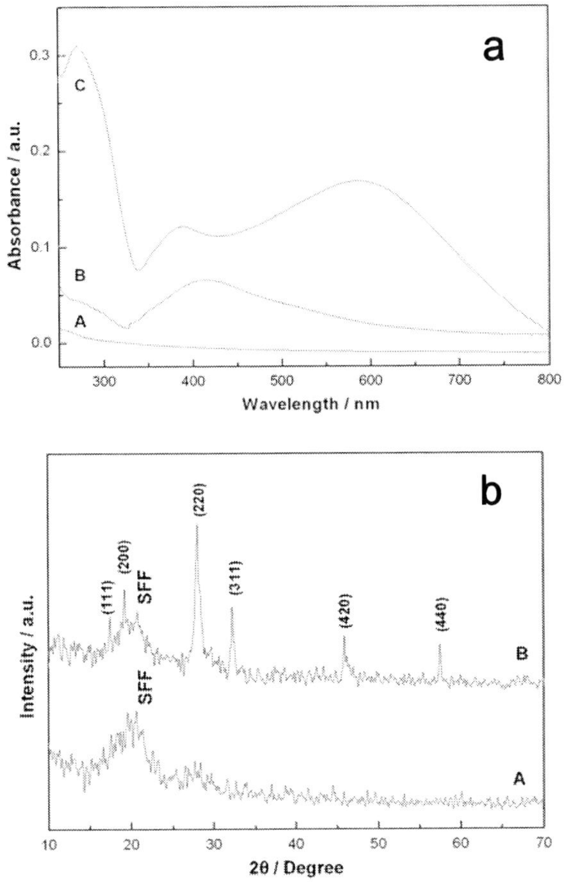

Figure 46. (a) UV-vis absorption spectra of the aqueous solutions of (A) SFF and (B, C) Ag/SFF with the additions of 20 ml and above 20 ml 0.01 mol·L-1 AgNO3 kept in the dark for 4 hours. (b) XRD patterns of (A) the original degummed silk and (B) silver nanoparticles on SFF supports.

UV-vis spectra clearly show that pure SFF does not absorb in the visible region, but the surface plasmon resonance band occurs at ca. 410 nm with addition of 20 mL AgNO3 solution kept 4 hours in the dark. Since silver nanoparticles are known to exhibit a size-dependent characteristic surface plasmon resonance band that can be measured, the characteristic absorption may be due to the redox of Tyr component in SFF in the case of the AgNO3 solution and the formation of silver nanoparticles, which is further proved by XRD results, and this absorption arising due to excitation of surface plasmons in the nanoparticles [30]. Actually it has been reported that the amino acid Tyr is an excellent reducing agent only under alkaline conditions [31]. However, it is not ambivalent with the current result owning to the pretreatment of SFF in Na2CO3 solution (a base medium) making for the ionization of the phenolic group of Tyr. Pretreated SFF thus has reduction activation and can *in situ* nucleate sliver nanocrystallites on SFF supports, regardless of rigorous pH value. Although the total content of Tyr only comes up to 5.17 mol% of silk fibroin, the Tyr is mainly located around SFF for its complex aromaticity, which may reduce silver ions around the membrane as the role of biotemplate. If the addition of AgNO3 solution is increased above 20 ml, the intensity of this plasmon peak changes very little with a blue shift, and a new peak appears at 570 nm. The two peaks, respectively, can be attributed to the transverse and longitudinal plasmon resonance of rod-shaped silver nanostructures [32]. However, no evident peaks can be tracked with addition of AgNO3 solution less than 1 mL due to the formation of few silver nanoparticles. It is considered that 4 hour is enough and the volume of 0.01 mol·L-1 AgNO3 solution shall range from 1 mL to 20 mL for the synthesis of silver nanocrystallites on SFF supports.

It can be seen from the XRD patterns that pure SFF containing plenty of amines, amides and carboxylic surface functional groups are nearly amorphous whereas Glycine–X repeats covering 94% of the silk sequence (X is Alanine in 65%, Serine in 23%, and Tyr in 9% of the repeats) [33] mainly form crystalline domains in silk fibers resulting in the peak around 20.64 degree [34]. A number of prominent Bragg reflections can be indexed as the face-centered cubic (fcc) structure of sliver with the corresponding diffraction peaks of (111), (200), (220), (311), (420) and (440) planes. The reflection peaks appear broader indicating as-prepared sliver nanoparticles with small dimensions, which is in agreement with UV-vis spectra results.

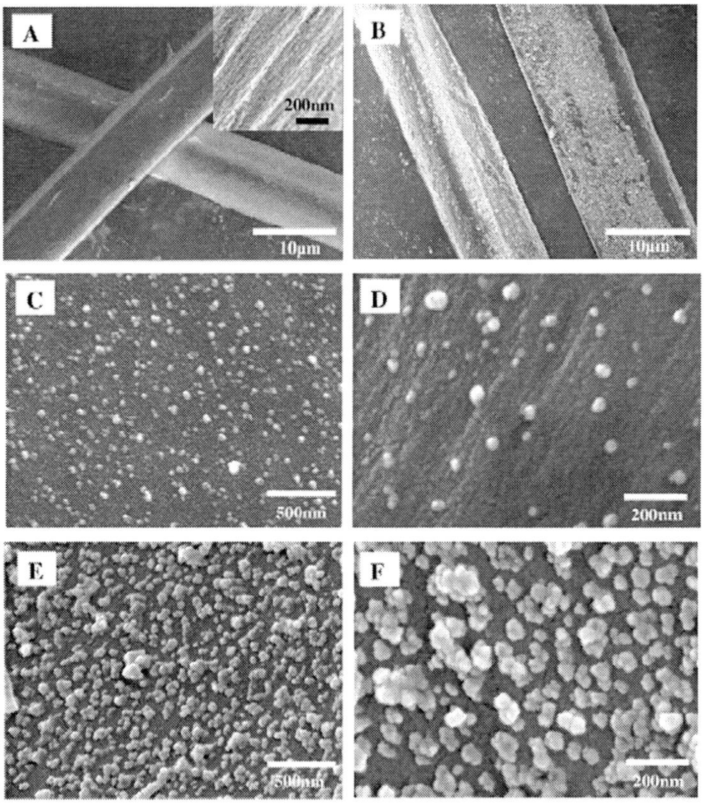

Figure 47. FESEM images of (A) the original SFF template, with the inset showing high magnification of the surface skin coat of aligned protein compared with the panorama of Ag/SFF composites, sliver nanoparticles on SFF supports with different dosages of 0.01 mol·L-1 AgNO3 as the reagents, (C) and (D) 1 ml, (B), (E) and (F) 20 ml. Images in Figure D and Figure F show high magnifications of Figure C and Figure E, respectively.

The characteristic FESEM images of original SFF supports seems as clean as a whistle under low magnification, as the surface skin coat of aligned protein obviously appears in the magnified images. It reveals that the surface has a regular rugged framework containing proteins chains with certain chemical residues. The typical images of silver nanoparticles depositing on SFF supports for 4 h, contrasting against original SFF supports, can be observed from the panorama that sliver nanoparticles distribute uniformly on SFF substrates. As a general rule, it is necessary to precisely control the reaction conditions, for example, the concentration of the silver ions, besides the base system during degumming process using Na2CO3 solution. As concerned as the images of Ag/SFF with 1 mL and 20 mL 0.01 mol·L-1 AgNO3 solution as the reactants, respectively, under low concentration (with 1mL AgNO3 added), as-prepared sliver nanoparticles distribute uniformly and present spherical shape with a mean size ca. 21 nm, few grow up exceptionally. As sliver ions concentration elevated (with 20 mL AgNO3 added), sliver nanocrystallites come into being ca. 30-50 nm clusters with the floriated figuration conjoined by about 5 nm small grains. It indicates that an increase in silver ions concentration leads to an obvious morphologic evolution of sliver nanostructures from dispersed grains to nanoclusters, which would be likely caused by the interactions of dissociative sliver ions and the chemical residues of SFF supports.

To further explore the *in situ* depositing process, ATR-FTIR measurements are carried out to identify possible biochemical radicals among biomolecules, which can be responsible for the reduction of Ag ions and capping silver nanoparticles. The degummed SFF supports are extracted from the solution, subsequently rinsed with distilled water and absolute alcohol, respectively, finally the fibers are collected for FTIR analysis. The FTIR spectrum of degummed silks and shows the diagnostic peaks at 1620, 1511, and 1224 (connected 1260) cm-1. These peaks are assigned to the amide I, amide II and amide III (crystal region, 1224 cm-1; amorphous zone, 1260 cm-1) bands of proteins released by SFF, respectively, which predominantly indicates β-sheet structure. The peaks at 1443 and 1372 cm-1 may be assigned to symmetric stretching vibrations of carboxylate groups of amino acid residues with free carboxylate groups and methylic groups of alanine in the proteins. The peaks at 1620 and 1160 cm-1 arises from carbonyl stretching vibration and phenolic groups of Tyr, respectively, which shows the carbonyl stretching vibration from the carboxylate ions and the hydroxyl stretching vibration from the phenolic ions in Tyr of the pure SFF. As for treated SFF, the peak of SFF at 1160 cm-1 disappears which indicates the phenolic residue exhausted, simultaneity showing the carbonyl stretching vibration peak at 1725 cm-1, which may be attributed to the formation of a quinone structure due to oxidation of phenolic group in Tyr. It has been reported that proteins can bind to metal nanoparticles through either free amine groups or cysteine residues in the proteins [35]. In the circumstances, oxidized tyrosine molecules bound Ag crystallites through amine groups so that Ag nanoparticles can stabilitate, which is also supported by the shift of the characteristic peaks. The amide I band at 1620 cm-1 has broadened, which also indicates the capping manner of silver nanoparticles by SFF macromolecules.

It has been further confirmed by MAS solid-state NMR analysis of pure SFF comparing with Ag/SFF. The resonances of phenyl group of Tyr is usually measured to be at 130-150 ppm, assigned to sp2 hybridized carbons of phenyl replaced by hydroxy. Due to the oxidation of Tyr components, phenyl group is converted into a semi-quinone; the chemical shift of quinone 13C is measured to be in the region of 180-190 ppm while the resonances at 160-170 ppm is just arose by the C of carbonyl in the carboxyl residues of SFF. In the case of Ag/SFF, a peak at 189 ppm shall be corresponding to semi-quinone hailing from the phenyl group during the reduction of Ag(I) to Ag(0), this peak is clearly missing in the NMR spectrum of pure SFF.

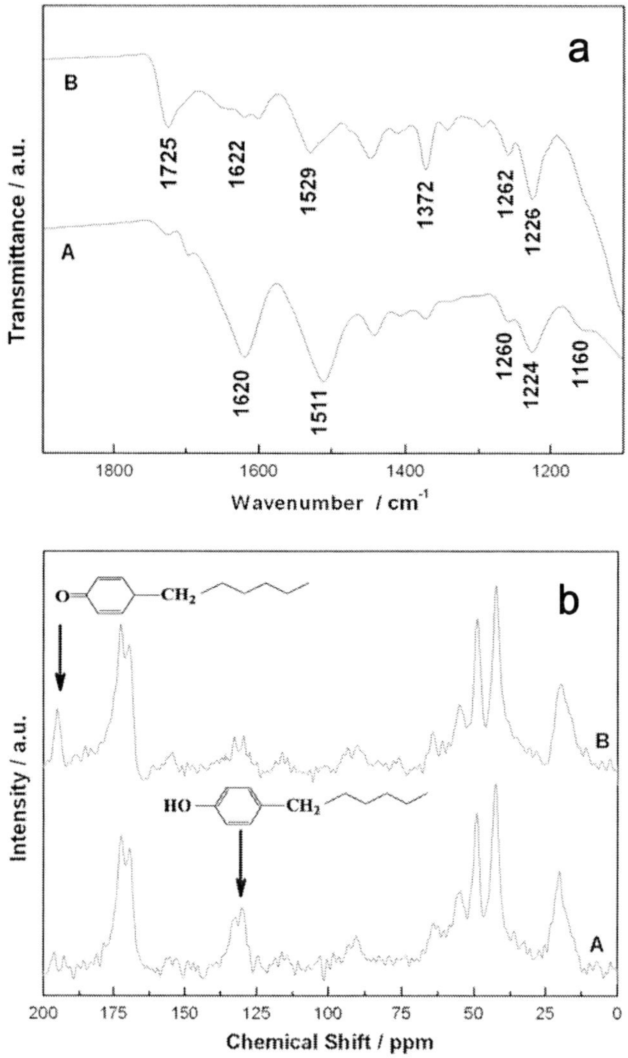

Figure 48. (a) ATR-FTIR spectra of (A) original degummed silk and (B) sliver nanoparticles on SFF supports. (b) MAS solid-state NMR spectra of (A) original degummed silk and (B) sliver nanoparticles on SFF supports. The inset shows a transformation of functional structure of Tyr component, which is a consequent on the reduction of sliver ions and formation of silver nanoparticles.

As proved above, it is ensured that the redox-active Tyr residues, mainly locating around the surface of SFF, can farthest reduce sliver ions on account the ionization of SFF during the pretreatment under alkaline conditions. A tentative degummed process substituting distilled water for 0.05 mol·L-1 Na2CO3 solution can not bring off ionization and there present no evident sliver grains, which indirectly proves that the reducing ability of Tyr achieved predominantly in alkaline system. Thus the reducing capability of Tyr is enhanced for phenolic group of Tyr at high pH. Consequently, the silver ions can be reduced by amines and amides, and interacted with carboxylic surface functional groups containing in those proteins chains around SFF substrates. The redox process shall occur in the present case between phenolic groups of Tyr exposed from SFF and captured Ag(I) ions, ultimately Ag nanoclusters can stabilize on SFF.

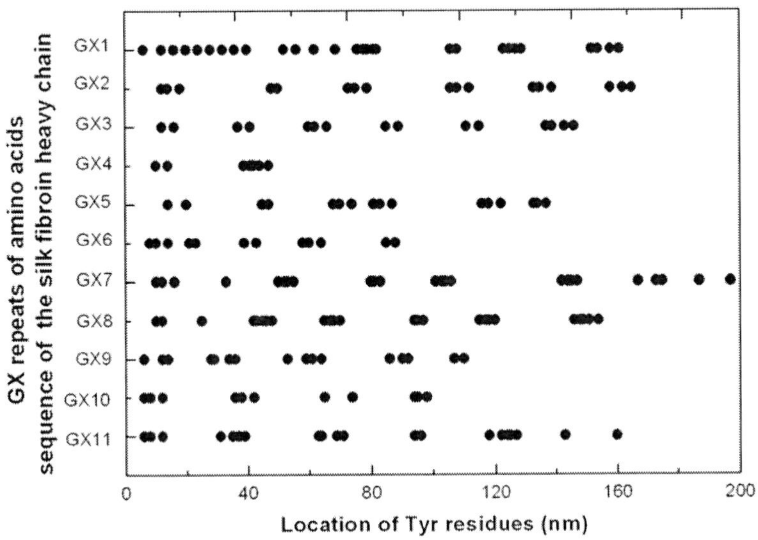

Figure 49. Distributing locations of Tyr residues according to the GX repeats of the amino acid sequence of the silk fibroin heavy chain.

The morphologies of silver nanoclusters reckon on special macromolecules structures of the biotemplate SFF as well as silver ions concentration. The fibroin is an insoluble protein containing up to 90% of the amino acids glycine, alanine, serine and tyrosine leading to significant content of antiparallel β-pleated sheet formation in the fibers. On examination of the primary sequence for the fibroin heavy chain, a pattern of hydrophobic and hydrophilic blocks has been identified that includes larger hydrophilic blocks at the chain ends, smaller hydrophilic blocks and hydrophobic blocks interval jointed. This pattern suggests the possibility to form micellar structures in liquid systems [36]. Fibroin molecules thus act as the hydrophilic–hydrophobic–hydrophilic polymers, during the formation of irregular sized micelles depending on chain folding and hydrophobic interactions. The smaller hydrophilic blocks can remain hydrate, and the larger terminal hydrophilic blocks at both the amino and carboxy termini define the outer edges of the micelles. Tyr residues are distributed according to the silk sequence of the amino acids and present all the way through the molecule, both in the hydrophilic and hydrophobic regions. Hereby, the organization does not give significant spatial ordering of the tyrosine. Distributing locations of Tyr residues are calculated according as the GX repeats formed the bulk of the polypeptide chain. Clearly, Tyr residues don't distributed well-proportioned but behave separated tufts at ca. 20~50 nm level. Through Coulombic interactions, the Ag(I) ions can be captured by phenolic groups of chemical residues. With this premise, the concentration of silver ions will affect the morphologies of as-prepared silver nanoparticles. Under higher concentration as-formed sliver crystallites present shapely spheres with the size of ca. 5 nm, due to homogeneous nucleation of ample silver ions. It is worth noting that sliver nanoparticles grow into nanoclusters. Tyr residues on SFF can contribute to the formation of silver nanoparticles in both reduction and location. Hence peculiar structures of Tyr residues and SFF shall be responsible for directing some morphological controls over the placement of silver nanocrystals.

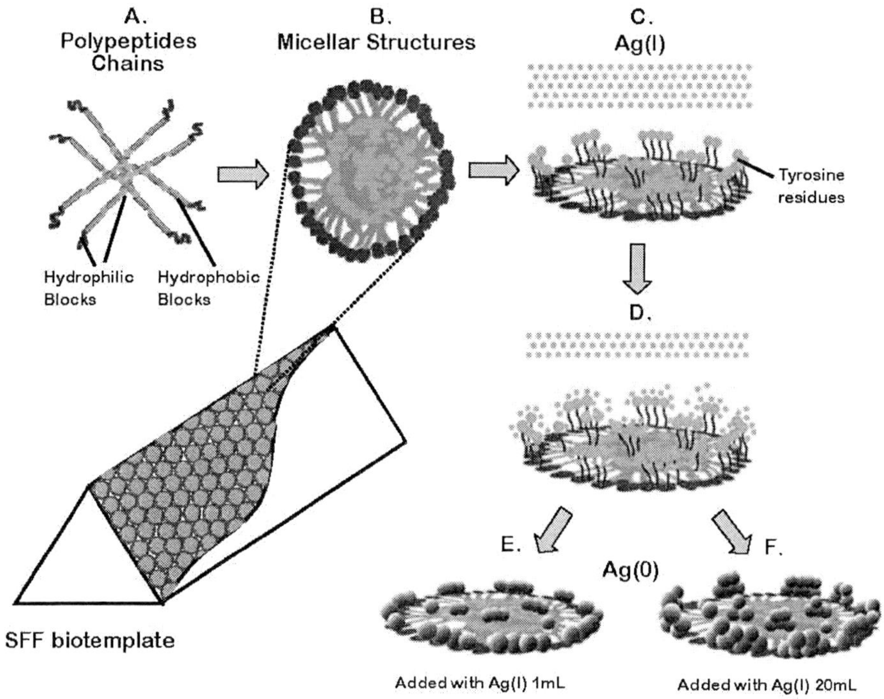

Figure 50. Illustration of in situ depositing Ag nanoparticles on SFF biotemplate for different morphologies and arrangements reckoning on the concentration of silver ions and macromolecules structures of SFF. It shows the model of (A) chain folding, (B) micelle formation, and (C, D) the reaction of sliver ions with chemical residues brings out two morphologies formed under different dosages of 0.01 $mol{\cdot}L^{-1}$ $AgNO_3$ as the reagents (E) 1 mL and (F) 20 mL, respectively.

The interaction of Tyr residues with silver nuclei provides a chemically reducing environment, thereby allowing further accelerated reduction of silver ions with sufficient metal ions at the interface between Tyr residues and sliver nanoparticles. It exists continuous growth by the diffusion of species from the solution with a great deal of silver ions towards the surfaces of primeval nuclei which build up into combined series leading to floriated clusters morphology.There are intervals between every two Tyr residues aggregations, which are consistent with the distributing locations of Tyr residues. Although the reductive reaction is the same as the former, lower concentration of silver ions generates single, long burst of nuclei, these nuclei can be bound by Tyr and slowly grow up. In some areas, a broad range of size distribution indicates the occurrence of multiple nucleation events [37]. Some sliver nanoparticles with the size of ca. 45 nm are formed anisomerously due to the transitory undulation of silver ions concentration. Other results also convincingly support the model. Sliver nanoparticles can present in interrupted arrays, in respect that amino acid residues such as arginine, cysteine, lysine and methionine are known to interact with silver ions, which leads to the faulty arrangement differing from the ideal state. In any case, the interactions among protein blocks, as well as distributing and spacing of these residues, play a critical and complicated role in the nucleation and growth of sliver nanocrystallites. Thus sliver nanoparticles present in arrays unconspicuously or in array out-of-order.

Representative TEM images of as-prepared silver nanoparticles separated from SFF and ones sticking on SFF proteins are shown in Figure 51A-B (with 20 mL silver nitrate solution added), respectively. It reveals that silver crystallites show itself nanoclusters. The SAED pattern reveals that the particles are polycrystalline in nature and the rings can be indexed as the fcc structure of crystalline silver which is consistent with the XRD results. As shown in Figure 51B, there are few spherical nanoparticles with the size of 10 nm on SFF proteins, and these special particles are indexed to single-crystalline silver. The uniformity in size for as-formed nanocrystallites can have been achieved through a self-sharpening growth process during which small particles grow more rapidly than larger ones.

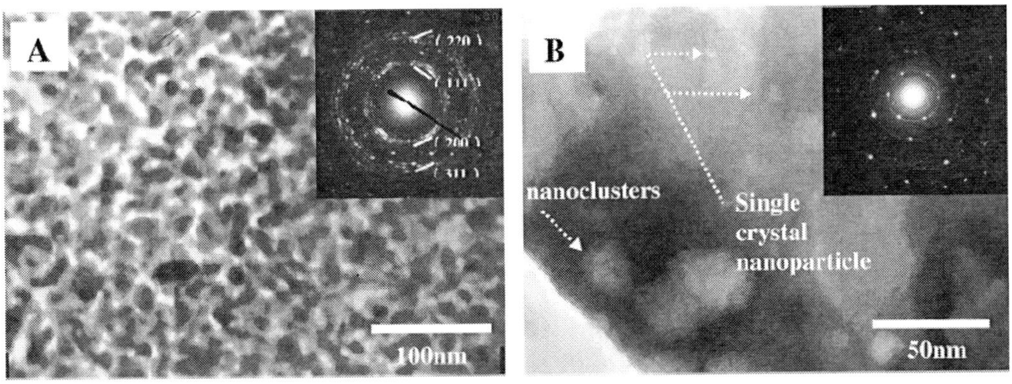

Figure 51. Representative TEM images of as-prepared Ag nanoclusters (added with 20 mL 0.01 mol·L^{-1} $AgNO_3$). (A) shows Ag nanoclusters separated from SFF substrates, the inset shows the selected area electron diffraction (SAED) pattern of the nanoclusters. (B) shows Ag particles sticking on SFF proteins, the inset shows the convergent beam electron diffraction patterns recorded by directing the electron beam perpendicular to the (111) facet.

In addition, it is found that sliver nanoparticles combine into larger swarms with ultraviolet radiation. A number of recent studies suggest that it shall rather be the conglomeration of some smaller subunits (or primary crystallites) than continuous growth by the diffusion of species from the solution towards the surfaces of nuclei. The as-prepared silver nanoclusters are rather stable on SFF supports over a period of several months, even under knocking on SFF substrates.

By the convenient biochemical technique with silk fibroin fibers as the biotemplate, support and reducer, silver nanoparticles/SFF composites has been successfully synthesized. The loading of silver nanoparticles is as high as 60% of the whole surface area, it is clear that the "biomaterial frameworks" provided highly effective sites for deoxidization and loading of noble metal nanoparticles, composing structures of macroscopic to nanoscopic dimensions. Our discovery that Tyr of SFF provides a redox-active framework to construct nanostructured silver crystallites is an important advance in combining biotemplate and chemical synthesis in the direction of organic-inorganic advanced materials synthesis.

CONCLUSION

Soft chemistry technique combined with materials science and biotechnology is presently developed to construct a variety of well-ordered functional nanostructures through appropriate bioinspired approaches.

A series of facile, green and versatile aqueous routes have been developed for the fabrication of hierarchical oxide films via different sol-gel soakage techniques combined relevant calcination treatment and involved the biomaterial eggshell membrane (ESM) as the template. The biomorphic replication of three diverse layers of ESM is systematically investigated by the control over synthesis conditions such as media pH value, dipping time, and calcination temperature. Herein, the biomaterial ESM serves both as the physical substrate and the functional macromolecule template to realize the precise biomorphic replication, by the rights of the interactions between ESM macromolecules (containing carboxyl, hydroxyl, amino groups, etc.) and relevant colloid ingredients. Moreover, some biomacromolecules of ESM components also act as the surfactants to obtain small-scaled and well-distributed oxide nanocrystallites, based on the strong bondage of short-chained amino acids within ESM glycoprotein with oxide nuclei.

Through in situ biotemplate soakage approaches at room temperature, a series of nanoclusters of chalcogenides and noble metal nanocrystallites have been successfully prepared and encrusted on degummed silk fibroin fibers (SFF) or eggshell membrane fibers (ESMF). Some amino acids of SFF or ESMF mainly provide both reactant and location functions under different solution conditions. The morphologies of these nanoclusters and the hierarchy of hybrid nanocomposites are mostly attributed to the concentration of relevant impregnant solutions as well as particular configurations and special functional groups of SFF or ESMF biomacromolecules. The biofibers not only serve as the substrates and templates, but also function as the surfactants, thus small-sized, well-distributed, and well-crystallized nanocrystallites can be organized into subtle hierarchical nanostructures combining organic biomaterials to form well-organized inorganic-organic hybrids.

This method and the relevant ideas offer an avenue to fabricate other functional nanomaterials with novel properties and valuable applications, on the basis of the integration among chemistry, materials sciences and biotechnologies.

REFERENCES

[1] Cook, G.; Timms, P. L.; Göltner-Spickermann, C. *Angew. Chem. Int. Ed.* 2003, *42*, 557.

[2] Cozzoli, P. D.; Kornowski, A.; Weller, H. *J. Am. Chem. Soc.* 2003, *125*, 14539.

[3] Lee, S. W.; Ichinose, I.; Kunitake, T. *Langmuir* 1998, *14*, 2857.

[4] Moss, J. M.; VanDamme, M. P. I.; Murphy, W. H.; Stanton, P. G.; Thomas, P.; Preston, B. N. *Arch. Biochem. Biophys.* 1997, *339*, 172.

[5] Liang, C. H.; Shimizu, Y.; Sasaki, T.; Koshizaki, N. *J. Phys.Chem. B* 2003, *107*, 9220.

[6] Dieguez, A.; Romano-Rodriguez, A.; Morante, J. R.; Kappler, J.; Barsan, N.; Gopel, W. *Sens. Actuators B* 1999, *60*, 125.

[7] a) Zhang, D. F.; Sun, L. D.; Jia, C. J.; Yan, Z. G.; You, L. P.; Yan, C. H. *J. Am. Chem. Soc.* 2005, *127*, 13492; b) Wang, Y.; Lee, J. Y.; Zeng, H. C. *Chem. Mater.* 2005, *17*, 3899–3903.

[8] Gnanasekar, K. I.; Jayaraman, V.; Prabhu, E.; Gnanasekaran, T.; Periaswami, G. *Sens. Actuators B* 1999, *55*, 170.

[9] Fu, M. G.; Wang, X. J.; Chen, J.; Wang, H. Y. *J. Zhengzhou Inst. Light Ind.* 1994, *28*, 160.

[10] Fang, G. J.; Liu, Z. L.; Zhang, Z. C. *J. Chin. Rare Earth Soc.* 1997, *15*, 10.

[11] Srivastava, D. N.; Chappel, S.; Palchik, O.; Zaban, A.; Gedanken, A. *Langmuir* 2002, *18*, 4160.

[12] Damen, T. C.; Porto, S. P. S.; Tell, B. *Phys. Rev.* 1966, *142*, 570.

[13] Hadjiev, V. G.; Iliev, M. N.; Vergilov, I. V. *J. Phys. C: Solid State Phys.* 1988, *21*, L199.

[14] van den Beucken, J. J. J. P.; Vos, M. R. J.; Thune, P. C.; Hayakawa, T.; Fukushima, T.; Okahata, Y.; Walboomers, X. F.; Sommerdijk, N. A. J. M.; Nolte, R. J. M.; Jansen, J. A. *Biomaterials* 2006, *27*, 691.

[15] Yi, F.; Guo, Z. X.; Zhang, L. X.; Yu, H.; Li, Q. *Biomaterials* 2004, *25*, 4591.

[16] Lu, W. G.; Fang, J.Y.; Stokes, K.L.; J. Lin, *J. Am. Chem. Soc.* 2004, *126*, 11798.

[17] Lee, S. M.; Cho, S.N.; Cheon, J. *Adv. Mater.* 2003, *15*, 441.

[18] Petroski, J. M.; Wang, Z. L.; Green, T. C.; El-sayed, M. A. *J. Phys. Chem. B* 1998, *102*, 3316.

[19] Iwahori, K.; Yoshizawa, K.; Muraoka, M.; Yamashita, I. *Inorg. Chem.* 2005, *44*, 6393.

[20] Smith, G. D.; Firth, S.; R. J.; Clark, H.; Cardona, M. *J. Appl. Phys.* 2002, *92*, 4375.

[21] Nanda, K. K.; Sahu, S. N.; Soni, R. K.; Tripathy, S. *Phys. Rev. B* 1998, *58*, 15405.

[22] Batonneau, Y.; Bremard, C.; Laureyns, J.; Merlin, J. C.; *J. Raman Spectrosc.* 2000, *31*, 1113.

[23] Krauss, T. D.; Wise, F. W. *Phys. Rev. B* 1997, *55*, 9860.

[24] Hayashi, S.; Sanda, H.; Agata, M.; Yamamoto, K. *Phys. Rev. B* 1989, *40*, 5544.

[25] Mlayah, A.; Brugman, A. M.; Carles, R.; Renucci, J. B.; Valakh, M. Ya.; Pogorelov, A. V. *Solid State Commun.* 1994, *90*, 567.

[26] Naik, R. R.; Stringer, S. J.; Agarwal, G.; Jones, S. E.; Stone, M. O. *Nat. Mater.* 2002, *1*, 169.

[27] Gardea-Torresdey, J. L.; Gomez, E.; Peralta-Videa, J. R.; Parsons, J. G.; Troiani, H.; Jose-Yacaman M. *Langmuir* 2003, *19*, 1357.

[28] Shankar, S. S.; Ahmad, A.; Sastry, M. *Biotechnol. Prog.* 2003, *19*, 1627.

[29] Hong, B. H.; Bae, S. C.; Lee, C. W.; Jeong, S.; Kim, K. S. *Science* 2001, *294*, 348.

[30] Kerker, M. *J. Colloid Interf. Sci.* 1985, *105*, 297.

[31] Selvakannan, P. R.; Swami, A.; Srisathiyanarayanan, D.; Shirude, P. S.; Pasricha, R.; Mandale, A. B.; Sastry, M. *Langmuir* 2004, *20*, 7825.

[32] Sun, Y. G.; Gates, B.; Mayers, B.; Xia, Y. N. *Nano Lett.* 2002, *2*, 165.

[33] Zhou, C. Z.; Confalonieri, F.; Jacquet, M.; Perasso, R.; Li, Z. G.; Janin, J. *Proteins* 2001, *44*, 119.

[34] Rathore, O.; Sogah, D. Y. *Macromolecules* 2001, *34*, 1477.

[35] Gole, A.; Dash, C.; Ramakrishnan, V.; Sainkar, S. R.; Mandale, A. B.; Rao, M.; Sastry, M. *Langmuir* 2001, *17*, 1674.

[36] Jin, H. J.; Kaplan, D. L. *Nature* 2003, *424*, 1057.

[37] Xia, Y. N.; Gates, B.; Yin, Y. D.; Lu, Y. *Adv. Mater.* 2000, *12*, 693.

In: Encyclopedia of Nanotechnology
Editor: Elwood D. Carlson

ISBN: 978-1-60692-079-4

Chapter 11

PREPARATION, CHARACTERIZATION AND THERMAL PROPERTIES OF NANOFLUIDS

***Hai-tao Zhu*[a], *Shi-quan Liu*[b], *Lei Xu*[a] *and Can-ying Zhang*[a]**

[a] College of Materials Science and Engineering,
Qingdao University of Science and Technology,
Qingdao 266042 China

[b] College of Materials Science and Engineering,
Jinan University, Jinan 250022 China

ABSTRACT

Nanofluids are fluids containing suspended solid nanoparticles. It is expected that the nanofluids will be the next generation of heat transfer fluids due to their unique thermal properties. The nanofluids may have potential applications in many important fields, such as micro-electronics, aerospace, transportation, and medicine, etc. In this chapter, a brief introduction of the background and concept of nanofluids is first presented, followed by reviewing the preparation, characterization and thermal properties of the nanofluids. In the preparation section, the stabilization mechanism of nanoparticles in liquids is described. Then several preparation methods are discussed. Emphasis is put on the stability of the nanofluids, which is influenced by the preparation methods and affects the properties of the nanofluids. Several typical characterization technologies of nanofluids are summarized in the characterization section. The thermal conductivity, convection, and boiling properties of nanofluids are depicted in the section of the thermal properties.

1. INTRODUCTION

The heating or cooling of fluids are of major importance to many industrial sectors, including transportation, chemical processes, energy supply and production, microelectronics, etc. The thermal conductivity of these fluids plays a vital role in the development of energy-efficient heat-transfer equipments. However, conventional heat-transfer fluids, such as water, oil and ethylene glycol, are inherently low-efficiency heat transfer fluids. There is a demanding need to develop advanced heat transfer fluids with significantly high thermal conductivities and improved heat transfer performances.

It is well known that crystalline solids have thermal conductivities with typically 1-3 orders of magnitude larger than those of fluids (Table 1). For example, the thermal conductivity of copper at room temperature is about 700 and 3000 times higher than that of water and engine oil, respectively. Therefore, fluids containing suspended solid particles are consequently expected to display significantly enhanced thermal conductivities, compared to traditional heat transfer fluids. It has been conducted numerous theoretical and experimental studies on the effective thermal conductivity of suspensions containing millimeter- or micrometer- sized solid particles. However, due to the large size and high density of the particles, there is no easy way to prevent the coarse particles from settling out of the suspension state. The lack of stability of such particle suspensions induces additional flow resistance and possible erosion. Hence, the idea of dispersing coarse particles in the traditional heat transfer fluids to enhance the thermal conductivity of fluids was rejected.

The development of nanotechnology revitalizes the pursuing to prepare fluids with high thermal conductivities. The thermal conductivities are enhanced by the solid particles added in the fluids. Modern nanotechnology makes it possible to produce and process materials with sizes below 50 nm. This offers an opportunity to apply the emerging nanotechnology to the established thermal energy engineering, Choi [1] put forward the concept of *nanofluid*. The nanofluid is a fluid containing dispersed nanometer-sized solid particles. Compared with the traditional fluids or suspensions containing coarse particles, nanofluids are expected to have superior properties. The small dimension and relatively large specific surface area of nanoparticles not only increase the stability of the suspensions, but also significantly improve the heat transfer capabilities of them. Furthermore, nanofluids can improve abrasive properties as compared to the conventional solid/fluid mixtures. Successful industrial application of the nanofluids will lead the current trend toward the component miniaturization, as a precursory role of the design of small and light heat exchanging systems.

Table 1. Thermal conductivities of various solids and liquids

	Materials	Thermal conductivity ($W \cdot m^{-1} \cdot K^{-1}$)
Matallic solids	Ag	429
	Cu	401
	Al	237
Nonmetallic solids	diamond	3300
	CNTs	3000
	Si	148
	Al_2O_3	40
	CuO	33
Metalic liquid	Na (664K)	72.3
Niometallic liquids	water	0.601
	ethylene glycol	0.257
	engine oil	0.145

The unique properties and potential applications of nanofluids in many industrial sectors provided a thrust necessity to the research of nanofluids. The interdisciplinary nature of nanofluids study also presents a great opportunity for the exploration and discovery at the frontier nanotechnology. In this chapter, the stabilization mechanism and preparation methods are discussed, followed by reviewing the characterization and thermal properties of the nanofluids.

2. Preparation of Nanofluids

2.1. Stabilization Mechanism of Nanoparticles in Liquids

Preparation of stable nanofluids is the key issue of nanofluid research. As mentioned above, a nanofluid is a fluid containing nanoparticles. The stability of nanofluids refers to the several aspects. 1) Thermodynamic stability. Nanofluids are multi-phase dispersion systems with high surface energies and are, therefore, thermodynamic unstable. 2) Kinetic stability. Nanoparticles dispersed in the nanofluids have strong Brownian movements. The mobility of the nanoparticles can offset their sedimentation caused by the gravity field. 3) Dispersion stability. Due to the aggregation of nanoparticles, the dispersion of nanoparticles in fluids may deteriorate with time. 4) Chemical stability. No chemical reactions either between the suspended nanoparticles or between the base fluid and nanoparticles are desired at the working conditions of the nanofluids. The last stability can only be evaluated upon the specific working environment. Therefore, there are two phenomena that are critical to the stability of a nanofluid, aggregation and sedimentation.

In stationary state, the sedimentation velocity of small spherical particles in a liquid follows the Stokes law [2]:

$$V = \frac{2R^2}{9\mu}(\rho_P - \rho_L)\cdot g \tag{1}$$

where V is the sedimentation velocity of particles; R is the radius of spherical particles; $\mu$is the viscosity of the liquid medium; ρ_P and ρ_L are the density of the particles and the liquid medium, respectively; g is the acceleration rate of the gravity.

This equation reflects a balance of the gravity, buoyant force, and frictional force that are acting on the suspended nanoparticles. According to the equation, the following measures can be taken to reduce the sedimentation velocity of nanoparticles in nanofluids, and hereafter to produce an increase of the stability of the nanofluids. 1) Reducing the size of the nanoparticles, *R;* 2) Increasing the viscosity of the base fluid, μ; 3) Minimizing the density difference between the nanoparticles and the base fluid, (ρ_P-ρ_L).

It is clear that reducing the particle size should remarkably decrease the sedimentation velocity of the nanoparticles and improve the stability of nanofluids, since **V** is proportional to the square of *R*. According to colloid chemistry theories, when the particle size reduces to a critical size, *Rc*, the sedimentation will "stop" (the equilibrium between sedimentation and diffusion) due to the Brownian motion of nanoparticles (diffusion). However, the smaller nanoparticles have higher surface energy, increasing the possibility of the aggregation of nanoparticles. Therefore, the key to prepare stable nanofluids is to use smaller nanoparticles and, simultaneously, to prevent the aggregation of the small nanoparticles.

The nanoparticles present in the dispersion medium (fluid) are always subjected to Brownian motion, resulting in frequent collisions to each other. Two basic interactions, attraction and repulsion, are involved in these collisions. The dispersion stability of the nanoparticles is thus determined by the interactions between the suspended nanoparticles. When the attraction dominates, a particle will adhere to the neighboring particles and agglomerate to form aggregates. If the attraction is overwhelmed by the repulsion, the nanoparticles can be stably dispersed.

Van der Waals force is the common attractive force acting on particles. The repulsion between particles mainly arises from two different sources. One is caused by the electrical double layers on the surface of the nanoparticles. The other might be generated by the adsorbed layers of polymer molecules on the surface of nanoparticles. The total potential energy [3], V_T, of a particle is the sum of the Van der Waals energy, V_A, and the repulsive energy, V_R. That is, $V_T = V_A + V_R$. The V_T, V_A, and V_R versus the distance between two particles are plotted in Figure 1. It can be seen that the attractive energy, V_A, increases rapidly as the particles are approaching to each other. While the repulsive energy, V_R, changes somewhat more slowly. As the particles approaching to each other, the total potential energy curve first reaches a maximum value point, V_{max}. This implies an energy barrier exists against the adherence of the particles. Once the energy barrier is overcome, the particles will attract strongly and the potential energy falls rapidly to the primary minimum value of the curve, V_{min}. Therefore, the smaller the energy barrier (V_{max}) is, the more possible the particles may aggregate.

From the above analysis, the aggregation of the nanoparticles in fluids can be avoided either by increasing the energy barrier, V_{max}, or by preventing the particles from approaching to one another in a distance where there is a significant attractive energy. The former can be realized by the formation of the electrical double layers on the surface of the nanoparticles. The latter can be achieved by constructing a physical steric barrier around the particles by adsorbing layers of nonionic materials. Accordingly, the stabilizations of the particles in fluids based these two routes are termed as electric and steric stabilization, respectively.

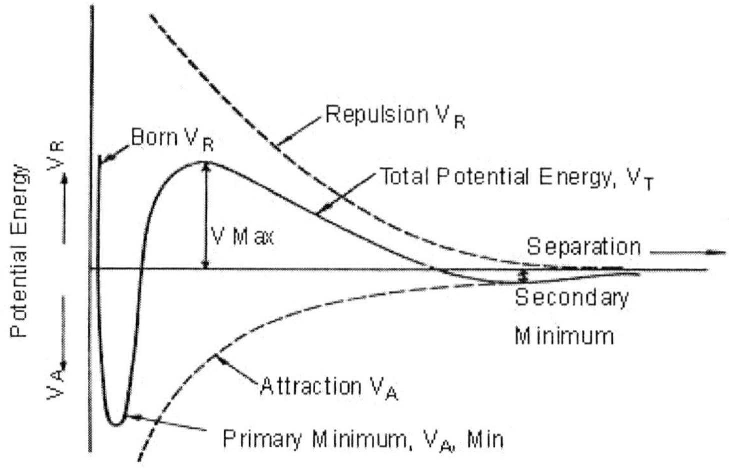

Figure 1. Plots of V_T, V_A, and V_R versus the distance between two particles.

It should be noted that, in the steric stabilization, the molecular weight and the amount of polymers affect the stability of suspension. For instance, the polymers in the liquid are at a low concentration. And the polymer molecule has more than two absorbable segments. Furthermore its dimension is long enough to be attached onto more than two particles. On this situation, two or more particles will be linked by one polymer molecule, and flocculation of particles takes place (see Figure 2a). If the concentration of polymer is notably high and the polymer chains are exceedingly long, the proportion of the coverage on particle surface by the adsorbed polymers is so high that the number of the adsorption sites on one particle becomes fairly small. On this situation, bridging flocculation still takes place (Figure 2b).

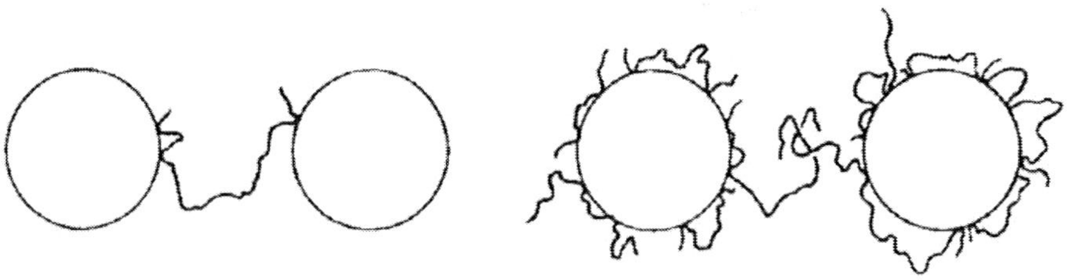

Figure 2. Bridging flocculation of particles. (a) two particles linked by one polymer molecule; (b) two particles linked by two separated adsorbed polymer molecules.

If polyelectrolytes (polymers with functional ionic groups, such as -COOH or $-SO_3H$,) are absorbed on the nanoparticle surfaces, the absorbed polymers, on the one hand, can cause a steric stabilization of the particles; on the other hand, the ionization of the functional groups can increase the surface electric charges of the nanoparticles, with a result of double electric layers on the nanoparticle surfaces. The combination of these two effects generates the electrosteric stabilization. When the nanoparticles are far away from each other, the repulsion force produced by the double electric layers dominates. However, the steric effect will play a main role if the nanoparticles are close to each other. Hence, the electrosteric stabilization will produce the best dispersion effect.

2.2. Preparation Methods of Nanofluids

Theoretically, all nanoparticles with high thermal conductivities could be used as dispersion in traditional fluids to prepare nanofluids. However, it is found that only a few oxides (Al_2O_3, CuO, TiO_2, and SiO_2), metals (Cu, Ag, and Au), carbides and carbon nanoparticles could be successfully used to produce nanofluids (Table 2-3). The used base fluids in the literature were water, oil, acetone, decene, and ethylene glycol, etc. The methods of preparing nanofluids are reviewed as follows.

Dispersion Method

Dispersion method is most commonly used to prepare nanofluids, especially at the early stage of the nanofluid investigations. Generally, the nanoparticles were produced as a dry powder form, and then dispersed into a fluid to obtain the nanofluids.

Table 2. Summary of nanofluids prepared by dispersion methods

Nanofluid	Preparation Condition	Volume fraction (vol%)	Stability and Observations
Al_2O_3/water, CuO/water, Al_2O_3/EG, CuO/EG [4]	mixing and shaking	1-5	some large agglomerates
Al_2O_3/water, CuO/water [5]	ultrasonication 3 hrs	2-10	several days
Fe/EG [6]	ultrasonication 10 - 70 mins	0.2-0.55	some clusters ultrasonic time affect stability
CuO/EG [7]	ultrasonication 1 - 30 hrs	0.001-5	some clusters ultrasonic time affect stability
Cu/oil Cu/water [8]	oleic acid, laurate salt as dispersant ultrasonication	2-5 2.5-7.5	1 week 30 hrs some clusters
Al_2O_3/water [9]	SDBS as dispersant ultrasonication 16-20 hrs	0.2-1.6	1 week
TiO2/water [10]	CTAB as dispersant ultrasonication 8-10 hrs	0.5-5	some clusters
CNTs/water [11,12]	SDS, CTAB as dispersant, ultrasonication 20-490 mins	0.6	ultrasonic time affect stability
Al_2O_3/water TiO2/water [13]	pH =3 pH= 10 mixing (10000rpm) 2 hrs	1-10	stable for more than 6 days
TiO2/water [14]	pH =3 ultrosonication and high-shear mixing	0.19-0.57	several weeks
Al_2O_3/water [15]	pH = 2-11.5 ultrasonication and magnetic agitation	2-5	
CNTs/water [16]	pH = 2-11 GA as dispersant ultrasonication and high-shear mixing	0.1-0.5	several months
graphite/water [17]	pH = 9-10 PVP as dispersant ultrasonication	0.5-2	2 months
CNTs/water CNTs/EG [18]	acid treatment ultrasonication	0.2-1.0	2 months
Fe_3O_4/water [19]	Fe_3O_4 wet precipitate dispersed in water ultrasonication	0.5-5	several months
CuO/water [20]	$Cu(OH)_2$ as precursor ACT as dispersant ultrasonic and microwave irradiation	0.1-10	several weeks

Table 3. Summary of nanofluids prepared by other methods

Method	Nanofluid	Preparation condition	Particle size (nm)	Volume fraction (vol%)	Stability and observations
Direct evaporation condensation method	Cu/EG[21]	Cu vapor condensed into nanoparticles by contacting with EG	10	0.1-0.3	little agglomeration
One-step chemical method	Cu/EG[22]	reducing $CuSO_4$ with NaH_2PO_2 in EG under microware irradiation using PVP as protective agent	10-20	<0.2	3 weeks Well dispersed
	Au/water Ag/water [23]	using citrate as reductant	10-20 60-80	0.0026 0.001	Well dispersed
	Cu/water [24]	reducing copper acetate with N_2H_4 in water without surfactant	75-100	<0.2	not stable
	Ag/oil [25]	thermal decomposition of silver lactate in mineral oil using Korantin SH as stabilized agent	9.5	0.001-0.3	1 month
SANSS method	CuO/water Cu_2O/wate+EG Cu/EG[26]	Cu bars were heated and vaporized by arc sparking, Cu aerosol was condensed to particles in the dielectric liquid	20×80 20-40 10		well dispersed
	Ni/water Ni/wate+EG Ni/EG[27]	Ni bar as electrode PVP as surfactant	30 10 20		well dispersed

The dispersion was usually performed under ultrasonic agitation or/and mechanical stirring. The aggregates of the nanoparticles, usually existing in dry powders, are broken down and separated in the process. For example, Lee *et al.* [4] prepared Al_2O_3/water, CuO/water, Al_2O_3/EG, and CuO/EG nanofluids by dispersing commercial Al_2O_3, CuO nanoparticles into water and ethylene glycol by the mechanical means of mixing and shaking. The volume fraction of nanoparticles is in the range of 1% ~ 5%. There were some large-size agglomerates of nanoparticles in those nanofluids. Li *et al.* [5] prepared Al_2O_3/water, CuO/water nanofluids by dispersing commercial Al_2O_3, CuO nanoparticles into water using 3-hour ultrasonic vibration. These nanofluids had been stable for several days. Generally, the processing time and intensity can also influence the dispersion effect significantly. Hong *et al.* [6] studied the effect of the ultrasonication on the stability of Fe/EG. With the increasing ultrasonic processing time, the size of nanoparticles decreases. Kwak *et al* [7] investigated the effect of the ultrasonic processing time on the stability of CuO/EG. When the time is less than 9 hrs, the measured size of the nanoparticles decreases with time. The prolonging time of the ultrasonication will lead to aggregation of nanoparticles.

Under high-intensity ultrasonic vibrating or shearing, the weak-bonding agglomerates in dry powders will be broken into the primary particle size. However, nanoparticles have a high tendency to re-agglomerate in dispersions due to the nature of attractive Van der Waals force among particles. According to the stabilization mechanism, addition of dispersant, or/and control of pH value of nanofluids, is usually used to prevent the re-aggregation of nanoparticles and therefore produce stable nanofluids. Xuan *et al.* [8] prepared Cu/oil and Cu/water nanofluids by mixing Cu nanoparticles with transformer oil and water under ultrasonication, respectively. To stabilize the suspension, oleic acid and laurate salt was selected as the dispersant of the oil and water system, respectively. The Cu/oil nanofluids can be stable 1 week long at a stationary state. Using sodium dodecyl benzene sulfonate(SDBS) as dispersant, Wen *et al.* [9] synthesized Al_2O_3/water nanofluids by dispersing Al_2O_3 nanoparticles into water under ultrasonication for 16 – 20 hrs. Nanofluids with less than 4 wt. % nanoparticles were found exceptionally stable and it can keep its stability for over a week. Using hexadecyltrimethyl ammonium bromide (CTAB) as the dispersant, Murshed *et al.* [10] prepared stable TiO_2/water nanofluids under ultrasonic vibration. Using sodium dodecyl sulfate (SDS), CTAB and nanosphere AQ as the dispersant, Assael *et al.* [11, 12] synthesized stable CNTs/water nanofluids with the help of ultrasonication.

By adjusting the pH value of the fluids to obtain electric stabilization, Pak et al. [13] synthesized TiO_2/water (pH=10) and Al_2O_3/water (pH=3) nanofluids under high-speed mixing. Those nanofluids can be stable for more than 6 days. In the same way, Wen *et al.* [14] prepared TiO_2/water (pH = 3) nanofluids under ultrasonication and following high-shear mixing. The nanofluids can last for at least a couple of weeks without visually observable sedimentation. Xie *et al.* [15] produced Al_2O_3/water nanofluids with different pH values under ultrasonication and subsequent magnetic force agitation.

Adding dispersant and adjusting pH value will produce enforced stable effect. By adjusting the pH value and using Gum Arabic (GA) as the steric stabilization agent, Ding *et al.* [16] prepared CNTs/water nanofluids under ultrasonic vibration and high shear homogenization. The prepared nanofluids were found to be very stable for months without visually observable sedimentation. With the help of ultrasonic vibrating, we [17] prepared graphite/water nanofluids by adjusting the pH value and adding polyvinylpyrrolidone (PVP) as the steric stabilization agent. The nanofluids can remain stable for more than 2 months with no visible precipitations.

In the case that the nanoparticles are incompatible to the base fluids, surface modification of the nanoparticles is necessary to obtain stable nanofluids. For example, carbon nanotubes (CNTs) are hydrophobic and can not be dispersed into the common heat transfer fluids without special treatment. Using an acid treatment on the CNTs, Xie [18] successfully prepared stable CNTs/water and CNTs/EG nanofluids with the help of intensive ultrasonication. The nanofluids can remain stable for more than 2 months.

The nanoparticles used for preparing nanofluids are usually dry powders. As mentioned above, under high-intensity ultrasonic vibrating or shearing, the weak-bonding agglomerates in the dry powders could be broken to the primary particle size. However, the hard agglomerates in the dry powders can not be crashed by the above means. To reduce the volume fraction of hard agglomerates in the powders, wet precipitates may be directly redispersed to prepared nanofluids. In our previous publications, Fe_3O_4 wet precipitate was first synthesized by a co-precipitation method. The wet precipitate was re-dispersed into distilled water under ultrasonication, and Fe_3O_4/water nanofluids with good stability were synthesized [19]. Similarly, $Cu(OH)_2$ precusor (Figure 3a) was synthesized by the reaction of $CuSO_4$ with NaOH in EG.

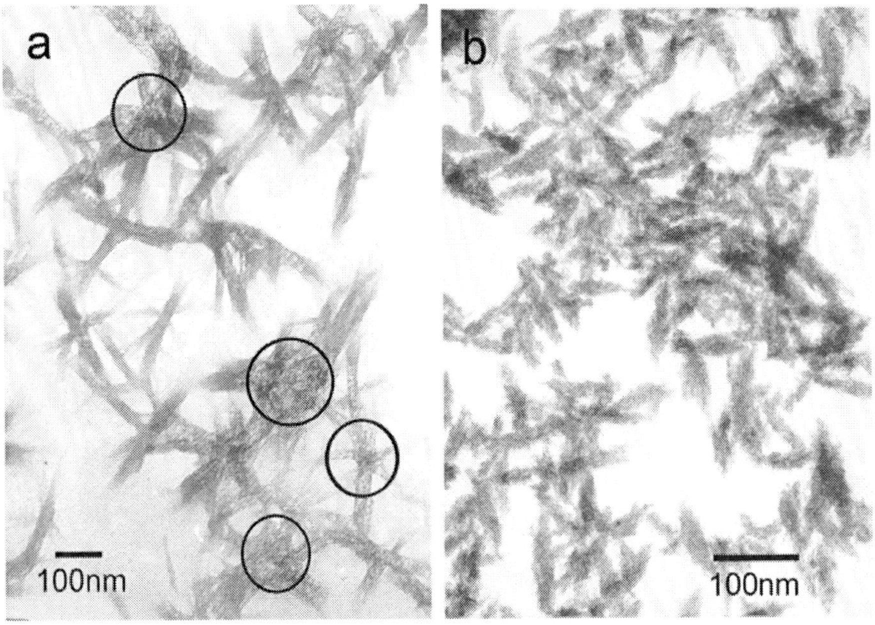

Figure 3. TEM images of the $Cu(OH)_2$ precursor (a), and the CuO nanofluids (b).

After filtrating and washing with ethanol and distilled water, the wet $Cu(OH)_2$ precursor was re-dispersed into distilled water. Using ammonium citrate (ACT) as the dispersant, the $Cu(OH)_2$ precursor suspension was transformed to CuO/water nanofluids (Figure 3b) under ultrasonication and microwave irradition. Different from the Fe_3O_4/water nanofluids, the final nanoparticles in the obtained nanofluids was CuO rather than the precursor, $Cu(OH)_2$. It was verified in the experiment that the precursor was unstable under irradiation of electromagnetic wave [20]. By change the ratio of the $Cu(OH)_2$ precursor to the distilled water, stable CuO nanofluids with solid volume fraction of 0.1% to 10% could be synthesized by this method.

Dispersion method is a two-step method. It has the advantage of preparing nanofluids with high solid volume fractions. The method is divided into the preparation of nanoparticles and the preparation of the nanofluids. If there are hard agglomerates in the dry powders, the stability of nanofluids is usually not ideal. In addition, due to the oxidation or the self-ignition of metallic nanoparticles, it is difficult to synthesize nanofluids containing metallic nanoparticles by this method.

Direct Evaporation Condensation Method

A direct evaporation condensation technique (Vacuum Evaporation onto a Running Oil Substrate) was originally used to produce nanoparticles. However, it was difficult to separate the particles from the fluids to obtain dry nanoparticles. Eastman *et al.* [21] modified this method to synthesize nanofluids. In preparation of Cu/EG nanofluids, solid bulk Cu was heated and vaporized, then Cu vapor was directly condensed into nanoparticles by contacting with a flowing ethylene glycol. The Cu/EG nanofluids containing Cu nanoparticles with an average size of less than 10 nm were obtained.

It is a one-step physical method that combines the preparation of nanoparticles with the preparation of nanofluids. The aggregation of nanoparticles in nanofluids was excessively reduced. This method provides an excellent control over the particle size and can produce particles for stable nanofluids without surfactant or electrostatic stabilizers. However, it is limited to the fluids which have sufficiently low vapor pressures to avoid the vaporization of the fluid and subsequent agglomeration of the nanoparticles.

One-Step Chemical Method

Most nanoparticles used in nanofluids can be synthesized using wet chemical method. By modifying the wet chemical method, it is possible to prepare nanofluids using the one-step chemical method that combines the preparation of nanoparticles and nanofluids.

For example, Cu/EG nanofluids were directly prepared by reducing copper sulfate with NaH_2PO_2 in EG under microware irradiation. Figure 4 is the TEM image of Cu nanoparticles. Most of the Cu nanoparticles are about 10 nm in diameter and are dispersed homogeneously in the fluid. The Cu/EG nanofluids are stable for more than 3 weeks at a stationary state and more than 2 weeks in the stationary state at 120^{0}C. It was found that the size of the nanoparticles in the nanofluids prepared by this method can be conveniently adjusted by altering the reaction conditions, such as $CuSO_4$ concentration, the amount of the protective agent (PVP) and the reductant NaH_2PO_2. The volume fraction of nanoparticle is less than 0.2%.

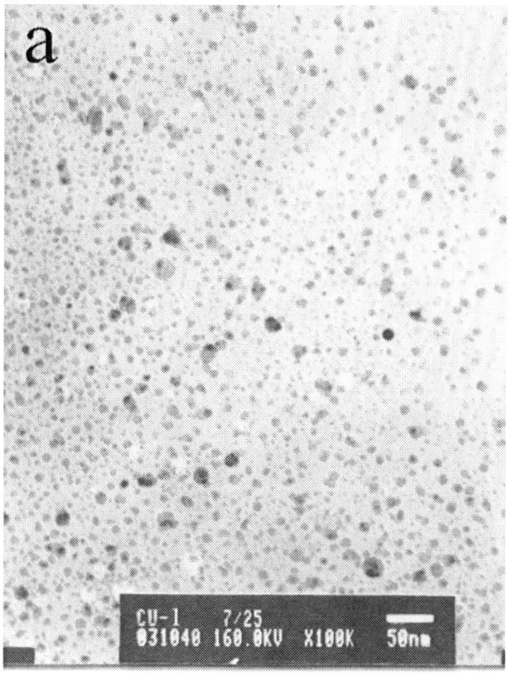

Figure 4. TEM image of Cu nanofluids.

Similarly, using citrate as the reductant, Patel *et al.* prepared Au/water and Ag/water nanofluids by this type of method [23]. The particle size of the obtained Au and Ag nanoparticles is 10 - 20 nm and 60 - 80 nm, respectively. The volume fraction of Au/water and Ag/water nanofluids is 0.00026% and 0.001%, respectively. Using hydrazine as the reducing agent and copper acetate as the precursor, Liu *et al.* [24] synthesized Cu/water nanofluids by a chemical reducing method. The volume fraction of nanoparticle is less than 0.2 %. Using Korantin SH as the stabilization agent, Bonnemann *et al.* [25] synthesized Korantin-stabilized silver/mineral oil nanofluids by the thermal decomposition pathway.

The one-step chemical method combines the synthesis of nanoparticles with the preparation of nanofluids. It has the advantages in terms of controlling the particle size, reducing the agglomeration of the nanoparticles, and producing nanofluids containing metallic nanoparticles. The disadvantage is that it is difficult to prepare nanofluids with a high volume fraction of solid.

SANSS (submerged arc nanoparticle synthesis system) method Lo [26] developed a submerged arc nanoparticles synthesis system to prepare Cu-based nanofluids. In his study, pure copper bars as the electrodes and the source of Cu nanoparticles were submerged in a dielectric liquid in a vacuum chamber. The copper bars were heated and vaporized by arc sparking. The copper aerosol was condensed to nanoparticles in the cooling media (dielectric liquid), forming nanofluids immediately. By changing the dielectric liquids (such as pure de-ionized water, de-ionized water containing different volumes of ethylene glycol, and pure EG) and the process parameters, nanofluids containing different nanoparticles (CuO , Cu_2O and Cu) were obtained. Using nickel bar as the electrode, they also successfully prepared Ni nanofluids [27].

3. CHARACTERIZATION OF NANOFLUIDS

3.1. Characterization of Nanoparticles in Nanofluids

A direct observation of nanoparticles can be conducted on a transmission or a scanning electron microscope. With the help of electron microscopes, the morphology and aggregation status of the nanoparticles can be revealed. Also, the particle size and its distribution can be estimated. The disadvantage of the SEM or TEM measurement is that only a few particles are taken as the representatives. This may not reflect the overall situations. In addition, the SEM/TEM measurements are facing the difficulties to judge whether the aggregated nanoparticles exist in the nanofluids or as a result of the evaporation of the nanofluids for SEM/TEM observations and analysis. This problem evokes more concerns in case of the nanofluids with high volume fractions of nanoparticles. Cryogenic transmission electron microscopy might provide a method to solve the problem. However, the structure of some nanofluids may alter during cryofixation.

Laser scattering method is commonly used to measure the particle size and distribution of the nanoparticles. Nanofluids with low volume fractions of nanoparticles can be measured without any pre-treatment. Dilution has to be made on the nanofluids with high concentrations of nanoparticles. This may also change the original configuration structure of the nanofluids. Therefore, it is necessary to develop new methods to characterize the size and distribution of nanoparticles in nanofluids, especially for high concentration nanofluids.

3.2. Evaluation of Stability of Nanofluids

The sediment weight or the sediment volume of nanoparticles, in a nanofluid under an external force field, is an indication of the stability of the characterized nanofluid. The stable nanofluid will have small sediment weight and sediment volume. Observation of the sediment volume is simple and commonly used method to evaluate the stability of nanofluids. For example, Figure 5 is the sedimentation image of CuO/water nanofluids with various amount of ammonium citrate as the dispersant [28]. It is clear that the stability of nanofluids is at optimum state when the amount of ammonium citrate is 0.3 wt. %. However, in some case, due to the discrepancy of sediment density for different sample, the veracity of this technique is low.

The sediment weight can be measured using a sedimentation balance immerged in the freshly prepared nanofluids [29]. The suspension fraction (Fs) of nanoparticles at a certain time was calculated by the following formula:

$$Fs = (W_0 - W)/W_0 \tag{2}$$

where W_0 is the total weight of all nanoparticles in the measured space; W is the weight of sediment nanoparticles at a certain time. Using this method, Zhu *et al.* [17] studied the influence of pH value on the stability of graphite/water nanofluids (Figure 6). When pH value is in the range of 9 - 10, the stability of nanofluids is the strongest.

Figure 5. Sedimentation (2 days) of CuO nanofluids with various amount of ammonium citrate as the dispersant (wt. %).

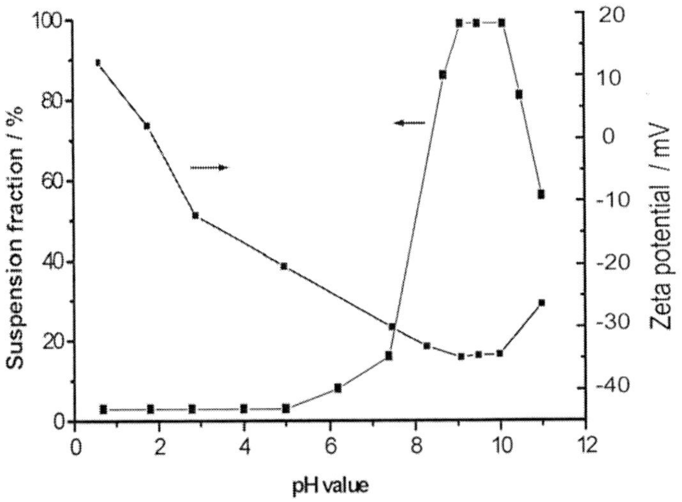

Figure 6. Influence of pH value on the stability and zeta potential of graphite/water nanofluids.

Gravity or centrifugation force field is often used in the sedimentation measurements. For the former, a long measuring time is needed. Sometimes, it may take a few days or even months. In addition, if the nanoparticles are too small, the Brownian movement of the nanoparticles is strongly active. The sedimentation caused by the gravity is negligible and thus the sedimentation measurement is not applicable.

In the measurement of the particle size stored at various time periods in a nanofluid, the corresponding particle size with the storing time is also an indication of the stability of the nanofluid. An increase of the particle size with the storing time may suggest that the nanofluid is not persistently stable. Using this type of assessment, Kwak *et al.* [7] evaluated the stability of CuO/EG nanofluids.

According to the stabilization theory of nanofluids, for electrically stabilized nanofluids, zeta potential is also an indication of the stability of the nanofluids. The experimental results of graphite/water nanofluids indicated it is a feasible judgment (Figure 6) [17]. When pH value is in the range of 9 - 10, the absolute value of zeta is highest and the stability of graphite/water nanofluids is at the maximum state.

Recently, a stability analyzer was invented to directly characterize the stability of the nanofluids [7]. The stability analyzer is a multiple light scattering system. In the measurement, a nanofluid was injected into a cylindrical cell. A vertical scan macroscopic analyzer consists of a reading head that moves vertically to scan the sample. The reading head contains a pulsed near infrared source and two synchronous detectors. One transmission detector monitors the light transmitted through the sample. Another backscattering detector receives the light backscattered by the product. The reading head acquires the transmitted and backscattered data along the length of sample cell. Results are represented by a curve showing the percentage of the backscattered or the transmitted light as a function of the sample length. The acquisition scan is then repeated over and again, and results are superimposed on a time scan graph. If the scan at different time overlaps, the product is stable, and vice versa. Using a TURBISCAN Lab stability analyzer, Kwak *et al.* [7] studied the effects of ultrasonication time on the stability of CuO/EG nanofluids. When ultrasonication time is 90 hrs, the CuO/EG nanofluids have higher stability.

3.3. Measurements of Thermal Conductivity of Nanofluids

The transient hot-wire method [30], temperature oscillation technique [31], steady-state parallel plate method [32], optical beam-deflection technique [33], and thermal lensing method [34] have been employed to measure the thermal conductivity of nanofluids. Among them, the transient hot-wire method is commonly used by many researchers. It has been proven to be one of the most accurate ways because it almost completely eliminates the effects of natural convection, whose presence can cause measurement errors. In addition, the data acquisition time in this method is fairly short relative to other traditional techniques.

A schematic of the transient hot-wire apparatus for measuring the thermal conductivity of nanofluids is shown in Figure 7. The main experimental cell is actually a part of the Wheatstone bridge circuit since the hot wire is used as one arm of the bridge circuit. A platinum wire is usually chosen as the hot wire because of its well-known resistance/temperature relationship over a wide temperature range. Since most nanofluids are electrically conductive, the wire and welded spot are usually coated with very thin electrical insulation layer that have high thermal conduction. The principle of the transient hot-wire measurement is based on the calculation of the transient temperature field around a hot wire [30], treated as a line source. An abrupt electrical pulse is supplied to the wire to generate the necessary temperature rise. The wire is surrounded by a sample nanofluid, whose thermal conductivity and thermal diffusivity are to be measured. The wire serves as both the heat source and the temperature sensor.

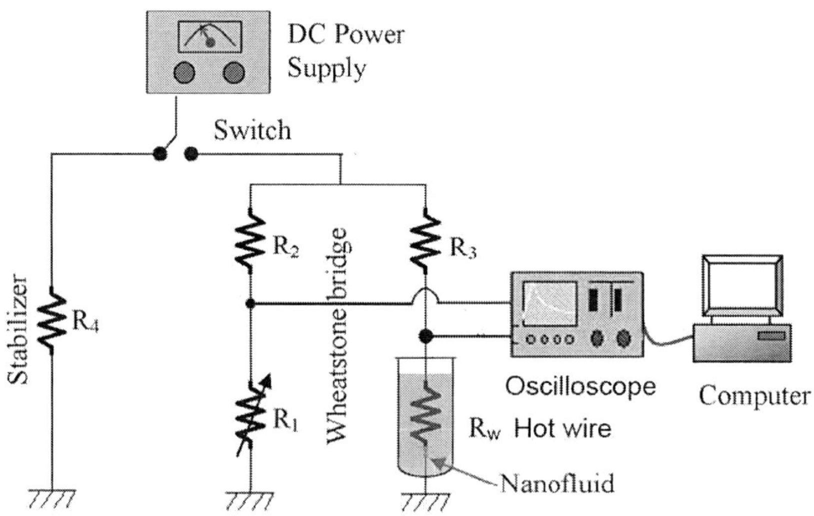

Figure 7. Schematic of the transient hot-wire apparatus for measuring the thermal conductivities of nanofluids.

The heat dissipated in the wire increases the temperature of the wire and also that of the sample. The temperature rise in the wire depends on the thermal conductivity of the sample through which the wire is inserted. A derivation of Fourier's law and temperature data were used to calculate the thermal conductivity.

$$T(t) - T_{ref} = \frac{q}{4\pi k} \ln\left(\frac{4K}{a^2 C} t\right) \quad (3)$$

where T(t) is the temperature of the wire in the fluids at time t, T_{ref} is the temperature of the cell, q is the applied electric power, k is the thermal conductivity, K is the thermal diffusivity of the fluid, a is the radius of the wire, and ln C = g, where g is Euler's constant.

4. Thermal Properties of Nanofluids

4.1. Experiments on Thermal Conductivity of Nanofluids

Thermal conductivity is an important parameter for heat transfer fluids. Many researchers investigated the thermal conductivity of nanofluids following the pioneer work of Choi [1]. Some conclusive results have been reported. For example, the thermal conductivity of nanofluids is evidently improved compared with the base fluids and micrometer-sized suspensions; the enhancement of thermal conductivity depends not only on the particle concentration, but also on the character of the base fluids and the nanoparticles; a nonlinear relationship between the thermal conductivity and the concentration is found for the nanofluids containing CNTs [35], TiO_2 [10] and CuO nanorods[20], etc.; the thermal conductivity of nanofluids is strongly temperature-dependent.

Oxide Nanofluids

The first materials prepared for nanofluids were oxide nanoparticles. The main reason is because they were easy to produce and chemically stable in solutions. The experimental studies on the thermal conductivities of ceramic nanofluids are listed in table 4. Using transient hot-wire method. Lee *et al.* [4] measured the thermal conductivity of nanofluids that contained Al_2O_3 and CuO nanoparticles in water and ethylene glycol (EG). The area weighted particle diameter of CuO and Al_2O_3 nanoparticles are 23.6 nm and 38.4 nm, respectively. The experimental results revealed that the thermal conductivities of Al_2O_3 and CuO nanofluids are substantially higher than those of the same liquids without the nanoparticles. The thermal conductivity of the nanofluids increases almost linearly with volume fraction of nanoparticles. The conductivity enhancements of EG nanofluids systems are always higher than those of water nanofluids systems, and the enhancements of CuO nanofluids are always higher than those of Al_2O_3 nanofluids. For example, an enhancement of 20% was observed for CuO/EG nanofluids with 4 vol. % CuO, while a 12% enhancement for CuO/water nanofluids with 3.5 vol. % CuO, and a 10% enhancement for Al_2O_3/water nanofluids with 4vol% Al_2O_3. But their result is smaller than that of Masuda *et al.* [36]. It was reported that a 30% enhancement for Al_2O_3/water nanofluids with 4.3vol% Al_2O_3 nanoparticles (13nm). The results indicated that the thermal conductivity of nanofluids depend on the particle size and thermal conductivity of both the base fluids and particles.

Wang's [32] and Xie's [15] experimental results have proved the effect of particles size on the effective thermal conductivity of nanofluids. Wang *et al.* [32] measured the thermal conductivity of nanofluids that contained Al_2O_3 (28nm) and CuO (23nm) nanoparticles in water, EG, pump oil (PO) and engine oil (EO). The measurements showed that the thermal conductivity of nanofluids increases with decreasing particle size. Results demonstrated 12% improvement of the effective thermal conductivity at 3 vol. % of nanoparticles as compared to 20% improvement reported by Masuda *et al.* [36] and 8% reported by Lee [4] at the same volume fraction of particles. Xie *et al.* [15] measured the thermal conductivity of Al_2O_3 nanofluids with even smaller particles (12.2 - 302 nm). They also observed the similar effect of particle size in addition to the effect of the base fluid. The thermal conductivity enhancement of Al_2O_3/water nanofluids is 20 at 5 vol. % of nanoparticles. However, Zhou and Wang [37] reported a surprising 17% increase in thermal conductivity for a loading of 0.4 vol. % CuO nanoparticles (50 nm) in water.

Table 4. Summary of experimental studies on the thermal conductivity of the oxide nanofluids

Investigator	nanofluids	Particle size (nm)	Enhancement	Influencing factor
Lee et al. [4]	CuO/water CuO/EG Al_2O_3/water Al_2O_3/EG	23.6 38.4	12% at 3.5vol% 20% at 4vol% 10% at 4vol% linear relationship	particle size volume fraction thermal conductivity of base fluid and particles
Wang et al. [32]	Al_2O_3/water Al_2O_3/EG CuO/EG CuO/water	28 23	12% at 3vol% 25% at 5vol% linear relationship	particle size volume fraction base fluid and particles dispersion method
Xie et al. [15]	Al_2O_3/water Al_2O_3/EG Al_2O_3/PO	12.2–302	20% at 5vol% 29% at 5vol% 38% at 5vol%	particle size (SSA) pH value base fluid
Zhou et al. [37]	CuO/water	50	17% at 0.4vol%	volume fraction
Murshed et al. [10]	TiO_2/ water	$\Phi 10 \times 40$ $\Phi 15$	33% at 5vol% 30% at 5vol% nonlinear relationship	Particle shape volume fraction
Zhu et al. [20]	CuO/water	shuttle-like 20×100	28% at 3vol% nonlinear relationship	Particle shape volume fraction
Zhu et al. [19]	Fe_3O_4/water	10	38% at 4vol% nonlinear relationship	cluster and alignments of nanoparticles volume fraction

Marshed *et al.* [10] measured the thermal conductivity of aqueous nanofluids of spherical and cylindrical shaped TiO_2 nanoparticles. It was found that the nanofluids containing spherical particles show slightly less enhancement than those with nanorods. Recently, Zhu et al. [20] studied the thermal conductivity of CuO/water nanofluids synthesized by the precursor dispersion technology. The CuO nanoparticles are shuttle-like shaped with an aspect ratio of about 5 (Figure 3b). The thermal conductivities (measured at room temperature) are larger than those reported results of CuO nanofluids containing nearly spherical CuO nanoparticles (Figure 8). This result comparison implies that using nanoparticles with high aspect ratio is helpful to enhance the thermal conductivity of nanofluids. The results are smaller than Das's [31] data at 51 °C. It can be accounted for as the temperature effect (it will discuss below).

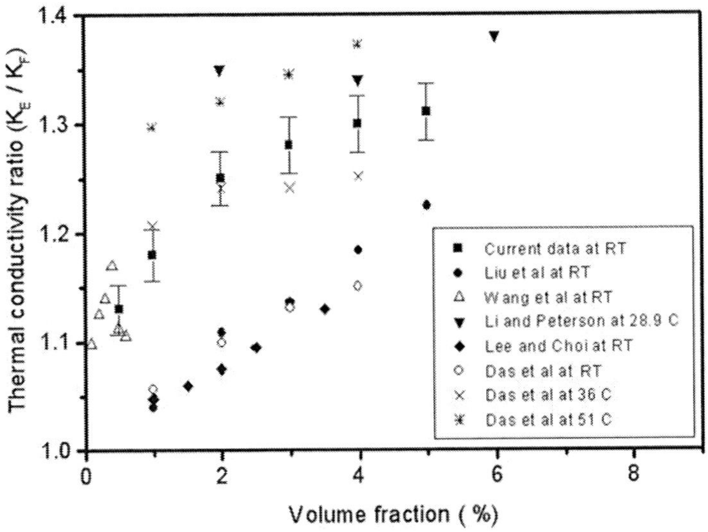

Figure 8. Comparison between the experimental data and the reported values of CuO nanofluid. KE and KF represent the thermal conductivity of nanofluids and the base fluids, respectively.

Zhu *et al.* [19] investigated the influences of clusters and alignments of nanoparticles on the thermal conductivity of Fe_3O_4/water nanofluids. The size of Fe_3O_4 nanoparticles was about 10 nm. Some loose clusters and alignments of Fe_3O_4 nanoparticles existed in nanofluids (Figure 9). Their results showed that the thermal conductivity of aqueous Fe3O4 nanofluids was larger than those of aqueous nanofluids containing the same volume fractions of Al2O3, CuO, and TiO2. A nonlinear dependence of thermal conductivity enhancement on particle concentration was also observed. The abnormal and nonlinear thermal conductivity of the aqueous Fe3O4 nanofluids were mainly caused by the nanoparticle alignment and clustering.

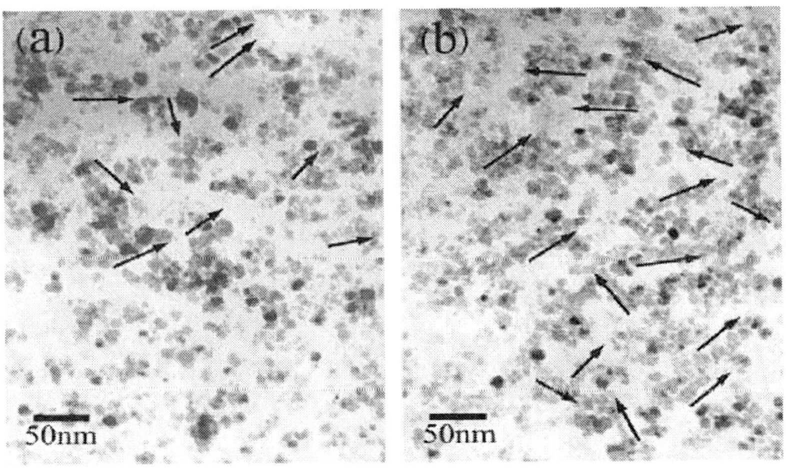

Figure 9. TEM image of Fe3O4/water nanofluids Fe3O4/water nanofluids. (a) 0.5 vol. %, (b) 1 vol. %.

The experimental results of ceramic nanofluids indicate that ceramic nanoparticles, which themselves do not exhibit high thermal conductivity, can enhance the thermal conductivity of fluids substantially, especially the rod-like nanoparticles, alignment and loose clustering of nanoparticles.

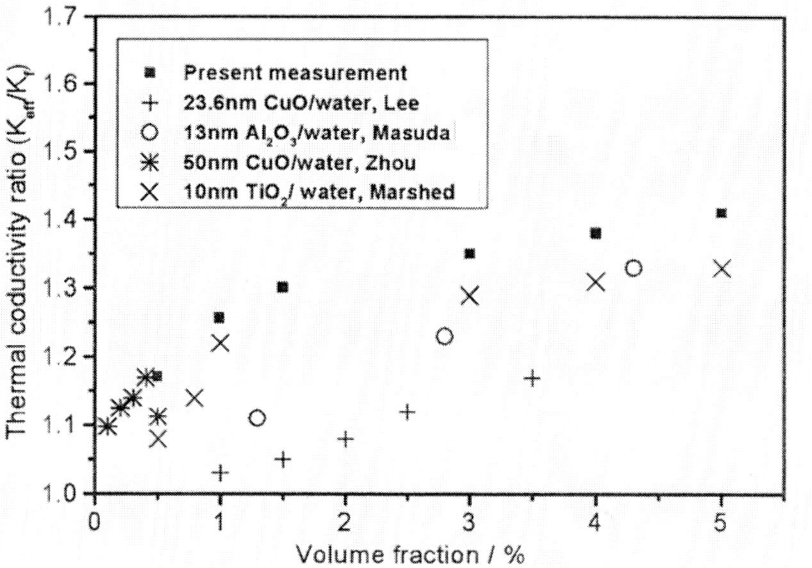

Figure 10. Thermal conductivity of Fe3O4 and other oxide aqueous nanofluids. Keff and Kf represent the thermal conductivity of the nanofluids and water, respectively.

Metallic Nanofluids

Although nanofluids containing metal nanoparticles are less studied than those containing ceramic nanoparticles, the results of metallic nanofluids have been encouraging (Table 5). Xuan and Li [8] prepared the Cu/water and Cu/transformer oil nanofluids by the ultrasonic dispersion of Cu nanoparticles (100 nm) into water and transformer oil. The effective thermal conductivity of the nanofluids increases with the volume fraction of the Cu nanoparticles. For the Cu/water nanofluids, though there are some agglomerates, the enhancement of thermal conductivity varies from 24% to 78% when the volume fraction of Cu nanoparticles increases from 2.5% to 7.5%. A more exciting result was reported by an ANL research group [21] that a 40% abnormal enhancement of conductivity with a concentration of 0.3% of 10 nm-sized copper particles suspended in EG. The Cu/EG nanofluid was synthesized by direct evaporation condensation method and only a few agglomerates of Cu nanoparticles were found in the nanofluids. Zhu *et al.* [22, 38] synthesized Cu/EG nanofluids using one-step chemical method. The Cu nanoparticles are about 10 nm in diameter and dispersed homogeneously in EG. The thermal conductivity of the Cu nanofluids is significantly greater than that of pure ethylene glycol. The effective thermal conductivity increases 9% at 0.1 vol. % volume fraction of Cu nanoparticles.

Patel *et al.* [23] studied the thermal conductivity of Au/water, Ag/water, and Au/toluene, using the previously mentioned transient hot-wire method. The particle size is 10 - 20 nm, 60

- 80 nm, and 3 – 4 nm, respectively. The most important observation in their study was a perceptible enhancement in thermal conductivity for vanishingly small concentrations.

Table 5. Summary of experimental studies on thermal conductivity of metal nanofluids

Investigator	nanofluids	Particle size (nm)	Enhancement	Influence factor
Xuan and Li [8]	Cu/water Cu/oil	100	78% at 7.5vol% 24% at 2.5vol% 43% at 7.5vol% linear relationship	volume fraction
Eastman et al. [21]	Cu/EG	<10	40% at 0.3vol% +acid 14% at 0.55vol% (fresh) linear relationship	volume fraction
Zhu et al. [22,38]	Cu/EG	10	9% at 0.1vol%	
Patel et al. [23]	Au/water Au/toluene Ag/water	10-20 3-4 60-80	3-5% for 0.0013-0.0026% 3-7% for 0.005-0.011% 3.2-4.5% for 0.01vol%	particle size temperature, chemical characteristics
Hong and Yang [6,39]	Fe/EG	10	18% at 0.55vol% nonlinear relationship	ultrasonic time cluster size

The enhancement of thiolate coated Au/toluene nanofluids is 3 - 7% for a volume fraction of only 0.005-0.011%; whereas the enhancement of naked Au/water nanofluids was 3.2 - 5% for a vanishingly small concentration of 0.0013 - 0.0026%. The effects of particle size, base fluids, and particle coating on the thermal conductivity of nanofluids were presented in the literature. Another important observation was that the relatively low thermal conductivity enhancement of Ag/water nanofluids, though Ag has higher thermal conductivity than Au and the conductivity of Ag/water nanofluid is an order of magnitude higher than that of Au/water nanofluid. This experimental finding indicates that the particle size can override the particle conductivity or concentration effect. However, Putnam *et al.* [33] did not observe significant enhancement in the thermal conductivity of the nanofluids with small volume fractions of Au nanoparticles.

Hong *et al.* [6] prepared Fe/EG nanofluids by ultrasonic dispersed Fe nanoparticles (10 nm) into EG. Some smaller clusters of Fe nanoparticles existed in the nanofluids. Their results showed that the ultrasonication time affected the thermal conductivity of the Fe/EG nanofluids. Though the thermal conductivity of bulk Fe is lower than that of bulk Cu crystal, the enhancement of Fe/EG nanofluids at same concentration is higher than that of Cu/EG nanofluids, in which the particle size of Cu nanoparticles is about 10 nm and little agglomeration is observed. Their result indicates that the material with a high thermal conductivity is not always the suitable candidate for the suspension to improve the thermal conductivity of base fluids. The nanofluids with small nanoparticle clusters are effectively modified on the thermal conductivity than that of individually dispersed nanoparticles. They have thoroughly investigated the effect of the clustering of Fe nanoparticles on the thermal conductivity of nanofluids [39]. They found that the thermal conductivity of nanofluids is directly related to the agglomeration of Fe nanoparticles. The nonlinear relation between the

Fe volume fraction and thermal conductivity of nanofluids is due to rapid clustering of nanoparticles in the condensed nanofluids.

Carbon Nanofluids

The large intrinsic thermal conductivity of carbon-based nanostructures, combined with their low densities, makes them attractive candidates for the application in nanofluids (Table 6). So far, the largest enhancement of thermal conductivity was observed in CNTs/oil nanofluids reported by Choi *et al.* [35]. The enhancement of the thermal conductivity is 160% with just 1% volume fraction of CNTs. The conductivity enhancement increased more rapidly than a linear dependence on nanotube loading. The reason for the abnormal enhancement and the nonlinear behavior is yet to be left unexplained. However, there are two facts that can be looked into. First, the thermal conductivity of CNTs is extremely high; second, the CNTs have a fairly high aspect ratio.

After treating CNTs with nitric acid, Xie et al. [18] synthesized CNTs/water, CNTs/EG and CNTs/decene (DE) nanofluids and measured the thermal conductivities of those nanofluids. The results showed that the thermal conductivity enhancement increases linearly with the increase of nanotude loading. But there was higher enhancement for the same volume fraction of the CNTs in the fluid with a lower thermal conductivity. Maximum enhancement is 19.6% for CNTs/decene with 1% volume fraction of CNTs. The enhancement is far smaller than what was achieved by Choi *et al.* [35]. This discrepancy would be attributed to the difference in preparation methods and the CNT specification. Liu *et al.* [40] measured the thermal conductivity of CNTs/EG and CNTs/engine oil (EO) nanofluids. The thermal conductivity enhancement increases almost linearly with the volume act fraction of CNTs.

Table 6. Summary of experimental studies on thermal conductivity of carbon nanofluids

Investigator	nanofluids	Particle size (nm)	Enhancement	Influence factor
Choi et al. [35]	MWCNTs/oil	Φ25nm × 50 μm	160% at 1vol% nonlinear relationship	volume fraction
Xie et al. [18]	CNTs/water CNTs/EG CNTs/DE	Φ15nm × 30 μm	7.0% at 1.0vol% 12.7% at 1.0vol% 19.6% at 1.0vol% linear relationship	volume fraction base fluids
Liu et al. [40]	CNTs/EG CNTs/EO	Φ20–30	12.4% at 1vol% 30% at 2vol% linear relationship	volume fraction base fluids
Assael et al. [11-12]	CNTs/water,	Φ130 × 3-50 μm	38% at 0.6vol%	concentration and type of dispersant, length of CNTs
Wen and Ding [41]	CNTs/water	Φ20–60 × 10 μm	23.7% and 31% at 0.84vol% for 20□ and 45□, respectively	volume fraction temperature
Zhu et al. [17]	graphite/water	200 × 10	34% at 2vol%	volume fraction

For CNTs/EO, the enhancement is 30% at a volume fraction of 2 vol. %. Assael *et al.* [11-12] prepared CNTs/water nanofluids with sodium dodecylsulfate as the dispersant under ultrasonic vibration. The maximum enhancement is 38% with 0.6 vol. % CNTs. The measurement results showed that the enhancement of the thermal conductivity increased with the length of CNTs.

Wen *et al.* [41] studied the thermal conductivity of CNTs/water nanofluids. The effective thermal conductivity increases with increasing CNTs concentration and the dependence is nonlinear even at extremely low concentrations. The enhancement also increases with increasing temperature and the dependence is also nonlinear. The temperature effect is more significant. The enhancement is close to those reported by Assael [11-12] and Xie [18], but much lower than that reported by Choi [35].

Zhu *et al.* [17] measured the thermal conductivity of graphite/water nanofluids using the transient hot-wire method. The enhancement increases with increasing graphite concentration (Figure 11). The enhancement of the thermal conductivity in suspension was 34% at 2.0 vol. % of graphite nanoparticles. The experimental data of graphite nanofluids is close to those reported data of the CNTs nanofluids by Xie [18] and Liu [40].

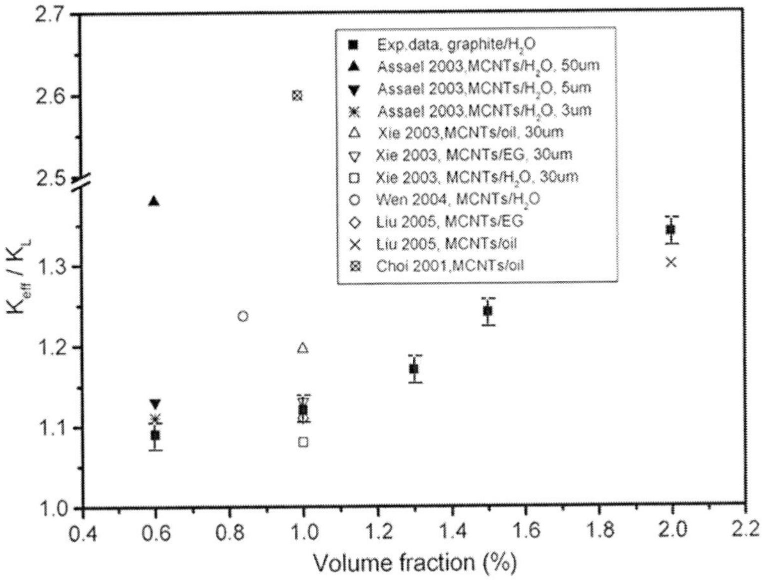

Figure 11. The thermal conductivity of graphite nanofluids and reported CNTs nanofluids. Keff and KL represent the thermal conductivity of nanofluids and base fluids, respectively.

Temperature Effect

A potentially important development in the nanofluids has been the observation by Das *et al.* [31] of strong temperature effects. Using the temperature oscillation technique, they measured the thermal conductivity of oxide nanofluids over the temperature range of 21 - 50□. The results revealed an almost two- to four-fold increase in the thermal conductivity enhancement for CuO and Al_2O_3 aqueous nanofluids. Patel *et al.* [23] studied the temperature effect on the thermal conductivity of the nanofluids containing Au and Ag nanoparticles at vanishingly small concentration. The thermal conductivity enhancement increase with increasing temperature (30 - 60□). Wen *et al.* [41] measured the thermal conductivity of CNTs/water nanofluids at the temperature range of 10 - 70□. The conductivity enhancement increases with increasing temperature and the dependence is nonlinear. At the temperature lower than 30□ , there is an approximately linear dependence of the thermal conductivity enhancement on the temperature. But this dependence tends to level off above 30□. Chon *et al.* [42] reported the effect of temperature on the thermal conductivity of Al_2O_3/water nanofluids. The enhancement of Al_2O_3/water nanofluids increases with increasing temperature within the range of 21 - 70□. Those discoveries make nanofluids particularly attractive for applications at elevated temperatures. These results also open up the possibility that nanofluids could be used as "smart fluids", sensing hot spots and providing more rapid cooling in those regions.

From the experimental results of many researches, it is known that the thermal conductivity of nanofluids depends on many parameters including the thermal conductivities of the base fluid and the nanoparticles, the volume fraction, the size, the shape, clustering of the nanoparticles, and the temperature, etc. However, the influences of above factors are unclear at present. To seek for the cause of the so-called anomalous increase in thermal conductivity of nanofluids, many theories have been proposed.

4.2. Theories of Thermal Conductivity of Nanofluids

Traditional theories, such as Maxwell model [43], Hamilton and Crosser model [44], explained the thermal conductivity enhancement of usual suspensions containing large particles (micro- or mini-size) quite extensively. The classical models were derived from continuum formulations and included only the particle shape and volume fraction as the variables. The large enhancement of thermal conductivity in nanofluids defies Maxwell's theory as well as Hamilton and Crosser theory. It gave rise to hypotheses about the mechanism of heat transfer in nanofluids.

Keblinski and Eastman [45, 46] proposed a comprehensive explanation of four possible factors to understand the mechanism of heat transfers in nanofluids. They were the Brownian motion of particles, liquid nano-layering at liquid/particle interface, the nature of heat transport in the nanoparticle, and the effects of nanoparticle clustering. The investigation demonstrated that the movement of nanoparticles due to Brownian motion is very slow in transporting significant amounts of heat through a nanofluid. The liquid nano-layering at the liquid/particles interface could give a path for rapid conduction. The mechanism of ballistic heat transport gains significance because the phonon mean free path is of the order of the nanoparticle dimensions. The thermal conductivity enhancement of loosely packed clusters is higher than that of closely packed clusters.

Liquid Layering at the Liquid- Particle Interface

It is well known that the liquid molecules close to a solid surface form a layered solid-like structure. As the atomic structure of the liquid layer is significantly more ordered than that of bulk liquid, the liquid layer at the solid/liquid interface would be expected to possess a higher thermal conductivity than the bulk liquid and becomes a thermal duct between the nanoparticles and the surrounding fluid [46]. In addition, the thicker interfacial layer results in a larger effective volume fraction of the particle/layer/liquid structure. This tendency will enhance the thermal conductivity of the liquid. Yu and Choi [47] proposed a modified Maxwell model that accounts for the effect of the liquid nano-layer by replacing the thermal conductivity of solid particle in Maxwell formula with the modified equivalent thermal conductivity of equivalent particles. The new model takes the liquid nano-layer into considerations and predicts the presence of ultra-thin nano-layers with a thickness less than 10 nm. It also indicates that the addition of smaller (<10nm) particles would be better than increasing solid volume fraction with respect to the improvement of the thermal conductivity. They also proposed a modified Hamilton and Crosser model to include the interfacial nano-layer for non-spherical particles [48]. The modified Hamilton-Crosser model can predict the thermal conductivity of CNTs/oil nanofluids reasonably well. However, it fails to predict the nonlinear behavior of the effective thermal conductivity of oxide and metallic nanofluids.

Based on the Maxwell theory and the average polarization theory, Xue [49] developed a model for the effective thermal conductivity of nanofluids with the respect of the effect of the interfacial nano-layer. With two adjustable parameters (i.e., the thickness and thermal conductivity value of the liquid layer), the theory matches the measured values of CNTs/oil and Al_2O_3/water nanofluids. Similarly, Xue and Xu [50] presented a model that considers the temperature distribution function and the liquid layer. The model predicted the thermal conductivity of CuO/water and CuO/EG nanofluids.

Considering the interfacial nano-layer with a linear thermal conductivity distribution, Xie *et al.* [51] proposed an effective thermal conductivity model to account for the effects of the nano-layer thickness, nanoparticle size, volume fraction, and thermal conductivities of fluid, nanoparticles, and nano-layer. But they only have verified the model calculations from a few experimental results. Moreover, it was assumed in the model for a fixed nano-layer thickness of 2 nm, although in practice the nano-layer thickness may be expected to vary for different combinations of the liquid and the solid.

The major drawbacks of the above models are that the size of interfacial nano-layer is assumed to be fairly high (several nanometers), and that the thermal conductivity of the nano-layer is assumed as high as the thermal conductivity of the solid. Considering the effect of the interfacial layer and the discontinuous temperature gradients at the interfacial boundaries, Leong *et al.* [52] developed a new model for the thermal conductivity of nanofluids. The proposed model takes into account some additional effects, including volume fraction, thickness, thermal conductivity of the interfacial nano-layer and particle size. Among them, the interfacial nano-layer thickness is chosen as 1 nm and its thermal conductivity is chosen as 2 to 5 times as that of base fluid thermal conductivity. The results predicted by this model are in good agreement with the experimental data of Al_2O_3/water, Al_2O_3/EG, CuO/water, and CuO/EG nanofluids.

Effects of Nanoparticle Clustering

As mentioned earlier, the loosely packed clusters could significantly enhance the effective thermal conductivity of nanofluids since the clusters may occupy more space than the individual nanoparticles that make up the clusters, and thus increase the effective volume of nanoparticles [46]. Based on the effective medium approximation and the fractal theory that can favorably describe the disorder and stochastic process of clustering and polarization of nanoparticles within the mesoscale limit, Wang *et al.* [53] proposed a fractal model to explain the clustering effects on the effective thermal conductivity of nanofluids. The results indicate that, for CuO/water nanofluids, the fractal model shows good promise when adsorption is included in the analysis, but it underestimates the enhancement in the absence of adsorption.

Nanoparticle Motion

Even though it has been stated earlier that Brownian motion alone cannot account for the thermal conductivity enhancement of nanofluids, for Brownian motion is unable to transport significant amount of heat in a short time period. More recently, a new look at Brownian motion has been presented. Xuan *et al.* [54] proposed a model that primarily combines the concept of fractals and Brownian motion. The whole process has been assumed to have two additive parts: the usual static theory of suspension and the Brownian motion dominated dynamics. Naturally, the Brownian motion part is temperature-dependent and was proposed to be proportional to $T^{1/2}$. This model was proved to be very weak when compared with experimental data.

Jang and Choi [55] devised a theoretical model that involves four modes of energy transport: collisions between base fluid molecules; thermal diffusion in nanoparticles; a collision between nanoparticles with the base fluid. Brownian motion produced convection-like effects at nano-scale. The advantage of the model is to include the effects of concentration, temperature, and particle size. As the particle size is decreased, random motion becomes more evident, and convection-like effects become dominant. This model was able to predict a particle size and the temperature-dependent thermal conductivity of nanofluids.

Kumar *et al.* [56] presented a model that account for the dependence of the thermal conductivity on particle size, concentration, and temperature. The proposed model incorporates two aspects: a stationary model and a moving particle model. The stationary particle model accounts for the geometrical effect of an increase in surface area with a decrease in particle size. The moving particle model developed from the Stokes-Edison formula accounts for the temperature effect. The Kumar model shows that the thermal conductivity enhancement is proportional to the volume fraction and inversely proportional to the radius of the particle, the temperature dependence is attributed to the variation of Brownian motion velocity for the particles. Experimental results for Au/water, Al_2O_3/water and Au/toluene nanofluids support the prediction of this model. However, this model cannot not be applied to the condition of large concentrations of the particles in which interparticle interactions becomes prevalent. In addition, the constant used in the model is empirical and varies over several orders of magnitude for different combination of nanoparticle/fluid mixture.

Prashe *et al.* [57] proposed that the convection caused by Brownian of nanoparticles is primarily responsible for the thermal conductivity enhancement of nanofluids. By introducing the general correlation for the heat transfer coefficient, he modified the Maxwell model by including the convection of the liquid near the particle due to Brownian motion. For the first time, they considered the effect of multi-particle convection. The model well predicted the thermal conductivity of oxide nanofluids.

Bvattacharya et al. [58] developed a technique to compute the effective thermal conductivity of nanofluids using Brownian motion simulation. The model predicted the effective thermal conductivities of Cu/EG and Al_2O_3/EG nanofluids with a high-degree of accuracy.

Ren *et al.* [59] proposed a model that account for the kinetic theory-based micro-convection and the effects of liquid nano-layering. The model works well for ceramic particle nanofluids. They also considered a fixed nano-layer thickness of 2 nm and determined the thermal conductivity of the nano-layer as the volume-averaged thermal conductivity of the base liquid and particles. Recently, Prakash *et al.* [60] developed a model that combined the hydrodynamical approach (micro-convection effects) with the nano-layering structure (interfacial effects). In the model, the interfacial resistance effects are incorporated through a phenomenological parameter. The micro-convection of the nanoparticles is included through Reynolds and Prandtl numbers. The model predicted the dependence of the thermal conductivity on the particle size and concentration for copper nanofluids.

Considering the fact that CNTs possess a large aspect ratio, their thermal conductivity is more difficult to predict. Nan *et al.* [61, 62] presented a simple formula for the thermal conductivity enhancement in CNT composites that is derived from the Maxwell-Garmett model by the effective medium approach. The results within the Nan's model agree well with the experimental observation. However, this model cannot explain the nonlinear phenomena of the thermal conductivity of CNTs nanofluids with the volume fraction of CNTs.

Xue [63] presented a model of the thermal conductivity for CNTs composites by incorporating the interface thermal resistance with an average polarization theory. The model includes the effects of CNTs length, diameter, concentration, interface thermal resistance, and the thermal conductivities of CNTs and base fluid. The model predicted that the nanotube diameter has a neglected effect on the thermal conductivity enhancement of the CNTs nanofluids. It seems that the model can agree well with the data from Xie *et al.* [18].

Thus various models and mechanisms that depend on an extension of classical theory, liquid nano-layering, particle clusters, interfacial resistance, and particle movement have been developed in an effort to explain the thermal conductivity of nanofluids. However, the success of these models has been very limited, and no clear picture has emerged until recently.

4.3. Convection in Nanofluids

For the practical application of nanofluids, the thermal conductivity enhancement only creates a "necessary" condition. The "sufficient" condition comes from the better performance of nanofluids in convection environments. In the convective heat transfer, the heat transfer coefficient depends not only on thermal conductivity but also on other properties, such as the specific heat, density, and dynamic viscosity of nanofluids.

Viscosity of Nanofluids

At low volume fractions, the density and specific heat of nanofluids are expected to be very similar to those of the base fluid. Li *et al.* [64] measured the viscosity of CuO/water nanofluid using a capillary viscometer. Results showed that the apparent viscosity of nanofluids decreased with increasing temperature. Wang *et al.* [32] measure the relative viscosity of Al_2O_3/water and Al_2O_3/EG nanofluids. The relative viscosity increases with increasing volume fraction of solid particles. That means a desirable heat transfer increase may be offset by the undesirable increase in pressure drop. Das *et al.* [65] measured the viscosity of Al_2O_3/water and CuO/water nanofluids as a function of the shear rate and showed

Newtonian behavior of the nanofluids for a range of volume fraction between 1% and 4%. The viscosity of CNTs/water nanofluids as a function of shear rate was measured by Ding *et al.* [16]. The viscosity of nanofluids increases with increasing CNT concentration and decreasing temperature. Also, the shear thinning behavior was found by the authors. That means the nanofluids can provide better fluid flow performance due to the higher shear rate at the wall where the viscosity is low.

Natural Convection in Nanofluids

The natural convection heat transfer in nanofluids has begun to receive attention owing to the potential application of nanofluids in micro-electro- mechanical systems. Comparatively, the natural convection of nanofluids is driven by the unstable density distribution of liquid and the distribution of the particle concentration. Putra *et al.* [66] studied natural convection of Al_2O_3/water and CuO/water nanofluids inside a horizontal cylinder heated from one end and cooled from the other. Unlike the results of conduction, they found a systematic and definite deterioration of the natural convection heat transfer that was dependent on the particle density, concentration, and the aspect ratio of the cylinder. The deterioration increases with particle concentration and is more significant for CuO nanofluids. Recently, Wen and Ding [14] also investigated the natural convection heat transfer of TiO_2/water nanofluids in a vessel that is composed of two horizontal aluminum discs with 240 mm in diameter and thickness of 10 nm, separated by a 10 mm gap. Both the transient and steady heat transfer coefficient were measured. The results show a systematic decrease in the natural convective heat transfer coefficient with increasing particle concentration.

In contrast, Khanafer et al. [67] numerically investigated the heat transfer behavior of nanofluids in a two-dimensional horizontal enclosure. The nanofluid in the enclosure was assumed to be in a single phase. That is, both the fluid and nanoparticles are in a thermal equilibrium and flow at the same velocity. The results showed that the heat transfer rate increased as the volume fraction of the particles increased at any given Grashof number. Kim *et al.* [68] also proposed an analytical investigation to describe the natural convection heat transfer of nanofluids. The result showed that the heat transfer coefficients of nanofluids increase with increasing volume fraction of the particles. Why the results from the analytical approach of Kim and Khanafer were contrary to the experimental results of Putra and Wen? The possible reason may lie in the assumption of the models. Considering the limited studies on the natural convection heat transfer in nanofluids, more efforts are needed in the future work.

Forced Convection in Nanofluids

Xuan and Li [69] experimentally investigated the forced convective heat transfer characteristics of Cu/water nanofluids (0.3 - 2 vol. %) through a straight tube with a constant heat flux at the tube wall. The results showed that the nanofluids gave substantial enhancement of the heat transfer rate compared to pure water; the friction factor for the nanofluids at a low volume fraction did not produce extra penalty in the pumping power.

Wen *et al.* [9] reported experimental results for the convective heat transfer for Al_2O_3/water nanofluids (0.6-1.6 vol. %) flowing through a copper tube in a laminar regime. Viscosity of the nanofluids was not measured and was assumed to follow the Einstein equation. They found the local heat transfer coefficient increased with increasing Reynolds number and particle concentrations. The improvement of the heat transfer coefficient was especially large in the entrance region and decreased with the axial distance. This behavior implied that it might be possible to create a "smart entrance" region to meet the highest performance of nanofluids. They also studied the convective heat transfer of CNTs nanofluids in the laminar regime with a constant flux wall boundary [16]. At a given shear rate, the viscosity of nanofluids increased with an increasing CNTs concentration and decreasing temperature. The presence of CNTs increased the convective heat transfer coefficient significantly, and the increase was more pronounced at high CNTs concentrations. They associated the possible reasons with the improved thermal conductivity, shear-induced enhancement in flow, reduced boundary layer, particle re-arrangement, and high aspect ratio of CNTs.

However, there are inconsistent reports on nanofluid behavior in the forced convection. Yang *et al.* [70] investigated the convective heat transfer of graphite nanofluids in laminar flow through a horizontal tube heat exchanger. The particles they used are disc-like (the average diameter is 1 - 2 μm with a thickness around 20 - 40 nm). The results showed that the heat transfer coefficient increased with the Reynolds number and the particle volume fraction and decreased with the fluid temperature. Unexpectedly, the increase of the heat transfer coefficient of the nanofluids is much lower than the enhancement of the thermal conductivity itself. That means apart from the effective thermal conductivity, particle shape or aspect ratio (0.02 here) of the nanoparticles should be an important factor in determining the heat transfer performance of nanofluids. Pak and Cho [13] studied the convective heat transfer performance of Al_2O_3/water, and TiO_2/water nanofluids in tubes. The viscosities of the Al_2O_3/water and TiO_2/water nanofluids at 10% volume fraction were approximately 200 and 300 times greater than that of pure water, respectively. The results showed that the convective heat transfer coefficient of the nanofluids with a concentration of 3% was 12% smaller than that of pure water, compared under the condition of a constant average velocity. The possible reason is that the nanofluids have higher viscosity than that of pure water, especially at high volume fraction of the nanoparticles.

4.4. Boiling in Nanofluids

The enhancement of the thermal conductivity of nanofluids encouraged researchers to study the boiling characteristics of nanofluids. Das *et al.* [65, 71] studied the pool boiling characteristics of Al_2O_3/water nanofluids (1-4 vol. %) under atmospheric conditions in a tube. The effects of particle concentration, heater diameter and surface roughness of the heater on the boiling characteristics of nanofluids were studied. The results indicated that, with the addition of nanoparticles, the boiling performance of the water deteriorated by increasing the wall superheat for a given heat flux. The deterioration in boiling performance increased with increasing the particle concentration and surface roughness. The deterioration in a narrow tube was lower than that in a wide tube. They attributed it to the affected surface roughness during pool boiling of nanofluids. For higher particle concentration and higher surface roughness, the fluctuating surface can trap the particles more easily and make the surface flatter, causing the degradation of the boiling performance.

Bang and Chang [72, 73] studied the boiling heat transfer characteristics of Al_2O_3/water nanofluids. The results showed that these nanofluids have poor heat transfer performance compared to the pure water in the natural convection and nucleate boiling. But the critical heat flux (CHF) has been enhanced to 32% and 13% for the horizontal flat surface and vertical flat surface in the pool boiling, respectively.

You *et al.* [74] investigated the boiling curve and the CHF of Al_2O_3/water nanofluids in pool boiling with various nanoparticle concentrations ranging from 0.001 g/L to 0.05 g/L. In the nucleate boiling regime of the boiling curves of nanofluids, the heat transfer enhancement or degradation were not observed. However the CHF of nanofluids increased extensively. The increasing of CHF was about 200% higher than pure water when the particles volume fraction was great than 0.005 g/L.

Vassallo *et al.* [75] also confirmed that the CHF increased for SiO_2/water nanofluids. They conducted experiments for both nano- and micro-suspension at the same volume fraction of the solid on a 0.4 mm diameter horizontal NiCr wire at atmospheric pressure. The heat transfer enhancement was not found in the nucleate boiling regime, but the CHF was increased significantly for both nano- and micro- particles. Addition of nanoparticles leaded to a maximum heat flux of about three times as that of pure water and almost twice as that of water with micro-particles.

Wen and Ding [76] conducted experiments on pool boiling heat transfer using Al_2O_3/water nanofluids. They found that the presence of Al_2O_3 nanoparticles in water could enhance the boiling heat transfer significantly, by ~ 40% for a 12.5 wt. % concentration of the particles. They proposed some possible reasons, such as the extra thermal resistance to the boiling surface caused by the sedimentation of nanoparticles, the effect of the surfactant, and the interaction between the boiling surface and nanofluids.

Zhou [77] conducted experiments on heat transfer characteristics of Cu/acetone nanofluids (0-0.4g/L) with and without acoustic cavitation. The results showed that the copper nanoparticle and acoustic cavitation had a significant influence on the heat transfer in the nanofluids. However, the addition of nanoparticles did not affect the dependence of the heat transfer on acoustic cavitation and fluid sub-cooling. The pool boiling heat transfer did not reduce with increased particle volume fraction in absence of acoustic field. While in an acoustic field, the boiling heat transfer of nanofluids was enhanced and the boiling hysteresis disappeared.

The currently available experimental data on the boiling heat transfer of nanofluids are limited and some of them are inconsistent. Therefore, our understanding of the boiling heat transfer of nanofluids is still poor. Further investigations on the boiling performance of nanofluid are necessary in the future.

5. CONCLUSION

In the past decade, many important, complex and interesting phenomena involving nanofluids have been reported. The present review gives a comprehensive overview of the fascinating research progress made in the area of nanofluids. The conclusions based on the reviewed studies in this chapter can be drawn as follows.

Dispersion method is a common and simple technique to prepare nanofluids with a high volume fraction of solids. However, the major problem of this method is that the nanoparticles are prone to form agglomerates. The best way to produce stable nanofluids that exhibits true "nano" behavior, may be a single-step method where nanofluids, instead of nanoparticles, are directly produced. In this way, the chance of agglomeration is reduced. On the other hand, until now, only a few oxides (Al_2O_3, CuO, TiO_2, and SiO2), metals (Cu, Fe, Ag, and Au), carbides and carbon nanoparticles have been used to produce nanofluids. Many other nano-sized materials that have high thermal conductivity, high aspect ratio and low density should be used in experimental practice.

Ceramic nanoparticles can enhance the thermal conductivity of fluids substantially and the largest enhancement of the thermal conductivity was observed in CNTs/oil nanofluids. Metallic nanoparticle-based nanofluids may be more promising for their magnificent enhancement of the thermal conductivity with an extremely low concentration of nanoparticles. It opens the prospect of increasing the thermal conductivity without making large changes in viscosity. The thermal conductivity of nanofluids depends on many parameters including the thermal conductivities of the base fluid and the nanoparticles, the volume fraction, the size, the shape, clustering of the nanoparticles and the temperature, etc. However, there is a lack of agreement between experimental results from different groups. It may be due to the discrepancies of properties of nanoparticles and base fluids as well as the characterization techniques. Thus, the future direction should be to standardize these techniques for characterizing the properties of nanofluids, such as, particle size and distribution, interface thermal resistance, thermal conductivity, etc.

Various models and mechanisms that depend on an extension of classical theory, liquid nano-layering, particle clusters, and particle movement have been developed in an effort to explain the thermal conductivity of nanofluids. However, the success of these models has been limited to an extent, and no clear picture has emerged until recently. Further theoretical investigations are needed to understand the heat transfer characteristics of nanofluids.

Finally, the studies on the convective and boiling heat transfer of nanofluids have just begun. Experimental work in the convection and boiling of nanofluids is still scarce. The critical heat flux seems to be enormously enhanced. However, the physical phenomenon behind this is unclear. Most convective studies are performed with oxide nanoparticle-based nanofluids. The convective heat transfer of metallic nanofluids with low concentrations should be studied.

References

[1] S.U.S. Choi, Enhancing thermal conductivity of fluids with nanoparticles, in *Developments and Applications of Non-Newtonian Flows*, eds. D. A. Singer and H. P. Wang, vol. FED231, American Society of Mechanical Engineers, New York, 1995. pp. 99–105.

[2] P.C. Hiemenz, *Principles of colloid and surface chemistry*, Marcel Dekker inc., New York, 1986.

[3] T. Sato, R. Ruch, *Stabilization of colloid dispersion by polymer adsorption*, Marcel Dekker inc., New York, 1980.

[4] S. Lee, S.U.S. Choi, S. Li, J.A. Eastman, Measuring thermal conductivity of fluids containing oxide nanoparticles, *Journal of Heat Transfer* 121 (1999) 280–289.

[5] C.H. Li, G.P. Peterson, Experimental investigation of temperature and volume fraction variations on the effective thermal conductivity of nanoparticle suspensions (nanofluids), *Journal of Applied Physics* 99 (8) (2006) 084314.

[6] T.K. Hong, H.-S. Yang, C.J. Choi, Study of the enhanced thermal conductivity of Fe nanofluids, *Journal of Applied Physics* 97 (6) (2005) 1–4.

[7] K. Kwak, C. Kim, Viscosity and thermal conductivity of copper oxide nanofluid dispersed in ethylene glycol, *Korea-Australia Rheology Journal* 17(2) (2005) 35-40.

[8] Y. Xuan, Q. Li, Heat transfer enhancement of nanofluids, *International Journal of Heat and Fluid Transfer* 21 (2000) 58–64.

[9] D. Wen, Y. Ding, Experimental investigation into convective heat transfer of nanofluids at the entrance region under laminar flow conditions, *International Journal of Heat and Mass Transfer* 47 (24) (2004) 5181.

[10] S.M.S. Murshed, K.C. Leong, C. Yang, Enhanced thermal conductivity of TiO2–water based nanofluids, *International Journal of Thermal Sciences* 44 (4) (2005) 367–373.

[11] M.J. Assael, C.F. Chen, I.N. Metaxa,W.A.Wakeham, Thermal conductivity of suspensions of carbon nanotubes in water, *International Journal of Thermophysics* 25 (4) (2004) 971–985.

[12] M.J. Assael, I.N. Metaxa, J. Arvanitidis, D. Christofilos, C. Lioutas, Thermal conductivity enhancement in aqueous suspensions of carbon multiwalled and double-walled nanotubes in the presence of two different dispersants, *International Journal of Thermophysics* 26 (3) (2005) 647–664.

[13] B. Pak, Y. Cho, Hydrodynamic and heat transfer study of dispersed fluids with submicron metallic oxide particles, Experimental Heat Transfer 11 (2) (1998) 151–170.

[14] D. Wen, Y. Ding, Formulation of nanofluids for natural convective heat transfer applications, *International Journal of Heat and Fluid Flow* 26 (6) (2005) 855–864.

[15] H. Xie, J. Wang, T. Xi, Y. Liu, F. Ai, Q. Wu, Thermal conductivity enhancement of suspensions containing nanosized alumina particles, *Journal of Applied Physics* 91 (7) (2002) 4568–4572.

[16] Y. Ding, H. Alias, D. Wen, R.A. Williams, Heat transfer of aqueous suspensions of carbon nanotubes (CNT nanofluids), *International Journal of Heat and Mass Transfer* 49 (1–2) (2005) 240–250.

[17] H.T. Zhu, C.Y. Zhang, Y.M. Tang, J.X. Wang, B.Ren. Y.S. Yin, Preparation and thermal conductivity of suspensions of graphite nanoparticles, *Carbon*, 45 (1) (2007) 226-228.

[18] H. Xie, H. Lee, W. Youn, M. Choi, Nanofluids containing multiwalled carbon nanotubes and their enhanced thermal conductivities, *Journal of Applied Physics* 94 (8) (2003) 4967–4971.

[19] H.T. Zhu, C.Y. Zhang, S.Q. Liu, Y.M. Tang, Y.S. Yin, Effects of nanoparticle clustering and alignment on thermal conductivities of Fe3O4 aqueous nanofluids, *Applied Physics Letters* 89 (2) (2006) 023123.

[20] H.T. Zhu, C.Y. Zhang, Y.M. Tang, J.X. Wang, Novel synthesis and thermal conductivity of CuO nanofluid, *Journal Of Physical Chemistry* C 111 (4) (2007) 1646-1650.

[21] J.A. Eastman, S.U.S. Choi, S. Li,W. Yu, L.J. Thompson, Anomalously increased effective thermal conductivities of ethylene glycol-based nanofluids containing copper nanoparticles, *Applied Physics Letters* 78 (6) (2001) 718–720.

[22] H. Zhu, Y. Lin, Y. Yin, A novel one-step chemical method for preparation of copper nanofluids, *Journal of Colloid and Interface Science* 227 (2004) 100–103.

[23] H.E. Patel, S.K. Das, T. Sundararagan, A.S. Nair, B. Geoge, T. Pradeep, Thermal conductivities of naked and monolayer protected metal nanoparticle based nanofluids: Manifestation of anomalous enhancement and chemical effects, *Applied Physics Letters* 83 (2003) 2931–2933.

[24] M. Liu, M.C. Lin, C.Y. Tsai, C. Wang.Enhancement of thermal conductivity with Cu for nanofluids using chemical reduction method. *International Journal of Heat and Mass Transfer* 49 (2006) 3028–3033.

[25] H. Bonnemann, S. S. Botha, B. Bladergroen, V. M. Linkov Monodisperse copper- and silver-nanocolloids suitable for heat-conductive fluids. *Appl. Organometal. Chem.* 2005; 19: 768–773.

[26] C.-H. Lo, T.-T. Tsung, L.-C. Chen, Shape-controlled synthesis of Cu-based nanofluid using submerged arc nanoparticle synthesis system (SANSS), *Journal of Crystal Growth* 277 (1–4) (2005) 636–642.

[27] C.-H. Lo, T.-T. Tsung, L.-C. Chen, Ni nano-magnetic fluid prepared by submerged arc nano synthesis system (SANSS), *JSME International Journal*, Series B 48 (4) (2006) 750–755.

[28] C.Y. Zhang, H.T. Zhu, J.X. Wang, S.B. Chang, Preparation of CuO nanoparticles under ultrasonic and microwave irradiation, in: *14th Chinese Advanced Ceramics Conference*, Luoyang, China, 2006.

[29] J. Ren, J. Shen, S.C. Lu, *Science and technology of particles dispersion*, Chemical Industry Press, Beijing, 2005.

[30] Y. Nagasaka, A. Nagashima, Absolute measurement of the thermal conductivity of electrically conducting liquids by the transient hot-wire method, *Journal of Physics E: Scientific Instruments* 14 (1981) 1435–1440.

[31] S.K. Das, N. Putta, P. Thiesen, W. Roetzel, Temperature dependence of thermal conductivity enhancement for nanofluids, *ASME Trans. J. Heat Transfer* 125 (2003) 567–574.

[32] X. Wang, X. Xu, S.U.S. Choi, Thermal conductivity of nanoparticle–fluid mixture, *Journal of Thermophysics and Heat Transfer* 13 (4) (1999) 474–480.

[33] S. A. Putnam, D.G. Cahill, PV. Braun, Z. Ge, R. G. Shimmin, Thermal conductivity of nanoparticle suspensions, *Journal of Applied Physics* 99, (2006) 084308.

[34] R. Rusconi, E. Rodari, R. Piazza, Optical measurements of the thermal properties of nanofluids, *Applied Physics Letters* 89 (2006) 261916.

[35] S.U.S. Choi, Z.G. Zhang, W. Yu, F.E. Lockwood, E.A. Grulke, Anomalous thermal conductivity enhancement in nano-tube suspensions, *Applied Physics Letters* 79 (2001) 2252–2254.

[36] H. Masuda, A. Ebata, K. Teramae, N. Hishinuma, Alteration of thermal conductivity and viscosity of liquid by dispersing ultra-fine particles (dispersion of Al_2O_3, SiO_2, and TiO_2 ultra-fine particles), *Netsu Bus-sei* (Japan) 7 (4) (1993) 227–233.

[37] L.P. Zhou, B.X. Wang, Experimental researches on the thermophysical properties of nanoparticle suspensions using the quasi-steady state method, *Annu. Proc. Chinese Eng. Thermophys.* (in Chinese), Shanghai, 2002.

[38] H.T. Zhu, *Preparation, stabilization and thermal conductivity of nanofluids*, PhD thesis, Shandong University, 2005.

[39] K. Hong, T.-K. Hong, H.-S. Yang, Thermal conductivity of fe nanofluids depending on the cluster size of nanoparticles, *Applied Physics Letters* 88 (3) (2006) 31901.

[40] M.-S. Liu, M. Ching-Cheng Lin, I.T. Huang, C.-C. Wang, Enhancement of thermal conductivity with carbon nanotube for nanofluids, *International Communications in Heat and Mass Transfer* 32 (9) (2005) 1202–1210.

[41] D. Wen, Y. Ding, Effective thermal conductivity of aqueous suspensions of carbon nanotubes (carbon nanotube nanofluids), *Journal of Thermophysics and Heat Transfer* 18 (4) (2004) 481–485.

[42] C.H. Chon, K.D. Kihm, S.P. Lee, S.U.S. Choi, Empirical correlation finding the role of temperature and particle size for nanofluid (Al_2O_3) thermal conductivity enhancement, *Applied Physics Letters* 87(15) (2005)153107.

[43] J.C. Maxwell, A *Treatise on Electricity and Magnetism*, second ed., Clarendon Press, Oxford, UK, 1881.

[44] R.L. Hamilton, O.K. Crosser, Thermal conductivity of heterogeneous two component systems, *I and EC Fundam* 1 (1962) 182–191.

[45] P. Keblinski, S.R. Phillpot, S.U.S. Choi, J.A. Eastman, Mechanisms of heat flow in suspensions of nano-sized particles (nanofluids), *International Journal of Heat and Mass Transfer* 45 (2002) 855–863.

[46] J.A. Eastman, S.R. Phillpot, S.U.S. Choi, P. Keblinski, Thermal transport in nanofluids, *Annu. Rev. Mater. Res.* 34 (2004) 219–246.

[47] W. Yu, S.U.S. Choi, The role of interfacial layers in the enhanced thermal of nanofluids: a renovated Maxwell model, *Journal of Nanoparticle Research* 5 (1–2) (2003) 167–171.

[48] W. Yu, S.U.S. Choi, The role of interfacial layers in the enhanced thermal conductivity of nanofluids: a renovated Hamilton–Crosser model, *Journal of Nanoparticle Research* 6 (4) (2004) 355–361.

[49] L. Xue, P. Keblinski, S.R. Phillpot, S.U.S. Choi, J.A. Eastman, Effect of liquid layering at the liquid–solid interface on thermal transport, *International Journal of Heat and Mass Transfer* 47 (19–20) (2004) 4277–4284.

[50] Q. Xue, W.-M. Xu, A model of thermal conductivity of nanofluids with interfacial shells, *Materials Chemistry and Physics* 90 (2–3) (2005) 298–301.

[51] H. Xie, M. Fujii, X. Zhang, Effect of interfacial nanolayer on the effective thermal conductivity of nanoparticle–fluid mixture, *International Journal of Heat and Mass Transfer* 48 (14) (2005) 2926–2932.

[52] K.C. Leong, C. Yang, S.M.S. Murshed, A model for the thermal conductivity of nanofluids– the effect of interfacial layer, *Journal of Nanoparticle Research* 8 (2006) 245–254.

[53] B.-X. Wang, L.-P. Zhou, X.-F. Peng, A fractal model for predicting the effective thermal conductivity of liquid with suspension of nanoparticles, *International Journal of Heat and Mass Transfer* 46 (2003) 2665–2672.

[54] Y. Xuan, Q. Li, W. Hu, Aggregation structure and thermal conductivity of nanofluids, *AIChE Journal* 49 (4) (2003) 1038–1043.

[55] S.P. Jang, S.U.S. Choi, Role of Brownian motion in the enhanced thermal conductivity of nanofluids, *Applied Physics Letters* 84 (2004) 4316–4318.

[56] D.H. Kumar, H.E. Patel, V.R.R. Kumar, T. Sundararajan, T. Pradeep, S.K. Das, Model for heat conduction in nanofluids, *Physical Review Letters* 93 (14) (2004) 144301.

[57] R. Prasher, P. Bhattacharya, P.E. Phelan, Thermal conductivity of nanoscale colloidal solutions (nanofluids), *Physical Review Letters* 94 (2) (2005) 025901.

[58] P. Bhattacharya, S.K. Saha, A. Yadav, P.E. Phelan, R.S. Prasher, Brownian dynamics simulation to determine the effective thermal conductivity of nanofluids, *Journal of Applied Physics* 95 (11) (2004) 6492–6494.

[59] Ren, Y., Xie, H., and Cai, A., Effective Thermal Conductivity of Nanofluids Containing Spherical Nanoparticles, *Journal of Physics D: Applied Physics* 38 (2005) 3958–3961.

[60] M. Prakash, E.P. Giannelis, Mechanism of heat transport in nanofluids, *Journal of Computer-Aided Materials Design* 14(1) (2007) 109-117.

[61] C.W. Nan, Z. Shi, Y. Lin, A simple model for thermal conductivity of carbon nanotube-based composites, *Chemical Physics Letters* 375 (5–6) (2003) 666–669.

[62] C.W. Nan, G. Liu, Y. Lin, M. Li, Interface effect on thermal conductivity of carbon nanotube composites, *Applied Physics Letters* 85 (2004) 3549–3551.

[63] Q.Z. Xue, Model for thermal conductivity of carbon nanotube-based composites, *Physica B: Condensed Matter* 368 (1–4) (2005) 302–307.

[64] J.M. Li, Z.L. Li, B.X. Wang, Experimental viscosity measurements for copper oxide nanoparticle suspensions, *Tsinghua Sci. Tech.* 7 (2) (2002) 198–201.

[65] S.K. Das, N. Putra,W. Roetzel, Pool boiling characteristics of nano-fluids, *International Journal of Heat and Mass Transfer* 46 (5) (2003) 851–862.

[66] N. Putra, W. Roetzel, S.K. Das, Natural convection of nano-fluids, *Heat and Mass Transfer* 39 (8–9) (2003) 775–784.

[67] K. Khanafer, K. Vafai, M. Lightstone, Buoyancy-driven heat transfer enhancement in a two-dimensional enclosure utilizing nanofluids, *International Journal of Heat and Mass Transfer* 46 (2003) 3639–3653.

[68] J. Kim, Y.T. Kang, C.K. Choi, Analysis of convective instability and heat transfer characteristics of nanofluids, *Physics of Fluids* 16 (7) (2004) 2395–2401.

[69] Y. Xuan, Q. Li, Investigation on convective heat transfer and flow features of nanofluids, *Journal of Heat Transfer* 125 (2003) 151–155.

[70] Y. Yang, Z.G. Zhang, E.A. Grulke, W.B. Anderson, G. Wu, Heat transfer properties of nanoparticle-in-fluid dispersions (nanofluids) in laminar flow, *International Journal of Heat and Mass Transfer* 48 (6) (2005) 1107–1116.

[71] S.K. Das, N. Putra, W. Roetzel, Pool boiling of nano-fluids on horizontal narrow tubes, *International Journal of Multiphase Flow* 29 (8) (2003) 1237–1247.

[72] I.C. Bang, S.H. Chang, Boiling heat transfer performance and phenomena of Al_2O_3–water nanofluids from a plain surface in a pool, in: *Proceedings of ICAPP*, Pitterburgh, US, 2004.

[73] I.C. Bang, S.H. Chang, Boiling heat transfer performance and phenomena of A12O3–water nano-fluids from a plain surface in a pool, *International Journal of Heat and Mass Transfer* 48 (12) (2005) 2420–2428.

[74] S.M. You, J.H. Kim, K.H. Kim, Effect of nanoparticles on critical heat flux of water in pool boiling heat transfer, *Applied Physics Letters* 83 (2003) 3374–3376.

[75] P. Vassallo, R. Kumar, S.D. Amico, Pool boiling heat transfer experiments in silica–water nano-fluids, *International Journal of Heat and Mass Transfer* 47 (2004) 407–411.

[76] D. Wen, Y. Ding, Experimental investigation into the pool boiling heat transfer of aqueous based γ-alumina nanofluids, *Journal of Nanoparticle Research* 7 (2) (2005) 265–274.

[77] D.W. Zhou, Heat transfer enhancement of copper nanofluid with acoustic cavitation, *International Journal of Heat and Mass Transfer* 47 (2004) 3109–3117.

In: Encyclopedia of Nanotechnology
Editor: Elwood D. Carlson

ISBN: 978-1-60692-079-4

Chapter 12

III-NITRIDE NANOWIRES RESEARCH

***Ho-Young Cha*[1] *and Huaqiang Wu*[2]**
[1] School of Electronic and Electrical Engineering
Hongik University, Seoul, 121-791, South Korea
[2] School of electrical and Computer Engineering
Cornell University, NY 14850, USA

ABSTRACT

Semiconductor nanowires have great potential for highly integrated nanoscale systems, high efficiency sensors, and advanced optical and electronic applications due to their reduced dimensionality incorporated with inherent doping capability and defect free material quality. Due to their unique properties, intensive research efforts have been made on III-nitride nanowires. Research progress has been impressive, both n-type and p-type doped III-nitride nanowires have been fabricated, and heterostructure nanowires have also been successfully synthesized. This chapter reviews various methods of III-nitride nanowire synthesis, and outstanding research on optical and electronic device demonstration.

I. INTRODUCTION

III-nitride semiconductors (GaN, AlN, InN, and their alloys) have been the subject of intensive research for the past decade. The breakthrough in III-nitrides research was first achieved by optical devices – III-nitride based light emitting diodes (LEDs) from the ultra violet (UV), to the green [1], [2] wavelength range. These LEDs have a wide range of applications, including traffic signals, outdoor displays, UV biological imaging and sensing, and water sterilization. Together with LEDs, blue lasers using GaN/InGaN systems have also become commercially available and have enormous impact on high density information storage. In regard to electronic devices, AlGaN/GaN based material systems are very attractive for high power electronics, especially in wireless communication applications. Since the first AlGaN/GaN high electron mobility transistor (HEMT) was demonstrated in early 1990s [3], very impressive progress has been made successively in both material growth and device development. The output power densities reported are already an order of magnitude higher compared to conventional Si and GaAs based devices and large signal operation up to 40 GHz has been demonstrated [4], [5]. The potential application areas are wireless communication infrastructures, satellite communications, broadband wireless networks, radars, etc. Other recent research interests using III-nitrides include solar cells, photodetectors, chemical/gas/bio sensors, actuators, and harsh environment MEMs and electronics. Such interests are obviously due to unique material properties that conventional semiconductors do not have. Figure 1 shows the relationship between energy bandgap and lattice constant of various materials and Table I shows comparison of material parameters. InN, GaN, and AlN can easily form alloys making it possible to tune the bandgap in a wide range from 0.8 eV to 6.2 eV. This bandgap modulation capability with a direct bandgap property is very attractive for developing novel light emitting and detecting devices in wide spectral range from the infrared to the deep UV. The large critical electric field together with the high mobility and saturation velocity in $Al_xGa_{1-x}N$ materials results in high breakdown voltage, high operating frequency, and high current density, which in turn produce high output power density. In addition, other characteristics such as radiation hardness, chemical stability, and biocompatibility make III-nitride semiconductors a promising candidate for harsh environment and biosensor applications.

One-dimensional nanostructures such as nanotubes and nanowires are popular research subjects because they offer substantial promise for future integrated nano systems when the current semiconductor technologies reach their physical limits. Due to the unique properties of III-nitride semiconductors, III-nitrides nanowires have received great attention from researchers in the field of nanotechnology.

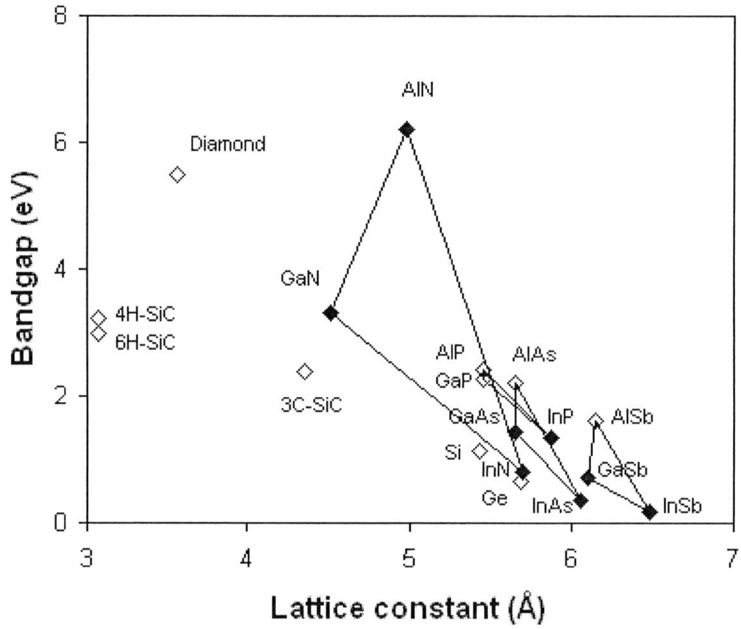

Figure 1. Energy bandgap versus lattice constant of various semiconductors.

Table I. Material parameters of various semiconductors [6], [7]

Parameters	Si	Ge	GaAs	InP	4H-SiC	6H-SiC	3C-SiC	GaN	InN	AlN	C
Bandgap (eV) D: direct, I: indirect	1.12 I	0.66 I	1.42 D	1.34 D	3.23 I	2.86 I	2.36 I	3.28 D	0.8 D	6.2 D	5.46-5.6 I
Relative dielectric constant	11.7	16.2	12.9	12.5	9.7	9.7	9.7	9	13.1	8.5	5.7
Electron mobility (cm^2/Vs)	≤1400	≤3900	≤8500	≤5400	≤900	≤400	≤800	≤1000	50 - 250	300	≤2200
Hole mobility (cm^2/Vs)	≤450	≤1900	≤400	≤200	≤120	≤90	≤320	≤350	-	14	≤1800
Saturation velocity ($\times 10^7$ cm/s)	1	0.7	1	1	2	2	2	2	2.5	-	2
Critical electric field (MV/cm)	0.3	0.1	0.4	0.5	3.2	2.4	1.3	5	-	1.2-1.8	1-10
Thermal conductivity (W/cmK)	1.3	0.58	0.55	0.68	3.7	4.9	3.6	1.3	0.45	2.85	6-20
Thermal expansion ($\times 10^{-6}$/K)	2.6	5.9	5.73	4.6	5.12	5.12	2.77	3.17	3.8	4.15-5.27	0.8

Like other semiconductor nanowires, high quality and defect free III-nitride nanowires can be grown on any substrates without using expensive epitaxial growth techniques or nanolithography tools and the entire semiconductor doping and heterostructure technology can be easily applied to one-dimensional nanowires. It is well known that epitaxial growth of thin film III-nitrides generally produces high dislocations and defects due to lattice mismatched substrates such as sapphire and SiC (GaN bulk substrates are not commercially available) [8]. Such defects would create leakage paths, cause trap-induced instability, and degrade optical characteristics, limiting emission spectra and efficiency. The growth of defect-free material can significantly reduce bulk and surface scattering of carriers and non-radiative recombination. Though current status of III-nitride nanowire research is still immature to compete with thin film technologies as other nanowires are, it is expected that cost effective and improved performance of III-nitride semiconductor devices can be ultimately achieved by utilizing nanowires. This chapter is devoted to reviewing research progress in synthesis and device development of III-nitride nanowires.

II. III-Nitride Nanowire and Nanorod Synthesis

Since the first GaN nanorods were demonstrated using a carbon nanotube (CNT) confined reaction in 1997 [9], numerous groups have successfully produced GaN, AlN, and InN nanowires and nanorods employing various techniques. Chen et al. [10] gave a detail review on III-Nitride nanowires growth in 2002. Many new techniques have been developed since then. In this section, different synthesis and growth techniques are reviewed. A large portion of the section focuses on the catalyst-assisted growth, e.g. vapor-liquid-solid (VLS) growth, as it has been widely used by many research groups. However, since the catalyst usually remains in the nanowires and nanorods, which may affect their intrinsic properties, removal of the catalyst may be required. This limits the application of VLS growth technique. Therefore, catalyst-free approaches including vapor-solid (VS) growth, sublimation, and metalorganic chemical vapor deposition (MOCVD) will also be discussed in this section. Because InN nanowire synthesis techniques have been explored widely in recent years, a special subsection is devoted to InN nanowire synthesis methods.

1. Confined Chemical Reaction Growth

This technique has been applied to AlN, GaN, and InN nanowires growth. CNTs [9], anodic alumina membrane (AAM) [11], and a patterned silicon nitride layer [12] have been utilized to provide a template of nanowire growth with confined spacers in nanometer.

1.1. CNT Template

Dai et al. [13] first demonstrated the use of CNTs as a template to grow nanostructures in which the nanotubes were converted into carbide (MC) nanorods by reaction with a volatile oxide species. Furthermore, using Ga_2O reaction with NH_3, Han et al. [9] successfully synthesized GaN nanorods with diameters between 4 - 50 nm and lengths up to 25 μm. A 4:1 molar ratio Ga-Ga_2O_3 powder mixture was used as the starting material to generate Ga_2O vapor through the following reaction:

$$Ga_2O_3\ (s) + 4Ga(l) \rightarrow 3Ga_2O\ (g) \quad (1)$$

The pressure of Ga_2O vapor from this reaction is quite high according to Frosch's report [14]. In Han's experiments, the mixture of Ga-Ga_2O_3 was placed in an alumina crucible and covered with a porous alumina plate with 3 - 5 μm diameter channels. CNTs were placed on top of the porous alumina plate. It is expected that the Ga_2O gas from Ga-Ga_2O_3 powder mixture would flow up toward the region of carbon nanotubes through the porous plate and react with the nanotubes and the NH_3 gas. The main reaction can be expressed as:

$$2Ga_2O\ (g) + C\ (\text{nanotubes}) + 4NH_3 \rightarrow 4GaN\ (\text{nanorods}) + \\ + H_2O + CO + 5H_2 \quad (2)$$

Similar to GaN nanorod growth, AlN nanowires were synthesized using Al and Al_2O_3 powders as the reactants and CNTs as the confined template [15]. InN nanowires synthesis through CNT templates has not been reported yet.

1.2. Anodic Alumina Membrane (AAM) Template

AAM has very wide applications in molecular filtration, drug separation, and selective ion-transport. Recently, AAM with nanosized pores has been used to synthesize nanowires [16-18]. The AAM possesses an ordered porous structure with a nanochannel diameter ranging from 10 to 100 nm, a typical channel length of tens of microns, and a channel density of 10^{10} cm^{-2}. It has been suggested that the capillary effect of the anodic nanochannels facilitates the formation of nanowires. Having these unique nanochannels, the AAM has been used to fabricate various kinds of ordered nanowires arrays including III-nitride nanowires.

In Cheng's GaN nanowire synthesis report [19], AAM was fabricated by means of the electrochemical anodization at a constant potential of 26V. The metallic indium was evaporated to form the nano-sized indium particles at the bottom of the uniform nanochannels of AAM. The gas reaction of Ga_2O and NH_3 at 1000 °C with the In catalyst produced GaN nanowires; nano-sized In particles served as the catalyst for vapor-liquid-solid (VLS) growth. Because of the highly ordered nanochannels of AAM, the as-grown GaN nanowires prepared by this method were also highly ordered. While the template-based technique is effective in producing nanowires with controllable size, the inability to isolate the product and then maintain the orientation alignment after removing the membrane prevents wider application of this technique.

1.3. Other Template Approaches

CNT and AAM are two popular templates for confined nanowire and nanorod growth. With recent developments in lithography, some research groups utilized patterned layers to serve as the template for nanowire growth.

Hersee et al. [12] reported highly ordered GaN nanowires using a patterned silicon nitride layer as the template. In this case, a 30nm silicon nitride growth-mask layer was deposited by low-pressure chemical vapor deposition (LPCVD) onto 600nm GaN films that had been previously grown by MOCVD on sapphire, silicon carbide, and silicon (111) substrates. Then, the LPCVD silicon nitride layer was patterned by lithography and dry etching was carried out to form a hexagonal array of aperture with a 500 nm pitch. The apertures were approximately circular and had an average diameter of 221 nm. This patterned silicon nitride layer then served as an MOCVD growth mask that defined the position each GaN nanowire. The scanning electron microscope (SEM) image of a highly ordered GaN nanowire array is shown in Figure 2.

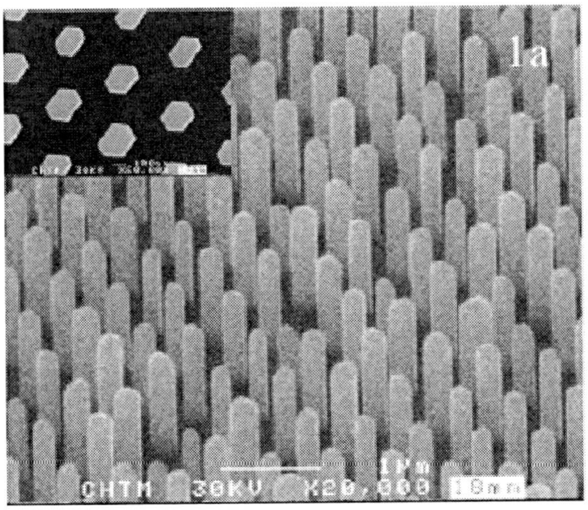

Figure 2. Scanning electron micrograph of a GaN nanowire array synthesized using a silicon nitride template consisting of 1 μm GaN nanowires (inset shows plan view and reveals the hexagonal symmetry of the nanowires). Reprinted with permission from [12]. Hersee, S. D. et al. *Nano Lett.* 2006, 6, 1808. Copyright (2006) American Chemical Society.

Once the GaN nanowires emerged from the growth mask, lateral growth occurred and the nanowire geometry was lost. However, the authors mentioned that introducing a pulsed MOCVD growth mode before the nanowires emerged from the growth mask, the nanowires maintained their geometry after they emerged from the mask. This technique could be used to grow high quality GaN nanowires and uniform nanowire arrays with precise control of the position and diameter of each nanowire. Goldberger et al. [20] reported hollow GaN nanotube synthesis using ZnO nanowires as the template. ZnO nanowire arrays were prepared on (110) sapphire wafers using a vapor deposition process. These ZnO nanowire arrays were placed inside a reaction tube for GaN CVD. Trimethylgallium and NH_3 were used as precursors. The deposition temperature was set at 600-700°C. After the GaN synthesis, the sample was treated at 600 °C with 10 % H_2 in argon to remove the ZnO nanowire template, resulting in GaN nanotubes. The synthesized GaN nanotubes are shown in Figure 3 (a) and the growth procedure is illustrated in Figure 3 (b).

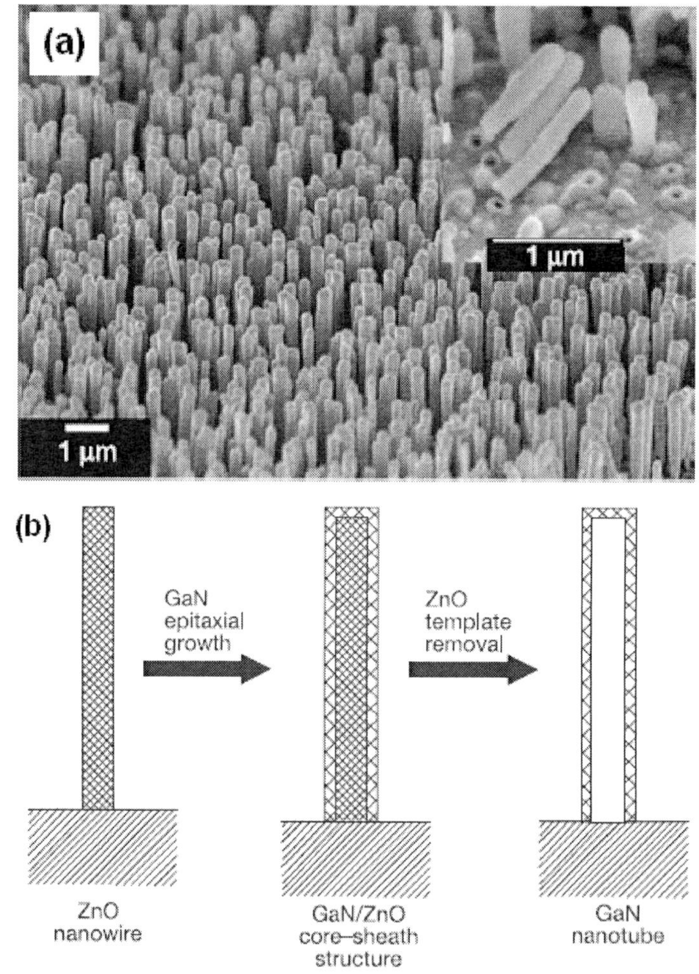

Figure 3. (a) SEM image of arrays of GaN nanotubes. The inset shows the fractured interface between the GaN nanotubes and the substrate. (b) Schematic illustration of the 'epitaxial casting' process for making single-crystal GaN nanotubes. Reprinted by permission from Macmillan Publishers Ltd: *Nature*, [20] Goldberger, J. et al., 2000, 422, 599, copyright (2000).

2. Catalyst Assisted Growth

Catalyst assisted growth is a very popular technique for growing various nanowires and nanotubes. From the name "catalyst assisted growth", this technique refers to a process wherein a catalyst is involved or assists the nanowire growth but does not incorporate into the final nanowire product. A popular synthesis mechanism in catalyst assisted growth is VLS growth. On the other hand, catalyst assisted growth also includes laser ablation, solution-liquid-solid where the source material is solution instead of vapor.

2.1. VLS Growth Technique

A typical experimental setup for GaN nanowire growth using VLS growth technique is shown in Figure 4 [21]. Group III elements (Al, Ga or In) are in the form of metal or powder. We have described the VLS growth process in Ref. [13]. In this case, about 2 grams of pure metal Ga were put in a small boron nitride boat, which was placed at the bottom of the quartz liner. The substrate was placed downstream from the metallic Ga. The quartz tube was evacuated by a mechanical pump capable of pumping the system down to a base pressure of 5 mTorr. 99.9999% NH_3 was cleaned by a purifier to remove residual water vapor and oxygen before being introduced into the growth system. H_2 was used as a carrier gas, and a mass flow controller was used to maintain the gas flow rate. During growth, the total system pressure was kept at one atmosphere; the furnace temperature was maintained at 900 ^oC while flowing 200 sccm NH_3 and 200 sccm H_2. The synthesized GaN nanowires are shown in Figure 5.

There are many ways to introduce a catalyst into the growth system. For CVD based growth techniques, catalyst coated substrates are a widely used method. One method to introducc the catalyst is to disperse the solution containing the catalyst onto the substrate. After the solvent evaporates, the catalyst stays on the substrate surface. Alternatively, the substrate could be pre-coated with a thin layer of catalytic metal film from vapor, via sputtering, or by evaporation. Introducing the catalyst by vapor deposition onto a substrate patterned by lithography offers the opportunity for selective area growth of the desired materials and paves the way for future device integration [21], [22].

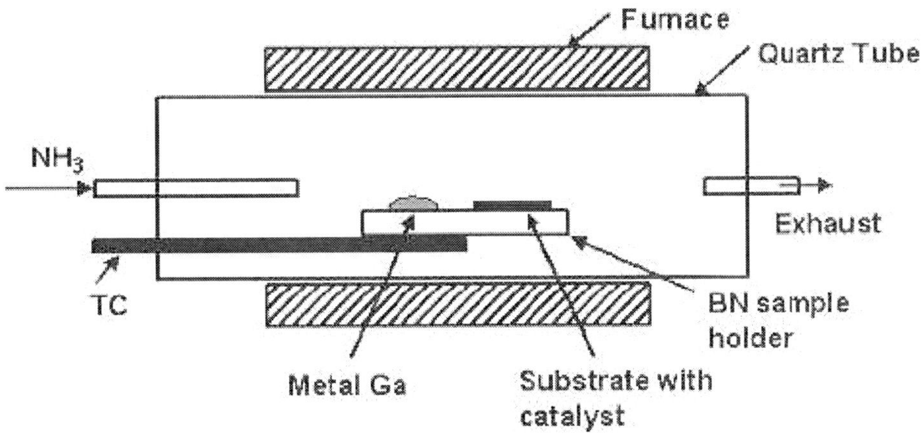

Figure 4. Schematic diagram of GaN nanowire growth via a VLS growth mechanism. Reprinted with permission from [21] Wu, H. et al., *J. Electon. Mater.* 35, 670, 2006, Copyright TMS (2006).

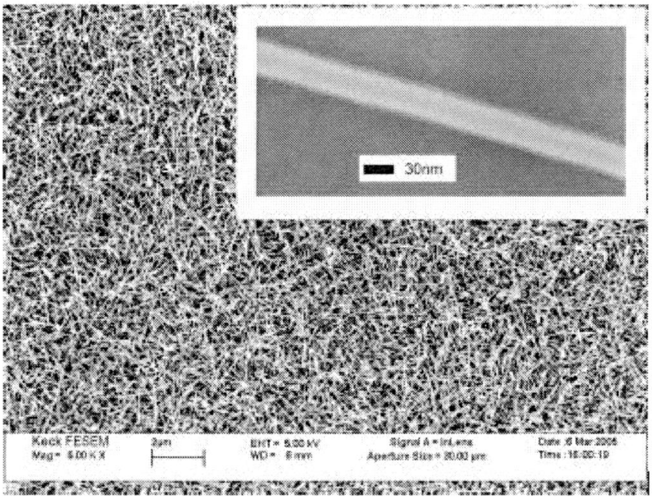

Figure 5. SEM image of dense and uniform GaN nanowires. The inset is a high magnification SEM image of a GaN nanowire. Reprinted with permission from [21] Wu, H. et al., *J. Electon. Mater*. 35, 670, 2006, Copyright TMS (2006).

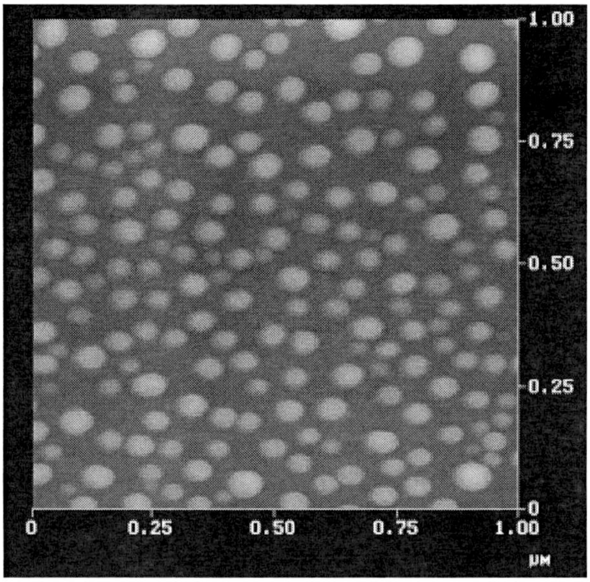

Figure 6. AFM image of Ni nano-droplets served as the catalyst of GaN nanowire growth via a VLS mechanism. Reprinted with permission from [21] Wu, H. et al., *J. Electon. Mater*. 35, 670, 2006, Copyright TMS (2006).

Our group has tried various catalysts including $InCl_3$, Co, Au, and Ni. Both solution and evaporation methods have been explored to prepare catalyst pre-coated substrates. Ni was found to be the best catalyst in our growth system. A 2 nm Ni layer was deposited by an e-beam evaporation system. The 2 nm Ni layer was continuous under SEM and AFM inspection. Rapid Thermal Annealing (RTA) near the eutectic temperature was used to break up the continuous Ni layer into Ni nano-droplets. The catalyst layer was annealed at 1000 ^{o}C for 2 min using RTA. The atomic force microscope (AFM) image of the Ni nano-droplets formed after annealing is shown in Figure 6. It is clear that the continuous Ni layer was broken up into many Ni nano-droplets with sizes ranging from 20-50 nm. These nano-droplets were suitable for nanowire growth via a VLS mechanism.

2.2. Laser Assisted Catalytic Growth

Duan et al. [23] utilized a laser ablation cluster formation method to synthesize GaN nanowires. The key feature of this method is that the catalyst can be selected from phase diagram data and/or knowledge of chemical reactivity. The catalyst should form a miscible liquid phase with GaN but should not form a more stable solid phase than GaN under the growth conditions. A GaN target with Fe catalyst was placed in the center of a furnace. The furnace temperature was set as 900 °C and 80 sccm NH_3 flow was maintained at the pressure of 250 Torr. A pulsed Nd:YAG laser was used to vaporize the target and catalyst with an ablation duration of 5 min. The resulting liquid nanoclusters were formed at elevated temperature, which directed the growth and defined the diameter of crystalline nanowires through a VLS growth mechanism. The synthesis process of the laser assisted catalytic growth combined with VLS is illustrated in Figure 7. The GaN nanowires from this growth technique had diameters on the order of 10 nm and lengths longer than 1 μm.

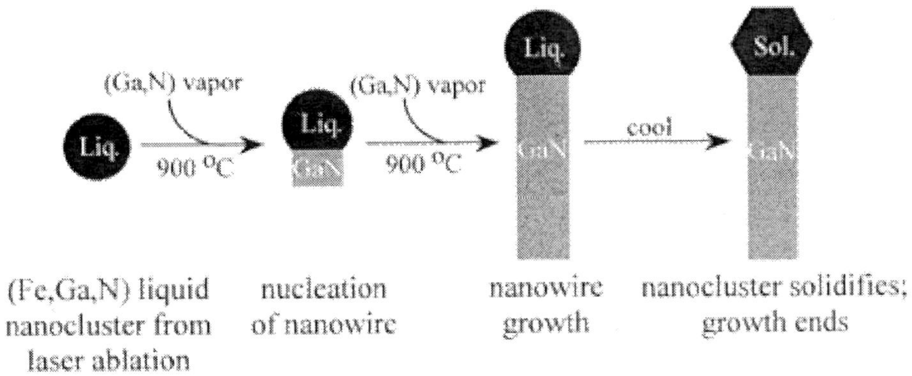

Figure 7. Schematic diagram for VLS growth mechanism. Reprinted with permission form [23] Duan, X. et al., *J. Am. Chem. Soc.* 122, 188, 2000. Copyright (2000) American Chemical Society.

3. Direct Reaction without Catalysts

He et al. [24] reported synthesis of GaN nanowires by a direct reaction between Ga and NH_3. The experimental setup is similar to Figure 4 without the substrate and catalysts. The key process parameters to control the form of product and diameter of nanowires are temperature and a NH_3 flow rate. Figure 8 is the summary graph from their report. It was observed that the growth rate and nanostructures were influenced by temperature and flow rate. Nanowires were synthesized at temperatures from 825 °C to 925 °C. The diameters were ranged from 20 to 150 nm depending on the NH_3 flow rate. In general, the diameter of nanowires increased with temperature and NH_3 flow rate. In the report, the growth process was divided into three stages.

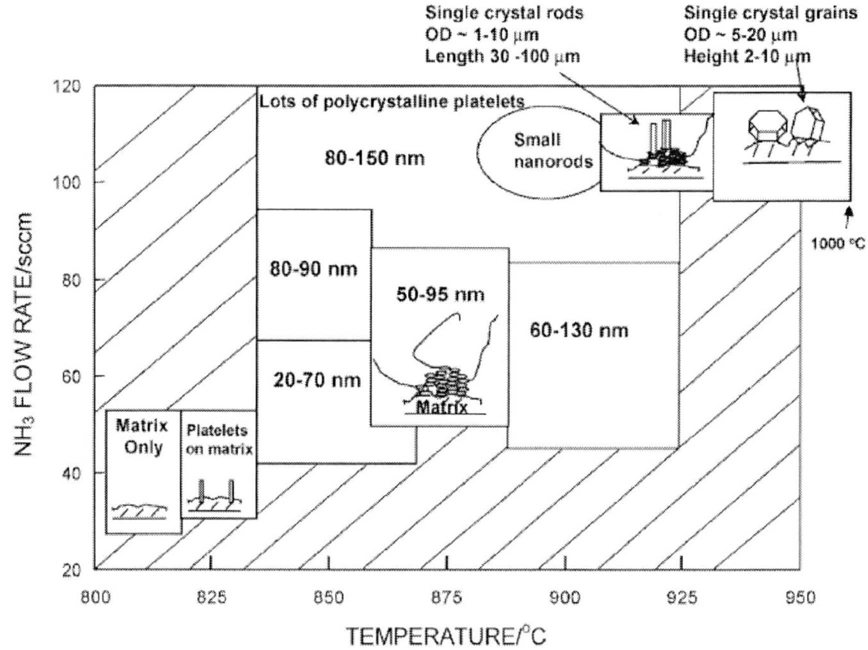

Figure 8. Map showing how nanowires diameter depends on temperature and NH3 flow rate for growth at a total pressure of 15 Torr. Reprinted from [24] J. Cryst. Grow, 231, He, M. et al., "Growth of GaN nanowires by direct reaction of Ga with NH3", 357, 2001, Copyright (2001), with permission from Elsevier.

During the initial stage, a nearly amorphous Ga-rich GaN matrix was formed when the temperature was lower than 750 ^{o}C. In the second stage, polycrystalline hillocks were formed with further heating. In the final stage, nucleation and growth of nanowires took place, and the diameter of the nanowire was controlled by the temperature and NH_3 flow rate.

Figure 9 from this report illustrates the various pathways for Ga and N atoms to arrive on the surface of the crystal platelets in the polycrystalline hillock. NH_3 molecules and Ga atoms arrive at the basal plane as indicated by path A. The NH_3 can decompose stepwise yielding N atoms, which react with the Ga to form GaN. In any case, molecules on the basal plane have four pathways. At low temperatures, the number of sites for crystal plane growth is low and the average molecular diffusion length before desorption is short. Such conditions would favor the formation of amorphous GaN (pathway B) or further small crystal platelets on the basal plane, but not incorporated into it (pathway C). At the highest temperatures (~950°C), the diffusion length of the molecules would be relatively long, allowing reaction and then be incorporated into the existing crystal (pathway D). Another possibility is that, the molecules and atoms would desorb rapidly from sites where they are not strongly bound to the existing crystal (pathways E and F). Except at the highest temperature, growth of the crystal platelets in the [001] direction is very slow. Instead, edges provide more nucleation sites for further growth. SEM investigations did show that the nanowires were not formed as separate objects but were emerging from the side of the crystal platelets. In this case, the nanowire diameters were the same as the thickness of the crystal platelets from which they grew. Since at higher temperature and NH_3 flow rates platelets are usually larger and thicker than their counterparts at lower temperature and NH_3 flow rates, the average diameter of the nanowires increases with both temperature and NH_3 flow rate.

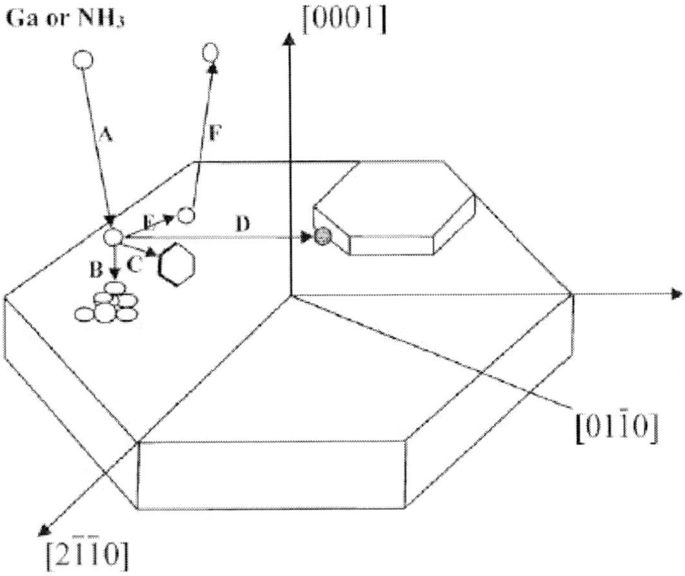

Figure 9. A model of one of the crystal platelets found in the polycrystalline hillock which can be used to discuss the growth of the hexagonal platelets in the [0001] direction. Reprinted from [24] *J. Cryst. Grow*, 231, He, M. et al., "Growth of GaN nanowires by direct reaction of Ga with NH_3", 357, 2001, Copyright (2001), with permission from Elsevier.

4. Other Growth Techniques

Besides the VLS, direction reaction and template assisted growth, many other techniques have been explored to grow GaN nanowires. Some of them are reviewed here briefly.

4.1. Molecular Beam Epitaxy (MBE) Growth

Calleja et al. [25] reported GaN nanocolumn growth using MBE. A cryogenically cooled RF plasma source supplied the active nitrogen and the growth was carried out at 770 °C. GaN nanocolumns were produced under N-rich conditions regardless of the substrates used, Si (111) substrate with or without AlN buffer, and c-sapphire with a GaN buffer. Dense nanocolumns were produced under N-rich conditions and the average diameter of nanocolumns ranged from 60 – 150 nm. Similar GaN nanocolumn growth on Si substrates via MBE has been reported by others [26], [27]. Bertness et al. [27] observed qualitative agreement with Calleja's results but used different growth conditions, such as a higher growth temperature (= 820 °C). The authors found that the growth morphology was strongly affected by the presence of Al and Be as well as V:III ratios. Based on experiments, the authors proposed that differential surface diffusion was more likely to be the main driving force to produce the isolated nanocolumns, which is different from a VLS mechanism proposed in Ref. 18. The images of GaN nanowire columns grown via MBE are shown in Figure 10.

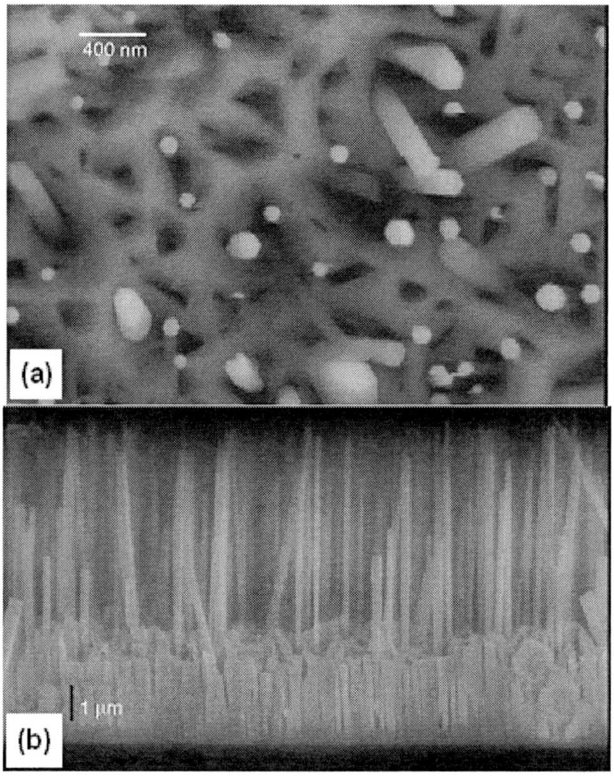

Figure 10. (a) Plan and (b) cross-sectional views of typical GaN nanowire grown on Si. The nanowires have hexagonal cross-sections. Reprinted from [27] *J. Cryst. Growth,* 287, Bertness, K. A. et al., "Spontanesouly grown GaN and AlGaN nanowires", 522, 2001, Copyright (2001), with permission from Elsevier.

4.2. MOCVD growth

Kuykendall et al. [28] studied GaN nanowire growth using a MOCVD method. Trimethlygallium and NH_3 were used as the precursor of Ga and N respectively. A 2 - 10 nm Ni, Fe, or Au layer was deposited on Si or sapphire substrates. The reaction was carried out at 800 - 1000 ^oC at atmosphere pressure. H_2 and NH_3 were supplied at a total flow rate of 155 sccm. It was found that the substrate location relative to the organic precursor outlet was not important. The reaction time was from 5 to 10 min. The SEM and transmission electron microscope (TEM) images of nanowires grown via are shown in Figure 11. The nanowires had a triangular cross-section with diameters ranging from 15 to 100 nm and the lengths between 1 and 5 μm. Various III-nitride nanowire radial heterostructures have been successfully demonstrated using MOCVD based growth methods, e.g. GaN/AlGaN [29], n-GaN/InGaN/p-GaN [30], GaN/AlN/AlGaN [31].

"Oxide-assisted growth", "chloride-assisted growth", "combustion method", and "sublimation method" to synthesize GaN nanowires have also been reported. Chen et al. [10] have a good review on those various techniques.

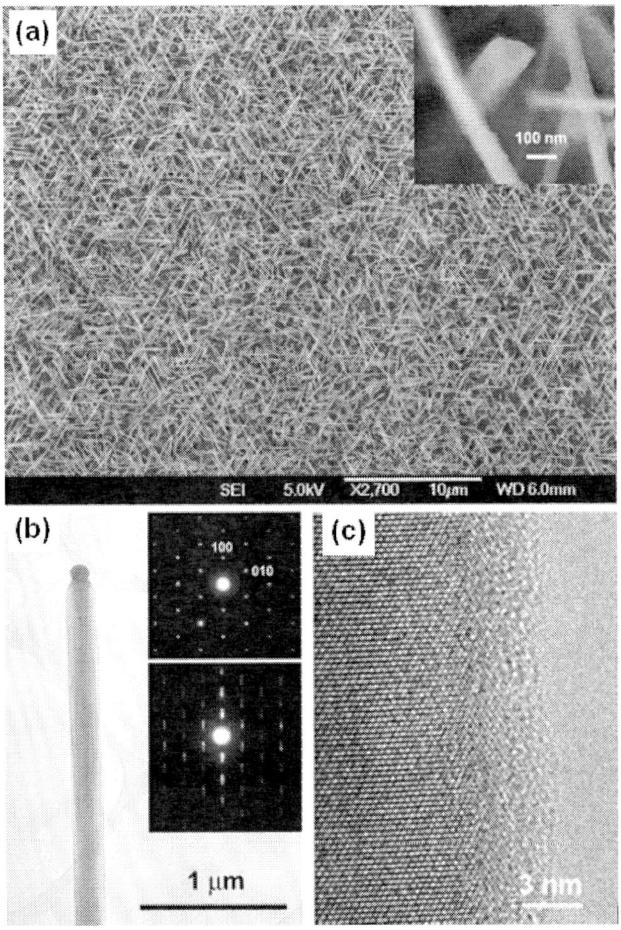

Figure 11. (a) FESEM image of the GaN nanowires grown on a gold-coated c-plane sapphire substrate. The inset shows a nanowire with its triangular cross section. (b) TEM image of a GaN nanowire with a gold metal alloy droplet on its tip. Insets are electron diffraction patterns taken along the [001] zone axis. The lower inset is the same electron diffraction pattern but purposely defocused to reveal the wire growth direction. (c) Lattice-resolved TEM image of the nanowire. Reprinted with permission from [28]. Kuykendall, T. et al. *Nano Lett.* 2003, 3, 1063. Copyright (2003) American Chemical Society.

5. InN Nanowire Growth

InN is a direct bandgap semiconductor and is a promising material for optoelectronic devices, low cost solar cells with high efficiency, optical coatings, and sensors [32]. However, due to low dissociation temperature of InN, there are fewer reports of InN nanowire growth compared to GaN nanowire growth. In this section, a detailed review on InN nanostructure growth is provided.

5.1. Au Catalyst Assisted VLS Growth

Liang et al. [22] reported the selective-area growth of InN nanowires on Au patterned Si (100) substrates. P-type Si (100) substrates were patterned with a 10 nm thick Au film by sputtering deposition through a shadow mask, which contained circle shaped openings of micrometer size. A pure In foil placed in an alumina boat served as the In vapor source upon heating. The boat was covered with a Au patterned Si substrate and then loaded into the center of a half-inch quartz tube, which was housed in a conventional tube furnace. After degassing and purging with NH_3 (40 sccm), growth was carried out at 500 °C for 8 hours. The SEM images of nanowires grown on a Au patterned Si substrate is shown in Figure 12. The image [see Figure 12 (a)] confirmed that the InN nanowires were grown only in the Au catalyst areas. InN nanowires had diameters ranging from 40 nm to 80 nm, which were related to the size of the aggregated Au cluster caused by heating the substrate during growth. Typical lengths of these nanowires were up to 5 μm.

Figure 12. SEM images of InN nanowires on a Si substrate. (a) Image taken at low magnification. The bright circles are InN nanowires grown on patterned Au film. The array of gold circles was formed by sputtering through a shadow mask. (b) Top view of the InN nanowires taken at high magnification. Reprinted with permission from [22] Liang, C. H. et al., *Appl. Phys. Lett.* 2006, 81, 22. Copyright 2006, American Institute of Physics.

5.2. Oxide Assisted VS Growth

Zhang et al. [33] reported InN/In_2O_3 coaxial nanocables synthesized via VS mechanism. A 4:1 molar ratio of In and In_2O_3 powder was used as the starting material and placed in an alumina crucible. For the fist stage, the reaction temperature was kept at 700 °C for 2 hours under a constant flow of NH_3 at 400 sccm. In the second stage, the temperature was still kept at 700 °C while the NH_3 flow was shut off and argon/oxygen (20:1) was introduced into the quartz tube. After 1 - 2 min, the mixture gases were turned off and NH_3 was introduced back to the quartz tube with a flow rate of 200 sccm. After the furnace was slowly cooled down to room temperature, the NH_3 flow was turned off. A light-yellow wool-like product was found on the surface and walls of the crucible. The SEM and high-resolution TEM (HRTEM) images of the synthesized InN nanocables are shown in Figure 13.

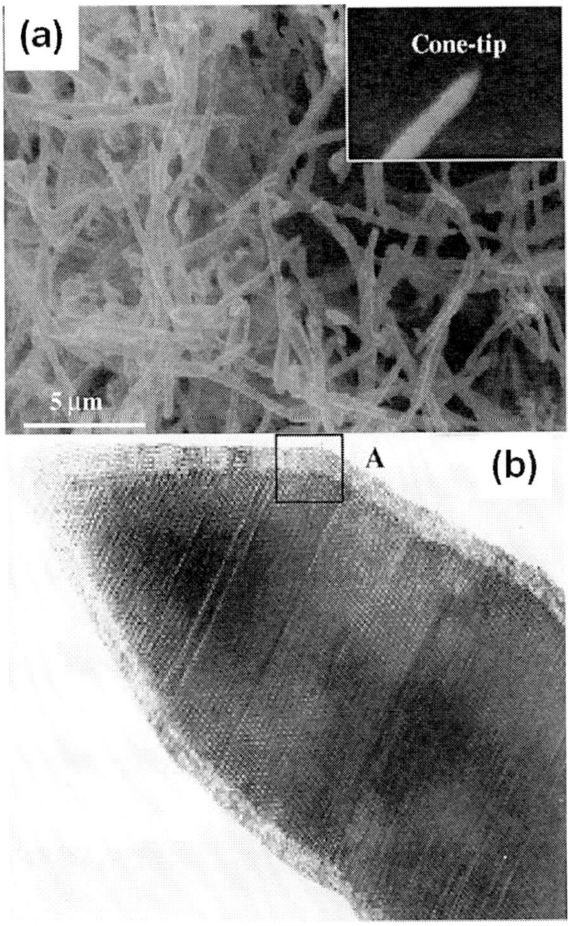

Figure 13. (a) A typical SEM image of the as-prepared wire-like product. The inset is a magnified SEM image of an individual nanowire with a cone-tip end. (b) An HRTEM image of the InN/In_2O_3 nanocable. The dark region is InN whereas the bright outer layer is In_2O_3. Reprinted from [33] *Mater. Lett.* 60, Zhang, J. et al., "Growth and structural characterization of InN/In_2O_3 coaxial nanocables", 2153, 2006, Copyright (2006), with permission from Elsevier.

It was suggested that InN nanocables were grown through the reaction between In_2O and N ions via two-stage VS mechanism as follows.

$$4In(s) + In_2O_3(s) \rightarrow 3In_2O(g) \quad (3)$$

$$In_2O(g) + 2NH_3 \rightarrow 2InN(nanowire) + H_2O + 2H_2 \quad (4)$$

During the cool down stage, the surface of the InN nanowires was dissociated and reacted at oxygen atmosphere to form In_2O_3 coated-layer. The authors claimed the core InN nanowire could avoid dissociation because In_2O_3 protecting layer had a very high melting point (at 1920°C). Lan et al. [34] also reported InN nanowire growth assisted by a small amount of oxygen. Their studies suggested that even a minute amount of oxygen from the residual oxygen in the growth environment and/or native oxide on the Si helps the growth of InN nanorods. The Si substrates were first coated with a 3 nm thick Au film by ion beam sputtering. Pure In powders were placed in an alumina boat and served as the In vapor source upon heating. The Au coated Si substrate was put face-down on top of the alumina boat and then loaded into the central zone of a one-inch quartz tube. The growth temperature was set at 500 °C at a pressure of 1 atm for 3 hours. 50 sccm NH_3 was introduced into the tube as the nitrogen source.

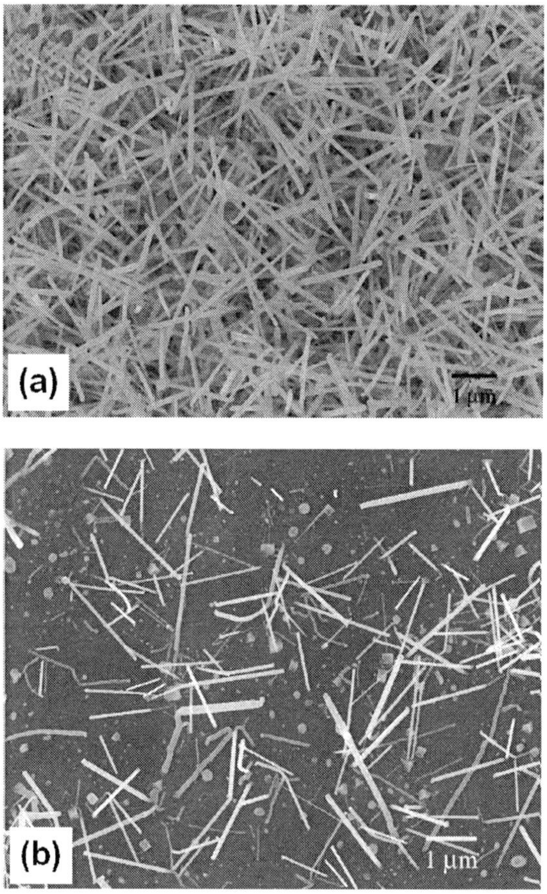

Figure 14. SEM images of catalytic-CVD grown InN nanorods produced at the (a) upstream and (b) downstream regions. Reprinted from [34] *J. Cryst. Growth,* 269, Lan, Z. H. et al., "Growth mechanism, structure and IR photoluminescence studies of indium nitride nanorods", 87, 2004, Copyright (2004), with permission from Elsevier.

It was observed that more dense nanorods with smaller diameters (30 – 50 nm) were formed at the upstream portion of the NH_3 flow and less dense nanorods with relatively larger diameters (50 – 100 nm) were formed at the downstream part. The InN nanorods ceased to form on the substrate when placed too far downstream. The SEM images for upstream and downstream regions are shown in Figure 14.

The authors also proposed a growth model to explain the synthesis process. Figure 15 illustrates the schematic diagram of this growth model. First, ultra-fine Au-In liquid alloy droplets are formed below 500 °C [Figure 15 (a)]. However, due to the presence of a minute amount of residual oxygen or the native oxide on Si, i.e. InO_x, which is energetically more favorable than InN to form, covers the surfaces of Au-In alloy droplets, resulting in core-shell Au@InO_x nanoclusters [Figure 15 (b)]. When the temperature is increased to 500 °C, it is speculated that N from the NH_3 source would then dissolve in the Au@InO_x nanoclusters [Figure 15 (c)] presumably with the formation of In-O-N as an intermediate state, due to the low miscibility of N and Au, before further reaction of N with In to form the InN nanorods [Figure 15 (d)].

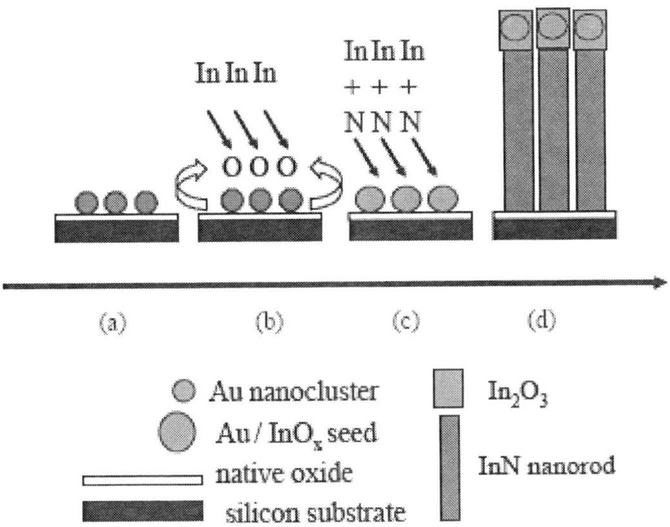

Figure 15. Schematic diagram of the growth mechanism for InN nanorods. (a) Au nanoclusters are formed at first; (b) Oxygen and In vapor dissolve in Au nanoclusters to form Au/InO_x nanoclusters in the early growth stage; (c) In vapor and N from the decomposition of NH_3 are absorbed in InO_x inducing the growth of InN nanorods; (d) The In_2O_3 encapsulated Au catalyst resides at the cap of InN nanorods after cooling. Reprinted from [34] *J. Cryst. Growth,* 269, Lan, Z. H. et al., "Growth mechanism, structure and IR photoluminescence studies of indium nitride nanorods", 87, 2004, Copyright (2004), with permission from Elsevier.

5.3. Reactive Gas Transport to Form InN Nanowires

Vaddiraju et al. [35] reported direct reaction of vapor phase In and N to synthesis InN nanowires. The synthesized InN nanowires exhibited a band gap of 0.8 eV, whereas the mixed phase of InN and In_2O_3 nanowires exhibited a peak at ~1.9 eV in addition to that at 0.8 eV. It was found that cleaning the native oxide layer on an as-received In source was a critical step in preventing the formation of any undesired indium oxide phase in the samples. In this study, the In source was cleaned by a 10 vol % nitric acid solution, followed by cleaning with distilled water and acetone to remove any native oxide layer present on the surface. The vapor transport experiments were performed using 140 sccm NH_3 at a total reactor pressure of 250 mTorr. The In source was heated to 650 – 850 °C while the corresponding substrate temperatures ranged from 400 - 480 °C. Figure 16 shows InN nanowires synthesized on quartz and polycrystalline AlN substrates at a heater temperature of 750 °C and substrate temperature of 450 °C.

Figure 16. SEM images of InN nanowires synthesized at a substrate temperature of 450 °C on (a) amorphous quartz substrate and (b) polycrystalline AlN substrates. The presence of crystals at the tip of the nanowires can be observed clearly in the micrograph. Reprinted with permission from [35]. Vaddiraju, S. et al. *Nano Lett.* 2005, 5, 1625. Copyright (2005) American Chemical Society.

The synthesized nanowires were straight, approximately 100 nm in diameter, and a few micrometers long. On the other hand, experiments performed at higher In vapor pressures by increasing the heater temperature to 850 °C (substrate temperature = 480 °C) resulted in multiple nucleation and growth of InN nanowires. Two growth mechanisms were proposed to explain different nanowire products at lower and higher In vapor, which are illustrated in Figure 17.

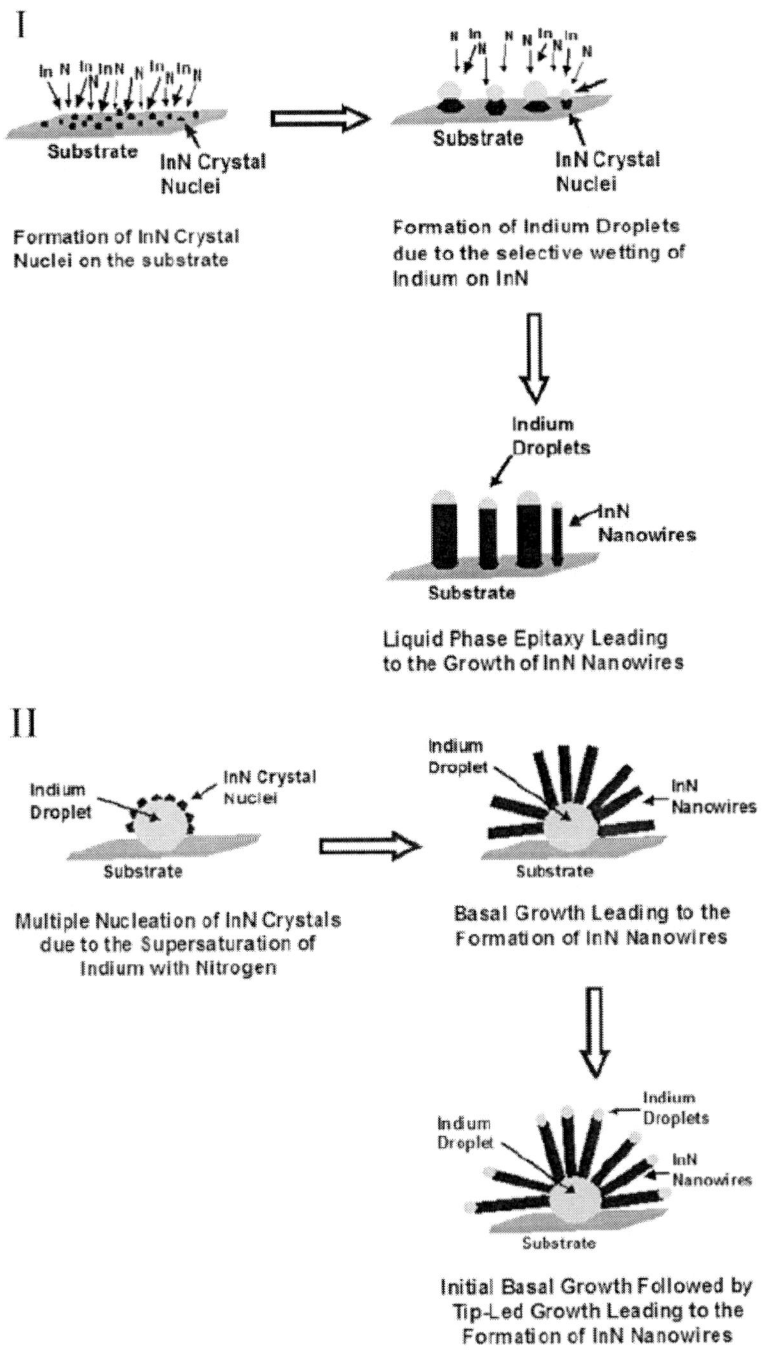

Figure 17. Schematic diagram of the growth mechanism proposed for the growth of InN nanowires synthesized by the vapor transport of In in the presence of NH_3. (I) The formation of InN crystal nuclei, followed by the formation of In droplets on the crystal nuclei. The nitridation of the droplets leads to the 1D growth and growth of InN nanowires by a tip-led growth route. (II) Multiple nucleation and growth of InN nanowires in a basal growth mode because of the supersaturation of In droplets with nitrogen dissolving from the gas phase. Reprinted with permission from [35]. Vaddiraju, S. et al. *Nano Lett*. 2005, 5, 1625. Copyright (2005) American Chemical Society.

For the lower In vapor under the lower heater temperature, the nucleation and growth of InN nanowires is inferred to occur as follows (see I in Figure 17): (a) formation of InN crystal nuclei because of the reaction between In vapor and N (from NH_3 decomposition at the heater) transported onto the substrate; (b) formation of In droplets on the surface of InN nuclei because of the selective wetting of InN by In molten metal in the presence of H_2 and NH_3 in the gas phase; and (c) selective growth into 1D nanowires through In droplets by liquid-phase epitaxy. For the higher In vapor under the higher heater temperature (see II in Figure 17), large In droplets are formed due to the high flux of In vapor during initial stages, without any InN crystals; Atomic nitrogen (from decomposed NH_3) gets dissolved into these large In droplets. Because the solubility of nitrogen in In is small at the temperature used (480 ^{o}C), the direct nitridation and supersaturation of In droplets with nitrogen led to the multiple nucleation of InN crystals on the surface of In.

III. Research Progress on III-Nitride Nanowire Devices

III-nitride nanowires were found have high unintentional n-type doping. The cause for such extremely high n-type doping is not clearly understood, but is speculated to be due to N-vacancies or oxygen related doping effects in the nanowires. The typical doping concentration of as-grown GaN nanowires is in the range from 10^{17} to 10^{19} /cm^3. Due to extremely small dimensions, surface scattering plays an important role in the carrier transport. The mobility of GaN nanowires, which is usually estimated using simple electrostatic models, ranges from a few tens cm^2/Vs to several hundreds cm^2/Vs. Considering the doping concentration of the nanowires, these values are at least comparable to the electron mobility of bulk GaN.

High n-type doping of nanowires makes it relatively easy to form ohmic contacts even without an additional annealing process that is generally required in thin film technology. On the other hand, it was also reported that both ohmic and Schottky contacts could be formed even with the same metallization. Such a behavior was due to the non-uniform surface barrier height of nanowires [36]. Most metal contact schemes for III-nitride nanowires employed Ti or Ni [37-40]. Hwang et al. [37] studied the effect of Ti thickness of Ti/Au contacts on GaN nanowires synthesized by a laser-assisted catalytic method. As the Ti thickness increased from 0 to 30 nm, the resistance reduced and became saturated for the thickness of > 20 nm. The specific contact resistivities reported in GaN nanowires were in a wide range from 10^{-5} Ω-cm^2 to 10^{-2} Ω-cm^2 [37], [40].

Since n-type and p-type incorporation is a key advantage of semiconductor nanowires, considerable research efforts have been made to realize homogeneous doping of III-nitride nanowires. Complementary crossed P-N junction diodes consisting of two individual p-type and n-type GaN nanowires exhibited the turn-on voltage of 3.5 V, which is consistent with the bandgap of bulk GaN [41]. Cheng et al. [42] demonstrated single GaN nanowires with an axial p-n junction. The first half way growth was carried out without dopant introduction, which led to an n-type GaN nanowire synthesis. In order to obtain the p-type growth of the other half, the synthesis was stopped and magnesium nitride was introduced into the furnace before resuming the growth process. The resulting nanowires exhibited rectifying characteristics of a p-n junction. The estimated doping concentrations in the p-type and n-type were 10^{17} - 10^{18} /cm^3 and 10^{17} – 10^{19} /cm^3, respectively. Qian et al. [30] synthesized radial p-n junction nanowires using radial growth heterostructures. The n-type GaN core nanowires were synthesized first, followed by sequential radial growth of InGaN and p-type GaN shells. Si and Mg were utilized for n-type and p-type dopants. Blue light emission was observed for the n-GaN/InGaN/p-GaN core/shell/shell nanowire heterostructures. Whereas the n-type GaN core nanowires showed a sharp peak at 367 nm consistent with band-edge emission from GaN, the heterostructure nanowire exhibited a dominant emission peak at 448 nm consistent with band-edge emission from InGaN with 18 % In composition. The electroluminescence (EL) spectrum showed strong blue light emission visible with the naked eye. The EL spectrum and image of the nanowire heterostructure device is shown in Figure 18.

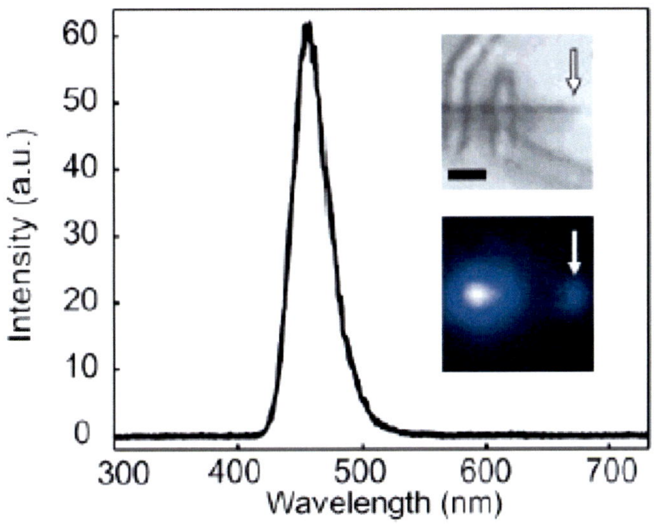

Figure 18. EL spectrum observed from as n-GaN/InGaN/p-GaN core/shell/shell nanowire device. The insets are bright field (upper) and EL (lower) images of the device. Two electrodes are seen in the upper inset image. The arrows point out the free-end of the nanowire. Strong emission is observed between two electrodes. Reprinted with permission from [30] Qian, F. et al., *Nano Lett.* 2004, 4, 1975. Copyright (2004) American Chemical Society.

Johnson et al. [43] observed UV blue lasing action from single GaN nanowires. The nanowires were synthesized using a Ni catalyst via direct reaction between Ga metal and NH_3 at 900 °C. Nanowires with a diameter of ~300nm and a length of ~ 40 nm were optically pumped by the fourth harmonic of a pulsed optical parametric amplifier. This excitation produced a plasma of electrons and holes in the nanowire, which then recombine across the bandgap to produce UV photons with wavelengths between 360 and 400 nm. Figure 19 shows lasing emission from a single GaN nanowire. The localized bright emission shown at the ends of the nanowire is the evidence of strong waveguiding behavior. The ends of the nanowire acted as mirrors to provide the feedback for stimulated emission. It was suggested that cavity modes were Fabry-Perot waveguide modes rather than whispering gallery modes. Emission spectra both below and above threshold are compared in Figure 19(b). A strong red shift in the emitted wavelength was observed as the excitation energy increases, which indicates that the electron-hole plasma mechanism is primarily responsible for the gain at room temperature.

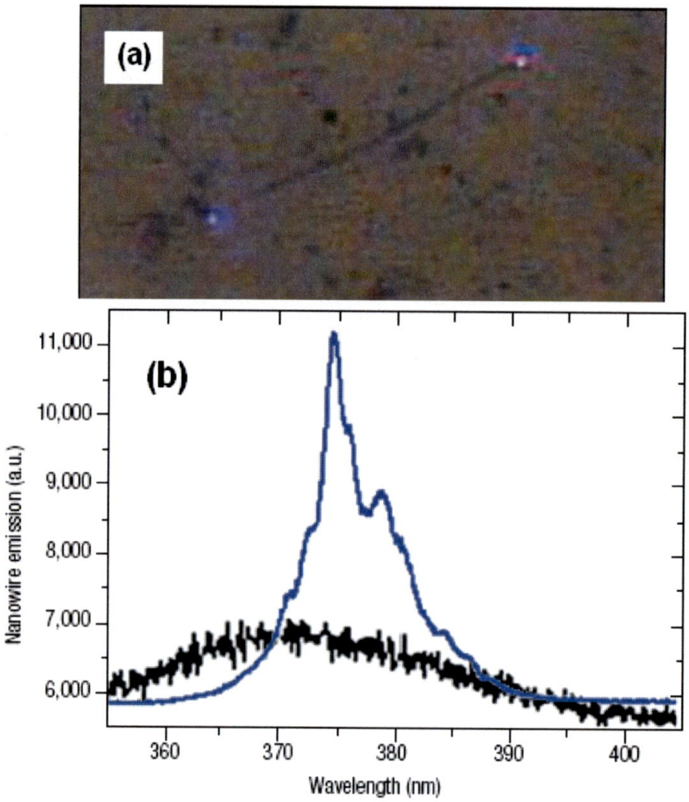

Figure 19. (a) Image of lasing emission from an optically pumped single GaN nanowire. (b) Spectrum of photoluminescence (dark) using 1 mW CW excitation and lasing (light) using ~ 1μJcm^{-2} pulsed excitation. The threshold was about 700 nJcm^{-2} in this case. Reprinted by permission from Macmillan Publishers Ltd: Nat. Mater., [43] Johnson, J. C. et al., 2002, 1, 106, copyright (2002).

Gradečak et al. [44] reported a low lasing threshold of 22 kW/cm^2 using MOCVD grown n-type GaN nanowires which were intentionally doped and grown in a nonpolar <11-20> direction. The MOCVD grown nanowires had triangular cross-sections with diameters of 100 – 300 nm. Power dependent PL images and spectra are shown in Figure 20. Whereas spontaneous emission is observed from both the body and ends of the nanowire at low excitation power densities (see Figure 20(a) and (b)), the emission intensity at nanowire ends becomes dominant with increasing the power level (see Figure 20(c)), indicating strong waveguiding properties. A redshift was observed in emission spectra from the nanowire ends at higher excitation densities (see Figure 20(d)), suggesting Fabry-Perot cavity modes. It was suggested that nonpolar growth direction, Si doping, and morphological uniformity contributed to the low threshold power density for stimulated emission.

Greytak et al. [45] demonstrated electric-field modulation of lasing in GaN nanowire lasers using parallel-plate electrodes that were placed both at the top and bottom of the nanowire. The electrodes were separated from the nanowire by SiO_2 layers. The bias dependent emission spectrum above and below lasing threshold is shown in Figure 21 (a) and modulation versus applied voltage below threshold is plotted in Figure 21 (b). The laser spectrum exhibited a dominant peak at 373 nm and the peak position did not change with applied voltages.

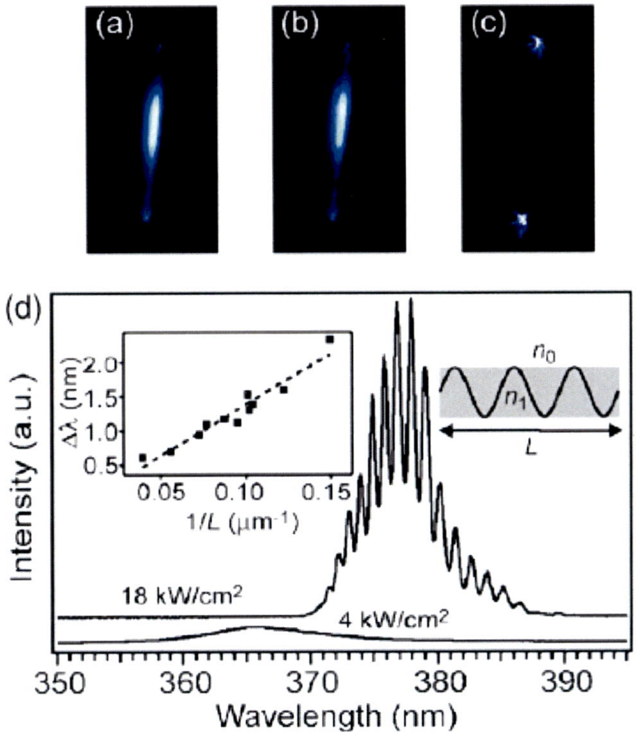

Figure 20. Power-dependent PL studies of an MOCVD grown n-type GaN nanowire. Images of a single nanowire excited with (a) 4 kW/cm^2, (b) 17 kW/cm^2, and (c) 66 kW/cm^2. (d) PL spectra recorded at the end of the nanowire for 4 kW/cm^2 and 18 kW/cm^2 power densities. Right inset, schematic representation of a longitudinal nanowire cavity with length L whose index of refraction, n_1, is larger than that of surrounding air, n_0. Left inset, mode spacing vs inverse nanowire length. Reprinted with permission from [44] Gradečak, S. H. et al., *Appl. Phys. Lett.* 2005, 87, 173111. Copyright 2005, American Institute of Physics.

The emission intensity below the lasing threshold was linearly proportional to the voltage applied over the ±45 V range and modulation of at least 20 % was achieved above and below the threshold. Further study suggested that electric field induced absorption caused the modulation.

GaN nanowire based field-effect transistors (FETs) have been reported from many different labs, mostly based on bottom-gate configurations that used a conductive substrate as the gate electrode. We have demonstrated both bottom-gate and top-gate GaN nanowire FETs using a "pre-aligned" process [39]. The device schematic diagrams of the bottom-gate and top-gate devices are shown in Figure 22 (a) and (b), respectively. The GaN nanowires were synthesized by direct reaction of metallic gallium and NH_3 at 900 °C. The typical diameter of the as-grown nanowires ranged from 20 nm to 40 nm. Both the bottom-gate and top-gate FETs exhibited good ohmic behavior of the source and drain contacts and gate modulation with pinch-off. The current-voltage characteristics of both devices are compared in Figure 23. The larger effective area of the nanowire surface covered by the top-gate structure contributed to the improved modulation with a low pinch-off voltage for the top-gate device. Under the pinch-off conditions, the top-gate devices had higher breakdown voltages ($BV_{ds} > 60$ V at $V_{gs} = -15$ V) than those of the bottom-gate devices ($BV_{ds} < 15$ V at $V_{gs} = -20$ V).

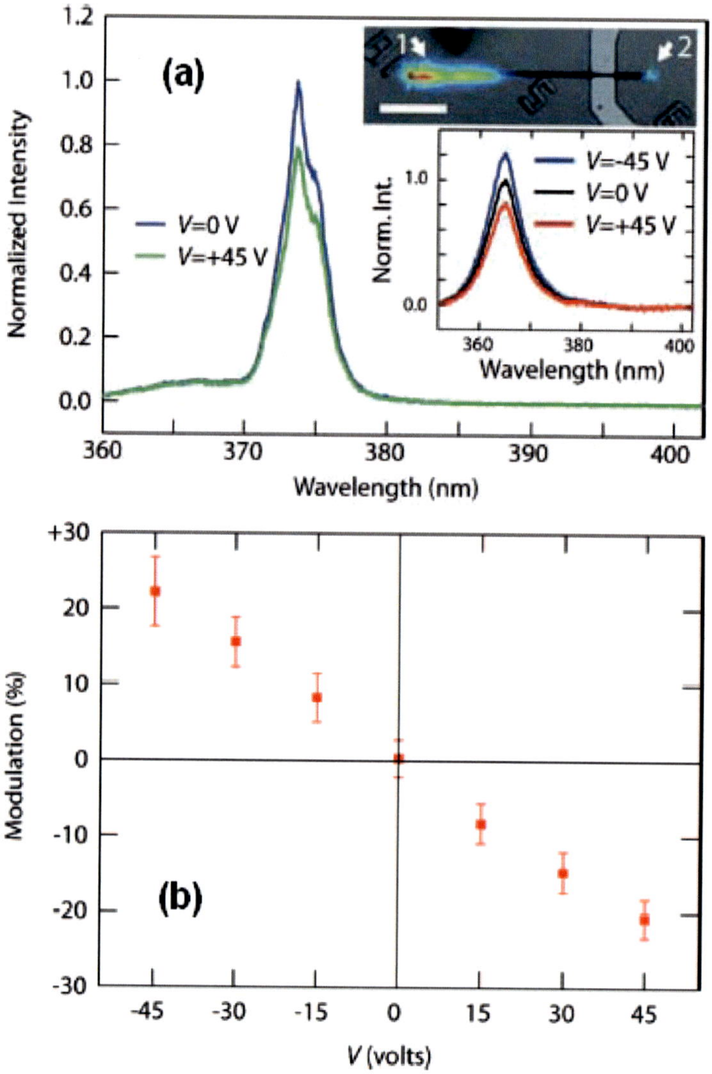

Figure 21. (a) Emission spectrum above lasing threshold with and without bias applied. Upper inset, superimposed PL and white-light image of device. Lower inset, bias dependent end emission spectra below threshold. (b) Modulation vs bias below threshold. Reprinted with permission from [45] Greytak, A. B. et al., *Appl. Phys. Lett.* 2005, 87, 151103. Copyright 2005, American Institute of Physics.

In the bottom-gate device, a high electric field is applied across the oxide between the bottom-gate and the drain, which causes oxide breakdown to occur prior to nanowire breakdown. Due to this early failure, it is unlikely to see current saturation characteristics for the bottom-gate devices especially with negative gate biases. On the other hand, clear current saturation characteristics were observed for the top-gate devices due to the higher breakdown voltage. The top-gate device operation was not limited by the oxide breakdown between the gate and the drain but rather governed by an inherent tunneling mechanism in the metal-oxide-semiconductor (MOS) gate region.

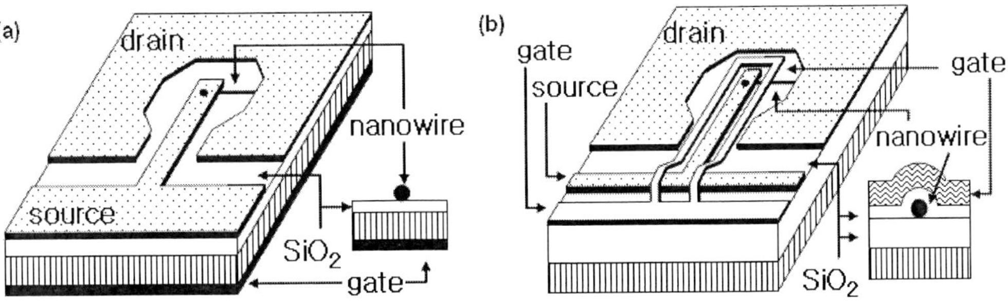

Figure 22. Schematic diagrams of (a) bottom-gate GaN nanowire FET and (b) top-gate GaN nanowire FET. Cross-sectional views across the nanowire channel are shown together. Reprinted with permission from [39] Cha, H.-Y. et al., "Fabrication and characterization of pre-aligned gallium nitride nanowire field-effect transistors", *Nanotechnology* 2006, 17, 1264. Copyright (2006) Institute of Physics.

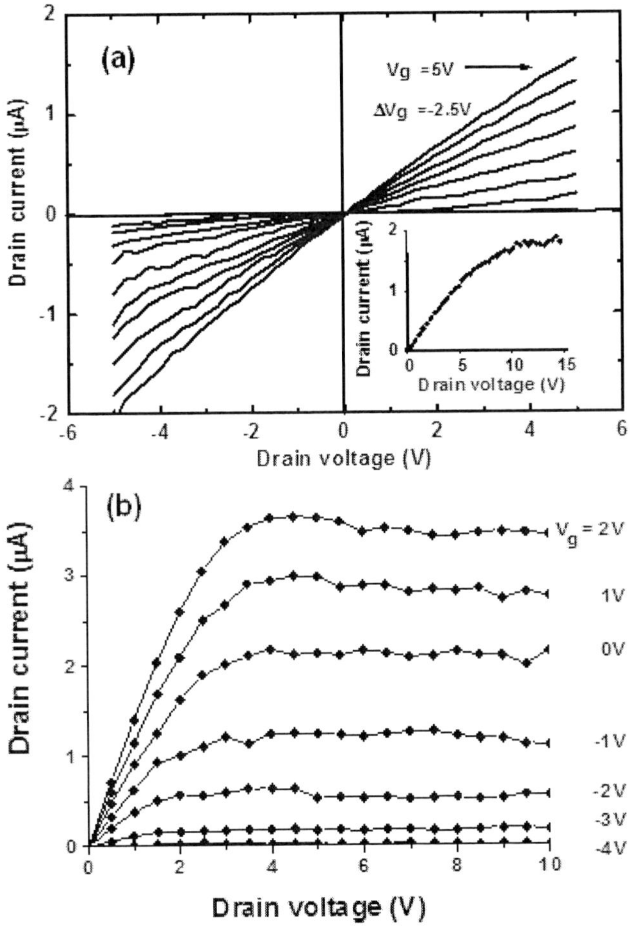

Figure 23. Current-voltage characteristics of (a) bottom-gate GaN nanowire FET and (b) top-gate GaN nanowire FET. The inset in (a) is the current-voltage characteristics without the gate bias. Reprinted with permission from [39] Cha, H.-Y. et al., "Fabrication and characterization of pre-aligned gallium nitride nanowire field-effect transistors", *Nanotechnology* 2006, 17, 1264. Copyright (2006) Institute of Physics.

Because of the small dimensionality of nanowires, channel depletion due to surface potential has to be taken into account to design and characterize nanowire devices. A common electrostatic model used in GaN nanowire FETs is the model developed for carbon nanowire transistors, which assumed a metallic cylindrical structure. With respect to carbon nanowires, GaN nanowires have much larger diameters and relatively lower carrier concentrations and thus it is questionable to adopt the same model. We have developed an electrostatic model taking into account of potential drop in the nanowire channel and investigated the effects of surface depletion [39]. Figure 24 shows the minimum required carrier concentration of nanowires versus the nanowire diameter to avoid full channel depletion. The surface barrier height of 0.7 V was used, which resulted in the best fit to our experimental data [39]. If the nanowires are within the dark regime, the whole nanowire will be depleted by surface barrier without an applied gate bias and result in no current flow with zero gate bias. For example, a nanowire with a diameter of 20 nm needs at least 2 × 19 /cm^3 in order to avoid full depletion with zero gate bias. However, appropriate surface treatments, e.g. surface passivation, can lower the surface potential at the nanowire surface and enhance the device current.

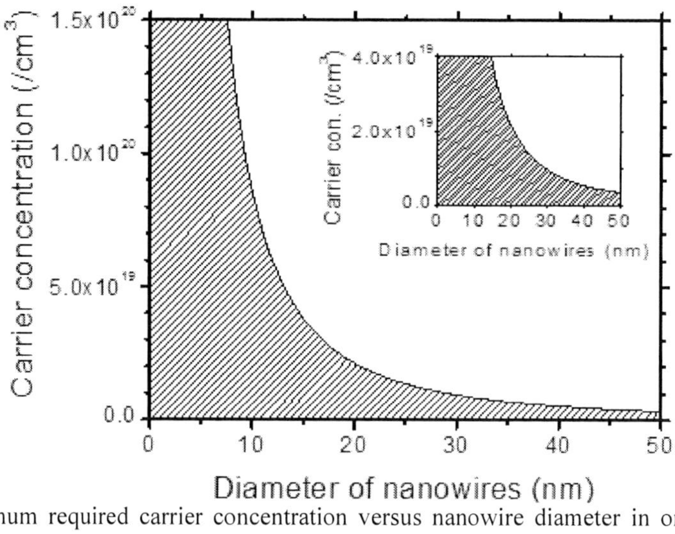

Figure 24. Minimum required carrier concentration versus nanowire diameter in order to avoid full channel depletion. Reprinted with permission from [39] Cha, H.-Y. et al., "Fabrication and characterization of pre-aligned gallium nitride nanowire field-effect transistors", *Nanotechnology* 2006, 17, 1264. Copyright (2006) Institute of Physics.

Using the top-gate configuration, we demonstrated non-volatile GaN nanowire devices [46]. The physical mechanism of the write and erase modes are governed by the locally enhanced electric field at the edge of the top-gate. The localized high electric field between the gate edge and the oxide induces Fowler-Nordheim tunneling of electrons for both write and erase modes while the relatively lower electric field under the gate center region and at the other side interface between the oxide and the nanowire does not contribute to the charge transport. The energy band profiles at the gate center and edge for the write and erase modes are shown in Figure 25 (a) and (b), respectively. Whereas the electric field under the gate center region is lower than 3 MV/cm with a gate bias of +10/-10 V, the localized enhanced electric field at the gate edge between the gate metal and the oxide is several MV/cm, which can cause the energy band profile to bend and generates an electron tunneling process. In the write mode, charges injected from the gate edge into the oxide continue to flow toward the nanowire channel due to ohmic conduction mechanisms until they loose energy and/or are captured in traps in the low field regime, depleting the nanowire channel. In the erase mode, the locally enhanced electric field at the gate edge allows the trapped charges to return back to the gate but no channel carriers are injected to the oxide due to the lower electric field between the nanowire and the oxide. This prototype device exhibited the memory effect lasting for several hours. The bias dependent memory effects and retention characteristics are shown in Figure 26 (a) and (b), respectively.

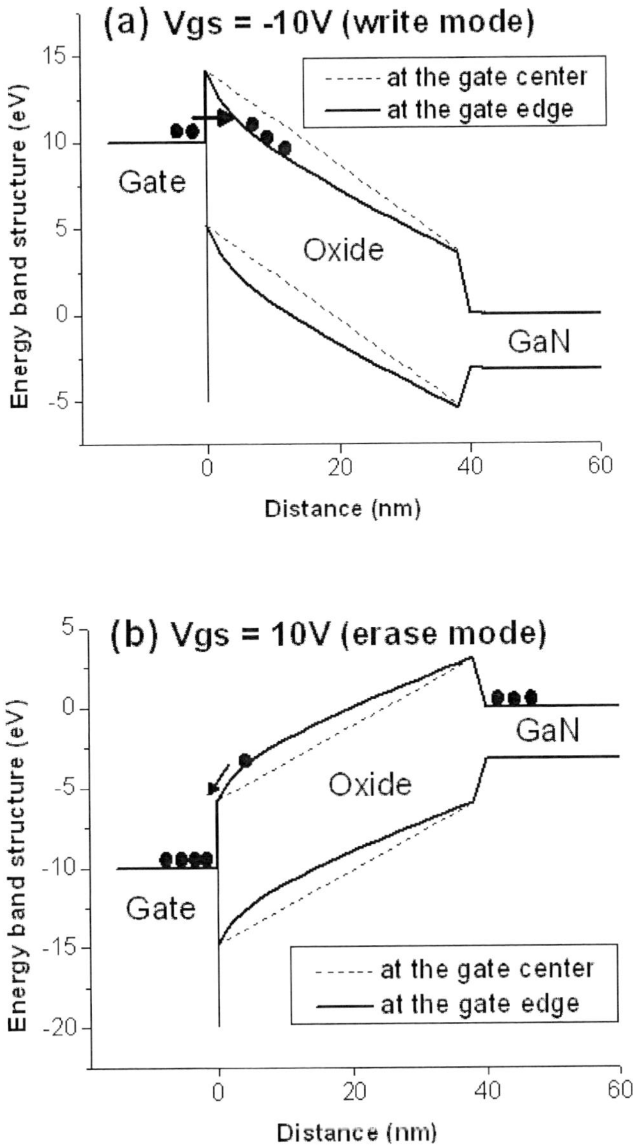

Figure 25. Energy band profiles for (a) write and (b) erase modes. The localized high electric field causes the energy band bending at the gate edge.

Li et al [31] recently demonstrated a GaN/AlN/AlGaN nanowire HEMT incorporating a 6 nm thick high-k ZrO_2 dielectric layer as a gate oxide. The ZrO_2 layer was formed by 50 cycles deposition at 110 °C using tetrakis(dimethylamino)-zirconium as the precursor. A very high mobility was obtained resulting in high transconductance and current density. The radial heterostructure had a triangular cross-section with a diameter of 100 nm. Figure 27 shows the image of cross-sectional high-angle annular darkfield scanning transmission electron microscopy.

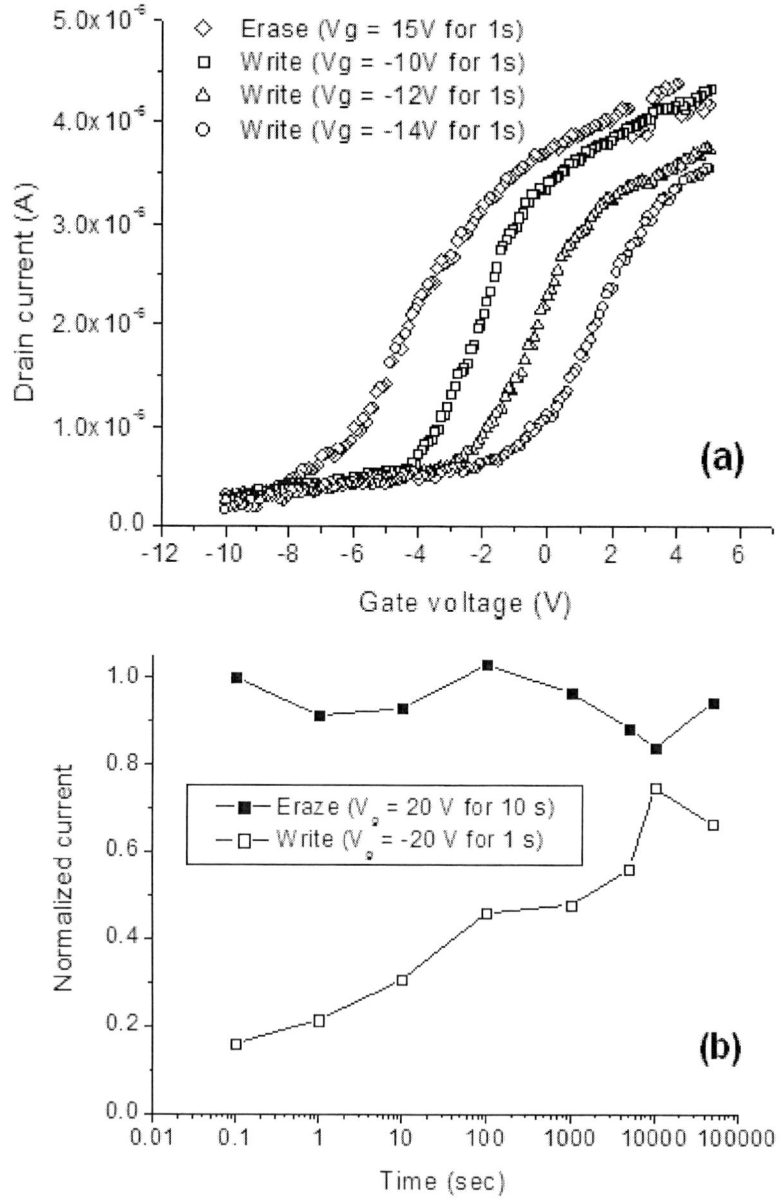

Figure 26. (a) Bias dependent memory effects and (b) retention characteristics of a GaN nonvolatile memory device. Reprinted with permission from [46] Cha, H.-Y. et al., *J. Appl. Phys.* 2006, 100, 024307. Copyright 2006, American Institute of Physics.

The mobility estimated from transconductance characteristics exhibited 2300 cm^2/Vs at room temperature, which is significantly higher than those reported in n-type GaN nanowires and comparable to those in thin film AlGaN/GaN heterostructures. The estimated extrinsic mobility and extracted intrinsic mobility of the radial heterostructure at different temperatures are shown in Figure 28.

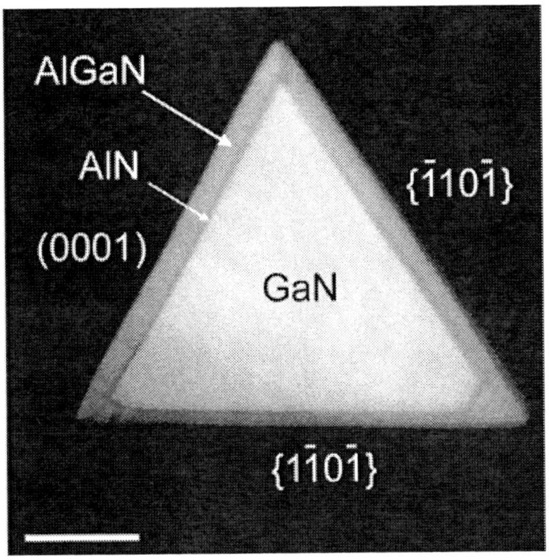

Figure 27. High angle annular dark field scanning transmission electron microscopy image of a GaN/AlN/AlGaN nanowire cross-section. Scale bar is 50 nm. Reprinted with permission from [31] Li, Y. et al., *Nano Lett.* 2006, 6, 1468. Copyright (2006) American Chemical Society.

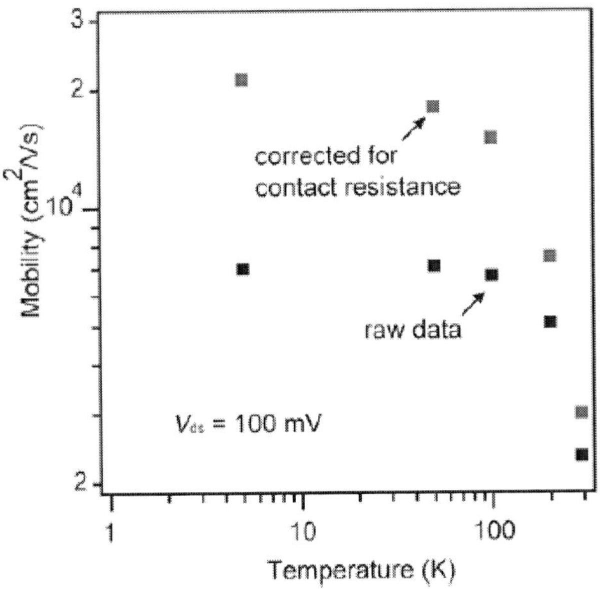

Figure 28. Measured (dark symbols) and intrinsic (bright symbols) electron mobility of GaN/AlN/$Al_{0.25}Ga_{0.75}N$ nanowire at different temperatures, where the intrinsic values were obtained after the correction for contact resistance. Reprinted with permission from [31] Li, Y. et al., *Nano Lett.* 2006, 6, 1468. Copyright (2006) American Chemical Society.

As the temperature decreases from room temperature, the mobility is increased sharply and then saturated because of reduced phonon scattering. The scaled transconductance and current density of this nanowire HEMT was 420 mS/mm and 500 mA/mm, respectively, which are comparable to those measured for thin film AlGaN/GaN HEMTs.

IV. Conclusion

III-nitride nanowires are an attractive material system for future nanoscale systems. N-type and p-type doped nanowires as well as nanowire heterostructures have been successfully synthesized. Many prototype nanowire devices have been fabricated and characterized to explore the full potential of III-nitrides. Much higher wafer costs associated with relatively immature growth techniques compared to other conventional semiconductors, especially Si, is a major drawback limiting commercialization of the current III-nitride thin film technology. However, III-nitride nanowires can be grown on foreign substrates and their research has been developed very actively parallel to other nanotechnologies, suggesting that the drawback mentioned above will not be a problem in III-nitride nanowires. However, like other nanowires and nanotubes, III-nitride nanowires also face common challenges including alignment, assembly, and control of doping and dimension. Since the nanowire dimension is extremely small, any change in dimension or surface defects will have significant influence on carrier transport and surface band bending. Therefore, uniform growth technique and appropriate surface passivation will also be crucial for device development. Once these issues are resolved, III-nitride nanowires will have a revolutionary impact on future photonics and electronics.

References

[1] Nakamura, S.; Mukai, T.; Senoh, M. *Jpn. J. Appl. Phys.* 1991, 30, L1998-L2001.

[2] Nakamura, S.; Senoh, M.; Iwasa, N.; Nagahama, S.-I. *Jpn. J. Appl. Phys.* 1995, 34, L797-799.

[3] Khan, M. A.; Bhattarai, A.; Nuznia, J. N.; Olson, D. T. *Appl. Phys. Lett.* 1993, 63, 1214-1215.

[4] Wu, Y.-F.; Saxler, A.; Moore, M.; Smith, R. P.; Sheppard, S.; Chavarkar, P. M.; Wisleder, T,; Mishra, U. K.; Parikh, P. *IEEE Electron Dev. Lett.* 2004, 25, 117-119.

[5] Boutros, K.; Regan, M.; Rowell, P.; Gotthold, D.; Birkhahn, R.; Brar, B. *IEEE IEDM Tech. Digest* 2003, 12.5.1-12.5.2.

[6] http://www.ioffe.ru/SVA/NSM/Semicond/

[7] Hudgins, J. L.; Simin, G. S.; Khan, M. A. *Proc. IEEE Power Electronics Specialists* 2002, 1747-1752.

[8] Davis, R. F.; Roskowski, A. M.; Preble, E. A.; Speck, J. S.; Heying, B.; Freitas, J. A. Jr.; Glaser, E. R.; Carlos, W. E.; *Proc. IEEE* 2002, 90, 993-1004.

[9] Han, W.; Fan, S.; Li, Q.; Hu, Y. *Science* 1997, 277, 1287-1289.

[10] Chen, L. C.; Chen, K. H.; Chen, C. C. *Nanowires and Nanobelts – Materials, Properties and Devices* Vol. 1; Kluwer Academic Publisher, 2003, 257.

[11] Cheng, G. S.; Zhang, L. D.; Zhu, Y.; Fei, G. T.; Li, L.; Mo, C. M.; Mao, Y. Q. *Appl. Phys. Lett.* 1999, 75, 2455-2457.

[12] Hersee, S. D.; Sun, X.; Wang, X. *Nano Lett.* 2006, 6, 1808-1811.

[13] Dai, H.; Wong, E. W.; Lu, Y. Z.; Fan, S.; Lieber, C. *Nature* 1995, 375, 769-772.

[14] Frosch, F. J.; Thurmond, C. D. *J. Phys. Chem.* 1962, 66, 877-878.

[15] Zhang, Y.; Liu, J.; He, R.; Zhang, Q.; Zhang, X; Zhu, J.; Chem. Mater. 2001, 13, 3899-3905.

[16] Nishizawa, M.; Menon, V. P.; Martin, C. R. *Science* 1995, 268, 700-702.

[17] Jirage, K. B.; Hulteen, J. C.; Martin, C. R. *Science* 1997, 278. 655-658.

[18] Lee, S. B.; Mitchell, D. T.; Trofin, L.; Nevanen, T. K.: Soderlund, H.; Martin, C. R. *Science* 2002, 296, 2198-2200.

[19] Cheng, G. S.; Chen, S. H.; Zhu, X. G.; Mao, Y. Q.; Zhang, L. D. *Mater. Sci. and Eng. A* 2000, 286, 165-168

[20] Goldberger, J.; He, R.; Zhang, Y.; Lee, S.; Yan, H.; Choi, H. J.; Yang, P. *Nature* 2003, 422, 599-602.

[21] Wu, H.; Cha, H. Y.; Chandrashekhar, M.; Spencer, M. G.; Koley, G *J. Electronic Mater.* 2006, 35, 670-674.

[22] Liang, C. H.; Chen, L. C.; Hwang, J. S.; Chen, K. H.; Hung, Y. T.; Chen, Y. F. *Appl. Phys. Lett.* 2002, 81, 22-24.

[23] Duan, X.; Lieber, C. M. *J. Am. Chem. Soc.* 2000, 122, 188-189.

[24] He, M.; Zhou, P.; Mohammad, S. N.; Harris, G. L.; Halpern, J. B.: Jacobs, R.; Sarney, W. L.; Salamanca-Riba, L. *J. Cryst. Grow.* 2001, 231, 357-365.

[25] Calleja, E.; Sanchez-Garcia, M. A.; Sanchez, F. J.; Calle, F.; Naranjo, F. B.; Munoz, E.; Jahn, U.; Ploog, K. *Phys. Rev. B* 2000, 62, 16826-16834.

[26] Calarco, R.; Marso, M.; Meijers, R.; Richter, T.; Aykanat, A. I.; Stoica, T.; Luth, H. *Proceedings of the 5th International Conference on Advanced Semiconductor Devices and Microsystems*, Smolenice Castle, Slovakia, 2004, 9-12.

[27] Bertness, K. A.; Roshko, A.; Sanford, N. A.; Barker, J. M.; Davydov, A. V. *J. Cryst. Grow.* 2006, 287, 522-527.

[28] Kuykendall, T.; Pauzauskie, P.; Lee, S. K.; Zhang, Y.; Goldberger, J.: Yang, P.; *Nano Lett.* 2003, 3, 1063-1066.

[29] Choi, H. J.; Johnson, J. C.; He, R. R.; Lee, S. K.; Kim, F.; Pauzauskie, P.; Goldberger, J.; Saykally, R. J.; Yang, P. D. *J. Phys. Chem. B* 2003, 107, 8721-8725.

[30] Qian, F.; Li, Y.; Gradečak, S.; Wang, D.; Barrelet, C. J.; Lieber, C. M. *Nano Lett.* 2004, 4, 1975-1979.

[31] Li, Y.; Xiang, J.; Qian, F.; Gradečak, S.; Wu, Y.; Yan, H.; Blom, D. A.; Lieber, C. M.; *Nano Lett.* 2006, 6, 1468-1473.

[32] Strite, S.; Morkoc, H. *J. Vac. Sci. Technol. B* 1992, 10, 1237-1266.

[33] Zhang, J.; Jiang, F.; Yang, Y.; Xu, B.; Li, J.; Wang, X.; Wang, S. *Mater. Lett.* 2006, 60, 2153-2157.

[34] Lan, Z. H.; Wang, W. M.; Sun, C. L.; Shi, S. C.; Hsu, C. W.; Chen, T. T.; Chen, K. H.; Chen, C. C.; Chen, Y. F.; Chen, L. C.; *J. Cryst. Growth* 2004, 269, 87-94.

[35] Vaddiraju, S.; Mohite, A.; Chin, A.; Meyyappah, M.; Sumanasekera, G.; Alphenaar, B. W.; Sunkara, M. K. *Nano Lett.* 2005, 5, 1625-1631.

[36] Koley, G.; Lakshmanan, L.; Wu, H.; Cha, H.-Y. *Phys. Stat. Sol. (a)* 2007, 204, 1123-1129.

[37] Hwang, J. S.; Ahn, D.; Hong, S. H.; Kim, H. K.; Hwang, S. W.; Jeon, B.-H.; Choi, J.-H.; *Appl. Phys. Lett.* 2004, 85, 1636-1638.

[38] Stern, E.; Cheng, G.; Young, M. P.; Reed, M. A.; *Appl. Phys. Lett.* 2006, 88, 053106.

[39] Cha, H.-Y.; Wu, H.; Chandrashekhar, M.; Choi, Y. C.; Chae, S.; Koley, G.; Spencer, M. G.; *Nanotechnology* 2006, 17, 1264-1271.

[40] Ham, M.-H.; Choi, J.-H.; Hwang, W.; Park, C.; Lee, W.-Y.; Myoung, J.-M.; *Nanotechnology* 2006, 17, 2203-2206.

[41] Zhong, Z.; Qian, F.; Wang, D.; Lieber, C. M.; *Nano Lett.* 2003, 3, 343-346.

[42] Cheng, G.; Kolmakov, A.; Zhang, Y.; Moskovits, M.; Munden, R.; Reed, M. A.; Wang, G.; Mosses, D.; Zhang, J.; *Appl. Phys. Lett.* 2003, 83, 1578-1580.

[43] Johnson, J. C.; Choi, H. J.; Knutsen, K. P.; Schaller, R. D.; Yang, P. D.; Saykally, R. J.; *Nat. Mater*. 2002, 1, 106-110.

[44] Gradečak, S.; Qian, F.; Li, Y.; Park, H.-G.; Lieber, C. M.; *Appl. Phys. Lett.* 2005, 87, 173111.

[45] Greytak, A. B.; Barrelet, C. J.; Li, Y.; Lieber, C. M.; *Appl. Phys. Lett*. 2005, 87, 151103.

[46] Cha, H.-Y.; Wu, H.; Chae, S.; Spencer, M. G.; *J. Appl. Phys.* 2006, 100, 024307.

In: Encyclopedia of Nanotechnology
Editor: Elwood D. Carlson

ISBN: 978-1-60692-079-4

Chapter 13

NANOSTRUCTURES GROWN VIA ELECTROCHEMICAL TEMPLATE METHODS: SYNTHESIS, MEASUREMENTS AND APPLICATIONS

Stefan Mátéfi-Tempfli
Université Catholique de Louvain,
Physical-Chemistry and Physics of Materials,
Place Croix du Sud 1, bat. Boltzmann A1,
B-1348 Louvain-la-Neuve, Belgium

ABSTRACT

Nanoscaled materials often exhibit very different properties from their macroscopic and even micrometer sized counterparts. This occurs when the size of an object is reduced down to the atomic level such that one or more of its dimensions become smaller than some characteristic lengths, i.e. mean free paths, coherence lengths and crystallite sizes. In this case new properties show up where no more scattering on defects and statistical averaging of various fundamental phenomena are dominant. Below these characteristic lengths quantizations of different parameters appear. The possibility of tuning the properties of nanomaterials by changing their sizes is very interesting on both fundamental and applied aspects. Nanostructures are even more attractive because there is virtually an infinity of possibilities to combine different nanomaterials together, thus creating new and smart nano devices and structures.

One of the great challenges of today is to find reliable techniques for the fabrication of such nanomaterials and nanostructures. Methods based on template synthesis and on self organization are the most promising due to the easiness and low cost of fabrication. This chapter will focus on various electrochemical synthesis methods of nanowires and nanostructured nanowires using nanoporous host materials as templates. A short overview of the templates and their fabrication methods will be presented. Among them, the nanoporous aluminum oxide shows a self-ordering of the nanopores in a honeycomb structure. It stands high temperatures and is insoluble in organic solvents. A recent development of a supported nanoporous aluminum oxide on silicon will be described. It will be shown that this new approach solves the handling problems associated with the brittle nanoporous aluminum oxide thin layers and allows its integration in silicon processes for microelectronic manufacturing.

Combining electrochemical template synthesis with nanolithography and patterning techniques opens up new ways for the fabrication of nanowires-based complex structures and devices. Two approaches will be presented. The first is aiming at contacting and measuring a large amount of single nanowires and compositionally modulated nanowires while the second is showing a localized growth of bundles or single nanowires for contacting them in a high density matrix arrangement.

Several results of measurements and applications will illustrate the presented methods and their use. Finally, concluding remarks and prospects for the electrochemical template synthesis will close the chapter.

1. INTRODUCTION

Homogeneous nanowires could be realized by patterning thin films using advanced nanolithographic techniques such as electron and ion beam lithography and nanoimprint methods. Despite of the flexibility offered by these techniques in defining various shapes and geometries, they do not permit the realization of nanowires with compositional modulation along their axis. They are also time-consuming and expensive methods.

Electrochemical reduction of various materials directly inside the nanopores of various host materials gives, due to their confinement, nanowires with geometries defined by the nanopores dimensions. Such host materials include nanoporous anodic aluminum oxide (AAO) films [1-5], nuclear track-etched mica [6-8] and polymer membranes [9-11], diblock copolymers [12], nanochannel array glasses [13] and SiO_2 nanocapillary arrays [14]. Using these nanoporous materials as templates for the electrochemical synthesis of nanowires allows the fabrication of a large number of nanowires in parallel (see Figure 1) and, by varying the electrodeposition conditions (i.e. voltage, current, electrolyte), the realization of multilayered nanowires along their growth axis.

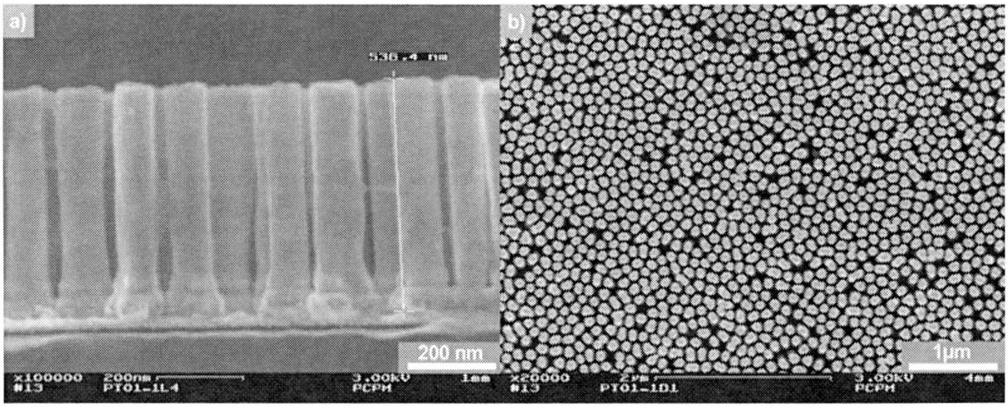

Figure 1. SEM micrographs showing Platinum nanowires electrodeposited inside the nanopores of a supported (on Au/Nb/Si substrate) aluminum oxide template after the removal of the alumina host. Sectional view (a) and top view (b).

An essential requirement for producing multilayered structures with the individual layers thicknesses in the nanometer range is the use of an atomistic deposition process. Several techniques, such as molecular beam epitaxy, sputter deposition and e-beam evaporation satisfy quite well this requirement. However, these thin film depositions techniques have some drawbacks, such as high cost, low deposition rates and very poor compatibility with template synthesis methods. Alternative atomistic deposition and cost saving electrochemical deposition techniques were developed for producing thin multilayers, namely the single bath technique (SBT) [15,16] and the dual-bath technique (DBT) [17-21]. Actually the DBT is a particular case of a multiple-bath technique (MBT) widely used in industrial processes for deposition of thick (i.e. micrometer range) layers.

2. Single Bath Technique

The SBT makes possible to fabricate periodic stacking of metal and alloy layers. For this technique, the electrochemical bath contains ions of two or more metals separable in two groups, more noble (more electro positive) and less noble metal ions, see Figure 2.

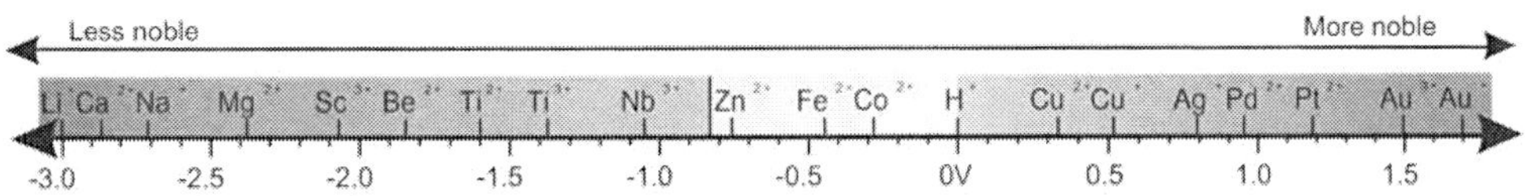

Figure 2. Reduction Potentials relative to Standard Hydrogen Electrode (SHE)

Alternate deposition in a controlled way of these metals is usually performed by varying the deposition potential or current [15,16]. The concentration of the nobler ions in the bath is usually much lower than the concentration of the less noble ions in order to reduce undesirable codeposition effects. We should note also that electrochemical reduction of ionic species at the working electrode (cathode) is taking place when the electrode potential is more negative than its standard reduction potential. As an example, let us consider the typical case of Co^{2+} and Cu^{2+} containing bath used frequently for the realization of Co/Cu multilayers which were found important in studying electron spin related transport phenomena [22]. For the considered example the reduction reactions at the working electrode are:

$$Co^{2+} + 2e^{-} = Co \qquad \text{for } U_{we} < -0.28 \text{ V (vs. SHE, 25°C, 1M } Co^{2+}) \qquad (1)$$

$$Cu^{2+} + 2e^{-} = Cu \qquad \text{for } U_{we} < +0.3419 \text{ V (vs. SHE, 25°C, 1M } Cu^{2+}) \qquad (2)$$

It is evident that applying for example -0.1 V/SHE, only Cu reduces, while switching, for example to -0.7 V/SHE, both Co and Cu will be deposited. Choosing sufficiently large differences in concentrations, i.e. 1M Co^{2+} : 2mM Cu^{2+} and alternatively switching between the two voltages mentioned above, we can realize multilayers of Cu/Co(<1% Cu)/ Cu/Co(<1% Cu)/.

The SBT can be used to deposit up to thousands of bilayers of various metals and alloys with individual thicknesses in the nanometer range. It is highly compatible with the template synthesis method, allowing the realization of multilayered nanowires. Unfortunately, SBT has also several limitations such as: the impossibility to deposit two different metals when their electrochemical reduction potentials are close to each other; the less noble deposited metal always contains a significant amount of nobler metal due to codeposition; the periodic stacking of more than two different materials is very problematic; the difficulty to find a unique electrolyte for each case, especially when the reduction of one of the ionic species is prohibited from aqueous solutions, i.e. its reduction potential is more negative than -0.8277 V/SHE (H2O decomposition).

3. Multiple Bath Technique

In contrast, the MBT does not suffer from these limitations. It permits to stack thin layers from various materials: metals, semimetals, alloys, semiconductors and oxides. This technique requires the use of a different bath for each type of deposited layer. The drawback of much longer processing time, due to cleaning and transfer between baths after each deposited layer, limits in fact the maximal number of the deposited layers to a few hundreds. Furthermore, the undesirable surface reactions appearing during the cleaning and transfer of the substrate between the different baths are unacceptable for nanometer-range thin layer deposition. Due to these problems, the MBT was not used for ultra-thin multilayers deposition. Some attempts using DBT was made using both manual [20,21] and automated [17,18] transfer of the substrate between different baths. However, the main problems of undesirable surface reactions and cross contamination between different baths remained unsolved.

We have developed a method involving a multiple-bath electrochemical system for the electrodeposition of a large variety of thin layers on a substrate [23], which was successful in solving the major problems associated with the MBT for nanometer-scale depositions. A block diagram of the modular multiple-bath electrodeposition system is shown in Figure 3. The main parts of the system are an electrochemical subsystem where electrodeposition of thin multilayers takes place and transfer subsystems 1, 2, 3... The number of transfer subsystems is not limited, i.e. the system is modular, and corresponds to the maximal number of different materials to be deposited. After each deposition step, the electrochemical subsystem is cleaned and rinsed using a cleaning subsystem.

In our method, electrodeposition takes place inside a closed electrochemical cell protected by inert gas, while the substrate remains on the same place attached to a working electrode during all the deposition steps. In that case, the substrate is not transferred from one bath to another. Different solutions are instead transferred from different chemical tanks to the electrochemical cell using separate transfer subsystems for each tank as depicted in Figure 4.

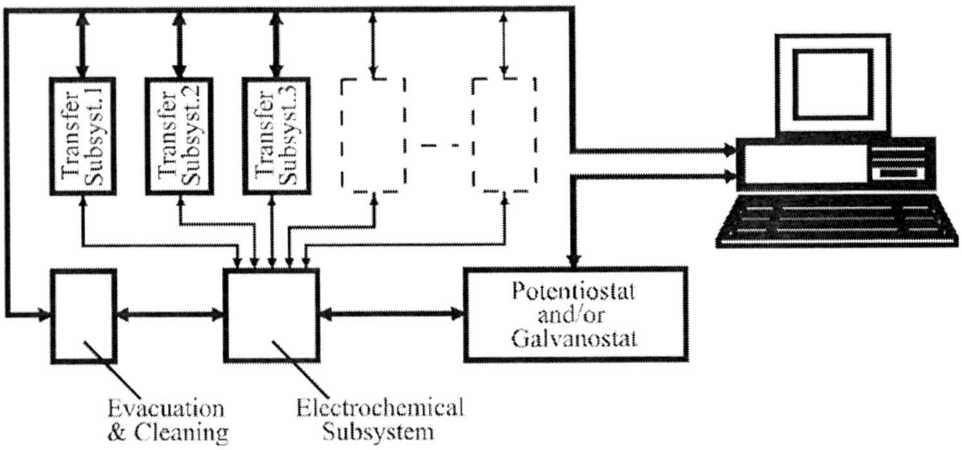

Figure 3. Block diagram of the modularly designed multiple bath electrodeposition system.

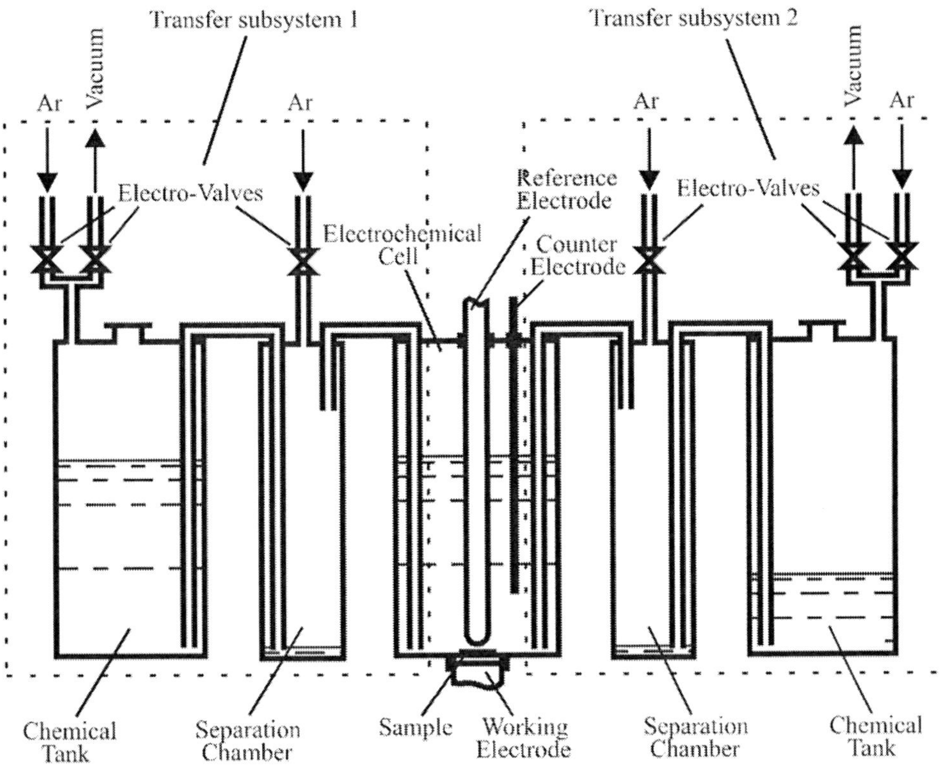

Figure 4. Schematic diagram of the system in which, for the sake of clarity, only two of the transfer subsystems are represented.

The deposition process includes the following steps: a) the substrate is placed and fixed to the working electrode inside the electrochemical cell; b) the chemical solution is transferred from the selected tank to the electrochemical cell; c) the computer-controlled electrodeposition is carried out on the substrate; d) the chemical solution is transferred back from the electrochemical cell to the chemical tank for future reuse; e) the electrochemical cell is cleaned by ultra pure water (or organic solvent); f) steps (b) to (e) are repeated until all the desired layers are deposited; g) the substrate is removed from the electrochemical cell.

The chemical solution transfer is realized using pressure differences between the tanks and the electrochemical cell applying an overpressure of inert gas (purified argon) and/or a moderate vacuum during transfer. In this procedure, the chemical solution flows through pipes and a separation chamber located between the chemical tank and the electrochemical cell as shown in Figure 4. The cross contamination between different chemical solutions is eliminated using these separation chambers by applying an overpressure of inert gas to separation chambers of all the transfer subsystems where no transfer operates. The cleaning procedure between two consecutive electrodepositions is made using ultra pure water (and/or solvent) including no dissolved oxygen or other chemically active gas in order to eliminate the undesirable surface reactions.

Pictures taken from a multiple-bath electrodeposition system implemented with three transfer subsystems could be seen in Figure 5.

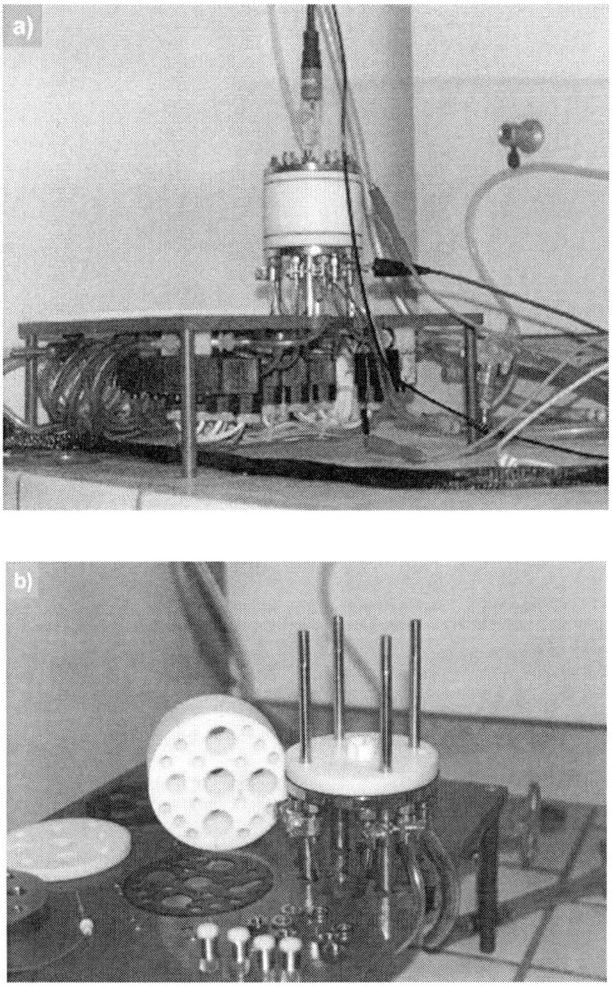

Figure 5. Photographs from the multiple bath electrodeposition system.

The control software was implemented in IgorPro, a scientific data analysis program [24], in a way that permits to mix the MBT and the SBT in the same electrochemical system.By overcoming the limitations of the classical MBT together with the versatility offered by our electrodeposition system, our procedure allows the realization of successive nanolayers of virtually any material that could be deposited electrochemically from aqueous or non aqueous baths (excluding high temperature molten salts).

The MBT like the SBT is of course highly compatible with the template synthesis of nanowires giving a very powerful tool for the fabrication of the nanostructured nanowires.

4. TEMPLATES

Among the various existing nanoporous materials track etched polymers, diblock copolymers with hexagonal ordering and nanoporous aluminas show a nanochanneled structure with hollow channels parallel to each other. They are the most interesting candidates as templates for nanowires synthesis.

4.1. Track Etched Polymers

When an energetic particle crosses a certain material it leaves a trail of radiation damage which shows up as a track. Such tracks were first observed in the 50^s in mica exposed to fission fragments from uranium [25]. If a sample of this material is etched in a suitable reagent it has been found that the tracks are very selectively attacked and hollow channels are formed along the particle paths, while the remaining material is mainly unaffected [6]. Various inorganic and organic materials were tested and studied [26]. It has been found that this effect could be efficiently used for charged particles detection [27].

It was only much later on, in the mid 90^s, that it was recognized that track etched polymers could be used also as templates for nanowires fabrication [22,28,29]. Such track etched templates are fabricated by irradiating polymer membranes such as bisphenol-A polycarbonate (PC) with heavy ions [30] produced by a particle accelerator in the MeV – GeV range. The ion irradiated PC film is then UV irradiated to increase the selectivity of the subsequent chemical etching. Etching is performed with a NaOH aqueous solution to form nanopores (nanochannels) with a desired diameter. A typical PC nanoporous membrane could be seen in Figure 6.

The main characteristics of this kind of templates are: random pore distribution; pore diameters from 10 nm to a few μm tunable by the etching time; and pore densities ranging from 10^5 to 10^{10} cm^{-2} controlled by the irradiation flux and time. At high densities of more than 10^9 cm^{-2} double and multiple shot defects appear leading to coalesced nanopores.

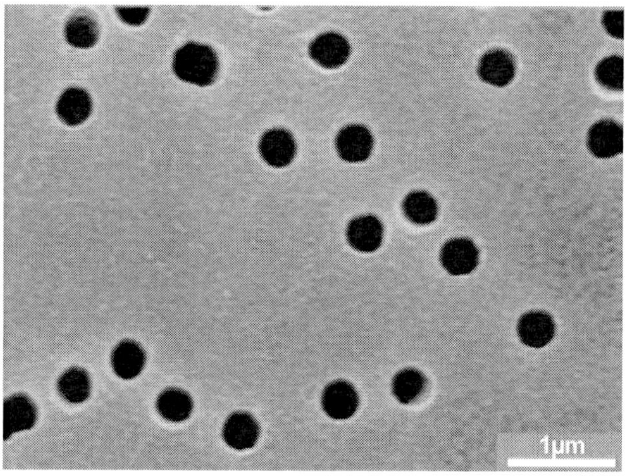

Figure 6. SEM micrograph from a track etched PC membrane surface.

4.2. Diblock Copolymers

Diblock copolymers have two constitutive elements (polymers) named A and B blocks. These blocks are immiscible but mix homogeneously in a disordered phase above certain temperatures. When lowering the temperature, segregation of the two phases from the disordered phase occurs in various microstructures [31-37]. The phase-diagram [38] that could be seen in Figure 7, which represents the Flory-Huggins parameter [39,40], χN, versus the concentration of A, shows clearly that different phases appear according to the relative amount of the constituent blocks A and B. The main microstructures observed are lamellar (LAM), hexagonally packed cylinders (HEX), a body centered cubic array of spheres (BCC) and a gyroid structure (GYR).

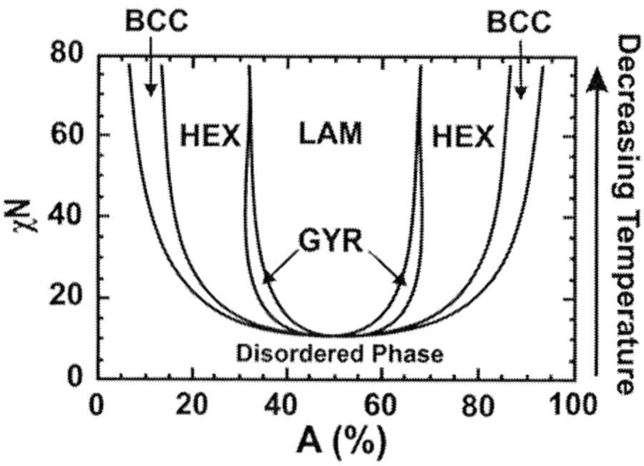

Figure 7. Mean-field phase diagram for conformationally symmetric diblock (A+B) melts. Phases are labeled LAM (lamellar), HEX (hexagonal cylinders), BCC (bcc spheres) and GYR (gyroidal). The χN designates the Flory-Huggins parameter [39,40].

Amongst these structures the HEX is the most interesting for use as template for high density nanowire arrays growth [12]. Applying an electric field during segregation of the A block in hexagonally ordered cylinders from the B block (in which they are embedded) their axes could be aligned parallel to the applied field and perpendicular to a conducting substrate. By selective removal of the constituent A, a hexagonally ordered nanoporous template of B constituent could be obtained.

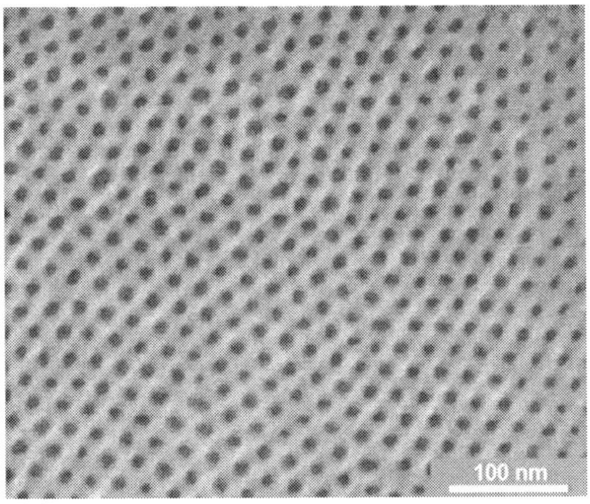

Figure 8. SEM micrograph from a polystyrene (PS) nanoporous template obtained by selective removal of the polymethylmethacrylate (PMMA) component of a PMMA/PS diblock copolymer segregated in hexagonally packed cylinders.

The main advantages of this template are the mild preparation conditions and the ordered arrangement of nanopores in a high density structure of typically $10^{11}...10^{12}$ pores/cm^2 (see Figure 8). Unfortunately, pore diameters and interpore distances are linked and can be tuned only by changing the A and/or B constituents and/or their molecular weights.

4.3. Nanoporous Alumina

Electrochemical oxidation of aluminum is used since almost a century for protective and/or decorative coating of aluminum products. It has been found that certain electrolytes used for the anodization of the aluminum produce compact oxide layers while others produce porous ones. The first characterization of these oxide layers dates back to the early 50^s [1]. A nanoporous structure is formed when a competitive equilibrium is established between electrochemical oxide growth and chemical dissolution of the oxide in certain acidic electrolytes [2-4]. Pores growth is accompanied by a phenomenon of self-ordering [5,41] tending to a perfectly ordered honeycomb nanoporous structure, as represented in Figure 9.

Systematic studies on this systems have shown that the degree of ordering could be influenced by the preparation conditions, i.e. anodization time, electrolyte type and its pH, applied voltage and temperature (see for example [42]). Perfect regularity could be obtained by prepatterning the aluminum surface before anodization [43] with various techniques like e-beam lithography or nanoindentation. Figure 10 shows some nanoporous alumina templates obtained in different preparation conditions.

The main futures of the alumina template are: the ordered arrangement of nanopores in a high density structure of typically $10^9 \ldots 10^{12}$ pores/cm^2; interpore distances controlled by the anodization voltages. Minimal pore diameter is also related to the anodization voltage but it can be increased by pore enlargement (etching) of the nanopores. This can range between 5 and 400 nm in function of the preparation conditions.

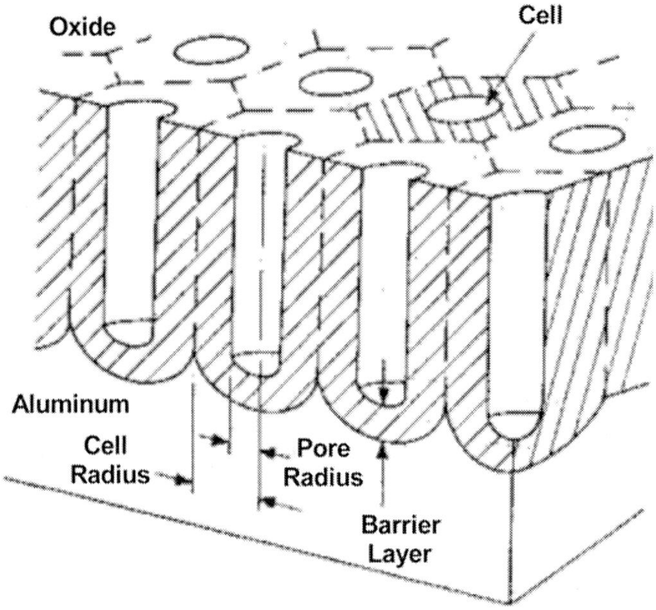

Figure 9. Schematic model of ideally ordered porous anodic alumina in a close-packed array of columnar hexagonal cells.

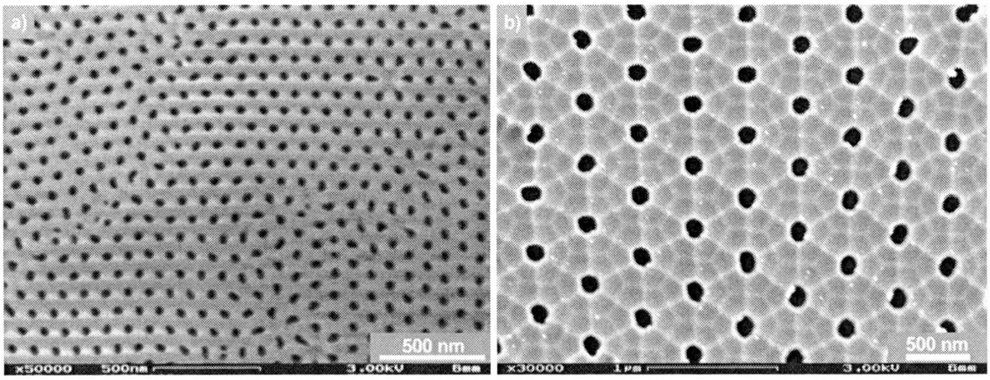

Figure 10. SEM micrographs from nanoporous alumina templates prepared using long anodization times to promote self ordering of the nanopores. Electrochemical oxidation of aluminum foils were carried out in oxalic acid (a) and phosphoric acid (b) solutions.

Common alumina membranes obtained by anodic oxidation of aluminum foils could be supported on the remaining non oxidized aluminum substrate or self-supported if they are removed from this substrate (Figure 11). The membrane can be removed, either by selective chemical etching of aluminum, or by diminishing stepwise the anodization voltage [4], or by reversing the applied voltages [44].

Despite the high melting point of Al_2O_3 (~2050 °C), the highest working temperature for such nanoporous membranes is limited by the melting point of aluminum (~658 °C) when the alumina membrane is supported on aluminum (AOA). It is also limited by the recrystallization temperature of the nanoporous alumina (>400 °C) and its phase transformations (>820 °C) [45] which cause mechanical deformation of the self-supported membranes.

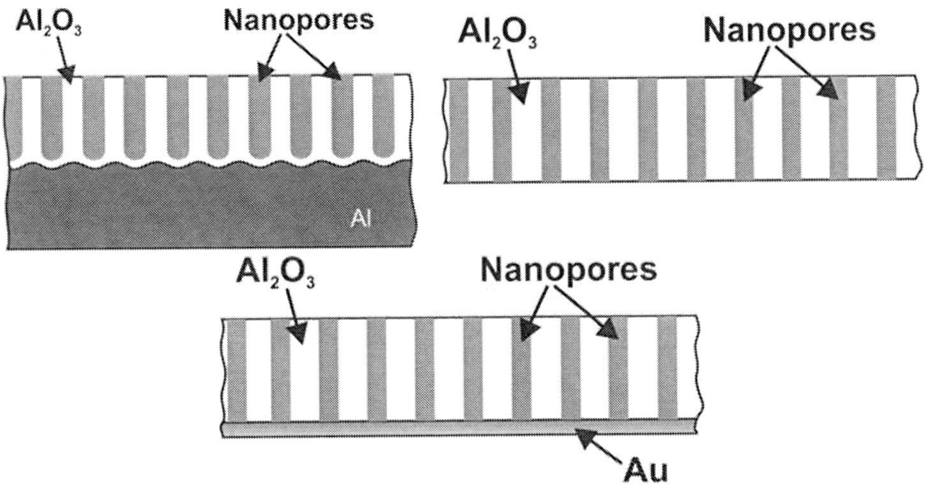

Figure 11. Nanoporous aluminum oxide membranes obtained by electrochemical oxidation of Al foils: alumina membrane naturally supported on aluminum (left); after removal from this substrate (middle), and with one side covered with a conducting layer (Au) serving as electrode for further electrodeposition inside the nanopores (right).

These drawbacks limit the use of nanoporous alumina as templates for chemical vapor depositions (CVD) where high temperatures are needed. They also restrain thermal treatments of the nanowires arrays which are synthesized by electrochemical methods thus limiting severely the advantages of this low temperature synthesis when these treatments are needed. The minimal thickness of the self-supported membrane is limited to ~20-50 μm due to its brittleness. Only AC electrodeposition could be used in case of AOA due to an oxide barrier layer between the nanopores and the aluminum substrate.

5. Supported Nanoporous Aluminum Oxide

Some attempts for supported alumina fabrication have been reported. One of them permits DC electrochemical synthesis of nanowircs inside nanopores for AOA by barrier layer removal [46].

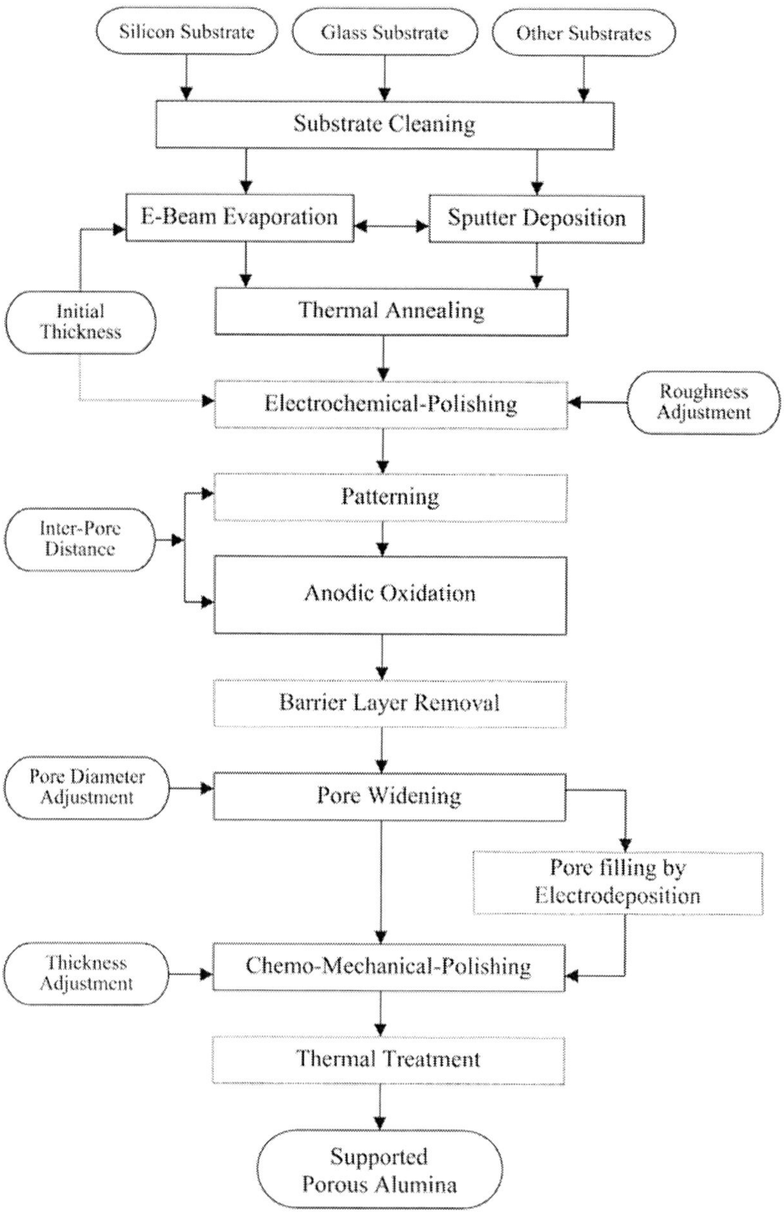

Figure 12. Flowchart showing the different steps of the supported nanoporous alumina fabrication.

Other attempts use conducting underlayers i.e. ITO [47,48], Pt [49,50], between an Al layer and a solid substrate like Si or glass. Unfortunately, none of these attempts succeeded to solve all problems and/or produce such supported membranes in a reliable manner and on large areas.

We have developed a fabrication procedure of a supported nanoporous aluminum oxide on silicon (or other substrates) that overcomes the above-mentioned major drawbacks and opens a way for its integration in electronic circuits. A flowchart of this fabrication procedure can be seen in Figure 12.

Each step has been optimized, some of them being critical to obtain good and reliable results. The process starts with a physical vapor deposition on a carefully cleaned substrate, typically Si, SiO_2, sapphire, MgO, glass, etc., of a conducting underlayer (see Figure 13). This conducting layer has a double role of serving as an anodization barrier during electrochemical oxidation of the aluminum layer deposited on it, and also as contacting electrode both during anodization and further electrochemical growth of nanowires inside the pores of the alumina. This conducting layer should not oxidize under anodization conditions, typically being a layer of gold or platinum. Intermediate adherence layers are used between the substrate and the conducting layer, and between the conducting layer and the subsequent aluminum layer.

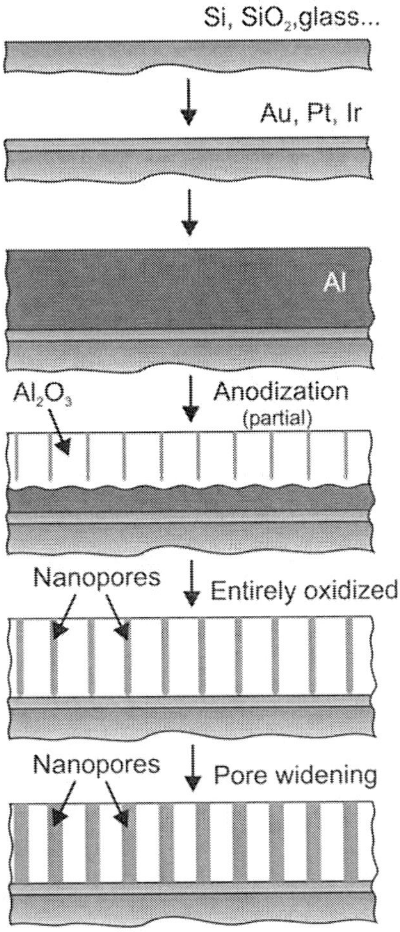

Figure 13. Schematic representation of the nanoporous alumina fabrication.

Aluminum layer deposition is a critical step in sense that small thicknesses (up to a few hundred nm) are easy to realize but thicker ones need special deposition conditions. The temperature of the substrate and deposition rate are key factors for compact aluminum layers deposition without hillock and pitting defects [51,52]. At substrate temperatures higher than

~70 °C, see Figure 14(a) and also at very low temperatures, Figure 14(b), such defects are dominant.

Figure 14. SEM micrographs from aluminum films surfaces, deposited by e-beam evaporation at a rate of 60 Å/min where: (a) substrate temperature was not kept constant but it was allowed to vary naturally from room temperature (300 K) to ~400 K during deposition for a total thickness of 2 μm; (b) substrate was cooled to the boiling point of liquid nitrogen (~77 K), deposited thickness 1 μm.

When the temperature of the substrate is kept constant at about 300 K for deposition rates larger than 300 Å/min, e-beam deposited Al layers are compact without defects. The deposit is polycrystalline with grain sizes (and surface roughness) decreasing with increasing deposition rate. Experimental investigations show that high rates, up to 6000 Å/min at room temperature, permit the realization of e-beam deposited Al layers with thicknesses up to 10 μm. On the other hand, using magnetron sputtering, the Al layer thickness can reach up to 2 μm for deposition rates of 700-1000 Å/min.

Depending on the final objectives, thermal treatment at 400-450 °C / (20 min - 4 h) in Ar gas and/or electrochemical polishing in perchloric acid ($HClO_4$) containing electrolytes at constant current densities of 100-200 mA/cm^2, can follow the deposition steps to increase grain sizes and decrease surface roughness.

Anodic oxidation of the supported thin aluminum layer is performed in similar conditions as for the case of classical aluminum foils anodization using sulfuric, oxalic or phosphoric acid solutions. In order to diminish oxidation reaction speeds and for a better control of thin layers anodization, low working temperatures of about 0°C are used. By controlling the anodization current at the end of the process, when no more metallic aluminum remains on the substrate, we obtain nanopores down to the conducting underlayer, see Figure 15(a). This allows the growth of nanowires inside nanopores by DC electrodeposition on large surface areas (Figure 15(b)).

Pore widening to desired values follows the anodization step using mainly hydrofluoric, phosphoric or sulfuric acid solutions at temperatures ranging between 30-40 °C. This step terminates, in principle, the supported nanoporous alumina fabrication. The template is well suited for the growth of nanowires or nanostructured nanowires inside nanopores by electrochemical or CVD methods. By using adequate substrates and underlayers it resists to high temperatures of more than 1000 °C allowing thermal treatments [53].

Figure 15. SEM sectional view of a supported nanoporous alumina template (a) and top view of an array of gold nanowires electrodeposited inside the nanopores after the removal of the alumina host (b).

By controlling the preparation conditions, pore diameters can be adjusted between 5 - 350 nm and pore lengths between 150 nm - 10 µm. Due to the possibility for thick Al layers deposition, a two step anodization is feasible especially for applications where an ordered arrangement of the nanopores (Figure 16), is important [54,55], this latter being promoted during first anodization step by a self ordering phenomenon [5].

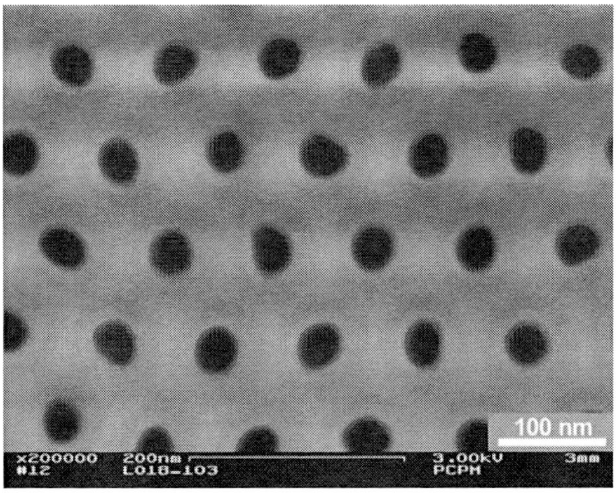

Figure 16. SEM surface image from a supported nanoporous alumina template obtained by a two step anodization process. The starting Al layer thickness was 7 μm and a special first anodization step [56] was conducted to a depth of 5 μm to promote ordering of the nanopores.

6. Contacting and Measuring a Large Amount of Single Nanowires and Compositionally Modulated Nanowires

The supported nanoporous aluminum oxide enables not only the fabrication of nanowires and nanostructured nanowires but also the combination of electrochemical template synthesis with nanolithography and patterning techniques, opening up new ways for the fabrication of nanowires-based complex structures and devices. The combination of these methods for microelectronic integration is under development with very promising preliminary results [57,58].

As an example we could mention a new method enabling electrical measurements on single nanowires [57] prepared by electrodeposition in supported alumina templates. This technique is based on electrically controlled nanoindentation of a thin insulating resist deposited on the top face of the template filled by the nanowires. A schematic representation of the fabrication steps could be seen in Figure 17.

Supported nanoporous alumina templates with thickness of about 750 nm, pore diameter of ~85 nm and interpore distance of about 140 nm were prepared. Then Co(14nm)/Cu(7nm) multilayered nanowires were electrodeposited within nanopores using a single bath method by switching alternatively between the deposition potentials of the two constituents [22]. As the growth rate is slightly different in different pores, only few pores show a complete filling followed by hemispherical metallic caps growth at the end of the deposition process. Prior to the nanolithography step, the template was thinned by mechanical polishing using colloidal silica (syton®), homogenizing in this manner nanowires length.

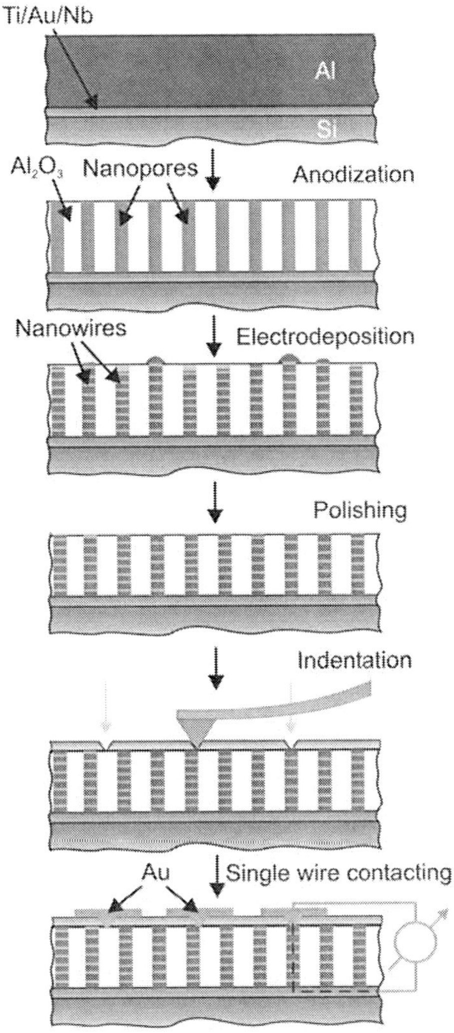

Figure 17. Schematic representation of the fabrication steps for single wire contacting

The fabrication process continues by depositing first a thin (60 nm) insulating polymer resist and then a second thick (1.2 μm) photoresist is spin-coated and patterned by classical optical lithography creating several openings of 30x30 μm^2. In each of these openings a CT-AFM is used to create a hole by indentation of the 60 nm resist until a good electrical contact between the tip and the underlying nanowire is obtained [59]. The bottom of the hole is enlarged and cleaned from resist residues by a soft O_2-Ar plasma etching. Then a 100 nm Au layer is sputtered to fill these access points to the nanowires in order to create the nanocontacts.

For a sample area of about 1 cm^2 we realized about 15 nanocontacts. The whole process allows connecting separately individual nanowires at different locations on the same sample and provides an easy way to perform single wire electrical measurements.

Magneto-transport measurements were performed at room temperature using dc signals. In the inset of Figure 18, we show an I(V) curve obtained on a single Co/Cu nanowire. A linear behavior without an observable joule effect is found for current densities up to 2.5 10^8

A/cm^2 that corresponds to a current of 12 mA through the single nanowire. Two giant-magnetoresistance-curves (GMR) measured with two different currents (100 μA and 15 mA respectively) are represented in Figure 18. In all cases the magnetic field used was perpendicular to the wire axis. The first effect of degradation as evidenced by irreversible change of the wire resistance, appears for a current density of 6 10^8 A/cm^2. Finally, the burn out of the nanowire was observed at 1.8 10^9 A/cm^2.

To exemplify how big these current densities are, it should be noted that the admissible current densities for copper wires used in designing coils and transformers with natural heat evacuation are 300-500 A/cm^2. Also, common fuses blow-out at ~10^4 A/cm^2, laser diodes supporting ~2 10^4 A/cm^2 and critical current densities of superconductors are in the range of 10^5-10^6 A/cm^2.

A possible explanation for the exceptional resistance of the nanowires to high current densities is that heat is evacuated by conduction through the alumina matrix surrounding the nanowires. This evacuation is favored also by the neighboring wires that increase the thermal conductivity of the Al_2O_3/metal nanocomposite system.

Our approach allows the electrical connection of one single nanowire in a large assembly of wires embedded in a supported nanoporous alumina template. The high current density permitted by this system is very promising to investigate magnetization reversal by spin polarized currents [60,61] for magnetic memory applications. This novel approach promises to be of strong interest for subsequent fabrication of phase-locked arrays of spin transfer devices [62] i.e. nano-oscillators with increased output power for microwave applications [63].

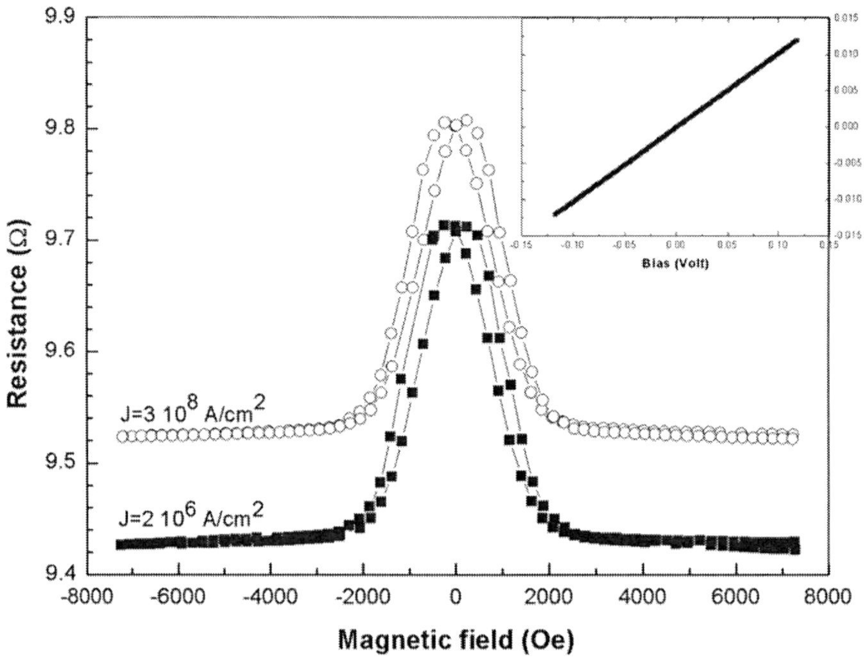

Figure 18. Resistance of a single Co/Cu multilayered nanowire versus magnetic field at 300 K, with magnetic field applied perpendicular to the wire axis. The inset shows an I(V) curve.

7. Localized Growth of Bundles or Single Nanowires

Arrays of vertically aligned nanowires that could be realized using the supported nanoporous alumina template show large surfaces. This is very promising for applications where high active surfaces are needed, i.e. sensors. Such a biosensor is under development [64] where a simple localization of the nanowires growth was realized by using an insulating substrate and by patterning the conductive underlayer. During a subsequent electrochemical deposition, nanowires will grow localized only on the conducting electrodes, see Figure 19.

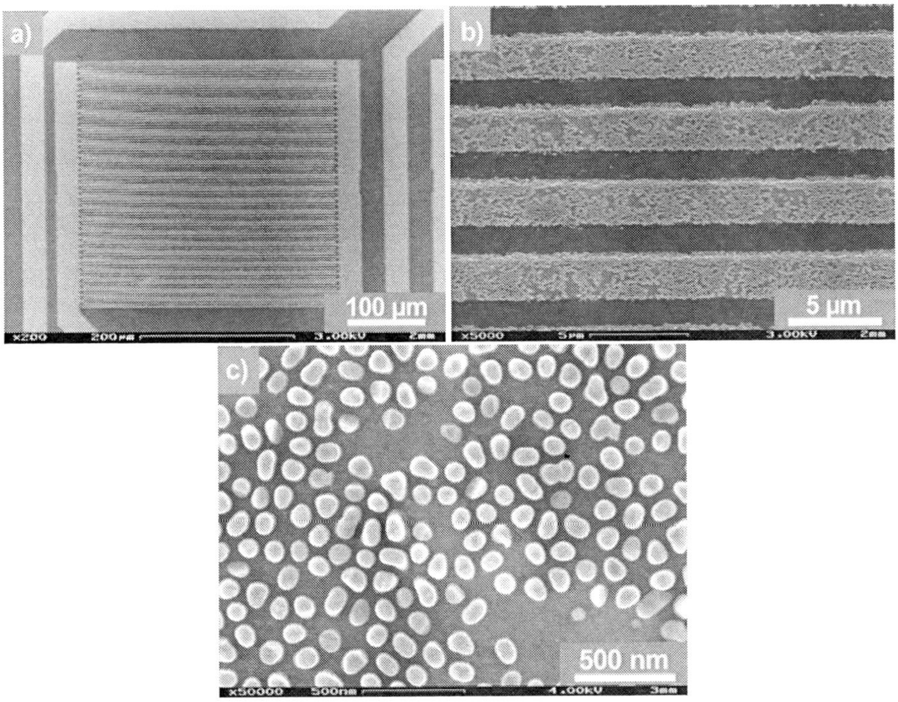

Figure 19. SEM Micrographs from localized nanowires on interdigited electrodes. Supported nanoporous alumina was prepared using a conducting underlayer of gold on SiO_2/Si substrate patterned by optical lithography then Au nanowires were grown by electrodeposition inside nanopores. SEM images were taken with increasing magnification from (a) to (c), after the removal of the alumina host.

A completely different approach for the localized growth of nanowires was adopted [58] by masking the top surface of the alumina template with a patterned SiN layer. In this manner, for a subsequent electrochemical deposition, only certain regions are exposed to the electrolyte with nanowires growing only in these regions (Figure 20).

After forming the nanopores by anodic oxidation on the supported template, a 100 nm thick SiN film is deposited by plasma enhanced chemical vapor deposition on top of the alumina layer. Then the sample is coated with a 200 nm thick layer of polymethylmethacrylate (PMMA) and patterned using *e*-beam lithography with different geometries. After resist development, the pattern is transferred to the SiN film using reactive ion etching followed by an oxygen plasma step to strip-out the PMMA layer.

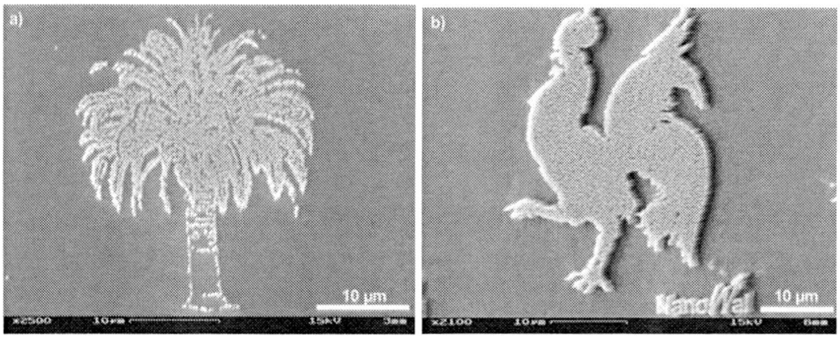

Figure 20. SEM micrographs from localized nanowires in artistic patterns by masking the supported nanoporous alumina surface using e-beam lithography.

By carefully adjusting the dimension of the openings in the SiN mask and, possibly, the interpore distance in the alumina template we are able to grow arrays of single nanowires or bunches of few nanowires, see Figure 21.

The pattern presented in Figure 21 is particularly interesting for crossbar nanowire circuits where each crossing point should be connected by a single nanowire. Work is in progress [65] to obtain such electrically connected single nanowires in a matrix arrangement combining the two patterning methods presented above. Taking into account the accuracy of modern nanolithography, less than 50 nm for the *e*-beam lithography, highly dense circuits including nanowires can be envisaged using this approach. In addition, the present technique allows the realization of contacts without dissolving the alumina host, thus protecting the nanowires from oxidation and allowing an efficient heat evacuation through the alumina matrix [57].

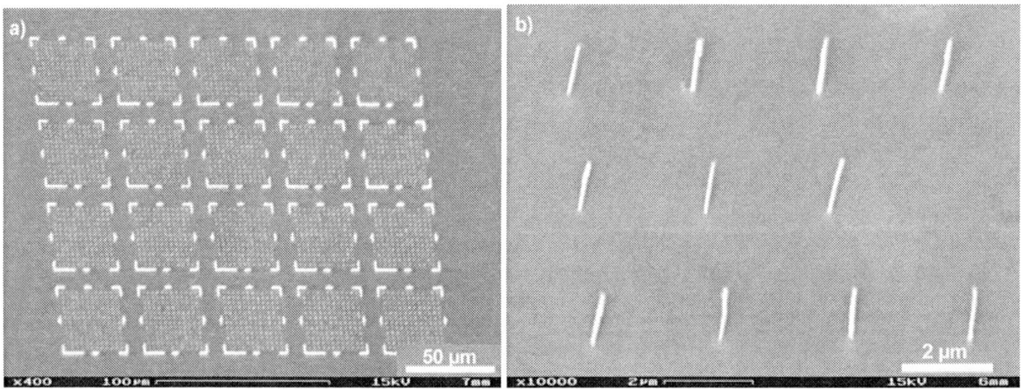

Figure 21. SEM micrographs from localized arrays of single nanowires in a matrix arrangement masking the supported nanoporous alumina surface using e-beam lithography.

CONCLUSION

Electrochemical synthesis using nanoporous materials as templates allows the realization of nanostructures that are difficult to obtain using conventional techniques. Among these templates the supported nanoporous alumina is the most promising. Its particular properties i.e. mechanical stability, good thermal conductivity and high temperature resistance, the possibility of using DC-electrochemical deposition inside nanopores, the easily tunable geometrical parameters (pore diameters and interpore distances) and the possibility for a high density ordered arrangement of nanopores, makes it an ideal candidate for further integration into large-scale fabrication of various nanowires based devices.

We have shown that combining nanolithographic techniques and electrochemical template synthesis, using supported nanoporous alumina, a powerful method was developed that opens a way for complex devices and circuits fabrication exploiting particular properties of nanowires and nanostructures.

Work is now in progress to go further both for realizing specific nanostructures for fundamental studies and for applications, such as spintronics, magnetic recording, sensors and biosensors, etc.

ACKNOWLEDGEMENT

I would like to thank Mátéfi-Tempfli Mária, B.Nagy Ottó and Jean-Paul Issi for a critical review of this chapter and also all collaborators who contributed to the work presented.

REFERENCES

[1] Keller, F.; Hunter, M. S.; Robinson, D. L. *J. Electrochem. Soc.* 1953, *100*, 411-419.
[2] O'Sullivan, J. P.; Wood; G. C. *Proc. R. Soc. London, Ser. A* 1970, *317*, 511-543.
[3] Thompson, G. E.; Furneaux, R. C.; Goode, J. S.; Wood, G. C.; Richardson, J. A. *Nature* 1978, *272*, 433-435.
[4] Furneaux, R. C.; Rigby, W. R.; Davidson, A. P. *Nature* 1989, *337*, 147-149.
[5] Masuda, H.; Fukuda, K. *Science* 1995, *268*, 1466-1468.
[6] Price, P. B.; Walker, R. M. *J. Appl. Phys.* 1962, *33*, 3407-3412.
[7] Martin, C. R. *Science* 1994, *266*, 1961-1966.
[8] Parthasarathy, R.; Martin, C. R. *Nature* 1994, *369*, 298-301.
[9] Fleischer, R. P.; Price, P. B.; Walker, R. M.; Hubbard, E. L. *Phys. Rev.* 1967, *156*, 353-355.
[10] Kautek, W.; Reetz, S.; Pentzien, S. *Electrochim. Acta* 1995, *40*, 1461-1468.
[11] Ferain, E.; Legras, R. *Nucl. Instrum. Methods Phys. Res., Sect. B* 2001, *174*, 116-122.
[12] Thurn-Albrecht, T.; Schotter, J.; Kästle, G. A.; Emley, N.; Shibauchi, T.; Krusin-Elbaum, L.; Guarini, K.; Black, C. T.; Tuominen, M. T.; Russell, T. P. *Science* 2000, *290*, 2126-2129.
[13] Tonucci, R. J.; Justus, B. L.; Campillo, A. J.; Ford, C. E. *Science* 1992, *258*, 783-785.
[14] Rajendra Kumar, R. T.; Badel, X.; Víkor, G.; Linnros, J.; Schuch, R. *Nanotechnology* 2005, *16*, 1697-1700.
[15] Yahalom, J.; Zadoc, O. *J. Mater. Sci.* 1987, *22*, 499-503.
[16] Lashmore, D. S.; Dariel, M. P. *J. Electrochem. Soc.* 1988, *135*, 1218-1221.
[17] Menezes, S.; Anderson, D. *174th Electrochem. Soc. Mtg. Chicago. Abstr.* 1988, *342*, 503.

[18] Goldman, L. M.; Ross, C. A.; Ohashi, W.; Wu, D.; Spaepen, F. *Appl. Phys. Lett.* 1989, *55*, 2182-2184.

[19] Barral, G.; Maximovitch, S. *Colloque de Physique* 1990, *C4*, 291-297.

[20] Celis, J. P.; Haseeb, A.; Roos, J. R. *Trans. Inst. Met. Finish.* 1992, *70*, 123-128.

[21] Haseeb, A. S. M. A.; Celis, J. P.; J.R. Roos, J. R. *J. Electrochem. Soc.* 1994, *141*, 230-237.

[22] Piraux, L.; George, J. M.; Despres, J. F.; Leroy, C.; Ferain, E., Legras, R.; Ounadjela, K.; Fert, A. *Appl. Phys. Lett.* 1994, *65*, 2484-2486.

[23] Mátéfi-Tempfli, S.; Piraux, L. (2002). Multiple bath electrodeposition, EP1256639 A1. Method, apparatus and system for electrodeposition of a plurality of thin layers on a substrate, WO02092883; US2006243597.

[24] http://www.wavemetrics.com/products/igorpro/igorpro.htm.

[25] Silk, E. C. H.; Barnes, R. S. *Philos. Mag.* 1959, *4*, 970-972.

[26] Katz, R.; Kobetich, E. J. *Phys. Rev.* 1968, *170*, 401-405.

[27] Medveczky, L.; Somogy, G. *Proceedings of the International Topical Conference on Nuclear Track Registration in Insulating Solids and Applications, Clermont-Ferrand (France)* 1969, *2*, 65-81.

[28] Blondel, A.; Meier, J. P.; Doudin, B.; Ansermet, J.-Ph. *Appl. Phys. Lett.* 1994, *65*, 3019-3021.

[29] Liu, K.; Nagodawithana, K.; Searson, P. C.; Chien, C. L. *Phys. Rev. B: Condens. Matter* 1995, *51*, 7381-7384.

[30] Ferain, E.; Legras, R. *Nucl. Instrum. Methods Phys. Res., Sect. B* 1997, *131*, 97-102.

[31] Aggarwal, S. L. *Polymer* 1972, 17, 938-956.

[32] Thomas, E. L.; Alward, D. B.; Kinning, D. J.; Martin, D. C.; Handlin, D. L.; Fetters, L. J. *Macromolecules* 1986, *19*, 2197-2202.

[33] Hajduk, D. A.; Harper, P. E.; Gruner, S. M.; Honeker, C. C.; Kim, G.; Thomas, E. L.; Fetters, L. J. *Macromolecules* 1994, *27*, 4063-4075.

[34] Schulz, M. F.; Bates, F. S.; Almdal, K.; Mortensen, K. *Phys .Rev. Lett.* 1994, *73*, 86-89.

[35] Hamley, I. W.; Koppi, K. A.; Rosedale, J. H.; Bates, F. S.; Almdal, K.; Mortensen, K. *Macromolecules* 1993, *26*, 5959-5970.

[36] Foerster, S.; Khandpur, A. K.; Zhao, J.; Bates, F. S.; Hamley, I. W.; Ryan, A. J.; Bras, W. *Macromolecules* 1994, *27*, 6922-6935.

[37] Khandpur, A. K.; Foerster, S.; Bates, F. S.; Hamley, I. W.; Ryan, A. J.; Bras, W.; Almdal, K.; Mortensen, K. *Macromolecules* 1995, *28*, 8796-8806.

[38] Matsen, M. W.; Bates, F. S. *Macromolecules* 1996, *29*, 1091-1098.

[39] Flory, P. J. *J. Chem. Phys.* 1941, *9*, 660-661.

[40] Huggins, M. L. *J. Chem. Phys.* 1941, *9*, 440.

[41] Masuda, H.; Hasegwa, F.; Ono, S. *J. Electrochem. Soc.* 1997, *144*, L127-L130.

[42] Nielsch, K.; Choi, J.; Schwirn, K.; Wehrspohn, R. B.; Gösele, U. *Nano Lett.* 2002, *2*, 677-680.

[43] Masuda, H.; Yamada, H.; Satoh, M.; Asoh, H.; Nakao, M.; Tamamura, T. *Appl. Phys. Lett.* 1997, *71*, 2770-2772.

[44] Tian, M.; Xu, S.; Wang, J.; Kumar, N.; Wertz, E.; Li, Q.; Campbell, P. M.; Chan, M. H. W.; Mallouk, T. E. *Nano Lett.* 2005, *5*, 697-703.

[45] Mardilovich, P. P.; Govyadinov, A. N.; Mukhurov, N. I.; Rzhevskii, A. M.; Paterson, R. *J. Membr. Sci.* 1995, *98*, 131-142.

[46] Nielsch, K.; Müller, F.; Li, A.-P.; Gösele, U. *Adv. Mater.* 2000, *12,* 582-586.
[47] Chu, S. Z.; Wada, K.; Inoue, S.; Todoroki, S. *J. Electrochem. Soc.* 2002, *149*, B321-B327.
[48] Chu, S. Z.; Wada, K.; Inoue, S.; Todoroki, S. *Electrochim. Acta* 2003, *48*, 3147-3153.
[49] Rabin, O.; Hertz, P. R.; Lin, Y-M.; Akinwande, A. I.; Cronin, S. B.; Dresselhaus, M. S. *Adv. Funct. Mater.* 2003, *13*, 631-638.
[50] Yasui, N.; Imada A.; Den, T. *Appl. Phys. Lett.* 2003, *83*, 3347-3349.
[51] Hale, G. J. *Thin Solid Films* 1979, *63*, 95-105.
[52] Mattsson, L.; Le Page, Y. H.; Ericson, F. *Thin Solid Films* 1991,*198*, 149-156.
[53] Mallet, J.; Yu-Zhang, K.; Mátéfi-Tempfli, S.; Mátéfi-Tempfli, M.; Piraux, L. *J. Phys. D: Appl. Phys.* 2005, *38*, 909-914.
[54] Vinckx, W.; Vanacken, J.; Moshchalkov, V. V.; Mátéfi-Tempfli, S.; Mátéfi-Tempfli, M.; Michotte, S.; Piraux, L. *Eur. Phys. J. B: Condens. Matter and Compl. Syst.* 2006, *53*, 199-203.
[55] Vinckx, W.; Vanacken, J.; Moshchalkov, V. V.; Mátéfi-Tempfli, S.; Mátéfi-Tempfli, M.; Michotte, S.; Piraux, L.; Ye, X. *Physica C: Supercond.* 2007, in press.
[56] Mátéfi-Tempfli, S. et al, unpublished.
[57] Fusil, S.; Piraux, L.; Mátéfi-Tempfli, S.; Mátéfi-Tempfli, M.; Michotte, S.; Saul, C. K.; Pereira, L. G.; Bouzehouane, K.; Cros, V.; Deranlot, C.; George, J.-M. *Nanotechnology* 2005, *16*, 2936-2940.
[58] Vlad, A.; Mátéfi-Tempfli, M.; Faniel, S.; Bayot, V.; Melinte, S.; Piraux, L.; Mátéfi-Tempfli, S. *Nanotechnology* 2006, *17*, 4873-4876.
[59] Bouzehouane, K.; Fusil, S.; Bibes, M.; Carrey, J.; Blon, T.; Le Dû, M.; Seneor, P.; Cros, V.; Vila, L. *Nano Lett.* 2003, *3,* 1599-1602.
[60] Katine, J. A.; Albert, F. J.; Buhrman, R. A.; Myers, E. B.; Ralph, D. C. *Phys. Rev. Lett.* 2000, *84*, 3149-3152.
[61] Grollier, J.; Cros, V.; Hamzić, A.; George, J. M.; Jaffrès, H.; Fert, A.; Faini, G.; Ben Youssef, J.; Legall, H. *Appl. Phys. Lett.* 2001, *78*, 3663-3665.
[62] *Piraux, L.; Renard, K.; Guillemet, R.; Mátéfi-Tempfli, S.; Mátéfi-Tempfli, M.; Antohe, V. A.; Fusil, S.; Bouzehouane, K.; Cros, V.* Submitted to Nano Lett. 2007.
[63] Kiselev, S.; Sankey, J. S.; Krivorotov, I. N.; Emley, N. C.; Schoelkopf, R. J.; Buhrman, R. A.; Ralph, D. C. *Nature* 2003, *425*, 380-383.
[64] Mátéfi-Tempfli, S. et al, unpublished.
[65] Vlad, A. et al, unpublished.

In: Encyclopedia of Nanotechnology
Editor: Elwood D. Carlson

ISBN: 978-1-60692-079-4

Chapter 14

SYNTHESIS OF EMBEDDED GOLD AND AGAU AND CUAU ALLOY NANOCLUSTERS IN SODA-LIME GLASS BY ION-EXCHANGE: A NOVEL ROUTE

***S. Mohan*[1*], *D. Manikandan*[1, 2], *K. G. M. Nair*[2] and *Sujin P. Jose*[3]**

[1]Raman School of Physics, Pondicherry University,
Pondicherry-605 014, India
[2]Materials Science Division, Indira Gandhi Center for Atomic Research,
Kalpakkam- 603 102, India
[3]Annai Fathima College, Madurai 625 706, India

ABSTRACT

Gold nanoclusters were formed in the soda-lime glass matrix by ion-exchange technique. The ion-exchanged glasses were subjected to heat treatment for different temperatures in air and the effect of it on the optical absorption spectra was analyzed. Theoretical optical absorption was calculated using the Mie's and Maxwell Garnett's effective medium theories. The simulated spectra matched with the experimentally recorded spectra confirming the presence of gold clusters in the glass matrix. As a novel route, hetero system nanoclusters like AgAu and CuAu were formed by ion-exchange method. The optical absorption measurements showed signatures of an incomplete alloy formation between the respective metal species.

Keywords: Embedded, alloy nanocluster, ion-exchange, soda-lime glass.

* E-mail address: smoh14@rediffmail.com

1. Introduction

One of the promising techniques for doping glasses with metal ions in a controllable way is the ion-exchange process which forms the first step in the multistep process of metal nanocluster composite glass preparation. Ion-exchange method has been extensively used for the fabrication of passive integrated-optical devices [1] and in the chemical strengthening of glasses [2]. Composites made of metal nanocluster in dielectric matrix have drawn much interest due to their enhanced nonlinear optical properties and for their magnetic applications [3-6]. Optical absorption of copper and silver ion-exchanged and annealed soda-lime glasses were discussed in our earlier work [7].

The present study is aimed to synthesize gold nanoclusters by the extensive ion-exchange technique and fire the metal doped glasses at different temperatures, to study their optical absorption properties. An attempt has been made to form complex nanoclusters of AgAu and CuAu in the same dielectric matrix through a combined ion-exchange method.

2. Experimental

Precleaned commercial soda-lime glasses (composition, wt. %, 72.90 SiO_2, 14.50 Na_2O, 6.72 CaO, 4.10 MgO, 1.40 Al_2O_3, 0.03 SO_3, and 0.08 Fe_2O_3) of 1 mm thickness are dipped in the molten salt bath solution of $HAuCl_4$ and $NaNO_3$ taken in the 1 : 5 ratio. In the bath, Au^+-Na^+ ion-exchange takes place with the possible formation of Au clusters inside the glass matrix. As for the AgAu ion-exchange, $HAuCl_4$: $AgNO_3$: $NaNO_3$ mixture in 0.3 : 4 : 2 ratio, was taken and the glass substrates were dipped in the melt for the same time for Ag and Au exchange. The Au and AgAu ion-exchanged samples were then cleaned and annealed in air for about 1 hour at three different temperatures (450, 550 and 650 °C). In the synthesize of complex CuAu nanoclusters, the glass slides are dipped into the molten salt bath mixture of $CuSO_4$: Na_2SO_4 : $HAuCl_3$ taken in the 5 : 5: 1 ratio for different dipping times viz., 1, 5, 10, 15 and 20 mins. During the complex ion-exchange, both Cu and Au ions enter into the glass matrix in case of CuAu exchange and Ag and Au ions in the case of AgAu exchange, in replacement for sodium.

Room temperature optical absorption spectra of the ion-exchanged and annealed samples were recorded in the 200-1500 nm wavelength region, using a UV-VIS-NIR dual beam spectrophotometer. A virgin soda-lime glass was kept as the reference for recording the spectra.

3. Results and Discussions

Figure 1.(a) shows the experimentally recorded and theoretically (Mie theory) calculated optical absorption spectra of gold nanoclusters in soda-lime glass. The experimentally recorded spectra exhibited a peak at 531 nm which corresponds to the surface plasmon resonance (SPR) of gold particles as reported in the case of implanted gold in silica [8]. For clusters smaller than the wavelength λ of incident light, the theoretical spectra was calculated using Mie's formula [9] which gives the absorption coefficient α of a material of refractive index n occupied with metal spheres of volume fraction V as follows

$$\alpha = \frac{18\pi V n^3}{\lambda} \frac{\varepsilon_2}{(\varepsilon_1 + 2n^2)^2 + \varepsilon_2^{\,2}} cm^{-1} \qquad (1)$$

where ε_1 and ε_2 are the real and imaginary components of the metal sphere. The equation has a maximum value at the surface plasmon resonance frequency ω_p, where $\varepsilon_1(\omega_p)+2n^2 = 0$. The Figure 1(b) shows the experimentally recorded and theoretically calculated spectra using Maxwell Garnett's effective medium theory, where the effective dielectric constant for spherical metal sphere embedded in a matrix is given below [10]

$$\frac{\varepsilon - \varepsilon_m}{\varepsilon + 2\varepsilon_m} = f \frac{\varepsilon_p - \varepsilon_m}{\varepsilon_p + 2\varepsilon_m} \qquad (2)$$

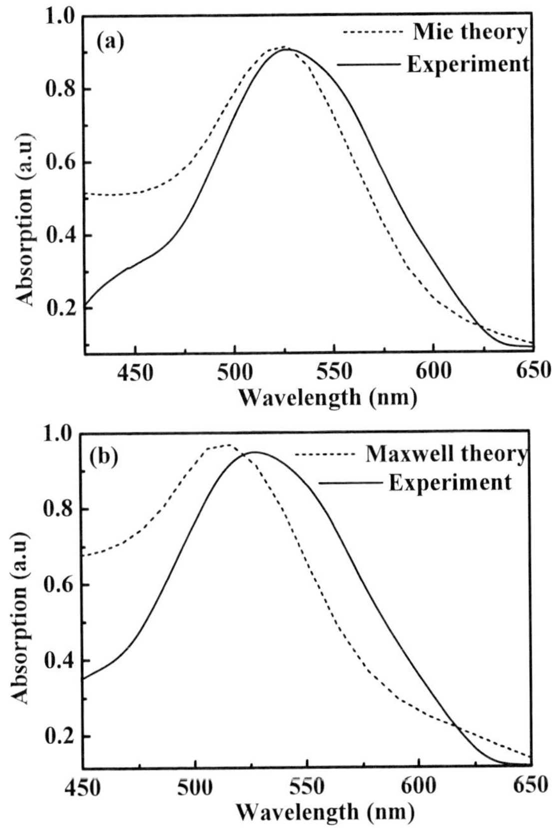

Figure 1. Optical absorption Vs wavelength for Au ion-exchanged and 1(a) Mie's theory and 1(b) Maxwell Garnett's theory.

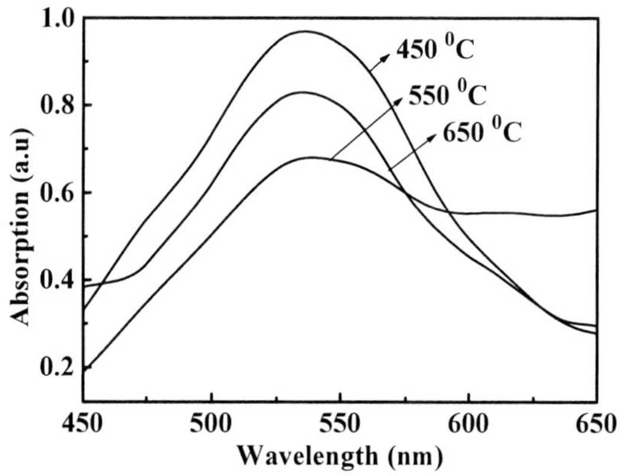

Figure 2. Optical absorption Vs wavelength for the Au ion-exchange and different temperature annealed soda-lime glass.

The values of ε_m and ε_p are the dielectric constants of the matrix and the particles. The values of the dielectric constants were derived from the optical constants of the respective bulk metal and matrices [11] after taking into account the size corrections [12]. The Figure 2 shows the different temperature annealed Au ion-exchanged glasses, with the spectra showing peaks due to absorption of gold particles with out any significant shift in the peak position.

The average cluster radii R of the embedded nano particles are calculated from the full width at half maximum Δλ (FWHM) of the optical absorption peaks using the formula given below

$$R = \frac{V_f \lambda_p^{\ 2}}{2\pi C \Delta\lambda} \tag{3}$$

V_f is the Fermi velocity of the electrons in bulk metal (Gold = $1.40X10^8$ cm/s), $\Delta\lambda$ is the full width at half maximum of the absorption band and λ_p is the characteristic wavelength at which surface plasmon resonance occurs. Both λ_p and $\Delta\lambda$ depend on the substrate and size of the metal nanoclusters forming the composite.

The calculated average radii of the embedded gold clusters ranged from 2.19 – 3.1 nm. It is observed that the gold clusters grew with increase in annealing temperatures and can be well understood from the Figure 3(a). The inverse relation of the FWHM to the cluster radius is followed in Figure 3(b), thus proving the applicability of the above equation (3).

Figure. 4 shows the SPR peaks of copper, silver and gold clusters embedded in soda-lime glass matrix formed by ion-exchange and annealing techniques.

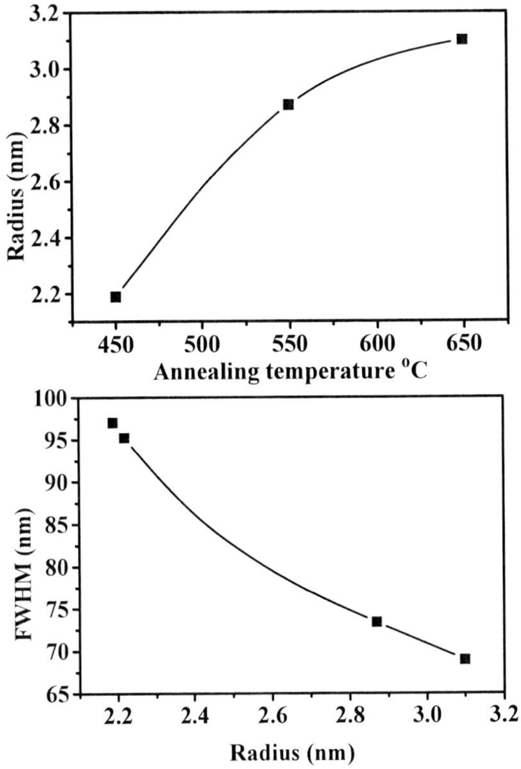

Figure 3. Average cluster radius Vs annealing temperatures (a) and FWHM of SPR peak Vs average radius of gold clusters (b).

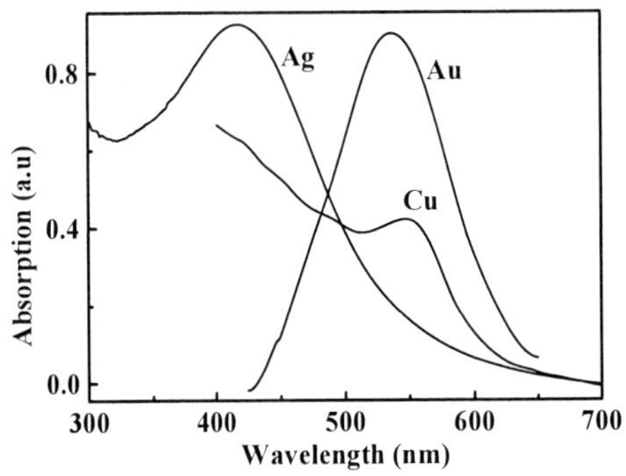

Figure 4. SPR band of respective copper, silver and gold clusters in soda-lime glass matrix.

3.1. AgAu System

We have formed complex clusters of Cu/Ag in soda-lime glass by sequential ion-exchange and ion irradiation methods [13] with the metals forming separate aggregates without any alloy formation between them. In the present work we have formed Ag and Au clusters in the same glass matrix by the combined ion-exchange method. The Figure 5 shows the absorption wavelength plot for AgAu ion-exchange and annealed samples. The as exchanged samples showed no peaks corresponding to either of the metal species but on annealing the samples in air at 400 and 600 °C resulted in the evolution of spectra with two separate well resolved SPR peaks at 421 and 511 nm for 400 °C annealed and around 428 and 494 nm for 600 °C annealed samples respectively. The peaks centered at 421 and 428 nm respectively were in consistent with the absorption of silver nano clusters in soda-lime glass, formed by us as shown in Figure 4. But the peaks around at 511 and 494 nm in Figure 5 shows a significant shift from the SPR position of pure gold (ie., 535 nm). These peaks can be attributed to the incomplete Au rich alloy clusters formation as documented by earlier researchers [14]. Further annealing of the samples at higher temperatures was restricted due to the softening of soda-lime glass above 650 °C and any oxidation of Ag or Au was also not evidenced from the absorption spectra. This clearly indicates the weak interaction of the two metal species with the oxygen of the matrix.

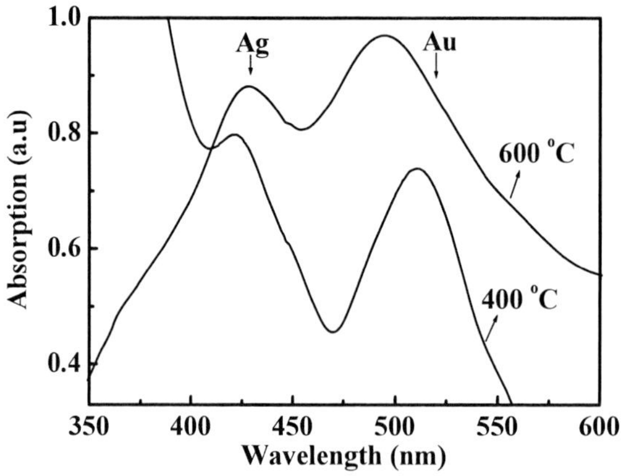

Figure 5. Absorption Vs wavelength of combined AgAu ion-exchanged and annealed soda-lime glass.

3.2. CuAu System

Figure. 6 shows the optical absorption spectra of the combined CuAu ion-exchanged glasses for different dipping times. The spectra of all samples dipped for 1,2,3,4 and 5 minutes showed peaks due to copper clusters only (example Figure 6 (5 min)). But the samples dipped for higher processing times like 10 and 20 minutes, each of the spectra exhibited two peaks centered at 554 and 606 nm and around at 555 and 601 nm respectively. The peaks at 554 and 555 nm are attributed to the CuAu alloy clusters with little concentration of gold. A prominent shift of the copper peak to higher wavelength side (ie., 606 and 601 nm) in the 10 and 20 min ion-exchanged samples are due to the possible formation of the Cu_2O as reported elsewhere [15]. The increased reactivity of copper with oxygen of the glass matrix is very well observed from the oxidation of the copper along with the Cu rich alloy formation.

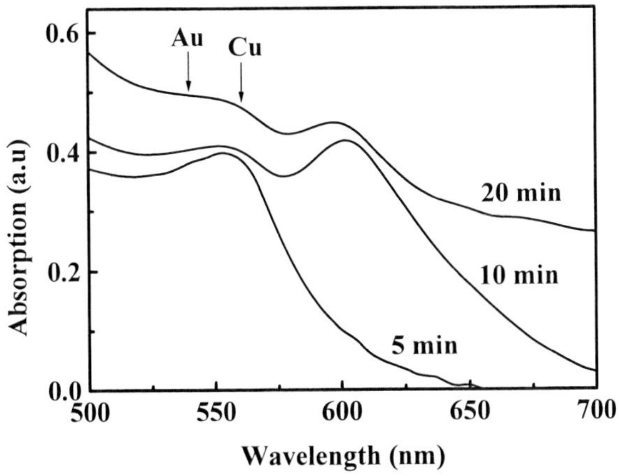

Figure 6. Absorption Vs wavelength of combined CuAu ion-exchanged soda-lime glass of different processing times.

4. Conclusion

The experimental results revealed that gold clusters are formed in soda-lime glass for the first time by the binary ion-exchange method with the clusters growing with increase in heat treatment temperatures. Complex clusters of CuAu and AgAu are also formed by combined ion-exchange and sequential combined ion-exchange followed by annealing. The absorption spectra showed signatures of alloying between the AgAu and CuAu metal species. In contrast to Ag, the Cu not only involved in alloy formation but also interacted with the oxygen of the matrix to form copper oxide. From this observation, it is found that the Cu has a strong interaction with the dielectric matrix when compared to Ag.

Acknowledgements

The authors thank the IUC-DAEF, Indore, for its financial assistance in carrying out this work. We also extend our sincere thanks to Dr. Narayan Kalkura and Mr. M. Ashok, Crystal Growth Centre, Anna University, Chennai, for recording the optical absorption spectra.

References

[1] R.V. Ramaswamy and R. Srivastava, *IEEE J.* Lightwave Technol. LT-6, 984 (1988).

[2] R.F. Bartholomew and H.M. Garfinkel, *"Glass: Science and Technology"*, ed. by. R. Uhlman and N.J. Kreidl), Vol. 5, 218, (Academic Press, New York, 1980).

[3] P. Mazzoldi, G.W. Arnold, G. Battaglin, F. Gonella, R.F. haglund Jr., J. Nonlin. *Opt. Phys. Mater.* 5, 285 (1996).

[4] F. Gonella, P. Mazzoldi, Metal nanocluster composited glasses, ed. by H.S. Nalwa, *Handbook of nanostructured Materials and nanotechonology*, Vol. 4, 81(Academic Press, San Diego, 2000).

[5] G. De, M. Gusso, L. Tapfer, M. Catalano, F. Gonella, G. Mattei, P. Mazzoldi, G. Battaglin, *J. Appl. Phys*. 80, 6734 (1996).

[6] F. Gonella, G. Mattei, P. Mazzoldi, C. Sada, G. Battaglin, E. Cattaruzza, *Appl. Phys. Lett.* 75, 55 (1999).

[7] D. Manikandan, S. Mohan, P. Magudapathy, K.G.M. Nair, *Physica B* (2002) (In press).

[8] G. Battaglin, E. Cattaruzza, F. Gonella, G. Mattei, P. Mazzoldi, C. Sada, *Nucl. Instr. and Meth. B* 166-167, 857 (2000).

[9] G. Mie, ann. *Physik* 25, 377 (1908).

[10] D.E. Aspnes, *Thin Solid Films* 89, 249 (1982).

[11] *Handbook of Optical Constants of Solids*, by EDWARD D. PALIK, (Academic Press, Inc. 1985).

[12] *Optical Properties of Metal Clusters, ed. by* M. Vollmer, U. Kreibig, *Springer Series in Materials Science*. (Springer-Verlag Berlin Heidelberg, 1995).

[13] D. Manikandan, S. Mohan, P. Magudapathy, K.G.M. Nair, *Nucl. Instr. and Meth. B* (2002) (In press).

[14] G. Battaglin, E. Cattaruzza, C. de Julian Fernandez, G. De marchi, F. Gonella, g. Mattei, C. Maurizio, P. Mazzoldi, A. Miotello, C. Sada, F. D'Acapito, *Nucl. Instr. and Meth. B* 175, 410 (2001).

[15] M. Mennig, M. Schmitt, B. Kutsch, H. Schmidt, *SPIE Proc*. 2288, 120 (1994).

In: Encyclopedia of Nanotechnology
Editor: Elwood D. Carlson

ISBN: 978-1-60692-079-4

Chapter 15

INVESTIGATION OF P-TYPE ZNO LAYERS OBTAINED BY RBQE METHOD

T. V. Butkhuzi*, T. G. Khulordava, M. M. Sharvashidze, N. G. Gapishvili, N. G. Bukhsianidze, L. T. Trapaidze, E. E. Kekelidze, G. G. Natsvlishvili, L. G. Aptsiauri and L. N. Gapishvili

Dep. of Physics, Tbilisi State University,
3, Chavchavadze Ave., 0128, Tbilisi, Georgia

ABSTRACT

We have investigated p-type ZnO layers obtained at T=400°C by Radical Beam Quasiepitaxy (RBQE). Hall Effect measurements performed at T=77K and 300K show resistivity $\rho=5\times10^{-2}$ and 3.1×10^{-3} Ωcm accordingly. At T=77K the hole concentration and mobility was $5\times10^{18}cm^{-3}$ and 22 cm^2/Vs. At T=300K the hole concentration and mobility was $8\times10^{18}cm^{-3}$ and 250 cm^2/Vs. Following bands in PL spectrum of p-type ZnO at T=77 K, λ=369.1 nm, λ=374.5 nm, λ=383.5 nm, λ=392.5 nm and λ=401 nm, were observed and identified. It is shown that RBQE technology enables the obtainment of high monocrystallinity ZnO samples with the low concentration of uncontrollable impurities.

Keywords: II-semiconductor, p-type ZnO, RBQE, PL spectra; PACS Codes: 71.55.Gs, 78.55.Cr, 73.61.Ga, 78.66.Hf.

---

* Corresponding Author: e-mail: tamazbutkhuzi@yahoo.com, Phone: +995 93 256194; 19-21a, Mtsxeta st., Ap. 12, 0179 Tbilisi, Georgia

1. Introduction

Wide gap II-VI compounds are characterized by n-type conductivity. The obtainment of stoichiometric and p-type samples is a serious problem. This is associated with the strong compensation and self-compensation caused by residual impurities and point defects (Kroger [1], Look [2], Nurmiko [3]). The obtainment of intrinsic defective p-n junction in wide gap semiconductors is very difficult (Tizemen [4]). ZnO is a direct gap (3.37 eV at room temperature) semiconductor with properties similar to GaN (Chen [5]). Obtainment of intrinsic defective hole conductivity in ZnO is especially problematic due to the native strong tendency to self-compensation (Sekiguchi [6], Tsukazaki [7]). To solve this problem original method Radial Beam Quasi Epitaxy (RBQE) has been developed (Butkhuzi [8]).

In this work we investigated electrical and optical properties of ZnO samples obtained by RBQE. P-type ZnO growth process was conducted at 400°C. For electrical and optical investigation we used Photoluminescence, thermo-electromotive force and Hall Effect measurements. We used X-ray diffraction to investigate structural characterisation of ZnO.

2. Method of Radical Beam Quasi Epitaxy

RBQE enables us to grow new quasi-epitaxial layers on the surface of binary compound such as ZnO, ZnS, ZnSe. The difference between RBQE and well-known gaseous-phase epitaxy is that during the formation of new layers, the non-metal component comes from the gaseous phase and the metal-component from the basic crystal. In our case the concentration of atomic oxygen is increased due to the external influence. The second significant difference is that the basic crystal is placed in the area of maximal temperature of the reactor. Temperature in other areas of the reactor is much less. This profile of temperature distribution promotes movement of residual impurities towards colder areas of the reactor. It should be mentioned that formation of new layers takes place in the wide temperature range (300 to 1100 K). Experiment shows that p-type ZnO has been obtained during treatment at the temperatures bellow 650°C (Butkhuzi [9]).

Single crystalline layers of ZnO were grown by RBQE on the basic crystal of ZnO but generally it can be used for other binary compounds. Figure 1 shows the scheme of experiment. During RBQE, the basic crystal (3) undergoes treatment in the atmosphere of singlet oxygen radicals obtained by RF discharge (6). The RF oscillator converts the molecular oxygen stream into plasma containing ions, electrons etc. in addition to oxygen radicals. Due to their activity, such oxygen radicals interact with the basic crystal and the growth of new layers is carried out by non-metal components adsorbed from the gas phase and metals extracted from the basic crystal. Singlet Oxygen radicals are highly active when they reach the surface of the basic crystal and they react with the basic crystal. During this process, oxygen radicals stimulate extraction of Zn atom from the basic crystal and the creation of new ZnO quasiepitaxial layer on the surface. Meanwhile, a number of Zn vacancies are created in the surface area of the basic crystal. The narrow profile of temperature distribution (11) is maintained with a water-cooling system in upper (1) and lower (2) parts of the reactor. The magnetic and electric fields (7) control the distance between the plasma and basic crystal. The experiment shows that the effective distance is $d \approx 1$

to 2.5 cm. At d<1 cm the crystal evaporation (ion etching) takes place; at d>2.5 cm, radicals can unify as molecules reducing the efficiency of layer growth. Concentration of oxygen radicals is about $n_o \approx 10^{14}$-$5\times10^{15}cm^{-3}$ (Elias [10]).

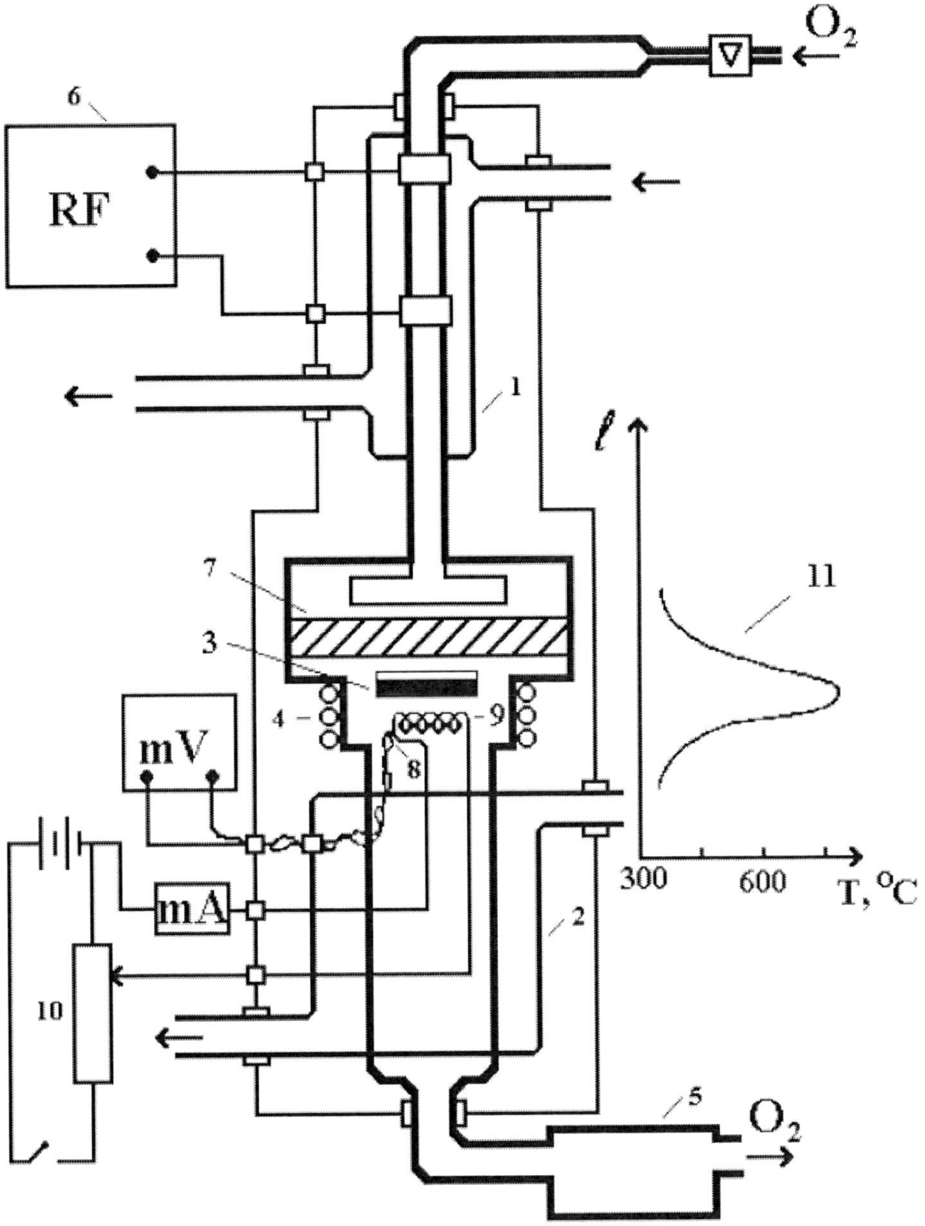

Figure 1. Scheme of RBQE technology: 1) and 2) cooling systems, 3) basic crystal, 4) heater, 5) vacuum pump, 6) RF device, 7) electric and magnetic fields, which separates only oxygen radicals, 8) thermocouple, 9) platinum coil, 10) potentiometer, 11) temperature distribution in the reactor.

According to the RBQE specifics, increased temperature increases the concentration of Zn atom extracted from the basic crystal. At the fixed concentration of oxygen radicals the intensity of metal extraction decreases with a decrease in the growth temperature.

Consequently metal concentration in ZnO layers decreases as well. This causes the raise of the concentration of V_{Zn} in basic crystal. As the temperature of treatment processing is rather high, oxygen vacancies mainly must be ionised, consequently the concentration of V_o is low. Due to the increase of V_{Zn} the Fermi level goes down to the valence band edge and transfer of electrons from V_o^+ to V_{Zn} takes place. As a result V_{Zn}^- and V_o^{+2} are created. With the increased concentration of V_{Zn}^- and V_o^{+2} increased the probability of their association at the expense of Coulomb interaction (V_{Zn}^--V_o^{+2}). That why we think that predominant defects are V_{Zn}^-, V_o^{+2} and (V_{Zn}^--V_o^{+2}). For interpretation of the peaks observed in PL spectra we use this mechanism of defects creation in RBQE. By means of RBQE method it is possible to obtained n-type, p-type and stoichiometric layers of ZnO. RBQE enable to obtained layers with thicknesses from few atomic layers to 25 μm.

3. EXPERIMENTAL

Monocrystalline layers of ZnO were grown on the basis of ZnO crystals by RBQE. As an initial crystal we used high quality n-type ZnO (ρ=9.4×10^{-2} Ωcm) monocrystals samples of wurtzite structure (produced by Cermet, Inc.). In order to obtain p-type ZnO epitaxial layers the basic crystal (n-type ZnO) has been treated at T=400°C in the atmosphere of activated oxygen obtained by RF-discharge with 80 W. Duration was 4 hours, and concentration of oxygen radicals was 10^{14}-10^{15} cm^{-3}. According to thermo-EMF measurement epitaxial ZnO layers showed p-type conductivity. We have studied Photoluminescence (PL) spectra of initial n-type ZnO crystals and new epitaxial p-type ZnO layers obtained by RBQE method. PL measurements were carried out at 70 K and 300K temperatures with a FS920 luminescence spectrometer from Edinburgh Instruments. A Xenon tube was used as an excitation source (λ_{Ex}=313 nm).

Figure (2) represents the PL spectra of initial n-type and obtained p-type samples recorded at T=300 K. In PL spectrum of initial n-type sample maximum at λ=378.5 nm were observed Figure (2-b). In PL spectrum of p-type ZnO layers λ=378.5 nm and λ=394.5 nm were observed Figure (2-a).

In PL spectrum (recorded at T=70K) of initial n-type ZnO λ=369.1 nm, λ=374.5 nm, λ=383.5 nm and λ=392.5 nm peaks were observed Figure (3-a). In PL spectrum of new quasiepitaxial p-type ZnO layers λ=369.1 nm, λ=374.5 nm, λ=383.5 nm, λ=392.5 nm and λ=401 nm peaks were observed but intensities were different than observed in initial crystal Figure (3-b). Figure (4-a) and (b) represent PL spectra of p-type ZnO layers recorded at T=300K and T=70K. Bands recorded at T=70K shifted at short wavelength area by 20nm.

For structural characterization of ZnO, new layers were checked with a Siemens D5000 XRD (X-Ray diffraction) spectrometer (CuKα line, step size 0.02°) Figure (5). Van-der-Pauw method was used to measure Hall Effect. The measurement was performed at T=300K and T=70K Table (1).

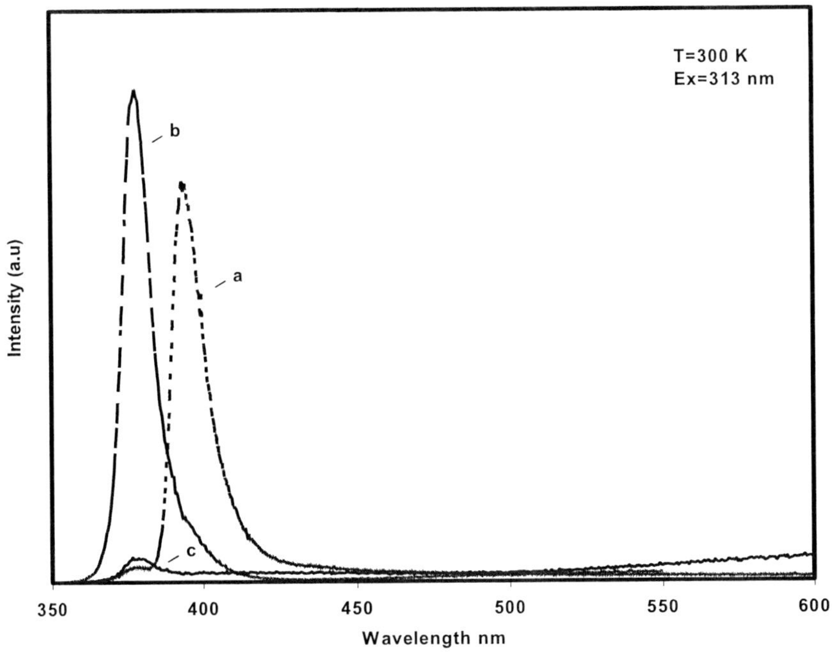

p-type ZnO treated RBQE; n-type initial ZnO; ZnO treated RBQE

Figure 2. Photoluminescence of Zno at T=300 K.

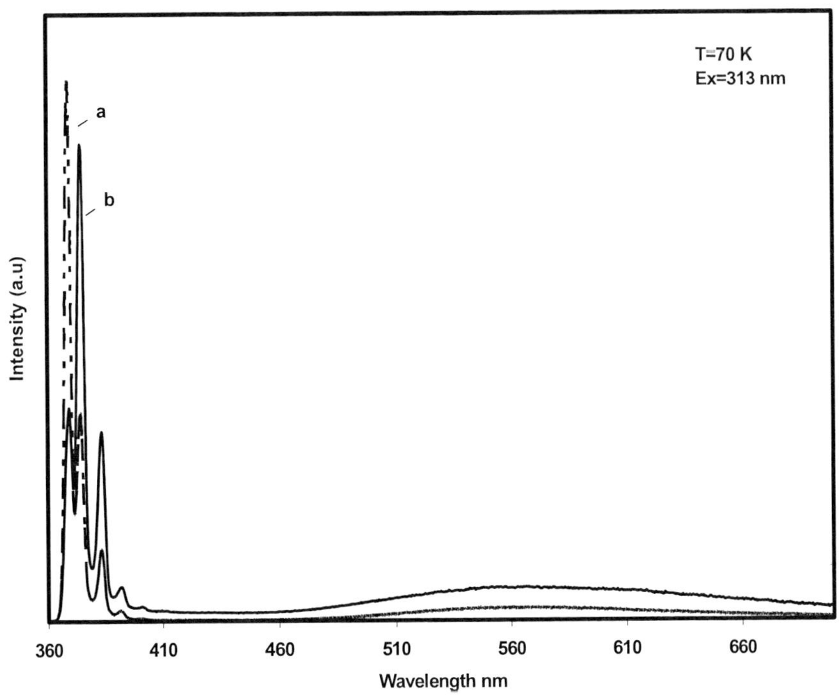

n-type ZnO. p-type ZnO

Figure 3. Photoluminescence of Zno at T=70 K.

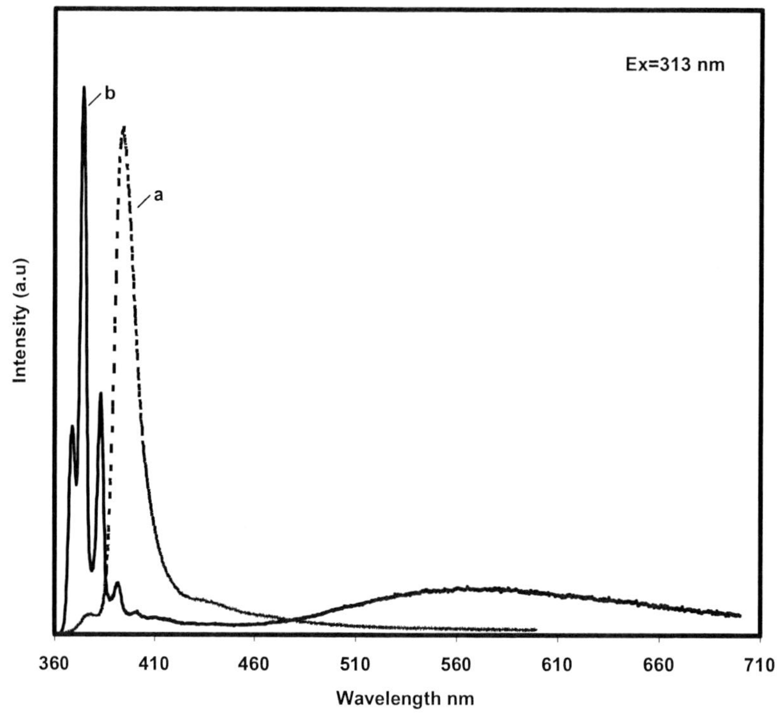

p-type ZnO, 300 K. p-type ZnO, 70 K.

Figure 4. Photoluminescence of p-type Zno.

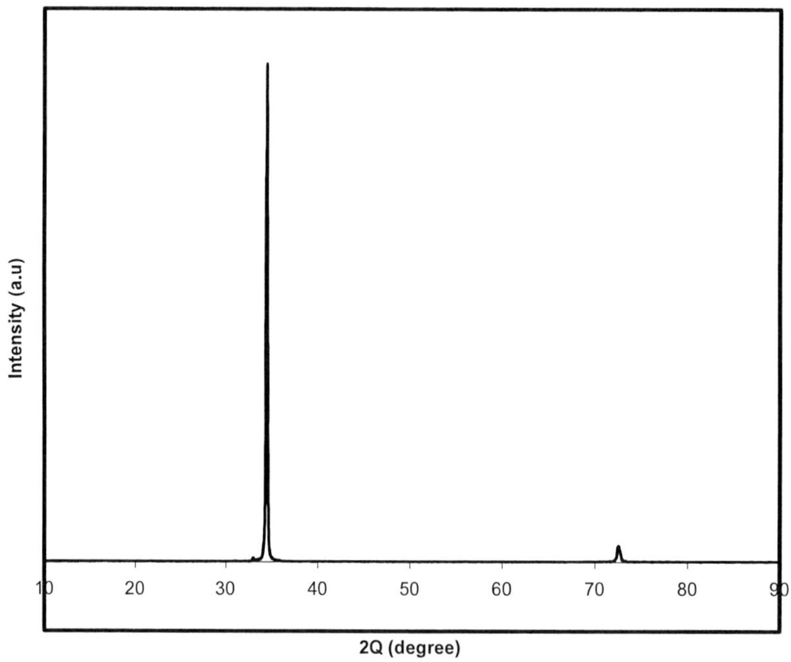

Figure 5. X-Ray diffraction of the ZnO.

Table 1.

Material	T (K)	Carrier mobility, μ, (cm^2/Vs)	Carrier concentration (cm^{-3})	Resistivity, ρ, (Ωcm)
p-type ZnO	77	22 ÷ 25	5 ÷ 6 × 10^{18}	5 ÷ 4.2 × 10^{-2}
p-type ZnO	300	230 ÷250	7 ÷ 8 × 10^{18}	3.9 ÷ 3.1 × 10^{-3}

4. Results and Discussion

In order to identify the defects responsible for the bands observed in the PL spectra we are taking into account the specific mechanisms of defects creation in ZnO under RBQE.

Taking into account the specific nature of RBQE, the maximum λ=401 nm band in PL spectrum at T=70K must be connected to V_{Zn}^-. The low intensity of this line may be explained with the low electron capture cross-section (Kohan [11]). The λ=392.5 nm band could be connected to the V_o^{2+}. The λ=383.5 nm band in the PL spectrum is connected to the associate (V_{Zn}^--V_o^{2+}) (Ursaki [12]). As it appears in the samples with p-type conductivity, it should be connected with acceptor centres. In p-type ZnO samples high intensity λ=374.5 nm band is observed and corresponds to acceptor bound exciton Figure (3-a). The peak at λ=369.1 nm in PL spectrum of initial n-type ZnO samples is more intensive than the one at λ=374.5 nm Figure (3-a). The peaks positions at λ=369.1 nm and λ=374.5 nm corresponds to donor bound excitons (Teke [13], Chichibu [14]) and acceptor bound excitons (Ashrafi [15]) respectively. Observation of the acceptor bound exciton in p-type ZnO layers obtained by RBQE testifies a high purity and perfect structure and significantly reduced number of residual defects, which is one of the most important issues.

According to the X-ray diffraction new layers of ZnO are characterized by the hexagonal structure. The X-ray diffraction study shows agreement of "a" and "c" parameters with known parameters of wurtzite ZnO (a=3.24982 Å; c=5.20661Å). X-ray diffraction analysis indicated that the prepared ZnO thin films had the preferred c-axis orientation of crystallites. Peak positiond at 34.4° could be observed for the films grown at the RBQE, which corresponded to the (0002) plane diffraction of ZnO with hexagonal wurtzite structure Figure (5). The intensity of the (0002) peak decreased when increasing the distance between oxygen plasma and crystal surface. This fact suggested that the crystalline quality of the ZnO thin films degraded from a single-oriented crystal to polycrystalline with the increase of the Zn content. With an increase of the distance between oxygen plasma and crystal surface O_2 molecules created by association of oxygen radicals, and concentration of oxygen in obtained surface layers decreased. This causes the degradation of the obtained ZnO crystalline quality. PL spectrum recorded at 300 K of the samples grown by RBQE with longer distance (>3 cm) between plasma and crystal surface shows weak maximum at 379.5 nm Figure (2c).

Thus, the study of the X-ray and PL spectra has shown that by RBQE method ZnO layers with wurtzite structure are obtained on the ZnO.

Hall Effect and thermo-EMF measurements were performed as well, which showed that we obtained ZnO layers with hole conductivity. Hall measurement results were summarized in Table (1).

With increasing temperature from 77 K to 300 K carrier mobility is increasing in one order of magnitude. It is known that carrier scattering occurs on the lattice defects (self and impurity defects) and phonons. At 300 K scattering on the phonons are predominated and directly proportional to the temperature. At lower temperatures carrier scattering mostly occurs on defects. Considering the specific nature of defect creation during RBQE, carrier scattering in obtained ZnO samples occur on charged point defects at low temperatures (the concentration of noncontrollable impurity defects are very low). Similar effect of temperature vs. mobility dependent (50-350 K) has been observed in p-type ZnO:P samples (Xiu [16]). From 250 to 350 K, the halls scattering on the phonons are dominated, while below 120 K scattering on ionized impurities take place. Between 120 and 250 K the both mechanisms coexist.

A similar relationship was observed by Schmidt [17]. He has investigated electrical properties of ZnO:Cu where mobility of the samples treated in vacuum are changing curiously. Particularly mobility increases by one order of magnitude from 10K to 350 K. Author suppose that this can be effect of impurity diffusion in the bulk of the crystal at low temperatures. Thus, in ZnO with low concentration of impurities in the temperature range 10 K to 350 K direct proportional relationship of mobility vs. temperature takes place.

In case of RBQE observed change of mobility can be explained by the fact that there is low concentration of residual impurities due to technological specifics of RBQE.

Thus, the RBQE method enables the control of the component concentration ratio and consequently electrical and optical properties of basic crystal as well as the new grown layers. Observation of acceptor bound exciton in p-type ZnO layers obtained by RBQE exhibits high purity and perfect structure, which is one of the most important issues in the production of wide band gap II-VI materials.

Acknowledgements

This work was supported by Cermet Inc., USA and Department of Science and Technology of Georgia.

References

[1] Kroger F., Khimia Nesovershennix Kristallov, *Mir*, M, Russian, 1969.

[2] Look D., Reynolds D., *Appl. Phys. Lett.* 81, 1830, 2002.

[3] Nurmiko A. and Gunshor R.L.: SST, 12, 1337, 1997.

[4] Tizemen S., Gang Xiong, Mishuck Braini, *Physica B*: 308, 1197, 2001.

[5] Chen Y.F., Bangall D.M., Koh H., Park K., *J. Appl. Phys*. 84, 3912, 1998.

[6] Sekiguchi T., Ohashi N. and yamane H., *Solid State Phenomena*, 63-64, 171, 1998.

[7] Tsukazaki A., Saiko H., Tamura K., Ohtani M., *Appl. Phys. Lett.* 81, 236, 2002.

[8] Butkhuzi T., Khulordava T., Tsekvava B., *Phys.Rev*. B Vol. 58, 16, 10692, 1998.

[9] Butkhuzi T., et. all. *Proc. of the Int. Conf. on Sol. State Crystals* Epilayers and Heterostructures in Optoelectronics and Semiconductor Technology, pp. 53-57, 1998.

[10] Elias L., Ogryzlo E.A. and Schiff H.J., *Can. Chem.* 37, 1680, 1959.

[11] Kohan A., Van de Walle C., *Phys. Rev.* B Vol. 61, N 22, 1519, 2000.
[12] Ursaki V.V., et.al. *Semicond. Sci. Technol.* v. 19 p. 851-854, 2004.
[13] Teke A., Ozgur U., Dogan S., *Phys. Rev.*, B 70, 1, 2004.
[14] Chichibu S.F. et. al. *J. Appl. Phys.* v. 93, p. 759, 2003.
[15] Ashrafi A.B.M.A. et. al. *Jpn. J. Appl. Phys. Part 2* 41, LI281, 2002.
[16] Xiu F. X., at. all. *Applied Physics Letters* 88, 052106, 2006.
[17] Schmidt O. at. all. *Procc. of E-MRS Spring Meeting*, Symposium G, Strasbourg, G.11.05, 2005.

In: Encyclopedia of Nanotechnology
Editor: Elwood D. Carlson
ISBN: 978-1-60692-079-4

Chapter 16

LOW-DIMENSIONAL ZNO NANOSTRUCTURES: DESIGN, STRUCTURES AND PROPERTIES

Guozhen Shen[a,1], Di Chen[b] and Cheol Jin Lee[c]

[a]Nanoscale Materials Center, National Institute for Materials Science (NIMS), Namiki 1-1, Tsukuba, Ibaraki 305-0044, Japan

[b]Photocatalytic Materials Center, National Institute for Materials Science (NIMS), Sengen 1-2-1, Tsukuba, Ibaraki 305-0047, Japan

[c]School of Electrical Engineering, Korea University, 5-1 Anam-Dong, Seongbuk-Gu, Seoul 136-701, Korea

ABSTRACT

Zinc oxide (ZnO) based one-dimensional (1-D) nanoscale materials, as important functional oxide nanostructures, have received increasing attention over the past few years due to their potential applications in optoelectronic switches, high efficiency photonic devices, near-UV lasers, and for assembling complex three-dimensional nanoscale systems. Various kinds of ZnO nanostructures have been realized, such as nanodots, nanorods, nanowires, nanohelixes, seamless nanorings, mesoporous nanowires and so on. This paper presents a brief review of our recent research achievements on low-dimensional ZnO nanostructures. By using a metal vapor deposition process, a group of unique nanoarchitectures of ZnO are synthesized, including several types of hierarchical ZnO nanostructures, vertically aligned nanowires, microtubes, nanonails, nanoribbons-nanowires junctions, etc. ZnO-based heterostructures are also briefly discussed in this paper. Structures of these nanostructures are discussed in detail using several techniques including scanning electron miscroscopy, high-resolution transmission electron microscopy and several structure models are presented. Finally, structural related photoluminescence properties are demonstrated. The synthesized ZnO nanostructures show excellent field emission properties. These novel nanoarchitectures could be unique for many applications in nanotechnology.

[1] E-mail: SHEN.Guozhen@nims.go.jp; gzshen@ustc.edu; Fax: +81-29-851-6280.

1. Introduction

Zinc oxide, with a wide band gap of 3.37 eV and a large excitation binding energy of 60 meV at room temperature is an important semiconducting and piezoelectric material. Wurtzite ZnO has a hexagonal structure, which can be simply described as a number of alternating planes composed of tetrahedral coordinated O^{2-} and Zn^{2+} ions stacking alternately along the *c* axis [1-3]. The tetrahedral coordination in ZnO results in noncentral symmetric structure and consequently piezoelectricity and pyroelectricity. In wrutzite ZnO crystal, the oppositely charged ions produce positively charged (0001)-Zn and negatively charged $(000\bar{1})$-O polar surfaces, which result in a normal dipole moment and spontaneous polarization along the c-axis as well as a divergence in surface energy (shown in Figure 1). It is a smart material and has versatile applications in the field of optoelectronics, piezoelectric sensors, transducers, and resonators. The wide band gap and large excitation binding energy make ZnO suitable for short wavelength optoelectronic applications and ensure an efficient excitonic emission at room temperature. ZnO is transparent to visible light and can be made highly conductive by doping. Element zinc is one of the absolutely necessary elements for human beings and ZnO is also biosafe and biocompatible and it can be directly used for biomedical applications without coating.

Since Yang and his co-workers reported ZnO nanowire array nanolasers in 2001 [4], synthesis and assembly of ZnO nanostructures, such as nanowires, nanobelts, nanotubes, have been widely explored in recent years and many kinds of methods based on chemical solution process and vapor phase approaches, such as thermal evaporation process, have been developed [5-40]. In principle, the thermal evaporation technique is a simple process in which the source materials are vaporized at elevating temperature and the resultant vapor phases condense under proper experimental conditions to form the desired products.

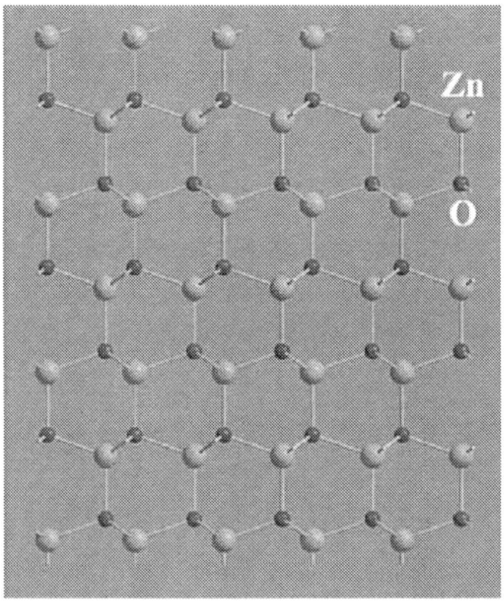

Figure 1. The projected structure of wurtzite structured ZnO along a-axis, showing the {0001} polar surfaces.

The processes are usually carried out in a horizontal tube furnace, which is composed of a horizontal tube furnace, a quartz tube, and a gas supply and control system. In the thermal evaporation process, ZnO is the commonly used source material. Many kinds of low-dimensional ZnO nanostructures have been obtained. However, temperature higher than 850 ^{o}C is usually required with the use of ZnO as the source material. Recently, it was found that metallic zinc can also be used as the source material to synthesize low-dimensional ZnO nanostructures [41-50] in vapor phase processes. With the use of metallic zinc, high temperature are avoid and the process are mainly performed at 400-600 ^{o}C since the Zn vapor pressure can reach up to 10.8 Pa at a temperature even down to 400 ^{o}C.

Low-dimensional ZnO nanostructures synthesized from ZnO powders have been extensively studied and several key reviews could be found on this topic. So we will not discuss it and in this work, we mainly focused on the synthesis of low-dimensional ZnO nanostructures using the low-temperature metal vapor deposition (MVD) process, in which metallic zinc powder is used as the source material. By using the MVD process, a diverse group of low-dimensional ZnO nanostructures are synthesized, such as hollow urchins, microtrepangs, microbelts of nanobowling-pin arrays, nanoflowers, nanocombs, nanowheels of ZnO nanoarrows, nanofans, nanowire-nanobelt junctions, microtubes by coalescence of ZnO nanowires and nanobarriers [51-60]. The objective of this article is to review these unique nanostructures and their corresponding growth processes. The properties, such as optical properties, field emission properties, of different ZnO nanostructures will also be reviewed.

2. Results and Discussion

2.1. Experimental Setup

Typical MVD process used in the work involves an experimental setup consisting of a horizontal tube furnace (one heating-zone furnace was used when pure zinc powder was selected as the only source materials and two heating-zone furnace when some additives were introduced), a quartz tube and a gas controlling system. The process was performed under atmospheric pressure and reaction temperature was set to 500-750 ^{o}C. The deposited products were characterized by scanning electron microscopy (SEM, Hitachi S-4700) and transmission electron microscopy (TEM, Tecnai F20) and selected area electron diffraction (SAED). The X-ray diffraction spectrum (XRD) was obtained on a Rigaku DMAX 2500 diffractometer. Photoluminscence (PL, 325 nm He-Cd laser) and Raman (Labram-HR) were used to check their optical properties.

2.2. Several Kinds of ZnO Nanostructures

2.2.1. ZnO Nanowires

Nanowire is one of the most common low-dimensional structures of ZnO. ZnO nanowires were firstly reported by Zhang et al. by using aAAM template-based method, which involves the generationg of AAM template, electrodeposition of pure metal zinc into the channel and the oxidation of zinc [61]. After that, many groups tried to synthesize ZnO nanowires using different methods. Using the MVD process, we could get uniform ZnO nanowires at low temperature. Figure 2 is the SEM image of ZnO nanowires deposited on silicon substrate by thermal evaporation of zinc powder at 500 °C. The nanowires have diameters of about 100-200 nm and lengths of several micrometers. They located randomly on the substrate. Higher-magnification SEM inset shows they are of perfect hexagonal cross-sections. All the ZnO nanowires have the preferred (0001) growth directions. Since no catalyst is used in the process, the ZnO nanowires are obviously formed by the vapor-solid mechanism.

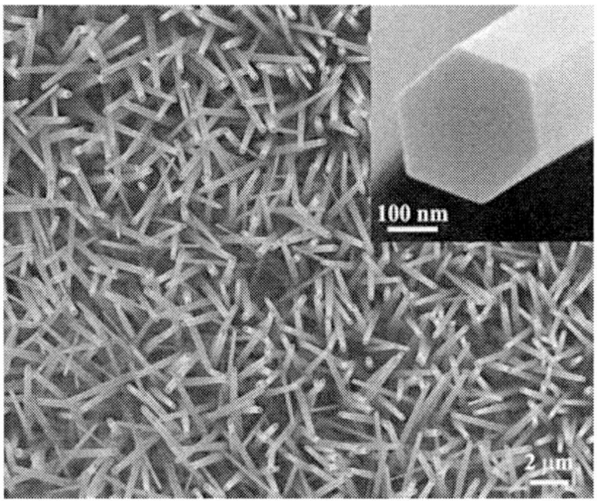

Figure 2. ZnO nanowires deposited on silicon substrate without catalyst.

2.2.2. ZnO Nanobelts

In the MVD process, the morphology of the ZnO product is greatly influenced by the experimental parameters, such as temperature, substrate position, gases flow, etc. Figure 3 shows the SEM images of ZnO nanobelts synthesized from zinc powders at 700 °C. These nanobelts are of rectangular cross-sections compared with the common ZnO nanowires. Their thickness lies in several to several tens of nanometers.

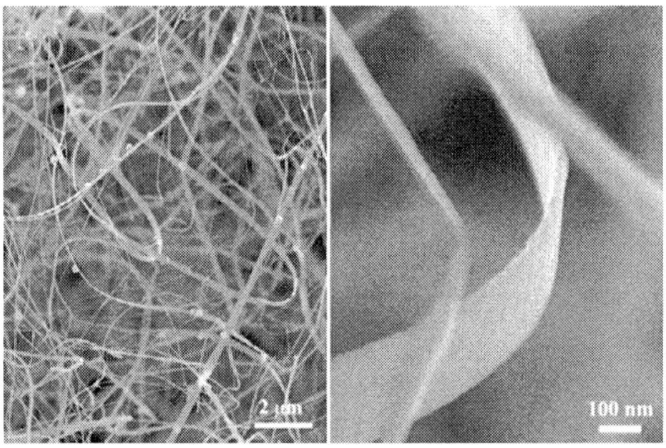

Figure 3. SEM images of the ZnO nanobelts from zinc powders without catalyst.

2.2.3. ZnO Nano/Microtubes

Inspired by the work of Iijima in 1991 [62], great interests have been paid to inorganic nanotubes with hollow cavities. A variety of them based on layered, or pseudo-layered materials have been prepared, which include various families such as transition-metal chalcogenide (for example, MX_2, M: Mo or W, X: S or Se; and others), $NiCl_2$, BN, vanadium oxide, Bi, etc [63-80]. Using a solvothermal method, polycrystalline ZnO nanotubes with diameters of about 450 nm could be obtained [81]. Single crystalline ZnO nanotubes were also synthesized by thermal reduction routes at high temperature using ZnO powders as the source material [82,83]. Other methods were also used to synthesize ZnO nanotubes [84-90]. MVD process is also useful for the synthesis of ZnO nanotubes. By thermal evaporation of Zn powder at 475 °C with a constant flow of Ar (95 sccm) and 0.25 sccm O_2, ZnO nanotubes with inerratic hexagonshaped cross section were grown perpendicular to the surface of silicon substrate [91]. Conic ZnO nanotubes could be fabricated at low pressure and low temperature by the MVD process without catalyst [92]. Very recently, it was found that if catalysts were introduced into the reaction system and the MVD process was performed at high temperature (950 °C), single-crystalline ZnO microtubes formed by coalescence of ZnO nanowires could be obtained at suitable temperature conditions [93]. The ZnO microtubes were deposited on NiO particles coated Al_2O_3 substrate. The reaction gas and carrier gas are O_2 (5 sccm) and Ar (500 sccm).

Figure 4 are the SEM images of the obtained ZnO microtubes. They reveal that the ZnO microtubes have outer diameters in the range of 03.-2 μm, wall thickness in the range of 100-500 nm, and lengths about 10 μm. The magnified SEM images show that most ZnO microtubes have a hollow cavity and a faceted hexagonal shape.

Figure 4. Typical SEM images of the synthesized ZnO microtubes: (a,b) low-magnification SEM images of the microtubes; (c,d) high-magnification SEM images revealing geometrical morphology of the microtubes with a faceted hexagonal shape. (Copyright American Chemical Society, reproduced with permission).

In addition, we found that there were many incomplete ZnO microtubes, which showed the special morphology assembled with several nanowires (Figure 5). Based on our observations, we proposed a possible growth mechanim for the formation of ZnO microtubes. The whole growth process involves the deposition of thin ZnO layer on NiO catalyst particles, generation of nucleation sites on large ZnO grains, vertical growth of ZnO nanowires, formation of incomplete ZnO tubular structures and formation of perfect ZnO microtubes by the coalenscence of neighboring nanowires.

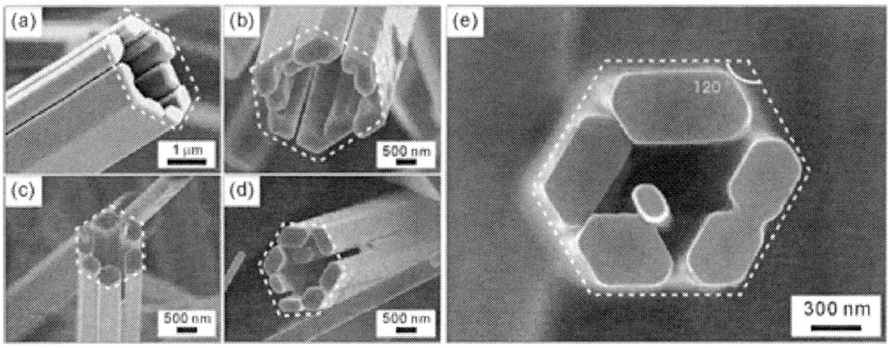

Figure 5. SEM images of the ZnO microtubes with the morphology assembled with several nanowires. (Copyright American Chemical Society, reproduced with permission).

2.2.4. Hierarchical ZnO Nanostructures

In recent years, the self-assembly and higher-order organization of nanostructured building blocks into two- or three-dimensional complex architectures have been of intense interest in current materials synthesis and nanodevice fabrication [94-100]. Using pure zinc powder as the source material, nanowires, nanobelts and nano/microtubes can be synthesized by the MVD process at convenient conditions. Simply modifying the reaction system, it was found that the morphology of the grown ZnO nanostructures can be dramatically changed. The most distinguished one is hollow ZnO urchins.

To get hollow ZnO urchins, a one heating zone tube furnace with very short heating zone was used and zinc powder is selected as the only source material. The reaction temperature was set to about 750 ^{o}C. Figure 6 shows the typical SEM images of hollow ZnO urchins deposited on silicon substrate after thermal evaporation for 60 min. The diameters of these hollow ZnO urchins range from several to several tens of micrometers. They clearly indicate interesting hollow structures and the cavities of the urchins can be clearly seen from Figure 6c. On each urchin, numerous one-dimensional ZnO nanowires with very high density grow pointing toward the center of the sphere. Each nanowire have diameter of about 40 nm and length of 4 μm, respectively (Figure 6d).

The formation of hollow ZnO urchins involves the vaporization of Zn powder, solidification of liquid droplets, surface oxidation, sublimation and self-catalytic growth of one-dimensional nanowires based on our experimental results. The whole process can be illustrated as shown in Figure 7.

During the synthesis of ZnO nanostructures, it was found that modifying the composition of the source materials can drastically change the morphology of the oxide nanostructure grown. When nickel sulfide powder was introduced into the source material, comb-like ZnO nanostructures were found deposited on silicon substrate. Figure 8 shows the SEM images of the synthesized ZnO nanocombs by thermal evaporation of Zn and Ni_2S powder in a two-heating zone furnace. The synthesized ZnO nanocombs are composed of aligned ZnO comb teeth growing along the conventional [0001] direction on [01-10] grown ZnO nanobelts. The comb teech and nanobelts can be clearly seen from Figure 8b and 8c.

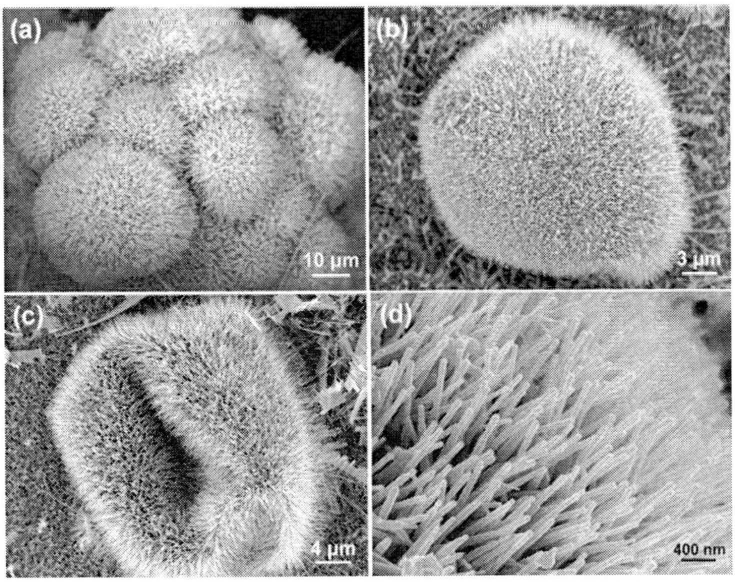

Figure 6. Typical SEM images with different magnifications of hollow ZnO urchins, which indicate high density ZnO nanowires radially aligned on the sphere surface. (Copyright American Chemical Society, reproduced with permission).

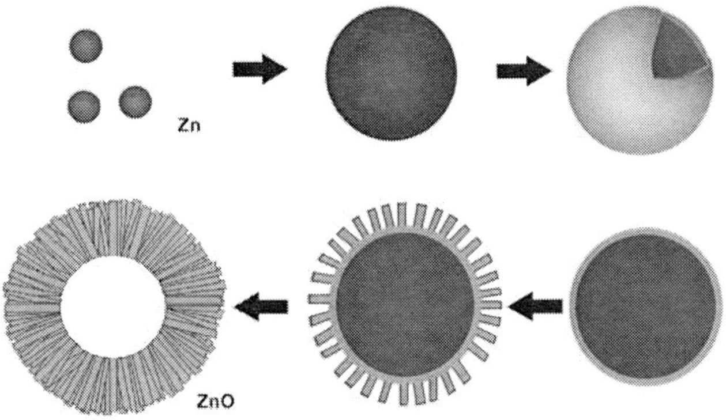

Figure 7. Schematic illustration of the formation process of hollow ZnO urchis. (Copyright American Chemical Society, reproduced with permission).

It is well known that the oppositely charged ions in wurtzite ZnO crystal can produce positively charged (0001)-Zn and negatively charged $(000\bar{1})$-O polar surfaces. The growth of asymmetric ZnO combs can be described based on this concept. It was induced by the polar surfaces in wurtzite ZnO crystals. During the growth process, ZnO nanobelts are firstly formed in very short time. The positively charged (0001)-Zn surface in the nanobelts tends to have tiny Zn clusters at the growth front. The Zn clusters as well as the local enrichment in

Zn at the growth front tend to initiate the growth of ZnO analogous to the vapor-liquid-solid (VLS) growth process. Therefore, the self-catalyzed effect of the Zn clusters results in the growth of the nanowires on one side of the nanobelts.

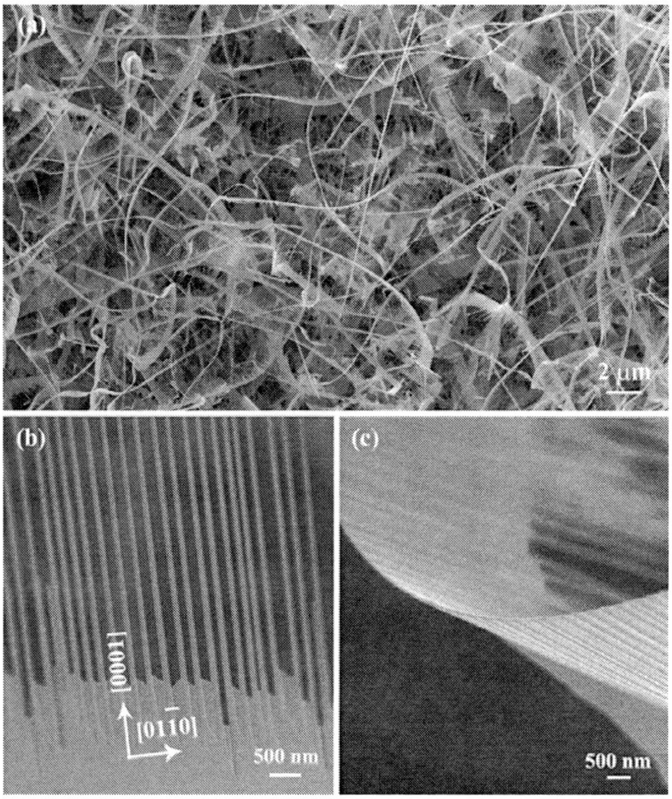

Figure 8. SEM images of the ZnO nanocombs, which are composed of aligned [0001] grown ZnO nanowires on one side of the [01-10] grown ZnO nanobelts.

In fact, the morphology of the branched teech in the nanocombs can be efficiently tuned by changing the conditions. Figure 9 shows the SEM images of a series of ZnO teeth in the synthesized nanocombs. The teech can vary from commn nanowires with uniform diameters along the whole length, sharp nanoneedles, nanopencils, plate nanonails, cubic nanowires, to nanonails.

When a mixture of In/In_2S_3 was introduced into the MVD process, other kinds of three-dimensional hierarchical ZnO nanostructures were obtained. Figure 10a shows the low-magnification SEM image of interesting 3D trepang-like ZnO nanostructures, which were synthesized by thermal evaporation of Zn powder and the mixture of In/In_2S_3 with a weight ratio of 1:2 (In to In_2S_3). The reaction temperatures for Zn powder and the mixture of In/In_2S_3 were set to 550 ^oC and 900 ^oC, respectively. The SEM image reveals that the ZnO trepangs have lengths of 3-20 μm and uniform diameters of 1.5 μm. High-magnification SEM images shown in Figure 10b demonstrates that, on each ZnO trepang, there are a number of uniform triangular ZnO nanocones radially oriented along the main trunk. The tip of each ZnO nanocones ranges from 10 to 30 nm. The top view image shown in Figure 10c also confirms the assembly of ZnO nanocones to form microtrepangs. Besides the straight trepangs, some

T-shaped microtrepangs were also found in the product as indicated in Figure 10d. TEM studies depict that all the branched ZnO nanocones growed along the [0001] directions.

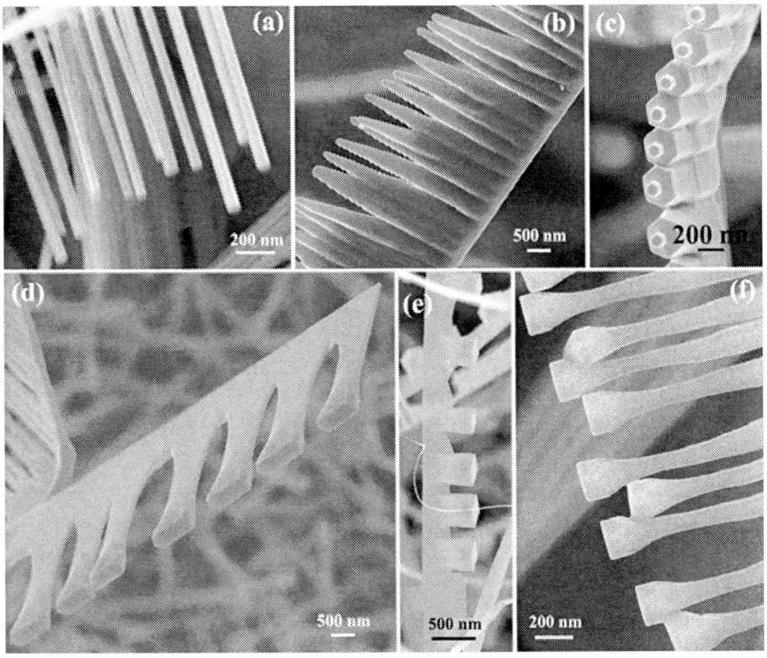

Figure 9. SEM images of the MVD synthesized ZnO combs with different teech structures.

Figure 10. SEM images of the ZnO microtrepangs composed of ZnO nanocones: (a) low-magnification image, (b) side-view image, (c) top-view image, and (d) side view of a T-shaped microtrepang. (Copyright American Chemical Society, reproduced with permission).

In this process, it was found that the weight ratio of In to In_2S_3 has great influence on the morphology of the final products. With the change of weight ratio of In to In_2S_3, several kinds of other 3D were obtained without the change of any other experimental parameter. Figure 11 shows the SEM images of the microbelts of ZnO nanobowling-pin arrays fabricated when the weight ratio of In to In_2S_3 was changed to 1:1. Figure 11a shows the general morphology of the product. Belt-like crystals with typical diameters of 1 μm and length up to tens of micrometers are deposited on the silicon substrate. Magnified SEM images shown in Figure 11b-d clearly indicate that two sets of ZnO nanobowling-pin arrays have perpendicular growth directions to the lengths of the ZnO microbelts. Each nanobowling-pin is composed of a hexagonal prism cap and a main trunk connected with a smaller diameter neck. The diameter of the hexagonal cap is 30 nm and the typical height is 100 nm. Compared with the synthsis of ZnO microtrepangs of nanocones, it can be seen that by only adjusting the weight ratio of In to In_2S_3, not only the morphologies of the products changed greatly from microtrepangs to microbelts, but also their construct units also changed from nanocones to nanobowling pins.

Further modifying the weight ratio of In to In_2S_3 to 3:2 led to the formation of nanoflower structures of ZnO nanobottles. Each ZnO nanoflower is about 2-5 μm in size with many petals radially grown from the center with spherical symmetry. Each petal is of bottle-like morphology of 200-500 nm in height and 100 nm in diameter.

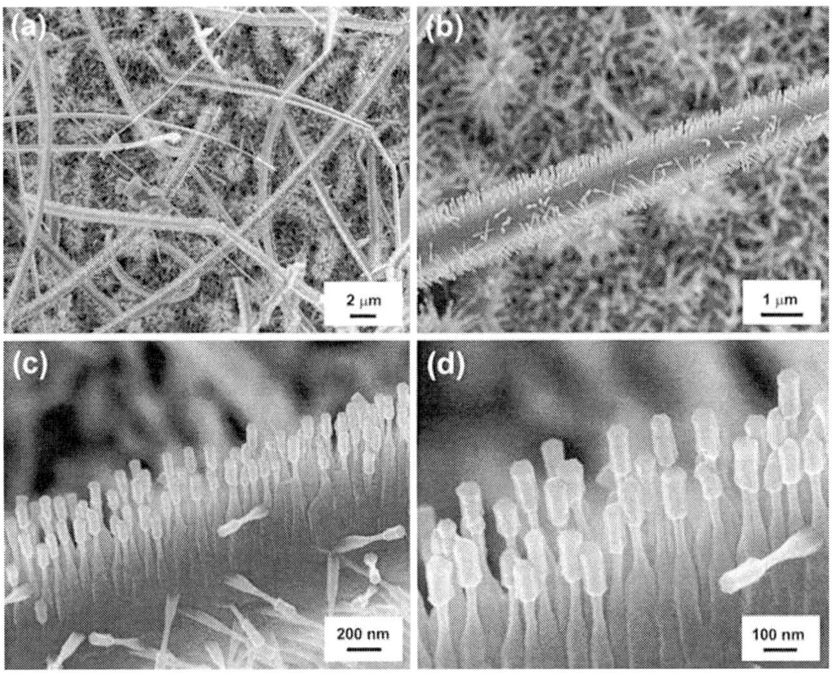

Figure 11. SEM images with different magnifications of microbelts of ZnO nanobowling-pin arrays. (Copyright American Chemical Society, reproduced with permission).

During the modification of weight ratio of In to In_2S_3, a novel delicate wheel-like structures by rolling up ZnO nanocombs are observed. Figure 12 shows the SEM images of this striking structure.

Figure 12. SEM images of novel nanowheel composed of ZnO nanoarrows.

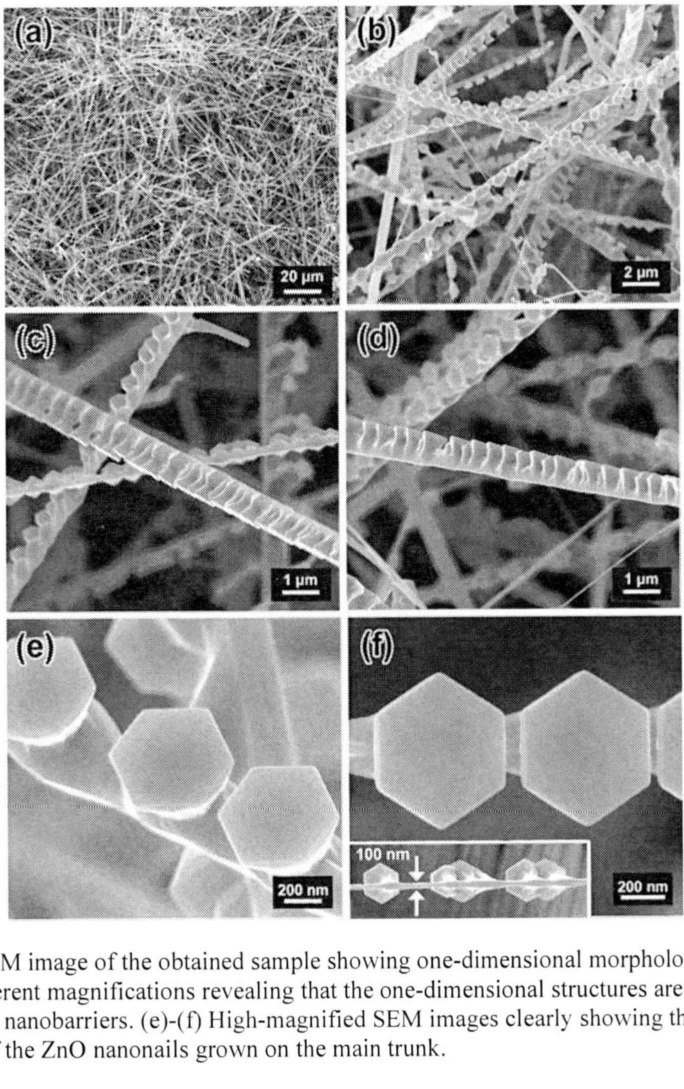

Figure 13. (a) SEM image of the obtained sample showing one-dimensional morphology. (b)-(d) SEM images with different magnifications revealing that the one-dimensional structures are actually hierarchical ZnO nanobarriers. (e)-(f) High-magnified SEM images clearly showing the perfect hexagonal cap of the ZnO nanonails grown on the main trunk.

It can be seen that the nanowheel is folded from a one-sided nanocomb. The composed branches are ZnO nanoarrows with lengths of 1 μm and the diameters of the trunks are 50 nm. According to Wang's work on single crystal ZnO nanorings, the formation of ZnO nanowheels can also be explained from the view point of long range electrostatic interactions. During the formation of ZnO nanowheels, the one-sided nanocomb structure first is formed, and then the alignment of the ZnO nanoarrow arrays on only one side of the nanocomb destroys the balance of the electrostatic interaction. As a result, the nanocomb tends to fold itself as its length gets longer to minimize the area of the polar surface and finally the nanocomb turns into the nanowheel nanostructure.

ZnO nanobarriers are of novel hierarchical structures consisted of numerous hexagonal ZnO nanonails periodly grown on one side of ZnO nanobelts. They were formed by thermal evaporation of Zn powder covered gallium film at 550 ^{o}C. In a typical hierarchical nanobarrier, numerous faceted nail-shaped nanowires (nanonails) (as the branches) grown periodically on a single nanobelt (as the trunk) with typical length of several to several tens of micrometers (Figure 13a-d). The ZnO nanonails are about several hundred nanometers in length with bigger caps on their tips. Figure 13e and 13f are the tilt- and top-view SEM images showing the caps of the nanonails periodically grown on typical hierarchical ZnO nanobarriers. These caps are of well faceted hexagonal structure. Each hexagonal cap is about 300 nm in length of side. A SEM image taken from the back side of a single hierarchical ZnO nanobarrier shown in Figure 13f inset reveals that the ZnO nanobelt is 100 nm in thickness. In fact, the as-obtained ZnO nanobarriers can be described as another kinds of comb-like structures. That is to say, the ZnO nanobarriers are also induced by the polar surfaces of wurtzite ZnO crystals.

2.2.5. Aligned ZnO Nanostructures

To improve the properties of ZnO nanostructures, such as field emission properties, patterned growth of aligned 1-D ZnO nanostructures is of particularly importance. Different methods have been reported for the fabrication of arrays of well-aligned ZnO nanowires including template methods, thermal evaporation, vapor transports and condensation methods, metal organic chemical vapor deposition methods, buffer layer pre-coated methods, solution methods, and laser ablation methods.

Without the use of catalyst, vertically aligned ZnO nanowires can be deposited on silicon substrate via the MVD process at temperature ranging from 400-800 ^{o}C. Figure 14a shows the SEM image of the synthesized vertically aligned ZnO nanowires. It indicates that the obtained ZnO nanowires are quite straight and have uniform diameters along their lengths. The diameters are about 70 nm in accordance to the SEM observations. Figure 14b is the lattice-resolved HRTEM image of a single ZnO nanowire. The lattice fringe of the ZnO nanowire is about 0.52 nm, corresponding to the (001) fringes perpendicular to the growth direction, which is consistent with that of the bulk wurtzite ZnO crystal. Since no metal catalyst was employed in our synthesis, the growth mechanism of vertically aligned ZnO nanowires cannot be explained using the conventional vapor-liquid-solid (VLS) model. During the process, zinc powders are evaporated to generate zinc vapors and they are transferred to the low temperature region by Ar gas and deposit on silicon substrate to form a thin layer of Zn film. And then small ZnO nanorods homogeneously epitaxially grow on the newly formed ZnO film and finally resulted in the formation of vertically aligned ZnO nanowires.

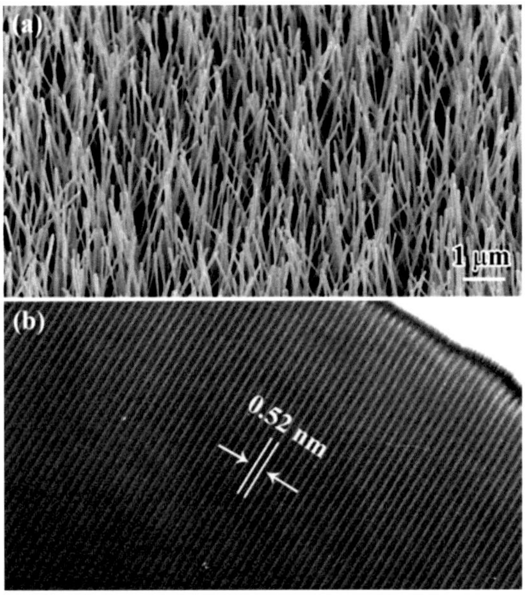

Figure 14. (a) SEM image and (b) HRTEM image of the synthesized vertically aligned ZnO nanowires on Si substrate.

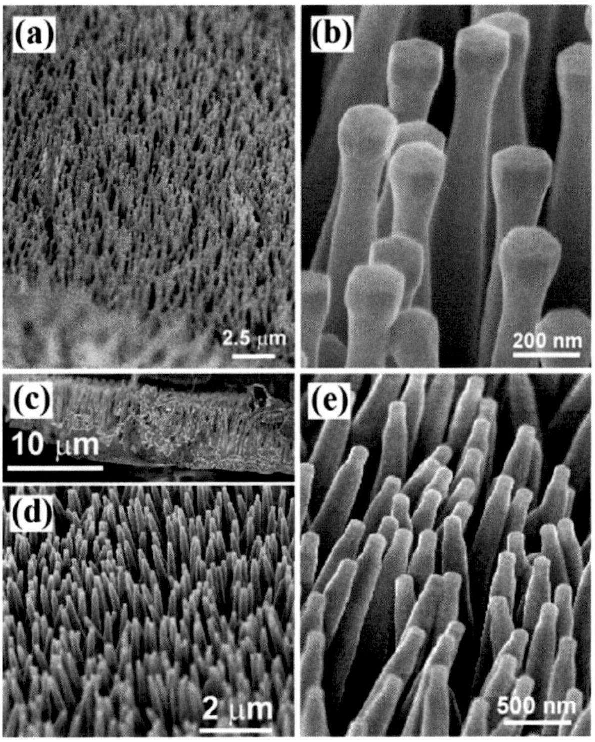

Figure 15. (a,b) SEM image of the vertically aligned ZnO nanonails synthesized at 600 °C. (c-e) SEM images of the vertically aligned ZnO nanopencils synthesized at 700 °C. (Copyright Wiley-VCH, reproduced with permission).

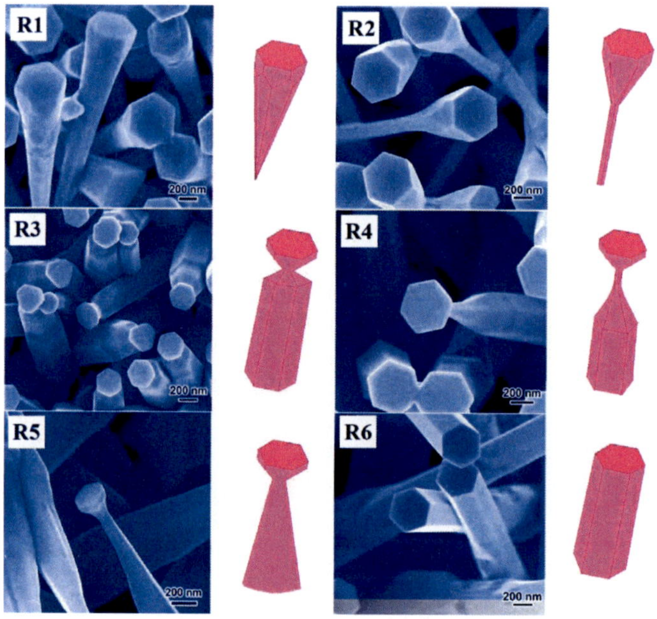

Figure 16. SEM image and structure models of the synthesized ZnO nanostructures from one experimental round to another. (Copyright American Chemical Society, reproduced with permission).

If the synthesis of ZnO structures from zinc powder was introduced a modified MVD process, where an adiabatic layer was set in the furnace, vertically ZnO nanostructures with varied tip structures can be synthesized on Si substrate at different temperature. Figure 15a shows the SEM image of the ZnO nanostructure synthesized at 600 °C. From the tilt-view image (Figures 15b), it can be seen that large-scale vertically-aligned one-dimensional ZnO nanostructures with high density were uniformly grown over the entire surface of the silicon substrate. Interestingly, the diameters of the one-dimensional ZnO nanostructures are no longer uniform along their length but gradually decrease from the base to the top and end with a big diameter hexagonal cap, composing nail-shaped nanowires. That is to say, each ZnO nanonail is consisted of a hexagonal cap attached to a cone-shaped stem.

When the reaction temperature was set to 700 °C, another interesting ZnO nanostructures, ZnO nanopencils, are obtained. Figure 15c shows the image of the cross section of the silicon substrate deposited with ZnO product, which clearly indicates the thickness of the product of about 9 μm. The tilt-view image (Figure 15d,e) also show the formation of vertically-aligned ZnO nanostructures with high density on the whole substrate. Compared with the ZnO nanonails obtained at 600 °C, the sample obtained at 700 °C exhibits the pencil-like morphology. Each ZnO nanopencil is about 8 μm in length and composed of a hexagonal stem with a small diameter hexagonal tip. The diameter of the stem is about 300 nm and that of the tip is about 50 nm.

The tip morphology of the synthesized vertically aligned ZnO nanostructures can be further tuned with the introduce of additives, such as metallic indium powders or thin films. With the transfer from one experimental round to another, the obtained nanostructures are ZnO nanocones with a diameter gradually increasing from about 50 nm at the root to 300 nm at the tip-ends possessing perfect atomically-flat hexagonal-shaped morphology from the first round, to ZnO nanonails with length of about 4 μm and each ZnO nanonail has a stem with a diameter of about 100 nm with a large diameter hexagonal-shaped cap attached from the second round, to the ZnO nanoprisms with uniform diameters along the whole length from the sixth experimental round or later.

Among various approaches to catalyst templates, nanosphere lithography (NSL) is a classical one because of its low cost and high efficiency compared with others. This technique works nearly perfectly on Si, glass and sapphire, and has been widely used for structuring substrates with a hexagonal array of various nanomaterials, including ferroelectric, magnetic, and catalytic metal dots for growth of 1-D nanomaterials. Recently, Fan and his co-workers obtained ZnO nanowires which exhibit truly vertical alignment and well separation on GaN substrate, and two types of patterns corresponding to a mono- and bilayer of spheres, respectively [47]. This is highly desirable for size-specific individual nanowire arrays on conductive substrates as needed for device applications such as sensor arrays or field emitter arrays. In such a process, it is worth mentioning that the use of a small-volume reaction tube and the low substrate temperature is essential. This sounds a minor point but actually turns out to be a key issue. Small-sized gold nanoparticles are very mobile at elevated temperature, which thus will strongly destroy the aligned pattern. The small tube in the big reaction chamber maintains a high vapor concentration above the substrate, which effectively reduces the temperature required for nanowire growth.

2.2.6. ZnO Based Heterostructures

In recent years, great efforts have been paid to the synthesis and characterization of nanoscale materials and the fabrication of devices made out of them. The manipulation with a given nanoscale object presumes not only the generation of pure nanomaterials with novel properties, but also the realization of multicomponent nanosystems (heterostructures) with diverse functions. One-dimensional (1-D) heterostructures with modulated compositions and interfaces are of particular interest with respect to the potential applications in unique types of nanoscale building blocks for future optoelectronic devices and systems.

In 2004, a group led by Prof. Park reported the synthesis of various kinds of heterostructures based on ZnO nanorods via a two-step MVD process [101]. In their process, various 1D nanostructures, including carbon nanotubes, GaN nanowires, GaP nanowires, SiC nanowires, SiC/C core/shell nanowires, were first synthesized using different methods, then these nanowires introduced into the MVD process which was performed at 500 ^{o}C. The results show that high-density ZnO nanorods deposit on pregrown 1D nanostructures. The growth direction of the ZnO nanorods is uniformly [0001]. Symmetric alignment of the ZnO nanorods was observed for the GaN-ZnO heterostructures. The growth of ZnO nanorods can be monitored as a function of reaction time.

Hierarchical ZnO/ZnS heterostructures were synthesized by our group using a one-step MVD process. In the process, Zn and SnS are used as source materials in a two heating zone furnace. After synthesis, it was found that the product consists of numerous one-dimensional saw-like nanostructures, which have one side flat and the other side with small teeth. Typical saw-like nanostructures have diameters of ~100 nm and lengths ranging from several tens to hundreds of micrometers as shown in Figure 17a. Its corresponding selected area electron diffraction (SAED) pattern is shown in Figure 17a inset. The diffraction spots in this pattern can be indexed as the diffraction along $[2\bar{1}\bar{1}0]$ zone axis of both ZnO (space group *P6₃mc*, a=b=3.249 Å, c=5.206 Å) and ZnS (space group *P6₃mc*, a=b=3.820 Å, c=6.257 Å). The

SAED pattern shows that the ZnO and ZnS have a crystalline orientation relationship as follows.

$$[2\bar{1}\bar{1}0]_{ZnO}//[2\bar{1}\bar{1}0]_{ZnS} \quad (1)$$

$$(0001)_{ZnO}//(0001)_{ZnS} \quad (2)$$

$$(01\bar{1}0)_{ZnO}//(01\bar{1}0)_{ZnS} \quad (3)$$

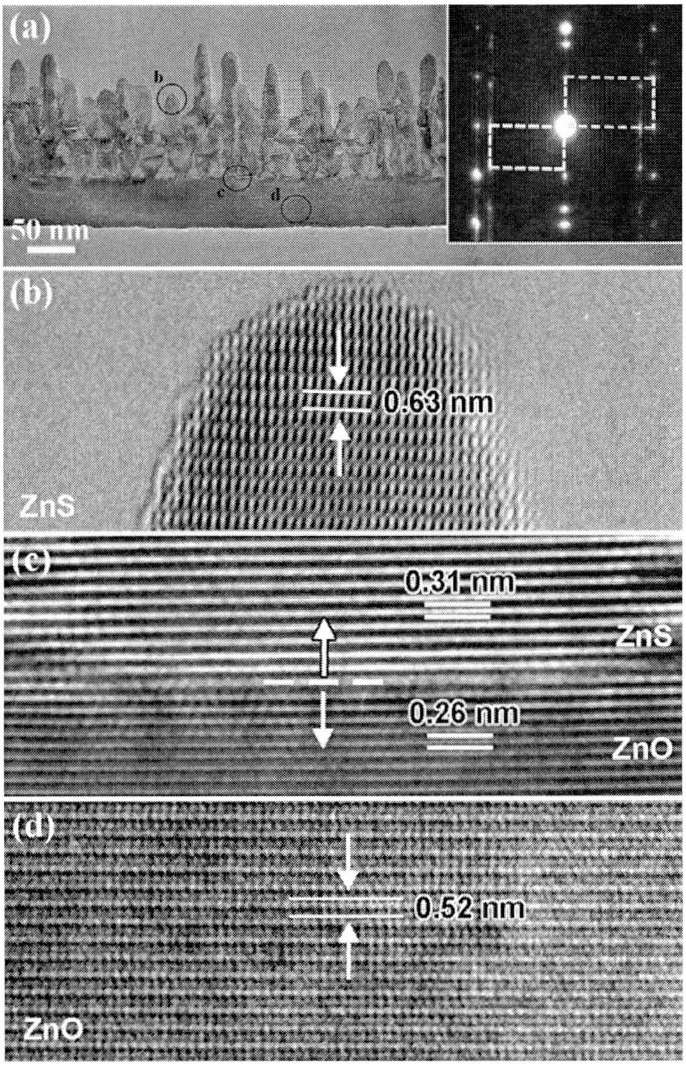

Figure 17. (a) TEM image and SAED pattern of a saw-like ZnO/ZnS heterostructure. (b-d) HRTEM image of the branched ZnS, interface, and the ZnO stem, respectively. (Copyright American Chemical Society, reproduced with permission).

Figure 17b shows the lattice-resolved HRTEM image taken from the ZnS nanowire framed in Figure 17a. The observed lattice fringe is separated by the 0.63 nm distance corresponding to the (0001) lattice plane spacing of wurtzite ZnS, indicating that the preferred growth direction of the ZnS nanowire is along the [0001] direction. Figure 17c shows the HRTEM image of the interface between ZnO and ZnS (framed in Figure 17a). The clearly resolved fringes are 0.31 nm for wurtzite ZnS and 0.26 nm for wurtzite ZnO, corresponding to the (0002) lattice plane spacing of wurtzite ZnS and ZnO, respectively. The interface between the ZnO and ZnS fragments is homogeneous and atomically sharp. The HRTEM image of the ZnO nanobelt is shown in Figure 17d. The resolved lattice fringe is 0.52 nm, corresponding to the (0001) spacing of wurtzite ZnO. Combined with the SAED pattern in Figure 17a, it can be seen that the ZnO nanobelt is a single crystal with the growth direction along the $[01\bar{1}0]$ direction.

Saw-like heterostructures synthesized presently are formed via a two-step growth process: the first step is the fast growth of ZnO nanobelts along the $[01\bar{1}0]$ direction at low temperature (550 ^{o}C) by thermal oxidation of zinc powder; the second step is the slow decomposition of SnS to Sn and S and then the reaction of Zn and S to subsequent growth of ZnS nanowires along the (0001) of ZnO, which can be expressed as the following reactions. SnS → Sn + S, and Zn +S → ZnS. As we have known, wurtzite ZnO and ZnS crystals can be described schematically as a number of alternating planes composed of 4-fold tetrahedral-coordinated O^{2-} (or S^{2-}) and Zn^{2+} ions, stacked alternatively along the c axis (Figure 1). The oppositely charged ions produce positively charged (0001)-Zn and negatively charged $(000\bar{1})$-O [or $(000\bar{1})$-S] polar surfaces. It is thought that the formation of saw-like ZnO/ZnS heterostructures here is also induced by the polar surfaces. In the first step, ZnO nanobelts are formed. The positively charged (0001)-Zn surface along the ZnO nanobelts is chemically active and tiny Zn clusters may exist at the growth front. In the second step, these Zn clusters react with sulfur and initiate the growth of ZnS teeth, forming the saw-like ZnO nanobelt/ ZnS nanowires heterostructures.

2.3. Properties of Low-Dimensional ZnO Nanostructures

2.3.1. Optical Properties

As is mentioned, ZnO is a wide bandgap semiconductor with a large exciton binding energy at room temperature. As a consequence, ZnO is recognized as a promising photonic material in the blue-UV region.

A photoluminescence (PL) of the vertically aligned ZnO nanowires shown in Figure 14a is shown in Figure 18. It exhibits a strong UV emission at 374 nm and a weak and broad emission band centered at 500 nm. The near-UV emission at 374 nm agrees well with the band gap of bulk ZnO, which is originated from the recombination of free excitons [102-104]. Careful studies found that the green emission at 500 nm is asymmetric (as shown in Figure 18 inset) and the intensity of this emission decreased with the increase of oxygen content. As previous reports, the green emission is related to the singly ionized oxygen vacancy and it results from the recombination of a photo-generated hole with a singly ionized charge state of the specific defects [105-110].

The PL properties of ZnO nanostructures can be tuned by introducing dopents. The PL of ZnO nanostructures synthesized in the round-to-round MVD process was shown in Figure 19. For comparison, the PL spectra of the ZnO nanoprisms (S1) synthesized from a pure zinc powder are also shown in this figure.

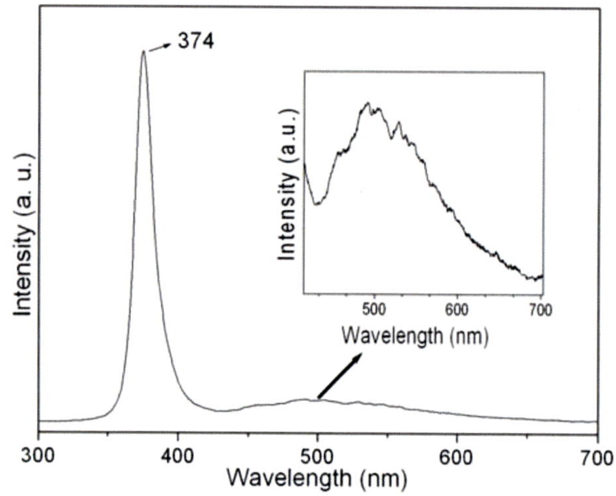

Figure 18. Room temperature photoluminescence of vertically aligned ZnO nanowires.

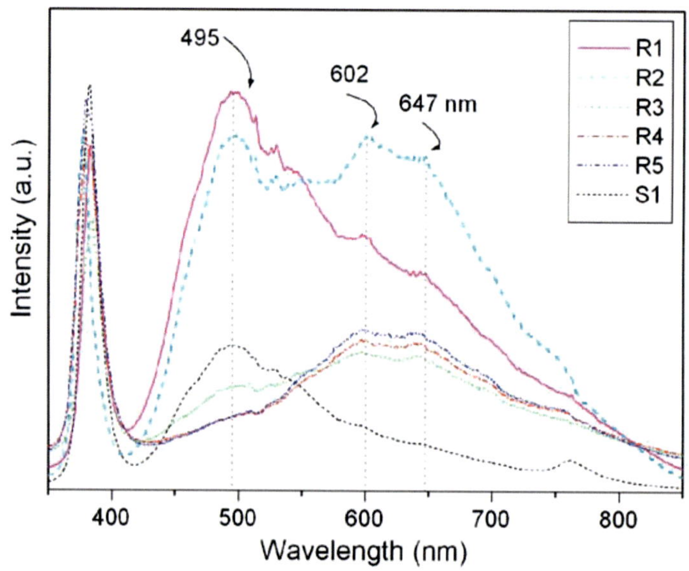

Figure 19. Room temperature PL spectra of the ZnO products obtained during the round-to-round MVD process as indicated in Figure 16. (Copyright American Chemical Society, reproduced with permission).

It is clear that all the spectra of one-dimensional ZnO nanostructures obtained from different experimental rounds, as well as the ZnO nanoprisms from pure zinc powder, exhibit a sharp UV emission at around 380 nm and broad green emissions ranging from 450 to 750 nm. Careful analysis shows that the broad green emissions ranging from 450 to 750 nm consist of three emissions centered at 495 nm, 602 nm and 647 nm. According to the literature, the UV emission of the ZnO nanostructures synthesized in the round-to-round MVD process, Figure 19, with a peak at ~380 nm corresponds to the near-band-edge (NBE) transition. The latter is responsible for the recombination of free excitons in ZnO. The exact origins of the green emissions are still unclear. However, it should be stressed that they are obviously defects-related emissions according to the previous reports. From the PL spectra, it was found that the relative intensity of the three emissions among the green emissions centered at 495 nm, 602 nm and 647 nm are different for different products. For the product obtained in R1, the peak at 495 nm is the strongest peak and it is much stronger than the other two peaks, while for other products obtained in R2, the peaks at 495 nm, 602 nm and 647 nm have almost the same peak intensity. For the product obtained in R3-R5, the intensity of peak at 495 nm is much weaker than the other two peaks at 602 nm and 647 nm, which have similar peak intensities, as shown in Figure 19. Besides the difference in the relative peak intensity in the green emissions, the relative intensity of the NBE band to the green emission band also changes with the experimental round number. From the PL spectra, the relative intensity of NBE to green emission is 0.86 for R1, 1.0 for R2, 1.98 for R3, 2.34 for R4, and 2.40 for R5. For the ZnO nanoprisms (S1) synthesized from pure zinc powder it shows the strongest relative intensity of 2.82. It indicates that with a decrease of the indium concentration, the product photoluminescence properties are improved.

PL properties of S doped ZnO nanostructures were also studied [111-114]. With the introduce of S into the ZnO nanostructures, PL spectra clearly show two feathers: first, the UV emission shifted to blue region. Second, the intensity of the green emission increased greatly. It is well known that when ZnO is doped with S element, the excess carriers supplied by the impurities contribute to enhance electrical conductivity of the ZnO nanostructures and thus lead to a blue shift of optical band-to-band transitions, known as the Burstein-Moss effect. Since S has a larger Bohr radius than O, the incorporation of S into ZnO crystal lattice may introduce lattice distortion in the ZnO nanostructures. After all, S doping into ZnO nanostructures influences the energy structure of ZnO and thus new band structure may be formed, which should result in changes of the PL intensity of green light emission. With the increase of S content in the S-doped ZnO nanostructures, the intensity of surface oxygen vacancies may also increase due to the insufficiency of oxygen in the reaction system, which also results in the increase of green emission.

2.3.2. Field Emission Properties

Field emission, as one of the applications of nanostructural materials, is of great commercial interest in display and other electron devices. In the past decade, researches on field emission have been mainly focused on carbon-based materials as well as 1-D nanostructures because of their low work function, high mechanical stability, high aspect ratio, and high conductivity [115-121]. In principle, low-dimensional ZnO nanostructures are very good field emitters because of their favorable aspect ratios. In 2001, we reported the field emission properties from ZnO nanowires grown on Co catalyst particles in which the turn-on field was about 6.0 V/μm at 0.1 $\mu A/cm^2$, and the emission current density was about

1.0 mA/cm^2 at 11.0 V/μm. Recently, Jo *et al.* showed similar field emission properties from thin films of ZnO nanowires grown on Au catalyst nanoparticles, which indicated the current density of about 1 mA/cm^2 at the electric field of about 15 V/μm [122]. On the other hands, Zhu *et al.* showed that ZnO nanoneedle arrays achieved the low turn-on field of about 2.4 V/μm and the high emission current density of about 2.4 mA/cm^2 at 7 V/μm from [123]. They suggested that good field emission properties was caused by the small geometry of ZnO nanowires.

Figure 20 shows the FE plot of the synthesized ZnO nanorods shown inset. The turn-on (corresponding to the current density of 0.1 μA/cm^2) electric field was about 6.2 V/μm and the emission current density was about 1.0 mA/cm^2 at an applied field of 15.0 V/μm. The Fowler-Nordheim plot usually presented in the literature is also shown in the inset of Figure 20. It exhibits two linear behaviors in the measurement range.

The field enhancement factor β can be calculated from the slope of Fowler-Nordheim plot if the work function of the emitter is known. The measurement value for the work function of ZnO nanowires was about 5.3 eV. In the case of high applied electric field, β value revealed about 1,500 from the inset of Figure 20. On the other hand, it showed much lower β value at lower applied electric field, resulting from unstable local field and emission current from ZnO nanowires during sweeping an applied voltage.

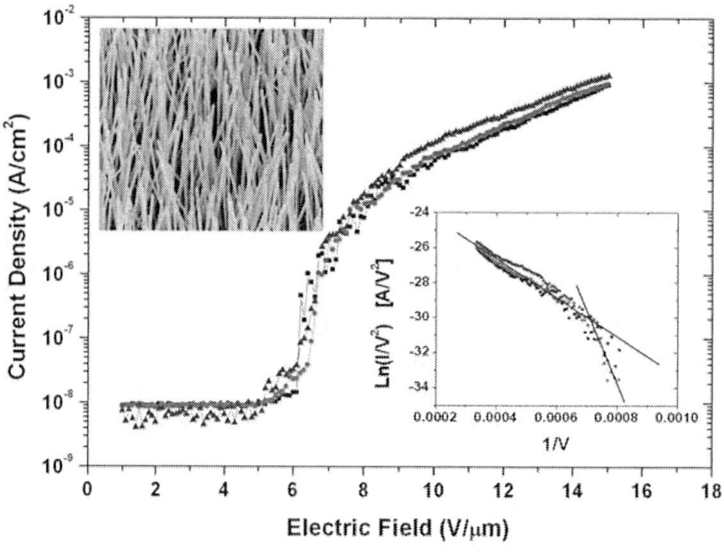

Figure 20. Field emission properties from ZnO nanowires grown on n-Si substrate. The turn-on electric field was about 6.2 V/μm, and the emission current density reached about 1 mA/cm2 at 15.0 V/μm. (Copyright Elsevier, reproduced with permission).

It is well known that field emission mainly depends on tip morphology and density of nanowires. ZnO nanostructures with different tip structures may show different FE properties. Figure 21 shows the FE properties of ZnO nanonails and nanopencils synthesized in the MVD process. A turn-on field of 7.9 V/μm was obtained for vertically-aligned ZnO nanonails and 7.2 V/μm for vertically-aligned ZnO nanopencils. Since vertically-aligned ZnO nanopencils have much small tips compared with the vertically-aligned ZnO nanonails, it is reasonable that the nanopencils show smaller turn-on field than nanonails. The emission characteristics were also analyzed using Fowler-Nordheim theory, which has been proved to be useful in describing FE from electron emittors. The FN law relates the emitted current and the local field E at the emitter surface,

$$J = (E^2\beta^2/\Phi)\exp(-B\Phi^{3/2}/E\beta)$$

Where J is the emission-current density, E is the local applied field (where $E=Vd^{-1}$, and d is the anode-sample distance), Φ is the work function of the material, β is the enhancement factor, and B is a constant.

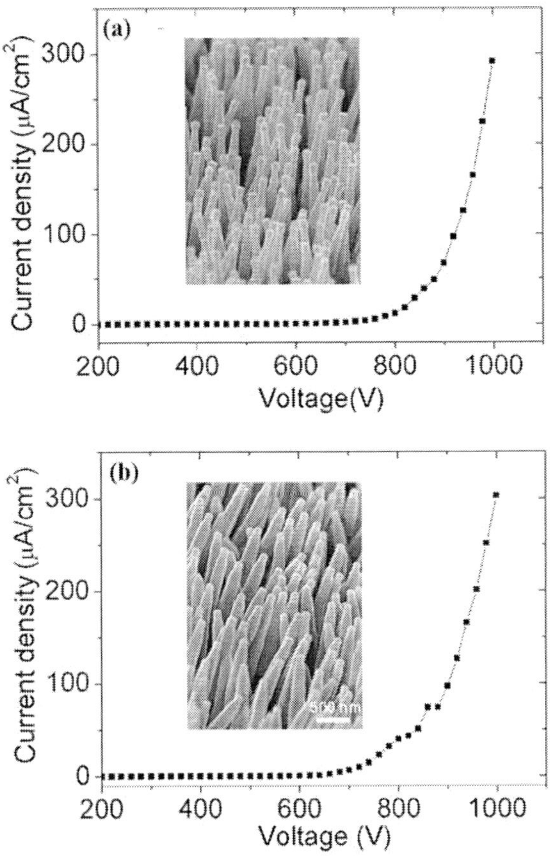

Figure 21. Field emission properties from (a) ZnO nanonails and (b) ZnO nanopencils grown on Si substrate. (Copyright Wiley-VCH, reproduced with permission)

3. Conclusion

This article reviews low-dimensional ZnO nanostructures grown by MVD process in different research groups, especially in our group. A diverse group of ZnO nanoarchitectures have been discussed. Properties, such as optical properties and field emission properties, of the synthesized low-dimensional ZnO nanostructures are also discussed. These novel nanoarchitectures could be unique for many applications in nanotechnology. ZnO has great potential as a useful material for nanoscale devices. Research investigation on this topic needed to be encouraged and there are much innovating and developing space and a lot of issues to be investigated, for example, to precisely control the size, size distribution, shape, alignments of low-dimensional ZnO nanostructures, to propose some theoretical mechanism for the formation of low-dimensional ZnO nanostructures, to fulfill real applications of the low-dimensional ZnO nanostructures, etc.

REFERENCES

[1] Wu, Y. Y.; Yan, H. Q.; Huang, M. H.; Messer, B.; Song, J. H.; Yang, P. D. *Chem. Eur. J.* 2002, *8*, 1261.

[2] Pan, Z. W.; Dai, Z. R.; Wang, Z. L. *Science* 2001, *291*, 1947.

[3] Wu, J. J.; Liu, S. C. *Adv. Mater.* 2002, *14,* 215.

[4] Huang, M. H.; Mao, S.; Feick, H.; Yan, H. Q.; Wu, Y. Y.; Kind, H.; Weber, E.; Russo, R.; Yang, P. D. *Science* 2001, *292*, 1897.

[5] Yang, P. D.; Yan, H. Q.; Mao, S.; Russo, R.; Johnson, J.; Saykally, R.; Morris, N.; Pham. J.; He, R. R.; Choi H. J. *Adv. Funct. Mater.* 2002, *12*, 323.

[6] Park, W. I.; Kim, D. H.; Jung, S. W.; Yi, G. C. *Appl. Phys. Lett.* 2002, *80*, 4232.

[7] Yao, B. D.; Chan, Y. F.; Wang, N. *Appl. Phys. Lett.* 2002, *81*, 757.

[8] Zhang, J.; Sun, L. D.; Yin, J. L.; Su, H. L.; Liao, C. S.; Yan, C. H. *Chem. Mater.* 2002, *14*, 4172.

[9] Park, W. I.; Yi, G. C.; Kim, M. Y.; Pennycook, S. J. *Adv. Mater.* 2002, *14*, 1841.

[10] Guo, L.; Ji, Y. L.; Xu, H. B.; Simon, P.; Wu, Z. Y. *J. Am. Chem. Soc.* 2002, *124*, 14864.

[11] Hu, J. Q.; Bando, Y. *Appl. Phys. Lett.* 2003, *82*, 1401.

[12] Vayssieres, L. *Adv. Mater.* 2003, *15*, 464.

[13] Tang, Q.; Zhou, W. J.; Shen, J. M.; Zhang, W.; Kong, L. F.; Qian, Y. T. *Chem. Commun.* 2004, 712.

[14] Zhang, J.; Sun, L. D.; Liao, C. S.; Yan, C. H. *Chem. Commun.* 2002, 262.

[15] Liu, B.; Zeng, H. C. *J. Am. Chem. Soc.* 2003, *125*, 4430.

[16] Kong, Y. C.; Yu, D P.; Zhang, B.; Fang, W.; Feng, S. Q. *Appl. Phys. Lett.* 2001, *78*, 407.

[17] Hu, J. Q.; Li, Q.; Wong, N. B.; Lee, C. S.; Lee, S. T. *Chem. Mater.* 2002, *14*, 1216.

[18] Greyson, E. C.; Babayan, Y.; Odom, T. W. *Adv. Mater.* 2004, *16*, 1348.

[19] Tseng, Y. K.; Huang, C. J.; Cheng, H. M.; Lin, L. N.; Liu, K. S.; Chen, I. C. *Adv. Funct. Mater.* 2003, *13*, 811.

[20] Lao, J. Y.; Wen, J. G.; Ren, Z. F. *Nano Lett.* 2002, *2*, 1287.

[21] Lao, J. Y.; Huang, J. Y.; Wang, D. Z.; Ren, Z. F. *Nano Lett.* 2003, *3*, 235.

[22] Hu, P. A.; Liu, Y. Q.; Wang, X. B.; Fu, L.; Zhu, D. B. *Chem. Commun.* 2003, 1304.

[23] Gao, P. X.; Wang, Z. L. *J. Phys. Chem. B* 2002, *106*, 12653.

[24] Chang, P. C.; Fan, Z. Y.; Wang, D. W.; Tseng, W. Y.; Chiou, W. A.; Hong, J.; Lu, J. G. *Chem. Mater.* 2004, *16*, 5133.

[25] Kong, X. Y.; Ding, Y.; Yang, R. S.; Wang, Z. L. *Science* 2004, *303*, 1348.

[26] Kong, X. Y.; Wang, Z. L. *Nano Lett.* 2003, *3*, 1625.

[27] Gao, P. X.; Wang, Z. L. *J. Am. Chem. Soc.* 2003, *125*, 11299.

[28] Yan, H. Q.; He, R. R.; Pham, J.; Yang, P. D. *Adv. Mater.* 2003, *15*, 402.

[29] Gao, P. X.; Wang, Z. L. *Appl. Phys. Lett.* 2004, *84*, 2883.

[30] Gao, P. X.; Wang, Z. L. *J. Appl. Phys.* 2005, *97*, 044304.

[31] Wang, Z. L.; Kong, X. Y.; Ding, Y.; Gao, P. X.; Hughes, W. L.; Yang, R. S.; Zhang, Y. *Adv. Funct. Mater.* 2004, *14*, 943.

[32] Wang, Z. L.; Kong, X. Y.; Zuo, J. M. *Phys. Rev. Lett.* 2003, *91*, 185502.

[33] Wang, Z. L. *Mater. Today*, 2004, *7*, 26.

[34] Pan, Z. W.; Mahurin, S. M.; Dai, S.; Lowndes, D. H. *Nano Lett.* 2005, *5*, 723.

[35] Ng, H. T.; Li, J.; Smith, M. K.; Nguyen, P.; Cassell, A.; Han, J.; Meyyappan, M. *Science* 2003, *300*, 1249.

[36] Dai, Y.; Zhang, Y.; Li, Q. K.; Nan, C. W. *Chem. Phys. Lett.* 2002, *358*, 83.

[37] Choy, J. H.; Jang, E. S.; Won, J. H.; Chung, J. H.; Jang, D. J.; Kim, Y. W. *Appl. Phys. Lett.* 2004, *84*, 287.

[38] Tian, Z. R.; Voigt, J. A.; Liu, J.; Mckenzie, B.; Medenmott, M. J. *J. Am. Chem. Soc.* 2003, *124*, 12954.

[39] Wang, X. D.; Summers, C. J.; Wang, Z. L. *Nano Lett.* 2004, *4*, 423.

[40] Jiang, Z. Y.; Xu, T.; Xie, Z. X.; Lin, Z. W.; Zhou, X.; Xu, X.; Huang, R.; Zheng, L. S. *J. Phys. Chem. B* 2005, *109*, 23269.

[41] Fan, H. J.; Bertram, F.; Dadgar, A.; Christen, J.; Krost, A.; Zacharias, M. *Nanotechnology*, 2004, *15*, 1401.

[42] Fan, H. J.; Fleischer, F.; Lee, W.; Nielsch, K.; Scholz, R.; Zacharias, M.; Gosele, U.; Dadgar, A.; Krost, A. *Supperlattices Microstruct.* 2004, *36*, 95.

[43] Fan, H. J.; Lee, W.; Scholz, R.; Dadgar, A.; Krost, A.; Nielsch, K.; Zacharias M. *Nanotechnology*, 2005, *16*, 913.

[44] He, F. Q.; Zhao, Y. P. *Appl. Phys. Lett.* 2006, *88*, 193113.

[45] Liao, L.; Li, J. C.; Liu, D.; Liu, C.; Wang, D.; Song, W.; Fu, Q. *Appl. Phys. Lett.* 2005, *86*, 083106.

[46] Fan, H. J.; Lee, W.; Hauschild, R.; Alexe, M.; Le Rhun, G.; Scholz, R.; Dadgar, A.; Nielsch, K.; Kalt, H.; Krost, A.; Zacharias, M.; Gosele, U. *Small* 2006, *2*, 561.

[47] Fan, H. J.; Fuhrmann, B.; Scholz, R.; Syrowatka, F.; Dadgar, A.; Krost, A.; Zacharias, M. *J. Cryst. Growth* 2006, *287*, 34.

[48] Wang, R. C.; Liu, C. P.; Huang, J. L.; Chen, S. J. *Appl. Phys. Lett.* 2005, *86*, 251104.

[49] Cheng, H. M.; Hsu, H. C.; Tseng, Y. K.; Lin, L. J.; Hsieh, W. F. *J. Phys. Chem. B* 2005, *109*, 8749.

[50] Meng, X. Q.; Zhao, D. X.; Zhang, J. Y.; Shen, D. Z.; Lu, Y. M.; Liu, Y. C.; Fan, X. W. *Chem. Phys. Lett.* 2005, *407*, 91.

[51] Lee, C. J.; Lee, T. J.; Lyu, S. C.; Zhang, Y.; Ruh, H.; Lee, H. J. *Appl. Phys. Lett.* 2002, *81*, 3648.

[52] Lyu, S. C.; Zhang, Y.; Lee, C. J. *Chem. Mater.* 2003, *15*, 3294.

[53] Ham, H.; Shen, G. Z.; Cho, J. H.; Lee, T. J.; Seo, S. H.; Lee, C. J. *Chem. Phys. Lett.* 2005, *404*, 69.

[54] Lyu, S. C.; Zhang, Y.; Ruh, H.; Lee, H. J.; Shim, H. W.; Suh, E. K.; Lee, C. J. *Chem. Phys. Lett.* 2003, *363*, 134.

[55] Shen, G. Z.; Bando, Y.; Lee, C. J. *J. Phys. Chem. B* 2005, 109, 10779.

[56] Shen, G. Z.; Bando, Y.; Lee, C. J. *J. Phys. Chem. B* 2005, 109, 10578.

[57] Shen, G. Z.; Chen, D.; Lee, C. J. *J. Phys. Chem. B* 2006, *110*, 15146.
[58] Shen, G. Z.; Bando, Y.; Chen, D.; Liu, B. D.; Zhi, C. Y.; Golberg, D. *J. Phys. Chem. B* 2006, *110*, 3973.
[59] Shen, G. Z.; Bando, Y.; Liu, B. D.; Golberg, D.; Lee, C. J. *Adv. Funct. Mater.* 2006, *16*, 410.
[60] Shen, G. Z.; Cho, J. H.; Lee, C. J. *Chem. Phys. Lett.* 2005, *401*, 414.
[61] Li, Y.; Meng, G. W.; Zhang, L. D. *Appl. Phys. Lett.* 2000, *76*, 2011.
[62] Iijima, S. *Nature*, 1991, *354*, 56.
[63] Shen, G. Z.; Bando, Y.; Ye, C. H.; Yuan, X. L.; Sekiguchi, T.; Golberg, D. *Angew. Chem. Int. Ed.* 2006, *45*, 7568.
[64] Shen, G. Z.; Chen, D. *J. Am. Chem. Soc.* 2006, *128*, 11762.
[65] Tenne, R.; Margulis, L.; Genut, M.; Hodes, G. *Nature* 1992, *360*, 444.
[66] Feldman, Y.; Wasserman, E.; Srolovitz, D. J.; Tenne, R. *Science* 1995, *267*, 222.
[67] Tenne, R. *Prog. Inorg. Chem.* 2001, *5*, 269.
[68] Nath, M.; Rao, C. N. R. *J. Am. Chem. Soc.* 2001, *123*, 4841.
[69] Ye, C. H.; Meng, G. W.; Jiang, Z.; Wang, Y. H.; Wang, G.; Zhang, L. D. *J. Am. Chem. Soc.* 2002, *124*, 15180.
[70] Yang, J.; Liu, Y. C.; Lin, H. M.; Chen, C. C. *Adv. Mater.* 2004, *16*, 713.
[71] Spahr, M. E.; Bitterli, P.; Nesper, R.; Muller, M.; Krumeich, F.; Nlssen, H. U. *Angew. Chem. Int. Ed.* 1998, *37*, 1263.
[72] Li, Y. D.; Wang, J. W.; Deng, Z. X.; Wu, Y. Y.; Sun, X. M.; Yu, D. P.; Yang, P. D. *J. Am. Chem. Soc.* 2001, *123*, 9904.
[73] Golgberger, J.; He, R.; Zhang, Y.; Lee, S.; Yan, H. Q.; Choi, H. J.; Yang, P. D. *Nature* 2003, *422*, 599.
[74] Liu, Z. Q.; Zhang, D. H.; Han, S.; Li, C.; Lei, B.; Lu, W. G.; Fang, J. Y.; Zhou, C. W. *J. Am. Chem. Soc.* 2005, *127*, 6.
[75] Hu, J. Q.; Bando, Y.; Liu, Z. W.; Zhan, J. H.; Golberg, D.; Sekiguchi, T. *Angew. Chem. Int. Ed.* 2004, *43*, 63.
[76] Hu, J. Q.; Bando, Y.; Zhan, J. H.; Golberg, D. *Angew. Chem. Int. Ed.* 2004, *43*, 4606.
[77] Shen, G. Z.; Bando, Y.; Golberg, D. *J. Phys. Chem. B* 2006, *110*, 23170.
[78] Shen, G. Z.; Bando, Y.; Zhi, C. Y.; Golberg, D. *J. Phys. Chem. B* 2006, *110*, 10714.
[79] Shen, G. Z.; Bando, Y.; Golberg, D. *Appl. Phys. Lett.* 2006, *88*, 123107.
[80] Nath, M.; Rao, C. N. R. *Angew. Chem. Int. Ed.* 2002, *41*, 3451.
[81] Zhang, J.; Sun, L. D.; Liao, C. S.; Yan, C. H. *Chem. Commun.* 2002, 262.
[82] Hu, J. Q.; Bando, Y. *Appl. Phys. Lett.* 2003, *82*, 1401.
[83] Hu, J. Q.; Li, Q.; Meng, X. M.; Lee, C. S.; Lee, S. T. *Chem. Mater.* 2003, *15*, 305.
[84] Xing, Y. J.; Xi, Z. H.; Xue, Z. Q.; Zhang, X. D.; Song, J. H.; Wang, R. M.; Xu, J.; Song, Y.; Zhang, S. L.; Yu, D. P. *Appl. Phys. Lett.* 2003, *83*, 1689.
[85] Zhang, X. H.; Xie, S. Y.; Jiang, Z. Y.; Zhang, X.; Tian, Z. Q.; Xie, Z. X.; Huang, R. B.; Zheng, L. S. *J. Phys. Chem. B* 2003, *107*, 10114.
[86] Kong, X. H.; Sun, X. M.; Li, X. L.; Li, Y. D. *Meter. Chem. Phys.* 2003, *82*, 997.
[87] Zhang, B. P.; Binh, N. T.; Wakatsuki, K.; Segawa, Y.; Yamada, Y.; Usami, N.; Kawasaki, M.; Koinuma, H. *Appl. Phys. Lett.* 2004, *84*, 4098.
[88] Li, Q. C.; Kumar, V.; Li, Y.; Zhang, H. T.; Marks, T. J.; Chang, R. P. H. *Chem. Mater.* 2005, *17*, 1001.

[89] Xu, W. Z.; Ye, Z. Z.; Ma, D. W.; Lu, H. M.; Zhu, L. P.; Zhao, B. H.; Yang, X. D.; Xu, Z. Y. *Appl. Phys. Lett.* 2005, *87*, 093110.

[90] Sun, Y.; Fuge, G. M.; Fox, N. A.; Riley, D. J.; Ashfold, M. N. R. *Adv. Mater.* 2005, *17*, 2477.

[91] Zhang, X. H.; Zhang, Y.; Xu, J.; Wang, Z.; Chen, X. H.; Yu, D. P.; Zhang, P.; Qi, H.; Tian, Y. J. *Appl. Phys. Lett.* 2005, *87*, 123111.

[92] Wang, R. C.; Liu, C. P.; Huang, J. L.; Chen, S. J. *Nanotechnology* 2006, *17*, 753.

[93] Jeong, J. S.; Lee, J. Y.; Cho, J. H.; Suh, H. J.; Lee, C. J. *Chem. Mater.* 2005, 17, 2752.

[94] Shi, H.; Qi, L. M.; Ma, J.; Cheng, H.; Zhu, B. *Adv. Mater.* 2003, *15*, 1647.

[95] Yan, H.; He, R.; Johnson, J.; Law, M.; Saykally, R. J.; Yang, P. D. *J. Am. Chem. Soc.* 2003, *125*, 4728.

[96] Shi, H.; Qi, L.; Ma, J.; Cheng, H. *J. Am. Chem. Soc.* 2003, *125*, 3450.

[97] Shen, G. Z.; Bando, Y.; Tang, C. C.; Golberg, D. *J. Phys. Chem. B* 2006, *110*, 7199.

[98] Colfen, H.; Mann, S. *Angew. Chem. Int. Ed.* 2003, *42*, 2350.

[99] Liu, B.; Zeng, H. C. *J. Am. Chem. Soc.* 2004, *126*, 8124.

[100] Cao, M.; Hu, J.; Liang, H.; Wan, L. J. *Angew. Chem. Int. Ed.* 2005, *44*, 4391.

[101] Bae, S. Y.; Seo, H. W.; Choi, H. C.; Park, J. *J. Phys. Chem. B.* 2004, *108*, 12318.

[102] Wong, E. M.; Searson, P. C. *Appl. Phys. Lett.* 1999, *74*, 2939.

[103] Stikant, V.; Clarke, D. R. *J. Appl. Phys.* 1998, *83*, 5447.

[104] Tseng, Y. K.; Chia, C. T.; Tsay, C. Y.; Lin, L. J.; Cheng, H. M.; Kwo, C. Y.; Chen, I. C. *J. Electrochem. Soc.* 2005, *152*, G95.

[105] Li, Y.; Cheng, G.; Zhang, L. D. *J. Mater. Res.* 2000, *15*, 2305.

[106] Konenkamp, R.; Word, R. C.; Schlegel, C. *Appl. Phys. Lett.* 2004, *85*, 6004.

[107] Wu, X.; Siu, G.; Fu, C.; Ong, H. C. *Appl. Phys. Lett.* 2001, *78*, 2285.

[108] Reynolds, D. C.; Look, D. C.; Jogai, B. *J. Appl. Phys.* 2001, *89*, 6189.

[109] Liu, M.; Kitai, A. H.; Mascher, P. *J. Lumin.* 1992, *54*, 35.

[110] Zhao, Q.; Yu, X. Y.; Song, X. F.; Zhang, X. Z.; Yu, D. P.; Li, C. P.; Guo, L. *Appl. Phys. Lett.* 2006, *88*, 033102.

[111] Na, C. W.; Bae, S. Y.; Park, J. *J. Phys. Chem. B* 2005, *109*, 12785.

[112] Shen, G. Z.; Cho, J. H.; Yoo, J. K.; Yi, G. C.; Lee, C. J. *J. Phys. Chem. B* 2005, *109*, 5491.

[113] Geng, B.; Wang, G. Z.; Jiang, Z.; Xie, T.; Sun, S. H.; Meng, G. W.; Zhang, L. D. *Appl. Phys. Lett.* 2003, *82*, 4791.

[114] Bae, S. Y.; Seo, H. W.; Park, J. *J. Phys. Chem. B* 2004, *108*, 5206.

[115] Shen, G. Z.; Cho, J. H.; Jung, S. I.; Lee, C. J. *Chem. Phys. Lett.* 2005, *401*, 529.

[116] Xu, N. S.; Chen, Y.; Deng, S. Z.; Chen, J.; Ma, X.; Wang, E. G. *J. Phys. D: Appl. Phys.* 2001, *34*, 1597.

[117] Liu, C.; Hu, Z.; Wu, Q.; Wang, X. Z.; Chen, Y.; Sang, H.; Zhu, J.; Deng, S. Z.; Xu, N. S. *J. Am. Chem. Soc.* 2005, *127*, 1318.

[118] Tang, C. C.; Bando, Y. *Appl. Phys. Lett.* 2003, *83*, 659.

[119] Bonard, J. M.; Weiss, N.; Kind, H.; Stokli, T.; Forro, L.; Kern, K.; Chatelain, A. *Adv. Mater.* 2001, *13*, 184.

[120] Zhu, W.; Bower, C.; Zhou, O.; Kochanski, G.; Jin, S. *Appl. Phys. Lett.* 1999, *75*, 873.

[121] Shen, G. Z.; Bando, Y.; Ye, C. H.; Liu, B.; Golberg, D. *Nanotechnology* 2006, *17*, 3468.

[122] Jo, S. H.; Lao, J. Y.; Ren, Z. F.; Farrer, R. A.; Baldacchini, T.; Fourkas, J. T. *Appl. Phys. Lett.* 2003, *83*, 4821.

[123] Zhu, Y.W.; Zhang, H.Z.; Sun, X.C.; Feng, S.Q.; Zhang, Q.; Xiang, B.; Wang, R. M; Yu, D. P. *Appl. Phys. Lett.* 2003, *83*, 144.

In: Encyclopedia of Nanotechnology
Editor: Elwood D. Carlson

ISBN: 978-1-60692-079-4

Chapter 17

STRUCTURE AND OPTO-ELECTRONIC BEHAVIOR OF DIAMONDOIDS, WITH APPLICATIONS AS MEMS AND AT THE NANOSCALE LEVEL

***G. Ali Mansoori*[1], *Thomas F. George*[2], *Guoping Zhang*[3] and *Lahsen Assoufid*[4]**

[1]Departments of Bio and Chemical Engineering,
University of Illinois at Chicago, Chicago, IL 60612, USA
[2]Office of the Chancellor and Center for Molecular Electronics
Departments of Chemistry and Biochemistry and Physics and Astronomy
University of Missouri–St Louis, St Louis, MO 63121, USA
[3]Department of Physics, Indiana State University, Terre Haute, IN 47809, USA
[4]Argonne National Laboratory, Argonne, IL 60439, USA

ABSTRACT

Diamondoids have been of great interest in recent years due to their role in nanotechnology, drug-delivery and medicine. Due to their six or more linking groups, they have found major applications as templates and as molecular building blocks in nanotechnology, polymers synthesis, drug delivery, drug targeting, DNA-directed assembly, DNA-amino acid nanostructure formation, and in host-guest chemistry. In this paper, the molecular nature of diamondoids, their molecular specificities, their intermolecular interactions and their opto-electronic properties are introduced. These will help in the understanding of the structure-property relations and self-assembly of diamondoids, which is essential for designing functional molecular gears for micro-electro-mechanical systems (MEMS) and sensitive bionanosensors, and for developing new nanodrugs, just to name a few.

DIAMONDOID MOLECULES

Diamondoid molecules are cage-like, ultra-stable, saturated hydrocarbons. These molecules are ringed compounds, which have a diamond-like structure consisting of a number of six-member carbon rings fused together (see Figure 1). More explicitly, they consist of repeating units of ten carbon atoms forming a tetra-cyclic cage system [1-3]. They are called "diamondoids" because their carbon-carbon framework constitutes the fundamental repeating unit in the diamond lattice structure. The diamondoid carbon skeleton is a basic portion of diamond as shown in Figure 2. This structure was first determined in 1913 by Bragg and Bragg [4] using X-ray diffraction analysis.

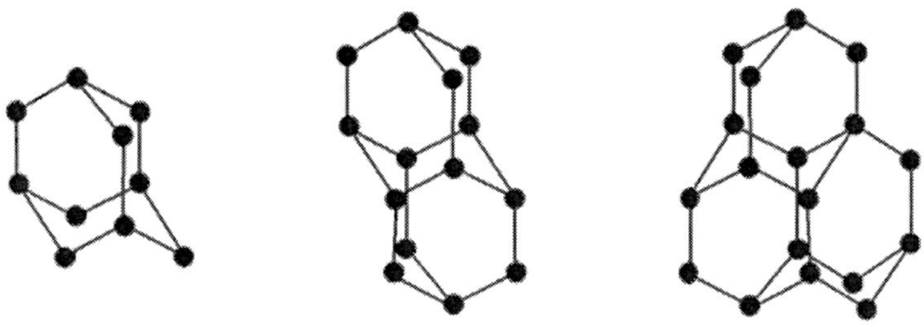

Figure 1. Molecular structures of adamantane, diamantane and trimantane (the smaller diamondoids) with chemical formulas C10H16, C14H20 and C18H24 from left to right.

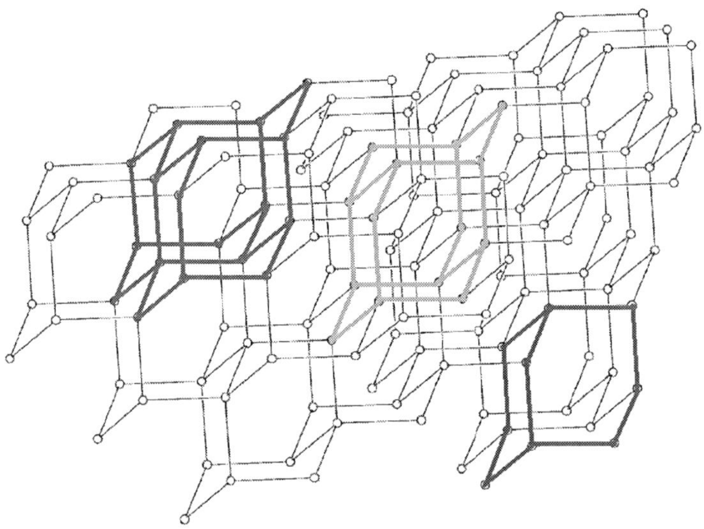

Figure 2. Structures of adamantane, diamantane and triamantane superimposed over the three-dimensional diamond network.

The first and simplest member of the diamondoids group, adamantane, is a tricyclic saturated hydrocarbon (tricyclo[3.3.1.1]decane). Adamantane is followed by its polymantane homologues (adamantologues): diamantane, tria-, tetra-, penta- and hexamantane. Figure 1 illustrates the smaller diamondoid molecules, with the general chemical formula $C_{4n+6}H_{4n+12}$:

adamantane ($C_{10}H_{16}$), diamantane ($C_{14}H_{20}$) and triamantane ($C_{18}H_{24}$). Each of these lower adamantologues has only one isomer.

Depending on the spatial arrangement of the adamantane units, higher polymantanes ($n \geq 4$) can have several isomers and non-isomeric equivalents. There are three possible tetramantanes, all of which are isomeric. They are depicted in Figure 3 as iso-, anti- and skew-tetramantane, respectively. Anti- and skew-tetramantanes each possess two quaternary carbon atoms, whereas iso-tetramantane has three.

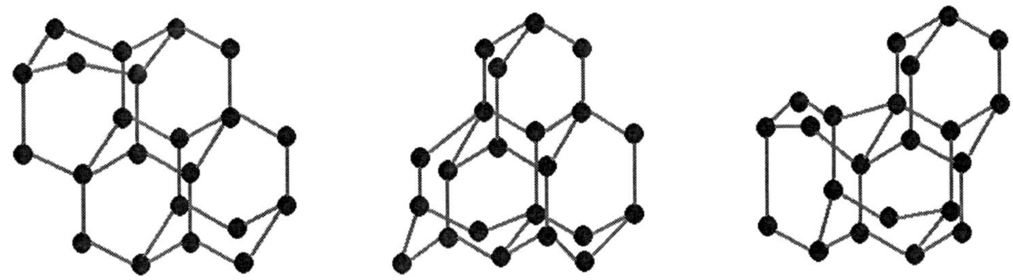

Figure 3. The three possible tetramantanes, all of which are isomeric, as anti-, iso- and skew-tetramantane from left to right. Anti- and skew-tetramantanes each possess two quaternary carbon atoms, whereas iso-tetramantane has three quaternary carbon atoms. The number of diamondoid isomers increases appreciably after tetramantane.

With regards to the remaining members of the diamondoids group, there are seven possible pentamantanes, six of which are isomeric ($C_{26}H_{32}$) and obeying the molecular formula of the homologous series, and one is non-isomeric ($C_{25}H_{30}$). For hexamantane, there are 24 possible structures: 17 are regular cata-condensed isomers with the chemical formula ($C_{30}H_{36}$), six are irregular cata-condensed isomers with the chemical formula ($C_{29}H_{34}$), and one is peri-condensed with the chemical formula ($C_{26}H_{30}$), as shown in Figure 4.

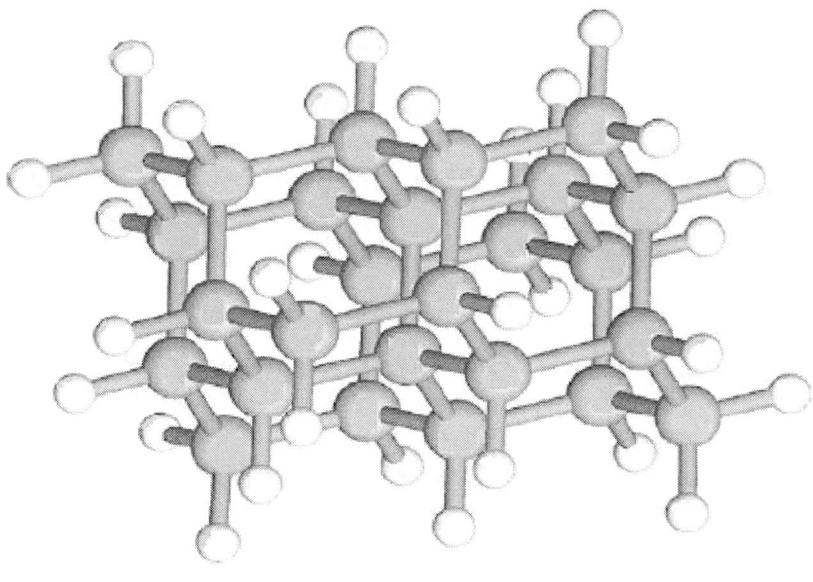

Figure 4. Molecular structure of (peri-condensed) cyclohexamantane (C26H30). The darker spheres represent carbon atoms, while the lighter spheres are hydrogen atoms.

Diamondoids, when in the solid state, melt at much higher temperatures than other hydrocarbon molecules with the same number of carbon atoms in their structure. Since they also possess low strain energy, they are more stable and stiff, resembling diamond in a broad sense. They contain dense, three-dimensional networks of covalent bonds, formed chiefly from first- and second-row atoms with a valence of three or more. Many of the diamondoids possess structures rich in tetrahedrally-coordinated carbon. They are materials with superior strength-to-weight ratios.

It has been found that adamantane crystallizes in a face-centered cubic lattice, which is extremely unusual for an organic compound. The molecule therefore should be completely free from both angle strain (since all carbon atoms are perfectly tetrahedral) and torsional strain (since all C-C bonds are perfectly staggered), making it a very stable compound and an excellent candidate for various applications, as will be discussed later.

At the initial point of growth, crystals of adamantane show only cubic and octahedral faces. The effects of this unusual structure on physical properties are interesting [5].

Many of the diamondoids can be brought to macroscopic crystalline forms with some special properties. For example, in its crystalline lattice, the pyramidal-shape [1(2,3)4]pentamantane has a large void in comparison to similar crystals. Although it has a diamond-like macroscopic structure, it possesses the weak, non-covalent, intermolecular van der Waals attractive forces involved in forming a crystalline lattice [6,7]. Another example is the crystalline structure of 1,3,5,7-tetracarboxy adamantane, which is formed via carboxyl hydrogen bonds of each molecule with four tetrahedral nearest-neighbors. The similar structure in 1,3,5,7-tetraiodoadamantane crystal would be formed by I…I interactions. In 1,3,5,7-tetrahydroxyadamantane, the hydrogen bonds of hydroxyl groups produce a crystalline structure similar to inorganic compounds, like the cesium chloride (CsCl) lattice [8] (see Figure 5).

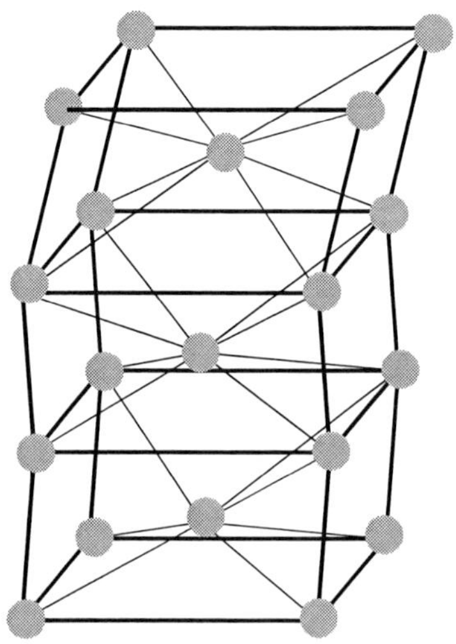

Figure 5. Quasi-cubic units of the crystalline network for 1,3,5,7- tetrahydroxy-adamantane, taken from [8]. The molecules have been shown as blue spheres and hydrogen bonds as solid linking lines. This crystalline structure is similar to that of CsCl.

The presence of chirality is another important feature in many derivatives of diamondoids. It should be pointed out that tetramantane [123] is the smallest of the lower diamondoids to possess chirality [6].

MOLECULAR STRUCTURE OF DIAMONDOIDS

Diamondoid molecules show unique properties due to their exceptional atomic arrangements. Adamantane consists of cyclohexane rings in 'chair' conformation. These molecules are chemically and thermally stable, strain-free and non-biodegradable.

Adamantane {(CAS No: 281-23-2) 1-tricyclo[3.3.1.$1^{3,7}$]decane}

Diamantane {(CAS No: 2292-79-7) pentacyclo[7.3.1.$1^{4.12}$.$0^{2.7}$.$0^{6.11}$]tetradecane} also known as {decahydro-3,5,1,7-[1.2.3.4]-butanetetraylnaphtalene }

Each successively-higher diamondoid family shows increasing structural complexity and varieties of molecular geometries. *Sui generis* properties of diamondoids have provoked an extensive range of inquiries in different fields of science and technology [9], including exploring these as molecular building blocks in nanotechnology.

Intermolecular Interactions

Knowledge about intermolecular interactions is very critical to designing and fabricating organic nanostructure-based devices such as organic thin film transistors, organic light-emitting devices, organic semiconductor nanorods, composites, etc. However, the nature and role of intermolecular interactions in such organic nanostructures as diamondoids is very challenging and not well understood [10-13,14].

The molecular structural characteristics of macroscopic systems based on the knowledge of statistically-averaged (effective) pair-interparticle interactions have already been defined and formulated in the thermodynamic limit [15-23]. As a result, the available atomic and molecular interaction database for macrobodies by means of effective-pair-intermolecular potentials is only an approximation [10-13]. Since the number of particles are finite in organic nanostructures, those statistical-averaging tools are not well-suited for this purpose. As a result, the theory of intermolecular potential functions of such systems and their relationship with their structural characteristics need to be formulated [10-13]. It is important to develop a phenomenological as well as mathematical foundation of intermolecular interactions for diamondoids which are presently non-existent, and to develop connections between intermolecular interactions and structural characteristics of diamondoid organic nanoscale systems [24-28].

It should be pointed out that structural characteristics in nanoscale systems are a dynamic phenomenon, not the static equilibrium of macroscopic phases. Coexistence of phases is expected to occur over bands of temperature and pressure, rather than at just sharp points

[10,24-26]. The Gibbs phase rule loses its meaning, and many phase-like forms may occur for nanoscale systems that are unobservable in the macroscopic counterparts of those systems.

For atoms and rather simple molecules, quantum mechanical *ab initio* calculational methods [29-33] have been successful in producing accurate intermolecular potential functions. For complex molecules and macromolecules, computations are prohibitively difficult to produce accurate potential data. Therefore, a joint theoretical and experimental effort is necessary. Experimentally, we use atomic force microscopy (AFM) and microelectrophoresis measurements to study the intermolecular interactions and structural characteristics in diamondoid-based organic nanosystems. These two different tools complement each other and measure levels of interactions on two different scales: The microelectrophoresis measurements reveal the information on a microscopic scale, while AFM investigates the interaction between atoms and the tip of the apparatus (for details, see the next section).

We have recently performed a first-principles simulation of the interaction between adamantane and an atomic force microscope tip made up of gold [34]. Our simulation results demonstrate quantitatively how an AFM tip interacts with adamantane. It is found that the AFM tip is able to detect the sharp potential change in adamantane. For instance, along different directions, the detailed morphology of the molecular structure can be probed directly. If the tip scans across the [111] direction, the potential change directly follows the sharp structural difference in two kinds of carbon atoms. This is an excellent starting point for our experimental effort, as one can see a large force change in the vicinity those atoms. In addition, if the scan is close to the tetragonal direction, a four-fold symmetry emerges. For the scan along the [111] direction, we estimate the force constant to be about 1.5 times that of C_{60}, which is an indication of the excellent mechanical properties of adamantane. We expect real experiments can probe these detailed changes. However, if the experimental tip contains more atoms, then the experimental results would be an average of the theoretical results.

To be more specific about our theoretical work, we directly compute the *xy*-scan image by scanning across two different surfaces, where one is hexagonal and the other is square surface. The probed atoms are the outside hydrogen atoms. We first hold our tip above adamantane and move along the *x*-direction. At each of the point, we do three radial scans and then use the finite difference method to compute the forces at each point.

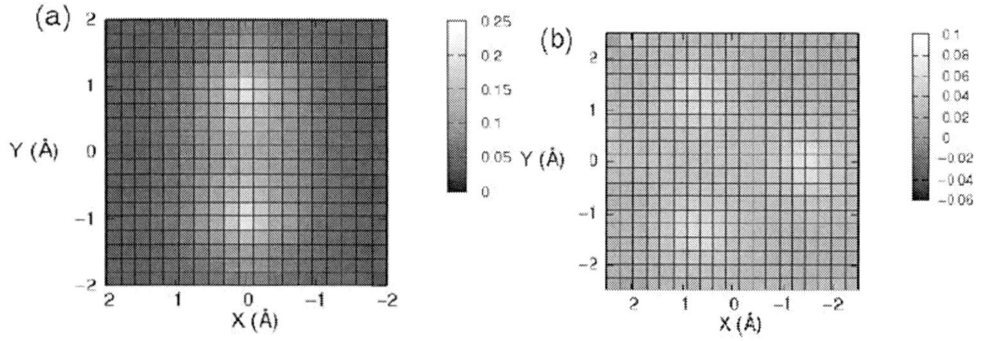

Figure 6. xy-scan image for (a) square surface and (b) hexagonal surface. The bright spots represent where the hydrogen atoms are located. For (a), the tip is 1.2 Å away from the center carbon atom; for (b), the tip is 1.8 Å away from the center of the triangle formed by the three outermost carbon atoms.

The number of calculations is huge. Figure 6(a) shows the force along the z-direction, where we use the intensity to show the magnitude of the force. One sees that there are two bright spots, which we know from the geometry of adamantane to correspond to the two outermost hydrogen atoms. Figure 6(b) shows a scan over the hexagaonal surface. Different from the square scan, there are three bright spots in this picture, corresponding to the three hydrogen atoms. These distinctive features are expected to show up in real experiments.

DIAMONDOIDS AS MOLECULAR BUILDING BLOCKS IN NANOTECHNOLOGY

Nanotechnology is the branch of engineering that deals with the manipulation of individual atoms, molecules and systems smaller than 100 nanometers. Two different methods are envisioned for nanotechnology to build nanostructured systems, components and materials: one method is named the "top-down" approach, and the other method is called the "bottom-up" approach. In the top-down approach, the idea is to miniaturize the macroscopic structures, components and systems towards the nanoscale. In the bottom-up approach, the atoms and molecules constituting the building blocks are the starting point to build the desired nanostructure [36-38].

Various illustrations are available in the literature depicting the comparison of the top-down and bottom-up approaches [37,38]. In the former, a macro-sized material is reduced in size to reach the nanoscale dimensions. The photolithography used in the semiconductor industry is an example of the top-down approach. The bottom-up nanotechnology is the engineered manipulation of atoms and molecules in a user-defined and repeatable manner to build objects with certain desired properties. To achieve this goal, a number of molecules are identified as the molecular building blocks (MBBs) of nanotechnology, among which diamondoids are the most important ones owing to their unique properties [6,39-44]. Diamondoids can be divided into two major clusters based upon their size: lower diamondoids (1-2 nm in diameter) and higher diamondoids (> 2 nm in diameter).

The building blocks of all materials in any phase are atoms and molecules. Their arrangements and how they interact with one another define many properties of the material. The nanotechnology MBBs, because of their sizes of a few nm's, impart to the nanostructures created from them new and possibly preferred properties and characteristics heretofore unavailable in conventional materials and devices. These nanosize building blocks are intermediate in size, lying between atoms and microscopic and macroscopic systems, and contain a limited and countable number of atoms. They constitute the basis of our entry into new realms of bottom-up nanotechnology [37,38].

The controlled and directed organization of MBBs and their subsequent assembly into nanostructures is one fundamental theme of bottom-up nanotechnology. Such an organization can be in the form of association, aggregation, arrangement or synthesis of MBBs through non-covalent van der Waals forces, hydrogen bonding, attractive intermolecular polar interactions, electrostatic interactions, hydrophobic effects, etc. [37].

The ultimate goal of assemblies of nanoscale molecular building blocks is to create nanostructures with improved properties and functionality heretofore unavailable to conventional materials and devices. As a result, one should be able to alter and engineer materials with desired properties. For example, ceramics and metals produced through controlled consolidation of their MBBs are shown to possess properties substantially improved and different from materials with coarse microstructures. Such different and improved properties include greater hardness and higher yield strength in the case of metals and better ductility in the case of ceramic materials [42].

Considering that nanoparticles have much higher specific surface areas, in their assembled forms there are large areas of interfaces. One needs to know in detail not only the structures of these interfaces, but also their local chemistries and the effects of segregation and interaction between MBBs and their surroundings. The knowledge of ways to control nanostructure sizes, size distributions, compositions and assemblies are important aspects of the bottom-up nanotechnology [37].

In general, nanotechnology MBBs are distinguished for their unique properties. They include, for example, graphite, fullerene molecules made of various numbers of carbon atoms (C_{60}, C_{70}, C_{84}, C_{240}, etc.), carbon nanotubes, nanowires, nanocrystals, amino acids and diamondoids [37]. All these molecular building blocks are candidates for various applications in nanotechnology.

One of the properties used to distinguish MBBs from one another is the number of their available linking groups. MBBs with three linking groups, like graphite, can only produce planar or tubular structures. MBBs with four linking groups can form three-dimensional diamond lattices. MBBs with five linking groups can create three-dimensional solids and hexagonal planes. The ultimate possibility at present would be MBBs with six or more linking groups. Adamantane with six linking groups (see Figure 7) and higher diamondoids are of the latter category, which can be used to construct many complex three-dimensional structures [41]. Such MBBs can have numerous applications in nanotechnology, and they are of major interest in designing shape-targeted nanostructures, including the synthesis of supramolecules with manipulated architectures [45-49].

In addition to possessing six or more linking groups, diamondoids have high strength, toughness and stiffness when compared to other known MBBs. They are tetrahedrally-symmetric stiff hydrocarbons. Strain-free structures of diamondoids give them high molecular rigidity, which is quite important for an MBB. High density, low surface energy and oxidation stability are some other preferred properties of diamondoids as MBBs. Diamondoids have noticeable electronic properties [50]. In fact, they are H-terminated diamond and the only semiconductors which show a negative electron affinity [6].

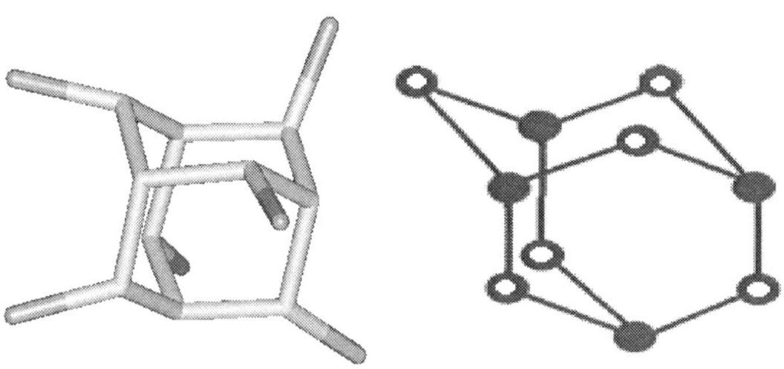

Figure 7. Demonstration of the six linking groups of adamantane.

Since diamondoids possess the capability for derivatization, they can be utilized for reaching suitable molecular geometries needed for molecular building blocks of nanotechnology. Functionalization by different groups can produce appropriate reactants for desired reactions, microelectronics, optics, employing polymers, films and crystal engineering. Over 20,000 variants of diamondoids have been identified and synthesized, and even more are possible [42], providing a rich and well-studied set of MBBs.

Adamantane can be used in molecular studies and preparation of fluorescent molecular probes [51,52]. Because of its incomparable geometric structure, the adamantane core (adamantyl) can impede interactions of fluorophore groups, and self-quenching would diminish due to steric hindrance. Hence, mutual quenching would be diminished, allowing the introduction of several fluorescent groups to the same molecular probe in order to amplify their signals.

Due to their demanding synthesis, diamondoids are helpful models to study structure-activity relationships in carbocations (carbonium ions, which have a positively-charged carbon atom) and radicals, to develop empirical computational methods for hydrocarbons, and to investigate orientational disorders in molecular crystals [5,35].

Atomic force microscope (AFM) is a powerful nanotechnology tool for molecular imaging and manipulations. One major factor limiting resolution in AFM to observe individual biomolecules such as DNA is the low sharpness of the AFM tip that scans the sample. Nanoscale 1,3,5,7-tetrasubstituted adamantane is found to serve as a molecular tip for AFM and may also find application as chemically well-defined objects for calibration of commercial AFM tips [53].

Opto-Electronic Properties of Diamondoids

Opto-electronics, a sub-field of photonics, refers to the study and applications of electronic devices that interact in general with invisible light (such as gamma ray, x-ray, ultraviolet and infrared). Optoelectronic devices are electrical-to-optical or optical-to-electrical transducers, or instruments that use such devices in their operation.

Diamondoids are generally optically transparent in visible light and have high electrical insulating properties as does diamond. In a recent study, it is found that the large gaps between the highest occupied molecular orbital and lowest unoccupied molecular orbital (HOMO-LUMO gaps) in diamondoids are the molecular counterparts of a large fundamental gap in diamond, which is responsible for its optical transparency in the visible range and its insulating properties [41]. The difference between the HOMO and LUMO energy levels is called the band gap energy, which can serve as a measure of the excitability of a molecule: the larger the energy, the harder it will be for excitation to occur. The highest occupied molecular orbital (HOMO) level for organic semiconductors is similar to what the valence band is to inorganic semiconductors. The same analogy is used between the lowest unoccupied molecular orbital (LUMO) level and the conduction band.

The bond lengths and angles in the carbon skeletal structure of diamondoids are essentially the same as in bulk diamond (see Figure 2).

The average bond length of C-C bonds in diamondoids is around 1.545 Å, and the average value of C-H bond lengths is about 1.10 Å. The former matches the C-C bond length in bulk diamond quite well. Upon an increase of the diamondoid size, the average C-C and C-H bond lengths will increase slightly, whereas the average bond angle decreases as the diamondoid size increases. Overall, the average bond angles of most diamondoids are less than the bond angle 109.47° of C-C-C in the bulk diamond structure. This is because the C-H bonds existing in the diamondoids make contributions of smaller bond angles of C-C-H or/and H-C-H to the average bond angles. It is also found that the similarity in bonding between diamondoids and diamond extends also to their bond strength and resistance to deformations [40,54].

The electronic density distribution in diamondoids is important to study for theoretical as well as general chemistry purposes. An explanation and prediction of the reactivity (including reaction mechanisms) and possible existence and stability of related compounds are based on the knowledge of the valence electron density distribution in the diamondoids molecular framework.

Some opto-electronic properties of adamantane, the first and the smallest of diamondoid molecules, has been studied in recent years. For example the geometry and electronic structure of adamantane and related molecular cations has been studied with the emphasis on the electron density distribution on the face planes and in the carbon cage on the basis of *ab initio* molecular orbital calculations. Figure 8 provides charts for the distribution electron density in adamantane across its C1–C7–C10 and C1–C2–C3 planes and along the C1–C4 line as calculated by Korolkov and Sizova [55].

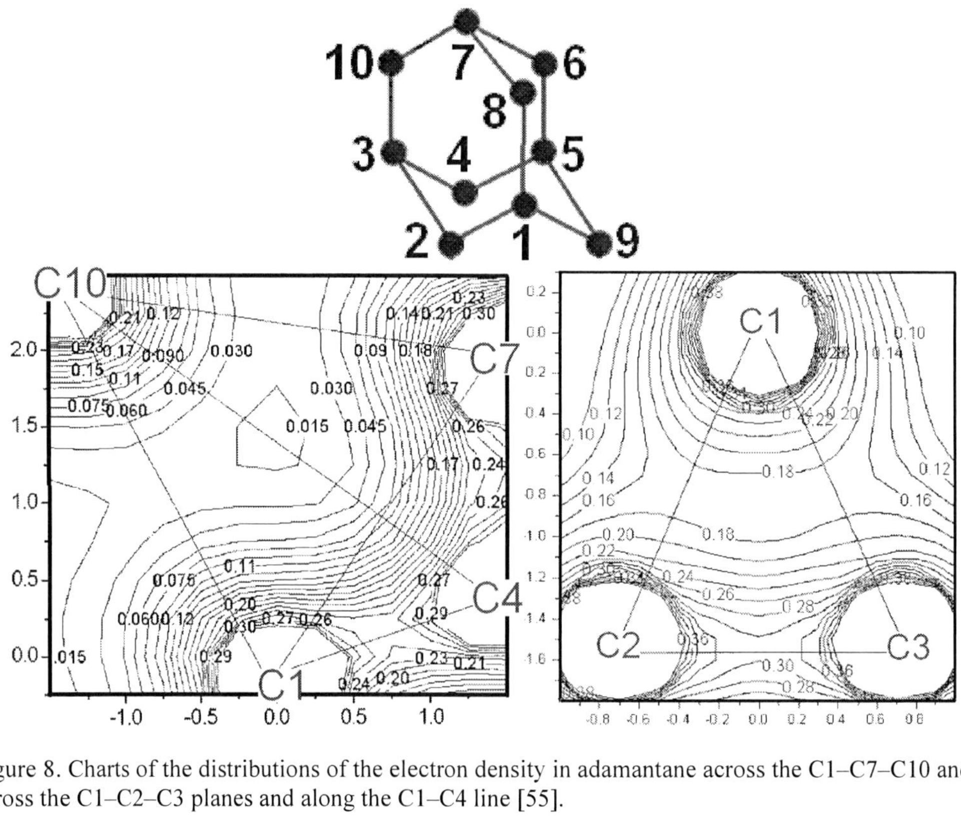

Figure 8. Charts of the distributions of the electron density in adamantane across the C1–C7–C10 and across the C1–C2–C3 planes and along the C1–C4 line [55].

These calculations [55] indicate very low electronic density in the inner parts of the adamantane molecule and its increase in the corresponding adamantly cations. It is believed that the differences in the electronic density distribution in saturated hydrocarbon–adamantane, and in polyhedral boron hydrides are caused by their different abilities to form additional weak bonds. The carbon atoms of saturated hydrocarbons have no vacant 2*p* orbitals in their valence structure, and that makes any additional interactions with the electron pairs on bonding orbitals impossible. As a result, the valences of the carbon atoms in saturated adamantane hydrocarbon are close to the usual value of four [55].

A group of investigators in Germany [56] have measured synchrotron-based carbon K-edge x-ray absorption (XAS) of diamondoids (starting from adamantane and continuing to hexamantane) together with that of a bulk diamond reference as it is shown by Figure 9. Near-edge x-ray absorption is a useful spectroscopy tool to probe electronic and structural properties of fine structure of thin organic films and liquids [57].

According to this figure, practically all the diamondoids possess the general features of the diamond band structure. The larger diamondoid sizes exhibit band structures much closer to that of bulk diamond as the second absolute gap at 303 eV emerges.

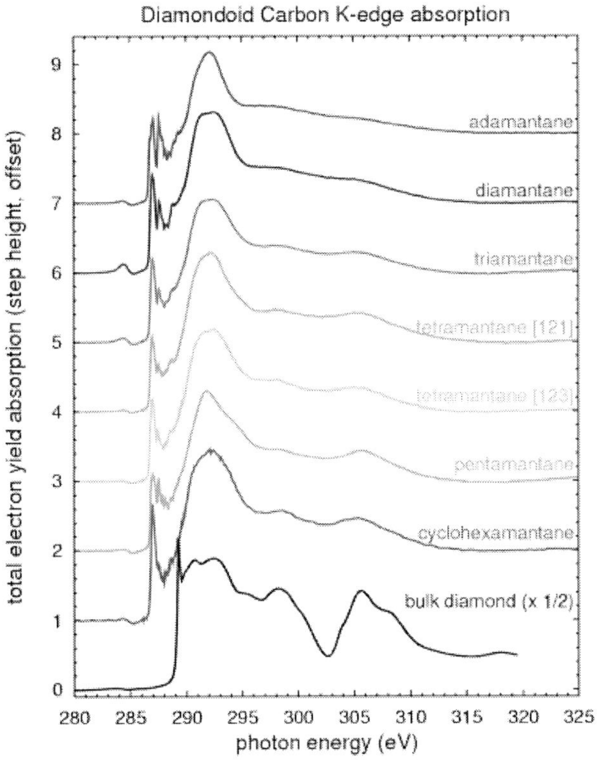

Figure 9. Carbon K-edge x-ray absorption of diamondoids (from adamantane to hexamantane) and of a bulk diamond.

Adamantane is considered an ideal molecule for the study of basic NMR spectroscopy mechanisms [58]. In solid-state NMR spectroscopy, adamantane is used as a common standard for chemical shift referencing (see, e.g., [59]). In dye lasers (i.e., which use an organic dye as a lasing medium), adamantane may be used to prolong their life. This is because adamantane cannot be photoionized since it lacks ultraviolet absorption bands [60]. In another study, adamantane molecule has been used in the design of a novel photoswitching single molecular tip for chemical identification through noncontact atomic force microscopy [61]. The synthesis of novel thermally reversible photochromic spiro is yet another interesting recent photonic application of adamantane [62]. In another study the electronic and vibrational properties of small diamondoids are investigated using an *ab initio* molecular orbital method [54].

Diamondoids can be used to produce electric, electronic, and magnetic functional complexes and polymers (see, e.g., [63] and also [64,65]). In recent years, a few patents have been issued for the opto-electronic applications of diamondoids including a patent for application of admantane as dye etching masks [66] and diamondoid-containing field emission devices [67].

Applications of Diamondoids in Micro-Electro-Mechanical Systems (MEMS)

Diamondoids are expected to be quite useful in the design of MEMS devices, as they have been also envisioned to be used in nanotechnology because of their stability, rigidity and strength. Due to the availability of multiple attachment sites (Figure 7), they are envisioned for the accurate design and building of durable and robust MEMS precision devices [68]. The exceptional hardness, fracture strength and inertness of diamondoids make them unique materials for miniaturized MEMS, such as rotors, propellers, ratches, gears, toothed cogs and cantilevers [69]. It has been estimated that MEMS constructed out of diamondids should last many times longer than current silicon-based MEMS. Since diamondoids are chemically benign, they would not promote allergic reactions in biomedical applications.

Crystal Engineering: The branch of MEMS and nanotechnology called "crystal engineering" is a new concept through which the power of non-covalent intermolecular forces is used in the solid-state to design new nanomaterials with desired functions. The approach here is to learn from known crystalline structures of, for example, minerals in order to design compounds with desired properties. Crystal engineering is considered to be a key new technology with applications in pharmaceuticals, catalysis and materials science. The structures of adamantane and other diamondoids have received considerable attention in crystal engineering due to their molecular stiffness, derivatization capabilities and their six or more linking groups (see Figure 10 [70-74].

Molecular Manufacturing: A concept called "molecular manufacturing," which was originally proposed in 1992 by K. E. Drexler [39] has attracted the attention of some investigators [40,75-78]. Molecular manufacturing is defined as "the production of complex structures via non-biological mechanosynthesis (and subsequent assembly operations)" [39]. Chemical synthesis controlled by mechanical systems operating on atomic scale, and with direct positional selection of reaction sites performed by atomic-precision manipulation systems, is known as mechanosynthesis.

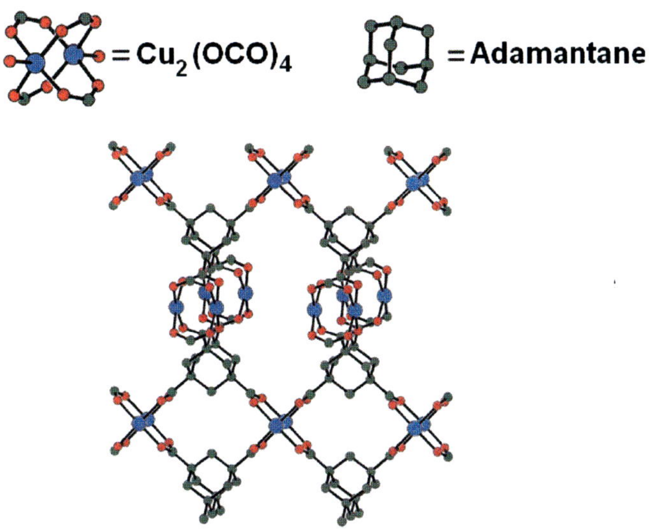

Figure 10. Crystal engineering example: design of a decorated framework made up of the Cu2(OCO)4 cluster (Cu, blue; C, green; O, red) and adamantane [74].

Due to their strong stiff structures containing dense, three-dimensional networks of covalent bonds, diamondoids are one of the favorite sets of molecules considered for molecular manufacturing as originally proposed, described and analyzed in "nanosystems." If diamondoid materials could be synthesized as proposed, they could be quite strong and yet light. For example, diamondoids are being considered to build stronger but lighter rockets and other space components and a variety of other earth-bound articles for which the combination of weight and strength is a consideration [39,40,78-80].

Some of the applications of molecular manufacturing based on diamondoids are: the design of an artificial red blood cell called respirocyte, nanomotors, nanogears, molecular machines and nanorobots [42,80-83]. As an example, the proposed design of nanogears is shown in Figure 11. Other potential applications of molecular manufacturing of diamondoids are in the design of molecular capsules and cages for various applications including drug delivery.

Figure 11. Low-friction nano-bearing assembly. In this design, two diamondoid rings replace small segments of a carbon nanotube, providing a lock for a third, larger ring. The larger ring includes a stitch-work of oxygens to create an electron-rich interior whose effective circular van der Waals packing just touches that of the nanotube framework [image taken from www.somewhereville.com].

For the concept of molecular manufacturing to become successful, among other requirements, a systematic study of the fundamental intermolecular interaction theory of the covalent and non-covalent molecular processes involved is needed [77].

DNA-Diamondoid Nanostructures: Due to the ability of adamantane and its derivatives to attach to DNA, it is possible to construct well-defined nanostructures consisting of DNA fragments as linkers between adamantane cores. This could be a powerful tool to design DNA-directed nanostructured self-assemblies [84]. A unique feature of such DNA-directed self-assemblies is its site-selective immobilization, which makes possible the construction of well-defined nanostructures, as shown in Figure 12.

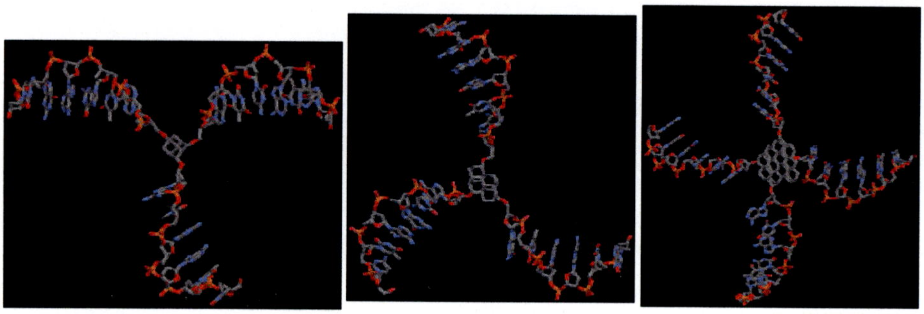

Figure 12. DNA fragments/diamondoid nanostructures left to right: (i) adamantane nucleus and three hexanucleotide linkers; (ii) tetramantane nucleus and three hexanucleotide linkers; (iii) undecamantane nucleus and four hexanucleotide linkers. The hydrogen atoms are removed for clarity [85].

On the other hand, the possibility of the introduction of a vast number of substitutes (like peptidic sequences, nucleoproteins, hydrophobic hydrocarbon chains, etc.) to an adamantane core (adamantyl) makes such a process capable of designing steric colloidal and supramolecular conformations via setting hydrophobic/hydrophilic and other interactions. In addition, the rigidity of adamantane structure can provide strength and rigidity to such self-assemblies [86].

Discussion

Diamondoids are a peculiar class of organic molecules (hydrocarbons) with unique structures and properties. They possess typical characteristics of diamond face-fused cages with hydrogen terminated dangling bonds. They may be used as molecular building blocks of complex functional nanostructures to be found in future NEMS devices. Diamondoids could have applications in self-assembly, immobilization, thin films, synthesis and characterization of MEMS and NEMS devices, ion-selective electrodes and cantilevers for scanning probe microscopy.

Diamondoids can be employed to construct molecular rods, cages and containers, and also for utilization in different methods of self-assembly. In fact, through the development of self-assembly approaches and utilization of diamondoids in these processes, it should be possible to design and construct novel nanostructures.

Diamondoids offer the possibility of producing a variety of MEMS and nanostructural devices, including molecular-scale and micro-scale components of machinery such as rotors, propellers, ratches, gears, toothed cogs, etc. We expect them to have potential for even more possibilities for applications in molding and cavity formation due to their organic nature and their sublimation properties. The diverse geometries and possibility of six or more attachment sites (linking groups) in diamondoids provide an extraordinary potential for the production of shape-derivatives in nanotechnology.

Despite many extensive investigations into properties of diamondoid, their opto-electronic properties are poorly studied. Our future work in this regard will involve joint experimental and theoretical studies of the opto-electric properties of diamondoids. These will be based on density functional theory (DFT) and *ab initio* quantum mechanical computations, collision-induced light scattering (CILS) and refractive index measurements.

ACKNOWLEDGMENTS

This work was supported by the U.S. Army Research Office under Contract W911NF-04-1-0383, the U.S. Department of Energy under Contract DE-FG02-06ER46304, and by a Promising Scholars grant from Indiana State University.

REFERENCES

[1] R. C. Fort, Adamantane: *The Chemistry of Diamond Molecules* (Marcel Dekker, New York, 1976).

[2] G. A. Olah, Ed., *Cage Hydrocarbons* (Wiley, New York, 1990).

[3] A. P. Marchand, *Science* 299, 52-53 (2003).

[4] W. H. Bragg and W. L. Bragg, *Nature* 91, 554-556 (1913).

[5] G. J. Kabo, A. V. Blokhin, M. B. Charapennikau, A. G. Kabo and V. M. Sevruk, *Thermochimica Acta,* 2000, 345, 125-133.

[6] J. E. Dahl, S. G. Liu, and R. M. K. Carlson, *Science* 299, 96-99 (2003).

[7] G. A. Mansoori, L. Assoufid, T. F. George and G. P. Zhang, "Measurement, Simulation and Prediction of Intermolecular Interactions and Structural Characterization of Organic Nanostructures," *Abstracts of the 2003 Nanotechnology Conference* (San Francisco, 2003).

[8] G. R. Desiraju, *J. Molecular Structure* 374,191-198 (1996).

[9] H. Ramezani and G. A. Mansoori, "*Diamondoids as Molecular Building Blocks for Nanotechnology, Drug Targeting and Gene delivery,*" in preparation.

[10] H. Rafii-Tabar and G. A. Mansoori, G. A., "*Interatomic Potential Models for Nanostructures,*" in press (2003). Preprint can be downloaded from http://www.uic.edu/labs/trl/NanostructurePotentialModels.pdf.

[11] W. Gans and J. C. Boeyens, Eds., *Intermolecular Interactions* (Plenum, New York, 1998).

[12] A. Stone, *The Theory of Intermolecular Forces*, (Oxford University Press, Oxford, 1997).

[13] K. Terakura and H. Akai, Eds., Interatomic Potential and Structural Stability: *Proceedings of the 15th Taniguchi Symposium* (Springer, Berlin, 1993).

[14] G. Maroulis and D. Xenides, *J. Chem. Phys.*115, 7957-7967 (2001).

[15] E. H. Benmekki and G. A. Mansoori, *Fluid Phase Equil.* 41, 43 (1988).

[16] A. P. Alivisatos, P. F. Barbara, A. W. Castleman, J. Chang, D. A. Dixon, M. L. Klein, G. L. McLendon, J. S. Miller, M. A. Ratner, P. J. Rossky, S. I. Stupp and M. E. Thompson, *Angew. Chemie, Adv. Mat.* 10, 1297-1336 (1998).

[17] F. Firouzi, H. Modarress and G. A. Mansoori, Europ. Polymer J. 34, 1489-1498 (1998).

[18] E. Z. Hamad and G. A. Mansoori, *J. Phys. Chem.* 94, 3148-3152 (1990).

[19] Z. Huang, A. A. Acero, N. Lei, S. A. Rice, Z. Zhang and M. L. Schlossman, *Faraday Trans.*, Issue 4, 545 (1996).

[20] D. Kinderlehrer, R. D. James and M. Luskin, Eds., Microstructure and Phase Transition, *IMA Volumes in Mathematics and Its Applications* (Springer-Verlag, New York, 1993), pp. 1-25.

[21] G. A. Mansoori, G. A., *Fluid-Phase Equil.* 4, 61-69 (1980).

[22] H. Modarress, E. Ahmadnia and G. A. Mansoori, *J. Chem. Phys.* 111, 10236-10241 (1999).

[23] P. Pourgheysar, G. A. Mansoori and H. Modarress, H., *J. Chem. Phys.* 105, 9580-9587 (1996).

[24] D. H. E. Gross, "Microcanonical Thermodynamics: Phase Transitions in 'Small' Systems," *World Science Lecture Notes in Physics* 66 (2001).

[25] T. L. Hill, Thermodynamics of Small Systems, Parts I and II, Reprint Edition (Dover, Mineola, New York, 1994).

[26] T. L. Hill, *Nano Lett.* 1, 273-275 (2001).

[27] Mansoori, G. A., "Organic Nanostructures and Their Phase Transitions," *Proceedings of the First Conference on Nanotechnology – The Next Ind. Rev.* 2, 345-367 (March 2002).

[28] G. A. Mansoori, "*Advances in Atomic and Molecular Nanotechnology*," Nanotechnology: Emerging Cutting-Edge Technology, U.N. Tech Monitor, Special Issue (September/October 2002).

[29] J. M. Haile and G. A. Mansoori, Eds., "Molecular-Based Study of Fluids," *Advances in Chemistry Series* (American Chemical Society, Washington, D.C., 1983).

[30] M. J. Frisch, G. W. Trucks, H. B. Schlegel, G. E. Scuseria, M. A. Robb, J. R. Cheeseman, V. G. Zakrzewski, J. A. Montgomery, Jr., R. E. Stratmann, J. C. Burant, S. Dapprich, J. M. Millam, A. D. Daniels, K. N. Kudin, M. C. Strain, O. Farkas, J. Tomasi, V. Barone, M. Cossi, R. Cammi, B. Mennucci, C. Pomelli, C. Adamo, S. Clifford, J. Ochterski, G. A. Petersson, P. Y. Ayala, Q. Cui, K. Morokuma, D. K. Malick, A. D. Rabuck, K. Raghavachari, J. B. Foresman, J. Cioslowski, J. V. Ortiz, A. G. Baboul, B. B. Stefanov, G. Liu, A. Liashenko, P. Piskorz, I. Komaromi, R. Gomperts, R. L. Martin, D. J. Fox, T. Keith, M. A. Al-Laham, C. Y. Peng, A. Nanayakkara, M. Challacombe, P. M. W. Gill, B. Johnson, W. Chen, M. W. Wong, J. L. Andres, C. Gonzalez, M. Head-Gordon, E. S. Replogle and J. A. Pople (Gaussian, Inc., Pittsburgh, 1998)

[31] MOLPRO is a package of ab initio programs written by H.-J. Werner and P. J. Knowles, with contributions from J. Almlf, R. D. Amos, M. J. O. Deegan, S. T. Elbert, C. Hampel, W. Meyer, K. Peterson, R. Pitzer, A. J. Stone, P. R. Taylor, R. Lindh, M. E. Mura and T. Thorsteinsson.

[32] R. L. Rowley, Y. Yang and T. A. Pakkanen, *J. Chem. Phys.* 114, 6058-6067 (2001).

[33] M. Edalat, S. S. Lan, F. Pang and G. A. Mansoori, *Int. J. Thermophys.* 1, 177-184 (1980).

[34] G. P. Zhang, Thomas F. George, Lahsen Assoufid, and G. Ali Mansoori, *Phys. Rev.* B 75, 035413-1-7 (2007).

[35] M. A. McKervey, *Tetrahedron* 36, 971-992 (1980).

[36] R. W. Siegel, E. Hu and M. C. Roco, Eds., *Nanostructure Sci. and Tech. - A Worldwide Study*. Prepared under the guidance of the IWGN, NSTC (WTEC, Loyola College in Maryland, 1999).

[37] G. A. Mansoori, *Principles of Nanotechnology: Molecular Based Study of Condensed Matter in Small Systems* (World Scientific, New York, 2005).

[38] G. A. Mansoori, T. F. George, G. P. Zhang and L. Assoufid, Editors, Molecular Building Blocks for Nanotechnology: From Diamondoids to Nanoscale Materials and Applications (Springer, New York, 2007), *Topics in Applied Physics*, in press.

[39] K. E. Drexler, *Engines of Creation* (Anchor Press/Doubleday, Garden City, New York, 1986).

[40] K. E. Drexler, *Nanosystems, Molecular Machinery, Manufacturing, and Computation* (Wiley, New York, 1992).

[41] G. C. McIntosh, M. Yoon, S. Berber and D. Tománek, *Phys. Rev.* B 70, 045401-1-8 (2004).

[42] R. C. Merkle, Trends in Biotechnology 17, 271-274 (1999); ibid., *Nanotechnology* 11, 89-99 (2000).

[43] G. A. Mansoori, "Advances in Atomic and Molecular Nanotechnology," United *Nations Tech Monitor Special Issue* (2002), pp. 53-59.

[44] R. Rawls, *Chem. and Eng. News* 80, 13 (2002).

[45] V. Kovalev, E. Shokova, A. Khomich and Y. Luzikov, *New J. Chem.* 20, 483-492 (1996).

[46] D. N. Chin, D. M. Gordon and G. M. Whitesides, *J. Amer. Chem. Soc.* 116, 12033-12044 (1994).

[47] O. Ermer and L. Lindenberg, *Helvetica Chimica Acta* 74, 825-877 (1991).

[48] O. Ermer and L. Lindenberg, *Helvetica Chimica Acta* 71, 1084-1093 (1988).

[49] E. Shokova, V. Tafeenko and V. Kovalev, *Tetrahedron Lett.* 43, 5153-5156 (2002); K. A. Hirsch, S. R. Wilson and J. S. Moore, *Chemistry-A European J.* 3, 765-771 (1997).

[50] N. D. Drummond, A. J. Williamson, R. J. Needs and G. Galli, *Phys. Rev. Lett.* 95, 096801-1-4 (2005).

[51] P. R. Seidl and K. Z. Leal, J. Mol. Structure: *THEOCHEM* 539, 159-162 (2002).

[52] V. V. Martin, I. S. Alferiev and A. L. Weis, *Tetrahedron Lett.* 40, 223-226 (1999).

[53] Q. Li, A. V. Rukavishnikov, P. A. Petukhov, T. O. Zaikova and J. F. W. Keana, *Org. Lett.* 4, 3631-3634 (2002)

[54] A. J. Lu, B. C. Pan and J. G. Han, *Phys. Rev.* B 72, 035447-1-9 (2005).

[55] D. Korolkov and O. Sizova, *Int. J. Quantum Chem.* 88, 606-613 (2002).

[56] T. M. Willey, C. Bostedt, T. van Buuren, J. E. Dahl, S. G. Liu, R. M. K. Carlson, L. J. Terminello and T. Moller, Phys. Rev. Lett. 95, 113401-1-4 (2005). See also S.G. Liu, J.D. Wind, J.E. Dahl, R. Gat and R.M.K. Carlson, "*Diamondoid-Containing Films for*

Optical Applications" in Molecular Diamond Technologies, ChevronTexaco Technology Ventures, LLC, US.

[57] G. Hähner, *Chem. Soc. Rev*. 35, 1244-1255 (2006).

[58] S. M. De Paul, M. Tomaselli, A. Pines, M. Erns and B. H. Meier, *J. Chem. Phys*. 108, 826-829 (1998).

[59] T. Asakura, Y. Nakazawa, E. Ohnishi and F. Moro, *Protein Sci*. 14, 2654-2657 (2005).

[60] R. Akaba, H. Sakuragi and K. Tokumaru , *J. Chem. Soc., Perkin Trans*. 2, 291-297 (1991).

[61] D. Takamatsu, Y. Yamakoshi and K. Fukui, *J. Phys. Chem*. 110, 1968-1970 (2006).

[62] Y. Yokoyama, H. Nagashima, S. M. Shrestha, Y. Yokoyama and K. Takada, *Bull. Chem. Soc. Jpn*. 76, 355-361 (2003).

[63] Y. Watanabe, Y. Shibasaki, S. Ando and M. Ueda, *Polymer J*. 37, 270-276 (2005).

[64] V. Rüdiger, A. Eliseev, S. Simova, H.-J. Schneider, M. J. Blandamer, P. M. Cullis and A. J. Meyer, *J. Chem. Soc., Perkin Trans*. 2, 2119-2124 (1996).

[65] V. B. Nazarov, V. G. Avakyan, M. V. Alfimov and T. G. Vershinnikova, *Russian Chem. Bull*. 52, 916-922 (2003).

[66] K. Watanabe et al. (2001), "*Resist Composition and Pattern Forming Process*," United States Patent Application 20010006752. Bandwidth Market, Ltd. Retrieved on 14 October, 2005.

[67] United States Patent 7160529; http://www.freepatentsonline.com/7160529.html. Inventors: J. E. Dahl, R. M. Carlson and S. Liu. Application number 10/784,885; filing date 24 February 2004; publication date 9 January 2007.

[68] J.-Y. Raty and G. Galli, "Structural and Electronic Properties of Isolated Nanodiamonds: A Theoretical Perspective," in *Proceedings of the UltraNanoCrystalline Diamond Conference* (St. Petersburg , Russia, 2004).

[69] A .R. Krauss, O. Auciello, D. M. Gruen, A. Jayatissa, A. Sumant, J. Tucek, D. C. Mancini, N. Moldovan, A. Erdemir, D. Ersoy, M. N. Gardos, H. G. Busmann, E. M. Meyer and M. Q. Ding, *Diam. Rel. Mater*. 10, 1952-1961 (2001).

[70] O. R. Evans, R.-G. Xiong, Z. Wang, G. K. Wong and W. Lin, Angew. Chem. Int. Ed. Engl. 38, 536-538 (1999); O. R. Evans and W. B. Lin, *Chem. Mater*. 13, 2705-2712 (2001).

[71] R. Vaidhyanathan, S. Natarajan and C. N. R. Rao, in *Encyclopedia of Supramolecular Chemistry*, edited by J. A. Atwood and J. W. Steed (Marcel Dekker, New York, 2004).

[72] M. J. Zaworotko, *Chem. Soc. Rev*. 23, 283-288 (1994).

[73] D. S. Reddy, D. C. Craig and G. R. Desiraju, *J. Amer. Chem. Soc*. 118, 4090-4093 (1996).

[74] B. Chen, M. Eddaoudi, T. M. Reineke, J. W. Kampf, M. O'Keeffe and O. M. Yaghi, *J. Am. Chem. Soc*. 122, 11559-11560 (2000).

[75] C. Phoenix, *J. Evolution and Tech*. 13 (2003).

[76] R. A. Freitas, Jr. and R. C. Merkle, *Kinematic Self-Replicating Machines* (Landes Bioscience, Georgetown, Texas, 2004).

[77] A. Herman, Modelling Simul. *Mater. Sci. Eng*. 7, 43-58 (1999).

[78] J. Peng, R. A. Freitas, Jr. and R. C. Merkle, *J. Comp. Theor. Nanosci*. 1, 62-70 (2004).

[79] K. E. Drexler, C. Peterson, G. Pergamit, *Unbounding the Future: The Nanotechnology Revolution* (William Morrow, New York, 1991).

[80] K. E. Drexler, *Trends in Biotechnology* 17, 5-7 (1999).

[81] K. Bogunia-Kubik and M. Sugisaka, *Biosystems* 65, 123-138 (2002).

[82] R. A. Freitas, *Nanotechnology* 2, 8-13 (1996).

[83] R. A. Freitas, Jr., Artificial Cells, *Blood Substitutes and Biotechnology* 26, 411-430 (1998).

[84] I. Habus, Q. Zhao and S. Agrawal, *Bioconjugate Chem.* 6, 327-331 (1995); M. Manoharan, K. L. Tivel and P. D. Cook, *Tetrahedron Lett.* 36, 3651-3654 (1995).

[85] H. Ramezani, G. A. Mansoori and M.-R. Saberi, *J. Comput. Theor. Nanosci.* 4, 1-11 (2007).

[86] C. L. D. Gibb, X. Li and B. C. Gibb, *Proc. Nat. Acad. Sci.* (USA) 99, 4857-4862 (2002); S. Aoki, M. Shiro and E. Kimura, Chemistry-A European J. 8, 929-939 (2002); S. B. Copp, S. Subramanian and M. J. Zaworotko, *J. Amer. Chem. Soc.* 114, 8719-8720 (1992).

In: Encyclopedia of Nanotechnology
Editor: Elwood D. Carlson

ISBN: 978-1-60692-079-4

Chapter 18

THREE-DIMENSIONAL CONTROLLABLE SUBMICRON GOLD FILMS: SULFHYDRYL-CONTAINING MOLECULAR ASSEMBLY AND CONTROLLABLE SURFACE PLASMON RESONANCE AND ELECTROCATALYSIS PROPERTIES

Jianqiang Hu, Juan Liu, Xianbo Lu, Wu Lu, Jianhua Zhou and Jinghong Li*

Department of Chemistry, Key Laboratory of Bioorganic Phosphorus Chemistry and Chemical Biology, Tsinghua University, Beijing 100084, P. R. China

ABSTRACT

Surfactant-free capped gold nanoparticles (SFGNPs) of about 60 nm in diameter were synthesized through using a seed-induced route and subsequently self-assembled on sulfhydryl-containing ITO glass slices. Two-layer or multi-layer submicron gold films could be obtained by repeatedly self-assembling the freshly prepared SFGNPs on the as-prepared one-layer gold nanofilm in the function of p-aminothiophenol (p-ATP) molecules, and the self-assembled thickness or layer number of which was tunable. The thickness control in the assemblies was demonstrated by the atomic force microscopy (AFM), scanning electron microscopy (SEM), and UV-visible absorption spectroscopy measurements. Moreover, UV-visible absorption spectroscopy, surface-enhanced Raman scattering (SERS) spectroscopy, and electrocatalysis property studies indicated that surface plasmon resonance (SPR) and electrocatalysis properties of the self-assembled films prepared thus were enhanced along with the increase in their layer-numbers. These controlled self-assembled films are not only helpful to study physicochemical properties of the uniform and controllable assemblies, but also are expected to explore applications in many fields such as chemical and biological sensors, transmission surface plasmon resonance transducer, SERS, biological detection, and electrochemistry.

* Email: jhli@mail.tsinghua.edu.cn

INTRODUCTION

Self-assembled nanocrystal superlattices of metal nanoparticles exhibit interesting electrical, optical, and catalytic properties that are distinct from their respective bulk properties and can be designed on a molecular level.[1] The characteristics of assembled films can be tuned by varying not only the size or shape of the particles as well as the distance between them but also the three-dimensional (3D) spatial arrangement. The novel electrical, optical, and catalytic properties of such assemblies hence provide a unique possibility for optimizing the properties of these films for potential applications in areas such as electronics, [2] optics, [3] catalysis, [4] sensors, [5] and biology [6]. At present, various sizes and shapes of gold nanoparticles with the best biocompatibility have been successfully prepared [7] and extensively employed to assemble into 1D or 2D ordered nanoarrays through all sorts of self-assembled strategies such as AFM nanolithography, [8] template-assisted assembly, [9] water/oil interface,[10] liquid-air interface, [11] and molecular assembly with a bifunctional compound acting as a linker unit for electrostatic or covalent attachment of pure nanoparticles to their functionalized surfaces or to each other [12]. Recent interest in this field has focused on the exploration of the collective properties of assembled 3D arrays of these nanoparticles, which leads to the formation of new and fascinating materials with tunable, designer-specified optical, electronic, and catalytic properties.[1c, 13, 14] Two factors for constructing a macroscopic 3D assembly or superlattice are of key importance: (i) the preparation of available building blocks of nanoparticles with uniform particle sizes and shapes and (ii) the selection of a suitable medium to link nanoparticles.

3D self-assembly of gold nanoparticles can, to the best of our knowledge, date back to the middle 1990s, which was easily carried out by alkanethiolate monolayer adsorbing gold nanoclusters [15]. Subsequently, gold nanoparticles were also assembled into 3D superlattices with nanoporous alumina membranes,[16] dialysis method, [17] soluble silica, [18] water-in-oil microemulsion process, [19] hydrogen-bonding mediation, [14] and dithio or bis-dithiocarbamate derivatives etc..[1c] However, the above-mentioned gold nanoarrays are, strictly speaking, not "real" 3D supperlattices but belong to 2D assemblies because of their exhibiting only monolayer and incomplete two-layer structures. Recently, "real" 3D self-assembly of gold nanoparticles was successfully acquired through using the template (*e.g.*, styrene microspheres etc.) and dipping etc. methods [20]. Nevertheless, a promising and "real" 3D self-assembly of gold nanoparticles is the fabrication of 3D multilayer film of citrate-stabilized gold nanoparticles self-assembled on the surface of a gold electrode using a layer by layer (LBL) approach and 1,4-benzenedimethanethiol as a covalently cross linker.[13] Regretfully, in the 3D assembly, a little shortcoming seems to only prefer assembling gold nanoparticles with small sizes (*e.g.*, less than 10 nm). Therefore, the "real" 3D self-assembly of gold nanoparticles with large sizes (*e.g.*, more than 10 nm) has still been rarely reported to date although gold nanoparticles of 16-170 nm have been self-organized into 2D hexagonal close-packed arrays at the air-water interface [21].

In this article, we utilized surfactant-free capped gold nanoparticles (SFGNPs) of *about* 60 nm as building blocks to obtain 3D submicron gold films with controllable thickness or layer numbers. In this process, SFGNPs were self-assembled on the surface of indium-tin-oxide (ITO) glass slices through using LBL technique and p-aminothiophenol (p-ATP) as a cross linker for electrostatically adsorbing and covalently bonding gold nanoparticles. Due to the fact that the electrostatic and covalent-bonding interaction between p-ATP and gold nanoparticles effectively worked at the ITO surface, the p-ATP molecules adsorbed onto gold nanoparticles through their sulfur atoms were standing-up relative to the ITO surface, [22] and thus that the specific binding (standing-up) mode finally resulted in the formation of multilayer gold films with thickness-dependend surface plasmon resonance and electrocatalysis properties.

EXPERIMENTAL SECTION

Reagents and Materials

Chloroauric acid ($HAuCl_4$, 99%) and p-aminothiophenol (p-ATP, 90%) were purchased from Aldrich Chem. Co.. All other reagents (analytical grade) were obtained from Beijing Chem. Co.. All reagents were used as received. Milli-Q water (>18.0 MΩ cm^{-1}) was used to prepare all solutions except p-aminothiophenol with absolute ethanol. All glassware used was washed with aqua regia and rinsed with 18 MΩ cm^{-1} water prior to use. 0.1 M phosphate buffer solution (PBS, pH=7.0) was prepared by an equal volume mixture of 0.078 M sodium dihydrogen phosphate (NaH_2PO_4) and 0.122 M disodium hydrogen phosphate (Na_2HPO_4).

Preparation of SFGNPs

First, spherical gold seeds of about 12 nm in diameter were prepared by citrate reduction method [7a, 23]. Then, 0.4 mL of the freshly synthetic 12 nm-seed and 10 mL of 1 mM $HAuCl_4$ were respectively added to round-bottom flask and diluted up to 60 mL. 40 mL of freshly prepared ascorbic acid (0.4 mM) was dripped into the solution (10 mL/min) while the diluted solution was stirring. Finally, surfactant-free gold nanoparticles of about 60 nm in diameter were obtained, the surface of which is advantageous and easy to be modified or functionalized with organic molecules (*e.g.*, sulfhydryl- or amino-containing compounds).

Three-Dimensional (3D) Assembly of SFGNPs

3D gold nanofilms were assembled by sulfhydryl-containing molecules according to the follow procedure: (1) ITO glass slices were cleaned respectively with washing agent, acetone, and Milli-Q water. Then these slices were removed and dried by N_2 (the following steps have the same operation). And 100 mL of as-prepared 60 nm-size colloidal solution was concentrated into 10 mL by centrifugation. (2) The cleaned ITO slices were dipped into 1 mM p-ATP in absolute ethanol solution for 6 h to obtain sulfhydryl-containing interface. Wherein, a drip (about 0.05 mL) of 1 M HCl was added in the solution (20 mL) to make amino group protonized. (3) The sulfhydryl-containing slices were placed in concentrated colloidal solution to allow them self-assemble through Au-S bonds for 12 h. Thus, uniform and ordered one-layer gold nanofilms in ITO slices were obtained. (4) Repeatedly performing (2) and (3), multi-layer gold submicron films could be successfully acquired.

Preparation and Electrochemical Measurements of SFGNPs/ITO Electrodes

According to the procedure of the SFGNPs assembly, after the ITO slices were assembled with three-layer and six-layer gold nanoparticles, the slices were taken out and covered by a beaker so that water could evaporate slowly in air. The two dried SFGNPs/ITO electrodes were stored at 4 □ in refrigerator prior to use.

Electrochemical experiments were performed at room temperature using a CHI 802B workstation (CH Instruments, Inc., Austin, USA). The measurements were based on a three-electrode system with the as-prepared ITO electrode as the working electrode, a platinum wire as the auxiliary electrode, and a saturated Ag/AgCl electrode as the reference electrode.

Results and Discussion

TEM Characterization of SFGNPs

Figure 1 shows typical TEM images of gold seeds prepared by the citrate reduction and SFGNPs prepared through using the ascorbic reduction, respectively. Well-monodisperse and uniform gold seeds, which had the sizes of 12 ± 2 nm, could be easily synthesized by the reduction of $HAuCl_4$ with $Na_3C_6H_5O_7$ (Figure 1A). Using different seed concentrations, the sizes of gold nanoparticles could be well controlled in the range of sizes from 20 to 80 nm by the ascorbic reduction and had good monodispersity in the absence of surfactant. Figure 1B gives representative TEM image of 0.4-mL-seed prepared SFGNPs with an average diameter of 60 ± 9 nm. Wherein, 60 nm-size gold nanoparticles were selected to prepare and subsequently self-organize into 3D films as building blocks because 60 nm-size gold nanoparticles could induce the strongest surface electric field for surface-enhanced Raman scattering (SERS) among all sizes of spherical gold nanoparticles [24].

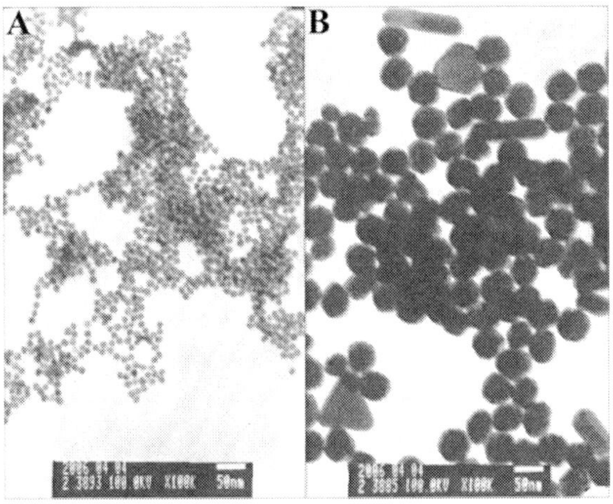

Figure 1. TEM images of (A) 12 nm-size gold seeds prepared by the citrate reduction and (B) surfactant-free capped gold nanoparticles prepared through using the ascorbic reduction, respectively. Scale bar: 50 nm.

It is very interesting to note that non-surfactant capped nanoparticles are usually endowed with a giant superiority in connecting other functionalized molecules or polymers and further self-assembling into ordered 1D, 2D, and even 3D superlattices because of eliminating steric hindrance in comparison with surfactant capped nanoparticles.

3D Assemblies and AFM and SEM Characterization of SFGNPs on ITO Glass Slices

Scheme 1 illustrates the strategy for the 3D controllable assembly of surfactant-free gold nanoparticles prepared thus. A cleaned ITO which surface contains copious hydroxyl groups was immersed in protonized p-ATP solution to afford a sulfhydryl-containing ITO slice through electrostatic interaction. According to Cumberland and Strouse,[25] gold nanoparticles could readily complex with various ligands containing oxygen, nitrogen, and sulfur etc. donor atoms. When the sulfhydryl-containing ITO was dipped into freshly prepared SFGNPs, ordered one-layer gold nanofilm assembled on the ITO slice was successfully obtained through the Au-S bonds. When the one-layer gold nanofilm was again immersed into the protonized p-ATP solution, the p-ATP molecules were again absorbed onto the gold nanofilm through their sulfur atoms and were standing-up relative to the substrate surface [22]. And the resulting one-layer nanofilm containing copious amido groups was sunk into negatively charged gold colloidal solution prepared thus to afford two-layer SFGNPs through electrostatic interaction between the positively charged amido end and negatively charged gold nanoparticles.

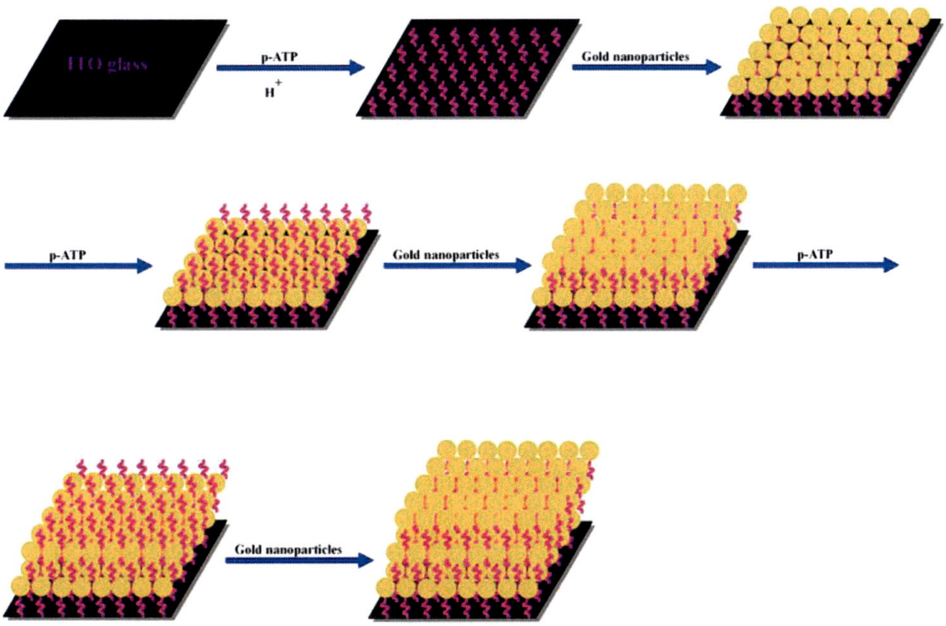

Scheme 1. Strategy for the three-dimensional controllable assembly of surfactant-free gold nanoparticles prepared thus.

For assembling multilayer gold nanofilms, it was only necessary to repeatedly immerse the two-layer film into protonized p-ATP and freshly prepared and negatively charged gold colloidal solution, respectively. Figures 2B, 2C, and 2D show typical AFM images of layer-by-layer (LBL) assembled one-, two-, and three-layer gold films using p-ATP, respectively. Their corresponding heights were 62 ± 6.6 nm, 114 ± 9.8 nm, 165 ± 13.4 nm, respectively. Figure 3 gives typical SEM image of the assembled three-layer gold submicron film and its theoretically assembled schematic plot. The SEM image of the three-layer nanofilm inferred potentially that these assembled films formed hexagonally close-packed nanoparticle arrays along with three-dimensional directions.

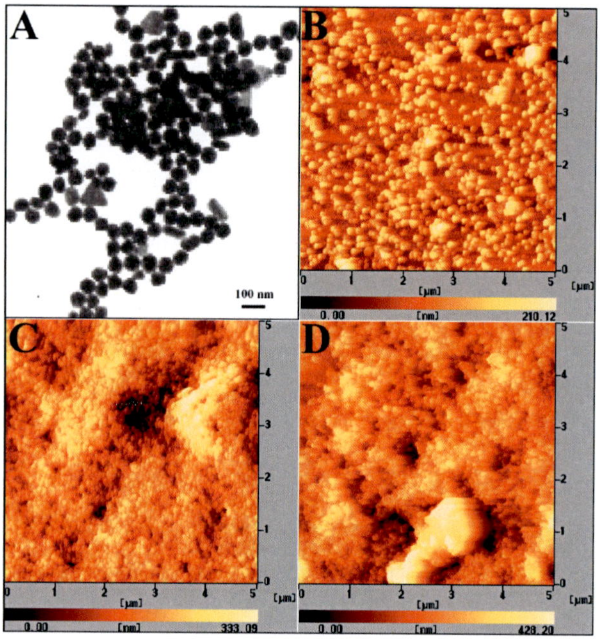

Figure 2. A) TEM image of surfactant-free capped gold nanoparticles prepared thus and B, C, D) AFM images of layer-by-layer (LBL) assembled one-, two-, and three-layer gold films using p-aminothiolphenol, respectively.

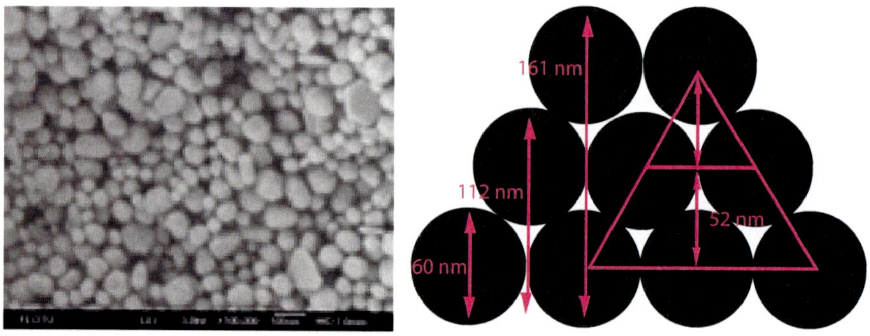

Figure 3. SEM image of the assembled three-layer gold submicron films and its theoretically assembled schematic plot.

The hexagonally close-packed mode could be clearly seen in the right schematic plot of Figure 3, which was consistent with that reported by Halas and his co-workers [26]. The distance between two layers of about 52 nm obtained through theoretical calculation could be well consistent with the results determined from AFM images.

UV-Visible Surface Plasmon Resonance (SPR) Absorption Property Investigation

It is well known that surface plasmon absorption spectroscopy can be used to effectively examine size and shape-controlled isolated colloidal nanoparticles [27]. Recently, UV-vis spectroscopy has been employed widely to monitor self-assembled gold nanofilms [1c]. Figure 4 shows UV-visible absorption spectra of different layer-number gold films prepared using the LBL self-assembly technology, and the insert plot of which is the dependent plot of assembled layer number vs. peak intensity. For all self-assembled gold films produced thus, the absorption peaks shifted to the red and their full-widths at half-maximum (FWHM) broadened as compared with those (peak position: 538 nm; FWHM: 54 nm) of isolated gold colloidal solution of about 60 nm in diameter (see Figure 5). The observed differences between self-assembled gold film and isolated gold nanoparticles in the colloidal solution could be due to differences in the local dielectric environment (*i.e.*, their dielectric constants of the surrounding medium are 1.00 (air) and 1.77 (water), respectively) [28] the interparticle spacing (close-packed and isolated, respectively) [29] and the chemical nature of the interface (with or without p-ATP as a linker) [30].

Along with the increase of the assembled layer numbers, the absorption peaks of these submicron films shifted to the higher wavelength and their FWHM trended toward the broader (see Table 1). Interestingly, the absorption peak intensity increased linearly with the number of self-assembly layers (see the inset of Figure 4). Increasing one layer gold nanoarray, the absorption intensity of about 0.007 a.u. would be increased, indicating that UV-visible surface plasmon resonance absorption property of self-assembled gold films depended on not only the size or shape of single gold nanoparticles but also the thickness of the gold films (*i.e.*, possessing thickness-dependent SPR property). The UV-visible absorption thickness-dependent inclination shown in Figure 4 is consistent with reported spectra for assembled multiplayer gold nanoparticles [31]. The distinct red-shift and broadening of the plasmon bands of the increased layer-number gold films could attribute to the increase of dielectric constant of the assembling matrix although they all possessed the same local dielectric environment, the interparticle spacing, and the chemical nature of the interface [30a]. These plasmon band characteristics could be explained in terms of the chemical interface-damping model [30b]. Excited electrons transferred among the gold SFGNPs through the p-ATP adsorbate inducing resonant states, and disturbed the coherence of the surface plasmon as a result of inelastic scattering processes and therefore led to the red-shift and broadening of the plasmon band. The distance dependence of the surface plasmon resonance absorption of the gold submicron films should have a promising application in chemical and biological sensing through binding an analyte to a selective receptor layer on the gold film at a certain distance from the metal surface. The resulting distance sensitivity to self-assembled layer thickness also provides an opportunity and basis for further development of transmission surface plasmon resonance (T-SPR) transducers based on receptor-coated Au island films [32].

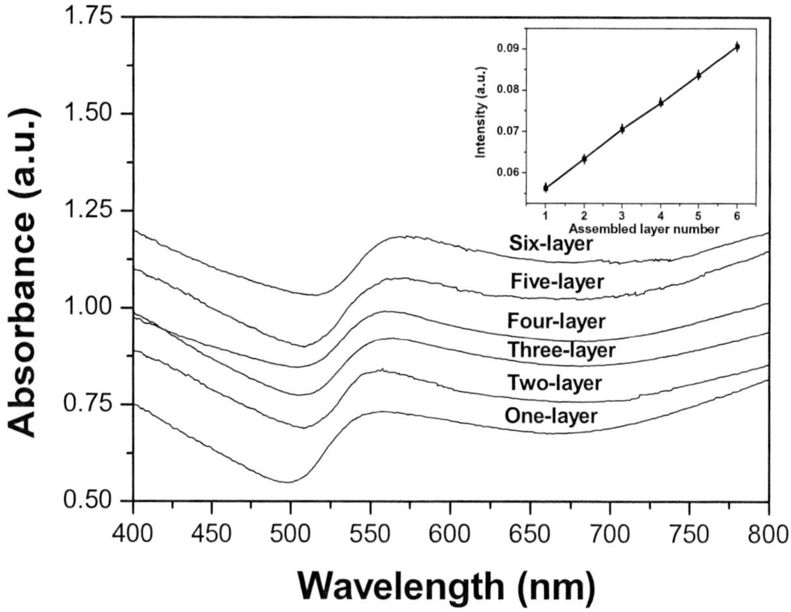

Figure 4. UV-visible spectra of LBL assembled gold films, the insert plot of which is the dependent plot of assembled layer number vs. peak intensity.

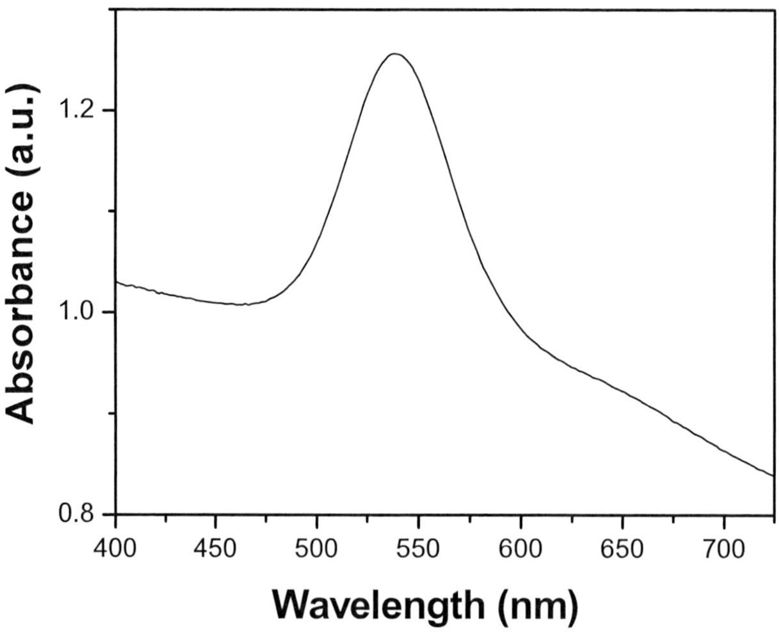

Figure 5. UV-visible absorption spectrum of spherical gold nanoparticles taken from the solution that was used for electronic microscopy studies in Figure 1B.

Table 1. UV-visible spectrum peaks and their corresponding FWHM and intensities of LBL self-assembled gold submicro films

Layer number	Peak position (nm)	FWHM (nm)	Peak intensity (a. u.)
1	555 ± 3	67 ± 8	0.0563 ± 0.002
2	557 ± 3	68 ± 8	0.0634 ± 0.002
3	560 ± 3	76 ± 8	0.0706 ± 0.002
4	562 ± 3	78 ± 8	0.0769 ± 0.002
5	567 ± 3	83 ± 8	0.0837 ± 0.002
6	573 ± 3	93 ± 8	0.0907 ± 0.002

*Freshly prepared surfactant-free gold nanoparticles were assembled through using p-aminothiolphenol on ITO glass. The UV-vis peak intensity was calculated by subtracting the adsorption peak intensity of the higher baseline from the UV-vis peak intensity.

Surface-Enhanced Raman Scattering (SERS) Property Investigation

Surface-enhanced Raman scattering (SERS) technique has gained a wide reputation in the detection of many adsorbate molecules with a wide range of concentrations down to the single molecule detection limit because of its extremely high surface sensitivity [33]. In contrast to electrochemically roughened noble metal electrodes, corresponding noble metal colloids intimately associated with their sizes and shapes seem to have the same and/or even higher detection sensitivity as a SERS substrate. It is generally believed that SERS arise at certain "hot sites", where the optical interactions are locally enhanced due to excitation of SPR.[34] Such sites could be located at junctions between closely spaced nanoparticles, at sharp edges or in crevices, at certain positions in fractal aggregates, etc.[35]. Therefore, the preparation and self-assembly of ordered nanostructured surfaces with rich "hot sites" will be optimal paths for obtaining highly SERS-active substrates [36]. Figure 6 shows typical SERS spectra of p-ATP adsorbed on assemblies of gold submicron films respectively with 1-6 layers, and the frequencies and intensities of their corresponding assigned vibration bands are listed in Table 2. Figure 6 and Table 2 reveal two key features and tendencies among the six spectra. The one is that, with the increase of the assembled layer numbers, the SERS spectrum peaks of every corresponding vibration band presented slightly the blue-shifted, which could be attributed to substantially different adsorption behavior of p-ATP on different assembled layers [37]. The other was the intensity increase of the seven bands located at about 1007, 1082, 1143, 1181, 1390, 1437, and 1587 cm^{-1} along with gradually increasing the self-assembled layer numbers. When SFGNPs were assembled up to the second and third layers, the intensities of p-ATP absorbed at the two submicron films had about one and two orders of magnitude enhancement as compared with those absorbed at the nanofilm assembled with one-layer SFGNPs, respectively. However, the assembled layer numbers changed from three layers to four, five, and six layers, these SERS intensities gradually increased only at the same order of magnitude.

Above underlying change in the SERS intensities could be ascribed to different electromagnetic (EM) and charge-transfer (CT) enhancements of the six assemblies.

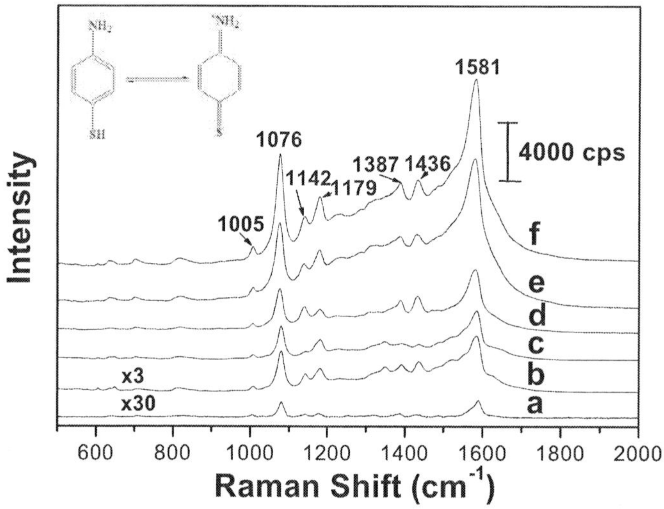

Figure 6. SERS spectra of p-ATP adsorbed on assemblies of one-layer (a), two-layer (b), three-layer (c), four-layer (d), five-layer (e), and six-layer (f) gold films, respectively. The wavelength of various bands and the scale bar of scattered intensity are shown in the figure.

Table 2. Thickness-depended SERS band frequencies and intensities of p-ATP adsorbed on LBL self-assembled gold submicron films

	SERS band frequencies (cm-1)							
Layer number	1	2	3	4	5	6	7[a]	SERS band intensities (cps)[b]
1	1007	1082	1143	1181	1390	1437	1587	31 ± 2
2	1008	1080	1142	1182	1392	1438	1586	755 ± 11
3	1007	1079	1142	1182	1393	1435	1585	1913 ± 53
4	1007	1076	1141	1181	1389	1434	1583	2231 ± 35
5	1006	1075	1141	1180	1388	1434	1582	3765 ± 34
6	1005	1075	1141	1179	1388	1434	1582	5717 ± 54

*The Raman and SERS spectra were performed with a microscopic confocal Raman spectrometer (Renishaw, RM 2000) operated with a He-Ne laser (632.8 nm), laser intensity (about 4.7 mW), and an objective len (20 X). [a] SERS vibrational bands of p-ATP located at about 1007, 1082, 1143, 1181, 1390, 1437, and 1587 cm^{-1} were assigned to CC and CCC bending (18a(a1)), CS stretching (7a(a1)), CH bending (9b(b1)), CH bending (9a(a1)), CH bending and CC stretching (3(b2)), CC stretching and CH bending (19b(b2)), and CC stretching (8a(a1)), respectively [37] [b] SERS band intensities were calculated from the SERS band frequency of about 1075 cm^{-1}. Laser line: 632.8 nm.

According to EM theory, the SERS intensity was determined by the orientation of the adsorbed p-ATP molecules, which were bound, due to their high affinities to gold, not only on the open surface of gold nanoparticles but also on sites located in gaps and junctions between neighboring gold nanoparticles (*i.e.*, "hot sites"). These "hot sites" induced extremely intense local electromagnetic fields, resulting in the main SERS enhancements. Therefore, the intensities of the vibrational bands, for example, the C-S stretching band (1082 cm^{-1}) enhanced only by an electromagnetic effect, [38] increased with increasing assembled layer numbers in the present study because of the increase of the available "hot sites" numbers. It is well known that the electromagnetic field of the light will exponentially attenuate with leaving the surface [39]. Thus, along with the increase of the assembled layer numbers, the available "hot sites" numbers gradually increased but reached a saturated value in a certain assembled layer, and accordingly the SERS intensities nonlinearly increased and reached a maximum intensity when the assembled layers were added. In contrast to the EM mechanism, the CT mechanism meaning the charge-transfer process between metal and adsorbate required the direct proximity between adsorbed molecules and metal surface. These bands around 1143, 1390, and 1437 cm^{-1} in the present study, which had been attributed to the CT of the metal to the adsorbed molecules in the previous studies, [40] were also enhanced along with the increase of the assembled layer numbers. The CT enhancement could be explained probably due to the charge-transfer among layers based on p-ATP served as an electronic circuitry (see Figure 6). The results presented herein suggested that these SERS intensities of p-ATP adsorbed on different self-assembled films intimately depended on the thickness or layer numbers of the assemblies.

Electrocatalysis Property Investigation

Previous studies have demonstrated that catalytic activity of gold nanoparticles depended on their sizes and shapes [41]. In the present study, we studied electrocatalysis properties of self-assembled films with different layers of gold nanoparticles. Figure 7 shows typical cyclic voltammograms of different assembled films modified electrodes in 50 mM pH 7.0 PBS.

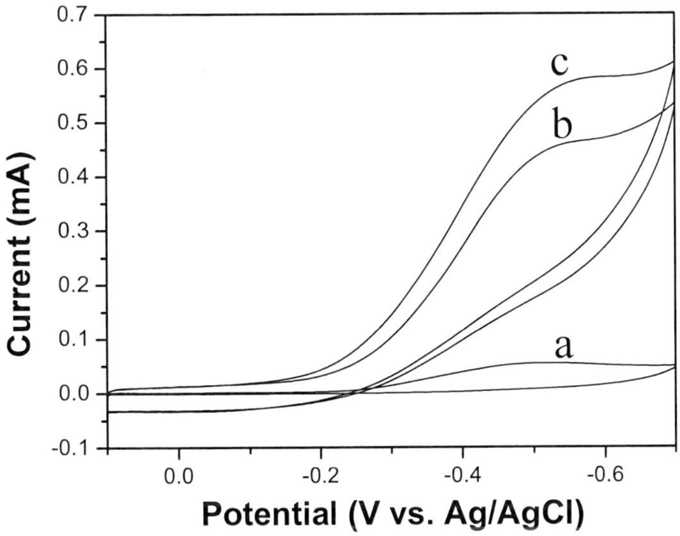

Figure 7. Cyclic voltammograms of assembled multilayer SFGNPs/ITO electrodes in oxygen-saturated pH 7.0 PBS with the scan rate of 100 mV/s. (a) naked ITO electrode, (b) ITO electrode assembled with three-layer SFGNPs, and (c) ITO electrode assembled with six-layer SFGNPs.

A stable and well-defined reduction peak of O_2 was all observed at these electrodes. The intensities of the pair reduction peak current generated from the two ITO electrodes assembled with multilayer SFGNPs were greatly increased as compared with that from the naked electrode and even reached one order of magnitude enhancement, indicating that assembled multilayer gold submicron films had highly catalytic activity. With the increase of the assembled layers, the effective surface areas of the three-dimensional gold film electrode were also increased, which resulted in obviously enhanced catalytic current of the submicron films (see Figure 7b and 7c). The new 3D submicron gold films as relatively high catalytic activity are expected to explore applications in bioanalysis and sensors.

Conclusion

In summary, we synthesized surfactant-free capped gold nanoparticles through using a seed-induced method, using which submicron gold films with different layer numbers could be successfully self-assembled on sulfhydryl-containing ITO glass slices by p-aminothiophenol molecules. The thickness or layer numbers of these assembled films could be controlled. The controllable assemblies were characterized by AFM, SEM, and UV-visible absorption, SERS, and electrocatalysis measurements. The results indicated that the surface plasmon resonance and electrocatalysis properties of the self-assembled films were gradually enhanced along with their layer-number increase. Therefore, the thickness-controlled self-assembled films may be helpful to study physicochemical properties of the uniform and controllable assemblies, and are also expected to explore applications in many fields such as chemical and biological sensors, SERS, biological detection, and electrochemistry.

Acknowledgement

This work was financially supported by the National Natural Science Foundation of China (No. 20435010, No. 20675044) and Postdoctoral Science Foundation of China (023205035).

References

[1] (a) Daniel, M. C.; Astruc, D.; Green, M. *Chem. Rev.* 2004, *104*, 293. (b) Schmid, G.; Simon, U. *Chem. Comm.* 2005, 697. (c) Wessels, J. M.; Nothofer, H. G.; Ford, W. E.; von Wrochem, F.; Scholz, F.; Vossmeyer, T.; Schroedter, A.; Weller, H.; Yasuda, A. *J. Am. Chem. Soc.* 2004, *126*, 3349. (d) Zheng, L. Z.; Yao, X.; Li, J. H. *Curr. Anal. Chem.* 2006, *2*, 279.

[2] Collier, C. P.; Saykally, R. J.; Shiang, J. J.; Henrichs, S. E.; Heath, J. R. *Science* 1997, *277*, 1978.

[3] Chen, J.; Wang, W.; Klemic, J.; Reed, M. A.; Axelrod, B. W.; Kaschak, D. M.; Rawlett, A. M.; Price, D. W.; Dirk, S. M.; Tour, J. M.; Grubisha, D. S.; Bennett, D. W. *Mol. Electron. II* 2002, *960*, 69.

[4] Jaramillo, T. F.; Baeck, S. H.; Cuenya, B. R.; McFarland, E. W. *J. Am. Chem. Soc.* 2003, *125*, 7148.

[5] Nolting, B.; Yu, J. J.; Liu, G. Y.; Cho, S. J.; Kauzlarich, S.; Gervay-Hague, J. *Langmuir* 2003, *19*, 6465.

[6] Phadtare, S.; Kumar, A.; Vinod, V. P.; Dash, C.; Palaskar, D. V.; Rao, M.; Shukla, P. G.; Sivaram, S.; Sastry, M. *Chem. Mater.* 2003, *15*, 1944.

[7] (a) Frens, G. *Nature* 1973, *241*, 20. (b) Jana, N. R.; Gearheart, L.; Murphy, C. J. *J. Phys. Chem. B* 2001, *105*, 4065. (c) Hu, J. Q.; Zhang, Y.; Liu, B.; Liu, J. X.; Zhou, H. H.; Xu, Y. F.; Jiang, Y. X.; Yang, Z. L.; Tian, Z. Q. *J. Am. Chem. Soc.* 2004, *126*, 9470. (d) Lu, L. H.; Kobayashi, A.; Tawa, K.; Ozaki, Y. *Chem. Mater.* 2006, *18*, 4894.

[8] (a) Fresco, Z. M.; Fréchet, J. M. J. *J. Am. Chem. Soc.* 2005, *127*, 8302. (b) Li, Q. G.; Zheng, J. W.; Liu, Z. F. *Langmuir* 2003, *19*, 166.

[9] (a) Deng, Z. X.; Tian, Y.; Lee, S. H.; Ribbe, A. E.; Mao, C. D. *Angew. Chem. Int. Ed.* 2005, *44*, 3582. (b) Wang, H.; Levin, C. S.; Halas, N. J. *J. Am. Chem. Soc.* 2005, *127*, 14992.

[10] Reincke, F.; Kegel, W. K.; Zhang, H.; Nolte, M.; Wang, D. Y.; Vanmaekelbergh, D.; Möhwald, H. *Phys. Chem. Chem. Phys.* 2006, *8*, 3828.

[11] (a) Cheng, W. L.; Dong, S. J.; Wang, E. K. *J. Phys. Chem. B* 2005, *109*, 19213. (b) Hu, J. W.; Han, G. B.; Ren, B.; Sun, S. G.; Tian, Z. Q. *Langmuir* 2004, *20*, 8831. (c) Hu, J. W.; Zhao, B.; Xu, W. Q.; Fan, Y. G.; Li, B.; Ozaki, Y. *Langmuir* 2002, *18*, 6839.

[12] (a) Zhao, Y.; Pérez-Segarra, W.; Shi, Q. C.; Wei, A. *J. Am. Chem. Soc.* 2005, *127*, 7328. (b) Kiely, C. J.; Fink, J.; Brust, M.; Bethell, D.; Schiffrin, D. *Nature* 1998, *396*, 444.

[13] Abdelrahman, A. I.; Mohammad, A. M.; Okajima, T.; Ohsaka, T. *J. Phys. Chem. B* 2006, *110*, 2798.

[14] Yao, H.; Kojima, H.; Sato, S.; Kimura, K. *Langmuir* 2004, *20*, 10317.

[15] Hostetler, M. J.; Stokes, J. J.; Murray, R. W. *Langmuir* 1996, *12*, 3604.

[16] Hanaoka, T. A.; Kormann, H. P.; Kröll, M.; Sawitowski, T.; Schmid, G. *Eur. J. Inorg. Chem.* 1998, *6*, 807.

[17] Adachi, E. *Langmuir* 2000, *16*, 6460.

[18] Fan, H. Y.; Yang, K.; Boye, D. M.; Sigmon, T.; Malloy, K. J.; Xu, H. F.; López, G. P.; Brinker, C.J. *Science* 2004, *304*, 567.

[19] Fan, H. Y.; Leve, E.; Gabaldon, J.; Wright, A.; Haddad, R. E.; Brinker, C. J. *Adv. Mater.* 2005, *17*, 2587.

[20] (a) Lu, L. H.; Randjelovic, I.; Capek, R.; Gaponik, N.; Yang, J. H.; Zhang, H. J.; Eychmüller, A. *Chem. Mater.* 2005, *17*, 5731. (b) Tessier, P. M.; Velev, O. D.; Kalambur, A. T.; Rabolt, J. F.; Lenhoff, A. M.; Kaler, E. W. *J. Am. Chem. Soc.* 2000, *122*, 9554.

[21] Kim, B.; Tripp, S. L.; Wei, A. *J. Am. Chem. Soc.* 2001, *123*, 7955.

[22] Baia, M.; Toderas, F.; Baia, L.; Popp, J.; Astilean, S. *Chem. Phys. Lett.* 2006, *422*, 127.

[23] Jana, N. R.; Gearheart, L.; Murphy, C. J. *Chem. Mater.* 2001, *13*, 2313.

[24] Schatz, G. C.; Zeman, E. J. *J. Phys. Chem.* 1987, *91*, 634.

[25] Cumberland, S. L.; Strouse, G. P. *Langmuir* 2002, *18*, 269.

[26] Wang, H.; Levin, C. S.; Halas, N. J. *J. Am. Chem. Soc.* 2005, *127*, 14992.

[27] (a) S. Link, M. A. El-sayed, *J. Phys. Chem. B* 1999, *103*, 8410. (b) Hu, J. Q.; Chen, Q.; Xie, Z. X.; Han, G. B.; Wang, R. H.; Ren, B.; Zhang, Y.; Yang, Z. L.; Tian, Z. Q. *Adv. Funct. Mater.* 2004, *14*, 183.

[28] Underwood, S.; Mulvaney, P. *Langmuir* 1994, *10*, 3427.

[29] (a) Schmitt, J.; Mächtle, P.; Eck, D.; Möhwald, H.; Helm, C. A. *Langmuir* 1999, *15*, 3256. (b) Link, S.; El-Sayed, M. A. *Int. Rev. Phys. Chem.* 2000, *19*, 409. (c) Mulvaney, P. *Langmuir* 1996, *12*, 788. (d) Hövel, H.; Fritz, S.; Hilger, A.; Kreibig, U. *Phys. Rev. B* 1993, *48*, 18178.

[30] (a) Kreibig, U.; Gartz, M., Hilger, A. *Ber. Bunsen-Ges. Phys. Chem.* 1997, *101*, 1593. (b) Quinten, M.; Kreibig, U. *Surf. Sci.* 1986, *172*, 557.

[31] Brust, M.; Bethell, D.; Kiely, C. J.; Schiffrin, D. J. *Langmuir* 1998, *14*, 5425.

[32] Doron-Mor, I.; Cohen, H.; Barkay, Z.; Shanzer, A.; Vaskevich, A.; Rubinstein, I. *Chem. Eur. J.* 2005, *19*, 5555.

[33] (a) Kneipp, K.; Kneipp, H.; Itzkan, I.; Dasari, R. R.; Feld, M. S. *Chem. Rev.* 1999, *99*, 2957. (b) Kneipp, K.; Wang, Y.; Kneipp, H.; Perelman, L. T.; Itzkan, I.; Dasari, R. R.; Feld, M. S. *Phys. Rev. Lett.* 1997, *78*, 1667. (c) Nie, S.; Emory, S. R. *Science* 1997, *275*, 1102.

[34] (a) Moskovits, M. *Rev. Mod. Phys.* 1985, *57*, 78. (b) Kneipp, K.; Kneipp, H.; Itzkan, I.; Dasari, R. R.; Feld, M. S. *Chem. Rev.* 1999, *99*, 2957.

[35] (a) Xu, H. X.; Aizpurua, J.; Käll, M.; Apell, P. *Phys. Rev. E* 2000, *62*, 4318. (b) Stockman, M. I.; Shalaev, V. M.; Moskovits, M.; Botet, R.; George, T. F. *Phys. Rev. B* 1992, *46*, 2821.

[36] Tian, Z. Q.; Ren, B.; Wu, D. Y. *J. Phys. Chem. B* 2002, *106*, 9463.

[37] (a) Yamada, H.; Yamamoto, Y. *Surf. Sci.* 1983, *134*, 71. (b) Ren, B.; Huang, Q. J.; Cai, W. B.; Mao, B. W.; Liu, F. M.; Tian, Z. Q. *J. Electroanal. Chem.* 1996, *415*, 175. (c) Wu, D. Y.; Xie, Y.; Ren, B.; Yan, J. W.; Mao, B. W.; Tian, Z. Q. *Phys. Chem. Commun.* 2001, *18*, 1.

[38] Cao, L. Y.; Diao, P.; Tong, L. M.; Zhu, T.; Liu, Z. F. *Chem. Phys. Chem.* 2005, *6*, 913.

[39] Kennedy, B. J.; Spaeth, S.; Dickey, M.; Carron, K. T. *J. Phys. Chem. B* 1999, *103*, 3640.

[40] (a) Hu, X.; Cheng, W.; Wang, T.; Wang, Y.; Wang, E.; Dong, S. *J. Phys. Chem. B* 2005, *109*, 19385. (b) Zheng, J.; Zhou, Y.; Li, X.; Ji, Y.; Lu, T.; Gu, R. *Langmuir* 2003, *19*, 632.

[41] (a) Haruta, M. *Catal. Today* 1997, *36*, 153. (b) Rashid, M. H.; Bhattacharjee, R. R.; Kotal, A.; Mandal, T. K. *Langmuir* 2006, *22*, 7141.

In: Encyclopedia of Nanotechnology
Editor: Elwood D. Carlson
ISBN: 978-1-60692-079-4

Chapter 19

NANOCRYSTALLINE ZINC OXIDE THIN FILMS PREPARED BY CHEMICAL SOLUTION DEPOSITION FROM A ZINC NAPHTHENATE PRECURSOR

***Seung Hwangbo*[a] *and Kyu-Seog Hwang*[b,1]**

[a]Major in Photonic Engineering, Division of Electronic & Photonic Engineering, Honam University, 59-1 Seobong-dong, Gwangsan-gu, Gwangju 506-714, Republic of Korea

[b]Department of Applied Optics and Institute of Photoelectronic Technology, Nambu University, 864-1 Wolgye-dong, Gwangsan-gu, Gwangju 506-824, Republic of Korea

ABSTRACT

Nanocrystalline ZnO thin films on silica glass substrates were prepared by spin coating-pyrolysis process from a zinc naphthenate precursor. As-deposited films were heat treated at 500°C ~ 1000°C. The ZnO thin films were analyzed by X-ray diffraction, field emission-scanning electron microscope, scanning probe microscope, and ultra violet-visible-near infrared spectrophotometer. Only (002) peak is observed for all samples on silica glass substrates. With the increase of heat-treatment temperature, the peak intensity of (002) plane increases. No evidently aggregated particles are present. From SPM analyses, three-dimensional grain growth, which is thought to be due to inhomogeneous substrate surface and *c*-axis oriented grain growth of the ZnO phase, is independent on heat treatment temperature. The most homogeneous surface of the highly oriented ZnO film was observed at the heat-treatment temperature of 800°C. All the films exhibit a high transmittance (>80%) in the visible region except film heat-treated at 1000°C and show a sharp fundamental absorption edge at about 0.38 ~ 0.40 μm. The estimated energy band gap for all the films are within the range reported for ZnO films and single crystal. ZnO films, consisting of densely packed grains with smooth surface morphology, were obtained by heat treatment at 600 ~ 800°C, expected to be ideal for practical usage.

[1] E-mail address: khwang@nambu.ac.kr, Phone: +82-62-970-0110, Fax: +82-62-972-0288

INTRODUCTION

High transmittance and good electrical-conductivity thin conducting oxides (TCOs) have potential use in a variety of technological applications, including plasma display panels (PDPs), electroluminescent devices, flat cathode ray tube (CRTs) and solar cells [1-4].

In recent years tin-doped indium oxide [5] (ITO) and tin oxide [6] (SnO_2) have been the most used materials for such applications. Recently, zinc oxide (ZnO) thin films have emerged as most promising oxide materials owing to their optical and electrical properties, together with their high chemical and mechanical stability. ZnO is a wide band gap semiconducting, piezoelectric and photoconducting material, which is transparent in the visible wavelength range and therefore used in a wide variety of applications, ranging from a transparent electrode material to gas sensors and surface acoustic wave (SAW) guides [7, 8]. Furthermore, ZnO has attracted considerable attention because it is a potential suitable candidate for ultra violet (UV) and blue optical applications [9]. ZnO not only has a wide direct band gap (3.37 eV) at room temperature but also shows a large exciton binding energy (60 meV) that is much larger than that of ZnSe (20 meV) and GaN (25 meV) [9]. ZnO generally emits three luminescence bands in the UV, green, and red regions. The UV emission is considered due to exciton-related activities. However, for the visible emissions, various mechanisms have been proposed including the involvement of zinc interstitials, oxygen interstitials, zinc vacancies, and oxygen vacancies [9 – 12].

Several important properties based on zinc oxide materials, such as electroluminescence and photoluminescence (PL), are strongly related to the defect formation. To provide an intuition for improving the PL properties in ZnO thin films, Lin et al [9] reported that the role of the predominant point defects in the as-grown ZnO films on Si substrates.

A variety of methods have been reported for the preparation of ZnO thin films. Physical processes such as metalorganic chemical vapor deposition (MOCVD), sputtering and pulsed laser deposition (PLD) have been adopted by various workers. The high-quality *c*-axis oriented growth of ZnO coatings on 6H-SiC(0001) substrates had been studied by Ashrafi et al [13]. The effect of thermal annealing was investigated on ZnO thin films grown on Si(001) by low-pressure MOCVD. From Zhang's report [14], annealing caused coalescence among different grains, which resulted in the formation of a dense surface and XPS confirmed that ZnO thin films changed from Zn-rich to O-rich due to more zinc atoms being bound to oxygen atoms after annealing. Zhao et al. [15] reported that preparation and optimization of *c*-axis oriented ZnO thin films on single-crystal diamond substrate by MOCVD.

Kim et al. [16] reported that magnetron sputtering growth and characterization of high quality single crystal ZnO thin films on (0001) sapphire substrates. The films showed structural and optical properties comparable with previously reported ZnO thin films grown by MOCVD, when grown at high temperature and O^2-rich atmosphere.

Furthermore, Hirata et al. [17] reported on gallium-doped ZnO transparent-conducting thin films grown on glass at different substrate temperatures by PLD. A widening in the optical band gap and a good gallium-doping efficiency were observed in the films when the substrate temperature was raised from 150°C to 300°C, as determined from optical and electrical measurements.

A chemical solution deposition (CSD) is another attractive technique for obtaining thin films and has the advantages of easy control of the film composition and easy fabrication of a large-area thin film with low cost [18 - 28]. However, the CSD method is relatively new and requires a greater understanding of the relationship between processing and defects in order to optimize film quality. In CSD, although the film showed high orientation or epitaxy to substrate with low fluctuation of alignments, the morphology of the film confirmed by microscope was similar to typical polycrystalline films. Conventional CSD processes utilize a polymeric precursor solution usually prepared from alkoxides, nitrates, or acetates [29, 30] in an organic solvent. A precursor solution is deposited onto a substrate via spin or dip coating. The as-deposited film will sequentially be prefired and heat treated to the desired oxide during heat treatment above 500°C. However, conventional CSD methods have revealed serious problems such as the occurrence of film cracking and pores brought about by stresses arising and volatilization of organics during the thermal conversion of organic-based gel to a ceramic film and instability by starting materials in air.

Thus, recently, interest in finding solutions to overcome the drawbacks has been increasing. In particular ZnO films with preferential orientation along the *c*-axis have been shown to work as SAW devices because of large piezoelectric constant. Ohyama et al. [31] prepared ZnO films on silica glass substrates from zinc acetate sol containing 2-methoxyethanol and monoethanolamine (MEA) with extremely preferred orientation along (002) plane. They described the effects of the heat-treatment conditions, the withdrawal speed of the substrate and the film thickness on the crystal orientation of the ZnO films as well as the relative density and morphology, in order to obtain more detailed insight on the conditions for the formation of highly oriented ZnO films.

Ohya et al. [32] reported transparent, electrically conducting ZnO thin films with polycrystalline structure from zinc acetate and alkoxide. Recently Wessler et al. [33] reported textured ZnO thin films on (0001) sapphire (SAP) by spin-coating a solution of acetate and ethanolamine. As far as we know, however, there is little information on preparing highly oriented ZnO thin films on amorphous substrates by CSD using a metal naphthenate as precursor.

In the present study, the preparation of ZnO thin films with a highly preferred orientation on amorphous silica substrates from a zinc naphthenate precursor is reported. The heat-treatment among many parameters on CSD process had a strong effect on the structural and optical behavior of ZnO thin film. The thin films were characterized by X-ray diffraction (XRD) θ-2θ scans, field emission-scanning electron microscope (FE-SEM), scanning probe microscope (SPM), and ultra violet (UV) – visible - near infrared (NIR) spectrophotometer. The results enabled us to develop guidelines for preparing highly oriented ZnO thin films on silica substrate from a metal naphthenate precursor.

Experimental Procedure

A coating solution was prepared using zinc naphthenate (Nihon Kagaku Sangyo Co., Ltd., Japan) and by diluting the solution with toluene (concentration: 4 wt% metal / 100 mL sol) to achieve an appropriate viscosity for the deposition of smooth and crack-free thin films.

Metal naphthenate which consist of cyclopentanes of cyclohezanes, methylene chains, carboxylates and metals mainly have the following structure (see Fig. 1):

$[(\text{cyclopentane})\text{-}(CH_2)_n\text{-}COO]^-_m\text{-}M^{m+}$

<R-O-M structure> M: metal

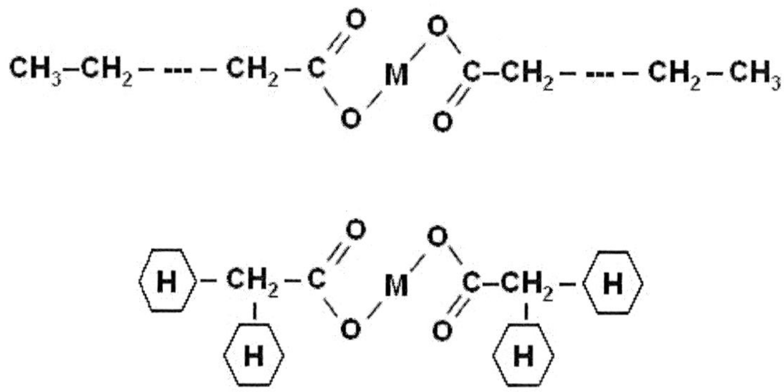

Figure 1. Structure of naphthenic acids.

Metal naphthenates are used in solid form or in organic solution with mineral spirit, toluene, xylene, naphthenic oil or other solvents. The properties mainly depend on the quality of naphthenic acids. Heating decomposition take place above 150 ~ 180°C. In presence of mineral acids are easily decomposed to naphthenic acids, solvent and inorganic salts. Alkaline solutions decomposed metal naphthenates to metal hydroxide and alkaline solution of the naphthenic acids. Metal naphthenates are more advantageous than metal alkoxides as starting materials, in terms of cost, stability in air and ease of handling. Preparation of the coating solution using zinc naphthenate was easily performed by the addition of toluene, while complicated procedures were needed for the preparation of sol from zinc alkoxide [32]. Furthermore, it should be noted that vaporization of the additives, such as alcohol, water and catalyst etc, during prefiring and heat treatment might cause unwanted cracks and pores in the coating [34]. On the other hand, in our paper, prefiring is the process for pyrolytic conversion of zinc naphthenate into ZnO, while the final heat treatment is the process for solid-state reaction. In our previous reports [18, 19] using metal naphthenates, it was observed the heat-treated films were smooth and it was difficult to identify cracks and pores, while defects were easily observed at the surface of metal alkoxide-derived films. In opto-electronic applications, the partially deteriorated film may present some disadvantages.

Prior to coating process, silica glass substrates were cleaned in deionized water, immersed in H_2O_2 solution and finally rinsed in acetone. The starting solution was spin-coated onto the cleaned substrate at 1500 rpm for 10 sec in air. The as-deposited film was prefired at 500°C for 10 min in air to vaporize organic components in coating solution. The coating process was repeated five times to prepare a thick coating of ZnO. Then the final annealing was performed in air at 500°C, 600°C, 700°C, 800°C, 900°C, and 1000°C for 30 min by directly inserting the samples into preheated tube-type furnace, which was followed by fast cooling.

The thickness of the heat-treated ZnO thin film was approximately 0.65 ~ 0.7 μm, as determined by observations of fracture-cross-sections using a FE-SEM (S-4700, Hitachi Co., Japan) and F20 (Filmetric Inc., San Diego, CA, U. S. A.) using reflection spectrum. The crystal structure, crystallinity, and orientation of the thin films were investigated by XRD (D-Max-1200, Rigaku Co., Japan) in θ-2θ geometry. A CuK_{α} (λ = 1.54056 Å) source was used, and the scanning range was between 2θ = 20° and 70°. The surface morphology of the thin films was evaluated using the FE-SEM micrographs. The microstructure of the samples was observed by FE-SEM cross-sectional view. The growth mechanism and surface roughness of the thin films were studied by a SPM (XE-200, PSIA Co., South Korea).

When one is evaluating surface roughness, the power spectral density (PSD) function is used increasingly as a valuable tool for topography description of extremely flat surfaces, such as epitaxial silicon or optical thin films. Average roughness (R_a) is one of the most commonly used roughness statistics. The illustration as shown in Fig. 2 represents two

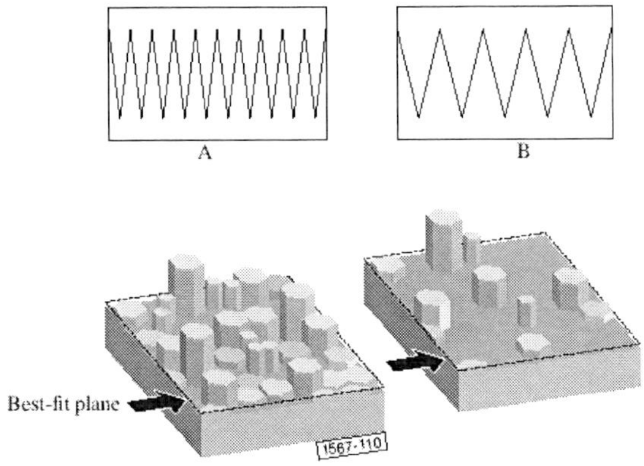

Figure 2. Two surfaces having the same "average roughness"[35].

surfaces having the same "average roughness". Similarly, a number of other roughness values are based upon least-squares calculations [e.g., root mean square (RMS) roughness, or R_q], and their algorithms are more concerned with a best fit of all height points than with the spatial frequency of features. Image "A" surface is represented as having a high frequency profile of features. Image "B" represents a separate surface having the same average feature height, but distributed at wider (lower-frequency) intervals. In terms of average and RMS roughness, both surfaces are "equally rough". What would better distinguish them apart is a method for characterizing each surface's feature frequency or wavelength. For more complex surfaces, moreover, it would be helpful to know which wavelengths occur most often and which impart the greatest influence or "power" to the surface's topography. This is presented in a PSD analysis [35].

Any randomly rough surface can be considered as a Fourier series of sinusoidal waves with different amplitudes, periods, and phases. The grating equation shows that a single grating with spacing d causes scatter into the angle according to sin $q = l/d$, where l is the wavelength of light. Hence, d can be considered as one spatial wavelength present on the surface, or accordingly, $f = l/d$ as one spatial frequency. For a randomly rough surface such as our thin film component, many different spatial frequencies are present. This is quantitatively expressed by the PSD curve, giving the relative strength of each roughness component of a surface microstructure as a function of the spatial frequency [36], whereas the root mean square (RMS) roughness as a single parameter of surface description reflects the standard deviation of all height values within the considered surface and does not provide any information on the topographical details of the surface.

Consider the following image (see Fig. 3),

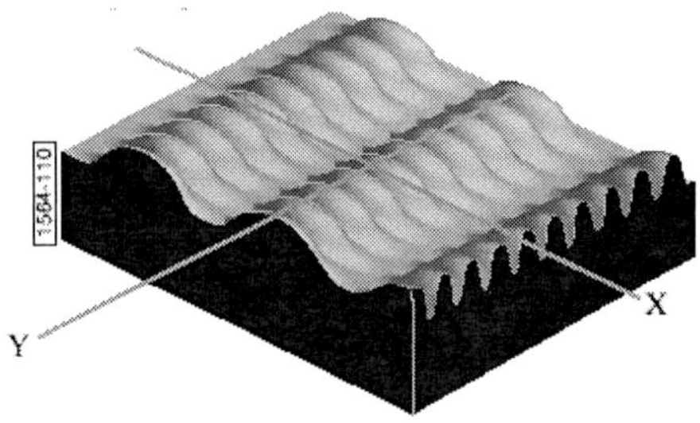

Figure 3. Synthetic surface consists of two dominant wave forms [35].

This synthetic surface consists of two dominant wave forms: a long period waveform along the X-axis, and a shorter period waveform along the Y-axis. A 2-dimentional PSD plot will consist of two dominant spikes (one or each dominant wavelength), plus some number of extra wavelengths inherent within the image. Because of the sinusoidal nature of the composited wave form, a relatively small composite of spectral frequencies suffice to describe the entire surface. By contrast, an image comprised of angular (saw-toothed of square) wave forms will generally yield a much broader spectral plot. This is largely because of the mathematical nature of PSDs, relying as they do upon Fourier transforms. Because PSDs transform the surface's real spatial elements (X-Y-Z dimensions) into an array of spatial frequency constructs (sine waves of differing spatial frequencies), a trend in surface features emerges: surfaces consisting of regular, rounded features yield PSD plots comprised of relatively fewer composite frequencies; surfaces with sharp, angular features yield PSD plots encompassing relatively more composite frequencies [35].

As shown in Fig. 4, this surface is comprised of "terraces" formed from the material's natural lattice structure; each terrace is one atomic monolayer thick and is spaced at fairly regular intervals. This degree of flatness is handily evaluated with PSD, as the terraces will produce a spectral spike corresponding to their spacing wavelength. A PSD plot for this type of surface generally takes the form shown in Fig. 5. This tapered PSD plot is characteristic of flat, isotropic surfaces. Longer wavelengths are present up to the scan width, and are accompanied by uniformly decreasing powers of shorter wavelengths down to 2 pixels. On the plot shown in Fig. 5 a spike stands out, corresponding to the wavelength spacing of the terraced features. Depending upon the qualitatively standards of the person evaluating such a plot, this spike may exceed a threshold standard of flatness [35].

PSD curves used throughout this work were calculated from selected atomic force microscope (AFM, NanoScope Multimode™ SPM, Digital Instruments, U. S. A.) scans with NanoScope off-line analysis software. In order to cover an extended range of spatial frequencies, we combined the PSD functions from $1 \times 1\ \mu m^2$ and $10 \times 10\ \mu m^2$ scans. All the SPM and AFM measurements were performed in air using the tapping mode with silicon probes having ~ 250 kHz resonant frequency.

Transmittance in the visible wavelength range was measured by a UV – visible - NIR spectrophotometer (CARY 500 Scan, VARIAN Co., Australia). The transmittance was automatically calibrated against a bare glass substrate as a reference sample, and the absorption coefficient was obtained from the transmittance curve.

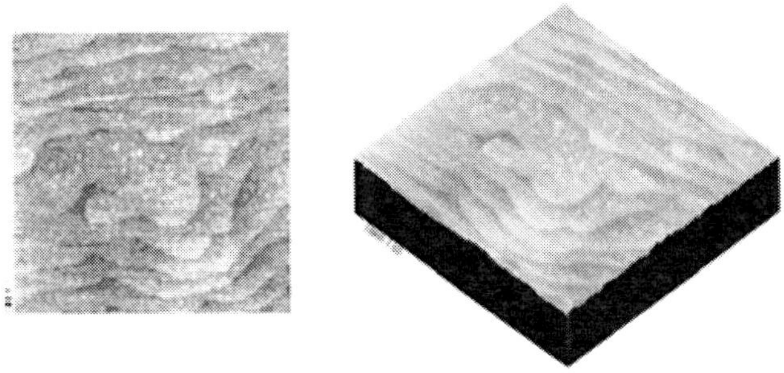

Figure 4. AFM Image of epitaxial gallium arsenide [35].

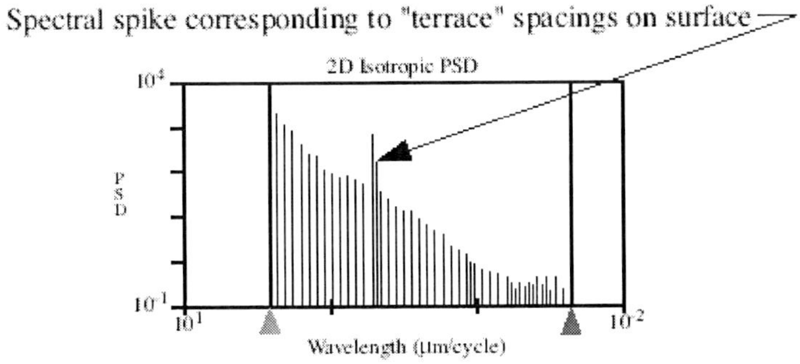

Figure 5. 2D isotropic PSD of epitaxial gallium arsenide [35].

RESULTS AND DISCUSSION

Figure 6 shows the XRD curves of ZnO thin films deposited on the silica glass substrates. Only the (002) ZnO peak at $2\theta \sim 34°$ is observed for all samples. This indicates the ZnO thin films prepared by CSD from a zinc naphthenate show a good *c*-axis orientation, i.e., the vertical growth to the substrate surface. The *c*-axis orientation in ZnO films by physical dry method can be understood by the 'grain boundary movement' model proposed by Loder et al. [37]. According to that model, at the very first stage of film growth, certain grains with particular orientation start to grow. By thermodynamical coalescence of crystallites during film growth, orientational growth (*c*-axis orientation for ZnO film) is achieved.

From Ohyama's work [34], for ZnO films prepared by CSD using zinc acetate – 2-methoxyethanol - MEA solution, removal of the solvent and organic substances produced by acetate decomposition prior to crystallization may be one of the key factors that provide oriented crystal growth. The vaporization of the solvents, the decomposition of the zinc acetate and the crystallization of the zinc oxide may occur almost simultaneously when the heating rate is high. Since the structural relaxation of the gel film, which is induced by the solvent vaporization and acetate decomposition, can take place only before the crystallization, the simultaneous vaporization, decomposition and crystallization may give the film less chance to be structurally relaxed [34].

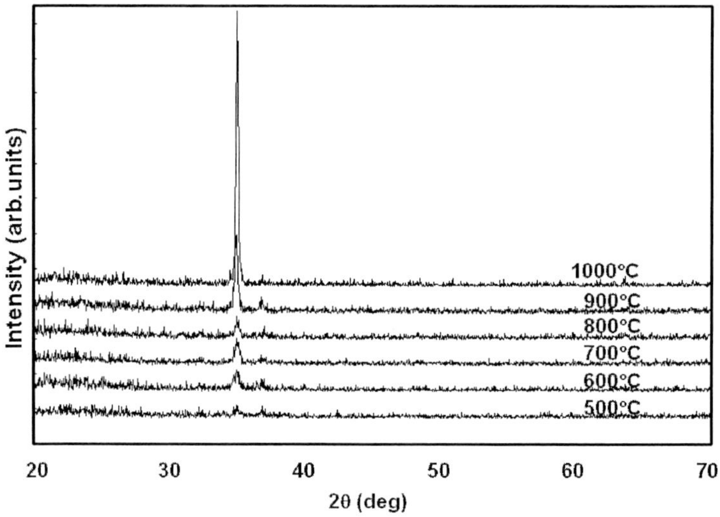

Figure 6. XRD spectra of ZnO films on silica glass substrates as a function of heat treatment temperature.

When the heating rate is low, the gel film is given an enough time to be structurally relaxed before crystallization, resulting in denser ceramic films. Moreover, when the prefiring temperatures too high (>300°C), the vaporization of the solvents, and the thermal decomposition of the zinc acetate may take place abruptly and simultaneously with the crystallization, disturbing the unidirectional crystal growth. On the contrary, when the preheating temperature is too low (<300°C), the complete vaporization and the thermal decomposition of the zinc acetate do not occur at the preheating step but do occur at the post heating step at post heating temperatures over 500°C. In such a case, abrupt solvent vaporization and acetate decomposition occur at the post-heating step, which may also disturb the unidirectional crystal growth. Thus, the preheating temperature seems to be the most important factor for obtaining ZnO films with preferred orientation, affecting the solvent vaporization, zinc acetate decomposition and zinc oxide crystal growth. Removal of the solvent and organic substances produced by acetate decomposition prior to crystallization may be one of the key factors that provide oriented crystal growth [34].

However, in this work, it should be noted here that highly *c*-axis oriented ZnO thin films were obtained by prefiring at high temperature, 500°C, and final heat-treatment with a rapid thermal annealing by directly inserting the samples into preheated furnace. This result suggests that a necessary condition using in Ohyama's work to obtain strongly preferred oriented ZnO thin film such as, usage of solvents having relatively high boiling points of about 200°C, heat-treatment of the film at lower rates and pyrolysis at lower temperatures of 200°C ~ 300°C, is unnecessitated in this work. The boiling point of toluene in our work is much lower, 110.6°C, than those of 2-methoxyethanol (125°C) and MEA (170°C) used in Ohyama's work [34]. Conclusively, the use of toluene, solvent of lower boiling point, resulted in strongly preferred orientation of ZnO crystals as in the present work.

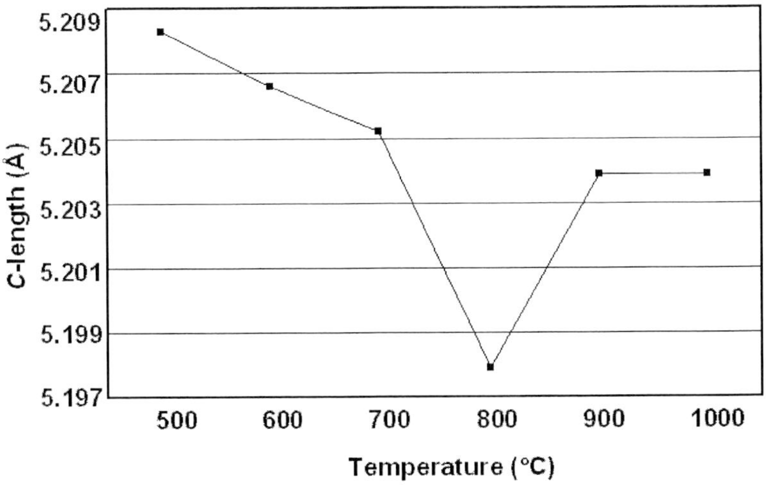

Figure 7. Variation of *c*-length of ZnO films as a function of heat treatment temperature.

It is also observed that with an increase in heat-treatment temperature, the peak intensity of (002) plane increases, as shown in Fig. 6. The lattice parameter, full width at half maximum (FWHM), and peak intensity of (002) reflection as shown in Figs. 7, 8, and 9. The diffraction angle of 2θ ranges from 34.41° to 34.49°. The value of the FWHM ranges from 0.194° to 0.830°. The diffraction 2θ peak position increase from 34.41° at 500°C up to 34.49° at 800°C and then decreases to 34.44° at 900°C and 1000°C. The FWHM decreases from 0.83° at 600°C to 0.194° at 1000°C. The FWHM at >700°C was comparable to those of highly-textured ZnO thin films prepared by the physical coating process [1]. The results indicate that the ZnO thin film using zinc naphthenate heat treated at >700°C is expected to have the best crystallinity.

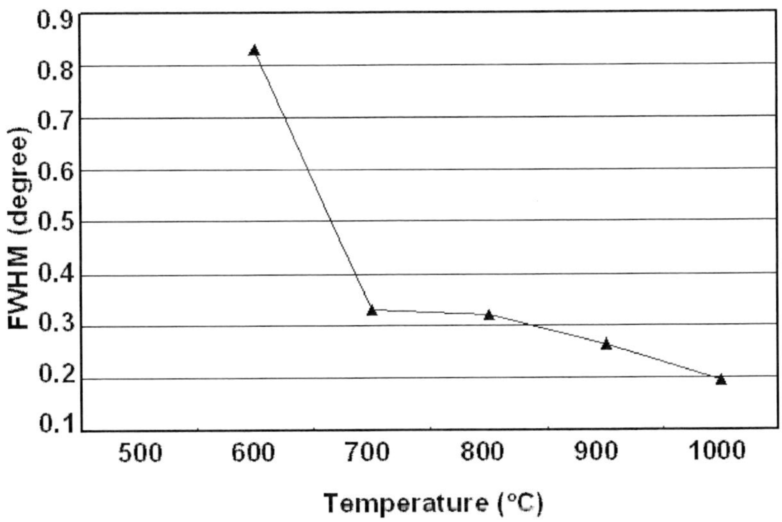

Figure 8. Variation of FWHM of ZnO films as a function of heat treatment temperature.

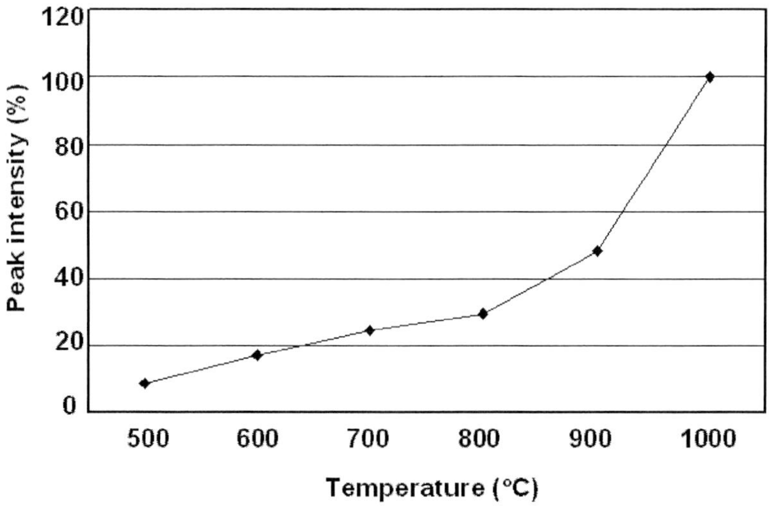

Figure 9. Peak intensity of (002) plane of ZnO films as a function of heat treatment temperature.

On the basis of XRD data, the lattice c parameter has been estimated to be in the range of 5.1979 to 5.2083Å, as shown in Fig. 7, which is similar to the JCPDS value of 5.2066 Å for bulk ZnO (JCPDS File 36-1451). The larger value of lattice constant for the film heat treated at 500°C compared to the standard powder value shows that the unit cell is elongated along the c-axis, and compressive forces act in the plane of the ZnO film. These compressive forces disappear as the heat treatment temperatures increases.

In order to more exactly investigate the structural properties of ZnO film, we calculated the stress in the film. The calculation of the film stress is based on the biaxial strain model [38].

The strain $\varepsilon = [(c_{bulk} - c_{film}) / c_{bulk}]$ in the direction of the c-axis, *i.e.*, perpendicular to the substrate surface, was measured by XRD. To derive the film stress of film parallel to the film surface, the following formula has been used [38], which is valid for a hexagonal lattice;

$$\sigma_{film} = [2c_{13}^{2}/2c_{13} - c_{33}(c_{11}+c_{12})/2c_{13}][(c_{film}-c_{bulk})/c_{bulk}] \qquad (1)$$

The elastic constant c_{ij} of single crystalline ZnO have been used: $c_{11} = 208.8$, $c_{33} = 213.8$, $c_{12} = 119.7$, and $c_{13} = 104.2$ GPa [39]. Stress in the film can be estimated using Eq. (1) and are plotted as a function of heat treatment temperature in Fig. 10. The negative sign for the film heat treated at 500°C indicates that the lattice constant c is elongated as compared to unstressed powder and therefore film is state of elongation. The film becomes completely stress free at about 600°C annealing temperature. After further heat treatment (>600°C) a film goes into state of compression as compared to the normal lattice state of bulk ZnO. Previous works [40] observed a state of compression in their films prepared by *rf* sputtering and annealed at 400°C. The stress observed in sputtered ZnO films by previous workers has been associated with interstitial oxygen. Their results can be qualitatively explained by assuming that, after annealing, the interstitial oxygen is probably removed and the defects in the lattice are annealed out. We assume that high stress relief temperatures (SRT), >600°C in this work, could be related with the volatilization of organics during heat treatment. Thus, in this work, the pyrolysis of organic components during annealing caused more defects in ZnO lattice, thus increasing the SRT of the film.

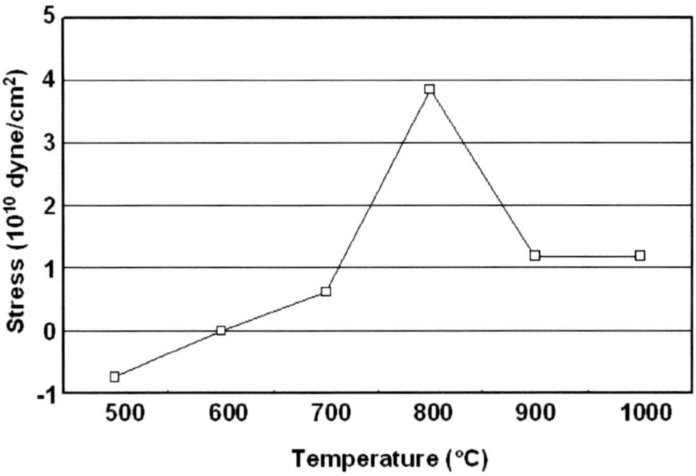

Figure 10. Variation of induced stress on the ZnO films as a function of heat treatment temperature.

Figure 11 shows the FE-SEM photographs of the ZnO thin films after heat treatment at 500°C (a), 600°C (b), 700°C (c), 800°C (d), 900°C (e), and 1000°C (f). Particulate structure is evident in all films except films heat treated at < 600°C. The particle size increases with increasing heat treatment temperature, and grain out-growth of the particles is observed in the films heat treated at 800°C and 900°C. It is noted that no evidently aggregated particles are present and that nano-sized particles were obtained in all the films except the film heat treated at 1000°C. Generally, in the CSD method, pores and cracks are easily recognized in the product films. On the contrary, the surface of the thin film produced in this work was relatively flat and smooth below 800°C.

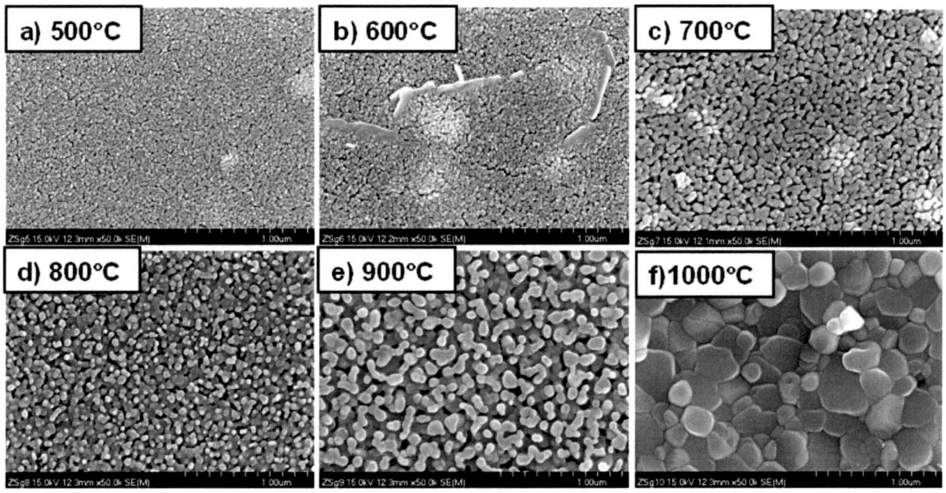

Figure 11. FE-SEM images of the ZnO films after heat treatment at various temperatures.

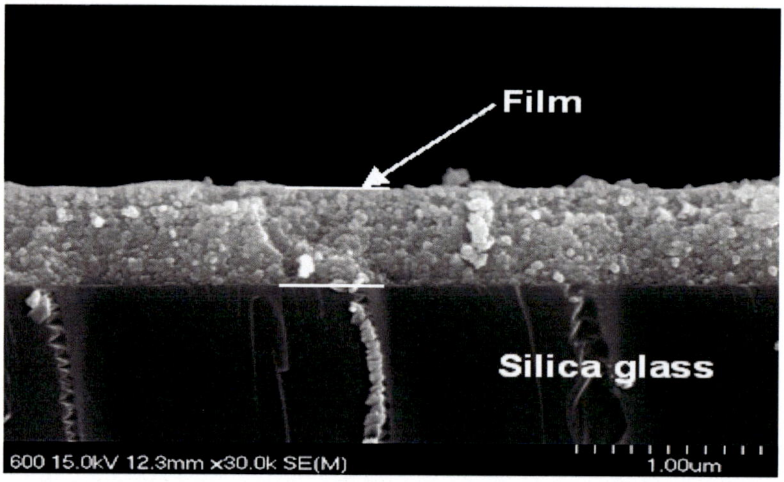

Figure 12. Cross-sectional image of the ZnO film heat treated at 600°C.

Figure 12 shows the cross-sectional morphology of the ZnO films hate treated at 600°C. The fractured cross-section of the ZnO thin film with an approximately 0.67 ~ 0.7 μm thickness appears dense and uniform. This suggests that the bonding strength between the substrate and the film is sufficiently strong.

To evaluate the surface roughness and morphology of the films according to the heat-treatment temperatures, SPM analysis was performed. Figure 13 shows the SPM images (5 × 5 μm^2) of ZnO films heat treated at various temperatures.

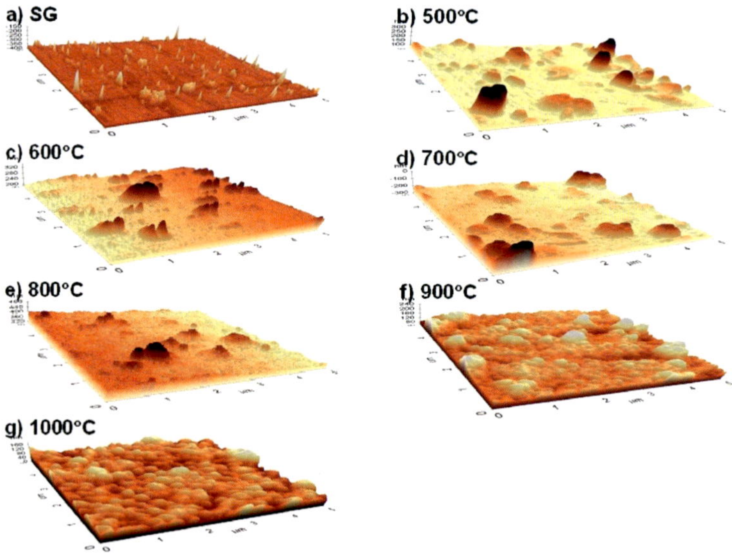

Figure 13. SPM photographs (5 × 5 μm^2) of the ZnO films after heat treatment at various temperatures.

The surface of polished silica glass substrate has a relatively high-surface roughness with some needle-like grain outgrowth. This surface is not smooth enough for automatically textured ZnO film. Due to a needle-like grain outgrowth on substrate surface, the film has three-dimensional island-like features, as shown in Fig. 13. It is observed that with increasing heat-treatment temperature to 800°C, the grain gradually converted to round-shaped peak. Above 900°C, it is easy to identify large grain growth, such as island growth on the two-dimensionally grown layer, although the grain size uniformly increased with increasing heat treatment temperatures to 900°C and 1000°C. In all the samples, three-dimensional grains were observed, although the previous XRD crystallographic results indicate a high orientation. The above results imply that the growth of three-dimensional grains identified in all the samples, which are thought to be due to the silica glass substrate naturally containing some needle-like grain outgrowth, depends on the heat treatment temperature.

Figures 14 and 15 show SPM top-view images and surface roughness of ZnO thin films as a function of heat-treatment temperatures. Generally, in previous works using a SPM, most surface roughness were obtained from line profiles. In this work, however, in order to obtain more exact roughness data, we performed line profiles and area scanning (5 × 5 μm^2) for the surfaces of the ZnO thin films. ZnO thin films on silica glass substrate having relatively flat surface structure annealed to 500°C showed high RMS roughness. We assume that 3-dimensional needle-like grain growth in bare glass substrate may increase surface roughness of ZnO after heating. With an increase in the annealing temperature to 800°C, the RMS roughness decreased to 12.524 or 8.558 nm. At 900°C and 1000°C, the surface roughnesses of the ZnO films were increased due to larger grain growth, resulting in higher RMS roughnesses as shown in Figs. 14 and 15. The most homogeneous film (RMS= 12.524 or 8.558 nm) having a preferred orientation was observed at a heat-treatment temperature of 800°C.

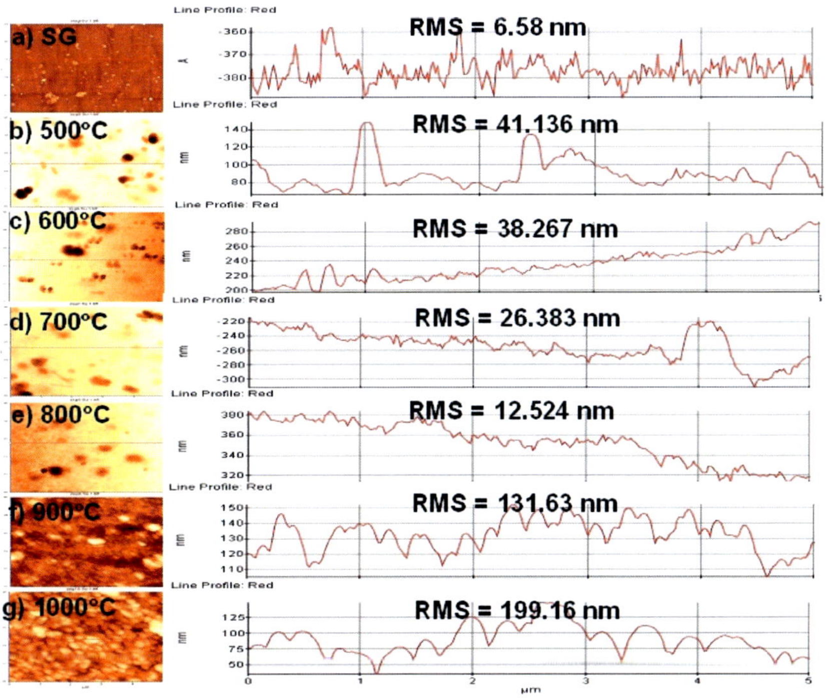

Figure 14. Line profiles (5 × 5 μm^2) for the surfaces of the ZnO films after heat treatment at various temperatures.

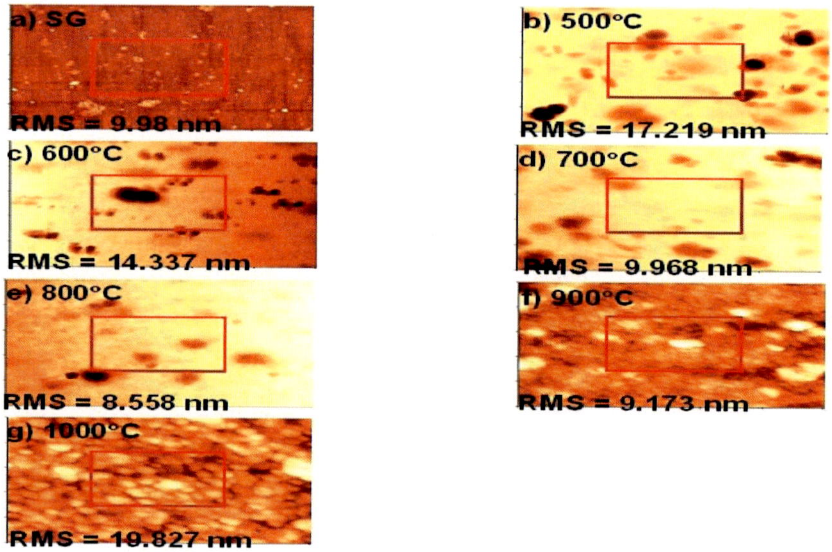

Figure 15. Area scanning (5×5 μm^2) for the surfaces of the ZnO films after heat treatment at various temperatures.

Furthermore, we studied the surface roughness characteristics of ZnO thin films as a function of heat treatment temperature by using PSD curves obtained by AFM measurements. In most cases where optical surfaces are to be evaluated, the parameter commonly used for describing the roughness characteristics of a surface of thin film is the RMS roughness. However, when specific roughness properties have to be investigated in detail, spectral roughness analysis should be performed. Because of PSDs transform the surface's real spatial elements (X-Y-Z dimensions) into an array of spatial frequency constructs (sine waves of differing spatial frequencies), a trend in surface features emerges. Determining the PSD provides valuable information not only on the height deviation of the roughness profile, but also on its lateral distribution, and hence it gave a more general description than the RMS roughness alone [18]. PSD is used increasingly as a metrology tool for evaluating extremely flat surfaces, while in the surfaces represented by RMS roughness, peaks and valleys are equidistant from the mean surface centerline.

Figure 16 shows the results of PSD functions for the ZnO thin films according to the heat treatment temperature. Similar to that of SG substrate, the PSD function of the ZnO thin films heat treated at 500°C and 600°C displays no inflated shape generally due to increase in grain size. In addition, PSD curve with a gentle grade indicating that the layer structure has grown uniformly both in height and width. At 800°C, the PSD curve of the ZnO film display inflated shape at spatial frequency around 10 ~ 100 μm^{-1}, which may be due to uniformly formed small grains. For the ZnO film heat treated at 1000°C, however, lower frequency range at a spatial frequency around 0.1 ~ 1 μm^{-1} was increased. At 1000°C, PSD curve of ZnO thin films revealed rough surface morphology, which could be attributed to the grain out growth as previously shown in Figs. 14 and 15.

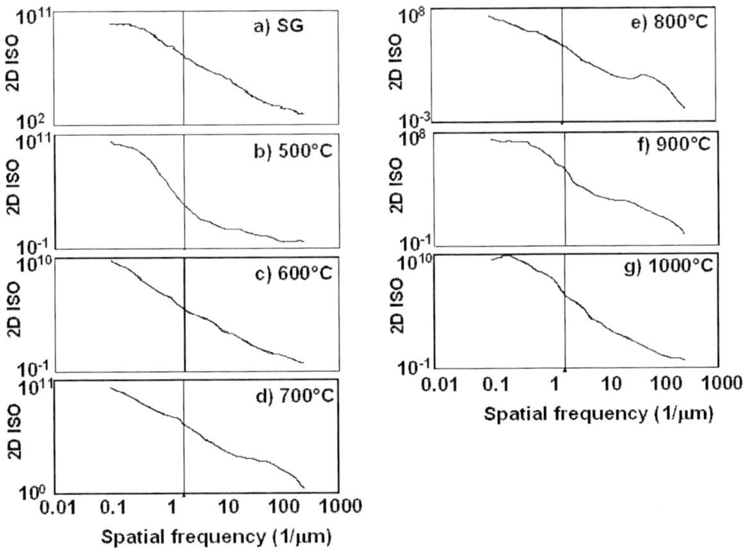

Figure 16. PSD curves of the ZnO thin films as a function of heat treatment temperature.

It is important that the surface of the deposited films should be smooth and flat to exhibit ideal properties for device applications. However, few textured films having excellent surface structure were obtained by the CSD process. Although the films showed epitaxy with the substrate or high orientation with low fluctuation of alignments, the morphology of the film generally exhibited porosity, roughness, and three-dimensional defects. This suggests that further investigation is required for the CSD method to obtain a great understanding of the relation between processing and defects in order to prepare highly *c*-axis oriented ZnO thin film on silica glass substrate without 3-dimensional grain outgrowth.

Figure 17 shows transmission spectra of the ZnO thin films heat treated at different temperatures.

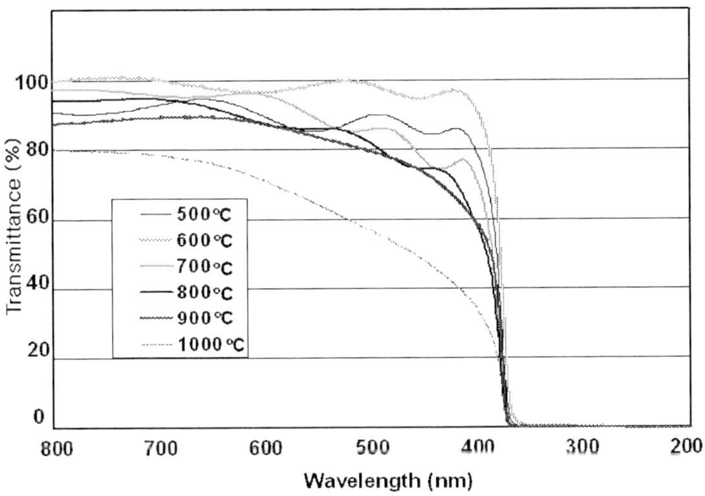

Figure 17. Transmittance of the ZnO films after heat treatment at various temperatures.

All the films exhibit a high transmittance (>80%) in visible region except for the film heat treated at 1000°C, and show a sharp fundamental absorption edge at about 0.38 ~ 0.4 μm, which is very close to the intrinsic band-gap of ZnO (3.2ev). The decrease in transmittance up to 900°C and 1000°C may be attributed to slight increase of surface roughness. For the films annealed below 800°C, the variation of transmittance with wavelength is principally due to the interference phenomenon, and the envelope of fringes starts to shrink after heat treatment the film above 900°C, as show in Fig. 17. The increase in surface roughness may be responsible for such shrinking in the fringes [41]. Briefly, the visible transmittance of ZnO thin film is quite comparable with those of film prepared by physical dry process [42].

The optical absorption coefficient, α, is defined as,

$$I = I_0 e^{-\alpha t} \quad (2)$$

where I is the intensity of transmitted light, I_0 is the intensity of the incident light, and t is the thickness of the ZnO film. As the transmittance is defined as I/I_0, we obtain α from Eq. (2). In the direct transition semiconductor, α and optical energy band gap (E_g) are related by [42, 43],

$$\alpha = (h\nu - E_g)^{1/2} \qquad (3)$$

where, h is Plank's constant, and ν is the frequency of the incident photon. Figure 18 shows the plot of α^2 *vs.* $h\nu$. The linear dependence of α^2 on $h\nu$ indicates that the ZnO films are direct-transition-type semiconductors. The photon energy at the point where α^2 is zero is E_g. Thus E_g can be determined by using an extrapolation method [42]. The optical band gap, E_g, is in the range between 3.288 eV and 3.315 eV. The estimated values of the band gap for all the films are within the range reported for such films [30, 31].

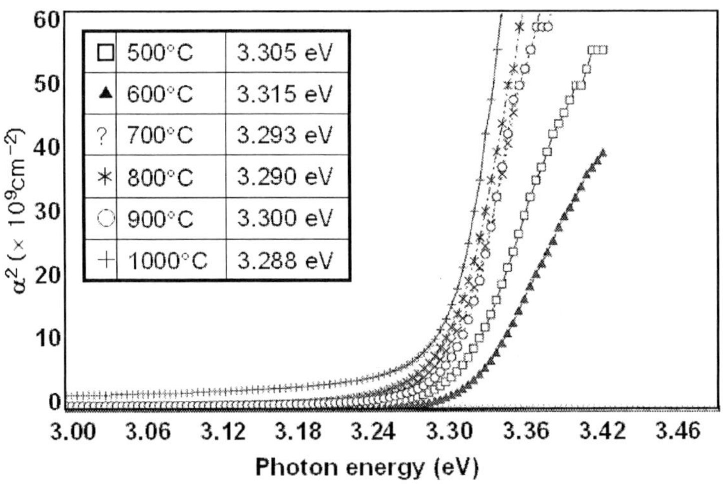

Figure 18. Square of the absorption coefficient as a function of photon energy for the ZnO films after heat treatment at various temperatures.

The refractive index of the ZnO film was calculated by Filmetrics F20, thin film measurement system. The F20 measures thin film characteristics by either reflecting or transmitting light through the sample, and then analyzing this light over a range of wavelengths. Because of its wave-like properties, light reflected from the top and bottom interfaces of a thin film can be in-phase so that reflections add, or out-of-phase so that reflections subtract. Whether the reflections are in- or out-of-phase (or somewhere in between) depends on the wavelength of the light, as well as the thickness and properties of the film [e.g., reflections are in-phase when $\lambda = (2^{nd})/I$, where λ is the wavelength, n is the reflective index, d is the film thickness, and I is an integer.]. The result is characteristic intensity oscillations in the reflectance spectrum. Figure 19 shows refractive index of the ZnO films as a function of heat treatment temperature. The refractive index increases with increasing heat-treatment temperature up to 800°C. At 800°C, the value of refractive index is close to the value of single crystal (n = 2.004). Above 800°C, the refractive index starts decreasing.

The variation of refractive index may be attributed to the change in the packing density with heat treatment temperatures. Tansley et al. [44] assumed a linear dependence of refractive index on packing density and reported a value of 0.95 for packing density for the sputtered ZnO film. Many attempts have been made to correlate the packing density with refractive index of the materials [40, 45]. Macleod [45] reported the relation based on Lorentz-Lorentz polarization theory, as below

$$n_f^2 = \frac{(1-p)(n_s^2+2)n_v^2 + p(n_v^2+2)n_s^2}{(1-p)(n_s^2+2) + p(n_v^2+2)} \tag{4}$$

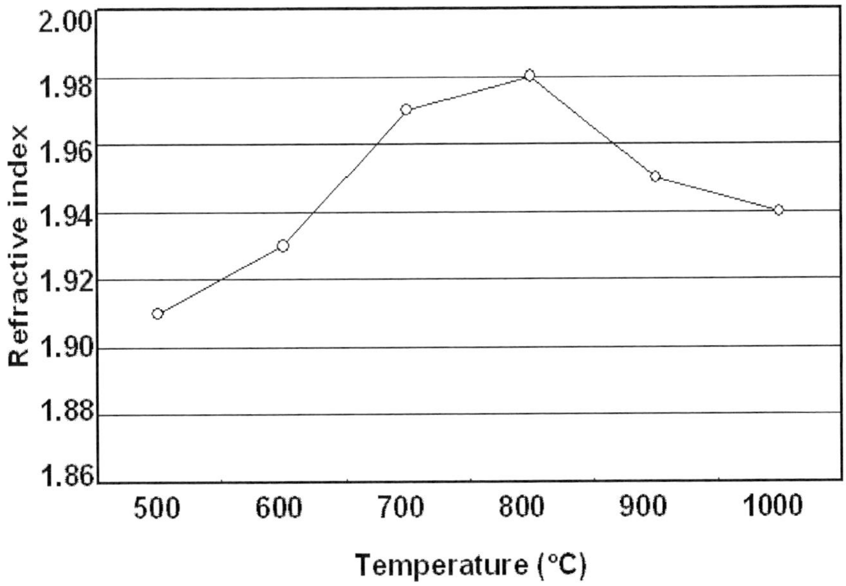

Figure 19. Refractive index of the ZnO films after heat treatment at various temperatures.

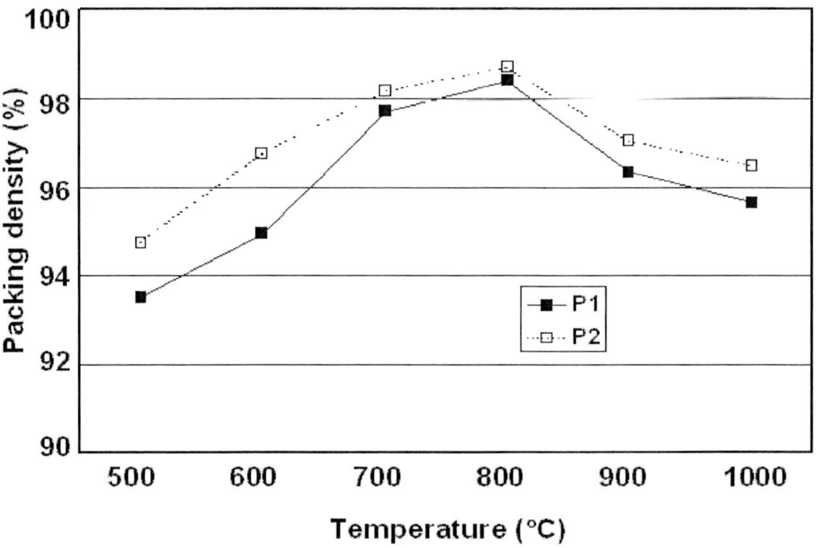

Figure 20. Estimated packing densities from two different formulas as a function of heat treatment temperature.

Where n_f is the refractive index of film (containing voids and columns), n_s is the index of the solid material of the film (single-crystal value), n_v is the index of the voids in the film (equal to one for air), and p is the packing density [40]. Eq. 4 is basically derived by assuming that the grains of the film are of spherical form. This approach is not strictly valid because of the possibility that grains of the film might depart from the spherical form and assuming cylindrical particles expression is [45]

$$n_f^2 = \frac{(1 - p)n_v^4 + (1 + p)n_v^2 n_s^2}{(1 + p)n_v^2 + (1 - p)n_s^2} \qquad (5)$$

The values of packing densities of the ZnO film as a function of heat treatment temperature are determined using Eqs. 4 and 5. The variation of packing densities is shown in Fig. 20. The values P_1 and P_2 are the packing densities calculated from the Eqs. 4 and 5, respectively. The maximum packing density was obtained at 800°C. This shows that the film heat treated at 800°C have minimum numbers of voids and pores. On the other hand, at higher temperature the pores owing to larger grain growth, as evident from FE-SEM studies, may be responsible for the decrease in packing density and therefore the refractive index. The FE-SEM images show round-shape grains of uniform size. Therefore Eq. 4, which is based on the assumption of spherical grains, seems to be more appropriate for calculation the packing density of the previous film [40].

CONCLUSION

Highly *c*-axis oriented ZnO thin films on silica glass substrates were prepared by spin coating-pyrolysis from metal naphthenate precursor. Structural relaxation of the precursor gel before crystallization, by employing solvents having high boiling points about 200°C, by heating the film at low rates or by prefiring at low temperatures of 200°C ~ 300°C, is unessential for obtaining oriented ZnO film from a zinc naphthenate precursor. The most homogeneous film having a preferred orientation was observed at 800°C. All the films exhibit a high transmittance (>80%) in visible region except film at 1000°C and show a sharp fundamental absorption edge at about 0.38 ~ 0.4 μm.

\

REFERENCES

[1] Kim, K.; Song, J.; Jung, H.; Choi, W.; Park, S.; Song, J.; Lee, J.; *J. Vac. Sci. Tech.* A 2000, 18(6), 2864-2868.

[2] Kim, Y.; Kim, Y.; Yang, H.; Park, J.; Han, J.; Lee, Y.; Kim, H.; *J. Vac. Sci. Tech.* A 1997, 15(3), 1103-1107.

[3] Ohtomo, A.; Kawasaki, M.; Ohkubo, I.; Koinuma, H.; Yasuda, T.; Segawa, Y.; *Appl. Phys. Lett.* 1999, 75(7), 980-982.

[4] Bundesmann, C.; Ashkenov, N.; Schubert, M.; Spemann, D.; Butz, T.; Kaidashev, E. M.; Lorenz, M.; Grundmann, M.; *Appl. Phys. Lett.* 2003, 83(10), 1974-1976.

[5] Shigesato, Y.; Takaki, S.; Haranoh, T.; *J. Appl. Phys.* 1992, 71, 3356-3364.

[6] Dominguez, J. E.; Fu, L.; Pan, X. Q.; *Appl. Phys. Lett.* 2002, 81(27), 5168-5170.

[7] Matsubara, K.; Fons, P.; Yamada, A.; Watanabe, M.; Niki, S.; *Thin Solid Films* 1999, 347, 238-240.

[8] Zhu, S.; Su, C.; Lehoczky, S.; George, M.; Lowndes, D.; *J. Mater. Res.* 2000, 15(5), 1125 - 1130.

[9] Lin, C. C.; Hsiao, C. S.; Chen, S. Y.; Cheng, S. Y.; *J. Electrochem. Soc.* 2004, 151, G285 - G288.

[10] Cho, S.; Ma, J.; Kim, Y.; Wang, G. K. L.; Ketterson, J. B.; *Appl. Phys. Lett.* 1999, 75, 2761 - 2763.

[11] Zhang, Y.; Du, G.; Liu, D.; Wang, X.; Ma, Y.; Wang, J.; Yin, J.; Yang, X.; Hou, X.; Yang, S.; *J. Cryst. Growth* 2002, 243, 439 - 443.

[12] Jin, B. J.; Im, S.; Lee, S. Y.; *Thin Solid Films* 2000, 366, 107 - 110.

[13] Ashrafi, A. B. M. A.; Zhang, B.; Binh, N. T.; Wakatsuki, K.; Segawa, Y.; *Jpn. J. Appl. Phys.* 2004, 43, 1114 - 1117.

[14] Zhang, Y.; Du, G.; Yang, X.; Zhao, B.; Ma, Y.; Yang, T.; Ong, H. C.; Liu, D.; Yang, S.; *Semicond. Sci. Technol.* 2004, 19, 755 - 758.

[15] Zhao, B.; Yang, H.; Du, G.; Fang, X.; Liu, D.; Gao, C.; Liu, X.; Xie, B.; *Semicond. Sci. Technol.* 2004, 19, 770 - 773.

[16] Kim, I. S.; Jeong, S. H.; Kim, S. S.; Lee, B. T.; *Semicond. Sci. Technol.* 2004, 19, L29 - L31.

[17] Hirata, G. A.; McKittrick, J.; Cheeks, T.; Siqueiros, J. M.; Diaz, J. A.; Contreras, O.; Lopez, O. A.; *Thin Solid Films* 1996, 288, 29 - 31.

[18] Hwang, K.; Park, Y.; *J. Mater. Res*. 2001, 16(9), 2519 – 2525.

[19] Hwang, K.; Yun, Y.; Kim, Y.; Ryu, H.; Kang, B.; *Jpn. J. Appl. Phys*. 2002, 41, 795 – 799.

[20] Hwang, K.; Yun, Y.; Kang, B.; Kim, S.; Jang, S.; Kim, C.; Oh, J.; *J. Mater. Sci.: Mater. in Med*. 2003, 14, 761 - 765.

[21] Hwang, K.; Manabe, T.; Yamaguchi, I.; Kumagai, T.; Mizuta, S.; *Jpn. J. Appl. Phys*. 1997, 36, 5221 – 5225.

[22] Hwang, K.; Ahn, J.; Kim, B.; *J. Mater. Sci*. 2006, 41, 3151 – 3153.

[23] Hwang, K.; Jeong, J.; Jeon, Y.; Jeon, K.; Kim, B.; *J. Sol-Gel Sci. & Tech*. 2005, 35, 237 – 243.

[24] Kim, B.; An, J.; Jeon, Y.; Jeong, J.; Kang, B.; Hwang, K.; *J. Mater. Sci*. 2005, 40, 237 – 239.

[25] Hwang, K., Kang, B.; Jeon, Y.; An, J.; Kim, B.; Nishio, K.; Tsuchiya, T.; *Surf. & Coatings Tech*. 2005, 190, 331 – 335.

[26] Ryu, H.; Park, J.; Hong, K.; Hwang, K.; Kang, B.; Oh, J.; *Surf. & Coatings Tech*. 2003, 173, 19 –23.

[27] Kim, B.; An, J.; Jeon, Y.; Kang, B.; Hwang, K.; *J. Kor. Phys. Soc*. 2004, 44(4), 1012 – 1014.

[28] Jeong, J.; Ahn, J.; Kim, B.; Jeon, Y.; Jeon, K.; Hwang, K.; *J. Kor. Phys. Soc*. 2005, 46(2), 559 – 561.

[29] Jiménez-González, A.; Soto Urueta, J.; Suárez-Parra, R.; *J. Crys. Growth* 1998, 192, 430 – 438.

[30] Shimono, D.; Tanaka, S.; Torikai, T.; Watari, T.; Murano, M.; *J. Ceram. Proc. Res*. 2001, 2(4), 184 - 188.

[31] Ohyama, M.; Kozuka, H.; Yoko, T.; *Thin Solid Films* 1997, 306, 78 – 85.

[32] Ohya, Y.; Saiki, H.; Takahashi, Y.; *J. Mater. Sci*. 1994, 29, 4099 – 4103.

[33] Wessler, B.; Lange, F. F.; Mader, W.; *J. Mater. Res*. 2002, 17(7), 1644 – 1650.

[34] Ohyama, M.; Kozuka, H.; Yoko, T.; Sakka, S.; *J. Ceram. Soc. Jpn*. 1996, 104(4), 296 – 300.

[35] *NanoScope Command Reference Manual*, Ver. 4.42 (Digital Instruments, Inc., Santa Barbara, CA, 1999), pp. 12.19 – 12.45.

[36] Ryu, H. W.; Hwang, K. S.; *J. Kor. Phys. Soc*. 2002, 40, 493 - 496.

[37] Loder, J. C.; Wielinga, T.; Worst, J.; *Thin Solid Films* 1983, 101, 61 - 73.

[38] Chen, M.; Pei, Z.; Wang, X.; Sun, C.; Wen, L.; *J. Mater. Res*. 2001, 16(7), 2118 – 2123.

[39] Cebulla, R.; Wendt, R.; Ellmer, K.; *J. Appl. Phys*. 1998, 83, 1087 - 1095.

[40] Gupta, V.; Mansingh, A.; *J. Appl. Phys*. 1996, 80(2), 1063 – 1073.

[41] Swanpoel, R.; *J. Phys. E* 1984, 17, 896 - 903.

[42] Park, K.; Ma, D.; Kim, K.; *Thin Solid Films* 1997, 305, 201 - 209.

[43] Ziegler, E.; Heinrich, A.; Oppermann, H.; Stover, G.; *Phys. Status Solidi A* 1981, 66, 635 - 648.

[44] Tansley, T. L.; Neely, D. F.; Foley, C. P.; *Phys. Status Solidi A* 1983, 77, 491 - 496.

[45] Macleod, H. A., *J. Vac. Sci. Tech. A* 1986, 4, 418 - 422.

In: Encyclopedia of Nanotechnology
Editor: Elwood D. Carlson

ISBN: 978-1-60692-079-4

Chapter 20

FABRICATION OF CORE-SHELL STRUCTURED PEROVSKITE NANOPARTICLES WITH HETEROGENEOUS OXIDE SHELLS AND THEIR CERAMICS

***Jian Quan Qi*[*, 1], *Helen Lai Wah Chan*[2], *Hu Yong Tian*[2], *Wan Ping Chen*[2] and *Long Tu Li*[1]**

[1] Department of Materials Sciences and Engineering and
State Key Laboratory of Fine Ceramics and New Processing,
Tsinghua University, Beijing 100084, China

[2] Department of Applied Physics and Materials Research Center,
The Hong Kong Polytechnic University, Hong Kong, China

ABSTRACT

The perovskite nanoparticles such as BaTiO3, SrTiO3, BaZrO3 and their solid solutions can be prepared using a new method, DSS, near room temperature. Hydrogen interstitials are introduced into the lattice of perovskite nanoparticles synthesized by DSS method so that the nanoparticles take positive charges and adsorb the hydroxyl group with negative charges on the surface. Hydroxyl groups adsorbed on the surface of nanoparticles can be ionized freely, and the nanoparticles exhibit an alkalescent property. When this kind of perovskite nanopaticles disperses in a certain solution of inorganic salt such as Mg(NO3)2•6H2O, Zn(NO3)2•6H2O, Al(NO3)3•9H2O, Y(NO3)3•6H2O etc., the hydroxide related to these salts spontaneously deposit on the surface of the nanoparticles, as a result, a perovskite structure nanoscaled core materials are successfully self-wrapped by a non-ferroelectric hydroxide derived from a nitrate salt solution. After annealing at high temperature, core-shell structured perovskite nanoparticles with heterogeneous oxide shells can be synthesized. When the core-shell structured perovskite nanoparticles with heterogeneous oxide shells are composited with doped shell oxide materials, the pinning effect during sintering can be avoided, as a result, the sintering temperature are decreased and dense ceramics are obtained. The core-shell structures are analyzed by a

---

* E-mail: jianquanqi@mail.tsinghua.edu.cn

conjunction of XRD, HRTEM and FE-SEM. The lattice match of shell and core in our ceramics of MZO wrapped BST nanoparticles probably follows the second match mode. The tunability and FOM values of the ceramics of MZO wrapped BST nanoparticles have been improved significantly compared with that of pure BST ceramics. The ceramics of MZO wrapped BST nanoparticles are promising materials for tunable microwave device applications and high energy density ceramic capacitors for their distinct dielectric loss improvement.

Core-shell structured nanoparticles received much attention for their unique physical and chemical properties, as well as their technological applications.[1-7] With oxide shells, they are also interesting from a number of perspectives.[8-11] The shells offer protection to the cores as well as introduce new properties to the hybrid structures.[12] This kind of hybrid structures can have advantages of both the core and the shell. Perovskite oxides such as BaTiO3, SrTiO3, PbTiO3, BaZrO3, PbZrO3 etc. is a series of important functional materials with a formula of ABO3, and most of them can form solid solution with each other while keeping a perovskite structure. The perovskite ceramics are important for their ferroelectricity and anti-ferroelectricity, which find a wide variety of applications in nonlinear optics, thin-film capacitors, pyroelectric detectors, optical memories, and electrooptic modulators [13]. However, with the reduction in grain size, the ratio of grain boundary to grain of the ceramics will increase and degenerate their electrical properties. In order to improve the properties of the ceramics, introduction of heterogeneous oxide shells which have outstanding properties, such as very low dielectric loss, is a wonderful choice. Self-wrapping MgO shell to BST nanoparticles is a successful example which can be applied for microwave materials.[14-15] Fabrication of this kind of core-shell structured perovskite nanoparticles with heterogeneous oxide shells and their ceramics will be discussed here and self-wrapping MgO shell to $Ba_{1-x}Sr_xTiO_3$ (BST) nanoparticles will be used as an example.

1. Synthesis of Perovskite Nanoparticles from Solution

In order to fabricate core-shell structured perovskite nanoparticles with heterogeneous oxide shells, it is a key to synthesize perovskite nanoparticles with suitable physical and chemical properties. However, most perovskite oxides are still prepared by conventional solid-state reactions between the corresponding oxides or oxides and carbonates at temperatures above 1000oC. Aiming at a better understanding of the structure-property relationship and the development of novel electronic devices, the study of the nano-scaled ferroelectric materials has attracted immense attention in recent years. Novel synthesis techniques have been developed. Among them, wet-chemical methods provide promising alternatives because they can be better controlled from the molecular precursor to the final material to give highly pure and homogeneous materials, and allow low reaction temperatures to be used, the size and morphology of the particles to be controlled, and metastable phases be prepared. There are two kinds of wet-chemical routes for the synthesis of perovskite nanocrystals. One is precipitation method which produces precursors such as carbonate, hydroxide, oxalate, dry gel etc. by solution method and then calcines them to obtain the perovskite nanocrystals. Although the methods are very simple and the reaction in solution can be even operated at room temperature, agglomeration and grain-growth are inevitable during the drying and calcination processes. The other one is direct synthesis of perovskite nanostructures in solution and much effort has been made on it in recent years. Hydrothermal method is one of the most popular approaches to obtain the perovskite nanostructures directly from solution, but the synthesis processes are often conducted at elevated temperatures (typically 100-280oC) and/or under relatively high pressure to improve the crystallinity of the products.[16-21] To further simplify the process and lower the processing temperature and pressure is very important and rather challenging from the technical point of view. Fortunately, some most important perovskite oxides have been directly synthesized from solution successfully near room temperature in recent years. This kind of method, which was called direct synthesis from solution (DSS),[22] was developed to synthesize BaTiO3,[23] SrTiO3,[24] BaZrO3[25] and their solid solutions near room temperature. The nanoparticles synthesized by DSS method are very suitable for the fabrication of core–shell structured perovskite nanoparticles with heterogeneous oxide shells.

1.1. Synthesis of $BaTiO_3$ Nanoparticles from Solution

The synthesis of fine ceramic powders has become an important part of modern ceramic research, so that there is a strong trend toward the application of chemical method for powder synthesis in electronic ceramics. Using fine ceramic powders can lower the sintering temperature, achieve uniform fine ceramic grain size and save energy. BaTiO3 is one of the most typical and important perovskite materials which has been widely used in multilayer ceramic capacitors (MLCC) and positive temperature coefficient of resistivity (PTCR) ceramics.[26-32] Recent advances in microelectronics and communications have led to the miniaturization of MLCCs. Higher capacitance in smaller case sizes requires a reduction of the thickness of the ceramic layers below 2-3 μm and the increase of the number of layers (>200).[33-34] To achieve this goal, powders with improved quality with small and uniform size are required. The synthesis of BaTiO3 nanostructure is also a basic method that many other perovskite nanostructures can be obtained similarly. There are several methods to synthesize BaTiO3 nanostructures from solution. For example, the precursors, barium titanium isopropoxide (BaTi[OCH(CH3)2]6), were used to synthesize BaTiO3 nanostructures at temperatures over 100oC by solution-phase decomposition in the presence of coordinating ligands followed by heat-treating at about 280oC.[16-17] Another synthesis protocol is based on the dissolution of alkali or alkaline earth metals in benzyl alcohol and subsequent reaction with transition metal alkoxides at relatively low temperatures of 200-220oC to synthesize BaTiO3 nanoparticles.[21] However, DSS is one of the most convenient methods to obtain BaTiO3 nanoparticles in a large-scale because the synthesis temperature can be as low as near room temperature, such as 45oC.[22-23] It may be a good replacement of the well known oxalate method.[35] Furthermore, improved DSS method can be operated even at room temperature or lower to synthesize BaTiO3 nanoparticles directly.

1. BaTiO3 Nanoparticle Synthesis Process

The chart of DSS method is shown in Figure 1. A typical method to synthesize BaTiO3 nanoparticles using DSS method is as follows: The reagents Ba(OH)2•8H2O and tetrabutyl titanate are adopted as starting raw materials to prepare nanoscaled BaTiO3 powders.

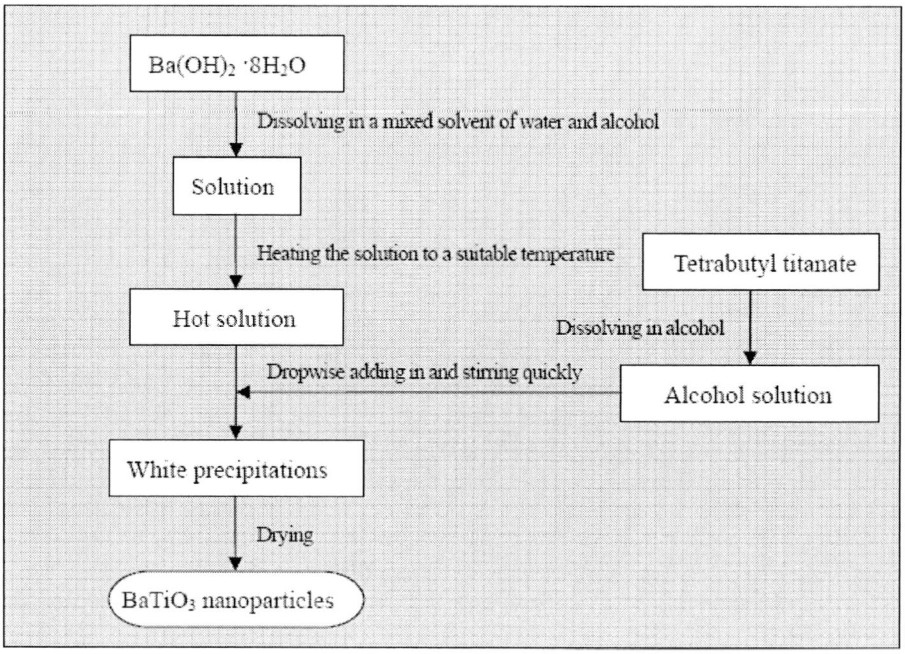

Figure 1. The chart of DSS method.

The base solution is prepared by dissolving Ba(OH)2•8H2O in deionized water (DIW) with the concentration of 0.4M ethanol. Tetrabutyl titanate is dissolved in ethanol absolute to obtain a solution with concentration of 0.4M ethanol. Different reaction temperatures of 45, 60, 70 and 85oC are adopted in order to study the temperature effect on the properties of prepared powders. When the base solution reaches the desired reaction temperature, the same amount of tetrabutyl titanate solution is added drop by drop to the base solution under normal pressure and is stirred quickly. An instantaneous formation of white precipitation can be observed. After the reaction has been completed, the white precipitation is filtered and dried at 60oC for 24h to obtain the powder.

2. Characterization of Nanoparticles

Modern analytic instruments such as X-ray diffraction (XRD), scanning electron microscopy (SEM), and transmission electron microscopy (TEM) etc can be used to characterize the features of the as-prepared nanoparticles. The powders prepared by conventional chemical method often are hardened easily due to agglomeration during the drying period. Fortunately, if the precipitations have been crystallized, agglomeration can be suppressed. The powders prepared by DSS method are all soft after drying at 60oC when the synthesis temperature is lower than or equal to 70oC. The powders synthesized at 85oC are hardened slightly, but it also can be crushed into soft powders easily. All of the samples are crystalline BaTiO3 powders as shown in the XRD spectra (Figure 2).

The line broadening effect of XRD profile is caused by the size reduction of the grains. The crystallite size of BaTiO3 can be estimated by the Scherrer's equation: [23]

$$d_x = K\lambda/(\beta\cos\theta) \quad (1)$$

where d_x is crystallite size, K a constant and usually taken as 0.89, λ the wavelength of diffraction X-ray (0.15418 nm for CuKα), β the line width at half-maxiumum height after subtraction of broadening caused by equipment and θ is half the diffraction angle of the center of the peak (in degrees). The XRD profiles of (111) plane of the samples synthesized at different temperatures are shown in Figure 3. The XRD peak shift with the synthesis temperature can be observed from this figure. It is revealed that the powders have different crystallite lattice parameters when synthesized at different temperatures. The lattice parameter and crystallite size of the powders synthesized at different temperatures are shown in Figure 4.

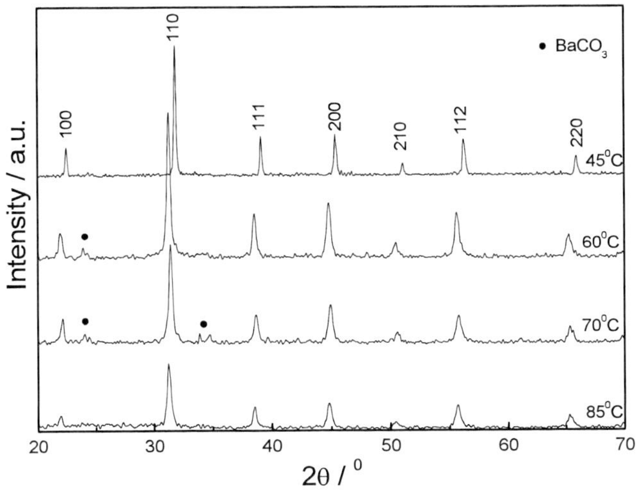

Figure 2. XRD profiles of $BaTiO_3$ powders synthesized at different temperatures [23].

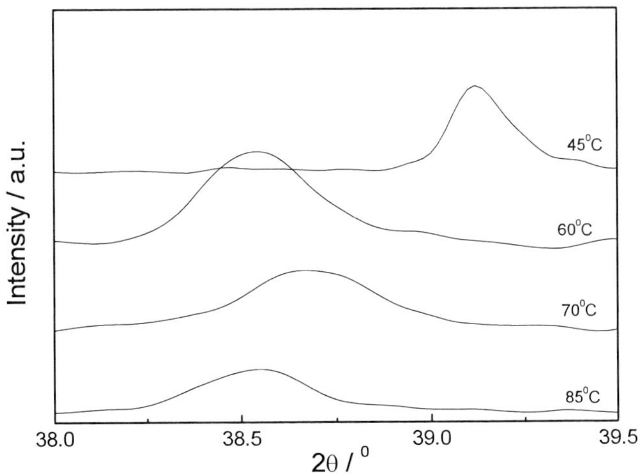

Figure 3. XRD profiles of (111) plane of $BaTiO_3$ powders synthesized at different temperatures [23].

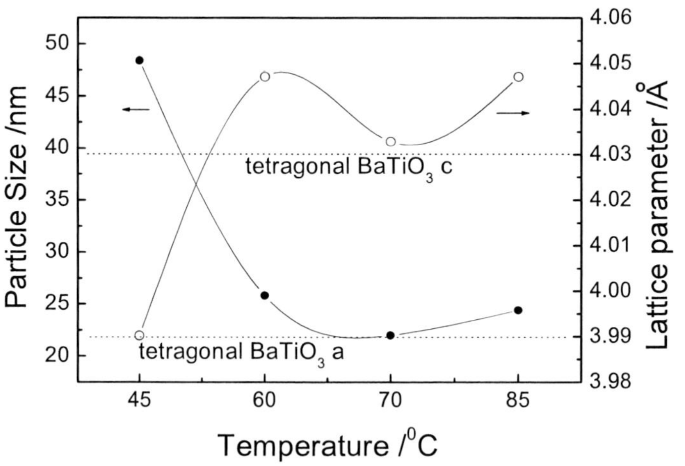

Figure 4. The lattice parameter and crystallite size of the $BaTiO_3$ powders synthesized at different temperatures [23].

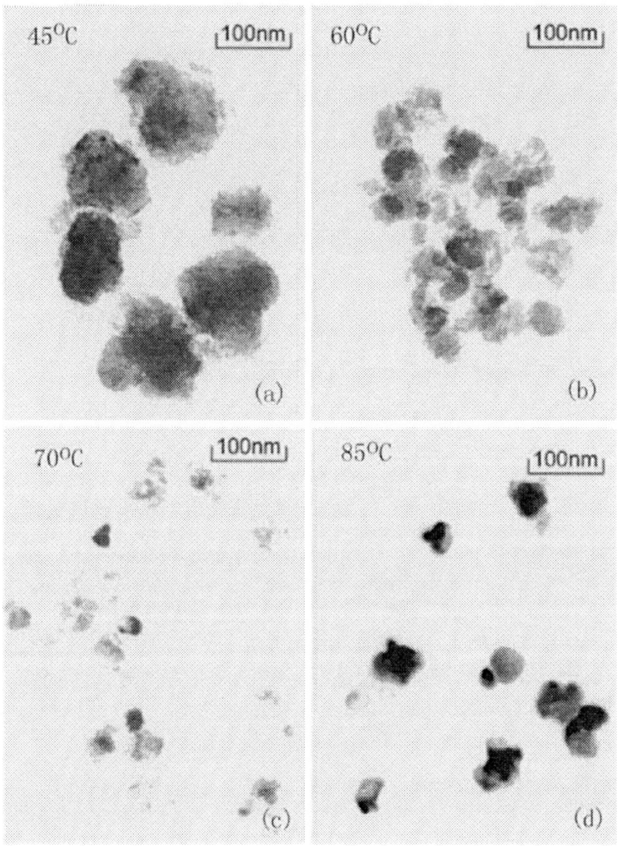

Figure 5. The TEM images of the samples [23].

The crystallite size of as-prepared powders decreases rapidly from 50nm to 20nm with the increase in the synthesized temperature from 45oC up to 70oC. After that, the crystallite size increases slightly. The TEM images of the samples are shown in Figure 5. The change tendency of the particle size with the synthesis temperature is similar to that of the crystallite size measured by XRD. Figure 6 shows the image of a single crystal diffraction spots obtained through focusing the transmission electron beam on one particle shown in Figure 5. Moreover, the shape of the particles in Figure 5 is almost spherical. Therefore, one particle is a single-crystal particle and the crystallite size is the average particle size of the powders.

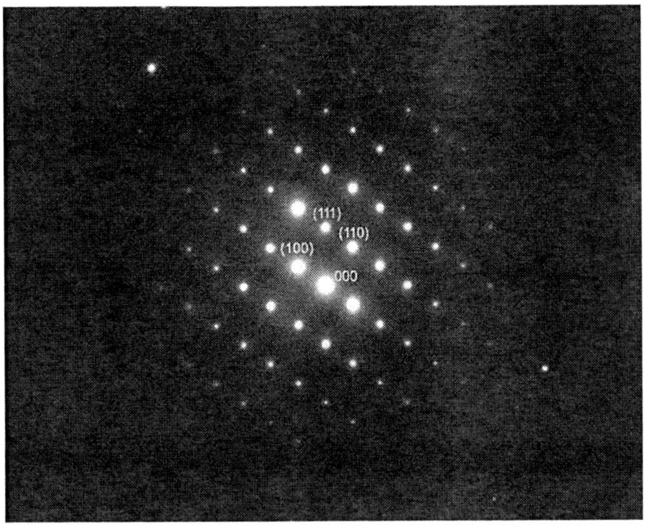

Figure 6. Transmission electron diffraction spot images of a single particle [22].

1.2. Synthesis of $SrTiO_3$ and BST Nanoparticles

Both SrTiO3 and Ba1-xSrxTiO3 (BST) are important functional materials. SrTiO3 ceramics can be used in grain boundary layer capacitor and voltage varistor.[36-38] The Curie temperature (TC) of the bulk BST varies almost linearly with the composition, x, in the solid solution Ba1−xSrxTiO3. For a particular application temperature, the composition is adjusted to move the ferroelectric phase transitions to the range of the device application temperature so that BST ceramics can be also used as low temperature PTCR and high dielectric constant capacitors.[39-40] BST films are being extensively investigated as the charge-storage dielectric for Gbit dynamic random access memory and other integrated capacitor structures, and more recently as electric-field tunable elements for high frequency devices, and infrared detectors.[41-51] The nano-scaled BST is also a good candidate for the fundamental research of nano-ferroelectricity because of its excellent compositional and structural modifiability.

The method of synthesis of SrTiO3 nanoparticles is very similar to that of BaTiO3 nanoparticles because of the very similar chemical properties between Ba and Sr, such as both Ba(OH)2 and Sr(OH)2 have strong alkalescence in water solution. However, their solubilities are quite different and Sr(OH)2 has a lower solubility than Ba(OH)2 especially at low temperature as shown in Figure 7.

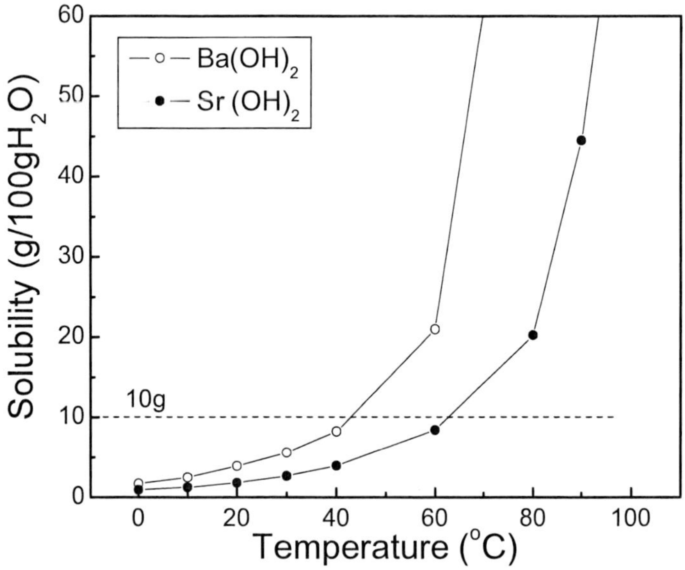

Figure 7. Solubility of $Ba(OH)_2$ and $Sr(OH)_2$, the number of grams of substance which can be dissolved in 100g of water at a certain temperature [22].

For example, the solubility of Ba(OH)2 reaches 10g at 43oC, while the solubility of Sr(OH)2 can only reach that number at 63oC. Thus, synthesis of the same amount of SrTiO3 nanoparticles needs a higher temperature than that of BaTiO3 nanoparticles in one batch so that the alkaline earth metal hydroxide can be dissolved completely at the beginning of the reaction. Therefore, the lower solubility of Sr(OH)2 than Ba(OH)2 should be noted during synthesis of Sr-included perovskite nanoparticles.

The process of DSS method for synthesizing BST nanoparticles is the same as that of synthesizing BaTiO3 and SrTiO3 nanoparticles as shown in Figure 1 except that the starting materials, Ba(OH)2·8H2O, is replaced with Ba(OH)2·8H2O and Sr(OH)2·8H2O according to a designated ratio in BST. The synthesis process may be divided into two steps.[24] The first step is to separately prepare a base solution and a titanium solution. The base solution is prepared by dissolving Ba(OH)2·8H2O (28.392 g) and Sr(OH)2·8H2O (2.658 g) into DIW (500 ml). The concentration of the base solution is ~ 0.2M. The solution was then heated to 60°C in a water bath. The titanium solution was prepared by dissolving Ti(OBu)4 (34.0 g) into absolute ethanol (250 ml). These solution concentrations are for the composition of (Ba0.9Sr0.1)TiO3; for other compositions, the solution concentration can be changed accordingly. The second step is to allow the reaction to take place by dripping the titanium solution into the base solution (60°C) under vigorous stirring. An instantaneous formation of white precipitation can be observed. After the reaction has been completed and followed by an ageing for 3 h, the white precipitation is obtained by filtration and baked at 60°C for 24 h in an oven.

Figure 8 shows the representative XRD pattern of the as-prepared powder that has been synthesized and baked at 60°C. Except for one small peak appearing at about 24°, all other peaks are identified as from perovskite BST. The peak sharpness and intensity indicate that the BST is well crystallized. The particles are found to have a cubic symmetry with lattice parameter ~0.402 nm, which is different from the symmetry of bulk BST.

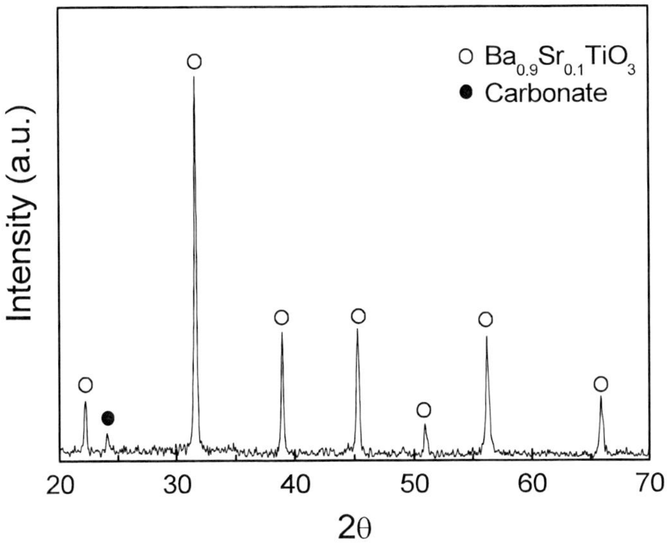

Figure 8. X-ray diffraction pattern of BST(90/10) nanoparticles [24].

In literature, BST ceramic of the same composition has a tetragonal phase at room temperature with lattice parameters a=0.399 nm and c=0.401 nm.[52] This symmetry change is a typical size effect that has been observed in other nano-scaled ferroelectric materials as well. The peak at 24° in the XRD pattern is identified as from barium carbonate. The formation of barium carbonate is believed to be a result of the reactions between carbon dioxide dissolved into the solutions from air and reacted with the alkaline earth hydroxides during the process. The impure phase can either be avoided by undertaking the whole process in a protective atmosphere (such as argon or nitrogen) or be removed by washing the powders using dilute acids.

Another microstructural factor that possibly affects the lattice symmetry is interstitial hydrogen. Synthesized from an aqueous solution, the particles are expected to contain hydrogen that may exist in the interstitial sites of BST lattices. Despite direct observation of such interstitial hydrogen is not yet available, some indirect evidences can support the assumption, such as the as-synthesized BST particles have a larger lattice than bulk BST because the interstitial hydrogen is responsible for such slight lattice expansion in the as-synthesized BST. For the particles subjected to post-synthesis thermal treatment, the interstitial hydrogen may be partially/completely removed.

The DTA-TG analysis has helped to study the release of hydrogen from the lattices. Figure 9 shows the weight and DSC of the as-synthesized powders as a function of temperature. From room temperature to about 220°C, there is a weight loss of ~1.5%. Such weight loss is mainly due to the removal of residual alcohol and adsorbed water in the particles. From 220°C to ~830°C, another weight loss, up to ~3%, is observed. This weight loss is believed to be corresponding to the release of interstitial hydrogen and the adsorbed hydroxyl groups. At higher temperatures, the sample weight is almost a constant value, meaning that the release of the interstitial hydrogen and hydroxyl groups is completed at ~830°C (i.e., the particles subjected to >830°C thermal treatment are stoichiometric BST). In the DSC curve, no sharp endothermic or exothermal peak is observed. This is because the as-synthesized particles are already well crystallized. A rather broadened exothermic curve is observed, which is likely corresponding to the crystallite size increase at higher temperatures.

The microstructure of the as-synthesized BST particles has been further characterized by means of scanning electron microscopy (SEM) and transmission electron microscopy (TEM). A representative SEM image of the as-prepared BST nanoparticles is shown in Figure 10. It is observed that the particles are uniform, with ~30 nm in diameter. The result is in good agreement with the XRD estimation.

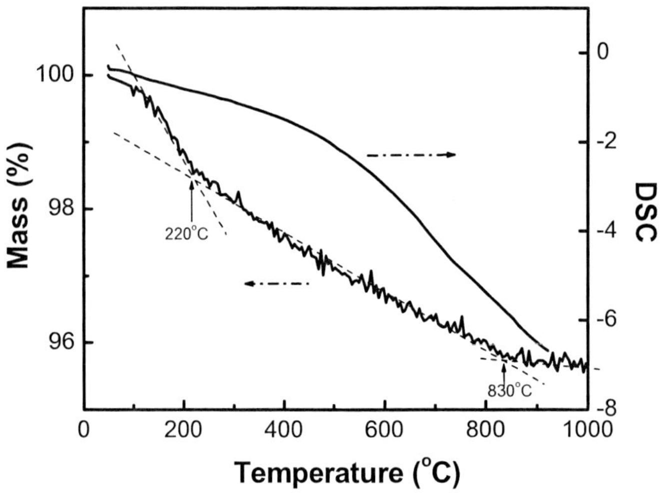

Figure 9. DTA-TG analysis of as-prepared BST(90/10) powder [24].

Figure 10. BST(90/10) nanoparticles observed under a field-emission scanning electron microscope (FE-SEM) [24].

The TEM observation results are shown in Figure 11. Figure 11(a) shows the morphology of the powder. Consistent with the SEM observation, the particle size is found under TEM to be ~50 nm in diameter. A selected-area electron diffraction (SAED) pattern with indexed atomic planes is shown in the inset of Figure 11(a).

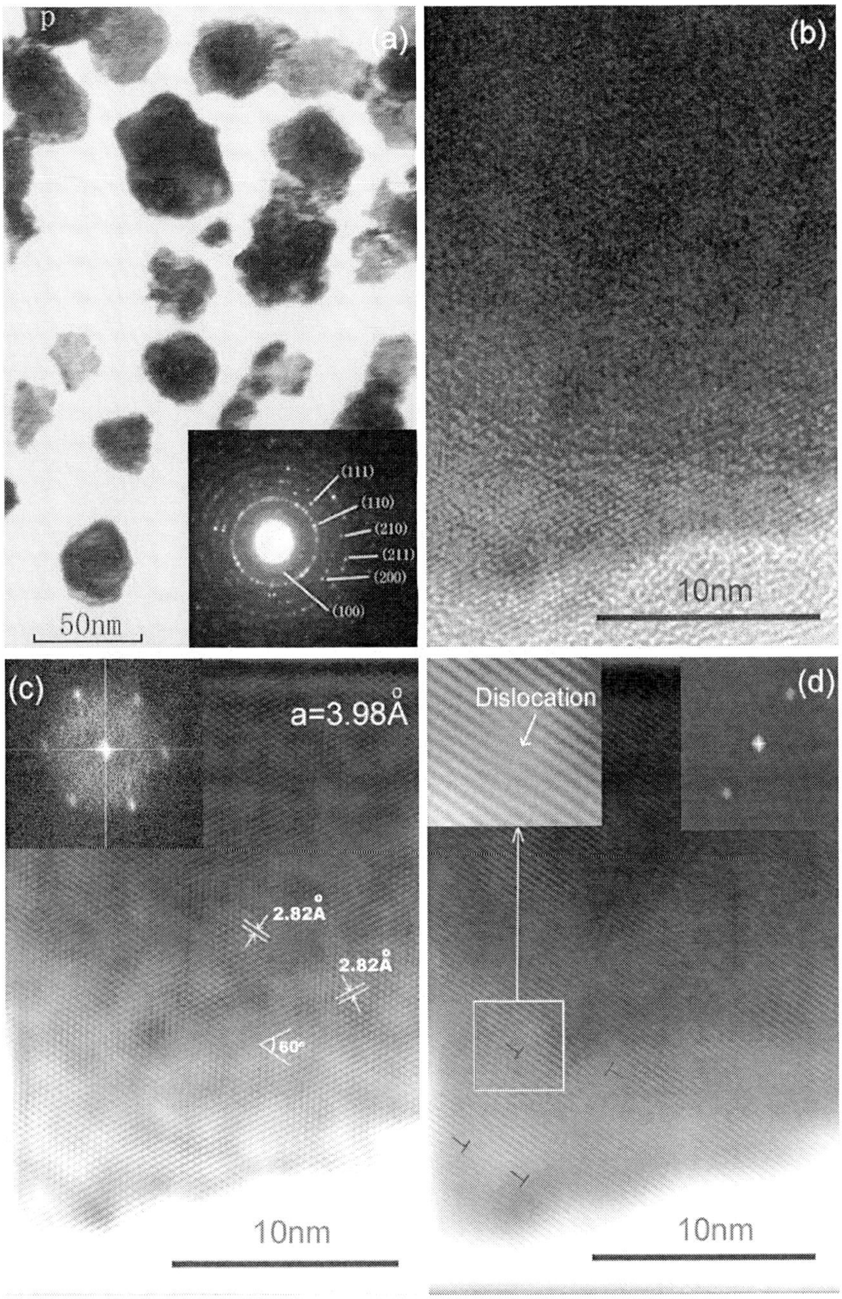

Figure 11. TEM observation of BST(90/10) nanoparticles: (a) Morphology, (b) High resolution image, (c) FFT image indicating relationship of atomic planes and (d) FFT image revealing the dislocations in the nano-structure [24].

Figure 11(b) is a high-resolution TEM image of "particle p" in Figure 11(a). Regularly arranged patterns can be observed in the darker region of the photo, indicating that the particles under observation are crystallized. A Fast Fourier Transform Algorithm (FFT) has yielded structural information of the nano-particles, as shown in Figures 11(c) and 11(d). The lattice parameter of the material is found to be ~0.398 nm. Dislocations have also been identified in the FFT image shown in Figure 11(d). Obviously, the low synthesis temperature is the main reason for the formation of the dislocations.

Further experiments have indicated that the particle size can be controlled through modifying the solution concentration, the synthesis temperature, the stirring rate and/or the drying temperature. For example, finer particles can be obtained in thinner solution. Although, as a typical example, the synthesis in this paper has been conducted at 60°C, it does not mean that crystallized nanoparticles could not be obtained via this technique at other temperatures. As a matter of fact, BST nanoparticles with acceptable crystallinity have been obtained via the same process conducted at several different temperature points (~25°C, 40°C, 80°C etc). Another point to mention is that this synthesis technique is applicable to BST with any Ba:Sr ratio, despite only the composition of Ba : Sr = 0.9 : 0.1 is discussed here.

It is difficult to enlarge the production scale of nanoparticles for many synthesis methods, especially for hydrothermal method in an autoclave. With DSS method, products can be synthesized in a large scale easily. In the laboratory, using a 2000 ml beaker as reaction container, about 100 g of nanoparticles can be obtained in one batch. Therefore, the method is ideal for industrial production.

The BST nanoparticles prepared by the above process are very useful for making high-quality ceramics and composites. The microstructural features are very small in size (hence large specific area) and already crystallized without any high temperature treatment mean a better sinterability of the BST nano-particles than powders prepared via regular solid-state reactions. Based on this, we have managed to prepare BST ceramics by a simplified ceramic processing in which the calcination step was omitted and the sintering temperature was lowered.

1.3. Synthesis of $BaZrO_3$ and BZT Nanoparticles

Barium zirconate (BZ) is another typical of perovskite oxide that can form solid solution with barium titanate completely, which is expressed as BZT generally. The substitution of small amounts of titanium by zirconium in barium titanate modifies the ferroelectric properties e.g., the thermal stability of the dielectric properties which is desirable for many technological applications [53, 54]. Also, the BZT ceramic [Ba(Zr, Ti)O3] shows excellent bulk properties of a high dielectric material (dielectric constant >20,000)[55, 56]. BZT ceramics have been established as one of the most important compositions for dielectrics in multilayer ceramic capacitors (MLCCs) of the EIA Y5V specification.[56-58] With the change of Zr/Ti ratio, the microstructure and properties of the material change, as already observed in BZT ceramics.[52] Due to the excellent and easily modifiable dielectric properties, BZT is useful for multilayer capacitors and frequency-agile microwave devices.[59] In the literature, the synthesis of perovskite BZT nano-powders is usually through the sol-gel method which requires post-annealing, typically at 400°C or above, to crystallize the powders.[51,61] In DSS method, however, by employing appropriate chemical precursors (barium hydroxide, zirconium isopropanol and tetrabutyl titanate, etc), perovskite BZT nanoparticles can be prepared under much milder conditions.

The synthesis process of BZT and BZ is very simple and similar to that of BST and BT [25] as shown in Figure 1. However, some starting materials are different, such as isopropanol zirconium is adopted to obtain the element zirconium in BZ and BZT. Because isopropanol zirconium is sensitive to moisture and even can form precipitates in absolute ethanol, the solvent of the alcohol solution has to be replaced by isopropanol. More details of our experiments are as follows: Firstly, a base solution is prepared by dissolving barium hydroxide into warm water (60°C). The concentration of the solution is usually 0.4 M. After dissolving, a small amount of ammonia is added into the solution. In the meantime, a solution is prepared by dissolving tetrabutyl titanate and zirconium isopropanol into isopropanol. In this solution, the concentration of titanium and zirconium has been adjusted in terms of the desired composition of the final products. To prepare Ba(Zr0.50Ti0.50)O3 powder, for example, the Ti and Zr concentration are both 0.2 M. Then the organic solution is slowly added into the base solution (~60°C) under vigorous stirring. An instantaneous formation of white precipitation can be observed. After the reaction has been completed, the white precipitation is filtered out of the solution and baked at ~60°C for 24h in an oven. Dried white powders obtained via the above process are the final products.

X-ray diffraction spectra of the BZT powders of different compositions (x = 0, 0.1, 0.2, 0.5, 0.7, and 1.0) are shown in Figure. 12.

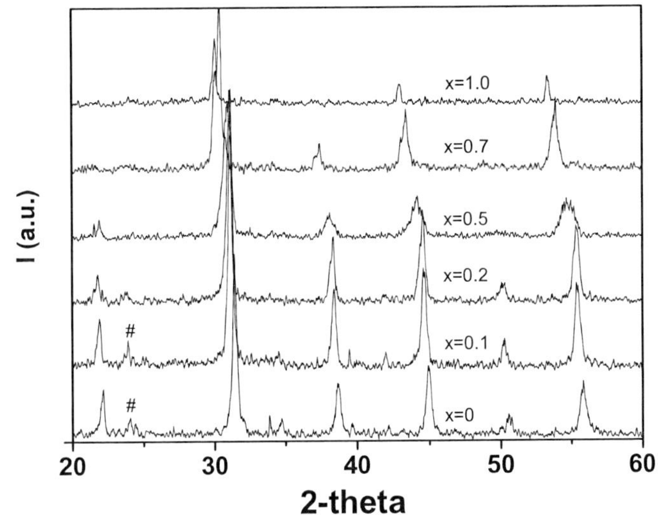

Figure 12. X-ray diffraction spectra of $Ba(Zr_xTi_{1-x})O_3$ powders. The peak marked with "#" is identified as $BaCO_3$ peak. All other peaks are perovskite peaks [25].

All the powders have been prepared at 60°C. Except for minor barium carbonate peaks (located at about 25°) occasionally appear in the plot, all major peaks shown in the patterns can be identified as from perovskite phase. With the increase in Zr/Ti ratio, the peak positions exhibit a shift towards lower angles, indicating an increase in the lattice constant. This trend is in good consistency with that in bulk BZT solid solutions,[52] suggesting a good compositional control in the as-prepared powders.

Figure 13 shows a TEM micrograph of the Ba(Zr0.50Ti0.50)O3 powders prepared in 60°C solutions. It is found that the powders are quite uniform in size, with an average diameter of about 40 nm. TEM observations and X-ray diffractions on the powder samples prepared under other conditions reveal that the particle size may vary as the processing parameters change. Typically, lowering the solution concentration and raising the solution temperature are helpful for making smaller-sized powders. Nevertheless, under the conditions we have investigated (i.e., solution temperature from room temperature to 80°C and concentration from 0.1 M to 1 M), the average diameter value of the powders is seldom beyond the range of 20 nm to 120 nm.

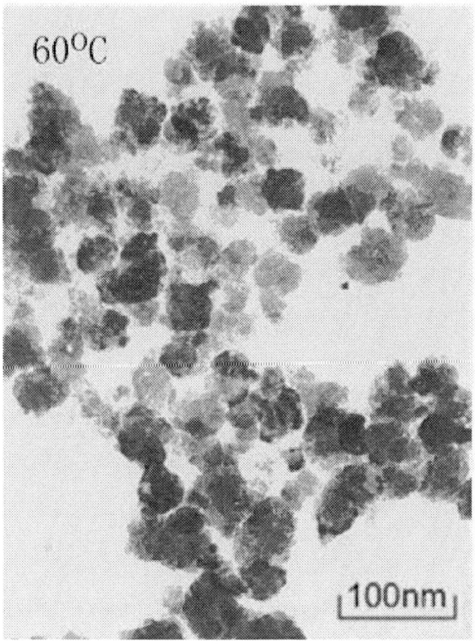

Figure 13. TEM observation of $Ba(Zr_{0.50}Ti_{0.50})O_3$ powders [25].

2. Synthesis of Core-Shell Structured Perovskite Nanoparticles with Heterogeneous Oxide Shells

Core-shell structured perovskite nanoparticles with heterogeneous oxide shell may have advantages of both the core and the shell. This kind of hybrid structure has potentials for some important applications, such as microwave devices and high energy density ceramic capacitors. Paraelectric phase Ba1-xSrxTiO3 (BST) at room temperature can be obtained by adjusting the ratio of Ba and Sr and/or doping foreign ions to substitute A or B sites in the ABO3 perovskite system, which is a promising material for electric tunable devices such as phase shifters, delay lines, tunable capacitors, and oscillators [62,63]. Driven by the applications of electric materials in telecommunications, the optimum material parameters have been focused on a low dielectric constant (ε), large tunability, a low loss dissipation factor (tanδ), and low temperature dependence of ε. Composite materials, including a ferroelectric BST phase and non-ferroelectric oxides with low loss (MgO, Ta2O3, Al2O3, etc.), have attracted much attention [63-67]. To obtain the size of ferroelectric materials in the nanometer scale, more attention has also been paid to chemical routes, like sol-gel procedures

[68], to the thermal decomposition of precursors [69], the hydrothermal process [70], selected coprecipitation techniques [71], as well as to the solvothermal method [72]. As noted previously, it is a key to synthesize perovskite nanoparticles with suitable physical and chemical properties. The special feature of the nanoparticles by DSS method is ideal in this respect, making them a favorable candidate for self-wrapping heterogeneous oxide shell during synthesis of core-shell structured perovskite nanoparticles. This strategy for creating the core-shell BST@MgO particles is described in Section 2.2. The nanoscaled BST powders synthesized by DSS method are used as core materials and the core-shell structure is formed using the chemical solution method, by adding magnesium nitrate solution into the BST base solution with an ultrasonic dispersion process.

2.1. The Special Feature of Perovskite Nanoparticles Prepared by DSS Method

The special feature of perovskite nanoparticles prepared by DSS method is that they can carry positive charge and attract free hydroxyl group on their surfaces. The hydroxyl group can often exist in perovskite lattice of compound such as BaTiO3, to form Ba1-$_x$[VBa'']$_x$TiO3-2$_x$[OHO•]2$_x$ when the BaTiO3 powders are synthesize from water solution[23]. Hydroxyl group in lattice can also be considered as interstitial hydrogen ion as shown in Equation (2)

$$V_O^{\bullet\bullet} + H_2O \rightarrow 2H_i^{\bullet} + O_O^{\times} \quad (2)$$

or

$$H_2O \rightarrow 2H_i^{\bullet} + O_O^{\times} + V_{Ba}^{''} \quad (3)$$

For a low synthesis temperature, it seems difficult to form vacancy defects and Equations (3) and (2) are not quite reasonable. A more reasonable equation may be revised as:

$$H_2O \rightarrow H_i^{\bullet} + OH^{-}(free) \quad (4)$$

The hydroxyl group with negative charge should locate on the surface of the nanoparticles and can be ionized freely. Because the perovskite nanoparticles by DSS method are synthesized from solution near room temperature, hydrogen interstitials could be introduced into the perovskite lattice and carry positive charges. When hydrogen interstitials are introduced into perovskite lattice, the lattice cell of the powders will expand and change the crystal symmetry. In our experiments, all the lattice cell parameters of the nanoparticles synthesized by DSS method are larger than that of the bulk materials and have cubic symmetry. This indicates that the lattice cell has expanded and hydrogen interstitials have been introduced into the perovskite lattice.

It is reasonable to expect that the hydrogen interstitial will introduce positive charges into the lattice and therefore, the particles have to adsorb negatively charged hydroxyl groups to compensate and neutralize such positive charges as shown in Equation (4). If such particles are dispersed in water, then the solution must show an alkalescent property. On the other hand, if stoichiometric perovskite particles are suspended in water, the solution should keep neutral. In Figure 14, DSS-prepared BaTiO3 nanoparticles have been dispersed in water and the pH value is checked. Alkalescent property of the suspension is proved and this means the hydrogen interstitials in perovskite lattice are actually compensated by surface-charged hydroxyl groups. It should be noted that around the nanoparticles, alkalinity is much stronger than far away from them. This is due to the adsorption effect of hydrogen interstitials with positive charges in the lattice which attract hydroxyl group with negative charges on the surface.

Another test has been adopted to prove this assumption. BST nanoparticles prepared by DSS method have been divided into three parts and subjected to different annealing processes. They are i) the as-synthesized, ii) the 500°C-annealed and iii) the 1000°C-annealed powders, respectively. With dispersion of the particles in water, the three suspensions have been found to have three different pH values: i) 10, ii) 8 and iii) 7. While samples i) and ii) are obviously the cases discussed above, sample ii) actually corresponds to a case where the interstitial hydrogen has been partially removed by the thermal treatment.

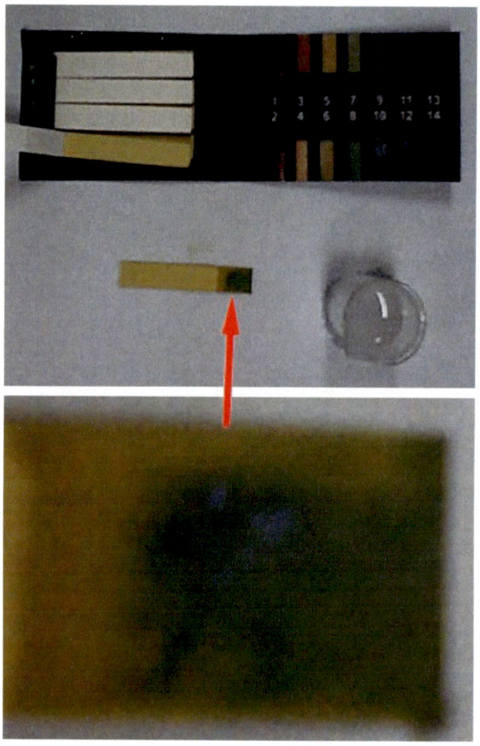

Figure 14. Checking the pH value of DSS-prepared $BaTiO_3$ nanoparticles dispersed in water.

The hydrogen interstitials have been removed in sample iii). All of these experiments are guided by the DSC-TG experiments of BST shown in Figure 9. Here we should note that hydrogen interstitials can be introduced into the lattice of perovskite nanoparticles synthesized by DSS method and behave as alkalescence. This can also happen in oxide particles prepared by other chemical routes such as hydrothermal method. However the alkalescence of perovskite nanoparticles by DSS method is much stronger than those by other methods because DSS method operates at much lower temperature compared to other methods.

2.2. Synthesis of Core-shell Structured BST Nanoparticles with MgO Shells

The perovskite nanoparticles prepared by DSS method behave as alkalescence, and can react with the solution of many salts. Because hydrogen interstitials will introduce positive charges into the lattice of perovskite nanoparticles obtained directly from solution and therefore the particles have to absorb negatively charged hydroxyl groups to compensate and neutralize such positive charges as shown in Equation (4). When this kind of perovskite nanopaticles are dispersed in a certain solution of inorganic salt such as Mg(NO3)2•6H2O, Zn(NO3)2•6H2O, Al(NO3)3•9H2O, Y(NO3)3•6H2O etc., the hydroxide related to these salts can spontaneously deposit on the surface of the nanoparticles. As a result, a perovskite structured nanoscaled core material will be successfully self-wrapped by a non-ferroelectric hydroxide derived from a nitrate salt solution. After annealing at high temperature, core-shell structured perovskite nanoparticles with heterogeneous oxide shells can be synthesized. Here nitrate salts are proposed in order to avoid some impurity ions, such as Cl-, SO42-, which are difficult to remove from the final products.

The flow chart of processing of self-wrapped core-shell structured perovskite nanoparticles with heterogeneous oxide shell is shown in Figure 15. Here we use BST(50/50) nanoparticles and magnesium nitrate hexahydrate [Mg(NO3)2·6H2O] as an example of perovskite nanoparticles and nitrate salt to illustrate the process. The BST nanoparticles prepared by DSS method is dispersed in ethanol by ball milling for 2 h. Mg(NO3)2·6H2O is dissolved into deionized water to form Mg-solution. This solution is added into the suspension and then subjected to ultrasonic agitation for 1 h. Ammonia is subsequently added to adjust the acidity of the mixture to near PH~11. The resulting MgO-wrapped BST mixture is filtered out, carefully washed several times by deionized water, and then dried and calcined at different temperatures.

A simultaneous thermal analyzer (NETZSCH STA 449C) was used for thermogravimetric analysis and differential scanning calorimetry (TGA and DSC). The weight of the sample is about 15 mg. The range in which the temperature increases is 30-1100oC. The measurements are made under oxygen (20 ml/min) with protective argon gas (10 ml/min) at a heating rate of 10oC/min. A phase analysis of the powder has been performed by X-ray diffraction (XRD) using a Rigaku D/Max-B diffractometer with CuKα radiation.

The powder sample is pressed on a glass plate and mounted vertically on the sample table, and diffraction patterns are recorded using CuKα (0.15406 nm) radiation. Data are collected in the 2θ range of 20o to 100o in steps of 0.02o using a scintillation detector. The morphology is observed by field emission scanning electron microscopy (SEM, JSM-6335F) and transmission electron microscopy (TEM, Philips CM20) working at 200 kV. The Fourier transform infrared spectroscopy (FTIR, Magna 760) has been used to determine the phase transitions that occur qualitatively during the syntheses. These spectra are recorded in the range of 400~4000 cm-1 at room temperature.

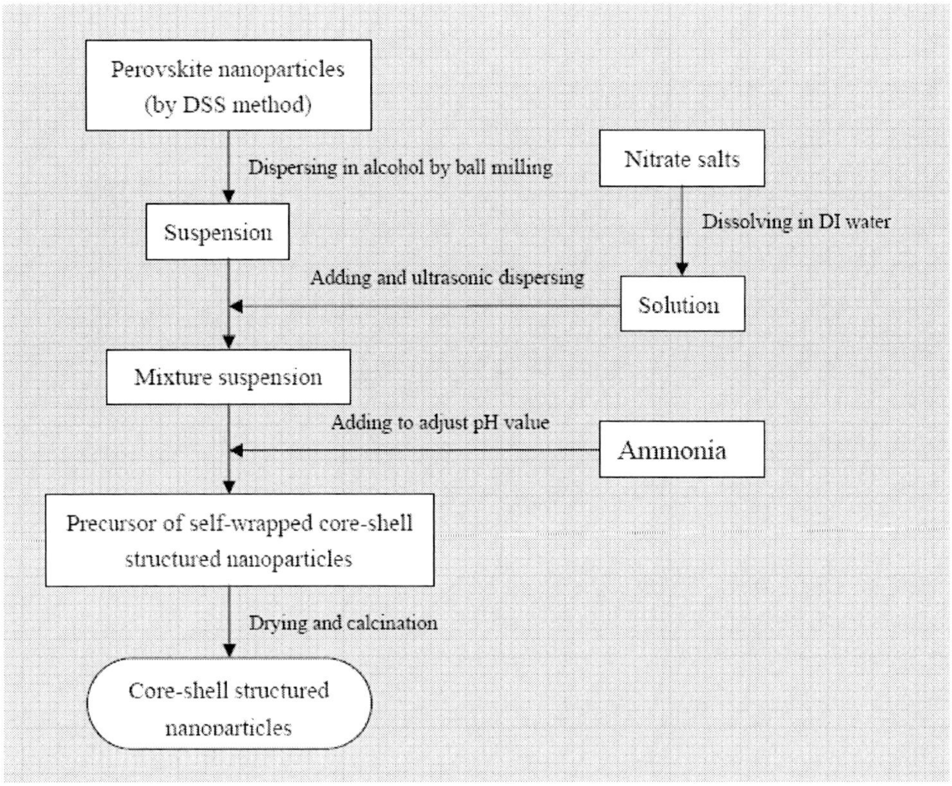

Figure 15. Processing of self-wrapped core-shell structured perovskite nanoparticles with heterogeneous oxide shell.

TGA and DSC are often employed as analysis of the temperature point of chemical reaction or phase transformation of a certain system. Figure 16 shows the TGA and DSC curves of as-prepared BST@MgO powder in flowing oxygen. The TGA trace indicates a three-step loss in the sample mass, corresponding to the evaporation of water and organic solvents, the reaction between metal ions and the surface of the BST core, and decomposition and crystallization of precursors. During the first step, the weight loss is close to 26.13%, which can be attributed to the evaporation of residual water and a small amount of organic solvent. Water can be found in the powder, even in the powder that has been dried at a higher temperature, probably because atmospheric H2O is trapped at the surface of the particles. It is reasonable to get the FTIR spectra belonging to the vibration bands of hydrogen bonding in the 500oC dried BST@MgO powder, although the spectra should disappear after the mixture is heated to over 100oC. It is speculated that the water does not only come from the process of synthesis. A sharp endothermic peak near 230.8oC is found in the DSC curve, which corresponds to the relatively large loss of mass in this step.

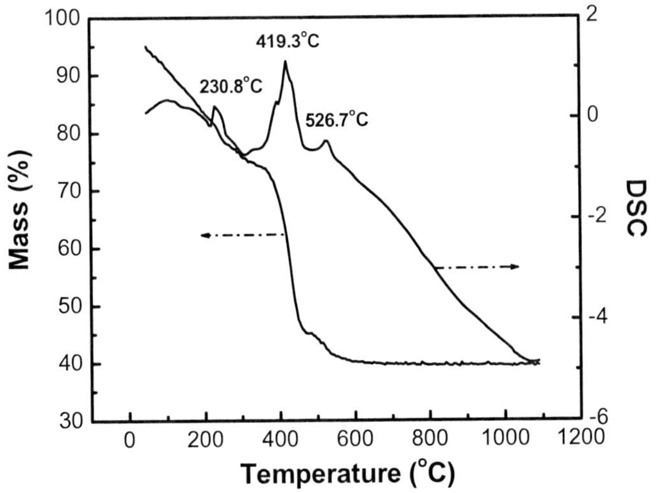

Figure 16. TGA and DSC curves of as-prepared BST(50/50)@MgO powder in flowing oxygen [15].

There is a sharp loss of mass from the onset temperature of 409oC, with the loss of mass being close to 29.09% in the narrow temperature range of 30oC. There is a sharp peak around 419.3oC, accompanied by two shoulder peaks in the DSC curve. We believe that the main chemical reactions occur in this temperature region. During synthesis process of core-shell structure in the suspension, the core with alkalescence can react with the acidic salt solution, and the hydroxide precursor can be self-wrapped onto the surface of the core. The shell will form after heating. As the temperature increases, one exothermic peak around 488.6oC and a weak endothermic peak around 526.7oC are shown in the DSC curve in the third step corresponding to a mass loss of 4.93%. This is attributed to decomposition and crystallization in the forming of MgO as a final shell.

Figure 17 shows the FTIR reflectivity spectra of: (a) as-prepared BST, (b) as-prepared BST@MgO, and (c) 500oC annealed BST@MgO between 4000~400 cm-1 at room temperature. The main bands are clearly shown in the spectra as follows: Metal-oxygen (M-O) bonding between 400 and 800 cm-1; Carbon-oxygen bonding between 1300 and 1700 cm-1 and around 2400 cm-1; C-H stretching vibrations between 2800 and 3000 cm-1 and weak nitrate peaks at about 1774 cm-1; and hydrogen bonding at around 3500 cm-1. The broad peak at 3522 cm-1 and one intensive peak at 1502 cm-1, shown in Figure 17(a), are due to the excitation of stretching and to the deformation vibration modes of hydroxide ions. These peaks are found at around 3536 cm-1, 3330 cm-1, 1502 cm-1, and 3499 cm-1, 1466 cm-1, 1643 cm-1 in the as-prepared BST@MgO powder and BST@MgO powder annealed at 500oC, as shown in Figure 17 (b) and (c). These peaks are assigned to the hydrogen bond between H2O and ethanol, which will become much weaker as the temperature increases because of the evaporations of the solvents. In Figure 17 (b) and (c), the absorption peak at 1678, 1774 cm-1 in curve (b), and only one peak at 1774 cm-1 in curve (c) are assigned to the nitrates group ($O-O-NO_2$ or $O-N=O$) because a small amount of residue of nitrate radicals has not been thoroughly removed after the mixtures have been washed several times.

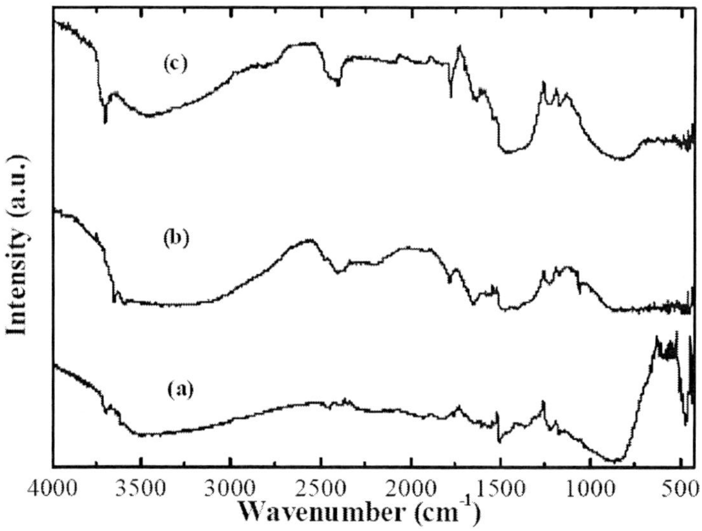

Figure 17. FTIR reflectivity spectra of the samples. (a) as-prepared BST(50/50), (b) as-prepared BST(50/50)@MgO, and (c) 500°C annealed BST(50/50)@MgO [15].

The bands at 2800-3000 cm-1 are due to C-H stretching vibrations, whereas the intensities are weaker in the as-prepared BST powder and as-prepared BST@MgO powder compared to those in the 500oC annealed BST@MgO powder. This is an intriguing phenomenon because the intensity should decrease as the ethanol and other organic solvents in these mixtures evaporate. Carbon-oxygen bonding presents between 1300 and 1700 cm-1, and entrapped atmospheric CO2 at 2455 cm-1 in curve (a), 2357 cm-1 in curve (b) and 2405 cm-1 in curve (c). The peaks 847 cm-1 in curve (a), 802 cm-1 in curve (b), and 825 cm-1 in curve (c) are typical deformation and stretching vibration modes of carbonate ions [73,74], although Liu et al. assigned these peaks to the N-O bond vibrations of NO3-[75]. It is reasonable to rule out the N-O bond vibrations of NO3- in this range, and N-O bonding vibrations should not be found in the as-prepared BST powder and BST@MgO powder after being annealed at 500oC. Naturally, these peaks around 802~847cm-1 should be caused by other bonding groups.

XRD investigations have shown that the as-precipitated BST powder is crystallized in a single phase (Ba0.5Sr0.5TiO3), as shown in Figure 18(a). No other phases, such as BaO, SrO, and TiO2 have been detected except for a weak peak of (Ba,Sr)CO3. BaCO3 is also found in the nano-BaTiO3 powder prepared using the DSS method [23]. This result is consistent with the FTIR spectra, a peak at 2455 cm-1 for the as-prepared BST powder, which is assigned to C-O bonding by entrapping atmospheric CO2 in the mixtures. The composition of the powder has been checked by energy dispersive X-ray spectroscopy (EDX) during the SEM and TEM measurements. The EDX analysis of the powder shows no traces of other ions, as the stoichiometry of as-prepared BST is Ba:Sr:Ti=10:9.16:19.22, near (Ba+Sr)/Ti=1 within the allowed measuring errors. From XRD patterns, the lattice parameters of BST decrease as the treating temperature increases. This is partly because of the release of hydrogen ions from the lattice and/or formation of vacancy defects. The lattice parameter of the as-prepared BST is about 3.9535 Å, decreasing to 3.9443 Å for BST in the 500oC annealed BST@MgO. Other results for sintering discs are derived from the as-prepared BST@MgO powder at higher temperatures, from 1100oC to 1300oC. The lattice parameters decrease from 3.9457 Ǻ (1100oC) to 3.9363 Ǻ (1300oC) (they are not shown here). In Figure 18(b), some peaks of NH4NO3 are presented in the XRD pattern. It is obvious that the nitrates are present, although the precipitations are washed several times with water. Naturally, we find the vibration bands of the nitrate group in the FTIR spectra (as shown in Figure 17 (b,c)).

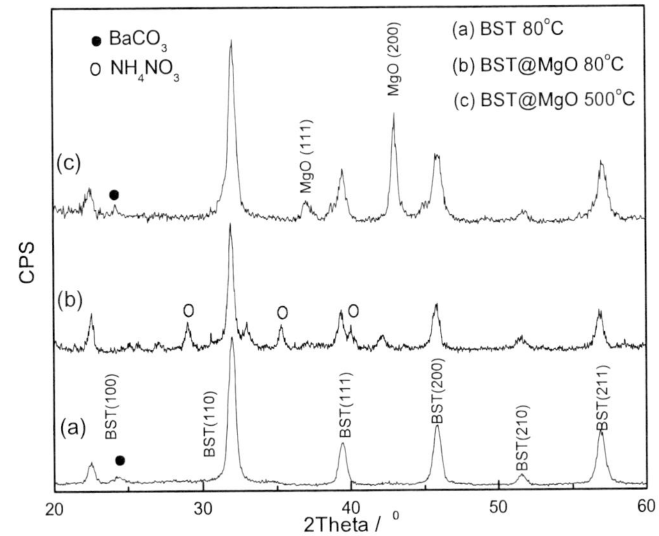

Figure 18. XRD patterns of : (a) as-prepared BST(50/50), (b) as-prepared BST(50/50)@MgO, and (c) 500°C annealed BST(50/50)@MgO [15].

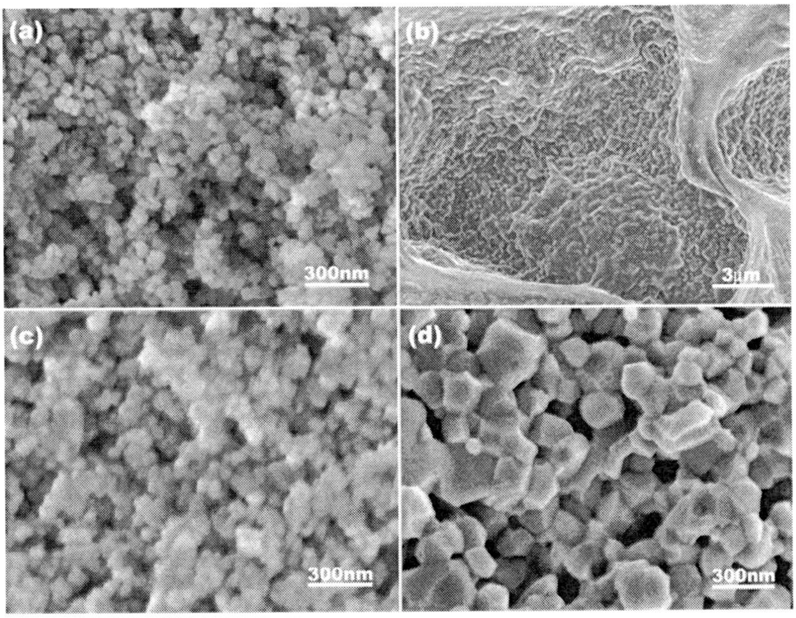

Figure 19. FE-SEM image, (a) BST(50/50) nanoparticles prepared at 70°C, (b) Interim composite of MgO precursor wrapped BST nanoparticles, (c) MgO wrapped BST(50/50) nanopaticles annealed at 500°C, and (d) MgO wrapped BST(50/50) nanopaticles annealed at 1100°C.

These peaks will become much weaker, and even disappear when the powders are heated at higher temperatures. MgO phase will become much stronger as the powders are heated at higher temperatures. Figure 19 shows the surface morphology of the nanoscaled BST powder and MgO-wrapped BST heat treated at different temperatures. The SEM observation shows that the as-prepared BST powder has an average particle size of 20 nm (Figure 19(a)). The particle size of BST powder calculated by the Scherrer's equation (Equation (1)) is 17.3 nm (as-prepared BST powder), while the particle size of BST in the composites, i.e., MgO-wrapped BST, shows a slight decrease from 17.9 nm (as-prepared BST@MgO) to 16 nm (BST@MgO annealed at 500oC). However, the size of the particle in the BST@MgO ceramic sintered at 1100oC increases to 22.6 nm. The ambiguous structure of the as-prepared BST@MgO powder is shown in Figure 19(b), possibly due to the amorphous structure of MgO caused by hydrolyzation and some organic residues or nitrates in it. Crystallized particles are shown to be embedded in an amorphous shell [65]. As temperature increases, all solvents and organic residues disappear, and well-crystallized MgO-wrapped BST will form. As shown in Figure 19(c) and (d), aggregated particles of about 100~150 nm (500oC) and several hundred nanometers (1100oC) consisting of more than one core-shell structure will form as crystalline growth is driven by the thermodynamic effect.

The fine details of the particles and their morphology have been investigated using a Philips TEM. The bright field electron micrograph of the BST powder is shown in Figure 20 (a). The average particle diameter is found to be approximately 20 nm. Figure 20 (b) shows the TEM images of BST@MgO annealed at 500oC. The particles from these images are not homogeneous in size, being between 40~200 nm. A big particle and its surroundings have been chosen to measure the EDX spectra to confirm the composition and its variations from core to shell. The spots are marked by "I," "II," and "III," as shown in Figure 20 (c). The results of the atomic percentage of the elements in BST@MgO are listed in Table I.

Table 1. Atomic percentage of the elements in BST@MgO[15]

Selected spot	Atomic %				Molar ratio of Mg/Ti
	Mg	Ti	Sr	Ba	
I	21.83	0.01	0.03	0.01	Without BST
II	13.9	10.4	4.2	7.4	1.34
III	8.12	19.22	10	9.16	0.42

According to the molar ratio of these spots, there is a radial chemical gradient from the outer spot (marked "I") to the boundary (marked "II") and then to the core of the grains (marked "III"). It is clear that a ferroelectric nanoscaled core, BST, in a thin surrounding shell, MgO, has been prepared successfully. As mentioned above, the outer shell consists of MgO. Figure 20(d) shows the SAED pattern recorded from an area containing a core-shell structure (BST@MgO) such as spot (III) in Figure 20(c). Two sets of diffraction patterns are found, which correspond to MgO and BST, respectively. There is a spot of diffraction plane of MgO beside that of the same diffraction plane of BST, such as labeled as M(220) for MgO and P(220) for BST, respectively in the figure, showing that the lattice cells are matched between the core and the shell. Therefore, BST nanoparticles synthesized by DSS method can be self-wrapped by the precursor of MgO such as amorphous Mg(OH)2 in magnesium nitrate solution and then form core-shell structures after annealed at 500oC.

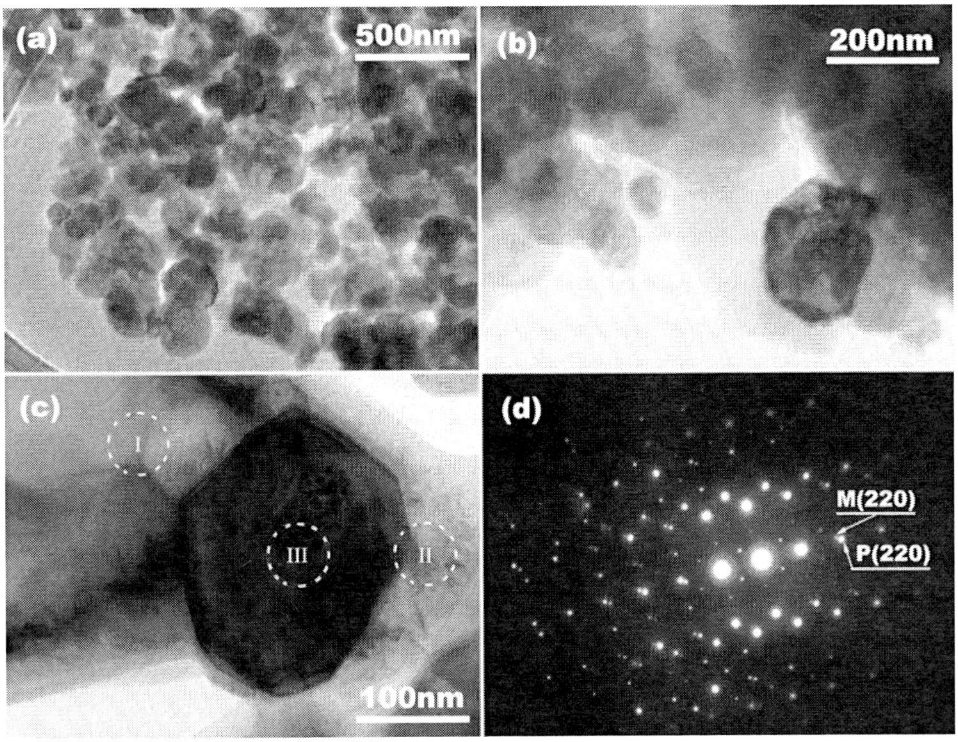

Figure 20. TEM image (a) BST(50/50) nanoparticles, (b) BST@MgO annealed at 500°C, (c) BST@MgO (a big particle and its surroundings), (d) SEAD of the BST@MgO (focus on (III) in (c)).

2.3. Other Patterns of Core-shell Structured Perovskite Nanoparticles with Heterogeneous Shells

Other patterns of core-shell structured perovskite nanoparticles with heterogeneous shell can be synthesized similar to that of BST@MgO. If the core is changed to BT or BZT, all the processing is the same as shown in Figure 15. If we use Mg(NO3)2 and Zn(NO3)2 as starting nitrates, the heterogeneous shell will change to Zn-doped MgO (MZO) and this kind of shell will improve the sintering characteristics of MgO. Thus, it can be expected that BT@MgO, BZ@MgO, ST@MgO, BZT@MgO and BST@MZO etc. can be synthesized by this method. If Mg(NO3)2 and Al(NO3)3 are used as the starting nitrates, spinel heterogeneous shell can be formed. Figure 21 shows the XRD pattern of BST@MgAl2O4 calcined at 1100oC. It is clearly revealed that spinel MgAl2O4 shell has been synthesized after calcined at 1100oC. There are two minor impurity phases indexed in this system, one is BaAl2O4 and the other is MgO. These two phases do not affect the electrical properties of the system and can be eliminated after a higher temperature annealing.

The shell with a spinel phase is not as sensitive to moisture as pure MgO. If the devices are fabricated using this kind of materials, much higher reliability can be expected than MgO wrapped materials.

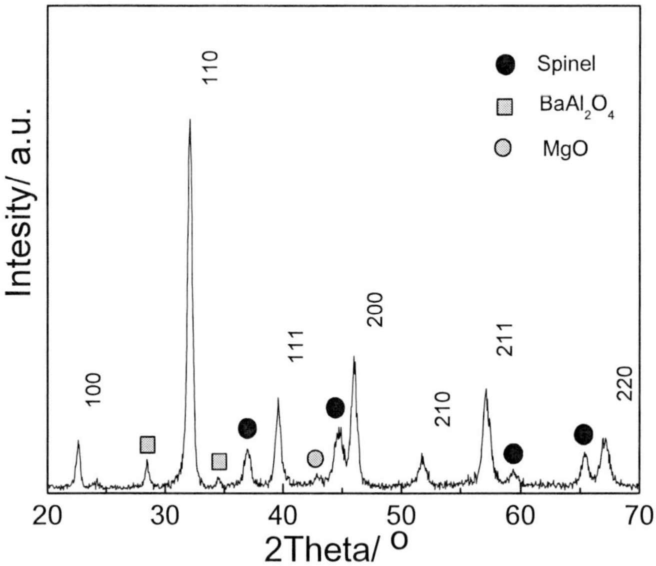

Figure 21. XRD pattern of BST(50/50)@$MgAl_2O_4$ calcined at 1100°C.

3. Ceramics of Core-Shell Structured Perovskite Nanoparticles with Heterogeneous Oxide Shells

It is very important to tailor either the grain size of the ferroelectric perovskite phase or the shell thickness of heterogeneous oxide shell. The ceramic materials with this kind of core-shell structures can be obtained by sintering of core-shell structured perovskite nanoparticles with heterogeneous oxide shells and can be applied to many electrical devices. However, it is difficult to obtain the dense ceramics if the powders with this kind of structures are directly sintered. Because inert oxide such as MgO etc. is often used as the heterogeneous shell in order to obtain very low dielectric loss. However, this kind of inert oxides with high melting point is very difficult to be sintered. In the conventional composite, it is well known that the sintering characteristics depend on the easily sintered phase, e.g., in the MgO-BST system, BST can be easily sintered at much lower temperatures than MgO. In MgO-wrapped BST, unfortunately, the sintering behavior can be dominated by MgO, making it difficult to be sintered at a low temperature like BST. Another reason of making the core-shell structured materials difficult to be sintered is that nanoparticles exert a pinning effect during sintering and thus the mass transfer can be suppressed from shell to shell. BST core in MgO can be considered as inclusions and increase the energy necessary for the movement of a grain boundary and inhibit grain growth during sintering. Here some efforts are focused on decreasing the sintering temperature and obtaining good ceramics.

Core-shell structured 40%wt MgO wrapped BST nanoparticles have been prepared following procedures shown in Figure 15. The powders of MgO wrapped BST nanoparticles have been pressed into disks with 1 mm thickness and 11.78 mm diameter under the pressure of 1000 kg/cm2. Then the samples are sintered at different temperature from 1250 to 1450oC to study their sintering characteristics. The firing shrinkages are listed in Table 2 and all of them are less than 10%. All the samples are porous and can be easily crushed. Their strength is bad.

Table 2. Firing shrinkage of samples (diameter of green ware is 11.78mm)

Sintering temperature	1250°C	1350°C	1450°C
Diameter of sintered samples	11.10mm	10.92mm	10.74mm
Firing shrinkage of samples	5.8%	7.3%	8.8%

In general, sintering aids will help in the densification of difficult-to-sinter ceramics by introducing SiO2 related secondary phase inclusion during sintering as reactive liquid. However, the results are not as good as expected when we use SiO2 as sintering aid in the system of core-shell structured MgO wrapped BST nanoparticles. The sintering has not been improved after adding the sintering aids which results in frothy samples. Firing shrinkage is only improved slightly. Another strategy is to improve the sintering characteristics of MgO shell. We try to substitute MgO with doped MgO. Of course the dopants should be restricted to a certain range.

3.1. ZnO Doped MgO

MgO and ZnO have rocksalt and wurtzite structures, respectively.[76] Since they have different coordination numbers and crystal symmetry, it is expected that Zn will replace Mg without changing the original structure up to a certain amount when alloying. According to the phase diagram, MgO allows a maximum of 56% ZnO (wt %, 40 at.%) at 1600 oC, and maintains its NaCl structure with the lattice constant staying close to that of pure MgO (4.208 Å). The bandgaps of ZnO doped MgO are listed in Table 3.[77] Being a kind of material with a wide bandgap, the slight doping will not change the electrical properties greatly. Thus, substitute of MgO by ZnO doped MgO can be accepted as a methodology for our applications.

Table 3. Bandgaps of ZnO doped MgO [77]

Mg atomic fraction	Bandgap energy (eV)
1	~7.7
0.88	>5.31
0.85	5.31
0.65	4.70
0.5	4.00
0.45	3.87
0.38	3.70
0.30	3.42
0.15	3.37
0	3.30

The conventional ceramic processing has been adopted to fabricate MgO solid solution. We use x=0.1 and 0.3, calcination temperature is 800oC. From Figure 22, we can see that x=0.1 at 800oC, only the pure rocksalt phase can be obtained, but the solution is not completed when x=0.3 at this temperature.

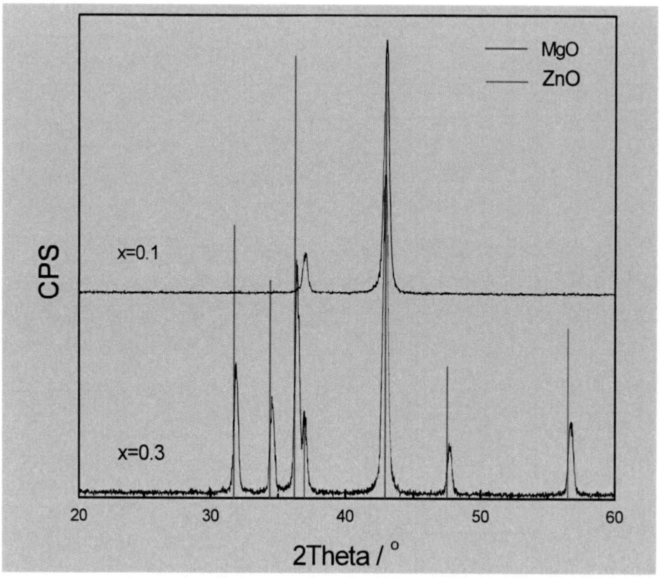

Figure 22. ZnO doped MgO calcined at 800°C.

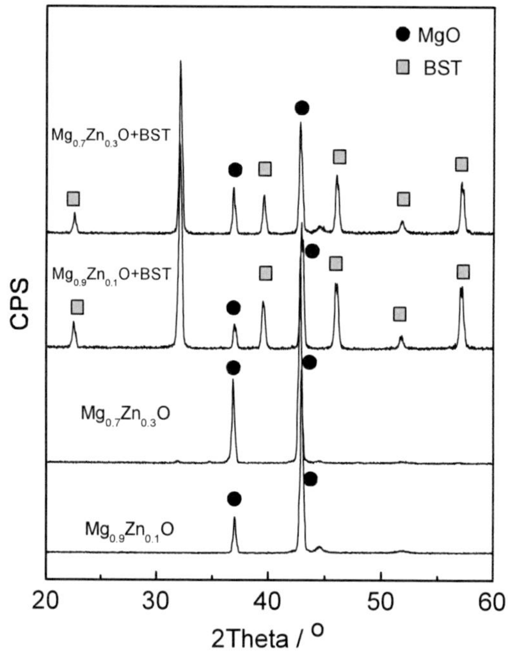

Figure 23. MZO and MZO-BST(50/50) composite ceramics sintered at 1250°C.

A long calcination time and a high temperature will promote the formation of solid solution. When the temperature increases to 1250oC, we obtain almost pure rocksalt MZO. There is no ZnO phase left. MZO is inert to BST, so that they do not react at high temperature. Composite of MZO and BST can be sintered into dense ceramics and keep their own phase respectively as shown in Figure 23. Firing shrinkage is listed in Table 4 and all the samples are dense.

Table 4. Firing shrinkage of MZO and composite of MZO-BST

Temperature	$Mg_{0.9}Zn_{0.1}O$	$Mg_{0.7}Zn_{0.3}O$	$60\%Mg_{0.9}Zn_{0.1}O+40\%BST$
1250°C	9.22mm/21.7%	8.97mm / 23.9%	9.64mm / 18.2%
1350°C	9.05mm / 23.2%		

3.2. Ceramics with Core-Shell Structured BST Nanoparticles with ZnO Doped MgO Shells

The advantages of core shell structures are that their properties can be conveniently tuned and tailored by designing the chemical compositions as well as sizes of the core and the shell. A strategy is synthesis of MgO wrapped BST core-shell structured nanoparticles using nanotechnology and obtaining their ceramics or films to improve the properties. In order to improve the sintering characteristics of this kind of materials, ZnO dopant has been added into the shell.

The flow chart of fabrication of MZO wrapped BST nanoparticles is similar to that of MgO wrapped one as shown in Figure 15. The BST nanoparticles are dispersed in distilled water by ball-grinding for 2 h. The solutions of magnesium nitrate hexahydrate (Mg(NO3)2·6H2O) and zinc nitrate hexahydrate (Zn(NO3)2·6H2O) with certain ratio to BST are added to the BST suspension under vigorous stirring, and then subject to ultrasonic agitation for 1 h. Ammonia is subsequently added to adjust the acidity of the mixture to near pH~11. The resulting MZO-wrapped BST mixture is filtered out, carefully wash several times by deionized water, and then dry at 80oC. In order to keep high resistivity and low electric loss of the shell but improve sintering distinctly, the fraction of ZnO in MgO is adopted as low as possible (here just 10%). The powders with the core of BST nanoparticles and the shell of MZO precursors are annealed at different temperatures from 500~1000oC in order to obtain the crystallized MZO shell. The MZO wrapped BST nanoparticle is pressed into disks with 1 mm thickness and 10 mm diameter and sintered at 1350oC to obtain the ceramics.

Table 5 lists the firing shrinkage of the samples of MZO wrapped BST nanoparticles with different ratio of MZO:BST. The sintering characteristics of MZO is improved distinctly after doped with ZnO, but that of core-shell structured powders is improved slightly (see Table 3). Here BST nanoparticles exert a pinning effect during sintering and thus the mass transfer can be suppressed from shell to shell. BST core in MZO can be considered as inclusions and increases the energy necessary for the movement of a grain boundary and inhibit grain growth during sintering.

In order to improve the sintering characteristics of this kind of materials, a strategy as shown schematically in Figure 24 has been adopted. When the core-shell structured MgO wrapped BST nanoparticles are composited with MZO powders as shown in Figure 24(b), the mass transfer can proceed from MZO to the shell MgO (or MZO) of BST nanoparticles without pinning effect as shown in Figure 24(c). Thus, dense ceramics can be obtained under this strategy as shown in Figure 24(d). Here is a typical example, core-shell structured MZO wrapped BST nanoparticles (5%wtMZO) are composited with 35%wt MZO conventional powders, firing shrinkage reach 25% when the sintering temperature is 1350oC (diameter shrinks to 8.87 from 11.78 mm) and thus obtained dense ceramics. At the same time, when all MZO are wrapped on the BST nanoparticles as shells, firing shrinkage is only 11% as shown in Table 5.

Table 5. Firing shrinkage of MZO(90/10) and its wrapped BST(50/50) nanoparticles(sintering at 1350oC, diameter of green disk is 11.78 mm)

Compositions	Pure MZO	20%wt MZO+80%BST	40%wtMZO+60%BST
Diameter/mm	9.05	10.35	10.50
Firing shrinkage	23%	12%	11%

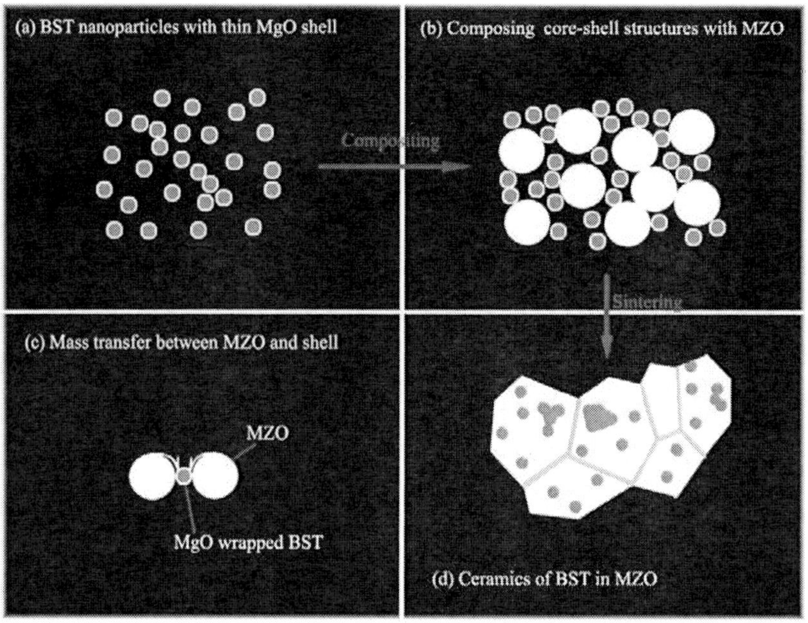

Figure 24. Schematic diagram showing the sintering process in MZO wrapped BST nanoparticles.

3.3. Analyzing the Core-Shell Structured BST Nanoparticles with ZnO Doped MgO Shells

The FE-SEM fractograph image of the ceramics of MZO wrapped core-shell structures prepared using the strategy shown in Figure 24, is shown in Figure 25. Both fractured styles of fracture along grain boundaries and fracture through grains, can be observed in the image. Some un-grown BST nanoparticles which keep their size similar to the as-prepared ones can also be observed in fractured grains, but on the boundary, it can hardly be observed. This confirmed our sintering strategy for MZO wrapped BST nanoparticles. On the other hand, it is also confirmed that the inert MZO shell can protect the core of BST nanoparticles and inhibit the grain growth.

Figure 25. Fractograph of the ceramics of MZO(90/10) wrapped BST(50/50) core-shell structures sintered at 1350°C [81].

The phases of the sample can be checked by XRD analysis as shown in Figure 26. In order to check the MgO phase clearly, The XRD profile of halite structured MZO ceramics is also given in Figure 26 (a). As shown in Figure 26 (b), because the MZO precursor is amorphous, there are no halite phase peaks in the as-prepared powders. Many other peaks are indexed as ammonia nitrate except for BST peaks After annealing or sintering at high temperature, the MZO precursor crystallized in halite structure as shown in Figure 26 (b), (c), (d), and (e), respectively.

The lattice cell parameters can be estimated from the peak positions and for a cubic crystal,

$$a = d\sqrt{h^2 + k^2 + l^2} \tag{5}$$

where a is lattice parameter and d is the distance between a certain diffraction crystalline plane and (h k l) crystalline plane index d can be calculated from Bragg's equation:

$$\lambda = 2d\sin\theta \tag{6}$$

where λis X-ray wavelength and θ is the diffraction angle.

The crystallite size of the powders can be estimated from the broadening of corresponding X-ray spectral peaks by the Scherrer's formula as shown in Equation (1). In the calculation the XRD peak of (200) plane of halite structured MZO (2θ~43) and (200) plane of perovskite structured BST (2θ~46) have been used.

The XRD profiles of (200) plane of two phases in different samples annealed at different temperatures, which have been regressed by Gaussian fitting, are shown in Figure 27. The annealing temperature dependence of the lattice cell parameters and crystallite size estimated by XRD data from Figure 27 are shown in Figures 28 and 29, respectively.

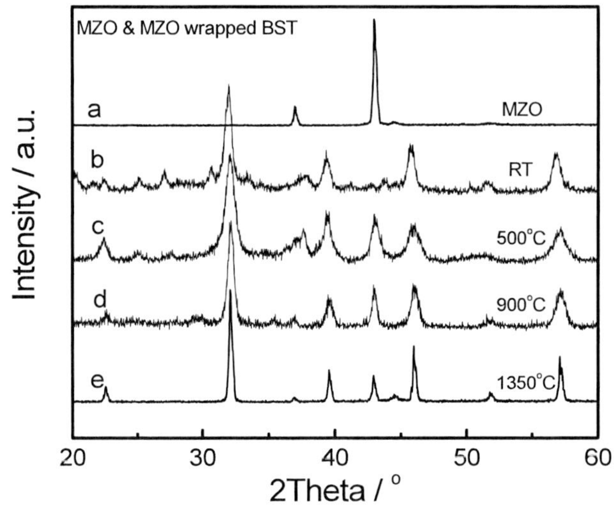

Figure 26. XRD profile of MZO(90/10) wrapped BST(50/50) nanoparticles annealing at different temperatures, (a)MZO ceramics, (b) as-prepared MZO precursor wrapped BST nanoparticles, (c) MZO wrapped BST nanoparticles annealed at 500°C, (d) MZO wrapped BST nanoparticles annealed at 900°C, (e) Ceramics of MZO wrapped BST nanoparticles sintered at 1350°C [81].

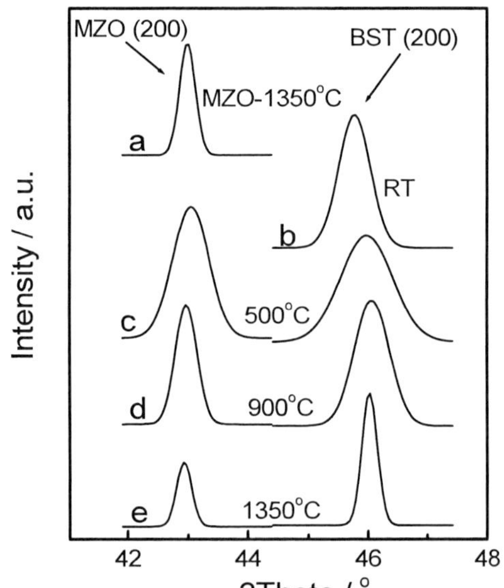

Figure 27. XRD profiles of (200) crystalline plane of (a) MZO(90/10) ceramics, (b) perovskite peak in MZO(90/10) precursor wrapped BST(50/50) nanoparticles, (c) MZO(90/10) wrapped BST(50/50) naoparticles annealed at 500°C (d) MZO(90/10) wrapped BST(50/50) nanoparticles annealed at 900°C, (e) Ceramics of MZO(90/10) wrapped BST(50/50) nanoparticles sintered at 1350°C [81].

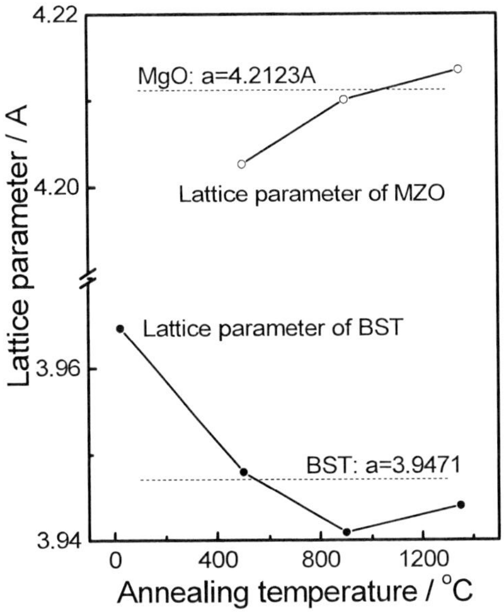

Figure 28. Lattice parameters of BST(50/50) core and MZO(90/10) shell as a function of annealing temperature.[81]

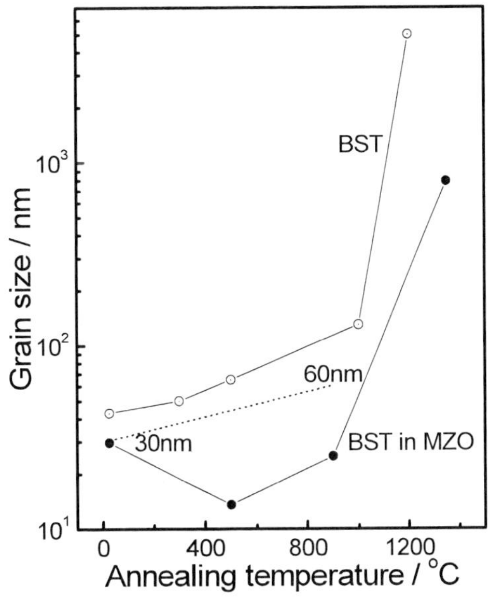

Figure 29. Annealing temperature dependance of nominal crystallite size of BST(50/50) as core in MZO(90/10) shell (solid circles) and that of pure BST(50/50) nanoparticles (hollow circles) [81].

The lattice cell parameter of MZO, aMZO, increases with annealing temperature. Because Zn ion is bigger (0.74A) than the Mg ion (0.72A) and cannot be dissolved into the MgO lattice completely at low temperature, the lattice cell parameter is small at 500oC. When the annealing temperature increases, more Zn ions are introduced into the lattice, so that the lattice cell of MZO expands. It is noted that the BST lattice cell is smaller than that in JCPDS cards (#39-1395) at high temperature because the lattice cell of BST core is subjected to inner stress caused by the MZO shell crystallization from amorphous to halite (rocksalt) structure.

It is easily understood that the crystallite size of pure BST increases with the annealing temperature for the grain growth even at moderately high temperature because nanoparticles have much greater chemical activity than conventional powders as shown in Figure 29. However, it should be noted that the nominal crystallite size of BST as core in MZO estimated from XRD data decreases before 500oC and then increases with annealing temperature and the size for the sample annealed at 900oC is smaller than the initial value. After the BST nanoparticles have been wrapped in MZO and the MZO crystallized from amorphous to halite structure, the lattice mismatch can cause inner stress and lead to a variation of lattice cell parameter which can cause XRD peaks broadening. Therefore, we should consider this factor when we estimate the crystallite size by XRD peak broadening effects. Different mathematical methods exist (e.g. Langford,[78] Balzar and Ledbetter,[79] Warren-Averbach[80]), which take into account strain in the analysis of XRD data and these methods use the integral breadth or the entire peak profile. However, here we focus much attention on the lattice parameter variation and the half width by Gaussian peak fits is used to estimate it in order to simplify the model. Given a variation of the lattice cell parameter δa, value of d changes by δd. According to Bragg's equation (Equation (6)),θ can change by δθ as Equation (7).

$$\delta d = -\frac{\lambda \cos\theta}{2\sin^2\theta}\delta\theta \quad \textbf{(7)}$$

Here, the minus sign implies that when the lattice cell is subjected to inner stress and shrank, the peak position shifts to higher angles. Because a cubic lattice cell parameter can be calculated using Equation (8):

$$\frac{1}{d^2} = \frac{h^2 + k^2 + l^2}{a^2} \quad \textbf{(8),}$$

the variation of lattice cell parameter can be given in Equation (9):

$$\delta a = \delta d\sqrt{h^2 + k^2 + l^2} \quad \textbf{(9).}$$

Using Equation (7), the broadening of XRD peak caused by lattice cell variation can be estimated using Equation (10):

$$\delta\theta = -\frac{2\sin\theta \, tg\theta}{\lambda\sqrt{h^2 + k^2 + l^2}}\delta a \quad \textbf{(10)}$$

or the variation of lattice cell parameter can be given by Equation (11):

$$\delta a = -\frac{\lambda\sqrt{h^2 + k^2 + l^2}}{2\sin\theta \, tg\theta}\delta\theta \quad \textbf{(11)}$$

where $\delta\theta$ is the line width at half-maxiumum height of the XRD peak after subtraction of equipment broadening and nano effect broadening β (refer to Scherrer's formula, Equation (1)). Because in our core-shell structure, the MZO shells form inert barriers around the BST nanoparticles, it prohibits the grain growth of BST nanoparticles before sintering started at low annealing temperatures, e.g., <1000°C. This conclusion can be confirmed by our observation of HRTEM. Therefore, the change of XRD peak broadening of BST caused by grain growth at low annealing temperature can be ignored. The variation of the lattice cell parameters is responsible for the increase in broadening of XRD peak compared with the initial as-prepared BST nanoparticles so that we can estimate it according to Equation (11) [81] as shown in Figure 30.

The maximum lattice cell variation is estimated to be 0.056Å, almost about 1.5% of 3.958Å. It means that the BST lattice parameter of some lattice cell in the nanosized BST core can decrease to 3.902Å from 3.958Å (calculated by the XRD peak position, refer to Figure 30). Therefore, the lattice mismatch in MZO wrapped BST nanoparticles is great and cause inner stress leading to shrinkage and variation in the lattice cell of BST core, which can be estimated by the XRD peak position and broadening effect.

There are two possible modes of lattice matching between MgO and BST lattice as shown in Figure 31. The bright bars in the Figure represent a certain match plane. The first mode, (I), is that the BST lattice matches MgO lattice one by one and insert dislocations when mismatch is great enough. The BST cell will expand in this mode. The second mode, (II), is that the BST lattice mismatches the MgO lattice and inserts dislocation to match MgO. The BST cell should shrink in this mode. From XRD analyzing results, the core-shell structure probably follows the second mode because the lattice cell parameter of BST nanoparticles decreases after wrapped by MZO.

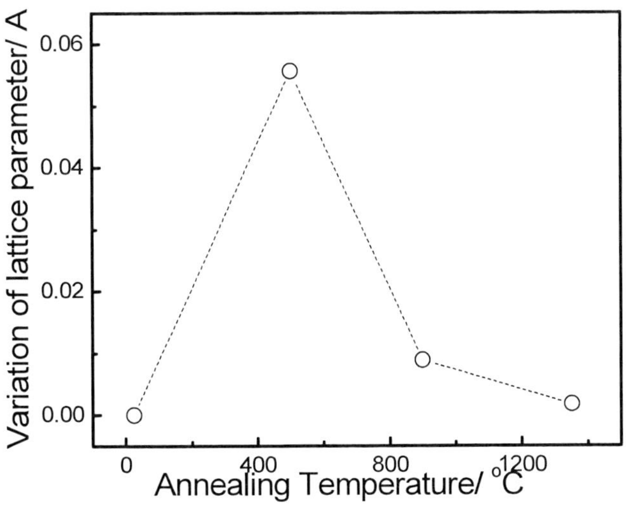

Figure 30. Variation of lattice cell parameters of BST core as a function of annealing temperature [81].

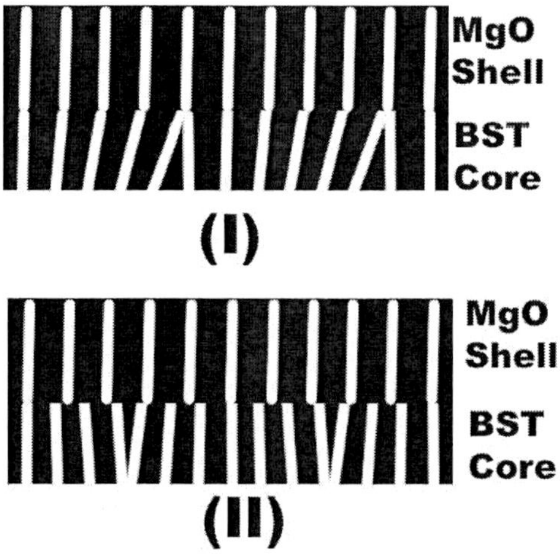

Figure 31. The lattice match modes between BST core and MZO shell [81].

The interface of BST core and MZO shell is particularly important because of its sharpness, lattice mismatch, and chemical gradient are critical for electron transfer and coupling. An epitaxial orientation relationship between the core and the shell is favorable, but an epitaxial growth is determined by their crystal structures.[1] The HRTEM image of MZO wrapped BST core-shell structure is shown in Figure 32. The interface of the core-shell structure is clearly shown in Figure 32(a). The lattice structure of MZO shell and BST core framed in Figure 32(a) are shown in Figure 32(b) and (c), respectively.

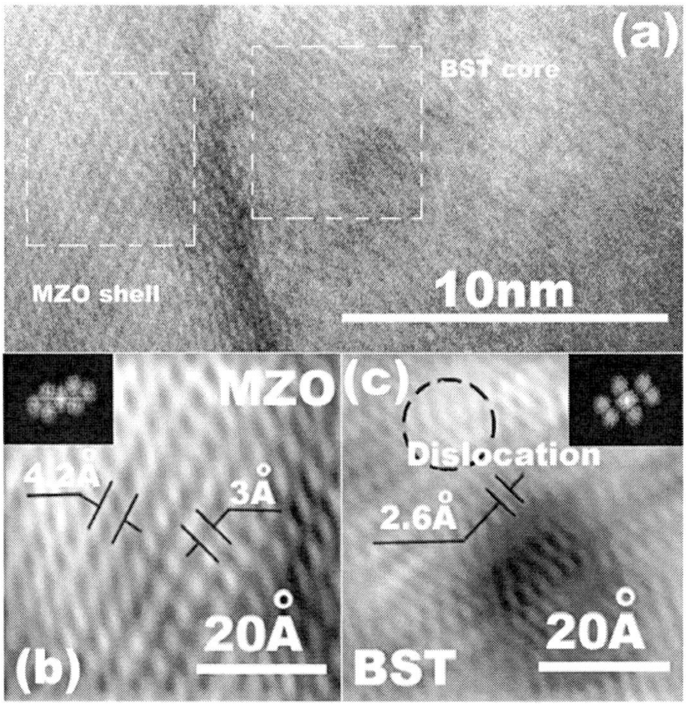

Figure 32. HRTEM image of MZO(90/10) wrapped BST(50/50) core-shell structure. (a) Interface of core-shell structure, (b) lattice of MZO shell, (c) lattice of BST core [81].

Figures 32(b) and (c) are obtained by fast Fourier transform (FFT) and smoothes the edge using a mask (as shown in insets of Figure 32(b) and (c), respectively) and then inverse FFT. It can be observed clearly that the MZO shell and BST core match along their (110) planes. The BST lattice cell is smaller than the normal one especially near the match boundary. There are many dislocations in BST lattice cell, and the integrity of MZO lattice is better than that of BST core. The distance between (110) planes is estimated as 2.6Å and the lattice cell parameter of BST is only 3.7Å. Therefore, the lattice cell has been depressed by the MZO shell as confirmed by XRD analysis.

3.4. Characterizing Electrical Properties of the Ceramics of the Core–shell Structured BST Nanoparticles with ZnO Doped MgO Shells

Relative permittivity and losses can be measured using a Stanford Research Systems model SR 720 LCR meter with the bias input by a Keithley 6517A electrometer/high resistance meter. In our experiments, the sample was fixed in an Oxford Cryostat with an intelligent temperature controller from 77 K to 400 K. High-frequency dielectric properties were measured using HP 4291B (1 MHz ~ 1.86 GHz) analyzer fitted with HP 16453 dielectric material testing fixture [14].

Core-shell structured perovskite nanoparticles with heterogeneous oxide shell may have advantages of both the core and the shell and have potentials for some important applications, such as microwave devices and high energy density ceramic capacitors. For high-frequency tunable device applications, a high dielectric constant is not desired for device impedance matching purposes, which leads to less efficient transfers of power to the device, thereby degrading the performance of the device. Moreover, a high dielectric loss also results in inferior device performance. Thus, the field dependence of the dielectric constant and loss should be evaluated. The tunability is expressed as:

$$T = \frac{\varepsilon(0) - \varepsilon(V)}{\varepsilon(0)} \times 100\% \quad \textbf{(12)}$$

where ε(V) and ε(0) denote the field-dependent and zero-field dielectric constants, respectively. The tunability as a function of temperature is shown in Figure 33. The figure of merit (FOM) is an important factor to demonstrate the quality of the ceramics. The FOM (K) is calculated based on the expression given below [82]:

$$K = \frac{T}{\tan\delta(0)} \quad \textbf{(13)}$$

where $\tan\delta(0)$ denotes the dielectric loss without a biased field on the samples.

In order to determine the temperature dependence of dielectric constant, we measured the dielectric constant and loss of the samples at 100 kHz from -200 °C to 50 °C under a biased field of 5 kV/cm. The tunability and FOM have also been calculated using Equations. (12) and (13). Figure 33 shows the tunability and FOM as functions of temperature. Relatively higher tunability has been obtained below the Curie temperature (T_c). The tunability is between 0.2% and 3.2% at room temperature (20~50°C), and the FOM value is between 0.8 and 12.8. The highest value in the tunability is close to 32% at -195°C, and the highest FOM value is close to 75 around the Curie temperature, as shown in Figure 33(a). The tunability of BST(50/50)-40%MZO(90/10) is very stable except for a sharp variation around the Curie temperature. The tunability is between 5.2% and 6.2% at room temperature, and the FOM value is between 12 and 13.9. The highest value of the tunability is close to 28%, below-100°C, and the highest FOM value is close to 65 around the Curie temperature, as shown in Figure 33(b). It is seen that the tunability and FOM of BST(50/50)-40%MZO(90/10) have improved compared with the pure BST(50/50) ceramic. Considering the applications in the microwave devices, a tunability larger than 10% is preferred. The tunability of BST(50/50)-40%MZO(90/10) is only 5.2~6.2% at room temperature (20~50°C), although it shows improved dielectric properties. Another composition, $Ba_{0.75}Sr_{0.25}TiO_3$ [BST(75/25)], has been chosen as the core material because its Curie temperature is closer to room temperature.

The temperature dependence of the dielectric constant and loss of the BST(75/25) ceramic is shown in Figure 34. The sample has been sintered at 1250oC for 2 h and measured at 100 kHz from 20oC to 120oC. The Curie temperature is near 55oC, and the loss is about 0.03 around the Curie temperature. The tunability is between 3.40% and 37.47% in the whole range of test temperatures with the highest value of 37.47% at 47.36oC, as shown in Figure 34(c). It increases from the lowest value of 3.43% at 20.25oC to 37.47% at 47.36 oC. The FOM value is between 1.75 and 36.41, and the largest value is 36.41 around 72.95oC. The FOM value is only ~7.5 at 30oC because of the high dielectric loss of ~0.024.

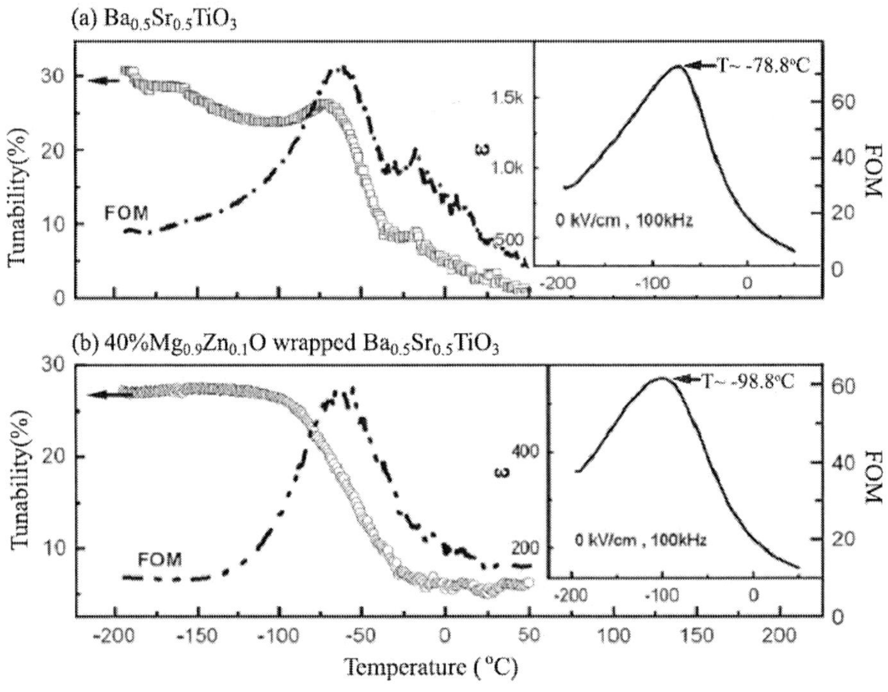

Figure 33. The tunability and the figure of merit as functions of measuring temperature. The insets show that the Curie temperatures are -78.8°C for BST(50/50) and -98.8°C for BST(50/50)-40%MZO(90/10), respectively [14].

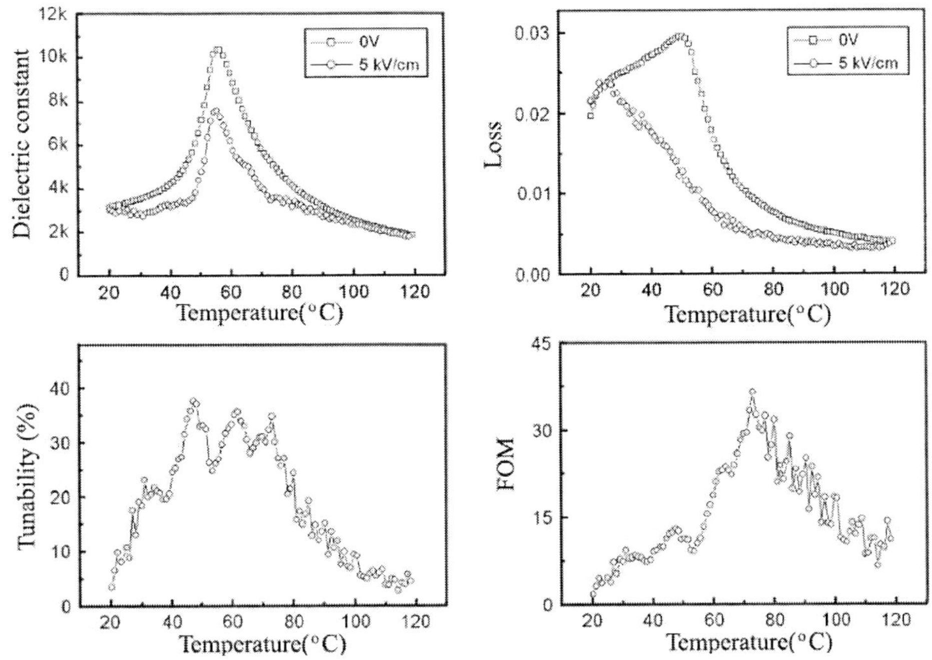

Figure 34. The dielectric properties of BST(75/25) ceramic sintered at 1250°C for 2 h.[14].

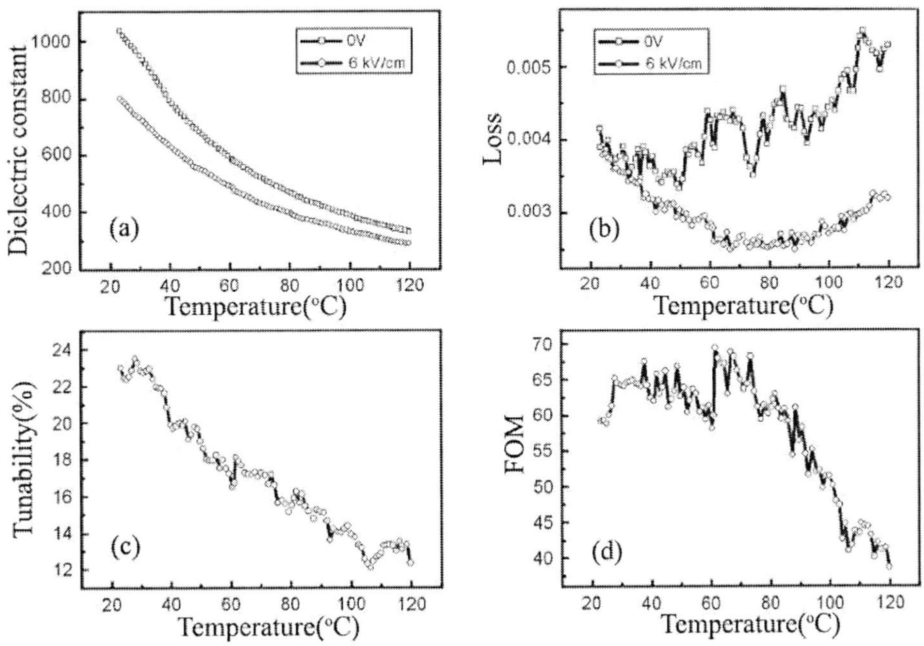

Figure 35. The dielectric properties of 40%$Mg_{0.9}Zn_{0.1}O$ wrapped $Ba_{0.75}Sr_{0.25}TiO_3$ ceramic sintered at 1350°C for 2h.[14].

The temperature dependence of the dielectric constant and loss of BST(75/25)-40%MZO(90/10) ceramic sintered at 1350oC for 2 h and measured at 100 kHz from 20oC to 120oC is shown in Figure 35. The dielectric constant decreases as the temperature increases, showing an obvious nonlinear dependence with regard to temperature. This meets an exponential decay law. For the sample without a bias:

$$\varepsilon=1370.78 \exp(-T/44.03)+243.97 \quad (14)$$

and for the sample with a bias of 5 kV/cm:

$$\varepsilon=961.62 \exp(-T/48.45)+210.20 \quad (15).$$

Here T is test temperature. As shown in Figure 35(b), the dielectric loss decreases significantly compared with that of pure BST(75/25), being between 0.002 and 0.005. The tunability has a monotonic decrease from 23.49% at 20 °C to 12.12% at 120 °C.

It is obvious that the Curie temperature of BST(75/25)-40%MZO shifts to a lower temperature than 25°C and cannot be observed in the measured temperature range. The FOM value is relatively stable between 60 and 70 from 20 °C to 80 °C, and decreases to 38.66 at 120 °C. High-frequency dielectric properties have also been measured by an HP 4291B (1 MHz ~ 1.86 GHz) analyzer at room temperature. The dielectric constant and loss of MZO(90/10) wrapped-BST and pure BST have been characterized and listed in Table 6.

Table 6. Room-temperature dielectric constant and loss at 100 MHz and 10 kHz

Samples	100MHz		10kHz	
	ε	Tanδ(%)	ε	Tanδ(%)
BST(50/50)	1994	2.0	1748	0.6
BST(50/50)-20%MZO	458	1.3	408	3.5
BST(50/50)-40%MZO	203	0.5	181	4.2
BST(75/25)	4655	8.0	4090	4.0
BST(75/25)-5%MZO	3220	1.2	2937	0.2
BST(75/25)-40%MZO	850	0.3	1001	0.4

It can be seen that the tunability and FOM value of the MZO wrapped-BST ceramics have been improved. These results show that BST-based composite ceramics derived from the core-shell structured nanoparticles can improve the dielectric properties, making it possible for use in tunable microwave device applications.

SUMMARY

In summary, perovskite nanoparticles such as BaTiO3, SrTiO3, BaZrO3 and their solid solutions can be fabricated using DSS near room temperature. When synthesized from solution directly, hydrogen interstitials will be introduced into the lattice of perovskite nanoparticles and carry positive charges, and they adsorb the hydroxyl group with negative charges on the surface of the nanoparticles. Because hydroxyl groups adsorbed on the surface of nanoparticles can be ionized freely, this kind of DSS nanoparticles exhibits an alkalescent property. When this kind of perovskite nanopaticles disperses in a certain solution of inorganic salt such as Mg(NO3)2•6H2O, Zn(NO3)2•6H2O, Al(NO3)3•9H2O, Y(NO3)3•6H2O etc., the hydroxide related to these salts spontaneously deposit on the surface of the nanoparticles, as a result, a perovskite structured nanoscaled core materials are successfully self-wrapped by a non-ferroelectric hydroxide derived from a nitrate salt solution. After annealing at high temperature, core-shell structured perovskite nanoparticles with heterogeneous oxide shells can be synthesized.

Because the sintering behavior are dominated by the inert oxide shells such as MgO and they often have high melting point, the core-shell structured perovskite nanoparticles with heterogeneous oxide shells are very difficult to be sintered directly at a low temperature like perovskite nanoparticles. Another reason of making the core-shell structured materials difficult to be sintered is that nanoparticles exert a pinning effect during sintering and thus the mass transfer can be suppressed from shell to shell. Thus it is difficult to obtain their dense ceramics using conventional method. When the core-shell structured perovskite nanoparticles with heterogeneous oxide shells are composited with doped shell oxide materials, the sintering temperature are decreased and pinning effect during sintering can be avoided. A good example is, when the core-shell structured MgO wrapped BST nanoparticles are composited with MZO powders, the mass transfer proceed from MZO to the shell MgO (or MZO) of BST nanoparticles. Thus, dense ceramics can be obtained.

From XRD analysis, the lattice cell parameters and crystallite size and the variation of lattice parameter can be estimated. When hydrogen interstitials are introduced into perovskite lattice, the lattice cell of the powders will expand and change the crystal symmetry. The lattice mismatch between core and shell causes the variation and shrinkage of lattice cell parameter of the perovskite core. The details of lattice match of shell and core can be observed and analyzed by HRTEM and SAED. In our ceramics of MZO wrapped BST nanoparticles, the core-shell structure probably follows the second match mode (Figure 31) because the lattice cell parameter of BST nanoparticles decreases after being wrapped by MZO. The morphology of ceramics can be observed by FE-SEM. Un-grown BST nanoparticles can also be observed in fractured grains of the ceramics of core-shell structured MZO wrapped BST nanoparticles.

The Curie temperatures of BST-MZO composite ceramics shift to a lower temperature compared with that of pure BST ceramics. The dielectric constant and loss decrease sharply depending on the percentage of the shell material (MZO) in the composites. The tunability and FOM values have been improved significantly compared with that of pure BST ceramics. It is thus seen that the ceramics of MZO wrapped BST nanoparicles are promising materials for tunable microwave device applications and high energy density ceramic capacitors for their distinct electric loss improvement.

REFERENCES

[1] J. Li, H. Zeng, S. H. Sun, J. P. Liu, Z. L. Wang, *J. Phys. Chem. B*, 108, 14005(2004).

[2] M Danek, K. F. Jensen, C. B. Murray, M. G. Bawendi, *Chem. Mater.*, 8, 173(1996).

[3] X. Peng, M. C. Schlamp, A. V. Kadavanich, A. P. Alivisatos, *J. Am. Chem. Soc.*, 119, 7019(1997).

[4] M. A. Malik, P. O'Brien, N. Revaprasadu, *Chem. Mater.*, 14, 2004(2002).

[5] P. Reiss, J. Bleuse, A. Pron, *Nano Lett.*, 2, 781(2002).
[6] V. Skumryev, S. Stoyanov, Y. Zhang, G. Hadjipanayis, D. Givord, J. Nogues, *Nature*, 423, 850(2003).
[7] H. Zeng, J. Li, Z. L. Wang, J. P. Liu, S. H. Sun, *Nano Lett.*, 4, 187(2004).
[8] N. Kakuta, K. H. Park, M. F. Finlayson, A. Veno, A. J. Bard, A. Campion, M. A. Fox, S. E. Webber, J. M. White, *J. Phys. Chem.*, 89, 3828(1985).
[9] C. Nasr, S. Hotchandani, W. Y. Kim, R. H. Schmehl, P. V. Kamat, *J. Phys. Chem. B*, , 101, 7480(1997).
[10] S. S. Davis, *Trends. Biotechnol.*, 15, 217(1997).
[11] V. Suryanarayanan, N. A. Sreekumaran, R. T. Tom, T. Pradeep, *J. Mater. Chem.*, 14, 2661(2004).
[12] I. Pastoriza-Santos, D. S. Koktysh, A. A. Mamedov, M. Giersig, N. A. Kotov, L. M. Liz-Marzan, *Langmuir* 16, 2731(2000).
[13] M. E. Lines, A. M. Glass, *Principles and Applications of Ferroelectrics and Related Materials*, Oxford University Press, Oxford, 2001.
[14] H. Y. Tian, J. Q. Qi, Y. Wang, H. L. W. Chan, C. L. Choy, *Progress in Solid State Chemistry* 33, 207(2005)
[15] H. Y. Tian, J. Q. Qi, Y. Wang, J. Wang, H. L. W. Chan, C. L. Choy, *Nanotechnology*, 16, 47(2005).
[16] J. J. Urban, W. S. Yun, Q. Gu, H. K. Park, *J. Am. Chem. Soc.*, 124, 1186 (2002).
[17] S. O'Brien, L. Brus, C. B. Murray, *J. Am. Chem. Soc.*, 123, 12085 (2001).
[18] B. A. Hernandez, K. S. Chang, E. R. Fisher, P. K. Dorhout, *Chem. Mater.*, 14, 480 (2002).
[19] L. A. Perez-Maqueda, M. J. Dianez, F. J. Gotor, M. J. Sayagues, C. Real, J. M. Criado, *J. Mater. Chem.*, 13, 2234 (2003).
[20] J. J. Urban, J. E. Spanier, L. Ouyang, W. S. Yun, H. K. Park, *Adv. Mater.*, 15, 423 (2003).
[21] M. Niederberger, N. Pinna, J. Polleux, M. Antonietti, R. *Angew. Chem., Int. Ed.*, 43, 2270(2004).
[22] J. Q. Qi, "Direct Synthesis Perovskite Nanoparticles from Solution", Editor: George V. Karas, "*New Developments in Crystal Growth Research*", 2006, ISBN: 1-59454-539-1, Nova Science Publishers, NY
[23] J. Q. Qi, L. T. Li, Y. L. Wang, Z. L. Gui, *J. Crys. Growth*, 260, 551(2004).
[24] J. Q. Qi, Y. Wang, W. P. Chen, L. T. Li, H. L. W. Chan, *J. Sol. State Chem.*, 178, 279(2005).
[25] J. Q. Qi, Y. Wang, W. P. Chen, L. T. Li, H. L. W. Chan, *J. Nanoparticle Res.* 8, 959(2006).
[26] U. Weber, G. Greuel, U. Boettger, *J. Am. Ceram. Soc.*, 84, 759 (2001).
[27] J. Q. Qi, W. P. Chen, Y. Wang, H. L. W. Chan, L. T. Li, *J. Appl. Phys.*, 96, 6937(2004).
[28] J. Tennery, R. L. Cook, *J. Am. Ceram. Soc.*, 44, 187(1961).
[29] Q. Li, J. Q. Qi, Y. L. Wang, Z. L. Gui, L. T. Li, *J. Eu. Ceram. Soc*, 21, 2217(2001).
[30] W. Li, J. Q. Qi, Y. L. Wang, L. T. Li, Z. L. Gui, *Mater. Lett.*, 57,1(2002).
[31] J. Q. Qi, Z. L. Gui, W. Li, Y. L. Wang, Y. J. Wu, L. T. Li, *Mater. Lett.*, 56, 507(2002).
[32] P. W. Haayman, R. W. Dam, H. A. Klasens, Method of Producing a Semiconding Materials, German Pat. No. 929,350, June 23, 1955.
[33] S. Venigalla , *Am. Ceram. Soc. Bull.*, 6, 63(2001).

[34] M. Viviani, M. T. Buscaglia, A. Testino, V. Buscaglia, P. Bowen, P. Nanni, *J. Eu. Ceram. Soc.* 23, 1383(2003).

[35] R. T. Kristóf, *Építôanyag* 58, 34(2006).

[36] P. E. C. Franken, M. P. A. Viegers, A. P. Gehring, *J. Am. Ceram. Soc.* 64, 687(1981).

[37] P. A. Fang, H. Gu, H. Shen, *J. Eu. Ceram. Soc.* 24, 2509(2004).

[38] J. Y. Li, S. T. Li, F. Y. Liu, *J. Mater. Sci.-Mater. in Electron.* 14, 483(2003).

[39] V. Hornebecq, C. Huber, M. Maglione, *Adv. Funct. Mater.* 14, 899(2004).

[40] J. Q. Qi, W. P. Chen, Y. J. Wu, L. T. Li, *J. Am. Ceram. Soc.*, 81, 437(1998).

[41] S. R. Summerfelt, in Ferroelectric Thin Films, edited by R. Ramesh (Kluwer Academic, Netherlands, 1997), Chap. 1, p. 1.

[42] Kingon, S. K. Streiffer, C. Basceri, S. R. Summerfelt, *Mater. Res. Bull.* 21, 18(1995).

[43] J. M. Ponds, S. W. Kirchoefer, W. Chang, J. S. Horwitz, and D. B. Chrisey, *Integr. Ferroelectr.* 22, 317(1998).

[44] F. A. Miranda, F. W. Van Keuls, R. R. Romanofsky, and G. Subramanyam, *Integr. Ferroelectr.* 22, 269(1998)

[45] C. S. Hwang, *Mater. Sci. Eng. B,* 56, 178 (1998).

[46] W. J. Kim, W. Chang, S. B. Qadri, J. M. Pong, S. W. Kirchoefer, D. B. Chrisey, J. S. Honwitz, *Appl. Phys. Lett.*, 76, 1185 (2000).

[47] J. B. L. Rao, D. P. Patel, V. Krichevsky, *IEEE Trans Annte. Prog.*, 47, 458 (1999).

[48] P. Padmini, T. R. Taylor, M. J. Lefevre, A. S. Nagra, R. A. York, J. S. Speck, *Appl. Phys. Lett.*, 75, 3186 (1999).

[49] Y. Gim, T. Hudson, C. Kwon, A. T. Findikoglu, B. J. Gibbons, B. H. Park, Q. X. Jia, *Appl. Phys. Lett.*, 77, 1200 (2000).

[50] Kingon, J. P. Maria, S. K. Streiffer, *Nature*, 406, 1032 (2000).

[51] E. Ngo, P. C. Joshi, M. W. Cole, C. W. Hubbard, *Appl. Phys. Lett.*, 79, 248 (2001).

[52] Landolt-Bornstein Numerical Data and Functional Relationship in Science and Technology, Springer-Verlag Berlin 1981.

[53] T. R. Armstrong, L. E. Morgens, A. K. Maurice, and R. C. Buchanan, *J. Am. Ceram. Soc.* 72, 605 (1989).

[54] D. M. Tahan, A. Safari, and L. C. Klein, *J. Am. Ceram. Soc.* 79, 1593 (1996).

[55] N. Kamehara, M. Tsukada, J. S. Cross, and K. Kurihara, *J. Ceram. Soc. Jpn.* 105, 801 (1997).

[56] J. Q. Qi, Z. L. Gui, Y. L. Wang, Q. Li, L. T. Li, *J. Mater. Sci. Lett.*, 21, 405(2002).

[57] D. Hennings, A. Schnell, and G. Simon, *J. Am. Ceram. Soc.* 65, 539 (1982).

[58] H. Shizuno, S. Kusumi, H. Saito, and H. Kishi, *Jpn. J. Appl. Phys.* 32, 4380 (1993).

[59] Z. Yu, C. Ang, R. Guo, and A.S. Bhalla, *Appl. Phys. Lett.*, 81, 1285(2002).

[60] M. Veith, S. Mathur, N. Lecerf, V. Huch, and T. Decker, *J. Sol-gel Sci Tech.*, 17,145(2000).

[61] T. Tohma, H. Masumoto, and T. Goto, *Jpn. J. Appl. Phys.*, 41[11B] 6643(2002).

[62] A. Outzourhit, J. U. Trefny, T. B. Kito Yarar, *J. Mater. Res.* 10, 6(1995).

[63] X. F. Liang, W. B. Wu, Z. Y. Meng *Mater. Sci. and Eng. B* 99, 366(2003).

[64] W. Chang, L. Sengupta, *J. Appl. Phys.,* 92, 3941(2002).

[65] C. Huber, M. Treguer-Delapierre, C. Elissalde, F. Weill, M. Maglione, *Ferroelctrics* 294 13(2003).

[66] C. Huber, M. Treguer-Delapierre, C. Elissalde, F. Weill, M. Maglione, *J. Mater. Chem.* 13 650(2003).

[67] K. B. Chong, L. B. Kong, L. F. Chen, L. Yan, C. Y. Tan, C. K. Ong, T. Osipowicz, *J Appl. Phys.,* 95, 1416(2004).
[68] L. H. Hench, J. K. West, *Chem. Rev*. 90, 33(1990).
[69] H. P. Baldus, M. Jansen, *Angew. Chem.,* 109, 338(1997).
[70] S. Wada, H. Chikamori, T. Noma, T. Suzuki, *J Mater. Sci. Lett.* 19, 245(2000).
[71] W. E. Rhine, R. B. Hallock, W. M. Davis, W. Wong-Ng, *Chem. Mater*. 4, 1208(1992).
[72] J. F. Bocquet, K. Chhor, C. Pommier, *Mater. Chem. Phys*., 57, 273(1999).
[73] C. H. Chen, W. G. Zhu, T. Yu, X. F. Chen, X. Yao, R. G. Krishman, *Surf. and Coating Technology* 167, 245(2003).
[74] K. Nakamoto, 1986 *Infrared and Raman Spectra of Inorganic and Coordination Compounds* (Wiley, New York) p106.
[75] J. B. Liu, H. Wang, Y. D. Hou, M. K. Zhu, H. Yan, M. Yoshimura, *Nanotechnology,* 15, 777(2004).
[76] C. Kim, *Phys. Rev. B*, 66, 113404 (2002).
[77] G. J. Fang, *Phys. Stat. Sol.,* 2, 200(2003).
[78] M. Rand, J. I. Langford, A. Drake, J. S. Abell, *Cryogenics,* 33, 291(1993).
[79] D. Balzar, H. Ledbetter, *Advances in X-ray Analysis*, 38, 397(1995).
[80] B. E.Warren, B. L. Averbach, *J. Appl. Phys.,* 21, 595(1950).
[81] J. Q. Qi, H. Y. Tian, Y. Wang, G. K. H. Pang, L. T. Li, H. L. W. Chan, *J.Phys. Chem. B*, 109, 14006 (2005).
[82] X. X. Xi, H. C. Li, W. D. Si, A. A. Sirenko, I. A. Akimov, J. R. Fox, *J. Electroceram.,* 4, 393(2000).

In: Encyclopedia of Nanotechnology
Editor: Elwood D. Carlson
ISBN: 978-1-60692-079-4

Chapter 21

WEAR DEPENDENT FRICTIONAL PROPERTIES OF ALKANETHIOLS SELF-ASSEMBLED FILMS

Mario D'Acunto[*]
Department of Physics, University of Pisa, Largo B. Pontecorvo 3, I-56127, Pisa, Italy
Department of Chemical Engineering and Materials Science, University of Pisa, Italy

ABSTRACT

Self-assembled monolayers (SAMs) present a 2D regular lattice, and the lateral force measured with an atomic force microscopy (AFM) scanning probe tip displays the typical stick-slip behavior. In this paper, we study the possible connection between an AFM induced damaged SAM surface and irregular behavior of the sticking phases.

In fact, the repeated passages of an AFM probe tip on an SAM surface can produce wear stripping off radicals from the carbon chain backbones, and irregularities of the SAM lattice take place. As a consequence, the frictional properties are changed because new extra space is made available to the final alkanethiol chain groups and the stick and slip phases are not time invariant and hysteresis between the forward and backward scanning direction is altered. The paper discusses if and how such friction irregularities can monitor the progressive modifications of the SAM worn surface.

Keywords: *Self-assembled monolayers, atomic force microscopy, friction, wear, stick-slip.*

INTRODUCTION

Self-assembled monolayers (SAMs) are very useful for the systematic modification of the physical, chemical and structural properties of a surface by varying the chain length, tail group and composition [1]. Many of these properties can be studied making use of the atomic

[*] E-mail: m.dacunto@ing.unipi.it

force microscopy (AFM), and the interaction between the AFM probe tip and the SAMs can also be considered an excellent reference to study the fundamentals properties of dissipation phenomena and onset wear for viscoelastic materials on the nanoscale [2-7], figure 1. It is well established that complete alkanethiol monolayers formed by [-SH(CH2)n-1CH3] molecules on Au(111) adopt a commensurate (√3×√3)R30° structure, with a nearest neighbour spacing of 0.497nm, which corresponds to a surface area of 0.2nm2 per chain [8]. AFM sliding friction experiments on Cn-alkanethiols SAMs showed a friction behaviour dependent on the chain length [9]. The coefficient of friction decreased as the chain length was increased. The low friction of methylic group terminated SAMs is due largely to the low surface energy coming from their hydrophobic tail groups and the high load bearing capacity is due to their compact structure minimizing the van der Waals interaction energy.

The first observation of an atomic stick-slip process was made by Mate *et al.* [10]. Making use of a tungsten tip sliding over graphite, they found a sawtooth-like modulation of the lateral force with the periodicity of the graphite's honeycomb structure. Tománek et al. gave an accurate description based on Tomlinson model for molecular friction [11]. Such description can be summarized as follows: the AFM tip is dragged over the periodic potential defined by the atomic structure of the surface and the contact. The lateral force is measured by means of a cantilever spring holding the tip. The tip sticks to a certain surface position until the force exerted by a moving support of the spring is high enough to activate a slip of the tip to the next atomic position. The stick-slip behavior shows up as the sawtooth-like modulation of the lateral force. Stick-slip movements in sliding friction are generally characterized by alternating periods of energy storage and period of energy dissipation in the driving cantilever spring.

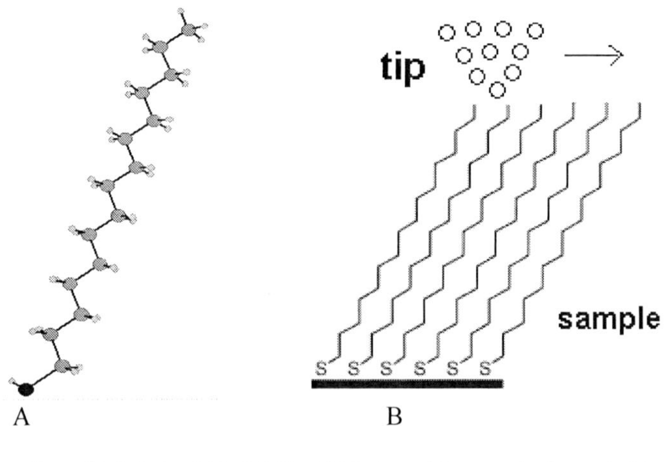

Figure 1. Single C_{15}-alkanethiol chain, [-SH(CH2)$_{14}$CH$_3$], the sulfur atom is drawn in black, figure 1A. Schematic sketch of the AFM probe tip-SAM system, figure 1B.

At any time, the total power can be considered the summation of the storage power and the dissipated power, where dissipated power refers to all power that is not stored in the driving spring and that can be converted to heat or to an increase of the free energy of the substrate. Measuring slip time is rather complicated for a number of reasons. First of all, drift and noise limit the smallest value of velocity that can reliably be measured. Uncertainties in cantilever stiffness as well as in substrate compliance add complications in ruling out that the tip is moving at extremely low speed with respect to the surface.

In this paper, we focus the attention on the frictional response at the contact between probe tip and damaged SAM surface when single AFM induced defects correspondent to methyl group stripped off are progressively produced. Stick-slip dynamics of such damaged surface present irregularities for both stick and slip scale times. Moreover, reduction of hysteresis between left-to-right (forward) and right-to-left (backward) is observed, making possible in some case, the overlap between the forward and backward scanning direction. We propose an explanation of such overlap of lateral force hysteresis loop based on the spatial re-orientation of methyl groups confining to defects. Such re-orientation is meanly increased in proximity of defects, because the van der Waals energy cohesion between chains is reduced and the free space available to the methyl group is larger respect to primitive undamaged profiles.

RESULTS AND DISCUSSIONS

The motion of a probe tip scanning a SAM lattice is characterized by two different timescales. The resonant frequency of a cantilever is around 105Hz, while the atomic vibration frequency is of the order of 1013Hz. As a consequence, in the contact area we have two kinds of dynamics, the slow dynamics of the cantilever and the fast vibrations of atoms. At the time scale of picoseconds, we encounter thermal vibrations of atoms, while the cantilever is stationary. On contrary, at the timescale of a microsecond, we observe the scanning vibrations of the cantilever in contact with a cluster of atoms. The energy modes between the cantilever and the atomic substrate can be decoupled, because the relaxation time of atomic cluster in contact with the probe tip is smaller than the cantilever timescale.

We focus the attention on the changes of the frictional properties produced by a damaged SAM lattice when end methyl groups can be stripped off by the progressive passage of an AFM probe tip at low applied loads. To do this, we use a one-dimensional Langevin equation describing an AFM probe tip sliding on a periodic potential, eq. (1):

$$M\ddot{x}(t) + M\gamma\dot{x}(t) + \frac{\partial U(x, x_{tip})}{\partial x} = \eta(t) \tag{1}$$

where M=3.0x10-11Kg, the damping factor is γ=8.9×105Hz, $\eta(t)$ is a random noise satisfying the fluctuation-dissipation relation described by $<\eta(t)\eta(t')>=2M\gamma k_B T\delta(t-t')$.

The eq. (1) describes the motion of the center of mass of the probe tip moving along the sample potential profile. It does not include parameters such as the dimension of the probe tip, and for a complete description of the influence of the AFM induced stresses also an equation along the z-dimension should be addressed in order to describe the balance of the applied load and deformation of the SAM surface [5-6]. The effective potential $U(x, x_{tip})$ can be written in absence of lattice defects as:

$$U(x, x_{tip}) = -U_0 \cos\left(2\pi \frac{x}{\lambda}\right) + \frac{k_x}{2}\left(x_{tip} - x\right)^2 \tag{2}$$

where the lattice potential barrier is U_0=0.6632×10-19 J, at room temperature the thermal energy is k_BT=0.041×10-19 J, λ is the lattice period =0.497nm, k_x =2.5N/m corresponds to an effective spring constant given by the combination of the stiffness of the cantilever and the lateral stiffness of the contact, *x*tip=*vt* is the instantaneous position of the probe tip, where the tip velocity is *v*=100nm/s. The parameters have been considered as effective in the load range of nearly 1nN, in the sense that their precise value may depend on the load applied and they can have a weak dependence on temperature [12]. The sinusoidal term in eq. (2) describes the regular lattice potential acting on the tip due to the interaction with the sample and the second term describes the elastic energy stored in the cantilever.

The actual interest is to reproduce the instantaneous lateral friction force that is given by the force $F(t)=k_x(vt-x)$. The stick-slip motion and continuous sliding are the two main regimes observed. Transitions between the two regimes as a function of the effective spring constant can be observed. The movement of the tip from one minimum to the next can be continuous or jumping, depending on the relation between corrugation U_0 and elastic energy which can be described by $\approx U_0/k_x\lambda^2$. When this ratio is greater than 1, then the stick-slip regime is observed [13]. The stick-slip dynamical behavior and the continuous regime for the probe tip sliding on a rigid substrate as described by eq. (1) are summarized in figure 2 for a forward and a backward scan. Hard-spring cantilever constants imply continuous regime.

Figure 2 demonstrates that to observe irregular stick-slip transitions, we need to work with soft spring cantilever constant. For regular periodicity, the saw-tooth modulation has the periodicity of the SAM lattice. The maximum lateral force is linearly related to the potential height corrugation of the SAM surface. The area enclosed in this hysteresis loop is the energy dissipated in one cycle, composed by the forward and the backward tip motion on a length of 2.5nm, including five methyl groups. For a pattern including defects and deformations, the potential as described by eq. (2) has to be modified locally accounting such defects and enclosed in the Langevin equation (1). As a consequence, when the SAM profile is progressively damaged by the probe tip, we should re-calculate a potential $U_{eff}(x)$ describing the SAM lattice profile to be inserted in eq.(1). This means that the effective potential can now be written as

$$U_{eff}\left(x, x_{tip}\right) = -U_w(x) + \frac{k_x}{2}\left(x_{tip} - x\right)^2 \tag{3}$$

where the subscript is used to distinguish such damaged profile potential respect to the potential used in eq. (2). Uw(x) represents the local damaged resultant potential describing the wear induced irregular SAM one-dimensional profile. The paper deals out the numerical results obtained by the Langevin, eq. (1), when an effective potential, as given by eq. (3) is used instead of the undamaged potential, eq. (2).

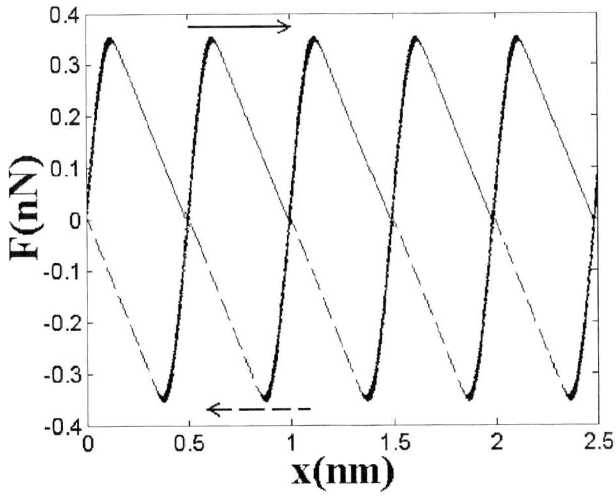

a

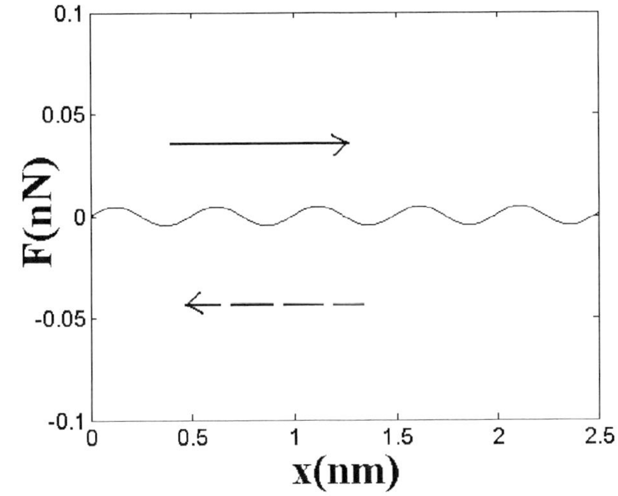

b

Figure 2. Numerical results for the lateral force F acting on a tip described by eq. (1) sliding on a C15 SAM surface for two different spring constants. Figure 2A shows the hysteresis between the forward (continue line) and backward (dashed line) scanning direction for a tip sliding on a regular period lattice for a soft effective spring constant k_x=2.5N/m. Figure 2B shows the continuous sliding regime for the lateral force as described by eq. (1) for a hard effective spring constant k_x=118N/m. The applied load, L, is =1nN and the tip velocity equals to 100nm/s, in both the figures.

Figure 3 shows a typical 2D lattice C18 SAM surface progressively damaged by the repeated passage of an AFM probe tip. The description of such wearing process is performed making two general assumptions, the first one is that the wear process takes place only stripping off the chain end methyl groups and the second assumption is that the n-1 backbone chain C carbons are locked, so that only the final chain carbon atom can be subject to be stripped off. It has been proposed that the progressive damaging starts from initial surface defects following a nucleation process [14]. The nucleation of wear domains starting by the initials SAM surface defects can be explained by the fact that defect involves a lowering of the van der Waals energy cohesion between neighboring chains and a consistent local change in the lateral force in correspondence of such defects. The AFM probe tip sliding on the SAM surface shows a typical stick-slip dynamics, where timescales for stick and slip phases are invariant and hysteresis between the forward and backward scanning direction is maintained. When the SAM surface is progressively damaged, as showed in figure 3, the probe tip moves on an irregular path, and the irregularity of the path reflects on the friction force.

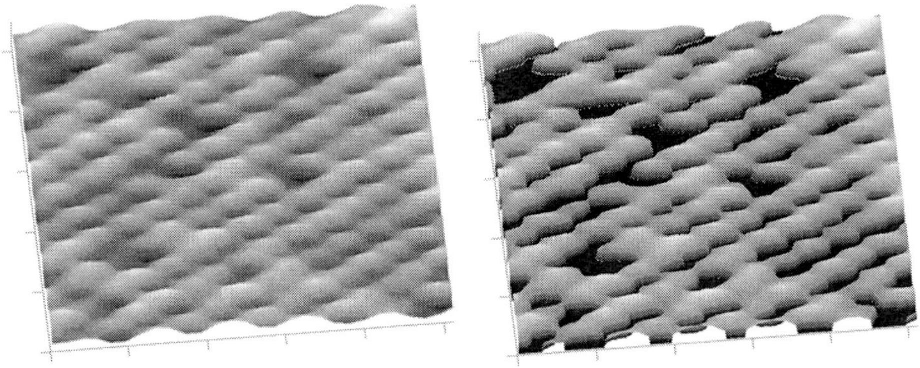

Figure 3. C18-chain progressive damage of simulated methyl group end surface. The initial image (left) is based on an AFM experimental acquired data on a C18 sample, the typical commensurate c(4×2) superlattice of ($\sqrt{3}\times\sqrt{3}$)R30° arrangement is slightly altered by the applied load [11]. The final image (right) shows that nearly 5% of methyl groups have been stripped off by the AFM tip after 50 scans with an applied load, L, of 1nN. The black portions represent the wear damaged domains.

The effective potential $U_{eff}(x)$, as given by eq. (3), describes the upper level of the alkanethiol chains. When an end methyl group is stripped off, the potential profile should take in account that the upper level is lowered to the subsequent chain carbon of a distance of nearly 0.153nm, that is the C-C distance in the alkanethiol chain, figure 1. When stressed by the probe tip, the carbon backbone flexibility is generally modeled by rigid bonds connecting the C-C atoms. The total energy of a given chain can be written as a superposition of a short-range valence or internal terms such as bond stretch, bond-angle bending and angle torsion terms, and a long-range non bonded interaction led by van der Waals intermolecular interaction [7], details of the hybrid molecular approach based on a Montecarlo code are included in [14]. The one-dimensional trajectory for the probe tip described by the Langevin eq. (1) implies that when we calculate the effective potential $U_{eff}(x)$ we have to project the effects of the lattice surrounding carbon atoms along the scanning direction of the probe tip. The lateral force calculated making use of the Langevin eq. (1) with the effective potential $U_{eff}(x)$ presents the results summarized by the figure 4. Changes of lattice periodicity accomplished by vacancy involve irregular paths for the probe tip and consequently, irregular stick-slip motions, figure 4A. Again, the variation of the potential walls involves a growth in the lateral friction force. An unexpected result is the drastic reduction of the hysteresis loop of the lateral force, figure 4B. The reduction allows an overlap of the lateral force between the forward and backward scanning directions. Moreover, the symmetry of the lateral force between the forward and backward scanning directions is lost, and the two directions present different stick phases. The reduction of hysteresis loop and the alteration of the lateral force loop symmetry should be a consequence of re-orientation of chains during forward and the backward scanning directions, i.e., the effective potential changes due to made easier re-orientations of the end C atoms. In fact, the van der Waals energy cohesion ε_{vdW} responsible for the chain packaging, generally falls around 1×10-20J, but is can be lowered of about 50% at the edge of a defect, because the extra space available due to the presence of larger defects.

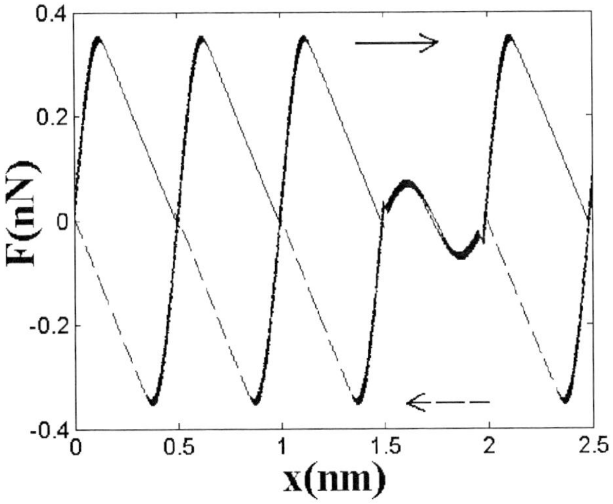

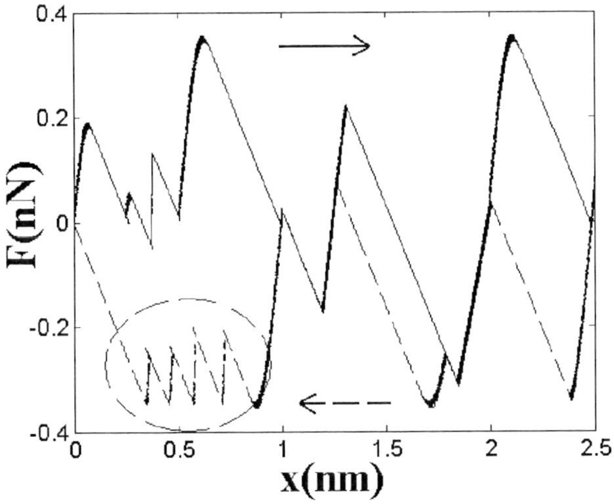

Figure 4. The figures represent the lateral force F for worn lattice profiles for a soft spring k_x=2.5N/m. The lateral force is increased with respect to the regular SAM periodicity and the directional symmetry is lost. As in figure 3, the applied load, L, is =1nN, the tip velocity equal to 100nm/s. The circle in figure 4B shows a regular stick-slip dynamics with a short periodicity respect to undamaged lattice period.

Figure 4 displays the typical stick-slip profiles for SAM surface where onset damages involving end chain groups taking place. In the left figure, the stick-slip profile corresponds to a 5 final methyl group where the 4-th methyl group has been previously stripped off by the AFM probe tip. It should be noted the corresponding frictional force has not been reduced, in fact while the other end methyl carbon atoms produce a stick phase starting by zero to nearly 0.35nN, the stick phase correspondent to the vacancy site of approximately 0.153nm starts with a zero force that has its initial value located below respect to the previous zero force. The maximum stick force must be calculated in this case considering that the stick value starts from a hole and the energy barrier meets by the scanning probe tip is $U_w > U_0$, where Uw is the energy barrier correspondent to a defect. Generally, for hole defects limited to a single stripped off methyl carbon atoms, we have a sticking force that should be twice respect to regular pattern stick force as confirmed by the numerical results reproduced in figure 4.

In the right figure, the potential represents a strong damaged surface and it is displayed in figure 5. The extra available space is now enhanced. When the probe tip meets a carbon atom, the extra space available to the carbon atoms permits movements of the carbon atoms that can change the stick force. In combination with the effects of the surrounding carbon atoms, limited domains of sawtooth-like modulated stick-slip dynamics can be observed, as highlighted by the circle in figure 4B.

The time scale for the atoms movements is rather fast falling in the range of picoseconds, while the collective probe movements fall in the milliseconds range. This implies that the two types of motion can be decoupled and the probe tip slides on a freeze potential configuration, which morphology is determined by the equilibrium positions of the atoms. Such equilibrium positions can be modified by internal rearrangements of end chain carbons atoms, or by the AFM probe tip induced stress causing surface damaging and friction.

It remains difficult to observe a direct connection between the effective potential and the different stick resultant phases. In our simulation, the stripping off process is limited to end chain methyl group carbon atoms and this is still an unrealistic approach. The physical situation is more difficult, nevertheless, we hope that such preliminary results can help to highlight the challenges presented in this paper, i.e., the possible monitoring of damaged surfaces making use of the sticking phases of lateral force generated at the contact between an AFM probe tip and a viscoelastic surface.

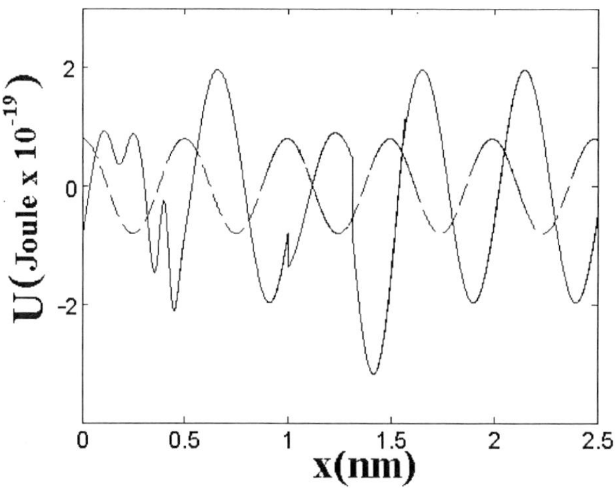

Figure 5. Damaged resultant potential $U_{eff}(x)$ correspondent to the stick-slip loop displayed in figure 4B. Such potential is the effective potential *seen* by sliding the probe tip. The dashed line represents the undamaged potential expressed by the sinusoidal term in the potential as in eq. (2).

CONCLUSIONS

Viscoelastic materials can exhibit irregular stick-slip dynamics when the regular lattice is progressively damaged by wear mechanisms. This irregularity could be used to monitor the damaged surface of a self-assembled monolayers (SAM). In order to demonstrate such behaviour, we consider an atomic force microscopy (AFM) probe tip sliding on a regular lattice of a SAM and progressively damaging the substrate. The collective coordinate of probe tip is described by a Langevin equation, where the potential describing the underlying surface is continuously recalculated as the damages take place. If we focus the attention on the not invariant time stick phase, it should be possible, almost in principle, to create a correspondence between the different stick phases and the corrugated potential describing the positions of the final SAM carbon atoms. In turn, worn SAM profiles can be also associated to anomalous hysteresis loops of frictional force between the forward and the backward sticking dynamics.

REFERENCES

[1] A. Ulman, *Introduction to Ultrathin Organic Films: From Langmuir-Blodgett to Self-Assembly*, Academic Press, San Diego, CA, 1991.

[2] V.N. Bliznyuk, M.P. Everson and V.V. Tsukruk, *J. of Tribology*, 120, 489 (1998).

[3] E. Barrena, S. Kopta, D.F. Ogletree, D.H. Charych, and M. Salmeron, *Phys. Rev. Lett.*, 82, 2880 (1999).

[4] B. Bhushan and H. Liu, *Physical Review* B, 63, 245412 (2001).

[5] Y.S. Leng and S Jiang, *J. Chem. Phys.*, 113, 8800 (2000).

[6] Y.S. Leng and S Jiang, *Phys. Rev. B*, 63, 193406 (2001).

[7] S. Jiang, *Molecular Physics*, 100, 2261 (2002).

[8] J. Hautman and M.L. Klein, *J. Chem. Phys*. 91, 4994 (1989).

[9] A. Lio, D.H. Charych and M. Salmeron, *J. Phys Chem*. B, 101, 3800 (1997).

[10] C.M. Mate, G.M. McClelland, R. Erlandsson, and S. Chang, *Phys. Rev. Lett.*, 59, 1942 (1987).

[11] D. Tomanek, W. Zhong, and H. Thomas, *Eurphys. Lett.* 15, 887 (1991).

[12] Y. Sang, M. Dubè, and M. Grant, *Phys. Rev. Lett.* 87, 174301 (2001).

[13] A. Sokoliuc, R. Bennewitz, E. Gnecco and E. Meyer, *Phys. Rev. Lett.*, 92, 134301 (2004).

[14] M. D'Acunto, *Nanotechnology*, 17, (2006).

In: Encyclopedia of Nanotechnology
Editor: Elwood D. Carlson

ISBN: 978-1-60692-079-4

Chapter 22

PREPARATION AND CHARACTERIZATION OF NANOPARTICLES AND NANOCRYSTALLINE BULKS OF PURE RARE-EARTH METALS

Xiaoyan Song

College of Materials Science and Engineering,
Key Laboratory of Advanced Functional Materials,
Ministry of Education, Beijing University of Technology,
Beijing 100022, China

ABSTRACT

Because of the extremely high chemical reactivity and hence the considerably rigorous requirements for the equipments to preserve a high purity of the product, the preparation and characterization of nanostructured pure rare-earth metals are still a big challenge in nanoscience and nanotechnology. In this chapter, followed a thermodynamic description of the energy state and the thermal stability of rare-earth nanoparticles and nanocrystalline bulks, the practical preparation and characterization of nanoscale rare-earths are depicted in detail. Different from a number of existing techniques that have been developed to synthesize nanocrystalline bulk materials, a home-designed configuration of the "oxygen-free" in-situ synthesis system, where the inert-gas condensation is combined with the SPS technology into an entirely closed and vacuumized system, is mainly introduced. With the technically designed route of sequentially performed processes, i.e. amorphization of nanoparticles, nucleation and growth of the short-range ordered "clusters" inside the nanoparticle, and the complete nanocrystallization, the ultrafine nanocrystalline bulk can be obtained with the grain size much smaller than the initial nanoparticle size. Microstructures and properties of some typical members of the rare-earth family are presented ordinally. The prominently changed thermal, physical and mechanical properties imply the promising applications of the nanocrystalline rare-earth metals and alloys in the fields of functional and structural materials. Finally, although the chapter is focused on rare-earth nanoparticles and nanocrystalline bulks, the proposed thermodynamic models and preparation techniques are potentially applicable to a large variety of nano metal materials, hence may establish a new way for studies on microstructures and performances of metallic nanomaterials, especially of those highly reactive in the air.

1. INTRODUCTION

Rare-earths are a group of 17 metal elements with atomic numbers 21, 39, and 57–71; the name lanthanides is reserved for the elements 58–71. The unique configuration of unpaired 4f electrons and the rich structures of the energy levels enable the rare-earth metals to possess many particular physical and chemical properties that are of world-wide interests, such as high electrical conductivity, large magnetic moment, and very high complexation reactivity [1,2]. Based on these properties, the rare-earth elements and compounds have been applied in many advanced materials such as high-performance magnets, fluorescent materials [3], chemical sensors, autocatalysts [4], and high-temperature superconductors [5]. For example, Nd2Fe14B and SmCo5/Sm2Co17 are powerful rare-earth permanent magnets, and Sm-Co permanent magnets have the highest resistance to demagnetization of any known magnetic material and very high intrinsic coercive force [6,7]. There is an increasing demand for high purity rare-earth metals and oxides particularly for use in heat storage and dispersal [8], corrosion resistance [9], as well as for environment-favorable applications, such as in pigments for paint and plastics [10] to replace heavy metal (e.g. Pb, Cr, Cd and Hg) pigments; in cement manufacture to reduce the temperature of calcination and help save energy [11]; and in refrigeration components as a potential substitute for chlorofluorocarbon (CFC) [12].

It is known that materials at nano-scale have properties significantly different from those at larger scales. There are two principal reasons for this: first, nanomaterials have a highly increased relative surface or boundary area, this can make materials more chemically reactive and affect their strength or electrical properties; second, quantum effects can begin to dominate the behavior of matter at the nano-scale, which affect the optical, electrical and magnetic behavior of materials.

The three-dimensional nanoscale materials, i.e. nanoparticles, classified as particles less than 100nm in diameter, and nanocrystalline metals, made up of nanometer-sized grains, are in the category that is discussed in this chapter. For the nanoscale rare-earth particles, due to a much greater surface area per unit mass hence a higher proportion of atoms at the surface compared with those of larger particles, some new or enhanced size-dependent properties such as chemical reactivity and optical behavior may exhibit. For the nanocrystalline rare-earth metals, as the grain size decreases, the interface area within the material greatly increases; this can significantly affect mechanical, electrical and magnetic properties of rare-earths. For example, it has been shown that magnets made of nanocrystalline Sm-Co grains possess unusual magnetic properties [13], as high coercivity can be obtained because magnetization flips cannot easily propagate past the extremely increased grain boundaries. This could lead to applications in motors, analytical instruments like magnetic resonance imaging (MRI), used widely in hospitals, and microsensors.

Since remarkably improved functional and structural properties can be expected for the nanoscale rare-earth materials, the related investigations have attracted growing interests in multiple interdisciplinary fields. However, because of the extremely high chemical reactivity and hence the considerably rigorous requirements for the equipments to preserve a high purity of the product, the preparation and characterization of nanoscale pure rare-earth metals are still a big challenge in nanoscience and nanotechnology. To synthesize nanoscale rare-earths is not simply another step in the miniaturization of materials with some existing 'top-down' and 'bottom-up' techniques, the chemical activity and physical properties of rare-earths, as well as the impact of those on the preparation route, have to be given careful considerations. Consequently, due to the confinement of the preparation technique, many important features of the nanoscale rare-earth metals such as the physical, the chemical, the thermal, and the mechanical characteristics have been rarely reported so far. Whereas, the corresponding research topics are of great importance both for the development of nanoscience and nanotechnology and for extending the applications of the rare-earth metals.

Although a number of reviews and books have been published on the nanoscience and nanotechnology, especially, there have been numerous reports on preparation and characterization of novel nanostructured materials, very few of them comprehensively deal with the features of nanoparticles and nanocrystalline bulks of the rare-earth family. In this consideration, this chapter is an effort to depict to the maximum extent possible the features of rare-earth nanoparticles and nanocrystalline bulks, presenting more detailedly the progress of our research specialization. The author tried to include as many as possible details in the following sections, however, it should be understood that the content of the chapter is still far from an exhaustive description of all the characteristic aspects of nanoscale rare-earths.

In this chapter, focused on the nanoscience and nanotechnology of nanoscale rare-earths, the content is outlined as follows: in Section 2, the energy states and the thermodynamic features of nanoparticles and nanocrystalline bulks are described by the developed thermodynamic models, with calculations performed on nanoscale rare-earth metals; then preparation and microstructure analysis of rare-earth nanoparticles and nanocrystalline bulks are presented in Section 3 and Section 4, respectively; characterization of thermal, physical and mechanical properties of nanocrystalline rare-earth bulks and comparison with those of conventional polycrystalline rare-earth metals are expounded in Section 5.

2. Thermodynamic Analysis of Nanoparticles and Nanocrystalline Bulks

It has been demonstrated in literature [14,15] that for nanoscale materials, owing to the significantly increased total surface area (e.g. nanoparticles) or the grain boundary area (e.g. nanocrystalline bulks), many new features in the crystal structure, interface, thermodynamics and phase transitions were along with introduced. Since the effects of the surface and the grain boundary on the total thermodynamic properties of materials were not emphasized or even neglected in the conventional thermodynamic models, the thermodynamic studies on nanomaterials, which are aimed to comprehend the thermal features of a material system including its surface or interface as well as the bulk, are needed extraordinarily to be carried out. In this section, the dependence of the fundamental thermodynamic functions on the nanoparticle size or the nano-grain size will be developed, based on which the thermal and physical properties, as well as the phase-transition features and thermal stability of the nanomaterials, may be quantitatively described.

2.1. Gibbs Energy of Nanoparticles

The thermodynamic equilibrium between solid and liquid phases in a small particle system was firstly considered nearly a century ago by Pawlow [16]. In 1954, Takagi [17] observed that the melting temperatures of pure Pb and Bi decreased with decreasing the size of the particles. From then on, several studies have been carried out to explain the dependence of the melting points of metals and alloys on the metallic particle size [18-23]. In recent years, Tanaka and his group [24-26] have evaluated the surface properties as well as phase diagrams of a few binary alloys in the small particle system on the basis of thermodynamic databases. The present thermodynamic model of the energy state of metal nanoparticles has been developed based on the Tanaka model of Gibbs energy in small particle systems.

For a single solid nanoparticle with its radius r, the total Gibbs energy of the particle is the sum of the surface energy and the Gibbs energy in the internal bulk of the particle, i.e.

$$G^{Total,Sol} = G^{Surface,Sol} + G^{Bulk,Sol} \tag{1}$$

The surface energy of the particle can be calculated as

$$G^{Surface,Sol} = (2\sigma^{Sol} V_{mol} / r) \cdot V / V_{mol} \tag{2}$$

where σSol is the surface tension of the solid particle; Vmol and V are the mole volume and the volume of the solid particle, respectively. For the surface tension of pure solid metals at temperature T, the estimation in the Tanaka model [24,25] is used, i.e.

$$\sigma^{Sol} = 1.25T_{mp} / 1000 - 0.0001 \cdot (T - T_{mp}) \tag{3}$$

where T_{mp} is the melting temperature of the material in the conventional bulk form.

The Gibbs energy of the internal bulk of the particle can be calculated by the classic thermodynamic model of the metal bulk material with a general function as

$$G^{Bulk,Sol} = G_{Sol}(T) \cdot V / V_{mol} \tag{4}$$

GSol(T) is the Gibbs free energy of the material in the conventional bulk form, which is a function of the temperature and can be obtained from the thermodynamic databases, e.g. SGTE. As the nanoparticle with a spherical shape is assumed, the volume of the particle is $V=4/3\pi r^3$. From eqs. (1)-(4), the total Gibbs energy of the single-phased solid nanoparticle is expressed as

$$G^{Total,Sol} = 4/3\pi r^3 / V_{mol} \cdot G_{Sol}(T) + 4/3\pi r^2 \cdot (0.0027T_{mp} - 0.0002T) \tag{5}$$

Thus, the Gibbs energy of the nanoparticles can be evaluated by the temperature and the particle size, and the intrinsic material parameters such as its bulk melting temperature and the specific heat capacity (from which GSol(T) can be deduced).

The Gibbs energy of the rare-earth nanoparticles were calculated by eq. (5). Figure 1 shows as examples the Gibbs energy of nanoparticles in the rare-earth family such as Sm, Nd, Gd and La with given crystal structures. It can be seen that the Gibbs energy of nanoparticles changes with both the temperature and the particle size. For a certain size class of the particles, the Gibbs energy reduces with increasing the temperature. At a given temperature, the Gibbs energy reduces with the increase of the nanoparticle size. The above changing features of the Gibbs energy of particles result in the spontaneous tendencies of segregation of fine nanoparticles and rapid coarsening of nanoparticles with increasing the temperature.

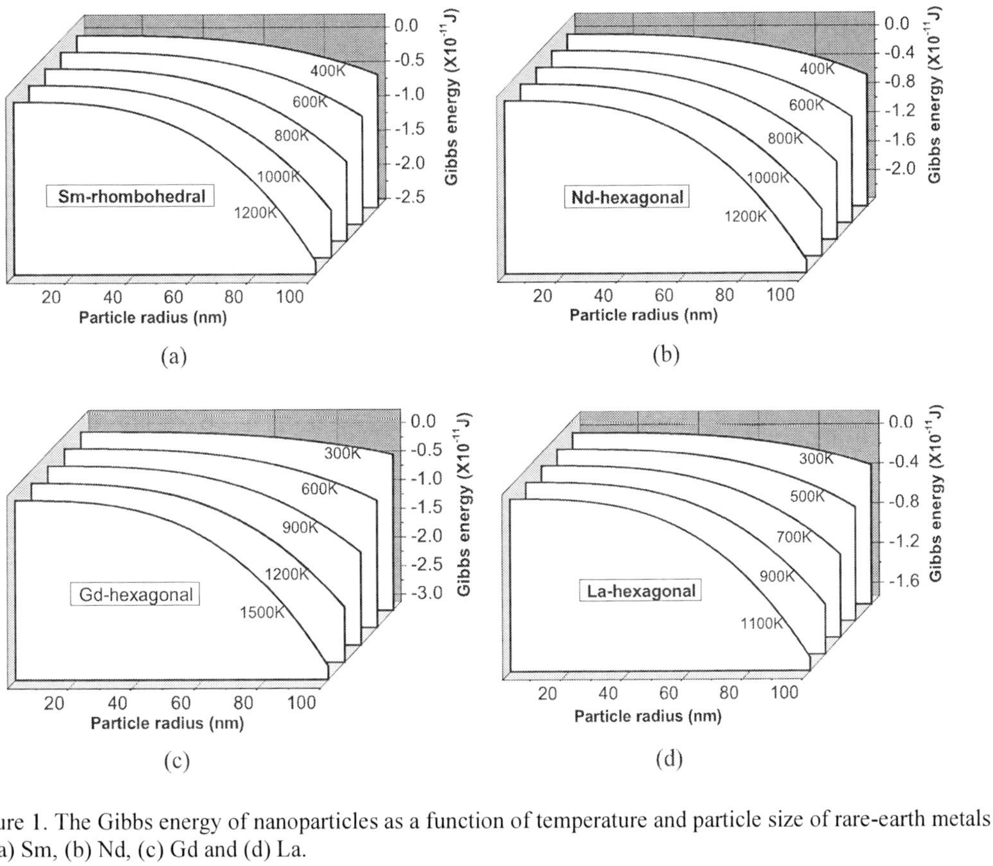

Figure 1. The Gibbs energy of nanoparticles as a function of temperature and particle size of rare-earth metals of (a) Sm, (b) Nd, (c) Gd and (d) La.

2.2. Melting Temperatures of Rare-Earth Nanoparticles

In the condition of phase equilibrium, the melting point of the particle is the temperature at which the Gibbs energy of the solid particle is equal to that of the particle liquid, i.e. $GLiq=G^{Total, Sol.}$ The Gibbs energy of the liquid state of the particle is estimated as

$$G^{Liq} = G_{Liq}(T) \cdot V / V_{mol} \tag{6}$$

where GLiq(T) is the Gibbs free energy at the liquid state of the material evaluated by the conventional thermodynamic model, it is a function of the temperature and can also be obtained from the thermodynamic databases. Vmol and V are the mole volume and the volume of the liquid phase, respectively, here the possible tiny change of the volume while the particle melts is neglected. Equalize eqs. (5) and (6), one obtains

$$T = 13.5T_{mp} - 5 \cdot 10^3 (G_{Liq}(T) - G_{Sol}(T)) \cdot r / V_{mol} \tag{7}$$

which is an implicit function equation between the melting temperature of the nanoparticle and the particle size, containing the Gibbs free energy functions of the solid and liquid phases, respectively.

Figure 2 shows as examples the size-dependence of the melting temperatures of rare-earth nanoparticles of Sm, Nd, Gd and La with given crystal structures, which are estimated by eq. (7).

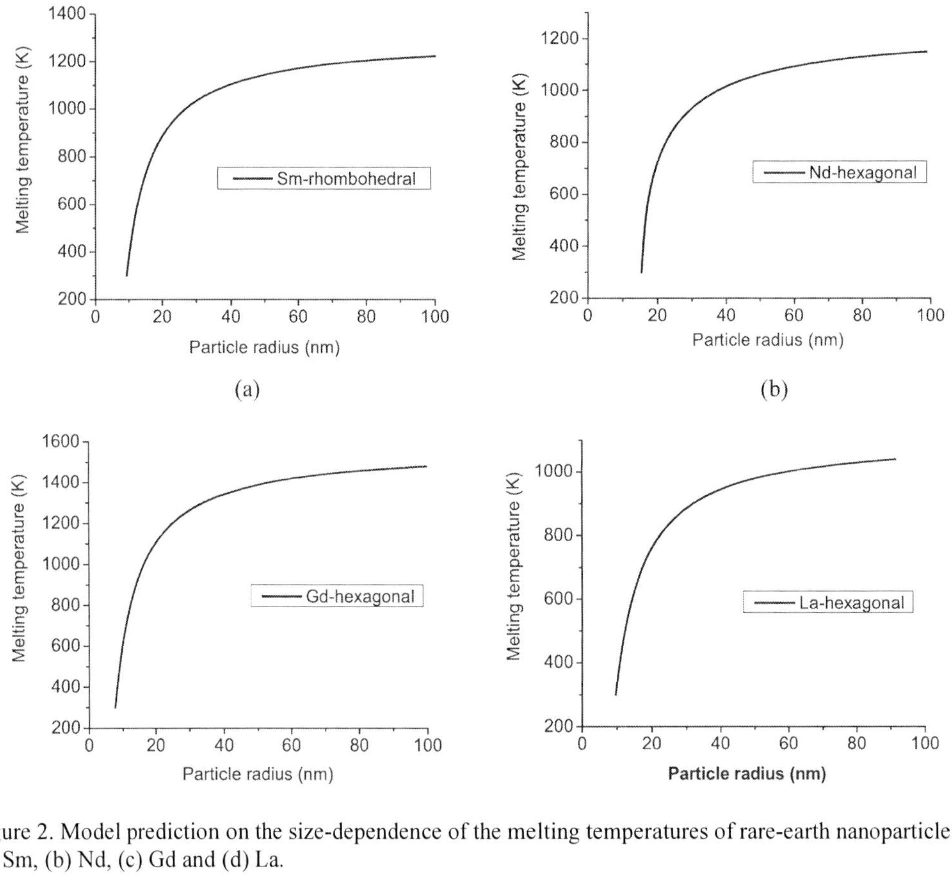

Figure 2. Model prediction on the size-dependence of the melting temperatures of rare-earth nanoparticles of (a) Sm, (b) Nd, (c) Gd and (d) La.

It can be seen that the rare-earth particles have the similar reduction tendency of the melting points with the decrease of the particle size. Generally, when the particle size is reduced to below 30 nm, there is a sharp decrease of the melting temperature of the nanoparticle. As the particle size decreases, the number of atoms at the surface increases, e.g. a particle of size 10 nm has 20% of its atoms on its surface, and at 3 nm 50% of its atoms. Thus nanoparticles have a much greater surface area per unit mass compared with larger particles. Due to the vibration relaxation of atoms at the surface, as the amplitude of vibration reaches 10-20% lattice constants, the melting takes place starting from the surface.

As is known, manufactured nanoparticles are typically not products in their own right, but generally serve as raw materials, ingredients or additives in other materials. For most applications, nanoparticles will be fixed, for example, consolidated into a bulk of the same material or within a composite. Based on the model prediction of the melting temperatures of the rare-earth nanoparticles, the processing parameters of consolidation can be optimized for the methods of powder metallurgy. Since the melting points of nanoparticles are greatly reduced as compared with those of the conventional bulk material and the coarse particles, the consolidation of nanoparticles can be performed at lower sintering temperatures.

2.3. Thermodynamic Model of Nanocrystalline Metal Bulk

As soon as the nanocrystalline materials are produced, the stability of nanostructures becomes most importance, namely, the nanoscaled grain structure should stay sufficiently stable over a useful period of temperature and/or time to retain the attractive properties of the nanocrystalline material. Rapid grain growth is the principal process of the destabilization of nanocrystalline structures. Nano-grain growth behaviors have been studied by various experimental techniques such as TEM [27-29], DSC [30], X-ray diffraction [31,32], and Raman [14] and Mössbauer spectroscopy [33], as well as Monte-Carlo [34] and molecular dynamics simulations [35]. Very recently, a few studies have attempted to investigate the grain growth kinetics in nanocrystalline materials with a thermodynamic point of view [36-38].

With reducing the grain size, the volume fraction of grain boundaries is significantly increased in the nanocrystalline materials. Structurally, a nanocrystalline material consists of two components: nanometer-sized crystallines composing of atoms that locate in the regular crystal lattice, and boundary regions composing of atoms that are partially ordered or disordered with a lower density [39]. In other words, a nanocrystalline material is inherently heterogeneous, or thermodynamically non-equilibrium on a nanometer scale. As compared to the conventional coarse-grained materials, the thermal stability of nanocrystalline materials is much more strongly affected by the grain boundaries. From the thermodynamic point of view the atomically disordered nano-grain boundaries influence the configurational and vibrational entropy as well as the enthalphy of the system [40]. Due to the fact that nanocrystalline materials contain a large volume fraction of grain boundaries, the thermal properties cannot be sophisticatedly described by the classic thermodynamic functions that have been derived for the polycrystalline materials where the contribution of the grain boundaries to the thermal features of the material is generally neglected.

Based on a "dilated crystal model", Fecht [41] and Wagner [42] established the thermodynamic basis for describing the thermal properties of interfaces in nanocrystalline metals. Fecht calculated some thermodynamic parameters of nano-grain boundaries of Cr, Pd and Hf in terms of a universal equation of state (EOS) [41]. Whereas, Wagner calculated the thermal properties of nano-grain boundaries of Cu, Pd and Ir using the quasiharmonic Debye approximation (QDA) [42]. Later Lu [43] and Meng et al [44] applied QDA to analyze the nano-crystallization of NiP and Ni from the amorphous state and the nanocrystalline γ-Fe→α-Fe phase transformation, respectively. Both the EOS and QDA approaches showed good agreement between the theoretical calculations and the experiments [41-44]. However, these models concentrated attention on interpreting the anomalies in the phase transformation behaviors of nanocrystalline materials as compared with those of polycrystalline specimens, but were not applied to analyze the characteristics of nano-grain growth. Considering that the thermal stability of the nano-grain boundary is remarkably influenced by its inherent non-equilibrium properties, the grain boundary migration is probably controlled by its thermodynamic energy state. Thus, the conclusions concerning the nano-grain growth kinetics without correlating with the thermodynamic features of nano-grain boundaries, are likely to be incomplete or unsubstantiated. Therefore, a model that has been recently developed in our group [37] which correlated the thermodynamics of nano-grain boundaries and the grain growth kinetics in nanocrystalline materials will be depicted in this section. The model can be used to describe the thermodynamic properties, as well as to predict the thermal stability of nanocrystalline metal bulks.

2.3.1. Thermodynamic Functions of Nanocrystalline Interfaces

To describe the structure state of the interfaces in nanocrystalline materials, Fecht [41] and Wagner [42] proposed a dilated crystal model, where a nano-grain boundary is equivalent to a dilated crystal with same crystallographic structure as the grain interior, but it has an expanded unit cell in which the nearest-neighbor separation between atoms is larger than the equilibrium interatomic distance as in a perfect crystal. Thus, a concept of 'excess volume', when compared to that of the perfect crystal, was introduced to characterize the features of the nano-grain boundary structure, which was defined as [41-44]:

$$\Delta V = \frac{V_b}{V_0} - 1 \quad (8)$$

where Vb is the atomic volume in the 'dilated crystal' at the nano-grain boundary, V0 is the atomic volume of perfect crystal in the equilibrium state. Distinctly, ΔV characterizes the deviation extent of the structure of nano-grain boundary with respect to the crystal. The excess volume ΔV, which is directly related to the nano-grain size (there is no conclusive relationship between the grain size and ΔV so far, though, it is recognized that ΔV increases with the decrease of grain size), is the most significant intrinsic parameter that affects the thermal properties of nano-grain boundary, as have been shown theoretically in Ref. [41-44]. Furthermore, it is a key parameter that correlates the thermodynamics of nano-grain boundary with the behavior of nano-grain growth, as will be shown in the following content.

Based on the dilated crystal model, the EOS and QDA approaches need to be combined in order to derive the fundamental thermodynamic functions for the nano-grain boundaries, where the nano-scale effects can be elaborated quantitatively. Assuming that the thermal features of the nano-grain boundary is equivalent to those of the dilated crystal with the same excess volume, the fundamental thermodynamic functions of the nano-grain boundary, including the excess enthalpy ΔH, the excess entropy ΔS, and the excess free energy ΔG, which are generated if compared with the grain interior, are given by [41]

$$\Delta H = E + PV_b \tag{9}$$

$$\Delta S = C_T \gamma \ln(V_b / V_o) \tag{10}$$

$$\Delta G = \Delta H + C_T (T - T_R) - T[\Delta S + C_T \ln(T - T_R)] \tag{11}$$

where E is the binding energy of atoms at the nano-grain boundary [45,46], and can be calculated using the following equation:

$$E = E_0 E^*(a^*) \tag{12}$$

E0 is the binding energy of atoms at the equilibrium state, and can be estimated [47] by

$$E_0 = \frac{(3.45K_B)^2}{12\pi r_0^3 \lambda_0^2 B_0} \tag{13}$$

where λ_0 and B_0 are the zero-pressure linear expansion coefficient and the modulus at the equilibrium state, respectively. K_B is the Boltzman constant. r_0 is the interatomic distance at the equilibrium state (the equilibrium Weigner-Seitz cell radius [48]). $E^*(a^*)$ is a universal function [45] with the expression

$$E^*(a^*) = (-1 - a^* - 0.05a^{*3})e^{-a^*} \tag{14}$$

where

$$a^* = \frac{r_b - r_0}{l} \tag{15}$$

Here rb is the interatomic distance in the dilated crystal at the nano-grain boundary, l is a scaling length characteristic of the metal [49].

P in eq. (9) is a negative pressure developed at the grain boundary as a result of the reduction of the grain size, and can be approximated as a function of the atomic volume at the grain boundary and the absolute temperature [49],

$$P(V_b, T) = \frac{3B_0}{(V_b / V_0)^{2/3}}(1 - (V_b / V_0)^{1/3}) \exp[\eta_0 (1 - (V_b / V_0)^{1/3})] + \alpha_0 B_0 (T - T_R) \tag{16}$$

α_0 is the zero-pressure bulk expansion coefficient at the reference temperature T_R. η_0 is a function of the derivative of the bulk modulus with respect to the pressure at the zero point, and is expressed as]1)/[(2/3$_{00}$−∂∂==PPBη.

γ in eq. (10) is the Grüneisen parameter of the dilated crystal at the nano-grain boundary, and is expressed by eq. (17) as suggested by Wagner [42], CV in eqs. (10)-(11) is the specific heat capacity at constant volume. In EOS, CV was a constant value of 3kB for each atom in the harmonic approximation at high temperature and is independent of ΔV [41]. To include the effect of ΔV, the Debye temperature Θ and Debye function D(Θ/T) [42],

$$\gamma = -1 - \frac{V_b}{2}\left[\frac{\partial^2 P/\partial V_b^2 - 10P/9V_b^2}{\partial P/\partial V_b + 2P/3V_b}\right] \tag{17}$$

which are both influenced by ΔV, are introduced to describe C_V:

$$C_V = 3R\left[12\left(\frac{T}{\Theta}\right)^3 \int_0^{\Theta/T} \frac{x^3}{e^x - 1}dx - \frac{3\left(\frac{\Theta}{T}\right)}{e^{\Theta/T} - 1}\right] \tag{18}$$

where R is the gas constant. The specific heat capacity at constant pressure C_P can be obtained from C_V by the relationship:

$$C_P - C_V = -T\frac{\left(\frac{\partial P}{\partial T}\right)_{V_b}^2}{\left(\frac{\partial P}{\partial V_b}\right)_T} = \frac{\alpha_0^2 B_0^2 V_b T}{B(V_b)} \tag{19}$$

where . $B(V_b) = -V_b(\partial P/\partial V_b)_T$.

The Debye temperature depends on the interatomic distance rb at the nano-grain boundary and is written as [42]

$$\Theta(r_b) = \left[\frac{\varphi'(r_b)}{\varphi''(r_0)}\right]^{1/2} \Theta_0 \tag{20}$$

with $\varphi'(r_b) = \partial\varphi/\partial r_b,\ \varphi''(r_0) = (\partial^2\varphi/\partial r^2)_{r=r_0}$. Θ_0 is the Debye temperature of the crystal in the grain interior at the equilibrium state, ϕ is the Morse function [50] and expressed as

$$\varphi(r_b) = D\{\exp[-4\gamma_0(r_b/r_0-1)] - 2\exp[-2\gamma_0(r_b/r_0-1)]\} \tag{21}$$

D is a fitting parameter, γ_0 is the Grüneisen parameter of the crystal at the equilibrium state. Thus, the dependence of the Debye temperature on the excess volume is derived as

$$\Theta(\Delta V) = \Theta_0 \left\{2\exp[-4\gamma_0((1+\Delta V)^{1/3} - 1)] - \exp[-2\gamma_0((1+\Delta V)^{1/3} - 1)]\right\} \quad \textbf{(22)}$$

Now, the thermodynamic parameters, including specific heat capacities C_V and C_P, the excess enthalpy ΔH, the excess entropy ΔS, and the excess free energy ΔG, are obtained as the functions of the excess volume ΔV and the temperature T from the equations (9)-(19).

2.3.2. Thermodynamic Properties of Nano-Grain Boundaries

To investigate the thermodynamic properties of nanocrystalline metals, the α-Sm (Sm-rhombohedral) bulk with nanocrystalline structure was selected from the rare-earth family as an example. The initial parameters that are needed for the thermodynamic calculations include: r0 = 1.9930Å [46], α0 = 31.2· 10-6/K [51], B0 = 100.5 GPa at TR =300K [52], and Θ0 ≈ 200K as estimated from experimental measurements in our group.

The thermodynamic parameters of nanocrystalline Sm-rhombohedral are shown in Figs. 3-4 as a function of the excess volume at different temperatures. CV approaches to a constant value of 3NkB (N, the Avogadro's number) and is independent of ΔV at high temperatures (Figure 3a). Whereas, there is an abrupt increase of CP above a certain excess volume (Figure 3b). For the thermodynamic functions of nano-grain boundaries, as shown in Figure 4, the excess enthalpy ΔH increases with ΔV (Figure 4a), indicating an increase of the heat storage capacity of the nano-grain boundaries with decreasing the grain size. With the increase of ΔV and hence that of the fraction of atoms at the grain boundaries, the excess entropy ΔS increases (Figure 4b) as a result of expanded volume of the disordered array of atoms. For the excess free energy ΔG as a function of ΔV, it is found that at relatively high temperatures there exists a maximum corresponding to a critical volume, and this critical value decreases with increasing the temperature (as indicated by arrows in Figure 4c). This implies that there will be a distinct change of the thermal stability of the nano-grain structure at the critical volume (or the critical nano-grain size) for a given temperature.

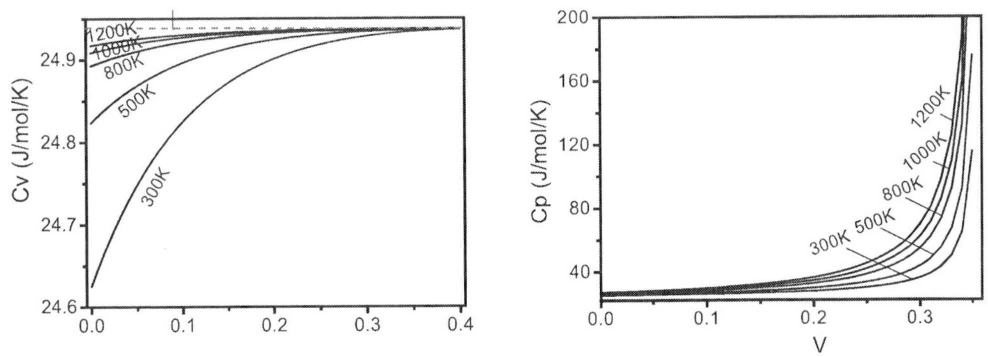

(a) (b)

Figure 3. Calculated dependence of (a) the specific heat capacity at constant volume CV and (b) the specific heat capacity at constant pressure CP on the excess volume ΔV at different temperatures. Nanocrystalline Sm-rhombohedral is used as an example.

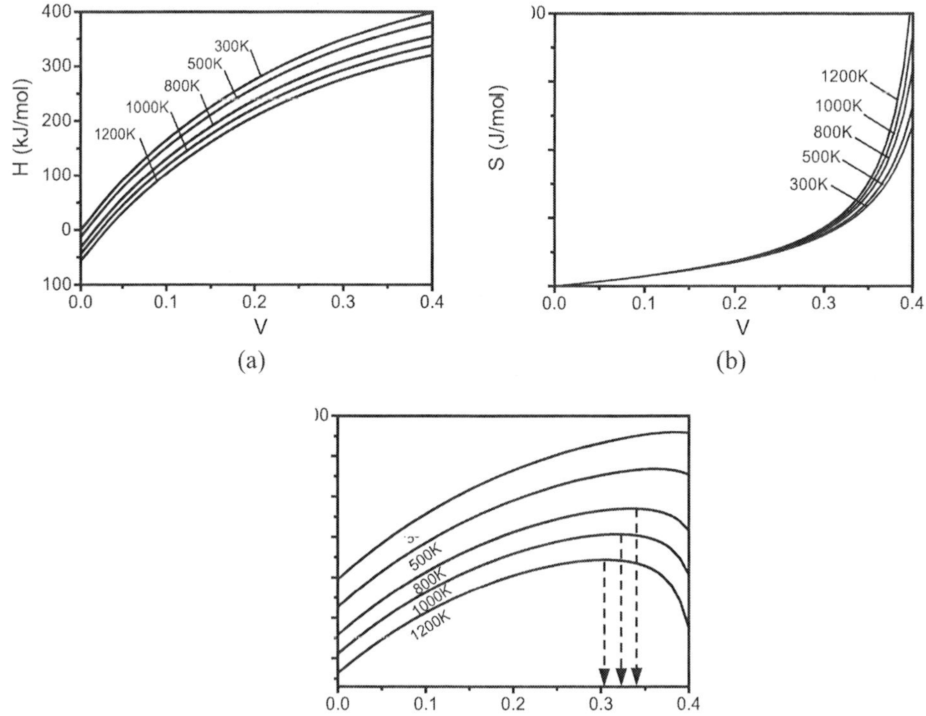

Figure 4. Calculated changes of (a) the excess enthalpy ΔH, (b) the excess entropy ΔS and (c) the excess free energy ΔG of nano-grain boundaries with the excess volume ΔV at different temperatures. Nanocrystalline Sm-rhombohedral is used as an example.

2.4. Thermal Stability of Nanocrystalline Metal Bulk

The grain growth behavior in nanocrystalline metals and alloys has been investigated rather extensively in experiments. It has been found that grains grow very slowly at low homologous temperatures in the nanocrystalline sample while rapidly above a certain temperature [28,29,53,54], and that grain growth becomes negligible for longer annealing times [29,55]. The empirical time exponent n, which is usually used to describe the isothermal grain growth, has been found to have variations of 2, 3 or 4 with the annealing time in nanocrystalline Cu [27], indicating deviation from the conventional polycrystalline metals, where n is close to 2 and is basically independent of time. According to the characteristics of nano-grain growth in the available reports, it is conjectured that grain growth in nanocrystalline materials occurs in a manner different from that in polycrystalline materials, thus is hard to be reasonably described by the conventional theories for the normal coarse-grain growth.

The thermodynamic model that describes the metastable energy state of nano-grain boundaries is more appropriate to explain the grain growth behavior in nanocrystalline metals from the viewpoint of the driving force. It is predicted by the thermodynamic model (Section 2.3) that at relatively high temperatures there exists a maximum in the excess free energy ΔG corresponding to a critical volume ΔV_C.

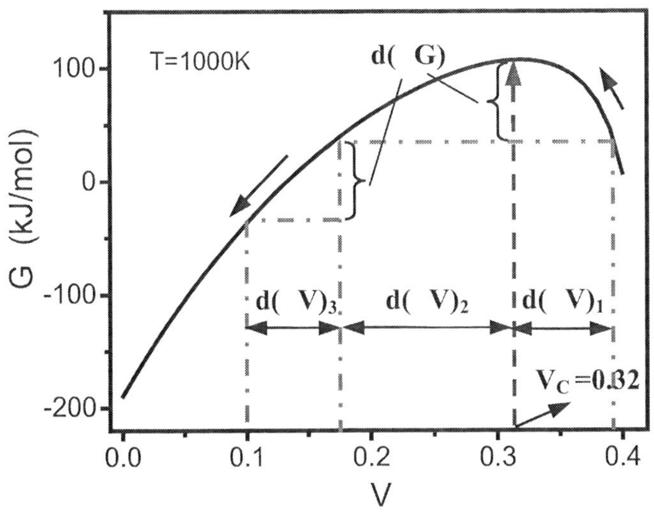

Figure 5. The excess free energy of nano-grain boundaries as a function of the excess volume, calculated for the nanocrystalline Sm-rhombohedral at T=1000K, showing the destabilization possibility in certain conditions.

As an example shown in Figure 5, at a temperature of T=1000K, when ΔV is comparatively smaller or larger than ΔV_C, e.g. $\Delta V = \Delta V_C - d(\Delta V)_2$, or $\Delta V = \Delta V_C + d(\Delta V)_1$, ΔG is lower than the maximum, implying that the system has a relatively more metastable energy state in both coarser- and finer-grain structures than in that with a critical nano-grain size. In the case of $\Delta V > \Delta V_C$, due to the fact that the excess free energy is relatively lower than those in the nano-grain structures with coarser grain sizes, i.e.

$$\Delta G_{(\Delta V = \Delta V_C + d(\Delta V)_1)} \leq \Delta G_{((\Delta V_C - d(\Delta V)_2) \leq \Delta V \leq \Delta V_C)},$$

the nanostructure with small grain size below the critical value may have higher thermal stability than the coarser nano-grain structures. This means that an energy barrier has to be overcome, e.g. by thermal activation, in order to initiate the nano-grain growth, namely, the evolution of the metastable nano-grain structure from very fine grain size towards the more stable coarser-grain structures. Although experimental data on the growth behavior of small grains in a range of a few nanometers are very scarce, a fairly high stability of the nanocrystalline metals with a mean grain size of 5-15 nm has been found in a very recent study [56].

2.4.1. Temperature-Varying Nano-Grain Growth

As indicated in Figure 5, the excess free energy of the nano-grain boundaries deviates from the progressive increase above a critical excess volume ΔVC, which implies that the thermal stability of the nanocrystalline material will change abruptly at a certain critical nano-grain size that is corresponding to ΔVC. This means that for a given nano-grain structure with an excess volume ΔV>ΔVC, the thermodynamically metastable state will be destabilized when external energy is supplied, for example by thermal activation, resulting in a critical temperature TC, at which the nanocrystalline structure will lose the thermal stability due to the high energy state. Except the thermal activation, the energy stored in the nanocrystalline structure, e.g. from the preparation history of the nanocrystalline material, may have the similar effect as supplying energy to trigger the thermodynamic instability in the nanostructure with grain size below the critical value.

It has been reported systematically [54] that for the pure nanocrystalline Co, in a series of one-hour annealing experiments performed in a wide range from room temperature to 1173K, rapid discontinuous grain growth occurred in a narrow intermediate temperature range between 773 and 873K with the mean grain size increased by approximately eight times. Similar behaviors of discontinuous nano-grain growth, on which the effects of solute segregation [28,57,58] and/or pinning of the precipitates [53,55,58] are eliminated, have been observed in nanocrystalline Al [59], the electrodeposited nanocrystalline Ni-Fe alloys with a fcc single phase [29], and the single-phased nanocrystalline Fe–Si and Fe–Si–Al alloys [60], respectively. According to the conditions in which the rapid grain growth occurred [29,54,59,60], it is confirmed that the discontinuous nano-grain growth is driven by the excess free energy of nano-grain boundaries in combination with the thermal energy or the stored energy from the preparation process of the nanocrystalline material.

2.4.2. Isothermal Nano-Grain Growth

As shown in Figure 4(c) and Figure 5, if $\Delta V<\Delta VC$, i.e. the nano-grain size is larger than the critical value, the excess free energy of nano-grain boundaries reduces monotonically with the increase of grain size, implying that the grains will grow continuously from the initial size scale. According to the dependence of ΔG on ΔV when $\Delta V<\Delta VC$, for the same reduction amount of ΔG, the release of the excess volume will decrease with the grain growth process, i.e. $d(\Delta V)2>d(\Delta V)3$ (see Figure 5), indicating that with the grain growth the increase in grain size becomes smaller. More specifically, for the same $d(\Delta G)$ at T=1000K, as marked in Figure 5, in the nano-grain structure having $\Delta V=0.32$ as a starting value, the nano-grains will grow into a grain size corresponding to $\Delta V=0.18$ (with $d(\Delta V)2=0.14$), whereas in a coarser nano-grain structure with $\Delta V=0.18$, the grain growth under the equivalent driving force will lead to a grain size corresponding to $\Delta V=0.10$ (with $d(\Delta V)3=0.08$). Therefore, it is predicted thermodynamically that at a constant temperature, once the nano-grain growth is initiated, the nano-grain grows faster in the stage with smaller grain size than in later stages, thus the rate of nano-grain growth is decreasing monotonically with the increase of grain size, i.e. the isothermal nano-grain growth has a slowing-down kinetics.

The experimental results of isothermal nano-grain growth in the pure nanocrystalline Co at T=803K was demonstrated in reference [37]. It was observed that the finer grains have a higher growing rate than the coarser ones. The grain growth exponent, n, which was estimated by the measurement of grain sizes at different times, was found to vary approximately in a range of 1.1~ 4.4. As exhibited in experiments, for the isothermal nano-grain growth, the growth rate decreases monotonically with time (or grain size), indicating that at the critical temperature the rapid nano-grain growth takes place at an early stage of anneal holding. The experimental finding of the slowing-down kinetics of isothermal nano-grain growth in the nanocrystalline metal bulk agrees well with the thermodynamic predications.

To summarize, the high energy states of the nanoparticles and the nanocrystalline bulks can be described by the thermodynamic models developed by aiming at features of nanostructures. The derived dependence relationships of the free energy and the melting temperature on the particle size provide a guide for the safekeeping and characterization of metal nanoparticles, as well as the design of the consolidation processing for the preparation of nanocrystalline bulks. The predicted thermodynamic properties and the thermal stability of the nanocrystalline metals, help to control the nano-grain growth and to investigate the physical and mechanical properties of nanocrystalline metal bulks as well as their nano-scale dependence. In the following sections, the preparation and characterization of nanoparticles and nanocrystalline bulks of rare-earth metals will be described.

3. Preparation and Characterization of Nanoparticles of Rare-Earth Metals

Studies on synthesis technologies and the related nanoparticle-based metal nanostructures have attracted great interests with their increasingly wide applications in the fields of information, energy, environmental and medical technologies. Unagglomerated nanoparticles with narrow size distribution are preferred for practical applications and technologies starting from nanoparticles. A number of techniques for the preparation of nanoparticles that satisfy this requirement have been developed mainly including liquid and gas phase processes. There are various liquid phase methods developed for preparing metal nanoparticles [61], such as the chemical reduction, sol-gel, reverse micelle process, pyrolysis, and spray pyrolysis methods. For the gas phase methods, condensation of the corresponding metal vapors is one of the predominant technologies for synthesis of metal nanoparticles [14,15,61]. At present characterization of nanoparticles is concentrated on measurements of particle diameter and particle size distribution, observations on morphology/shape and surface characteristics.

For the particularity of the extremely high chemical activity of pure rare-earth particles, hence the requirement of high purity for the safekeeping and the further usage of rare-earth nanoparticles, a technique of inert-gas condensation that applies processes of solid $\leftrightarrow$ gas phases transformation in an enclosed system will be introduced in this section for the preparation of rare-earth nanoparticles.

3.1. Preparation of Rare-Earth Nanoparticles

The inert-gas consolidation method was first generated by Gleiter and coworkers for preparing the materials that demonstrated the exciting properties of nanostructured materials [14]. In their method, a metal was typically vaporized into a low density gas by electrical heating, then metal vapors migrate from the hot source into a cooler inert gas. The decreasing temperature leads to a rapid reduction in the equilibrium vapor pressure and correspondingly a high supersaturation state. Subsequently the vapors rapidly nucleate, forming a large number of nanoparticles.

The inert-gas condensation for preparing rare-earth nanoparticles was carried out with a home-made instrument in the author's group, which was appended with forced inert-gas flow and surrounding cooling of the evaporation chamber with cyclic cold-water, as compared with some previously reported inert-gas condensation instruments [62], in order to enhance the deposition efficiency and the dispersing degree of the produced nanoparticles. It was found that the inert-gas pressure, the applied voltage, the input current and the distance between the cathode and the vaporizing metal, as well as the vapor pressure of the metal, all influence to different contents the resultant particle size and the production rate of the nanoparticles. In the highly purified inert (Ar or He) gas at a pressure of ~10-1 Torr, the mean size of the condensed particles could be controlled in the range of 20-80 nm when the applied voltage was adjusted from 10 to 20V, the input current from 80 to 200A, and the evaporation time from 0.5 to 3 hours. When the other conditions are fixed, as the inert-gas pressure increases, both the mean size and the production rate of the rare-earth particles increase as a result of the higher density of plasma generated between the two electrodes, hence the higher collision frequency of extremely small particles that leads to coagulation and growth of particles. Increasing the voltage and/or the current intensity result in a higher production rate and probably coarser particle size as a reason of the temperature increase in the electrode zones. With reducing the distance between the cathode and the vaporizing metal, the evaporation process becomes faster, and the production rate is increased.

A series of experiments showed that the production rate depends strongly on the vapor pressures of rare-earth metals. Generally, the larger the difference between the melting and the boiling temperatures of the rare-earth metal, the higher is the transition temperature for the sharp increase of the vapor pressure of the rare-earth, and the lower is the production rate of the rare-earth nanoparticles by the inert-gas condensation method. A figure demonstrating this rule is shown in Figure 6 including most rare-earths. In principle, any metals whose transition temperatures for the sharp increase of the vapor pressures are not very high can be possibly used to prepare the nanoparticles by the inert-gas condensation method. Those rare-earth metals that nanoparticles have been made in our work are shown with bold lines in Figure 6. Table 1 shows correspondingly the production rate of the rare-earth nanoparticles made by inert-gas condensation.

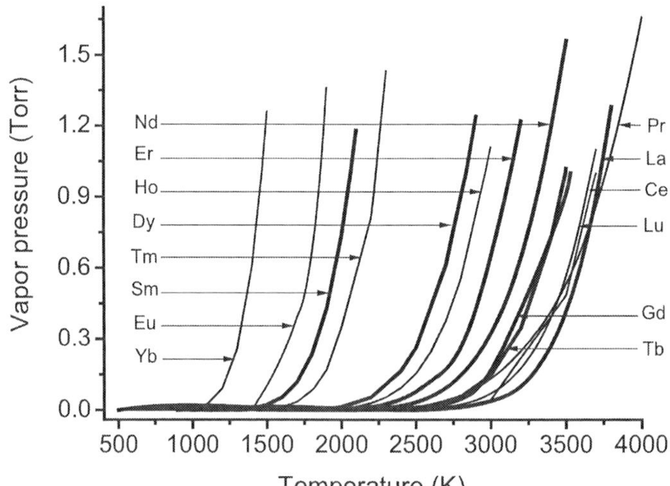

Figure 6. The vapor pressures of rare-earth metals via temperature.

Table 1. The production rates of rare-earth nanoparticles prepared by inert-gas condensation, together with the melting and boiling points of the rare-earths

Rare-earth metals	Sm	Dy	Er	Nd	Tb	Gd	La
Melting point (K)	1345	1682	1795	1289	1630	1585	1193
Boiling point (K)	2064	2835	3136	3341	3496	3539	3730
Production rate (g/h)	300	200	150	80	50	50	5

3.2. Microstructure of Rare-Earth Nanoparticles

Since the pure rare-earth nanoparticles are so active that become burning immediately when they are exposed in the air, the prepared rare-earth nanoparticles were preserved in the glove box filled with highly purified argon gas. Also because of their extremely high chemical activity, at present the rare-earth nanoparticles cannot be detected by X-ray diffraction (XRD) method yet. The transmission electron microscopy (TEM) was chosen as a master characterization method for the shape, size and size distribution, and crystal structure of rare-earth nanoparticles. The nanoparticles were put into the acetone while in the glove box and subsequently fetched onto the copper-grid film for the TEM observation.

The TEM images of the prepared rare-earth nanoparticles are shown in Figure 7, using Sm, Nd, Dy and Tb as examples, together with the high-resolution TEM (HRTEM) images of single rare-earth particles, selected area electron diffraction (SAED) patterns and its indexing of the particles, and the particle size (diameter) distributions. It is observed from Figure 7 that the inert-gas condensed rare-earth nanoparticles have a comparatively homogeneous size distribution with mean diameters of 25-40nm. The nanoparticles keep basically the same crystal structure of the raw polycrystalline rare-earth metal, as indicated by the indexing of the SAED patterns of the particles. In an overview, each nanoparticle is a single-crystal.

4. Preparation of Nanocrystalline Bulks of Rare-Earth Metals

4.1. Preparation Route of Nanocrystalline Bulks

In the past two decades, a number of techniques have been developed to synthesize nanocrystalline bulk materials, such as inert-gas condensation and consolidation [63], electrodeposition [64], severe plastic deformation [65], crystallization of amorphous solid [66], surface mechanical attrition [67], and powder metallurgy [68-70]. However, it is hard to produce nanocrystalline bulk materials with controllable grain sizes in a wide range below 100 nm. Particularly, in conventional powder metallurgy processes, the rapid coarsening of the particles occurs very often, even with the nanoparticles as the starting powder, grains in the scale of submicro- or even micrometer may form in the resultant bulk. Thus, in the preparation of bulk materials by the conventional powder-metallurgical methods, the grain size in the synthesized bulk is generally larger than the initial particle size [68-70].

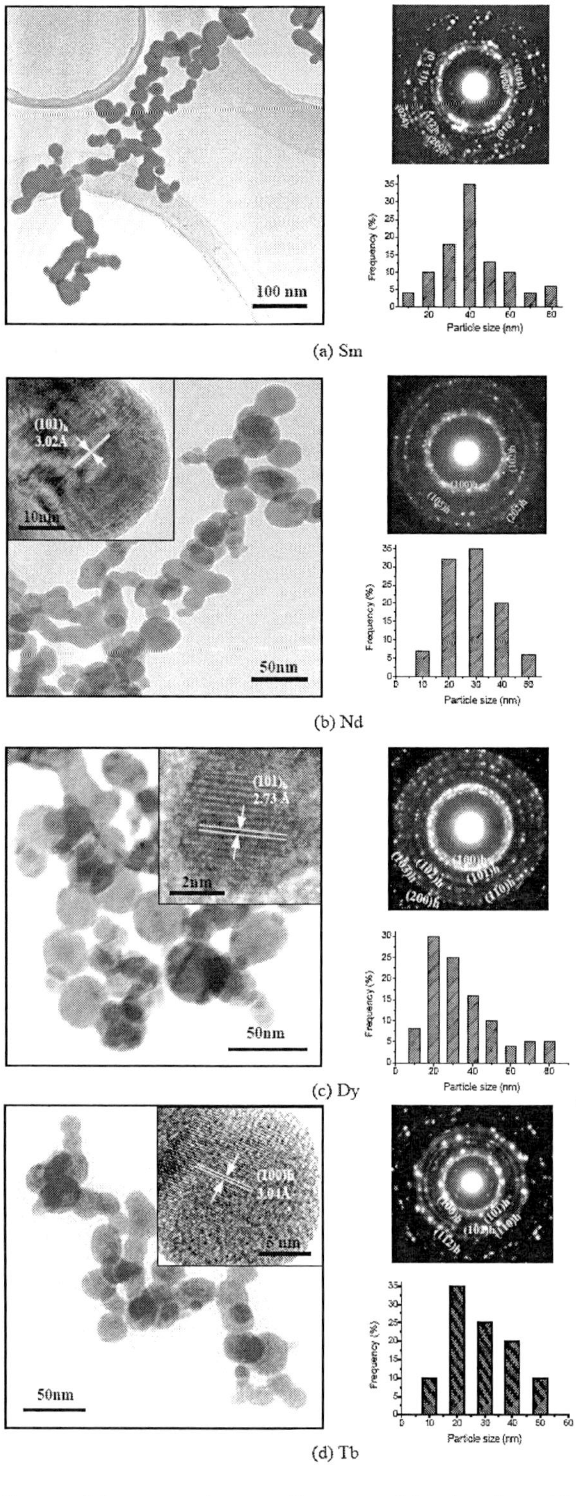

Figure 7. TEM images of some of the prepared rare-earth nanoparticles, together with the HRTEM images of single rare-earth particles (inset in each TEM image), SAED patterns and its indexing (upper right in each partial figure), and size distributions (lower right in each partial figure) of the nanoparticles of (a) Sm, (b) Nd, (c) Dy and (d) Tb, respectively.

An advanced powder metallurgical technology, so-called as spark plasma sintering (SPS), has been fast developed in recent years. It is a solid consolidation process that has the key advantage of the very high degree of densification obtained at low sintering temperatures and within a short time, compared with the conventional sintering methods. The SPS process applies simultaneously high-intensity pulsed direct current and uniaxial pressure, which offers the possibilities of using rapid heating rates (e.g. achieving 1000°C/min) and very short holding times (e.g. 3-5min) to obtain highly dense samples [71-73]. As a reason, grain growth in the sintering samples can be effectively restrained. Up to now the SPS technology has been applied to fabricate various materials including metal and alloys, compounds, ceramics, composites, especially, some new-type materials such as bulk amorphous and nanocrystalline materials [74], multi-scaled structure and function graded materials [75] and etc.

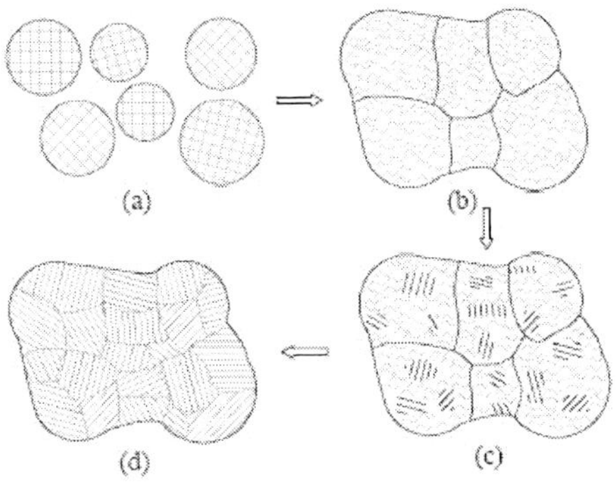

Figure 8. Schematic diagram of preparing metal bulk with the ultrafine nanocrystalline structure with (a) single-crystal nanoparticles, (b) amorphization of nanoparticles in SPS consolidation process, (c) nucleation and growth of the short-range ordered "clusters" inside the nanoparticle, and (d) complete nanocrystallization.

A novel technique combining inert-gas condensation with the SPS technology has been recently established in our group [52,56]. Taking into account the extremely high chemical activity of rare-earth nanoparticles, we integrated the configuration of inert-gas condensation with SPS equipment into an entirely closed and vacuumized system, named as "oxygen-free" in-situ synthesis system where the oxygen concentration is controlled below 0.5ppm. According to the special mechanisms of SPS consolidation, which were proposed [71,72] and developed [76,77] in the literature, when the pulsed current passes through the powder of rare-earth metals that has a high electrical conductivity while a relatively low heat conductivity, a high temperature can be reached rapidly in the powder particles. Based on the model calculations [76], at local sites of the sintering particles, the temperature induced by the Joule-heating can be much higher than that measured by the thermocouple placed in the wall of the graphite die, and there is a sharp increase of the temperature from the particle body to the particle-to-particle interface. Since the temperatures of phase transition of the inert-gas condensed nanoparticles are reduced due to the nano-scale effect as referred to Section 2.1, the high temperature induced by the Joule-heating at the local site of each nanoparticle can exceed its glass transition temperature, and can even reach the melting point at the interfaces between the nanoparticles. As a result, the nanoparticles are likely to have a metallic-glass (amorphous) body and a melted boundary surface. Enlightened by the experimental finding that a metastable amorphous state exists in the compact powder sample in SPS process under certain conditions of sintering parameters and nanoparticle size classes, a novel preparation scheme for metal bulks with ultrafine nanocrystalline structures was proposed for the first time by the author's group. The procedure is composed of sequentially arranged processes of: amorphization of nanoparticles, nucleation and growth of the short-range ordered "clusters" inside the nanoparticle, and the complete nanocrystallization. The diagram of the above route is shown in Figure 8.

4.2. Rare-Earth Bulks with Ultrafine Nanocrystalline Structure

Using the route introduced in Section 4.1, pure metal bulks of a series of typical members in the rare-earth family, e.g. Sm, Gd, Nd, Dy, Tb and Er, have been prepared with the ultrafine nanocrystalline structures.

For different rare-earth metals, the processing parameters in the preparation procedure were modified depending on the particle size and the melting temperature of the rare-earth nanoparticles. For the modification, the prediction on the size dependence of the melting temperatures of rare-earth nanoparticles, which was proposed in Section 2.2, was used as the foremost reference data. In a glove box connected with the powder reservoir of the inert-gas condensation apparatus, the prepared rare-earth powder was fed into the high-strength graphite die, which was subsequently sent to the SPS equipment (SPS-5.40-IV/ET, Sumitomo Coal Mining Co., Ltd) by a sliding rail. An external pressure was then applied to the powder through the upper and lower punches. In the sintering process, the external pressure was kept constant, and the temperature, the heating rate and the holding time were adjusted in order to achieve the consolidated bulks with nearly full relative density and desired nanocrystalline microstructures. The details in the step sintering processes to obtain the amorphous matrix and the final nanocrystallized structure can be found in references [52] and [56] by using the preparation of nanocrystalline Sm as an example. The microstructures of the prepared nanocrystalline rare-earth bulks, using some members of the rare-earth family as examples, i.e. Sm, Nd, Dy and Tb, are shown in Figure 9.

It is observed that compact ultrafine nano-grains formed in the rare-earth bulks with the grain sizes of 5-15 nm, an example displaying by arrows the different orientations of ultrafine nano-grains is given in Figure 9(a). The crystal structures of the prepared nanocrystalline rare-earth bulks were kept the same as those of the raw polycrystalline metals at the room temperature, as analyzed by the SAED patterns. Especially, the grain sizes in the resultant rare-earth bulks (ca. 10-15nm on average) are distinctly smaller than the initial nanoparticle sizes (ca. 40-60nm on average), which may be the first demonstration that the traditionally accepted relationship, that the grain size in the sintered bulk is larger than the initial nanoparticle size, can be changed. Applying the process of nucleation and growth of the short-range ordered "clusters" in the amorphous matrix inside the nanoparticle is the most significant advantage of the preparation route proposed in Section 4.1. In virtue of this technique, nanocrystalline bulk materials with controllable grain sizes can be obtained, which will improve the material properties in various aspects by applying the scale effects; moreover, the prepared ultra-fine nanocrystalline bulks can be used to study some special features of nanostructured materials.

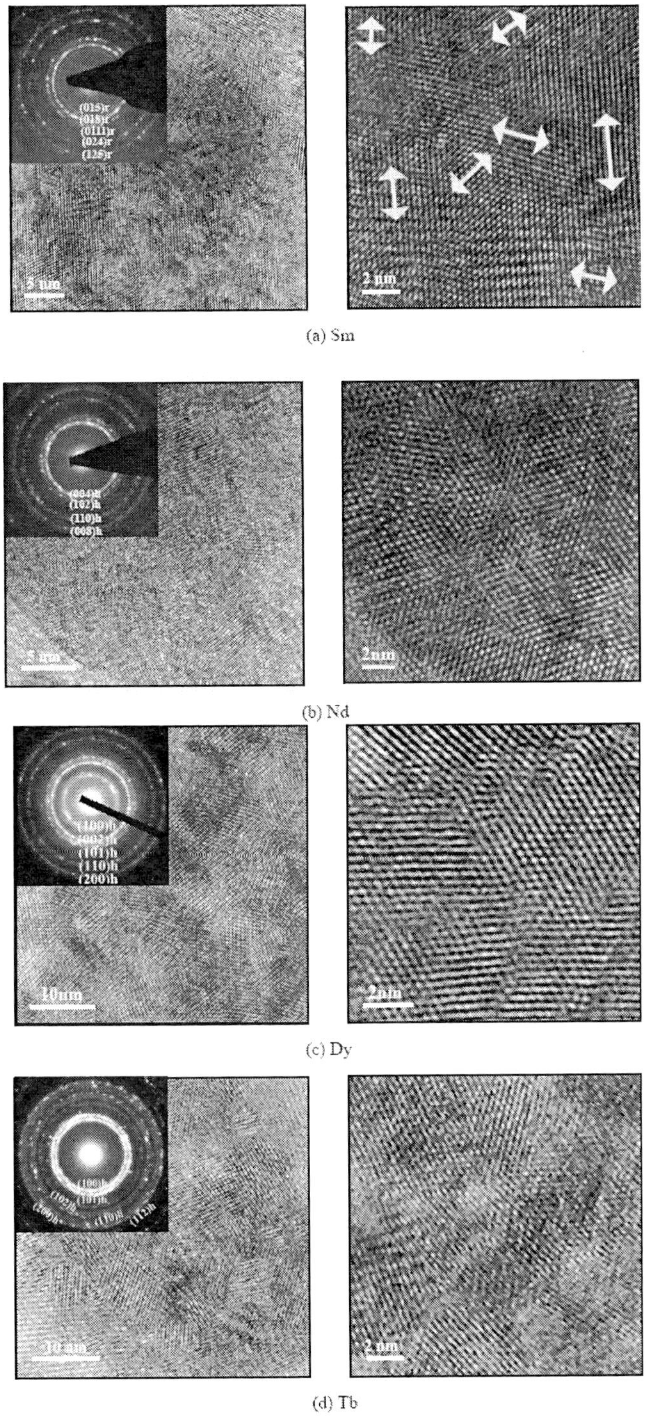

Figure 9. The nanocrystalline microstructures of rare-earth metals prepared by combining inert-gas condensation and SPS, shown with HRTEM images (left in each partial figure), SAED patterns and its indexing (upper left in each partial figure), and locally enlarged areas revealing ultrafine nano-grains (right in each partial figure) of (a) Sm, the arrows indicating the orientations of nano-grains, (b) Nd, (c) Dy and (d) Tb, respectively.

5. Properties Characterization of Nanocrystalline Rare-Earth Metals

It is considered that as the nanocrystalline microstructures of rare-earth metals are obtained, especially when the ultrafine nano-grain structures are achieved, the rare-earth metals would present different properties from those of the polycrystalline rare-earths. In order to characterize the properties of nanocrystalline rare-earth metals, the thermal, physical and mechanical properties of the prepared nanocrystalline bulks were measured for the typical members in the rare-earth family. For the sake of comparison, the measurements of properties on the polycrystalline rare-earth metals were performed correspondingly. To keep the purity of the samples, all these measurements were performed under highly pure argon or helium gas to protect from oxidization or reactions.

The remarkably changed thermal properties (e.g. specific heat capacity), physical properties (e.g. electrical resistivity and thermal conductivity) and mechanical properties (e.g. micro-hardness and bulk modulus of elasticity) of nanocrystalline Sm, and magnetic properties (e.g. Curie transition temperature and magnetic moment) of nanocrystalline Gd, have been reported in literature [56,78]. Here a detailed demonstration of the properties of nanocrystalline rare-earth metals is given by using Nd and Dy as examples.

5.1. Thermal Properties

5.1.1. Specific Heat Capacity

The isobaric heat capacities (Cp) were measured on both the prepared nanocrystalline and the raw polycrystalline rare-earth bulks by two apparatuses of DSC 404C, PEJASUS and DSC Q1000, TA Instruments Inc. respectively for each sample, in order to verify the data repeatability.

The curves of the specific heat capacities of Nd and Dy bulks are shown in Figure 10 for the nanostructured samples and the raw polycrystalline metals, respectively. It is noted that the specific heat capacity of the nanocrystalline bulk has a stronger increasing tendency with the temperature than that of the polycrystalline metal. Furthermore, compared with the conventional polycrystalline metals, the specific heat capacities of the ultrafine nanocrystalline rare-earth bulks are apparently larger in the whole range of the testing temperature. For example, compared with the Cp values of the raw polycrystalline bulks, which are ca. 32.3 J/mol/K for Nd and 28.5 J/mol/K for Dy at 350oC, the Cp values of the ultrafine nanocrystalline Nd and Dy bulks are ca. 47.0 and 35.4 J/mol/K at the same temperature, which have been increased approximately by 50% and 20%, respectively. As is known that the heat capacity of a material is closely related to its vibrational and configurational entropies of atoms [79,80], the increase of the heat capacity in the ultrafine nanocrystalline bulks reflects the substantial contribution of the highly dense nano-grain boundaries to the thermodynamic features of the material. of (a) Nd and (b) Dy, together with the comparison with those of the corresponding raw polycrystalline metals.

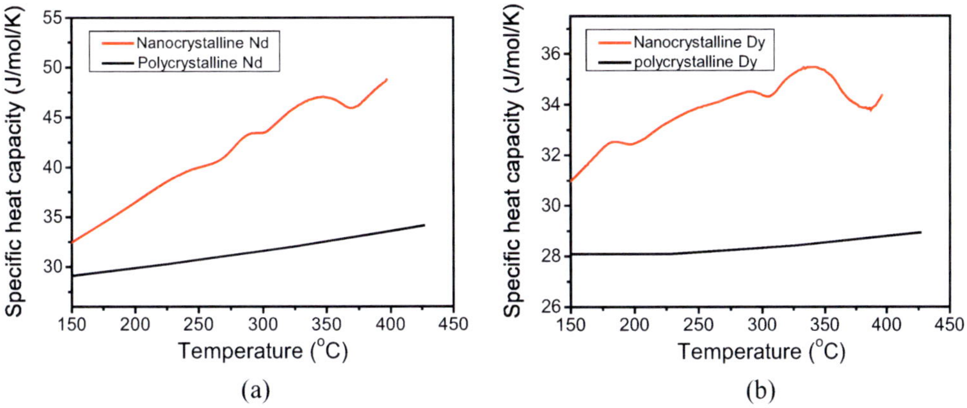

Figure 10. The specific heat capacities of the prepared ultrafine nanocrystalline bulks of (a) Nd and (b) Dy, together with the comparison with those of the corresponding raw polycrystalline metals.

5.1.2. Thermal Expansion Coefficient

The thermal expansion coefficients (the linear expansion, αL) were measured for the prepared nanocrystalline and the raw polycrystalline rare-earth bulks with the NETZSCH DIL 402EP apparatus, respectively.

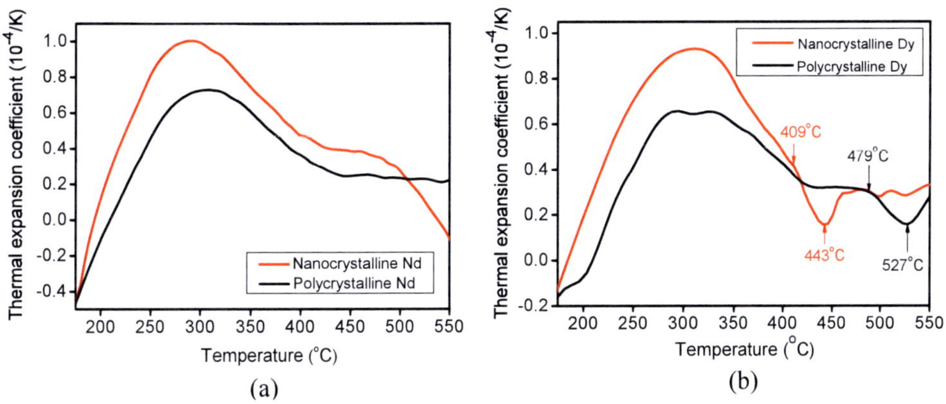

Figure 11. The thermal expansion coefficients of the prepared ultrafine nanocrystalline bulks of (a) Nd and (b) Dy, together with the comparison with those of the corresponding raw polycrystalline metals.

The variations of the thermal expansion coefficients of Nd and Dy bulks with the temperature are shown in Figure 11 for both the nanocrystalline samples and the raw polycrystalline metals, respectively. In the range of the testing temperature, the αL value of the ultrafine nanocrystalline bulk is generally larger than that of the polycrystalline metal, except some points corresponding to phase transformations. The appearance of sharp wave in the αL curve reveals phase transformation at the corresponding temperature. It is observed in Figure 11(b) that, the starting temperature of phase transformation from hexagonal to rhombohedral crystal structure is about 479°C in the polycrystalline Dy, and the temperature corresponding to the wave hollow is about 527°C (which is consistent with the value in thermodynamic database). However, these two temperatures are about 409°C and 443°C in the ultrafine nanocrystalline Dy bulk, respectively. Therefore, the starting temperature for phase transformation in the ultrafine nanocrystalline Dy bulk is reduced by 70°C from that of the polycrystalline metal. Similarly, the starting temperature for hexagonal → rhombohedral phase transformation in the ultrafine nanocrystalline Tb bulk is found reduced by about 65°C from that of the polycrystalline Tb.

5.2. Physical Properties

5.2.1. Electrical Resistivity

The electrical resistivities were measured for the prepared nanocrystalline and the raw polycrystalline rare-earth bulks with the Model ZEM-2, ULVAC apparatus, respectively.

Figure 12 shows the changes of the electrical resistivities of Nd and Dy bulks with the temperature for the nanocrystalline samples and the raw polycrystalline metals, respectively. As shown in the figure, the electrical resistivities of all the rare-earth samples increase with temperature as expected in normal electrical property of metal materials. However, at the same temperature, the electrical resistivities of the prepared ultrafine nanocrystalline rare-earth bulks are higher than those of the raw polycrystalline metals, which can be explained by the "quantum confinement" [14,15], which is essentially due to the changes in the atomic structure as a result of the direct influence of the ultra-small length scale on the energy band structure. The characteristic electronic properties of the nanocrystalline structures are a result of tunneling currents, purely ballistic transport and the coulomb blockade effects [61].

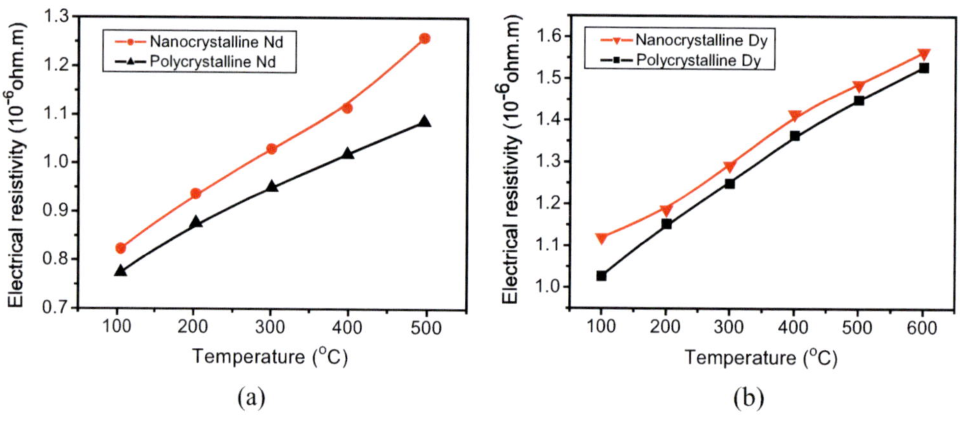

Figure 12. The electrical resistivities of the prepared ultrafine nanocrystalline bulks of (a) Nd and (b) Dy, together with the comparison with those of the corresponding raw polycrystalline metals.

5.2.2. Thermal Conductivity

The thermal conductivities were measured for the prepared nanocrystalline and the raw polycrystalline rare-earth bulks with the TC-7000H apparatus, respectively.

Figure 13 shows the changes of the thermal conductivities of Nd and Dy bulks with the temperature for the nanocrystalline samples and the raw polycrystalline metals, respectively. It is noted that the thermal conductivities of Nd samples vary little with the temperature, while those of Dy samples increase with the temperature. For the same rare-earth, the thermal conductivity of the prepared ultrafine nanocrystalline bulk is apparently lower than that of the polycrystalline metal, which is contrary to the changing tendency of the electrical resistivity as the microstructure transforms from polycrystalline to nanocrystalline. As compared with the raw polycrystalline metal, the thermal conductivity of the ultrafine nanocrystalline bulk is reduced by ca. 78% on average for Nd and by 27-31% for Dy.

In general, the thermal conductivity of a material contains two portions of contributions [2,81], one is generated from the phonon scattering, another is from the current carrier. In the ultrafine nanocrystalline bulk, the larger fraction of the nano-grain boundaries increases apparently the probability of phonon scattering, hence reduces the thermal conductivity from the phonon scattering effect. On the other hand, the high density of the nano-grain boundaries increases the resistance to the motion of the current carrier and reduces the migration rate, thus increases the thermal conductivity from the contribution of the current carrier. However, as the portion of the thermal conductivity from the phonon scattering effect has a much stronger impact on the total thermal conductivity of the material, the thermal conductivity of the ultrafine nanocrystalline bulk is lower than that of the raw polycrystalline metal.

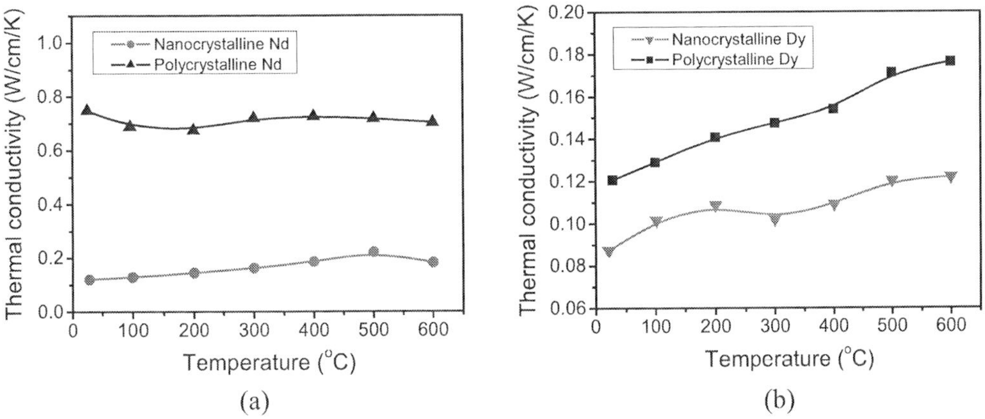

Figure 13. The thermal conductivities of the prepared ultrafine nanocrystalline bulks of (a) Nd and (b) Dy, together with the comparison with those of the corresponding raw polycrystalline metals.

5.3. Mechanical Properties

The mechanical properties of the nanocrystalline rare-earth bulks are characterized by the micro-hardness and the modulus of elasticity, measured by the nano-indentation method [82] with the Hysitron TriboIndenter apparatus.

5.3.1. Micro-Hardness

The measurements of the micro-hardness on the nanocrystalline and the raw polycrystalline Nd and Dy bulks, respectively, are shown in Figure 14. With the loads in a range of 500-4000 μN, the values of the micro-hardness become stable when the load is increased to above 2500μN. It can be seen that the micro-hardness of the prepared ultrafine nanocrystalline rare-earth bulks are apparently higher than those of the raw polycrystalline metals. For example, the micro-hardness are 3.1 GPa for ultra-fine nanocrystalline and 1.3 GPa for polycrystalline Nd bulks, respectively, which is approximately 1.4 times increased.

For Dy, the micro-hardness of the ultra-fine nanocrystalline bulk is approximately 2.5 times as high as that of the raw polycrystalline metal.

5.3.2. Modulus of Elasticity

Figure 15 shows the measured modulus of elasticity of the nanocrystalline and the raw polycrystalline Nd and Dy bulks, respectively. The similar feature as that of micro-hardness is found, i.e. the modulus of elasticity of the prepared ultrafine nanocrystalline rare-earth bulks are much higher than those of the raw polycrystalline metals. For example, the modulus of elasticity are 64.5 GPa for ultra-fine nanocrystalline and 28.0 GPa for polycrystalline Nd bulks, respectively, which is increased approximately as 2.3 times high. For Dy, the modulus of elasticity of the ultra-fine nanocrystalline bulk is approximately 2.0 times as high as that of the raw polycrystalline metal.

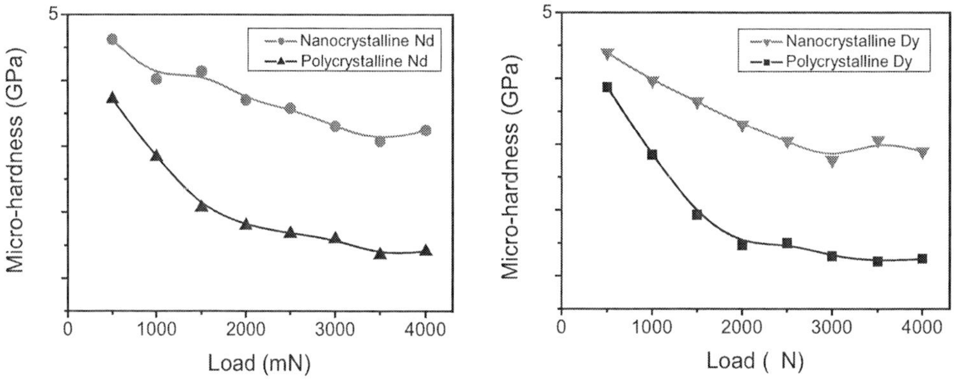

(a) (b)

Figure 14. The micro-hardness of the prepared ultrafine nanocrystalline bulks of (a) Nd and (b) Dy, together with the comparison with those of the corresponding raw polycrystalline metals.

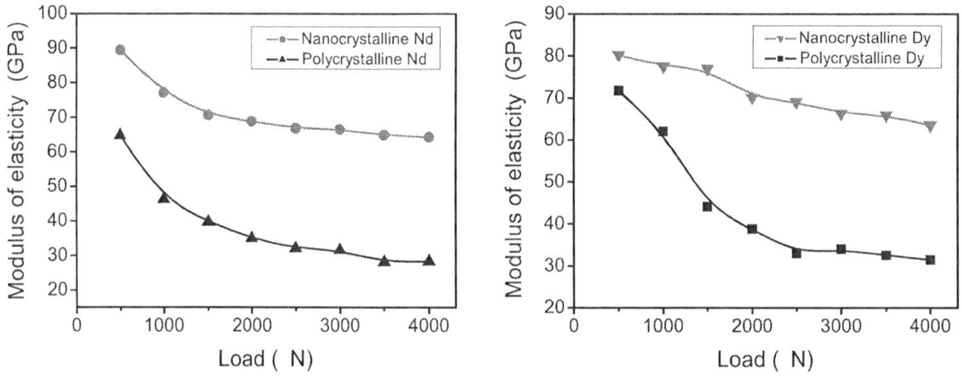

(a) (b)

Figure 15. The modulus of elasticity of the prepared ultrafine nanocrystalline bulks of (a) Nd and (b) Dy, together with the comparison with those of the corresponding raw polycrystalline metals.

As demonstrated in Section 5, compared with the conventional polycrystalline rare-earth metals, the prominent improvement of the thermal, physical and mechanical properties of the prepared ultrafine nanocrystalline rare-earth bulks have been found, which indicates promising applications of the nanocrystalline rare-earth metals and alloys in functional materials, such as heat-storage and thermal-electric materials, as well as in structural materials, such as rare-earth-light metal alloys, nanocrystallized rare-earth-based metallic glass, and rare-earth doped composites.

6. Conclusion

Since the nanostructured rare-earth metals are significant in fundamental research and technological applications, whereas there are very few journal reviews or text books that comprehensively deal with the features of nanoparticles and nanocrystalline bulks of the rare-earth family, an effort has been made in this chapter to depict the energy state, thermal stability, microstructure and property of rare-earth nanoparticles and nanocrystalline bulks, respectively, which are research fields of our specialization.

The fundamental thermodynamic functions of metal nanoparticles and nanocrystalline bulks deduced by our group were introduced firstly, which describe the high energy state and the dependence of thermodynamic properties on both the temperature and the nanoparticle size or the nano-grain size. The predicted melting temperatures of the rare-earth nanoparticles as a function of the particle size help to design the consolidation processing for preparing the nanocrystalline rare-earth bulks. Based on calculations of the excess free energy, the thermal stability of nanocrystalline bulks, including the temperature-varying and the isothermal nano-grain growth behavior, can be quantitatively described. The possible occurrence of discontinuous rapid grain growth in the intermediate temperature range for the temperature-varying case, and the monotonically decreasing nano-grain growth rate with the annealing time for the isothermal case, were predicted for rare-earth nanocrystalline bulks. The thermodynamics-based description of the thermal stability provides a guide for the control of nano-grain growth in rare-earth nanocrystalline bulks in the situation of hot-working treatment on them after preparation.

The preparation of nanoparticles and nanocrystalline bulks of pure rare-earth metals was then introduced. A home-made configuration of combining the inert-gas condensation with the SPS technology in an entirely closed and vacuumized system, where the oxygen concentration is controlled below 0.5ppm, has been established in our group by taking into account the extremely high chemical activity of nanoscale rare-earths. The structures, as well as the nano-grain sizes in the consolidated rare-earth bulks, can be tailored either by changing the nanoparticle size or by modifying the sintering parameters. Ultrafine nanocrystalline bulk can be obtained with the grain size much smaller than the initial nanoparticle size using sequentially performed processes, i.e. amorphization of nanoparticles, nucleation and growth of the short-range ordered "clusters" inside the nanoparticle, and the complete nanocrystallization. This technique has a primary advantage of an "oxygen-free" in-situ synthesis of the pure rare-earth bulks with controllable nanostructures over the conventional power metallurgy techniques for nanomaterials preparation, it can be potentially used to prepare any nanocrystalline bulk materials if the metal precursors can be thermally evaporated without much difficulty.

The properties of the prepared nanocrystalline rare-earth bulks were introduced subsequently. Particularly, the thermal properties (e.g. specific heat capacity), physical properties (e.g. electrical resistivity and thermal conductivity) and mechanical properties (e.g. micro-hardness and modulus of elasticity) of the ultrafine nanocrystalline rare-earth bulks were characterized and compared with those of the corresponding raw polycrystalline rare-earth metals, respectively. The prominently changed thermal, physical and mechanical properties imply the promising applications of the nanocrystalline rare-earth metals and alloys in the fields of functional and structural materials. Accordingly, the advanced properties and important applications of nanoscale rare-earth metals will stimulate increasingly great research efforts in the future, and the challenge may lie in the integration of nanostructured rare-earths into the devices and components.

Finally, although the focus of this chapter are rare-earth nanoparticles and nanocrystalline bulks, the proposed thermodynamic models and preparation techniques are applicable to a large variety of nano metal materials, hence establish a new route for the studies on microstructures and performances of metallic nanomaterials, especially of those highly reactive in the air.

ACKNOWLEDGEMENTS

Our nanoscience and nanotechnology research is supported by the National Natural Science Foundation of China (50401001, 50671001), the Program for New Century Excellent Talents in University (NCET 2006), the Key Project of Science and Technology Innovation Engineering, Chinese Ministry of Education (705004), and the Program of Beijing New Star of Science and Technology (2004B04). The author sincerely acknowledges Prof. Jiuxing Zhang, from Beijing University of Technology, and Markus Rettenmayr, from Friedrich-Schiller-University Jena, for their valuable comments and suggestions. The author would also like to acknowledge her collaborators and co-researchers.

REFERENCES

[1] Gschneider, K. A. In Book *The Rare Earth*; Spedding, F. H.; Daane, A. H. Ed.; 1st Ed.; J Wiley Press: New York, 1961, pp 5-20.

[2] Gschneider Jr., K. A.; Eyring, L. *Handbook on the Physics and Chemistry of Rare Earths;* ISBN 0 444 85022 8; North-Holland Publishing: Amsterdam, 1978; Vol. 1, pp 1-15; Vol. 2, pp 164-170; Vol. 16, pp 107-120.

[3] Uda, T.; Jacob, K. T.; Hirasawa, M. *Science* 2000, 289, 2326-2331.

[4] Harrison, P. G.; Bailey, C.; Azelee, W. J. *Catalysis*. 1999, 186, 147-159.

[5] Laurent, C.; Zoltán, A.; Gál, T. P.; Braun, F. J.; DiSalvo, B. B.; Jürgen, H. M. *J. Sol. St. Ch.* 2001, 162, 90-95.

[6] Legrand, B. A.; Perrier de la Bathie, R.; Tournier, R.; Chateigner, D. *J. Magn. Magn.* 1997, 173, 20-28.

[7] Kerschl, P.; Handstein, A.; Khlopkov, K.; Gutfleisch, O.; Eckert, D.; Nenkov, K.; Téllez-Blanco, J.-C.; Grössinger, R.; Müller K.-H.; Schultz, L. *J. Magn. Magn.* 2005, 290-291, 420-423.
[8] Yamaguchi, K.; Itagaki, K.; Kim, D.-Y.; Ohtsuka, M. *J. Alloy. Com.* 1995, 221, 161-168.
[9] Amadeh, A.; Pahlevani, B.; Heshmati-Manesh, S. *Corros. Sci.* 2002, 44, 2321-2331.
[10] Fray, D. J. *Science* 2000, 289, 2295-2300.
[11] Worrell, E.; Martin, N.; Price, L. *Energy* 2000, 25, 1189-1214.
[12] Dai, W.; Shen, B. G.; Li, D. X.; Gao, Z. X. *J. Alloy. Com.* 2000, 311, 22-25.
[13] Sellmyer, D. J.; Skomski, R. *Scrip. Mater.* 2002, 47, 531-535.
[14] Gleiter, H. *Acta Mater.* 2000, 48, 1-29.
[15] Tjong, S.C.; Chen, H. *Mater. Sci. Eng. R.* 2004, 45, 1-88.
[16] Pawlow, P. Z. *Phys. Chem.* 1909, 65, 1-8.
[17] Takagi, M. *J. Phys. Soc. Jpn.*1954, 9, 359-365.
[18] Wronski, C. R. M.; *Br. J. Appl. Phys.* 1967, 18, 1731-1737.
[19] Couchman, P. R.; Jesser, W. A. *Nature*, 1977, 269, 481-486.
[20] Borel, J. P. *Surf. Sci.* 1981, 106, 1-7.
[21] Saka, H.; Nishikawa, Y.; Imura, T. *Phil. Mag.* A. 1988, 57, 895-906.
[22] Sasaki, K.; Saka, H. *Phil. Mag.* A. 1991, 63, 1207-1221.
[23] Lee, J.G.; Mori, H.; Yasuda, H. *Phys. Rev.* B. 2002, 66, 012105-1-4.
[24] Tanaka, T.; Hara, S. Z. *Metallkd.* 2001, 92, 467-472.
[25] Tanaka, T.; Hara, S. Z. *Metallkd.* 2001, 92, 1236-1241.
[26] Tanaka, T.; Lee, J.; Hirai, N. In Book *Chemical Thermodynamics for Industry*; Letcher, T. Ed.; ISBN 0 85404 591 0; Thomas Graham House: Cambridge, 2004, 209-218.
[27] Ganapathi, S. K.; Owen, D. M.; Chokshi, A. H. *Scrip. Metall. Mater.* 1991, 25, 2699-2704.
[28] Malow, T. R.; Koch, C. C. *Acta Mater.* 1997, 45, 2177-2185.
[29] Ebrahimi, F.; Li, H. *Scrip. Mater.* 2006, 55, 263-266.
[30] Ivanisenko, Y.; Wunderlich, R. K.; Valiev, R. Z.; Fecht, H. *J. Scrip. Mater.* 2003, 49, 947-952.
[31] Lu, K. *Scrip. Metall. Mater.* 1991, 25, 2047-2052.
[32] Winning, M.; Rollett, A. D. *Acta Mater.* 2005, 53, 2901-2907.
[33] Duhaj, P.; Šveca, P.; Sitek, J.; Janickovic, D. *Mater. Sci. Eng.* A. 2001, 304, 178-186.
[34] Liu, Z. J.; Shen, Y. G. *Acta Mater.* 2004, 52, 729-736.
[35] Millett, P. C.; Selvam, R. P.; Molecular, A. S. *Acta Mater.* 2006, 54, 297-303.
[36] Millett, P. C.; Selvam, R. P.; Saxena, A. *Acta Mater.* 2007, 55, 2329-2336.
[37] Song, X. Y.; Zhang, J. X.; Li, L. M.; Yang, K. Y.; Liu, G. Q. *Acta Mater.* 2006, 54, 5541-5550.
[38] Liu, F.; Chen, Z.; Yang, W.; Yang, C. L.; Wang, H. F.; Yang, G.C. *Mater. Sci. Eng. A.* 2007, 457, 13–17.
[39] Song, X. Y.; Gao, J. P.; Zhang, J. X. *Acta Phys. Sinca.* 2005, 54, 1313-1319.
[40] Gartner, F.; Bormann, R.; Birringer, R.; Tschope, A. *Scrip. Mater.* 1996, 35, 805-810.
[41] Fecht, H. J. *Phys. Rev. Lett.* 1990, 65, 610-613.
[42] Wagner, M. *Phys. Rev. B.* 1992, 45, 635-639.
[43] Lu, K. *Phys. Rev. B.* 1995, 51, 18-27.
[44] Meng, Q.; Zhou, N.; Rong, Y.; Chen, S.; Hsu, T. Y. *Acta Mater.* 2002, 50, 4563-4570.

[45] Rose, J. H.; Ferrante, J.; Smith, J. R. *Phys. Rev. B*. 1983, 28, 1835-1840.
[46] Rose, J. H.; Smith, J. R.; Guinea, F.; Ferrante, *J. Phys. Rev. B*. 1984, 29, 2963-2969.
[47] Guinea, F.; Rose, J. H.; Smith, J. R. *Appl. Phys. Lett*. 1984, 44, 53-55.
[48] Vinet, P.; Ferrante, J.; Smith, J. R.; Rose, J. H. *J. Phys. C*. 1986, 19, L467-L473.
[49] Vinet, P.; Smith, J. R.; Ferrante, J.; Rose, J. R. *Phys. Rev. B*. 1987, 35, 1945-1953.
[50] Gerifalco, L. A.; Weizer, V. G. *Phys. Rev*. 1959, 114, 687-692.
[51] Villars, P.; Calvert, L. D. Pearson's *Handbook of Crystallographic Data for Intermetallic Phases;* ISBN 0 87170 219 3; American Society for Metal: Metals Park, Oh, 1985, Vol. 2, pp 5277-5278.
[52] Song, X. Y.; Zhang, J. X.; Li, E. D.; Lu, N. D.; Yin, F. X. *Nanotechnology*. 2006, 17, 5584-5589.
[53] Driver, J. H. *Scrip. Mater*. 2004, 51, 819-823.
[54] Song, X. Y.; Yang, K. Y.; Zhang, J. X. *J. Nanosci. Nanotechnol*. 2005, 5, 2155-2160.
[55] Xun, Y.; Lavernia, E. J.; Mohamed, F. A. *Mater. Sci. Eng. A*. 2004, 371, 135-140.
[56] Song, X. Y.; Zhang, J. X.; Yue, M.; Li, E. D.; Zeng, H.; Lu, N. D.; Zhou, M. L.; Zuo, T. Y. *Adv. Mater*. 2006, 18, 1210-1215.
[57] Kirchheim, R. *Acta Mater*. 2002, 50, 413-419.
[58] Choi, P.; Silva, M.; Klement, U.; Al-Kassab, T.; Kirchheim, R. *Acta Mater*. 2005, 53, 4473-4481.
[59] Gianola, D. S.; Petegem, S. V.; Legros, M.; Brandstetter, S.; Swygenhoven, H. V.; Hemker, K. J. *Acta Mater*. 2006, 54, 2253-2263.
[60] Zuo, B.; Saraswati, N.; Sritharan, T.; Hng, H. H. *Mater. Sci. Eng. A*. 2004, 371, 210-216.
[61] Kuchibhatla, S.; Karakoti, A. S.; Bera, D.; Seal, S. *Prog. Mat. Sci.* 2007, 52, 699-913.
[62] Hai, N. H.; Lemoine, R.; Remboldt, S.; Strand, M.; Shield, J. E.; Schmitter, D.; Kraus, R. H.; Espy, M.; Leslie-Pelecky, D. L. *J. Magn. Magn*. 2005, 293, 75-79.
[63] Gleiter, H. *Prog. Mater. Sci*. 1989, 33, 223-315.
[64] Erb, U.; El-Sherik, A. M.; Palumbo, G.; Aust, K. T. *Nanostruct. Mater*. 1993, 2, 383-390.
[65] Valiev, R. Z.; Korznikor, A. V.; Mulyukov, R. R. *Mater. Sci. Eng. A*. 1993, 168, 141-148.
[66] Yoshizawa, Y.; Oguma, S.; Yamauchi, K. *J. Appl. Phys*. 1988, 64, 6044-6046.
[67] Zhang, H. W.; Hei, Z. K.; Liu, G.; Lu, J.; Lu, K. *Acta Mater*. 2003, 51, 1871-1881.
[68] Dogan, C.; Rawers, J.; Govier, D.; Korth, G. *Nanostruct. Mater*. 1994, 4, 631-644.
[69] Lillo, T. M.; Korth, G. *Nanostruct. Mater*. 1998, 10, 35-43.
[70] Zhang, H. W.; Gopalan, R.; Mukai, T.; Hono, K. *Scrip. Mater*. 2005, 53, 863-868.
[71] Tokita, M. In *Proceedings of the 2000 Powder Metallurgy World Congress*; Kosuge, K.; Nagai, H. Ed.; Japanese Society of Powder and Powder Metallurgy: Kyoto, 2001, pp 729-732.
[72] Mamedov, V. *Powd. Metall*. 2002, 45, 322-328.
[73] Shen, Z. J.; Johnson, M.; Zhao, Z.; Nygren, M. *J. Am. Ceram. Soc*. 2002, 85, 1921-1927.
[74] Shearwood, C.; Fu, Y. Q.; Yu, L.; Khor, K. A. *Scrip. Mater*. 2005, 52, 455-460.
[75] Tokita, M. in *Proceedings of the Seventh International Symposium on Functionally Graded Materials*; Beijing, 2002, pp 1-8.
[76] Song, X. Y.; Liu, X. M.; Zhang, J. X. *J. Am. Ceram. Soc*. 2006, 89, 494-500.

[77] Vanmeensel, K.; Laptev, A.; Hennicke, J.; Vleugels, J.; Biest, O. *Acta Mater.* 2005, 53, 4379-4388.
[78] Yue, M.; Zhang, J. X.; Zeng, H.; Wang, K. *J. Appl. Phys. Lett.* 2006, 89, 232504-232508.
[79] Sun, N. X.; Lu, K. *Phys. Rev. B.* 1996, 9, 54-57.
[80] Rupp, J.; Birringer, R. *Phys. Rev. B.* 1987, 15, 36-40.
[81] Tritt, T. M.; Nolas, G. S.; Slack, G. A.; Cohn, J. S. *J. Appl. Phys.* 1996, 11, 8412-8415.
[82] Mirshams, R. A.; Parakala, P. *Mater. Sci. Eng. A*, 2004, 372, 252-260.

In: Encyclopedia of Nanotechnology
Editor: Elwood D. Carlson

ISBN: 978-1-60692-079-4

Chapter 23

REFLUXING POLYOL SYNTHESIS OF TRANSITION METAL CHALCOGENIDES NANOCRYSTALLITES: A GENERAL PROCESS

Guozhen Shen[a,b1] ***and Di Chen***[b]

[a] Nanoscale Materials Center, National Institute for Materials Science (NIMS), Namiki 1-1, Tsukuba, Ibaraki 305-0044, Japan

[b] Nano-materials and Nano-chemistry, Hefei National laboratory for Physical Sciences at Microscale, University of Science and Technology of China, Hefei 230026, P. R. China

ABSTRACT

Transition metal chalcogenides, including sulfides, selenides, and tellurides are important optoelectronic, luminescent, and semiconducting materials. With regard to all of the applications of chalcogenides, highly crystalline particles with almost monodisperse size distribution and regular morphology are required. The synthesis of chalcogenides nanocrystallites under mild conditions is of great interest in both basic research and engineering applications. In this paper, the synthesis of chalcogenides nanocrystallites, such as nanoparticles, nanoplatelets, nanorods, nanowires, and their assembled patterns, via the simple, rapid and convenient refluxing polyol method was reviewed based on our works. Several techniques including powder X-ray diffraction (XRD), scanning electron microscopy (SEM), transmission electron microscope (TEM), X-ray photoelectron (XPS), Raman, photoluminescence (PL), etc. have been used to characterize the produced nanocrystallites. Possible growth mechanism of the refluxing polyol technique to transition metal chalcogenides nanocrystallites is discussed. Combining with other techniques, the refluxing polyol method could be extended to synthesize other important inorganic materials nanocrystallites. Given to the generality of this simple refluxing polyol method, coating and modification of the surfaces of transition metal chalcogenide nanocrystallites may also be fulfilled.

---

[1] E-mail address: SHEN.Guozhen@nims.go.jp; gzshen@ustc.edu; Fax: +81-29-851-6280.

1. INTRODUCTION

The development of novel materials is a fundamental focal point of chemical research, and this interest is mandated by advancements in all areas of industry and technology. Over the past decade, there has been a dramatic increase of interest in both the preparation and properties of nanocrystalline materials.[1-14] This interest has been fueled by the unique properties that such materials possess when compared to bulk phases, as well as the potential they hold for such varied applications as electronics, catalysis, and biological labeling, etc. Transition metal chalcogenides including sulfides, selenides and tellurides are often found in minerals and have attracted considerable attention in recent decades due to the interesting properties they possess.[15-17] They have a number of commercial applications in pigments, semiconductors, fluorescence devices and even superconductors.[18,19] With regard to all of the applications mentioned, highly crystalline particles with almost monodisperse size distribution and regular morphology are required. In fact, many different methods of preparation, liquid-based as well as gas phase-based, are aiming at the fulfillment of these requirements.[20-32]

The polyol method has been intensively investigated for the synthesis of metals, alloys, and oxides nanocrystallites, in which the reducing properties of a high-boiling alcohol (e.g. glycerol, glycol) towards a suitable metal precursor were utilized.[33-36] In the present work, we briefly review our recent progress in the refluxing polyol preparation of metal chalcogenide nanomaterials. The refluxing polyol process can also be combined with microwave irradiation technique.

2. APPARATUS AND CHARACTERIZATION

All reagents used for the refluxing polyol synthesis were of analytical purity and purchased from different chemical companies and used without further purification. The experimental setup for the refluxing polyol synthesis can be illustrated in Figure 1, which includes a heating system, a refluxing system, and a cooling system. The source materials are put into the round-bottom flask and heated until the finish of the reaction.

The obtained precipitates were washed with absolute alcohol and distilled water for several times and dried in air.

The phase purity of the products was examined by X-ray powder diffraction (XRD) using a Japan Rigaku D/max γA X-ray diffractometer equipped with graphite monochromatized Cu $K\alpha$ radiation (λ = 1.54178 Å). To obtain further information about stoichiometry of some key compounds, X-ray photoelectron spectra was recorded on a VGESCALAB MKII X-ray photoelectron spectrometer using non-monochromatized Mg $K\alpha$ as the excitation source and choosing C_{1s} (284.60 eV) as the reference line. Scanning electron microscope (SEM) measurements were carried out with a field-emission microscope (JEOL, 7500B) operated at an acceleration voltage of 10 kV. Transmission electron microscopy (TEM) investigation was made on a Hitachi H-800 transmission electron microscope with an accelerating voltage of 200 kV.

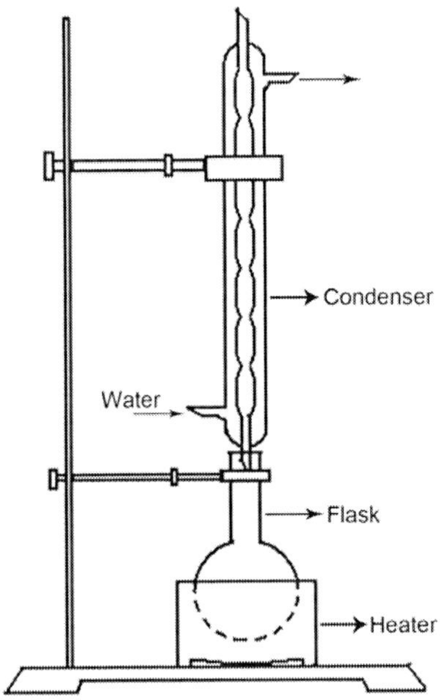

Figure 1. Schematic illustration of the apparatus used in the refluxing polyol process.

3. Results and Discussion

3.1. Preparation of Binary Sulfides Nanocrystallites by the Refluxing Polyol Process

To get transition metal sulfide nanocrystallites, appropriate amount of metal salts and thiourea (Tu) were placed in a round-bottomed flask. Then the flask was filled with glycol solution. Stirred for a few minutes, the solution was heated and refluxed for 0.5-2 h. After the mixture had been cooled, the solid was filtered and washed several times with carbon bisulfide, absolute alcohol and distilled water. Then vacuum dried at 60 ^{o}C for 3 h. Lots of binary sulfides can be obtained by this technique.[37-39] The detail experimental results are shown in Table 1.

Figure 2 shows the SEM image of the synthesized Bi_2S_3 product in the refluxing polyol process. It can be seen that urchin-like patterns consisted of Bi_2S_3 nanorods are synthesized on a large scale. Each urchin-like Bi_2S_3 pattern has a diameter of ~1 μm with acicular crystallites radiating from the center in a uniform size distribution. The nanorods grown from the center have almost uniform diameters of about 50 nm.

The analysis by TEM accompanied with selected area electron diffraction (SAED) gives further information about this product. As shown in Figure 3a, all samples dispersed on the TEM grids show spherical aggregates with diameters of ~1 μm, consistent with the SEM results. Figure 3b is a TEM image of a single Bi_2S_3 nanorod radiating from the center. The corresponding SAED pattern shows its single crystal nature and can be indexed to the reflections of orthorhombic Bi_2S_3 (Figure 3b inset). These unusual urchin-like Bi_2S_3 patterns can be used as self-sacrificing templates for the synthesis of $Bi_2(Se,S)_3$ nanorods and Bi_2Se_3 flakes with similar urchin-like morphology.[40]

Table 1. Results of the products from the reactions between metal salts and thiourea by the simple polyol process

Reactants	Reactive time	Products	Morphology	JCPDS Card No.
$Bi(NO_3)_3 \cdot 5H_2O$+Tu	0.5 h	Bi_2S_3	urchins	17-230
$NiCl_2 \cdot 6H_2O$+Tu	1.5 h	NiS	nanoparticles	75-0613
CuCl+Tu	1 h	$Cu_{1.8}S$	nanoparticles	23-0962
$CuCl_2 \cdot 2H_2O$+Tu	1 h	CuS+$Cu_{1.8}S$[a]	nanoparticles	/
$CuSO_4 \cdot 5H_2O$+Tu	1 h	$Cu_{1.8}S$+CuS[a]	nanoparticles	/
$PbCl_2$+Tu	0.5 h	PbS	nanoparticles	5-0592
$Pb(NO_3)_2$+Tu	0.5 h	PbS	nanoparticles	5-0592
AgCl+Tu	0.5 h	Ag_2S	nanoparticles	14-0072
$AgNO_3$+Tu	0.5 h	Ag_2S	nanoparticles	14-0072
$InCl_3 \cdot 4H_2O$+Tu	2 h	In_2S_3	nanoflakes	25-390
$FeCl_3 \cdot 6H_2O$+Tu	1.5 h	Fe_3S_4	nanoparticles	16-0713
$SnCl_2$+Tu	1.5 h	SnS	flake-like	39-354
$SnCl_4$+Tu	1.5 h	SnS_2	flake-like	23-677
$AgNO_3$+$SbCl_3$+Tu	2 h	Ag_3SbS_3	nanoparticles	21-1173
$Cu(NO_3)_2$+$SbCl_3$+Tu	2 h	Cu_3SbS_3	nanoparticles	75-1574
$CuCl_2 \cdot 2H_2O$+$InCl_3 \cdot 4H_2O$+Tu	1 h	$CuInS_2$	foam-like	75-106
$AgNO_3$+$Bi(NO_3)_3 \cdot 5H_2O$+Tu	1.5 h	$AgBiS_2$	nanorods	21-1178

[a] trace product

Figure 2. SEM images of as-prepared product, which shows large-scale urchin-like patterns of three-dimensional aligned Bi_2S_3 nanorods [Reprinted from Ref. 37, Copyright 2003, with permission from Elsevier].

Based on our experimental results, we found that the polyol used has great effect on the final products. For comparison, different polyols are tested. The results reveal that the viscosity and boiling point of polyol play an important role and ethylene glycol is the best one. During the experiment, the ethylene glycol provides a uniform reaction temperature and makes the reactants mixing homogeneously, which is important to the formation of the homogeneously distributed nanoscale crystallines.

Figure 4a is the XRD pattern of the synthesized NiS product by using $NiCl_2 \cdot 6H_2O$ and thiourea as the reactants in ethylene glycol for 1.5 h. All the peaks in this pattern can be indexed to hexagonal phase NiS (JCPDS Card No. 75-0613). No impurities such as Ni, NiO_x or intermediary phase nickel sulfides are detected in the XRD pattern. According to Scherrer's formula, the average size of the prepared NiS is about 25 nm. Figure 4b is the corresponding TEM image of NiS product, which reveals the formation of nanoparticles with average diamters of about 35 nm.

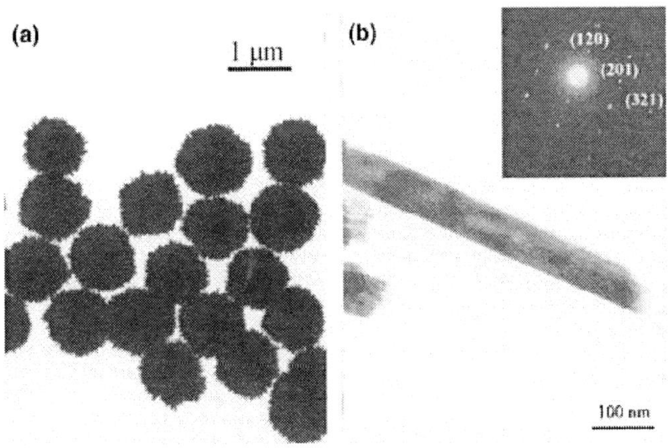

Figure 3. (a) TEM image of Bi_2S_3 urchin-like patterns; (b) TEM image of a single Bi_2S_3 nanorod (inset: the corresponding ED pattern) [Reprinted from Ref. 37, Copyright 2003, with permission from Elsevier].

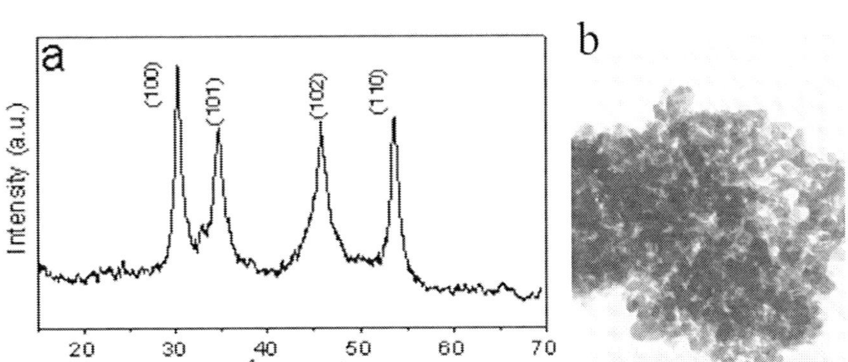

Figure 4. (a) XRD pattern of NiS; (b) TEM image of NiS [Reprinted from Ref. 38, Copyright 2003, with permission from Elsevier].

In fact, in the refluxing polyol process, it was found that the metal salts used have no obviously influence on the morphology and structure of the product. Figure 5 is the XRD patterns of produced PbS by using $PbCl_2$ and $Pb(NO_3)_2$ as the metal source, respectively. All peaks in both patterns can be indexed to PbS with cubic cell. The phases and morphologies of these two samples are the same. TEM observation shows that they are of similar morphology and size, namely, the average size of both PbS products is about 50~120 nm.

Table 2. The relationship between viscosity of different solvents and the morphology of the products [Reprinted from Ref. 37, Copyright 2003, with permission from Elsevier]

Solvent	Viscosity ($mN{\cdot}s{\cdot}m^{-2}$)/25°C	Morphology
Distilled water	0.89	Aggregates of nanorods
Mixture of water and EG	0.89~26.09	Nanorods and urchin-like patterns
Ethylene glycol (EG)	26.09	Uniform urchin-like patterns
Polyol ethylene glycol 200	>26.09	Ununiform urchin-like patterns

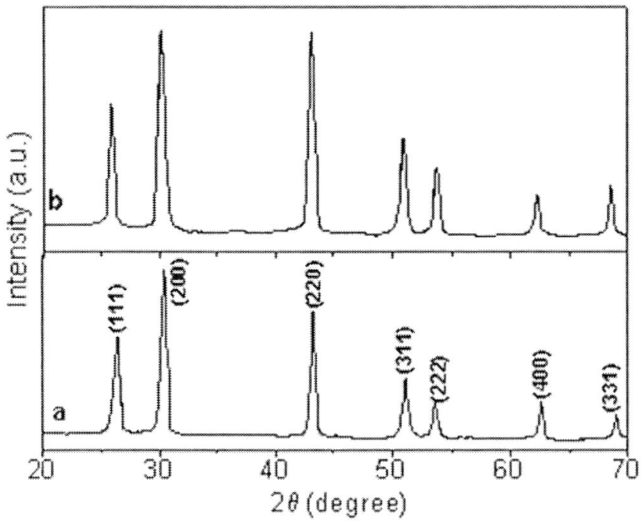

Figure 5. XRD patterns of PbS obtained (a) from $PbCl_2$; (b) $Pb(NO_3)_2$ [Reprinted from Ref. 38, Copyright 2003, with permission from Elsevier].

Besides the analysis of XRD, EDX analysis was also used to determine the composition of these products. Typical EDX pattern of as-obtained PbS sample is shown in Figure 6. The results show that the PbS nanoparticles are composed of the elements Pb and S and the ratio of Pb:S is 1:1.07, in agreement with the expected value (the Cu peaks arise from the copper grid). EDX analyses of other products have the similar results, which are also in good agreement with the expected values of final products.

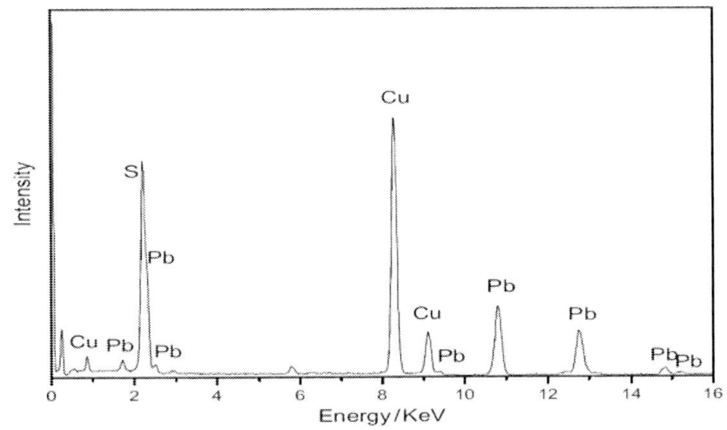

Figure 6. EDX spectrum of the PbS sample [Reprinted from Ref. 38, Copyright 2003, with permission from Elsevier]

3.2. Preparation of Ternary Sulfides Nanocrystallites by the Simple Polyol Process

By carefully controlled the experimental parameters, the refluxing polyol process was found to be very useful for the synthesis of ternary sulfides nanocrystallites.[41-44]

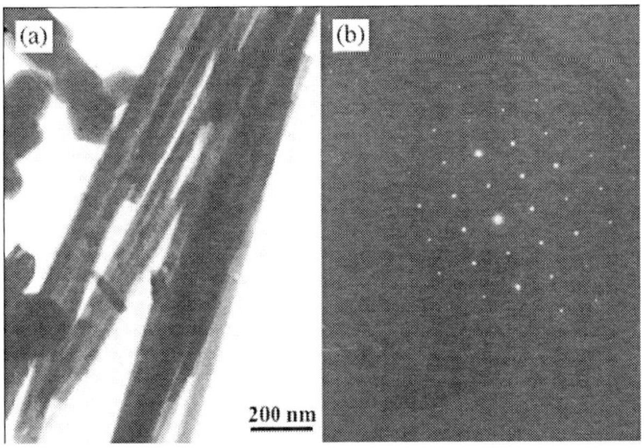

Figure 7. (a) TEM image of the synthesized $AgBiS_2$ nanorods. (b) the corresponding SAED pattern [Reprinted from Ref. 42, Copyright 2003, with permission from Elsevier].

Ternary metal chalcogenide $AgBiS_2$ belongs to the class of I–V–VI semiconducting materials, which are technologically important in linear, nonlinear, optoelectronic, and thermoelectric devices as well as optical recording media.[45] The TEM image (Figure 7a) shows the $AgBiS_2$ samples obtained in the present refluxing polyol process consist of nanowires with ~40 nm in diameter and length up to 10 μm. The corresponding SAED pattern shown in Figure 7b indicates their single-crystal nature.

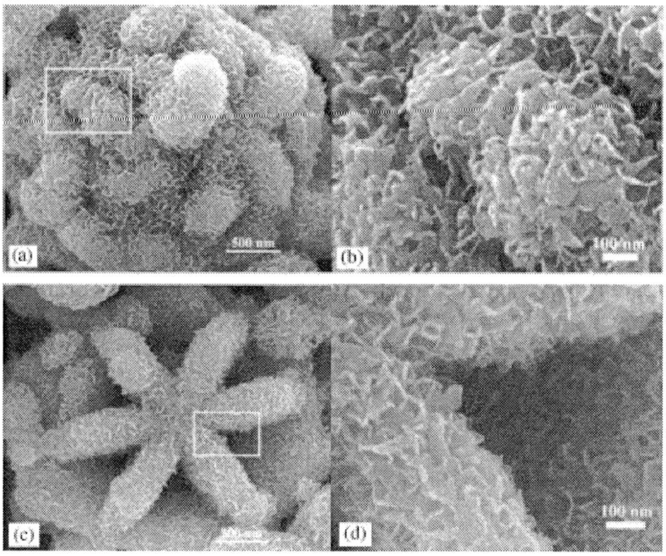

Figure 8. SEM images of the as-prepared $CuInS_2$ product, which clearly shows the porous foam-like microstructures [Reprinted from Ref. 43, Copyright 2003, with permission from Elsevier].

Due to its direct energy gap of 1.5 eV, ternary compound $CuInS_2$ with the chalcopyrite structure has emerged as a leading material for high-efficiency and radiation-hard solar cell applications. In addition, $CuInS_2$ is also a candidate for photochemical applications because its band gap matches well the solar spectrum.[46,47] Through the reaction of $CuCl_2$, $InCl_3$ and thiourea in ethylene glycol at 195 ^oC, porous foam-like $CuInS_2$ nanocrystallines were obtained. The low-magnification SEM in Figure 8a shows the formation of a porous and sponge-like structure with an extended network of features separated by some pores-like structures. The fine texture of the foam features can be clearly seen in Figure 8b. No existence of grainy matter was observed in the product. The structure resembles a network of interconnected flakes forming irregular polygonal cavities with different sizes and the wall thickness is very thin and less than 10 nm. We also observed $CuInS_2$ flower- and hexagram-like crystals composed of foam-like microstructure (Figure 8c, 8d).

The growth process for binary or ternary metal sulfides in the refluxing process can be expressed as follows. When metal salts mixed with thiourea in ethylene glycol, metal-thiourea complexes can form according to literatures.[48,49] In the polyol route, the relatively high solubility of metal salts and thiourea in polyol solution and the change of solution color also reveal the strong complex action between metal ions and thiourea. So the formation of nanoscale metal sulfides crystallines in glycol may be through the following steps: first, metal ions can combine with thiourea molecules to form metal-thiourea complexes in the solution. Second, at a given temperature, the stability of metal-thiourea complexes decreases and undergoes thermal decomposition to form metal sulfides. Meanwhile, it is well known that the thiourea takes part in the following reversible decomposition reaction in solution:

$$NH_2CSNH_2 \leftrightarrow CH_2N_2 + H_2S \qquad (1)$$

The formed CH_2N_2 reacts with H_2O at high temperature to produce NH_3 and CO_2:

$$CH_2N_2 + H_2O \leftrightarrow NH_3 + CO_2 \quad (2)$$

So the decomposition of NH_2CSNH_2 can be expressed as:

$$NH_2CSNH_2 + H_2O \leftrightarrow NH_3 + CO_2 + H_2S \quad (3)$$

In this reaction, the trace water (coming from analytical polyol or hydrated metal salts in our experiments) plays a critical role in the formation of metal sulfides crystals, since it can promote the decomposition of NH_2CSNH_2 and the formation of metal sulfides. In the synthesis of ternary metal sulfide nanocrystallites, the molar ratio of metal salts should be well controlled; otherwise, binary metal sulfide would be produced among with the ternary sulfide products.

3.3. Preparation of Selenides Compounds by the Rapid Ethylenediamine-Assisted Polyol Process

Ethylenediamine (EN) is strongly basic and elemental chalcogen is activated by a nucleophilic attack of the amine to form E^{2-} (E=S, Se) in the polyol route. So ethylenediamine-assisted polyol process was designed to synthesize metal selenide nanocrystallites.[50-52] Table 3 shows the products obtained from the ethylenediamine(en)-assisted polyol route.

Table 3. Results of the products from the rapid ethylenediamine-assisted polyol process

Reactants	Reactive time	Products	Morphology	JCPDS Card No.
$SbCl_3$+S+DEG+EN	1.5 h	Sb_2S_3	nanowires	6-474
$SbCl_3$+Se+DEG+EN	1.5 h	Sb_2Se_3	nanowires	72-1184
$ZnCl_2$+Se+EG+EN	1 h	ZnSe	spheres	80-0021
$CdCl_2$+Se+EG+EN	1 h	CdSe	flowers	77-0021
$SnCl_2 \cdot 2H_2O$+Se+DEG+EN	1.5 h	SnSe	nanowires	32-1382

SnSe belongs to the interesting class of isomorphic materials that are in many ways between two-dimensional systems and three-dimensional crystals. It has been widely used in infrared optoelectronic devices, memory switching devices and holographic recording devices.[53-55] As shown in Figures 9a and 9b, the as-prepared product are SnSe nanowires having a narrow size distribution with a mean diameter of ~50 nm (Figure 9c). The observed average aspect ratio of these SnSe nanowires is about 150. The electron diffraction pattern (Figure 9b inset) taken from the selected area of SnSe nanowires shown in Figure 9b shows that the as-prepared SnSe nanowires are well crystallized.

Energy-dispersive X-ray (EDX) analysis was used to determine the composition of these nanowires. It (Figure 9d) shows that these nanowires are composed of the elements Sn and Se and the ratio of Sn:Se is 1:1.02, in agreement with the expected values (the Cu peaks arise from the copper grid).

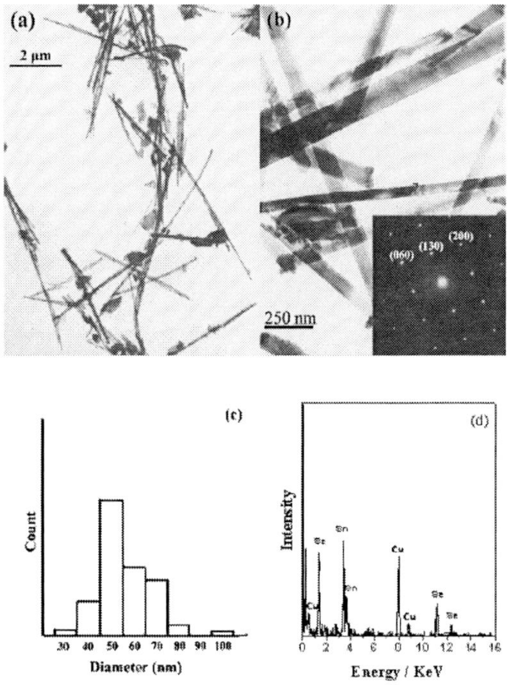

Figure 9. (a,b) TEM images of the as-grown SnSe nanowires. (c) the diameter distribution of SnSe nanowires. (d) EDX spectrum of the SnSe nanowrie [Reprinted from Ref. 51, Copyright 2003, with permission from the Chemical Society of Japan].

Figure 10. SEM image of (a) Sb_2S_3 and (b) Sb_2Se_3 nanowires obtained in the ethylenediamine(en)-assisted polyol route [Reprinted from Ref. 50, Copyright 2003, with permission from Elsevier].

Group V–VI binary compounds (Sb_2S_3, Sb_2Se_3), are highly anisotropic semiconductor with a layered structure parallel to the growth direction and with orthorhombic phase crystal structure, which is known to adopt a number of packing structures resulting in either trigonal prismatic or octahedral coordination of the metals within a layered matrix of chalcogens.[56,57] The compounds have a great deal of attention due to its switching effects and excellent photovoltaic and thermoelectric power (TEP), which allow possible applications in photo-electrochemical devices, optical and thermoelectric cooling devices, decorative coatings and solar selective.[58] In the ethylenediamine-assisted polyol process, Sb_2S_3, and Sb_2Se_3 nanowires are prepared by the reaction of $SbCl_3$ and S (or Se) powder. The SEM images are shown in Figure 10a and 10b. They showed that the obtained nanowires are with typical lengths in the range of several to several tens of micrometers. While the lengths of as-prepared Sb_2Se_3 nanowires range from 1 to 15 μm.

As a direct band gap material, zinc selenide, ZnSe, can be widely used in many fields. It exhibits great potential for various optoelectronic and high-speed applications like blue–green light emitting diodes, photoluminescent and electroluminescent devices, lasers and thin film solar cells. Lots of methods of synthesizing ZnSe have been developed, including molecular beam epitaxy (MBE), metalorganic chemical vapor deposition (MOCVD), solvothermal or hydrothermal route, organometallic vapor phase epitaxy (OMVPE) and some other methods. Figure 11 is the SEM image of the synthesized ZnSe product. It can be seen that the as-prepared sample composed of relatively uniform spheres crystals with a diameter of about 1 μm on average. Besides the spheres, some nanorods are also detected, which have a diameter of about 50 nm and length up to 10 μm. The ratio of this structure is about 5%. XRD analysis reveals that the ZnSe spheres obtained in the present enthylenediamine-assisted polyol process are of cubic structures.

Figure 11. FE-SEM image of as-prepared sample, which shows the formation of ZnSe spheres with a diameter of about 1 μm [Reprinted from Ref. 52, Copyright 2003, with permission from Elsevier].

It is well known that solvent plays an important role in determining the crystal morphology. Solvents with different physical and chemical properties can influence the solubility, reactivity, and diffusion behavior of the reactants, in particular, the polarity and coordinating ability of the solvents can influence the crystal morphology of the final product. EN is strongly basic and elemental E is activated by a nucleophilic attack of the amine to form E^{2-} (E=S, Se) in the polyol route. On the other hand, the N-chelating behavior makes EN a strong donor ligand, and it plays an important role in the formation of one-dimensional nanowires. EN can react with metal ions to form relatively stable antimony complexes, which may serve as molecular templates in control of the crystal growth. We deduced the growth mechanism of metal selenide nanocrystallites as follows. (1) metal salts dissolved in EN to form the chain type structure of metal ions complexes under the template effect of EN; (2) Elemental E is activated by EN to form E^{2-}; (3) At appropriate temperature, E^{2-} attacks the metal ions complexes and the ligands are lost gradually to form the final metal selenide nanocrystallites.

3.4. Preparation of other Chalcogen-Related Compounds by the Simple Refluxing Polyol Process

As important members of the semiconductor family, selenium and tellurium show unique physical and chemical properties, which provide them with the chances to widen the range of applications. The polyol synthesis of Se and Te nanorods is carry out under atmosphere and the chemical reaction we employed can be formulated as the following equations:[59]

$$Na_2SeO_3+(NH_4)_2S_2O_3+2H_2O \rightarrow Se+Na_2SO_4+(NH_4)_2SO_4+2H_2 \quad (1)$$

$$Na_2TeO_3+(NH_4)_2S_2O_3+2H_2O \rightarrow Te+Na_2SO_4+(NH_4)_2SO_4+2H_2 \quad (2)$$

Figure 12. (a) TEM image of submicron Se rod-like crystals; (b) TEM image of a single submicron Se rod-like crystal; (c) SAED pattern of (b); (d) SEM image of submicron Te rod-like crystals [Reprinted from Ref. 59, Copyright 2004, with permission from Elsevier].

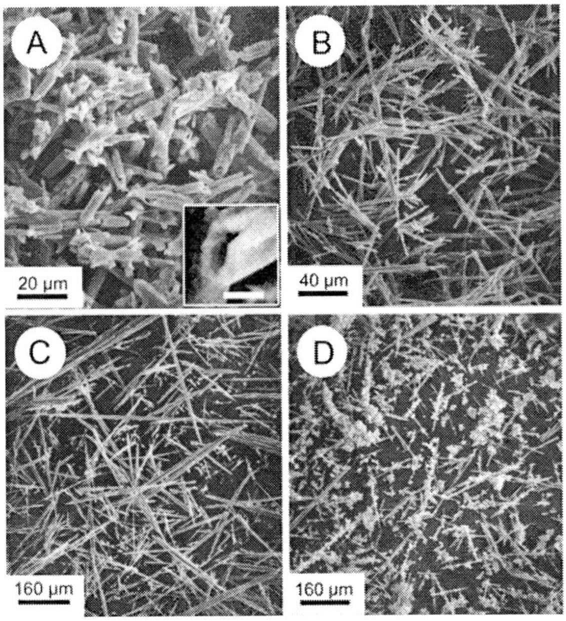

Figure 13. SEM images of the as-obtained Se crystals under different reaction conditions: (a) 150 °C, 0.5 h; (b) 195 °C, 0.5 h; (c) 180 °C, 5 h; (d) 180 °C, without stirring [Reprinted from Ref. 59, Copyright 2003, with permission from Elsevier].

Figure 12a is the SEM image of as-obtained Se nanorods with diameter of ~500 nm and length up to 15 µm. TEM result (Figure 12b) shows the long Se rod-like crystal with diameter of about 450 nm, consisting with the SEM results. The corresponding selected area electron diffraction (SAED) pattern shows its single crystal nature (Figure 12c). SEM image of as-obtained Te sample shown in Figure 12d also reveals that the obtained Te sample is consisted of uniform rod-like crystals. The average diameter of these nanorods is ca. 400 nm and length ranges from 2 mm to 10 µm.

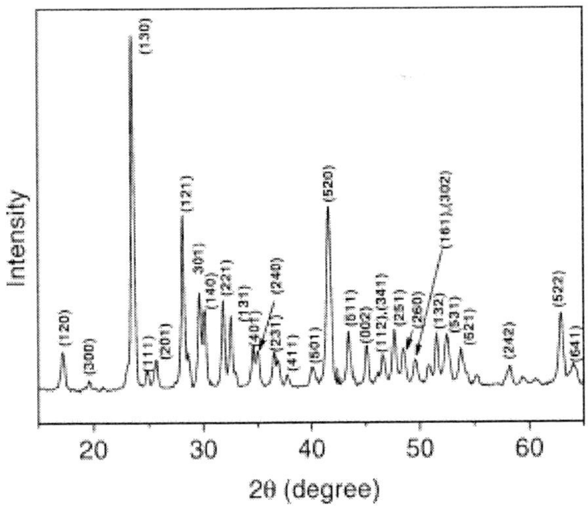

Figure 14. XRD pattern of the synthesized $(Bi(Bi_2S_3)_9I_3)_{0.667}$ needlelike crystals [Reprinted from Ref. 61, Copyright 2003, with permission from Elsevier].

For the formation of submicron Se and Te rod-like crystals, the feature of the present polyol route is the high boiling point of ethylene glycol, which crystallizes the as-obtained products at relatively low temperature and short reaction time. Both reaction time and reaction temperature have influences on the morphology of the obtained products. If the reaction was performed at low temperature, the yield of Se product decreased dramatically and they are of tube-like morphology instead of rod-like morphology (Figure 13a). High temperature and long reaction time are better for the crystallinity of the products but would result in the increase of diameters to the microscale (Figure 13).

BiSI and its related chalcogenide halide compounds belonging to the V–VI–VII group ternary compounds are ferroelectric, electro-optic, electric-mechanical and photo-electric semiconductors. Many studies in these compounds have focused on the synthesis and characterization of BiSI. However, few reports could be found on the synthesis and characterization of complexed $(Bi(Bi_2S_3)_9I_3)_{0.667}$ due to the difficulty in obtaining stoichiometric compounds since it was first reported by Miehe in 1971.[60] Here, $(Bi(Bi_2S_3)_9I_3)_{0.667}$ needlelike crystals were obtained by refluxing in ethylene glycol solution.[61] We used $Bi(NO_3)_3{\cdot}6H_2O$, $(NH_2)_2CS$ and NaI as the source materials. As shown in Figure 14, the XRD pattern can be indexed to hexagonal cell of $(Bi(Bi_2S_3)_9I_3)_{0.667}$, with lattice parameters a=15.66 Å, c=4.02 Å. The SEM image indicates that the needles have diameter of 400 nm and length up to 15 μm (Figure 15).

Figure 15. SEM image of the $(Bi(Bi_2S_3)_9I_3)_{0.667}$ needlelike crystals [Reprinted from Ref. 61, Copyright 2003, with permission from Elsevier].

We have reported the synthesis of cubic phase ZnSe with sphere morphology in the enthylenediamine-assisted polyol process when refluxing the source material in ethylene glycol solution. However, we found that when we added some polyethylene glycol 200 into the solution, $(ZnSe)_2{\cdot}EN$ will be obtained as shown in Figure 16a.[62] If the $(ZnSe)_2{\cdot}EN$ precursor is annealed at 500 °C in N_2 gas, XRD result shows the formation of hexagonal ZnSe product (Figure 16b).

SEM images of the as-prepared $(ZnSe)_2{\cdot}EN$ precursor are shown in Figures 17a,b. The low-magnification image (Figure 17a) shows that large-scale disk-like crystals are obtained when PEG-200 was added to the present route. Most of the disk-like crystals are of uniform hexagonal morphology. The typical SEM image of the hexagonal disk-like crystal is shown in Figure 17b, which clearly indicates that the crystal is about 1.5 μm in length of side. And the thickness of these disk-like crystals is about 100 nm on average. Figure 17c and 17d are the SEM images of the ZnSe product after annealing the $(ZnSe)_2$·EN precursor. A low-magnified image (Figure 17c) shows that annealed at 500 °C, the sample is also with the hexagonal-shaped morphology. But from a high-magnified image (Figure 17d), one can see that this morphology is quite different with the simple aggregates of individual nanoparticles, for that each hexagonal-shaped crystals are composed with little nanocrystals connected with each other and we called it the quasi-network structure. And the pores in the quasi-network may be caused by the release of EN from the $(ZnSe)_2$·EN precursor under high temperature.

In the present polyol process, no $(ZnSe)_2$·EN precursor could be obtained without the addition of PEG-200. When we substituted PEG-200 with other polyols, from oligomeric polyols, such as DEG, TEG, to higher polyethylene glycols, such as PEG400, PEG 600, PEG20000, no $(ZnSe)_2$·EN precursor could be obtained. XRD pattern showed that the product in all these solvents is cubic phase ZnSe. It has been well known that the non-ionized surfactant PEG forms chainlike structure due to the assembly in the reaction system. Surfactant PEG with lower molecular weights gets more flexible and looser aggregate.[63]

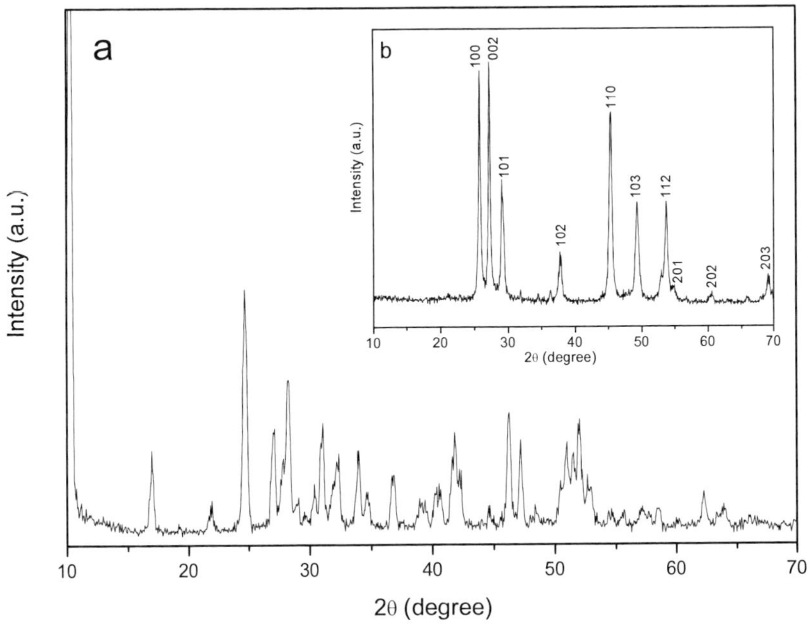

Figure 16. (a) XRD pattern of the synthesized $(ZnSe)_2$·EN precursor. (b) XRD pattern of the annealed sample, indicating the formation of hexagonal ZnSe product [Reprinted from Ref. 62, Copyright 2004, with permission from MRS].

Therefore, we can imagine that the loose and flexible aggregate of PEG-water (coming from EN and EG solution) divided reaction space into linear reaction fields when PEG and a small quantity of water were mixed together. The best polyol in the synthesis of (ZnSe)·2EN in the present process is PEG200 just as proved by numerous experiments.

Figure 17. (a,b) SEM image of the obtained $(ZnSe)_2 \cdot EN$ precursor. (c,d) SEM image of the annealed samples [Reprinted from Ref. 62, Copyright 2004, with permission from MRS].

3.5. Preparation of Sulfides Compounds by the Rapid Microwave Polyol Process

Since 1986, microwave irradiation as a heating method has found a number of applications in chemistry.[64] The microwave synthesis, which is quite fast, simple and environmentally friendly, has been widely used in various fields including molecular sieve preparation, plasma chemistry, analytical chemistry catalysis, inorganic complexes synthesis and organic reactions, etc. More importantly, it is a rapidly developing area of research, which has attracted much attention as a new method for preparing nanoscale inorganic compounds.[65] If the microwave technique could be combined with the simple refluxing polyol process, transition metal chalcogenide nanocrystallites may be prepared at much lower cost. Manipulations and reactions were carried as follows: appropriate amount of metal salts and thiourea were used to obtain metal sulfides (Table 4).[66-68]

For example, IV–VI intermetallic compounds, SnS and SnS_2 are increasingly important because of their specially semiconducting properties. For example, flaky SnS with the orthorhombic phase is a narrow band gap semiconductor with an optical band gap of 1.08 eV, which is a kind of ideal material for photoelectric energy conversion. And SnS thin films have also exhibited conversion efficiency similar to the one found for silicon films (up to 25%) in photovoltaic devices.[69,70] On the other hand, layered SnS_2 with the layer CdI_2-type structure has a layer band gap of about 2–3 eV.Broad band gap leads to photoconductions, and makes it potential to be a candidate in solar cells and opto-electronic devices.[71-73]

Table 4. Results of the products from the reactions between metal salts and thiourea by the microwave polyol process

Reactants	Products	Phase	Morphology
$CdCl_2 2.5H_2O$+Tu	CdS	Hexagonal	nanoparticles
$Zn(NO_3)_2{\cdot}6H_2O$+Tu	ZnS	Hexagonal	nanoparticles
$CoCl_2{\cdot}6H_2O$+Tu	CoS	Hexagonal	nanoparticles
$PbCl_2$+Tu	PbS	Cubic	dendrite
$CuCl_2{\cdot}2H_2O$+Tu	CuS, $Cu_{2-x}S$	/	nanoparticles
$Bi(NO_3)_3{\cdot}5H_2O$ +Tu	Bi_2S_3	Orthorhombic	radiate Nanorods
$SbCl_3$+Tu	Sb_2S_3	Orthorhombic	nanorods
$AgNO_3$+Tu	Ag_2S	Monoclinic	nanoparticles
$SnCl_2{\cdot}2H_2O$+Tu	SnS	Orthorhombic	flakes
$SnCl_4$+Tu	SnS_2	Hexagonal	layer-rolled

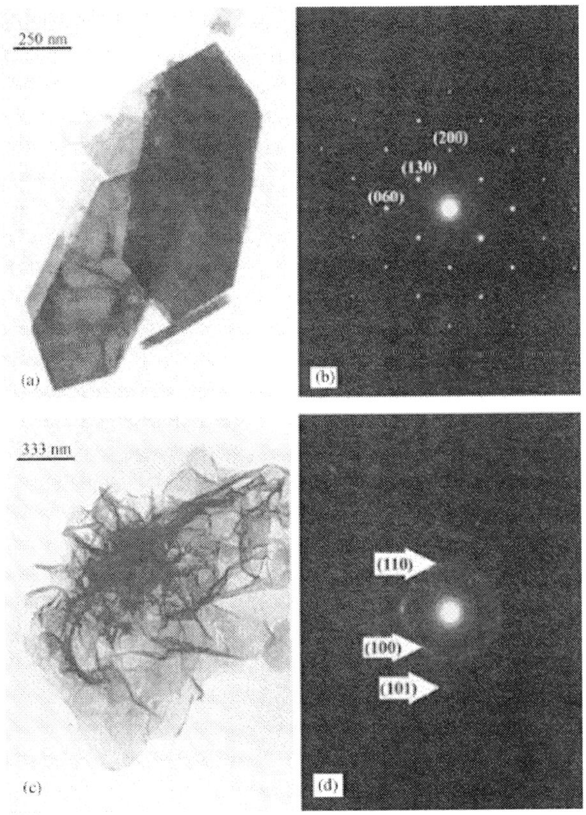

Figure 18. TEM image (a) and ED pattern (b) of SnS; TEM image (c) and ED pattern (d) of SnS_2 [Reprinted from Ref. 66, Copyright 2004, with permission from Elsevier].

Figure 18 shows the different morphologies of the as-prepared SnS and SnS_2 products. From Figure 18a, one can see that the orthorhombic phase SnS crystallines consist of nanoflakes with hexagonal shapes. The corresponding SAED pattern (Figure 18b) shows that the flaky SnS is single-crystal. Figure 18c shows the layer-rolled crystals morphology of hexagonal phase SnS_2, which is different from that of the obtained SnS. The corresponding

SAED pattern shown in Figure 18d shows the relatively good crystallinity of the SnS_2 layer-rolled crystals.

The proposed mechanism of microwave-assisted ethylene glycol synthesis of metal sulfides nanocrystallites may be summarized as below: firstly, the strong complexation between M^{n+} and thiourea lead to the formation of M-Thiourea complexes in the microwave synthesis, which prevent the production of a large number of free S^{2-} in the solution and will be favorable for the formation of the products. Secondly, the M-Thiourea complexes undergo thermal decomposition under the microwave irradiation to produce metal sulfides.

In this microwave synthesis, ethylene glycol (EG) and microwave irradiation play key roles for the preparation of metal sulfides. EG acting as both reaction media and dispersion media can efficiently adsorb and stabilize the surface of the particles and favor producing monodispersed metal sulfides crystals with good dispersivity. Meanwhile, EG with the high permanent dipole is an excellent susceptor of the microwave irradiation, which can take up the energy from the microwave field and get the polar reaction solution heated up to high temperature instantaneously.[74,75] All these help to the decomposition of M-Thiourea complexes and the formation of the products. Secondly, microwave irradiation as a heating method affords a uniform growth environment for the formation of nanosized metal sulfides. With microwave irradiation of reactants in polar solution, temperature and concentration gradients can be avoided, providing a uniform environment for the nucleation and the growth of the products. So it can be assumed that microwave irradiation not only provides the energy for the decomposition of the M-Thiourea complexes, but also greatly accelerates the formation of metal sulfides as a result of the fast and homogenous heating effects of microwaves.

4. CONCLUSION

Transition metal chalcogenides nanocrystallites are highly interesting materials from many points of views. They are important optoelectronic, luminescent, and semiconducting materials. In this chapter, we reviewed the synthesis and characterization of transition metal chalcogenides by the refluxing polyol process. The rapid refluxing polyol process is proved to be a general process for the synthesis of chalcogenide nanocrystallites with many kinds of morphologies, such as nanoparticles, nanorods, nanowires etc. Transition metal chalcogenides nanocrystallites can also be synthesized by combining the refluxing polyol process with the microwave techniques.

REFERENCES

[1] Iijima, S. *Nature* 1991, 354, 56.
[2] Golbstein, A. N.; Echer, C. M.; Alivisatos, A. P. *Science* 1992, 256, 1425.
[3] Alivisatos, A. P. *Science* 1996, 271, 933.
[4] Murray, C. M.; Norris, D. J.; Bawendi, M. G. *J. Am. Chem. Soc.* 1993, 115, 8706.
[5] Weller, H. *Angew. Chem.* 1996, 35, 1079.

[6] Li, W. Z.; Xie, S. S.; Qian, L. S.; Chang, B. H.; Zou, B. S.; Zhou, W. Y.; Zhao, R. A.; Wang, G. *Science* 1996, 274, 1701.

[7] Wang, Z. L. *Adv. Mater.* 1998, 10, 13.

[8] Rao, C. N. R.; Muller, A.; Cheetham, A. K. *the Chemistry of Nanomaterials: Synthesis, Properties and Applications*, Wiley-VCH, 2004.

[9] Zhang, J. Z.; Wang, Z. L. *Self-Assembled Nanostructures*, Kluwer Academic Publisher, 2002.

[10] Lauhon, L. J.; Gudiksen, M. S.; Lieber, C. M. *Phil. Trans. R. Soc. Lond. A.* 2004, 362, 1247.

[11] Xia, Y. N.; Yang, P. D.; Sun, Y. G.; Wu, Y. Y.; Mayer, B.; Gates, B.; Yin, Y. D.; Kim, F.; Yan, H. Q. *Adv. Mater.* 2003, 15, 353

[12] Law, M.; Goldberg, J.; Yang, P. D. *Annu. Rev. Mater. Res.* 2004, 34, 83.

[13] Jun, Y. W.; Lee, J. H.; Choi, J. S.; Cheon J. *J. Phys. Chem. B* 2005, 109, 14795.

[14] Shen, G. Z.; Tang, K. B.; Qian, Y. T.; Chen, D. *Int. J. Nanotechnology* 2004, 1, 366.

[15] S.M. Stuczynski, Y.U. Know, M.L. Steigerwald, J. *Organomet. Chem.* 1993, 449, 167.

[16] Feng, P. Y.; Bu, X. H.; Zheng, N. F. *Accounts Chem. Res.* 2005, 38, 293.

[17] Rajamathi, M.; Seshadri, R. *Current Opin. Solid State and Mater. Sci.* 2002, 6, 337.

[18] Greenwood, N. N.; Earnshaw, E. A. *Chemistry of the Elements*, Pergamon, Oxford, 1990.

[19] Nickless, G. *Inorganic Chemistry of Sulfur*, Elsevier, London, 1968, p. 670.

[20] Qi, L. M.; Colfen, H.; Antoniette, M. *Nano Lett.* 2001, 1, 61.

[21] Walton, R. I. *Chem. Soc. Rev.* 2002, 31, 230.

[22] Gorer, S. Penner, S. M. *J. Phys. Chem. B.* 1999, 103, 5750.

[23] Turco Lierri, V.; Rossi, M.; Arrigo, G. D.; Manno, D.; Micocci, G. *Appl. Phys. A.* 1999, 69, 369.

[24] Zhu, J. J.; Liu, S.; Palchik, O.; Koltypin, Y.; Gedanken, A. *J. Solid State Chem.* 2003, 153, 342.

[25] Shen, G. Z.; Cho, J. H.; Yoo, J. K.; Yi, G. C.; Lee, C. J. *J. Phys. Chem. B* 2005, 109, 9294.

[26] Shen, G. Z.; Lee, C. J. *Cryst. Growth and Design* 2005, 5, 1085.

[27] Jiang, Y.; Meng, X. M.; Liu, J.; Hong, Z. R.; Lee, C. S.; Lee, S. T. *Adv. Mater.* 2003, 15, 1195.

[28] Li, Q.; Wang, C. R. *Appl. Phys. Lett.* 2003, 83, 359.

[29] Parak, W. J.; Gorion, D.; Pellegrino, T.; Zanchet, D.; Micheel, C.; William, S. C.; Boudreau, R.; Le Gros, M. A.; Larabell, C. A.; Alivisatos, A. P. *Nanotechnology*, 2003, 14, R15.

[30] Aldana, J.; Lavelle, N.; Wang, Y. J.; Peng, X. G. *J. Am. Chem. Soc.* 2005, 127, 2496.

[31] Yu, S. H.; Yang, J.; Han, Z. H.; Zhou, Y.; Yang, R. Y.; Qian, Y. T.; Zhang, Y. H. *J. Mater. Chem.* 1999, 9, 1283.

[32] Li, B.; Xie, Y.; Huang, J. X.; Liu, Y.; Qian, Y. T. *Chem. Mater.* 2000, 12, 2614.

[33] F. Fievet, J.P. Lagier, M. Figlarz, *MRS Bull.* 1989, 29.

[34] L.K. Kurihara, G.M. Chow, P.E. Schoen, *Nanostruct. Mater.* 1995, 5, 607.

[35] P. Toneguzzo, G. Viau, O. Acher, F. Fievet-Vincent, F. Fievet, *Adv. Mater.* 1998, 10, 1032.

[36] S. Sun, C.B. Murray, D. Weller, L. Folks, A. Moser, *Science* 2000, 287, 1989.

[37] Shen, G. Z.; Chen, D.; Tang, K. B.; Li, F. Q.; Qian, Y. T. *Chem. Phys. Lett.* 2003, 370, 334.
[38] Shen, G. Z.; Chen, D.; Tang, K. B.; Liu, X. M.; Huang, L. Y.; Qian, Y. T. *J. Solid State Chem.* 2003, 173, 232.
[39] Shen, G. Z.; Chen, D.; Tang, K. B.; Huang, L. Y.; Qian, Y. T.; Zhou, G. E. *Inorg. Chem. Commun.* 2003, 6, 178.
[40] Shen, G. Z.; Chen, D.; Tang, K. B.; Qian, Y. T. *Nanotechnology* 2004, 15, 1530.
[41] Chen, D.; Shen, G. Z.; Tang, K. B.; Jiang, X.; Huang, L. Y.; Jin, Y.; Qian, Y. T. *Mater. Res. Bull.* 2003, 38, 509.
[42] Shen, G. Z.; Chen, D.; Tang, K. B.; Qian, Y. T. *J. Cryst. Growth* 2003, 252, 199.
[43] Shen, G. Z.; Chen, D.; Tang, K. B.; Fang, Z.; Sheng, J.; Qian, Y. T. *J. Cryst. Growth* 2003, 254, 75.
[44] Shen, G. Z.; Chen, D.; Tang, K. B.; Qian, Y. T. *J. Cryst. Growth* 2003, 257, 293.
[45] Snmanta, L.K.; Chatterjee. S. *Phys. Stat. Sol. B* 1994, 182, 85.
[46] Birkmire, R.W.; Eser. E. *Annu. Rev. Mater. Sci.* 1997, 27, 625.
[47] Thiel; F.A. *J. Electrochem. Soc.* 1982, 129, 1570.
[48] Karanjai, M.K.;Dasgupta; D. *Thin Solid Films* 1987, 155, 309.
[49] Yu, S.H.; Shu, L.; Yang, J.; Han, Z.H.; Qian, Y.T.; Zhang, Y. H. *J. Mater. Res.* 1999, 14, 4157.
[50] Shen, G. Z.; Chen, D.; Tang, K. B.; Jiang, X.; Qian, Y. T. *J. Cryst. Growth* 2003, 252, 350.
[51] Shen, G. Z.; Chen, D.; Jiang, X.; Tang, K. B.; Liu, Y. K.; Qian, Y. T. *Chem. Lett.* 2003, 32, 426.
[52] Shen, G. Z.; Chen, D.; Tang, K. B.; Qian, Y. T. *J. Cryst. Growth* 2003, 257, 276.
[53] Maier, H.; Daniel, D. R. *J. Electron. Mater.* 1971, 6, 693.
[54] Baxter, C. R.; Melennam, M. *J. Vac. Sci. Technol.* 1975, 12, 110.
[55] Valiukonis, G.; Guseinova, D. A.; Krivaite, G.; Sileika, A. *Phys. Status Solidi B* 1986, 139, 299.
[56] Zheng, X.W.; Xie, Y.; Zhu, L. Y.; Jiang, X. C.; Jia, Y. B.; Song, W. H.; Sun. Y. P. *Inorg. Chem.* 2002, 41, 455.
[57] Hulliger, F. *Structural Chemistry of Layered-Type Phase.* , Reidel, Dordrecht, 1976.
[58] Rajapure, K.; Lokhande, C.; Bhosele, C. *Thin Solid Films* 1997, 311 114.
[59] Shen, G. Z.; Chen, D.; Tang, K. B.; Qian, Y. T. *Mate. Res. Bull.* 2004, 39, 2077.
[60] Miehe, G.; Kupcik V. *Naturwissenschaften* 1971, 58, 219.
[61] Shen, G. Z.; Chen, D.; Tang, K. B.; Huang, L. Y.; Qian, Y. T. *J. Cryst. Growth* 2003, 249, 331.
[62] Shen, G. Z.; Chen, D.; Tang, K. B.; Qian, Y. T. *J. Mater. Res.* 2004, 19, 1367.
[63] Wang, W. Z.; Zhan, Y. J.; Wang, G. H. *Chem. Commun.* 2001, 727.
[64] Mingos; D.M.P. *Res. Chem. Intermediat.* 1994, 20, 85.
[65] Saskia, A.G. *Chem. Soc. Rev.* 1997, 26, 233.
[66] Chen, D.; Shen, G. Z.; Tang, K. B.; Lei, S. J.; Zheng, H. G.; Qian, Y. T. *J. Cryst. Growth* 2004, 260, 469.
[67] Chen, D.; Shen, G. Z.; Tang, K. B.; Jiang, X.; Huang, L. Y.; Jin, Y.; Qian, Y. T. *Inorg. Chem. Commun.* 2003, 6, 710.
[68] Chen, D.; Shen, G. Z.; Tang, K. B.; Sheng, J.; Fang, Z.; Liu, X. M.; Zheng, H. G.; Qian, Y. T. *Mater. Chem. Phys.* 2003, 82, 206.

[69] Nair, M. T. S.; Nair, P. K. *Semicond. Sci. Technol.* 1991, 6, 132.
[70] Prince, M. B. *J. Appl. Phys.* 1955, 26, 534.
[71] Lokhande, C. D. *J. Phys. D: Appl. Phys.* 1990, 23, 703.
[72] Domingo, G.; Itoga, R. S.; Cannewurf, C. R. *Phys. Rev.* 1966, 143, 536.
[73] Chen, D.; Shen, G. Z.; Tang, K. B.; Liu, Y. K.; Qian, Y. T. *Appl. Phys. A* 2003, 77, 747.
[74] Tu, W. X.; Liu, H. F. *Chem. Mater.* 2000, 12, 564.
[75] Rao, K. J.; Vaidhyanatham, B.; Ganguli, M.; Ramakrishnan, P. A. *Chem. Mater.* 1999, 11, 882.

In: Encyclopedia of Nanotechnology
Editor: Elwood D. Carlson

ISBN: 978-1-60692-079-4

Chapter 24

MOLECULAR DYNAMICS SIMULATION FOR MECHANICAL BEHAVIOR OF NANOCRYSTALS

Wei Yang*, Xiaoyan Li, Xinling Ma and Zengcai Guo
Department of Engineering Mechanics, Tsinghua University,
Beijing 100084, China

1. INTRODUCTION

1.1. Micro- and Nano-Scale Phenomena in Mechanics

The endeavor toward the understanding of micro- and nano-scale phenomena in mechanics is powered by three technology drives. The first drive comes from the field of material sciences, accompany with the endless search for stronger, lighter, tougher and multi-functional materials[1-3]. The second drive emerges from the rapid growth in information technology whose products involving microelectronic packaging, MEMS and NEMS technologies and nano-electronics[3-7]. Nano-structured information materials can offer more information storage capacity. The third, but more recent technology drive is formed upon the emergence of nano-science and engineering, where an understanding for nano-scale failure and deformation is in urgent demands[8-10].

The characterization of the mechanical properties, such as the Young's modulus, the flow stress and the hardness, of materials in a nanometer scale becomes an emerging research land as the field of nanotechnology unfolds. Thin films have small dimensions whose properties might differ from those of a bulk material of the same compositions[11,12]. Fundamental understanding of nanoscale friction and wear processes calls for the knowledge of contact nanomechanics[13,14]. Instruments with atomic resolution, such as the atomic force microscopy, high resolution electron microscopy and nano-moire method, enable the investigation of surface and bulk material properties at a nanoscale[15,16].

* E-mail: yw-dem@tsinghua.edu.cn

1.2. Nanocrystals

Nanocrystals find increasing use in nowadays technology. Since the German material scientist Gleiter sintered the metallic powders of 6nm diameter to bulk samples and proposed the concept of nanocrystalline materials [17], nanostructured materials have nourished throughout the world [18,19]. A great deal of studies on the mechanical properties of nanocrystals, such as their elasticity, hyper-plasticity and abnormal Hall-Petch relation are being pursued from different perspectives. It was shown [20] that the incorporation of the configuration entropy softened the stress strain response in a sub-micron regime.

1.2.1. Nano-Grained Bulk Materials

The nano-grained bulk samples can be synthesized by electro-deposition [21], ball milling, alloying, consolidation and severe plastic deformation (SPD). When a grain shrinks below a critical size of dislocation depletion that ranges from 15nm to 20nm for various metals, the available experimental evidence suggests that dislocations seldom present within the grain. When dislocations were observed, they were primarily in grains of larger sizes or in immobile or locked configuration. Nanocrystals exhibit peculiar deformation behavior[22-24]. Molecular dynamics (MD) simulations indicated that metal of an average grain size of several nanometers deformed mainly by many small, independent slip events in the grain boundaries[25-27]. Though dislocations were seldom observed within the grains, they did observe noticeable traces of stacking faults due to the passage of partial dislocations across the grains. The MD calculations, due to a typical high strain rate (108~109/s) imposed during the simulation, overestimate the threshold stress for creep by more than one order of magnitude [28,29], in contrast to many creep experiments that were actually conducted at a very slow strain rate (10^{-7}/s)[30].

Absence of interior dislocations does not cripple the capability of plastic deformation of nanocrystalline metals. Instead, material scientists had great expectation [31] on the super-ductility behavior of nano-grained materials at room temperature. Recently, Lu et al. [21] reported a stunning experiment of rolling nanocrystalline copper (about 30nm grain size) at room temperature (0.23 of homologous temperature) to an elongation of 5100%. No texture seemed to form during rolling, and the deformed sample exhibited random orientations of the equiaxed nano-grains, with about the same grain size and slightly increased dislocation density as that of the as-deposited sample.

When the grain sizes fall within the range of about 10nm, they can be regarded as crystallized grain cores glued together by viscous interlayers with atoms of short-range order. At a low extending rate, nano-grained metals of such small grain sizes deform mainly by neighboring grains sliding pass each other with insignificant distortion in the grain shape [32]. The crystallized cores move in trajectories of quasi-uniform deformation superimposed by oscillations when they switch the neighbors. Yang and Hong [33] adopted a structural evolution algorithm to simulate the process. Accommodation mechanisms such as grain boundary diffusion, sliding and grain rotation control the kinetics of the process. Actual deformation minimizes the plastic dissipation and stores strain energy for representative steps of grain neighbor switching. Numerical simulations based on these principles are conducted for a representing cell composed of 200 non-uniform grains of an average diameter of 30nm. The log-log plot of creep stress versus grain diameter reveals the dependence of the creeping threshold on the grain size. A negative power of 2.61 is found from the simulation. A typical cycle of deformation is composed of an insertion stage and a rotation stage. The latter can be induced through the motion of dipoles of disclinations, as suggested by Ovid'ko [34] and Murayama et al.[35].

1.2.2. Surface Self-Nanocrystallization

Among various methods of SPD, high frequency and multi-directional shot peen is an efficient method for surface self-nanocrystallization [36]. The technique is later generalized to several patents of the surface mechanical attrition treatment (SMAT) as comprehensively discussed by Lu and Lu [37]. The nano-structured surface layers block the movement of dislocations that are nucleated within the specimen, and consequently strengthen the material in a proportion far exceeding the mixture estimate [37,38]. A material with nanostructured surface layer is often resilient to the fatigue crack growth from the surface. The fast formation of fine grained passivation layer is beneficial to resist stress corrosion. Moreover, recent investigation leads to an opportunity of low temperature nitriding due to SMAT [39].

One treatment of high frequency and multi-directional shot-peening [36] consists of pure iron samples with an initial grain size of about 50 microns. The velocity distribution of the shots ranged from 5 to 20m/s. A nano-grained surface layer was formed after shot-peens of 450 seconds [36]. Substantial multi-directional slips fragmented the grains, and the surface self-nanocrystallized layer contained grains of the size about 10-20nm. Subsequent experiments on stainless steel [40] and low carbon steel [41] verified the formation of nano-grains under ultrasonic shot peening, as well as the co-existence of subgrains and grains. It was anticipated that subgrains formed first, then evolved under consecutive passages of trans-grain dislocations to stable nano-grains.

SMAT of this kind involves a deformed volume exceeding $10mm^3$, and a dislocation density in the order of $10^{11}\,mm^{-2}$. Even a two-dimensional calculation would encounter trillions of discrete dislocations. The huge scale of the system precludes a direct dislocation interaction calculation. A dislocation density approach is adopted instead [42]. Under a loading condition similar to that of Lu and Lu [36], a dislocation density as high as $2.17\times10^{11}\,mm^{-2}$ is simulated, and that is sufficient to form nano-grained iron.

1.3. Nanoindentation

Nanoindentation for measuring thin film mechanical properties is probably the most popular yet ill-understood method due to its inherent complexities [43]. Because of the test's simplicity and ability to mechanically probe into a nano-scale, it is becoming increasingly applied. The nanometer-ranged indentation depth can be utilized to probe thin films materials whose thickness or grain size is in a nano-scale. For a nano-indenter, the accuracy of the loading control reaches the range of nano-newton and that of the displacement control of 0.1nm. For relatively large indenting depth, the depth-sensing approach of Oliver and Pharr [44] is frequently adopted. Bahr et al. [45] used nanoindentation to perform the hardness measurement with indent less than 50nm. Adams [46] investigated the fracture mechanism of fine-grained thin film via nanoindentation. Nanoindentation can also be used to measure the residual stress in the surface layers of materials, the interface strength of composites, the hardness variation near grain boundaries, and the dislocation structures in semiconductors such as silicon.

As opposed to the burst pressure or micro-tensile tests of lithographed structures, where relatively uniform stress fields may be generated, the indentation-induced stress gradients can produce unique challenges. Possible stages of deformation include Hertzian elasticity to film delamination[47] and double buckling. Current limitations of indentation-induced delamination measures of toughness involve large scatter associated with sensitivity of the fracture radius to the contact radius ratio [43].

A key issue for nanoindentation is the size effect. The shallower the indentation is, the higher the hardness appears. Based on the concept of geometrically necessary dislocations introduced during nanoindentation, Nix and Gao [48] discovered that the size effect in nanoindentation can be well characterized under Mechanism-based Strain Gradient (MSG) plasticity theory. The existing data for (100) tungsten with four different conical tip radii varying from 85 to 5000 nm and the data for four conical tips (R = 0.5 to 20μm) into (100) aluminum, however, indicated a somewhat different trend. Both single crystals exhibit a decrease about a factor of two in hardness with increasing depth, irrespective of either increasing or decreasing average strain gradients. Based on these data, Tymiak et al. [49] discussed implications to strain gradient plasticity and indentation size effect interpretations at very shallow depths.

For the extremely shallow nanoindentation stage, the indenting force relates to the indenter displacement by a step-like curve. That is mainly due to the incubation and the burst of the first batch of dislocation loops [50]. The bubble raft simulation [51] indicated that a dislocation loop nucleates at a location of about 0.78 of the indenting half-width below the indenter. The bubble raft experiment is also capable of producing the step-like relation between the indenting force and indenter displacement.

Yet another interesting issue is the nanoindentation for nano-grained thin films. The ratio between the grain size D and the indentation depth h becomes a critical parameter. When $D/h >> 1$, the situation resembles the nanoindentation to a single crystal, the mechanical behavior is featured by the step-like response due to the incubation and burst of the first batch of dislocations. When D/h is in the neighborhood of unity, the blocking effect of grain boundaries becomes important. When $D/h << 1$, the other extreme, a deep indentation for nano-grained materials, takes place, the linear viscous constitutive law [52] should be used for analyses.

2. Basic Molecular Dynamics

2.1. Fundamentals of Molecular Dynamics

Computer simulation of material structure and properties can be divided naturally into two categories: continuum and atomistic formulations. In a continuum model, the material concerned is regarded as a continuous medium. Most continuum simulations use finite-element method. In contrast, in an atomistic model simulation, the material concerned is regarded as an aggregate of atoms, and each atom is treated on an individual basis. Two atomistic simulation methods used the most are the MD method and Monte Carlo (MC) method [53].

Nanoscale simulation can be conducted in a variety of methods, such as computation based on quantum mechanics [4-7], massive MD simulation and multiscale simulation. The available computation models suit for a spectrum of problems.

The continuum mechanics simulation suits for a spatial scale larger than nanometers, and a time scale longer than nanoseconds. On the other hand, MD suits for a computation scale less than or equal to giga-atoms and a simulation period shorter than nanoseconds. The ab initio calculation is only applicable for problems involving dozens of atoms such that near a dislocation core [54]. For the energy band calculation, however, the approach of effective electron mass can be adopted to large opto-electronics problems [7]. Whenever simulating defect-related behavior such as surface, grain boundaries, nano-grains and nanocracks, one can only rely on the multi-scale simulation that links continuum mechanics and MD [2,55-58].

Powered by the Moore-type multiplication in the computing capacity in recent years, the MD method emerges as an ideal tool for studying the mechanics and mechanisms in a nanometer-length scale [59,60]. The simulated phenomena range from nanojets [61,62], ice-water transition [63,64], silicon fracture [65,66], supersonic fracture [67,68], dislocation evolution [69-72], and the thermodynamic properties of nanocrystals [73]. During these simulations, the response of a material subjected to an external force is evaluated by following the dynamics of every atom in the simulation sample. Atomistic simulation provides opportunity for the exact control over the parameters of the numerical experiment. The information stimulates new insights when compared to the experimental data. The MD simulations are typically run on scaled-down systems compared to experiments but the results capture many of the features observed experimentally.

Given the instantaneous position and velocity vectors of the atoms, the dynamical history of the assembly is generated by numerically integrating the equations of Newtonian mechanics. The energy and forces of the system are calculated, while atoms in the assembly are given for the Cartesian coordinate components in different conditions of thermodynamics and kinetics. Computer simulation process leads to fluctuation on an equilibrium system that has minimal interatomic energy and force through development of suitable inter-atomic potentials.

The atoms follow the classical equations of motion,

$$m_i \dot{\boldsymbol{v}}_i = \boldsymbol{F}_i = -\frac{\partial V_{\text{tot}}}{\partial \boldsymbol{r}_i} \tag{1}$$

where the total inter-atomic potential V_{tot} consists of an interior part V_{I} and an exterior part V_{E}. The most commonly used the integrate algorithm is the following Verlet leapfrog algorithm [74],

$$\boldsymbol{v}_i(t+\frac{\Delta t}{2}) = \boldsymbol{v}_i(t-\frac{\Delta t}{2}) + \frac{\boldsymbol{F}_i}{m_i}\Delta t$$
$$\boldsymbol{r}_i(t+\Delta t) = \boldsymbol{r}_i(t) + \boldsymbol{v}_i(t+\frac{\Delta t}{2})\Delta t$$
$$\boldsymbol{v}_i(t) = \frac{1}{2}\left[\boldsymbol{v}_i(t+\frac{\Delta t}{2}) + \boldsymbol{v}_i(t-\frac{\Delta t}{2})\right] \quad (2)$$

where m_i, $\boldsymbol{v}_i(t)$ and $\boldsymbol{r}_i(t)$ denote the mass, the velocity and the position of the i-th atom, respectively. The time step for this integration is in the order of femtosecond (fs), which is a small fraction (less than one percent) of the inter-atomic vibration period.

The MD simulation comprises two phases: initial relaxation and loading. In the first phase, atoms were positioned according to an assumed lattice sites and their velocities were randomized to satisfy the Maxwell-Boltzmann distribution at prescribed temperature [74]. In order to reach the equilibrium state, the substrate atoms were relaxed according to Eq.(1) in the absence of any external forces. For the simulations presented in the sequel, the relaxation phase was completed within 5000 time steps by controlling the substrate temperature at a 300K using NVT ensemble [74]. In the second phase, the appropriate loading is applied by tiny increments, composed of a deterministic part compatible with the loading and a stochastic part. Thousands of relaxation steps will be engaged after each loading increment.

2.2. Ab Initio Molecular Dynamics

Empirical or semi-empirical interatomic potentials used in MD simulations fail to describe complex chemical, solid-state physical and bio-physical processes. Therefore, physicists started to work from the fundamental principles of quantum mechanics to search for more accurate and transportable interatomic potential. As a result, a new methodology, which is so-called ab initio molecular dynamics (AIMD) [75-79], is established by combining finite temperature MD with forces acting on the nuclei from electronic structures calculations performed 'on the fly' as the MD simulation proceeds. Because the system is regarded as a collection of nuclei and electrons in AIMD simulations, many-body effect, electronic polarization and bond-forming and breaking events are reflected within the accuracy of the electronic structure representation [80]. This leads to realistic simulations that are performed without adjustable parameters, using only the values of the fundamental constants: the electronic charge and mass and Planck's constant [78]. Shortly speaking, AIMD not only enhances reliability and predictive power of atomic-scale simulations, but also opens an unbiased way along which one is likely to observe the microscopic details of the real-world.

Originally AIMD was developed in the solid state community, but its scope has become particularly widespread in material science, chemistry and biology. Although AIMD has better computing accuracy and transportability than the classical MD, a significant increase in computational cost is also obvious. In contrast to the classical MD, the length and time scales that are accessible in AIMD are much smaller. Currently, AIMD calculations can be applied routinely to systems consisting of just hundreds of atoms and access timescales on the order of tens of picoseconds (ps). Moreover, the accuracy of AIMD calculations is limited by the accuracy of the electronic structure method employed [80]. Presently, AIMD simulations commonly use the Car-Parrinello approach in which the electronics structure is described via the Kohn-Sham formulation [81] of the density functional theory, and the Kohn-Sham orbitals are expanded in a plane-wave basis set.

2.3. Interatomic Potential

2.3.1. Pair Potential

The selection of interatomic potentials for the system is crucial in gaining accurate simulation results. The pair potential gives a simple description for the pair-like interaction between atoms[74]. It suits for the covalent bonds between atoms, as well as the ionic crystal with insignificant many-body effect. In the popular 12-6 LJ potential formula, the interior inter-atomic potential obeys the following law:

$$U_{\mathrm{I}} = 4\varepsilon_0\left(\left(\frac{\sigma}{r}\right)^{12} - \left(\frac{\sigma}{r}\right)^{6}\right) \tag{3}$$

In the case of Cu, the reference energy ε_0 is chosen as 0.583eV and the atom diameter σ is 2.27 angstrom.

During the simulation, one may need to compute the stress field of the system. The macroscopic stress tensor is the derivative of the free energy of the system with respect to the strain tensor. In an atomistic scale, that can be translated to a suitable derivative of the energy with respect to the inter-atomic distances [82],

$$\sigma_{i,\alpha\beta} = \frac{1}{V_i}\left(-\frac{p_{i,\alpha}p_{i,\beta}}{m_i} + \frac{1}{2}\sum_{j\neq i}\frac{\partial U_I}{\partial r_{ij}}\frac{r_{ij,\alpha}r_{ij,\beta}}{r_{ij}}\right) \tag{4}$$

where $\sigma_{i,\alpha\beta}$ is the (α, β) component of the stress tensor for atom i, V_i is the volume assigned to atom i (subjected to $\sum_i V_i = V$ with V being the total volume of the system), m_i is the mass of atom i, $p_{i,\alpha}$ is the α-th component of its momentum and $r_{ij,\alpha}$ is the α-th component of the position vector from atom i to atom j. The atomic stress tensor defined in Eq. (4) is based on the Virial Theorem. When the atomic stress is averaged over a spatial region, the atomic stress quickly converges to a macroscopic stress field.

The use of pair potential, however, leads to poor representation for the deformation of solids involving shear, bending and torsion. For metallic crystals, pair potential (that inevitably leads to Cauchy relations among elastic constants) fails to quantify the components C_{12} and C_{44} of the elasticity tensor.

2.3.2. Many-Body Potential

The frequently used many-body corrections include Embedded Atom Method (EAM) [83-87], Effective Medium Theory [25] and Finnis-Sinclair model [88]. Exemplified here is the inter-atomic potential [88] for iron, featuring the widely used multi-body potential proposed by Finnis and Sinclair based on the Embedded Atoms Method. The total energy of an assembly of atoms at positions $\{\boldsymbol{r}_i\}$ is given by

$$U_{\mathrm{I}} = U_{\mathrm{N}} + U_{\mathrm{P}} \tag{5}$$

where $U_{\mathrm{N}} = -A\rho^{1/2}$ is the N-body term and $U_{\mathrm{P}} = \sum U(\boldsymbol{r}_i)/2$ a conventional central pair-potential summation. The charge density at the *i*-th atomic core-cite is given by $\rho = \sum \phi(\boldsymbol{r}_i)$, and the inter-atomic potential is given by

$$\begin{aligned}
&\phi(r) = (r-d)^2 + \beta(r-d)^3/d \qquad (r \le d)\\
&\phi(r) = 0 \qquad (r > d)\\
&a < d < \sqrt{2}a \qquad a \text{ is the lattice constant}\\
&U(r) = (r-c)^2(c_0 + c_1 + c_2 r^2) \qquad (r \le c)\\
&U(r) = 0 \qquad (r > c)
\end{aligned} \tag{6}$$

The parameters $c, d, A, \beta, c_0, c_1, c_2$ are given in Finnis and Sinclair[88].

The EAM potentials [84-87] provide a simple framework for accurate description of noble transition metals and their alloys. In EAM scheme the total energy of a mono-atomic system is represented as,

$$U_{\mathbf{tot}} = \frac{1}{2}\sum_{ij}\phi(r_{ij}) + \sum_i F(\overline{\rho}_i) \tag{7}$$

Here $\phi(r_{ij})$ is a pair potential as a function of the distance r_{ij} between atoms *i* and *j*, and *F* is the embedding energy as a function of the host electron density $\overline{\rho}_i$ at site *i*, which is induced by the other atoms in the system as,

$$\overline{\rho}_i = \sum_{j \ne i} \rho(r_{ij}) \tag{8}$$

where $\rho(r_{ij})$ is the electron density function. The second term in Eq. (7) is volume dependent and represents, in an approximate manner, many-body interactions in the system. Thus many-body effects are considered by introducing the embedding energy function, and the computation efficiency is comparable to simple pair potential models.

2.3.3. Indenter Potential

For the case of nano-indentation, the atomic nature of the indenter tip may be ignored. The effect of indenter can be modeled as a frictionless spherical tip having no chemical interaction with the sample surface by using a repulsive spherical potential of the form [82]:

$$U_{\mathrm{E}} = A_{\mathrm{E}} \cdot (R-r)^{p} \cdot H(R-r) \tag{9}$$

where $H(r)$ is a standard Heaviside step function, R is the indenter radius, r is the distance from the center of indenter sphere; A_{E} and p are constants reflecting the hardness of an indenter tip. For the indentation on coppers, one may choose A_{E} as 3.0 nN/Å^2 and p as 3[82].

The position and the velocity of each dynamic atom in the model are calculated by simultaneously solving a system of equations of atomic motion given by

$$m\ddot{r}_{i} = -\sum_{j=1}^{n} \frac{\partial U_{\mathrm{I}}(r_{ij})}{\partial r_{ij}} \boldsymbol{n}_{ij} - \frac{\partial U_{\mathrm{E}}(r_{iK})}{\partial r_{iK}} \boldsymbol{n}_{iK} \tag{10}$$

where n is the total number of atoms in the substrate, r_{ij} the distance between substrate atoms i and j, r_{iK} the distance between substrate atom i and the tip sphere center labeled by K, $\boldsymbol{n}_{ij}$ the unit vector with direction from substrate atom j to atom i, and $\boldsymbol{n}_{iK}$ the unit vector with direction from tip sphere center K to substrate atom i. Function $U_{\mathrm{I}}(r_{ij})$ is the inter-atomic potential.

2.4. Multiple Time Step Integration Algorithms

Over the past decades, researchers have paid much attention to expanding the time scale of MD simulation. Strett et al. [89] introduced a multiple time step (MTS) integration method via Taylor expansion. Other MTS methods such as Verlet-I [90] and reversible reference system propagator algorithm [91] (rRESPA) were also suggested. To overcome the linear instability of Verlet-I and rRESPA methods, the mollified impulse method [92,93] (MOLLY) has been developed, as a result of achieving the longer time step size. MTS methods can run at much faster speeds than do the standard algorithms. The rRESPA method has implemented successfully in various complex systems [94-97] with speedup factor of 4 to 15. Furthermore, the rRESPA method can be also carried out to accelerate simulation for systems with different time scales [98]. The related theory of the RESPA method are outlined as follows.

Consider a system with N degrees of freedom, the corresponding Hamilton's equations can be cast in the following form,

$$\dot{\boldsymbol{q}}_i = \frac{\partial H}{\partial \boldsymbol{p}_i}, \ \dot{\boldsymbol{p}}_i = -\frac{\partial H}{\partial \boldsymbol{q}_i} \quad i = 1, 2, \cdots, N \tag{11}$$

where H is the Hamiltonian of the system, $\boldsymbol{q}_i$ and $\boldsymbol{p}_i$ indicate the i-th generalized coordinate and conjugated momentum, respectively. The Liouville operator L for such system is defined as

$$iL \equiv \{\cdots, H\} = \sum_{i=1}^{N} \left(\dot{\boldsymbol{q}}_i \frac{\partial}{\partial \boldsymbol{q}_i} + \dot{\boldsymbol{p}}_i \frac{\partial}{\partial \boldsymbol{p}_i} \right) = \sum_{i=1}^{N} \left(\frac{\partial H}{\partial \boldsymbol{p}_i} \frac{\partial}{\partial \boldsymbol{q}_i} - \frac{\partial H}{\partial \boldsymbol{q}_i} \frac{\partial}{\partial \boldsymbol{p}_i} \right) \tag{12}$$

Let $\Gamma(t)$ denote the set of particle positions and momenta at time t such as $\Gamma(t) = \{\boldsymbol{q}_i(t), \boldsymbol{p}_i(t)\}$. With the help of the Liouville operator, one obtains the system state $\Gamma(t)$ expressed as

$$\Gamma(t) = e^{iLt}\Gamma(0) \equiv G(t)\Gamma(0) \tag{13}$$

in which $G(t)$ is called the classical propagator, and $\Gamma(0)$ the initial state of the system. For Cartesian coordinates, Eq. (12) can be changed into

$$iL = \sum_{i=1}^{N} \left(\dot{\boldsymbol{r}}_i \frac{\partial}{\partial \boldsymbol{r}_i} + \boldsymbol{F}_i \frac{\partial}{\partial \boldsymbol{p}_i} \right) \tag{14}$$

where $\boldsymbol{r}_i$, $\boldsymbol{F}_i$ are the position and the force of the i-th degree of freedom.

The rRESPA method is to generate a discretized propagator $G(\Delta t)$ which reversibly integrates different components of the system under two or more different time steps [99]. Accordingly, the force $\boldsymbol{F}(\boldsymbol{r})$ on the system is decomposed into the "slow" force $\boldsymbol{F}_s(\boldsymbol{r})$ and the "fast" force $\boldsymbol{F}_f(\boldsymbol{r})$,

$$\boldsymbol{F}(\boldsymbol{r}) = \boldsymbol{F}_s(\boldsymbol{r}) + \boldsymbol{F}_f(\boldsymbol{r}) \tag{15}$$

In the process of integrating equations of motion, the fast force is computed after each time step δt, and the slow force is determined every n time steps $\Delta t \equiv n\delta t$. To accomplish the decomposition of the force, a switching function $S(\boldsymbol{r})$ is introduced so that

$$\boldsymbol{F}_s(\boldsymbol{r}) = [1 - S(\boldsymbol{r})]F(\boldsymbol{r}), \ \boldsymbol{F}_f(\boldsymbol{r}) = S(\boldsymbol{r})\boldsymbol{F}(\boldsymbol{r}) \tag{16}$$

A typical expression of the switching function is [100]

$$S(\boldsymbol{r})=\begin{cases}1 & r_c-\lambda>|\boldsymbol{r}| \\ 1+R^2(2R-3) & r_c-\lambda\leq|\boldsymbol{r}|\leq r_c \\ 0 & r_c<|\boldsymbol{r}|\end{cases} \tag{17}$$

where

$$R=\frac{|\boldsymbol{r}|-r_c+\lambda}{\lambda} \tag{18}$$

r_c is the cutoff distance, and λ is a parameter used to optimize the computing time. Based on the force spilt Eq. (15), the Liouville operator is divided into two parts such that

$$iL=\dot{\boldsymbol{r}}\frac{\partial}{\partial \boldsymbol{r}}+\boldsymbol{F}_f(\boldsymbol{r})\frac{\partial}{\partial \boldsymbol{p}}+\boldsymbol{F}_s(\boldsymbol{r})\frac{\partial}{\partial \boldsymbol{p}}=iL_f+\boldsymbol{F}_s(\boldsymbol{r})\frac{\partial}{\partial \boldsymbol{p}} \tag{19}$$

Using the symmetrical Totter factorization[101], as well as Eq. (13) and the initial state $\{\boldsymbol{r}(0),\boldsymbol{p}(0)\}$, the dynamics trajectory within a time step can be obtained [91],

$$\begin{aligned}\boldsymbol{r}(\Delta t)&=\boldsymbol{r}_f\left\{\Delta t;\boldsymbol{r}(0),\dot{\boldsymbol{r}}(0)+\frac{\Delta t}{2m}\boldsymbol{F}_s\left[\boldsymbol{r}(0)\right]\right\},\\ \dot{\boldsymbol{r}}(\Delta t)&=\dot{\boldsymbol{r}}_f\left\{\Delta t;\boldsymbol{r}(0),\dot{\boldsymbol{r}}(0)+\frac{\Delta t}{2m}\boldsymbol{F}_s\left[\boldsymbol{r}(0)\right]\right\}+\frac{\Delta t}{2}\boldsymbol{F}_s\left[\boldsymbol{r}(\Delta t)\right]\end{aligned} \tag{20}$$

The implement of rRESPA consists of iterations in terms of integration scheme Eqs. (20). Such iterations can reduce the computing time spent in evaluating intermolecular forces, and accelerate the integration of the equations of motion. In addition, the propagator $G(t)$ is a unitary operator, i.e., $G(-t)=G^{-1}(t)$, which ensures that the rRESPA method is truly time reversible.

2.5. Ordering Parameters and Visualization

To identify the location and type of defects, we introduce three methods to analyzing the local atomic ordering.

2.5.1. Centrosymmetry Disorder

The first method may be called as the method of centrosymmetry disorder. A centrosymmetric material (such as copper or other FCC metals) will remain centrosymmetric under homogeneous elastic deformation [102,103]. In a centrosymmetric material, each atom has pairs of equal and opposite bonds among its nearest neighbors. As the material is distorted, these bonds will change direction and/or length, but they will remain equal and opposite. If there is a defect nearby, however, this equal and opposite relation no longer holds for all of the nearest-neighbor pairs. Thus a centrosymmetry parameter is defined which is zero for a centrosymmetric material under homogeneous elastic deformation but nonzero for plastic deformation of the material. The centrosymmetry parameter for each atom, denoted as a dimensionless parameter Δ, is defined as follows [102,103]:

$$\Delta = \frac{1}{6d_0^{\,2}} \sum_{i=1,6} \left| \boldsymbol{R}_i + \boldsymbol{R}_{i+6} \right|^2 \tag{21}$$

where $\boldsymbol{R}_i$ and $\boldsymbol{R}_{i+6}$ are the vectors or bonds corresponding to the six pairs of opposite nearest neighbors in an FCC lattice, and d_0 stands for the atom diameter. In a perfect bulk FCC lattice, each atom has 12 nearest-neighbor vectors. If the given atom has nearest neighbors other than 12 or in a highly distorted local environment, this set may contain duplicates or non-nearest neighbors.

The following modifications are proposed [103] to resolve the highly-distorted case. Let us start from any atom, termed the center-atom, within the deformed lattice, and assume a centrosymmetry approximation around it. All atoms of distance less than $1.2d_0$ from the center-atom are regarded as the set of the nearest neighbors. Select two roughly opposed atom sites from the set. If the sum of two vectors can be bounded within the offset range of $d_0/3$, the atom-pair will be included in the family of the regular pairs associated with the center-atom, and the value of $\left|\boldsymbol{R}_i + \boldsymbol{R}_{i+6}\right|^2$ will be added to the Δ parameter. Otherwise, the pair will be discarded and we will search for the next possible pair. After the selection of all regular pairs, the left-over atom, if any, is regarded as the irregular atom site, and whose distance from the center-atom will be squared to include in Δ as penalty. The final sum of Eq. (21) provides a dimensionless and squared measure for the departure from centrosymmetry that highlights irregular atoms, in the immediate vicinity of the selected atom.

2.5.2. Local Crystalline Order

The second method is termed the method of local crystalline order. That method examines the coordination number and type of each atom in the atomistic configuration using the topological analysis developed by Honeycutt and Andersen [104]. According to this method, the state of an atom is characterized by four indices *i*, *j*, *k* and *l*. Index *i* denotes the bond relation of two atoms. Two atoms are bonded if $i = 1$ and not bonded if $i = 2$. Index *j* indicates the number of common nearest neighbors between the two bonding atoms. Index *k* provides the number of bonds among the common nearest neighbor atoms. The last index *l* indicates the number of bonds in the longest continuous path that goes through the common neighbors. The above procedure provides a distinction between FCC and HCP structures, taking into account only the nearest-neighbor pairs. An arbitrary atom in a perfect FCC structure forms only 1421-type quadru-pairs with its 12 nearest neighbors, while in the case of the HCP lattice one would find six 1421 and six 1422 quadru-pairs. Using this analysis, one may assigns different colors to atoms for visualization of defects.

2.5.3. Radial Distribution Function

The method of radial distribution function (RDF) utilizes the following dual statistic distribution function,

$$g(r) = \rho(r)/\rho_0 \tag{22}$$

where ρ_0 denotes the mean atomic density of the material, and $\rho(r)$ the mean atomic density in a spherical surface of radius r centered at a given atom core. The dimensionless ratio reflects the atom distribution at different radii from a given atom center. The RDF can be calculated from the dual statistic distribution function $g(r)$ as $4\pi r^2\rho_0 g(r)$, or equivalently as $4\pi r^2\rho(r)$. It represents the average number of atoms located on a spherical surface of radius r from the atom center.

Either the dual statistic distribution function $g(r)$ or the RDF can describe the order-disorder behavior of materials [105]. For gaseous matters, $g(r)$ behaves like a step function that rises suddenly after a characteristic radius of the repulsive corc; the other extreme of crystalline matters has a $g(r)$ curve featured by many peaks; the intermediate cases of liquids and amorphous solids have gradually decayed peaks in the $g(r)$ curves, characterizing the short-range order.

2.6. Parallel Algorithms

The use of massive parallel clusters leads to the birth of massive molecular dynamics simulation, with giga-atom capacity [68,71,106]. For simulation involving Giga-atoms, hundreds of 100G processors and 10Gbytes DRAM have to be used at each time step. If the cluster stores the atom coordinates every 10 seconds during the simulation, a steady data flow of 250Mb/s will be generated.

A major parallel algorithm for short-range MD interaction is named Link-Cell-List algorithm [107-111]. Its basic principle is to divide, in computation space, the entire model into multiple cells of equal or non-equal sizes. Each cell corresponds to a computation process. The algorithm is devoted to optimizing the cell division such that the computing burdens in different cells are evenly distributed and the data exchange among cells is minimized. Such an algorithm may achieve scalability and transportability for parallel computation. After allocating all the atoms into different processes, the information such as atom coordinates can be communicated through neighboring cells. The Neighbor-List algorithm[74] for serial MD will be implemented simultaneously. For each cell, one may finds its neighbor-list according to the interior atom coordinates within the cell and the atomic positions of the neighboring cells logged into the processor. The inter-atomic force on each atom within the cell can be calculated by this neighbor-list. Consequently, one may use the classical Verlet leapfrog algorithm for series computation to update the velocities and positions of every atom. Figure 1 outlines the flowchart of the parallel algorithm.

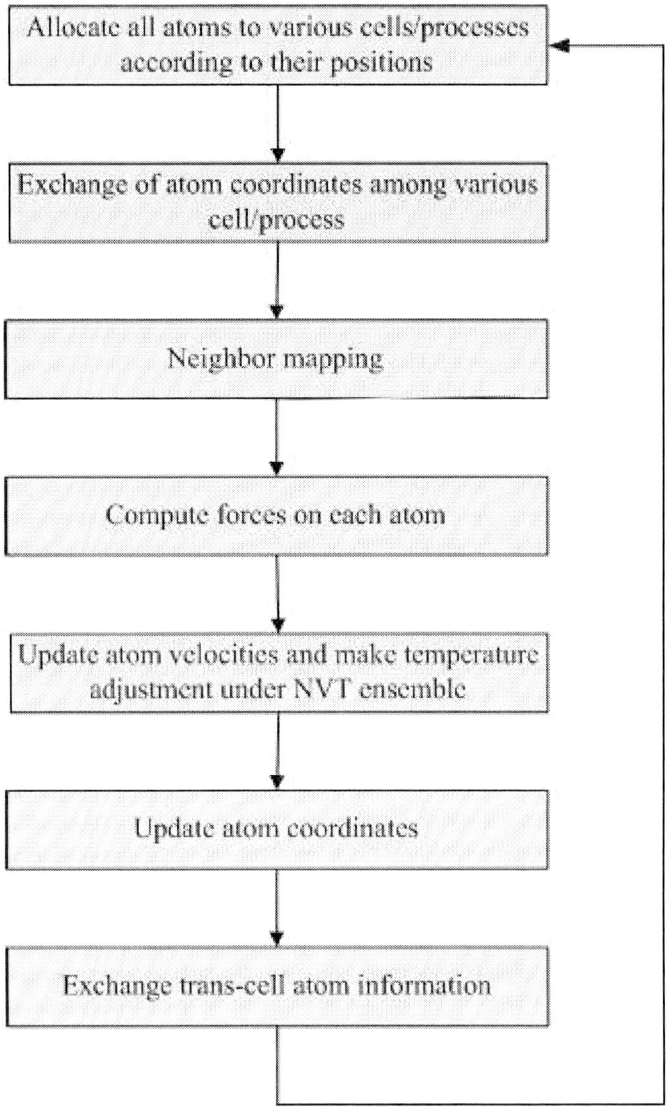

Figure 1. Flowchart of parallel MD algorithm.

In data transmission among different cells, it is critical to reduce idle waiting times of processes and to prevent dead-locking. One may assign an integer number to each cell, and set the sequence of data communication in different directions according to the numbering characteristics such as odd or even. Consider the three-dimensional case where one cubic cell is surrounded by 26 cells. Set the cells odd numbers to transmit data to cells of even numbers in the positive x direction at first, and then conduct data transmission from even to odd after the first group of messages were received and processed. Accordingly, the data from the center cell are communicated to 9 cells in the positive x direction. The same sequence is repeated to the 9 cells in the negative x direction. One then conducts similar data communication to the cells (three each) in the positive and negative y directions. The 2 cells in the positive and negative z directions are communicated last to complete data exchange with all 26 neighboring cells. The process can repeat itself for data transmission to other cells. The sequence designed in this way enforces data transmission for all cells at every data transmission round. After 26 rounds of data transmission, all cells attain the updated information of the neighboring cells.

In the MD simulation under short-range inter-atomic potential, only the atomic interactions within a cut-off distance are of interest. In the Link-List method, one only tabulates the neighbors for those atoms within a distance, labeled as Rlist, several times of the cut-off distance from the cell under consideration. Since the dimension of a cell is much larger than Rlist, the number of tabulated neighbors for each cell is significantly reduced. Accordingly, only the atom information in the layer measured by Rlist needs to be exchanged to the neighboring cells, so that the costs for both neighboring table generation and data transmission are considerably reduced. This method is called atom-caching in the literatures[107-111]. Its schematics are shown in Figure 2.

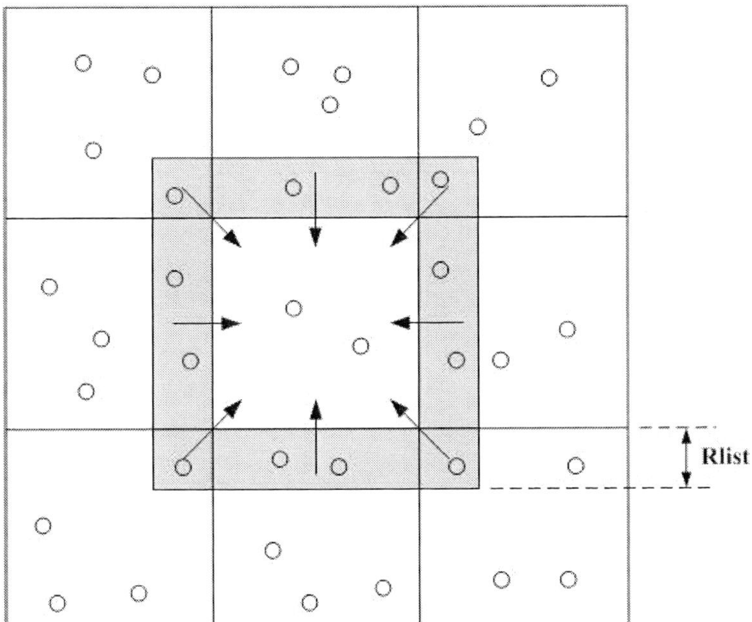

Figure 2. Schematics for method of atom-caching.

From a simulation perspective, the computational time becomes prohibitive beyond the first few nanoseconds, even for the state-of-the-art scalable parallel machines. To overcome these limitations, a way of connecting the MD simulation results to other simulation methods such as rate theory [112] or kinetic MC simulations [113-115] is required. In particular, MC simulations appear very promising because they provide the ability to perform atomic-level simulations of the defect kinetics and microstructural evolution over relevant length and time scales. A method to extend atomistic simulations of solids beyond the nanosecond time scale is the multiple time method [116]. The method was used to simulate crystal growth on the time scale of laboratory experiments. Transitions involving concerted motion of multiple atoms on the crystal surface are found to lead to remarkably smooth growth of pure metals such as Al and Cu.

3. MPM/MD Handshaking Method

3.1. Multiscale Modeling and Simulation

Many physical phenomena, such as crack propagation, turbulence, and high energy cluster impacts, occur in multiple length scales. Different length scales interact strongly to generate complicated behaviors. These phenomena pose notorious difficulty in computation. To study such multiscale problems, some multiscale methods for the modeling of solids and fluids have come forth. In recent years, the multiscale simulation approaches have made amazing progress in many fields including material science, fluid mechanics, chemistry and biology. Especially towards researches of nanoscale systems, multiscale simulations successfully bridge the gap in the understanding of mechanical properties and processes at length scales ranging from 100 nm to a micron [117]. It is widely recognized that most multiscale techniques are hybrid numerical methods that employs models at different physical scale in combination to build a comprehensive and accurate description of multiscale systems. This section aims to review some existing multiscale methods and introduce a new multiscale scheme called MPM/MD handshaking method [118].

3.1.1. Lattice Static Simulation

Using lattice static approach can dramatically downsize the scale of simulation. The earlier method, by Thomson and coworkers [119], used the lattice Green function and molecular statics. In the lattice Green function method, atoms at the lattice sites interact with their neighbors through short-range nonlinear potential. That resolves a static defect structure of the lattice. The calculation is proceeded in a hierarchical manner: the lattice Green function for a perfect lattice is calculated first, then leads to the defect containing lattice, and finally iterates to the structure under nonlinear interaction.

3.1.2. Quasi-Continuum Simulation

The quasi-continuum method was proposed by Tadmor et al. [120], Ortiz and Phillips [57] and Phillips [58]. It has three features: (1) the potential of the solid is written in terms of a collection of representing "atoms" through appropriate triangularization and they condense the action of a collection of the nearby atoms; (2) a sensitivity parameter is introduced, the number of the representing atoms can change adaptively according to this parameter; and (3) the method is based on lattice statics rather than MD since the kinetic energy associated with the representation atom is hardly distinguishable with the thermally related phonon scattering.

3.1.3. Coupled Continuum-MD Method

According to the domain decomposition strategy, the simulation domain is decomposed into atomistic and continuum regions where the MD and continuum models are used, respectively. And then, the matching condition at the continuum-MD interface will be a crucial issue. Sinclair et al. [121] proposed the flexible boundary conditions in atomistic dislocation modeling. For general linear problems, Cai et al. [122] found exact boundary conditions which become a conceptual starting point. Further theoretic work has been provided by Wagner et al. [123], and corresponding practices have been carried out by E and Huang [124,125]. In a somewhat different approach, Kohlhoff et al.[126] applied a hybrid finite element (FE)-MD method to crack propagation in bcc crystals. The link between atomistic and FE regions is implemented through a transition zone made up of combination of atoms and bulk elements. By introducing an overlapping layer with double identities, Yang and his collaborators [127,128,129] also developed a combined FE-MD method which has been applied to the evolution of crack tip and interfacial fracture. In the overlapping layer, an element node associates with a set of atom points. The nodal force is distributed to those atoms, and the atom movements are averaged to give the nodal motion. During the simulation for nanoscale fracture, the MD under EAM potential is used to simulate the nanoscale behavior near the crack tip, and the nanoscale region is embedded within a micron size annulus described by elasticity and discrete dislocations whose speed is dictated by an empirical law of dislocation dynamics. The continuum and atomistic regions are hand-shaken by an overlapping layer, and this permeable layer allows the trans-scale connection of defects such as dislocations. If the atomistic region generates dislocations under the prescription of interatomic potential, these dislocations will be converted into continuum dislocations when they pass through the overlapping layer. Applying the foregoing method, Tan and Yang [129] simulated the emission of 13 dislocations from an aluminum-alumina interface crack tip within 30 ps, and discussed the effect of a zigzag interface on dislocation emission under different loading modes.

3.1.4. Lattice Material Point Method

Tan and Nairn [130] proposed the lattice material point method (LMPM) that combines the adaptive material point method (MPM) and MD method. A lattice material possesses dual presentations of a continuum material point and a lattice structure formed by the associated atoms. The MPM has two descriptions: a material formulation where the material particles are arranged under a hierarchical structure, and a grid formulation which map the material points to a 3D space at each time step. LMPM provides a smooth transition for two formulations and projects them in a background mesh. The method is able to provide a seamless and adaptive link between MD and finite element continuum computation.

3.1.5. Coupled Quantum Mechanics-MD Method

To investigate chemical reactions which involve large molecules and chemical bond formation or breaking, the quantum-chemistry community has extensively developed coupled quantum mechanics-MD methods. Warshel and Levitt [131] firstly presented the basic concepts and algorithms of QM-MD approaches. Subsequently, there has been comparatively much effort devoted to studies [132,133] for such methods. In order to take advantage of the applicability and accuracy of the ab initio QM methods for chemical reactions and the computational efficiency of the MD method, a couple QM and MD approach uses the QM method in the active region of chemical reaction, but the molecular mechanics method elsewhere [132].

3.1.6. Macro Atomistic AB Initio Dynamics

It is quite inspiring that Abraham et al. [134] and Broughton et al. [135] propound the macro atomistic ab initio dynamics (MAAD) with a seamless coupling of quantum to statistical to continuum mechanics. This approach makes the finite element (FE), MD and semi-empirical tight-binding (TB) run concurrently with minimal coupling, which guarantees physical consistency of the simulative systems [136]. At present, the approach has been applied to simulating the rapid brittle fracture of a silicon slab with a microcrack at its center. In the course of simulation, the continuum region away from the crack tip is treated by FE; the atomistic statistical region around the crack tip is implemented by MD; the region at the crack tip is dominated by mean-field quantum mechanics represented by TB.

3.1.7. Coarse-Grained MD

The coarse-grained molecular dynamics (CGMD) [137] is an attempt to extend atomistic simulation to much larger systems than amenable to MD alone. In CGMD, equations of motion are derived directly from finite temperature MD through a statistical coarse graining procedure. It is guaranteed that the generalized FE can pass smoothly into MD as the mesh size is reduced to the atomic spacing [136]. The main step of this methodology is to construct a microscopic Hamiltonian for a set of coarse-grained variables by averaging coarse-grain energy of the canonical ensemble. As a technique for simulation of the mechanics of microscale systems, CGMD has been illustrated and validated by using the example of the dynamics simulation of MEMS resonators [138].

3.1.8. Coupled Continuum-Atomic-Scale Finite Element Method

Considering the principle of energy minimization, Liu et al. [139] proposed an order-N atomic-scale finite element method (AFEM) which is as accurate as MD simulations. Furthermore, they discussed the stability and convergence of AFEM in detail, and then applied it to simulating the vibration of a carbon nanotube. Since AFEM and FEM are in the same theoretical framework, their linkage can provide a seamless multiscale computation method. By adopting two computational schemes, Liu et al. [139] calculated an armchair carbon nanotube with two fixed ends under a displacement loading. It is found that combined AFEM/FEM method not only reduces the CPU time but also saves the computer memory. This numerical example demonstrates that the AFEM/FEM method is suitable for large scale static problems relating to multiple atomistic interactions.

3.1.9. Heterogeneous Multiscale Method

The heterogeneous multiscale method (HMM) [140] is a unified framework for designing efficient multiscale methods that couple macro- and micro-scale computational models. The full philosophy of HMM is to start with a macroscale solver and find the missing macroscale data by performing local simulation of the microscale models constrained to be consistent with the local macroscale state of system [141]. The macroscale models in HMM are expressed in terms of conservation laws, whereas the microscale models are described by the discrete atomistic methods such as MD or MC. Now HMM has successfully dealt with a wide variety of problems, encompassing complex fluids [142], shock propagation [141], thermal expansion [141] and twin boundary dynamics.

3.2. MPM/MD Handshaking Method

3.2.1. Material Point Method

Motion of a continuum is governed by conservation equations of mass, momentum and energy. If one lets ρ be the mass density, $\boldsymbol{v}$ the velocity, $\boldsymbol{\sigma}$ the Cauchy stress, $\boldsymbol{b}$ the specific body force and $\boldsymbol{a}$ the acceleration, the conservation equations for mass and momentum are,

$$\dot{\rho} = -\rho \nabla \cdot \boldsymbol{v} \tag{23}$$

$$\nabla \cdot \boldsymbol{\sigma} + \rho \boldsymbol{b} = \rho \boldsymbol{a} \tag{24}$$

Though not listed here, one has to bear in mind that the consideration of thermal effects may involve the conservation of energy. In MPM, a piece of material is discretized into a collection of material points (particles) as illustrated in Figure 3(a).

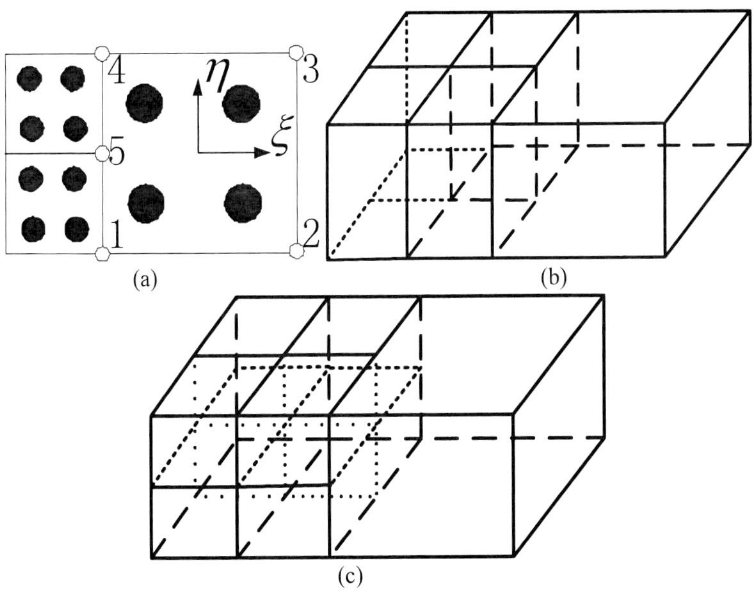

Figure 3. (a) A schematic view of two-dimensional (2-D) MPM calculation. The solid lines show a regular hierarchical grid. The black dots are material points. Four material points in each element are located at the reduced coordinates ($\pm 0.5, \pm 0.5$) and this arrangement is usually adopted in MPM. Graphs (b) and (c) delineate two kinds of three-dimensional (3-D) hierarchical grids. Four and eight finer background elements are condensed to one big element.

Each material point is assigned a mass based on the volume it occupies and the initial mass density. Eq. (23) is satisfied if the mass of each material point is kept fixed throughout the computation. To provide a Lagrangian description of material points, other properties, such as position, velocity, acceleration, stress, strain and constitutive parameters are also assigned to each material point. The Lagrangian scheme usually results in mesh lockup and entanglement if the material is seriously distorted. Yet in MPM the conservation equation of momentum is solved on a background Eulerian mesh and thus the trouble of mesh entanglement is avoided. For clarity, in this section all variables with subscript i or j are associated with Eulerian description, representing values on background grids. Variables associated with Lagrangian description, representing material point values, are labeled by subscript p. The discretization of Eq. (24) takes the following form [143],

$$\sum_{j=1}^{N_g} m_{ij}\boldsymbol{a}_j = \boldsymbol{f}_i^{\text{int}} + \boldsymbol{f}_i^{\text{ext}}, \quad i = 1,\cdots,N_g \tag{25}$$

where m_{ij} is the consistent mass matrix, $\boldsymbol{a}_j$ is the acceleration on node j, $\boldsymbol{f}_i^{\text{int}}$ and $\boldsymbol{f}_i^{\text{ext}}$ are the internal and external forces on node i, and N_g is the number of grid nodes. By introducing interpolation shape functions, the mass matrix, internal and external forces are defined by [143],

$$m_{ij} = \sum_{p=1}^{N_p} m_p N_i(\boldsymbol{x}_p) N_j(\boldsymbol{x}_p) \tag{26}$$

$$\boldsymbol{f}_i^{\text{int}} = -\sum_{p=1}^{N_p} V_p \boldsymbol{\sigma}(\boldsymbol{x}_p)\cdot\nabla N_i(\boldsymbol{x}_p), \qquad \boldsymbol{f}_i^{ext} = \hat{\boldsymbol{\tau}}_i + \sum_{p=1}^{N_p} m_p \boldsymbol{b}(\boldsymbol{x}_p) N_i(\boldsymbol{x}_p) \tag{27}$$

where m_p and V_p are the mass and volume of the *p-th* material point, $N_i(\boldsymbol{x})$ is the interpolation shape function on node i, $\hat{\boldsymbol{\tau}}_i$ is the surface traction associated with node i, and N_p is the number of material points. Usually the consistent mass matrix is replaced by a diagonal mass matrix for computation efficiency. The expense is a slight amount of numerical dissipation[144]. By doing so, Eq. (25) takes the reduced form,

$$m_i \boldsymbol{a}_i = \boldsymbol{f}_i^{\text{int}} + \boldsymbol{f}_i^{\text{ext}}, \quad i = 1,\cdots,N_g \tag{28}$$

where the diagonal mass matrix is

$$m_i = \sum_{p=1}^{N_p} m_p N_i(\boldsymbol{x}_p), \quad i = 1,\cdots,N_g \tag{29}$$

The discrete momentum conservation equation (Eq. (28)) is solved on a background grid. The obtained grid accelerations are used to update the state of each material point. To update the stress state on each material point, a constitutive relation is required. The Lagrangian description of MPM follows with material points and thus easily accommodates a history dependant constitutive relation. Here our attention is focused on the coupling of different regions, i.e. the MPM/MD regions. The coupling needs the match of constitutive relations in the two regions. Their difference in formulations dictates that the matching can only be enforced in the elastic regime when atoms or material points in the coupling layer are vibrating from their equilibrium sites in small amplitude. For linear elastic materials, the relation between the stress and strain in the MPM region can be simplified as,

$$\dot{\boldsymbol{\sigma}} = \boldsymbol{C} : \dot{\boldsymbol{\varepsilon}} \text{ where } \boldsymbol{\varepsilon} = \frac{1}{2}[(\nabla \boldsymbol{u}) + (\nabla \boldsymbol{u})^T] \tag{30}$$

The symbol $\boldsymbol{C}$ denotes the fourth-rank elasticity tensor and $\boldsymbol{u}$ the displacement field. For nonlinear elastic materials, $\boldsymbol{C}$ is a tangential stiffness tensor and varies with the strain of the material. After updating the information on material points, the background mesh can be discarded which enables adaptive meshing[130] or reused for the next time step. Details of time-integration algorithm can be found in Ref. [143]. Several kinds of modified algorithm[145,146] are employed to improve the accuracy. Here we adopt the algorithm described in Ref.[145]. A fixed Eulerian background mesh is used here for simplicity.

For most MPM calculations, standard finite element (FE) shape functions [147] are adopted to map information between material points and background grids [130]. For the hierarchical grid as illustrated in Figure 3(a), the standard FE shapc function on node 5 is,

$$N_5(\xi,\eta) = \frac{1}{2}(1-\xi)(1-\eta^2) \tag{31}$$

Consider the special case of a uniformly stressed state, the internal force (Eq. (27)) simplifies to,

$$\boldsymbol{f}_i^{\text{int}} = -\boldsymbol{\sigma} \cdot \sum_{p=1}^{N_p} V_p \nabla N_i(\boldsymbol{x}_p) \text{ where } \boldsymbol{\sigma} = \boldsymbol{\sigma}(\boldsymbol{x}_p) \;\; \forall p \tag{32}$$

For the hierarchical grid in Figure 3(a), $\boldsymbol{f}_i^{\text{int}}$ is not zero (for i=1, 4 and 5). We call this non-zero $\boldsymbol{f}_i^{\text{int}}$ the artificial forces. To suppress its occurrence, we modify the interpolation shape functions. The square terms in the standard FE shape functions are replaced with their absolute value. For example, the interpolation shape function on node 5 in Figure 3(a) is modified to the following bilinear form,

$$N_5(\xi,\eta) = \frac{1}{2}(1-\xi)(1-|\eta|) \tag{33}$$

Similar modification can be exercised for three-dimensional (3-D) background grids, i.e. replacing the square terms with their absolute value. By doing so, the internal forces vanish for the special case in Figure 3(a). It's also true for the 3-D cases as illustrated in Figures 3(b) and 3(c) if eight material points are located at the reduced coordinates ($\pm 0.5, \pm 0.5, \pm 0.5$). After modification, the interpolation shape functions are linear in each direction, so that the modification can suppress artificial forces on hierarchical nodes. A step stress wave is used to test the modified interpolation shape functions in hierarchical MPM scheme. The stress wave propagates along $[\bar{1}00]$ direction. Sliding boundary conditions are employed in directions perpendicular to [100]. The 3-D background mesh in Figure 3(c) is employed and each contains eight material points, that are initially located at the reduced coordinates ($\pm 0.5, \pm 0.5, \pm 0.5$). Figure 4 shows the snapshots of the stress wave propagating along $[\bar{1}00]$.

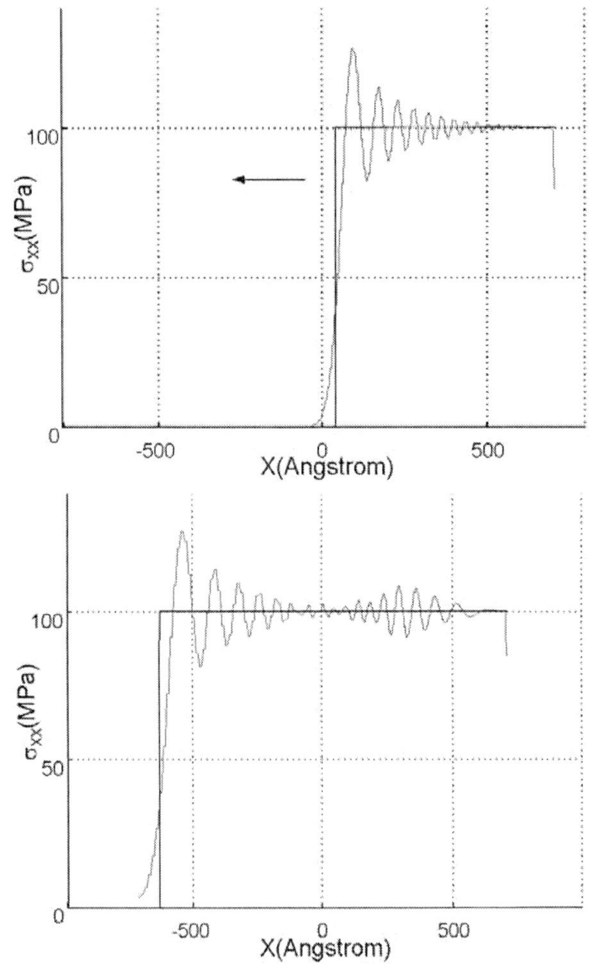

Figure 4. Amplitude of a step stress wave propagating from a finer MPM region to a coarser MPM region. The two regions are separated by $x = 0$. The step line is the corresponding analytical stress wave with amplitude of 100MPa. (a) Profile of the stress wave in the MPM finer region. (b) Profile of the stress wave that has passed into the coarser MPM region.

It can be seen that along the boundary of hierarchical background grids there are no visible artificial forces. The simulated stress wave agrees well with the analytical stress wave. It can also be seen that some small short wave length phonons are not supported by the coarse MPM region and part of the stress wave is reflected on the boundary. Proper damping [148] on the boundary is expected to reduce the amplitude of the reflected waves.

One drawback of the linear interpolation shape functions is that the functions are not smooth across adjacent cells. The lack of smoothness can produce numerical noise in MPM. The generalized interpolation MPM [149] can remove numerical artifacts inherent in the MPM formulation when material points fail to register in a self-similar fashion on the background computational grid.

3.2.2. Handshaking of MPM/MD Region

The handshaking scheme described here spatially divides the physical system into the MPM/MD regions (Figure 5). An imaginary surface (interface) is drawn between the MPM and MD regions. The handshaking domain is composed of regions I, II and III. The width of each region should be no less than a potential cutoff. In regions II and III, material points are initially located at the ideal lattice sites. Displacement boundary conditions are employed to decouple two regions [126]. Regions II and III supply the boundary conditions for the MD region as illustrated in the top diagram in Figure 5(b). Conversely, as illustrated in the bottom diagram of Figure 5(b), atoms in region I supply the boundary conditions for the MPM continuum region. At each time step in the MD calculation, particles in regions II and III are viewed as atoms (Figure 5(b)). Position of each atom in regions II and III is conveyed from the position of the corresponding particle. Forces on each atom in the MD region are obtained from the total potential energy by taking derivatives with respect to the atomic coordinates. Particles in regions II and III have different bearings in the MD and the continuum calculations. In the MD calculation, each particle is assigned an electron cloud density. The electron density on each particle in region II is partly induced by particles in region III so that the latter serves to provide an atomic atmosphere to particles in region II. Similarly, the electron density on each atom in region I is partly induced by particles in region II, and region II serves to provide an atomic atmosphere to atoms in region I. Accordingly, an electron density corresponding to the deformation throughout the simulation is correctly obtained in regions I and II. At each time step in the MPM calculation, atoms in region I are viewed as particles (Figure 5(b)), and each particle receive information, such as mass, volume, position, velocity and stress that needed for MPM calculations, from the corresponding atoms. The background mesh is extended to region I for the MPM calculation. Each particle evolves under the discrete momentum equation (Eq. (28)). In the MPM calculation, atoms in region I are different from the atoms in the MD calculation. They evolve under the discrete momentum equation which is solved on background grids. Thus the mass, velocity and internal force on the background grids near the MPM/MD interface are partly induced from the "particles" in region I.

After the two regions decouple as illustrated in Figure 5(b), each region evolves under their governing equations concurrently with a suitable time step. For MPM, the critical time step is proportional to the smallest element size and inversely proportional to the wave speed of the material [150].

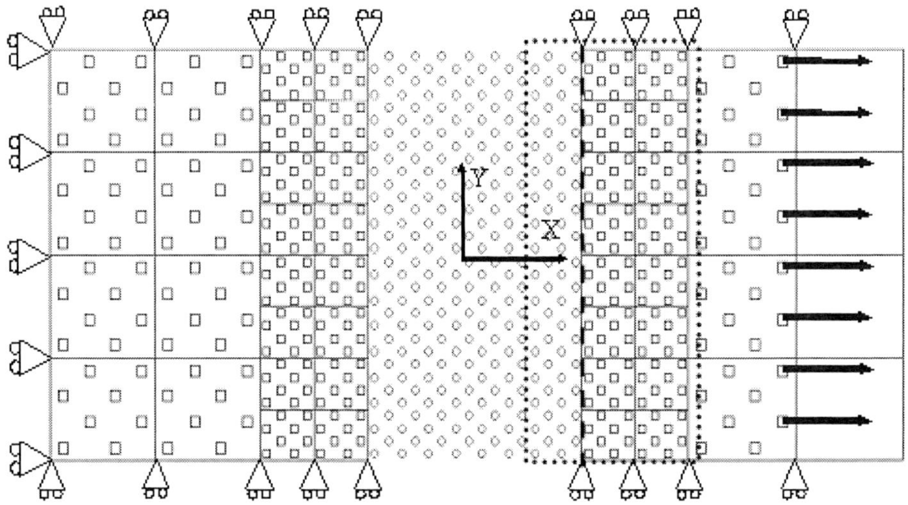

(a)

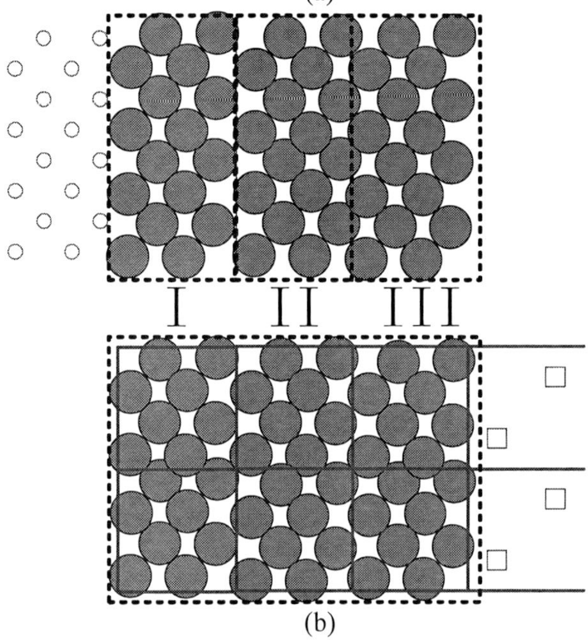

(b)

Figure 5. (a) Schematic view of a bar under tension. Only a slice of the bar (FCC crystal) is shown. Big squares represent heavy material points. Small squares represent light material points. Circles represent atoms. The solid lines delineate the background mesh. Dotted boxes denote the right part of the handshaking region. The dashed line represents the interface between the MPM and MD regions. Right arrows symbolize boundary forces under 500MPa theoretical stress. Graph (b) enlarges the handshaking region in (a). Filled circles in region I represent atoms. Filled circles in regions II and III are material points. The top diagram in (b) elucidates the MD calculations and the bottom one the MPM calculations. Background mesh is extended to region I for MPM calculations.

The decouple scheme ensures equality of the displacement field in the MD/MPM regions. Since forces are explicitly avoided in the decouple scheme, the stress equilibrium between two regions is not established *a priori*. As pointed out by Kohlhoff et al. [126], the stress equilibrium in the two regions requires the equal elastic constants between the continuum/MD regions. Also the matching of elastic constants in the MD/MPM regions is a necessary condition. For simplicity, only elastic constants are considered here, which corresponds to linear elasticity. For cubic crystals under linear deformation, there are three elastic constants C_{11}, C_{12} and C_{44}. These constants can be derived from the atomic potential and then are used in the MPM region. Thus the MPM region only sustains elastic deformation. The dividing of the MD/MPM regions should be based on physical conditions. When plastic deformation approaches the handshaking region, the MD region should be extended to ensure the covering of the inelastic region throughout the simulation. Also when there is no need of MD description, that region should be reversed into the MPM region. The hierarchical, adaptive, MPM developed by Tan and Nairn [130] is expected to achieve this goal. Here in our simulation the MD region is large enough and the MPM region only sustains linear elastic deformation. Thus a fixed background mesh is adopted for simplicity.

Leaving the handshaking region, material points are enlarged to represent many atoms, rendering the computational efficiency of MPM. We generalize the hierarchical background mesh [130] to the three dimensional case. For bulk materials, the background mesh can be enlarged in three directions, i.e. eight elements condense to one big element (Figure 3(c)). Similarly for plates and bars, the background mesh can be enlarged in two (Figure 3(b)) or one directions, i.e. four or two elements condense to one element. Material points are enlarged in proportion to the background mesh. The mass of each material point is proportional to the volume it covers. The coarser the mesh is, the heavier the material points initially placed in that element. The reduced coordination of each material point is the same after the enlargement, i.e. the material points and background mesh are enlarged proportionally. The enlargement results in missing degrees of freedom. As the size of background mesh increases, a small portion of short wavelength phonons are not supported by the coarser mesh, as shown in Figure 4. The largest spacing depends on the shortest wavelength phonons expected to propagate unimpeded in the MPM region.

3.2.3. Validation of the MPM/MD Handshaking Method

The multiscale method developed here handshakes the MD/MPM regions. The MPM region provides an environment for the MD region. The seamlessness of the method has two meanings. First, the outer load applied in the MPM region can pass into the MD region without obvious hindering. Second, the disturbance in the MD region, arising from the bond breaking at the crack tip [151] or from energetic cluster impacts, can propagate into the MPM region without obvious reflection. Usually the former is easy to be implemented, while the latter is more difficult to be achieved since the continuum region usually is coarser and may not support some short wavelength phonons. Thus the validation here is divided into two parts, validating the two meanings separately.

The first example tests the transmission ability from the MPM region to the MD region. An undisturbed FCC crystal lattice was embedded in the MPM region as illustrated in Figure 5(a). The bar is 3-D but only a slice of which is shown. A uniform tensile step force is applied at time zero at the right end of the bar. Hierarchical mesh (of rather large mesh size) and material points are employed in the MPM region. In the handshaking region material points are placed at the ideal crystal lattice sites. In the coarse MPM region, eight finer elements and lighter material points condense to one big element and one heavier material point (Figure 3(c)). Forces on boundary are applied on material points along [100]. In Y, Z and –X directions the sliding boundary conditions are imposed. Periodic side boundary conditions are applied in the MD region for the consistency with the MPM region. The EAM potential [152] for Cu is employed here. The initial temperature is set at zero. The definition of the stress on an atom is based on Virial theorem. Here the formulations of calculating the stress are used by those in Schiøtz et al. [83]. Note that the bar in Figure 5(a) is only a schematic view and the true region of simulation is much larger.

Figure 6 shows the passing of stress wave from the MPM to the MD region. The analytical solution is obtained by using the elastic constants in the MPM region. It can be seen that the simulation results agree well with the analytical solution. No visible wave reflection is observed in the handshaking region. Note that the stress configuration fluctuates along the analytical solution. The fluctuation in the MPM region is partly due to discretization since a continuum of infinite degrees of freedom is replaced by material points of limited degrees of freedom. Another reason is the lack of smoothness of the interpolation shape functions. Contrasting with that in MPM region, the fluctuation in MD region is intrinsic since atoms are discrete.

The quasi-static performance of this MPM/MD handshaking method is also tested. A damping force is used to quiet the bar. At long times, all velocities of material points and atoms approach zero and the final solution corresponds to a static loading of an end-loaded bar. The stress distribution is shown in Figure 7. The distribution of σ_{xx} is not shown on the right MPM force boundary since solutions are usually not very accurate on the boundary just like FE method. The stress in the MD region is slightly higher and the stress in the MPM region is slightly lower. The maximum deviation amounts to 0.36% of the true value, a scatter regarded acceptable for applications.

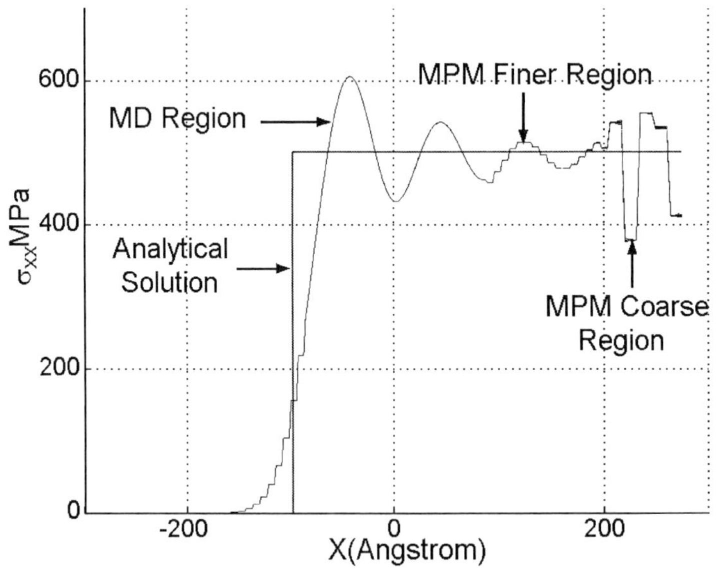

Figure 6. Stress wave propagates from the MPM region to the MD region. The analytical solution corresponds to a 500MPa step stress wave. Different regions are denoted in the figure.

We also test the method for Lennard-Jones potential, EAM potential for bcc crystals [83] and a highly optimized empirical potential for silicon [153]. The results are similar and thus encouraging. From above one can see that the load applied in the MPM region can pass into the MD region almost without impeding.

Like Moseler et al. [148], we use Gaussian wave packets with small standard deviations to test the transmission ability of short wave length fluctuations produced in the MD region. The simulation configuration is similar with the configuration in Figure 5, but the background mesh is uniform and refined, providing each particle occupies a mesh. The disturbance is first initiated in the MD region. Two Gaussian wave packets, with a standard deviation of one and a half lattice constant of copper, are initiated in the MD region and they propagate in the opposite directions (Figure 8).

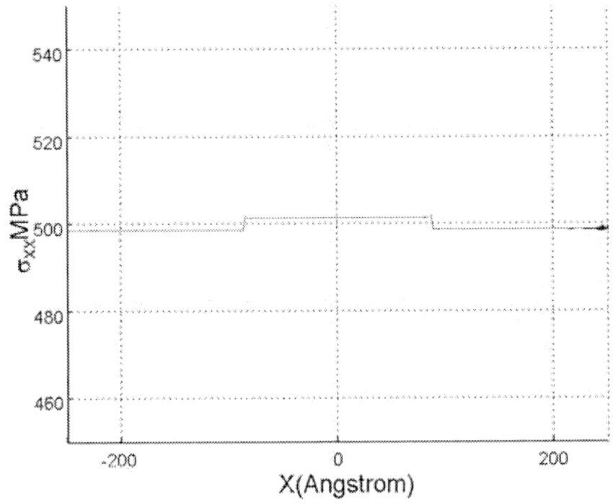

Figure 7. Profile of the stress component σ_{xx} along the bar for static loading.

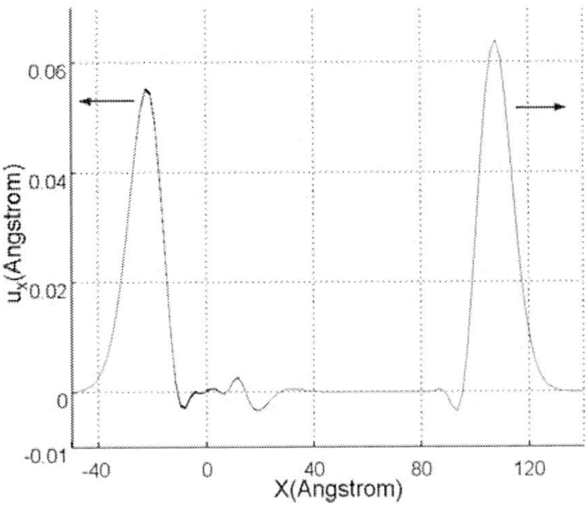

Figure 8. Amplitude of two Gaussian wave packets. The region of x > 0 is the MD region. The region of x < 0 is the MPM region.

The leftward propagating wave passes from the MD region to the MPM region, while the rightward propagating wave is a reference and propagates always in the MD region. It can be seen that only a small fraction of wave is reflected back into the MD region. Even when the standard deviation is as small as one lattice constant, more than half of the energy passes into the MPM region. Moseler et al found that a proper damping on the P-Q boundary could drastically reduce the amplitude of the reflected waves on the P-Q boundary [148]. Here damping is not employed in the handshaking region but the transmission ability is still better than the conventional Langevin MD [154] for Gaussian wave packets with standard deviation greater than one lattice constant.

4. Simulations for Mechanical Behavior of Nanocrystals

4.1. Formation of Nanocrystals after Severe Plastic Deformation

After SPD such as SMAT, the accumulation of highly dense dislocations may trigger their reorganization to form low angle grain boundaries, as described by Ma et al. [42].

To verify the nano-grain formation, MD simulation was carried out [42]. The simulation starts from a perfect pure iron lattice. A dislocation density of $2.17\times10^{11}\,\mathrm{mm}^{-2}$ is introduced to simulate the shot-peening process of Lu and Lu [36]. For simplicity, the dislocations are generated along the closely packed plane. The number of dislocations is given by $n = 2.17\times10^{11}\times A_{\mathrm{cell}}$, with A_{cell} being the cell area for the simulation. The locations and the orientations of the initial dislocations are generalized randomly via computer algorithm. Either inserting or removing a semi-array of atom creates a dislocation. The process of dislocation generation is controlled by mass conservation so that the number of the removed atoms approximately equals to that of the inserted atoms. The two-dimensional cell consists of nearly 10,000 atoms.

The MD calculation is conducted at NVT ensemble fixed at a temperature of 313K. During the computation, the velocities of atoms are rescaled at each time step ensuring the temperature not to deviate from the specified one [74]. Periodic boundary condition is described. The relaxation is conducted for 10,000 MD steps, with each step having a time elapse of 0.5fs.

The simulation result is depicted in Figure 9 in terms of the Δ parameter. The process of relaxation reveals a clue for the formation of subgrains triggered by large amount of dislocation injection, and the evolution of subgrains via thermal fluctuation to stable grains. Figure 9(a) depicts the simulation result after 2000 relaxation steps. The image clearly indicates the formation of very small crystallites (about 5nm in diameter) generated by large influx of dislocation injection. Further relaxations 4,000 and 6,000 steps, witness the merge of subgrains to larger grains of larger grain boundary mis-orientations, as shown in Figure 9(b) and (c). The stabilized grain shape is shown in Figure 9(d), after 8,300 relaxation steps. The simulated final grain sizes range from 5nm to 20nm, in agreement with the experiment of Lu and Lu [36].

The spontaneous formation of organized nanocrystals in semiconductors has been observed during heteroepitaxial growth and chemical synthesis. Grom et al. [155] reported the self-organization of Si nanocrystals larger than 80 angstroms into brick-shaped crystallites oriented along the [111] crystallographic direction. The results should aid the development of maskless, reproducible Si nanofabrication techniques.

Nam et al. [156] studied the freezing behavior of gold nanoclusters by employing MD simulations based on the semi-empirical EAM. Investigations of the gold nano-clusters revealed that, just after freezing, ordered nano-surfaces with a five-fold symmetry were formed with interior atoms remaining in the disordered state. Further lowering of temperatures induced nanocrystallization of the interior atoms that proceeded from the surface towards the core region, finally leading to an icosahedral structure. These dynamic processes explain why the icosahedral cluster structure is dominantly formed in spite of its energetic metastability.

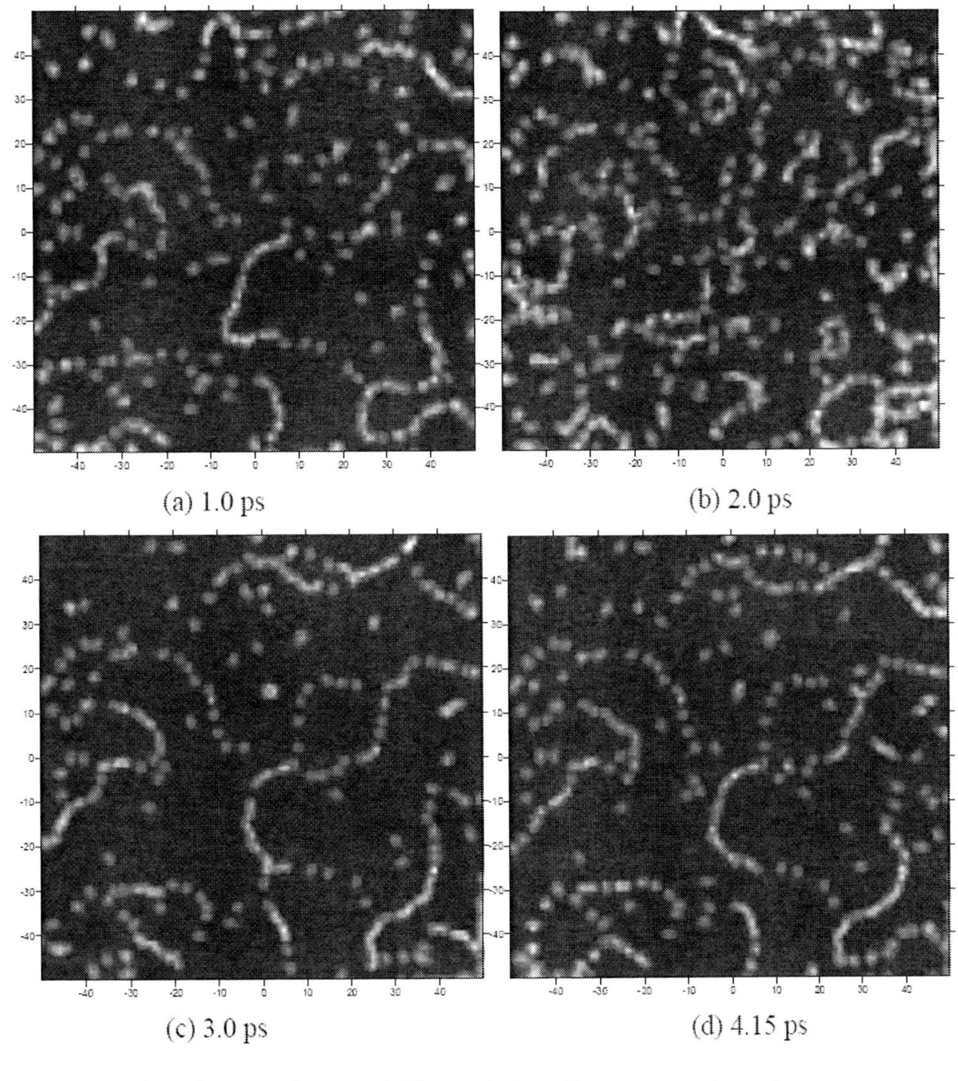

Figure 9. Formation of nano-grains revealed by evolution of centrosymmetry order.

A new growth mechanism featuring grain-rotation-induced grain coalescence was recently discovered by Moldovan et al. [157]. Based on a simple model employing a stochastic theory and using computer simulations, the coarsening of a polycrystalline microstructure due solely to the grain-rotation coalescence mechanism was revealed. The study demonstrates that this mechanism exhibits power-law growth with a universal scaling exponent. The value of this universal growth exponent is shown to depend on the assumed mechanism by which the grain rotations are accommodated.

4.2. Nanoindentation

Nanometer-scale contact experiments and simulations demonstrate the potential to probe incipient plasticity, namely the onset of permanent deformation, in crystals. Many indentation simulations have been reported for both the elastic and the plastic behaviors of materials in recent years [82,158-163], though their attention, as preliminary studies, was rather focused on simulating nanoindentation of single crystals. Li et al. [164] explored the atomistic mechanisms governing elastic limit and incipient plasticity in single crystals. They proposed a position-sensitive criterion based on elastic stability determines the location and character of homogeneously nucleated defects, and validated it at both the atomistic and the continuum levels. That permits an interpretation of the experimentally observed sequence of displacement bursts to elucidate the role of secondary defect sources operating locally at stress levels considerably smaller than the ideal strength required for homogeneous nucleation. Until quite recently, Feichtinger et al. [165] simulated the nanoindentation of nanocrystalline-Au of 5nm and 12nm grain sizes. Multiple sources of grain boundary sliding and dislocation movement were identified to contribute the plasticity of nanocrystals under indentation.

Ma and Yang [103] simulated the processes of nanoindentation via parallel MD. The indenter is applied to the nanocrystalline grain boundary (GB) and to the grain interior, with result shown in Figure 10. The intended initial contact points are marked by the red dots "P" in Figure 10. The profiles of the centrosymmetry disorder Δ in the vertical cross-section of $x = 0$ are plotted. The sites where local FCC order is reserved are colored as black, while the degree of whitening indicates the deviation from the ordered FCC state. The initial configurations, namely graphs (a) and (c) in Figure 10 have nearly perfect polycrystalline structures, with disorder only appearing near the grain boundaries. As the indenter moves downward, one observes the bursts of defects (in the disorder form of stacking faults) within the grains nearest to the indenters. The propagating stacking faults are arrested by the grain boundaries. Further downward movements of the indenters trigger more bursts and subsequent arrests of stacking faults. The pile-ups of the stacking faults near the grain boundaries would induce the burst of new stacking faults in the grains next to the indenting grain, shown in graphs (c) and (d), but not any further for a total indenting depth of 38.4 angstrom. The simulation suggests that the plasticity of nano-grained metals under nanoindentation is dominated by the bursts and arrests of stacking faults. A comparison between graph (b) and graph (d) of Figure 10 indicates that more stacking faults are induced if the indenter is loaded on GBs than on the grain interiors. When indented on the grain boundary, the two grains are crushed apart gradually. When indented on the grain, on the other hand, the contacted grain is pushed deeper into the nanocrystalline aggregate. Intergranular motion is also observed as a result of grain boundary sliding.

The case of nanoindentation to a perfect single-crystal leads to a different situation, as shown in Figure 11 for cross-sectional view of Δ, while the top surface coincides to (111) plane of the lattice. Partial dislocations nucleate below the indenter and propagate into the crystal along different sets of $\{111\}$ planes, until reaching to the opposite surface setting in the simulation. The absence of grain boundaries eliminates the arrest and re-nucleation process.

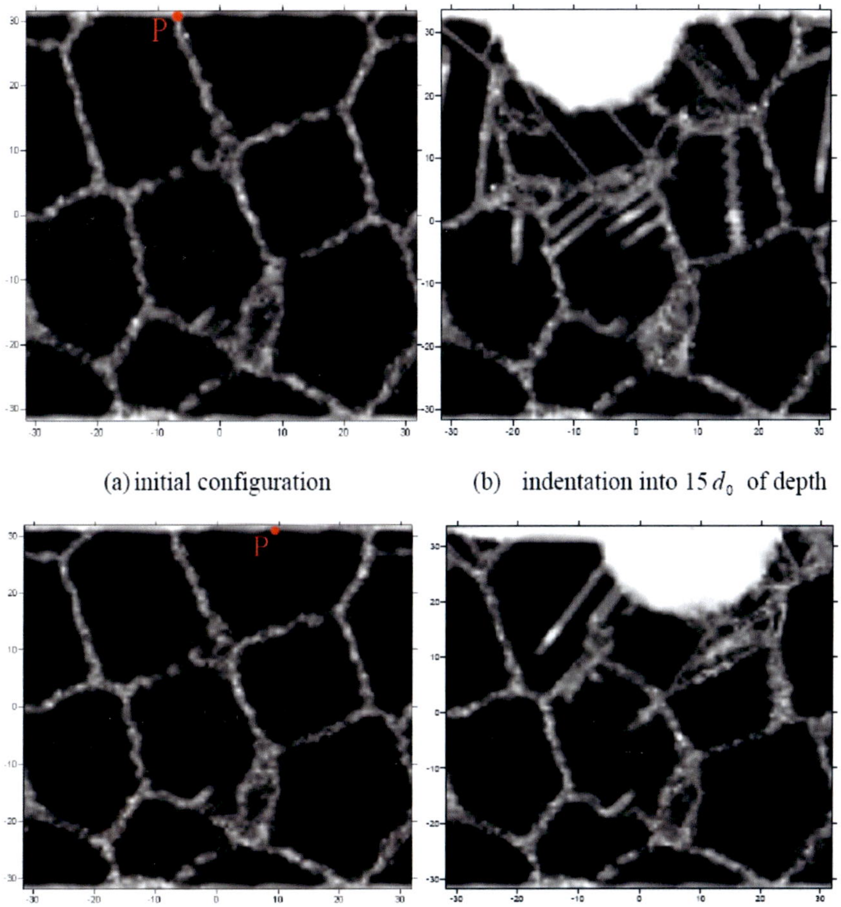

(a) initial configuration (b) indentation into $15\,d_0$ of depth

(c) initial configuration (d) indentation into $15\,d_0$ of depth

Figure 10. Profiles of Δ for nanoindentation applied to the grain boundary, see graphs (a) and (b), and to the grain interior, see graph (c) and (d).

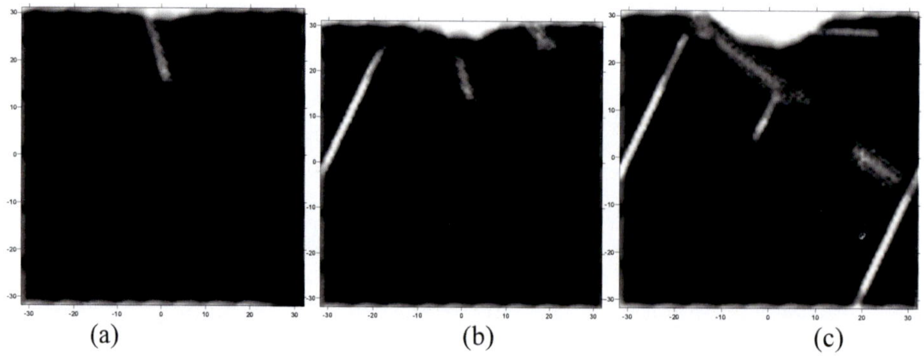

Figure 11. Profiles of Δ for nanoindentation applied to a single crystal. Indentation into (a) $4\,d_0$ of depth; (b) $5\,d_0$ of depth; and (c) $7\,d_0$ of depth.

To further illustrate the defect evolution under nanoindentation, we plot in Figure 12, for the nanocrystalline and single-crystal Cu cases respectively, the results of local crystalline order simulated under the LJ potential.

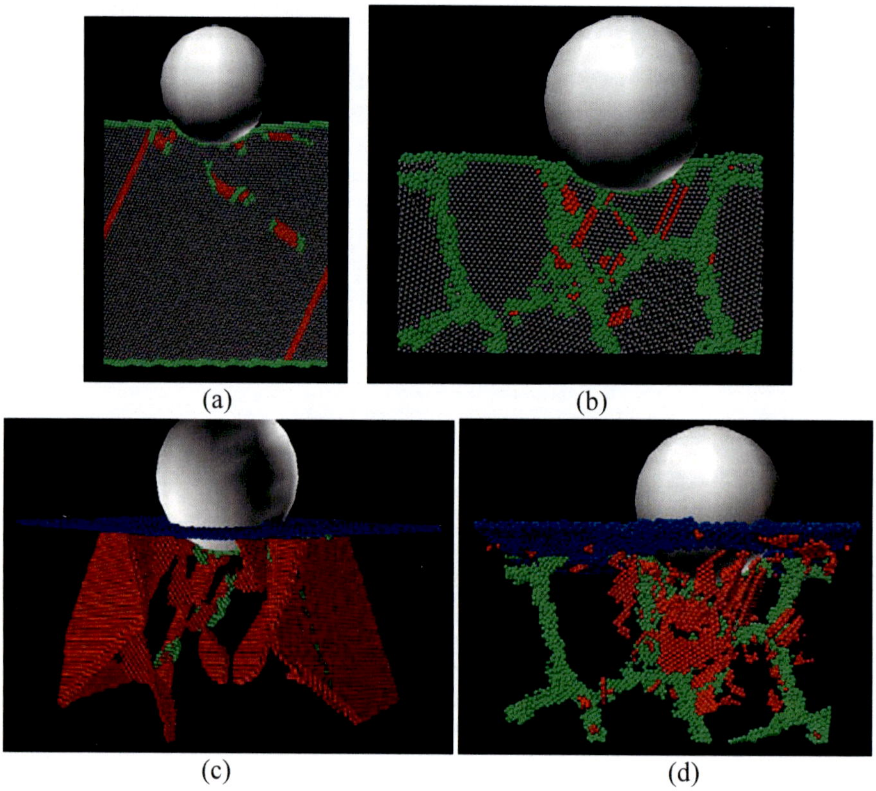

Figure 12. Profiles of local coordinate order for nanoindentation of $7\,d_0$ to single crystal (graphs (a) and (c)) and nanocrystal at the grain boundary (graphs (b) and (d)). Top graphs presents a cross-sectional view where grey, green and red atoms denotes regular, disordered and stacking fault sites, respectively. Bottom graphs present three dimensional views. The blue, green and red colors mark atoms on surface or dislocation cores, grain boundaries and stacking faults, respectively.

The top graphs of these figures present the cross-sectional view, the arrest of the stacking faults by grain boundaries in the nanocrystal case provides a clear contrast to the unconstrained but sparse bursts of stacking faults in the single crystal case.

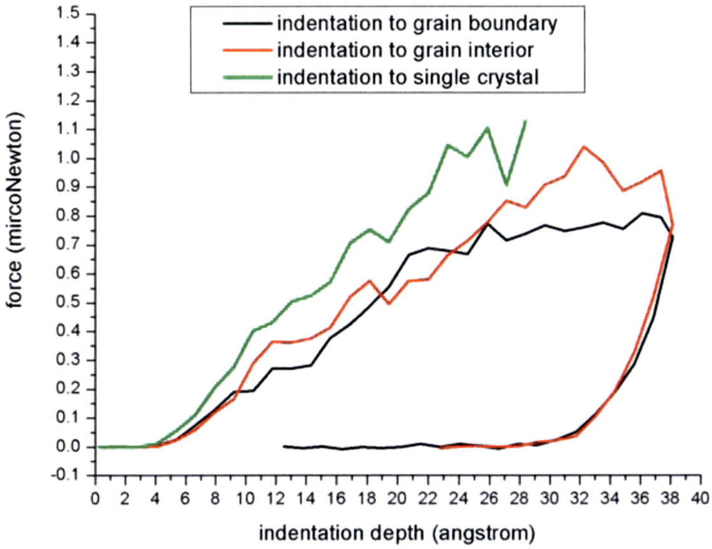

(a) LJ potential

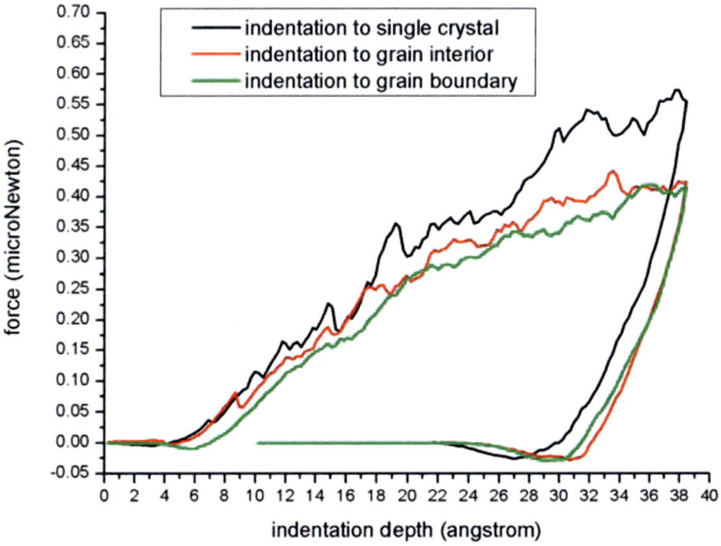

(b) EAM potential

Figure 13. The contact force vs indentation depth curves of indentation under LJ and EAM potentials. Black, red and green curves refer to indentations on single crystal, on grain interior and on grain boundary of nanocrystal, respectively.

Two bottom graphs in Figure 15 present three dimensional views. The stacking faults, especially for the case of single crystal nanoindentation, are clearly shown in the [111] orientations. Similar trend is observed under a simulation based on EAM with detailed parameters for copper given by Johnson [87].

Three force versus indentation depth curves are compared for nanocrystal indented on grain boundary, nanocrystal indented on grain interior, and nanoindentation on a single crystal. The simulation based on LJ potential are plotted in Figure 13(a), while that under EAM potential in Figure 13(b). Regardless the potential adopted, the case of indentation on a single-crystal encounters the stiffest and the most jagged response, followed by the indentation to the grain interior of nanocrystal, and the indentation on the grain boundary of nanocrystal gives the most compliant response. The usage of LJ potential leads to roughly twice the indentation force of that of EAM potential.

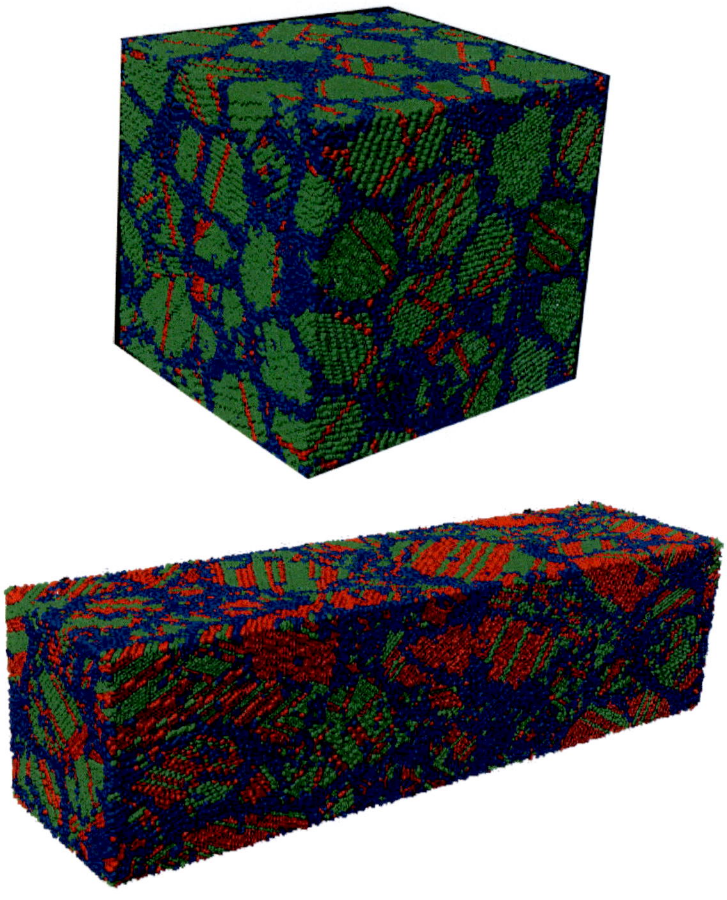

Figure 14. Local crystalline order of nanocr0ystalline Cu under uniform stretching of (a) 6% and (b) 91% logarithmic strains. The atoms in FCC stacking sequence are referred to as the perfect lattice and colored green; the atoms of HCP stacking sequence are regarded as the stacking fault and colored red, while the atoms with other crystalline orders are thought as grain boundary and painted blue.

4.3. Fast Stretching of Nanocrystals

Now consider the macroscopically uniform deformation of nanocrystals. The MD simulation is only capable of the case of fast stretching. There were several publications [21,25,26,59,83,166-172] concerning on this subject, but none of them is conducted to the regime of large deformation. The simulation has been conducted under different scales, from about 50,000 to 1,000,000 atoms, under LJ and EAM potentials. Here we only show a simulated example for the nanocrystal containing a million copper atoms under EAM potential [87]. According to the corresponding results, we introduce a new mechanism so called dislocation-assisted grain growth [173].

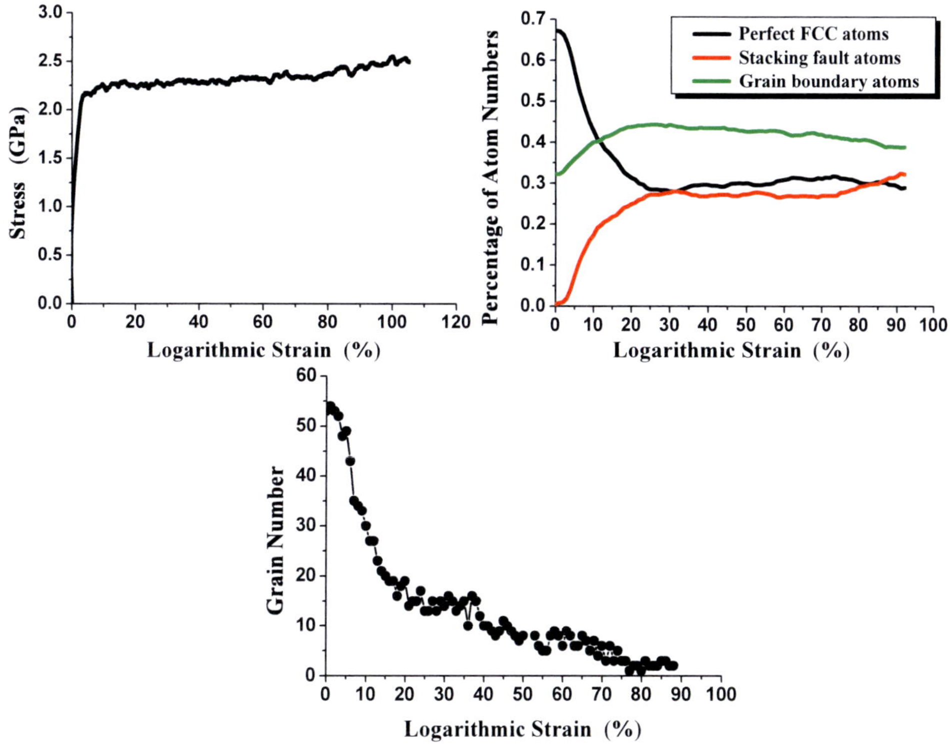

Figure 15. (a) The curve for the stress versus logarithmic strain; (b) the percentage variations of atoms in perfect FCC, stacking faults and grain boundaries; (c) the variation of the grain number during the deformation process.

In this simulation, three-dimensional Voronoi construction [103] was used to generate the nanocrystalline model for computation. The sample configuration involves 1,048,645 atoms and 64 randomly oriented grains with an average grain size about 7nm. Periodic side boundaries were maintained throughout the simulation. The isochoric stretching of a nanocrystalline copper box was conducted by parallel molecular dynamics simulations under the Linked-Cell-List algorithm [108]. The initial MD simulation carried 3000 annealing relaxation steps at 300K, using Verlet leapfrog algorithm [74] with MD time step of 2fs. The loading was imposed under a quasi-homogeneous manner. A homogeneous affine deformation of 0.1% stretching and 0.05% lateral contraction was enforced. Then 500 relaxing steps were executed at 300K by maintaining a NVT ensemble. The stretching rate was $10^9 s^{-1}$ and the total simulation time was 1.006ns.

Figures 14(a) and 14(b) showed the atom configurations of 6% and 91% logarithmic strains respectively. The atoms in grain boundaries, stacking faults or inside grains were identified by local crystalline order [104,174]. Comparing the two configurations in Figure 14, one finds the grain boundaries were widened and the disordered atoms increased after a large deformation. In order to maintain the plastic deformation, more and more partial dislocations emitted from grain boundaries or triple junctions and passed through grains. Partial dislocations were left behind the stacking faults. At the stage of both small (Figure 14(a)) and large deformation (Figure 14(b)), only a few slip systems, generally less than three in each grain, were activated. The strain compatibility could not be met along the grain boundaries if the plasticity was only mediated by movements of partial dislocations. The atoms near grain boundaries adjusted their locations locally under the high flow stress, in the form of a large number of small sliding events. These small slipping facilitated the grain rotation process. It was found in the simulation that the grains tended to rotate to the tensile direction, with a grain rotation rate less than 4 degree per 1% logarithmic strain. As the deformation continued, the grain boundaries widened and triple junctions enlarged to accommodate the accumulated misfit strain among grains.

The flow stress, shown in Figure 15(a), was kept about 2.25GPa after about 5 to 6% logarithmic strain. However, a stable flow stress does not necessarily imply a steady state. The atom configuration kept evolving as the deformation proceeded. Figure 15(b) showed the percentage variations of perfect FCC, stacking fault and grain boundary atoms. The regular FCC atoms steadily decline from the initial percentage of about 70% to merely 30% at 30% logarithmic strain. At the same strain level, the atoms percentage in stacking faults shoots up to nearly 30%. The atom percentage in grain boundaries modulates from the initial percentage of 30% to 40%. After 30% logarithmic strain, these percentages remained steady with the deformation up to 100% logarithmic strain. The increases on the percentages of the stacking fault and grain boundary atoms indicated the disorder of the nanocrystalline sample was enhanced at large deformation. Hidden behind the steady percentages of perfect FCC, stacking fault, and grain boundary atoms, the changes in the grain configuration were still enormous in the large stretch regime. To reveal this change, one may classify atoms of the sample into three categories: the “grain boundary atoms” unrecognized by the local crystalline order method; the “pseudo-boundary atoms” locating within three layers thickness between the perfect FCC atoms and the grain boundary ones; and the remaining “pure grain atoms”. This classification obtains clean isolated grains and eliminates the scattering clusters of a few atoms which observe the FCC crystalline order. The number of grains which is larger than the average one decreases sharply at the initial 20% logarithmic strain. Then the trend became slow with the stretching strain, as plotted in Figure 15(c). At the end of the deformation, most of the grains merged into a few super grains and others were cut into small colonies, containing only several thousands of atoms.

The grain size distribution became inhomogeneous in the regime of large deformation, as delineated in Figure 16. The grain size distribution became bimodal after 10% of deformation (Figure 16(b)). A few grains grew up obviously. When the deformation reached 50% logarithmic strain, the first peak moved toward a smaller size and the second peak became flat (Figure 16(c)). Two grains grew to rather large sizes. After 91% logarithmic strain, the size distribution of the grains in the sample became very dispersive (Figure 16(d)). The largest grain had the number of atoms about 3.8×10^5 and an approximate size of 18nm. The upper right inset showed that the number of the smaller grains also decreased. The reason might lie in the merging process among grains. The process of grain refinement is unlikely since the grain boundary sliding, rather than the movements of partial dislocations, would dominate the deformation mechanisms for a small grain size of 4 or 5 nm.

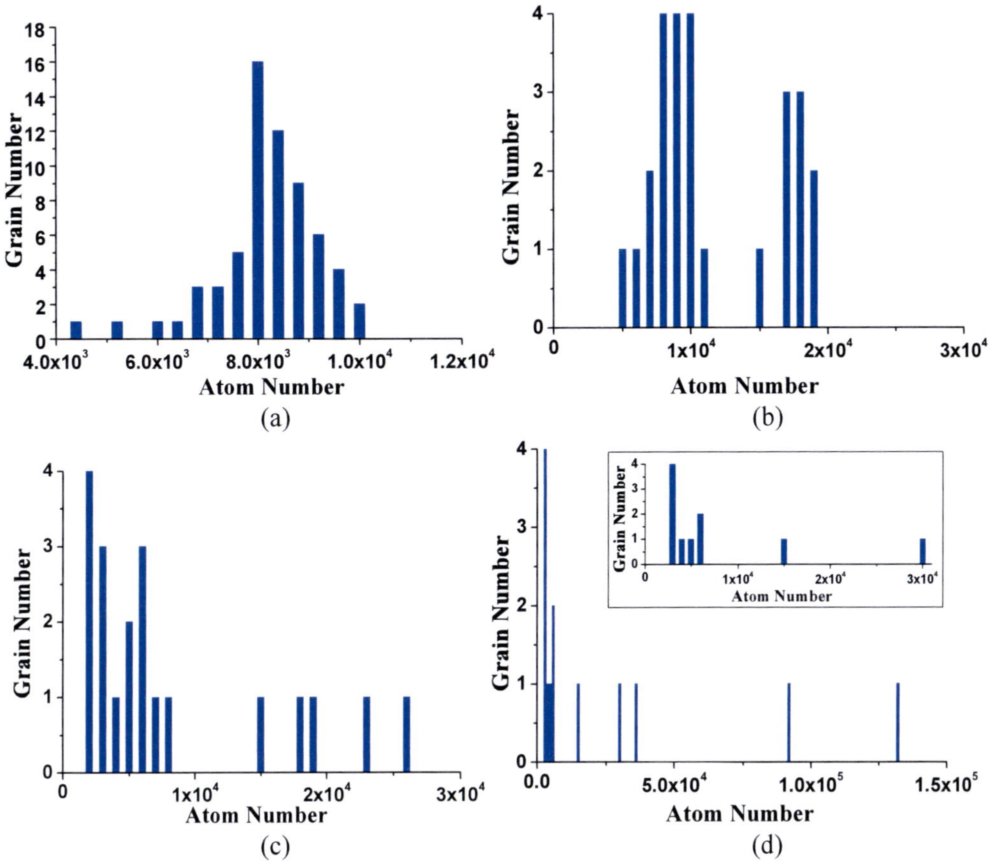

Figure 16. Size distribution of grains at (a) 0%, (b) 10%, (c) 50% and (d) 91% logarithmic strains. Only the grains containing more than 1000 atoms are considered.

Two modes of grain combination processes were revealed: penetration of the stacking faults through low angle grain boundaries at small deformation (Figure 17(a)) and the formation of twin-like structures at large deformation (Figure 17(b)). The two mechanisms were different from the observations of MD simulation by Hasnaoui et al. [175] and Haslam et al. [176]. In Ref. [175] the sample of the nanocrystalline nickel was deformed at $0.46T_\mathrm{m}$ and in Ref. [176] that of the nanocrystalline palladium at $0.66T_\mathrm{m}$, with T_m being the melting temperature. For the relative high temperature, grain boundaries diffusion gained higher mobility to replace the role of partial dislocations. In both works, low-angle grain boundaries were found to disassociate into partial dislocations, to propagate into grains and to lead to the grain coalescence. In Ref. [176], grain boundary migration assisted by diffusion accounted for the most grain growth.

The simulation was undertaken at rather low temperature, *i.e.* $0.22T_m$. Two modes of grain growth concerned with the kinematics of stacking fault propagation and the grain rotation. In the small deformation regime, grain merge gave priority to those neighboring grains whose mis-orientations are below 10 degrees.

(a)

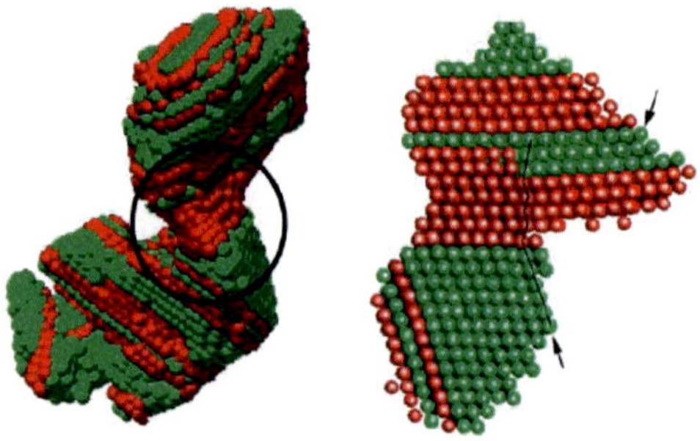

(b)

Figure 17. (a) Stacking faults vertically pass through small angle grain boundary (from the left to the right). Three stages of the stacking fault penetration are shown: onset of stacking fault passing (left); partial merge of neighboring grains (middle) and complete merge (right). (b) Configuration of a large pure grain under 40% stretching strain. The connection area has a twin-like structure manifested by stacking faults parallel to the previous grain boundary. The cross-section view gives the orthomorphic projection of the area circled in the left.

The stacking faults burst through the boundary between the neighboring grains vertically, see Figure 17(a) from the left to the right. The mis-orientation of two neighboring grains would be eliminated gradually, which was presented as the partial merge (the middle graph in Figure 17(a)) and the final consolidation (the right graph of Figure 17a)) processes. The larger grain formed by such mode was stable in the later evolution. Another mode dominated the merge in the large strain regime for the neighboring grains with large mis-orientations. The merge process was featured by parallel (instead of vertical in the small strain regime) passing of stacking faults through the grain boundary. Figure 17(b) showed the configuration of a large grain under 40% logarithmic strains. The grain was evolved by combining two small ones situated at the top and the bottom of the figure. The central connection neck is composed of stacking faults which are parallel to the previous grain boundary. A cross-section view of the orthomorphic projection of the area circled in the left was delineated in the right of Figure 17(b). The red neck-like HCP configuration area can be viewed as the twin-type grain boundary layer for the adjacent grains (in green color), as evidenced by the twin-type orientations marked by the arrows. The zigzag path indicates a linking route through the closest atoms to depict the twin-like structure. As suggested by Derlet et al. [177] and Kumar et al. [178], this twin-like structure was formed at large deformations and was caused by emission of a second partial dislocation on an adjacent slip plane. The merged grain was comparatively unstable and might separate again.

Careful observations on the structural evolution showed no textures were formed in our simulation. Initial deformation was dominated by partial dislocations. As the deformation continued, the disordered atoms and the formation of twin-like structures encumbered the activities of partial dislocations. The twin-like structures, formed at large deformations, are effective barriers to the movements of dislocations and thus disturb and hinder the texture evolution among large grains. The grain merge process gives rise to inhomogeneous grain size distribution in the sample. The extremely small grains would facilitate the grain boundary activities that also served for the reduction of textures. This nature of nanocrystalline metals provides a mechanism to maintain their physical properties isotropic even after large deformations.

4.4. Impact of High Energy Nano-Cluster

We use the coupled MPM/MD multiscale method to model high energy cluster impacts. Normal impacts of Cu clusters flying with a velocity of 10km/s are simulated. The configuration (illustrated in Figure 18) is arranged as follows. Cu Clusters are formed from a spherical cutoff of an FCC lattice and are initially placed symmetrically above the target surface. The top surface (001) of the single crystal substrate is kept free and the rest five surfaces are fixed throughout the simulation. The time step is chosen to ensure that the changing of atom positions at one step is at least one magnitude smaller than the equilibrium atom spacing. The temperature is initially set at zero but allowed to evolve freely. For all cases the simulation time is up to 10 ps and the simulation region is chosen large enough to avoid the reflected waves from the fixed boundary to the MD region. For computational efficiency, the background mesh is not the finest. The background mesh in Figure 18 can let a Gaussian wave packet with a standard deviation greater than three times lattice constant transmit the handshaking region without considerable reflection. For the impact of a cluster with 1052 atoms, the target is represented by a bulk region of dimension 277.6×277.6×138.8 nm^3, containing over 2.3 million atoms and 0.5 million material points. Were the total region modeled with atomistic description, about 900 million atoms would be needed. Material points are arranged in 5 hierarchical layers, with each material point in the outside layer condensates 8 material points (binary condensation in each direction) in the inner layer.

For the visualization technique of impacted atomistic system, we still make use of the aforementioned method in which atoms with non-FCC structure is distinguished from the local crystalline order [104,174]. Different colors are assigned to different structures to provide a clear image of defects: atoms with local FCC order are referred to as a perfect lattice and are colored transparent, atoms on and above target surface are colored white denoting surface and sputtering atoms, atoms with local HCP order are colored red and thus denoting stacking faults, atoms with other 12-coordinated combinations are colored green, and atoms with none 12-coordinated combinations are colored yellow. For impacts of Si cluster, those with none 4-coordinated atoms are colored green. Surface atoms are also colored white.

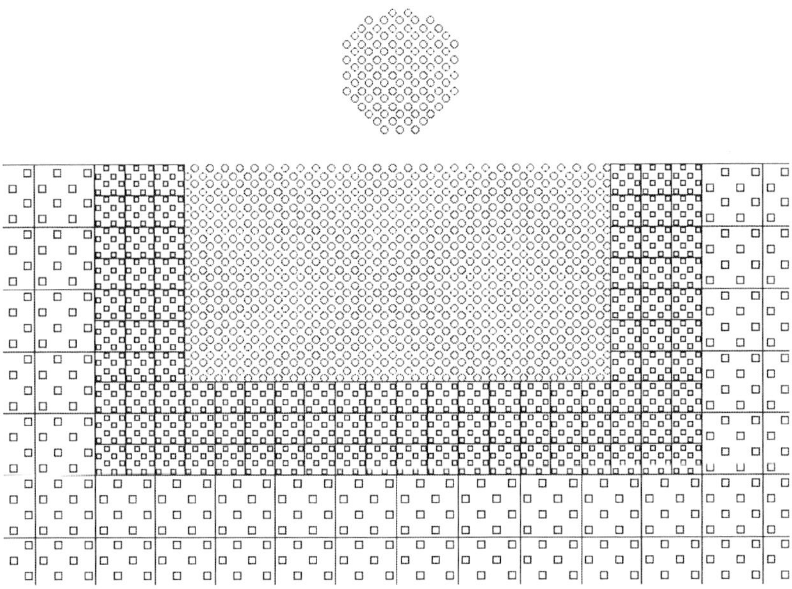

Figure 18. A schematic view of high energy cluster impacts. Only a slice of material points and atoms are shown. Circles represent atoms. Small squares represent light material points, and big squares represent heavier material points. Solid lines denote the background mesh. Only one layer hierarchical grids are shown.

Graphs in Figure 19-Figure 21 give snapshots of impact for clusters with 14, 128, and 1052 atoms. It can be seen that damage first spreads out almost in an isotropic manner (graph (a) of Figure 19-Figure 21). Afterward, disordered atoms prefer to propagate along the four directions $[10\bar{1}]$, $[\bar{1}0\bar{1}]$, $[01\bar{1}]$ and $[0\bar{1}\bar{1}]$ (graphs (b) and (c) of Figure 19-Figure 21). This damage evolution process is made clearer through the bottom-view in Figure 19(f). These four directions are denoted in Figure 22(a) as thick black lines. Damage spread along the four intersection lines of the four closely packed slip planes $\{111\}$. As rarefaction waves come from the surface, the number of disordered atoms diminishes (Figure 19(d-e), Figure 20(d-f) and Figure 21(d)). For the cluster with 14 atoms, hardly any disordered atoms are left after several ps . For the cluster with 128 atoms, rarefaction waves let the disordered atoms labeled by "A" rotate inward, as shown in Figure 20(c-d). Stacking faults appeared in the later stage of impact and at $10ps$ two partial dislocations interact with each other as shown in Figure 20(f). For the cluster with 1052 atoms, the anisotropy is not evident compared with the clusters having fewer atoms. In the late stage of impact, many stacking faults appeared. The stacking faults propagate along the four close packed planes $\{111\}$. Most of them intersect on the four directions $[10\bar{1}]$, $[\bar{1}0\bar{1}]$, $[01\bar{1}]$, $[0\bar{1}\bar{1}]$, and the rest on the two directions $[110]$ and $[1\bar{1}0]$ (Figure 21(c-f)). Dislocation loops can be observed in Figure 21(c). It can also be seen that the four common slip directions of slip planes (illustrated in Figure 22(a)) are the most favorable propagation paths of displaced atoms.

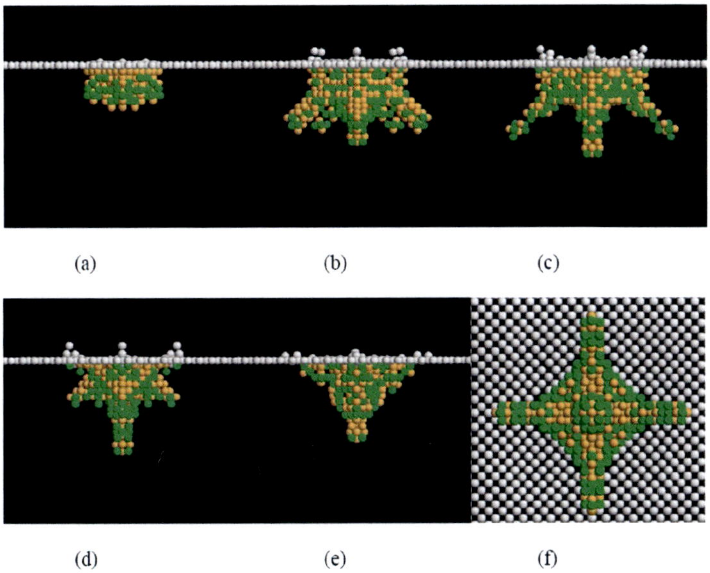

Figure 19. (a)-(e) Frontal views (view from [100]) of local crystalline orders for 0.2ps, 0.4ps, 0.5ps, 0.7ps and 1.0ps after impact for the cluster with 14 atoms. Graph (f) shows the bottom view for graph (c). Only atoms with non-FCC crystalline order are shown.

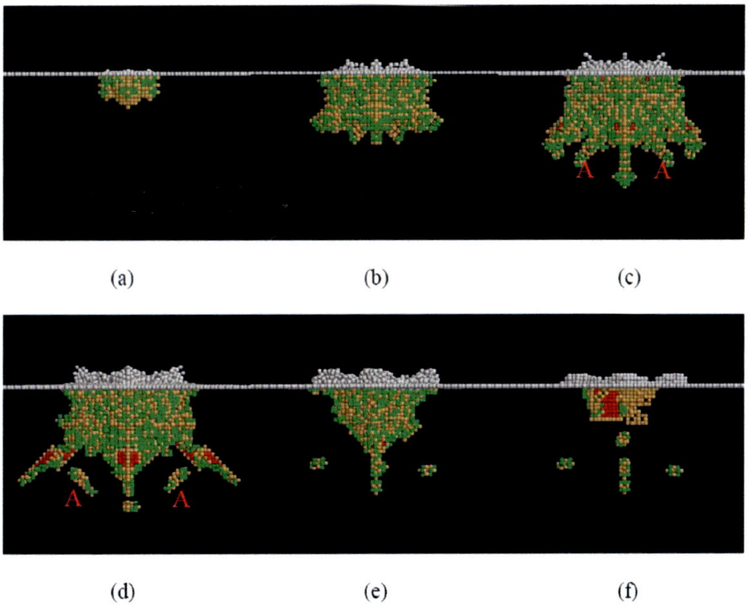

Figure 20. (a)-(f) Frontal views of local crystalline orders for 0.2ps, 0.4ps, 0.7ps, 1.2ps, 3.1ps and 10ps after impact for the cluster with 128 atoms.

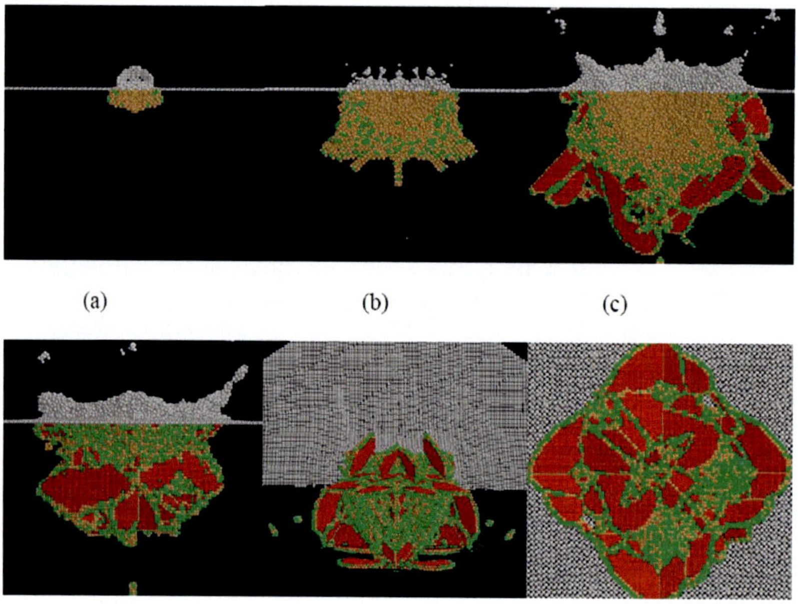

(a) (b) (c)

(d) (e) (f)

Figure 21. (a)-(d) Frontal views of local crystalline orders for 0.2ps, 0.6ps, 4.2ps and 6.2ps after impact for the cluster with 1052 atoms. For clarity, the frontal half atoms of (a-c) are removed. Graph (e) is the view from [-1 1 -1] of graph (c), and graph (f) is the bottom view of graph (c).

For clusters with 14 and 128 atoms, hardly any sputtering yield is observed; while many monomers and small clusters are sputtered out for the cluster with 1052 atoms.

The evolution of crater shapes also indicates that the FCC lattice structure plays an important role in the impact process. For the cluster containing as few as 14 atoms, no observable crater is left after 10*ps*. For the cluster with 128 atoms, the crater's top view evolves from a circle (Figure 23(a)) to a square (Figure 23(b)), and then back to a circle (Figure 23(c)). For the cluster with 1052 atoms, the crater's top view evolves from a circle (Figure 24(a)) to a square (Figure 24(b)) which is rotated about 45^0 from that in Figure 23(b). In the impact process, damage first spreads out in an isotropic manner and leads to a circular top view of the crater. Then the disordered atoms prefer to propagate along the four directions illustrated in Figure 22(a). Since the region of displaced atoms is small, actually four small planes propagate along the four directions. The deformation rate along the intersection lines is higher, changing the crater's top view from a circle to a square, as shown in Figure 23(b). In the late stage of impact, there is not enough energy to drive the small planes in Figure 22(b) outward and the protrusion of the central parts gradually changes the crater to a circular top view (Figure 23(c)). Because the anisotropy is not evident for the cluster with 1052 atoms, the crater's top view keeps a circle until stacking faults burst (Figure 21(c-f)), which makes damage to propagate along the four close packed planes $\{111\}$. The deformation rate in their intersection directions is larger (illustrated in Figure 25(c)) and the top view of the crater evolves from a circle to a square rotated 45 degrees (Figure 23(b)). The same configuration persists at 10ps as shown in Figure 23(c).

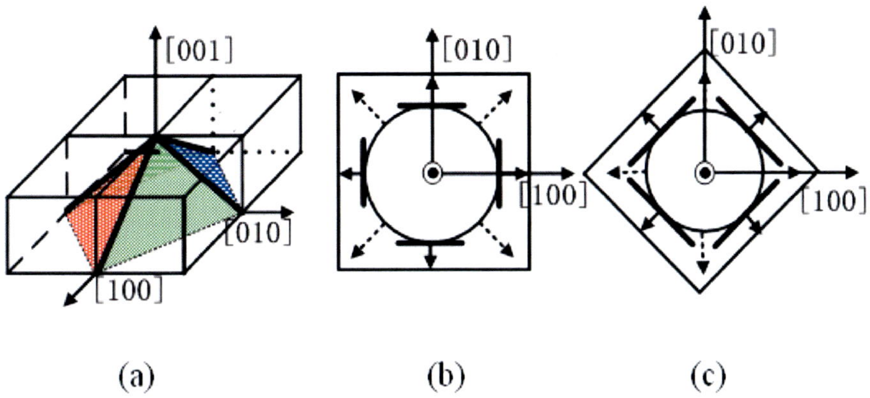

Figure 22. Illustration of the deformation process. (a) Thick black lines denote the four intersection lines of four slip planes {111}. Damage prefers to propagate along the four thick black lines. Graphs (b) and (c), corresponding to the cases of small and large cluster impacts respectively, show the deformation rate in different directions of the crater's top view. Thick back lines denote the planes that propagate in the directions shown by short solid arrows. This propagation leads to higher deformation rates in the directions of dotted long arrows.

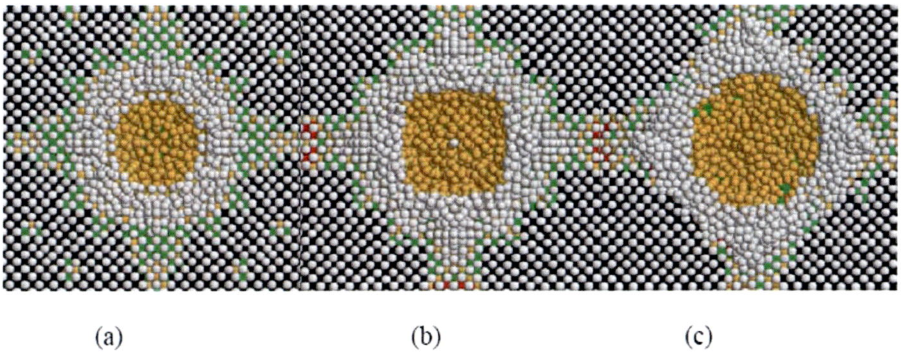

Figure 23. (a)-(c) Top view of the local crystalline orders for 0.5ps, 0.8ps, 1.7ps after the impact with a cluster of 128 atoms.

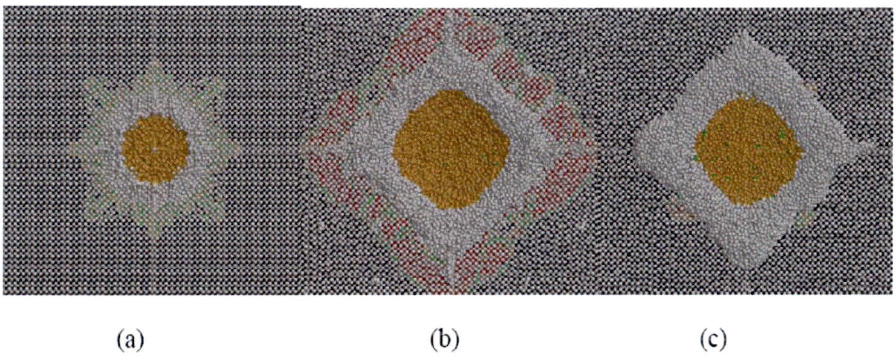

Figure 24. (a)-(c) Top view of the local crystalline orders for 0.6ps, 4.2ps and 10ps after the impact with a cluster of 1052 atoms.

It can be seen that energetic cluster impacts differ from nanoindentation in several aspects. Nanoindentation of MD simulations [82,164,103] aims to mimic quasi-equilibrium experiments, while the process of high energy cluster impacts is far away from the equilibrium state, characterized with high deformation rates. The impact case is intrinsically thermo-mechanically coupled and proceeds with possible state changes. The propagations of shock waves [179] as well as the front of the disordered atoms dominate the initial stage of the impact. The formation of the stacking faults and dislocation bursts, as frequently observed in the initial stage of nano-indention [82,164,103], only occurs in the late stage of cluster impact. The behavior of incipient plasticity also depends on the impact speed.

For the purposes of comparison, impacts of Si clusters on Si surface are also simulated. We adopt the potential proposed in Ref.[153]. The potential combines the Stillinger and Weber (SW) formulation [180] and the EAM form [181]. That potential provides a reasonable description of the energetics for all atomic coordinations Z, from the dimmer (Z=1) to FCC and HCP (Z=12). It accurately predicts formation energies and geometries of interstitial complexes: small clusters, interstitial-chain and planar {311} defects. One possible drawback of the potential is that the domain of the functions which comprise the potential is set by the extent of the fitting data. When evaluation of the potential leads to arguments beyond their domain, one has to use extrapolations. We regard that as a minor deficiency and ignorable for the purpose of qualitative comparison. Various graphs in Figure 25 are snapshots of frontal and top views of local crystalline orders for a Si-Si cluster impact.

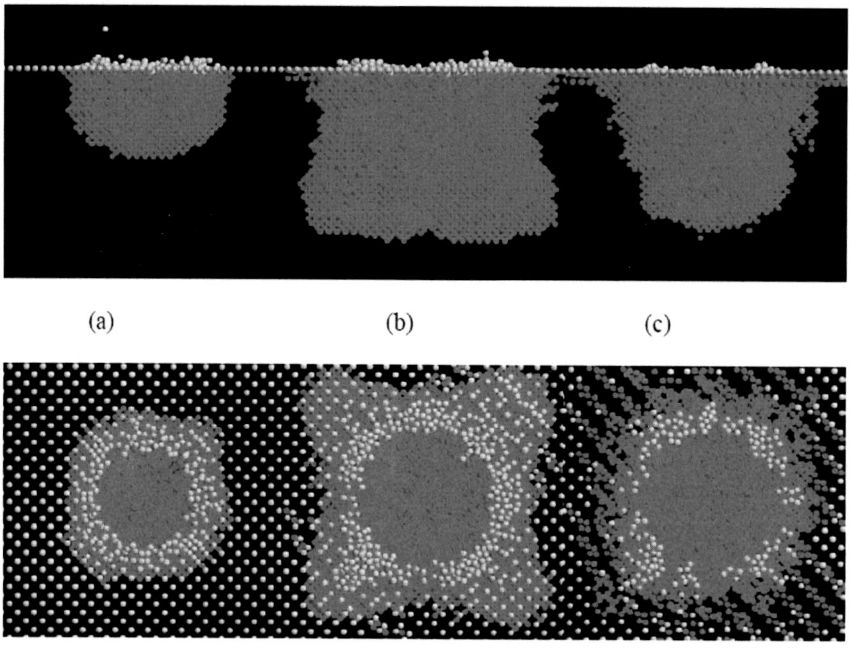

(a) (b) (c)

(d) (e) (f)

Figure 25. (a)-(c) are frontal views of the local crystalline orders at 0.4ps, 1.3ps, and 2.5ps after a Si cluster impacting on a Si substrate. The Si cluster contains 260 atoms. Graphs (d-f) correspond to the top view of graphs (a-c). The white spheres represent surface atoms, and the green ones represent atoms with none 4-coordinated combinations.

It can be seen that the target first deforms in an isotropic manner in the initial stage of impact, as shown in Figure 25(a) and (d). Anisotropic damage in the target emerges subsequently, as indicated in Figure 25(b) and (e). More damage occurs in $[110]$ and $[1\bar{1}0]$ directions, which is quite distinct from that of a Cu-Cu cluster impact. At the late stage of impact, the target recovers to an almost isotropic profile (Figure 25(c) and (f)), in contrast to the anisotropic profile for the Cu-Cu impact (Figure 21(c-f)).

For hypervelocity impacts of macroscopic bodies, the crater is almost hemispherical with volume proportional to the impacting energy of the macro cluster, provided the impacting velocity is high enough [182]. This volume dependence on the impacting energy is valid for micro cluster impacts at least for the following three cases. Case one is the simulated impacts of an Ar cluster (containing atoms from 13 to 3000) on a Si (001) surface [183], for total cluster energies of 0.5-55keV. Case two is the simulated impacts of an Ar cluster (containing atoms from 200 to 800) on a Cu (001) surface [184], for total cluster energies of 6-20keV. Case three is an experiment result featuring impacts of an Ar cluster (containing 3000 atoms) on a Au (111) surface [184], for total cluster energies between 20-150keV. These cases indicate that the peculiar damage formation process is caused by an isotropic energy transportation process [183]. Here in our simulations the total Cu cluster energies vary from 0.5 to 35keV and the total Si cluster energy is 4keV. The simulation results here show that the damage formation process, created by a cluster impact, possesses certain anisotropy which is in accordance with the results obtained by Nordiek et al. [185]. They found that the shape and size of craters depended on the surface orientation. They explain this anisotropy from two aspects. One is that the propagation velocity of disturbances relies on different crystallographic directions. The other is that the shear mode, a stretch along the [100] axis and a compression of the same amount along the [010] axis, costs the lowest energy. Thus the crater formation is enhanced in directions <100> (Figure 24(b)). Since dislocations manifests macroscopically as shear deformation, the burst of stacking faults (Figure 24(b)) is the underlying cause of the enhanced deformation in directions <100>. So the anisotropy of crystals dictates the anisotropic evolution of craters under the energetic cluster impacts. The displaced atoms propagate along preferable paths. Accordingly, the energy transportation process, characterized with multi-interactions among the cluster and surface atoms, is not an isotropic process. The underlying anisotropic lattice structure plays an important role in the impacting process and makes this transportation process anisotropic. For Cu cluster impacts on Cu surface, displaced atoms prefer to propagate along the intersection directions of slip places.

4.5. Interaction between Dislocations and Low-Angle Grain Boundaries

To study interaction between dislocations and low-angle GBs, we establish a 2D model of Cu bicrystal with low-angle interface. Because the interaction of dislocations and GBs is a long-time dynamical process, the MTS method is applied to simulating this bicrystal model [186]. Here the shorter simulation time step is taken as 1.0 fs, and the longer time step is 3.0 fs. The initial configuration of the model is composed of about 10,000 atoms. The mis-oriented bicrystal interface is set along the direction $[010]$. The lattice on the right and left sides of the interface are rotated 5° clockwise and anti-clockwise with respect to the [010] axis, respectively. The model is then relaxed for 4,000 steps at the temperature of about 0K. Thus the symmetrical tilt grain boundary with a misorientation 10° is created. To identify the type and the location of defects induced by deformation, we make the potential energy of atom corresponding to RGB in the color-map. If an atom has lower potential energy, it is likely to be colored blue. On the contrary, a higher energy atom is possibly red. In Figure 26, the low-angle GB of the bicrystal is formed by distributed arrays of edge dislocations [187].

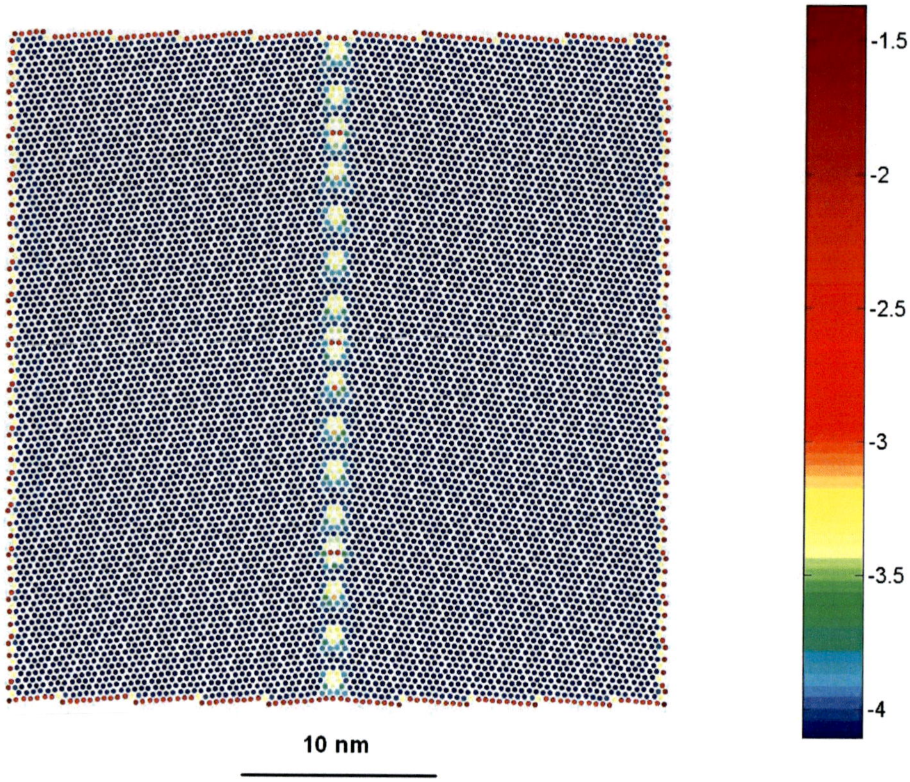

Figure 26. Initial configuration of Cu bicrystal with low-angle grain boundary.

We remove a row of atoms as shown in Figure 27(a), Figure 28(a), Figure 29(a) and Figure 30(a), respectively. The top and bottom sides of the model are imposed on the constant tension stress of 3.64GPa, while the right and left sides keep free. The simulation maintains a NVT ensemble fixed at nearly 0K. For the above four cases, all of the simulations last 100 ps.

Since the given defects appear in the different places, Figure 27-Figure 30 show in turn snapshots of interactions between the dislocations and the GBs. At the initial stage, the artificial defect collapses and dissociates into two partial dislocations with a stacking fault between them. Then the stacking fault vanishes by forming two perfect dislocations. The previous process is shown as Figure 27(b) and (c), and the same as that of Kuronen et. al. [188]. Such process also occurs in other three cases. The two perfect edge dislocations glide along the slip plane due to the resolved shear stress. One of them moves toward the right surface, while the other toward the GB. The surface approaching dislocation reaches the free surface, and generates a step there. The GB approaching dislocation, however, may have a spectrum of outcomes when interacting with the low-angle GB.

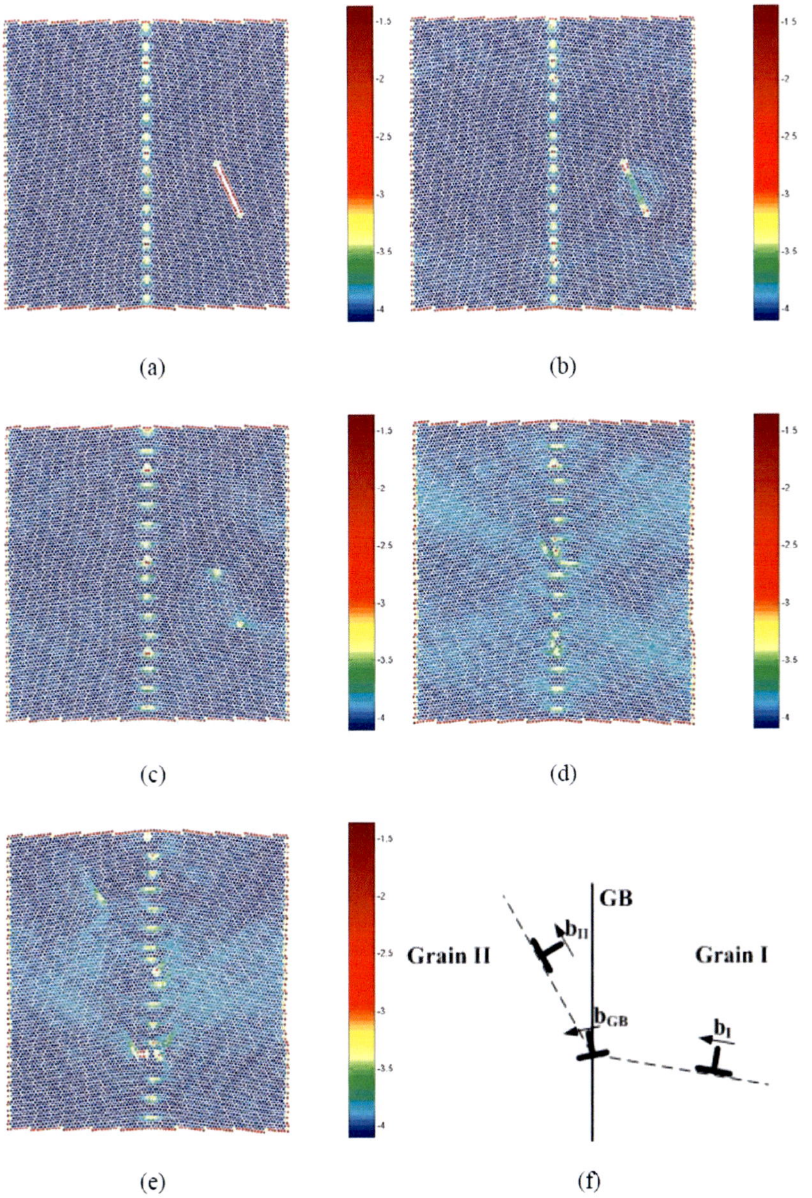

Figure 27. (a)-(e) Snapshots of interaction between lattice dislocation and GB for 0ps, 2.0ps, 3.4ps, 8.0ps and 10.9ps (Case I). Graph (f) shows related mechanism.

In Figure 27(d) and (e), it can be seen that a lattice dislocation interacts with GB dislocations, and decomposes into a transmitted partial dislocation in the neighboring grain, leaving a residue dislocation in the GB. The corresponding dislocation reaction, as delineated in Figure 27(f), is $\boldsymbol{b}_I \rightarrow \boldsymbol{b}_{II} + \boldsymbol{b}_{GB}$, where $\boldsymbol{b}_I$, $\boldsymbol{b}_{II}$ and $\boldsymbol{b}_{GB}$ are Burger vectors of incident lattice dislocation, transmitted dislocation and residue dislocation, respectively. In addition, the transmitted dislocation moves towards the top surface. Figure 28(b-e) show that a lattice dislocation is absorbed by the GB, and interact with GB dislocations, as a result of emitting a lattice dislocation to the incident grain and creating a ledge on the GB. The explanation concerned with this process is illustrated in Figure 28(f). The relevant dislocation reaction is $\boldsymbol{b}_I \rightarrow \boldsymbol{b}'_I + \boldsymbol{b}_{GB}$, where $\boldsymbol{b}_I$, $\boldsymbol{b}'_I$ and $\boldsymbol{b}_{GB}$ are Burger vectors of incident lattice dislocation, emitted dislocation and residue dislocation, respectively.

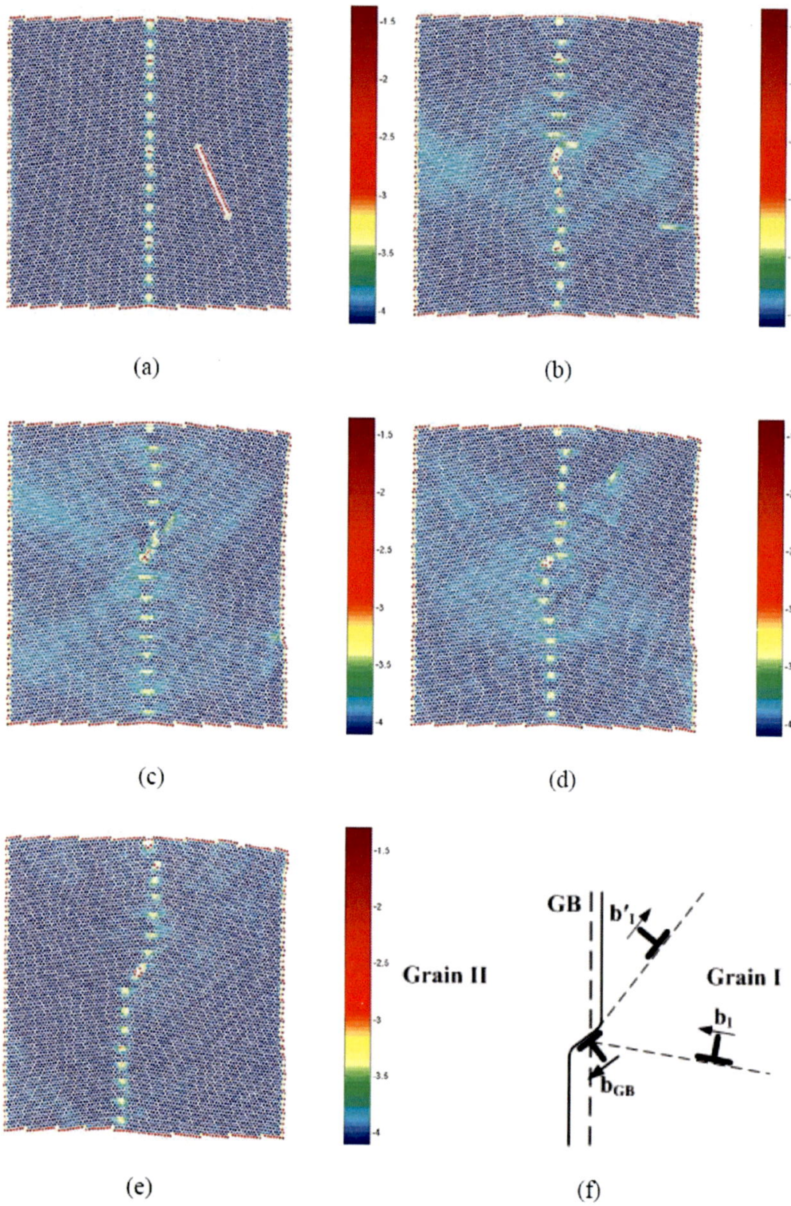

Figure 28. (a)-(e) Snapshots of interaction between lattice dislocation and GB for 0ps, 5.5ps, 7.8ps, 9.5ps and 14.5ps (Case II). Graph (f) shows related mechanism.

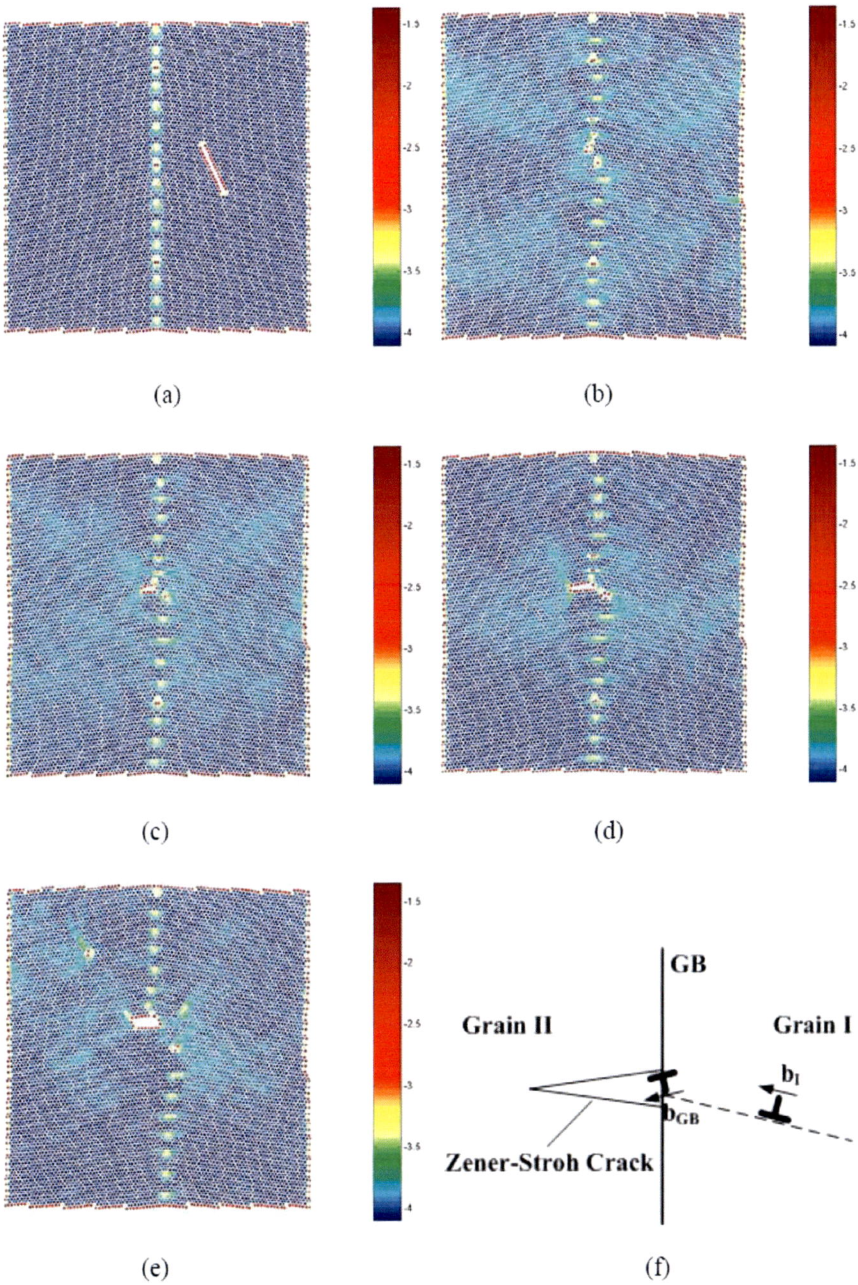

(a) (b) (c) (d) (e) (f)

Figure 29. (a)-(e) Snapshots of interaction between lattice dislocation and GB for 0ps, 8.2ps, 9.4ps, 10.9ps and 28.0ps (Case III). Graph (f) shows related mechanism.

When the artificial defect is placed as Figure 29(a), one may observe some interesting phenomena as shown in Figure 29(b-e). A lattice dislocation enters the GB, and interacts with the GB dislocations around it. Thereby a concentrated stress field forms. The GB dislocations on two intersecting slip planes can coalesce into a so-called Zener-Stroh crack [189-192]. Dislocations are emitted from the microcrack tip and consequently blunt the crack tip.

Because of the tension stress, the microcrack evolves as a quadrangular hole. The related mechanism is illustrated in Figure 29(f).

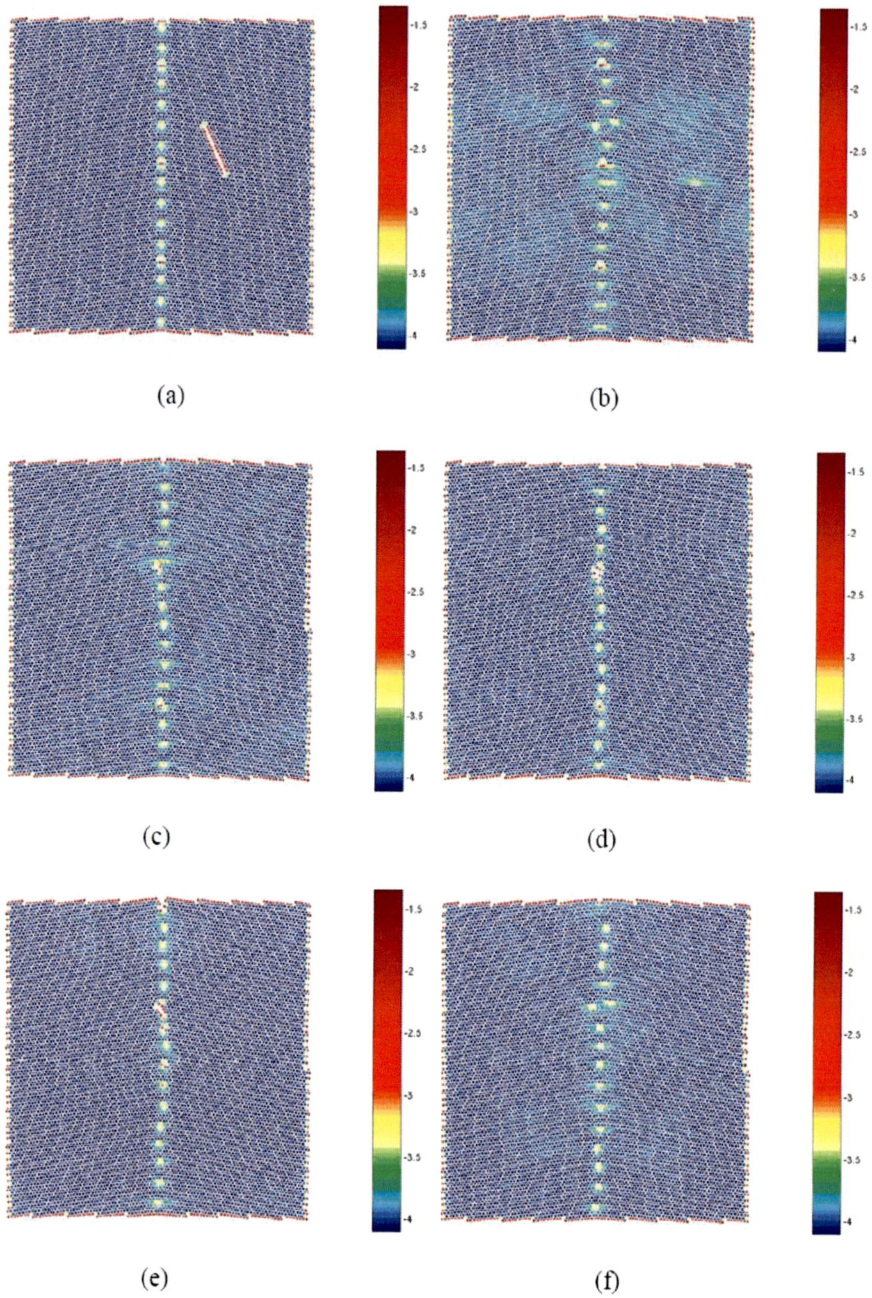

Figure 30. (a)-(f) Snapshots of interaction between lattice dislocation and GB for 0ps, 6.0ps, 13.1ps, 18.7ps, 32.3ps and 46.8ps (Case IV).

If the model is initialized as that of Figure 30(a), a different outcome appears. A lattice dislocation is captured by the GB, and incorporates a GB dislocation into a micro-hole. After a period of time, this micro-hole decomposes into the two dislocations which confront each other. Later on, the two dislocations combine again to a micro-hole. Then the micro-hole decomposes into dislocations. Such a process, shown in Figure 30(b-f), may repeat through the simulation, indicating a waiting period for the lattice dislocation to overcome the energy barrier and to transmit through the GB.

Through the above simulations, one realizes the complexity of the interaction between a lattice dislocation and a GB. The GB may act as sources, sinks or traps for lattice dislocations.

5. Discussion and Outlook

In this brief review, current methodological developments in MD are summarized, with complementary computer simulations for mechanical behavior of nanocrystals. From aforementioned simulations, the formation and deformation processes of nanocrystals are depicted at atomic level. Our discussions will be focused on two aspects: (1) dislocation-assisted grain growth in nanocrystalline copper under fast stretching, and (2) the MPM/MD handshaking method applied to impacts of high energy nanocluster.

In the simulation for fast stretching of nanocrystals, we observed that the movements of partial dislocations that dominate the deformation process did not lead to texture formation. Hence, we present a new mechanism characterized as the dislocation-assisting grain coalescence, to distinct it from conventional mechanism of textures induced by dislocation plasticity. Whether this mechanism relies on the rate or the mode of loadings is the main content of future works. Moreover, this mechanism may exhibit a certain length-scale effect.

In the simulation for impacts of high energy nanocluster, the temperature is initialized at 0K, like that in Refs. [126, 135, 151]. For finite temperature simulations, velocity of material points should be thermalized to a Maxwellian distribution [135]. For different temperatures, elastic constants and lattice spacing vary in the MD region. Thus the elastic constants in the MPM region and lattice spacing in the handshaking region should change in accordance with that in the MD region. Another problem is the dissipation in the MPM region. If the elasticity theory is used in the MPM region, vibration modes will propagate undamped. In order to thermalize short-wavelength phonons and also to allow energy to be dissipated in the MPM region, the MPM degrees of freedom should be weakly coupled to a Brownian heat bath whose dynamics are set to the temperature at which the simulation is being performed [135]. Thus Newton's equations of motion are replaced by Langevin equation [193,194]. In this case, an external force, which combines a random force and a frictional dissipation term, is prescribed on each material point. A proper damping [148] is expected to reduce the reflection waves on the MPM/MD boundary.

So far MD has not only been a standard tool in academic researches, but also becomes increasingly attractive to industrial researchers. With the help of MD, those researchers are engaging in advanced explorations, such as design and fabrication for novel nanoscale systems. Although MD has made considerable success in academic and industrial investigations, it must be faced with one serious challenge: How to bridge the gap in the different computational models at length and time scales? Namely, the atomistic calculations need to find the way to overcome their natural barrier of length and time scales. For the length-scale problem, the multiscale and hierarchical modeling techniques look very promising. But most multiscale methods by now remain at the level of testing concepts. Few have gone through the rigorous procedure of verifying the results [195]. Nevertheless, it is our belief that the multiscale methods will become gradually perfect in the next decade.

For the time-scale problem, how to enhance the sampling efficiency of MD simulations is a longstanding issue of research. For the sake of this purpose, transition path sampling scheme [196, 197] has been developed to overcome both the long timescale problem and the lack of prior knowledge about transition mechanisms. In addition, Voter et al. [198] have presented three new TST-based dynamics methods including hyper-dynamics [199], parallel replica dynamics [200] and temperature-accelerated dynamics [201]. These dynamics methods are only suitable for infrequent-event systems. By now MD simulations are limited to tens of nanoseconds, in vast contrast to realistic time scales of dynamical processes that often extend into the microsecond or the second range. Nevertheless, it is possible to go beyond the time scales accessible by today's MD and to reach simulation times an order of magnitude longer. In the future, new developments in hierarchical methodologies coupling a wide range of temporal and spatial scales will provide a drive force to achieve the above goal. Furthermore, the rapid advance in supercomputer power will also facilitate the goal to be accomplished.

REFERENCES

[1] W Yang, W. B. Lee. *Mesoplasticity and Its Applications*. Berlin: Springer-Verlag, 1993.

[2] W Yang. *Macroscopic and Microscopic Fracture Mechanics*. National Defence Industry Press, (in Chinese), 1995.

[3] W Yang. *Mechatronic Reliability*. TUP-Springer-Verlag, Berlin, 2002.

[4] LB Freund. The mechanics of electronic materials. *International Journal of Solids and Structures*., 2000, 37:185-196.

[5] LB Freund, HT Johnson. Influence of strain on functional characteristics of nanoelectronic devices. *Journal of the Mechanics and Physics of Solids*, 2001, 49:1925-1935.

[6] HT Johnson, LB Freund, CD Akyuz, A Zaslavsky. Finite element analysis of strain effects on electronic and transport properties in quantum dots and wires. *Journal of Applied Physics*, 1998, 84:3714-3725.

[7] HT Johnson, V Nguyen, AF Bower. Simulated self-assembly and optoelectronic properties of InAs'GaAs quantum dot arrays. *Journal of Applied Physics*, 2002, 92:4653-4663.

[8] KE Drexler. *Nanosystems: Molecular Machinery, Manufacturing and Computation*. New York: John Wiley and Sons Inc., 1992.

[9] W Yang, XL Ma, HT Wang. Advances in nanomechanics. *Advances in Mechanics*, 2002, 32:161-174 (in Chinese).

[10] WW Gerberich and W Yang eds., Interfacial and Nanoscale Failure. In: I Milne, RO Ritchie and B Karihaloo eds. *Comprehensive Structural Integrity*, Vol. 8. Oxford: Elsevier Ltd, 2003, Chapter 8.01.

[11] B Freund, S Suresh. *Thin Film Materials*. Cambridge: Cambridge University Press, 2003.

[12] F Liu, MH Huang, PP Rugheimer, DE Savage, MG Lagally. Nanostressors and the nanomechanical response of a thin silicon film on an insulator. *Physical Review Letters,* 2002, 89 (13): art. no. 136101.

[13] B Bhushan. Handbook of Micro/Nano Tribology. *CRC Series of Mechanics and Materials Science*. Boca Raton, Florida: CRC Press, 1999.

[14] G He, MH Muser, MO Robbins. Adsorbed layers and the origin of static friction. *Science*, 1999, 284:1650-1652.

[15] I Chasiotis, W Knauss. Experimentation at the Micron and Submicron Scale, In: WW Gerberich and W Yang eds. *Interfacial and Nanoscale Failure*, Oxford: Elsevier Ltd, 2003, Chap. 8.02.

[16] FL Dai, YM Xing. Nano-moire method. *Acta Mechanica Sinica*, 1999, 15:283-288.

[17] H Gleiter. Nanocrystalline materials. Progress in Materials Science, 1989, 33:223-315.

[18] H Gleiter. Nanostructured materials: basic concepts and microstructure. *Acta Materialia*, 2000, 48:1-29.

[19] F Celarie, S Prades, D Bonamy, L Ferrero, E Bouchaud, C Guillot, C Marliere. Glass breaks like metal, but at the nanometer scale. *Physical Review Letters*, 2003, 90(7): art. no. 075504.

[20] Q Yang, W Yang. Three-dimensional analysis of scale dependence of sub-micron polycrystals due to configuration entropy. *Acta Mechanica Sinica*, 2001, 17(2): 172-182.

[21] L Lu, ML Sui, K Lu. Superplastic extensibility of nanocrystalline copper at room temperature. *Science*, 2000, 287(5457):1463-1466.

[22] X McFadden, RS Mishra, RZ Valiev, AP Zhilyaev, AK Mukherjee. Low-temperature superplasticity in nanostructured nickel and metal alloys. *Nature*, 19999, 398 (6729):684-686.

[23] PM Derlet, R Meyer, LJ Lewis, U Stuhr, H Van Swygenhoven. Low-frequency vibrational properties of nanocrystalline materials. *Physical Review Letters*, 2001, 87 (20): art. no. 205501.

[24] Y Champion, C Langlois, S Guerin-Mailly, P Langlois, JL Bonnentien, MJ Hytch. Near-perfect elastoplasticity in pure nanocrystalline copper. *Science*, 2003, 300 (5617):310-311.

[25] J Schiotz, FD Di Tolla, KW Jacobsen. Softening of nanocrystalline metals at very small grain sizes. *Nature,* 1998, 391:561-563.

[26] H Van Swygenhoven. Polycrystalline materials: Grain boundaries and dislocations. *Science*, 2002, 296:66-67.

[27] M Samaras, PM Derlet, H Van Swygenhoven, M Victoria. Computer simulation of displacement cascade in nanocrystalline Ni. *Physical Review Letters*, 2002, 88 (12): art. no. 125505.

[28] J Schiøtz. Simulation of nanocrystalline metals at the atomics scale. What can we do? What can we trust? In: Dinesen AR etal. Eds. *Proceedings of the 22nd International Symposium on Materials Science:* Science of Metastable and Nanocrystalline Alloys-Structure, Properties and Modeling, Roskilde: Risø National Laboratory, Sep 2001, 127-139.

[29] H Liang, X Wang, H Wu and Y Wang. Molecular dynamics simulation of length scale effects on tension nanocrystalline copper wire. *Acta Mechanica Sinica*, 2002, 34(2):208-215 (in Chinese).

[30] B Cai, QP Kong, L Lu, K Lu. Interface controlled diffusional creep of nanocrystalline pure copper. *Scripta Metallurgica*, 1999, 41:755-759.

[31] J Karch, R Birringer, H Gleiter. Ceramics ductile at low temperature. *Nature*, 1987, 330:556-558.

[32] MF Ashby, RA Verrall. Diffusion-accommodated flow and superplasticity. *Acta Metallurgica,* 1973, 21:149-163.

[33] W Yang, W Hong. Numerical simulation for deformation of nano-grained metals. *Acta Mechanica Sinica*, 2002, 18:506-515.

[34] IA Ovid'ko. Materials science: deformation of nanostructures. Science, 2002, 295:2386-2386.

[35] M Murayama, JM Howe, H Hidaka, S Takaki. Atomic-level observation of disclination dipoles in mechanically milled, nanocrystalline Fe. *Science*, 2002, 295 (5564): 2433-2435.

[36] K Lu, J Lu. Surface nanocrystallization (SNC) of metallic materials-presentation of the concept behind a new approach. *Journal of Materials Science and Technology*, 15(1999) 3: 193-197.

[37] J Lu, K Lu. Surface Nanocrystallization (SNC) of Materials and Its Effect on Mechanical Behavior. In: WW Gerberich and W Yang eds, *Interfacial and Nanoscale Failure*, chap 8.14, Oxford: Elsevier Science, 2003, 495-528.

[38] NR Tao, ZB Wang, WP Tong, W Sui, J Lu, K Lu. An investigation of surface nanocrystallization mechanism in Fe induced by surface mechanical attrition treatment. *Acta Materialia*, 2002, 50(18):4603-4616.

[39] WP Tong, NR Tao, ZB Wang, J Lu, K Lu. Nitriding iron at lower temperatures. *Science*, 2003, 299, (5607):686-688.

[40] G Liu, J Lu, K Lu. Surface nanocrystallization of 316L stainless steel induced by ultrasonic shot peening. *Materials Science and Engineering* A, 2000, 286(1): 91-95.

[41] G Liu, SC Wang, XFLou, J Lu, K Lu. Low carbon steel with nanostructured surface layer induced by high-energy shot peening. *Scripta Materialia*, 2001, 44(8-9):1791-1795.

[42] XL Ma, W Wang, W Yang. Simulation for surface self-nanocrystallization under shot peening. *Acta Mechanica Sinica*, 2003, 19(2):172-180.

[43] WW Gerberich, DE Kramer, NI Tymiak, AA Volinsky, DF Bahr, MD Kriese. Nanoindentation-induced defect-interface interactions: Phenomena, methods and limitations. *Acta Materialia*, 1999, 47:4115-4123.

[44] WC Oliver, GM Pharr. An improved technique for determining hardness and elastic-modulus using load and displacement sensing indentation experiments. *Journal of Materials Research*, 1992, 7:1564-1583.

[45] DF Bahr, DE Kramer, WW Gerberich. Nonlinear deformation mechanisms during nanoindentation. *Acta Materialia*, 1998, 46:3605~3617.

[46] MJ Adams. Nanoindentation of particulate coating. *Journal of Materials Research*, 1999, 14:2344~2350.

[47] T Chang, W Guo, G Li. Analysis of microbucking for monomolecular layers adhering to a substrate. *Acta Mechanica Sinica*, 2002, 18(6):608-620.

[48] WD Nix, H Gao. Indentation size effects in crystalline materials: a law for strain gradient plasticity. *Journal of the Mechanics and Physics of Solids*, 1998, 46:411-425.

[49] NI Tymiak, DE Kramer, DF Bahr, TJ Wyrobek, WW Gerberich. Plastic strain and strain gradients at very small indentation depths. *Acta Materialia*, 2001, 49:1021-1034.

[50] WW Gerberich, JC Nelson, ET Lilleodden, P Anderson, JT Wyrobek. Indentation induced dislocation nucleation: The initial yield point. *Acta Materialia*, 1996, 44:3585-3598.

[51] A Gouldstone, KJ Van Vliet, S Suresh. Simulation of defect nucleation in a crystal. *Nature*, 2001, 411:656-656.

[52] W Yang, XL Ma, HT Wang, W Hong. Advances in nanomechanics. *Advances in Mechanics*, 2002, 32 (2):161-174 (in Chinese).

[53] DW Hermann. *Computer Simulation Methods in Theoretic Physics*, 2nd Edition, Berlin: Springer-Verlag, 1990.

[54] G Cicero, L Pizzagalli, A Catellani. Ab initio study of misfit dislocations at the SiC/Si(001) interface. *Physical Review Letters*, 2002, 89 (15): art. no. 156101.

[55] W Yang, HL Tan. Chaotic atom motion excited by fracture. *Material Scientific Research International*, 1996, 2:1-12.

[56] V Bulatov, FF Abraham, L Kubin, B Devincre, S Yip. Connecting atomistic and mesoscale simulations of crystal plasticity. *Nature*, 1998, 391(6668): 669-672.

[57] M Ortiz, R Phillips. Nanomechanics of Defects in Solids. *Advances in Applied Mechanics*, 1999, 36:1-79.

[58] R Phillips. *Crystals, Defects and Microstructures – Modeling across Scales*. Cambridge: Cambridge Press, 2001.

[59] SR Phillpot, D Wolf, H Gleiter. Molecular-dynamics study of the synthesis and characterization of a fully dense, three-dimensional nanocrystalline material. *Journal of Applied Physics,* 1995, 78(2):847-861.

[60] J Wass. 3-D Molecular Dynamics version 2.5. *Science*, 2000, 288(5469): 1191-1191.

[61] M Moseler, U Landman. Formation, stability and breakup of nanojets. *Science*, 2000, 289 (5482):1165-1169.

[62] J Eggers. Dynamics of liquid nanojets. *Physical Review Letters*, 2002, 89 (8): art. no. 084502.

[63] PJ Feibelman. Partial dissociation of water on Ru(0001). *Science*, 2002, 295 (5552):99-102.

[64] M Matsumoto, S Saito, I Ohmine. Molecular dynamics simulation of the ice nucleation and growth process leading to water freezing. *Nature*, 2002, 416 (6879):409-413.

[65] YM Xing, FL Dai, W Yang. Experimental study about nano-deformation field near quasi-cleavage crack tip. *Science in China* A, 2000, 43:963-968.

[66] JG Swadener, MI Baskes, M Nastasi. Molecular dynamics simulation of brittle fracture in silicon. *Physical Review Letters*, 2002, 89 (8): art. no. 085503.

[67] FF Abraham, H Gao. How fast can cracks propagate? *Physical Review Letters*, 2000, 84:3113-3116.

[68] FF Abraham, R Walkup, H Gao, et al. Simulating materials failure by using up to one billion atoms and the world's fastest computer: Brittle fracture. *Proceedings of National Academy of Sciences of the USA*, 2002, 99:5777-5782.

[69] SJ Zhou, DL Preston, PS Lomdahl, DM Beazley. Large-scale molecular dynamics simulations of dislocation intersection in copper. *Science,* 1998, 279 (5356):1525-1527.

[70] J Christiansen, K Morgenstern, J Schiotz, KW Jacobsen, KF Braun, KH Rieder, E. Logsgaard, F Besenbacher. Atomic-Scale Structure of Dislocations Revealed by Scanning Tunneling Microscopy and Molecular Dynamics. *Physical Review Letters*, 2002, 88: art. No. 206106.

[71] FF Abraham, R Walkup, H Gao, et al. Simulating materials failure by using up to one billion atoms and the world's fastest computer: work-hardening. *Proceedings of National Academy of Sciences of the USA*, 2002, 99:5783-5787.

[72] S Li, K Gao, L Qiao, W Chu, F Zhou. Molecular dynamic simulation of the role of dislocations in microcrack healing. *Acta Mechanica Sinica*, 2000, 16 (4): 366-373.

[73] JQ Ding, ZY Chen. Molecular dynamics calculation of thermodynamic properties of nanocrystaline α-iron. *Acta Mechanica Sinica*, 2000, 32(6): 739~743 (in Chinese).

[74] MP Allen, DJ Tildesley. Computer Simulation of Liquids, Oxford: Clarendon, 1989.

[75] R Car, M Parrinello. Unified approach for molecular dynamics and density-functional theory. *Physical Review Letters*, 1985, 55:2471-2474.

[76] MC Payne, MP Teter, DC Allan, TA Arias, JD Joannopoulos. Iterative minimization techniques for ab initio total-energy calculations: molecular dynamics and conjugate gradients. *Reviews of Modern Physics*, 1992, 64:1045-1097.

[77] ME Tuckerman, PJ Ungar, T Rosenvinge, ML Kein. Ab initio molecular dynamics simulations. *Journal of Physical Chemistry*, 1996, 100(31):12878-12887.

[78] MJ Cillan. The virtual matter laboratory. *Contemporary Physics*, 1997, 38(2):115-130.

[79] M Parrinello. From silicon to RNA: the coming of age of ab initio molecular dynamics. *Solid State Communication*, 1997, 102:107-120.

[80] ME Tuckerman. Ab initio molecular dynamics: basic concepts, current trends and novel applications. *Journal of Physics: Condensed Matter*, 2002, 14:R1297-1355.

[81] W Kohn, LJ Sham. Self-consistent equations including exchange and correlation effects. *Physical Review*, 1965, 140:A1133-A1138.

[82] CL Kelchner, SJ Plimpton and JC Hamilton. Dislocation nucleation and defect structure during surface indentation. *Physical Review* B, 1998, 58:11085-11088.

[83] J Schiøtz, T Vegge, FD Di Tolla, KW Jacobsen. Atomic-scale simulations of the mechanical deformation of nanocrystalline metals. *Physical Review* B, 1999, 60:11971-11983.

[84] SM Foiles, MI Baskes, MS Daw. Embedded-atom-method functions for the fcc metals Cu, Ag, Au, Ni, Pd, Pt and their alloys. *Physical Review* B, 1986, 33: 7983-7991.

[85] DJ Oh, RA Johnson. Simple embedded atom method model for fcc and hcp metals. *Journal of Materials Research*, 1988, 3:471-478.

[86] MI Baskes, JS Nelson, AF Wright. Semiempirical modified embedded-atom potentials for silicon and germanium. *Physical Review* B, 1989, 40:6085-6100.

[87] RA Johnson. Alloy models with the embedded-atom method. *Physical Review* B, 1989, 39:12554-12559.

[88] MW Finnis, JE Sinclair. A simple empirical N-body potential for transition metals. *Philosophical Magazine* A, 1984, 50(1): 45-55.

[89] WB Streett, DJ Tildesley, G Saville. Multiple time-step methods in molecular dynamics. *Molecular Physics*, 1978, 35(3):639-648.

[90] H Grubmüller, H Heller, A Windemuth, K Schulten. Generalized Verlet algorithm for efficient molecular dynamics simulations with long-range interactions. *Molecular Simulation,* 1991, 6:121-142.

[91] M Tuckerman, BJ Berne , GJ Martyna. Reversible multiple time scale molecular dynamics. *Journal of Chemical Physics*, 1992, 97(3):1990-2001.

[92] BG Archilla, JM Sanz-Serna, RD Skeel. Long-time-step methods for oscillatory differential equations. *SIAM Journal on Scientific Computing*, 1998, 20:930-963.

[93] JA Izaguirre, S Reich, RD Skeel. Longer time steps for molecular dynamics. *Journal of Chemical Physics*, 1999, 110(20):9853-9864.

[94] RH Zhou , BJ Berne. A new molecular dynamics method combining the reference system propagator algorithm with a fast multipole method for simulating proteins and other complex systems. *Journal of Chemical Physics*, 1995, 103(21):9444-9459.

[95] DD Humphreys, RA Friesner, BJ Berne. A multiple time step molecular dynamics algorithm for macromolecules. *Journal of Physical Chemistry*, 1994, 98: 6885-6892.

[96] M Watanabe, M Karplus. Simulation of macromolecules by multiple time-step methods. *Journal of Physical Chemistry*, 1995, 99:5680-5697.

[97] P Procacci, M Marchi. Taming the Ewald sum in molecular dynamics simulations of solvated proteins via a multiple time step algorithm. *Journal of Chemical Physics*, 1996, 104(8):3003-3012.

[98] S Reich. Multiple time scales in classical and quantum-classical molecular dynamics. *Journal of Computational Physics*, 1999, 151:49-73.

[99] SJ Stuart, RH Zhou, BJ Berne. Molecular dynamics with multiple time scales: the selection of efficient reference system propagators. *Journal of Chemical Physics*, 1996, 105(4):1426-1436.

[100] ME Tuckerman, BJ Berne. Molecular dynamics algorithm for multiple time scales: systems with long rang forces. *Journal of Chemical Physics*, 1991, 94(10): 6811-6815.

[101] HF Trotter. On the product of semi-groups of operators. *Proceedings of the American Mathematical Society*, 1959, 10(4): 545-551.

[102] A Richter, R Ries, R Smith, M Henkel and B Wolf. Nanoindentation of diamond, graphite and fullerene films. *Diamond and Related Materials*, 2001, 9:170-184.

[103] XL Ma, W Yang. Molecular dynamics simulation on burst and arrest of stacking faults in nanocrystalline Cu under nanoindentation. *Nanotechnology*, 2003, 14:1208-1215.

[104] JD Honeycutt and HC Andersen. Molecular dynamics study of melting and freezing of small Lennard-Jones clusters. *Journal of Physical Chemistry*, 1987, 91:4950-4963.

[105] R Zallen. The Physics of Amorphous Solids. New York: Wiley-Interscience , 1983.

[106] FF Abraham. Atomic simulations of crack propagation. In: The Society of Materials Sciences of Japan ed. *Materials Science for the 21st Century, Invited papers*, vol. A, May 2001, 195-202.

[107] DM Beazley, PS Lomdahl. Message-passing multi-cell molecular dynamics on the connection machine. *Parallel Computing*, 1994, 20:173-195.

[108] S Plimpton. Fast parallel algorithms for short-range molecular dynamics. *Journal of Computational Physics*, 1995, 117:1-19.

[109] R Murty, D Okunbor. Efficient parallel algorithms for molecular dynamics simulations. *Parallel Computing*, 1999, 25:217-230.

[110] H Schwichtenberg, G Winter and H Wallmeier. Acceleration of molecular mechanic simulation by parallelization and fast multipole techniques. *Parallel Computing*, 1999, 25:535-546.

[111] ZY Mo, JL Zhang. Parallelization and optimization of two-dimensional molecular dynamics program (MDP). *Computational Physics*, 2000, 17:193-198 (in Chinese).

[112] WJ Phythian, RE Stoller, AJE Foreman, AF Calder, D J Bacon. A comparison of displacement cascades in copper and iron by molecular dynamics and its application to microstructural evolution. *Journal of Nuclear Materials*, 1995, 223:245-261.

[113] HL Heinisch. Simulating the production of free defects in irradiated metals. *Nuclear Instruments and Methods in Physics Research Science* B, 1995, 102:47-50.

[114] HL Heinisch and BN Singh. Stochastic annealing simulation of differential defect production in high energy cascades. *Journal of Nuclear Materials*, 1996, 232:206-213.

[115] M Jaraiz, GH Gilmer, JM Poate, TD de la Rubia. Atomistic calculations of ion implantation in Si: Point defect and transient enhanced diffusion phenomena. *Applies Physics Letters*., 1996, 68:409-411.

[116] G Henkelman, H Jonsson. Multiple time scale simulations of metal crystal growth reveal the importance of multiatom surface processes. *Physical Review Letters*, 2003, 90 (11): art. no. 116101.

[117] A Nakano, ME Bachlechner, RK Kalia, E Lidorikis, P Vashishta, GZ Voyisajis, TJ Campbell, S Ogata, F Shimojo. Multiscale simulation of nanosystems. *Computing in Science and Engineering*, 2001, 3(4):56-66.

[118] Z Guo, W Yang. MPM/MD handshaking method for multiscale simulation and its application to high energy cluster impacts. *International Journal of Mechanical Science*, 2005, in press.

[119] R Thomson, SJ Zhou, AE Carlsson, BK Tewary. Lattice imperfections studied by use of lattice Green's functions. Physical Review B: *Condens. Matter*, 1992, 46: 10613-10622.

[120] EB Tadmor, M Ortiz, R Phillips. Quasicontinuum analysis of defects in solids. *Philosophical Magazine* A, 1996, 73:1529-1563.

[121] J Sinclair, PC Gehlen, RG Hoagland, JP Hirth. Flexible boundary conditions and nonlinear geometric effects in atomic dislocation modeling. *Journal of Applied Physics*, 1978, 49:3890-3897.

[122] W Cai, M de Koning, VV Bulatov, S Yip. Minimizing boundary reflections in coupled-domain simulations. *Physical Review Letters*, 2000, 85:3213-3216.

[123] GJ Wagner, GK Eduard, WK Liu. Molecular dynamics boundary conditions for regular crystal lattice. *Computer Methods in Applied Mechanics and Engineering*, 2004, 193:1579-1601.

[124] W E, Z Huang. Matching conditions in atomistic-continuum modeling of materials. *Physical Review Letters*, 2001,87:135501-135504.

[125] W E, Z Huang. A dynamics atomistic-continuum method for the simulation of crystalline materials. *Journal of Computational Physics*, 2002, 182:234-261.

[126] S Kohlhoff, P Gumbsch, HF Fischmeister. Crack propagation in bcc crystals studied with a combined finite-element and atomistic model. *Philosophical Magazine* A, 1991, 64:851-878.

[127] W Yang, HL Tan, TF Guo. Evolution of crack tip process zone. *Modeling and Simulation in Material Science and Engineering,* 1994, 2:767-782.

[128] HL Tan, W Yang. Atomistic/continuum simulation of interfacial fracture, Part I: Atomistic simulation. *Acta Mechanica Sinica*, 1994a, 10:150-161.

[129] HL Tan, W Yang. Atomistic/continuum simulation of interfacial fracture, Part II: Atomistic/dislocation/continuum simulation. *Acta Mechanica Sinica*, 1994b, 10:237-249.

[130] HL Tan, JA Nairn. Hierarchical, adaptive, material point method for dynamic energy release rate calculations. *Computer Methods in Applied Mechanics and Engineering*, 2002, 191:2095-2109.

[131] A Warshel, M Levitt. Theoretical studies of enzymic reactions. *Journal of Molecular Biology*, 1976, 103:227-249.

[132] YK Zhang, TS Lee, WT Yang. A pseudobond approach to combining quantum mechanical and molecular mechanical methods. *Journal Chemical Physics*, 1999, 110:46-54.

[133] M Eichinger, P Tavan, J Hutter, M Parrinello. A hybrid method for solutes in complex solvents: Density functional theory combined with empirical force fields. *Journal of Chemical Physics*, 1999, 110:10452-10467.

[134] FF Abraham, JQ Broughton, N Bernstein, E Kaxiras. Spanning the continuum to quantum length scales in a dynamics simulation of brittle fracture. *Europhysics Letters*, 1998, 44(6):783-787.

[135] JQ Broughton, FF Abraham, N Bernstein, E Kaxiras. Concurrent coupling of length scales: Methodology and application. *Physical Review* B, 1999, 60(4):2391-2402.

[136] RE Rudd, JQ Broughton. Concurrent coupling of length scales in solid state systems. *Physica Stattus Solidi* (B), 2000, 217:251-291.

[137] RE Rudd, JQ Broughton. Coarse-grained molecular dynamics and the atomic limit of finite elements. *Physical Review* B, 1998, 58:R5893-R5896.

[138] RE Rudd, JQ Broughton. Atomistic simulation of MEMS resonators through the coupling of length scales. *Journal of Modeling and Simulation of Microsystems*, 1999, 1:29-38.

[139] B Liu, Y Huang, H Jiang, S Qu, KC Hwang. The atomic-scale finite element method. *Computer Methods in Applied Mechanics and Engineering*, 2004, 193:1849-1864.

[140] W E, B Engquist. The heterogeneous multiscale methods. *Communications in Mathematical Science*, 2003, 1:87-132.

[141] XT Li, W E. Multiscale modeling of the dynamics of solids at finite temperature. *Journal of the Mechanics and Physics of Solids*, 2005, 53:1650-1685.

[142] WQ Ren, W E. Heterogeneous multiscale method for the modeling of complex fluids and micro-fluidics. *Journal of Computational Physics*, 2005, 204:1-26.

[143] D Sulsky, Z Chen, HL Schreyer. A particle method for history-dependent materials. *Computer Methods in Applied Mechanics and Engineering*, 1994, 118:179-196.

[144] D Burgess, D Sulsky, JU Brackbill. Mass matrix formation of the FLIP particle-in-cell method. *Journal of Computational Physics*, 1992, 103:1-15.

[145] D Sulsky, SJ Zhou, HL Schreyer. Application of a particle-in-cell method to solid mechanics. *Computer Physics Communications*, 1995, 87:236-252.

[146] SG Bardenhagen. Energy conservation error in the material point method for solid mechanics. *Journal of Computational Physics*, 2002, 180:383-403.

[147] T Hughes. *The Finite Element Method*. Prentice Hall, 1987.

[148] M Moseler, J Nordiek, H Haberland. Reduction of the reflected pressure wave in the molecular dynamics simulation of energetic particle-solid collisions. *Physical Review* B, 1997, 56:15439-15445.

[149] SG Bardenhagen, EM Kober. The generalized interpolation material point method. *Computer Modeling in Engineering and Science*, 2004, 5: 477-496.

[150] D Sulsky, JU Brackbill. Cluster: A numerical method for suspension flow. *Journal of Computational Physsics,* 1991, 96:339-368.

[151] FF Abraham, JQ Broughton, N Bernstein, E Kaxiras. Spanning the length scales in dynamics simulation. Computers in Physics, 1998, 12:538-546.

[152] M Doyama, Y Kogure. Embedded atom potentials in fcc and bcc metals. *Computational Materials Science*, 1999, 14:80-83.

[153] TJ Lenosky, B Sadigh, E Alonso et al. Highly optimized empirical potential model of silicon. *Modeling and Simulation in Materials Science and Engineering*, 2000, 8:825-841.

[154] AE DePriesto, H Metiu. Molecular dynamics simulations of energy flow at a solid surface: New methods using a small number of atoms. *Journal of Chemical Physics*, 1989, 90(2):1229-1236.

[155] GF Grom, DJ Lockwood, JP McCaffrey, HJ Labbe, PM Fauchet, B White, J Diener, D Kovalev, F Koch and L Tsybeskov. Ordering and self-organization in nanocrystalline silicon. *Nature*, 2000, 407 (6802):358-361.

[156] HS Nam, NM Hwang, BD Yu and JK Yoon. Formation of an icosahedral structure during the freezing of gold nanoclusters: Surface-induced mechanism. *Physical Review Letters,* 2002, 89 (27): art. no. 275502.

[157] D Moldovan, V Yamakov, D Wolf and S R Phillpot. Scaling behavior of grain-rotation-induced grain growth. *Physical Review Letters*, 2002, 89 (20): art. no. 206101.

[158] J Belak, DB Boercher and IF Stowers. Simulation of nanometer-scale deformation of metallic and ceramic surfaces. *MRS Bulletin*, 1993, 5:55-60.

[159] K Komvopoulos and W Yan. Molecular dynamics simulation of single and repeated indentation. *Journal of Applied Physics*., 1997, 82:4823-4830.

[160] P Walsh, RK Kalia, A Nakano and P Vashishta. Amorphization and anisotropic fracture dynamics during nanoindentation of silicon nitride: a multimillion atom molecular dynamics study. *Applied Physics Letters*, 2000, 77:4332-4334.

[161] A Gannepalli and SK Mallapragada. Molecular dynamics studies of plastic deformation during silicon nanoindentation. *Nanotechnology*, 2001, 12:250-257.

[162] D Christopher, R Smith and A Richter. Atomistic modelling of nanoindentation in iron and silver. *Nanotechnology,* 2001, 12:372-383.

[163] D Christopher, R Smith and A Richter. Nanoindentation of carbon materials. *Nuclear Instruments and Methods in Physics Research B*, 2001, 180:117-124.

[164] J Li, KJ Van Vliet, T Zhu, S Yip, S Suresh. Atomistic mechanisms governing elastic limit and incipient plasticity in crystals. *Nature,* 2002, 418:307-310.

[165] D Feichtinger, PM Derlet and H Van Swygenhoven. Atomistic simulation of spherical indentaions in nanocrystalline gold. *Phys. Rev*. B, 2003, 67:024113.

[166] SR Phillpot, J Wang, D Wolf and H Gleiter. Computer simulation of the structure and dynamics properties of grain boundaries in a nanocrystalline model material. *Materials Science and Engineering,* 1995, A204:76-82.

[167] H Van Swygenhoven and A. Caro. Molecular dynamics computer simulation of nanophase Ni: structure and mechanical properties. *NanoStructured Materials*, 1997, 9:669-672.

[168] H Van Swygenhoven and A. Caro. Plastic behavior of nanophase Ni: a molecular dynamics computer simulation. *Applies Physics Letters*, 1997, 71(12):1652-1654.

[169] H Van Swygenhoven and A. Caro. Plastic behavior of nanophase metals studied by molecular dynamics. *Physical Review* B, 1998, 58:11246-11251.

[170] V Yamakov, D Wolf, M Salazar, SR Phillpot and H Gleiter. Length-scale effects in the nucleation of extended dislocations in nanocrystalline Al by molecular dynamics simulation. *Acta Materialia*, 2001, 49:2713-2722.

[171] L Lu, SX Li and K Lu. An abnormal strain rate effect on tensile behavior in nanocrystalline copper. *Scripta Materialia*, 2001, 45:1163-1169.

[172] PM Derlet and H Van Swygenhoven Length scale effects in the simulation of deformation properties of nanocrystalline metals. *Scripta Materialia*, 2002, 47: 719–724.

[173] XL Ma, HT Wang, W Yang. *Dislocation-assisted grain Growth in nanocrystalline copper under large deformation*. Submitted to Nanotechnology, 2005.

[174] H Van Swygenhoven, D Farkas and A Caro. Grain boundary structures in polycrystalline metals at the nanoscale. *Physics Review* B, 2000, 62:831-838.

[175] A Hasnaoui, H Von Swygenhoven, PM Derlet. Cooperative process during plastic deformation in nanocrystalline fcc metals: A molecular dynamics simulation. *Physical Review* B, 2002, 66:184112-1-184112-8.

[176] AJ Haslam, D Moldovan, V Yamakov, D Wolf, SR Phillpot, H Gleiter. Stress-enhanced grain growth in a nanocrystalline material by molecular-dynamics simulation. *Acta Materialia*, 2003, 51:2097-2112.

[177] PM Derlet, A Hasnaoui, H Van Swygenhoven. Atomistic simulations as guidance to experiments. *Scripta Materialia*, 2003, 49:629-635.

[178] KS Kumar, H Van Swygenhoven, S Suresh. Mechanical behavior of nanocrystalline metals and alloys. *Acta Materialia*, 2003, 51:5743-5774.

[179] XL Ma, W Yang. Supersonic wave propagation in Cu under high speed cluster impact. *Nanotechnology*, 2004, 15:449-456.

[180] F Stillinger, T Weber. Computer simulation of local order in condensed phases of silicon. *Physical Review* B, 1985, 31:5262-5271.

[181] MS Daw, MI Baskes. Semiempirical, quantum mechanica caluculation of hydrogen embrittlement in metals. *Physical Review Letters*, 1983, 50:1285-1288.

[182] W Herrmann, JS Wilbeck. Review of hypervelocity penetration theories. *International Journal of Impact Engineering*, 1987, 5:307-322.

[183] T Aoki, J matsuo, Z Insepov, I Yamada. Molecular dynamics simulation of damage formation by cluster ion impact. *Nuclear Instruments and Methods in Physics Research* B, 1997, 121:49-52.

[184] Z Insepov, I Yamada. Surface processing with ionized cluster beams: computer simulation. Nuclear Instruments and Methods in Physics Research B, 1999, 153:199-208.

[185] J Nordiek, M Moseler, H Haberland. Energetic impact of Cu-clusters on Cu-surfaces. *Radiation Effects and Defects in Solids*, 1997, 142:27-38.

[186] XY Li, W Yang. Multiple time step molecular dynamics simulation for interaction between dislocations and grain boundaries. *Acta Mechanica Sinica*, 2005, 21:371-379.

[187] AP Sutton, RW *Balluffi. Interfaces in Crystalline Materials*. New York: Oxford University Press, 1995.

[188] A Kuronen, K Kaski, L Perondi, J Rintala. Atomistic modeling of interaction between dislocations and misfit interface. *Europhysics Letters*, 2001, 55:19-25.

[189] AN Stroh. The formation of cracks as a result of plastic flow I. *Proceedings of the Royal Society London A,* 1954, 223:404-414.

[190] AN Stroh. The formation of cracks as a result of plastic flow II. *Proceedings of the Royal Society London A,* 1955, 223:548-560.

[191] AH Cottrell. Theory of brittle fracture in steel and similar metals. *Transaction of the Metallurgical Society of the AIME*, 1958, 212:192-203.

[192] GP Cherpanov. Interface microcrack nucleation. *Journal of the Mechanics and Physics of Solids,* 1994, 42:665-680.

[193] SA Adelman, JD Doll. Generalized Langevin equation approach for atom/solid-surface scattering: General formulation for classical scattering off harmonic solids. *Journal of Chemical Physics,* 1976, 64(6):2375-2387.

[194] DJ Diestler, ME Riley. Tests of approximations for inelastic molecule-surface scattering dynamics. *Journal of Chemical Physics*, 1987, 86(9):4885-4894.

[195] W E, XT Li, EV Eijnden. Some recent progress in multiscale modeling. In: S Attinger, P Koumoutsakos eds, *Multiscale modeling and simulation*, Berlin: Springer-Verlag, 2000, 1-21.

[196] C Dellago, PG Bolhuis, FS Csajka, D Chandler. Transition path sampling and the calculation of rate constants. *Journal of Chemical Physics*, 1998, 108:1964-1977.

[197] C Dellago, PG Bolhuis, D Chandler. Efficient transition path sampling: Application to Lennard-Jones cluster rearrangements. *Journal of Chemical Physics*, 1998, 108:9236-9245.

[198] AF Voter, F Montalenti, TC Germann. Extending the time scale in atomistic simulation of materials. *Annual Review of Materials Research*, 2002, 32:321-346.

[199] AF Voter. A method for accelerating the molecular dynamics simulation of infrequent events. *Journal of Chemical Physics*, 1997, 106(11):4665—4677.

[200] AF Voter Parallel replica method for dynamics of infrequent events. *Physical Review* B, 1998, 57(22):13985-13988.

[201] MR Sørensen, AF Voter. Temperature-accelerated dynamics for simulation of infrequent events. *Journal of Chemical Physics*, 2002, 12(21):9599-9606.

In: Encyclopedia of Nanotechnology
Editor: Elwood D. Carlson

ISBN 978-1-60692-079-4

Chapter 25

CARBON NANOTUBES BASED ELECTROCHEMICAL SENSORS

Chengguo Hu and Shengshui Hu*
Department of Chemistry
Wuhan University, Wuhan 430072 China

ABBREVIATIONS

AA	ascorbic acid
AChE	acteylcholinesterase
ADH	alcohol-dehydrogenase
o-AP	o-aminophenol
APTES	3-aminopropyltriethoxy-silane
ARS	3,4-dihydroxy-9,10-dioxo-2- anthracenesulfonic acid, sodium salt
ASV	anodic stripping voltammetry
ATCh	acteylthiocholine
[bmim]PF_6	1-butyl-3-methylimidazolium hexafluorophosphate
bppg	basal-plane pyrolytic graphite
CCNNEs	ceramic-carbon nanotubes nanocomposite electrodes
CDs	cyclodextrins
CE	capillary electrophoresis
CFNE	carbon fiber nanoelectrode
CHIT	chitosan
ChOx	cholesterol oxidase
CNTPEs	carbon nanotube paste electrodes
CNTPMEs	carbon nanotube powder microelectrodes
CNTs	carbon nanotubes
CNTs-IE	carbon nanotubes-intercalated graphite electrode

* Department of Chemistry, Wuhan University, Wuhan 430072 China, +86-27-8721 8904 (O). +86-27-6875 4067 (Fax). sshu@whu.edu.cn

CoTMPyP	[tetrakis(N-methylpyridyl)porphyrinato] cobalt
CP	carbon paste
CPEs	carbon paste electrodes
Ct	catalase
CTAB	cetyltrimethylammonium bromide
CV	cyclic voltammetry
CVD	chemical vapour deposition
CySH	cysteine
Cyt c	cytochrome c
Da	dalton
DA	dopamine
DAB	diaminobenzidine
DCC	dicyclohexyl carbodiimide
DET	direct electron transfer
DHB	dihydroxybenzaldehyde
DHP	dihexadecyl hydrogen phosphate
DL	detection limit
DMF	N,N-dimethylformamide
dsDNA	double-strand DNA
ssDNA	single-strand DNA
DNR	daunomycin
DPV	differential pulse voltammetry
$E^{\circ\prime}$	formal potential
EB	ethidium bromide
ECD	electrochemical detection
EDC	1-ethyl-3-(3-dimethylaminopropyl)carbodiimide
EIS	electrochemical impedance spectroscopy
ΔE_p	redox peak separation
EP	epinephrine
ESEM	environmental scanning electron microscopy
ET e	lectron transfer
FAD	flavin adenine dnucleotide
Fc-SWNTs	ferrocenyl-derivatised SWNTs
FDH	D-fructose dehydrogenase
FePc	iron-phthalocyanine
FIA	flow injection analysis
GCE	glassy carbon electrode
GDH	glucose dehydrogenase
GDI	glutaraldehyde
GOx	glucose oxidase
GSH	glutathione
GSSG	glutathione disulfide
Hb	hemoglobin
HcySH	homocysteine
HOPG	highly oriented pyrolytic graphite
HPLC	high performance liquid chromatography

HRP	horseradish peroxidase
5-HT	serotonin
IR	infrared spectra
LBL	layer-by-layer
Mb	myoglobin
MB	methylene blue
MCMB	mesocarbon microbeads
M-DNA	metallized double-stranded DNA
MWNTs	multi-walled carbon nanotubes
MP-11	microperoxidase
NAD^+	β-nicotinamide adenine dinucleotide
NADH	reduced β-nicotinamide adenine dinucleotide
NE	norepinephrine
NEAs	nanoelectrode arrays
NMP	N-methyl-2-pyrrolidone
NSA	N-succinimidyl acrylate
OCNTs	ordered carbon nanotubes
$OMIMPF_6$	1-octyl-3-methyl imidazolium hexafluorophosphate
OP	organophosphorus
PABS	poly(aminobenzenesulfonic acid)
PA-FTIR	photoacoustic Fourier transform infrared
PANT	polyaniline
PB	Prussian blue
PBS	phosphate buffer solution
PDAB	poly(3,3′-diaminobenzidine)
PDDA	poly(diallyl dimethyl ammonium) chloride
PG	pyrolytic graphite
PMEs	powder microelectrodes
POx	putrescine oxidase
PPD	poly(o-phenylenediamine)
PPy	polypyrrole
PQCI	piezoelectric quartz crystal impedance
RTILs	room-temperature ionic liquids
RSH	mercapatoacetic acid
SAMs	self-assembled monolayers
SCE	saturated calomel electrode
SDS	sodium dodecyl sulphate
SEM	scanning electron microscopy
S/N	signal-to-noise ratio
SSB	Escherichia coli single-strand binding protein
sulfo-NHS	N-hydroxysulfo-succinimide
SWNTs	single-walled carbon nanotubes
TBO	Toluidine Blue O
TEM	transmission electron microscopy
TGA	thermogravimetric analysis
TM-AFM	tapping-mode atomic force microscopy

TPP	tetra-phenyl-porphyrin
UA	uric acid
Xa	xanthine
XOx	xanthine oxidase
XPS	X-ray photoelectron spectroscopy

1. INTRODUCTION

Carbon nanotubes (CNTs) have become the subject of intense researches in the last decades because of their unique properties and promising applications in any aspect of nanotechnology. As a consequence of their small dimensions and unique structural properties, CNTs display fascinating electronic and optical properties that are distinct from other carbonaceous materials and nanoparticles of other types. CNTs are currently being used in electronic and optoelectronic, biomedical, pharmaceutical, energy, catalytic, analytical and materials applications. Specifically, the properties of small dimensions, functional surfaces, good conductivity, excellent biocompatibility, modifiable sidewall and high reactivity make CNTs ideal candidates for constructing sensors with high performance. As an example, CNTs have been widely employed in constructing various electrochemical sensors. Compared with the conventional-scale materials and other types of nanomaterials, the special nanostructural properties make CNTs have some overwhelming advantages in fabricating electrochemical sensors, including:

i. the larger specific area that produces higher sensitivity;
ii. the tubular structure and the chemical stability allowing the fabrication of ultrasensitive sensors consisting of only one nanotube;
iii. the better biocompatibility that is suitable for constructing electrochemical biosensors, especially for facilitating the electron transfer of redox proteins and enzymes;
iv. the modifiable ends and the sidewalls providing a chance for fabricating multi-functioned electrochemical sensors via the construction of functional nanostructures;
v. the possibility of achieving miniaturization;
vi. the possibility of constructing ultrasensitive nano arrays.

In fact, since the first application of CNTs in electrochemistry by Britto et al. [1], numerous works have dealt with the construction of electrochemical sensors by CNTs and the sequential applications. Generally, the replacement of ordinary materials by CNTs can effectively improve the redox currents of both inorganic and organic species and reduce the redox overpotentials. The electron transfer and the direct electrochemistry of redox proteins at CNTs-based electrochemical sensors were also widely reported. The well-defined structure and the chemistry stability as well as the electrocatalytic activity towards many substances, CNTs are also extensively used as the carrier platforms for construction various electrochemical sensors. The promising applications of CNTs have been reviewed by several authors [2-6]. Instead of covering all aspects of CNTs applications in electrochemistry, this chapter is restricted to electrochemical sensors based on common electrochemical methods (e.g. voltammetry and amperommetry) and doses not cover other methods like potentiometry, impedance spectroscopy and piezoelectricity. It not only pays attention to the principle, fabrication and application of various CNTs-based electrochemical sensors but also attempts to focuses on the fundamental knowledge of the CNTs, such as electrochemical properties and electrocatalytic activity.

2. Electrochemical Properties of Carbon Nanotubes

The electrochemical qualities of carbon materials-based electrodes are significantly dependent on the surface properties as the creation of specific surface functional groups (especially oxygen-containing groups) can increase the rate of electron transfer (ET) considerably [7]. Based on their specific structures, two distinct surface regions exist in carbon nanotubes (CNTs): the sidewalls and the ends. Since carbon nanotubes can be seen as the graphene sheets rolled into tubes, the electrochemical properties of carbon nanotubes are comparable to the basal planes of pyrolytic graphite (bppg). For intact carbon nanotubes, the defect-free structure makes the whole tubes possess almost the same properties to that of bppg except that the cap regions may be more reactive due to the much higher curve strain than the sidewall. The opening of the ends by physical/chemical treatments on carbon nanotubes produces a variety of oxygen-containing groups, which possess the properties similar to the edge places of bppg [8].

Due to the simple and well-defined responses at carbon materials, the $Fe(CN)_6^{3-}$/ $Fe(CN)_6^{4-}$ couple has been widely used to characterize the surface properties of all kinds of carbon electrodes. Similarly, the electrochemical properties of carbon nanotubes-based electrodes are generally investigated by $Fe(CN)_6^{3-}/Fe(CN)_6^{4-}$ as the probe. Nugent et al. [9] have compared the electrochemical behaviors of aligned bundles of carbon nanotubes with other carbon electrodes with similar structures. The results indicated $Fe(CN)_6^{3-}$/ $Fe(CN)_6^{4-}$ showed an ideal redox peak separation (ΔE_p) of 59 mV at the aligned multi-walled carbon nanotubes (MWNTs). In comparison, they reported ΔE_p of more than 100 mV for basal planes of highly oriented pyrolytic graphite (HOPG) with electrochemical pretreatments and greater than 700 mV for basal planes without electrochemical pretreatments. Whereas, ΔE_p of about 70 mV was observed at edge planes of HOPG. In contrast to the ideal response reported by Nugent et al. [9], Li et al. [8] observed much larger ΔE_p at single-walled carbon nanotubes (SWNTs) papers (ΔE_p = 96 mV) and at aligned MWNTs with heat pretreatment to remove impurities like amorphous carbon and catalyst (ΔE_p = 230 mV). Particularly, for aligned MWNTs, the apparent electron transfer rate (as indicated by ΔE_p) was found to correlate to both the area of the exposed side-walls (with graphite basal-plane-like properties) and the density of graphite edge-plane-like defects, i.e. the electrochemistry of the probe was also influenced by the sidewalls of the nanotubes. Li et al. also attributed the small ΔE_p at the SWNTs paper to the promoted electron transfer by the oxygen-containing defects produced during the acid purification step for preparing the papers. Recently, the fundamental electrochemical properties of inherently hydrophobic MWNTs electrodes and acid-treated MWNTs were reported by Papakonstantinou et al. using AC impedance spectroscopy and cyclic voltammetry (CV) [10]. Slow electron transfer kinetics with ΔE_p larger than 200 mV and an electron transfer resistance of about 370 Ω were observed on pristine MWNTs, close to the results reported by Li et al. [8] at aligned MWNTs. However, acid treatment was found to significantly improve the electron transfer kinetics of $Fe(CN)_6^{3-}/Fe(CN)_6^{4-}$ redox couple, approaching almost reversible ET kinetics. They believed that acid treatment modified the charge carrier density at the surface by introducing surface states and serves to facilitate electron transfer. The better wetting properties of functionalized CNTs and the high local density of states as compared with untreated MWNTs were regarded to be responsible for favoring faster ET kinetics.

In a word, the intact CNTs sidewalls resemble the basal planes of pyrolytic graphite and can be regarded to be electrochemically inert to the substrates. The apparent improved electron transfer at the intact nanotube sidewalls are attributed to possessing the higher stains than the basal planes. As for the opened caps, the presence of defects and oxygen-containing functional groups makes them possess similar electrochemical properties to those of edge planes of pyrolytic graphite. The introduction of edge-like defect sites and oxygen-containing functional groups at both the caps and the sidewalls by chemical or physical treatments can significantly improve the electrochemical properties of CNTs by changing the electronic structures, the surface states and the wettability of the sidewall for the substrates, reflected by the apparently facilitated electron transfer kinetics.

The critical roles of defect sites and oxygen-containing groups on the electrochemical performances of CNTs-based electrodes have been proved by several fundamental research works [11-14]. Compton et al. [11] compared the oxidative behaviors of reduced β-nicotinamide adenine dinucleotide (NADH), epinephrine (EP) and norepinephrine (NE) that were widely used to verify the so-called 'electrocatalytical properties' of CNTs at CNTs and graphite powder-modified basal plane pyrolytic graphite electrodes. They found both nanotube and graphite powder-modified electrodes shown electrocatalytical activities towards these species, with enhanced currents and reduced peak-to-peak separations in comparison with naked basal plane pyrolytic graphite. Based on this, they recommended that the catalytic properties should be cautiously assigned to carbon nanotubes. Further, they explored the origin of catalytic properties of CNTs by comparing the voltammetric behaviors of ferricyanide and epinephrine at different modified pyrolytic graphite electrodes [12-13]. The results indicated that these species exhibited similar behaviors at C_{60} film-modified bppg electrode and at naked bppg electrode [13]. Whereas, the response at CNTs film-modified electrode was close to that at edge plane pyrolytic graphite electrode [12-13] (Figure 1).

They concluded that the basis of the electrocatalytic nature of MWNTs was suggested to reside in electron transfer from the edge-plane-like defect sites at the open ends of nanotubes, which structurally resembled the behavior of edge plane (as opposed to basal plane) graphite. Latter, Gooding et al. [14] demonstrated the critical role of oxygen-containing groups on the electrochemical properties of CNTs-based electrodes using randomly orientated nanotubes films by a simple casting method and vertically aligned nanotubes by a self-assembled method.

In both cases, the electrochemical properties of CNTs were favored with the increase of the cutting time by sonication in a 3:1 v/v mixture of concentrated sulfuric acid (98%) and concentrated nitric acid (70%) with 8 h. The X-ray photoelectron spectroscopic (XPS) data indicated the content of the oxygenated species increased with the cutting time. Thus, the favorable electrochemical properties of CNTs were unambiguously correlated with the oxygen-containing groups, consistent with the electrocatalytic activity at the edge-like defect sites on CNTs. In fact, a pair of redox couple located at around 0.0 V in neutral media was usually observed at acid-treated CNTs-based electrodes, which was attributed to the redox of the carboxyl groups by Li. et al. using XPS and infrared spectra (IR) [15]. In addition, the electrode process involved four electrons while the rate-determining step was a one-electron reduction.

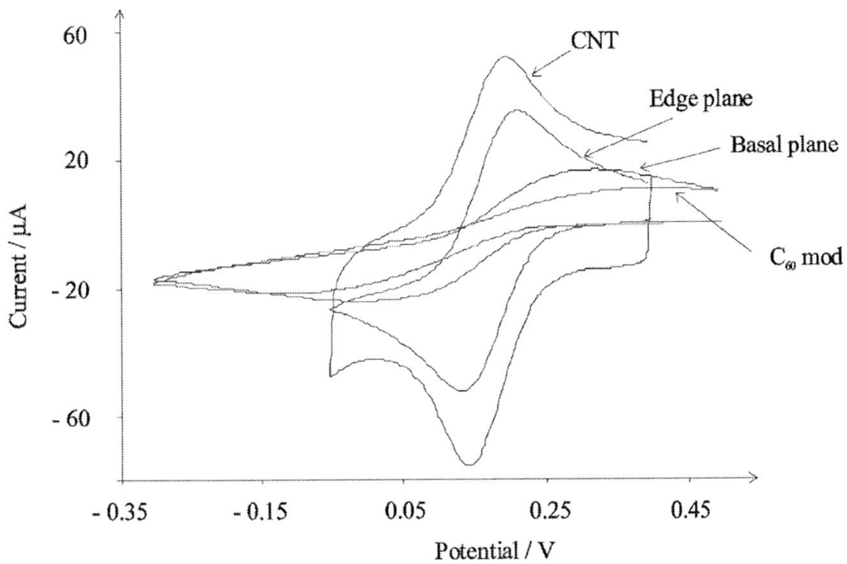

Figure 1. Cyclic voltammograms for the reduction of 1 mM ferricyanide for CNTs- and C_{60} film-modified bppg electrodes. Also shown is the response of a bare bppg electrode and an edge plane pyrolytic graphite electrode. All scans at a rate of 100 mV/s. [Reprinted with permission from C. E. Banks et al. [12], *Chem. Commun.*, 16, 1804 (2004). Copyright 2004, Royal Society of Chemistry].

As a result of the strong intertube interactions, CNTs generally existed as highly tangled ropes and were insoluble in all solvents, which greatly hindered their practical applications in electrochemical sensors. In addition, as-produced SWNTs, regardless of the production procedure, usually contain amorphous carbon, carbon nanoparticles, and residues from the metal catalysts. To overcome these limitations and to create defect-like sites and oxygen-containing groups on CNTs that favored the electrochemical properties of CNTs, as-received CNTs were usually treated by chemical methods using oxidative, such as refluxing in dilute nitric acid or refluxing/sonication in a concentrated H_2SO_4/HNO_3 mixture [16]. These treatments, by generating surface defects and sometimes resulting in tube shortening, can provide abundant carboxylated sites along the nanotube surface and shortened tube ends [16-18]. During these processes, nitric acid molecules or NOx residues can intercalate the SWNTs [19] and lead to changes in the Raman spectra of SWNTs by doping [20], which can be de-intercalated by thermal annealing treatments in vacuum or argon along with the removal of most defect-like sites on CNTs [21]. A typical example of the annealing treatment effect on the defect sites of CNTs was reported by Kuznetsova et al. using transmission IR spectra (Figure 2) [22].

It is clear that the oxygen-containing groups on the defect sites can be effectively eliminated with annealing treatment. Whereas, the defect sites cannot be removed by this method, i.e. the annealing treatment cannot restore the perfect structure of CNTs.

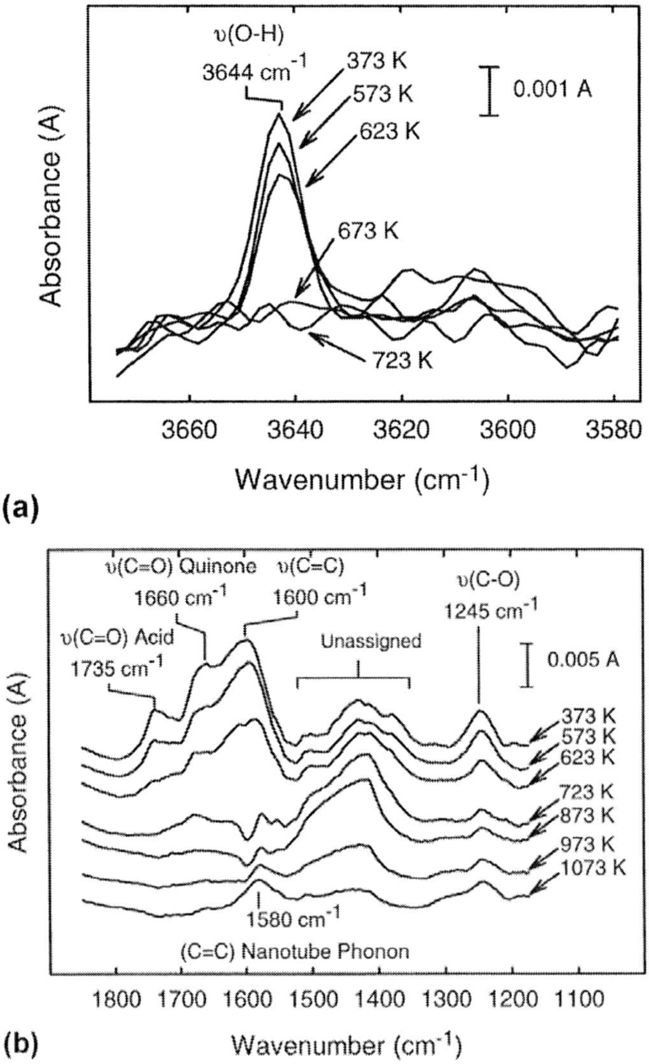

Figure 2. Decomposition of oxygen related functionalities on o-SWNTs. Transmission IR spectra of o-SWNTs following heating in vacuum to the indicated temperatures. The spectra indicate that carboxylic acid and quinone groups are thermally destroyed in vacuum above about 600 K. Other unassigned surface functionalities are also observed to disappear upon heating. Removal of these functionalities reveals the IR transition associated with a 1580 cm-1 phonon mode of nanotubes. [Adapted with permission from A. Kuznetsova et al. [22], Chem. Phys. Lett., 321, 292 (2000). Copyright 2000, Elsevier Science].

The annealing effects on the electrochemical properties of CNTs-based electrodes were investigated by Barisci et al. [23-24]. They found that thermal annealing produced significant changes in a range of properties of the material including increased hydrophobicity and elimination of electroactive surface functional groups. The activation of CNTs was also achieved by electrochemical methods. Wang et al. [25] has explored the electrochemical activation of MWNTs from different sources by anodization treatment. They found that while the anodic pretreatment resulted in a dramatic improvement in the electrochemical reactivity of the ARC-produced CNTs, CNTs produced by chemical vapour deposition (CVD) appeared to be resistant to the anodic activation. These differences in the effect of the electrochemical pretreatment were attributed to the anodic preanodization effectively 'breaking' the basal-plane end caps of ARC-CNTs thereby exposing edge plane defects, similar to those already present in the open-end caps of CVD-CNTs. The electrochemical activation of as-grown CNTs ensembles in 1.0 M NaOH was also achieved by Lin et al. [26] by applying a bias at 1.5 V for 90 s. After electrochemical treatment, some functional groups (e.g., carboxylic acid) were created at the CNTs ends, which were further used to immobilize proteins using standard water-soluble coupling agents 1-ethyl-3- (3-dimethylaminopropyl)carbodiimide (EDC) and N-hydroxysulfo-succinimide (sulfo-NHS) by forming amide linkages between their amine residues and carboxylic acid groups on the CNTs ends.

3. Fabrication of Carbon Nanotubes-Based Electrochemical Sensors

Since their first application in electrochemistry by Britto et al. [1], numerous papers dealing with the use of CNTs as electrode have been reported. The major problem on the promising applications of CNTs in electrochemical sensors is the immobilization of activated CNTs on the electrode surface because CNTs generally exist as highly tangled ropes and are insoluble in all solvents, which greatly hinder their capacity of forming uniform and stable films on the electrode surface. To overcome this deficiency, CNTs are first dispersed or dissolved in various solutions or suspensions and immobilized on the surfaces of various substrates by physical or chemical methods. This section focuses on the introduction of some typical immobilization of CNTs on electrode surfaces that are widely used in constructing CNTs-based electrochemical sensors.

3.1. Solvent Dispersion and Casting Immobilization

The most widely used methods for fabricating of CNTs-based electrochemical sensors are the approaches that involve in the dispersing of CNTs in a certain solvent with sonication after their purification and activation pretreatments, and the following immobilization by dropping the resultant suspensions on the electrode surfaces and allowing to dry (i.e. the casting methods). Among the used solvents, N,N-dimethylformamide (DMF) is the most extensively used polar solvent and more than half of the papers deal with CNTs-based electrochemical sensors using DMF as the dispersing solvent [15,27-40]. There are also some solvents used to prepare CNTs suspensions, like water [41-46], acetone [47-51], ethanol [52-54], and even toluene [55-56]. However, compared with the other solvents, DMF has some overwhelming advantages for dispersing CNTs, including much higher solubility, stability and exfoliation efficiency. The debundling and dissolution of SWNTs in DMF and N-methyl-2-pyrrolidone (NMP) has been systematacially investigated by Furtado et al. [57]. In the case of the improved debundling of CNTs, they suggested that the amide adsorption on the nanotube wall, stabilized by the donation of π-electrons from the aromatic rings to the N-based electronegative solvents, should be the dominant interaction that drives the tube debundling and that polar groups added on the nanotube walls, for example, COOH groups, might provide a second mechanism for the debundling process. As for the high solubility, a dipolar interaction between the solvent and the oxygen-containing groups on CNTs was believed to provide a significant benefit to the dissolution of CNTs in these solvents. Compared with other approaches, the solvent-dispersing methods of CNTs inevitably suffer from some disadvantages, such low solubility, low stability and low exfoliation efficiency, due to the rather weak interactions between these solvents and CNTs. The predominance of these methods is their simplicity, convenience and no need of other additives, which makes them suitable for using as the foregoing step in multi-step fabrications of complicated CNTs-based electrochemical sensors.

3.2. Additive-Assisted Dispersion and Immobilization

To improve the solubility and stability of CNTs in their suspensions, various additives are added to solvents to assist the dispersion of CNTs, such as surfactants and water-soluble polymers. The structures of these species are shown in Figure 3.

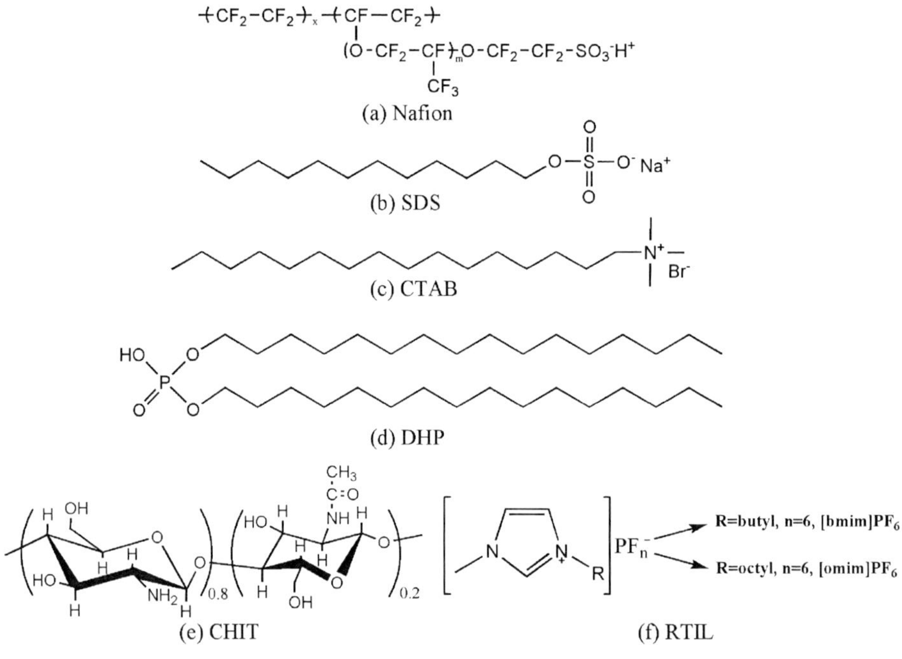

Figure 3. Molecular structures of dispersing reagents of CNTs.

3.2.1. Nafion

As a typical cationic ion exchange resin, Nafion films have been used extensively for the modification of electrode surfaces and for the construction of amperometric biosensors due to their unique ion-exchange, discriminative, and biocompatibility properties [58-59]. It is clear from Figure 3a that Nafion contains two different regions: the hydrophobic polymer backbone and the ionized hydrophilic sulfonate groups outside the hydrophobic region. This special amphiphilic structure render Nafion bear the capacity of combining with CNTs by hydrophobic interactions between the hydrophobic backbone of Nafion and the sidewall of

CNTs and dispersing them in solutions by the hydrophilic groups. The dispersion of CNTs in Nafion solutions with sonication was firstly reported by Wu et al. and used to the selective determination of dopamine (DA) via the elimination of the interferences from negatively charged uric acid (UA) and ascorbic acid (AA) by the Nafion film [60]. Later, Wang et al. systematacially studied the dispersion of CNTs in Nafion solutions and used this system to construct glucose electrochemical biosensors based on the electrocatalytic action towards hydrogen peroxide (H_2O_2) [61]. Since then, many works on construction of CNTs-based electrochemical sensors were reported based on this method, including the direct determination of DA [62-63], nitrophenol [64-65], cadmium ion [66], carbohydrates [67] and homocysteine (HcySH) [68], the fixation of glucose oxidase (GO_x) [69], copper (Cu) [70] and platinum (Pt) nanoparticles [71] for constructing glucose electrochemical sensors, and the electrodeposition of nickel (Ni) for analysis of amino acids in flow injection analysis (FIA) [72].

3.2.2. Surfactants

3.2.2.1. Water-Soluble Surfactants

Since the report by Abatemarco [73], sodium dodecyl sulphate (SDS) (Figure 3b) has been widely used to prepare stable suspension of purified SWNTs in water with the aid of sonication. The mechanism for the dissolution of SWNTs in SDS aqueous solutions has been well investigated. The common agreement is that the encapsulation of SWNTs in SDS micelle and the repulsive interactions between negatively charged SDS micelles account for the stable suspension of SWNTs in SDS aqueous solution. The reports on the using of SDS dispersed CNTs in electrochemical sensors are rare, due to the rather weak adsorption of SDS and CNTs at the smooth surfaces of conventional solid electrodes and the unstable films prepared from this suspension. However, Chen et al. prepared a SWNTs modified carbon fiber nanoelectrode (CFNE) for the sensitive determination of DA from the suspension SWNTs in SDS aqueous solution [74]. The achievement of this successful preparation might arise from the strong adsorption of SWNTs on the porous and hydrophilic surface of CFNE. The interaction between the oxygen-containing groups on acid treated SWNTs and burning pretreated CFNE in air might also contribute to the strong attachment of SWNTs on CFNE. The suspension of CNTs in cetyltrimethylammonium bromide (CTAB) (Figure 3c) aqueous solution was also used by Cai et al. to prepare CNTs-based hemoglobin (Hb) electrochemical biosensors [75]. The suspension of CNTs was mixed with the phosphate buffer solution (PBS) of Hb and fixed on the surface of a glassy carbon electrode (GCE) by a Nafion coating, resulting in the direct electrochemistry of Hb. The direct electrochemistry of GOx was also achieved at similar electrochemical biosensors and used for the determination of glucose [76].

3.2.2.2. Water-Insoluble Surfactants

Besides the Nafion dispersing system of CNTs, Wu et al. also developed another CNTs dispersing system on the basis of the stable dispersion of CNTs in the aqueous suspension of dihexadecyl hydrogen phosphate (DHP) (Figure 3d) [77]. Different from SDS and CTAB, DHP consists of two hydrophobic tails and a dissociable phosphate group. This special structure enables DHP to form stable and homogeneous suspensions in water, which can be cast on the electrode surfaces to prepare stable and uniform films. Based on the hydrophobic interactions between the hydrophobic tails of DHP and the sidewall of the nanotubes as well as the possible interactions between the phosphate groups on DHP and the oxygen-containing groups on acid-treated CNTs, MWNTs were dispersed in the aqueous suspension of DHP with sonication, resulting in the formation of a stable and homogeneous aqueous suspension of DHP and CNTs. The casting of the suspension of DHP and MWNTs on the electrode surface produced a electrochemical CNTs-based sensors that showed excellent electrochemical activity towards many species, such as hormone [78-80], drugs [81-91], biomolecules [77,92-97], and heavy metal ions [98-99].

3.2.3. Chitosan

Chitosan (CHIT) is a polysaccharide biopolymer (Figure 3e), which displays excellent film-forming ability, high water permeability, good adhesion, and susceptibility to chemical modifications due to the presence of reactive amino and hydroxyl functional groups. Jiang et al. firstly reported the stable dispersion of CNTs in the acidic aqueous solutions of CHIT with sonication and used to the simultaneous determination of DA and UA [100]. The special interaction between CHIT and CNTs was characterized by Zhang et al. [101]. They demonstrated by thermogravimetric analysis (TGA) that CHIT might adsorbed onto CNTs and form a special CHIT-CNTs system, which can be precipitated from the solutions by the addition of concentrated salts or the adjustment of the solution acidity. The selective interaction between CHIT and CNTs also provided a possible approach for separating of CNTs from carbonaceous impurities. Based on the derivation of the reactive groups on CHIT, they also developed a CNTs-based glucose electrochemical biosensor, which might be applied to a large group of dehydrogenase enzymes for designing of a variety of bioelectrochemical devices (e.g., sensors, biosensors, biofuel cells). The solubilization of CNTs in aqueous solutions with polysaccharides has also been reported previously by Star et al. using starch as the dispersing agent [102]. They found that that common starch, provided it is activated toward complexation by wrapping itself helically around small molecules, would transport SWNTs competitively into aqueous solutions, and that the process was sufficiently reversible at high temperatures to permit the separation of SWNTs in their supramolecular starch-wrapped form by a series of physical manipulations from amorphous carbon. Based on their previous work [100], Jiang et al. applied the CHIT-CNTs system for the direct determination of nitrite [36], the simultaneous detection of UA and NE [103], and the selective determination of DA in the presence of AA [104]. Another marvelous work on the simultaneous electrodeposition of CHIT-CNTs on gold electrodes was reported by Chen et al. [105]. The typical electrodeposition procedure was expressed as follows: a pair of polished and cleaned gold electrodes (diameter 2.0 mm, separation of about 0.5 cm) was connected to a direct current power supply (3.0 V) and dipped into the CNTs CHIT solution (pH 5.0); H^+ in the solution was reduced to H_2 at the cathode, and the pH near the cathode surface gradually increased; as the solubility of CHIT was pH-dependent, when the pH exceeded the pKa of CHIT (about 6.3), CHIT became insoluble and the CHIT entrapped CNTs would deposit onto the cathode surface as a result. Based on this, they developed a glucose electrochemical biosensor through the codeposition of GOx with CHIT-CNTs. This simple and controllable electrodeposition method overcomes the major obstacle for preparing CNTs-based biosensor systems and expands the scope of electrochemical devices based on CNTs. However, this method might also suffer from some inevitable disadvantages, e.g. the using of high potential bias (3.0 V) might influence the quality of the substrate electrodes and the bubbling of H_2 produced during the electrolysis process might influence the uniformity of the deposited films.

3.2.4. Sel-Gel

The sol-gel and hydrogel have been widely used in recent years to immobilize biomolecules (e.g. enzymes) for the fabrication of the electrochemical biosensors because of their easy fabrication, chemical inertness, thermal stability and good biocompatibility. However, the lack of electrochemical reactivity and the poor conductivity of these materials, especially for silica sol-gel, greatly hinder their promising applications. A typical procedure for preparing CNTs-based sol-gel consists of the dispersion of CNTs in solvents, the mixing of the CNTs suspensions with the sols and the casting of the resultant mixture the electrode surfaces. Recently, Gavalas et al. prepared CNTs aqueous sol-gel composites, which were used as enzyme-friendly platforms for the development of stable biosensors [106] (Figure 4).

These CNTs-based sol-gel electrochemical biosensing platforms were demonstrated to possess the both the electrochemical characteristics of CNTs and the role of sol-gel for eliminating by-products. Tan et al. also reported a new type of amperometric cholesterol biosensor based on sol-gel CHIT/silica and MWNTs organic-inorganic hybrid composite materials [107]. They found that the MWNTs acted as both nanometer conducting wire and catalyst, which can effectively promote electron transfer between enzyme and surface of the electrode. Gong et al. also developed a new kind of ceramic-carbon nanotubes nanocomposite electrodes (CCNNEs) prepared by doping MWNTs into silicate gel [108]. In contrast to the conventional sol-gel or CNTs-based electrochemical sensors, the electrochemical response of these electrodes can be conveniently tuned from that conventional-scale electrodes to that of microelectrodes by just varying the content of MWNTs in the composites. Besides silica-based sol-gel, the preparation of CNTs-redox polymer hydrogel composite films was also reported recently by Joshi et al. [109]. By incorporation of enzyme modified SWNTs into PVP-Os polymer hydrogel, the sensor's current output was increased for 2-3 times.

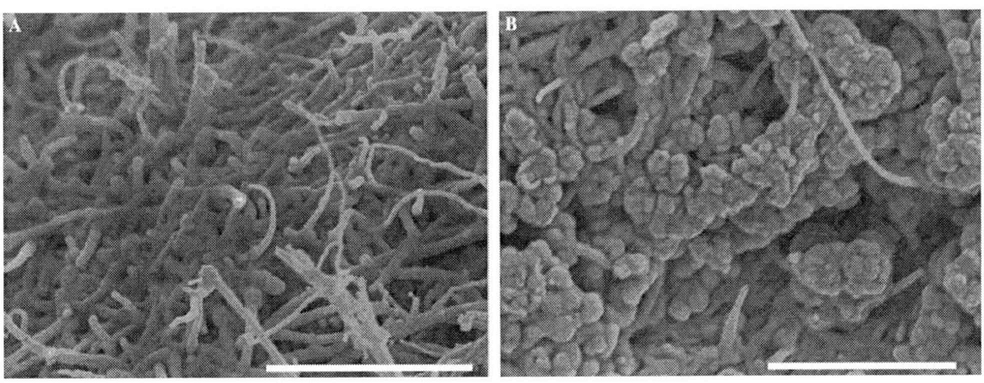

Figure 4. SEM pictures of the surface of the CNTs sol-gel composite (A) without and (B) with 0.25 mg colloidal silica particles per mg of MWNTs. Bar, 5 μm. [Reprinted with permission from V. G. Gavalas et al. [106], *Anal. Biochem.*, 329, 247 (2004). Copyright 2004, Elsevier Science].

3.2.5. Room-Temperature Ionic Liquids

Room-temperature ionic liquids (RTILs) are thermally stable salts, liquid at room temperature, constituted by an organic cation and either an organic or an inorganic anion. Unlike traditional inorganic molten salts such as NaCl, $NaAlF_6$ or the eutectic mixture LiCl-KCl, room temperature ionic liquids present a high degree of asymmetry that inhibits crystallisation at room temperatures. RTILs have some marvelous properties, including nonvolatile nature or non-measurable vapour pressure, low melt point (even as low as -90 °C), wide liquidus range (up to 200 °C), strong electrostatic field, wide electrochemical potential range (i.e. RTILs remain stable at potentials even higher than 5 V), relative air and water stability, high polarity, favorable viscosity and density as solvents, high thermal stability (i.e. decomposition temperature might be higher than 400 °C), and the ability of solvating a wide range of species including organic, inorganic, and organometallic compounds by varying the R group in RTILs. These special properties make them have promising applications in green chemistry (including synthesis, catalysis and biocatalysis), functional materials (e.g. photoelectrical materials and lubricate materials), energy science, environment science, and so on.

Recently, Zhao et al. [110] proposed a novel strategy for investigating the electrical-ionic properties of RTILs and carbon composite materials formed by mixing a water-insoluble RTILs (1-butyl-3-methylimidazolium hexafluoro- phosphate, $[bmim]PF_6$) (Figure 3f) and carbon materials of two types: one was MWNTs with a tube shape and the other was mesocarbon microbeads (MCMB) with a bead shape (Figure 5). The hybrid MWNTs/RTILs and MCMB/RTILs materials showed a different conductivity mechanism determined by ac impedance technology. The RTILs and carbon composite materials can also be used to act as modifiers in the direct electrochemistry of protein and catalyze the reduction of O_2 and H_2O_2. Later, Zhang et al. [111] developed a RTILs supported three-dimensional network SWNTs electrode. In that work, large quantities of SWNTs were considerably untangled in RTILs so as to greatly increase the effective area of the electrode. N-succinimidyl acrylate (NSA), as a model monomer, was dissolved in the supporting RTILs and was electrografted onto SWNTs (SWNTs-poly-NSA). Then, GOx was directly covalently anchored on the SWNTs-poly-NSA assembly.

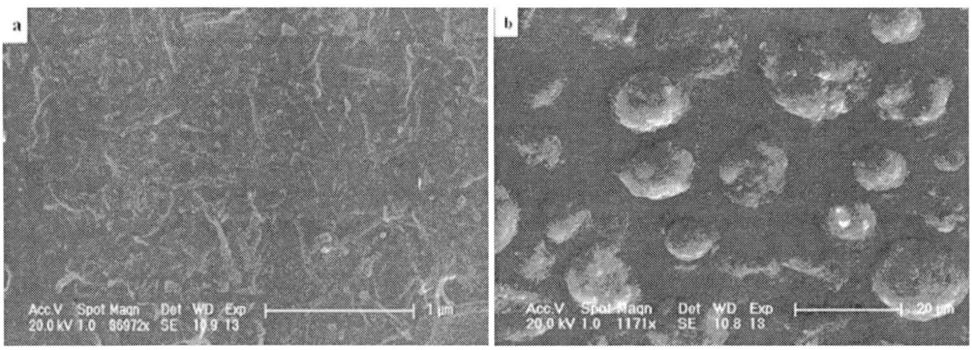

Figure 5. Environmental scanning electron microscopic (ESEM) images of (a) MWNTs/[bmim]PF6 and (b) MCMB/[bmim]PF6, The scale bar is 1μm in (a) and 20μm in (b). The accelerating voltage is 20 kV. [Adapted with permission from F. Zhao et al. [110], *Anal. Chem.*, 76, 4960 (2004). Copyright 2004, American Chemical Society].

More recently, Zhao et al. [112] prepared a RTILs-CNTs modified GCE by grinding the mixture of a RTILs (1-octyl-3-methyl imidazolium hexafluorophosphate, $OMIMPF_6$) (Fig. 3f) and MWNTs on certain weight ratio and casting the resulting black gel on the electrode surface, which was successfully used to the selective determination of DA in the presence of UA and AA.

3.2.6. Cyclodextrins

Cyclodextrins (CDs) are crystalline, water soluble, cyclic oligosaccharides built up of glucopyranose units (glucose units) and contain a relatively hydrophobic central cavity and hydrophilic outer surface. The soft cutting of SWNTs by CDs was firstly reported by Chen et al. [113] and then characterized by Chambers et al. [114] using adsorption and Raman spectra as well as differential scanning calorimetery (DSC). The results confirmed conclusively that γ-CDs and SWNTs did interact with each other in a fashion similar to that originally suggested by Chen et al. [113], who proposed that the CDs were absorbed at the surface of nanotube ropes by van der Waals forces. This result foretold the use of CDs for the chemical manipulation and processing of CNTs. Based on this, Luo et al. developed CDs-CNTs composite films modified electrodes by dispersing CNTs in the aqueous solutions of either α- [115-117] or β-CD [118-119] with sonication, which was used for the determination of NE [119], UA [118] and thymine [115], the selective determination of p-nitrophenol [116], the simultaneous determination of DA and EP [117].

3.3. Self-Assembling Immobilization

3.3.1. Polyelectrolyte Assisted Self-Assembly

The applications of CNTs in electrochemical sensors by self-assembled methods were mainly achieved through the attachment of water-soluble polyelectrolytes on the sidewall of CNTs. The molecular design of strong SWNTs/polyelectrolyte multilayer composites has been reported previously [120]. The typical polyelectrolyte used for preparing CNTs-based electrochemical sensors by self-assembled methods was a cationic polyelectrolyte, poly(diallyl dimethyl ammonium) chloride (PDDA). The interaction between PDDA and CNTs has been systematacially investigated by Yang et al. [121] using XPS and photoacoustic Fourier transform infrared (PA-FTIR) spectroscopies. They found that the mild sonication of MWNTs in aqueous PDDA resulted in a significant improvement of CNTs dispersibility and greatly enhanced their adhesion to Au and Si substrates.

The MWNTs-PDDA interaction was due to the presence of an unsaturated contaminant in the PDDA chain, as confirmed by both XPS and PA-FTIR, which entered into a π-π interaction with the CNTs. Electrostatic group repulsions of the coated CNTs then provided the dispersibility and adhesion. Guo et al. firstly reported the electrostatic assembly of calf thymus DNA on MWNTs-modified gold electrode via PDDA and used this electrode to study the interaction between chlorpromazine chloride and DNA [42]. Soon, they also prepared a CNTs-based amperometric cholesterol biosensor through layer-by-layer (LBL) deposition of PDDA and cholesterol oxidase (ChOx) on MWNTs-modified gold electrode, followed by electrochemical generation of a nonconducting poly(o-phenylenediamine) (PPD) film as the protective coating [122]. Zhang et al. fabricated multilayer films of shortened MWNTs that were homogeneously and stably assembled on GCE with the LBL method based on electrostatic interaction of positively charged PDDA and negatively charged and shortened MWNTs (Figure 6), which were further applied to the electrocatalytic reduction of O_2 [123] and the selective determination of DA [124].

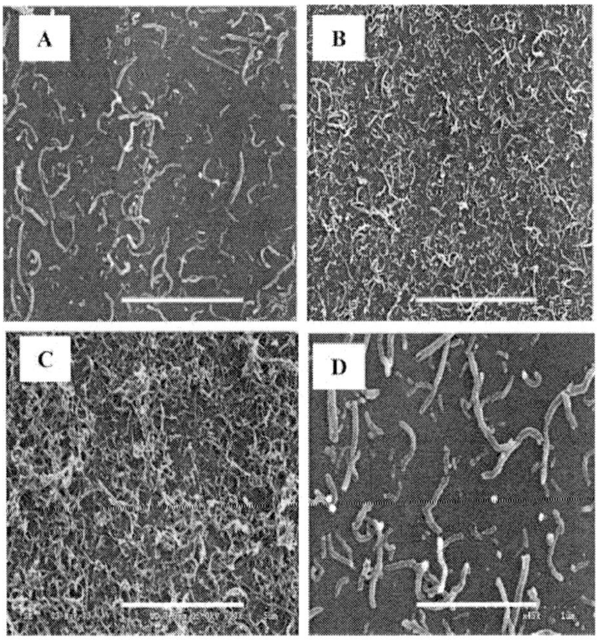

Figure 6. Representative SEM images of {PDDA/MWNTs}1 (A), {PDDA/MWNTs}3 (B), and {PDDA/MWNTs}5 (C) assembled on a silicon wafer. The scale bar in A-C was 5 μm. (D) represents the SEM image of {PDDA/MWNTs}1 with a high amplification with a scale bar of 1 μm. [Reprinted with permission from M. Zhang et al. [123], Langmuir, 20, 8781 (2004). Copyright 2004, American Chemical Society].

Recently, He et al. introduced a new fabrication of DNA-CNTs particles using the LBL technique on SWNTs by the alternative deposition of PDDA and DNA on the water-soluble oxidized SWNTs [125]. The electrodes modified by the DNA/PDDA/SWNTs particles were used as DNA sensors for the investigation of DNA damage by nitric oxide. The self-assembly of negatively charged poly(aminobenzenesulfonic acid) (PABS) on SWNTs was also proposed by Liu et al. [126], which was used for the assembly of a conducting polymer, polyaniline (PANT), on the surface of SWNTs. The obtained PANI/PABS-SWNTs multilayer films were very stable and showed high electrocatalytic ability toward the oxidation of NADH at a much lower potential (about +50 mV vs. Ag|AgCl). Lenihan et al. proposed a protocol for immobilizing alkaline phosphatase on the surface of MWNTs utilizing a LBL methodology. They first incubated CNTs with streptavidin to form a protein layer on the surface of the nanotubes and then attached biotinylated alkaline phosphatase to streptavidin, anchoring the sensing protein onto the surface of CNTs, which was used to the determination of 1-naphthyl phosphate. Besides macromolecules, small conjugated molecules were also used for the construction of CNTs-based electrochemical sensors. Yan et al. described and characterized the adsorption of electroactive methylene blue (MB) dye onto SWNTs to form an electrochemically functional nanostructure [127]. They found that MB essentially interacted with SWNTs through charge-transfer and hydrophobic interactions, leading to the formation of a MB-SWNTs adsorptive nanostructure, which exhibited distinct electrochemical properties from those of MB adsorbed onto a GCE. The stable adsorption of water-soluble and positively charged MB molecules onto the SWNTs was further demonstrated to be able to solubilize the formed nanostructure in water quite well and to fabricate a functional nanocomposite by LBL assembling of the formed narrostructure on a solid substrate.

3.3.2. Direct Self-Assembly

The negatively charged carboxyl groups on CNTs produced during the acid treatments provides two approaches for direct self-assembling immobilization of CNTs on the electrode surfaces: one is the covalent bonding of CNTs to the self-assembled monolayers (SAMs) of cysteamine modified gold electrodes via the reaction of carboxyl groups on CNTs and amino groups on cysteamine SAMs in the presence of coupling reagents, and the other is the attachment of CNTs to the electrode surfaces via the electrostatic interactions between negatively charged carboxyl groups on CNTs and the positively charged species on the electrode surfaces. The self-assembly of CNTs on a gold electrode was firstly reported by Liu et al. [128].They first cut as-grown nanotubes into short pipes and thiol-derivatized at the open ends by chemical methods.

The ordered assembly of SWNTs was then made by their spontaneous chemical adsorption to gold via Au-S bonds. They found that the nanotubes were organized on gold, forming a self-assembled monolayer structure with a perpendicular orientation, and the adsorption kinetics of the nanotubes was very slow in comparison to conventional alkanethiols. Different from the work by Liu et al. [128], Gooding et al. fabricated an aligned CNTs array for the direct electrochemistry of microperoxidase MP-11 via a three-step procedure [129] (Figure 7). First, a polycrystalline gold electrode was cleaned electrochemically in 0.05 M sulfuric acid and placed in an aqueous ethanol solution containing 1 mM cysteamine for 5 h to give a cysteamine self-assembled monolayer (SAM). Next, 0.2 mg of shortened SWNTs was dispersed in 1 ml DMF with 0.5 mg of dicyclohexyl carbodiimide (DCC) to convert the carboxyl groups at the ends of the shortened SWNTs into active carbodiimide esters. Finally, the cysteamine-modified gold electrode was placed in the nanotube solution for 4 h, while the amines at the terminus of the SAM formed amide bonds with one end of the tubes. The further immobilization of MP-11 on the other end of the aligned shortened SWNTs arrays on the electrode surface achieved the direct electrochemistry of MP-11. Later, they applied this method to GOx and also accomplished the direct of the direct electron transfer between the electrode and the redox active centre of GOx, flavin adenine dnucleotide (FAD) [130].

Recently, Xu et al. fabricated a MWNTs monolayer modified gold electrode by a similar method and also achieved the direct electrochemistry of horseradish peroxidase (HRP) that was physically adsorbed on the surface of MWNTs [131]. In contrast to the covalent methods for attaching CNTs to cysteamine modified gold electrodes, Qu et al. [132] attached acid-treated SWNTs to cysteamine modified gold electrodes via electrostatic adsorption, i.e. the cysteamine modified gold electrode was treated with 0.1 M HCl aqueous solution for 5 min to bear positive charge and the resulting electrode was placed into 0.5 mg/ml SWNTs aqueous solution for 2 h to attach SWNTs on the electrode surface by electrostatic adsorption. Distinguished from the cysteamine method, Chattopadhyay et al. also developed another method for preparing aligned CNTs arrays on the surface of ordinary pyrolytic graphite (PG) electrodes from randomly dispersed CNTs [133].

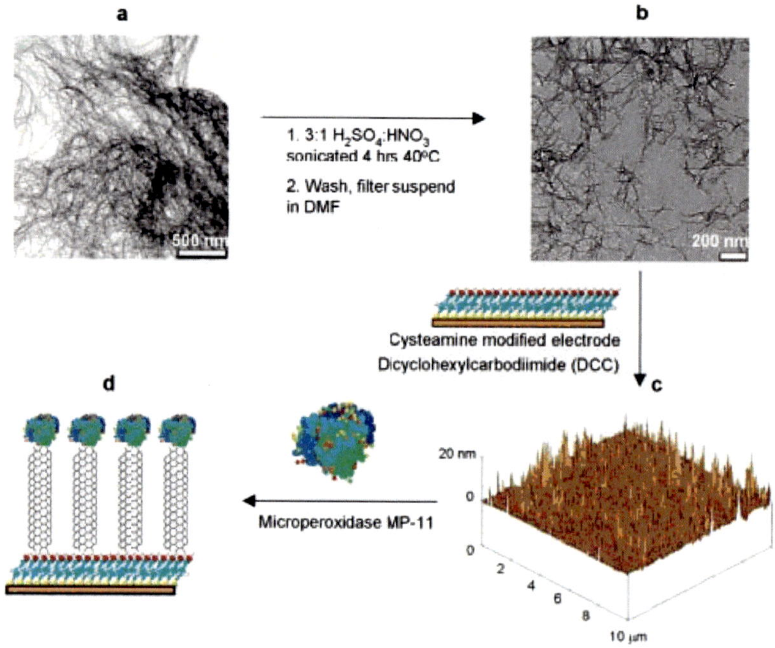

Figure 7. A schematic showing the steps involved in the fabrication of aligned shortened SWNTs arrays for direct electron transfer with enzymes such as microperoxidase MP-11. [Reprinted with permission from J. J. Gooding et al. [129], J. Am. Chem. Soc., 125, 9006 (2003). Copyright 2003, American Chemical Society].

A typical procedure of this method was described as follows: SWNTs forests were constructed on basal plane pyrolytic graphite disk electrodes that were first abraded on 400 grit SiC paper, then ultra-sonicated in water for 1 min. The SWNTs were carboxyl functionalized and shortened by sonication in a mixture of HNO_3 and H_2SO_4 for 4 hours at 70 °C. The electrodes were dipped in 1 mg/ml Nafion solution in 9:1 (v/v) methanol- water for 15 min. This was quickly washed with water and then placed in 5 mg/ml of $FeCl_3 \cdot 6H_2O$ for 15 min. After washing with water the electrodes were then placed in a freshly sonicated DMF suspension of SWNTs for 30 min, washed with methanol and dried in vacuo for 18 h. Based on the successful fabrication of the CNTs arrays on the electrode surfaces, proteins like HRP were either covalently attached [134] or physically adsorbed [135] onto the end of the CNTs arrays and achieved their electroanalytical applications.

3.4. Carbon Nanotubes Paste Electrodes

CNTs paste electrodes (CNTPEs) refer to the paste electrodes made up of liquid binder and CNTs. CNTPEs retain the properties of the classical carbon paste electrodes (CPEs) such as the feasibility to incorporate different substances, the low background currents, the wide potential range, the easy renewal and composite nature. CNTPEs also preserve the unique electrochemical properties of CNTs, like the ability to promote electron-transfer reactions of various species, especially for biomolecules. Compared with other methods for preparing CNTs-based electrochemical sensors, the CNTPEs don't need additional CNTs treatments, e.g. purification, modification or shortening treatment, and is suitable for all kinds of CNTs. Britto et al. [136] firstly reported a dramatic improvement in the electrochemical behavior of DA with ΔE_p of 30 mV at nonactivated carbon nanotubes electrodes constructed by using bromoform as binder. Davis et al. [137] further explored the possibility of using CNTs as the electrode materials. They achieved for the first time the protein direct electrochemistry at a CNTPE that was prepared from protein decorated CNTs by physical adsorption. The formal appellation 'CNTs paste' or 'CNTP' electrode was firstly reported by Valentini et al. in 2003 [138]. In that work, they firstly prepared CNTPEs by mixing SWNTs with mineral oil. Studies on the composition of the SWNTs paste electrodes showed a different behavior, as compared to graphite paste electrodes (CPEs). In the case of CPEs, the treatment of graphite powder with HNO_3 oxidation hardly changed the electrochemical properties of CPEs. As for CNTPEs, the HNO_3 oxidation treatment significantly improved the electrochemical activity of CNTPEs towards many electroactive substances, especially for small biomolecules. Later, Valentini et al. [139] further explored the various advantages of CNTPEs over conventional carbon paste (CP), Pt and glassy carbon electrodes, such as a very low capacitance (background current) and fast electron transfer rates towards various redox couples, especially two-electron quinonic structure species. Since then, more and more works focuses on the preparation and the applications of CNTPEs in electrochemical sensors, including the determination of amitrole [140], HcySH [141] and DNA [142] at CNTPEs, the determination of carbohydrates at a copper particles incorporated CNTPE [143], the determination of lactate, phenols, catechols, alcohols or glucose at enzyme incorporated CNTPEs [144-145], and the electrocatalytic oxidation of NADH at a 3,4-dihydroxybenzaldehyde (3,4-DHB) modified CNTPE [146-147]. The using of solid binders in preparing CNTPEs was also reported. Wang et al. prepared CNTs/Teflon composite electrodes by the dispersion of CNTs within a Teflon binder [148]. The resulting CNTs/Teflon material brought new capabilities for electrochemical devices by combining the advantages of CNTs and 'bulk' composite electrodes, including the accelerated electron transfer, the minimization of surface fouling and surface renewability. The electrocatalytic properties of CNTs were not impaired by their association with the Teflon binder. Later, Chen et al. prepared a CNTs-epoxy composite microdisc electrode by mixing CNTs powder with epoxy resin/hardener and packing the resulting CNTs-epoxy composite into the opening end of capillary [149].

3.5. Abrasive Immobilization

3.5.1. Carbon Nanotubes Powder Microelectrodes

Powder microelectrodes (PMEs) have been proved an effective technology for the purpose. PMEs, which are originated in Liu's laboratory, are a convenient technique for both characterization of powder materials and electroanalytical applications [150]. In fact, the so-called PMEs is a combination of porous and microelectrode. Besides, the PMEs have the properties of the thin layer cell to some extent. No binder is need for preparing a PME; this not only makes the preparation simple but also can prevent impurities and keep the powder material in its pristine state [150]. Liu et al. [151] firstly reported the immobilization of CNTs on PMEs by filling CNTs in the microcaves of PMEs through a simple rubbing operation. The resultant CNTs-based PMEs (CNTPMEs) were further modified by $Os(bpy)_3^{2+}$ and used for the electrocatalytic determination of nitrite. Then, CNTPMEs were used for the determination of hydrazine [152] and cysteine (CySH) [153], the direct electrochemistry of HRP [154] by Zhao et al. and for the studies on electrochemical properties and scavenge of superoxide anion in aprotic media by Wei et al. [155].

3.5.2. Carbon Nanotubes Abrasively Modified Conventional-Scale Electrodes

The immobilization of CNTs on the surface of conventional-scale electrodes was firstly reported by Luo et al. [156]. In that work, a CNTs-intercalated graphite electrode (CNTs-IE) was fabricated by grinding with a suitable amount of CNTs powder. A thin CNTs layer was formed on the graphite surface by intercalating the CNTs into the soft graphite layers with the aid of mechanical force and also the action of chemical and physical adsorption. This CNTs-IE can effectively separate the oxidation potential of DA and AA, which was used for the simultaneous determination of DA and AA [156] and the selective determination of DA and serotonin (5-HT) in the presence of AA [157]. The CNTs-IE was further employed by Shi et al. to prepare an amperometric cholesterol biosensor by immobilizing of ChOx in sol-gel on a Pt-decorated CNTs-IE [158] and a Pt-decorated CNTs intercalated a waxed graphite electrode [159].

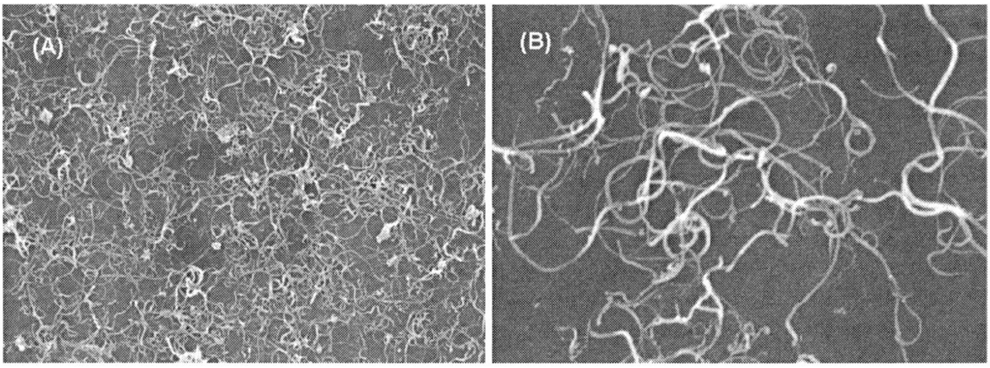

Figure 8. SEM images of CNTs used to fabricate the CNTs modified bppg electrode, magnification (A) 10000 and (B) 33000. [Reprinted with permission from A. Salimi et al. [160], *Analyst*, 129, 225 (2004). Copyright 2004, Royal Society of Chemistry].

Salimi et al. also developed a CNTs intercalated bppg electrode by gently rubbing the electrode surface of bppg on fine qualitative filter paper containing CNTs for 1 min, which was used for the determination of EP [160] (Figure 8). Based on this electrode, they also prepared a glucose electrochemical biosensor by coating the CNTs intercalated bppg electrode with a sol-gel composite film containing GOx [161]. Later, Salimi et al. produced a CNTs intercalated GCE by immobilizing MWNTs on a GCE that was preheated for 5 min at 50 °C through gentle rubbing of the electrode surface on a filter paper supporting CNTs, which was applied to the determination of morphine [162], and thiols like thiocytosine, L-CySH and glutathione (GSH)[163].

3.6. Electropolymerization Immobilization

The polymerization of various monomers in the presence of dispersed CNTs in solutions by electrochemical methods has been employed for the immobilization of CNTs on the electrode surface. In this process, CNTs were enwrapped in the polymers during the electropolymerization process in the form of counter ions or dopants. Wu et al. reported a marvelous method for the dispersion and the immobilization of CNTs on the electrode surface by water-soluble alizarin red S (ARS, 3,4-dihydroxy-9,10-dioxo-2-anthracenesulfonic acid, sodium salt) [164]. In that work, the open-ended and cut MWNTs (10 mg) were added into 50 ml of 2 mg/ml aqueous ARS solution. The resulting mixture solution was stirred for 12 h with a magneting stirring bar, and then left undisturbed for 6 h. A homogeneous, well-distributed solution of the MWNTs-ARS complex was then obtained. The formation of the complex was qualitatively indicated by the color change of the solution from orange to black.

This solution was quite stable as even after three months the MWNTs did not precipitate from the ARS solution. The MWNTs-ARS film can be electrochemically deposited on the surface of a GCE by 40 successive cyclic potential sweeps at a scan rate of 100 mV/s between 0.0 and 2.2 V (vs. saturated calomel electrode (SCE)) in the MWNTs-ARS solution. Pan et al. prepared a CNTs-based glucose electrochemical biosensor by the simultaneous immobilization of CNTs and GOx on the gold electrode through the electropolymerization of o-aminophenol (o-AP) in 0.2 M acetate buffer solution (pH 5.0) containing GOx, o-AP monomer and fresh CNTs dispersion [165]. Cheng et al. fabricated a MWNTs/polypyrrole (PPy) modified electrode through cyclic sweeping from 0.0 to 0.7 V at a GCE with the scan rate 0.1 V/s for eight circles in the solution with 0.1 M pyrrole, 0.2 mg/ml MWNTs and 0.2 M KCl in phosphate buffer (pH 2.0) [166]. It is clear that in these two works, CNTs were used as the dopants in the polymers. Wang et al. also reported a one-step preparation route of amperometric enzyme electrodes based on incorporating a CNTs dopant and the biocatalyst (e.g. GOx) within an electropolymerized polypyrrole film [167]. They found that the entrapment of the CNTs had little effect upon the electropolymerization rate and redox properties of the resulting film, and the CNTs dopant retained its electrocatalytic activity to impart high sensitivity and selectivity.

3.7. Direct Growth

Despite the randomly dispersed CNTs, well-aligned CNTs were also used for constructing electrochemical sensors. These CNTs-based sensors were generally constructed from CNTs ensembles directly grown from CVD methods. Based on their different density, aligned CNTs electrodes have the electrochemical properties resembling conventional-scale electrodes at high density or the microelectrode arrays at low density. This section will introduce the fabrication of aligned CNTs electrochemical sensors made up of these two kinds.

3.7.1. Dense Carbon Nanotubes Forests

3.7.1.1. Unmodified Carbon Nanotubes

The unmodified aligned CNTs electrodes were typically constructed by two steps: the CVD synthesis and the transfer of the aligned CNTs ensembles to various substrates. Sometimes, the prepared aligned CNTs electrodes were further activated by electrochemical methods. Ye et al. developed an aligned CNTs electrode for the selective determination of UA [168]. They found that compared to GCE, the CNTs electrode catalyzed oxidation of UA and L-AA, reducing the overpotentials by about 0.028 and 0.416 V, respectively, and resolved the overlapping voltammetric response of UA and L-AA into two well-defined voltammetric peaks in applying both CV and differential pulse voltammetry (DPV). Chen et al. developed a simple method for preparing ordered CNTs (OCNTs) in the nanopores of an alumina template membrane [169]. The catalyst was then removed by 0.1 M HCl solution and the alumina was dissolved away in 3 M NaOH solution to obtain the desired freestanding OCNTs membrane, which was further fixed on the surface of a GCE by a Nafion film. The OCNTs modified GCE (OCNTs/GC) showed potent electrocatalytic activity toward the electrochemical oxidation of NADH in PBS (pH 6.8) with a diminution of the overpotential of 645 mV. Ye et al. also reported a nonenzymatic glucose electrochemical sensor using aligned CNTs [170]. They first synthesized the aligned CNTs on a tantalum (Ta) plate and then attached the CNTs ensembles to the surface of a GCE using conductive silver paint. The CNTs-based electrode was used without any further treatment. Compared to GCE, a substantial (+400 mV) decrease in the overpotential of the glucose oxidation reaction was observed at this electrodes with oxidation starting at ca. +0.10 V (vs. 3 M KCl-Ag|AgCl).

3.7.1.2. Modified Carbon Nanotubes

The modified aligned CNTs electrochemical sensors were usually fabricated through the immobilization of functional species (e.g. biocatalysts like enzymes) by various methods. Soundarrajan et al. developed a simple pyrolytic method for large-scale production of aligned CNTs arrays perpendicular to the substrate [171]. These aligned CNTs arrays can be transferred onto various substrates of particular interest (e.g., polymer films for organic optoelectronic devices) in either a patterned or non-patterned fashion. They further prepared aligned coaxial nanowires by electrochemically depositing a concentric layer of an appropriate conducting polymer onto the individual aligned CNTs, which were used for electrochemical sensing applications. On the basis of electropolymerization of some organic monomers, GOx were immobilized on the aligned CNTs to prepared CNTs-based glucose electrochemical biosensors by Gao et al. using pyrrole [172-173], Loh et al. using 3,3'-diaminobenzidine (DAB) [174] and Ye et al. using o-aminophenol [175] as the monomers. The physical adsorption of GOx on the naked [176] or the Pt decorated [177] surfaces of aligned CNTs were also employed for constructing glucose electrochemical biosensors. Recently, Ye et al. developed a novel oxygen sensor for working at a relatively low potential by the physical adsorption of hemin on an aligned CNTs electrode [178]. They found that CV of the hemin-modified MWNTs electrode in pH 7.4 PBS clearly showed the dioxygen reduction peaks close to 0.0 V (vs. Ag|AgCl).

3.7.2. Carbon Nanotubes Nanoelectrode Arrays

At low density, the CNTs ensembles possessed the electrochemical properties resembled the microelectrodes, which were denoted as CNTs nanoelectrode arrays (NEAs). To make each nanotube work as an individual nanoelectrode, the spacing needs to be sufficiently larger than the diameter of the nanotubes to prevent diffusion layer overlap with the neighboring electrodes. In contrast to the conventional-scale aligned CNTs ensembles at high density, the size reduction of each individual electrode and the increased total number of electrodes result in improvements in both the signal-to-noise ratio and detection limits (DL). A typical procedure for preparing CNTs-based NEAs included the direct CVD growth of well aligned CNTs with low density, the coating of the CNTs layer by insulating materials (e.g. epoxy) and the removing of the protruding parts of the CNTs by polishing. Sometimes, the CNTs-based NEAs were further electrochemical activated by applying a potential bias 1.0 M NaOH for a certain period to obtain functional groups like carboxyl groups. The major works regarding the construction of CNTs-based NEAs were done by Ren et al. [26,179-181] and Meyyappan et al. [182-184] these two groups. Ren et al. first fabricated CNTs-based NEAs from low site density aligned CNTs by plasma-enhanced CVD on Ni nanoparticles made by electrochemical deposition [179] (Figure 9).

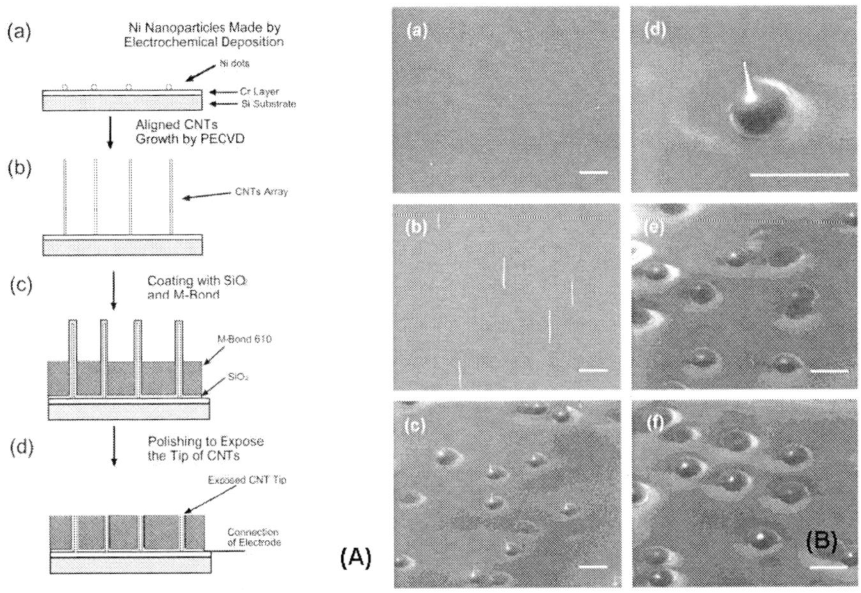

Figure 9. Fabrication scheme of the NEAs (A): (a) Ni nanoparticles electrodeposition; (b) aligned carbon nanotube growth; (c) coating of SiO2 and M-Bond; and (d) polishing to expose CNTs. Scanning electron microscope images of the NEAs (B): (a) Top view; (b-f) from a 45° side view. All of the scale bars represent 5 μm. (a) Electrodeposited Ni nanoparticles; (b) low site density aligned CNTs array; (c) CNTs array coated with SiO2 and an epoxy layer, (d) close-up look at a single half-embedded CNTs; (e) CNTs after polishing; and (f) second electrodeposition of Ni nanoparticles on the broken CNTs only. [Adapted with permission from Y. Tu et al. [179], Nano Lett., 3, 107 (2003). Copyright 2003, American Chemical Society].

They found that each nanotube was separated from the nearest neighbor by several micrometers and NEAs of 1 cm^2 consisted of up to millions of individual nanoelectrodes, each with a diameter of 100 nm. Based on this method, they further constructed a glucose electrochemical biosensor through the covalent immobilization of GOx on the ends of CNTs NEAs that were activated in 1.0 M NaOH at 1.5 V for 90 s beforehand [26]. The CNTs-based NEAs were also used successfully by Ren et al. for the voltammetric detection of trace concentrations of lead (II) [180-181] and cadmium(II) [181] at ppb level. Meyyappan et al. mainly focused on the construction of ultrasensitive DNA sensors from CNTs-based NEAs. Based on the signal change of the electrochemical indicator, $Ru(bpy)_3^{2+}$, before and after the hybridization of the oligonucleotide probes that were covalently attached to the open ends of CNTs with the target oligonucleotides, ultrasensitive determinations of DNA/RNA were achieved with the detection limit of nM level [182-184].

4. Carbon Nanotubes-Based Electrochemical Sensors for Electroanalysis

Although possessing the electrochemical properties similar to edge plane pyrolytic graphite electrodes, CNTs still have some overwhelming advantages for fabricating carbon materials-based electrochemical sensors, including the good conductivity, good biocompatibility, the large specific area, the well-defined nanotube structure, the modifiable surface, the functional groups on the surface with pretreatments and the capacity of being solubilized or dispersed in solutions. Based on these unique properties, numerous works have been dealt with CNTs-based electrochemical sensors, which can be mainly divided into two kinds: the nonenzymatic and the enzymatic CNTs-based electrochemical sensors.

4.1. Carbon Nanotubes-Based Nonenzymatic Electrochemical Sensors

4.1.1. Carbon Nanotubes-Based Nonenzymatic Electrochemical Sensors for Biomedical Analysis

4.1.1.1. Thiols

Aminothiols are of physiological importance as biological agents and metabolites. The levels of these compounds in biological matrices such as plasma and urine are valuable biomarkers in a number of clinical situations. Increased levels of HcySH and CySH are associated with risk of cardiovascular diseases. The ratio of GSH to glutathione disulfide (GSSG) indicates the redox status of cells. In addition, N-acetylcysteine is an important mucolytic agent used to reduce the viscosity of pulmonary secretions in respiratory diseases. A substantial challenge to the development of electrochemical methods for thiol detection was that the direct oxidation of thiols at solid electrodes is slow and usually large overpotentials are required [149]. Recently, Cao et al. reported the direct determination of L-CySH and GSH at a MWNTs modified electrode [185]. They found that the CNTs modified electrode exhibited efficiently electrocatalytic oxidation for L-CySH and GSH with relatively high sensitivity, stability and long-life. Coupled with high performance liquid chromatography (HPLC), this electrode was utilized for amperometric detection of the thiols. The peak currents of L-CySH and GSH were linear to their concentrations ranging from 3.0×10^{-7} to 1.0×10^{-3} M with the calculated detection limit (signal-to-noise ratio (S/N) = 3) of

1.2×10^{-7} and 2.2×10^{-7} M, respectively. The method had been successfully applied to assess the contents of L-CySH and GSH in rat striatal microdialysates. This work paved the way to determine thiols by CNTs-based electrochemical sensors. Since then, many works regarding the determination of thiols by electrochemical methods used CNTs to construct the working electrodes. In general, these CNTs-based electrochemical sensors were constructed from two type of sensing materials: CNTs and Pt-CNTs composite.

4.1.1.1.1. Unmodified Carbon Nanotubes-Based Electrochemical Sensors

The electrochemical behavior of CySH at CNTs powder microelectrodes (CNTPMEs) were investigated by Zhao et al. [153]. In contrast to the absence of signals at a GCE and a graphite powder modified powder microelectrode, CySH exhibited a sensitive oxidation peak at around 0.58 V (vs. SCE) in 0.2 M H_2SO_4 at a CNTPMEs, indicating the strong ability of CNTs to improve the electrochemical response of CySH. Chen et al. reported the fabrication of a CNTs-based microdisc electrode by mixing MWNTs with epoxy for a specially designed miniaturized capillary electrophoresis (CE)-amperometric detection system for the separation and detection of several bioactive thiols [149]. The end-channel CNTs amperometric detector offered favourable signal-to-noise characteristics at a relatively low potential (0.8 V) for detecting thiol compounds. Four thiols (HcySH, CySH, GSH, and N-acetylcysteine) were separated within 130 s at a separation voltage of 2000 V using a 20 mM phosphate running buffer (pH 7.8). Highly linear response was obtained for HcySH, CySH, GSH, and N-acetylcysteine over the range of 5-50 μM with detection limits of 0.75, 0.8, 2.9, and 3.3μM, respectively. Lawrence et al. reported an effective means for the determination of HcySH using a CNTP electrode [141]. At the CNTP electrode, HcySH showed a well defined electrochemical signal emerging at +0.28 V and a current plateau at +0.64 V. In contrast, the HcySH response obtained at the carbon paste electrode showed only a slight increase in the oxidation current at +0.40 V. The sensitive determination of HcySH was also achieved by Gong et al. at a CNTs modified GCE [68]. Cyclic voltammetric results clearly showed that CNTs, especially those pretreated with HNO_3, possessed an excellent electrocatalytic activity toward the oxidation of HcySH at a low potential (0.0 V versus Ag|AgCl). The remarkable catalytic property of the acid-pretreated CNTs was believed to be essentially associated with oxygen-containing moieties introduced on the tube surface. Zhang et al. also studied the electrochemical behaviors of GSH and GSSG at an acid-treated MWNTs modified GCE [40]. They found that the functionalized CNTs exhibited efficiently electrocatalysis on the current responses of GSH and GSSG. Recently, Salimi et al. reported the electrochemical responses of three thiols, thiocytosine, GSH and L-CySH, at a CNTs modified GCE constructed by abrasive methods [163]. They found that the overpotentials of these species were apparently reduced along with a dramatic increase in the peak currents in comparison to bare GCE. Furthermore, the thiols amperometric responses of the coated electrodes were extremely stable, with more than 95% of the initial activity after 30 min stirring of 0.1 mM thiols.

4.1.1.1.2. Platinum-Carbon Nanotubes Composites-Based Electrochemical Sensors

Besides the unmodified CNTs, Pt decorated CNTs were also used for the sensitive determination of thiols. Xian et al. prepared a MWNTs/Pt microparticles nanocomposite by electrodepositing Pt microparticles onto the MWNTs matrix [186]. The GCE modified with this nanocomposite was used for the measurement of thiols, such as L-CySH and GSH. Compared with the MWNTs or Pt microparticles modified electrode, the nanocomposite modified electrode exhibited high sensitivity and good stability for detection of thiols, which was applied successfully to the HPLC-electrochemical detection (ECD) system for the determination of L-CySH and GSH in rat striatum. The electrochemical behavior of L-CySH on Pt/CNTs electrode was also investigated by Fei et al. using cyclic voltammetry [187]. In that work, CNTs were grown directly on graphite disk by CVD and Pt was electrochemically deposited on the activated CNTs/graphite electrode by electroreduction of Pt(IV) complex ion on the surface of CNTs. Among graphite, CNTs/graphite, and Pt/CNTs electrodes, improved electrochemical behavior of CySH oxidation was found with Pt/CNTs electrode.

4.1.1.2. Carbohydrates

Electrocatalytic oxidation of carbohydrates is of great interests from the points of view ranging from medical applications of the blood glucose sensing to ecological approaches, such as waste-water treatment in food industry. Most previous studies on this subject involved the use of the enzyme GOx, which catalyzes the oxidation of glucose to gluconolactone. However, due to the intrinsic nature of enzymes, such enzyme-based sensors suffer from the stability problem. Thus, great efforts have made to achieve direct determination of carbohydrates. Ye et al. recently reported the direct electrocatalytic oxidation of glucose in alkaline medium at well-aligned MWNTs electrodes [170]. Compared to GCE, a substantial (+400 mV) decrease in the overpotential of the glucose oxidation reaction was observed at MWNTs electrodes with oxidation starting at ca. +0.10 V (vs. 3 M KCl-Ag|AgCl). The only problem of using this electrode for the direct determination of glucose was the serious interferences from some electroactive biomolecules like UA and AA, which might coexist in real biological samples. Deo et al. also reported on the electrocatalytic oxidation of carbohydrates at glassy-carbon electrodes modified with SWNTs coatings [67]. They found that the oxidation processes of four carbohydrates, including galactose, fructose, mannose and xylose, started around +0.25 V in 0.1 M NaOH solution. Moreover, the amperometric responses of these species at the CNTs modified electrodes were extremely stable, with no loss in sensitivity over a continuous 2 h operation.

Besides unmodified CNTs, the composites made up of copper particles and CNTs were also used by several works to achieve the electrooxidation of carbohydrates. Recently, Wang et al. prepared a copper modified CNTPE by incorporating copper powder in CNTPE [143]. This new composite electrode was used for the improved amperometric detection of carbohydrates following their capillary electrophoresis (CE) microchip separations. The CNTs/Cu composite electrode detector displayed enhanced sensitivity compared to detectors based on copper or CNTs alone. The marked catalytic action of the CNTs/Cu composite material permitted effective low potential (+0.5 V vs. Ag|AgCl) amperometric detection, and was coupled to the renewability, bulk modification and versatility advantages of composite electrodes. The CNTs/Cu composite surface also led to a greater resistance to surface fouling compared to that observed at the copper electrode. The method used by Male et al. for constructing copper decorated CNTs-based carbohydrate electrochemical sensors was much different [70]. In that work, Copper (Cu) nanoparticles (4-8 nm in diameter), prepared by using sodium borohydride to reduce copper dodecyl sulfate (Cu(DS)2), were used with SWNTs to fabricate electrochemical sensors. Nafion was used to solubilize SWNTs and displayed interactions with Cu nanoparticles to form a network, connecting Cu nanoparticles to a glassy carbon (GC) or Cu electrode surface. The composite of SWNTs/Cu_{nano} gave a much enhanced current response at +0.65 V with a relatively low starting oxidation potential

(+0.3 V vs. Ag|AgCl (3 M NaCl)) in comparison to the GC electrode modified by either Cu nanoparticles or SWNTs alone. Similar results were reflected from amperometric detection when 0.1 mM glucose was detected at +0.65 V. The current responses for other saccharides (100 μM) were also analyzed and compared to the glucose response. The monosaccharides, galactose (120%), arabinose (94%), xylose (109%), fructose (99%), and mannose (97%) gave similar responses as glucose. However, the current signal was lower for the dissacharides, maltose (50%), lactose (56%), sucrose (39%), and trehalose (28%), as well as the trisaccharide raffinose (56%).

4.1.1.3. Electroactive Microbiomolecules

The determination of neurotransmitters and other important microbiomolecules is of considerable significance in biochemical and clinical diagnosis. Generally, these species are electroactive, providing a simple but sensitive way for their determination by electrochemical methods. However, owning to the overlapping of the oxidation potentials of these species, it is difficult to selectively determine them at bare electrodes. To overcome this obstacle, various materials have been used to modify the bare electrodes. Among these modifiers, CNTs have become the most outstanding material and most recent works on the determination of the microbiomolecules employ CNTs as the sensing materials.

4.1.1.3.1. Unmodified Carbon Nanotubes-Based Electrochemical Sensors

The work regarding the enhanced electrochemical response of some microbiomolecules at CNTs electrodes was firstly reported by Luo et al. [15]. In contrast to the poor response at bare GCE, DA exhibited a pair of well-defined redox peaks at SWNTs modified GCE. Similar enhanced signals of AA and NE were also observed at the modified electrodes. Since then, many works focused on the selective or simultaneous determination of these species at CNTs modified electrodes (table 1).

Table 1. Sensitive determination of electroactive microbiomolecules at CNTs-based sensors

Targets	Electrodes	Supporting electrolyte	Detection potential/V	Linear range/μM	DL/μM	Ref.
UA	SWNTs/Au	0.1 M HAc- NaAc pH 5.0	0.451	4.0 ~ 700	1.0	[188]
Thymine	MWNTs-CD/CE	$NaHCO_3$- Na_2CO_3 pH 10.8	0.87	25 ~ 1800	5.0	[115]
NE	SWNTs/GCE	B-R pH 5.72	0.261	10 ~ 1100	6.0	[189]
Xanthine (Xa)	MWNTs/GCE	HAc-NaAc pH 5.20	0.8	0.2 - 20	0.004	[190]
DA 5-HT	MWNTs-IE	0.05 M PBS pH 7.4	0.13 0.32	0.5 ~ 10 1 ~ 15	0.1 0.2	[157]
Hypo-xanthine	MWNTs-DHP/GCE	0.1 M PBS pH 7.3	1.06	0.5 ~ 200	0.2	[92]
DA AA	MWNTs-IE	50 mM NaH_2PO_4 pH 5.0	0.4 0.13	0.5 ~ 10 80 ~ 1360	0.1 20	[156]
DA NE 5-HT	MWNTs/GCE	0.2 M PBS pH 5.0	0.24	0.0005 ~ 10 0.0005 ~ 10 0.0007 ~ 10	0.00025 0.00025 0.00035	[191]*

Table 1. Continued

Targets	Electrodes	Supporting electrolyte	Detection potential/V	Linear range/μM	DL/μM	Ref.
DA				0.1 ~ 100	0.0077	
EP	SWNTs/CFNE	25 mM Tris-HCl pH 7.4	___	0.3 ~ 100	0.038	[74]
NE				0.5 ~ 100	0.042	
UA	Aligned MWNTs	PBS pH 7.4	0.295	0.2 ~ 80	0.1	[168]
Dopa	SWNTs/GCE	0.1 M PBS pH 5.0	0.27	0.5 ~ 20	0.3	[28]
EP	MWNTS/bppg	0.1 M PBS pH 4.3	0.18	0.1 ~ 100	0.02	[160]
DA				0.0005 ~ 10	0.0002	
NE	MWNTs/GCE	0.2 M PBS pH 5.0	___	0.0004 ~ 10	0.0002	[192]*
5-HT				0.0007 ~ 10	0.0003	

* The results were obtained from an electrochemical detection (ED) system for liquid chromatography (LC).

4.1.1.3.2. Carbon Nanotubes Composites-Based Electrochemical Sensors

The selective determination of the electroactive microbiomolecules was also performed at CNTs composite modified electrodes. Compared with the CNTs, the combination of CNTs with other components not only provides higher stability and better uniformity of the composite films but also endows the composite films with the unique properties of both components. A typical example is that the composite films made up of CNTs and Nafion can effectively eliminate the interference of negatively charged AA on the determination of other species due to the exclusion of AA from the electrode surface by negatively charged Nafion film. Table 2 lists the recent works on the determination of the electroactive microbiomolecules at CNTs composite films modified electrodes.

Table 2. Sensitive determination of electroactive microbiomolecules at CNTs composite film modified electrodes

Targets	Electrodes	Supporting electrolyte	Detection potential/V	Linear range/μM	DL /μM	Ref.
UA	CNTs-β-CD/PG	0.2 M HAc-NaAc pH 4.5	0.45	0.5 ~ 50	0.2	[118]
DA	MWNTs-Nafion/GCE	0.1 M PBS pH 5.5	0.235	0.05 ~ 1.0	0.0025	[60]
XA UA	MWNTs-DHP/GCE	0.1 M PBS pH 5.5	0.77 0.38	0.02 ~ 20 0.1 ~ 100	0.04 0.01	[77]
DA 5-HT	MWNTs-DHP/GCE	0.1 M PBS pH 7.0	0.18 0.36	0.05 ~ 5.0 0.02 ~ 5.0	0.011 0.005	[94]
NE	CNTs-β-CD/PG	0.1 M PBS pH 6.0	0.25	1.0 ~ 300	0.5	[119]
DA AA	MWNTs-CHIT/GCE	0.1 M PBS pH 7.2	0.144 -0.029	0.5 ~ 100 5.0 ~ 1000	0.2 2.0	[100]
DA	MWNTs-Nafion/GCE	0.1 M PBS pH 6.0	0.235	0.01 ~ 10	0.0025	[62]
UA EP	MWNTs-CHIT/GCE	0.1 M PBS pH 5.5	0.484 0.329	2.0 ~ 100 5.0 ~ 250	0.75 1.0	[103]
DA	MWNTs-Nafion/CFME	0.05M PBS pH 7.5	0.145	2.0 ~ 20	0.07	[63]
DA	MWNTs-RTIL/GCE	0.1 M PBS pH 7.0	0.16	1.0 ~ 100	0.1	[112]

4.1.1.4. Amino Acids

The amino acids detected by electrochemical methods are generally electroactive and usually contain phenol moieties. Xu et al. prepared a MWNTs film modified electrode and coupled this electrode to an ion chromatography system for the simultaneous determination of oxidizable amino acids, including CySH, tryptophane and tyrosine [193]. The results show that when the concentrations range from 1.5×10^{-6} to 2.5×10^{-4} M for CySH, 5.5×10^{-7} to 1.5×10^{-4} M for tryptophane and 9.0×10^{-7} to 3.5×10^{-4} M for tyrosine, respectively, the peak currents responded linearly to the concentrations of these analytes. The detection limits were 7.0×10^{-7} M for CySH, 2.0×10^{-7} M for tryptophane and 3.5×10^{-7} M for tyrosine at a S/N of 3. Wu et al. investigated the electrochemical behavior of tryptophan at MWNTs/GCE [49]. Compared with a bare electrode, the oxidation peak current of tryptophan was obviously increased, and the peak potential negatively shifted. Under the chosen conditions, the DPV peak current was linear to the concentration of tryptophan in the range of $2.5\times10^{-7} \sim 1.0\times10^{-4}$ M, and the detection limit was 2.7×10^{-8} M. The voltammetric response of tryptophan at a SWNTs modified GCE was also studied by Huang et al. [95]. In pH 2.5 Na_2HPO_4-citric acid buffer, tryptophan yielded a well-defined and very sensitive oxidation peak at about 1.08 V at the SWNTs-film coated GCE. The oxidation peak current increased greatly and the peak potential shifted toward negative direction at the SWNTs-modified GCE in contrast to that at the bare GCE. Under optimized conditions, the oxidation peak current was proportional to the concentration of tryptophan over the range from 4×10^{-8} to 1×10^{-5} M. The detection limit was 1×10^{-8} M at 3 min of accumulation.

The electrochemical determination of traditionally 'non-electroactive' amino acids was also performed by Deo et al. at a SWNTs modified electrode that was coated by a nickel hydroxide film [72]. The determination process was achieved through the electrocatalytic oxidation of these amino acids at the $Ni(OH)_2$ layer that was electrodeposited on surface of SWNTs. In the presence of amino acid, the Ni-CNTs layer underwent an electrocatalytic process in which the amino acid reduces the newly formed NiO(OH) species. The results showed that under flow injection analysis conditions, a linear response was observed from 10 ~ 80 mM arginine with a detection limit of around 1 mM (based on S/N ~ 3). This electrode was also applied to the detection of other amino acids like histidine, lysine, asparagine, methionine and phenylalanine.

4.1.2. Carbon Nanotubes-Based Nonenzymatic Electrochemical Sensors for Environment Protection and Drug Analysis

The monitoring of some harmful substances in environment or in vivo has important significances in environment protection, clinical medicine and pharmacology. The determination of these species at CNTs-based electrochemical sensors has been performed by many workers due to their enhanced responses at the CNTs sensors.

4.1.2.1. Metal Ions

Wu et al. firstly reported the simultaneous determination of trace levels of cadmium (II) and lead (II) by anodic stripping voltammetry (ASV) at a MWNTs-DHP composite modified GCE [98]. In pH 4.5 NaAc-HAc buffer, Cd^{2+} and Pb^{2+} first adsorbed onto the surface of a MWNTs film coated GCE and then were reduce at -1.20 V. During the positive potential sweep, reduced cadmium (II) and lead (II) were oxidized, and two well-defined stripping peaks appeared at -0.88 and -0.62 V. The addition of 0.02 M KI significantly enhanced the responses of cadmium (II) and lead (II), which was attributed to the synergistic adsorption of cadmium (II) and lead (II) at MWNTs via the adsorption of KI at MWNTs and the interactions between KI and these metal ions. Under optimal conditions, the striping peak currents changed linearly with the concentration of Cd^{2+} from 2.5×10^{-8} to 1×10^{-5} M and with that of Pb^{2+} from 2×10^{-8} to 1×10^{-5} M. The lowest detectable concentrations of Cd^{2+} and Pb^{2+} were estimated to be 6×10^{-9} and 4×10^{-9} M, respectively. This electrode was also employed by Yi for the sensitive determination of mercury (II) and the striping peak current (I_P) was proportional to the concentration of Hg^{2+} from 8×10^{-10} to 5×10^{-7} M with a low detection limit of 2×10^{-10} M [99]. On the basis of the cationic ion exchange ability of Nafion, Cai et al. proposed an anodic stripping method for the determination of cadmium (II) at a MWNTs-Nafion composite modified GCE and achieved a linear range of anodic stripping response from a solution of 1 ~ 4 μM Cd^{2+} in 0.1 M acetate buffer with a detection limit of about 76 nM [66]. Besides the randomly oriented CNTs-based conventional-scale electrodes, nanoelectrode arrays (NEAs) that are based on low-site density CNTs were also used by Ren et al. for the sensitive determination of lead (II) [180,181] and cadmium [181]. The spin-coating of an epoxy resin on the electrode surface provided a new way to create the electrode passivation layer effectively reducing electrode capacitance and current leakage, which not only helped improve the signal-to-noise ratio but also reduce the response of potential species that might interfered the determination process.

4.1.2.2. Phenolic Compounds

Phenol and substituted phenols are widely distributed in industrial and natural waste. These components have toxic effects on human beings, animals, and plants. They give an undesirable taste and odor to drinking water, even in very low concentration. The determination of phenolic compounds at CNTs-based electrochemical sensors has been performed by many workers. Generally, the electrochemical responses of these compounds at CNTs electrodes are greatly enhanced compared to the bare electrodes and the selectivity are usually improved (Table 3).

Table 3. Sensitive determination of phenolic compounds at CNTs modified electrodes

Targets	Electrodes	Supporting electrolyte	Detection potential/V	Linear range/μM	DL/μM	Ref.
2-Nitrophenol 4-Nitrophenol	MWNTs-Nafion/GCE	0.1 M PBS pH 4.0	-0.8 -1.0	0.05 ~ 10 0.1 ~ 10	0.01 0.04	[64]
Estradiol	MWNTs/GCE	0.1 M PBS pH 2.0	0.82	0.25 ~ 50	0.01	[194]
4-Aminophenol	MWNTs-Nafion/GCE	0.1 M citrate- HCl pH 3.0	0.42	0.005 ~ 2.0	0.0008	[65]
3,4-Dihydroxy-phenylacetic acid	SWNTs/GCE	0.1 M HAc- NaAc pH 4.4	0.32	1.0 ~ 120	0.4	[195]
Catechol	MWNTs/GCE	0.05 M PBS pH 7.0	0.215	20 ~ 1200	10	[196]
Bisphenol A	MWNTs-DHP/GCE	____	0.58	0.05 ~ 20	0.02	[197]
Hydroquinone Catechol Resorcinol	MWNTs/GCE	0.1 M HAc- NaAc pH 5.5	0.152 0.256 0.677	2.0 ~ 100 2.0 ~ 100 5.0 ~ 80	0.6 0.6 1.0	[35]
Hydroquinone Catechol	MWNTs/GCE	0.20 M HAc- NaAc pH 4.5	0.126 0.264	1.0 ~ 100 0.6 ~ 100	0.75 0.2	[198]

4.1.2.3. Drugs

The development of sensitive analytical techniques of drugs has drawn much attention due to the promising applications in environment protection, pharmacology and biomedical studies. However, due to the chemical stability and the large bodies, the responses of these substances were generally weak at common bare electrodes.

Table 4 .Sensitive determination of drugs at CNTs modified electrodes

Targets	Electrodes	Supporting electrolyte	Detection potential/V	Linear range/μM	DL /μM	Ref.
6-Mercaptopurine	MWNTs/GCE	0.1 M PBS pH 7.0	0.53	0.4 ~ 100	0.2	[199]
Daunomycin	MWNTs-DHP/GCE	0.1 M PBS pH 5.0	-0.564	0.02 ~ 10	0.008	[81]
6-Benzylamino-purine	CNTs/GCE	NH3- NH4Cl	0.65	0.08 ~ 10	0.005	[200]
Chloramphenicol	MWNTs-DHP/GCE	0.1 M PBS pH 7.0	-0.65	0.3 ~ 12	0.045	[82]
Indole-3-acetic acid	MWNTs-DHP/GCE	0.1 M PBS pH 2.0	0.68	0.1 ~ 50	0.02	[78]
Vitamin B6	MWNTs-DHP/GCE	0.1 M PBS pH 6.0	0.8	0.5 ~ 100	0.2	[201]
Thyroxine	MWNTs-DHP/GCE	0.1 M HCl	0.8	15 ~ 400 μg/l	5μg/l	[79]
8-Azaguanine	MWNTs-DHP/GCE	0.1 M PBS pH 7.0	0.86	0.025 ~ 10	0.01	[85]
Insulin	MWNTs/GCE	0.1 M PBS pH 7.4 + 0.02% (v/v) Tween-80	0.8	0.1 ~ 1.0	0.014	[202]
Tinidazole	MWNTs-DHP/GCE	0.1 M B-R pH 10.0	-0.78	0.05 ~ 40	0.01	[203]
Metronidazole	MWNTs-DHP/GCE	0.1 M B-R pH 9.0	-0.71	0.025 ~ 10	0.006	[86]
Fluphenazin	MWNTs-MPS/Au	0.05 M HAc-NaAc pH 3.0	0.78	0.05 ~ 15	0.01	[204]
Thyroxine	SWNTs-DHP/GCE	0.1 M HCl	0.78	0.1 ~ 7.0	0.02	[80]
Reserpine	MWNTs-DHP/GCE	0.1 M PBS pH 6.0	0.64	0.02 ~ 10	0.0075	[88]
Mitomycin c	MWNTs-DHP/GCE	——	0.79	0.25 ~ 100	0.08	[89]
Epirubicin	SWNTs-DHP/GCE	0.1 M PBS pH 7.0	-0.74	0.05 ~ 10	0.02	[90]
Brucine	MWNTs/GCE	H_2SO_4 pH 1.0	0.405	1.0 ~ 100	0.2	[205]
Amitrole	CNTPE	0.05 M PBS pH 7.4	0.9	0.8 ~ 7.0	0.57	[140]
Simvastatin	MWNTs-DHP/GCE	0.1 M H_2SO_4	1.06	0.1 ~ 7.5	0.05	[91]
Morphine	MWNTs/GCE	0.1 M PBS pH 7.0	0.5	0.5 ~ 150	0.2	[162]
Prochlorperazine	MWNTs/DDT /Au	0.05 M PBS pH 5.8	0.69	0.5 ~ 15	0.1	[206]
Melatonin	MWNTs-DHP/GCE	0.1 M PBS pH 7.5	0.61	0.08 ~ 10	0.02	[207]

To improve the detection sensitivity, various nanomaterials have employed to fabricate modified electrodes with high performance. Among these materials, CNTs have also been widely to improve the sensitivity of some drugs. Compared with other nanomaterials, CNTs have some advantages, such as good conductivity and strong electrocatalytic activity. The enhanced adsorption of some water insoluble or conjugated drugs on the hydrophobic and π-conjugated sidewalls of CNTs contributes to the sensitive determination of these species at CNTs electrodes. Table 4 lists the applications of CNTs-based electrochemical sensors for drug analysis.

4.1.3. Carbon Nanotubes-Based Electrochemical Sensors for Electrocatalytic Determination

Due to their unique properties, CNTs are proved to exhibit electrocatalytic activity towards several small molecules like dioxygen [155,208-210], hydrogen peroxide [61,122], carbon monoxide [211] and nitrite [151]. At the same time, some overwhelming advantages, such as the good conductivity and chemical satiability, the modifiable surface, the large specific area and the strong adsorption of some macromolecules on the sidewall, make CNTs the ideal supporting materials of some catalysts, which have been employed by many researchers to fabricate highly efficient electrocatalytic systems of some important molecules. For example, the strong adsorption of some porphyrin-based compounds on CNTs has been used to prepare CNTs- porphyrin hybrids that showed enhanced electrocatalytic activity towards the reduction of dioxygen, including [tetrakis(N-methylpyridyl)porphyrinato] cobalt (CoTMPyP) [53,212], hemin [178] and tetra- phenyl-porphyrin (TPP) [30]. On the basis of the strong adsorption of positively charged PDDA on negatively charged and shortened MWNTs, Zhang et al. prepared a multilayer films of shortened MWNTs that are homogeneously and stably assembled on GCE by the LBL method [123]. Electrochemical studies indicate that the LBL assembled MWNTs films possess a remarkable electrocatalytic activity toward O_2 reduction in alkaline media.

4.2. Carbon Nanotubes-Based Enzymatic Electrochemical Sensors

4.2.1. Carbon Nanotubes-Based Enzymatic Electrochemical Biosensors for Electroanalysis

The incorporation of enzymes with a transducer allows one to fabricate enzyme electrodes which exhibit high selectivity, high-sensitivity and long-term stable response to bioanalytes. Many different materials and methods have been used to immobilize enzymes on the electrode surfaces, such as sol-gels, carbon pastes, conductive polymers and recently nanostructured materials. Of particular interest is the use of CNTs as an electrochemical transducer, due to the excellent electron transfer rate, the well-defined nanostructure, the good biocompatibility, and the modifiable surface allowing biomaterial immobilization of CNTs.

4.2.1.1. Glucose Oxidase

Among all the enzyme-based biosensors, glucose biosensor is the most widely studied because the diagnosis of diabetes mellitus requires an accurate monitoring of blood glucose levels. Most of the amperometric glucose biosensors are based on the glucose oxidase (GOx) enzyme, which catalyzes the oxidation of glucose to gluconolactone:

$$\text{Glucose} + O_2 \xrightarrow{\text{GOx}} \text{Gluconolactone} + H_2O_2$$

The quantification of glucose can be achieved via electrochemical detection of the enzymatically liberated H_2O_2. However, the overpotential necessary for the oxidation or reduction of H_2O_2 at solid electrodes is rather high. Recently, Wang et al. demonstrated that the overpotential of H_2O_2 can be greatly reduced at CNTs-based electrodes [61], indicating the promising applications of CNTs in constructing glucose electrochemical biosensors. The hybrids of CNTs and some noble metals nanoparticles, such as palladium, platinum, gold, copper and iridium have also been demonstrated to lower the H_2O_2 oxidation/reduction overpotential efficiently. Besides the materials, the immobilization methods effectively influence the activity and the quality of enzyme electrodes. Typical GOx immobilization methods include the electropolymerization embedment, composite blending, covalent bonding, electrodeposition and physical adsorption.

4.2.1.1.1. Electropolymerization Embedment

Electropolymerization represents an attractive well controlled one-step avenue for preparing amperometric enzyme electrodes. Thc main advantages of this immobilization avenue are the simple one-step preparation, exclusion of electroactive and surface-active interferences, control of the film thickness, and localization of biocatalysts onto tiny electrode surfaces [167]. The electropolymerization of some monomers on the surface of CNTs provided a simple but effective approach for immobilization of GOx on CNTs, including pyrrole, o-aminophenol and o-phenylenediamine. In a typical immobilization process, GOx and the monomers are placed together in the supporting electrolytes on certain ratio and GOx is immobilized on the electrode surface via the physical entrapment of GOx within the electropolymerized film during the electropolymerization process. Gao et al. reported a glucose electrochemical biosensor through the immobilization of GOx on well aligned CNTs via pyrrole electropolymerization [172,173]. They prepared the biosensor by immobilizing GOx in the electropolymerization film of pyrrole through electrochemically oxidizing pyrrole monomer (0.1 M) in a pH 7.45 buffer solution containing 2mg/ml GOx and 0.1 M $NaClO_4$ at a constant potential of 1.0 V for 1 min at 10 °C. The oxidation potential of H_2O_2 at this electrode (0.45 V vs. Ag|AgCl) was found to be apparently lower than that at a gold electrode (0.65 V vs. Ag|AgCl), providing a selective approach for glucose determination. Callegari et al. proposed a convenient and versatile strategy for the construction of glucose amperometric biosensors [213]. They prepared the sensors by the co-immobilization of GOx and a suitable redox mediator, namely ferrocenyl-derivatised SWNTs (Fc-SWNTs) within a thin polypyrrole film adsorbed onto the GCE surface. They found that Fc-SWNTs might provide preferential routes for the electrical wiring of the enzyme to the electrode, likely via the aligned ferrocenyl groups, thus illustrating the potentialities of such redox-active CNTs for the elaboration of mediated biosensors. The co-immobilization of GOx on iron-phthalocyanine (FePc) functionalized well aligned MWNTs by the electropolymerization of nonconducting polymers was also performed by Ye et al. to prepare a glucose biosensor [175]. The CNTs modifier FePc was used to electrocatalyze the oxidation of H_2O_2 that was liberated by the enzymatic reaction of glucose. Dai et al. prepared an amperometric glucose biosensor via the immobilization of GOx on a MWNTs modified GCE by electrochemical deposition of PPD film [209]. This electrode showed enhanced electrocatalytic activity toward oxygen at a potential of -0.40 V (vs. Ag|AgCl) in neutral medium and the steady-state amperometric response to glucose was determined at a selected potential of -0.30 V by means of the reduction of dissolved oxygen consumed by the enzymatic reaction. Wang et al. recently proposed a one-step method for the construction of glucose biosensors by applying a fixed potential (usually 0.7V) onto a GCE that was placed beforehand in a aqueous solution containing CNTs, GOx and pyrrole monomer [167]. They found that the CNTs dopant retained their electrocatalytic activity toward hydrogen peroxide to impart high sensitivity upon entrapment within the PPy network. The simultaneous incorporation of CNTs and GOx thus imparted biocatalytic and electrocatalytic properties onto amperometric transducers and represented a simple and effective route for preparing enzyme electrodes. A similar method was also used by Pan et al. to fabricate glucose biosensors except that additional supporting electrolytes were added to the electropolymerization solution [165].

4.2.1.1.2. Blending

4.2.1.1.2.1. Composite Films

Blending of enzymes and some polymeric materials that are capable of forming uniform and stable films on the electrode surfaces provides a convenient noncovalent approach for the immobilization of enzymes while maintaining the native properties and reactivities of enzymes. This makes the composite blending technique a potential tool for the development of new electrochemical biosensors. Leaching of the entrapped enzymes does not occur or occurs very slowly because these species usually have large bodies and possible interactions (e.g. electrostatic and hydrophobic interactions) between enzymes and the film matrices might exist. Several types of film matrices have been employed for constructing glucose electrochemical biosensors, including sol-gels, Nafion, CHIT. However, due to their poor conductivity and film compactness, the performance of the biosensors fabricated from these species and enzymes were usually poor. This deficiency can be completely overcome by the incorporation of CNTs, which leads to the improvement of the conductivity and the diffusion process of the composite films.

On the basis of the stable dispersion of CNTs in Nafion solutions, Wang et al. firstly studied the electrochemical behavior of H_2O_2 at CNTs-based electrodes [61]. In contrast to the absence of redox activity at a bare GCE over a wide potential range (-0.4 ~ 1.0 V), the CNTs-coated electrode was found to exhibit significant oxidation and reduction currents of H_2O_2 with a rather low starting potential (around +0.20 V), providing a highly selective low-potential biosensing of glucose when GOx was immobilized in the CNTs-Nafion composite. The excellent analytical performance of the CNTs/Nafion/GOx electrode was further demonstrated by Tsai et al. [69]. They found that under optimal conditions, the glucose biosensor displayed a sensitivity of 330 nA/mM, a linear range of up to 2 mM, a detection limit of 4 μ M, and a response time of < 3 s. Zhu et al. fabricated an amperometric glucose biosensor based on the enzyme electrode with CNTs/Pt nanoparticle modified on GCE surface (CNTs-Pt/GCE) [214]. GOx in gelatin was cross-linked by the use of glutaraldehyde (GDI). Compared with normal GOx/Pt electrode, the analytical property of GOx/CNTs-Pt/GCE electrode was apparently improved, reflected by the lowered detection limit, overpotential and response time. Salimi et al. developed a glucose biosensor by immobilizing GOx into a sol-gel composite at the surface of a bppg electrode modified with MWNTs [161]. The CNTs modified electrode was found to offer excellent electrocatalytic activity toward reduction and oxidation of hydrogen peroxide liberated in the enzymatic reaction between GOx and glucose, enabling sensitive determination of glucose. The amperometric detection of glucose carried out at 0.3V (vs. SCE) in 0.05 M PBS (pH 7.4) showed a good linear response range of 0.2 ~ 20 mM glucose with a low detection limit of 50 μM (S/N = 3). In addition, the response time of the electrode was less than 5 s when it was stored dried at 4 °C, and the sensor showed almost no change in the analytical performance after operation for 3 weeks. Recently, Joshi et al. described the construction of amperometric biosensors based on the incorporation of SWNTs modified with GOx into a redox polymer hydrogel [109]. The composite films were constructed by first incubating GOx in a SWNTs solution and then cross-linking within a poly[(vinylpridine) Os(bipyridyl)$_2$Cl$^{2+/3+}$] polymer film. Incorporation of SWNTs, modified with GOx, into the redox polymer films resulted in a 2-10-fold increase in the oxidation and reduction peak currents during cyclic voltammetry, while the glucose electrooxidation current was increased 3-fold to similar to 1 mA/cm^2 for glucose sensors.

4.2.1.1.2.2. Carbon Pastes

As a typical method, carbon pastes have been extensively used to prepare enzyme-based electrochemical biosensor because this method has some unique advantages, such as easy fabrication, low cost, wide potential window, high signal-to-noise ratio and retaining of enzyme activity. Recently, Rubianes et al. studied the electrochemical response of H_2O_2 at a CNTPE by the hydrodynamic voltammetric method. Similar to that at the MWNTs-Nafion/GCE [61], the signals of H_2O_2 were significantly improved at CNTPE in comparison with common carbon paste electrodes and decreases of 300 mV in the oxidation overpotential and 400 mV in the reduction overpotential were obtained for hydrogen peroxide, allowing the development of a highly selective and sensitive glucose biosensor without using any metal, redox mediator or anti-interference membrane. Ricci et al. also prepared a Prussian blue (PB) modified CNTPE, which showed a high sensitivity towards hydrogen peroxide with a detection limit of 7.4×10^{-6} M. Further they fabricated a glucose biosensor by incorporating GOx into the paste, creating a sensitive tool for the detection of glucose in the range between 0.1 and 50 mM.

4.2.1.1.3. Covalent Bonding

Although the noncovalent methods for fabricating CNTs-based electrochemical biosensors have some advantages like easy fabrication and low cost, the poor stability and leaching of enzymes sometimes significantly influence the analytical performance of the biosensors. To overcome this limitation, great efforts have made to covalently attach enzymes to the electrode surfaces via various methods and reagents. Xue et al. described a simple method for fabricating glucose biosensors by covalently attaching GOx onto a SWNTs modified Pt electrode [215,216]. The Pt/SWNTs electrode was firstly activated with EDC in PBS with pH 5.5 for 1 h, and then immersed in the PBS containing 5 mg/ml GOx at 4 °C for

12 h. The resulting enzyme electrode was washed thoroughly with PBS, and stored at -10 °C. This biosensor showed a stable amperometric response at a low applied potential (0.4 V vs. SCE), a wide response range of pH (5.5-8.0) and good stability (kept 90% activity after 4 months) [216].

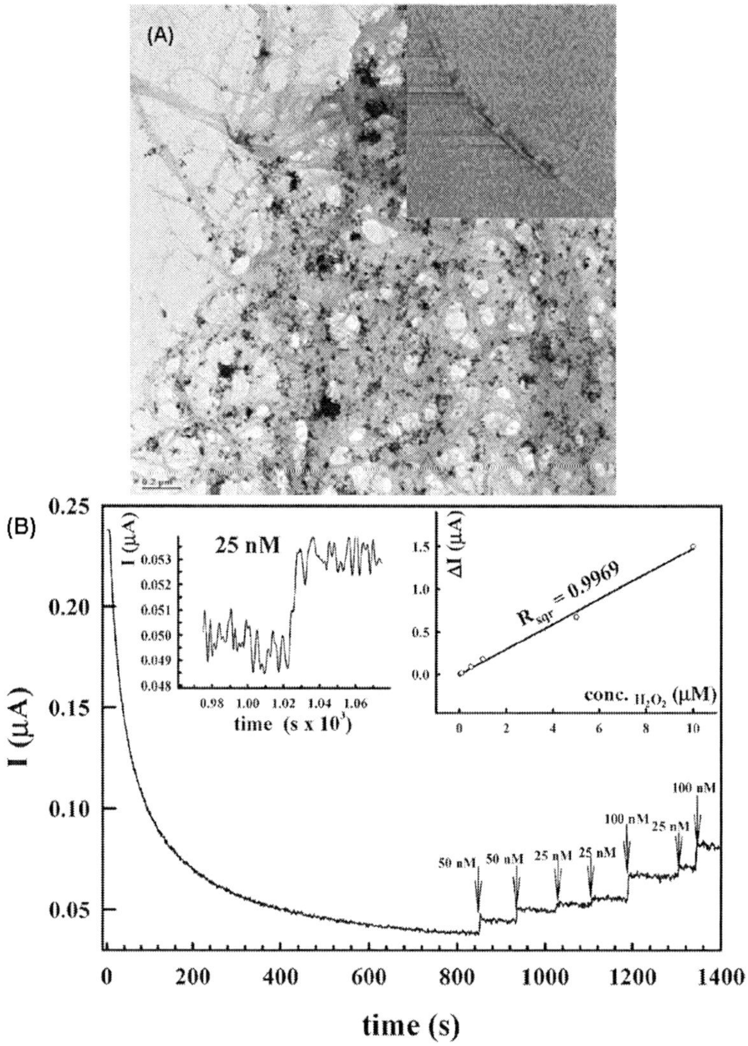

Figure 10. (A). TEM micrograph of SWNTs in the presence of Pt nanoparticles. Inset: AFM tapping-mode phase image (size, 1 μm×1 μm; data scale, 20 nm) of one SWNTs in the presence of Pt nanoparticles. (B). Performance of GC/CNT + Ptnano electrode in amperometric detection of low concentrations of hydrogen peroxide at +0.55 V vs. Ag|AgCl (3 M NaCl). The inset (left) shows detection limit of 25 nM with S/N=3. Inset (right) shows calibration curve for H2O2 concentrations between 25 nM and 10 μM. [Adapted with permission from S. Hrapovic et al. [71], Anal. Chem., 76, 1083 (2004). Copyright 2004, American Chemical Society].

As mentioned above, Gooding et al. also covalently immobilized MP-11 on the open end of a SWNTs array that was fabricated self-assembling methods with the aid of EDC [129]. Hrapovic et al. [71] recently developed a highly sensitive hydrogen peroxide sensor by the incorporation of CNTs and Pt nanoparticles into the Nafion films on a GCE, which exhibited a linear range from 25 nM to 10 μM (R^2=0.997, sensitivity of 3.57 A M^{-1} cm^{-2}), and also from 100 μM to 2 mM (R^2 = 0.996, sensitivity of 1.85 A M^{-1} cm^{-2}), with a rather low detection limit of 25 nM (S/N=3) (Figure 10). They further prepared a glucose biosensor by cross-linking the GOx films that were dropped and dried on the CNTs/Pt-modified GC electrode with 2.5% glutaraldehyde (GDI). This nanocomposite biosensor exhibited a rapid response (3s), a low detection limit (0.5μM, S/N=3) and a wide linear range from 0.5 μM up to 5 mM with a sensitivity of 2.11 μA/mM.

There was also an interesting work performed by Zhang et al. for the covalent immobilization of GOx on CNTs composites using RTIL as the CNTs dispersing reagent and N-succinimidyl acrylate (NSA) as both the surface modifier of CNTs and the functional sites for GOx covalent immobilization [111].

4.2.1.1.4. Other Methods

Some other methods have also been proposed by several workers for immobilizing of GOx on CNTs-based electrodes to prepare glucose electrochemical biosensors, such as physical adsorption and co-electrodeposition. Wang et al. developed a novel biosensor, comprised of electrode of gold/MWNTs-GOx (Au/MWNTs-GOx) [176]. MWNTs were grown on n-type silicon substrates using an ASTeX microwave plasma system. After growth of MWNTs on silicon substrate, a thin gold film was E-beam evaporated on the top surface of MWNTs. Then, the silicon substrate was etched away using a mixture acid solution of HCl and HNO_3. Finally, the electrode modified with MWNTs (Au/MWNTs) was put into the 0.1 M PBS containing GOx, and the GOx was immobilized by MWNTs. The biosensor showed a linear response to glucose concentration within the range from 0.05 to 13 mM with a low detection limit of less than 0.01 mM for glucose, suggesting MWNTs enhanced efficiency of electron transfer between GOx and electrode greatly, resulting in high sensitivity in detecting glucose. Loh et al. prepared a glucose biosensor by the physical adsorption of GOx on the poly(3,3′-diaminobenzidine) (PDAB) film modified electrode [174]. They first grew MWNTs (10-mm thick, erected) in a quartz furnace using ethylenediamine on 1-mm diameter tantalum substrates seeded with cobalt nanoparticles, which was further functionalized with DAB by the electropolymerization of DAB using a CV method and a self-limiting, very thin layer of polymer with NH_2 functional groups can be introduced onto the CNTs surface in one easy step. For immobilizing the GOx on the DAB-modified electrode surface, a second round of CV was then performed with GOx in the buffer solution to attach the GOx on the DAB-modified CNTs. They believed that the immobilization of GOx probably proceeded via entrapment within the highly porous wiring matrix of CNTs. This glucose biosensor had some overwhelming advantages, such as the low detection potential due to the lowered overpotential of H_2O_2 at the CNTs electrode and the absence of the concomitant oxidation of interferent molecules like AA and UA. Tang et al. fabricated a new amperometric biosensor, based on adsorption of GOx at the Pt nanoparticle-modified CNTs electrode [177]. The CNTs/graphite electrode (0.4×0.4 cm, geometry area 0.16 cm^2) was prepared by the CVD method and used as the working electrode. Using a standard three-electrode cell, Pt nanoparticles were electrodeposited on the CNTs electrode by the potential step deposition method from N_2-saturated 7.7 mM H_2PtCl_6 plus 0.5 M HCl aqueous solution. GOx immobilization was achieved by incubation for 12 h in a GOx solution (500U/ml) at less than 4 °C. The obtained GOx/Pt/CNTs electrode was washed carefully with double-distilled water and dried at less than 4 °C, and then 0.5wt% Nafion solution was cast on the surface of the resulting GOx/Pt/CNTs electrode to avoid the loss of GOx in determination and to improve the anti-interferent ability. The excellent electrocatalytic activity and special three-dimensional structure of the enzyme electrode resulted in good characteristics such as a large determination range (0.1 ~ 13.5mM), a short response time (within 5 s), a large current density (1.176 mA/cm^2), and high sensitivity (91 mA M^{-1} cm^{-2}) and stability (73.5% remains after 22 days). Lim et al. employed an electrodeposition method for the co-deposition of GOx and palladium nanoparticles onto a Nafion-solubilized CNTs film [56]. The co-deposited Pd-GOx-Nafion CNTs bioelectrode retained its biocatalytic activity and offered an efficient oxidation and reduction of the enzymatically liberated H_2O_2, allowing for fast and sensitive glucose quantification. The combination of Pd-GOx electrodeposition with Nafion-solubilized CNTs enhanced the storage time and performance of the sensor. An extra Nafion coating was used to eliminate common interferents such as UA and AA. The fabricated Pd-GOx-Nafion CNTs glucose biosensor exhibited a linear response up to 12 mM glucose and a detection limit of 0.15 mM (S/N = 3).

4.2.1.2. Other Enzymes

4.2.1.2.1. Cholesterol Oxidase

The determination of cholesterol is of great significance in clinical diagnosis of coronary heart disease, arteriosclerosis, myocardial infarction and brain thrombosis. Among the various biosensors of cholesterol, cholesterol oxidase (ChOx)-based electrochemical biosensors are especially attractive because they offer a simple, rapid and inexpensive means of quantifying cholesterol. Similar to glucose electrochemical biosensors, most of the amperometric cholesterol biosensors are based on the ChOx enzyme, which catalyzes the oxidation of cholesterol to 4-cholesten-3-one:

$$\text{Cholesterol} + O_2 \xrightarrow{\text{ChOx}} \text{4-Cholesten-3-one} + H_2O_2$$

The quantification of cholesterol can be achieved via electrochemical detection of the enzymatically liberated H_2O_2. Guo el al. proposed a CNTs-based amperometric cholesterol biosensor through LBL deposition of PDDA and ChOx on MWNTs-modified gold electrode, followed by electrochemical generation of a nonconducting PPD film as the protective coating [122]. Due to the strong electrocatalytic properties of MWNTs toward H_2O_2 and the low permeability of PPD film for electroactive species, such as AA, UA and acetaminophen, this biosensor exhibited high sensitivity and good anti-interferent ability in the detection of cholesterol with a linear range up to 6.0 mM and a detection limit of 0.2 mM. Tan et al. developed an amperometric cholesterol biosensor based on sol-gel CHIT/silica and MWNTs organic-inorganic hybrid composite material [107]. They first prepared a Prussian blue-modified GCE (PB/GCE) by electrodeposition, followed by the casting of the aqueous mixture of MWNTs, sol-gel silica, CHIT and ChOx on the surface of PB/GCE. The analytical characteristics and dynamic parameters of the biosensors with and without MWNTs in the hybrid film were compared, and the results showed that analytical performance of the biosensor can be improved greatly after introduction of the MWNTs. Shi et al. recently fabricated a novel H_2O_2 sensor by intercalating Pt nanoparticles decorated CNTs (CNTs-Pt) that were prepared using a chemical reduction method on the surface of a waxed graphite electrode [159]. The results showed that the Pt nanoparticles at a waxed graphite electrode exhibited high catalytic activity for the reduction of hydrogen peroxide. Based on this, ChOx was immobilized with sol-gel on the CNTs-Pt base electrode to construct a cholesterol biosensor, which was found to have excellent analytical performance for cholesterol detection with a rapid response (< 20 s), a wide linear range from 4.0×10^{-6} -1.0×10^{-4} M and a low detection limit of 1.4×10^{-6} M.

4.2.1.2.2. Dehydrogenases

Similar to the oxidases, some dehydrogenases are also employed to construct various electrochemical biosensors, including alcohol-dehydrogenase (ADH), D-fructose dehydrogenase (FDH) and glucose dehydrogenase (GDH). Generally, the response of dehydrogenase-based electrochemical biosenors is a current due to the electrooxidation of NADH that that is produced in the enzymatic reaction

$$\text{Substrate} + \text{NAD}^+ \xrightarrow{\text{Dehydrogenase}} \text{Product} + \text{NADH}$$

In this process, NAD^+ behaves as the enzymatic mediator. This section will first focus on the development of CNTs-based NADH electrochemical sensors and then on the construction of dehydrogenases-based electrochemical biosensors.

4.2.1.2.2.1. Carbon Nanotubes-Based NADH Electrochemical Sensors

The direct electrochemical oxidation of NADH is becoming more and more interesting since many dehydrogenases need it as cofactors to catalyze redox reaction. However, the direct electrochemical oxidation of NADH at a bare electrode takes place at high overpotential, e.g. the reported overpotential at pH 7.0 is about 1.1 V at carbon and 1.3 V at Pt electrode. Many efforts have been made to reduce the overpotential of NADH electrooxidation. Musameh et al. [217] firstly reported the electrocatalytic oxidation of NADH at a GC electrode modified with disordered CNTs coatings with decreasing of 490 mV in overpotential. Since then, many works dealt with the electrocatalytic oxidation of NADH at various CNTs-based electrodes. Chen et al. studied the electrochemical behavior of NADH at an OCNTs/GC electrode [169,218]. They found that the direct electrocatalytic oxidation of NADH occurred at a much lower potential (at about 0.0 V) with high stability at an OCNTs/GC electrode than at disordered CNTs modified electrode. A reduction of the overpotential of more than 600 mV was achieved, and the anodic peak currents increased linearly with NADH concentration from 2×10^{-5} to 1×10^{-3} M with a low detection limit of about 5×10^{-7} M. Recently, Zhang et al. [219] developed the CHIT-CNTs system by integrating the redox mediator Toluidine Blue O (TBO) and CNTs in CHIT and used for the determination of NADH. In that work, they firstly covalently attached TBO to CHIT via a two-step procedure and then solubilized CNTs in the aqueous solution of CHIT-TBO. The resulting mixture was cast on the surface of a GCE to form thin CHIT-TBO/CNTs surface films. As compared to CHIT-TBO, the CHIT-TBO/CNTs films displayed large amplification of a current due to the TBO-mediated oxidation of NADH at -0.10 V. This was discussed in terms of the TBO/CNTs synergy that resulted in the improved charge propagation through the

CHIT-TBO/CNTs matrix. The electrocatalytic oxidation of NADH was also achieved by Liu et al. at PANI/PABS-SWNTs multilayer films [126].

4.2.1.2.2.2. Dehydrogenase-Based Electrochemical Biosensors

Based on the electrocatalytic oxidation of NADH at CNTs electrodes, several dehydrogenase- based electrochemical biosensors have been constructed. Wang et al. reported on a MWNTs-based alcohol biosensor, based on the incorporation of alcohol-dehydrogenase /NAD^+ within a three-dimensional MWNTs/Teflon matrix [220].

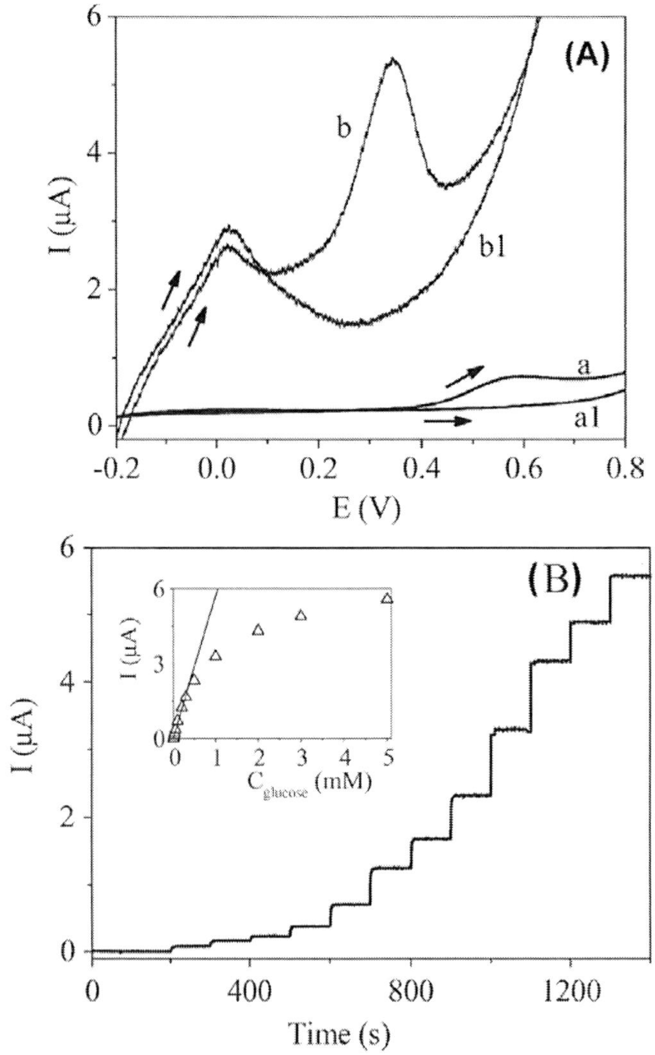

Figure 11. (A). Linear scan voltammograms recorded at (a) GC/CHIT and (b) GC/CNT-CHIT film electrodes in 0.10 mM NADH solution. Traces a1 and b1 were recorded in a background electrolyte solution. For the clarity of presentation, traces b and b1 were shifted downward by 6 μA along the y-axis. Scan rate, 50 mV/s. (B). Amperometric response of the GC/CNT-CHIT-GDI-GDH film electrode to additions of glucose aliquots (10 μM - 5.0 mM) into a stirred solution of pH 7.40 phosphate buffer that contained 0.10 mM NAD+. Insets: corresponding calibration plots. Potential, 0.400 V. [Adapted with permission from M. Zhang et al. [101], Anal. Chem., 76, 5045 (2004). Copyright 2004, American Chemical Society].

The MWNTs component offered a marked decrease in the overpotential for the oxidation of the liberated NADH along with minimal surface fouling effects, hence allowing convenient low-potential stable detection of the ethanol. Antiochia et al. developed a new mediated CNTP amperometric biosensor of fructose [221]. The biosensor was formed by a CNTP electrode modified with an electropolymerized film of 3,4-dihydroxy benzaldehyde (3,4-DHB) and was based on the activity of a commercial available D-fructose dehydrogenase (FDH) immobilized on an immobilon membrane placed on the top of the electrode surface. The biosensor response current was directly proportional to D-fructose concentration from 5×10^{-6} to 2×10^{-3} M with a detection limit of 1×10^{-6} M. On the basis of the electrocatalytic oxidation of NADH at a CNTs-CHIT composite film modified GCE, Zhang et al. recently proposed a simple but useful approach for the fabrication of glucose biosensors by the incorporation of glucose dehydrogenase (GDH) into the CNTs-CHIT films through the tethering of GDH to CHIT with GDI [101] (Figure 11). The stability and sensitivity of the GC/CNTs-CHIT-GDI-GDH biosensor allowed for the interference-free determination of glucose in the physiological matrix (urine). In pH 7.40 PBS, linear least-squares calibration plots over the range 5 ~ 300 μM glucose (10 points) had slopes 80 mA $M^{-1}cm^{-2}$ and a correlation coefficient 0.996. The detection limit was 3 μM glucose (S/N = 3). Another advantage of such an electrode system was that the oxygen did not influence its analytical performance. This was in contrast to oxidase-based electrodes that were prone to interferences from changing levels of oxygen in the samples. Rubianes et al. also studied the performance of enzymatic biosensors based on the immobilization of different dehydrogenase within a CNTPE prepared by dispersion of MWNTs and mineral oil [145]. The strong electrocatalytic activity of CNTs towards the oxidation of NADH allowed an effective low-potential amperometric determination of ethanol, in connection with the incorporation of alcohol dehydrogenase/NAD^{+}, within the composite matrix. Compared to the analogous enzymatic CPEs, a great enhancement in the response was observed at the enzymatic CNTPEs.

4.2.1.2.3. Other Enzymes

There were also some other enzymes used by several workers to prepare electrochemical biosensors, such as acetylcholinesterase, alkaline phosphatase, organophosphorus hydrolase, urease and tyrosinase. Based on the reversible Nernstian response of a SWNTs modified electrode to analytes associated with solution pH change, Xu et al. constructed voltammetric urease and acetylcholinesterase biosensors by immobilizing the enzymes with a sol-gel hybrid material that was cast on the surface of the SWNTs modified electrode [29], which were used for the determination of urea and acetylthiocholine according the potential shift of the anodic peak of SWNTs. The methods for preparing sensor and biosensor were simple and reproducible and the range of analytes could be extended to substrates of other hydrolyases and esterases. Lenihan et al. proposed a new approach to the modification of CNTs with biomolecules for the development of nanoscale biosensors [222]. Alkaline phosphatase was immobilized on the surface of MWNTs utilizing a LBL methodology. CNTs were incubated with streptavidin, resulting in the formation of a protein layer on the surface of the nanotubes. Biotinylated alkaline phosphatase was then allowed to bind to streptavidin, anchoring the sensing protein onto the surface. Electrochemical biosensors were constructed by using carbon nanotubes compacted into pellets. Electrodes constructed in this manner were observed to generate an electrochemical signal that was a function of substrate concentration.

Trojanowicz et al. prepared a pesticide paraoxon biosenors by the physical adsorption of organophosphorus hydrolase on the surface of MWNTs modified working graphite ink electrode of the three-electrode screen-printed sensing stripe [223]. Joshi et al. reported the fabrication of a disposable biosensor based on acetylcholinesterase-CNTs modified thick film strip electrode for organophosphorus (OP) insecticides [39]. The degree of inhibition of the enzyme acteylcholinesterase (AChE) by OP compounds was determined by measuring the electrooxidation current of the thiocholine generated by the AChE catalyzed hydrolysis of acteylthiocholine (ATCh). The large surface area and electro-catalytic activity of CNTs lowered the overpotential for thiocholine oxidation to 200 mV (vs. Ag|AgCl) without the use of mediating redox species and enzyme immobilization by physical adsorption. The biosensor detected as low as 0.5 nM (0.145 ppb) of the model organophosphate nerve agent paraoxon with good precision, reproducibility and stability. HRP was also employed by several workers to prepare hydrogen peroxide biosensors. Xu et al. used MWNTs used as a coimmobilization matrix for the incorporation of HRP and electron transfer mediator MB onto a GCE surface [224]. Cyclic voltammetry and amperometric measurements demonstrated the feasibility of MB as an electron carrier between the immobilized peroxidase and the surface of GCE. The amperometric response of this resulting biosensor to H_2O_2 showed a linear relation in the range from 4 μM to 2 mM with a detection limit of 1 μM (S/N=3). Yamamoto et al. reported a hydrogen peroxide biosensor based on the immobilization of HRP on the surface of a MWNTs modified electrode via the cross-linking of HRP and BSA composite film by GDI [27]. The MWNTs-HRP modified electrode showed excellent response of reduction current for the determination of H_2O_2 at the potential of -300 mV (vs. Ag|AgCl) with a linear concentration range of $3.0\times10^{-7} \sim 2.0\times10^{-4}$ M and a low detection limit of 1.0×10^{-7} M (S/N = 3).

4.2.2. Carbon Nanotubes-Promoted Direct Electron Transfer of Redox Proteins and Enzymes

The direct electron transfer of redox proteins and enzymes at electrode surface has attracted more and more attention recently. These works are of great importance for studying the biological redox process, which can help us understand the energy transform and material metaboly in life action. But unfortunately, it is difficult for these macrobiomolecules to achieve direct exchange electron with electrode surface because they usually have big and complex structures with the redox centers deeply immersing in the bodies and the adsorption of these species at the electrode surface often results in the change of their native configuration and the consequent denature. Earlier, mediators were employed by some groups to improve the electron contact between enzymes and electrodes. However, the using of mediators generally makes the biosensors complicated and leads to slow response. The development of the mediatorless biosensors, thus, becomes the goal of extensive researches. The first work regarding the direct electron transfer of proteins at CNTs electrodes was reported by Davis et al. in 1997 by the physical adsorption of horse heart cytochrome c (Cyt c) on CNTs [137]. Since then, many works focused on the achievement of the direct electron transfer of various redox proteins and enzymes at CNTs electrodes, including hemoglobin (Hb) [44,54,75,225], myoglobin (Mb) [48,50,226, 227], Cyt c [41,51,132,228-230], peroxidases (e.g. HRP [110,154,231,232], catalase (Ct) [32,233] and MP-11 [110,129,234]), oxidases (e.g. GOx [76,130,235-238], putrescine oxidase (POx) [239,240] and xanthine oxidase (XOs) [241]), and so on.

4.2.2.1. Direct Electron Transfer of Redox Proteins and Enzymes at Carbon Nanotubes Electrodes

4.2.2.1.1. Heme Proteins

4.2.2.1.1.1. Myoglobin

Myoglobin (Mb) is a small size heme protein with a molar mass of approximately 17 500, and it functions in both transport and short-term storage of oxygen. Due to its commercial availability and a known structure, Mb is considered to be an idea model protein for the study of heme proteins or enzymes.

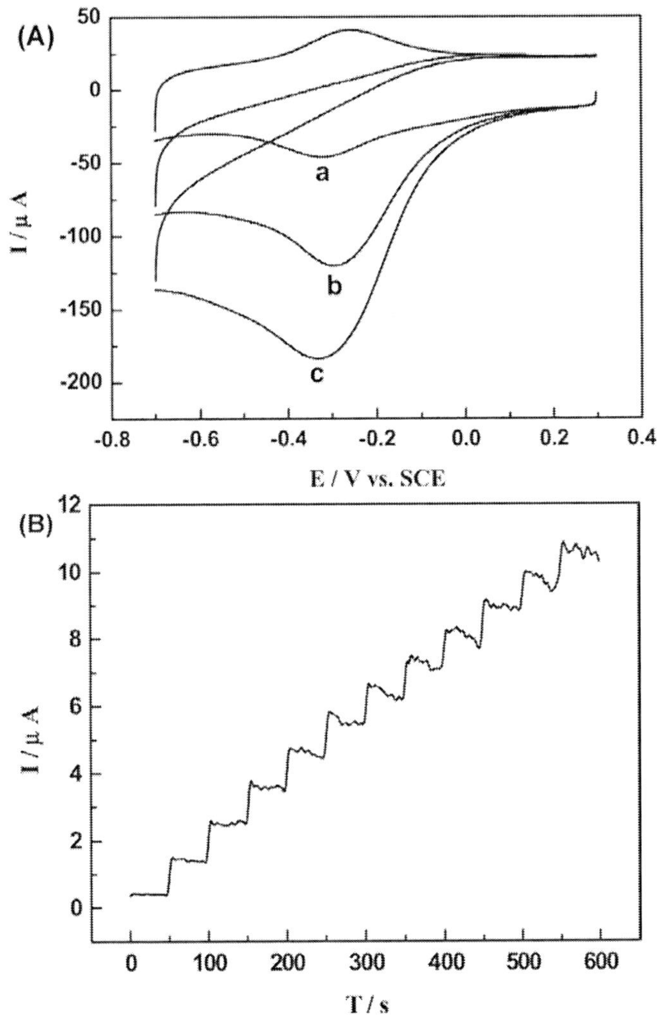

Figure 12. (A). CV of Mb/MWNTs/GCE in the presence of H2O2: (a) 0mM, (b) 0.4mM, and (c) 0.8mM; scan rate: 200 mV/s. (B). Amperometric response of H2O2 at Mb/MWNTs/GCE in 0.1 M PBS (pH 4.0); each addition was 30 µmol/l H2O2; operation potential -0.14 V. [Adapted with permission from G. Zhao et al. [50], Anal. Biochem., 329, 160 (2004). Copyright 2004, Elsevier Science].

Thus, the achievement of the direct electron transfer (DET) of Mb not only provides a convenient approach for fabricating Mb-based biosensors but also creates nice model systems for studying the biological functions of the proteins. Similar to other nanoscale materials, the good conductivity, the nanosized tubular structural and the functional groups make CNTs act as molecular wires to allow the electrical communication between the underlying electrodes and Mb. Recently Zhao's group developed a simple but effective method for the immobilization and the direct electron transfer of Mb at CNTs-based electrodes [48,50,226,227]. In a typical procedure, Mb was immobilized on the surface of a MWNTs modified electrode by immersing the CNTs electrode in an acetate buffer solution of Mb for a long period (e.g. 72 h), followed by thoroughly washing with double-distilled water to remove loosely bonded enzyme molecules. The resultant electrode exhibited a pair of well-defined redox peaks with cathodic and anodic peak potentials at 0.298 and 0.198 V (vs. SCE), respectively. The shapes of the cathodic and anodic peak were nearly symmetric, and the reduction and oxidation peaks have the same heights, indicating the reversible process of Mb at this electrode. This biosensor was further used for the sensitive detection of nitric oxide (NO) [48,227], O_2 [226] and H_2O_2 [50] (Figure 12).

4.2.2.1.1.2. Hemoglobin

Hemoglobin (Hb), functioning physiologically in the storage and transport of molecular oxygen in the blood of vertebrates, has a molecular weight of 64,000 to 67,000, and comprises four polypeptide subunits (two α and two β subunits), each of which has an iron-bearing heme within molecularly accessible crevices and has a similar structure. Although Hb does not play a role as an electron carrier in biological systems, it has been shown to possess enzyme-like catalytic activity. Similar to Mb, it is also an ideal model molecule for the study of direct electron transfer reactions of heme enzymes because of its commercial availability and a known and documented structure. The direct electron transfer of Hb at CNTs electrodes has been achieved by Cai et al. based on different immobilization methods [44,75]. Based on the facilitated dispersing of CNTs in water by CTAB, Cai et al. fabricated a CNTs-based Hb electrochemical biosensor by first casting the mixture of CNTs aqueous suspension and Hb phosphate buffer solution and fixing the resultant composite film with a Nafion coating [75]. Cyclic voltammetric results showed a pair of well-defined redox peaks, which corresponded to the direct electron transfer of Hb, with the formal potential ($E^{0'}$) at about 0.343 V (vs. SCE) in the PBS (pH 6.8). Further experiments demonstrated that the immobilized Hb retained its bioelectrocatalytic activity to the reduction of H_2O_2 and the electrocatalytic current was proportional to the concentration of H_2O_2 at least up to 20mM. Cai et al. also reported an adsorption method for the immobilization of Hb on a CNTs electrode by incubating the electrode into the protein solution (5 mg/ml in 0.1 M PBS, pH 5.5) for 1 h at 4 °C [44]. Jia et al. also reported the using of novel bamboo-like nitrogen- doping CNx nanotubes for the immobilization of Hb by dipping the CNTs electrode into a 0.2 M acetate buffer solution (pH 6.0) containing 10 mg/ml Hb over 72 h at 4 °C. The transmission electron microscopy (TEM) clearly showed that Hb could be strongly adsorbed on the surface of CNx nanotubes. The CV results of Hb/CNx nanotubes modified electrode showed a pair of well-defined, quasi reversible redox peaks centered at approximate 0.33 V (vs. SCE) in pH 7.0 PBS.

4.2.2.1.1.3. Cytochrome c

Cytochrome c (Cyt c) is a basic redox heme protein with 12,384 g/mol molecular weight and approximately spherical shaper with 34 A° diameter. It plays an important role in the biological respiratory chain, whose function is to receive electrons from Cyt c reductase and deliver them to Cyt c oxidase. As mentioned above, the direct electron transfer of Cyt at CNTs electrodes was first reported by Davis et al. in 1997 [137]. The electrochemistry of Cyt c was later studied by Wang et al. using cyclic voltammetry at a GCE modified with SWNTs [228]. A pair of well-defined redox waves was obtained in Cyt c aqueous solution at an activated SWNTs film-modified electrode. The electrode reaction of Cyt c was found to be a diffusion-controlled process. The peak current increased linearly with the concentration of Cyt c in the range from $3.0\times10^{-5} \sim 7.0\times10^{-4}$ M with a detection limit of 1.0×10^{-5} M. Chen et al. successfully immobilized MWNTs on the surface of Pt electrode by mixing with DNA [41]. The experimental data indicated that Cyt c can strongly adsorb on the surface of the modified electrode, and forms an approximate monolayer. The immobilized MWNTs can promote the redox of Cyt c, which gave reversible redox peaks with a formal potential of 81 mV vs. SCE. Recently, Qu et al. [132] characterized the SWNTs sub-monolayer film that was prepared by simply electrostatically adsorbing nanotubes onto a 2-aminoethanethiol self-assembled monolayer (SAM) on a gold bead electrode using tapping-mode atomic force microscopy (TM-AFM). The results showed that the orientation of SWNTs on the SAM was horizontal and the surface coverage was quite low. However, this electrode was demonstrated to be capable of facilitating the direct electron transfer of Cyt c in solution. Zhao et al. also studied the direct electron transfer of Cyt c on a MWNTs modified electrode [51]. In contrast to the time-consuming methods by immersing the CNTs electrodes in the protein or enzyme solutions for a long period [48,50,226,227], they fabricated the Cyt c modified CNTs electrodes by placing a MWNTs modified electrode in a Cyt c solution (0.09 mmol/l, pH 7.0 PBS) and then performing a consecutive cyclic potential scans in the potential range from 0.3 to -0.8 V with a scan rate 50 mV/s up to obtain a stable CV curve. The results from electrochemical impedance spectroscopy (EIS) and CV suggested that Cyt c could be tightly adsorbed on the surface of MWNTs and MWNTs showed an obvious promotion for the direct electron transfer between Cyt c and electrode. A couple of well-defined and quasi-reversible CV peaks of Cyt c can be observed in a PBS (pH 7.0). Cyt c adsorbed on MWNTs exhibited a remarkable electrocatalytic activity for the reduction of hydrogen peroxide (H_2O_2).

4.2.2.1.2. Peroxidases

4.2.2.1.2.1. Horseradish Peroxidase

Horseradish peroxidase (HRP), functioning physiologically as the biocatalysis of H_2O_2, has been extensively investigated, due to its commercial availability and well known structure. HRP has been used as a typical model for studying the structural, kinetic and thermodynamic properties of peroxidases. The first work on the direct electron transfer of HRP was reported by Zhao et al. using CNTs-based powder microelectrodes (CNTPME) [154]. The immobilization of HRP was achieved by mixing 1 mg CNTs powder with 200 ml 0.1 M PBS solution containing 1 mg/ml HRP and packing the dry mixture into the microcaves of CNTPME. The resultant electrode exhibited a pair of reversible redox peaks -0.7 and - 0.6 V in 0.1 M PBS (pH 7.0). It was found that the adsorbing HRP still kept its catalytic activity to H_2O_2 upon the immobilization on the CNTs electrode. Cai et al. [231] also developed a HRP biosensor based on the method used for constructing Mb biosensors [75]. It was found that HRP underwent effective and stable direct electron transfer reaction and retained its catalytic activity towards H_2O_2. On the basis of the stable dispersion of CNTs in RTILs, Zhao et al. prepared a HRP biosensor by first dispersing HRP and MWNTs in a RTIL, then grinding the mixture to form a black gel and casting the gel on the electrode surface with spinning [110]. HRP coated by RTILs-modified gold and glassy carbon electrodes allowed efficient electron transfer between the electrode and the protein and also catalyzed the reduction of O_2 and H_2O_2.

4.2.2.1.2.2. Catalase

Catalase (Ct) is a redox enzyme, present in almost all aerobic organisms. It catalyses the disproportionation of hydrogen peroxide into oxygen and water without the formation of free radicals, so it serves in part to protect cells from the toxic effects of hydrogen peroxide. Catalase contains four equal subunits, and each of them has a molecular weight of 57,000 daltons, is equipped with a Fe (III)-protoporphyrin-IX. The direct electron transfer of Ct was mainly accomplished through physical adsorption approaches. Wang et al. recently studied the direct electron transfer of Ct at a gold electrode modified with SWNTs [32]. They found that a pair of well-defined redox peaks was obtained for Ct with the reduction peak potential at -0.414 V and a peak potential separation of 32 mV in 0.05 M PBS (pH 5.9). Both reflectance FT-IR spectra and the dependence of the reduction peak current on the scan rate revealed that Ct adsorbed onto the SWNTs surfaces. Compared to other types of carbonaceous electrode materials (e.g., graphite and carbon soot), the electron transfer rate of Ct redox reaction was greatly enhanced at the SWNTs-modified electrode. The peak current was found to increase linearly with the Ct concentration in the range of $8\times10^{-6} \sim 8\times10^{-5}$ M used for the electrode preparation. The catalytic activity of Ct adsorbates at the SWNTs appears to be retained, as the addition of H_2O_2 produced a characteristic catalytic redox wave. Similar direct electron transfer of Ct at a SWNTs modified GCE was also studied by Zhao et al. [233].

4.2.2.1.2.3. Microperoxidase

Microperoxidase (MP-11) is an oligopeptide consisting of eleven amino acids and a covalently linked FeIII- protoporphyrin IX heme site. The direct electron transfer between MP-11 and electrodes can be used in biosensors and biofuel cells. In biofuel cells, the MP-11 modified electrodes are used as cathode materials to catalyze the reduction of oxygen. The direct electron transfer of MP-11 has recently been studied by Dong et al. in different CNTs-based electrode systems [110,234]. Based on the immobilization of enzymes in the RTIL-MWNTs composite films, Dong et al. also studied the direct electron transfer of MP-11 at this film modified electrodes [110]. In 0.1 M PBS, a pair of well-defined redox waves were obtained, corresponding to the active FeIII/II center of MP-11 at $E_{1/2}$ [$(E_{pa} + E_{pc})/2$ (vs. Ag|AgCl)] = -0.25 V with a peak separation of 61 mV at a scan rate of 100 mV/s. This result was different from the report by Willner et al. that $E_{1/2}$ = -0.4 V (vs. SCE), which was attributed to the fact that the redox active center in MP-11, an iron protoporphyrin IX, was not shielded by a polypeptide and was easily affected by the environments of the ionic liquids compared with other systems. They also studied the direct electron transfer of MP-11 by cyclic voltammetry at a GCE modified with MWNTs. In that work, MP-11 was immobilized by immersing the CNTs electrode in 0.1 M PBS containing 1 mg/ml MP-11 at 4 °C for about 10 h. The resulting electrode showed a pair of well-defined redox peaks with a formal potential at about -0.26 V (vs. Ag|AgCl) in PBS (pH 7.0). The experimental results also demonstrated that the immobilized MP-11 retained its bioelectrocatalytic activity to the reduction of H_2O_2 and O_2.

4.2.2.1.3. Oxidases

4.2.2.1.3.1. Glucose Oxidase

Glucose oxidase (GOx), from Aspergillus or Penicillium, is a homodimer with a molecular weight of about 150 ~ 180 kDa containing two tightly bound flavine adenine dinucleotide (FAD) cofactors and catalyzes the electron transfer from glucose to oxygen accompanying the production of gluconic acid and hydrogen peroxide. Industrially, it has been used in the production of gluconic acid. The most important application is in biosensors for the quantitative determination of glucose in body fluids, foodstuffs, beverages, and fermentation liquor. As mentioned above, although CNTs have been widely used to fabricate the GOx-based glucose electrochemical biosensors due to the excellent electrocatalytic activity of CNTs towards H_2O_2, few works accomplished the direct electron transfer between GOx and the underlying electrodes. Xu et al. recently studied the electrochemical behavior of GOx at a MWNTs modified electrode by cyclic voltammetry [235]. They found that the MWNTs electrode exhibited two pairs of redox peaks with formal potentials of -0.45 and -0.55 V (vs. SCE) when immersed into 0.1 M PBS containing 2 mg/ml GOx for 2 h, which were attributed to the responses from GOx in the solution and adsorbed on the MWNTs surface, respectively. In addition, with the increase of scan time, the currents of both redox peaks decreased except that the redox peaks at -0.45 V disappeared after about 1 h , and the redox peaks at -0.55 V tended to a stable value. The reservation of adsorbed GOx at MWNTs foretells the possibility of immobilization of GOx on CNTs by physical adsorption. Wang et al. proposed a simple approach for the immobilization and the direct electron transfer of GOx at CNTs electrodes [237]. The GOx modified SWNTs (GOx/SWNTs) electrode was prepared by just dropping 5μL 1.0×10^{-4} M GOx solution onto the surface of the electrode surface and allowing to dry under ambient conditions. The resultant GOD/SWNTs electrode was rinsed completely with water and the working buffer solution prior to any measurements. A pair of well-defined redox peaks was obtained for GOx with the reduction peak potential at -0.465 V and a peak potential separation of 23 mV in 0.05 M PBS. The specific enzyme activity of GOx adsorbates at the SWNTs was proved to be retained by spectral methods. Based on the facilitated dispersing of CNTs in water by CTAB [75], Cai et al. also fabricated a CNTs-based GOx biosensor by first casting the mixture of CNTs aqueous suspension and GOx phosphate buffer solution and fixing the resultant composite film with a Nafion coating [76]. Cyclic voltammetric results showed a pair of well-defined redox peaks, which corresponded to the direct electron transfer of GOx, with a midpoint potential of about -0.466 V (vs. SCE) in the PBS (pH 6.9). The experimental results also demonstrated that the immobilized GOx retained its bioelectrocatalytic activity for the oxidation of glucose, suggesting that the electrode may find use in biosensors (for example, it may be used as a bioanode in biofuel cells). More recently, Luong et al. proposed a simple procedure for the fabrication of CNTs-based glucose electrochemical sensors [221]. 3-Aminopropyltriethoxy-silane (APTES) was used as a solubilizing agent for CNTs and an immobilization matrix for GOx to construct a mediatorless biosensor. The resultant GOx biosensor exhibited a pair of well-defined redox peaks at around 0.45 V (vs. Ag|AgCl).

4.2.2.1.3.2. Putrescine Oxidase and Xanthine Oxidase

Besides GOx, the direct electron transfer of some other oxidases like putrescine oxidase (POx) and xanthine oxidase (XOx) was also investigated. Similar to other oxidases, POx is rather big molecule (>200 kDa) with the redox center deeply inlaid in its structure, which reacts with biogenic amines (e.g. putrescine, cadaverine and histamine) with a certain degree of substrate specifically as follows:

$$H_2N\text{-}(CH_2)_n\text{-}NH_2 + O_2 + H_2O \xrightarrow{POx} H_2N\text{-}(CH_2)_{n-1}\text{-}CHO + NH_3 + H_2O_2$$

Although feasible, the electrochemical measurement of H_2O_2 at positive potentials usually suffers from the serious interferences from some electroactive micromolecules like UA and AA, even for CNTs electrodes. The direct electron transfer of POx at CNTs electrodes was mainly accomplished by Luong et al. [239,240]. In their earlier work, POx was immobilized on the electrode surface by the deposition of a mixture of MWNTs, PDDA, APTES, and POx on the electrode surface and the coating of the resulting electrode by a Nafion film [239]. The use of APTES permits the adsorption of GOx on CNTs. They believed that nanoscale 'dendrites' of MWNTs were reasoned to form a network, projecting outward from the electrode surface acting like bundled ultra-microelectrodes, thereby permitting access to the active site and facilitating direct electron transfer to the immobilized enzyme. This was demonstrated by the appearance of a pair of redox peaks for POx with an oxidation peak near -0.45 V and a reduction peak near -0.49 V (vs. Ag|AgCl, 3 M NaCl) in 0.05 M PBS (pH 7.0). The direct electron transfer permitted the detection of putrescine at negative potentials, circumventing the interference of endogenous ascorbic and uric acids. Compared with the most common interfering species, such as spermine, spermidine, cadaverine, and histamine, a detection limit of 5 μM and a response 20 times greater were found for putrescine. Later, Luong et al. further fabricated a POx-based biosensor through a covalent method [83]. In that work, MWNTs were dissolved in a mixture of APTES, Nafion and ethanol with sonication to get uniformly dispersed CNTs. The resulting mixture was dropped on the electrode surface and allowed to dry. Upon the enzyme preparation, a drop of the enzyme solution was dried on the CNTs/APTES modified electrode and a certain volume of GDI was dropped on the resulting electrode to crosslink the enzyme with APTES. Similar to the noncovalent immobilization method [239], POx exhibited a pair of well-defined peaks at -0.52 V (vs. Ag|AgCl) in 0.05 M borate buffer (pH 8.35), providing a selective and sensitive method for putrescine determination. Similar to the immobilization of GOx on CNTs electrodes [237], Wang et al. also reported the noncovalent immobilization of XOx on CNTs electrodes by just dropping the XOx solution onto the surface of the electrode surface and allowing to dry under ambient conditions. The resulting electrodes were completely rinsed with water prior to any measurements, which exhibited a pair of well-defined redox peaks at around -0.46 V with a small peak separation of 28 mV in 0.05 M phosphate buffer (pH 7.0).

4.2.2.2. Fundamentals of Direct Electron Transfer of Redox Proteins and Enzymes at Carbon Nanotubes Electrodes

Based on the above discussions, it is clear that CNTs electrodes can effectively facilitate the direct electron transfer of various redox proteins and enzymes. However, few works gain insights into the fundamental information of this important process. Guiseppi-Elie et al. studied the direct electron transfer between SWNTs and the redox center of adsorbed GOx [242]. It was found that both FAD and GOx spontaneously adsorbed to annealed CNTs and displayed quasi-reversible one-electron transfer. They believed that the nanotubes become positioned within a tunneling distance of the cofactors with little consequence to denaturation. Willner's group proposed an approach for the covalent attachment of reconstituted GOx on the end of aligned SWNTs [243]. The electrons were transported along distances higher than 150 nm and the transport rate was controlled by the nanotube length, interfacial electron transfer rate constants for 50 and 100 nm SWNTs were 42 and 19 s^{-1}, respectively. Such plugging of CNTs into enzymes was also reported by Gooding et al. by two different ways [130]. In the first method, native GOx was covalently attached to the ends of the aligned tubes, which allowed close approach to FAD and direct electron transfer to be observed with a rate constant of 0.3 s^{-1}. The second strategy was similar to the method proposed by Willner et al. [243], i.e. FAD was attached to the ends of the tubes and the enzyme reconstituted around the surface immobilized FAD. This latter approach allowed more efficient electron transfer to the FAD with a rate constant of 9 $^{-1}$. The direct electron transfer of both native and reconstituted GOx at aligned CNTs also demonstrated the molecular wire role of CNTs in promoting the electron transfer of redox proteins and enzymes. Rusling et al. [134] investigated the electrochemical behaviors of HRP and Mb covalently bound to the ends of SWNTs forest electrodes. Their data indicated that the "trees" in the nanotube forest behaved electrically similar to a metal, conducting electrons from the external circuit to the redox sites of the enzymes. The peroxidase activity of these enzymes was demonstrated, with detection limits for hydrogen peroxide in buffer solutions of ~100 nM. The direct electron transfer of MP-11 covalently attached to aligned SWNTs arrays was also examined by Gooding et al. [129]. It was found that although the enzyme also adsorbed along the hydrophobic walls of the nanotubes, the majority of the electron transfer was dominated by proteins immobilized on the electroactive ends of the nanotubes, which further supported the excellent electrochemical properties of the ends of the nanotubes. In addition, despite different distributions of tube lengths, the rate constants for heterogeneous electron transfer to the enzyme were essentially the same, demonstrating the efficiency of the shortened nanotubes as molecular wires. Jiang et al. recently studied the electrochemical redox processes of Cyt c at a SWNTs modified electrode by in situ UV-vis and circular dichroism (CD) spectroelectrochemistry [37]. It was found that SWNTs can effectively decrease the effect of electric field on the configurational changes of Cyt c and stabilize the Met80-Fe bond of adsorbed Cyt c, and that electron transfer induced a more opening configurational change at SWNTs/GCE than that at the bare electrode, which might be responsible for the direct electron transfer of enzymes on CNTs electrodes and the retaining of their biocatalytic activity. The CNTs-induced configurational change of enzymes has been verified by Gooding et al. [244]. Rate constants for direct electron transfer to GOx at CNTs-modified electrodes and graphite electrodes suggest the enzyme was partially denatured from its native configuration, thus bringing the redox-active center of the enzyme closer to the electrode and allowing appreciable electron transfer.

4.3. Carbon Nanotubes-Based DNA Electrochemical Sensors

Sequence-specific DNA detection has been a topic of significant interest due to its promising application in the diagnosis of pathogenic and genetic diseases. Electrochemical DNA biosensors are currently receiving considerable attention because of their high sensitivity, portability, low-cost and rapid response. Such devices rely on the conversion of the nucleic acid recognition processes into a useful electrical signal. Similar to other electrochemical sensors, the DNA electrochemical sensors benefit much from the use of CNTs based on the enhanced oxidation of the oxidable bases, the improved responses of the indicaters or the excellent properties of CNTs as the carrier platforms.

Fang's group has reported several CNTs-based DNA electrochemical sensors for the determination of sequence-specific DNA over the last few years [166,245-247,]. In an earlier work, they a novel and sensitive electrochemical DNA biosensor based on MWNTs functionalized with carboxylic acid groups for the covalent DNA immobilization and the enhanced hybridization detection [245]. A MWNTs modified GCE was fabricated from a simple casting method and oligonucleotides with the 5'-amino group were covalently bonded to the carboxyl groups of CNTs with the aid of EDC. The hybridization process on the electrode was monitored by DPV analysis using an electroactive intercalator daunomycin (DNR) as an indicator. The sensitive response of DNR at CNTs electrode led to a significant increase of the hybridization signal of DNA compared to DNA sensors with oligonucleotides directly incorporated on carbon electrodes. Similar improved hybridization responses of DNA covalent immobilized on CNTs electrodes were also observed by Wang et al. [248] and Kerman et al. [249] using MB and Escherichia coli single-strand binding protein (SSB) as the indicator, respectively. Wang et al. proposed an effective method for amplifying electrical detection of DNA hybridization based on CNTs carrying a large number of CdS particle tracers [250]. Such use of CNT amplification platforms was combined with an ultrasensitive stripping voltammetric detection of the dissolved CdS tags following dual hybridization events of a sandwich assay on a streptavidin modified 96-well microplate. Fang et al. also reported a new indicator free biosensing strategy for direct electrochemical detection of DNA hybridization by AC impedance measurement [246]. They first immobilized the oligonucleotide probes on a MWNTs modified GCE through the entrapping of DNA as the sole counter anion during the electropolymerization growth of the PPy films. Before and after hybridization reaction with the complementary DNA sequences, a decrease of impedance values was observed as a result of the reduction of the electrode resistance: Hybridization amounts of the one-base, two-base and three-base mismatched sequences were obtained only in a 51%, 18% and 8.2% response when compared to that for the complementary matched sequence. Such unique response was attributed to the concomitant conductivity changes of the PPy-polymerized CNTs, and offered great promise for reagentless DNA hybridization analysis. Later, they further improved this method by adding zinc ions (Zn^{2+}) to the hybridization solution, resulting in the formation of metallized double-stranded DNA (M-DNA) that can significantly change the electrochemical impedance response [247]. Recently, Fang et al. reported a new sensitive electrochemical biosensor based on magnetite nanoparticle for monitoring DNA hybridization by using MWNTs/PPy-modified glassy carbon electrode [166]. In such detection system, mercapatoacetic acid (RSH)-coated magnetite nanoparticles, capped with 5'-(NH2) oligonucleotide, was used as DNA probe to complex 29-base polynucleotide target (a piece of human porphobilinogen deaminase PBGD promoter from 170 to 142). Target sequence hybridized with the probe resulted in the decrease of the reduction peak current of DNR connected with probe. As mentioned above, Meyyappan et al. reported a novel approach for the fabrication of nanoelectrode arrays using vertically aligned MWNTs embedded within a SiO_2 matrix [182,183] (Figure 13). Based on the a low background and a small cell time constant of these nanoelectrodes, they prepared a highly sensitive DNA sensors by covalently attaching the DNA probes on the ends of the CNTs and monitoring the electrochemical response of $Ru(bpy)_3^{2+}$ mediator amplified guanine oxidation.

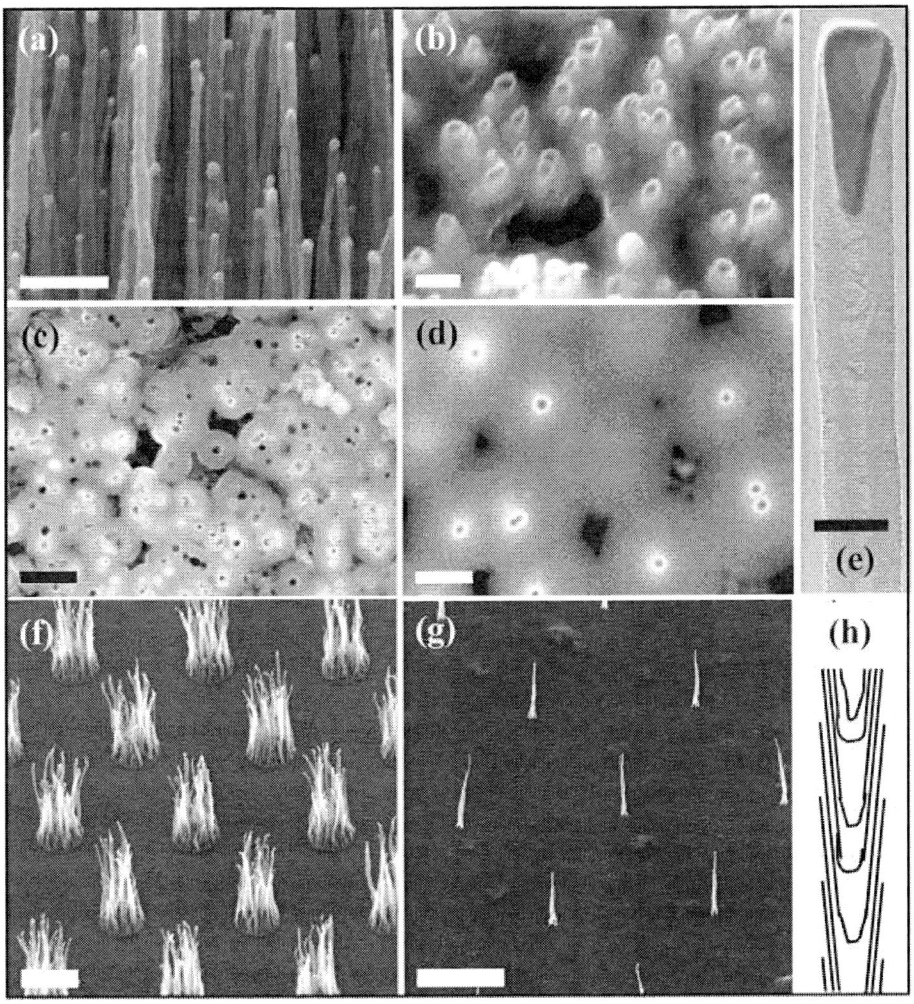

Figure 13. SEM images of (a) an as-grown MWCNT array, (b) the surface of a polished MWCNT array embedded in an SiO2 matrix, (c) a high-density MWCNT nanoelectrode array (hd, ~2×109 electrodes cm-2), (d) a low-density one (ld-1, ~7×107 electrodes cm-2), (f) an as-grown MWCNT array on 2 mm Ni spots patterned with UV lithography, (g) an as-grown MWCNT array on ~100 nm diameter Ni spots patterned with e-beam lithography, and (h) a schematic of the defective bamboo-like structure. (a), (b), (f), and (g) are 45° uperspective views while (c) and (d) are top views. (e) is the TEM view of a single MWCNT. The scale bars in (a)–(g) are 500 nm, 200 nm, 500 nm, 500 nm, 50 nm, 2 μm, and 5 μm, respectively. [Reprinted with permission from J. Koehne et al. [183], J. Mater. Chem., 14, 676 (2004). Copyright 2003, Royal Society of Chemistry].

Wu et al. has studied the electrochemical behaviors of DNA, guanine and adenine at a MWNTs-DHP composite film-modified electrode (MWNTs-DHP/GCE) [93]. Compared with the undiscerned responses at the bare electrode, adenine and guanine exhibited a sensitive oxidation peak at 0.73 and 1.02 V on MWNTs-DHP/GCE in 0.1 M PBS (pH 7.0), respectively. These responses were also observed for the hydrolyzed product of calf thymus DNA, providing a simple but sensitive method for DNA measurement. The enhanced oxidation of guanine and adenine residues in DNA was also observed by Wang et al. at a SWNTs modified GCE [251]. On the basis of the enhanced oxidation of guanine on pretreated CNTPEs, Pedano et al. recently proposed an adsorptive stripping potentiometric method for the measurement of trace levels of nucleic acids [142]. Trace (μg/l) levels of the oligonucleotides and polynucleotides can be readily detected following short accumulation periods with detection limits of 2.0 μg/l for a 21 bases oligonucleotide and 170 μg/l for calf thymus dsDNA. The interaction between nucleic acids and CNTPE was demonstrated to be mainly hydrophobic.

Besides the application in DNA detection, CNTs were also used as the carrier platforms of DNA by several workers. Guo et al. examined the electrochemical characteristics of the covalent immobilization of calf thymus DNA molecules on MWNTs by EDC [43]. By studying the CV and electrochemical impedance spectroscopy (EIS) of $Fe(CN)_6^{3-}/Fe(CN)_6^{4-}$, they found that most of calf thymus DNA are covalently immobilized on MWNTs via diimide-activated amidation between the carboxylic acid groups on CNTs and the amino groups on DNA bases, though the direct adsorption of the DNA molecules on MWNTs can be also observed. The studies on the interaction between DNA molecules and small biomolecules ethidium bromide (EB) implied that DNA molecules immobilized at the surface of MWNTs still had the ability to interact with small biomolecules. Based on the strong adsorption of PDDA on CNTs, Guo et al. also proposed a LBL method for the immobilization of DNA on the electrode surface and monitored the electrostatic assembly of DNA on PDDA modified CNTs by the piezoelectric quartz crystal impedance (PQCI) technique [42]. It was found that PDDA played a key role in the attachment of DNA to MWNTs and acted as a bridge to connect DNA with MWNTs, though the direct adsorption of DNA on MWNTs has been observed. Recently, He et al. introduced a new fabrication of DNA-CNTs particles using the LBL technique on SWNTs [125]. PDDA, a positively charged polyelectrolyte, and DNA as a negatively charged counterpart macromolecule were alternatively deposited on the water-soluble oxidized SWNTs. Pure DNA/PDDA/SWNTs particles can be prepared and separated by simple unltracentrifugation. An electrode modified by the DNA/PDDA/SWNTs particles shows a dramatic change of the electrochemical signal in solutions of $(Ru(bpy)_3)^{2+}$ as a reporting redox probe. A preliminary application of the DNA-modified carbon nanotubes in the development of DNA sensors used in the investigation of DNA damage by nitric oxide was presented.

5. Conclusion and Perspectives

As a new type of carbonaceous materials, carbon nanotubes (CNTs) possess some unique properties that are much different from either the conventional-scaled materials or other types of nanoparticles, such as the special and well-defined tubular structure of nanosizes, the excellent chemical stability, the modifiable surface and the strong electrocatalytic activity. These special properties foresee their promising applications in electroanalytical chemistry for constructing CNTs-based sensors. The fact is that the modification of electrodes with CNTs has been observed to apparently improve the responses of substrates from small H_2O_2 molecules to huge redox proteins. Most of current works mainly focused on the fabrication of randomly dispersed CNTs-based electrochemical sensors and few reports deal with the fabrications of well organized nanostructures of CNTs. As a tubular nanomaterials, the key advantages of CNTs are their small diameter and huge length to diameter ratio that allow them to be used as molecular wires for facilitating electron transfer between proteins and electrodes or to be organized in a low density manner for using as a single nanoelectrode or nanoelectrode arrays with ultra-sensitivity. Since the direct growth by CVD methods inevitably surfers from the need of rigorous conditions, development of simple physical or chemical immobilization methods that can control the orientation of CNTs on electrode surfaces has become a matter of great urgency.

ACKNOWLEDGEMENTS

The research is supported by the National Natural Science Foundation of China (Nos. 30370397 and 60571042).

REFERENCES

[1] P.J. Britto, K.S.V. Santhanam, P. M. Ajayan, *Bioelectrochem. Bioenerg.* 1996, 41, 121.

[2] Q. Zhao, Z. H. Gan, Q. K. Zhuang, *Electroanalysis* 2002, 14, 1609.

[3] N. Q. Li, J. X. Wang, M. X. Li, *Rev. Anal. Chem.* 2003, 22, 19.

[4] J. Li, J. E. Koehne, A. M. Cassell, H. Chen, H. T. Ng, Q. Ye, W. Fan, J. Han, M. Meyyappan, *Electroanalysis* 2005, 17, 15.

[5] J. Wang, *Electroanalysis* 2005, 17, 7.

[6] J. J. Gooding, *Electrochim. Acta* 2005, 50, 3049.

[7] R.L. McCreery, *Electroanalytical Chemistry*, vol. 17, Dekker, New York, 1991.

[8] J. Li, A. Cassell, L. Delzeit, J. Han, M. Meyyappan, *J. Phys. Chem. B* 2002, 106, 9299.

[9] J. M. Nugent, K. S. V. Santhanam, A. Rubio, P. M. Ajayan, *Nano Lett.* 2001, 1, 87.

[10] P. Papakonstantinou, R. Kern, J. Irvine, E. McAdams, J. McLaughlin, T. McNally, *Fullerenes Nanotubes and Carbon Nanostructures* 2005, 13, 275.

[11] R. R. Moore, C. E. Banks, R. G. Compton, *Anal. Chem.* 2004, 76, 2677.

[12] C. E. Banks, R. R. Moore, T. J. Davies, R. G. Compton, *Chem. Commun.* 2004, 6, 1804.

[13] C. E. Banks, T. J. Davies, G. G. Wildgoose, R. G. Compton, *Chem. Commun.* 2005, 7, 829.

[14] A. Chou, T. Bocking, N. K. Singh, J. J. Gooding, *Chem. Commun.* 2005, 7, 842.

[15] H. X. Luo, Z. J. Shi, N. Q. Li, Z. N. Gu, Q. K. Zhuang, *Anal. Chem.* 2001, 73, 915.

[16] J. Liu, A. G. Rinzler, H. Dai, J. H. Hafner, R. K. Bradley, P. J. Boul, A. Lu, T. Iverson, K. Shelimov, C. B. Huffman, F. Rodriguez-Macias, Y. S. Shon, T. R. Lee, D. T. Colbert, R. E. Smalley, *Science* 1998, 280, 1253.

[17] H. Hu, P. Bhowmik, B. Zhao, M. A. Hamon, M. E. Itkis, R. C.Haddon, *Chem. Phys. Lett. 2001*, 345, 25.

[18] D. B. Mawhinney, V. Naumenko, A. Kuznetsova, J. T. Yates, J. Liu, R. E. Smalley, *Chem. Phys. Lett.* 2000, 324, 213.

[19] C. Bower, A. Kleinhammes, Y. Wu, O. Zhou, *Chem. Phys. Lett.* 1998, 288, 481.

[20] A. Kukovecz, C. Kramberger, M. Holzinger, H. Kuzmany, J. Schalko, M. Mannsberger, A. Hirsch, *J. Phys. Chem. B* 2002, 106, 6374.

[21] M. T. Martinez, M. A. Callejas, A. M. Benito, M. Cochet, T. Seeger, A. Anson, J. Schreiber, C. Gordon, C. Marhic, O. Chauvet, W. K. Maser, *Nanotechnology* 2003, 14, 691.

[22] A. Kuznetsova, D. B. Mawhinney, V. Naumenko, J. T. Yates, J. Liu, R. E. Smalley, *Chem. Phys. Lett.* 2000, 321, 292.

[23] J. N. Barisci, G. G. Wallace, D. Chattopadhyay, F. Papadimitrakopoulos, R. H. Baughman, *J. Electrochem. Soc.* 2003, 150, E409.

[24] J. N. Barisci, G. G. Wallace, R. H. Baughman, *J. Electroanal. Chem.* 2000, 488, 92.

[25] M. Musameh, N. S. Lawrence, J. Wang, *Electrochem. Commun.* 2005, 7, 14.

[26] Y.H. Lin, F. Lu, Y. Tu, Z. F. Ren, *Nano Lett.* 2004, 4, 191.

[27] K. Yamamoto, G. Shi, T. S. Zhou, F. Xu, J. M. Xu, T. Kato, J. Y. Jin, L. Jin, *Analyst* 2003, 128, 249.

[28] X. X. Yan, D. W. Pang, Z. X. Lu, H. Q. Lu, H. Tong, *J. Electroanal. Chem.* 2004, 569, 47.

[29] Z. Xu, X. Chen, X. H. Qu, J. B. Jia, S. Dong, *Biosens. Bioelectron.* 2004, 20, 579.

[30] Zhao Q., Gu Z. N., Zhuang Q. K. Electrochem. Commun. 2004, 6: 83.

[31] J. Wang, S. B. Hocevar, B. Ogorevc, *Electrochem. Commun.* 2004, 6, 176.

[32] L. Wang, J. X. Wang, F. M. Zhou, *Electroanalysis* 2004, 16, 627.

[33] Z. H. Gan, Q. Zhao, Z. N. Gu, Q. K. Zhuang, *Anal. Chim. Acta* 2004, 511, 239.

[34] N. S. Lawrence, R. P. Deo, J. Wang, *Anal. Chim. Acta* 2004, 517, 131.

[35] Y. P. Ding, W. L. Liu, Q. S. Wu, X. G. Wang, *J. Electroanal Chem.* 2005, 575, 275.

[36] L. Y. Jiang, R. X. Wang, X. M. Li, L. P. Jiang, G. H. Lu, *Electrochem. Commun.* 2005, 7, 597.

[37] X. Jiang, Z. L. Zhang, H.Y. Bai, X.H. Qu, J. G. Jiang, E. K. Wang, S. J. Dong, *Spectrochim. Acta Part A* 2005, 61, 943.

[38] Q. Zhao, L. H. Guan, Z. N. Gu, Q. K. Zhuang, *Electroanalysis* 2005, 17, 85.

[39] K. A. Joshi, J. Tang, R. Haddon, J. Wang, W. Chen, A. Mulchandani, *Electroanalysis* 2005, 17, 54.

[40] W. Zhang, F. L. Wan, W. Zhu, H. H. Xu, X. Y. Ye, R. Y. Cheng, L. T. Jin, *J. Chromatogr. B* 2005, 818, 227.

[41] G. Wang, J. J. Xu, H. Y. Chen, *Electrochem. Commun.* 2002, 4, 506.

[42] M. L. Guo, J. H. Chen, L. H. Nie, S. Z. Yao, *Electrochim. Acta* 2004, 49, 2637.

[43] M. L. Guo, J. H. Chen, D. Y. Liu, L. H. Nie, S. Z. Yao, *Bioelectrochemistry* 2004, 62, 29.

[44] J. Chen, C. X. Cai, *Chin. Chem. Lett.* 2004, 15, 813.

[45] Y. J. Zhang, Y. Wen, Y. Liu, D. Li, J. H. Li, *Electrochem. Commun.* 2004, 6, 1180.

[46] M. L. Guo, J. H. Chen, J. Li, B. Tao, S. Z. Yao, *Anal. Chim. Acta* 2005, 532, 71.
[47] F. H. Wu, G. C. Zhao, X. W. Wei, *Electrochem. Commun.* 2002, 4, 690.
[48] G. C. Zhao, L. Zhang, X. W. Wei, Z. S. Yang, *Electrochem. Commun.* 2003, 5, 825.
[49] F. H. Wu, G. C. Zhao, X. W. Wei, Z. S. Yang, *Microchim. Acta* 2004, 144, 243.
[50] G. C. Zhao, L. Zhang, X. W. Wei, *Anal. Biochem.* 2004, 329, 160.
[51] G. C. Zhao, Z. Z. Yin, L. Zhang, X. W. Wei, *Electrochem. Commun.* 2005, 7, 256.
[52] C. G. Hu, W. L. Wang, S. X. Wang, W. Zhu, Y. Li, *Diamond and Related Materials* 2003, 12, 129.
[53] J. Y. Qu, Y. Shen, X. H. Qu, S. J. Dong, *Chem. Commun.* 2004, 1, 34.
[54] N. Q. Jia, L. J. Wang, L. Liu, Q. Zhou, Z. Y. Jiang, *Electrochem. Commun.* 2005, 7, 349.
[55] S. Lefrant, M. Baibarac, I. Baltog, J. Y. Mevellec, L. Mihut, O. Chauvet, *Synth. Metals* 2004, 144, 133.
[56] S. H. Lim, J. Wei, J. Y. Lin, Q. T. Li, J. KuaYou, *Biosens. Bioelectron.* 2005, 20, 2341.
[57] C. A. Furtado, U. J. Kim, H. R. Gutierrez, L. Pan, E. C. Dickey, P. C. Eklund, *J. Am. Chem. Soc.* 2004, 126, 6095.
[58] G. Fortier, M. Vaillancourt, D. Belanger, *Electroanalysis* 1992, 4, 275.
[59] Z. Fan, D. J. Harrison, *Anal. Chem.* 1992, 64, 1304.
[60] Y. Y. Sun, K. B. Wu, S. S. Hu, *Chemical Journal Of Chinese Universities-Chinese* 2002, 23, 2067.
[61] J. Wang, M. Musameh, Y. H. Lin, *J. Am. Chem. Soc.* 2003, 125, 2408.
[62] K. B. Wu, S. S. Hu, *Microchim. Acta* 2004, 144, 131.
[63] S. R. Hocevar, J. Wang, R. P. Deo, M. Musameh, B. Ogorevc, *Electroanalysis* 2005, 17, 417.
[64] W. S. Huang, C. H. Yang, S. H. Zhang, *Anal. Bioanal.Chem.* 2003, 375, 703.
[65] W. S. Huang, W. B. Hu, J. C. Song, *Talanta* 2003, 61, 411.
[66] Y. C. Tsai, J. M. Chen, S. C. Li, F. Marken, *Electrochem. Commun.* 2004, 6, 917.
[67] R. P. Deo, J. Wang, *Electrochem. Commun.* 2004, 6, 284.
[68] K. P. Gong, Y. Dong, S. X. Xiong, Y. Chen, L. Q. Mao, *Biosens. Bioelectron.* 2004, 20, 253.
[69] Y. C. Tsai, S. C. Li, J. M. Chen, *Langmuir* 2005, 21, 3653.
[70] K. B. Male, S. Hrapovic, Y. L. Liu, D. S. Wang, J. T. Luong, *Anal. Chim. Acta* 2004, 516, 35.
[71] S. Hrapovic, Y. L. Liu, K. B. Male, J. H. T. Luong, *Anal. Chem.* 2004, 76, 1083.
[72] R. P. Deo, N. S. Lawrence, J. Wang, *Analyst* 2004, 129, 1076.
[73] T. Abatemarco, J. Stickel, J. Belfort, B. P. Frank, P. M. Ajayan, G. Belfort, *J. Phy.s Chem. B* 1999, 103, 3534.
[74] R. S. Chen, W H. Huang, H. Tong, Z. L. Wang, J. K. Cheng, *Anal. Chem.* 2003, 75, 6341.
[75] C. X. Cai, J. Chen, *Anal. Biochem.* 2004, 325, 285.
[76] C. X. Cai, J. Chen, *Anal. Biochem.* 2004, 332, 75.
[77] Y. Y. Sun, J. J. Fei, K. B. Wu, S. S. Hu, *Anal. Bioanal. Chem.* 2003, 375, 544.
[78] K. B. Wu, Y. Y. Sun, S. S. Hu, *Sens. Actuators B* 2003, 96, 658.
[79] K. B. Wu, X. B. Ji, J. J. Fei, S. S. Hu, *Nanotechnology* 2004, 15, 287.
[80] F. Wang, J. J. Fei, S. S. Hu, C*olloids Surf. B* 2004, 39, 95.

[81] S. F. Lu, *Anal. Lett.* 2003, 36, 2597.
[82] S. F. Lu, X. P. Dang, K. B. Wu, S. S. Hu, *J. Nanosci. Nanotech.* 2003, 3, 401.
[83] Y. H. Wu, X. B. Ji, S. S. Hu, *Bioelectrochemstry* 2004, 64, 91.
[84] S. F. Lu, *Chin. Anal. Chem.* 2004, 32, 412.
[85] S. F. Lu, *Microchem. J.* 2004, 77, 37.
[86] S. F. Lu, K. B. Wu, X. P. Dang, S. S. Hu, *Talanta* 2004, 63, 653.
[87] S. F. Lu, *Russ. J. Electrochem.* 2004, 40, 750.
[88] H. J. Zhang, K. B. Wu, *Microchim. Acta* 2005, 149, 73.
[89] W. Y. Qu, H. Wang, H. B. Wu, *Collection Of Czechoslovak Chemical Communications* 2005, 70, 178.
[90] H. J. Zhang, *J. Nanoparticle Research* 2004, 6, 665.
[91] H. J. Zhang, C. G. Hu, S. J. Wu, S. S. Hu, *Electroanalysis* 2005, 17, 749.
[92] S. F. Lu, *Anal. Sci.* 2003, 19, 1309.
[93] K. B. Wu, J. J. Fei, W. Bai, S. S. Hu, *Anal. Bioanal. Chem.* 2003, 376, 205.
[94] K. B. Wu, J. J. Fei, S. S. Hu, *Anal. Biochem.* 2003, 318, 100.
[95] W. S. Huang, G. Mai, Y. X. Liu, C. Yang, W. Y. Qu, *J. Nanosc. Nanotech.* 2004, 4, 423.
[96] S. H. Zhang, W. Y. Qu, W. S. Huang, Y. Y. Wu, *J. Nanosci. Nanotech.* 2004, 4, 553.
[97] Y. Z. Wang, Q. Li, S. S. Hu, *Bioelectrochemistry* 2005, 65, 135.
[98] K. B. Wu, S. S. Hu, J. J. Fei, W. Bai, *Anal. Chim. Acta* 2003, 489, 215.
[99] H. C. Yi, *Anal. Bioanal. Chem.* 2003, 377, 770.
[100] L. Y. Jiang, C. Y. Liu, L. P. Jiang, Z. Peng, G. H. Lu, *Anal. Sci.* 2004, 20, 1055.
[101] M. G. Zhang, A. Smith, W. Gorski, *Anal. Chem.* 2004, 76, 5045.
[102] A. Star, D. W. Steuerman, J. R. Heath, J. F. Stoddart, *Angew. Chem. Int. Ed.* 2002, 41, 2508.
[103] G. H. Lu, L. Y. Jiang, F. Song, C. Y. Liu, L. P. Jiang, *Electroanalysis* 2005, 17, 901.
[104] L. Y. Jiang, C. Y. Liu, L. P. Jiang, G. H. Lu, *Chin. Chem. Lett.* 2005, 16, 229.
[105] X. L. Luo, J. J. Xu, J. L. Wang, H. Y. Chen, *Chem. Commun.* 2005, 16, 2169.
[106] V. G. Gavalas, S. A. Law, J. C. Ball, R. Andrews, L. G. Bachas, *Anal. Biochem.* 2004, 329, 247.
[107] X. C. Tan, M. J. Li, P. X. Cai, L. J. Luo, X. Y. Zou, *Anal. Biochem.* 2005, 337, 111.
[108] K. P. Gong, M, N. Zhang, Y. M. Yan, L. Su, L. Q. Mao, Z. X. Xiong, Y. Chen, *Anal. Chem.* 2004, 76, 6500.
[109] P. P. Joshi, S. A. Merchant, Y. D. Wang, D. W. Schmidtke, *Anal. Chem.* 2005, 77, 3183.
[110] F. Zhao, X. Wu, M. K. Wang, Y. Liu, L. X. Gao, S. J. Dong, *Anal. Chem.* 2004, 76, 4960.
[111] Y. J. Zhang, Y. F. Shen, J. H. Li, L. Niu, S. J. Dong, A. Ivaska, *Langumuir* 2005, 21, 4797.
[112] Y. F. Zhao, Y. Q. Gao, D. P. Zhan, H. Liu, Q. Zhao, Y. Kou, Y. H. Shao, M. X. Li, Q. K. Zhuang, Z. W. Zhu, *Talanta* 2005, 66, 51.
[113] Chen J., Dyer M. J., Yu M. F. *J. Am. Chem. Soc.* 2001, 123, 6201.
[114] G. Chambers, C. Carroll, G. F. Farrell, A. B. Dalton, M. McNamara, M. Panhuis, H. J. Byrne, *Nano Lett.* 2003, 3, 843.
[115] Z. H. Wang, Y. M. Wang, G. Luo, *Electroanalysis* 2003, 15, 1129.

[116] Z. H. Wang, G. A. Luo, S. F. Xiao, G. Y. Wang, *Chemical Journal Of Chinese Universities- Chinese* 2003, 24, 811.
[117] G. Y. Wang, X. J. Liu, G. A. Luo, Z. H. Wang, *Chin. J. Chem.* 2005, 23, 297.
[118] Z. H. Wang, Y. M. Wang, G. Luo, *Analyst* 2002, 127, 1353.
[119] Wang G. Y., Liu X. J., Yu B., Luo G. J *J. Electroanal. Chem.* 2004, 567: 227.
[120] A. A. Mamedov, N. A. Kotov, M. Prato, D. M. Guldi, J. P. Wicksted, A. Hirsch, *Nature mater.* 2002, 1, 191.
[121] D. Yang, J. Rochette, E. Sacher, *J. Phys. Chem. B* 2005, 109, 4481.
[122] M. L. Guo, J. H. Chen, J. Li, L. H. Nie, S. Z. Yao, *Electroanalysis* 2004, 16, 1992.
[123] M. N. Zhang, Y. M. Yan, K. P. Gong, L. Q. Mao, Z. X. Guo, Y. Chen, *Langmuir* 2004, 20, 8781.
[124] M. N. Zhang, K. P. Gong, H. W. Zhang, L. Q. Mao, *Biosens. Bioelectron.* 2005, 20, 1270.
[125] P. G., He, M., Bayachou, *Langmuir* 2005, 21, 6086.
[126] J. Y. Liu, S. J. Tian, and W. Knoll, *Langmuir* 2005, 21, 5596.
[127] Y. M. Yan, M. N. Zhang, K. P. Gong, L. Su, Z. X. Guo, L. Q. Mao, *Chem. Mater.* 2005, 17, 3457.
[128] Z. F. Liu, Z. Y. Shen, T. Zhu, S. F. Hou, L. Z. Ying, Z. J. Shi, Z. N. Gu, *Langmuir* 2000, 16, 3569.
[129] J. J. Gooding, R. Wibowo, J. Q. Liu, W. R. Yang, D. Losic, S. Orbons, F. J. Mearns, J. G. Shapter, D. B. Hibbert, *J. Am. Chem. Soc.* 2003, 125, 9006.
[130] J. Q. Liu, A. Chou, W. Rahmat, M. N. Paddon-Row, J. J. Gooding, *Electroanalysis* 2005, 17, 38.
[131] J. J. Xu, G. Wang, Q. Zhang, X. H. Xia, H. Y. Chen, *Chin. Chem. Lett.* 2005, 16, 523.
[132] X. H. Qu, Z. Q. Peng, Y. L. Wang, S. J. Dong, *Electroanalysis* 2005, 17, 59.
[133] D. Chattopadhyay, I. Galeska, F. Papadimitrakopoulos, *J. Am. Chem. Soc.* 2001, 123, 9451.
[134] X. Yu, D. Chattopadhyay, I. Galeska, F. Papadimitrakopoulos, J. F. Rusling, *Electrochem. Commun.* 2003, 5, 408.
[135] M. O'Connor, S. N. Kim, A. J. Killard, R, J. Forster, M. R. Smyth, F. Papadimitrakopoulos, J. F. Rusling, *Analyst* 2004, 129, 1176.
[136] P. J. Britto, K. S. V. Santhanam, P. M. Ajayan, *Bioelectrochem. Bioenerg.* 1996, 41, 121.
[137] J. J. Davis, R. J. Coles, H. A. O. Hill, *J. Electroanal. Chem.* 1997, 440, 279.
[138] F. Valentini, A. Amine, S. Orlanducci, M. L. Terranova, G. Palleschi, *Anal. Chem.* 2003, 75, 5413.
[139] R. Antiochia, I. Lavagnini, F. Magno, T. Valentini, G. Palleschi, *Electroanalysis* 2004, 16, 1451.
[140] M. Chicharro, E. Bermejo, M. Moreno, A. S. Sanchez, A. Zapardiel, G. Rivas, *Electroanalysis* 2005, 17, 476.
[141] N. S. Lawrence, R. P. Deo, J. Wang, *Talanta* 2004, 63, 443.
[142] M. L. Pedano, G. A. Rivas, *Electrochem. Commun.* 2004, 6, 10.
[143] J. Wang, G. Chen, M. Wang, M. P. Chatrathi, *Analyst* 2004, 129, 512.
[144] M. D. Rubianes, G. A. Rivas, *Electrochem. Commun.* 2003, 5, 689.
[145] M. D. Rubianes, G. A. Rivas, *Electroanalysis* 2005, 17, 73.
[146] R. Antiochia, I. Lavagnini, P. Pastore, F. Magno, *Bioelectrochemstry* 2004, 64, 157.

[147] R. Antiochia, I. Lavagnini, F. Magno, *Anal. Bioanal. Chem.* 2005, 381, 1355.
[148] J. Wang, M. Musameh, *Anal. Chem.* 2003, 75, 2075.
[149] G. Chen, L. Y. Zhang, J. Wang, *Talanta* 2004, 64, 1018.
[150] C. S. Cha, C. M. Li, H. X. Yang, P. F. Liu, *J. Electroanal. Chem.* 1994, 368, 47.
[151] P. F. Liu, J. H. Hu, *Sens. Actuators B* 2002, 84, 194.
[152] Y. D. Zhao, W. D. Zhang, H. Chen, Q. M. Luo, *Talanta* 2002, 58, 529.
[153] Y. D. Zhao, W. D. Zhang, H. Chen, Q. M. Luo, *Sens, Actuators B* 2003, 92, 279.
[154] Y. D. Zhao, W. D. Zhang, H. Chen, Q. M. Luo, S. F. Y. Li, *Sens. Actuators B* 2002, 87, 168.
[155] Y. L. Wei, X. B. Ji, X. P. Dang, S. S. Hu, *Bioelectrochemistry* 2003, 61, 51.
[156] Z. H. Wang, J. Liu, Q. L. Liang, Y. M. Wang, G. Luo, *Analyst* 2002, 127, 653.
[157] Z. H. Wang, Q. L. Liang, Y. M. Wang, G. A. Luo, *J. Electroanal. Chem.* 2003, 540, 129.
[158] Q. C. Shi, T. Z. Peng, J. Y. Cheng, *Chin. J. Anal. Chem.* 2005, 33, 329.
[159] Q. C. Shi, T. Z. Peng, Y. N. Zhu, C. F. Yang, *Electroanalysis* 2005, 17, 857.
[160] A. Salimi, C. E. Banks, R. G. Compton, *Analyst* 2004, 129, 225.
[161] A. Salimi, R. G. Compton, R. Hallaj, *Anal. Biochem.* 2004, 333, 49.
[162] A. Salimi, R. Hallaj, G. R. Khayatian, *Electroanalysis* 2005, 17, 873.
[163] A. Salimi, R. Hallaj, *Talanta* 2005, 66, 967.
[164] K. B. Wu, S. S. Hu, *Carbon* 2004, 42, 3237.
[165] D. W. Pan, J. H. Chen, S. Z. Yao, W. Y. Tao, L. H. Nie, *Anal. Sci.* 2005, 21, 367.
[166] G. F. Cheng, J. Zhao, Y. H. Tu, P. A. He, Y. H. Fang, *Anal. Chim. Acta* 2005, 533, 11.
[167] J. Wang, M. Musameh, *Anal. Chim. Acta* 2005, 539, 209.
[168] J. S. Ye, Y. Wen, W. De Zhang, L. M. Gan, G. Q. Xu, F. S. Sheu, *Electroanalysis* 2003, 15, 1693.
[169] J. Chen, J. C. Bao, C. X. Cai, T. H. Lu, *Anal. Chim. Acta* 2004, 516, 29.
[170] J. S. Ye, Y. Wen, W. D. Zhang, L. M. Gan, G. Q. Xu, F. S. Sheu, *Electrochem. Commun.* 2004, 6, 66.
[171] P. Soundarrajan, A. Patil, L. M. Dai, *J. Vacuum Sci. Tech. A* 2003, 21, 1198.
[172] M. Gao, L. M. Dai, G. G. Wallace, *Electroanalysis* 2003, 15, 1089.
[173] M. Gao, L. Dai, G. G. Wallace, *Synth. Metals* 2003, 137, 1393.
[174] K. P. Loh, S. L. Zhao, W. De Zhang, *Diamond and Related Materials* 2004, 13, 1075.
[175] J. S.Ye, Y. Wen, W. D. Zhang, H.F. Cui, G. Q. Xu, F. S. Sheu, *Electroanalysis* 2005, 17, 89.
[176] S. G. Wang, Q. Zhang, R. L. Wang, S. F. Yoon, J. Ahn, D. J. Yang, J. Z. Tian, J. Q. Li, Q. Zhou, *Electrochem. Commun.* 2003, 5, 800.
[177] H. Tang, J. H. Chen, S. Z. Yao, L. H. Nie, G. H. Deng, Y. F. Kuang, *Anal. Biochem.* 2004, 331, 89.
[178] J. S. Ye, Y. Wen, W. De Zhang, H. F. Cui, L. M. Gan, G. Q. Xu, F. S. Sheu, *J. Electroanal. Chem.* 2004, 562, 241.
[179] Y. Tu, Y. H. Lin, Z. F. Ren, *Nano Lett.* 2003, 3, 107.
[180] Y. Tu, Y. H. Lin, W. Yantasee, Z. F. Ren, *Electroanalysis* 2005, 17, 79.
[181] G. D. Liu, Y. H. Lin, Y. Tu, Z. F. Ren, *Analyst* 2005, 130, 1098.
[182] J. Li, H. T. Ng, A. Cassell, W. Fan, H. Chen, Q. Ye, J. Koehne, J. Han, M. Meyyappan, *Nano Lett.* 2003, 3, 597.

[183] J. Koehne, J. Li, A. M. Cassell, H. Chen, Q. Ye, H. T. Ng, J. Han, M. J. Meyyappan, *Mater. Chem.* 2004, 14, 676.

[184] J. E. Koehne, H. Chen, A. M. Cassell, Q. Yi, J. Han, M. Meyyappan, J. Li, *Clin. Chem.* 2004, 50, 1886.

[185] X. N. Cao, L. Lin, Y. Z. Xian, W. Zhang, Y. F. Xie, L. T. Jin, *Electroanalysis* 2003, 15, 892.

[186] Y. Z. Xian, Y. Y. Zhou, H. T. Wang, L. H. Zhou, F. Liu, L. T. Jin, *J. Chromatogr. B* 2005, 817, 239.

[187] S. D. Fei, J. H. Chen, S. Z. Yao, G. H. Deng, D. L. He, Y. F. Kuang, *Anal. Biochem.* 2005, 339, 29.

[188] J. X. Wang, M. X. Li, Z. J. Shi, N. Q. Li, Z. N. Gu, *Microchem. J.* 2002, 73, 325.

[189] J. X. Wang, M. X. Li, Z. J. Shi, N. Q. Li, Z. N. Gu, *Electroanalysis* 2002, 14, 225.

[190] G. Zhao, S. Q. Zang, K. Z. Liu, S. Lin, J. Liang, X. Y. Guo, Z. J. Zhang, *Anal. Lett.* 2002, 35, 2233.

[191] W. Zhang, Y. F. Xie, S. Y. Ai, F. L. Wan, J. Wang, L. T. Jin, J. Y. Jin, *J. Chromatogr. B* 2003, 791, 217.

[192] W. Zhang, F. L. Wan, Y. F. Xie, J. Gu, J. Wang, K. Yamamoto, L. T. Jin, *Anal. Chim. Acta* 2004, 512, 207.

[193] J. M. Xu, Y. P. Wang, Y. Z. Xian, L. T. Jin, K. Tanaka, *Talanta* 2003, 60, 1123.

[194] Y. Y. Sun, K. B. Wu, S. S. Hu, *Microchim. Acta* 2003, 142 , 49.

[195] J. X.Wang, M. X. Li, Z. J. Shi, N. Q. Li, Z. N. Gu, *Electrochim. Acta* 2001, 47, 651.

[196] Z. Xu, X. Chen, X. H. Qu, S. J. Dong, *Electroanalysis* 2004, 16, 684.

[197] S. F. Lu, J J. Fei, Q. He, S. S. Hu, *Chemia Analityczna* 2004, 49, 607.

[198] H. L. Qi, C. X. Zhang, *Electroanalysis* 2005, 17, 832.

[199] X. N. Cao, L. Lin, Y .Y. Zhou, G. Y. Shi, W. Zhang, K. Yamamoto, L. T. Jin, *Talanta* 2003, 60, 1063.

[200] G. Zhao, K. Z. Liu, S. Lin, J. Liang, X. Y. Guo, Z. J. Zhang, *Microchim. Acta* 2003, 143, 255.

[201] W. Y. Qu, K. B. Wu, S. S. Hu, *J. Pharm. Biomed.Anal.* 2004, 36, 631.

[202] J. Wang, M. Musameh, *Anal. Chim. Acta* 2004, 511, 33.

[203] C. H. Yang, *Anal. Sci.* 2004, 20, 82.

[204] B. Z. Zeng, F. Huang, *Talanta* 2004, 64, 380.

[205] S. F. Wang, Q. Xu, *Anal. Lett.* 2005, 38, 657.

[206] F. Zhao, F. Huang, Q. P. Yan, B. Z. Zeng, *Microchim. Acta* 2005, 150, 179.

[207] W. Y. Qu, F. Wang, S. S. Hu, D. Cui, *Microchim. Acta* 2005, 150, 109.

[208] Y. Q. Chu, C. A. Ma, Y. H. Zhu, *Acta Phys. Chim. Sin.* 2004, 20: 331.

[209] Y. Q. Dai, K. K. Shiu, *Electroanalysis* 2004, 16, 1697.

[210] F. Wang, S. S. Hu, *J. Electroanal. Chem.* 2005, 580, 68.

[211] J. B. He, C. L. Chen, J. H. Liu, *Sens. Actuators B* 2004, 99, 1.

[212] J. Y. Qu, S. Yan, X. H. Qu, S. J. Dong, *Electroanalysis* 2004, 16, 1444.

[213] A. Callegari, S. Cosnier, M. Marcaccio, D. Paolucci, F. Paolucci, V. Georgakilas, N. Tagmatarchis, E. Vazquez, M. J. Prato, *Mater. Chem.* 2004, 14, 807.

[214] Y. N. Zhu, T. Z. Peng, J. P. Li, *Chin. Anal. Chem.* 2004, 32, 1299.

[215] H. G. Xue, W. L. Sun, B. J. He, Z. Q. Shen, *Chin. Che. Lett.* 2002, 13, 799.

[216] H. G. Xue, W. L. Sun, B. J. He, Z. Q. Shen, *Synthetic Metals* 2003, 135, 831.

[217] M. Musameh, J. Wang, A. Merkoci, Y. Lin, *Electrochem. Commun.* 2002, 4, 74.

[218] J. Chen, J. C. Bao, C. X. Cai, T. H. Lu, *Chin. Chem. Lett.* 2003, 14, 1171.
[219] M. G. Zhang, W. Gorski, *J. Am. Chem. Soc.* 2005, 127, 2058.
[220] J. Wang, M. Musameh, *Anal. Lett.* 2003, 36, 2041.
[221] R. Antiochia, I. Lavagnini, F. Magno, *Anal. Lett.* 2004, 37, 1657.
[222] J. S. Lenihan, V. G. Gavalas, J. Q. Wang, R. Andrews, L. G. Bachas, *J. Nanosci. Nanotech.* 2004, 4, 600.
[223] M. Trojanowicz, A. Mulchandani, M. Mascini, *Anal. Lett.* 2004, 37, 3185.
[224] J. Z. Xu, J. J. Zhu, Q. Wu, Z. Hu, H. Y. Chen, *Electroanalysis* 2003, 15, 219.
[225] Y. P. Li, H. B. Cao, Y. Zhang, *Acta Phys. Chim. Sin.* 2005, 21, 187.
[226] L. Zhang, G. C. Zhao, X. W. Wei, Z. S. Yang, *Chem. Lett.* 2004, 33, 86.
[227] L. Zhang, G. C. Zhao, X. W. Wei, Z. S. Yang, *Electroanalysis* 2005, 17, 630.
[228] J. X. Wang, M. X. Li, Z. J. Shi, N. Q. Li, Z. N. Gu, *Anal. Chem.* 2002, 74, 1993.
[229] F. L. Cheng, S. Du, B. K. Jin, *Chin. J. Chem.* 2003, 21, 436.
[230] Y. H. Wu, S. S. Hu, *Colloids Surf. B* 2005, 41, 299.
[231] C. X. Cai, J. Chen, *Acta Chim. Sin.* 2004, 62, 335.
[232] W. Zheng, Q. F. Li, Y. M. Yan, L. Su, L. Q. Mao, *Indian J. Chem. Section A* 2005, 44, 950.
[233] Q. Zhao, L. H. Guan, Z. N. Gu, Q. K. Zhuang, *Chin. Chem. Lett.* 2005, 16, 501.
[234] M. K. Wang, Y. Shen, Y. Liu, T. Wang, F. Zhao, B. F. Liu, S. J. Dong, *J. Electroanal. Chem.* 2005, 578, 121.
[235] J. Z. Xu, J. J. Zhu, Q. Wu, Z. Hu, H. Y. Chen, *Chin. J. Chem.* 2003, 21, 1088.
[236] C. X. Cai, J. Chen, T. H. Lu, *Science in China Series B-Chemistry* 2004, 47, 113.
[237] W. Liang, Z. B. Yuan, *Sensors* 2003, 3, 544.
[238] J. H. T. Luong, S. Hrapovic, D. Wang, F. Bensebaa, B. Simard, *Electroanalysis* 2004, 16, 132.
[239] J. F. Rochette, E. Sacher, M. Meunier, J. H. T. Luong, *Anal. Biochem.* 2005, 336, 305.
[240] Luong J. H T., Hrapovic S., Wang D. S. Electroanalysis 2005, 17: 47.
[241] L. Wang, Z. B. Yuan, *Anal. Sci.* 2004, 20, 635.
[242] A. Guiseppi-Elie, C. Lei, R. Baughman, *Nanotechnology* 2002, 13, 559.
[243] F. Patolsky, Y. Weizmann, I. Willner, *Angew. Chem. Int. Ed.* 2004, 43, 2113.
[244] J. Q. Liu, M. N. Paddon-Row, J. J. Gooding, *J. Phys. Chem. B* 2004, 108, 8460.
[245] H. Cai, X. N. Cao, Y. Jiang, P. G. He, Y. Z. Fang, *Anal. Bioanal. Chem.* 2003, 375, 287.
[246] H. Cai, Y. Xu, P. G. He, Y. Z. Fang, *Electroanalysis* 2003, 15, 1864.
[247] Y. Xu, Y. Jiang, H. Cai, P. G. He, Y. Z. Fang, *Anal. Chim. Acta* 2004, 516, 19.
[248] S. G. Wang, R. L. Wang, P. J. Sellin, Q. Zhang, *Biochem. Biophys. Res. Commun.* 2004, 325, 1433.
[249] K. Kerman, Y. Morita, Y. Takamura, E. Tamiya, *Anal. Bioanal. Chem.* 2005, 381, 1114.
[250] J. Wang, G. D. Liu, M. R. Jan, Q. Y. Zhu, *Electrochem. Commun.* 2003, 5, 1000.
[251] J. X. Wang, M. X. Li, Z. J. Shi, N. Q. Li, Z. N. Gu, *Electroanalysis* 2004, 16, 140.

ENCYCLOPEDIA OF NANOTECHNOLOGY VOLUME 2

ENCYCLOPEDIA OF NANOTECHNOLOGY
VOLUME 2

ELWOOD D. CARLSON
EDITOR

Nova Science Publishers, Inc.
New York

For permission to use material from this book please contact us:
Telephone 631-231-7269; Fax 631-231-8175
Web Site: http://www.novapublishers.com

NOTICE TO THE READER

The Publisher has taken reasonable care in the preparation of this book, but makes no expressed or implied warranty of any kind and assumes no responsibility for any errors or omissions. No liability is assumed for incidental or consequential damages in connection with or arising out of information contained in this book. The Publisher shall not be liable for any special, consequential, or exemplary damages resulting, in whole or in part, from the readers' use of, or reliance upon, this material.

Independent verification should be sought for any data, advice or recommendations contained in this book. In addition, no responsibility is assumed by the publisher for any injury and/or damage to persons or property arising from any methods, products, instructions, ideas or otherwise contained in this publication.

This publication is designed to provide accurate and authoritative information with regard to the subject matter covered herein. It is sold with the clear understanding that the Publisher is not engaged in rendering legal or any other professional services. If legal or any other expert assistance is required, the services of a competent person should be sought. FROM A DECLARATION OF PARTICIPANTS JOINTLY ADOPTED BY A COMMITTEE OF THE AMERICAN BAR ASSOCIATION AND A COMMITTEE OF PUBLISHERS.

LIBRARY OF CONGRESS CATALOGING-IN-PUBLICATION DATA

Encyclopedia of nanotechnology / editor, Elwood D. Carlson.
p. cm.
ISBN 978-1-60692-079-4 (hardcover)
1. Nanotechnology. I. Carlson, Elwood D.
T174.7.E53 2009
620'.503--dc22

2009004646

Published by Nova Science Publishers, Inc. ✝ New York

Errata for *Encyclopedia of Nanotechnology*

Editor:Elwood D. Carlson

ISBN 978-1-60692-079-4

Due to a misprint, the following chapters page numbers were missing:

Preface		**xi**
Expert Commentary	Nanostructured Polymers: Developments and Trends, Highlights and Shadows *Maria Vittoria Russo and Ilaria Fratoddi*	**1**
Chapter 1	TiO2 Nanostructured Materials: Design, Characterization and Applications *Aicheng Chen, Xinsheng Peng and Peter Holt-Hindle*	7
Chapter 2	Molecular Foundations for Onset Wear. A Review *Mario D'Acunto*	**37**
Chapter 3	Application of Coordination Chemistry in the Fabrication of Inorganic Anostructures *Yi Xie, Yujie Xiong and Zhengquan Li*	**65**
Chapter 4	1-D Zinc Sulfide Nanostructures: Synthesis and Characterization *Soumitra Kar and Subhadra Chaudhuri*	**85**
Chapter 5	Advance and Application Prospect of Dendrimers *Bifeng Pan and Daxiang Cui*	**113**
Chapter 6	Thermodynamic and Kinetic Theories of Nanowire Growth *G. W. Yang*	**201**
Chapter 7	Novel Manufacturing and Processing Technologies of Nanoporous Silicon *Jia-Chuan Lin and Wei-Chih Tsai*	**237**
Chapter 8	MOVPE Growth of High Density InAs/InGaAs/InP Quantum Dots for Mid-Infrared Emission *Tang Xiaohong and Yin Zongyou*	**277**
Chapter 9	Self-Assembled Hierarchical Architectures by Liquid-Phase Pulsed Laser Ablation *Li Yang, Paul W. May and Lei Yin*	**307**
Chapter 10	Formation and Assembly of Hierarchical Nanoarchitectures by Bioinspired Techniques *Huilan Su, Di Zhang and Qun Dong*	**329**
Chapter 11	Preparation, Characterization and Thermal Properties of Nanofluids *Hai-tao Zhu, Shi-quan Liu, Lei Xu and Can-ying Zhang*	**393**

Chapter 12	III-Nitride Nanowires Research *Ho-Young Cha and Huaqiang Wu*	**427**
Chapter 13	Nanostructures Grown via Electrochemical Template Methods: Synthesis, Measurements and Applications *Stefan Mátéfi-Tempfli*	**459**
Chapter 14	Synthesis of Embedded Gold and AgAu and CuAu Alloy Nanoclusters in Soda-Lime Glass by Ion-Exchange: A Novel Route *S. Mohan, D. Manikandan, K. G. M. Nair and Sujin P. Jose*	**483**
Chapter 15	Investigation of P-Type ZnO Layers Obtained by RBQE Method *T. V. Butkhuzi, T. G. Khulordava, M. M. Sharvashidze, N. G. Gapishvili, N. G. Bukhsianidze, L. T. Trapaidze, E. E. Kekelidze, G. G. Natsvlishvili, L. G. Aptsiauri and L. N. Gapishvili*	**491**
Chapter 16	Low-Dimensional ZnO Nanostructures: Design, Structures and Properties *Guozhen Shen, Di Chen and Cheol Jin Lee*	**501**
Chapter 17	Structure and Opto-Electronic Behavior of Diamondoids, with Applications as MEMs and at the Nanoscale Level *G. Ali Mansoori, Thomas F. George, Guoping Zhang and Lahsen Assoufid*	**527**
Chapter 18	Three-Dimensional Controllable Submicron Gold Films: Sulfhydryl-Containing Molecular Assembly and Controllable Surface Plasmon Resonance and Electrocatalysis Properties *Jianqiang Hu, Juan Liu, Xianbo Lu, Wu Lu, Jianhua Zhou and Jinghong Li*	**547**
Chapter 19	Nanocrystalline Zinc Oxide Thin Films Prepared by Chemical Solution Deposition from a Zinc Naphthenate Precursor *Seung Hwangbo and Kyu-Seog Hwang*	**561**
Chapter 20	Fabrication of Core-Shell Structured Perovskite Nanoparticles with Heterogeneous Oxide Shells and Their Ceramics *Jian Quan Qi, Helen Lai Wah Chan, Hu Yong Tian, Wan Ping Chen and Long Tu Li*	**581**
Chapter 21	Wear Dependent Frictional Properties of Alkanethiols Self-Assembled Films *Mario D'Acunto*	**623**
Chapter 22	Preparation and Characterization of Nanoparticles and Nanocrystalline Bulks of Pure Rare-Earth Metals *Xiaoyan Song*	**633**
Chapter 23	Refluxing Polyol Synthesis of Transition Metal Chalcogenides Nanocrystallites: A General Process *Guozhen Shen and Di Chen*	**665**
Chapter 24	Molecular Dynamics Simulation for Mechanical Behavior of Nanocrystals *Wei Yang, Xiaoyan Li, Xinling Ma and Zengcai Guo*	**687**
Chapter 25	Carbon Nanotubes Based Electrochemical Sensors *Chengguo Hu and Shengshui Hu*	**749**

CONTENTS

Volume 1

Preface Error! Bookmark not defined.

Expert Commentary Nanostructured Polymers: Developments and Trends, Highlights and Shadows **1**
Maria Vittoria Russo and Ilaria Fratoddi

Chapter 1 TiO2 Nanostructured Materials: Design, Characterization and Applications Error! Bookmark not defined.
Aicheng Chen, Xinsheng Peng and Peter Holt-Hindle

Chapter 2 Molecular Foundations for Onset Wear. A ReviewError! Bookmark not defined.
Mario D'Acunto

Chapter 3 Application of Coordination Chemistry in the Fabrication of Inorganic Anostructures Error! Bookmark not defined.
Yi Xie, Yujie Xiong and Zhengquan Li

Chapter 4 1-D Zinc Sulfide Nanostructures: Synthesis and CharacterizationError! Bookmark not define
Soumitra Kar and Subhadra Chaudhuri

Chapter 5 Advance and Application Prospect of DendrimersError! Bookmark not defined.
Bifeng Pan and Daxiang Cui

Chapter 6 Thermodynamic and Kinetic Theories of Nanowire GrowthError! Bookmark not defined.
G. W. Yang

Chapter 7 Novel Manufacturing and Processing Technologies of Nanoporous Silicon Error! Bookmark not defined.
Jia-Chuan Lin and Wei-Chih Tsai

Chapter 8 MOVPE Growth of High Density InAs/InGaAs/InP Quantum Dots for Mid-Infrared Emission Error! Bookmark not defined.
Tang Xiaohong and Yin Zongyou'

Chapter 9 Self-Assembled Hierarchical Architectures by Liquid-Phase Pulsed Laser Ablation Error! Bookmark not defined.
Li Yang, Paul W. May and Lei Yin

Chapter 10 Formation and Assembly of Hierarchical Nanoarchitectures by Bioinspired Techniques Error! Bookmark not defined.
Huilan Su, Di Zhang and Qun Dong

Chapter 11 Preparation, Characterization and Thermal Properties of NanofluidsError! Bookmark
Hai-tao Zhu, Shi-quan Liu, Lei Xu and Can-ying Zhang

Chapter 12 III-Nitride Nanowires Research Error! Bookmark not defined.
Ho-Young Cha and Huaqiang Wu

Chapter 13 Nanostructures Grown via Electrochemical Template Methods: Synthesis, Measurements and Applications Error! Bookmark not defined.
Stefan Mátéfi-Tempfli

Chapter 14 Synthesis of Embedded Gold and AgAu and CuAu Alloy Nanoclusters in Soda-Lime Glass by Ion-Exchange: A Novel RouteError! Bookmark not defined.
S. Mohan, D. Manikandan, K. G. M. Nair and Sujin P. Jose

Chapter 15 Investigation of P-Type ZnO Layers Obtained by RBQE MethodError! Bookmark no
T. V. Butkhuzi, T. G. Khulordava, M. M. Sharvashidze, N. G. Gapishvili, N. G. Bukhsianidze, L. T. Trapaidze, E. E. Kekelidze, G. G. Natsvlishvili, L. G. Aptsiauri and L. N. Gapishvili

Chapter 16 Low-Dimensional ZnO Nanostructures: Design, Structures and Properties Error! Bookmark not defined.
Guozhen Shen, Di Chen and Cheol Jin Lee

Chapter 17 Structure and Opto-Electronic Behavior of Diamondoids, with Applications as MEMs and at the Nanoscale LevelError! Bookmark not defined.
G. Ali Mansoori, Thomas F. George, Guoping Zhang and Lahsen Assoufid

Chapter 18 Three-Dimensional Controllable Submicron Gold Films: Sulfhydryl-Containing Molecular Assembly and Controllable Surface Plasmon Resonance and Electrocatalysis Properties Error! Bookmark not defined.
Jianqiang Hu, Juan Liu, Xianbo Lu, Wu Lu, Jianhua Zhou and Jinghong Li

Chapter 19 Nanocrystalline Zinc Oxide Thin Films Prepared by Chemical Solution Deposition from a Zinc Naphthenate Precursor Error! Bookmark not defined.
Seung Hwangbo and Kyu-Seog Hwang

Chapter 20 Fabrication of Core-Shell Structured Perovskite Nanoparticles with Heterogeneous Oxide Shells and Their CeramicsError! Bookmark not defined.
Jian Quan Qi, Helen Lai Wah Chan, Hu Yong Tian, Wan Ping Chen and Long Tu Li

Chapter 21 Wear Dependent Frictional Properties of Alkanethiols Self-Assembled Films Error! Bookmark not defined.
Mario D'Acunto

Chapter 22 Preparation and Characterization of Nanoparticles and Nanocrystalline Bulks of Pure Rare-Earth MetalsError! Bookmark not defined.
Xiaoyan Song

Chapter 23 Refluxing Polyol Synthesis of Transition Metal Chalcogenides Nanocrystallites: A General Process Error! Bookmark not defined.
Guozhen Shen and Di Chen

Chapter 24 Molecular Dynamics Simulation for Mechanical Behavior of Nanocrystals Error! Bookmark not defined.
Wei Yang, Xiaoyan Li, Xinling Ma and Zengcai Guo

Chapter 25 Carbon Nanotubes Based Electrochemical SensorsError! Bookmark not defined.
Chengguo Hu and Shengshui Hu'

Volume 2

Chapter 26 Synthesis of Carbon Nanotubes Using Novel Alloy Hydride Catalysts by Thermal CVD and their Energy-Related Applications **809**
Sundara Ramaprabhu and Mancheri Manikoth Shaijumon

Chapter 27 Properties, Synthesis, Purification, and Integration of Carbon Nanotubes for the Electronic Device Applications **845**
Wanjun Park and Yo-sep Min

Chapter 28 Field Enhanced Thermal Conductivity in Heat Transfer Nanofluids with Magnetic Coating on Carbon Nanotubes **885**
Brian Wright , Dustin Thomas, Haiping Hong, Lori Groven, Jan Puszynski, Edward Duke, Xiangrong Ye , Sungho Jin, Pauline Smith, Walter Roy

Chapter 29 Synthesis and Investigation of Magnetic Properties of GD - Substituted Mn - Zn Ferrite Nanoparticles as a Potential Low Curie Temperature Agent for Magnetic Fluid Hyperthermia **895**
F. S. Bayburtskiy, Yu. V. Baldokhin, D. B. Korman, A. E. Chekanova and O. A. Shlyakhtin

Chapter 30 Lipid Nanotubes as Scaffolds to Create Structured Inorganic Nanomaterials **905**
Yong Zhou and Xiaqin Wang

Chapter 31 Multifunctional Reaction Platform Based on Carbon Nanotubes **921**
Wenwen Li, Yu Zhang and Chao Gao

Chapter 32 Super Carbon Nanotube: Its Concept, Geometric Conservation Laws and Mechanical Properties **943**
Yuli Chen, Yajun Yin, Bin Liu and Keh-Chih Hwang

Chapter 33	Improving Dielectric Properties of Polymer Nanocomposites through Nanoparticle Surface Modification *Dongling Ma, Linda S. Schadler, Richard W. Siegel, Nianqiang Wu, Diana F. Rogers,and Brian C. Benicewicz*	**961**
Chapter 34	Formation of Nanoparticles Under Laser Ablation of Solids in Liquids *G. A. Shafeev*	**993**
Chapter 35	Carbon Nanoparticles as Substrates for Cell Adhesion and Growth *Lucie Bacakova', Lubica Grausova, Marta Vandrovcova, Jiri Vacik', Aneta Frazcek, Stanislaw Blazewicz, Alexander Kromka', Bohuslav Rezek, Milan Vanecek, Milos Nesladek', Vaclav Svorcik, Vladimir Vorlicek and Milan Kopecek*	**1029**
Chapter 36	Organic-Shell Inorganic-Core Hybrid Nanoparticles with Advanced Functions Designed by Wet Process *Mami Yamada*	**1097**
Chapter 37	Highly Stabilized Gold Nanoparticles Synthesized and Modified by PEG-b-polyamine *Daisuke Miyamoto and Yukio Nagasaki*	**1131**
Chapter 38	Nanoparticles and Quantum Dots as Biomolecule Labels for Electrochemical Biosensing *Martin Pumera*	**1153**
Chapter 39	Photoluminescence Behavior of Water-Soluble Thiol-Capped CdTe Quantum Dots in Living Cells *Pei-Nan Wang and Ji-Yao Chen*	**1159**
Chapter 40	Hybrid Nanoparticle Based on Silica *Shaojun Guo, Erkang Wang*	**1187**
Chapter 41	Heat Transfer of Nanoparticle Suspensions (Nanofluids) *S. M. Sohel Murshed, Kai Choong Leong and Chun Yang*	**1217**
Chapter 42	Recent Developments in the Effective Thermal Conductivity of Nanoparticle Suspensions (Nanofluids) Research *Calvin H. Li and G. P. Peterson*	**1259**
Chapter 43	Enhancement and Temperature Variation in the Thermal Conductivity of Nanofluids *Pei Tillman, Miccal T. Matthews and James M. Hil*	**1293**
Chapter 44	Nanotecnology: Obtaining of Nanoparticles and Nanocomposites and their Use in Food and Drug Packaging *A. Edwin Moncada'*	**1323**
Chapter 45	The Applications of Nanoparticles in Electrochemistry *Ruhai Tian and Jinfang Zhi*	**1351**

Chapter 46 Polymeric Nanoparticles for Oral Delivery of Protein Drugs **1379**
Ji-Shan Quan, Hu- Lin Jiang, Jia-Hui Yu, Ding-Ding Guo, Rohidas Arote, Yun-Jai Choi, Chong-Su Cho

Chapter 47 Fabrication of Micro/Nano-Composite Structures and Their Photovoltaic Behaviors **1393**
Jin Zhai, Xiangliang Sheng, Yong Zhao and Liping Heng

Chapter 48 Charge Separation and Transport in Organic/Inorganic Nanocomposites and Its Applications in Photovoltaic Devices **1403**
Chun-Wei Chen, Yun-Yue Lin, Yu-Ting Lin, Tsung-Wei Zeng, Chia-Hao Chang and Wei-Fang Su

Chapter 49 Polymeric Nanocomposites, Stabilized Organic Derivatives of Five-Valent Phosphorus **1429**
A. Kh. Shaov, Kh. Kh. Gurdaliev and A. M. Kharaev

Chapter 50 Polymer/Silicate Nanocomposites Based on Organomodified Clays **1443**
A. K. Mikitaev, O. B. Lednev, A. Yu. Bedanokov and M. A. Mikitaev

Chapter 51 Fundamental Aspects of Filling of Nanocomposites with High-Elasticity Matrix: Fractal Models **1459**
Georgi V. Kozlov, Yurii G. Yanovskii and Gennadi E. Zaikov

Chapter 52 Structure Formation of Polymer Nanocomposites Based on Polypropylene **1469**
Ahmed Kh. Malamatov, Georgi V. Kozlov and Eugeni M. Antipov

Chapter 53 The Interrelation of Elasticity Modulus and Amorphous Chain's Tightness for Nanocomposites Based on the Polypropylene **1475**
Georgi V. Kozlov, Ahmed Kh. Malamatov, Eugeni M. Antipov and Abdulah K. Mikitaev

Chapter 54 Thermal Degradation and Combustion of Polypropylene Nanocomposite **1481**
S. M. Lomakin, I. L. Dubnikova, S. M. Berezina, G. E. Zaikov, R. Kozlowski, Gyeong-Man Kim and G. H. Michler

Chapter 55 Biomedical Application of Nanotechnology in Cancer Early Diagnosis **1501**
Min Song and Xuemei Wang

Chapter 56 Gold Nanorods: Synthesis, Optical Properties, Assembly and Bioapplications **1523**
Zhanfang Ma and Chungang Wang

Chapter 57 Nanotechnology Based Gas Sensors **1547**
John T.W. Yeow and Yun Wang

Index **1565**

In: Encyclopedia of Nanotechnology
Editor: Elwood D. Carlson

ISBN 978-1-60692-079-4

Chapter 26

SYNTHESIS OF CARBON NANOTUBES USING NOVEL ALLOY HYDRIDE CATALYSTS BY THERMAL CVD AND THEIR ENERGY-RELATED APPLICATIONS

Sundara Ramaprabhu* ***and Mancheri Manikoth Shaijumon***

Alternative Energy Technology Laboratory
Department of Physics,
Indian Institute of Technology Madras,
Chennai - 600 036,
India

ABSTRACT

This chapter presents the recent trends in the carbon nanotubes growth by chemical vapour deposition (CVD) technique. The use of CVD technique and novel alloy hydride catalysts for the synthesis of multiwalled carbon nanotubes (MWNTs) by the pyrolysis of acetylene is presented. The advantages of this novel approach to catalyst preparation using hydrogen decrepitation with reference to the increase in the catalytic reactivity and active sites for the formation of CNTs are highlightened. A growth process of CNTs with alloy hydride catalysts by vapour-liquid-solid (VLS) mechanism is proposed. The results of the different characterization of the as-grown and purified MWNTs are described. The dependence of the yield and the purity of the MWNTs synthesized on the alloy hydride catalysts are discussed. The various experimental techniques of the energy- related applications of MWNTs are presented.

* Corresponding author. Tel: +91-44-22574862; Fax: +91-44-22570509. E-mail: ramp@iitm.ac.in [S. Ramaprabhu]

1. INTRODUCTION

The discovery of fullerenes [1] has significantly raised the general interest in carbon research activity. Carbon nanotubes (CNTs) are currently the focus of intense research worldwide because of their unique properties that could impact various areas of science and technology. Recent experimental studies have shown that CNTs have high mechanical strength suggesting their potential for advanced composites [2, 3]. The remarkable electronic properties offer them great potential for novel applications including various nano-devices [4-6]. Because of their small diameter involving only a small number of carbon atoms, and due to their large aspect ratio, CNTs are classified as one-dimensional (1D) carbon systems and most of the theoretical studies on CNTs emphasize their 1D properties [7]. The most interesting of these theoretical developments was the prediction that single walled carbon nanotubes (SWNTs) could be either semiconducting or metallic depending on their geometrical characteristics, namely their diameter and the orientation of the hexagons with respect to the nanotube axis [8]. Though predicted in 1992, these remarkable electronic properties were experimentally confirmed only in 1998 [9, 10]. Extensive effort has been taken to study the structural, electrical, mechanical and chemical properties of CNTs in order to explore the potential applications of these novel materials. The properties and applications of CNTs have been extensively reviewed by several authors [11-14] and many books have been published [15-18]. The recent developments in this field have generated great excitement in the area of nano scale science and technology.

CNTs can be used individually or as assembly for its various applications such as to build nano-electronic devices such as field effect transistors and rectifying electrodes [5]. Bundles of nanotubes have been used for field emission based flat panel displays [19, 20] and other applications such as CNT based sensors and filters [21, 22]. Individual nanotubes have been used as tips for scanning probe microscopy [23]. Bulk quantities of CNTs have also been used in other applications such as hydrogen storage media and composite materials with improved mechanical properties [2 -3, 24].

For realizing the possible applications of CNTs, controlled and optimized growth of nanotubes is very significant. The nano-electronics applications require controlled and selective growth of CNTs on substrates. Most of the applications demand high quality nanotubes in high yield. Even though lots of work has been carried out in recent years towards overcoming these issues, simple, efficient and inexpensive methods for large scale production of CNTs are still a major challenge. In recent years, chemical vapor deposition (CVD) method has been shown to be promising for producing nanotubes with different morphologies in large quantities which is essential for the various applications.

In this chapter, we present the progress made in CNTs growth by CVD. Synthesis of MWNTs by pyrolysis of acetylene over alloy hydride catalysts [25], prepared by hydrogen decrepitation technique, will be focussed. The proposed growth mechanism of CNTs produced by this technique and the various characterization methods employed are also discussed. Further, some of the applications of MWNTs with an emphasis on energy-related applications such as catalyst support material in PEM Fuel cells, supercapacitor electrode materials and for hydrogen storage media are discussed in detail.

2. Synthesis of Carbon Nanotubes

There are several methods to produce CNTs which can be broadly divided into physical and chemical depending on the technique applied to release carbon atoms from carbon-containing precursor molecules. In physical methods such as arc discharge [26] and laser ablation [27], high-energy input is used to release carbon atoms. Chemical methods such as CVD depend mainly on the catalytic decomposition of carbon precursors on the surface of transition metal particles [28].

In arc discharge method, CNTs are grown by a dc arc discharge between two carbon electrodes in a chamber filled with an inert gas. The nanotubes thus produced are generally bound together by strong van der Waals interactions and form bundles. SWNTs are produced using carbon electrodes that contained Fe, Ni or Co catalyst and Ar or He as the inert gas in the chamber [29]. However this method suffers from a number of disadvantages. The yield is rather low, since most of the evaporated carbon is deposited on the walls of the chamber than on the cathode and the materials that are formed in the deposit generally contains a mixture of components including fullerenes, amorphous carbon and catalytic impurities.

The laser evaporation technique [30] produces CNTs in high-yield. In this method a carbon target doped with powdered Ni or Co as catalyst is heated to 1200°C in vacuum and then bombarded with intense laser pulses. During laser ablation, an inert gas is passed through the chamber, which carries the grown nanotubes to a colder end from which they are collected. Various attempts have been made to improve the quality and yield of nanotubes produced by this method [31-33].The use of high powered lasers makes this method expensive.

CVD of hydrocarbon over metal catalyst has been a successful technique to produce CNTs. Various forms of carbon fibers and filaments have been synthesized by this method in the past [34, 35]. The CVD method for the production of CNTs is essentially a two-step process, which involves a catalyst preparation step followed by the growth of nanotubes [28]. The growth process involves heating a catalyst material to high temperatures (500-1100°C) in a tubular furnace, and flowing a hydrocarbon gas through the reactor over a period of time. The catalytic species are typically transition metal nanoparticles that are formed on support materials such as alumina, silica and zeolites [36-39]. Iron, nickel or cobalt nanoparticles are often used as catalysts. For MWNTs growth, most of the CVD methods employ either acetylene or ethylene as the carbon feed stock and the growth temperature is typically in the range of 550-750°C. MWNTs were produced by catalytic CVD of acetylene over Ni nanoparticles obtained from Ni/Cr HT-type anionic clay catalysts [40]. SWNTs growth by CVD generally uses methane as carbon feed stock with various supported catalysts in the reaction temperature range 850-1000°°C [41-45]. A single-step process has been developed to synthesize SWNTs and alloy nanowire encapsulated MWNTs by thermal CVD using certain Zr based AB_2 alloy hydride catalysts [46]. The encapsulation creates dimensionally confined nanowires inside the CNTs, leading to CNT encapsulated composite nanostructures, which have potential applications as catalysts, electronic devices and magnetic tapes [47]. The pyrolysis of metallocenes and metallocene-benzene mixture has been shown to yield CNTs [48]. The nature and yield of carbon nanostructures produced by thermal CVD by the pyrolysis of hydrocarbons over metallocenes depend on the nature of catalyst, growth temperature, growth time, gas flow rates and quartz reactor dimensions. The effect of these influential parameters on the growth of various carbon nanostructures has been systematically studied by the pyrolysis of acetylene with ferrocene at 1050°C, using quartz reactors with varying dimensions and gas flow rates in a double-stage furnace thermal CVD setup [49]. MWNTs growth on patterned SiO_2 and Si substrates was studied using the vapour phase catalyst delivery method [50]. The potential of the CVD and gas phase techniques of CNTs growth for satisfying the technological requirements has been recognized since 1995 [51]. Aligned and ordered nanotubes could be grown on patterned substrate surfaces using CVD technique [42, 50, 52-55]. Figure 1 shows one example of organized nanotube patterns grown on pre-selected substrate sites using photolithography by Ajayan's group [52]. Vertically aligned CNTs have been grown with plasma enhanced and microwave CVD [56-62]. Thus, CVD of hydrocarbons over transition metal catalysts has been shown to be most suitable for the production of CNTs. Although CVD has various advantages over other methods, appropriate selection of feeding gas and catalysts with high reactivity is essential for controlled growth of CNTs in large quantities. Synthesis of MWNTs by thermal CVD using novel alloy hydride catalysts and the role of these new catalysts for producing MWNTs in high yield are discussed in detail in the following section.

3. Synthesis of MWNTs Using Novel Alloy Hydride Catalysts

As the diameters and morphology of CNTs and hence their physical properties are expected to be dependent on the size and shape of catalytic nanoparticles [63], controlled growth of CNTs would require an efficient and reproducible route for catalyst preparation.

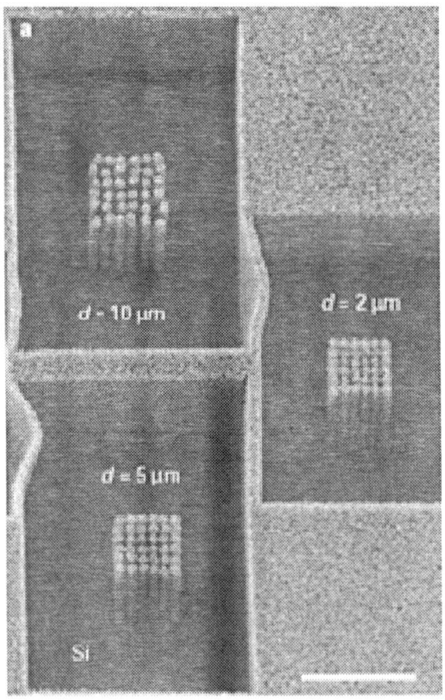

Figure 1. SEM images of organized nanotube patterns on pre-selected substrates sites. Scale bars, 100 μm (Adapted from [52] with permission).

Hence it is important to prepare effective catalysts with active Fe, Co, or Ni metal particles with appropriate sizes. Zr based C 14 type AB_2 alloys with Fe, Co, or Ni have been shown to have large hydrogen decrepitation and about 20 % volume expansion upon several cycles of hydrogen absorption and desorption, during which the breakdown of particles into smaller sizes with fresh surfaces occurs due to plastic deformation of the alloy [64]. These particles are catalytically very active, due to the presence of transition metals such as Fe, Co or Ni and are free from being oxidized. In this method, MWNTs have been synthesized by the catalytic decomposition of acetylene over Zr based AB_2 alloy hydride, prepared by hydrogen decrepitation method [25]. Mischmetal (Mm) and Fe/Ni based C 15 type AB_2 and hexagonal $CaCu_5$ type AB_5 alloy hydrides have also been prepared by hydrogen decrepitation route and have been used as catalysts for the growth of MWNTs as they show high hydrogen intake capacity, large hydrogen decrepitation, and are inexpensive [65].

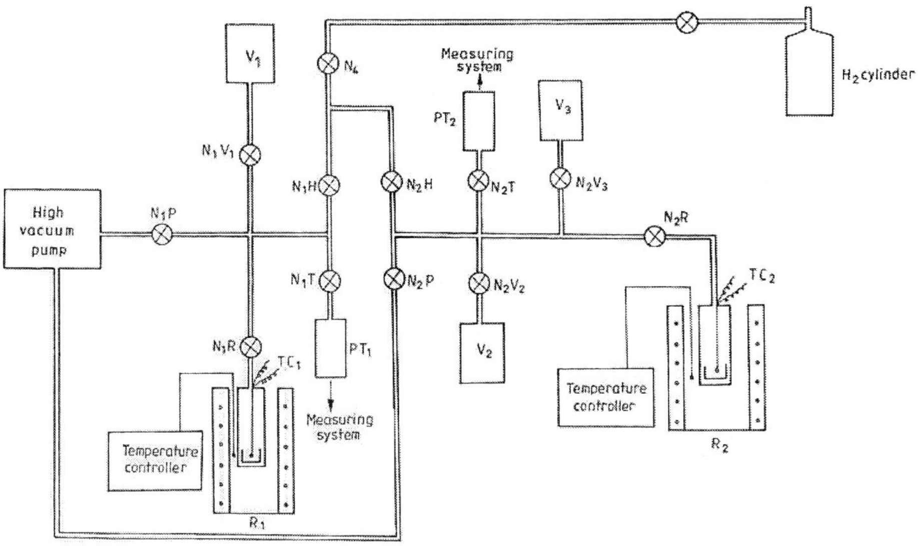

Figure 2. Schematic diagram of high pressure experimental facility for the hydrogen adsorption studies.

3.1. Preparation of Alloy Hydride Catalysts

The alloys were prepared by arc melting the constituent elements under argon atmosphere (99.995 % purity). Each alloy was hydrogenated to a maximum storage capacity using a high pressure Seiverts apparatus (Figure 2). Prior to hydrogenation, the sample is activated by degassing at 200°C in high vacuum (10^{-5} torr), using a diffusion pump for about 3 h in order to remove any dissolved gases in the alloy. The sample is then cooled to 100°C, and hydrogen gas at a pressure of around 30 bar was allowed to react with it, after which, it is slowly cooled to room temperature. For the dehydrogenation process, the hydride is degassed at 200°C for about 3 h in a vacuum of 10^{-5} torr, cooled to 100°°C and hydrogen gas is allowed to react with it, after which it is cooled to room temperature. This hydriding-dehydriding cycles are repeated several times and fine powders of alloys are obtained after about 4-5 cycles of hydrogenation/dehydrogenation. Figures 3 and 4 show respectively the powder X-ray diffractograms of Zr based AB_2 alloy and SEM image of its hydrides. SEM image clearly reveals the alloy hydride particle size in the range of 5-20 μm. These powders are directly used as catalysts for the growth of MWNTs.

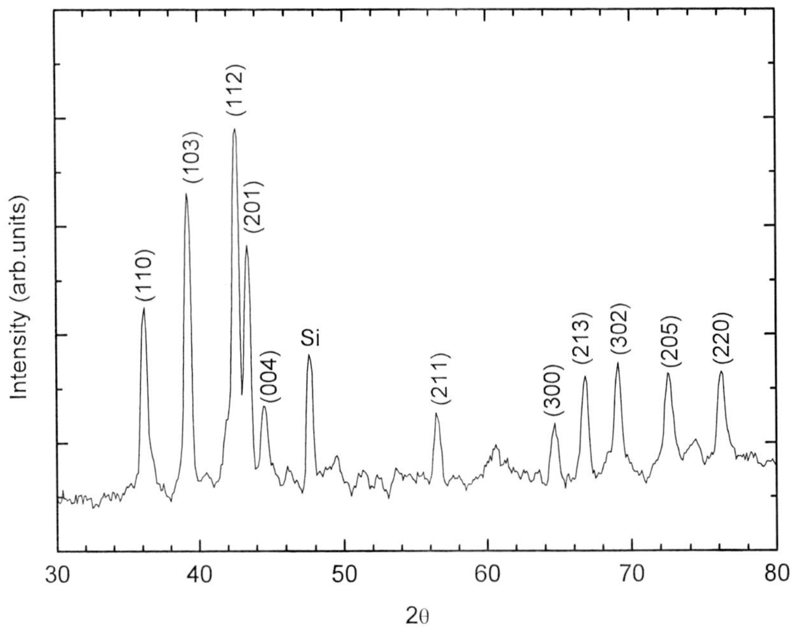

Figure 3. Powder X-ray diffractogram of Zr based AB_2 alloy taken with Cu K_α radiation.

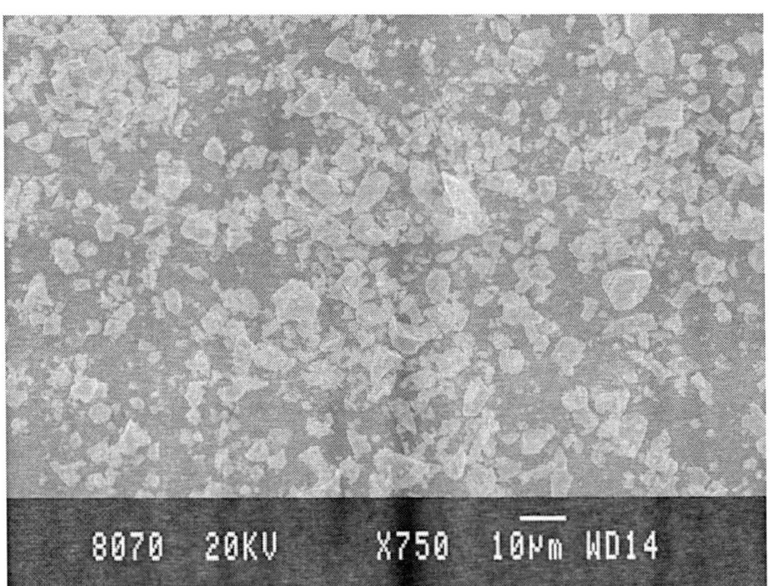

Figure 4. SEM image of fine powders of Zr based AB_2 alloy hydride.

3.2. MWNTs Growth

The thermal CVD set up designed and developed for the catalytic synthesis of MWNTs using alloy hydride catalysts is schematically shown in Figure 5. The apparatus consists of a single-stage furnace system fitted with a quartz tube (Figure 6). For the experiment, the catalysts were placed in a quartz boat and kept at the center of the furnace in a quartz tube reactor. Here, argon is used as the carrier gas, while acetylene serves as the carbon feed stock. MWNTs were synthesized by the pyrolysis of acetylene over alloy hydride catalysts. Flushing the catalyst with hydrogen for 30 min removes the oxide layer and makes their surface active. Acetylene gas is passed through the quartz tube reactor for 30 min. The reactor is cooled to room temperature in argon flow and the quartz boat with the carbon deposits is removed. Figure 7 shows the photograph of the nanotubes deposit in the quartz boat and alloy hydride catalyst precursor in the quartz boat before the reaction.

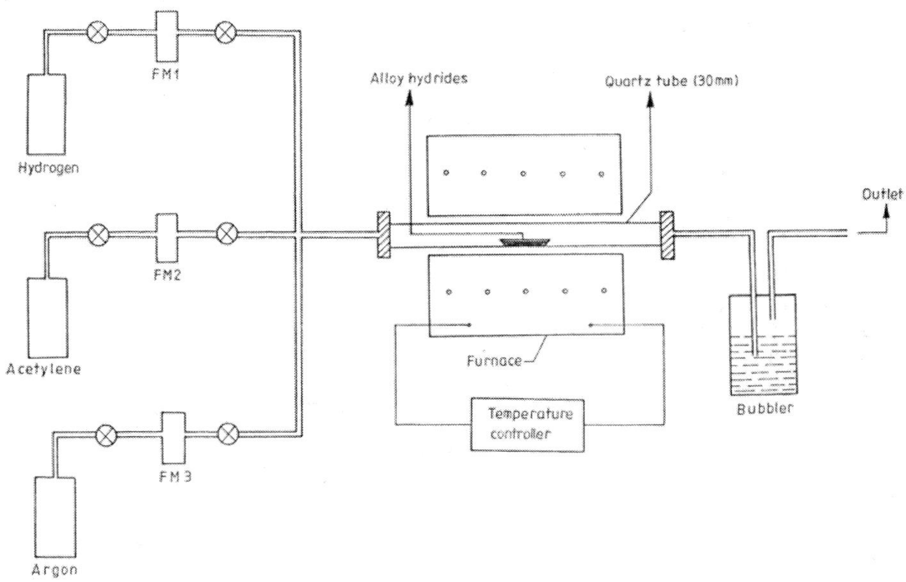

Figure 5. Schematic diagram of a single-stage thermal CVD facility employed for the synthesis of MWNTs.

The yield of the graphitic carbon deposited using different alloy hydride catalysts, measured using the following equation gives a weight gain of ~300 %.

$$Carbon\,yield\,(\%) = \frac{(m_{tot} - m_{cat})}{m_{cat}} \tag{1}$$

where m_{cat} is the initial amount of the catalyst (before reaction) and m_{tot} is the total weight of the sample after reaction.

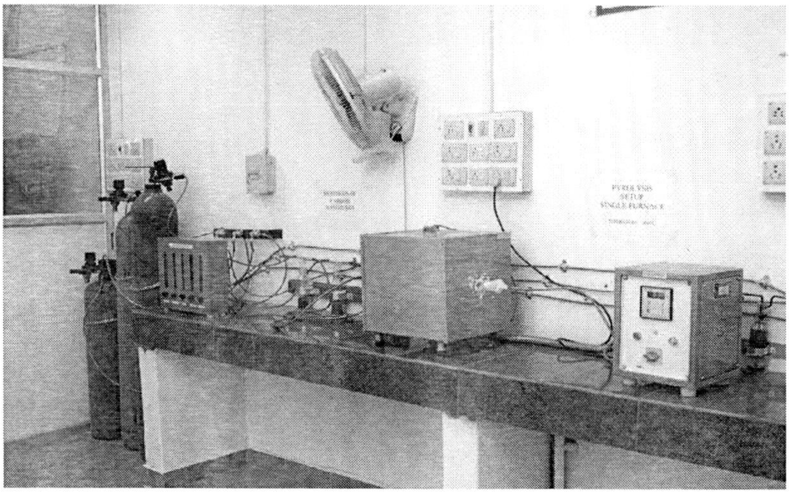

Figure 6. Photograph of the single-stage furnace thermal CVD facility developed for the synthesis of MWNTs.

Figure 7. Photographs of (a) quartz boat with Mm and Ni based AB_2 alloy hydride catalyst before deposition and (b) quartz boat with MWNTs deposits with a reaction time of 30 min.

3.3. Purification of MWNTs

Pure and defect free CNTs are expected to have significant properties and hence the use of nanotubes for various applications will require the elimination of defects and other reaction products such as amorphous carbon and catalyst particles. Various methods have been developed to purify the as-grown CNTs [66, 67]. Of all the purification methods, oxidation followed by acid treatment appears to be effective, wherein the acid dissolves the metallic oxides exposed by the oxidation treatment [67]. Here, the as-prepared carbon samples were heated in air at 350°C for 2 h to remove the carbonaceous impurities. Sample was spread in a quartz boat to ensure uniform heating and placed at the center of a quartz tube, which was inserted horizontally in a furnace. The air-oxidized samples were further purified by refluxing in concentrated HNO_3 at 60°C in a round bottom flask for 24 h to remove the catalytic impurities. The resulting samples were washed with de-ionized water several times and dried in vacuum at 100°C for 1 h.

3.4. Characterization of CNTs

A wide range of characterization tools have been used to reveal the properties of CNTs [68]. Techniques based on electron microscopy such as scanning electron microscopy (SEM), transmission electron microscopy (TEM), scanning tunneling microscopy (STM), atomic force microscopy (AFM) and those based on spectroscopy such as Raman spectroscopy and Infrared spectroscopy (IR) have been widely used. In addition, X-ray diffraction (XRD, X-ray photoelectron spectroscopy (XPS and thermal methods such as thermogravimetric analysis (TGA) and temperature programmed desorption (TPD) have been used. Powder X-ray diffractograms of as-prepared and purified MWNTs grown by catalytic decomposition of acetylene using Zr based AB_2 alloy hydride catalyst (Figure 8) show the graphitic nature of the samples. The complete removal of impurities upon purification has also been revealed. TGA is used to monitor the changes in weight of the sample with increasing temperature to estimate the yield and purity of CNTs. The TG curves of as-grown and purified MWNTs prepared by the catalytic decomposition of acetylene using Mm and Fe based AB_2 alloy hydride catalyst is shown in Figure 9. The slight weight loss for the as-grown MWNTs up to around 450°C is due to the burning of amorphous carbon. Weight loss between ~ 500-700°C is attributed to the oxidation of MWNTs. Final residual weights of around 52 % and 7 % were obtained respectively for as-grown and purified MWNTs synthesized using Mm and Fe based AB_2 hydride catalyst, revealing a purity of about 93 % for the purified sample. Variability of oxygen exposure of the samples during the initial oxidative heat treatment could also retain the catalytic impurities, as they are not fully exposed to acid treatment due to the amorphous carbon adhered on its surface. The yield of MWNTs is defined as the ratio of weight loss between 500°C and 800°C to the weight remained at 850°C, both measured from TGA. This is a measure of the ratio of the weight gain by MWNTs to the weight of the catalytic powder, assuming that the MWNTs get oxidized completely in the temperature range 500-800°C. A maximum yield of 60 % has been obtained for MWNTs synthesized using Mm and Ni based AB_2 alloy hydride catalyst. Figure 10 shows the SEM images of MWNTs grown over Mm based AB_5 and Zr based AB_2 alloy hydride catalysts by catalytic decomposition of acetylene using thermal CVD. These images show the presence of tangled tubes with high packing density.

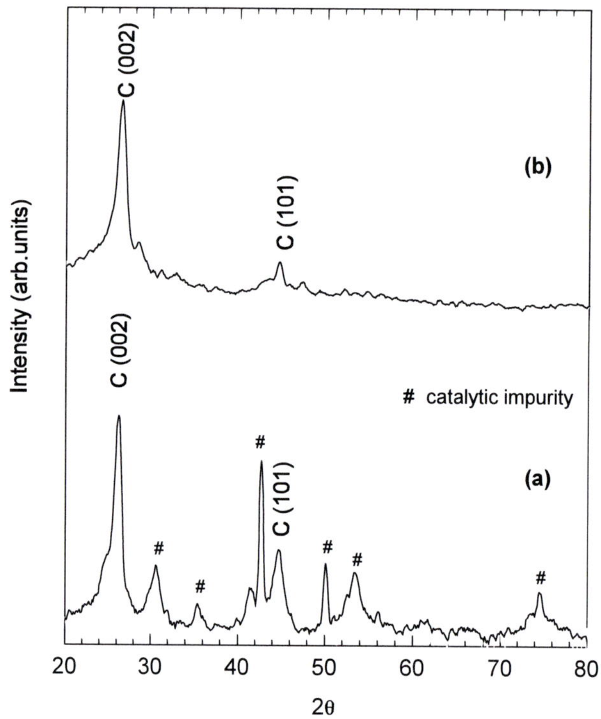

Figure 8. Powder X-ray diffractograms of (a) as-grown and (b) purified MWNTs synthesized by the pyrolysis of acetylene over Zr based AB_2 alloy hydride catalyst.

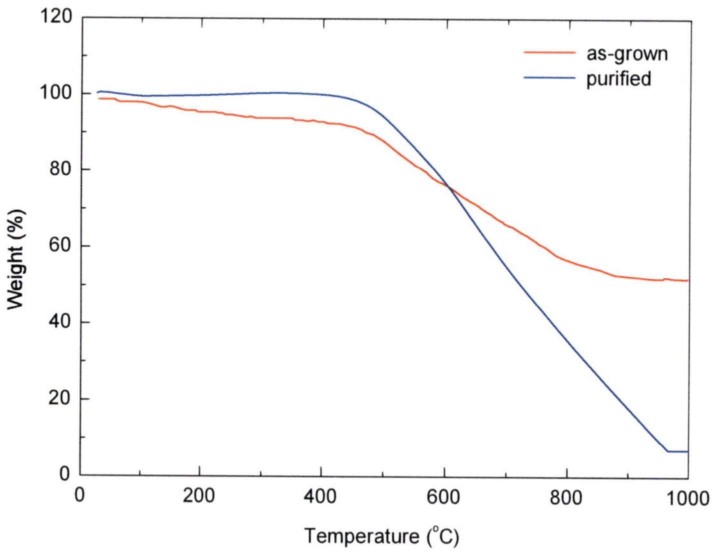

Figure 9. Thermogravimetry (TG) curves of MWNTs grown over Mm and Fe based AB_2 hydride catalyst.

TEM is one of the best tools to characterize nanotubes and related nanostructures. Both MWNTs [26] and SWNTs [69] were discovered on the TEM micrographs of the cathode deposit formed using arc-discharge method. TEM images of MWNTs synthesized by catalytic decomposition of acetylene over Zr based AB_2 alloy and Mm and Fe based AB_2 hydride catalyst are shown in Figure 11. The tubes are of approximately 20-30 nm diameters and with a hollow core of approximately 6-8 nm. MWNTs with opened tips are seen in the TEM image as shown in Figure 11 a, which would have resulted from the preferential oxidation of the tube ends during purification.

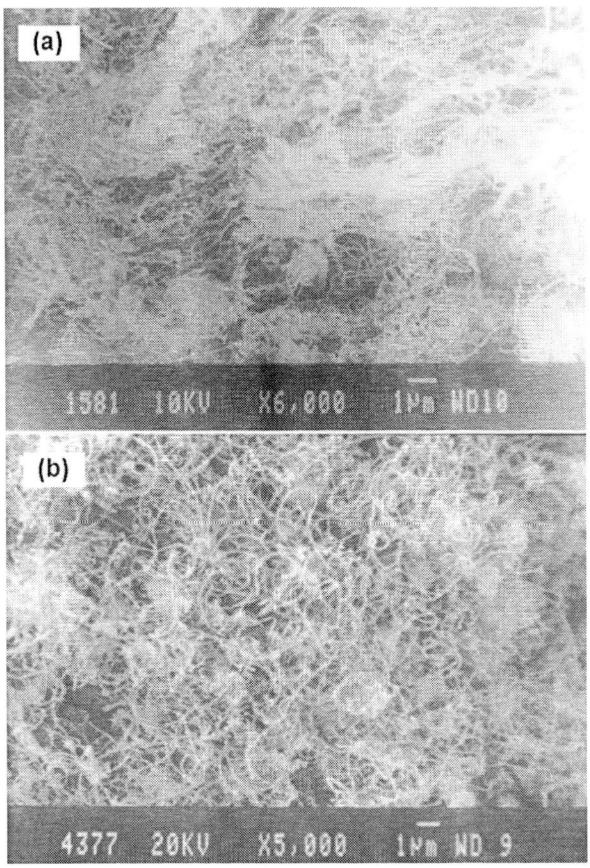

Figure 10. SEM images of MWNTs obtained from pyrolysis of acetylene over (a) Zr based AB_2 alloy hydride catalyst, (b) Mm based AB_5 alloy hydride catalyst.

High resolution TEM (HRTEM) images of purified MWNTs prepared with Mm and Ni based AB_2 alloy hydride catalyst show clearly the graphite layers (Figure 12). Raman spectroscopy has been used to investigate the vibrational properties of the carbon samples. Over the years, Raman scattering has been shown to be a useful technique to probe the various carbon based materials [70]. Raman scattering in CNTs results from the inelastic scattering of light from the nanotubes, leading to an increase or decrease of energy of the incident light due to an emission or absorption of a phonon present in the nanotube [68].

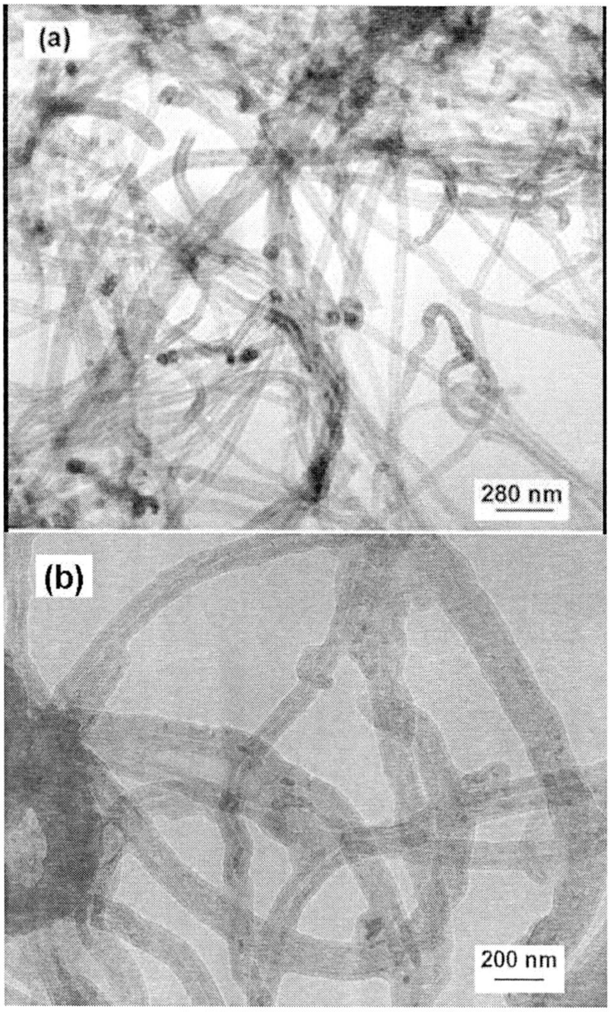

Figure 11. TEM images of (a) as-grown MWNTs synthesized by the pyrolysis of acetylene over Mm and Fe based AB_2 alloy and (b) Zr based AB_2 alloy hydride catalysts.

Figure 13 shows the Raman spectra of purified MWNTs prepared with Mm and Ni based AB_2 hydride catalysts. The spectrum shows two strong peaks at 1365 cm^{-1} and 1590 cm^{-1}, which are the typical high frequency vibrational modes of MWNTs. The peak at 1590 cm^{-1} (G-band) corresponds to the Raman allowed optical mode E_{2g} of 2- dimensional graphite, while the peak at 1365 cm^{-1} (D-band) is mainly derived from disordered carbon defects of the MWNTs [71]. The smaller value of the ratio of intensities of the two peaks (I_D/I_G), reveals that the sample is of sufficiently high purity [72].

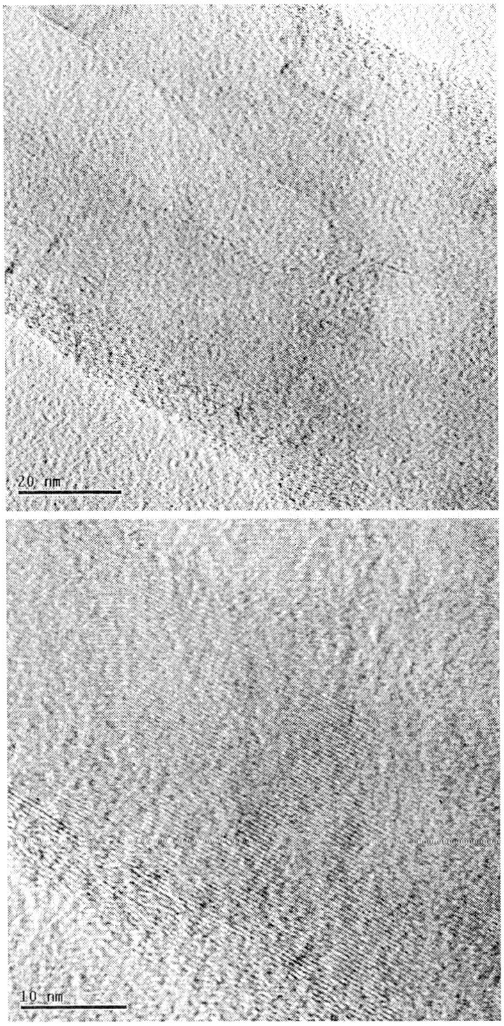

Figure 12. HRTEM images of MWNTs synthesized from pyrolysis of acetylene over Mm and Ni based AB_2 alloy hydride catalyst.

3.5. Growth Mechanism of MWNTs

A thorough understanding of the formation mechanism of nanotubes is important for controlling the various growth conditions. The growth process of CNTs by thermal CVD with a catalyst is proposed to include three steps; (i) dissociation of hydrocarbon gas on the surface of the catalyst, (ii) diffusion of carbon atoms on the surface or through the bulk of the catalyst and (iii) growth of nanotubes [73]. These mechanisms are illustrated schematically in Figure 14 [74]. It has also been proposed that the form of graphite that is produced is closely related to the physical dimensions of the metal catalyst particles and the most effective metals have been shown repeatedly to be iron, nickel and cobalt [75].

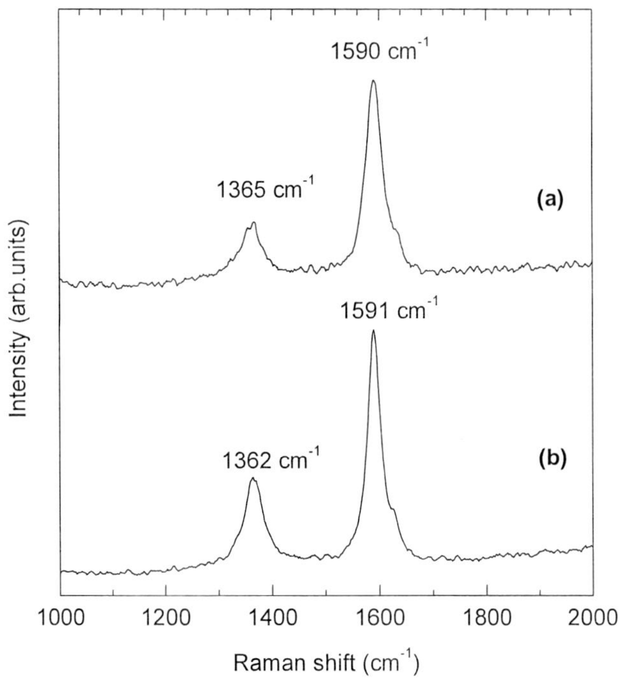

Figure 13. Raman spectra of MWNTs synthesized by pyrolysis of acetylene using (a) Mm based AB_2 and (b) Mm based AB_5 alloy hydride catalyst.

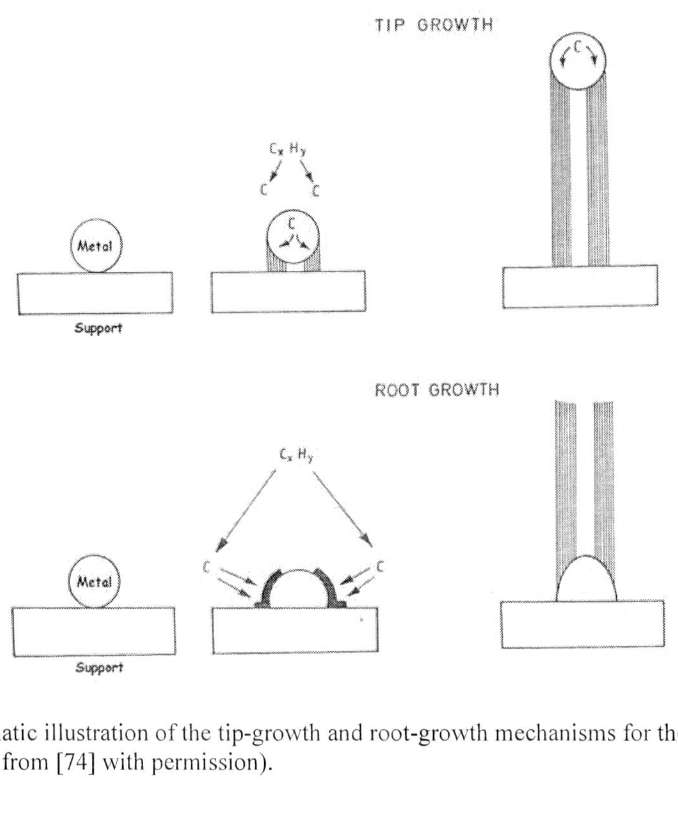

Figure 14. Schematic illustration of the tip-growth and root-growth mechanisms for the carbon filament growth (Adapted from [74] with permission).

It has been reported that transition metal carbide particles are fluid like during the catalytic process, which could be due to the over saturation of the metal particles with carbon, leading to the formation of metastable carbides [76].

Since the average size of the alloy hydride catalyst particles before deposition is in the range of 5-20 μm (Figure 4), the possible MWNTs growth mechanism in this new process could be through the precipitation of carbon in the form of MWNTs from the molten catalytic particles. The melting temperatures of the alloy-C system are lower than those of the metal-C system. Further, reduction in particle size results in lowering of melting temperature [77]. Hence, the vapour–liquid–solid (VLS) mechanism [78] might be responsible for the growth process of MWNTs using alloy hydride catalyst.

The novel approach to catalyst preparation using hydrogen decrepitation ensures increase in total surface area by providing fresh surfaces, which further enhance the catalytic reactivity and active sites for the formation of CNTs. The alloy hydride catalysts with fresh surfaces which are already in the metallic from are very reactive and are free from getting oxidized. Furthermore the presence of transition metals such as Fe, Co or Ni makes these alloy hydrides effective catalysts for the growth of MWNTs in high yield and with low cost.

4. Energy-Related Applications of CNTs

The extraordinary structure and properties of CNTs which includes nanometer-size, large surface area, high aspect ratio, high mechanical strength, flexibility, high electron conductivity, low mass density, high thermal conductivity and chemical stability, impart them a great potential for use in a number of applications [79]. The remarkable electronic properties of CNTs and its applications as quantum wires [4] have inspired the design of several components for nanoelectronics such as field effect transistors and rectifying diodes [80-82]. CNTs have also been recognized as one of the most promising electron field emitters [83-85]. CNTs due to its high aspect ratio, mechanical strength and flexibility make potentially ideal structures for use as tips in scanning probe microscopes [86-88]. CNTs have the ideal combination of mechanical strength and aspect ratio for use as fiber reinforcing elements in structural composites [2, 3]. CNT- based sensor devices make use of the strong electronic response of nanotubes to changes in the local environment [79]. It has been shown that nanotubes can be used as advanced miniaturized chemical sensors [41]. Miniaturized gas ionization sensors using CNTs have been fabricated by [21]. CNT filters have been recently demonstrated with freestanding monolithic uniform macroscopic hollow cylinders having radially aligned CNT walls [22].

Several studies have shown the superiority of nanotubes to other carbon-based electrodes for applications in batteries and capacitors [90-92]. Some of the energy-related applications of CNTs are reviewed in the following section. The properties of MWNTs synthesized using the new process discussed in the previous section, as catalyst support material for PEM Fuel cell, supercapacitor electrode materials and hydrogen storage media are discussed in detail.

4.1. Lithium Intercalation

The basic principle of rechargeable lithium batteries is electrochemical intercalation and de-intercalation of lithium in both electrodes [93]. The cathode materials are transition metal oxides (Li_xMO_2, where M- transition metal), while carbon materials are used as anodes. CNTs have been investigated for lithium storage by various groups [94-95]. High reversible and irreversible capacities have been obtained for SWNTs [96]. Figure 15 shows the electrochemical intercalation data for MWNT sample using an electrochemical cell with a MWNT film and lithium foil as two working electrodes [96]. However the high divergence between insertion and extraction excludes it from any practical application in lithium-ion batteries. Hence, further studies are required to overcome some of these barriers such as irreversibility and large voltage hysteresis in order that nanotubes be used for practical applications.

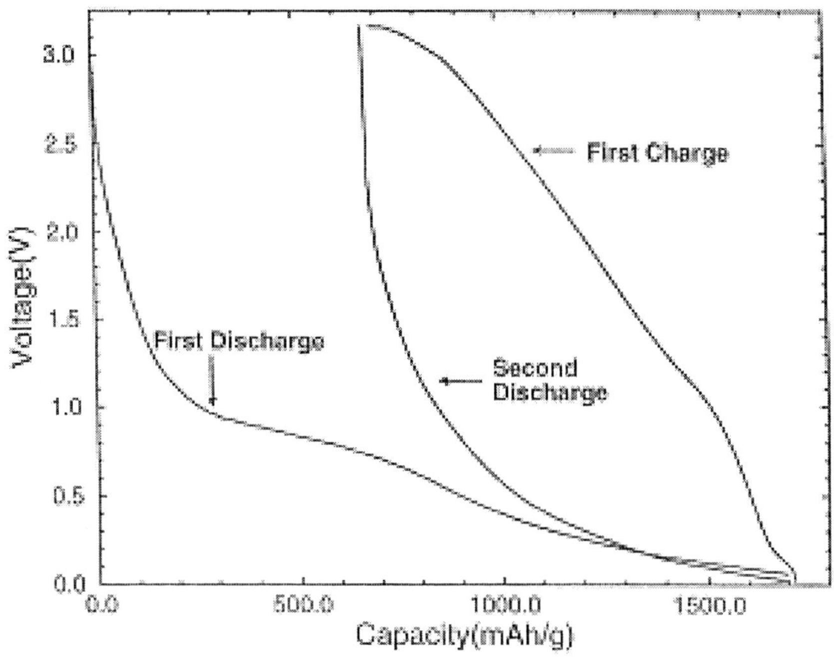

Figure 15. Charge-discharge data of purified SWNTs. Reversible capacity is 1000 mAh/g (Adapted from [96] with permission).

4.2. CNTs as Catalyst Support Material in Proton Exchange Membrane Fuel Cells (PEMFC)

Proton Exchange Membrane Fuel Cells (PEMFCs) have gathered tremendous amount of attention, as a result of their applicability to transportation systems and stationary and portable power applications [97]. They use a solid ion exchange polymer as an electrolyte and porous carbon containing platinum catalyst as electrodes, with hydrated protons as the charge carriers. PEMFCs are typically fueled with pure hydrogen supplied from storage tanks or on-board reformers. The morphology and size of CNTs make them very significant in catalytic applications due to their ability to disperse catalytically active metal particles [98, 99]. The high accessible surface area, interesting electronic properties and high stability suggest that CNTs are suitable materials for electrodes and catalyst support in PEMFCs [100], which effectively reduces the Pt catalyst loading and improves the fuel cell performance. Functionalization and oxidation of CNTs have been shown to be effective for better loading and dispersion of the metal particles [101, 102]. Various methods have been followed in the literature to prepare metal-loaded CNTs [103-105]. Pt nanoparticles have been electrodeposited on MWNTs by potentiostatic method and its electrocatalytic activity for oxygen reduction has been studied [106, 107]. The electrocatalytic activity of Pt depends on many factors such as high surface area and good electronic activity of the conductive catalyst support [108, 109].

MWNT-supported Pt (Pt/MWNT) electrocatalysts were prepared by chemical reduction method using a Pt salt and pre-treated MWNTs, synthesized by pyrolysis of acetylene over Mm and Ni based AB_2 alloy hydride catalysts [110]. The pre-treated MWNTs are functionalized with carboxylic acid groups attached. This would act as anchoring sites for better adherence of Pt nanoparticles onto the MWNTs surface, there by giving a better performance at higher current densities. XPS spectrum of pre-treated MWNTs (C_{1S} spectrum) shown in Figure 16 appears to be composed of graphitic carbon (284.5 eV). The peak at 288.9 eV shows that the MWNTs surfaces are covered with carboxylic groups.

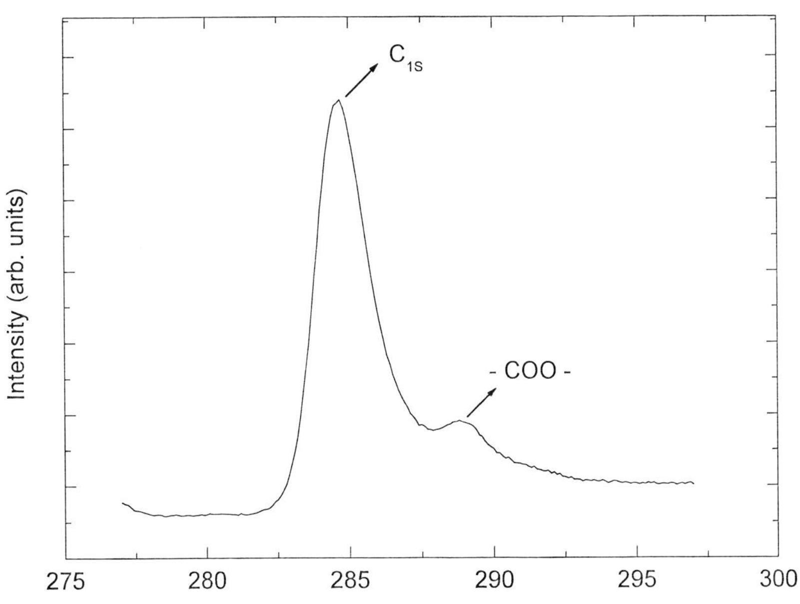

Figure 16. X-ray photoelectron spectra for C_{1S} state of HNO_3 - treated MWNTs, prepared by pyrolysis of acetylene over Mm and Ni based AB_2 alloy hydride catalyst.

The powder XRD patterns for Pt/MWNTs electrocatalysts, prepared using MWNTs synthesized with Mm and Ni based AB_2 alloy hydride catalysts (Figure 17) show the characteristic patterns of Pt face centred cubic diffraction along with the hexagonal reflections from graphitic carbon. Figure 18 shows the SEM images of purified MWNTs and Pt-loaded MWNTs, synthesized with Mm and Ni based AB_2 alloy hydride catalyst. The CNT walls are decorated by Pt nanoparticles with a particle size distribution in the range 3-15 nm. EDX measurements showed Pt contents of 19.6 % (Figure 19).

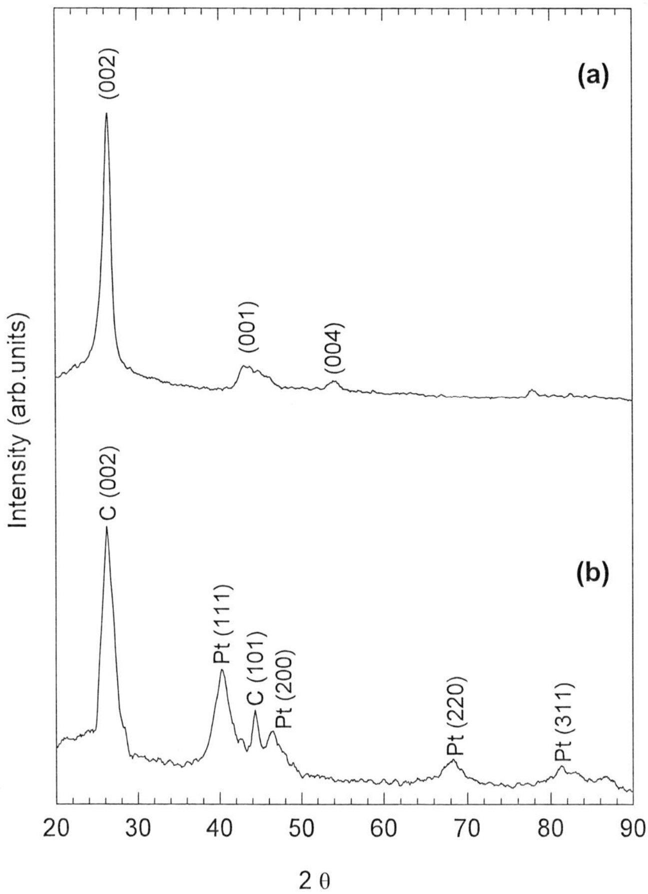

Figure 17. Powder X-ray diffractograms of (a) Purified MWNTs prepared by pyrolysis of acetylene over Mm and Ni based AB_2 alloy hydride catalyst and (b) Pt-loaded MWNTs.

TEM images of MWNT-supported Pt electrocatalysts prepared using pre-treated MWNTs synthesized with Mm and Ni based AB_2 alloy and Mm based AB_5 alloy hydride catalysts are shown in Figures 20 (a), (b) respectively. A uniform and high dispersion of Pt nanoparticles are seen on the MWNTs surface with a particle size distribution of about 3-5 nm.

The basic performance measure of a fuel cell is the voltage output as a function of the electrical current density drawn, or polarization curve. The current-voltage characteristics of a PEMFC fabricated with MWNTs supported Pt electrocatalysts as cathode catalysts were studied using a PEMFC test station comprising of a single stack fuel cell. The polarization curves were all obtained from the single stack PEMFC using same type of anode.

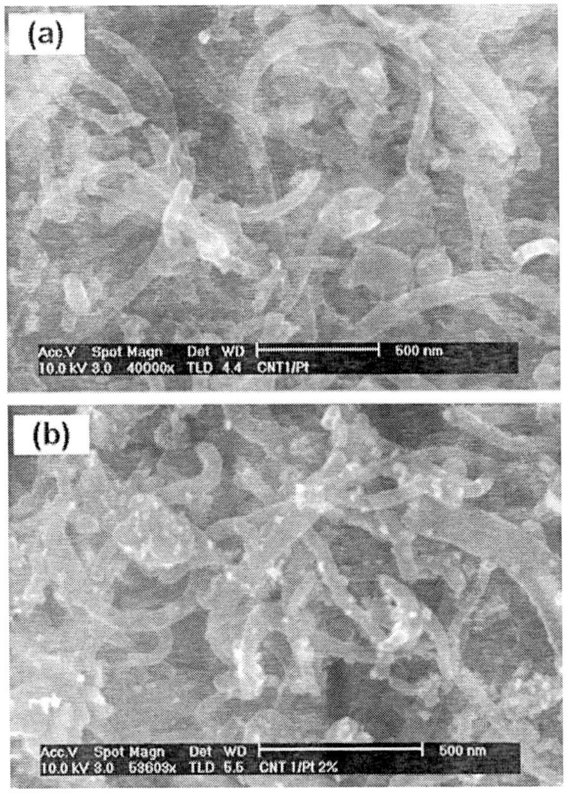

Figure 18. SEM images of (a) purified MWNTs and (b) Pt-loaded MWNTs prepared using pre-treated MWNTs synthesized by pyrolysis of acetylene over Mm and Ni based AB_2 alloy hydride catalyst.

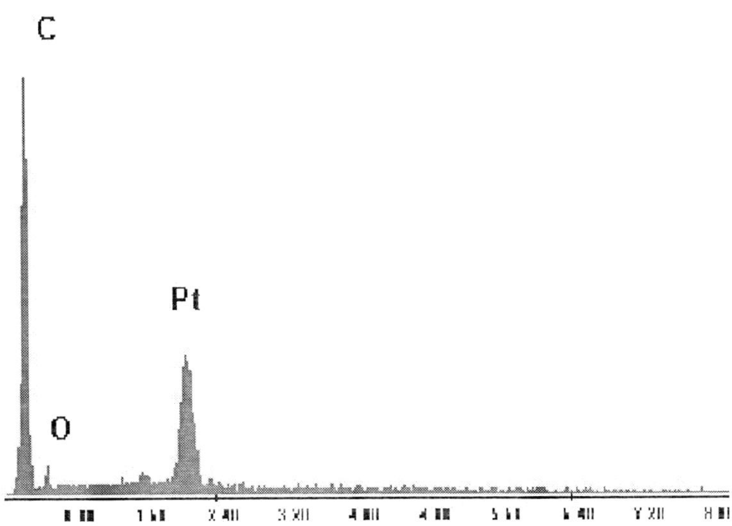

Figure 19. Energy dispersive analysis of Pt/MWNTs electrocatalysts prepared using pre-treated MWNTs synthesized from pyrolysis of acetylene over Mm and Ni based AB_2 alloy hydride catalysts.

A commercial 20 wt %. Pt on carbon (Pt/C) obtained from E-TEK Inc. was used as the anode with a Pt loading of 0.25 mg cm^{-2}. To study the electrocatalytic activity of Pt/MWNTs on the performance of PEMFC, active material of composition of x wt % Pt/MWNTs and y wt % Pt/C, with $x + y = 100$, where $x = 0$, 25, 50, 75 and 100 %, were taken as cathode catalyst. Different cathode catalyst composition was studied with Pt/MWNTs content in the active material varying from 0 to 100 % with a Pt loading of 0.5 mg cm^{-2} and the performance was systematically studied.

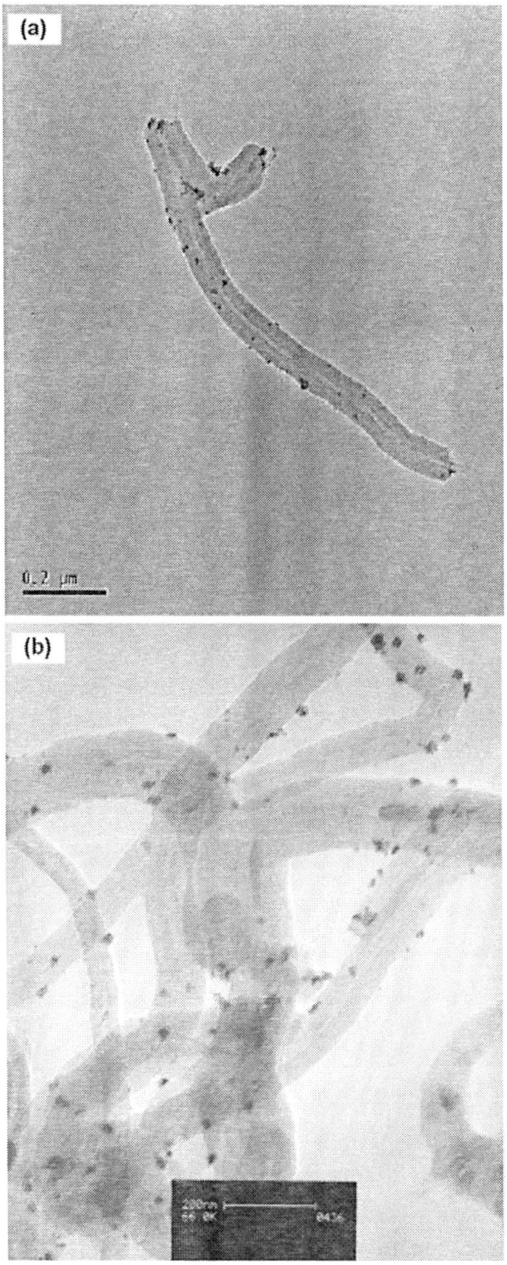

Figure 20. TEM images of Pt-loaded MWNTs prepared using pre-treated MWNTs synthesized by pyrolysis of acetylene over (a) Mm and Ni based AB_2 and (b) Mm based AB_5 alloy hydride catalysts.

A comparison of the fuel cell performance with cathode catalyst containing mixture of different compositions of Pt/MWNTs, prepared using MWNTs synthesized over Mm and Ni based AB_2 alloy hydride catalyst, and Pt/C is shown in Figure 21 [111]. In the low current density region, the voltage drop in the potential-current curve, generally known as activation polarization, reflects the sluggish kinetics intrinsic to the oxygen reduction reaction at the cathode surface.

The linear decrease in voltage in the mid to high current density range, or ohmic polarization, arises from limitations in proton transport through the electrolyte membrane from anode to cathode and/or limitations in electron flow in the electrode materials. Better performance of PEMFC was observed for cathode catalysts with Pt/MWNTs content compared to those containing commercial 20 % Pt/C, which could be attributed to the higher catalytic reactivity of smaller Pt particles with uniform sizes decorated on the MWNTs. The PEMFC performance also depends on the electrical conductivity of the MWNTs support, and its ability to transport electrons to the current collector of MEA. In the present work, it was found that the performance of the PEMFC passes through a maximum for cathode catalyst with 50 wt % of Pt/MWNTs. PEMFC with cathode catalyst containing 50 wt % Pt/MWNTs shows the maximum performance, with a potential of 540 mV at a current density of around 535 mA cm^{-2} and thus seems to be the optimal composition for better performance of PEMFC, under the present study.

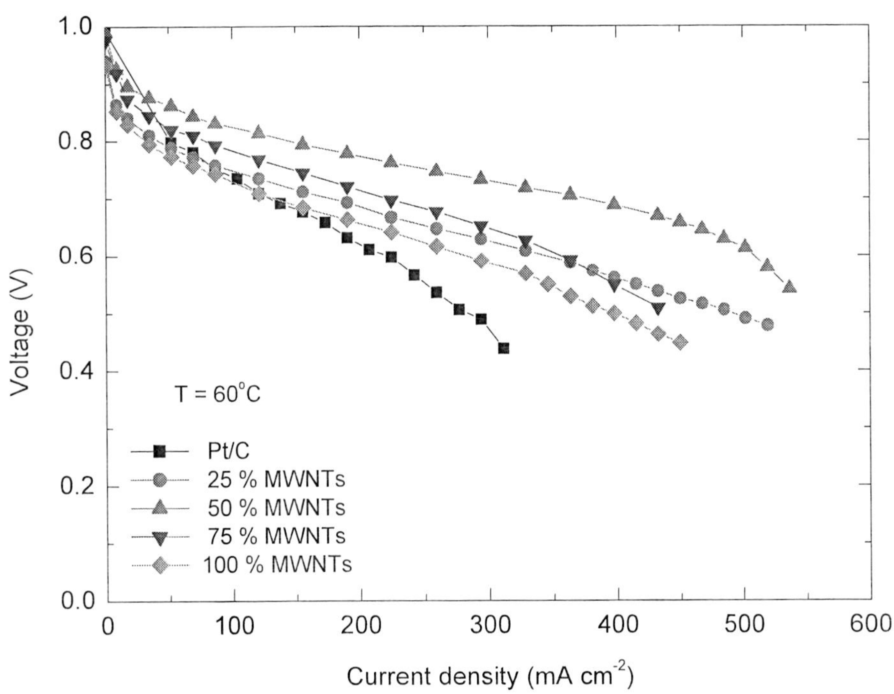

Figure 21. Polarization curves of PEMFC with cathode catalyst containing a mixture of Pt/C and Pt/MWNTs electrocatalysts (prepared using purified MWNTs synthesized by pyrolysis of acetylene over Mm and Ni based AB_2 alloy hydride catalyst).

Thus the composition of Pt/MWNT plays the key role in the overall performance of the PEMFC. The present study on MWNTs synthesized using alloy hydride catalyst indicates that the accessibility of MWNTs and hence the Pt electrocatalysts in the mixture for the oxygen reduction reaction, may be the determining factor in deciding the performance of the cell.

4.3. CNTs Based Electrochemical Supercapacitors

Due to their high specific surface area, relatively low cost, availability and established production technology, use of activated carbon materials as the electrode materials in supecapacitors has been intensively investigated during the last decades [112, 113-116]. The experimentally measured capacitance of the activated carbons is not in good agreement with the theoretical values, since a significant part of the surface area remains in the micropores, which are not accessible to the electrolyte ions [112]. Therefore, pore size distribution together with surface area is important for the determination of electric double layer capacitance [117]. Recently there have been considerable attempts to use CNTs as electrode materials for supercapacitors because of its narrow distribution of mesopore sizes, highly accessible surface area, low resistivity and high stability [117-119].

Figure 22. Photograph of a supercapacitor test cell based on MWNTs electrodes.

Purified MWNTs synthesized by the pyrolysis of acetylene using various alloy hydride catalysts have been used as supercapacitor electrode materials and the electrochemical performance have been investigated [121]. Figure 22 shows the supercapacitor test cell that has been fabricated with two MWNT electrodes separated by a thin polymer in 1 M H_2SO_4 solution. The specific capacitance for a single electrode has been measured from the cyclic voltammogram, galvanostatic charge-discharge curve and complex plane impedance plots and the values calculated from these three techniques are consistent with each other.

The capacitance values were obtained from the CV curves (Figure 23) according to the equation:

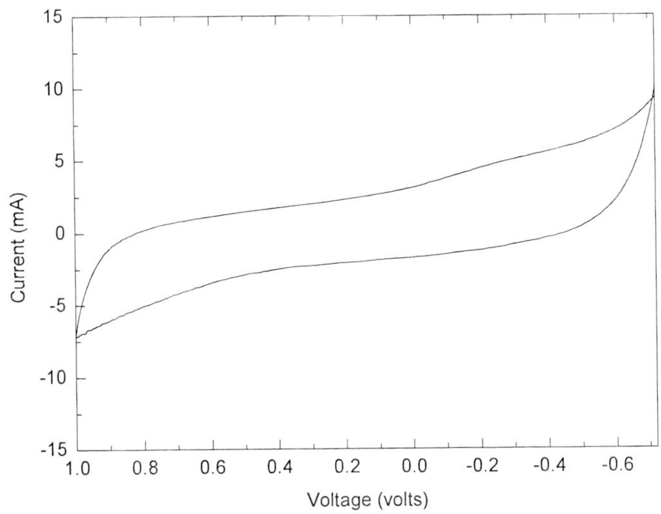

Figure 23. Cyclic voltammograms of MWNTs electrodes, prepared using purified MWNTs synthesized by pyrolysis of acetylene over Mm and Ni based AB_2 alloy hydride catalysts (scan rate of 10 mV s^{-1} in 1 M H_2SO_4).

$$C = \frac{i}{s} \tag{2}$$

where i is the average cathodic current, which was determined by integration of the negative sweep of the CV curves and s is the potential sweep rate. C_{sp} is the specific capacitance measured in F/g, given by Eqn. 3,

$$C_{sp} = \frac{C}{m} \tag{3}$$

where m is the amount of MWNTs in one electrode. The specific capacitance of the MWNTs electrodes was also determined by galvanostatic charge-discharge technique. Galvanostatic cycling of supercapacitor cells was performed at a constant current of 2 mA/10 mA between 0 and 1 V. The typical constant current charge-discharge (V-t) curves for the MWNT electrodes prepared with purified MWNTs synthesized using Mm and Ni based AB_2 alloy hydride catalysts is shown in Figure 24. The discharge curves are almost linear in the total range of potential with constant slopes, showing perfect capacitive behaviour. The capacitance of the cell was deduced from the slope of the V- t curves using the equation,

$$C = \frac{I}{(dV/dt)} \tag{4}$$

The capacitance of the MWNT electrode is also calculated from the low frequency data of the impedance spectra (Figure 25), according to the formula,

$$C = \frac{1}{2\pi f(-Z_{im})} \tag{5}$$

where f is the frequency at the point and Z_{im} represents the imaginary part of the complex impedance at this frequency. A maximum specific capacitance of 26 F/g has been achieved for MWNT electrodes prepared using purified MWNTs synthesized from the pyrolysis of acetylene over Mm and Ni based AB_2 alloy hydride catalyst.

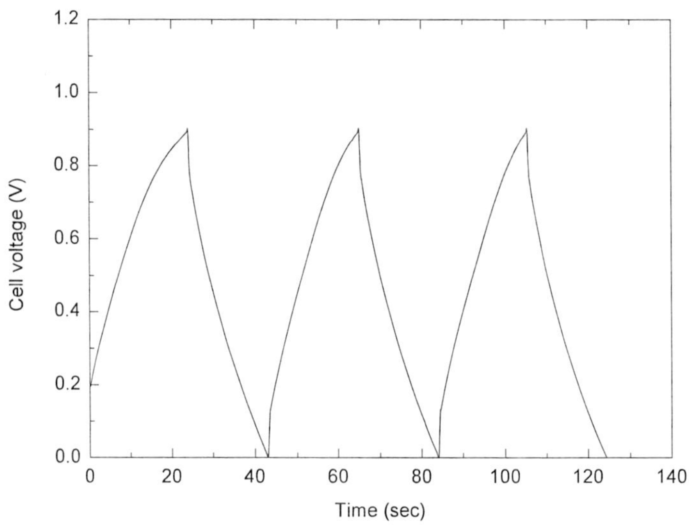

Figure 24. Galvanostatic charge-discharge characteristics of MWNTs electrodes prepared using purified MWNTs synthesized by pyrolysis of acetylene over Mm and Ni based AB_2 alloy hydride catalysts.

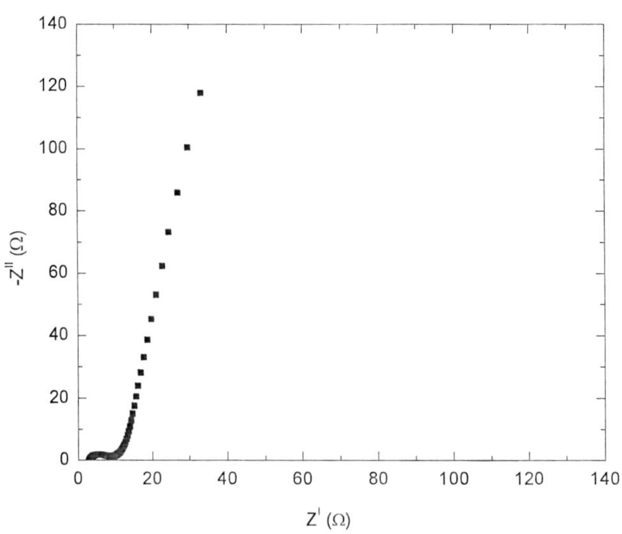

Figure 25. Complex plane impedance plots at an AC voltage amplitude of 10 mV for MWNTs electrodes prepared using purified MWNTs synthesized by pyrolysis of acetylene over Mm and Ni based AB_2 alloy hydride catalysts.

4.4. Hydrogen Adsorption Studies of MWNTs

Hydrogen provides a clean, reliable and affordable source of energy for meeting the growing energy needs, while protecting the environment and ensuring energy security. An economically viable hydrogen storage medium is an essential component of the hydrogen fueled transportation systems and there is an on-going research for advanced hydrogen storage materials. Since the initial findings of hydrogen adsorption in SWNTs [122], lot of research has been carried out to study the hydrogen storage capacity of CNTs as well as other carbon materials in several ways. Theoretical studies have predicted that interaction of hydrogen with CNTs is through physisorption of hydrogen on the exterior and possibly the interior surfaces [123-126]. Inter nanotube space between densely aligned CNTs has been shown to contribute to the effective uptake of molecular hydrogen [127] Studies on the investigation of interaction of hydrogen with SWNTs highlight the advancement in the understanding of the phenomenon of hydrogen adsorption in CNTs [128, 129]. Numerous other promising and comparable results were obtained [130, 131], while discrepancies were also shown in some reports. Apart from the different synthesis and purification methods that could lead to different results, the measuring methods are also very important [132]. The low bulk density of CNTs and the limited amount used could result in large uncertainties in the experiments, especially for high-pressure gas phase storage. Various other factors, such as the unavoidable leaks at high pressures and the influence of temperature variations on pressure also bring about wide range of values for hydrogen uptake in CNTs. Various comparative works have been presented on hydrogen adsorption in carbon materials [133, 134]. Hence there is a need for a careful and systematic study of the hydrogen sorption measurements in CNTs taking into account the various factors that are generally encountered such as unavoidable leaks in the system, temperature effect on the pressure, proper calibration of the sample volume and the low mass density of CNTs.

Figure 26. Photograph of the experimental facility developed for the high-pressure hydrogen adsorption studies based on pressure reduction method.

Hydrogen adsorption studies have been carried out for the as-grown and purified MWNTs synthesized by pyrolysis of acetylene using Zr based AB_2, Mm and Ni/Fe based AB_2, and Mm based AB_5 alloy hydride catalysts, in the pressure range 10-100 bar and at 298 K [25, 65], using a high pressure Seiverts apparatus (Figure 26). Various precautions were taken during the adsorption measurements, as all these factors affect the adsorption capacities [132]. The low sample density of MWNTs was also taken into account while calculating the adsorption capacity.

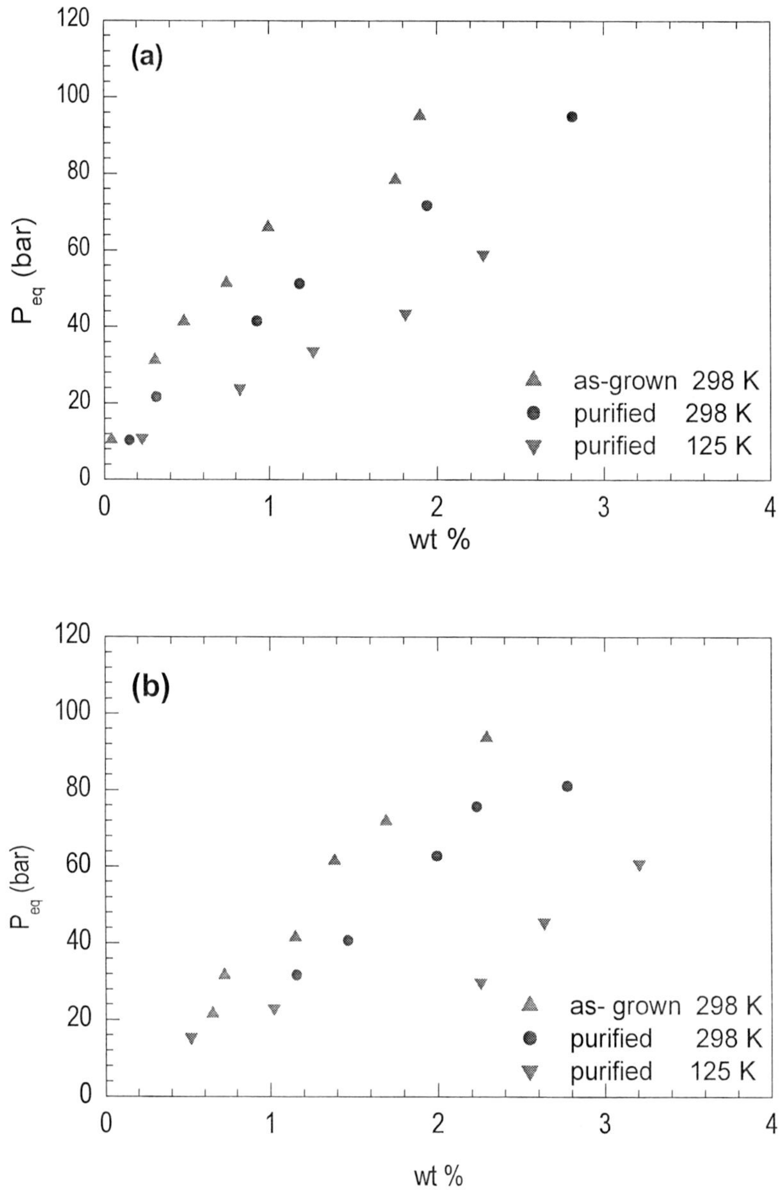

Figure 27. Hydrogen adsorption isotherms of MWNTs synthesized by pyrolysis of acetylene over (a) Mm and Ni based AB_2 and (b) Mm and Fe based AB_2 alloy hydride catalysts.

The hydrogen molecules could be physically adsorbed on the external nanotube walls or stay in the micro channels between the nanotube bundles. In our experiments, maximum storage capacities of 2.4 wt. % and 3.3 wt. % were obtained at 298 K and at around 100 bar respectively for as-grown and purified MWNTs synthesized by pyrolysis of acetylene using Mm based AB_5 alloy hydride catalyst. Figure 27 shows the hydrogen adsorption isotherms for MWNTs synthesized by Mm and Ni/Fe based AB_2 alloy hydride catalysts. The purified MWNTs which were subjected to air oxidation and acid treatment, which could open their ends effectively, showed increased hydrogen sorption capacity, as hydrogen could have entered nanotubes through their ends.

Hence for the purified MWNTs, the increase in storage capacity of the adsorbed hydrogen would have resulted from the condensation of hydrogen inside the cavity of the tubes. Hydrogen adsorption studies performed at 125 K for purified MWNTs showed an enhancement in the storage capacity, which is due to the low temperature condensation effect of hydrogen, as CNTs adsorb a hydrogen molecule by physical van der Waals type forces [124, 125].

5. SUMMARY

An overview of various synthesis methods for carbon nanotubes has been presented. It is shown that CVD methods are promising for producing large quantities of CNTs. MWNTs have been synthesized by catalytic decomposition of acetylene over alloy hydride catalysts obtained through hydrogen decrepitation route and this new process has been shown to be a simple and cost effective method to produce good quality MWNTs in high yield. Various characterization techniques have been employed to demonstrate the quality and yield of MWNTs produced using the high reactive alloy hydride catalysts. Vapour-liquid-solid (VLS) mechanism is thought to be responsible for the formation of MWNTs using alloy hydride catalysts. The enormous potential of CNTs in Science and Technology has been illustrated by giving a short review of the several possible applications of CNTs with emphasis on their energy-related applications. Various experimental techniques for the studies of some of the energy-related applications such as catalyst support material in PEMFC, supercapacitor electrode materials and hydrogen storage media, have been discussed in detail.

ACKNOWLEDGEMENTS

The authors acknowledge Indian Institute of Technology Madras, India. Defence Research and Development Organization (DRDO) India, Department of Science and Technology (DST) India, Ministry of Non-conventional Energy Sources (MNES) India, Ministry of Human Resource Development (MHRD) India, are gratefully acknowledged for their financial support. We thank Dr. N. Rajalakshmi, CFCT, Chennai, India for fruitful discussions.

REFERENCES

[1] Kroto, H.W., J.R. Heath, S.C. O'Brien, R.F. Curl, and R.E. Smalley (1985) C_{60}: Buckminsterfullerene. *Nature,* 318, 162-163.

[2] Ajayan, P.M., L.S. Schadler, C. Giannaris, and A. Rubio (2000) Mechanical response of singlewalled carbon nanotubes in polymer nanocomposites. *Adv. Mater.*, 12, 750.

[3] Peigney, A. (2003) Tougher ceramics with carbon nanotubes. *Nature Mater.*, 2, 15-16.

[4] Tans, S.J., M.H. Devoret, H. Dai, A. Thess, R.E. Smalley, L.J. Geerligs, and C. Dekker (1997) Individual single-wall carbon nanotubes as quantum wires. *Nature,* 386, 474 – 477.

[5] Tans, S.J., A.R.M. Verschueren, and C. Dekker (1998) Room-temperature transistor based on a single carbon nanotube. *Nature*, 393, 49-52.

[6] Martel, R., T. Schmidt, H.R. Shea, T. Hertel, and P. Avouris (1998) Single and multi-wall carbon nanotube field effect transistors. *Appl. Phys. Lett.*, 73, 2447-2449.

[7] Hamada, N., S. Sawada, and A. Oshiyama (1992) New one-dimensional conductors: graphitic microtubules. *Phys. Rev. Lett.,* 68(10), 1579–1581.

[8] Dresselhaus, M.S., G. Dresselhaus, and R. Saito (1992) Carbon fibers based on C_{60} and their symmetry. *Phys. Rev. B,* 45, 6234-6242.

[9] Wildöer, J.W.G, L.C. Venema, A.G. Rinzler, R.E. Smalley, and C. Dekker (1998) Electronic structure of atomically resolved carbon nanotubes. *Nature,* 391, 59-61.

[10] Odom, T.W., J.L. Huang, P. Kim, and C.M. Lieber (1998) Atomic structure and electronic properties of single-walled carbon nanotubes. *Nature, 391*, 62-64.

[11] Ebbesen, T.W. (June 1996) Carbon nanotubes. *Physics Today*, 26-32.

[12] Ajayan, P.M. (1999) Nanotubes from carbon, *Chem. Rev.,* 99, 1787-1799.

[13] Dai, H. (2002) Carbon nanotubes: opportunities and challenges. *Surface Science,* 500, 218-241.

[14] Popov, V.N. (2004) Carbon nanotubes: properties and application. *Mater. Science and Engineering R,* 43, 61-102.

[15] Saito, R., M. Fujita, G. Dresselhaus, and M.S. Dresselhaus (1992) Electronic structure of chiral graphene tubules. *Applied Physics Letters,* 60(18), 2204–2206.

[16] Dresselhaus, M.S., G. Dresselhaus, and P.C. Eklund *Science of fullerenes and carbon nanotubes* Academic press, San Diego, 1996.

[17] Reich, S. and C. Th, *Carbon nanotubes: Basic Concepts and Physical Properties.* Wiley-VCH, Darmstadt, 2004.

[18] Meyyappan, M., (2004) Carbon nanotube growth by chemical vapour deposition. 1, 581-589, *In Encyclopedia of Nanoscience and Nanotechnology*, H. S. Nalwa (ed.).

[19] Rinzler, A.G., J.H. Hafner, P. Nikolaev, L. Lou, S.G. Kim, D. Tomanek, D. Colbert, and R.E. Smalley (1995) Unraveling nanotubes: field emission from an atomic wire, *Science,* 269, 1550-1553.

[20] de Heer, W.A., A. Chatelain, and D. Ugarte (1995) A Carbon nanotube field-emission electron source, *Science*, 270, 1179-1180.

[21] Modi, A., N. Koratkar, E. Lass, B.Q. Wei, and P.M. Ajayan (2003) Miniaturized gas ionization sensors using carbon nanotubes. *Sciecne*, 424, 171-174.

[22] Srivastava, A., O.N. Srivastava, S. Talapatra, R. Vajtai, and P.M. Ajayan (2004) Carbon nanotube filters. *Nature Materials,* 3, 610-614.

[23] Hafner, J.H., C.L. Cheung, T.H. Oosterkamp, and C.M. Lieber (2001) High-yield assembly of individual single-walled carbon nanotube tips for scanning probe microscopies. *J. Phys. Chem. B*, 105(4) 743–746.

[24] Tibbetts, G.G., G.P. Meisner, and C.H. Olk (2001) Hydrogen storage capacity of carbon nanotubes, filaments, and vapor-grown fibers. *Carbon*, 39, 2291-2301.

[25] Shaijumon M M and Ramaprabhu S (2003) Synthesis of carbon nanotubes by pyrolysis of acetylene using alloy hydride materials as catalysts and their hydrogen adsorption studies. *Chem. Phys. Lett.* 374, 513 - 520.

[26] Iijima, S. (1991) Helical microtubules of graphitic carbon. *Nature,* 354, 56- 58.

[27] Guo, T., P. Nikoleev, A. Thess, D.T. Colbert, and R.E. Smally (1995) Catalytic growth of single-walled manotubes by laser vaporization. *Chem. Phys. Lett.,* 243 (1-2), 49-54.

[28] Yacaman, M.J., M. Miki-Yoshida, L. Rendon, and J. G. Santiesteban (1993) Catalytic growth of carbon microtubules with fullerene structure. *Appl. Phys. Lett.,* 62, 202-204.

[29] Iijima, S. and T. Ichihashi (1993) Single-shell carbon nanotubes of 1-nm diameter. *Nature,* 363(6430), 603–615.

[30] Thess, A., R. Lee, P. Nikolaev, H. Dai, P. Petit, J. Robert, C. Xu, Y.H. Lee, S.G. Kim, A.G. Rinzler, D.T. Colbert, G. E. Scuseria, D. Tom′anek, J.E. Fischer, and R.E. Smalley (1996) Crystalline ropes of metallic carbon nanotubes. *Science,* 273(5274), 483–487.

[31] Bolshakov, A.P., S.A. Uglov, A.V. Saveliev, V.I. Konov, A.A. Gorbunov, W. Pompe, and A. Graff (2002) A novel CW laser–powder method of carbon single-wall nanotubes production. *Diamond. Rel. Mater.,* 11, 927-930.

[32] Eklund, P.C., B.K. Pradhan, U.J. Kim, Q. Xiong, J.E. Fischer, A.D. Friedman, B. C. Holloway, K. Jordan, and M.W. Smith (2002) Large-scale production of single-walled carbon nanotubes using ultrafast pulses from a free electron laser. *Nano Lett.,* 2(6), 561-566.

[33] Moravsky, A.P., E.M. Wexler, and R.O. Loutfy *Growth of carbon nanotubes by arc discharge and laser ablation.* pp. 65-97. *In* M. Meyyappan (ed) *Carbon nanotubes, Science and Applications*. CRC Press, Florida, (2004).

[34] Baker, R.T.K. (1989) Catalytic growth of carbon filaments. *Carbon,* 27, 315-323.

[35] Tibbetts, G.G. (1989) Vapor-grown carbon fibers: status and prospectus. *Carbon,* 27, 745-757.

[36] Li, W.Z., S.S. Xie, L.X. Qian, B.H. Chang, B.S. Zou, W.Y. Zhou, R.A. Zhao, and G. Wang (1996) Large-scale synthesis of aligned carbon nanotubes. *Science,* 274 (5293), 1701-1703.

[37] Kim, W., H.C. Choi, M. Shim, Y. Li, D. Wang, and H. Dai (2002) Synthesis of ultralong and high percentage of semiconducting single-walled carbon nanotubes. *Nano Lett.,* 2, 703-708.

[38] Harutyunyan, A.R., B.K. Pradhan, U.J. Kim, G. Chen, and P.E. Eklund (2002) CVD synthesis of single wall carbon nanotubes under "Soft" conditions. *Nano Lett.,* 2, 525-530.

[39] Choi, H.C., S. Kundaria, D. Wang, A. Javey, Q. Wang, M. Rolandi, and H. Dai (2003) Efficient formation of iron nanoparticle catalysts on silicon oxide by hydroxylamine for carbon nanotube synthesis and electronics. *Nano Lett.,* 3, 157-161.

[40] Shaijumon M M, Bejoy N, and Ramaprabhu S (2005) Catalytic growth of carbon nanotubes over Ni/Cr hydotalcite – type anionic clays and the effect of its purification on their hydrogen storage properties *Appl. Surf. Science* 242 192 - 198.

[41] Kong, J., A.M. Cassell, and H. Dai (1998) Chemical vapor deposition of methane for single-walled carbon nanotubes. *Chem. Phys. Lett.,* 292, 567-574.

[42] Kong, J., H.T. Soh, A.M. Cassell, C.F. Quate, and H. Dai (1998) Synthesis of individual single-walled carbon nanotubes on patterned silicon wafers. *Nature,* 395(6705), 878–888.

[43] Flahaut, E., A. Govindaraj, A. Peigney, Ch. Laurent, A. Rousset, and C.N.R. Rao (1999) Synthesis of single-walled carbon nanotubes using binary (Fe, Co, Ni) alloy nanoparticles prepared in situ by the reduction of oxide solid solutions. *Chem. Phys. Lett.,* 300, 236-242.

[44] Su, M., B. Zheng, and J. Liu (2000) A scalable CVD method for the synthesis of single-walled carbon nanotubes with high catalyst productivity. *Chem. Phys. Lett.*, 322, 321-326.

[45] Colomer, J.F., C. Stephan, S. Lefrant, G. Van Tendeloo, I. Willems, Z. Kónya, A. Fonseca, Ch. Laurent, and J.B. Nagy (2000) Large-scale synthesis of single-wall carbon nanotubes by catalytic chemical vapor deposition (CCVD) method. *Chem. Phys. Lett.,* 317, 83-89.

[46] Ramaprabhu, S., M. M. Shaijumon (2005) Carbon nanotubes and their synthesis. *Patent filed* (No:1129/CHE/05)

[47] Che, G., B. B. Lakshmi, C. R. Martin, and E. R. Fisher (1999) Metal-Nanocluster-Filled Carbon Nanotubes: Catalytic Properties and Possible Applications in Electrochemical Energy Storage and Production. *Langmuir.,* 15, 750–758.

[48] Sen, R., A. Govindaraj, and C.N.R. Rao (1997) Carbon nanotubes by the metallocene route. *Chem. Phys. Lett.,* 267, 276-280.

[49] Shaijumon M M and Ramaprabhu S (2005) Studies of yield and nature of carbon nanostructures synthesized by pyrolysis of ferrocene and hydrogen adsorption studies of carbon nanotubes.*Int. J. Hydrogen Energy* 30 311- 317.

[50] Jung, Y.J., B.Q. Wei, R. Vajtai, P.M. Ajayan, Y. Homma, K. Prabhakaran, and T. Ogino (2003) Mechanism of selective growth of carbon nanotubes on SiO_2/Si patterns. *Nano Lett*., 3, 561-564.

[51] Dai, H., J. Kong, C. Zhou, N. Franklin, T. Tombler, A. Cassell, S. Fan, and M. Chapline (1999) Controlled chemical routes to nanotube architectures, physics and devices. *Journal of Physical Chemistry B,* 103(51), 11246–11255.

[52] Zhang, Z.J., B.Q. Wei, R. Vajtai, G. Ramanath, and P.M. Ajayan (2000) Substrate-site selective growth of aligned carbon nanotubes. *Appl. Phys. Lett.,* 77, 3764-3766.

[53] Wei, B.Q., R. Vajtai, Y.Y. Choi, P.M. Ajayan, H. Zhu, C. Xu, and D. Wu (2002) Structural characterizations of long single-walled carbon nanotube strands. *Nano Lett*., 2, 1105-1107.

[54] Cao, A., B. Wei, Y. Jung, R. Vajtai, P.M. Ajayan, and G. Ramanath (2002) Growth of aligned carbon nanotubes on self-similar macroscopic templates. *Appl. Phys. Lett.,* 81, 1297.

[55] Li, Y.H., C.L. Xu, B.Q. Wei, X.F. Zhang, M.X. Zheng, J. Liang, D.H. Wu, and P.M. Ajayan (2002) Self-organized ribbons of aligned carbon nanotubes. *Chemistry of Materials*, 14(2), 483-485.

[56] Ren, Z.F, Z.P. Huang, J.W. Xu, J.H. Wang, P. Bus, M.P. Siegal, and P.N. Provencio (1998) Synthesis of large arrays of well-aligned carbon nanotubes on glass. *Science,* 282, 1105-1107.

[57] Chen, Y., D.T. Shaw, and L.P. Guo (2000) Field emission of different nanotubes. *Appl. Phys. Lett.,* 76, 2469-2471.

[58] Yabe, Y., Y. Ohtake, T. Ishitobi, Y. Show, T. Izumi, and H. Yamauchi (2004) Synthesis of well-aligned carbon nanotubes by radio frequency plasma enhanced CVD method. *Diamond and Related Mater.,* 13, 1292-1295.

[59] Song, I.K., W.J. Yu, Y.S. Cho, G.S. Choi, and D. Kim (2004) The determining factors for the growth mode of carbon nanotubes in the chemical vapour deposition process. *Nanotechnology,* 15, S590- S595.

[60] Qin, L.C., D. Zhou, A.R. Krauss, and D.M. Gruen (2000) Growing carbon nanotubes by microwave plasma-enhanced chemical vapor deposition. *Appl. Phys. Lett.,* 72, 3437-3439.

[61] Bower, C., O. Zhou, W. Zhu, D.J. Werder, and S. Jin (2000) Nucleation and growth of carbon nanotubes by microwave plasma chemical vapor deposition. *Appl. Phys. Lett.,* 77, 2767-2769.

[62] Taniguchi, M., H. Nagao, M. Hiramatsu, Y. Ando, and M. Hori (2005) Preparation of dense carbon nanotube film using microwave plasma-enhanced chemical vapor deposition. *Diamond Relat. Mater.,* 14, 855-858.

[63] Kukovitsky, E.F., S.G. L'vov, N.A. Sainov, V.A. Shustov, and L.A. Chernozatonskii (2002) Correlation between metal catalyst particle size and carbon nanotube growth. *Chem. Phys. Lett.,* 355, 497-503.

[64] Mani N and Ramaprabhu S (2004) Effect of substitutional elements on hydrogen absorption properties in Mm based Ab_5 alloys. *Journal of Alloys and Compounds,* 363, 275-295.

[65] Shaijumon M M, Rajalakshmi N, Hojin Ryu, and Ramaprabhu S (2005) Synthesis of multi walled carbon nanotubes in high yield using Mm based AB_2 alloy hydride catalysts and the effect of purification on their hydrogen adsorption properties. *Nanotechnology* 16, 518-524.

[66] Strong, K. L., D.P. Anderson, K. Lafdi, and J.N. Kuhn (2003) Purification process for single-wall carbon nanotubes. *Carbon,* 41, 1477-1488.

[67] Chiang, I.W., B.E. Brinson, R.E. Smalley, J.L. Margrave, and R.H. Hauge (2001) Purification and characterization of single-wall carbon nanotubes *J. Phys. Chem. B,* 105, 1157-1161.

[68] McGuire K and Rao A M, (2004) Characterization techniques in carbon nanotubes research. 1, 117-136, *In Carbon nanotubes-Science and Applications* Meyyappan M (ed.).

[69] Iijima, S. and T. Ichihashi (1993) Single-shell carbon nanotubes of 1-nm diameter. *Nature,* 363(6430), 603–615.

[70] Dresselhaus M S and Eklund P C (2000) Phonons in carbon nanotubes, *Advances in Physics*. 49, 705-814.

[71] Saito, R., G. Dresselhaus, and M.S. Dresselhaus *Physical Properties of Carbon nanotubes,* Imperial College press, London, 1998.

[72] Reich, S. and C. Th, *Carbon nanotubes: Basic Concepts and Physical Properties.* Wiley-VCH, Darmstadt, 2004.

[73] Charlier, J.C. and S. Iijima *Growth mechanisms of carbon nanotubes*. pp. 29-53. In M.S. Dresselhaus, G. Dresselhaus and P. Avouris (eds.), *Carbon nanotubes*, Springer, Berlin, 2001.

[74] Baker, R.T.K. (1989) Catalytic growth of carbon filaments. *Carbon,* 27, 315-323.

[75] Sinnott, S.B., R. Andrews, D. Qian, A.M. Rao, Z. Mao, E.C. Dickey and F. Derbyshire (1999) Model of carbon nanotube growth through chemical vapor deposition. *Chem. Phys. Lett.,* 315, 25-30.

[76] Parmon V.N. (1996) Fluidization of the active component of catalysts in catalytic formation of carbon assisted by iron and nickel carbides. *Catal. Lett.,* 42,195-199.

[77] Moisala, A.A., G. Nasibulin, and E.I. Kauppinen (2003) The role of metal nanoparticles in the catalytic production of single-walled carbon nanotubes - a review. *J. Phys: Condens. Matter,* 15, S3011-S3035.

[78] Leu, I.C., M.H. Hon, and Y.M. Lu (1999) Chemical Vapor Deposition of Silicon Carbide Whiskers Activated by Elemental Nickel. *J. Electrochem. Society,* 146, 184-188.

[79] Ajayan, P.M. and O. Zhou *Applications of carbon nanotubes*. pp. 391-425 *In* M.S. Dresselhaus, G. Dresselhaus and Ph. Avouris (eds.) *Carbon Nanotubes.* Springer, 2001.

[80] Lambin, P., A. Fonseca, J.P. Vigneron, J.B. Nagy, and A.A. Lucas (1995) Structural and electronic properties of bent nanotubes, *Chem. Phys. Lett.,* 245, 85-89.

[81] Chico, L., V.H. Crespi, L.X. Benedict, S.G. Louie, and M.L. Cohen (1996) Pure carbon nanoscale devices: Nanotube heterojunctions, *Phys. Rev. Lett.,* 76, 971-974.

[82] Yao, Z., H.W.C. Postma, L. Balents, and C. Dekker (1999) Carbon nanotube intramolecular junctions. *Nature*, 402, 273-276.

[83] Cheng, Y. and O. Zhou (2003) Electron field emission from carbon nanotubes. *C.R. Physique,* 4, 1021-1033.

[84] Bonard, J.M., H. Kind, T. Stöckli, and L.O. Nilsson (2001) Field emission from carbon nanotubes: the first five years. *Solid State Electronics,* 45, 893-914.

[85] Lim, S.C., H.J. Jeong, K.H. An, D.J. bae, Y.H. Lee, Y.M. Shin, and Y.C. Choi *Carbon nanotube- based field emitters.* pp. 611-624. In H. S. Nalwa (Eds) *Encyclopedia of Nanoscience and Nanotechnology*, 2004.

[86] Dai, H., E.W.Wong, and C.M.Lieber (1996) Probing Electrical Transport in Nanomaterials: Conductivity of Individual Carbon Nanotubes. *Science*, 272, 523-526.

[87] Dai, H., J.H. Hafner, A.G. Rinzler, D.T. Colbert, and R.E. Smalley (1996) Nanotubes as nanoprobes in scanning probe microscopy. *Nature,* 384, 147–150.

[88] Wong, S.S., E. Joselevich, A.T. Woolley, C.L. Cheung, and C.M. Lieber (1998) Covalently functionalized nanotubes as nanometre-sized probes in chemistry and biology. *Nature,* 394, 52–55.

[89] Kong, J., N.R. Franklin, C. Zhou, M.C. Chapline, S. Peng, K. Cho, and H. Dai (2000) Nanotube Molecular Wires As Chemical Sensors, *Science,* 287, 622-625.

[90] Britto, P.J., K.S.V. Santhanam, and P.M. Ajayan (1996) Carbon nanotube electrode for oxidation of dopamine. *Bioelectrochem. Bioenergetics*, 41, 121-125.

[91] Britto, P.J., K.S.V. Santhanam, A. Rubio, A. Alonso, and P.M. Ajayan (1999) Improved charge transfer on carbon nanotube electrodes, *Adv. Mater.*, 11, 154-157.

[92] Planeix, J.M., N. Coustel, B. Coq, V. Brotons, P.S. Kumbhar, R. Dutartre, P. Geneste, P. Bernier, and P.M. Ajayan (1994) Application of carbon nanotubes as supports in heterogeneous catalysis, *J. Am. Chem. Soc.*, 116, 7935-7936.

[93] Wakihara, M. and O. Yamamoto (Eds) *Lithium ion batteries-fundamentals and performance.* Tokyo, Weinheim: Kodansha, Wiely-VCH, Vol. 3, No. 4, 1998.

[94] Frackowiak, E., S. Gautier, H. Gaucher, S. Bonnamy, and F. Beguin (1999) Electrochemical storage of lithium in multiwalled carbon nanotubes. *Carbon,* 37, 61-69.

[95] Beguin, F., K. Metenier, R. Pellenq, S. Bonnamy, and E. Frackowiak (2000) Lithium insertion in carbon nanotubes. *Mol. Crys. Liq. Cryst.*, 340, 547-552.

[96] Gao, B., A. Kleinhammes, X.P. Tang, C. Bower, Y. Wu, and O. Zhou (1999) Electrochemical intercalation of single-walled carbon nanotubes with lithium. *Chem. Phys. Lett.*, 307, 153- 157.

[97] Costamagna, P. and S. Srinivasan (2001) Quantum jumps in the PEMFC science and technology from the 1960s to the year 2000: Part I. Fundamental scientific aspects, Part II. Engineering, technology development and application aspects. *J. Power Sources*, 102, 242-269.

[98] Planeix, J.M., N. Coustel, B. Coq, V. Brotons, P.S. Kumbhar, R. Dutartre, P. Geneste, P. Bernier, and P.M. Ajayan (1994) Application of carbon nanotubes as supports in heterogeneous catalysis, *J. Am. Chem. Soc.*, 116, 7935-7936.

[99] Che, G., B.B. Lakshmi, C.R. Martin, and E.R. Fisher (1999) Metal-nanocluster- filled carbon nanotubes: catalytic properties and possible applications in electrochemical energy storage and production. *Langmuir,* 15(3), 750–758.

[100] Liu, Z., X. Lin, J.Y. Lee, W. Zhang, M. Han, and L.M. Gan (2002) Preparation and characterization of platinum-based electrocatalysts on multiwalled carbon nanotubes for proton exchange membrane fuel cells. *Langmuir,* 18, 4054-4060.

[101] Michelson, E.T., I.W. Chiang, J.L. Zimmermann, P.J. Boul, J. Lozano, J. Liu, R. E. Smalley, R.H. Hauge, and J.L. Margrave (1999) Solvation of Fluorinated Single-Wall Carbon Nanotubes in Alcohol Solvents. *J. Phys. Chem. B,* 103, 4318-4322.

[102] Satishkumar, B.C., E.M. Vogl, A. Govindaraj, and C.N.R. Rao (1996) The decoration of carbon nanotubes by metal nanoparticles. *J. Phys. D.: Appl. Phys.,* 29, 3173.

[103] Lordi, V., N. Yao, and J. Wei (2001) Method for supporting platinum on single-walled carbon nanotubes for a selective hydrogenation catalyst. *Chem. Mater.,* 13, 733-737.

[104] Choi, H.C., M. Shim, S. Bangsaruntip, and H. Dai (2002) Spontaneous reduction of metal ions on the sidewalls of carbon nanotubes. *J. Am. Chem. Soc.,* 124, 9058-9059.

[105] Liu, Z., X. Lin, J.Y. Lee, W. Zhang, M. Han, and L.M. Gan (2002) Preparation and characterization of platinum-based electrocatalysts on multiwalled carbon nanotubes for proton exchange membrane fuel cells. *Langmuir,* 18, 4054-4060.

[106] Cui, H.F., J.S. Ye, W.D. Zhang, J. Wang, and F.S. Sheu (2005) Electrocatalytic reduction of oxygen by a platinum nanoparticle/carbon nanotube composite electrode. *J. Electroanal. Chem.,* 577, 295-302.

[107] Tang, H., J.H. Chen, Z.P. Huang, D.Z. Wang, Z.F. Ren, L.H. Nie, Y.F. Kuang, and S.Z. Yao (2004) High dispersion and electrocatalytic properties of platinum on well aligned carbon nanotube arrays. *Carbon*, 42, 191-197.

[108] Kabbabi, A., F. Gloaguen, F. Andolfatto, and R. Durand (1994) Particle size effect for oxygen reduction and methanol oxidation on Pt/C inside a proton exchange membrane. *J. Electroanal. Chem.*, 373, 251–254.

[109] Yu, J.S., S. Kang, S.B. Yoon, and G. Chai (2002) Fabrication of ordered uniform porous carbon networks and their application to a catalyst supporter. *J. Am. Chem. Soc.*, 124, 9382-9383.

[110] Rajalakshmi N, Hojin Ryu, Shaijumon M M, and Ramaprabhu S (2005) Performance of Polymer Electrolyte Membrane Fuel Cells with Carbon Nanotubes as Oxygen Reduction Catalyst Support Material *Journal of Power Sources* 140, 250 –257.

[111] Shaijumon M M, Rajalakshmi N, Ramaprabhu S (2006), Pt/MWNT - Pt/C composites as electrocatalysts for oxygen reduction reaction in PEMFC *(under review in Appl. Phys. Lett).*

[112] Kötz, R. and M. Carlen (2000) Principles and applications of electrochemical capacitors. *Electrochim. Acta*, 45, 2483-2498.

[113] Gamby, J., P.L. Taberna, P. Simon, J.F. Fauvarque, and M. Chesneau (2001) Studies and characterizations of various activated carbons used for carbon/carbon supercapacitors. *J. Power sources*, 101, 109-116.

[114] Frackowiak, E. and F. Beguin (2001) Carbon materials for the electrochemical storage of energy in capacitors. *Carbon,* 39, 937-950.

[115] Frackowiak, E., K. Jurewicz, S. Delpeux, and F. Beguin (2001) Nanotubular materials for supercapacitors. *J. Power Sources*, 97-98, 822-825.

[116] Kibi, Y., T. Saito, M. Kurata, J. Tabuchi, and A. Ochi (1996) Fabrication of high-power electric double-layer capacitors. *J. Power Sources*, 60, 219-224.

[117] Lee, Y.H., K.H. An, J.Y. Lee, and S.C. Lim (2004) Carbon nanotube-based supercapacitors, in: H.S. Nalwa (ed) *Encyclopedia of Nanoscience and Nanotechnology*, 1, 625-634.

[118] Niu, C., E.K. Sichel, R. Hoch, D. Moy, and H. Tennet (1997) High power electrochemical capacitors based on carbon nanotube electrodes. *Appl. Phys. Lett.*, 70, 1480 – 1482.

[119] Frackowiak, E., K. Metenier, V. Bertagna, and F. Beguin (2000). Supercapacitor electrodes from multiwalled carbon nanotubes. *Appl. Phys. Lett.*, 77, 2421-2423.

[120] An, K.H., W.S. Kim, Y.S. Park, J.M. Moon, D.J. Bae, S.C. Lim, Y.S. Lee, and Y.H. Lee (2001) Electrochemical properties of high-power supercapacitors using single-walled carbon nanotube electrodes. *Adv. Funct. Mater.*, 11, 387-392.

[121] Shaijumon M M and Ramaprabhu S (2006) Supercapacitor electrodes from multiwalled carbon nanotubes synthesized using alloy hydride catalysts obtained through hydrogen decrepitation route *(under review in J. Power Sources).*

[122] Dillon, A.C., K.M. Jones, T.A. Bekkedahl, C.H. Kiang, D.S. Bethune, and M.J. Heben (1997) Storage of hydrogen in single-walled carbon nanotubes. *Nature,* 386, 377-379.

[123] Rzepka, M., P. Lamp, and M.A. de la Casa-Lilo (1998) Physisorption of hydrogen in microporous carbon and carbon nanotubes. *J. Phys. Chem. B*, 102, 10894-10898.

[124] Ma, Y., Y. Xia, M. Zhao, R. Wang, and L. Mei (2001) Effective hydrogen storage in single-wall carbon nanotubes. *Phys. Rev. B*, 63, 115422/1 – 115422/6.

[125] Yin, Y.F., T. Mays, and B. McEnaney (2000) Molecular simulations of hydrogen storage in carbon nanotube arrays. *Langmuir*, 16, 10521-10527

[126] Stan, G., and M.W. Cole (1998) Hydrogen adsorption in nanotubes. *J. Low Temp. Phys.*, 110, 539-544.

[127] Cao, A., H. Zhu, X. Zhang, X. Li, D. Ruan, C. Xu, B.Q. Wei, J. Liang, and D. Wu (2001) Hydrogen storage of dense-aligned carbon nanotubes. *Chem. Phys. Lett.*, 342, 510-514.

[128] Arellano, J.S., L.M. Molina, A. Rubio, M.J. Lopez, and J.A. Alonso (2002) Interaction of molecular and atomic hydrogen with (5,5) and (6,6) single-wall carbon nanotubes. *J. Chem. Phys.,* 117, 2281-2288.

[129] Shiraishi, M., T. Takenobu, and Ata (2003) Gas-solid interactions in the hydrogen/single-walled carbon nanotube system. *Chem. Phys. Lett.,* 367, 633-636.

[130] Zhang, H., Y. Chen, S. Li, X. Fu, Y. Zhu, S. Yi, X. Xue, Y. He, and Y Chen (2003) Hydrogen storage for carbon nanotubes synthesized by the pyrolysis method using lanthanum nickel alloy as the catalysts. *J. Appl. Phys.,* 94, 6417-6422.

[131] Lee, S.M. and Y.H. Lee (2000) Hydrogen storage in single-walled carbon nanotubes, *Appl. Phys. Lett.,* 76, 2877-2879.

[132] Kiyobayashi, T., H.T. Takeshita, H. Tanaka, N. Takeichi, A. Zuttel, L. Schlapbach, and N. Kuriyama (2002) Hydrogen adsorption in carbonaceous materials- How to determine the storage capacity accurately. *J. Alloys. Compd.*, 330-332, 666-669.

[133] Anson, A., M. Benham, J. Jagiello, M.A. Callejas, A.M. Benito, A.K. Maser, A. Züttel, P. Sudan, and M.T. Martinez (2004) Hydrogen adsorption on a single-walled carbon nanotube material: a comparative study of three different adsorption techniques. *Nanotechnology*, 15, 1503-1508.

[134] Ströbel, R., L. Jörissen, T. Schliermann, V. Trapp, W. Schütz, K. Bohmhammel, G. Wolf, and J. Garche (1999) Hydrogen adsorption on carbon materials. *J. Power Sources,* 84, 221-224.

In: Encyclopedia of Nanotechnology
Editor: Elwood D. Carlson

ISBN 978-1-60692-079-4

Chapter 27

PROPERTIES, SYNTHESIS, PURIFICATION, AND INTEGRATION OF CARBON NANOTUBES FOR THE ELECTRONIC DEVICE APPLICATIONS

Wanjun Park[1] ***and Yo-sep Min***
Samsung Advanced Institute of Technology, Yong-In, Kyeongki-Do, 449-712, Korea

ABSTRACT

Since the discovery of carbon nanotubes (CNTs), extensive studies on this new carbon allotrope have formed a new epoch in various fields of science and technology with their excellent physical and electrical properties, and chemical functionality. In particular, understanding of their electrical properties has aroused many research groups in academy and industry to explore the nano-scale electronics based on CNTs such as transistor, memory, interconnect, logic array, and so on. Although the CNT-based electronics is regarded as a strong potential candidate for the next generation electronic devices, many technological barriers are still remained.

First of all, pure nanotubes should be separated from a mixture of single-walled, multi-walled nanotubes, and carbonaceous impurities or should be grown without impurities for reliable device performance. Another critical barrier to integrate nanotubes in devices is uncertainty and uncontrollability in positioning nanotubes on a right region. In addition, transport mechanism in CNTs should be more understood to modulate band gap of them to fabricate various device elements.

In this article, we present recent efforts to overcome the barriers focusing on transistors based on single-walled carbon nanotubes (SWNTs). Experimental and theoretical studies to control the CNT electronic structure are covered for CNT-field effect transistors. The large scale fabrication of CNT electronic elements are also given for the selective growth of CNT that is extendable to device integration. For the purity issue of nanotubes, sono-chemical route to single-walled carbon nanotube and purification method of nanotubes integrated in transistors via sulfidation are demonstrated. For the positioning problems, direct photolithographic method using a mixture of catalytic precursors and conventional resists as a catalytic resist is introduced with other positioning methods of nanotubes.

[1] E-mail address: wanjun@samsung.com

1. INTRODUCTION

Since the discovery of carbon nanotubes (CNT) [1], their various aspects have been explored in different fields of science and technology. In particular, the unique one-dimensional geometry and the excellent transport property have made CNT the most promising building blocks for the next generation electronic devices [2, 3]. After finding of the current modulation by the gate bias with CNT transistor structure [4], many extensive works have been tried to realize the nano-scale electronics with various types of CNT electronic elements –resistor, capacitor, memory cell, logic array, and so on.

Although carbon nanotube can be a strong candidate for the next-era technology based on nanotechnology after the current semiconductor electronics with its nano-sized feature and unique electronic properties, there are many technology barriers such as control of individual CNTs in the array form to build the IC systems as well as control of CNT placement on the right position. The first hurdle for nanotube electronics is in synthesis or growth of nanotubes [5, 6]. To utilize nanotubes as an electronic material, pure nanotubes should be separated from a mixture of single-walled, multi-walled nanotubes, and carbonaceous impurities [7] or should be grown without impurities for reliable device performance. The electrical nature of nanotubes due to their chirality and diameter should be controlled to be semiconducting or metallic for particular applications. Another critical barrier for nanotube electronics is uncertainty and uncontrollability in positioning nanotubes on a right region to integrate as a device element. One more concern for device integration is the growth temperature should be lowered to be compatible to current CMOS process. Fundamental point of view, transport mechanism in CNTs should be more understood to modulate band gap of them to fabricate various device elements.

In this article, we shall discuss the recent progresses in the carbon nanotube electronics toward a view point of integration.

In section 2, experimental and theoretical studies on carbon nanotube transistors are discussed. The general current-voltage characteristics of CNTFETs in intrinsic nanotube are expressed with a ballistic nature in CNT and the Schottky barrier at electrode and nanotube junction. Then molecular adsorption is also discussed with a special emphasis on suppression of ambipolar behavior that is attributed to the Schottky barriers.

In section 3, low temperature of single walled CNT growth is reviewed and discussed. For high purity and uniform distribution of single walled nanotubes, sonochemical route is introduced. On the other hand, a remote plasma enhanced chemical vapor deposition on substrates is discussed as an essential technique for the CNT integration.

In section 4, a direct photolithographic method using a mixture of catalytic precursors and conventional resists is introduced as a fundamental processing technology for large scale device integration with other positioning methods of nanotubes. Purification method of nanotubes integrated in transistors via sulfidation which is useful for removal of undesired electrical leakage path is also demonstrated.

Finally, we summarize this article in section 5.

2. ELECTRIC PROPERTIES OF CNTFET

The atomic structure of a single-wall carbon nanotube (SWNT) is specified by the chiral vector, $C_h = na_1+ma_2$, that can be expressed by the real space unit vectors a_1 and a_2 of the hexagonal lattice. The electronic structure of SWNT is well developed with a simple tight-binding calculation since its one-dimensional nature. Indices (n, m) of the chiral vector indicate either metallic or semiconducting, i.e. if n-m is multiple numbers of 3 including zero, nanotube is metallic, otherwise, nanotube is semiconductor with the energy gap opening in electronic density of states (DOS).

The ballistic nature of carbon nanotubes is attributed to the potential applications for the electronic devices with high driving current capability and less heat dissipation that are essential for the trend of scaling down in either the active or the passive device elements of the integrated circuit systems. Details of fundamental properties for electronic structure and transport of carbon nanotubes are elsewhere [8].

Of special interest is the carbon nanotube field effect transistor (CNTFET) that is expected as the most important building block for the future electronic system. CNTFET has an analogy of MOSFET in device structure and I-V characteristics but operational basics are different. While early devices have shown poor device characteristics, improvements were achieved by using doped CNTs [9] or high-k materials [10]. The contact between metal and CNT can be of ohmic [11] or Schottky type [12, 13]. CNTFETs operate by modulating the transmission coefficient of the Schottky barriers at the contact between the metal and the CNT.

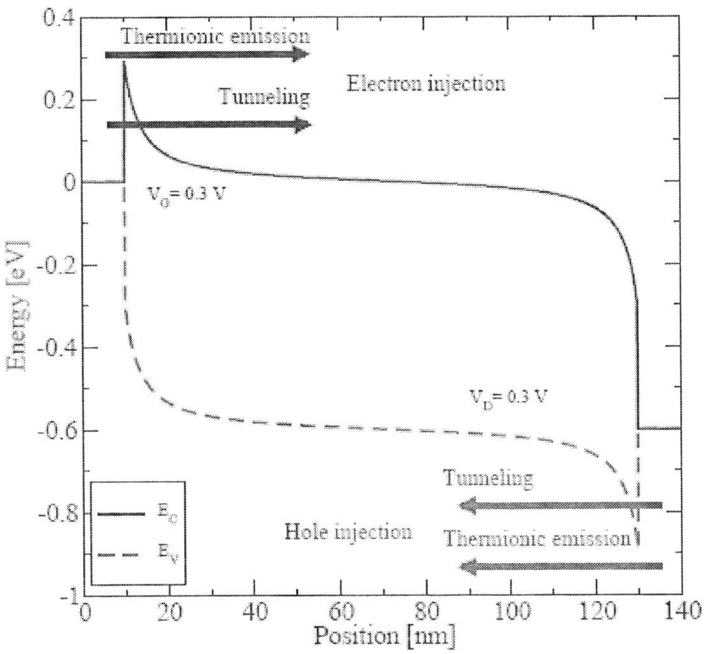

Figure 2.1. Band edge profile with the Schottky barriers at the CNT/metal junctions for source and drain of the CNTFET.

2.1. Intrinsic Characteristics of CNTFET

Without external charges, the band edge profile for CNTFET is shown in Figure 2.1. Positive gate voltages near the source increase the tunneling current of electrons, which is desirable for n-type devices. By decreasing the gate voltage the tunneling current of electrons decreases, but the thermionic emission current of electrons does not vary. If the gate voltage decreases further to negative values the thermionic emission current of electrons also decreases. On the other hand by applying positive voltages higher than the gate voltage to the drain, the Schottky barrier near the drain is suppressed and consequently hole injection at the drain increases, an undesirable phenomenon for an n-type device. Especially in the off regime this would result in an intolerably high off-current.

The resulted current-voltage characteristics are shown in Figure 2.2 where the current is symmetric with respect to the gate voltage and the gate voltage of valley depends on both of metal work function and the drain voltage. This ambipolar behavior of Schottky barrier CNTFETs limits the performance of the CNTFET since the I_{on}/I_{off} ratio is limited [14, 15].

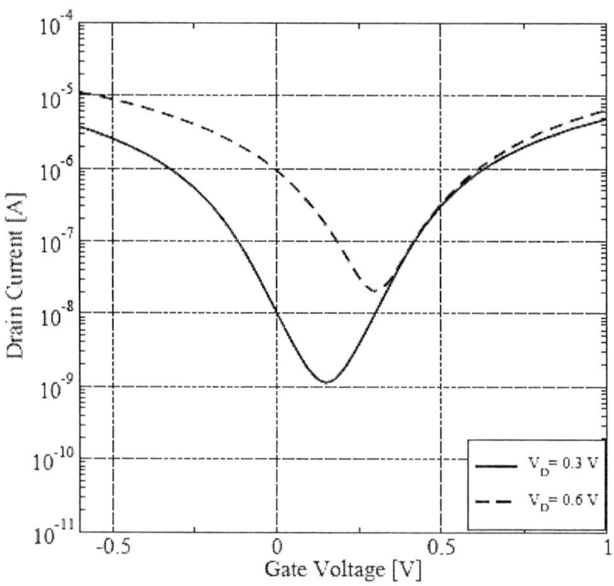

Figure 2.2. Current-voltage characteristics of the CNTFET without external charge.Assuming ballistic transport, the drain current was calculated using theLandauer-Büttiker formula with equilibrium Fermi functions at the source anddrain contacts and the transmission coefficient through the device. The twofold band and twofold spin degeneracy were used. This work was focused on ambipolar devices, where the metal Fermi level is located in the middle of the CNT band gap at each contact. The transmission coefficient was evaluated using the WKB approximation and an idealized band structure. The lattice constant of CNT was taken as 0.246 nm and the band gap energy set here to 0.6 eV corresponding to a CNT of a diameter of 1.4 nm. (See Ref. [16] for more details).

In order to suppress the ambipolar behavior, an asymmetric structure of CNTFET has been proposed, where the gate is not extended to the drain and the control of the gate over carrier injection from the drain is reduced. As a result the ambipolar behavior decreases to some extent.[15] A double-gate (DG) device is also proposed [17], where the first gate controls carrier injection at the source contact and the second one independently controls carrier injection at the drain contact. In this device structure the band-edge profiles near the source and drain are controlled separately by two different gate biases. In order to suppress the tunneling current of holes at the drain it is necessary that the voltage of the drain side gate be equal or higher than the voltage of the drain. In consequence the parasitic tunneling current of holes is suppressed and the parasitic current is limited to thermionic emission of holes over the Schottky barrier at the drain, as shown in Figure 2.3. While electron injection at the source contact can be controlled through the first gate (source side), the second gate (drain side) suppresses parasitic hole current at the drain contact. The same discussion holds for *p*-type devices. By applying negative voltages to the first gate, hole injection at the source contact can be controlled and the second gate suppresses parasitic electron current at the drain contact. the thermionic emission current of holes at the drain contact will decrease exponentially and consequently, a lower parasitic current can be achieved This work showed that by applying a proper voltage range to the second gate, parasitic carrier injection at the drain can be avoided, and improved device characteristics are achieved.

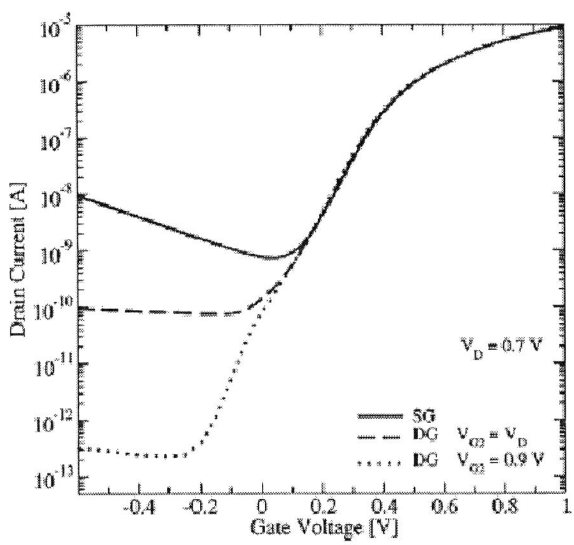

Figure 2.3. Calculated transfer characteristics for the single gate CNTFET (solid line), the DG-CNTFET with $V_{G2}=V_D$ (dashed line), and the DG-CNTFET with V_{G2}=0.9 Volt (dotted line). For both devices V_D=0.7 Volt [17].

2.2. Oxygen Molecule Effects as an External Charge on Electronic Structure of Carbon Nanotube

We studied the carrier suppression on CNT with an asymmetric structure where the modulating potentials on each Schottky barriers of source and drain are independently controlled. This geometric asymmetry is useful for understanding fundamental I-V characteristics of CNT-FET but limited to build the actual nano-scale devices. One dimensional constituent such as nanotube or nano-wire is sensitive to the external charge sources because of large surface area. Specially, CNT transport is extremely affected because of tube geometry.

Common observations of the *p*-type behaviors of the nanotube devices have been attributed to the doping by the adsorbed oxygen from the ambient air [18, 19, 20]. It was shown theoretically [21] and experimentally [22] that electronic properties of CNT-FET are sensitive to the absorption of oxygen molecules attributed to charge transfer and gas induced charge fluctuation. Meanwhile, Derycke et al. have suggested the main effect of the oxygen adsorption lies in the adjustment of the Schottky barrier rather than the bulk doping of the CNT body [9, 23], which is consistent with subsequent studies showing a negligible charge transfer between oxygen adsorbator and the nanotube wall [24, 25, 26]. Despite this apparent controversy, bundle or networks of the nanotube may react differently from the individual nanotube upon the oxygen adsorptions. The thermoelectric power measurement proved that the networks of CNT could be doped as *p*-type upon oxygen adsorption, irrespective of the contact metal [18, 19]. On the other hand, the transfer character of the transistor with a single and short nanotube could sensitively depend on the Schottky barrier rather than doping [9, 23].

Irrespective of whether the binding energy of the O_2 on the wall is strong or not, *ab initio* calculation should predict the existence of the LUMO level of O_2 just above the valence band edge of the semiconducting nanotube [27, 28]. Here we performed *ab initio* density-functional calculations using the Vienna simulation package (VASP) [29, 30, 31]. The spin-polarized gradient-corrected approximation is adopted for the exchange-correlation potential [32]. The ionic potentials are described by the projector augmented wave (PAW) method as provided with VASP [33], and the energy cutoff for the plane-wave basis is set to 400eV [34].

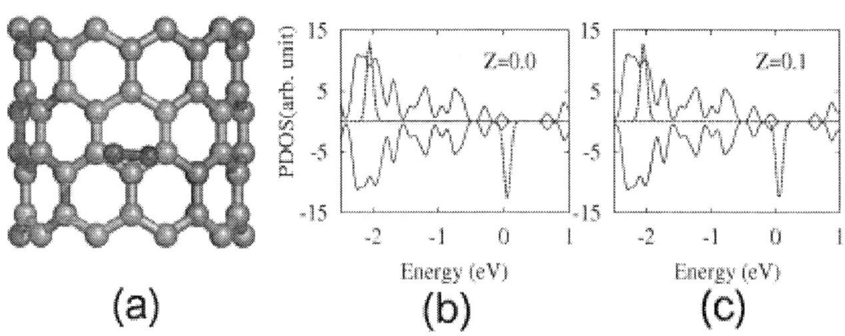

Figure 2.4. (a)Adsorption geometry of oxygen molecule on the wall of the (10,0) nanotube. The electronic density of states for (b) the O_2 adsorbed carbon nanotube and (c) that with excess charge of 0.1 electron. Upper panels in (b) and (c) are for the spin-up electrons, while lower panels are for the spin-down electrons. Dotted lines and solid lines in (b) and (c) are the projected density of states for oxygen molecule and the carbon nanotube, respectively.

Figure 2.4 presents the adsorption geometry of oxygen molecule on the (10, 0) nanotube and the electronic density of states. To simulate an isolated single nanotube, we assumed a large two-dimensional lattice in the perpendicular plane to the tube axis, where the distance between the nanotube and its replica is larger than 10 Å. The unit-cell along the axial direction is twice the minimal unit-cell of the zigzag carbon nanotube (≈8.49 Å). The atomic positions are relaxed with residual forces smaller then 0.05eV/ Å with the given unit-cell parameters. The upper panels in Figure 4(b) and 4(c) are for the majority-spin states and lower panels are for the minority-down states. The dotted peaks in the upper panel at about -2 eV and that in the lower panel at just above 0 eV shows the triplet ground state of the oxygen molecule. The LUMO level of O_2, which is the $pp\pi^*$ with the minority-spin state, is just above the valence band maximum of the (10, 0) nanotube. In other word, the Fermi level of the oxygen adsorbed nanotube sits at the valence band edge rather than at the midgap as shown in Figure 2.5.

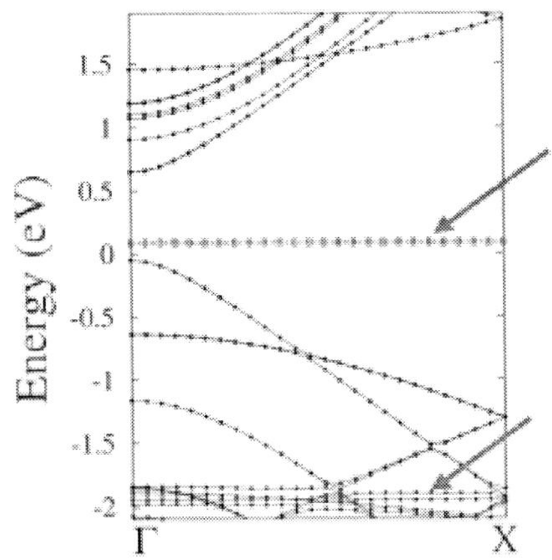

Figure 2.5. Adsorption geometry of oxygen molecule on the wall of the (10,0) nanotube and *ab initio* band structure with the oxygen adsorbed nanotube system. Arrows indicate oxygen molecular states. The LUMO level of O_2 adsorbators could provide the charge trapping centers, irrespective of the possibility of a practical hole-doping by the oxygen adsorption. These charge trapping centers could screen the electric field from the positively biased gate, resulting in the suppression of electron channel formation [22].

We increased the electron occupation to have 0.1 excess electrons in the unit cell (80 carbon atoms and 2 oxygen atoms). We find that the excess electron occupy the LUMO of O_2 rather the conduction band of the carbon nanotube. Thus, the turn-on voltage of the *n*-channel current will be increased by the presence of the oxygen molecule on the nanotube wall. This shift of the Fermi level, and the consequent unease in turning on the *n*-channel current, should be interpreted as p-type doping of the nanotube, regardless of whether the charge transfer is strong or not. Moisture can be considered as another source of charge transfer for CNT under ambient air. However it is found that the LUMO level of H_2O is about 4 eV above the Fermi level of the carbon nanotube. Thus, it is believed that the effect of water vapors on the aforementioned suppression of the n-channel formation is insignificant. Since the HOMO-LUMO gap in the DFT is not so rigorous, TDDFT calculations are necessary for the H_2O effect using the Gaussian package. The HOMO-LUMO gap of H_2O is found to be 6.4 and 7.5 eV in DFT and TDDFT, respectively. Thus, the DFT artifact would not affect our model discussed here.

2.3. Current-Voltage Characteristics of CNTFET

Figure 2.6 shows that a typical p-channel behavior in I_{ds}-V_{gs} curves of the CNT-FET measured with various source and drain voltages (V_{ds}) under the exposure of ambient air without passivation. It could be described as a device operating in p-channel depletion mode with an analogue of MOSFET, whose turn-on voltage (V_{th}) is about V_{gs} = 4V. This turn-on voltage depends on the gate dielectric thickness. This asymmetric change in the I_{ds}-V_{gs} curve, inducing less current for one polarity of gate bias and more current for the opposite polarity is attributed to the charge sources in ambient air. Intrinsic transport was investigated after removing the possible charge sources such as N_2, O_2, H_2O, and so on.

Figure 2.7 shows complete conversion of transport property of CNTFET from unipolar to ambipolar at vacuum of 3×10^{-6} Torr as discussed above for transport of CNTFET without the external charge. These external charge sources are volatile at vacuum even without the extra heat treatment since the overall conversion process can be reversible and can reproduce the patterns.

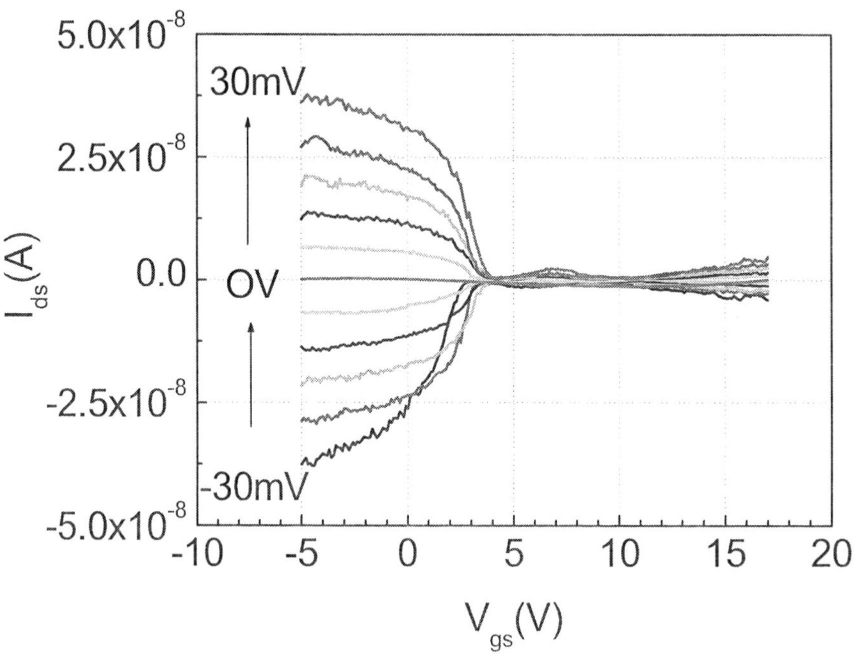

Figure 2.6. I_{ds}-V_{gs} curves of the CNTFET under the influence of ambient air condition with various source-drain voltages of 5mV steps. DWNTFET was chosen to maximize ambipolar behavior since it is mainly ascribed to the smaller band gap (V_g=0.2eV) arising from the large diameter (2nm) of outer shell. Purified DWNTs were suspended in a solvent and spin coated on the SiO_2 grown on heavily doped p-type Si substrate. Scanning Probe Microscopy (SPM) locates individually separated nanotubes on the substrate. After generating patterns for metal contact by electron beam lithography, metal electrodes with 100nm of Pd were defined by the lift-off process. Typical system resistances without gate bias were in the range of several hundreds of KΩ to MΩ. [22].

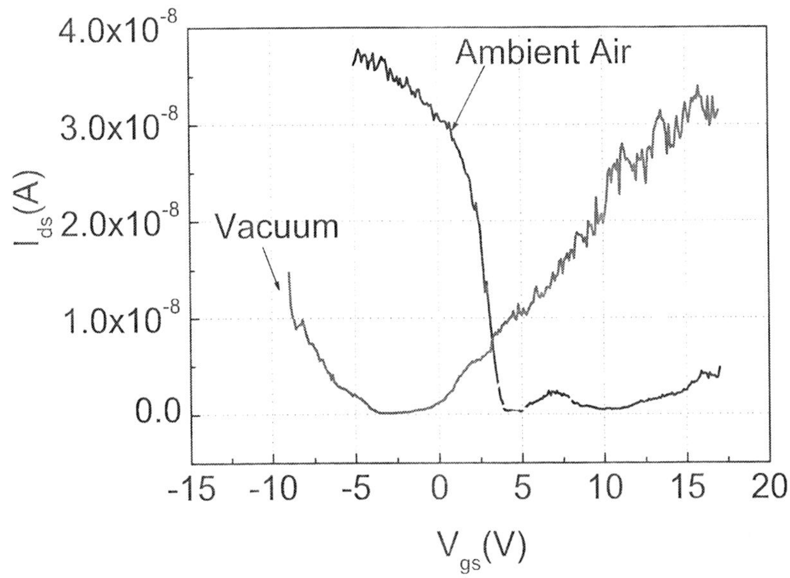

Figure 2.7. Transfer characteristics of CNTFET under ambient air, vacuum of 3×10^{-6} Torr, The n-channel starts to emerge and the FET shows an ambipolar behavior after vacuum environment with complete conversion of transport property. The overall conversion process is reversible and reproducible.

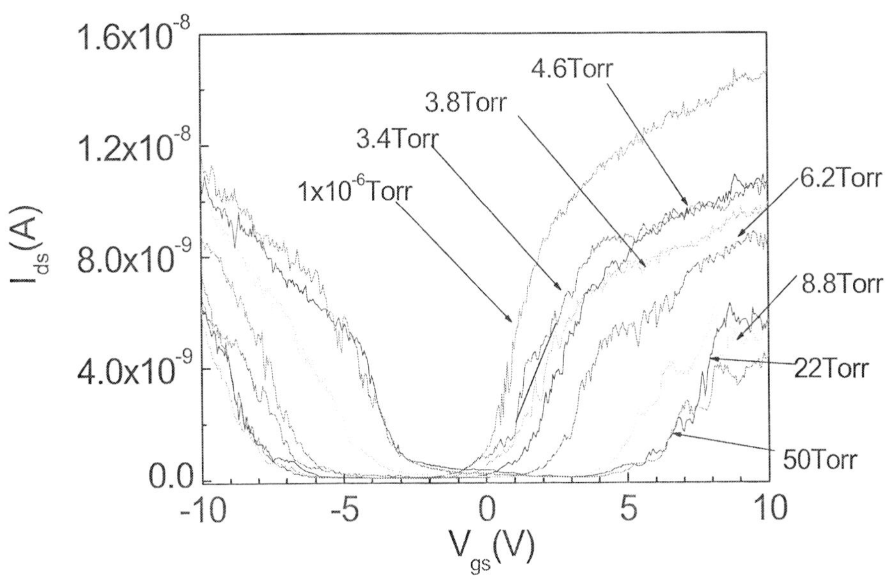

Figure 2.8. The I_{ds}-V_{gs} curves of the CNTFET under oxygen. Single-walled carbon nanotube was grown by the patterned catalyst method on the thermally grown SiO_2 of 500nm. CH_4 gas was introduces in a thermal CVD furnace at 905oC for 10 min with hydrogen gas after loading the wafer. After defining the metal contact patterns with the conventional photolithography technique, the Ti/Au electrodes were deposited on the patterns by using thermal evaporation. The second lift-off process was carried out to form metal contacts with CNT. The distance source and the drain electrode was fixed at 10·m [35].

The absorption of oxygen molecules in ambient air is dominant to electronic properties of CNT as discussed by ab initio calculation. However it also leads the work function variation in the electrode and may result the ambipolar behavior [23]. To investigate the effect of O_2 adsorption on the transport properties, a long individual single-walled carbon nanotube is studied for the transport measurement. I_{ds}-V_{gs} curves shown in Figure 2.8 indicate that main effect on the transport is O_2 adsorption on the CNT channel. Under the high vacuum of $1x10^{-6}$ Torr, the n-channel almost turns on at $V_g = 0$. This could be ascribed to the fact that the Fermi level of Ti aligns with the conduction band edge of the nanotube. The n-channel current is more suppressed, while the p-channel current is enhanced as the O_2 pressure is higher. These observations bring important discussions about the origin of the change whether doping-like effect on CNT or metal work function change in metal electrode. The slope of the I_{ds}-V_{gs} curves change upon the Schottky barrier variation, while the curves shift with negligible slope change upon the doping [23]. In this regard the observed I_{ds}-V_{gs} curves from the oxygen exposure are rather close to doping-like effect on CNT than the modulation of Schottky barrier.

The comparison of the curves of 1×10^{-6} Torr and 6.2 Torr shows that the n-channel current is reduced while the p-channel current is almost unchanged. That means there has been no clear demonstration of the O_2-induced *p*-type doping of an individual carbon nanotube and should be ascribed to the n-channel current suppression by the trapped negative charges on the LUMO level of the O_2 molecules nearby nanotube surfaces or the surface of the gate dielectric.

The oxygen exposure of the CNT-FET could induce the work function increase in the metal surface, *p*-channel current suppression, as well as the hole-doping like effect of the CNT body. However, such a doping effect would be dominant as the length of the nanotube increases.

The unipolar behavior of CNT-FET in ambient air needs more comprehensible arguments. It is reasonable to assume that there are significant molecular adsorptions on the wall of CNT as well as on the surface of metal electrode. Provided that the LUMO levels of the adsorbators sit somewhere in the gap of the carbon nanotube, electrons would occupy the molecular levels under a positive gate voltage, as shown in Figure 2.9(a) and 2.9(b). As a result, the electrostatic potential generated by the negatively charged adsorbators screens the gate field and, consequently, suppresses the electron channel formation. The nanotube bands in the channel region would resist to down-shift because of the presence of charged adsorbators. Two upward arrows in Figure 2.9(b) describe the effect of such an electrostatic potential generated by the charged adsorbators. On the other hand, conduction bands of a clean nanotube shift down allowing electron current under such a positive gate voltage, as shown in Figure 2.9(c).

To find a molecule for suppressing holes to build the n-type CNT-FET is still challenging issue for non-volatile adsorption to be acceptable for integration. Actually, for real device application for the CMOS-like logic, carrier suppression by a molecular adsorption may not be sufficient since the molecules on CNT surface is not expected to be stable under the operational environment. Atomic doping in CNT may be a key solution. But there are fundamental issues for degradation of ballistic nature of CNT since dopants act as source of defects as well as for the synthetic method of doped SWNT. Encapsulation is also considered but the technology is very early stage for device application.

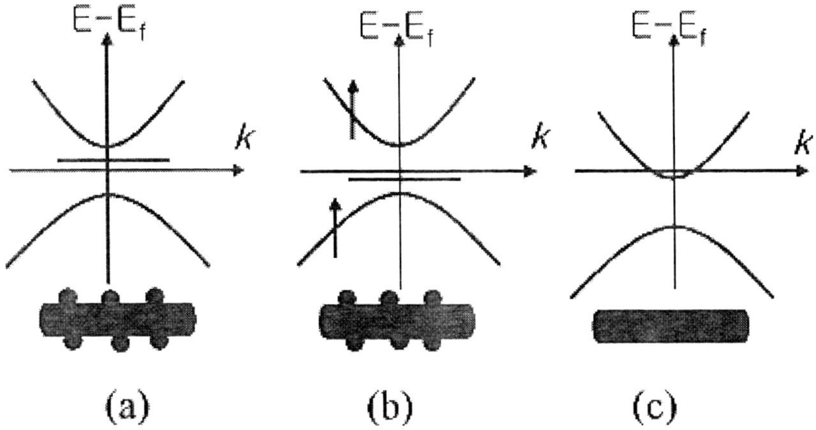

Figure 2.9. Schematic band structure of the carbon nanotube and the LUMO level of the adsorbators, with respect to the Fermi level.

3. Growth of Single Walled Carbon Nanotubes for the Electronic Device Applications

To fabricate nanotube-based FETs, SWNTs should be grown without multi-walled carbon nanotubes (MWNTs) and carbonaceous impurities. However, the growth of SWNTs is generally accompanied by the undesired carbon materials and MWNTs that make electrical short and unexpected leakage. In this section, we will discuss two different ways to obtain high purity SWNTs.

One way to build the CNT array for IC fabrication is to place the individual nanotubes on the desired positions from nanotube powder with an appropriate dispersion method [36]. Syntheses of single-walled carbon nanotube (SWNT) powders have been achieved by several methods such as arc discharge, laser ablation, and HiPCo process [38, 39]. SWNTs synthesized by these techniques should be subsequently purified to fabricate nanotube-based device elements. We introduce a simple way to obtain high purity SWNT powder in this article to overcome the various CNT purifying steps that are inevitable for the previous synthetic methods.

On the other hand, controlled positioning of SWNTs on a defined region has been regarded as a crucial step for device integration [40]. SWNTs grown by chemical vapor deposition (CVD) can be more easily integrated in devices than the synthesized nanotubes, since nanotubes are directly and selectively grown on catalysts such as iron, cobalt, and nickel [41]. Before the SWNT growth, the catalyst is pre-patterned as films or nanoparticles on a defined region by various techniques such as lithographic lift-off, soft lithography, offset printing, template methods, and direct lithography [41-45].

In these two synthetic approaches to SWNTs, there are common efforts to lower the synthesis temperature. For the SWNT powder synthesis, the growth temperature was recently lowered to room temperature by a sonochemical route. For the direct growth of SWNTs on substrates by CVD, remote plasma-enhanced chemical vapor deposition (PECVD) has shown a clue for the low temperature growth of SWNTs.

3.1. Sonochemical Synthesis of SWNTs at Room Temperature

It is well known that cavitation bubbles produced in liquid solution during sonication can instantaneously generate local spots of several thousand degrees Celsius of temperature and several thousand atmospheres of pressure [46]. Inspired by the chemical effects of ultrasound, the sonochemical route to SWNTs was achieved in a liquid solution at atmospheric pressure and room temperature [47]. In addition, this method offers a facile route to high-purity SWNTs without specialized equipment and dose not require multi-step purification process to eliminate the amorphous carbon and multi-walled carbon nanotubes.

The typical process of this method for SWNT growth is depicted in Figure 3.1. The procedure begins with the preparation of a mixture solution of ferrocene and *p*-xylene. Silica powder was then added in this solution. Ferrocene was chosen as a precursor of Fe catalyst for nanotube growth. The carbon source for SWNT growth was provided by *p*-xylene as well as by ferrocene [48], while silica powder acted as a nucleation site for SWNT growth. Ultrasonication was performed under ambient conditions to grow SWNTs. Finally, the resulting powder was collected on filter membrane. Initially, 2 grams of silica powder (diameters of 2.5 mm) were added to 0.01 mol % ferrocene-dissolved *p*-xylene solution (50 mL). The solution was then sonicated by a 1/2 inch titanium tip 200-W probe pulsed 65 % at atmospheric pressure and room temperature for 20 minutes. During this period, the color of silica powder turned into gray, indicating a carbon deposition on its surface. It was also found that silica powder broke into small pieces below several hundreds of μm. This could be attributed to interparticular collisions of silica powder triggered by ultrasound irradiation [49]. The resulting powder, collected on a filter membrane, shows that fiber-like structures exist on the surfaces of silica powder as shown in Figure 3.2.

After filtration to remove the silica powder by using HF solution, fiber-like structures remained on the filter membrane as shown in Figure 3.3(a). This morphology is similar to purified SWNT bundles produced by the laser ablation method [50]. High-resolution transmission electron microscope (HR-TEM) confirmed that these structures mainly consisted of SWNT bundles as shown in Figure 3.3(b). The absence of multi-walled carbon nanotubes was also confirmed by HR-TEM observations. The average diameter of SWNTs based on TEM data was 1.5 ± 0.1 nm as shown in an inset image of Figure 3.3(b).

Control experiments were carried out to determine if the presence of silica powder is essential for producing SWNTs. For a sample sonicated without silica powder, no SWNTs were shown in the resulting HR-TEM images. Concentration effect of ferrocene in *p*-xylene solution was also studied to achieve the best conditions for the SWNT growth. It was found that high-purity SWNTs were obtained at relatively low concentration of ferrocene (0.01 mol %), showing no noticeable evidence for the existence of multi-walled carbon nanotubes and amorphous carbon. Once Fe nanoparticles are produced by decomposition of ferrocene during sonication, there would be two possible competing reactions; SWNT growth by catalytic action of fine Fe catalyst and Fe particle growth by agglomeration of fine Fe catalyst. However, it is expected that the former reaction occur dominantly under our experimental conditions, reducing the chance to form the larger Fe particles, which may result in multi-walled carbon nanotube growth. These are probably caused by very low concentration of ferrocene in our experiment.

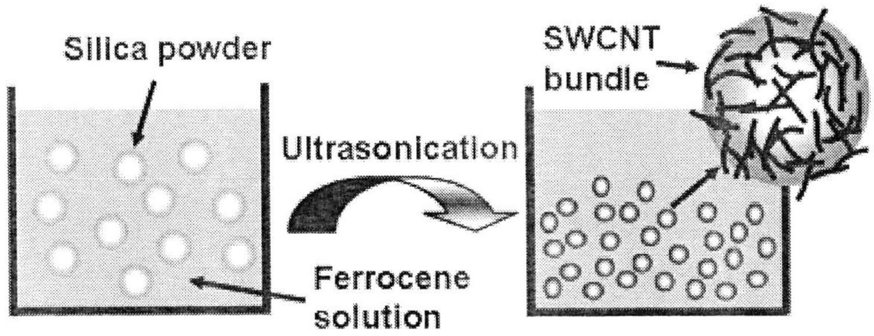

Figure 3.1. Schematic diagram of sonochemical route to SWNTs. Silica powder is immersed in the solution (ferrocene-xylene mixture). Ultrasonication then was done for 20 min at room temperature and atmospheric pressure. Sonication produces high-purity SWNTs on the surface of silica powder [47].

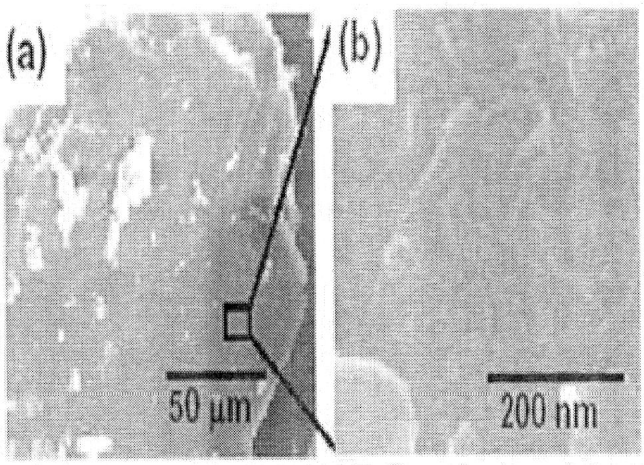

Figure 3.2. SEM images of silica powder after the sonochemical synthesis.

Raman spectroscopy has also confirmed sample homogeneity as shown in Figure 3.4. Spectra were recorded at 10 randomly selected positions and averaged. It shows the typical spectral characteristics of SWNT structure, as characterized by the strong G band (tangential mode) and the presence of the sharp radial breathing mode (RBM) in the lower wave number region (100-300 cm^{-1}). RBM spectra were further analyzed to give the diameter distribution of SWNTs. According to the equation: $\omega_{RBM} = 235/d + 9$ for bundled SWNTs [51], where d (nm) is the diameter and ω (cm^{-1}) is the Raman shift of a SWNT, the radial breathing mode (RBM) region in Figure 3.4(a) yields SWNT diameters of 1.1-2.1 nm with an average diameter of 1.55 nm. The diameters of most SWNTs (>90%) were in the range of 1.3.1.8 nm. These estimates agree well with the measured values from HR-TEM images. A weak peak at 1340 cm^{-1} of D-band (related to disordered graphite or amorphous carbon) and strong peak at 1572 cm^{-1} of G-band (tangential mode) are shown in Figure 3.4(b). The relatively low intensity of D band relative to the G-band indicates that only a small amount of amorphous carbon or few defects exist in these SWNTs [52].

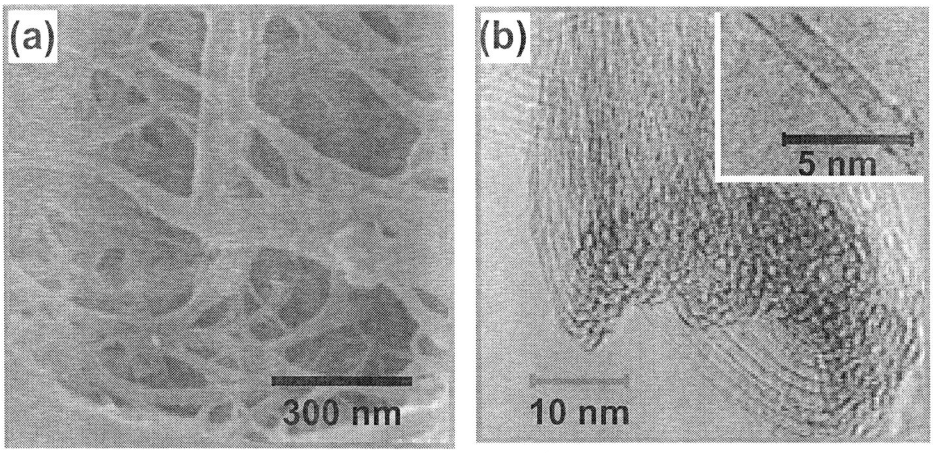

Figure 3.3. (a) SEM image of SWNT bundles on polycarbonate filter membrane. (b) HR-TEM images of SWCNTs. Most of SWNTs exist in bundles. For the measurements of diameters of SWNTs, individual SWCNTs (inset) are also imaged by HR-TEM [47].

Although details of SWNT growth mechanism are still under investigation, we believe that these sonochemical treatments both facilitate decomposition of ferrocene and *p*-xylene, and provide the required energy to synthesize SWNTs. Unlike previous methods for producing SWNTs [37-39, 41], which require severe conditions such as high temperature, high vacuum, or high voltage arc discharge, our method provides a facile route to high-purity SWNTs under ambient conditions.

3.2. Low Temperature Growth of SWNTs by Remote PECVD

Despite the promise of nanotubes grown by CVD, the high growth temperature (> 800 °C) of CNTs is a barrier for device fabrication. Since lower growth temperatures significantly enhance the compatibility of CNT growth with current complementary metal-oxide-silicon (CMOS) technology for CNT-based electronics, plasma enhanced chemical vapor deposition (PECVD) has been recently utilized to grow nanotubes at lower temperatures [53.57]. Unfortunately, the grown nanotubes at lower temperatures by PECVD are generally multi-walled carbon nanotubes (MWNTs) [53, 54]. Using the microwave plasma chemical vapor deposition method, MWNTs were grown at ~ 450 °C by Wilson and his coworkers [54]. Kato and coworkers used radio-frequency magnetron PECVD to grow SWNTs at 550 °C with zeolites as a supporting template for the Fe/Co catalyst [55]. Dai's group demonstrated that SWNTs can be grown at 600 °C without any template [56, 57]. Recently, it was demonstrated that SWNTs can be grown even at 400 ~ 450 °C without any template by remote PECVD [58, 59]. The key process parameter to enable the low temperature growth is low plasma power, which should be held at a minimum.

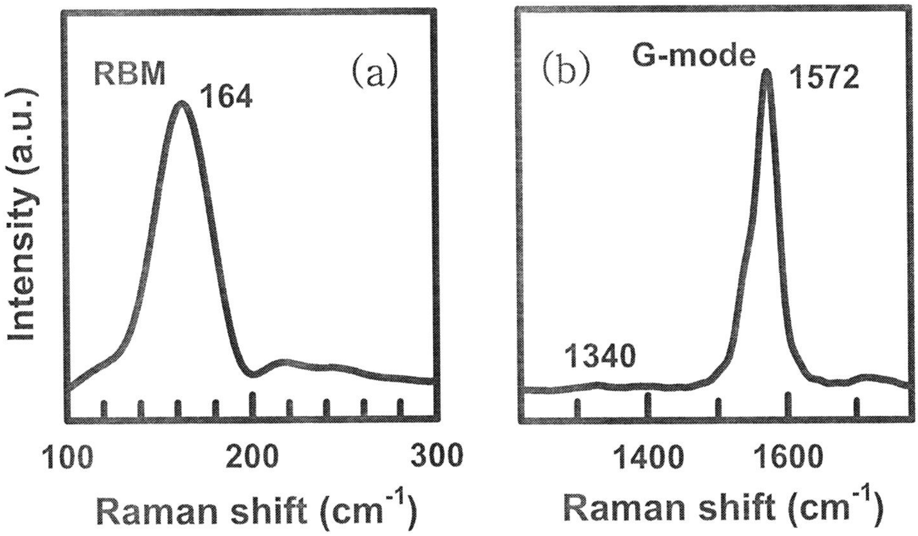

Figure 3.4. Raman spectra (Ar^+ laser: λ=514.5 nm) of synthesized SWNTs. (a) RBM and (b) G-line mode (right graph) are presented for the average of 10 spectra recorded at random points.

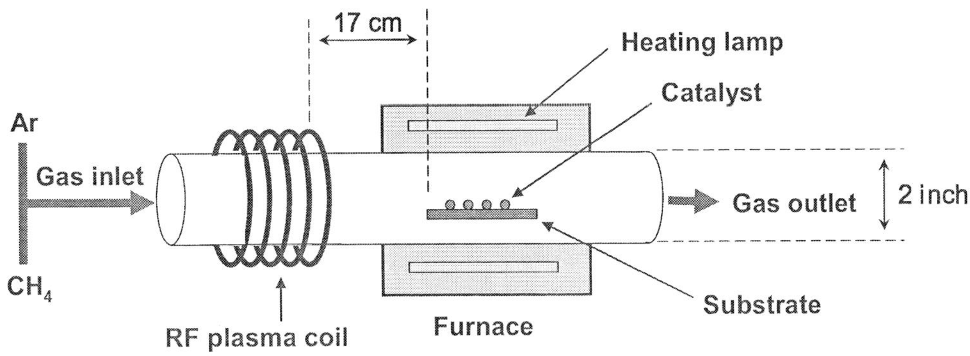

Figure 3.5. A schematic diagram of the remote PECVD system.

Figure 3.5 shows a schematic diagram of the homemade radio frequency (rf, 13.56 MHz) remote PECVD system. A quartz tube with a 2-inch diameter was surrounded by a copper coil in a helical manner from the gas inlet to the sample holder which was located at a distance of 17 cm from the end of the copper coil. Substrates were heated by halogen lamp and the growth temperature was calibrated with a thermocouple-implanted quartz holder. The calibrated temperature was not influenced by the remote plasma ignition.

Catalyst-deposited substrates were placed on a quartz holder and then heated up to the growth temperature for 300 sec in an argon atmosphere (100 sccm). After an interval for the stabilization of substrate temperature, methane (60 sccm) was introduced with argon (15 sccm) in the quartz tube and the subsequent plasma was ignited to grow CNTs. During the CNT growth (10 min), the working pressure was maintained at ~ 0.4 Torr.

To optimize the process condition for SWNT growth, we monitored the presence of radial breathing mode (RBM) peaks due to SWNTs and the D-band to G-band (I_D/I_G) ratio in Raman spectra. Since the G-band and D-band are due to tangential stretch modes of nanotubes and disordered graphite or amorphous carbons, respectively, the ratio generally increases with an increasing amount of disordered or amorphous carbons [60].

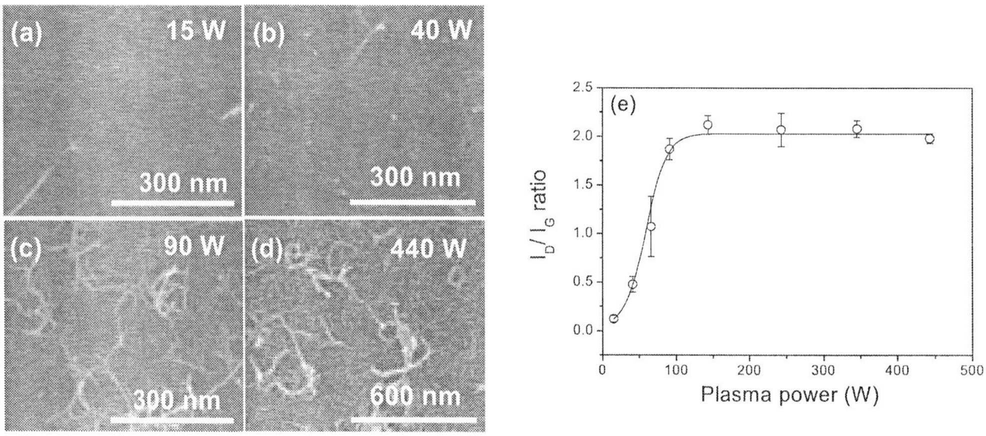

Figure 3.6. SEM images of nanotubes grown at 450 °C with plasma powers of 15 (a), 40 (b), 90 (c), and 440 W (d). (e) I_D/I_G ratio variations with plasma power [58].

The plasma power effect on nanotube growth was also shown with nanotubes grown at 450 °C in Figure 3.6. In the low power region (< 40 W, Figure 3.6(a-b)), CNTs are sparsely grown, but in the high power region, nanotubes with larger diameters are densely grown with short nanotubes or carbonaceous particles as shown in Figure 3.6(c-d). The plasma effects are more clearly revealed in Figure 3.6(e). The I_D/I_G ratio dramatically decreases from ~ 1.9 at 90 W to ~ 0.1 at 15 W as the plasma power decreases. However, the ratio is nearly constant in the high power region. It should be noted that the plasma lowers the growth temperature of nanotubes but the power should be held at a minimum to avoid the increase of I_D/I_G ratio.

Figure 3.7(a) shows I_D/I_G ratio as a function of growth temperature for nanotubes grown with 15 W, 40 W and 65 W plasma powers. Solid circles denote that RBM peaks were obtained from all of scans. Process conditions showing RBM peaks from some scans positions were designated with open circles. Cross symbols represent no RBM peaks. Figure 3.7(b) clearly reveals that the plasma power should be low to obtain SWNTs showing RBM peaks with low I_D/I_G ratio at low temperature. The lowest growth temperature showing RBM peaks from all of scans was 425 °C but their I_D/I_G ratios were higher than those of nanotubes grown at 450 °C. From the nanotubes grown at 400 °C with plasma powers of 15 W and 40W, the RBM peaks respectively appeared from three and two positions of five different scans. SWNT growth may be suppressed by high plasma power due to the bombardment effect [61]. The importance of low plasma power is also supported by a previous report by Delzdit. In their work, MWNTs were grown at low power levels and nanofibers were formed at high power levels, of which transition occurred at 30 ~ 40 W [53].

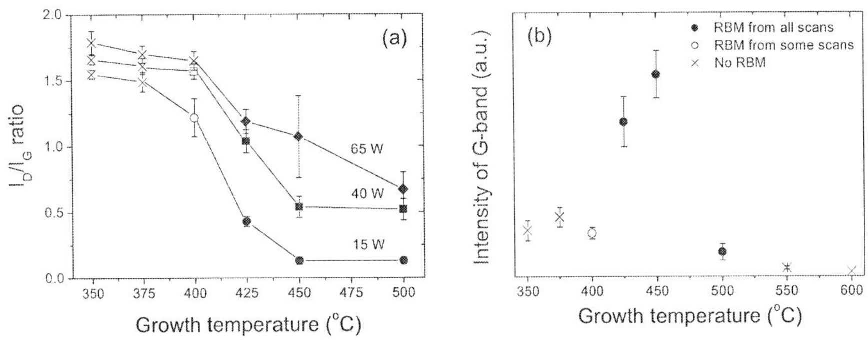

Figure 3.7. (a) ID/IG ratio variation with growth temperature. Solid or open circles designate that RBM peaks appear from all or a few of five different scans, respectively. Cross symbols indicate that RBM peaks were not observed from several scans. (b) Intensity variation of G-band with growth temperature. The nanotubes were grown at 450 °C with a plasma power of 15 W.

Temperature effect on the intensity of G-band was investigated with nanotubes grown by using a low plasma power of 15 W. As the growth temperature increases, the intensity of G-band shows an increasing tendency up to 450 °C and then decreases at higher temperatures than 450 °C as shown in Figure 3.7(b). Eventually, the G-band disappears at 600 °C wherein any peaks due to the carbon materials are not observed in Raman spectra. The disappearance of the Raman peaks may be attributed to an etch effect by plasma. To verify the etch effect by the weak plasma power; we treated nanotubes at 450 °C under argon plasma (15 W) without any flow of methane gas. As the plasma treatment time increased, G-band was gradually weakened and nearly disappeared after 10 min. Therefore, such a temperature dependency of the G-band in Figure 3.7(b) can be explained by a competition between growth and etch reactions of sp^2 carbon structures such as SWNTs, MWNTs, and graphene sheets. As the growth temperature increases, the contribution of the etch reaction may increase. When the etch effect is small, the G-band shows the increasing tendency with the temperature. However, as the etch effect increases, the G-band intensity gradually decreases with the growth temperature (> 450 °C). At 600 °C, the growth of sp^2 carbons is eventually overcame by the etch effect, to result in the disappearance of the G-band. It should be noted that the RBM peaks with a strong G-band are obtained from the specimens grown at 425 ~ 450 °C wherein the etch effect of sp^2 carbons is considerably high. This means that the etch effect by the low plasma power plays a critical role for SWNT growth.

Although the remote PECVD using a low power of argon plasma gives SWNTs at 450 °C as shown in Figure 3.8, the observed nanotubes detached from the substrates for TEM analysis are mainly bundles and defective in their surface. In addition, MWNTs or double-walled carbon nanotubes (DWNTs) were also observed with SWNTs. Figure 3.8(c) shows a DWNT with an outer diameter of 1.33 nm and an inner diameter of 0.61 nm. The interlayer distance of the DWNTs is ~ 0.36 nm, which is similar with reported values [62]. Figure 3.8(d) shows an image of SWNTs with carbonaceous materials.

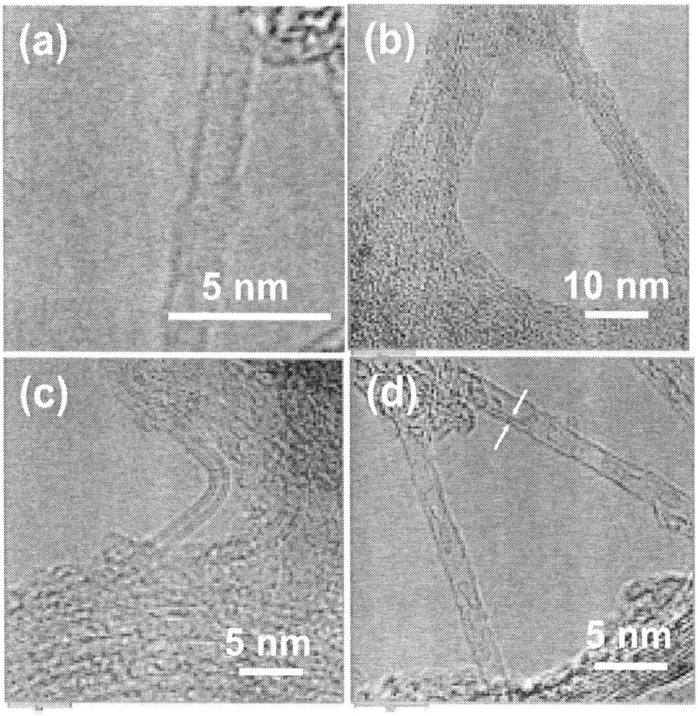

Figure 3.8. HRTEM images of nanotubes grown by remote argon PECVD. (a) Singlewalled nanotuebs. (b) Bundles. (c) Double-walled nanotubes. (d) Single-walled nanotubes with carbonaceous materials inside. Nanotubes were grown at 450 °C with a plasma power of 40. Dark lines correspond to the sidewalls of nanotubes [58].

To fabricate nanotube-based FETs, SWNTs should be grown without multi-walled carbon nanotubes (MWNTs) and carbonaceous impurities. However, the growth of SWNTs is generally accompanied by the undesired carbon materials, especially during chemical vapor deposition (CVD) process, which is the preferred growth method of nanotubes for integrated electronic devices. Although sulfidative purification of nanotubes was recently reported to selectively remove carbonaceous impurities from nanotubes integrated in devices, the undesired MWNTs would be problematic for reliable transistor performance. Therefore, preferential growth of SWNTs without both carbonaceous impurities and MWNTs is essential in CVD for device applications.

The preferential growth of SWNTs over MWNTs and carbonaceous impurities can be achieved by water-plasma CVD. Water plasma lowers the growth temperature down to 450 °C as argon plasma does, and the grown nanotubes are pure SWNTs without MWNTs.

Several groups have widely investigated an effect of water on purification and synthesis of carbon nanotubes. Tohji, et al. reported that SWNTs could be purified with water by hydrothermal treatment [63]. Yoshimura et al. demonstrated that MWNTs were synthesized in the absence of metal catalyst by hydrothermal treatment of amorphous carbon in pure water [64]. Wang et al. introduced one-step water-assisted synthesis of MWNTs only by dipping hot graphite rod into cool water in the absence of catalyst [65]. Hata, et al. showed a thermal CVD synthesis of SWNTs where the activity and lifetime of the catalysts were enhanced by water. The most common finding is that all of the nanotubes prepared from the mentioned approaches are pure without carbonaceous impurities. Inspired by the water effect, water plasma was chosen as both an oxidant for high purity of nanotubes and an energy source for low temperature growth.

Iron catalyst nanoparticles were deposited on SiO_2 (400 nm)/Si substrates with a mixed solution of ferrocene and conventional photoresist as described in the section 4. A homemade radio frequency (13.56 MHz) remote PECVD system was used for nanotube growth. Water was evaporated from a canister and carried to the quartz tube by the vacuum without any carrier gas. The amount of water vapor was regulated by a metering valve and monitored by pressure increase in the quartz tube.

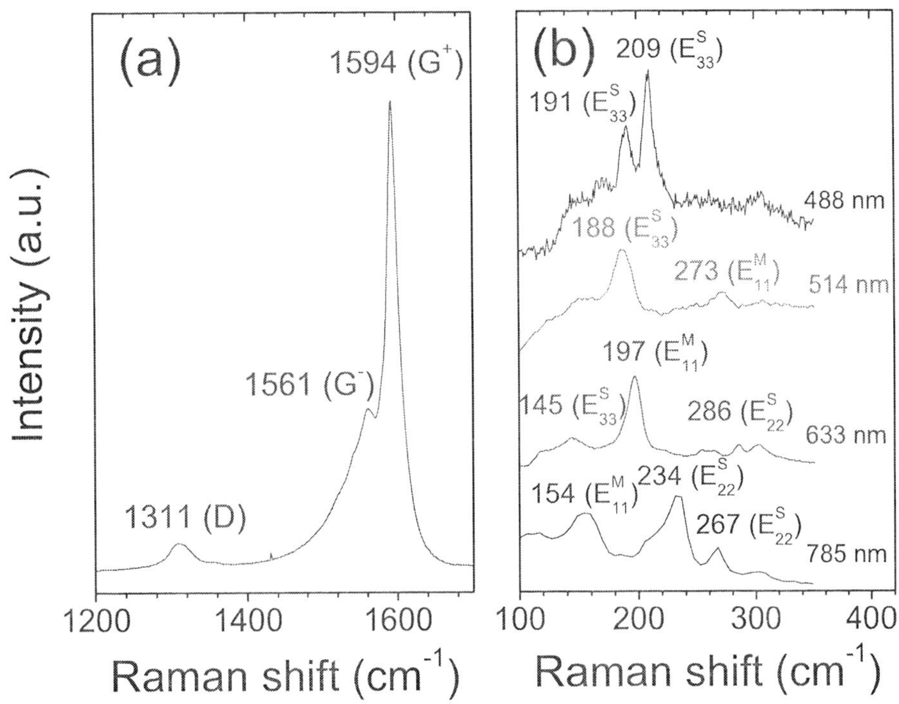

Figure 3.9. Raman spectra of the grown nanotubes. (a) G-band and D-band obtained from 633 nm laser excitation. (b) RBM peaks from 488, 514, 633, and 785 nm laser excitations. All spectra were measured on 5 different positions and averaged. Raman shifts were calibrated with a silicon peak at 521 cm^{-1}.

SWNTs were grown on catalyst-deposited substrates at 450 ^{o}C with methane gas (60 sccm) and water plasma (15 W) in the remote PECVD system. Raman spectra from laser excitations of various wavelengths (488, 514, 633 and 785 nm) reveals that the grown nanotubes were single-walled. Figure 3.9(a) and (b) show the typical Raman spectra of SWNTs, as characterized by the strong G band (tangential stretch mode) and the presence of the sharp radial breathing mode (RBM). Most RBM peaks appear in frequency region between 120 and 250 cm^{-1}, which roughly corresponds to the diameter range of 1 ~ 2 nm.

According to the Kataura plot (transition energies as a function of nanotube diameter) [66], the grown nanotubes are a mixture of metallic and semiconducting tubes as designated in Figure 3.9(b). All spectra were measured on 5 different positions and averaged. Raman shifts were calibrated with a silicon peak at 521 cm^{-1}.

To evaluate the quality of nanotubes from the amount of disordered and amorphous carbons, we monitored a Raman intensity ratio (I_D/I_G) of D-band to G-band as growth time increased. The ratio of the nanotubes grown by water plasma is compared to that of the nanotube by argon plasma (15 W, methane 60 sccm) as shown in Figure 3.10. Since water plasma with a weak power (15 W) plays a role of oxidant which removes carbonaceous impurities and disordered carbons, SWNTs grown by water plasma shows lower I_D/I_G ratio than those by argon plasma over all growth time. The I_D/I_G ratio decreases in the initial stage of nanotube growth with the growth time and then keeps nearly constant during a period and subsequently increases possibly due to deformation of nanotubes by active species such as radical or ions generated by plasma. It should be noted that growth and etch reactions of nanotubes compete more severely in plasma than in thermal CVD.

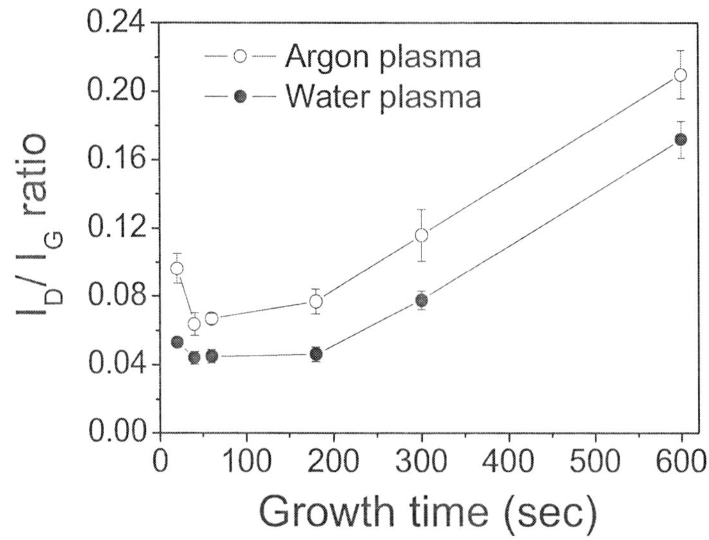

Figure 3.10. Variations of I_D/I_G ratios with growth time. Open and solid circles were plotted with Raman intensity ratio (633 nm) of SWNTs grown by argon and water plasma, respectively.

Figure 3.11(a) shows HRTEM image of the grown SWNTs of which diameters are mainly in a range of 1 ~ 2 nm (Figure 3.11(b)) as expected in Raman spectra. The observed nanotubes were single-walled or bundled. MWNTs were rarely observed as the growth time is short. Among the nanotubes grown for 180 seconds, we could not find MWNTs in HRTEM images. Considering that SWNTs were less frequently observed than MWNTs among the nanotubes grown by argon plasma CVD, the water plasma effect on the growth of pure SWNTs without MWNTs is obvious.

To investigate electrical nature of SWNTs grown by water plasma, we fabricated FETs with random network of SWNTs as a channel. The source-drain channel length was 5 μm and the channel width was 40 μm. The number of SWNTs per unit area was ~ 30/μm^2.

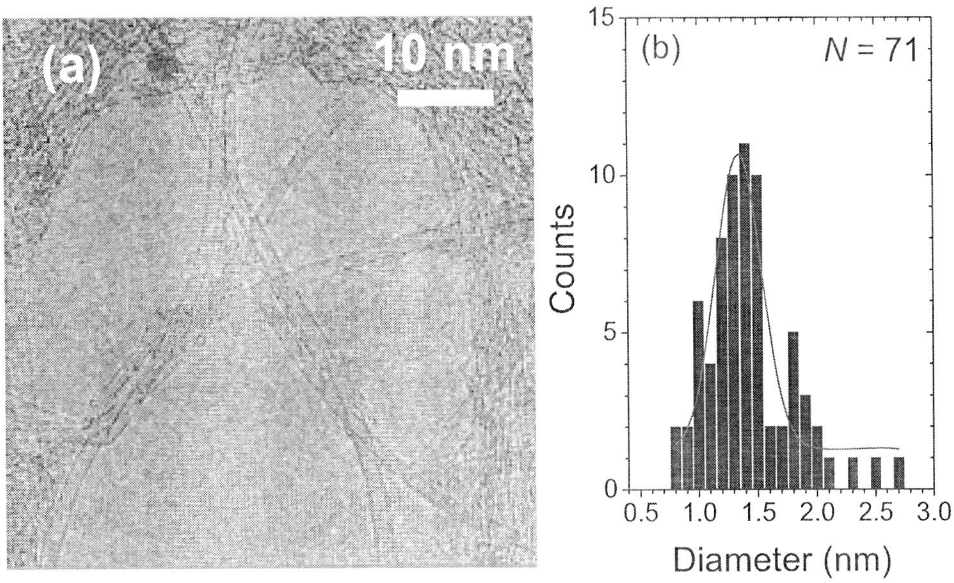

Figure 3.11. (a) HRTEM image of SWNTs grown at 450 oC with a water plasma power of 15 W. (b) The histogram of the diameter of SWNTs determined from HRTEM images [59].

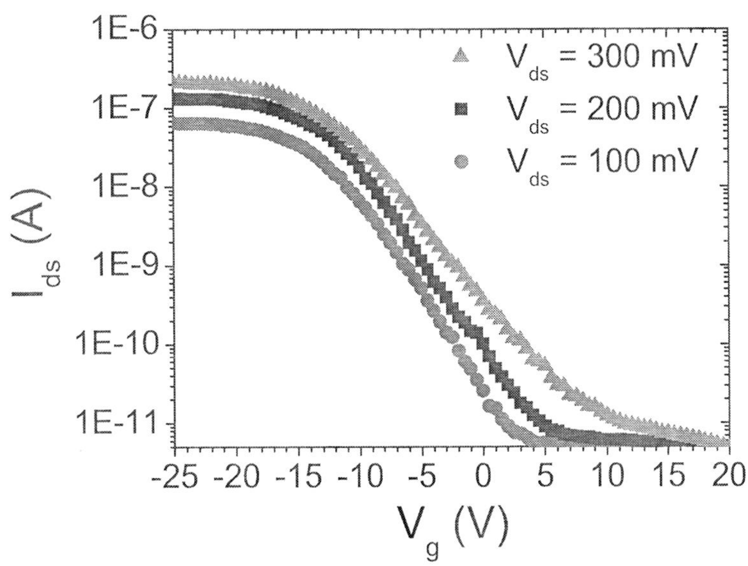

Figure 3.12 Drain-source current (Ids) versus gate voltage (Vg) characteristics of a FET with random network of SWNTs [59].

It is reported that, for SWNTs with densities exceeding a percolation threshold, FETs in the random network are electrically conducting [67]. Above the percolation threshold, which is $0.3/\mu m^2$ for nanotubes with a length of 1~3 μm, the SWNTs will intersect with one another and form continuous conduction path with metallic tubes. However, even though the nanotube density is much higher than the percolation threshold, the FET with network of SWNTs grown by water plasma shows typical p-type semiconducting characteristics with an On/Off ratio of ~ 10^4 (Figure 3.12). This can be explained by some higher portion of semiconducting nanotubes over metallic tubes than the 2:1 ratio expected for random chirality distribution. Dai et al. reported that nearly 90 % of the grown nanotubes by remote PECVD are semiconducting, while the HiPco and laser ablation nanotubes consist of 61 and 30 % semiconductors, respectively [56, 57].

The semiconducting nature of the network FETs also supports the preferential growth of SWNTs over MWNTs. Since MWNTs are generally metallic due to their large diameters, the network transistors should show metallic nature unless the preferential growth was used.

This water plasma CVD method for the low temperature growth of pure SWNTs facilitate to fabricate nanotube-based devices for fundamental characterizations and industrial applications.

4. Carbon Nanotube Placement and Fabrication for the Electronic Devices Integration

Since carbon nanotubes (CNT) are one-dimensional systems which can be metals or semiconductors depending on its chirality, these new carbon allotropes have been considered as the material to replace conventional silicon-based electronics in future. However, the uncertainty and uncontrollability of CNT positioning is one of the most critical barriers for CNT integration. To successfully build upon an integrated circuit (IC) with CNT-based active and passive elements, CNTs should be selectively and individually grown or located only on a defined region.

From a methodological point of view, there are two pathways for positioning CNTs on a defined region: top-down and bottom-up approaches. For an example of the top-down approach, a self-assembly method for CNT positioning has been attempted with an interacting force between CNTs and organic molecular marks pre-patterned on a surface [46]. Currently, this approach is in the early stages of building ICs over a large area and has difficulties in extending the organic molecules to various surfaces of oxides or metals for real applications.

On the other hand, many bottom-up approaches have been tried using pre-patterned catalysts before a subsequent chemical vapor deposition of CNTs, since it is well known that nanoparticles or thin films of some transition metals, such as Fe, Co and Ni, play the role of catalyst for CNT growth [68]. The catalyst pre-patterns have been obtained from various techniques such as lithographic lift-off [41], soft lithography [42], and offset printing [69]. However, these pre-patterning methods are not compatible with current microelectronic processing for IC fabrication. For an example of the bottom-up approach, a pre-patterning method using a catalyst-dispersed solution with nanoparticle supporters is the most popular way to obtain CNTs with high electrical performance [41]. However, since the dispersion solution of catalyst and supporting particles are not uniform in concentration, this method is not reliable for mass production.

Recently, several groups demonstrated direct photolithographic methods for selective growth of CNTs using photoresists or electron beam resists containing catalyst precursors [45]. The modified resist plays the role as both catalyst and resist for growth and alignment of

CNTs, respectively. The modified resist patterns are obtained by following general lithographic procedures. After removal of organics from the resist pattern, catalytic particles are deposited on the defined region and then CNTs are selectively grown on the pre-patterned catalyst. Huang, et al used poly (vinyl alcohol) as a polymer resin with metal nitrate precursors [45]. Hinderling et al. demonstrated that iron-containing block copolymer polystyrene-poly (ferrocenylsilane) can be used for CNT growth [70]. Ishida, et al. reported electron beam lithographic method using a calixarene resist modified with iron(III) acetylacetonate ($Fe(acac)_3$) [71]. However, those modified resists were not successful to grow single-walled carbon nanotubes (SWNTs).

Here, we report a combination for the direct photolithographic route to selective growth of SWNTs using a simple mixture (Cat-PR) of ferrocene and a conventional photoresist (AZ 5214E, Clariant Co) [72]. From the Cat-PR pattern, we obtained catalytic nanoparticles of which diameters are in the range of 3 – 10 nm and are mainly smaller than 5 nm. Nanotubes with diameters of 1.0 ~ 1.5 nm are grown on the catalytic nanoparticles as individual SWNTs or bundles of them. It is also demonstrated that nanotubes suspended between three-dimensional structures could be obtained by utilizing the Cat-PR pattern as an etch mask.

4.1. Selective Deposition of Catalytic Nanoparticles

Ferrocene was chosen as the catalyst precursor in creating a homogeneously dissolved and chemically stable solution with commercially available PR (AZ 5214E). Since ferrocene has a sandwich structure consisting of two cyclopentadienyl and an iron ion, it is moderately stable and soluble to various organic solvents [73]. The addition of ferrocene to the PR does not significantly deteriorate the photo-active properties of PR. Furthermore, ferrocene has been recently utilized as a catalyst to grow CNTs in gas phase or liquid phase [47, 74].

Different amounts of ferrocene were dissolved in AZ 5214E to obtain various concentrations of Cat-PR solutions. Standard photolithographic procedures were then followed.[17] The Cat-PR was spin-coated on substrates and baked, followed by the exposure step with a mask aligner in order to define where the CNTs will be positioned. After the photolithographic development step, the Cat-PR pattern with ferrocene was obtained as shown in Figure 4.1.

Before the growth of CNTs, the burning step is added to general photolithographic procedures in order to remove organics from the Cat-PR pattern. The burning temperature was determined to be 550 ^{o}C which is higher than the temperature of a strong exothermic peak in differential scanning calorimetric thermogram obtained in air. The Cat-PR shows a strong exothermic pick at ~ 460 ^{o}C accompanying weight loss due to the removal of organics, and the residue at 550 ^{o}C is around 0.8 %. Supposing that the residue is Fe_2O_3 without any organic substance, the residual weight should be theoretically 0.73 %. The good agreement between the experimental and the theoretical values of the residue supports that nanoparticles were formed without organic residue by the burning process. This complete removal of organics from the Cat-PR pattern is crucial to obtain catalytically active nanoparticles. The exothermic peak and weight loss in TG-DSC should appear in order to remove the organics and to deposit catalyst particles on the defined regions during the burning process. For example, a mixture of polymethylmethacrylate (PMMA) and ferrocene shows no exothermic reactions. Consequently, catalysts are not deposited and CNTs are not grown from the combination.

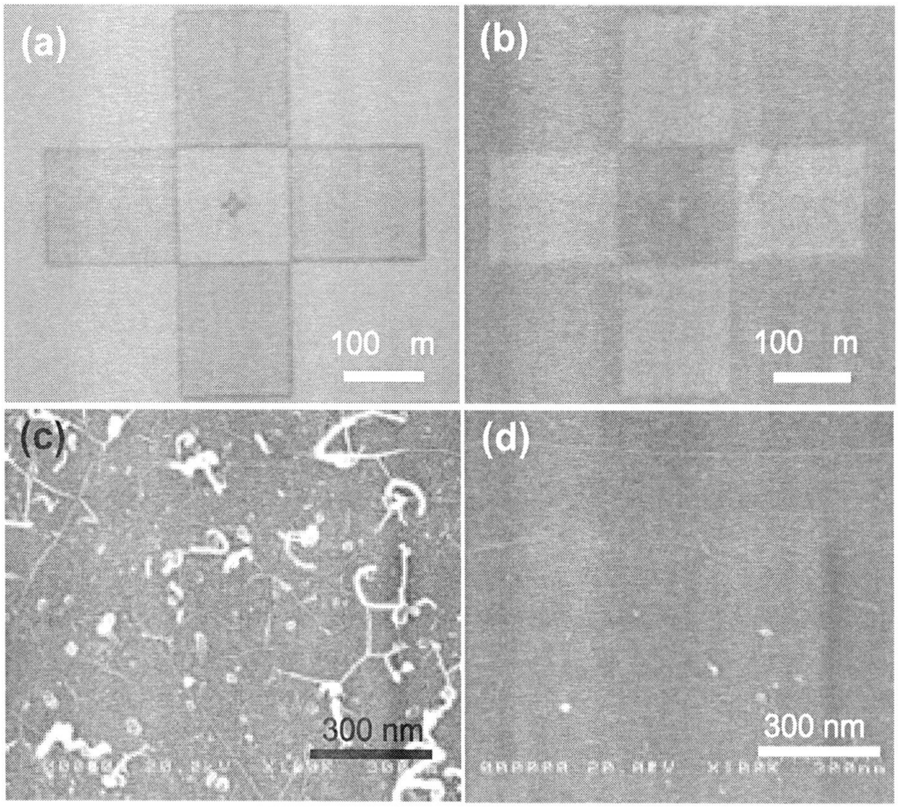

Figure 4.1. CNT selective growth by the direct photolithography using a simple mixture (Cat-PR) of ferrocene and conventional photoresist (AZ 5214E). (a) Optical microscopic image of Cat-PR patterns. (b) SEM image after CNT growth on catalytic particles prepared from 0.3 M Cat-PR solution. (c) SEM image of CNTs grown on catalyst particles prepared from 0.3 M Cat-PR solution. (d) SEM image of CNTs grown on catalyst particles prepared from 0.05 M solution [72].

Figure 4.2 shows low- and high-resolution TEM images of catalytic nanoparticles deposited with 0.1 M Cat-PR. Most particles are randomly and separately distributed on the surface of SiO_2 substrate and well crystallized. However, it was also observed that a few of them were coalesced together (Figure 4.2(a)). Diameters of catalytic particles were in the range of 3 – 10 nm and were mainly smaller than 5 nm (Figure 4.2(b)). Root-mean-square (rms) roughnesses of the catalyst deposited substrate were determined to be 0.8 nm and 1.3 nm from 0.05 M and 0.3 M Cat-PRs, respectively, by atomic force microscope (Seiko, SPA500). The rms roughness increased with the concentration of ferrocene in the Cat-PR solution.

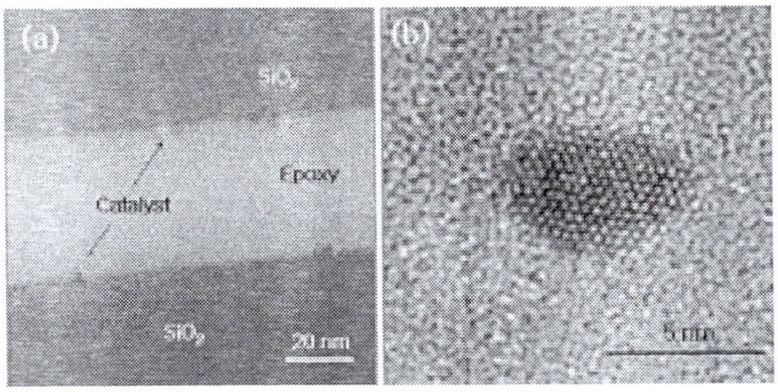

Figure 4.2. TEM micrographs of catalytic nanoparticles prepared from 0.1 M Cat-PR solution after the burning process. (a) Cross-sectional image. (b) High-resolution image of a catalyst particle.

To investigate the chemical state of the deposited catalytic nanoparticles after the burning process, we analyzed the catalytic particles with XPS without removing adventitious carbons of which peak (284.8 eV) was utilized as an internal reference. Fe 2p peaks (Figure 4.3) from catalytic particles are located at 711.5 eV for a $2p_{3/2}$ component and at 724.9 for $2p_{1/2}$. The binding energy of Fe $2p_{3/2}$ peak is similar with the literature values of iron oxyhydroxide (FeOOH) or iron oxide (Fe_2O_3) of which $2p_{3/2}$ peaks are located at 711.6 eV and 710.9 eV, respectively [75]. Although the O 1s region of the oxyhydroxide would be expected to show a doublet with one peak due to oxide and the other for hydroxide [76], since the catalytic particles are located on SiO_2 underlying layer, the O 1s region could not be used to distinguish FeOOH and Fe_2O_3.

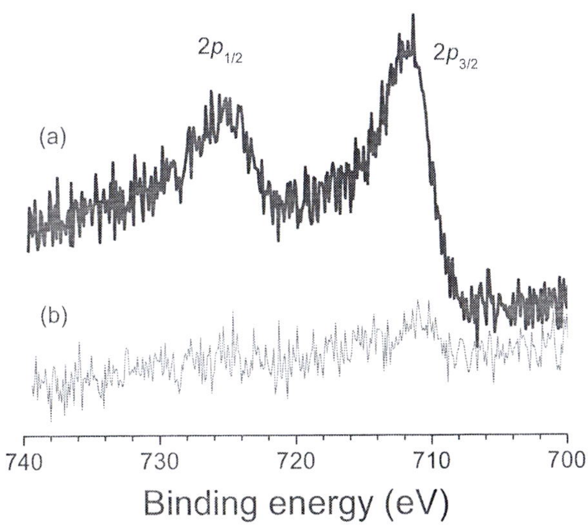

Figure 4.3. Fe $2p$ core level photoemission spectra. (a) From the specimen with catalytic nanoparticles. (b) From the specimen of which catalytic particles are removed by exposal and development processes. The binding energies were referenced to C_{1s} at 284.8 eV.

Ishida et al. indicated that their catalyst particles, after removal of residual carbons, formed a mixture of α-Fe, Fe_2O_3, and Fe_3C.[11] However, the $2p_{3/2}$ peak in Figure 4.3 reveals that metallic iron (707.0 eV) or Fe_3C (708.1 eV) are not main components of catalyst particles [75, 77]. It is believed that the catalytic nanoparticles mainly consist of oxyhydroxide or oxide forms of iron without organic residues, which is possibly reduced to iron particles by hydrogen gas during the subsequent CVD process.

For comparison, a specimen coated with the Cat-PR was exposed all over the area without photomask and then developed to remove catalyst. From the catalyst-removed specimen, Fe 2p peaks (Figure 4.3(b)) were not distinguishable from noises, since the iron concentration was very low. This means that catalyst particles can be pre-patterned by the Cat-PR method.

4.2. SWNT Growth and Characterization

CVD for nanotube growth was performed with acetylene gas on the catalyst pre-pattern, which was formed by the burning process. As shown in Figure 4.1(b), CNTs were grown only on the defined region by the Cat-PR pattern (Figure 4.1(a)). Figure 4.1(c) and (d) show SEM images of CNTs grown from 0.3M and 0.05M solutions, respectively. The density of the grown CNTs increases with the ferrocene concentration of the Cat-PR. As shown in Figure 4.1(c) and (d), particles and nanotubes with large diameters more frequently appear as the concentration of Cat-PR increases.

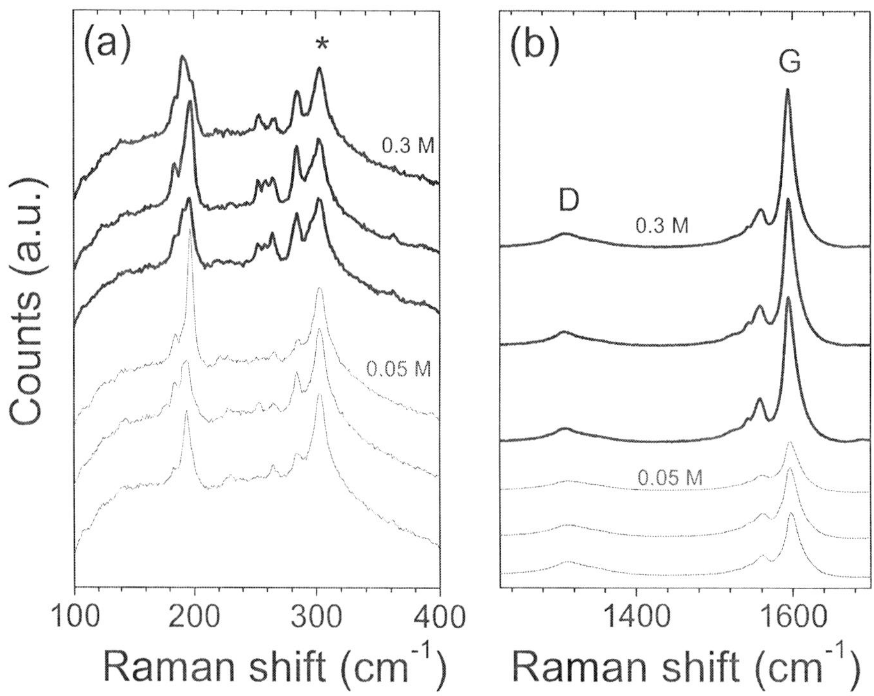

Figure 4.4. Raman spectra of SWNTs from 633 nm laser excitation. (a) Radial breathing modes. (b) Tangential modes. Raman spectra were obtained from 3 different positions of each specimen which were grown with 0.05 M (thin lines) and 0.3 M (thick lines) Cat-PR, respectively. All Raman spectra were calibrated with a peak at 303 cm^{-1} denoted with an asterisk due to the SiO_2/Si substrate [72].

The diameters of nanotubes grown on SiO_2/Si substrates were determined by Raman spectroscopy using 633 nm (Figure 4.4) and 785 nm (Figure 4.5) laser excitations. Figure 4.4 shows the typical Raman spectra of SWNTs, as characterized by the strong G band (tangential stretch mode) and the presence of the sharp radial breathing mode (RBM) in the low frequency region (100 – 300 cm^{-1}) [78]. The diameter (d_t, nm) of SWNTs can be calculated from the frequency (ω_{RBM}, cm^{-1}) of the RBM by the equation, $\omega_{RBM} = A/d_t, + B$ where A and B are experimentally found to be 223.5 $cm^{-1}nm$ and 12.5 cm^{-1} for SWNTs. In the RBM mode from 633 nm laser excitation (Figure 4.5(a)), several peaks appear in a wide range of 170 – 290 cm^{-1} with the main peaks in the frequency range of 170 – 210 cm^{-1}, regardless of the concentration of the Cat-PR or the laser-exposed position. These intensities are comparable to the substrate peak at 303 cm^{-1}. The corresponding diameter of the main peaks is 1.13 – 1.42 nm. However, the intensities of the minor peaks increase with the increase in concentration of the Cat-PR.

A weak peak at 1310 cm^{-1} of the D-band in Figure 4.4(b) is related to disordered graphite and amorphous carbon. The intensity ratios of D-band relative to G-band (I_D/I_G) are 0.38±0.06 and 0.21±0.02 in the CNTs grown with 0.05 M and 0.3 M Cat-PR solution, respectively. The higher frequency component (ϖ_G^+) at 1594 cm^{-1} and the lower frequency component (ϖ_G^-) at 1560 cm^{-1} of the G-band (Figure 4.4(b)) are attributed to vibrations along the direction of the nanotube axis and along the circumferential direction, respectively. The sharp Lorentzian features of the peaks indicate that the grown SWNTs are mainly semiconducting. For metallic SWNTs, the lower frequency component generally shows a very broad Breit-Wigner-Fano line [60]. The G band components, ϖ_G^+ and ϖ_G^-, are related to the equation, $\varpi_G^- = \varpi_G^+ - c/d_t^2$ where c = 47.7 $cm^{-1}nm^2$ for semiconducting nanotubes and c = 79.5 $cm^{-1}nm^2$ for metallic nanotubes. Substituting the diameter of 1.21 nm attributed to the strongest RBM peak at 197 cm^{-1} in the equation, the calculated (ϖ_G^-) value for the semiconducting nanotube is 1561 cm^{-1}, which agrees well with the observed value (1560 cm^{-1}). If the grown nanotubes have random chirality distribution, the ratio of semiconducting to metallic nanotubes should be 2:1. The position (1560 cm^{-1}) and the sharp Lorentzian features of the lower frequency component (ϖ_G^-) of the G-band also reveal that nanotubes are mainly semiconducting. However, the strongest peak at 197 cm^{-1} seems to be attributed to an E_{11} transition of metallic nanotubes, according to the Kataura plot (transition energies versus diameter) [66].

In recent studies on the transition energies of nanotubes as a function of diameter [79], it was reported that the first metallic band in the visible can barely be distinguished from the higher-lying band. Therefore, we used 785-nm-laser to characterize the grown nanotubes. The overall tendency of the Raman spectra from 785 nm is similar to that from 633 nm. However, the main peaks of the RBM mode from 785 nm (Figure 4.5(b)) are located in the higher frequency range of 210 – 240 cm^{-1} than those from 633 nm. The intensities are stronger than those of the substrate peak due to the strong coupling between electrons and phonons of nanotubes with 785 nm laser excitation. In Figure 4.5(a), the empirical Kataura plot reported by Weisman et al. was shown as a function of Raman shift.[25] Supposing a resonant window by 785-nm-laser is 1.58 ± 0.1 eV as designated by horizontal lines in Figure 4.5(a) [80], most RBM peaks in Figure 4.5(b) are attributed to E_{22} transitions of semiconducting nanotubes of which frequency range is designated by vertical lines. The strongest peak at 233 cm^{-1} is due to (11, 3) nanotubes, and the diameter is 1.01 nm. This supports that the grown SWNTs are mainly semiconducting, although it is not clear if the semiconducting portion is higher than the ratio of 2:1 from random chirality distribution. The corresponding diameter of the main peaks (210 – 240 cm^{-1}) is 0.98 – 1.13 nm. Considering the frequency range of the main peaks from 633 and 785 nm laser excitation, the diameter of the grown SWNTs is roughly 1.0 – 1.5 nm which is much smaller than the diameters (3 ~ 10 nm) of nanoparticles deposited from the Cat-PR by the burning process.

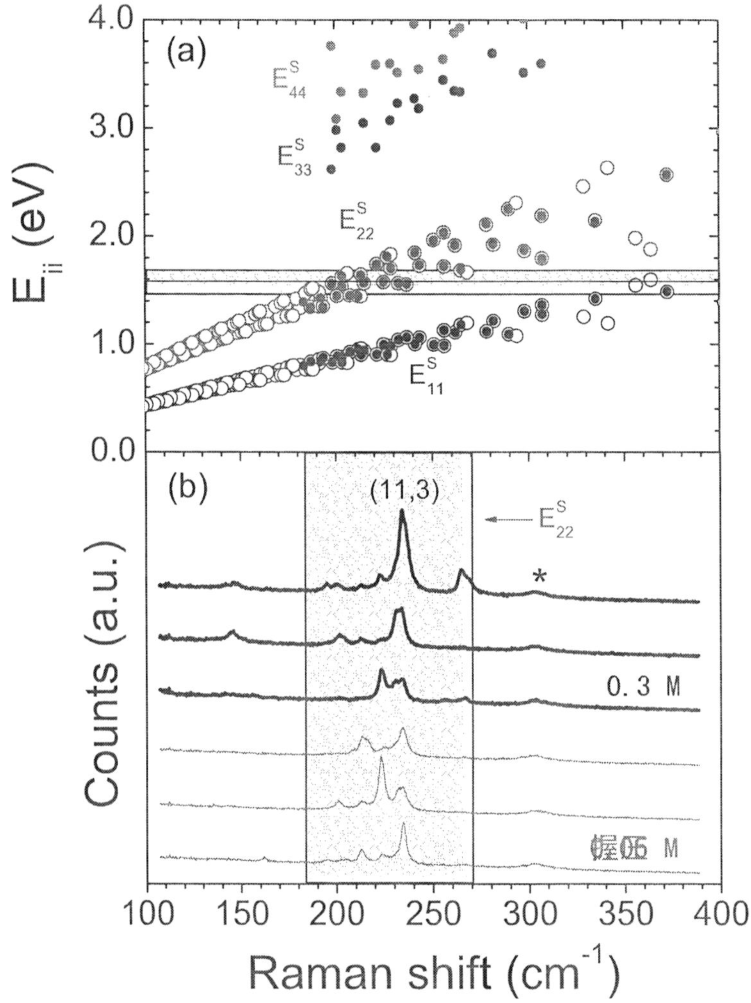

Figure 4.5. (a) Optical transition energy versus Raman shift for semiconducting SWNTs which was re-plotted from the Ref 25. Solid symbols are experimental data; open symbols are from the empirical fitting functions. The horizontal lines designate the resonant window by 785-nm-laser (1.58 eV). (b) Radial breathing modes of SWNTs from 785 nm laser excitation. Raman spectra were obtained from 3 different positions of each specimen which were grown with 0.05 M (thin lines) and 0.3 M (thick lines) Cat-PR, respectively. Vertical lines designate the frequency range due to the E_{22} transition of semiconducting nanotubes. All Raman spectra were calibrated with a peak at 303 cm^{-1} denoted with an asterisk due to the SiO_2/Si substrate.

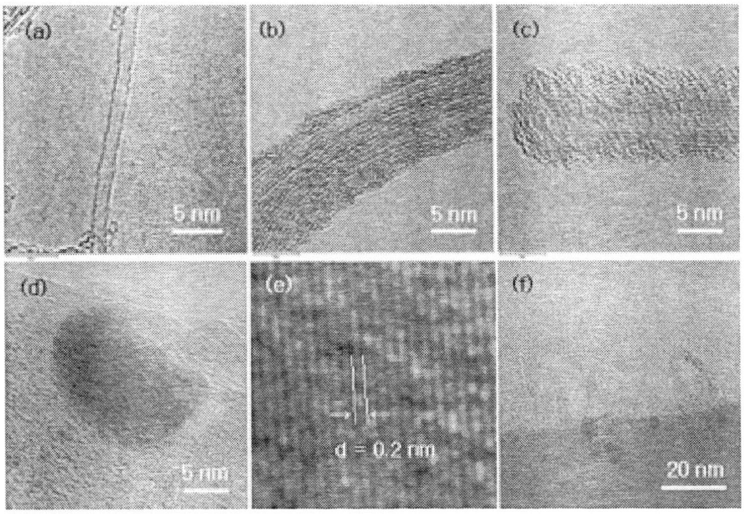

Figure 4.6. High and low resolution TEM images of CNTs and catalysts. (a) An individually separated SWNT with a diameter of 1.3 nm. (b) A bundle of SWNTs. (c) A MWNT with a closing cap. (d) A catalyst nanoparticle after the CNT growth. (e) A magnified image of the catalyst in Figure 7d. (f) Cross-sectional TEM image of nanotubes and nanoparticles after the growth.

The structure and the diameter of nanotubes were confirmed by high resolution TEM. We observed SWNTs as expected from Raman spectroscopy. However, the grown nanotubes were a mixture of individual SWNTs, bundles of SWNTs and multi-walled carbon nanotubes as shown in Figure 4.6. The bundles were most frequently observed and SWNTs were less frequently observed than MWNTs from individually separated nanotubes in the TEM specimen. Figure 4.6(d) shows a catalytic nanoparticle surrounded by graphene sheets after nanotube growth. From a magnified image in Figure 4.7(e), we determined the interplanar spacing to be ~ 0.2 nm. The catalyst particles after nanotube growth were qualitatively analyzed by energy-dispersive X-ray (EDX) spectroscopy, in which iron and carbon were observed without oxygen. This supports the oxidized form of iron is reduced during the growth. However, it is not clear whether the reduced form is metallic iron or iron carbide [81-83].

It should be noted that catalytic nanoparticles are partially or fully embedded in the underlying oxide layer as shown in Figure 4.2(a) and 4.6(f). The depth of the nanoparticles embedded in SiO_2 layer is much deeper after the CNT growth (Figure 4.6(f)) than after the burning process (Figure 2(a)). The embedment may decrease the effective size of catalyst nanoparticles. On the other hand, after the nanotube growth (Figure 4.6(d)), nanoparticles larger than 5 nm are more frequently observed than after the burning process, possibly due to coalescence of small particles. Therefore, the effective nanoparticle size will be influenced by both the embedment and the coalescence of nanoparticles. For the controlled growth of SWNTs, the effective size of catalytic particles should be further optimized by adjustment of the growth temperature and the Cat-PR concentration, since the diameters of nanotubes are determined by the diameters of catalytic nanoparticles [84, 85].

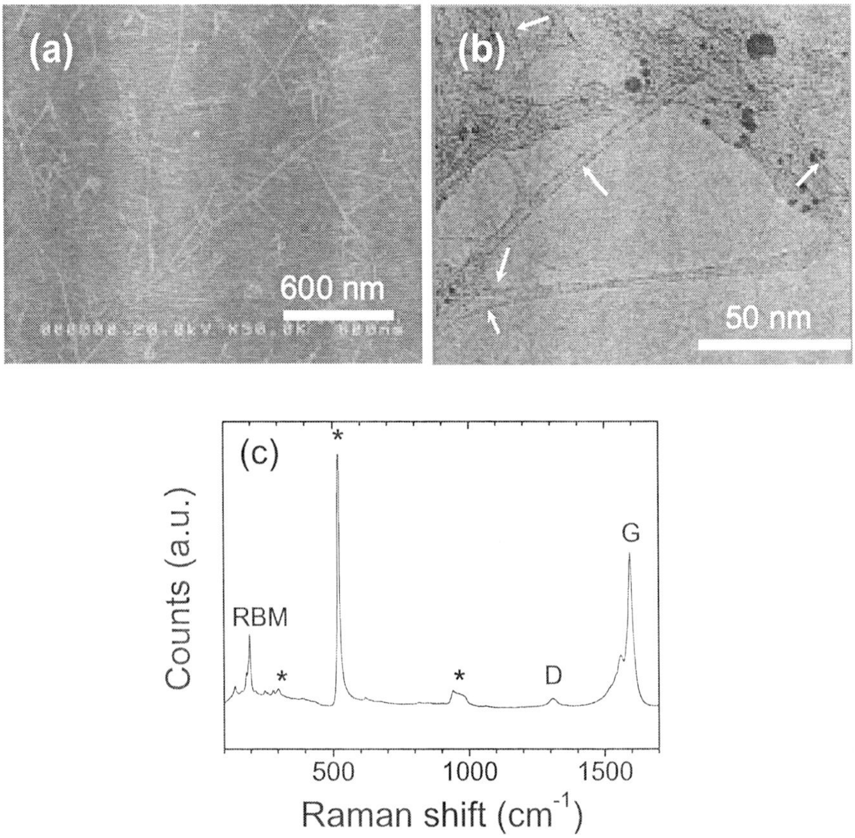

Figure 4.7. SWNTs grown on catalytic nanoparticles (0.1 M Cat-PR) by supplying water vapor during the CVD. (a) SEM image. (b) TEM. Arrows indicates SWNTs with various diameters. (c) Raman spectra (633 nm).

The burning process to remove organics from the Cat-PR pattern should be carefully performed in direct photolithographic methods, because some residues on catalytic particles poison catalytic activity. In contrast, Ishida et al. obtained RBM peaks from only three points among one hundred different scans over the patterned area [71], Raman scan on nanotubes grown with the Cat-PR shows RBM peaks from most points over all area. We believe that the difference originates from the carbonaceous residue remained on catalytic nanoparticles after the removal of organics.

Recently, Hata, et al. reported a thermal CVD synthesis of impurity-free SWNTs by using water vapor where the activity and lifetime of the catalyst nanoparticles were enhanced by water, since water plays a role of a mild oxidant to remove carbon impurities on the catalyst surface. In order to improve the activity of catalytic nanoparticles, CNT growth by using water vapor were also carried out on the catalyst nanoparticles deposited from 0.1 M Cat-PR. The nanotubes grown with water show better quality and higher density than those grown without water, as shown in SEM and TEM images of Figure 4.7(a-b). The high quality of SWNTs by water is verified by much lower I_D/I_G ratio (~ 0.06) in Raman spectra than that (0.2 ~ 0.4) of nanotube grown without water. Due to the higher density of SWNTs, the RBM peaks in Figure 4.7(c) are much stronger than those in Figure 4.4(a) as comparable to the substrate peak at 520 cm^{-1}. The improvement of the quality and the density of SWNTs by water also supports that the activation of catalyst nanoparticles is critical to obtain SWNTs in this direct lithographic method.

Suspended SWNTs. Fabrication of suspended nanotubes between three-dimensional structures is obviously a suitable way for fundamental characterization or electronic device applications of CNT, since nanotubes can be isolated from the adjacent CNTs and the interfaces with substrates [86]. The standard etching process, which is performed with conventional PRs, can be also applied to the Cat-PR. This facilitates the fabrication of the suspended nanotubes, because the Cat-PR pattern can be used as a etch mask during the etching process. A square shape pattern was formed on a SiO_2 layer with Cat-PR (0.05 M) as described above. The oxide layer was subsequently etched with a conventional oxide etching solution (buffered oxide etch 6:1, J.T. Baker Co). After burning the organics from the pattern and growing CNTs, we successfully obtained suspended-CNTs on the SiO_2 stands. Figure 4.8(a) and 4.8(b) show suspended nanotubes grown over four stands and two stands, respectively. Nanotubes grown from the spacing between stands are not observed due to the absence of catalytic particles by etching. Most of suspended nanotubes were connected to the nearest-neighbor stands. It should be further investigated that the direction of the growth can be controlled by changing dimensions of stands and spacing between stands.

The direct photolithographic method for the selective growth of SWNTs by using a simple mixture of ferrocene and conventional photoresist is a promising and practical pathway for large-scale applications of ICs with nanotube-based electronic elements such as transistors, resistors, and interconnects.

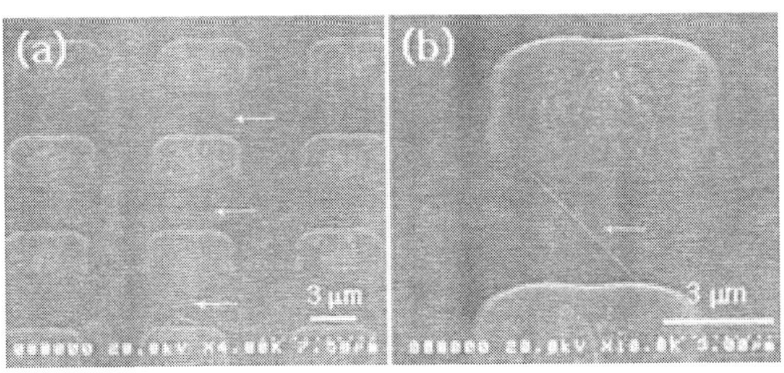

Figure 4.8. Suspended nanotubes obtained by using the Cat-PR pattern as a etch mask. (a) Over four stands. (b) Over two stands [72].

4.3. Sulfidative Purification of Carbon Nanotubes Integrated in Transistors

To obtain any intrinsic properties from CNTs, they should be pure without carbonaceous impurities such as amorphous carbons on their surface. Unfortunately, the growth of CNTs is inevitably accompanied by carbonaceous impurities [87]. Regardless of the preparation method of CNTs, which is mainly arc-discharge, laser ablation, chemical vapor deposition (CVD) or high pressure CO (HiPco) process, the carbonaceous materials on their surface are the most common impurities formed during the growth [88]. Therefore, the separation of pure nanotubes is essential for fundamental research and for device applications.

Since the carbonaceous particles are more susceptible to oxidation, in contrast to the perfect nanotube walls [89], CNTs have been purified by oxidative treatments in gas phase or liquid phase [87, 90-93]. Nowadays, nanotubes for device applications are directly grown on pre-patterned catalyst by CVD due to its compatibility with conventional integration processes and capability on large areas. However, the oxidative treatments are not suitable for the purification of CNTs integrated in a device because CNTs can be lost and deformed during the oxidation process. Therefore, CNTs grown by CVD are generally used without the subsequent purification process in order to avoid damage from the oxidative treatment. Recently, several groups have grown CNTs without surface carbon impurities by using water vapor, ammonia, or hydrogen gas or by using rapid growth process. Nevertheless the appearance of carbonaceous impurities is inevitable during CVD. Although many purification methods have been intensively studied, to our knowledge, there is no report regarding the removal of carbonaceous impurities from *CNTs integrated in a device*.

Recently, we demonstrated a simple purification method of CNTs integrated in field-effect transistors (FET) by using the sulfidation reaction of carbon, $C + 2S \rightarrow CS_2$, to selectively remove carbonaceous impurities. The sulfidative purification of CNTs integrated in FETs results in a dramatic improvement of switching characteristics due to removal of carbonaceous impurities.

It is well known that the sulfidation reaction of charcoal gives carbon disulfide. Until about 1950, the primary industrial production method of carbon disulfide was by heating charcoal to 750-900 °C in the presence of vaporized sulfur [94]. We chose sulfidation reaction to selectively remove carbonaceous impurities from CNTs because sulfidation is less reactive to carbon than oxidation, considering that molar Gibbs free energies of formation of gaseous CO_2 and CS_2 are –394.4 and 67.1 kJ/mol at room temperature, respectively [95]. Furthermore, sulfur shows high volatility in elemental form and the reaction product (CS_2) is easily removable. Carbon disulfide is a colorless liquid, which evaporates at room temperature.

The sulfidation of carbonaceous impurities was performed at 300 °C to enhance the reactivity difference between carbonaceous impurities and CNTs. At first, the occurrence of sulfidation reaction at lower temperatures was proven with simultaneous thermogravimetry-differential scanning calorimetry-mass spectrometry (TG-DSC-MS) analysis. A mixture of single-walled carbon nanotubes (NanoCarb Inc., arc-discharge, purity 60 %, diameter of 1.2.1.4 nm) and elemental sulfur (Aldrich, 99.998 %) was loaded in an alumina crucible and heated to 300 °C with a heating rate of 20 °C/min. The temperature was maintained at 300 °C for 30 min in vacuum (~ 0.01 Torr). During the heating (see the temperature profile in Figure 4.9(a)), TG-DSC-MS was simultaneously analyzed with a Netzsch STA-MS-Skimmer instrument. In the DSC curve (data not shown), two endothermic peaks owing to the melting and vaporization of sulfur are observed at 120 °C and 250 °C, respectively, and an exothermic peak from the sulfidation of carbonaceous impurities appears at 290 °C. The open circles in Figure 4.9(a) show that the mass intensity of carbon disulfide (76 amu) synchronously increases from 150 °C to 290 °C with the exothermic peak, and subsequently reaches its maximum intensity at 306 °C. It then decreases due to the exhaustion of sulfur. This reveals that sulfur reacts to carbonaceous impurities on the surface of CNTs by the sulfidation reaction, $C + 2S \rightarrow CS_2$, at 300 °C in vacuum (~0.01 Torr).

The selectivity of the sulfidation of the carbonaceous impurities is evaluated with Raman spectroscopy (514 nm laser) by comparing the spectra before and after the sulfur treatment as shown in Figure 4.9(b). Spectra were obtained at ten positions and averaged. The overall tendency of the Raman spectra of the sulfur treated CNTs are nearly same with that of the pristine. However, the ratio (I_D/I_G) of the disorder-induced D-band to the tangential G-band (derived from the graphite-like in-plane mode) decreases from 0.05 to 0.035 after the sulfur treatment [60]. This means that the nanotube walls are not sulfidized and not deformed, but carbonaceous impurities are selectively removed by the sulfur treatment at 300 °C. It is also proven by transmission electron microscopic (TEM) images before (Figure 4.10(c)) and after (Figure 4.10(b)) the treatment that the carbonaceous impurities were removed by sulfidation.

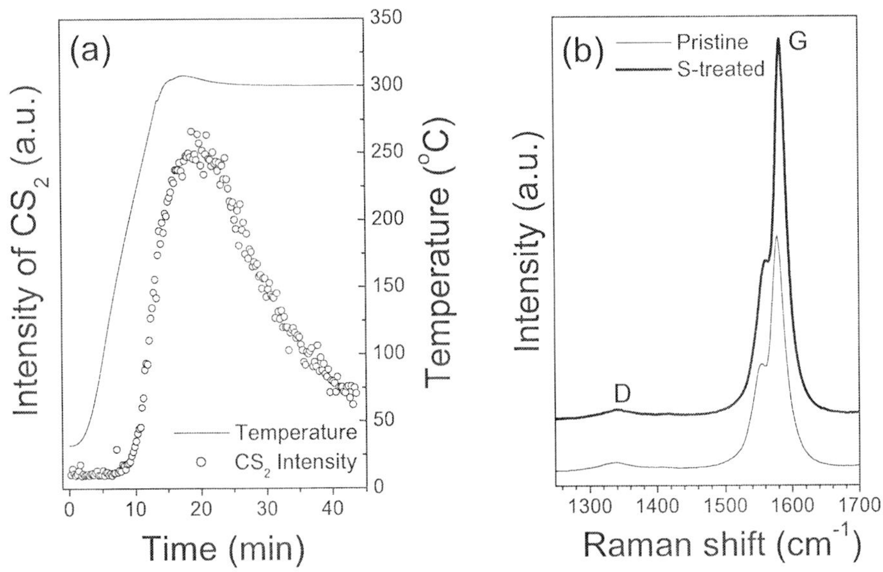

Figure 4.9. Sulfidation of carbon nanotubes with carbonaceous impurities. (a) Mass intensity variation of CS_2 (76 amu) evolved during the heating of a mixture of nanotubes and sulfur. (b) Raman spectra of nanotubes before and after the sulfidation [96].

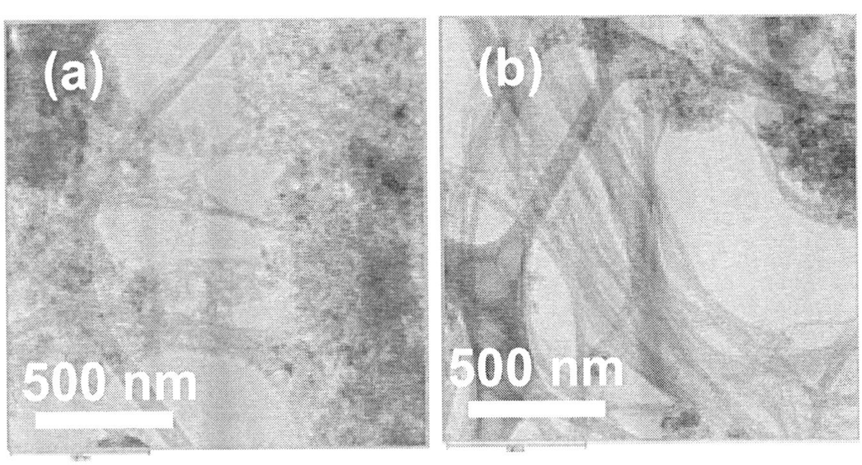

Figure 4.10. TEM images of nanotubes (c) before and (d) after the sulfidation [96].

The purification method by sulfidation can be applied to *CNTs integrated in devices* since nanotube walls are impervious to the sulfidation as proven with Raman spectroscopy. We compared electrical properties of CNTs integrated in a field effect transistor (FET) before and after the sulfur treatment. Figure 4.11(a) shows a typical atomic force microscopic (AFM) image of CNT-FET studied in this work (see Supporting Information for the detailed procedure of FET fabrication).

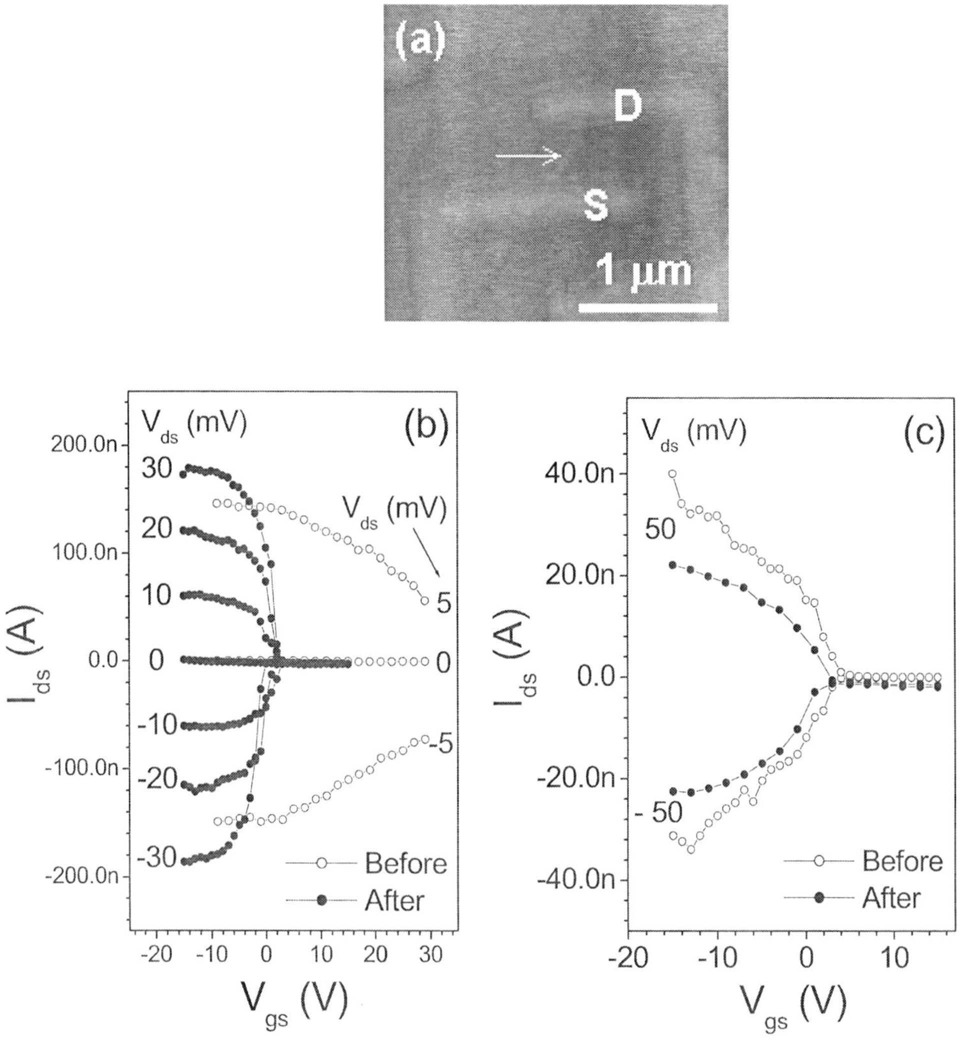

Figure 4.11 (a) AFM image of a back-gated CNT-FET. S and D denote source and drain, respectively, and CNT is designated by the arrow. (b-c) Drain-source current (I_{ds}) versus gate-source (V_{gs}) voltage characteristics of CNT-FETs before and after sulfur treatment. (b) The transistor with high off-state current before the sulfur treatment turns on and off at 3V after treatment. (c) The normally operating transistor before the sulfur treatment shows the same threshold voltage at 3 V after the treatment [96].

Among the CNT-FETs fabricated, we selected a transistor, which shows poor switching characteristics as shown in the drain-source current (I_{ds}) versus gate-source voltage (V_{gs})

curve of Figure 4.11(b) (open circles). The transistor with a resistance of 40 kΩ does not show a clear threshold voltage (V_{th}), which indicates the transition voltage from off-state to on-state. There are two possibilities for the transistor to show such poor switching performance. Firstly, if the CNT integrated in the transistor is metallic or semiconducting with narrow band gap, high off-state current can appear [67]. Secondly, the carbonaceous impurities on the outer surface of the CNT may act as a conduction path between source and drain, consequently increasing the off-state current in the transistor.

Sulfur treatment for the nanotube integrated in the FET was as follows. The CNT-FET specimen was placed on a vacuum chamber with a base pressure of 11.7 mTorr, together with sulfur contained in an Al crucible. The specimen was then heated at 300 ^{o}C for 30 min without any carrier gas. During the sulfur treatment, working pressure gradually decreased from 15.3 mTorr to the base pressure as sulfur was evaporated and exhausted. After cooling down, the specimen was unloaded from the chamber.

The solid circles in Figure 4.11(b) show I_{ds}-V_{gs} characteristics after sulfur treatment. Different from the device before the sulfidation, the sulfur treated CNT-FET shows clear on/off switching with a threshold voltage (V_{th}) of 3V and with an on-/off-state current ratio of two orders (see Supporting Information). For comparison, we investigated another CNT-FET, which had normally operated as a transistor before sulfur treatment. As shown with open circles in Figure 4.11(c), the FET shows a typical p-FET I_{ds}-V_{gs} curve with a threshold voltage of ~3V. After sulfur treatment, the FET still operates as a p-FET with the same threshold voltage.

If the sulfur treatment gave rise to the sulfidation of nanotube walls, the threshold voltage should shift due to the widened band gap by the deformation of the nanotube wall. However, as expected from the Raman spectra, there is no shift in V_{th} with sulfur treatment. This reveals that the high off-state current is due to the carbonaceous impurities on the surface of the nanotube, which provide conduction paths between source and drain, and the impurities were selectively removed from the nanotube through the sulfidation reaction [96].

The purification of CNTs by sulfidation can be used to remove carbonaceous impurities that may remain after initial purifying procedures by other methods, and be more useful for nanotubes integrated in devices for fundamental characterizations or industrial applications.

5. SUMMARY

Electronic and transport properties of carbon nanotubes have been discussed theoretically and experimentally for ambipolar behavior on CNT-FET with intrinsic CNT and carrier suppression by molecular adsorption where oxygen molecules act as a main factor for p-type behavior with suppression of electron current. Although the molecular adsorption can be a method for choosing the carrier type of the actual CNT-FET by controlling the CNT electronic structure, as discussed in this article, seeking the more stable methods is the future challenging issue for CNT-FETs to be acceptable for integration and application.

Single wall carbon nanotube with narrow windows of diameter and high purity is essential for the CNT electronics. The sono-chemical route is an efficient method for nanotube synthesis for this point. On the other hand, direct growth of CNT on substrate makes easier way to position the individual CNTs for nanotube arrays. Remote-PECVD technique allows achieving the low temperature growth of single wall carbon nanotube as low as 450 °C so far. Water assisted remote-PECVD is effective for the CNT growth with defect reduction and density increase. The catalytic photo-resist is employed for growing the individual CNT on the pre-defined area with large scale capability. Carbonaceous impurities followed by CVD growth frequently make the undesired electrical leakage paths on the CNT channel. Sulfur based purification is effective to remove this impurities without degradation of CNT electrical characteristics.

ACKNOWLEDGEMENT

The authors would like to thank Dr. H. Kosina for modeling of CNTFET, Dr, N. Park for *ab initio* calculation, Dr. J. Park for HRTEM analysis, and Dr. S. Jeong, Dr. D. Kang and E. Bae for SWNT growth and transport measurements.

This work was supported by TND (Tera-level Nano Devices), a Korean National Research Program

REFERENCES

[1] S. Iijima, *Nature* **354**, 56 (1991).

[2] Javey, H. Kim, M. Brink, Q. Wang, A. Ural, J Guo, P. Mcintyre, P. Mceuen, M. Lundsrtrom, and H. Dai, *Nature Materials* **1**, 241 (2002).

[3] S. Frank, P. Poncharal, Z. Wang, and W. Heer, *Science* **280**, 1744 (1998)..

[4] Dekker, *Phys. Today* **5**, 22 (1999)..

[5] Y. Ando, X. Xhao, T. Sugai, and M. Kumar, *Materials Today* (2004, Oct. 22).

[6] M. Meyyappan, L. Delzeit, A. Cassell, and D. Hash, *Plasma Source Sci. Technol.* **12**, 205 (2003).

[7] R. Krupke and F. Hennrich, *Adv. Func. Mater*. **7**, 111 (2005).

[8] R. Saito, G. Dresselhaus, and M. S. Dresselhaus, *Physical Properties of Carbon Nanotube, (Imperial College Press, London*, 1998).

[9] S. Heinze, J. Tersoff, R. Martel, V. Derycke, J. Appenzeller, and P. Avouris, *Phys. Rev. Lett.* **89**, 106801 (2002).

[10] M. Kim, T. Brintlinger, E. Cobas, H. Zheng, M. Fuhrer, Z.Yu, R. Droopad, J. Ramdani, and K. Eisenbeiser, *Appl.Phys.Lett*. **84**, 1946 (2004).

[11] Javey, J. Guo, Q. Wang, M. Lundstrom, and H. Dai, *Nature* **424**, 654 (2003).

[12] R. Martel, V. Derycke, C. Lavoie, J. Appenzeller, K. Chan, J. Tersoff, and P. Avouris, *Phys. Rev. Lett.* **87**, 256805 (2001).

[13] J. Appenzeller, J. Knoch, V. Derycke, R. Martel, S.Wind, and P. Avouris, *Phys. Rev. Lett.* **89**, 126801 (2002).

[14] M. Radosavljevic, S. Heinze, J. Tersoff, and P. Avouris, *Appl. Phys. Lett.* **83**, 2435 (2003).

[15] S. Heinze, J. Tersoff, and P. Avouris, *Appl. Phys. Lett.* **83**, 5038 (2003).

[16] M. Pourfath, E. Ungersboeck, A. Gehring, B.H. Cheong, W. Park, H. Kosina, and S. Selberherr, *34th European Solid-State Device Research Conference (2004).*

[17] M. Pourfath, A. Gehring, E. Ungersboeck, H. Kosina, S. Selberherr, B. H. Cheong, and W. J. Park, J. *Appl. Phys.* **97**, 106130 (2005).
[18] P. G. Collins, K. Bradley, M. Ishigami, and A. Zettle, *Science* **287**, 1801 (2000).
[19] G. U. Sumanasekera, C. K. W. Adu, S. Fang, and P. C. Eklund, *Phys. Rev. Lett.* **85**, 1096 (2000).
[20] S.-H. Jhi, S. G. Louie, and M. L. Cohen, *Phys. Rev. Lett.* **85**, 1710 (2000).
[21] J. Jhao, A. Buldum, J. Han, and J. P. Lu, *Nanotechnology* **13**, 195 (2002).
[22] Kang, N. Park, E. Bae, J. Ko, J. Kim, and W. Park, *Appl. Phys. Lett.* **86**, 093105 *(2005).*
[23] V. Derycke, R. Martel, J. Appenzeller, and Ph. Avouris, *Appl. Phys. Lett.* **80**, 2773 (2002).
[24] J. Kong, H. T. Soh, A. M. Cassell, C. F. Quate, and H.Dai, *Nature* **395**, 878 (1998).
[25] T. Shimada, T. Sugai, Y. Ohno, K. Kishimoto, T. Mizutani, H. Yoshida, T. Okazaki, and H. Shinohara, *Appl. Phys. Lett.* **84**, 2412 (2004).
[26] W. J. Wildoer, L. G. Venema, A. G. Rinzler, R. E. Smalley, and C. Dekker, *Nature* **391**, 59 (1998).
[27] S. Dag, O. Gulsere, T. Yildirim, and S. Ciraci, *Phys. Rev.* **B 67**, 165424 (2003).
[28] P. Giannozzi, R. Car and G.. Scoles, *J. Chem. Phys.* **118**, 1003 (2003).
[29] P. Hohenberg and W. Kohn, *Phys. Rev.* **136**, B864 (1964).
[30] W. Kohn and L. J. Sham, *Phys. Rev.* **140**, A1133 (1965).
[31] G. Kresse and J. Furthmüller, *Phys. Rev.* **B 54**, 11169 (1996).
[32] J. P. Perdew, K. Burke, and M. Ernzerhof, *Phys. Rev. Lett.* **77**, 3865 (1996).
[33] G. Kresse and D. Joubert, *Phys. Rev.* **B 59**, 1758 (1999).
[34] J. Ihm, A. Zunger, and M. L. Cohen, *J. Phys.* **C 12**, 4409 (1979).
[35] Kang, N. Park, J. Ko, E. Bae, and W. Park, *Nanotechnology* **16**, 1048 (2005)
[36] S. G. Rao, L. Huang, W. Setyawan, and S. Hong, *Nature* **425**, 36 (2003).
[37] Journet, W. K. Maser, P. Bemier, A. Loiseau, M. L. Chapelle, S. Lefrant, P. Deniard, R. Lee, and J. E. Fischer, *Nature* **388**, 756 (1997).
[38] T. Guo, P. Nikolaev, A. Thess, D. T. Colbert, and R. E. Smally, *Chem. Phys. Lett.*, **243**, 49 (1995).
[39] P. Nikolaev, M. J. Bronikowski, R. K. Bradley, F. Rohmund, D. T. Colbert, K. A. Smith, and Smalley, *R.E. Chem. Phys. Lett.* **313**, 91 (1999).
[40] H. Dai, *Acc. Chem. Res.* **35**, 1035 (2002).
[41] J. Kong, H. T. Soh, A. M. Cassell, C. F. Quate, and H.Dai, *Nature* **395**, 878 (1998).
[42] M. Cassell, N. R. Franklin, T. W. Tombler, E. M. Chan, J. Han, and Dai, H., *J. Am. Chem. Soc.* **121**, 7975 (1999).
[43] W. B. Choi, D. S. Chung, J. H. Kang, H. Y. Kim, Y. W. Jin, I. T. Han, Y. H. Lee, J. E. Jung, N. S. Lee, G. S. Park, and J. M. Kim, *Appl. Phys. Lett.* **75**, 3129 (1999).
[44] J. Bae, W. B. Choi, K. S. Jeong, J. W. Chu, G. S. Park, S. Song, and I. K. Yoo, *Adv. Mater.* **14**, 277 (2002).
[45] S. Huang, L. Dai, and A. H. M. Mau, *Adv. Mater.* **14**, 1140 (2002).
[46] K. S. Suslick, MRS *Bull.* **20**, 29 (1995).
[47] S. H. Jeong, J. H. Koh, J. B. Park, and W. Park, *J. Am. Chem. Soc.* **126**, 15982 (2004).
[48] R. Andrews, D. Jacques, A. M. Rao, F. Derbyshire, D. Qian, X. Fan, E. C. Dickey, and C. Chen, J. Chem. *Phys. Lett.* **303**, 467 (1999).
[49] S. J. Doktycz and K. S. Suslick, *Science* **247**, 1067 (1999).

[50] G. Rinzler, J. Liu, H. Dai, P. Nikolaev, C. B. Huffman, F. J. Rodriguez-Macias, P. J. Boul, A. H. Lu, D. Heymann, D. T. Colbert, R. S. Lee, J. E. Fischer, A. M. Rao, P. C. Eklund, and R. E. Smalley, *Appl. Phys.* **A 67**, 29 (1998).
[51] R. Pfeiffer, H. Kuzmany, C. Kramberger, C. Schaman, T. Pichler, H. Kataura, Y. Achiba, J. Kürti, and V. Zólyomi, *Phys. Rev. Lett.* **90**, 225501 (2003).
[52] S. Lefrant, I. Baltog, M. Baibarac, J. Schreiber, and O. Chauvet, *Phys. Rev.* **B 65,** 235401 (2002).
[53] L. Delzeit, I. MaAninch, B. A. Cruden, D. Hash, B. Chen, J. Han, and M. Meyyappan, J. *Appl. Phys.* **91**, 6027 (2002).
[54] J. I. B. Wilson, N. Scheerbaum, S. Karim, N. Polwart, P. John, Y. Fan, and A. G. Fitzgerald, *Diam. Relat. Mater.* **11**, 918 (2002).
[55] T. Kato, G. H. Jeong, T. Hirata, R. Hatakeyama, K. Tohji, and K. Motomiya, *Chem. Phys. Lett.* **381**, 422 (2003).
[56] Y. Li, D. Mann, M. Rolandi, W. Kim, A. Ural, S. Hung, A. Javey, J. Cao, D. Wang, E. Yenilmez, Q. Wang, J. F. Gibbons, Y. Nishi, and H. Dai, *Nano Lett.* **4**, 317 (2004).
[57] Y. Li, S. Peng, D. Mann, J. Cao, R. Tu, K. J. Cho, and H. Dai, *J. Phys. Chem.* **B 109**, 6968 (2005).
[58] E. J. Bae, Y. S. Min, D. Kang, J. H. Ko, and W. Park, *Chem. Mater.*17, 5141 (2005).
[59] Y. S. Min, E. J. Bae, B. S. Oh, D. Kang, and W. Park, *J. Am Chem. Soc.***127**, 12498 (2005).
[60] M. S. Dresselhaus, G. Dresselhaus, A. Jorio, A. G. Souza Filho, and R. Saito, R., *Carbon* **40**, 2043 (2002).
[61] K. Song, Y. S. Cho, G. S. Choi, J. B. Park, and D. J. Kim, *Diam. Relat. Mater.* **13**, 1210 (2004).
[62] J. Wei, B. Jiang, X. Zhang, H. Zhu, and D. Wu, *Chem. Phys. Lett.* **376**, 753 (2003).
[63] K. Tohji, T. Goto, H. Takahashi, Y. Shinoda, N. Shimizu, B. Jeyadevan, and I. Matsuoka, *Nature* **383**, 679 (1996).
[64] J. M. C. Moreno and M. Yoshimura, *J. Am. Chem.* Soc. **123**, 741 (2001).
[65] Z. Kang, E. Wang, L. Gao, S. Lian, M. Jiang, C. Hu, and L. Xu, *J. Am. Chem. Soc.* **125**, 13652 (2003).
[66] Kataura, Y. Kumazawa, Y. Maniwa, I. Umezu, S. Suzuki, Y. Ohtsuka, and Y. Achiba, *Synth. Met.* **103**, 2555 (1999).
[67] E. S. Snow, J. P Novak, P. M. Campbell, and D. Park, *Appl. Phys. Lett.* 82, 2145 (2003).
[68] W. Q. Deng, X. Xu, and W. A. Goddard, *Nano Lett.* **4**, 2331 (2004).
[69] Ago, K. Murata, M. Yumura, J. Yotani, and S. Uemura, *Appl. Phys. Lett.* **82**, 811 (2003).
[70] Hinderling, Y. Keles, T. Stockli, H. F. Knapp, T. Arcos, P. Oelhafen, I. Korczagin, M. A. Hempenius, J. Vancso, R. Pugin, and H. Heinzelmann, *Adv. Mater.* **16**, 876 (2004).
[71] M. Ishida, H. Hongo, F. Nihey, and Y. Ochiai, Jpn. *J. Appl. Phys.* **43**, L1356 (2004).
[72] Y. S. Min, E. J. Bae, J. B. Park, and W. Park, *Nanotechnology* **17**, 116 (2006).
[73] G. Wilkinson, M. Rosenblum, M. C. Whiting, and R. B. Woodward, *J. Am. Chem. Soc.* **74**, 2125 (1952).
[74] W. Zhu, C. L. Xu, D. H. Wu, B. Q. Wei, R. Vajtai, and P. M. Ajayan, *Science* **296**, 884 (2002).
[75] J Tan, K. J. Klabunde, and P. M. A. Sherwood, *Chem. Mater.* **2**, 186 (1990).

[76] D. Welsh and P. M. A. Sherwood, *Phys. Rev.* **B 40**, 6386 (1989).
[77] N. Shabanova and V. A. Trapeznikov, *J. Electron Spectrosc. Relat. Phenom.* **6**, 297 (1975).
[78] S. M. Bachilo, M. S. Strano, C. Kittrell, R. H. Hauge, R. E. Smalley, and R. B. Weisman, *Science* **298**, 2361(2002).
[79] R. B. Weisman and S. M. Bachilo, *Nano Lett.* **3**, 1235 (2003).
[80] Jorio, R. Saito, J. H. Hafner, C. M. Lieber, M. Hunter, T. McClure, G. Dresselhaus, and M. S. Dresselhaus, *Phys. Rev. Lett.* **86**, 1118 (2001).
[81] Y. Homma, Y. Kobayashi, T. Ogino, D. Takagi, R. Ito, Y. Jung, and P. M. Ajayan, J. *Phys. Chem.* **B 107**, 12161 (2003).
[82] H. Hou, A. K. Schaper, Z. Jun, F. Weller, and A. Greiner, *Chem. Mater.* 15, 580 (2003).
[83] N. Y. Jin-Phillipp and M. Ruhle, *Phys. Rev.* **B 70,** 245421 (2004).
[84] Y. J. Jung, B. Wei, R. Vajtai, P. M. Ajayan, Y. Homma, K. Prabhakaran, and T. Ogino, *Nano Lett.* **3**, 561 (2003).
[85] Y. J. Jung, Y. Homma, T. Ogino, Y. Kobayashi, D. Takagi, B. Wei, R. Vajtai, and P. M. Ajayan, *J. Phys. Chem.* **B 107**, 6859 (2003).
[86] B. Farmer and R. G. Gordon, Electrochem. *Solid State Lett.* **8**, G89 (2005).
[87] T. W. Ebbesen, P. M. Ajayan, H. Hiura, and K. Tanigaki, *Nature* **367**, 519 (1994).
[88] T. W. Ebbesen, In Carbon Nanotubes: Preparation and Properties, edited by T. W. Ebbesen (*CRC, Boca Raton,* 1996), *Chapter 4.*
[89] T. Colbert, J. Zhang, S. M. McClure, P. Nikolaev, Z. Chen, J. H. Hafner, D. W. Owens, P. G. Kotula, C. B. Carter, J. H. Weaver, A. G. Rinzler, and R. E. Smalley, *Science* **266**, 1218 (1994).
[90] T. Jeong, W. Y. Kim, and Y. B. Hahn, *Chem. Phys. Lett.* **344**, 18 (2001).
[91] H. Hiura, T. W. Ebbesen, and K. Tanigaki, *Adv. Mater.* **7**, 275 (1995).
[92] M. Bonard, T. Stora, J. P. Salvetat, F. Maier, T. Stockli, C. Duschl, L. Forro, W. A. Heer, and A. Chatelain, *Adv. Mater.* **9,** 827 (1997).
[93] Y. Q. Xu, H. Peng, R. H. Hauge, and R. E. Smalley, *Nano Lett.* **5**, 163 (2005).
[94] A. Lowenheim and M. K. Moran, *Keyes and Clark's Industrial Chemicals (Wiley-Interscience, New York,* 1978).
[95] R. Lide, CRC *Handbook of Chemistry and Physics (CRC, Boca Raton, 2004).*
[96] Y. S. Min, E. J. Bae, and W. Park, *J. Am. Chem. Soc.* **127**, 8300 (2005)

In: Encyclopedia of Nanotechnology
Editor: Elwood D. Carlson
ISBN 978-1-60692-079-4

Chapter 28

FIELD ENHANCED THERMAL CONDUCTIVITY IN HEAT TRANSFER NANOFLUIDS WITH MAGNETIC COATING ON CARBON NANOTUBES

Brian Wright , Dustin Thomas, Haiping Hong,
Department of Material and Metallurgical Engineering,
South Dakota School of Mines and Technology, Rapid City, SD 57701, USA
Lori Groven, Jan Puszynski,
Department of Chemical and Biological Engineering,
South Dakota School of Mines and Technology, Rapid City, SD 57701, USA
Edward Duke,
Department of Geology and Geological Engineering,
South Dakota School of Mines and Technology, Rapid City, SD 57701, USA
Xiangrong Ye , Sungho Jin,
Department of Mechanical & Aerospace Engineering, University of California,
San Diego, La Jolla, CA 92093, USA
Pauline Smith, Walter Roy
Army Research Lab, Aberdeen Proving Ground, MD 21005, USA

ABSTRACT

In this paper, we report that the thermal conductivity (TC) of heat transfer nanofluids containing Ni coated single wall carbon nanotube (SWNT) can be enhanced by applied magnetic field. A reasonable explanation for these interesting results is that Ni coated nanotubes form aligned chains under applied magnetic field, which improves thermal conductivity via increased contacts. On longer holding in magnetic field, the nanotubes gradually move and form large clumps of nanotubes, which eventually decreases the TC. When we reduce the magnetic field strength and maintain a smaller field right after TC reaches the maximum, the TC value can be kept longer compared to without magnetic field. We attribute gradual magnetic clumping to the gradual cause of the TC decrease in the magnetic field. We also found that the time to reach the maximum peak value of TC is increased as the applied magnetic field is reduced. Scanning Electron Microscopy (SEM) images show that the Ni coated nantubes are aligned well under the influence of a magnetic field. Transmission Electron Microscopy (TEM) images indicate that nickel remains attached onto the nanotubes after the magnetic field exposure.

Keywords: Thermal conductivity, Carbon nanotube, Ni coated, Magnetic field.

1. Introduction

The discovery of carbon nanotubes (CNT) has instigated tremendous research efforts in recent years, due to their promising thermal, electrical, mechanical, and functional properties. For example, it has been reported that single wall carbon nanotubes exhibit a thermal conductivity (TC) value as high as 2000 – 6000 W/m.K.[1] under an ideal circumstance. By contrast typical heat transfer fluids like water and oil have TC values of only 0.6W/m.K. and 0.2 W/m.K. respectively.

We prepared a composite fluid containing high-TC materials of CNTs and magnetic-field-responsive magnetic nanoparticles with an aim to increase the thermal conductivity of the fluid. Such fluids with enhanced thermal conductivity are useful for a variety of applications such as heat transfer coolants and lubricants. [2,3] The fluids containing carbon nanotubes (called "nanofluids" hereafter) should exhibit substantially improved TC values. [4-6] However, a simple composite structure turned out to be not effective in enhancing the TC of nanofluids. At low nanotube percentage loading, no significant improvements in TC were reported. At the loading of 1 vol% CNTs (~1.4wt %), there is about 10-20% TC increase reported by different groups.[7,8] However, at such a high concentration, the fluid already lost most of its fluidity and became mud-like, thus making the fluid much less useful for coolant & lubricant applications.

A possible explanation for this lack of sufficient TC increase through compositing with carbon nanotubes may be attributed to a lack of alignment and orientation in the fluids when the carbon nanotubes are irregularly positioned in the fluids with only a random and infrequent chance for them to contact each other. Therefore, only very high concentrations of CNTs produce noticeable TC improvements. Therefore, there is a need to introduce nanotube-nanotube physical contacts in the fluids in order to increase the thermal (and electrical) conductivity significantly, while maintaining the nanotube concentration relatively low to ensure a desirably low fluid viscosity.

We introduce a new concept of incorporating magnetically sensitive metal or metal oxide nanoparticles in a carbon-nanotube-containing fluid. Magnetic particles in a liquid medium can have a variety of configurations depending on the nature of magnetic particles and the field strength.[9-10] Under a relatively strong magnetic field, small magnetic particles form connected networks and also tend to get somewhat oriented toward the field direction, which also moves the carbon nanotubes nearby and induce more physical contacts. Thus improved thermal conductivity is anticipated. We incorporated carbon nanotubes and magnetic-field-sensitive nanoparticle Fe_2O_3 into water under a magnetic field and the results confirmed this assumption. [11]

Based on the above results, we estimated that the metal encapsulated carbon nanotubes [12] dispersed in water, water/ethylene glycol or oil with the help of appropriate chemical surfactant, could also significantly increase the TC value under a magnetic field if the coating metal is magnetic. The coated layer configuration of the magnetic material could be more beneficial than the presence of isolated magnetic nanoparticles as small magnetic particles tend to be superparamagnetic with a weak response to applied magnetic field.

In this paper, we investigated nanofluid systems containing 0.01 and 0.02 wt% of Ni coated, magnetic nanotubes in water under different magnetic field strengths. Understanding of thermal conductivity behavior of such a composite material system in magnetic field would be valuable in analysis and synthesis of nanofluids.

2. EXPERIMENTAL

Single wall carbon nanotubes (SWNT) were purchased from Helix Material Solutions Inc (Helix, Richardson, Texas). Chemical surfactant sodium dodecylbenzene sulfonate (SDBS) was purchased from Sigma Aldrich.

Ni coated SWNTs were prepared using the following methodology:

1. Oxidation: HNO_3: H_2SO_4 (3:1) reflux at 110 °C for 6 hrs.
2. Sensitization and Activation: Immersion in 0.1 M $SnCl_2$-0.1 M HCl for 30 min, immersion in 0.0014 M $PdCl_2$-0.25 M HCl for 30 min.
3. Electroless Plating: Introduction into electroless bath for 25 min with composition as listed in Table 1.

The Ni coated SWNTs were put into water together with the appropriate amount of chemical surfactant sodium dodecylbenzene sulfonate (SDBS). Sonication was performed using a Branson Digital Sonifier, model 450. Magnetic field was provided by a pair of spaced apart Ba-ferrite magnet plates (4 x 6 x 1 inch dimension) and placing the sample in the middle of the gap between the magnets.

Table 1. Electroless bath conditions

Chemical	Concentration (g/L)
$NiSO_4 \cdot 6H_2O$	25
$Na_3C_6H_5O_7 \cdot 2H_2O$	5
$NaH_2PO_2 \cdot H_2O$	15
NH_4Cl	60
$Pb(NO_3)_2$	2.5
Bath temperature	20°C
pH (adjusted w/ NH_4OH)	~8.0

The microstructure was taken using an Olympus VANOX-T microscope. The magnification was 500X. Scanning electron microscopy (SEM) images were acquired using the backscattered electron detector on a Zeiss Supra40VP variable pressure system. Transmission electron microscopy (TEM) images were acquired with a Hitachi H-7000 FA.

The pH values of the fluid containing the Ni coated SWNTs were measured using a Denver instrument UP-10 pH/mV meter. The thermal conductivity data was obtained by the Hot Disk™ thermal constants analyzer [13], using the following parameters: measurement depth 6 mm, room temperature, power 0.012W, measurement time 15s, sensor radius 3.189mm, TCR 0.0471/K, disk type kapton, temperature drift rec yes. The magnetic field intensity was recorded by F.W. Bell Gaussmeter Model 5060.

Dispersion stability of the Ni coated nanotubes in fluid was observed with visual inspection.. We put the nanofluids in a see-through glass beaker and observed if there was any precipitation at the edge and/or bottom of the glass beaker.

3. Results and Discussion

Figure 1 shows the optical microstructure of 0.01wt % of Ni-coated carbon nanotube in DI water. Because of the heavy Ni, 0.01 wt % of Ni-free nanotubes will contain several times larger number of nanotubes than 0.01 wt % of Ni-coated nanotubes. Relatively large dark spots which may be attributed to the agglomerated nanotubes or other particles can barely be observed. This indicates that the Ni coated nanotubes are dispersed quite well in the water with the help of the chemical surfactant used.

The thermal conductivity (TC) versus time plot for various concentrations of Ni coated nanotubes (0.01 wt% and 0.02 wt%) in magnetic fields (two different intensities) are shown in Figure 2. Without a magnetic field, the TC value of Ni coated SWNT loaded nanofluids is around 0.63 ~ 0.64 W/m.K. and stays mostly constant versus the time. Because of very few contacts between Ni coated carbon nanotubes, the TC value is essentially the value for the DI water itself. Figure 2a shows the TC versus time at a 51mm gap separation between the two magnets (0.62 kOe magnetic field) with two concentrations (0.01 wt% and 0.02 wt%), and Figure 2b shows the TC versus time at a 91mm gap (0.38 kOe) with the same concentrations. Under the magnetic field, thermal conductivity shows very interesting behavior. TC initially increases with time but eventually reaches a peak. On longer time holding in magnetic field, the Ni coated nanotubes gradually move and form bigger clumps of CNTs, thus a decrease in the TC. The maximum TC value is around of 1.10 W/m.K. in 51mm gap (0.62 kOe) at 0.02 wt% and 1.01 W/m.K. in 91mm gap (0.38 kOe) at 0.02wt%, about 75% and 60% higher than the value of DI water. For both the 51mm gap and the 91mm gap, the TC of 0.02 wt% nanofluids is obviously higher than that of 0.01 wt% sample, and the peak of maximum TC is slightly left-shifted versus time.

These results strongly support our hypothesis that a good alignment and orientation of conducting materials in a fluid are critical and essential to the enhancement of TC of the composite fluid. However, in the case of nano particle containing composite fluid, the increase ratio of TC is not as significant due to the thermal contact resistance in the nanofluids [14].

Figure 1. Optical microscope picture of 0.01wt % of Ni coated carbon nanotubes in DI water.

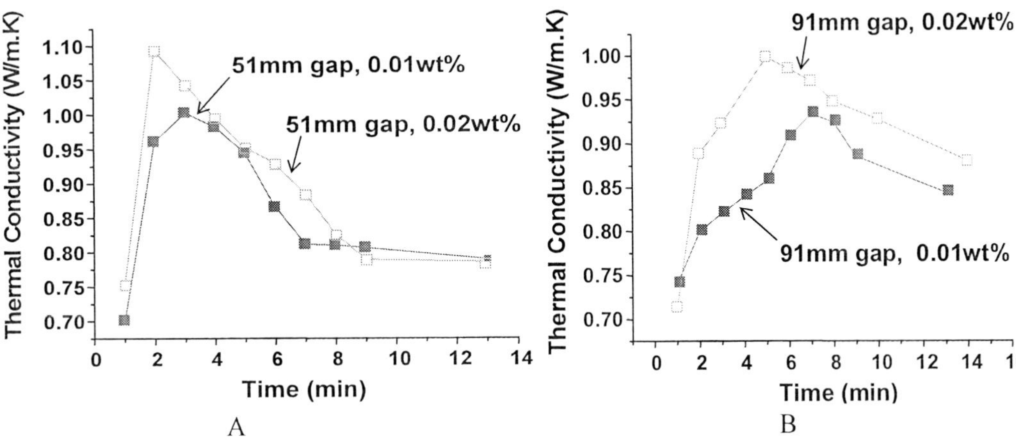

Figure 2. Thermal conductivity versus time in different concentrations at two different magnetic field intensities. A) 51mm gap (0.62 kOe) . B) 91mm gap (0.38 kOe).

The next step was to determine what happens to the TC value of nanofluids containing Ni coated single wall carbon nanotube if the external magnetic field is reduced and maintained at a much smaller field once the TC reaches the maximum. Figure 3 is a plot of thermal conductivity versus time in a nanofluid containing 0.01wt% Ni coated SWNTs.. A) after placing the nanofluid in a 0.62 kOe magnetic field for 2-3 minutes to reach the maximum TC, the magnetic field was reduced to 0.18 kOe and maintained. B) after placing the nanofluid in a 0.62 kOe magnetic field for 2-3 minutes to reach the maximum TC, removed the magnetic field completely. C) placed the nanofluid in a 0.18 kOe magnetic field all the time. It is clearly seen from Figure 3 that the TC of nanofluids under somewhat reduced magnetic field after TC reaches the maximum is higher than that for removed magnetic field after TC reaches the maximum. The results indicate that the gradual magnetic clumping was the cause of the TC decrease. To better understand this, measurements of TC versus time in a constant weak magnetic field (0.18 kOe) were conducted as for the sample C in Figure. 3, the result of which was a relative low TC value. Further optimization or time-dependent programming of the applied magnetic field is likely to yield more desirable thermal conductivity as a function of time.

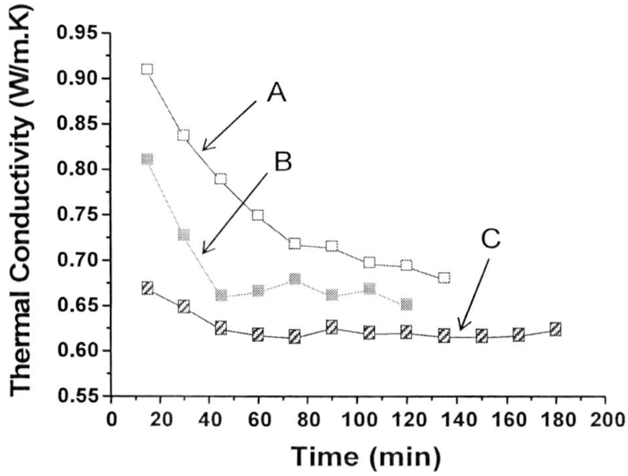

Figure 3. Thermal conductivity versus times in 0.01wt% Ni-coated SWNT loading nanofluids. A) after putting fluids in 0.62 kOe magnetic field for 2-3 mins to reach the maximum TC, reduce the magnetic field to 0.18 kOe and maintain. B) after putting fluids in 0.62 kOe magnetic field for 2-3 minutes to reach the maximum TC, withdraw the magnetic field. C) put the fluids in 0.18 kOe magnetic field all the time.

It is interesting to note that if one extrapolates the curve A and B to the zero time, two curves would come to the similar starting point, around 1.00 W/m.K. Actually this more or less coincides with the maximum TC value of the nanofluid under 0.62 kOe magnetic field.

Table 2. Magnetic field dependence of the time to reach peak TC and the value of TC maximum at the concentration of 0.02 wt% Ni coated SWNT

Distance between two magnets (mm)	Magnetic field (kOe)	Time to reach TC maximum (min)	TC maximum value (W/m.K)
35	0.86	<2	N/A
51	0.62	2	1.10
91	0.38	5	1.01
141	0.18	7	0.95

Also, it is interesting to note that different magnetic field strength influences the maximum TC value, the time to reach the maximum, and the shape of the TC vs time curve. Figure 4 shows thermal conductivity versus time at different magnetic field strengths (as adjusted by changing the distance between the pair of magnets as shown in Table 2). , A) 51mm gap. B) 91mm gap. C) 141mm gap. It is seen that the shorter the gap distance (i.e., the higher the applied magnetic field is), the shorter the time to reach the maximum TC. At 35mm gap, it is impossible to observe the maximum TC value. This is as anticipated since the shorter distance means a stronger magnetic field, which enables to very quickly assemble and align the nanotubes at a rate quicker than we are able to measure the thermal conductivity with our current equipment.

The quantitative data correlating the magnetic field intensity, time to reach TC maximum and TC value at different gap distances between the two magnets are listed in Table 2.

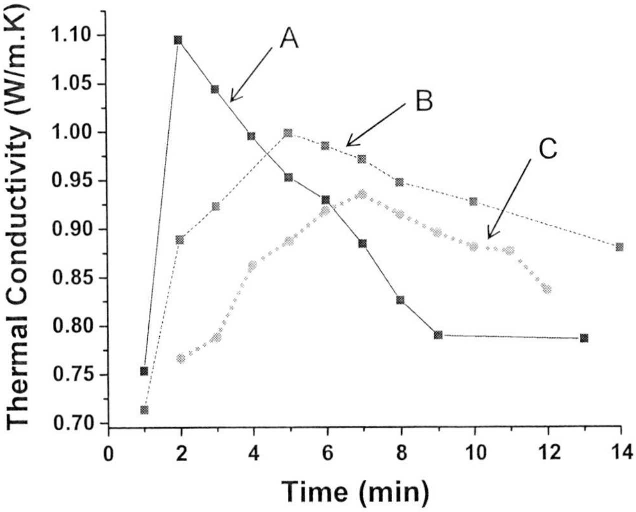

Figure 4. Thermal conductivity versus time at different gap distances between two magnets (different magnetic field intensity) at 0.02 wt% concentration. A) 51mm gap. B) 91mm gap. C) 141mm gap.

From Figure 5, it is clearly seen that under the magnetic field, the nanotubes align well in the magnetic field direction due to the magnetic moment of the Ni coating. The red arrow represents the magnetic field direction. The black material represents the carbon nanotube. Small bright particles are images from some portion of the coated Ni metal. Since Ni is much heavier than carbon, Ni appears brighter in the backscattered electron image. Energy-dispersive X-ray analysis showed that the bright particles contain Ni element.

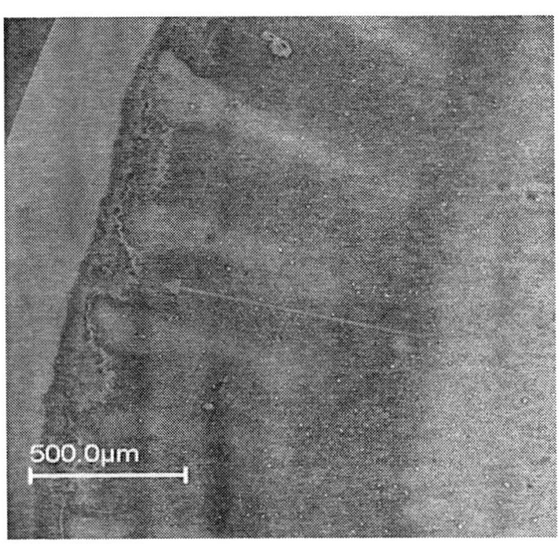

Figure 5. Backscattered electron SEM image of 0.01wt% Ni coated single wall carbon nanotubes aligned with magnetic field. Scale bar is 500μm. Red arrow is magnetic field direction.

To prepare the SEM sample, drops of nanofluids containing Ni coated nanotubes were placed on the SEM sample holder (usually made by Al) and left to dry under the magnetic field, see Figure. 5. Since long time exposure to a magnetic field will reduce the TC value, the samples were only placed under a magnetic field for a short time (2-4 mins). During this time, the samples could reach substantially improved TC values, indicating that nanotubes and nano particles align very well, this is evident from Figure 5. No significant change was observed in the thermal conductivity of the fluids after the magnetic field was removed, i.e., after ~10 minutes.

In order to make sure that all these interesting results are solely due to the Ni-coated SWNT, and not Ni particles detached from the nanotubes, several Transmission Electron Microscopy (TEM) pictures were taken. Figure 6 shows TEM images of carbon nanotube. A) As received, uncoated Ni. B) Ni coated before the experiment. C) Ni coated after the experiment. It is clearly seen that Ni is still attached to the nanotube after the experiment. All the results reflect the magnetic nature of Ni coated SWNT.

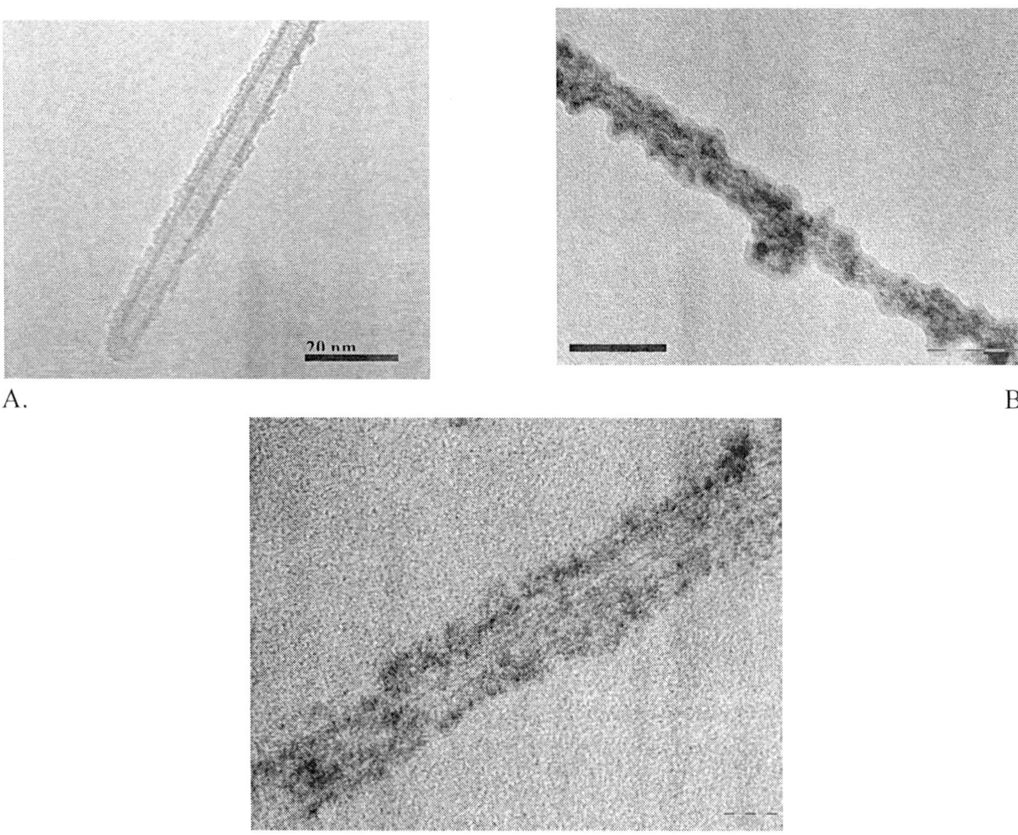

Figure. 6. TEM images of carbon nanotube. A) As received, uncoated Ni. B) Ni coated before the experiment. C) Ni coated after the experiment. Scale bar is 20nm.

4. Conclusion

In summary, we have successfully demonstrated that the thermal conductivity (TC) of the heat transfer nanofluids containing Ni-coated single wall carbon nanotubes (SWNTs) can be significantly enhanced in the presence of applied magnetic field. The explanation for these interesting results is that Ni coated nanotubes form aligned chains under an applied magnetic field, resulting in improved contacts and hence improved thermal conductivity. On longer holding in magnetic field, the nanotubes gradually move and form large clumps of nanotubes, thus decreasing the TC. Reducing the magnetic field strength and maintaining a much smaller field right after TC reaches the maximum value enables to keep the TC value longer compared to the case of complete magnetic field removal.

We also found that the time to reach the maximum peak value of TC is increased as the applied magnetic field is reduced by increasing the gap between two magnets. Scanning Electron Microscopy (SEM) pictures show that the Ni coated nantubes are aligned well under the influence of a magnetic field.

ACKNOWLEDGMENTS

H. Hong would like to thank Army Research Lab (Cooperative agreement DAAD19-02-2-0011) for financial support. Stimulating discussion with Prof. G.P. Peterson of University of Colorado, Boulder is acknowledged.

REFERENCES

[1] S. Berber, Y. K. Kwon, and D. Tománek, *Phys. Rev. Lett.* 84, 4613 (2000).

[2] H. Hong, J. Wensel, F. Liang, W. E. Billups, W. Roy, J. Thermophys. *Heat Trans.* 21, 234 (2007).

[3] P. Keblinski, J. A. Eastman, D. G. Cahill, *Materials Today*, 8(6), 36 (2005).

[4] X. Wang, X. Xu, S. Choi, J. Thermophys. *Heat Trans.* 13(4), 474 (1999).

[5] S. Choi, Z. Zhang, W. Yu, F. E. Lockwood, E. A. Grulke, *Applied Phys Letters,* 79(14), 2252 (2001).

[6] BH. Kim, GP. Peterson, *J. Thermophys. Heat Trans.* 21, 451 (2007).

[7] H. Xie, H. Lee, W. Youn, M. Choi, *Journal of Applied Physics,* 94 (8), 4967 (2003).

[8] H. Hong, J. Wensel, S. Peterson, W. Roy, *Polymeric Materials: Science & Engineering,* 95, 1076 (2006).

[9] S. Jin and M. McCormack. *J. Electronic Materials* 23, 8 (1994).

[10] X. R. Ye, C. Caraio, C. Wang, and J. B. Talbot, S. Jin, *J. Nanosci. and Nanotechnol* 6, 852 (2006).

[11] H. Hong, B. Wright, J. Wensel, S. Jin, X. Ye, W. Roy, *Synthetic Metal,* 157, 437 (2007).

[12] Y. Zhang, N. W. Franklin, R. J. Chen, and H. Dai, *Chem. Phys. Lett.* 331, 35 (2000).

[13] Detail information see www.hotdisk.se.

[14] B. Wright , D. Thomas, H. Hong , L. Groven, J. Puszynski, E. Duke, X. Ye , S. Jin, *Applied Physics Letter,* 91, 173116 (2007).

In: Encyclopedia of Nanotechnology
Editor: Elwood D. Carlson

ISBN 978-1-60692-079-4

Chapter 29

SYNTHESIS AND INVESTIGATION OF MAGNETIC PROPERTIES OF GD - SUBSTITUTED MN - ZN FERRITE NANOPARTICLES AS A POTENTIAL LOW CURIE TEMPERATURE AGENT FOR MAGNETIC FLUID HYPERTHERMIA

F. S. Bayburtskiy[1], Yu. V. Baldokhin[1], D. B. Korman[1], A. E. Chekanova[2] and O. A. Shlyakhtin[2]
[1]N. M. Emanuel Institute of Biochemical Physics, RAS, Moscow
[2]M. V. Lomonosov Moscow State University, Moscow

ABSTRACT

Gd-substituted Mn–Zn ferrite nanoparticles of different compositions were synthesized by a chemical co-precipitation method. To study the reduction of the Curie temperature (T_C) for different samples, their magnetic properties in dependence to the composition and cationic distribution were investigated. An attempt to lower the T_C of superparamagnetic particles to the optimal temperature required in magnetic fluid hyperthermia (44–47 °C) was made.

Key words: Ferrites; Nanoparticles; Magnetic hyperthermia; Curie temperature (T_C).

---

[1] N. M. Emanuel Institute of Biochemical Physics, RAS, Moscow, 119991, Kosygina street, 4, E – mail: Chembio@sky.chph.ras.ru, Bayburt@mail.ru.

[2] M. V. Lomonosov Moscow State University, Moscow, 119899, Vorobyovy gory, Department of Chemistry. E – mail: Head@inorg.chem.msu.ru, Annash@sky.chph.ras.ru

Introduction

Magnetic fluid hyperthermia (MFH) of malignant tumors is a method in cancer therapy which seems to be very prospective, especially in view of the rapid development of nanotechnology's achievements [1 – 8]. Technical problems in MFH still to be conquered include controlling the target temperature and reaching homogeneous temperature distributions. One approach, and the subject of interest in this paper, is to evaluate biocompatible superparamagnetic nanoparticles sized 6 – 12 nm in diameter and prepare them with a Curie temperature in the range of 44 – 47°C that allows self-regulated temperature control in a variable magnetic field This temperature range is the therapeutically used temperature for cancer therapy. In this way the problem of temperature control could be solved partially. Currently, all magnetic particles used in MFH are biocompatible, stabilized magnetic fluids based on Fe_3O_4 nanoparticles with a Curie temperature 585°C [1 – 8]. There is currently no suitable magnetic material for the preparation of ferrofluids with a Curie temperature ideal for MFH. But various ferrite-spinel systems with different, composition-depending properties and the simplicity of the synthesis using chemical co-precipitation method seem to be a suitable for investigation. Recently, a ferrofluid prepared for technical applications and based on Gd-substituted Mn–Zn ferrite nanoparticles was synthesized by a co-precipitation method [9 – 10] with a comparatively low T_C (100°C). There is, therefore a good chance to further develop such «advanced» ferrite nanoparticles with the desirable magnetic properties.

To approach the development of such materials, the subject of our work was to synthesize substituted ferrite nanoparticles to investigate their magnetic properties and to determine their T_C values in dependence of composition and cation distribution. The starting point was the Gd-substituted Mn–Zn ferrite system ($Mn_{1-X}Zn_X[Fe_{2-Y}Gd_Y]O_4$). In this paper, we report the magnetic data of the synthesized ferrites.

Sample Preparation

$Mn_XZn_Y[Fe_{2-Z}Gd_Z]O_4$ and $Mn_XZn_YFe_2O_4$ particles of different ionic composition (Table 1) were synthesized using chemical co-precipitation method. Fe_2O_3 dissolved in HCl, $MnCl_2$ $4H_2O$, $ZnSO_4$ $7H_2O$ and $Gd(NO_3)_3$ $6H_2O$ were used to obtain Fe^{3+}, Zn^{2+}, Mn^{2+} and Gd^{3+} ions in aqueous solutions The solutions containing these ions were mixed in an appropriate molar proportion in distilled water, filtered and added to 8 M NaOH solution at the temperature 80°C (pH of the medium 12). After stirring by the same temperature during 10 min, the obtained precipitate was washed off by adding of appropriate portions of boiling water on permanent magnet for several times, up to pH = 7. Next the samples were dried up to powder-like state.

Table 1. Composition of the different $Mn_XZn_Y[Fe_{2-Z}Gd_Z]O_4$ and $Mn_XZn_YFe_2O_4$ particles synthesized by a chemical co-precipitation method and their crystal structure and magnetic moment properties

№	Formula	a (Å)	Linear T_C (°K)	T3/2-Law T_C (°K)	Magnetic moment per particle (μB)
1	Fe_3O_4	8.394			
2	$Mn_{0.36}Zn_{0.64}Fe_2O_4$	8.480	385.4	383.2	5631.86
3	$Mn_{0.59}Zn_{0.41}Fe_2O_4$	8.402	509.3	489.5	13819.70
4	$Mn_{0.54}Zn_{0.46}[Fe_{1.90}Gd_{0.10}]O_4$	8.456	385.0	383.5	19971.50
5	$Mn_{0.86}Zn_{0.14}[Fe_{1.74}Gd_{0.26}]O_4$	8.400	379.3	377.5	8837.69
6	$Mn_{0.50}Zn_{0.50}[Fe_{1.90}Gd_{0.10}]O_4$	8.435	389.9	386.8	10007.40
7	$Mn_{0.67}Zn_{0.33}[Fe_{1.83}Gd_{0.17}]O_4$	8.439	391.8	389.7	8794.37
8	$Mn_{0.77}Zn_{0.23}[Fe_{1.72}Gd_{0.28}]O_4$	8.405	373.6	372.1	4852.07
9	$Mn_{0.53}Zn_{0.47}[Fe_{1.80}Gd_{0.20}]O_4$	8.416	364.9	363.6	5112.00
10	$Mn_{0.60}Zn_{0.40}[Fe_{1.72}Gd_{0.28}]O_4$	8.408	377.3	375.7	6758.24
11	$Mn_{0.62}Zn_{0.38}[Fe_{1.90}Gd_{0.10}]O_4$	8.454	395.8	393	13819.70
12	$Mn_{0.74}Zn_{0.26}[Fe_{1.80}Gd_{0.20}]O_4$	8.421	372.8	371.4	11523.70
13	$Mn_{0.48}Zn_{0.52}Fe_2O_4$	8.415	502.3	484.3	12780.00

The structural parameters derived from the fit are the particle size Dx = 60 Å, lattice parameter, a = 8.453 Å.

RESULTS AND DISCUSSION

Structural Characterization

The structure of the particles was characterized using a $Co_{K\alpha}$ radiation (λ = 1.789 Å DRON UM-1 X-ray diffractometer). The analysis confirms the formation of single-phase spinel structure. The structural parameters derived from the X-ray (typical X-ray for $Mn_XZn_Y[Fe_{2-Z}Gd_Z]O_4$ – Figure 2) are the lattice parameter a (Table 1) and the particle size (9 nm).

Quantitative Analysis

The cationic distribution was derived from quantitative spectrum analysis (ICP-AES) and is given in Table 1.

Transmission Electron Microscopy (TEM)

The nanosize (9 nm) of the ferrites obtained, derived from the X-ray, was also confirmed by TEM using a Philips EM (Figure 1 and Figure 2).

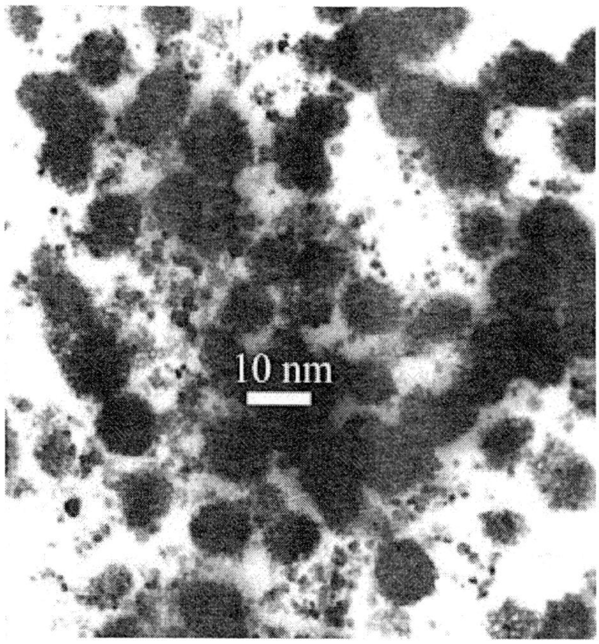

Figure 1. TEM of the $Mn_{0.62}Zn_{0.38}[Fe_{1.90}Gd_{0.10}]O_4$ sample's aqueous suspension confirmed the nanosize of the particles.

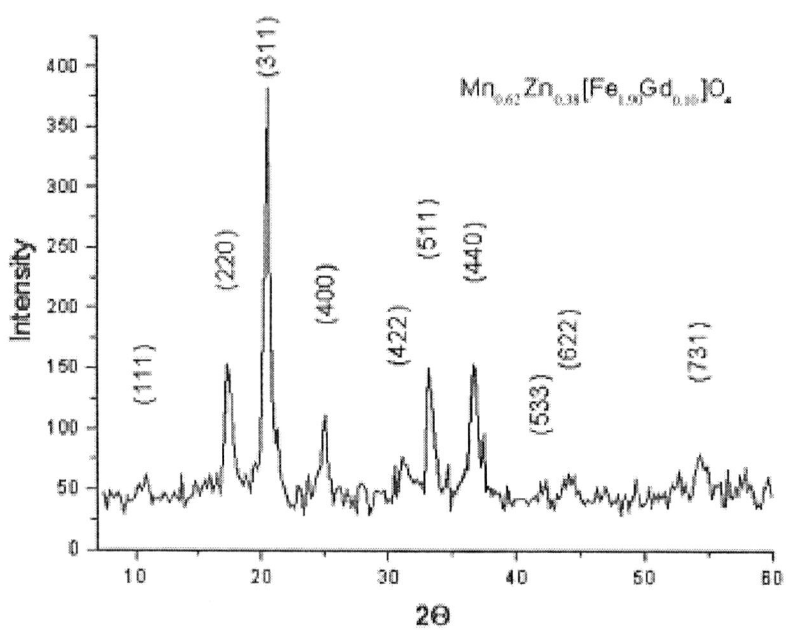

Figure 2. X-ray of (Mn0.62Zn0.38[Fe1.90Gd0.10]O4) – confirms the formation of single-phase spinel structure.

Magnetic Measurements

Temperature and field dependencies of the magnetic moment were measured with VSM PARC-155 (Princeton Applied Research, USA), SQUID and express-magnetometer. Figure 3 shows what a typical MH curve for all ferrite samples evaluated looks like, indicating their superparamagnetic behavior.

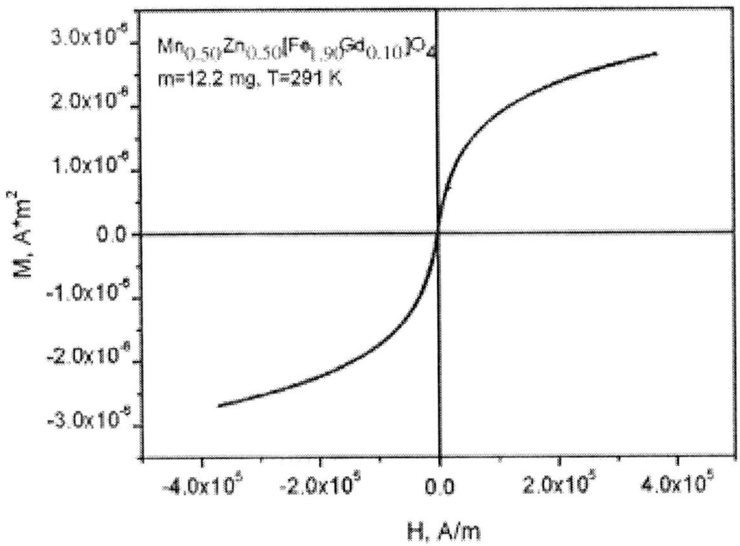

Figure 3. Typical magnetic field dependence of specific magnetic moment demonstrates the super-paramagnetic behavior of the ferrites samples synthesized $Mn_{0.50}Zn_{0.50}[Fe_{1.90}Gd_{0.10}]O_4$

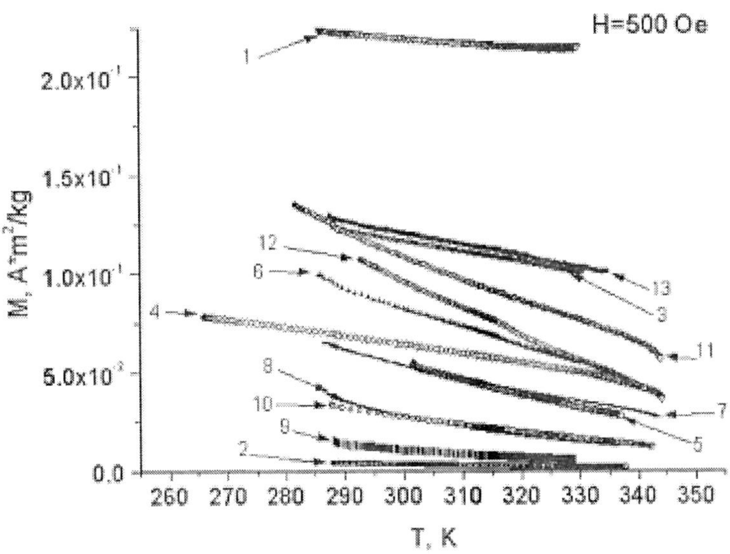

Figure 4. Temperature dependencies of specific magnetic moment for ferrite samples of different compositions.

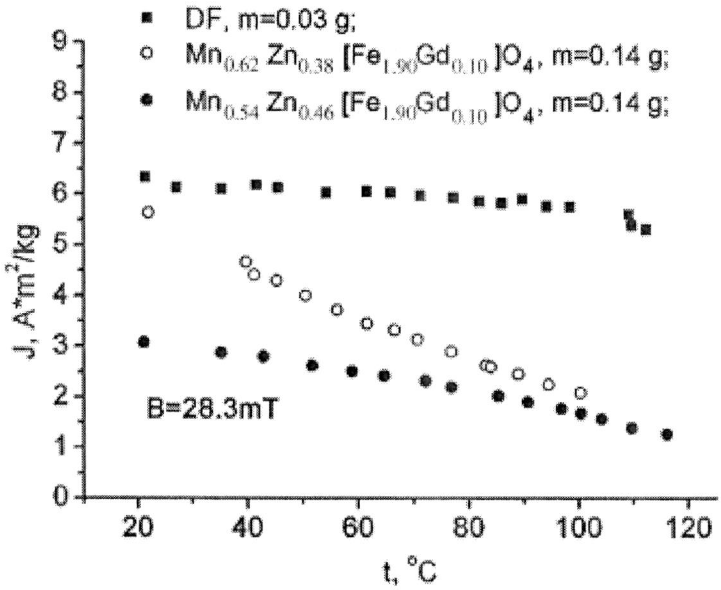

Figure 5. Temperature dependencies of specific magnetic moment for two ferrite samples compared with dextran magnetite ($Mn_{0.54}Zn_{0.46}[Fe_{1.90}Gd_{0.10}]$ O_4 and $Mn_{0.62}Zn_{0.38}[Fe_{1.90}Gd_{0.10}]O_4$).

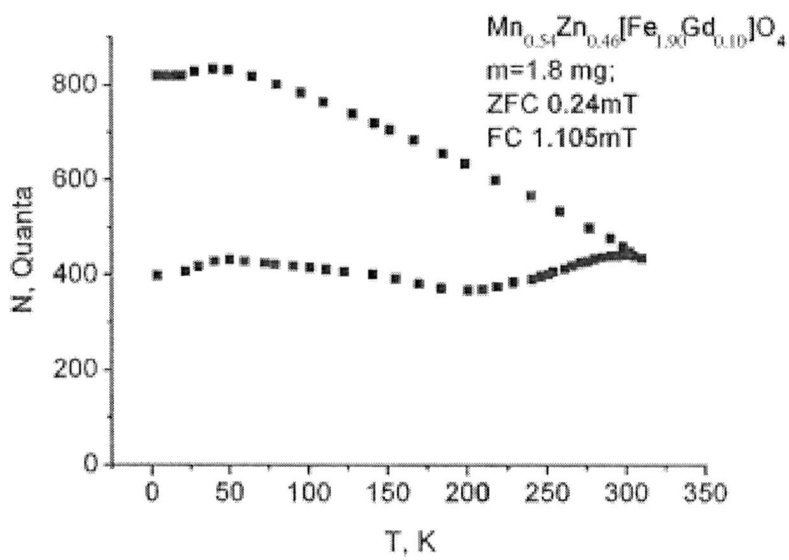

Figure 6. Temperature dependencies of magnetic moment for Mn0.54Zn0.46[Fe1.90Gd0.10]O4 in the low-temperature range.

The magnetic moment's field dependencies of all samples are in a good accordance with the Langevin's law, so the magnetic moment of the particle was calculated. The magnetic moments of the particles are presented in Table 1. The average magnetic moment is Mperpart.10 500 μB. Figure 4 shows an evolution of the temperature dependencies of specific magnetic moments for the samples of different composition at temperatures ranging from 290 to 363°K. With increasing temperature, the magnetic moment of the $Mn_{1-X}Zn_X[Fe_{2-Y}Gd_Y]O_4$

samples decreases according to linear (close to Blokh T3/2) law. The results of the linear and the Blokh's approximations are presented in Table 1. The most promising results in regards of T_C and magnetic moment are shown by samples № 11, № 12, № 6 and № 4 in Figure 4.

Figure 5 shows temperature dependencies of specific magnetic moment for two $Mn_XZn_Y[Fe_{2-Z}Gd_Z]O_4$ ferrites samples compared with the same of dextran magnetite in the range of 20–118°C.

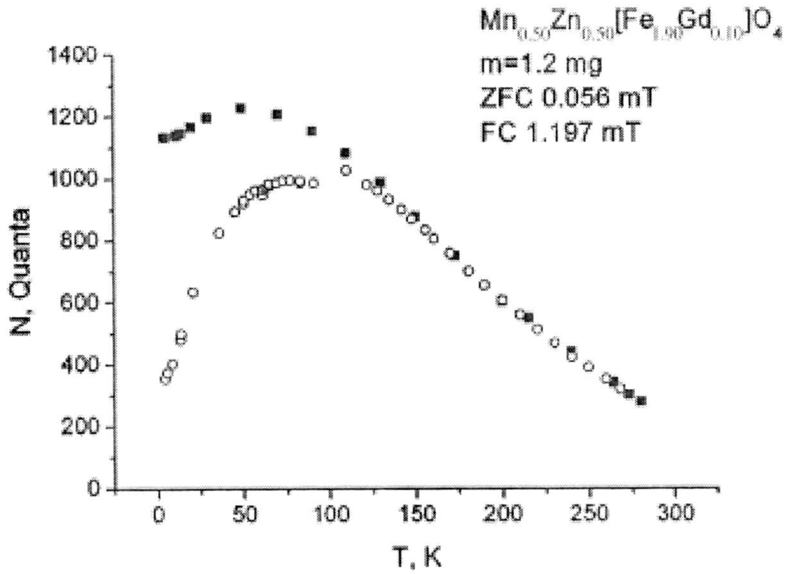

Figure 7. Temperature dependencies of magnetic moment for $Mn_{0.50}Zn_{0.50}[Fe_{1.90}Gd_{0.10}]O_4$ in the low-temperature range.

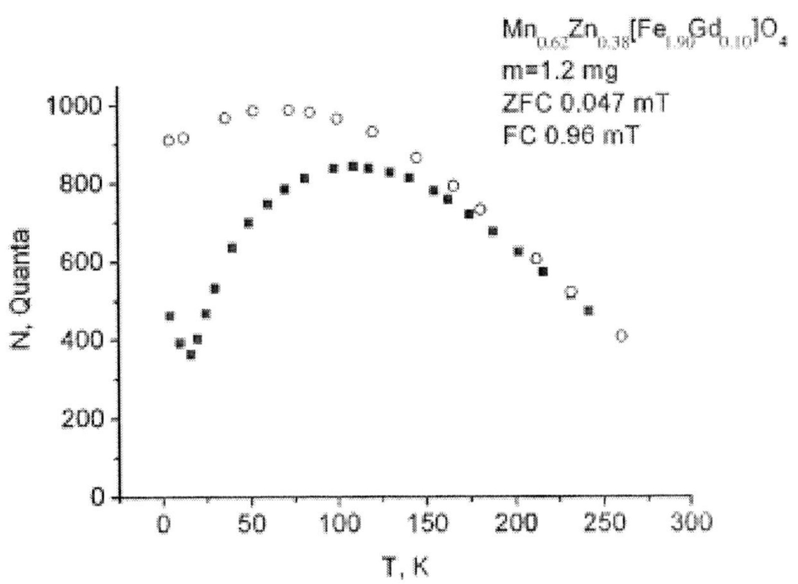

Figure 8. Temperature dependencies of magnetic moment for $Mn_{0.62}Zn_{0.38}[Fe_{1.90}Gd_{0.10}]O_4$ in the low-temperature range.

Figures 6 – 9 show temperature dependencies of magnetic moment for $Mn_XZn_Y[Fe_{2-Z}Gd_Z]O_4$ ferrites samples in the low-temperature range.

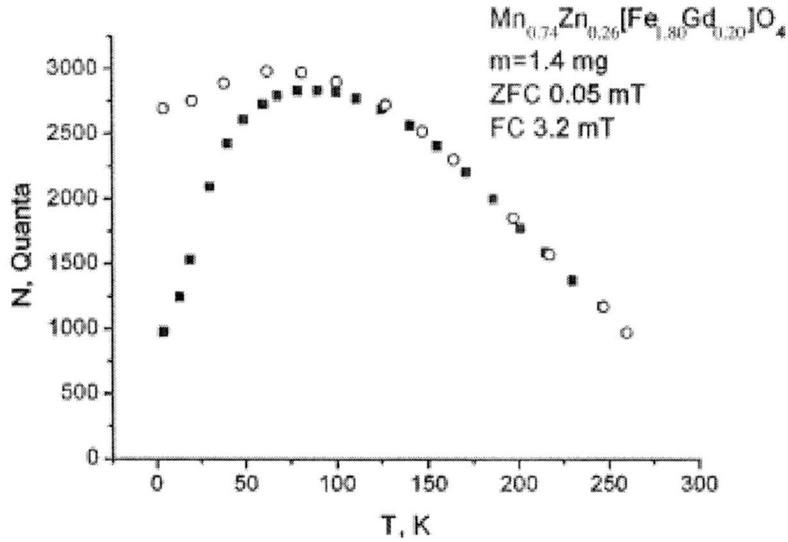

Figure 9. Temperature dependencies of magnetic moment for Mn0.74Zn0.26[Fe1.80Gd0.20]O4 in the low-temperature range.

Conclusion

The $Mn_XZn_Y[Fe_{2-Z}Gd_Z]O_4$ and $Mn_XZn_YFe_2O_4$ ferrite superparamagnetic particles of different ionic distribution were synthesized and their magnetic properties investigated. The T_C of all $Mn_XZn_Y[Fe_{2-Z}Gd_Z]O_4$ and $Mn_XZn_YFe_2O_4$ samples proved to be too high for the auto-regulation of temperature in the desirable range Apparently, it is not possible to get the ferrites of this composition with the T_C lower than 70–100°C. It is quite possible that there is a need to include additional components, such as Mg or Cu (II), and try to then substitute Fe (III) further, as possible for Gd(III) in the new ferrite-system to achieve the lower T_C. From our experience, such a substitution in $Mn_XZn_Y[Fe_{2-Z}Gd_Z]O_4$ would only be possible up to Z = 2, because no spinel single-phase would form above this value.

From a low T_C point of view, there are some samples (№ 12, № 6, № 4 and № 11), which exhibit the promising magnetic behavior and show a sufficiently high magnetic moment (14000 μB/particle). In addition, these nanoparticles do not contain any biologically highly toxic elements, and further investigations with them seem warranted.

Acknowledgement

Scientific investigation is executed under the Grant RFBR № 05 – 01 – 04001 (DFG).

References

[1] Jordan, R. Scholz and P. Wust et al., *J. Magn. Magn. Mater*. 201 (1999), p. 413.

[2] Jordan, R. Scholz and K. Maier-Hauff et al., *J. Magn. Magn. Mater*. 225 (2001), p. 118.

[3] M. Babincova, P. Cicmanec and V. Altanerova et al., *Bioelectrochemistry* 55 (2002), p. 17.

[4] Hilger, R. Hiergeist and R. Hergt et al., *Invest. Radiol*. 37 (2002), p. 580.

[5] M. Babincova, V. Altanerova and C. Altaner et al., *Med. Phys*. 31 (2004), p. 2219.

[6] R. Hergt, R. Hiergeist and I. Hilger et al., *J. Mag. Magn. Mater*. 270 (2004), p. 345.

[7] Bayburtskiy F . S., N. A. Brusentsov et al., *J. Mag. Magn. Mater*. 252 (2004), pp. 2350–2351.

[8] Gogosov V. V., Bayburtskiy F. S., Brusentsov N. A. *J. Magn. Magn. Mater*. 225 (2001), p. 113.

[9] Parekh. Kinnari, R.V. Upadhyay and R.V. Mehta, *J. Magn. Magn. Mater*. 252 (2002), p. 35.

[10] R.V. Upadhyay, R.V. Mehta and Kinnari. Parekh. *J. Magn. Magn. Mater*. 201 (1999), p. 129.

In: Encyclopedia of Nanotechnology
Editor: Elwood D. Carlson

ISBN 978-1-60692-079-4

Chapter 30

LIPID NANOTUBES AS SCAFFOLDS TO CREATE STRUCTURED INORGANIC NANOMATERIALS

***Yong Zhou*[1] *and Xiaqin Wang*[2]**

[1] SORST, Japan Science and Technology Agency (JST),
Tsukuba Central 5, 1-1-1 Higashi, Tsukuba, Ibaraki 305-8565, Japan
and Nanoarchitectonics Research Center (NARC), National Institute of Advanced Industrial Science and Technology (AIST),
Tsukuba Central 5, 1-1-1 Higashi, Tsukuba, Ibaraki 305-8565, Japan
[2] College of Materials Science and Engineering, Donghua University,
Shanghai 200051, P. R. China

ABSTRACT

Harnessing nature's amazing ability to form self-assembled structures for nanotechnology applications is an attractive alternative to conventional fabrication methods. Lipid nanotubes (LNTs) are especially intriguing assemblies, which represent a potentially powerful architecture generated through noncovalent self-assembly of amphiphilic molecules. They can be considered among the largest self-organized non-living structures yet observed. LNTs have hydrophilically internal and external membrane surfaces, and can provide the wide scope for chemical modifications, in sharp contrast to carbon nanotubes. These unique properties make themselves as ideal candidates for a variety of applications in chemistry, biochemistry, materials science and medicine. The lipid headgroups can function as templates for the nucleation, growth and deposition of inorganic substances on the external and internal surfaces of preformed LNTs, and the linear hollow cylinders and channels inside the nanotubes may also supply specific nanospaces for nanomaterials reaction and orrientation. This chapter assesses recent progress in application of LNTs as scaffolds for synthesis of structured inorganic nanomaterials, aiming to present the significant importance of the key technology of the 21st century, so-called "nanobiotechnology" by combination of biotechnology and nanoscicence for material fabrication.

INTRODUCTION

Lipid is the basic building blocks of biological membrane [1]. In aqueous media, lipid molecules self-assemble into diverse aggregate morphologies, depending on the molecular shape and solution condition such as lipid concentration, electrolyte concentration, pH, and temperature. Many lipid molecules can self-assemble into open ended, hollow cylindrical structures, named LNTs, which are composed of rolled-up bilayer membrane wall [2]. This potentially powerful architecture can be considered among the largest self-organized non-living structures yet observed. Such tubular structures are of interesting hydrophilic internal and external membrane surfaces, in sharp contrast to carbon nanotubes, which make themselves as ideal candidates for controlled release, chemical reaction and selective filling of nanomaterials and so on [2–5].

In recent years, benefiting from the specific properties of lipid molecules like highly order architecture and precise molecular recognition, there is increasing interest in LNTs for templating the growth of a large variety of inorganic nanomaterials. The headgroups of the LNT can modulate the nucleation, growth and deposition of inorganic substances on the external and inner surface of preformed LNTs [2–5]. This chapter represents an overview of recent promising applications of LNTs as scaffolds for inorganic synthesis by beginning with the brief introduction of the formation mechanism of LNTs.

FORMATION OF LNTS

Yager *et al* first reported hollow cylindrical structures of lipid derived from 1,2-bis (tricosa-10,12-diynoyl)-*sn*-glycero-3-phospho-choline, **1**(*8,9*) in 1984 when they studied the properties of the diacetylenic monomeric lipid [6]. **1**(*8,9*) was dispersed in distilled water above its chain melting transition temperature (T_m) and subsequent cooling. When aqueous dispersions were cooled below the gel-to-liquid crystalline phase transition temperature, the **1**(*8,9*) lipid bilayer transited from the chain disordered L_α phase into an ordered $L_{\beta'}$ gel phase, and the tubular structure was spontaneously formed from fusion of lipid vesicles. These tubules appear to be similar in shape to a paper soda straw and have an average diameter of the order of 0.5 μm with lengths varying from a few to several hundred micrometers and wall thickness of several to many bilayers (10 to 60 nm or more). The conversion from tubules to vesicles is reversible; when a tubule suspension is heated above the T_m, transformation to vesicles occurs quite instantly.

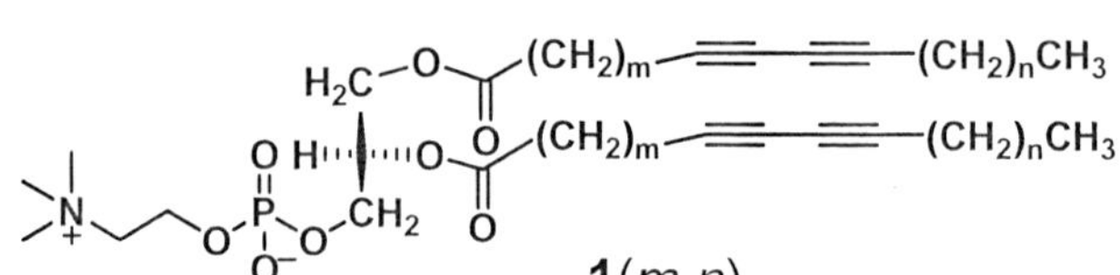

1(*m,n*)

For diacetylenic LNT, the tubules can be formed by different processes. In addition to formation in aqueous solutions, a more easy and convenient process for large amounts of the LNT was developed with ethanol/lipid/water solution [7]. The process involves dissolving the lipids in alcohol and then mixing with water above their T_m and slowly cooling the mixture. By this route longer and more robust LNTs can be produced. Aging under suitable conditions may obtain tubules in excess of 1.2 mm in length. Nevertheless, bulk water is not a prerequisite for tubule formation. The tubules may also precipitate from acetonitrile in absence of water [8]. This finding may have important technological implications in the development of these unique microstructures for promising use in organic solvents and under controlled environmental conditions.

Since the discovery of the LNTs, considerable research has been carried out to understand their formation mechanism including hollow morphology and dimension [9–11]. A variety of theoretical models have been proposed to explain the growth of these high-curvature structures. Among the models, the most reasonable one which is well consistent with the experiments is based on the molecular chiral packing, first proposed by Helfrich [12, 13]. According to the theory, chiral interactions cause constituent molecules to pack at a nonzero angle with respect to their nearest neighbors in a solid bilayer membrane. This situation leads to a spontaneous torsion and twisting of the edges of the bilayer membrane and subsequently formation of a cylindrical hollow, as shown in Figure 1. Chiral self-assembly involves a helically coiled ribbon structure as an intermediate from fusion of vesicles after hot aqueous dispersions of nanotube-forming lipids are cooled, which has been evidenced with experimental observation [14]. The ribbon then twists to form tubular structure. There are two routes for the tubular formation. One route proceeds with shortening of the helical pitch of the ribbon and maintaining a constant tape width (Figure 1a) [15]. The other involves widening of the tape width and maintaining a constant helical pitch (Figure 1b) [16]. The latter route is more common in literatures than the former one. In addition to the twisting-induced LNT, there is another route based on packing directed self-assembly [17]. In this route, wedge-shaped lipid molecules have a tendency to directly assemble into hollow cylindrical structures without forming helically twisted or coiled ribbons during the course of the reaction.

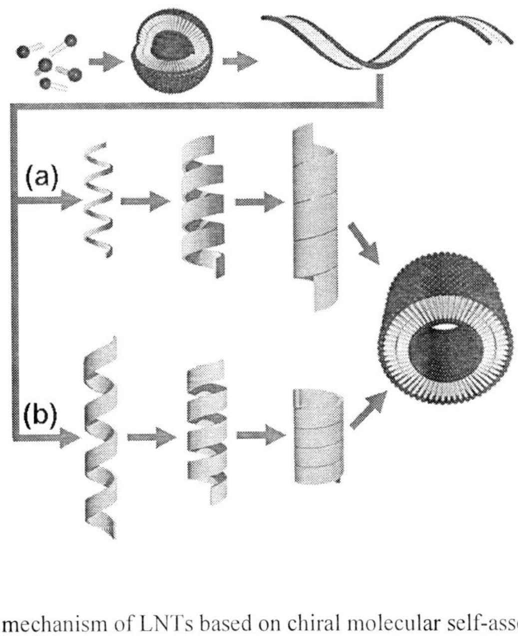

Figure 1. Possible formation mechanism of LNTs based on chiral molecular self-assemble.

Although the optimum mechanism behind the formation of the tubules has been identified to explain current experimental results, several fundamental questions on tubules have still risen such as: (1) what kind of materials will form these structures? (2) what is the molecular basis for the formation of the tubule structures? As a result, it is necessary to develop a rational understanding of the relation between the individual component molecule and the tubular microstructure derived from it. It has been revealed that the lipid molecular conformation, a variety of functionalities necessary for aggregation, and the location and orientation of those functionalities play crucial roles in determining the tubular self-assembling behavior [18–20]. In general, tubular formation depends on some typical parameters such as the degree of hydration of the lipid head group, the crystallinity of the membrane and the chirality as well as presence of amide bond and hydrogen bonding [21–25]. Singh *et al* has synthesized a number of molecules **1**(*m,n*) with *m* varying form 4 to 15 and *n* varying from 6 to 17 to assess the effect of variation of these features on the self-ssembly of cylindrical microstructures [26]. The most prominent results of these studies were: (1) The nondiacetylenic version of these molecules did not form tubules. (2) Molecules with nonsymmetrical positions of the diacetylenes in the acyl chain did not form tubules (m≠n). (3) The position of the diacetylene group had little or no effect on the observed structure of tubules formed from these substances. (4) The acyl chains had very high trans-conformational order. Singh *et al* also studied a number of diacetylenic lipids in which the chirality, size, and charge of the head groups were systematically varied [27]. They found that a nonchiral head group attached to similar acyl tails, as in the **1**(*8,9*) molecule, did not form tubules but did form sheet-like structures.

The ability to precisely control the inner and outer diameters of self-assembled LNTs directly determines their suitability for technological applications. Understanding how structural variation and experimental conditions affect nanotube dimensions at the molecular level would facilitate more efficient and systematic approaches to generating rationalized tubular libraries. Thomas and co-workers found that variations on the chemical structure can lead to tubules with tunable diameters [28]. They synthesized an analogue of **1**(*8,9*) where the phosphate linkage between the choline headgroup and glycerol backbone was replaced by a phosphonate. The tubules formed under similar conditions had twice the diameter of **1**(*8,9*) tubules. Mixtures of **1**(*8,9*) with the shorter-chain saturated lipid 1,2-bis(dinonanoyl)-*sn*-glycero-3-phosphocholine (DNPC) give tubules with outer diameters as small as 50 nm, approximately an order of magnitude less than those of pure **1**(*8,9*) microtubules [29]. Changes in experimental conditions such as temperature, concentrations, and the particular solvents used in the precipitation process afford considerable control over the diameters and the lengths of the resultant molecular assemblies. For instance, the average lengths of the tubules formed were found to depend on the alcohol used and be a function of the alcohol-water ratio [30, 31]. Nanotubes precipitated from a methanol/water solution of **1**(*8,9*) consist of single bilayers with a wall thickness of approximately 8 nm, whereas nanotubes obtained from ethanol/water or water consist of multiple (more than five) bilayers. Increase of alcohol-water ratio leads to the reduced length of the LNT. Meanwhile, decreasing the cooling rate through T_m has been shown to greatly increase the average length of tubules and decrease the wall thickness in acohol/water solutions [32]. For controlling diameter distribution, Guo *et al* reported a promising way to prepare a dimension-controlled glycolipid nanotube assembled from *N*-(11-*cis*-octadecenoyl)-β-D-glucopyranosyl-amine (**2**) by successive use of vesicle extrusion and porous templates [33]. They extruded an aqueous dispersion of **2** at 90 °C through a polycarbonate filter to produce uniform vesicles with thin walls. The extruded vesicles were filled into the nanopores of an anodic aluminum oxide (AAO) membrane filter and then self-assembled into tubular structures within the nanopores on cooling. This method can narrow the distribution of outer diameters, inner diameters, and wall thickness of the resultant LNTs compared to the solution-based preparative procedure. While so far all the research on length control has focused on changing the conditions of the self assembly process, such as the solvent composition, the cooling rate, and the addition of metal salts, a simple approach has been reported for obtaining intermediate length and short LNTs using a mechanical stirring system [34]. A suspension of **2**-based LNT can be cut shorter under mechanical stirring so that the LNT lengths could be controlled by varying the stirring time and the stirring rate.

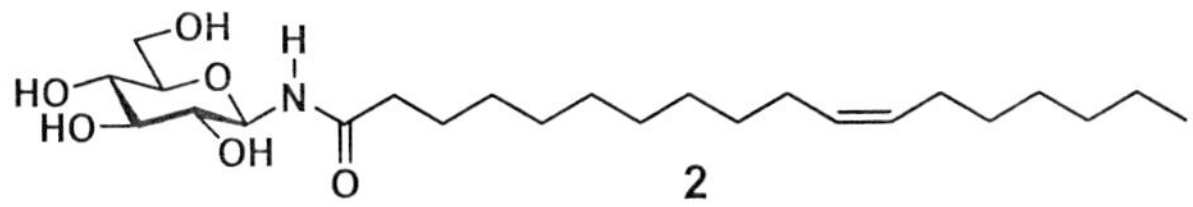

2

TEMPLATE FOR NANOSTRUCTURES

Harnessing nature's amazing ability to form self-assembled structures for nanotechnology applications is an attractive alternative to conventional fabrication methods. Employment of various biomolecules of linear morphology such as tobacco mosaic virus, protein microtubules and DNA to create 1-D array of nanoparticles, tubular and wire-like nanostructures has been widely developed for many advanced nanodevices. Headgroups of LNTs can function as templates for the nucleation, growth, and deposition of inorganic substances on their surface, hollow cylinder and bilayer membrane wall to form different functional nanostructures. The early report about this application of LNTs is that the LNTs acted as an excellent template to create metal nanotubes by coating the organic nanotube with metal precursors with an elegant electroless plating process [35–38]. The electroless plating, developed by Brenner [39] and Shipley [40], is a particularly rapid and effective means of producing thin, reasonably uniform metal coatings on the exterior and interior surfaces of the tubules. The coating procedure mainly involves inexpensive reagents and do not require complicated or expensive processing equipment. With the electroless plating, the thickness of the metal layer can be varied from 20 to 200 nm. The metallization procedure of the electroless plating is substantially compatible with the chemical and mechanical constraints of the lipid tubular microstructures so that the tubular morphology can be preserved. However, rapid changes in pH, osmolarity, and alcohol concentrations, and the generation of large quantities of hydrogen gas during the plating procedure often results in significant reduction of the length of the produced metal microcylinders with an aspect ratio less than the employed LNTs. Reduction of the aspect ratio of the metal microcylinders is also attributed to the mechanical stress on the tubules from the centrifugation and resuspension procedures [41]. The hollow metal microcylinders thus produced have potential applications in microwave [42] and long-term sustained release systems in agriculture, environment aquality and medicine [43].

Besides the eletroless plating technique, metal nanoparticles can also been controllably deposited on the surface of the LNT [44] by other soft chemical routes. Mastui *et al* report homogeneous coating carboxylic acid-thiol capped Au nanocrystals on a glycyglycine bolaamphiphiles, bis(*N*-α-amidoglycylglycine)-1,7-heptane dicarboxylate-based tubule templates (**3**) [45], which was firstly synthesized by Shimizu and Kogiso *et al* [46]. Hydrogen bonding between the amide groups of the tubules and the carboxylic acid groups of the nanocrystals was used as a driving force for this assembly process. In addition, by anchoring of sequenced peptides on the tubular surface of **3** which can cooperate with metal ions, a serial of monodispersed metal nanocrystals in control of particle size [47], shape [48], and even coating density [49] have been uniformly formed on the tubular surface. This method can also regulate growth of the wurtzite crystalline phase of coating ZnS nanocrystals at room temperature [50].

Lipid self-assembled tubular nanostructures can also serve as templates for mineral nucleation and deposition of metal oxide precursors with sol-gel routes to yield metal oxide nanotubes. The deposition or coating of inorganic substances occurs concurrently at both the inner and the outer surfaces. Shimizu and Ji *et al* reported direct sol-gel replication from a peptidic LNT self-assembling from **4** with a single bilayer wall in aqueous solution [51]. Different from traditional sol-gel reaction requiring active solution catalysts such as H^+, OH^-, minimal amount of positive charges of wall of the performed LNT can lead to slow hydrolysis and condensation of the silica precursor, tetraethyl orthosilicate (TEOS), in the absence of any catalyzes, and subsequently form the silica nanotube by calcinations. The wall thickness of thus-produced silica nanotube can be controlled within 4-nm precision by tuning amount ratio of the LNT to TEOS [52].

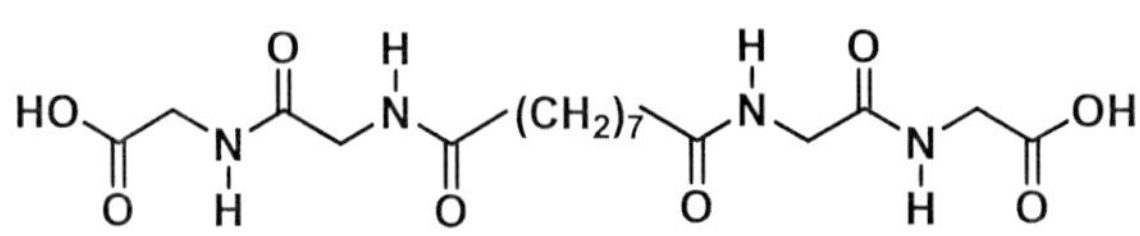

3

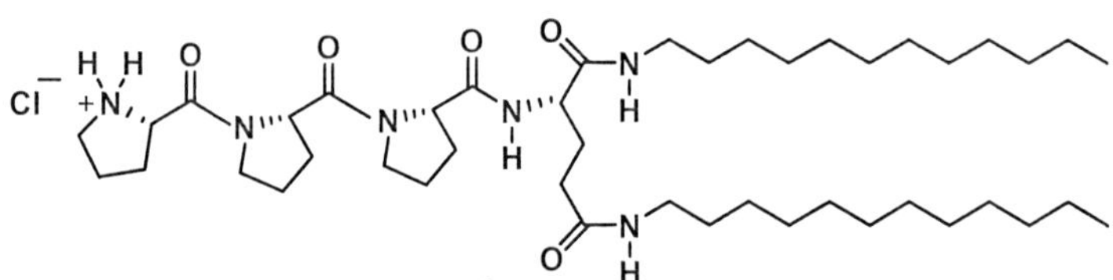

4

The same peptidic LNT was also effective for templated synthesis of transition metal oxide nanotubes including titanium oxide, tantalum oxide and vanadium oxide with a modified synthetic procedure by freezing the LNT to slow hydrolysis rate of the metal precursors [53]. In detail, the aqueous dispersion of the LNT was treated with short-time freezing in liquid nitrogen to form so-called iced LNT. It is proposed that there is unfrozen water layer around both inner and outer surfaces of the iced LNT. When the iced LNT began to melt, the ethanol solution of the precursors would permeate into this unfrozen water layer and initiate hydrolysis to produce tubular structures of metal oxides by subsequent calcinations. Interestingly, when examining the self-assembly of **4** in ethanol/water mixtures, it was unexpectedly found that the self-assembling morphologies were highly sensitive to the ethanol/ water volume fractions, changing from tubular structures with different diameters to spherical ones [54]. Therefore, using the obtained self-assembled structures as templates, tubular silica structures with different diameters of 80, 50, and 30 nm and hollow spheres can be obtained.

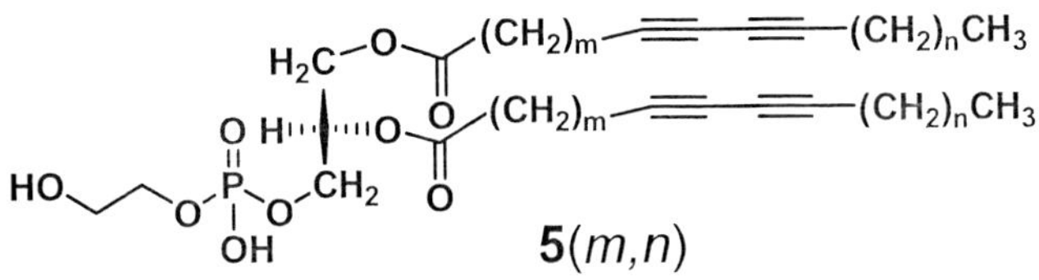

5(*m,n*)

One of the most impressive features with LNTs as template for inorganic nanostructures is to produce helical architectures. It is well known that DNA and RNA in nature self-assemble into fascinating highly ordered superstructures such as single-, double- and triple-helical structures. As shown above, the LNT was generally formed through helical winding of flat ribbons self-assembled from chiral molecular packing. In many cases, helical markings (or helical defect lines) were preserved on the tubules as reminiscence of winding [55]. The lipid molecules residing at the edge sites along the helical marking have different binding strength from those sited within the membrane wall of the LNT. These helical lines and marks can be replicated and serve as a chemically patterned organic surface for the helical organization of nanoparticles, which have potential utilization as asymmetric reaction catalysts [56, 57], helical sensors [58], optical materials [59] and so forth. Schnur *et al* gave an earlier report about the helical winding of Pd/Ni catalyst nanoparticles around the tubular cylinder in 1993 [60]. Later on Burkett and Mann reported the observation that helical-pattern arrays of gold nanoparticles by in situ reduction of $HAuCl_4$ were formed on the performed LNT of **1**(*8,9*), following the underlying helical ribbon structure of the tubule [61]. Lvov *et al.* also made the beautiful helices of SiO_2 nanoparticles winding around the interior of the tubules of **1**(*8,11*) , mixed with 2% of the charged lipid **5**(*8,9*), as shown in Figure 2 [62]. By the multistage layer-by-layer (LBL) absorption process of $PEI^+/PSS^-/PEI^+$ [PEI: poly(ethylenimine), PSS: poly(styrenesulfonate)] to induce positive charge on the tubules, the negatively charged SiO_2 nanoparticles absorbed along charged line defect of the interior of the tubules. This finding gives a new insight that the charged lipids are concentrated along helical defect lines that are not otherwise visible. With the similar LBL route, a metallic Cu spiral/helical nanostructure was also prepared *via* selectively electroless metallization [63]. In that preparation procedure, the negatively charged Pd catalyst nanoparticle was bound to the terminal PEI cationic layer of the multilayer film and initiated selective template metallization to form the helical Cu nanostructure. Helically-patterned arrays of Pd nanoparticles was also formed on a helical ribbon self-assembled from a cholesterol derivative doped with diaminoethylene glycol moiety as a binding site of Pd^{2+}, followed by reduction with ascorbic acid into Pd element [64].

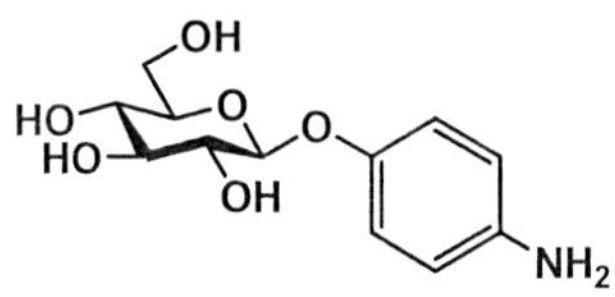

6

Another striking example is the formation of helical arrays of CdS nanoparticles tracing on a functionalized chiral template of glycolipid nanotubes from the binary components, **2** and aminophenyl-β-D-glucopyranoside (**6**) as an additive [65]. The glycolipid nanotube of **2** was functionalized by incorporation of **6** through the self-assembly. This functionalization process created active binding sites, which trace the chiral molecular packing of the nanotube. Figure 3a-3c show the time resolved high-resolution FE-SEM images that trace the formation of helical arrays of CdS nanoparticles on the functionalized LNT (*f*-LNT) surface. Figure 3a revealed that at the initial stage CdS selectively nucleated and grew into nanoparticles on the *f*-LNT. These nanoparticles connected one by one in a 1-D pattern, twining the curved surface of the *f*-LNT scaffold, as marked by solid arrows. The existence of a gap between the sections (shown as dot arrows) indicates that there are many nucleation sites in the 1-D growth track along the *f*-LNT. With reaction time, the discontinuous connection of the nanoparticles evolved into chain- and nanowire-like architectures along the scaffold (Figure 3b). These nanowires and chains arrange almost in a parallel fashion to one another with tilting by approximately 45° with respect to the long axis of the *f*-LNT. After 1 day's incubation, the helical array of the CdS nanoparticles on the *f*-LNT formed (Figure 3c). Based on the theoretical model for chiral self-assembly of tubules proposed by Helfrich and Prost [9, 10], it can be deduced that the nanotube structure from **2** has the molecular tilt optimally oriented at an angle of 45°. Thus, it is expected that **6** is also distributed in a helical fashion within the LNT membrane, as indicated with the dot curves in Figure 4a. As a result, Cd^{2+} will helically nucleate on the surface of the *f*-LNT through preferential affinity of Cd^{2+} to the amino group of **6**. Interaction with S^{2-}, slowly released by decomposition of thioacetamide, allows the heterogeneous nucleation of CdS along the surface of the *f*-LNT (Figure 4b). Increase in the incubation time promoted the growth of the nuclei into nanoparticles (Figure 4c) and eventually the formation of chain- or wire-like architecture helically twinning the *f*-LNT (Figure 4d). The left-handed helical sense of the CdS chains on the *f*-LNT suggests the identical helical sense of the chiral molecular packing. Consequently, the helical nucleation and growth of the CdS nanoparticles took place on the template surfaces.

Lipid tubular self-assembly can provide a hollow cylindrical environment to serve as a confined reaction field whose dimensions range from several to tens of nanometers. This nanospace with a high axial ratio favors the production of confined 1-D organization of appropriate nanostructures with dimensions in the single-nanometer range [66]. Shimizu and Yang *et al* developed a so-called "ship-in-bottle" fashion to introduce $HAuCl_4$ aqueous solutions into a vacant hollow cylinder of **2** by capillary force and succeeded in subsequent photoreduction of Au(III) into Au(0) by ultraviolet ray irradiation to get 1-D arrangement of the nanoparticles [67]. Complete removal of the LNT organic shell from the LNT-gold nanocomposite by firing in air can allow the nanoparticles to remain with being confined and eventually result into 1-D gold nanowires. The same hollow cylinder also encapsulated semiconductor nanoparticles and led to the formation of semiconductor nanowires [68]. Gazit *et al.* found the self assembly of a very short peptide (NH_2-phe-phe-COOH) into nanotubes

[69]. By dilution of this dipeptide organic solution into the aqueous, the peptide molecules show a rapid assembly into order semicrystalline structures within seconds, which were confirmed to be rather stiff nanotubular structures. The hollow section of the formed peptide nanotube was demonstrated as an ideal mold for nanocasting Ag nanowires.

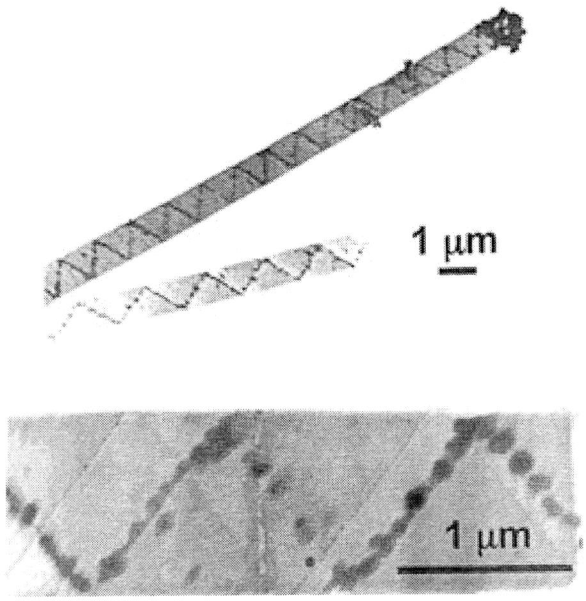

Figure 2. TEM images of nanoparticle helices inside the tubules of **1**(*8,11*)+2% **5**(*8,9*) after $PEI^+/PSS^-/PEI^+/(45\ nm\ SiO_2)$ treatment, tubule diameter 1 μm. Reproduced with permission from Ref. [62]. Copyright 2000 *Am. Chem. Soc.*

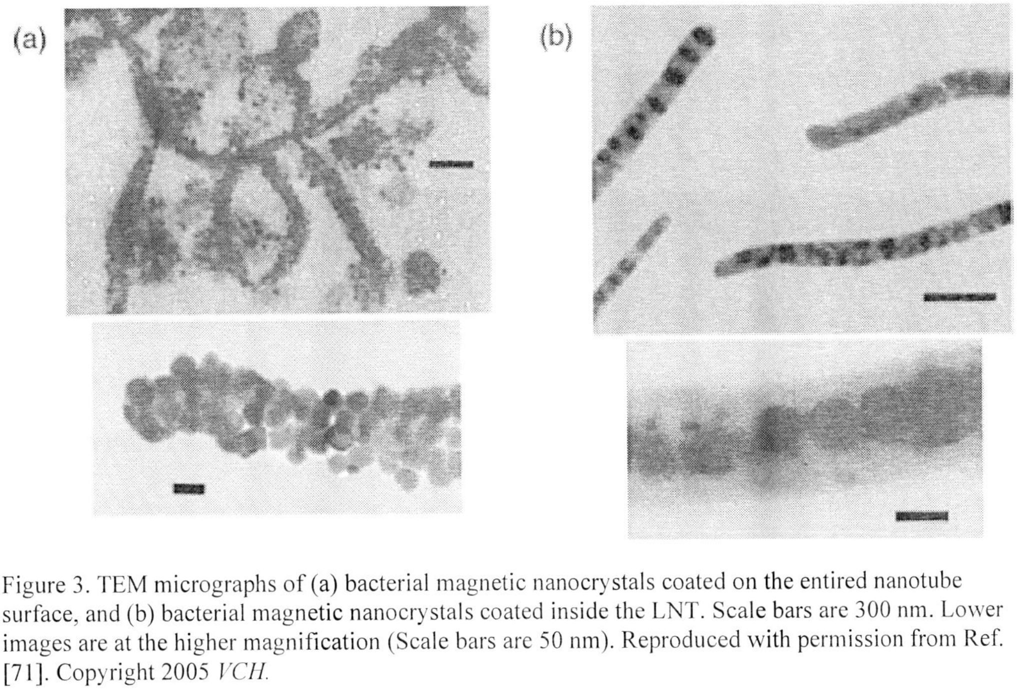

Figure 3. TEM micrographs of (a) bacterial magnetic nanocrystals coated on the entired nanotube surface, and (b) bacterial magnetic nanocrystals coated inside the LNT. Scale bars are 300 nm. Lower images are at the higher magnification (Scale bars are 50 nm). Reproduced with permission from Ref. [71]. Copyright 2005 *VCH.*

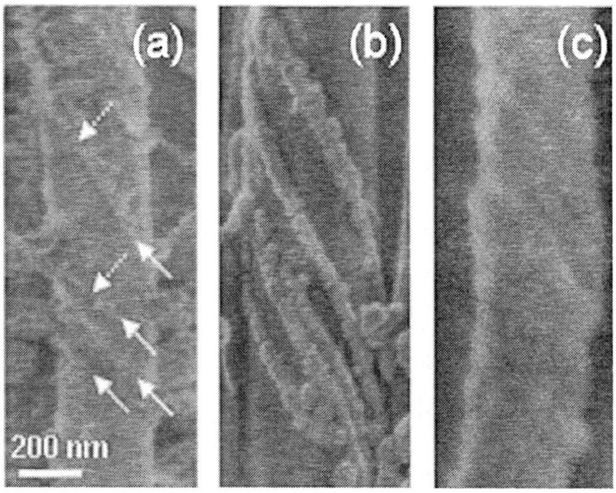

Figure 4. Time-resolved high-resolution FE-SEM images that show the formation of the helical array of CdS nanoparticles on the f-LNT surface for (a) 3 h, (b) 5 h and (c) 1 day. Reproduced with permission from Ref. [65]. Copyright 2006 Am. Chem. Soc.

By proteolytic lysis of the peptide with a proteinase K enzyme, individual silver nanowires about 20 nm in diameter can be obtained. In their further work, trilayered coaxial nanocables (metal/insulator/metal) using the same self-assembled peptide nanotube as the scaffold was fabricated by utilization of designed peptides as linkers to attach gold nanoparticles onto the Ag-filling peptide nanotube surfaces, in which the diphenylalanine motif at one end of the designed peptide molecule can interact with the peptide nanotube surface by weak nocovalent bonds, and the cysteine residue at the other end provides the reactive thiol group for the covalent binding of gold nanoparticles [70]. The attached gold nanoparticles were ideal for site specific initiation of electroless deposition of more gold nanoparticles over the peptide nanotube surface. Such coaxial geometry may give rise to useful and unique electromagnetic properties, which can be used as potential capacitors. Matsui *et al* reported fabrication of magnetic LNTs with the aid of magnetic bacteria [71]. The fabrication process consists of three steps: firstly magnetite (Fe_3O_4) nanocrystals was produced by a bacteria [72], then the magnetic nanocrystals thus produced were extracted from the cells, and incorporated into the hollow cylinder of a LNT produced by self assembly of **3**. Figure 5 shows a typical TEM image of the bacterial magnetic nanocrystals coated on the entired nanotube surface (a) and inside the LNT (b). The bacterial magnetic nanocrystals can be selectively deposited only to the interior or to the exterior surfaces of the nanotubes when the concentration of the nanocrystal in the nanotube solution is optimized. At high concentration, the nanocrystal was found to deposit on the outer surface of the nanotube, and low concentration resulted in selective incorporation of the magnetic nanocrystals substantially inside the nanotube to form a linear chain. The produced linear chain of the magnetic nanoparticles introduces a uniaxial magnetic anisotropy, not present in the magnetic particles or assemblies commonly known. In addition to metal, semiconductor and magnetic nanoparticles, ferritins composed of 24 polypeptide similar subunits joining each other through noncovalent interactions to form a 12-nm spherical protein cage surrounding a ferrihydrite ($5Fe_2O_3{\cdot}9H_2O$) core of 6 nm [73] were encapsulated into hollow cylinder of LNTs via either capillary force [74] or electrostatic interactions [16b].

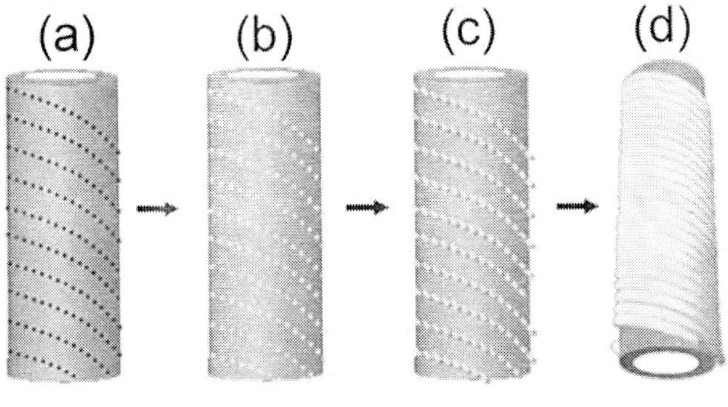

Figure 5. Illustration of the formation mechanism for the helical arrays of CdS nanoparticles on the *f*-LNT surfaces. Reproduced with permission from Ref. 65. Copyright 2006 Am. Chem. Soc.

Besides the surface and hollow nanospace of the LNT, the bilayer membrane wall of the unique tubular nanostructure can also serve as matrices to homogeneously embed nanoparticles. Compared with Gazit's work [69, 70], the NH_2-phe-phe-COOH peptide molecules can also self-assemble into tubular structure with nanoporous wall by a simplified method, *i.e.*, just gradually cooling the heated peptide aqueous solution from 65 °C to the room temperature without using organic solvent [75]. The porous wall would allow free diffusion of metal ions and reductant agent to all regions within the walls, resulting in the formation of metal-encapsulated tube wall of peptide nanocomposite. Fluorescent nanotubes consisting of CdS-embedded bilayer membranes of a synthetic peptide lipid the sodium salt of 2-(2-(2-tetradecanamidoacetamido) acetamido) acetic acid (**7**) have been prepared [76]. The preparation procedure of the CdS-LNT is schematically illustrated in Figure 6. **7** can self-assemble in aqueous solutions into a hollow cylindrical structure in the presence of proton (H^+) (abbreviated as a H-LNT) or a series of transition metal cations [77]. In the CdS-LNT case, coordination of Cd^{2+} to two negatively charged COO^- groups of **7** allows the lipid **7** to form a Cd-complexed LNT (abbreviated as a Cd-LNT). Upon exposure to H_2S vapor, the Cd^{2+} in the Cd-LNT were released as a result of competitive binding of the proton to the COO^- group, resulting in the formation of H-LNT. The released Cd^{2+} subsequently reacted with S^{2-} to initiate CdS nuclei, and finally grew into the CdS nanodots (NDs) in all over the lipid bilayer membranes of **7**. Figures 7 (a) and 7 (b) show a typical STEM image for the formed CdS-LNT and the enlarged image of the selected area in (a), respectively. The assigned CdS NDs clearly seen as dark dots in the image are homogenously dispersed in the bilayer membranes of the LNT and separated from one another without any aggregation. The tubular nanocomposites clearly exhibited fluorescence originating from electronic transition of the CdS NDs. The relative homogeneity of the fluorescence signal also implies that the CdS NDs were uniformly dispersed in the membrane wall of the LNT, in consistent with the STEM observation. The fluorescence of the CdS-LNT is easily distinguishable and resistant to photobleaching compared to other organic moiety-based fluorescence, and enables ones to visualize for long time and to trace the localization in biological systems. The fluorescent CdS-LNT has proved to successfully act as a supramolecular nanotube host to encapsulate ferritin and gold nanoparticles, which is applicable for biomolecule delivery.

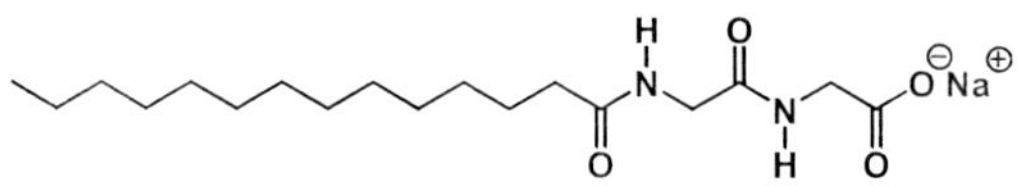

7

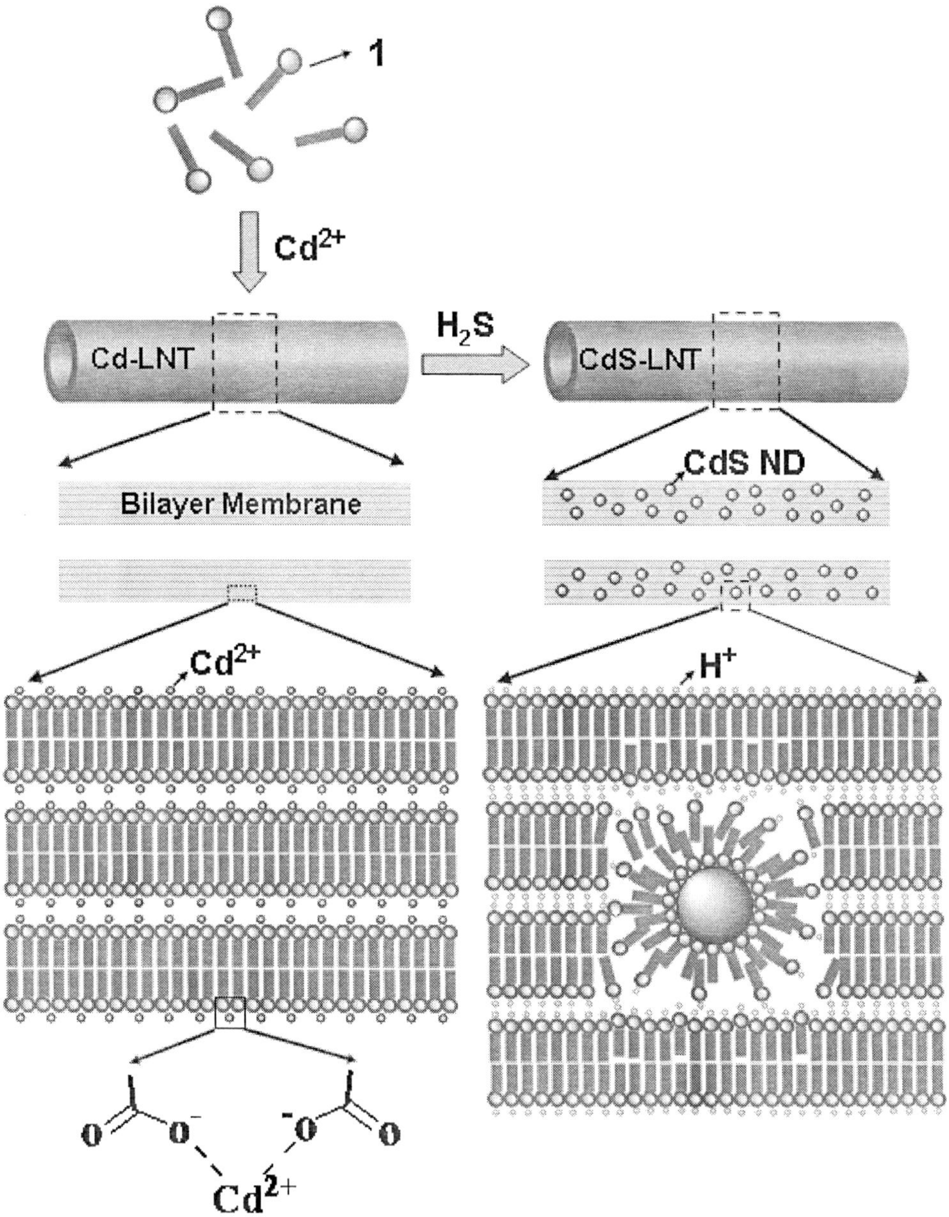

Figure 6. Schematic illustration for the preparation procedure of CdS-LNTs via the formation of Cd-LNT. For simplicity, the number of layers and tilting of constituting molecules are neglected. Reproduced with permission from Ref. [76]. Copyright 2007 VCH.

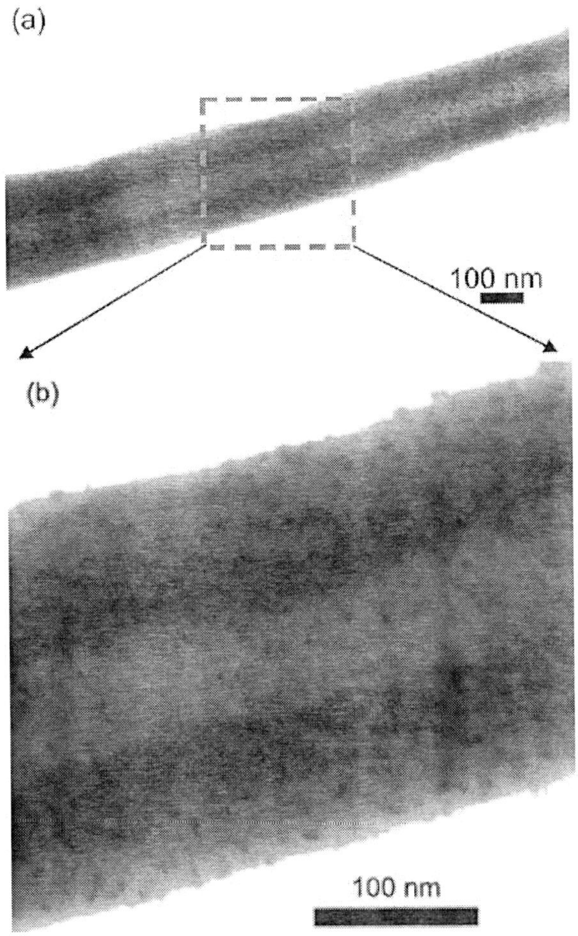

Figure 7. (a) STEM image of the formed CdS-LNT where CdS NDs are embedded in the bilayer membrane walls of the LNT and (b) enlarged STEM image of the selected area in (a). Reproduced with permission from Ref. [76]. Copyright 2007 VCH.

CONCLUSION

This chapter has outlined recent advances on the LNT as template directing growth of structured nanomaterials. The produced helical patterns on the surfaces of the LNT, one-dimensional arrays in the hollow cylinder, and nanotubes with controlled wall thickness within nano-scale precision demonstrate the LNT superior to other templates. Several fundamental features of the unique LNT make the formed nanostructures attractive for future microelectronic applications. The connection of the nanostructures to the external nanodevices is one of key issues for the application. Despite that a variety of serious technique obstacles remain to be solved, it is believed that the world-wide interdisciplinary activities from physicists, chemists, engineers, biologists, and material scientists will achieve this goal in near future and explore this system with more exciting functions.

ACKNOWLEDGMENTS

Dr. Toshimi Shimizu, NARC, AIST, Japan, is acknowledged to direct the related work. The Japan Science and Technology Agency (JST) is acknowledged for financial support of the CREST and SORST projects. Partial financial support from State Key Lab for Modification of Chemical Fibers and Polymer Materials, Donghua University, P. R. China, is also thanked.

REFERENCES

[1] Kunitake, T. *ibid*, 1992, *31*, 709.
[2] Shimizu, T.; Masuda, M.; Minamikawa, H. *Chem. Rev.* 2005, *105*, 1401.
[3] (a) Zhou, Y. *Curr. Nanosci.* 2006, *2*, 123. (b) Zhou, Y. *Recent Patents on Nanotechnology* 2007, *1*, 21.
[4] (a) Schnur, J. M. *Science* 1993, *262*, 1669. (b) Spector, M. S.; Price, R. R.; Schnur, J. M. *Adv. Mater.* 1999, *11*, 337.
[5] Zarif, L. *J. Controlled Release* 2002, *81*, 7.
[6] Yager, P.; Schoen, P. E. *Mol. Cryst. Liq. Cryst.* 1984, *106*, 371.
[7] Georger, J. H.; Singh, A.; Price, R. R.; Schnur, J. M.; Yager, P.; Schoen, P. E. *J. Am. Chem. Soc.* 1987, *109*, 6169.
[8] Rudolph, A. S.; Calvert, J. M.; Ayers, M. E.; Schnur, J. M. *J. Am. Chem. Soc.* 1989, *111*, 8516.
[9] Helfrich, W. *J. Chem. Phys.* 1986, *85*, 1085.
[10] Selinger, J. V.; Schnur, J. M. *Phys. Rev. Lett.* 1993, *71*, 4091.
[11] Chappell, J. S.; Yager, P. *Chem. Phys.* 1991, *150*, 73.
[12] Helfrich, W. *J. Chem. Phys.* 1986, *85*, 1085.
[13] Helfrich, W.; Prost, J. *Phys. Rev. A* 1988, *38*, 3065.
[14] Cheng, Q.; Yamamoto, M.; Stevens, R. C. *Langmuir* 2000, *16*, 5333.
[15] Chung, D. S.; Benedek, G. B.; Konikoff, F. M.; Donovan, J. M. *Proc. Natl. Acad. Sci. USA* 1993, *90*, 11341.
[16] Nakashima, N.; Asakuma, S.; Kunitake, T. *J. Am. Chem. Soc.* 1985, *107*, 509.
[17] (a) Masuda, M.; Shimizu, T. *Langmuir* 2004, *20*, 5969. (b) Kameta, N.; Masuda, M.; Minamikawa, H.; Goutev, N. V.; Rim, J. A.; Jung, J. H.; Shimizu, T. *Adv. Mater.* 2005, *17*, 2732.
[18] Spector, M. S.; Easwaran, K. R. K.; Jyothi, G.; Selinger, J. V.; Singh, A.; Schnur, J. M. *Proc. Natl. Acad. Sci. USA* 1996, *93*, 12943.
[19] Spector, M. S.; Price, R. R.; Schnur, J. M. *Adv. Mater.* 1999, *11*, 337.
[20] Spector, M. S.; Selinger, J. V.; Schnur, J. M. In *Topics in Stereochemistry*, Green, M. M.; Nolte, R. J. M.; Meijer, E. W., Eds.; Wiley; Germany, 2003; Vol. 24.
[21] Fuhrhop, J. H.; Schnieder, P.; Rosenberg, J.; Boekema, E. *J. Am. Chem. Soc.* 1987, *109*, 3387.
[22] Schnur, J. M.; Ratna, B. R.; Selinger, J. V.; Singh, A.; Jyothi, G.; Easwaran, K. R. K. *Science* 1994, *264*, 945.

[23] Rudolph, A. S.; Singh, B. R.; Singh, A.; Burke, T. G. *Biochim. Biophys. Acta* 1988, *943*, 454.
[24] Selinger, J. V.; Mackintosh, F. C.; Schnur, J. M. *Phys. Rev. E* 1996, *53* 3804.
[25] (a) Nakashima, N.; Asakuma, S.; Kunitake, T. *J. Am. Chem. Soc.* 1985, *107*, 509. (b) Kunitake, T.; Nakashima, N.; Shimomura, M.; Okahata, Y.; Kano, K.; Ogawa, T. *J. Am. Chem. Soc.* 1980, *102*, 6642.
[26] Ihara, H.; Fukumoto, T.; Hirayama, C.; Yamada, K. *Polym. Commun.* 1986, *27*, 282.
[27] Singh, A.; Marchywka, S. *Polym. Mater. Sci. Eng.* 1989, *61*, 675.
[28] Thomas, B. N.; Corcoran, R. C.; Cotant, C. L.; Lindemann, C. M.; Kirsch, J. E.; Persichini, P. J. *J. Am. Chem. Soc.* 1998, *120*, 12178.
[29] Svenson, S.; Messersmith, P. B. *Langmuir* 1999, *15*, 4464.
[30] Georger, J. H.; Singh, A.; Price, R. R.; Schnur, J. M.; Yager, P.; Schoen, P. E. *J. Am. Chem. Soc.* 1987, *109*, 6169.
[31] Ratna, B. R.; Baral-Tosh, S.; Kahn, B.; Schnur, J. M.; Rudolph, A. S. *Chem. Phys. Lipids* 1992, *63*, 47.
[32] Thomas, B. N.; Safinya, C. R.; Plano, R. J.; Clark, N. A. *Science* 1995, *267*, 1635.
[33] Guo, Y., Yui, H.; Minamikawa, H.; Yang, B.; Masuda, M.; Ito, K.; Shimizu, T. *Chem. Mater.* 2006, *18*, 1577.
[34] Yang, B.; Kamiya, S.; Yui, H.; Masuda, M.; Shimizu, T. *Chem. Lett.* 2003, *32*, 1146.
[35] Schnur, J. M.; Georger, J. H.; Price, R.; Yager, P.; Singh, A.; Schoen, P. E. Invention Disclosure NC 70173, 1986.
[36] Schnur, J. M.; Price, R.; Schoen, P.; Yager, P.; Calvert, J. M.; Georger, J.; Singh, A. *Thin Solid Films* 1987, *152*, 181.
[37] Matsui, H.; Pan, S.; Gologan, B.; Jonas, S. *J. Phys. Chem. B* 2000, *104,* 9576.
[38] Krebs, J. J.; Rubinstein, R.; Lubitz, P.; Harford, M. Z.; Baral, S.; Shashidhar, R.; Ho, Y. S.; Chow, G. M.; Qadri, S. *J. Appl. Phys.* 1991, *70*, 6404.
[39] Brenner, A.; Ridell, B. E. *Proc. Am. Electrochem. Soc.* 1946, *33*, 23.
[40] Shipley, C. R. *Plating and surface Finishing* 1984 *71*, 92.
[41] Zabetakis, D. *J. Mater. Res.* 2000, *15*, 2368.
[42] Chow, G. M.; Stockton, W. B.; Price, R.; Baral, S.; Ting, A. C.; Ratna, B. R.; Shoen, P. E.; Schnur, J. M.; Bergeron, G. L.; Czarnaski, M. A.; Hickman, J. J.; Kirkpatrick, D. A. *Mater. Sci. Eng. A* 1992, *158*, 1.
[43] (a) Price, R. R.; Patchan, M. *J. Microencaps*. 1991, *8*, 301. (b) Ahl, P. L.; Price, R. R.; Smuda, J.; Gaber, B. P.; Singh, A. *Biochim. Biophys. Acta* 1990, *1028*, 141.
[44] Gao, X.; Djalai, R.; Haboosheh, A.; Samson, J.; Nuraje, N.; Matsui, H. *Adv. Mater.* 2005, *17*, 1753.
[45] Matsui, H.; Pan, S.; Douberly, Jr. G. E. *J. Phys. Chem. B* 2001, *105,* 1683.
[46] Kogiso, M.; Ohnishi, S.; Yase, K.; Masuda, M.; Shimizu, T. *Langmuir* 1998, *14*, 4978.
[47] (a) Banerjee, I. A.; Yu, L.; Matsui, H. *Proc. Natl. Acad. Sci. USA*, 2002, *99*, 6451. (b) Yu, L.; Banerjee, I. A.; Shima, M.; Rajan, K.; Matsui, H. *Adv. Mater*. 2004, *16*, 709.
[48] Yu, L.; Banerjee, I. A.; Matsui, H. *J. Am. Chem. Soc*. 2003, *125*, 14837.
[49] Djalali, R.; Chen, Y.; Matsui, H. *J. Am. Chem. Soc*. 2002, *124*, 13660. (b) Dajali, R.; Chen, Y.; Matsui, H. *J. Am. Chem. Soc*. 2003, *125*, 5873.
[50] Banerjee, I. A.; Yu, L.; Matsui, H. *J. Am. Chem. Soc.* 2005, *127*, 16002.
[51] Ji, Q.; Iwaura, R.; Kogiso, M.; Jung, J. H.; Yoshida, K.; Shimizu, T. *Chem. Mater.* 2004, *16,* 250.

[52] Ji, Q.; Iwaura, R.; Shimizu, T. *Chem. Lett.* 2004, 504.
[53] Ji, Q.; Shimizu, T. *Chem. Commun.*, 2005, 4411.
[54] Ji, Q.; Iwaura, R.; Shimizu, T. *Chem. Mater.* 2007, *19,* 1329.
[55] (a) Goldstein, A. S.; Lukyanov, A. N.; Carlson, P. A.; Yager, P.; Gelb, M. H. *Chem. Phys. Lipids* 1997, *88*, 21. (b) Lee, K. C.; Carlson, P. A.; Goldstein, A. S.; Yager, P.; Gello, M. H. *Langmuir* 1999, *15*, 5500.
[56] Kogiso, M.; Yoshida, K.; Yase, K.; Shimizu, T. *Chem. Commun.* 2002, 2492.
[57] Jung, J. H.; Ono, Y.; Hanabusa, K.; Shinkai, S. *J. Am. Chem. Soc.* 2000, *122*, 5008.
[58] Cornell, B. A.; Braach-Maksvytis, V. L. B.; King, L. G.; Osman, P. D. J.; Raguse, B.; Wieczorek, L.; Pace, J. *Nature* 1997, *87*, 580.
[59] Hodgekinson, I.; Wu, Q. H. *Adv. Mater.* 2001, *13*, 889.
[60] Selinger, J. V.; Schnur, J. M. *Phys. Rev. Lett.* 1993, *71*, 4091.
[61] Burkett, S. L.; Mann, S. *Chem. Commun.* 1996, *1996*, 321.
[62] Lvov, Y. M.; Price, R. R.; Selinger, J. V.; Singh, A.; Spector, M. S.; Schnur, J. M. *Langmuir* 2000, *16*, 5932.
[63] Price, R. R.; Dressick, W. J.; Singh, A. *J. Am. Chem. Soc.* 2003, *125*, 11259.
[64] Jung, J. H.; Rim, J. A.; Lee, S. J.; Lee S. S. *Chem. Commun.* 2005, 468.
[65] Zhou, Y.; Ji, Q.; Masuda, M.; Kamiya, S.; Shimizu, T. *Chem. Mater.* 2006, *18*, 403.
[66] Shimizu, T. *J. Poly. Sci. part A-Poly. Chem.* 2006, *44*, 5137.
[67] (a) Yang, B.; Kamiya, S.; Yoshida, K.; Shimizu, T. *Chem. Commun.* 2004, 500. (b) Yang, B.; Kamiya, S.; Shimizu, Y.; Koshizaki, N.; Shimizu, T. *Chem. Mater.* 2004, 2826.
[68] Zhou, Y.; Kamiya, S.; Shimizu, T. Unpublished results.
[69] Reches, M.; Gazit, E. *Science* 2003, *300*, 625.
[70] Carny, O.; Shalev, D. E.; Gazit, E. *Nano lett.* 2006, *6*, 1594.
[71] Banerjee I. A; Yu L.; Shima, M.; Yoshino, T.; Takeyama, H.; Matsunaga, T.; Matsui, H. *Adv. Mater.* 2005, *17*, 1128.
[72] Yang, C. D.; Takeyama, H.; Tanaka, T. ; Matsunaga, T. *Enzyme Microb Technol* 2001, *29*, 13.
[73] Proulx-Curry, P. M.; Chasteen, N. D. *Coord. Chem. Rev.* 1995, *144*, 347.
[74] Yui, H.; Shimizu, Y.; Kamiya, S.; Masuda, M.; Yamshita, I.; Ito, K.; Shimizu, T. *Chem. Lett.* 2005, *34*, 232.
[75] Song, Y.; Challa, S. R.; Medforth, C. J.; Qiu, Y.; Watt, R. K.; Peña, Miller, J. E.; van Swol, F.; Shelnutt, J. A. *Chem. Commun.* 2004, 1044.
[76] Zhou, Y.; Kogiso, M.; He, C.; Shimizu, Y.; Koshizaki, N.; Shimizu*, T. Adv. Mater.* 2007*,* 19, *1055.*
[77] Kogiso, M.; Zhou, Y.; Shimizu, T. *Adv. Mater. 2007,* 19*, 242.*

In: Encyclopedia of Nanotechnology
Editor: Elwood D. Carlson

ISBN: 978-1-60692-079-4

Chapter 31

MULTIFUNCTIONAL REACTION PLATFORM BASED ON CARBON NANOTUBES

Wenwen Li, Yu Zhang and Chao Gao*
College of Chemistry and Chemical Engineering,
Shanghai Jiao Tong University, 800 Dongchuan Road,
Shanghai 200240, P. R. China

ABSTRACT

Functionalization of carbon nanotubes (CNTs) opens a versatile avenue for the preparation of soluble CNTs and novel CNT-based hybrid nanomaterials and nanodevices. In this work, we present a multifunctional reaction platform based on CNTs, poly(glycidyl methacrylate)-grafted multiwalled carbon nanotubes (MWNT-g-PGMA). The reaction platform of MWNT-g-PGMA was achieved by surface-initiated atom transfer radical polymerization (ATRP). The polymerization kinetics was investigated. The epoxy groups of MWNT-g-PGMA can be converted into azido groups by reaction with NaN_3. The "click" reaction between azido groups and alkyne modified coumarin 343 affords luminescent CNTs. The epoxy groups can also be readily converted into hydroxyl and amino groups in the presence of hydrochloric acid and 1,2-ethylenediamine, respectively. Further reaction of the hydroxyl groups with succinic anhydride results in the formation of multicarboxylic acid polymer-functionalized CNTs. The chemical structure, morphology and the grafted quantities of epoxy, hydroxyl, amino and carboxylic acid groups functionalized MWNTs were determined by FTIR, ^{1}H NMR, SEM, TEM and TGA measurements. Magnetic CNTs and silver nanocrystals-decorated CNTs were fabricated through subsequent assembly of magnetic nanoparticles (e.g., Fe_3O_4) and in situ reduction of Ag^+ in aqueous solution using the amino and carboxylic groups functionalized CNTs as templates, respectively.

Keywords: *ATRP, carbon nanotubes, click chemistry, functionalization, poly(glycidyl methacrylate).*

---

* Corresponding author. Tel: +86-21-5474 2676. Fax: +86-21-5474 1297. E-mail: chaogao@sjtu.edu.cn

INTRODUCTION

Carbon nanotubes (CNTs) [1] based nanocomposites and nanodevices which have potential application in the fields of chemistry, material, biology, nanoscience and nanotechnology have been extensively studied over the past several years [2-7]. For instance, carbon nanotubes were reported to be modified by kinds of compounds including polymers [8-33], organic molecules [34-39], metal [41-47] and nanoparticles [23, 48-57] though covalent linkage, adsorption, mechanical mixing or other methods. Many CNT-based nanocomposites were successfully prepared through these methods. Therein preparation of polymer functionalized CNTs nanohybrids is one of the most fascinating fields.

Two main methodologies, "grafting to" and "grafting from", have been used to covalently attach polymer chains to the surface of nanotubes. In the "grafting to" route, as-prepared polymers are attached to CNTs by the reaction between their terminal groups and the functional groups of CNTs. On the contrast, in the "grafting from" strategy, polymers are growing from the surfaces of CNTs by in situ polymerization of monomers in the presence of CNT-based macro-initiators. Up to now, many polymers including polystyrene (PS) [8,9], poly(sodium 4-stryrenesulfonate) [10,11], poly(*tert*-butyl acrylate) (P*t*BA) [10,12], poly(acrylicacid) (PAA) [10], poly(methyl methacrylate) (PMMA) [13-17], poly(2-hydroxyethyl methacrylate) (PHEMA) [17], poly(n-butyl methacrylate) (PnBA) [18], poly(2,3-dihydroxypropyl methacrylate) [19], poly(2-dimethylaminoethyl methacrylate) (PDMAEMA) [20], poly(2-diethylaminoethylmethacrylate) (PDEAEMA) [21], temperature-switching poly(N-isopropyl acrylamide) (PNIPAA) [22,23], poly(2-vinylpyridine) [24], poly(4-vinylpyridine) [25], polyimide [26], poly(N-vinylcarbazole) [27], poly(vinyl alcohol) (PVA) [28], hyperbranched poly(amidoamine) [29], hyperbranched poly(3-ethyl-3-hydroxymethyloxetane) (PEHO) [30], poly(2-((bromobutyryl)oxy)ethyl acrylate) (PBBEA) [31], poly(3-*O*-Methacryloyl-1,2:5,6-di-*O*-isopropylidene-D-glucofuranose) (PMAIG) [32] and some other copolymers [14,26,32,33] were employed to functionalize CNTs through either "grafting to" or "grafting from" method. For the polymer and CNTs nanohybrids, the inherent excellent properties of CNTs could be kept, the solubility of CNTs could be increased and the conventional properties of polymers can be improved, which opens up a very useful way for the preparation of novel CNT-based nanomaterials and nanodevices.

Based on the polymer-coated CNTs, magnetic nano-particles, quantum dots (QDs), such as CdTe and CdS, and metal particles Au, Ag, Pt were loaded to the surfaces of carbon nanotubes [23,47,57]. For instance, polyamine- and amino-functionalized MWNTs were then used as templates to load CdTe quantum dots (QDs) and Fe_3O_4 magnetic nanoparticles by the electrostatic self-assembly approach [57]. Poly(carboxylic acid)-modified MWNTs were used to sequester metal ions such as Ag^+, Co^{2+}, Ni^{2+} and Au^{3+}, forming MWNT-polymer/metal hybrid nanocomposites [47]. One defect for the aforementioned preparation of the nanocomposites is that the polymers attached to the surfaces of CNTs should contain a special functional group for the following reaction with nanoparticles or metals. So it will be more convenient for us to fabricate CNTs-based nanocomposites, if we can prepare a multifunctional reactive platform based on CNTs. A valid strategy for the preparation of a multifunctional reaction platform is functionalization of CNTs with polymers containing groups that can be easily transformed into kinds of functional groups. Poly(glycidyl methacrylate)(PGMA) is a kind of such polymer and the epoxy groups of PGMA can be easily converted into other functional groups, such as hydroxyl, amino and azido groups. The multifunctional polymer-functionalized CNTs offer a versatile platform for tailoring and fabricating novel hybrid nanomaterials and nanodevices.

On the other hand, the "click chemistry" popularized by Sharpless groups also attracted considerable attention recently, and the copper catalyzed azide-alkyne [2+3] cycloaddition is the most popular click chemistry reaction [58, 59]. It is proved to be efficient due to its high yield, stereo selectivity and excellent functional group compatibility and has already used to prepare functional polymers [60-66], dentrimers [67,68], affinity chromatography supports [69,70], biomolecules [71-73]. Thus it is of great interest to introduce this reaction to the functionalization of CNTs, preparing multifunctional CNTs.

In this work, PGMA was grown from the surfaces of multi-walled carbon nanotubes (MWNTs) by *in situ* atom transfer radical polymerization (ATRP), resulting in a versatile reactive platform of PGMA-grafted MWNTs (MWNT-*g*-PGMA). The epoxy groups of PGMA grafts are then converted into hydroxyl, amino, carboxyl, and azido- groups subsequently. With the derivative polymers-functionalized CNTs as templates, other CNT-based nanocomposites such as MWNT/Fe_3O_4 and MWNT/Ag nanohybrids are prepared successfully and the azido groups are used to introduce some fluorescent molecules to the surfaces of CNTs by the "click" chemistry.

EXPERIMENTAL SECTION

Materials. The multiwalled carbon nanotubes (MWNTs) were purchased from Tsinghua-Nafine Nano-Powder Commercialization Engineering Centre in Beijing. Glycidyl methacrylate (GMA) was purchased from Aldrich and purified by vacuum distillation. Cu(I)Br was purchased from Aldrich and purified by extraction with glacial acetic acid overnight and then washed with alcohol and ether. The chemicals of 2-bromo-2-methylpropionyl bromide (α-bromoisobutyryl bromide), ethyl 2-bromoisobutyrate, propargyl alcohol, *N,N,N',N",N"*-pentmethyldiethylenetriamine (PMDETA), coumarin 343 and dicyclohexylcarbodiimide (DCC) were purchased from Aldrich and used as received. Tetrahydrofuran (THF), *N,N*-dimethylformamide (DMF), dimethyl sulfoxide (DMSO), acetone, methanol, chloroform, 1,2-ethylenediamine, succinic anhydride, NaN_3 and all other reagents or solvents were obtained from Shanghai Reagents Company. MWNT-supported macroinitiators (MWNT-Br) were prepared according to previous procedures [17]. Calculated from corresponding TGA weight loss data, the content of Br of MWNT-Br is 5.4 wt %, which corresponds to 0.67 mmol per gram MWNT-Br.

Instrumentation. Thermal gravimetric analysis (TGA) was measured on a PE TGA-7 instrument with a heating rate of 20°C/min in a nitrogen flow (20 mL/min). Fourier-transform infrared (FTIR) spectra were recorded on a PE Paragon 1000 spectrometer. Hydrogen nuclear magnetic resonance (^1H NMR) spectra were measured with a Varian Mercury Plus 400 MHz spectrometer. Absorption and fluorescence spectra were recorded at room temperature on a Lambda 20 UV-visible spectrophotometer (Perkin-Elmer) and Varian Cary Fluorescence Spectrometer. Transmission electron microscopy (TEM) analysis was performed on a JEOL JEL2010 electron microscope at 200 kV. Scanning electron microscopy (SEM) images were recorded using a LEO 1550VP field-emission microscope, and the samples were loaded on the silicon surface, previously sputter-coated with a homogenous gold layer for charge dissipation during the SEM imaging.

Preparation of PGMA-grafted MWNTs without free initiator [47,74]. Here MWNT-*g*-PGMA was synthesized by *in situ* atom transfer radical polymerization (ATRP) with only MWNT-Br as the initiator. Typically, 50.0 mg of MWNT-Br (0.034 mmol Br), 14.5 mg (0.10 mmol) of CuBr, 17.3 mg (0.10 mmol) of PMDETA, and 0.5 mL of diphenyl ether were placed in a 10 mL dry flask. The flask was evacuated and filled thrice with N_2, and then 0.14 g of GMA (0.98 mmol) was injected into the flask. The mixture was stirred at 50 ^{o}C for 48 h. The resulting sample was diluted with THF and separated by centrifugation. The product was washed with THF for eight times and MWNT-*g*-PGMA (CP1b) was obtained after drying overnight under vacuum.

Preparation of MWNT-g-PGMA with MWNT-Br and ethyl 2-bromoisobutyrate as co-initiators [47,74]. Herein the sacrificial initiator of ethyl 2-bromoisobutyrate was mixed with MWNT-Br to investigate the polymerization of GMA from the surfaces of MWNTs. Typically, 30.0 mg of MWNT-Br (0.02 mmol Br), 14.5 mg (0.10 mmol) of CuBr, 17.3 mg (0.10 mmol) of PMDETA, 26.6 mg of ethyl 2-bromoisobutyrate (0.136 mmol) and 0.6 mL of diphenyl ether were placed in a 10 mL dry flask. The flask was evacuated and filled thrice with N_2, and then 0.2 g of GMA (1.4 mmol) was injected into the flask. The mixture was stirred at 50 ^{o}C for 48 h. The resulting sample was diluted with THF and vacuum- filtered. The filtrate was purified by passing through a neutral alumina column. The solution was concentrated by a rotary distiller and then precipitated in hexane. PGMA (P1) was obtained after drying overnight under vacuum. On the other hand, the filtered solid was washed with THF several times and separated by centrifugation and MWNT-*g*-PGMA (CP2a) was obtained after drying overnight under vacuum.

Preparation of amino derivatizated polymer-functionalized MWNTs (MWNT-g-PGMA-NH_2, CP3a) [75]. The epoxy groups of PGMA were further derivatizated into amino groups in the presence of amino chemicals. Typically, 20 mg of CP1a, 2.4 g of ethylene diamine and 10 mL of DMF were added to a 25 mL flask. The mixture was stirred at room temperature for 10 hours. The resulting sample was precipitated in acetone and separated by centrifugation. MWNT-*g*-PGMA-NH_2 (CP3a) was obtained after drying overnight under vacuum.

Preparation of magnetic MWNT-polymer/Fe_3O_4 nanohybrids [57]. Typically, 30.0 mg of MWNT-*g*-PGMA-NH_2 was dispersed in 100 mL water, and the solution was adjusted to weak acidity (pH 4.0) with aqueous HCl. Then 30 mL of Fe_3O_4 aqueous solution (4.31 mmol/L) was added to the flask. The mixture was stirred at room temperature overnight. The product was separated by centrifugation and washed with deionized water. After drying overnight under vacuum, the sample of magnetic MWNT-polymer/Fe_3O_4 nanohybrids was obtained.

Preparation of hydroxyl derivatizated polymer-functionalized MWNTs (MWNT-g-PGMA-OH, CP3b) [76]. The epoxy groups of PGMA can also be converted into hydroxyl groups in the presence of hydrochloric acid. Typically, 30 mg of CP1a, 2 mL of THF and 0.5 mL of methanol were added to a 10 mL flask. Then 0.3 mL of 6M HCl was added to the mixture under continuous stirring. Another 1.5 mL of methanol was injected to the flask as the reaction continued. The mixture was stirred at room temperature for 3 hours. The resulting product was precipitated in ether and separated by centrifugation. MWNT-*g*-PGMA-OH (CP3b) was obtained after drying overnight under vacuum.

Preparation of carboxyl derivatizated polymer-functionalized MWNTs (MWNT-g-PGMA-COOH, CP3c). The hydroxyl groups of CP3b were partially converted to carboxylic acid groups in the presence of succinic anhydride, triethylamine and DMAP [47]. Typically, 30 mg of CP1a, 50 mg of succinic anhydride, 1 mL of triethylamine, 100 mg of DMAP and 30 mL of DMF were added to a 100 mL flask and the mixture was stirred at 0 °C for 2 hours. The temperature was then increased to 40 °C and the mixture was stirred at 40 °C for another 24 hours. The reaction mixture was separated by centrifugation and then precipitated in acetone. MWNT-*g*-PGMA-COOH (CP3c) was obtained after drying overnight under vacuum.

Silver nanocrystals-decorated MWNTs were fabricated according to previous procedures with MWNT-*g*-PGMA-COOH (CP3c) as the templates (data not shown) [47].

Preparation of azido derivatizated polymer-functionalized MWNTs (MWNT-g-PGMA-N_3, CP3d) [77]. Typically, 100.0 mg of CP1b was dispersed in 10 mL DMF in a 50 mL flask, and the mixture was sonicated in a bath (40 kHz) for 10 min. Then 5 mL of NaN_3 aqueous solution (500 mg of NaN_3) was added into the flask. The flask was kept at 50 °C for 20 hours. The resulting sample was washed with water and THF, and then separated by centrifugation. The sample of MWNT-*g*-PGMA-N_3 (CP3d) was obtained after drying overnight under vacuum.

Preparation of alkyne modified coumarin 343 [65]. Typically, 57 mg of coumarin 343 (0.2 mmol), 56 mg of propargyl alcohol (1.00 mmol) and 2 mL CH_2Cl_2 were added in a 25 mL flask. Then 42 mg of DCC and 24 mg of DMAP dispersed in 2 mL CH_2Cl_2 was inject into the flask in N_2 atmosphere. The flask was immersed in an ice-water bath and kept at 0 °C for 2 h. and then the mixture was stirred at ambient temperature for 20 hours. The *N, N'*-dicyclodexylurea was filtered off and the filtrate was washed with CH_2Cl_2. The solvent was evaporated under reduced pressure to give alkyne modified coumarin 343.

Preparation of coumarin 343 modified MWNTs by click reaction [65,77]. Typically, 15.0 mg of alkyne modified coumarin 343 and 6 mg of CuBr were dispersed in 10 mL DMF in a 25 mL flask in N_2 atmosphere, and the mixture was sonicated in a bath (40 kHz) for 4 min. Then 30 mg of CP3d was added into the flask and the mixture was stirred at room temperature for 2 days. The resulting product was separated by centrifugation and washed with THF and methanol several times. Coumarin 343-modified MWNTs was obtained after drying overnight at 40 °C under vacuum.

RESULTS AND DISCUSSION

Design of Reactive Nanoplatform

Glycidyl methacrylate (GMA) is a kind of functional monomer containing epoxy groups. Herein, it is polymerized from the surfaces of multiwalled carbon nanotubes (MWNTs) by *in situ* atom transfer radical polymerization (ATRP). The epoxy groups were further converted to hydroxyl, amino, carboxyl, and azido groups, resulting in series of functional polymers-coated MWNTs. Scheme 1 shows the polymerization of GMA initiated by MWNT-Br and the further derivatization of the polymers.

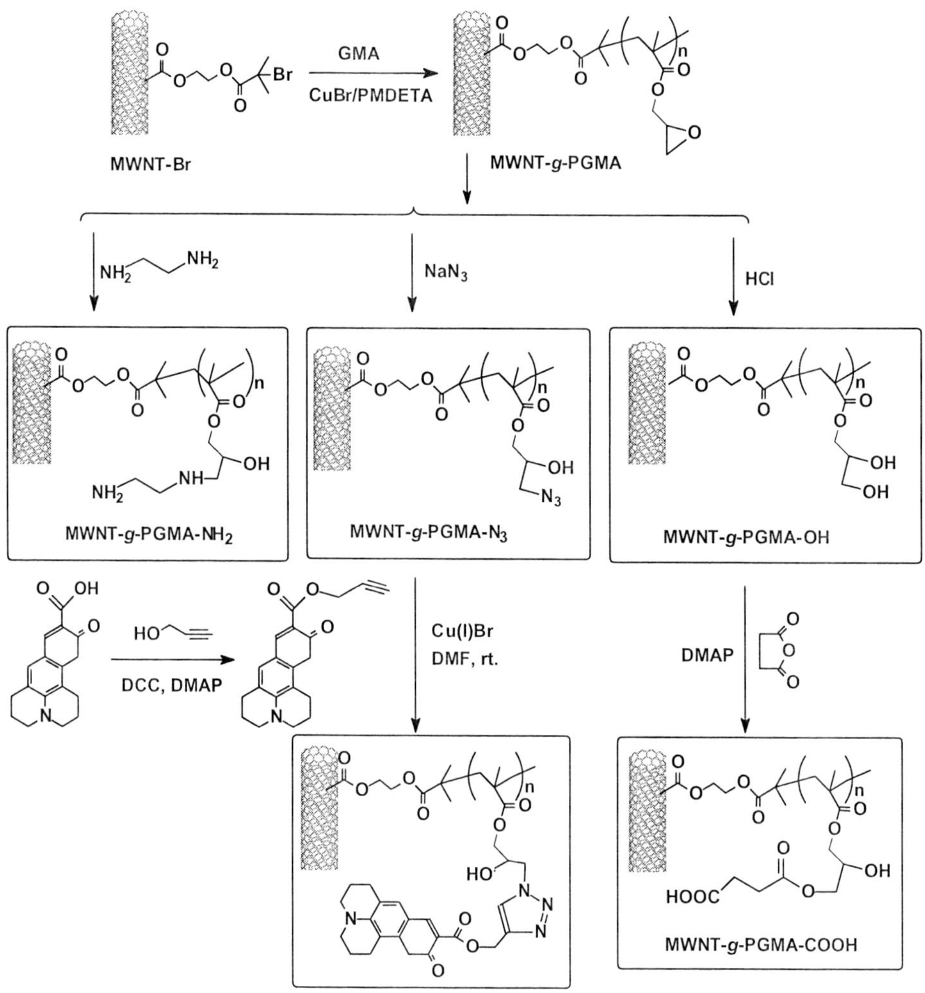

Scheme 1. Preparation of PGMA grafted-MWNT (MWNT-g-PGMA) by atom transfer radical polymerization (ATRP) and further modification of the reactive platform of MWNT-g-PGMA.

Table 1. Polymerization of glycidyl methacrylate (GMA) initiated with multiwalled carbon nanotube-based macroinitiator (MWNT-Br) without free initiator

Sample	R_{weight} [a]	R_{mole} [b]	f_{wt}% [c]	$M_{n,\ TGA}$ [d]
CP1a	1.9:1	20:1:3:3	64	2650
CP1b	2.8:1	29:1:3:3	70	3480
CP1c	6.0:1	62:1:3:3	77	5000
CP1d	10:1	105:1:3:3	82	6800

[a] The weight ratio of monomer/MWNT-Br.

[b] The mole ratio of monomer/initiator point/CuBr/PMDETA.

[c] The weight fraction of grafted PGMA calculated from TGA data.

[d] The average molecular weight of the grafted PGMA calculated from TGA data: $M_{n,TGA} = f_{wt} / [(1-f_{wt}) * 0.67 * 10^{-3}]$; herein, 0.67 represents the concentration of initiating sites per gram of MWNTs (mmol/g).

Table 2. Polymerization of glycidyl methacrylate (GMA) initiated with multiwalled carbon nanotube-based macroinitiator (MWNT-Br) and ethyl 2-bromoisobutyrate

Sample	R_{mole} [b]	$M_{n,\ theory}$	f_{wt}%	$M_{n,\ TGA}$ [c]	$M_{n,\ GPC}$	M_w/M_n
CP2a	8.4/1	1200	57	1900	6400	1.63
CP2b	16.5/1	2300	61	2300	7000	1.48
CP2c	67.0/1	9500	80	5900	8900	1.46

[a] The mole ratio of /MWNT-Br/ethyl 2-bromoisobutyrate /CuBr/PMDETA is 1/6.8/5/5.

[b] The mole ratio of monomer to total initiating points.

[c] The average molecular weight calculated from TGA results: $M_{n,TGA} = f_{wt} / [(1-f_{wt}) * 0.67 * 10^{-3}]$, herein, 0.67 represents the concentration of initiating sites per gram of MWNTs (mmol/g).

In this paper both MWNT-Br and the mixture of MWNT-Br with a co-initiator of ethyl 2-bromoisobutyrate were used to initiate the polymerization of GMA. The reaction conditions and results were summarized in table 1 and 2. The products of MWNT-*g*-PGMA were characterized by TGA, FTIR, ^{1}HNMR, SEM and TEM.

Characterization of MWNT-PGMA

Figure 1 shows the TGA curves of the obtained MWNT-*g*-PGMA with different feed ratio of monomer to initiator (R_{feed}) in the absence (a) or presence (b) of free initiators. The quantities of the coating polymers can be calculated through these TGA curves. The amount of PGMA increased from ca. 64% to 82% with the increasing monomer/MWNT-Br feed ratio in the absence of free initiator (see Table 1). Similar conditions can be observed when ethyl 2-bromoisobutyrate was used as a co-initiator: when the mole ratio of monomer to total initiating points (MWNT-Br and ethyl 2-bromoisobutyrate) increased from 8.4/1 to 67.0/1, the grafted PGMA quantities also increased from 57% to 80% (see Table 2).

On the other hand, the number average molecular weight of PGMA coated on the surfaces of MWNTs which can be calculated from the TGA results ($M_{n,TGA}$) increased from 2650 to 6800 as the monomer/initiator molar feed ratio varied from 20 to 105 (see Table 1) without free initiator. For the MWNT-*g*-PGMA and PGMA obtained in the presence of free initiator, $M_{n,TGA}$ of the grafted PGMA increased from 1900 to 5900 when the monomer/initiator molar feed ratio varied from 8.4 to 67. At the same time the molecular weight of free PGMA ($M_{n,GPC}$) increased from 6400 to 8900 (see Table 2).

From these data we can conclude that PGMA chains still grows from MWNTs surfaces in the presence of free initiator and MWNTs with high amounts of PGMA coated on the surfaces can be obtained in the absence/presence of the free initiator. These results also confirmed the previous conclusion that: the quantities of grafted polymer can be controlled to some extent by adjusting the R_{feed} [12,17].

FTIR was used to determine the chemical structure of as-prepared MWNT-*g*-PGMA and figure 2 shows the FTIR spectra of CP1a, CP2b and neat PGMA. The characteristic adsorption peaks of PGMA can be clearly observed in these representative spectra of MWNT-g-PGMA, which indicated that PGMA have been grown from the surfaces of MWNTs. The peak at 1100 cm^{-1} can be attributed to C-O-C groups, 1700 cm^{-1} for C=O stretching and 2900 cm^{-1} for C-H stretching.

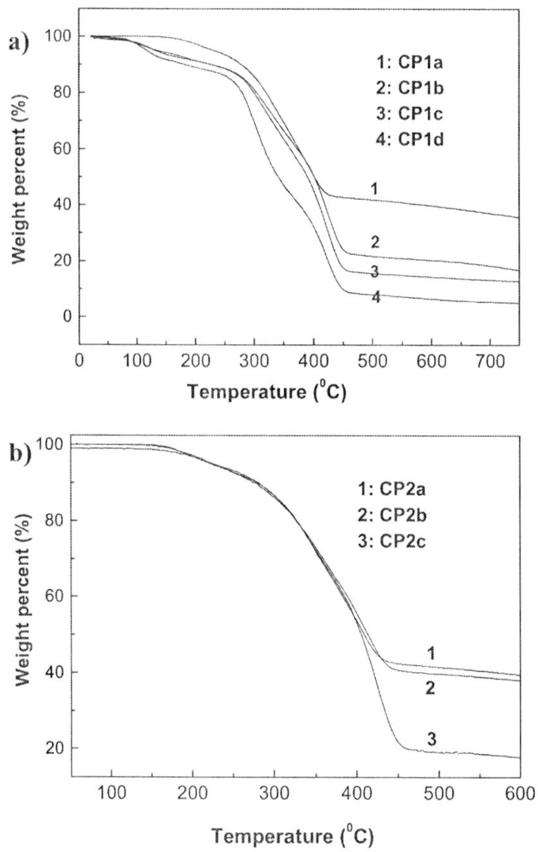

Figure 1. TGA curves of poly(glycidyl methacrylate)-grafted MWNTs (MWNT-g-PGMA) in the absence (a: CP1 series) and presence of free initiator (b: CP2 series).

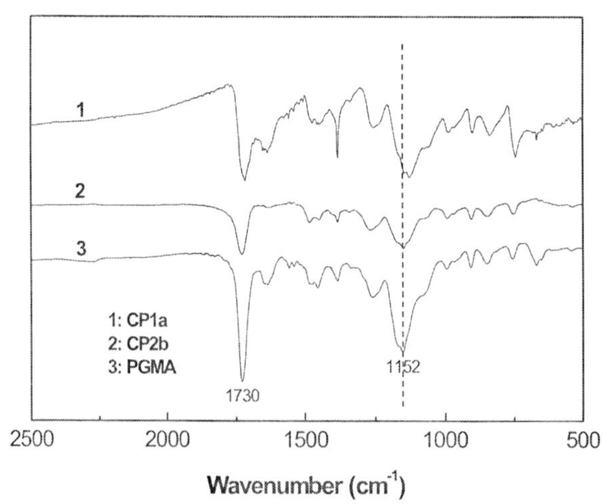

Figure 2. Fourier-transform infrared (FTIR) spectra of CP1a, CP2b and neat poly(glycidyl methacrylate) (PGMA) (PGMA was prepared by ATRP initiated by ethyl 2-bromoisobutyrate with number average molecular weight Mn = 17900, PDI= 1.43).

The morphology of the resulting MWNT-*g*-PGMA was observed under high-resolution TEM and SEM. Figure 3 shows the typical TEM images of MWNT-g-PGMA samples (CP1b and CP2b). A core-shell structure of the polymer-functionalized MWNTs can be clearly observed in these images with the polymer layer of ca. 5 nm thickness (figure 3a and b for CP1b). The polymer layers coated evenly and thoroughly on the surfaces of MWNTs. For the MWNT-g-PGMA samples prepared in the presence of free initiator, a similar structure was found: MNWTs with a even polymer layer coated on the surfaces.

SEM was also used to observe the morphology of the resulting MWNT-*g*-PGMA. Figure 4 displays the SEM images of CP1a. The tube like structure can be clearly viewed. Furthermore, some functionalized carbon nanotubes tangled together in some places probably due to the interaction between the polymer chains and the high polymer grafting density. These SEM observations together with TEM images intuitionisticly proved that the polymers have been successfully attached to MWNTs surfaces.

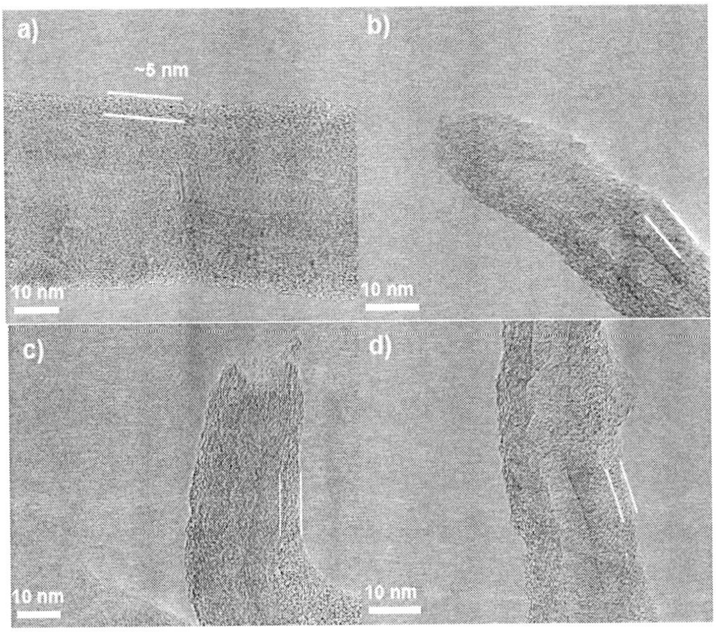

Figure 3. TEM images of CP1b (a~b) and CP2b (c~d).

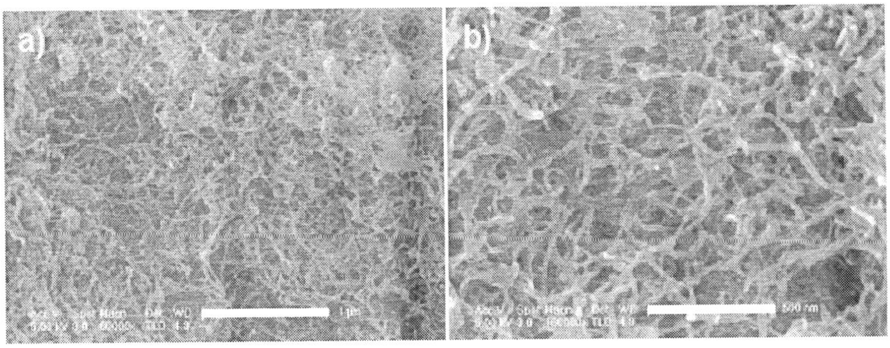

Figure 4. SEM images of CP1a.

Kinetics of the Preparation of MWNT-*g*-PGMA

The polymerization processes of MWNT-*g*-PGMA in the presence or absence of free initiator were monitored by ^{1}H NMR spectra in the experiments [32]. The samples taken from the reaction system were used to detect monomer conversions with the peak at δ 6.6-6.8 ppm of anisole as the reference. On the other hand, these samples were also purified to obtain MWNT-*g*-PGMA or neat PGMA: for the MWNT-*g*-PGMA prepared with only MWNT-Br as the initiator, the grafted polymer fraction (f_{wt}%) was obtained from corresponding TGA curves. For the MWNT-*g*-PGMA prepared with the co-initiators of MWNT-Br and ethyl 2-bromoisobutyrate, the molecular weight of free PGMA was measured by GPC and the grafted polymer fraction (f_{wt}%) of MWNT-*g*-PGMA was also obtained from corresponding TGA curves.

For the polymerization system with only MWNT-Br as the initiator, the value of ln([M_0]/[M]) increases linearly with the reaction time (Figure 5) ([M_0], the initial concentration of monomer; [M], the concentration of monomer at a given conversion). This is in good agreement with kinetics of normal ATRP in a homogeneous system [74]. The amount of the grafted polymer also increased with the reaction time: ca. 58wt% polymer coated on carbon nanotubes after 1 h of reaction and the polymer fraction increased to ca. 71% after 3 h of reaction.

For the reaction system with the co-initiators of MWNT-Br and ethyl 2-bromoisobutyrate, the grafted polymer fraction on MWNTs increased from 35 wt% after 0.5 h, 45 wt% after 1 h, 57 wt% after 5 h to 63 wt% after 7 h, implying that the polymer content can be well controlled in a wide range by the polymerization time. At the same time, for the free PGMA collected in the filtrate, linear kinetics plot between monomer conversion and molecular weight of the free polymer was also observed (Figure 6).

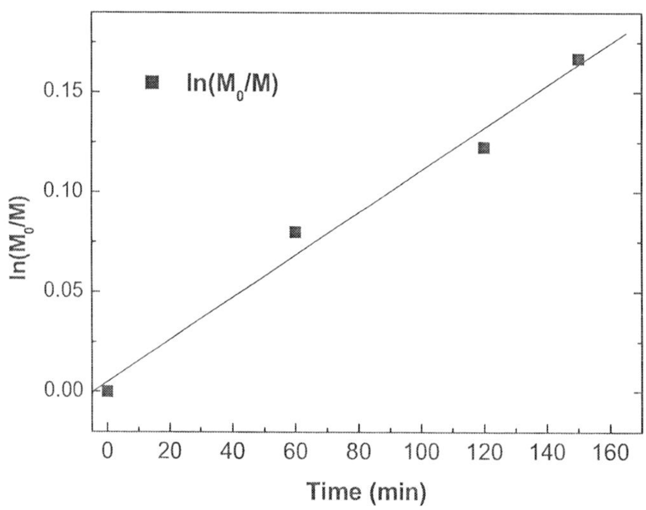

Figure 5. The ln([M0]/[M]) of the polymerization system as a function of reaction time during the preparation of MWNT-g-PGMA initiated only by MWNT-Br: herein M0 and M represent the initial concentration of the monomer and the concentration at a given time, respectively.

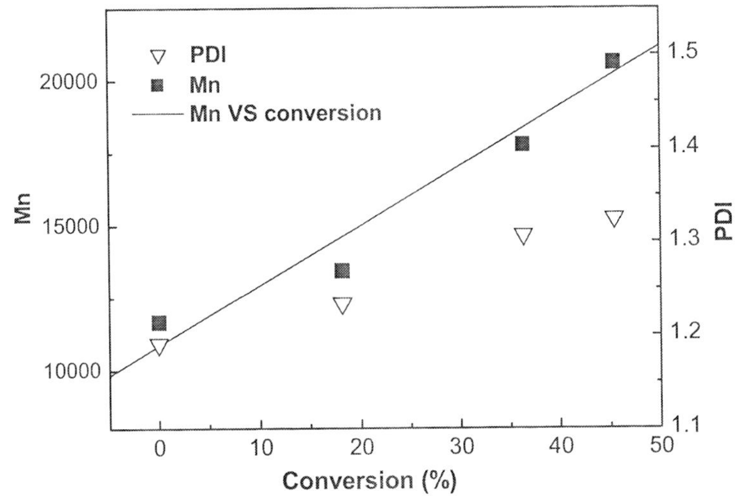

Figure 6. The number-average molecular weight and polydispersity index (PDI) of the free polymer obtained by grafting PGMA from surfaces of MWNTs initiated by MWNT-Br and ethyl 2-bromoisobutyrate, as a function of monomer conversion.

Based on these results we can draw the following conclusions: 1. MWNT-Br is a good initiator for the polymerization of GMA by ATRP both in the presence and absence of ethyl 2-bromoisobutyrate as the co-initiator; 2. the polymerization process can be controlled very well with MWNT-Br as the initiator, MWNTs with different amounts of PGMA coated on the surfaces can be obtained if we terminated the polymerization at different reaction time. Similar results were observed in our pervious work: the polymerization process can be controlled very well either initiated with MWNT-Br in the absence or presence of free initiator [32].

Preparation of MWNT-*g*-PGMA-NH_2 and MWNT-*g*-PGMA-NH_2/Fe_3O_4 Nanohybrids

Figure 7 shows the fourier-transform infrared (FTIR) spectra of MWNT-*g*-PGMA (CP1a) and MWNT-*g*-PGMA-NH_2 (CP3a). Compared to the FTIR spectra of MWNT-*g*-PGMA, two other typical adsorption peaks at 1570 cm^{-1} and 1230 cm^{-1} appeared in the spectra of MWNT-*g*-PGMA-NH_2, which can be assigned to -NH_2 groups. The quantities of the amino groups can be estimated through the TGA curves. As shown in Table 1, the grafted polymer fraction increased from ca. 64 wt% for MWNT-*g*-PGMA to 69 wt% for MWNT-*g*-PGMA-NH_2. This data suggests that ca. 70 % of epoxy groups are converted to amino groups.

The as-prepared MWNT-*g*-PGMA-NH_2 was then used to adsorb Fe_3O_4 magnetic nanoparticles by the electrostatic self-assembly approach, resulting in MWNT-*g*-PGMA-NH_2/Fe_3O_4 nanohybrids [21,57]. Figure 8 shows the typical TEM images of the MWNT-*g*-PGMA-NH_2/Fe_3O_4 nanohybrids. We can clearly observe in both images that a large amount of nanoparticles were assembled onto the MNWTs surfaces. Nanoparticles dispersed along the tubes and some particles accumulated together in some places.

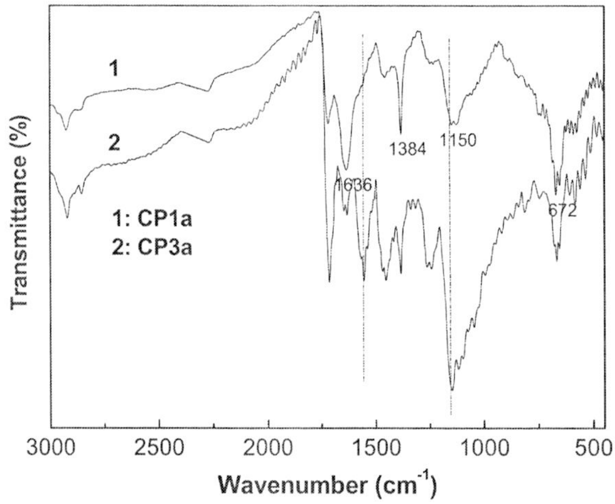

Figure 7. Fourier-transform infrared (FTIR) spectra of MWNT-g-PGMA (CP1a), MWNT-g-PGMA-NH2 (CP3a).

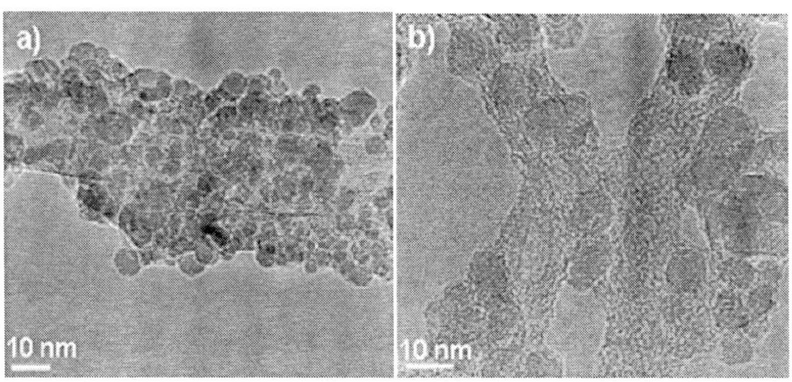

Figure 8. TEM images of MWNT/Fe3O4 nanohybrids.

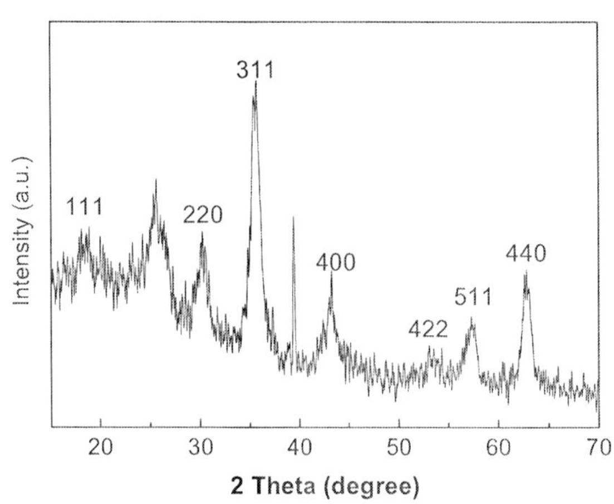

Figure 9. XRD patterns of MWNT/Fe3O4 nanohybrids.

In the XRD patterns, the characteristic reflection peaks of Fe_3O_4 at the 2θ angles of 30.2°, 35.5°, 43.6°, 53.6°, 57.2°, and 62.7°, assigning to *hkl* 220, 311, 400, 422, 511 and 440 respectively, all can be found (Figure 9). These TEM observations and XRD measurements further illuminated that amino groups were successfully introduced to the polymer chains and then interacted with negatively charged Fe_3O_4 nanoparticles forming magnetic MWNTs /Fe_3O_4 nanohybrids.

Preparation of MWNT-*g*-PGMA-OH and MWNT-*g*-PGMA-COOH

The grafted polymer fraction increased from 64% to 69%, then to 79% when the epoxy groups were converted to hydroxyl groups and then to carboxylic acids groups, implying the success of the derivatization (Figure 10). In the FTIR spectra of the resulting samples, the strong adsorption peaks at 1730 cm^{-1} illuminated that carboxyl groups have been successfully attached to the polymer chains. Carboxyl polymers-coated MWNTs can also be used to sequester metal ions such as Ag^+, forming MWNT-polymer/metal nanocomposites (data not shown) [47].

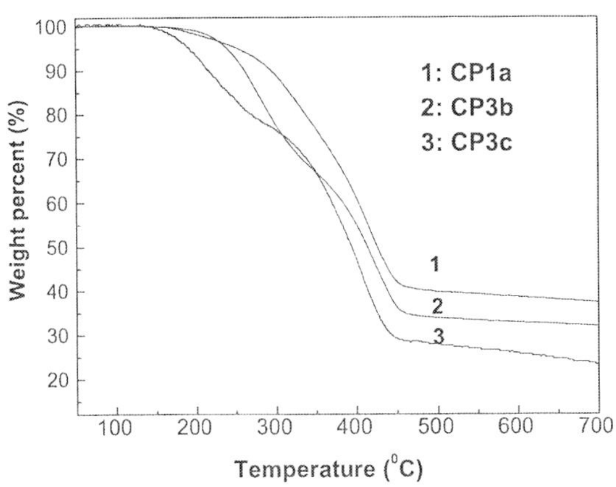

Figure 10. TGA curves of MWNT-g-PGMA (CP1a), MWNT-g-PGMA-OH (CP3b) and MWNT-g-PGMA-COOH (CP3c).

Preparation of MWNT-*g*-PGMA-N_3 and coumarin-modified MWNTs

Scheme 1 also shows the preparation steps of coumarin modified MWNTs. Alkyne-modified coumarin was synthesized by reaction of coumarin 343 with propargyl alcohol in the presence of DCC and DMAP. The epoxy groups of MWNT-*g*-PGMA were converted to azido groups by reaction with NaN_3. The "click" reaction between MWNT-*g*-PGMA-N_3 and alkyne modified coumarin 343 gave rise to luminescent MWNTs.

In the FTIR spectrum of MWNT-*g*-PGMA-N_3, the characteristic peak of azido groups at 2100 cm^{-1} was clearly observed, confirming the existence of azido groups in MWNT-*g*-PGMA-N_3. On the other hand all the characteristic peaks of coumarin 343 can be found in the

FTIR spectrum of coumarin-modified MWNTs, and characteristic peak of azido groups at 2100 cm^{-1} still exists but gets smaller, which illuminates that only parts of the azido groups were consumed in the click reaction and the coumarin 343 was successfully attached to the surfaces of carbon nanotubes. Hence, the residual azido groups can be further used to react with other functional molecules, producing multifunctional nanohybrids.

The emission property of the resulting coumarin-modified MWNTs was checked through the luminescence spectra. Figure 12 shows the luminescence spectra of coumarin-modified MWNTs, coumarin 343 and alkyne-modified coumarin in THF. For coumain 343, the emission peak appears at *ca.* 485 nm if excited at 437 nm, whereas for alkyne-modified coumarin, the emission peak shifts to *ca.* 460 nm likely due to the influence of propargyl groups. When coumarin was attached to the surfaces of carbon nanotubes, the emission peak was still at *ca.* 460 nm at the same exciting wavelength, but the emission intensity was much weaker. This is probably caused by the small quantity of coumarin attached on the CNTs surfaces.

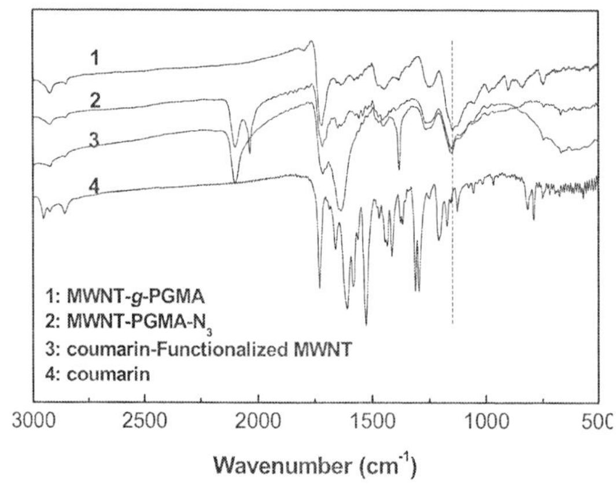

Figure 11. FTIR spectra of coumarin-functionalized MWNT, coumarin and alkyne modified Coumarin.

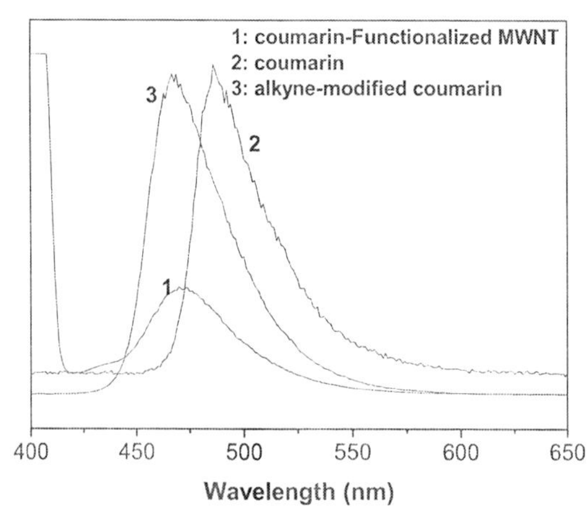

Figure 12. Luminescence spectra of coumarin-functionalized MWNT, coumarin and alkyne modified Coumarin in THF solution with a concentration of 0.1 mg/mL (λex = 437 nm).

CONCLUSION

A versatile reactive platform based on poly(glycidyl methacrylate) (PGMA)-functionalized multiwalled carbon nanotubes has been successfully prepared by *in situ* atom transfer radical polymerization (ATRP). The function of the MWNT-*g*-PGMA nanoplatform was certified by derivatization of the epoxy groups into hydroxyl, amino, carboxyl, and azido groups. Moreover, the resulting cationic amino- and anionic carboxyl-modified CNTs were employed as templates to adsorb magnetic nanoparticles and sequester metal nanocrystals, providing magnetic nanotubes and metal-decorated nanotubes. The PGMA-functionalized CNTs promises a multifunctional efficient platform for the synthesis of a series of functional nanocomposites and nanohybrids.

ACKNOWLEDGEMENTS

This work was financially supported by the National Natural Science Foundation of China (No. 50473010), the Foundation for the Author of National Excellent Doctoral Dissertation of China (No. 200527), Fok Ying Tung Education Foundation (No. 91013), the Program for New Century Excellent Talents in University, and the Opening Research Foundation of the Key Laboratory of Molecular Engineering of Polymers of Ministry of Education (Fudan University).

REFERENCES

[1] Iijima, S. Helical microtubules of graphitic carbon. *Nature,* 1991, *354*, 56-58.

[2] Dai, L; Mau, A. W. H. Controlled synthesis and modification of carbon nanotubes and C60: carbon nanostructures for advanced polymeric composite materials. *Adv. Mater.*, 2001, *13*, 899-913.

[3] Niyogi, S; Hamon, MA; Hu, H; Zhao, B; Bhowmik, P; Sen, R; Itkis, ME; Haddon, RC. Chemistry of single-walled carbon nanotubes. *Acc. Chem. Res.*, 2002, *35*, 1105-1113.

[4] Dai, H. Carbon nanotubes: synthesis, integration, and properties. *Acc. Chem. Res.*, 2002, *35*, 1035-1044.

[5] Hirsch, A. Functionalization of single-walled carbon nanotubes. *Angew. Chem., Int. Ed.*, 2002, *41*, 1853-1859.

[6] Bianco, A; Prato, M. Can carbon nanotubes be considered useful tools for biological applications? *Adv. Mater.*, 2003, *15*, 1765-1768.

[7] Lin, Y; Taylor, S; Li, H; Fernando, S; Qu, L; Wang, W; Gu, L; Zhou, B; Sun, Y.-P. Advances toward bioapplications of carbon nanotubes. *J. Mater. Chem,* 2004, *14*, 527-541.

[8] Qin, SH; Qin, DQ; Ford, WT; Resasco, DE; Herrera, JE. Functionalization of single-walled carbon nanotubes with polystyrene via grafting to and grafting from methods. *Macromolecules,* 2004, *37*, 752-757.

[9] Liu, YQ; Yao, ZL; Adronov, A. Functionalization of single-walled carbon nanotubes with well-defined polymers by radical coupling. *Macromolecules,* 2005, *38*, 1172-1179.

[10] Kong, H; Luo, P; Gao, C; Yan, D. Polyelectrolyte-functionalized multiwalled carbon nanotubes: Preparation, characterization and layer-by-layer self-assembly. *Polymer*, 2005, *46*, 2472-2485.

[11] Qin, SH; Qin, DQ; Ford, WT; Herrera, JE; Resasco, DE; Bachilo, SM; Weisman, RB. Solubilization and purification of single-wall carbon nanotubes in water by in situ radical polymerization of sodium 4-styrenesulfonate. *Macromolecules,* 2004, *37*, 3965-3967.

[12] Kong, H; Gao, C; Yan, D. Constructing amphiphilic polymer brushes on the convex surfaces of multi-walled carbon nanotubes by in situ atom transfer radical polymerization. *J. Mater. Chem.*, 2004, *14*, 1401- 1405.

[13] Koshio, A; Yudasaka, M; Zhang, M; Iijima, S. A simple way to chemically react single-wall carbon nanotubes with organic materials using ultrasonication. *Nano Lett.,* 2001, *1*, 361-363.

[14] Baskaran, D; Mays, JW; Bratcher, MS. Polymer-grafted multiwalled carbon nanotubes through surface-initiated polymerization. *Angew. Chem., Int. Ed.,* 2004, *43*, 2138-2142.

[15] Park, SJ; Cho, MS; Lim, ST; Cho, HJ; Jhon, MS. Synthesis and dispersion characteristics of multi-walled carbon nanotube composites with poly(methyl methacrylate) prepared by in-situ bulk polymerization. *Macromol. Rapid Commun.*, 2003, *24*, 1070-1073.

[16] Yao, Z; Braidy, N; Botton, GA; Adronov, A. Polymerization from the surface of single-walled carbon nanotubes--preparation and charac- terization of nanocomposites. *J. Am. Chem. Soc.*, 2003, *125*, 16015-16024.

[17] Kong, H; Gao, C; Yan, D. Controlled functionalization of multiwalled carbon nanotubes by in situ atom transfer radical polymerization. *J. Am. Chem. Soc.*, 2004, *126*, 412-413.

[18] Qin, S; Qin, D; Ford, WT; Resasco, DE; Herrera, JE. Polymer brushes on single-walled carbon nanotubes by atom transfer radical polymerization of n-butyl methacrylate. *J. Am. Chem. Soc.*, 2004, *126*, 170-176.

[19] Gao, C; Duan Vo, C; Jin, YZ; Li, W; Armes, SP. Multihydroxy polymer-functionalized carbon nanotubes: synthesis, derivatization, and metal loading, *Macromolecules*, 2005, *38,* 8634-8648.

[20] Li, W; Kong, H; Gao, C; Yan, D. pH-responsive poly(2-diethylaminoethyl methacrylate)-functionalized multiwalled carbon nanotubes. *Chinese Science Bulletin*, 2005 *50*, 2276-2280.

[21] Gao, C; Li, W; Morimoto, H; Nagaoka, Y; Maekawa, T. Magnetic carbon nanotubes: synthesis by electrostatic self-assembly approach and application in biomanipulations. *J. Phys. Chem. B,* 2006, *110,* 7213-7220.

[22] Kong, H; Li, W; Gao, C; Yan, D; Jin, YZ; Walton, DRM; Kroto, HW. Poly(N-isopropylacrylamide)-coated carbon nanotubes: temperature-sensitive molecular nanohybrids in water. *Macromolecules,* 2004, *37*, 6683-6686.

[23] Sun, T.; Liu, H.; Song, W.; Wang, X.; Jiang, L.; Li, L.; Zhu, D. Responsive aligned carbon nanotubes. *Angew. Chem., Int. Ed.,* 2004, *43*, 4663-4666.

[24] Lou, XD; Detrembleur, C; Pagnoulle, C; Jerome, R; Bocharova, V; Kiriy, A; Stamm, M. Selective deposition, and decoration of the nanotubes. *Adv. Mater.*, 2004, *16*, 2123-2127.

[25] Qin, SH; Qin, DQ; Ford, WT; Herrera, JE; Resasco, DE. Grafting of poly(4-vinylpyridine) to single-walled carbon nanotubes and assembly of multilayer films. *Macromolecules,* 2004, *37*, 9963-9967.

[26] Qu, LW; Lin, Y; Hill, DE; Zhou, B; Wang, W; Sun, XF; Kitaygorodskiy, A; Suarez, M; Connell, JW; Allard, LF; Sun, Y-P. Polyimide-functionalized carbon nanotubes: synthesis and dispersion in nanocomposite films. *Macromolecules,* 2004, *37*, 6055-6060.

[27] Wu, W; Zhang, S; Li, Y; Li, JX; Liu, LQ; Qin, YJ; Guo, ZX.; Dai, LM; Ye, C; Zhu, DB. PVK-modified single-walled carbon nanotubes with effective photoinduced electron transfer. *Macromolecules,* 2003, *36*, 6286-6288.

[28] Lin, Y; Zhou, B; Fernando, KAS; Allard, LF; Sun, Y -P. Polymeric carbon nanocomposites from carbon nanotubes functionalized with matrix polymer *Macromolecules*, 2003, *36*, 7199-7204.

[29] Cao, L; Yang, W; Yang, J; Wang, C; Fu, S. Hyperbranched poly(amidoamine)-modified multi-walled carbon nanotubes via grafting-from method. *Chem. Lett.*, 2004, *33*, 490-491.

[30] Xu, YY; Gao, C; Kong, H; Yan, DY; Jin, YZ; Watts, PCP. Growing multihydroxyl hyperbranched polymers on the surfaces of carbon nanotubes by in situ ring-opening polymerization. *Macromolecules*, 2004, *37*, 8846-8853.

[31] Hong, C-Y; You,Y-Z; Wu, D; Liu, Y; Pan, C-Y. Multiwalled carbon nanotubes grafted with hyperbranched polymer shell via SCVP. *Macromolecules*, 2005, *38*, 2606-2611.

[32] Gao, C; Muthukrishnan, S; Li, W; Yuan, J; Xu, Y; Müller, Axel H. E. Linear and Hyperbranched Glycopolymer-Functionalized Carbon Nanotubes: Synthesis, Kinetics, and Characterization. *Macromolecules*, 2007, *40*, 1803-1815.

[33] Zhao, B; Hu, H; Haddon, HC. Synthesis and properyies of a water-soluble single-walled carbon nanotube-poly(m-aminobnzene sulfonic acid) graft copolymer. *Adv. Func. Mater.* 2004, *14*, 71-76.

[34] Zhu, YQ; Hsu, WK; Kroto, HW; Walton, DRM. Carbon nanotube template promoted growth of NbS2 nanotubes/nanorods. *Chem. Commun.*, 2001, 2184-2185.

[35] Zhao, W; Song, C; Pehrsson, PE. Water-soluble and optically pH-sensitive single-walled carbon nanotubes from surface modification. *J. Am. Chem. Soc.*, 2002, *124*, 12418-12419.

[36] Chen, RJ; Zhang, Y; Wang, D; Dai, H. Noncovalent sidewall functionalization of single-walled carbon nanotubes for protein immobilization. *J. Am. Chem. Soc.*, 2001, *123*, 3838-3839.

[37] Kong, J; Chapline, M; Dai, H. Functionalized carbon nanotubes for molecular hydrogen sensors. *Adv. Mater.*, 2001, *13*, 1384-1386.

[38] Azamian, BR; Davis, JJ; Coleman, KS; Bagshaw, CB; Green, MLH. Bioelectrochemical single-walled carbon nanotubes. *J. Am. Chem. Soc.*, 2002, *124*, 12664-12665.

[39] Terrones, M; Redlich, P; Grobert, N; Trasobares, S; Hsu, WK; Terrones, H; Zhu, YQ; Hare, JPC; Reeves, L; Cheetham, AK; Ruhle, M, Kroto, HW; Walton, DRM. Carbon nitride nanocomposites. *Adv. Mater.*, 1999, *11*, 655-658.

[40] Zhao, LP; Gao, L. Novel in situ synthesis of MWNTs-hydroxyapatite composites. *Carbon*, 2004, *42*, 423-426.

[41] Jiang, K; Eitan, A; Schadler, LS; Ajayan, PM; Siegel, RW. Selective attachment of gold nanoparticies to nitrogen-doped carbon nanotubes. *Nano Lett.,* 2003, *3*, 275-3277.

[42] Azamian, BR; Coleman, KS; Davis, JJ; Hanson, N; Green, MLH. Directly observed covalent coupling of quantum dots to single-wall carbon nanotubes. *Chem. Commun.,* 366–367.

[43] Zanella, R; Basiuk, EV; Santiago, P; Basiuk, VA; Mireles, E; Puente-Lee, I; Saniger, JM. Deposition of gold nanoparticles onto thiol-functionalized multiwalled carbon nanotubes. *J. Phys. Chem. B*, *109*, 16290-16295.

[44] Xing, YC. Synthesis and electrochemical characterization of uniformly-dispersed high loading Pt nanoparticles on sonochemically-treated carbon nanotubes. *J. Phys. Chem. B*, *108*, 19255-19259.

[45] Quinn, BM; Dekker, C; Lemay, SG. Electrodeposition of noble metal nanoparticles on carbon nanotubes. *J. Am.Chem. Soc.* 2006, *127*, 6146–6147.

[46] Lee, Y; Song, HJ; Shin, HS; Shin, HJ; Choi, HC. Spontaneous formation of transition-metal nanoparticles on single-walled carbon nanotubes anchored with conjugated molecules. *Small , 1*, 975–979.

[47] Gao, C; Li, W; Jin, YZ; Kong, H. Facile and large-scale synthesis and characterization of carbon nanotube/silver nanocrystal nanohybrids. *Nanotechnology*, 2006, *17*, 2882–2890.

[48] Korneva, G; Ye, H; Gogotsi, Y; Halverson, D; Friedman, G; Bradley, J-C; Kornev, KG. Carbon nanotubes loaded with magnetic particles. *Nano Lett.*, 2005, *5*, 879-884.

[49] Georgakilas, V; Tzitzios, V; Gournis, D; Petridis, D. Attachment of magnetic nanoparticles on carbon nanotubes and their soluble derivatives. *Chem. Mater.*, 2005, *17*, 1613-1617.

[50] Lee, SW; Sigmund, WM. Formation of anatase TiO2 nanoparticles on carbon nanotubes. *Chem. Commun.*, 2003, 780-781.

[51] Huang, Q; Gao, L. Synthesis and characterization of CdS/multiwalled carbon nanotube heterojunctions. *Nanotechnology*, 2004, *15*, 1855-1860.

[52] Cao, J; Sun, J-Z; Hong, J; Li, H-Y; Chen, H-Z; Wang, M. Carbon nanotube/CdS Core-shell nanowires prepared by a simple room-temperature chemical reduction method. *Adv. Mater.*, 2004, *16*, 84-87.

[53] Banerjee, S; Wong, SS. Situ quantum dot growth on multiwalled carbon nanotubes. *J. Am. Chem. Soc.*, 2003, *125*, 10342-10350.

[54] Banerjee, S; Wong, SS. Synthesis and characterization of carbon nanotube-nanocrystal heterostructures. *Nano Lett.*, 2002, *2*, 195-200.

[55] Ravindran, S; Chaudhary, S; Colburn, B; Ozkan, M; Ozkan, CS. Covalent coupling of quantum dots to multiwalled carbon nanotubes for electronic device applications. *Nano Lett.*, 2003, *3*, 447-453.

[56] Azamian, BR; Coleman, KS; Davis, JJ; Hanson, N; Green, MLH. Directly observed covalent coupling of quantum dots to single-wall carbon nanotubes. *Chem. Commun.*, 2002, 366-367.

[57] Li, W; Gao, C; Qian, H; Ren, J; and Yan, D. Multiamino-functionalized carbon nanotubes and their applications in loading quantum dots and magnetic nanoparticles. *J. Mater. Chem.*, 2006, *16*, 1852–1859.

[58] Kolb, HC; Finn, MG; Sharpless, KB. Click chemistry: diverse chemical function from. a few good reactions. *Angew. Chem., Int. Ed.* 2001, *40*, 2004-2021.

[59] Rostovtsev, VV; Green, LG; Fokin, VV; Sharpless, KB. A stepwise Huisgen cycloaddition process: copper(I)-catalyzed regioselective "ligation" of azides and terminal alkynes. *Angew. Chem., Int. Ed.*, 2002, *41*, 2596-2599.

[60] Binder, WH; Kluger, C. Combining ring-opening metathesis polymerization (romp) with sharpless-type click reactions: An easy method for the preparation of side chain functionalized poly(oxynorbornenes). *Macromolecules*, 2004, *37*, 9321-9330.

[61] Lutz, J-F; Boerner, HG; Weichenhan, K. Combining Atom Transfer Radical Polymerization and click chemistry: A versatile method for the preparation of end-functional polymers. *Macromol. Rapid Commun.* 2005, *26*, 514-518.

[62] Mantovani, G; Ladmiral, V; Tao, L; Haddleton, DM. One-pot tandem living radical polymerisation-Huisgens cycloaddition process ("click") catalysed by N-alkyl-2-pyridylmethanimine/Cu(I)Br complexes. *Chem. Commun.*, 2005, 2089-2091.

[63] Sumerlin, BS; Tsarevsky, NV; Louche, G; Lee, RY; Matyjaszewski, K. Highly Efficient "Click" Functionalization of Poly(3-azidopropyl methacrylate) Prepared by ATRP. *Macromolecules*, 2005, *38*, 7540-7545.

[64] Gao, H; Louche, G; Sumerlin, BS; Jahed, N; Golas, P; Matyjaszewski, K. Dihydroxypolystyrene synthesized via ATRP and click chemistry. *Macromolecules,* 2005, *38*, 8979-8982.

[65] Ladmiral, V; Mantovani, G; Clarkson, GJ; Cauet, S; Irwin, JL; Haddleton, DM. Synthesis of neoglycopolymers by a combination of "click chemistry" and living radical polymerization. *J. Am. Chem. Soc.,* 2006, *128*, 4823-4830.

[66] Gao, H; Matyjaszewski, K. Synthesis of star polymers by a combination of ATRP and the "click" coupling method. *Macromolecules*, 2006, *39*, 4960-4965.

[67] Wu, P; Feldman, AK; Nugent, AK; Hawker, CJ; Scheel, A; Voit, B; Pyun, J; Frechet, JMJ; Sharpless, KB; Fokin, VV. Efficiency and fidelity in click chemistry route to triazole dendrimers via the Cu(I)-catalyzed ligation of azides and alkynes. *Angew. Chem., Int. Ed.*, 2004, *43*, 3928-3932.

[68] Wu, P; Malkoch, M; Hunt, JN; Vestberg, R; Kaltgrad, E. Finn, MG; Fokin, VV; Sharpless, KB; Hawker CJ. Multivalent, bifunctional dendrimers prepared by click chemistry. *Chem. Commun.*, 2005, 5775–5777.

[69] Guo, Z; Lei, A; Liang, X; Xu, Q. Click chemistry: a new facile and efficient strategy for preparation of functionalized HPLC packings. *Chem. Commun.*, 2006, 4512 – 4514.

[70] Mariano, O-M; Javier, L-J; Fernando, H-M; Francisco, S-G. Synthesis of glyco-silicas by Cu(I)-catalyzed "click-chemistr" and their applications in affinity chromatography. *Advanced Synthesis and Catalysis* 348, *16-17*, 2410 – 2420.

[71] Zeng, Q; Li, T; Cash, B; Li, S; Xie, F; Wang, Q. Chemoselective derivatization of a bionanoparticle by click reaction and ATRP reaction. *Chem. Commun.*, 2007, 1453–1455.

[72] Hsu, T-L; Hanson, SR; Kishikawa, K; Wang, S-K; Sawa, M; Wong, C-H. Alkynyl sugar analogs for the labeling and visualization of glycoconjugates in cells. *Proc. Natl. Acad. Sci. USA.* 2007, *104*, 2614-2619.

[73] Gallant, ND; Lavery, KA; Amis, EJ; Becker ML. Universal gradient substrates for "click" biofunctionalization. *Adv. Mater.*, 2007, *19*, 965-969.

[74] Cañamero, PF; Fuente, JL de la; Madruga, EL; Fernández-García, M. Atom transfer radical polymerization of a functional monomer, glycidyl methacrylate. *Macromol. Chem. Phys.*, 2004, *205*, 2221-2228.

[75] Yu, WH; Kang, ET; Neoh KG. Controlled grafting of well-defined epoxide polymers on hydrogen-terminated silicon substrates by surface-initiated ATRP at ambient temperature. *Langmuir*, 2004, *20,* 8294-8300.
[76] Li, Z; Liu, G. Water-dispersible tetrablock copolymer synthesis, aggregation, nanotube preparation, and impregnation. *Langmuir*, 2003, *19,* 10480-10486.
[77] Slater, M; Snauko, M; Svec, F; Fréchet, Jean MJ. "Click Chemistry" in the preparation of porous polymer-based particulate stationary phases for *i*-HPLC separation of peptides and proteins *Anal. Chem.*, 2006, *78*, 4969 -4975.

In: Encyclopedia of Nanotechnology
Editor: Elwood D. Carlson

ISBN: 978-1-60692-079-4

Chapter 32

SUPER CARBON NANOTUBE: ITS CONCEPT, GEOMETRIC CONSERVATION LAWS AND MECHANICAL PROPERTIES

Yuli Chen, Yajun Yin*, Bin Liu and Keh-Chih Hwang
FML, Department of Engineering Mechanics,
Tsinghua University, 100084, Beijing, China

ABSTRACT

Based on the integral theorems for mean curvature and Gauss curvature, geometric conservation laws for spontaneous Y-branched carbon nanotubes are presented. From these laws the concept of "super carbon nanotube" is proposed. It may be proved that such a super structure might be of the lowest energy both locally and globally. To make clear the response of the super carbon nanotube under loading, the mechanical properties of the fundamental cell in super carbon nanotube are studied by atomic-scale finite element method.

Keywords and phrases: *Super carbon nanotubes, geometric conservation laws, mechanical properties.*

1. INTRODUCTION

The discovery of carbon nanotubes in 1991 [1] opened up a new research field. In 1995, Zhou et al. observed branched carbon nanotubes in experiments, such as L, Y, and T patterns [2]. These nanostructures have been applied as multi-terminal electronic devices and circuits [3, 4]. Among them, the Y-branched carbon nanotubes have drawn intense attentions [5-14],

* Corresponding author: Prof. Yajun YIN, E-mail: yinyj@mail.tsinghua.edu.cn, Tel: +86-10-62795536, Fax: +86-10-62781824.

because their symmetrical structures make them possible to act as new nanotube-based networks [15, 16].

On the basis of symmetric Y-branched carbon nanotubes, a novel carbon nanostructure called "super carbon nanotube" is published recently [17, 18]. Such beautiful carbon nanostructures have drawn the attentions of researchers. Pugno [19] studied the mimicking nacre with super-nanotubes for producing optimized super-composites from the viewpoint of fractal structures. Wang and Qiu et al. [20] simulated the mechanical properties of the super-graphite (from which super carbon nanotubes may be made) through elastic thin shell theory and finite element method. For their unique configurations and potential applications in industries, super carbon nanotubes will attract more and more researchers without doubt. Hence to make clear their concepts, geometries and mechanical properties may be of essential importance.

The paper is organized as follows. In section 2, the geometric conservation laws for perfect Y-branched carbon nanotubes are developed. Then, section 3 presents the possible applications of the geometric regulations to the design of super carbon nanostructures. Most of the details in sections 2 and section 3 are cited from the work of YIN et al. in reference [18]. Section 4 shows the predicted mechanical properties of super carbon nanotubes by atomic-scale finite element method.

2. Geometric Conservation Laws and Physical Principle for Perfect Y-Branched Carbon Nanotubes

Recently, geometric conservation laws [21] for 3-way nanotube junctions [22, 23] are revealed. A 3-way nanotube junction, with three lipid vesicles interconnected by three membrane nanotubes intersecting at a point (figure 1), may be formed through dynamic self-organizations of lipid molecules [22, 23]. Geometrically, this beautiful bionano structure has been proved to observe very simple geometric regulations [21]. Surprisingly, such kind of geometric structures exist not only in biology but also in physics and material sciences.

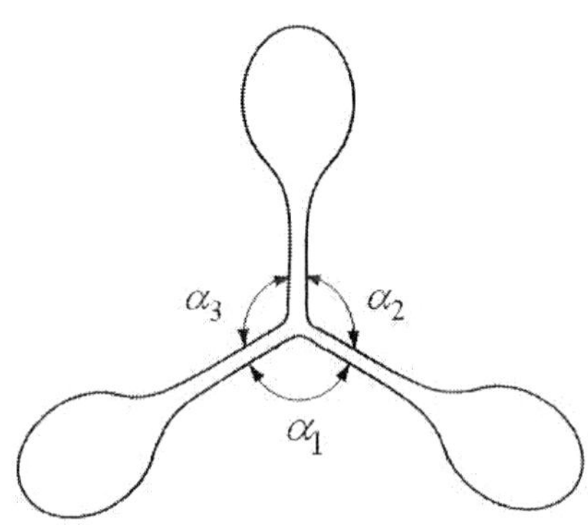

Figure 1. Smooth and closed curved surface formed from three vesicles and a 3-way nanotube junction.

A very typical example is the Y-branched carbon nanotubes [5-14]. Hence a question may be asked: Do Y-branched carbon nanotubes also obey geometric conservation laws or what geometric regulations control their growth? To answer this question, similar ideas and procedures in reference [21] will be used.

2.1. Geometric Conservation Laws for Perfect Y-Branched Carbon Nanotubes

In this paper, perfect Y-branched carbon nanotubes will be studied. Here "perfect" means that there are no defects such as dislocations or misconnected bonds. Geometrically such perfect structures may be idealized as globally smooth curved surfaces or 2D Riemann manifolds. There are two reasons for this idealization: (a) It coincides with existing evidence [5] —— Seamless and smooth transition has been found at the connecting locations between neighboring branches of a perfect Y-branched carbon nanotube. (b) It makes differential geometry applicable and meaningful information about Y-branched carbon nanotubes available. To deal with the perfect Y-branched carbon nanotube correctly, a general smooth curved surface in figure 2 is considered first. Let $\boldsymbol{n}$ be the outward unit normal of the surface and C be any smooth and closed curve drawn on the surface. At any point of the curve, let $\boldsymbol{t}$ be the unit tangent to the curve and along the positive direction of the curve. Let $\boldsymbol{m}$ be the unit vector tangential to the surface and normal to the curve, drawn outward from the region enclosed by C. Unit vectors $\boldsymbol{t}$, $\boldsymbol{n}$ and $\boldsymbol{m}$ form a right-handed system on curve C (figure 2) with the relation $\boldsymbol{m} = \boldsymbol{t} \times \boldsymbol{n}$ satisfied. On such a surface, integral theorems about two characteristic geometric quantities, i.e. the mean curvature $H = (c_1 + c_2)/2$ and Gauss curvature $K = c_1 c_2$ with c_1 and c_2 the two principle curvatures, will be specially focused. The first integral theorem in differential geometry is about the mean curvature H:

$$\oint_C d\boldsymbol{s} = \iint_A 2H d\boldsymbol{A} \tag{1}$$

The second one is about the Gauss curvature K [24]:

$$\oint_C \left(k_n \boldsymbol{m} + \tau_g \boldsymbol{t}\right) ds = \iint_A 2K d\boldsymbol{A} \tag{2}$$

Here $d\boldsymbol{s} = \boldsymbol{m} ds$ is the vector element with ds the length element along the curve C. $k_n = \frac{d\boldsymbol{t}}{ds} \cdot \boldsymbol{n}$ and $\tau_g = \frac{d\boldsymbol{n}}{ds} \cdot \boldsymbol{m}$ are respectively the normal curvature and the geodesic torsion at a point on the curve C. $d\boldsymbol{A} = \boldsymbol{n} dA$ is the element area vector in the normal direction of the curved surface and A is the area enclosed by C. Equations (1) and (2) lay the foundation of the conservation laws for perfect Y-branched carbon nanotubes.

Once equations (1) and (2) are applied to the perfect Y-branched carbon nanotube, a problem may not be avoided: How to select properly the boundary curve C along which the line integrals are carried out? Here three principles are suggested. First, the line integrals should be easy enough to calculate. Second, they should include enough information about the geometric characteristics of the Y-branched carbon nanotube. Third, they should lead to simple conservation laws. According to these principles, we cut the three branches respectively along their cross sections (figure 3a).

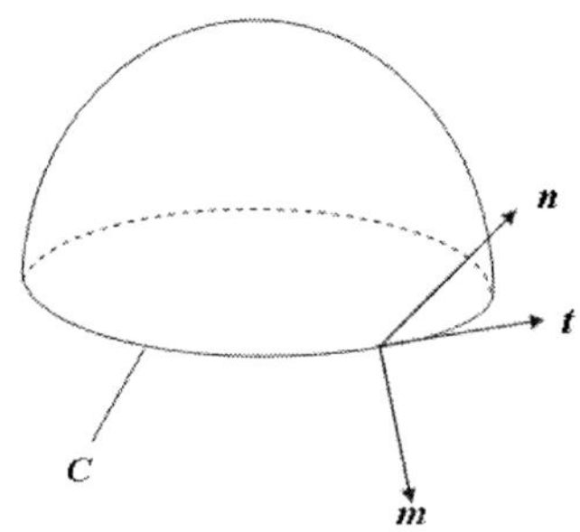

Figure 2. Schematic of curved surface with unit vectors $\boldsymbol{m}$, $\boldsymbol{t}$ and $\boldsymbol{n}$ at its boundary.

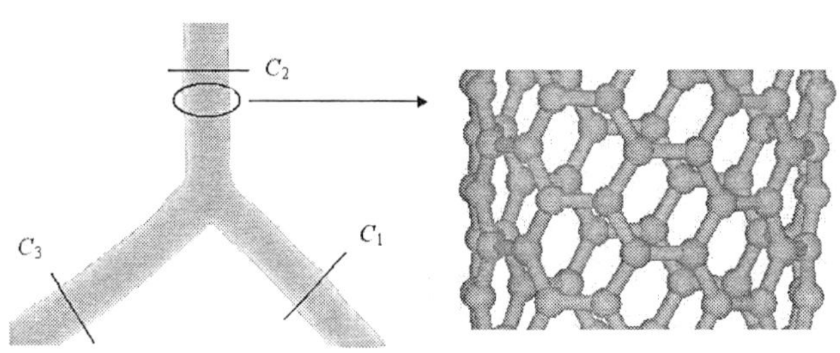

Figure 3a. Perfect Y-branched carbon nanotube with the three branches cut respectively along their cross sections.

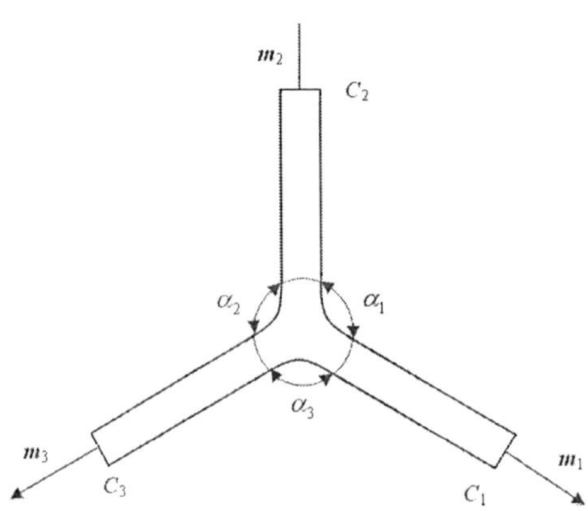

Figure 3b. Smooth and open curved surface with three boundaries, abstracted from the perfect Y-branched carbon nanotube in figure 3a.

Thus a smooth and open curved surface A with three boundaries C_i $(i=1,2,3)$ may be obtained (figure 3b). If the location of C_i is far enough from the junction, then the branch's shape near C_i may be treated as part of a circular cylindrical surface. Thus the unit vector $\boldsymbol{m}_i$ on C_i may be parallel to the axis of the branch and the line integrals will become:

$$\oint_{C_i} d\boldsymbol{s} = \boldsymbol{m}_i \int_0^{2\pi} r_i d\theta = 2\pi r_i \boldsymbol{m}_i \tag{3}$$

$$\oint_{C_i} (k_n \boldsymbol{m} + \tau_g \boldsymbol{t}) ds = \boldsymbol{m}_i \oint_{C_i} k_n ds = -\boldsymbol{m}_i \int_0^{2\pi} d\theta = -2\pi \boldsymbol{m}_i, \ (\tau_g)_i = 0 \tag{4}$$

Here r_i is radius of C_i. At last equations (1) and (2) may be changed into

$$\iint_A H dA = \frac{1}{2}\sum_{i=1}^{3} \oint_{C_i} d\boldsymbol{s} = \pi \sum_{i=1}^{3} (r_i \boldsymbol{m}_i) \tag{5}$$

$$\iint_A K dA = \frac{1}{2}\sum_{i=1}^{3} \oint_{C_i} (k_n \boldsymbol{m} + \tau_g \boldsymbol{t}) ds = -\pi \sum_{i=1}^{3} \boldsymbol{m}_i \tag{6}$$

Equations (5) and (6) are the geometric conservation laws for perfect Y-branched carbon nanotubes. Of course these laws may not be valid for imperfect ones, because the basis for the assumption of globally smooth curved surface is lost. Equation (5) means that the integral of the mean curvature H on the Y-branched carbon nanotube is determined not only by the directions but also by the radii of branches. Equation (6) implies that the integral of the Gauss curvature K on the Y-branched carbon nanotube is just related to the directions of branches.

2.2. Symmetric Regulations for Spontaneous Y-Branched Carbon Nanotubes

Technically there are different ways to produce branched carbon nanotubes, e.g. "forced" branching [7] in templates with branched nano-channels, spontaneous branching [8-14] and electron beam welding [25]. To evaluate the reasonability of equations (5) and (6), spontaneous Y-branched carbon nanotubes will be carefully investigated. From experiments [6, 8, 9, 13], it is found that a spontaneous Y-branched carbon nanotube usually has very regular geometry and the radius of each branch looks very uniform along the branch. Thus every branch may be further abstracted as a circular cylindrical surface A_i on which constant mean curvature ($H_i - Const.$) and zero Gauss curvature ($K_i = 0$) are kept. If all the three branches are long enough, then the following integrals in which the contribution of the junction is omitted may be approximated:

$$\iint_A H dA \approx \sum_{i=1}^{3}\left(\iint_{A_i} H_i dA\right) = \sum_{i=1}^{3}\left(H_i \iint_{A_i} dA\right) = \mathbf{0} \tag{7}$$

$$\iint_A K dA \approx \sum_{i=1}^{3} \iint_{A_i} K_i dA = \mathbf{0} \tag{8}$$

Equations (5) ~ (8) together will give

$$\sum_{i=1}^{3}\left(r_i \boldsymbol{m}_i\right) = \mathbf{0} \tag{9}$$

$$\sum_{i=1}^{3} \boldsymbol{m}_i = \mathbf{0} \tag{10}$$

Equations (9) and (10) may lead to detailed information about the geometry of the spontaneous Y-branched carbon nanotube. Here equation (10) will be discussed first. Mechanically equation (10) may be equivalent to the equilibrium equation for three unit forces. In mechanics, it is well known that the necessary and sufficient condition for this equilibrium is: (a) the three unit forces intersecting at a point should keep on the same plane; and (b) the angle between two neighboring forces should be 120°, i.e.

$$\alpha_1 = \alpha_2 = \alpha_3 = 120^\circ \tag{11}$$

Similarly, equation (9) may also be regarded as an equilibrium equation for three forces. With equation (11) satisfied, the necessary and sufficient condition for the equilibrium in equation (9) is that the radii of the three branches should be equal:

$$r_1 = r_2 = r_3 \tag{12}$$

Equations (11) and (12) are the geometric regulations for spontaneous Y-branched carbon nanotubes. From experiments, it may be confirmed that forced Y-branched nanotubes [9] have non-equal angle between branches but spontaneous ones possess structures with branches orientated at 120° [6, 8, 9, 13]. Therefore equation (11) coincides strictly with existing experiments. Besides, TEM images have shown that the radii of the three branches of a spontaneous Y-branched carbon nanotube are roughly the same [6, 12,13]. Hence equation (12) also agrees well with experiments. Such coincidences support the validity of the geometric conservation laws in equations (5) and (6).

2.2. Physical Principle for Perfect Y-Branched Carbon Nanotubes

A spontaneous Y-branched carbon nanotube with equations (11) and (12) satisfied corresponds to an equilibrium state with both minimum energy and symmetric geometry as well. This judgment may be interpreted in detail as follows. For simplicity (but without losing generality), an arbitrary Y-branched carbon nanotube with junction S is supposed to grow in such a way that its three branches SA_i ($i = 1, 2, 3$) are connected just right with three fixed points A_1, A_2 and A_3 on the plane (figure 4). The total elastic curvature energy of this Y-branched carbon nanotube may be written as:

$$E = \sum_{i=1}^{3} E_i \tag{13}$$

E_i is the ith branch's elastic curvature energy. Through a continuum limit treatment [26] to the potential given by Lenosky et al. [27], the curvature elastic energy of a single layer curved graphite has been proved to be:

$$E_s = \iint_A \left[\frac{1}{2} k_c (2H)^2 + \bar{k}_1 K \right] dA \tag{14}$$

k_c and $\bar{k}_1$ are the bonding elastic constants. Once equation (14) is applied to the Y-branched carbon nanotube with $H_i = -1/2r_i$ and $K_i = 0$, one has

$$E_i = \frac{\pi k_c l_i}{r_i} \tag{15}$$

Here l_i is the length of the ith branch. Suppose $r_1 = r_2 = r_3 = r$, then equations (13) and (15) will lead to

$$E = \frac{\pi k_c}{r} \sum_{i=1}^{3} l_i = \frac{\pi k_c l}{r} \tag{16}$$

$l = \sum_{i=1}^{3} l_i$ is the total length of the three branches. The total potential energy F for an arbitrary Y-branched carbon nanotube is determined by $F = E + F_p$ with F_p the potential of external forces. However, for a spontaneous Y-branched carbon nanotube external forces do not exist and $F = E$ is met. Thus the equilibrium of the spontaneous Y-branched carbon nanotube will render F a minimum value $F_{\min} = E_{\min}$:

$$E_{\min} = \frac{\pi k_c l_{\min}}{r} \tag{17}$$

Equation(17) clearly reveals that the spontaneous Y-branched carbon nanotube, at equilibrium and with minimum energy, is of minimum total length $l_{\min}$. According to the theory of Steiner minimal tree [28] in geometry, the necessary and sufficient condition for $l_{\min}$ is to set S as the Steiner point S' at where $\angle A_1S'A_2 = \angle A_2S'A_3 = \angle A_3S'A_1 = 120^\circ$ (figure 4). Therefore equation (17) includes the following information: (a) For a spontaneous Y-branched carbon nanotube at equilibrium, minimum energy means symmetric geometry, and vice versa. (b) Once a Y-branched carbon nanotube is grown spontaneously through dynamic self-organization, it tends to form automatically into a perfect geometry with low energy and high symmetry. In another word, the symmetric geometry described by equations (11) and (12) is a "natural" one to which a spontaneous branching process will approach with priority. (c) Any Y-branched carbon nanotube that diverges from the symmetric geometry may possess higher energy. In fact, any divergence needs external forces. For example, once a Y-branched carbon nanotube is grown through a template, the external forces acted by the template will force the nanotubes to grow along the channels. At last the forced Y-branched carbon nanotube not only possesses non-symmetric structure such as curved branches, non-equal radii and non-equal angles, but also stores residual stresses or energies inside. In short, spontaneous Y-branched carbon nanotube at equilibrium, minimum energy and symmetric geometry are unitary.

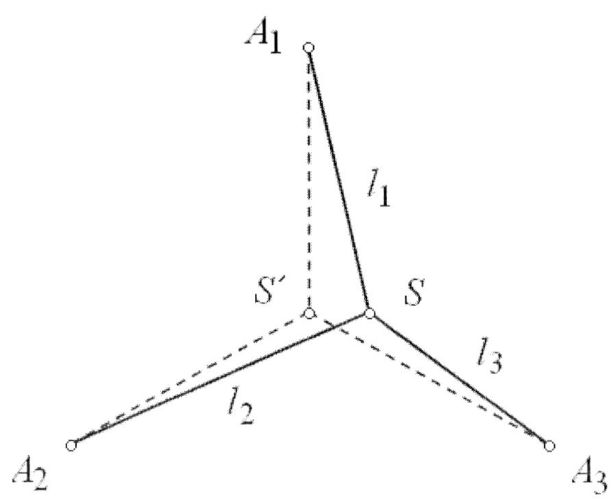

Figure 4. Y-branched carbon nanotube with three branches connecting with three fixed points.

To one's surprise, although the derivation processes are different, both the conservation laws in equations (5) and (6) and the geometric regulations in equations (11) and (12) for Y-branched carbon nanotubes are similar to that for 3-way membrane nanotube junctions [21]. Why do these similarities exist? There are three reasons. First, Y-branched carbon nanotubes and 3-way membrane nanotube junctions have similar geometries. Second, the major premises for the derivation processes, i.e. the integral theorems in equations (1) and (2), are the same. Third, the philosophies or beliefs for the two studies are the same: i.e. 2D Riemann manifolds instead of Euclidian spaces should be the natural choice of both the carbon nano world and the bionano world.

3. Concept of Super Carbon Nanotubes

Equations (11) and (12) may be of potential applications in nanoscience and nanotechnology —— They may provide inspirations for probing new carbon nanostructures.

3.1. Conformation of Super Carbon Nanotubes

First, a possible and simple carbon nanostructure is the periodic network of carbon nanotubes called "super graphite". As mentioned above, spontaneous branching has symmetry but no controllability, whereas forced branching has controllability but no symmetry. This may stimulate such an inspiration: Is it possible to design a forced branching process with both controllability and symmetry? The answer is positive: If the template with symmetric Y-branched nano-channel is prepared (figure 5a), then the controllable forced branching with equations (11) and (12) satisfied will be equivalent to a symmetric spontaneous one. Extending this inspiration one can have two symmetric Y-branched nano-channels interconnected together (figure 5b). If four symmetric Y-branched nano-channels are connected in the way shown in figure 5c, a hexagonal nano-channel network may be formed. Repeating the above procedures one can get a nano-channel network covered periodically with "hexagonal cells" (figure 5d).

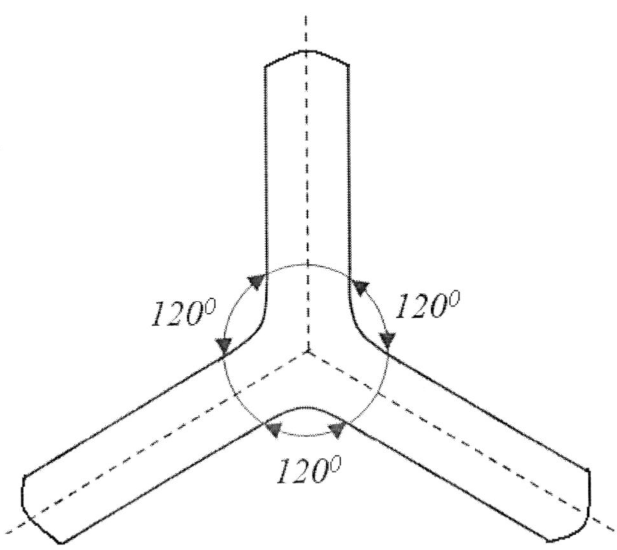

Figure 5a. Template with one symmetric Y-branched nano-channel.

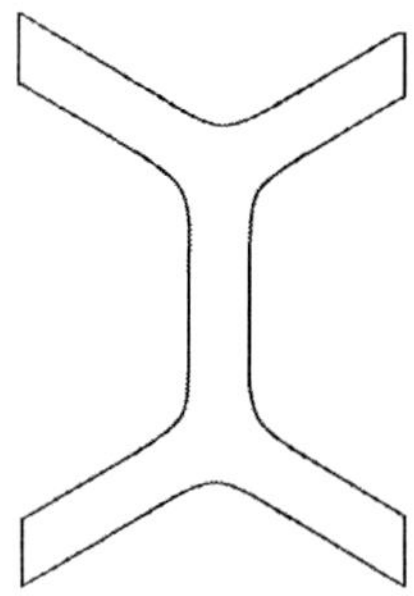

Figure 5b. Template with two symmetric Y-branched nano-channels.

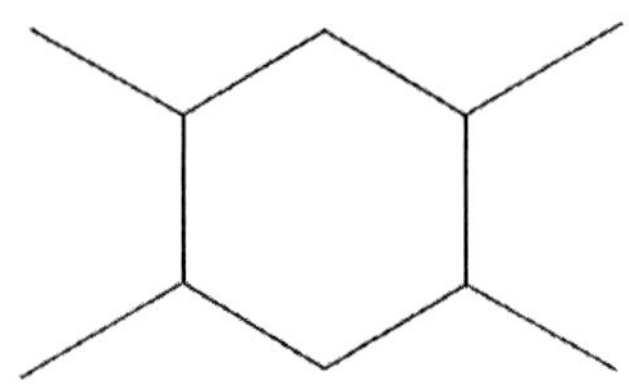

Figure 5c. Template with network formed by four symmetric Y-branched nano-channels.

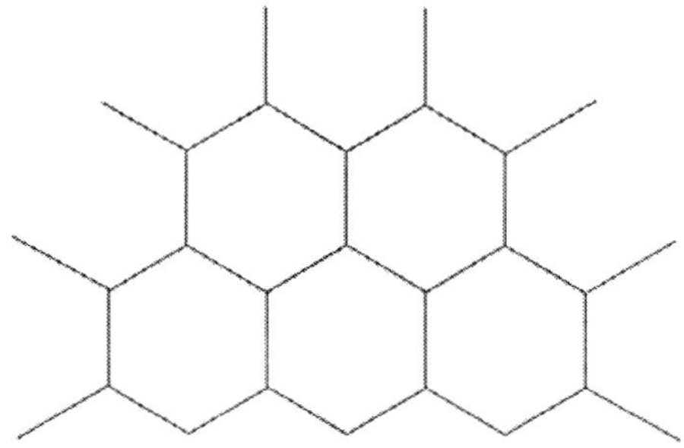

Figure 5d. Template of nano-channel network covered periodically with hexagonal cells.

From this template, a periodic hexagonal network constructed by carbon nanotubes may be grown (figure 6a). This is an interesting structure: It possesses low energy and high symmetry not only locally but also globally. Besides, it includes the first-level self-similarity —— The similarity between the network of carbon nanotubes and the network of carbon atoms (or single layer graphite).

Then, the "super carbon nanotube" can be made from the "super graphite". In carbon nano science and technology, it is well known that a common carbon nanotube may be "theoretically manipulated" by rolling up a single layer graphite into a tube. Similarly, a super carbon nanotube may also be "theoretically manufactured" by curling up the super graphite into a tube (figure 6b). Figure 6b also displays an interesting structure: It includes the second-level self-similarity —— The similarity between the super carbon nanotube and the common carbon nanotube.

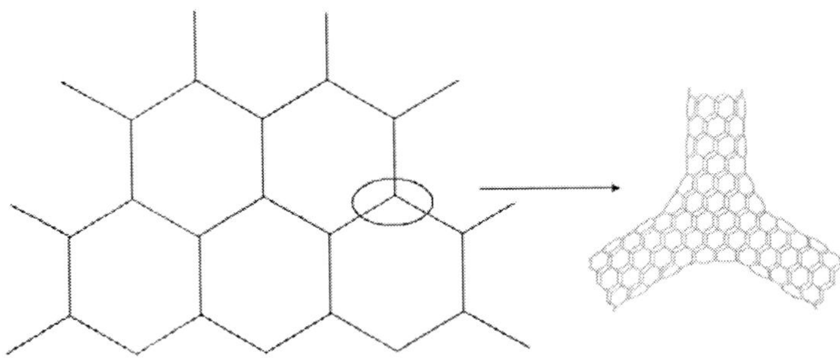

Figure 6a. Periodic network of hexagonal cells constructed by carbon nanotubes, grown from the template in figure 5d.

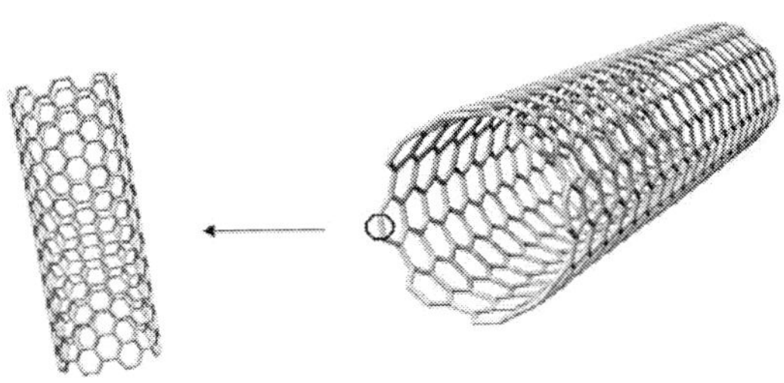

Figure 6b. Super carbon nanotube theoretically manufactured from the network of carbon nanotubes in figure 6a.

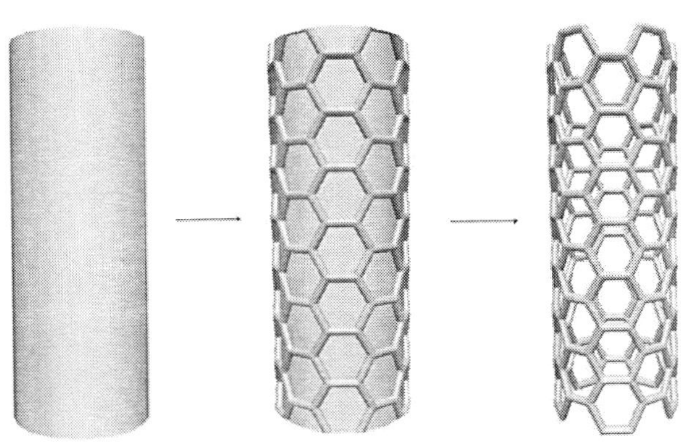

Figure 6c. Potential techniques for manufacturing super carbon nanotubes through circular cylindrical template (left: a circular cylindrical template; middle: nano channel networks carved on the surface of the template; right: super carbon nanotube).

In the above proposed manipulation process for super carbon nanotubes, the term "theoretically" is used, because two technical difficulties exist: One is to roll up the planar network of nanotubes into a cylindrical one, and another is to "weld" the cylindrical network into a seamless tube. How to overcome these difficulties? The answer is to give up the concept of "planar template" and introduce the concept of "circular cylindrical template". In differential geometry, a circular cylindrical surface is a developable one that may be unrolled into a plane after it is cut along its generatrix. Hence, the nano-channel network in the planar template in figure 5d may be manipulated conformably on the surface of a circular cylinder, and then a circular cylindrical template may be made (figure 6c). From this template, the super carbon nanotube may be grown directly and "practically".

3.2. Possible Topologies for Super Carbon Nanotubes

It is well known that geometrically there are three types of carbon nanotubes —— zigzag, armchair and chirality. Theoretically it is reasonable to believe that there are also three types of super carbon nanotubes, each of which might be constructed by three types of common carbon nanotubes called "fundamental elements". Hence, there are possibly nine types of topologies for super carbon nanotubes, among them the armchair-armchair and zigzag-zigzag topologies are the most symmetric and simple.

Early in 1992, Scuseria [5] proposed a kind of symmetric Y-branched junction with armchair topology, which composed of heptagons at the connected junction (figure 7b). This symmetric Y-branched junction has been proven to be a stable structure, which means that this structure is both geometrically and physically realizable. Now, if the carbon nanotubes are connected with the Scuseria-type Y-branched junctions, the super carbon nanotube (figure 7a) with armchair-armchair topology may be formed.

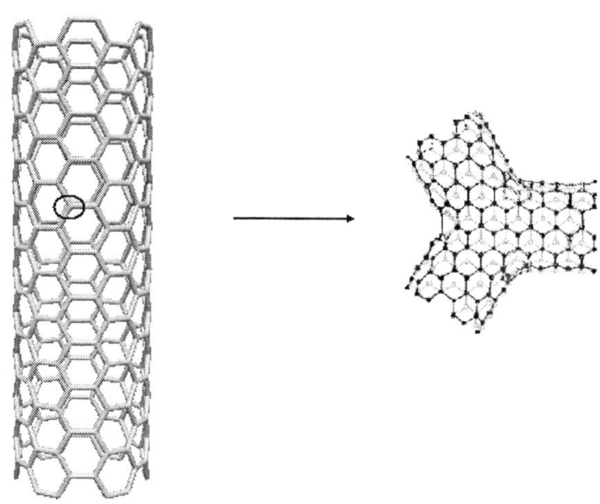

(a) (b)

Figure 7. The super carbon nanotube in (a) includes symmetric Y-branched carbon nanotube junction in (b) with armchair-armchair topology (the junction in (b) is cited from Scuseria in reference [5]).

Another kind of symmetric Y-branched carbon nanotube junction may possess zigzag topology (figure 8b) in which heptagons are also involved to realize the smooth transitions at the junction. If such junctions are included in the structure (figure 8a), hopeful super carbon nanotubes with zigzag-zigzag topology may be created. For simplicity, super carbon nanotubes with armchair-armchair topology will be focused below.

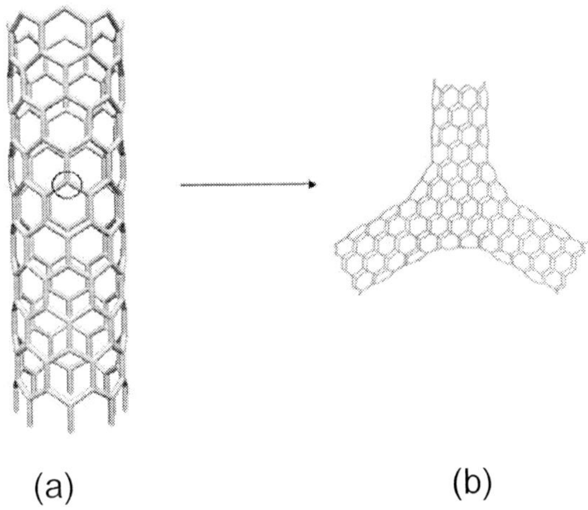

Figure 8. The super carbon nanotube includes symmetric Y-branched carbon nanotube junction with zigzag-zigzag topology.

4. Mechanical Properties of Super Carbon Nanotubes with Armchair-Armchair Topologies

From the geometric conservation laws for spontaneous Y-branched carbon nanotubes, we have proved in the above sections that the super carbon nanotube (figure 6b) might be of the lowest energy both locally and globally. Besides, the concept of circular cylindrical template makes it possible to manufacture super carbon nanotubes. Now further question may be asked: What mechanical properties may these beautiful carbon nanostructures posses? We will try to answer this question below.

4.1. Numerical Method and Model

Both super graphite and super carbon nanotubes are atomic systems that totally consist of C-C bonds and Van der Waals interactions. Hence all kinds of theoretical and numerical methods developed for carbon nanotubes can be used to investigate these systems. In 2004, Liu et al. [29, 30] proposed an order-N atomic-scale finite element method (AFEM). The AFEM may involve all the inter-atomic potentials and solve the equilibrium configuration of atoms based on the minimal energy principle. This method is of the same accuracy as the molecular dynamics, but is much faster than the standard conjugate gradient method widely used. Here the AFEM is adopted to calculate the mechanical properties of the super structures.

As mentioned above, super graphite (figure 9b) may be made through repeating Y-branched carbon nanotubes (figure 9c) periodically.

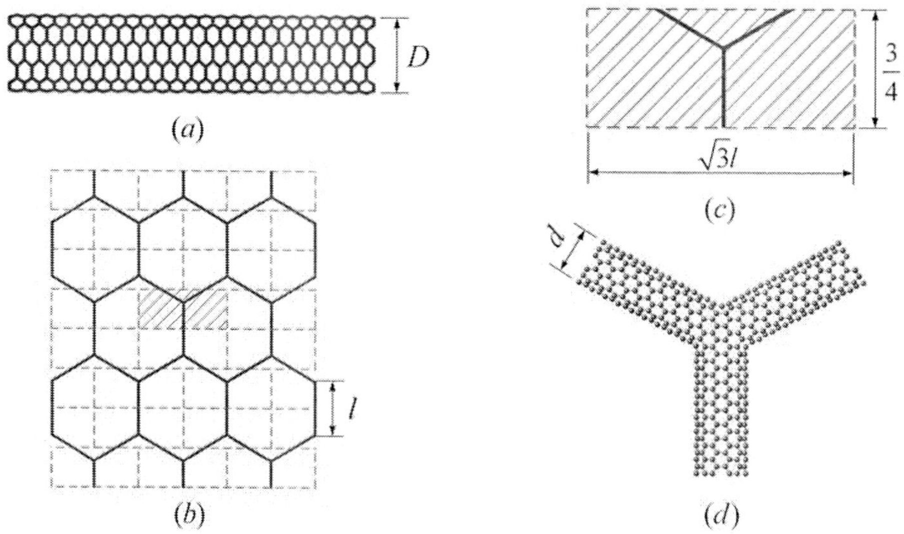

Figure 9. The cell and calculation model of the super carbon nanotube with armchair-armchair topology.

In another word the Y-branched carbon nanotube may be considered as the fundamental cell for the super graphite, and we can get the mechanical properties of super graphite from the fundamental cell. Wang [31] reported that single-walled nanotube with large radius has the same tension modulus with the graphite sheet, while the one with relatively small radius has the slightly reduced modulus. Analogously, when super graphite is wrapped into the super tube (figure 9a), the tension modulus of the super tube may be very near to that of the super graphite. Therefore, if the behavior of Y-branched carbon nanotube (figure 9d) is made clear, the mechanical properties of the super nanotubes may be roughly evaluated.

The following section will just focus on axial tensile modulus of armchair-armchair super carbon nanotubes constructed by Y-branched carbon nanotubes with different aspect ratios (l/d).

4.2. Mechanical Properties Predicted by Atomic-Scale Finite Element Method

The Y-branched carbon nanotubes are constituted by Scuseria-type junctions and (6, 6) carbon nanotubes. The aspect ratios (l/d) are taken as 3.0, 4.2, 6.0, 12 and 24 respectively. The second-generation Brenner potential [32] and Lennard-Jones 6-12 potential are selected to describe the energies of C-C covalent bonds and Van der Waals interactions respectively.

Figure 10 shows the predicted relation between tension F and strain ε of the Y-branched carbon nanotube. The following information is revealed. (a) When the strain is less than 10%, the tension increases along with the strain almost linearly.

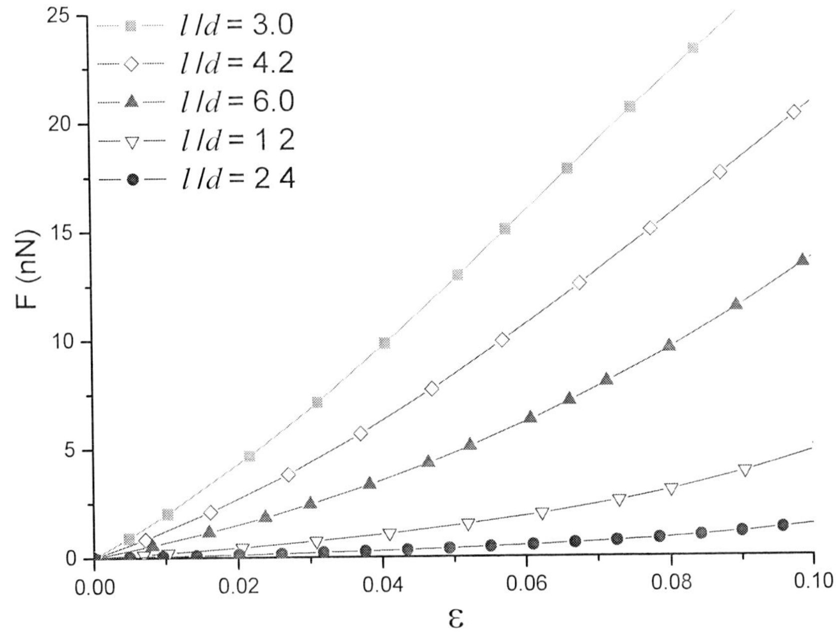

Figure 10. The uniaxial tension-strain curve of the Y-branched carbon nanotubes with different aspect ratios.

This means the structure is flexible enough to endure large stain (up to 10%) without yielding or buckling. (b) The larger the aspect ratio is, the smaller the slope of the curve is. This implies that the structure with larger aspect ratios seems to be softer. When the super tube are formed by the Y-branched carbon nanotubes, the thickness of its wall should be given by

$$\delta = \frac{D_{CC}}{2} + d + \frac{D_{CC}}{2} \tag{18}$$

Here d is the diameter of the carbon nanotube. D_{CC} is the equilibrium distance between graphite sheets and is usually taken as 0.34 nm. Then the stress in the cross-section of the super carbon nanotube may be expressed as

$$\sigma = F_{\mathrm{ST}} / (\pi D \delta) = \frac{2F}{3l\delta} \tag{19}$$

Here F_{ST} and D are the axial force and the diameter of the super carbon nanotube respectively, and l is the arm length of the Y-branched carbon nanotube, just as shown in figure 9b. Thus the axial tensile modulus E of the super carbon nanotube may be calculated as

$$E = \mathrm{d}\sigma / \mathrm{d}\varepsilon \tag{20}$$

Here ε is the axial tensile strain of the super carbon nanotube. The predicted stress-strain curves in figure 11 seem to be linear but not strict linear. The modulus E in equation (20) is tangent modulus of the curves in figure 11. The $E-\varepsilon$ curve is presented in figure 12.

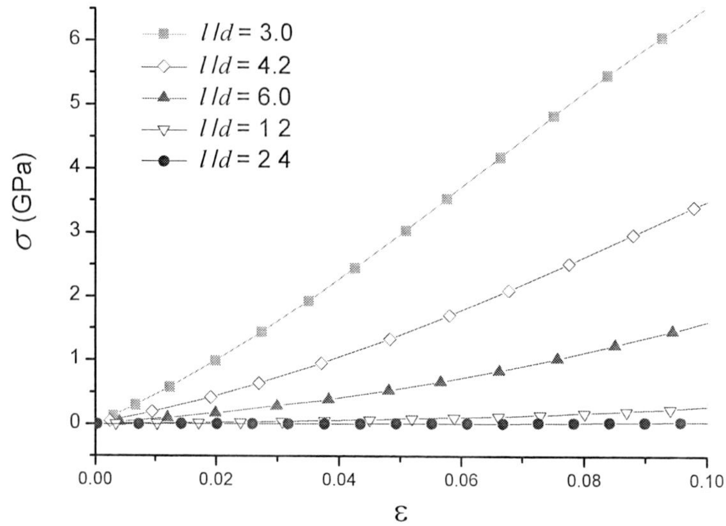

Figure 11. The axial stress-strain curve of the super carbon nanotubes with different aspect ratios.

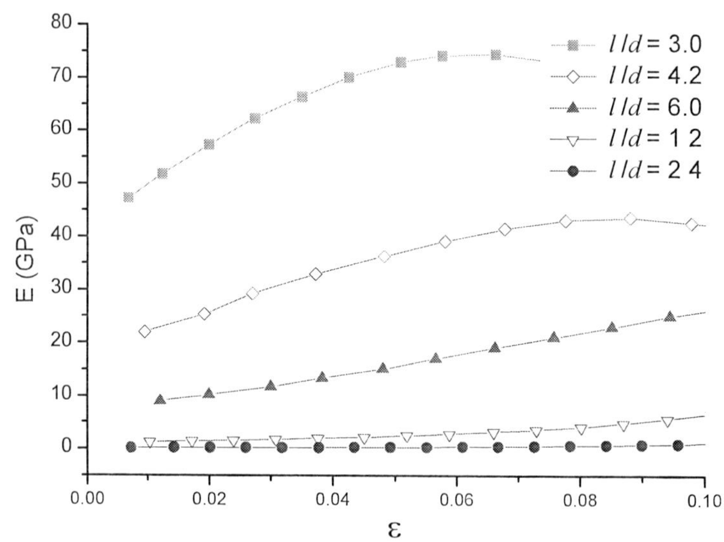

Figure 12. The axial modulus-strain curve of the super carbon nanotubes with different aspect ratios.

Figure 12 indicates that the axial tensile modulus of the super carbon nanotube is very sensitive to the aspect ratio of the carbon nanotube. The larger the aspect ratio is, the smaller the modulus is. Therefore, to adjust the aspect ratio may be the effective method to improve the axial tensile modulus of the super carbon nanotube.

5. Conclusion

Spontaneous Y-branched carbon nanotubes should observe geometric conservation laws and simple geometric regulations, i.e. the angle between two neighboring branches should be 120° and the radii of the three branches should be equal. From the symmetric Y-branched carbon nanotubes the concept of "super carbon nanotube" is proposed. Such a super structure might be of the lowest energy. Besides, with the aid of the concept of "circular cylindrical template", super carbon nanotube is possible to be manufactured.

Preliminary investigation to the mechanical properties of super carbon by using atomic-scale finite element method indicates that the axial tensile modulus of the super carbon nanotube is very sensitive to the aspect ratio of the carbon nanotube.

Acknowledgements

Supports by the Chinese NSFC under Grant No.10572076 are gratefully acknowledged.

References

[1] Iijima S. 1991 *Nature* 354 56.
[2] Zhou D. and Seraphin S. 1995. *Chem. Phys. Lett.* 238 286.
[3] Chico L, Crespi V. H., Benedict L. X., Louie S. G. and Cohen M. L. 1996. *Phys. Rev. Lett.* 76 971.
[4] Terrones M., Banhart F, Grobert N., Charlier J.-C., Terrones H. and Ajayan P. M. 2002 *Phys. Rev. Lett.* 89 075505.
[5] Scuseria G. E. 1992. *Chem. Phys. Lett.* 195 534.
[6] Biró L. P., Ehlich R., Osváth Z., Koós A., Horváth Z. E., Gyulai J. and Nagy J. B. 2002. *Diamond Relat. Mater.* 11 1081.
[7] Li J., Papadopoulos C. and Xu J. M. 1999. *Nature* 402 253.
[8] Nagy P., Ehlich R., Biró L. P. and Gyulai J. 2000. *Appl. Phys.* A 70 481.
[9] Satishkumar B. C., Thomas P. J., Govindaraj A. and Rao C N R 2000. *Appl. Phys. Lett.* 77 2530.
[10] Gan B. et al 2000. *Diamond Relat. Mater.* 9 897.
[11] Gan B. et al 2000. *Mater. Lett.* 45 315.
[12] Deepak F. L., Govindaraj A. C. and Rao N. R. 2001. *Chem. Phys. Lett.* 345 5.
[13] Li W. Z., Wen J. G. and Ren Z. F. 2001. *Appl. Phys. Lett.* 79 1879.
[14] Ting J. M. and Chang C. C. 2002. *Appl. Phys. Lett.* 80 324.
[15] Terrones M. 2003. *Annu. Rev. Mater. Res.* 33 419.
[16] Dag S., Senger R. T. and Ciraci S. 2004. *Phys. Rev.* B 70 205407.
[17] Coluci V. R., Galvão D. S. and Jorio A. 2006. *Nanotechnology* 17 617.
[18] Yin Y. J., Chen Y. L., Yin J. and Huang K. Z. 2006. *Nanotechnology* 17 4941.
[19] Nicola M. Pugno 2006. *Nanotechnology* 17 5480.
[20] Wang M., Qiu X. and Zhang X. 2007. *Nanotechnology* 18 No 7 075711 (6pp).
[21] Yin Y. and Yin J., J. 2006. *Nanobiotechnology* 4 : 6.

[22] Karlsson A., Karlsson R., Karlsson M., Cans A., Stromberg A., Ryttsen F. and Orwar O. 2001. *Nature* 409 150-152.

[23] Karlsson M., Sott K., Davidson M., Cans A-S., Linderholm P., Chiu D. and Orwar O. 2002. *Natl. Acad. Sci.* 99 11573-11578.

[24] Yin Y. 2005. *Tsinghua Science and Technology*, 10 373-377; Yin Y, Yin J. and Wu J., 2007, *Applied Mathematical Sciences*, 1 1465-1484.

[25] Terrones M., Banhart F., Grobert N., Charlier J. C., Terrones H. and Ajayan P. M. 2002. *Phys. Rev. Lett.* 89 075505.

[26] Ou-Yang Z. C., Su Z. B., and Wang C. L. 1997. *Phys. Rev. Lett.* 78 4055.

[27] Lenosky T., Gonze X., Teter M., and Elser V. 1992. *Nature* (London) 355 333.

[28] Gilbert E. N. and Pollak H. O. 1968. *SIAM J. Appl. Math.* 16 1.

[29] Liu B., Huang Y., Jiang H., Qu S. and Hwang K. C. 2004. *Comput. Methods Appl. Mech. Engrg.* 193 1849–1864.

[30] Liu B., Jiang H., Huang Y., Qu S. and Yu M-F. 2005. *Phys. Rev.* B 72(3) 035435/1-8.

[31] Wang L. F., Zheng Q. S., Liu J. Z., and Jiang Q. 2005. *Phys. Rev. Lett.* 95 105501.

[32] Brenner D. W., Shenderova O. A., Harrison J. A., Stuart S. J., Ni B. and Sinnott S. B., 2002. *J. Phys.: Condens. Matter* 14 783-802.

In: Encyclopedia of Nanotechnology
Editor: Elwood D. Carlson
ISBN: 978-1-60692-079-4

Chapter 33

IMPROVING DIELECTRIC PROPERTIES OF POLYMER NANOCOMPOSITES THROUGH NANOPARTICLE SURFACE MODIFICATION

Dongling Ma,[a,1] Linda S. Schadler,[a] Richard W. Siegel,[a] Nianqiang Wu,[b] Diana F. Rogers,[c] and Brian C. Benicewicz[c]

[a] Department of Materials Science and Engineering and Rensselaer Nanotechnology Center, Rensselaer Polytechnic Institute, Troy, NY 12180, USA

[b] Keck Interdisciplinary Surface Science Center, NU*ANCE*, Northwestern University, IL 60208, USA

[c] Department of Chemistry and NYS Center for Polymer Synthesis, Rensselaer Polytechnic Institute,Troy, NY 12180, USA

ABSTRACT

In this study, the potential of using nanoparticle surface chemistry to tune the dielectric properties of polymer nanocomposites is explored. Composites with 5 wt% of titania (TiO_2) nanoparticles and low-density polyethylene (LDPE) were fabricated by melt mixing. Both polar (N-(2-aminoethyl)3-aminopropyl-trimethoxysilane (AEAPS)) and nonpolar (decyltrichlorosilane (DCS)) silane coupling agents were used to modify the nanoparticle surface to yield either hydrophilic or hydrophobic surfaces. Two coating methods were employed and the resulting surface structures were characterized in detail. It was found that the surface structure was controlled primarily by the experimental conditions of the coating process. In addition, a previously un-described surface reaction that resulted in the formation of Si-CH_3 bonds was proposed, which may be due to the reaction of AEAPS with oxygen vacancies on the TiO_2 nanoparticle surface. A large observed change in the chemical shift (over 40 ppm) in the ^{29}Si NMR and the –2.3 ppm chemical shift in ^{13}C NMR supported this conclusion. The method leading to a "dry" compact monolayer surface structure was chosen for preparing TiO_2 nanoparticles with either polar or non-polar surfaces for composite sample preparation and subsequent electrical studies. In both cases, surface modification had a positive effect on dielectric breakdown strength although the possible underlying mechanisms were different. Compared to the uncoated TiO_2 filled samples, AEAPS-TiO_2 and DCS-TiO_2 filled samples showed a 35% and 43% increase in the short-term Weibull breakdown strength, respectively. The space charge distribution patterns were also clearly different among the

[1] E-mail address: ma@emt.inrs.ca. Phone: 1-450-929-8120. Fax: 1-450-929-8102. Corresponding author.

samples. The microstructure characteristics of the composite samples, associated with different TiO_2 surfaces, were studied and correlated to the corresponding dielectric properties.

INTRODUCTION

Polymers play an important role in electrical insulating technology. Polyethylene (PE) is one of the most attractive cable insulating materials due to its unique characteristics such as high dielectric strength, ease of fabrication, low cost and simple maintenance and, thus, has long been of research interest [1-5]. There are problems, however, related to the application of PE as an insulating material that still have not been resolved successfully. First is the inherent dependence of PE resistivity on temperature and electrical field [1]; the other is the reliability and long-term life of PE. Degradation [6,7], which can be thermal, oxidative, photo-oxidative or electrical, results in material failure.

Additives/fillers are mixed with polymer matrices to modify their high field stability, mechanical and thermal–mechanical properties, and resistance to degradation [8-10]. But in most cases, the introduction of additives/fillers has a significant negative effect on dielectric breakdown strength due to increased field distortion. Therefore, one important goal is to improve other properties without negatively affecting or even improving the electrical properties. To achieve this goal, it is important to carry out detailed investigations of the electrical properties and structure of the composite materials in order to gain sufficient understanding about the system. Most of the studies so far have focused on conventional micron-size additives/fillers, while in this work, the focus is on nanoparticles.

Nanomaterials are attracting more and more attention because their small size of their constituent domains leads to unique properties. The constituent domains of nanomaterials have at least one dimension less than 100 nm in size. They are intermediate in size between molecular and microscopic (micron-size) structures [11]. In this regime, material properties are dominated by quantum and other confinement behavior that makes their characteristics fundamentally different from their micron-scale counterparts [12-17]. In addition, nanoscale structures have high surface area and high surface-volume ratio. As the particle size decreases, the surface area to volume ratio increases dramatically when the length scale falls below 100 nm. Therefore, at the same particle loading, nanocomposites possess much more interfacial area than microcomposites. This large surface area, in combination with enhanced surface reactivity of nanoparticles, suggests a great potential of tailoring the macroscopic properties of nanocomposites by modifying the nanoparticle surface at the nanoscale. To be specific, for example, it is known that particle filler dispersion, interfacial compatibility and matrix morphology are all possible factors in affecting the dielectric breakdown strength. For a specific polymer, the interfacial compatibility is solely determined by the surface chemistry of the nanoparticles and the other two factors are also closely related to the surface chemistry of the nanoparticles. Therefore, the surface chemistry must have an important effect on the dielectric breakdown behavior. Despite this large potential, only recently have studies emerged on the effects of surface modification of the nanoparticles on the dielectric properties of the nanocomposites [18-20]. In our previous work [18,21], we have demonstrated that the surface states of TiO_2 nanoparticles had non-negligible effects on the dielectric breakdown strength and space charge distribution in polyethylene nanocomposites.

In the present work, we further study the effect of different surface modifications of the TiO_2 nanoparticles, with either polar or non-polar silane coupling agents, on the permittivity, dielectric breakdown strength and space charge distribution of LDPE. TiO_2 was chosen because it is becoming increasingly important in composite applications.

EXPERIMENTAL

Materials

Low-density (0.922 g/cc) polyethylene (LDPE) DOW 6811 from Dow Chemical was used as the matrix, with the number-average and weight-average molecular weight being 14000 and 70000, respectively. TiO_2 nanoparticles with a surface area of 42 m^2/g were provided by Nanophase Technologies Corporation. The average particle diameter was 23 nm as determined by transmission electron microscopy (TEM) as shown in Figure 1.

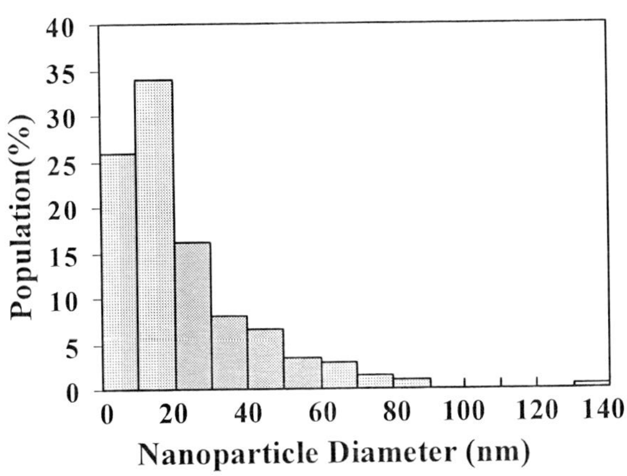

Figure 1. TiO_2 nanoparticle size distribution determined from TEM.

The particle diameters ranged from less than 10 nm to over 100 nm, with about 70% of the nanoparticles smaller than 30 nm in diameter. X-ray diffraction confirmed that the particles were principally anatase. N-(2-aminoethyl)3-aminopropyl-trimethoxysilane (AEAPS) ($(CH_3O)_3Si(CH_2)_3NH(CH_2)_2NH_2$) and decyltrichlorosilane (DCS) ($CH_3(CH_2)_9SiCl_3$) were purchased from Gelest, Inc. and used as received. Toluene (water content <0.03%), calcium hydride (CaH_2, purity 93%) and potassium bromide (purity 99+%) were purchased from Mallinckrodt Baker Inc., Acros Organics and Aldrich Chem. Co., respectively.

Coating Procedures

The coating procedures were derived from those described by Tsubokawa *et al* [22]. In coating method I, 10 g of nanoparticles were dispersed in toluene via sonication (Sonics Vibracell, 750 W) at 70% power for 5 minutes. The solid full wave probe was 13 mm in diameter and 254 mm in length. To avoid a sharp temperature rise of the suspension during sonication, a beaker containing the suspension was surrounded by ice. 0.015 mol of silane coupling agents was added slowly to the nanoparticle-toluene mixture and stirred for one hour at room temperature. Then the mixture was refluxed for 48 hrs under dry air atmosphere. The resultant slurry was centrifuged at 6,000 rpm for 10 minutes and the deposited particles were washed with fresh toluene. The centrifuge-toluene wash procedure was repeated two more times and the silane-coated particles were dried in *vacuo* for 24 hrs at 30 °C.

Coating method II was similar to method I with two modifications. The toluene was dried over CaH_2 powder in a nitrogen atmosphere, and then purified by distillation. The TiO_2 nanoparticles were dried in *vacuo* at 195 °C for 24 hrs prior to their immediate use in the coating procedures.

Sample Preparation

5 wt% of as-received or surface modified TiO_2 nanoparticles were melt-mixed into LDPE at 130 °C using a Haake System 90 Buchler Mixer for 10 minutes. Pellets for permittivity and thin films (10 to 40 μm in thickness) for dielectric breakdown measurement were formed by hot compression at 160 °C in a Carver Press (Model 3912). Polytetrafluoroethylene films (McMaster-Carr Supply Company) were used as backing. The diameter of these prepared samples was 58 mm. Processing details can be found in ref [23]. The pressure was changed slightly among the samples to accommodate the viscosity of the mixtures. Each sample was kept in a desiccator for at least one day for relaxation before any measurements were carried out. The neat LDPE went through exactly the same process to exclude the factor of different processing on the properties of pure polymer and filled polymer samples.

Plaque samples with a thickness of 2 mm for space charge measurements were pre-press molded, then pressed against two 1 mm-thick semicon electrodes [18]. The semicon electrodes were composed of polyethylene copolymer and carbon black particles, and were also prepared by hot compression. All the press moldings were performed at 130 °C and polyethylene terephthalate films were used as backing during these processes.

Structural characterizations were made with thin film samples prepared for dielectric breakdown testing except for those used for atomic force microscopy (AFM), which required an extra step. AFM samples were prepared by placing a pre-pressed film on a hot, clean Si wafer in the mold at 160 °C without pressure. The fully melted film was slowly cooled to below 50 °C in the mold. In this case, Kapton (DuPont) was used as backing films. The free surface of thus pressed films was examined by tapping mode AFM.

Characterization Techniques

Solid-state ^{29}Si and ^{13}C cross-polarization magic-angle-spinning (CPMAS) nuclear magnetic resonance (NMR) measurements were made on a Chemagnetics CMX 360 instrument operating at a frequency of 71.25 and 90.53 MHz, respectively. The ^{29}Si spectra were obtained using a 4.5 μs 90° pulse width, a contact time of 6 ms, a 5 s recycle delay, and a spinning speed of 3 kHz; tetrakis (trimethylsilyl) silane was used as an external reference. The ^{13}C spectra were obtained using a 4.5 μs 90° pulse width, a contact time of 1 ms, a 5 s recycle delay, and a spinning speed of 3 kHz; hexamethyl benzene was used as an external reference.

Fourier-transform infrared spectra (FTIR) were obtained on a Magna-IR 560 spectrometer in transmission mode. Potassium bromide (KBr) pressed pellets were prepared by first drying the KBr in an oven at 200 °C for 3 hrs followed by mixing with about 3 wt% of coated TiO_2 nanoparticles. The spectrum of pure AEAPS silane was obtained by placing a couple of drops of the coupling agent between two pressed pure KBr pellets. The spectra of composite samples were obtained by using as-molded thin films directly. The signal was averaged over 512 scans with a resolution of 1 cm^{-1}.

Thermogravimetric analysis (TGA) was performed on a Mettler TGA/SDTA 851e at a heating rate of 20 °C/min in air for coated particles, and in nitrogen for water content determination of uncoated particles.

The particle dispersion in the composite samples was observed with a JEOL JSM-6330F field emission scanning electron microscope (FE-SEM). The samples were broken in liquid nitrogen and the cross section was examined after being sputter-coated with either gold or platinum.

Time-of-flight secondary ion mass spectroscopy (SIMS) analysis was performed with a Physical Electronics PHI TRIFT III, which is equipped with a pulsed Ga^+ liquid ion gun operated at 15 kV. The ion source was operated with a current of 600 pA. A pulsed low-energy electron flood gun neutralized the charging effect. The secondary ions were accelerated to ±3 kV by applying a bias on the sample.

Differential scanning calorimetry (DSC) experiments were carried out on a Mettler DSC 822e under N_2 atmosphere with a ramp rate at 10 °C/min. The weight of samples ranged from 3 to 6 mg. The original DSC curves were normalized to the net weight of LDPE and then used for analysis. The degree of crystallinity was estimated using $W_c^{DSC} = \Delta H_f/\Delta H^o$, where ΔH_f is the heat of fusion obtained after integrating the area under the normalized melting curve and $\Delta H^o = 293$ J/g is the reference heat of fusion for 100 % crystalline polyethylene.

X-ray diffraction patterns (XRD) were recorded at room temperature with a Scintag XDS 2000 diffractometer using a wavelength of 0.154 nm. The signal was obtained in step mode with the step size of 0.02° and the collection time of 1 s. The peak analysis was done by using a mixed Gaussian and Lorentzian function. For comparison, the original diffraction patterns were normalized to their own (110) reflection intensities.

A NanoScope III (Digital Instruments Veeco Metrology Group) AFM was operated in tapping mode at room temperature to visualize composite morphology. Both topographic and phase images were simultaneously obtained using Si cantilevers with a spring constant of 40 N/m.

Permittivity measurements were carried out at room temperature with a HP 4192A LF Impedance Analyzer. The fixture had two parallel disk electrodes with the top electrode having a guard ring. During the testing, the top electrode was grounded and the bottom electrode was connected to a high potential. Sample plates were inserted between the two electrodes without an air gap. Capacitance was read at 1 kHz for the calculation of permittivity.

To measure the dielectric breakdown strength, a sphere-plane geometry was employed. A stainless steel sphere with a diameter of ¼ ″ was connected to high potential while the plane electrode was grounded. The test was performed with a Glassman High Voltage FC60P02 at room temperature using a stepwise increased voltage with steps of 100 V and application periods for each step of approximately 1 second. Between steps the high voltage electrode was grounded and then the next higher voltage was applied. The process proceeded until a sharp current increase was observed indicating electrical failure of the sample. Dielectric breakdown strength was calculated as the electrical voltage causing the sample failure divided by the sample thickness.

A pulsed electro-acoustic (PEA) technique was used to measure the space charge distribution. The tests were carried out at 50 ^{o}C under the voltage of +40 kV (20 kV/mm). The space charge distribution was measured at 0, 15, 30, 60, 180, and 360 minutes after the voltage application and a 2.2 kV pulse voltage was used. A calibration run was done at low electrical field, where no space charge build-up was assumed to be present, and used as a reference signal. The relative space charge density is presented here for a comparative evaluation. Details of the basic PEA measurements can be found elsewhere [24].

PARTICLE SURFACE MODIFICATION

Two similar coating methods I and II were employed to coat TiO_2 nanoparticles using the AEAPS silane. The resultant coating structures were characterized in detail using TGA, FTIR, ^{29}Si NMR and ^{13}C NMR. Based on the coating structure analysis, and accordingly the expectation for the influence of coating structure on the insulating properties, coating method II was chosen for chemical modification of the nanoparticles by DCS.

TiO_2 surfaces contain water and hydroxyl groups, both of which are involved in the coating process. Surface water can hydrolyze the silane, while hydroxyl groups contribute directly to the formation of coating structures by reacting with other hydrolyzed silane [25]. Similarly, the hydrolyzed product of the silane coupling agent may hydrogen bond and/or covalently bond to the surface [25]. Hydrogen bonds are formed between hydrolyzed silane and the TiO_2 surface via surface hydroxyl groups or adsorbed molecular water. Further condensation reactions may produce Ti-O-Si bonds, in which the silane is categorized as covalently bonding to the surface. Also, $-OCH_3$ groups of the silane can exchange with surface hydroxyl groups to form a covalently bound coating structure. The bonding status and organization of silane on the surface affect the performance of the composite materials [25]. In this study, a covalent bond was preferred. Detailed structure characterization is very important towards achieving controlled synthesis.

Characterization of Nanoparticle Surfaces Coated with AEAPS

The water content and hydroxyl group density of the particles before the reaction were approximated from the TGA thermograms reported previously in ref [21]. For as-received TiO_2 nanoparticles, approximately 13 $\mu mol/m^2$ of water and 4.5 $\mu mol/m^2$ of surface hydroxyl groups were present on the surface. The drying process had efficiently removed most of the surface water. However, the density of the hydroxyl groups (approximately 12 $\mu mol/m^2$) on the dried particles was more than double that of as-received TiO_2 nanoparticles, which may be due to dissociation of trapped water before it evaporates. A similar surface –OH content of

13.7 μmol/ m^2 on dry TiO_2 has also been reported [26]. The calculation was based on the surface area value (42 m^2/g) provided by the producer of the nanoparticles.

The TGA thermograms for coated particles (Figure 2) showed a significant mass loss, indicating the presence of coupling agent. Approximately, 4.9 and 2.8 wt%, or 12.1 and 6.8 μmol/m^2 surface coverage, of AEAPS silane were present on the particles after coating with methods I and II, respectively. Drying had a significant effect on the amount of coating on the nanoparticle surfaces: more silane was present for coating method I compared to method II. The amount of silane present on the surface was more than the hydroxyl group density, indicating possible silane multi-layers resulting from method I. It was also noted that particles prepared by coating method I showed a sudden and more evident decrease in weight at about 210 °C than those prepared by coating method II, which may have been due to the loss of trapped water.

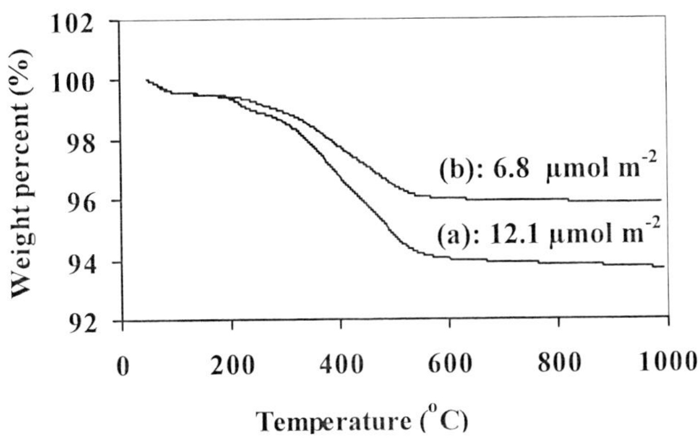

Figure 2. Thermogravimetry of AEAPS silane modified TiO_2 by (a) coating method I, and (b) coating method II.

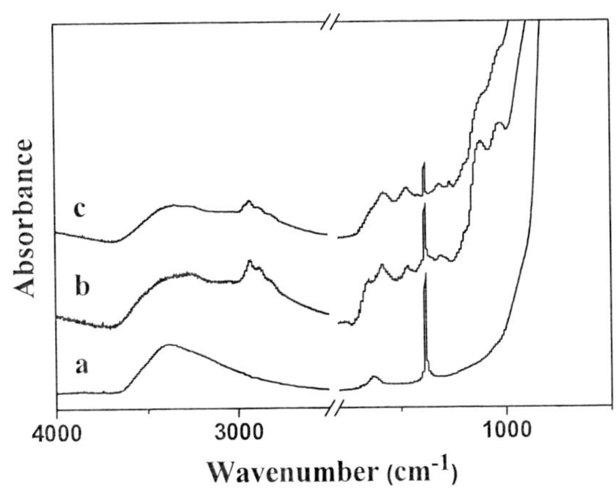

Figure 3. IR spectra of (a) as-received TiO_2, AEAPS silane modified TiO_2 by (b) coating method I, and (c) coating method II.

The FTIR spectra for untreated and silane coated TiO_2 particles are shown in Figure 3. The broad O-H stretching band from 2800-3700 cm^{-1} arises from surface hydroxyl groups and the weak 1640 cm^{-1} peak from molecular water on pure TiO_2 [27,28]. Upon surface modification, the molecular water band shifted to 1660 cm^{-1} for method I, and was evidently diminished for method II. This peak should not be confused with two nearby peaks at 1480 and 1590 cm^{-1}, which were present for both coating methods I and II and were due to the reaction product of $-NH_2$ of AEAPS silane with CO_2 in the air [29].

A broad and strong peak of the coated particles ranging from 2500 to 3700 cm^{-1} was again due to the formation of the amine bicarbonate [30]. The symmetric and asymmetric C-H stretches for both coating methods are observed at 2940 and 2910 cm^{-1}. The 1390 cm^{-1} peak in as-received TiO_2 was believed to be due to organic contaminants of the particles during packaging and most of them can be removed by the drying, as evidenced by SIMS. Three sharp bands for $Si-O-CH_3$ [30] at 2840, 1200, and 1090 cm^{-1} present for pure AEAPS coupling agent were not present for either coating method, indicating that the methoxy groups had been hydrolyzed in coating method I or exchanged in method II. Two broad peaks centered at 1040 and 1130 cm^{-1}, respectively, indicate the formation of a long or branched Si-O-Si structure [30] and were much stronger in coating method I than II. An interesting peak at 1270 cm^{-1} was present for modified TiO_2 prepared by coating method II. The band was attributed to the Si-C stretching mode of a silane methyl group [31,32].

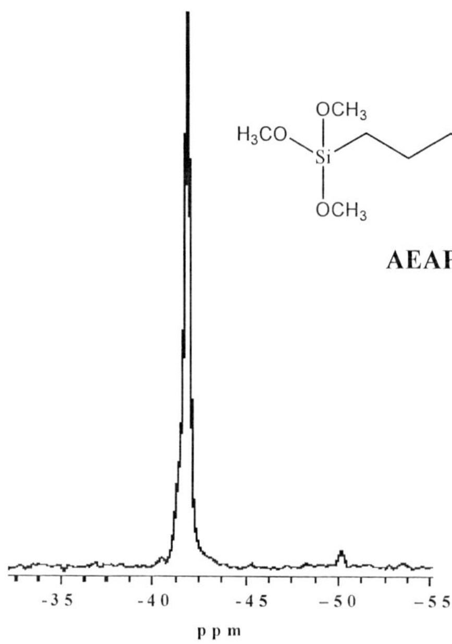

Figure 4. ^{29}Si NMR of AEAPS liquid coupling agent.

^{29}Si NMR was employed to determine the chemical environment around the silicon atoms. The ^{29}Si NMR spectrum of as-received AEAPS silane coupling agent, with a major peak at –41 ppm and a minor peak at –50 ppm, is shown in Figure 4. The latter peak was due to the hydrolysis of the AEAPS silane during storage and the amount of hydrolysis was estimated to be less than 2%. Figure 5 shows the spectra of AEAPS silane coated TiO_2 particles. A broader resonance had been observed for coated particles than that of pure AEAPS silane, indicating reduced mobility and surface attachment. The lower spectrum (coating method I) displayed a single peak at –67 ppm, which was attributed to the anchoring of all three oxygen atoms in the AEAPS silane with the TiO_2 surface or with nearby silane molecules [33-35]. The upper spectrum (coating method II) clearly contained two distinct resonances. As described earlier, the peak at –67 ppm is ascribed to the coupling of all three oxygen atoms of the silane molecule with the TiO_2 surface or other silane molecules. In addition, a shoulder at –49 ppm was present, which was assigned to the mono-coupling of the silane with the TiO_2 surface or other silane molecules. The resonance at –19 ppm, however, was unexpected.

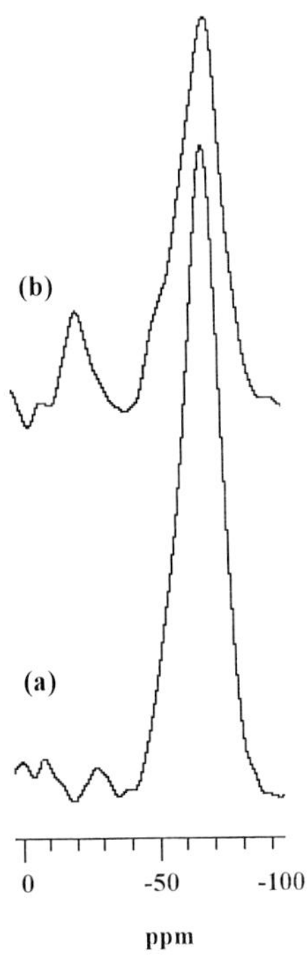

Figure 5. ^{29}Si NMR of AEAPS silane treated TiO_2 (a) prepared by method I, and (b) prepared by method II.

This peak could not be attributed to the pure coupling agent, which had a chemical shift of –41 ppm, nor could it be attributed to hydrolyzed forms of the molecule according to our ^{29}Si NMR observation of hydrolyzed AEAPS silane. Hydrolyzed AEAPS silane (not shown here) showed chemical shifts at –41, –49, –58 and –67 ppm; they are assigned to $RSi(OR')_3$, $RSi(OSi{\equiv})(OR')_2$, $RSi(OSi{\equiv})_2OR'$, and $RSi(OSi{\equiv})_3$, respectively, where R represents the organic group in the AEAPS silane and R′ is H or $-CH_3$. In fact, the ^{29}Si chemical shift at –19 ppm was not consistent with silicon having three oxygen atoms as its nearest neighbors. The chemical shift was, however, consistent with silicon bound by two oxygen atoms and two carbon atoms [36].

The ^{13}C NMR spectra are shown in Figure 6. The chemical shifts were labeled for each carbon atom of the AEAPS silane according to [34] and the additional peak at 168 ppm is attributed to the amine bicarbonate salt formed from reaction of the primary amine with carbon dioxide [37]. The sharp resonance at –2.3 ppm in the ^{13}C NMR spectrum for the silane coated particles prepared by method II is attributed to Si-CH_3 where Si was bound by two oxygen atoms and two carbon atoms directly [36].

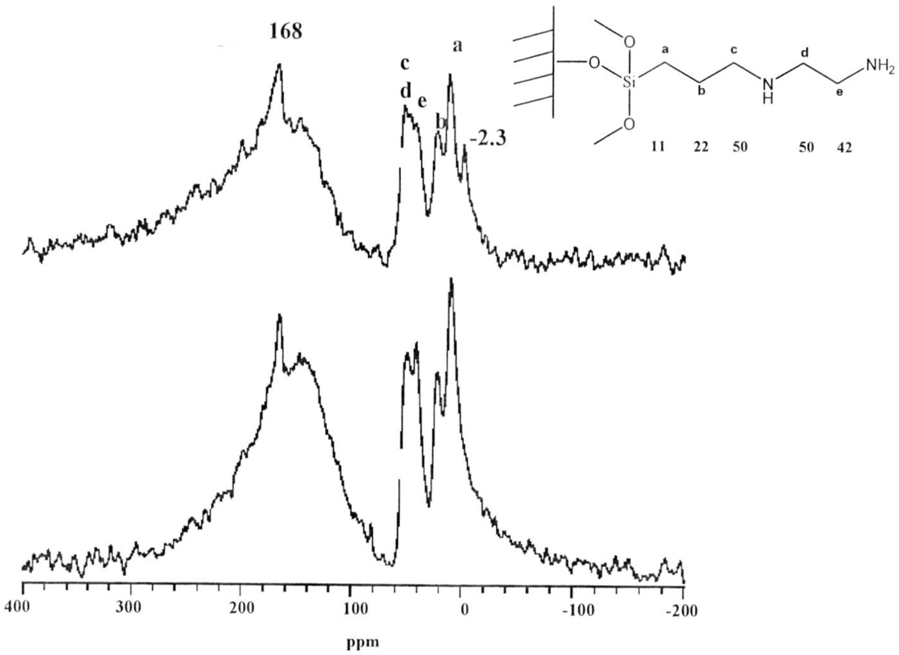

Figure 6. ^{13}C NMR of AEAPS silane treated TiO_2 (a) prepared by method I, and (b) prepared by method II.

Determination and Comparison of AEAPS Coating Structures

It is crucial to determine whether the silane was covalently bound (i.e., forming Si-O-Ti) or hydrogen-bound (i.e., forming Si-OH---OH-Ti) to the particle surface because the former provides a stronger interface and is preferred. In the case of coating method I, both water on the nanoparticle surface and in the as-received toluene can hydrolyze the silane. Therefore, hydrolysis of the Si-OCH_3 bond proceeds prior to condensation with either surface hydroxyls or other silanols. The resulting Si-OH can form a hydrogen bond with the Ti-OH or surface water on the TiO_2 nanoparticles or further condensation may occur to form Si-O-Ti covalent bonds at the interface.

As for coating method II, the reaction mechanism could differ from that for method I because of the absence of water in the system. Most of the surface moisture on the TiO_2 surface was removed during the drying treatment and also most of water in toluene was removed. Thus, theoretically, most of the AEAPS silane was initially available to form Si-O-Ti chemical bonds via direct alcoxolation. It is still possible that small amounts of Si–OH, either hydrolyzed during storage (Figure 4) or by small amounts of water left in the anhydrous toluene, form hydrogen bonds with the Ti-OH on the nanoparticle surface.

Mass spectrometry (MS) is commonly used to study interfacial bonding. According to a previous report [38], silicon-containing fragments could be detected by MS during the heating process if the silane was hydrogen-bound to the surface. These authors also reported that silicon-containing fragments would not be detected if the silane was covalently bonded to the surface. However, this method is only suitable for silanes with relatively low boiling points. During the heating process, two reactions at the interface between particles and a silane could occur and are competitive: one is evaporation of the hydrolyzed (totally or partially) silane and the other is the condensation reaction between Si-OH and surface hydroxyls. Thus, if the silane has a much higher boiling point than the corresponding condensation temperature, a condensation reaction may occur first and covalently bond the silane to the particles. In this case, the absence of silicon-containing fragments is not conclusive proof that the original interfacial bonding is covalent.

Here we employed ^{29}Si NMR, which is a very useful technique to identify interfacial structure and to investigate the nature of bonding at the interface. According to Maciel *et al.* [33], the chemical shift due to hydrolysis of Si–OCH_3 to Si-OH could not be resolved by ^{29}Si solid-state NMR. However, reaction of Si-OCH_3 or Si-OH to form a Si-O-Si bond resulted in a chemical shift of approximately 10-ppm upfield. The upfield shift of 10 ppm for the formation of Si-O-Si bonds was consistent with other literature reports [34,35] as well as our data of the hydrolyzed AEAPS. As noted earlier, the chemical shifts at –41, -49, -58, and –67 ppm corresponded to $RSi(OR')_3$, $RSi(OSi{\equiv})(OR')_2$, $RSi(OSi{\equiv})_2OR'$, and $RSi(OSi{\equiv})_3$, respectively. For coated particles prepared by method I, the absence of -41, -49 and –58 ppm shifts (Figure 5(a)) indicated that Si-OH and Si-OCH_3 were not present in the coating structure, which was also supported by the FTIR results. Therefore, these results confirmed the covalent bonding of AEAPS coupling agent to the particle surface, as well as condensation reactions resulting in Si-O-Si bonds. For coated particles prepared by method II, the presence of the small shoulder at –49 ppm in the ^{29}Si NMR spectrum (Figure 5(b)) indicated a small amount of Si-OCH_3 or Si-OH was present in particles prepared by coating method II, which was not detected by FTIR. However, the major peak at –67 ppm in the ^{29}Si NMR spectrum indicated that most of the AEAPS coupling agent bonded either to the particle surface covalently forming Si-O-Ti bonds, or other silane molecules, as evidenced by the weak Si-O-Si FTIR absorbances, which appeared as two shoulders in the region of 1000-1170 cm^{-1}. The difference between Si-O-Si and Si-O-Ti could not be resolved by ^{29}Si NMR.

Coating coverage was calculated from the TGA results (Figure 2). It can be seen that nearly twice as much silane was coated onto the particle surfaces using coating method I compared to coating method II. By comparing the surface hydroxyl concentration (4.5 $\mu mol/m^2$) of the as-received TiO_2 nanoparticles with the silane coverage (12.1 $\mu mol/m^2$) coated with method I, it was concluded that silane multilayers were formed in coating method I. This would result from an excess of Si-OH groups formed by the reaction of the silane coupling agent with readily available water in both toluene and on the TiO_2 surface. In addition, the silane molecules in the layer nearest to the TiO_2 surface may only form a single Ti-O-Si bond. Subsequent condensation between neighboring coupling agents that are only singly bonded to the TiO_2 would result in the formation of a Si-O-Si network, which has been suggested by two broad peaks centered at 1040 and 1130 cm^{-1} (Figure 3(b)). Again due to the high concentration of Si-OH groups, the final coating coverage would be restricted primarily by spatial hindrance due to the large organic group, and secondarily by the availability of the hydrolyzed groups. As for the particles modified with method II, because the concentration of surface hydroxyls (12 $\mu mol/m^2$) was nearly twice that of the coating coverage (6.8 $\mu mol/m^2$), it could be concluded that a compact monolayer was formed and, on average, each silicon formed two covalent bonds with the nanoparticle surface. It may be argued that a “patchy” multilayer surface structure could be formed in the above situation, instead of a relatively uniform monolayer structure. However, the FTIR results did not support this argument. The two Si-O-Si shoulders in the FTIR spectra were quite weak and inconsistent with a patchy multilayer surface structure. Thus, the anhydrous conditions of method II resulted in lower coating amounts as compared to method I, and a predominantly monolayer structure.

The sudden weight loss at approximately 210°C was likely due to water trapped by the silane-layered structure. MS at that temperature showed no difference for coated particles prepared by both methods. Since MS only can detect ions with molecular weights higher than 20 and our expected silane fragments have a molecular weight higher than 20, the loss could not be due to an organic fragment. FTIR results supported this conclusion where particles prepared by coating method I showed a broad peak at 1660 cm^{-1}, which was assigned to residual water and a weak contribution from the amine bicarbonate salt [29]. This result further supported the conclusion that coating method I yielded a multi-layered structure with some water trapped between silane layers and the particle surface. FTIR and TGA studies did not detect any trapped water in the case of coating method II.

A previously unreported [29] Si NMR resonance appearing at –19 ppm was observed for particles prepared by coating method II (Figure 5(b)), which suggested the occurrence of a previously unobserved reaction during the coating process. The resonance was consistent with silicon bonded only to two oxygen atoms. Based on this analysis, the following reaction mechanism was proposed. After the particle drying treatment, the TiO_2 surface could become oxygen deficient. In addition, it is also known that oxygen vacancies are present even in as-received TiO_2 nanoparticles [39], which are preferred sites of impurity or gas adsorption. Drying then exposes these "old" vacancies to the silane by removing small molecule impurities. Thus, when the surface was treated with the silane, a silane methoxy group may fill the surface oxygen vacancy thereby breaking a silicon-oxygen bond. Therefore, the number of first nearest neighbors of Si changed from three to two oxygen atoms. The silicon atom may then react with a methyl group from either a neighboring silane compound or the surface methoxy group producing a Si-CH_3 bond in place of the silane methoxy group, consistent with the –19 ppm shift in ^{29}Si NMR. Additionally, the presence of the –2.3 ppm resonance in ^{13}C NMR and 1270 cm^{-1} in FTIR supported this hypothesis.

In summary, the surface coating structures were largely controlled by the experimental conditions of the coating process. The presence of water at the interface in particular was found to exert a large effect on the nature of the surface structure. Coating method I led to a multilayer silane structure on the TiO_2 nanoparticle surface with residual water trapped under the silane layers. The multilayer structure formed covalent bonding (Ti-O-Si) to the titania surface, as well as Si-O-Si bonding within the multilayer. For coating method II, a primarily compact monolayer surface structure was formed with each silane molecule forming, on average, two primary bonds to the surface. Two different structures were formed based on the reactivity with adsorbed water on the nanoparticle surfaces. In addition, a new reaction was detected on the titania nanoparticle surface in the coating method II. The formation of Si-CH_3 bonds possibly resulted from the previously unreported reaction of the silane with oxygen vacancies on the dried nanoparticle surface.

Both coating methods yielded covalent-bonded coating structures. Compared to hydrogen-boned coating structures, these structures are more stable and more resistant to water attack. But for the purpose of modifying the nanoparticle/polymer interface, generally, monolayer coating, yielded through coating method II, is sufficient. In addition, trapped water was not detected for the particles coated by method II, while it was detected for those coated by method I. As known, the presence of water at the interface can electrically degrade the insulation. Therefore, coating method II was more suitable for surface modifying the nanoparticles for insulating application. Accordingly, the TiO_2 particles modified with the AEAPS silane (AEAPS-TiO_2) via the coating method II were chosen for composite sample preparation and characterization.

Characterization and Determination of DCS Coating Structure

As stated above, coating method II was preferred for coating nanoparticles. Therefore, only coating method II was applied to coating the nanoparticles with DCS ($CH_3(CH_2)_9SiCl_3$). TGA of the DCS coated particles (DCS-TiO_2) is shown in Figure 7. The amount of DCS coating was 1.8 wt%, or 3.1 μmol/m^2. Similar to the AEAPS, the DCS, which was coated onto the particle surface by coating method II, presented a monolayer coating structure. The lower coverage of DCS compared to the AEAPS was believed to result from the large spatial hindrance caused by its long hydrocarbon chain.

^{29}Si NMR results of neat DCS silane and DCS coated particles are shown in Figure 8. The DCS silane exhibited a single peak at –13 ppm indicating that no observable hydrolyzation or condensation occurred during storage. The spectrum of the DCS coated particles contained two nearby peaks; and as expected, the resolved two peaks were centered at the same positions as the AEAPS coated particles, i.e., –58 and –67 ppm, respectively. This can be easily understood because the chemical component difference in the long side chains in both silanes began from the fourth atom away from the central Si atom; thus, the first different atom was too far away to influence the ^{29}Si NMR spectra of the coated particles. Based on what was described earlier about the correlation of the ^{29}Si NMR peaks with chemical bonding in the section on AEAPS coated nanoparticles, it was determined that the DCS was primarily covalently bonded to the TiO_2 surface, or with itself. No trapped water was detected from TGA or FTIR.

In summary, similar to the nanoparticles coated with the AEAPS silane via method II, the DCS silane formed a "dry", mainly covalently bound monolayer structure on the TiO_2 nanoparticles.

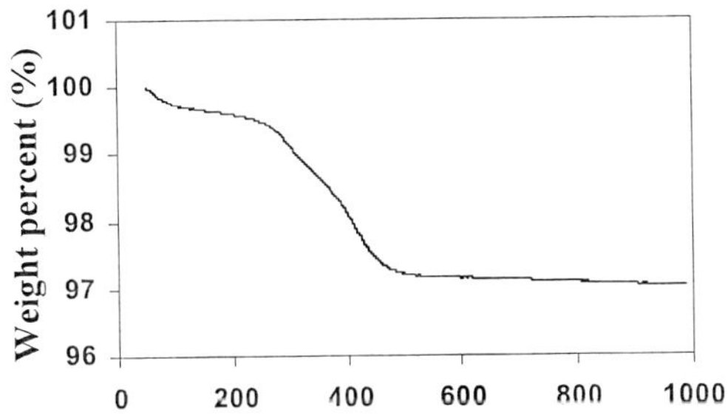

Figure 7. Thermogravimetry of DCS silane modified TiO_2 by the coating method II.

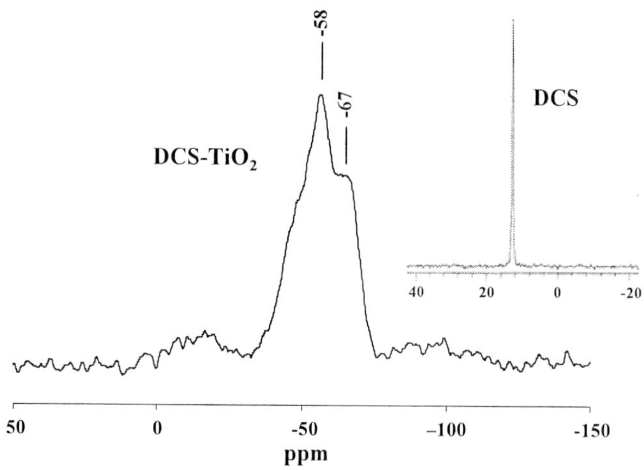

Figure 8. ^{29}Si NMR of the DCS silane and DCS modified TiO_2 nanoparticles.

MICROSTRUCTURE CHARACTERIZATION OF POLYMER NANOCOMPOSITES

Dispersion of nanoparticles in a polymer and the morphology of resulting composites have non-negligible effects on dielectric properties. For example, it is known that particles act as local field enhancing defect centers in filled polymers [1,40,41], which causes a decrease in the dielectric breakdown strength as compared to the neat polymer. The extent of the local field enhancement, and hence the dielectric breakdown strength, is directly related to the size of the particles (or agglomerates) and their dispersion in the matrix. In addition, the motion of charges could be impeded at the interface between the particles and the matrix, and hence the space charge distribution could also be affected by the particle dispersion. On the other hand, polymer morphology can affect the dielectric properties. LDPE is a semi-crystalline polymer, so it consists of an ordered crystalline phase and a disordered non-crystalline phase comprising a partially ordered layer adjacent to the crystalline phase [42] and a completely disordered amorphous region. The degree of crystallinity, the crystallites' size and their distribution, and the organization of the crystallites with respect to the non-crystalline phase govern various physical properties [42], and have long been of research interest. In order to gain a fundamental understanding of the dielectric properties of interesting systems, it is crucial to carry out a thorough microstructural investigation of the prepared nanocomposites. In our previous work [23,43], we have combined a variety of methodologies to characterize the crystalline structure of polyolefin nanocomposites, including a deep exploration of the use of small-angle light scattering (SALS) in the nanocomposites. Herein, we will include some of our previous work and also additional work with a different nanocomposite system.

Dispersion

In order to fully characterize the dispersion of the nanoparticles, FE-SEM and SIMS were employed together. Figure 9 (a) and (b) show FE-SEM micrographs of the bulk dispersion of the as-received TiO_2 (AR-TiO_2) and AEAPS-TiO_2 nanoparticles in the LDPE, respectively. The dispersion of the DCS-TiO_2 was similar to that of the AEAPS-TiO_2. Qualitatively, the as-received nanoscale TiO_2 exhibited a much better dispersion in the matrix than did the coated nanoparticles. Except for individual particles, agglomerates were observed in all samples, but the coated TiO_2 nanoparticles formed more and larger agglomerates in the matrix. Some agglomerates can even be seen without the aid of a microscope. They appeared as "chunks" under polarized reflection optical microscopy. To complement the dispersion characterization, the time-of-flight SIMS technique was used (Figure 10). One advantage of SIMS over FE-SEM to observe the dispersion is that it will not distort the dispersion, which possibly occurs during preparation of the SEM samples. It can be seen that although the dispersion of the nanoparticles in the matrix was not very homogeneous at the nanoscale and agglomerates were also present, "globally", the particles were uniformly distributed in all three composites. Converging all the above observations, it was clear that the nanoparticles were dispersed in the composites at different length scales, from nano to micro, in the form of individual nanoparticles, small aggregates and large agglomerates, respectively. The coated particle filled samples, no matter whether the particle surface was more hydrophilic or hydrophobic as compared to the AR-TiO_2, had more and larger agglomerates than the uncoated particle filled composites. Generally, the dispersion was closely related to the compatibility of particles with the polymer matrix. The increasing order of surface hydrophobicity of all the particles studied was AEAPS-TiO_2 < AR-TiO_2 < DCS-TiO_2 expected from the silane chemical structure. Because the LDPE molecules were hydrophobic, the compatibility between the particles and the matrix would order in the same way, and thus the dispersion would be expected to improve in the same order. The inconsistency arose from the coating process, which formed some agglomerates that could not be broken during shear mixing. Agglomerates may also form during melt mixing because of enhanced incompatibility in the AEAPS-TiO_2 / LDPE.

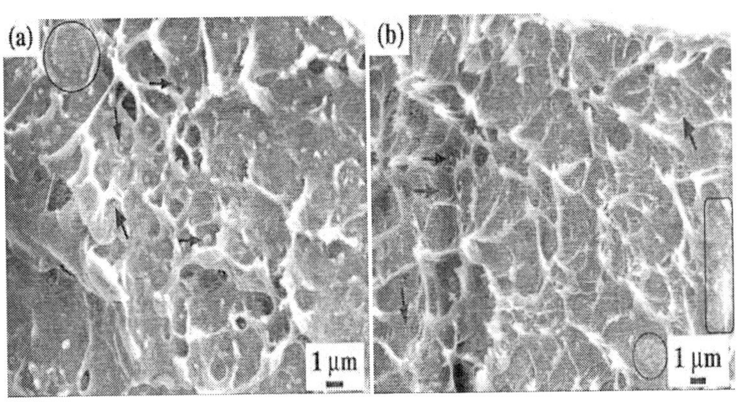

Figure 9. FE-SEM images showing the dispersion of (a) as-received TiO_2 nanoparticles and (b) AEAPS coated TiO_2 nanoparticles in LDPE. Circles and a rounded rectangle indicate agglomerates and arrows point to particles. Reprinted with permission from [18]. Copyright (2005) IOP Publishing Ltd.

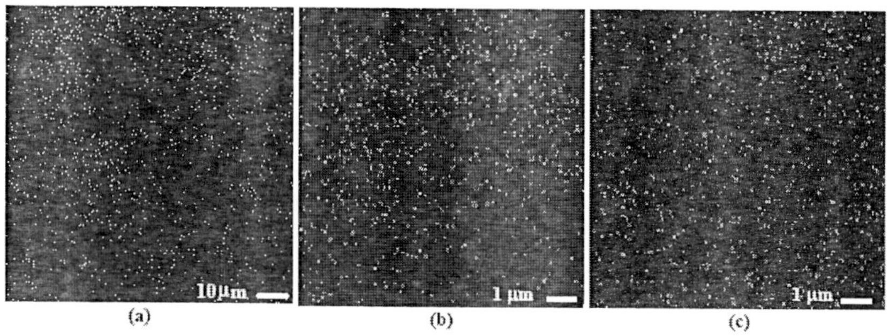

Figure 10. SIMS images showing Ti^+ map in the LDPE composite samples filled with (a) as-received, b) AEAPS coated, and (c) DCS coated nanoparticles.

Morphology

DSC was employed to measure the degree of crystallinity. The values calculated from normalized DSC thermograms (Figure 11) varied from 45 to 48% among all the samples, indicating that the incorporation of the nanoparticles and their different surface states did not evidently affect the degree of crystallinity of the polymer matrix. The same conclusion was made from XRD although the absolute values obtained from this technique were systemically lower than those from DSC, mainly due to the sensitivity difference of each technique to complex crystalline morphology [44].

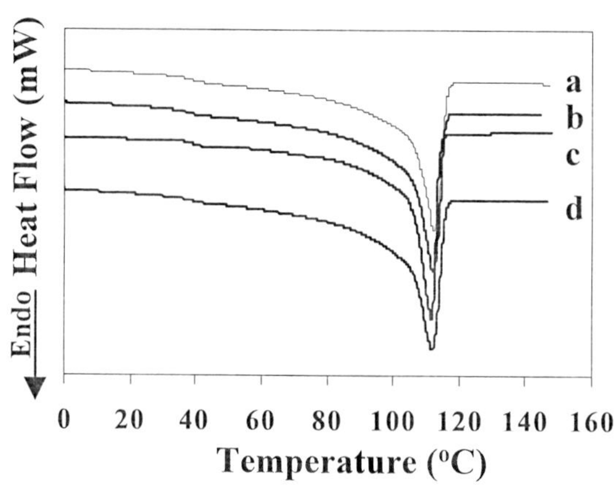

Figure 11. DSC melting curves for (a) neat LDPE, (b) AR-TiO_2 / LDPE, (c) AEAPS-TiO_2 / LDPE, and DCS-TiO_2 / LDPE.

The unit cell parameters were determined from XRD. All normalized diffraction patterns for the samples studied are shown in Figure 12. The amorphous scattering appeared as a broad peak under the strong and sharp diffraction pattern of the crystallites. The diffraction peaks centered near 22° and 24° are characteristic of the (110) and (200) planes in the crystalline region of the LDPE. Among all the samples, the shifts of crystalline peak positions were within experimental errors, which meant that the unit cell dimension was not altered by the nanoparticles and their surface modification. However, the widths of the (110) and (200) diffraction peaks narrowed in the nanocomposites (except for the (200) of the DCS-TiO_2 / LDPE) as compared to those of the neat LDPE. Therefore, the lateral dimension of the PE lamellae might increase with the incorporation of nanoparticles. It should be pointed out, however, that the above discussion was based on an assumption that the peak width is only related to crystallite size. The internal strain and degree of perfection of the crystallites could also alter the peak width. Although it is still not clear which was the most direct reason for the changes of the peak widths, it is likely that both nanoparticles and their different surface features contributed to this variation. The differences were not significant, however.

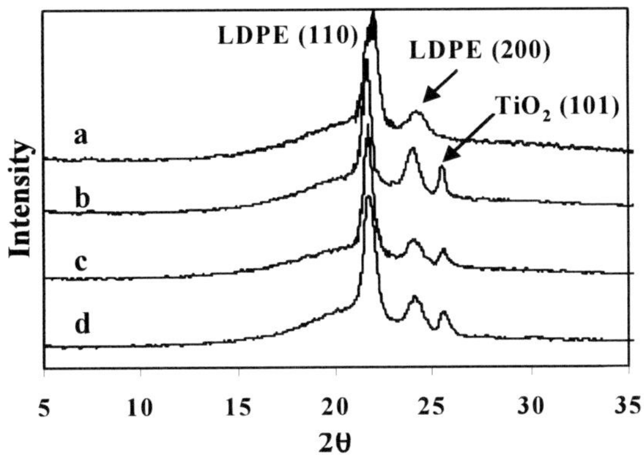

Figure 12. XRD scans of (a) neat LDPE, (b) AR-TiO_2 / LDPE, (c) AEAPS-TiO_2 / LDPE and (d) DCS-TiO_2 / LDPE.

The average lamellar thickness (i.e., the third dimension of the lamellae) and thickness distribution were studied by AFM. Representative phase AFM images of the neat LDPE and the AR-TiO_2 / LDPE are presented in Figure 13. The width of the linear patterns, which is a projection of the lamellae width, can represent the lamellar thickness for relative comparison. All lamellar thickness as observed by AFM varied in the range of 20-40 nm, with the most probable thickness around 25 nm. Although the AFM observations were not able to provide absolute values of lamellar thickness, it still can be reasonably concluded that the average lamellar thickness did not change significantly with the addition of nanoparticles. In addition, DSC analysis also yielded useful information about the lamellar thickness. All DSC thermograms revealed similar melting shape, peak melting temperatures and final melting temperatures, suggesting similar lamellar size distribution. Furthermore, the same final melting temperatures indicated the same thickness of the largest crystals (lamellae) formed in all samples according to the Thompson-Gibbs equation [45,46].

The macroscopic lamellae arrangement (i.e., the suprastructure) was studied by cross-polarized light microscopy. All samples showed similar fine crystallite structure (not shown here) [18], which appeared as bright spots with similar sizes. Small-angle light scattering (SALS) confirmed that they were spherulites with different degrees of perfection [23]. AFM was employed to directly demonstrate the lamellar arrangement. Figure 14 shows the phase images of the neat LDPE film, the AR-TiO_2 / LDPE, the AEAPS-TiO_2 / LDPE, and the DCS-TiO_2 / LDPE, respectively.

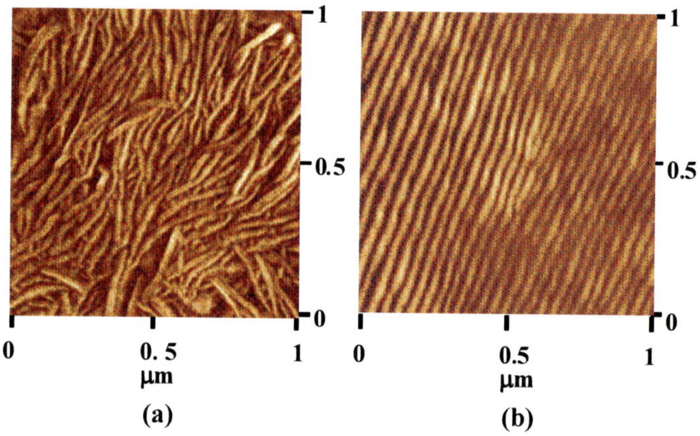

Figure 13. AFM phase images showing lamellae in (a) neat LDPE and (b) AR-TiO_2 / LDPE. Reprinted with permission from [23]. Copyright (2005) Wiley Periodicals, Inc.

It was clear that the neat LDPE exhibited a well-defined impinging spherulite texture (Figure 14(a)). The interspherulitic boundaries could be resolved easily and the diameter of the spherulites ranged from several microns to over 10 microns, consistent with that observed from optical microscopy. Curved individual lamellae or lamellar-bundles were observed. All the nanocomposites exhibited coarser spherulitic texture, increasing the complexity of resolving individual spherulites, lamellar bundles or individual lamellae. Despite this difficulty, it was still clear that all the nanocomposites exhibited textures characteristic of poorly developed spherulites, with the AEAPS-TiO_2 / LDPE displaying the most severely disrupted spherulites with the highest disorder of internal organization of lamellae or their bundles. In addition, micrometer-scale nanoparticle aggregates were easily observed surrounding lamellar bundles. Not like AFM, which only yields surface texture information, SALS studies the bulk crystalline structure and has been carried out on the neat LDPE, the AR-TiO_2 / LDPE, and the AEAPS-TiO_2 / LDPE [23]. From the SALS, relatively quantitative information about the degree of spherulite perfection was achieved. It was found that among these three samples, the AEAPS-TiO_2 / LDPE possessed the worst developed spherulites. This finding was in a good agreement with what was observed at the surface in AFM, confirming that the surface and the bulk had similar crystalline structure.

Based on the above investigations about morphology, the TiO_2 nanoparticles, with various surface conditions, did not evidently change the degree of crystallinity, the unit cell dimension, the average lamellar thickness, and the average spherulite size. However, they definitely affected the degree of perfection of the supramolecular structure by influencing the relative arrangement and orientation of the lamellae within the spherulites and the particle concentrations at the interspherulite boundaries. The AEAPS-TiO_2 / LDPE exhibited the most disrupted and poorly developed spherulite structure. This is associated with surface treatments. The surface of the AEAPS-TiO_2 was more hydrophilic as compared to those of the AR-TiO_2 and the DCS-TiO_2, and thus less compatible with the polymer matrix. This incompatibility led to repelling of particle aggregates to boundary regions and also caused most severe interference to spherulite organization.

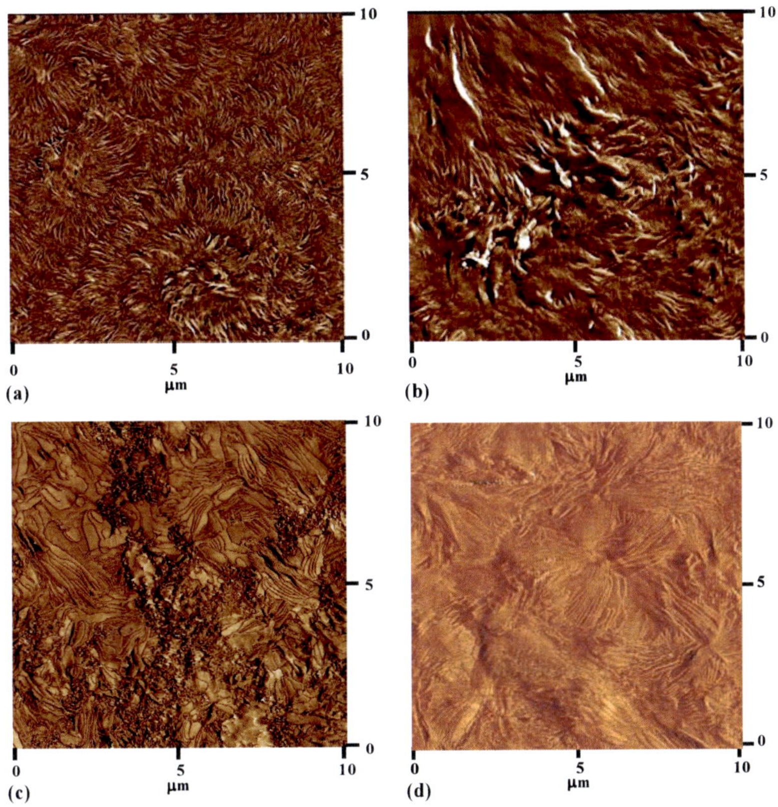

Figure 14. AFM phase images of (a) neat LDPE, (b) AR-TiO_2 / LDPE, (c) AEAPS-TiO_2 / LDPE and (d) DCS-TiO_2 / LDPE. (a), (b) and (c) are reprinted with permission from [23]. Copyright (2005) Wiley Periodicals, Inc.

DIELETRIC BEHAVIOR

In this section, the dielectric properties of the nanocomposites filled with the TiO_2 nanoparticles with different surfaces are presented. To be specific, permittivity, dielectric breakdown strength, and space charge distribution were measured and analyzed. It was found that tailoring the nanoparticle surface by silane modification had a significant effect on the dielectric breakdown strength and space charge distribution, while having little effect on permittivity.

Permittivity

The permittivity of the neat LDPE and LDPE nanocomposite samples is shown in Figure 15. The permittivity of the neat LDPE is 2.26 and all the other data fall into the same region between 2.0 to 2.5. The experimental error was 10 %. The difference in permittivity in those samples was not significant. It was because the nanoparticle loading, 5 wt% (about 1.1 vol %), was too low for the nanoparticles to show their influence on the permittivity of the composite samples. When samples filled with the as-received TiO_2 nanoparticles at different loadings were examined, however, it was found that the overall permittivity increased nearly linearly with increasing TiO_2 volume fraction (Figure 16). This increase could be easily understood because TiO_2 nanoparticles have a much higher permittivity than neat LDPE. This linear relationship indicated that the composite permittivity was mainly affected by the bulk property, instead of by the surface property, of the nanoparticles. Otherwise, the permittivity would have shown a linear relationship with d^2, instead of d^3 (proportional to the volume fraction), where d is the average diameter of the nanoparticles. For comparison, the permittivity of the microcomposite samples filled with as-received TiO_2 microparticles was also measured, which yielded nearly the same permittivity value as the nanocomposites. The above results indicate that there is not much opportunity to change permittivity by doing surface modification in our studied system.

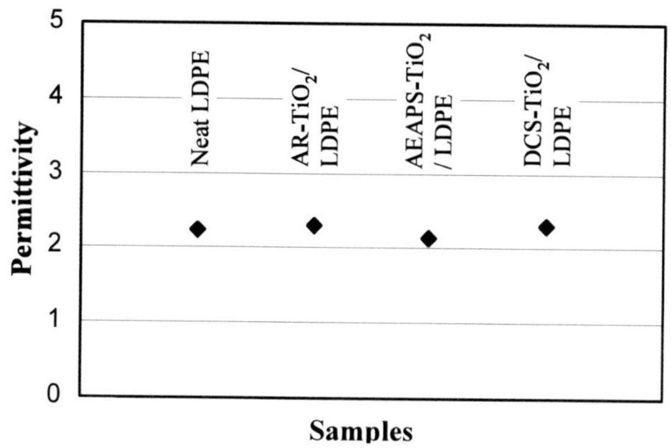

Figure 15. Permittivity of neat LDPE and LDPE samples filled with as received (AR), AEAPS coated, and DCS coated TiO_2 nanoparticles.

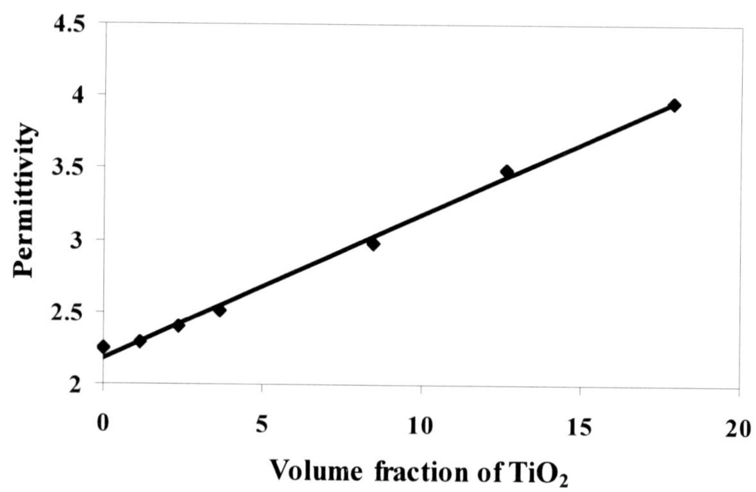

Figure 16. Permittivity linearly increased with increasing volume fraction of as-received TiO_2 nanoparticles. A trendline is shown.

Dielectric Breakdown Strength

Dielectric breakdown strength is defined as the electric field intensity that causes an "insulator-to-conductor" transition in a material and is accompanied by a sharp increase in current. It is known that even if variables such as morphology, additives and test conditions are under exact control, scatter in the dielectric breakdown is inevitable, which increases the difficulty of data analysis. Therefore various statistical methods are used to treat data. A Weibull distribution [1,47] is used in this work to present breakdown strength data. The cumulative probability P of the electrical failure takes the form of [1,47]:

$$P = 1 - \exp[-(\frac{E}{E_0})^{\beta}] \quad (1)$$

where β is a shape parameter that represents the data scatter and E_0 is a scale parameter that represents the breakdown strength at the cumulative failure probability of 63.2%. E_0 is also called Weibull breakdown strength and frequently used to evaluate the dielectric strength of different materials. P is calculated from [47]:

$$P_i = \frac{i - 0.5}{n + 0.25} \quad (2)$$

where i is the i-th result when the values of E are sorted in ascending order and n is the number of measurements. Both β and E_0 were obtained from linear regression of the linearized Equation (1) by taking the logarithm to it. Figure 17(a) shows the Weibull plots of the dielectric breakdown strength of neat LDPE, the AR-TiO_2 / LDPE and the AEAPS-TiO_2 / LDPE samples. The scattered data of the DCS-TiO_2 / LDPE fell in between those of the AEAPS-TiO_2 / LDPE and the neat LDPE. In order to avoid too crowded data and to highlight the difference in the Weibull strength E_0 of different samples more clearly, the Weibull plot of the DCS-TiO_2 / LDPE was not included in Figure 17(a), but shown in Figure 17(b) separately. For comparison, the Weibull plot of the neat LDPE is also shown in Figure 17(b).

The data points were not aligned very linearly. The rapid decrease at relatively low breakdown strength values are thought to be due to the presence of agglomerates [18,21]. Despite this, the Weibull plots were sufficient for comparing differences between our samples and identifying general trends in the results. Neat LDPE had an E_0 of 3.8±0.4 MV/cm. With the incorporation of 5 wt% of AR-TiO_2, E_0 decreased by about 40% to 2.3±0.3 MV/cm. The large negative impact of small amounts of additive particles to the dielectric breakdown strength has been reported previously. For example, Khalil *et al.* [48] showed that the addition of 1 wt% of TiO_2 caused a 9% decrease in step dc breakdown strength of LDPE. Ueki *et al.* [47] reported that the incorporation of 2.5 wt% of carbon black N220 into high-density polyethylene led to a reduction of about 50% in dc breakdown strength. After particle surface modification, E_0 of the nanocomposite samples was still lower than that for the neat LDPE. However, both coated samples showed evidently higher dielectric breakdown strength than that for the AR-TiO_2 / LDPE. The dielectric breakdown strength of the AEAPS-TiO_2 / LDPE (3.1±0.3 MV/cm) was nearly 40% higher than that for the AR-TiO_2 / LDPE. The DCS-

TiO_2 / LDPE showed an even higher Weibull breakdown strength of 3.3±0.4 MV/cm. Evidently, the nanoparticle surfaces affected the dielectric breakdown strength significantly. Furthermore, it appeared that both the hydrophilic surface (coated with polar AEAPS) and the more hydrophobic surface (coated with nonpolar DCS) showed an enhanced Weibull breakdown strength compared to the nanocomposites filled with the uncoated TiO_2 nanoparticles. It is important to understand this interesting phenomenon by investigating the microstructure and correlating it with dielectric breakdown fundamentals.

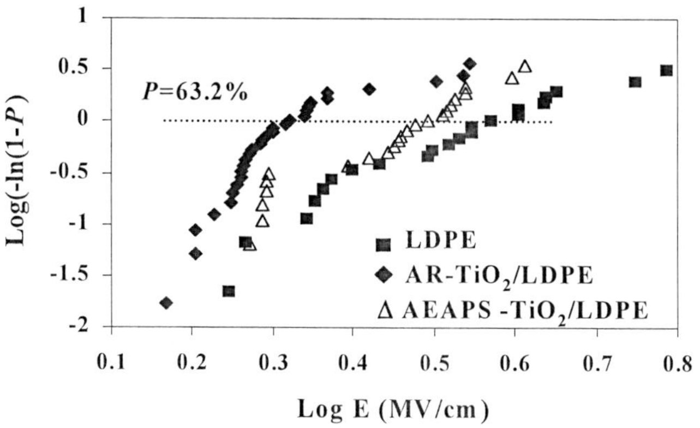

a)

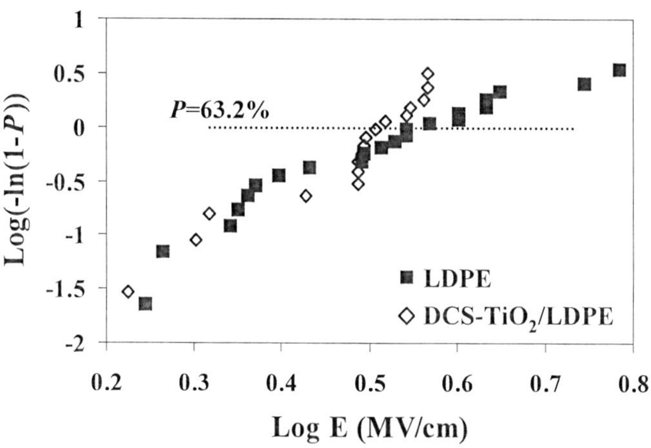

b)

Figure 17. Weibull plots of dielectric breakdown strength of (a) neat LDPE, AR-TiO_2 / LDPE and AEAPS-TiO_2 / LDPE, and (b) neat LDPE and DCS-TiO_2 / LDPE. (a) is adapted with permission from [18]. Copyright (2005) IOP Publishing Ltd.

It is known that dielectric breakdown in polymers is a complicated process, and could be dominated by more than one mechanism [49-52]. The process is influenced by many factors such as temperature [53], molecular weight [53], degree of crystallinity [53], and spherulite size and structure [54]. With the introduction of the nanoparticles, the complexity is further increased. Nanoparticles could not only influence the polymer morphology, but also could act as defect centers electrically and introduce many unique interfacial areas as well. Therefore, in addition to those factors in the neat polymer, the nature of the particles, their dispersion and surface features all could directly affect the dielectric breakdown process. In our case, for example, the reported conductivity of nanocrystalline TiO_2 (35 nm) is on the order of 10^{-6} to 10^{-7} $(\Omega\text{-cm})^{-1}$ at 500 °C and under atmospheric pressure [55], while that for neat polyethylene at room temperature is in the range of 10^{-15}-10^{-17} $(\Omega\text{-cm})^{-1}$ [56]. The difference between these two conductivities exceeds 9 orders of magnitude. Therefore, electric field distortion around the TiO_2 nanoparticles is inevitable. If two particles happen to be quite close to each other so that their distorted field superimposes, a higher field enhancement will occur and the polymer between the two particles will suffer a much higher local electrical stress compared to the bulk polymer. Therefore, the polymer between the two particles will fail at a lower external electric field first and then this local failure will initiate macroscopic breakdown. Thus, the TiO_2 nanoparticles can act as electrical defect centers in the composites and their poor dispersion would significantly reduce the dielectric breakdown strength further.

In our previous work [18], we analyzed the effects of the AEAPS modification on the dielectric breakdown strength in detail. Herein, in order to compare how different surface modification influenced the dielectric breakdown strength with different mechanisms, it is necessary to give a brief summary. The AEAPS-TiO_2 / LDPE exhibited a worse particle dispersion, had a more incompatible interface with the LDPE matrix and revealed spherulites with less organized internal structure, as compared to those of the AR-TiO_2 / LDPE. All these facts would theoretically lead to lowering of the dielectric breakdown strength. On the contrary, the AEAPS-TiO_2 / LDPE had Weibull breakdown strength nearly 40% higher than that of the AR-TiO_2 / LDPE. The improved dielectric breakdown strength was associated with modification of electrical features at the interface between the coated nanoparticles and the matrix: the polar groups could act as charge scattering centers, traps or excitation sites. Therefore, the incorporation of polar groups can decrease the acceleration of electrons, thus decreasing the chance of them initiating dielectric breakdown.

This mechanism cannot explain the behavior for the DCS-TiO_2 / LDPE as no polar groups are presented. The DCS silane has a 10-C hydrocarbon chain, and thus the presence of it on the nanoparticle surface could change the nanoparticle surface from hydrophilic to hydrophobic. Therefore, the interfacial compatibility between the nanoparticles and the hydrophobic LDPE matrix should be improved as compared to that of the AR-TiO_2 / LDPE and the AEAPS-TiO_2 / LDPE. The improved compatibility in the DCS-TiO_2 / LDPE was evidenced by using solution mixing and sonication, which were able to break the agglomerates formed during the coating process. These agglomerates were not broken up by melt-mixing. In this case, the dispersion of the nanoparticles would theoretically be a direct reflection of the interfacial compatibility. It was found that the DCS-TiO_2 / LDPE revealed a nice dispersion of the nanoparticles, while the AEAPS-TiO_2 / LDPE showed inhomogeneous dispersion of the nanoparticles after solution mixing.

The DCS-TiO_2 / LDPE prepared by melt mixing also had worse dispersion than the AR-TiO_2 / LDPE. In characterizing the morphology, the DCS-TiO_2 / LDPE and the AR-TiO_2 / LDPE had a similar degree of crystallinity, unit cell parameters, and average lamellar thickness and degree of perfection of spherulite structure. Therefore, the over 40% increase in the dielectric breakdown strength of the DCS-TiO_2 / LDPE over the AR-TiO_2 / LDPE was mainly due to improved interfacial compatibility despite the worse dispersion in the former sample.

Based on the above discussion, it could be predicted that there was still a possibility to further enhance the Weibull breakdown strength by improving the particle dispersion in the DCS-TiO_2 / LDPE. In order to test this hypothesis, breakdown measurements were carried out on solution mixed DCS-TiO_2 / LDPE and compared with that of the melt-mixed DCS-TiO_2 / LDPE. As expected, the Weibull strength of the solution-mixed DCS-TiO_2 / LDPE was increased by 11% and, thus, quite close to that of the neat LDPE. Investigation of the microstructure of the solution processed samples did not show any significant differences in crystallinity and spherulite structure compared to the melt mixed DCS-TiO_2/LDPE composite, indicating that the main difference was the nanoparticle dispersion between the DCS-TiO_2 / LDPE samples prepared by melt mixing and solution mixing. Therefore, by carrying out particle surface modification and adjusting the processing conditions, the Weibull breakdown strength of the nanocomposites approached that of the neat LDPE.

On the other hand, β is an important parameter and need to be taken into account. β values were 3.5, 5.1, 3.3 and 5.1 for the neat LDPE, AR-TiO_2 / LDPE, AEAPS-TiO_2 / LDPE and DCS-TiO_2 / LDPE, respectively. β has been used to evaluate the data scatter and, in polymer composites, the data scatter is closely related to the size distribution of the defects or filler particles. Higher β values indicate a narrower data distribution. Worse particle dispersion in the AEAPS-TiO_2 / LDPE caused relatively greater data spread in this sample than in the AR-TiO_2 / LDPE. However, it is not clear at this moment why the DCS-TiO_2 / LDPE showed the much decreased data scatter than the AEAPS-TiO_2 / LDPE and the similar data scatter as the AR-TiO_2 / LDPE although the AR-TiO_2 / LDPE exhibited better particle dispersion. Better interface in the DCS-TiO_2 / LDPE could be a reason.

Space Charge Distribution

Space charge accumulation is a common phenomenon in polymer materials and has been extensively studied [24,57-60]. The build-up of space charge can ruin both short and long term performance of polymers in electrical applications [1,61,62]. It is imperative to include a study of space charge distribution, if one would like to evaluate a polymer insulating material comprehensively. It should be pointed out, however, because space charge distribution is much affected by the electrode materials [24,57], the results and discussion of the space charge distribution will be only restricted to the specific measuring system investigated. The underlying meaning of this statement is that both material properties of a sample and interaction between the sample and electrodes account for the space charge distribution.

The space charge distribution of the neat LDPE, the AR-TiO_2 / LDPE, the AEAPS-TiO_2 / LDPE and the DCS-TiO_2 / LDPE is presented in Figure 18(a), (b), (c), and (d), respectively. Figure 18(a) illustrates development of the space charge distribution versus time in the neat LDPE sample. The building-up of a small amount of hetero space charges, i.e. the charges of the opposite polarity to the nearby electrodes, was completed within the first 30 minutes. Then, the space charges were nearly constant with time. The distribution of the space charge was symmetric with positive and negative space charges being of the same density near the two electrodes, indicating little or no charge injection from either electrode [62]. These hetero space charges possibly originated from impurity dissociation and migration under the field.

Figure 18(b) shows the space charge distribution in the AR-TiO_2 / LDPE. Its space charge distribution was completely different from that of the neat LDPE. The nature of the charges changed. The space charges were of the same polarity as the nearby electrodes, i.e. primarily homo. This indicates enhanced charge injection from the electrodes with the introduction of the nanoparticles. It was also noted that the charge distribution in the sample was not symmetric. The accumulation of positive charges near the anode was more evident than that of the negative charges near the cathode, which spread out evenly, indicating that the electrons/negative charge carriers injected from the cathode had a higher mobility than that of holes/positive charge carriers injected from the anode. The space charge became stabilized after 30 minutes. The above observations were different from those previously reported [63]. In that work, LDPE revealed homo space charge distribution, but the TiO_2 / LDPE sample revealed negative space charges all through the sample. Therefore, the space charge distribution is affected significantly by the nature of both the matrix and the filler. Specifically, it could be influenced by the surface state and crystal structure of the filler.

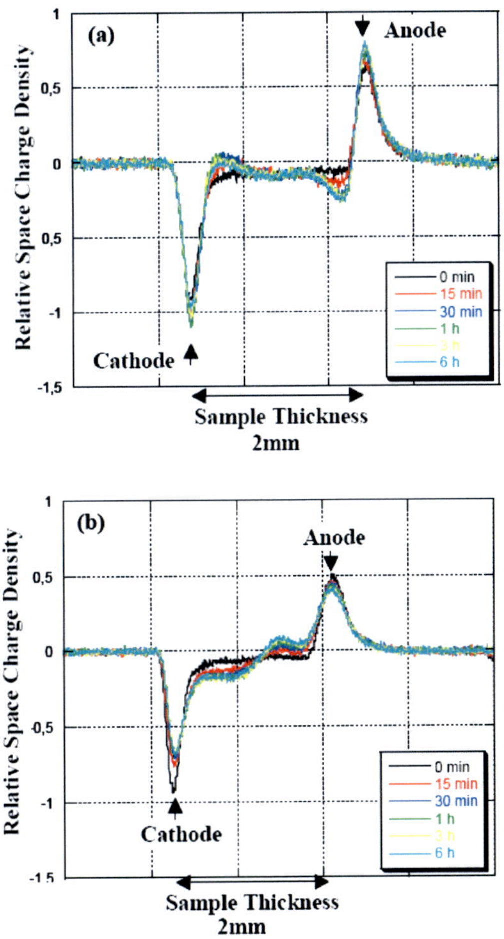

Figure 18. continued on next page.

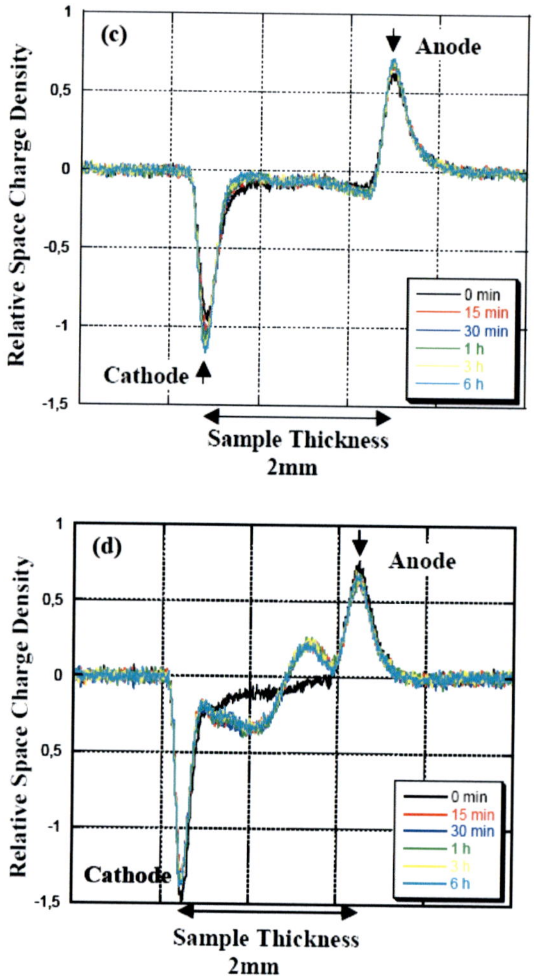

Figure 18. Space charge distribution in (a) neat LDPE, (b) AR-TiO_2 / LDPE, (c) AEAPS-TiO_2 / LDPE, and (d) DCS-TiO_2 / LDPE at different times after application of an electric field of 20 kV/mm. (a), (b) and (c) is reprinted with permission from [18]. Copyright (2005) IOP Publishing Ltd.

Figure 18(c) demonstrates the clearly suppressed space charge distribution in the AEAPS-TiO_2 / LDPE. Only a small number of homo space charges were observed in the entire sample and during the observation period of 6 hr, no appreciable increase in the space charge was found. A small number of space charge together with a constant density (flat curve) in the entire composite are beneficial to both short and long term breakdown strength. The evident improvement of the space charge distribution in this sample was the result of introduction of polar groups at the interface between the nanoparticles and the polymer matrix, as explained in our previous work [18].

The space charge distribution of the DCS-TiO_2 / LDPE is revealed in Figure 18(d). The homo space charge increased remarkably during the initial 15 min, then remained fairly stable. Both the positive and negative charges formed the peaks near the two electrodes. No "flat" region was observed in this sample. The DCS-TiO_2 / LDPE had the largest net intensity of the space charge among all the samples studied here.

The variation of the space charge distribution among four samples was easily observed. As shown in Figure 18, the nature of the space charge distribution changed from hetero to homo with the incorporation of 5 wt% of the AR-TiO_2 nanoparticles into the LDPE. Charge suppression and charge amplification were observed for the AEAPS-TiO_2 / LDPE and the DCS-TiO_2 / LDPE, respectively. Moreover, the nature of charges was different: hetero in the AEAPS-TiO_2 / LDPE, homo in the DCS-TiO_2 / LDPE. Clearly, the introduction of the nanoparticles and their surface modification both had significant effects on the space charge distribution.

Space charge distribution is determined by comprehensive trapping/detrapping and transport behavior of charges in insulating polymers. Additive particles could affect the space charge distribution directly by annihilating the trap centers or by introducing new traps [62,64,65], or indirectly by modifying the polymer morphology [59,64,66], and thus varying trap density, trap depth distribution or charge carrier mobility. However, the microstructure of the polymer had perhaps second-order effects on the space charge distribution, while electrical features of the filler particles themselves played a more important role [24,62]. In view of the observation that the morphology difference in our samples was even less than what was previously reported [24,62], here we discuss the difference in the space charge distribution from the aspect of electrical features of the particles and particle/electrode interaction, disregarding the morphology variations in the composite samples due to different particle surface conditions.

The reason that the AR-TiO_2 / LDPE showed a completely opposite space charge distribution pattern to that of the neat LDPE is probably because the TiO_2 nanoparticles changed the surface chemistry of the electrodes, allowing a massive charge injection to form a homo space charge distribution [62]. It also could not be excluded that ionic conduction of Ti ions contributed to the overall space charge distribution. The DCS-TiO_2 / LDPE showed a similar homo space charge pattern. However, the space charge density was much higher indicating more charge injection from electrodes related to the surface modified nanoparticles. Moreover, the more symmetric positive and negative charge distribution suggested similar mobility of both charges in the DCS-TiO_2 / LDPE, while in the AR-TiO_2 / LDPE the mobility was different for positive and negative charges. Therefore, DCS surface modification apparently not only enhanced the electrode charge injection, but also altered the charge mobility.

The AEAPS-TiO_2 differed from the AR-TiO_2 and DCS-TiO_2 because of the presence of polar molecules on the surface. These polar molecules or dipoles at the particle/polymer interface may affect charge carrier transport and trapping. The observed tiny number of hetero space charges was from deep traps because the charge build-up seen in Figure 18(c) was still present for over 6 hr after discharge. The suppression of the space charge accumulation therefore indicates a reduced number of charges trapped in deep traps in this sample. This could be due to a reduced number of deep traps or an increased number of shallow traps formed when introducing polar groups [18], and hence the decreased chance of charges being trapped by deep traps. Except for positive effects on the space charge accumulation, it should also be noted that the nature of the space charge changed from homo to hetero through modifying the TiO_2 surface with the AEAPS. This finding suggests reduced or no charge injection from the electrodes as compared to the samples filled with the AR-TiO_2 and DCS-TiO_2 nanoparticles. This was quite likely due to the change of interaction between the sample and the electrode materials after the nanoparticle surface modification with the polar AEAPS.

Another possibility was that the AEAPS surface modification of the nanoparticles encouraged local charge transfer at the polymer/electrode interface [67]. The energy of the donor states could be raised due to the existence of a large polarization energy related to the polar group [67]. Therefore, polar groups on the nanoparticle surface may play a multiple role in the mechanism of space charge generation.

In summary, the surface modification of the nanoparticles had different effects on the different dielectric properties of the nanocomposites. Permittivity seems not to be influenced by the particle surface modification appreciably at the low loading. Both polar and non-polar particle surface modification improved the dielectric breakdown strength significantly as compared to the composite filled with the uncoated nanoparticles. On the other hand, the surface modification influenced the space charge build-up in a complicated way; not only could the amount of charges, but also the nature of charges could be affected.

CONCLUSION

Surface modifications of the nanoparticles were carried out by using two silanes, which had quite different features. They were N-(2-aminoethyl)3-aminopropyl-trimethoxy-silane(AEAPS) having polar groups, and decyltrichlorosilane(DCS) having a relatively long hydrocarbon chain, which changed the nanoparticle surface from hydrophilic to hydrophobic. Surface coating structures were very sensitive to coating conditions. By using a dry nanoparticle / dry toluene reflux coating process, a desired coating structure was obtained. It was found that particle surface modification played a very important role in determining the dielectric breakdown strength. The AEAPS-TiO_2 / LDPE had Weibull breakdown strength 35% higher than that of the AR-TiO_2 / LDPE, associated with modification of electrical features at the interface between the coated nanoparticles and the matrix: the polar groups could act as charge scattering centers, traps or excitation sites in complicated electrical processes. The dielectric breakdown strength of the DCS-TiO_2 / LDPE was further increased, as compared to the AEAPS-TiO_2 / LDPE, mainly attributed to the improved interfacial compatibility between the particle fillers and the LDPE matrix. Although both polar and nonpolar surface modification had beneficial effects on the dielectric breakdown strength, their effects on the space charge distribution were quite different. The AEAPS-TiO_2 / LDPE exhibited hetero space charge distribution, with evidently suppressed charges, while the DCS-TiO_2 / LDPE revealed similar homo space charge distribution to that of the AR-TiO_2 / LDPE, with a largely enhanced charge accumulation. The evident improvement in the space charge formation in the AEAPS-TiO_2 / LDPE was again due to modification of electrical features at the interface between the coated nanoparticles and the matrix, and hence modification of the trap distribution in this material. The increased homo charges in the DCS-TiO_2 / LDPE were due to enhanced charge injection from the electrode because of particle/electrode interaction. This work has clearly demonstrated that the dielectric properties of polymer nanocomposites can be tuned by doing particle surface modification, a finding that can be used in insulating material optimization for industrial applications.

ACKNOWLEDGEMENTS

The work was supported by the ABB Group Service Center and the Nanoscale Science and Engineering Initiative of the National Science Foundation under NSF Award Number DMR-0117792. We would also like to acknowledge receipt of the Mettler-Toledo Thermal Analysis Education Grant 2001, and the TGA and DSC equipment. In addition, we would like to gratefully acknowledge Anna Christerson and Dmitri V. Zagorevski for their work on space charge distribution and mass spectroscopy, respectively. Finally, we would like to give our thanks to Eva Mårtensson, Carina Önneby, Benjamin J. Ash, and Tom Apple for their very useful discussions of this work.

REFERENCES

[1] Khalil, M. S. *IEEE T. Dielect. El. In.* 2000, 7, 261.
[2] Zhu, Y.; Yoon, H. G.; Suh, K. S. *IEEE T. Dielect. El. In.* 1999, 6, 164.
[3] Tanaka, Y.; Ohnuma, N.; Katsunami, K.; Ohki, Y. *IEEE T. Electr. Insul.* 1991,26, 258.
[4] Morshuis, P. H. F.; Kreuger, F. H.; Leufkens, P. P. *IEEE T. Electr. Insul.* 1988, 23, 1051.
[5] Wintle, H. J. *IEEE T. Electr. Insul.* 1990, 25, 27.
[6] Ramu, T. S. Conference Record of the 1990 *IEEE international Symposium on Electrical Insulation*, Toronto, Canada, 1990, 21.
[7] Peacock, A. J. Handbook of Polyethylene: Structures, Properties and Applications, *Marcel Dekker*: New York, NY, 2000.
[8] Ashcraft, A. C.; Eichhorn, R. M.; Shaw, R. G. Conference Record of the 1976 IEEE *International Symposium on Electrical Insulation*, Montreal, Canada, 1976, 213.
[9] Gächter, R.; Müller, H. (Editor) *Plastics Additives Handbook,* Hanser Publishers: Munich, Germany, 1990.
[10] Ryfšavý, D.; Tkadlečková, H. Polym. *Degrad. Stabil.* 1992, 37, 19.
[11] Siegel, R. W.; Hu, E.; Roco, M. C. (Editor) WTEC Panel Report on Nanostructure Science and Technology: Rand D Status and Trends in Nanoparticles, Nanostructured Materials, and Nanodevices, *International Technology Research Institute*: Dordrecht, Baltimore; Kluwer Academics Publishers: Boston, 1999.
[12] Buuren, T. V.; Dinh, L. N.; Chase, L. L.; Siekhaus, W. J.; Terminello, L. *J. Phys.Rev. Lett.* 1998, 80, 3803.
[13] Ren, S. Y.; Dow, J. D. *Phys. Rev.* 1992, B 45, 6492.
[14] Frenkel, J.; Dorfman, *J. Nature* 1930, 126, 274.
[15] Mahoney, W.; Schaefer, D. M.; Patil, A.; Andres, R. P.; Reifenberger, *R. Surf. Sci.* 1994, 316, 383.
[16] Rastogi, A. C.; Sharma, S. N.; Kohil, S. *Semicond. Sci. Technol.* 2000, 15, 1011.
[17] Rao, C. N. R.; Kulkarni, G. U.; Thomas, P. J.; Edwards, P. P. *Chem Eur. J.* 2002, 8, 28.
[18] Ma, D.; Hugener, T. A.; Siegel, R. W.; Christerson, A.; Mårtensson, E.; Önneby, C.; Schadler, L. S. *Nanotechnology* 2005, 16, 724.

[19] Roy, M.; Nelson, J. K.; Reed, C. W.; MacCrone, R. K.; Keefe, R. J.; Zenger, W.; Schadler, L. S. *IEEE T. DIELECT. EL. IN.* 2005, in press.

[20] Roy, M.; Reed, C. W.; MacCrone, R. K.; Schadler, L. S.; Nelson, J. K.; Keefe, R.; Zenter, W. 2005 *International Symposium on Electrical Insulating Materials*, Tokyo, Japan, 2005.

[21] Ma, D.; Siegel, R. W.; Hong, J-I.; Schadler, L. S.; Mårtensson, E.; Önneby, C. *J. Mater. Res.* 2004, 19, 857.

[22] Tsubokawa, N.; Kogure, A.; Maruyama, K.; Sone, Y.; Shimomura, M.; Polym. *J. 1990*, 22, 827.

[23] Ma, D.; Akpalu, Y. A.; Li, Y.; Siegel, R. W.; Schadler, L. S. J. Polym. Sci. B: Polym. Phys. 2005, 43, 488.

[24] Tanaka, Y.; Chen, G.; Zhao, Y.; Davies A. E.; Vaughan, A. S.; Takada, T. 2003 *IEEE Trans. Dielectr. Electr. Insul.* 2003, 10, 148.

[25] Plueddemann, E. P. Silane Coupling Agents, Plenum: New York, 1982.

[26] Munuera, G.; Rives-Arneau, V.; Saucedo, A. *J. Chem. Soc. Faraday Trans.* 1979, 75, 736.

[27] Finklea, H. O.; Vithanage, R. *J. Phys. Chem.* 1982, 86, 3621.

[28] Maira, A. J.; Coronado, J. M.; Augugliaro, V.; Yeung, K. L.; Conesa, J. C.; Soria, J.; *J. Catal.* 2001, 202, 413.

[29] Culler, S. R.; Ishida, H.; Koenig, *J. L. Appl. Spectrosc.* 1984, 38, 1.

[30] Launer, P. J. Infrared Analysis of Organosilicon Compounds, Spectra-structure Correlations, in: *Silicone Compounds Register and Review,* Petrarch Systems Catalogue, or www.gelest.com.

[31] Flett, M. St. C. Characteristic Frequencies of Chemical Groups in the Infra-Red, *Elsevier Publishing Company*: Amsterdam, London, New York, 1963.

[32] Kim, Y.; Kim, H. J. *J. Korean Phys*. Soc. 2002, 40, 94.

[33] Yang, J. J.; El-Nahhal, I. M.; Maciel, G. E. *J. Noncryst. Solids* 1996, 204, 105.

[34] Yang, J. J.; El-Nahhal, I. M.; Chuang, I-S.; Maciel, G. E. *J. Noncryst. Solids* 1997, 209, 19.

[35] da Fonseca, M. G.; Airoldi, C. *J. Mater. Chem.* 1999, 9, 1375.

[36] Sindorf, D. W.; Maciel G. E. *J. Am. Chem. Soc.* 1983, 105, 3767.

[37] Zaper, A. M.; Koenig, *J. L. Polym. Compos.* 1985, 6, 156.

[38] Abboud, M.; Turner, M.; Duguet, E.; Fontanille, M. *J. Mater. Chem.* 1997, 7, 1527.

[39] Parker, J. C.; Siegel, R. W. *J. Mater. Res.* 1990, 5, 1246.

[40] Chen, G.; Davies A. E. *IEEE Trans. Dielectr. Electr. Insul.* 2000, 7, 401.

[41] Fujita, S.; Ruike M.; Baba, M. 1996 *Annual Report of the Conference on Electrical Insulation and Dielectric Phenomena, Millbrae, CA*, 1996, 2, 738.

[42] Peacock, A. J. Handbook of Polyethylene: Structures, Properties and Applications, *Marcel Dekker:* New York, 2000.

[43] Akpalu, Y. A.; Ma, D.; Siegel, R. W.; Schadler L. S. PMSE Preprints, *American Chemical Society, Division of Polymeric Materials: Science and Engineering,* 2004, 91, 918.

[44] Mark, J. E.; Eisenberg, A.; Graessley, W. W.; Mandelkern, L.; Samulski, E. T.; Koenig, J. L.; Wignall, G. D. *Physical Properties of Polymers, American Chemical Society:* Washington DC, 1993.

[45] Sanchez, I. C.; Eby, R. K. *Macromolecules* 1975, 8, 638.

[46] Crist, B.; Howard, P. R. *Macromolecules* 1999, 32, 3057.
[47] Ueki, M. M.; Zanin, M. *IEEE T. Dielec. El.* In. 1999, 6, 876.
[48] Khalil, M. S.; Henk, P. O.; Henriksen, M. Conference Record of the 1990 *IEEE International Symposium on Electrical Insulation*, Toronto, Canada, 1990, 268.
[49] Blythe, A. R. *Electrical Properties of Polymers*, Cambridge University Press: Cambridge, London, New York, Melbourne, 1979.
[50] Wagner, K. W. *AIEE Trans.* 1922, 41, 288.
[51] Stark, K. H.; Garton, C. G. *Nature* (London) 1955, 176, 1225.
[52] Artbauer, J.; Kolloid Z. und Z. *Polymere* 1965, 15, 202.
[53] Ieda, M. *IEEE T. Electr. Insul.* 1980, 15, 206.
[54] Kolesov, S. N. Vysokomol. *Soed.* 1968, 10, 582.
[55] Knauth, P.; Tuller, H. L. *J. Appl. Phys.* 1999, 85, 897.
[56] Callister, W. D. Materials Science and Engineering: an Introduction, John Wiley and Sons, Inc.: New York, Chichester, Brisbane, Toronto, Singapore, 1990.
[57] Chen, G.; Tay, T. Y. G.; Davies, A. E.; Tanaka, Y.; Takada, T. *IEEE Trans. Dielectr. Electr. Insul.* 2001, 8, 867.
[58] Matsui, K.; Tanaka, Y.; Takada, T.; Fukunaga, K.; Maeno, T. *Proceedings of the 7th International Conference on Properties and Applications of Dielectric Materials*, Nagoya, Japan, 2003, 867.
[59] Fleming, R. *J. Brazilian J. Phys.* 1999, 29, 280.
[60] Zhang, C.; Mizutani, T.; Kaneko, K.; Ishioka, M. J. Phys. D: *Appl. Phys.* 2002, 35 1875.
[61] Laurent, C.; Massines, F.; Mayoux, C. *IEEE Trans. Dielectr. Electr. Insul.* 1997, 4, 585.
[62] Cartwright, G. A.; Davies, A. E.; Swingler, S. G.; Vaughan, A. S. *IEE Proceedings Science, Measurement and Technology* 1996, 143, 26.
[63] Kahali, M. S.; Zaky, A. A.; Hansen, B. S. 1985 *Annual Report of the Conference on Electrical Insulation and Dielectric Phenomena*, Buffalo, NY, 1985, 143.
[64] Khalil, M. S.; Montanari, G. C.; Fabianl, D.; Palmieri, F. *XIIIth International Symposium on High Voltage Engineering,* Delft, Netherlands, 2003, 1.
[65] Perlman, M. M.; Haridoss, S. *IEEE Trans. Electr. Insul.* 1987, 22, 9.
[66] Li, X. ; Du, Q.; Kang, J.; Tu, D. *J. Polym. Sci. B: Polym. Phys.* 2002, 40, 365.
a. Lewis, T. *J. IEEE Trans. Electr. Insul.* 1984, 19, 210.

In: Encyclopedia of Nanotechnology
Editor: Elwood D. Carlson

ISBN: 978-1-60692-079-4

Chapter 34

FORMATION OF NANOPARTICLES UNDER LASER ABLATION OF SOLIDS IN LIQUIDS

G. A. Shafeev
Wave Research Center, Prokhorov General Physics Institute,
Russian Academy of Sciences, Moscow, Russia

ABSTRACT

The process of nanoparticle formation under laser ablation of solids in liquids is described. Critical parameters are discussed that govern the properties of nanoparticles ejected into the surrounding liquid. These parameters are laser wavelength, pulse duration, interaction of individual nanoparticles with laser beam inside the liquid. A review of previous results is presented on the properties of nanoparticles of noble metals. Recent data on laser-assisted generation of other metals is given, including the formation of alloyed nanoparticles. Micro- and nanostructuring of the target upon its laser ablation in liquid environment is discussed. Examples are given of the influence of the surrounding liquid on the chemical composition of generated nanoparticles. The laser control over the size distribution of nanoparticles in liquids is demonstrated either by spatial profiling of laser beam intensity or proper tuning of laser wavelength into plasmon resonance of nanoparticles. Recent results are given on excitation of high energy levels of media under laser exposure to laser pulses of picosecond range of duration.

INTRODUCTION

The advent of lasers opened a new branch of research of interaction of radiation with matter. The primary "eye-visible" effect of laser action on a solid target is removing of some material from the target surface within the laser spot. This process was called "laser ablation" from a Latin word *ablatio*, which means removal.

The process of laser ablation of solids in liquids has attracted much attention of researchers during the last decade. This is due mainly to the simplicity of the experimental setup. Many modern laboratories (and not only physical ones) are equipped with lasers, and synthesis of nanoparticles is a strong temptation. The process proceeds in one step and results in immediate formation of nanoparticles in the liquid in which the target is immersed. The main feature of the process is that ideally the liquid contains only nanoparticles made of the target material and the liquid. There are no counter-ions or residuals of reducing agents left in the solution. For this reason laser ablation of solids in liquids can be considered as a method of nanoparticle synthesis, which is an alternative to chemical methods.

Commercially available laser sources are characterized by a number of parameters, such as peak power, average power, wavelength of emission, pulse repetition rate, etc. If the final purpose of ablation of a target immersed into a liquid is the synthesis of nanoparticles with desired properties, such as their chemical composition, size distribution, concentration, etc., then the above mentioned laser parameters are of different importance to the properties of desired nanoparticles. Also, the nature of the liquid plays a significant role in the final properties of nanoparticles generated under laser ablation.

The objective of this text is to outline the relative importance of various experimental parameters to the properties of nanoparticles synthesized by laser ablation of a solid target immersed into a liquid.

General Setup of Laser Ablation in Liquids

Figure 1 shows a generalized setup of experiments on laser ablation of solids in a liquid environment. This setup may vary from one research group to another, though their common features are the same.

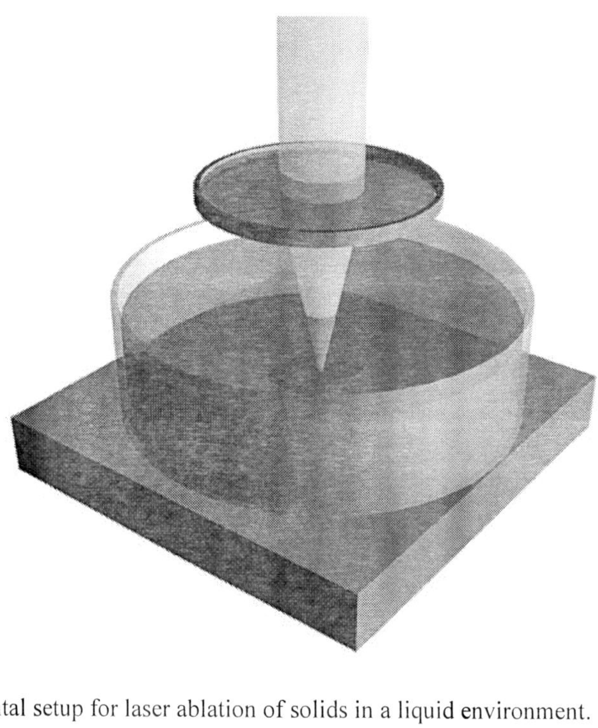

Figure 1. Experimental setup for laser ablation of solids in a liquid environment.

Laser radiation is focused onto a solid target immersed in a liquid. It is assumed that the liquid is transparent at the laser wavelength, otherwise the focalization of the beam would be problematic due to absorption of laser radiation in it. The simplest way is working with the free surface of the liquid, which allows avoiding additional reflection at the interface "covering glass/air". However, the use of volatile liquids, such as acetone, ethanol, etc., requires covering the liquid with a window that is transparent at the laser wavelength.

Experimental Technique

The experimental setup for laser ablation in a liquid environment is perfectly simple. It is usually assumed that the liquid that surrounds the solid target is transparent at laser wavelength while the laser radiation is absorbed by the target. A solid target is placed under a thin (several millimeters) layer of liquid and is exposed to laser radiation through this layer. Different pulsed laser sources can be used, e.g., a Nd:YAG laser at 1.06 μm output and its harmonics, a Cu vapor laser, a Ti:sapphire laser, etc. UV excimer lasers are less common, since most liquids and NP absorb in the UV region. The only necessary requirement is that the laser beam is sufficiently powerful to induce local melting of the target. Usually the laser beam is focused onto the target using an appropriate optics to a certain size of laser spot (see Figure 1). In some experiments the target is rotated under the laser beam to avoid the exposure of the same area. Some research uses a sealed-off cell to avoid oxidation of NP by air oxygen, but basically the setup is the same. It is the density of the laser energy (in J/cm^2), or so-called *fluence*, that determines the temperature of the target and the possibility to produce surface melting and eventual generation of NP. The irradiation of the metal surface results in fast removal of the material that is confined to the laser spot. The ejected nanoparticles remain in the liquid that surrounds the target, resulting in formation of a so-called colloidal solution. Unlike *real* solutions that contain ionic or molecular species, colloidal solutions also contain particles, e.g., NP and clusters. Due to accumulation of NP in the surrounding liquid, their prolonged interaction with laser radiation is possible. Therefore, the thickness of the liquid layer above the target is also an important experimental parameter that may influence the properties of generated NP.

Different laser parameters are of different importance to the efficiency of NP generation.

1. Pulse Duration

Usually, pulsed laser sources are used for generation of NP in liquids with pulse duration from hundreds of femtoseconds to hundreds of nanoseconds. The NP under laser ablation in liquids are formed owing to sputtering of the molten layer by the recoil pressure of the liquid that surrounds the target. Therefore, the necessary condition of NP synthesis is melting of the target material. In general, the temperature distribution under laser exposure of solids can be found solving a heat conduction equation with corresponding boundary conditions. However, in the case of short laser pulses of the duration mentioned above, the complicated problem of temperature calculation can be significantly simplified. This simplification is based on the fact that the heat diffusion length l_d from laser-exposed areas of the target during the duration of the laser pulse t_p is small compared to laser spot size d. Indeed, in a typical experiment on pulsed laser ablation d ~ 10 μm, while the heat diffusion length is much less. Assuming that the laser beam has a flat-top profile (typical of excimer lasers, metal vapor lasers, etc.) one may suggest that the absorbed laser energy is spent for heating of the target layer whose thickness is of order of $(at_p)^{1/2}$, where a stands for the heat diffusion coefficient of the target material, while the area of this layer coincides with the laser spot. Unlike to what one might believe, the presence of liquid around the target does not alter noticeably the temperature inside the laser spot. The reason is that the heat diffusion coefficient for liquids is even smaller than that of solids. Presence of the liquid may affect only the average temperature rise of the target if the repetition rate of laser pulses is elevated, e.g., of order of 1 kHz and more. Simple heat balance equation leads for the following expression of the temperature T within the laser spot:

$$T \approx \frac{Aj}{c\rho h},$$

where A stands for absorptivity of the target at the laser wavelength, A = 1- R, where R is the reflectivity coefficient at laser wavelength, c stands for heat capacity of the target material, □ is density of the target material, and h is the heat diffusion length inside the target. One can see that in a quite natural way the temperature is proportional to the energy density of the laser beam, or so called *fluence* j. The heat diffusion length h depends on the heat diffusivity of the target material:

$$h \propto \sqrt{at_p},$$

where in turn a = k/c□, where k is the heat conduction coefficient of the target, and t_p stands for laser pulse duration. The longer is the laser pulse t_p, the thicker is the layer of the material which is heated by absorbed laser energy. The above-made estimation of the temperature rise T is made assuming superficial absorption of laser radiation. If α is the coefficient of absorption of laser radiation, then this condition can be written as follows:

$$\alpha^{-1} << h$$

Of course some energy is consumed for heating and evaporation of a liquid adjacent to the laser spot, but this energy is small compared to the absorbed one due to low thermal conductivity of liquids.

In case of metal targets laser radiation is absorbed by free electrons that transfer their energy to the metal lattice within 3- 5 picoseconds. Virtually no heat exchange with the bulk of the solid target occurs, and the absorbed laser energy is spent for heating of the layer within the absorption depth α^{-1} of laser radiation.

The absorptivity A of the target surface is a complex parameter. For a smooth metallic surface it can be calculated using reference data on both real and imaginary part of a complex dielectric function of the material. However, as soon as the surface of the target is not flat and is characterized by a certain relief, the absorptivity of the target may largely deviate from its theoretical value. This is due to the dependence of absorptivity on the angle of incidence of radiation. Since the target material is dispersed into surrounding liquid as nanoparticles during laser ablation in liquids, the formation of the relief indeed occurs.

2. Laser Wavelength

As soon as laser ablation of metals is considered, any wavelength is appropriate. However, laser radiation can be absorbed by NP that are generated under ablation of the target. The majority of NP absorb in UV region, which imposes certain drawback on the use of excimer UV lasers for generation of NP.

3. Repetition Rate

Nanoparticles are ejected from the solid target at each pulse provided that the absorbed laser energy is sufficient to melt it. Therefore, the higher is the repetition rate of laser pulses, the higher is the rate of NP generation. However at high repetition rate the target may be screened from the beam by gas bubbles that have remained from previous pulses. This can be avoided using either a flow cell or elevated velocity of scanning (rotation) of the target.

Historical Review

Formation of nanoparticles under laser ablation of solids either in gas or in vacuum has been extensively explored during the last decade. Understanding of the mechanisms of cluster formation is needed to control the process of Pulsed Laser Deposition (PLD) that is widely used now for deposition of a large variety of compounds. Formation of nanoparticles under laser ablation of solids in liquid environment has been studied to a lower extent. A number of wet chemical methods of nanoparticles (NP) preparation are known up to date. Laser ablation of solids in liquid environment emerged as an alternative technique that is capable of producing "pure" NP without counter-ions and surface-active substances. Primary motivation for generation of NP by laser exposure of solid targets in liquids was the hope that pure NP would be more efficient in the process of Surface Enhanced Raman Scattering (SERS), since no other compounds can obscure the surface of a metallic NP [1-5]. Formation of both large variety of NP under laser ablation of corresponding solids has been reported during the last decade [6-17]. Further studies showed that the laser ablation in liquids is a complex process with a number of experimental parameters. In particular, chemical interaction of ejected nano-sized species with hot liquid vapors at high transient pressure results in large diversity of final particles that remain in the liquid. Also, surface tension at the interfaces is dominant force that governs the dynamics of interaction of molten target material with vapors of surrounding liquid on a nanometer scale. Under pulsed laser exposure the liquid also undergoes chemical changes that may be stipulated by ejected NP that act in this case as a catalyst. Initially liquid substance may become supercritical in the vicinity of the target, which alters its reactivity and opens new channels of chemical reactions especially in case of aqueous solutions.

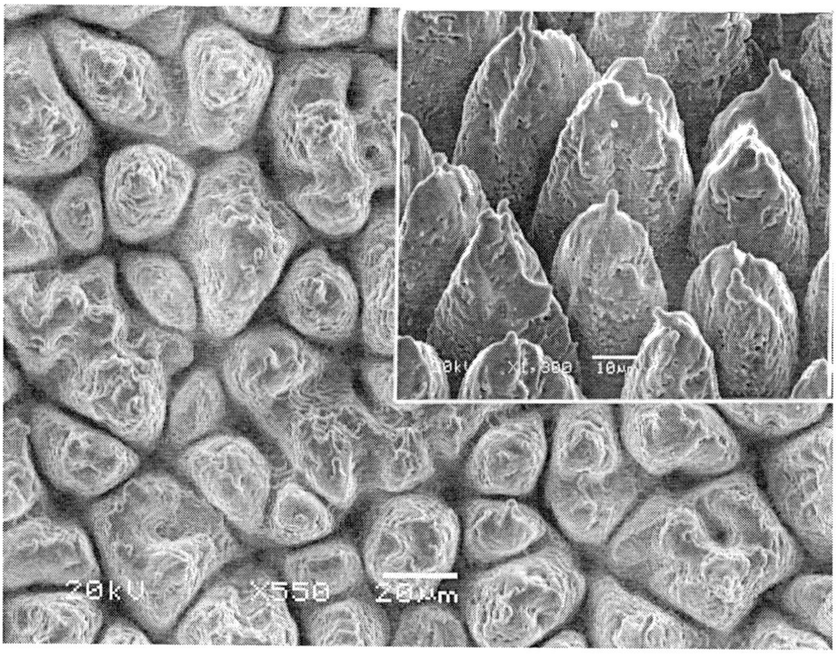

Figure 2. SEM top view of a brass target produced by scanning beam of a Cu vapor laser. Inset shows the enlarged view of the same surface tilted for 35°. Scale bar denotes 20 μm and 10 μm in inset.

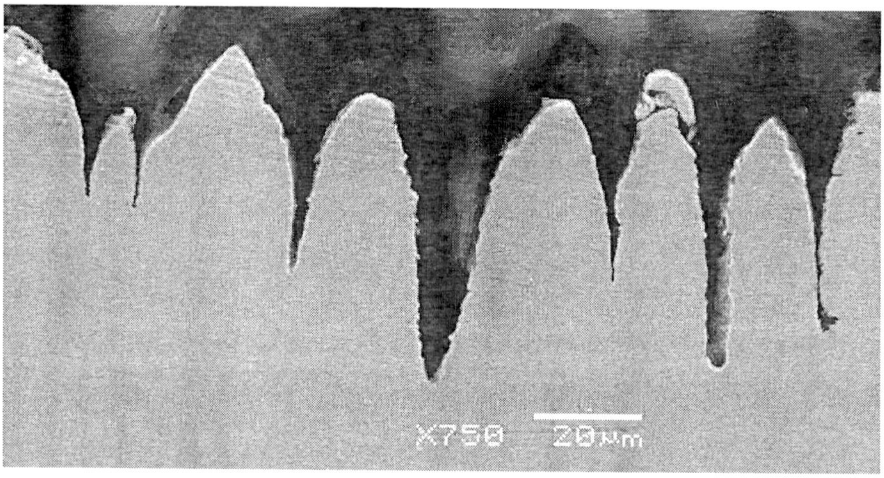

Figure 3. SEM view of a cross section of a brass target subjected to laser beam of a Cu vapor laser under the layer of ethanol. Space bar denotes 20 μm.

Interaction of metal nanoparticles with laser light proceeds via its absorption by free electrons. Free electrons exhibit plasmon resonance, whose position is determined by both their concentration, their effective mass (that depends on the solid) and particle size. NP of noble metals, e.g., Au or Ag, show a strong selective absorption in the visible, so the suspensions of these NP in liquids are colored. Plasmon resonances of other metallic NP are situated in the UV region [18, 19]. Though if the liquid is transparent at the onset of laser exposure of the target, the appearance of NP may change this situation, and the laser radiation is absorbed by the colloidal solution. This may lead to poor control of the laser spot size on the target due to either thermal lensing inside the solution or different non-linear phenomena induced by laser beam in NP [20-22]. So, the thickness of the liquid layer above the target is also one of the experimental parameters that are essential during ablation in liquids.

Depending on the dwell time of the laser beam on the target, a more or less significant amount of the target material is transferred into the surrounding liquid as NP. The target relief is therefore modified, and a crater is formed in the target in case of its exposure to a stationary laser beam. To avoid the formation of the crater the target is either rotated or is scanned under the laser beam. However, a periodic relief arises on the target surface at sufficiently high number of laser shots of order of 10^4. The structures are densely packed micro-cones, and their period linearly varies with the size of laser spot on the target as shown in Figure 2 [23].

These micro-cones are separated by long channels with high aspect ratio that run deep inside the target body. The formation of these micro-structures alters the properties of NP generated by laser ablation, since the conditions of vapor expansion inside the channels are different from those on a flat solid-liquid interface. In other words, the molten areas of the target are no longer on its surface but are situated mostly in the channels. Indeed, the micro-cones are oriented in a way to reflect the incident laser beam. Therefore, the properties of NP generated from a smooth target surface at the onset of laser exposure are different from those at later stages when the micro-relief has been formed.

The cross section of a brass target subjected to radiation of a Cu vapor laser in ethanol is presented in Figure 3. One can see deep channels that separate the microstructures formed in the target upon laser ablation. The melt is formed predominantly on the bottom of channels due to both increased angle of incidence of laser radiation compared to a flat surface and to reflection of laser light from side surface of micro-cones.

The molten layer is pushed along the cone surface and is partially dispersed into surrounding liquid as NP. Other part of melt solidifies on cone surface forming in some cases typical tips on their tops as can be seen in Figure 2.

Formation of periodic micro-relief on the target surface inexorably alters the character of temperature distribution along its surface during the laser pulse. The temperature distribution is smooth on initial flat target surface, and in case of a flat top profile of laser beam intensity the temperature profile is also almost flat except for side effects. As soon as periodic micro-relief is formed, the temperature profile consists of numerous "hot spots" located in the channels between micro-cones. The energy density (fluence) inside these spots exceeds by far the fluence on the initial surface without micro-relief. Since the size of NP depends on the fluence, the size distribution of NP generated via laser ablation of the micro-structured target should be different from that generated at the first stages of laser ablation when the target is still flat.

Unlike laser ablation in vacuum where synthesized NP rapidly leave the laser beam and never come back due to sticking on the chamber walls, NP synthesized by ablation in liquids remain in it and may return into the laser beam during their motion. The efficiency of coupling of radiation to nanoparticles depends on the proximity of laser wavelength λ to plasmon resonance of charge carriers. The energy from electrons to the lattice is transferred within 3-5 ps [24], and the temperature T of a nanoparticle of radius R can be estimated on the basis of conventional heat diffusion equation [25]. For $8\pi kR/l \ll 1$ one obtains $T \sim 2\pi I_0 kR^2/\lambda k_{liq}$, that is the temperature of the nanoparticle in the laser beam is proportional to its geometric cross-section. Here k_{liq} stands for the thermal conductivity of the surrounding liquid which is assumed to be constant during its evaporation for the sake of simplicity. Note that due to the small size of NP their temperature is proportional to the peak power density of the laser beam I_0 (Watts per square centimeter). The extinction coefficient κ under large detuning from the plasmon resonance is close to that of the bulk metal. However, in the vicinity of plasmon resonance $\kappa=\kappa(\lambda)$ shows resonant behavior, and the temperature T of the particle strongly depends on the laser wavelength.

The main difference between laser ablation in vacuum and in liquid environment is a short free path of ablated species in the latter case. Indeed, the formation of NP under laser irradiation of a metal target immersed into a liquid proceeds via local melting of the metal. The adjacent liquid layer is heated to almost the same temperature owing to heat transfer from the metal. The thickness h of this layer can be estimated using the heat diffusion coefficient of a liquid $a = 10^{-3}$ cm^2/s and the diffusion time equal to duration of the laser pulse $\tau_p = 20$ ns. Then the thickness of the liquid layer $h \sim (a\tau_p)^{1/2} = 0.3$ μm. This thin layer has a temperature much higher than boiling point of the liquid at normal pressure and is hence in the vapor phase. The pressure in this vapor layer can be roughly estimated as the vapor pressure of the liquid at temperature of the substrate. This means that in typical experimental conditions all liquids are in vapor phase at pressure of the order of hundreds atmospheres. The thickness of the molten layer on the target is about the length of heat diffusion into the metal, $h_m \sim (a_m\tau_p)^{1/2}$. $a_m = 1.7$ and 0.13 cm^2/s for Ag and Au, respectively, which gives the heat diffusion length of 1.9 μm for Ag and 0.5 μm for Au. This is the maximal thickness of the reservoir for generation of nanoparticles that is gained at fluence much higher than the melting threshold. That is why the size of nanoparticles is almost independent on the pulse width, from hundreds of nanosecond through hundreds of femtosecond. At lower fluence the thickness of the melt is smaller than h_m. Expanding vapors of the liquid splash this reservoir resulting in the removal of the molten layer. Note that formation of nanoparticles via evaporation of the metal is unlikely, since the pressure of metal vapor at a temperature close to melting is too low compared to vapor pressure of the surrounding liquid. Surface tension stabilizes the molten drop of the metal, while the pressure of surrounding vapor of the liquid tends to split it. As a first approximation one may suggest that the size of a stable drop of metal can be found from the following relation: 2□/R ~ p_{liq}, where □ is the surface tension coefficient of the metal, R is the radius of the drop, and p_{liq} is the pressure of the vapor of the liquid surrounding the target. Substitution of these parameters for Ag ablated in water for R = 30 nm and T = 1000° C gives $p_{liq} \sim 10^8$ Pa, which is a reasonable value. The melting temperature decreases with the size of the nanoparticles, so at sufficiently small R the particle melts under the laser pulse, but the vapor pressure of surrounding liquid is not high enough to cause its further splitting.

Laser Ablation of an Ag Target in Liquid Environment

Historically Ag NP were the first ones synthesized by laser ablation of a silver plate in water. Various laser sources have been successfully used for this synthesis ranging from a nanosecond Nd:YAG through Ti:sapphire femtosecond lasers [21,26,27]. Ablation of a metallic Ag immersed into water by radiation of a Cu vapor laser at wavelength of 510.6 nm and pulse duration of 20 ns results in visible coloration of the liquid; it takes on a yellowish color. The optical spectra of Ag NP generated by laser ablation in different liquids are presented in Figure 3. The maximum of absorption is at about 400 nm, which is typical for so called plasmon absorption of Ag nanoparticles in water [18,19]. The intensity of the plasmon band increases with irradiation time, while its position does not vary significantly. Ag NP are slowly oxidized by air oxygen dissolved in the liquid, and the position of plasmon resonance shifts to the red with time due to formation of oxide layer with high refractive index [22].

The Transmission Electron Microscope (TEM) view of Ag nanoparticles produced by ablation of an Ag target in acetone is shown in Figure 5. The average size of particles is around 15 nm. Moreover, they are flat disks, and their thickness is of order of few nanometers [15]. This is clearly seen in those areas where the TEM image of intersection of two or more Ag particles appears darker than each separate particle.

Smaller Ag NP are synthesized using sodium dodecyl sulphate (SDS) as a surface active substance [26]. Laser pulses of various durations have been used for synthesis of Ag NP in water ranging from ns to fs domain [21, 27].

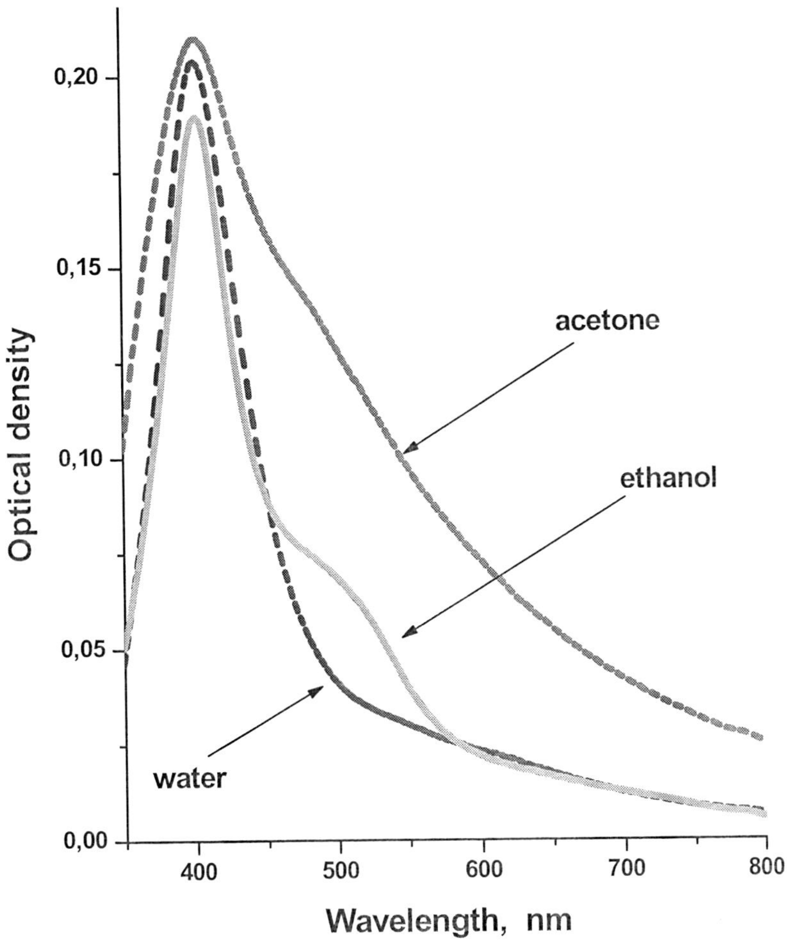

Figure 4. Absorption spectrum of Ag nanoparticles obtained by ablation of Ag in different liquids. The plasmon peak of Ag is well pronounced at ca 400 nm.

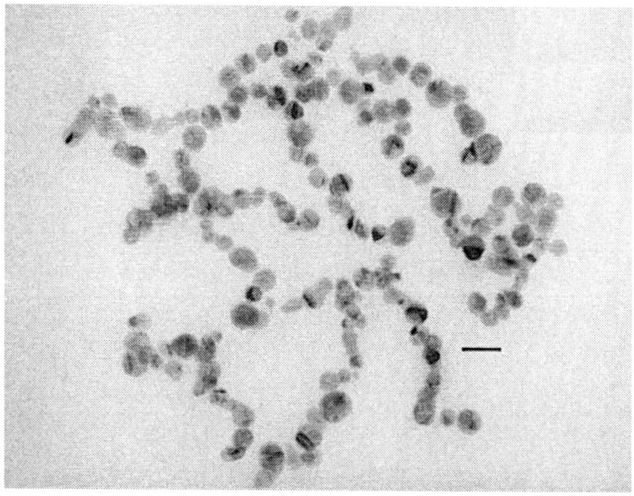

Figure 5. TEM view of Ag nanoparticles obtained by ablation of a bulk Ag in acetone. Scale bar denotes 50 nm.

LASER ABLATION OF AN AU TARGET IN LIQUID ENVIRONMENT

Gold NP exhibit a well-pronounced dependence of their absorption spectrum on their shape and size. In general, a metallic ellipsoid is characterized by three different plasmon resonances. In case of a spherical particle all three of them are degenerated, and Au NP show the maximum of absorption around 520 nm independent on the way of their preparation. Elongated Au NP (nanorods) are characterized by two absorption bands that correspond to so called transverse and longitudinal plasmon resonances. The position of the latter varies with aspect ratio of Au nanorods, and for sufficiently high aspect ratio the longitudinal resonance moves from the visible to the near IR range of the spectrum [28].

The optical spectra of water after laser exposure of metallic Au show well-pronounced absorption band at ca 520 nm that is characteristic of transverse plasmon resonance [18, 19]. At higher laser fluence elongated Au NP are also synthesized, which is accompanied by the appearance of a red wing. Au NP prepared in water in absence of surface-active agents are metastable against precipitation, and their significant precipitation is observed within several days after preparation. Precipitation of NP is accompanied by appearance of a wide red wing that indicates the presence of elongated Au NP with various aspect ratios. Synthesis of a stable colloidal solution of Au NP in water has been reported by several groups [29-32].

The coupling of laser radiation to the colloidal solution can lead to the modification of the size distribution of Au NP. This is confirmed by TEM view of Au particles as obtained after the ablation of Au in water. Laser radiation modifies not only the size of the particles, but also their shape. As-obtained particles are elongated, while those exposed to laser radiation are the nano-disks similar to Ag nanoparticles. Average size of Au as-obtained after ablation by a Cu vapor laser in H_2O in absence of surface-active substances is 20 nm, while with further laser exposure of the solution alone their size diminishes to ca 10 nm.

Au NP produced at high laser fluence on the target are characterized by elongated shape independent of the nature of the surrounding liquid. This effect has been observed for laser ablation in alkanes [33] as well as in water. Figure 6 shows the TEM view of Au NP synthesized by laser ablation of a gold target in water. The majority of NP has a spherical shape, while some NP have aspect ratio ranging from 2 to 10. Accordingly, the absorption spectrum of this solution is characterized by two maxima, one at ca 540-560 nm and the second having a wide red wing centered at longer wavelengths.

The rate of generation of Ag and Au NP depends on laser fluence. For instance, it is 0.6 and 1.2 mg/hour, for respective metals, with a laser spot of 30 μm and fluence of 30 J/cm^2.

Nanoparticle size can be drastically reduced by the use of aqueous solutions of surfactants, which cover the particles just after their ablation and thus prevent them from further agglomeration. Sodium dodecyl sulphate (SDS) is one of the most popular agents capable of reducing the mean size of Au nanoparticles down to 5 nm during nanosecond laser ablation of gold [29-31]. Ablation of a gold target with shorter laser pulses, e.g., femtosecond ones, results in synthesis of bimodal distribution function of nanoparticle size at elevated laser fluence and almost mono-dispersed Au NP at lower fluence of 50 J/cm^2 [32]. Another stabilizing agent for Au NP are cyclodextrins [34]. Their addition provides perfect stability of Au NP as small as 2-2.5 nm synthesized by laser ablation with femtosecond pulses.

NP of another noble metal Pt can also be synthesized by laser ablation of a metallic Pt target. In presence of SDS their size can be as small as 5 nm [35].

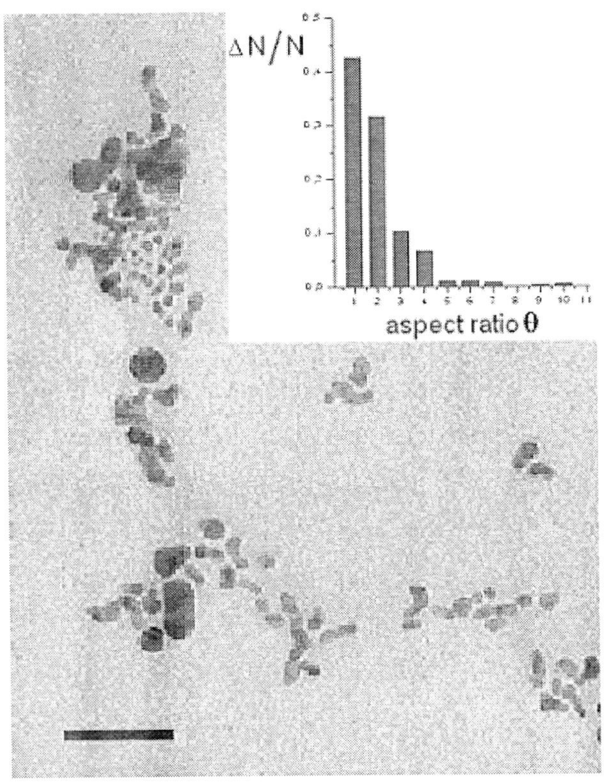

Figure 6. TEM view of Au NP obtained by ablation of a gold target immersed into water at elevated laser fluence of 90 J/cm^2 at 1.06 μm, pulse width of 130 ns.

Interaction of Nanoparticles with Laser Beam

Coupling of laser radiation to isolated NP proceeds via its absorption by free electrons. Then the absorbed energy is thermalized and transferred to NP material. Since these electrons are characterized by some plasmon resonance, the efficiency of this coupling strongly depends on the detuning of laser radiation from maximum of absorption. This interaction is not important in case of laser generation of NP in vacuum, since NP do not come back to the laser beam. In liquid environment NP remain in the liquid and do come back inside the beam due to convective flows. Repeated interaction of NP with laser beam leads to observation of some interesting phenomena described below.

Fragmentation of NP under Laser Exposure in Liquids

Fragmentation of NP under their laser exposure in liquids has been reported as early as the synthesis of NP itself [2, 6, 30]. The process manifests itself as a gradual decrease of the size of NP upon laser irradiation. As a result, the distribution function of particle size shifts to lower dimensions. The dependence of the NP size on the laser wavelength has been reported [36, 37] indicating the spectral dependence of the fragmentation process. Optical constants of metallic targets do not vary largely in the wavelength range of lasers used for NP synthesis, so fragmentation of NP should be ascribed to interaction of synthesized NP themselves with the laser beam inside the liquid.

The physical processes that lead to fragmentation of NP in the laser beam are still under discussion. The most probable reason leading to reduction of NP size are the instabilities that develop at the interface of molten NP with the vapors of surrounding liquid [25]. Indeed, a NP may absorb enough energy from the laser beam to undergo the phase transition into a liquid state. The temperature of NP exceeds at this moment the boiling temperature of surrounding liquid. The latter is therefore vaporized and forms a shell around the NP. The pressure of vapors in this shell is around 10^8 Pa, and its possible asymmetry would result in the break of molten NP inside it into smaller parts. The most probable process is splitting of the molten NP into two equal particles, since, first, this corresponds to lowest value of total surface energy, and second, is the lowest mode of perturbation of the surface of molten NP. This hypothesis is in qualitative agreement with the fact that the average size of NP exposed to laser radiation in various liquids decreases with the decrease of boiling temperature of the liquid. The balance is gained when the pressure inside the NP of radius R is of order of the pressure of the surrounding liquid p_{liq}: $2\sigma/R \sim p_{liq}$, where σ stands for the surface tension of molten NP.

Fragmentation of the NP under their laser exposure introduces *feedbacks* into the system "laser radiation – NP". Indeed, the average size of NP under their laser exposure is inversely proportional to the peak intensity of the laser beam inside the liquid. The size of NP decreases upon exposure till NP become so small that the energy absorbed from the laser pulse is not sufficient for melting. The system therefore demonstrates a negative feedback and auto-stabilization. On the other hand, the melting temperature of a NP decreases with its size, which is a positive feedback. Finally, small NP formed via fragmentation from bigger ones do not absorb enough energy from the laser beam to be molten, and the ensemble of NP acquires its upper limit of NP size.

SHAPE-SELECTIVE FRAGMENTATION

The fragmentation of NP under laser exposure of their suspension in a liquid can be shape-selective, if the spectrum of plasmon resonance depends on their shape. This can be illustrated by shape-selective fragmentation of Au NP having elongated shape (nanorods). Au nanorods are characterized by two absorption peaks, one of them corresponds to so called transverse resonance (TR), while the second is called longitudinal one (LR). These resonances correspond to oscillations of free electrons in the directions across the nanorods axis and along it, respectively [28]. The position of TR near 560-570 nm is common for Au nanorods with any aspect ratio, while the position of their LR is a linear function of the aspect ratio and moves to the red with its increase [38-41]. If laser radiation is tuned into TR, then all nanorods are molten and fragmented into spherical NP at sufficiently high peak power of the laser beam [40]. However, tuning the laser wavelength into LR allows selective fragmentation of Au nanorods with aspect ratios which are in resonance with laser radiation [42].

This fragmentation (splitting) is illustrated in Figure 7 in which the evolution of absorption spectrum of suspension of Au NP with wide distribution of aspect ratio is presented. Au NP are especially convenient for this type of experiments since their absorption spectrum depends on their shape. The radiation of a green line of a Cu vapor laser fits well the transverse plasmon resonance of Au NP, which is common for all Au NP and does not depend on their aspect ratio. On the contrary, the yellow line of a Cu vapor laser is absorbed by non-spherical (elongated) Au NP. Therefore, the evolution of the absorption spectrum is different if various laser outputs are used for NP exposure.

One can see gradual decrease of the optical density of the suspension in the red that corresponds to fragmentation of Au nanorods irrespective to their aspect ratio (Figure 7, 1). On the contrary, exposure of the suspension to radiation that is preferentially absorbed by Au nanorods with the aspect ratio 2 – 3 results in the decrease of the optical density just at the laser wavelength (Figure 7, 2). This decrease indicates the depopulation of the suspension with nanorods that are in resonance with laser radiation. Of course the laser fluence should not be too high otherwise nanorods with "non-resonant" aspect ratio will also be fragmented, since their absorption wings are very wide.

Au nanorods with aspect ratio about 10 absorb in the near IR region. Exposure of their suspension to radiation of a Nd:YAG laser operating at its fundamental output at 1.06 μm results in self-stabilization of fragmentation: no changes of absorption spectrum of the solution occur as soon as long nanorods are fragmented [42].

Similar fragmentation of Au NP is also observed in ethanol and acetone. Their size in these liquids is somewhat smaller than in water and is of 7 nm under the same laser fluence. As a consequence, these colloidal solutions are stable against sedimentation. Finally, the addition of a surface-active substance, e.g., poly(vinyl pyrrolidone) (PVP) with molecular mass of 10^4, to ethanol in which the ablation is carried out further decreases the average size of Au nanoparticles down to ca 4 nm at otherwise equal conditions.

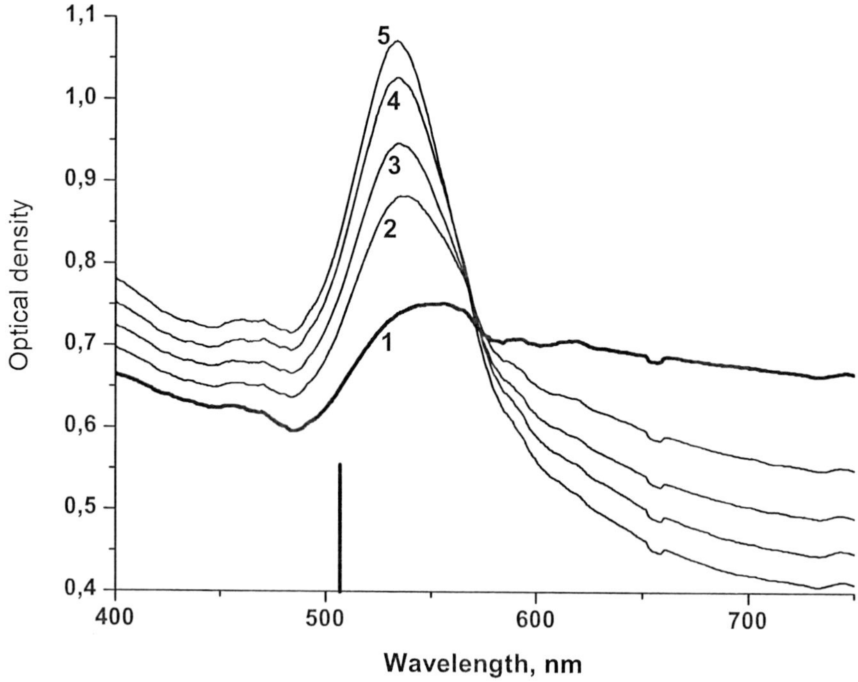

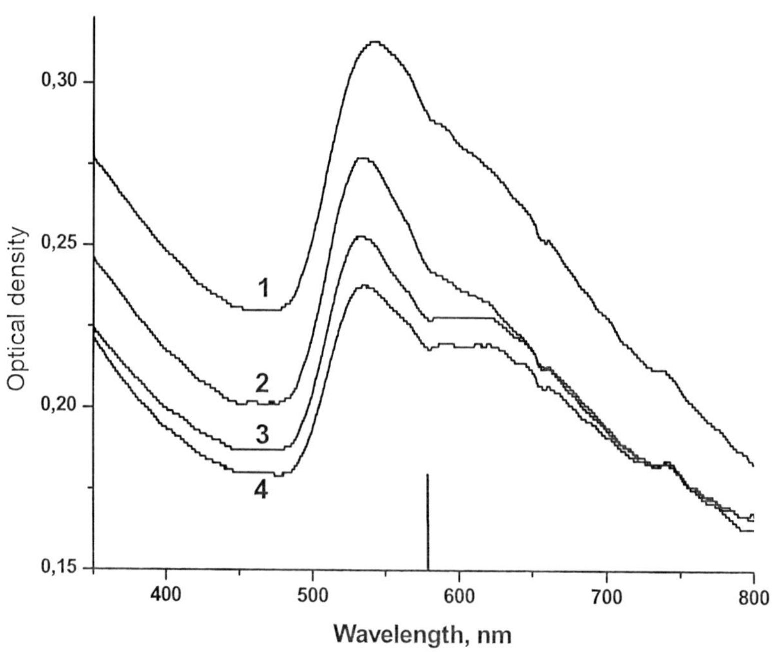

Figure 7. Evolution of optical density of suspension of Au nanorods in water under exposure to radiation of a Cu vapor laser at 510.6 nm (1) and 578.2 nm laser output. Vertical lines indicate the position of the laser wavelength. Curve (1) in both figures corresponds to the spectrum of initial colloidal solution.

FORMATION OF THE AU-AG ALLOY UNDER LASER IRRADIATION OF NANOPARTICLES

Direct interaction of NP with laser beam inside the liquid may lead to their melting during each laser pulse at sufficiently high laser peak intensity. If the colloidal solution contains NP of various materials, their absorption of laser radiation may be different. Laser exposure of a mixture of different NP may lead to formation of alloyed NP, and at least one kind of NP should be molten by laser beam. Recent publications [43-45] have reported the formation of gold-silver alloy nanoparticles in liquid environment under laser exposure of core-shell particles obtained by a chemical method. The spectrum of core shell (non-alloyed) nanoparticles is not a linear combination of the spectra of monometallic colloids [45]. On the contrary, the spectrum of a mixture of individual colloids is composed of spectra of monometallic particles. Upon laser exposure, these peaks disappear, and the single alloy peak arises. Laser exposure of Ag-Au core-shell NP may lead to both Ag shell removal from Au NP and their alloying depending on laser wavelength and its fluence [46]. Individual NP of various materials have to come into contact to form an alloy, and therefore the laser alloying of NP is concentration-dependent process. The position of plasmon resonance of alloyed NP is a linear combination of plasmon frequencies of individual NP. Formation of alloyed AuAg NP is accompanied by peculiar modification of the absorption spectrum indicating an intermediate phase of the Au-Ag alloy formation [47-49]. The temporal evolution of the optical density of the mixture is presented in Figure 8.

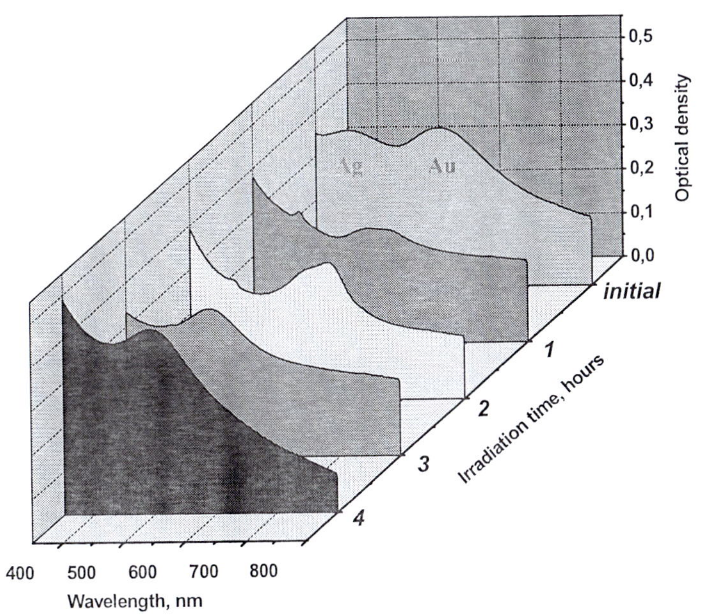

Figure 8. Evolution of the absorption spectrum of the mixture of both Au and Ag NP synthesized by laser ablation of respective metals in ethanol under laser exposure at 55 J/cm^2 of a Cu vapor laser radiation (wavelength of 510.6 nm).

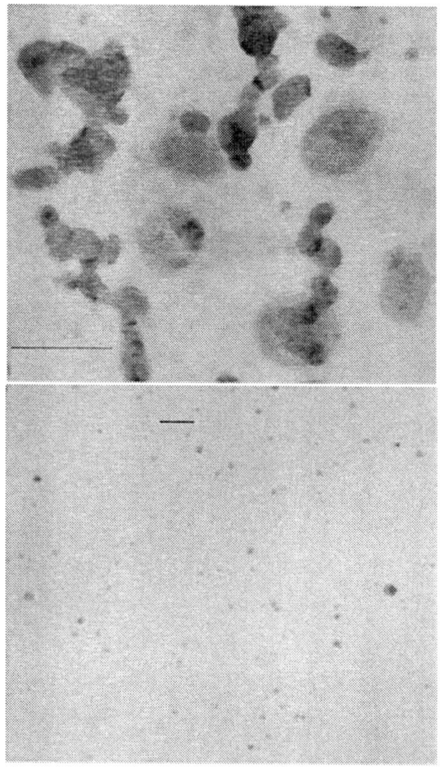

Figure 9. TEM view of Au-Ag nanoparticles after 2 hours of exposure (a). Concentration of PVP is 0.1 g/l, Cu vapor laser, laser fluence of 9 J/cm^2. Brighter particles are Ag while the darker ones are Au. Scale bar denotes 40 nm. Alloyed Au-Ag nanoparticles obtained by exposure of the mixture of individual colloids in ethanol during 4 hours (b). Scale bar denotes 100 nm. Average size of alloyed Au-Ag particles is 7 nm.

The formation of an intermediate phase of the alloy is observed only in case of sufficiently large initial NP of each metal (50 -70 nm). Smaller NP of 10-20 nm in diameter also show the deviation of the maximum of plasmon resonance from its stoichiometric position but this effect is weak.

The position of plasmon frequency of an Au-Ag alloy varies continuously with relative concentration of metals and is always in between the maxima of plasmon resonance at 400 nm and 520 nm of Ag and Au, respectively. In case of individual particles, however, the maximum of absorption falls out of this interval after certain exposure time and is significantly red-shifted after 1-2 hours of exposure. With further exposure the maximum returns to the interval corresponding to an Au-Ag alloy and remains there for even longer exposures indicating the formation of the alloy. The TEM view of NP at the stage of anomalous spectrum (Figure 8, spectrum 2) reveals multiple contacts between Au and Ag nanoparticles that are well observed due to different atomic mass of both metals.

With further exposure these hybrid particles are alloyed, and the final position of the absorption maximum corresponds to the Ag-Au alloy (Figure 8, spectrum 4). The size of alloyed nanoparticles is about 5 nm, which is much smaller than initial ones.

Formation of alloyed nanoparticles is sensitive to the presence of surface-active substances in the liquid. The rate of alloy formation increases under addition of PVP (10^{-5} M).

However, the alloying is inhibited at given laser intensity at elevated concentrations of PVP exceeding 5×10^{-5} M [48]. The rate of alloying decreases with dilution of the mixture, presumably due to the increase of the average distance between the nanoparticles. It is pertinent to note that the surface-active substance, e.g., PVP, apparently remains untouched under laser exposure of NP in spite of their elevated temperature. Raman analysis shows no variation in the glassy carbon content before and after laser exposure of suspensions of NP with PVP, though this should be so if the pyrolysis of the surfactant on hot NP took place. This might be explained by the preferential localization of PVP at the interface of a vapor bubble that is formed around the NP during their laser exposure.

The absorption spectrum of the Au-Ag alloyed NP is closer to the wavelength of a Cu vapor laser (510.6 nm) than the plasmon maxima of individual NP. Therefore, the efficiency of coupling of laser radiation increases upon formation of the alloy. As a result, the average size of alloyed Au/Ag NP is much smaller than that of individual ones [48, 49]. This is illustrated in Figure 9 where TEM images of NP are presented at different stages of Au-Ag alloy formation.

Similar behavior is observed under exposure of the mixture of colloidal solutions of Cu/Ag and Cu/Au. The alloyed NP are formed under sufficiently long laser exposure of the mixture, and the final position of the plasmon resonance depends on ratio of individual colloids in it.

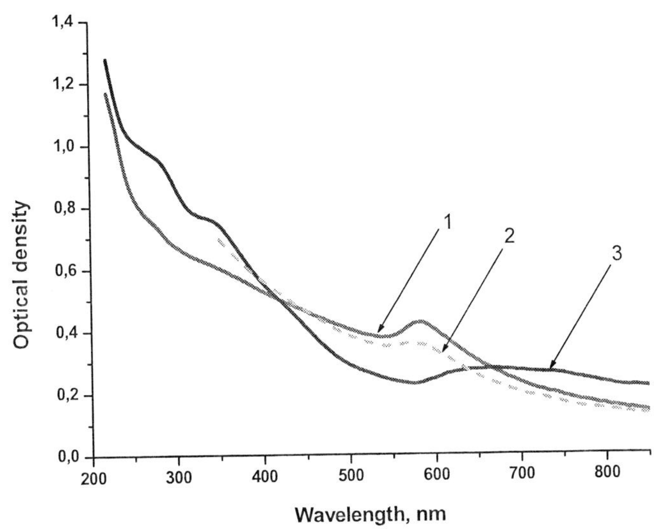

Figure 10. Absorption spectra of the liquids after laser ablation of a bulk Cu target. Ethanol (1), acetone (2), and water (3). Cu vapor laser, fluence of 30 J/cm^2.

NANOPARTICLES OF CU, BRASS, AND BRONZE

Copper is chemically active and easily reacts with vapors of surrounding liquid. The chemical interaction may take place both during the ablation and after it due to contact with air oxygen dissolved in the colloidal solution. Typically, the plasmon resonance of Cu nanoparticles lies in the visible. Cu NP synthesized by chemical means in anaerobic conditions show the absorption peaks from 570 to 590 nm [49, 50].

Laser ablation of Cu target in various liquids also leads to formation of NP that are characterized by absorption in the visible [51]. The peak position is around 590 nm, which agrees with previously reported data for Cu NP (see Figure 10). Ablation of a Cu target in water leads to formation of compounds containing only ionic species Cu^{+} without any plasmon peak. The stability of generated Cu NP to slow oxidation is different. Namely, Cu NP produced by ablation in ethanol are slowly oxidized. This is manifested by gradual disappearance of the peak of plasmon resonance (Figure 10). Remarkably, the initial plasmon peak can be restored by exposure of aged colloidal solution to the radiation of a Cu vapor laser at sufficiently high fluence of several tens of J/cm^2.

On the contrary, ablation of a Cu target in acetone results in a stable spectrum of the colloidal solution of NP. The difference in behaviour of Cu nanoparticles in these two liquids stems from their images obtained with a transmission electron microscope. Cu NP generated by ablation in ethanol are low-contrast entities with average size of 5 nm. Cu NP in acetone are metallic as one can conclude from their high contrast and well pronounced peak of plasmon resonance. However, in this instance they are embedded into some amorphous cloud [51]. Raman analysis of the dried suspension indicates that this cloud consists of glassy carbon. The layer of glassy carbon protects Cu NP from oxidation with air oxygen and preserves their plasmon resonance in the visible. Note that decomposition of acetone on Cu NP that occurs during laser ablation of a Cu target is highly chemically selective - it takes place on Cu nanoparticles but not on Ag ones in similar experimental conditions.

Cu NP can also be synthesized via exposure of a suspension of micro-particles of CuO to laser radiation [52, 53]. At first stages of the exposure the micro-particles are broken into smaller entities, and then their reduction occurs by surrounding 2-propanol. It is worth mentioning that Ag NP can also be produced by exposure of suspended Ag_2O micro-particles in an appropriate hydrocarbon liquid to laser radiation. In both cases, fragmentation of micro-particles into smaller particles occurs due to absorption of laser energy by inter-band transition of corresponding oxide, while the subsequent formation of metallic NP takes place due to chemical reduction of oxide by surrounding liquid that acts like a reducing agent.

Brass NP can be synthesized in a similar way. Unlike Cu NP they are stable to oxidation for at least several months. The position of the plasma resonance of brass nanoparticles depends slightly on the type of laser source used for their synthesis and lies near 510-515 nm. The presence of brass NP in the liquid is corroborated by the X-ray diffractometry of the evaporated suspension. Bronze NP synthesized by laser ablation of a bronze target do not show any well-defined absorption maximum, instead a wide plateau in the green range of spectrum is observed in the absorption spectrum of their colloidal solution. However the absorption spectrum is merely different from that of Cu NP. The mean size of NP of these Cu alloys is around 20-30 nm.

INTERNAL SEGREGATION OF BRASS NP

Apart from fragmentation described above brass NP undergo drastic modification under their laser exposure in a liquid. This is an internal segregation of a nanoparticle that is repulsion of one of its components to its periphery [54]. The material of the nanoparticle is subjected to additional pressure due to its small radius. For a liquid nanoparticle, the capillary pressure is given by the expression $p=2\sigma/R$, where σ is the surface tension coefficient and R is the particle radius, which is equal for Cu NP to 120 MPa for R=20 nm. In addition to this pressure, the nanoparticle during the laser pulse is subjected to the vapor pressure of the surrounding liquid, which reaches a maximum simultaneously with the particle temperature and decreases rapidly upon expansion of the vapor shell. The pressure of ethanol vapors surrounding the particle with a temperature of 1500 K is estimated at about 20 MPa from the equation of state of a real gas. Note that the pressure of nanoparticle-substance vapors at the melting temperature for most metals is negligibly low compared to the above values.

As synthesized Cu NP are not stable towards oxidation by air oxygen. The oxidation is accompanied by disappearance of the peak of plasmon resonance of Cu NP as shown in Fig. 11. It is pertinent to note that the spectra of NP of Cu alloys are more stable with time. This is due to the formation of a shell around them as described below.

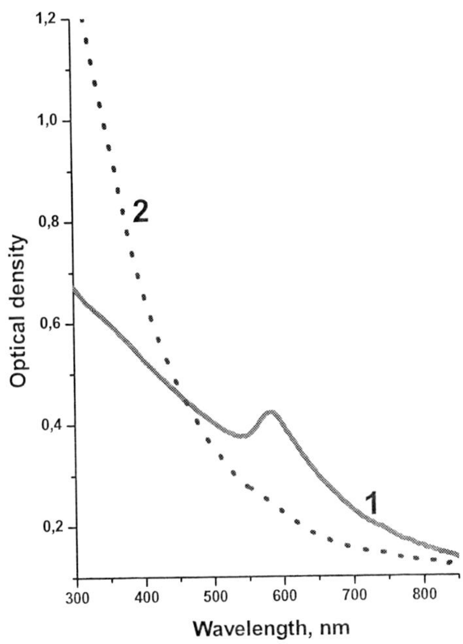

Figure 11. Evolution of absorption spectrum of Cu NP produced by laser ablation of a bulk Cu target in ethanol. As prepared (1), 6 months later (2). Ablation was carried out using a Cu vapor laser.

Laser radiation acting on the ensemble of brass NP in the liquid changes their absorption spectrum. The absorption spectrum of brass NP evolves to that of Cu NP (Figure 12).

With laser exposure the peak corresponding to Cu NP becomes noticeable. Upon further action on the colloidal solution, the peak corresponding to brass nanoparticles disappears, and only the plasmon resonance peak of Cu NP remains. Qualitative evidence of the effect is the absence of small brass nanoparticles (with a radius smaller than 5 nm) upon the laser ablation of a brass target in ethanol. Estimates show that the time of diffusion of zinc atoms to a distance of about the diameter of a nanoparticle in the liquid state is of order of several ns, which is comparable with duration of the laser pulses commonly used in experiments [54].

Small nanoparticles rapidly lose zinc upon laser irradiation and are transformed to copper NP oxidized by air oxygen.

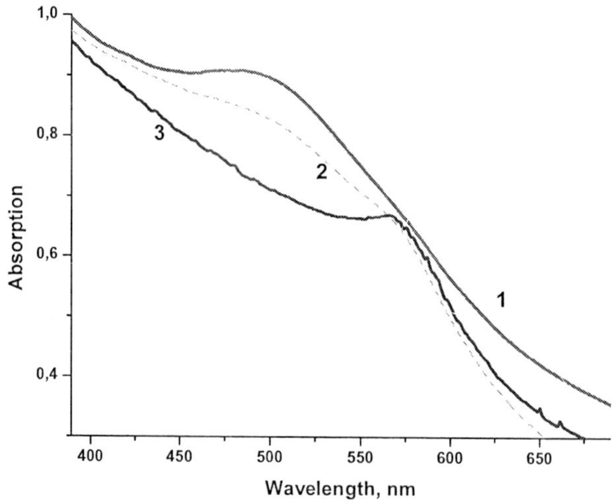

Figure 12. Evolution of the absorption spectrum of brass NP upon exposure of their colloidal solution in ethanol to radiation of a 130 ns Nd:YAG laser. As synthesized brass NP (1), 10 (2), and 90 min of exposure (3).

A similar shell is observed under laser exposure of bronze NP. In this case the shell thickness is only few nm due to lower content of the component with low melting temperature (Sn) (Figure 13). Plasmon resonance of bronze NP is less pronounced than that of brass NP though is merely different from both Cu and brass NP.

Small radius of NP is responsible for decrease of their melting temperature compared to bulk material. Also, it has been theoretically demonstrated that the melting temperature is different for different shapes of NP composed of the same material [55].

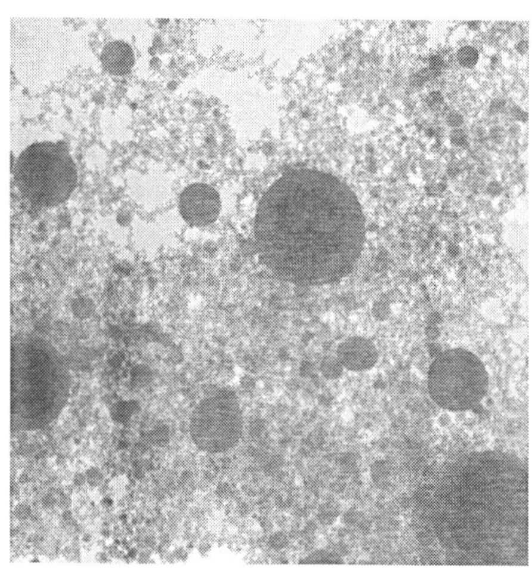

Figure 13. TEM view of core-shell bronze NP synthesized by laser ablation of corresponding targets (8% Sn, 92%Cu) and then subjected to laser radiation in suspension. The diameter of big NP is about 40 nm.

Internal segregation of NP under laser exposure leads to repulsion of the components with lower melting temperature. In this sense one may speak about purification of the core material under laser exposure.

SELF-INFLUENCE OF A FEMTOSECOND LASER BEAM

Remarkably, the initial average size of nanoparticles produced by ablation of metallic targets is almost independent on the duration of laser pulse from 100 fs to 100 ns and is of order of 10 nm. This implies that the size is determined rather by thermal properties of the target. Indeed, the layer of metal molten by a laser pulse is the only source from which the nanoparticles are formed. The difference in size becomes pronounced under sufficiently long laser exposures when the majority of generated nanoparticles pass through the laser beam in the liquid. In this case the key parameter that determines the efficiency of coupling is detuning of laser wavelength from the plasmon resonance. In certain cases a good coupling of laser radiation to the ensemble of NP is achieved via non-linear transformation of laser radiation by NP themselves. Indeed, the wavelength of a femtosecond Ti:sapphire laser (810 nm) is too far from the plasmon resonances of both Au and Ag NP. However, prolonged laser exposure of the colloidal solution leads to drastic difference in their average size (14 nm vs 4 nm) [21]. Also, there is a significant amount of even smaller Ag nanoparticles whose size is below the measuring limits of the microscope (< 2 nm). The ablation is accompanied by intense generation of the second harmonics of laser radiation at 405 nm. This wavelength exactly fits the plasmon resonance of Ag NP, while for Au nanoparticles it is significantly detuned. The effect of second harmonics generation (SHG) on small metal clusters is well documented in the literature [56, 57]. Despite too small size of metal clusters compared to the laser wavelength, SHG can be very effective due to their high density, so that the conditions of spatial synchronism are easily fulfilled. Several models account for high value of the second order non-linear susceptibility $\chi^{(2)}$ and for efficient generation of the second harmonics. This is assigned to quantum size effect for wave functions of electrons confined to small metal spheres, which gives rise to quadruple oscillations and, consequently, to non-zero $\chi^{(2)}$ even for nanoparticles with a center of symmetry. The observed difference in particle size is attributed to the self-influence of laser radiation upon generation of nanoparticles via laser ablation in liquids. Indeed, appearing nanoparticles double the frequency of laser radiation. Owing to exact matching to plasmon resonance of Ag NP, their temperature can be much higher than under exposure to initial laser radiation. This causes fragmentation of nanoparticles that are in resonance with the second harmonics that is Ag ones.

The third order non-linear susceptibility of NP $\chi^{(3)}$ is responsible for non-linear lensing of laser radiation inside the colloidal solution. As a result, the focusing conditions change during the accumulation of NP in the beam path, and the ablation rate decreases even though laser radiation is largely detuned from plasmon resonance of NP.

In general, the limiting factor of interaction of NP immersed in a liquid with a femtosecond laser beam is generation of a white continuum. This continuum consumes a significant part of laser energy and reduces the efficiency of interaction with NP. This is well pronounced for pulse durations less than 60 fs.

Influence of the Nature of the Liquid

Ablation of a Ti Target

The composition of NP produced by laser ablation of Ti was found to be dependent on the nature of the liquid [16] (see Figure 14). Namely, laser ablation of Ti in ethanol results in formation of Ti nanoparticles having the cubic structure. It should be reminded that the initial polycrystalline Ti plate has tetragonal structure. It is known that the cubic phase of Ti is metastable and exists only at elevated temperature > 600° C. At the same density of laser energy laser ablation of Ti target immersed into dichlorethane leads to the formation of nanoparticles of titanium carbide TiC. Four the most intense peaks of TiC can be distinguished on the X-ray diffractogram of the evaporated suspension. Finally, laser ablation of Ti in water results in the formation of nanoparticles having the composition of non-stoichiometric oxide TiO_x, where x = 1.04. The width of the peaks indicates the small size of the nanoparticles. All the diffractograms contains several unidentified peaks.

Formation of TiC via ablation of Ti in dichlorethane may be due to catalytic action of Ti nanoparticles, since no carbon-containing particles have been observed during ablation of either Si or Au in this liquid under otherwise equal experimental conditions. It is worthwhile to mention that the surface of Ti target after laser ablation in dichlorethane does not contain any detectable by X-ray diffraction amount of TiC. This indicates the formation of TiC nanoparticles via chemical reaction of ejected metal nanoparticles with liquid vapors.

Unlike to what is believed, NP generated by laser ablation in liquids are always crystalline, and the fraction of amorphous component is usually very low.

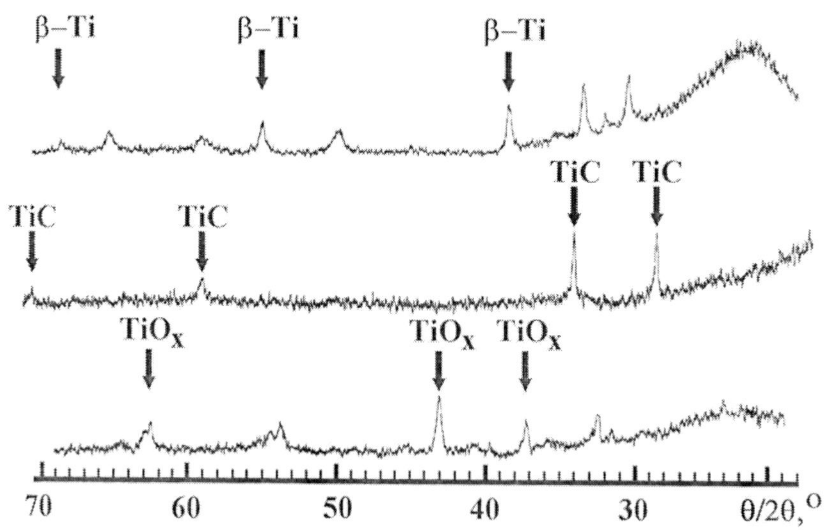

Figure 14. Difractograms of NP synthesized by laser ablation of metallic Ti target in different liquids. Ethanol (upper curve), dichlorethane (middle), and water (bottom curve). The size of nanoparticles calculated from broadening of diffraction peaks is 25, 35, and 25 nm, respectively. Cu vapor laser, fluence of 4 J/cm^2.

Oxidation state of NP material is strongly influenced by the nature of surrounding liquid even in case noble metals, e.g., Au. It was confirmed by ablation of Au in water, ethanol, and chloroform [58]. X-ray Photoelectron Spectroscopy (XPS) analysis of dried colloids showed different degree of oxidation of Au. The formation of a gold–chlorine compound is suggested in case of ablation of Au in chloroform with 5 ns laser shots at wavelength of 532 nm.

Another example of the influence of the liquid on the chemical composition of NP generated via laser ablation of a solid target is copper. In this case final NP are metallic only under ablation in acetone, while water and ethanol lead to non-metallic NP.

Ablation of Sn

Tin is a metal with low melting point, and laser fluence required for its ablation is much lower than for other metals mentioned above. A certain dependence of the particles composition on the concentration of a surface-active substance is found in this case [59].

Non-stoichiometric tin oxide SnO_{2-x} is formed under laser ablation of a tin target in aqueous solutions containing SDS [60]. Ablation of Sn in either water or ethanol at moderate laser fluence (less than 0.2 J/cm^2 at nanosecond pulse duration) leads to formation of Sn NP. This is corroborated both by X-ray diffraction of the dried suspension and its optical absorption spectra.

Since melting temperature of bulk Sn is low, Sn NP are liquid at room temperature. In general, this is due to high fraction of surface atoms in NP compared to their total number. However, evaporation of their colloidal solution in absence of surface-active substances leads to aggregation, and X-ray diffraction shows narrow peaks of Sn NP about 100 nm in diameter. A gold plate immersed into colloidal solution is rapidly covered by numerous layers of Sn. Once fixed on a large substrate, Sn NP are solidified and form a kind of nano-contacts to the substrate. Mixing Sn NP and Au NP in ethanol leads to wetting of Au NP by liquid Sn and therefore to specific core-shell NP. Their TEM view is presented in Figure 15.

The thickness of the shell depends on the volume of the Sn NP. Large Sn NP are solidified just after the contact with Au NP, and no shell is formed. Smaller Sn NP envelop Au NP resulting in a shell with homogeneous thickness. Diffuse entities are tentatively assigned to tin oxide.

At picosecond pulse duration the resulting NP consist mostly of tin oxide, which is confirmed by their characteristic absorption peak at 360 nm.

W and Mo NP

Other metals that might interact with surrounding liquid during ablation remain metallic and are oxidized very slowly with air oxygen that is dissolved in liquid. This concerns NP of either W or Mo upon their ablation in water. These metals with high melting temperature are dispersed into surrounding liquid as metallic NP, which is corroborated by both X-ray diffraction and optical spectra of corresponding colloidal solutions.

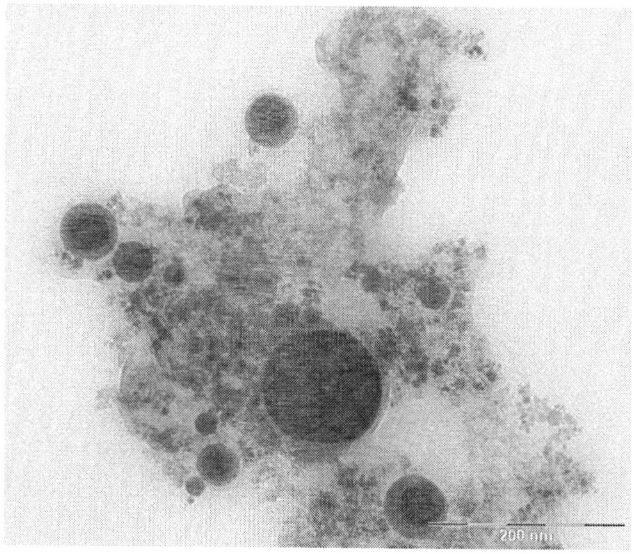

Figure 15. TEM view of Au NP wetting by Sn NP. Both colloids were synthesized by ablation of corresponding metallic targets in ethanol using radiation of a Cu vapor laser. Scale bar denotes 200 nm.

Their morphology alters during several months of storage at room temperature due to oxidation, which results in the formation of core-shell NP (see Figure 16). Like many other metals, the oxide layer prevents further oxidation of the metallic core. Similar result has been reported for W NP obtained by ablation of a W target in ethanol using high-repetition-rate copper bromide laser radiation [61]. The authors deduce the formation of non-stoichiometric oxide WO_{3-x} around W NP on the basis of absorption spectra of the colloidal solution.

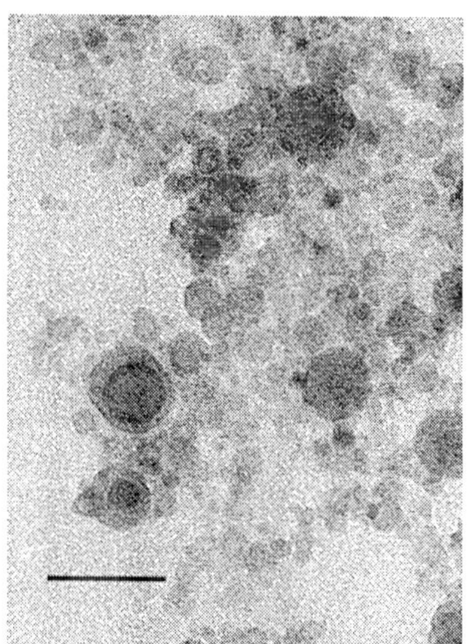

Figure 16. TEM view of W NP synthesized by ablation of a W target in ethanol by radiation of a Cu vapor laser. Scale bar denotes 50 nm.

Carbon NP can also be obtained by laser ablation of a graphite target immersed into liquid [62].

A peculiar feature of laser ablation in water is enhanced solubility of some compounds in it under high temperature and high transient pressure that are established during the laser pulse. This leads to so called hydrothermal synthesis of NP of unusual morphology, e.g., platelets or disks of $Pb(Zr,Ti)O_3$ [63]. Aqueous solutions at pressure and temperature that exceed its critical value take on the ability to dissolve solids that are not soluble in normal conditions. Synthesis of compounds via their dissolution in supercritical compounds is called the hydrothermal one. In nature this type of synthesis occurs in the vicinity of underwater volcano, which results in formation of various minerals. Hydrothermal synthesis is typical for non-metallic compounds, such as oxides of more complex ceramics. The material of the target is dispersed and dissolved in a supercritical medium during the laser pulse. Upon cooling the solubility drops, and the formation of NP occurs from oversaturated solution. Apparently, similar process may take place under laser ablation of single crystal rutile in an aqueous environment [64].

The specific surface of NP is very high, for instance, 1 ml of Au colloid prepared by laser ablation of a bulk target at typical density has the surface of 30 m^2. High surface favors catalytic reactions in liquids. Not only NP material but also the liquid itself may undergo chemical changes during laser ablation. In some cases these changes are catalyst-specific that is occur with NP of some specific metals. Laser exposure of most metallic targets (Sn, brass, etc.) in ethanol leads to its noticeable modification towards formation of products with higher molecular mass. Indeed, a wide absorption band centered at 350 nm is observed that correspond to a viscous non-volatile liquid that remains after evaporation of the colloidal solution synthesized by laser ablation. This product might be a low-molecular mass polyethylene, which may also form a kind of shell on the synthesized NP. Formation of this product occurs mostly on the stage of exposure of NP to laser radiation, and catalytic effects of its formation mediated by NP of various compositions are not excluded.

Modeling of Distribution Function

Modeling of distribution function of nanoparticles size can be performed by a numerical solution of kinetic equation on the basis of first principles. The following processes should be taken into account: generation of nanoparticles by laser ablation of a solid target immersed into liquid, aggregation of nanoparticles, their fragmentation and escape of NP adsorbed on the wall of a reactor [25, 65]. In the specific conditions of laser ablation in liquids, the last term can be negligibly small unlike laser ablation in vacuum. On the contrary, fragmentation of the particles in the laser beam is the main channel of modification of their distribution function. The peculiar feature of ablation in liquids is simultaneous action of all 4 processes, since the generated NP remain in the liquid and, in particular, on the optical path of laser beam. Formation of NP due to coagulation of individual metal atoms ejected from the target is unlikely, since the necessary condition of their synthesis is melting of the target. NP are ejected from the target at the very beginning of ablation.

Melting of NP and their subsequent fragmentation may occur only in those areas of the liquid where the peak power of laser radiation is sufficiently high. The model takes into account a factor proportional to the ratio of the volume of laser beam waist to total volume of the liquid. In typical experimental conditions this factor is about 10^{-4}, which explains the need of prolonged laser exposures required to alter the distribution function at measurable level.

Laser exposure of NP in liquid in absence of the metal target results in their fragmentation due to absorption of laser energy by individual particles. Under laser fluence typically used NP are molten during the laser pulse, and their splitting is due to hydrodynamic instabilities of the molten metal embedded into a vapor pocket. According to the model derived above, the size of the particles goes to 0 under long exposure. In reality for each value of laser fluence there is a certain minimal size of particles that are too small to absorb the energy sufficient for their melting from laser beam. The model that takes into account all above-mentioned considerations shows good qualitative agreement with the experimentally measured distribution function [25].

Influence of Intensity Distribution of the Laser Beam on the Shape of Nanoparticles

The influence of the laser beam profile onto the morphology of NP generated via laser ablation of solids in liquids is not studied so far and is often ignored in current scientific literature. The origin of this influence is a certain value of the energy density of the laser beam, which is needed to melt the target material. The size and shape of NP generated via laser ablation are sensitive to the laser fluence. Typically higher fluence leads to higher average size of generated NP. Also, in case of laser ablation of Au at elevated laser fluence the generated NP have elongated shape (see Figure 6). The term "elevated" should be understood in term of laser fluence needed to melt the target surface. One may suggest that this is due to specific way of interaction of the melt bath within the laser spot with surrounding medium. Indeed, sputtering of the melt is due to splashing of the molten layer on the target surface by high pressure vapors of surrounding liquid. However, the ejection of the melt in the central areas of the laser spot is different from that at its periphery. The melt ejected in the spot center has high probability to return to the target surface being pushed by the expanding vapor cloud. On the contrary, ejection of the melt from the periphery of the melt bath very likely results in the formation of NP, since the lateral dimensions of the expanding vapor pocket above the target are quite close to the laser spot size. This means that only the edges of the molten area contribute to NP formation, while the central part of the laser spot provides relatively small fraction of NP.

These considerations suggest the following experiment on NP generation in which the laser beam that causes ablation of a target immersed into liquid has another type of symmetry of its intensity distribution different from that of a flat-top or Gaussian profile. A good possibility is formation of a periodic intensity distribution in the plane of the target. Such periodic distribution may occur spontaneously due to, for example, interference of the laser beam with Surface Electromagnetic Wave excited in the solid by this beam. Another possibility is exposure of the target to the interference pattern of two coherent laser beams. In this case the laser fluence is a periodic function along one of coordinate, so does the temperature distribution as schematically shown in Figure 17.

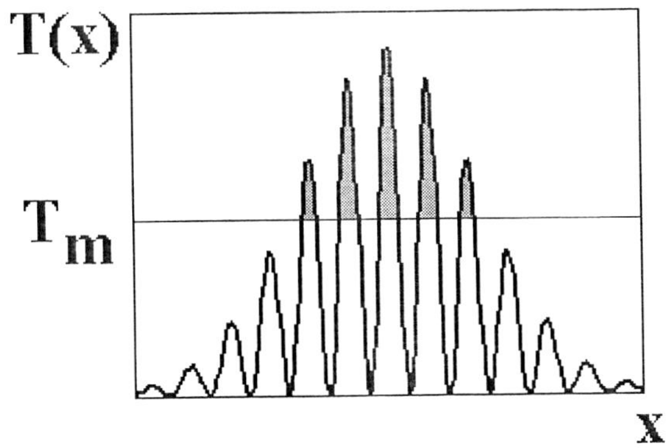

Figure 17. Temperature distribution on a target exposed to two coherent laser beams. Solid line $T(x)=T_m$ indicates melting temperature of the target.

If the fluence in the maxima of interference pattern are sufficient to melt the target, then the melt baths will have elongated shape on the target, which in turn may affect the shape of ejected NP.

The experimental setup is shown in Figure 18. A Cu vapor laser emits at two laser lines, either green (510.6 nm) or yellow (578.2 nm). In this particular experiment only one output was used, namely, the green one. The green laser beam was split into two beams of approximately equal intensities using dividing glass cube. These two beams were superimposed on the surface of a gold target placed into liquid ethanol. The angle between two beams corresponded to the interference pattern on the target surface with period of 4 µm. The cell with gold sample was moved under the laser beams in the direction parallel to the maxima of interference pattern.

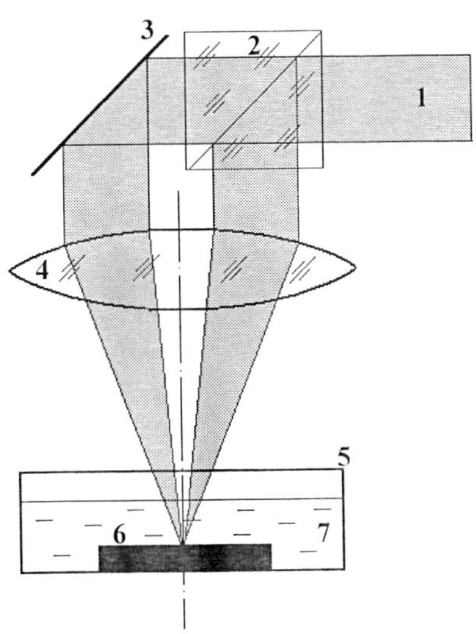

Figure 18. Experimental setup on laser-assisted synthesis of elongated Au NP. 1 – laser beam (Cu vapor laser, 510.6 nm laser output), 2 – splitting cube, 3 – dielectric mirror, 4 – focusing lens, 5 – cell, 6 – Au target, 7 – liquid ethanol.

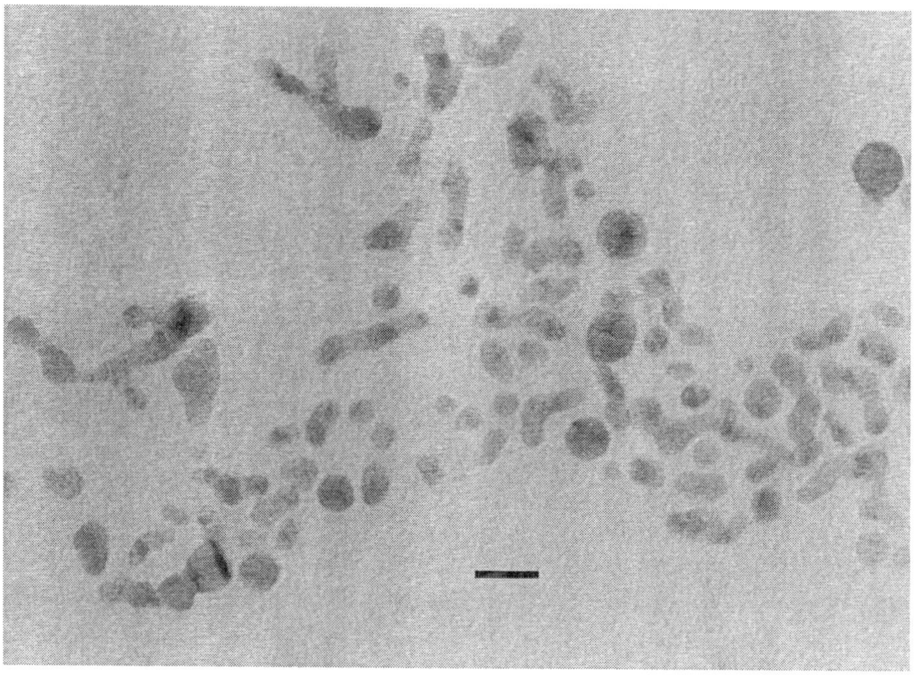

Figure 19. TEM view of Au nanoparticles obtained by ablation of an Au target in ethanol by two interfering laser beams of a Cu vapor laser. Scale bar denotes 50 nm.

Ablation of the Au target in these conditions results in visible coloration of the liquid, and the color of the solution has bluish tint typical of elongated Au NP. TEM examination of generated Au colloid confirms the presence of elongated Au NP, as demonstrated in Figure 19. Some Au NP are indeed elongated, and their aspect ratio (ratio of length to diameter) exceeds 10.

Nanostructuring of Solids under their Laser Ablation in Liquids

The total laser fluence absorbed by a target should be sufficient to melt it in order to produce NP in the liquids. Ideally smooth surface is characterized by definite threshold laser fluence needed for melting. In practice the surface of a target has certain micro-roughness, and this surface may not melt simultaneously. Indeed, any protrusion which is weakly thermally coupled to the bulk of the target is molten at lower fluence than the smooth target. Therefore, at certain fluence a rough surface will not be molten only in some protruding areas. The vapors of surrounding liquid that expands from the target will then carry away the molten fragments thus forming spikes in its surface. The rate of NP generation in the liquid is very low, so at threshold laser fluence the material tends to leave the target in the form of NP but remains on it solidified as nano-spikes or nano-bumps. Period of nano-spikes does not depend on the laser wavelength and is determined mostly by characteristics of the target material.

Formation of NS on the target is accompanied by modification of its absorption spectrum. Indeed, the electrons that oscillate within single NP are also confined, and therefore the spike edges act like the edge of NP as soon as their dimensions are compared with mean free path of electrons at Fermi level (see Figure 20).

This confinement causes the coloration of some metals with formed NS. For instance, Ag plate with NS looks yellow in appearance [66]. A macro view of Ag surface with NS is shown in Figure , a. Yellowish coloration in Ag is typical of silver NP.

Absorption spectrum of both initial Ag target and of Ag with NS is shown in Figure 21.

Initial spectrum corresponds to plasmon resonance of electrons in bulk Ag. Formation of NS manifests itself in broadening of this resonance. A new maximum appears near 360 nm. This maximum shifts to red with oxidation of Ag upon storage in air during several days. Similar modifications of absorption have also been observed in large variety of metals, e.g., Ti, W, Cu, etc., subjected to laser ablation in liquids at threshold laser fluence.

Typical morphology of both Ag and Ta target with NS formed by laser ablation in water is presented in Figure 16. One can see that the period of NS is about 50 nm for Ag and 300 nm for Ta, which is much smaller than laser spot. In this sense NS are self-organized. NS on Ta are characterized by high aspect ratio, while on Ag they are rather nano-bumps.

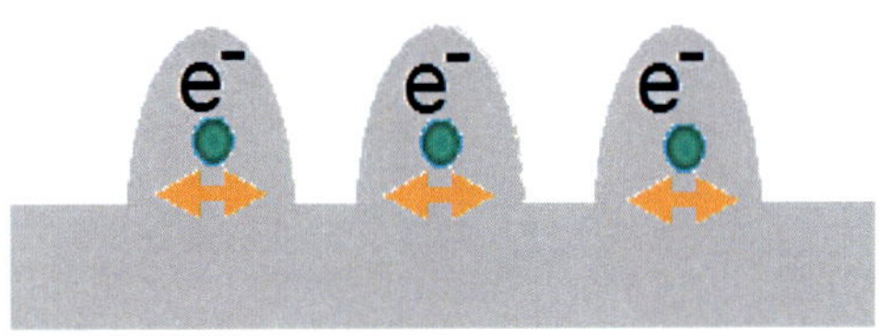

Figure 20. Schematic view of oscillations of free electrons in metallic nanostructures. Plasmon resonance of these electrons is close to that of nanoparticles of the same lateral dimensions.

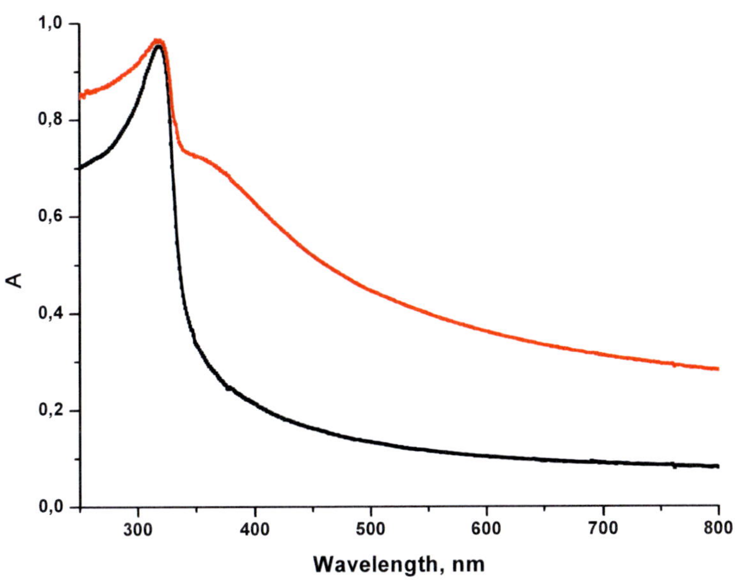

Figure 21. Absorption of an Ag target ablated under water layer with 5 ps laser pulses at wavelength of 248 nm. Black curve corresponds to initial Ag target while the red one to Ag with NS.

Formation of NS on metallic targets ablated in liquids requires sufficiently short laser pulses. Indeed, no NS are formed with nanosecond pulse duration. The longest tested pulses that produce NS are 350 ps delivered by a Nd:YAG laser at wavelength of 1.06 μm. Another necessary condition for formation of NS is certain initial roughness of the target surface. The local roughness required for NS formation should exceed 30-50 nm. These two conditions allow suggesting local melting of the target in the micro-protrusions of the target relief, which develops with further increase of the number of laser shots into an array of self-organized nano-spikes.

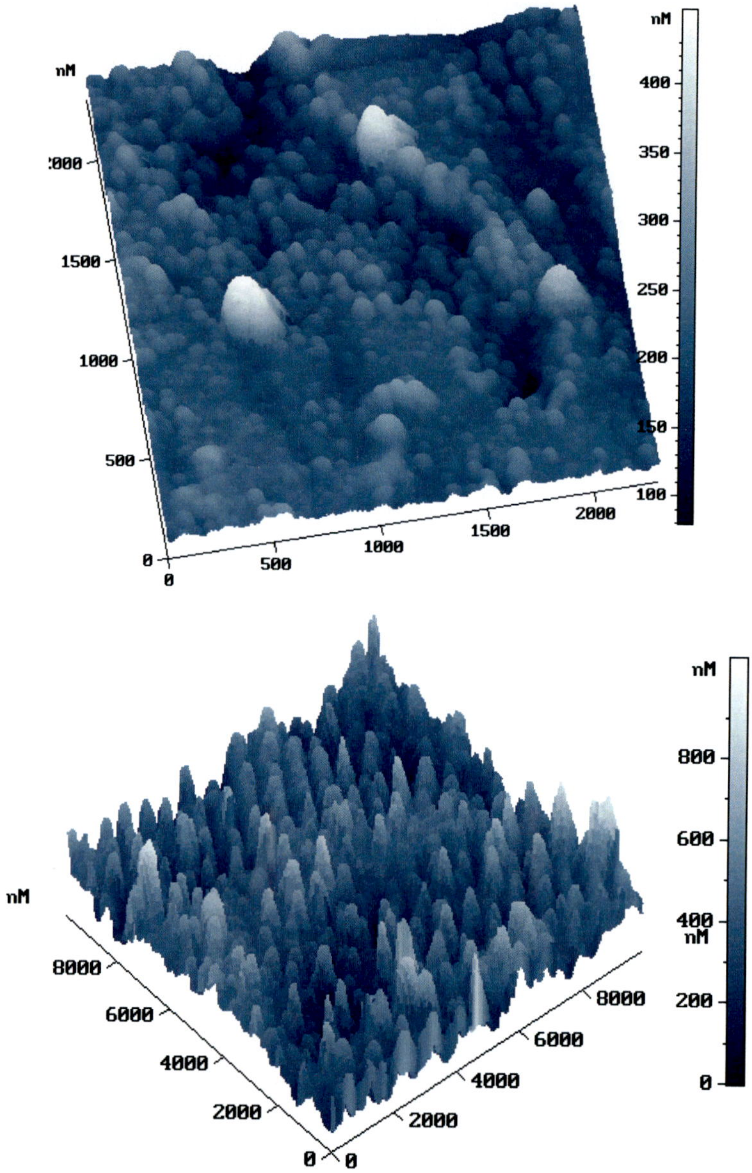

Figure 22. AFM view of NS on Ag (top) and Ta (bottom). NS were generated by ablation of corresponding metal target in water, pulse width of 350 ps at wavelength of 1.06 μm.

Figure 23. Schematic picture of detachment of nanodrops from the melt on the surface of a solid target.

The process of NS formation under laser ablation of solids immersed into liquids requires further studies. However the experimental data available so far indicate relation of NS to some instability that develops at the interface "melt-vapor of liquid" within the laser spot. In typical conditions the pressure of vapor of the liquid exceed by at least one order of magnitude the capillary pressure stipulated by curvature of the melt surface, so the recoil pressure of vapor that expands from the target plays dominant role.

Nanostructures on Ag generated by its laser ablation in water show the SERS effect of adsorbed organic molecules [67]. The estimated enhancement factor is about 10^5. Unlike Ag colloid in a liquid, Ag plate with nanostructures can be used for SERS measurements many times, since the NS do not degrade upon measurements.

Different stages of NS development are schematically shown in the Figure 23. These structures correspond to morphology that can be found within the laser spot on the target. At sufficiently high laser fluence the nano-drop of the melt is detached from the target surface into surrounding liquid as nanoparticles.

It is also clear that probe microscopes fail to image correctly the mushroom-like structures. This type of nanostructures is imaged by an AFM as a cone.

Small size of NS implies another mechanism of their formation, which should not involve capillary waves. Indeed, capillary waves with period of hundreds of nanometers would be suppressed since capillary pressure at this scale tends to smooth the melt.

Thus, formation of nanostructures under laser ablation of metals in liquid environment is a general conformity. It is characterized by several features. First, nanostructures are observed only in case of sufficiently short laser pulses. Second. Their period foes not depend on the laser wavelength. Nanostructures are merely different from periodic ripples that are generated owing to interference of a Surface Electromagnetic Wave with laser beam.

The reviewed results on laser-assisted generation of nanoparticles via laser ablation in liquids represent only one point of view. Recent review [69] and recent book [70] consider another processes and approaches that have been developed during last decade.

Excitation of High Energy Levels

In typical conditions of experiments on laser ablation of solids in liquid environment the nanoparticles (NP) sputtered into liquid are optically thin at the wavelength of most common lasers. Despite their small size (about 10 nm), the NP can efficiently absorb the laser radiation. The efficiency of this interaction is a function of numerous experimental parameters, such as particle size, detuning of laser frequency from the position of the plasmon resonance of NP, etc. However in most cases, NP are optically thin, that is almost transparent at the laser wavelength. Hence, the time required to reach their temperature under laser exposure is shorter than the duration of laser pulse as soon as the latter exceeds the characteristic time of electron-phonon relaxation. The temperature of NP inside the laser beam is proportional to the peak power of the laser radiation. Estimations show that the temperature of Au nanoparticles inside Cu vapor laser beam with peak intensity of 10^8 W/cm^2 is several kK, and this temperature scales linearly with laser intensity. Therefore, exposure of NP suspension in liquid environment is a novel approach that allows excitation of high energy levels of both the NP material and its close environment. The liquid surrounding the NP in this case turns into a vapor having high temperature and pressure, and eventually into plasma that perfectly surrounds it. The present paper describes several proofs that laser exposure of NP suspension in liquid may lead to excitation of high energy levels including those of nuclei and neutron release. Laser initiation of neutron emission may open new possibilities for transformation of elements, synthesis of desired isotopes, disposal of nuclear waste, etc. The suspension of Hg in D_2O is a model system for laser-assisted transmutation of Hg into Au since the neutron binding energy in D has the lowest value [71,72].

Two kinds of Hg were used for a model system of transmutation of Hg into Au: (i) Hg of analytical purity with natural isotopic composition or (ii) enriched Hg containing 55.6% of ^{196}Hg and 41.4 % of ^{199}Hg, with much lower content of other isotopes, obtained by selective photochemical reaction of Hg with oxygen [73].

Exposure of the colloidal solution of Au NP in water to radiation of a 350 ps Nd:YAG laser is accompanied by visible emission from the solution. The spectrum of this emission contains several spectral features. The first one corresponds to anti-Stokes Raman scattering of H_2O vapors of laser radiation excited at 1060 nm. The second broad peak coincides with the emission of atomic Au(I) at 627.8 nm during transition from the resonant to the meta-stable level. The peak at 530 nm coincides with the position of plasmon resonance of Au NP in H_2O. One can also see that the laser-produced plasma is characterized by intense UV radiation at least up to 200 nm where water itself and Au NP have strong absorption [74].

If the laser spark is placed in the vicinity of a Be window of the cell one can get an image of a metallic grid on Be using an X-ray photo film, though the exposure time has to be at least 2 hours. Therefore, both deep UV and X-ray radiation is produced during exposure of Au NP colloid in H_2O to laser radiation with rather moderate intensity. This result is in agreement with previous observations of X-ray emission under exposure of aqueous solutions to femtosecond laser radiation [75,76].

The energy of X-ray photons is sufficient to cause the modification of the NP environment. This has been confirmed in a model experiment on exposure of Au NP in D_2O to radiation of a 350 ps Nd:YAG laser at peak intensity of 10^{10} W/cm^2. Raman spectrum of D_2O with Au NP subjected to 2 hours laser exposure indicates the formation of HDO molecules.

Suspension of Hg in D_2O is characterized by an absorption peak in the range 270-290 nm even prior to laser exposure. This peak is assigned to the plasmon resonance of Hg NP, which is close to the theoretical position of 290 nm reported for 10 nm NP of Hg in H_2O [18]. Therefore, in the initial stages, exposure of Hg suspensions to laser radiation leads to the formation of Hg NP (size of the order of 10 nm). Using either a Cu vapor laser (8 hours of exposure) or a 100 femtosecond Ti:sapphire laser (2 hours of exposure), no Au content was detected within the accuracy of measurements.

Table 1. Analysis of Hg samples of natural isotope composition exposed to radiation of a 90 ps Nd:YAG laser, exposure time of 60 min

Sample	Hg in D_2O initial	Hg in D_2O sample 1	Hg in D_2O sample 2
Hg, mg/l	35	23	8
Au, mg/l	0.009	0.1	0.024
Au/Hg ratio	0.00026	0.0043	0.003

Table 2. Exposure of suspension of ^{196}Hg in D_2O to radiation of a 350 ps Nd :YAG laser. The results are averaged over 4 different exposures

Sample	$^{196}Hg/D_2O$ initial	$^{196}Hg/D_2O$ exposed Hg drop	$^{196}Hg/D_2O$ 4 hours of exposure (averaged over 4 samples)	$^{196}Hg/D_2O$ after sedimentation
Au, mg/l	0.0073	0.38	0.23	0.17
Hg, mg/l	20	8.94	2.31	12.6
Au/Hg	0.00036	0.0425	0.10	0.0135

After exposure to a 90 ps Nd:YAG laser radiation, the suspension precipitates very slowly, unlike a freshly prepared suspension in ultrasonic bath. After 1 day of sedimentation the liquid separated into two parts, the first close to the bottom and the other remaining in suspension. Analytic data presented in Table 1 show the formation of Au.

Experiments with a 350 ps Nd:YAG laser were carried out with two types of Hg samples, either of natural isotopic composition or enriched with ^{196}Hg. Results of analysis of Hg suspension of natural isotope composition in D_2O are similar to those obtained with a 90 ps laser. The formation of Au in exposed suspensions of Hg in D_2O is confirmed by analysis, though in general the Au/Hg ratio is lower.

The results of analysis for Hg enriched with ^{196}Hg are presented in Table 2. Note that Au content is found under laser exposure of a Hg drop immersed into D_2O for about 10 min in order to disperse the metal. Four hours of laser exposure increases the Au content to almost 10%.

The mechanism of transmutation is still under discussion. The effect is observed only under sufficiently long laser pulses of picosecond range. One may suggest that the beginning of a laser pulse ionizes the nanoparticles within the laser beam while the remaining laser energy is spent for acceleration of electrons as it is observed in the case of interaction of strong laser beams with plasma [77]. These electrons provide X-ray photons needed for release of a neutron from Deuterium.

CONCLUSION

The research on the processes of laser ablation in liquids is still in progress. There is a certain hope that the simplicity of the experimental setup on laser-matter interaction will allow developing new cost-effective technologies. Several specific features, such as the absence of a vacuum, may provide wide use of liquid-assisted generation of nanoparticles in medicine, catalysis, etc. The novel process of laser-assisted nanostructuring of solids via their laser ablation in a liquid environment is also of high interest for various applications in medicine, biology, aero- and hydrodynamics.

References

[1] J. Nedersen, G. Chumanov, and T.M. Cotton, *Appl. Spectrosc.*, 1993, *47*, 1959.

[2] M. S. Sibbald, G. Chumanov, and T.M. Cotton, *J. Phys. Chem.*, 1996, *100*, 4672.

[3] A. Fojtik, A. Henglein, *Ber. Busenges. Phys.Chem.*, 1993, *97*, 252.

[4] A. Henglein, *J. Phys.Chem.*, 1993, *97*, 5457.

[5] M. Procházka, Štepánek, B. Vickova, I. Srnova, P. Maly, *J. Mol. Struct.* 1997, *410/411*, 213.

[6] M. Procházka, P. Mojzeš, J. Štepánek, B. Vlèková, P.-Y. Turpin, *Anal. Chem.*, 1997, *69*, 5103.

[7] M. Procházka, J. Štepánek, B. Vlèková, I. Sronová, P. Malŷ, *J.Mol. Struc.*, 1997, *410*, 213.

[8] I. Srnová, M. Procházka, B. Vlèková, J. Štepánek, P. Malŷ, *Langmuir,* 1998, *14*, 4666.

[9] J.S. Jeon, C.H. Yeh, *J. Chin. Chem. Soc.*, 1998, *45*, 721.

[10] M.-S. Yeh, Y.-S. Yang, Y.-P. Lee, H.-F. Lee, Y.-H. Yeh, and C.-S. Yeh, *J. Phys.Chem.*, 1999, *B 103*, 6851.

[11] A. Takami, H. Kurita, and S. Koda, *J. Phys.Chem.,* 1999, *B103*, 1226.

[12] S. Link, C. Burda, B. Nikoobakht, and M.A. El-Sayed, *J. Phys.Chem.*, 2000, *B104*, 6152.

[13] F. Mafuné, J. Kohno, Y. Takeda, T. Kondow, H. Sawabe, *J. Phys. Chem.*, 2000, *B 104*, 9111.

[14] F. Mafuné, J. Kohno, Y. Takeda, T. Kondow, H. Sawabe, *J. Phys. Chem.*, 2001, *B 105*, 5114.

[15] A.V. Simakin, V.V. Voronov, G.A. Shafeev, R. Brayner, and F. Bozon-Verduraz, *Chem. Phys. Lett.*, 2001, *348*, 182.

[16] S.I. Dolgaev, A.V. Simakin, V.V. Voronov, G.A. Shafeev, F. Bozon-Verduraz, *Appl.Surf.Sci.*, 2002, *186*, 546.

[17] P. V. Kamat, M. Flumiani, and G.V. Hartland, *J. Phys.Chem.*, 1998, *B 102*, 3123.

[18] J.A. Creighton and D.G. Eadon, *J. Chem. Soc. Faraday Trans.*, 1991, *87*, 3881.

[19] P. Mulvaney, *Langmuir*, 1996, *12*, 788.

[20] R.A. Ganeev, A.I. Ryasnyansky, R.I. Tugushev, and T. Usmanov, *J. Opt. A: Pure Appl. Opt.* 2003, *5*, 409.

[21] G.A. Shafeev, E. Freysz, and F. Bozon-Verduraz, *Appl. Phys.*, 2003, *A78*, 307.

[22] V.A. Karavansky, A.V. Simakin, V.I. Krasovsky, and P.V. Ivanchenko, *Quantum Electronics*, 2004, *34(7)*, 644.

[23] P.V. Kazakevich, A.V. Simakin, and G.A. Shafeev, *Quantum Electronics*, 2005, *35*, 831-834.

[24] S. Link and M. El-Sayed, *J. Phys. Chem.,* 1999, *B103*, 8410.

[25] A.V. Simakin, V.V. Voronov, N.A. Kirichenko, and G.A. Shafeev, *Appl. Phys.*, 2004, *A79*, 1127-1132.
[26] F. Mafune, J. Y. Kohno, Y. Takeda and T. Kondow, *J.Phys. Chem.*, 2000, *104*, 8333.
[27] T. Tsuji, T. Kakita, and M. Tsuji, *Appl.Surf.Sci.*, 2003, *206*, 314.
[28] B. Nikoobakht and M.A. El-Sayed, *Chem. Mater.*, 2003, 15, 1957.
[29] F. Mafuné, J.-Y. Kohno, Y. Takeda, T. Kondow, and H. Sawabe, *J. Phys.Chem.*, 2000, *B104*, 9111.
[30] F. Mafuné, J.-Y. Kohno, Y. Takeda, T. Kondow, and H. Sawabe, *J. Phys.Chem.*, 2001, *B105*, 5144.
[31] Y.-H. Chen and C.-S. Yeh, *Colloids Surf.*, 2002,*197*, 133.
[32] A.V. Kabashin and M. Meunier, *J. Appl. Phys.*, 2003, *94*, 7941.
[33] G. Compagnini, A. A. Scalisi, and O. Puglisi, *J. Appl. Phys.*, 2003, *94(12)*, 7874.
[34] J.-P. Sylvestre, A. V. Kabashin, E. Sacher, M. Meunier, and J. H. T. Luong, *J. Am. Chem. Soc.*, 2004, *126*, 7176.
[35] F. Mafuné, J-Y. Kohno, Yo. Takeda, and T. Kondow, *J. Phys. Chem. B* 2003, 107, 4218-4223.
[36] J.S. Jeon and C.H. Yeh, *J. Chin. Chem. Soc.*, 1998, *45*, 721.
[37] T. Tsuji, K. Iryo, N. Watanabe, M. Tsuji, *Appl. Surf. Sci.*, 2002, *202*, 80.
[38] X.Y. Zhang, L.D. Zhang, Y. Lei, L.X. Zhao, and Y.Q. Mao, *J. Mat. Chem.*, 2001, *11*, 1732.
[39] B.M.I. van der Zande, M. R. Bohmer, L.G.J. Fokkink, and C. Schonenberger, *Langmuir*, 2000, *16*, 451.
[40] S.-S. Chang, C.-W. Shih, C.-D. Chen, W.-C. Lai, and C.R. C. Wang, *Langmuir*, 1999, *15*, 701.
[41] Y.-Y. Yu, S.-S. Chang, C.-L. Lee, and C.R. C. Wang, *J. Phys. Chem.,* 1997, *B101*, 6661.
[42] P.V. Kazakevich, A.V. Simakin, G.A. Shafeev, *Chem. Phys. Lett.*, 2006, *421*, 348-350. See also in: P.V. Kazakevich, A.V. Simakin, G.A. Shafeev, G. Viau, Y. Soumare, F. Bozon-Verduraz, *Appl. Surf. Sci.,* 2007, 253 7831–7834.
[43] S. Link and M. El-Sayed, *J. Phys. Chem.* 1999, *B103*, 8410.
[44] Y.-H. Chen, Y.-H. Tseng, and C. S. Yeh, *J. Mater Chem.*, 2002, *12*, 1419.
[45] Y. Kim, R.C. Johnson, J. Li, J.T. Hupp, G.C. Schatz, *J. Phys. Chem.,* 2002, *352*, 421.
[46] J.-P. Abid, H.H. Girault, and P.F. Brevet, *Chem.Comm.*, 2001, 829.
[47] A.T. Izgaliev, A.V. Simakin, and G.A. Shafeev, *Quantum Electronics*, 2004, *34(1)*, 47-52.
[48] A.T. Izgaliev, A.V. Simakin, and G.A. Shafeev, and F. Bozon-Verduraz, *Chem.Phys.Lett.*, 2004, *390*, 467.
[49] P.V. Kazakevich, A.V. Simakin, V.V. Voronov, and G.A. Shafeev, *Appl.Surf.Sci.,* 2006, *252*, 4373.
[50] P.V. Kazakevich, V.V. Voronov, A.V. Simakin, and G.A. Shafeev, 2004, *Quantum Electronics*, 34(10) 951.
[51] P.V. Kazakevich, V.V. Voronov, A.V. Simakin, and G.A. Shafeev, *Appl. Surf. Sci.,* 2006, 252 4373.
[52] Y.-H. Yeh, M.-S. Yeh, Y.-P. Lee, C.-S. Yeh, *Chemistry Letters*, 1998, 1183-1184.
[53] M.-S. Yeh, Y.-S. Yang, Y.-P. Lee, H.-F. Lee, Y.-H. Yeh, C.-S. Yeh, *J. Phys. Chem.* 1999, *B* 103 6851-6857.

[54] V. V. Voronov, P. V. Kazakevich, A. V. Simakin, and G. A. Shafeev, 2004, *JETP Letters,* 80(11) 684.

[55] M. Wautelet, J. P. Dauchot, and M. Hecq, *Nanotechnology* 2000, 11, 6–9; M. Wautelet, J.P. Dauchot, M. Hecq, *Materials Science and Engineering* 2003, C 23 187–190; R. Vall′ee, M. Wautelet, J. P. Dauchot, and M. Hecq, *Nanotechnology*, 2001, 12 68–74.

[56] K.Y.Lo, J.T. Lue, *Phys. Rev.*, 1995, *B51(4)* 2467.

[57] O. A. Aktsipetrov, P. V. Elyutin, A. A. Fedyanin, A. A. Nikulin, A. N. Rubtsov, *Surface Science*, 1995, *325(3)*, 343.

[58] G. Compagnini, A. A. Scalisi, and O. Puglisi, *Phys. Chem. Chem. Phys.*, 2002, 4, 2787–2791.

[59] T. Sasaki, C. Liang, W.T. Nichols, Y. Shimizu, N. Koshizaki, *Appl. Phys.* 2004, A79 1489.

[60] C. Liang, Y. Shimizu, T. Sasaki, and N. Koshizaki, *J. Phys. Chem.,* 2003, B107, 9220-9225.

[61] M.S.F. Lima, F.P. Ladàrio, and R. Riva, *Appl. Surf. Sci.*, 2006, 252, 4420–4424.

[62] G.X. Chen, M.H. Hong, Q. He, W.Z. Chen, H.I. Elim, W. Ji, T.C. Chong, 2004, *Appl. Phys.* A79 1079.

[63] A. Kruusing, *Optics and Lasers in Engineering*, 2004,41, 307.

[64] A. Iwabuchi, C.-K. Choo, and K. Tanaka, *J. Phys. Chem.,* 2004, B108, 10863.

[65] F. Bozon-Verduraz, R. Brayner, V.V. Voronov, N.A. Kirichenko, A.V. Simakin, and G.A. Shafeev, *Quantum Electronics*, 2003, 33(8), 714-720.

[66] E.V. Zavedeev, A.V. Petrovskaya, A.V. Simakin, and G.A. Shafeev, *Quantum Electronics*, 2006, 36(10) 978-980.

[67] S. Lau Truong, G. Levi, F. Bozon-Verduraz, A.V. Petrovskaya, A.V. Simakin, and G.A. Shafeev, *Appl.Phys.* 2007, A89 (2) 373 – 376.

[68] S. Lau Truong, G. Levi, F. Bozon-Verduraz, A.V. Petrovskaya, A.V. Simakin, G.A. Shafeev, *Applied Surface Science*, 2007, 254, 1236–1239.

[69] G.W. Yang, *Progress in Materials Science*, 2007, 52, 648–698.

[70] A. Kruusing, *Handbook of Liquids-assisted laser processing*, Elsevier, 2008, 454 p.

[71] G.A. Shafeev, F. Bozon-Verduraz, and M. Robert, *Preprint of GPI, No45*, Moscow, April 12, 2006.

[72] G.A. Shafeev, F. Bozon-Verduraz, and M. Robert, *Physics of Wave Phenomena*, 2007,15(3), 131–136.

[73] Yu.V. Viazovetsky and A.P. Senchenkov, *Journal of Technical Physics*, 1998, 68(1) 67.

[74] G.A. Shafeev, A.V. Simakin, F. Bozon-Verduraz, M. Robert, *Applied Surface Science* 2007, 254 1022–1026.

[75] K. Hatanaka, T. Miura, and H. Fukumura, *Appl.Phys.Lett.*, 2002, 80(21) 3925.

[76] K. Hatanaka and H. Fukumura, X-ray generation from optical transparent materials by focusing ultra-short laser pulses, Chapter 9 in: *Three-Dimensional Laser Microfabrication: Principles and Applications*, H. Misawa and S. Juodkazis, Eds. Wiley-VCH, (2006) (199-238).

[77] A. Pukhov and J. Meyer-ter-Vehn, *Appl. Phys.*, 2002, B74, 355.

In: Encyclopedia of Nanotechnology
Editor: Elwood D. Carlson

ISBN: 978-1-60692-079-4

Chapter 35

Carbon Nanoparticles as Substrates for Cell Adhesion and Growth

Lucie Bacakova[1,*], Lubica Grausova[1], Marta Vandrovcova[1], Jiri Vacik[2,†], Aneta Frazcek[3,‡], Stanislaw Blazewicz[3], Alexander Kromka[4,§], Bohuslav Rezek[4], Milan Vanecek[4], Milos Nesladek[5,||], Vaclav Svorcik[6,#], Vladimir Vorlicek[7&] and Milan Kopecek[1]

[1]Institute of Physiology, Academy of Sciences of the Czech Republic, Videnska 1083, 142 20 Prague 4-Krc, Czech Republic
[2]Nuclear Physics Institute, Academy of Sciences of the Czech Republic, 250 68 Rez near Prague, Czech Republic
[3]AGH University of Science and Technology, Faculty of Materials Science and Ceramics, Department of Biomaterials, Al. Mickiewicza 30, 30-059, Cracow, Poland
[4]Institute of Physics, Academy of Sciences of the Czech Republic, Cukrovarnicka 10, 162 53 Prague 6, Czech Republic
[5]CEA-LIST, Centre d'Etudes Saclay, Bat. 451, p. 84, 91191 Gif Sur Yvette, France
[6]Department of Solid State Engineering, Institute of Chemical Technology, Technicka 5, 166 28 Prague 6, Czech Republic
[7]Institute of Physics, Academy of Sciences of the Czech Republic, Na Slovance 2, 182 21 Prague 6, Czech Republic

---

[*] E-mail: lucy@biomed.cas.cz; grausova@biomed.cas.cz; vandrovcova@biomed.cas.cz; kopecek@biomed.cas.cz
[†] E-mail: vacik@ujf.cas.cz
[‡] E-mail: afraczek@op.pl; blazew@agh.edu.pl
[§] E-mail: kromka@fzu.cz; rezek@fzu.cz; vanecek@fzu.cz
[||] E-mail: Milos.NESLADEK@cea.fr
[#] E-mail: vaclav.svorcik@vscht.cz
[&] E-mail: vorlicek@fzu.cz

ABSTRACT

Carbon nanoparticles, such as fullerenes, nanotubes and nanodiamonds, have been considered as promising building blocks in nanotechnology for broadscale nanostuctured devices and materials, e.g., microchips, nanorobots and biosensors, carriers for controlled drug and gene delivery or tracers for novel imaging technologies. Soon after macroscopic production was established, fullerenes, as donors and acceptors of electrons, started to be targeted for use in photodynamic therapy against tumor and microbial cells and also for quenching oxygen radicals. It has been discovered that nanodiamonds can also act as antioxidant and anti-inflammatory agents.

The interaction of carbon nanoparticles with cells and tissues has been investigated mainly using suspensions of these particles in cell culture media or other fluids. Relatively little is known about the influence of layers consisting of carbon nanoparticles on cell-substrate adhesion. Carbon nanoparticles can be advantageously used for surface modification of various artificial materials developed for the construction of tissue replacements and other body implants. In the form of films deposited on the surface of implants, carbon nanoparticles can not only improve the mechanical and other physical properties of a body implant, but can also enhance the attractiveness of the implant for cell colonization. The latter effect is probably mediated by the surface nanostructure of the films, which, at least to a certain degree, mimics the architecture of physiological cell adhesion substrates, such as extracellular matrix molecules. On nanostructured substrates, the cell adhesion-mediating molecules are adsorbed in the appropriate spectrum, amount and spatial conformation that make specific sites on these molecules (e.g. amino acid sequences like RGD) accessible to cell adhesion receptors. In addition, the nanostructured surfaces have been reported to enhance the adsorption of vitronectin, preferred by osteoblasts for their adhesion. The electrical conductivity of carbon nanotube films has also had beneficial effects on the growth and maturation of osteoblasts. Thus, the nanoparticles deposited on the bone-anchoring parts of bone, joint or dental replacements can improve the integration of these devices with the surrounding bone tissue. In addition, carbon nanoparticles admixed into polymers, designed for the fabrication of three-dimensional scaffolds for bone tissue engineering, could decorate the walls of the pores in these materials, and thus promote the ingrowth of bone cells. The interaction of cells with substrates modified with carbon nanoparticles could be further intensified by functionalizing these particles with various chemical functional groups or biomolecules, including KRSR-containing adhesion oligopeptides, recognized preferentially by osteoblasts.

This interdisciplinary review, based mainly on our own results, deals with the interaction of bone-derived cells with glass, polymeric and silicon substrates modified with fullerenes C_{60}, nanotubes, nanocrystalline diamond as well as other carbon allotropes, such as amorphous hydrogenated carbon or pyrolytic graphite. In general, these modifications have resulted in improved adhesion, growth, viability and osteogenic differentiation of these cells. The use of carbon and/or metal-carbon composite nanoparticles for the construction of patterned surfaces for regionally-selective adhesion and directed growth of cells is also discussed in this work.

1. INTRODUCTION

Carbon nanoparticles, such as fullerenes, nanotubes and nanodiamonds, are promising building blocks for the construction of novel nanomaterials for emerging industrial technologies, such as molecular electronics, advanced optics, and storage of hydrogen as a potential source of energy. In addition, they are considered as promising materials for a wide range of biomedical applications, including photodynamic therapy against tumors and infectious agents, quenching oxygen radicals, construction of microchips, nanorobots and biosensors, advanced imaging technologies, controlled drug or gene delivery, and also in the simulation of cellular components such as membrane pores or ion channels.

Fullerenes are spheroidal molecules made exclusively of carbon atoms (e.g., C_{60}, C_{70}), present in both terrestrial and extraterrestrial materials [1]. From these materials, fullerenes can be purified by chromatographic methods, selective complexation and fractional crystallization. For large-scale industrial production, various more efficient methods have been developed, such as filtration of a fullerene extract through a thin layer of activated carbon [2]. Fullerenes were discovered, prepared and systemized by Kroto *et al.* [3, 4]. They exhibit a wide range of unique physical and chemical properties, e.g., extreme stability, ability to withstand high temperatures and pressures, a high proclivity to react with other species while maintaining their spherical geometry, the ability to entrap other smaller species such as helium, at the same time not reacting with the fullerene molecule. Potential technical applications for fullerenes include superconductors, lubricants, optical devices, chemical sensors, photovoltaics, polymer additives, polymer electronics such as Organic Field Effect Transistors (OFETS). Due to their high reactivity, fullerenes can serve as catalysts. They also represent an advantageous medium for hydrogen storage, because almost every carbon atom in C_{60} can absorb a hydrogen atom without disrupting the buckyball structure, which could lead to applications in fuel cells. Fullerenes have been also used as precursors to produce diamond films.

In addition, fullerene molecules display a diverse range of biological activity. Their unique hollow cage-like shape and structural analogy with clathrin-coated vesicles in cells support the idea of the potential use of fullerenes as drug or gene delivery agents [5-8]. Fullerenes are able to accept and release electrons. Thus, when irradiated with ultraviolet or visible light, they can convert molecular oxygen into highly reactive singlet (i.e., atomic) oxygen [8, 9]. Thus, they have the potential to inflict photodynamic damage on biological systems, including damage to cellular membranes, inhibition of various enzymes or DNA cleavage [10-13]. This harmful effect can be advantageously exploited for photodynamic therapy against tumors, viruses and bacteria resistant to multiple drugs [14, 15]. On the other hand, C_{60} is considered to be the world's most efficient radical scavenger. This is due to the relatively large number of conjugated double bonds in the fullerene molecule, which can be attacked by radical species. Thus, fullerenes would be suitable for applications in quenching oxygen radicals in medicine, as well as in cosmetics [9, 16-22].

In their pristine unmodified state, fullerenes are highly hydrophobic and water-insoluble. On the other hand, they are relatively highly reactive, which enables them to be structurally modified. Fullerenes can form complexes with other atoms and molecules, e.g. metals, porphyrins, nucleic acids, proteins, as well as other carbon nanoparticles, e.g., nanotubes [6, 7, 23-26]. In addition, fullerenes can be functionalized with various chemical groups, e.g., the hydroxyl, aldehydic, carbonyl, carboxyl, ester or amine groups, as well as amino acids and peptides. This usually renders them soluble in water and intensifies their interaction with biological systems [10, 11, 13, 27, 28]. For example, derivatized fullerenes decrease neuronal degeneration and death, protect the brain against ischemia [19, 20], attenuate ischemia-reperfusion-induced lung injury [22, 29], interact with receptors and inhibit various enzymes, such as HIV protease, HIV-reverse transcriptase, enzymes synthesizing cholesterol, redox enzymes or nitric oxide synthase [8, 17, 30]. Fullerenes C_{60} derivatized with monomalonic acid showed an inhibitory effect on nitric oxide-dependent relaxation of rabbit aortic smooth muscle [16].

Carbon nanotubes were discovered, synthesized and systemized by Iijima and Ichihashi [31-33]. These tubular structures are formed by a single cylindrically-shaped graphene sheet (single-wall carbon nanotubes, SWCNT or SWNT) or several graphene sheets arranged concentrically (multi-wall carbon nanotubes, MWCNT or MWNT). Their diameter is on the nanometer scale (e.g., from 0.4 nm to 2-3 nm in single-walled nanotubes), but their length can reach several micrometers. Related structures, with differences in shape but with a similar architectural principle, known as carbon nanorings, nanohorns, graphene nanoribbons or nanowires, have also been observed [23, 34, 35].

Carbon nanotubes can be prepared by arc-discharge and related methods [36], by laser ablation, laser evaporation or a laser beam pulse [37] or by chemical vapor deposition (CVD) methods [38]. The preparation of nanotubes usually requires the presence of metal catalysts, e.g., Fe, Co, Ni or Y, which can have adverse effects on cells and tissues [39-42]. In addition, in their pristine unmodified state, carbon nanotubes are hydrophobic, and therefore they are prone to form agglomerates in a water environment. To disperse these aggregates, various organic solvents or surfactants, such as Triton X and sodium dodecylbenzene sulfonate have been used [38, 43, 44], but these can alter biomolecules and damage cells. Thus, a more appropriate approach seems to be derivatization of nanotubes, especially with oxygen-containing chemical functional groups. Treatment of carbon nanotubes with acids (usually nitric and sulphuric, also hydrochloric or hydrofluoric acid), which attack the reactive ends of the carbon nanotubes, resulted in the creation of oxygen-containing groups, e.g. carboxylic. At the same time, this treatment purified the nanotubes from metal catalysts [36, 42, 45, 46]. However, Bottini *et al.* [47] and Magrez *et al.* [48] reported increased toxicity of nanotubes after oxidation in acids. Oxidized nanotubes were shorter and straighter than their pristine counterparts, thus they could penetrate more easily inside the cells. In addition, they were better dispersed in an aqueous solution, and therefore reached higher concentrations of free nanotubes at similar weight per volume [47].

Carbon nanotubes have unique mechanical, physical and chemical properties in comparison with conventional carbon fibers [49]. Due to their sp^2 bonds, the tensile strength of SWCNT is about one hundred times higher than that of steel, while their density (i.e., specific weight) is about six times lower [50, 33]. The excellent mechanical properties of carbon nanotubes could be utilized in hard tissue surgery, e.g., to reinforce bone implants, particularly scaffolds for bone tissue engineering made of synthetic polymers or chitosan [51, 52]. In addition, nanotubes can resemble nanofibres of collagen and other extracellular proteins of the bone, and together with other carbon nanoparticles, they can also resemble hydroxyapatite and other inorganic crystals in the bone [53, 54]. In addition, carbon nanotubes can be used effectively as substrates for growth and electrical stimulation of osteoblasts, which has been reported to promote their proliferation, differentiation, production of mineralized bone matrix and thus healing of the damaged bone [55-57]. Carbon nanotubes in the form of layers or as a component of 3D scaffolds can also be used as substrates for attachment, growth and electrical stimulation of neuronal cells [58]. For example, layers of multiwalled carbon nanotubes functionalized with 4-hydroxynonenal were successfully used as substrates for growing embryonic rat brain neurons *in vitro* [59].

Similarly as fullerenes, carbon nanotubes are relatively highly reactive, able to be functionalized, to form complexes with various atoms and molecules, and to emit photoluminescence. Due to these properties, both fullerenes and carbon nanotubes can be used in a wide range of advanced imaging, diagnostic and therapeutic strategies, such as transmission electron microscopy of DNA [60], labeling and tracer studies [1, 61-63], magnetic resonance [25, 26, 64], radiotherapy of tumors, including advanced methods of boron neutron capture therapy [65, 66], controlled drug and gene delivery [5-8, 67] and biosensor technology [23, 68-70]. The elongated shape of carbon nanotubes, resembling "tiny drinking straws", predisposes them to serve as tips for ultra-high-resolution atomic-force microscopy, and scanning tunneling microscopy [33, 71, 72]. In addition, carbon nanotubes and related materials, such as carbon nanohorns and nanotube-derived carbon foam, are promising substrates for storage of hydrogen as a potential source of energy (34, 73-75], and also for storage of methane, a key technology for producing natural gas vehicles [76].

Carbon nanotubes also represent a novel class of antimicrobial agents [77], ion (e.g., K^+) channel blockers [78], artificial ion and water channels and components of nanoporous membranes [79-81].

On the other hand, the reactivity of nanotubes, their ability to emit electrons from their ends and to generate free radicals, can induce oxidative stress, inflammation and alterations of various biomolecules and cellular structures, such as genes, transcription factors, enzymes, extracellular matrix molecules, mitochondria and cell membranes [40, 82-85], which should be taken into account in the design of nanotube-based materials for biomedical applications.

Nanocrystalline diamond (NCD) has also attracted considerable attention as a promising material for advanced biomedical applications. In the form of a powder, NCD has been prepared by detonation synthesis from a mixture of trinitrotoluene and hexogene [86]. Thin NCD films can also be deposited using microwave plasma-enhanced chemical vapor deposition (MW PECVD) from methane or fullerene C_{60} precursors [86, 87]. Similarly as fullerenes and nanotubes, NCD can also be relatively easily functionalized with various chemical groups and biomolecules (e.g., nucleic acids, enzymes, antibodies and other proteins) with strong covalent bonding [88-90]. At the same time, in comparison with fullerenes and nanotubes, NCDs are practically non-toxic and thus they are the most advantageous among all carbon nanoparticles for biomedical applications [88, 91]. In the form of a powder, NCD could serve as a carrier for controlled drug and gene delivery, for advanced imaging technologies or as an antioxidant and anti-inflammatory agent [91-93]. NCD films exhibit a wide range of unique physicochemical properties, such as mechanical hardness, chemical and thermal resistance, excellent optical transparency and controllable electrical properties (e.g., insulator/semiconductor/metal-like). Thus, biocompatible NCD films could markedly improve the mechanical and other physical properties of medical implants, particularly those designed for hard tissue surgery. Nanocrystalline diamond has already been used for coating the head and cup of artificial joint replacements, e.g., prostheses of temporomandibular joints [94]. In addition, NCD films could be applied in the bone-anchoring stems of articular prostheses or other permanent bone implants in order to improve their integration with the surrounding bone tissue. In studies *in vitro*, NCD provided excellent substrates for adhesion, growth, metabolic activity and maturation of human osteoblast-like MG 63 cells, human SAOS-2 osteoblasts, mouse MC3T3 osteoblasts, as well as other non-osteogenic cell types, such as human cervical carcinoma HeLa cells, rat pheochromocytoma PC12 cells or bovine pulmonary artery endothelial CPAE cells [57, 88, 95-97]. Type IIa diamond, microcontact-printed with laminin, supported adhesion of mouse cortical neurons and neurite outgrowth [98]. In studies *in vivo*, diamond layers, deposited by an MW CVD method on TiAl6V4 probes implanted into a rabbit femur, showed very high bonding strength to the metal base and also to the surrounding bone tissue, without any problems with corrosion [99]. Last but not least, similarly as fullerenes and nanotubes, NCDs are promising building blocks for constructing biomarkers, microchips, nanorobots or biosensors [88-91, 100].

Layers of *other carbon allotropes*, e.g., a-C:H, graphite or hydrocarbon on pristine and modified polymers, are also characterized by nanostructure (i.e., the presence of particles less than 100 nm), in addition to their atomic structure. These layers are represented by carbon structures formed on the surface of synthetic polymers after irradiation, e.g., with plasma, by an ion beam or laser beam, or after modification of the material surface by magnetron sputtering, vacuum evaporation or chemical vapor deposition (CVD). These carbon allotropes could be used for surface modification of various materials designed for the construction of body implants or tissue engineering, e.g., synthetic polymers, metals or carbon-based composites [101-103]. These allotropes can be enriched with metals in order to improve their mechanical properties and to enhance their bioactivity [103-105].

Among all these exciting findings, we have concentrated on the influence of carbon nanoparticles, particularly when arranged into layers and used for biomaterial coating, on cell-substrate adhesion, subsequent growth, differentiation and viability of cells, especially bone-forming cells. A very interesting finding has been reported that nanostructured surfaces can promote preferential adhesion and growth of osteoblasts over other "competitive" cell types, including fibroblasts, and thus they can prevent fibrous encapsulation and loosening of bone implants. It is considered that the underlying mechanism is higher adsorption to the nanostructured surfaces of vitronectin, an extracellular matrix protein preferred by osteoblasts. Therefore, it can be expected that carbon nanoparticles may serve as novel building blocks for creating artificial bioinspired nanostructured surfaces for bone tissue engineering [53, 54, 106-111]. The preferential adhesion of osteoblasts could be further enhanced by the functionalization of carbon nanoparticles by the amino acid sequence Lys-Arg-Ser-Arg (KRSR), a ligand for adhesion receptors on osteoblasts [112-115]. In addition, carbon nanoparticles could be admixed to synthetic polymers, chitosan and other matrices designed for the fabrication of three-dimensional porous scaffolds for tissue engineering. The nanoparticles would form prominences and create a nanopattern on the pore walls, and thus promote the ingrowth of cells inside the scaffolds [109, 116].

Therefore, in our interdisciplinary studies, the following types of materials modified with carbon nanoparticles were investigated:

(1) Continuous layers of fullerenes C_{60} deposited on microscopic glass coverslips or composites with a carbon matrix reinforced with carbon fibers, i.e., a material applicable in hard tissue surgery.
(2) Micropatterned layers created by deposition of fullerenes C_{60} through metallic grids on microscopic glass coverslips.
(3) Hybrid layers created by co-deposition of fullerenes C_{60} and metals on microscopic glass coverslips.

(4) Polymeric materials (i.e., a terpolymer of polytetrafluoroethylene, polyvinyldi fluoride and polypropylene, or polysulphone) mixed with single- or multi-walled carbon nanotubes.
(5) Nanostructured or hierarchically micro- and nanostructured nanocrystalline diamond layers deposited on silicon substrates.
(6) Layers of amorphous hydrogenated carbon (a-C:H), pyrolytic carbon, pyrolytic graphite as well as nanocomposite metal-hydrocarbon plasma polymer films.

The adhesion, growth, viability, metabolic activity and maturation of human osteoblast-like MG 63 cells were then studied in cultures on these materials. The cell behavior correlated with the physical and chemical properties of these materials, and also with results obtained by other authors.

2. Cell Growth on Fullerene C_{60} Layers

2.1. Continuous C_{60} Layers

In our experiments, fullerenes C_{60} (purity 99.5 %, SES Research, U.S.A.) were deposited on to microscopic glass coverslips (Menzel Glaser, Germany; diameter 12 mm) by evaporation of C_{60} in the Univex-300 vacuum system (Leybold, Germany) in the following conditions: room temperature of the substrates, C_{60} deposition rate $\leqslant$ 10 Å/s, temperature of C_{60} evaporation in the Knudsen cells about 450° C, time of deposition up to 50 minutes (1 A). The thickness was measured by atomic force microscopy (AFM). A scratch was made in the layer and its profile was measured in contact mode. The thickness of the layers increased proportionally to the temperature in the Knudsen cell and the time of deposition. We prepared two continuous films of different thicknesses, i.e., a thin layer of 505 ± 43 nm and a thick layer of 1090 ± 8 nm. Optical microscopy revealed that the fullerene layers were brownish in color (Figure 1 B), and the color intensity increased with layer thickness. Raman spectra of the deposited C_{60} films were measured in the back-scattering geometry at room temperature using a Ramascope 1000 Raman microscope (Renishaw, UK) with a 514.5 nm excitation wavelength Ar-ion laser. Immediately after deposition, the Raman spectra showed that the prepared fullerene films were of high quality, confirmed by a high peak $A_g(2)$ at wavenumber 1468 cm^{-1}, low peaks $H_g(7)$ and $H_g(8)$ and the absence of D (disorder, ~ 1350 cm^{-1}) and G (graphitic, ~ 1600 cm^{-1}) bands, which are signs of fragmentation and graphitization of C_{60}, respectively (Figure 2). The surface wettability of the films was estimated from the contact angle measured by a static method in a material-water droplet system using a reflection goniometer. The fullerene C_{60} films were relatively hydrophobic, having contact angles 100.6 ± 6.8° for thin and 96.7 ± 1.7° for thick continuous layers, respectively [117, 118].

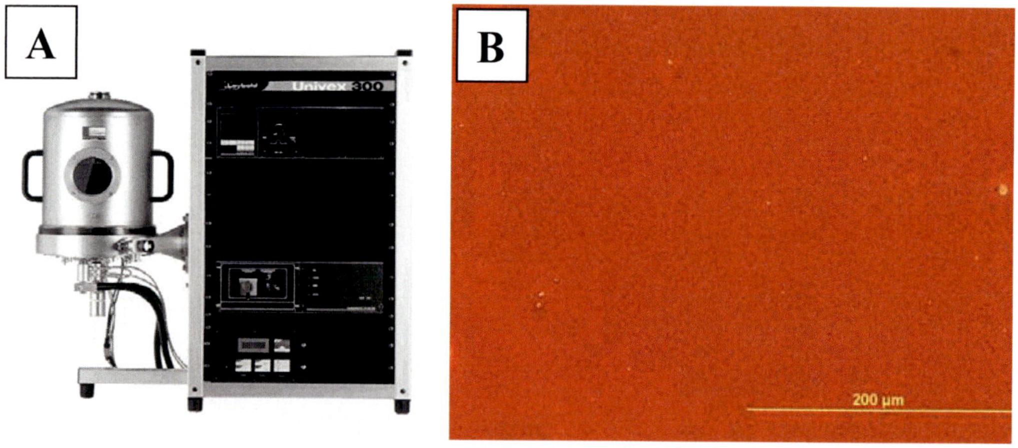

Figure 1. A: Univex-300 vacuum system (Leybold, Germany); B: continuous fullerene C_{60} layer (thickness of 1090 ± 8 nm) deposited on to a microscopic glass coverslip in this system [117, 118].

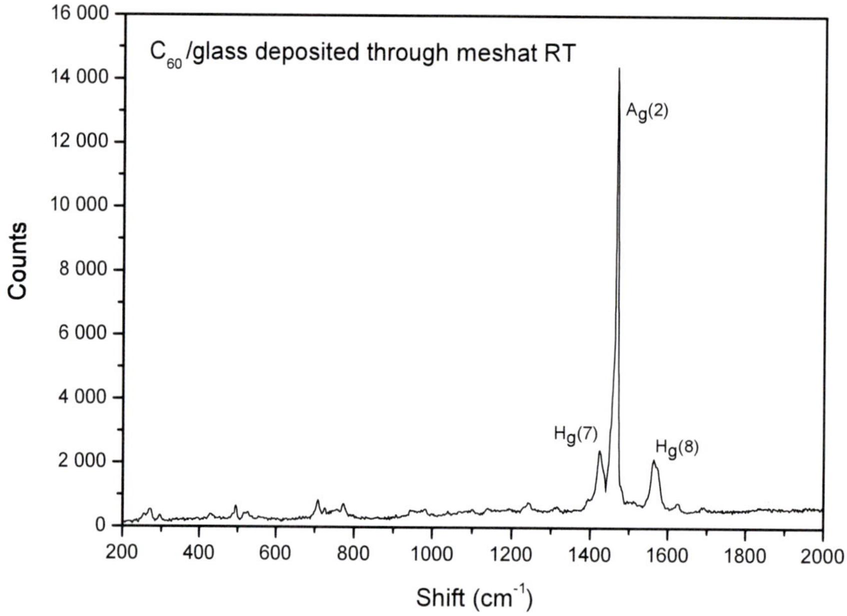

Figure 2. Raman analysis of a fullerene C_{60} layer immediately after deposition on glass coverslips. The quality of the fullerene layer is confirmed by a high peak $A_g(2)$ at wavenumber 1468 cm^{-1}, low peaks $H_g(7)$ and $H_g(8)$ and absence of D (disorder, ~ 1350 cm^{-1}) and G (graphitic, ~ 1600 cm^{-1}) bands, which are signs of fragmentation and graphitization of C_{60}, respectively [117, 118].

To cultivate the cells, the fullerene-coated glass coverslips were sterilized with 70% ethanol for 1 hour, inserted into 24-well polystyrene multidishes (TPP, Switzerland; diameter 15 mm) and seeded with human osteoblast-like MG 63 cells (European Collection of Cell Cultures, Salisbury, UK). Each dish contained 5,000 cells (i.e., about 2,830 cells/cm^2) and 1.5 ml of Dulbecco's modified Eagle's Minimum Essential Medium supplemented with 10% fetal bovine serum and gentamicin (40 μg/ml). The cells were cultured for 1, 3 and 5 or 7 days at 37° C in a humidified air atmosphere containing 5% of CO_2. On day 1 after seeding, the cells on both continuous thin and continuous thick fullerene layers adhered in similar numbers (3,420 ± 420 cells/cm^2 and 2,880 ± 440 cells/cm^2, respectively), which was comparable to the values found on standard cell culture substrates, represented by the tissue culture polystyrene dish (3,080 ± 290 cells/cm^2) and the microscopic glass coverslip (2,560 ± 310 cells/cm^2). In addition, the cells on the fullerene layers and on the control substrates were of similar morphology, i.e., mostly polygonal and spread over a similar area (Figure 3). From days 1 to 5 after seeding, the cells on both continuous C_{60} films, polystyrene and glass also proliferated with similar cell population doubling times ranging from 18.6 ± 0.3 h to 19.6 ± 0.5 hours, and on day 5 they reached similar cell population densities (from 91 600 ± 9500 cells/cm^2 to 112 700 ± 9800 cells/cm^2; Figure 4). In addition, as revealed by staining the adhered cells with calcein and ethidium homodimer using the LIVE/DEAD kit, the percentage of viable cells on days 1 to 7 (after seeding on all fullerene layers) ranged from 80 ± 10 % to 100 ± 15 %, and was similar to the values obtained on the polystyrene culture dishes (99 ± 24 % to 100 ± 15 %), and also on the microscopic glass coverslips (88 ± 21 % to 99 ± 26 %). The trypan-blue exclusion test (performed on the trypsinized cells while they were being counted in the ViCell Analyser) also showed similar proportions of viable cells on the fullerene layers (82 ± 11 % to 89 ± 13 %), as well as on standard cell culture surfaces (80 ± 13 % to 95 ± 13 %). The average viability of the cells tended to increase with time of cultivation.

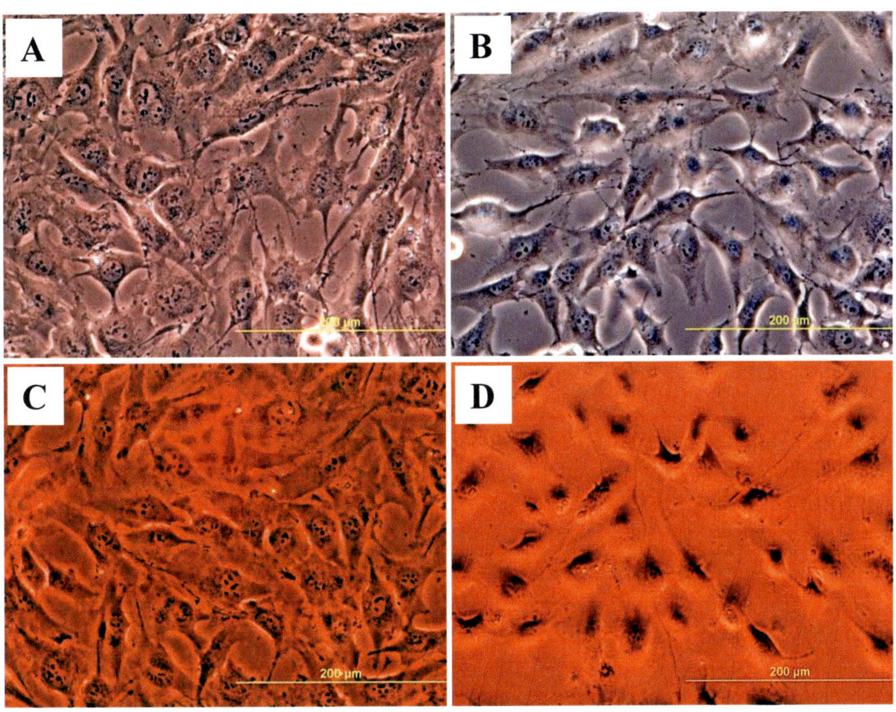

Figure 3. Morphology of human osteoblast-like MG 63 cells in 7-day-old cultures on a polystyrene dish (A) a microscopic glass coverslip (B), thin continuous (C) and thick continuous (D) fullerene C_{60} layers. Native cultures, Olympus IX 50 microscope, DP 70 digital camera, obj. 20x, bar = 200 µm [117].

Therefore, in our studies, fullerene C_{60} layers acted as good substrates for adhesion, growth and maturation of human osteoblast-like MG 63 cells, comparable to standard cell culture materials, such as tissue culture polystyrene and microscopic glass coverslips. These results are consistent with earlier findings on the supportive effects of fullerene C_{60} layers (formed on polystyrene culture dishes by the evaporation of methanol from colloidal C_{60}-methanol suspensions) on the adhesion, proliferation and assembly of an actin cytoskeleton in several lines of normal and malignant breast epithelial cells [119]. The latter cells, like those in our study, were well stained with calcein using a live-dead assay kit and only sparsely stained with ethidium homodimer, which indicated a lack of cytotoxic damage to the cell enzymatic (i.e., esterase) activity and membrane integrity. Grafting C_{60} molecules on to polyurethane surfaces also increased the number of adhered platelets and enhanced their activation [120].

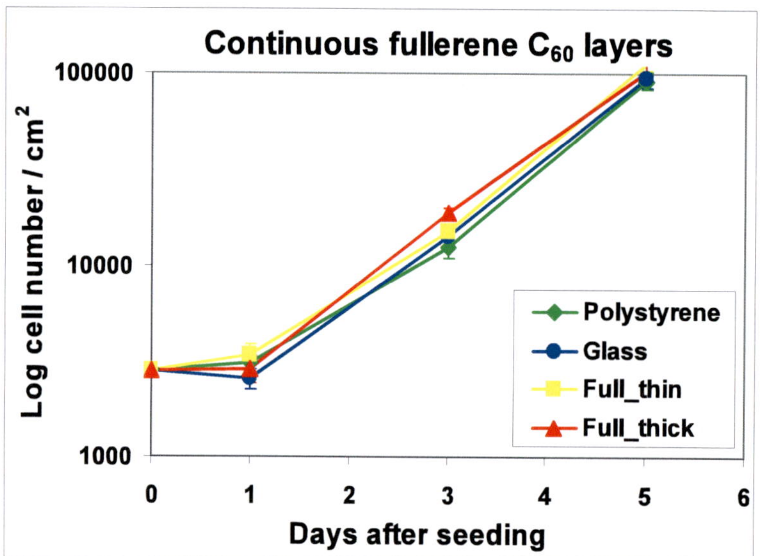

Figure 4. Growth curves of human osteoblast-like MG 63 cells in cultures on polystyrene dishes, microscopic glass coverslips or continuous thin and thick fullerene C_{60} layers. Mean ± S.E.M. (Standard Error of Mean) from 9 to 30 measurements. No statistically significant differences were found by ANOVA, Student-Newman-Keuls Method [117].

The supportive effects of fullerene C_{60} layers on cell colonization could be explained by their surface nanostructure, mimicking the nanoarchitecture of the natural extracellular matrix (ECM), basal lamina, as well as the cell membrane. The nanostructure of the cell adhesion substrate has been reported to enhance the adsorption of cell adhesion-mediating ECM molecules, such as fibronectin, collagen and especially vitronectin, provided by the serum of the culture medium or synthesized by the cells. The spatial conformation of the adsorbed molecules on our films may also be more appropriate for the accessibility of specific sites on these molecules by cell adhesion receptors, e.g., integrins [53, 54, 106-108, 110, 116, 121, 122]. As revealed by immunofluorescence staining of the MG 63 cells on our samples, β_1 integrins (i.e., receptors for collagen, fibronectin and vitronectin), and talin (an integrin-associated protein) were assembled in dot- or streak-like focal adhesion plaques, mainly visible on the cell periphery (Figure 5), and these processes were accompanied by the formation of a fine mesh-like β-actin cytoskeleton and by the presence of a considerable amount of osteopontin, a marker of osteogenic cell differentiation (Figure 6). All these events can be considered signs of active cell-substrate interaction, followed by signal transduction inside the cells. Moreover, β_1 integrins have been reported to enhance osteogenic cell differentiation, which is manifested by higher levels of osteocalcin, higher activity of alkaline phosphatase and a more pronounced response to 1,25-dihydroxyvitamin D_3 [111].

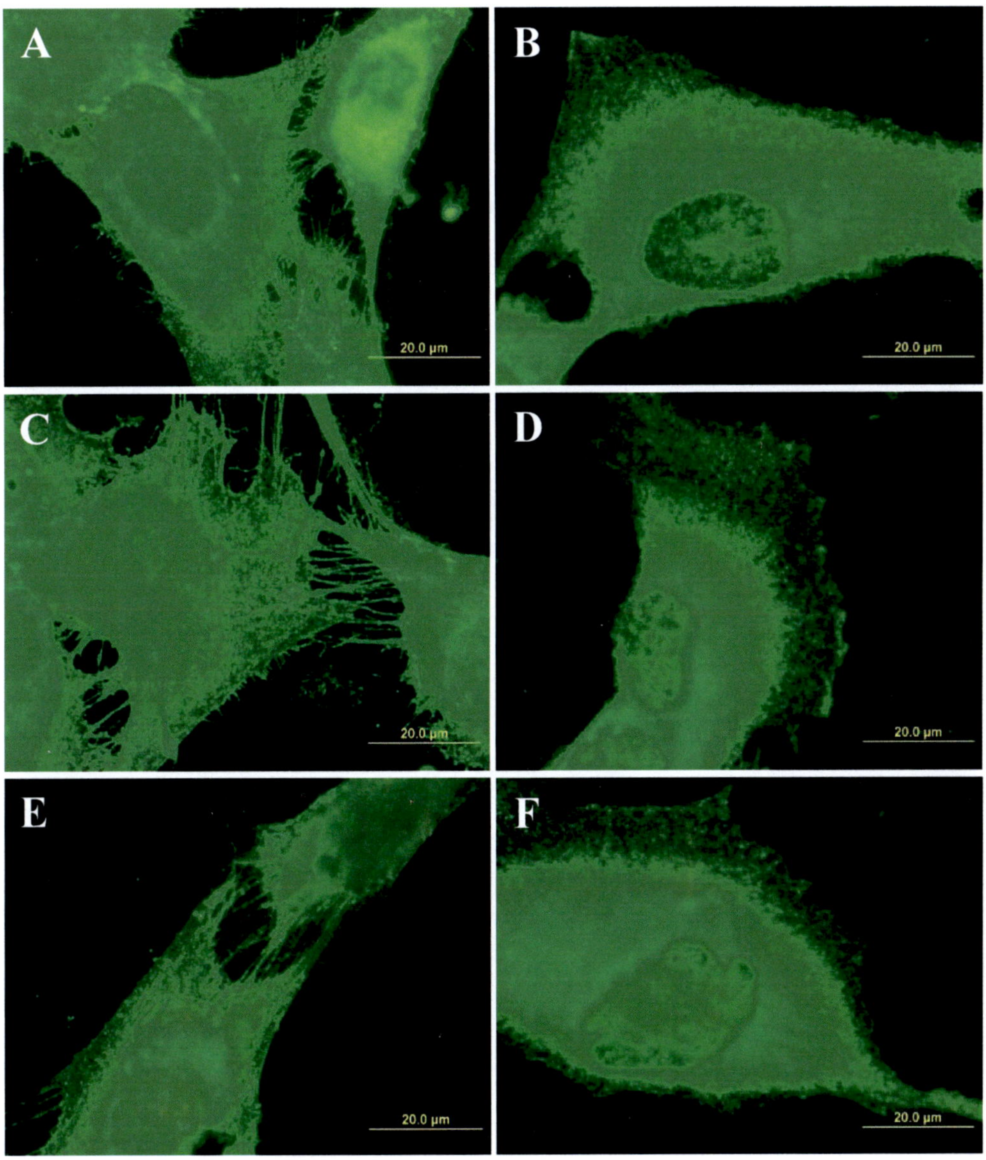

Figure 5. Immunofluorescence staining of β_1 integrins (A, C, E) and talin (B, D, F) in human osteoblast-like MG 63 cells on day 3 after seeding on microscopic glass coverslips (A, B) and thin (505 ± 43 nm; C, D) or thick (1090 ± 8 nm; E, F) continuous fullerene C_{60} layers. Olympus IX 50 epifluorescence microscope, DP 70 digital camera, obj. 100x, bar = 20 µm [117].

The proliferation and differentiation of chondrocytes (another cell type important in bone reconstruction) were also promoted by fullerenes. C_{60} added in cell culture enhanced the production of specific large proteoglycan ECM molecules, typical for cartilage, in rat embryonic limb bud cells in culture [123].

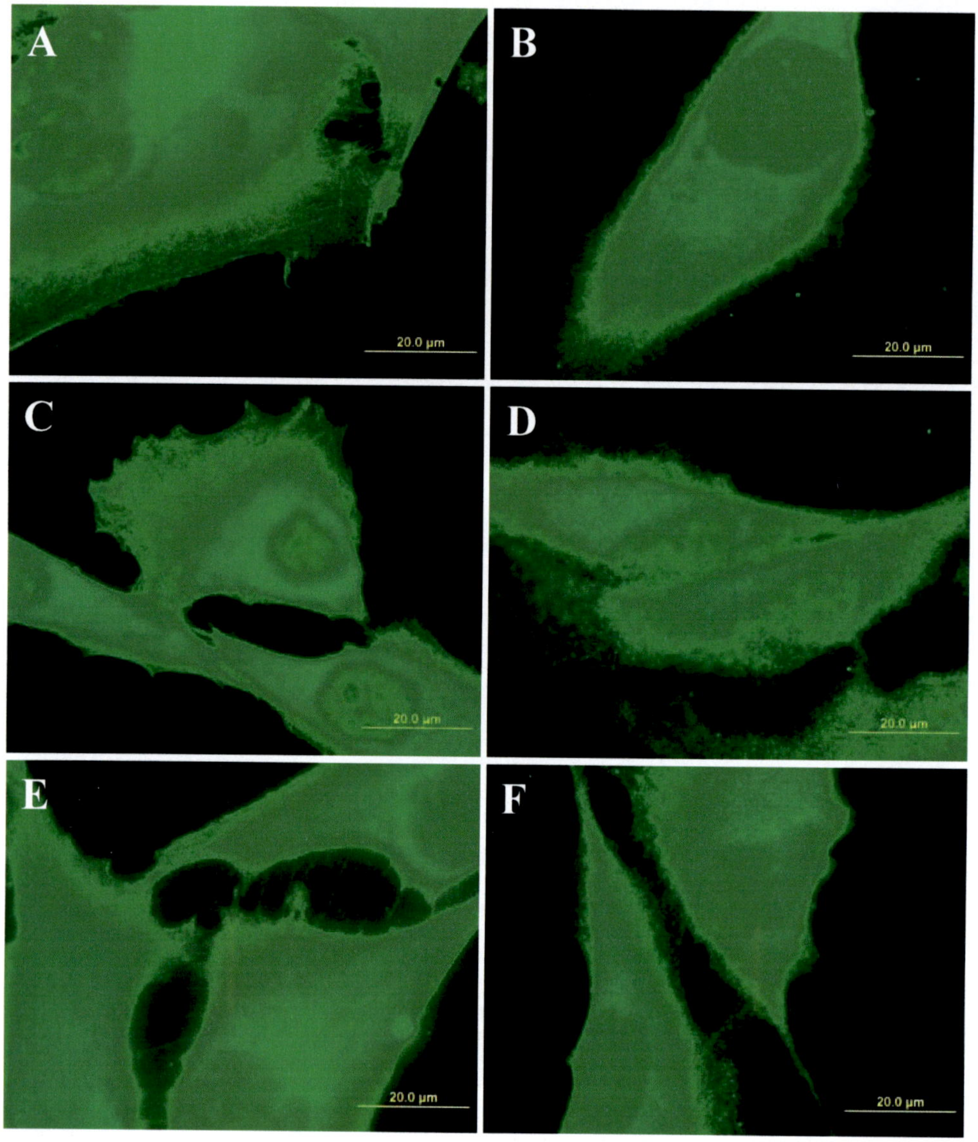

Figure 6. Immunofluorescence staining of β-actin (A, C, E) and osteopontin (B, D, F) in human osteoblast-like MG 63 cells on day 3 after seeding on microscopic glass coverslips (A, B) and thin (505 ± 43 nm; C, D) or thick (1090 ± 8 nm; E, F) continuous fullerene C_{60} layers. Olympus IX 50 epifluorescence microscope, DP 70 digital camera, obj. 100x, bar = 20 µm [117].

Similarly, C_3-fullero-*tris*-methanodicarboxyllic acid (a water-soluble fullerene derivative), supported the formation of focal adhesion plaques, an assembly of an actin cytoskeleton and cell spreading in human epidermoid carcinoma cells exposed to UV light, probably due to its antioxidative action [124]. The free radical scavenger activity of fullerenes and their derivatives (e.g., hydroxylated, carboxylated, butylated or sulphonated C_{60}) also protected keratinocytes from apoptosis induced by ultraviolet light B [27], minimized the oxidative damage of small bowel transplants [18], attenuated ischemia-reperfusion-induced lung injury [22], prevented focal cerebral ischemia and decreased neuronal degeneration and death [19, 20]. However, hydroxylated fullerene derivatives protected cells against oxidative stress only at lower concentrations (10-50 μM in cell culture media), while at higher concentrations (1 mM and 1.5 mM), they caused cell death [22]. Hydroxylated fullerenes usually induced cell death characterized by signs of apoptosis, such as increased caspase activity, DNA fragmentation and annexin V binding to the phosphatidylserine, exposed on the outer side of the cell membrane, without increased membrane permeability [28]. On the other hand, a fullerene derivative $C_{60}(OH)_{24}$ (concentrations from 1 to 100 μg/ml) did not seem to induce apoptosis but caused the accumulation of polyubiquitinated proteins and facilitated autophagic cell death in human umbilical vein endothelial cells [125]. Fullerene colloids in water, also called nano-C_{60}, i.e. fullerene aggregates that readily form when pristine C_{60} is added to water, mainly induced cell necrosis characterized by oxidative damage to the cell membrane (e.g., peroxidation of the membrane lipids) and loss of membrane integrity [10], though the genotoxicity of these agents has also been demonstrated on human lymphocytes [11]. The harmful effects of fullerenes C_{60}, usually potentiated by the light, especially ultraviolet light, could be utilized for antimicrobial therapy or photodynamic therapy against tumor cells [126, 127]. In addition, a nanocrystalline suspension of fullerenes C_{60} and C_{70} prepared by the solvent exchange method using tetrahydrofuran (THF/n$C_{60/70}$) augmented the cytotoxicity of the tumor necrosis factor [128, 129].

Continuous layers of fullerenes C_{60}, experimentally developed on glass, were also tested on composites with the carbon matrix reinforced with carbon fibers (i.e., carbon fiber-reinforced carbon composites, CFRC), which have been considered as promising materials applicable in hard tissue surgery [104, 130-133]. These composites were prepared at the Institute of Rock Structure and Mechanics, Acad. Sci. CR, Prague. Briefly, a commercially available woven fabric (made of Toray T 800 carbon fibers) was arranged in layers, infiltrated with a carbon matrix precursor (phenolic resin UMAFORM LE, Synpo Ltd., Pardubice, CR), pressed, cured, carbonized at 1000°C, and finally graphitized at 2200°C [104, 130-133]. Prior to the deposition of fullerenes, the CFRCs were ground using a metallographic paper of 4000 grade in order to decrease the material surface micro-scale roughness, which has been shown to be less appropriate for spreading and subsequent growth of bone-derived and also vascular smooth muscle cells [104, 130, 131]. This grinding lowered the surface roughness of CFRC about twice. As measured by a profilometer (Rank Taylor Hobson Ltd., England), the departures of the roughness profile from the mean line (i.e., R_a parameter) decreased from 6.5 ± 1.8 μm to 3.5 ± 0.6 μm, and the mean spacing of the adjacent local peaks (parameter S) lengthened from 38 ± 11 μm to 96 ± 49 μm. The fullerene coating did not significantly change this surface microroughness, but created a nanostructured pattern on the pre-existing microarchitecture of the CFRC surfaces. The fullerene C_{60} layers were deposited on the CFRC by evaporation of C_{60} in the Leybold Univex-300 vacuum system (room temperature of the substrates, deposition rate $\lesssim$ 1 Å/s, temperature of C_{60} evaporation in the Knudsen cells 450°C, time of deposition about 15 min, thickness of the fullerene layers < 100 nm). Similarly as in the continuous fullerene layers (deposited on microscopic glass coverslips), the Raman analysis confirmed again that the prepared fullerene films were of high quality and with no fragmentation or graphitization of C_{60} (Figure 7). One third of the CFRC samples (3 x 3 cm) were protected against fullerene coating by a mask in order to achieve regionally-selective growth of the fullerene layer. The other completely uncoated CFRC samples, as well as tissue culture polystyrene (TCPS) dishes were used as control materials [95].

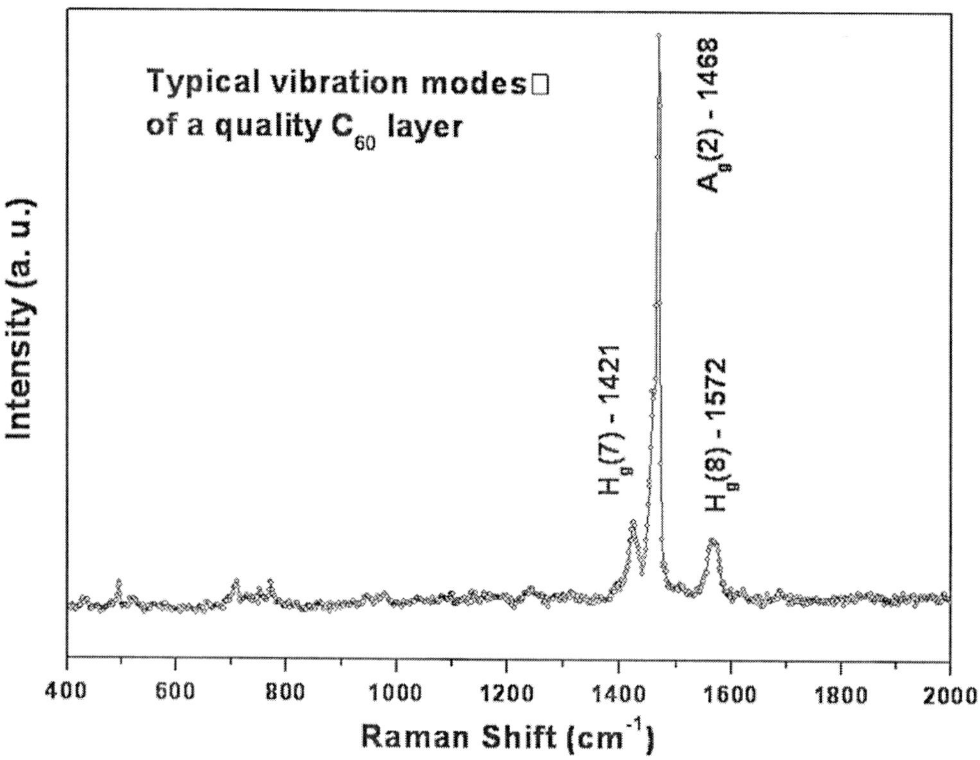

Figure 7. Raman analysis of a thin C_{60} layer deposited on a carbon-fibre reinforced composite (CFRC) substrate. The quality of the fullerene layer is confirmed by a high peak $A_g(2)$ at wavenumber 1468 cm^1, low peaks $H_g(7)$ and $H_g(8)$ and absence of D (disorder, ~ 1350 cm^{-1}) and G (graphitic, ~ 1600 cm^{-1}) bands, which are signs of fragmentation and graphitization of C_{60}, respectively [95].

The material was then seeded with human-osteoblast-like MG 63 cells. Unlike the fullerene layers on glass, the cell population density on day 2 after seeding (Figure 8 A) on C_{60}-coated CFRC was significantly lower (about 19,000 ± 700 cells/cm^2) than that on the control uncoated CFRC and polystyrene dishes (44,000 ± 4,000 cells/cm^2 and 67,000 ± 7,000 cells/cm^2, respectively). This finding could be attributed to a combination of at least two factors less favorable for cell adhesion, such as the remaining surface microroughness and relatively high hydrophobicity of the non-functionalized fullerenes (static water drop contact angle about 100°, see above). It is known that hydrophobic materials promote preferential adsorption of cell non-adhesive proteins from the serum of the culture media, such as albumin. In addition, the cell adhesion-mediating extracellular matrix proteins may be adsorbed in a relatively rigid state, thus their specific amino acid sequences were less accessible for cell adhesion receptors [121, 122]. On the other hand, the water drop contact angle, measured by a static method in a material-water droplet system using a reflection goniometer, was not able to reveal any hydrophobicity of the fullerene layer deposited on

CFRCs. The contact angle was non-measurable due to complete absorption of the water drop into the fullerene layer, which suggested a certain non-compactness or porosity of this layer on CFRC. The contact angle of the non-coated CFRC was 99.5 ± 1.0°.

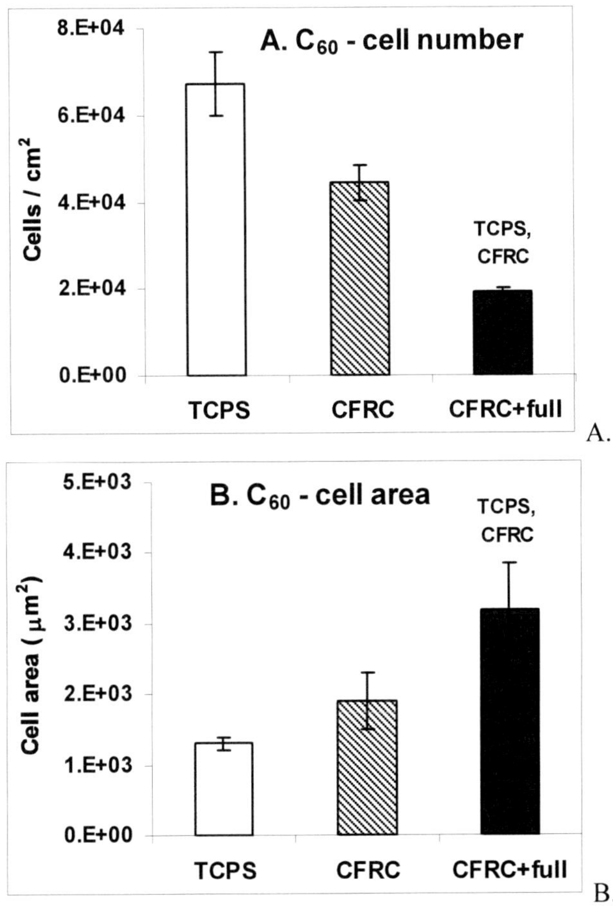

Figure 8. Population density (A) and adhesion area (B) of osteoblast-like MG 63 cells on day 2 after seeding on a tissue culture polystyrene dish (TCPS), carbon fibre-reinforced carbon composites (CFRC) and CFRC coated with a fullerene layer (CFRC+full). Mean ± S.E.M. (Standard Error of Mean) from 4-12 measurements, ANOVA, Student-Newman-Keuls method. Statistical significance: $^{TCPS, CFRC}$: $p \leq 0.05$ compared to the values on tissue culture polystyrene and pure CFRC [95].

The release of fullerenes into the culture media and their cytotoxic action seemed to be less probable in our experiments as an explanation of the lower number of cells adhered to C_{60}-coated CFRC. The cell attachment and spreading on the uncoated regions of the fullerene-modified CFRC, and also on the bottom of the polystyrene dishes containing fullerene-coated samples, were similar as in the control polystyrene dishes without fullerene samples (Figure 9). At the same time, the fullerene layer was resistant to mild wear, represented by swabbing with cotton, rinsing with liquids (water, phosphate-buffered saline, culture media) and exposure to cells and proteolytic enzymes (trypsin) used for cell harvesting. After these procedures and/or one-year storage at room temperature in a dark place, the Raman spectra did not change significantly. The peak $A_g(2)$ at wavenumber 1468

cm^{-1} still remained the highest, though some of the peaks $H_g(7)$ and $H_g(8)$ increased slightly, and additional small D and G bands appeared, which indicated some fragmentation, graphitization, polymerization or oxidation of C_{60} molecules.

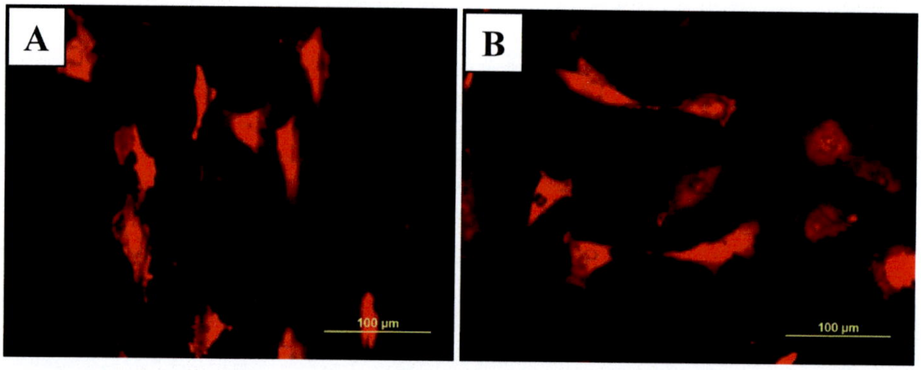

Figure 9. Human osteoblast-like MG 63 cells on day 1 after seeding on the uncoated region of CFRC partially coated by fullerenes (A) and the bottom of a polystyrene dish containing a fullerene-coated CFRC sample (B). Cells fixed with 70% ethanol and stained with propidium iodide. Olympus IX 50 microscope, DP 70 digital camera, obj. 20, bar=100 µm [95].

In addition, the fullerene-coated CFRC surfaces were stronger and less prone to release carbon particles, which is an important limitation to the potential biomedical use of CFRC [104, 130, 131]. Moreover, the spreading area of cells on the fullerene-coated samples amounted to 3,182 ± 670 μm^2, while on both control surfaces it was only 1,888 ± 400 and 1,300 ± 102 μm^2 (Figure 8 B). This could be explained by the low cell population density on the fullerene layer, which provided the cells with more space on which they could spread. On the other hand, the nanostructure of the fullerene layer might enhance the adsorption of vitronectin, i.e., an extracellular matrix protein promoting adhesion of osteoblasts [106-108]. The MG 63 cells on the fullerene-coated CFRC formed dot-like vinculin-containing focal adhesion plaques and a fine network of actin microfilaments (Figure 10), which is a sign of cell vitality and effective binding between cell adhesion receptors and extracellular matrix molecules adsorbed on the material surface [122].

2.2. Micropatterned C_{60} Layers

An interesting idea is that fullerenes might be used for constructing patterned surfaces for regionally-selective adhesion and directed growth of cells. Domains adhesive for cells on the material surface have been created using a wide range of physico-chemical and biochemical approaches. These approaches involve attachment of moderately hydrophilic molecules (e.g., acrylic acid, [134]) or attachment of various chemical functional groups, particularly those containing oxygen. The latter can also be created by irradiation of the material surface, especially the surface of synthetic polymers, by plasma, ions or ultraviolet light [121, 135-140]. In addition, the material surface can be functionalized by various biomolecules, particularly ligands for cell adhesion receptors, such as oligopeptides containing amino acid sequences recognized by all cell types (sequence RGD) or preferentially by a certain cell type (e.g., REDV by endothelial cells, VAPG by smooth muscle cells or KRSR recognized by osteoblasts) [112-114, 122, 141]. Regions repulsive to cells can be prepared using hydrophobic compounds, such as octadiene or polysiloxanes, or extremely hydrophilic molecules, e.g. polyethylene oxide [134, 141, 142].

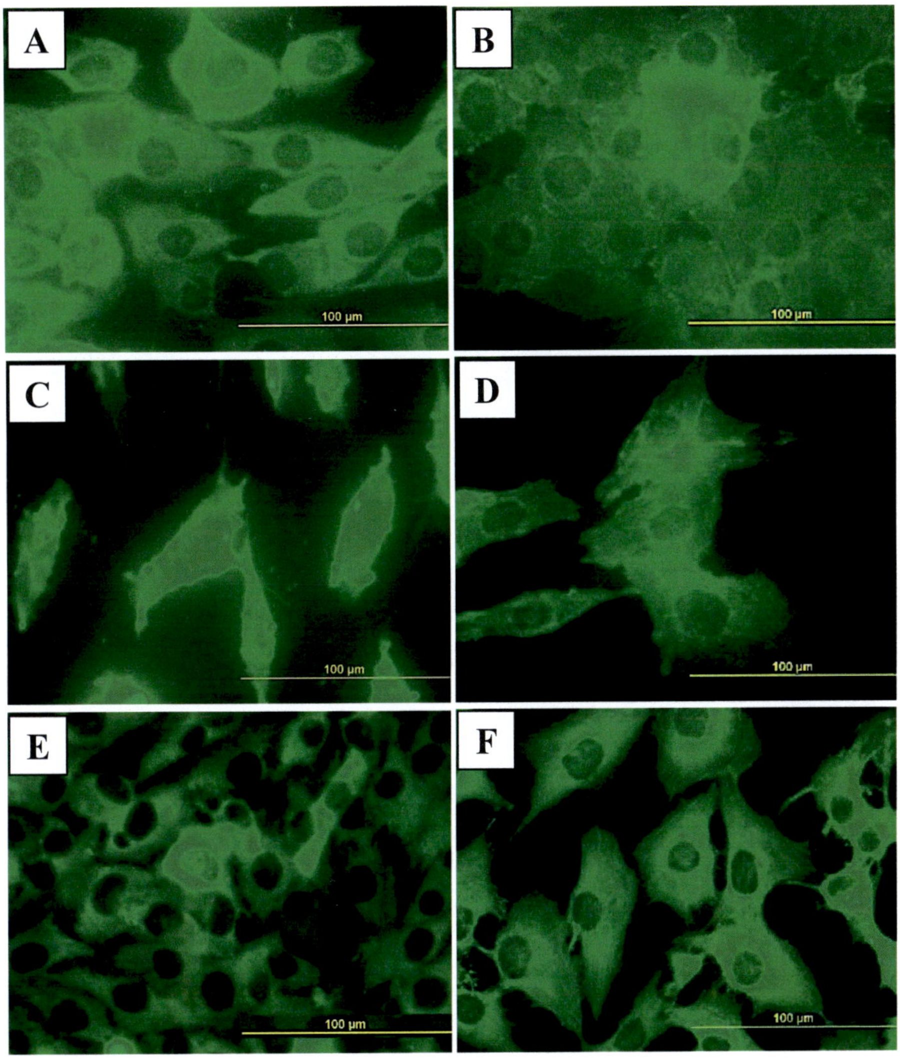

Figure 10. Immunofluorescence staining of vinculin (A, C, E) and β-actin (B, D, F) in osteoblast-like MG 63 cells on day 2 after seeding on CFRC (A, B), CFRC coated with a fullerene layer (C, D) and tissue culture polystyrene dish (E, F). Olympus IX 50 microscope, DP 70 digital camera, obj. 40, bar = 100 μm [95].

Patterned surfaces can also be created by changing the surface roughness and topography, i.e., by creating prominences and hollows of different size, shape and depth [116, 143]. In our experiments, micropatterned layers were prepared by deposition of fullerenes C_{60} on microscopic glass coverslips through contact metallic masks with rectangular holes with an average size of 128 ± 3 μm per 98 ± 8 μm (about 12 430 μm^2) and 50 μm in distance, or 470 x 440 μm (about 207,000 μm^2), distance 200 μm (Figure 11). The fullerenes formed bulge-like prominences below the openings of the metallic grid. Similarly as in the continuous fullerene C_{60} films, the thickness of these prominences increased proportionally to the temperature in the Knudsen cell and the time of deposition. As revealed by AFM, the thickness of the layers on sites underlying the openings of the grid was 128 ± 8 nm, 238 ± 3 nm, 484 ± 5 nm or 1043 ± 57 (data presented as mean ± standard deviation). Fullerene layers were also formed below the metallic part of the grid, where their thickness was either within the size of the standard deviations (the 1st, 2nd and 4th layer) or it amounted to 158 ± 5 nm (the 3rd layer). Therefore, in the 3rd layer, the effective height of the fullerene prominences was 326 ± 5 nm [117, 118].

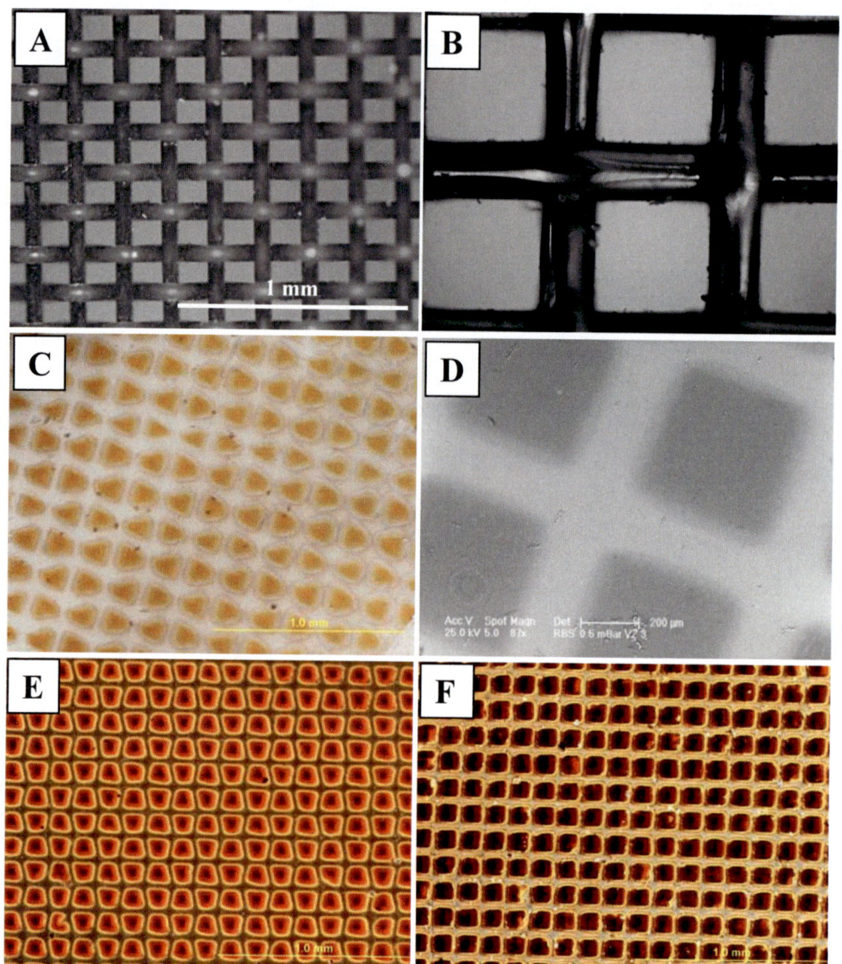

Figure 11. A, B: Metallic grids used for creating micropatterned fullerene C_{60} layers on microscopic glass coverslips. A: openings of 128 ± 3 μm per 98 ± 8 μm (about 12,500 μm^2) and 50 μm in distance, B: openings of 470 x 440 μm (about 207,000 μm^2), distance 200 μm. C-F: micropatterned layers created by the deposition of fullerenes C_{60} through a mesh with smaller (C, E, F) and larger (D) openings. The height of the fullerene protrusions was 128 ± 8 nm (C), 238 ± 3 nm (D), 484 ± 5 nm (E) and 1043 ± 57 nm (F). Olympus BX41 microscope (A, B), Olympus IX 50 microscope (C, E, F) and SEM-EDS scanning electron microscope (JOEL JSM-5600) (D). Objective 10x (A, B) or 4x (C, E, F); bar = 1mm (A, B, C, E, F) or 200 μm (D) [118].

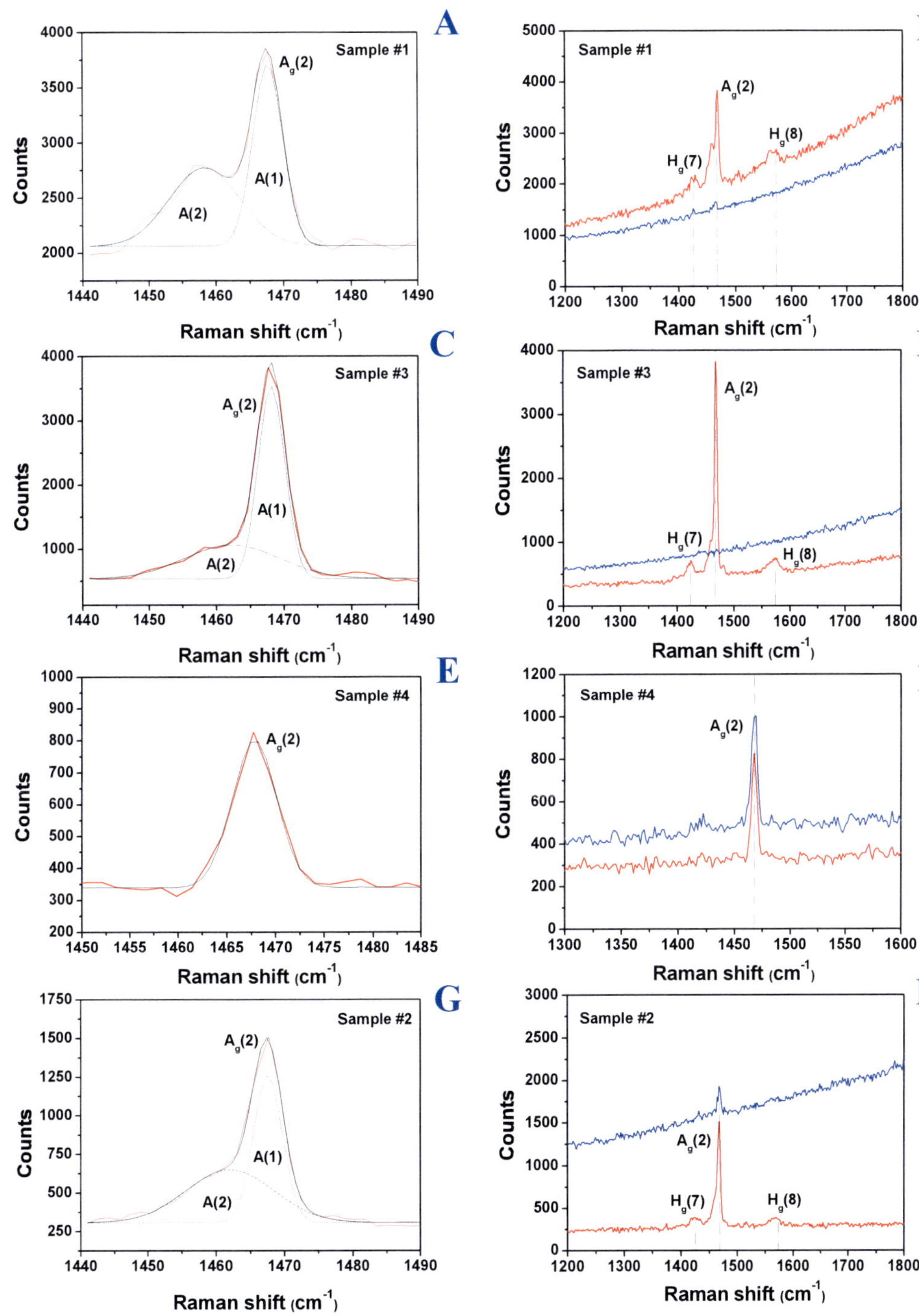

Figure 12. Raman analysis of microstructured C_{60} layers after 1-hour-sterilization with 70% ethanol. A, C, E, G: Gaussian peak fit analysis; B, D, F, H: Raman spectra below the openings (red line) or bars (blue line) of the metallic lattice. As revealed by AFM, the layer thickness below the openings and bars was 128 ± 8 and ≤ 8 nm, respectively, in the sample on B, 238 ± 3 nm and ≤ 3 nm on D, 484 ± 5 nm and 158 ± 5 nm on F, and 1043 ± 57 nm and ≤ 57 nm on H [117, 118].

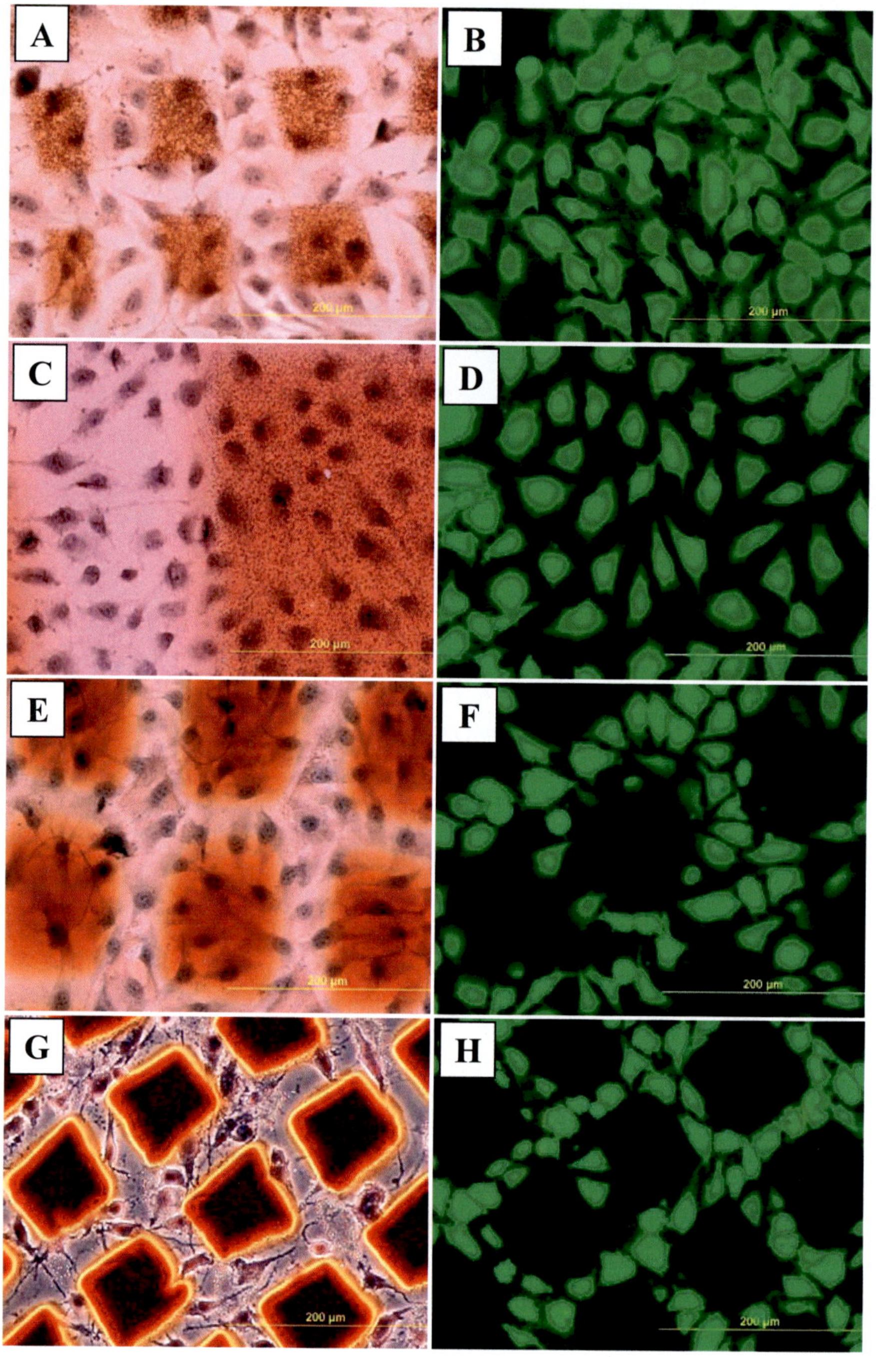

Figure 13. Human osteoblast-like MG 63 cells on day 7 after seeding on fullerene layers micropatterned with prominences 128 ± 8 nm in height (A, B), 238 ± 3 nm in height (C, D), 326 ± 5 nm in height (E, F) or 1043 ± 57 nm in height (G, H). Stained with hematoxylin and eosin (A, C, E, G) or LIVE/DEAD viability/cytotoxicity kit (B, D, F, H). Olympus IX 50 microscope, DP 70 digital camera, obj. 20x, bar = 200 μm [117, 118].

For cell culture, the fullerene-modified samples were sterilized by 70% ethanol in water (1 hour, at room temperature). Analysis of the vibration mode $A_g(2)$ on the Raman spectra showed that, after sterilization, some of the micropatterned fullerene layers were almost intact (the 3rd layer, Figure 12 E), whereas in the other, some of the C_{60} molecules reacted with oxygen or polymerized, as indicated by the satellite peak A(2). The proportion of C_{60} molecules involved in these chemical changes was determined by the ratio A(2)/A(1), and reached about 30% in the 2nd layer (Figure 12 C) and 50% in the 1st and 4th layer (Figure 12 A, G). Similarly as in AFM, fullerenes were also found below the metallic bars of the grid, though in the thinner layers 1 and 2, the amount was very low and barely detectable (Figure 12 B, D, F, H).

Reflection goniometry showed that similarly to the continuous fullerene C_{60} layers, all of the micropatterned C_{60} layers were relatively highly hydrophobic. All micropatterned layers had similar water drop contact angles, i.e., 93.6 ± 3.6° (layer 1), 94.5 ± 5.5° (layer 2), 95.3 ± 3.1° (layer 3) and 95.6 ± 3.8° (layer 4).

Optical microscopy revealed that below the openings of the grids, the fullerene prominences were brownish in color, and the color intensity increased with layer thickness. Up to a thickness of 484 ± 5 nm, the layers were relatively well transparent in a conventional light microscope (Figure 11). Thus, the cells growing on the layers were well observable in native or hematoxylin- and eosin-stained cultures, and the fullerene bulges and grooves were also well distinguishable. However, the fullerene prominences 1043 ± 57 nm in thickness were less transparent, thus the cells were better observable after fluorescence staining (Figure 13).

The micropatterned layers were seeded with human osteoblast-like MG 63 cells. From day 1 to 7 after seeding, the cells on the layers with fullerene prominences up to 326 ± 5 nm were distributed almost homogeneously over the entire material surface (Figure 13), thus the cell population densities on the prominences and in the grooves were similar (Figure 14). At the same time, the cells on these layers adhered and grew to an extent comparable to that on tissue culture polystyrene dishes and microscopic glass coverslips. Similarly as in continuous fullerene C_{60} layers, these beneficial effects on cell colonization could be explained by their surface nanostructure, mimicking the nanoarchitecture of the natural extracellular matrix [53, 54, 106-110]. Moreover, due to the micropatterning, the fullerene layers in our study were constructed with a hierarchically-organized micro- and nanostructure, i.e., with a similar architectural principle as natural tissues [116]. The presence of oxygen, suggested in some of our fullerene layers by Raman spectroscopy, might also contribute to the supportive effects of these films on cell adhesion and growth. Oxygen-containing chemical functional groups have been repeatedly shown to enhance the colonization of various materials with cells [121, 135-137, 144]. In the case of fullerene-grafted polyurethane, the presence of oxygen-containing and amine functional groups (due to pre-treatment of the polymer with oxygen plasma and the use of aminosilanes as coupling agents) was also attributed to the enhanced adhesion and activation of platelets [120].

However, the cells on the layers with prominences of 1043 ± 57 nm were found preferentially in the grooves (Figure 13). These grooves contained from 75.0 ± 13.4 % to 99.6 ± 0.3 % of cells, though they occupied only 44 ± 3 % of the material surface. The cell population density in the grooves was about 5 to 192 times higher than on the bulges, and these differences increased proportionally to the time of cultivation (Figure 14). Preferential growth of bone cells in grooves, pits and other types of hollows has been observed on various artificially microfabricated or primary rough polymeric, metallic or carbon-based materials [104, 116, 130, 143, 145-148]. However, the cells were also able to colonize the prominences on these surfaces, though the prominences were usually much higher (i.e., several micrometers or even tens of μm) than those on the thick micropatterned fullerene layers in our present study (only about 1 μm). Surprisingly, on our fullerene C_{60} layers, the MG 63 cells were not able to "climb up" relatively low prominences only about 1 μm in height, even at a relatively late culture interval of 7 days after seeding. This may be due to a synergetic action of certain physical and chemical properties of the fullerene bulges less appropriate for cell adhesion, such as their hydrophobicity, a steep rise of the prominences, as well as the tendency of spherical ball-like fullerene C_{60} molecules to diffuse out of the prominences [149].

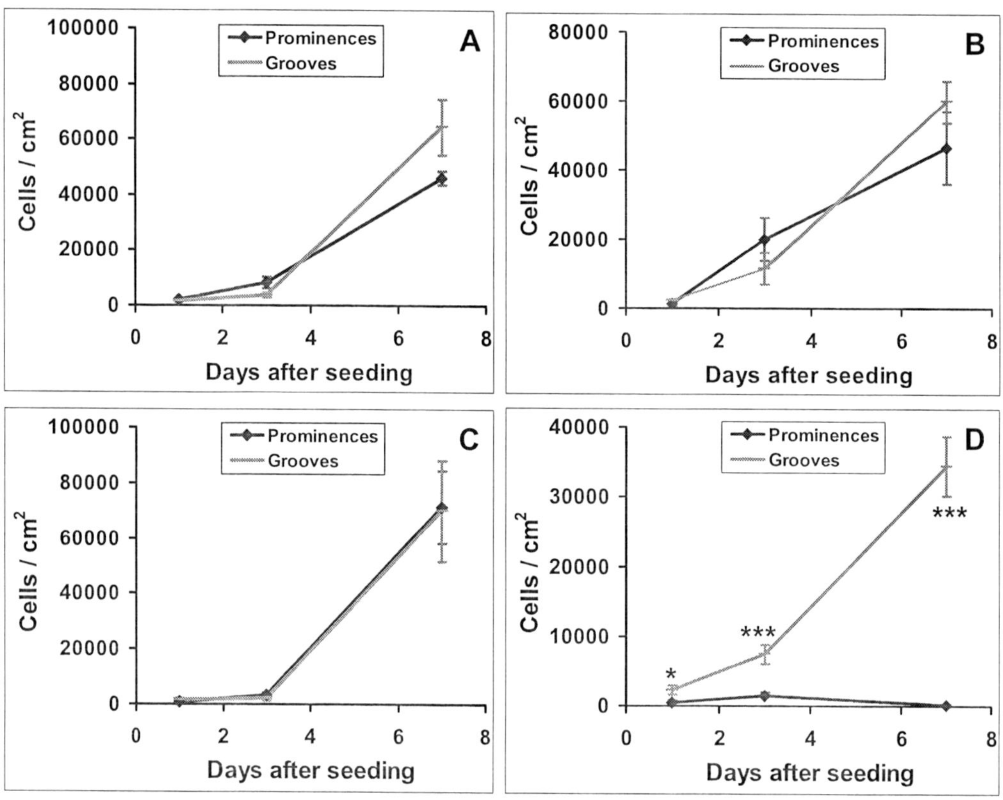

Figure 14. Population densities of MG 63 cells on prominences and grooves on fullerene layers micropatterned with prominences 128 ± 8 nm in height (A), 238 ± 3 nm in height (B), 326 ± 5 nm in height (C) or 1043 ± 57 nm in height (D). Mean ± Standard Error of Mean (S.E.M.) from 9 to 21 microphotographs, Student t-test, statistical significance: $*p \leq 0.05$, $***p \leq 0.001$ [118].

Fluorescence staining of cells using the LIVE/DEAD kit revealed that, on day 1 after seeding, the percentage of viable cells on all micropatterned fullerene layers ranged between 80 ± 10 % and 87 ± 20 %, and was similar to the values obtained on the control tissue culture polystyrene dishes (96 ± 21%) and microscopic glass coverslips (67 ± 16 %). It can be assumed that the presence of a certain number of dead cells was due to the stress to which the cells were submitted during the cell seeding procedure, particularly during trypsinization and keeping the anchorage-dependent cells in suspension, rather than due to a cytotoxic action of the fullerenes. Accordingly, the percentage of viable cells increased with time of cultivation, reaching 97 ± 25 % to 99 ± 15 % on day 7 after seeding on all tested surfaces.

Thus, micropatterned fullerene films could be used as templates for regionally-selective cell adhesion and growth. Moreover, osteogenic cells growing in hollows have been reported to be more active in the phosphorylation of various kinases and transcription factors, signal transduction and differentiation [143, 145-148]. From this point of view, the microstructured fullerene layers (if they adhere strongly to the underlying materials and are resistant to the release of fullerene molecules) could be used as a bioactive coating for bone implants in order to improve their integration with the surrounding bone tissue.

2.3. Fullerene C_{60} Layers Modified by Ion- or Laser-irradiation

Both continuous and micropatterned fullerene C_{60} layers were irradiated with ion beams (Au^+ ions with energy of 1.8 MeV and fluence up to 2 x 10^{15} cm^{-2}) with the aim (i) to convert C_{60} into another biocompatible carbon allotrope, and/or (ii) to induce pattern formation in various scales and dimensions (in order to promote bio-compatibility and bio-attractivity of the micro- and/or nano-patterened carbon substrates with respect to the inspected MG 63 cells).

Figure 15 A shows that using ion irradiation (1.8 MeV Au^+, 2 x 10^{14} cm^{-2}) both goals can be achieved – i.e., conversion of C_{60} into bio-compatible amorphous carbon and selforganization of the as-deposited carbon allotrope layer (i.e., alteration of the original array of rectangular bulges into a system of periodically repeating islands of trapezoidal shape). The form of the trapezoids changes in one direction as a mirror-reflection recurrence; in the perpendicular direction the change is more complex, again combining a mirror reflection with a flip-flop turn. The bulge array form alteration may be due to a coordinated stress relaxation that is accumulated in the system (of a thin C_{60} backing layer supporting the bulge prominences) during ion irradiation. The continuous transformation of C_{60} into amorphous carbon (during progressive ion irradiation) results in the gradual inner stress growth that has (by contrast) a tendency to be relieved. This tendency grows with the accumulation of stress energy, and it results in a sudden stress relaxation. The consequence is a different sample configuration with a distinct topography that has lower area (stress) energy. The observed trapezoidal form and repeating spatial arrangement of the bulge prominences is an interesting self-shaping pattern formation that deserves further attention.

A similar effect can be found when a continuous layer of fullerenes (a thin film fullerite) is irradiated with energetic ions. In Figure 15 B, a dramatic alteration of the originally smooth (as-deposited) C_{60} layer is demonstrated after 1.8 MeV Au^+ ion irradiation to the fluence 2 x 10^{14} cm^{-2}. It can be seen that ion irradiation leads to the formation of ridges and grooves in the surface morphology that could advantageously be controlled by a selected fluence of the ions. As mentioned above, C_{60} is converted during ion irradiation into amorphous carbon, which (due to a different molecular arrangement) requires expansion of the transformed layer into a larger area (than the original C_{60} film holds). This results in an interesting topography of a periodic-like corrugation of the amorphous carbon film. It might be expected that irradiation of the fullerite film through a metallic mask will create a periodic C_{60} - amorphous carbon system with original spatial patterns. Based on the effect of higher bio-compatibility of the micro- and nano-structured surface morphologies, micro-corrugated amorphous carbon films might become attractive for cell adhesion and growth.

The first experiments confirmed that the ion-irradiated fullerene layers enable homogeneous colonization with MG 63 cells to an extent similar to the non-irradiated layers, as well as reference cell culture substrates, represented by microscopic glass coverslips and standard polystyrene dishes. The vast majority of the cells were viable, of normal polygonal or spindle-shaped morphology, well-spread and with a well-assembled fine mesh-like beta-actin cytoskeleton (Figure 16).

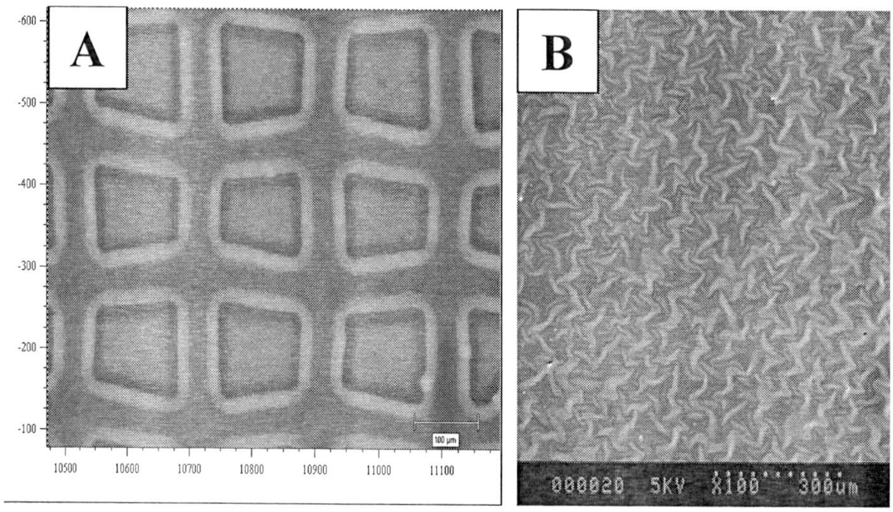

Figure 15. A. Fullerene C_{60} microdomains, created by deposition through a metallic mesh and irradiated with Au^+ ions (energy of 1.8 MeV and fluence of $2x10^{14}$ cm^{-2}). After irradiation, the original system of symmetrical (rectangular) C_{60} bulges was converted into a system of amorphous carbon bulges of trapezoid shapes varying (as a mirror image) periodically. B. A continuous fullerene C_{60} layer irradiated with Au^+ ions (1.8 MeV, $2x10^{15}$ cm^{-2}). The irradiation led to corrugation of the originally smooth fullerene layer.

2.4. Metal-fullerene C_{60} Composites

Another novel and interesting material with potential use in biotechnology and medicine is represented by binary metal/C_{60} composites. In our experiments, several different kinds of these hybrid (metallic-organic) systems, mainly thin films of Ni-C_{60} and Ti-C_{60}, have been inspected. These composites combine important properties of both phases and are assumed also to exhibit satisfying bio-compatible qualities.

In our earlier studies, nanocomposite Ti/hydrocarbon plasma polymer films, created by DC magnetron sputtering and containing amorphous hydrogenated carbon-like material with nanoclusters of Ti, supported the adhesion, growth and maturation of human bone-derived MG 63 cells to an extent comparable to that found on standard cell culture materials, represented by polystyrene dishes, microscopic glass slides or coverslips. In the case of vascular endothelial cells of the line CPAE, derived from the bovine pulmonary artery, the cell adhesion, growth and maturation was even improved in comparison with the standard culture materials and hydrocarbon plasma polymer films without Ti [103]. The beneficial effects of Ti on the colonization of carbon fibre-reinforced carbon composites (with MG 63 cells, as well as primary or low-passaged vascular smooth muscle cells, derived from the rat aorta) were also observed after the CFRC was coated with a carbon-titanium layer [104].

In the case of metal-C_{60} hybrids, both inspected systems (Ni-C_{60} and Ti-C_{60}) were deposited on various supports (e.g., Si and MgO; Figure 17). They exhibited a number of interesting properties, mainly a tendency toward spontaneous self-organization, either already during hybrid synthesis (i.e., simultaneous co-deposition of both metal and organic phases) or during post-deposition processing (e.g., thermal annealing). A typical example of pattern formation of an as-deposited Ni-C_{60} hybrid is shown in Figure 18 A. Under certain deposition kinetics a periodic system of stripe domains was formed. The domains consist of sub-half-micrometer large droplets encompassed with a thin C_{60}-based rind with a polymeric structure. The stripes are embedded in the double-layer platform falling into the epitaxial Ni and amorphous carbon thin films. It has been clarified that during co-deposition the C_{60} molecules partly disintegrate and transform toward amorphous carbon. Due to the accumulation of amorphous carbon (with limited solubility in Ni), thermodynamic instability arises in the system, and after it reaches a certain level it triggers spontaneous partitioning and self-organization of the deposited material. A sequential drift and coordinate release of the thermodynamic instability has been proposed as a principal pattern formation mechanism [150].

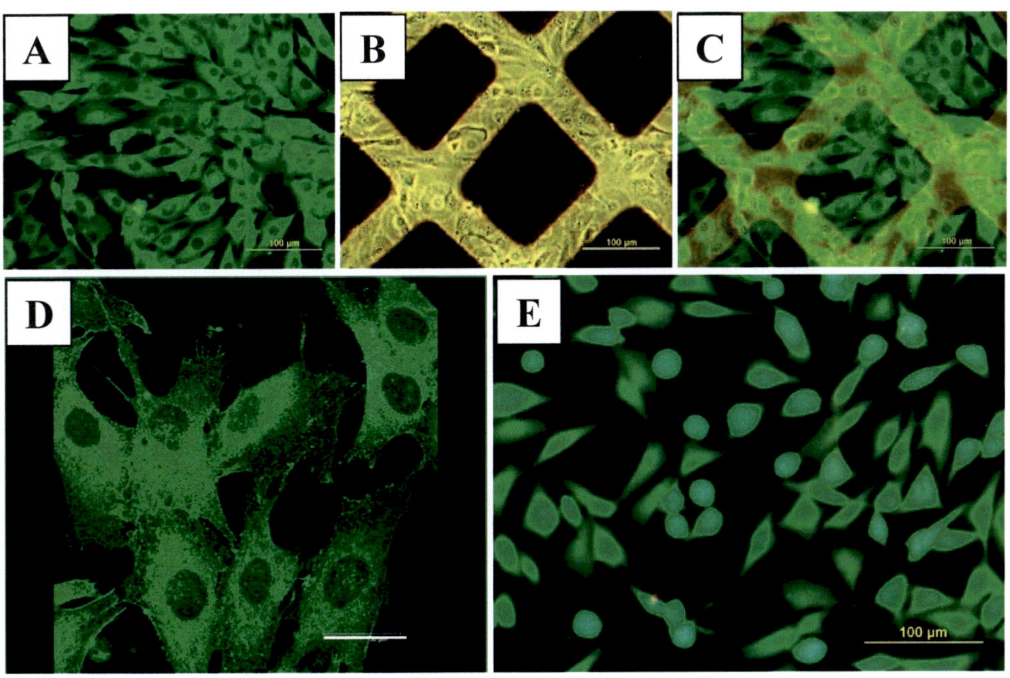

Figure 16. Human osteoblast-like MG 63 cells in cultures on a fullerene C_{60} layer micropatterned by deposition through metallic grid and then irradiated with Au ions of the energy of 1635 keV and fluence of $2x10^{14}$ cm^{-2}. A-D: immunofluorescence staining of beta-actin, day 3 after seeding; E: staining with LIVE/DEAD viability/cytotoxicity kit, day 1 after seeding. A, B, C, E: Olympus IX 50 microscope, DP 70 digital camera. A: fluorescence; B: conventional light; C: combination of both fluorescence and conventional light. Bar = 100 µm. D: Leica DM 2500 confocal microscope, obj. 100, bar = 30 µm.

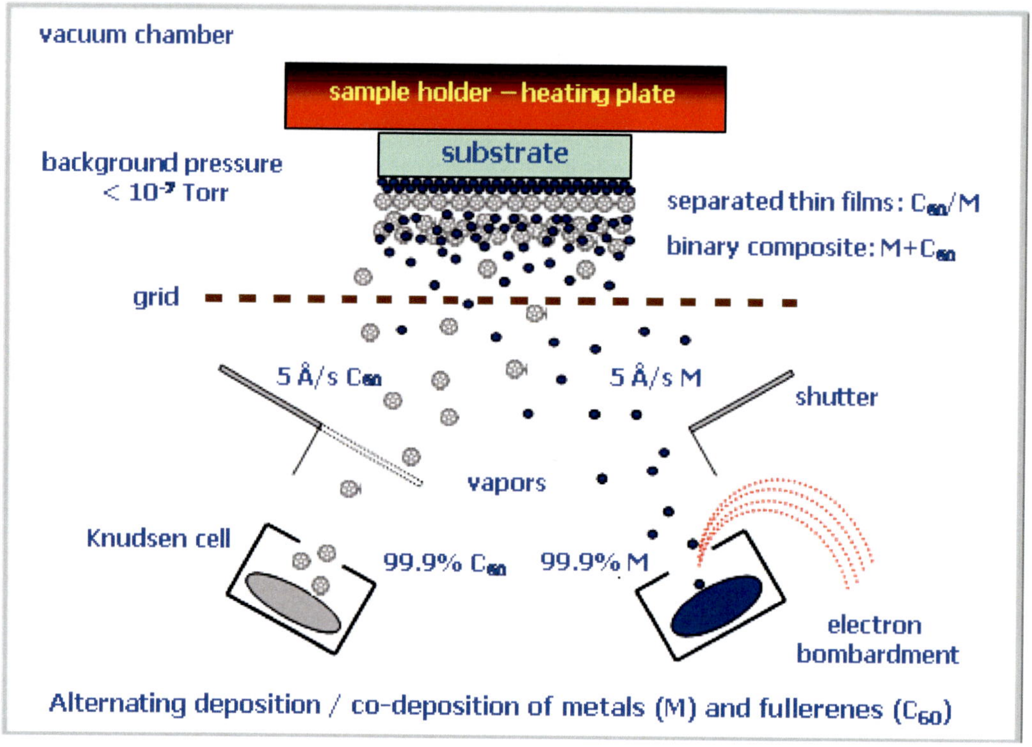

Figure 17. A scheme of hybridization of the metal/C_{60} composites in an Univex-300 vacuum system (Leybold, Germany; see Figure 1). Deposition rates: DR(M) = DR(C_{60}) ~ 5 Å/s; temperatures during deposition: 500°C (M), 120°C (C_{60}); total thickness of the hybrid composites 100 - 1000 nm.

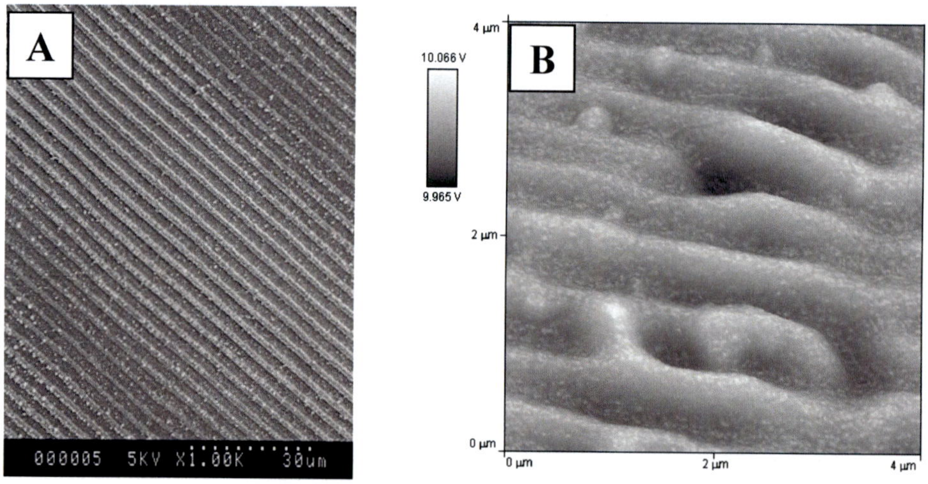

Figure 18. A. SEM micrograph of a stripe system spontaneously formed during co-deposition of Ni and C_{60} on the MgO (001) substrate. SEM-EDS (JOEL JSM-5600), bar = 30 μm. B. MFM (Magnetic Force Microscopy) analysis of the Ni+C_{60} composite synthesized at room temperature on the Si (001) substrate. Veeco CP-II microscope, scan area of 5 x 5 μm.

Another interesting manifestation of the pattern formation proclivity of the hybrid system is shown in Figure 18 B. Here, we observe an interesting property of Ni-C_{60} hybrids deposited on Si at RT - i.e., a buried self-organized morphology, consisting of periodic stripe-like magnetic domains, that has been revealed using the MFM (Magnetic Force Microscopy) method. The 'hidden' system of magnetic domains indicates that even at RT (when only a very small additional thermal energy may enter the system) a certain coordinated rearrangement (relaxation) of the stressed microstructure may appear [151].

An interesting issue is how these hybrid composites and their micro- and nano-pattern forms may affect the bio-activity of the cells. Preliminary results (obtained only recently) have demonstrated that the composites are bio-compatible and support adhesion and growth of the human osteoblast-like MG 63 cells (Figure 19 A-F).

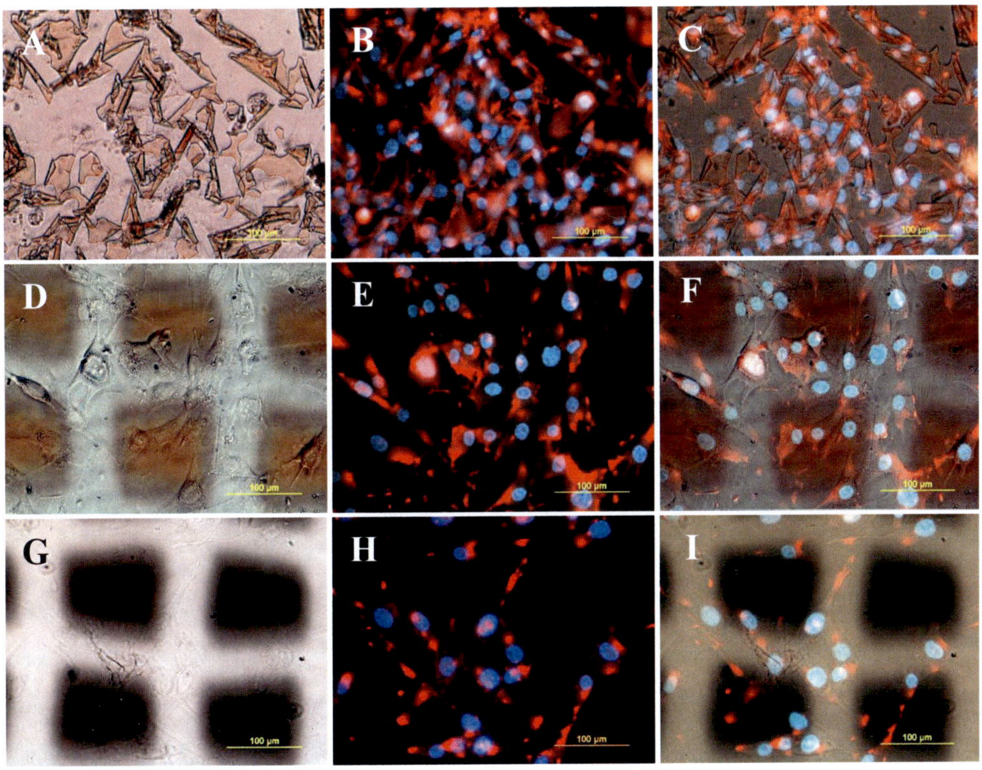

Figure 19. Human osteoblast-like MG 63 cells on day 4 after seeding on hybrid Ti-C_{60} composites deposited on microscopic glass coverslips in the form of continuous films (A-C) or films micropatterned by deposition through a metallic grid (D-F) and irradiated with Au^+ ions of the energy of 1635 keV and fluence of $2x10^{14}$ cm^{-2} (G-I). The cells were fixed with paraformaldehyde and stained with Hoechst 33342 and Texas Red C_2-maleimide fluorescence dyes. Olympus IX 50 microscope, DP 70 digital camera. A,D,G: conventional light; B,E,H: fluorescence; C,F,I: combination of both conventional light and fluorescence. Bar = 100 µm.

A study of ion-irradiated composites has shown that, similarly as in the case of plain fullerite films, also in the case of hybrids the fullerene molecules are converted to amorphous carbon. Because of the metallic phase (which keeps the integrity of the composites firm) no manifestation of the spatial geometry alteration has been observed during the $C_{60} \rightarrow$ amorphous carbon transformation (as it is in the case of the plain fullerites). A thorough analysis of the systems showed that the hybrids are (after ion irradiation) more stable than the original structures (which are highly stressed due mainly to low miscibility of the C_{60} and metallic phases). More significant activity of the phase components (e.g., phase separation) can be observed only when the system is exposed to an energy flow, e.g., during thermal annealing or laser illumination. The preliminary results of our study showed that colonization of the human osteoblast-like MG 63 cells may also be possible on ion-irradiated and/or thermally processed hybrids, especially of the Ti-C_{60} type (Figure 19 G-I).

3. Cell Growth on Polymer-Carbon Nanotube Composites

Our studies on cell interaction with nanotube-modified materials were carried out on polymer-nanotube composites, i.e., synthetic polymers mixed with single- or multi-walled carbon nanotubes. As the polymeric matrix, we used a terpolymer of polytetrafluoroethylene, polyvinyldifluoride and polypropylene (PTFE/PVDF/PP) or polysulphone (PSU), i.e., polymers which were considered to be promising for the construction of bone implants. The nanotubes were expected to reinforce the relatively soft and elastic polymeric material and thus to improve its mechanical properties for hard tissue surgery. The other desired function of nanotubes was the formation of prominences in nanoscale on the material surface, which could promote its colonization with bone cells.

3.1. Composites of Carbon Nanotubes and PTFE/PVDF/PP

For the preparation of composites consisting of carbon nanotubes and a PTFE/PVDF/PP terpolymer matrix, a commercially available terpolymer of polytetrafluoroethylene, polyvinyldifluoride and polypropylene (density of 1600 g/dm^3, Aldrich Chemical Co., U.S.A.) was dissolved in acetone at a concentration of 0.1 g/ml. Single-wall carbon nanohorns (SWNH) or high crystalline electric arc multi-wall nanotubes (MWNT-A; both from NanoCraft Inc., Renton, U.S.A.) were mixed with acetone in a sonicator for 5 minutes, and then with the terpolymer solution for 15 minutes to a concentration of 2, 4, 6 or 8 wt. %. The mixtures were then poured on to Petri dishes and left to evaporate the solvent [95].

Despite sonication, both SWNH and MWNT-A formed aggregates in the polymeric matrix, which created microscale irregularities on the material surface (Figure 20). The surface microscale roughness, measured by surface profilometry, increased proportionally to the nanoparticle concentration. In the pure terpolymer, the R_a parameter, i.e., the arithmetic mean of the departures of the roughness profile from the mean line, was only 0.20 ± 0.04 μm, whereas in the SWNH- and MWNT-A-modified material, the R_a parameter ranged from 0.41 ± 0.01 μm to 2.22 ± 0.36 μm in terpolymer samples with 2 to 8 wt. % of nanoparticles (Figure 21 A). The R_a parameter was usually similar in materials containing the same concentrations of SWNH or MWNT-A.

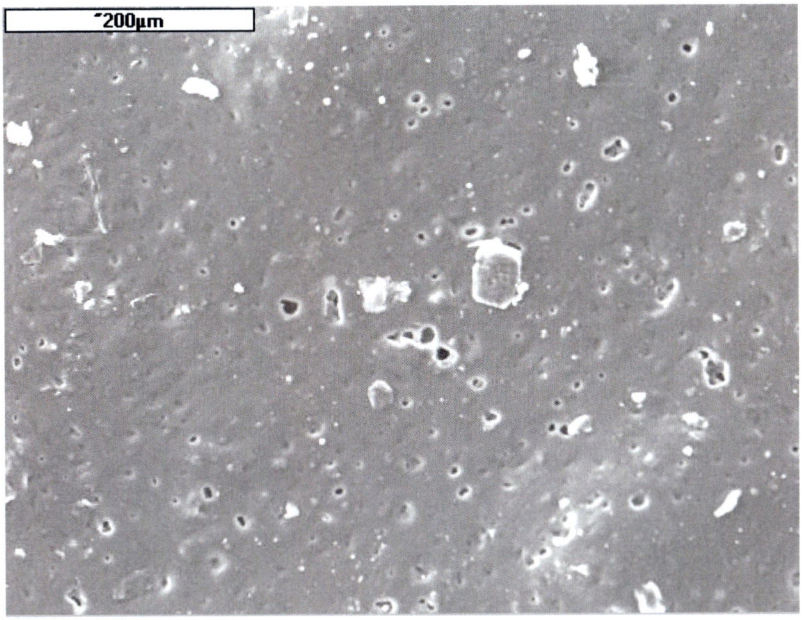

Figure 20. An example of the surface morphology of the terpolymer of polytetrafluoroethylene, polyvinyldifluoride and polypropylene mixed with 4 wt.% of single-wall carbon nanohorns. Jeol JSM 5400 Scanning Electron Microscope, obj. 2000x. Bar = 200μm [95].

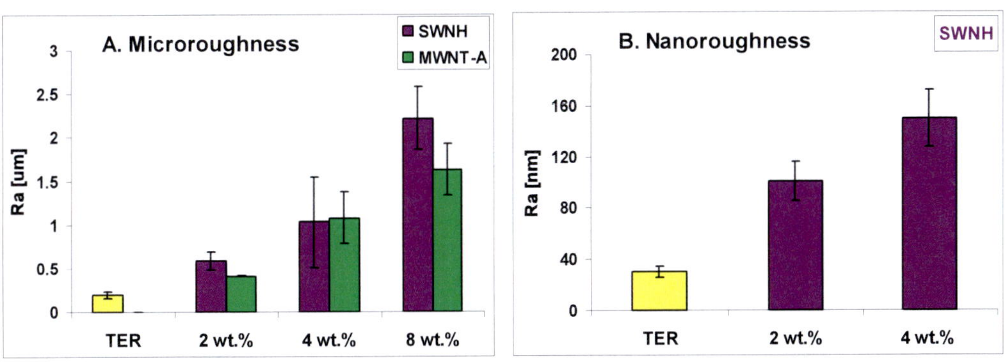

Figure 21. Surface roughness in microscale (A) and nanoscale (B) of unmodified terpolymer of polytetrafluoroethylene, polyvinyldifluoride and polypropylene (TER) and terpolymer samples mixed with 2, 4, 6 and 8 wt.% of single-wall carbon nanohorns (SWNH) or multi-wall carbon nanotubes (MWNT-A). R_a is an average deviation of the roughness profile from the mean line. Measured with a T500 Hommel Tester (Hommelwerke Co., Germany; A) or by atomic force microscopy using SPM LAB 5.01 software (B). Mean ± S.D. (Standard Deviation) from 3 measurements [95].

At the same time, the surfaces of the terpolymer-nanotube composites contained irregularities in nanoscale, as revealed by AFM (Figure 21 B). Therefore, in fact, similarly to the fullerene-coated CFRC and micropatterned surfaces (created by deposition of C_{60} through metallic grids), the surface of the carbon nanotube-polymer composites was of a hierarchically organized micro- and nanostructure.

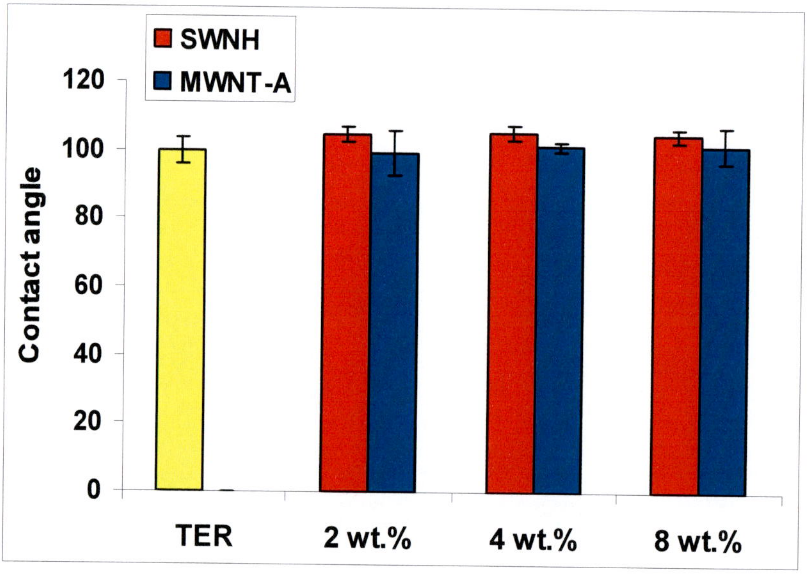

Figure 22. Static water drop contact angle measured on unmodified surface terpolymer of polytetrafluoroethylene, polyvinyldifluoride and polypropylene (TER) and terpolymer samples mixed with 2, 4, 6 and 8 wt.% of single-wall carbon nanohorns (SWNH) or multi-wall carbon nanotubes (MWNT-A). Measurement by the sessile drop method using the DSA 10 Mk2 automatic drop shape analysis system (Kruss, Germany). Mean ± S.D. (Standard Deviation) from 6-10 measurements [95].

The composites of PTFE/PVDF/PP and carbon nanotubes were relatively highly hydrophobic. The water drop contact angle, measured in a static water drop-material contact system, was in the range from 99.3 ± 6.56° to 105.2 ± 2.12°, and was similar in the unmodified and all carbon-nanoparticle-modified terpolymer samples (Figure 22). The surface energy, measured in terpolymers modified with 4 wt. % of SWNH or MWNT-A, was also similar in pure terpolymer and both types of composite samples. Only in the sample with 4 wt. % of SWNH was the polar component of the surface energy significantly lower (Table 1).

Table 1. The surface energy and its dispersion and polar components in pure terpolymer and terpolymer modified with 4 wt. % of SWNH or MWNT-A, measured by contact angle for polar liquid (water) and for non-polar liquid (diodomethane)

Sample name	Surface energy [mN/m]	Dispersion component [mN/m]	Polar component [mN/m]
Terpolymer	22.63 ± 0.61	21.18 ± 0.47	1.45 ± 0.13
4 wt. % SWNH	21.12 ± 0.43	20.47 ± 0.38	0.65 ± 0.04
4 wt. % MWNT-A	22.61 ± 1.08	21.50 ± 0.94	1.11 ± 0.14

Mean ± S.D. (Standard Deviation) from 12 measurements.

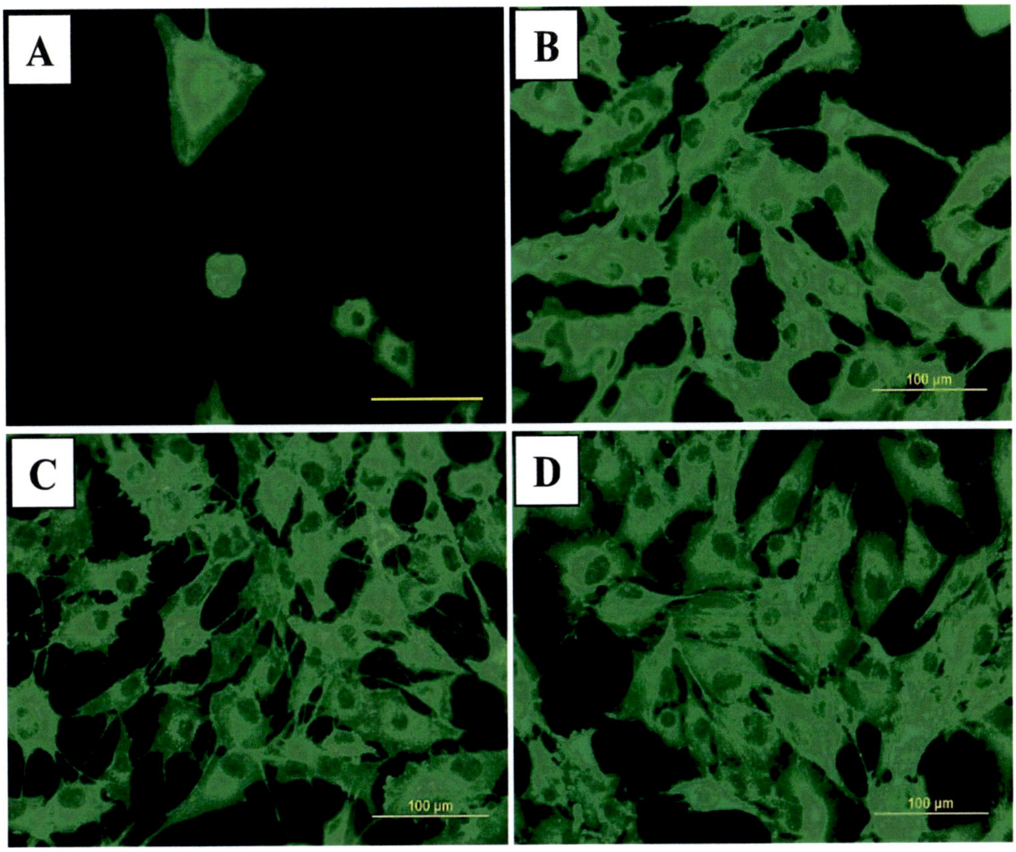

Figure 23. Immunofluorescence staining of β-actin in osteoblast-like MG 63 cells on day 3 after seeding on a terpolymer of polytetrafluoroethylene, polyvinyldifluoride and polypropylene (A), terpolymer mixed with 4 wt. % of single-wall carbon nanohorns (B) or 4 wt. % of multi-wall carbon nanotubes (C) and tissue culture polystyrene dish (D). Olympus IX 50 microscope, DP 70 digital camera, obj. 20, bar = 100 µm [95].

The addition of carbon nanotubes to the PTFE/PVDF/PP terpolymer markedly improved several parameters of cell adhesion, growth and maturation. The human osteoblast-like MG 63 cells in cultures on PTFE/PVDF/PP mixed with SWNH or MWNT-A were well-spread, polygonal, and contained distinct beta-actin filament bundles, whereas most cells on the pure terpolymer were less spread or even round and clustered into aggregates (Figure 23). The enzyme-linked immunosorbent assay (ELISA) revealed that the cells on the material with 4 wt. % SWNH contained a higher concentration of vinculin, a component of focal adhesion plaques, in comparison with the values in cells on the pure terpolymer and also on the tissue culture polystyrene (Figure 24 A). The concentration of talin, another important integrin-associated focal adhesion protein, was also higher in cells grown on terpolymer samples with 4 and 8 wt.% of SWNH (Figure 24 C). However, the presence of MWNT-A did not affect the concentration of vinculin and talin in cells cultured on these samples (Figure 24 B, D). The concentration of osteocalcin, a marker of osteogenic cell differentiation, was similar in cells on all SWNH-modified and pure terpolymer samples, and also TCPS (Figure 24 E). However, in cells on samples with 4 and 6 wt. % of MWNT-A, the concentration of osteocalcin was significantly lower than the concentration in cells on the pure terpolymers and TCPS (Figure 24 F). This could be explained by a higher proliferation activity of these cells (Figure 25), which may delay the process of cell differentiation [122]. Another factor affecting the osteogenic maturation of cells is the softness of the adhesion substrate. On softer substrates, human mesenchymal stem cells differentiated towards muscle or neuronal cells, whereas osteogenic differentiation required matrices of relatively high rigidity [152]. The terpolymer used in our studies, though reinforced with carbon nanoparticles, could still be considered as a relatively soft material in comparison with the metals and ceramics that are also used in bone tissue engineering. Nevertheless, the concentration of intercellular adhesion molecule-1 (ICAM-1), an adhesion molecule of the immunoglobulin superfamily and an important marker of cell immune activation, was not increased in cells on terpolymers modified with SWNH or MWNT-A (Figure 24 G, H), which suggests that the material could be tolerated by the host tissue.

In summary, the improved colonization of the PTFE/PVDF/PP terpolymer with human osteoblast-like MG 63 cells after it has been modified with carbon nanotubes could be attributed to changes in its surface roughness and topography, especially the creation of the surface nanostructure, rather than to its surface wettability, which remained unchanged and relatively low. Only the increased proliferation activity of cells on samples with 4 wt. % of MWNT-A (Figure 25) could be attributed to a favorable combination of the material surface nanostructure and a relatively high polar component of the free surface energy (Table 1) [153-156]. Such a combination was not observed on the other samples, which were either nanostructured (terpolymer with SWNH) or only had a higher polar component of the surface energy (pure terpolymer). On the other hand, the terpolymer modified with 4 wt. % of SWNH seemed to be the most appropriate substrate for cell adhesion, spreading and formation of focal adhesion plaques.

3.2. Composites of Carbon Nanotubes and PSU

The composites of polysulphone (PSU) and carbon nanotubes were prepared by a similar approach as the constructs with a PTFE/PVDF/PP matrix. The PSU (Aldrich Chemical Co., U.S.A.) was dissolved in dichloromethane at a concentration of 0.1 g/ml, mixed with 0.5, 1.0 or 2.0 wt.% of SWNH or MWNT-A, sonicated, poured on to Petri dishes and left to evaporate freely. The PSU-carbon nanotube composites also had a combined micro- and nanostructure of the material surface. The microroughness was similar in all nanotube-containing PSU samples, but was higher in nanotube-modified PSU than in unmodified polymer. The surface nanoroughness, measured by AFM, increased proportionally with the concentration of carbon nanotubes in the polymer matrix (Figure 26) [157, 158].

In comparison with the terpolymer PTFE/PVDF/PP, the PSU used in our studies was more wettable, having a sessile water drop contact angle of 84.8 ± 5.4^{o}. Unlike the terpolymer, PSU in its pristine unmodified state gave good support for the adhesion and growth of bone-derived MG 63 cells. PSU proved to be an appropriate growth substrate for human periodontal ligament fibroblasts [159], freshly isolated rat hepatocytes [160], as well as a suitable component of three-dimensional scaffolds for the construction of bioartificial liver assist devices [161]. After the addition of MWNT-A, the surface wettability of the composite remained similar (contact angle in the range from 85.0 ± 4.6^{o} to 90.1 ± 2.7^{o}), while after addition of SWNH, the wettability decreased slightly (contact angle from 93.2 ± 4.3^{o} to 91.2 ± 3.0^{o}).

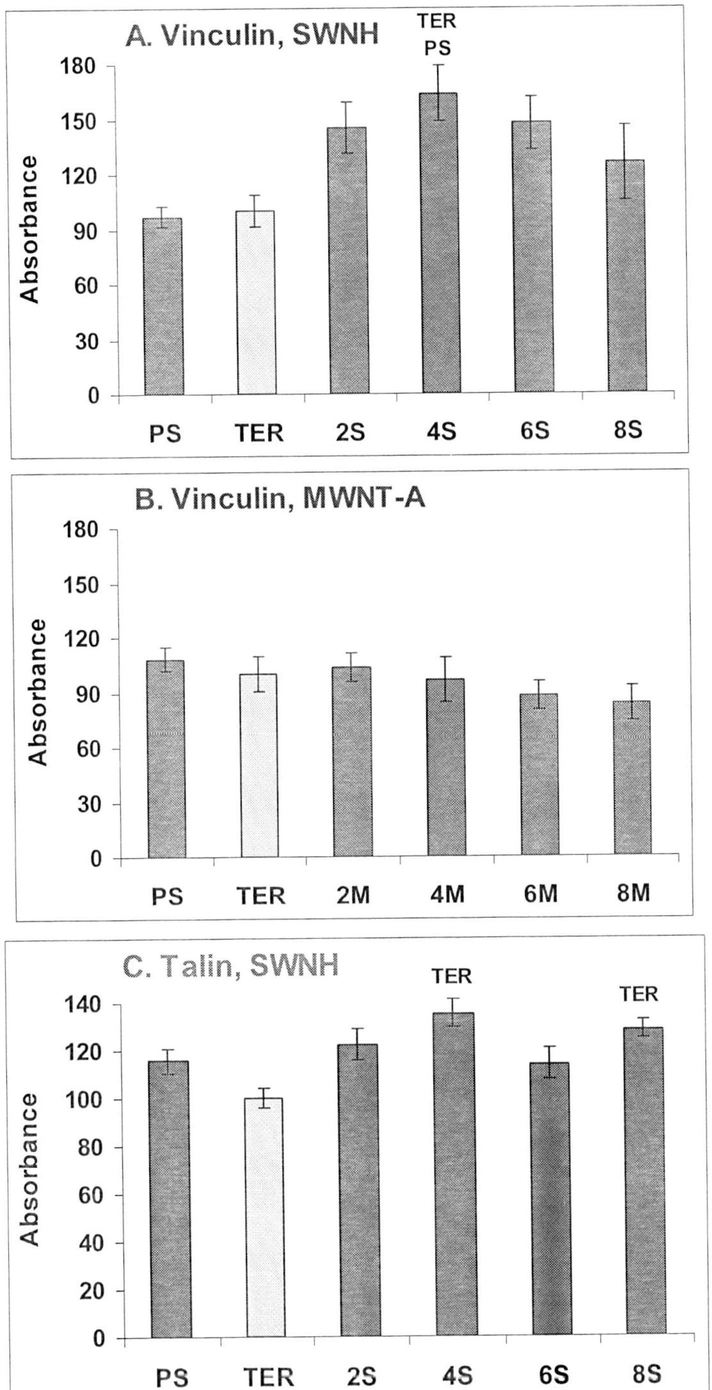

Figure 24. Continued.

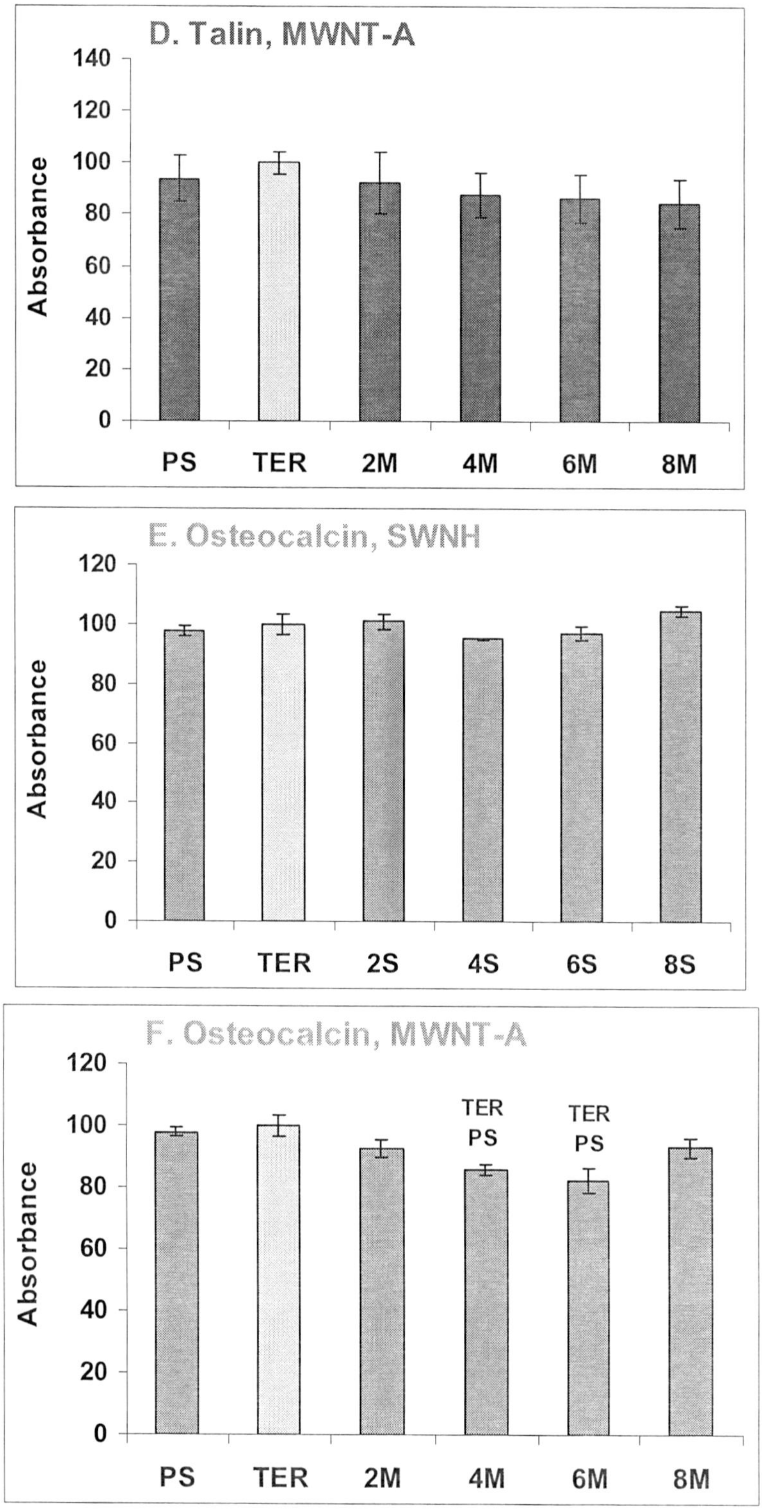

Figure 24. Continued.

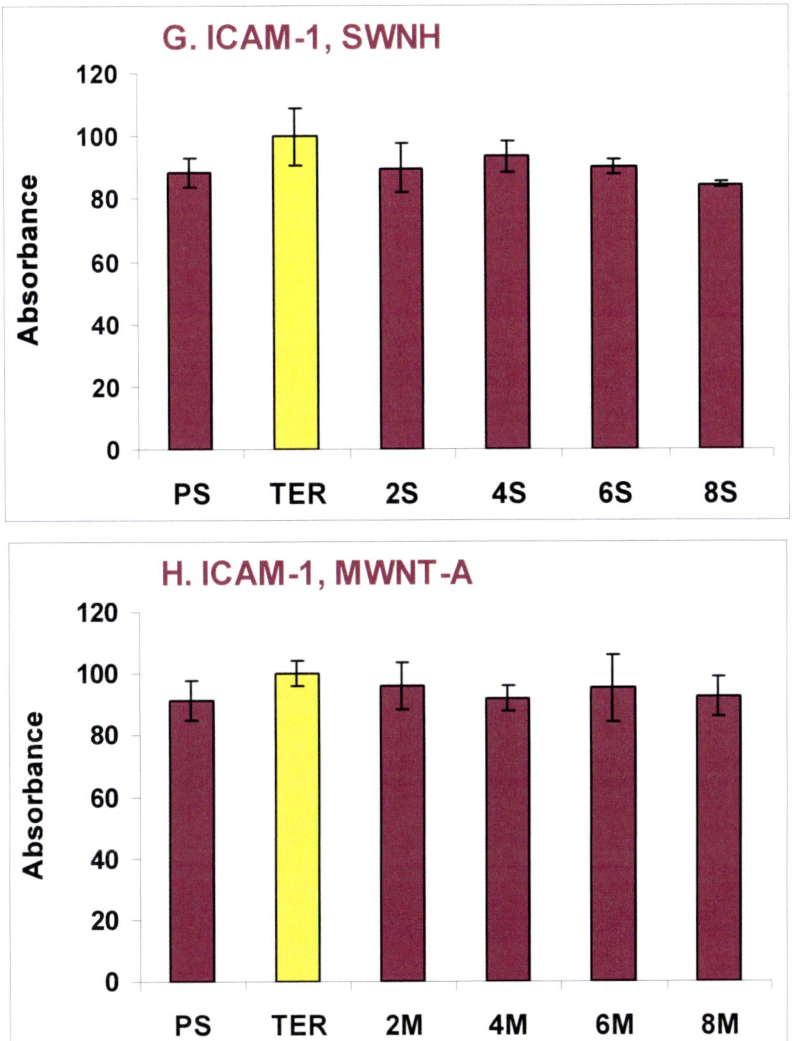

Figure 24. Concentration of vinculin (A, B), talin (C, D), and osteocalcin (E, F) in osteoblast-like MG 63 cells on day 8 after seeding on the pure terpolymer of polytetrafluoroethylene, polyvinyldifluoride and polypropylene (TER), terpolymer mixed with 2, 4, 6 or 8 wt.% (2S, 4S, 6S, 8S) of single-wall carbon nanohorns (SWNH) or 2, 4, 6 or 8 wt.% (2M, 4M, 6M, 8M) of multi-wall carbon nanotubes (MWNT-A) and tissue culture polystyrene (PS). Measured by ELISA per mg of protein; the absorbance values of cells from the modified terpolymers were expressed in % of the values obtained from the control cells grown on the unmodified terpolymer. Mean ± S.E.M. (Standard Error of Mean) from 4 measurements, ANOVA, Student-Newman-Keuls method. Statistical significance: $^{TER, PS}$: $p \leq 0.05$ compared to the values on the pure terpolymer and tissue culture polystyrene, respectively [95].

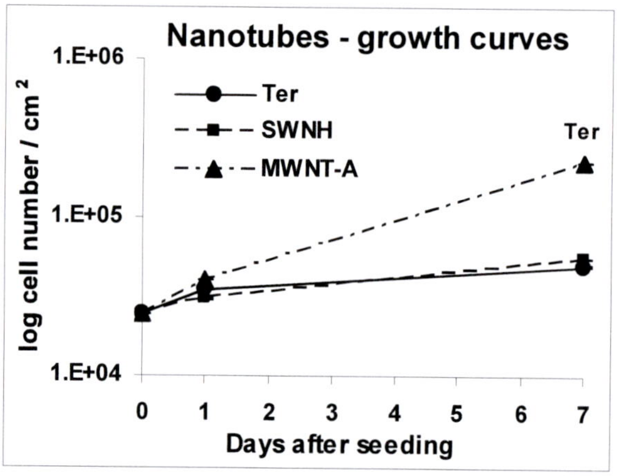

Figure 25. Growth curves of MG 63 cells on a terpolymer of polytetrafluoroethylene, polyvinyldifluoride and polypropylene (Ter), terpolymer mixed with 4 wt.% of single-wall carbon nanohorns (SWNH) or 4 wt.% of multi-wall carbon nanotubes (MWNT-A) [95].

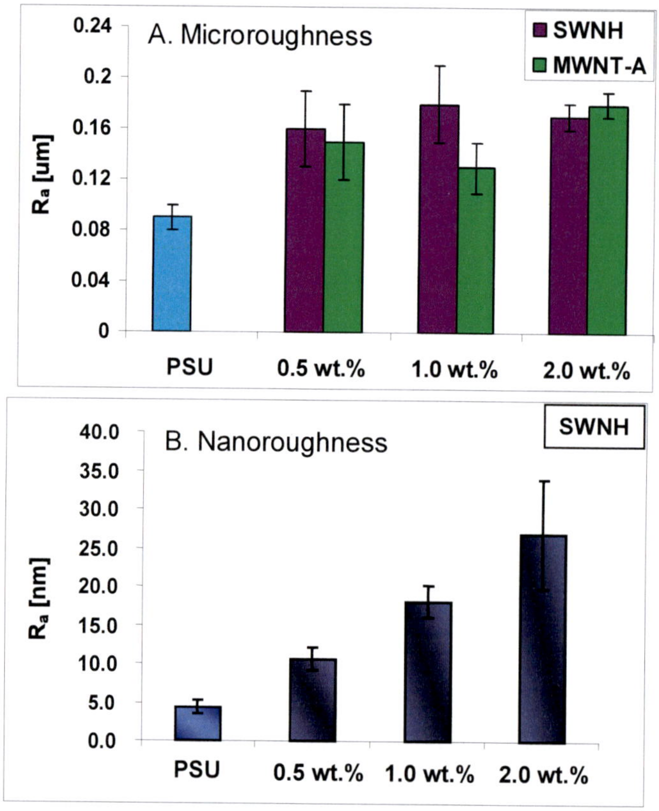

Figure 26. Surface roughness in microscale (A) and nanoscale (B) of unmodified polysulphone (PSU) and PSU mixed with 0.5, 1.0 or 2.0 wt.% of single-wall carbon nanohorns (SWNH) or multi-wall carbon nanotubes (MWNT-A). R_a is an average deviation of the roughness profile from the mean line. Measured using the T500 Hommel Tester (Hommelwerke Co., Germany; A) or by or by atomic force microscopy using SPM LAB 5.01 software (B). (B). Mean ± S.D. (Standard Deviation) from 6 measurements [157, 158].

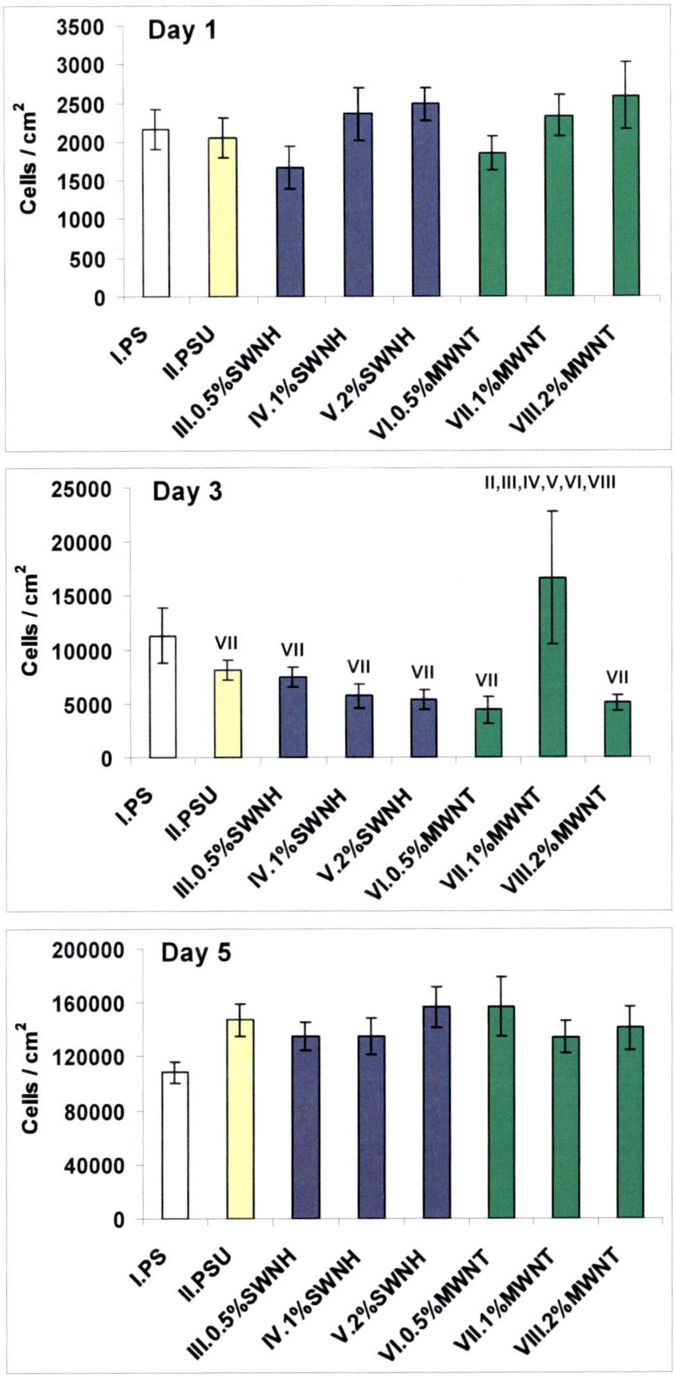

Figure 27. Cell population density (cells/cm^2) of human osteoblast-like MG 63 cells on day 1, 3 and 5 after seeding on cell culture polystyrene dishes (PS), polysulphone (PSU) and polysulphone mixed with 0.5, 1 or 2 wt.% of single-wall carbon nanohorns (SWNH) or multi-wall carbon nanotubes (MWNT). Mean ± S.E.M. (Standard Error of Mean) from 15-50 measurements. ANOVA, Student-Newman-Keuls method. Statistical significance: [I, II, III, IV, V, VI, VII, VIII]: $p \leq 0.05$ compared to the values on the sample labeled with the same number [157, 158].

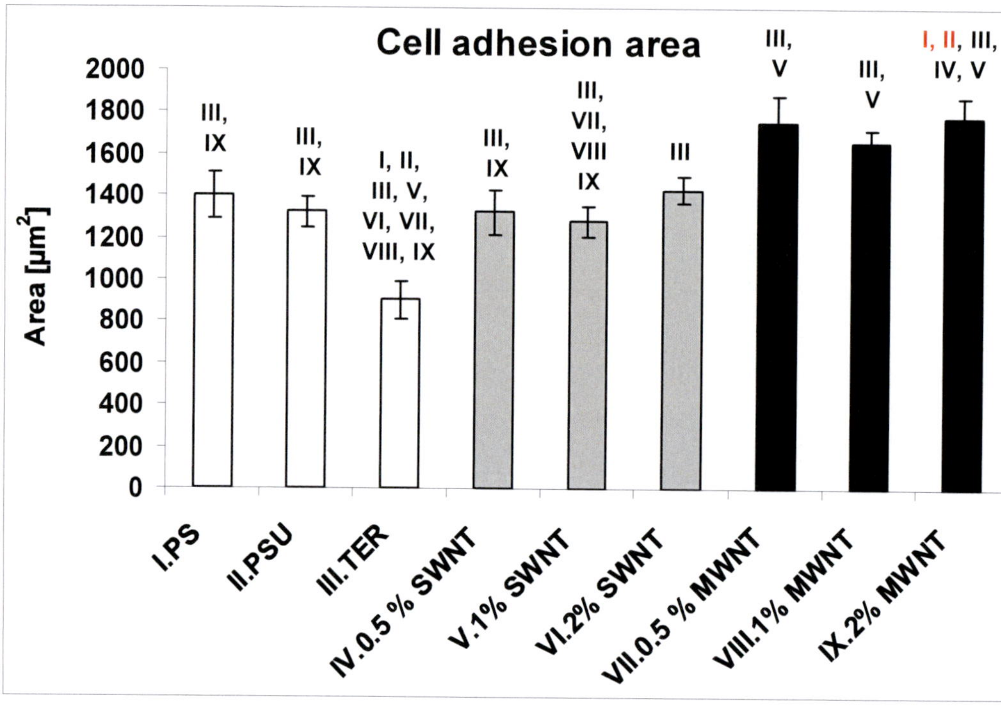

Figure 28. Cell spreading area (µm^2) of human osteoblast-like MG 63 cells on day 1 after seeding on cell culture polystyrene dishes (PS), polysulphone (PSU), terpolymer of polytetrafluoroethylene, polyvinyldifluoride and polypropylene (TER) and polysulphone mixed with 0.5, 1 or 2 wt.% of single-wall carbon nanohorns (SWNH) or multi-wall carbon nanotubes (MWNT). Mean ± S.E.M. (Standard Error of Mean) from 26-124 cells for each experimental group. ANOVA, Student-Newman-Keuls method. Statistical significance: I, II, III, IV, V, VI, VII, IX: p≤ 0.05 compared to the values on the sample labeled with the same number [157, 158].

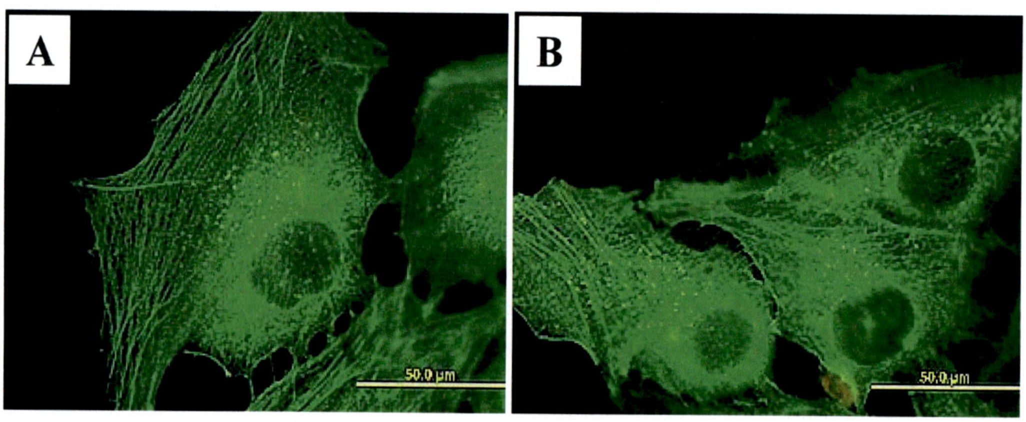

Figure 29. Immunofluorescence of beta-actin in human osteoblast-like MG 63 cells on day 3 after seeding on polysulphone mixed with 2 wt.% of single-wall carbon nanohorns (A) or 2 wt.% of multi-wall carbon nanotubes (B). Olympus IX 50 microscope, DP 70 digital camera, obj. 100, bar = 50 µm [157, 158].

As for the colonization with cells, the numbers of MG 63 cells achieved on PSU-nanotube composites were usually similar to those obtained on unmodified PSU and standard polystyrene cell culture dishes. Only on day 3 after seeding on PSU with 1 wt. % of MWNT-A was the cell population density significantly higher (Figure 27). However, the addition of carbon nanotubes improved the cell spreading. The cell adhesion area, i.e., the cell area projected on the material surface, and measured in cells stained against beta-actin using the immunofluorescence technique, was the largest in cells grown on PSU samples with 0.5, 1 and especially 2 wt. % of MWNT-A (Figure 28). Moreover, the cells on PSU-carbon nanotube composites developed thick beta-actin cables oriented in parallel and resembling those observed in cells of muscle type (Figure 29). However, similarly as in cells on the composites PTFE/PVDF/PP and MWNT-A, this phenotype was often associated with a loss of markers of osteogenic differentiation, particularly osteocalcin [157, 158].

3.3. Interaction of Cells with Carbon Nanotubes Suspended in the Culture Media

Carbon nanotubes can be added to polymers designed for the construction of three-dimensional scaffolds for tissue engineering. These nanoparticles would form nanoscale prominences on the pore walls and thus improve the ingrowth of cells inside the scaffolds. In addition, they would reinforce the scaffolds and improve their mechanical properties in order to be more appropriate for bone tissue engineering, and they would also enable electrical stimulation of cells, which has been reported to promote their growth and maturation, manifested by collagen synthesis and the formation of mineralized bone tissue matrix [55, 56]. An important question is what would happen with the nanotubes if the scaffolds were made of degradable polymers, such as polylactide, polyglycolide, polycaprolactone and their copolymers, which are often used in tissue engineering [162]. While in stable terpolymer PTFE/PVDF/PP, as well as PSU, the nanotubes were permanently bound in the polymeric matrix, they would be released from degradable materials. The nanotubes could then accumulate in adjacent and remote tissues, penetrate inside the cells and have adverse effects on the cells, e.g., a cytotoxic, immunogenic or even genotoxic, mutagenic and carcinogenic action [83-85]. On the other hand, a recent paper showed that there is a real chance of clearing the nanotubes from the organism relatively quickly by glomerular filtration [163].

In order to simulate less or more massive release of carbon nanotubes from the scaffolds and to evaluate their direct effects on cells, we suspended SWNH or MWNT-A in the culture media (DMEM, Sigma, Cat. No. D5648, supplemented with 10% of fetal bovine serum) at concentrations of 0.004, 0.04, 0.4, 4 and 40 mg/ml. This concentration range was chosen because it involves similar concentrations of nanotubes present in our polymer-nanotube composites, as well as concentrations lower or higher by several orders. The nanotube-containing media were then used for seeding human osteoblast-like MG 63 cells into standard polystyrene 24-well multidishes (TPP, Switzerland). Each well contained 30,000 cells (about 17,000 cells/cm^2 and 1.5 ml of the nanotube-containing culture medium). The medium without nanotubes served as a control sample [164]. The cell number and morphology on days 1, 3 and 7 after seeding were then evaluated.

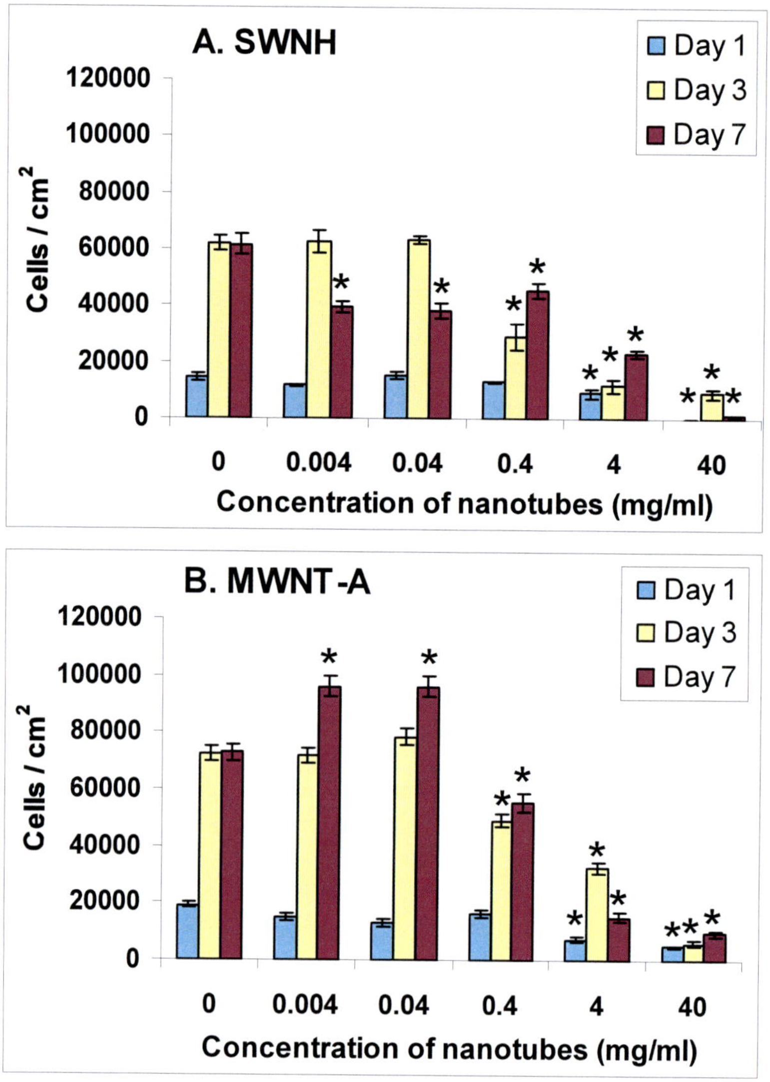

Figure 30. Cell population density (cells/cm^2) of human osteoblast-like MG 63 cells on day 1, 3 and 7 after seeding in the culture medium containing 0-40 mg/ml of single-wall carbon nanohorns (SWNH) or multi-wall carbon nanotubes (MWNT-A). Mean ± S.E.M. (Standard Error of Mean) from 3 samples for each experimental group. ANOVA, Student-Newman-Keuls method. Statistical significance: *: $p \leq 0.05$ compared to the corresponding values obtained in the medium without nanotubes [164].

On day 1 after seeding, the cells in both media adhered in numbers similar to the control value obtained in the medium without nanotubes up to a concentration of 0.4 mg/ml, where the numbers of initially adhering cells were significantly lower (Figure 30). This may be due not only to a cytotoxic action of carbon nanotubes but also to the fact that these nanoparticles, at higher concentrations (4 and 40 mg/ml), covered most of the bottom of the culture well and left only a limited space for cell attachment and spreading (Figures 31 and 32). On day 3 after seeding, the cell number in the media with nanotubes in concentrations of 0.004 and 0.04 mg/ml was similar (SWNH) or even higher (MWNT-A) than in the control nanotube-free medium (Figure 30). A significant decrease in the cell number was apparent from the concentration of 0.4 mg/ml of both types of nanotubes. However, on day 7 after seeding, the cell numbers in the media with SWNH were already significantly decreased from the lowest nanotube concentration of 0.004 mg/ml; whereas in the MWNT-A-containing media, the reduction was only from 0.4 mg/ml, i.e., a concentration 100 times higher (Figure 30). Therefore, it can be concluded from these experiments that the cell growth was inhibited more in the media containing single-walled nanotubes than in media containing multi-walled nanotubes. Similar results were also obtained in cultures of human fibroblasts, where single-wall carbon nanotubes induced stronger toxic effects, manifested by cellular apoptosis or necrosis, than other carbon nanoparticles represented by active carbon, carbon black, multi-wall carbon nanotubes and graphite [165]. In alveolar macrophages, the single-wall carbon nanotubes also impaired phagocytosis and induced cell degeneration and death to an extent more profound than multi-wall carbon nanotubes and fullerenes C_{60} [166]. In addition, single-wall carbon nanotubes stimulated the aggregation of human platelets *in vitro* and vascular thrombosis in rat carotid arteries *in vivo* more than their multi-wall counterparts [167].

The more pronounced adverse effects of single-wall carbon nanotubes on cells could be attributed to their relatively small size, which allows them to penetrate inside the cells more easily than bigger particles. In our experiments, the SWNH were only 2 to 3 nm in diameter and 30 to 50 nm in length, while the diameters of MWNT-A were about 5-20 nm and they were from 300 to 2000 nm in length. In addition, the SWNH had a relatively sharp closed horn-like end (19 degrees), which might further facilitate their penetration inside the cells. Recently, the more pronounced cytotoxicity of single-wall carbon nanotubes has been attributed to their relatively small surface area [165]. The cytotoxicity of both single- and multi-wall carbon nanotubes can be enhanced by the presence of impurities, particularly metal catalysts used during their preparation [39-42], the formation of big, stiff and solid agglomerates resembling asbestos [42, 84, 168] or some types of nanotube derivatization, e.g., oxidation by treatment in acids, as mentioned in the Introduction [47, 48]. On the other hand, other types of nanotube functionalization, such as by 1,3-dipolar cycloaddition reaction, oxidation/amidation treatment or side-wall fuctionalization with phenyl-SO_3H, SO_3Na or phenyl-$(COOH)_2$ has been reported to attenuate the toxicity of carbon nanotubes [44, 169].

In cultures of human embryo kidney HEK 293 cells, single-wall carbon nanotubes in concentrations from 0.001 mg to 0.25 mg/ml, i.e., a concentration range corresponding, at least partly, to that used in our studies, inhibited cell growth by alterations in the expression of cell-cycle associated genes, such as cdk2, cdk4, cdk6 and cyclin D3, leading to an arrest in the G1 phase of the cell cycle, down-regulation of several signal transduction-associated genes (mad2, jak1, ttk, pcdha9 and erk), as well as a lower expression of cell adhesion-associated proteins such as laminin, fibronectin, collagen IV, cadherin and focal adhesion kinase. High nanotube concentration of 0.25 mg/ml induced cell death within 24 hours [82]. In a concentration of 0.1 mg/ml, single-walled carbon nanotubes significantly decreased the number of rat aortic smooth muscle cells relative to the cultures in the control pure medium [168]. In human epidermal keratinocytes, 6-aminohexanoic acid (AHA)-derivatized single-wall carbon nanotubes, applied to the culture media in concentrations from 0.00005 to 0.05 mg/ml, decreased the activity of mitochondrial enzymes, measured by MTT assay. At the same time, the cells treated with 0.05 mg/ml of AHA-SWNTs increased their production of interleukins IL-6 and IL-8 [84]. On the other hand, in human lung A549 cells, the EC_{50} value after 24 hours of exposure to a medium with single-wall carbon nanotubes exceeded the concentration of 8 mg/ml [83]. Relatively safe concentrations of single-walled carbon nanotubes, recommended by Bottini *et al.* [47], did not exceed 40 μg (0.04 mg) per ml of cell culture media. As for the multi-wall carbon nanotubes, when these nanoparticles of high purity were applied to the culture media in concentrations of 16 μg/ml and 20 μg/ml, the growth rate, vitality and shape of Crandell feline kidney fibroblasts remained unchanged, though the nanotube aggregates showed high adhesion to the cell membrane, and were also observed beneath the cell membrane [41].

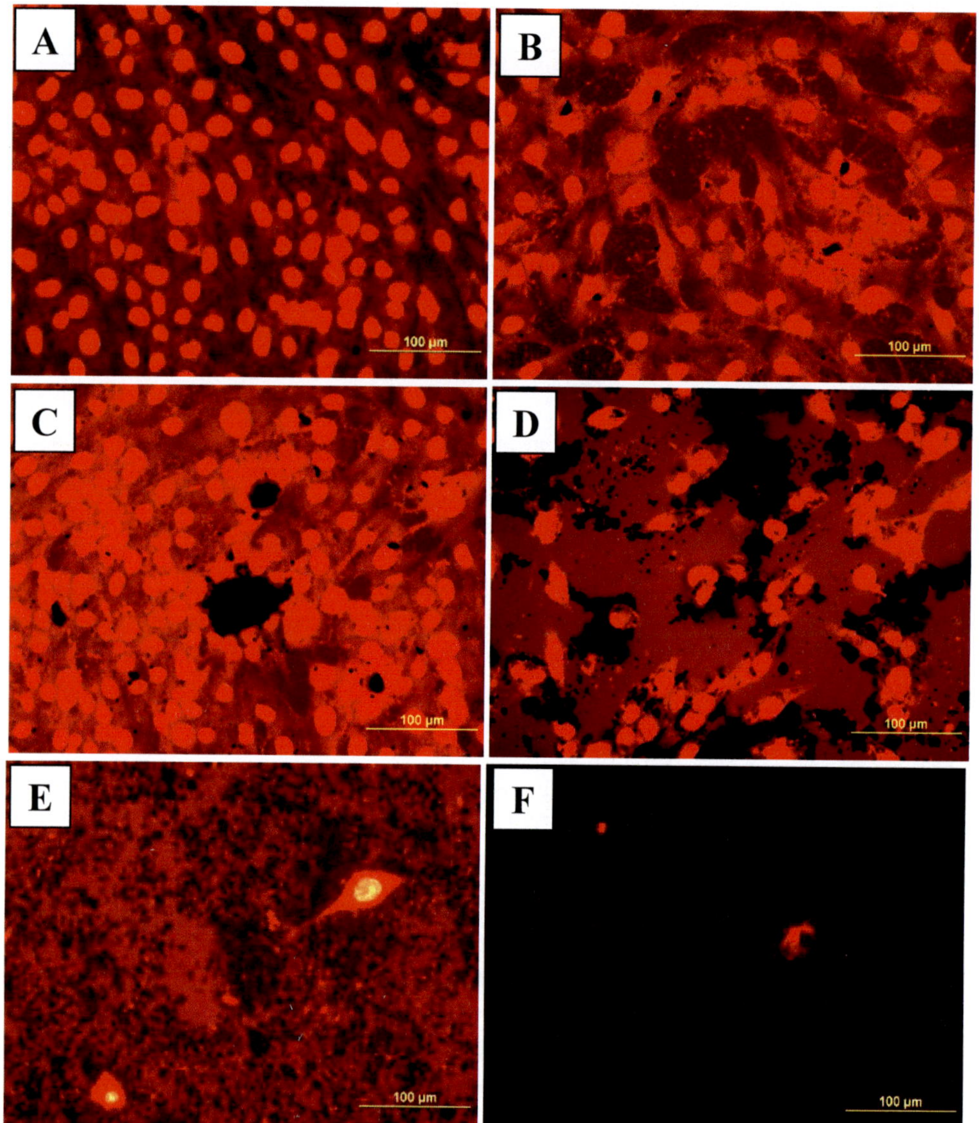

Figure 31. Morphology of human osteoblast-like MG 63 cells on day 3 after seeding in the culture medium containing single-wall carbon nanohorns in concentrations of 0 (A), 0.004 (B), 0.04 (C), 0.4 (D), 4 (E) or 40 (F) mg/ml. Cells fixed with 70% ethanol, stained with propidium iodide. Olympus IX 50 microscope, DP 70 digital camera, obj. 20, bar = 100 μm [164].

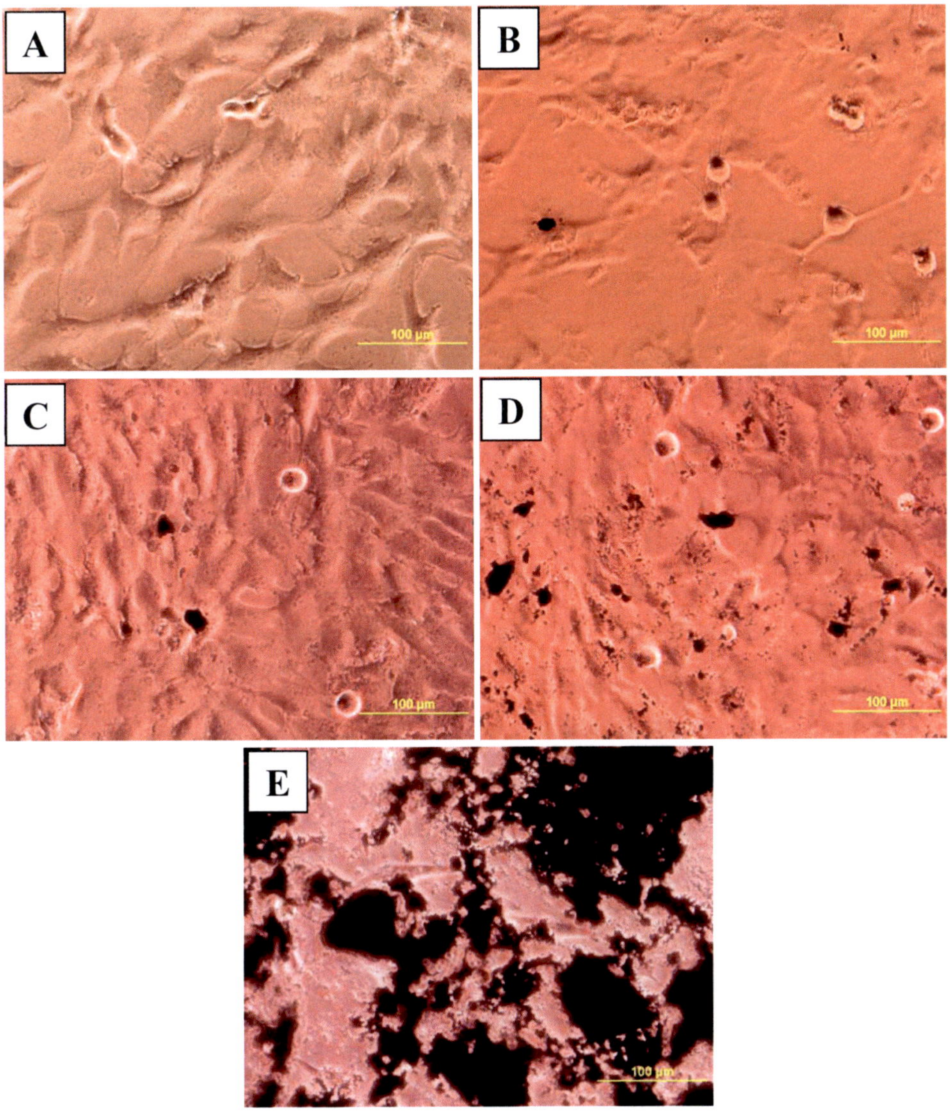

Figure 32. Morphology of human osteoblast-like MG 63 cells on day 3 after seeding in the culture medium containing multi-wall carbon nanotubes in concentrations of 0 (A), 0.004 (B), 0.04 (C), 0.4 (D) or 4 (E) mg/ml. Living cells in native cultures, Olympus IX 50 microscope, DP 70 digital camera, obj. 20, bar = 100 μm [164].

4. NANOCRYSTALLINE DIAMOND LAYERS

In our studies, nanocrystalline diamond (NCD) films were grown on silicon substrates by a microwave plasma enhanced chemical vapor deposition (MW PECVD) method in an ellipsoidal cavity reactor (Figure 33) [95-97, 170]. The silicon substrates were either mechanically lapped to root mean square (*rms*) roughness up to 300 nm or polished to atomic flatness (*rms* roughness about 1 nm). Prior to the deposition process, the substrates were mechanically seeded in an ultrasonic bath, using a 5 nm diamond powder for 40 minutes. The nucleation procedure was followed by PECVD growth using H_2/CH_4 gas mixtures. A constant methane concentration (1 % CH_4 in H_2), at a total gas pressure of 30 mbar was used. The substrate temperature was 860 °C, as measured by a two-color pyrometer working at wavelengths of 1.35 and 1.55 μm. The silicon substrates were overcoated with an NCD film on both sides, i.e., on the top and bottom side, respectively. Thus, hermetic sealing of the Si substrate minimized any unwanted biochemical reaction, as an opened Si area could result in pollution and disturbances of the subsequent cell experiments. As calculated from optical measurements [171], the film thickness was 330 nm on the atomically polished side.

Prior to the biological test, all NCD films were wet-cleaned in a solution of H_2SO_4 + KNO_3 for 30 min at 200 °C to remove any non-carbon phases from the surface of the deposited NCD films. Finally, the wet-cleaned NCD films were treated in oxygen plasma for 3 minutes at a total power of 300 W and gas pressure 10 mbar to enhance the hydrophilic character of the diamond surface. As mentioned above, moderately wettable surfaces have repeatedly been shown to promote adhesion and growth of cells [57, 121, 134-140]. The wetting angle, measured by a static method in a material-water droplet system using a reflection goniometer, was 35° for both types of NCD surfaces deposited on flat or rough Si substrates. Generally, the NCD surfaces were stable without significant changes of the wetting angle. Virgin silicon substrates treated in oxygen plasma also showed a low contact angle value (~17°), but this value was significantly less stable than for the NCD surface. A Si sample, standing in air for several hours showed local non-stabilities, and the wetting angle varied up to 42° [96, 97].

Figure 33. An ellipsoidal cavity reactor (AIXTRON - P6, Germany), characterized by the following parameters: microwave frequency 2.45 GHz, maximum power 6 kW, substrate size 2" or 3", rotating substrate stage, stable plasma at high pressure, fully PC controlled.

The Raman spectra of the deposited NCD films were measured using a 514.5 nm excitation wavelength laser, which enabled us to determine the diamond character of the deposited films (i.e., the sp^3 hybridization). The typical Raman spectrum of an NCD film displayed one dominant peak centered at wavenumber 1333 cm^{-1} (optical phonon in diamond, Figure 34) [172]. In addition, X-ray photoelectron spectroscopy (XPS) determined that the ratio of carbon in sp^3 hybridization, characteristic for diamond, to the sp^2 C was more than 95 % [173].

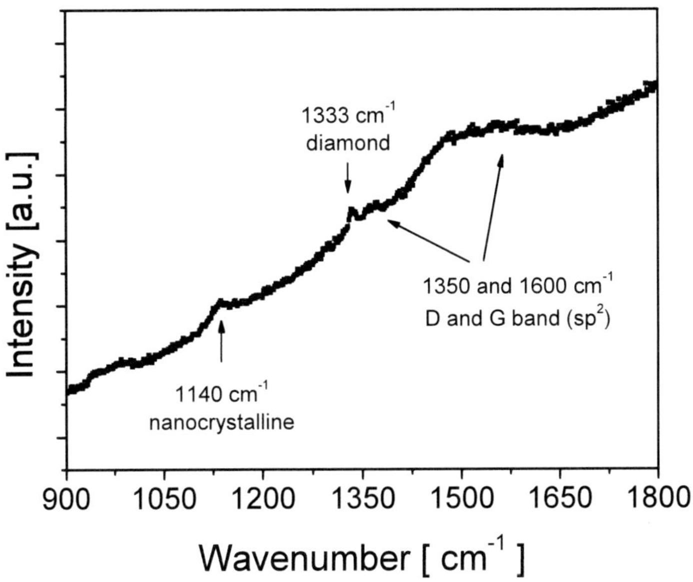

A

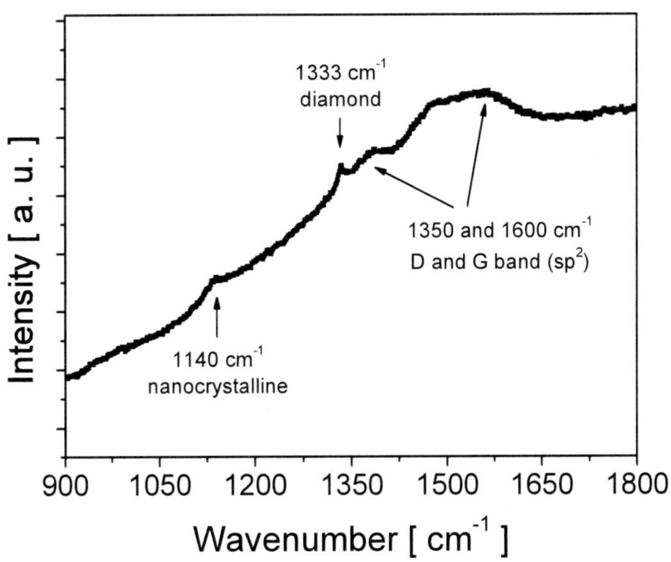

B

Figure 34. Raman spectroscopy of nanostructured (A) and hierarchically micro- and nanostructured (B) nanocrystalline diamond (NCD) films. The peak 1333 cm^{-1}, characteristic for diamond, and 1140 cm^{-1}, often attributed to NCD, are well visible in both types of films [96].

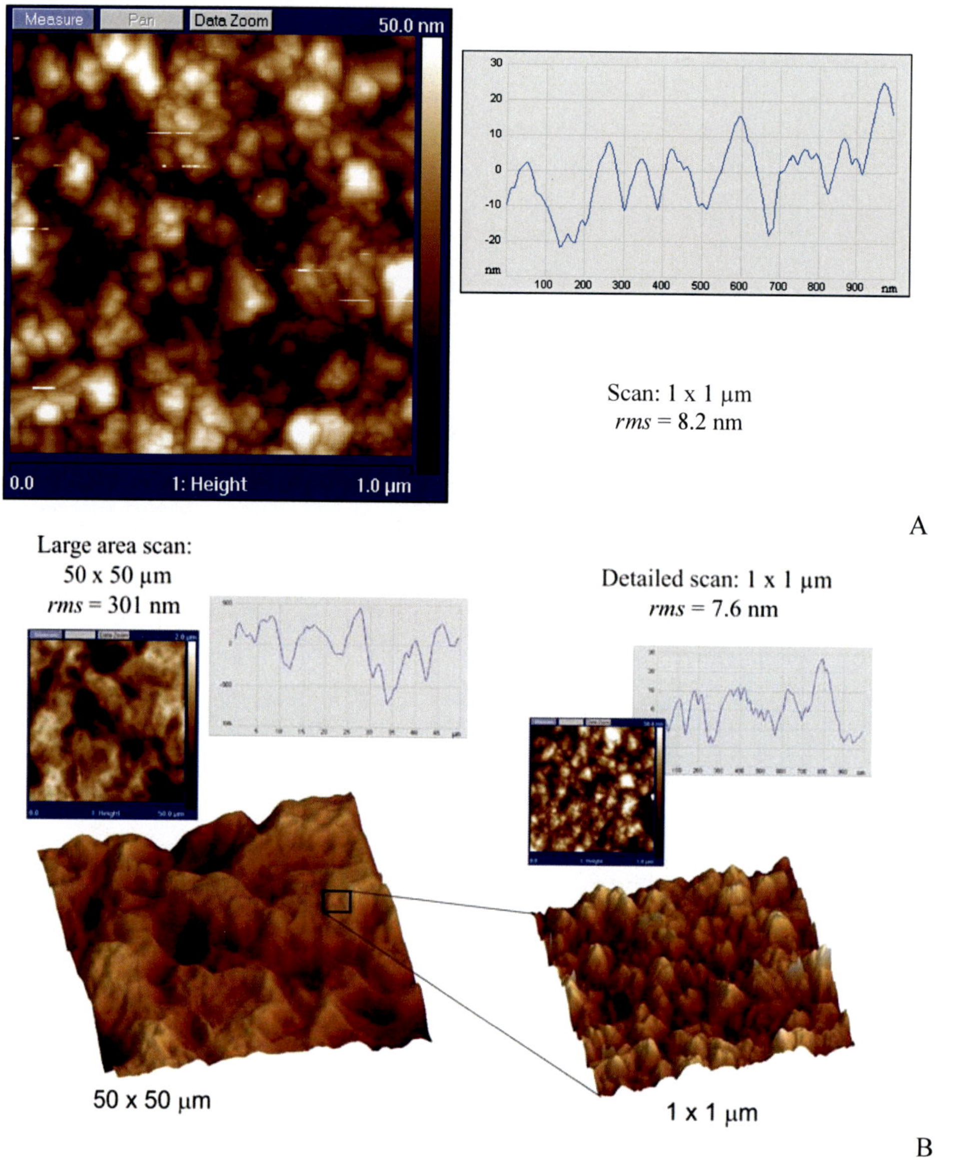

Figure 35. Atomic force microscopy of nanostructured (A) and hierarchically micro-nanostructured (B) NCD films. Root mean square (*rms*) roughness is 8.2 nm (A) or 301 nm and 7.6 nm (B). AFM Microscope Dimension 3100, Veeco [96].

As confirmed by AFM, two types of NCD layers were prepared. The first was nanostructured with a relatively flat surface (Figure 35A). On a detailed AFM scan of 1 x 1 µm, the development of crystal faceting was observable. This NCD film consisted of small crystals up to 30 nm in size. The calculated *rms* roughness was 8.21 nm. The surface features were similar and independent of the scanned area. The second type of layer was hierarchically micro- and nanostructured (Figure 35B). A scan over a large area (50 x 50 µm) exhibited features of relatively high, i.e. submicron, roughness up to ~300 nm. As shown by the detailed scan of 1 x 1 µm, the surface of these irregularities contained nano-scale features with *rms* roughness of 7.6 nm. Both nanostructured and hierarchically micro-nanostructured films thus differ mainly in surface roughness on a large scale [96, 97].

The reason for preparing the hierarchically micro- and nanostructured NCD films was that not only the nanoroughness of the material surface, but also its microstructure, is important for its colonization with osteoblasts and for regeneration of the well functioning bone tissue. Surface roughness on the submicron or micrometer scale (R_a parameter, i.e., departures of the roughness profile from the mean line, in the range approx. from hundreds of nanometers to several micrometers) has repeatedly been shown to enhance the strength of osteoblast adhesion, their spreading and differentiation [111, 154, 174-176], including the deposition of mineralized bone extracellular matrix [177], and the acquisition of osteoblast phenotype in mesenchymal and osteoprogenitor cells [178, 179]. At the same time, increasing surface microroughness reduced the activity of osteoblasts, and thus bone resorption [180]. Microstructured surfaces also modulated the number of attached cells, cell shape, presence and activity of cell adhesion receptors, the assembly of focal adhesion plaques and cytoskeleton, locomotion, proliferation, reactivity to hormones, vitamins and growth factors, as well as the production of various bioactive molecules in cells [111, 154, 174, 175, 177-182]. In addition, as mentioned several times, hierarchically-organized structures of graduated sizes represent the main architectural principle of natural tissues and organs [116].

NCD films were also deposited on microscopic glass slides using the MW PECVD method (26.7 mbar, 710 °C, 0.8-1% CH_4). After the growth, the films were chemically oxidized and subsequently hydrogenated. Some NCD films were also doped with boron (NCD-B; 3 000 ppm B:C; $\rho\sim 1\ 10^{-1}\ \Omega$ cm). The reason was that boron-doped NCD can be used in active sensing applications *in vivo* [175, 176, 183], and the presence of charged states of the ionized B acceptors on the diamond film surface can also influence cell colonization. The wetting angle of both types of films was 85-90°, and the root mean square roughness (*rms*), measured by AFM, was 18 nm.

All samples were sterilized in 70% ethanol, inserted in culture wells in polystyrene multidish plates and seeded with human osteoblast-like MG 63 cells. The number of initially adhered cells, evaluated 24 hours after seeding, was usually similar on all tested NCD layers, microscopic glass slides and cell culture polystyrene dishes. Only on silicon substrates was the number of initially adhering viable cells markedly lower, and this effect was more pronounced in CPAE cells than in MG 63 cells. On the other hand, the percentage of viable CPAE cells on silicon substrates improved with time of cultivation, though their absolute number still remained lower than on NCD-coated surfaces, glass slides and polystyrene wells. In MG 63 cells, the number of viable cells decreased with time of cultivation, and on day 5 after seeding, no cells were detected on Si substrates (Figure 36). It has been reported that the cytotoxicity of silica nanoparticles is dependent on the metabolic and growth activity of a given cell type [184]. For example, fibroblasts with relatively long doubling times were more susceptible to injury induced by silica exposure than tumor cells with short doubling times [184]. Silicon-containing materials can damage cells by the induction of reactive oxygen species, damage to cell membrane and mitochondrial metabolism, as well as inflammatory reactions [184, 185]. Lower adhesion and spreading on silicon substrates was also observed in human cervical carcinoma HeLa cells, rat pheochromocytoma PC12 cells or mouse MC3T3 osteoblasts [88]. Silicon substrates implanted into rabbit eyes behaved in a highly corrosive manner [186].

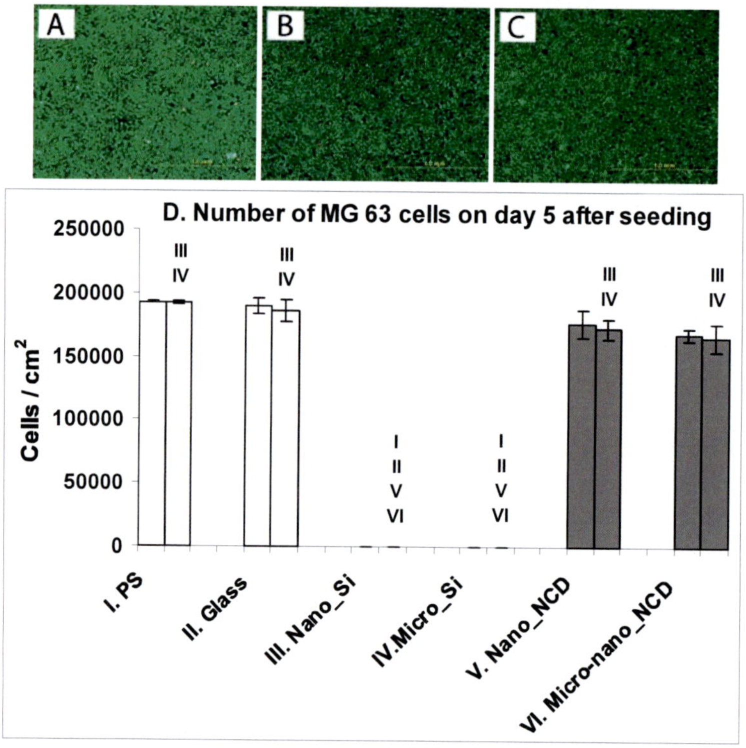

Figure 36. A-C: Fluorescence staining of MG 63 cells with a LIVE/DEAD kit on day 5 after seeding on a control tissue culture polystyrene dish (A), nanostructured NCD films (B) and hierarchically organized micro- and nanostructured NCD (C). D: Number of MG 63 cells on polystyrene culture dishes (PS), microscopic glass coverslips (Glass), nanostructured silicon substrates (Nano_Si), microstructured silicon substrates (Micro_Si), nanostructured diamond films (Nano_NCD) and diamond films with hierarchically organized micro- and nanostructure (Micro-nano_NCD). First column: total cell number, second column: number of viable cells. Mean ± S.E.M. (Standard Error of Mean) from 3 samples. ANOVA, Student-Newman-Keuls Method. Statistical significance: I, II, III, IV, V, VI: $p \leq 0.05$ in comparison with the sample of the same number [97].

The number of MG 63 and CPAE cells on days 3 and 5 after seeding on nanostructured and hierarchically micro-nanostructured O-terminated NCD was usually higher, or at least similar, in comparison with the values obtained on glass and polystyrene dishes (Figures 36, 37). Similarly as in the case of fullerene layers and nanotube-polymer composites, this favorable cell behavior was probably due to the surface nanostructure of both NCD layers. As mentioned above, the material nanostructure has been reported to be highly supportive for colonization with cells, preferentially osteoblasts. Nanostructured surfaces improve the adsorption of cell adhesion-mediating extracellular matrix molecules, e.g., fibronectin, vitronectin and collagen, from the serum of the culture medium or body fluids. These proteins are adsorbed in an appropriate amount and probably also in a favorable spatial conformation, enabling the accessibility of specific amino acid sequences by cell adhesion receptors, e.g., integrins. Preferential adhesion of osteoblasts over other cell types has been explained by preferential adsorption of vitronectin on nanostructured surfaces, due to its relatively small molecule, and its preferential recognition by osteoblasts [53, 54, 106-111, 122].

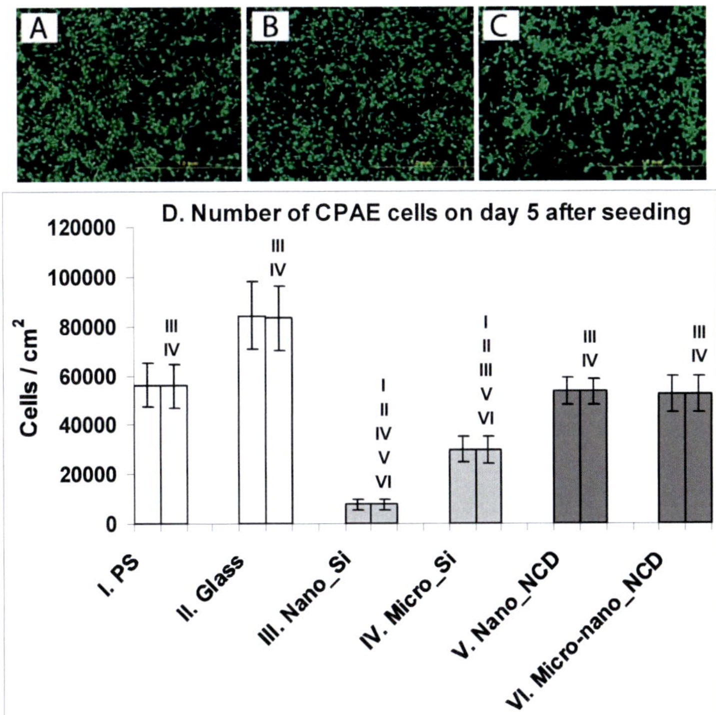

Figure 37. A-C: Fluorescence staining of CPAE cells with a LIVE/DEAD kit on day 5 after seeding on a control tissue culture polystyrene dish (A), nanostructured NCD films (B) and hierarchically organized micro- and nanostructured NCD (C). D: Number of MG 63 cells on substrates described in Figure 36. First column: total cell number, second column: number of viable cells. Mean ± S.E.M. (Standard Error of Mean) from 3 samples. ANOVA, Student-Newman-Keuls Method. Statistical significance: I, II, III, IV, V, VI: $p \leq 0.05$ in comparison with the sample of the same number [97].

In MG 63 cells, the microstructure of the NCD layers also seemed to have a positive effect on their adhesion. On hierarchically micro- and nanostructured surfaces, the MG 63 cells were better spread, i.e., they adhered over a significantly larger area than on the other tested substrates. This result is in good correlation with an earlier study reporting a supportive effect of the surface microroughness of dental titanium implants on the spreading of rat osteoblasts in primary cultures [181], as well as the development of a polygonal shape in osteogenic cells [182]. In addition, as mentioned above, the hierarchically organized micro- and nanostructure of a cell adhesion substrate is considered as a factor mimicking the architecture of natural tissues, and thus beneficial for cell colonization [116].

However, on micro- and nanostructured surfaces, the cell number tended to be lower than in purely nanostructured NCD, especially in endothelial CPAE cells. For example, on day 3 after seeding, the cell population density obtained on nanostructured NCD amounted to 44,000 ± 6,900 cells/cm^2, while on micro- and nanostructured NCD it was only 17,400 ± 2,400 cells/cm^2. A possible explanation is that on the latter surfaces, the spreading of CPAE cells, which are relatively large and have a flat polygonal morphology, and their subsequent growth, might be hampered by micro-sized surface irregularities [104, 130-133]. It can be hypothesized that the idea of a hierarchically-organized micro- and nanoarchitecture of biomaterials, resembling the arrangement of natural tissues [116], may be useful mainly for constructing three-dimensional porous bioartificial implants rather than for planar two-dimensional surfaces. For example, nanodiamonds could be admixed into a polymeric matrix designed for the production of scaffolds for bone tissue engineering. The nanodiamonds would then decorate the walls of the pores, usually hundreds of micrometers in diameter, and the resulting nanostructure of the pore walls would facilitate the ingrowth of osteoblasts and bone tissue regeneration.

The cells on NCD films, especially those that are purely nanostructured, also displayed higher metabolic activity. As revealed by the XTT test, the absorbance of the formazan dye produced by MG 63 cells on nanostructured and hierarchically micro-nanostructured diamond films was significantly higher (by 260 ± 4 % and 102 ± 2 %, respectively) than on the control polystyrene dishes (absorbance normalized to 100 ± 8 %). The absorbance was not detectable on the silicon substrates, probably due to the very low number of viable cells on these materials. In addition, the color of the XTT reagent solution was altered, which may have been due to some non-specific reaction with the silicon substrates. Higher absorbance of formazane from cells cultivated on the two diamond layers indicates higher activity of mitochondrial enzymes in these cells than in those on the control cell culture surfaces (i.e., glass and polystyrene), and is in good correlation with the cell numbers found on both types of NCD films [96].

On boron-doped NCD films, the numbers of MG 63 cells were often lower than on undoped films modified only by oxidation and hydrogenation, though these cells were well spread and of normal morphology. Similarly as on the other NCD films, the cells on boron-doped NCD assembled dot- or streak-like focal adhesion plaques containing integrins $alpha_v$ and $beta_1$, i.e. receptors for vitronectin, fibronectin and collagen, as well as integrin-associated proteins talin and vinculin. These cells also developed a mesh-like beta actin cytoskeleton (Figure 38). However, the presence of boron might affect their proliferation ability, e.g., through an alteration to the calcium metabolism of these cells. Incubation of MG 63 cells with 2-aminoethoxydiphenylborate (2-APB) resulted in a blockage of store-operated Ca^{2+} channels, followed by lengthening of the S and G2/M phases of the cell cycle [187]. On the other hand, supplementation of rats with boron in drinking water improved the mechanical properties of the bone tissue [188]. In accordance with this finding, the MG 63 cells grown on boron-doped NCD surfaces were more brightly stained for osteocalcin, an important extracellular matrix glycoprotein involved in bone tissue mineralization (Figure 38).

In summary, nanocrystalline diamond films give good support for the adhesion and growth of osteogenic and endothelial cells and can be used in tissue engineering applications and *in vivo* experiments. This is a great advantage over other carbon nanoparticles, such as fullerenes and nanotubes, which, as mentioned above, can have a cytotoxic behavior, especially in a free form dispersed in cell culture media [10, 11, 82, 47]. Virgin silicon substrates (typically used in the semiconductor industry) have been found to be cytotoxic and not suitable for biomedical applications, e.g., for biomaterial coatings or for *in situ* bio-sensor elements. Endothelial cells were usually more sensitive to surface roughness and their adhesion and growth was better on purely nanostructured than on hierarchically organized micro- and nanostructured NCD films. At the same time, MG 63 cells were more responsive to the chemical composition of the adhesion substrate, i.e., Si cytotoxicity [97].

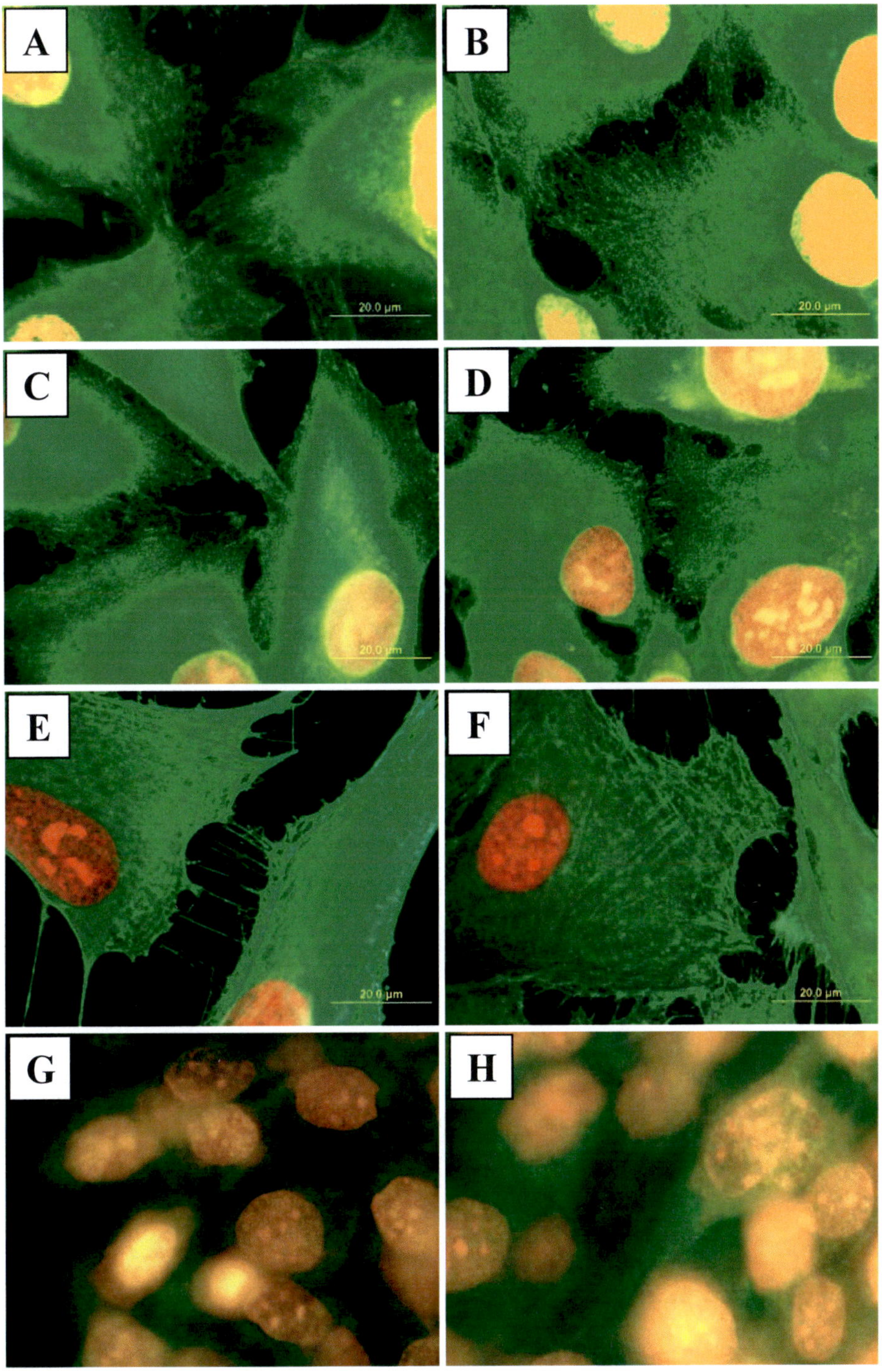

Figure 38. Immunofluorescence of alpha$_v$ integrins (A, B), talin (C, D), beta-actin (E, F) and osteocalcin (G, H) in human osteoblast-like MG 63 cells on day 1 (E, F), 3 (A-D) or 7 (G, H) after seeding on non-doped (A, C, E, G) and boron-doped (B, D, F, H) nanocrystalline diamond films deposited on microscopic glass slides. Olympus IX 50 microscope, DP 70 digital camera, obj. 100, bar = 20 µm.

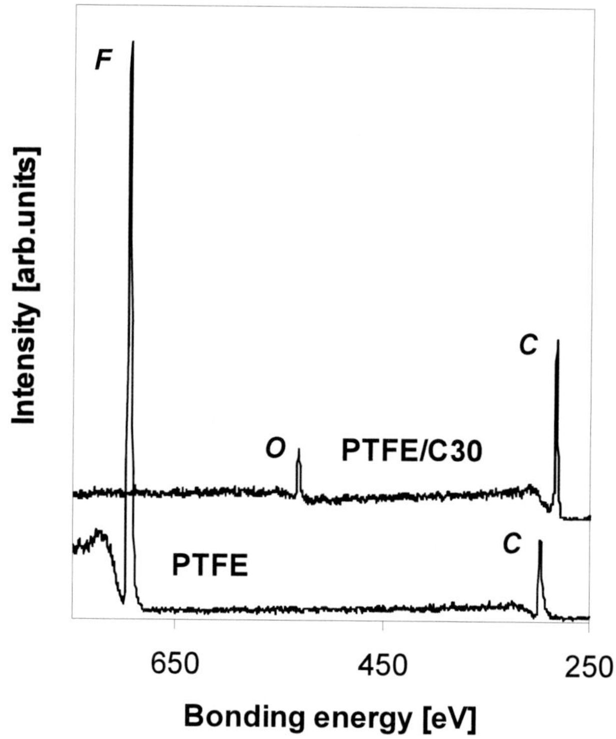

Figure 39. XPS spectra of the pristine polytetrafluoroethylene (PTFE) and PTFE with a carbon layer deposited for 30 min (PTFE/C30). A constant off-set is added to the latter spectrum [102].

5. Other Carbon-Based Layers for Potential Biomaterial Coating

These layers comprised films of amorphous hydrogenated carbon (a-C:H), deposited by by magnetron sputtering on carbon fiber-reinforced carbon composites (CFRC) to strengthen and smooth their surface in order to prevent a potential release of carbon particles from the material, and to enhance its colonization with cells. However, the adhesion and growth of MG 63 cells and primary vascular smooth muscle cells was not significantly increased in comparison with uncoated CFRC and reference cell culture polystyrene. This was most probably due to the relatively high hydrophobicity of these a-C:H films [189]. Similarly, a-C:H films have been used as bioinert coatings for blood-contacting implants and devices, as well as articular surfaces of joint replacements [103, 105, 122, 190, 191]. However, other a-C:H in our experiments, deposited on polyethyleneterephtalate (PET) by RF magnetron sputtering, enhanced adhesion and growth of mouse 3T3 fibroblasts [101]. Similarly, the a-C:H formed on polytetrafluoroethylene (PTFE) by chemical vapor deposition from acetylene induced by a UV-excimer lamp increased the colonization of the polymer with human umbilical vein endothelial cells (HUVEC, line EA.hy926) [102]. The explanation is that these a-C:H films contained additional oxygenic structures (Figure 39) and were moderately wettable (the static water drop contact angle decreased from 90° to approx. 65° with the deposition time, see Figure 40), which rendered them more attractive for cells. In addition, the material surface morphology was changed during the photo-deposition of carbon. The dependence of the surface morphology on the deposition time is shown in Figure 41 in AFM scans of 1×1 μm^2. In the first phase of deposition, the surface is covered by carbon and then the holes present in the material are filled. However, as is apparent from Figure 41, the surface nanoroughness of PTFE then increased up to 20 minutes of photo-deposition of carbon, after which it decreased. The increasing nanoscale surface roughness of the adhesion substrate has been shown to have a positive influence on cell functioning. Human osteoblast-like cells of the line SAOS-2 in cultures on nanocrystalline diamond layers showed increasing activity of cellular dehydrogenases with increasing surface nanoroughness (the root mean square roughness, rms, was in the range from 11 nm to 39 nm) [57]. Similarly, in our studies, the a-C:H layers photo-deposited on PTFE through a metallic mask for 20 minutes, were optimal for micropatterning the material surface with domains adhesive for cells, and thus for induction of regionally-selective cell adhesion and directed cell growth (Figure 42) [102].

The heterogeneous results obtained on a-C:H films indicate that the physical and chemical properties of these films, and their subsequent behavior in a biological environment can vary among different laboratories depending on the conditions of preparation, including the heterogeneity of the pristine substrates used for deposition. For example, due to the production process, the surfaces of pristine PTFE foils are typically not very homogeneous regarding surface roughnesses.

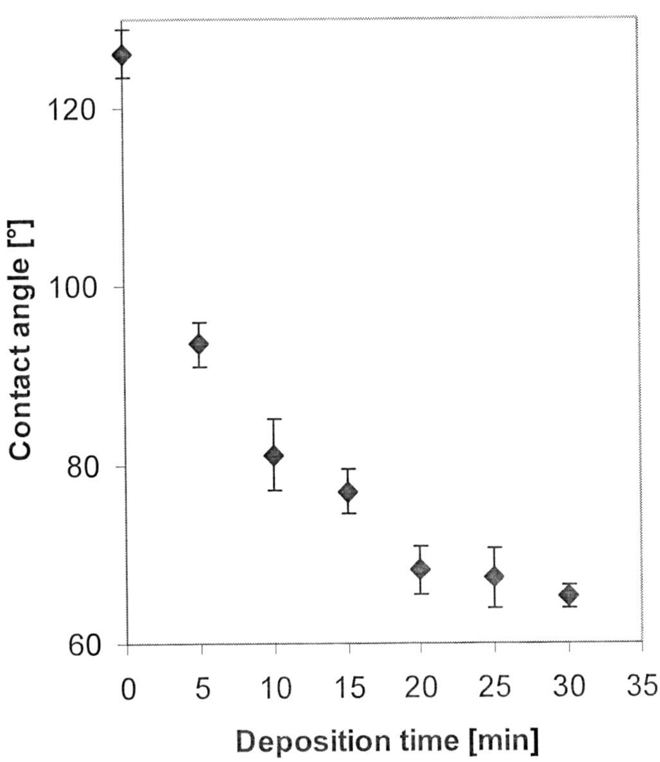

Figure 40. Dependence of water contact angle on CVD carbon deposition time on PTFE [102].

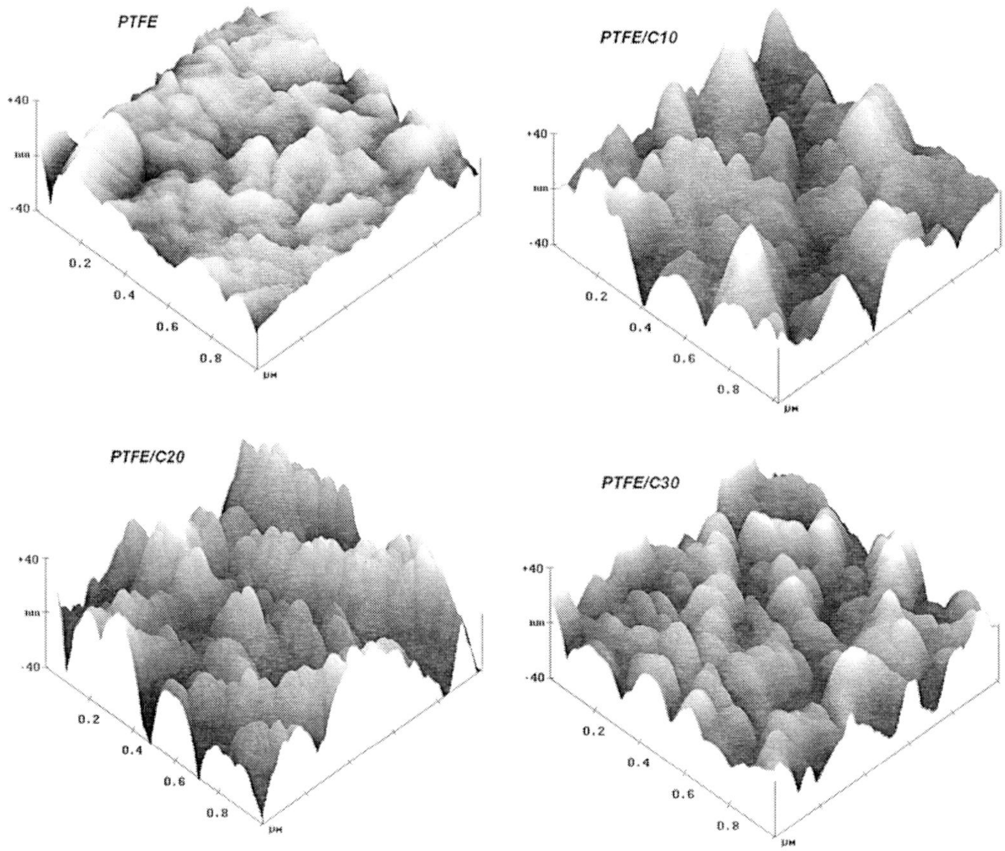

Figure 41. AFM images (scans of 1×1 μm^2) of the surface morphology of pristine polytetrafluoroethylene (PTFE) and PTFE with carbon layers deposited from 10 to 30 min (PTFE/C10 to PTFE/C30) [102].

Enrichment of amorphous carbon films with Ti also stimulated cell adhesion and growth. Coating CFRC with a carbon-titanium layer deposited by the plasma-enhanced physical vapor deposition method (PECVD), using unbalanced d.c. planar magnetron sputtering, stimulated attachment, spreading, proliferation, protein synthesis and an increase in cell volume in MG 63 and vascular smooth muscle cells [104]. Similarly, nanocomposite hydrocarbon plasma polymer films with 20 at. % of Ti, deposited on microscopic glass slides and polystyrene dishes, also had a supportive effect on adhesion, growth and maturation of MG 63 cells and especially bovine vascular endothelial cells of the line CPAE [103]. On nanocomposite hydrocarbon plasma polymer films with 3 at. % of silver, focal adhesion plaques in CPAE cells were more brightly immunostained against vinculin than those in the cells on control Ag-free hydrocarbon films, glass or polystyrene dishes. The endothelial CPAE cells on films adhered, grew, formed a filamentous beta-actin cytoskeleton, developed Weibel-Palade bodies containing the von Willebrand factor, a marker of endothelial cell maturation, and created a confluent layer in an extent similar to that observed in cells growing on control surfaces. At the same time, films with 3at. % of Ag displayed bacteriostatic activity against *Escherichia coli* [105].

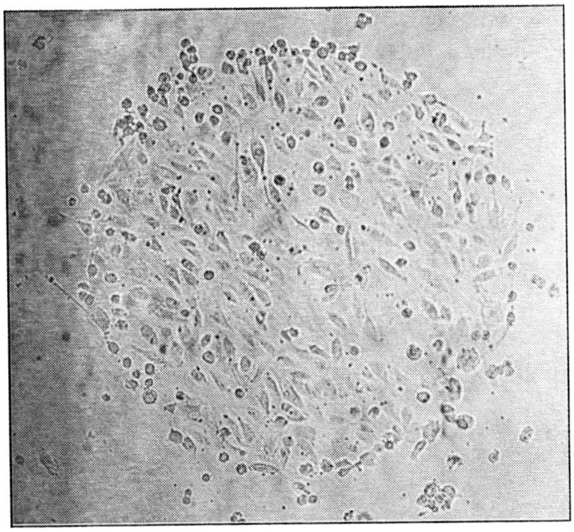

Figure 42. Phase-contrast micrographs of HUVEC (7 days after cell seeding) on a PTFE sample coated selectively with carbon for 20 min by photo-induced CVD through a contact mask (spot diameter 1.5 mm) [102].

Pyrolytic carbon deposited on CFRC in the form of isotropic layers created in a tumbling bed reactor also supported adhesion and growth of rat vascular smooth muscle cells, human osteoblast-like MG 63 cells and human embryonal lung LEP fibroblasts [155, 192, 193]. However, these effects were more pronounced after coating CFRC with pyrolytic graphite prepared by decomposition of butane in a mixture with N_2 (pressure 4 Pa, temperature 1900 ^{o}C, exposure 300–400 min), especially if this treatment was combined with polishing CFRC with metallographic paper and diamond paste [132, 133]. The MG 63 on these surfaces were better spread, grew more quickly, attained higher maximum population densities and contained higher concentrations of osteocalcin and osteopontin, i.e. indicators of osteogenic maturation. At the same time, the concentration of immunoglobulin molecules ICAM-1, markers of cell immune activation, in MG 63 cells remained similar as on reference culture substrates, represented by polystyrene dishes [131].

6. Conclusion and Further Perspectives

Fullerenes C_{60} deposited as continuous films or layers micropatterned with grooves and bulges clearly supported the adhesion, growth, viability and maturation of human osteoblast-like MG 63 cells to a similar extent as standard tissue culture polystyrene dishes and microscopic glass coverslips. In addition, layers with fullerene prominences more than 1000 nm in height promoted regionally-selective adhesion and growth of MG 63 cells in the grooves located among the prominences, and thus they could be used as templates for directed cell adhesion and growth. Composites of C_{60} and metals, particularly Ti, are also promising as growth substrates for bone-derived cells.

The terpolymer of polytetrafluoroethylene, polyvinyldifluoride and polypropylene (PTFE/PVDF/PP), or polysulphone mixed with single- or multi-wall carbon nanotubes also provided good support for the adhesion and growth of MG 63 cells. In the case of

PTFE/PVDF/PP, these parameters improved markedly in comparison with cells on the pure terpolymer. Carbon nanotubes could be added into polymeric matrices designed for the construction of porous 3D scaffolds for bone tissue engineering in order to stimulate cell ingrowth inside the material. However, the concentration of osteocalcin and osteopontin, indicators of osteogenic maturation, in cells on polymer-nanotube composites was often lower than on pure terpolymer or polystyrene dishes.

All nanocrystalline diamond films used in this study, showing different levels of surface organization, i.e., nanostructured (*rms* 8.2 nm or 18 nm) and hierarchically micro-nanostructured (*rms* 301 and 7.6 nm, respectively), as well as different chemical composition (i.e. hydrogenated, oxygen-terminated and boron-doped) provided good support for the adhesion, spreading, viability, growth and maturation of human osteoblast-like MG 63 cell in cultures on these materials. Thus, these NCD films could be used for the surface modification of bone implants (e.g., bone-anchoring parts of joint prostheses or bone replacements) in order to improve their integration with the surrounding bone tissue.

Other carbon allotropes and carbon-based materials, such as amorphous hydrogenated carbon, pyrolytic graphite, hydrocarbon plasma polymers, as well as their composites with metals can also serve for coating various materials in order to enhance their bioactivity.

The interactions of carbon nanoparticles with biological systems can be further intensified by the functionalization of these molecules with various atoms, chemical groups or molecules including ligands for cell adhesion receptors, such as RGD-containing oligopeptides or KRSR preferred by bone cells. However, at the same time, thorough studies on the potential cytotoxicity, immunogenicity, mutagenicity and carcinogenicity of carbon nanoparticles are needed for their potentional biomedical applications.

7. Acknowledgements

This study was supported by the Grant Agency of the Czech Republic (Grants No. 204/06/0225 and 101/06/0226), by the Ministry of Education of the Czech Republic (Research Program No. LC 06041), and the Acad. Sci. CR (Grants No. KAN400480701, KAN200100801, KAN 101120701 and KAN 400100701). Mr. Robin Healey (Czech Technical University, Prague) and Mrs. Sherryl Ann Vacik, Pangrac & Associates, Aransas, Texas, U.S.A. are gratefully acknowledged for their language revision of the manuscript. We also thank Dr. Vera Lisa and Mrs. Ivana Zajanova (Inst. Physiol., Acad. Sci. CR) for their excellent assistance in cell culturing and immunocytochemical methods.

References

[1] Poreda, RJ; Becker L. Fullerenes and interplanetary dust at the Permian-Triassic boundary. *Astrobiology*, 2003, *3*, 75-90.

[2] Komatsu, N; Ohe, T; Matsushige, K. A highly improved method for purification of fullerenes applicable to large-scale production. *Carbon*, 2004, *42*, 163–167.

[3] Kroto, HW; Heath, JR; O'Brien, SC; Curl, RF; Smalley, RE. C_{60}: Buckminsterfullerene. *Nature*, 1985, *318*, 162-163.

[4] Kroto, HW. Smaller carbon species in the laboratory and space. *Int. J. Mass Spectrometry and Ion Processes*, 1994, *138*, 1-15.

[5] Foley, S; Crowley, C; Smaihi, M; Bonfils, C; Erlanger, BF; Seta, P; Larroque, C. Cellular localisation of a water-soluble fullerene derivative. *Biochem. Biophys. Res. Commun.*, 2002, *294*, 116-119.

[6] Isobe, H; Nakanishi, W; Tomita, N; Jinno, S; Okayama, H; Nakamura, E. Gene delivery by aminofullerenes: structural requirements for efficient transfection. *Chem. Asian. J.*, 2006, *1*, 167-175.

[7] Isobe, H; Nakanishi, W; Tomita, N; Jinno, S; Okayama, H; Nakamura, E. Nonviral gene delivery by tetraamino fullerene. *Mol. Pharm.*, 2006, *3*, 124-134.

[8] Bakry, R; Vallant, RM; Najam-ul-Haq, M; Rainer, M; Szabo, Z; Huck, CW; Bonn, GK. Medicinal applications of fullerenes. *Int. J. Nanomedicine.*, 2007, *2*, 639-649.

[9] Guldi, DM; Prato, M. Excited-state properties of C60 fullerene derivatives. *Acc. Chem. Res.*, 2000, *33*, 695-703.

[10] Sayes, CM; Gobin, AM; Ausman, KD; Mendez, J; West JL; Colvin,VL. Nano-C60 cytotoxicity is due to lipid peroxidation. *Biomaterials*, 2005, *26*, 7587-7595.

[11] Dhawan, A; Taurozzi, JS; Pandey, AK; Shan, W; Miller, SM; Hashsham, SA; Tarabara, VV. Stable colloidal dispersions of C60 fullerenes in water: evidence for genotoxicity. *Environ. Sci. Technol.*, 2006, *40*, 7394-7401.

[12] Isakovic, A; Markovic, Z; Todorovic-Markovic, B; Nikolic, N; Vranjes-Djuric, S; Mirkovic, M; Dramicanin, M; Harhaji, L; Raicevic, N; Nikolic, Z; Trajkovic, V. Distinct cytotoxic mechanisms of pristine versus hydroxylated fullerene. *Toxicol. Sci.*, 2006, *91*, 173-183.

[13] Yamawaki, H; Iwai, N. Cytotoxicity of water-soluble fullerene in vascular endothelial cells. *Am. J. Physiol. Cell. Physiol.*, 2006, *290*, C1495-C1502.

[14] Yamakoshi, Y; Umezawa, N; Ryu, A; Arakane, K; Miyata, N; Goda, Y; Masumizu, T; Nagano, T. Active oxygen species generated from photoexcited fullerene (C_{60}) as potential medicines: O_2^- versus $^1O_2$. *J. Am. Chem. Soc.*, 2003, *125*, 12803-12809.

[15] Tang, YJ; Ashcroft, JM; Chen, D; Min, G; Kim, CH; Murkhejee, B; Larabell, C; Keasling, JD; Chen, FF. Charge-associated effects of fullerene derivatives on microbial structural integrity and central metabolism. *Nano Lett.* 2007, *7*, 754-760.

[16] Satoh, M; Matsuo, K; Kiriya, H; Mashino, T; Hirobe, M; Takayanagi, I. Inhibitory effect of a fullerene derivative, monomalonic acid C60, on nitric oxide-dependent relaxation of aortic smooth muscle. *Gen. Pharmacol.*, 1997, *29*, 345-351.

[17] Wilson, SR. *Fullerenes: Chemistry, Physics, and Technology*. Kadish KM and Ruoff RS (Eds.), John Wiley and Sons, 2000, 437-465.

[18] Lai, H-S; Chen, Y; Chen, W-J; Chang, K-J; Chiang, L-Y. Free radical scavenging activity of fullerenol on grafts after small bowel transplantation in dogs. *Transplant. Proc.*, 2000, *32*, 1272–1274.

[19] Huang, SS; Tsai, SK; Chih, CL; Chiang, LY; Hsieh, HM; Teng, CM; Tsai, MC. Neuroprotective effect of hexasulfobutylated C_{60} on rats subjected to focal cerebral ischemia. *Free Radic. Biol. Med.*, 2001, *30*, 643-649.

[20] Yang, DY; Wang, MF; Chen, IL; Chan, YC; Lee, MS; Cheng, FC. Systemic administration of a water-soluble hexasulfonated C(60) (FC(4)S) reduces cerebral ischemia-induced infarct volume in gerbils. *Neurosci. Lett.*, 2001, *311*, 121-124.

[21] Nakamura, E; Isobe, H. Functionalized fullerenes in water. The first 10 years of their chemistry, biology, and nanoscience. *Acc. Chem. Res.* 2003, *36*, 807-815.

[22] Chen, Y-W; Hwang, KC; Yen, C-C; Lai, YL. Fullerene derivatives protect against oxidative stress in RAW 264.7 cells and ischemia-reperfused lungs. *Am. J. Physiol. Regul. Integr. Comp. Physiol.*, 2004, *287*, R21-R26.

[23] Kawase, T; Tanaka, K; Seirai, Y; Shiono N; Oda, M. Complexation of carbon nanorings with fullerenes: Supramolecular dynamics and structural tuning for a fullerene sensor. *Angew. Chem. Int. Ed. Engl.*, 2003, *42*, 5597-5600.

[24] Bhattacharya, S; Tominaga, K; Kimura, T; Uno, H; Komatsu, N. A new metalloporphyrin dimer: Effective and selective molecular tweezers for fullerenes. *Chem. Phys. Lett.*, 2007, *433*, 395-402.

[25] Simon, F. Studying single-wall carbon nanotubes through encapsulation: from optical methods till magnetic resonance. J. *Nanosci. Nanotechnol.*, 2007, *7,* 1197-1220.

[26] Sitharaman, B; Tran, LA; Pham, QP; Bolskar, RD; Muthupillai, R; Flamm, SD; Mikos, AG; Wilson, LJ. Gadofullerenes as nanoscale magnetic labels for cellular MRI. *Contrast Media Mol. Imaging.*, 2007, *2*, 139-146.

[27] Fumelli, C; Marconi, A; Salvioli, S; Straface, E; Malorni, W; Offidani, AM; Pellicciari, R; Schettini, G; Giannetti, A; Monti, D; Franceschi, C; Pincelli, C. Carboxyfullerenes protect keratinocytes from ultraviolet-B-induced apoptosis. *J. Invest. Dermatol.*, 2000, *115*, 835-841.

[28] Isakovic, A; Markovic, Z; Todorovic-Markovic, B; Nikolic, N; Vranjes-Djuric, S; Mirkovic, M; Dramicanin, M; Harhaji, L; Raicevic, N; Nikolic, Z; Trajkovic, V. Distinct cytotoxic mechanisms of pristine versus hydroxylated fullerene. *Toxicol. Sci.*, 2006, *91*, 173-183.

[29] Lai, YL; Murugan, P; Hwang, KC. Fullerene derivative attenuates ischemia-reperfusion-induced lung injury. *Life Sci.*, 2003, *72*, 1271-1278.

[30] Wolff, DJ, Papoiu, AD; Mialkowski, K; Richardson, CF; Schuster, DI; Wilson, SR. Inhibition of nitric oxide synthase isoforms by tris-malonyl-C(60)-fullerene adducts. *Arch. Biochem. Biophys.*, 2000, *378*, 216-223.

[31] Iijima, S. Helical microtubules of graphitic carbon. *Nature*, 1991, *354*, 56-58.

[32] Iijima, S; Ichihashi, T. Cobalt-catalysed growth of carbon nanotubes with single-atomic-layer walls. *Nature*, 1993, *363*, 603-605.

[33] Iijima, S. Carbon nanotubes: past, present, and future. *Physica B*, 2002, *323*, 1–5.

[34] Fernandez-Alonso, F; Bermejo, FJ; Cabrillo, C; Loutfy, RO; Leon, V; Saboungi, ML. Nature of the bound states of molecular hydrogen in carbon nanohorns. *Phys. Rev. Lett.*, 2007, *98*, 215503.

[35] Lee, H; Cho, J. Sn(78)Ge(22)@carbon core-shell nanowires as fast and high-capacity lithium storage media. *Nano Lett.*, 2007, *7*, 2638-2641.

[36] Ando, Y; Zhao, X; Inoue, S; Iijima, S. Mass production of multiwalled carbon nanotubes by hydrogen arc discharge. *J. Crystal Growth*, 2002, *237–239*, 1926–1930.

[37] Yudasaka, M; Ichihashi, T; Komatsu, T; Iijima, S. Single-wall carbon nanotubes formed by a single laser-beam pulse. *Chemical Physics Letters*, 1999, *299*, 91–96.

[38] Okazaki, T; Saito, T; Matsuura, K; Ohshima, S; Yumura, M; Oyama, Y; Saito, R; Iijima, S. Photoluminescence and population analysis of single-walled carbon nanotubes produced by CVD and pulsed-laser vaporization methods. *Chem. Phys. Lett.*, 2006, *420*, 286–290.

[39] Kagan, VE; Tyurina, YY; Tyurin, VA; Konduru, NV; Potapovich, AI; Osipov, AN; Kisin, ER; Schwegler-Berry, D; Mercer, R; Castranova, V; Shvedova, AA. Direct and indirect effects of single walled carbon nanotubes on RAW 264.7 macrophages: role of iron. *Toxicol. Lett.*, 2006, *165*, 88-100.

[40] Pulskamp, K; Diabaté, S; Krug, HF. Carbon nanotubes show no sign of acute toxicity but induce intracellular reactive oxygen species in dependence on contaminants. *Toxicol. Lett.*, 2007, *168*, 58-74.

[41] Pensabene, V; Vittorio, O; Raffa, V; Menciassi, A; Dario, P. Investigation of CNTs interaction with fibroblast cells. *Conf. Proc. IEEE Eng. Med. Biol. Soc.*, 2007, *1*, 6620-6623.

[42] Wick, P; Manser, P; Limbach, LK; Dettlaff-Weglikowska, U; Krumeich, F; Roth, S; Stark, WJ; Bruinink, A. The degree and kind of agglomeration affect carbon nanotube cytotoxicity. *Toxicol. Lett.*, 2007, *168*, 121–131.

[43] Islam, MF; Rojas, E; Bergey, DM; Johnson, AT; Yodh, AG. High weight fraction surfactant solubilization of single-wall carbon nanotubes in water. *Nano Letters*, 2003, *3*, 269-273.

[44] Sayes, CM; Liang, F; Hudson, JL; Mendez, J; Guo, W; Beach, JM; Moore, VC; Doyle, CD; West, JL; Billups, WE; Ausman, KD; Colvin, VL. Functionalization density dependence of single-walled carbon nanotubes cytotoxicity in vitro. *Toxicol. Lett.*, 2006, *161*, 135-142.

[45] Dillon, C; Gennett, T; Jones, KM; Alleman, JL; Parilla, PA; Heben, MJ. A simple and complete purification of single-walled carbon nanotube materials. *Adv. Mater.*, 1999, 11, 1354-1358.

[46] Chattopadhyay, D; Galeska, I; Papadimitrakopoulos, F. Complete elimination of metal catalysts from single wall carbon nanotubes. *Carbon*, 2002, *40*, 985–988.

[47] Bottini, M; Bruckner, S; Nika, K; Bottini, N; Bellucci, S; Magrini, A; Bergamaschi, A; Mustelin, T. Multi-walled carbon nanotubes induce T lymphocyte apoptosis. *Toxicol. Lett.*, 2006, *160*, 121-126.

[48] Magrez, A; Kasas, S; Salicio, V; Pasquier, N; Seo, JW; Celio, M; Catsicas, S; Schwaller, B; Forro, L. Cellular toxicity of carbon-based nanomaterials. *Nano Lett.*, 2006, *6*, 1121-1125.

[49] Ebbesen, TW. Carbon nanotubes. *Phys. Today,* 1996, *49*, 26-32.

[50] Yakobson, B; Smalley, RE. Fullerene Nanotubes: C[sub1,000,000) and Beyond. *Am. Sci.*, 1997, *85*, 324-333.

[51] Shi, X; Sitharaman, B; Pham, QP; Liang, F; Wu, K; Edward Billups, W; Wilson, LJ; Mikos, AG. Fabrication of porous ultra-short single-walled carbon nanotube nanocomposite scaffolds for bone tissue engineering. *Biomaterials*, 2007, *28*, 4078-4090.

[52] Abarrategi, A; Gutiérrez, MC; Moreno-Vicente, C; Hortigüela, MJ; Ramos, V; López-Lacomba, JL; Ferrer, ML; del Monte, F. Multiwall carbon nanotube scaffolds for tissue engineering purposes. *Biomaterials*, 2008, *29*, 94-102.

[53] Price, RL; Waid, MC; Haberstroh, KM; Webster, TJ. Selective bone cell adhesion on formulations containing carbon nanofibers. *Biomaterials*, 2003, *24*, 1877-1887.

[54] Price, RL; Ellison, K; Haberstroh, KM; Webster, TJ: Nanometer surface roughness increases select osteoblast adhesion on carbon nanofiber compacts.*J Biomed. Mater. Res. A*, 2004, *70*, 129–138.

[55] Supronowicz, PR; Ajayan, PM; Ullmann, KR; Arulanandam, BP; Metzger, DW; Bizios, R. Novel current-conducting composite substrates for exposing osteoblasts to alternating current stimulation. *J. Biomed. Mater. Res.*, 2002, *59*, 499-506.

[56] Zanello, LP; Zhao, B; Hu, H; Haddon, RC. Bone cell proliferation on carbon nanotubes. *Nano Lett.*, 2006, *6*, 562-567.

[57] Kalbacova, M; Kalbac, M; Dunsch, L; Kromka, A; Vanecek, M; Rezek, B; Hempel, U; Kmoch, S: The effect of SWCNT and nano-diamond films on human osteoblast cells. *Phys. Stat. Sol.*, 2007, *(b)*, 1–4.

[58] Malarkey, EB; Parpura, V. Applications of carbon nanotubes in neurobiology. *Neurodegener. Dis.*, 2007, *4*, 292-299, 2007.

[59] Mattson, MP; Haddon, RC; Rao, AM. Molecular functionalization of carbon nanotubes and use as substrates for neuronal growth. *J. Mol. Neurosci.*, 2000, *14*, 175-182.

[60] http://www.cellnucleus.com/TEM.htm.

[61] Ohtsuki, T; Masumoto, K; Shikano, K; Sueki, K; Tanaka, T; Komatsu, K. Direct preparation of radioactive fullerenes as a tracer for applications. *Biol. Trace. Elem. Res.*, 1999, *71-72*, 489-498.

[62] Bottini, M; Cerignoli, F; Dawson, MI; Magrini, A; Rosato, N; Mustelin, T. Full-length single-walled carbon nanotubes decorated with streptavidin-conjugated quantum dots as multivalent intracellular fluorescent nanoprobes. *Biomacromolecules*, 2006, *7*, 2259-2263.

[63] Hoshino, A; Manabe, N; Fujioka, K; Suzuki, K; Yasuhara, M; Yamamoto, K. Use of fluorescent quantum dot bioconjugates for cellular imaging of immune cells, cell organelle labeling, and nanomedicine: surface modification regulates biological function, including cytotoxicity. *J. Artif. Organs.*, 2007, *10*, 149-157.

[64] Hartman, KB; Wilson, LJ. Carbon nanostructures as a new high-performance platform for MR molecular imaging. *Adv. Exp. Med. Biol.*, 2007, *620*, 74-84.

[65] Yinghuai, Z; Peng, AT; Carpenter, K; Maguire, JA; Hosmane, NS; Takagaki, M. Substituted carborane-appended water-soluble single-wall carbon nanotubes: new approach to boron neutron capture therapy drug delivery. *J. Am. Chem. Soc.*, 2005, *127*, 9875-9880.

[66] Escorcia, FE; McDevitt, MR; Villa, CH; Scheinberg, DA. Targeted nanomaterials for radiotherapy. *Nanomed.*, 2007, *2*, 805-815.

[67] Singh, R; Pantarotto, D; McCarthy, D; Chaloin, O; Hoebeke, J; Partidos, CD; Briand, JP; Prato, M; Bianco, A; Kostarelos, K. Binding and condensation of plasmid DNA onto functionalized carbon nanotubes: toward the construction of nanotube-based gene delivery vectors. *J. Am. Chem. Soc.*, 2005, *127*, 4388-4396.

[68] Yao, DS; Cao, H; Wen, S; Liu, DL; Bai, Y; Zheng, WJ. A novel biosensor for sterigmatocystin constructed by multi-walled carbon nanotubes (MWNT) modified with aflatoxin-detoxifizyme (ADTZ). *Bioelectrochemistry*, 2006, *68*, 126-133.

[69] Du, D; Huang, X; Cai, J; Zhang, A; Ding, J; Chen, S. An amperometric acetylthiocholine sensor based on immobilization of acetylcholinesterase on a multiwall carbon nanotube-cross-linked chitosan composite. *Anal. Bioanal. Chem.*, 2007, *387*, 1059-1065.

[70] Goyal, RN; Gupta, VK; Bachheti, N. Fullerene-C60-modified electrode as a sensitive voltammetric sensor for detection of nandrolone--an anabolic steroid used in doping. *Anal Chim Acta.*, 2007, *597*, 82-89.

[71] Woolley, AT. Biofunctionalization of carbon nanotubes for atomic force microscopy imaging. *Methods Mol. Biol.*, 2004, *283*, 305-319.

[72] Gibson, CT; Carnally, S; Roberts, CJ; Attachment of carbon nanotubes to atomic force microscope probes. *Ultramicroscopy*, 2007, *107*, 1118-1122.

[73] Ding, F; Lin, Y; Krasnov, PO; Yakobson, BI. Nanotube-derived carbon foam for hydrogen sorption. *J. Chem. Phys.*, 2007, *127*, 164703.

[74] Kowalczyk, P; Hołyst, R; Terrones, M; Terrones, H. Hydrogen storage in nanoporous carbon materials: myth and facts. *Phys. Chem. Chem. Phys.*, 2007, *9*, 1786-1792.

[75] Nikitin, A; Li, X; Zhang, Z; Ogasawara, H; Dai, H; Nilsson, A. Hydrogen storage in carbon nanotubes through the formation of stable C-H bonds. *Nano Lett.*, 2008, *8*, 162-167.

[76] Murata, K; Miyawaki, J; Yudasaka, M; Iijima, S; Kaneko, K. High-density of methane confined in internal nanospace of single-wall carbon nanohorns. *Carbon,* 2005, *43*, 2817–2833.

[77] Kim, JW; Shashkov, EV; Galanzha, EI; Kotagiri, N; Zharov, VP. Photothermal antimicrobial nanotherapy and nanodiagnostics with self-assembling carbon nanotube clusters. *Lasers Surg. Med.*, 2007, *39*, 622-34.

[78] Park, KH; Chhowalla, M; Iqbal, Z; Sesti, F. Single-walled carbon nanotubes are a new class of ion channel blockers. *J. Biol. Chem.*, 2003, *278*, 50212-50216.

[79] Hinds, BJ; Chopra, N; Rantell, T; Andrews, R; Gavalas, V; Bachas, LG. Aligned multiwalled carbon nanotube membranes. *Science*, 2004, *303*, 62-65.

[80] Majumder, M; Zhan, X; Andrews, R; Hinds, BJ. Voltage gated carbon nanotube membranes. *Langmuir*, 2007, *23*, 8624-8631.

[81] van Hijkoop, VJ; Dammers, AJ; Malek, K; Coppens, MO. Water diffusion through a membrane protein channel: a first passage time approach. *J. Chem. Phys.*, 2007, *127*, 085101.

[82] Cui, D; Tian, F; Ozkan, CS; Wang, M; Gao, H. Effect of single wall carbon nanotubes on human HEK293 cells. *Toxicol. Lett.*, 2005, *155*, 73-85.

[83] Davoren, M; Herzog, E; Casey, A; Cottineau, B; Chambers, G; Byrne, HJ; Lyng, FM. In vitro toxicity evaluation of single walled carbon nanotubes on human A549 lung cells. *Toxicol. In Vitro*, 2007, *21*, 438-448.

[84] Zhang, LW; Zeng, L; Barron, AR; Monteiro-Riviere, NA. Biological interactions of functionalized single-wall carbon nanotubes in human epidermal keratinocytes. *Int. J. Toxicol.*, 2007, *26*, 103-113.

[85] Kisin, ER; Murray, AR; Keane, MJ; Shi, XC; Schwegler-Berry, D; Gorelik, O; Arepalli, S; Castranova, V; Wallace, WE; Kagan, VE; Shvedova, AA. Single-walled carbon nanotubes: geno- and cytotoxic effects in lung fibroblast V79 cells. *J. Toxicko. Environ. Health.*, 2007, *A 70*: 2071-2079.

[86] Iakoubovskii, K; Adriaenssens, GJ; Meykens, K; Nesladek, M; Vul, AY; Osipov, VY. Study of defects in CVD and ultradisperse diamond. *Diamond Relat. Mater.*, 1999, *8*, 1476–1479.

[87] Qin, LC; Zhou, D; Krauss, AR; Gruen, DM. TEM Characterization of nanodiamond thin films. *NanoStructured Materials*, 1998, *10*, 649-660.

[88] Bajaj, P; Akin, D; Gupta, A; Sherman, D; Shi, B; Auciello, O; Bashir, R. Ultrananocrystalline diamond film as an optimal cell interface for biomedical applications. *Biomed. Microdevices*, 2007, *9*, 787–794.

[89] Hartl, A; Schmich, E; Garrido, JA; Hernando, J; Catharino, SC; Walter, S; Feulner, P; Kromka, A; Steinmüller, D; Stutzmann, M. Protein-modified nanocrystalline diamond thin films for biosensor applications. *Nat. Mater.*, 2004, *3*, 736-742.

[90] Christiaens, P; Vermeeren, V; Wenmackers, S; Daenen, M; Haenen, K; Nesládek, M; vandeVen, M; Ameloot, M; Michiels, L; Wagner, P. EDC-mediated DNA attachment to nanocrystalline CVD diamond films. *Biosens. Bioelectron.*, 2006, *22*, 170-177.

[91] Schrand, AM; Huang, H; Carlson, C; Schlager, JJ; Osawa, E; Hussain, SM; Dai, L. Are diamond nanoparticles cytotoxic? *J. Phys. Chem.*, 2007, *111*, 2-7.

[92] Bakowicz, K; Mitura, S. Biocompatibility of NCD. *J. Wide Bandgap Mater.*, 2002, *9*, 261.

[93] Fu, CC; Lee, HY; Chen, K; Lim, TS; Wu, HY; Lin, PK; Wei, PK; Tsao, PH; Chang, HC; Fann, W. Characterization and application of single fluorescent nanodiamonds as cellular biomarkers. *Proc. Natl. Acad. Sci. U. S. A.*, 2007, *104*, 727-732.

[94] Papo, MJ; Catledge, SA; Vohra, YK; Machado, C. Mechanical wear behavior of nanocrystalline and multilayer diamond coatings on temporomandibular joint implants. *J. Mater. Sci. Mater. Med.*, 2004, *15*, 773-777.

[95] Bacakova, L; Grausova, L; Vacik, J; Fraczek, A; Blazewicz, S; Kromka, A; Vanecek, M; Svorcik, V. Improved adhesion and growth of human osteoblast-like MG 63 cells on biomaterials modified with carbon nanoparticles. *Diamond Relat. Mater.*, 2007, *16*, 2133-2140.

[96] Grausova, L; Bacakova, L; Kromka, A; Potocky, S; Vanecek, M; Nesladek, M; Lisa, V. Nanodiamond as a promising material for bone tissue engineering. *J. Nanosci. Nanotechnol.,* 2008, in press.

[97] Grausova, L; Kromka, A; Bacakova, L; Potocky, S; Vanecek, M; Lisa, V. Bone and vascular endothelial cells in cultures on nanocrystalline diamond films. *Diamond Relat. Mater.*, 2008, in press.

[98] Specht, CG; Williams, OA; Jackman, RB; Schoepfer, R. Ordered growth of neurons on diamond. *Biomaterials*, 2004, *25*, 4073-4078.

[99] Rupprecht, S; Bloch, A; Rosiwal, S; Neukam, FW, Wiltfang, J. Examination of the bone-metal interface of titanium implants coated by the microwave plasma chemical vapor deposition method. *Int. J. Oral Maxillofac. Implants*, 2002, *17*, 778-785.

[100] Xiao, X; Wang, J; Liu,C; Carlisle, JA; Mech, B; Greenberg, R; Guven, D; Freda, R; Humayun, MS; Weiland, J; Auciello, O. In vitro and in vivo evaluation of ultrananocrystalline diamond for coating of implantable retinal microchips. *J. Biomed. Mater. Res. B Appl. Biomater.*, 2006, *77*, 273-281.

[101] Kubova, O; Bacakova, L; V. Svorcik, V. Biocompatibility of carbon layer on polymer. *Mater. Sci. Forum*, 2005, *482*, 247-250.

[102] Kubova, O; Svorcik, V; Heitz, J; Moritz, S; Romanin, C; Matejka, P; Mackova, A. Characterization and cytocompatibility of carbon layers prepared by photo-induced chemical vapor deposition. *Thin Solid Films*, 2007, *515*, 6765-6772.

[103] Grinevich, A; Bacakova, L; Choukourov, A; Boldyryeva, H; Pihosh, Y; Slavinska, D; Noskova, L; Skuciova, M; Lisa, V; Biederman, H. Nanocomposite Ti/hydrocarbon plasma polymer films from reactive magnetron sputtering as growth supports for osteoblast-like and endothelial cells. *J. Biomed. Mater. Res. Part A*, 2008, in press.

[104] Bacakova, L; Stary, V; Kofronova, O; Lisa, V. Polishing and coating carbon fiber-reinforced carbon composites with a carbon-titanium layer enhances adhesion and

growth of osteoblast-like MG63 cells and vascular smooth muscle cells *in vitro*. *J. Biomed. Mater. Res.*, 2001, *54*, 567-578.

[105] Bacakova, L; Koshelyev, H; Noskova, L; Choukourov, A; Benada, O; Mackova, A; Lisa, V; Biederman, H. Vascular endothelial cells in cultures on nanocomposite silver/hydrocarbon plasma polymer films with antimicrobial activity. *J. Optoelectronics and Advanced Materials*, 2008, in press.

[106] Webster, TJ; Ergun, C; Doremus, RH; Siegel, RW; Bizios, R. Enhanced functions of osteoblasts on nanophase ceramics. *Biomaterials*, 2000 , *21*, 1803-1810.

[107] Webster, TJ; Ergun, C; Doremus, RH; Siegel, RW; Bizios, R. Specific proteins mediate enhanced osteoblast adhesion on nanophase ceramics. *J. Biomed. Mater. Res.*, 2000, *51*, 475-483.

[108] Webster, TJ; Smith, TA. Increased osteoblast function on PLGA composites containing nanophase titania. *J. Biomed. Mater. Res. A*, 2005, *74*, 677-686.

[109] Woo, KM; Chen,VJ; Ma, PX. Nano-fibrous scaffolding architecture selectively enhances protein adsorption contributing to cell attachment. *J. Biomed. Mater. Res. A*, 2003, *67,* 531-537.

[110] Wei, G; Ma, PX. Structure and properties of nano-hydroxyapatite/polymer composite scaffolds for bone tissue engineering. *Biomaterials*, 2004, *25*, 4749-4757.

[111] Wang, L; Zhao, G; Olivares-Navarrete, R; Bell, BF; Wieland, M; Cochran, DL; Schwartz, Z; Boyan, BD. Integrin $beta_1$ silencing in osteoblasts alters substrate-dependent responses to 1,25-dihydroxy vitamin D_3. *Biomaterials*, 2006, *27*, 3716-3725.

[112] Dee, KC; Andersen, TT; Bizios, R. Design and function of novel osteoblast-adhesive peptides for chemical modification of biomaterials. *J. Biomed. Mater. Res.*, 1998, *40*, 371-377.

[113] Gobin, S; West, JL. Val-ala-pro-gly, an elastin-derived non-integrin ligand: Smooth muscle cell adhesion and specificity. *J. Biomed. Mater. Res. A*, 2003, *67*, 255–259.

[114] Shin, H; Jo, S; Mikos, AG. Biomimetic materials for tissue engineering. *Biomaterials*, 2003, *24*, 4353–4364.

[115] Bacakova, L; Noskova, L; Koshelyev, H; Biederman, H; Vascular endothelial cells in cultures on metal/C:H composite films. *Engineering of Biomaterials*, 2004, *7 [37]*, 18-20.

[116] Tan, J; Saltzman, WM. Biomaterials with hierarchically defined micro- and nanoscale structure. *Biomaterials*, 2004, *25*, 3593-3601.

[117] Bacakova, L; Grausova, L; Vacik, J; Svorcik, V. Human bone-derived cells in cultures on fullerene C_{60} layers of continuous and patterned morphology. *Conference NanoSMat 2007*, Algarve, Portugal, Abstract No. BBE20, p. 66.

[118] Grausova, L; Bilkova, P; Bacakova, L; Vacik, J; Svorcik, V.: Regionally-selective adhesion and growth of human osteoblast-like MG 63 cells on micropatterned fullerene C_{60} layers. *14th International Conference on Plasma Physics and Applications (14th CPPA 2007)*, Brasov, Romania, p. 127.

[119] Levi, N; Hantgan, RR; Lively, MO; Carroll, DL; Prasad, GL. C_{60}-fullerenes: detection of intracellular photoluminescence and lack of cytotoxic effects. *J. Nanobiotechnol.*, 2006, *4*, 14.

[120] Lin, JC; Wu, CH: Surface characterization and platelet adhesion studies on polyurethane surface immobilized with C_{60}. *Biomaterials*, 1999, *20*, 1613-1620.

[121] Bacakova, L; Walachova, K; Svorcik, V; Hnatowitz, V. Adhesion and proliferation of rat vascular smooth muscle cells (VSMC) on polyethylene implanted with O(+) and C(+) ions. *J. Biomater. Sci. Polym. Ed.*, 2001, *12,* 817-834.

[122] Bacakova, L; Filova, E; Rypacek, F; Svorcik, V; Stary, V. Cell adhesion on artificial materials for tissue engineering. *Physiol Res.*, 2004, *53 Supp* , S35-45.

[123] Tsuchiya, T; Yamakoshi, YN; Miyata, N. A novel promoting action of fullerene C_{60} on the chondrogenesis in rat embryonic limb bud cell culture system. *Biochem. Biophys. Res. Commun.*, 1995, *206*, 885-894.

[124] Straface, E; Natalini, B; Monti, D; Franceschi, C; Schettini, G; Bisaglia, M; Fumelli, C; Pincelli, C; Pellicciari, R; Malorni, W. C3-fullero-tris-methanodicarboxylic acid protects epithelial cells from radiation-induced anoikia by influencing cell adhesion ability. *FEBS Lett.*, 1999, *454*, 335-340.

[125] Yamawaki, H; Iwai, N. Cytotoxicity of water-soluble fullerene in vascular endothelial cells. *Am. J. Physiol. Cell. Physiol.*, 2006, *290*, C1495-C1502.

[126] Yamakoshi, Y; Umezawa, N; Ryu, A; Arakane, K; Miyata, N; Goda, Y; Masumizu, T; Nagano, T. Active oxygen species generated from photoexcited fullerene (C_{60}) as potential medicines: O_2^- versus $^1O_2$. *J. Am. Chem. Soc.*, 2003, *125*, 12803-12809.

[127] Tang, YJ; Ashcroft, JM; Chen, D; Min, G; Kim, CH; Murkhejee, B; Larabell, C; Keasling, JD; Chen, FF. Charge-associated effects of fullerene derivatives on microbial structural integrity and central metabolism. *Nano Lett.*, 2007, *7*, 754-760.

[128] Harhaji, L; Isakovic, A; Raicevic, N; Markovic, Z; Todorovic-Markovic, B; Nikolic, N; Vranjes-Djuric, S; Markovic, I; Trajkovic, V. Multiple mechanisms underlying the anticancer action of nanocrystalline fullerene. *Eur J Pharmacol.*, 2007, *568*, 89-98.

[129] Harhaji, L; Isakovic, A; Vucicevic, L; Janjetovic, K; Misirkic, M; Markovic, Z; Todorovic-Markovic, B; Nikolic, N; Vranjes-Djuric, S; Nikolic, Z; Trajkovic, V. Modulation of tumor necrosis factor-mediated cell death by fullerenes. *Pharm. Res.*, 2007, in press.

[130] Bacakova, L; Stary, V; Hornik, J; Glogar, P; Lisa, V; Kofronova O. Osteoblast-like MG63 cells in cultures on carbon fibre-reinforced carbon composites. *Inzynieria Biomateriałów-Engineering of Biomaterials*, 2001, *4* [17-19], 11-12.

[131] Bacakova, L; Stary, V; Glogar, P; Lisa, V. Adhesion, differentiation and immune activation of human osteogenic cells in cultures on carbon-fibre reinforced carbon composites. *Inzynieria Biomateriałów-Engineering of Biomaterials*, 2003, *6*, 8-9.

[132] Stary, V; Bacakova, L; Hornik, J; Chmelik, V. Bio-compatibility of the surface layer of pyrolytic graphite. *Thin Solid Films*, 2003, *433*, 191-198.

[133] Stary, V; Glogar, P; Bacakova, L; Hnilica, F; Chmelik, V; Korinek, Z; Gregor, J; Mares, V; Lisa, V. A study of surface properties of composite materials and their influence on biocompatibility. *Acta Montana AB*, 2003, *11*, 19-36.

[134] Haddow, DB; France, RM; Short, RD; MacNeil, S; Dawson, RA; Leggett, GJ; Cooper, E. Comparison of proliferation and growth of human keratinocytes on plasma copolymers of acrylic acid/1,7-octadiene and self-assembled monolayers. *J. Biomed. Mater. Res.*, 1999, *47*, 379-387.

[135] Bacakova, L; Mares, V; Bottone, MG; Pellicciari, C; Lisa, V; Svorcik, V. Fluorine-ion-implanted polystyrene improves growth and viability of vascular smooth muscle cells in culture. *J. Biomed. Mater. Res.*, 2000, *49*, 369-379.

[136] Bacakova, L; Mares, V; Lisa, V; Svorcik, V. Molecular mechanisms of improved adhesion and growth of an endothelial cell line cultured on polystyrene implanted with fluorine ions. *Biomaterials*, 2000, *21*, 1173-1179.

[137] Mikulikova, R; Moritz, S; Gumpenberger, T; Olbrich, M; Romanin, Ch; Bacakova, L; Svorcik, V; Heitz, J. Cell microarrays on photochemically modified PTFE, *Biomaterials*, 2005, *26*, 5572.

[138] Parizek, M; Bacakova, L; Lisa, V; Kubova, O; Svorcik, V; Heitz J. Vascular smooth muscle cells in cultures on synthetic polymers with adhesive microdomains. *Inzynieria Biomaterialow (Engineering of Biomaterials)*, 2006, *IX (58-60)*, 7-10.

[139] Parizek, M; Bacakova, L; Kasalkova, N; Svorcik, V; Kolarova, K. Improved adhesion and growth of vascular smooth muscle cells in cultures on Polyethylene modified by plasma discharge. *Inzynieria Biomaterialow (Engineering of Biomaterials)*, 2007, *X (67-68)*, 1-4.

[140] Kasalkova, N; Kolarova, K; Bacakova, L; Parizek, M; Svorcik, V. Cell adhesion and proliferation on modified polyethylene, *Mater. Sci. Forum*, 2007, *567-568*, 269-272.

[141] Bacakova, L; Filova, E; Kubies, D; Machova, L; Proks, V; Malinova, V; Lisa, V; Rypacek, F. Adhesion and growth of vascular smooth muscle cells in cultures on bioactive RGD peptide-carrying polylactides. *J. Mater. Sci. Mater. Med.*, 2007, *18*,1317-1323.

[142] Patrito, N; McCague, C; Norton, PR; Petersen, NO. Spatially controlled cell adhesion via micropatterned surface modification of poly(dimethylsiloxane). *Langmuir*, 2007, *23*, 715-719.

[143] Hamilton, DW; Chehroudi, B; Brunette, DM. Comparative response of epithelial cells and osteoblasts to microfabricated tapered pit topographies in vitro and in vivo. *Biomaterials*, 2007, *28*, 2281-2293.

[144] Heitz, J; Svorcik, V; Bacakova, L; Rockova, K; Ratajova, E; Gumpenberger, T; Bauerle, D; Dvorankova, B; Kahr, H; Graz, I; Romanin, C. Cell adhesion on polytetrafluoroethylene modified by UV-irradiation in an ammonia atmosphere. *J. Biomed. Mater. Res.*, 2003, *67A*, 130-137.

[145] Charest, JL; Bryant, LE; Garcia, AJ; King, WP. Hot embossing for micropatterned cell substrates. *Biomaterials*, 2004, *25*, 4767-4775.

[146] Kenar, H; Kose, GT; Hasirci, V. Tissue engineering of bone on micropatterned biodegradable polyester films. *Biomaterials*, 2006, *27*, 885-895.

[147] Hamilton, DW; Brunette, DM. The effect of substratum topography on osteoblast adhesion mediated signal transduction and phosphorylation. *Biomaterials*, 2007, *28*, 1806-1819.

[148] Papenburg, BJ; Vogelaar, L; Bolhuis-Versteeg, LA, Lammertink, RG; Stamatialis, D; Wessling, M. One-step fabrication of porous micropatterned scaffolds to control cell behavior. *Biomaterials*, 2007, *28*,1998-2009.

[149] Guo, S; Fogarty, DP; Nagel, PM; Kandel, SA: Scanning tunneling microscopy of surface-adsorbed fullerenes: C_{60}, C_{70}, and C_{84}. *Surf. Sci.*, 2007, *601*, 994-1000.

[150] Vacik, J; Naramoto, H; Narumi, K; Yamamoto, Y; Miyashita, K. Pattern formation induced by co-deposition of Ni and C60 on MgO.1*00*. *Journal of Chemical Physics*, 2001, *114 [20]*, 9115 -9119.

[151] Vacik, J; Lavrentiev, V; Hnatowicz, V; Yamamoto, S; Vorlicek, V; Stadler, H. Spontaneous Partitioning of the Ni+C_{60} Thin Film Grown at RT. *Journal of Alloys and Compounds* 2008, in press.

[152] Engler, AJ; Sen, S; Sweeney, HL; Discher, DE. Matrix elasticity directs stem cell lineage specification. *Cell*, 2006, *126*, 677-689.

[153] Hao, L; Lawrence, J; Chian, KS. Osteoblast cell adhesion on a laser modified zirconia based bioceramic. *J. Mater. Sci. Mater. Med.*, 2005, *16*, 719-726.

[154] Zhao, G; Schwartz, Z; Wieland, M; Rupp, F; Geis-Gerstorfer, J; Cochran, DL; Boyan, BD. High surface energy enhances cell response to titanium substrate microstructure. *J. Biomed. Mater. Res. A*, 2005, *74*, 49-58.

[155] Pesakova, V; Kubies, D; Hulejova, H; Himmlova, L. The influence of implant surface properties on cell adhesion and proliferation. *J. Mater. Sci. Mater. Med.*, 2007, *18*, 465-473.

[156] Dos Santos, EA; Farina, M; Soares, GA; Anselme, K. Surface energy of hydroxyapatite and beta-tricalcium phosphate ceramics driving serum protein adsorption and osteoblast adhesion. *J. Mater. Sci. Mater. Med.*, 2008, in press.

[157] Grausova, L; Bacakova, L; Fraczek, A; Blazewicz, S; Blazewicz, M; Pepka, M. Biological effects of polymers modified with carbon nanotubes on human osteoblast-like MG 63 cells. *Engineering of Biomaterials*, 2006, *9[58-60]*, 14-16.

[158] Grausova, L; Bacakova, L; Fraczek, A; Filova, E; Blazewicz M. Adhesion, proliferation and viability of human osteoblast-like MG 63 cells on carbon nanotube-polysulfone composites. *Eight International Conference on the Science ad Application of Nanotubes, „Nanotubes 2007"*, Ouro Preto, Minas Gerais, Brasil, p. 181.

[159] Mailhot, JM; Sharawy, MM; Galal, M; Oldham, AM; Russell, CM. Porous polysulfone coated with platelet-derived growth factor-BB stimulates proliferation of human periodontal ligament fibroblasts. *J. Periodontol.*, 1996, *67*, 981-985.

[160] Jozwiak, A; Karlik, W; Wiechetek, M; Werynski, A. Attachment and metabolic activity of hepatocytes cultivated on selected polymeric membranes. *Int. J. Artif. Organs*, 1998, *21*, 460-466.

[161] Hoque, ME; Mao, HQ; Ramakrishna, S. Hybrid braided 3-D scaffold for bioartificial liver assist devices. *J. Biomater. Sci. Polym. Ed.*, 2007, *18*, 45-58.

[162] Pamula, E; Bacakova, L; Filova, E; Buczynska, J; Dobrzynski, P; Noskova, L; Grausova, L: The influence of pore size on colonization of poly(L-lactide-glycolide) scaffolds with human osteoblast-like MG 63 cells in vitro. *J. Mater. Sci. Mater. Med.*, 2008, *19*, 425-435.

[163] McDevitt, MR; Chattopadhyay, D; Jaggi, JS; Finn, RD; Zanzonico, PB; Villa, C; Rey, D; Mendenhall, J; Batt, CA; Njardarson, JT; Scheinberg, DA. PET imaging of soluble yttrium-86-labeled carbon nanotubes in mice. *PLoS ONE 2*, 2007, e907.

[164] Fraczek, A; Menaszek, E; Czajkowska, B; Bacakova, L; Blazewicz, M. In vitro and in vivo biocompatibility of single wall carbon nanotubes. *Engineering of Biomaterials*, 2007, *10(69-72)*, 8-11.

[165] Tian, F; Cui, D; Schwarz, H; Estrada, GG; Kobayashi, H. Cytotoxicity of single-wall carbon nanotubes on human fibroblasts. *Toxicol. In Vitro*, 2006, *20*, 1202-1212.

[166] Jia, G; Wang, H; Yan, L; Wang, X; Pei, R; Yan, T; Zhao, Y; Guo, X. Cytotoxicity of carbon nanomaterials: single-wall nanotube, multi-wall nanotube, and fullerene. *Environ. Sci. Technol.*, 2005, *39*, 1378-1383.

[167] Radomski, A; Jurasz, P; Alonso-Escolano, D; Drews, M; Morandi, M; Malinski, T; Radomski, MW. Nanoparticle-induced platelet aggregation and vascular thrombosis. *Br. J. Pharmacol.*, 2005, *146*, 882-893.

[168] Raja, PM; Connolley, J; Ganesan, GP; Ci, L; Ajayan, PM; Nalamasu, O; Thompson, DM. Impact of carbon nanotube exposure, dosage and aggregation on smooth muscle cells. *Toxicol. Lett.*, 2007, *169*, 51-63.

[169] Dumortier, H; Lacotte, S; Pastorin, G; Marega, R; Wu, W; Bonifazi, D; Briand, JP; Prato, M; Muller, S; Bianco, A. Functionalized carbon nanotubes are non-cytotoxic and preserve the functionality of primary immune cells. *Nano Lett.*, 2006, *6*, 1522-1528.

[170] Potocky, S; Kromka, A; Potmesil, J; Remes, Z; Polackova, Z; Vanecek, M. Growth of nanocrystalline diamond films deposited by microwave plasma CVD system at low substrate temperatures. *Phys. Stat. Sol. (a)*, 2006, *203*, 3011-3015.

[171] Poruba, A; Fejfar, A; Remes, Z; Springer, J; Vanecek, M; Kocka, J; Meier, J; Torres, P; Shah, A. Optical absorption and light scattering in microcrystalline silicon thin films and solar cells. *J. Appl. Phys.*, 2000, *88*, 148-160.

[172] Buhlman, S; Blank, E; Haubner, R; Lux, B: Characterization of ballas diamond depositions, *Diamond Rel. Mater.*, 1999, *8*, 194-201.

[173] Zemek, J; Houdkova, J; Lesiak, B; Jablonski, A; Potmesil, J; Vanecek, M. Electron spectroscopy of nanocrystalline diamond surfaces. *J. Optoelectr. & Adv. Materials*, 2006, *8*, 2133.

[174] Kim, HJ; Kim, SH; Kim, MS; Lee, EJ; Oh, HG; Oh, WM; Park, SW; Kim, WJ; Lee, GJ; Choi, NG; Koh, JT; Dinh, DB; Hardin, RR; Johnson, K; Sylvia, VL; Schmitz, JP; Dean, DD.Varying Ti-6Al-4V surface roughness induces different early morphologic and molecular responses in MG63 osteoblast-like cells. *J. Biomed. Mater. Res.* 2005, *74*, 366-373.

[175] Zhao, G; Zinger, O; Schwartz, Z; Wieland, M; Landolt, D; Boyan, BD. Osteoblast-like cells are sensitive to submicron-scale surface structure. *Clin. Oral. Implants Res.*, 2006, *17*, 258-264.

[176] Zhao, W; Xu, JJ; Qiu, QQ; Chen, HY. Nanocrystalline diamond modified gold electrode for glucose biosensing. *Biosens Bioelectron*, 2006, *22*, 649-655.

[177] Boyan, BD; Bonewald, LF; Paschalis, EP; Lohmann, CH; Rosser, J; Cochran, DL; Dean, DD; Schwartz, Z; Boskey, AL. Osteoblast-mediated mineral deposition in culture is dependent on surface microtopography. *Calcif. Tissue Int.*, 2002, *71*, 519-529.

[178] Lohmann, CH; Bonewald, LF; Sisk, MA; Sylvia, VL; Cochran, DL; Dean, DD; Boyan, BD; Schwartz, Z. Maturation state determines the response of osteogenic cells to surface roughness and 1,25-dihydroxyvitamin D3. *J. Bone Miner. Res.*, 2000, *15*, 1169-1180.

[179] Schneider, GB; Zaharias, R; Seabold, D; Keller, J; Stanford, C. Differentiation of preosteoblasts is affected by implant surface microtopographies. *J. Biomed. Mater. Res. A*, 2004, *69*, 462-468.

[180] Lossdorfer, S; Schwartz, Z; Wang, L; Lohmann, CH; Turner, JD; Wieland, M; Cochran, DL; Boyan, BD. Microrough implant surface topographies increase osteogenesis by reducing osteoclast formation and activity. *J. Biomed. Mater. Res. A*, 2004, *70*, 361-369.

[181] Sammons, RL; Lumbikanonda, N; Gross, M; Cantzler, P. Comparison of osteoblast spreading on microstructured dental implant surfaces and cell behaviour in an explant model of osseointegration. A scanning electron microscopic study. *Clin. Oral. Implants. Res.*, 2005, *16*, 657-666.

[182] Sader, MS; Balduino, A; Soares, G de A; Borojevic, R. Effect of three distinct treatments of titanium surface on osteoblast attachment, proliferation, and differentiation. *Clin. Oral. Implants Res.*, 2005, *16,* 667-675.

[183] Mares, JJ; Hubik, P; Nesladek, M; Kristofik. Boron-doped diamond — Grained Mott's metal revealing superconductivity. *Diamond Relat. Mater.*, 2007, *16*, 921-925.

[184] Chang, JS; Chang, KL; Hwang, DF; Kong, ZL. *In vitro* cytotoxicitiy of silica nanoparticles at high concentrations strongly depends on the metabolic activity type of the cell line. *Environ. Sci. Technol.*, 2007, *41*, 2064-2068.

[185] Oner, G; Cirrik, S; Bakan, O. Effects of silica on mitochondrial functions of the proximal tubule cells in rats. *Kidney Blood Press Res.*, 2005, *28*, 203-210.

[186] Xiao, X; Wang, J; Liu, C; Carlisle, JA; Mech, B; Greenberg, R; Guven, D; Freda, R; Humayun, MS; Weiland, J; Auciello, O. In vitro and in vivo evaluation of ultrananocrystalline diamond for coating of implantable retinal microchips. *J. Biomed. Mater. Res. B Appl. Biomater.*, 2006, *77*, 273-281.

[187] Labelle, D; Jumarie, C; Moreau, R. Capacitative calcium entry and proliferation of human osteoblast-like MG-63 cells. *Cell Prolif.*, 2007, *40*, 866-884.

[188] Naghii, MR; Torkaman, G; Mofid, M. Effects of boron and calcium supplementation on mechanical properties of bone in rats. *Biofactors*, 2006, *28*, 195-201.

[189] Bacakova, L; Stary, V., Glogar P.: Adhesion and growth of cells in culture on carbon-carbon composites with different surface properties. *Engineering of Biomaterials*, 1998, *1 [2]*, 3-5.

[190] Lappalainen, R; Selenius, M; Anttila, A; Konttinen, YT; Santavirta, SS. Reduction of wear in total hip replacement prostheses by amorphous diamond coatings. *J. Biomed. Mater. Res. B Appl. Biomater.*, 2003, *66*, 410-413.

[191] Ma, WJ; Ruys, AJ; Mason, RS; Martin, PJ; Bendavid, A; Liu, Z; Ionescu, M; Zreiqat, H. DLC coatings: Effects of physical and chemical properties on biological response. *Biomaterials*, 2007, *28*, 1620–1628.

[192] Bacakova, L; Balik, K; Zizka, S; Adhesion and growth of vascular smooth muscle cells in cultures on carbon fibre-reinforced carbon composites covered with pyrolytic carbon. *Engineering of Biomaterials*, 1998, *1, [-]*, 19-22.

[193] Pesakova, V; Klezl, Z; Balik, K; Adam, A. Biomechanical and biological properties of the implant material carbon-carbon composite covered with pyrolytic carbon. *J. Mater. Sci.: Mater. Med.*, 2000, *11*, 793-798.

In: Encyclopedia of Nanotechnology
Editor: Elwood D. Carlson

ISBN: 978-1-60692-079-4

Chapter 36

ORGANIC-SHELL INORGANIC-CORE HYBRID NANOPARTICLES WITH ADVANCED FUNCTIONS DESIGNED BY WET PROCESS

Mami Yamada*

Department of Applied Chemistry, Tokyo University of Agriculture and Technology, 2-24-16 Nakacho, Koganei, Tokyo 184-8588: PRESTO/Japan Science and Technology Agency, 4-1-8, Honcho, Kawaguchi-shi, Saitama 332-0012, Japan

ABSTRACT

This chapter reviews the novel synthesis of inorganic nanoparticles stabilized by an organic shell layer with investigation into their specific characteristics. How to design organic-inorganic hybrid nanoparticles in chemical solution process effectively is indispensable to uncover their unprecedented natures due to the fusion of merits of both organic and inorganic substances. Expressly, the two synthetic parameters are a focus of this chapter: the physical construction of the inorganic core and the chemical constituents introduced. Concretely, the catalytic activities of nanoparticles are controllable by modifying the crystal faces and the surface area of their inorganic cores in connection with their shape and size. In another case, when functional units, as an organic shell, are attached to the surface of nanoparticles, external stimuli such as electric field or light are utilized to transform the characteristics of nanoparticles themselves. It is also interesting to bring metal coordination polymers into an inorganic core, leading to the first isolation of alkyl chain stabilized-metal coordination nano-polymers with ferromagnetism. The nanomaterials presented here are listed as Pt nano-cube, size-selected Au nanoparticles, metal hexacyanoferrate coordination nano-polymers, and metal nanoparticles functionalized by bifferocene, anthraquinone, porphyrin and triphenylene derivatives. The synthetic procedure and their remarkable characteristics with some application examples are demonstrated.

* Tel./Fax: +81-1-42-388-7379. Email: m-yamada@cc.tuat.ac.jp

INTRODUCTION

Nanometer-sized metal particles are generally of mesoscopic areas consisting of dozens to thousands atoms [1]. Since gold colloid sol was synthesized for the first time in the 17th century, the particular natures of metal nanoparticles, which are different from those of bulk metals and molecules, have attracted much attention from researchers. These unique properties of nanoparticles are derived from two characteristic effects—the quantum size effect and the surface effect, leading to the expectation of diverse applications [2] *e.g.* optoelectronic devices, molecular catalysts and chemical sensors. The preparation methods of metal nanoparticles are mainly divided into two categories: physical dry method and chemical wet process. Basically, the former means that bulk metals are downsized mechanically such as with the breaking grinder method, laser ablation, and plasma irradiation, while the latter comprises metal ions dissolved in solution and reduced to neutral atoms, gathering on a nanometer scale. Recently, the improvement of the chemical process is remarkable after Brust *et al.* reported in 1994 the large-scale preparation of mono-dispersed gold nanoparticles which are significantly air-stable in solvent-free form [3]. The nanoparticles obtained by the chemical method commonly possess "the metallic nanocore-organic shell hybrid structure" [4-6]. The organic shell supplies the high stability of nanoparticles, avoiding thermal aggregation among inorganic cores. The chemical process has some benefits over the physical method: the equipment is very convenient and easy (so to speak "synthesis in one vessel"), and the innumerable kinds of nanoparticles could be constructed just by adjusting appropriate synthetic parameters (starting materials, unit ratio, temperature, solvents, etc.) in thereaction mixture. These synthetic variations enable the unique design of nanoparticle materials with unprecedented physical and chemical structures. The other astonishing merit of nanoparticles synthesized by wet process is the re-dispersibility into solvent after purification, which gives the facile fabrication of nanoparticle films onto substrates for application usage. From these points, stable nanoparticle dispersions have been developed for industrial usage by many academic institutes and manufacturing companies.

Against the backdrop noted above, this chapter describes the latest nanoparticle research under the primary theme of 'creation and functionality of organic-shell inorganic-core hybrid nanoparticles by wet process'. There exist numerous patterns in selecting components for organic and inorganic composites. Generally, an organic material excels in the versatility of its physical structure, including flexibility, while an inorganic substance can accumulate multiple electrons in forming various electronic states. First, the creation of organic/inorganic hybrid nanoparticles should be pursued through effective selection of components. It should be also considered that functions of nanoparticles are affected substantially by their physical structure (shape and size) and composition ratio among the units selected. In other words, in order to build up hybrid nanoparticles with novel functions, the preparation method requires the precise control of at least three synthetic parameters of "unit selection", "physical structure" and "component ratio". Simultaneously, the promising development of the wet process would be promoted by exploring the effective synthetic procedure which accompanies the assembling process of the nanoparticles on a substrate for the purpose of further application. Through feedback on the experimental information gathered on the basis of these research concepts, the research presented in this chapter aim to show the preparation of novel organic/inorganic nanoparticles with functions including electrochemical, optical, catalytic, and magnetic properties. The synthesized nanoparticles have been classified into three groups: (1) functionalized organic molecules-attached metal nanoparticles, (2) shape- and size-controlled metal nanoparticles, and (3) nanometer-sized metal coordination polymers. These topics are demonstrated mostly in connection with my related publications.

1. Functionalization of Metal Nanoparticles by Organic Molecules

The first section is presented to prepare and investigate the functional-molecule modified noble metal nanoparticles which possess the prominent properties derived from the combination of the inorganic metal core and organic shell. Functionalization based on a metal core will open an unexpected interesting way to study intra- and intermolecular nanoparticle chemistry and to design new photoelectrochemical systems or multi-step, and electron donor/acceptor reactions using nanoparticles. The fundamental idea of functionalization of metal nanoparticles by thiol exchange reaction was advocated by Murray *et al,*. who proposed the preparation of functionalized gold nanoparticles stabilized by alkylthiolate by exchanging some thiolates on the particle surface with ferrocene thiol derivatives in toluene solution [7]. The method is facile and can be utilized for various nanoparticle functionalizations. In addition, it could create poly-functionalized metal nanoparticles by means of particles modified with mixtures of different thiols [8]. The reactivity of functionalized alkyl thiolate-stabilized metal nanoparticles should be investigated [9-12] for the further synthetic steps and the development of novel functions of nanoparticles. In this section, as functional molecules, redox-active units (biferrocene and anthraquinone), a chlomophore (phorphyrine) and a liquid crystal molecule (triphenylene) are selected. The characters of the prepared organic-inorganic nanoparticles could be controlled by physical and chemical states of the surrounding functional species with outside stimulation, *i.e.,* applied electronic field, light and dielectric strength around the particles.

1.1. Electrodeposition of Metal Nanoparticles Surrounded by Multiple Redox Units

The biferrocene (BFc) unit is one of multiple-redox species, which undergoes a two-step one-electron oxidative reaction of $BFc/BFc^{+}/BFc^{2+}$ in electrolyte solution [13]. The high charge accumulation on a particle/solution interface is expected by the applied electronic field when the BFc units are attached onto a particle surface. The BFc-modified gold nanoparticles (Au_n-BFc) were synthesized by a thiolate exchange between an alkyl thiolate-stabilized metal nanoparticles (Au_n-AT) and a BFc thiol derivative, (1-(9-thiononyl-1-one)-1', 1''-biferrocene (BFcS) (Scheme 1) [13-15]. Au_n-ATs were obtained by the reduction of Au (III) ions ($HAuCl_4$, $AuCl_3$ *etc.*) in solution mixture with a surface active agent (TOAB, DDAB *etc.*), a reducing agent ($NaBH_4$, Super Hydride *etc.*) and an alkyl thiolate with appropriate chain length [3, 16, 17]. The average diameter of gold core, d_{av}, was changed with the range of 1.5~8.0 nm by the synthetic parameters of Au_n-AT, mainly the molar ratio of Au ions to an alkyl thiolate. The number of the exchanged BFcS on the Au_n-AT surface, θ_{BFc}, was calculated based on the ratio of the integrals of the ^{1}H-NMR signals between BFc (2.9-4.7 ppm) and methyl protons (0.8-0.9 ppm), respectively. Figure 1 shows the transmission electron microscopy (TEM) image of Au_n-BFcs with d_{av} = 2.3 nm and θ_{BFc} = 7.5) [14]. It is seen that the nanoparticles are individually isolated by stable organic shells. Au_n-BFcs could be obtained as a black powdery form after the alcohol purification process, however, they were easily redispersed into several organic solvents, *e.g.*, CH_2Cl_2, chloroform, THF, hexane due to the hydrophobic nature of organic shell.

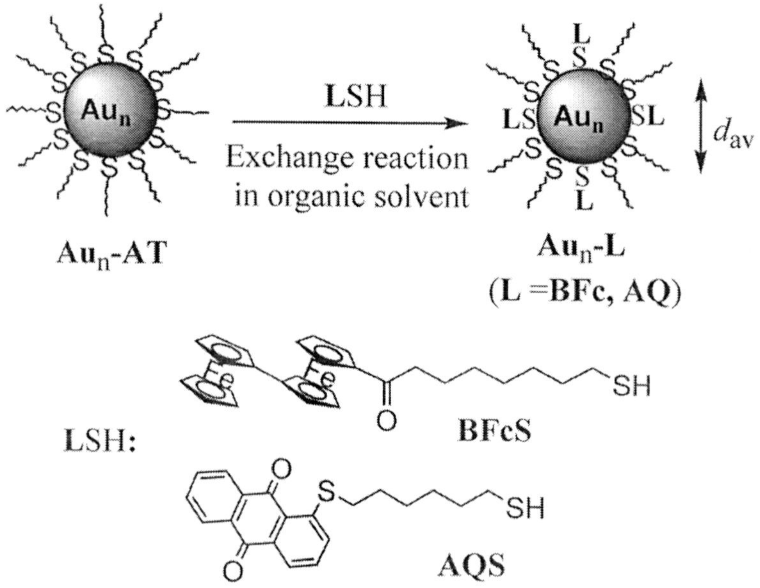

Scheme 1. Illustration of Au_n-BFc and Au_n-AQ.

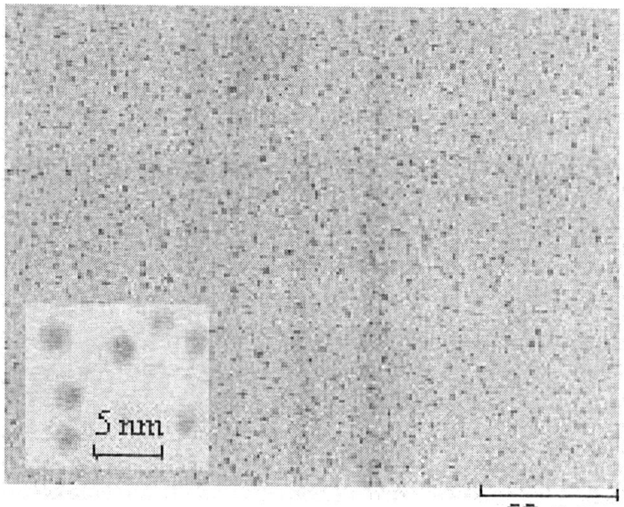

Figure 1. TEM image of Au_n-BFc (d_{av} = 2.3 nm, θ_{BFc} = 7.5).

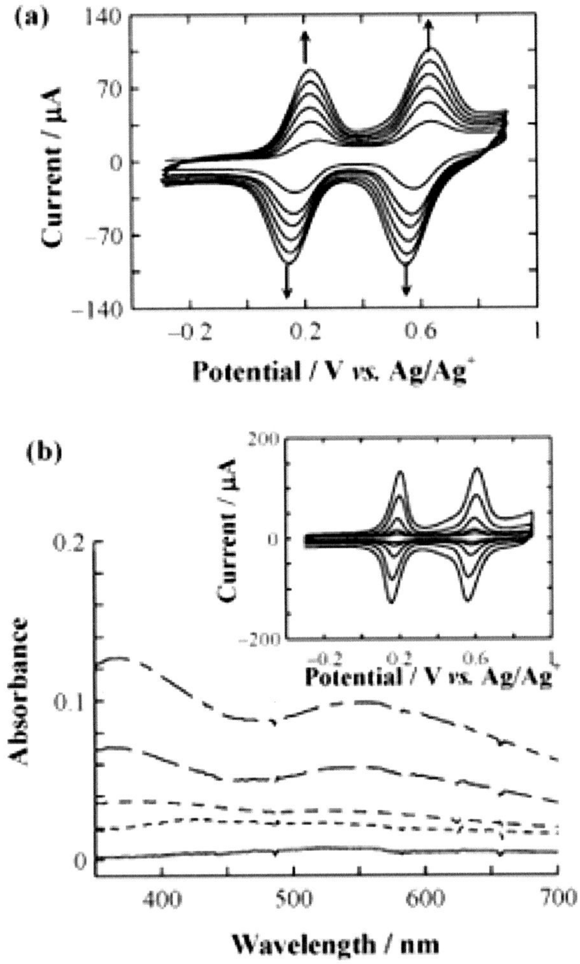

Figure 2. (a) Cyclic voltammograms of 1.7 μM of Au_n-BFc (d_{av} = 2.9 nm, θ_{BFc} = 20.8) at ITO in 0.1 M Bu_4NClO_4–CH_2Cl_2 at 100 mV/s between –0.3 and 0.9 V *vs.* Ag/Ag^+ in the positive direction at the 1st, 10th, 20th, 30th, 40th, and 50th cyclic scan, as shown from the bottom of the figures to the top. (b) UV-Vis spectra and (inset) cyclic voltammograms in 0.1 M Bu_4NClO_4–CH_2Cl_2 of the electrodeposited Au_n-BFc films prepared in the same condition of Figure 2a with the 3, 10, 25, 50, and 75 cyclic scans, as shown from the bottom of the figure to the top [15].

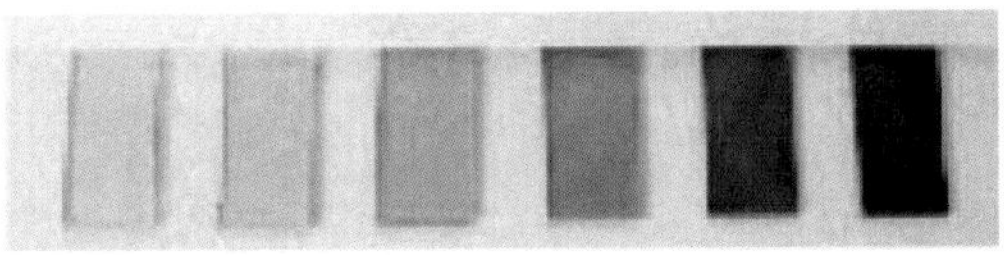

10 20 40 60 120 180

Figure 3. Photographs of electrodeposited Au_n-BFc films prepared in a solution of 5.0 μM Au_n-BFc (d_{av} = 2.3 nm, θ_{BFc} = 15) at ITO 0.1 M Bu_4NClO_4–CH_2Cl_2 at 100 mV/s between –0.3 and 0.9 V *vs.* Ag/Ag^+. The numbers in the figure are those of the cyclic scans [14].

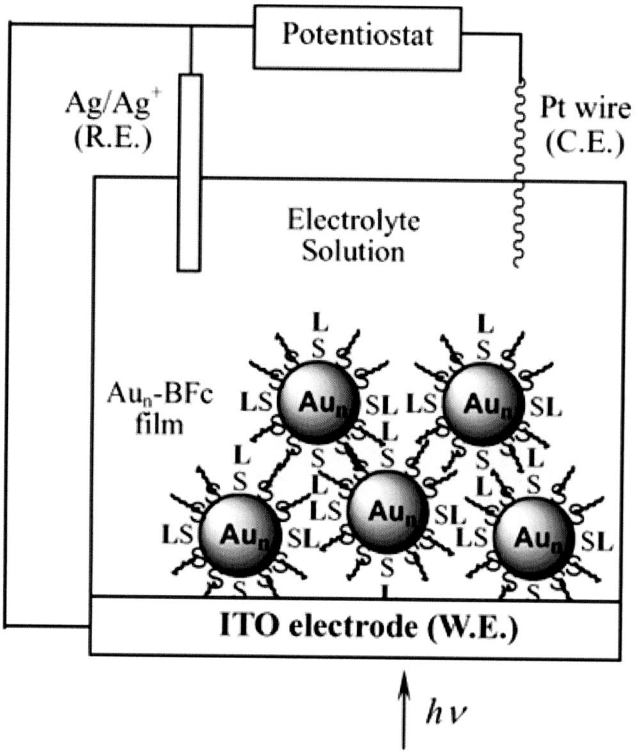

Scheme 2. Electro-optical measurement spectroelectrochemical cell of the Au_n-BFc film [21].

Figure 2a shows the typical cyclic voltammograms (CVs) of Au_n-BFcs, which is measured at an indium tin oxide (ITO) electrode in electrolyte solution of Bu_4NClO_4–CH_2Cl_2 [15]. The peak current increases gradually by consecutive potential scans between –0.3 and 0.9 V *vs.* Ag/Ag^+ where two-step $1e^-$ oxidation due to BFc units on a particle surface occurs at 0.20 and 0.61 V *vs.* Ag/Ag^+, suggesting that Au_n-BFc particles flock on an electrode/solution interface by 2 electron oxidation of BFc sites. UV-Vis spectra of Au_n-BFc films thus prepared exhibit broad absorption bands that grow in intensity with increasing in the number of potential scans (Figure 2b), meaning that the thickness of the film is controlled by changing the number of potential scans. Figure 3 displays the actual photographs of the Au_n-BFc films on ITO, which confirms that the black deposition due to Au_n-BFc is tightly attached onto a transparent ITO electrode [18]. CVs of the electrode films in pure electrolyte solution show two pairs of cathodic and anodic waves (Figure 2b, inset), the peak currents of which are proportional to the potential scan rate, indicating behavior of surface immobilized BFc species in the film. The construction of the Au_n-BFc films with a different core diameter was applicable by using this electrodeposition process, independent of the core size [15]. In turn, this oxidative deposition also occurred by introducing Pd in place of Au, BFc-modified palladium nanoparticles (Pd_n-BFc) (see below) [19].

As seen in Figure 2b, the Au_n-BFc film with relatively large diameter (2.3 nm~) exhibits the broad absorption band at *ca.*550 nm due to the collective surface plasmon (SP) band influenced by the dipole-dipole coupling among the adjacent particles [20]. This collective SP band in the formed film offered special electro-optical features different from a single isolated nanoparticle (Scheme 2) [21]. Figure 4 illustrates that the UV-Vis spectra of the Au_n-BFc film change with the λ_{max} of the SP band at 584 nm by applied potential. The λ_{max} value is extremely shifted to a shorter wavelength by 74 nm with increasing absorbance by potential in the negative direction to −1.6 V. On the contrary, the positive potential shift to 1.6 V gives the longer λ_{max} by 8 nm with decreasing absorbance. These spectral changes means that the direct charging of electrons in the gold core of Au_n-BFc is predominant comparing to the charge accumulation at the immobilized BFc units, because the λ_{max} shift in the negative potential is significantly larger than that in the positive direction where redox reaction of BFc moities occur.

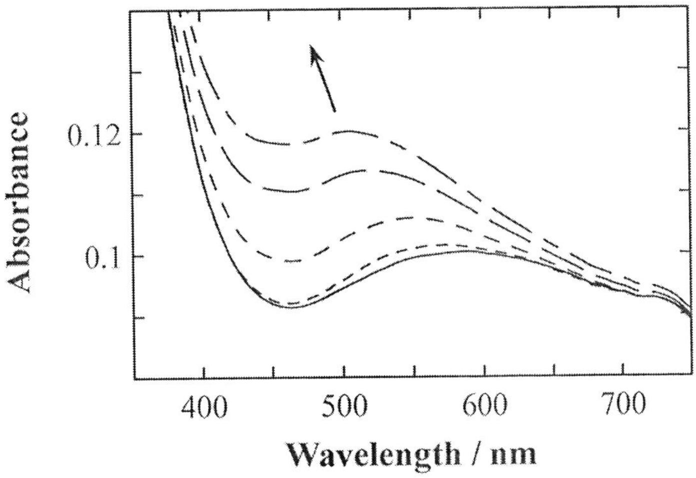

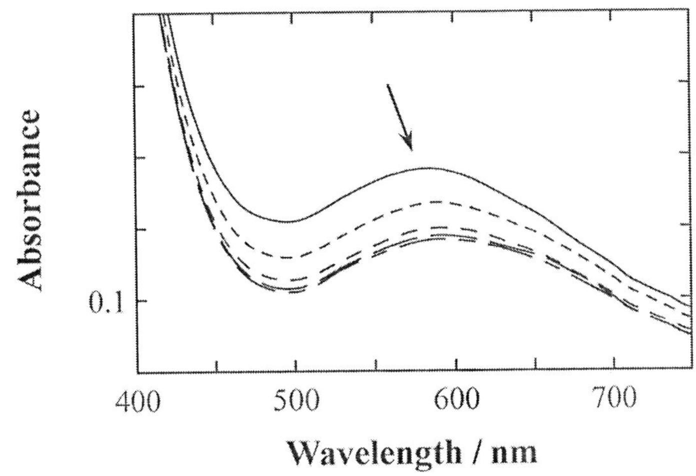

Figure 4. (Top) UV-Vis spectra of the Au_n-BFc film (d_{av} = 6.4 nm, θ_{BFc} = 4.6) on ITO at given potentials of 0, −0.4, −0.8, −1.2, and −1.6 V *vs.* Ag/Ag^+ in 0.1 M Bu_4NClO_4–CH_2Cl_2 in the negative direction, and (bottom) at given potentials of 0, 0.4, 0.8, 1.2, and 1.6 V *vs.* Ag/Ag^+ in 0.1 M Bu_4NClO_4–CH_2Cl_2 in the positive direction [21].

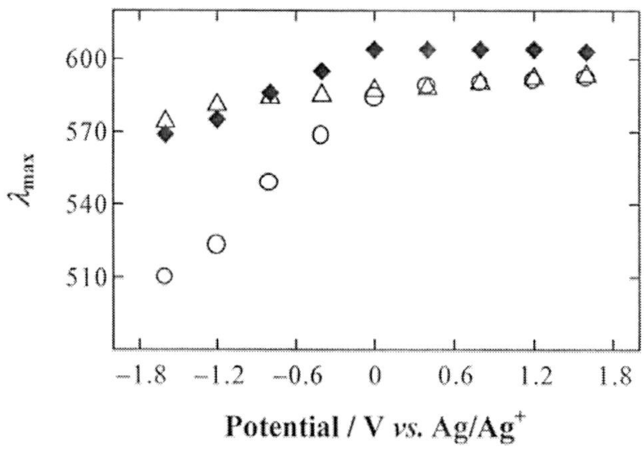

Figure 5. The plots of the absorption maximum (λ_{max}) of the Au$_n$-BFc film (d_{av} = 2.3 nm, θ_{BFc} = 15) (triangles), the Au$_n$-BFc film (d_{av} = 4.3 nm, θ_{BFc} = 11.3) (squares), and Au$_n$-BFc film (d_{av} = 6.4 nm, θ_{BFc} = 4.6) (circles) film *vs.* the applied potential [21].

Table 1. Spectroelectrochemical data of the Au$_n$-BFc films [21]

Particle Diameter / nm	C_{core} [a] / aF	ΔV [b] / V	Number of Injected Electrons to the Core [c]	λ_{init} [d] / nm	λ_{final} [d] / nm	$\lambda_{final.ideal}$ [e] / nm	$\Delta\lambda_{ex}$ [f] / nm	$\Delta\lambda_{ideal}$ [g] / nm
2.3	1.1	0.15	11e^-	587	574	577	13	10
4.3	2.8	0.057	28e^-	604	569	601	35	3
6.4	4.2	0.038	42e^-	584	510	582	74	2

a) The core capasitance: $C_{core} = 4\pi\varepsilon\varepsilon_0 r\ (r + d)/d$ where ε is the monolayer dielectric constant, r is the core radius, and d is the monolayer thickness. b) The potential spacing of consecutive single-electron transfer processes: $\Delta V = e/C_{core}$. c) Electron provision from 0 V to −1.6 V calculated by the value of ΔV. d) See Figure 20 at 0 V (λ_{init}) and −1.6 V (λ_{final}). e) The value calculated by $\lambda_{final.ideal} / \lambda_{init} = (N_{init} / N_{final})^{1/2}$ where N_{init} and N_{final} are the numbers of free electrons per metal core after and before the charging, respectively., referred to Table 1. f) $\Delta\lambda_{ex} = \lambda_{init} - \lambda_{final}$. g) $\Delta\lambda_{deal} = \lambda_{init} - \lambda_{final.ideal}$

Au$_n$-BFc films with a different core diameter could be prepared with the same electrochemical method. Notably, the shift of λ_{max} is more conspicuous for the Au$_n$-BFc film with larger core size in the negative potential region (Figure 5). The electroscopic data are summarized in Table 1, suggesting that the particle-particle interaction works more effectively to the Au$_n$-BFc film with larger particle size, considering that the shifts grow with increasing the core size of particles. It is interesting that the order of ideal amount of the λ_{max} shift is opposite to the experimental values; that is, the theoretical calculation [20] predicts that the SP band position of the isolated particles are stimulated more easily with smaller core size by applied potential. These results point out that the collective SP band of the Au$_n$-BFc films might be shifted particularly with interparticle interaction different to a single particle and/or some other parameters, which affects the SP resonance, *e.g,.* electronic state of the surface ligands [22], chemical adsorbates on the surface, and the number of layers, should be considered for understanding these electro-optical shifts. The properties of nanoaprticles provoked by their interparticle interactions are also filled with many potentialities for nanotechnological science.

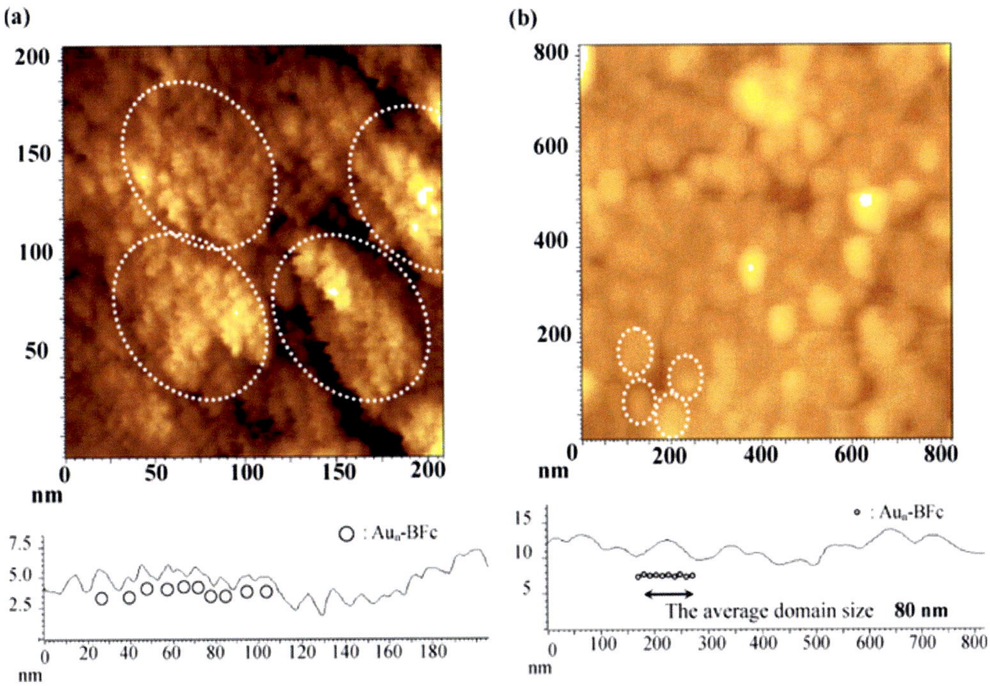

Figure 6. (a) STM and (b) AFM images of the Au$_n$-BFc film prepared in a solution of 5.2 μM Au$_n$-BFc (d_{av} = 2.3 nm, θ_{BFc} = 7.5) at HOPG with 5 cyclic scans in 0.1 M Bu_4NClO_4–CH_2Cl_2 at 100 mV/s between –0.3 and 0.9 V *vs.* Ag/Ag^+ in the positive direction, and the typical cross-sectional profile along the cross axis [14].

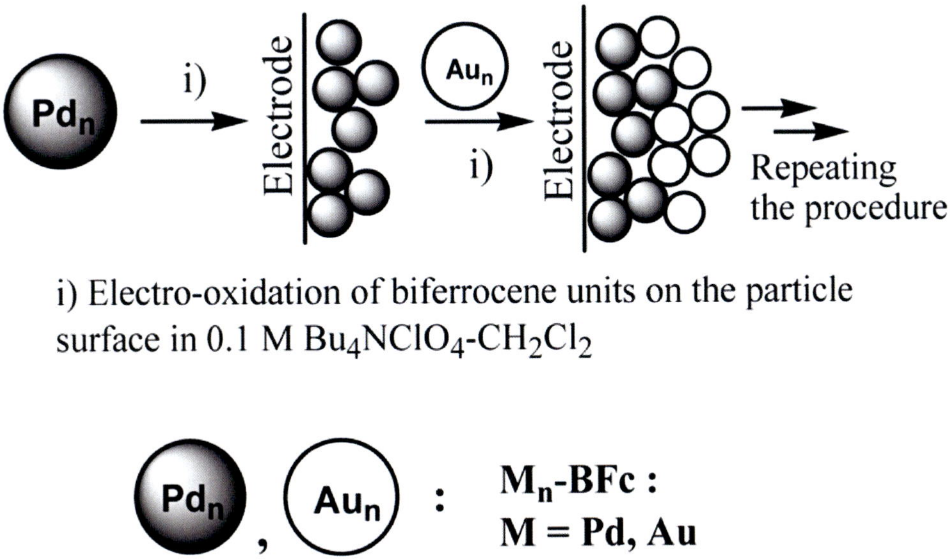

Scheme 3. Preparation of the Au$_n$-BFc and Pd$_n$-BFc composite film [27].

STM and AFM images of the Au$_n$-BFc film (d_{av} = 2.3 nm, θ_{BFc} = 7.5) are shown in Figure 6 [14]. The surface maintains a monolayer-level flatness seen in its cross sectional view, whereas it is apparent that domains of particles *ca.* 70–80 nm in diameter (encircled by a dotted line) are constructed. The AFM image of the same sample reveals this peculiar nanostructure of the Au_n-BFc film; round-shaped domains spreading on the whole surface shown in Figure 6b. About the Pd_n-BFc deposition, in Figure 7, tetragonal-like aggregation is observed similar to Au_n-AQ (see below). These specific morphological features testify that these characteristic assembling structures proceed in this deposition system.

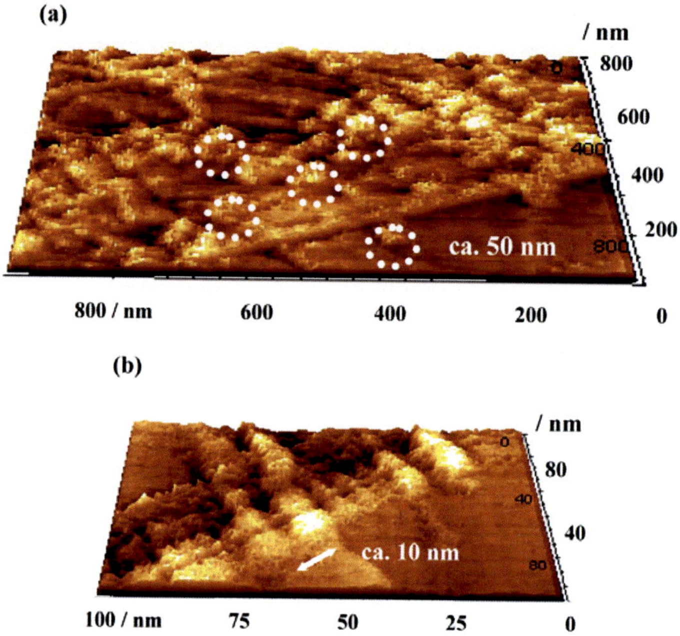

Figure 7. STM image of the Pd_n-BFc film (d_{av} = 3.8 nm, θ_{BFc} = 26.3)prepared in a solution of 1.9 μM Pd_n-BFc at HOPG with 60 cyclic scans in 0.1 M Bu_4NClO_4–CH_2Cl_2 at 100 mV/s between –0.3 and 0.9 V *vs.* Ag/Ag^+ in the area of (a) 800 × 800 nm^2 and (b) 100 × 100 nm^2 [19].

The other multiple redox unit, an anthraquinone (AQ) thiol derivatice 1-(1,8-dithiaoctyl)anthracene-9,10-dione (AQS) [23] is next introduced onto a gold nanoparticle to produce Au_n-AQ (Scheme 1) [24, 25]. BFc and AQ are both multiple redox species, however, they are contrastive. Namely, the former undergoes two-step one electron 'oxidative' reaction of $BFc/BFc^+/BFc^{2+}$, while the latter receives two-step one electron 'reductive' reaction of $AQ/AQ^-/AQ^{2-}$. From the results noted above, the 2 electron oxidation process of BFc units of Au_n-BFc led to the formation of a uniform BFc-active gold nanoparticle film on an electrode. On the contrary, Au_n-AQ is found to be aggregated by 2 electron "reduction" of AQ sites (Figure 8), clarifying that metal nanoparticles functionalized with multiple-redox molecules could assemble by charge accumulation of redox species on a particle/solution interface. Note that the multiple-electron system is indispensable for this deposition phenomenon since it is not observed for particles with a single redox species such as ferrocene [8]. The detailed mechanism of Au_n-BFc is revealed by electrochemical quartz crystal microbalance (EQCM) [26] and summarized in Figure 9 that (1) Au_n-BFcs are apt to aggregate at the electrode/electrolyte interface and adsorb to the electrode by the formation of the BFc^{2+} state on the particle surface, (2) significant desorption of softly assembled Au_n-BFcs from the electrode occurs after BFc sites are returned to the neutral state by reversible reduction, and (3) a small portion of the strongly adsorbed flocks of Au_n-BFcs remains on the electrode [14]. By repeating the potential scans, the Au_n-BFc film is gradually fabricated by the remaining Au_n-BFc domains observed in the STM and AFM images.

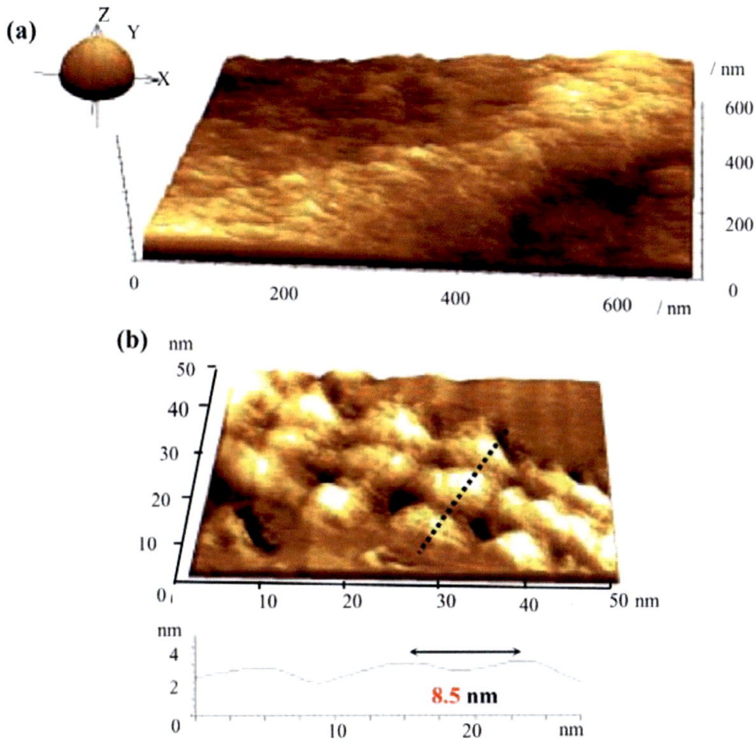

Figure 8. (a) STM image of a multilayer Au_n-AQ (d_{av} = 2.2 nm, θ_{AQ} = 26) film electrodeposited on HOPG in the area of 700 × 700 nm^2. (b) STM image of a submonolayer Au_n-AQ film on HOPG in the area of 50 × 50 nm^2 with the cross-sectional profile along the solid line [25].

About application experiments using this system, a combination of electro-oxidative deposition of Au_n-BFc and Pd_n-BFc forms a thin BFc-active composite film with a layered hybrid structure [27]. First, the electrodeposition of Pd_n-BFc (d_{av} = 3.8 nm) is performed by potential cyclic scans, which is followed by the electrodeposition of Au_n-BFc (d_{av} = 2.9 nm), thus forming the Pd_n-BFc/Au_n-BFc composite film (Scheme 3). The two-step hetero-electrodeposition procedure is repeated in order to increase the number of composite layers (Figure 10a). Figure 10b shows the XPS spectra of the composite film, displaying that the Au $4f_{5/2}$ (87.5 eV) and $4f_{7/2}$ (83.9 eV) peaks are not detected in the first layer of Pd_n-BFc, although they appear after deposition of the second layer of Au_n-BFc. In addition, the formation of the third layer of Pd_n-BFc shelters most of these Au peaks, suggesting that the prepared films build up an alternately layered structure of Pd_n-BFc and Au_n-BFc. It proves that enough electron transfer exists among heterogeneous interfaces of Pd_n-BFc and Au_n-BFc for developing the composite film in gaining the average 5 M_n-BFc (M = Pd, Au) layers [28] above the second layer.

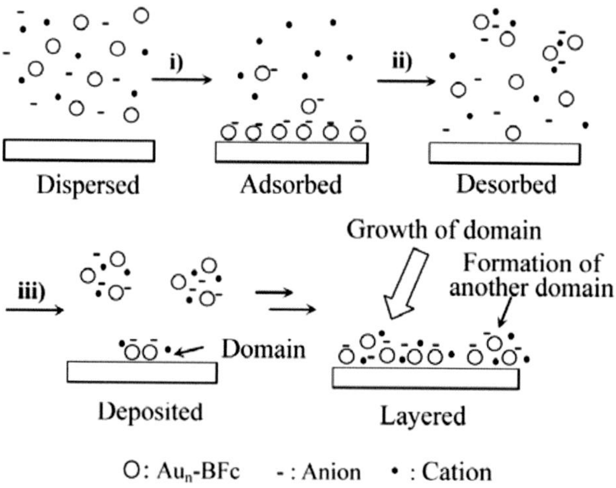

Figure 9. The illustration of the electrodeposition process in Au$_n$-BFc solution at the electrode interface. i) Two-electron oxidation of the BFc sites, ii) set back to the neutral state by two-electron reduction, and iii) repeating the potential sweep [14].

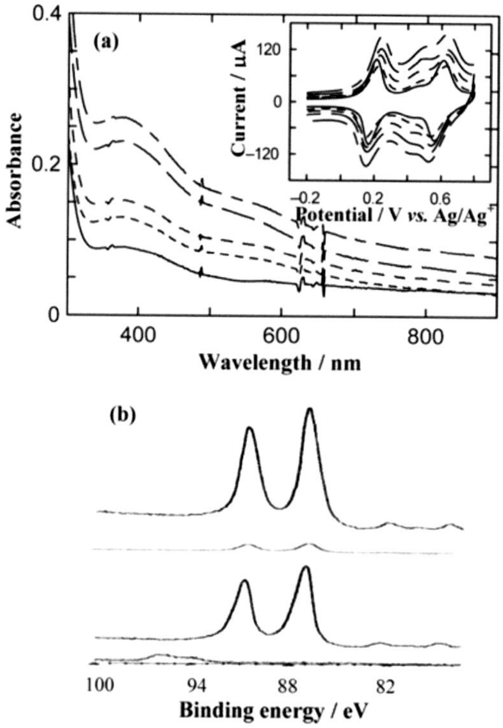

Figure 10. (a) The UV-Vis spectrum, (inset) the cyclic voltammogram, and (b) the XPS spectrum of Pd$_n$-BFc, Pd$_n$-BFc/Au$_n$-BFc, Pd$_n$-BFc/Au$_n$-BFc/Pd$_n$-BFc, [Pd$_n$-BFc/Au$_n$-BFc]$_2$, [Pd$_n$-BFc/Au$_n$-BFc]$_2$Pd$_n$-BFc films, as shown from the bottom of the figure to the top. The films were prepared with 25 cyclic potential scans between –0.3 V and 0.9 V *vs.* Ag/Ag$^+$ for each metal nanoparticle film in a solution of 3.2 μM Pd$_n$-BFc or Au$_n$-BFc, at ITO in 0.1 M Bu_4NClO_4–CH_2Cl_2 at 100 mV/s [27].

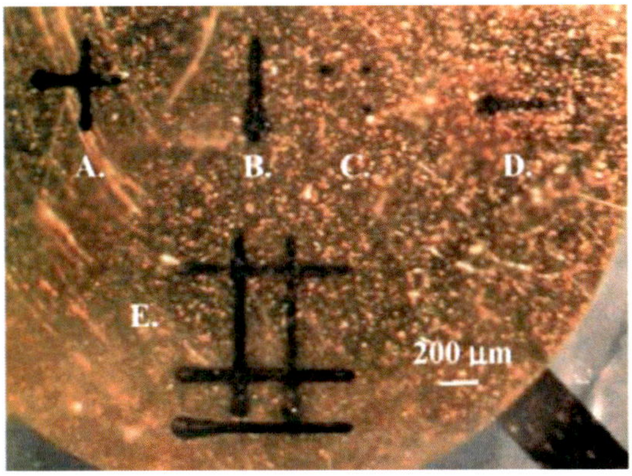

Figure 11. CCD image of the lithographic Au_n-BFc deposition on a gold electrode substrate prepared in 10 µM Au_n-BFc in 0.1 M Bu_4NClO_4-CH_2Cl_2 at the substrate potential 1.1 V by the SECM apparatus. Each line was deposited by moving the Pt counter electrode in a 7 µm diameter at 122.7 µm/s for 500 µm (A, B, and D) or 900 µm (E) with 10 scans. As for part D, the deposition time was maintained for 5 s without moving the counter electrode [18].

By utilizing scanning electrochemical microscopy (SECM) [29], dot-and line-shape lithographic assemblies of Au_n-BFc particles at a resolution of 50-100 nm were constructed [13]. When the substrate potential is kept at 1.1 V in Au_n-BFc solution where the electro-oxidative deposition of Au_n-BFc occurs, the electronic density due to the faradaic current tends to gather at the area of the substrate where the distance to the counter electrode is the closest. Figure 11 displays the CCD images of the lithographic deposition of Au_n-BFc by using this method under several conditions. We can clearly recognize the black precipitates deposited in a specific area. The simplest diagram is a "dot" described in Figure 11C, which is prepared by maintaining the deposition time for several seconds without moving the Pt counter electrode. The line drawing of the Au_n-BFc deposition is possible with *ca.* 100 µm width deposited by moving the counter electrode, displayed in Figure 11A, B, D, and E. Furthermore, it is expected that the variable parameters, such as the distance between a counter electrode and a substrate electrode, the scan speed, and the diameter of a counter electrode, could be applied to change the line width and the deposition thickness to produce more precisely controlled shapes. These assemblies systems would be also applicable by attempting to use STM and AFM instruments in place of SECM in order to achieve higher lithographic resolution.

1.2. Discotic Liquid Crystalline Molecule Modified Gold Nanoparticles with Controlling their Assemble Structure

Many studies related to the construction of highly-ordered metal nanoparticle assemblies have been reported [2]. Among them, from the viewpoint of electronic energy transport, researchers should especially focus on the one-dimensional (1D) arrangement of nanoparticles. The decrease in interparticle spacing can elicit a dramatic increase in particle-particle electronic interaction, leading to efficient 1D tunneling current as a role for nanowires [30]. Existing methods for fabricating the 1D arrangement of particles include the use of electron beam lithography, templates and self-assembly of nanoparticles. The self-assembly method is the most facile way, by which an appropriate volume of nanoparticle solution is spread over a substrate, spontaneously constructing the thermodynamically stable arrangement of nanoparticles accompanied by solvent vaporization. The development of the 1D assembly is still hard to achieve, unfavourable to that of the hexagonal closed packed (hcp) structure because of the low-symmetric structural isotropy. In order to accomplish the 1D assembly of nanoparticles, their structural design should be refined at best. As a secret, the discotic liquid crystal molecule, hexaalkoxy-substituted triphenylene (TP) [31], is introduced into an organic ligand surrounding a gold nanoparticle surface (Au-TP) (Figure 12) [32, 33]. TP molecules are known to self-assemble into 1D columnar mesophase via π-π interaction [34]. Consequently, it is expected that TP molecules would be a strong trigger for the 1D assembly of metal cores by introducing TP units onto the metal nanoparticle surface.

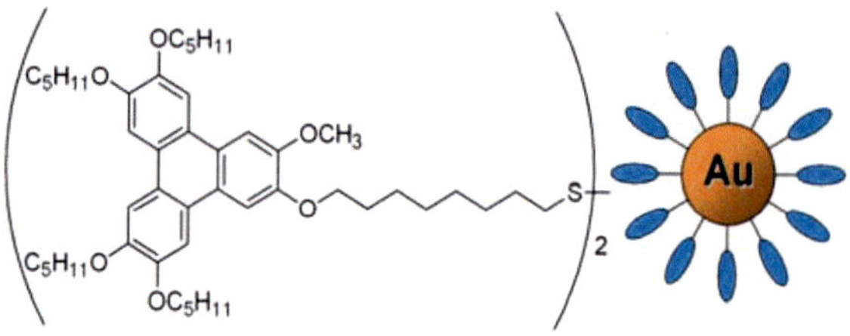

Figure 12. (Left) Structure of TPD. (Right) Schematic representation of Au-TP [32].

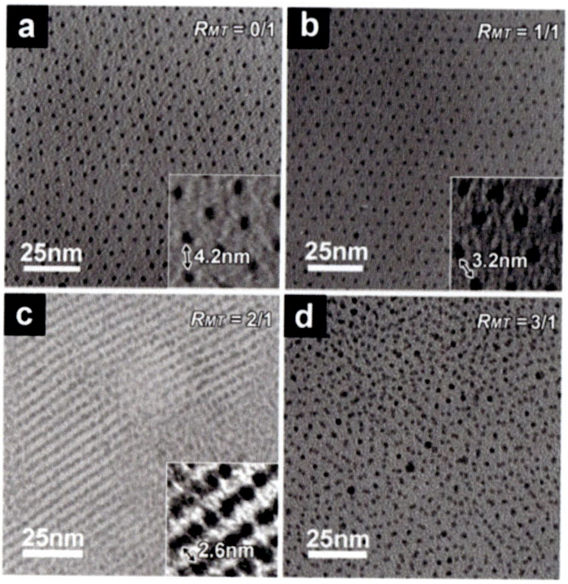

Figure 13. TEM images of Au-TP prepared in (a) R_{MT} =0/1, (b) 1/1, (c) 2/1, and (d) 3/1. The inset shows the enlarged image [32].

Au-TP with 2.4 nm diameter was synthesized in a homogeneous reduction of $HAuCl_4$ in DMF/water with TP thiol derivatice, [8-(3-methoxy-6, 7, 10, 11-tetrakis-pentyloxy-triphenylen-2-yloxy)-octanyl disulfide (TPD, Figure 12). Figure 13 shows the TEM images of the Au-TP assemblies prepared in a mixture of a polar (methanol) and a less polar (toluene) solvent. The solvent polarity is controlled by altering the mixed ratio of methanol to toluene (v/v), R_{MT}, from 0/1 to 3/1. The R_{MT} value is regarded as a kind of polarity index. When the solvent polarity is relatively low (R_{MT} = 0/1, 1/1), a clear hcp structure is formed with a particle spacing, d_{int}, of 4.2 and 3.2 nm, respectively (Figure 13a and 13b). The full molecular length of the TP ligand is *ca.* 2.4 nm calculated by MOPAC described in Figure 14d, where the C8 alkyl chain is 1.1 nm, the aromatic frame of the TP moiety is 0.6 nm, and the surrounding pentyloxy group of the TP moiety is 0.7 nm. These values indicate that the Au-TP in Figure 13a self-assemble by intercalating only the adjacent pentyloxy moiety around the TP ligands (Figure 14a). On the other hand, those in Figure 13b should accompany the *partial* π-π interaction of TP units (Figure 14b).

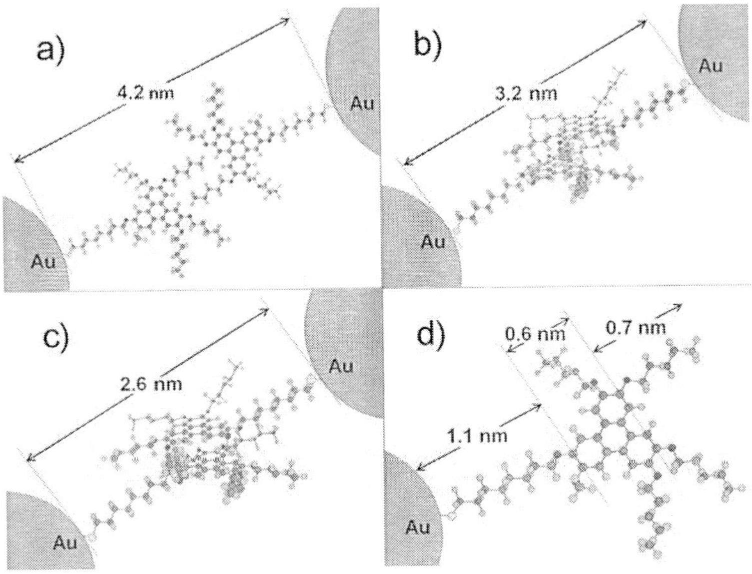

Figure 14. Schematic illustration of (a) adjacent TP ligands on a gold core surface with non-stacking (d_{int} = 4.2 nm), (b) with partial stacking (d_{int} = 3.2 nm), (c) with full stacking (d_{row} = 2.6 nm) of TP units, and (d) a single TP ligand on a gold core surface [32].

Interestingly, when the polarity of the solvent is further increased to R_{MT} = 2/1, the assembly structure is entirely converted from hcp to a 1D arrangement (Figure 13c). The interparticle spacing in a chain is *ca.* 0.7 nm. The row spacing, d_{row}, is 2.6 nm, a value which is consistent with the *full* stacking of the TP units interdigitated among the adjacent Au-TP particles (Figure 14c). The well-defined superstructure of Au-TP disappears with an increase of R_{MT} above 3/1 (Figure 13d). The whole results deduce that the strong interparticle π-π interaction of TP units in solution is inevitable for the 1D arrangement of Au-TPs, while that interaction is supportive, but not indispensable for the hcp ordering of Au-TPs.

The plausible mechanism is attributed to the specific nature of TP ligands that discotic liquid crystalline molecules tend to gather into a 1D column as mentioned at the beginning. When the *inter*particle π-π stacking is promoted in an appropriate mixed solvent, the small columnar phase of TP units will be constructed by the intercalation of TP ligands among Au-TPs. During the ordering process of the collected TP columnar phase, the partial area on a metal core surface, where the protection by TP ligands becomes weak due to the alkyl chain bending of TP ligands turning to the TP column direction, is inevitably produced. Accordingly, the van der Waals attraction among the metal cores is accelerated, causing the gradual aggregation of Au-TPs. At that time, the specific 1D nucleus of Au-TPs are compulsorily constructed on account of the surrounding TP column, resulting in the secondary growth of the 1D Au-TP arrays through directional motion of nanoparticles toward the ordered arrays, accompanied by the solvent evaporation process. The further investigation of the other parameters, *e.g.*, solvent molecular structure, boiling point, Au-TPs on their physical properties (core size, alkyl chain length of the TP ligand) produced more variable assemble structure of Au-TPs [33]. The general relations between the design of organic/inorganic metal nanoparticles and their self-assemble nature should been clarified in detail. From this point, this study is important which demonstrates definitely that both the appropriate physical architecture of nanoparticles and the parametric balance in self-assembly process are essential for the precise control of nanoparticle arrangement.

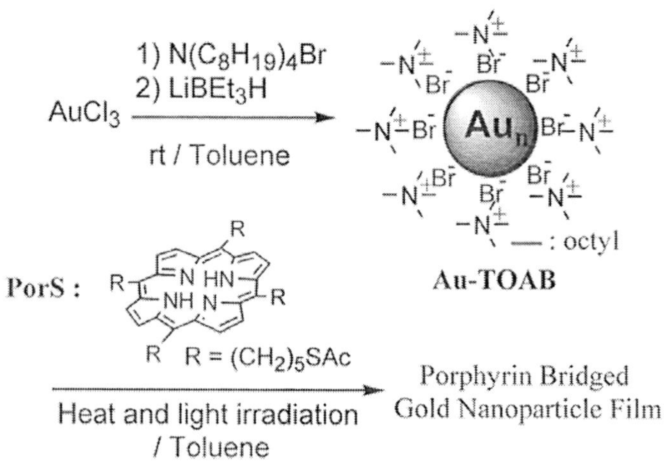

Scheme 4. Preparation of the PorS-bridged Au nanoparticle film [35].

1.3. Formation of a Porphyrin-Gold Nanoparticle Network

Next, the combination between metal nanoparticles and photo-functional molecules is presented. The system is made by the connection between octylammonium bromide-covered gold nanoparticle (Au-TOAB) and a porphyrin derivative, meso-tetra (5-thioacetylpentyl)porphyrin with four alkyl chains terminated with a thioacetate group (PorS) (Scheme 4) [35]. It is imaginable that the heat activation can produce aggregation of Au-TOAB since the adsorption force of the ammonium salt to gold nanoparticle surface is weak. Actually, heating a toluene solution of Au-TOAB caused enormously grown coagulation, resulting in precipitation at the bottom of the vessel. In this study, an extra perturbation of 'light irradiation' is introduced in addition with heat and PorS in order to control the cohesion process of Au-TOAB to form a rigid network film on a specific part of the substrate successfully.

A toluene solution of Au-TOAB was prepared by reducing $AuCl_3$ by lithium triethylborohydride dissolved in a micelle solution of tetraoctylammonium bromide (TOAB) in degassed toluene under N_2 inert atmosphere based on the literature [4]. The core diameter of the prepared gold particles was 3.8 ± 0.8 nm, and the nanoparticle solution was stable for a few weeks in a condition. The porphyrin derivative, PorS, equipped with four adsorptive thiolate hands around rigid planar porphyrin moiety to generate chemically tight combination with Au-TOAB, since the adsorption strength of thiolate is larger than that of TOAB. Figure 15a displays the time-resolved UV-Vis spectra of the mixed solution of Au-TOAB and PorS in toluene with an incident light of "D_2 + W" lamps simultaneous heating at 50 °C. A formation of a purple thin film on the inner wall of the used UV quartz cell is observed, which corresponds to a gradual increase of the absorption in the range of 500 nm to 900 nm (Figure 15a).The nanoparticle film thus prepared exhibits a red-shifted surface plasmon band of assembled gold nanoparticles at 576 nm, and a sharp Soret band due to the porphyrin moiety at 420 nm, confirming that the film is composed of PorS-gold nanoparticles (Figure 15a inset).

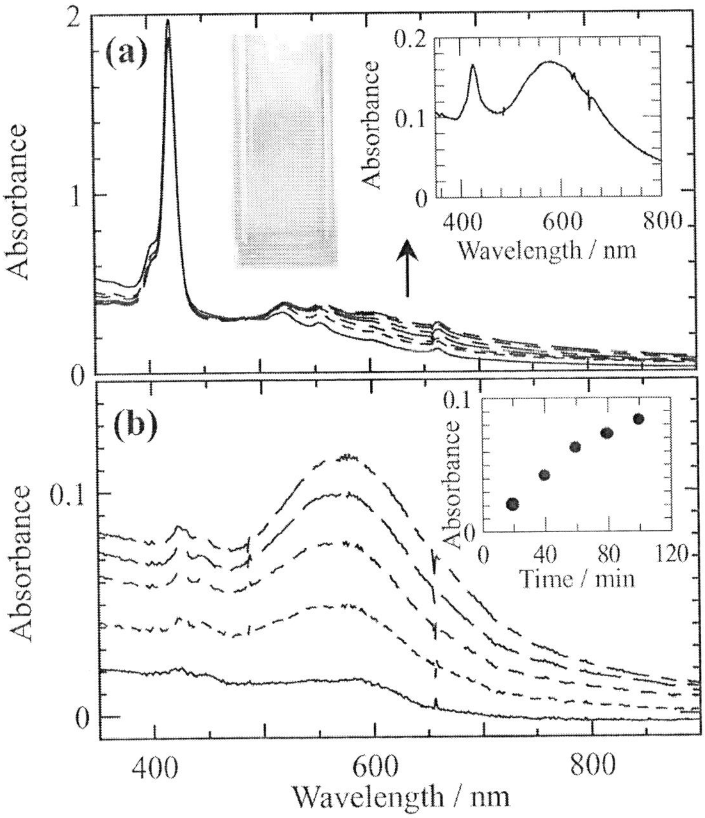

Figure 15. (a) UV-Vis spectral changes of the mixture of Au-TOAB and PorS with an incident light in every 20 min for 2 h with simultaneous heating at 50 °C from the bottom to the top. (Inset) The picture of the prepared film on a cell wall after the measurement and its UV-Vis spectrum. (b) The UV-Vis spectrum of the prepared PorS-Au nanoparticle film on the wall of the UV cell prepared with a Xe lamp from the bottom to the top for 20, 40, 60, 80, and 100 min with simultaneous heating at 50 °C in the mixture of Au-TOAB and PorS. (Inset) The correlation between the absorption intensity at 420 nm of the film and the irradiation time [35].

Figure 15b shows the UV-Vis spectra of the films stuck on the inside of the cell prepared using a "Xe" lamp with different irradiation time. The spectra present two bands derived from Au-TOAB and PorS similarly to Figure 15a, and the intensity of the absorption bands grows with increases in irradiation time almost linearly (Figure 15b inset). The molar ratio of Au-TOPB to PorS is 1: 7.5, indicating that the film contains less porphyrin moieties than that prepared using the D_2 + W lamps, in addition the film formation process depends on the light source. Additional experiments were carried out for the detailed elucidation of this networking mechanism, indicating that the main factor for this system is 'the thermal energy of molecular vibration' in substrate (here quartz of the UV cell) by infrared light which generates a specifically heated area of the inner wall. As a result, the isolated and/or partially gathered nanoparticles are stuck on the irradiated spot connected by a thioacetate group of PorS, and the nanoparticle networks grow gradually in three-dimensional direction.

This film formation is applicable for the larger gold nanoparticles in 5.6 nm core diameter as shown in Figure 16. It demonstrates that this technique would be possibly utilized for the film formation with appropriate metal nanoparticels in designed area. The photochemical properties of the film was also examined using the interdigitated array (IDA) electrode, which showed that photo-current response was switchable by light irradiation, involving the reliable increase of conductivity in 30 %. The excited electrons of the porphyrin units could induce the faster electron transfer through the film.

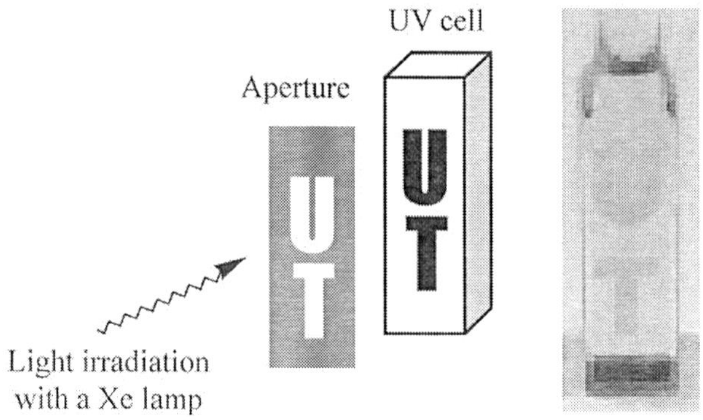

Figure 16. A photograph of porphyrin bridged gold nanoparticles on the inner wall of the UV cell and the illustration of the irradiation process.

2. Physical Structure Control of Metal Nanoparticles as Advanced Catalysts

A large number of researchers up to now have pointed out that the properties of a metal nanoparticle are greatly dependent on its physical structure. For instance, the absorption originated to the surface plasmon oscillation of gold nanoparticles is dramatically red-shifted when the nanoparticle shape changes from spherical structure to a rod like one [37]. In common, the catalytic activity of metal nanoparticles is accelerated with decreasing their diameter, since the ratio of surface atoms increases exceedingly [38]. The focus of this section is to develop advanced catalysts by controlling the shape and size of metal nanoparticles. It is not so simple to arrange their physical structure precisely in solution, since among various synthetic parameters, at least solution temperature, time, an organic ligand, and a free energy change (ΔG) should be precisely controlled in order to obtain the specific shape of nanoparticles [39, 40]. The intelligent synthetic system which controls the accurate crystal growth of metal nanoparticles both kinetically and thermodynamically should be invented by selecting appropriate synthetic parameters. Under consideration of these factors, highly selective synthesis of cubic Pt nanoparticles is accomplished by means of an ionic additive in mixture, and the preparation of gold nanoparticle catalyst with high monodispersity in size is practicable for carbon nanotube synthesis.

2.1. Synthesis and Size Control of Platinum Nanocubes with High Selectivity of Shape

The catalytic properties of Pt nanoparticles are greatly expected for the organic synthesis, the decomposition of exhaust gas, and the polymer electrolyte fuel cell (PEFC). Fuel cells possess high energy conversion efficiency originated from the chemical energy of the combustion reaction between hydrogen and oxygen, without emitting environmental pollutants including CO_2. Among several kinds of fuel cells, PEFC would become the most common source in the general community because of its low operating temperature (below 100 °C). However, to enable its practical usage, the cost of the Pt nanoparticles used as an electrode catalyst in PEFCs must be reduced, which can be carried out by developing Pt nanoparticles with more efficient catalytic activity. Their catalytic activity depends on their shape and size, and cubic Pt nanoarticles are anticipated to exhibit higher catalytic activity compared to spherical particles, since cubic particles are composed of only (100) facets, with more defective and active sites accompanied by dissolution and surface reconstruction than (111) facets [41]. As for particle size, around 5 nm is reported to produce the most efficient catalytic performance [42].

Here, described is the novel preparation and size control of Pt nanocubes stabilized by polyacrylic acid sodium salt (PAA-Pt) with selectivity of max. 84% by the additive effect of NaI [43]. NaI was selected as an additive to control the particle shape because I^- anions are known to adsorb onto a Pt crystal surface most strongly among halogen anions, and was thought to act as a good controller for the growth rate between {100} and {111} faces of PAA-Pt. PAA-Pt could be prepared by the reduction of K_2PtCl_4 with H_2 bubbling in the presence of PAA and NaI in aqueous solution.

Figure 17 shows the TEM image of the prepared PAA-Pt with and without NaI. From these images, it is obvious that the shape of PAA-Pt tends to become cubic when NaI coexists in the reaction mixture. The further investigation of the NaI effect on the physical structure (shape and size) of PAA-Pt was carried out by changing the molar ratio of NaI to K_2PtCl_4, as summarized in Figure 18. The ratio of the cubic nanoparticles to all of the produced particles (= cubic ratio) and the average particle size tend to increase with increasing NaI ratio and Pt concentration. This result indicates the formation mechanism of the cubic PAA-Pt: after a particle nucleus is generated at the first stage of the reaction, the particle crystal grows gradually, accompanied by the different growth ratios between the {111} and {100} faces under the additive effect of I^-.

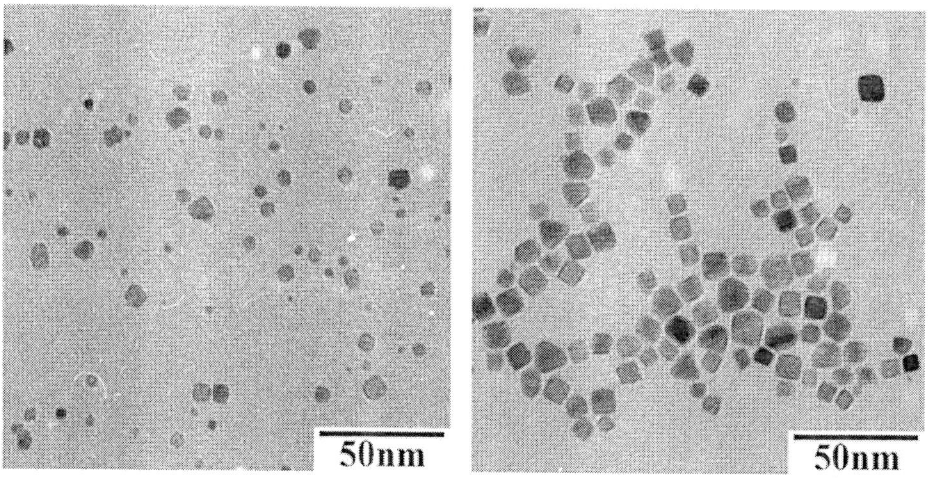

Figure 17. TEM image of PAA-Pt prepared under the reaction condition of NaI/K_2PtCl_4 = (right) 1.0 mM/1.0 mM and (left) 0 mM/1.0 mM.

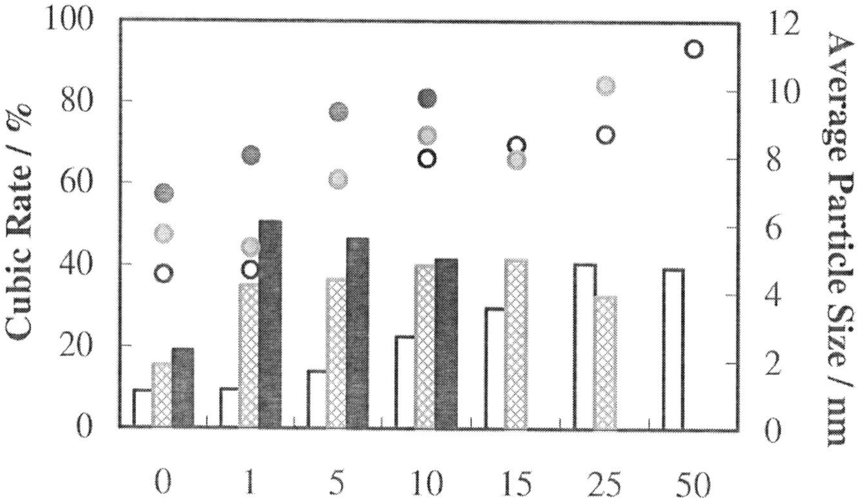

Figure 18. Effect of the NaI molar ratio to K_2PtCl_4 on the average particle size (circles) and the cubic selectivity (bars) of the prepared PAA-Pt. The concentration of K_2PtCl_4 was 0.1 (white), 0.5 (grey) and 1.0 mM (black), respectively [43].

Figure 19 shows the TEM images of PAA-Pt synthesized at a different reaction temperature, *T*, from 5 to 60 ° C. They demonstrate that the cubic PAA-Pt can be preferentially obtained between 20 and 60 ° C with high selectivity *ca.* 70-80 %, in addition, the cubic size is controllable between 7.5–10.5 nm by changing *T*. The particle size first increased with rise of *T* until 20 ° C, and then gradually decreased. In this system, the temperature was the most effective parameter to control the cubic size of PAA-Pt, keeping the high shape selectivity. It was preliminarily confirmed that the catalytic performance of the prepared PAA-Pt for the O_2 reduction was higher than the ordinary spherical Pt nanoparticles. Our group would report their detailed catalytic capacity in near future.

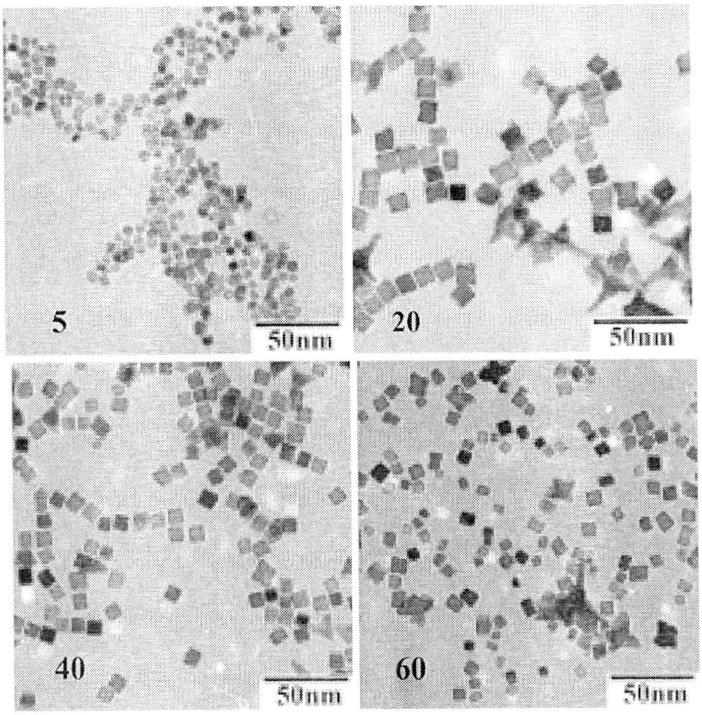

Figure 19. TEM image of PAA-Pt prepared with a different reaction temperature of which the value is noted in the images [43].

2.2. Synthesis and Diameter Control of Multi-walled Carbon Nanotubes over Gold Nanoparticle Catalysts

The characteristics of carbon nanotubes (CNTs) [38, 44] have been energetically studied for decades including electrical behavior and mechanical strength. In general, nanoparticles of 3d transition metals (Fe, Co, Ni *etc.*) are used as catalysts for the synthesis of CNTs in chemical vapor deposition (CVD) [38]. Accordingly, the studies of "noble" metal nanoparticles for a CNT catalyst attract great interest. Bulk gold has long been regarded as inert and less active as a catalyst, a condition attributed to its completely filled 5d shell and the relatively high value of its first ionization; however, gold begins to show interesting catalytic behavior, *e.g*,. CO oxidation, hydrocarbon oxidation, and NO reduction when its size is reduced into a nanoscale regime [45, 46]. In this section, the preparation of multi-walled carbon nanotubes (MWNTs) over a gold nanoparticle catalyst is discussed [47, 48]. The adopted gold nanoparticles have been stabilized by dodecanethiol (DT), used as a catalyst after supported on SiO_2-Al_2O_3.

DT-stabilized gold nanoparticle with a core diameter of 2.1 ± 0.5 nm (Au-DT1) was synthesized on the basis of reducing Au (III) ions in toluene/water solution with DT and tetraoctyl ammonium bromide (TOAB) [49]. The diameter control of Au-DT1 was carried out by the calcinations process of the crude solid of Au-DT1 containing both TOAB and DT, obtained before the purification process of methanol. The size selection of gold nanoparticles could be accomplished by the solid heating method, where the crude Au-DT1 was heated in an electric furnace at fixed temperatures of 150, 180, and 200 °C to produce Au-DT2 (d_{av} = 3.7 nm) Au-DT3 (d_{av} = 5.3 nm) and Au-DT4 (d_{av} = 7.0 nm), respectively [49, 50]. Seen in the TEM images of the prepared samples (Figure 20), the particles are well-assembled due to their very low size distribution. By this step, the particle size could be conveniently adjusted almost linearly to the applied heating temperature.

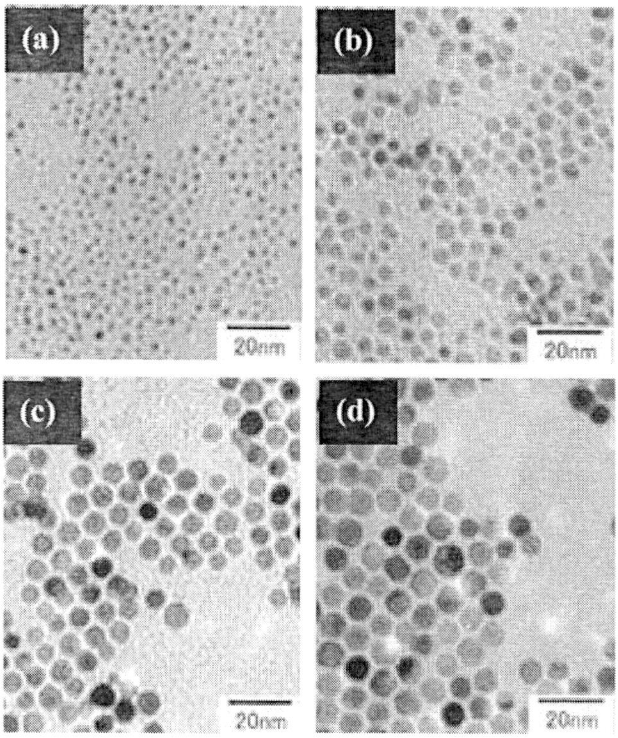

Figure 20. TEM image of (a) Au-DT1, (b) Au-DT2, (c) Au-DT3, and (d) Au-DT4 [47].

Each size-controlled Au-DT was supported to produce the Au-DT/SiO_2-Al_2O_3 catalyst (0.1 wt. % gold), obtained after the calcination process for 1 h min at 200 °C. The particle core diameter in Au-DT(1-4)/SiO_2-Al_2O_3 was measured at 3.0 ± 0.7 nm, 5.0 ± 0.7 nm, 8.3 ± 1.0 nm, and 10.0 ± 1.1 nm, respectively. Comparing the values of the particle size before and after being supported, an aggregation among adjacent particles in the SiO_2-Al_2O_3 matrix occurred, leading to the *ca.* 1.5 times crystal growth in size during the calcination. However, the particles remained separately dispersed in SiO_2-Al_2O_3, and reflected the size order of the used Au-DT as a starting particle. This confirms that this catalyst preparation using a size-selective metal nanoparticle prepared in solution is effective for managing the precise particle size in catalysts. Apparently, the particle size distribution is narrower than that prepared by general preparation of nanoparticle catalysts, such as an impregnation method [51] in which metal ions dispersed into a support are reduced in a solid state.

The influence of the reaction temperature and the size of Au-DT on the formation of MWNTs was investigated by using C_2H_2 as a carbon source over Au-DT(1-4)/SiO_2-Al_2O_3, summarized in Table 2. It is obvious that the outer diameter of the formed MWNTs becomes larger with an increase of the size of Au-DT introduced as a catalyst. The diameter of

MWNTs also tends to increase at higher reaction temperatures, which is explained by the considerations that (1) the size of the gold nanoparticle catalyst grows gradually by heat-induced aggregation during the reaction among adjacent particles, and (2) the catalytic reaction is accelerated at higher temperatures. Additionally, it should be mentioned that MWNTs can be generated at lower reaction temperatures with decreasing size of the Au-DTs, attributed to higher catalytic ability for smaller nanoparticles.

Table 2. Dependence of a used catalyst on the outer diameter of the MWNTs formed by using C_2H_2 as a carbon source [47]

Used Nanoparticle for a Catalyst	Outer Diameter of MWNTs / nm				
	Reaction Temperature / °C				
	450	500	550	600	700
Au-DT1	-[a]	9.1	9.5	10.3	14.0
Au-DT2	-	-	13.0	14.3	16.5
Au-DT3	-	-	23.0	17.0	20.2
Au-DT4	-	-	-	19.3	23.3

a) Dash indicates that no MWNTs were identified.

Figure 21 shows the relation between the size of Au-DT used as a catalyst and the outer diameter of MWNTs formed at 600 and 700 °C, which displays the discrepancy between C_2H_2 and C_2H_4 used as a carbon source. The diameter of MWNTs produced by C_2H_2 (MWNT-C_2H_2: Figure 22) increases almost proportionally with increasing the size of Au-DT. On the other hand, the reasonable dependence of the MWNT diameter synthesized by C_2H_4 (MWNT-C_2H_4) on Au-DT size cannot be recognized.

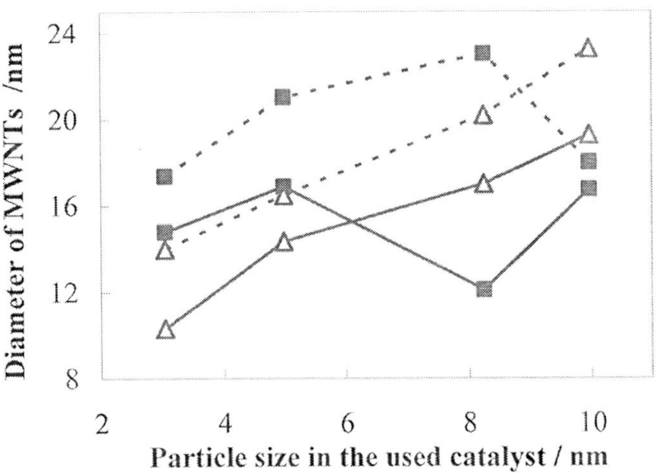

Figure 21. Relation between the particle size in the used catalyst and the outer diameter of MWNTs formed at 600 (solid line) and 700 (dotted line) °C by using C_2H_2 (triangles) and C_2H_4 (squares) as a carbon source [47].

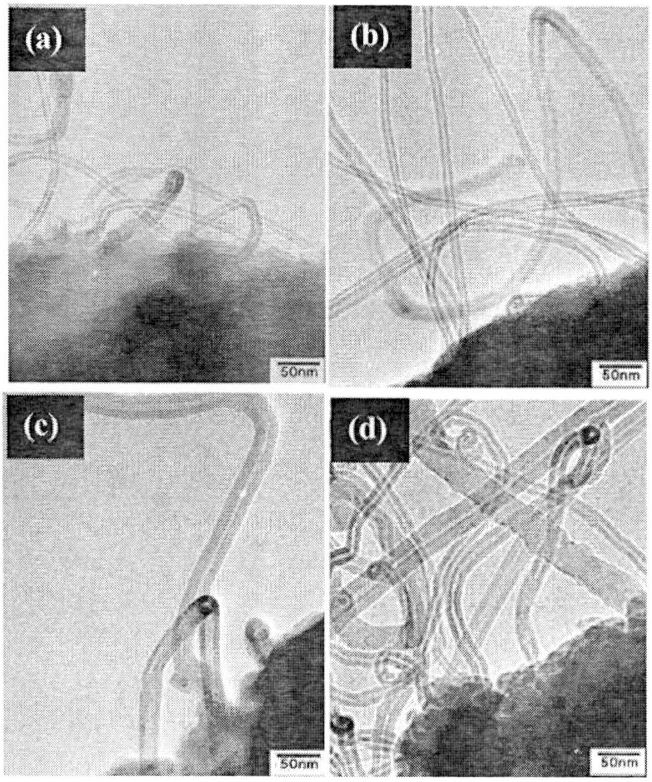

Figure 22. TEM image of MWNTs formed at 600 °C by using C_2H_2 as a carbon source over (a) Au-DT1, (b) Au-DT2, (c) Au-DT3, and (d) Au-DT4/SiO_2-Al_2O_3 [47].

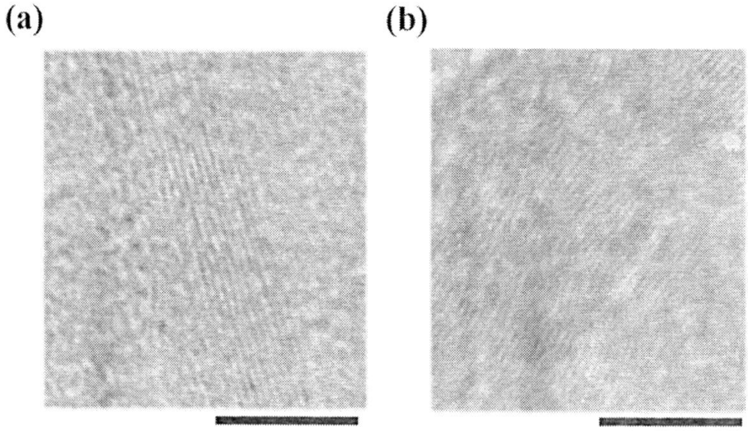

Figure 23. HRTEM images of MWNTS formed at formed at 600 °C by using (a) C_2H_2 and (b) C_2H_4 as a carbon source over Au-DT2/SiO_2-Al_2O_3 [47]. Scale bar : 5 nm.

To elucidate this phenomenon, the HRTEM images of MWNT-C_2H_2 and –C_2H_4 were taken, especially in the part of the tube inner wall (Figure 23); these demonstrate that consecutive carbon layers with interlayer ideal spacing of 0.38 nm [52] are observed in the image of MWNT-C_2H_2, while undulated layers covered by amorphous carbon are identified in that of MWNT-C_2H_4. Many unrevealed points remain concerning the mechanism of CNT formation, even over a general 3d metal catalyst; however, it is proposed that the tip-and base-growth models [53] can be also applied for a gold nanoparticle catalyst. After the nucleation of CNTs at the gold nanoparticle surface, the tube growth process begins. During the growth, the wall layers are created in a uniform axial direction by the introduction of C_2H_2 molecules, possessing a regular growth of the parameter values relating to the CNT diameter. On the other hand, the carbon wall grows more randomly into several axes with C_2H_4 molecules, leading to less dependence of the CNT configurations on the size of Au-DT.

3. Introduction of Metal Coordination Polymers into a Nanoparticle Core

The last section focused on metal coordination compounds as an inorganic core in place of metal. Metal coordination polymers, where some transition metal ions are connected in three-dimensional (3-D) nets by organic ligands, are fascinating organic-inorganic hybrid materials [54]. Variable geometries and connections have been created by selecting proper bridging ligands and metal ions in order to produce novel functional materials with useful wide-ranging properties [55, 56], *e.g.* electronic, magnetic, electrochromic, optical, catalytic, and adsorptive characteristics. The discussion about their specific natures has centered on bulk crystals (μm ~) for a long time. It has not been well clarified yet whether downsizing effects like metallic nanoparticles are observed for the properties of nanometer-sized metal coordination polymers, which we have named as metal coordination "nano"-polymers (MCNPs) [57-59]. For the first step, it is important to popularize the synthetic procedure of MCNPs in order to reveal the relations between the physical structure of MCNPs and their characters experimentally. As a metal coordination polymer, metal cyanometalates, M_1-CN-M_2 (M_1 = Fe, Ru, Co, Cr, Os, *etc.*, M_2 = Fe, Co, Ni, Mn, Cr, Eu, La, Sm, *etc.*), called Prussian blue (M_1 = M_2= Fe) [61] and its analogues, are selected [56]. Because of its convenient synthetic procedure compared to other metal coordination polymers, M_1-CN-M_2 is one of the most widely-explored metal coordination polymers with various outstanding features as a bulk crystal including, *e.g*,. magnetism (photo-induced magnetism), electrochromism, and conductivity, etc. M_1-CN-M_2 MCNPs would be a standard model case to demonstrate the possibility of MCNPs as novel nanomaterials.

3.1. Synthesis and Downsizing Effect of Metal Coordination Nano-polymers

The attempts to synthesize MCNPs have appeared by degrees in the past 5 years using, *e.g.*, the reverse micelles technique [62], the Langmuir-Bloddgett method [63], and template synthesis [64]. Among them, our group succeeded to isolate the Fe/Cr-CN-Co MCNPs (d_{av} = 5~7 nm) stabilized by organic shell of stearylamine (SA), utilizing reverse micelle solution of H_2O/cyclohexane/a nonionic surfactant, polyethylene glycol mono 4-nonylphenyl ether (NP-5: $HO(CH_2CH_2O)_nC_6H_4C_9H_{19}$) [57]. The complexation between $[Fe/Cr(CN)_6]^{3-}$ anions and Co^{2+} occurs in a water nano-droplet to generate nanometer-sized crystals of Fe/Cr-CN-Co MCNPs with narrow size distribution (Scheme 5). The merits of this synthesis deserving special mention is that the prepared Fe/Cr-CN-Co MCNPs can be obtained as an air-stable powdery form, and the excessive organic stabilizers of SA are sufficiently removed during the purification process. The key to this isolation is the stable and strongly coordinated protection by an amine group of SA onto the Co sites around the Fe/Cr-CN-Co inorganic core. The powder X-ray diffraction (XRD) patterns of the samples showed a typical face centered cubic (fcc) structure as the same as their bulk crystal, though the peaks were widely broadened due to the small crystal size [65]. In this synthesis, the metal elemental control among Fe, Cr and Co is also demonstrated by adjusting the ratio among the starting complexes in reaction mixture (Table 3). In particular, compound **1**, comprised of only Co and Fe, exhibits the typical cubic shape stacked together by self-assembly (Figure 24). The lattice constant, *a*, is in good accordance with that of a bulk form (*e.g.*, a = 10.06 and 10.54 Å for **1** and **5**, respectively). The value of *a* shifted almost linearly from **1** to **5** with increase of the Cr component, suggesting that the metal units were uniformly dispersed in the nano-polymers.

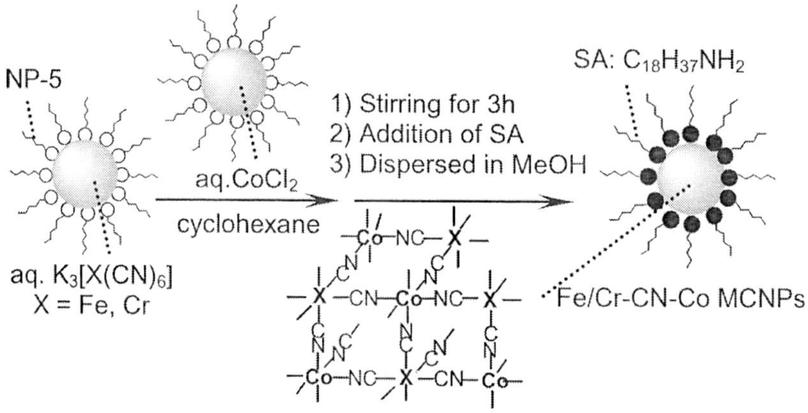

Scheme 5. Preparation of Fe/Cr-CN-Co MCNPs [57].

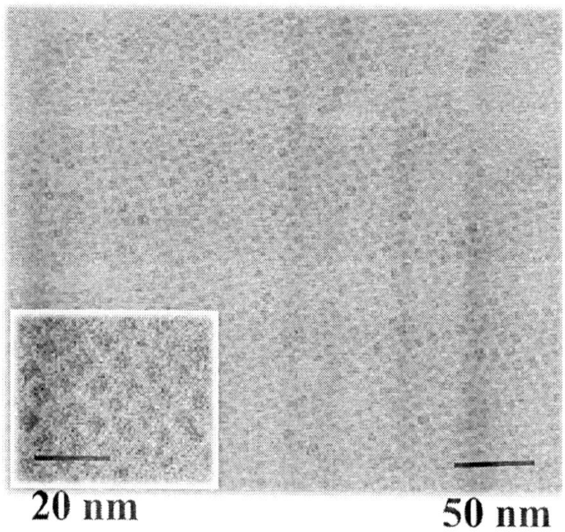

Figure 24. TEM image of the prepared Fe/Cr-CN-Co MCNPs (**1**).

Table 3. Formula of the prepared Fe/Cr-CN-Co MCNP [57]

Compound	Formula[a]
1	$K_{1.04}Co_{1.10}[Fe(CN)_6](SA)_{0.41}\cdot 2.1H_2O$
2	$K_{0.81}Co_{1.10}[Fe(CN)_6]_{0.72}[Cr(CN)_6]_{0.28}(SA)_{0.53}\cdot 3.0H_2O$
3	$K_{0.80}Co_{1.01}[Fe(CN)_6]_{0.62}[Cr(CN)_6]_{0.38}(SA)_{0.60}\cdot 2.4H_2O$
4	$K_{0.82}Co_{1.30}[Fe(CN)_6]_{0.32}[Cr(CN)_6]_{0.68}(SA)_{0.46}\cdot 3.1H_2O$
5	$K_{0.30}Co_{1.35}[Cr(CN)_6](SA)_{0.69}\cdot 4.3H_2O$

a) Determined by the thermogravinetric analysis (TGA), the induced couple plasma (ICP) analysis, and the common CHN analysis method. The ratio of the hexacyano complexes is normalized as 1.

All of the samples were redispersed in generally less polar solvents, *e.g.* CH_2Cl_2, $CHCl_3$, THF, pyridine, while their bulk form [66, 67] was insoluble in any solvent). Figure 25 displays the THF solution of **1-5,** which indicates that the solution color is dependent on the elemental ratio in Fe/Cr-CN-Co MCNPs. This color variation reflects the electronic state of metal components in MCNPs, affecting their properties. For example, about their magnetism (Figure 26), field-cooled magnetization versus temperature curves of **1-5** show that compounds **4** and **5**, containing a high ratio of Cr units, exhibit ferromagnetism with spontaneous magnetization at the Curie temperature (*T*c). These findings indicate that the isolated Fe/Cr-CN-Co MCNPs would become a potential substance for the field of coordination chemistry in a nano-scale range.

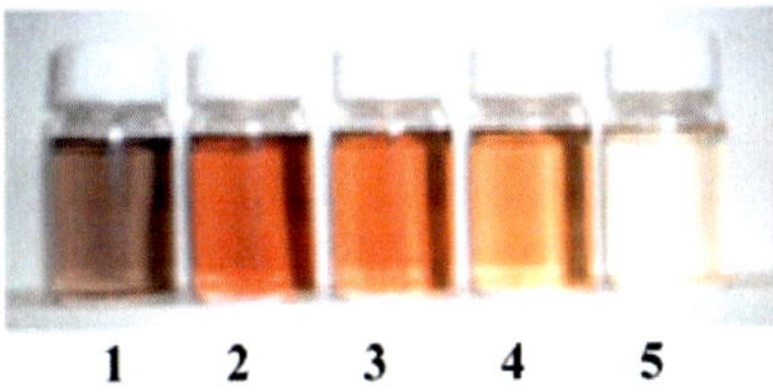

Figure 25. THF solution of the prepared Fe/Cr-CN-Co MCNPs (**1-5**). The number under the figure refers to that of the compound [57].

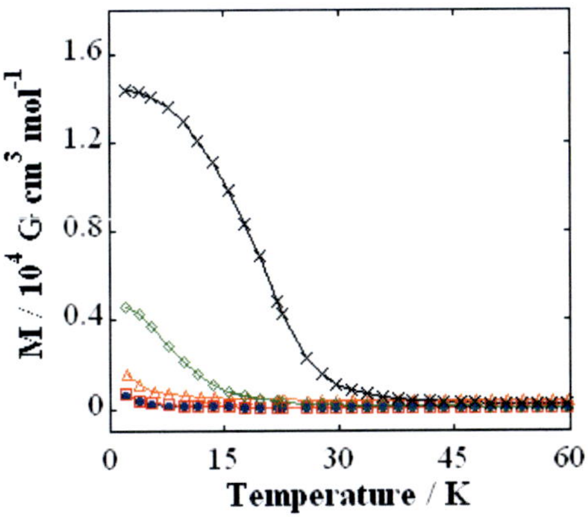

Figure 26. Field-cooled magnetization curves in an external magnetic field of 0.1 T of Fe/Cr-CN-Co MCNPs: blue (**1**), red (**2**), orange (**3**), green (**4**), and black (**5**) [57].

Figure 27. Photographs of Fe-CN-Co MCNPs prepared in the reverse micelle solution of CTAB during the reaction [59]. The reaction times are noted in the figures.

The other surfactants except for nonionic NP-5, of which the hydrophilic group is cationic or anionic, can be applied for the synthesis of MCNPs. Fe-CN-M_2 MCNPs (M_2 = Eu, La) were prepared in an anionic reverse micelle of sodium bis(2-ethylhexyl)sulfosuccinate (AOT) [68, 69]. The crystal size is *ca.* 50 ~ 800 nm, indicating that the size tends to be larger when rare earth metal ions are introduced into MCNPs. In another case, Fe-CN-Co MCNPs with the average diameter of *ca.* 3 nm in cationic reverse micelle solution of cetyltrimethyl ammonium halides [CTAX, X = B(bromide), C(chloride)] [59]. It is interesting to note that the material color dramatically changes from *red* to *green* with increasing reaction time when Fe-CN-Co MCNPs are synthesized in CTAB reverse micelles (Figure 27). This is attributed to the temporal evolution of elemental composition and crystal structure of Fe-CN-Co nanoparticle core, of which the phenomenon is not observed for its bulk crystal. Various measurement revealed that the driving force was the interaction between the Fe-CN-Co MCNP surface and cetyltrimethyl ammonium species in solution, since the ratio of surface area in MCNPs was exceedingly increased compared to bulk material. This is one piece of evidence that metal coordination polymers possess a characteristic size effect on their properties.

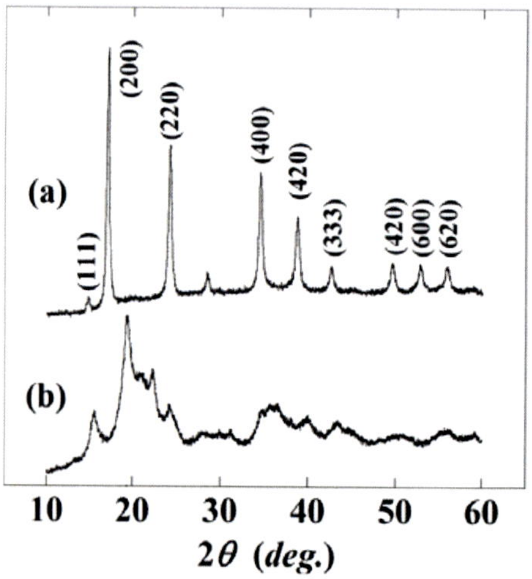

Figure 28. XRD patterns for (a) Cr-CN-Cu bulk crystals and (b) Cr-CN-Cu MCNPs [67].

In another experiment, these downsizing effects were also demonstrated by investigating the crystal structure and magnetic properties of Cr-CN-M_2 (M_2 = Co, Fe, Cu) MCNPs stabilized by pyridine derivatives [67]. For example, the XRD pattern of Cr-CN-Cu MCNPs with 15 nm diameter showed unexpected diffraction peaks, entirely different from the typical fcc structure of its bulk crystal originated from the complexation between a $[Cr(CN)_6]^{3-}$ anion and an octahedral Cu^{2+} cation (Figure 28). The crystal disorder of Cr-CN-Cu MCNPs is possibly attributed to some tetrahedral Cu sites which are provoked by the release of water molecules from Cr-CN-Cu crystals. This is also the surface effect of MCNPs derived from the interaction between "the organic species in solution" and "Cu sites in MCNPs". Numbers of unrevealed questions have been left about MCNPs. Further investigation is very significant not only for the coordination chemistry but also for the nanotechnology science.

3.2. Applicable Usage of Metal Coordination Nano-polymers

M_1-CN-M_2 MCNPs thus prepared become novel materials based on the original properties of their bulk crystal, *e.g.* nano-magnets, sensors (H_2O_2, glucose), electrochromism *etc.* For example, Fe-CN-Cr MCNPs can act as a nano-magnet material as noted above. In another case, the toluene dispersion of Fe-CN-Fe MCNPs [70] was spread over an ITO electrode to form a thin film of Fe-CN-Fe MCNPs. The film displays the reversible electrochromism between blue (+0.6 V *vs.* SCE) and coloress (–0.4 V *vs.* SCE) in 0.1 M potassium hydrogen phthalate [71]. Generally, the electrochromic film of Fe-CN-M_2 (M_2 = Fe, Co, Ni) are made by electrodeposition process of starting metal ions, while versatile electrochromic patterns can be fabricated more conventionally by utilizing the combination between Fe-CN-M_2 MCNP dispersion and appropriate film formation methods (*e.g.* ink jet printing, spincoating, LB, *etc.*)

From the other point, it is also perceptive to use MCNPs as a novel precursor of metal alloy or ceramic nano-materials [58, 60]. It is proposed that a novel synthetic approach to create metal alloy or ceramic nanoparticles by transforming MCNPs through the decomposition of the bridging organic ligand of MCNPs. Metal coordination polymers have a great advantage where the various metal elements are connected by an organic ligand with uniform composition [54]. In M_1-CN-M_2 MCNPs, M_1 and M_2 ions are bridged in advance by CN ligands with a regular elemental composition; therefore, the M_1M_2 alloy or M_1M_2 oxide nanoparticles obtained after the transformation reaction hold the uniform dispersion of the metal constituents with almost the same metal elemental ratio in the M_1-CN-M_2 MCNPs precursor. This method has the advantages, first, that the kinds of the metal elements can be accurately introduced before hand, and second, the electronic state of the nano-materials finally obtained is controllable just by altering the decomposition condition between reductive (H_2) and oxidative (O_2) atmosphere.

The transforming process of cobalt tetracyanoplatinate (Pt-CN-Co) MCNPs through a gas phase reduction in the H_2 atmosphere is demonstrated as an example. Pt-CN-Co MCNPs (Pt/Co = 0.9) with *ca.* 15 nm diameter were prepared by the complexation between $[Pt^{II}(CN)_4]^{2-}$ and Co^{2+} as the same as Fe/Cr-CN-Co MCNPs (Figure 29a). From thermal gravimetry analysis (TGA) of Pt-CN-Co MCNPs under H_2 atmosphere, the removal of the CN ligands occurred from 350 to 450 ° C with a weight loss of 24 %. The X-ray photoelectron spectroscopy (XPS) spectra of Pt-CN-Co MCNPs before and after the gas-phase reduction at 350 ° C show that when the reaction time is increased, the Pt4f signals at 73.1 eV and 76.4 eV, due to Pt^{II} sites in Pt-CN-Co MCNPs, decrease in intensity, while new signals appear at 71.2 eV and 74.5 eV assigned to the metal Pt^{0} (Figure 30, left). The intensity of the CN stretching signal of Pt-CN-Co MCNPs in an IR spectrum was reduced with increasing reaction time. It is understood that the transformation reaction of Pt-CN-Co MCNPs gradually occurs by the reduction of Pt and Co ions accompanied by the removal of the bridging CN ligands. When reaction time is fixed as 3 h, the higher reaction temperature in the range of 350 to 450 ° C promotes the metallization reaction of Pt^{II} (Figure 30, right).

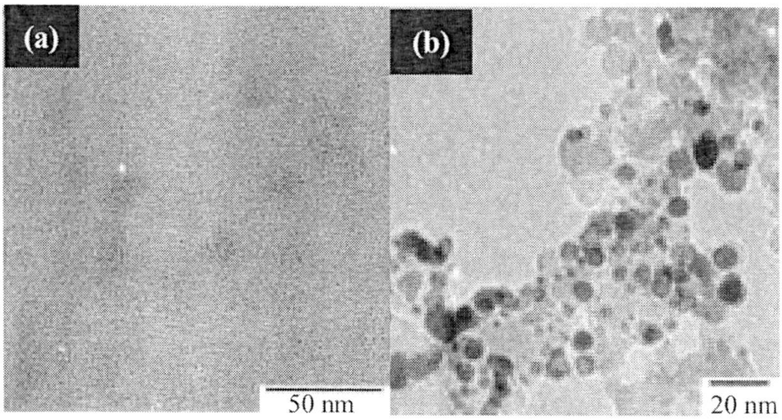

Figure 29. TEM image of Pt-CN-Co MCNPs (Pt/Co = 0.9) (a) before and (b) after the transforming reaction in H_2 atmosphere at 400 °C for 3h [58].

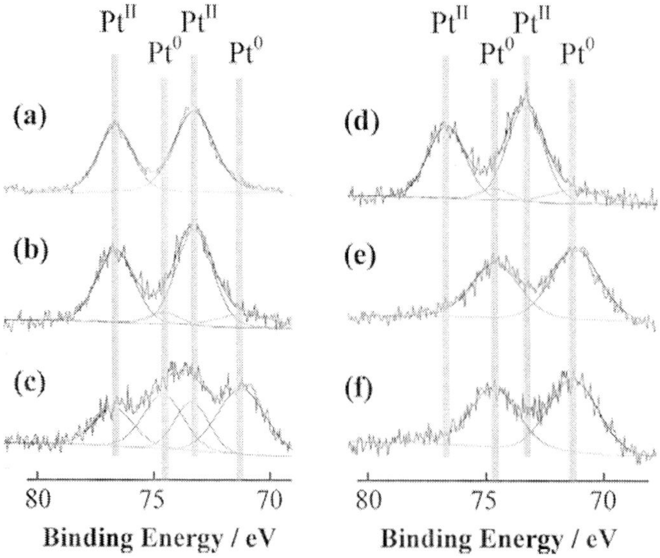

Figure 30. XPS curve in the Pt4f energy range of Pt-CN-Co MCNPs (Pt/Co = 0.9) before (a) and after the transformation reaction in H_2 atmosphere at 350 °C for 3h (b)(d), at 350 °C for 5h (c), at 400 °C for 3h (e), and at 450 °C for 3h (f) [58].

Table 4. Pt/Co ratio and the lattice constant of Pt-CN-Co MCNPs [58]

Pt/Co (cal.)[a]	Pt/Co (anal.)[b]	Lattice Constant, *a* /Å[c]
1	0.9	3.76
2	1.6	3.83
3	2.9	3.86
5	3.6	3.88

a) The mixed ratio of the starting complexes. b) Determined by EDX. c) Calculated from the XRD results.

The XRD pattern of Pt-CN-Co MCNPs after the transformation reaction at 400°C with 3h exhibits a fcc structure consistent with metallic PtCo. In the TEM image of Pt-CN-Co MCNPs after the reaction, most particles have a below 10 nm diameter (Figure 29b). The particle size is reduced compared to that of Pt-CN-Co MCNPs (Figure 29a). The size shrinkage happened by the removal of the bridging CN ligand between Pt and Co ions. The value of Pt/Co in Pt-CN-Co MCNPs is changeable by inserting $Pt^{IV}Cl_6^{2-}$ ions in place of Co^{2+} in reaction solution (Pt/Co = 0.9, 1.6 , 2.9, and 3.6). In the XRD patterns, the diffraction peaks are shifted to lower diffraction angles almost linearly, meaning that the lattice constant, *a*, calculated by the XRD patterns increases almost linearly with an increase of the Pt component in the compound, suggesting that the metal components are well dispersed in an PtCo alloy nanoparticle (Table 4). Using the same concept, GeFe nanoparticles with a $B8_2$ type structure (η phase) could be composed from iron tetraoxalategermanium MCNPs and exhibited the unique ferromagnetism [60]. The uniformity of metal units in MCNPs as a precursor plays an important role for the fabrication of nanomaterials.

CONCLUSION

The novel organic-shell inorganic-core hybrid nanoparticles were synthesized by introducing metal or metal coordination compounds as an inorganic core. In favor of withdrawing the advantage of the combination between organic substances and inorganic ones, the suitable design of their construction is inevitable by tuning the principal parameters of the unit selection, the physical structure and the constituent ratio between inorganic and organic units. In the first section, metal nanoparticles with added functionalities have been created by attaching molecules with functions including a multiple redox moiety (biferrocene and anthraquinone), a photo-active unit (porphyrin) and a liquid crystalline molecule (triphenylene) onto the surface core through coordination bonds by thiolate. External stimuli such as light and electric field were utilized to transform the properties of functionalized molecules. Second, the catalytic activities of noble metal nanoparticles were investigated from the points of controlling these activities by their physical structure of shape and size. PAA-protected Pt nanoparticles (PAA-Pt) with cubic shape were selectively prepared by addition of I^- anions to give a Pt nanocube surrounded by (100) facets with the high catalytic activity for O_2 reduction. The other experiment clarified the correlation between the diameter of an Au particle and that of the synthesized carbon nanotube (CNT) by changing the size of Au nanoparticle catalysts. The last section was allocated to the achievements with novel nanomaterials using metal coordination polymers as an inorganic core. The metal coordination nano-polymers (MCNPs) were stable in air, obtained as a solid powdery form by organic stabilizers, and they exhibited characteristic properties different from their bulk crystals derived from a size effect. Additionally, they could work as a novel intelligent precursor of an alloy or ceramic nanomaterial after the decomposition of organic ligands under an appropriate gas atmosphere. Nanoparticle science in wet process could progress further from the point of the development of the synthetic technique, the investigation of specialized natures, and the construction of their dimensional network, which could lead to practical widely-used nanoparticle materials in the future.

ACKNOWLEDGEMENTS

The described studies were partially supported by Grants-in-Aid for Precursory Research for Embryonic Science and Technology (PRESTO) from Japan Science and Technology Agency (JST) and for Scientific Research (No. 20710069) from the Ministry of Culture, Education, Sports, Science and Technology, Japan. I thank Prof. H. Nishihara in the University of Tokyo for the research contents.

REFERENCES

[1] Schimid, G. *Chem. Rev.* 1992, 92, 1709-1727.

[2] Daniel, M.-C.; Astruc, D. *Chem. Rev.* 2004, 104, 293-346.

[3] Brust, M.; Walker, M.; Bethell, D.; Schiffrin, D. J.; Whyman, R. *J Chem Soc, Chem. Commun.* 1994, 801-802.

[4] Quiros, I.; Yamada, M.; Kubo, K.; Mizutani, J.; Kurihara, M.; Nishihara, H. *Langmuir,* 2002, 18, 1413-1418.

[5] Kiely, C. J.; Fink, J.; Brust, M.; Bethell, D.; Schiffrin, D. J. Nature 1998, 396, 444-446.

[6] Chen, S.; Ingram, R. S.; Hostetler, M. J.; Pietron, J. J.; Murray, R. W.; Schaaff, T. G.; Khoury, J. T.; Alvarez, M. M.; Whetten, R. L. *Science* 1998, 280, 2098-2101.

[7] Hostetler, M. J.; Green, S. J.; Stokes, J. J.; Murray, R. W. J. Am. Chem. Soc. 1996, 118, 4212-4213.

[8] Ingram, R. S.; Hostetler, M. J.; Murray, R. W. *J. Am. Chem. Soc.* 1997, 119, 9175-9178.

[9] Templeton, A. C.; Hostetler, M. J.; Warmoth, E. K.; Chen, S.; Hartshorn, C. M.; Krishnamurthy, V. M.; Forbes, M. D. E.; Murray, R. W. *J. Am. Chem. Soc.* 1998, 120, 4845-4849.

[10] Gittins, D.; Bethell, D.; Schiffrin, D. J.; Nishols, R. J. Nature 2000, 408, 67-69.

[11] Templeton, A. C.; Cliffel, D. E.; Murray, R. W. *J. Am. Chem. Soc.* 1999, 121, 7081-7089.

[12] Kamat, P. V.; Barazzouk, S.; Hotchndani, S. Angew. *Chem. Int. Ed.* 2002, 41, 2764-2767.

[13] Horikoshi, T.; Itoh, M.; Kurihara, M.; Kubo, K.; Nishiahra, H. *J. Electroanal. Chem.* 1999, 473, 113-116.

[14] Yamada, M.; Tadera, T.; Kubo, K.; Nishihara, H. *J. Phys. Chem. B* 2003, 107, 3703-3711.
[15] Yamada, M.; Nishihara, H. *Eur. Phys. J. D* 2003, 24, 257-260.
[16] Hostetler, M. J.; Wingate, J. E.; Zhong, C.-J.; Harris, J. E.; Vachet, R. W.; Clark, M. R.; Londono, J. D.; Green, S. J.; Stokes, J. J.; Wignall, G. D.; Glish, G. L.; Porter, M. D.; Evans, N. D.; Murray, R. W. *Langmuir* 1998,14, 17-30.
[17] Lin, X. M.; Sorensen, C. M.; Klabunde, K. *J. Chem. Mater.* 1999, 11, 198-202.
[18] Yamada, M.; Nishihara, H. *Langmuir* 2003, 19, 8050-8056.
[19] Yamada, M.; Quiros, I.; Mizutani, J.; Kubo, K.; Nishihara, H. *Phys.Chem. Chem. Phys.* 2001, 3, 3377-3381.
[20] Templeton, A. C.; Pietron, J. J.; Murray, R. W.; Mulvaney, P. *J. Phys. Chem. B.* 2000, 104, 564-570.
[21] Yamada, M.: Nishihara, H. *ChemPhysChem* 2004, 5, 555-559.
[22] Alvarez, M. M.; Khoury, J. T.; Schaaff, T. G.; Shafigullin, M. N.; Vezmar, I.; Whetten, R. L. *J. Phys. Chem.* B 1997, 101, 3706-3712.
[23] Nishiyama, K.; Kubo, A.; Taniguchi, I., Yamada, M.; Nishihara, H. *Electrochemistry* 2001, 69, 980-983.
[24] Yamada, M.; Kubo, K.; Nishihara, H. *Chem. Lett.* 1999, 1335-1336.
[25] Yamada, M.; Tadera, T.; Kubo, K.; Nishihara, H. *Langmuir* 2001, 17, 2363-2370.
[26] Sauerbrey, G. Z. *Phys.* 1959, 155, 206-222.
[27] Yamada, M.; Nishihara, H. *Chem. Commun.* 2002, 2578-2579.
[28] The coverage of Au_n-BFc on an electrode, Γ_{BFc}, can be estimated from the optical measurements by equation: Γ_{BFc} = A/1000ε [mol cm^{-2}] where A is the absorbance at the SP band and ε is the molar extinction coefficient at the SP band for each particle (see ref. 22). The number of layers is estimated by assuming that the electrodeposited nanoparticles are hexagonaly packed in the film.
[29] Bard, A. J.; Denuault, G.; Lee, H.; Mandeler, D.; Wipf, D. O. *Acc. Chem. Res.* 1990, 23, 357-363.
[30] Berry, V.; Saraf, R. F. Angew. *Chem., Int. Ed.* 2005, 44, 6668-6611.
[31] D. Adam, D.; Schuhmacher, P.; Simmerer, J.; Haussling, L.; Siemensmeyer, K.; Etzbach, K. H.; Ringsdorf, H.; Haarer, D. *Nature* 1994, 371, 141-143.
[32] Yamada, M..; M.; Shen, Z.; Miyake, M. *Chem. Commun.* 2006, 2569-2571.
[33] Shen, Z.; Yamada, M.; Miyake, M. *J. Am. Chem. Soc.* 2007,129, 14271-14280.
[34] Markovitsi, D.; Germain, A.; Millie, P.; Lecuyer, P.; Gallos, L.; Argyrakis, P.; Bengs, H.; Ringsdorf, H. *J. Phys. Chem.* 1995, 99, 1005-1017.
[35] Yamada, M.; Kuzume, A.; Kurihara, M.; Kubo, K.; Nishihara, H. *Chem. Commun.* 2001, 2476-2477.
[36] Fink, J.; Kiely, C. J.; Bethell, D.; Schiffrin, D. *J. Chem. Mater.* 1998, 10, 922-926.
[37] Link, S.; El-Sayed, M. A. *J. Phys. Chem. B.* 1999, 103, 8410-8426.
[38] Saito, T.; Ohshima, S.; Xu, W.-C.; Ago, H.; Yumura, M.; Iijima, S. *J. Phys. Chem. B.* 2005, 109, 10647-10652.
[39] Burda, C.; Chen, X.; Narayanan, R.; El-Sayed, M. A. *Chem. Rev.* 2005, 105, 1025-1102.
[40] Shen, Z.; Yamada, M.; Miyake, M. *Chem. Commun.* 2007, 245-247.
[41] Narayanan, R.; El-Sayed, M. A. *J. Am. Chem. Soc.* 2004, 126, 7194-7195.

[42] Stonehart, P. *J. Appl. Electrochem.* 1992, 22, 995-1001.

[43] Yamada, M.; Kon, S.; Miyake, M. *Chem. Lett.* 2005, 34, 1050-1051.

[44] Tans, S. J.; Verschueren, A. R. M.; Dekker, C. Nature 1998, 393, 49–52.

[45] Okumura, M.; Akita, T.; Haruta, M. *Catal. Today* 2001, 74, 265–269.

[46] Jie, J.; Haraki, K.; Kondo, J. N.; Domen, K.; Tamaru, K. *J. Phys. Chem.* B 2000, 104, 11153-11156.

[47] Yamada, M.; Kawana, M.; Miyake, M. *Appl. Catal. A-Gen.* 2006, 302, 201-207.

[48] Lee, S. Y.; Yamada, M.; Miyake, M. *Carbon* 2005, 43, 2654-2663.

[49] T. Teranishi, S. Hasegawa, T. Shimizu, M. Miyake, *Adv. Mater.* 2001, 13, 1699-1701.

[50] Kyung-Hoon Kim, K. –H.; Yamada, M.; Park, D. –W.; Miyake, M. *Chem. Lett.* 2004, 33, 344-345.

[51] Merenyi, G.; Lind, J.; Shen, X.; Eriksen, T. E. *J. Phys. Chem.* 1990, 94, 748-752.

[52] Lyu, S. C.; Lee, T. J.; C.W. Yang, C. W.; Lee, C. *J. Chem. Commun.* 2003, 1404-1406.

[53] Amelinckx, S.; Zhang, X. B.; Bernaerts, D.; Zhang, X. F.; Ivanov, V.; Nagy, J. B. *Science* 1994, 265, 635-639.

[54] Veciana, J.; Rovira, C.; Amabilino, D. B. *Supramolecular Engineering of Synthetic Metallic Materials*; Kluwer Academic Publishers: Dordrecht,The Netherlands, 1998.

[55] Ferlay, S.; Mallah, T.; Ouahe´s, R.; Vellet, P.; Verdaguer, M. *Nature 1*995, 378, 701-703.

[56] Sato, O.; Hayami, S.; Einaga, Y.; Gu, Z.-Z. *Bull. Chem. Soc. Jpn.* 2003, 76, 443-470.

[57] Yamada, M.; Arai, M.; Kurihara, M.; Sakamoto, M.; Miyake, M. *J. Am. Chem. Soc.* 2004, 126, 9482-9483.

[58] Yamada, M.; Maesaka, M.; Kurihara, M.; Sakamoto, M.; Miyake, M. *Chem. Commun.* 2005, 4851-4853.

[59] Yamada, M.; Sato, T.; Miyake, M.; Kobayashi, Y. *J. Colloid. Intef. Sci.* 2007, 315, 369-375.

[60] Yamada, M.; Ohkawa, R.; Miyake, M. *IEEJ Trans.* 2007, 127, 1342-1343.

[61] Itaya, K.; Ataka, T.; Toshima, S. J. *Am. Chem. Soc.* 1982, 104, 4767-4772.

[62] Vaucher, S.; Fielden, J.; Li, M.; Dujardin, E.; Mann, S. *Nano Lett.* 2002, 2 225-229.

[63] Culp, J. T.; Park, J. H.; Benitez, I. O.; Huh, Y. D.; Meisel, Y. D.; Talham, D. R. *Chem. Mater.* 2003, 15, 3431-3436.

[64] D-Vera, J. M.; Colacio, E. *Inorg. Chem.* 2003, 42, 6983-6985.

[65] Cullity, B. D. *Elements of X-Ray Diffraction,* 2nd ed.; Addison-Wesley Publishers: Reading, MA, 1978.

[66] The bulk crystal of M_1-CN-M_2 is generally prepared by mixing diluted solution of M_1 cyano complexes and M_2 cations in water. See ref. 56.

[67] Arai, M.; Miyake, M.; Yamada, M. *J. Phys. Chem.* C. 2008, 112, 1953-1962.

[68] Kondo, N.; Nakajima, A.; Sasaki, Y.; Kurihara, M.; Yamada, M.; Miyake, M.; Mizukami M.; Sakamoto, M. *Chem. Lett.* 2006, 35, 1302-1303.

[69] Kondo, N.; Yokoyama, A.; Kurihara, M.; Sakamoto, M.; Yamada, M.; Miyake, M.; Ohsuna, T.; Aono, H.; Sadaoka, Y. *Chem. Lett.* 2004, 33, 1182-1183.

[70] Gotoh, A.; Uchida, H.; Ishizaki, M.; Satoh, T.; Kaga, S.; Okamoto, S.; Ohta, M.; Sakamoto, M.; Kawamoto, T.; Tanaka, H.; Tokumoto, M.; Hara, S.; Shiozaki, H.; Yamada, M.; Miyake, M.; Kurihara, M. *Nanotechnology* 2007, 18, 345609.

[71] Hara, S.; Tanaka, H.; Kawamoto, T.; Tokumoto, M.; Yamada, M.; Gotoh, A.; Uchida, H.; Kurihara, M.; Sakamoto, M. *Jpn. J. Appl. Phys.* 2007, 46, L945-L947.

In: Encyclopedia of Nanotechnology
Editor: Elwood D. Carlson

ISBN: 978-1-60692-079-4

Chapter 37

HIGHLY STABILIZED GOLD NANOPARTICLES SYNTHESIZED AND MODIFIED BY PEG-*B*-POLYAMINE

Daisuke Miyamoto*[1,2,3] *and Yukio Nagasaki*[1,2,3,4,5,6],
[1]Graduate School of Pure and Applied Sciences; Japan
[2]Tsukuba Research Center for Interdisciplinary Materials Science; Japan
[3]Center for The Tsukuba Advanced Research Alliance; Japan
[4]College of Engineering Sciences; Japan
[5]Master's School of Medical Sciences, Graduate School of Comprehensive Human Sciences, University of Tsukuba; Japan
[6]Satellite Laboratory, International Center for Materials Nanoarchitectonics, Japan National Institute of Materials Science, Japan

INTRODUCTION

Colloidal gold nanoparticles (GNPs) with a size range of several to hundreds of nanometers show a bright-pinkish color due to plasmon resonance and have been widely utilized for preparation of stained glasses for several hundred years. Recently, much attention has been given to gold nanoparticles (GNPs) because of their unique characteristics, and GNPs have been applied in many areas, such as biomedical materials science, optics and electronics [1].

In the 20th century, various methods for the preparation of gold colloids were reported and reviewed [2-8]. Most of these methods were based on the reduction of tetrachloroauric acid ($HAuCl_4$). The most popular one for a long time has been that using citrate reduction of $HAuCl_4$ in water [3]. It leads to GNPs of ca. 20 nm. To obtain GNPs of prechosen size between 16 and 147 nm via their controlled formation, a method was proposed where the ratio between the reducing and stabilizing agents was varied [4]. GNPs thus prepared are dispersed in the solution by ionic repulsion of adsorbed ions such as citric acid on their

* 1-1-1 Tennoudai, Tsukuba, Ibaraki 305-8573, Japan; E-mail: nagasaki@nagalabo.jp

surface [9-12]. Under physiological condition, however, the ionically stabilized GNPs tend to aggregate because of the charge shielding. To improve the dispersion stability of ionically stabilized GNPs, much kind of polymer has been used as GNPs-stabilizer [11].

Although there are a variety of ways to achieve nanoparticle-polymer composites[13,14], most of them divided two different approaches. The first one consist of the *in situ* synthesis of the nanoparticles in the polymer matrix either by reduction of the metal salts dissolved in that matrix [15], or by evaporation of the metals on the heated polymer surface [16]. The second one involves polymerization of the matrix around the nanoparticles [17-19]. Blending of *pre*made GNPs into a *pre*synthesized polystyrene (PSt) bound to a thiol group was also reported [20]. The physical process involving mechanical crushing or pulverization of bulk metals and arc discharge yielded large nanoparticles with a wide size distribution, while nanoparticles prepared by reduction of metal salts tends to give small and a narrow size distribution. For this reduction process, several kinds of reducing agents such as $NaBH_4$ [21] and alchol [22] are utilized. Reduction of metal ions in the presence of the polymer is the most often chosen because the complexation of the metal cations by the ligand atoms of the polymer is crucial before reduction. In particular, it dramatically limits the particles size [23]. The most important role of the stabilizing polymer is to protect the nanoparticles from coagulation.

To apply GNPs for biomedical materials science, it is important to modify the surface of GNPs in order to improve their biocompatibility and colloidal stability under physiological conditions. Thiol chemistry has been widely used for the modification on gold surfaces with chemical species ranging from small molecules [24-26] to biomacromolecules [27-30]; synthetic polymers [31,32], especially, thiol-ended PEG (PEG-SH), have been widely used [33-38], and they have shown the inhibition of the adsorption of nonspecific molecules. The S-Au linkage is strong enough (47 kcal/mol) [39] for the construction of a dense PEG brush on the colloid surface, however, the oxidative stability of thiolate species was limited [40-42], as well as exchanged with thiolated compounds inside the body. Thiolate-modified surfaces are also damaged by exposure to light, high temperature and oxygen. In this way, the stability of PEGylated GNPs via a thiol-gold bond is not always sufficient to maintain its functionality for long-term, especially *in vivo*. For versatile applications, alternative methodology must be developed in terms of surface modification chemistry.

One of the best ways to solve these problems is usage of a water soluble polymer possessing multiple-coordination ability with metals, such as poly(2-vinylpyridine) (PVP) and poly(ethleneimine) (PEI), as a stabilizer for metal colloids [43,44]. Although the N-Au linkage is not as strong (6 kcal/mol) [66] as the linkage, these polymers were strongly interacted with gold surface by the multiple-coordination. For example, Antonietti and co-workers reported the preparation of PEG-stabilized gold colloids through the mixing of $AuCl_3$ with PEI-PEG graft copolymers [44]. The amino groups in the PEI segment of the PEI-PEG graft copolymer was considered to have the ability to reduce auric cations to form a gold colloid in the $AuCl_3^-$ incorporated PEI/PEG micelles, however, the surface modified with PEG/polyamine graft copolymers was not stable [45,46].

In this study, we used PEG/polyamine block copolymers and highly stabilized PEGylated GNPs (PEG-GNPs) was facilely synthesized by autoreduction of $HAuCl_4$ using poly(ethylene glycol)-*b*-poly(2-(*N,N*-dimethylamino)ethyl methacrylate) (PEG-*b*-PAMA) without addition of the further reduction reagents. The condition of GNP synthesis, such as pH and the block copolymer concentration, was investigated. Furthermore, a function of target molecule recognition was added to the PEG-GNPs and the ability of the target molecule recognition was investigated. In order to investigate the interaction between the GNPs surface and PEG-*b*-PAMA, the technique of the GNPs stabilization was applied for the surface modification of GNPs using PEG-*b*-PAMA and commercial GNPs. The GNP modification conditions, such as pH, block copolymer concentration and PAMA chain length, were investigated, and the dispersion stability of the PEG-GNPs under various conditions was evaluated in detail.

2.1. Synthesis of α-acetal-poly(ethylene glycol)-*block*-poly(2-(*N*,*N*-dimethylamino)ethyl methacrylate) [47]

3,3-Diethoxy-1-propanol (1.00 mmol) and potassium naphthalene (1.00 mmol) were added to 45 mL of tetrahydrofuran (THF) in a 100 mL flask with a three-way stopcock under a nitrogen atmosphere [48], then, ethylene oxide (EO) was added to the solution. After the mixture was stirred for 2 days at room temperature, the appropriate amount of AMA was added in the solution and stirred for 20 min at 0 ℃. The reaction was stopped by adding 5 mL of methanol. The solution was poured to excess cold 2-propanol to precipitate the polymer. The precipitate was dissolved in methanol and precipitated again in cold 2-propanol. This procedure was repeated again. The precipitate were dried *in vacuo* and freeze-dried with benzene. The recovered polymers were protonated by hydrochloric acid and purified by soxhlet extraction for 2 days using THF to remove unreacted acetal-PEG. The white precipitate was dissolved water and the polymer was obtained by freeze-dry. The chemical structure of PEG-*b*-PAMA [49] was confirmed by ^{1}H-NMR (EX-400, JEOL, Japan), (400 MHz, $CDCl_3$): δ (ppm) 0.85-1.23 (d, 3H, α-CH_3), 1.81-1.87 (d, 2H, -C(COO-)(CH_3)-CH_2-), 2.52-2.57 (d, 6H, -N(CH_3)$_2$), 3.00-3.06 (d, 2H, -CH_2-CH_2-N-) 3.27 (d, 3H, -O-CH_3), 3.45-3.74 (d, 4H, -CH_2-CH_2-O-), and 4.15-4.21 (d, 2H, -CH_2-CH_2-N-). The polymerization results are summarized in Table 1.

Table 1. Results of acetal-poly(ethylene glycol)-*block*-poly(2-(*N*,*N*-dimethylamino)ethyl methacrylate)

Run	Code	Mn ($\times 10^{-3}$)[a]	
		PEG	PAMA
1	6k/16k	5.7	15.7
2	5k/4k	4.9	3.8

a) Determined by SEC data. Column: TSK-Gel SUPER HZ3000, HZ2500; carrier: THF with 0.5 wt% triethylamine; detecter: RID, flow rate: 0.35 mL/min; Temperature: 40 ℃.

2.2. Facile Synthesis of PEGylated Gold Nanoparticles [47]

Five mL of acetal-PEG-*b*-PAMA (Run 1, Table 1) aqueous solution (6.8 mg/mL) was added in the 1 mL of $HAuCl_4 \cdot 4H_2O$ aqueous solution (2.5 mg/mL) and stirred at room temperature. This reaction was monitored by the measurement of UV spectrum. A schematic representation of PEG-*b*-PAMA synthesis and preparation of PEG-GNPs is shown in Scheme 1.

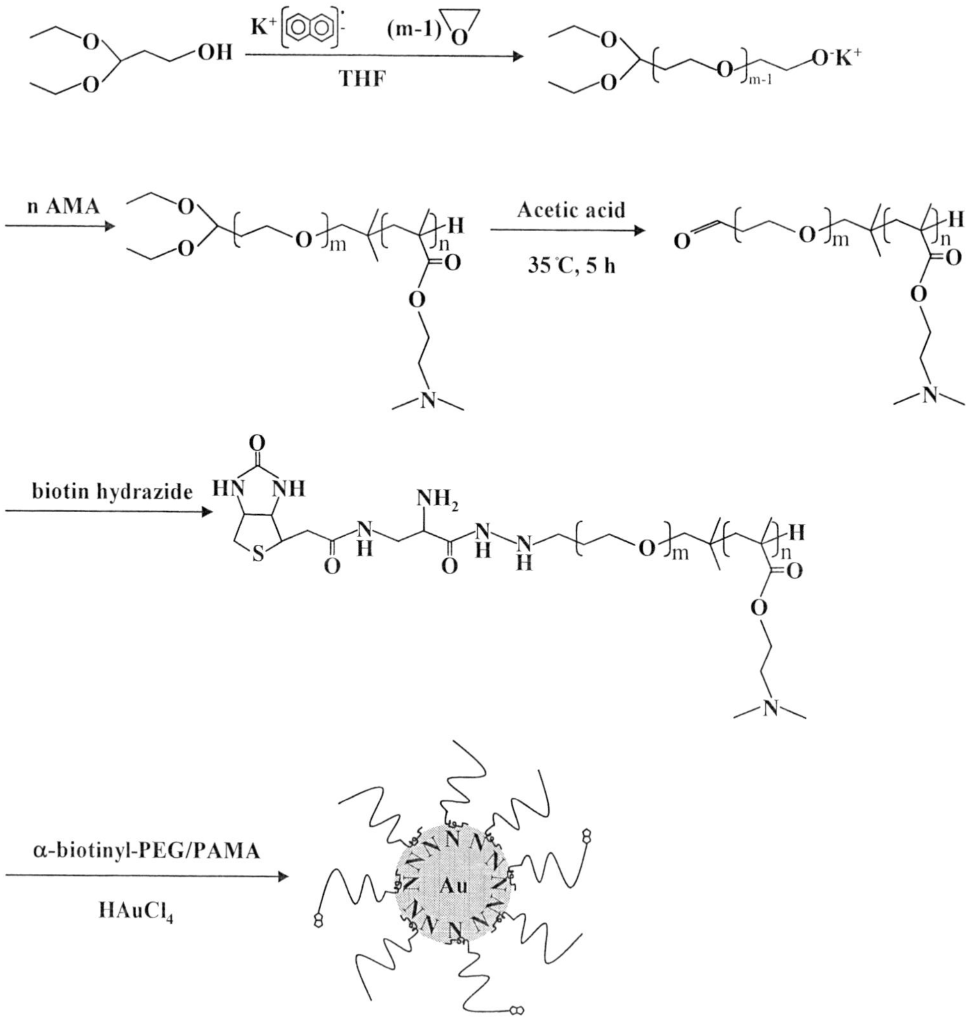

Scheme 1. Synthesis of PEG-*b*-PAMA and facile synthesis of PEGylated nanoparticles.

Since amino groups are known to show coordination ability on a metal surface, acetal-PEG-*b*-PAMA was expected to function as a stabilizer of GNPs. An unprecedented finding is that the block copolymer even facilitated autoreduction of the auric cations to obtain GNP without any additional reducing agent. The tetrachloroauric acid solution was changed from colorless to bright red due to the formation of GNPs when the block copolymer was added to the solution. The tertiary amino groups in the PAMA sgment play a crucial role in the reduction of auric cations as well as the anchoring of PEG on the surface of the GNPs. Figure 1 shows the change in the plasmon absorbance at 520 nm of tetrachloroaurate aqueous solution with different polymer addition. In the presence of acetal-PEG-OH at room temperature, no color change was observed. On the contrary, both PAMA homopolymer and acetal-PEG-*b*-PAMA induced a change in the color of the solution due to a gradual increase in the absorbance at 520 nm over several hours, indicating the reduction of auric cations to form GNPs. Nevertheless, the GNPs obtained by the PAMA homopolymer were rather unstable under a high salt concentration, and the characteristic absorption of GNPs was gradually reduced with time (data not shown). Worth nothing is that the reduction process finished within ca. 3 h in the case of acetal-PEG-*b*-PAMA at room temperature to form stable nanoparticles having a narrow size distribution. Figure 2 shows transmission electron microscope (TEM) image of (a) GNPs synthesized by acetal-PEG-*b*-PAMA, (b) GNP synthesized by PAMA homopolymer. The GNPs synthesized by acetal-PEG-*b*-PAMA showed homogeneous particles size. On the contrary, the GNPs synthesized by PAMA homopolymer showed large particle size distribution. The acetal-PEG-*b*-PAMA-stabilized GNPs have a substantially narrower distribution than those prepared by the PEI-PEG graft copolymers [44], PAMA homopolymer, and comparable to conventional method. This is obviously an advantage of using block copolymers as stabilizers because they can form a multimolecular micelle structure with a definite association number and well-defined core-shell architecture: an aurate-complex core segregated and surrounded by hydrophilic shell layer.

2.3. Characterization of Facilely Synthesized PEGylated Gold Nanoparticles [47]

GNPs were synthesized using the acetal-PEG-*b*-PAMA (Table 1, Run 2). The obtained GNPs were purified by centrifuge at 45,000 g, for 30 min at 20 ℃. The precipitated particles were resuspended in water or various buffer solutions with a wide range of pH. The centrifugations were repeated several times to remove free block copolymers. The average size and the distribution index (D_w/D_n) of the obtained GNPs were 11.8 nm and 1.01, respectively, which was determined by dynamic light-scattering (DLS) measurement.

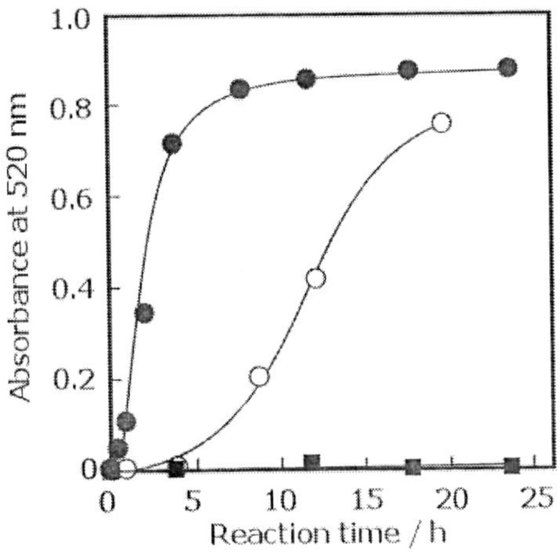

Figure 1. Change in the plasmon absorbance at 520nm of the tetrachloroaurate aqueous solution in the presence of R-acetal-PEG-OH: open circle, PAMA: closed triangle, and R-acetal-PEG-*b*-PAMA: closed square. Reproduced from ref. 61 by courtesy of publishers, American Chemical Society, U.S.A.

Commercially available GNPs, prepared by citrate, were dispersed in an aqueous solution through the ionic repulsion of the surface-adsorbed ions. Thus, the citrate-reduced gold nanoparticles have an appreciably negative ζ-potential (<-20 mV) at neutral pH, yet they undergo aggregation in the acidic region (pH <5) due to the neutralization of the surface negative charge. On the other hand, as can be seen in Figure 3, the ζ-potential of the GNPs prepared by the acetal-PEG-*b*-PAMA was close to zero regardless of the environmental pH. This result indicates that the PAMA segment in the block copolymer coordinates with the gold surface, allowing the PEG segment to be tethered from the surface into the aqueous exterior to shield the surface charge. The clear phase separation of the PEGlayer and the anchored PAMA segment should be an important factor in shielding the cationic charge of the PAMA segment.

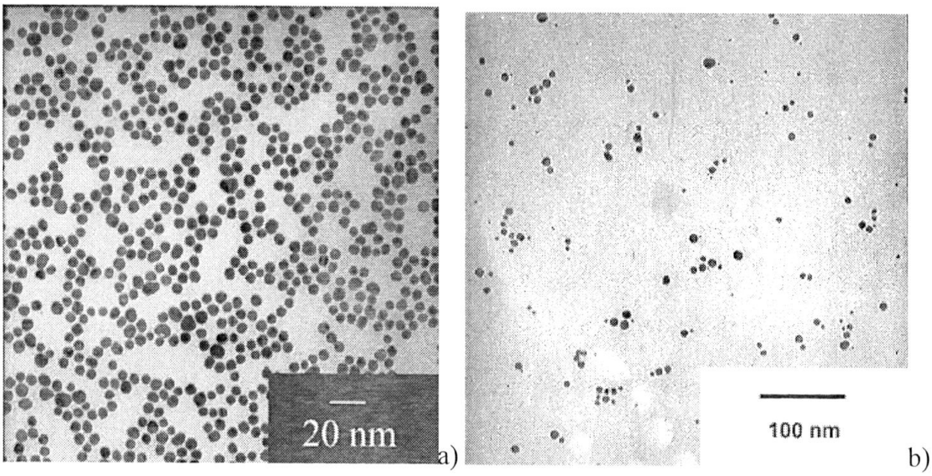

Figure 2. TEM image of gold nanoparticles prepared in the presence of (a) R-acetal-PEG-*b*-PAMA and (b) PAMA. Reproduced from ref. 61 by courtesy of publishers, American Chemical Society, U.S.A.

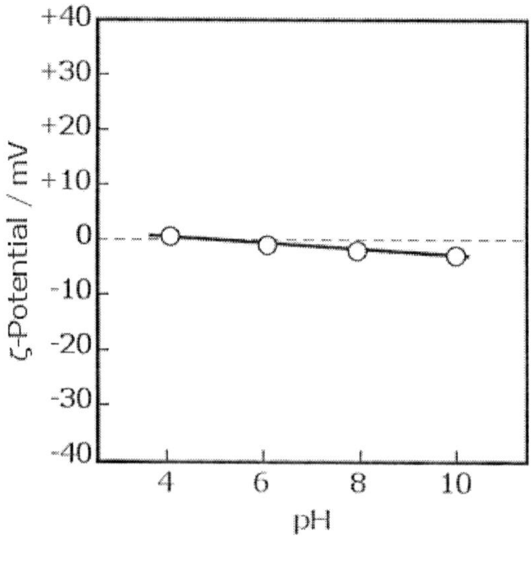

Figure 3. Change in ζ-potential of R-acetal-PEG-*b*-PAMA-anchored gold nanoparticles with pH ranging from 4 to 10 at 25 °C in 10mM NaCl solution. Reproduced from ref. 61 by courtesy of publishers, American Chemical Society, U.S.A.

Although the gold nanoparticles prepared by the acetal-PEG-*b*-PAMA showed an almost neutral surface charge, they maintain a fairly high dispersion character even after repeated centrifugation. To evaluate the stability of the block-copolymer-anchored nanoparticles, the time course of the change in the plasmon absorption was monitored under very high salt concentration (1.0 M NaCl). Note that the characteristic plasmon band at 520 nm is lowered, and an absorption band in a longer wavelength region appears when the GNPs are close to each other due to the electric dipole-dipole interaction between plasmons of neighboring particles[50-52]. Figure 4 shows the time course of plasmon absorbance at 520 nm as a function of time in 1.0 M NaCl. It should be noted that citrate stabilized GNPs immediately aggregate even under physiological salt concentration (0.15 M NaCl), while gold nanoparticles anchored with acetal-PEG-*b*-PAMA showed almost no decrease in the plasmon absorbance at 520 nm even under high ionic strength (1.0 M NaCl). The appreciable stability of the PEG-GNPs should be due to the steric stabilization effect of the hydrophilic PEG palisade tethered from the surface of the nanoparticles. The nanoparticles were also stable at a low pH region for several days even though the amino groups in acetal-PEG-*b*-PAMA should protonate completely under this condition.

The PEG density on the GNPs surface was also evaluated by thermal gravimetric analysis (TGA). GNPs was synthesized using the acetal-PEG-*b*-PAMA (Table 1, Run 2). After the solution of the PEG-GNPs was substituted with water through centrifugation and the PEG-GNPs was completely dried *in vacuo*, TGA was carried out. The average size was 8.16 nm, which was determined by TEM measurement.

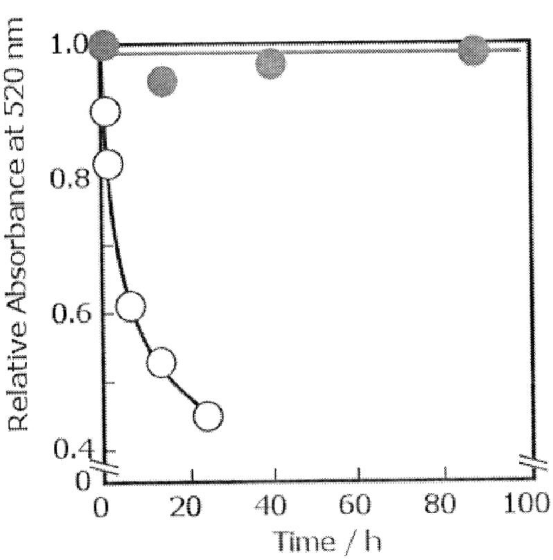

Figure 4. Change in the relative absorbance at 520nm of gold nanopaticles R-acetal-PEG-*b*-PAMA-anchored gold nanoparticles in 1.0 M NaCl solution: closed circle, citrate-stabilized gold nanoparticles in 0.1 M NaCl solution: closed square. Reproduced from ref. 61 by courtesy of publishers, American Chemical Society, U.S.A.

The amounts of PEG-*b*-PAMA on the GNPs are summarized in Table 2. PEG chain density on the GNP surface was 0.49 chains/nm^2. This PEG chain density was slightly low, comparing to the references [33, 53]. The reason was not clear. This result may be caused by the different chain length of terminal segment of PEG or the difference of the curvature of

GNPs. However, the PEG chain density was enough to disperse the PEG-GNPs under very high salt concentration (1.0 M NaCl).

2.4. Facile Synthesis of Biotinyl-PEGylated Gold Nanoparticles Using α-biotinyl-poly(ethylene glycol)-*block*-poly(2-(*N*,*N*-dimethylamino)ethyl methacrylate) [47]

One of the other important characteristics of these acetal-PEG-*b*-PAMA-anchored nanoparticles is the ligand installing site at the distal end of the PEG chain. The PAMA segment in the block copolymer is anchored on the surface of the GNPs, while the PEG segments extend into the aqueous exterior from the surface of the GNPs.

Table 2. Experimental characteristics of PEGylated gold nanoparticles

Mn of PEG derivatives	d_{core} (nm)	PEG chain density (chains/nm^2)	reference
5700/2800 (PEG/PAMA)	8.16	0.49	
5000 (MeO-PEG-SH)	2.8	2.86	33)
2100 (MeO-PEG-SH)	3.6	2.00	50)

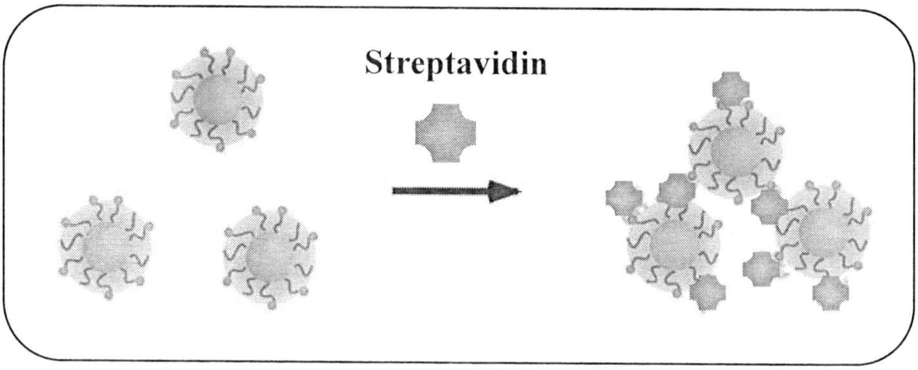

Scheme 2. Aggregation of biotinyl-PEG coated gold nanoparticles by streptavidin recognition.

To obtain aldehyde-PEG-*b*-PAMA, 1.0 g of acetal-PEG-*b*-PAMA was dissolved in acetic acid aqueous solution (10 vol %) and the mixture was stirred for 5 h at 35 °C. After the mixture was neutralization by adding 10 mol/L NaOH aqueous solution, the solution was dialyzed against excess water. After the pH of the solution was adjusted to 9 using NaOH aqueous solution, the aldehyde-PEG-*b*-PAMA was obtained by freeze-drying. One point seven mg of biotin hydrazide was added in the 10 mL of aldehyde-PEG-*b*-PAMA aqueous solution (10 mg/mL) and stirred for 5 h. Then, 1.5 mg of sodium cyanoborohydride was added in the solution three times per 15 min. The solution was dialyzed against excess water. The α-biotinyl-PEG-*b*-PAMA was obtained by freeze-drying. The chemical structure of biotyl-PEG-*b*-PAMA was confirmed by ^{1}H-NMR (EX-400, JEOL, Japan), (400 MHz, $CDCl_3$): δ (ppm) 0.85-1.23 (d, 3H, α-CH_3), 1.81-1.87 (d, 2H, -C(COO-)(CH_3)-CH_2-), 2.52-

2.57 (d, 6H, -N(CH_3)$_2$), 3.00-3.06 (d, 2H, -CH_2-CH_2-N-) 3.27 (d, 3H, -O-CH_3), 3.45-3.74 (d, 4H, -CH_2-CH_2-O-), 4.15-4.21 (d, 2H, -CH_2-CH_2-N-), 4.25-4.35 4.35-4.45 (d, 2H, -*CH*-NH-), 6.15-6.25 (d, 2H, -CH-*NH*-C) and 7.50-7.60 (d, 1H, -C(=O)-*NH*-NH-CH_2-). The biotinyl-PEG-GNPs was prepared by the same method of chapter 2.3.

Using biotin-PEG-GNPs thus prepared, molecular recognition experiment were carried out. Streptavidin, which forms an extremely stable complex with biotin ($K=10^{-15}$) [54-56], was used to investigate the functionality of the biotins installed on the periphery of the PEG-*b*-PAMA-anchored GNPs. Figure 5 shows a typical time course of the plasmon absorption change after the addition of 100 μg/mL streptavidin. The maximum wavelength of the absorption showed a bathochromic shift with time, indicating an aggregation of the GNPs via streptavidinbiotin interaction. At the same time, the absorption at the longer wavelength region gradually increased. Actually, the colloid solution underwent a visual color change gradually from pink to purple. On the other hand, no absorption change was observed when bovine serum albumin as a control was added to the biotinyl-PEG-GNPs solution. That is, the aggregation of biotinyl-PEG- GNPs induced by streptavidin is indeed due to the specific biotin/streptavidin interaction. A schematic representation of the molecular recognition of the biotinyl-PEG- GNPs is shown in Scheme 2.

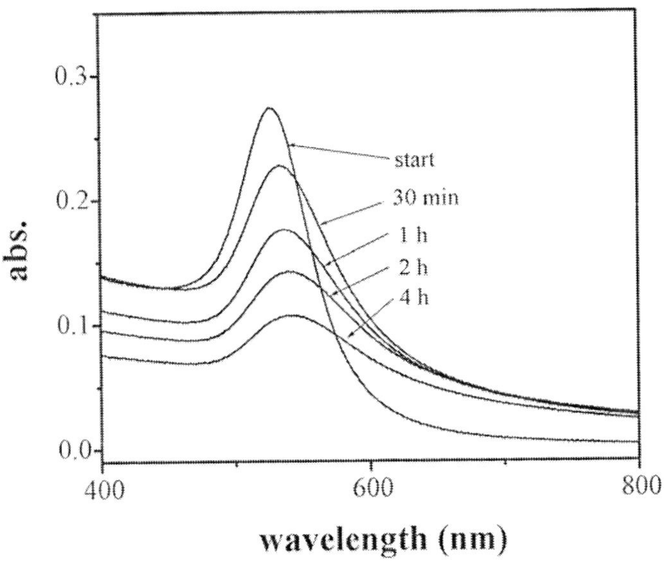

Figure 5. Time-dependent change in the absorption spectrum of R-biotinyl-PEG/PAMA-anchored gold nanoparticles after adding 100 μg/mL streptavidin. Reproduced from ref. 61 by courtesy of publishers, American Chemical Society, U.S.A.

2.5. Synthesis of Metoxy-poly(ethylene glycol)-*block*-poly(2-(*N*,*N*-dimethylamino)ethyl methacrylate) with a Variety of Polyamine Chain Length [57]

Our original oxyanion-initiated polymerization technique has been employed for the preparation of PEG-*b*-PAMA [49](chapter 2.1). This method has several advantages: fast polymerization reaction, narrow molecular weight distribution, controlling the molecular weight based on monomer/initiator ratio, and so on. In the preparation of PEG-*b*-PAMA having a short PAMA chain length, however, it is difficult to control the molecular weight because of the very fast polymerization reaction as compared to the initiation rate. In order to synthesize well-controlled oligomers, the control of the initiation reaction is crucial. Atom transfer radical polymerization (ATRP) can yield short polymer chains with controlled molecular weight and molecular-weight distribution because the propagation species is a stable copper complex, which converts to the active free form in an equilibrium manner [58,59]. If the deactivation rate constant (k_{deact}) is sufficiently larger than the activation rate constant (k_{act}), the polymerization proceeds step by step to result in the synthesis of polymers controlled from the beginning of the polymerization reaction.

Various molecular weights of PEG-*b*-PAMA were synthesized by ATRP, using methoxy-poly(ethylene glycol)-2-bromoisobutyrate (PEG-Br), copper bromide and 2,2'-bipyridine, and AMA as initiator, catalyst and monomer, respectively. The molecular weights of the block copolymer were controlled by reaction time and feed. A schematic representation of PEG-*b*-PAMA synthesis is shown in Scheme 3. The chemical structure of PEG-*b*-PAMA [49] was confirmed by ^{1}H-NMR (Avance 500 NMR Spectrometer, Bruker, Germany), (500 MHz, D_2O): δ (ppm) 0.85-1.23 (d, 3H, α-CH_3), 1.85 (d, 6H, -C(CH_3)-Br), 1.81-1.87 (d, 2H, -C(COO-)(CH_3)-CH_2-), 2.52-2.57 (d, 6H, -N(CH_3)$_2$), 3.00-3.06 (d, 2H, -CH_2-CH_2-N-) 3.27 (d, 3H, -O-CH_3), 3.45-3.74 (d, 4H, -CH_2-CH_2-O-), and 4.15-4.21 (d, 2H, -CH_2-CH_2-N-).

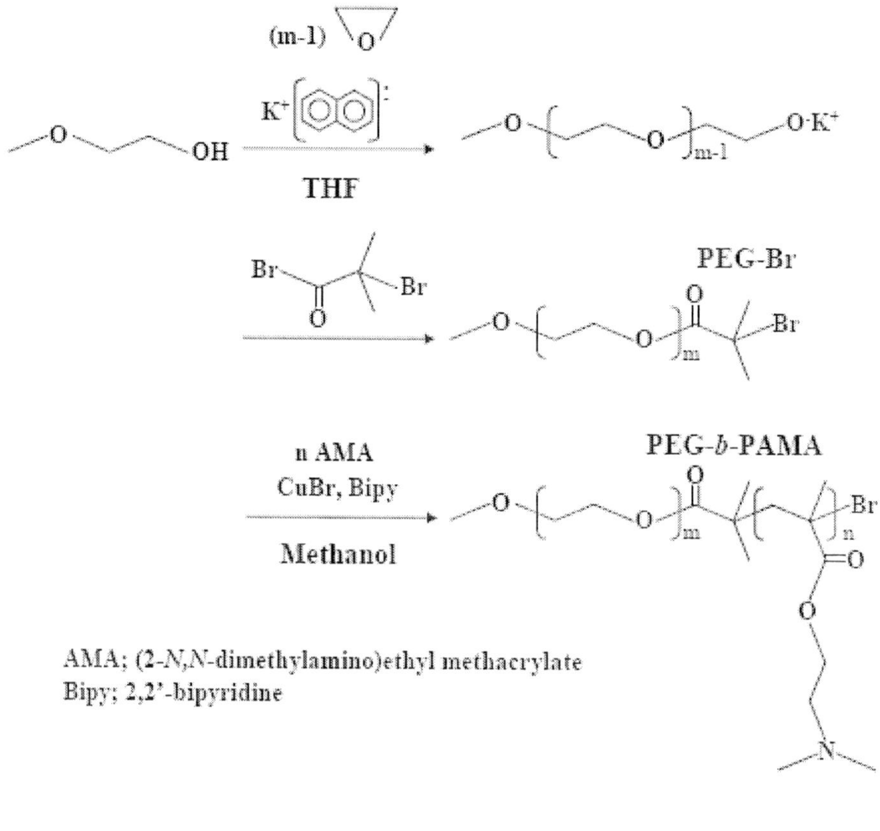

Scheme 3. Schematic representation of PEG-*b*-PAMA synthesis.

The results of the synthesis and characterization of the prepared PEG-*b*-PAMAs are summarized in Table 3. The molecular weights were well controlled, with narrow molecular weight distribution factors. Especially, PEG-*b*-PAMA possessing a very short PAMA length was clearly obtained by this method. The key step of this synthesis was to add fairly large amounts of AMA monomer to the reaction system in order to increase the initiation rate. PEG-*b*-PAMAs with short PAMA chain lengths were prepared by adjusting the polymerization time. If the polymerization was quenched to be for 3 h, the degree of polymerization of the PAMA segment was adjusted to 3 (Table 3, Run 1). Thus, PEG-*b*-PAMAs with a variety of PAMA chain lengths from DP=3 to 43 (DP: degree of polymerization) were precisely synthesized. These polymers were purified through repeated precipitations and ion exchange column fractionation techniques. Using PEG-*b*-PAMA thus prepared, GNPs surface modifications were carried out.

Table 3. Synthesis of MeO-PEG-*b*-PAMA block copolymers via living radical polymerization (ATRP) of 2-(*N.N*-dimethylamino)ethyl methacrylate(AMA) initiated with MeOPEGBr coupled with CuBr and 2,2'-bipyridine[1]

Run	Code	feed (mmol/L)		Time	Yield	10^{-3} x MW of Copolymer[2]			Number of amino groups[3]
		PEGBr	AMA	h	%	Mn	Mw	Mw/Mn	
1	MeOPEG-*b*-PAMA(3)	0.10	4.0	3.0	6.1	4.7	4.8	1.02	3
2	MeOPEG-*b*-PAMA(6)	0.10	4.0	3.5	10	5.3	5.5	1.03	6
3	MeOPEG-*b*-PAMA(15)	0.10	2.0	18	54	6.6	7.2	1.09	15
4	MeOPEG-*b*-PAMA(18)	0.10	3.5	18	52	7.4	8.1	1.09	18
5	MeOPEG-*b*-PAMA(24)	0.10	4.0	18	63	8.8	9.9	1.12	24
6	MeOPEG-*b*-PAMA(36)	0.10	5.0	18	60	11.4	13.0	1.14	36
7	MeOPEG-*b*-PAMA(43)	0.10	7.0	18	62	11.7	13.3	1.14	43

[1] MeOPEGBr (Mn=4,100, Mw=4,200, Mw/Mn=1.03) : 0.1 mmol/L; CuBr : 0.1 mmol/L; 2,2'-bipyridine : 0.2 mmol/L. polymerization temperature : 25 °C.

[2] Determined by SEC data. Column :Superdex 75 10/300 GL; carrier:10 mM, pH 7.0 phosphate buffer with 140 mM NaCl; detecter: RID, flow rate: 0.7 mL/min.

[3] Determined by ^{1}H NMR data.

2.6. Modification of Gold Nanoparticles Using PEG-b-PAMA(43) under Different pH Conditions [57]

In order to investigate the mechanism of GNPs surface modification with PEG-*b*-PAMA, the interaction between PEG-*b*-PAMA and commercial GNPs were investigated under several conditions. PEG-GNPs were prepared by mixing commercial GNPs with the PEG-*b*-PAMAs prepared in this study under a variety of conditions. One of the representative preparation procedures for PEG-GNPs is described as follows. Four hundred and fifty μL of commercial GNPs solution (7.1×10^{-3} wt%, pH 6.5) were added to 50 μL of PEG-*b*-PAMA(43) solution ([N]=960 μM) in 1.5 mL vial. One hundred mM phosphate buffer was used as the solvent of the block polymer for pH values below 8.5. For pH values above 8.5, 100 mM carbonate buffer was employed. The pH value was varied from 5.0 to 11.0, using these two buffers in these experiments. The solution was finally diluted to adjust the GNPs, buffer and block copolymer concentrations to be 6.4 wt%, 10 mM and 96 μM, respectively.

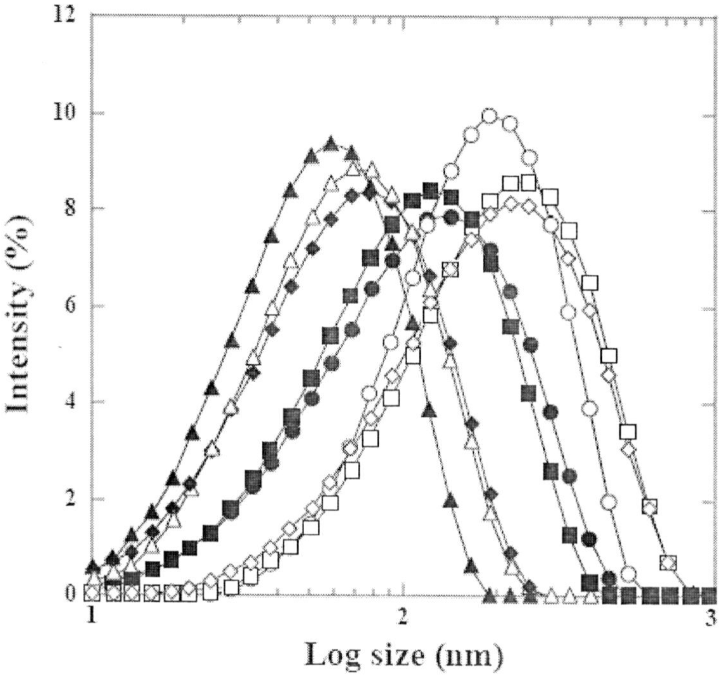

Figure 6. Particle size distribution of GNPs modified with PEG-*b*-PAMA(43) under various pH conditions in 10 mM phosphate buffer. Open circle, pH 5.0; open square, pH 5.5; open diamond, pH 6.0; closed circle, pH 6.5; closed square, pH 7.0; closed diamond, pH 7.5; open triangle, pH 8.0; closed triangle, pH 8.5.

The protonation degree of the amino groups in the block copolymer depended on the environmental pH [49]. The effect of the state of the amino groups on GNP PEGylation was investigated using PEG-*b*-PAMA(43) and varying the environmental pH. In order to confirm the dispersion stability, DLS measurement was carried out after the PEG-*b*-PAMA modification. Figure 6 shows the particle size distribution of the PEG-GNP solutions, which was kept for 18 h at room temperature after preparation under different pH conditions in 10 mM phosphate buffer (pH 5.0-8.5). The PEG-GNPs prepared under acidic conditions increased in size after 18 h at room temperature, indicating the coagulation of the particles in the solution. On the contrary, the particle size did not increase much under alkaline conditions in the same phosphate buffer (pH<8.5). However, in phosphate buffer solution at pH 8.5, a slight but definite increase in particle size was observed. In order to further investigate the preparation conditions in the higher pH region, the buffer of the preparation experiment was changed to carbonate buffer. Figure 7 shows the particle size distribution of the PEG-GNP solutions after 18 h at room temperature in 10 mM pH 8.5-11.0 carbonate buffer. Even in carbonate buffer at pH<10, a slight increase in particle size was observed. At pH>10, on the contrary, an almost constant size (ca. 37 nm) with narrow size distribution was obtained even after 18 h. The size of the PEG-GNPs prepared at pH 10, 37 nm, is reasonable for a PEG brush composed of PEG with a molecular weight of 4.2 kDa coating the GNP (18.4 nm), assuming a single GNP for each particle.

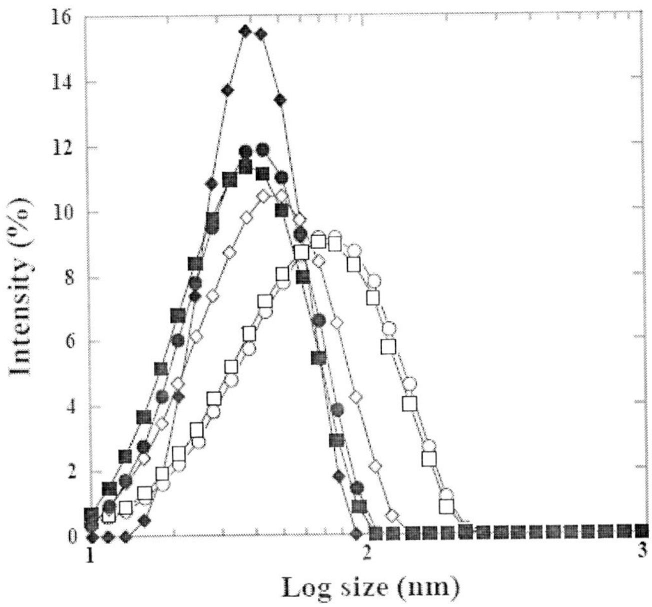

Figure 7. Particle size distribution of GNPs modified with PEG-*b*-PAMA(43) under various pH conditions in 10 mM carbonate buffer. Open circle, pH 8.5; open square, pH 9.0; open diamond, pH 9.5; closed circle, pH 10.0; closed square, pH 10.5; closed diamond, pH 11.0.

It has been reported previously that the pK_a value of the amino groups of PEG-*b*-PAMA is 7.0 [49, 60], which means that all the amino groups in the PAMA segment of the block copolymer are completely depronated at pH 10. This means that the GNPs are stabilized by the neutral PEG-*b*-PAMA, probably via the coordination of amino nitrogen atoms on the GNP surface, but not via electrostatic interaction between the protonated PAMA and the negative GNP surface, caused by the absorbed citric acid. As stated above, although the N-Au bond is not strong enough, unlike the S-Au bond, the polyvalency [61] coordination of the tertiary amino groups of PEG-*b*-PAMA might work effectively on the GNP surface.

2.7. Modification of Gold Nanoparticles Using PEG-*b*-PAMA(43) under Different Concentration [57]

Since commercial GNPs which was modified with PEG-*b*-PAMA at alkaline condition showed stable dispersion as well as the PEGylated GNPs prepared by autoreduction method, the effect of the environmental polymer concentration on GNP PEGylation was investigated using PEG-*b*-PAMA(43). PEGylated GNP was prepared by the same method as in chapter 2.6. The pH was fixed to 10.5. The concentration of the amino groups of PEG-*b*-PAMA was varied from 0.96 to 960 μM, which GNPs concentration was kept constant, 6.4 wt%.

Figure 8 shows the particle size distributions of the PEG-GNPs prepared with different polymer concentrations ([N] concentration: 0.96-960 μM). When the [N]/[GNP] ratio, which denotes the number of the tertiary amino groups to the number of GNPs, was 330, the particles size of the PEG-GNP was fairly large (75.7 nm), which indicates that the amount of polymer was not enough to stabilize the GNPs. When the [N]/[GNP] ratio was greater than 3,300, the particle sizes of the PEG-GNPs were almost equal (about 37 nm), and an [N]/[GNP] ratio of more than 3,300 was required to maintain the stability of their dispersion.

2.8. Modification of Gold Nanoparticles Using Various Molecular Weight of PEG-*b*-PAMAs [57]

The effect of the chain length of the polyamine segment was evaluated using various molecular weights of PEG-*b*-PAMAs, where the number of PAMA segments in the block copolymers was changed from 3 to 43, possessing the same PEG molecular weight (4.2 k). PEGylated GNP was prepared by the same method as in chapter 2.6. The concentration of the amino groups of PEG-*b*-PAMA and pH were fixed 96 μM and 10.5, respectively. All experiments were carried out at room temperature, and the total reaction time was fixed to 18 h. For the purification of PEG-GNPs thus prepared, the free excess polymers in the solutions was removed by repeated centrifugations (2.0×10^4 *g*, 4 times), and the solvent was substituted with pH 7.4 phosphate buffer saline. In order to confirm the dispersion stability of the PEG-GNPs in the protein solution, after the obtained PEG-GNPs were precipitated by centrifugation, the solvent was substituted with BSA or 95 % human serum solution by the same method as the one described above and these solutions were stored at 4 ℃ for 4 days.

Figure 9 shows the ζ-potential of the PEG-GNPs prepared from various block copolymers after the free polymer in the solution was removed and the solvent of the PEG-GNPs was changed from pH 10.5 carbonate buffer to pH 7.4 phosphate buffer saline by centrifuge purification. Since citric acid was adsorbed on the native GNPs, the ζ-potential was negative. Once the GNPs were modified by PEG-*b*-PAMA at an [N]/[GNP] ratio of 33,000, the ζ-potentials of the GNPs were almost completely shielded, regardless of the PAMA chain lengths. Actually, none of the ζ-potentials of the PEG-GNPs depended on PAMA length, and the ζ-potentials were almost 0 mV. During the immobilization of PEG-*b*-PAMAs on the GNP surface, it is not clear whether citric acid is adsorbed together with PEG-*b*-PAMAs or liberated. However, it is known that the slipping plane on the surface shifted offshore [62]. Thus, the adsorbed citric acid on the periphery of the gold surface should not affect the ζ-potential measurement in this case. The complete shielding of the surface charge indicates that PEG formed tetherlike structure when GNPs were modified by PEG-*b*-PAMA block copolymers.

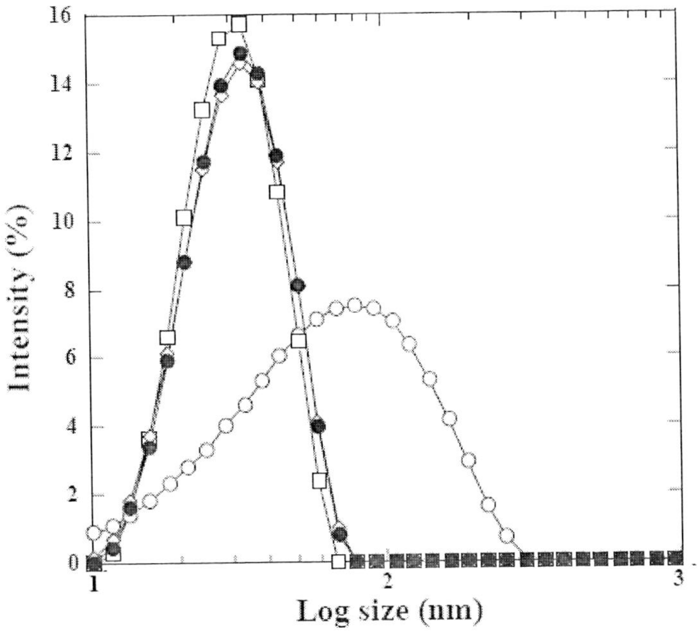

Figure 8. Particle size distribution of GNPs modified with PEG-*b*-PAMA(43) of various concentrations in 10 mM, carbonate buffer, pH 10.5 ([N] concentration, 0.96-960 μM). Open circle, 0.96 μM; open square, 9.6 μM; open diamond, 96 μM; closed circle, 960 μM.

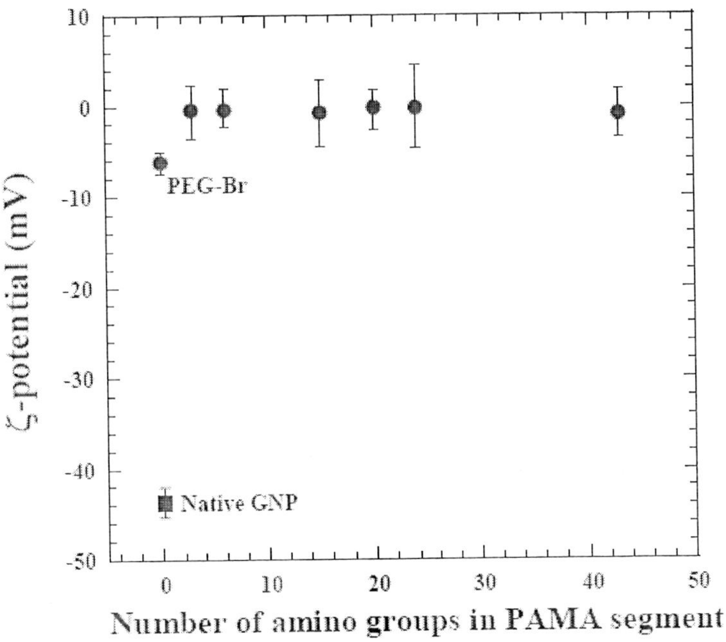

Figure 9. Relationship between the ζ-potential of the PEG-GNPs and number of amino group units in the PAMA segment of PEG-*b*-PAMA. Closed circle, PEG-GNPs(0-43); closed square, native GNPs.

As stated above, although an [N]/[GNP] ratio of 3,300 was enough to disperse the GNPs, not all the polymers in the solution interacted with the particles. When PEG-GNPs are administered *in vivo,* the free polymer in the solution must be removed. The dispersion stability of the PEG-GNPs after centrifuge purification was evaluated. Figure 10 shows the average diameter of various PEG-GNPs at an [N]/[GNP] ratio of 33,000 before and after purification. Immediately after the preparation of PEG-GNPs in the presence of excess free polymer, the PEG-GNPs showed the same average diameter regardless of the PAMA length. When the excess free polymer was removed completely from the solution by centrifugation, the particle size of the PEG-GNPs in PBS increased. The coagulation tendency was remarkable in PEG-*b*-PAMAs with long polyamine segments. When the GNPs were modified with PEG-*b*-PAMA having short AMA units (DP=3, 6), on the contrary, the PEG-GNPs showed excellent dispersion stability, which was maintained for 14 days at room temperature in pH 7.4 PBS. It is rather surprising that the purified PEG-GNPs (DP=3, 6) were stable at neutral pH even though they were prepared from deprotonated PEG-*b*-PAMA umder alkaline conditions (pH=10.5). This probably means that, once the neutral amino groups in PEG-*b*-PAMA coordinate on the GNP surface, they are not cleaved even in neutral pH. The pK_a value of the amino groups which coordinate on the gold surface might change [63]. When the GNPs were modified with PEG-*b*-PAMA having long polyamine segments, a part of the amino groups in the single polymer chain could not coordinate on the GNP surface even at high pH, due to the conformational restriction of the PAMA segments. The protonation of the free amino groups in this case may decrease the dispersion stability under physiological conditions. The dispersion stability of PEG-*b*-PAMA was further examined in the presence of BSA, as shown in Figure 10. The PEG-GNPs modified with PEG-*b*-PAMA possessing long PAMA segments aggregated further in 4.5 mg/mL BSA solution. However, the PEG-GNPs modified with PEG-*b*-PAMA(3) (PEG-GNP(3)) showed excellent dispersion even when it was stored for 4 days in BSA solution at 4 ℃.

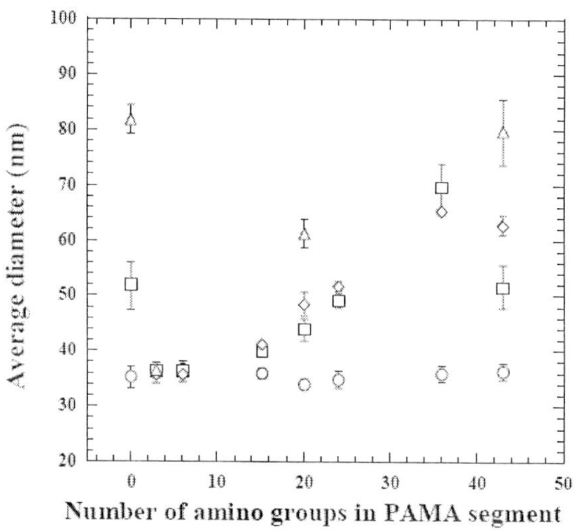

Figure 10. Average diameter of the PEG-GNPs before and after purification and in BSA solution. Open circle, before purification; open square, after purification; open diamond, after purification and stored for 14 days; open triangle, stored in BSA solution.

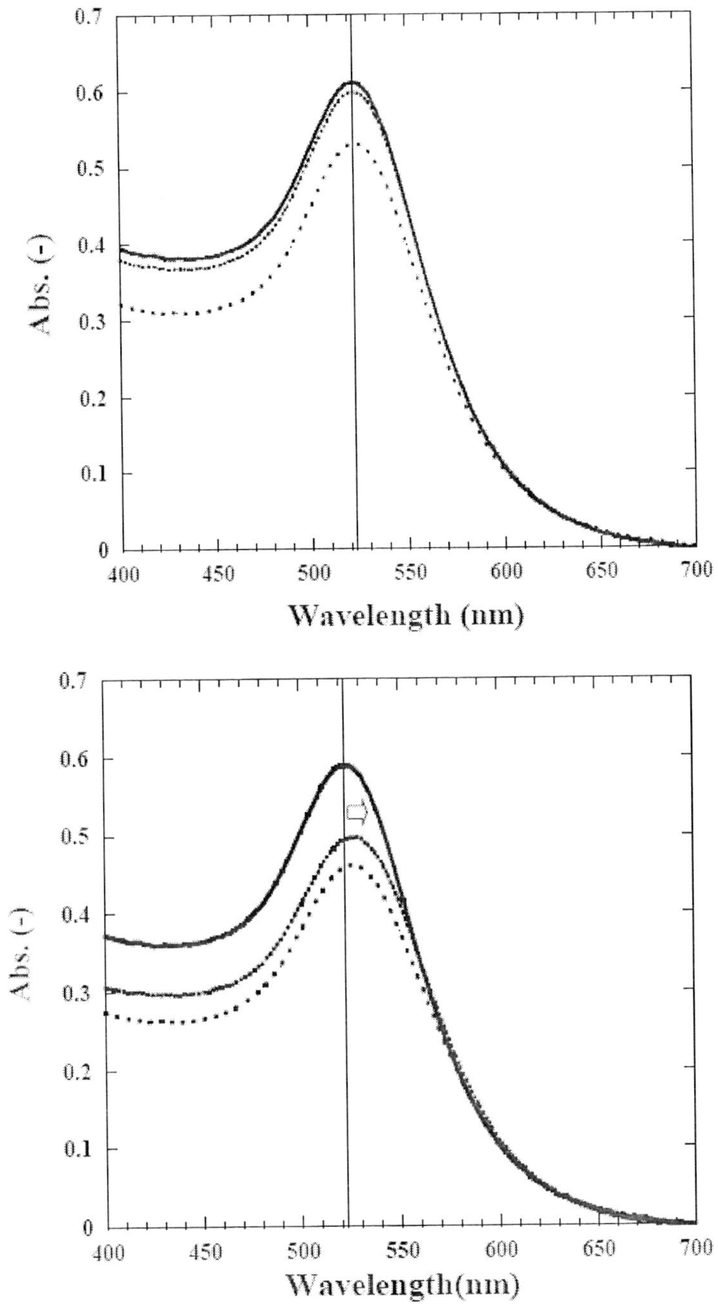

a)

b)

Figure 11. UV spectra of (a) PEG-GNP(3) and (b) PEG-GNP(43) before and after purification and stored for 14 days. Solid line, before purification; dashed line, after purification; dotted line, stored for 14 days after purification.

Figure 11 shows the UV spectra of GNPs modified with PEG-*b*-PAMA before and after purification. The surface plasmon peak was observed at 520.0 nm for the native GNPs. It shifted to the longer wavelength region upon modification of the GNP surface with PEG-*b*-PAMA. This means that the permittivity of the GNP surface environment was changed by the adsorption of the polymer on the GNP surface [64] and it is more proof of GNP surface modification by the block copolymer. The UV spectra of the PEG-GNPs after purification were then investigated. The UV spectra of PEG-GNP(3) after purification showed no change compared to those before purification. In the case of PEG-GNP(43), however, a slight but definite red shift was observed after purification. The red shift of the plasmon peak is known as signifying the coagulation of GNP [1]; and this is in good agreement with the DLS results. The observed bathochromic shift of the plasmon peak depended on the PAMA length of the block copolymers; *i.e.*, the shift increased with decreasing PAMA chain length. In addition to the protonation characteristics of PEG-*b*-PAMA on GNPs, as descrived above, the PEG density of the PEG-GNPs should influence on their dispersion stability. The high dispersion of the PEGylated GNPs was mainly governed by the entropic repulsion of the PEG. We assumed that the PEG density on the GNP surface depended on the PAMA chain length in the PEG-*b*-PAMA modification reaction. The PEG density of the PEG-GNPs was thus investigated by TGA.

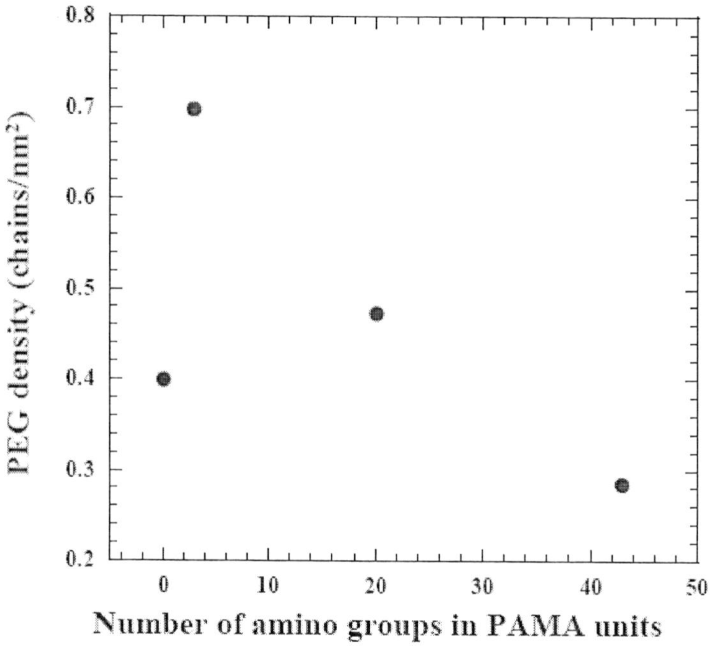

Figure 12. PEG chain density on GNPs modified with PEG-*b*-PAMA, as determined by TGA, assuming that the PAMA segments are located on the GNP surface to form a PEG brush.

2.9. Quantification of Adsorbed PEG-*b*-PAMA on GNPs [57]

After the solution of the PEG-GNPs was substituted with water through repeated centrifugations and the PEG-GNPs were completely dried *in vacuo*, thermal gravimetric analysis was carried out. Figure 12 shows the PEG brush density, as determined from the TGA data; it is a function of the number of amino groups in PEG-*b*-PAMA. As a control, PEG-Br was used as the GNP modification agent. Since the ether oxygen in the ethylene glycol units of PEG interacted weakly with gold surface, a limited amount of PEG was adsorbed on the GNP surface. The GNPs coupled with PEG-Br were dispersed in the aqueous medium when excess PEG-Br solution was used. After removing the excess free PEG-Br from the solution by centrifugation, however, the PEG-GNPs did not redisperse completely, probably due to the insufficient amounts of PEG on the GNPs.

The amount of modified PEG-*b*-PAMA on the GNP surface increased significantly, as compared to that of PEG-Br. Among the PEG-*b*-PAMAs used in this study, the shortest PAMA (PEG-*b*-PAMA(3)) had highest density, 0.7 chains/nm^2, assuming that the PAMA segments adsorbed on the GNP surface formed a PEG brush. With increasing PAMA chain length, the brush density gradually decreased. The effect of the PAMA chain length on the PEG chain density was in good agreement with the PEG-GNP dispersion stability (*i.e.*, the dispersion stability increased with increasing PEG chain density on the GNP surface) [65-67]. Fukuda, Tsujii and co-workers reported [68] that densely packed polymer brush effectively repelled protein adsorption. They explained that the large protein can not migrate in densely packed brush layer. The colloidal stability of PEGylated nanoparticles was also interpreted in the same way [69,70], *viz.*, the PEG brushes on GNPs surface can not migrate each other when the brush density becomes high enough. When the GNPs were modified with PEG-SH, whose molecular weight was the same as that of the PEG segment in PEG-*b*-PAMA, the PEG density was 1.3 chains/nm^2 under the same modification conditions. Although this value was larger than that of PEG-GNP(3), PEG-GNP(3) showed an extremely high dispersion stability under various conditions.

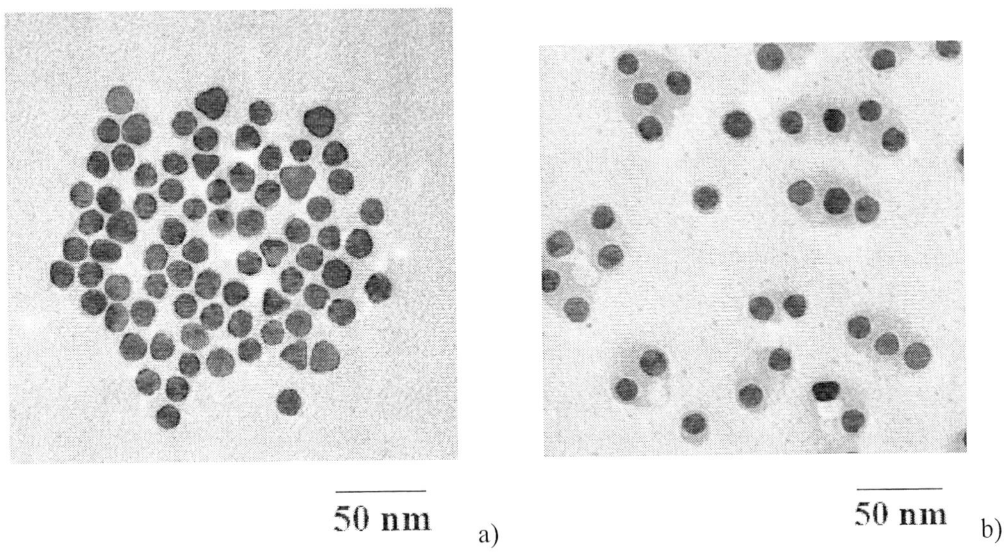

Figure 13. TEM image of PEG-GNP(3) after purifica tion (a) and PEG-GNP(3) kept in 95 % human serum for 4 days (b).

Finally, the dispersion stability of the PEG-GNPs was examined in the presence of 95 % human serum as model blood. Since direct DLS measurement of PEG-GNPs in human serum would be too difficult to conduct because of the very high protein concentration, the average diameter of the redispersed PEG-GNPs after separation of human serum proteins by centrifugation was evaluated by DLS. In the case of PEG-GNP(3), the particle size after separation of human serum proteins (ca. 43.8 nm) was almost the same as that without serum, indicating stable dispersion under high serum proteins conditions. The stable dispersion of PEG-GNP(3) in 95 % human serum was also confirmed by TEM measurement (Figure 13). On the contrary, PEG-GNP(43) in 95 % human serum precipitated on the bottom of sample tube and adsorbed on the tube surface after 4-day incubation. These results confirm the fact that PEG-GNPs(3) was completely dispersed even in 95 % human serum,, which strongly suggests that it is suitable for *in vivo* application.

Using PEG-*b*-PAMA modified GNPs thus prepared, further functionalization can be achieved. For example, PEG-*b*-PAMA and RNA-SH was able to be co-immobilized on GNPs surface (PEG/RNA-GNPs). The PEG/RNA-GNPs could thus prepared control the RNA release [71] and expected to apply for the high performance gene delivery systems.

3. Conclusion

Monodispersed GNPs were readily prepared by the simple addition of a ligand installed PEG-*b*-PAMA in $HAuCl_4$ aqueous solution at room temperature without any additional reducing reagent. The ligand-installed PEG layer locating on the outer surface of the obtained GNPs substantially contributed to sterically stabilize them even under very high salt concentrations. Furthermore, biotin as the ligand molecule installed at the end of tethered PEG chain showed specific recognition of streptavidin under physiological salt concentration indicating the high utility of these ligand-installed PEG/PAMA-gold nanoparticles as colloidal biosensor systems.

The polyvalency coordination of tertiary amino groups on the gold surface was confirmed to lead to stable modification with PEG-*b*-PAMA, and not electrostatic interaction between the protonated PAMA and the gold surface. PEG-GNPs modified with PEG-*b*-PAMA having a short PAMA chain length showed excellent dispersion stability under physiological conditions, even when kept in 95% human serum for 4 days. It was confirmed that the increased PEG chain density resulting from PEG-*b*-PAMA with a short PAMA chain length is one of the factors which improve the dispersion stability of GNPs, indicating that three amino nitrogens are enough to stabilize the PEG brush on the gold colloid surface via the polyvalency effect. The obtained PEG-GNPs may be useful in a variety of biological applications, for example, as carriers of gene delivery, model carriers of drug delivery systems, and contrast agents for *in vivo* X-ray computed tomography.

References

[1] Daniel, M. C.; Astruc, D. *Chem. Rev.* 2004, *104*, 293.
[2] Brown, D. H.; Smith, W. E. *Chem. Soc. Rev.*, 1980, *9*, 217.
[3] Turkevitch, J.; Stevenson, P. C.; Hiller, J. *Discuss. Faraday Soc.*, 1951, *11*, 55.
[4] Frens, G. *Nature: Phys. Sci.*, 1973, *241*, 20.
[5] Schmid, G. *Chem. Rev.*, 1992, *92*, 1709.
[6] Brust, M.; Walker, M.; Schiffrin, D.; Whyman, R. *J. Chem. Soc., Chem. Commun.*, 1994, 801.

[7] Han, M. Y.; Quek, C. H.; Huang, W.; Chew, C. H.; Gan, L. M. *Chem. Mater.*, 1999, *11*, 1144.

[8] Brown, K. R.; Walter, D. G.; Natan, M. *J. Chem. Mater.*, 2000, *12 (2)*, 306.

[9] Weiser, H. B. *Inorg. Colloid Chem.*, 1993, *1*, 21.

[10] Biggs, S.; Chow, M. K.; Grieser, F.; *J. Colloid Interface Sci.*, 1993, *160*, 511.

[11] Hayat, M. A. Colloidal Gold Principles, *Method and Applications, Academic Press: New York*, 1989.

[12] Roucoux, A.; Schulz, J.; Patin, H. *Chem. Res.* 2002, *102*, 3757.

[13] Ziolo, R. F.; Glannelis, E. P.; Weinstein, B. A.; OHoro, M. P.; Gamguly, B. M.; Mwhrota, V.; Russel, M. W.; Huffman, D. R. *Science*, 1992, *257*, 219.

[14] Jordan, R.; West, N.; Chou, Y. M.; Nuyken, O. *Macromolecules*, 2001, *34*, 1606.

[15] Selven, S. T.; Spatz, J. P.; Klock, H. A.; Moller, M. *Adv. Mater.*, 1998, *10*, 132.

[16] Sayo, K.; Deki, S.; Hayashi, S. *Eur. Phys. J. D.*, 1999, *9*, 429.

[17] Lee, J.; Sunder, V. C.; Heine, J. R.; Bawendi, M. G.; Jensen, K. F. *Adv. Mater.*, 2000, *12*, 1102.

[18] Teichoeb, J. H.; Forrest, J. A. *Phys. Rev. Lett.*, 2003, *91*, 016104-1.

[19] Raula, J.; Shan, J.; Nuoponen, M.; Niskanen, A.; Jiang, H.; Kauppinen, E. I.; Tenhu, H. *Langmuir*, 2003, *19*, 3499.

[20] Corbierre, M. K.; Cammeron, N. S.; Sutton, M.; Mochrie, S. G. J.; Lurio, L. B.; Ruhm, A.; Lennox, R. B. *J. Am. Chem. Soc.*, 2001, *123*, 10411.

[21] Kolb, U.; Quaiser, S. A.; Winter, M.; Reets, M. T. *Chem. Mater.*, 1996, *8*, 1889.

[22] Toshima, N.; Yonezawa, T. *New J. Chem.*, 1998, 1179.

[23] Schaaf, T. G.; Whetten, R. L. *J. Phys. Chem. B*, 2000, *104*, 2630.

[24] Templeton, A. C.; Wuelfing M. P.; Murray, R. W. *Acc. Chem. Res.* 2000, *33*, 27.

[25] Hong, R.; Han, G.; Fernandez, J. M.; Kim, B. J.; Forbes, N. S.; Rotello, V. M. *J. Am. Chem. Soc.* 2006, *128*, 1078.

[26] Barrientos, A. G.; Fuente, J. M.; Rojas, T. C.; Fernandez, A.; Penades, S. *Chem. Eur. J.* 2003, *9*, 1909.

[27] Mirkin, C. A.; Letsinger, R. L.; Mucic, R. C.; Storhoff, J. J. *Nature* 1996, *382*, 607.

[28] Kerman, K.; Saito, M.; Morita, Y.; Takamura, Y.; Ozsoz, M.; Tamiya, E. *Anal. Chem.* 2004, *76*, 1877.

[29] Balamurugan, S.; Obubuafo, A.; Soper, S. A.; McCarley, R. L.; Spivak, D. A. *Langmuir* 2006, *22*, 6446.

[30] Rosi, N. L.; Giljohann, D. A.; Thaxton, C. S.; Lytton-Jean A. K. R.; Han, M. S.; Mirkin, C. A. *Science,* 2006, *312*, 1029.

[31] Shimin, R. G.; Scohch, A. G.; Braun, P. V.; *Langmuir*, 2004, *20*, 5613.

[32] Kitano, H.; Kawasaki, A.; Kawasaki, H.; Morokoshi, S. *J. Coll. Int. Sci,.* 2005, *282*, 340.

[33] Wuelfing W. P., Gross S. M., Miles D. T., Murray R. W. *J. Am. Chem. Soc.*, 1998, *120*, 12696.

[34] Otsuka H., Akiyama Y., Nagasaki Y., Kataoka K. *J. Am. Chem. Soc.*, 2001, *123*, 8226.

[35] Kim, D. K.; Park, S. J.; Lee, J. H.; Jeong, Y. Y.; Jon, S. Y. *J. Am. Chem. Soc.*, 2007, *129*, 7661.

[36] Bergen, J. M.; Recum, H. A.; Goodman, T. T.; Massey, A. P.; Pun, S. H. *Macromol. Biosci.*, 2006, 6, 506.

[37] Zheng, M.; Davidson, F.; Huang, X. *J. Am. Chem. Soc.*, 2003, *125*, 7790.

[38] Niidome, T.; Yamagata, M.; Okamoto, Y.; Akiyama, Y.; Takahashi, H.; Kawano, T.; Katayama, Y.; Niidome. Y. *J. Cont. Rel.*, 2006, *114*, 343.

[39] Felice R. D., Selloni A. *J. Chem. Phy. B*, 2004, *120*, 4906.

[40] Bearinger, J. P.; Terrettaz, S.; MicheL, R.; Tirelli, N.; Vogel, H.; Textor, M.; Hubbell, J. A. *Nature materials,* 2003, *2*, 259.

[41] Cooper, E.; Leggett, G. J. *Langmuir,* 1998, *14*, 4795.

[42] Castner, D. G.; Hinds, K.; Grainger, D. W. *Langmuir,* 1996, *12*, 5083.

[43] Mayer, A. B. R,; Mark, J. *E. Eur. Polym. J.*, 1998, *34(1)*, 103.

[44] Bronstein, L. M.; Gourkova, S. N.; Sidorov, A. Y.; Valetsky, P. M.; Hartmann, J.; Breulmann,K.; Colfen, H.; Antonietti, M. *Inorg. Chim. Acta*, 1998, *280(1-2)*, 348.

[45] Harrisa L. G., Tosattib S., Wielandc M., Textorb M., Richardsa R.G. *Biomaterials*, 2004, *25*, 4135.

[46] Blattler T. M., Pasche S, Textor M., Griesser H. J., *Langmuir*, 2006, *22*, 5760 .

[47] Ishii, T.; Otsuka, H.; Kataoka, K.; Nagasaki, Y. *Langmuir,* 2004, *20*, 561.

[48] Akiyama Y., Harada A. Nagasaki Y., Kataoka K., *Macromolecules*, 2000, *33*, 5841.

[49] Kataoka K., Harada A., Wakebayashi D. Nagasaki Y. *Macromolecules*, 1999, *32*, 6892.

[50] Kreibig, U.; Genzel, L. *Surf. Sci.* 1985, *156*, 678-700.

[51] Collier, C. P.; Saykally, R. J.; Shiang, J. J.; Henrichs, S. E.; Heath, J. R. *Science,* 1997, *277*, 1978.

[52] Heath, J. R.; Knobler, C. M.; Leff, D. V. *J. Phys. Chem. B,* 1997, *101*, 189.

[53] Corvierre, M. K.; Cameron, N. S.; Lennox, R. B. *Langmuir*, 2004, *20*, 2867.

[54] Green, N. M. *Adv. Protein Chem.,* 1975, *28*, 85.

[55] Blankenburg, R.; Meller, P.; Ringsdorf, H.; Salesse, C. *Biochemistry,* 1989, *28*, 8214.

[56] Herron, J. N.; Muller, W.; Paudler, M.;Riegler, H.; Ringsdorf, H.; Suci, P. A. *Langmuir,* 1992, *8*, 1413.

[57] Miyamoto, D.; Oishi, M.; Kojima, K.; Yoshimoto, K.; Nagasaki, Y. *Langmuir*, in press.

[58] Tang, W.; Matyjaszewski, K. *Macromolecules,* 2006, *39*, 4953 .

[59] Matyjaszewski, K.; Xia, J. *Chem. Rev.,* 2001, *101*, 2921.

[60] Oishi, M.; Kataoka, K.; Nagasaki, Y. *Bioconjugate Chem*., 2006, *7*, 677.

[61] Mammen, M.; Choi ,S. K.; Whitesides G. M. *Angew. Chem. Int. Ed*., 1998, *37*, 2755.

[62] Koopal, L. K.; Hlady, V.; Lyklema, J. *J. Colloid Interface Sci.*, 1988, *121*, 49.

[63] Nagasaki Y., Ohishi M., Nakamura T., in preparation.

[64] Nagasaki Y., Yoshinaga K., Kurosawa K., Iijima M., *Coll. Polym. Sci.*, 2007, *285*, 563.

[65] Pasche, S.; Vo1ro1s, J.; Griesser, H. J.; Spencer, N. D.; Textor, M. *J. Phys. Chem. B*, 2005, *109*, 17545.

[66] Paul, S. M.; Falconnet, D.; Pasche, S.; Textor, M.; Abel, A. P.; Kauffmann, E.; Liedtke, R.; Ehrat, M. *Anal. Chem*., 2005, *77*, 5831.

[67] Unsworth, L. D.; Sheardown, H.; Brash, J. L. *Langmuir,* 2005, *21*, 1036.

[68] Yoshikawa, C.; Goto, A.; Tsujii, Y.; Ishizuka, N.; Nakanishi, K.; Fukuda, T. *J. Polym. Sci. Part A-Polym. Chem,.* 2007, *45*, 4795.

[69] Mclean, S. C.; Lioe, H.; Meagher, L.; Craig, V. S. J.; Gee M. L. *Langmuir*, 2005, *21*, 2199-2208.

[70] Romero-Cano, M. S.; Puertas, A. M.; Nieves, F. J. *J. Chem. Phys*., 2000, *112*, 8654-8659.

[71] Oishi, M.; Nakaogami, J.; Ishii, T.; Nagasaki, Y. *Chem. Lett*., 2006,*35*, 1046.

In: Encyclopedia of Nanotechnology
Editor: Elwood D. Carlson

ISBN: 978-1-60692-079-4

Chapter 38

Nanoparticles and Quantum Dots as Biomolecule Labels for Electrochemical Biosensing

*Martin Pumera**
National Institute for Materials Science, 1-1 Namiki, Tsukuba, Ibaraki, Japan

Abstract

The aim of this chapter is to discuss the advantages of use of nanoparticles and quantum dots as tags for electrochemical bioassays.

Nanoscale materials bring new opportunities for electrochemical biosensing. Quantum dots are very stable (comparing to enzyme labels), they offer high sensitivity (thousands of atoms can be released from one nanoparticle) and wide variety of nanoparticles brings possibility for multiplex detection of several biomolecules.

Metallic Nanoparticles

While gold nanoparticles have been known as "colloidal gold" for centuries, their use for electrochemical detection of biomolecules were introduced in 2001 by Limoges' [1] and Wang's groups [2]. The general scheme of utilization of nanoparticles and quantum dots in biomolecule electrochemical sensing is shown in Figure 1. Probe biomolecule is immobilized on surface, then the solution containing target molecule is introduced. After the specific biorecognition event (antigen-antibody reaction or DNA hybridization), the rest of the non-complementary molecules is washed away. Consequently, the second probe molecule tagged with quantum dot or nanoparticle is introduced, interacting with target molecule and labeling it for the final step, electrochemical detection. Originally, in Wang's scheme [2], hybridization of a target oligonucleotide to magnetic bead-linked oligonucleotide probes was followed by binding of the streptavidin-coated metal nanoparticles to the captured DNA,

---

* Fax: +81-29-860-4706; Email: martin.pumera@gmail.com

dissolution of the nanometer-sized gold tag by HBr/Br_2 solution, and potentiometric stripping measurements of the dissolved metal tag at single-use thick-film carbon electrodes [2]. Limoges and coworkers immobilized amplified 406-base pair DNA sequence to polystyrene microwell by passive adsorption. The assay was based on the hybridization of the single-stranded target DNA with an oligonucleotide-modified Au nanoparticle probe, followed by the release of the gold metal atoms anchored on the hybrids by oxidative metal dissolution, and the indirect determination of the HBr solubilized Au^{III} ions by anodic stripping voltammetry [1]. Analogous assay using Ag nanoparticles was developed, relying on hybridization of the target DNA with the silver nanoparticle–oligonucleotide DNA probe, followed by the release of the silver metal atoms anchored on the hybrids by oxidativemetal dissolution using nitric acid and the indirect determination of the solubilized Ag^{I} ions by anodic stripping voltammetry at a carbon fiber ultramicroelectrode [3]. However, the HBr/Br_2 or HNO_3 solution is highly toxic and therefore method based on direct electrochemical detection of nanoparticle tags, which would replace the chemical oxidation agent, was developed [4,5]. This scheme is using Cl^- chemistry where Au nanoparticles are electrooxidized at 1.2 V for short time (1-5 minutes) in 0.1 M HCl. As a result of this electrooxidation step in a highly acidic environment, $[AuCl_4]^-$ ions are produced and the biomolecules are oxidized and denatured on the electrode surface. The denaturation of the biomolecules allows an open surface for the $[AuCl_4]^-$ ions to be reduced at potential of 0.4 V (vs. Ag/AgCl) [4,5]. For DNA detection, probe DNA was immobilized on the paramagnetic beads surface via biotin-streptavidin interaction and target DNA was labeled with Au_{67} nanoparticle in ratio 1:1 preventing multiple DNA links between paramagnetic bead and nanoparticle (typical for above described Au nanoparticle-based assays), thus enhancing achievable detection limits. The hybridized paramagnetic beads were accumulated on the surface of a magnetic electrode and enabled the magnetically triggered direct electrochemical detection of gold quantum dot tracers without prior chemical dissolution of the Au quantum dots. Electrochemical magnetogenosensors for biomedical applications based on above described direct detection of gold nanoparticles was recently developed by our group [6]. The electrochemical detection of biomolecules via Au and Cl^- chemistry is gaining popularity for its non-toxicity and simplicity, and it was recently used i.e. for detection of kinase-catalyzed thiophosphorylation [7].

Biomolecule Multiplexing via Quantum Dots with Different Electrochemical Signature

Multiple target DNA/protein detection is very important feature of modern bioassays. The goal is not to detect just one biomolecule in one measurement, but several of them. This is possible to achieve using different quantum dots with different electrochemical properties (see Figure 2). Here we list some of examples of variety of nanoparticles for electrochemical detection of DNA: Core-shell Cu@Au where hybridization events between probe and target were monitored by the release of the copper metal atoms anchored on the hybrids by acidic oxidative metal dissolution and the indirect determination of the dissolved Cu^{2+} ions by anodic stripping voltammetry [8]. CdS quantum dots are another addition to the spectrum of nanoparticle tags for DNA hybridization detection [9,10]. PbS quantum dots also offer different voltammetric detection potential for following DNA hybridization [11].

Therefore it is possible to demonstrate real hybridization assays that permit simultaneous determination of multiple DNA targets. Three encoding nanoparticles (zinc sulfide, cadmium sulfide, and lead sulfide) have been used to differentiate the signals of three DNA targets in connection with a sandwich DNA hybridization assay and stripping voltammetry of the corresponding heavy metals [12]. Multiplexing capabilities of quantum dots were also demonstrated in connection to immunoassay. Electrochemical immunoassay protocol for the simultaneous measurements of proteins, based on the use of different inorganic nanocrystal tracers was described [13]. The multiprotein electrical detection capability was coupled to the amplification attribute of electrochemical stripping transduction (to give in *fmol* detection limits). The multianalyte electrical sandwich immunoassay involved a dual binding event, based on antibodies linked to the nanocrystal tags and magnetic beads. Carbamate linkage was used for conjugating the hydroxyl-terminated nanocrystals with the secondary antibodies. Each biorecognition event provided a distinct voltammetric peak, whose position and size reflected the identity and concentration of the corresponding antigen. The concept was demonstrated for a simultaneous immunoassay of β_2-microglobulin, IgG, bovine serum albumin, and C-reactive protein in connection with ZnS, CdS, PbS, and CuS colloidal crystals, respectively.

Soft Nanoparticles

However, not only metallic or binary nanoparticles can be used as tags for electrochemical biosensing of biomolecules. Very recently, there has been substantial research activity in development soft nanoparticle labels which are easy to synthesize and easy to detect by electrochemical means. Liu et al. developed apoferritin-templated phosphate nanoparticle labels and used them for electrochemical immunoassay [14,15] (see Figure 3). Apoferritin is a native protein composed of 24 polypeptide subunits that interact to form a hollow cage-like structure 12.5 nm in diameter; the interior cavity of apoferritin is about 8 nm in diameter and has an interior volume that can store several thousands of metal ions in form of phosphate salt. There is wide variability of metals which can be introduced in apoferritin cavity which paves the way for highly multiplexed assays. These novel soft nanoparticles brings new possibilities for electrochemical sensing of proteins and alternatively also for DNA sensing.

Outlook

Electrochemical nanobiosensors offer without doubts an important step towards development of selective, down to few target molecules sensitive biorecognition device for medical and security applications. There is high expectation that such devices will develop toward reliable point-of-care diagnostics of cancer and other diseases, and as useful tools for molecular biosystems, intra-operation pathological testing or proteome research.

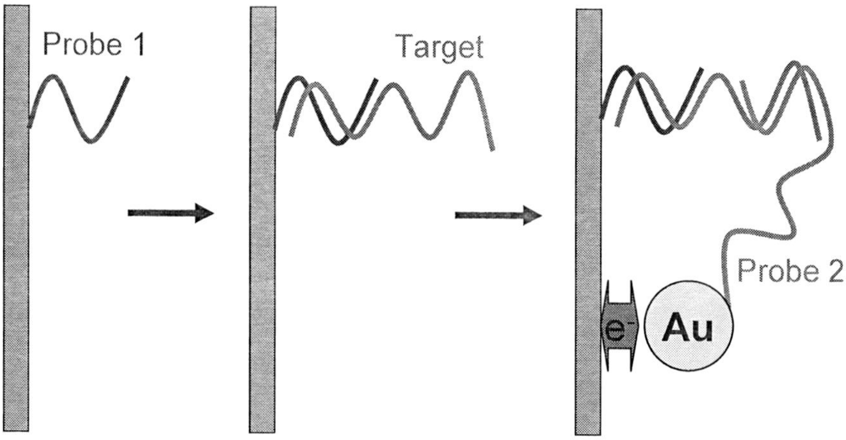

Figure 1. Schematic of sandwich bioassay employing gold nanoparticle as a electrochemical tag.

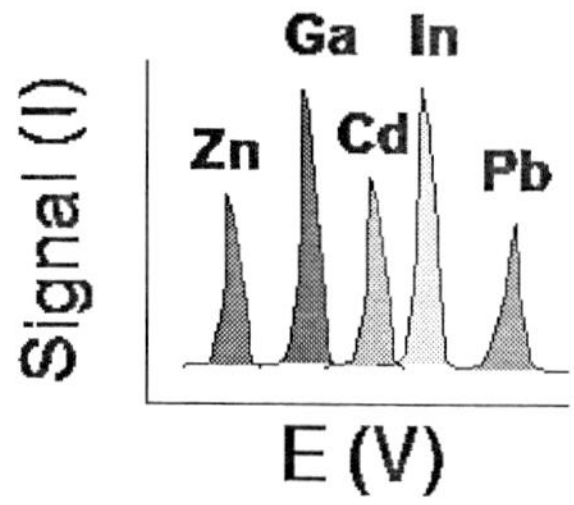

Figure 2. Schematic showing different electrochemical detection potentials of several metallic components of quantum dots.

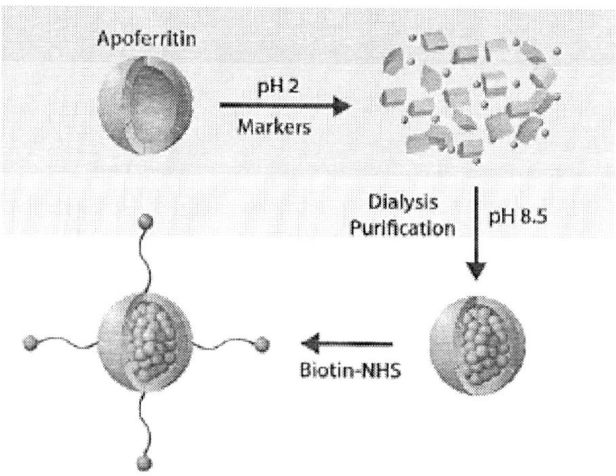

Figure 3. Preparation of biotin-functionalized appoferitin labels. Reprinted with permission from [15] G. Liu, J. Wang, H. Wu, Y. Lin, Anal. Chem. 78, 7417 (2006). Copyright (C) American Chemical Society.

REFERENCES

[1] L. Authier, C. Grossiord, P. Brossier, B. Limoges, *Anal. Chem.* 73, 4450 (2001).

[2] J. Wang, D. Xu, A.-N. Kawde, R. Polsky, *Anal. Chem.* 73, 5576 (2001).

[3] H. Cai, Y. Xu, N. Zhu, P. He, Y. Fang, *Analyst* 127, 803 (2002).

[4] M. Pumera, M. T. Castaneda, M. I. Pividori, R. Eritja, A. Merkoci, S. Alegret, *Langmuir* 21, 9625 (2005).

[5] M. Pumera, M. Aldavert, C. Mills, A. Merkoci, S. Alegret, *Electrochim. Acta* 50, 3702 (2005).

[6] M. T. Castaneda, A. Merkoci, M. Pumera, S. Alegret, *Biosens. Bioelectr.* 22, 1961 (2007).

[7] K. Kerman, H.-B. Kraatz, *Chem. Comm.* 5019 (2007).

[8] H. Cai, N. N. Zhu, Y. Jiang, P. G. He, Y. Fang, *Biosens. Bioelectron.* 18, 1311 (2003).

[9] J. Wang, G. Liu, R. Polsky, A. Merkoçi, *Electrochem. Commun.* 4, 722 (2002).

[10] N. N. Zhu, A. P. Zhang, P. G. He, Y. Z. Fang, *Analyst* 128, 260 (2003).

[11] N. N. Zhu, A. P. Zhang, Q. J. Wang, P. G. He, Y. Z. Fang, *Electroanal.* 16, 577 (2004).

[12] J. Wang, G. Liu, A. Merkoci, *J. Am. Chem. Soc.* 125, 3214 (2003).

[13] G. Liu, J. Wang, J., Kim, M. R. Jan, G. E. Collins, *Anal. Chem.* 76, 7126 (2004).

[14] G. Liu, H. Wu, J. Wang, Y. Lin, *Small* 2, 1139 (2006).

[15] G. Liu, J. Wang, H. Wu, Y. Lin, *Anal. Chem.* 78, 7417 (2006).

In: Encyclopedia of Nanotechnology
Editor: Elwood D. Carlson

ISBN: 978-1-60692-079-4

Chapter 39

PHOTOLUMINESCENCE BEHAVIOR OF WATER-SOLUBLE THIOL-CAPPED CDTE QUANTUM DOTS IN LIVING CELLS

***Pei-Nan Wang*[1,a] *and Ji-Yao Chen*[2]**

State Key Lab for Advanced Photonic Materials and Devices,
Departments of [1]Optical Science and Engineering [2]Physics,
Fudan University, Shanghai 200433, China

ABSTRACT

Water-soluble thiol-capped colloidal CdTe QDs were synthesized by the hydrothermal route. The PL wavelength was determined by the QD size due to the quantum confinement effect. The PL properties of the CdTe QDs were not changed upon aging in air for several months. When studying time-dependent PL spectroscopy of thiol-capped CdTe QDs in living cells, it was observed that the PL was first enhanced and then redshifted until it was fully quenched. This characteristic PL behavior was proved to be originated from the effect of photooxidation. Deoxygenation could effectively prevent the PL blueshift to occur. It was revealed that singlet oxygen is responsible for the PL blueshift. The singlet oxygen molecules were presumably formed from the oxygen molecules intercalated in the thiol layer at the QD core surface. The photostability of QDs in living cells was also studied. Photobleaching of QDs occurred in living cells. The higher irradiation dose and environment oxygen density and the lower QD concentration resulted in the higher photobleaching rate. It was proved that photobleaching resulted mainly from the photooxidation of the cellular QDs. The photostability of cellular QDs was well improved when the cells were deoxygenated by purging with nitrogen. However, the photostability of cellular QDs is better than that of the chlorophyll and green fluorescence protein (GFP) in living cells, and is much better than that of FITC when the local concentration of QDs is not too low.

---

[a] Corresponding author. E-mail: pnwang@fudan.edu.cn

INTRODUCTION

Semiconductor quantum dots (QDs), with the radius smaller than the Bohr radius, have strong size-dependent photoluminescence (PL) properties due to the quantum confinement effect of charge carriers [1]. Their size-dependent optical properties and many potential applications have been extensively studied in the past few decades [2-5]. Recently, their advantages have been exploited as potential candidates for optoelectronic devices, biological labeling, building block of a superstructure and so on. In particular, as biological labels [6-11], only the water-soluble QDs can be internalized into living cells. Thus, the preparation of water-dispersed semiconductor QDs with a high PL quantum yield (QY) becomes a significant research field.

Generally, the water-dispersed semiconductor QDs can be prepared either in aqueous solutions using thiols as stabilizing agents [12-14] or by the ligand exchange method [8,15-20]. Although semiconductor QDs prepared by an organometallic route possess excellent luminescent properties, the PL QY of QDs treated with the water-soluble ligands becomes rather low, and the PL even completely quenched after being transferred into water [17]. Via aqueous synthetic method, the PL QY of the QDs can be improved (up to 40% at room temperature) after post-preparative treatments [21,22] such as size-selective precipitation, selective photochemical etching, and surface modification. However, the typical PL QY of the as-prepared CdTe QDs lies between 3% and 10%. Recently, Zhang et al. developed the hydrothermal route to obtain water-dispersed semiconductor QDs with a narrow particle size distribution and an enhancement of PL QY (around 30% in the green-yellow emission window) under the condition of low precursor concentration [23]. This method utilized the high reaction temperature (180°C), which could accelerate the growth of CdTe QDs.

The success in producing water-soluble semiconductor QDs with high PL QY opened a novel approach of fluorescent biological labeling. Since Alivisatos [6] and Nie [9] first reported the cellular labeling with hydrophilic QDs, QDs have received widespread attention from the biological and medical fields [24-29]. Compared to the traditional cellular labeling fluorophores, semiconductor QDs have the advantages of emission tunability, superior photostability, longer PL decay times, higher PL efficiency, and narrow emissions with a broad excitation band. In recent years, along with the development of water-soluble QDs, there has been an increasing interest in their potential applications in labeling of cells, protein trafficking, DNA detection, etc. [30-35]. For those applications, it is of particular importance to study the photophysical properties of QDs in living cells. Phenomena such as photoenhanced luminescence, the photo-induced spectral blueshift, and the on/off behavior (blinking), have been observed in QD solutions as well as for single QDs [36-39]. However, few works concerning the PL behavior of QDs in living cells have been reported so far. The photo-induced optical properties of the intracellular QDs are far from being fully understood. In this chapter, the PL spectroscopy and the photostability of the thiol-capped CdTe QDs in living cells were studied and the mechanism of the photoactivation process was explored. To further improve the photostability and QY of QDs, the approach of surface treatment was explored.

PREPARATION OF WATER-SOLUBLE THIOL-CAPPED CDTE QDS

The water-soluble thiol-capped CdTe QDs used in this work were prepared by Guo et al. of the Department of Macromolecular Science, Fudan University, through a modified hydrothermal route, which was considered as a simple and highly efficient method [40]. Briefly, by a molar ratio of 2:1, sodium borohydride was used to react with tellurium in water to prepare sodium hydrogen telluride (NaHTe). Fresh solutions of NaHTe were diluted by N_2-saturated deionized water to 0.0467 M for further use. $CdCl_2$ (1 mmol) and thioglycolic acid (TGA) (1.2 mmol) were dissolved in 50 mL of deionized water followed by adjustment to pH = 9.0 in the ice-water bath. The initial mixtures of $CdCl_2$ and TGA produced white precipitates, and the pH of solution was measured as 2.3. By dropwise addition of NaOH solution, the precipitates did not dissolve until pH = 7.3. When TGA was replaced by 3-mercaptopropionic acid, the dissolution of the precursors of $CdCl_2$ and 3-mercaptopropionic acid took a long time. Continuous adding of NaOH solution adjusted the precursor solution to pH = 9, and it was consequently stored at 4 °C. Then, 0.096 mL of oxygen-free solution containing fresh NaHTe, cooled to 0 °C, was added to 10 mL of the above precursor solution and vigorously stirred. Finally, the solution with a faint yellow color was put into a Teflon-lined stainless steel autoclave with a volume of 15 mL. The autoclave was maintained at the reaction temperature (180-200 °C) for a certain time and then cooled to room temperature by a hydro-cooling process. The obtained CdTe concentration was calculated from the known concentration of NaHTe added to the precursor solution, assuming that the reaction yield of NaHTe was 100% with a large excess of Cd^{2+} ions.

The PL QY of CdTe QDs used in this work was estimated to be about 47% by comparison with Rhodamine 6G in ethanol at room temperature, assuming its PL QY as 95% [41].The absorption and PL spectra of QDs in aqueous solution are demonstrated in Figure 1.

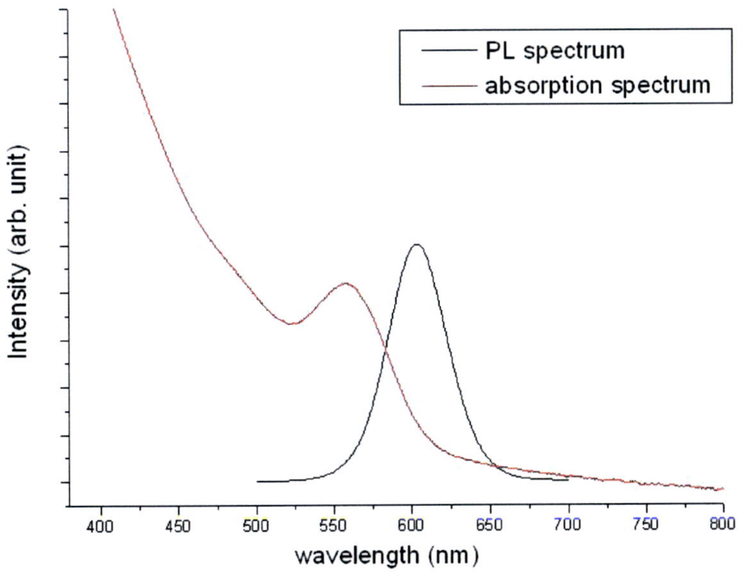

Figure 1. Absorption and PL spectra of thoil-capped CdTe QDs in aqueous solution.

The absorption spectrum was acquired using a spectrometer (Shimadzu Uv3101pc). PL spectrum was measured with a fluorescence spectrometer (HITACHI F-2500) using an output of a xenon lamp at 400 nm as the excitation source. As can be seen in the figure, the absorption band extends far into the UV region, while the PL spectrum is narrow and symmetric as compared to that from the conventional organic dyes. The observed first exciton absorption peak is at 558 nm and the exciton emission peak is at 603 nm. Comparing to the bulk band gap of CdTe (1.54 eV), the semiconductor core size is estimated to be around 3.5 nm in diameter according to the quantum confinement effect [42]. This size is small enough for the uptake of QDs into the living cells.

Time-Dependent PL blue shift of CdTe QDs in Living Cells

When studying the optical properties of the colloidal QDs, investigations have been focused on their PL behaviors. Phenomena such as photoenhanced luminescence, the photoinduced spectral blue shift, and the on/off behavior (blinking), have been observed in QD solutions as well as for single QDs [36-39]. Subsequently, many theories have been developed to explain these phenomena [43-45]. For biological and medical applications, it is of particular importance to study the photophysical properties of QDs in living cells. However, the photoinduced optical properties of the intracellular QDs are far from being fully understood. In this work, the PL spectroscopy of the thiol-capped CdTe QDs in living cells was studied. Time-dependent PL enhancement, spectral blue shift and photobleach of the intracellular QDs were observed and the mechanism of the photoactivation process was explored.

Uptake of QDs into Living Cells

The uptake of QDs into living cells was carried out as follows: The mouse myoblast cells (C2C12) or human primary liver cancer cells (PLC) were seeded onto a glass cover slip placed in a culture dish containing DMEM-H medium with 10% calf serum, 100 units/mL penicillin, 100 μg/mL streptomycin and 100 μg/mL neomycin. The cells were then cultured in a fully humidified incubator at 37 °C with 5% CO_2. When the cells adhered to the cover slip and reached 80% confluence with normal morphology, the CdTe QD aqueous solution was added into the culture dish to achieve a QD concentration of 100 μg/mL. These cells were then incubated for 4 h in an incubator. After incubation, the cover slip with the adhered living cells was washed with PBS (phosphate-buffered saline) for three times to remove the unbound QDs and was then sealed on a glass slide for microscopic measurement. With such a QD concentration in solution, no detectable damage to cells was observed.

Intracellular Distribution of QDs

The intracellular QD distributions were detected with a laser scanning confocal microscope (Olympus, FV-300, IX71) using a 405 nm semiconductor laser (Coherent) as the excitation source. The laser beam was focused by a 60× objective to a spot of about 1 μm in diameter. The QD distributions in C2C12 and PLC cells are depicted in Figures 2 and 3, respectively. Using the z-scan mode, the images in different layers were recorded to detect the distributions of QDs inside the cells. From those images, the three-dimensional intracellular distribution of QDs was obtained. In both Figures 2 and 3, A shows the fluorescence micrographs recorded with a 585-640 nm band-pass filter, where the main image exhibits the QD distribution in an X-Y plane inside the cell and the images displayed below and at the right side of the main one are the X-Z and Y-Z profiles obtained by z-scanning along the marked lines in the main image, showing the cross sections of the cell; B shows the differential interference contrast (DIC) micrograph to exhibit the cell morphology, which was acquired in a transmission channel simultaneously. C shows the merged image of A and B.

The QDs were found to distribute diffusely inside the cells according to the three dimensional micrographs (Figures 2A and 3A), indicating that the uptake of QDs into the living cells really occurred.

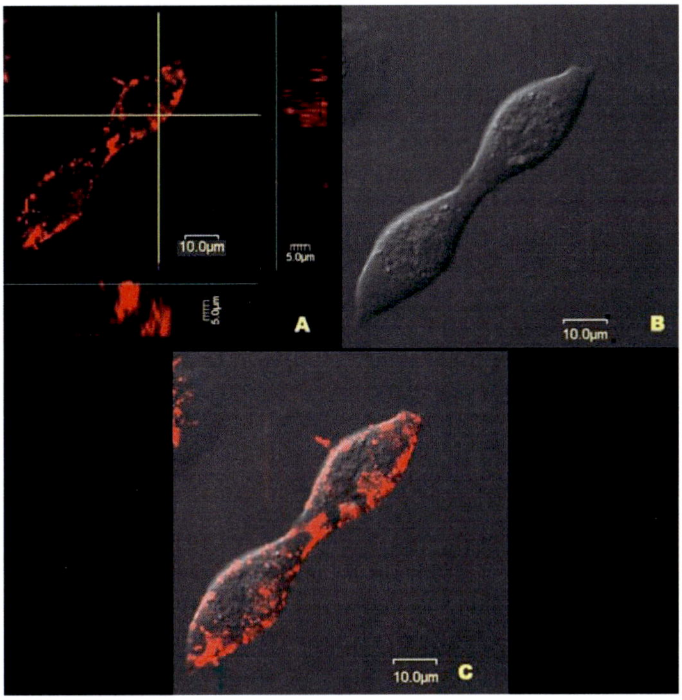

Figure 2. Micrographs of a QD labeled C2C12 cell. (A) Micrographs of QD distributions. The main image exhibits a QD distribution in an X-Y plane inside the cell. The images displayed below and on the right side of the main one are the X-Z and Y-Z profiles measured by z-scanning along the marked lines in the main image; (B) The DIC micrograph; (C) The merged image of (A) and (B). (Reproduced with permission from J. Am. Chem. Soc. 2006, 128, 13396-13401. Copyright 2006 American Chemical Society).

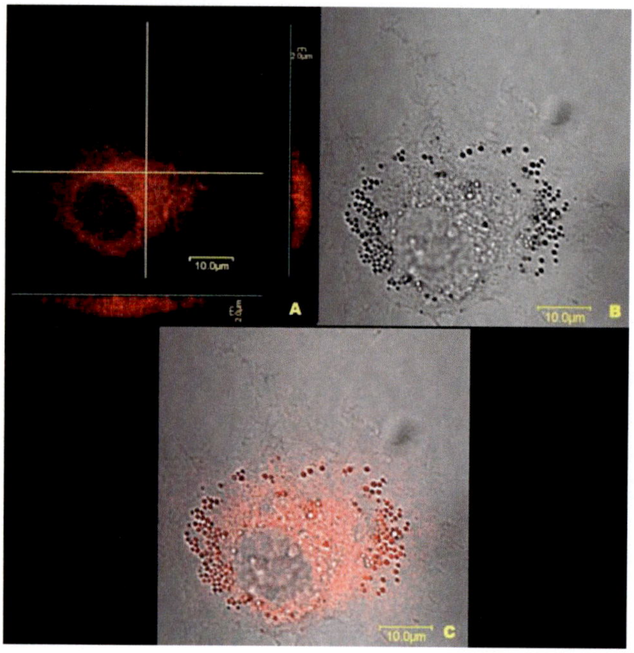

Figure 3. Micrographs of a QD labeled PLC cell. (A), (B) and (C) represent the images as described in Figure 2. (Reproduced with permission from *J. Am. Chem. Soc.* 2006, 128, 13396-13401. Copyright 2006 American Chemical Society).

Time-Dependent PL Behavior of QDs in Living Cells

On the basis of the obtained micrographs, we selected some parts in the cells to measure the micro-PL spectra of the intracellular QDs using the point-stay mode of the microscope. In this mode, the focused laser beam was fixed at the selected spot of the sample and the fluorescence spectrum was recorded using a photomultiplier tube (PMT) with the appropriate filter sets. The spot size of the laser beam focused by a 60× objective lens was about 1 μm in diameter. The laser power on the sample was measured with a laser-power meter (Coherent), and the irradiation power density could be calculated. The excitation power density used was about 250 W/cm^2. The spectra were measured using a spectrometer (Acton, Spectropro 2150i) equipped with a liquid-nitrogen-cooled CCD (Princeton, Spec-10:100B LN). The PL output from the side exit of the microscope was directly focused onto the entrance slit of the spectrometer.The PL spectra of the intracellular QDs with different irradiation times are demonstrated in Figures 4A and 4B for C2C12 and PLC cells, respectively. As an ensemble luminescence from the QD aggregates, the spectra are relatively broader as compared with that from single-dot [37,45]. In both kinds of cells, the spectra recorded at the later times were clearly blue shifted with respect to the initial wavelength of 602 nm (peak wavelength). The PL intensity was slightly enhanced first (within 10 s) and then decreased with time. A blueshhift of about 35 nm was observed for an irradiation time of 15 min before the emission was fully photobleached.

The PL blue shift and photobleach of QDs were previously observed in QD solutions as well as for single QDs and were believed to be attributed to the irreversible photooxidation of

QDs [37,45-47]. As a result of photooxidation, macroscopic oxide layers were created and the effective QD size was reduced with time. The quenching states were expected to be formed at the interface between CdTe and the surface oxide [45], and new surface defects were also possible to be generated by oxidation of the QD surface. Hence, the PL intensity decreased and finally the QDs were bleached [48]. On the basis of the well-known quantum confinement effect [42,49], the observed blue shift of approximately 35 nm in this work corresponds to a shrinkage in the diameter of about 0.33 nm. The PL enhancement of colloidal QDs was observed previously in pure QDs [50], aqueous and nonaqueous solutions [36,47], QD films between insulating layers [51] and QD monolayers [48]. Various mechanisms were proposed to explain this effect [36,46-48,52]. Regardless of different models with complex kinetics presented, the primary reason for photoenhancement is believed to be the passivation of the core surface.

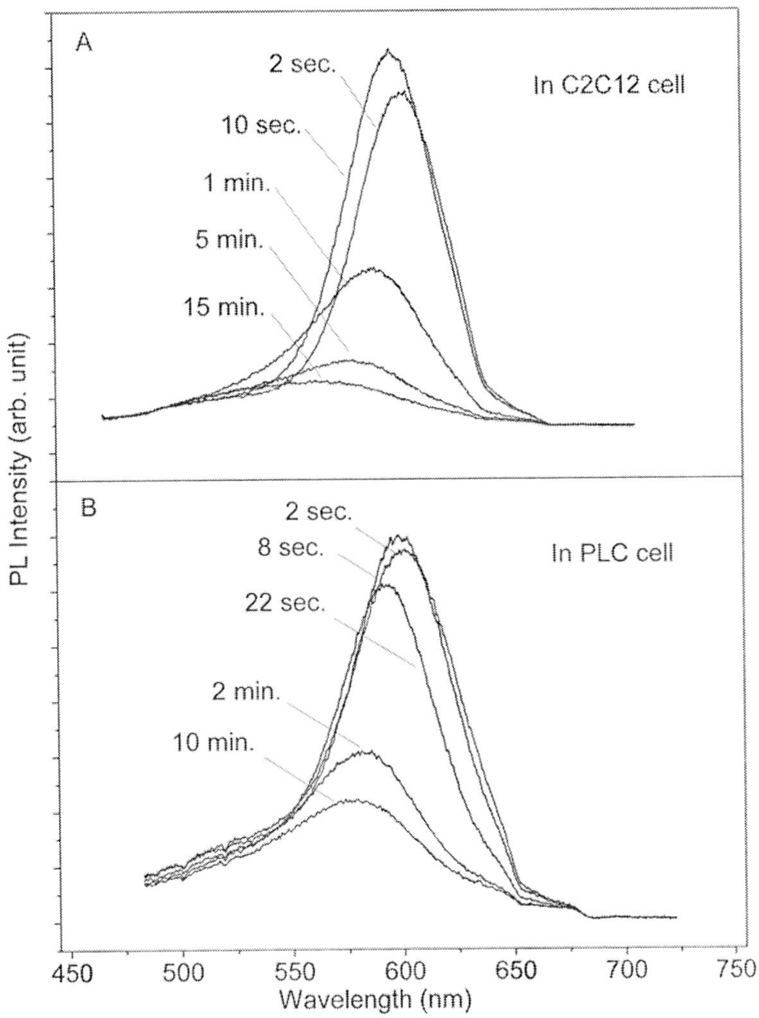

Figure 4. Time evolutions of the PL spectra of CdTe QDs in a C2C12 cell (A) and a PLC cell (B). (Reproduced with permission from J. Am. Chem. Soc. 2006, 128, 13396-13401. Copyright 2006 American Chemical Society).

Passivation of the surface states that are not blocked by the capping agent causes the PL enhancement, which does not change the shape and position of the PL peak [46,47]. Although the precise nature of the passivation process is not yet fully understood, with the effects of passivation and photooxidation, the time evolutions of the PL spectra observed in this work can be well explained. Passivation of surface traps enhances luminescence, and the oxidation of the QD cores reduces the QD size and introduces quenching states and new defects, leading to the blue shift and photobleach. It is the competition between the two processes that is responsible for the complex kinetics of the luminescence observed [46,48].

Origin of the PL Blue Shift

Although the time-dependent PL behavior observed in living cells can be explained by photooxidation, it is necessary to further confirm this effect directly by experiments. Therefore, the QD-loaded C2C12 cells were bubbled with nitrogen gas for 30 min to purge away oxygen molecules. The samples for microscopic measurement were then prepared in a glovebox filled with flowing pure nitrogen gas. A time evolution of the PL spectra from a deoxygenated C2C12 cell is shown in Figure 5. Contrary to the situation without deoxygenation (as shown in Figure 4 A), all the spectral peaks were almost at the same wavelength. No significant shift was detected. The PL intensity was enhanced slightly at the very beginning and then kept constant. These results provide unambiguous evidence that the photooxidation played a dominant role in the spectral blue shift and luminescence quenching. The photooxidation is likely a universal effect, whatever for single QDs [37,45], the monodisperse QDs in solutions [47], or aggregate QDs in living cells. Removing of oxygen molecules can therefore prevent the photooxidation of QDs effectively.

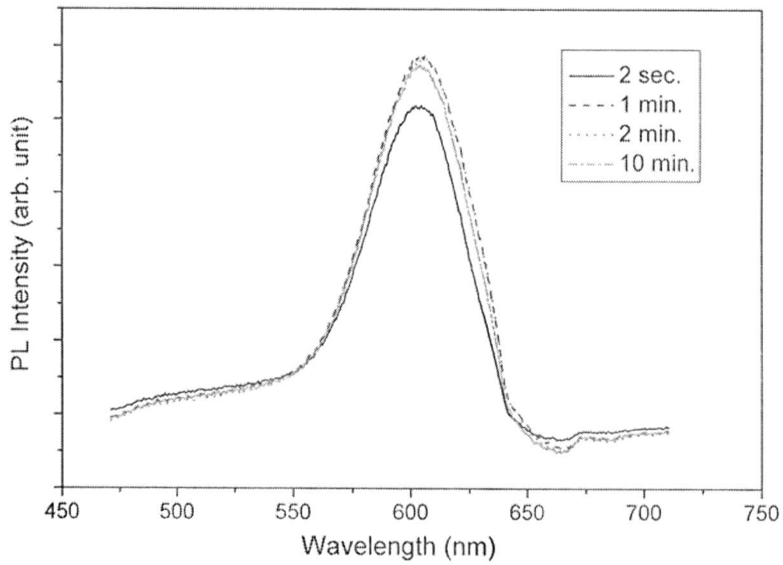

Figure 5. Time evolution of the PL spectra of CdTe QDs in a C2C12 cell deoxygenated by purging with nitrogen. (Reproduced with permission from J. Am. Chem. Soc. 2006, 128, 13396-13401. Copyright 2006 American Chemical Society).

However, the photoenhancement of PL in C2C12 cells was observed at the beginning of the illumination in either air or a nitrogen atmosphere (Figures 4 and 5). Similar results were reported in QD aqueous solutions by Wang et al. [47]. They suggested that the slight PL enhancement in N_2 was attributed to the presence of residual amount of oxygen.

To eliminate the influence of the complex environment in cells, the PL spectroscopy of QDs was studied in a deoxygenated aqueous solution. A QD aqueous solution (1mg/mL) was bubbled with nitrogen gas for 30 min to purge away oxygen and then sealed between two cover slips in a nitrogen atmosphere. The PL spectra of this prepared sample were also measured with the confocal microscope but at a higher power density of 3.7 kW/cm^2. The PL spectra as a time evolution are shown in Figure 6. As expected, no significant enhancement, blue shift or photobleach was observed. On the contrary, without purging with nitrogen (as can be seen later in Figure 13A), the PL spectra from the QD aqueous solution were remarkably blue shifted and photobleached with irradiation time. These results strongly support our above explanations.

To identify the nature of the products formed by photooxidation, the QD powders extracted from the solutions before and after laser irradiation were analyzed by X-ray photoelectron spectroscopy (XPS). The solutions were dripped on silicon wafers and dried in a vacuum chamber. The dried samples were carefully kept in a nitrogen atmosphere until XPS measurements were conducted. The measured high-resolution XPS spectra of the Te $3d_{5/2}$ and $3d_{3/2}$ peaks before and after laser irradiation are shown in Figure 7. It can be seen clearly in the figure that after laser irradiation, two additional peaks appeared at the binding energies of 576.0 and 586.4 eV, corresponding to the Te-O bonding states in $CdTeO_3$ [53].

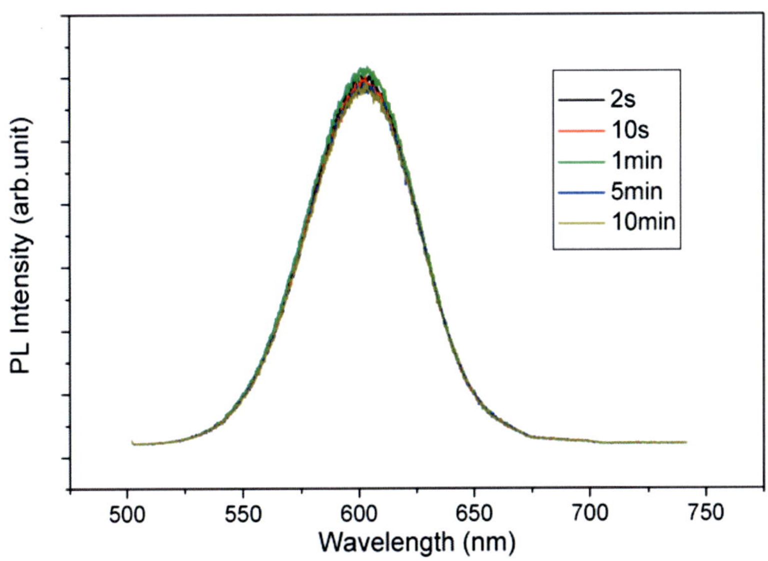

Figure 6. Time evolution of the PL spectra of CdTe QDs in aqueous solution deoxygenated by purging with nitrogen. (Reproduced with permission from J. Am. Chem. Soc. 2006, 128, 13396-13401. Copyright 2006 American Chemical Society).

This information further supports the above conclusion that photooxidation reaction occurred during laser irradiation. Since the difference of the binding energies between the Cd 3d peaks of CdTe and CdO is only 0.1 eV [53], with the limitation of the XPS resolution, this small chemical shift cannot give an explicit conclusion about the formation of CdO although we have in fact detected a 0.2 eV shift to the lower energy.

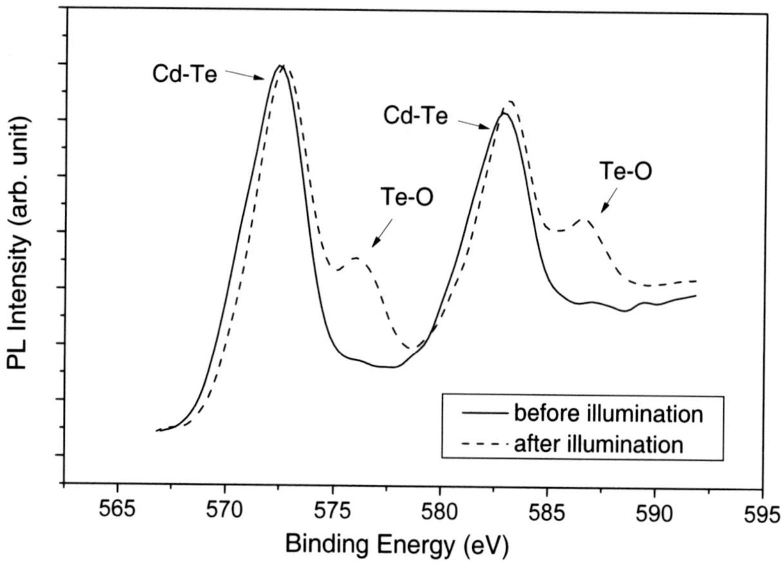

Figure 7. High-resolution XPS spectra of Te 3d5/2 and 3d3/2 peaks from CdTe QDs before (solid line) and after (dashed line) laser irradiation. (Reproduced with permission from J. Am. Chem. Soc. 2006, 128, 13396-13401. Copyright 2006 American Chemical Society).

However, since CdO and SeO_2 were proposed to be produced in the oxidation reactions of CdSe QDs [20,54,55], CdO and TeO_2 are therefore also possible products from the oxidation reactions of CdTe QDs.To gain further insight into the photooxidation kinetics, additional experiments were conducted with histidine, sodium azide (NaN_3) and deuterated water.

Histidine is known as a scavenger of singlet oxygen ($^1O_2$). It has been revealed that histidine can react directly with $^1O_2$ at a very high reaction rate [56-58]. The effect of histidine was first studied in QD aqueous solutions. A 10 mg amount of histidine powder was dissolved in 5 mL of water to make the histidine stock solution. A 2 mL amount of QD aqueous solution (1 mg/mL) was added with the histidine stock solution to reach a histidine concentration of 3.2 mM. The irradiation power density was 3.7 kW/cm^2. The PL spectra from the histidine-added solution are illustrated in Figure 8 as a time evolution. It is distinct that no blue shift was observed.

To prepare histidine-treated cell samples, histidine stock solution was added to the cell dish that containing QD-loaded C2C12 cells to reach a histidine incubation concentration of 3.2 mM. The cells were then incubated for 10 min. Figure 9 shows the time evolution of the PL spectra from the histidine-treated cells. Only a blue shift of 4 nm was observed for an illumination time of 100 s and no further shift occurred afterward. This small blue shift was likely due to the insufficient histidine-$^1O_2$ reactions inside the cell. Comparing with the blue shift of 35 nm in the C2C12 cells without histidine treatment (Figure 4A), these data (in cells as well as in solutions) reveal that the singlet oxygen played a dominant role in the photooxidation of QDs.

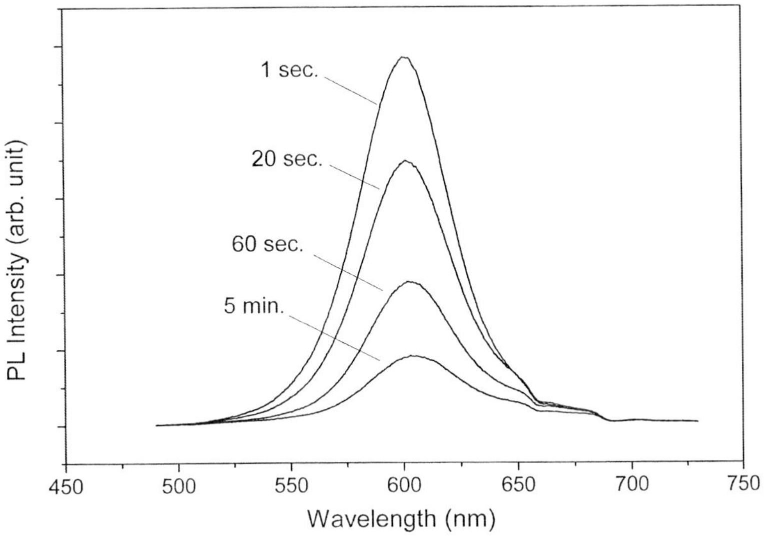

Figure 8. Time evolution of the PL spectra recorded from the QD aqueous solution with histidine added. (Reproduced with permission from J. Am. Chem. Soc. 2006, 128, 13396-13401. Copyright 2006 American Chemical Society).

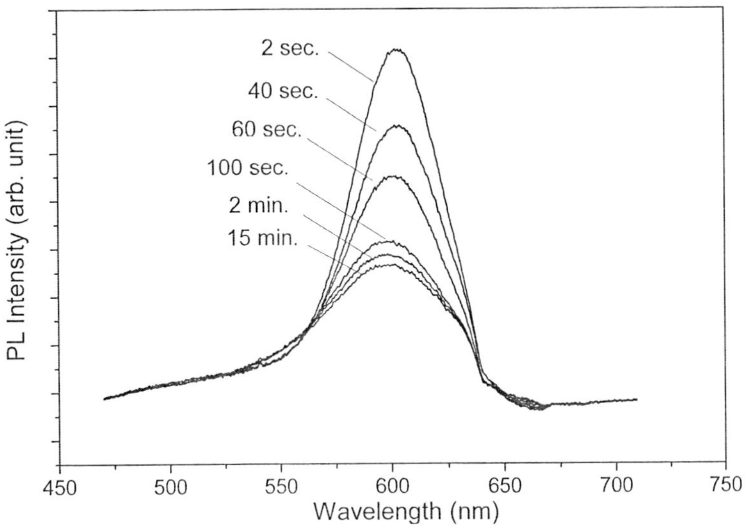

Figure 9. Time evolution of the PL spectra recorded from the QDs in a C2C12 cell treated with histidine. (Reproduced with permission from J. Am. Chem. Soc. 2006, 128, 13396-13401. Copyright 2006 American Chemical Society).

It was reported that semiconductor QDs alone could actually generate photoinduced $^1O_2$ species without a mediating molecule [59]. The QDs are strong absorbers due to their large transition dipole moment. As primary energy donors, they transfer their absorbed photoenergies to the nearby oxygen molecules to form reactive singlet oxygen species, which subsequently react with the QDs to form oxides [59]. When histidine is added in, it reacts rapidly with singlet oxygen and effectively hinders the oxidation of QDs to occur. Although the singlet oxygen quantum yield is low (only 5% for CdSe QDs as reported by Burda et al.

[59]), in our experiments, the laser beam was focused by a 60× objective and the excitation power densities used were about 250 W/cm^2 in cells and 3.7 kW/cm^2 in solutions, hence the generation of enough singlet oxygen for the localized photooxidation on the QD surface was reasonable.

However, in each of the histidine-treated samples, photobleach was still observed (Figures 8 and 9). To find out the origin of this un-blue-shifted photobleach, a QD aqueous solution was added with histidine and bubbled with nitrogen gas. In the measured PL spectra, no blue shift was observed but photobleach still occurred as shown in Figure 10. Note that this situation is entirely different from that shown in Figure 6 where the solution was also deoxygenated but histidine was not added and no photobleach was observed. It is likely that this photobleach resulted not from the photooxidation but from the histidine added. Histidine is a weak acid. The pH value of the QD solution was measured to vary from 7 to 6 when histidine was added in. As was reported, the pH value in the QD solution influenced the luminescence efficiency [47,60]. The acid in the solution caused the dissociation of organic ligands from the QD surface and therefore reduced the PL efficiency [60]. Laser irradiation might accelerate this process.

We must note that other potential reactive oxygen species (ROS) formed by a Type I photooxidation mechanism could also be responsible for the blue shift. These ROS would also react with histidine. To clarify this issue, the experiments with NaN_3 were conducted in solutions as well as in cells. NaN_3 is an efficient physical quencher of $^1O_2$ [61,62]. The presence of a physical quencher of singlet oxygen would not remove Type I intermediates. Figure 11 shows the time evolution of the PL spectra of a QD aqueous solution with a NaN_3 concentration of 20 mM. The irradiation power density applied was also 3.7 kW/cm^2. It is distinct that no blue shift was observed.

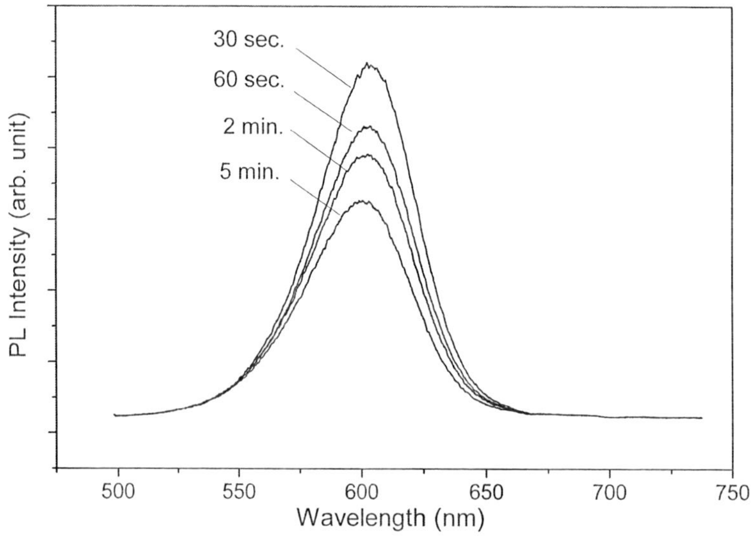

Figure 10. Time evolution of the PL spectra of the QD aqueous solution bubbled with nitrogen gas and added with histidine. (Reproduced with permission from J. Am. Chem. Soc. 2006, 128, 13396-13401. Copyright 2006 American Chemical Society).

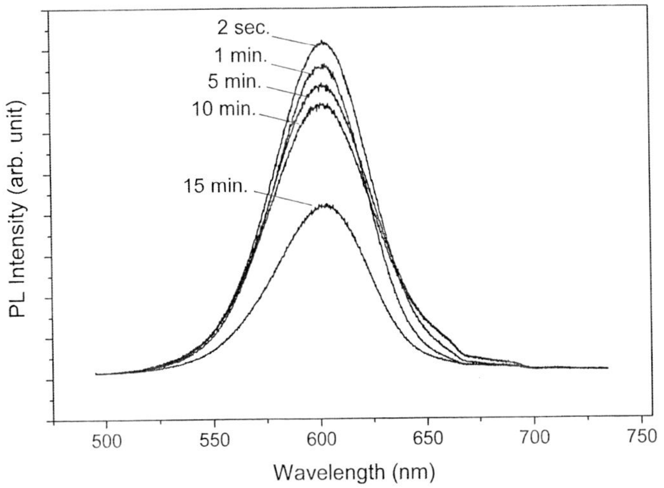

Figure 11. Time evolution of the PL spectra from the QDs in aqueous solution with NaN3 added. (Reproduced with permission from *J. Am. Chem. Soc.* 2006, 128, 13396-13401. Copyright 2006 American Chemical Society).

This result definitely supports our above conclusion that the blue shift resulted from the photooxidation with singlet oxygen. Nevertheless, the PL was quenched with time. Similar quenching was also observed when the NaN_3-added QD solutions was bubbled with nitrogen gas, suggesting that this kind of quenching was not associated with oxidation but was most probably due to the photoinduced reaction of the organic ligands with NaN_3 although the details are still unknown. The NaN_3 effect was also studied in cells. The C2C12 cells were incubated with QDs and NaN_3 for 4 h. The QD incubation concentration was 100 μg/mL. The incubation concentration of NaN_3 was 10 mM, which is commonly used for cells [63,64]. The time evolution of the PL spectra in a NaN_3-treated C2C12 cell is illustrated in Figure 12.

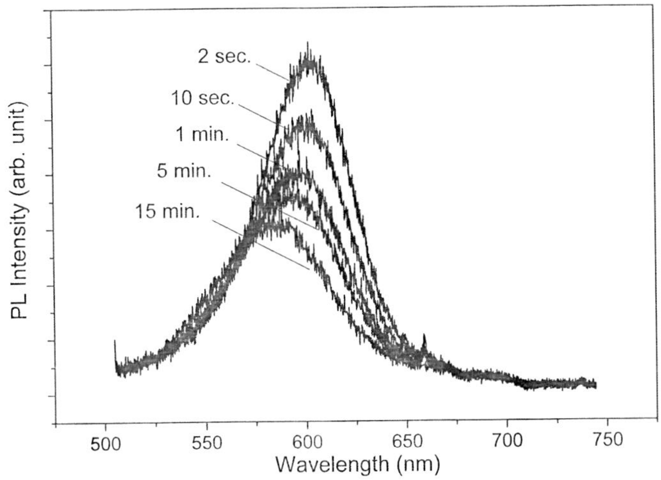

Figure 12. Time evolution of the PL spectra from the QDs in an NaN3-treated C2C12 cell. (Reproduced with permission from *J. Am. Chem. Soc.* 2006, 128, 13396-13401. Copyright 2006 American Chemical Society).

The blue shift was still showed but slowed as compared with the untreated sample (Figure 4A). The blue shift was 18 nm for an irradiation time of 15 min., while it was 35 nm when the cells were not treated with NaN_3 (Figure 4A). Since there was no blue shift observed in solutions, where the NaN_3 concentration was higher, these blue shifts were most likely due to the low intracellular concentration of NaN_3, although it is difficult to measure the actual NaN_3 concentrations inside the cells. Increasing the intracellular concentration of NaN_3 may inhibit the blue shift further, but that would cause the cytotoxic effect on cells.

On the other hand, it is known that the lifetime of the singlet oxygen in deuterated water (D_2O) is 10 times longer than that in water [65]. If the singlet oxygen molecules are generated directly on the QD core surface, the lifetime of singlet oxygen will not be an important factor for the oxidation efficiency, otherwise, the longer lifetime would increase the oxidation probability effectively. Figure 13 shows the time evolutions of the PL spectra from QDs in water solution (A) and in deuterated water solution (B), respectively.

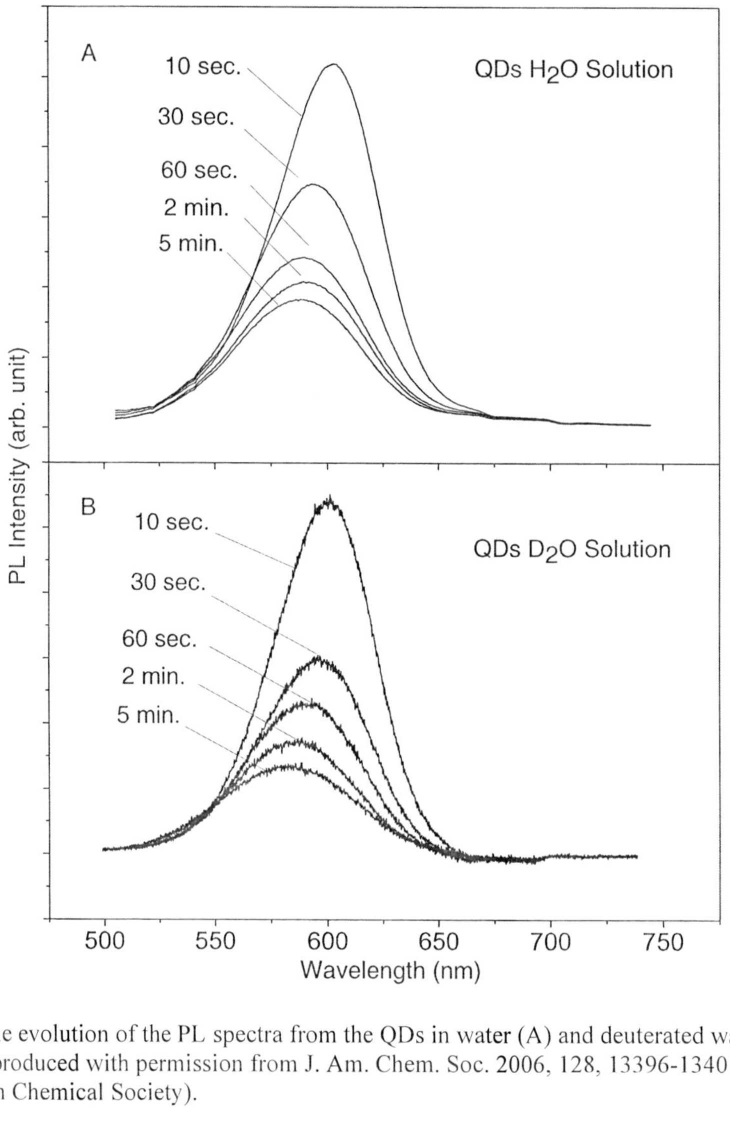

Figure 13. Time evolution of the PL spectra from the QDs in water (A) and deuterated water (B) solutions. (Reproduced with permission from J. Am. Chem. Soc. 2006, 128, 13396-13401. Copyright 2006 American Chemical Society).

The solutions were prepared by adding 0.1 mg of CdTe QD powders into 5 mL of H_2O and 5 mL of D_2O, respectively. It is found that the rate of the blue shift in D_2O is only a little faster than that in H_2O. These data support our above hypothesis that the singlet oxygen is formed on the QD surface by energy transfer from the QDs. The O_2 molecules involved in oxidation are likely to intercalate in the thiol layer at the core surface as suggested by Burda et al. [59].

Photostability of CdTe QDs in Living Cells

An important feature of the hydrophilic photo-luminescent semiconductor QDs is their photostability in biological applications. Although the photostability of semiconductor QDs was reported to be much better than Rhodamine and FITC [6,9,66], the photochemical instability of the thiol-capped QDs was also observed and the mechanism is not yet well understood [67,68]. Therefore, it is of great interest to study the mechanism of the photoinduced instability of the water-soluble semiconductor QDs. In this work, the photostabilities of thiol-capped CdTe QDs in two kinds of living cells were studied and compared with that of FITC, GFP and native chlorophyll. We found that the photostability of the cellular QDs was dependent on their local concentration as well as the irradiation intensity. The photonbleaching mechanism of the cellular QDs was shown to be the oxygen dependent photooxidation.

Incubation of Living Cells with QDs

Euglena gracilis 277 (EG 277) cells were cultured in normal culture conditions. Briefly, the cells were cultured in a 250 mL flask containing 100 mL liquid culture medium. The cell flask was exposed in 12 h irradiation (2300 lx)/12 h dark cycles at 25 °C for 48–72 h. To prepare the cell samples, the cells were centrifuged three times in deionized water to remove the culture medium. The washed cells were then re-suspended in water, and incubated with QDs (0.1–0.3 mg/mL) for one day. After incubation, the cell samples were further washed by three cycles of centrifugation to remove the unbound QDs in the solution. The QD treated cells were suspended in water again with a cell density of 106 /mL, and were ready for measurement. The tolerance of EG 277 cells to QDs is very strong. Even when incubated with QDs at a concentration of 0.5 mg/mL, no detectable damage to cells was found.

The HEK 293 cells, originally derived from a human embryonic kidney, were cultured in DEME medium containing 10% foetal calf serum, 100 units/mL penicillin, 100 μg/mL streptomycin, and 100 μg/mL neomycin in a fully humidified incubator at 37 °C with 5% of CO_2. The cells were subcultured in tissue culture flasks to keep them in an exponential growth phase for use in experiments.

To stain the cells with QDs, the cells in culture flasks were incubated with QDs (0.15 mg/mL) in DMEM medium for 4 h in an incubator. Then, these QD loaded cells were washed with PBS for three times and fixed with 10% formol (pH = 7.4) for 10 min at room temperature. After washing again, these cell samples were ready for the fluorescence experiments. The HEK 293 cells are fragile to the QDs of high incubation concentration (over

0.5 mg/mL). Under incubation conditions in this experiment, the cells were in the normal state.

For GFP labeling in 293 cells, the EGFPC1 (Clontech), an expression vector containing an enhanced green fluorescent protein coding region, was transfected into HEK 293 cells using liposomal transfection reagent in accordance with the standard protocol [69]. Transfection was carried out in a 35 mm culture dish by adding 1 μg of plasmid DNA and 3 μg of Fugene 6 liposomal transfection reagent (Boehringer) according to the manufacturer's instructions. The transfected cells were checked by a fluorescence microscope.

FITC labelling of cells was conducted with immunostaining. The HEK 293 cells adhered on slide glass were incubated with phalloidin-FITC for 30 min, and then washed and mounted with a drop of 50% glycerol.

Cellular Images

The three dimensional cellular images were acquired with a confocal laser scanning microscope as described above. The intracellular distributions of QDs, FITC or GFP could be obtained at the same time by recording their luminescence or fluorescence signals in different channels using a set of filters. From these images, the intracellular distribution of QDs and the relative luminescence intensities were obtained.

In EG 277 cells, the chlorophyll (mainly chlorophyll a- and chlorophyll ab-proteins [70]) is the native fluorophore, which emits fluorescence peaked around 685 nm, while the observed PL of cellular QDs in EG 277 is peaked around 605 nm. When the cells were excited with a 488 nm laser beam in a confocal microscope, the fluorescence images of cellular QDs and chlorophyll were recorded simultaneously in channel 1 with a 585–630 nm band-pass filter and in channel 2 with a 630 nm long-pass filter, respectively. Figure 14 shows the images obtained in the two channels as well as the DIC image acquired simultaneously in a transmission channel. The cells have been incubated with 0.2 mg/mL QDs for one day. Figure 14A demonstrates the intracellular distribution of QDs in an EG 277 cell, where the main image shows the cellular QDs in the x–y plane and the upper profile exhibits the QD distributions in the x–z plane along the marked line in the main image obtained by z-scanning. The QDs were found to distribute diffusely inside the cell, but no QDs were detected in the outer membrane of the cell.

The intracellular distribution of QDs in HEK 293 cells depicted in Figure 15 shows that the cellular uptake of QDs also occurred. Compared with EG 277 cells (Figure 15b), the distribution patterns are different. The QDs are more uniformly distributed in the EG 277 cells, but are patchy in appearance in the cytoplasm of HEK 293 cells, which might be due to their different cellular structures. Figure 15d shows the QD distribution in x–y and x–z planes in an HEK 293 cell, demonstrating that the QDs are not only bound to the plasma membrane but also localized in the cytoplasm of the cell. Note that the local brightness in the image is proportional to the local concentration of QDs. Under the same measuring conditions, the luminescence brightness in HEK 293 cells (Figure 15a) is much lower than that in EG 277 cells (Figure 15b), corresponding to a lower local concentration of QDs in HEK 293 cells.

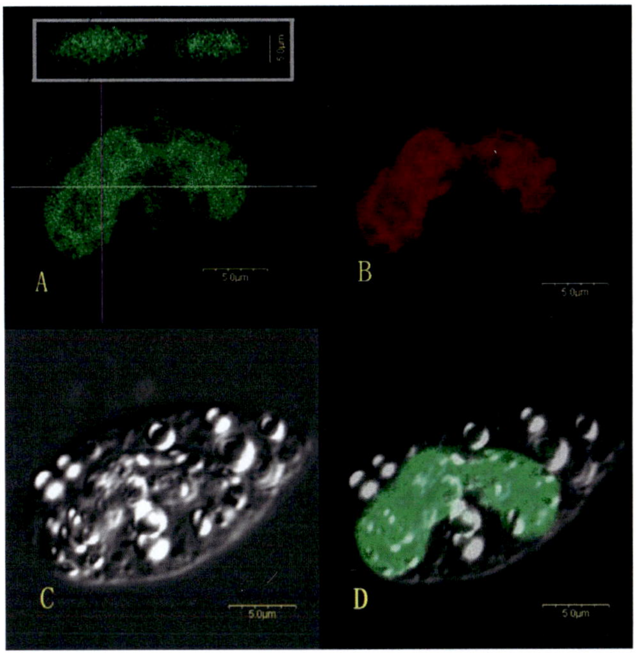

Figure 14. Micrographs of an EG 277 cell. (A) PL images of cellular QDs in x–y (main one) and x–z (upper one) planes. (B) Fluorescence image of chlorophyll; (C) the DIC image. (D) The merged image of (A) and (C).

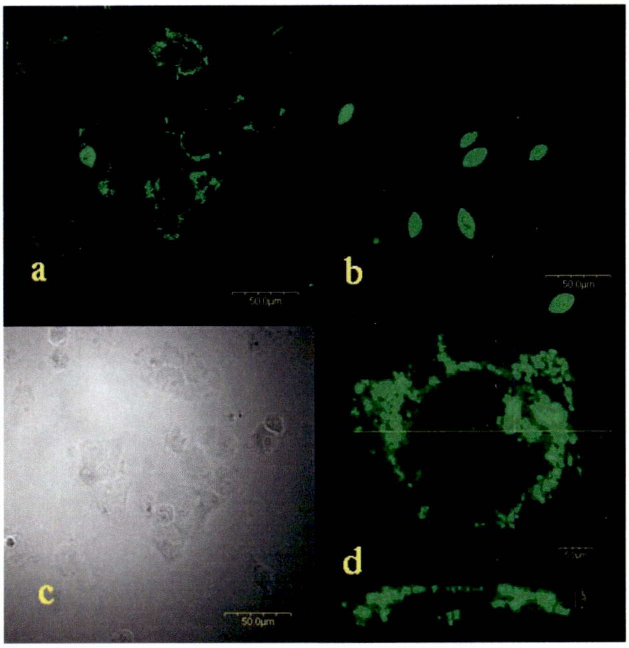

Figure 15. PL images of cellular QDs. (a) HEK 293 cells; (b) EG 277 cells; (c) DIC image of the HEK 293 cells in (a); (d) luminescence images in x–y (main one) and x–z (bottom one) planes from the cellular QDs in an HEK 293 cell. Excitation, 488 nm; filter, 585–630 nm.

The average cellular QD concentration was estimated by the method of alkali extraction as the following procedures. After incubation with QDs, the cells were washed three times to remove unbound QDs and then digested. The cell density in each sample suspension was counted in the microscope with the cell-counter. These cell samples were then incubated in NaOH solution (0.1 M) for 2 h to thoroughly destroy the whole cells as well as the cellular organelles and extract the cellular QDs into the solutions. These sample solutions were centrifuged, and the QD containing supernatant in each sample was collected, respectively, and the average amount of cellular QDs per cell can be calculated according to the cell density counted. Meanwhile, the PL spectra and the intensities at the spectral peak of these supernatants were measured by a fluorescence spectrophotometer (Hitachi, F-2500) using the excitation wavelength of 400 nm. A calibration curve of the peak intensity as a function of QD concentration was made as shown in Figure 16. With this calibration curve, the local concentration of QDs in a living cell can be determined by comparing the PL intensity with that in the curve.

Since the cell numbers of the samples in Figures 14 and 15 were already countered to be 1.2×10^6 cells/mL and 1.1×10^6 cells/mL for EG 277 and HEK 293 supernate, respectively, the cellular amounts of QDs were finally estimated to be 8.75×10^{-7} and 2.32×10^{-7} μg/cell (namely 2.3×10^6 and 6.1×10^5 QDs per cell) for EG 277 and HEK 293, respectively. The cellular QD amount in HEK 293 cells is much lower than that in EG 277 cells, but is comparable to that in Chinese hamster ovary cells [71]. The average volume of a single HEK 293 cell is about 1200 μm^3, which is larger than that of an EG 277 cell (470 μm^3). The average QD concentration in EG 277 cells was calculated to be 1.86 mg/mL, which was much higher than the incubation concentration of QDs (0.2mg/mL), indicating that the cells actively engulfed QDs during the incubation.

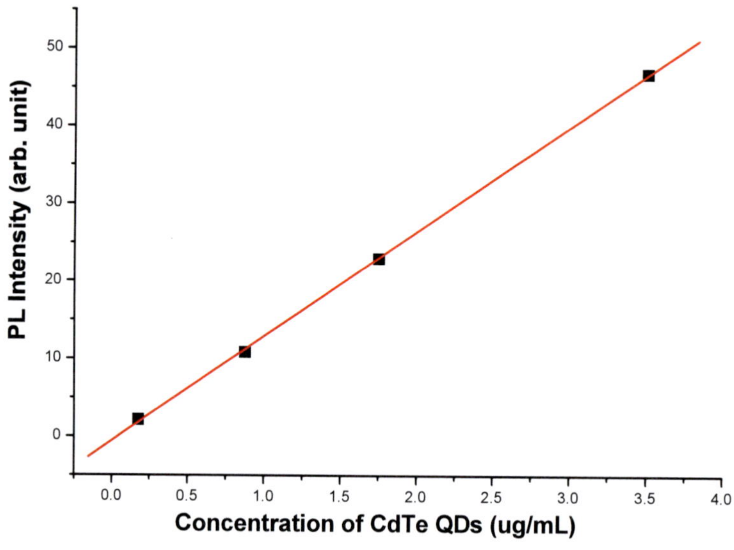

Figure 16. Calibration curve demonstrates the dependence of the luminescence intensity on the QD concentration.

This phenomenon is not so surprising since the EG cells tend to internalize the outside materials as reported previously [72]. The average QD concentration in HEK 293 cells was estimated to be 0.19 mg/mL. Compared with the QD incubation concentration of 0.15 mg/mL, the active internalization of QDs by HEK 293 cells might also have occurred. For these mammal cells, endocytosis was suggested as the mechanism of the internalization [73].

Photostability of Cellular QDs

Based on the images shown in Figures 14 and 15, we selected some parts in the cell to study photobleaching of QDs and cellular fluorophores. The photobleaching was measured with the aforementioned point-stay mode of the scanning microscope. The time-dependent fluorescence intensity was recorded by a photomultiplier tube with an appropriate filter set. The typical photobleaching curves of the chlorophyll and QDs in an EG277 cell, under an excitation power density of 8×10^4 W/cm^2, are shown in Figure 17. Both the cellular QDs and chlorophyll were photobleached as a function of irradiation time but the QDs seem more stable than the native chlorophyll.

The dependence of QD photobleaching in the EG 277 cells on the irradiation power density was shown in Figure 18. As can be seen, under a low power density of 8×10^3 W/cm^2, the QDs were very stable. When the power density was increased to 6.4×10^5 W/cm^2, the QDs were bleached in a few seconds, indicating that the irradiation power density plays an important role in photobleaching. Thus, for the imaging measurements, over-irradiation should be avoided.

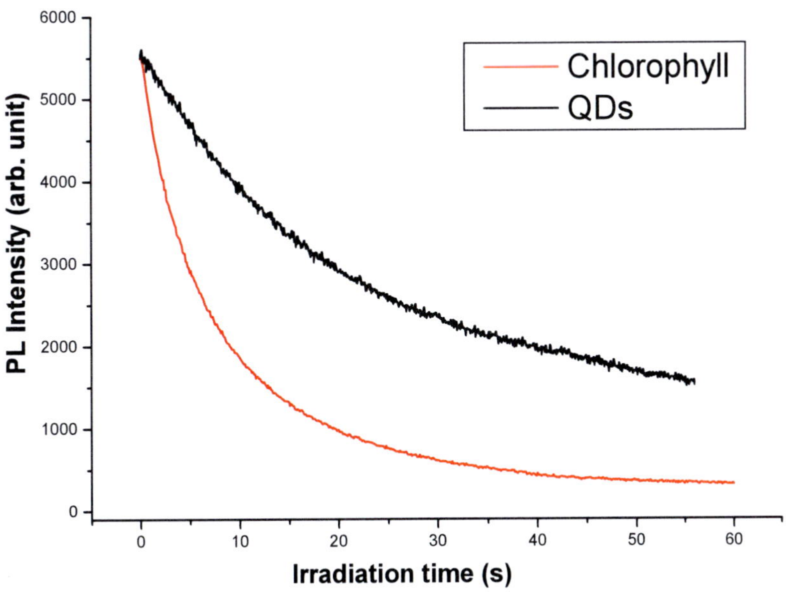

Figure 17. Photobleaching of QDs and Chlorophyll in an EG 277 cell as a function of irradiation time.

The dependence of photobleaching of cellular QDs on the QD local concentration was also studied in this work. The EG 277 cells with different cellular QD concentrations were obtained by incubating the cells with different QD concentrations. The relative cellular QD concentrations were determined by measuring the PL intensities for each cell group as shown in Table 1.

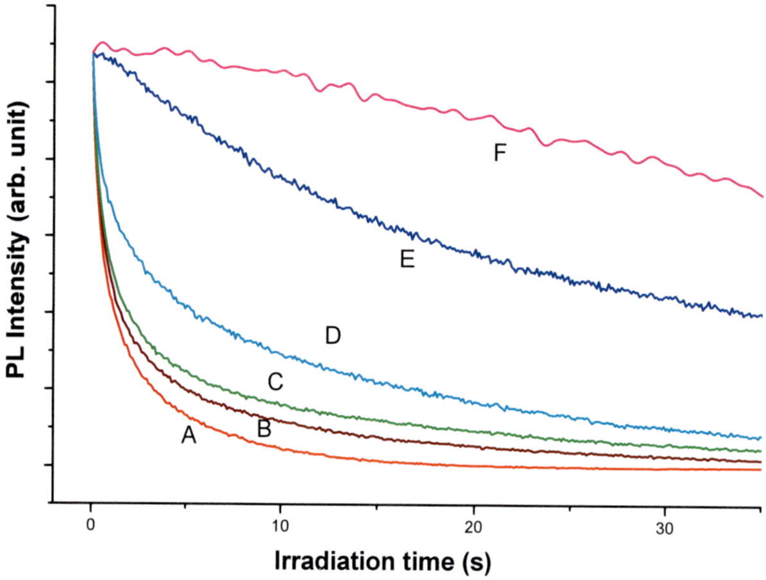

Figure 18. Dependence of photobleaching of QDs in EG 277 cells on the irradiation power density. The irradiation power densities (488 nm) are 6.4×105 (A), 4.8×105 (B), 3.2×105 (C), 1.6×105 (D), 8×104 (E), 8×103 (F) W/cm2.

The cellular QD concentration increased with the incubation concentration, but tended to saturate in an incubation concentration over 0.2 mg/mL. The photobleaching curves with different QD cellular concentrations in EG 277 cells are shown in Figure 19, demonstrating that the photobleaching was slower when the cellular QD concentration was higher. The photobleaching times (the time for PL intensity decreased to half) are also listed in Table 1, showing further that the average bleaching time in cells with higher cellular concentration is significantly longer than that of the cells with the lower cellular concentration (24.0 ± 7.7 to 7.1 ± 1.6 s). These results indicate that the photobleaching of the cellular QDs is cellular concentration dependent.

Table 1. The relative cellular QD concentrations and the photobleaching times (the time for PL intensity decreased to half) of QDs in EG 277 cells with different incubated concentrations. For each group, 10-20 cells were measured and the data were calculated statistically. The irradiation power density is 8×10^4 W/cm^2

Incubation cncentration of QDs (mg/mL)	Concentration of cellular QDs (relative)	Bleaching time (s)
0.3	1000±99	32.1±9.4
0.2	834±203	24. 0±7.7
0.1	402±75	7.1±1.6

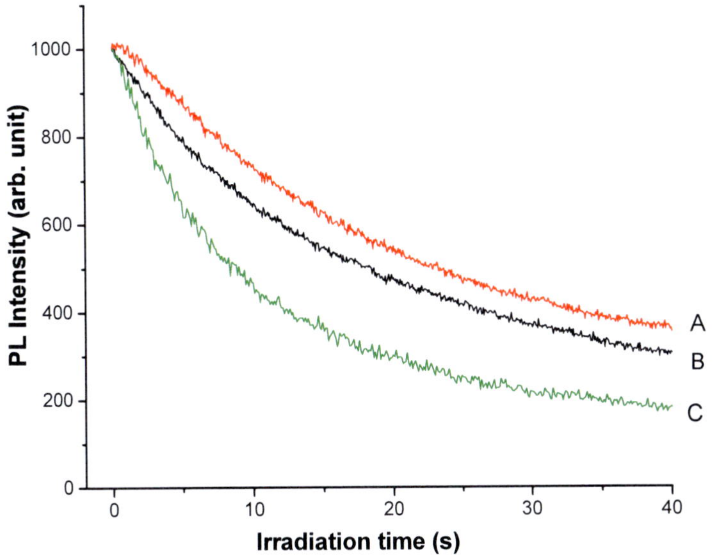

Figure 19. The photobleaching of cellular QDs in EG 277 cells incubated with different QD concentrations (curves have been normalized). (A) 0.3 mg/mL; (B) 0.2 mg/mL; (C) 0.1 mg/mL. The irradiation power density is 8×104 W/cm2.

The photobleaching in HEK 293 cells was also performed under the same irradiation conditions as that for EG 277 cells. The photobleaching curves of cellular QDs, FITC and GFP in HEK 293 cells are shown in Figure 20. Although all three species were photobleached, the bleaching of QDs is a little bit slower than that of GFP, and much slower than that of FITC. It should be noted that under the same irradiation power density the photobleaching of QDs in HEK293 cells (Figure 20) is much faster than that in EG 277 cells (Figure 17). The cellular QDs lost half of their luminescence intensities within 20 s in EG 277 cells, but within only 3 s in HEK 293 cells. What causes such a remarkable difference? The difference of cellular QD concentrations must be one of the reasons. As discussed above, the average concentration as well as the local concentration of QDs in EG277 cells is obviously higher than that in HEK293 cells. Hence, the cellular QD concentration must be an important factor for the photostability of QDs in cells.

As the distribution of QDs in HEK293 cells is not uniform (Figure 15a), these cells are suitable samples to study the dependence of photobleaching on the local concentration of QDs. The relationship between photobleaching and local concentration was studied by choosing different areas with different brightnesses in the same cell. The results are shown in Figure 21, where the bleaching curves were normalized and the value of the initial brightness of each curve at the start point was listed in the inset. These values correspond to the local concentrations of QDs at these irradiated spots. The lower the QD local concentration, the faster the bleaching, indicating that photobleaching is inversely proportional to the QD local concentration. This effect was confirmed by repeating measurements for about 30 cells.

Though QDs are believed to be relatively photostable, our results indicate that the thiol-capped QDs would still be photobleached, especially when the local concentration of cellular QDs is low. Therefore, when these QDs are used as the probes in cellular labelling, reaching a certain cellular concentration level is a critical condition.

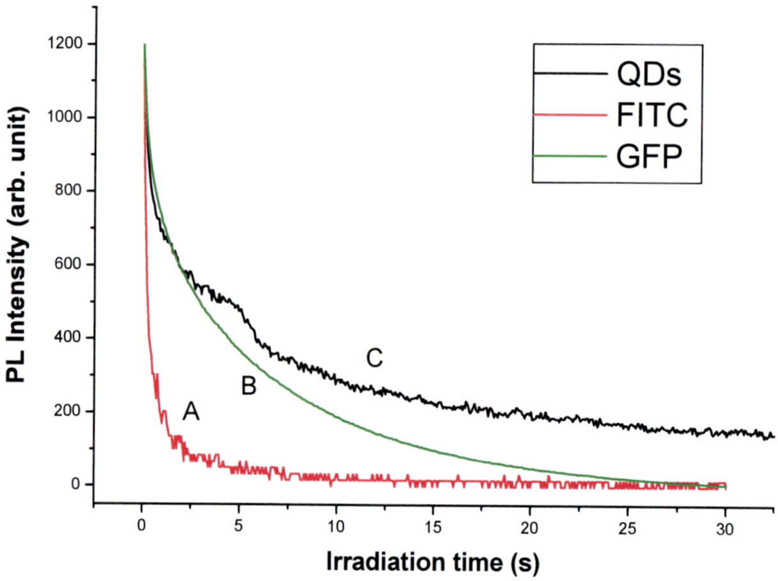

Figure 20. Photobleaching curves of (A) FITC, (B) GFP and (C) QDs in HEK 293 cells. The irradiation power density was 8×104 W/cm2, the same as that in Figure 17.

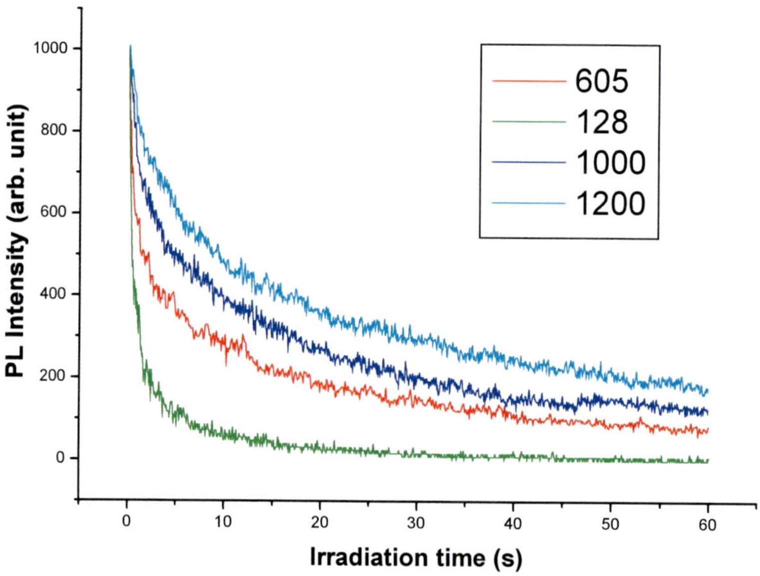

Figure 21. Photobleaching curves with different local QD concentrations in an HEK 293 cell. The curves were measured in the areas with different brightness and had been normalized. The values in the inset are the initial luminescent brightness of each curve. The power density of the irradiation at 488 nm is 8×104 W/cm2.

When the cellular concentration of QDs is too low, their photobleaching is comparable with or even worse than that of FITC. The advantage of their photostability would be lost.

As was proved above, the photobleaching of QDs results from the effect of photooxidation. To further confirm it, an experiment was conducted for EG 277 cells with nitrogen treatment. After incubation with QDs, the QD loaded EG 277 cell suspension was bubbled with nitrogen gas for 1 h. The cell suspension was then sealed between a piece of slide glass and a piece of cover slip for the photobleaching measurement. All the procedures were carried out in a chamber with nitrogen atmosphere. Meanwhile, the photobleaching of QD loaded cells without N_2 treatment was measured under the same conditions as a comparison. Figure 22 Demonstrates a comparison of the photobleaching for EG 277 cells, N_2 treated (curves 1 and 2) and untreated (curves 3 and 4). The differences between the two situations are obvious, indicating that the photooxidation must be the main reason for the photobleaching of cellular QDs. To avoid fast photobleaching, increasing QD concentration is necessary, especially for those subcellular organelles with high oxygen density. However, the high concentration of cellular QDs may cause cytotoxicity in cells [54,74]. The surface treatment or surface coating of the QDs with a shell can decrease the cytotoxicity in cells [75,76], and also decrease the QD photooxidation due to the shielding effect of the shell [37,77]. Thus, in the application of cellular labeling, the surface modification of QDs, such as coating the QD core or conjugation with biomolecules, should be considered.

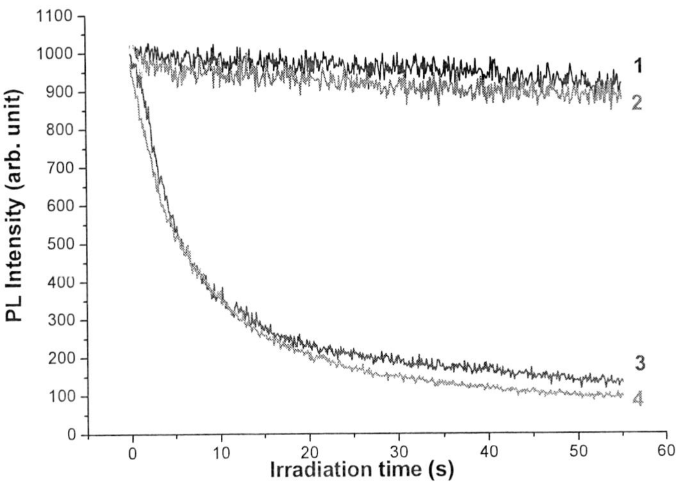

Figure 22. Photobleaching of cellular QDs in N2 treated (curves 1 and 2) and untreated (curves 3 and 4) EG 277 cells.

CONCLUSION

Water-soluble thiol-capped colloidal CdTe QDs have been synthesized by the hydrothermal route. Without any postpreparative treatment, the room temperature PL QY of the as-prepared CdTe QDs with an average diameter of 3.5 nm reaches 47% at the emission wavelength around 603 nm. The PL properties of the CdTe QDs are not changed upon aging in air for several months. PL behavior of thiol-capped CdTe QDs in living cells has been investigated. Strong evidences have been presented that singlet oxygen is responsible for the time-dependent PL blue shift of the thiol-capped CdTe Qds in living cells. The PL behavior is directly related to the existent oxygen molecules in the living cells. Deoxygenation can effectively prevent the PL blue shift from occurring. The singlet oxygen molecules are presumably formed by energy transfer from QDs to the oxygen molecules intercalated in the thiol layer at the QD core surface. The photobleaching of thiol-capped CdTe QDs inside living cells is QD cellular concentration dependent. Such bleaching is mainly due to the photooxidation of the cellular QDs. The rate of photobleaching of thiol-capped CdTe QDs inside living cells is dependent on the local concentration of cellular QDs and the irradiation intensity. Higher irradiation power density and oxygen density and lower QD concentration will result in a higher photobleaching rate. Such bleaching is mainly due to the photo-oxidation of the cellular QDs. To avoid overly rapid photobleaching in experiments, such as cellular imaging, the local concentration of cellular QDs should reach a certain level and the over-irradiation must be avoided.

ACKNOWLEDGMENTS

Financial support from the National Natural Science Foundation of China (60638010) is gratefully acknowledged. The authors are grateful to Wu-Li Yang and Chang-Chun Wang of the Department of Macromolecular Science, Fudan University for their help in preparing the QDs.

REFERENCES

[1] Murray, C. B.; Kagan, C. R.; Bawendi, M. G. *Annu. Rev. Mater. Sci.* 2000, *30*, 545-610
[2] Alivisatos A. P. *Science* 1996, *271*, 933-937.
[3] Weller, H. *Angew. Chem. Int. Ed. Engl.* 1993, *32*, 41-53.
[4] Klein, D.; Roth, R.; Lim, A.; Alivisatos, A.; McEuen, P. *Nature* 1997, *389*, 699-701.
[5] Weller, H. *Adv. Mater.* 1993, *5*, 88-95.
[6] Bruchez, M. P.; Moronne, M.; Gin, P.; Weiss, S.; Alivisatos, A. P. *Science* 1998, *281*, 2013-2016.
[7] Gaponik, N.; Radtchenko, I. L.; Sukhorukov, G. B.; Weller, H.; Rogach, A. L. *Adv. Mater.* 2002, *14*, 879-882.
[8] Mattoussi, H.; Mauro, J. M.; Goldman, E. R.; Anderson, G. P.; Sundar, V. C.; Mikulec, F. V.; Bawendi, M. G. *J. Am. Chem. Soc.* 2000, *122*, 12142-12150.
[9] Chan, W. C. W.; Nie, S. *Science* 1998, *281*, 2016-2018.
[10] Han, M.; Gao, X.; Su, J. Z.; Nie, S. *Nat. Biotechnol.* 2001, *19*, 631-635.
[11] Battersby, B. J.; Bryant, D.; Meutermans, W.; Matthews, D.; Smythe, M. L.; Trau, M. *J. Am. Chem. Soc.* 2000, *122*, 2138-2139.
[12] Rogach, A. L.; Katsikas, L.; Kornowski, A.; Su, D.; Eychmüller, A.; Weller, H. *Ber. Bunsen-Ges. Phys. Chem.* 1996, *100*, 1772-1778.
[13] Rogach, A. L.; Kornowski, A.; Gao, M.; Eychmuller, A.; Weller, H. *J. Phys. Chem. B.* 1999, *103*, 3065-3069.
[14] Gaponik, N.; Talapin, D. V.; Rogach, A. L.; Hoppe, K.; Shevchenko, E. V.; Kornowski, A.; Eychmüller, A.; Weller, H. *J. Phys. Chem. B* 2002, *106*, 7177-7185.

[15] Talapin, D. V.; Rogach, A. L.; Kornowski, A.; Haase, M.; Weller, H. *Nano Lett.* 2001, *1*, 207-211.

[16] Qu, L.; Peng, Z. A.; Peng, X. *Nano Lett.* 2001, *1*, 333-337.

[17] Talapin, D. V.; Rogach, A. L.; Mekis, I.; Haubold, S.; Kornowski, A.; Hasse, M.; Weller, H. *Colloids Surf. A* 2002, *202*, 145-154.

[18] Murray, C. B.; Norris, D. J.; Bawendi, M. G. *J. Am. Chem. Soc.* 1993, *115*, 8706-8715.

[19] Talapin, D. V.; Haubold, S.; Rogach, A. L.; Kornowski, A.; Haase, M.; Weller, H. *J. Phys. Chem. B* 2001, *105*, 2260-2263.

[20] Katari, J. E.; Colvin, V. L.; Alivisatos, A. P. *J. Phys. Chem.* 1994, *98*, 4109-4117.

[21] Talapin, D. V.; Rogach, A. L.; Shevchenko, E. V.; Kornowski, A.; Haase, M.; Weller, H. *J. Am. Chem. Soc.* 2002, *124*, 5782-5790.

[22] Gao, M.; Rogach, A. L.; Kornowski, A.; Kirstein, S.; Eychmuller, A.; Mohwald, H.; Weller, H. *J. Phys. Chem. B* 1998, *102*, 8360-8363.

[23] Zhang, H.; Wang, L.; Xiong, H.; Hu, L.; Yang, B.; Li, W. *Adv. Mater.* 2003, *15*, 1712-1715.

[24] Mitchell P. *Nat. Biotechnol.* 2001, *19*, 1013-1017.

[25] Klarreich E. *Nature* 2001, *413*, 450-452.

[26] Jovin T. M. *Nat. Biotechnol.* 2003, *21*, 32-33.

[27] Seydel C. *Science* 2003, *300*, 80-81.

[28] Taton T. A. *Nat.Mater.* 2003, *2*, 73-74.

[29] Uren R. F. *Nat. Biotechnol.* 2004, *22*, 38-39.

[30] Wang, Y.; Tang, Z. Y.; Kotov, N. A. *Materials Today* 2005, *8*, 20-31.

[31] Gao, X. H.; Cui, Y. Y.; Levenson, R. M.; Chung, L. W. K.; Nie, S. M. *Nature Biotechnol.* 2004, *22*, 969-976.

[32] Chan, W. C. W.; Maxwell, D. J.; Gao, X. H.; Bailey, R. E.; Han, M. Y.; Nie, S. M. *Curr. Opin. Biotechnol.* 2002, *13*, 40-46.

[33] Hasegawa, U.; Nomura, S. M.; Kaul, S. C.; Hirano, T.; Akiyoshi, K. *Biochem. Biophys. Res. Commun.* 2005, *331*, 917-921.

[34] Alivisatos, P. *Nature Biotechnol.* 2004, *22*, 47-52.

[35] Ho, Y. P.; Kung, M. C.; Yang, S.; Wang, T. H. *Nano Lett.* 2005, *5*, 1693-1697.

[36] Jones, M.; Nedeljkovic, J.; Ellingson, R. J.; Nozik, A. J.; Rumbles, G. *J. Phys. Chem. B* 2003,*107*, 11346-11352.

[37] Nirmal, M.; Dabbousi, B. O.; Bawendi, M. G.; Macklin, J. J.; Trautman, J. K.; Harris, T. D.; Brus, L. E. *Nature* 1996, *383*, 802-804.

[38] Tang, J.; Marcus, R. A. *J. Chem. Phys.* 2005, *123*, 1-12.

[39] van Sark, H. M. W. G. J.; Patrick, L. T.; Frederix, M.; Bol, A. A.; Gerritsen, H. C.; Meijerink, A. *ChemPhyChem.* 2002, *3*, 871-879.

[40] Guo, J.; Yang, W.; Wang, C. *J. Phys. Chem. B* 2005, *109*, 17467-17473.

[41] Demas, J. N.; Grosby, G. A. *J. Phys. Chem.* 1971, *75*, 991-1024.

[42] Yu, W. W.; Qu, L. H.; Guo, W. Z.; Peng, X. G. *Chem. Mater.* 2003, *15*, 2854-2860.

[43] Maenosono, S. *Chem. Phys. Lett.* 2003, *376*, 666-670.

[44] Nazzal, A. Y.; Qu, L. H.; Peng, X. G.; Xiao, M. *Nano Lett.* 2003, 3, 819-822.

[45] van Sark, H. M. W. G. J.; Patrick, L. T.; Frederix, M.; Van den Heuvel, D. J.; Gerritsen, H. C. *J. Phys. Chem. B* 2001, *105*, 8281-8284.

[46] Myung, N.; Bae, Y.; Bard, A. J. *Nano Lett.* 2003, *3*, 747-749.

[47] Wang, Y.; Tang, Z. Y.; Correa-Duarte, M. A.; Pastoriza-Santos, I.; Giersig, M.; Kotov, N. A.; Liz-Marza'n, L. M. *J. Phys. Chem. B* 2004, *108*, 15461-15469.

[48] Cordero, S. R.; Carson, P. J.; Estabrook, R. A.; Strouse, G. F.; Buratto, S. K. *J. Phys. Chem. B* 2000, *104*, 12137-12142.

[49] Sharma, S. N.; Kohli S.; Rastogi, A. C. *Physica E* 2005, *25*, 554-561.

[50] Korsunska, N. E.; Dybiec, M; Zhukov, L; Ostapenko, S.; Zhukov, T. *Semicond. Sci. Technol.* 2005, *20*, 876-881.

[51] Kimura, J.; Uematsu, T.; Maenosono, S.; Yamaguchi, Y. *J. Phys. Chem. B* 2004, *108*, 13258-13264.

[52] Shinya, M.; Noriko, E.; Yukio, Y. *J. Phys. Chem. B* 2003, *107*, 2645-2650.

[53] Castro-Rodrgueza, R.; Iribarrenb, A.; Bartolo-Perez, P.; Pena, J. L. *Thin solid films* 2005, *484*, 100-103.

[54] Derfus, A. M.; Chan, W. C. W.; Bhatia, S. N. *Nano Lett.* 2004, *4*, 11-18.

[55] Jang, E.; Jun, S.; Chung, Y. S.; Pu, L. S. *J. Phys.Chem.B* 2004, *108*, 4597-4600.

[56] Nilsson, R.; Merkel, P. B.; Kearns, D. R. *Photochem. Photobiol.* 1972, *16*, 109–116.

[57] Miskoski, S.; Garcia, N. A. *Photochem. Photobiol.* 1993, *57*, 447-452.

[58] Telfer, A.; Bishop, S. M.; Phillips, D.; Barber, J. *J. Bio. Chem.* 1994, *269*, 13244-13253.

[59] Samia, A. C. S.; Chen, X. B.; Burda, C. *J. Am. Chem. Soc.* 2003, *125*, 15736-15737.

[60] Aldana, J.; Lavelle, N.; Wang, Y. J.; Peng, X. G. *J. Am. Chem. Soc.* 2005, *127*, 2496-2504.

[61] Monroe, B. M. In *Singlet O_2*; Frimer, A. A., ed.; CRC Press: Boca Raton, FL., 1985; *Vol. I*, pp 177-224.

[62] Beghetto, C.; Renken, C.; Eriksson, O.; Joril, G.; Bernardi, P.; Ricchelli, F. *Eur. J. Biochem.* 2000, *267*, 5585-5592.

[63] Ding, X. M.; Xu, Q. Z.; Liu, F. G.; Zhou, P. K.; Gu, Y.; Zeng, J.; An, J.; Dai, W. D.; Li, X. S. *Cancer Lett.* 2004, *216*, 43–54.

[64] Srivastava, L. P.; Singh, R. P.; Raizada, R.B. *Food and Chem. Toxicology* 1999, *37*, 177-181.

[65] Gorman, A. A.; Rodgers M. A. J. *Chem. Phys. Lett.* 1978, *55*, 52-57.

[66] Gao X. H.; Nie S. M. *Trends in Biotechnology* 2003, *21*, 371-373.

[67] Hanaki K.; Momo A.; Oku T.; Komoto A.; Maenosono S.; Yamaguchi Y.; Yamamoto K. *Biochem. Bioph. Res. Co.* 2003, *302*, 496-501.

[68] Aldana J.; Wang Y. A.; Peng X. G. *J. Am. Chem. Soc.* 2001, *123*, 8844-8850.

[69] Xu Y.; Sun H. C.; Tian B.; Li Y.; Chen J.; Chen J.; Gao D. M.; Xue Q.; Tang Z. Y. *J.Cancer Res.Clin.Oncol.* 2004, *130*, 375-382.

[70] Brown J. S. *Biochimica et Biophysica Acta.* 1980, *591*, 9-21.

[71] Li Z. F.; Ruckenstein E. *Nano Letters* 2004, *4*, 1463-1467.

[72] Einicker-Lamas M.; Mezian G. A.; Fernandes T. B.; Silva F. L. S.; Guerra F.; Miranda K.; Attias M.; Oliveira M. M. *Environ. Pollut.* 2002, *120*, 779-786.

[73] Hoshino A.; Hanaki K.; Suzuki K.; Yamamoto K. *Biochem. Bioph. Res. Co.* 2004, *314*, 46-53.

[74] Kloepfer J. A.; Bradforth S. E.; Nadeau J. L. *Journal of Physical Chemistry B.* 2005, *109*, 9996-10003.

[75] Hoshino A.; Fujioka K.; Oku T.; Suga M.; Sasaki Y. F.; Ohta T.; Yasuhara M.; Suzuki K.; Yamamoto K. *Nano. Lett.* 2004, *4*, 2163-2169.

[76] Kirchner C.; Liedl T.; Kudera S.; Pellegrino T.; Javier A. M.; Gaub H. E.; Stolzle S.; Fertig N.; Parak W. J. *Nano. Lett.* 2005, *5*, 331-338.

[77] Spanhel L.; Haasse M.; Weller H.; Henglein A. *J. Am. Chem.Soc.* 1987, *109*, 5649-5655.

In: Encyclopedia of Nanotechnology
Editor: Elwood D. Carlson

ISBN: 978-1-60692-079-4

Chapter 40

Hybrid Nanoparticle Based on Silica

Shaojun Guo, Erkang Wang*

State Key Laboratory of Electroanalytical Chemistry, Changchun Institute of Applied Chemistry, Chinese Academy of Sciences, Changchun, 130022, Jilin, China. Graduate School of the Chinese Academy of Sciences, Beijing, 100039, P. R. China

Abstract

Inorganic nanoparticles have received great interests due to their attractive electronic, optical, and thermal properties as well as catalytic properties and potential applications in the fields of physics, chemistry, biology, medicine, and material science and their different interdisciplinary fields. Great deals of synthetic methods such as aqueous- and organic-phase approaches have been reported to synthesize high-quality inorganic nanoparticles, but the stability of colloid nanoparticles obtained is still a great problem. In addition, most of these methods for preparing nanoparticles were based on the use of organic ligand, surfactant or polymeric templates. However, the addition of organic ligand, surfactant or polymeric templates may introduce contamination in the final product due to the formation of undesired byproduct in the reaction medium, which makes them unsuitable for further application, especially in biology. On the other hand, fluorohores such as dyes has been employed as optical probe by many groups for carrying out bioassay such as detecting target DNA, protein, etc. However, because one DNA or other probes can only be labeled with one or a few fluorophores, the fluorescence signal is too weak to be detected when the target concentration is low. Furthermore, most organic dyes suffer serious photobleaching, often resulting in irreproducible signals for ultratrace bioanalysis. Silica coating of inorganic nanomaterial or fluorohores appears as an attractive alternative for providing enhanced colloidal or photostable stability and suitable functional groups for bioconjugation. Hybrid nanoparticles based on silica has been pursued vigorously in the last few years and in this review chapter we provide a perspective on the present status of the subject. Described approaches include encapsulating inorganic nanomaterials using SiO_2 sphere, fluorophore-doped silica nanoparticles, constructing muti-functional hybrid materials using SiO_2 sphere as supporting material, and some important applications of hybrid

* Email: ekwang@ciac.jl.cn; Fax: +86-431-85689711; Tel: +86-431-85262003

nanoparticle based on silica. Finally, summary and outlook of these hybrid nanoparticles are also presented.

Keywords: inorganic nanoparticle, hybrid nanomaterial, quantum dot, electrochemical sensor, magnetic nanoparticle, silica

1. Introduction

Fabrication of functional inorganic particles with nanometer-scale dimensions has been intensively studied in the last couple of decades due to their unusual chemical and physical properties compared with their bulk materials, which enable them to be promising in application over broad areas such as electronics [1], optics [2] optoelectronics [3], nanosensor [4], information storage [5], fuel cell [6], biomedicine [7], biological labeling [8], gene delivery [9], electrocatalysis [10] and surface enhanced Raman scattering (SERS) [11, 12]. Inspired by these important applications, many groups around the world have devoted their great efforts to synthesize functional inorganic nanomaterials via diversified approaches such as chemical (bottom-up) and physical (top-down) methods. Wet-chemical approach as a simple and effective method has been employed by many scientists to synthesize functional inorganic nanoparticles with different morphologies [12-35]. For example, metal nanoparticles with different novel morphologies as stylish functional nanomaterials have been widely researched by our group [12-30] and other famous groups [31-35]; Quantum dots (QDs) as new class of fluorescent probes have been facilely synthesized via organic or aqueous phase routes [36-40]; Superparamagnetic magnetic nanoparticles [41-45], which are of great interest for researchers from a wide range of disciplines, including biotechnology/biomedicine [46], and magnetic resonance imaging [47, 48], have been prepared by a number of suitable methods. However, most of these methods for preparing nanoparticles were based on the use of organic ligand, surfactant or polymeric templates. The addition of organic ligand, surfactant or polymeric templates may introduce contamination in the final product due to the formation of undesired byproduct in the reaction medium, which makes them unsuitable for further application [49], especially in biology. Additionally, inorganic nanoparticles as friendly-environmentally materials for application in biology need to be water-soluble, colloidally stable and have a tailored surface chemistry. Currently, most syntheses for high-quality, quasi-monodisperse nanoparticles of semiconductors [50], metals [51], metal oxides [52], and magnetic material [53] involved organic solvents and coating with monolayers of hydrophobic surfactants. Such nanoparticles could not be used directly for biofunctionalization as they were water-insoluble and did not have suitable functional groups for constructing nanoparticle-biomolecule conjugates. Although many groups synthesized water-soluble nanoparticles using some polymer and surfactant as protecting agents, these protecting agents usually brought potential toxicity and side-effect for the biosystem of studying. On the other hand, fluorohores such as dyes has been employed as optical probe by many groups for carrying out bioassay such as detecting target DNA, protein, etc. However, because one DNA or other probes can only be labeled with one or a few fluorophores, the fluorescence signal is too weak to be detected when the target concentration is low. Furthermore, most organic dyes suffer serious photobleaching, often resulting in irreproducible signals for ultratrace bioanalysis. Silica coating of inorganic nanomaterial or fluorohores appears as an attractive alternative for providing enhanced colloidal or photostable stability and suitable functional groups for bioconjugation. Particularly after discovering monodisperse silica sphere by stöber et al., many excellent papers have been reported to synthesize hybrid nanoparticle based on silica because silica has very important predominance for application in biology. 1) Silica is a stabilizer which not only prevents particle coalescence but also is chemically inert and optically transparent. 2) Silica has been reported to be good bio-compatible material. 3) Silica owns excellent surface chemistry, which enable silica coated inorganic nanoparticles or fluorophore-doped silica nanoparticles dissolvable in different solvents. 4) Hybrid nanoparticle through silica as supporting or coating material will exhibit superior multifunctional properties contributed by different inorganic nanomaterials (metal, semiconductor, fluorophore and magnetic material, etc.). Thus, hybrid nanoparticles based on silica has been pursued vigorously in the last few years and in this review chapter we provide a perspective on the present status of the subject. Described approaches include encapsulating inorganic nanomaterials using SiO_2 sphere, fluorophore-doped silica nanoparticles, constructing muti-functional hybrid materials using SiO_2 sphere as supporting material, and some important applications of hybrid nanoparticle based on silica. Finally, summary and outlook of these hybrid nanoparticles are also presented.

However, due to the explosion of publications in this field, we do not claim that this review includes all of the published work about hybrid nanoparticle based on silica. We apologize to the authors of much excellent work that due to the large activity in this field, we have unintentionally left out.

2. Encapsulating Inorganic Nanomaterials Using SiO_2 Sphere

2.1. Metal/Silica Core/Shell Nanostructure

Synthesis of metal nanomaterials such as gold and silver exhibiting particular optic, electronic, magnetic properties have received great intensity in the past decades. Particularly, the surface plasmon resonance (SPR) of metal nanomaterials (an important optical phenomenon of metal nanomaterials) can be tuned across the near-UV, visible, and near-infrared spectral range by varying the size, shape, and surrounding dielectric environment of particle. So, the SPR of silica coated metal nanomaterials can be facilely adjusted via changing the thickness of silica on the surface of metal nanomaterials. On the other hands, silica coated metal nanomaterials will commendably solve the bio-incompatible problem usually caused by protected ligands on the surface of metal nanomaterials. Most importantly, the stabilization and functionalization of metal nanomaterials with silica provides flexibility for a variety of applications, including bioassay, bioimaging and biosensor. In this part, we will discuss recent advances on encapsulating metal nanomaterials using SiO_2 sphere.

A reverse micelle and sol-gel technique has been employed to synthesize spherical Ag/SiO_2 composite particles [54, 55]. The size of the particles and the thickness of the coating can be controlled by manipulating the relative rates of the hydrolysis and condensation reactions of tetraethoxysilane (TEOS) within the microemulsion. For example,

Adair et al. [54] employed the above method to facilely prepare silica coated Ag nanoparticles. The process is performed in conjunction with hydrolysis of TEOS, followed by condensation in the water nanodroplets to form a coating on the nanosized silver with reverse micelle structure. Later, Asher's group [55] reported a microemulsion method to fabricate nanocomposite silica colloidal particles containing homogeneously dispersed silver quantum dots.

Simple sol-gel technique can also be employed to fabricate silica coated metal nanoparticles. But, the silica-coating procedures reported in the literature generally involve surfaces with a significant chemical or electrostatic affinity for silica. Gold or silver metal has very little affinity for silica because it does not form a passivating oxide film in solution. So, introducing a primer layer into the surface of metal is very necessary for obtaining well-defined silica coated metal nanoparticles. (3-aminopropyl) trimethoxysilane (APTMS) and (3-mercaptopropyl) trimethoxysilane (MPTMS) as bifunctional molecules have been widely used as primers to fabricate well-defined silica coated metal nanoparticles [56-60]. For instance, Liz-Marzán et al. [56] showed that gold colloids have been homogeneously coated with silica using the silane coupling agent, APTMS, as a primer to render the gold surface vitreophilic. After the formation of a thin silica layer in aqueous solution, the particles can be transferred into ethanol for further growth using the Stöber method. The thickness of the silica layer can be completely controlled, and (after surface modification) the particles can be transferred into practically any solvent. Figure 1 shows the transmission electron microscopy (TEM) images of silica-coated gold particles with different diameters produced during the extensive growth of the silica shell. It is found that silica has homogeneously coated on the surface of gold particles. Later, Murphy and coworkers [60] employed MPTMS as a coupling agent to fabricate silica coated gold nanorods. But, in this case, it is found that for small aspect ratio nanorods, the reproducibility is poor and particle aggregation (again, preferentially through the tips) is usually observed during the sodium silicate addition step. Poly-(vinylpyrrolidone) (PVP) [61] as a functional molecule can also be employed to fabricate silica coated metal nanoparticles because PVP can provide good link points for OH group. Graf et al. [61] developed a general method to coat colloids with silica. The amphiphilic, nonionic polymer, PVP, was adsorbed to various colloidal particles such as small gold colloids, gold-shell silica-core particles, small and large silver colloids. After this functionalization, the stabilized particles could be transferred to a solution of ammonia in ethanol and directly coated with smooth and homogeneous silica shells of variable thickness by addition of tetraethoxysilane in a seeded growth process. Figure 2 shows the typical TEM images of silica coated gold nanoparticles. From the different images, gold nanoparticles have been homogeneously coated with silica. Although the above method using PVP as a coupling agent was interesting and could be applied to many particle systems, some particles such as CTAB protected gold nanoparticles or nanorods coated through PVP capping have been found to fail for these particles. CTAB-coated nanoparticles can indeed be transferred into ethanol upon functionalization with the slightly negatively charged polymer PVP, however, controlled hydrolysis and condensation of TEOS on the nanoparticle surface could not be achieved, probably because the remaining CTAB promote the formation of silica [62]. In order to solve the above problem, Pastoriza-Santos et al. [62] developed a novel silica-coating procedure that has been devised for CTAB-stabilized gold nanorods but that can be readily extended to other CTAB-stabilized nanoparticles [62]. The method comprises a combination of the polyelectrolyte layer-by-layer (LBL) technique and the hydrolysis and condensation of

TEOS in a 2-propanol-water mixture and leads to homogeneous coatings with tight control on shell thickness. Figure 3 shows the effectiveness of this method for deposition of silica shells with accurate thickness control. For thin shells (Figure 3a),the silica surface is rough, probably because of the presence of polyelectrolyte chains on the nanoparticles, whereas upon subsequent additions of TEOS, the thickness of the silica shells increases (Figure 3b, c and d) and the surface quickly becomes smoother.

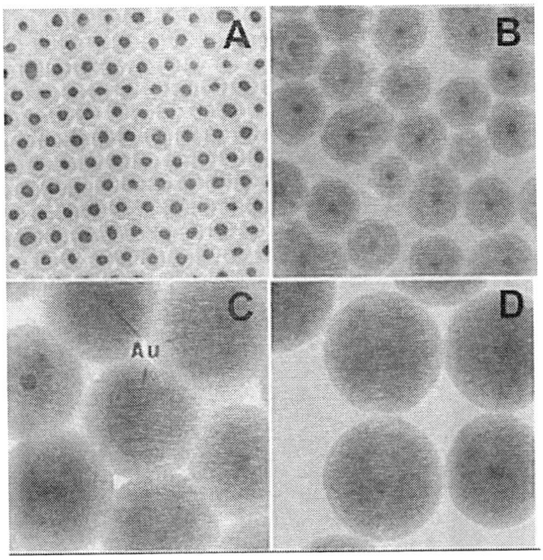

Figure 1. TEM images of silica-coated gold particles produced during the extensive growth of the silica shell around 15 nm Au particles with TEOS in 4:1 ethanol/water mixtures. The shell thicknesses are (A) 10 nm, (B) 23 nm, (C) 58 nm, and (D) 83 nm. Reprinted with permission from Ref. 56, L. M. Liz-Marzán, Langmuir 12, 4329 (1996), Copyright American Chemical Society (1996).

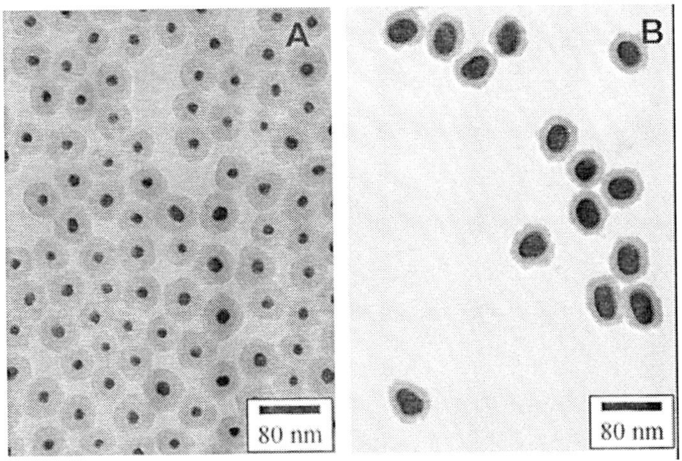

Figure 2. TEM images of gold particles coated with silica. (A) Gold particles (7nm, polydispersity 15%) with an 18nm shell (total polydispersity 9%) grown by two additions of TEOS. (B) Gold particles (20 nm radius, polydispersity 12%) with 12 nm shell (total polydispersity 9%). Reprinted with permission from Ref. 61, C. Graf, Langmuir 19, 6693 (2003), Copyright American Chemical Society (2003).

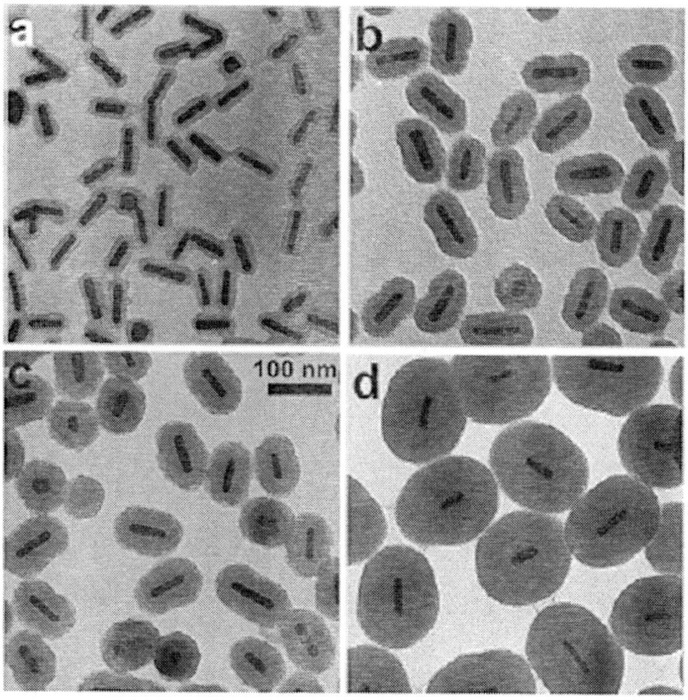

Figure 3. TEM images of silica-coated gold nanorods, with silica shell thickness increasing from a to d. The scale is the same for all images. Reprinted with permission from Ref. 62, Pastoriza-Santos, Chem. Mater. 18, 2465 (2006). Copyright American Chemical Society (2006).

Preparation of silica coated metal nanoparticles via sol-gel technique without the addition of primer have received great interests in recent years because it will greatly reduce the cost of materials and avoid the toxicity cause by the introduction of primers. Several excellent papers [63-66] have reported to fabricate the silica coated metal nanoparticles without the use of primer as coupling agent. Han et al. [63] developed a facile synthetic route to prepare monodisperse Au@SiO_2 nanoparticles with a homogeneous silica shell via well-established silica surface chemistry with readily available silane coupling agents. This simple preparation process is very fast, allowing the direct and efficient coating of the pretreated citrate-stabilized gold nanoparticles with silica, without the use of any specific, surface-coupling silane agents or large stabilizers (PVP). Typical TEM images of three Au@SiO_2 samples with different silica-shell thicknesses (~35, 75, and 90 nm) are shown in Figure 4. It is observed that monodisperse Au@SiO_2 nanoparticles with 50 nm gold cores and uniform thicknesses of silica shell were successfully prepared.

2.2. Semiconductor/Silica Core/Shell Nanostructure

Colloidal semiconductor nanocrystals have attracted extensive interest because of their variety of size- and shape-dependent optical and electrical properties and various applications such as in optoeletronic devices, photovoltaic devices, and biological fluorescence labeling. Over the past two decades, many methods for synthesizing high-quality semiconductor nanocrystals via water- or organic-phase routes have been well developed. However, semiconductor nanocrystals obtained are usually toxic, which caused by heavy metal ions effect (Cd^{2+}, Hg^{2+}). Thus, their application as in-vitro or in-vivo bioimaging agents will receive tremendous challenge, which is an agonizing problem puzzling many scientists around the word. On the other hand, although QDs have shown higher sensitivity and better photostability and chemical stability than conventional fluorophore markers [67], their photophysical behavior is usually affected by the use of solvents, ligands, environments and photoionization of QDs. In order to improve the photostability of QDs, they need to be encapsulated within a rigid matrix. Silica is an ideal choice, and it can be applied for a coating using a versatile sol-gel process. Thus, silica can act as a robust, inert layer against the degradation of optical properties and also imparts water solubility. In this part, we will discuss recent advances on encapsulating semiconductor nanomaterials (soluble in water or organic phase) using SiO_2 sphere.

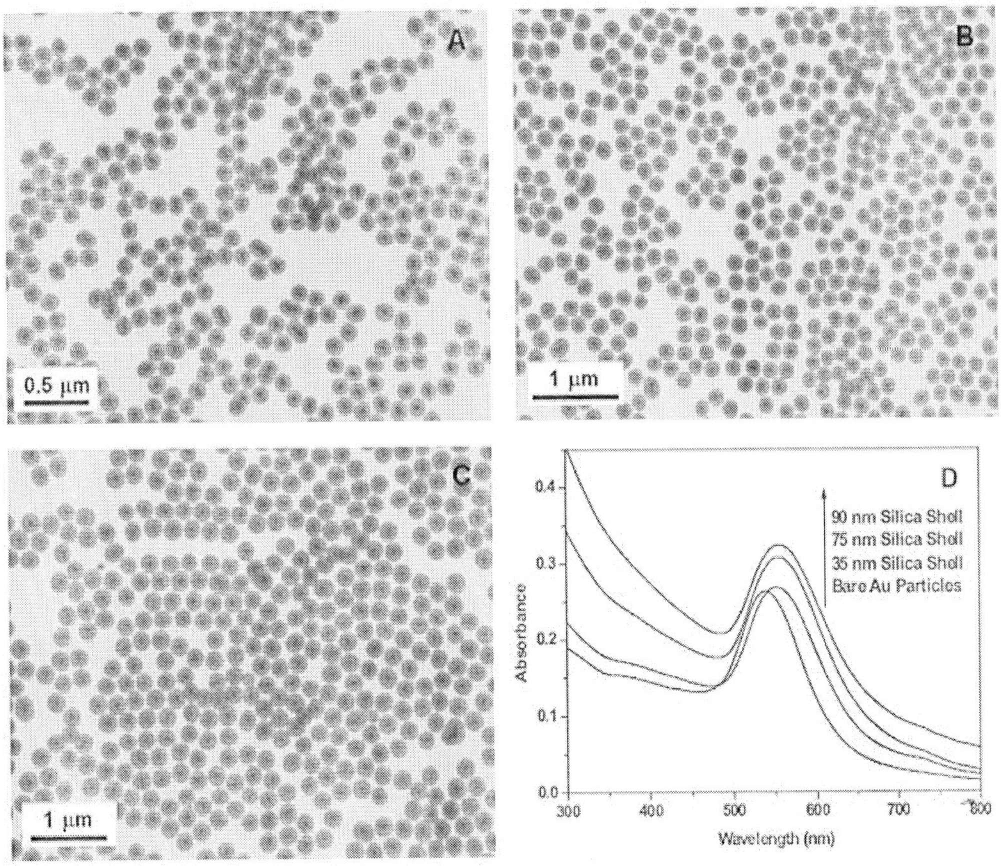

Figure 4. TEM images of Au@SiO_2 nanoparticles with silica-shell thickness of A) 35, B) 75, and C) 90 nm. D) UV-vis spectra of citrate-stabilized 50 nm gold nanoparticles with their corresponding Au@SiO_2 nanoparticles with silica-shell thickness of 35, 75, and 90 nm. Reprinted with permission from Ref. 63, M. Han, Adv. Funct. Mater. 15, 961 (2005), Copyright Wiley-VCH (2005).

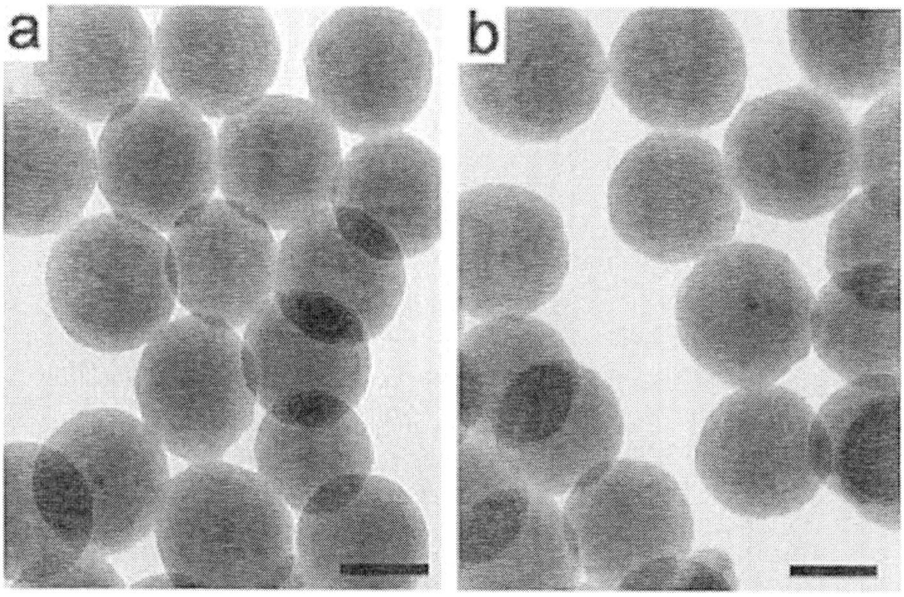

Figure 5. TEM images of core-shell-structured $CdTe@SiO_2$ particles prepared under different initial concentrations of CdTe QDs: a) 1.35×10^{-5} M; b) 5.40×10^{-6} M (with respect to CdTe QDs). The scale bars correspond to 50 nm. Reprinted with permission from Ref. 69, M. Gao, Adv. Mater. 17, 2354 (2005), Copyright Wiley-VCH (2005).

Semiconductor nanoparticles with water solubility coated with homogeneous silica can also be facilely prepared through the above methods for preparing silica coated metal nanoparticles. The combination of microemulsion and sol-gel technique has been developed to fabricate CdS and CdTe nanoparticles [68, 69]. For instance, Gao et al. [69] reported the preparation of $CdTe@SiO_2$ core/shell structured fluorescent spheres via the above method. CdTe nanocrystals stabilized by thioglycolic acid and 1-thioglycerol were first synthesized. Under optimal conditions, highly fluorescent silica spherical particles with single CdTe nanocrystal cores were obtained. Figure 5 shows the typical TEM images of two different $CdTe@SiO_2$ samples with different shell thicknesses. In this system, the thickness of silica can be easily controlled via changing the initial concentration of CdTe nanocrystals used. Furthermore, some primers such as PVP and MPTMS can also be used as coupling agents for water-soluble semiconductor nanocrystals to prepare silica coated fluorescent hybrid nanospheres. Zhang et al. [70] demonstrated that PVP-stabilized $NaYF_4$ nanocrystals could be directly coated with a uniform layer of silica. Kotov and coworkers [71] reported that CdTe nanowires with bristled SiO_2 coatings resembling nanoscale centipedes were produced in a modified Stöber process when mercaptosuccinic acid and MPTMS were used as a stabilizer and coupling agent, respectively. In addition, coating CdTe nanocrystals with homogeneous silica have also successfully demonstrated by Zhang et al. [72] and Weller et al. [73] via derivatived-silane as coupling agents.

Coating hydrophobic semiconductor nanoparticles with silica will have tremendous significance because high fluorescent quantum dots prepared via organic-phase route for application in biology usually renders them amenable to soluble in water. So, in the past years, a great deal of work has been reported to prepare silica coated quantum dots (organic solubility). In general, there exist two approaches to prepare silica coated hydrophobic semiconductor nanoparticles. One approach involves the use of a surface primer (a silane coupling agent) before introducing TEOS, ammonium and water into QD solution. Several excellent papers [74-83] have been mentioned to prepare silica coated QDs using MPTMS or APTMS as a coupling agent. For instance, Alivisatos et al. [77] reported that hydrophobic CdSe/ZnS core/shell nanocrystals with a core size between 2 and 5 nm are embedded in a siloxane shell through adding MPTMS as a coupling agent. Later, Nann and coworkers [81] extended the above method to directly prepare photostable, highly luminescent, water-soluble core-shell CdSe/ZnS/SiO_2 hybrid nanoparticles. To our great interest, this novel method could coat single luminescent QDs with silica shells. Figure 6 shows high-resolution TEM images of single QDs encapsulated in silica. The size of the particles is approximately 30 nm and the great majority contain single QDs at the center. In addition, the diameter of the final silica particles can be tuned from approximately 30 nm to 1 mm by means of seeded growth. More recently, Ying et al. [82] have developed a novel silica-coating method for QDs in toluene using commercially available silanes. Particularly, this silica-coating procedure can be facilely suitable for various hydrophobic inorganic nanoparticles, including Au, Ag, Fe_3O_4, and ZnS-CdSe nanoparticles. Although the success has been achieved through the above Stöber-based method and additionally introducing coupling agents, the above Stöber-based method have major disadvantages such as high requirements on purity of the reactants, the difficulty and multiplicity of the preparation steps, and the fact that it is not possible to directly coat nanoparticles with nonpolar ligands. The W/O microemulsion system in conjunction with the Stöber synthesis and silane coupling method can well solve the above problem and be used for the preparation of silica-coated semiconductor nanocrystals [84-87]. Ying et al. [84] described a simple strategy for making plain CdSe quantum dot water soluble by coating with silica. In this microemulsion system, the introducing a surfactant (Igepal CO-520) is very necessary. The surfactant interaction prior to silica coating allows the hydrophobic QDs to be encapsulated within the aqueous domains of the reverse microemulsion. Figure 7a shows typical TEM image of homogeneously dispersed 40 nm silica nanoparticles prepared by the Igepal microemulsion method. Single QDs (Figure 7b) or multiple QDs (Figure 7c, d) could be facilely encapsulated within each silica nanoparticles. Ying et al. [87] further extended this method to prepare silica coating of near-Infrared-emitting PbSe QDs. In addition, Nann and coworkers [86] found that introduction an polymer, poly (ethylene glycol) nonylphenyl ether into a microemulsion system could also be used to prepare spherical PbSe@SiO_2 core/shell nanoparticles. These core/shell nanoparticles obtained have a spherical shape with narrow size distribution and smooth surfaces. Importantly, the size of the particles and the thickness of the shells can be easily controlled by manipulating the relative rates of the hydrolysis and condensation reactions of TEOS within the microemulsion.

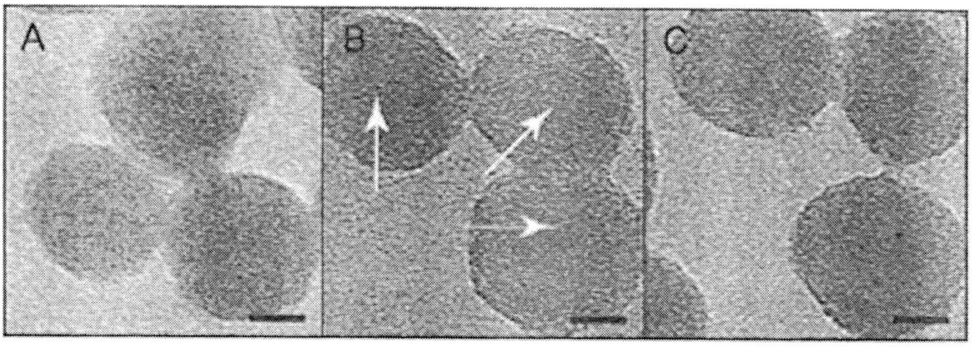

Figure 6. HRTEM images of single QDs encapsulated within a silica shell. A) QD/SiO_2 spheres of approximately 30 nm diameter. B) QDs are misarranged with respect to the electron beam. The arrows indicate the positions of the QDs. C) SiO_2 sphere with two QDs incorporated (lower right). Scalebars: 10 nm. Reprinted with permission from Ref. 81, T. Nann, Angew. Chem. Int. Ed. 43, 5393 (2004), Copyright Wiley-VCH (2004).

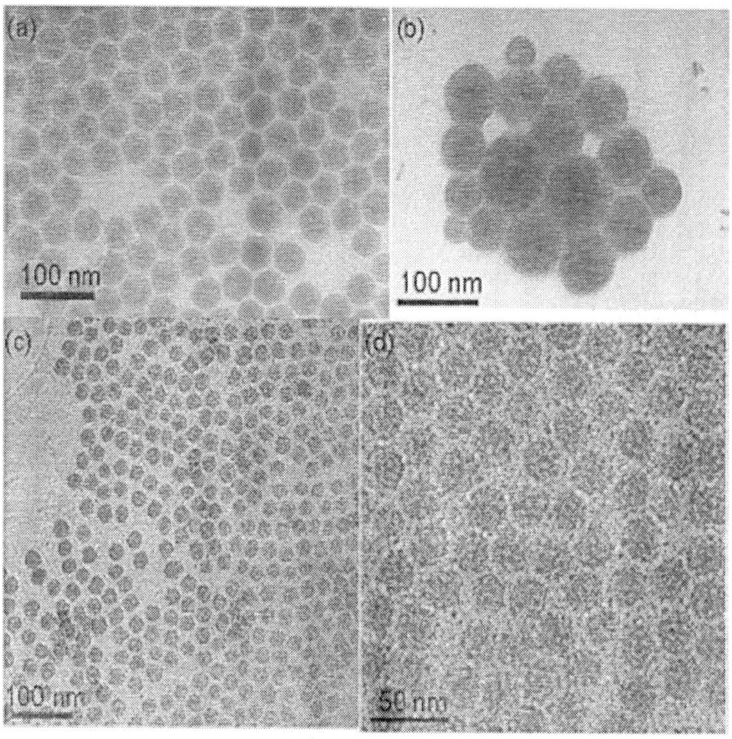

Figure 7. TEM image (a) of 40 nm SiO_2 nanoparticles (control) prepared by the Igepal microemulsion method. b) TEM image of 5 nm CdSe QDs individually encapsulated within SiO_2 nanoparticles. c, d) TEM images of multiple 5 nm CdSe QDs encapsulated within 25 nm SiO_2 nanoparticles. Unencapsulated QDs were also observed when the QD concentration was high. Reprinted with permission from Ref. 84, J. Y. Ying, Adv. Mater. 17, 1620 (2005), Copyright Wiley-VCH (2005).

2.3. Magnetic Nanoparticle/Silica Core/Shell Nanostructure

Magnetic nanoparticles have been the focus of intense research for researchers from a wide range of disciplines such as magnetic fluids, catalysis, biotechnology/biomedicine, magnetic resonance imaging, data storage, and environmental remediation [88]. Particularly, synthesis of superparamagnetic magnetic nanoparticles, the size of which is below a critical value (typically around 10-20 nm) has gained more attention due to their special features and important applications in biomedicine. A number of suitable methods have been developed for the synthesis of magnetic nanoparticles of various different compositions. However, monodisperse magnetic nanoparticles, which are intensively desirable for biomedicine, are usually synthesized via organic phase route at high temperature. A great deal of organic ligands existed on the surface of magnetic nanoparticles will greatly limit the bio-application of magnetic materials themselves. On the other hand, small magnetic particles (synthesized via aqueous phase route) usually tend to form agglomerates to reduce the energy associated with the high surface area to volume ratio of the nanosized particles. Moreover, naked magnetic nanoparticles are chemically highly active, and are easily oxidized in air, resulting generally in loss of magnetism [88]. In order to solve the above problem, a great deal of methods have been developed to design novel magnetic nanoparticle/silica core/shell nanostructure. A silica shell does not only protect the magnetic cores, but can also prevent the direct contact of the magnetic core with additional agents linked to the silica surface, thus avoiding unwanted interactions. Most importantly, the protecting shells can also be used for further functionalization with other functional nanoparticles or various ligands, depending on the desired applications. In this part, we will discuss recent advances on encapsulating magnetic nanoparticles using SiO_2 sphere.

Two different approaches have been explored to generate silica coatings on the surfaces of magnetic particles. The first method relied on the well-known Stöber process, in which silica was formed in-situ on the surface of magnetic nanoparticles, particularly iron oxide particles, through the hydrolysis and condensation of a sol-gel precursor. For magnetic oxide, because the iron oxide surface has a strong affinity toward silica, no primer was required to promote the deposition and adhesion of silica. This method was originally applied to ferromagnetic rod-like nanoparticles [89], then to micrometer-sized hematite colloids by Matijevic and co-workers [90], and later extended to other magnetic nanoparticles by a number of research groups [91-100]. For instance, Xia et al. [93] developed a sol-gel approach for the coating of superparamagnetic iron oxide nanoparticles with uniform shells of amorphous silica. The thickness of silica coating could be conveniently controlled in the range of 2-100 nm by changing the concentration of the sol-gel solution. Figure 8 shows the typical TEM images of iron oxide nanoparticles whose surfaces had been coated with silica shells using different TEOS concentrations. When the thickness of silica coating was increased, the core-shell nanoparticles became more monodispersed because of a reduction in the relative size distribution. Figure 8D shows the HRTEM image of a silica-coated iron oxide nanoparticle. This image clearly indicates the single crystallinity of the iron oxide core and the amorphous nature of the silica shell. Later, Gao and coworkers [97] extended this simple method to prepare silica coated big iron oxide nanosphere. It is found that homogeneous silica can still coat on the surface of magnetic nanoparticles. Another important advance on silica coated magnetic nanoparticles was focused on the work reported by Salgueiriño-Maceira et al. [101, 102] and Luis M. Liz-Marzán et al. [103]. Particularly,

Salgueiriño-Maceira et al. [101] highlighted the preparation of stable silica-coated cobalt nanoparticles through silica coating on the surface of cobalt nanoparticles, which is in-situ produced via the reduction of Co^{2+} in the presence of $NaBH_4$ and citrate. This method is simple, rapid and good reproducibility. Depending on the experimental conditions, both the diameter of the magnetic core and the thickness of the dielectric shell can be easily tuned. Figure 9 shows the typical TEM images of spherical cobalt particles coated by a uniform silica shell. It can be seen that the magnetic core varied from 32 (Figure 9a and b) to 95 (Figure 9c) nm. Carefully observing the above experimental results, it is found that it is hard to reach the good result that silica can be homogeneously coated on the surface of magnetic nanoparticles. Combining microemulsion synthesis and the stöber method probably provides a state-of-art strategy for obtaining well-defined magnetic nanoparticle/silica core/shell nanostructure because the microemulsion synthesis can be used to confine and control the coating of silica on core nanoparticles. Similar to the above metal or semiconductor coating process, the microemulsion procedure for obtaining silica coated magnetic nanoparticles still needs to introduce some surfactants into the solution. For instance, Igepal CO-520 [104-107], Triton X-100 [104] and Brij-97 [104] have been employed by several groups to prepare silica coated magnetic nanoparticles. The noticeable example is the work reported by Korgel et al. [106], who demonstrated that colloidal FePt nanocrystals, 6 nm in diameter, could be coated with silica shells in the microemulsion system. Figure 10a shows a TEM image of the oleic acid/oleylamine-capped FePt nanocrystals prior to coating with silica. The nanocrystals were relatively size-monodisperse with an average diameter of 6 nm. After coating with silica, monodisperse silica coated FePt nanoparticles can be facilely (Figure 10b and c). The SiO_2 shell thickness could be controlled from 7 to 23 nm by adjusting the TEOS concentration and the ratio of TEOS to FePt nanocrystals used during the coating step.

Herein, another important hybrid nanoparticles based on silica containing magnetic nanoparticles is mesoporous silica coated magnetic nanoparticles. This hybrid nanostructure is very important for magnetic induced catalytic application, which is caused by the properties of magnetic nanoparticles and mesoporous silica as individual component. Inspired by this, several excellent papers have mentioned to prepare this hybrid nanostructrure [108-112]. For instance, Shi et al. [111] reported a novel scheme to fabricate uniform magnetic nanospheres with a magnetic core/mesoporous silica shell structure. The synthesis strategy is presented in Figure 11. A thin and dense silica layer was first deposited on the surface of hematite particles of desired thickness in order to protect the iron oxide core from leaching into the mother system under acidic circumstances. Then, the mesoporous silica shell was formed from simultaneous sol-gel polymerization of tetraethoxysilane (TEOS) and n-octadecyltrimethoxysilane (C18TMS) followed by removal of the organic group. Finally, the hematite cores of the nanospheres were reduced in a flowing gas mixture of H_2 and N_2 to produce the final hybrid nanospheres. In addition, the thickness of could be easily tuned by changing the concentration of the TEOS/C18TMS mixture. This novel hybrid material may be applicable in targeted drug delivery, catalysis and multiphase separation in the future.

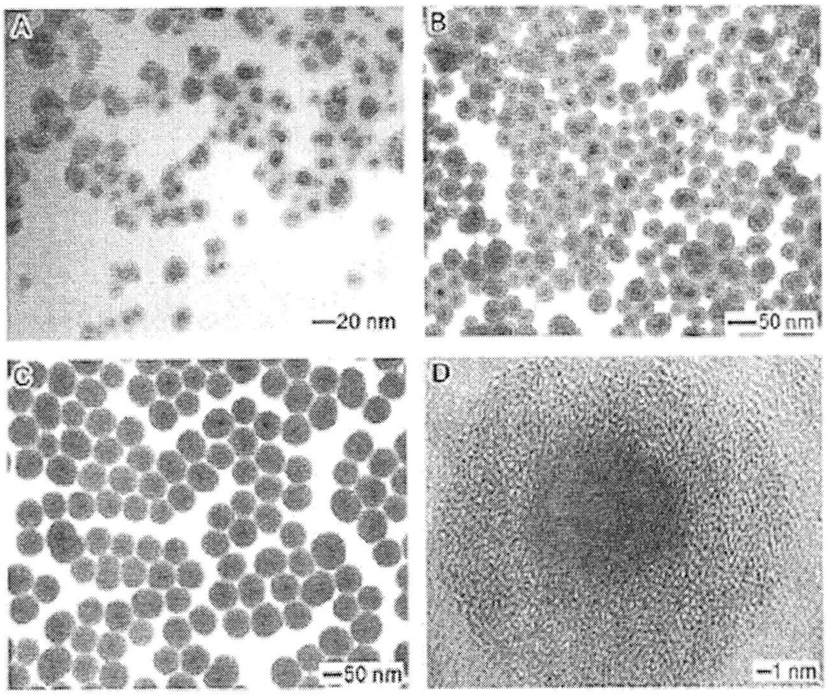

Figure 8. (A-C) TEM images of iron oxide nanoparticles whose surfaces have been coated with silica shells of various thicknesses. In this case, the thickness of silica coating could be adjusted by controlling the amount of precursor added to the solution: (A) 10, (B) 60, and (C) 1000 mg of TEOS to 20 mL of 2-propanol. (D) A HRTEM image of the iron oxide nanoparticle whose surface has been uniformly coated with 6 nm of amorphous silica shell. Reprinted with permission from Ref. 93, Y. Xia, Nano. Lett. 2, 183 (2002), Copyright American Chemical Society (2002).

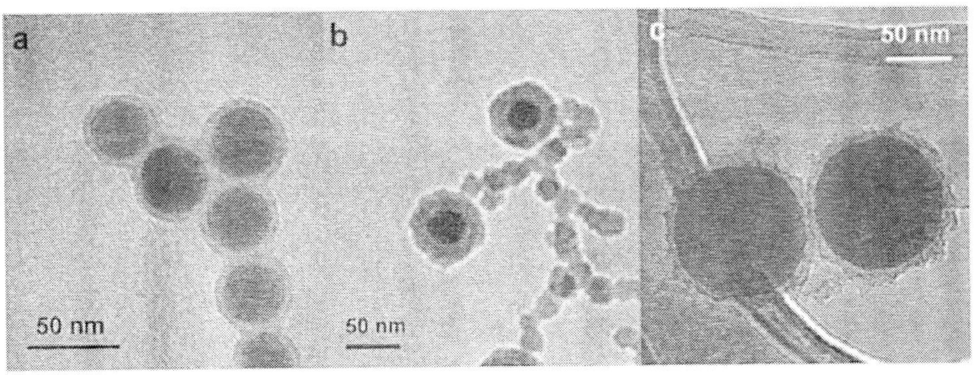

Figure 9. TEM images of silica coated cobalt nanoparticles with 32 (a and b) and 95 (c) nm cores. Reprinted with permission from Ref. 101, V. Salgueiriño-Maceira, J. Mater. Chem.16, 3593 (2006), Copyright Royal Society of Chemistry (2006).

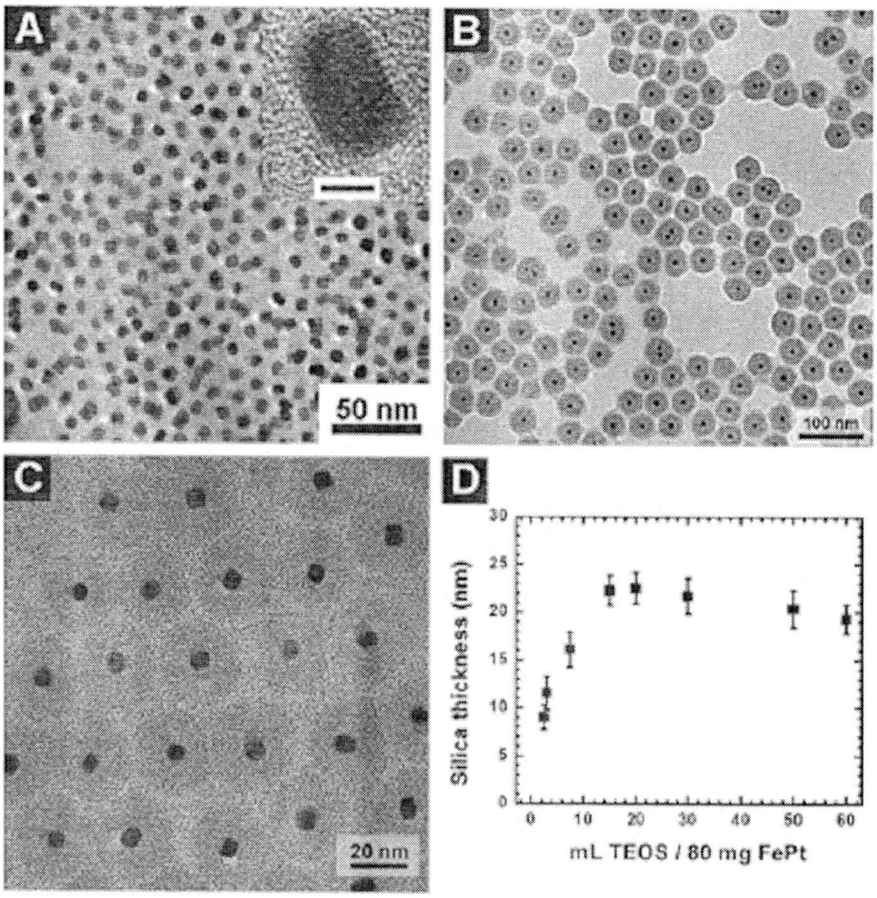

Figure 10. TEM images of FePt nanocrystals (A) prior to and (B and C) after coating with SiO_2. (The inset scale bar in part A is 3 nm.) (D) The SiO_2 shell thickness obtained as a function of TEOS concentration. Reprinted with permission from Ref. 106, B. A. Korgel, J. Phys. Chem. B 110, 11160 (2006), Copyright American Chemical Society (2006).

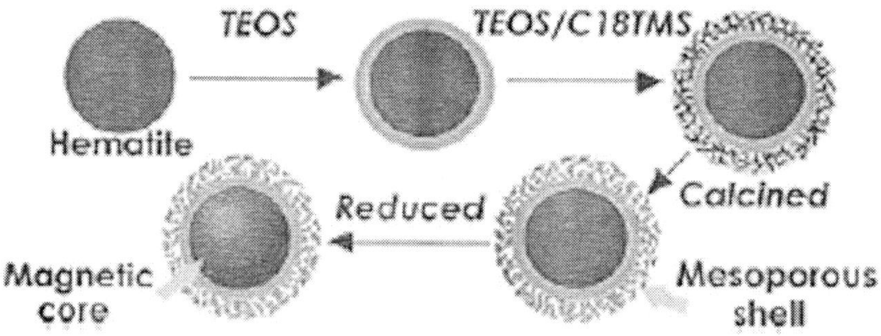

Figure 11. Illustration of synthesis of uniform magnetic nanospheres with a magnetic core/mesoporous silica shell structure. Reprinted with permission from Ref. 111, J. Shi, J. Am. Chem. Soc. 127, 8916, (2005), Copyright American Chemical Society (2005).

In summary, we have reviewed recent advances on preparation of silica coated inorganic nanoparticles via different coating techniques. On one hand, It should be noted that searching a general, simple and rapid method to preparation of metal (or semiconductor, magnetic nanoparticle)@silica core/shell nanoparticles is still great challenge. On the other hand, currently, the as-prepared inorganic nanoparticles generally own different surface properties such as negative or positive charges and ligands. We have to choose the best method for the as-prepared inorganic nanoparticles and render them amenable to facilely functionalize with homogeneous silica.

3. Fluorophore-Doped Silica Nanoparticles

Although silica coated semiconductor nanoparticles could be facilely obtained and used as bioprobe, an unavoidable question is the fact that the each hybrid nanoparticles own one or several semiconductor nanoparticles, which leads to low quantum yield. On the other hands, fluorohore such as dyes has been employed as optical probe by many groups for carrying out bioassay such as detecting target DNA, protein, etc. However, because one DNA or other probes can only be labeled with one or a few fluorophores, the fluorescence signal is too weak to be detected when the target concentration is low. In addition, the poor photostability of many fluorophores is anothor changlleging question. Most organic dyes suffer serious photobleaching, often resulting in irreproducible signals for ultratrace bioanalysis. Fluorophore-doped silica nanoparticles developed recently, which has a silica matrix that entraps a large number of fluorophores, can be used as alternative substitutes for solving the above existed questions because of their major advantages such as strong fluorescence signal, highly photostable performance benefiting from the silica matrix shielding effect, easy surface modifications for bioconjugation, and size uniformity and tunability.

Thus, many efforts have been directed towards synthesizing fluorophore-doped silica nanoparticles and further exploring their advanced applications [113-132]. For example, inorganic dyes such as tris(2,2'-bipyridyl)dichlororuthenium(II) hexahydrate ((Ru(bpy)) doped silica hybrid nanoparticles have been facilely obtained by several groups via the microemulsion methods [113-124]. Particularly, Tan's groups [119] studied the formation of Ru(bpy) dye-doped silica nanoparticles by ammonia-catalyzed hydrolysis of TEOS in water-in-oil microemulsion in details. It is found that the fluorescence spectra, particle size, and size distribution of Ru(bpy) dye-doped silica nanoparticles could be easily controlled via manipulating some parameters such as reactant concentrations (TEOS and ammonium hydroxide), nature of surfactant molecules, and molar ratios of water to surfactant and cosurfactant to surfactant. This optimization study of the preparation of dye-doped silica nanoparticles provides a fundamental knowledge of the synthesis and optical properties of Ru(bpy) dye-doped silica nanoparticles. But, the fluorescence intensity of inorganic dye doped silica hybrid nanoparticle is still limited due to the relatively low quantum yield of inorganic dyes. To achieve high signal amplification in ultra-sensitive bioanalysis, silica nanoparticles doped with highly fluorescent organic dyes are in great demand as organic dyes ususlly have higher quantum yield. Several important results have been reported by Tan et al. to synthesize organic dyes doped silica nanoparticles [125-132]. For instance, they have developed highly fluorescent, extremely photostable, and biocompatible organic-dye doped silica nanoparticles using a reverse microemulsion [125]. Their strategy was to link a hydrophilic dextran molecule to an organic dye molucule (hydrophobic) and use acidic conditions in the water core of a reverse microemulsion system to create a strong attractive force between the dextran linked dye and the silica matrix. In addition, the size of hybrid nanoparticles can be easily tuned by varying the microemulsion conditions. Furthermore, this method can be also extended to obtain mutil-fluorophores doped silica nanoparticles [120, 127, 129, 132]. Besides inorganic or organic dye doped silica matrix, some other inorganic molecules such as polyoxometalates (POMs) could also be doped into silica matrix to form hybrid system [133, 134]. Green et al. [133] immobilized the POMs and the enhancing agent (amine-rich species) in silica spheres, which confines the POM to the core of the particle while increasing the emission of the material. The above fluorophore doped silica nanoparticles will be probably useful in widespread fields such as bioimaging, bioassay and electrochemical detection, etc..

4. Constructing Muti-functional Hybrid Materials Using SiO_2 Sphere as Supporting Material

In the above part, we have simply reviewed the case that many scientists around the world have developed diverse methods for synthesizing stylish functional inorganic nanoparticles with particular optical, electronic and magnetic properties. With the development of nanoscience and nanotechnology, the clever combination of different nanoscale materials can probably lead to the development of multifunctional nano-assembly system with optical, electronic and magnetic properties simultaneously. Silica, as a high stable supporting material, can commendably complete the above intention due to its excellent surface chemistry. In recent years, several groups have devoted their momentous efforts to design multifunctional advanced nanoarchitectures using biocompatible SiO_2 spheres as supporting materials.

Magnetic fluorescent nanoparticles using silica as supporting material have received great attention because of their bifunctional properties [97, 135-141]. Recently, magnetic fluorescent nanoparticles with magnetic nanoparticles located at the interior of silica shell have been facilely obtained [97, 135-136]. For instance, a new class of highly fluorescent, photostable, and magnetic core/shell nanoparticles in the submicrometer size range has been synthesized by Salgueiriño-Maceira et al. [135] from a modified Stöber method combined with the layer-by-layer (LbL) assembly technique. In this synthesis system, the luminescent magnetic nanoparticles were prepared via two main steps. The first step involves controlled addition of tetraethoxysilane to a dispersion of $Fe_3O_4/\gamma\text{-}Fe_2O_3$ nanoparticles, which are thereby homogeneously incorporated as cores into monodisperse silica spheres. The second step involves the LbL assembly of polyelectrolytes and luminescent CdTe quantum dots onto the surfaces of the silica-coated magnetite/maghemite particles, which are finally covered with an outer shell of silica. Nanoparticles with such a core/shell architecture have the added benefit of providing a robust platform from silica for incorporating diverse functionalities into a single nanoparticle. In addition, silica coated magnetic fluorescent nanospheres have been well-defined demonstrated by Ying et al. [138]. For example, a novel water-soluble hybrid material consisting of QDs and magnetic nanoparticles encapsulated in a silica shell has been facilely obtained via the microemulsion method [138]. Furthermore, silica nanospheres can also be used as temples for obtaining magnetic and fluorescent nanoparticles supported on the surface of silica simultaneously [139]. Hyeon and coworkers [139] designed a muti-step strategy for obtaining multifunctional composite nanoparticle assemblies by subsequent assembly of nanoparticles of CdSe/ZnS, Au or Pd on the magnetite nanoparticle-bearing silica spheres. The general synthetic procedure for multifunctional nanoparticle/silica sphere assemblies is shown in Figure 12. First, uniformly sized silica spheres were prepared by the Stöber method and then functionalized with amino groups by treatment with APTMS. Second, the Fe_3O_4 nanoparticles were assembled on the surfaces of the amino-functionalized silica spheres. Third, functional nanoparticles of CdSe/ZnS, Au or Pd were additionally assembled on the uncovered surface of SiO_2 spheres to give multifunctional assemblies. The synthesized multifunctional silica spheres exhibited a combination of magnetism and luminescence (CdSe/ZnS), surface plasmon resonance (Au) or catalysis (Pd). Silica coated magnetic nanoparticles and dye molecules can also provide muti-functional properties [140-141] besides the above muti-functional properties provided by magnetic and fluorescent nanoparticles. Mou and coworkers [140] developed a new strategy for the synthesis of uniform magnetic/luminescent functional mesoporous silica nanoparticles with well-ordered porous structure and good aqueous dispersity through the muti-step strategy. In the above muti-functional nanostructure, the dye molecule (FITC) was employed to supply the fluorescent function of hybrid spheres.

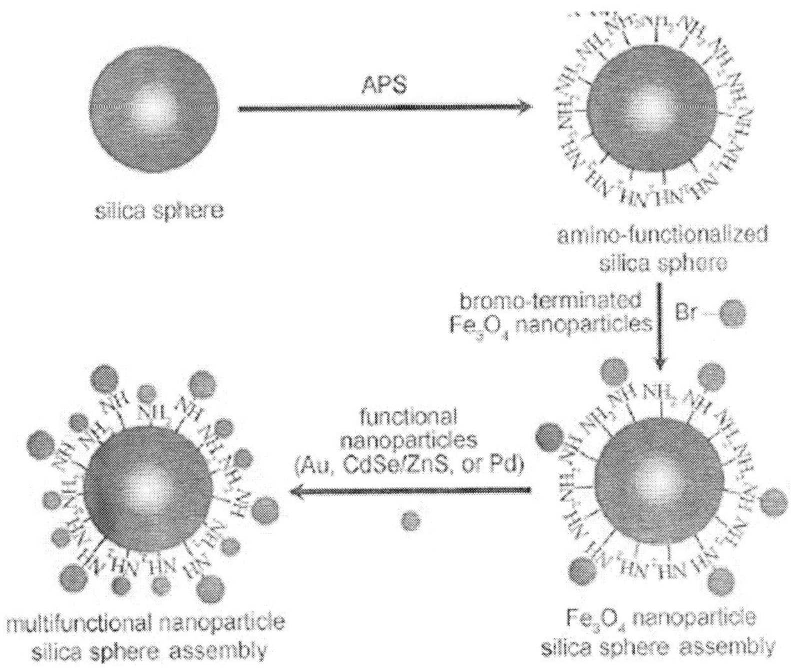

Figure 12. Synthetic procedure to obtain multifunctional nanoparticle/silica sphere assemblies. Reprinted with permission from Ref. 139, T. Hyeon, Angew. Chem. Int. Ed. 45, 4789 (2006), Copyright Wiley-VCH (2006).

Similarly, magnetic optical or magnetic catalytic hybrid nanospheres have also received great intensity in the recent years due to their applications in photothermal therapy, magnetic resonance imaging and catalysis. Several successful examples included the work reported by Salgueiriño-Maceira [142], Mirkin [143], Cho [144] and Ying [145]. For example, Salgueiriño-Maceira et al. [142] developed a method for obtaining bifunctional magnetic and optically tunable nanoparticles with a structural design involving a magnetic iron oxide core (Fe_3O_4/γ-Fe_2O_3) surrounded by a thick silica shell and further covered with an outer shell of gold. The whole preparation procedure is followed. The first step was carried out by using magnetic nanoparticles, which were coated with a silica shell using a sol-gel process based on the hydrolysis of TEOS. In this step, no primer was required because the iron oxide surface has a strong affinity for silica. The second step involved the assembly gold nanoparticles on the surface of the silica coated magnetic nanoparticles. Finally, gold shell was form on the surface of above assembling system. These obtained gold-coated magnetic silica spheres will take advantage of the strong resonance absorption gold shells show in the visible or near-infrared (NIR) range and can be controlled by using an external magnetic field, which makes them very promising in biomedical applications. Cho et al. [144] also found that magnetic and optical nanomaterials can also coated on the surface of monodisperse silica nanospheres through an advanced strategy. Figure 13a shows the synthetic approach for typical preparation process. First, 100 nm silica spheres were synthesized using the Stöber method, and then the surfaces of the particles were modified with 3-aminopropyltrimethoxysilane (Figure 13 b). Second, the Fe_3O_4 nanoparticles were then covalently attached to the amino-modified silica spheres (Figure 13c). Third, gold seed nanoparticles of 1-3 nm were attached to the residual amino groups of the silica spheres (Figure 13d). Finally, a complete 15-nm-thick gold shell with embedded Fe_3O_4 nanoparticles was formed around the silica spheres (Figure 13e). In addition, magnetic catalytic hybrid nanospheres have been well developed by Ying et al. [145], which reported on the synthesis of a nanocomposite of Pd nanoclusters supported on silica coated Fe_2O_3 nanoparticles. It is found that these bifunctional nanocomposites provided excellent catalytic reactivity and reusability in the hydrogenation of nitrobenzene.

5. Important Applications of Hybrid Nanoparticles Based on Silica

It is well known that ultimate target of nanomaterials will commendably complete their applications in many fields such as electronics, optics, optoelectronics, catalysis, sensor, biomedicine and others. Many groups have gained great benefits from studying nanoscience and nanotechnology, especially exploring the applications of different nanomaterials. Every year, excellent papers have still been published in great quantity for reporting the important applications caused by unique properties of nanomaterials. Unexceptionally, in this part, we will simply discuss recent advances on several important applications of hybrid nanoparticles based on silica.

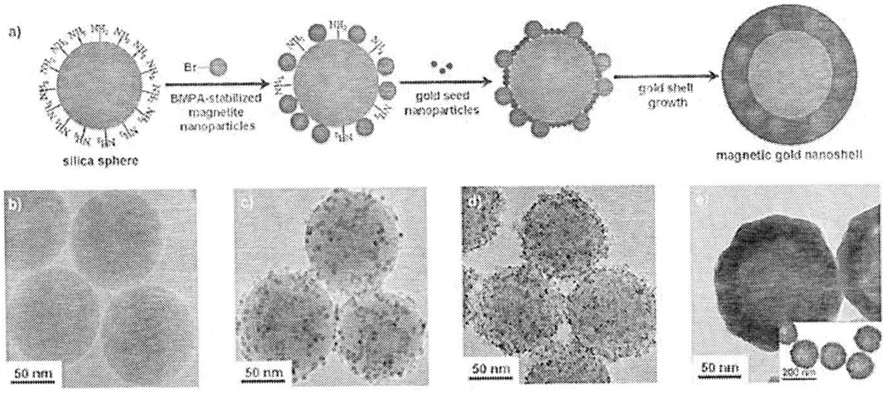

Figure 13. a) Synthesis of the magnetic gold nanoshells (Mag-GNS). TEM images of b) amino-modified silica spheres, c) silica spheres with Fe_3O_4 (magnetite) nanoparticles immobilized on their surfaces, d) silica spheres with Fe_3O_4 and gold nanoparticles immobilized on their surfaces, and e) the Mag-GNS. Reprinted with permission from Ref. 144, M. H. Cho, Angew. Chem. Int. Ed. 45, 7754 (2006), Copyright Wiley-VCH (2006).

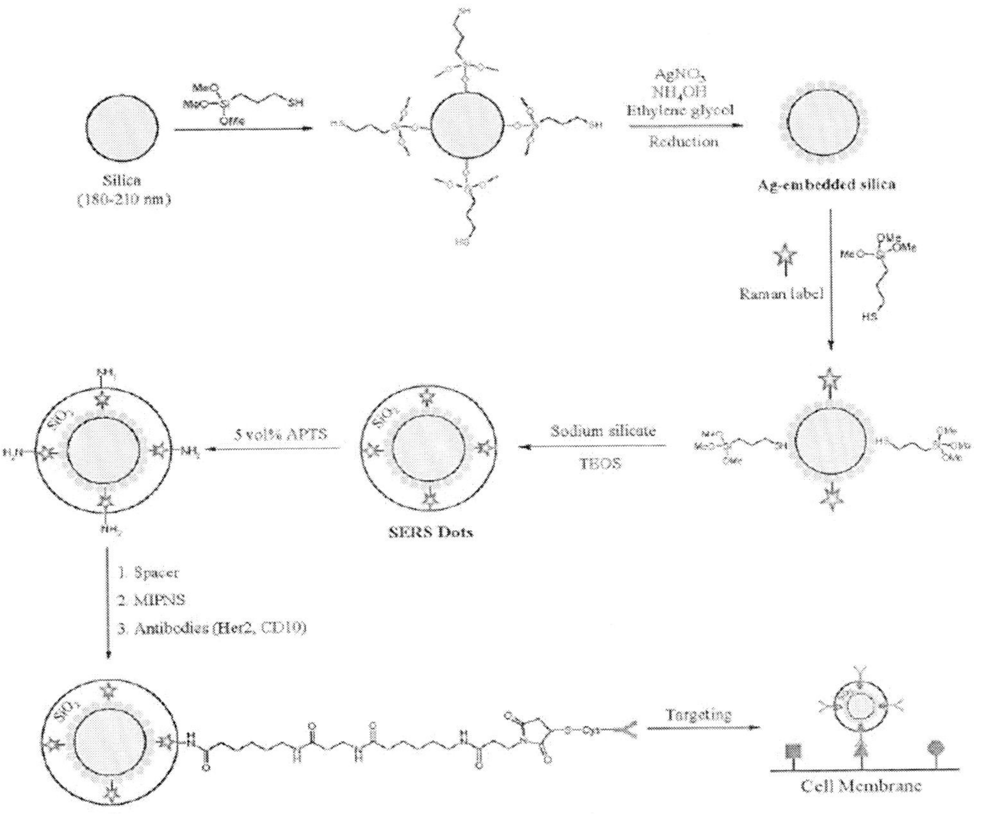

Figure 14. Schematic illustration of SERS dots synthesis and targeting of cellular cancer markers. Reprinted with permission from Ref. 147, Y. S. Lee, Anal. Chem. 78, 6967 (2006), Copyright American Chemical Society (2006).

Fluorescent or magnetic fluorescent nanoparticles based on silica have initially been used as in the field of biological and biomedical imaging [136, 137, 140, 146]. For instance, Wang et al. [136] designed a poly(N-isopropylacrylamide) (PNIPAM)-coated luminescent/magnetic silica microspheres. Further, the author demonstrated that PNIPAM-covered luminescent/magnetic microspheres could also be able to be taken up by CHO cells without inducement of any specific ligands. Ying and coworkers [137] demonstrated that Fe_2O_3/CdSe heterodimers could be coated with a thin homogeneous silica shell and used for the labeling of different live cell membranes through a simple bioconjugation method. Another important application of hybrid nanoparticles based on silica was focused on bioanalyis, biosensor and biodection [63, 66, 143, 147-148]. For instance, Ma et al. [66] designed a Au nanorod@SiO_2 nanostructure via an improved Stöber methods. These hybrid nanostructures can be easily deposited on the solid surface and further functionalized with goat anti-human-immunoglobulin G (anti-h-IgG). Finally, these anit-h-IgG modified hybrid nanostructures was successfully employed for the colorimetric detection (changes in the absorption of light) of h-IgG in a model reaction based on the specific binding affinity between the proteins. Significantly, the protein-recognition event on the Au rod@SiO_2 films could not only be monitored by changes in the absorption of light, but also was observable by the naked eye. In addition, construction of biosensors based on surface-enhanced Raman scatting (SERS) using hybrid nanoparticle based on silica is a promising direction for biodetection. For instance, Lee et al. [147] developed biocompatible, photostable, and multiplexing-compatible surface-enhanced Raman spectroscopic tagging material (SERS dots) composed of silver nanoparticle-embedded silica spheres and organic Raman labels for cellular cancer targeting in living cells. A typical preparation procedure for SERS dots synthesis and targeting of cellular cancer markers is shown in Figure 14. Finally, the obtained antibody-conjugated SERS dots were successfully demonstrated to be applied to the targeting of HER2 and CD10 on cellular membranes and exhibited good specificity for the targets. Furthermore, hybrid nanoparticles based on silica can also be used to some other important fields such as bioseparation, phototherapy and catalysis. Prominent examples include template-assisted fabrication of magnetic mesoporous silica-magnetite nanocomposite [149] and its potential for application in magnetic bioseparation (its ability to bind and elute DNA and extract RNA from bacterial cells) and preparation of magnetic gold nanoshells supported silica nanospheres for magnetic resonance imaging and photothermal therapeutic applications [150] and synthesis of magnetic nanocomposite catalysts for application in the catalytic hydrogenation of nitrobenzene with ease of recycling via magnetic separation [145].

Dye doped silica nanoparticles can be easily functionalized with different sensing probes and useful in many important fields such as biomedicine, nanobiology, and nanoelectrochemistry. Typical examples include leukemia cell identification using Ru(bpy) doped silica as an opticl probe [116, 117], parallel and high-throughput signaling of biomolecules by employing dual-dye based silica nanoparticles [120], multiplexed bacteria monitoring [127], staining probes for affymetrix genechips [128], and ultrasensitive DNA detection [130]. More recently, Ru(bpy) doped silica nanoparticles have been employed as advanced immobilization matrix for application in solid electrochemluminescence (ECL) system [113-115, 121, 123, 124]. For instance, Dong's group [113] reported a novel ECL sensor based on Ru(bpy) doped silica nanoparticles conjugated with a biopolymer chitosan membrane. Figure 15A shows cyclic voltammograms (CVs) of Ru(bpy) in the absence (dotted line) and presence (solid line) of 3×10^{-5} M tripropylamine (TPA) at the scan rate of

100 mV/s in PBS (pH 7.5). Since the TPA is quite small, it could permeate into the silica nanoparticle through the pores of the nanoparticles to react with the oxidized Ru(bpy). The presence of TPA made the oxidation current of Ru(bpy) increase clearly while the reduction current decreased. Figure 15B shows corresponding ECL intensity-time curve in the presence of 7.5 × 10^{-6} TPA under continuous potential scanning for ten cycles, which was no detectable change for ECL intensity. Furthermore, ECL intensity had good linearity with the TPA concentration and the linear range was wide, extending from 8.5 × 10^{-9} to 8.1 × 10^{-5} M with a remarkable detection limit (S/N=3) of 2.8 nM. Our group further extended to study the interactions between the Ru(bpy) dope silica nanoparticles and some biomacromolecules (BSA, lysozyme and ctDNA) [114]. The ECL behaviour of the RuSi nanoparticles was investigated after deposition with biomolecules through LBL self-assembly. When the surface of RuSi nanoparticles was modified using biomolecules, an ECL decrease with surface modification was observed. The more polyelectrolytes and biomolecules deposited on the RuSi nanoparticles, the more obvious ECL decrease was observed. The ECL decrease varied with different polyelectrolytes and biomolecules, and such decrease could be attributed to the changes of steric hindrance, limited diffusion of the coreactant and electrostatic interaction induced by surface modifications.

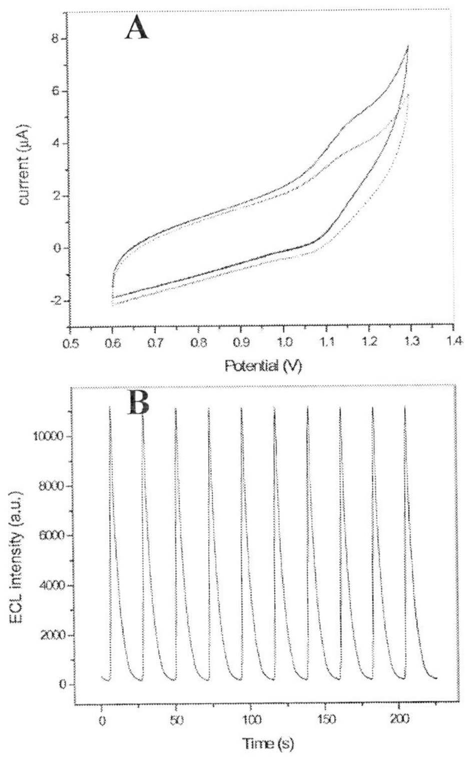

Figure 15. Cyclic voltammograms (A) of Ru(bpy) immobilized in composite film electrode in the absence (dashed line) and presence (solid line) of 3 × 10^{-5} M TPA in PBS (pH 7.5) at the scan rate of 100 mV/s; ECL intensity (B) of composite film electrode in PBS (pH 7.5) containing 7.5 × 10^{-6} M TPA under continuous CVs for 10 cycles with the scan rate of 100 mV/s. Reprinted with permission from Ref. 113, Y. S. Lee, Anal. Chem. 78, 5119 (2006), Copyright American Chemical Society (2006).

6. SUMMARY AND OUTLOOK

The design of hybrid nanoparticle based on silica, covering a wide range of functional materials, has made substantial progress, especially over the past decade. However, synthesis of high-quality inorganic nanoparticles using disverse ligands with OH groups as protective agents is still challenges to be faced in the coming years. Although silica coated inorganic nanoparticles make great progress, silica coated inorganic nanomaterials with different morphologies (rod, wire, tube, plate, cube and other novel shapes) will a promising research field. This is because these nanomaterials with novel morphologies have been reported to exhibit more peculiar physical and chemical properties and many important applications in many fields. Furthermore, in the case of silica coated inorganic nanoparticles, it should be noted that it is difficult or impossible to achieve a fully dense and nonporous silica coating, and it is thus difficult to maintain high stability of these nanoparticles under harsh conditions. So, we must explore novel synthetic methods to ensure the stability of inorganic nanoparticles at other conditions. Thus, carbon or polymer coated inorganic nanoparticle as a good substitute developed recently, provides good opportunities for scientists engaging in coating materials.

Complex, multifunctional nanoparticle systems based on silica with designed active sites, including nanoparticles, ligands, enzymes, chiral catalysts, drugs, and other species, seem to be promising for a variety of applications. Optional combination of the above functional components will bring enormous opportunities for scientists engaging in physics, chemistry, biology, medicine, material science and their different interdisciplinary fields. On the other hand, it is noted that constructing muti-functional hybrid materials using SiO_2 sphere as supporting material usually needs complex and tedious assembly process, which usually needs long time and meantime has poor reproducible. Industrial applications of such systems can not be achieved based on current case. Recent developed bifunctional dimer nanoparticles [137,151-162] probably provide good opportunities for solving the above problem. Finally, whether industrial applications of such systems can be achieved and how far we can go in all those areas will greatly depend on our ability to synthesize hybrid nanoparticles based on silica in a simple, economical and scalable fashion.

ABOUT THE AUTHORS

Erkang Wang obtained his BS at the Univ. of Shanghai in 1952 and his PhD under Prof. J. Heyrovsky at the Czechoslovak Academy of Sciences, Prague, in 1959. He has received many national and international awards for his research, which involve electro- and bioelectrochemistry and environmental analysis. He was Visiting Prof. at many institutions and has presented over 150 seminars and 70 plenary and keynote lectures. He is the Member of Chinese Academy of Sciences (CAS) and the Third World Academy of Sciences (TWAS). His current research interests are bioelectrochemistry including biosensors and biomembranes, capillary electrophoresis, synthesis, assembly and application of micro/ nanomaterials and nanoelectrochemistry, etc.

Shaojun Guo is studying for his PhD under the supervision of Prof. Erkang Wang at Changchun Institute of Applied Chemistry, Chinese Academy of Sciences. He received her BS degree (2005) in college of chemistry from Jilin University. In 2005, he joined Prof. Wang's group in the Changchun Institute of Applied Chemistry, Chinese Academy of Sciences. His current scientific interests are focused on synthesis, design, assembly and application of functional micro/nanomaterials.

Suggested Reviewer

Weihong Tan. Center for Research at the Bio/Nano Interface, Department of Chemistry, McKnight Brain Institute, University of Florida, GainesVille, Florida 32611-7200. E-mail: tan@chem.ufl.edu

ACKNOWLEDGMENT

This work was supported by the National Science Foundation of China. (Nos. 20575064, 20675076 and 20735003)

REFERENCES

[1] El-Sayed, M. A. *Acc. Chem. Res.* 2001, 34, 257-264.
[2] Murphy, C. J.; Jana, N. R. *Adv. Mater.* 2002, 14, 80-82.
[3] Kamat, P. V. J. Phys. Chem. B 2002, 106, 7729-7744.
[4] Zhang, J.; Liu, J.; Peng, Q.; Wang, X.; Li, Y. *Chem. Mater.* 2006, 18, 867-871.
[5] Lu, A.; Salabas, E. L.; Schüth, F. Angew. *Chem. Int. Ed.* 2007, 46, 1222-1244
[6] Srivastava, R.; Mani, P.; Hahn, N.; Strasser, P. Angew. *Chem. Int. Ed.* 2007, 46, 8988-8991.
[7] Chen, J.; Wang, D.; Xi, J.; Au, L.; Siekkinen, A.; Warsen, A.; Li, Z.-Y.; Zhang, H.; Xia, Y.; Li, X. *Nano Lett.* 2007, 7, 1318-1322.
[8] Jailswail, J. K.; Mattoussi, H.; Mauro, J. M.; Simon, S. M. *Nat. Biotechnol.* 2003, 21, 47-51.
[9] Huh, Y.; Lee, E.; Lee, J.; Jun, Y.; Kim, P.; Yun, C.; Kim, J.; Suh, J.; Cheon, *J. Adv. Mater.* 2007, 19, 3109-3112.
[10] Guo, S.; Fang, Y.; Dong, S.; Wang, E. *J. Phys. Chem.* C 2007, 111, 17104-17109.
[11] Doering, W. E.; Piotti, M. E.; Natan, M. J.; Freeman, R. G. *Adv. Mater.* 2007, 19, 3100-3108.
[12] Guo, S.; Wang, L.; Wang, E. *Chem. Commun.* 2007, 3163-3165.
[13] Guo, S.; Wang, E. *Inorg. Chem.* 2007, 46, 6740-6743.
[14] Guo, S.; Wang, E. *J. Colloid Interf. Sci.* 2007, 315, 795-799.
[15] Guo, S.; Wang, E. *Anal. Chim. Acta* 2007, 598, 181-192.
[16] Guo, S.; Wang, L.; Wang, Y.; Fang, Y.; Wang, E. *J. Colloid Interf. Sci.* 2007, 315, 363-368.
[17] Zhang, L.; Sun, X.; Song, Y.; Jiang, X.; Dong, S.; Wang, E. *Langmuir* 2006, 22, 2838-2843.
[18] Sun, X.; Dong, S.; Wang, E. *Polymer* 2004, 45, 2181-2184.
[19] Sun, X.; Jiang, X.; Dong, S.; Wang, E. *Macromol. Rapid Commun.* 2003, 24, 1024-1028.
[20] Cheng, W.; Dong, S.; Wang, E. *Chem. Mater.* 2003, 15, 2495-2501.
[21] Cheng, W.; Dong, S.; Wang, E. *Anal. Chem.* 2002, 74, 3599-3604.
[22] Cheng, W.; Dong, S.; Wang, E. *Langmuir* 2002, 18, 9947-9952.
[23] Cheng, W.; Dong, S.; Wang, E. *Langmuir* 2003, 19, 9434-9439.
[24] Cheng, W.; Wang, E. *J. Phys. Chem.* B 2004, 108, 24-26.
[25] Cheng, W.; Dong, S.; Wang, E. *J. Phys. Chem.* B 2004, 108, 19146-19154.
[26] Cheng, W.; Dong, S.; Wang, E. *Angew. Chem. Int. Ed.* 2002, 42, 449-452.
[27] Cheng, W.; Dong, S.; Wang, E. *J. Phys. Chem.* B. 2005, 109, 19213-19218.
[28] Sun, X.; Dong, S.; Wang, E. *Langmuir* 2005, 21, 4710-4712.
[29] Sun, X.; Dong, S.; Wang, E. *Angew. Chem. Int. Edit.* 2004, 43, 6360-6363.
[30] Sun, X.; Dong, S.; Wang, E. *Macromolecules* 2004, 37, 7105-7108.
[31] Xiong, Y.; McLellan, J. M.; Chen, J.; Yin, Y.; Li, Z.-Y.; Xia, Y. *J. Am. Chem. Soc.* 2005, 127, 17118-17127.
[32] Xiong, Y.; Cai, H.; Wiley, B. J.; Wang, J.; Kim, M. J.; Xia, Y. *J. Am. Chem. Soc.* 2007, 129, 3665-3675.
[33] Sun,Y.; Xia, Y. *Science* 2002, 298, 2176-2179.

[34] Tian, N.; Zhou, Z. Y.; Sun, S. G.; Ding, Y.; Wang, Z. L. *Science* 2007, 316, 732-735.
[35] Rao, C. N. R.; Vivekchand, S. R. C.; Biswas, Kanishka; Govindaraja, A.; *Dalton Trans.,* 2007, 3728-3749.
[36] Zheng, R.; Guo, S.; Dong, S. *Inorg. Chem.* 2007, 46, 6920-6923.
[37] Bao, H.; Wang, E.; Dong, S. *Small* 2006, 2, 476-480.
[38] Zheng, Y.; Gao, S.; Ying, J. Y. *Adv. Mater.* 2007, 19, 376-380.
[39] Pradhan, N.; Peng, X. *J. Am. Chem. Soc.* 2007, 129, 3339-3347.
[40] Pradhan, N.; Battaglia, D. M.; Liu,Y.; Peng, X. *Nano Lett.* 2007, 7, 312-317.
[41] Han, M.; Liu, Q.; He, J.; Song, Y.; Xu, Z.; Zhu, J. *Adv. Mater.* 2007, 19, 1096-1100.
[42] Ge, J.; Hu, Y.; Yin, Y. *Angew. Chem. Int. Ed.* 2007, 46, 7428-7431.
[43] Deng, H.; Li, X.; Peng, Q.; Wang, X.; Chen, J.; Li, Y. *Angew. Chem. Int. Ed.* 2005, 44, 2782 -2785.
[44] Ge, J.; Hu, Y.; Biasini, M.; Beyermann, W. P.; Yin, Y. *Angew. Chem. Int. Ed.* 2007, 46, 4342-4345.
[45] Chen, H. M.; Liu, R.; Li, H.; Zeng, H. *Angew. Chem. Int. Ed.* 2006, 45, 2713-2717.
[46] Gupta, A. K.; Gupta, M. *Biomaterials* 2005, 26, 3995-4021.
[47] Mornet, S.; Vasseur, S.; Grasset, F.; Verveka, P.; Goglio, G.; Demourgues, A.; Portier, J.; Pollert, E.; Duguet, E. *Prog. Solid State Chem.* 2006, 34, 237.
[48] Li, Z.; Wei, L.; Gao, M. Y.; Lei, H. *Adv. Mater.* 2005, 17, 1001-1005.
[49] Guo, S.; Fang, Y.; Dong, S.; Wang, E. *Inorg. Chem.* 2007, 46, 9537-9539.
[50] Li, J. J.; Wang, Y. A.; Guo, W.; Keay, J. C.; Mishima, T. D.; Johnson, M. B.; Peng, X. *J. Am. Chem. Soc.* 2003, 125, 12567-12575.
[51] Sun, S.; Murray, C. B.; Weller, D.; Folks, L.; Moser, A. *Science* 2000, 287, 1989-1992.
[52] Sun, S.; Zeng, H. *J. Am. Chem. Soc.* 2002, 124, 8204-8205.
[53] Park, J.; An, K. J.; Hwang, Y. S.; Park, J. G.; Noh, H. J.; Kim, J. Y.; Park, J. H.; Hwang, N. M.; Hyeon, T. *Nat. Mater.* 2004, 3, 891-895.
[54] Li, T.; Moon, J.; Morrone, A. A.; Mecholsky, J. J.; Talham, D. R.; Adair, J. H. *Langmuir* 1999, 15, 4328-4334.
[55] Wang, W.; Asher, S. A. J. *Am. Chem. Soc.* 2001, 123, 12528-12535.
[56] Liz-Marzán, L. M.; Giersig, M.; Mulvaney, P. *Langmuir* 1996, 12, 4329-4335.
[57] Ung, T.; Liz-Marzán, L. M.; Mulvaney, P. *Langmuir* 1998, 14, 3740-3748.
[58] Kobayashi, Y.; Correa-Duarte, M. A.; Liz-Marzán, L. M. *Langmuir* 2001, 17, 6375-6379.
[59] Salgueiriňo-Maceira, V.; Caruso, F.; Liz-Marzán, L. M. *J. Phys. Chem.* B 2003, 107, 10990-10994.
[60] Obare, S. O.; Jana, N. R.; Murphy, C. *J. Nano. Lett.* 2001, 1, 601-603.
[61] Graf, C.; Vossen, D. L. J.; Imhof, A.; van Blaaderen, A. *Langmuir* 2003, 19, 6693-6700.
[62] Pastoriza-Santos, I.; Pérez-Juste, J.; Liz-Marzán, L. M. *Chem. Mater.* 2006, 18, 2465-2467.
[63] Liu, S.; Han, M. *Adv. Funct. Mater.* 2005, 15, 961-967.
[64] Yin, Y.; Lu, Y.; Sun, Y.; Xia, Y. *Nano. Lett.* 2002, 2, 427-430.
[65] Lu, Y.; Yin, Y.; Li, Z.; Xia, Y. *Nano. Lett.* 2002, 2, 785-788.
[66] Wang, C.; Ma, Z.; Wang, T.; Su, Z. Adv. *Funct. Mater.* 2006, 16, 1673-1678.
[67] Wu, X.; Liu, H.; Liu, J.; Haley, K. N.; Treadway, J. A.; Larson, J. P.; Ge, N. F.; Peale, F.; Bruchez, M. P. *Nat. Biotechnol.* 2003, 21, 41-46.

[68] Chang, S.; Liu, L.; Asher, S. A. *J. Am. Chem. Soc.* 1994,116, 6139-6744.
[69] Yang, Y.; Gao, M. *Adv. Mater.* 2005, 17, 2354-2357.
[70] Li, Z.; Zhang, Y. *Angew. Chem. Int. Ed.* 2006, 45, 7732-7735.
[71] Wang, Y.; Tang, Z.; Liang, X.; Liz-Marzán, L. M.; Kotov, N. A. Nano. Lett. 2004, 4, 225-231.
[72] Wolcott, A.; Gerion, D.; Visconte, M.; Sun, J.; Schwartzberg, A.; Chen, S.; Zhang, J. Z. *J. Phys. Chem.* B 2006, 110, 5779-5789.
[73] Schroedter, A.; Weller, H.; Eritja, R.; Ford, W. E.; Wessels, J. M. *Nano. Lett.* 2002, 2, 1363-1367.
[74] Rogach, A. L.; Nagesha, D.; Ostrander, J. W.; Giersig, M.; Kotov, N. A. *Chem. Mater.* 2000, 12, 2676-2685.
[75] Parak, W. J.; Gerion, D.; Zanchet, D.; Woerz, A. S.; Pellegrino, T.; Micheel, C.; Williams, S. C.; Seitz, M.; Bruehl, R. E.; Bryant, Z.; Bustamante, C.; Bertozzi, C. R.; Alivisatos, A. P. *Chem. Mater.* 2002, 14, 2113-2119.
[76] Bruchez, M. J.; Moronne, M.; Gin, P.; Weiss, S.; Alivisatos, A. P. *Science* 1998, 281, 2013-2016.
[77] Gerion, D.; Pinaud, F.; Williams, S. C.; Parak, W. J.; Zanchet, D.; Weiss, S.; Alivisatos, A. P. *J. Phys. Chem.* B 2001, 105, 8861-8871.
[78] Mokari, T.; Sertchook, H.; Aharoni, A.; Ebenstein, Y.; Avnir, D.; Banin, U. *Chem. Mater.* 2005, 17, 258-263.
[79] Chan, Y.; Zimmer, J. P.; Stroh, M.; Steckel, J. S.; Jain, R. K.; Bawendi, M. G. *Adv. Mater.* 2004, 16, 2092-2097.
[80] Sorensen, L.; Strouse, G. F.; Stiegman, A. E. *Adv. Mater.* 2006, 18, 1965-1967.
[81] Nann, T.; Mulvaney, P. *Angew. Chem. Int. Ed.* 2004, 43, 5393-5396.
[82] Jana, N. R.; Earhart, C.; Ying, J. Y. *Chem. Mater.* 2007, 19, 5074-5082.
[83] Rogach, A. L.; Kotov, N. A.; Koktysh, D. S.; Susha, A. S.; Caruso, F. Colloid Surf. A: Physicochem. *Eng. Aspects* 2002, 202, 135-144.
[84] Selvan, S. T.; Tan, T. T.; Ying, J. Y. *Adv. Mater.* 2005, 17, 1620-1625.
[85] Darbandi, M.; Thomann, R.; Nann, T. *Chem. Mater.* 2005, 17, 5720-5725.
[86] Darbandi, M.; Lu, W.; Fang, J.; Nann, T. *Langmuir* 2006, 22, 4371-4375.
[87] Tan, T. T.; Selvan, S. T.; Zhao, L.; Gao, S.; Ying, J. Y. *Chem. Mater.* 2007, 19, 3112-3117.
[88] Lu, A.; Salabas, E. L.; Schüth, F. *Angew. Chem. Int. Ed.* 2007, 46, 1222-1244.
[89] James, R. O.; Bertucci, S. J.; Oltean, *G. L. U.S. Patent* 5, 217, 804, 1993.
[90] Ohmori, M.; Matijevic, E. J. *Colloid Interface Sci.* 1992, 150, 594-598.
[91] Tang, D.; Yuan, R.; Chai, Y.; An, H. *Adv. Funct. Mater.* 2007, 17, 976-982.
[92] Correa-Duarte, M. A.; Giersig, M.; Kotov, N. A.; Liz-Marzán, L. M. *Langmuir* 1998, 14, 6430-6435.
[93] Lu, Y.; Yin, Y.; Mayers, B. T.; Xia, Y. Nano. *Lett.* 2002, 2, 183-186.
[94] Philipse, A. P.; van Bruggen, M. P. B.; Pathmamanoharan, C. *Langmuir* 1994, 10, 92-99.
[95] Aliev, F. G.; Correa-Duarte, M. A.; Mamedov, A.; Ostrander, J. W.; Giersig, M.; Liz-Marzán, L. M.; Kotov, N. A. *Adv. Mater.* 1999, 11, 1006-1010.
[96] Cannas, C.; Musinu, A.; Peddis, D.; Piccaluga, G. Chem. Mater. 2006, 18, 3835-3842.
[97] Chen, M.; Gao, L.; Yang, S.; Sun, J. *Chem. Commun.* 2007, 1272-1274.

[98] Xu, X.; Deng, C.; Gao, M.; Yu, W.; Yang, P.; Zhang, X. *Adv. Mater.* 2006, 18, 3289-3293.
[99] Deng, Y.; Qi, D.; Deng, C.; Zhang, X.; Zhao, D. 2008, 130, 28-29.
[100] Ge, J.; Hu, Y.; Zhang, T.; Yin, Y. *J. Am. Chem. Soc.* 2007, 129, 8974-8975.
[101] Salgueiriño-Maceira, V.; Correa-Duarte, M. A. *J. Mater. Chem.* 2006, 16, 3593–3597.
[102] Salgueiriño-Maceira, V.; Correa-Duarte, M. A.; Farle, M.; López-Quintela, M. A.; Sieradzki, K.; Diaz, R. *Langmuir* 2006, 22, 1455-1458.
[103] Kobayashi, Y.; Horie, M.; Konno, M.; Rodriguez-González, B.; Liz-Marzán, L. M. *J. Phys. Chem.* B 2003, 107, 7420-7425.
[104] Yi, D. K.; Lee, S. S.; Papaefthymiou, G. C.; Ying, J. Y. *Chem. Mater.* 2006, 18, 614-619.
[105] Vestal, C. R.; Zhang, Z. J. *Nano. Lett.* 2003, 3, 1739-1743.
[106] Lee, D. C.; Mikulec, F. V.; Pelaez, J. M.; Koo, B.; Korgel, B. A. *J. Phys. Chem. B* 2006, 110, 11160-11166.
[107] Santra, S.; Tapec, R.; Theodoropoulou, N.; Dobson, J.; Hebard, A.; Tan, W. *Langmuir* 2001, 17, 2900-2906.
[108] Meng, Y.; Chen, D.; Jiao, X. *J. Phys. Chem.* B 2006, 110, 15212-15217.
[109] Wu, P.; Zhu, J.; Xu, Z. Adv. *Funct. Mater.* 2004, 14, 345-351.
[110] Sen, T.; Sebastianelli, A.; Bruce, I. J. *J. Am. Chem. Soc.* 2006, 128, 7130-7131.
[111] Zhao, W.; Gu, J.; Zhang, L.; Chen, H.; Shi, J. *J. Am. Chem. Soc.* 2005, 127, 8916-8917.
[112] Lu, A. H.; Li, W. C.; Kiefer, A.; Schmidt, W.; Bill, E.; Fink, G.; Schüth, F. *J. Am. Chem. Soc.* 2004, 126, 8616-8617.
[113] Zhang, L.; Dong, S. *Anal. Chem.* 2006, 78, 5119-5123.
[114] Wei, H.; Liu, J.; Zhou, L.; Li, J.; Jiang, X.; Kang, J.; Yang, X.; Dong, S.; Wang, E. *Chem. Eur.* J. 2008, accepted.
[115] Wei, H.; Wang, E. *Chem. Lett.* 2007, 36, 210-211.
[116] Santra, S.; Zhang, P.; Wang, K.; Tapec, R.; Tan, W. *Anal. Chem.* 2001, 73, 4988-4993.
[117] Herr, J. K.; Smith, J. E.; Medley, C. D.; Shangguan, D.; Tan, W. *Anal. Chem.* 2006, 78, 2918-2924.
[118] Wang, X.; Zhou, J.; Yun, W.; Xiao, S.; Chang, Z.; He, P.; Fang, Y. *Anal. Chim. Acta* 2007, 598, 242–248.
[119] Bagwe, R. P.; Yang, C.; Hilliard, L. R.; Tan, W. *Langmuir* 2004, 20, 8336-8342.
[120] Wang, L.; Yang, C.; Tan, W. *Nano. Letter.* 2005, 5, 37-43.
[121] Chang, Z.; Zhou, J.; Zhao, K.; Zhu, N.; He, P.; Fang, Y. *Electrochim. Acta* 2006, 52, 575–580.
[122] Santra, S.; Wang, K.; Tapec, R.; Tan, W. *J. Biomed. Optics* 2001, 6, 160-166.
[123] Zhang, L.; Dong, S. *Electrochem. Commun.* 2006, 8, 1687-1691.
[124] Zhang, L.; Liu, B.; Dong, S. *J. Phys. Chem.* B 2007, 111, 10448-10452.
[125] Zhao, X.; Bagwe, R. P.; Tan, W. *Adv. Mater.* 2004, 16, 173-176.
[126] Zhou, X.; Zhou, *J. Anal. Chem.* 2004, 76, 5302-5312.
[127] Wang, L.; Zhao, W.; O'Donoghue, M. B.; Tan, W. *Bioconjugate Chem.* 2007, 18, 297-301.
[128] Wang, L.; Lofton, C.; Popp, M.; Tan, W. *Bioconjugate Chem.* 2007, 18, 610-613.
[129] Kim, S. H.; Jeyakumar, M.; Katzenellenbogen, J. A. *J. Am. Chem. Soc.* 2007, 129, 13254-13264.
[130] Zhao, X.; Tapec-Dytioco, R.; Tan, W. *J. Am. Chem. Soc.* 2003, 125, 11474-11475.

[131] Montalti, M.; Prodi, L.; Zaccheroni, N.; Battistini, G.; Marcuz, S.; Mancin, F.; Rampazzo, E.; Tonellato, U. *Langmuir* 2006, 22, 5877-5881.

[132] Wang, L.; Tan, W. *Nano. Lett.* 2006, 6, 84-88.

[133] Green, M.; Harries, J.; Wakefield, G.; Taylor, R. *J. Am. Chem. Soc.* 2005, 127, 12812-12813.

[134] Qi, W.; Li, H.; Wu, L. *Adv. Mater.* 2007, 19, 1983-1987.

[135] Salgueiriño-Maceira, V.; Correa-Duarte, M. A.; Spasova, M.; Liz-Marzán, L. M.; Farle, M. *Adv. Funct. Mater.* 2006, 16, 509–514.

[136] Guo, J.; Yang, W.; Wang, C.; He, J.; Chen, *J. Chem. Mater.* 2006, 18, 5554-5562.

[137] Selvan, S. T.; Patra, P. K.; Ang, C. Y.; Ying, J. Y. *Angew. Chem. Int. Ed.* 2007, 46, 2448 -2452.

[138] Yi, D. K.; Selvan, T.; Lee, S. S.; Papaefthymiou, G. C.; Kundaliya, D.; Ying, J. Y. *J. Am. Chem. Soc.* 2005, 127, 4990-4991.

[139] Kim, J.; Lee, J. E.; Lee, J.; Jang, Y.; Kim, S.W.; An, K.; Yu, J. H.; Hyeon, T. *Angew. Chem. Int. Ed.* 2006, 45, 4789-4793.

[140] Lin, Y. S.; Wu, S. H.; Hung, Y.; Chou, Y. H.; Chang, C.; Lin, M. L.; Tsai, C. P.; Mou, C. Y. *Chem. Mater.* 2006, 18, 5170-5172.

[141] Ma, D.; Guan, J.; Normandin, F.; Dénommeé, S.; Enright, G.; Veres, T.; Simard, B. *Chem. Mater.* 2006, 18, 1920-1927.

[142] Salgueiriño-Maceira, V.; Correa-Duarte, M. A.; Farle, M.; López-Quintela, A.; Sieradzki, K.; Diaz, R. *Chem. Mater.* 2006, 18, 2701-2706.

[143] Stoeva, S. I.; Huo, F.; Lee, J. S.; Mirkin, C. A. *J. Am. Chem. Soc.* 2005, 127, 15362-15363.

[144] Kim, J.; Park, S.; Lee, J. E.; Jin, S. M.; Lee, J. H.; Lee, I. S.; Yang, I.; Kim, J. S.; Kim, S. K.; Cho, M. H.; Hyeon, T. *Angew. Chem. Int. Ed.* 2006, 45, 7754-7758.

[145] Yi, D. K.; Lee, S. S.; Ying, J. Y. *Chem. Mater.* 2006, 18, 2459-2461.

[146] Santra, S.; Zhang, P.; Wang, K.; Tapec, R.; Tan, W. *Anal. Chem.* 2001, 73, 4988-4993.

[147] Kim, J. H.; Kim, J. S.; Choi, H.; Lee, S. M.; Jun, B. H.; Yu, K. N.; Kuk, E.; Kim, Y. K.; Jeong, D. H.; Cho, M. H.; Lee, Y. S. *Anal. Chem.* 2006, 78, 6967-6973.

[148] Zhang, L.; Liu, B.; Dong, S. *J. Phys. Chem.* B 2007, 111, 10448-10452.

[149] Sen, T.; Sebastianelli, A.; Bruce, I. J. *J. Am. Chem. Soc.* 2006, 128, 7130-7131.

[150] Kim, J.; Park, S.; Lee, J. E.; Jin, S. M.; Lee, J. H.; Lee, I. S.; Yang, Il.; Kim, J. S.; Kim, S. K.; Cho, M. H.; Hyeon, T. *Angew. Chem. Int. Ed.* 2006, 45, 7754-7758.

[151] Choi, J. S.; Jun, Y. W.; Yeon, S. I.; Kim, H. C.; Shin, J. S.; Cheon, J. *J. Am. Chem. Soc.* 2006, 128, 15982-15983.

[152] Chiang, I. C.; Chen, D. H. *Adv. Funct. Mater.* 2007, 17, 1311-1316.

[153] Yu, H.; Cen, M.; Rice, P. M.; Wang, S. X.; White, R. L.; Sun, S. *Nano Lett.* 2005, 5, 379-382.

[154] Pellegrino, T.; Fiore, A.; Carlino, E.; Giannini, C.; Cozzoli, P. D.; Ciocarella, G.; Respaud, M.; Palmirotta, L.; Cingolani, R.; Manna, L. *J. Am. Chem. Soc.* 2006, 128, 6690-6698.

[155] Xu, Z.; Hou, Y.; Sun, S. *J. Am. Chem. Soc.* 2007, 129, 8698-8699.

[156] Li, Y. Q.; Zhang, Q.; Nurmikko, A. V.; Sun, S. *Nano Lett.* 2005, 5, 1689-1692.

[157] Gu, H. W.; Zheng, R. K.; Zhang, X. X.; Xu, B. *J. Am. Chem. Soc.* 2004, 126, 5664-5665.

[158] Gu, H. W.; Yang, Z. M.; Gao, J. H.; Chang, C. K.; Xu, B. *J. Am. Chem. Soc.* 2005, 127, 34-35.

[159] Teranishi, T.; Inoue, Y.; Nakaya, M.; Oumi, Y.; Sano, T. *J. Am. Chem. Soc.* 2004, 126, 9914-9915.

[160] Shi, W.; Zeng, H.; Sahoo, Y.; Ohulchanskyy, T. Y.; Ding, Y.; Wang, Z. L.; Swihart, M.; Prasad, P. N. *Nano Lett.* 2006, 6, 875-881.

[161] Kudera, S.; Carbone, L.; Casula, M. F.; Cingolani, R.; Falqui, A.; Snoeck, E.; Parak, W. J.; Manna, L. *Nano Lett.* 2005, 5, 445-449.

[162] Kwon, K. W.; Shim, M. *J. Am. Chem. Soc.* 2005, 127, 10269-10275.

In: Encyclopedia of Nanotechnology
Editor: Elwood D. Carlson

ISBN: 978-1-60692-079-4

Chapter 41

HEAT TRANSFER OF NANOPARTICLE SUSPENSIONS (NANOFLUIDS)

S. M. Sohel Murshed, Kai Choong Leong and Chun Yang
School of Mechanical and Aerospace Engineering
Nanyang Technological University, Republic of Singapore

ABSTRACT

Nanofluids, a dispersion of nanoparticles in conventional heat transfer fluids, promises to offer enhanced heat transfer performance. Nanofluids are a new, innovative class of heat transfer fluids and represent a rapidly emerging field where nanoscale science and thermal engineering meet. In recent years, nanofluids have attracted great interest from researchers of multi-disciplines because of their superior thermal properties and potential applications in important fields such as microelectronics, microfluidics, transportation, and biomedical. Published research works have shown that nanofluids possess higher effective thermal conductivity and effective thermal diffusivity compared to their base fluids and the magnitudes of these properties increase remarkably with increasing nanoparticle volume fraction. Particle size and shape as well as fluid temperature are also found to influence the enhancement of thermal conductivity of the nanofluids. Despite numerous theoretical studies on model development for nanofluids, there is no widely accepted model available due to inconclusive heat transfer mechanisms of nanofluids. There are also many inconsistencies in reported experimental results and controversies in the proposed mechanisms for enhanced thermophysical properties of nanofluids. In this chapter, the current state-of-research in nanofluids including synthesis, potential applications, experimental and analytical studies on the effective thermal conductivity, thermal diffusivity, and viscosity of nanofluids are critically reviewed. Results from the authors' extensive theoretical and experimental studies on thermophysical and electrokinetic properties of nanofluids are also summarized.

1. INTRODUCTION

1.1. Development and Concept of Nanofluids

With an ever-increasing thermal load due to smaller features of microelectronic devices and more power output, cooling for maintaining desirable performance and durability of such devices is one of the most important technical issues in many high-tech industries. The conventional method to increase the cooling rate is to use extended heat transfer surfaces. However, this approach requires an undesirable increase in the size of the thermal management system. In addition, the inherently poor thermophysical properties of traditional heat transfer fluids such as water, ethylene glycol (EG) or engine oil (EO) greatly limit the cooling performance. Thus, these conventional cooling techniques are not suitable to meet the demand of these high-tech industries. There is therefore, a need to develop advanced cooling techniques and innovative heat transfer fluids with improved heat transfer performance.

It is well known that at room temperature, metals possess at least an order-of- magnitude higher thermal conductivity than fluids. For example, the thermal conductivity of copper at room temperature is about 700 times greater than that of water and about 3000 times greater than that of engine oil (Figure 1). Therefore, the thermal conductivities of fluids that contain suspended metallic or nonmetallic (oxide) particles are expected to be significantly higher than those of conventional heat transfer fluids. As thermal conductivity of a fluid plays a vital role in the development of energy-efficient heat transfer equipment, numerous theoretical and experimental studies on increasing the thermal conductivity of liquids by suspending small particles have been conducted since the treatise by Maxwell more than a century ago [1]. However, all such studies on thermal conductivity of suspensions have been confined to millimeter- or micrometer-sized particles. The major problems of such suspensions are the rapid settling of these particles, clogging of flow channels, and increased pressure drop in the fluid. Even if the fluid is kept circulating rapidly enough to prevent much settling, the micro-particles would damage the walls of the heat transfer devices (e.g. pipes and channels) and wear them thin. In contrast, nanoparticles in a liquid can maintain better suspension and thereby reduce erosion and clogging.

Over the last several decades, scientists and engineers have attempted to develop fluids, which offer better cooling or heating performance. However, it is only in 1995 that Choi and his co-workers at Argonne National Laboratory of USA [3] developed the novel concept of a "nanofluid" to meet the cooling challenges facing many high-tech industries. This new class of heat transfer fluids (nanofluids) is engineered by dispersing nanometer-sized solid particles, rods or tubes in traditional heat transfer fluids. From past investigations, nanofluids were found to exhibit significantly higher thermophysical properties, particularly thermal conductivity than those of base fluids [4-12]. Thus, nanofluids have attracted great interest from the research community due to their potential benefits and applications in numerous important fields such as microelectronics, microfluidics, transportation, manufacturing, medical, and so on.

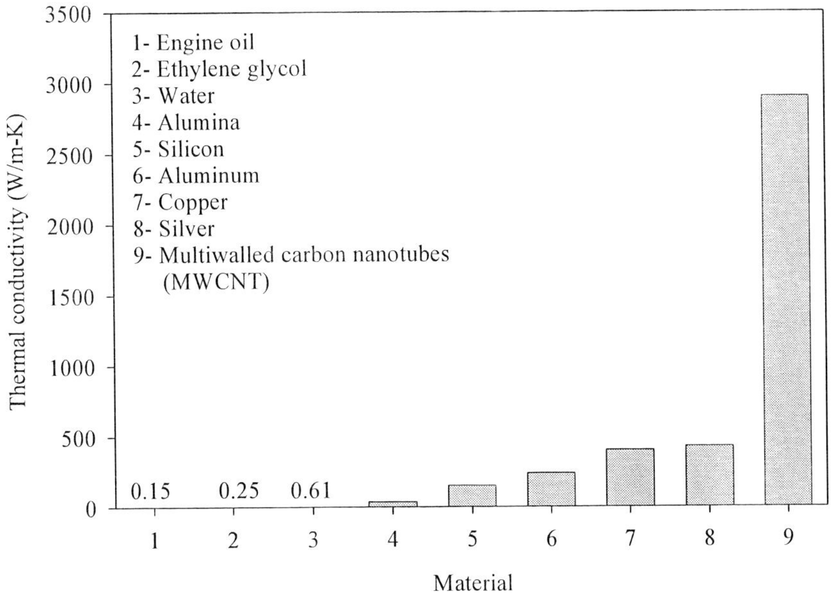

Figure 1. Thermal conductivity of typical materials (solids and liquids) at 300 K [2].

1.2. Impact and Potential Benefits of Nanofluids

The impact of nanofluids is expected to be great considering that heat transfer performance of heat exchangers or cooling devices is vital in many industries. For example, the transport industry has a strong incentive to reduce the size and weight of vehicle thermal management systems and nanofluids can increase thermal transport of coolants and lubricants. When the nanoparticles are properly dispersed, nanofluids offer numerous benefits [2] besides the anomalously high effective thermal conductivity. These benefits include:

1.2a. Improved Heat Transfer and Stability

Because heat transfer takes place at the surface of the particles, the much larger relative surface areas of nanoparticles compared to micro-particles, provide significantly improved heat transfer capabilities. In addition, particles finer than 20 nm carry 20% of their atoms on their surface, making them instantaneously available for thermal interaction [2].

1.2b. Microchannel Cooling without Clogging

Nanofluids will not only be a better medium for heat transfer in general, but they will also be ideal for microchannel applications where high heat loads are encountered. The combination of microchannels and nanofluids will provide both highly conducting fluids and a large heat transfer area. This cannot be attained with microparticles because they can easily clog the microchannels.

1.2c. Miniaturized Systems

Nanofluid technology can support the current industrial trend toward component and system miniaturization by enabling the design of smaller and lighter heat exchanger systems.

Miniaturized systems will reduce the inventory of heat transfer fluids and will result in cost savings.

The better stability of nanofluids will prevent rapid settling and reduce clogging in the walls of heat transfer devices. The high thermal conductivity of nanofluids translates into higher energy efficiency, better performance, and lower operating costs. They can reduce energy consumption for pumping heat transfer fluids. Thermal systems can be smaller and lighter. In vehicles, smaller components result in better gasoline mileage, fuel savings, lower emissions, and a cleaner environment [13].

1.3. Potential Applications of Nanofluids

With the aforementioned highly desired thermal properties and potential benefits, nanofluids are thought to have a wide range of applications, which are elaborated here.

1.3a. Nanofluids in Transportation

Because engine coolants, engine oils, automatic transmission fluids, and other synthetic high temperature fluids currently possess inherently poor heat transfer capabilities, they could benefit from the high thermal conductivity offered by nanofluids. Nanofluids would allow for smaller, lighter engines, pumps, radiators, and other components.

1.3b. In Micromechanics

Micro-electromechanical systems (MEMS) generate a lot of heat during operation. Conventional coolants do not work with MEMS because they do not have enough cooling capability. Moreover, even if solid particles are added to these coolants to enhance their thermal conductivity, they still could not work, because the particles would be too big to flow smoothly in the extremely narrow cooling channels required by MEMS. Since nanofluids can flow in microchannels without clogging, they would be suitable coolants. They could enhance cooling of MEMS under extreme heat flux conditions.

1.3c. In Electronics and Instrumentation

Demand for ultra-high-performance cooling in this area has been increasing, and conventional enhanced surface techniques have reached their limit with regards to improving heat transfer. Since nanoparticles are much smaller than the diameter of microchannels, smooth –flowing nanofluids could support this demand.

1.3d. In Medical Applications

Magnetic nanoparticles in biofluids can be used as delivery vehicles for drugs or radiation, providing new cancer treatment techniques. Nanoparticles are more adhesive to tumor cells than normal cells. Thus, magnetic nanoparticles excited by an AC magnetic field are promising for cancer therapy. The combined effect of radiation and hyperthermia is due to the heat-induced malfunction of the repair process right after radiation-induced DNA damage. In the near future, nanofluids can be used as advanced drug reduction fluids.

2. SYNTHESIS OF NANOFLUIDS

Preparation of nanofluids is the first key step to investigate their heat transfer performance and novel characteristics. Modern fabrication technologies provide great opportunities to process materials at the nanometer scale. Nanostructured materials exhibit new or enhanced properties, which are not exhibited by the bulk solids. There are mainly two techniques for synthesizing nanofluids: the two-step process, and the direct evaporation technique or single-step process [2-4].

In the two-step process, dry nanoparticles are first produced by an inert gas condensation method and they are then dispersed into a fluid. Figure 2 shows TEM images of several types of commercially available nanoparticles. This method may result in a large degree of nanoparticle agglomeration. Thus, proper dispersion techniques and small volume fraction of nanoparticles are important to produce stable nanofluids by this technique. An advantage of the two-step process in terms of eventual commercialization of nanofluids is that the inert gas condensation technique can produce large quantities of nanopowders.

The direct evaporation technique synthesizes nanoparticles and disperses them into a fluid in a single step. As with the inert gas condensation technique, this technique involves the vaporization of a source material under vacuum conditions. An advantage of this process is that nanoparticle agglomeration is minimized. The disadvantages are that the liquid must have a very low vapor pressure and that this technique can produce very limited amounts of nanofluids. At present, most researchers used the two-step process to produce nanofluids by dispersing commercial or self-produced nanoparticles in a liquid. The optimization of thermal properties of nanofluids requires stable nanofluids, which can be ensured by proper synthesis and dispersion procedures.

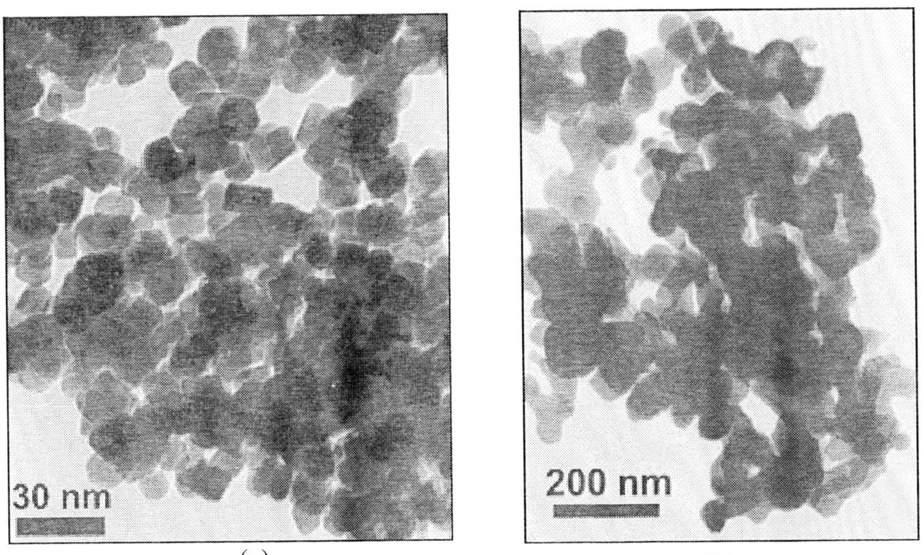

(a) (b)

Figure 2. TEM photographs of nanoparticles from Nanostructured and Amorphous Materials, Inc., USA – (a) TiO_2 (15 nm) and (b) Al_2O_3 (80 nm).

Figure 3. Ultrasonic dismembrator probe inside sample nanofluids.

3. Preparation and Characterization of Sample Nanofluids

A nanofluid does not mean a simple mixture of liquid and solid nanoparticles. Techniques for good dispersion of nanoparticles in liquids or directly producing stable nanofluids are crucial. To prepare nanofluids by the two-step process i.e. by suspending nanoparticles into base fluids, proper mixing and stabilization of the particles are essential.

Like most of the reported nanofluids, by dispersing commercially available nanoparticles in base fluids sample nanofluids were prepared for the present study. This is the two-step process of preparing nanofluids. Several types of nanoparticles were purchased from Nanostructured and Amorphous Materials, Inc., USA. These are spherical-shape titanium dioxide (TiO_2) nanoparticles of Φ15 nm and rod-shape of Φ10 nm × L 40 nm (here Φ is the diameter and L is the length), spherical-shape alumina (Al_2O_3) nanoparticles of Φ80 nm and Φ150 nm, and spherical-shape aluminum (Al) nanoparticles of Φ80 nm. Deionized water, ethylene glycol, and engine oil were used as base fluids.

After different volumetric loading of these nanoparticles (0.1 to 5%) into base fluids, their proper mixing was ensured by 8-10 hours of homogenization using an Ultrasonic Dismembrator (Fisher Scientific Model 500) as shown in Figure 3. Ultrasonication is very important in breaking the larger clusters into smaller clusters [14]. Figure 4 depicts the procedure for the preparation and characterization of sample nanofluids. A Transmission Electron Microscope (TEM) and a Particle Size Analyzer (PSA) were then used to characterize the dispersion, clustering and morphology of nanoparticles in base fluids, which can influence heat transfer performance of nanofluids.

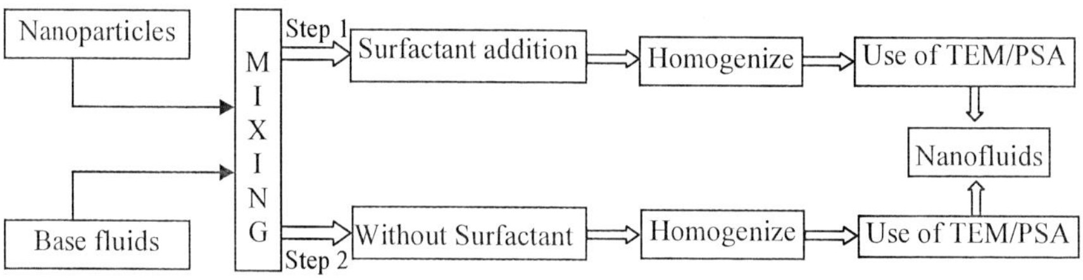

Figure 4. Process flow chart for sample preparation of nanofluids.

Primarily, the particles were neither uniformly dispersed in the base fluids nor a satisfactory stability of the nanofluids were observed. There is, therefore, a need to use surfactants to break down the particle agglomeration in suspension. Cetyl Trimethyl Ammonium Bromide (CTAB) surfactant was used to ensure better stability and proper dispersion without affecting the thermophysical properties and single-phase heat transfer performance of the nanofluids. After addition of CTAB, all types of nanoparticles were found to be well dispersed in their base liquids and formed stable suspensions. Almost no sedimentation of nanoparticles was observed within four to five days after sample preparation. Moreover, engine oil-based nanofluids were found to be stable for about few weeks. This is not surprising because engine oil is a highly viscous liquid compared to water. Based on Stokes' sedimentation law [15], the rate of particle sedimentation or Stokes settling velocity is low for small particles and highly viscous liquids. Before characterizing the thermophysical or other properties of nanofluids, proper sample preparation and its stability are very important.

Table 1. Thermophysical properties of base fluids at 300 K [16]

Base fluids	Thermal conductivity, k (W/m·K)	Thermal diffusivity, α (m^2/s)	Density, ρ (kg/m^3)	Specific heat, c_p (kJ/kg·K)	Kinematic viscosity, υ (m^2/s)
Deionized water (DIW)	0.607	14.55×10^{-8}	998	4.2	9.2×10^{-7}
Ethylene glycol (EG)	0.255	9.385×10^{-8}	1111	2.4	18.1×10^{-6}
Engine oil (EO)	0.145	8.740×10^{-8}	884	1.9	9.44×10^{-4}

Table 2. Thermophysical properties of several nanoparticles at 300 K [16, 17]

Nanoparticles	Thermal conductivity, k (W/m·K)	Thermal diffusivity, α (m^2/s)	Density, ρ (kg/m^3)	Specific heat, c_p (kJ/kg·K)
TiO_2	8.04	2.9×10^{-6}	4000	0.711
CuO	17.65	5.17×10^{-6}	6500	0.525
Al_2O_3	39	11.9×10^{-6}	3970	0.775
Al	237	97.1×10^{-6}	2700	0.877
Cu	401	117×10^{-6}	8933	0.385

4. Properties of Base Fluids and Nanoparticles

The thermophysical properties of both the nanoparticle and the base fluid are important in preparing sample nanofluids as well as determining the thermophysical properties such as thermal conductivity and viscosity of nanofluids. Therefore, the thermophysical properties of commonly used base fluids and nanoparticles are given in Tables 1 and 2.

5. Experimental Studies on Thermal Conductivity of Nanofluids

5.1. Measurement Method

Since the transient hot-wire (THW) method was first suggested by Stalhane and Pyk in 1931 to measure the absolute thermal conductivity of powders [14], many researchers have modified the method to make it more accurate. With modern electronic instrumentation and development of a proper theoretical basis, this method has become an accurate means of determining the thermal conductivity of fluids. Attempts were made by several researchers to extend the transient hot-wire method to measure the thermal conductivity of electrically conducting media. Nagasaka and Nagashima [19] performed thermal conductivity measurements of electrically conducting liquids by considering the electrical insulation layer effect on the heat transfer system in the rig. The transient hot–wire method has proved to be one of the most accurate techniques of determining the thermal conductivity of fluids [18-19]. The advantage of this method lies in its elimination of natural convection effects which will cause measurement errors. In addition, this method is very fast compared to other techniques. The conceptual design of this hot-wire apparatus is also simple when compared to other techniques. Although several studies reported the use of the steady-state technique [7, 20], temperature oscillation technique [9], and the 3ω-wire method [21] to measure the effective thermal conductivity of nanofluids, these techniques are not as accurate as the transient hot–wire method. The temperature oscillation technique measures the thermal diffusivity and the thermal conductivity is then calculated using the volumetric specific heat of the sample. Similar to the hot-wire method, the 3ω-wire method uses a metal wire suspended in a liquid. A sinusoidal current at frequency ω is passed through the metal wire and generates a heat wave at frequency 2ω, which is deduced by the voltage component at frequency 3ω. The 3ω-wire method may be suitable to measure temperature-dependent thermal conductivity.

Most researchers used the transient hot-wire method to measure the thermal conductivity of nanofluids. A schematic of the transient hot-wire apparatus used in the authors' study is shown in Figure 5. The entire experimental setup comprises several major units including experimental (hot-wire) cell, power supply, Wheatstone bridge circuit, data acquisition and control system (A/D converter and computer). Details of the transient single hot-wire setup and procedures are discussed in the paper by Murshed *et al.* [10].

Table 3. Summarized results for thermal conductivity of different types of nanofluids

Researchers	Nanofluids [Particle (size in nm) /base fluid]	Measurement technique	Observed significant enhancement of thermal conductivity at particle volume%
Eastman *et al.* [4]	Al_2O_3 (33)/water CuO (36)/water Cu (35)/oil	THW method	29% for 5 vol.% 60% for 5 vol.% 44% for 0.052 vol.%
Lee *et al.* [6]	Al_2O_3 (38)/water/EG CuO (23.6)/water/EG	THW method	CuO/EG: 22% for 4 vol.% Al_2O_3 /EG: 18% for 5 vol.%
Wang *et al.* [7]	Al_2O_3 (28)/water Al_2O_3 (28)/EG	steady-state method	12% for 3 vol.% 26% for 5 vol.%
Xuan and Li [8]	Cu (100)/water	THW method	Cu/water: 54% for 5 vol.%
Eastman *et al.* [5]	Cu (<10)/EG CuO (35)/EG	THW method	40% for 0.3 vol.% 22% for 4 vol.%
Xie *et al.* [22]	Al_2O_3 (60.4)/water/EG	THW method	Al_2O_3 / EG: 30% for 5 vol.%
Wang *et al.* [23]	Al_2O_3 (29)/EG TiO_2 (40)/EG	steady-state parallel plate method	18% for 4 vol.% 13% for 5 vol.%
Wang *et al.* [20]	CuO (50)/DIW	Quasi-steady state	17% for 0.4 vol.%
Xuan and Li [24]	Cu (10)/water	THW method	70% for 3 vol.%
Patel *et al.* [25]	Au (10-20)/toluene/ water	THW method	Au/toluene: 5.5% for 0.008 vol.% at 30 °C
Kumar *et al.* [26]	Au (4)/water/toluene	THW method	Au /water: 20% for 0.00013 vol.% at 30 °C
Hong *et al.* [11]	Fe (10)/EG	THW method	18% for 0.55 vol.%
Kwak and Kim [27]	CuO (12)/EG	THW method	6% for 1 vol.%
Li and Peterson [12]	CuO (29)/water Al_2O_3 (36)/water	steady-state method	52% for 6 vol.% at 34 °C 30% for 10 vol.% at 34 °C
Zhu *et al.* [28]	Fe_3O_4 (10)/water	THW method	38% for 4 vol.%
Hwang *et al.* [29]	CuO (35.4)/water/EG	THW method	CuO/EG: 9 % for 1 vol.%
Xuan *et al.* [30]	Cu (35.4)/water	THW method	24% for 2 vol.%
Liu *et al.* [31]	CuO (29)/EG	THW method	23% for 5 vol.%
Krishnamurthy *et al.* [32]	Al_2O_3 (20)/water	unspecified	16% for 1 vol.%
Wen and Ding [33]	TiO_2 (34)/water	THW method	6% for 0.66 vol.%
Putnam *et al.* [34]	Au (4)/ethanol	Optical beam deflection technique	1.3% $\pm$ 0.8% for 0.018 vol.%
Kang *et al.* [35]	Diamond (30-40)/EG Ag (8-15)/water	THW method	75% for 1.32 vol.% 11% for 0.4 vol.%

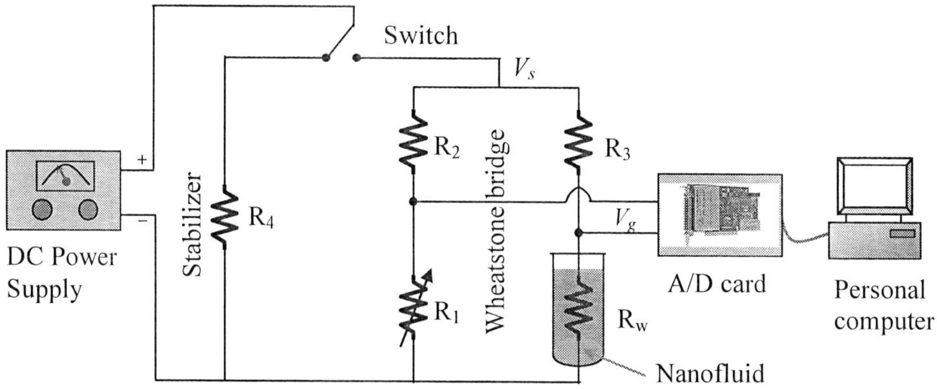

Figure 5. Schematic of transient hot-wire experimental setup [10].

5.2. Effect of Particle Volume Fraction and Base Fluids

Since Maxwell's model [1] was developed more than a century ago, it is known that the effective thermal conductivity of suspensions containing metallic or nonmetallic particles increases with the volume fraction of the solid particles. However, the enhancement of thermal conductivity of a mixture (solid/liquid) may vary significantly particularly in the case of nanofluids.

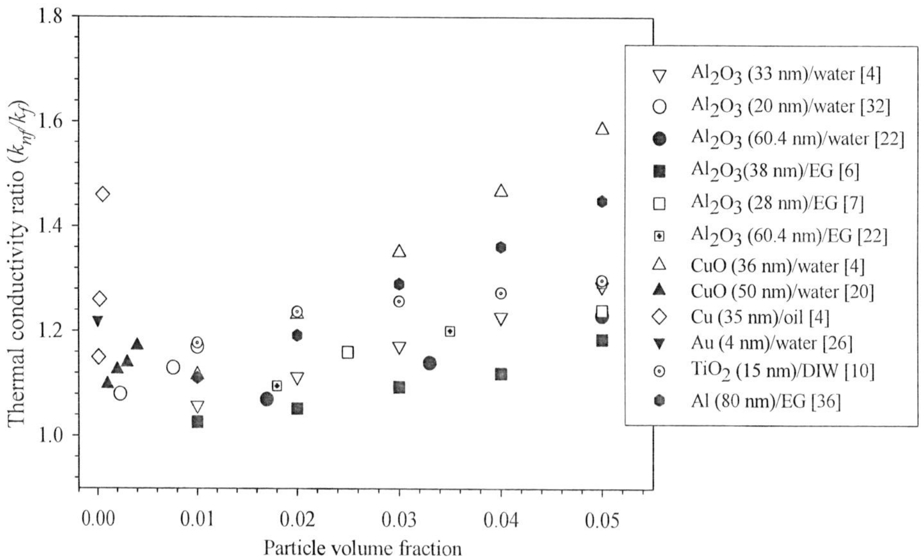

Figure 6. Comparison of experimental results of the enhanced thermal conductivity of nanofluids.

5.2.1. Results from the Literature

In the last several years, many experimental investigations on the effective thermal conductivities of nanofluids containing different concentrations, materials and sizes of nanoparticles dispersed in different base fluids have been reported. The published significant results on the thermal conductivity of nanofluids at room temperature are summarized in Table 3.

Some key results of the effective thermal conductivity of nanofluids as a function of nanoparticle volume fraction from various research groups are also shown in Figure 6.

From the reported results shown in the Table 3 and Figure 6, it is clear that nanofluids exhibit much higher thermal conductivities than their base fluids even when the concentrations of suspended nanoparticles are very low and they increase significantly with nanoparticle volume fraction. However, it is found that even for the same nanofluids different groups reported different enhancements. Several representative studies on the thermal conductivity of nanofluids are elaborated here.

The first results on the enhanced effective thermal conductivity of nanofluids were reported by Eastman *et al.* [4]. By dispersing Al_2O_3 and CuO nanoparticles in water, the reported increase in thermal conductivity were 29% and 60%, respectively for a nanoparticle volumetric loading of 5%. Surprisingly, in the case of Cu/oil-based nanofluids, the thermal conductivity was increased by about 44% by dispersing only 0.052 volume % of Cu nanoparticles in HE-200 oil. The same group [6] later showed a moderate enhancement of thermal conductivity for the same ceramic nanoparticles i.e. Al_2O_3 and CuO dispersed in water and ethylene glycol. For instance, a 20% enhancement in the thermal conductivity of ethylene glycol was observed for 4.0 volume % of CuO nanoparticles. Wang *et al.* [17] reported a surprising 17% increase in the thermal conductivity for a loading of 0.4 volume % of 50 nm sized same nanoparticles (CuO) in water. Recently, Li and Peterson [12] measured the thermal conductivity of the same ceramic nanofluids i.e. CuO (29 nm)/water and Al_2O_3 (36 nm)/water by the steady-state method. Their results were also astounding as both the CuO and Al_2O_3 nanoparticles increased the thermal conductivity of water by 52% and 22%, respectively at a volume fraction of just 6% at 34°C.

By using the steady-state parallel plate method, thermal conductivities of several types of nanofluids were measured by Wang *et al.* [21]. The Al_2O_3/EG-based nanofluids showed 18% increase in thermal conductivity at particle volume % of 4. In contrast, Xie *et al.* [20] observed about 30% increase in thermal conductivity for 5 volume % of Al_2O_3 (60.4 nm) nanoparticles in the same base fluid. Although the particle size used by Xie *et al.* was double that of the particles of Wang *et al.*, their results showed much higher thermal conductivity than those of Wang *et al.* for this nanofluid. This discrepancy could be due to different measurement methods and adjustment of pH values of the nanofluids used by Xie *et al.*

For Fe (10 nm)/EG-based nanofluids, a large increase in thermal conductivity was reported by Hong *et al.* [11]. They obtained an enhancement of 18% with just 0.55 volume % of Fe nanoparticles. They also noticed that sonication had significant effect on the thermal conductivity of nanofluids. Nonetheless, their observed enhancement for this nanofluid was even much higher than that of the Cu/EG-based nanofluids obtained by Eastman *et al.* [5]. This indicates that a suspension of high conductivity materials does not always result in an increase in the thermal conductivity of nanofluids. By using the co-precipitation method, Zhu *et al.* [26] prepared Fe_3O_4 (10 nm)/water-based nanofluids and measured their thermal conductivity values by the hot-wire method. They found a 38 % increase in the thermal conductivity for the nanoparticle volume fraction of 0.04.

Putnam *et al.* [33] measured the thermal conductivity of Au (4 nm)/ethanol-based nanofluids by using the optical beam deflection technique. For the first time, their results showed no anomalous enhancement in the thermal conductivity of nanofluids with very low particle volume fractions. Their observed maximum enhancement in thermal conductivity was 1.3%±0.8% for 0.018% volumetric loading of 4 nm-sized Au particles in ethanol. This result is directly in conflict with the anomalous result of Patel *et al.* [23] for the same nanofluid.

From a comparison of the reported studies, the increments of thermal conductivities are different for different types of nanofluids. The thermal conductivity of nanofluids varies with the size, material of nanoparticles as well as the base fluids. For instance, nanofluids with metallic nanoparticles were found to have a higher thermal conductivity than nanofluids with non-metallic (oxide) nanoparticles. The smaller the particle size, the larger the thermal conductivities of nanofluids. A few studies [11, 14] also reported that highly-conductive nanoparticles are not always effective in enhancing the thermal conductivity of nanofluids.

The particle size is important because shrinking it to nanoscale dimensions not only increases the surface area relative to volume but also generates some nanoscale mechanisms in the suspensions [14, 36-40]. Experimental [41] and theoretical evidence [37-38, 40] indicate that the effective thermal conductivity of nanofluids increases with decreasing particle size.

5.2.2. Results by the Authors

By using transient hot-wire method the effective thermal conductivities of several types of nanofluids are measured by the authors [10, 36]. Similar to the most studies reported in the literature, our results show a substantial increase in thermal conductivity with particle volume fraction. The choice of base fluids also influences this enhancement. From Figure 7, it can be seen that the effective thermal conductivities of TiO_2 (15 nm)/deionized water/ethylene glycol-based nanofluids increase substantially with increasing volumetric loading of TiO_2 (15 nm) nanoparticles. The thermal conductivity of TiO_2 (15 nm)/water nanofluids has a maximum enhancement of 29.7% for a particle volume fraction of 0.05 while the maximum increase of thermal conductivity of TiO_2 (15 nm)/ethylene glycol-based nanofluids is 18% for the same particle volume fraction of 0.05. Figure 7 also shows that water-based nanofluids exhibit much higher thermal conductivity compared to the ethylene glycol-based nanofluids for this nanoparticles. This is mainly due to higher value of the thermal conductivity of water compared to ethylene glycol (Table 1).

Figure 8 illustrates that the effective thermal conductivities of Al_2O_3 (80 nm)/deionized water/ethylene glycol-based nanofluids also increase significantly with increasing volumetric loading nanoparticles. By suspending 0.05 volume fraction of Al_2O_3 (80 nm) nanoparticles in deionized water and in ethylene glycol, the maximum enhancements of thermal conductivities were found to be 23% and 18%, respectively. This indicates that water is a better base fluid compared to ethylene glycol for these nanofluids.

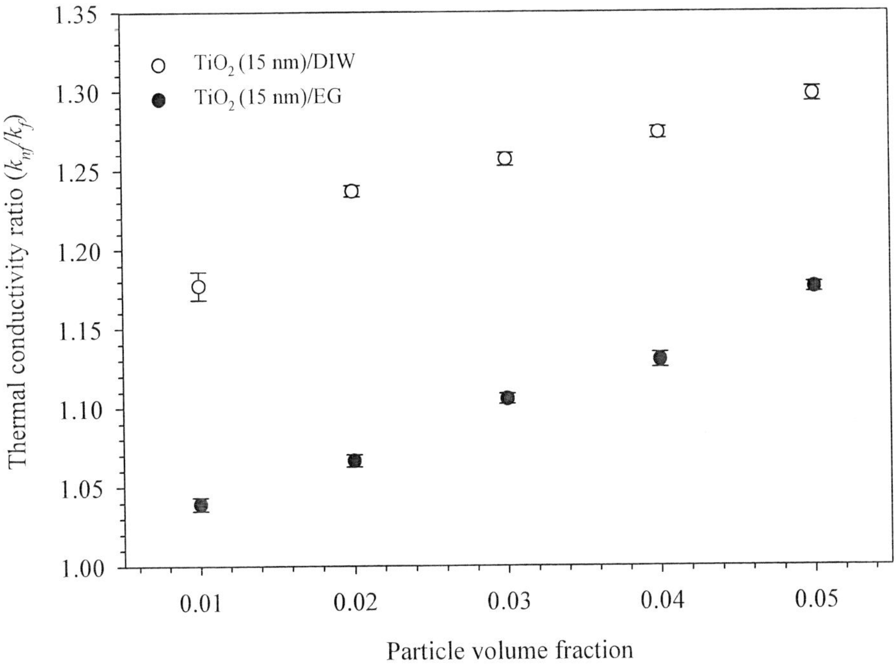

Figure 7. Enhancement of the effective thermal conductivity of TiO_2 (15 nm)/DIW/EG-based nanofluids with particle volume fraction [10, 36].

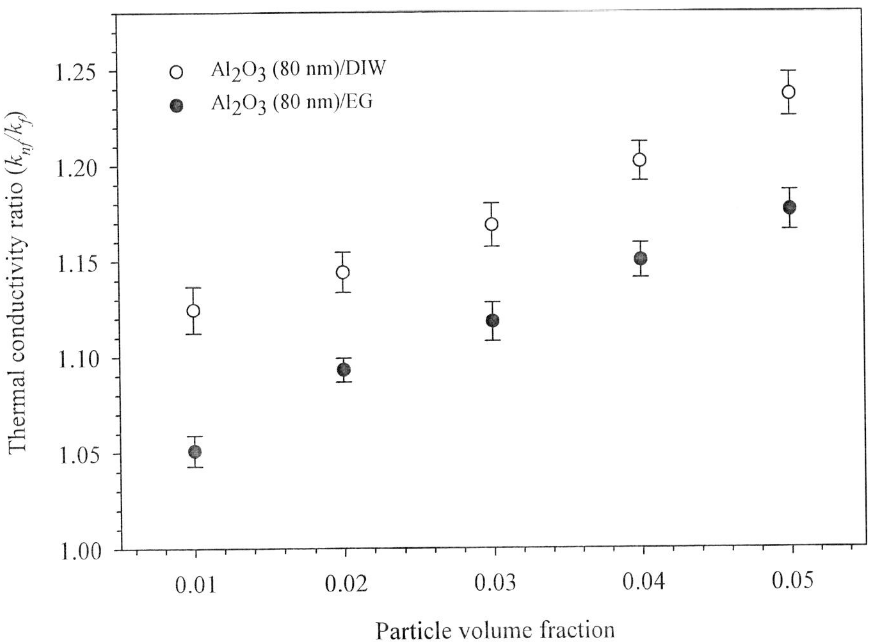

Figure 8. Results of the authors for the effective thermal conductivity enhancement of Al_2O_3 (80 nm)/DIW/EG-based nanofluids with particle volume fraction.

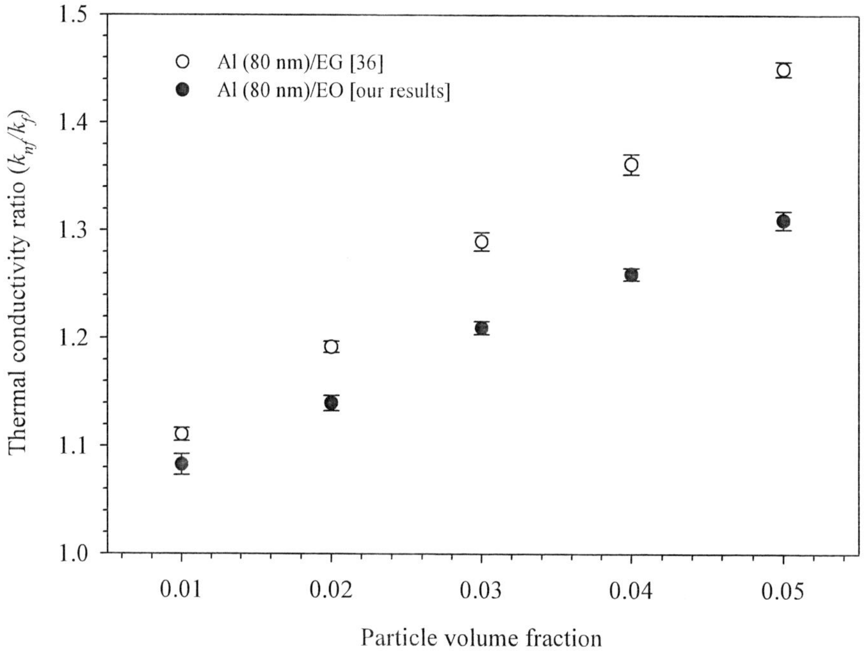

Figure 9. Enhancement of the effective thermal conductivity of Al (80 nm)/ethylene glycol /engine oil-based nanofluids with particle volume fraction.

The thermal conductivity of Al (80 nm)/ethylene glycol-based nanofluids has a maximum enhancement of 45% for a particle volume fraction of 0.05 while the maximum increase of thermal conductivity of Al /engine oil-based nanofluids is 31% for the same particle volumetric loading (i.e. 5%) as shown in Figure 9. Similar to the previous results (Figures 7 and 8), it is seen that ethylene glycol-based nanofluids show much higher thermal conductivities than those of engine oil-based nanofluids.

From the above results, it can be concluded that suspending a small volume percentage (1 to 5%) of nanoparticles in base fluids significantly increases the effective thermal conductivities of nanofluids. The thermal conductivity increases almost linearly with increasing the volumetric loading of nanoparticles. However, for lower volume fraction (< 1%) and very small-sized particles, the enhancement of thermal conductivity may not always be linear. Besides the nanoscale mechanisms such as the size of nanoparticle, nanolayer at the particle/liquid interface and Brownian motion of nanoparticles, the dissociation of surfactant may play a significant role in the enhancement of thermal conductivity. The thermophysical properties of base fluids are also found to have impact on the thermal conductivity of nanofluids. The higher the thermal conductivity of the base fluid, the larger the enhancement of thermal conductivity of nanofluids. The reason may also lie in the dispersion behavior of nanoparticles in the base fluids as well as the effect of surfactant. A small amount of CTAB surfactant was added in all sample nanofluids for a better dispersion of nanoparticles.

5.3. Effect of Particle Size and Shape

Although the thermal conductivities of nanofluids reported by different researchers were shown to be particle size dependent, very few experimental works have been reported [41] on the effect of particle size on the thermal conductivity of nanofluids. Other than results for carbon nanotubes, no experimental study of the effect of nanoparticle shape has been reported in the literature. The thermal conductivities of nanofluids containing different sizes and shapes of nanoparticles are measured by the authors and the results are presented in this section. Due to limited availability of different sizes of the same type of nanoparticles, TiO_2 nanoparticles of spherical (15 nm) and rod shape (10×40 nm) and spherical shape Al_2O_3 nanoparticles of 80 nm and 150 nm were used in our experiments.

Our results for TiO_2/deionized water-based nanofluids (Figure 10) show a nonlinear relationship between thermal conductivity and particle volume fraction at lower volumetric loading (0.005-0.02) and a linear relationship at higher volumetric loading (0.02-0.05). This nonlinear behavior of nanofluids at lower volume fractions of nanoparticles may be due to the influence of the CTAB surfactant, long time of sonication, and hydrophobic surface forces in the nanofluids. Figures 10 and 11 illustrate that for nanofluids with rod-shaped nanoparticles (TiO_2), the increase in thermal conductivity is larger than that of the nanofluids with spherical-shaped nanoparticles (TiO_2). Due to the larger shape factor and alignment, the thermal conductivity of nanofluids composed of rod-shape nanoparticles is larger than that of the nanofluids with spherical-shaped nanoparticles.

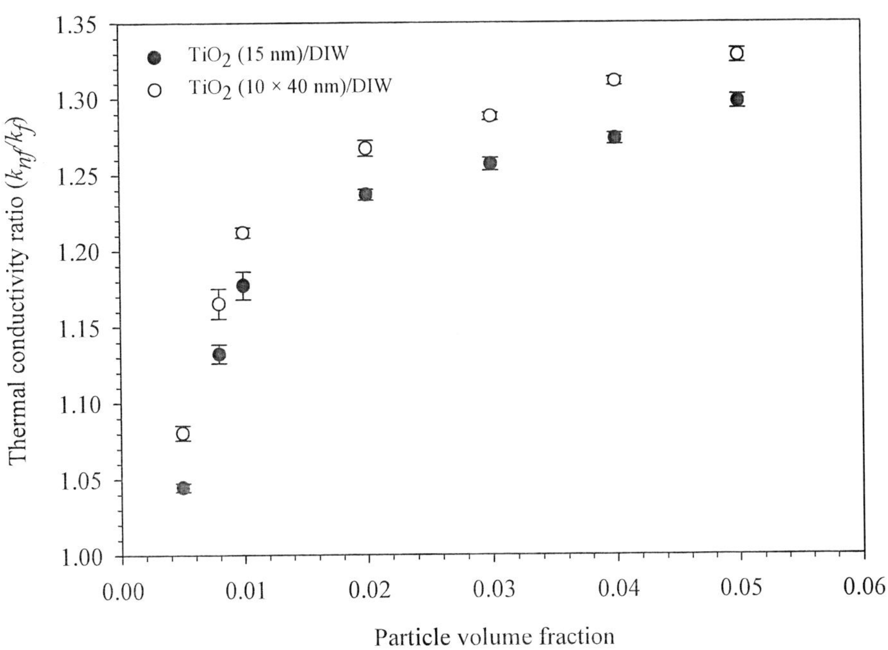

Figure 10. Enhancement of the effective thermal conductivity of TiO_2/DIW-based nanofluids [10].

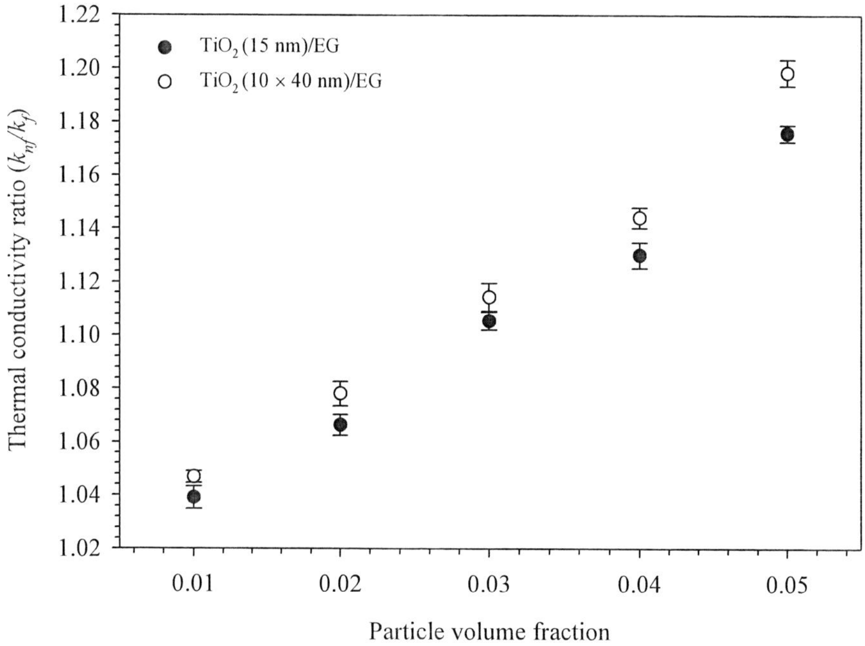

Figure 11. Results of the authors for the effective thermal conductivity enhancement of TiO_2/EG-based nanofluids.

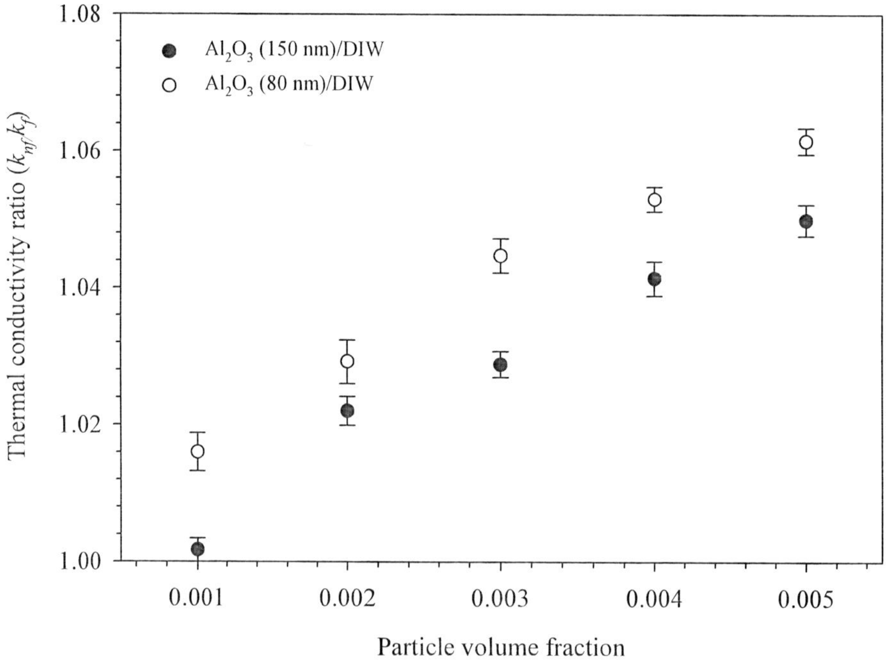

Figure 12. Results of the authors for the effective thermal conductivity enhancement of Al_2O_3/DIW-based nanofluids.

Figure 12 illustrates that for the same base fluid, smaller size (80 nm) Al_2O_3 nanoparticles show higher enhancement of thermal conductivity compared to larger-sized (150 nm) nanoparticles. Chon and Kihm [41] also found that nanofluids having smaller-sized Al_2O_3 particles have higher thermal conductivities than similar nanofluids with smaller-sized nanoparticles. The reason is that the smaller sized particles provide better dispersion and can also enhance the contribution of other nanoscale mechanisms such as Brownian motion of particles compared to larger-sized particles.

The above results demonstrate that the smaller the particle size, the higher the effective thermal conductivity of nanofluids. It is also concluded that the particle shape influences the thermal conductivity of nanofluids.

5.4. Effect of Fluid Temperature

Fluid temperature may play an important role in enhancing the effective thermal conductivity of nanofluids. Very few studies were performed to investigate the temperature effect on the effective thermal conductivity of nanofluids.

5.4.1. Results from the Literature

A summary of the published results on temperature-dependent thermal conductivity of various nanofluids is provided in Table 4 followed by brief discussion on several representative studies.

Table 4. Temperature-dependent thermal conductivity of nanofluids

Researchers	Nanofluids [Particle (size in nm)/base fluid]	Measurement technique	Observed maximum enhancement of thermal conductivity at temperature
Das *et al.* [9]	Al_2O_3(38.4)/water CuO (28.6)/water	temperature oscillation technique	for 4 vol.%: 16% at 36 °C and 25% at 51 °C for 1 vol.%: 22% at 36 °C and 30% at 51 °C
Patel *et al.* [25]	Au (10-20)/water	THW method	for 0.00026 vol.%: 5% to 21% at temperature range of 30 - 60 °C
Chon and Kihm [41]	Al_2O_3 (47)/water Al_2O_3 (150)/water	THW method	6% at 31 °C and 11% at 51 °C 3% at 31 °C and 8.5% at 51 °C [for 1 volume%]
Li and Peterson [12]	Al_2O_3 (36)/water	steady-state method	for 2 vol.%: 7% at 27.5 °C and 23% at 36 °C
Yang and Han [21]	Bi_2Te_3 (20× 170) /FC72/oil	3ω wire method	for Bi_2Te_3 /FC72 at 0.8 vol.%: 8% at 10 °C and 7% at 40 °C
Venerus *et al.* [42]	Au (22) /water Al_2O_3 (30)/ Petroleum oil	Forced Rayleigh scattering	no enhancement but slight decrease with the temperature in the range of 25-75 °C

Das *et al.* [9] investigated the effect of temperature on the thermal conductivity of nanofluids containing Al_2O_3 (38.4 nm) and CuO (28.6 nm) nanoparticles in water. A temperature oscillation technique, which determines thermal conductivity through measuring the thermal diffusivity of sample medium, was used in their study. They reported a two- to four-fold increase in thermal conductivity enhancement for these nanofluids over a temperature range of 21°C to 51°C. Das *et al.* suggested that the strong temperature dependence of thermal conductivity was due to the motion of nanoparticles. Using the transient hot-wire technique, the same group [25] performed experiments for water-based nanofluids containing 10-20 nm-sized Au particles. Nanofluids with citrate stabilization showed thermal conductivity enhancement of 5-21% in the temperature range of 30 to 60°C at a very low volumetric loading of 0.00026%. Chon and Kihm [41] studied the effects of particle size and fluid temperature on the thermal conductivity of nanofluids. They observed a moderate enhancement of thermal conductivity with respect to temperature. The thermal conductivity of Al_2O_3 (47 nm)/water-based nanofluids increased by 6% to 11% when the fluid temperature was increased from 31°C to 51°C. They identified Brownian motion of the nanoparticles as the possible mechanism for increased thermal conductivity with fluid temperature. However, results obtained by Li and Peterson [12] demonstrated a surprising increase in thermal conductivity of nanofluids with a small increase of fluid temperature. For example, for Al_2O_3 (47)/water-based nanofluids at 2 volume %, an increase in fluid temperature to 36°C resulted in an enhancement of three times the thermal conductivity at 27.5°C. Yang and Han [21] measured the thermal conductivity of suspensions of Bi_2Te_3 (20×170 nm) nanorods in FC72 and in oil (hexadecane) by using the 3ω-wire method. Interestingly, they observed a slightly decrease in thermal conductivity with increasing temperature, which is in contrast to the trend observed in nanofluids containing spherical nanoparticles. The contrary trend was claimed to be due mainly to the particle aspect ratio.

Venerus *et al.* [42] used the Forced Rayleigh Scattering technique to measure the thermal conductivity of Au (22) /water and Al_2O_3 (30) /petroleum oil-based nanofluids in the range of 25 −75°C. In contrast to all other reported results, they found the level of thermal conductivity enhancement for these nanofluids to be independent of temperature.

Other than Yang and Han [21] and Venerus *et al.* [42], all published results demonstrate that the thermal conductivity of nanofluids increases significantly with the fluid temperature. However, there is a clear lack of consistency among the reported results of different research groups.

5.4.2. Results by the Authors

The effect of the temperature on the enhancement of effective thermal conductivity of several types of nanofluids was also investigated in this study. The transient hot-wire system with a fluid heating facility was used to measure the thermal conductivity of nanofluids at different temperatures ranging from 20 to 60°C. A refrigerating/heating circulator was used to the hot-wire system to generate different temperatures of nanofluids in the experiments. The measurement procedure and detailed description of the transient hot-wire technique are similar to experiments conducted at room temperature [10]. The experimental apparatus was calibrated by measuring the effective thermal conductivity of the base fluids.

Figure 13 shows that at a temperature of 60°C, the effective thermal conductivity of Al_2O_3 (80 nm)/EG based nanofluids increases by about 9% and 12% (compared to base fluid) for nanoparticle volumetric loadings of 0.5% and 1%, respectively. Whereas for the same temperature, the enhancement of effective thermal conductivity of Al (80 nm)/EO nanofluids are 20% and 37% for volumetric loadings of 1% and 3% nanoparticles in the base fluid, respectively. It is also seen from Figure 13 that dependence of temperature on thermal conductivity of nanofluids with metallic nanoparticles is more significant than that of nanofluids with nonmetallic nanoparticles.

Figures 14 and 15 compare the present results for Al_2O_3/ water based nanofluids at 1 % volumetric loading of nanoparticles with the data reported by Das *et al.* [9] and Chon and Kihm [41]. Our results are consistent with the experimental data of Das *et al.* and Chon and Kihm although the size of the Al_2O_3 nanoparticles used in our study is much larger than those of Das *et al.* and Chon and Kihm (Figure 14). For the same size (150 nm) of Al_2O_3 nanoparticles (Figure 15), our results for thermal conductivity of nanofluids were found to be higher than those reported by Chon and Kihm [41]. The observed differences could be due to difference in preparing sample nanofluids and different measurement methods used.

As shown in Figures 13 to 15, the experimental thermal conductivity values of different nanofluids increase significantly with fluid temperature. A linear increase in the effective thermal conductivity of nanofluids with temperature was also observed. This is because the high fluid temperature intensifies the Brownian motion of nanoparticles and also decreases the viscosity of the base fluid. With an intensified Brownian motion, the contribution of micro-convection in heat transport increases, thereby intensifying the enhancement in the thermal conductivity of nanofluids [7, 43].

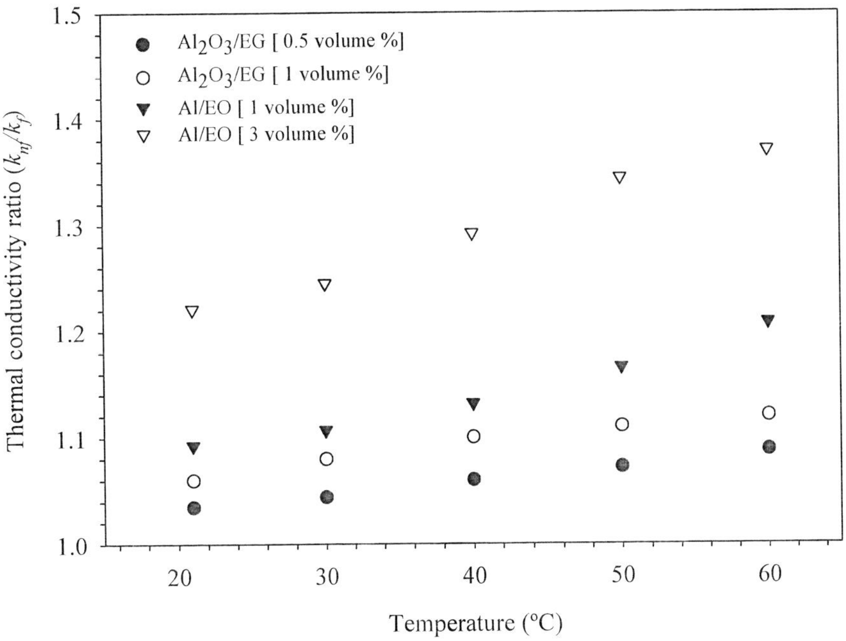

Figure 13. Thermal conductivity enhancement with temperature for Al_2O_3 (80 nm)/ethylene glycol and Al (80 nm)/engine oil-based nanofluids [36].

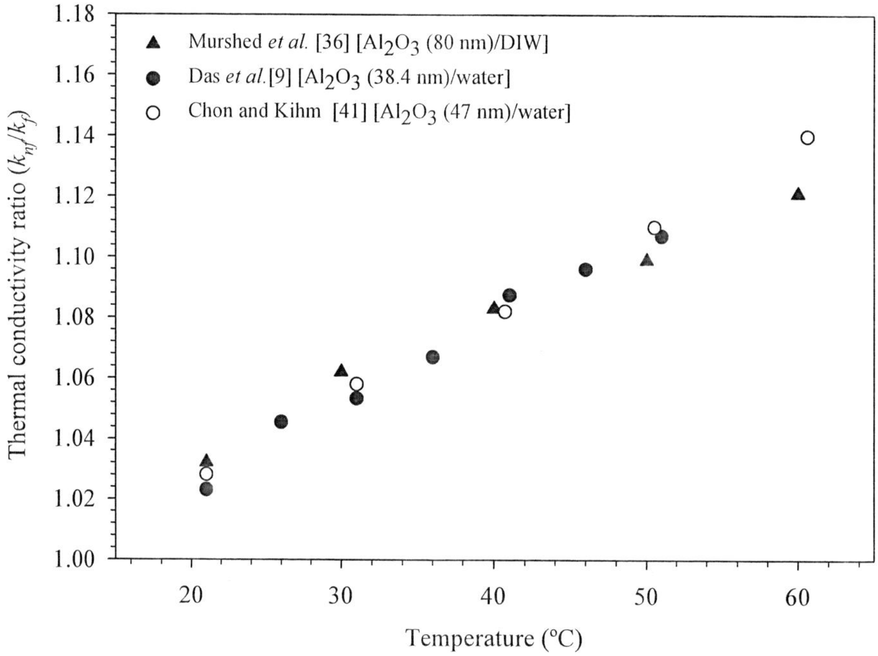

Figure 14. Thermal conductivity enhancement with temperature for Al_2O_3 (different sizes)/water-based nanofluids.

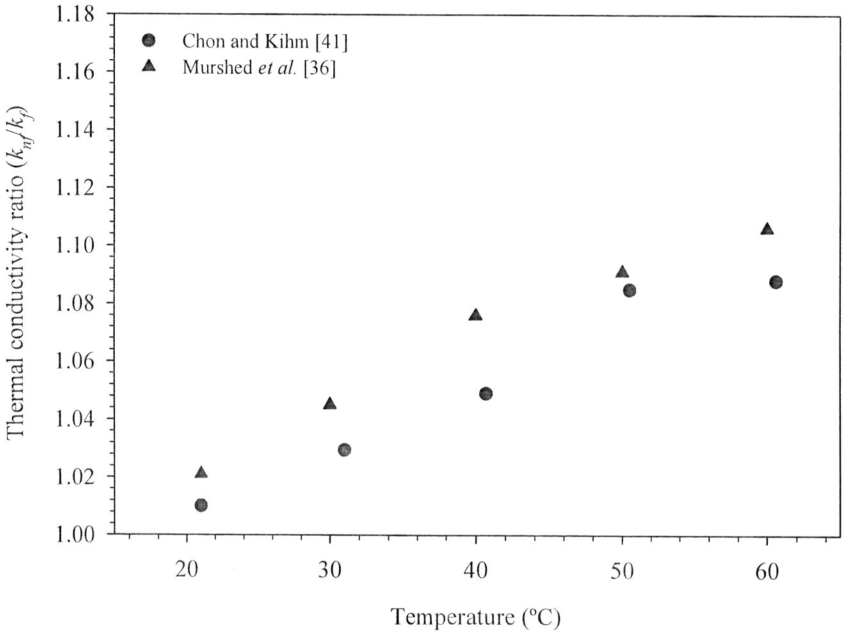

Figure 15. Thermal conductivity enhancement with temperature for Al_2O_3 (150 nm)/water based-nanofluids.

6. THEORETICAL STUDIES ON NANOFLUIDS

6.1. Mechanisms for the Enhanced Thermal Conductivity of Nanofluids

Experiments have shown that nanofluids exhibit anomalously high thermal conductivity which cannot be predicted by classical models. To explain the observed enhanced thermal conductivity of nanofluids, Wang *et al.* [7] and Keblinski *et al.* [37] proposed various mechanisms, which are not considered by classical models. Wang *et al.* [7] suggested that the microscopic motion of nanoparticles, surface properties, and the structural effects might cause enhanced thermal conductivity of nanofluids. In nanofluids, the microscopic motion of the nanoparticles due to van der Waals force, stochastic force causing Brownian motion and electrostatic force can be significant. Wang *et al.* however, showed that Brownian motion does not contribute significantly to energy transport in nanofluids. They indicated that the electric double layer and van der Waals force could have strong electrokinetic effects on the nanoparticles. The surface properties and structural effects were not confirmed as potential mechanisms in their study. Keblinski *et al.* [37] later elucidated four possible mechanisms for the anomalous increase in nanofluids heat transfer which are i) Brownian motion of the nanoparticles, ii) liquid layering at the liquid/particle interface, iii) nature of the heat transport in the nanoparticles, and iv) effect of nanoparticle clustering.

Due to the Brownian motion, particles move randomly through the liquid and may collide, thereby enabling direct solid-solid transport of heat from one to another, which can increase the effective thermal conductivity. However, by a simple analysis, Keblinski *et al.* [37] showed that the thermal diffusion of nanoparticles is much faster than Brownian diffusion and thus the contribution of Brownian motion to energy transport in nanofluids is not significant.

When the size of the nanoparticles in a nanofluid becomes less than the phonon mean-free path, phonons no longer diffuse across the nanoparticle but move ballistically without any scattering. However, it is difficult to envision how ballistic phonon transport could be more effective than a very-fast diffusion phonon transport, particularly to the extent of explaining anomalously high thermal conductivity of nanofluids. No further work or analysis has been reported on the ballistic heat transport nature of nanoparticles. Instead, the continuum approach was adopted in all reported works [20, 36, 38, 44].

Liquid layering around the particle i.e. nanolayer was offered as another mechanism responsible for higher thermal properties of nanofluids. The basic idea is that liquid molecules can form a layer around the solid particles and thereby enhance the local ordering of the atomic structure at the interface region. Hence, the atomic structure of such liquid layer is significantly more ordered than that of the bulk liquid. Given that solids, which have much ordered atomic structure, exhibit much higher thermal conductivity than liquids the liquid layer at the interface would reasonably have a higher thermal conductivity than the bulk liquid. Thus the nanolayer is considered as an important factor enhancing the thermal conductivity of nanofluids.

The effective volume of a cluster is considered much larger than the volume of the particles due to the lower packing fraction (ratio of the volume of the solid particles in the cluster to the total volume of the cluster) of the cluster. Since heat can be transferred rapidly within such clusters, the volume fraction of the highly conductive phase (cluster) is larger than the volume of solid, thus increasing its thermal conductivity [11, 37]. However, in general, clustering may exert a negative effect on heat transfer enhancement, particularly at a low volume fraction by settling small particles out of the liquid and creating a large region of "particle free" liquid with a high thermal resistance [37].

Besides these mechanisms, the effects of particle surface chemistry and particles interaction for nanometer-sized particles could be significant in enhancing the thermal conductivity of nanofluids.

6.2. Models for the Effective Thermal Conductivity

To predict the effective thermal conductivity of solid particle suspensions, several models have been developed since the treatise by Maxwell [1]. As mentioned previously, these classical models such as those attributed to Maxwell [1], Hamilton-Crosser [45], and Bruggeman [Hui *et al.* [46]] were developed for predicting the effective thermal conductivity of a continuum medium with well-dispersed solid particles. The Maxwell model [1] predicts the effective electrical or thermal conductivity of liquid-solid suspensions for low volumetric loading of spherical particles. Hamilton and Crosser [45] modified Maxwell's model to determine the effective thermal conductivity of non-spherical particles by applying a shape factor. For spherical particles, the Hamilton and Crosser (HC) model reduces to the Maxwell model. The Bruggeman model (BGM) [Hui *et al.* [46]] is another well-known model for determining the effective thermal conductivity of mixtures and composites. In this model, the mean field approach is used to analyze the interactions among the randomly distributed particles.

The classical models have been found to be unable to predict the anomalously high thermal conductivity of nanofluids. This is because these models do not include the nanoscale effects of particle size such as the interfacial layer at the particle/liquid interface, and motion of particles, which are considered as important factors for enhancing thermal conductivity of nanofluids [7, 37-40, 47-48,]. Recently, many theoretical studies have been carried out to predict the anomalously increased thermal conductivity of nanofluids. Several models have been proposed to consider various mechanisms. A detailed summary of all classical and recently developed models for the prediction of effective thermal conductivity of nanofluids is provided in review articles [49-50]. Therefore, these models are not reproduced here. Instead recently developed models are categorically briefed.

All the reported models can be categorized into two general groups, which are

(i) *static models* which assume stationary nanoparticles in the base fluid in which the thermal transport properties are predicted by conduction-based models such as those of Maxwell and Hamilton-Crosser.
(ii) *dynamic models* which are based on the premise that nanoparticles have lateral, random motion in the fluid. This motion is believed to be responsible for transporting energy directly through collision between nanoparticles or indirectly through micro liquid convection that enhances the transport of thermal energy.

Researchers in the first group [20, 35, 36, 38, 40, 47-48, 51, 52-55] use the concept of liquid/solid interfacial layer to develop models and to explain the anomalous improvement of the thermal conductivity in nanofluids. Approaches adopted in these studies can be classified under the static model category.

Other groups of researchers emphasize the contribution of dynamic part related to particle Brownian motion. Although Wang *et al.* [7] and Keblinski *et al.* [37] showed that the contribution of Brownian motion to energy transport in nanofluids is not significant, other researchers [26, 43, 56] held contrary views.

Most recently, the authors have developed an improved model to predict the effective thermal conductivity of nanofluids containing spherically shaped nanoparticles [57]. Compared to existing models, this model shows better agreement with experimental results. This was attributed to the inclusion of both static and dynamic mechanisms such as particle size, nanolayer, particle movement, particle surface chemistry and interaction potential.

7. Studies on Thermal Diffusivity and Specific Heat of Nanofluids

7.1. Thermal Diffusivity of Nanofluids

Despite the fact that thermal diffusivity is an important thermophysical property especially in convective heat transfer applications, very little work have been performed on the determination of the effective thermal diffusivity of nanofluids. Xuan and Roetzel [58] was the first to discuss the effective thermal diffusivity tensor for flowing nanofluids under both laminar and turbulent flow conditions. However, neither experimental nor theoretical result for the effective thermal diffusivity of nanofluids was provided in their paper. Wang *et al.* [59] measured the thermal conductivity and specific heat of some nanofluids and thereby calculated their effective thermal diffusivity. Their calculated results were also found to fluctuate severely with volume fraction. The accurate measurement of thermal diffusivity is much more complex for composite fluids such as nanofluids compared to solids or gases. Although there are several methods to measure the thermal diffusivity of solid, gases or liquid, no precise technique has been developed to measure the thermal diffusivity of composite fluids. Among all existing methods of measuring the effective thermal diffusivity of composite materials, the most common one is the flash method discussed by Parker *et al.* [60]. However, this method is mainly used for solid composite materials. To measure the thermal diffusivity of gases and liquids, several techniques such as the thermal–wave cavity technique [61] and the temperature oscillation technique [62] have been proposed. In the temperature oscillation technique, the thermal diffusivity is deduced by measuring and evaluating the amplitude attenuation and the phase shift between the temperature oscillations at the surface of the liquid specimen and at a specific position inside the reference specimen. The principle of the thermal–wave cavity technique consists in measuring the temperature fluctuations in a sample as a result of nonradiative deexcitation processes, which take place following the absorption of intensity-modulated radiation. The accuracy of these methods are not as good compared to the transient hot-wire method. Nevertheless, the applications of these methods are very limited due to complexities and difficulties to achieve accurate results particularly for composite fluids. They are therefore not suitable for the convenient and accurate determination of the thermal diffusivity of solid-liquid mixtures such as nanofluids.

The authors have developed a transient technique termed as the *double hot-wire (DHW)* method for the direct and precise measurement of the effective thermal diffusivity of nanofluids. Figure 16 shows the schematic of the double hot-wire experimental setup used in this study. The advantages of the double hot-wire technique are its simplicity and high measurement accuracy. Besides thermal conductivity, this technique can be used to determine the thermal diffusivity more accurately than that of the conventional hot-wire method. The theoretical basis for calculating the effective thermal diffusivity of nanofluids and detailed description of the experimental technique is provided in [63].

No investigation has been reported in the literature on the determination of the specific heat of nanofluids. Thus, the specific heat of the nanofluids is also determined from the measured thermal conductivity and thermal diffusivity. Our results for the specific heat of nanofluids are analyzed and compared with calculated values obtained from the volume fraction mixture rule (see subsection 7.2).

The effective thermal diffusivity of several types of nanofluids at different volume percentages (1 to 5%) of nanoparticles are measured by using double hot-wire technique. The effective thermal diffusivities of TiO_2/EG-based of nanofluids are shown in Figure 17. It is seen that for maximum 5% volumetric loading of TiO_2 nanoparticles of 15 nm and 10×40 nm in ethylene glycol, the maximum increase in effective thermal diffusivity was observed to be 25% and 29%, respectively. These results indicate that along with the particle volume fraction, particle size and shape have effect on the enhancement of thermal diffusivity of the nanofluids. This is probably because the particle size and shape affect the effective thermal conductivity of nanofluids [40-41] and the dispersion of particles in the base fluids, which eventually influence the effective thermal diffusivity of the mixture.

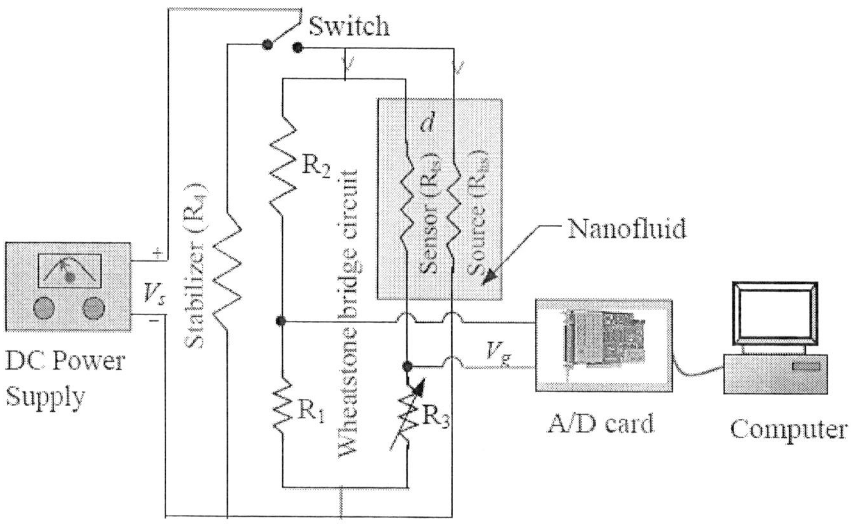

Figure 16. Schematic of the double hot-wire (DHW) experimental system [63].

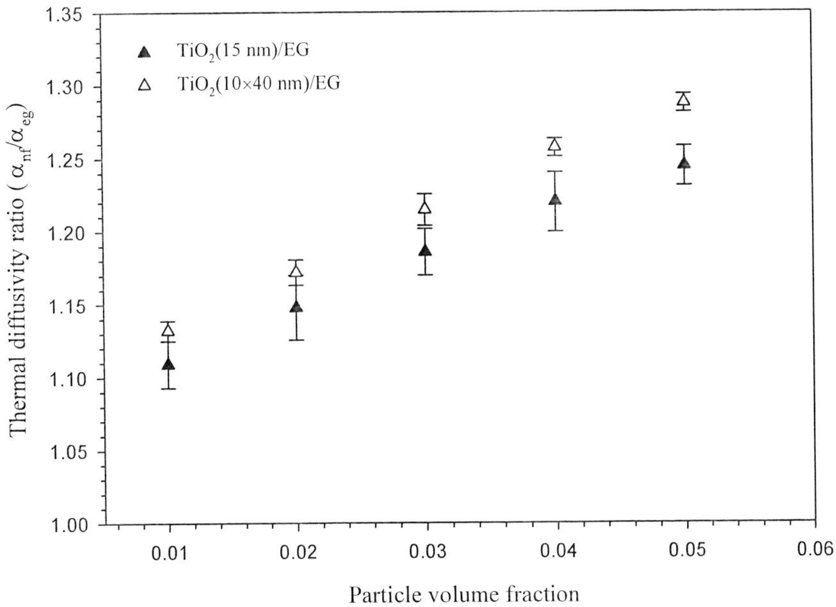

Figure 17. Thermal diffusivity of TiO_2/ethylene glycol-based nanofluids with particle volume fraction [63].

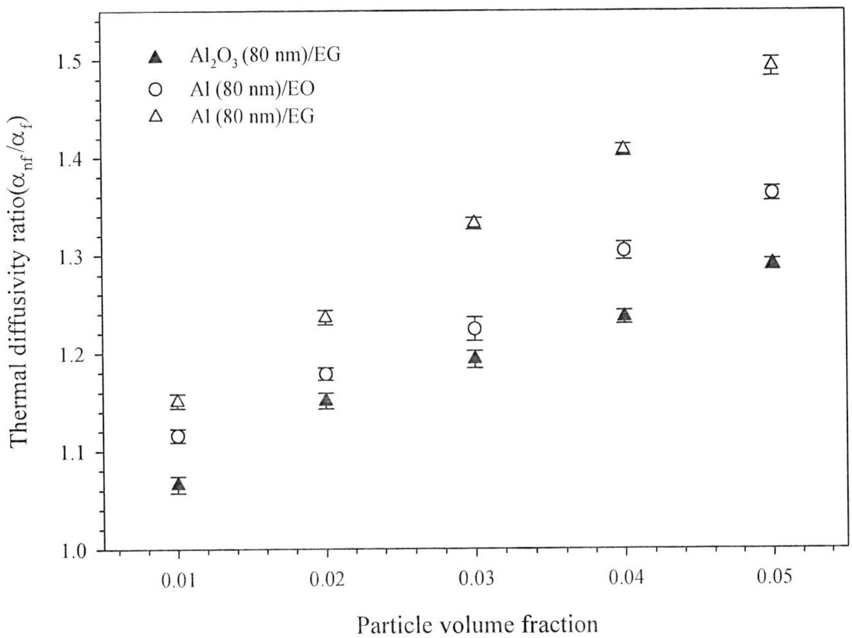

Figure 18. Effects of particle volume fraction, particle material, and base fluid on effective thermal diffusivity of nanofluids [63].

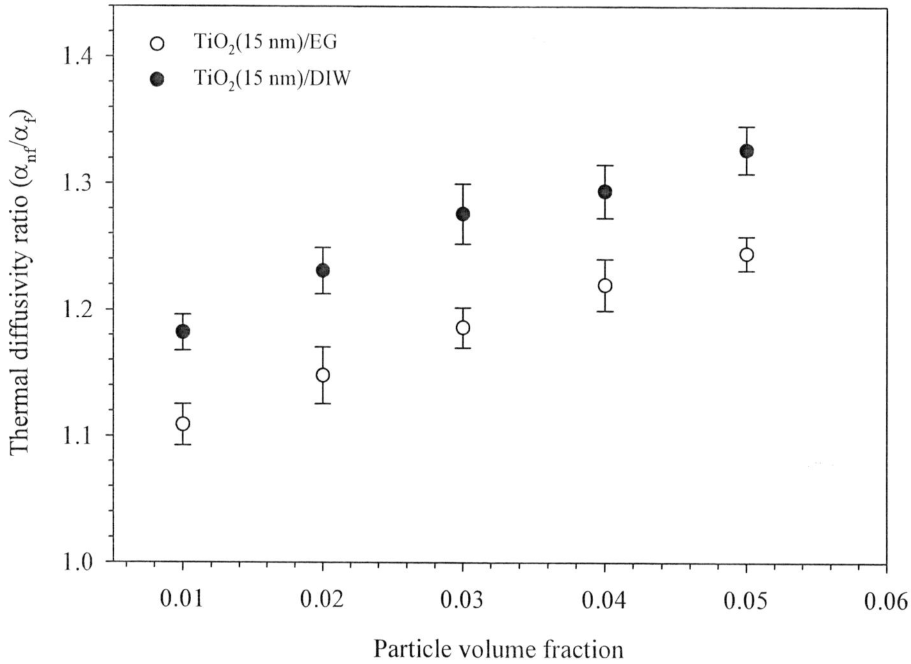

Figure 19. Thermal diffusivities of TiO_2/DIW and EG-based nanofluids with particle volume fraction [63].

Figure 18 shows that nanofluids with aluminum nanoparticles in ethylene glycol and engine oil exhibit substantial enhancement of thermal diffusivity i.e. maximum 49% and 36%, respectively compared to their base fluids. The ethylene glycol-based nanofluid was found to have a higher thermal diffusivity than that of the engine oil-based nanofluids due to the higher thermal diffusivity of ethylene glycol as compared to the engine oil. It is also seen that the nanofluids with Al nanoparticles in ethylene glycol exhibit significantly higher thermal diffusivity than that of Al_2O_3 nanoparticles in same base fluid (EG). This is expected because the thermal diffusivity of Al (97.1×10^{-6} m^2/s) at 300 K is almost an order of magnitude higher than that of Al_2O_3 (11.9×10^{-6} m^2/s) at 293 K [16]. Figure 18 also demonstrates the impact of the particle material and the base fluid on the enhancement of the thermal diffusivity of nanofluids.

Figure 19 illustrates the effect of base fluids on the effective thermal diffusivity of nanofluids. It is found that TiO_2 (15 nm)/DIW-based nanofluids have higher thermal diffusivities compared to that of ethylene glycol-based nanofluids. As mentioned previously, this is because of much higher thermal diffusivity of water compared to the ethylene glycol (Table 1).

Figures 17 to 19 show that nanofluids exhibit significant increase in thermal diffusivity with the nanoparticle volume fraction. This is mainly because the effective thermal conductivity of nanofluids was found to be substantially higher than the value for the base fluids and increases with the particle volume fraction [4-10, 22]. The dissociation of surfactant (i.e. CTAB) may play a significant role in intensifying the enhancement of thermal properties (e.g. thermal conductivity and diffusivity) of nanofluids by inducing electrostatic repulsion among the suspended nanoparticles resulting in better dispersion and stabilized suspensions. Moreover, with increased volumetric loading of nanoparticles, the specific heat of nanofluids decreases which results in an increase in the effective thermal diffusivity of the mixture (nanofluids). The calculated results of thermal diffusivity (i.e. by using the measured thermal conductivity and specific heat) of CuO (50 nm)/water-based nanofluids reported by Wang *et al.* [59] also showed significant enhancement of the thermal diffusivity of nanofluids. For example, for 0.1% volumetric loading of CuO nanoparticles in water, the enhancement of the thermal diffusivity of nanofluids was calculated to be 40%. The authors' above results also demonstrate that besides the particle volume fraction, base fluid and nanoparticle material influence the enhancement of the thermal diffusivity of nanofluids.

7.2. Determination of Specific Heat of Nanofluids

7.2.1. Theoretical Model for Specific Heat

The specific heat of sample nanofluids can be determined by using measured thermal conductivity and thermal diffusivity values and the definition of specific heat given as

$$c_{p-nf} = \frac{k_{nf}}{\alpha_{nf}\rho_{nf}} \tag{1}$$

where the density of mixture can be obtained from

$$\rho_{nf} = \phi_p\rho_p + (1-\phi_p)\rho_f \tag{2}$$

Since there is no model available to determine the effective specific heat of nanofluids, the volume fraction mixture rule can be used to formulate an expression. The effective volumetric specific heat nanofluids can be expressed as [24, 39]

$$(\rho c_p)_{eff} = [\phi_p\rho_p c_{p-p} + (1-\phi_p)\rho_f c_{p-f}] \tag{3}$$

where ϕ is the particle volume fraction and subscripts p and f stand for nanoparticle and base fluid, respectively.

From Equations (2) and (3), the effective specific heat of nanofluids can be expressed as

$$c_{p-nf} = \frac{\phi_p\rho_p c_{p-p} + (1-\phi_p)\rho_f c_{p-f}}{\phi_p\rho_p + (1-\phi_p)\rho_f} \tag{4}$$

The experimentally calculated specific heats of nanofluids will be compared with theoretically calculated results from Equation (4).

7.2.2. Results and Comparisons

Applying Equation (1), the effective specific heats of several types of nanofluids were determined from the measured thermal conductivity and thermal diffusivity obtained by double hot-wire method. As can be seen from Figures 20 to 22, the specific heats of nanofluids significantly decrease with increasing volume fraction of nanoparticles. Because of the much smaller specific heat of solid compared to that of liquid, it is obvious that the specific heat of a mixture will decrease with an increase of volumetric loading of solid particle.

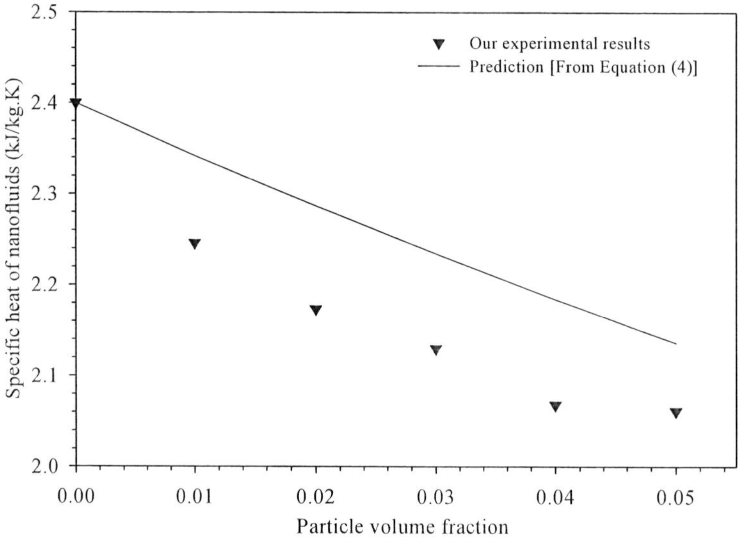

Figure 20. Our measured and theoretically calculated specific heats of TiO_2 (15 nm)/EG-based nanofluids with particle volume fraction.

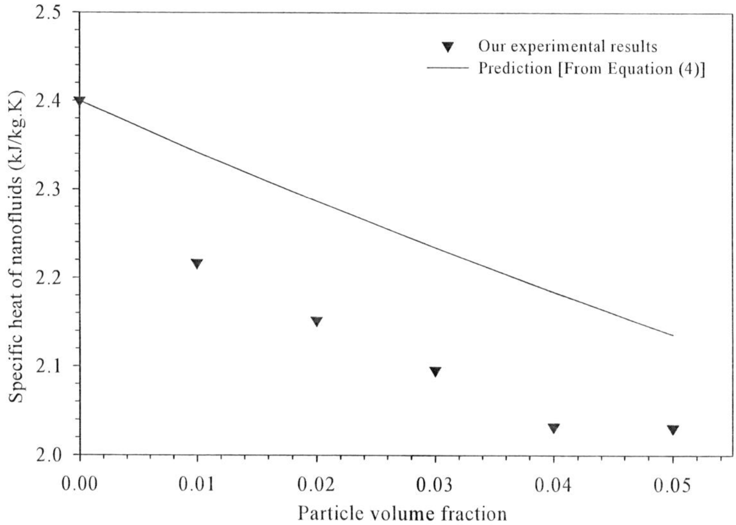

Figure 21. Measured and theoretically calculated specific heats of TiO_2 (10 × 40 nm)/EG-based nanofluids with particle volume fraction (authors' results).

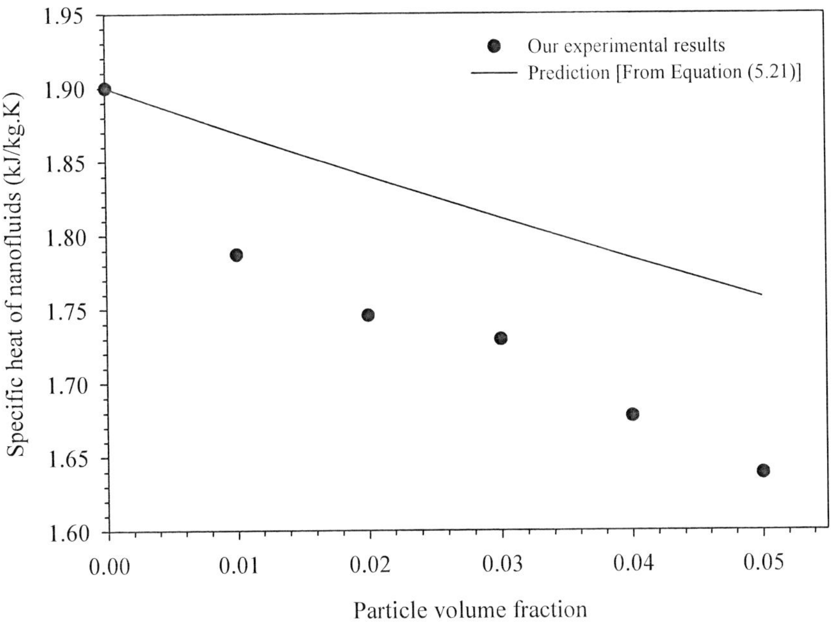

Figure 22. Measured and theoretically calculated specific heats of Al (80 nm)/EO-based nanofluids (authors' results).

Since no experimental data for the effective specific heats of nanofluids are available in the literature, no comparison of present results can be made. However, our results were compared with the calculated values obtained from the volume fraction mixture rule (Equation (4)). Figures 20 to 22 show that the effective specific heats of nanofluids determined from measured thermal conductivity and thermal diffusivity values are much lower than those calculated from Equation (4). This indicates that the volume fraction mixture rule is not suitable for the determination of the specific heats of nanofluids. The reason is unclear at the moment and requires further study.

7.3. Summary of Results

The experimental results presented in this section show that nanofluids containing a small amount of nanoparticles have much higher effective thermal diffusivity values than those of base fluids. The thermal diffusivity of nanofluids increases significantly with increasing volume fraction of nanoparticles. Moreover, the experimental results of this study demonstrate that the particle size and shape together with particle material and base fluid have effect on the effective thermal diffusivity of nanofluids. The specific heats of the nanofluids are also determined from the measured thermal conductivity and thermal diffusivity. With increasing volume fraction of nanoparticles, the specific heat of nanofluids is much lower and decreases more rapidly than the results calculated from the volume fraction mixture rule.

8. Studies on Viscosity of Nanofluids

8.1. Reported Studies in Literature

Although several articles [39, 64-65] emphasized the significance of investigating the viscosity of nanofluids, very few studies on this property were reported. It is believed that viscosity is as critical as thermal conductivity in engineering systems that employ nanofluids convective heat transfer where fluid flow is present. Pumping power is related to the pressure drop, which in turn is dependent on viscosity. In laminar flow, the pressure drop is directly proportional to the viscosity.

Masuda *et al.* [66] measured the viscosity of suspensions of dispersed ultra-fine particles in water. They found that TiO_2 (27 nm) particles at a volumetric loading of 4.3% increased the viscosity of water by 60%. Wang *et al.* [7] observed that the effective viscosity of Al_2O_3 (28 nm)/distilled water-based nanofluids was increased by about 86% for a 5% volume concentration of nanoparticles. In their case, a mechanical blending technique was used for the dispersion of Al_2O_3 nanoparticles in distilled water. They also found an increase of about 40% in viscosity of ethylene glycol at a volumetric loading of 3.5% of Al_2O_3 nanoparticle. Their results indicate that the viscosity of nanofluids depends on dispersion methods. In contrast, at 10% volume concentration of nanoparticles, the viscosities of Al_2O_3 (13 nm)/water and TiO_2 (27 nm)/water-based nanofluids used by Pak and Cho [67] were several times greater than that of water. This large discrepancy could be due to differences in dispersion techniques and differences in size of particles. Pak and Cho also used adjusted pH values and employed an electrostatic repulsion technique. As expected, the viscosity of nanofluids depends on the methods used to disperse and stabilize the nanoparticle suspension. Their viscosity results were significantly larger than the predictions from the classical theory of suspension rheology such as Einstein's model [68].

Das *et al.* [69] and Putra *et al.* [70] measured the viscosity of Al_2O_3/water and CuO/water-based nanofluids as a function of shear rate and showed Newtonian behavior of the nanofluids for a range of volume percentage between 1% and 4%. For Al_2O_3/water-based nanofluids, Das *et al.* [69] also observed an increase in viscosity with an increase of particle volume fraction.

At a given shear rate, Ding *et al.* [71] observed that the viscosity of nanofluids increased with an increasing CNT concentration and decreasing temperature. Shear thinning was observed at all concentrations. This suggests that nanofluids can offer better fluid flow performance due to the higher shear rate at the wall, which results in low viscosity.

Prasher *et al.* [72] reported results on the viscosity of alumina-based nanofluids for various shear rates, temperature, and particle volume fraction. Their results demonstrate that viscosity is independent of shear rate showing that the nanofluids are Newtonian in nature. It also is shown that with increasing nanoparticle volume fraction, the viscosity increases. On the other hand, it is found that viscosity is independent of temperature. This is contrary to the nature of liquid viscosity variation with the temperature.

For TiO_2 (34 nm)/water-based nanofluids, Wen and Ding [33] reported about 20% increase in the effective viscosity for a concentration of 2.4 weight % of particles. However, a much higher viscosity increase was observed under low share rate conditions i.e. 25-100 1/s.

All reported results show that the viscosity of nanofluids is increased anomalously and cannot be predicted by classical models such those attributed to Einstein [68], Krieger and Dougherty [73], Nielsen [74], and Batchelor [75]. No firm conclusion can be drawn from the above random data of several nanofluids. Thus, more extensive investigations need to be carried out on the effective viscosity of nanofluids.

8.2. Models for the Effective Viscosity

On top of very few experimental studies, no established model is available for the prediction of the effective viscosity of nanofluids. The Einstein model is commonly used to predict the effective viscosity of suspensions containing a low volume fraction of particles (usually < 1 volume %). The effective viscosity (η_{eff}) of dilute suspensions is commonly determined by Einstein's viscosity equation [68] expressed as

$$\frac{\eta_{eff}}{\eta_f} = 1 + 2.5\phi_p \tag{5}$$

Batchelor [75] considered interactions between particles in dilute suspension and modified the Einstein equation to

$$\frac{\eta_{eff}}{\eta_f} = 1 + 2.5\phi_p + 6.2\phi_p^2 \tag{6}$$

where ϕ_p is the volume fraction of the dispersed phase and η_f is the viscosity of the base fluid.

Since the nanofluids used in most of the studies are not dilute suspensions, the power law-based models [73, 74] are more appropriate for the prediction of the effective viscosity compared to Einstein's and Batchelor's models. For simple, hard sphere systems, the relative viscosity i.e. the ratio of the effective viscosity of suspension, η_{eff} to that of the suspending medium, η_f increases with particle volume fraction. To determine this relative viscosity, a semi-empirical equation formulated by Krieger and Dougherty (K-D) [73] expressed as

$$\frac{\eta_{eff}}{\eta_f} = (1 - \frac{\phi_p}{\phi_m})^{-[\eta]\phi_m} \tag{7}$$

is used.

A decade later, a generalized equation for the relative elastic moduli of composite materials, which is also widely used for relative viscosity, was proposed by Nielsen [74]. For low concentrations of dispersed particles, Nielsen's equation can be simplified as

$$\frac{\eta_{eff}}{\eta_f} = (1+1.5\phi_p)e^{\phi_p/(1-\phi_m)} \quad (8)$$

where ϕ_m is the maximum packing fraction and $[\eta]$ is the intrinsic viscosity ($[\eta]$ = 2.5 for hard spheres). For randomly mono-dispersed spheres, the maximum close packing fraction is approximately 0.64 [73].

8.3. Present Studies by the Authors

In this study, the effective viscosity of nanofluids was measured by a controlled rate rheometer (Contrases Low Shear 40 Model) and the results were compared with those predicted by the existing models as well as experimental data obtained from the literature.

Figure 23 shows that the viscosities of suspensions of TiO_2 and Al_2O_3 in water are independent of shear rates tested. This demonstrates that these nanofluids are Newtonian in nature. Figure 24 compares the relative viscosity data for TiO_2/water-based nanofluids from this study and those from the literature. The effective viscosity was found to increase with the particle volume fraction. For 15 nm TiO_2 nanoparticles in water, the maximum increase in viscosity was found to be 86% for a volume fraction of 0.05. For the same volume fraction of cylindrical shape (10 × 40 nm) TiO_2 nanoparticles, the viscosity of water is increased by 58%. This indicates that the shapes of the particles may influence the suspension viscosity. Through the surface charge, particle shape influences the electrostatic repulsive force. This repulsive force plays a significant role in altering the viscosity of nanofluids through the dispersion and sedimentation of nanoparticles. The particle shape affects clustering and adsorption, which may also change the viscosity of nanofluids. In fact, TiO_2 (10 × 40 nm) nanoparticles in base fluids were found to be more stable than TiO_2 (15 nm) nanoparticles. The authors' results for these nanofluids are found to be higher than those of Masuda *et al.* [66] and Pak and Cho [67]. Although both of them used the same size TiO_2 nanoparticles of 27 nm and adjusted to a high pH value (pH = 10) of their suspensions, Pak and Cho observed much higher viscosity values than those of Masuda *et al.* Figure 24 also depicts that the existing models are unable to predict the viscosity of these nanofluids.

Figure 25 presents measured viscosities of Al_2O_3 (80 nm)/water-based nanofluids with nanoparticle volume fraction and their comparisons with the models of Einstein [68], Batchelor [75], and Nielsen [74]. Al_2O_3 (80 nm) nanoparticles were found to increase the viscosity of water by nearly 82% for the maximum volumetric loading of 5%. A similar increment (86%) of the effective viscosity of Al_2O_3 (28 nm)/distilled water-based nanofluids was also observed by Wang *et al.* [7] for the same volumetric loading of 5%. In spite of their smaller particle size, Masuda *et al.'s* [66] results showed much larger increases in viscosity for the same nanofluids. For example, the viscosity of Al_2O_3 (13 nm)/water-based nanofluids at a volume % of 4.16 was about 2.68 times that of the viscosity of water in their study. Interestingly, the viscosity of the same nanofluid i.e. Al_2O_3 (13 nm)/water prepared by Pak and Cho [67] was three times higher than that of water for a 3 volume % of nanoparticles. However, a larger increase in viscosity was also observed in the present study for 150 nm sized Al_2O_3 nanoparticles. The maximum increase in viscosity was about 180% for the maximum volumetric loading of 5%. The present results for Al_2O_3/DIW-based nanofluids demonstrate the effect of particle size on the viscosity of suspensions and the larger the particle size, the higher the viscosity. Figure 25 also shows that the measured viscosities are severely under-predicted by the existing models.

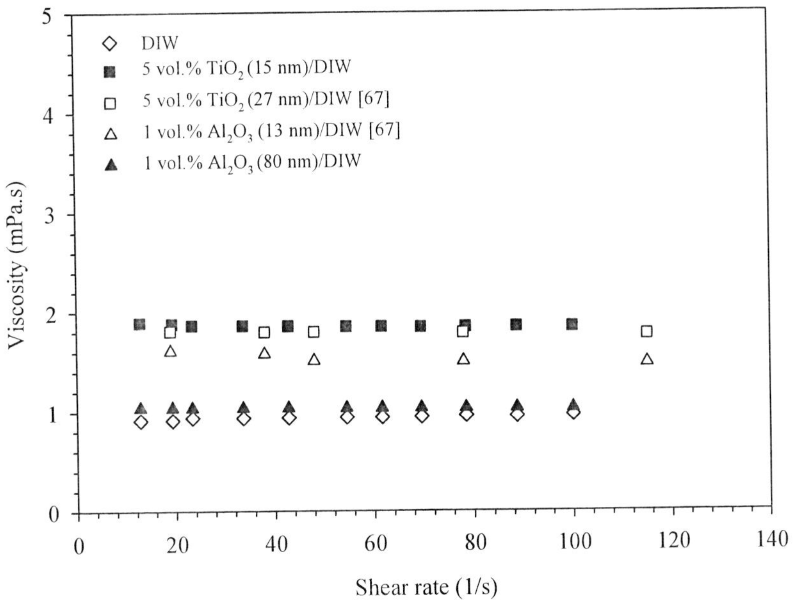

Figure 23. Comparison of authors' results and data obtained from the literature on the viscosity of nanofluids with shear rate.

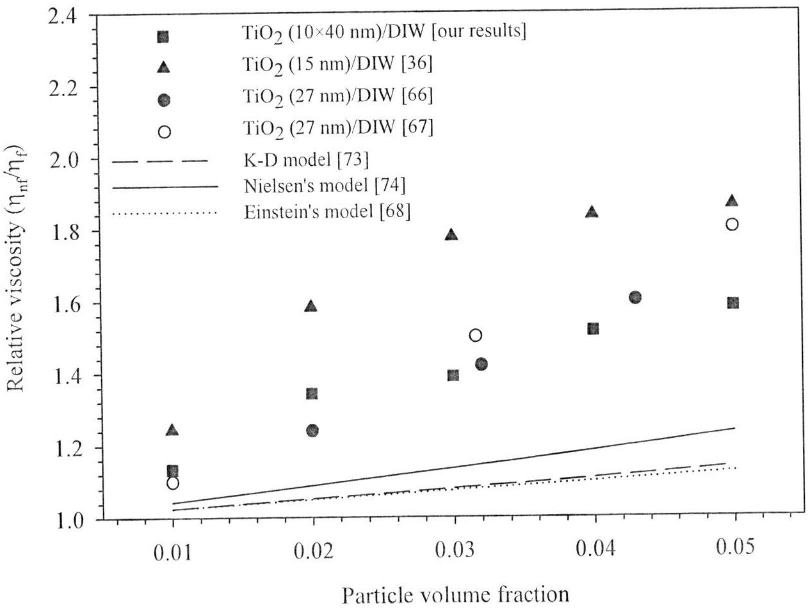

Figure 24. Experimental and predictions of relative viscosity of TiO_2/water-based nanofluids.

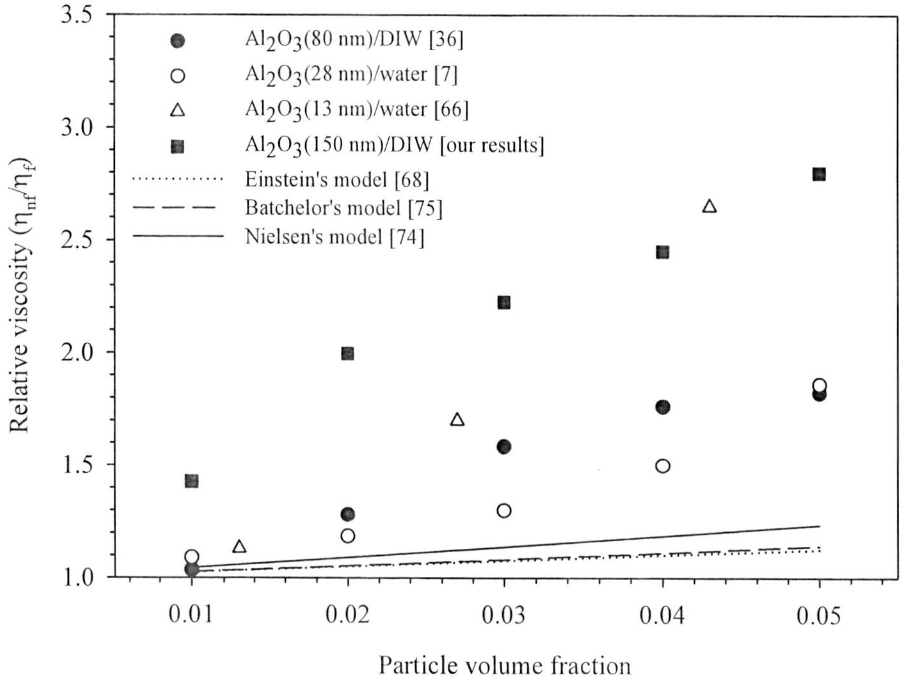

Figure 25. Experimental and predictions of relative viscosity of Al_2O_3/water-based nanofluids.

8.4. Summary

From Figures 24 and 25, it is found that the viscosity of nanofluids increases significantly with particle volume fraction. However, there are discrepancies between the present results and the reported data from the literature. The reasons for the differences could be the difference in the size of the particle clusters, differences in the dispersion techniques, and the use of surfactants. Indeed, the viscosity of nanofluids greatly depends on the methods used to disperse and stabilize the nanoparticle suspension [7]. Existing models were also found to be unable to predict the effective viscosity of nanofluids. This is probably because these models considered only particle volume fraction, whereas the nanoparticles in fluids can easily form clusters and can experience surface adsorption. Clustering and adsorption increase the hydrodynamic diameter of nanoparticles leading to the increase of relative viscosity. Besides the particle volume fraction and size, the nature of the particle surface, ionic strength of the base fluid, surfactants, pH values, inter particle potentials such as repulsive (electric double layer force) and attractive (van der Waals force) forces may play significant roles to alter the viscosity of nanofluids. However, such enhancement of viscosity may reduce the potential benefits of nanofluids.

9. Electrokinetic Phenomena of Nanofluids

9.1. Reported Studies in the Literature

Electrochemical effects certainly influence the thermal conductivity of nanofluids through stability and interparticle interactions. For example, the electrostatic repulsive force, which is described by the zeta potential, is important to avoid agglomeration and sedimentation of nanoparticles. Although some review articles [39, 76] pronounced that the surface chemistry of nanoparticles could be an important factor influencing the thermal conductivity of nanofluids, very few studies have been reported to investigate its effect on the enhanced thermal conductivity of nanofluids. Xie *et al.* [77] showed that simple acid treatment of carbon nanotubes enhanced the suspension stability of carbon nanotubes in water. It was attributed to hydrophobic-to-hydrophilic conversion of the surface nature due to the generation of a hydroxyl group. Patel *et al.* [25] reported that 4 nm gold nanoparticles with a coating of a covalent chain in toluene were about 50 times less effective for heat transport than uncoated 10-20 nm gold particles in water. This is contrary to the findings of the nanoparticle size effect on the effective thermal conductivity of nanofluids [26, 40]. Lee *et al.* [78] investigated the effect of zeta potential and particle hydrodynamic size (i.e. the mobility equivalent size) on the thermal conductivity of nanofluids. The hydrodynamic diameter of CuO nanoparticles and thermal conductivity of the nanofluids at different pH values were measured in their study. Their results demonstrated that the pH value affects the thermal performance of the nanofluids. As the solution pH value departs from the iso-electric point of particles, the colloidal particles become more stable and eventually alter the thermal conductivity of the fluid. They identified that the surface charge state is primarily responsible for the enhancement of thermal conductivity of the nanofluids. However, no further research was reported. Prasher *et al.* [79] studied the effect of aggregation on the effective thermal conductivity of nanofluids. They claimed that the observed thermal conductivity of nanofluids can be explained by the aggregation kinetics. It was also reported that the colloidal chemistry plays a significant role in deciding the thermal conductivity of nanofluids.

9.2. Studies by the Authors

The characterization of electrokinetic properties of nanofluids and their effects on thermal conductivity [80] are investigated. The effects of pH value, surfactant, and electrolyte concentration on zeta potential and thermal conductivity of TiO_2 (15 nm)/water–based nanofluids are discussed here.

9.2.1. Effects of pH value and Electrolyte Concentration on Zeta Potential

Figure 26 presents the variation of zeta potential for TiO_2 (15 nm)/water–based nanofluids with solution pH. The iso-electric point (IEP), a specific pH value at which the zeta potential is zero, was found to be between 4.9 and 5.2. This value is consistent with that reported in the literature (IEP = 5.1) for this type of nanoparticles [81]. It can also be seen from the figure that the presence of surfactant increases the magnitude of zeta potential, which indicates better stability of the nanofluids because the larger the magnitude of zeta potential, the stronger the repulsive forces. In addition, the surfactant can be adsorbed at the charged particle surface and modify the charges distribution in the electrical double layer (EDL) resulting a change in zeta potential. Due to the presence of CTAB, the IEP also decreased slightly from 5.2 to 4.9.

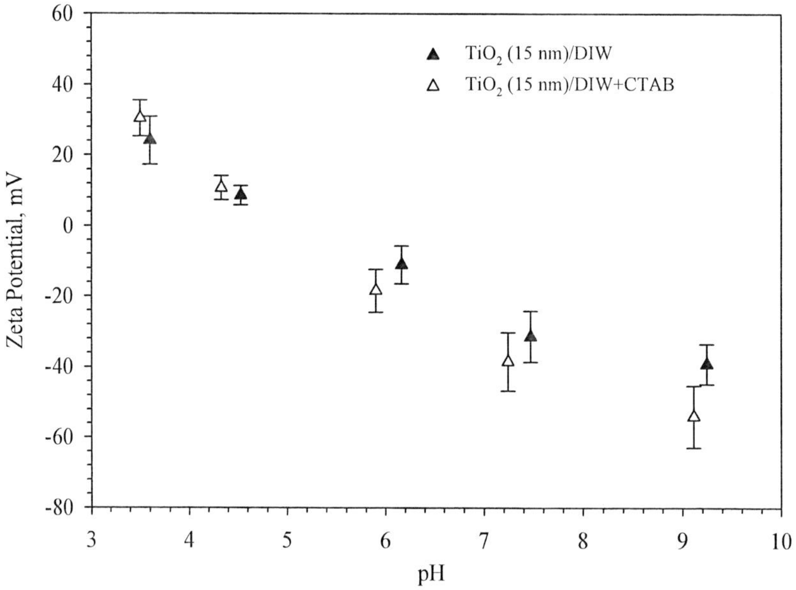

Figure 26. Effect of pH value on zeta potential of nanoparticles in deionized water [80].

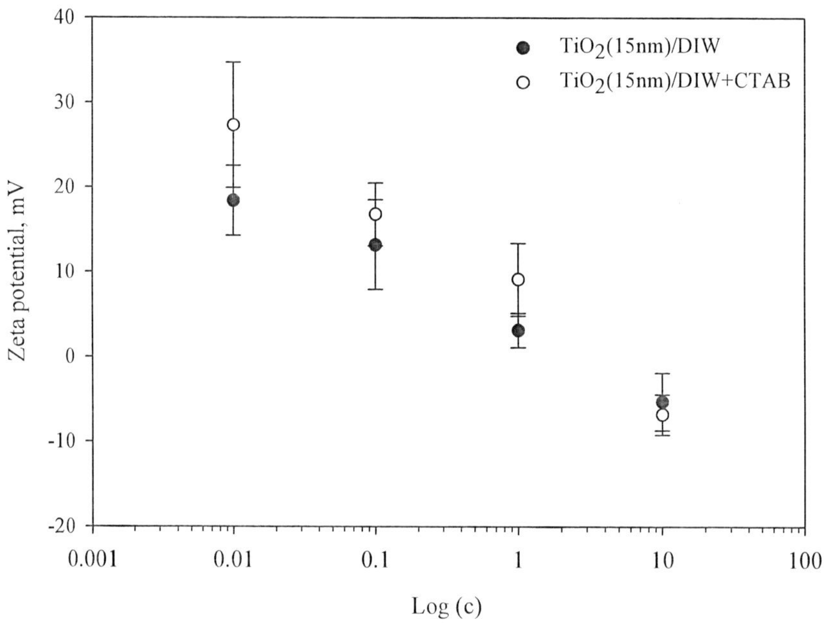

Figure 27. Effect of electrolyte concentration on zeta potential of nanoparticles [80].

The effect of electrolyte concentration (c, in mM) on the zeta potential is shown in Figure 27. Sodium chloride (NaCl) electrolyte of various concentrations was used. It is found that the ionic concentration of the suspensions has significant impact on the zeta potential. Similar to Figure 26, the addition of CTAB surfactant results in larger magnitudes of the zeta potential. The higher the concentration of the salt, the smaller the zeta potential. It is noted that the pH value at each NaCl concentration was remained the same i.e. about 6.2.

9.2.2. Effects of pH value and Electrolyte Concentration on Thermal Conductivity

Figure 28 shows that the enhanced thermal conductivity of this nanofluid decreases with an increase in pH value. For 0.2 volume % of TiO_2 nanoparticle, the thermal conductivity decreased from 5.5% to 2.55% when the pH value was increased from 3.4 to 9. A larger decrease in the enhanced thermal conductivity of Al_2O_3/water-based nanofluids was observed by Xie *et al.* [22] when the pH value was increased from 2 to 12. The trend of decreasing thermal conductivity with pH value until the IEP is reached can be explained by the DVLO theory. When the pH value is close to or equal to the IEP (zero repulsive force), the particle suspension becomes unstable and large-scale agglomerations occur which can lead to a decrease in effective thermal conductivity of nanofluids. However, when the pH value is further increased from the IEP, the effective thermal conductivity should increase gradually due to less agglomeration of particles, which was not observed in the authors' study [80]. Nevertheless, the decrease in effective thermal conductivity in this study was within 3% when the pH value changes from 3.4 to 9.

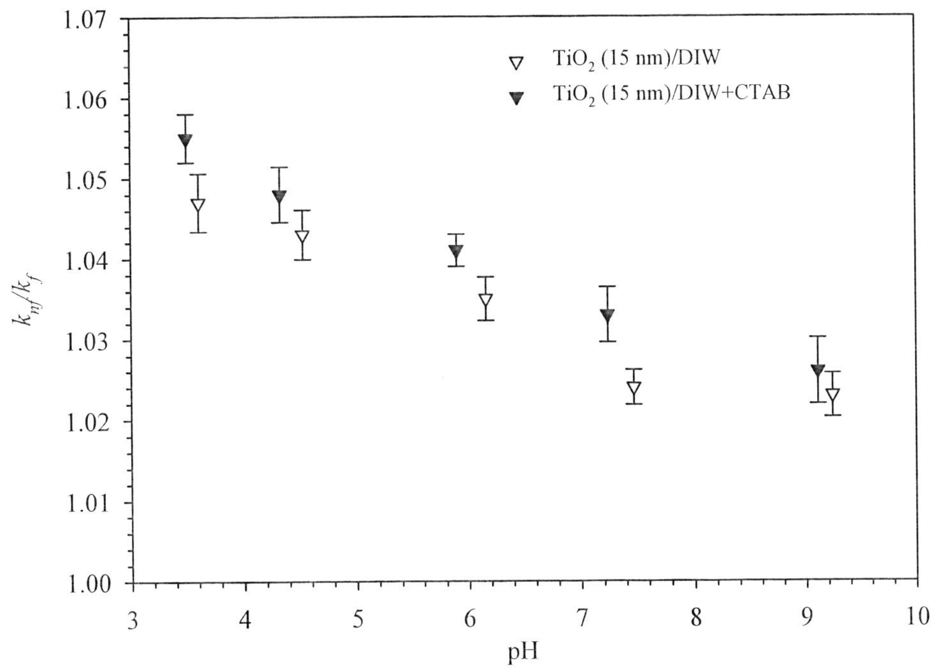

Figure 28. Effect of pH value on the effective thermal conductivity of nanofluids [8].

The influence of electrolyte concentration on the effective thermal conductivity of nanofluids is shown in Figure 29. The effective thermal conductivity was found to decrease with increasing NaCl concentration. However, the decrease of thermal conductivity was only within 1.5 to 2% when NaCl concentration was increased from 0.01 mM to 10 mM. This is consistent with the results of zeta potential (Figure 27) which decreases with increasing electrolyte concentration.

The positive effect of CTAB surfactant on the effective thermal conductivity of nanofluids can also observed from Figures 28 and 29. Surfactants promote to enhance the effective thermal conductivity of nanofluids through better particle dispersion i.e. less sedimentation/ agglomeration [82] and enhancing the interactions between components of nanofluids [83]. Such positive effect of surfactant on the enhancement of the effective thermal conductivity of Cu/ethylene glycol-based nanofluid was also reported by Yu and Choi [38]. It should be pointed out that the thermal conductivity of the base fluid is independent of pH value and electrolyte concentration. Since no other studies were reported in the literature regarding the influences of pH value and electrolyte concentration on the effective thermal conductivity for this nanofluid, no comparison of the present results can be made.

9.3. Summary

The pH value and the electrolyte concentration were found to have a significant effect on the zeta potential. The presence of a surfactant causes improved particle dispersion behavior, which resulted in better stability with less agglomeration and slightly larger effective thermal conductivity compared to nanofluids without surfactant. The enhanced thermal conductivity of this nanofluid was found to decrease with increasing pH value or electrolyte concentration.

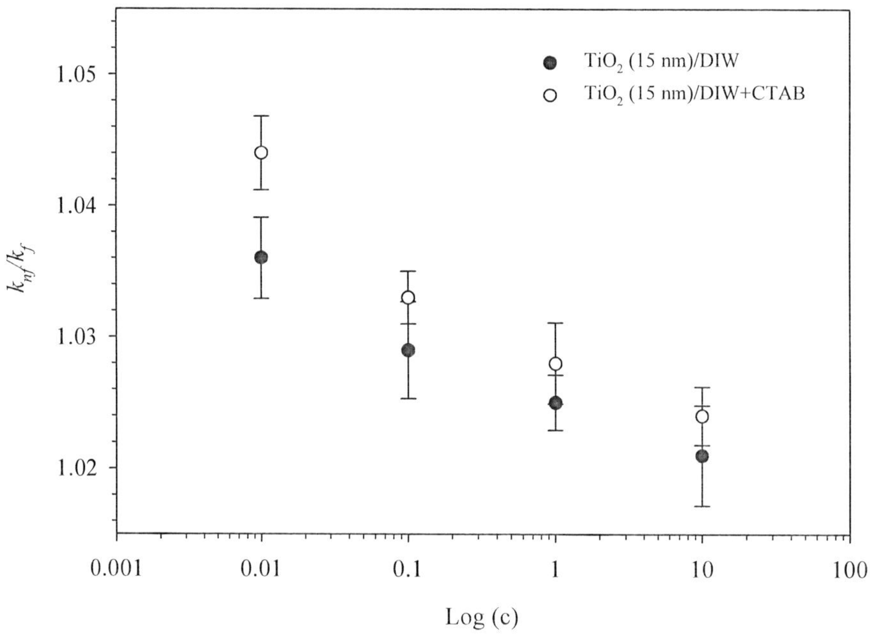

Figure 29. Effect of electrolyte concentration on the effective thermal conductivity [80].

10. CONCLUSIONS AND FUTURE WORK

From the review and presented studies by the authors, it can be summarized that nanofluids exhibit enhanced effective thermal conductivity, which increases with increasing volumetric loading of nanoparticles. Review of experimental studies clearly shows the lack of consistency in the reported results from the various research groups. The effects of several important factors such as particle size and shapes, clustering of particles, temperature of the fluid, and dissociation of surfactant on the effective thermal conductivity of nanofluids have not been investigated adequately in the literature. In order to confirm the effects of these factors on the effective thermal conductivity of nanofluids, the present results are of great significance.

It is found that that the existing classical models cannot explain the observed enhanced thermal conductivity of nanofluids. On the other hand, most of the recently developed models are not based on first principles, and thus failed to explain the observed thermal conductivity of nanofluids. These models also cannot be validated with a wide range of experimental data. There is, therefore a need to develop new models, which can explicitly explain the enhanced thermal conductivity of nanofluids. Besides particle volume fraction, particles size, temperature, particles dispersions, particle interaction and clustering should be taken into account in developing model for the effective thermal conductivity of nanofluids.

Despite the importance of the effective thermal diffusivity, very little work has been reported on the determination of the effective thermal diffusivity of nanofluids. The present results show that nanofluids containing a small amount of nanoparticles have much higher effective thermal diffusivity values than those of base fluids. The thermal diffusivity of nanofluids increases significantly with increasing volume fraction of nanoparticles. The particle size and shape together with particle material and base fluid are also observed to have effect on the effective thermal diffusivity of nanofluids. With increasing volume fraction of nanoparticles, the specific heat of nanofluids is found to be much lower and decreases more rapidly than that of calculated from the volume fraction mixture rule. Effective viscosity of nanofluids is found to increase substantially with volumetric loading of nanoparticles. The existing models are also found to be unable to predict the viscosity of nanofluids. Thus, the theoretical model for the effective viscosity of nanofluids is to be developed. Furthermore, studies on characterization of electrokinetic properties and convective heat transfer of nanofluids are rather scarce. More extensive studies in these areas need to be performed for the potential applications of nanofluids in thermal management systems.

Although many studies have been performed on the heat transfer behavior of nanofluids, there has been very little agreement in the data from different research groups. Hence, proper sample preparation and repeatable and more systematic experimental studies on thermophysical properties, particularly thermal conductivity, thermal diffusivity, convective heat transfer coefficients and viscosity of nanofluids are worthwhile.

Focusing on the potential applications of nanofluids in advanced cooling techniques for more efficient cooling of electronics and microelectromechanical systems (MEMS), investigations on microchannel flow of nanofluids are of great interest.

References

[1] Maxwell, J. C. *A Treatise on Electricity and Magnetism*, Clarendon Press: Oxford, U.K., 1891.

[2] Choi, S. U. S.; Zhang, Z. G.; Keblinski, P. In *Encyclopedia of Nanoscience and Nanotechnology;* Nalwa, H. S.; Ed.; American Scientific Publishers: Los Angeles, CA, USA, 2004; Vol. 6, pp 757-773.

[3] Choi, S. U. S. In *Developments and applications of non-Newtonian flows*; Siginer, D. A.; Wang, H. P.; Eds.; ASME Publishing: New York,USA,1995; FED-Vol. 231/MD-Vol. 66, pp 99-105.

[4] Eastman, J. A.; Choi, S. U. S.; Li, S.; Thompson, L. J. In *Proceedings of the Symposium on Nanophase and Nanocomposite Materials II*; Komarneni, S.; Parker, J. C.; Wollenberger, H.; Eds.; Materials Research Society: Boston, USA, 1997; Vol. 457, pp 3-11.

[5] Eastman, J. A.; Choi, S. U. S.; Li, S.; Yu, W.; Thompson, L. J. *Appl. Phys. Lett.* 2001, 78, 718-720.

[6] Lee, S.; Choi, S. U. S.; Li, S.; Eastman, J. A. *J. Heat Transfer* 1999, 121, 280-289.

[7] Wang, X.; Xu, X.; Choi, S. U. S. *J. Thermophys. Heat Transfer* 1999, 13, 474-480.

[8] Xuan, Y.; Li, Q. *Int. J. Heat Fluid Flow* 2000, 21, 58-64.

[9] Das, S. K.; Putra, N.; Thiesen, P.; Roetzel, W. *J. Heat Transfer* 2003, 125, 567-574.

[10] Murshed, S. M. S.; Leong, K. C.; Yang, C. *Int. J. Therm. Sci.* 2005, 44, 367-373.

[11] Hong, T.; Yang, H.; Choi, C. J. *J. Appl. Phys.* 2005, 97, 064311-1-064311-4.

[12] Li, C. H.; Peterson, G. P. *J. Appl. Phys.* 2006, 99, 084314-1 – 084314-8.

[13] Choi, S. U. S.; Xu, X.; Keblinski, P.; Yu, W. In *Proceedings of 20th Symposium on Energy Engineering Science*, Illinois, USA, 2002.

[14] Hong, K. S.; Hong, T.; Yang, H. *Appl. Phys. Lett.* 2006, 88, 031901-1-031901-3.

[15] Steinour, H. H. *Ind. Eng. Chem.* 1944,36, 618-624.

[16] Kaviany, M. *Principles of Heat Transfer*, John Wiley & Sons, Inc.: New York, 2002.

[17] Bolz, R.; Tuve, G. *Handbook of Tables for Applied Engineering Science*, The Chemical Rubber Co.: Boca Raton, USA, 1973.

[18] Horrocks, J. K.; McLaughlin, E. *Proc. R. Soc. Lond. A.1963,* 273, 259-274.

[19] Nagasaka, Y.; Nagashima, A. *J. Phys. E: Sci. Instrum.* 1981, 14, 1435-1440.

[20] Wang, B.-X.; Zhou, L.-P.; Peng, X.-F. *Int. J. Heat Mass Transfer* 2003, 46 2665-2672.

[21] Yang, B.; Han, Z. H. *Appl. Phys. Lett.* 2006, 89, 083111-1–083111-3.

[22] Xie, H.; Wang, J.; Xi, T.; Liu, Y.; Ai, F.; Wu, Q. *J. Appl. Phys.* 2002, 91, 4568-4572.

[23] Wang, Y.; Fisher, T. S.; Davidson, J. L.; Jiang, L. In *Proceedings of 8th AIAA/ASME Joint Thermophysics and Heat Transfer Conference*, USA, 2002, AIAA 2002-3345.

[24] Xuan, Y.; Li, Q. *J. Heat Transfer* 2003, 125, 151-155.

[25] Patel, H. E.; Das, S. K. Sundararajan, T.; Nair, A. S.; George, B.; Pradeep, T. *Appl. Phys. Lett.* 2003, 83, 2931-2933.

[26] Kumar, D. H.; Patel, H. E.; Kumar, V. R. R.; Sundararajan, T.; Pradeep, T.; Das, S. K. *Phys. Rev. Lett.* 2004, 93, 4301-4304.

[27] Kwak, K.; Kim, C. *Kor.- Aus. Rheol. J.* 2005, 17, 35-40.

[28] Zhu, H.; Zhang, C.; Liu, S.; Tang, Y.; Yin, Y. *Appl. Phys. Lett.* 2006, 89, 023123-1–023123-3.

[29] Hwang, Y. J.; Ahn, Y. C.; Shin, H. S.; Lee, C. G.; Kim, G. T.; Park, H. S.; Lee, J. K. *Curr. Appl. Phys.* 2006, 6, 1068-1071.
[30] Xuan, Y.; Li, Q.; Zhang, X.; Fujii, M.; *J. Appl. Phys.* 2006, 100, 043507-1– 043507- 6.
[31] Liu, M. S.; Lin, M. C.-C.; Huang, I. T.; Wang, C.-C. *Chem. Eng. Technol.* 2006, 29, 72-77.
[32] Krishnamurthy, S.; Bhattacharya, P.; Phelan, P. E.; Prasher, R. S. *Nano Lett.* 2006, 6, 419– 423.
[33] Wen, D.; Ding, Y. *IEEE Trans. Nanotechnology* 2006, 5, 220-227.
[34] Putnam, S. A.; Cahill, D. G.; Braun, P. V.; Ge, Z.; Shimmin, R. G. *J. Appl. Phys.* 2006, 99, 084308-1–084308-6.
[35] Kang, H. U.; Kim, S. H.; Oh, J. M. *Exp. Heat Transfer* 2006, 19,181-191.
[36] Murshed, S. M. S.; Leong, K. C.; Yang, C. *Int. J. Therm. Sci.* 2008, 47, 560-568.
[37] Keblinski, P.; Phillpot, S. R.; Choi, S. U. S.; Eastman, J. A. *Int. J. Heat Mass Transfer* 2002, 45, 855-863.
[38] Yu, W.; Choi, S. U. S. *J. Nanoparticle Res.* 2003, 5, 167-171.
[39] Eastman, J. A.; Phillpot, S.R.; Choi, S. U. S.; Keblinski, P.; *Ann. Rev. Mater. Res.* 2004, 34, 219-246.
[40] Jang, S. P.; Choi, S. U. S. *Appl. Phys. Lett.* 2004, 84, 4316-4318.
[41] Chon, C. H.; Kihm, K. D. *J. Heat Transfer* 2005, 127, 810.
[42] Venerus, D. C.; Kabadi, M. S.; Lee, S.; Perez-Luna, V. *J. Appl. Phys.* 2006, 100, 094310-1– 094310-5.
[43] Koo, J.; Kleinstreuer, C. *J. Nanoparticle Res.* 2004, 6, 577-588.
[44] Murshed, S. M. S.; Leong, K. C.; Yang, C. *Int. J. Nanosci.* 2006, 5, 23-33.
[45] Hamilton, R. L.; Crosser, O. K. T *I&EC Fundamentals* 1962, 1, 187-191.
[46] Hui, P. M.; Zhang, X.; Markworth, A. J.; Stroud, D. *J. Mater. Sci.* 1999, 34, 5497-5503.
[47] Yu, W.; Choi, S. U. S. *J. Nanoparticle Res.* 2004, 6, 355-361.
[48] Xue, Q.-Z. *Phys. Lett. A.* 2003, 307, 313-317.
[49] Wang, X. Q.; Mujumdar, A. S., *Int. J.Therm. Sci.* 2007, 46, 1-19.
[50] Murshed, S. M. S.; Leong, K. C.; Yang, C. *Appl. Therm. Eng.* 2008 (in press, doi:10.1016/j.applthermaleng.2008.01.005).
[51] Xie, H.; Fujii, M.; Zhang, X. *Int. J. Heat Mass Transfer* 2005, 48, 2926-2932.
[52] Leong, K. C.; Yang, C.; Murshed, S. M. S. *J. Nanoparticle Res.* 2006, 8, 245-254.
[53] Yu, W.; Choi, S. U. S. *J. Nanosci. Nanotechnol.* 2005, 5, 580-586.
[54] Sabbaghzadeh, J.; Ebrahimi, S. *Int. J. Nanosci.* 2007, 6, 45-49.
[55] Ren, Y.; Xie, H.; Cai, A., *J. Phys. D: Appl. Phys.* 2005, 38, 3958-3861.
[56] Xuan, Y.; Li, Q.; Hu, W. *AIChE J.* 2003, 49, 1038-1043.
[57] Leong, K. C.; Yang, C.; Murshed, S. M. S., *Presented in Nanofluids: Fundamentals and Applications Conference*, Copper Mountain, Colorado, USA, September 2007.
[58] Xuan, Y.; Roetzel, W.; *Int. J. Heat Mass Transfer* 2000, 43, 3701-3707.
[59] Wang, B-X.; Zhou, L.-P.; Peng, X.-F. *Prog. Nat. Sci.* 2004, 14, 922-926.
[60] Parker, W. J.; Jenkins, R. J.; Butler, C. P.; Abbott, G. L. *J. Appl. Phys.* 1961, 32,1679-1684.
[61] Balderas-Lopez, J. A.; Mandelis, A., *Rev. Sci. Instrum.* 2001, 72, 2649-2652.
[62] Czarnetzki, W.; Roetzel, W., *Int. J. Thermophys.* 1995, 16, 413-422.

[63] Murshed, S. M. S.; Leong, K. C.; Yang, C. *J. Phys. D: Appl. Phys.* 2006, 39, 5316-5322.

[64] Keblinski, P.; Eastman, J. A.; Cahill, D.G. *Materials Today* 2005, 8, 36-44.

[65] Das, S. K.; Choi, S. U. S.; Patel, H. E.; *Heat Transfer Eng.* 2006, 27, 3-19.

[66] Masuda, H.; Ebata, A.; Teramae, K.; Hishinuma, N. *Netsu Bussei* 1993, 4, 227-233.

[67] Pak, B. C.; Cho, Y. I. *Exp. Heat Transfer* 1998, 11, 151-170.

[68] Einstein, A. *Investigations on the Theory of the Brownian Movement*, Dover Publications, Inc.: New York, 1956.

[69] Das, S. K.; Putra, N.; Roetzel, W. *Int. J. Heat Mass Transfer* 2003, 46, 851-862.

[70] Putra, N.; Roetzel, W.; Das, S. K. *Heat Mass Transfer* 2003, 39, 775-784.

[71] Ding, Y.; Alias, H.; Wen, D.; Williams, A. R. *Int. J. Heat Mass Transfer* 2006, 49, 240-250.

[72] Prasher, R.; Song, D.; Wang, J.; Phelan, P.E. *Appl. Phys. Lett.* 2006, 89, 133108-1–133108-3.

[73] Krieger, I. M.; Dougherty, T. J. *Trans. Soc. Rheology* 1959, 3, 137-152.

[74] Nielsen, L. E. *J. Appl. Phys.* 1970, 41, 4626-4627.

[75] Batchelor, G. K. *J. Fluid Mech.* 1977, 83, 97-117.

[76] Kabelac, S.; Kuhnke, J. F. In *Proceedings of International Heat Transfer Conference*, Sydney, Australia, August 2006.

[77] Xie, H.; Lee, H.; Youn, W.; Choi, M. *J. Appl. Phys.* 2003, 94, 4967-4971.

[78] Lee, D.; Kim, J. W.; Kim, B. G. *J. Phys. Chem. B* 2006, 110, 4323-4328.

[79] Prasher, R.; Phelan, P. E.; Bhattacharya, P. *Nano Lett.* 2006, 6, 1529-1534.

[80] Murshed, S. M. S.; Leong, K. C.; Yang, C. *J. Nanosci. Nanotechnol.* 2008, 8, 1-6..

[81] Kosmulski, M.; *J. Colloid Interface Sci.* 2006, 298, 730-741.

[82] Kim, S. H.; Choi, S. R.; Kim, D. *J. Heat Transfer* 2007, 129, 298-307.

[83] Assael, M. J.; Metaxa, I. N.; Kakosimos, K.; Constantinou, D. *Int. J. Thermophys.* 2006, 27, 999-1017.

In: Encyclopedia of Nanotechnology
Editor: Elwood D. Carlson

ISBN: 978-1-60692-079-4

Chapter 42

RECENT DEVELOPMENTS IN THE EFFECTIVE THERMAL CONDUCTIVITY OF NANOPARTICLE SUSPENSIONS (NANOFLUIDS) RESEARCH

Calvin H. Li
Department of Mechanical, Industrial, and Manufacturing Engineering,
University of Toledo, Toledo, Ohio 43606, USA
G. P. Peterson
Department of Mechanical Engineering, University of Colorado
Boulder, Colorado 80309, USA

ABSTRACT

Since the first report in 1995 that indicated that the addition of nanometer copper metal particles to traditional heat transfer liquids, such as water and Ethylene Glycol, could result in an enhanced effective thermal conductivity which was considerably higher than that predicted by traditional mean-field theory, tremendous experimental studies have been conducted. Most of these support this conclusion, but there is some variation in the degree of the level of the enhancement. These investigations include experiments on metal nanoparticle/liquid suspensions, metal oxide nanoparticle/liquid suspensions, liquid nanoparticle/liquid suspensions, and carbon nanotube/liquid suspensions, and cover a wide range of nanoparticle shapes and volume fractions. These investigations include variations in the size of the nanoparticle, the type of nanoparticle material, the type of base fluid, the effect of surfactants, the effect of pH value of the suspension, the temperature of the suspension, and the ultrasonic vibrating time. Moreover, a number of experimental techniques and methods have been adopted to measure the effective thermal conductivity of different types of suspensions, including steady-state cut-bar methods, transient hot-wire methods and other less well known methods.

The theoretical exploration of the fundamental mechanisms for enhanced effective thermal conductivity of nanoparticle suspensions has resulted in a number of theoretical models. There are basically two different mechanisms proposed for these variations, the Brownian motion effect and the agglomeration effect. In addition to these, several other alternatives, such as the effect of the adsorption layer of the fluid molecules on the surface of the nanoparticles and the thermophoresis of the nanoparticles inside the suspension have been proposed.

The Brownian motion mechanism in nanoparticle suspensions is a relatively new concept, which has grown in acceptance with the addition of additional experimental evidence. While the agglomeration mechanism is not new, it assumes that the agglomeration of the nanoparticles replaces the micron size and/or millimeter size particles in the mean-field theory. Both major mechanisms are supported by the comparisons between predictions of models and reported experimental results. Moreover, there is almost equal support for both from numerical simulation studies and experimental observations.

An extensive review of the available experimental reports and theoretical developments indicates that a majority of the experiments have confirmed the enhanced effective thermal conductivities of nanoparticle suspensions and indicated that these are greater than the values predicted from mean-field theory models. In addition, the mean-field theory can not explain why the effective thermal conductivity of nanoparticle suspensions could have this unusually high enhancement, or why it increases with increases in the bulk temperature, and with decreases in the nanoparticle size. Finally, the evolution process of newly developed Brownian motion models and agglomeration models is discussed in detail.

As a result, what is clear from the available research is that it is necessary to illuminate why effective thermal conductivities of nanoparticle suspensions are higher than the predictions of mean-field theory models. Extensive study on the contributions of the mechanisms of Brownian motion, agglomeration and correlation of both is critically needed. This review will help the development of a more robust understanding of the new phenomenon and the further application of effective thermal conductivity of nanoparticle suspensions in the future.

1. INTRODUCTION

The term nanofluids is used to describe in general terms 1-100 nm diameter nanoparticle/liquid suspensions, which are used to achieve enhanced thermal conductivity for heat transfer and thermal management purposes. The concept was first proposed in 1995 when it was observed that by dispersing metallic nanoparticles into traditional heat transfer fluids, the effective thermal conductivity of copper nanoparticle/water suspensions could be dramatically enhanced and the pumping power significantly decreased [1, 2]. Based on the theoretical model of Hamilton and Crosser [3], it was estimated that the Cu nanoparticle with a sphericity of 0.3 could dramatically enhance the effective thermal conductivity of nanoparticle suspensions by as much as 3.5 times that of pure water, at a volume fraction of 20 %, and Masuda et al.'s experimental result of the effective thermal conductivity of γ Al_2O_3 nanoparticle/Distilled water suspensions indicated that a 30 % enhancement could be achieved at a nanoparticle volume fraction of 4.3 % [4].

Adding high thermal conductivity particles into a low thermal conductivity matrix to achieve increased thermal conductivities while retaining the other desirable physical properties of the matrix is far from a novel idea. This kind of work has been extensively studied for many centuries. Due to the limitations of technology, however, the size of the dispersed particles has been restricted to the order of millimeter and microns, such as the experimental results reported by Hamilton and Crosser [3] and by Meredith and Tobias [5]. Previous theoretical research can be traced even earlier than the first experimental investigations, and many successful mean-field theory models have been developed and have been used to accurately predict the effects of the thermal conductivity of solid/solid mixtures, solid/liquid mixtures, and liquid/liquid mixtures. A number of these models have been used to predict the effective thermal conductivities of nanoparticle suspensions, but deviations of experimental data from the predictions of those mean-field theory models were reported, and meanwhile this lack of correlation between the measured and predicted values required further study of the fundamental mechanisms of the effective thermal conductivity of nanoparticle suspensions.

2. THE MEAN-FIELD THEORY AND MODELS

The mean-field theory has a long history which can be traced back as far as the 18th century. The first mean-field model capable of predicting the effective thermal conductivity of solid particles and fluid mixtures was initially developed by Maxwell [6] in 1873 who studied conduction through heterogeneous media. In this initial work, the potentials at the interface were assumed to be continuous and equal, thus the current passing through the interface of the two media would be the same. Examples of two and three different physical properties with interfaces that were spherical were conducted, and the mathematical treatment was applied to introduce the potential expanded in solid body harmonics and the surface harmonics. By analyzing the algebraic sign of the constants in the harmonics, it became clear that if the conduction of one of the materials were higher than the rest, the material with the higher conduction would tend to equalize the potential throughout the material. Otherwise, the current was prevented from reaching the inner of the lower conduction materials. By examining the case of n spheres of radius a_1 and resistance R_1, placed in a medium of resistance material, R_2, where all the spheres and media were contained in a sphere of radius a_2, and assuming no interaction between the spheres or medium, the effective resistance of the sphere of radius a_2 with a volume fraction of ϕ could be calculated as follows:

$$R = \frac{2R_1 + R_2 + \phi(R_1 - R_2)}{2R_1 + R_2 - 2\phi(R_1 - R_2)} R_2 \tag{1}$$

This equation serves as the original formula, which was subsequently modified by other researchers and has been extensively used to predict the behavior of these types of systems, with the limitation that the volume fraction, ϕ, should be very small in order to avoid the interaction between spheres and that the magnitude of $\frac{R_1 - R_2}{2R_1 + R_2}$ should not be large.

Based on Maxwell's work, a number of models were developed, and only the models which were used to compare the predictions and experimental data in the nanoparticle suspension study will be introduced in this chapter.

2.1. Models with Consideration of the Shape Influence of Dispersed Particle

In 1924, Fricke [7] developed the mean-field theory of Maxwell's by including the effect of particle shape to the effective thermal conductivity. For other particle geometries, such as ellipsoid or spheroid particles, the geometrical arrangement of the suspended particles was introduced. For suspensions of non-spherical particles, the effective thermal conductivity could be calculated using a form somewhat analogous to the equation for a suspension with spherical particles as Maxwell's equation,

$$(k_e / k_1 - 1)/(k_e / k_1 + x) = \phi(k_2 / k_1 - 1)/(k_2 / k_1 + x) \tag{2}$$

where k_e, k_1, and k_2 are the specific conductivities of the suspension, the suspending medium, and the suspended spheroids, ϕ is the volume concentration of the suspended spheroids. The influence of the geometric factors of the suspended particles was considered through a parameter, x, which is a function of the ratio k_2/k_1 and the ratio a/b of the axis of symmetry of the spheroids to the other axis. Fricke compared the result of this equation with experimental data for the conductivity of blood (k_2=0, a/b=1/4.25, x=1.05), which showed excellent agreement for concentrations from 10 % to 90 %. For a constant volume concentration, the conductivity of a suspension was shown to be independent of the size of the suspended particles and also nearly independent of the form of the particles when the difference between the conductivities of the dispersed particles and the continuous matrix was not very large, especially with suspensions of prolated spheroids that are less conducting than the suspending medium. This was a pure mathematical treatment, which gave the hints to calculate the effective thermal conductivity of suspensions in the later theoretical study.

Following this lead, a significant theoretical work was performed and resulted in the development of equations to calculate effective thermal conductivity. R. L. Hamilton and O. K. Crosser (1962) [3] proposed an equation to predict the effective thermal conductivity of heterogeneous systems consisting of a continuous matrix with dispersed particles within it. This was based on the work of Maxwell and Frick and introduced a new constant n, which was dependent on the shape of the particle, and independent of the ratio of the thermal conductivities of the dispersed particles and the continuous matrix. When the particle was a sphere, n would not demonstrate any dependence at all. Thus, in the development of this equation, the effect of the particle shape was referred to the spherical shape, and a shape factor was added into the equation, which was based on the experiments of spherical particles and non-spherical particle mixtures. A coefficient ψ was then introduced and defined as the ratio of the surface area of a sphere to the surface area of a non-spherical particle having the same volume as the spherical particle. For spherical particles, n=3/ψ; for prolate ellipsoids, n=3/ψ^2; and for oblate ellipsoids n=3/$\psi^{1.5}$. The equation for the effective thermal conductivity was

$$k_e = k_1\left[\frac{k_2 + (n-1)k_1 - (n-1)\phi(k_1 - k_2)}{k_2 + (n-1)k_1 + \phi(k_1 - k_2)}\right] \tag{3}$$

With this equation, it was found that if the continuous matrix had a higher conductivity than that of the dispersed particles, the particle shape would have little effect on the effective thermal conductivity of the mixture. However, if the dispersed particles had a higher conductivity, there was a strong shape effect. When using Eq. 3 with n in terms of sphericity and the conductivity of the dispersed particles at least 100 times larger than the conductivity of the continuous matrix, equation 3 agreed with the experimental results of silastic rubber and variously shaped particles of aluminum and balsa at selected volume compositions at 95 ^{o}F, in which the diameters of particles were in the magnitude of a millimeter. Finally, it was also pointed out that the agglomeration might change the shape, size, and orientation of particles, which would make the prediction of the equation deviate from the experimental data.

Another significant theoretical development on the shape effect was accomplished one year later after Hamilton and Crosser's model, J. B. Keller (1963) [8] considered the extreme situation in which the spheres and circular cylinders imbedded in a continuous matrix of conductivity k_1 have perfect conduction with a dense cubic array, and nonconducting cylinders in a medium with square array. For the perfectly conducting spherical particles, it was determined that the maximum value of the fractional volume is π/6. With this maximum value of fractional volume, the equation to calculate the effective conductivity was derived as follows:

$$k_e / k_1 = -(\pi/2)\log[(\pi/6)-\phi]+\cdots(\pi/6)-\phi << 1 \qquad (4)$$

When ϕ=0.5161, the result yielded k_e/k_1=7.65, which agreed with the measured result of k_e/k_1=7.6 in the experiment of Merdith and Tobias [5].

For the perfectly conducting circular cylinders, the maximum value of the fractional volume is π/4. With the same procedure, the equation for medium containing perfectly conducting circular cylinders was obtained as Eq. 5 for the square array of cylinders, which agreed with the numerical results obtained by J. B. Keller [8].:

$$k_e / k_1 = -(\pi^{\frac{3}{2}}/2)\log[(\pi/4)-\phi]^{\frac{1}{2}}+\cdots(\pi/4)-\phi << 1 \qquad (5)$$

This investigation also considered a situation in which the cylinders contained in the continuous matrix were non-conducting, and there was a resistance of each gap between two adjacent cylinders, which was $R=1/k_e c$. Here, c was the distance between the axes of the two adjacent cylinders. It was found that the equation of effective thermal conductivity for nonconducting cylinders and the equation for perfectly conducting cylinders were the reciprocal of each other, and had the following relationship:

$$\frac{k_e^{\infty}}{k_1}=\frac{k_1}{k_e^{0}} \qquad (6)$$

where k_e^{∞}, k_e^{0} were the effective thermal conductivities of perfectly conducting cylinders and of nonconducting cylinders. Here, k_1 was the conductivity of a continuous medium.

A. Rocha and A. Acrivos (1973) [9, 10] made the assumption that no significant interactions existed in the dilute suspension as previous researchers had, and got a general expression from Fourier's law for the heat conduction in statistical homogeneous suspension as Eq. 7. Their principal contribution is that the effective thermal conductivity of suspensions was studied not only with spherical particles but with arbitrary shaped particles as well.

$$q + k_1 \frac{\partial T}{\partial x_i} = (k_1 - k_2) c \int_{S_p} T n_i dS \tag{7}$$

where q was the heat flux, k_1 was the thermal conductivity of continuous matrix, k_2 was the thermal conductivity of dispersed particle, c was the number of dispersed particles per unit volume, S_p was the surface of a particle and n_i was the normal on the surface. The effective thermal conductivity of the suspension with spherical particles was then derived as

$$\frac{k_e}{k_1} = \{1 + 3[(\alpha - 1)/(\alpha + 2)]\phi\}\delta \tag{8}$$

where α was the ratio of thermal conductivity of dispersed particles to that of the continuous matrix, ϕ was the volume fraction of particles, and δ was the Kronecker delta.

When the particles were slender or ellipsoids, the effective thermal conductivity was not isotropic, but had different values along different particle axes direction in the suspension. Finally, the effective thermal conductivity equations for suspensions with arbitrary shaped particles, axially symmetric shaped particles, and slender particles were obtained under the steady state and non-interaction condition. The general equation was

$$\frac{k_e}{k_1} = \delta_{ij} + 2[(\alpha - 1)/(\alpha + 1)]\phi \times \left\{ \delta_{ij} + v_2^{-1} \sum_{m=1}^{\infty} (2\pi)^m [(\alpha - 1)/(\alpha + 1)]^m \int_{\Omega} \right\} \tag{9}$$

It was pointed out that the effective thermal conductivity of the suspension would still be isotropic if the function was truncated to the first order of (α-1)(α+1), and reduce to only a function of volume fraction of dispersed particles.

I. C. Kim and S. Torquato (1993) [11] applied computer simulation to calculate the effective thermal conductivity of aligned spheroidal particle suspensions with first passenger time algorithm. The dispersed particle had a wide range of aspect ratio from 0.1 to 10 and the ratio of the thermal conductivities of dispersed particle and the continuous matrix was from zero to infinite. This method was mainly concerned with the Brownian motion of dispersed particles and was a totally different approach to calculating the effective thermal conductivity of suspensions from the previously dominant thought, which held that the derivative of the Maxwell's equation should be conducted to the high order of volume ratio to include the far and near range interactions between two or more particles. This approach was from the viewpoint of motion instead of the statistic force field to consider the mechanism of the enhancement of effective thermal conductivity of suspension.

2.2. Models with Consideration of Interaction between Particles

Due to the nature of the atoms and molecules, when the dispersed particles are close to each other, their interaction would influence the effective thermal conductivity of the suspension, too.

G. K. Batchelor [12] reviewed the study made on the transport properties of heterogeneous mixtures that were composites of two phases that separately were homogeneous, and pointed out that the use of probability theory should be very important and essential for the study on the bulk properties of these two-phase dispersed systems. Batchelor started from the Fourier theory that $F=K\bullet G$, where F was the flux, K was the bulk transport coefficient characteristic of the medium, and G was the gradient of intensity. Then a table of the physical phenomena that represented the problem of determining an effective transport coefficient was given, and the theoretical analysis was conducted based on the common principles of these phenomena.

Table 1. Effective transport coefficient for two-phase disperse systems [12]

Cases of two-phase disperse systems with random structure in which the problem is to determine an effective transport coefficient

Nature of the two uniform media of which the two-phase disperse system is composed	Quantity represented by F	Quantity represented by G	Transport coefficient (or scalar components in case of isotropic structure)	Local differential equation satisfied in each phase (in steady state)
(1) Thermal conductor	Heat flux	Temperature gradient	Thermal conductivity	$\mathbf{F}=\mathbf{K}\cdot\mathbf{G}$ $\nabla\cdot\mathbf{F}=0$
(2) Electrical conductor	Electric current	Electric field	Electrical conductivity	
(3) Electrical insulator	Electric displacement	Electric field	Dielectric constant	
(4) Dia- or para-magnetic material	Magnetic induction	Magnetic field	Magnetic permeability	
(5) Small fluid or rigid particles sedimenting through Newtonian incompressible fluid (a "fluidized bed")	Flux of particle number relative to zero-volume-flux axes	Gravitational force on particles in unit volume of mixture (minus the buoyancy force)	Mobility	$\nabla p=\mu\nabla^2\mathbf{u}$ $\nabla\cdot\mathbf{u}=0$ where $\mathbf{u}$ = velocity p = pressure μ = viscosity
(6) "Porous medium" consisting of a fixed array of small rigid particles through which incompressible Newtonian fluid is passing	Force on particles in unit volume of mixture (= pressure gradient calculated from pressure drop between distant parallel planes)	Flux of fluid volume relative to particles	Permeability (Darcy constant) divided by μ	
(7) Small fluid or rigid particles suspended in incompressible Newtonian fluid	Deviatoric stress	Rate of strain	Shear viscosity	As immediately above; or $\mathbf{F}=2\mu\mathbf{G}$, $\nabla\cdot\mathbf{F}=0$ $G=\text{trace}\,\mathbf{G}=0$
(8) Elastic inclusions embedded in an elastic matrix	Stress	Strain	Lamé constants (or rigidity and bulk moduli)	$\mathbf{F}=2\mu\mathbf{G}+\lambda G\mathbf{I}$ $\nabla\cdot\mathbf{F}=0$ where μ,λ = local Lamé constants

This table discussed the applicability of the equations and theories to all the volume fractions and gave the upper and lower limits of effective thermal conductivity for arbitrary volume fractions found by Z. Hashin and S. Shtrikman [12] for the effective magnetic permeability of multiphase materials. Then, it discussed the calculations on the mean flux F

in two-phase statistic homogeneous mixtures, in which the local heat flux always had a linear relation to the local intensity gradient. Here, a new parameter was introduced: the dipole strength *S*, which was a measure of the result when the matrix material was replaced by the particle material, and a statistical property of microstructure of the mixture representing the interaction between particles.

Three kinds of interacting conditions of particles that dilute suspensions were discussed: (1) non-interaction between particles, (2) first order of interaction between particles, and (3) strong interaction in small volume fraction suspensions. The corresponding expression for the effective thermal conductivity of the first condition was given as

$$k_e = k_1\left\{1+\left[2(\alpha-1)/(\alpha+2)\right]\phi+\left[3(\alpha-1)^2/(\alpha+2)^2\right]\phi^2\right\} \tag{10}$$

where k_e was the effective thermal conductivity, k_1 was the thermal conductivity of continuous matrix, α was the ratio of the thermal conductivity of particle material to that of the matrix material, and ϕ was the volume fraction of particles.

For the last two conditions, a broader analysis considered the situation in which there was a bulk movement in the suspensions. This was a more complicated situation which included the interactions between particles plus the influence of other external forces.

D. J. Jefferey (1973) [13] suggested that the probability $P(r|o)$ for one sphere only in the distance $O(r)$ of the reference sphere was $O(\phi)$ and the probability $P(r|o)$ for two spheres in the distance $O(r)$ was $O(\phi^2)$. The choice for the probability $P(r|o)$ satisfied

$$\begin{aligned} P(r|o) &= 0, r \le 2a \\ P(r|o) &= n, r >> a \end{aligned} \tag{11}$$

where a is the radius of the particle and r is the distance of two particles.

This could be interpreted as following. The first condition was that the spheres did not overlap and the second condition was that all members of spheres were far away from the reference sphere. When solving the problems involving two spheres, the twin spherical expansions were applied. After obtaining the expressions for the heat flux and temperature within and out of the spheres, the author took the interactions between pairs of spheres into consideration through a second order of the volume ratio between particles and fluid. This equation depends on the way pairs of spheres are distributed with respect to each other. As to the convergence, this work was done on the basis of Batchelor's method in 1972 [12] for the average interactions between spheres to obtain bulk properties of suspensions. The equation developed is shown as Eq. 12.

$$k_e / k_1 = 1 + 3\beta\phi + 3\beta\phi^2\left(\beta + \sum_{p=6}^{\infty}\frac{B_p - 2A_p}{(p-3)2^{p-3}}\right) \tag{12}$$

where β=(α-1)(α-2) and α=k_2/k_1, volume fraction ϕ =4/3πa^3n and a is the radius of the particle, P is the probability density for a second sphere being at location r to the center of the reference sphere and A_p and B_p are known coefficients.

R. H. Davis (1986) [14], fully aware of the difficulty of the integral convergence encountered by early researchers, considered a situation that inside the composite material, the undisturbed temperature gradient varied inversely with the square of the distance from the heated body. Thus, the correction to the dipole strength of the reference sphere varied as $(a/r)^5$ instead of $(a/r)^3$ due to the presence of a second sphere, and only convergent integrals were encountered. Finally, Davis proposed an equation (Eq. 13) based on Green's theorem to calculate the rate of heat transfer from the heated body to the composite material, which was similar to Jeffrey's model [13], shown as below.

$$k_e / k_1 = 1 + \frac{3(\alpha - 1)}{\left[\alpha + 2 - (\alpha - 1)\phi\right]}\left\{\phi + f(\alpha)\phi^2 + 0(\phi^3)\right\} \tag{13}$$

here $f(\alpha) = \sum_{p=6}^{\infty} \frac{B_p - 3A_p}{(p-3)2^{p-3}}$ was used to represent the decaying temperature field in the process of developing this equation, while early researchers had considered it a composite material of infinite extent on which an undisturbed linear temperature field was imposed. The $O(\phi^2)$ in the equation contributing to the effective thermal conductivity was not given entirely by the solution for two spheres placed in the undisturbed temperature field, but the actual environment of these two spheres was the undisturbed temperature plus an $O(\phi)$ term from all of the remaining spheres. It was also pointed out that contributions of $O(\phi^3)$ and higher orders were even more difficult to obtain and not generally needed for the dilute suspension cases. Here, the maximum volume ratio is 0.62 in order to prevent the spheres from touching one another.

J. F. Brady et al. (1988) [15] presented a method to calculate hydrodynamic interactions among particles in infinite suspensions with infinite small particle Reynolds number. The numerical calculation applied to the spatially periodic suspension problem was the Stokesian-dynamics method.

Based on this work, R. T. Bonnecaze and J. F. Brady [16, 17], in order to avoid the convergent problem in integration over the infinite extent of matrix, developed a method for determining the individual dipoles for a finite number of particles, which included the particles interactions in both long range and short range fields. The individual particle potentials and the potential gradient were related to the sum of particle moments to generate a "potential" matrix. Then, by inverting this potential matrix, a capacitance matrix was formed. The potential invert produced the many-body reflections of the particle moments that were included in its formulation, and the potential invert included the far field interactions of the particles. Plus, the exact two-body interactions added into the potential invert, the accurate capacitance matrix was obtained. With this capacitance matrix, the dipole strength of particles could be calculated directly under the given external potential field. The equation developed was

$$k_e = k_1 - n(S / G^E) \tag{14}$$

where S was the dipole strength and G^E was the potential gradient applied.

This method was applied to the infinite suspensions by using the O'Brien's method [16] and periodic boundary conditions which reduced the system to a finite number of linear equations. With this method, the effective thermal conductivities of the suspensions with cubic arrays of spheres which had infinite thermal conductivity, 10 times the thermal conductivity of the matrix's, and 0.01 times the thermal conductivity of the matrix's, were calculated.

S. Y. Lu and S. T. Kim 1990 [18] studied the relationship between the microstructure of the dispersed particles and the effective thermal conductivity of the anisotropic composite, consisting of homogeneous matrix with spheroidal particles dispersed within it. The systems with fixed aligned dispersed particles were the only ones considered, and only the variation of the particles' positions influenced the effective thermal conductivity in this study. The dipole strength and renormalization approach in the integration were used to calculate the dipole strength as early researchers had done. The model that included the assumption that dispersed particles would take non-overlapping positions with equal probability was calculated to show the flaw of the absence of potential energy existing between particles. Then, the hard-sphere fluid model was adopted, in which the pair distribution function was a function only of the distance between two axes. This model showed a better agreement with the former work and the experimental data.

$$k_e = \widehat{I} + \left(-\frac{M}{V_2 r_1} \right) \phi + \left(\frac{-\Xi + \Omega}{V_2 r} \right) \phi^2 \tag{15}$$

where $\widehat{I}$ was the unit tensor, V_2 was the volume of dispersed particles, M was the second-order tensor for one-body contribution, Ξ was the second-order tensor for two-body contribution, and Ω was the renormalization quantity.

S. Y. Lu and H. C. Lin (1996) [19] developed an expression for the effective conductivity tensor to calculate the effective conductivity in the form of the viral expansion of the dispersed particle volume fraction ϕ which was truncated at the $O(\phi^2)$ term. The particle shape discussed was ellipsoid and its mathematical expression was

$$\frac{x^2}{a^2} + \frac{y^2}{b^2} + \frac{z^2}{c^2} = 1, a \geq b \geq c \tag{16}$$

In the study, pair interaction, the second-order tensor for one-body contribution M, the second-order tensor for two-body contribution Ξ, and the renormalization quantity Ω for the two aligned spheroids, were evaluated by solving a boundary-value problem involving two aligned spheroids with a boundary collocation scheme. This boundary collocation scheme was derived based on the pre-limit that the governing equation was the Laplace equation and the system was a two-particle system. The temperature fields outside the particles and inside two particles were expressed, first, with the general solution for the Laplace equation, then forced to satisfy the boundary conditions at the particle surfaces, and finally the relevant thermal dipole strength. The series ratio inferred from the viral expansion of the equivalent inclusion estimate served as a reasonably good validity test parameter for the second-order virial approximation. The magnitude of this series ratio was also a reflection of intensity of thermal interactions within the system. Two models for the pair distribution function were studied; one was the well-stirred mode and the other one was the hard spheroid model. The equivalent inclusion estimate method of Hatta and Taya was applied here and derived as

$$\frac{k_e}{k_1} = 1 + \frac{\alpha - 1}{(\alpha - 1)h_i + 1}\phi + \frac{(\alpha - 1)^2 h_i}{[(\alpha - 1)h_i + 1]^2}\phi^2 + \frac{(\alpha - 1)^3 h_i}{[(\alpha - 1)h_i + 1]^3}\phi^3 + \ldots \tag{17}$$

where $\alpha = k_2/k_1$, and the equation would converge with the condition $\frac{\alpha - 1}{(\alpha - 1)h_i + 1} < 1$.

2.3. Models with the Consideration of Interfacial Resistance

Based on the experimental results of spherical nickel dispersions in a sodium borosilicate glass matrix and carbon fiber-reinforced glass ceramic, D. P. H. Hasselman and L. F. Johnson (1987) [20] pointed out that the thermal barrier resistance at the interface between the dispersed particle and the continuous matrix would have an important influence on the effective thermal conductivity of the composite, which the previous study did not take into consideration in the development of the equation or in the calculation of the effective thermal conductivity. The idea that potential was continuous and the current was the same at the two sides of interface between dispersed particle and continuous matrix, was still adopted in this study, just as Maxwell had proposed.

$$k_1\left(\frac{\partial T_1}{\partial r}\right) = k_2\left(\frac{\partial T_2}{\partial r}\right) \tag{18}$$

$$T_1 - T_2 = -\frac{k_2}{h_c}\frac{\partial T_2}{\partial r} \tag{19}$$

where T_1 was the temperature in the continuous matrix out of the dispersed particle, T_2 was the temperature inside of the dispersed particle, k_1 was the thermal conductivity of the continuous matrix, k_2 was the thermal conductivity of the dispersed particle, r was the radius of the interfacial surface, and h_c was the interfacial thermal barrier resistance. If the interfacial thermal barrier resistance was zero, then $T_1 = T_2$ at the interface, which agreed with the solutions of the Maxwell equations [6].

Physically the interfacial thermal barrier resistance h_c was not zero, so the effective thermal conductivity of the composite was affected by the cumulative effect on T_1, and the radius of the large sphere, b, which included n dispersed spheres of radius a. When the large sphere was large enough, $b \to \infty$, the equation for the effective thermal conductivity was

$$k_e = k_1 \frac{\left[2\left(\frac{k_2}{k_1} - \frac{k_2}{ah_c} - 1\right)\phi + \frac{k_2}{k_1} + \frac{2k_2}{ah_c} + 2\right]}{\left[\left(1 - \frac{k_2}{k_1} + \frac{k_2}{ah_c}\right)\phi + \frac{k_2}{k_1} + \frac{2k_2}{ah_c} + 2\right]} \qquad (20)$$

The equations for two kinds of particles, circular cylinders and flat plate, were conducted as:

$$k_e = k_1 \frac{\left[\left(\frac{k_2}{k_1} - \frac{k_2}{ah_c} - 1\right)\phi + \frac{k_2}{k_1} + \frac{2k_2}{ah_c} + 1\right]}{\left[\left(1 - \frac{k_2}{k_1} + \frac{k_2}{ah_c}\right)\phi + \frac{k_2}{k_1} + \frac{2k_2}{ah_c} + 1\right]} \qquad (21)$$

$$k_e = \frac{k_2}{\left[\left(1 - \frac{k_2}{k_1} + \frac{2k_2}{ah_c}\right)\phi + \frac{k_2}{k_1}\right]} \qquad (22)$$

respectively, and they all agreed with the previous reported solutions when the interfacial thermal barrier resistance was zero. The equation (20) was the most widely used equation, beside the Hamilton-Crosser model [3], in the later nanoparticle suspension effective thermal conductivity study.

The original contribution of work is that it did not only consider the influence of the interfacial resistance to the effective thermal conductivity, but included the particle size in the model as well.

2.4. Models of Liquid-Liquid Mixture

In 1961, Meredith and Tobias [5] found that the Maxwell's equation [6], the Fricke equation [7] and the Bruggeman equation [5] all failed to predict the experimental results of effective thermal conductivity of the liquid-liquid emulsions when the volume fraction is greater than 20 %, by comparing their experimental data with the predictions of the models. Hence they developed a new expression to predict the effective conductivity of the water-oil emulsion suspensions which have both a wide range of size distribution of dispersed spherical oil particles in a water matrix and a very narrow size distribution of dispersed spherical water particles in an oil matrix, which is shown as Eq. 23 below,

$$\frac{k_e}{k_1} = \left(\frac{2+2X\phi}{2-X\phi}\right)\left(\frac{2+(2X-1)\phi}{2-(X+1)\phi}\right) \tag{23}$$

where $X = \frac{k_2/k_1 - 1}{k_2/k_1 + 2}$. Here, an assumption was made that spherical shape liquid particles were dispersed in the other liquid matrix.

L. G. Leal [21] analyzed the effective conductivity of a dilute suspension in a simple shear flow and studied the role of the mechanical and thermal properties of the particles in contributing to the bulk thermal transport characteristics of the suspension. The bulk heat flux of the suspension was calculated by the averaging of the microscale heat flux. The local temperature field was considered under the assumption that the suspension was experiencing a simple shear flow with a linear bulk temperature field across the shear gradient in which the Reynolds number and the Peclet number were very small. In this investigation, it was found that the effective conductivity was largely determined by the thermal conductivity of the suspended particles and the shear flow disturbance contribution. Further, it was pointed out that the particle thermal conductivity might have either a positive or a negative contribution, according to the conductivity ratio between the particle and base fluid, while the flow always contributed positively. The study gave people another perspective of the enhancement of the effective thermal conductivity of suspensions, stating that not only the physical and thermal properties of the suspended particles would change the effective thermal conductivity of suspension, but their motion could also result in a considerable change to the effective thermal conductivity as well. However, because this work lacked information on the effect of the particle size, shape, volume fraction, and other factors, the resulting expression was not capable of providing good predictions on the experimental data.

$$k_e = k_1[1+\phi\{\frac{3(k_2-k_1)}{k_2+2k_1} + (\frac{1.176(k_2-k_1)^2}{(k_2+2k_1)^2} + \frac{2\mu_1+5\mu_2}{\mu_1+\mu_2}(0.12\frac{2\mu_1+5\mu_2}{\mu_1+\mu_2} - 0.028\frac{k_2-k_1}{k_2+2k_1}))Pe_1^{3/2} + 0(Pe_1^2)\}] \tag{24}$$

Here μ is the viscosity, and subscript 1 represents the basefluid; 2 represents the particle if the particle was formed by another fluid.

All of the above discussed models are well constructed and have been successfully used to predict the effective thermal conductivity of many kinds of mixtures. However, the application to effective thermal conductivity of nanoparticle suspensions was not quite well. One problem is that the predictions of those models are significantly lower than the experimental results of nanoparticle suspension effective thermal conductivities at room temperature; Second, those models could not predict the elevation of nanoparticle suspension effective thermal conductivity with an increase of temperature; Third, only few of those models considered the nanoparticle size and could not predict that the effective thermal conductivity would increase with a decrease of nanoparticle size. Hence, a new theory, which could well describe the new phenomena of the effective thermal conductivity variation of nanoparticle suspensions, is critical needed.

3. Development of Nanoparticle Suspension Study and Two Major Possible Mechanisms: Brownian Motion and Agglomeration

In the following several years after the first proposal that indicated that adding nanoparticle into the traditional heat transfer fluid media could obtain much higher effective thermal conductivity than that of the base fluids, Choi's prediction was confirmed by several experimental reports that nanoparticle suspensions could obtain unusually high effective thermal conductivity compared to that of base fluids [22-24]. In those first experimental reports, both a transient hot-wire method and a parallel plates method were employed to measure the effective thermal conductivity of suspensions consisting of Cu, CuO, and Al_2O_3 nanoparticles in Distilled (DI) water, Ethylene glycol and engineer oil [22-24].

3.1. Experimental Reports on Effective Thermal Conductivity of Nanoparticle Suspensions

Eastman et al. [22] reported that a 36 nm diameter CuO nanoparticle/DI water suspension could obtain 60 % enhancement on thermal conductivity at a volume fraction of 5 % over that of DI water. And a 33 nm diameter Al_2O_3 nanoparticle/DI water suspension could achieve around 30 % enhancement on thermal conductivity at 5 % volume fraction over that of DI water. Interestingly, the 18 nm diameter Cu nanoparticle/HE-200 oil suspension could have as high as 43 % enhancement over the thermal conductivity of HE-200 oil at a volume fraction as small as around 0.05 %, which is two orders less in volume fraction than that of CuO and Al_2O_3 nanoparticle/DI water suspensions at the same magnitude of enhancement in thermal conductivity [4]. However, in another later experimental report by the same group of researchers [23], the enhancements on thermal conductivity were much lower than previously reported. The 23.6 nm diameter CuO nanoparticle/DI water suspension had an enhancement above 20% at the volume fraction of 4%, and the 38.4 nm diameter Al_2O_3 nanoparticle/DI water suspension had an enhancement around 12 % at the volume fraction of 4 %, and an enhancement around 20 % at the volume fraction of 5 %. The Hamilton and Crosser model was used to compare the predictions and the experimental results. It was illustrated that this model could well predict the effective thermal conductivity of Al_2O_3 nanoparticle suspensions, while predictions of the CuO nanoparticle suspension model were much smaller.

The new experiment of the Cu nanoparticle suspensions showed that, with the Cu nanoparticles of diameter less than 10 nm, the Cu/ethylene glycol suspension could have an enhancement above 40 % at the volume fraction of 0.3 %, which is much less than the 43 % enhancement achieved at around 0.05 % volume fraction of 18 nm Cu/HE-200 oil suspension [4].

In 1999, Wang and Xu et al., [24] employed a steady-state parallel-pate method to measure the effective thermal conductivities of suspensions with 20 nm diameter Al_2O_3 and CuO nanoparticles in DI water, ethylene glycol, engine oil, and vacuum pump fluid. The experimental results were compared with a series of mean-field models, such as Maxwell's model [6], Jeffrey's model [13], Davis' model [14], Lu and Lin's model [19], Bonnecase and Brady's model [16, 17], and Hamilton-Crosser's model [3]. The comparisons between the experimental results and predictions on Al_2O_3 nanoparticle/ethylene glycol suspensions pointed out that those available models all failed to give predictions that were large enough to match the experimental results of thermal conductivity enhancements. Specifically, it was pointed out that the microscopic motion and the particle structure could be the reasons which induced the unusually high effective thermal conductivity of nanoparticle suspensions. This report is the first theoretical analysis of the two extensively debated mechanisms of enhanced thermal conductivity, Brownian motion effect and agglomeration effect, on which researchers have been focusing in the later investigations.

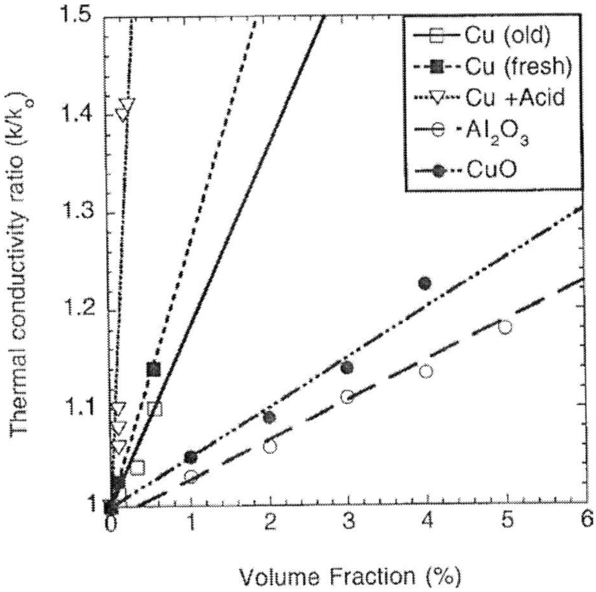

Figure 1. Effective thermal conductivities of metallic and oxide nanoparticle suspensions [22].

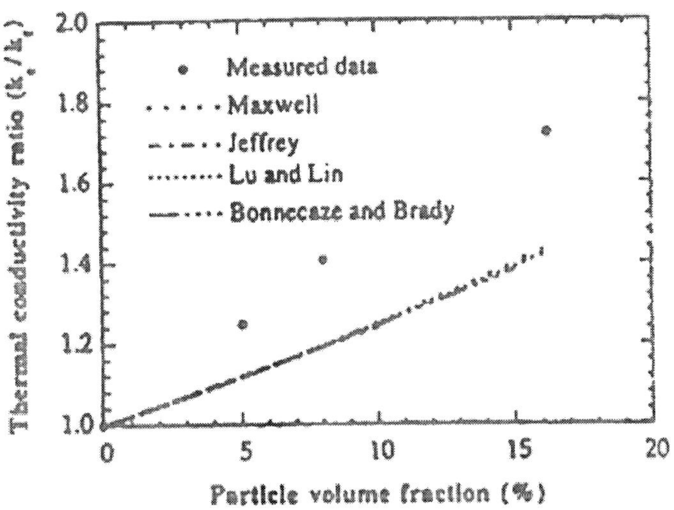

Figure 2. Measured thermal conductivities of Al_2O_3/ethylene glycol mixtures vs. effective thermal conductivities calculated from theories [24].

Since 2001, there has been an explosive growth in interest in this new heat transfer liquid media, both experimentally and theoretically [25-83]. The experimental data dramatically enriched the database of all kinds of effective thermal conductivity tests of nanofluids. Therefore, it is possible to understand the behavior of this new heat transfer fluid media and to find out the relationships between the effective thermal conductivity and many factors, such as the nanoparticle shape, nanoparticle material, nanoparticle volume fraction, nanoparticle size, base fluid material, and bulk temperature. In 2002, Xie et al., [25, 26] measured the effective thermal conductivity of suspensions with a range of diameter (5 nm, 25 nm, 58 nm, 101 nm, 122 nm and 124 nm) and two crystalline phase Al_2O_3 nanoparticles (α crystalline and γ crystalline) in pump oil, DI water at different pH value, and ethylene glycol, as the base fluids. There were several important discoveries in this report. First, it illustrated that the thermal conductivity enhancement of Al_2O_3 nanoparticle suspensions decreased with an increase of pH value in the DI water base fluid. Second, it showed that the crystalline phase of Al_2O_3 did not have an influence on the effective thermal conductivity of suspensions. Third, the nanoparticle diameter had an influence on the effective thermal conductivity and the best enhancement of effective thermal conductivity happened to be that of 25 nm diameter Al_2O_3 nanoparticle suspension. And last, the enhancement of suspension effective thermal conductivity increased with a decrease of base fluid thermal conductivity.

The same year, another two experimental reports brought one more parameter, bulk temperature, which could influence the effective thermal conductivity of nanoparticle suspensions [27, 28]. The first experimental report introduced a temperature oscillation technique to obtain the effective thermal conductivity through measuring the thermal diffusivity of the nanoparticle suspension [27].

It was demonstrated that the effective thermal conductivities of both 38.4 nm diameter Al_2O_3/DI water and 28.6 nm diameter CuO/DI water nanoparticle suspensions would increase with an increase of bulk temperature [27]. Additionally, it pointed out that the Hamilton-Crosser model could not predict the elevated effective thermal conductivity of nanoparticle suspensions with the increase of temperature, although the effective thermal conductivity of Al_2O_3/DI water nanoparticle suspension generally agreed with the prediction of the Hamilton-Crosser model at room temperature. The second report confirmed this temperature dependence relation by investigating nanoparticle suspensions with 60-80 nm diameter Ag nanoparticle, 10-20 nm diameter Au nanoparticle and 3-4 nm diameter Au nanoparticle covered with a monolayer of Octadecanethiol in water and toluene base fluids. The 3-4 nm diameter Au nanoparticle covered with a monolayer of Octadecanethiol in toluene base fluid at 0.0079 % volume fraction showed a thermal conductivity enhancement of 4.5 % at 30 ^{0} C and 6.3 % at 60 ^{0} C. The 10-20 nm diameter Au nanoparticle/water suspensions at 0.00026 % volume fraction had a thermal conductivity enhancement of 4.6 % at 30 ^{0} C and 8.3 % at 60 ^{0} C. The 60-80 nm diameter Ag nanoparticle/water at 0.001 % volume fraction had a thermal conductivity enhancement of 3.2 % at 30 ^{0} C and 4.5 % at 60 ^{0} C [28].

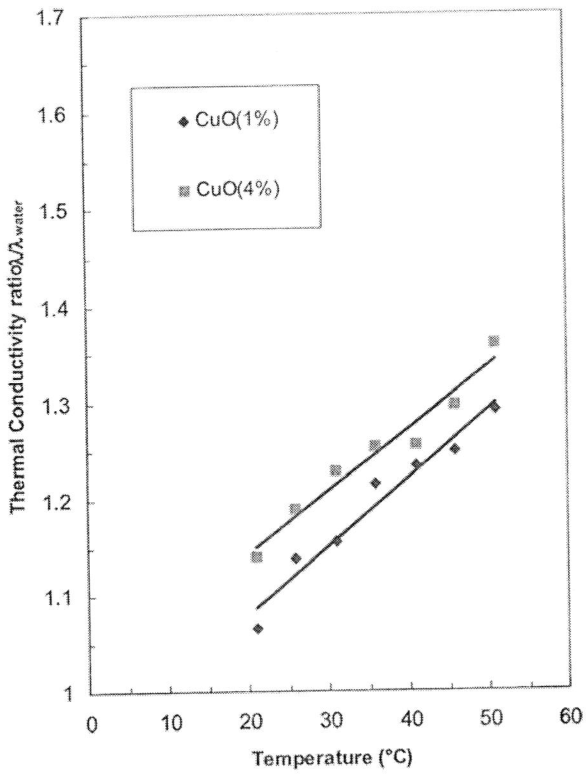

Figure 3. Temperature dependence of thermal conductivity enhancement for Al_2O_3/water nanoparticle suspensions [27].

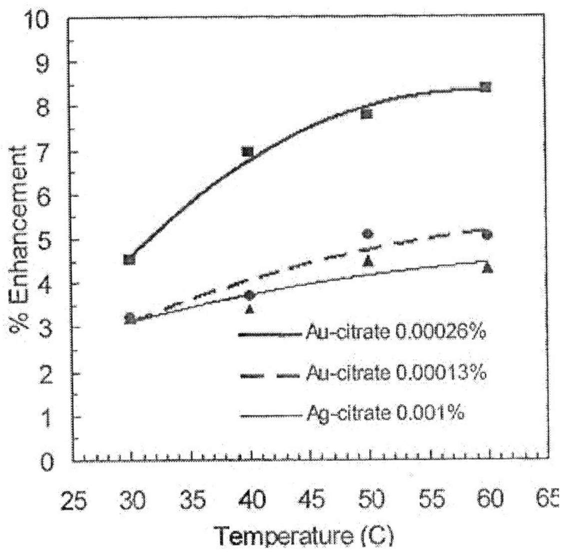

Figure 4. Percentage of enhancement in thermal conductivity vs. the temperature of Au-citrate and Ag-citrate with reference to the conductivity of blank water (with 5 nM trisodium citrate) at each temperature [28].

In 2006, a visual experiment of enhanced mass transport in nanoparticle suspensions was reported, which indirectly proved the Brownian motion effect toward the effective thermal conductivity of the nanoparticle suspension [29]. It was observed that the same dye in water and in nanoparticle suspensions showed different transport abilities. The distances dye transported in nanoparticle suspensions were much greater than the distance dye transported in pure water. And there is an optimal volume fraction, 0.5 %, in which the dye transported best. As a result, it was explained that nanoparticle Brownian motion could create a velocity disturbance field and this disturbance field enhanced the transport ability of dye inside nanoparticle suspensions.

A new experimental study by Putnam et al. [30] introduced an optical beam deflection technique to measure the thermal diffusivity of C_{60}-C_{70} fulleren nanoparticle/tolune suspensions, and alkanethiolate-protected Au nanoparticle/ethanol and toluene suspensions. It was stated that in the dilute volume fraction (0.01 % to 1 %), the enhancement was not as great as it was previously reported by [28]. However, there was another interesting discovery in this report that the mixture of 2 % volume fraction water in ethanol could have an effective thermal conductivity enhancement of 6 %, two folds higher than the prediction of 3 % by mean-field model.

When researchers focused on the Brownian motion effect of the nanoparticle, some experimental reports brought the attention to other factors, such as the agglomeration effect [31-39].

A cryogenic transmission electron microscope technique was employed to determine the morphology of nanoparticles and agglomerations in the suspensions [33]. In this method, the suspension specimen, which could not be studied under a conventional transmission electron microscope, was frozen with liquid nitrogen below the devitrification temperature to reserve the original structure of the nanoparticles/agglomerations inside the suspension. Then, it was fractured into two parts in the middle, which gave a cross-section surface. On this surface, it had the original size distribution and morphology information of nanoparticles/ agglomerations in the suspension. Then, the carbon and platinum ions beams were sprayed on the surface at an angle of 45 ^{0} and formed a thin layer membrane of 1.5 ~ 2.0 nm thickness. This thin layer membrane could duplicate the information of size distribution and morphology of nanoparticles/agglomerations on the surface, and be studied under a transmission electron microscope. The picture of 25 nm diameter nanoparticle in water is shown below.

Hong and Yang et al. [34, 35] experimentally investigated the dependence of effective thermal conductivity of 10 nm diameter Fe nanoparticle suspensions on the Fe nanoparticle cluster size. It was reported that the effective thermal conductivity of nanoparticle suspensions was not necessarily dependent on the thermal conductivity of nanoparticles by comparing this experimental results with that of Cu nanoparticle suspensions reported previously [22]. It was also shown in another paper [35] that the effective thermal conductivity of Fe nanoparticle suspensions would decrease with an increase of the time after ultrasonication was stopped, and the size of the Fe nanoparticle cluster increased with an increase in the time after ultrasonication was stopped. These experimental results clearly demonstrated that, first, the mean-field theory could not be applied to predict the effective thermal conductivity of nanoparticle suspensions because the prediction of effective thermal conductivity by mean-field theory is always dependent on the thermal conductivity of nanoparticles; and second, the effective thermal conductivity of Fe nanoparticle suspensions decreased with an increase in the Fe nanoparticle cluster size, which confirmed the previous conclusion that the smaller the nanoparticle size is, the better the effective thermal conductivity will be. But here, it also attributed the better enhancement of Fe nanoparticle suspensions over that of Cu nanoparticle suspensions as a result of clustering of Fe nanoparticles. While another experimental report by Liu and Wang et al. [36] showed that the effective thermal conductivity of Cu/DI water nanoparticle suspensions decreased with an increase of time after the measurement started. Moreover, it was found that Cu nanoparticle would agglomerate into 50-100 nm diameter spherical and square shaped clusters. The ratio between the effective thermal conductivity of the nanoparticle suspension and the thermal conductivity of base fluid, DI water, would decrease quickly in the first ten minutes and then kept a relatively stable value which was slightly higher than unity due to the severe agglomeration of the Cu nanoparticles.

Peterson, Li and Kim [40-43] noticed the possible contribution of nanoparticle clustering and tried to experimentally verify this by measuring the effective thermal conductivities of the CuO nanoparticle, the Al_2O_3 nanoparticle, and the single-walled carbon nanotube and multiwalled carbon nanotube suspensions. The relationships between the effective thermal conductivity of metal oxide nanoparticles and carbon nanotube suspensions and volume fraction, temperature, and base fluid were confirmed. It was pointed out that for the carbon nanotube suspensions, the effective thermal conductivity might be well predicted with the extended mean-field theory by choosing the correct shape factors [40]. The work [40] corresponded to the other experimental work by Choi et al. [31], Xie and Choi et al. [32], and Liu et al. [36], which all reported that both single-walled and multi-walled carbon nanotube suspensions could enhance the effective thermal conductivity much greater than the predictions of mean-field models.

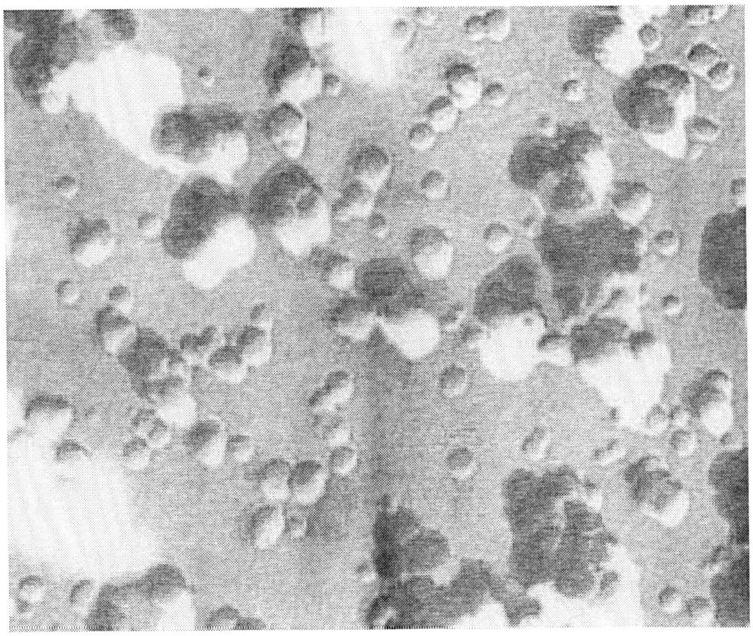

Figure 5. Cry-TEM pictures of SiO_2 nanoparticles/agglomerations in suspension [33].

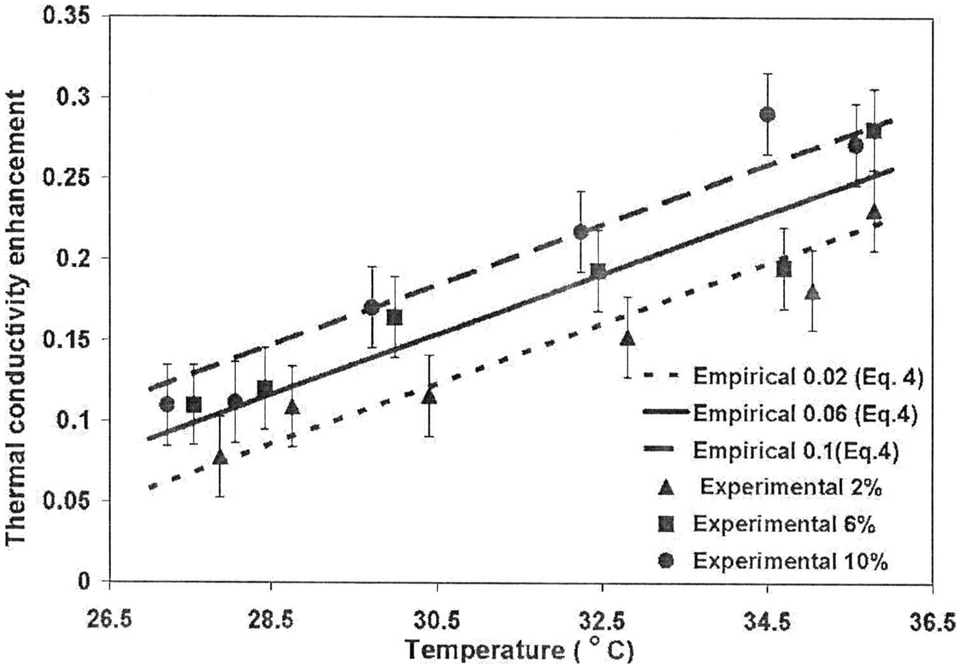

Figure 6. Thermal conductivity enhancement of Al_2O_3/DI water suspensions vs. temperature [40].

The Al_2O_3 nanoparticle suspensions were further studied by measuring the effective thermal conductivities of different diameter Al_2O_3 nanoparticle suspensions, and it once more confirmed that nanoparticle size would influence the enhancement of the effective thermal conductivity [41, 42]. Li and Peterson et al. [43] also conducted a comparing experiment of steady state cut bar method and transient hot wire method to respond to the questioning on the validity of employing the transient hot wire method to measure the effective thermal conductivity of nanoparticle suspensions. The experiment well illustrated that the transient hot wire method was fully capable of measuring the enhancement of the effective thermal conductivity with very good accuracy.

Zhu and Yin et al. [44] measured the effective thermal conductivity of 9.8 nm diameter Fe_3O_4 nanoparticle/DI water suspensions, and found that the effective thermal conductivity of Fe_3O_4 nanoparticle/DI water suspensions was higher than that of other metal oxide nanoparticle suspensions, such as Al_2O_3, CuO, and TiO_2, and significantly higher than the predictions of Maxwell's model [6], the renovated Maxwell's model [45], and Das' Brownian motion model [46]. It is interesting to note in this report that the cluster size of the Fe_3O_4 nanoparticle agglomerations increased with an increase of volume fraction. For example, at a volume fraction of 0.5 %, only few loose clusters were found with a size of 40 nm, at volume fraction of 1 %, the cluster size increased to around 60 nm, and at volume fraction of 3 %, the cluster size dramatically increased to several hundreds nanometers. The alignment of Fe_3O_4 nanoparticles was observed to be random in all the samples.

Chopkar and Kumar et al. [47] introduced another new experimental method of a modified thermal comparator to measure the effective thermal conductivities of 15 nm diameter Al_2Cu and 18 nm diameter Ag_2Al nanoparticles in DI water and ethylene glycol base fluids. It was once again confirmed that the effective thermal conductivity of nanoparticle suspensions increased with an increase of volume fraction and the enhancements of this experiment were much greater than the predictions of the Maxwell model [6] and the Hamilton-Crosser model [3]. A new kind of needle-like 30 nm CuO nanoparticle/DI water suspensions were measured by Jwo et al. [48]. This work reported that for the needle-like shape CuO /DI water nanoparticle suspension, the enhancement of the effective thermal conductivity could be even greater than that of spherical shape CuO/DI water nanoparticle suspensions. At the volume fraction of 0.8 %, the enhancement of effective thermal conductivity could be as high as 5.9 % at 5 ^{0} C to 38.2 % at 40 ^{0} C. Even at the volume fraction of 0.2 %, the enhancement of effective thermal conductivity could be 14.7 % at 40 ^{0} C.

Kim et al. [49] experimentally studied the size dependence of the effective thermal conductivity of suspensions with 38 nm diameter Al_2O_3 nanoparticle, 10 nm, 30 nm and 60 nm diameter ZnO nanoparticles, and 10 nm, 34 nm, and 70 nm diameter TiO nanoparticles in DI water and ethylene glycol base fluids. The results of the 38 nm diameter Al_2O_3 nanoparticle in both DI water and ethylene glycol suspensions confirmed the previous reported data by Lee et al. [23], and the results of both ZnO nanoparticle suspensions and TiO nanoparticle suspensions demonstrated that the smaller size could induce greater enhancement in effective thermal conductivity. The experimental data were compared with both a mean-field model, the Hamilton-Crosser model [3], and a Brownian motion model, Jang-Choi's model [50, 51]. There was no satisfactory agreement achieved. This report also presented the concern of the elevation of the effective thermal conductivity by a laser beam shooting on the ZnO nanoparticle suspensions.

Except for the measurement of effective thermal conductivities of the solid/liquid suspensions, liquid/liquid suspensions were also measured. Yang and Han [52, 53] adopted a 3-ω method to measure the effective thermal conductivity of water in FC72 nanoemulsion suspensions, and found a remarkable enhancement of 126 % at a volume fraction of 12 %. This work extended the selection of nanoparticle suspensions from solid/liquid matches towards the liquid/liquid matches.

3.2. The Simulation Reports on the Effective Thermal Conductivity of Nanoparticle Suspensions

In 2004, Bhattacharya and Phelan et al. [54] adopted a Brownian dynamics simulation technique with the equilibrium Green-Kubo method to compute the effective thermal conductivity of nanoparticle suspension. This simulation work confirmed the experimental reports [4] that Al_2O_3/ethylene glycol nanoparticle suspension and Cu/ethylene glycol nanoparticle suspension had much higher effective thermal conductivities compared to the predictions of the Hamilton-Crosser model. Another simulation work of Lattice Boltzmann simulation agreed with the results of the Brownian dynamics simulation [55], and discussed the role of the irregular motion of the nanoparticle and its contribution toward the effective thermal conductivity of nanoparticle suspension.

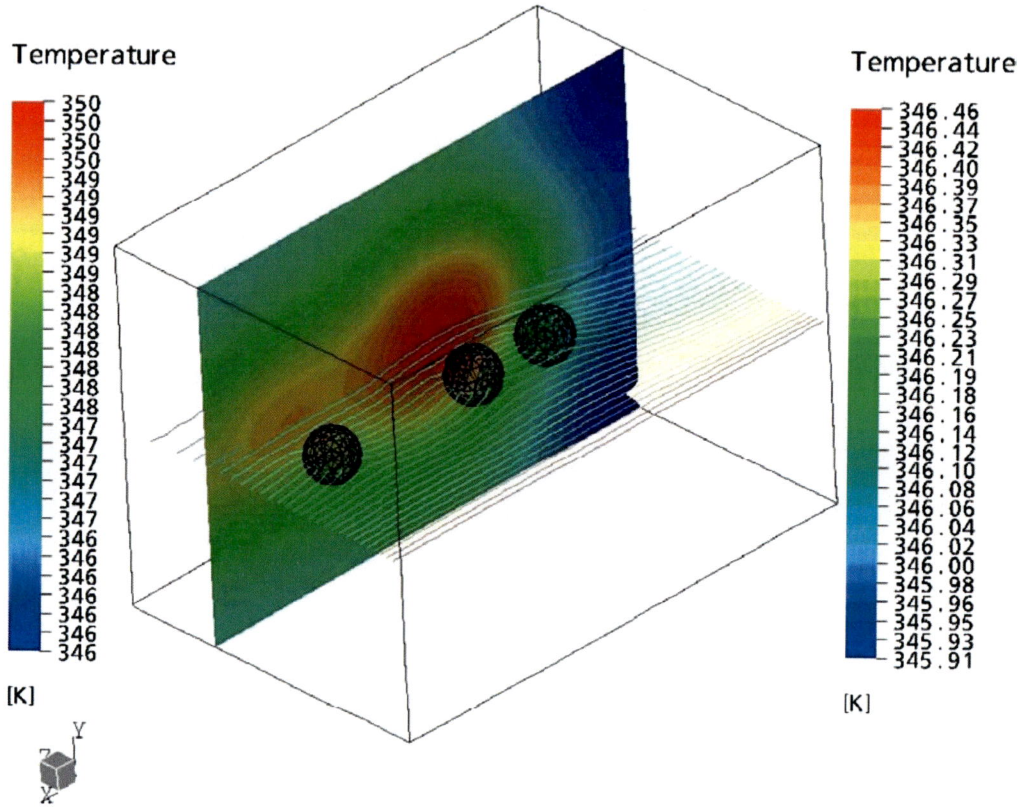

Figure 7 The comparison for simulated temperature field on vertical plane (left, the temperature scale of perpendicular contour; right, the temperature scale of horizontal contour) [56].

Li and Peterson [56] demonstrated in a simulation that the Brownian motion of nanoparticle could significantly change the temperature field in the suspensions and smooth the temperature gradient not only along the moving direction of the Brownian motion, but the temperature gradient perpendicular to the moving direction of the Brownian motion through the base fluid microconvection introduced by the nanoparticle Brownian motion. A comparison of one single nanoparticle case and a two-adjacent nanoparticles case showed that the hydrodynamic interaction between nanoparticles would dramatically enhance the influence on the temperature gradient and field compared to the effect of two single non-interacting nanoparticles. This simulation work demonstrated that the microscale hydrodynamic interaction of Brownian motion could explain the reason that macroscale effective thermal conductivity would increase with an increase of temperature and an increase of volume fraction. Moreover, Keblinski and Thomin [57] studied the hydrodynamic field around a nanoparticle with molecular dynamics simulation, which was conducting Brownian motion. The simulation showed that time-integrated response of the Brownian motion velocity field around a nanoparticle should be the same as that when the nanoparticle was having a constant velocity, which might imply that the convective heat transfer of a nanoparticle conducting Brownian motion could be the same as that when it was having a constant velocity.

As researchers had focused on adopting the Brownian motion effect to explain the effective thermal conductivity of nanoparticle suspensions, some criticizing voices on the Brownian motion effect have been reported, too. In 2006, Evans and Keblinski et al. [58] employed molecular dynamics simulation to study the Brownian motion effect of nanoparticles with 296 atoms of each in fluid. The interaction potential between solid atoms was FENE potential, $U_{FENE} = -5.625\varepsilon \cdot \ln[1-(r/1.5\sigma)^2]$, and the interaction potential between base fluid molecules was the standard Lennard-Jones potential, $U_{LJ} = 4\varepsilon[(\sigma/r)^{12} - (\sigma/r)^6]$, where ε is the unit of energy, σ is the unit of length, and r is the distance between a pair of atoms or molecules. The interaction potential between base fluid molecule and the solid atom on the surface of nanoparticle is still a Lennard-Jones potential, but the cutoff distance was set to be 1.5σ, which led to attractive forces between base fluid molecule and the solid atom. Moreover, the unit of energy between the base fluid molecule and solid atom was set to be three values, $\varepsilon_{SF}=0.25\varepsilon$, $\varepsilon_{SF}=1.25\varepsilon$, and $\varepsilon_{SF}=2.25\varepsilon$ for different wetting conditions between the nanoparticles and the base fluids. It concluded that only minor differences of thermal conductivity compared to that of the pure fluid was observed in this simulation, and the thermal conductivity increases were rather small and could not correspond to spectacular increases in experiments.

Eapen et al. [59, 60] employed molecular dynamics simulation to study the mechanisms of heat transfer in nanoparticle suspensions. The microscale heat flux is separated into the self-correlations and cross-correlations between the kinetic, potential, and collision modes. It showed in a demonstrating picture that the total effective thermal conductivity of Pt/Xenon nanoparticle suspension would be the sum of the contributions for all correlations, and much higher than the prediction of the Maxwell mean-field model [6]. Moreover, it was pointed out later that the Brownian motion models gave too high predictions on effective thermal conductivity of nanoparticle suspensions over the corresponding experimental results, and the enhanced effective thermal conductivity would be the reason of chain-like agglomeration of nanoparticles [60]. This simulation work was echoed by another molecular dynamics simulation report by Vladkov and Barrat [39], which stated that the enhanced thermal conductivity was caused by the collective effects of confinement, nanoparticle mass, Brownian motion, nanoparticle agglomeration and percolation.

However, the Brownian motion effect was also assured by a molecular dynamics simulation work by Sarkar and Selvam [61] that the effective thermal conductivity was much higher than the prediction of Hamilton-Crosser model. An equilibrium molecular dynamics simulation was conducted with the Green-Kubo method to study the effective thermal conductivity of different volume fraction Cu/argon nanoparticle suspensions. The mean square displacement analysis showed that both the enhancement of effective thermal conductivity and the enhancement of diffusivity had the similar mechanism, which was caused by the base fluid molecule microconvection induced by the nanoparticle Brownian motion.

3.3. Brownian Motion Models of Effective Thermal Conductivity of Nanoparticle Suspensions

Based on the previous experimental and simulating investigations on effective thermal conductivity of nanoparticle suspensions, there were several Brownian motion models proposed [46, 50, 51, 62,-65]. The first model took consideration of nanoparticle size, volume fraction, and bulk temperature, and based its velocity on the Stokes-Einstein formula [46], shown as in Eq. 25. This model believes that nanoparticles would carry thermal energy from hot place to cold place by mass diffusion, and enhance the effective thermal conductivity of the suspension as represented by the parameter, *c,* in the equation.

$$\frac{k_e}{k_1} = 1 + c \cdot \overline{u}_p \frac{\eta r_m}{k_1(1-\phi)r_p} \tag{25}$$

where k_e is the effective thermal conductivity of the nanoparticle suspension, k_l is the thermal conductivity of the base fluid, c is a constant, $\overline{u}_p = \frac{2k_b T}{\left(\pi\eta d_p^2\right)}$ is the average nanoparticle velocity, k_b is Boltzmann's constant, η is the base fluid dynamic viscosity, d_p=$2r_p$ is the diameter of nanoparticle, r_m is the radius of base fluid molecule, and ϕ is the volume fraction of nanoparticle. Moreover, the constant is estimated as $c = \frac{nlc_v}{3}$, where n is the number concentration of nanoparticle, l is the mean free path of the nanoparticle and c_v is the specific heat of nanoparticle.

The second Brownian motion model included one more factor, the adsorption layer of base fluid molecules by Jang and Choi [50], as show in Eq. 26. The first two parts on the right side of the equation are the contributions of the thermal conductivities of the nanoparticle and base fluid, and the third part is the contribution of Brownian motion and adsorption layer of base fluid molecules.

$$k_e = k_1(1-\phi) + k_2\phi + \phi h \delta_T \tag{26}$$

where k_e is the effective thermal conductivity, k_l is the thermal conductivity of base fluid, ϕ is the volume fraction of nanoparticle, k_2 is the thermal conductivity of the nanoparticle, h is the convective heat transfer coefficient, δ_T is the thickness of the thermal boundary layer. The third part on the right side of the equation was further developed as $3C_1 \frac{d_f}{d_p} k_1 \mathrm{Re}_{dp}^2 \mathrm{Pr}\,\phi$, where $h = \frac{k_1}{d_p}\mathrm{Re}_{dp}^2 \mathrm{Pr}^2$ is the heat transfer coefficient for flow past nanoparticles, $\delta_T = \frac{\delta}{\mathrm{Pr}} = \frac{3d_f}{\mathrm{Pr}}$, and $\mathrm{Re}_{dp} = \frac{Cd_p}{\upsilon}$. Here $C = \frac{D}{l_f}$ is the Brownian motion velocity of nanoparticle, D is the nanoparticle diffusion coefficient, l_f is the mean free path of a base fluid molecule, and υ is the kinematic viscosity of base fluid.

Jang and Choi [51] also tried to clarify some doubts of their Brownian motion model later and explained the assumptions made during the building up of the model that the nanoparticles would carry thermal energy by the nanoparticle mass diffusion through different temperature regions. This effort was made to ensure the Brownian motion effect to the effective thermal conductivity, whose intensity was decided by the bulk temperature, nanoparticle size, and base fluid properties.

The third Brownian motion model [62] included the base fluid effect that was carried by the Brownian motion of nanoparticle, instead of the adsorption layer [50, 51] shown in Eq. 27. It argued that it was the base fluid flow carried by the nanoparticle, instead of the nanoparticle's mass movement, that helped to transfer the thermal energy in the suspension.

$$k_e = 5\times10^4 \beta\alpha_p\rho_f c_f\sqrt{\frac{kT}{\rho_p D}}f(T,\alpha_p,etc.) \tag{27}$$

where ρ_f and ρ_p are the density of the base fluid and the nanoparticle, c_f is the specific heat of the base fluid, k is the Boltzmann constant, T is bulk temperature, D is the diameter of nanoparticle, α_p=NV_d/V, N is the total number of nanoparticles in a cell, V_d is the volume of a nanoparticle, V is the volume a nanoparticle sweeping in a certain time period, β is the fraction of the base fluid volume travelling with nanoparticles, and $f(T,\alpha_p,\text{etc})$ is a factorial function of temperature, T, α_p, and the physical properties of the base fluid and nanoparticle materials, which depends on the properties of the intervening base fluid and the nanoparticle interactions.

This model attributed the temperature dependence of effective thermal conductivity to the long impact range of the interparticle potential, and illustrated that this temperature dependence was not due to the thermophoresis or osmophoresis of nanoparticles [62]. The thermophoresis influence on effective thermal conductivity was represented with Eq. 28, and the osmophoresis influence was represented with Eq. 29, as below. It was clearly demonstrated through the comparison of these three influences (Eq. 27-29) that the impact of the Brownian motion is much more significant than that of thermophoresis or osmophoresis effect, and the thermophoresis and osmophoresis motion effects were independent of nanoparticle size.

$$k_{thermoph} = \frac{1}{6\pi}\frac{\alpha_p k}{\mu_f}\frac{3k_1}{k_2+2k_1}(1\times10^5\rho_f c_f)\nabla T \tag{28}$$

$$k_{osmoph} = \frac{1}{3\pi}\frac{\alpha_p^2 k}{\mu_f}\frac{3k_1}{k_2+2k_1}(1\times10^5\rho_f c_f)\nabla T \tag{29}$$

where α_p is the same as in Eq. 27, k is the Boltzmann constant, μ_f is the dynamic viscosity, k_1 and k_2 are the thermal conductivities of base fluid and nanoparticle, ρ_f and c_f are density and specific heat of base fluid, and ∇T is the temperature gradient.

Still another Brownian motion, the forth one, was proposed with an order-of-magnitude analysis of various possible mechanisms [63, 64]. This model combined the Maxwell-Garnett mean-field model [20] and a Brownian motion convection model together, and would reduce to only the Maxwell-Garnett model [20] once the nanoparticle size is large enough. Due to the Maxwell-Garnett mean-field model [20], the interfacial thermal boundary resistance was considered in this model, too. In this report, the heat transport in nanoparticle suspensions was compared with the particle-to-fluid heat transfer in fluidized beds in the way that convective currents from various particle interactions in fluidized bed enhanced the heat transfer. Henceforth, a convective heat transfer model in the Stokes regime was adopted for the heat transfer enhancement of Brownian motion effect of the nanoparticles. The model was presented as Eq. 30.

$$\frac{k_e}{k_1} = (1 + A \cdot \mathrm{Re}^m \, \mathrm{Pr}^{0.333} \, \phi)\left[\frac{(1+2\alpha) + 2\phi(1-\alpha)}{(1+2\alpha) - \phi(1-\alpha)}\right] \tag{30}$$

where k_e is the effective thermal conductivity of the nanoparticle suspension, k_1 is the thermal conductivity of base fluid, Re is the Reynolds number based on the nanoparticle Brownian motion velocity, u, and diameter, d, Pr is the Prandtl number of the base fluid, ϕ is nanoparticle volume fraction, $\alpha = 2R_{\mathrm{interfacial}} k_1/d$, $R_{interfacial}$ is the interfacial thermal resistance, and A and m are constants.

In 2005, Chon and Kihm et al. [66] measured the effective thermal conductivities of Al_2O_3/DI water suspensions with diameter size at 11 nm, 47 nm, and 150 nm and bulk temperature ranging from 21 °C to 71 °C. This report clearly illustrated that for the same nanoparticle material, Al_2O_3, and the same base fluid, DI water, the effective thermal conductivity of nanoparticle suspension increased with an increase of bulk temperature, and increased with a decrease of nanoparticle diameter. Based on the experimental results, an empirical equation was generated via a linear Buckingham-Pi theorem regression, Eq. 31. With this equation, it confirmed the importance of the Brownian motion on the effective thermal conductivity, and concluded that the elevation of the effective thermal conductivity at high temperature was a primary result of the intensity increase of nanoparticles Brownian motion. This finding was also supported by another experimental report later [67].

$$\frac{k_e}{k_1} = 1 + 64.7 \cdot \phi^{0.746} \left(\frac{d_f}{d_p}\right)^{0.369} \left(\frac{k_2}{k_1}\right)^{0.7476} \times \mathrm{Pr}^{0.9955} \, \mathrm{Re}^{1.2321} \tag{31}$$

where ϕ is the volume fraction of nanoparticles, d_f and d_p are the diameter of the base fluid molecule and nanoparticle, k_2 and k_1 are the thermal conductivity of the nanoparticle and base fluid, Pr is the Prandtl number of the base fluid, and Re is the Reynolds number based on nanoparticle Brownian motion velocity and diameter.

3.4. Related Research on the Agglomeration Mechanisms

Wang and Li et al. [68, 69] measured the effective thermal conductivity of the 25 nm diameter SiO_2 nanoparticle/DI water suspensions and demonstrated the clustering of SiO_2 nanoparticles inside the suspension with TEM pictures. It was proposed that the microscope motion should be the microconvection introduced by the nanoparticle Brownian motion and the structure should be the agglomeration of nanoparticles. The microconvection and agglomeration could be the major mechanisms for the enhanced thermal conductivity of nanoparticle suspensions, as well as the adsorption fluid layer on nanoparticles. In the same year, Keblinski et al. [70] broke further down the proposed mechanisms of the microscope motion and the nanoparticle structure, and reported the theoretical analysis results on the Brownian motion effect, liquid layering effect, ballistic transport inside of nanoparticles effect, and the nanoparticle clustering effect. In this report, the role of Brownian motion was argued to produce the nanoparticle agglomeration. It was concluded that the agglomeration produced by the nanoparticle Brownian motion with the assistance of liquid layering effect and ballistic transport inside nanoparticles served as a fast lane for heat flux, and the agglomeration of nanoparticles was the mechanism of the unusual high thermal conductivity of nanoparticle suspension [69]. In 2003, Wang et al. [71], considering the fact of nanoparticles clustering inside the suspensions, further developed a fractal model to predict the effective thermal conductivity of nanoparticle suspensions, which tried to include the size effect, the fractal of nanoparticle agglomerations and the adsorption layer of the base fluid.

Because the agglomerations of nanoparticles are relatively loose compared to the same size of solid particles, the base fluid is filled in the void of the cluster to connect the nanoparticles together as a whole body. If the agglomeration of nanoparticles could serve as a fastline for the heat flux insider the suspensions, the adsorbed base fluid inside the cluster should play an important role. The effect of the adsorption layer was extensively studied with molecular dynamics simulation by Xue and Keblinski et al. [72-75]. A simple Lennard-Jones interaction potential was applied to the liquid, solid and the interface of the liquid-solid in the simulation. And the binding energy parameter of the liquid-solid interaction was set as proportional to the binding energy parameter of liquid-liquid interaction, $\varepsilon_{ls}=\alpha\varepsilon_{ll}$, where α was chosen from 0.2 to 5 to represent the non-wetting, weakly-wetting, and the wetting interaction of the liquid-solid interface. It was concluded that the adsorption layer could not play an important role in the effective thermal conductivity even in the limit of nanometer size particles because the non-wetting/weakly-wetting liquid-solid interface showed a high thermal resistance, as well as that the wetting liquid-solid interface had a much smaller but still large thermal resistance, which would be equal to that of liquid-liquid situation over a distance of around 1 nm [72-74]. It was also found that the interfacial resistance of weakly coupled between carbon nanotube and octane liquid was relatively larger and the adsorbed liquid layer did not result in an observable effect in the heat transfer [73]. Despite the possible negative effect from simulation results, Xue and Choi [74] considered the possible positive effect of adsorbed base fluid layer and tried to modify the Hamilton-Crosser model to include the base fluid interfacial layer. Xie and Zhang et al. [76] also proposed that the adsorbed base fluid layer could contribute to the effective thermal conductivity of nanoparticle suspension in a theoretical analysis. Prakash and Giannelis [77] tried to incorporate the interfacial resistance effect of the adsorbed base fluid layer with the Brownian motion of the nanoparticles together, to explain the enhanced effective thermal conductivity of nanoparticle suspension.

Vadas [78] used both the DuPhlag solution and the Fourier solution to interpret the experimental data reported previously, and showed that the solid-fluid interfacial heat transfer mechanism should be accounted to reach identical results. Lee [79] introduced the electrical double layer concept and calculated the thermal conductivity and thickness of interfacial layer. It was concluded that the interfacial layer could enhance the overall effective thermal conductivity. Evans and Keblinski [80] studied the effective thermal conductivity of water molecules with different orientational and translational orders induced by an electric field. It was found that the electric-field induced crystallization and translational order of water molecules could bring a three fold increase in thermal conductivity compared to that of the base water. This nonequilibrium molecular dynamics simulation confirmed the possibility of the contribution from the adsorbed base fluid layer, but also pointed out that it might not be possible to have the crystallinity at room temperature in nanoparticle suspensions. Tillman and Hill [81] analyzed the adsorbed base fluid layer with mathematical approach and disagreed any possibility of contribution from the adsorbed layer to the effective thermal conductivity.

Based on the experimental reports on agglomeration of nanoparticles inside nanoparticle suspensions, and the reported experimental relationship between the effective thermal conductivity of nanoparticle suspensions and the agglomeration sizes, Prasher et al. [82, 83] also tried to modify the mean-field theory by including the fractal clusters, and to use the modified model to predict the contribution of agglomeration to the effective thermal conductivity enhancement. It was concluded in this effort that the morphology of nanoparticle clusters might be very important when using mean-field theory to predict the effective thermal conductivity.

4. Discussion and Conclusion

From all of the experimental reports on the effective thermal conductivity of nanoparticle suspensions, it is for certain that the effective thermal conductivity of nanoparticle suspensions will increase with an increase of volume fraction, bulk temperature, and a decrease of nanoparticle size. The stable nanoparticle suspensions could be used in many areas in which the thermal control and heat transfer are critical issues, such as the automobile industry, manufacturing industry, microelectronic industry, and many other cutting-edge scientific research fields. In order to guide the discovery of high effective thermal conductivity nanoparticle suspensions to satisfy the urgent need of better heat transfer liquid media for the rapidly growing heat flux cooling demands, many models were proposed based on the Brownian motion effect and the agglomeration, as well as other possible mechanisms. However, the general acceptance of any one of the models has not yet been reached and how those mechanisms play a role in the effective thermal conductivity enhancement is still not clear. Before the new heat transfer media, nanoparticle suspensions, could be massively used in the daily cooling of devices and other applications, there is still a long way to go in the development of a fundamental understanding of the microscale thermophysical transport phenomena.

REFERENCES

[1] Choi, S. U. S., Enhancing thermal conductivity of fluids with nanoparticles. In *''Developments and Applications of Non-Newtonian Flows''* (D.A. Siginer and H.P. Wang, eds.). American Society of Mechanical Engineers, New York, 1995.

[2] S. P. Lee and U. S. Choi, Application of metallic nanoparticle suspensions in advanced cooling sytems, Recent advances in solids/structures and application of metallic materials, eds. Y. Kwon, D. Davis, and H. Chung, *The American Society of Mechanical Engineers,* New York, PVP-Vol 342/MD-Vol 72, pp. 227-234, 1996.

[3] Hamilton, R. L., and Crosser, O. K., *Thermal conductivity of heterogeneous Two component Systems, I&EC Fundamentals,* Vol. 1 (3), pp. 187–191, 1962.

[4] Eastman, J. A., Choi, S. U. S., Li, S. Thompson, L. J., and Lee, S., Enhanced thermal conductivity through the development of nanofluids. In *''Material Research Society Symposium Proceedings''* (S. Komarnei, J.C. Parker and H.J. Wollenberger, eds.), Vol. 457, pp. 9–10, Warrendale, PA, 1997.

[5] Meredith, R. E., and Tobias, C. W., *Conductivities in emulsions.* J. Electrochem. Soc, Vol. 108(3), pp. 286–290, 1961.

[6] Maxwell, J. C., *A Treatise on Electricity and Magnetism,* 3rd edn. Oxford University Press, Oxford, 1873.

[7] Fricke, H., *A mathematical treatment of the electric conductivity and capacity of disperse systems. Physical Review,* Vol. 24, pp. 575–587, 1924.

[8] Keller, J. B., Conductivity of a medium containing a dense array of perfectly conducting spheres or cylinders or nonconducting cylinders. *Journal of Applied Physics,* Vol. 34(4), pp. 991–993, 1963 .

[9] Rocha, A. and Acrivos, A., On the effective thermal conductivity of dilute dispersions: highly conducting inclusions of arbitrary shape. *Quarterly journal of mechanics and applied mathematics,* Vol. 26(4), pp. 441-455, 1973.

[10] Rocha, A. and Acrivos, A. On the effective thermal conductivity of dilute dispersions. *Quarterly journal of mechanics and applied mathematics,* Vol. 26(2), pp. 217-233, 1973.

[11] Kim, I. C. and Torquato, S. Effective conductivity of composites containing spheroidal inclusions: comparison of simulations with theory. *Journal of applied physics,* Vol. 74(3), pp. 1844-1854, 1993.

[12] Batchelor, G. K., Transport properties of two-phase materials with random structure. *Annual review of fluid mechanics,* Vol. 6, pp. 227-255, 1974.

[13] Jeffrey, D. J., Conduction through a random suspension of spheres. *Proceedings of the royal society of London,* series a (mathematical and physical sciences), Vol. 335(1602), pp. 355-367, 1973.

[14] Davis, R. H., The effective thermal conductivity of a composite material with spherical inclusions. *International journal of thermophysics,* Vol. 7(3), pp. 609-620, 1986.

[15] Brady, J. F., Phillips, R. J., Lester, J. C., and Bossis, G. Dynamic simulation of hydrodynamically interacting suspensions. *Journal of fluid mechanics,* Vol. 195, pp. 257-280, 1988.

[16] Bonnecaze, R. T. and Brady, J. F. A method for determining the effective conductivity of dispersions of particles. *Proceedings of the royal society of London: Mathematical and physical sciences,* Vol. 430(1879), pp. 285-313, 1990.

[17] Bonnecaze, R. T. and Brady, J. F. The effective conductivity of random suspensions of spherical particles. *Proceedings of the royal society of London, series a* (mathematical and physical sciences), Vol. 432, pp. 445-465, 1991.

[18] Lu, S. Y. and Kim, S. T. Effective thermal conductivity of composites containing spheroidal inclusions. *AIChE journal,* Vol. 36(6), pp. 927-938, 1990.

[19] Lu, S. Y. and Lin, H. C. Effective conductivity of composites containing aligned spheroidal inclusions of finite conductivity. *Journal of applied physics,* Vol. 79(9), pp. 6761- 6769, 1996.

[20] Hasselman, D. P. H., and Johnson, L. F., Effective thermal conductivity of composites with interfacial thermal barrier resistance. *Journal of composite materials,* Vol. 21, pp. 508-515, 1987.

[21] Leal, L., G. On the effective conductivity of a dilute suspension of spherical drops in the limit of low particle peclet number. *Chemical engineering communication,* Vol. 1, pp. 21-31, 1973.

[22] Eastman J. A., Choi S. U. S., Li S., Anomalously increased effective thermal conductivities of ethylene glycol-based nanofluids containing copper nanoparticles, *Applied Physics Letters, V*ol. 78 (6), pp. 718-720, 2001 .

[23] Lee, S., Choi, S. U. S., Li, S., and Eastman, J. A. Measuring thermal conductivity of fluids containing oxide nanoparticles. *Transactions of the ASME,* Vol. 121, pp. 280-289, 1999.

[24] Wang, X. and Xu, X. Thermal conductivity of nanoparticle-fluid mixture. *Journal of thermophysics heat transfer,* Vol. 13(4), pp. 474-480, 1999.

[25] Xie, H., Wang, J., Xi, T., Liu, Y., and Ai, F. Dependence of the thermal conductivity of nanoparticle-fluid mixture on the base fluid. *Journal of materials science letters,* Vol. 21, pp. 1469-1471, 2002.

[26] Xie, H., Wang, J., Xi, T., Liu, Y., and Ai, F. Thermal conductivity enhancement of suspensions containing nanosized alumina particles. *Journal of applied physics,* Vol. 91(7), pp. 4568-4572 , 2002.

[27] Das, S. K., Putra, N., Thiesen, P. and Roetzel, W. Temperature dependence of thermal conductivity enhancement for nanofluids. *Journal of heat transfer,* Vol. 125, 567-574, 2003.

[28] Patel, H. E., Das, S. K., and Sundararajan, T. Thermal conductivities of naked and monolayer protected metal nanoparticle based nanofluids: Manifestation of anomalous enhancement and chemical effects. *Applied physics letters,* Vol. 83(14), pp. 2931-2933, 2003.

[29] Krishnamurthy, S., Bhattacharya P., Phelan P. E., and Prasher R., Enhanced mass transport in nanofluids, *Nano letters,* Vol. 6, pp. 419-423, 2006.

[30] Putnam, S. A., Cahill, D. A., and Braun, P. V., Thermal conductivity of nanoparticle suspensions, *Journal of Applied Physics,* Vol. 99 (8), pp. 084308, 2006.

[31] Choi, S. U. S., Zhang, Z. G., Yu, W., Lockwood, E. A., Grulke, E. A., Anomalous thermal conductivity enhancement in nanotube suspensions, *Applied Physics Letters,* Vol. 79, pp. 2252-2254, 2001.

[32] Xie, H., Lee, H., Youn, W., and Choi, M. Nanofluids containing multiwalled carbon nanotubes and their enhanced thermal conductivities. *Journal of applied physics,* Vol. 94(8), pp. 4967-4971, 2003.

[33] Peterson, G. P., and Li, C. H., An Overview of Heat and Mass Transfer in Fluids with Nano-particle Suspensions" , *Advances in Heat Transfer,* Vol. 39, J. P. Hartnett and T. F. Irvine (eds.), Academic Press, New York, NY, pp. 261-392, 2006.

[34] Hong T. K., Yang H. S., Choi C. J., Study of the enhanced thermal conductivity of Fe nanofluids, *Journal of Applied Physics,* Vol. 97 (6), pp. 064311, 2005.

[35] Hong T. S., Hong T., K, and Yang H. S., Thermal conductivity of Fe nanofluids depending on the cluster size of nanoparticles, *Applied Physics Letters,* Vol. 88, pp. 031901, 2006.

[36] Liu M. S., Lin M. C. C., Tsai C. Y., and Wang, C., Enhancement of thermal conductivity with Cu for nanofluids using chemical reduction method, *International Journal of Heat and Mass Transfer,* Vol. 49 (17-18), pp. 3028-3033, 2006.

[37] Jana S., Salehi-Khojin A., Zhong W. H., Enhancement of fluid thermal conductivity by the addition of single and hybrid nano-additives, *Thermochimica Acta,* Vol. 462 (1-2), pp. 45-55, 2007.

[38] Wensel, J., Wright, B., Thomas, D., Douglas, W., Mannhalter, B., Cross, W., Hong, H., Kellar, J., Smith, P., and Roy, W., Enhanced thermal conductivity by aggregation in heat transfer nanofluids containing metal oxide nanoparticles and carbon nanotubes, *Applied Physics Letters,* Vol. 92, pp. 023110, 2008.

[39] Vldakov, M., and Barrat, J., Modeling transient absorption and thermal conductivity in a simple nanofluid, *Nano Letters,* Vol. 6, pp. 1224-1228, 2006.

[40] Kim, B. H., and Peterson, G. P., Effect of morphology of carbon nanotubes on thermal conductivity enhancement of nanofluids, *Journal of Thermophysics and Heat Transfer,* Vol. 21, pp. 451-459, 2007.

[41] Li, C. H. and Peterson, G. P., "Experimental investigation of temperature and volume fraction variations on the effective thermal conductivity of nanoparticle suspensions (nanofluids)", *Journal of applied physics,* Vol. 99(8), pp. 084314, 2006.

[42] Li, H. C. and Peterson, G. P., "The effect of particle size on the effective thermal conductivity of Al_2O_3-water nanofluids", *Journal of Applied Physics,* Vol. 101(4), pp. 044312, 2007.

[43] Li, H. C., Williams, W., Hu, L., Buonqiorno, J. and Peterson, G. P., "Transient and steady-state experimental comparison study of effective thermal conductivity of Al_2O_3/water nanofluid", *Journal of Heat Transfer,* Vol. 130(4), 2008, in print.

[44] Zhu H. T., Zhang C. Y., Liu, S. Q., Tang Y. M., and Yin, Y. S., Effects of nanoparticle clustering and alignment on the thermal conductivities of Fe_3O_4 aqueous nanofluids, *Applied Physics Letters,* Vol. 89, pp. 023123, 2006.

[45] Yu W., Choi S. U. S., The role of interfacial layers in the enhanced thermal conductivity of nanofluids: A renovated Maxwell model, *Journal of Nanoparticle Research,* Vol. 5 (1-2), pp. 167-171, 2003.

[46] Kumar, H. D., Patel, H. E., Kumar, V. R. R., Sundararajan, T., Pradeep, T., and Das, S. K., Model for heat conduction in nanofluids, *Physical Review Letters,* Vol. 93, pp. 144301, 2004.

[47] Chopkar M., Kumar S., Bhandari D. R., Development and characterization of Al_2Cu and Ag_2Al nanoparticle dispersed water and ethylene glycol based nanofluid, *Materials*

Science and Engineering B-Solid State Materials for Advanced Technology, Vol. 139 (2-3), pp. 141-148, 2007.

[48] Jwo C. S., Chang H., Teng T. P., A study on the effects of temperature and volume fraction on thermal conductivity of copper oxide nanofluid, *Journal of Nanoscience and Nanotechnology,* Vol. 7 (6), pp. 2161-2166 Sp. Iss. SI, 2007.

[49] Kim S. H., Choi, S. R., Kim D., Thermal conductivity of metal-oxide nanofluids: Particle size dependence and effect of laser irradiation, *Journal of Heat Transfer-Transactions of The ASME,* Vol. 129 (3), pp. 298-307, 2007.

[50] Jang, S. P., and Choi, S. U. S., Role of Brownian motion in the enhanced thermal conductivity of nanofluids, *Applied Physical Letters,* Vol. 84, pp. 4316-4318, 2004.

[51] Jang S. P., Choi S. U. S., Effects of various parameters on nanofluid thermal conductivity, *Journal of Heat Transfer-Transactions of The ASME,* Vol. 129 (5), pp. 617-623, 2007.

[52] Yang, B. and Han, Z. H., Temperature-dependent thermal conductivity of nanorod-based nanofluids, *Applied Physics Letters,* Vol. 89, pp.083111, 2006.

[53] Han, Z. H., and Yang, B., Thermophysical characteristics of water-in--FC72 nanoemulsion fluids, *Applied Physics Letters,* Vol. 92, pp. 013118, 2008.

[54] Bhattacharya P., Phelan P. E., and Prasher R. Brownian dynamics simulation to determine the effective thermal conductivity of nanofluids, *Journal of Applied Physics,* Vol. 95, pp. 6492-6494, 2004.

[55] Xuan, Y., and Yao, Z., Lattice Boltzmann model for nanofluids, *Heat Mass Transfer,* Vol. 41, pp. 199-205, 2005.

[56] Li, H. C. and Peterson, G. P., "Mixing effect on the enhancement of thermal conductivity of nanoparticle suspensions (nanofluids)", *Int. J. Heat and Mass Transfer,* Vol. 50, pp. 4668-4677, 2007.

[57] Keblinski, P., and Thomin, J., Hydrodynamic field around a Brownian particle, *Physical Review E,* Vol. 73, pp. 010502, 2006.

[58] Evans W., Fish J., Keblinski P., Role of Brownian motion hydrodynamics on nanofluid thermal conductivity, *Applied Physics Letters,* Vol. 88 (9), pp. 093116, 2006.

[59] Eapen J., Li J., Yip S., Mechanism of thermal transport in dilute nanocolloids, *Physical Review Letters,* Vol. 98 (2), pp. 028302, 2007.

[60] Eapen J., Williams W. C., Buongiorno J., Mean-field versus microconvection effects in nanofluid thermal conduction, *Physical Review Letters,* Vol. 99 (9), pp. 095901, 2007.

[61] Sarkara S., Selvam R. P., Molecular dynamics simulation of effective thermal conductivity and study of enhanced thermal transport mechanism in nanofluids, *Journal of Applied Physics,* Vol. 102 (7), pp. 074302, 2007.

[62] Koo J., Kleinstreuer C., A new thermal conductivity model for nanofluids, *Journal of nanoparticle research,* Vol. 6, pp. 577-588, 2004.

[63] Koo J., Kleinstreuer C., Impact analysis of nanoparticle motion mechanisms on the thermal conductivity of nanofluids, *International Communications in Heat and Mass Transfer,* Vol. 32 (9), pp. 1111-1118, 2005.

[64] Prasher R., Bhattacharya P., Phelan P. E., Thermal conductivity of nanoscale colloidal solutions (nanofluids), *Physical Review Letters,* Vol. 94, pp. 025901, 2005.

[65] Prasher R., Bhattacharya P., Phelan P. E., Brownian-motion-based convective-conductive model for the effective thermal conductivity of nanofluids, *Journal of Heat Transfer-Transactions of The ASME,* Vol. 128 (6), pp. 588-595, 2006.

[66] Chon C. H., Kihm K. D., Lee S. P., Empirical correlation finding the role of temperature and particle size for nanofluid (Al2O3) thermal conductivity enhancement, *Applied Physics Letters,* Vol. 87 (15), pp. 153107, 2005.

[67] Chon, C. H., and Kihm, K. D., Thermal conductivity enhancement of nanofluids by Brownian motion, *Journal of Heat Transfer,* Vol. 127, pp. 810, 2005.

[68] Wang, B.X., Li, H., Peng, X.F., Research on the Heat-Conduction Enhancement for Liquid with Nano-Particle Suspensions, *Journal of Thermal Science,* Vol. 11(3), pp. 214-219, 2002.

[69] Wang, B.X., Li, H., Peng, X.F., Effect of Surface Adsorption on Heat Transfer Enhancement for Liquid with Nano-particle Suspensions, *Journal of Engineering Thermophysics,* Vol. 24(4), pp. 664-666, 2003.

[70] Keblinski, P., Phillpot, S. R., Choi, S. U. S., and Eastman, J. A. Mechanisms of heat flow in suspensions of nano-sized particles (nanofluids). *International journal of heat and mass transfer,* Vol. 45, pp. 855-863, 2002.

[71] Wang B. X., Zhou L. P., Peng X. F., A fractal model for predicting the effective thermal conductivity of liquid with suspension of nanoparticles, *International Journal of Heat and Mass Transfer,* Vol. 46 (14), pp. 2665-2672, 2003 .

[72] Xue, L., Leblinski, P., Phillpot, S. R., Choi, S. U. S., and Eastman, J. A. Two regimes of thermal resistance at a liquid-solid interface. *Journal of chemical physics,* Vol. 118(1), pp. 337-339, 2003.

[73] Shenogin, S., Xue, L., Ozisik, R., and Keblinski, P. Role of thermal boundary resistance on the heat flow in carbon-nanotube composites. *Journal of applied physics,* Vol. 95(12), pp. 8136-8144, 2004.

[74] Xue, l., Keblinski, P., Phillpot, S. R., Choi, S. U. S., and Eastman, J. A., Effect of liquid layering at the liquid-solid interface on thermal transport, *International Journal of Heat and Mass Transfer,* Vol. 47, pp. 4277-4284, 2004.

[75] Evans, W., Fish, J., and Keblinski, P., Thermal conductivity of ordered molecular water, *The Journal of Chemical Physics,* Vol. 126, pp. 154504, 2007.

[76] Xie H. Q., Fujii M., Zhang X., Effect of interfacial nanolayer on the effective thermal conductivity of nanoparticle-fluid mixture, *International Journal of Heat and Mass Transfer,* Vol. 48 (14), pp. 2926-2932, 2005.

[77] Prakash M., Giannelis E. P., Mechanism of heat transport in nanofluids, *Journal of Computer-Aided Materials Design,* Vol. 14 (1), pp. 109-117, 2007.

[78] Vadasz P., Heat conduction in nanofluid suspensions, *Journal of Heat Transfer-Transactions of The ASME, Vol.* 128 (5), pp. 465-477, 2006.

[79] Lee, D., Thermophysical properties of interfacial layer in nanofluids, *Langmuir,* Vol. 23, pp. 6011-6018, 2007.

[80] Evans, W., Fish, J., and Keblinski, P., Thermal conductivity of ordered molecular water, *The Journal of Chemical Physics,* Vol. 126, pp. 154504, 2007.

[81] Tillman P., Hill J. M., Determination of nanolayer thickness for a nanofluid, International Communications In *Heat and Mass Transfer,* Vol. 34 (4), pp. 399-407, 2007.

[82] Prasher R., Phelan P. E., Bhattacharya P., Effect of aggregation kinetics on the thermal conductivity of nanoscale colloidal solutions (nanofluid), *Nano Letters,* Vol. 6 (7), pp. 1529-1534, 2006.

[83] Prasher R., Evans, W., Meakin, P., Phelan P. E., and Keblinski, P., Effect of aggregation on thermal conduction in colloidal nanofluids, *Applied Physics Letters,* Vol. 89, pp. 143119, 2006.

In: Encyclopedia of Nanotechnology
Editor: Elwood D. Carlson

ISBN: 978-1-60692-079-4

Chapter 43

ENHANCEMENT AND TEMPERATURE VARIATION IN THE THERMAL CONDUCTIVITY OF NANOFLUIDS

Pei Tillman, Miccal T. Matthews and James M. Hill
School of Mathematics and Applied Statistics,.
University of Wollongong, Wollongong, Australia

ABSTRACT

One of the most important findings for nanofluids is the discovery of a significant enhancement and a strong temperature effect in their thermal conductivity. Research confirms that the standard rules for mixtures is not capable of describing the nanoparticle's contribution to the thermal conductivity, and some researchers believe that the Brownian motion is the key mechanism affecting the variation of thermal conductivity with temperature. In this paper, the authors review existing experimental results and explore different approaches to estimating the effective thermal conductivity of nanofluids from recent theoretical models. Focus is primarily on an explanation of the temperature-dependence on the thermal conductivity of nanofluids. It is found that the mobility of nanoparticles is theoretically conjectured as having a key role in determining the temperature effect on the thermal conductivity of a nanofluid. The limitations of existing models and future research directions are discussed.

1. INTRODUCTION

The development of thermal energy-efficient heat transfer fluids is required in many industrial and commercial applications. Since solid materials typically have a thermal conductivity several orders of magnitude larger than commonly-used liquids, many heat transfer fluids containing millimeter or micrometer-sized particles were developed for improved heat exchange systems. However, these so-called "charged" fluids had several drawbacks, most commonly due to particle settling and cogging of machine components. With the emergence of nanotechnology, a new class of heat transfer fluids was developed by suspending nanoparticles and carbon nanotubes in these liquids. At present, three types of nanofluids containing single- and multi-wall carbon nanotubes, vapor grown carbon fibres, amorphous carbon and copper, CuO and Al2 O3 nanoparticles have been developed, manufactured, and tested [1]. These are:

1. heat transfer nanofluids or nanocoolants based on three base fluids: water, water/ethylene glycol mixtures, water/antifreeze mixtures,
2. nanolubricant fluids: commercial 15W-40 oil, BP Amoco DS-1666 Durecene oil (synthetic poly-α-olefin oil), military specification-based fluids,
3. nanolubricant grease, based on fluids and greases: BP 1666 Durecene oil, military specification-based greases.

Typical applications of these types of nanofluids are vehicle radiator coolants, air conditioning systems, cutting fluids, quenching fluids and high-stress contact gears for rotary machines. The enhanced thermal conductivity of the nanofluids tends to remove unwanted heat from the engines while providing lubrication and temperature control. The anticipated benefits of nanofluids are lower operation costs, improved thermal transfer properties, minimal clogging, suspension stability and long-term performance life.

The thermal conductivities of nanofluids are significantly higher than those of the base fluid, even when the loaded nanoparticle concentration is less than 1 % by volume. The main reason for this is that the suspended nanoparticles remarkably increase the thermal conductivity of the fluid, primarily as a result of the high surface-to-volume ratio of the nanoparticles (for example, the surface-to-volume ratio of a 10 nm particle is 1,000 times larger than that of a 10 μm particle). In nanoparticle-fluid mixtures, other effects such as the microscopic motion of particles, particle structure and surface properties may cause additional heat transfer mechanisms, which are not well understood at present. Traditional multiphase effective medium theories such as those of Maxwell [2] and Hamilton and Crosser [3] cannot account for the thermal conductivity enhancement achieved in nanofluids. The effective medium theories predict no change in the conductivity enhancement with temperature, since they are only dependent on the volume fraction of the solid particles and the thermal conductivity of the pure components. The model of Hamilton and Crosser [3] is dependent on the volume fraction and the sphericity of the particles, which is defined as the ratio of the surface area of a particle with a perfectly spherical shape to that of a non-spherical particle with the same volume. These, and other effective medium theories, predict no effect of particle size and only a weak effect of particle conductivity.

Various attempts have been made in the past ten years to theorize this phenomenon by taking into account the effects of particle radius and shape, concentration, properties of the nanolayer surrounding the nanoparticle, nanoparticle aggregation, nanoparticle motion, ballistic phenomena, dynamic viscosity of the liquid and the temperature of the medium. However, many models do not have any temperature-dependence and cannot be treated as a general theoretical framework. The purpose of this article is to present a review of the progress of research in the important aspect of the temperature-dependence on the thermal conductivity enhancement in nanofluids. Our review of the temperature effect on the thermal conductivity behavior of materials begins with single-phase systems, such as solids and liquids, then multiphase systems, in particular nanofluids. We examine the physical reasoning behind the models adopted, which correlate the thermal conductivity of nanofluids as

a function of temperature, particle size and volume fraction. Limitations of the existing models and some future research directions are discussed.

2. Thermal Conductivity of Liquids

Accurate knowledge of the thermal transport properties of pure liquids is essential for the development of a theory of the transport properties of nanofluids. In heat conduction, heat is transferred due to an energy exchange between heat carriers in the collision process. It is believed that liquids mainly conduct heat through the vibration of single molecules, while other methods of transfer such as molecular convection and radiation impart a small contribution to the overall heat transfer mechanism. Based on statistical mechanics (in particular kinetic gas theory) and correlated with experimental data, Avsec and Oblak [4] extended the Chung-Lee-Starling model (CLS) [5,6] and developed equations to calculate the transport properties of pure fluids.

For most simple liquids (except water) the thermal conductivity is generally characterized as decreasing linearly with increasing temperature. The thermal conductivity of liquids can generally be expressed as a function of temperature by

$$k_f = A_1 + A_2T + A_3T^2 \ (Wm^{-1}K^{-1}), \quad (1)$$

where the subscript f represents the fluid, the temperature T is in Kelvin (K) and A_1, A_2 and A_3 are constants in units of $Wm^{-1}K^{-1}$, $Wm^{-1}K^{-2}$ and $Wm^{-1}K^{-3}$, respectively. For water, $A_1 = -0.76760$, $A_2 = 7.5390 \times 10^{-3}$, $A_3 = -9.8250 \times 10^{-6}$; for toluene ($C_7H_8$), $A_1 = 0.22050$, $A_2 = -3.000 \times 10^{-4}$, $A_3 = 0$; for the hexadecane oil ($C_{16}H_{34}$), $A_1 = 0.1963$, $A_2 = -1.9 \times 10^{-4}$, $A_3 = 0$ [7]. Figure 1 displays the variation of the thermal conductivities of water, toluene, hexadecane oil, and pure perfluoro-n-hexane (FC72) with temperature, as measured by Yang and Han [8].

3. Thermal Conductivity of Solids

The thermal conductivity of solids (metals and semiconductors) has a more complex behavior than those of liquids. Starting from T = 0 K, the thermal conductivity rises to a maximum k_{max} in the range 10 to 50 K in the case of metals, and then decreases rapidly, finally leveling off at about 300 K for typical metals. The maximum value generally occurs at a higher temperature for non-metals, since this behavior is highly dependent on the purity of the material, and the maximum thermal conductivity may change by an order-ofmagnitude with a small change in the impurity level. It is found that the temperature T_{max} corresponding to the thermal conductivity peak k_{max} can be significantly increased by the impurity level of the material [9], hence a lower T_{max} corresponds to a material with a higher purity. Thus, it is difficult to treat the low-temperature thermal conductivity as a true property of the material.

Heat conduction in pure metals actually results from the random motion of free electrons in the system, which carry thermal energy from one location to another. In dielectric solid materials (such as insulators and semiconductors) heat is conducted via the vibration of atoms, and the atoms are bonded to each other through interatomic force interactions.

Each atom in a solid vibrates around its equilibrium position, and the atoms near the hot side will have large vibrational amplitudes, which will affect the atoms on the other side of the system. The vibration of any one atom can cause the vibration of the entire system by creating lattice waves. A phonon is defined as a quasiparticle with an energy equal to that of all of the vibration modes of the lattice [10]. Heat in a dielectric crystal is conducted by such a phonon gas, similar to that in a box of gas molecules. The collision of phonons is due to the interaction of phonon waves. In impure metals or disordered alloys, the electron mean free path is reduced by collisions with impurities, and the phonon contribution may be comparable with the electronic contribution. In semiconductors at low temperatures where they are not electrical conductors, heat is transported only by phonon thermal conduction.

A phonon gas model of heat conduction is adopted in insulators or semiconductors in the lower temperature range. When the temperature is above the Debye temperature, defined by

$$T_\theta = \frac{\hbar v_s}{k_B}\left(\frac{6\pi^2 N}{V}\right)^{\frac{1}{3}}, \tag{2}$$

where V is the volume of the specimen, k_B is the Boltzmann constant (1.38×10^{-23} JK^{-1}), N is the number of atoms in the specimen and v_s is the speed of sound, semiconductors start to generate current and phonon thermal conduction is accompanied by a significant fraction of electron thermal conduction. Therefore, the thermal conductivity of semiconductors is generally evaluated by use of the additive law

$$k = k_e + k_{ph}, \tag{3}$$

where k_e and k_{ph} are the thermal conductivities due to electron and phonon heat transport, respectively. The unified temperature-dependence of thermal conductivity for metals and insulators or semiconductors is deduced from the well-known Debye formula for thermal conductivity as [9]

$$k_e = \frac{1}{3} C_v \ell_e v_e B_1 B_2 = \frac{1}{3} C_v \tau_e v_e^2 B_1 B_2, \tag{4}$$

where C_v is the heat capacity per unit volume, u_e is the velocity of the electrons, ℓ_e is the mean free path of the electrons, τ_e is the relaxation time of the electrons, B_1 is a function of the electron scattering by the sample boundary and B_2 is a function of the electron scattering by defects and other electrons. An identical expression for k_{ph} may also be applied, with an obvious change in notation.

The electron/phonon mean free path ℓ is determined by two processes; geometrical scattering and scattering by other electrons or phonons [11]. When the mean free path becomes comparable with the size of the nanoparticles at low temperature, the size effect dominates which causes a reduction of the thermal conductivity of the solid.

Kittel [11] explains the temperature-dependence of the phonon thermal conductivity from the theory of anharmonic coupling. If the force between atoms was purely harmonic, there would be no mechanism for collisions between different phonons, and the mean free path is solely based on collisions between a phonon and a crystal boundary or lattice imperfections. There are situations where these effects are dominant. At temperatures higher than the Debye temperature T_ϑ, the determining factor for the thermal conductivity is phonon- phonon scattering. With anharmonic lattice interactions, there is a coupling between different phonons which limits the value of the mean free path. This theory predicts that £ is proportional to 1/T at high temperatures, in agreement with many experiments [11]. At low temperatures, the only temperature-dependent term on the right hand side of Eq. (3) is C_v, the heat capacity per unit volume, which varies as T^3 [11].

Table 1. Thermal conduction mechanism in solids.

Temperature range	Electron/phonon scattering by boundary (B_1)	Electron/phonon scattering by defect (B_2)	Electron/phonon scattering by electron/phonon
$0 < T < 10K$	√	-	-
$10 < T < T_{ph}$	-	√	√
$T > T_{ph}$	-	-	√

Avsec and Oblak [4] express the phonon contribution to the thermal conductivity of an isotropic solid as the integral of the product of the phonon mean free path £, heat capacity per unit volume of a single phonon mode C_v and phonon density of states F over the acentric factor w via

$$k_{ph} = \frac{N v_g}{3V} \int_0^{\omega_{max}} \ell(\omega) C_v(\omega) F(\omega) d\omega, \tag{5}$$

where w_{max} is the phonon spectrum width, v_g is an average phonon group velocity, and the ratio N/V is the number of atoms per volume. For the low and high temperature region where the temperature is lower or higher than the Debye temperature T_ϑ, they used the following solution for the thermal conductivity of an isotropic solid

$$k_{ph} = \begin{cases} k_0 \exp\left(-\frac{T_\theta}{T}\right), & \text{if } T < T_\theta, \\ \frac{BM\Omega_a^{1/3} k_B^3 T_\theta^3}{(2\pi)^3 \hbar^3 \gamma_G^2 T}, & \text{if } T > T_\theta, \end{cases} \tag{6}$$

where B is a dimensionless constant, Ω_a is atomic volume, M is the molecular mass and γ_G is the Grüneisen constant.

Okhotin *et al.* [9] introduce a characteristic temperature T_{ph}, defined as a certain temperature above which electron scattering by defects of any kind exert practically no effect on the thermal conductivity. They outlined the thermal conduction mechanisms in different temperature ranges, as summarized in Table 1. The Dulong-Petit law is a chemical law which offers fairly good predictions for the specific heat capacity of solids with relatively simple crystal structures at high temperatures. It states that the dimensionless heat capacity $C_v/(Nk_B)$ is equal to 3, where N is the number of identical oscillators. Above T_{ph}, scattering by defects of any kind exerts practically no effect on the thermal conductivity because the product of the heat capacity ($C_v = 3k_BN$ by the law of Dulong and Petit [11]) and the terms allowing for election/phonon scattering by boundaries and defects (B_1 and B_2) disappear. Their ideal gas model for each temperature-dependant factor in Eq. (4) is listed in Table 2. Table 2 represents Okhotin *et al*'s complete theory; from the temperature ranges given in Table 1, the expressions may be simplified [9].

Table 2. Temperature-dependent factors in Okhotin *et al.*'s model [9]

	Phonon gas model	Electron gas model
C_v	$9R\left(\frac{T}{T_\theta}\right)^3\int_0^{\frac{T_\theta}{T}}\frac{x^4e^x}{(e^x-1)^2}dx$	$9R\left(\frac{T}{T_\theta}\right)^3\int_0^{\frac{T_\theta}{T}}\frac{x^4e^x}{(e^x-1)^2}dx$
B_1	$1-\exp(-aT^b)$, $a=(4\sim 50)\times 10^{-3}$, $b=1.5\sim 3.25$	$1+a_1T^{-b_1}\exp\left[-c_1\left(\frac{T}{T_{ph}}\right)^3\right]$. The values of a_1, b_1 & c_1 for some materials can be found in [9]
B_2	$\exp(a\pm bT+cT^2)$, $a=0.2\sim 0.7$, $b=(5\sim 10)\times 10^{-3}$, $c=(2.5\sim 10)\times 10^{-4}$	$\left\{\exp\left[a_2+b_2\left(\frac{T}{T_{ph}}\right)+c_2\left(\frac{T}{T_{ph}}\right)^2\right]\right\}^{-1}$. The values of a_2, b_2 & c_2 for some materials can be found in [9]
τ	$\frac{\left(\frac{k_BT}{\hbar}+\frac{k_BT_{max}}{\hbar}\right)^{-1}}{\exp\left(\frac{T_{max}}{T}\right)-1}$	$\frac{\left(\frac{k_BT}{\hbar}+\frac{k_BT_{max}}{\hbar}\right)^{-1}}{\exp\left(\frac{T_{max}}{T}\right)+1}$
u	$\sqrt{\frac{v_s(6\pi^2)^{\frac{1}{3}}}{r_c m_{ph}}}$	$\frac{k_BTm_e}{\pi\hbar^3\rho v_s^2}\left[\frac{3}{4m_e}\left(\frac{\hbar}{r_c}\right)^2\left(\frac{T}{T_{ph}}\right)^{-n}\right]^2$ For most metals $n=0.3$.

A reduction in material dimensionality can cause a change in the phonon dispersion, phonon density of states, phonon speed and phonon relaxation time. These factors may lead to significant changes in the magnitude and temperature variation of the thermal conductivity. For example, a carbon nanotube is considered a rolled-up graphene sheet, and the high thermal conductivity of a nanotube is believed to be associated with the large phonon mean free paths in the system [12], which is limited by scattering from sample boundaries, point defects and phonon-phonon umklapp scattering processes [13]. This is obvious from Eq. 4, which shows that $k \sim C_v u \ell$. Figure 2 shows the temperature-dependence on the thermal conductivity of some metals, semiconductors [14,15], a (10,10) carbon nanotube calculated from molecule dynamics simulation [12] and a multiwall carbon nanotube (MWCNT) bundle [13]. The calculation for the (10,10) carbon nanotube by Berber *et al.* [12] suggests that at room temperature the thermal conductivity of the nanotube is 6,600 $Wm^{-1}K^{-1}$, exceeding the thermal conductivity of 3,320 $Wm^{-1}K^{-1}$ for near-isotropic pure diamond. Kim *et al.*'s [13] study indicates that the thermal conductivity of a MWCNT bundle decreases with the diameter of the MWCNT bundle, with a thermal conductivity of about 3,000 $Wm^{-1}K^{-1}$ at room temperature.

Studies of the thermal conductivity of nanofluids are normally performed in a temperature range of 278 K to 323 K, although data over a much larger temperature range is valid for the pure components of the nanofluid. Figure 3 shows the variation of the thermal conductivity of nanoparticle materials normally used in nanofluids (Al_2O_3, Au and CuO) with temperature [7, 15]. It is obvious that the thermal conductivity of the nanoparticles decreases with increasing temperature in the temperature range 200 K to 500 K.

4. EXPERIMENTAL DATA OF NANOFLUIDS

An important finding concerning nanofluids was the discovery of a significant enhancement and a strong temperature effect on their thermal conductivity. Experimental investigations on the effective thermal conductivity of nanofluids have examined a wide variety of different solvent-particle systems.

For a 1 %-loaded Al_2O_3/water nanofluid at 21 °C there is a 2 % enhancement, while at 51 °C there is a 10.8 % enhancement; for a 4 %-loading the enhancement goes from 9.4 % to 24.3 %. For a CuO/water nanofluid the results were an enhancement from 6.5 % to 29 % and 14 % to 36 % corresponding to a 1 %-load and 4 %-loading, respectively [16]. A similar increase in the thermal conductivity of CuO and Al_2O_3 nanofluids was also observed by Li and Peterson [17]. Patel *et al.* [18] observe a 5 % - 21 % increase in thermal conductivity for a temperature range of 30-60 °C at a loading of 0.00026 % Au in water. There have been several other experimental studies on various types of nanofluids [8, 19–24], and the maximum enhancements obtained are summarized in Table 3.

Kim *et al.* [24] investigate the effect of the morphology of carbon nanotubes (CNT) on the effective thermal conductivity of suspensions as a function of the volume fraction, aspect ratio (defined as the ratio of the length to the diameter of the nanotubes) and temperature. The measured data was classified into three different groups: CNT A (functionalized SWCNT/H_2O), B (SWCNT/H_2O) and C (MWCNT/H_2O). Their findings indicate that the degree of thermal conductivity enhancement increases with the aspect ratio of the carbon nanotube, volume fraction and temperature. Wen and Ding's [23] finding concerning the effect of tempearture on the thermal conductivity of carbon nanotube-nanofluids is similar to Kim *et al.* 's results

[24]; the effective thermal conductivity of nanofluids increases with temperature, but they discovered a nonlinear dependence on temperature and nanotube concentration.

In contrast to Kim *et al.* [24] and Wen and Ding's [23] findings concerning non-spherical nanoparticle suspensions, Yang and Han [8] experimentally found a thermal conductivity enhancement for Bi_2Te_3 nanorods in FC72 and in oil nanofluids which decreases with increasing temperature, and proposed the relation $k_{eff} \propto T^{-a}$, where $a = 1.68 - 1.95$. They analyzed the thermal diffusion coefficient of the nanorods and the thermal conductivity predicted by the effective medium theory versus the particle aspect ratio. They found that the diffusive conduction benefits from the increased aspect ratio of the nanoparticles. Although the long fibres are less mobile than spherical particles in a nanolfluid, and hence the effects of Brownian motion are reduced, they can promote heat transfer by providing a rapid and longer heat flow path. Their analysis suggests that the conduction-based mechanism should be primarily responsible for the thermal conductivity enhancement in nanofluids containing nanorods.

Table 3. Experimental studies of the temperature effect on the thermal conductivity of nanofluids. Only the maximum enhancements obtained are shown.

Author	Nanofluid	Temp. (°C)	Enhancement
Das *et al.* [16]	CuO/H_2O	51	36 % at $\phi = 4$ %
Patel *et al.* [18]	Au thiolate/H_2O	60	21 % at $\phi = 0.00026$ %
Kumar *et al.* [19]	Au/H_2O	50	9 % at $\phi = 0.00026$ %
Chon *et al.* [20]	Al_2O_3/H_2O	71	29 % at $\phi = 4$ %
Zhang *et al.* [21]	Al_2O_3/H_2O	10, 30, 50	20 % at $\phi = 15$ %
Zhang *et al.* [22]	CNT/H_2O	23	40 % at $\phi = 0.9$ %
Li *et al.* [17]	CuO/H_2O (distilled)	34	52 % at $\phi = 6$ %
Yang & Han [8]	Bi_2Te_3/FC72	3	7.7 % at $\phi = 0.8$ %
Wen *et al.* [23]	MWCNT/H_2O	45	31 % at $\phi = 0.84$ %
Kim *et al.* [24]	MWCNT/H_2O	44	37 % at $\phi = 1$ %

5. Present Models to Account for the Temperature Effect

From the classic effective medium theories the definition for the thermal conductivity of a two-phase mixture is

$$\frac{k_{eff}}{k_f} = 1 + \frac{n(k_p - k_f)\phi}{k_p + (n-1)\,k_f - (k_p - k_f)\phi}, \tag{7}$$

where n is an empirical constant for mixtures containing particles of arbitrary shape. For Maxwell's formula $n = 3$, while for that of Hamilton and Crosser we have $n = 3/\Phi$ where Φ is the sphericity defined as the ratio of the surface area of a sphere, with a volume equal to that of a particle, to the surface area of the particle.

The Hamilton and Crosser model yields excellent results for particles larger than 13 nm, but for smaller particles the theory breaks down with a deviation of more than 100 % compared to experimental results. This is primarily because the theory possesses no dependence on particle size and surface phenomena.

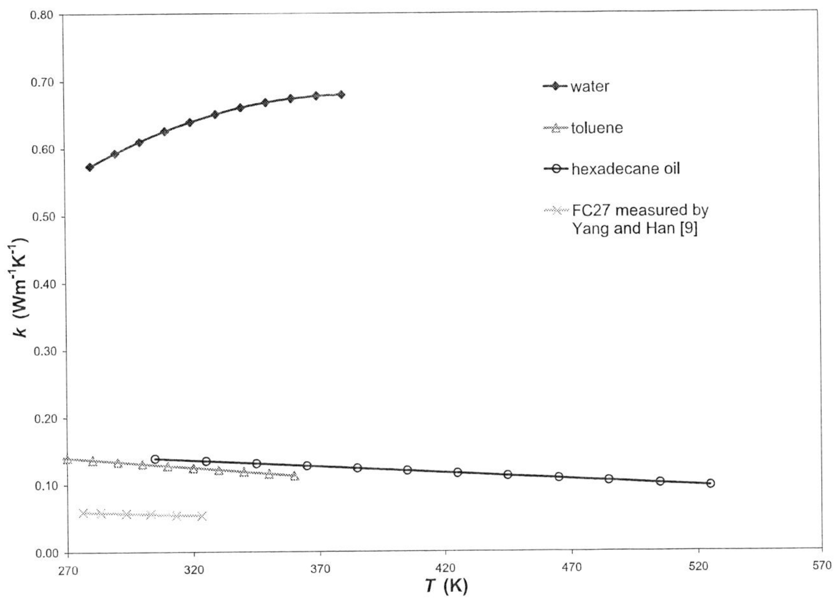

Figure 1. Thermal conductivity of pure fluids as a function of temperature [7,8].

It is important to note that the bulk thermal conductivity of metals and ceramics such as Au, Al_2O_3 and CuO and fluids such as toluene and oil decrease with increasing temperature at 0 °C to 100 °C, but water does not (see Fig. 1). If we use a linear function $k_p(T) = A_p - B_pT$ to represent the thermal conductivity of a nanoparticle and $kf(T) = Af + BfT$ for that of the base fluid, Maxwell's formula implies that the effective thermal conductivity of nanofluids is a monotonically decreasing function with respect to T for volume fraction $\varphi \leq 10$ %. This behavior is in contrast to most of the experimental findings shown in Figs. 4 and 5. As a result, it is clear that the effective thermal conductivity of nanoparticle suspensions is not a simple function of the combined thermal properties of the base fluid and the nanoparticle. Maxwell's rule of mixtures is not adequate to describe the nanoparticle contribution to the thermal conductivity.

Various approaches have been developed in the past ten years based on the phenomenology and morphology of the agglomerates of observed nanoparticles. Several theoretical descriptions have been proposed that incorporate a variety of fundamental mechanisms as explanations of the thermally-driven particle flow in liquids. Many researchers believe that the Brownian motion of the nanoparticles constitutes a key mechanism for the thermal conductivity enhancement with increasing temperature and decreasing nanoparticle size.

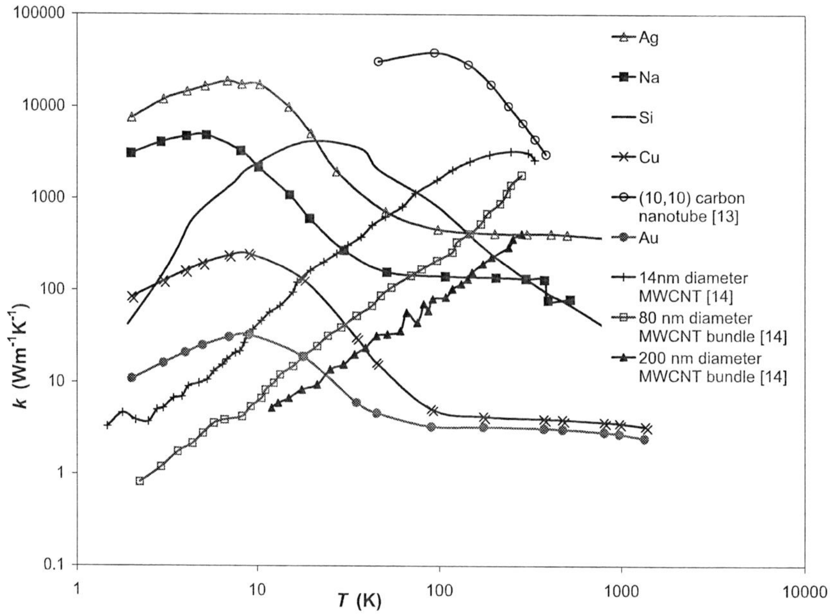

Figure 2. Thermal conductivity of some metals and semiconductors [12–15].

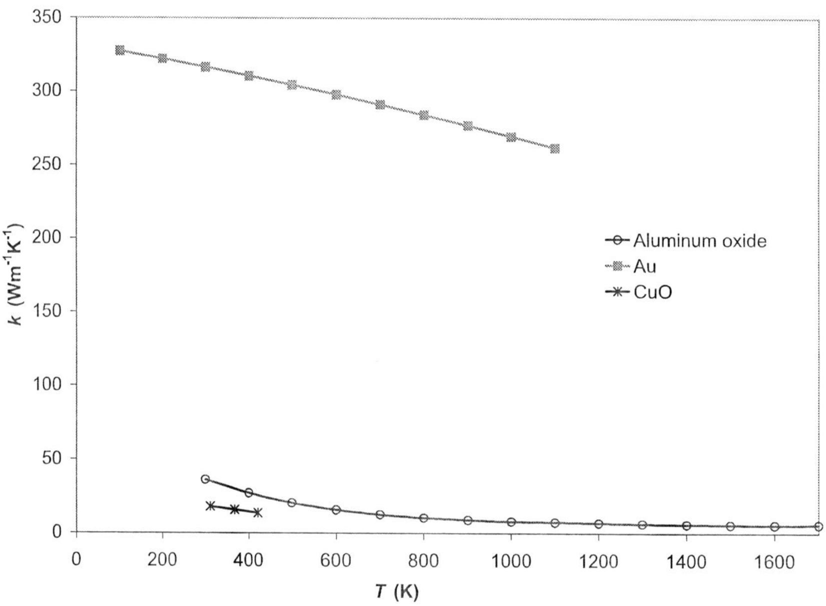

Figure 3. Thermal conductivity of Au, Al_2O_3 and CuO [7, 15].

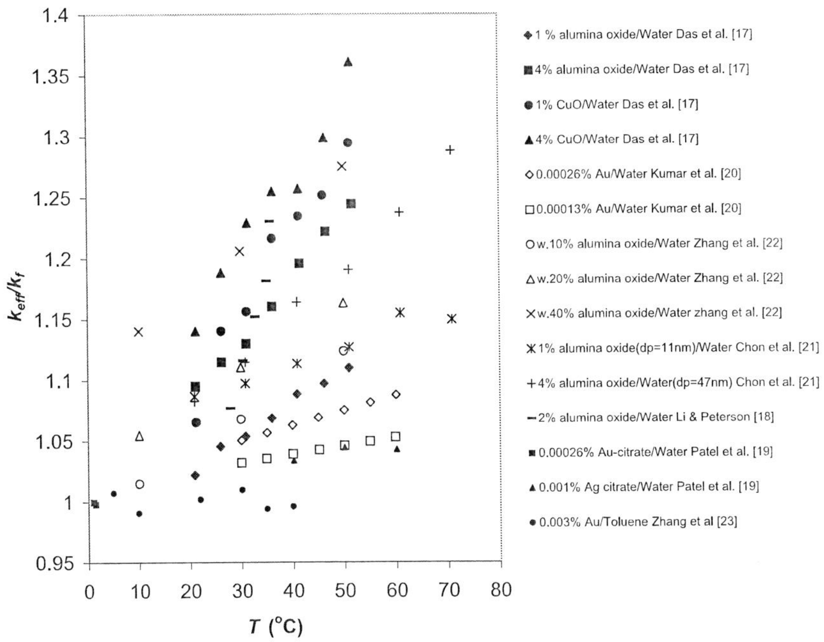

Figure 4. Measured thermal conductivity of spherical particle suspensions versus temperature.

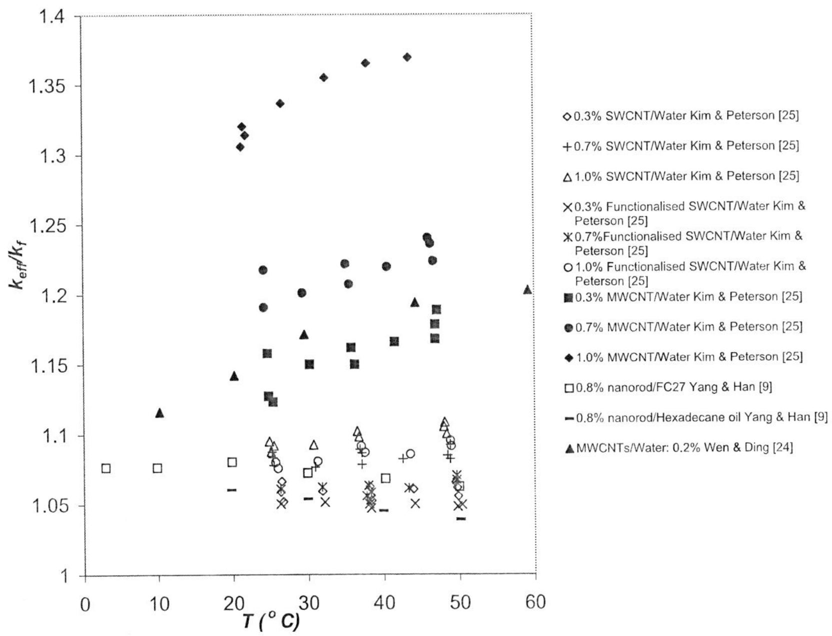

Figure 5. Measured thermal conductivity of non-spherical particle suspensions versus temperature [8,24].

The effects of Brownian motion on small particles are more pronounced at high temperature, as has been observed experimentally [16–21]. Based on the theory of Brownian motion of particles suspended in a fluid, many researchers propose a microscopic convective-current model which takes into account the dependence of temperature on the effective thermal conductivity of nanofluids, along with the traditional effective medium theories. The motion of particles inside a fluid is modelled as low Reynolds number flow past a sphere. In the postulated micro-convection model, four parameters are used: the Reynolds number Re = $u_p d_p/v$ which represents the mobility of the nanoparticles, the Prandtl number Pr = v/αf = μCp,f/kf, Nusselt number Nu = hd_p/kf, and Péclet number Pe = updp/αf = RePr. The classical solution for flow past a sphere is given by [25]

Nu = 2.0 + 0.5RePr + O(Re2Pr2) = 2.0 + 0.5Pe + O(Pe2). (8)

Table 4. Overview of development of models including temperature effects for evaluation of the thermal conductivity of nanofluids.

Model type	Formula
Analytical Wang *et al.* [36]	Nanosphere (1999) $k_{Brownian} = k_r + k_t$ $k_r = \left[1.176\left(\frac{k_p - k_f}{k_p + 2k_f}\right)^2 + 5\left(0.6 - 0.028\frac{k_p - k_f}{k_p + 2k_f}\right)\right]\phi k_f Pe_f^{1.5}$ $k_t = (0.0556Pe_t + 0.1649Pe_t^2 - 0.0391Pe_t^3 + 0.0034Pe_t^4)k_f$ $Pe_f = \frac{r_p^2 \gamma \rho_f C_{p,f}}{k_f}$ $Pe_t = \frac{u_p \rho_f C_{p,f} L \phi^{0.75}}{k_f}$ $L = r_p\left(\frac{4\pi}{3\phi}\right)^{\frac{1}{3}}$
Analytical Xuan *et al.* [42]	Nanocluster (2003) $k_{eff} = k_{static}(\text{Maxwell}) + \frac{1}{2}\rho_p c_p \phi \sqrt{\frac{k_B T}{3\pi r_c \mu}}$
Analytical Jang *et al.* [27,28]	Nanosphere (2004) $k_{eff} = (1-\phi)k_f + \phi k_p + \frac{3c_1 \phi d_f k_f Re_p^2 Pr^2}{d_p}$ $Re_p = \frac{u_p d_p}{\nu}$ $u_p = \frac{k_B T}{3\pi\mu d_p \ell}$ $C_1 = 18 \times 10^6$
Analytical Kumar *et al.* [19]	Nanosphere (2004) $k_{eff} = k_f + \frac{c u_p \phi r_f}{(1-\phi) r_p}$ $u_p = \frac{2k_B T}{\pi\mu d_p^2}$ $c \approx \frac{N\ell\hat{C}_v}{3}$

Table 4. Continued

Model type	Formula
Analytical Ren *et al.* [29]	Nanosphere (2005) $k_{eff} = k_f \left[1 + F(Pe) + 3\Theta\phi_T + \frac{3\Theta^2\phi_T^2}{1-\Theta\phi_T}\right]$ $F(Pe) = 0.0556Pe + 0.1649Pe^2 - 0.0391Pe^3 + 0.0034Pe^4$ $Pe = \frac{u_p L \phi_T^{0.75}}{\alpha_f}$ $L = (r_p + d)\sqrt[3]{\frac{4\pi}{3\phi_T}}$ $\Theta = \frac{\beta_{lf}\left[\left(1+\frac{d}{r_p}\right)^3 - \beta_{pl}\beta_{fl}\right]}{\left(1+\frac{d}{r_p}\right)^3 + 2\beta_{lf}\beta_{pl}}$ $\beta_{ij} = \frac{k_i - k_j}{k_i + 2k_j}$ $u_p = \sqrt{\frac{3k_B T}{m_c}}$ $\phi_T = \phi\left(1 + \frac{d_l}{r_p^3}\right)$
Analytical Xu *et al.* [37]	Nanoparticle cluster (2006) $k_{eff} = k_{static}$ (Hamilton & Crosser) $+ k_{Brownian}$ $k_{Brownian} = \frac{ck_f Nu(2-D_f)\, d_f \left[\left(\frac{d_{p,max}}{d_{p,min}}\right)^{1-D_f} - 1\right]^2}{Pr d_{p,min}(1-D_f)\left[\left(\frac{d_{p,max}}{d_{p,min}}\right)^{2-D_f} - 1\right]}$ $D_f = 2 - \frac{\ln(\phi)}{\ln\left(\frac{d_{p,min}}{d_{p,max}}\right)}$ $c = O(10^2)$
Analytic Sabbaghzadeh *et al.* [30]	Cylindrical nanoparticles (2007) $k_{eff} = k_f[1 - \phi(1+M)] + \phi(k_p + k_l M) + \frac{k_f d_f \phi(1+M)}{2Pr_f(r_p+d)}\left(0.35 + 0.56Re^{0.52}\right)Pr_f^{0.3}$ $Re = \frac{u_p L}{\nu_f}$ $u_p = \sqrt{\frac{3k_B T}{m_p}}$ $L = \sqrt[3]{\frac{\pi r_p^2 \ell M^2}{\phi}}$ $M = \left(\frac{d}{r_p} + 1\right)^2 - 1$

Table 4. Continued

Model type	Formula
Semi-empirical Shukla & Dhir [38]	Nanosphere (2005) $k_{keff} = k_{static}(\text{Maxwell}) + \frac{c\phi(T - T_0)}{\mu r_p^4}$ c, T_0 are experimentally determined constants.
Semi-empirical Prasher [31]	Nanosphere (2005) $k_{eff} = k_f \left(1 + A Re^m Pr^{0.333}\phi\right) \left[\frac{(1+2a) + 2\phi(1-a)}{(1+2a) - \phi(1-a)}\right]$ $a = \frac{2R_b k_f}{d_p}$ $u_p = \sqrt{\frac{3k_B T}{m_p}}$ $A = 40,000, m = 2.5 \pm 15\ \%$
Semi-empirical Patel *et al.* [26]	Nanosphere (2005) $k_{eff} = k_f + \frac{k_p d_f \phi}{d_p(1-\phi)} \left(1 + \frac{c u_p d_p}{\alpha_f}\right)$ $u_p = \frac{2k_B T}{\pi \mu d_p^2}$ $c = 25,000$
Analytic and Correlation Koo *et al.* [39]	Nanosphere (2004) $k_{eff} = k_{static}(\text{Maxwell}) + k_{Brownian}$ $k_{Brownian} = (5 \times 10^4)\, \beta_1 \phi \rho C_v \sqrt{\frac{k_B T}{\rho d_p}}$ $\times [(-6.04\phi + 0.4705)T + (1722.3\phi - 134.63)]$ $u_p = \sqrt{\frac{18 k_B T}{\pi \rho d_p^3}}$ For Au-citrate, Ag-citrate & CuO: $\beta_1 = 0.0137(100\phi)^{-0.8229}, \phi < 1\ \%$ For CuO: $\beta_1 = 0.0011(100\phi)^{-0.7272}, \phi > 1\ \%$ For Al_2O_3: $\beta_1 = 0.0017(100\phi)^{-0.0841}, \phi > 1\ \%$
Correlation Chon *et al.* [20]	Nanosphere(2005) $\frac{k_{eff}}{k_f} = 1 + c\left(\frac{1}{d_p}\right)^{0.369} \left(\frac{T^{1.2321}}{10^{\frac{2.4642B}{T-C}}}\right)$ For water: $B = 247.8$, $C = 140$ c is an empirical constant

In the analysis of the Stokes flow regime, there are two approaches to evaluate the average velocity of the nanoparticles, u_p. One approach is estimated from the Stokes-Einstein diffusion theory, assuming that a nanoparticle freely moves over a distance equal to the mean-free path of a base fluid molecule ℓ_f. Thus $u_p = D_0/\ell_f = k_B T/(6\pi\mu d_p \ell_f K)$, where D_0 is the particle thermal diffusion coefficient and K is a shape factor (K = 1/2 for spheres). Another approach is the root-mean-square velocity from the kinetic theory of gases, u_p = $3k_BT/m_p$, where m_p is the mass of a nanoparticle. Finally, a direct expression for the effective thermal conductivity of nanofluids, k_{eff} as a function of T, is obtained via the mathematical relation $u_p = u_p(T)$. Table 4 gives a chronological account of some of the prediction methods developed over the past decade concerning the temperature-dependence on the thermal conductivity enhancement. These existing theoretical models can be categorized into three groups: analytical; semi-empirical; correlation.

5.1. 1D Parallel Process-Based Heat Flow Paths

One group of researchers [19,26–31] begins their analysis by considering one-directional heat transfer over the base fluid and the solid. Fourier's law of heat conduction, and the heat transfer micro-convection from the particles is applied to the liquid and the solid. The effective thermal conductivity of the nanofluid is subsequently derived by substituting relevant parameters from thermal statistical theory and Stokes flow theory. These sort of models involve process-based techniques. This technique is quite useful for examining what physical processes cause heat flow in a nanofluid medium.

In Jang and Choi's model [27,28] based on kinetics, Kapitza resistance and convection, the whole process of energy transport is assumed to consist of three additive modes; (a) collisions between base fluid molecules; (b) thermal diffusion in nanoparticles in fluids; (c) collision of nanoparticles with each other due to Brownian motion; and (d) thermal interaction of nanoparticles with the base fluid (nanoconvection). In the first mode, the energy carriers (the base fluid molecules) travel freely only over the mean free path ℓ_f before they collide, and the net energy flux q across a plane at z is given by

$$q = -\frac{1}{3}\ell_f C_{v,f} u_f (1-\phi)\frac{dT}{dz} = -k_f(1-\phi)\frac{dT}{dz}, \tag{9}$$

In the second mode, the net energy transport is carried by electrons or phonons in the suspended nanoparticles, so that the energy flux is given by

$$q = -\frac{1}{3}\ell_{nano} C_{v,nano} u_{nano}\phi\frac{dT}{dz} = -k_{nano}\phi\frac{dT}{dz}, \tag{10}$$

where k_{nano}, ℓ_{nano} and u_{nano} are the thermal conductivity of the suspended nanoparticles, the mean free path and the mean speed of the electrons or phonons, respectively. The Kapitza resistance is considered in order to evaluate the thermal conductivity of the suspended nanoparticle and the effect of nanoparticle size, and is given by

$$k_{nano} = \beta k_p, \tag{11}$$

where $\beta \sim O(10^{-2})$ is a constant related to the Kapitza resistance per unit area.

The third mode concerns the collision of nanoparticles with each other via Brownian motion. The translational speed of the of the nanoparticle may be calculated from kinetic theory, and from an order-of-magnitude analysis it may be deduced that this mode is smaller than the other three modes and may be neglected.

The last mode of the energy transport is the thermal interactions of the randomly moving nanoparticles with the base fluid molecules by thermally-induced fluctuations. They postulate that the Brownian motion of the nanoparticles in nanofluids produces convection- like effects at the nanoscale. Hence the Brownian-motion-induced nanoconvection energy flux for the last mode may be defined by

$$q = -h\delta_T \phi \frac{dT}{dz}, \tag{12}$$

where δ_T is the thickness of the thermal boundary layer, estimated from $\delta_T \sim \delta/Pr \sim d_f/Pr$, where δ is the the hydrodynamic boundary layer thickness. For typical nanofluids, the Reynolds and Prandtl numbers are of the order of 1 and 10, respectively, and the Nusselt number for the flow past a sphere, Eq. (7), is simplified to $Nu \sim (RePr)^2$. The heat transfer coefficient for the flow past nanoparticles is then defined by

$$h \sim \frac{k_f}{d_p}(RePr)^2. \tag{13}$$

The expression for the thermal conductivity of nanofluids k_{eff} is derived as a function of k_p, k_f, φ, d_p, d_f, Re and Pr, as shown in Table 4. As the particle size is decreased or the temperature is increased, the random motions become larger in magnitude and convection-like effects dominate. This model is able to predict a particle size- and temperature-dependent thermal conductivity.

Sabbaghzadeh *et al.* [30] follow the same approach as Jang and Choi's model [27,28] and in addition consider the effects of liquid layering and the nanolayer structure [32] for the case of a cylindrical nanoparticle suspension. They assume a linear distribution of the thermal conductivity of the nanolayer and a fixed nanolayer thickness. The heat transfer coefficient h is given by Fand [33] for a flow direction perpendicular to the cylinder axes where $10^{-1} < Re < 10^5$ as

$$h = \frac{k_f}{2(r_p + d)}(0.35 + 0.56Re^{0.52})Pr^{0.3}, \tag{14}$$

where d is the nanolayer thickness, $Re = u_p L/v$, L is the specific length of the cylindrical shape calculated from $L = \sqrt[3]{\pi r_p^2 l (r_p + d)^2/(r_p^2\phi)}$ and l is the length of the cylinder. This process links k_{eff} with temperature through the Reynolds number. Their model successfully shows that the effective thermal conductivity increases with decreasing nanoparticle diameter if the nanolayer thickness increases, and increases with the volume fraction of the nanoparticle. However, they didn't plot the variation of the effective thermal conductivity with temperature. We found that the contribution from the thermal interaction of dynamic complex nanoparticles with the base fluid molecules is $O(10^{-5})$ for their model, which is very small compared to the other processes. As a result, the relationship between k_{eff} and temperature T is insignificant.

Kumar *et al.* [19] divide the heat transfer in nanofluids into two portions only, namely by the fluid and the particles. From the one-dimensional heat flux of the parallel model, they found

$$q = -k_p A_p \left(\frac{dT}{dx}\right)_p - k_f A_f \left(\frac{dT}{dx}\right)_f, \tag{15}$$

Assuming that the nanofluid system is in thermodynamic equilibrium such that

$$\left(\frac{dT}{dx}\right)_p = \left(\frac{dT}{dx}\right)_f = \left(\frac{dT}{dx}\right), \tag{16}$$

and that the ratio of the heat transfer areas A_p/A_f is proportional to the total surface area of the nanoparticles and the liquid species per unit volume of the suspension, the effective thermal conductivity of the nanofluid is expressed as

$$q = -k_{eff} A_m \frac{dT}{dx}, \tag{17}$$

where

$$k_{eff} = k_f \left[1 + \frac{k_p \phi r_f}{k_f (1-\phi) r_p}\right]. \tag{18}$$

Kumar *et al.* [19] also postulated that the effective thermal conductivity of the particle k_p is directly proportional to its mean velocity according to kinetic theory, $k_p = cu_p$, where c is a constant. The velocity of the particles u_p is taken to be that due to Brownian motion at a given temperature of the suspension and may be calculated using the Stokes-Einstein formula $u_p = 2k_B T/(\pi \mu d_p^2)$. After substituting $k_p = cu_p$ into Eq. (17), a comprehensive model is derived which explains the enhancement of the thermal conductivity of a nanofluid with respect to variations in particle size, particle volume fraction and temperature. However, the assumption $k_p = cu_p$ results in a zero thermal conductivity of a nanoparticle when the mean velocity of the particle is zero - this is clearly incorrect, since the thermal conductivity of the particle should be equal to its own property value.

To circumvent this error, Patel *et al.* [26] improved Kumar *et al*'s model [19] by introducing the concept of micro-convection. They assumed that heat flows through the nanofluid via conduction through both the solid and liquid and advection from the Brownian motion of the particles. The motion of particles inside the liquid is modelled as flow past a sphere, and from the 1D parallel model, the overall heat transfer is written as

$$\frac{k}{k_f} = (1 + CRe^m Pr^{0.333}\phi)\left[\frac{(1+2a)+2\phi(1-a)}{(1+2a)-\phi(1-a)}\right]. \tag{19}$$

As the size of the the particles is of the order of 10^{-7} to 10^{-8} m and the Brownian motion velocity is also very low, Pe^2 is negligible compared to Pe. As the conduction through the liquid and the particle conduction is already separately accounted for in q_f and q_p, Patel *et al.* [26] simplified the classical solution of the Nusselt number for the flow past a sphere Eq. (8) via

$$Nu = 0.5Pe. \quad (20)$$

They postulate that the thermal conductivity of the fluid is $k_f = ck_p$, where c is constant, which circumvents the zero thermal conductivity of the nanoparticles at T = 0 K in Kumar *et al*'s [19] model. The Nusselt number is then defined as

$$Nu = 0.5Pe. \quad (21)$$

The new semi-empirical model of the effective thermal conductivity is obtained by deriving h from Eqs. (20) and (21) and substituting all relevant parameters for the situation of micro- heat convection of flow over a sphere, yielding

$$\frac{k_{eff} - k_f}{k_f} = \frac{k_p d_f \phi}{k_f d_p (1-\phi)} \left(1 + \frac{c u_p d_p}{\alpha_f}\right), \quad (22)$$

where α_f is the thermal diffusivity of the fluid and the constant c is determined experimentally. The Brownian velocity of the particles u_p, which causes the convection, is calculated the same way as in Kumar *et al.* [19], that is from the Stokes-Einstein formula using the particle diameter d_p as the characteristic length. In their work, c = 25, 000 gives accurate predictions for a wide range of experiments.

Ren *et al.* [29] presented a model that considered four heat transport modes; heat flux through the nanoparticle; heat flux through the nanolayer; heat flux through caused by micro-convection; and heat flux through the base fluid - $q = q_p + q_l + q + q_f$. They assumed a linear linear function for the thermal conductivity of the nanolayer between the values for the nanoparticle and the base fluid, the thickness of which is a variable. Expressions for q_f, q_p and q_l were based on classical theories, while they derived an expression for the heat transfer from micro-convection. In their model q is described as a function of the modified Péclet number $Pe = u_p L / \alpha_f \phi_T^{0.75}$ via

$$q_c = -k_f \nabla T F(Pe), \quad (23)$$

where the function F is given by

$$F(Pe) = 0.0556Pe + 0.1649Pe2 - 0.0391Pe3 + 0.0034Pe4. \quad (24)$$

The specific length L is calculated from L = (r_p + dl) 3 4π/(3 T), and the total volume fraction of the original nanoparticle and the nanolayer is given by T = $(1 + d_l/r_p)^3$. The particle velocity is estimated from the kinetic theory of a monatomic gas molecule, $u_p = \sqrt{3k_B T/m_c}$, where m is the mass of the complex nanoparticle. The results fit well for ceramic particle suspensions. The model reveals that the enhancement in the effective thermal conductivity increases with the decreasing nanoparticle radius and temperature, due primarily to the contribution of the particle motion.

Prasher *et al.* [31] analyzed convection in the liquid where a sphere is imbedded in a semi-infinite medium. They considered the effects of possible mechanisms for the thermal energy transfer in nanofluids, and an order-of-magnitude analysis demonstrated that local convection caused by the Brownian motion of the nanoparticles is the only important mechanism to explain the thermal conductivity enhancement. The heat transfer coefficient h is calculated from Eq. (8) as

$$h = \frac{k_f}{r_p}\left(1 + \frac{RePr}{4}\right). \tag{25}$$

If a sphere is imbedded in a semi-infinite medium of thermal conductivity k_m, then Nu based on the radius of the particle ($r_p = d_p/2$) is given to be 1, that is $h = k_m/r_p$ according to [34]. Substituting this h into the left side of Eq. (25) gives the new effective thermal conductivity k_m of the fluid due to the micro-convection caused by the movement of a single sphere as

$$k_m = k_f\left(1 + \frac{RePr}{4}\right). \tag{26}$$

The root-mean-square velocity of a Brownian particle from the kinetic theory of gases is used as a particle velocity for modelling the thermal conductivity of nanoparticles, $u_p = \sqrt{3k_BT/(m_p)}$. Substituting the estimated k_m due to the convection contribution into the Maxwell-Garnett model, the theory is extended to predict the thermal conductivity k_{eff} of a nanofluid which includes the conduction contribution of the particles, the thermal interfacial resistance and the Brownian convection contribution from a single nanoparticle, and is given by

$$\frac{k}{k_f} = \left(1 + \frac{RePr}{4}\right)\left[\frac{(1+2a) + 2\phi(1-a)}{(1+2a) - \phi(1-a)}\right], \tag{27}$$

where $a = 2R_bk_f/d_p$ and R_b is the interfacial resistance between the fluid and a nanoparticle. However, their analysis showed that even at very small values of φ the flow field due to two particles will interact, and Eq. (27) needs to be modified to reflect this. Heat transport in a nanofluid systems is similar to the particle-to-fluid heat transfer in fluidized beds. They followed the findings of Brodkey *et al.* [35] for the Nu-correlation for the particle-to-fluid heat transfer in fluidized beds, and proposed a general correlation for h due to the presence of multiple nanoparticles. With two introduced empirical coefficients C and m, this modification leads to

$$\frac{k}{k_f} = (1 + CRe^mPr^{0.333}\phi)\left[\frac{(1+2a) + 2\phi(1-a)}{(1+2a) - \phi(1-a)}\right]. \tag{28}$$

They found that m = 2.5 ± 15 % is the best value for water-based nanofluids, and C = 40, 000 matched experimental data well.

5.2. Models of the Additive Type

In a quiescent suspension, nanoparticles move randomly and carry a relatively large volume of the surrounding fluid with them, which may result in a heat flux between hot and cold regions. A group of researchers [36–3 9] propose that the effective thermal conductivity of nanofluids k_{eff} is composed of the particle's conventional static part k_{static} and a Brownian motion part $k_{Brownian}$, that is

$$k_{eff} = k_{static} + k_{Brownian}. \quad (29)$$

The stationary effective thermal conductivity models assume no bulk motion of the fluid, and k_{static} is predicted from traditional effective medium theories such as those of Maxwell [2] and Hamilton and Crosser (HC) [3]. In a nanoparticle-fluid mixture, microscopic forces can be significant; forces such as the Van der Waals force, the electrostatic force resulting from the electric double layer at the particle surface, the stochastic force that gives rise to the Brownian motion of particles, and of course hydrodynamic forces. The motion of the particles and the fluid are induced and affected by the collective effect of these forces. At present, these forces have not been calculated accurately, and only $k_{Brownian}$ due to the convective effect from the particle Brownian motion have been studied in detail.

Wang *etal.* [36] propose that $k_{Brownian}$ consists of the rotational and translational motions of a spherical particle k_r and k_t, respectively. They can be estimated from provided functions of k_p, k_f, φ and the modified Peclet numbers Pe_f and Pe_t. Pe_f is defined as $Pe_f = r^2_p \gamma \rho_f C_{p,f}/k_f$, where γ is the velocity gradient calculated from the mean Brownian motion velocity and the average distance between particles, and ρ_f and $C_{p,f}$ are the base fluid density and specific heat, respectively. Pe_t is defined as $Pe_t = (u_p L \rho_f C_{p,f}/k_f)\varphi^{3/4}$, where $L = (r_p/\varphi^{1/3})(4\pi/3)^{1/3}$. Their calculation reveals that the thermal conductivity increase by Brownian motion is less than 0.5 % up to a volume fraction of 10 % for an Al_2O_3-liquid mixture. They conclude that Brownian motion does not contribute significantly to the energy transport in nanofluids. Thus the variation of the thermal conductivity of nanofluids with temperature is insignificant based on their model.

Xu *et al.* [37] presented a new model by taking into account the fractal distribution of nanoparticle sizes and the heat convection between the nanoparticles and the fluid due to the Brownian motion of the nanoparticles. Combined with the HC model [3] for stationary nanoparticles, they proposed a heat transfer model by micro-convection of all the particles in the fluid. The heat flux due to the Brownian motion of the nanoparticles is expressed as an integral of the heat transfer by a single particle assuming thermal equilibrium, that is

$$\begin{aligned} q_c &= -\int_{d_{p,min}}^{d_{p,max}} hA_p(T_p - T_f)dN \\ &= -\frac{\Delta T}{\delta_T}\int_{d_{p,min}}^{d_{p,max}} hA_p\delta_T dN, \end{aligned} \quad (30)$$

where $d_{p,min}$ and $d_{p,max}$ are the minimum and maximum diameters of the nanoparticle, respectively, h_p is the heat transferred by convection from a single nanoparticle moving in the liquid, A_p is the surface area of a particle and the thickness of the thermal boundary layer is $\delta_T \sim \delta/Pr = 3d_f/Pr$. The fractal power law for the size distribution of the nanoparticles in nanofluids is introduced in the system via

$$-dN = D_f d_{p,max}^{D_f} d_p^{-(D_f+1)} d(d_p) > 0, \tag{31}$$

$Nu = 2.0 + 0.5RePr + O(Re^2Pr^2)$ 2.0 is applied to a single particle, which implies $h = k_f Nu/d_p$. Substituting expressions for h and dN into Eq. (30), the equivalent thermal conductivity contributed by heat convection is approximated by

$$k_{Brownian} = \frac{q_c}{-A\frac{\Delta T}{\delta_T}} = \frac{\int_{d_{p,min}}^{d_{p,max}} hA_p\delta_T dN}{\int_{d_{p,min}}^{d_{p,max}} A_p dN}, \tag{32}$$

where A is the total area of all the nanoparticles. By introducing an empirical constant c for the thickness of the thermal boundary layer such that $\delta_T = c\delta/Pr$ and a viscosity given by $\mu = \mu_0 10^{B/(T-C)}$ into the Prandtl number $Pr = \mu C_p/k_f$, the model shows that the thermal conductivity of a nanofluid due to heat convection by Brownian motion of the nanoparticles is proportional to the temperature and the reciprocal of the nanoparticle size d_p^{-1}, via

$$\frac{k_{Brownian}}{k_f} = c\frac{d_f k_f Nu(2-D_f)\left[\left(\frac{d_{p,max}}{d_{p,min}}\right)^{1-D_f} - 1\right]^2}{10^{\frac{B}{T-C}} A C_p (1-D_f) d_{p,min}\left[\left(\frac{d_{p,max}}{d_{p,min}}\right)^{2-D_f} - 1\right]}. \tag{33}$$

It is found that c is around $O(10^2)$. Their model suggests that the maximum value of the contribution from the heat convection due to the nanoparticle's Brownian motion is about 21 % and 2.1 % for nanoparticles with an average diameter of 5 and 60 nm, respectively. There is a critical concentration of nanoparticles at which $k_{Brownian}$ reaches a maximum value, below which $k_{Brownian}$ increases with concentration and above which $k_{Brownian}$ decreases with concentration. It is found that this critical concentration is 12.6 % for all nanoparticles. It is believed that at higher concentrations, the nanoparticles may easily agglomerate to form clusters, and the strength of the Brownian motion is thus weakened, leading to the decrease in heat convection.

Shukla and Dhir [38] describe the Brownian motion of the nanoparticles using Langevin dynamics. They derived an explicit expression for the thermal conductivity contribution $k_{Brownian}$ for N spherical Brownian particles suspended in a system occupying volume V at an equilibrium temperature T using the Green-Kubo relation

$$k_{Brownian} = \frac{V}{k_B T^2}\int_0^{\infty}\langle J_x(\tau)J_x(0)\rangle d\tau = C\frac{\phi(T-T_0)}{\mu r_p^4}, \tag{34}$$

where C and T_0 are constants to fit the individual solid/liquid combination. T_0 represents a reference temperature, below which the contribution of Brownian motion to the effective thermal conductivity is negligible. In their work $C = 7.0 \times 10^{-36}$ and $T_0 = 294$ K for an alumina/water nanofluid, and $C = 1.5 \times 10^{-33}$ and $T_0 = 273$ K for an gold/water nanofluid. Reasonable agreement with experimental data is obtained for the variation of the effective thermal conductivity enhancement with temperature Koo and Kleinstreuer [39] propose a model where k_{static} is calculated from Maxwell theory. To derive $k_{Brownian}$, they consider two nanoparticles with translational time- averaged Brownian velocities u_p in two different temperature fields of extent l, where l is the average distance a particle travels in one direction without changing its direction due to the Brownian motion of the other particles. The relationship between the translational time-average speed u_p and temperature T is described by the real-gas conduction law

$$u_p = \sqrt{\frac{18k_BT}{\pi\rho d_p^3}}, \tag{35}$$

where ρ is the particle density and d_p is its diameter. They introduce a probability p for a particle to travel along any direction, and assume that each of the two particle cells are in thermal equilibrium at temperatures T_1 and T_2. The energy the particles carry across the interface from multiple particles moving to neighboring cells is deduced to be

$$q = \frac{\Delta Q}{A\Delta t} \approx -p\phi\rho C_v u_p d_p \nabla T = -k_{Brownian}\nabla T. \tag{36}$$

They argue that each Brownian particle generates a velocity field in the surrounding fluid described by Stokes flow past a sphere. The ability of a large volume of fluid dragged by the nanoparticle to carry a substantial amount of heat is represented by the fraction of liquid volume /3₁ which travels with a particle and a factorial function f(T,). Together with Eq. (36), and the expected value of the probability for particles to move in one direction as p = 0.197, the model is described by

$$k_{Brownian} = \left(5 \times 10^4\right)\beta_1\phi\rho C_p f(T,\phi)\sqrt{\frac{k_BT}{\rho D}}. \tag{37}$$

Both /3₁ and f(T,) are determined from experimental data. The function f(T,) is given by

$$f(T,\phi) = (-6.04\phi + 0.4705)T + (1722.3\phi - 134.63), \tag{38}$$

while /3₁ has the form

$$\beta_1 = A\phi^{-B}, \tag{39}$$

where A, B > 0 are experimentally determined constants.

5.3. Models of the Correlation Type

The last group of models considered are of the correlation type [17,20], which are fitted to experimental data using classical linear regression techniques. Chon *et al.* [20] provide an empirical correlation to describe the effect of individual physical parameters on the thermal conductivity of a nanofluid. The thermal conductivity of a nanofluid is derived as a function of Pr and Re, which are dependent on temperature, and is given by

$$\frac{k_{eff}}{k_f} = 1 + 64.7\phi^{0.7460}\left(\frac{d_f}{d_p}\right)^{0.3690}\left(\frac{k_p}{k_f}\right)^{0.7476} Pr^{0.9955} Re^{1.2321}, \tag{40}$$

where the temperature-dependence of the viscosity $j = j_0 10^{B/(T-C)}$ is included in the estimation of Pr and Re, and the Brownian velocity of the nanoparticles is based on Einstein's diffusion theory by using the mean free path of liquid molecules as a characteristic length. Thus, the empirical correlation of k_{eff} can explicitly show the effect of nanofluid temperature and size as

$$\frac{k_{eff}}{k_f} = 1 + c\left(\frac{Pr\,(T)^{0.9955}\,T^{1.2321}}{d_p^{0.369}\,k_f\,(T)^{0.7476}\,\mu\,(T)^{2}}\right) = 1 + c\left(\frac{1}{d_p}\right)^{0.369}\left(\frac{T^{1.2321}}{10^{\frac{2.4842B}{T-C}}}\right), \tag{41}$$

where c is an empirical coefficient. The model is fitted to the thermal conductivity measurements from three different batches of fairly mono-dispersed Al_2O_3 samples. The most important finding of their work is the effect of nanoparticle mobility, or u_p, is a dominant function of temperature for a given particle size, and that Re dominates the temperature dependence whereas Pr shows a slightly decreasing temperature dependence.

Li and Peterson [17] applied a two-factor linear regression analysis in the measured effective thermal conductivity of CuO and Al_2O_3 nanoparticle suspensions to examine the effects of variations in temperature and volume. For the Al_2O_3/water nanofluid they found

$$\frac{k_{eff} - k_f}{k_f} = 0.764481464\phi + 0.018688867T - 0.462147175,\ R^2 = 0.9171, \tag{42}$$

where the temperature T is measured in °C. For the CuO/water nanofluid they found

$$\frac{k_{eff} - k_f}{k_f} = 3.761088\phi + 0.0017924T - 0.30734,\ R^2 = 0.9078. \tag{43}$$

It may be noted that the coefficient for the volume fraction is much larger than that for temperature for both correlations, indicating the relative importance of these two parameters.

6. Conclusions

Various models and mechanisms that depend on an extension of classical theory involving the movement of the particles has been developed in an effort to explain the enhancement and temperature-dependence on the thermal conductivity of nanofluids. The measurements show that in a nanofluid containing spherical nanoparticles or MWCNT bundles with a high aspect ratio, the thermal conductivity enhancement increases rapidly by two-to-four orders of magnitude with increasing temperature over a temperature range of 20 °C to 50 °C. In contrast, the thermal conductivity of a nanorod nanofluid as examined by Yang and Han [8] decreases with increasing temperature. For a low volume fraction of SWCNT suspensions [24], the thermal conductivity enhancement was found to be nearly insensitive to temperature.

As shown in Sections 2 and 3, the thermal conductivity of liquids such as toluene and oil usually decrease with increasing temperature, while the thermal conductivities of nanoparticles, both metallic and metallic oxide, decrease with temperature. If we simply use a linear function $kf = Af \cdot BfT$ or $k_p = A_p \cdot B_pT$ to represent temperature-dependence of the thermal conductivity of the nanoparticles and the base fluid over the temperature range of 5 °C to 50 °C and substitute it into any traditional effective medium theory, such as those of Maxwell [2] and Hamilton and Crosser [3] and calculate the weighted average volume, the effective thermal conductivity of a nanofluid always appears as a decreasing function of temperature, even for an aqueous suspension when the volume fraction of the nanoparticles is less than 10 %. Although $kf = Af + BfT$ for water and a small loading of nanoparticles (< 1 % volume fraction), keff of an aqueous suspension still appears as a decreasing function of temperature. As a result, it is clear that the effective thermal conductivity of nanofluids is not a simple function of the combined properties of the base fluid and the nanoparticles. This analysis indicates that the traditional effective medium theories cannot explain the phenomena of thermal conductivity variation with temperature.

Many authors believe that Brownian motion is the key mechanism which affect the variation of the thermal conductivity with temperature. The existing models are able to show this phenomena, that is, the higher the temperature and the smaller the average size of the nanoparticles in a fluid, the higher the velocity of the nanoparticle's Brownian motion. Xu *et al.* [37] found that the contribution from the heat convection due to a nanoparticle's Brownian motion is between 2.1 % and 21 % for nanoparticles with an average diameters of 5 and 60 nm, respectively. The intensive Brownian particle motion causes a high heat flux along with micro-convention currents, thus it improves the heat transfer performance. A nanoparticle's Brownian velocity u_p is considered to be the single dominating function describing the temperature-dependence of the thermal conductivity of a nanofluid, dominating over other parameters such as kf and Pr. It has been noticed that models lack consistency in the formulation of the nanoparticle Brownian velocity, u_p. Some authors estimate u_p from the kinetic theory of dilute gases, which is $u_p \times T$- the validity of this estimate for dense liquids is questionable. Others use the Stokes-Einstein formula to estimate $u_p \times T/d_p$. However, in the postulated micro-convection model, to account for the heat transfer between the particles and the fluid the Reynolds number, defined as $Re = u_pd_p/v$, becomes independent on the particle diameter based on u_p estimated from the Stokes-Einstein formula. However, Re is linearly dependent on the particle diameter in the correlation for the flow past a solid sphere.

Some researchers believe that the interaction of the flow field due to the movement of multiple nanoparticles is important even at small values of [31], but others believe that it is insignificant in dilute cases where < 1 % due to retardation effect on the inter-particle potential [39], and can be neglected.

In the postulated micro-convection model to account for the heat transfer between particles and fluid, based on the theory of Tomotika *et al.* [25], the Nusselt number can be expressed as $Nu = 2.0 + 0.5RePr + O(Re^2Pr^2)$. We have noted that each investigator applies it in their own model in different ways. Some assume $Re << 1$ and use $Nu = 2$ [37], others $Nu = 0.5RePr$ with the assumption $Pe^2 << Pe$ [26], and even others assume the Reynolds number and the Prandtl number are of the order 1 and 10 and use Nu " Re^2 Pr^2 [27,28]. The validity of these simplified Nusselt number correlations is questionable.

Several authors have argued that the large thermal conductivity increase is due to the hydrodynamic effects of the large volume of fluid dragged by the Brownian motion of the nanoparticles [36,40,41]. The calculation by Wang *et al.* [36] shows that up to a volume fraction of 10 % the thermal conductivity increase by Brownian motion is less than 0.5 % for an Al_2O_3-liquid mixture. They concluded that Brownian motion does not contribute significantly to the energy transport in a nanoparticle-fluid mixture. Evans *et al.* [40] uses a kinetic theory-based analysis of heat flow in nanofluids. Their results demonstrate that the hydrodynamics effects associated with Brownian motion have a only minor effect on the effective thermal conductivity of nanofluids. Their analysis is also supported by the results of molecular dynamics simulations of heat flow in a model of a nanofluid. Keblinski *et al.* [41] compares the time scale of a particles motion with that of heat diffusion in the liquid. The comparison shows that the thermal diffusion is much faster than the Brownian diffusion, even in the limit of extremely small particles. Their analysis demonstrates that the movement of nanoparticle due to Brownian motion is too slow to transport a significant amount of heat through a nanofluid, a conclusion supported by the results of their molecular dynamic simulations.

As mentioned above, several arguments presented in these models are neither consistent nor convincing. It is obvious that success of these models has been very limited, and no concrete conclusions have been reached. Due to the complexities in the convective currents arising from a system comprising multiple nanoparticles, most existing extended models for the prediction of the effective thermal conductivity of a nanofluid are semi-empirical in nature. Based on Yang and Han's analysis [8], the thermal conductivity enhancement of a nanofluid containing a moderate aspect ratio is attributed to the combined effects of the particles Brownian motion and diffusive heat conduction. According to Sabbaghzade *et al.*'s [30] model for cylindrical nanoparticles, we found that the contribution from the heat convection is about 0.004 % for a volume fraction of $\varphi = 1$ % of carbon nanotubes (diameter = 25 nm, length = 50 μm). The nanoparticles with a cylindrical shape are generally bulky and less mobile than spherical nanoparticles. A more comprehensive study is needed to explore the effects of particle shape on the Brownian motion in nanofluids.

Despite the theoretical effort on the subject, a generally accepted theory for the temperature-dependence of k_{eff} has yet to be established. A unified theoretical model is needed for the temperature-dependence of the thermal conductivity of nanofluids which matches well with experimental results. Accounting for Brownian motion is the predominant mechanism for spherical nanoparticles, while the diffusive heat conduction mechanism will become dominant as the particle aspect ratio increases. To improve the accuracy and versatility of the temperature-dependency of the effective thermal conductivity of nanofluids, more experimental data sets, in particular for non-spherical particle suspensions, are required.

ACKNOWLEDGMENTS

The authors would like to acknowledge the support of the Discovery Project scheme of the Australian Research Council. JMH is grateful to the Australian Research Council for provision of an Australian Professorial Fellowship.

NOMENCLATURE

a	constant
A	surface area, m^2/constant
B	constant
c	constant
C	constant
C_p	specific heat capacity at constant pressure, $Jg^{-1}K^{-1}$
C_v	specific heat capacity at constant volume, $Jm^{-3}K^{-1}$
d	diameter/thickness, m
D_f	fractal dimension
D_0	nanoparticle thermal diffusion coefficient, m^2/s
F	defined function
$F(\omega)$	phonon density of states
h	convective heat transfer coefficient, $Wm^{-2}K^{-1}$
$\hbar$	Planck's constant, 1.05459×10^{-34} Js
J_x	heat current vector in x-direction
k	thermal conductivity of material, $Wm^{-1}K^{-1}$
k_B	Boltzmann constant, 1.38×10^{-23} JK^{-1}
K	shape factor
£	mean free path, m
L	length/specific length, m
m	mass, g
M	molecular mass
n	constant
N	number of particles or atoms
Pe	Péclet number
Pr	Prandtl number
q	heat flux
r	radius, m
R_b	thermal resistance, Km^2W^{-1}
Re	Reynolds number

T	temperature, K
T_θ	Debye temperature, K
u	velocity, m/s
v_s	speed of sound, m/s
V	volume, m^3

Greek and symbols

a	thermal diffusion coefficient of fluid, m^2/s
β	constant related to the Kapitza resistance
γ	velocity gradient
γ_G	Grüneisen constant
δ	hydrodynamic boundary layer thickness, m
δ_T	thermal boundary layer thickness, m
v	kinematic viscosity of fluid, m^2/s
μ	dynamic viscosity of fluid, Pa.s
ω	angular frequency, s^{-1}
Ω_a	atomic volume
φ	volume fraction of loaded particles
φ_T	volume fraction of loaded particles and nanolayer
Φ	sphericity of particles
ρ	density of material, g/cm^3
τ	relaxation time, s
h	Planck's constant, 1.05459×10^{-34} Js

Subscripts

adv	advective component
c	cluster/convestion
Brownian	Brownian motion component
e	electron
eff	effective thermal conductivity
f	base fluid
l	nanolayer
max	maximum
min	maximum
nano	nanolayer
p	nanoparticle
ph	phonon
r	rotational component
s	sound
static	static component
t	translational component

References

[1] Marquis, E. D. S & Chibante, L. P. F. Improving the heat transfer of nanofluids and nanolubricants with carbon nanotubes. *The Journal of the Minerals, Metals & Materials Society*. 2005, vol. 57, pp. 32-43.

[2] Maxwell, J. C. A *Treatise on Electricity and Magnetism*. 3rd ed.; Oxford: Clarendon Press, 1892, vol. 1, pp. 440-435.

[3] Hamilton, R. L. & Crosser, O. K. Thermal conductivity of heterogeneous two-component systems. *Industrial & Engineering Chemistry Fundamentals*. 1962, vol. 1,pp. 187-191.

[4] Avsec, J. & Oblak, M. The calculation of thermal conductivity, viscosity and thermodynamic properties for nanofluids on the basis of statistical nanomechanics. *International Journal ofHeat and Mass Transfer*. 2007, vol. 50, pp. 433 1-4341.

[5] Chung, T., Ajlan, M., Lee, L. L. & Starling, K. E. Generalized multiparameter correlation for nonpolar and polar fluid transport properties. *Industrial & Engineering Chemistry Research*. 1988, vol. 27, pp. 67 1-679.

[6] Chung, T., Lee, L. L. & Starling, K. E. Applications of kinetic gas theories and multiparameter correlation for prediction of dilute gas viscosity and thermal conductivity. *Industrial & Engineering Chemistry Fundamentals*. 1984, vol. 23, pp. 8-13.

[7] Lide, D. R. & Kehiaian, H. V. CRC *Handbook of therm ophysical and thermochemical data*. ISBN-0849301971. Boca Raton: CRC Press: USA, FL, 1994; pp. 417-419.

[8] Yang, B. & Han, Z. H. Temperature-dependent thermal conductivity of nanorod-based nanofluids. *Applied Physics Letters*. 2006, vol. 89, 083111.

[9] Okhotin, A. S., Zhmakin, L. I. & Ivamyuk, A. P. The temperature dependence of thermal conductivity of some chemical elements. *Experimental Thermal and Fluid Science*. 1991, vol. 4, pp. 289-300.

[10] Chen, G. *Nanoscale energy transport and conversion;* ISBN-10:019515942X; Oxford University Press; USA, 2005, pp18-22.

[11] Kittel, C. *Introduction to Solid State Physics,* Eighth edition; ISBN 0-471-41 526-X; John Wiley & Sons, Inc: USA, NJ, 2005, pp121-128.

[12] Berber, S., Kwon, Y. K. & Tománek, D. Unusually high thermal conductivity of carbon nanotubes. *Physics Review Letters*. 2000, vol. 84, pp. 4613-4616.

[13] Kim, P., Shi, L., Majumdar, A. & McEuen, P. L. Thermal transport measurement of individual multiwalled nanotubes. *Physics Review Letters*. 2001, vol. 87, 215502.

[14] Ho, G.Y.; Power, R.W.; Liley, P.E. Thermal conductivity of the elements: A Comprehensive Review. *J. Phys. Chem. Res. Data*. 1974, vol.4, suppl. 1, 340.

[15] Shackelford, J.F.; Alexander, W. *CRC Material Science and Engineering Handbook*; ISBN-0849326966; Boca Raton: CRC Press: USA, FL, 2001:252.

[16] Das, S. K., Putra, N., Thiesen, P. & Roetzel, W. Temperature dependence of thermal conductivity enhancement for nanofluids, *Journal of Heat Transfer*. 2003, vol. 125, pp. 567-574.

[17] Li, C. H. & Peterson, G. P. Experimental investigation of temperature and volume fraction variations on the effective thermal conductivity of nanoparticle suspensions (nanofluids). *Journal ofApplied Physics*. 2006, vol. 99, 084314.

[18] Patel, H. E., Das, S. K., Sundararajan, T., Nair, A. S., George, B. & Pradeep, T. Thermal conductivities of naked and monolayer protected metal nanoparticle based nanofluids: Manifestation of anomalous enhancement and chemical effects. *Applied Physics Letters*. 2003, vol. 83, pp. 293 1-2933.

[19] Kumar, D. H., Patel, H. E., Kumar, V. R. R., Sundararajan, T., Pradeep, T. & Das, S. K. Model for heat conduction in nanofluids. *Physics Review Letters*. 2004, vol. 93, 144301.

[20] Chon, C. H., Kihm, K. D., Lee, S. P. & Choi, S. U. S. Empirical correction finding the role of temperature and particle size for nanofluid (Al2 O3) thermal conductivity enhancement. *Applied Physics Letters*. 2005, vol. 87, 153107.
[21] Zhang, X., Gu, H. & Fujii, M. Experimental study on the effective thermal conductivity and thermal diffusivity of nanofluids. *International Journal of Thermophysics*. 2006, vol. 27, pp. 569-580.
[22] Zhang, X., Gu, H. & Fujii, M. Effective thermal conductivity and thermal diffusivity of nanofluids containing spherical and cylindrical nanoparticles. *Journal of Applied Physics*. 2006, vol. 100, 044325.
[23] Wen, D. & Ding, Y. Effective thermal conductivity of aqueous suspensions of carbon nanotubes (carbon nanotube nanofluids). *Journal of Therm ophysics and Heat Transfer*. 2004, vol. 18, pp. 48 1-485.
[24] Kim, B. H. & Peterson, G. P. Effect of morphology of carbon nanotubes on thermal conductivity enhancement of nanofluids. *Journal of Therm ophysics and Heat Transfer*. 2007, vol. 21, pp. 45 1-459.
[25] White, F.M. *Viscousfluidflow*; ISBN 0-07-069710-8; McGraw-Hill Inc.: USA, 1974; pp.212-214.
[26] Patel, H. E., Sundararajan, T., Pradeep, T., Dasgupta, A., Dasgupta, N. & Das, S. K. A micro-convection model for thermal conductivity of nanofluids. Pramana - *Journal ofPhysics*. 2005, vol. 65, pp. 863-869.
[27] Jang, S. P. & Choi, S. U. S. Role of Brownian motion in the enhanced thermal conductivity of nanofluids. *Applied Physics Letters*. 2004, vol. 84, pp. 4316-4318.
[28] Jang, S. P. & Choi, S. U. S. Effects of various parameters on nanofluid thermal conductivity. *Journal ofHeat Transfer*. 2007, vol. 129, pp. 617-623.
[29] Ren, Y., Xie, H. & Cai, A. Effective thermal conductivity of nanofluids containing spherical nanoparticles. *Journal of Physics D: Applied Physics*. 2005, vol. 38, pp. 3958-3961.
[30] Sabbaghzadeh, J. & Ebrahimi, S. Effective thermal conductivity of nanofluids containing cylindrical nanoparticles. *International Journal of Nanoscience*. 2007, vol. 6, pp. 45-49.
[31] Prasher, R. Thermal conductivity of nanoscale colloidal solutions (Nanofluids), *Physics Review Letters*. 2005, vol. 94, 025901.
[32] Tillman, P., & Hill, J. M. Determination of nanolayer thickness for a nanofluid. *International Communications on Heat and Mass Transfer*. 2007, vol. 34, pp. 3 99-407.
[33] Fand, R. M. Heat transfer by forced convection from a cyclinder to water in crossflow. *International Journal ofHeat and Mass Transfer*. 1965, vol. 8, pp. 995-1010.
[34] Incropera, F.P.; DeWitt, D.P. *Fundamentals ofheat and mass transfer*, Second edition; ISBN 0-471-51728-3; John Wiley & Sons Inc.; NY , 1990, pp357-409.
[35] Brodkey, R.S.; Kim, D.S.; Sidner, W. Fluid to particle heat transfer in a fluidized bed and to a single particle. *Int. J. Heat Mass transfer* 1991, vol.34, 2327-2337.
[36] Wang, X., Xu, X. & Choi, S. U. S. Thermal conductivity of nanoparticle-fluid mixture. *Journal of Thermophysics and Heat Transfer*. 1999, vol. 13, pp. 474-480.
[37] Xu, J., Yu, B., Zou, M. & Xu, P. A new model for heat conduction of nanofluids based on fractal distributions of nanoparticles. *Journal ofPhysics D: Applied Physics*. 2006, vol. 39, pp. 4486-4490.
[38] Shukla, R. K. & Dhir, V. K. Study of the effective thermal conductivity of nanofluids. *Proceedings ofIMECE2005, 2005 ASME International Mechanical Engineering Congress and Exposition*. IMECE05-8028 1.
[39] Koo, J. & Kleinstreuer, C. A new thermal conductivity model for nanofluids. *Journal ofNanoparticle Research*. 2004, vol. 6, pp. 577-588.
[40] Evans, W., Fish, J. & Keblinski, P. Role of Brownian motion hydrodynamics on

nanofluid thermal conductivity. *Applied Physics Letters.* 2006, vol. 88, 093116.

[41] Keblinski, P., Phillpot, S. R., Choi, S. U. S. & Eastman, J. A. Mechanisms of heat flow in suspensions of nano-sized particles (nanofluids). International *Journal ofHeat and Mass Transfer*. 2002, vol. 45, pp. 855-863.

[42] Xuan, Y., Li, Q. & Wu, W. Aggregation structure and thermal conductivity of nanofluids. *AIChE Journal of Thermodynamics*. 2003, vol. 49, pp. 1038-1043.

In: Encyclopedia of Nanotechnology
Editor: Elwood D. Carlson

ISBN 978-1-60692-079-4

Chapter 44

NANOTECNOLOGY: OBTAINING OF NANOPARTICLES AND NANOCOMPOSITES AND THEIR USE IN FOOD AND DRUG PACKAGING

A. Edwin Moncada[*]
Braskem S.A. III Polo Petroquímico, Via Oeste Lote 5 - Passo Raso,
Braskem. S.A. Triunfo, RS, Brazil

ABSTRACT

This work aims to review the immense influence that nanotechnology has having along the last years in all areas of knowledge, as will as, the method of obtaining of nanoparticles, trough of clays modification and Sol-gel synthesis and the used of the nanoparticles obtained in the formation of nanocomposites with polyolefin, also is discussed the interaction nanoparticle-polymer, and the used of a new compatibilizer agent in the properties of nanocomposites. Is presented one of its main applications and the main capital investment from the governments is the packaging industry of food and pharmaceutical products. This is due, among many other factors, to the possibility of using nano-sensors that when combined with the package communicate to the consumer/producer, as well an in control situations, the characteristics of package material such as: microbiological quality, storage conditions (humidity, temperature, light, etc), time-temperature treatment and shelf-life, among others. Finally is shows some trends for the application of nanoparticles.

INTRODUCTION

The development of materials has been fundamental to modify the behavior and human culture throughout the time. It is possible to describe the historical evolution periods of human race through advance, manipulation and knowledge of chemistry and physics of

* Phone: 55.51.3721.8251, Fax:55.51.3457.1084; E-mail address: edwin.moncada@braskem.com.br; moncada.edwin@gmail.com

materials. That is, how its development can be categorized in different ages: Stone age, Bronze age, Iron age, Steel age (industrial revolution), Polymers age, and Silicone and Silicon age (the revolution of telecommunication). For this reason, materials have been so important to man and the researches in this area present advance in a rate, every time, more accelerated.

This century has been characterized for getting huge advances in every area of research. In the last years have had a growth phenomenon that has centered the attention of scientist all over the world, leading the government to make huge investments in this area, we are talking about: nanotechnologies. These are characterized to allow the manipulation of the matter structure to a smaller scale, in the order of nanometers (nm, thousandth of micrometers, millionth of mm or thousand-millionth of meters), generating materials and structures with different characteristics from those, which are normally used.

The nanotechnologies, today, are like an innovative and radical jump, that will affect the whole economic sectors. The spread effect with extensive dissemination of nanotechnologies is principally due to; the substantial changes caused in the field of material that are used in the whole manufacturing and service sectors, conditioning many time its development. By means of these technologies it is possible to modify the properties of known materials much more radically that what has been achieved until now, as well as creating completely new materials. On the other side, nanotechnologies allow to adjust materials in a scale better reduced than the actual manufacture of microprocessors.

Everything looks like to indicate that we are now in the initial stages of an expansive wave equivalent to that initiated in the first years of 70´s in correlation with the introduction of microprocessors.

The nanotechnology word was presented in 1974, by the researcher Nori Tanoguchi of the Tokio University, who showed the differences between microelectronic and the new engineering field in a submicrometric scale. [1] According to the National Science Foundation (NSF) and the National Nanotechnology Initiative (NNI) of EUA, nanoscience and nanotechnology are defined like the understanding, control, and exploration of materials and systems whose structures and components show physical, chemical and biologist properties and phenomena that are, significantly, new and modified due to their nanometric scale (1-100 nm).

Some questions emerge, such as: Is Nanotechnology a radical technology that involves the construction of macroscopic products through direct manipulation of software? Or are we near to a decade of virtual utopian technology where the materials can be constructed through molecular manufacture? These questions were asked on the Nottingham University Symposium (August 26th 2005), where they were extensively discussed.[2,3] One important point was that: the concept of molecular manufacture was established (enunciated originally by Drexler in 1981 [4]) in which, later on, it was proposed as a molecular nanotechnology (MNT) by Drexler and others.[5,6,7,8]

In the figure 1, it is possible to see the evolution of nanoscience and nanotechnology through the time. The indexed annual publications (where the word nano is on the title of the article) (Web of Science) and international patents found in Internet (Derwent). In the graph, it is possible to see an exponential growth in publications (since 1993, with 955 articles on indexed magazines) and international patents (initiating on 1999, with 296 registered patents). In the year 2006, 26967 international publications were found in indexed magazines and 5369 concessions of international patents.

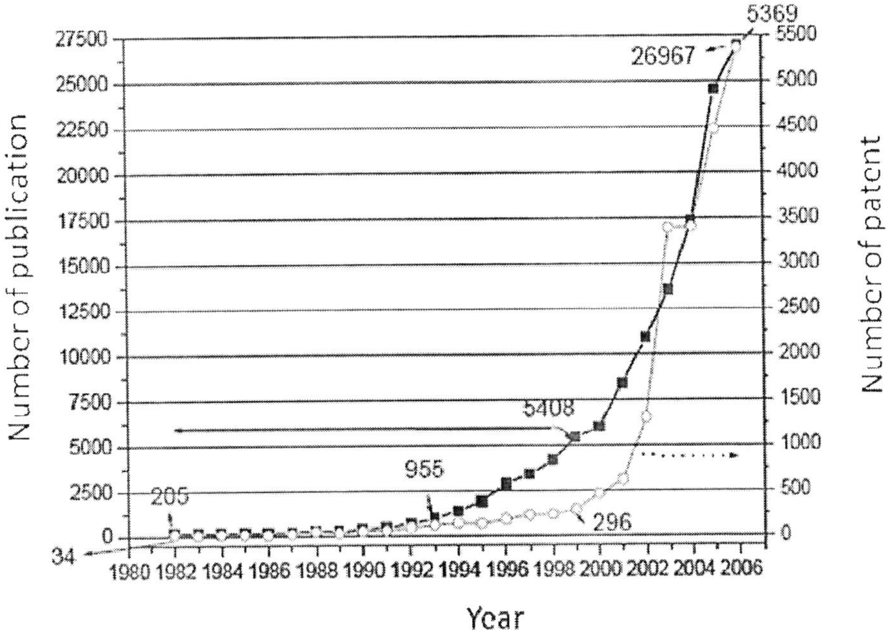

Figure 1. Numbers of scientific publication per year, where the word nano is found on the title of the published work and the international patents of the last 15 years. The search was done on dice base "web of science" and "Derwent" of Thomson Corporation on April 2007 [9].

Source: Scioence citation index and Derwent, 2007.

What said before gives an idea of the great increase in researches, measured by the quantity of articles, industrial applications, quantity of patents and how it will be seen through this chapter, the tendency for the following years continue increasing the amount of publication and patents. Besides the governments invest every time more money to finance, specifically, nanoscience and nanotechnology.

Inside of investigations in this scale of physical greatness, Public Universities, Institutes of Governmental Investigation and some of their departments in several countries in the world, it was used around US$ 4 billions in the year 2005, in the construction of "facilities" and in the establishment of the research and development (P&D) lines. On the other hand companies of P&D are investing more than nine times of used resource in other field of research. In 2005, 1331 bound companies to 76 industries, second to LUX Research Inc/EUA, invested US$ 3,2 billions in nanotechnology (NT) and sold approximately US$ 32 billions in products that used some way of NT. Or CEO, of "General Electric", Jeffrey Immelt fixed nanotechnology (NT), personalized medicine and renewable energy, like the most priority of P&D. [10] According to the published article on February 2006 in the magazine "Science", or amount that is being estimated to 2015, it is of a global market for Nanotechnology (NT) and Nanoscience (NC), around of US$1,0 trillion (NEL, 2006). To May 16th 2007, the products stocktaking to the consumer, organized by the "Woodrow Wilson International Center for Scholars", it already was of 475 produced products, or in production lines, coming from nanotechnology.

INDUSTRIAL APPLICATIONS OF NANOTECHNOLOGY

The applications to produce nanostructure materials and nanotechnologies are being developed with extreme speed and a simple list of limited number of applications only gives a small idea of its potentialities. Since long time, there are techniques that allow working in a nanostructural level, like some areas of condensed matter, science of colloid and the growth of superficial coating among others. The great development, in molecular biology fields and biotechnologies since 80´s, has increased its expansion towards every kind of materials: metallic, non-metallic, plastics and compound materials, and through them towards the most diverse fields: scientific, technological and industrials.

There are a lot of areas in science that has been, are and will be implicated in nanotechnologies, such as: medicine, biology, pharmacology and materials with all theirs applications in the engineering sectors like: civil, construction, electronic, mechanics, chemistry and foods engineering among others.

Some examples of innumerable applications of nanotechnology, already present at the market or near to its commercialization are as follows. [11]

- Sensors: They are destined to detect the presence of specific compounds in different environments; closed and opened or aromas that characterize the quality of beverages and food products (electronic nose, intelligent packaging).
- Photovoltaic systems of high efficiency for the conversion of the solar energy.
- New materials with an elevated relation between resistance and mass to airspace, biomedical and transport applications.
- Packaging of food products with better barrier characteristics for gases penetration and capacity to indicate the state of conservation (intelligent packaging).
- Diagnosis techniques based on a system named "lab-on-a-chip" to make the clinical and genetics analysis with minimum sample and on real time.
- Cosmetics, especially to protect against the solar radiation.
- Materials for the filtration and hydrocarbons catalysis and other substances.
- Superficial coatings with resistance to corrosion scratched and notably improved ragged.
- Cut tools with highest tenacity and reduced fragility.
- More functional and lighter screen videos based on electronics of polymers.
- New prosthesis and implant to placement in vivo.
- Work techniques of pieces to micromechanics and microelectronic in a 100 nm scale.

The nanostructured materials can be gotten in two ways: one named "top down" in which the nanostructures are sculptured on a block of material and the other called "bottom up" where the nanostructured materials are gotten through nanoparticles. The "top down" techniques have similarity with the actual technique of productions of electronic microprocessors. "Bottom up" techniques are based on similar processes used in technology of materials and can get powders, compact objects or thin coatings with great changed properties with regard to the same materials gotten by conventional technologies.

Nanoparticles refers to particles where at least one of its space dimensions is in nanometric scale (smaller than 100 nm) and it can be found nanospheres, nanotubes, nanosheet, nanofiber, etc. A possible works listed where the word nanoparticles appears, would be too extensive to be named, but between the more used are those coming from clays, inorganic oxides and concerning ones to the carbon.

It will be presented in the following pages, related concepts to nanotechnology and nanoparticles, some methodologies of obtaining nanoparticles, specifically by means of clays treatments and Sol-gel synthesis method, besides, the possibility to form sensors with these particles and their use in the industry of food and pharmaceutical products, specifically in food and drug packaging.

NANOPARTICLES OBTAIN BY CLAYS

Clay is a mineral coming from the decomposition of rocks originated in a natural process during thousands of years. Physically it is considered a colloid of small particle and flat surface. Chemically it is a hydrated alumina silicate which general formula is $Al_2O_32SiO_22H_2O$. It is, also, called aluminosilicatos or philosilicates due to their structure in sheets. The main chemical component of clay is the silicate minerals (SiO4) depending on the structural organization and cations presented by the family of generated clay. The most commonly utilized clays for obtaining nanoparticles are those coming from aluminosilicates 2:1. These structures are called TOT due to their tetrahedral-octahedral-tetrahedral layers. The thickness of the sheet is 0.96 nm. A scheme of the structure of these compounds (aluminosilicates) is presented in the figure 2.

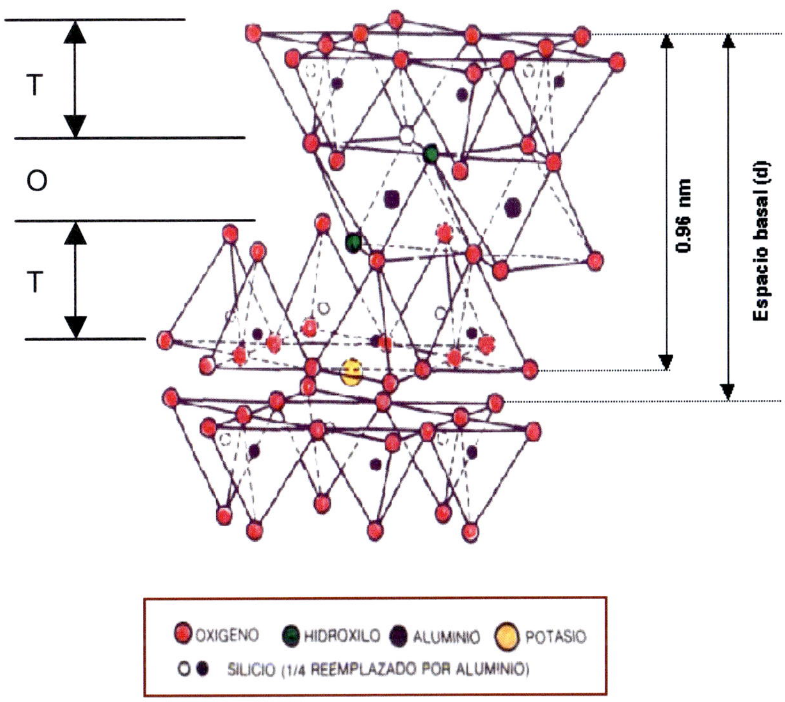

Figure 2. Structure in philosilicates layer 2:1 [12].

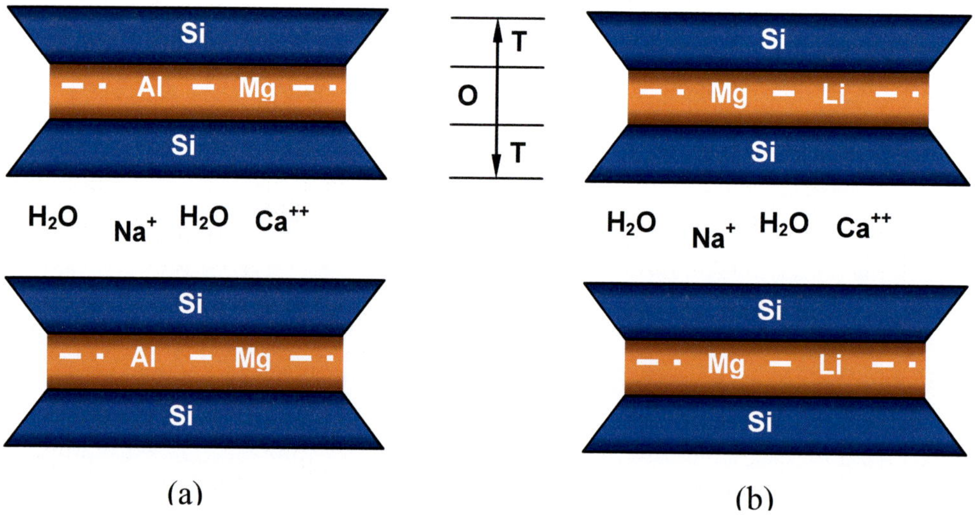

Figure 3. Schematic representation of (a) Montmorillonite clays (b) Hectorite clays.

But if in the octahedral layer, the magnesium atoms predominate and they present isomorphic substitutions for lithium, this clay is named Hectorite (HTA). The differences between these two important clays are shown in the figure 3, where it is appreciated the characteristic ions of each one of the layers (TOT) of the MMT and HTA.

Among clays type TOT, commonly named smectite, there is a great variety depending on the layer where the isomorphic replacement is carried out (tetrahedral, octahedral or both). If in the octahedral layer, 1/6 of the aluminum atoms are replaced for magnesium, and in the tetrahedral layer there are not substitutions, this clay is named montmorillonite (MMT).

One of the principal philosilicates 2:1 characteristics is its capacity to exchange its interlayer cations with other compounds; this process is named cationic exchange capacity (CEC) expressed in meq cation/100g clay. This property gives an endless of applications due to the possibility of including great variety of compounds between the philosilicates sheets that modifies their properties.

Clay's Modification

In general, many substances, such as oils, pollutants [13], amines, pesticides [14] and others [15,12] can be exchanged or inserted into the clay interlayer space modifying their characteristics. [16,17] The intercalations with alquil-amine compounds inside the clay increase the interlayer spacing and also modify its polarity transforming clay in a more hydrophobic compound, due to the alquil group. A scheme of this reaction is shown in the Figure 4, where it is observed how the protonated amine displaces cations through cationic exchange reaction present in the interlayer of clay. This is due, mainly, to the effect of added clorhydic acid, since the cations of interlayer space form very stable and soluble chlorides, allowing the cations to leave the interlayer, and the protonated amine penetrates on the interlayer replacing cations.

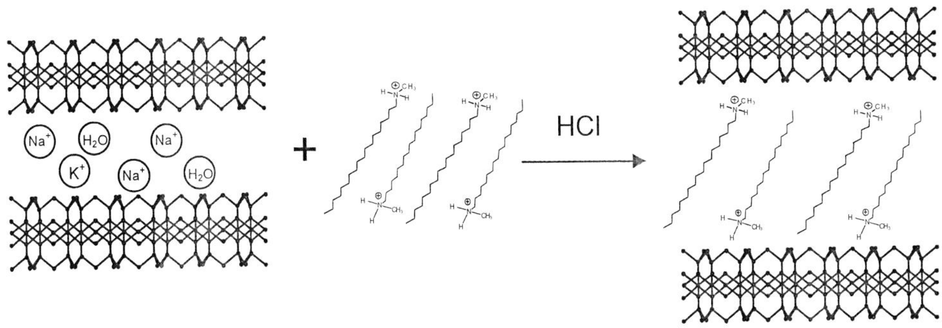

Figure 4. Schematic representation of the modification of clays with octadecylamine.

Studies, in this area, have shown that when the alquil-amine presents 18 atoms of carbon the biggest increases are gotten in the interlayer distance. [18,19] This is the vital importance to many applications and in particular for this research paper, since it wants to disperses sheets of clay in a nonpolar polymer like polypropylene. For this reason, in this chapter some variable of clay cationic exchange reactions with octadecylamine (ODA) are studied. Octadecylamine is a primary amine that has a C18 chain and accepts a proton, in an acid environment, remaining positively charged (ODA+). In this chapter, time and temperature reaction are variables taking into consideration to exchange cationic reaction.

Nanoparticulas Obtained by the Sol-gel Method

Among the principal difficulties in nanocomposites formation with clays, is its reduced interlayer space (1,0 nm), since, it makes difficult the polymer insertion between their sheets to generate inserted and exfoliated states. Another important characteristic is the lack of clays homogeneity and the presence of some ions like impurities that give coloration to nanocomposites, for example iron; or give toxicity like Cd and Cr. [20] For this reason, techniques to obtain synthetic clays or inorganic compounds with laminate structures [21,22] have been looked for, techniques that can modify the polymer properties when they are dispersed on the polymer, but without influencing their properties negatively. Here, the method of synthesis Sol-gel becomes one of the most important tools to obtain laminate structures of high purity. Certainly, the grade of structure organizations and their properties depend on the nature of their components, such as: organic, inorganic or hybrid systems in order to generating synergic interactions. [23,24,25] This synthesis method, also, allows modifying the morphology of particles that can be obtained such as: sheets, spheres, threads, etc. when template is used. [24]

Brinker and Sherer defined the Sol-gel method as the obtaining of materials through the preparation of a sol and the obtaining f gel by the solvent elimination. But, Since and Kolbe were who previously discovered the formation of silica mono-dispersed particles through hydrolysis and condensation of tetraethyl ortosilicate (TEOS) in ethanol with ammonia as catalytic compound. Since that moment numerous modifications have been developed and new variables have been introduced trying to have control on the size and the morphology of particles.

Breck and their collaborators in Mobil Oil Corporation were the pioneers in using surfactants as structural patternmakers (templates) for silicates synthesis and ordered aluminosilicates structurally. These materials were synthesized in a basic solution and were used surfactants of long alquilic chain, which were removed, later on, to obtain mesoporic silica. There are many "templates" used. In general, they are organic molecules and can be cationic, anionic or neuters. Also, copolymer with functionalities of different polarity can be used. When template is changed, the surfactant interaction (S) with the inorganic species (I) are modified (S^+I, $S^+X^-I^+$, S^0I^0, S^-I^+) affecting the properties of resulting material such us; pore wall thickness, superficial concentration of silanol groups, thermal stability and also the size of the pore. [24]

Many works (research papers) present the nanoparticles obtaining by the Sol-gel method and its later use in the nanocomposites formation [26,27,28,29,30], in the Moncada et, al, work, [31], a comparison between the works mentioned previously is made.

The possibility to obtain nanoparticles with different morphologies by means of the Sol-gel method makes it so much versatile for many applications, as it was mentioned previously. In figure 5, different morphologies are shown for the nanoparticles obtained by means of the Sol-gel method, the authors present the spherical nanoparticles formation (SNP), hybrid spherical nanoparticles (HSNP), and hybrid layer nanoparticles (HLNP), by means of variation in the synthesis conditions of the template agent like the polar and no-polar compounds are in general.

In order to generate silica particles with spherical morphology (SNP or HSNP), a modified version of the classical method of Stober [32] was applied. Instead of performing the sol–gel process of TEOS by adding it to an ethanol/NH4OH solution, prior to this step TEOS was slightly hydrolyzed in an ethanol/water azeotrope (4.4%H2O), forming short siloxane oligomers. Furthermore, template was added in order to obtain hybrid silica nanoparticles (HSNP) and to control the size and aggregation of the nanoparticles state. Figure 5 shows TEM images of particles obtained by the sol–gel method. For SNP, a spherical morphology with an approximate size of 10 nm can be seen with philiform aggregate formation. This unusual form could be made due to slightly prehydrolyzed TEOS before the addition to ammonia. The addition of a small amount of Template in the synthesis of HSNP allowed dispersed nanoparticles (10 nm) with a low degree of agglomeration to be obtained. Although also some discreet philiform aggregates (2–10 nanoparticles) are seen, they are dispersed without forming agglomerates. The probable way in which the template acts is by joining to the surface of the particles through R–NH2 ↔ HO–Si → R–NH+3 O–Si hydrogen bonds.

In this way the particles are capped with a layer allowing the particles to remain largely separated and dispersed. The template molecules attached to the surface of the silica nanospheres, surrounding them, give them the hybrid particles character. When greater amounts of template were added, the formation of a laminar micelle in the solution allowed the synthesis of laminar silica structures like HLNP. Figure 5 shows TEM images of these laminar particles in which the approximate size of a 300 nm × 300 nm sheets can be determined, but whose thickness cannot be measured exactly in this image.

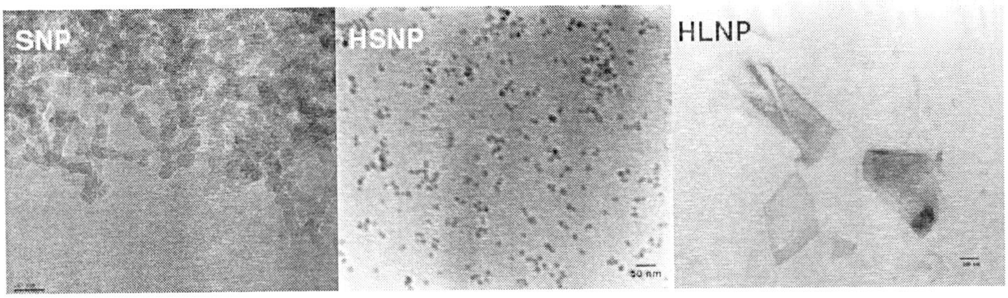

Figure 5. TEM images of nanoparticles obtained by Sol-Gel Method.
SNP= Spherical nanoparticles, HSNP= hybrid spherical nanoparticles, HLNP= Hybrid layer nanoparticles.

Besides, nanoparticles previously named can have hybrid characteristic (organic-inorganic) such as HSNP or HLNP, and not hybrid characteristic such as SNP. This is presented in the figure 6. Where in the FT-IR spectrum can be observed the characteristic bands such inorganic elements (synthesis Sol-gel) as organic elements (template).

In the Table 1 are presented the assignments of the bands frequency.

The spectra of SNP and HSNP are characteristics of silica samples synthesized by sol–gel [33,34,35], but in the case of HSNP the spectrum also shows a shoulder at 2920 cm−1 corresponding to the C–H absorption from the template present in the sample. The broad band at 3000–3700 cm−1 is due to water and silanol groups engaged in hydrogen bonds. On the other hand, the spectrum of HLNP corresponds to a typical aluminosilicates where strong peaks at 2856 and 2920 cm−1 are attributed to template, which is present in considerable amounts in the particles. For more details of these particles see [31].

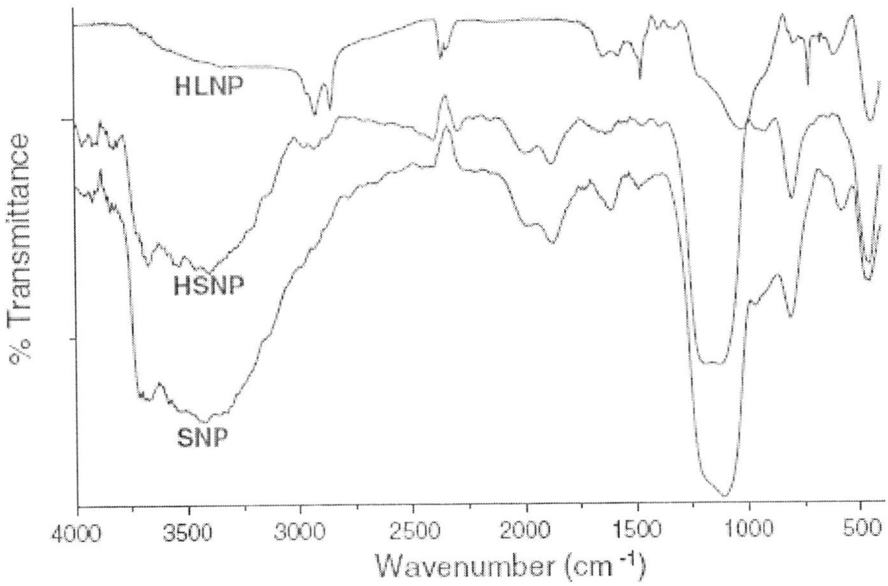

Figure 6. FT-IR of nanoparticles obtained by Sol-Gel method.

Table 1. Characteristic band of hybrid and unhybrid material obtained by Sol-Gel synthesis

Position of the band (cm^{-1})	Assignment			Reference
	SNP	HSNP	HLNP	
450	Bending Si-O-Si ____			[36,37,38]
720	------	------	"Rocking" -(CH_2)n-	[35,40]
800	____Flexion O-Si-O ____			[39,36]
960	Flexion Si-OH		-----	[39,41,44]
1000-1200	____Tension Si-O-Si ____			[39,40,41,44]
1210	----	----	Tension Al-O	[39,45]
1465	----	----	Flexion -C-H	[9]
1640			-NH_3+	[39]
1640	Flexion OH of the water			[39,44]
2918	----	-CH_2-____		[39,40,42]
3000-3700	OH of water and silica			-

Nanoparticles Obtain by Sol-gel Compared with Nanoparticles Obtained by Treatment of Clays

Nanoparticles obtained by Sol-gel method, like it was mentioned previously, show advantages in comparison with nanoparticles obtained by means of clay treatment, these are some of them:

- High purity and uniformity of chemical structure, because they are obtained by synthesis methods.
- Approval by legislation for direct contact with alive creatures (non toxicity), presenting it applications in such exigent industries with the legislation, for example: health areas, medicine, foods and drug.
- Obtaining of hybrid nanoparticles (organic-inorganic), it increases more the applications range.

Nanocomposites

Nanocomposite is the dispersion of nanometric size particles (nanoparticles) in a matrix, one of the first research to obtain nanocomposites was made by the TOYOTA company between the end of 80`s and the beginning of 90`s [40,41,42], with a clays exfoliation in a naylon-6 matrix. These results showed significative increments in an extensive range of properties to polymers reinforcement [43,44].

Later on to this discovery, a great explosion about nanocomposite researches has been developed, and researches were made in order to obtaining them with a great majority of polymers, such as: polypropylene [45,46,47], polyethylene [48], polystyrene [49], polyvinylclorure [50],acrilonitrile butadiene styrene copolymer (ABS) [51], polymetilmetacrylate [52] polyethylenterefthalate (PET) [53],ethylene-vinyl acetate copolymer (EVA) [54],polyacrilonitrile [55], polycarbonate [56], polyethylen oxide (PEO) [57],epoxy resin [58], polyamides [59], polycaprolactone [60], phenolic resin [61], rubers [62], polyurethanes [63], polyvinyl piridine [64]

The developed methods for the production of these nanocomposites are mainly three: mixed in solution, polymerization in-situ and mixed in fused.

Mixed in solution.[65] It consists in dissolving clay and polymer in an appropriate solvent and agitate the system, so that the polymer can introduce itself between the clay sheets. Then, the solvent is evaporated until obtaining a solid sample, in which is hoped to obtain the nanocomposite.

Polymerization In-Situ. [66] It consists in inserting the necessary catalytic system to the polymerization reaction in the inter-laminar region of clays, introducing the new system to the polymerization reactor and making the reaction. As the polymer chain grows, it will go separating the sheets until dispersing them.

Mechanic mixed in melt state. [67] It consists in using a mechanical mixer, which heats the polymer until the fused point. The mixer has molars that apply a force. When the clay is added the polymer is introduced between the clay's sheets.

There are three terms to describe the dispersion grade of clay in nanocomposites, as follows:

Nanocomposites Obtained Using Clays Modified

Where clays are used for the generation of nanoparticles, three possible morphological states can be found at the nanocomposites.

Tactoide state: When the polymer surrounds the clay structure.

Intercalated sate: when the polymer has the possibility to introduce itself between the clay's sheets given a highly ordered state that contain the clay's sheets and the polymer.

Exfoliated state: In contrast, this state is the dispersion of the clay's sheets in the polymeric matrix.

A graphic of these states is shown in the Figure 7, where its characteristics can be appreciated.

Depending on the characteristics of this mixture, it is necessary to reduce the repulsion in interface if the polymers are polar (nylon) or non-polar (PE, PP, etc) in order to minimize the repulsion force. For this, a third component named compatibilizer agent is used in the mixture in order to help with the compatibility between the two other materials. The compatibilizer agent more used has been maleic anhydride implanted in polypropylene and, in general, the most efficient mass relationship between this and the quantity of load has been 3/1. [68,69,70,71] This material has great importance because the modification in the properties not only depends of the particle size but also of an effective interaction between matrix and nanoparticles.

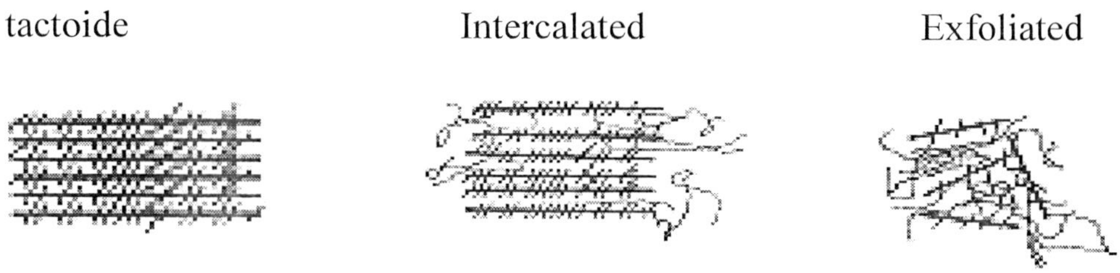

Figure 7. Different states of nanoparticles dispersion in a nanocomposites.

Compatibilizer Agent

Surface modification of hydrophilic clays through ion exchange reaction, normally with long chain alkyl ammonium ions increases the hydrophobic character as well as the interlayer spacing of these materials. [72,73,74] Therefore, the organically modified clays (O-clays) would be more compatible with non-polar polymers such as PP. However, in the case of polyolefin (PE, PP) the main difficulty encountered in preparing nanocomposites with O-clays is the lack of sufficient interaction between the hydrophobic polyolefin matrix and still relatively hydrophilic O-clays. [75,76,77] However, depending on the hydrophilic/hydrophobic balance of a polymer, the polymer chains could enter into the interlayer space of the clay and form either tactoid, intercalated or exfoliated nanocomposites. In the tactoid hybrid structure, the polymer chains mainly surround the stacks of the clay resulting in a material with properties similar to those of micro composites. In the intercalated hybrid structure, extended polymer chains are introduced between the silicate sheets, resulting in a multilayered structure consisting of alternating polymer and inorganic clay layers. In the exfoliated hybrid structure, the clay nanolayers are dispersed as single layers in the polymer matrix. However, the challenge is to obtain exfoliated nanocomposites since these materials show superior properties than those of intercalated or tactoid hybrid materials. Owing to the strong hydrophilic nature of the clays, low adhesion between clays and non-polar polymers causes a considerable decrease in the mechanical properties of the nanocomposites, as the interface becomes a weak point in the hybrid material. To overcome this limitation and in order to adequately disperse the O-clays in PP matrix, functionalized PP materials such as PP grafted with maleic anhydride has been used as compatibilizer agent.[78] Recent study carried out by Lertwimolnun et.al. [79] On the degree of dispersion of montmorillonite in PP nanocomposites, prepared by melt intercalation, showed that the dispersion of the clay in PP matrix was improved by addition of a modified PP with maleic anhydride as compatibilizer. However, this improvement was obtained for concentrations of the compatibilizer higher than 10 weight percent and up to 25 weight percent. In a more detailed study on this subject, the effect of the molecular weight and content of the maleic anhydride of PP grafted with this monomer have been studied by Wang et. al.[80] The results showed that not only the extent of grafting and the molecular weight of the compatibilizer but also the mixing temperature and processing conditions affected the structure of PP nanocomposites obtained with octadecylamine-modified montmorillonite. It was found that the exfoliation depended mainly on the intercalation capability of the compatibilizer as well as its composition in the nanocomposites. Other compatibilizer agents are study, as acrylic acid, [81] itaconic acid and its derivatives. [82,83] Itaconic acid is non-oil based dicarboxylic acid monomer. It is obtained through large-scale fermentation of agricultural wastes such as molasses, a sub-product of sugar industry. Due to its double functionality, itaconic acid and its derivatives offer interesting possibilities as polar functional monomers for the modification of polyolefin. The scheme of grafted reaction of itacónico acid in polypropylene is shown in the figure 8.

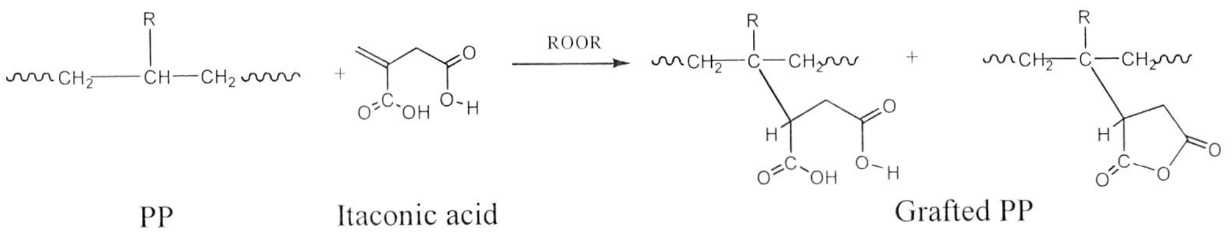

Figure 8. Schematic representation of the grafted reaction of itaconic acid in polypropylene. [46]

NANOCOMPOSITES OBTAINED USING SOL-GEL NANOPARTICLES

Nanocomposites obtained using polypropylene with two different clays Montmorillonite (MMT) and Hectorite (HTA) and two different compatibilizer agents like compatibilizer commercial maleic anhydride (PP-g-MA) and the patented compatibilizer by this group: itaconic acid (PP-g-AI). They presented remarkable increments in mechanical nanocomposites properties in comparison with the polymer alone. [11]

In the figure 9, nanocomposites transmission microphotographs of polypropylene are shown using clays type Montmorillonite and Hectorite organically modified. TEM images show an effective nanocomposites formation and an effect of itaconic acid incorporation percentage (PP-g-$AI_{1,8}$ = 0.7% of grafting). They also show a better dispersion of clays nano sheets in the polymeric matrix when the clay type Hectorite was used, in comparison with the clay type Montmorillonite. This effect is more important in the mechanical properties, which will be shown below.

Table 2 shows mechanical properties of different nanocomposites obtained with polypropylene, two different clays (MMT and HTA) and itaconic acid implanted in polypropylene with different implant percentages. It is proved the nanocomposites formation by increasing the mechanical properties in comparison with the polymer without load. It is, also, observed an effect of itaconic acid implant percentage in the values of mechanical properties and consequently in the nanocomposites formation. [46]

When clay type MMT was used, it was observed a decrease of the mechanical property with the decrease of itaconic acid implant percentage. When clay type HTA was used the behavior was on the contrary, when the itaconic acid implant percentage was decreased, it was observed an increase in mechanical properties. The authors explain this tendency through different cationic exchange capacities of clays and their chemical interaction with the compatibilizer, which are shown in the Table 2.

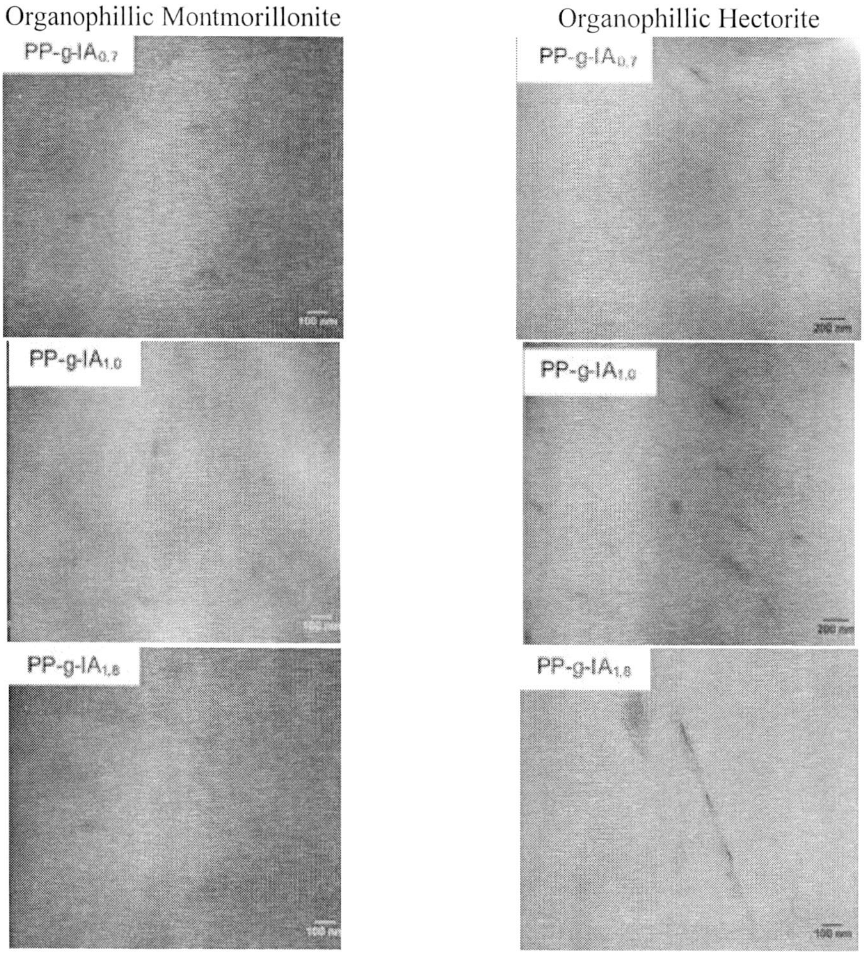

Figure 9. TEM images of nanocomposites obtained by using organically modified Montmorillonite/Hectorite and polypropylene grafting itaconic acid with different percentages of grafting.

Table 2. Mechanical properties of nanocomposites prepared by used PP, 1 wt -% of different chalice and 3% of functionalized PP grafted with different percentages of itaconic acid. (And, Modulus, (and, tensile strength; (, elongation at break)

PP-g-AI Wt.-%	Montmorillonite			Hectorite		
	E (MPa)	σ_y (MPa)	ε (%)	E (MPa)	σ_y (MPa)	ε (%)
1,8	2117 (±35)	40 (±1)	17	1791 ((±40)	49 (±2)	9
1,0	1941 (±33)	38 (±2)	8	2075 (±35)	48 (±1)	8
0,7	1900 (±42)	38 (±2)	8	2137 (±28)	51 (±2)	9

Unmodified PP: E= 1092 (±45) MPa, σ_y=30 (±2) MPa, ε= 20%.

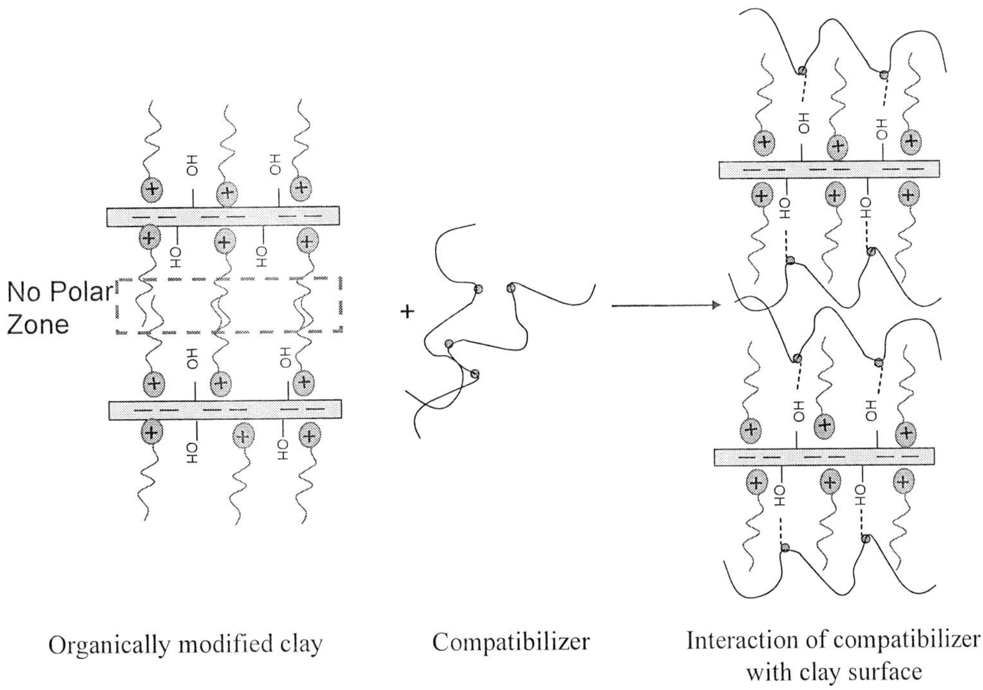

Figure 10. Model showing interaction of the organically modified clay with grafted PP used as compatibilizer.

Interaction of PP-g-IA with Nanoparticles

Grafted PP used as compatibilizer interacts with nanoparticles, for example clays, the interaction are through hydrogen bonding with OH groups on the surface of the clays as shown in Figure 10. As seen from this scheme, the polar groups of grafted PP interact with clay surface and therefore the interlayer space of the clay increases as the result of this interaction, generating a less polar region. This facilitates the entry of non polar PP chains into the clay gallery. [46]

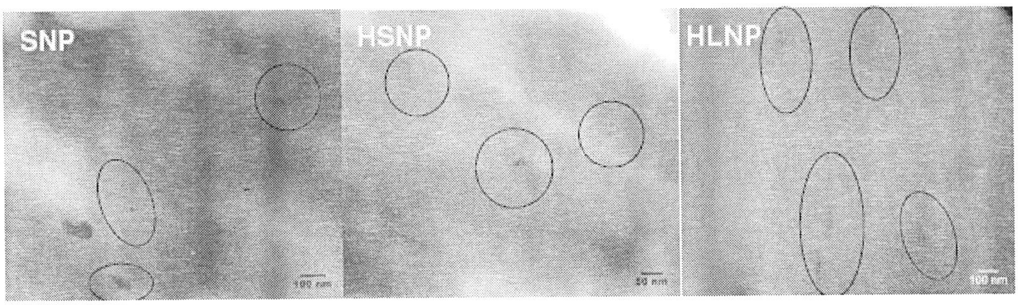

Figure 11. TEM images de nanocomposites obtained with 1% sol-gel nanoparticles, 3% de PP-g-MA and Polypropylene.

Nanocomposites Obtained Using Sol-Gel Nanoparticles

Recently, Chastek et al prepared silicate sheets with covalently attached C16-alkyl chains [84] and made nanocomposites with polystyrene samples of various molecular weights. A dramatic increase in Young's modulus of the nanocomposites compared with neat polystyrene was found, especially in the case of the polymers with the higher molecular weight. On the other hand, polypropylene nanocomposites with superficially treated silica nanoparticles were reported by Wu et al [85], showing remarkable improvements in the mechanical properties of the nanocomposites, pointing to the importance of the nanoparticles/matrix interface. Jain et al [86] worked on the modification of the surface of silica nanoparticles containing vinyl triethoxysilane groups that were dispersed in polypropylene, but the mechanical properties of the resulting materials are not mentioned in the paper. Many matrices have been used for the dispersion and evaluation of the modification of the properties of silica nanoparticles, such as polypropylene (PP) [87], high-density polyethylene (HDPE) [88], acrylic latex [89], polyethylene terephthalate (PET) [90] and polyethersulfone (PES) [91].

Nanocomposites generated with nanoparticles obtained previously (SNP, HSNP and HLNP) are going to be shown as follows and then it will be done a comparison with nanocomposites obtained with clays showed in the first section.

The figure 11 shows de TEM images of nanocomposites obtained with: 1% of Sol-Gel nanoparticles and 3% of compatibilizer agent (PP-g-MA) in a polypropylene matrix.

It is possible to see in the TEM images that spherical particles without hybrid composition (SNP) show small amassed of particles until 100 nm; these particles show sizes approximately of 10 nm (as it was shown previously). The nanocomposites obtained with the hybrid spherical particles (HSNP) show a better dispersion that the non-hybrid ones (SNP) with sizes approximately of 10 nm. This is the original size when the particles are alone. It shows the efficiency of the particles dispersion in the resin and so the contribution of the zone hybrid particle in the dispersion.

For the hybrid laminate particles (HLNP), an effective dispersion of laminates nanoparticles is shown, with dimensions of 10 nm and until 200nm (thickness and largess). The above show the efficiency of the nanoparticles obtained by Sol-gel. These nanoparticles without hybrid composition show good nanocomposites formation. When the hybrid function is established in nanoparticles, the dispersion in the resin is significantly improved; it changes of having nano-aggregates of nanoparticles of 100 to nanoparticles totally dispersed of 10 nm.

Mechanical Properties

The Table 3 shows the mechanical properties of nanocomposites obtained by Sol-Gel nanoparticles, grafted polypropylene with maleic anhydride and polypropylene as matrix.

It is evident the nanocomposites formation with all Sol-gel nanoparticles with increases in the module as the follows: since 40% for non-hybrid spherical nanoparticles (SNP), 48% for hybrid spherical nanoparticles (HSNP), until 62% for hybrid laminate nanoparticles (HLNP). Tensile strength (σ_y) has showed bigger increases: 57% for SNP, 73% for HSNP and 90% for HLNP.

Table 3.Mechanical properties of nanocomposites using, 1 wt.-% of nanoparticles obtained by Sol-Gel and 3% of PP-g-MA and different PPs. (Note: E, Modulus; σ_y, tensile strength; ε, elongation at break)

MFI PP	SNP			HSNP			HLNP		
	E (MPa)	σ_y (MPa)	ε (%)	E (MPa)	σ_y (MPa)	ε (%)	E (MPa)	σ_y (MPa)	ε (%)
13	1523(±22)	47(±2)	9	1620 (±22)	52(±1)	90	1770(±28)	57(±1)	10

Unmodified PP: E= 1092 (±45) MPa, σ_y=30 (±2) MPa, ε= 20%.

These values not only show the big efficiency in the nanocomposites formation generated by nanoparticles obtained by the Son-Gel method and the influence of the formation of hybrid systems, but also the big influence that presents the morphology of the nanoparticles in the mechanical properties of nanocomposites. It show that for laminate morphologies are more effective the increase of the mechanical properties.

Wu et al [13] in your publication are showing different concentrations of functionalized spherical silica nanoparticles were added to polypropylene and we compared them with hybrid nanoparticles of similar morphology (HSNP), presented by Moncada et, al. [REF] in the formation of nanocomposites in isotactic PP of MFI = 3. The higher changes in percentage of the mechanical properties for a 1% of load obtained in that work were approximately 17% for Young's modulus (E), where for pristine polymer E = 1390 MPa and polymer nanocomposites E = 1620 MPa and 5% for yield stress (σy), where for pristine polymer σy = 37 MPa and for polymer nanocomposites σy = 39 MPa. The changes percentage of the mechanical properties obtained in our work were 46% for Young's modulus (E), where for pristine polymer E = 1090 MPa and polymer nanocomposites E = 1590 MPa and 83% for yield stress (σy) where for pristine polymer σy = 30 MPa and for polymer nanocomposites σy = 55 MPa, clearly greater increases of these properties with silica nanoparticles obtained in our work was obtained. As to elongation at break (ε), in the work of Wu et al there is approximately a 300% increase where for pristine polymer ε = 90% and polymer nanocomposites ε = 320%, while in our work we found a 36% decrease of this property, where for pristine polymer ε = 250% and polymer nanocomposites ε = 160%. This shows the importance of the superficial modification of the nanoparticles on the mechanical properties of the nanocomposites, because it has a direct influence on the interfacial zone of the nanocomposites and on the energy transfer mechanism between the matrix and the nanoparticle.

The thermal stability of these nanocomposites also shows big increases for the nanocomposites compared with the polymer without load, in the polypropylene the temperature of decomposition is 319°C measured by DSC, and authors [REF] have reported values in nanocomposites of 350°C. This, amplifiers the applications range of this polymer so used at the moment and with a great expansion.

Table 4. Comparative mechanical properties of nanocomposites using 1% w/w of HLNP and Montmorillonite clays, 3% w/w PP-g-MA, and PP

MFI PP	HLNP			MMT[a]		
	E (MPa)	σ_y (MPa)	ε (%)	E (MPa)	σ_y (MPa)	ε (%)
3[b]	1607 (±23)	50 (±2)	42	1310 (±31)	34 (±1)	8
13[b]	1770 (±28)	57 (±1)	10	1415 (±23)	48 (±3)	5
3[c]	1711 (±35)	52 (±2)	50	1452 (±41)	43 (±1)	7

Note: E = Young's modulus, σy = elastic limit, ε = elongation at break.
Unmodified PP: E= 1092 (±45) MPa, σ_y=30 (±2) MPa, ε= 20%Values.
[a] taken from [45].
[b] Ziegler-Natta Polypropylene.
[c] Metallocene Polypropylene.

Comparison between Nanocomposites Obtained with Clays and Sol-gel Particles

Like it has been presented through this chapter, nanoparticles obtained by Sol-gel methods show advantages in the structure, composition and purity compared with nanoparticles obtained by clays.

In the Table 4, it is show a comparison between the properties of nanocomposites obtained by Sol-gel and clay nanoparticles, doing emphasis in mechanical properties.

For this, is done a comparison between laminates particles obtained by Sol-gel (HLNP) and the ones obtained by clays, principally with the Montmorillonite clay (MMT), because it is the most used in the studies. The obtainment conditions of these two nanocomposites were the same and were done by the same research group. In general, bigger increases are observed in the values of the mechanical properties in comparison with polymer without load, for nanocomposites generated with nanoparticles obtained by the Sol-gel method.

Comparison between PP-g-IA and PP-g-MA as Compatibilizer Agents in Nanocompositos Formation

It has been mentioned the big modifications in the mechanical properties of obtaining nanocomposites when the compatibilizer agent itaconic acid grafted in polypropylene (PP-g-IA) was used, now is shown a comparison of this one with the compatibilizer agent maleic anhydride grafted in polypropylene (PP-g-MA), that is the most used, conserving a similar percentage of grafted for the two products.

The table 5, shows the different values in the mechanical properties for obtaining nanocomposites, using MFI=26 Polypropylene as matrix, organically modified Montmorillonite and Hectorite as nanocomposites source, and the two different compatibilizer agents with similar percentage of grafted, itaconic acid grafted in polypropylene (0,7 % of graft) and maleic anhydride grafted in polypropylene (0,6% of graft).

Table 5. Mechanical properties of nanocomposites prepared by used PP of MFI=26, 1 wt.-% of different clays and 3% of PP grafted with 0,7% of itaconic acid (IA) or PP grafted with 0,6% of maleic anhydride (MA). (E, Modulus; σy, tensile strength; ε, elongation at break)

PP-g-X	Montmorillonite (OMMT)			Hectorite (OHTA)		
	E (MPa)	σ_y (MPa)	ε (%)	E (MPa)	σ_y (MPa)	ε (%)
[(a)]PP-g-$IA_{0,7}$	1900 (±42)	38 (±2)	8	2137 (±28)	51 (±2)	9
[(b)]PP-g-$MA_{0,6}$	1415 (±27)	38 (±3)	10	1607 (±32)	43 (±2)	10

Unmodified PP: E= 1092 (±45) MPa, σ_y=30 (±2) MPa, ε= 20%.
[(a)] Reference [31].
[(b)] Reference [46].

Considering the values for OM, and PP-g-$AI_{0.7}$ of E=1900 MPa, σy=38 MPa, and compared with the values of E=1415 MPa, σy=38 MPa when PP-g-$AM_{0,6}$ was used with the same clay. It is possible to appreciate the big difference in the obtained values. When grafted polypropylene with itaconic acid as compatibilizer agent was used, it presented the biggest increases. When OMMT clay was changed for OHTA and PP-g-$AI_{0.7}$ was used, values were: E=2137 MPa and σy=51 MPa, it compared with E=1607 MPa and σy=43 MPa for PP-g-$MA_{0.6}$. Again, the biggest increases were with PP-g-AI. These values show a bigger efficiency in the increment of the mechanical properties when grafted with itaconic acid polypropylene was used. It shows that in nanocomposites formation the itaconic acid grafted in polypropylene (PP-g-IA) is more effective that the maleic anhydride grafted in polypropylene (PP-g-MA).

This effect with better properties as compatibilizer agent of PP-g-AI in comparison with PP-g-AM can be explained with the chemical structures of the two compatibilizer agents studied yet. PP-g-AI show two groups free carbonic, with which can be done linkages bridges type of hydrogen with clay sheets. The oxygen of the compatibilizer with the hydrogen of the clay and vice versa (the oxygen of the clay with the hydrogen of the compatibilizer) are showing in the figure 12 generating six possible linkages points, in the other hand, PP-g-AM has an anhydride group that allows it to generate three possible linkage points by bridge of hydrogen between the compatibilizer oxygen and clay.

Therefore, when the number of possibilities of generated hydrogen bridges is bigger, the interaction of the compatibilizer agent with the clay is also bigger and better interface among the different materials (clay and PP) and also better compatibility, this is going to help, like it has been mentioned previously, to a bigger energy and stress transfer between the nanocomposites matrix and its load. All the above-mentioned is increased in great measure due to the great superficial area that nanoparticles have.

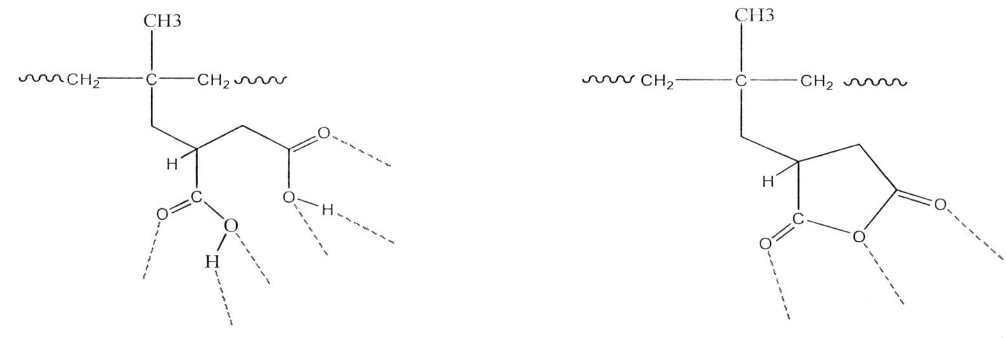

Figure 12. Interactions bridge type of hydrogen generated by the different compatibilizer with the clay surface.

Nanoparticles Used as Sensor

The nanoparticles obtained in this chapter, can be functionalized to increase its applications and to meet the needs of the industry. The functionalization is possible using the Sol-gel technique, enabling the generation of hybrid materials and from this, to obtained materials with characteristic chemical that depending of functional group added to the structure of the nanoparticles can be generate new materials. For example materials with more bio-compatibility, smart and intelligent particles that can do a molecular identification of compound (toxical compound and viruses etc). These compounds are called chemical sensor, other sensors can be obtained by polymer [92,93] and oxides [94,95].

Sol-gel modification usually, the firs step un functionalizing nanoparticles involves building a self-ensambly monolayer (SAM) on the particles surface. In this approach, bifunctional molecules containing a hydrophilic head group and hydrophobic tail group absorb onto a substrate or an interface as closely packed monolayer, mostly thought the attraction between hydrophobic chains. The tail group and the head group can be chemical modified to contain functional group. SAMs are widely explored for engineering the surface and interfacial properties of materials such to wetting adhesion and friction. [96,97] These monolayer are also used to mediate the molecular recognition processes and to direct oriented crystal growth. [98] Chemical and physical modification of solid surface is extensively used in chromatography, chemistry analysis, catalysis, electrochemistry, electronic industry and the more actually used are for sensing and electronic devise.

Depending of substrate surface, two kinds of SAMs are commonly encountered. On metal surface such a gold and silver thiolated alkyl chains are often used, and the oxides surface such as silica, alkyl silanes are usually used. The silane and group hydrolyze first, and then become covalently bound to the substrate through a condensation reaction. The SAMs molecules may also be covalently linked with one another through siloxane group as a result of the hydrolyses of the silane group and their subsequent condensation. Because of the disordered nature of the substrate surface, SAMs on the oxide surface are usually disordered [99].

USE OF NANO SENSOR IN THE INDUSTRY

These sensors generated with nanometer size, can be to present applications in the broadest sectors of the industry, as medicine, agriculture, environmental science, construction, aerospace, even packing. Below will be presented few details of its application in the food and drug packaging industry.

FOOD AND DRUG PACKAGING

More than 200 companies around the world are developing nanomaterials in order to be used in food and pharmaceutical products packaging. United States is the main country in research, next Japan and China. In 2006 these investments increased in dollars, of $2.6 bn to $7.0 bn. It is expected that in 2010 this value can increase to $20.4 bn (49) and for the pharmacists industry in 2011 can be of $11 bn. It is predicted that in 2010 Asia will be the great market for these products, with the 50% of the world population. [100]

On the other hand, parameters like increment in the demand of surer feeding products (bio-terrorism), better relationship cost-efficiency, bigger control (producer) and opportune and effective information to the consumer, are taking to the development of packaging materials for foods and more demanding pharmaceutical products. Investigations in packaging are being carried out that allow the consumer to detect the possibility of microbiology, chemistry or physic contamination; and allow the producer better controls in their systems of production, transport and storage. This is possible to get with the called intelligent packaging, which, through sensors detect the parameters before mentioned by means of packaging and immediately, without analysis in laboratories. These sensors can be obtained using diverse technologies [101,102,103,104], among them, nanotechnology is predicted like more efficient and with more easiness to disperse it in the polymeric resins. But, this concept of intelligent packaging (IP) is not alone; it is ligature directly to other concepts like active packaging (AP) and smart packaging (SP) that will be enunciated next.

Generally, in literature it is possible to find some divergences among the packaging definitions above mentioned, (smart packaging SP, active packaging AP, and intelligent packaging IP). As follows: For Clarke (2001) IP is packaging that presents logical capacity, SP is one that communicates and AP is one that controls. For Brody and others (2001) IP is packaging systems that indicator and communicate and SP is the packaging that has the capacity of both AP and IP. For Rijk (2002) IP is one that monitories the conditions of packed food and gives information about the food quality during the transport and storage. The ambiguity showed in these definitions can cause difficulties in their interpretation and it is for this reason that the definitions of Kil L.Yan (Investigator of the Department of Foods Science of the University of Rutgers [105] (USA)) is considered the most appropriate, where SP are the packaging that gather AP and IP characteristics. IP are the packaging systems able to carry out intelligent functions (to detect, to indicator, to record, to communicate, with a scientific logic) in order to facilitate the decision of extending useful life, increasing security, improving quality, providing information and noticing about possible food problems. For AP, increasing food protection is a function.

This is, in a total packaging system, IP is responsible component to indicator the food environment and to give the information. AP is responsible component to make some action to protect the food (like an antimicrobial compound). It is important to say that IP and AP are not mutually excluding, some packaging systems can be classified as IP and others as AP or maybe both as SP. In an appropriate situation IP, AP and the traditional function of packaging are going to work in association to provide a total packaging solution. They always should depend on necessities and characteristic of packed product. The table 6 shows the main applications of each one of mentioned packaging.

The previous information's can be collected in the bars code, which allows to the producer knows always the quality of his product and to the consumer has more information about the product that wants to buy.

With the above-mentioned, it is possible to say that the traditional function of packaging is changing. It has been simply the material that contained the food and protects it of the external atmosphere. And it is becoming in a functional material that allows that as producer as consumer can verify the product they are selling or buying and with a nutritious quality and appropriate microbiological.

INTELLIGENT PACKAGING BY SUPPLY CHAIN

The packaging technologies using systems RFID (intelligent packaging), are being presented as the most suitable for the identification and communication of the characteristics of the food, allowing know through the supplement chain of food, the state of preservation, as well as the location both on the shelf and in transport. The communication could be done at any time allowing for continuous monitoring of the food through network connection. In the future, the supplement chain will be principally of intelligent and active packaging integrates with communication system (RFID).

Table 6. Main applications of the active, intelligent and smart packaging [106]

Active Packaging AP	Intelligent Packaging IP	Smart parking SP
Antimicrobial	Microbial growth Indicator	Antimicrobial with indicator integrated
Oxygen absorbers	Time-temperature indicator	Oxygen absorbers with indicator integrated
Ethylene absorbers	Nutritional attributes Indicator	Ethylene absorbers with indicator integrated
Humidity absorbers	Indicator of gas concentration in packaging of modified atmosphere	Barrier plus with indicator integrated
Absorbers of off flavors, amines and aldehydes	Indicator of impact occurrence	Self - warming / cooling with indicator integrated
UV-light absorbers	Radio Frequency Indicator RFID	

The packages will protect the food without additives, inform about the product quality and history in every stage of the logistic chain, guide the journey of the package, reduce product loss, and will give real-time information to the consumer about the properties /quality/use of the product.

FUTURE TRENDS IN THE WORLD OF NANOPARTICLES

The nanoparticles regardless of their provenance (clay, Sol-Gel, etc.) should see each needs more momentous and more demanding both chemical and selectivity of compound identification of organic substances such as viruses and bacteria live, or recognition of specific molecules as cancer cells, DNA, etc.. They must meet the requirements of interaction with the human body through functionalizations organic-inorganic enabling good communication and acceptance by living cells. Another application the need is of new physical functions as a high thermal and mechanical resistance, good compatibility with anywhere materials etc.

REFERENCES

[1] Medeiros ES, Mattoso LH, Nanotecnologia. In: Duran N, Mattoso LHC, Morais PC. (Org.). Nanotecnologia: Introdução, preparação e caracterização de nanomateriais e exemplos de aplicação. São Paulo: Artliber, 2006;1: 1.

[2] Moriarty P, Nanotechnology: Radical new science or plus ça change?. Nanotechnology Perceptions. 2005;1(3):115-118.

[3] Scott F, Nanotechnology: Radical new science or plus ça change?- the debate. Nanotechnology Perceptions. 2005;1(3):119-146.

[4] Drexler KE. Molecular Engineering: An approach to the development of general capabilities for molecular manipulation. Proc. Nad. Acad. Sci. 1981; 78(9): 5275-5278.

[5] Merkle RC. Convergent assembly. Nanotehcnology 1997; 8(1): 18-22.

[6] Merkle RC, Parc X, Casing and assembler. Nanotechnology 1999;10:315-322.

[7] Merkle RC, Robert A, Freitas Jr. Theoretical analysis of a carbon-carbon dimer placement tool for diamond mechanosynthesis. J. Nanosci. Nanotechnol. 2003; 3: 319-324.

[8] Hall JS, Architectural considerations for self-replicating manufacturing systems. Nanotechnology 1999; 10(3): 323-330.

[9] Herrmann PSP, Mattoso LH, A Nanotecnología no agronegócio, uma realidade na embrapa. XXXVI Congreso Brasilero de Engenharia agrícola, 2007.

[10] Kellehr K. http://www.thestreet.com/tech/internet/10279136.html, visited in December 2007.

[11] Moncada E, PhD thesis, Universidad de Chile, 2006.

[12] http://www.uclm.cs/uscrs/higucras/yymm/Arcillas.htm, Visited in Julio 2004.

[13] Ogawa M, Chem Mater. 1998; 10: 1382.

[14] Wang L.Q, Liu J, Exarhos G.J, Flanigam K.Y and Bordia R, J. Phys. Chem. B. 2000; 104: 2810.

[15] Kunyima B, Viaene K, Hassan Khalil M, Schoonheydt R.A, Langmiur, 1990; 6: 482.
[16] Ray S.S, Okamoto M, Prog. Polym. Sci, 2003; 28: 1539.
[17] Zanneti M, Camino G, Reichert P, Miilhaupt R, Macromol. Rapid Comun, 2001; 22: 176.
[18] http//www. nanomat.de Visited April 2007
[19] Alexandre M, Dubois P, Materials Science and Engineering, 2000; 28: 1.
[20] Carrado K.A, Applied Clay Science, 2000;17:1.
[21] Kim D.W, Blumstein A, Kumar J, Tripathy S.K, Chem. Mater., 2001;13: 243.
[22] Shimojima A, Umeda N, Kuroda K, Chem. Mater., 2001;13: 3610.
[23] Hench Larry L, West Jon K, Chem. Rev., 1990; 90; 33.
[24] Galo J. de A. Llia-Soller G.J, Sanchez C, Lebeau B, Chem. Rev., 2002; 102: 4093.
[25] Hong-Ping L, Chung-Yuan M, Acc. Chem. Res., 2002; 35: 927.
[26] Soler-Illia G. J, Sanchez C, Lebeau B and Patarin J, Chem. Mater. 2002; 102: 4093.
[27] Ogawa M, Langmuir, 1997; 13: 1853.
[28] Yun S.K and Maier J, Inorg. Chem., 1999; 38: 545.
[29] Chastek T.T, Stein A and Macosko C, Polymer, 2005; 46: 4431.
[30] Wu C.L, Zhang M.Q, Rong M.Z and Friedrich K, Compos. Sci. Technol. 2005; 65: 635.
[31] Moncada E, Quijada R, Retuert J. Nanoparticles prepared by the Sol-Gel method and their use in the formation of nanocomposites with polypropylene, Nanotechnology 2007; 18: 335606.
[32] Stober W, Fink A and Bohn E, J. Colloid Interface Sci., 1968; 26: 62.
[33] Ellis T.S and D'Angelo J.S, J. Appl. Polym. Sci., 2003; 90: 1639.
[34] Hernandez C and Pierre A.C, Langmuir, 2000; 16: 530.
[35] Gao Y, Choudhury N.R, Dutta N, Matisons J, Reading M and Delmotte L, Chem. Mater., 2001; 13: 3644.
[36] Hernandez C, and Pierre A.C, Langmuir, 2000; 16: 530.
[37] Transfetetti B.C and Davanzo C.U, Macromolecules, 2004; 37: 459.
[38] Nakanishi K, Infrared Absortion Spectroscopy, Ed. Holden-Day, San Francisco, 1967. 39.
[39] Lopez T, Asomoza M, Gómez R, Journal of Non-Crystalline Solids, 1992; 147&148: 769.
[40] Okada A, Fukushima Y, Kawasumi M, Inagaki S, Usuki A, Sugiyama S, et al. Inventors 1986. Composite material and process for manufacturing same. United Estate Patent 4,739,007 19 Abril 1988.
[41] Kawasumi M, Kohzaki M, Kojima Y, Okada A, Kamigaito O. Inventors 1988 Process for producing composite material. United Estate Patent 4,810,734. 7 March 1989.
[42] Usuki A, Kawasumi M, Okada A. Synthesis of nylon 6-clay hybrid, J. Mater. Res. 1993; 8(5): 1179-1184.
[43] Yang F, Yingar R, Nelson G.L, Flammability of polymer-clay and polymer-silica nanocompositos. Journal of fire Sci, 2005; 23: 209-226.
[44] Kojima A, Yamada Ki, Fujii T, Hirata M. Improvement in adhesive-free adhesion by the use of electrostatic interactions between polymer chains grafted onto polyethylene plates. J. Appl. Polym. Sci. 2006;101(4): 2632-2638.

[45] Moncada E, Quijada R, Retuert J. Comparative effect of metallocene and Ziegler-Natta polypropylene on the exfoliation of Montmorillonite and Hectorite clays to obtain nanocompositos. J. Appl. Polym. Sci. 2007; 103; 698 -706.

[46] Moncada E, Quijada R, Yazdani-Pedram M. Usse of PP grafted with itaconic acid as new compatibilizer for PP/Clays nanocomposites. Macromol. Chem. Phys. 2006; 207: 1376-1386.

[47] Moncada E, Quijada R, Yazdani-Pedram M, Inventors 2006. Patent PCT/IB2007/003008

[48] Zhao C, Qin H, Gong F, Feng M, Zhang S and Yang M, Mechanical, thermal and flammability properties of polyethylene/Clay nanocompositos. Polym. Degrad. Stabil 2005; 87(1): 183-189.

[49] Mohanty S, Nayak S.K. Melt blended polystyrene/layered silicate nanocomposites: effect of Clay modification on the mechanical, thermal, morphological and viscoelastic behavior. J. Thermoplast. Compo. Mater, 2007; 20(2): 175-193.

[50] Xu W, Zhou Z, Ge M, Pan W. Polyvinyl chloride/Montmorillonite Nanocomposites: glass transition Temperature and mechanical properties. Journal of Thermal Analysis and Calorimetry, 2004;78 (1): 91-99.

[51] Alma P, Patiño S, Ramos LF. Morphological and thermical properties of ABS/Montmorillonite nanocomposites using ABS with different AN content, Macromol. Mater. Eng, 2007; 299(3): 302-309.

[52] Guo-An Wang, Cheng-Chien Wang, Chuh-Yung Chen. Preparation and characterization of cayered double hydroxides – PMMA nanocomposites by solution polymerization, Journal of Inorganic and Organometallic Polymers and Materials, 2005; 15(2): 239-251.

[53] Kráčalík M, Mikešová J, Puffr R, Baldrian J, Thomann R, Friedrich C. Effect of 3D structures on recycled PET/organoclay nanocomposites. Polymer Bulletin, 2007; 58(1): 313-319.

[54] Pasanovic-Zujo V, Gupta R.K, Bhattacharya S.N. Effect of vinyl acetate content and silicate loading on EVA nanocomposites under shear and extensional flow. Rheologica Acta, 2004; 43(2): 99-108.

[55] Tianshi Yu, Jiaping Lin, Jiafu Xu, Tao Chen, Shaoliang Lin. Novel polyacrylonitrile nanocomposites containing Na-Montmorillonite and nano SiO2 particle. Polymer, 2005; 46(15): 5695-5697.

[56] Yong Gao, Peng He, Jie Lian, Lumin Wang, Dong Qian, Jian Zhao, et al. Improving the mechanical properties of polycarbonate nanocomposites with plasma-modified carbon nanofibers. Journal of Macromolecular Sciences, Part B: Physics, 2006; 45: 671–679.

[57] Marc X, Reinholdt, R, Kirkpatrick J, Pinnavaia T. Montmorillonite-poly(ethylene oxide) nanocomposites: interlayer mlkali Metal behavior. Phys. Chem. B, 2005; 109 (34): 16296 -16303.

[58] Lijia P, Pingsheng H, Gang Z, Dazhu C. PbS/epoxy resin nanocomposite prepared by a novel method. Materials Letters, 2004; 58(1-2):176-178.

[59] Contreras V, Cafiero M, Da Silva S, Rosales C, Perera R, Matos M. Characterization and tensile properties of ternary blends with PA-6 nanocomposites. Polymer Engineering & Science, 2006; 46(8): 1111-1120.

[60] Khalid Saeed, Soo-Young Park. Preparation and properties of multiwalled carbon nanotube/polycaprolactone nanocompositos, Journal of Applied Polymer Science, 2007; 104(3): 1957-1963.

[61] Zhang Y, Lee SH, Mitra Y, Toghiani H, Pittman C. Phenolic resin/octa(aminophenyl)-T8-polyhedral oligomeric silsesquioxane (POSS) hybrid nanocomposites: synthesis, morphology, thermal and mechanical properties. Journal of Inorganic and Organometallic Polymers and Materials, 2007; 17(1): 159-171.

[62] Valentini L, Biagiotti J, Kenny J.M , López Manchado M.A. Physical and mechanical behavior of single-walled carbon nanotube/polypropylene/ethylene-propylene-diene rubber nanocompositos, Journal of Applied Polymer Science, 2003; 89(10): 2657-2663.

[63] Pattanayak A, Jana S. Properties of bulk-polymerized thermoplastic polyurethane nanocompositos, Polymer, 2005; 45(10): 3394-3406.

[64] Reiko Saito, Tadakuni Tobe. Synthesis of poly(vinyl pyridine)-silica nanocomposites using perhydropolysilazane, J. Appl. Polym. Sci., 2004; 93(2): 740-757.

[65] K. Yano, J. Polym. Sci., Part A: Polym. Chem.,1993; 31(10): 2493.

[66] Heinemann J, Reichert P, Thomas R, Mulhaupt R, Macromol. Rapid. Commun., 1999; 20: 423.

[67] Vaia R, Giannelis A, Chem. Mater., 1993; 5(12): 1694.

[68] Hasegawa N, Kawasumi M, Kato M, Usuki A, Okada A, J. Appl. Polym. Sci., 1998; 67: 87.

[69] Hasegawa N, Kawasumi M, Kato M, Usuki A, Okada A, Macromolecules, 1997; 30: 6333.

[70] Ye W, Feng-B C, Yann-C L, Kai-C W, Composites Part B: engineering, 2004; 35: 111.

[71] Ye W, Feng-B C, Kai-C W, Journal of Applied Polymer Science, 2005; 97: 1667.

[72] Suprakas R, Okamoto M, Prog. Polym. Sci.,2003; 28: 1539.

[73] Silva C.R, Fonseca M.G, Barone J.S, and Airoldi C, Chem. Mater., 2002; 14: 175.

[74] Ishida H, Blackwell S.C and Blackwell J, Chem Mater., 2000; 12: 1260.

[75] Zhu J, Uhl F.M, Morgan A.B and Wilkie C.A, Chem. Mater., 2001; 13: 4649.

[76] Koo C.M, Ham H.T, Kim S.O, Wang K.H, Macromolecules, 2002; 35: 5116–5122.

[77] Zanneti M, Camino G, Reichert P, Miilhaupt R, Macromol. Rapid Comun., 2001; 22: 176.

[78] Sinha S, Okamoto M, Prog. Polym. Sci., 2003; 28: 1539.

[79] Lertwimolnun W, Vargas B, Polymer, 2005; 46: 3462.

[80] Wang Y, Chen F.B, Wu K.C, J. Appl. Polym. Sci., 2005; 97: 1667.

[81] Moad G, Prog. Polym. Sci., 1999; 24: 81.

[82] Yazdani-Pedram M, Vega H, Quijada R, Macromol. Chem.Phys., 1998; 199: 2495.

[83] Yazdani-Pedram M, Vega H, Quijada R, Polymer, 2001; 42(10): 4751.

[84] Chastek T T, Stein A and Macosko C, Polymer, 2005 46 4431.

[85] Wu C.L, Zhang M.Q, Rong M.Z and Friedrich K, Compos.Sci. Technol., 2005; 65: 635.

[86] Jain H, Goossens F, Picchioni P, Magusin B, Mezari M and Van D, Polymer, 2000; 46: 6666.

[87] Wu C.L, Zhang M.Q, Rong M.Z and Friedrich K, Compos.Sci. Technol., 2002; 62: 1327.

[88] Zhang M.Q, Rong M.Z, Zhang H.B and Friederich K, Polym. Eng. Sci., 2003; 43: 490

[89] Nicolas L and Narkis M Polym. Eng. Sci. 1971; 11: 194.

[90] Kr´a˘cal´ık M, Mike˘scov´a J, Puffr R, Baldrian J, Thomann R and Friedrich C, Polymer Bulletin, 2007; 58: 313.

[91] Jana S.C and Jain S, Polymer, 2001; 42: 6897.

[92] Riul A, Malmegrim R.R, Fonseca F.J, Matosso L.H, Nano-Assembled Films for taste sensor application, Artf. Organs. 2003; 27:(5), 469-472.

[93] Dos Santos D, Riul A, Fonseca F.J, Mattoso L.H, A layer –by-layer film of chitosan in a taste sensor application, Macromol. Biosci., 2003; 3: 591-595.

[94] Pacquit A, Tong Lau, Doamond D, Smart packaging for monitoring of fish freshness, Proc. of SPIE, 2005; 5826: 545-550.

[95] Seok-In B, Gravure-printed colour indictor for monitoring kimchi fermentation as a novel intelligent packaging, Packag. Technol Sci., 2002; 15: 155-160.

[96] Whitesides G.M, Self-assembling materials, Sci. Am. 1995; 273: 146-149.

[97] Ulman A, Formation and structure of self-assembled monolayer, Chem. Rev., 1996; 96: 1533-1554.

[98] Bunker B.C, Rieke P.C, Tarasevich B.J, Cambell A.A, Fryxell G.E, Ceramic thin films formation on functionalized interface through biomimetic processing, Sicence, 1994; 264: 48-55.

[99] Self-Assembled Nanostructues: Nanostruture Science and Technology, (David J. Lockwood) New York, United States of America. Jin Zhang, Zhong-lin Wang, Jun Liu, Shaowei Chem, and Gang-yu Liu. ISBN:0-306-47299-6.

[100] Moraru, C, Panchapakesan C, Takhistov K. Nanotechnology: new frontier in food science. Food Technology., 2003; 57: 25-28.

[101] Das R, Selke S, Harte J. Development of electronic nose method for evaluation of HDPE flavour characteristics, correlated with organoleptic testing. Packaging Technology and Science, 2007; 20(2): 125-136.

[102] Seok-In Hong. Gravure-printed color indicators for monitoring Kimchi fermentation as a novel intelligent packaging. Packag. Technol. Sci., 2002;15:155-160.

[103] O'Mahony F, O'Riordan T, Papkovskaia N, Ogurtsov V, Kerry J, Papkovsky D. Assessment of oxygen levels in convenience-style muscle-based sous vide products through optical means and impact on shelf-life stability. Packaging Technology and Science, 2004; 17(4): 225-234.

[104] Bodenhamer WT. Inventor 2002. Toxin Alert. United Estate Patent 6,376,204. 18 April 2004.

[105] Yam K.L. Takhistov P.T. and Miltz J. Intelligent packaging: concepts and applications. Journal of Food Science. 2005; 70(1): R1-10.

[106] Butler P. Smart packaging – Intelligent packaging for food, beverages, pharmaceuticals and household products: http://www.azom.com/details.asp? ArticleID=2152. visited April 2007.

In: Encyclopedia of Nanotechnology
Editor: Elwood D. Carlson
ISBN 978-1-60692-079-4

Chapter 45

THE APPLICATIONS OF NANOPARTICLES IN ELECTROCHEMISTRY

Ruhai Tian, Jinfang Zhi
Technical Institute of Physics and Chemistry, Chinese Academy of Sciences, No.2, Beiyitiao, Zhong-guan-cun, Haidian District, Beijing, 100080, P. R. China

ABSTRACT

Combination of Nano-sciences & techniques and electrochemistry makes electrochemical techniques are promising in many aspects, such as electroanalysis, bio-analysis, power & energy, etc. At the same time, the introduction of Nano-sciences & techniques in electrochemistry also expands the unknown filed of the interdiscipline.

In this comment paper, we will discuss the applications of nanoparticles (NPs) in electrochemistry including the following threads 1) why scientists introduce the NPs into electrochemistry; in this part, we summarize the develop of Nano-sciences & techniques; mainly discuss some basic theories and concepts about NPs and electrochemistry and how scientists use NPs in electrochemistry; 2) progresses of NPs application in electrochemistry , in this part, we summarize the achieved marvelous achievements in this area; 3) challenge of the NPs application in electrochemistry, in this part, we summarize the challenges that electrochemists are encountering with and bring up our opinions and strategies.

1. INTRODUCTION

Electrochemical methods are widely adopted techniques because of their essential significance not only in scientific research but also in industry. [1] Actually, the field of electrochemistry is a very large domain, encompassing fundamental theories (e.g. electron transfer, electrolyte diffusion etc), phenomena (e.g., electrophoresis and corrosion), technologies and practical applications (e.g. electrochromics, electrolysis, sensors, batteries, fuel cells, supercapacitors, etc). Electrochemistry related publications such as journals, monographs, patents and conferences are largely published.

On the other hand, nano-sciences & techniques have been developed a lot since Dr Richard P. Feynman's famous prophetic lecture "There's Plenty of Room at the Bottom" in 1959. Along with the development and advancement of nano-related theories and techniques, the programmable and reproduceable fabrications of nano-materials and nano-structures are becoming more and more reliable. Nano-sciences & techniques are dramatically infiltrating to experimental sciences, such as chemistry, physics, biology, materials, electronics, etc. To this day, more and more organizations and scientists are paying more and more attentions on nano-sciences & techniques. Nano-sciences & techniques are opening new doors for not only sciences & techniques but also industries.

Nano-sciences & techniques is one of the developing areas at top speed due to a wide variety of potential applications in biomedical, optical, material, and electronic fields. Nano-related organizations, journals were established largely and quickly after 90s of last century and numerous of nano-related achievements have been published. "Nanoparticle" is a descriptive name of small particles with at least one dimension is less than 100 nm, which consists of nanospheres, nanoreefs, nanoboxes, nanocrystals, nanosheets, etc. The unconscious application of NPs started from 9th century in pottery and art so as to get glittering effect or beautiful colors (this is obviously based on the optical properties of NPs). As early as 1857, Faraday consciously synthesized very stable gold nanoparticle colloidal solution through solution process but people didn't know the details of "nano" at that time. After the development of advanced instruments (Scanning electronic microscope/SEM, Transmission electron microscopy/TEM, X-ray Photoelectron Spectroscopy/XPS, atomic force microscopy/AFM, dynamic light scattering/DLS, powder x-ray diffractometry/XRD, etc.) in the early of 20th century, scientists got the tools to develop nano-related sciences and technologies.

Today, we know that the small size of NPs results in some fancy properties (electronic band structure change, crystal structure change etc) and size-effects (quantum confinement effect in semiconductor particles, surface plasmon resonance in some metal particles especially on the surface and superparamagnetism in magnetic materials, etc). No other reasons than these fancy properties attracted more and more scientists to focus on nano-related researches.

In this chapter, we mainly discuss the applications of NPs in electrochemistry. It is not only impossible to cover all the NPs-related electrochemistry in this small chapter, but also unnecessary because a lot of reviews about NPs related electrochemistry have been published.

Reference [2,3] summarized the applications of NPs in electroanalysis.

Reference [4] summarized the applications of Carbon nanotubes (CNTs) in the DNA sensor.

Reference [5] reviewed the applications of CNTs in electrochemistry.

Reference [6] reviewed nanowire-based electrochemical biosensors.

Reference [7] reviewed the application of NPs in energy fields.

Moreover, some special topics were also published , e.g. prof. Erkang Wang reviewed the advances in synthesis and electrochemical applications of gold nanoparticles (AuNPs). [8]

We hope this chapter about the NPs applications in electrochemistry could be helpful to those interested researchers. At the same time, we are looking forward to see more achievements and publications in this area. In this chapter, we discuss two parts of contents: 1) the fabrication of NPs 2) applications of NPs in electrochemistry. The second part is the focal point we want to discuss. We will just discuss the applications of NPs in electrochemical analysis and energy.

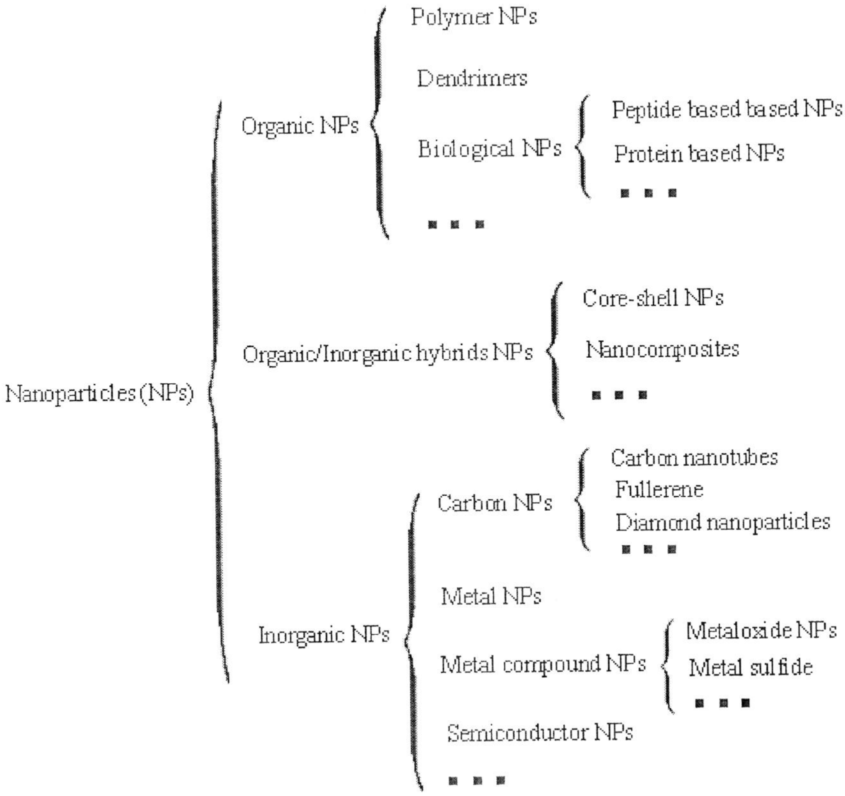

Chart 1. The classification of NPs.

2. THE NPS FABRICATION

Synthesis and applications of NPs have witnessed tremendous growth during the past decades. Here, we can just simply classify NPs as shown in chart 1. Inorganic NPs are most widely used nanoelements in electrochemistry because of their excellent conductivity, catalytic abilities and stabilities. In this chapter, we just discuss the applications of inorganic NPs in electrochemistry. The applications of inorganic NPs mainly focus on metal NPs; carbon NPs; semiconductor NPs; other nanostructures materials.

2.1. 0-Dimensional Inorganic Nanostructure Materials (Metal NPs)

The main approaches used to fabricate 0-dimensional metal NPs for the applications in electrodes were either the evaporation or direct electrochemical deposition of metal onto an electrode surface or bottom-up self-assembly of metal NPs by solution processes.

Solution Processes

Fabrication of NPs through solution process is most popular because of good reproducibility, easy control, high dispersion and narrow size distribution. Frens [9] method and phase-transfer [10,11] method are widely used in solution synthesis of metal NPs. The difference between these two methods is that metal NPs synthesized by Frens method are surface charged but NPs synthesized by phase-transfer are more surface-functional because of the capping organic layer. Synthesis of colloids AuNPs [9,10,12-19], cobalt NPs (CoNPs) [20] via chemical reduction methods has been reported. Multifarious solution process, such as microemulsions [21], reverse micelles and organometallics [22] were also developed to synthesize silver NPs. Selective structured platinum NPs were also synthesized in solution with different reagent as reduction agent [23,24]

Electrochemical Methods

Electrochemical method is another common method used in the fabrication of metal NPs, especially nanostructured transition metal clusters [25]. The advantage of this method is the particles size, shape, surface properties and product quantity could be easily controlled [26-28]. For electrochemistry research, metal NPs can be directly deposited onto electrode surface without more subsequent immobilization procedures. Manfred T. Reetz reported the synthesis of tetra-alkyl-ammonium-stabilized Pd and Ni clusters by a simple electrochemical process [26]. The size of particles could be controlled by variation of current density and the processes were high yield and absence of side products. Reference [27] reported the electrochemical synthesis of gold nanorods with different mean aspect ratio. L. Rodrı´guez-Sa´nchez reported the synthesis of 2-7nm silver NPs with silver as sacrificial anode, and tetrabutylammonium bromide as stabilizing salt [28]. Direct deposition of silver NPs (AgNPs) [29,30], platinum NPs (PtNPs) [31-35], nickel NPs (NiNPs) [36], ruthenium composites (RuNPs) [37], chromium NPs (CrNPs) [38], palladium NPs (PdNPs) [26,39], CdS NPs [40] on electrode surface was also studied.

Other Methods

Many other methods are also used for the fabrication of metal NPs such as radio-frequency sputtering [41-45], electron beam lithography [46], sonochemical synthesis [47], chemical vapor deposition [48], laser evaporation [49,50] Sonoelectrochemical method [51], etc.

2.2. 1-Dimensional Inorganic Nanostructure Materials

The most widely used 1-D nanostructure materials in electrochemistry are CNTs, carbon wires, silicon nanowire (SiNWs), conducting polymer wires (CP-NWs), etc. The basic construction methods of 1-D nanostructure materials including:

1. Morphology template auxiliary direct formation (porous Al_2O_3, DNA etc)
2. Symmetric growth of seed at liquid/solid, vapor/solid, vapor/liquid/solid interface
3. Selective etching according to the intrinsically properties of different crystalline face
4. Capping agents kinetically controlled growth of seed

5. Dip-pen nanolithography (AFM, STM tips)
6. Other methods

Reference [6] reviewed the fabrication, assembly/alignment and surface functionalization of 1-D nanostructures (CNTs, SiNWs, CP-NWs). As one of the 1-D nanostructures, because of their remarkable mechanical, chemical, thermal and electrical properties [52-54], CNTs have been developed a lot since its discovery in 1991 [55,56]. CNT structure can be thought as rolled-up rectangular sheet of graphite and can be divided into single-walled (SWNTs) and multi-walled (MWNTs). MWNTs are coaxial CNTs with 0.34nm of interlayer space. The diameters of CNTs range from 0.4nm to 3nm for SWNTs and from 1.4nm to 100 nm for MWNTs. The excellent properties of CNTs results from their tube-shape rolled graphite structure and their symmetric and electronic structure. CNTs electrodes can be fabricated through 1) polymer confining 2) paste conglutination 3) layer by layer assembly 4) matrix growth and sol-gel matrix assembly [5].

3. The Application Fields of NPs in Electrochemistry

3.1. Electrochemical Analysis

Electrochemical analysis is one of the most sensitive analytical methods. [1] The applications of NPs in electrochemical analysis mainly focus on the utilization of the nano-effects of NPs: catalysis abilities, large specific surface area (surface area to volume), etc. In Reference [3], Christine M. Welch and Richard G. Compton reviewed the applications of metal NPs (gold, silver, platinum, nickel, copper, iron, etc) in electroanalysis. Here, we discuss the applications of NPs in electrochemical analysis in two aspects:

3.1.1. As Nanoelectrode or Arrays

Nanoelectroces (NEs) and nanoelectrodes arrays (NEAs) which possess 1-100nm dimension range show great advantages with respect to traditional macro electrodes. The large mass transport mode of electrolyte on NEs&NEAs results in the low Faradic capacitance, fast electrochemical response and high signal-to-noise ratio, which makes them very suitable in the trace analysis and ultra fast electron exchange kinetic research [57,58]. Moreover, the small size of NEs&NEAs makes them suitable in the analysis of small volume sample and easy to assemble them as chips. Based on these fancy and unique properties, NEs&NEAs can be used as electrochemical detector in high-performance liquid chromatography (HPLC) and capillary electrophoresis (CE) [59,60].

NPs are widely used as elements for the fabrication of ultramicro-electrode and arrays even the nanoelectrode and arrays. Comparing with macroelectrode, ultramicroelectrode, ultramicroelectrode arrays, NEs&NEAs possess great advantages, such as lower ohmic drop of potential, shorter time to get the steady-state signal, enhancement of signal-to-noise, etc. [58] NEs&NEAs based on the NPs immobilization and assembly have been proved to be ultra-sensitive and catalytic. [61] Here, the definition of nanoelectrode is electrode with a critical dimension in the range of nanometers, where the critical dimension is not an exact number. According to the agreed consensus in nanoscience and technologies, we define the critical dimension of nanoelectrode is less than 100nm.

3.1.1.1. Gold Nanoparticles (AuNPs) as NEs&NEAs Element

In metal NPs, AuNPs were fabricated very early [62] , in 1857, (benzene was discovered in 1825 and its structure was given in 1865) and studied almost most clearly [16]. Allen J. Bard etc reported the fabrication and characterization of self-assembled spherical gold ultramicroelectrodes in the size of 1-30μm with 8nm AuNPs as basic elements through the bottom-up approach. [63] Figure 1A shows the fabrication procedures of gold ultramicroelectrodes based on the self-assembly of individual AuNPs. Figure 2B is the SEM images of different sizes of gold ultramicroelectrodes constructed. They found the ultramicroelectrode geometries and sizes were perfectly controllable and reproducible. The ultramicroelectrode they made had the electrochemical properties of metallic gold and showed ideal microelectrode behavior both in aqueous and acetonitrile electrolyte solution.

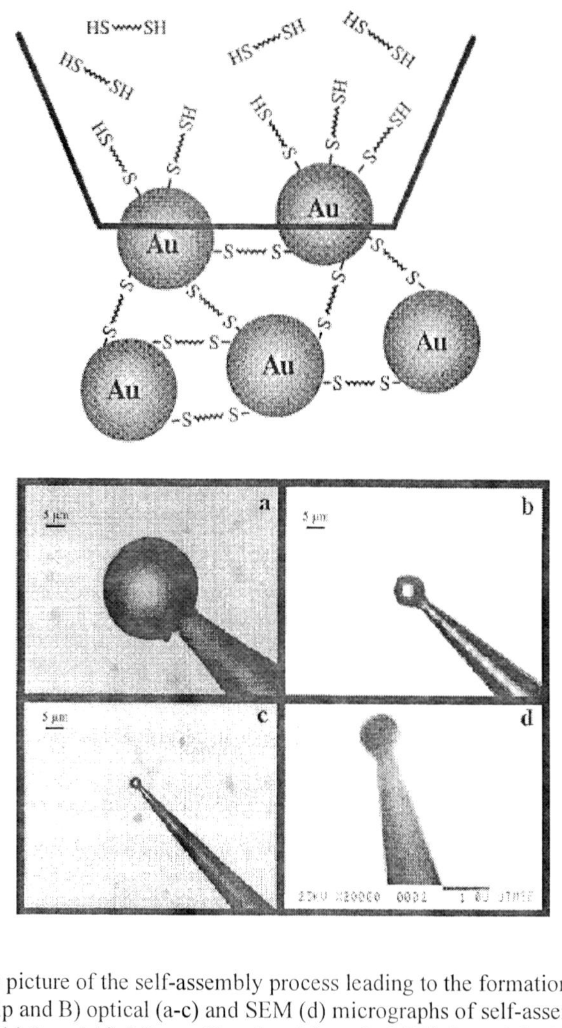

A.

B.

Figure 1. A) Schematic picture of the self-assembly process leading to the formation of microspheres at the end of the micropipet tip and B) optical (a-c) and SEM (d) micrographs of self-assembled UMEs. Electrode diameter, (a) 26, (b) 8, (c) 3, and (d) 0.9 μm. (Reprinted from Ref [63] Copyright 1997 American Chemical Society).

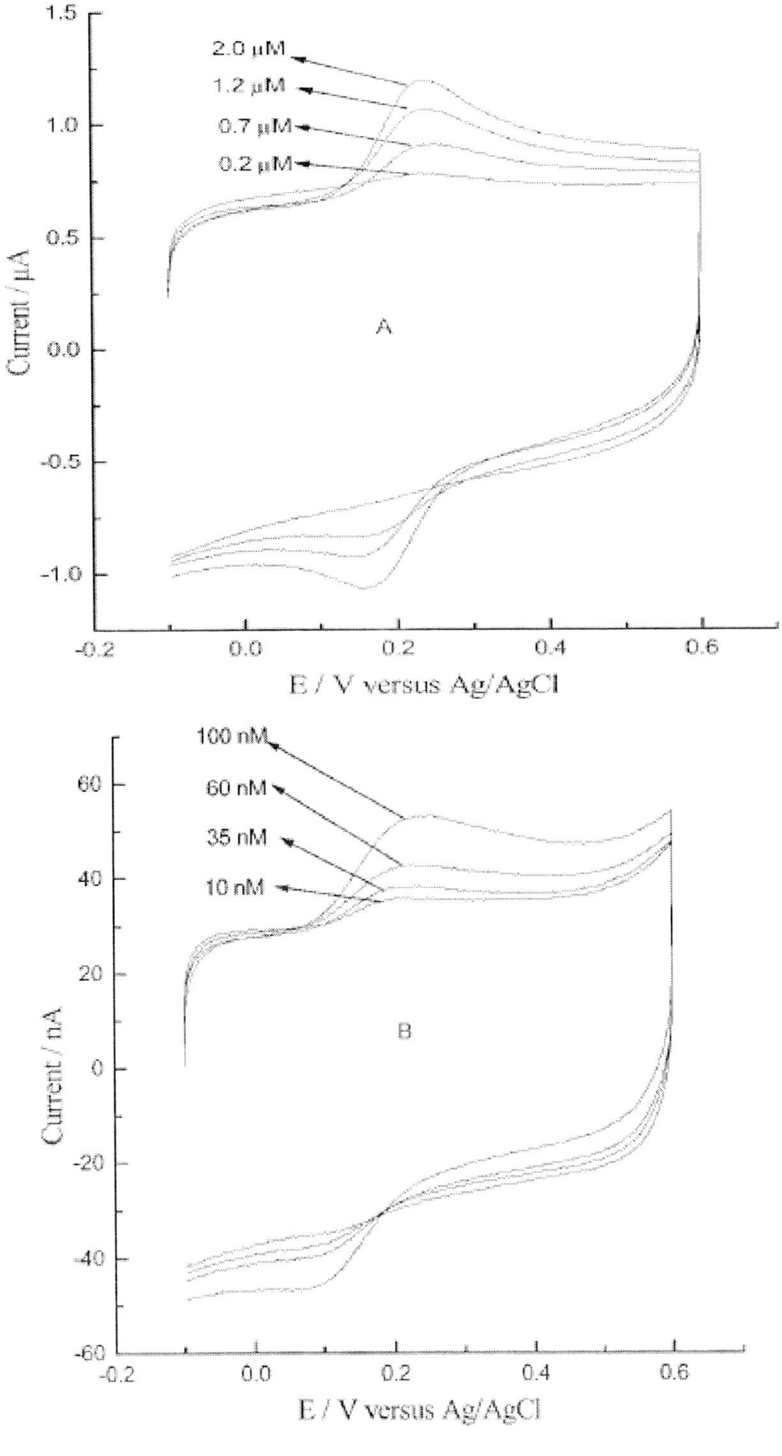

Figure 2. Cyclic voltammograms at 50 mV/s in aqueous $Fe(CN)_6^{4-}$ at (A) a gold macroelectrode in 50 mM KNO_3 and (B) the NEEs obtained by self-assembly of AuNPs for 9minutes in 50 mM KNO_3. $Fe(CN)_6^{4-}$ concentrations are as indicated. (Reprinted from Ref [64] Copyright 1997 American Chemical Society).

Erkang Wang et al reported the fabrication of gold nanoelectrode arrays through the assembly of colloidal AuNPs on (3-Mercaptopropyl)-trimethoxysilane (MPTMS) functionalized ITO glass surface. [64] They found that the assembly processes were high controllable and the size, shape and interspacing of as-prepared NEAs are highly flexible. Their electrochemical analysis experiments indicated that the detection limit of $Fe(CN)_6^{4-}$ on NEAs obtained by AuNPs assembly for 9 minutes were 20 times lower than at gold macroelectrode (Figure 2). Jinfang Zhi and Ruhai Tian reported the modification of AuNPs on functional boron doped diamond (BDD) electrode surface and investigated its basic electrochemical properties [65]. They found the AuNPs surface charge strongly affect the electron exchange between electrode and electrolyte and the electrode characterizations could be controlled by the AuNPs size and modification density [65,66].

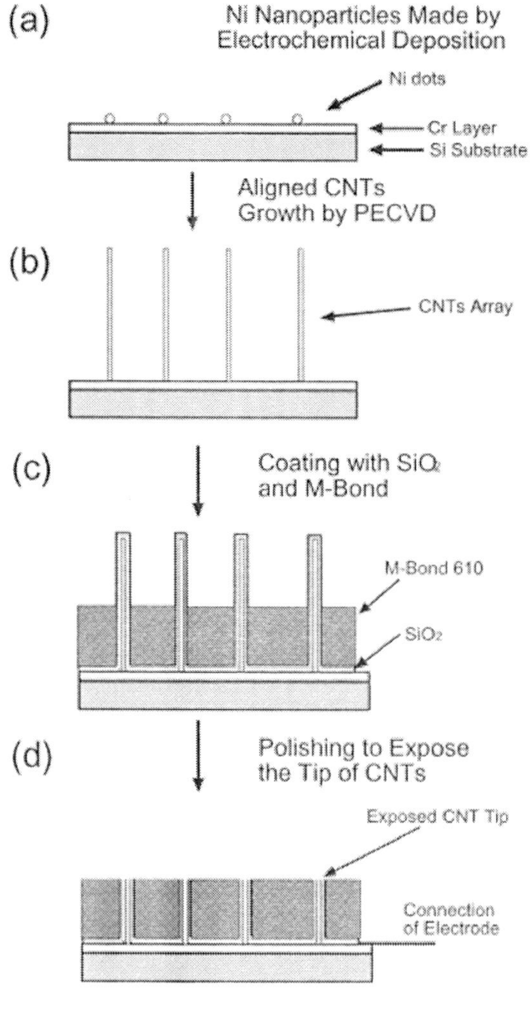

Figure 3. Fabrication procedures of CNTs electrode arrays a) Ni NPs catalysis electrodeposition; b) aligned CNTs growth; c) coating of SiO_2 on aligned CNTs array d) polishing the surface to expose CNTs (Reprinted from Ref [69] Copyright 2000 American Chemical Society).

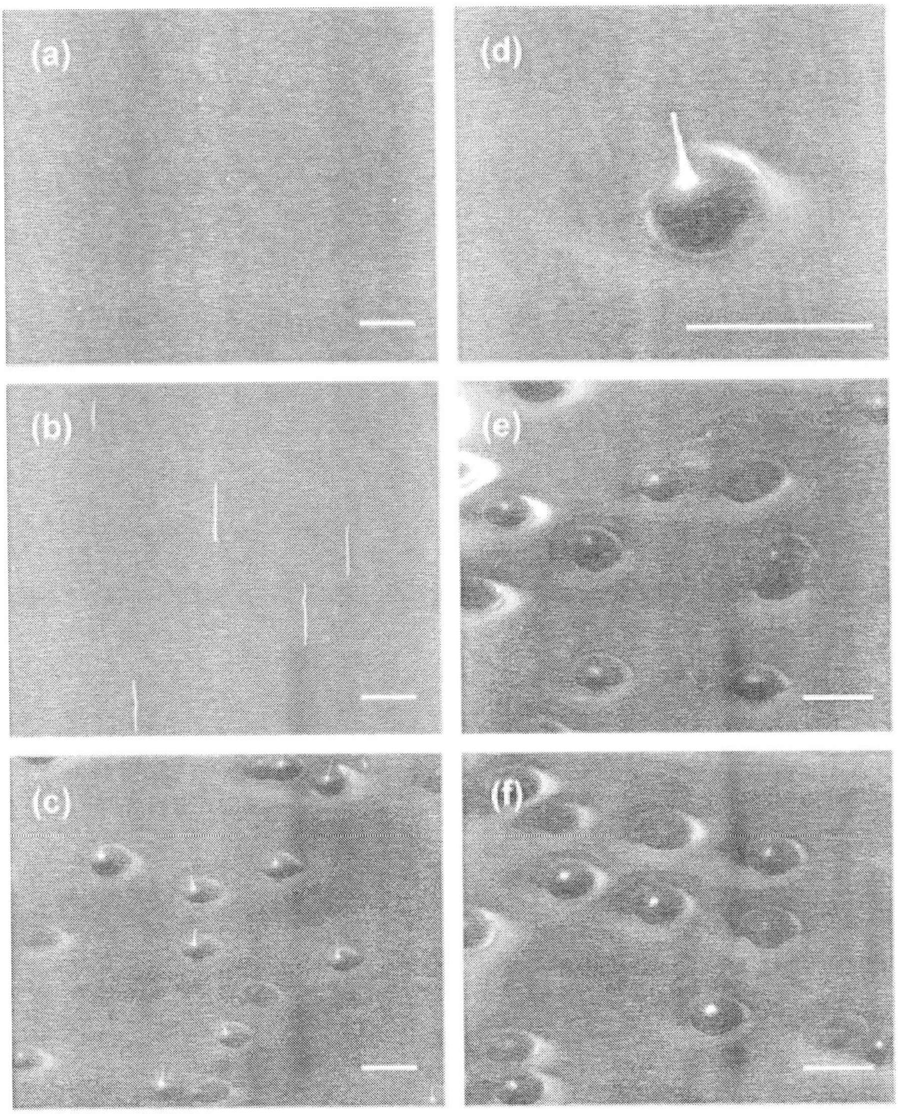

Figure 4. Scanning electron microscope images of a) Electrodeposited Ni nanoparticles; b) low site density aligned CNTs array; c) CNTs array coated with SiO_2 and an epoxy layer, d) close-up look at a single half-embedded CNT; e) CNTs afterpolishing; and f) second electrodeposition of Ni nanoparticles onthe broken CNTs only. (Reprinted from Ref [69] Copyright 2000 American Chemical Society).

3.1.1.2. Carbon Naontubes (CNTs) as NEs and NEAs Element

CNT is another kind of important element used in the NEs&NEAs construction [67,68]. Yuehe Lin and coworkers reported the fabrication of CNTs nanoelectrode array with vertically aligned single wall CNT as assembly element [69,70]. They fabricated highly dispersed and well aligned CNTs arrays as NEAs. Figure 3 shows the fabrication procedures of CNTs-NEAs on Chromium (Cr) coated silicon substrate. First, Cr layer was deposited onto the silicon surface as contact metal; then Ni-NPs were electrochemically deposited onto contact metal layer surface as CNTs growth catalyst. After the growth of aligned CNTs, the block material, SiOx or epoxy was coated onto the substrate surface to stuff the CNTs free areas. Finally, the substrate surface was polished to make the CNTs tips exposed. Figure 4 shows the SEM images of NEAs fabrication steps corresponding to the scheme shown figure 3. Moreover, they also studied the immobilization of glucose oxidase on the exposed CNT tips through the covalent amide linkages [71] (Figure 5). The glucose sensor they fabricated performed a selective electrochemical analysis of glucose in the presence of common interferents (e.g. ascorbic acids, uric, acetaminophen, etc). CNTs NEAs are also promising in the ultra-trace analysis of NADH, hydrogen peroxide, thiol-containing homocysteine, organophosphorus compound, alcohol, etc [72].

Jun Li and coworkers also reported the fabrication of integrated CNTs arrays on contact metals coated silicon surface with metal as CNTs growth catalyst [59]. By this way, they fabricated well-regulated CNTs bundle microelectrode arrays (Figure 6A, Figure 6B) and the electrode arrays typically performed as microelectrode arrays (Figure 6C). The detection limit of electroactive species could reach as low as a few nM on these electrodes. Otherwise, if the CNTs tips are functionalized with oligonucleotide probes or enzymes, this platform can be practicably applicable in DNA/RNA and bio-molecules analysis [73].

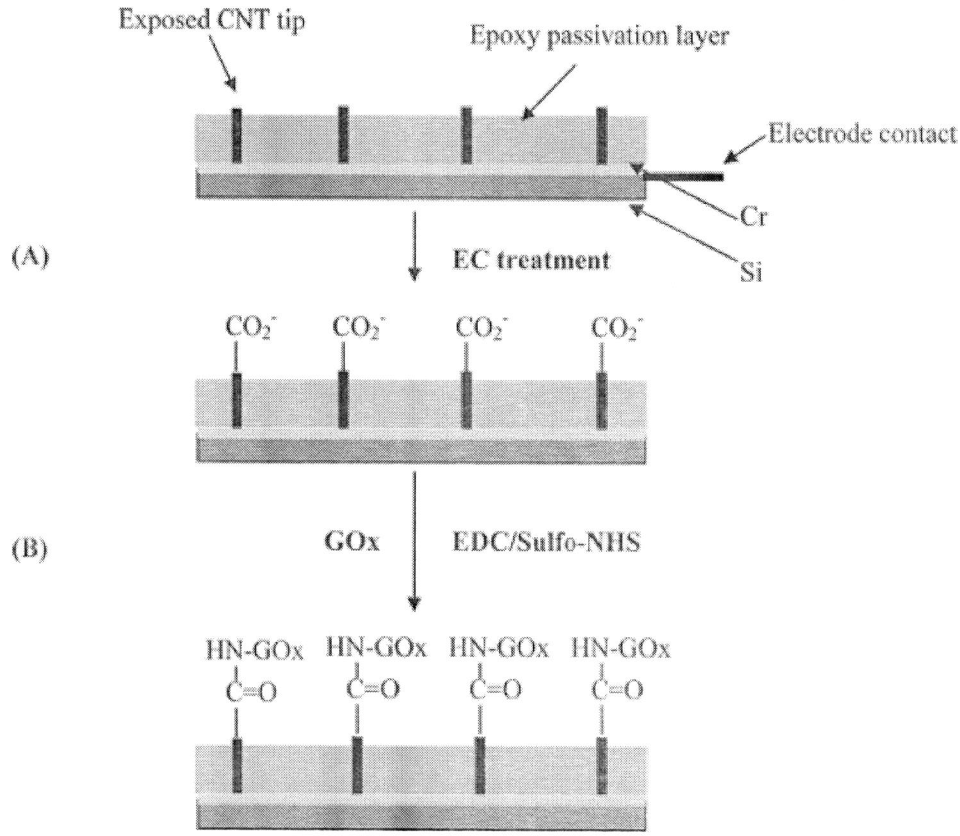

Figure 5. Fabrication of a glucose biosensor based on CNTs-NEAs (Reprinted from Ref [71] Copyright 2000 American Chemical Society).

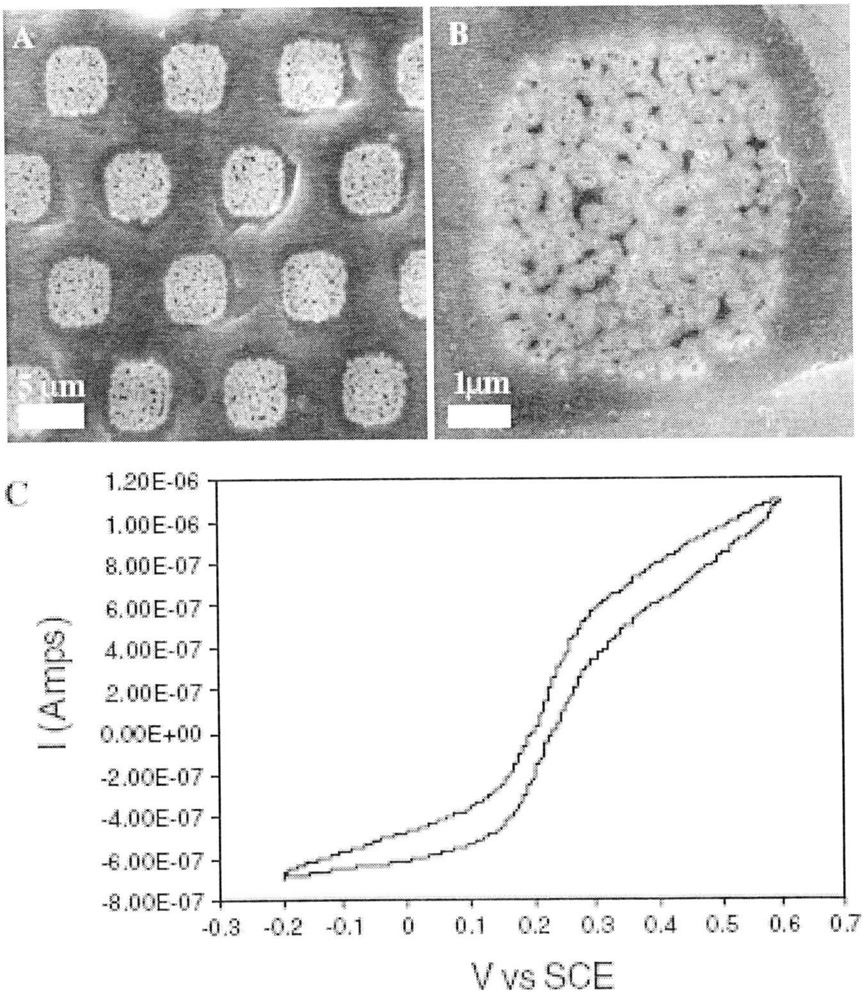

Figure 6. The top view of a CNTs bundle array (A) on silicon surface with 6 µm single-bundle-electrode diameter (B), the electrode array performs as microelectrode array, (C) CV scan of the CNTs bunch array in 1 mM $K_4Fe(CN)_6$/1 M KCl. Scan rate −10 mV s^{-1}. (Reprinted from Ref [59] Copyright 2004 Nanotechnology).

3.1.2. As Electrochemical Catalyst and Bio-label

One of the most significant applications of NPs in electroanalysis field is adopting them as electrochemical catalyst and bio-label. Applications of nanomaterials to biology or medicine include fluorescent biological labels; drug and gene delivery; bio detection of pathogens; detection of proteins; probing of DNA structure; tissue engineering; tumour destruction via heating (hyperthermia); separation and purification of biological molecules and Cells; MRI contrast enhancement; phagokinetic studies, etc [74]. The applications of NPS in bio and the combination of bio-techniques with electrochemistry techniques lead to the generation of some more precise and rapid analysis methods. This is one of the applications of NPs in electrochemistry.

3.1.2.1. Gold Nanoparticles (AuNPs)

AuNPs can be employed as a catalytic element in electrochemical analysis. Table 1 summarizes the applications of AuNPs in electroanalysis. AuNPs showed the excellent catalytic effect to the oxidation of dopamine. [75] Another research indicated that the electrochemical sensitivity of AuNPs modified carbon electrode to dopamine could be increased when some albumin was added during the preparation of AuNPs [19]. The selective detection of dopamine with AuNPs modified electrodes was also reported by many other groups [76-79].

The indium tin oxide (ITO) electrode modified by AuNPs (stabilized by dimethylamino pyridine) enables detection of nitric oxide. [80] Liu S et.al. reported the detection of phenol [81], glucose [82], hydrogen peroxide and nitrite [12,83,84] by the using of AuNPs modified electrodes. They found AuNPs could tune the electron transfer between electrode and myoglobin or hemoglobin which was immobilized onto the AuNPs surface.

S. Bharathi [85] fabricated the glucose oxidase enzyme/AuNPs/silicate network on the ITO glass surface by the co-deposition of these three elements. The glucose oxidase enzyme can catalyze the oxidation of glucose and AuNPs in the network can catalyze the oxidation and reduction of H_2O_2, the by-product of the enzymatic reaction. They also found that this kind of glucose sensor can eliminate the interference from acetaminlphen, ascorbic acid, dopamine etc because of the low operating potential. The electrochemical catalysis and detection of uric acid with AuNPs and gold nanorods modified ITO electrode were also studied [86,87].

J.M. Pingarrón constructed colloidal-gold/cysteamine modified carbon paste electrode (AuNPs-Cyst-CPE) and their studies showed that to the voltammetric determination of thio-compounds such as methionine, as well as of methionine-based peptides, AuNPs-Cyst-CPE exhibited improved electroanalytical characteristics when compared with colloidal-gold cysteamine-modified Au disk electrodes (AuNPs-Cyst-AuE) [88].

Table 1. Applications of AuNPs in electrochemical analysis

Compound	Nitric oxide	Hydrogen peroxide	Dopamine	phenol	Glucose
Detection limit	0.01mM [80] 0.7 μM [12]	4.6μM12 9.1×10^{-7} M. [83]	2×10^{-7} M [19,79] 0.13 μM [75]	6.1 nM [81]	0.01mM [82] 0.04ng [89]
Compound	Fructose	Lactose	Sucrose	Sorbitol	Arabinose
Detection limit	10ng [89]	9ng [89]	12ng [89]	9ng [89]	6ng [89]
Compound	Xylitol	Ascorbic acid	Serotonin (5-HT)	SO_2	Cholesterol
Detection limit	6ng [89]	[87]	8×10^{-7} M [79]	[90]	[91]
Compound	Norepinephrine	Uric acid	Arsenic-(III)	Methionine	
Detection limit	[87]	[86,87]	1 nM [92]	5.9×10^{-7}M [88]	

* In Flow injection and Liquid Chromatography.

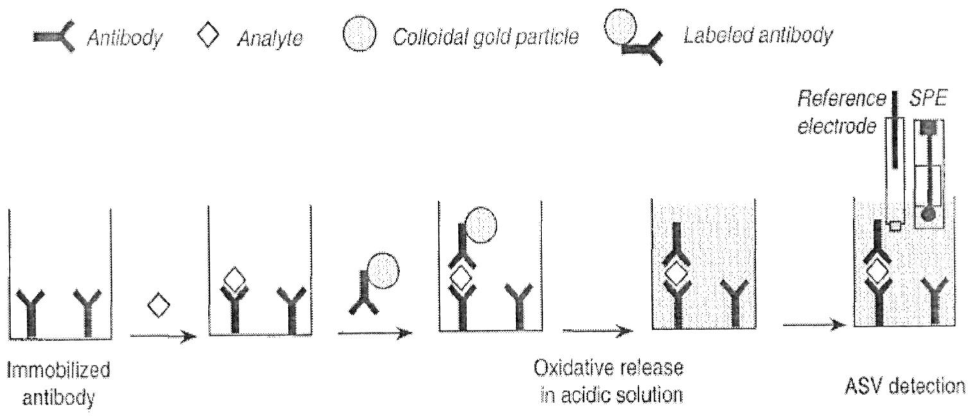

Figure 7. Schematic representation of noncompetitive heterogeneous electrochemical immunoassay based on a colloidal gold label. (Reprinted from Ref [93] Copyright 2000 American Chemical Society).

AuNPs were also used in bio-label and bio-electrochemical-analysis. Benoit Limoges group reported the sensitive electrochemical immunoassay which is based on the anodic stripping voltammetric detection of dissolved gold-label [93]. Figure 7 shows the principle of the procedures: arm-antibodies were absorbed onto the polystyrene micro-wells and goat IgG analyte was then captured by the antibody. The gold labeled antibody was employed to sandwich the goat IgG.. Finally, the gold was dissolved into acid solution for the subsequent electrochemical detection. By this method, 3 pM of immunoglobulin was determined. The exact idea was also used in the electrochemical detection of DNA [93].

Yuzhi Fang et.al. reported the colloidal AuNPs enhanced detection of ssDNA [94]. Reference [95,96] reported detection of DNA hybridization, based on the deposition of silver on AuNPs tags and a subsequent electrochemical stripping detection of dissolved silver. This method avoids the need of enzymes or fluorescence detection. The similar ideas were also used into the gene-analysis [97,98]

Reference [99] reported the reversible electrochemistry of horse heart cytochrome c at SnO_2 electrode modified with colloidal AuNPs. The electrochemical behavior of gold colloid and colloidal gold labeled anti-human IgG (immunogold) on carbon paste electrode was also studied [100]. Based on the electrochemical response of colloidal gold, low detection limits were achieved. And this methodology has been used in pharmacology research [101].

3.1.2.2. Carbon Nanotubes (CNTs)

CNT is a kind of widely used nanomaterial in electrochemistry because of its excellent and fancy electronic structure properties. Modification of CNTs onto traditional electrode surface has been widely reported. CNT plays very important role in the electron transfer between electrode and electrolyte. In most cases, CNT performs as electron tunnel and electrocatalyst. [5] High electrocatalysis of CNTs result from the surface defect sites and ends of nanotube [102,103]. CNTs modified electrodes and the nanocomposites consists of CNTs were employed in electrochemical applications such CNT/metal (oxide), CNT/organic layer, CNT/biomaterials [5].

Table 2 summarizes the application of CNTs in the electrochemical analysis of some inorganic and organic molecules.

Table 2. Applications of CNTs in electrochemical analysis

Compound	NPs type	Media	Detection limit
Nitric oxide	MWNT [104,105]	PBS (pH 7.0)	2×10^{-8}M [104] 8×10^{-8}M [105]
Hydrogen peroxide	MWNT [106-108]	PBS (pH 7.4)	50 μM [108]
Dopamine	SW/MWNT [108-112]	PBS (pH 7.4) [109] PBS (pH 7.1) [112] PBS (pH 5.0) [113]	0.5 μM [108] 1×10^{-7}M [110] 3×10^{-7}M [112] 2.5×10^{-10}M [113]
Serotonin	MWNT [110]	PBS (pH 7.0)	1×10^{-7}M [110]
Glucose	MWNT [109,114]	PBS (pH 7.4) [109] 0.1M NaOH [114]	0.6 mM [109] 1 μM [114]
Uric acid	SWNT [115]	PBS (pH 7.0)	
Cysteine	MWNT [116,117]	PBS (pH 7.0) [116] PBS (pH 7.8) [117]	2μM [116] 0.8μM [117]
N-acetylcysteine	MWNT [117]	PBS (pH 7.8) [117]	3.3μM [117]
Homocysteine	MWNT [117,118]	PBS (pH 7.8) [117] PBS (pH 7.0) [118]	0.75μM [117]
Epinephrine	SWNT [112]	PBS (pH 7.1) [112]	
Ascorbic acid	MWNT [119]	PBS (pH 7.0) [119]	
Glutathione	MWNT [117-119]	PBS (pH 7.8) [117] PBS (pH 7.0) [118]	2.9μM [117] 0.06μM [118]
NADH	MWNT [107,108]	PBS (pH 7.4)	2 μM [108]
Oxygen	MWNT [120]	0.1M NaOH	
Guanine	MWNT [121,122]	Acetate buffer (pH 5.0)	
Insulin	MWNT [122]	PBS (pH 7.4)	14 nM [122]
3,4-dihydroxyphenylaceticacid (DOPAC)	MWNT [113]	PBS (pH 5.0)	3×10^{-10}M [113]
Norepinephrine	MWNT [113]	PBS (pH 5.0)	2.5×10^{-10}M [113]
3-methoxy-4-hydroxyphenylglycol (MHPG)	MWNT [113]	PBS (pH 5.0)	5×10^{-10}M [113]
5-hydroxytryptamine (5-HT)	MWNT [113]	PBS (pH 5.0)	3.5×10^{-10}M [113]
homovanillic acid (HVA)	MWNT [113]	PBS (pH 5.0)	1.25×10^{-9}M [113]
5-hydroxyindoleacetic acid (5-HIAA)	MWNT [113]	PBS (pH 5.0)	6×10^{-10}M [113]
Catechol	MWNT [123,124]	PBS (pH 7.0)	

Because CNTs have large specific surface area, wide electrochemical window, flexible surface chemistry, conductivity, good stability, biocompatibility, they are also promising in the electrochemical DNA-sensors based on the electrochemistry. [4] Especially, the flexible surface chemistry make CNTs are attractive because their surface could be activated easily and versatilely, which facilitates the immobilization of bio-molecules on carbon surface and can also increase the immobilization yield. [125-127] Moreover, CNTs are very suitable in the applications of bioelectrochemistry and biosensors because the electron transfer can happen smoothly between CNTs and enzymes/proteins (glucose oxidase/GOx, hemeproteins, GRP, hemoglobin/Hb, myoglobin/Mb). The electron transfer between CNTs and biomolecules are mainly through the CNTs surface defect [128] and π -π communication [129]. The electrochemical detection limit of DNA with CNTs modified electrode can reach 1fg/ml [130] and 2.3×10^{-14} mol L^{-1131}. Table 3 summarizes the possible application of CNTs in bio analysis.

3.1.2.3. Nickel Nanoparticles (Ni NPs)\

Nickel electrode has been widely utilized in carbohydrate electrochemical analysis with alkaline solution as electrolyte. [132] Highly dispersed nickel-NPs were also modified onto graphite-like carbon film electrode surface for electroanalysis. This nickel NPs modified electrode had excellent electrocatalytic ability and good stability compared with a Ni-bulk electrode with regard to the electrooxidation of sugars. [43] The improved detection limits for the investigated sugars, namely, 20, 25, 50, and 37 nM for glucose, fructose, sucrose, and lactose, respectively (table 4).

Table 3. Electron transfer properties of proteins on CNTs (Reprinted from Ref [5] Copyright 2005 The Japan Society for Analytical Chemistry)

Protein	CNT electrode	$E^{0\prime}$/V	ET rate/s^{-1}	Analyte
Cyt. *c*	CNT paste	*ca.* 0.06	NA	NA
HRP	Poly-L-lysine/SWNT	NA	NA	H_2O_2
Hb	CNTPME	−0.278 (SCE)	0.062	H_2O_2
Hb	CNT film	−0.343 (SCE)	1.25 ± 0.25	H_2O_2
Mb	MWNT film	NA	NA	H_2O_2
Mb	MWNT film	−0.248 (SCE)	5.4	NO
MB	Oriented SWNT	−0.21 (SCE)	NA	H_2O_2
HRP	Oriented SWNT	−0.25 (SCE)	NA	H_2O_2
HRP	CNTPME	−0.643 (Ag/AgCl)	2.48 ± 0.03	H_2O_2
Cyt. *c*	SWNT film	NA	NA	Adenine
Catalase	SWNT film	−0.414 (SCE)	NA	H_2O_2
MP-11	MWNT film	−0.26 (Ag/AgCl)	38	H_2O_2,O_2
MP-11	Aligned SWNT/Au	−0.390 (Ag/AgCl)	2.8 - 3.9	NA
GOx	CTAB-CNT	−0.466 (SCE)	1.53 ± 0.45	NA
GOx	Aligned SWNTs	−0.422 (Ag/AgCl)	0.3	NA
GOx	Aligned SWNTs	−0.45 (SCE)	NA	Glucose
GOx	CNTPME	−0.659 (Ag/AgCl)	1.61 ± 0.3	NA
GOx	SWNT film	−0.441 (Ag/AgCl)	1.7	Glucose

$E^{0\prime}$, formal potential; NA, not available, MP-11, microperoxidase 11.

Table 4. Applications of Ni NPs in electrochemical analysis

Compound	Glucose	Fructose	Sucrose	Lactose
Detection limit	20nM [43]	25nM [43]	50nM [43]	37nM [43]

Table 5. Applications of copper NPs in electrochemical analysis

Compound	Hydrogen peroxide	o-diphenols (catechol, dopamine, pyrogallo)	Glucose	Oxygen	Nitric acid
Detection limit	0.97 μM [134]	Catechol, 3μM [135] Dopamine, 5μM [135]	40nM 0.8pmol [136]*	1ppm [133]	1.5 μM [137]

*in flow injection system

3.1.2.4. Copper Nanoparticles (Cu NPs)

Copper NP is another commonly used NPs in electrochemistry. The priority of CuNPs is its versatile surface state and surface reaction. The Cu_2O/CuO component generated on copper surface always participates in the electrochemical reaction through the hydrogen bond or complex interaction. Otherwise, because Cu_2O is a good p-type semiconductor, sometimes, photoelectrochemical reaction happens between excited Cu_2O and electrolyte. [133] Table 5 lists some examples of the applications of CuNPs in electrochemical analysis.

Table 6. detection of α amino acids at 100nm CuNPs modified screen-printed carbon electrode (Reprinted from Ref [138] Copyright 2004 The Royal Society of Chemistry)

Amino acid	Linear range/μM	*R*	Sensitivity/ nA μM^{-1}	D_L (S/N = 3) /nM	D_L (S/N = 3) /ng
Ala	5—500	0.9990	7.29	24	0.04
Arg	5—300	0.9960	5.97	587	2.47
	5—500	0.9894	5.05	695	2.93
Asn	5—300	0.9998	11.68	94	0.28
	5—500	0.9901	9.78	112	0.34
Asp	5—500	0.9951	5.98	2693	7.17
Cys	10—300	0.9955	6.41	875	2.12
Glu	5—500	0.9901	4.90	624	1.84
Gln	5—500	0.9962	7.40	221	0.65
Gly	5—500	0.9996	2.83	161	0.21
His	5—500	0.9961	12.07	44	0.14
Iso-Leu	5—500	0.9976	4.40	297	0.78
Leu	5—300	0.9978	7.44	231	0.61
Lys	5—500	0.9994	4.30	1890	7.23
Met	5—500	0.9929	8.99	102	0.30
Phe	5—500	0.9991	7.62	472	1.56
Pro	5—300	0.9988	3.70	2545	5.86
Ser	5—500	0.9914	7.58	1393	2.93
Thr	5—500	0.9954	8.46	429	1.02
Trp	5—500	0.9993	6.66	389	1.59
Tyr	5—500	0.9999	7.17	548	1.98
Val	5—500	0.9992	4.21	574	1.34

Copper NPs modified electrodes are also used in amino-acid analysis. 100nm copper NPs modified screen-printed carbon electrode was used in the detection of a-amino acids based on two step complexation between amino acid and copper surface [138]. They found the $Cu^{II}O$-amino acid complex on the 100nm copper surface played a very important role in the surface complexation processes. The reversible adsorption (on $Cu^{II}O$)/desorption (on $Cu^{I}O$) make 100nm Cu NPs modified screen-printed carbon electrode more stable and versatile in the amino-acid analysis. All 20 underivatized amino acids could be sensitively determined in buffer solution (table 6).

3.1.2.5. Other Nanoparticles

Platinum NPs (PtNPs), silver NPs (AgNPs) modified electrodes were also used in electroanalysis of some organic and inorganic molecules. [139,140,141] Table 7 gives some examples of the applications of other NPs in electroanalysis.

Table 7. Applications of other NPs in electrochemical analysis

Compound	Nitric oxide	Hydrogen peroxide	Dopamine	Glucose	Oxygen
Detection limit	$3\times^{-7}$M [141]	0.01μM [41,42] 25nM [88]	3×10^{-8} M [142]	0.5 μM [88]	
NPs	Silver NPs [141]	Platnium NPs [41,42,88] Iron NPs [44]	Fe_3O_4 NPs [142]	Platnium NPs [88]	Iridium NPs [143]

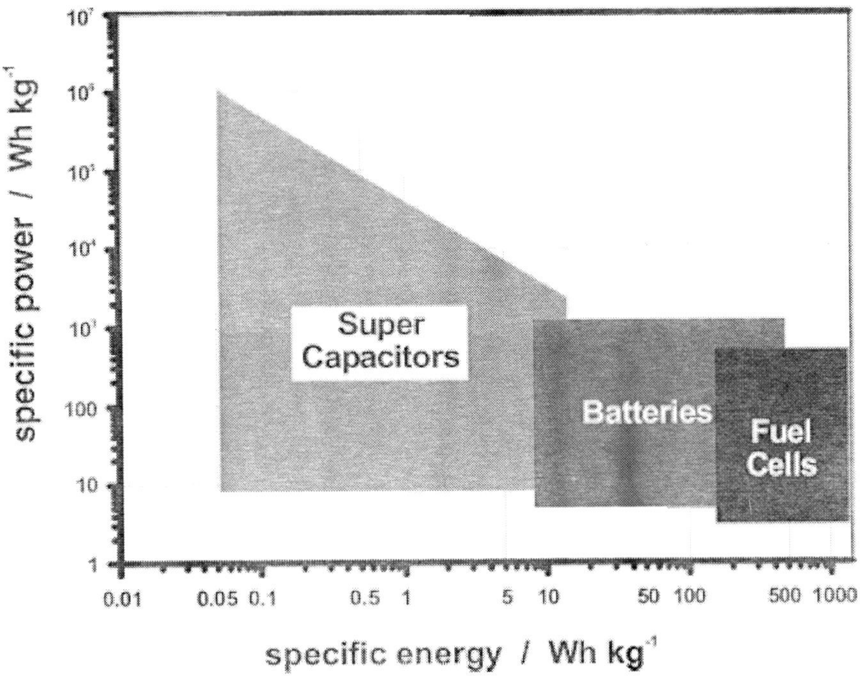

Figure 8. the different merits of different electrochemical power supplying system.

3.2. Power and Energy

Electrochemical energy storage and conversion include batteries, fuel cells, and electrochemical capacitors (ECs), etc [144]. Batteries and fuel cells are based on the chemical energy conversion to electrical energy and electrochemical capacitors are based on capacitors result from the generation and destruction of electric double layer or solid surface reaction. The different working mechanisms lead to the different characterizations of these three power supplying systems. Figure 8 shows the difference: batteries and fuel cells can provide long-time power but supercapacitors can provide high power energy. Single electrochemical power source can't supply both high energy and high power but the best system with high power and high energy can be achieved when the available electrochemical power systems are combined.

The application of NPs in power and energy is one of the most promising and active fields. Reference [7] reviewed the application of NPs in power and energy. The authors gave the fundamental principles and some very good example about the application of NPs in power and energy. The advantage of NPs in power and energy field is its catalysis and ultra-high specific surface area. All the researches of this field are all based on these two fundaments. The studies in this area focus on 1) trying to increase the activity per unit area 2) decreasing the necessary amount of expensive catalyst 3) exploiting the new process to get new materials 4) engineering design.

3.2.1. Fuel Cell

The applications of NPs in fuel are mostly based on their high catalytic abilities. Some catalytic reactions just happen on the specific crystal face, surface step sites or lattice defects. NPs possesses large specific surface area (surface area to the volume), numerous free valence positions to the cluster, which makes NPs are very suitable for catalyst. NPs-catalysis has electronic effects, geometric effects and support effects [7]. Electronic effects arise from the electronic structure compactness. For metal NPs, especially when the size of nanoparticle was smaller than 5nm, nanoparticle shows totally different electronic structure comparing with bulk metal. The small number of atoms involved in the electron bond formation leads to strong valence electrons localization. The heavy electrons localization can result in the smaller lattice constant with respect to bulk metal. The smaller lattice constant results in the high surface catalytic ability [145,146]. The more the nanoparticle is small, the more the electronic structures are different. The geometric effect results from the morphological properties of nanoparticle which strongly affect the catalytic activity during the reaction process. The supporting material always affects the catalytic activity of NPs because the interaction of catalytic NPs with supporting material may lead to the electronic properties change of those atoms close to the supporting surface [147].

3.2.1.1. PtNPs

The most widely used catalysts in fuel cell is platinum containing metal or alloys [148]. PtNPs were well-known catalytic to the reduction of oxygen in both acid and alkaline media [149-155] and platinum alloys were also employed and developed as highly-active catalyst. Pt_xCr_y, Pt_xCo_y, Pt_xNi_y alloy and PtV/C composite were proved highly active to the reduction of oxygen [37,156-158]

Platinum metal is also used as catalyst in the oxidation of methanol. The deadly drawback of platinum material as methanol oxidizing electrode is the irreversible adsorption of monoxide carbon which results in the electrode poisoning. A way to resolve this problem is to synthesize the alloy NPs [159] . To solve this problem, gold-platinum NPs were studied as catalyst for the oxidation of methanol [160]. Other platinum containing alloy NPs (Pt-Sn, Pt-Ru) were also studied in the oxidation of methanol. [37] Ruthenium, tin, molybdenum were co-alloyed with platinum to make CO tolerate catalyst [161,162]. Research indicated the metal oxide (RuO_x) played very important roles for the tolerance of CO) Moreover, platinum-alloy NPs were also employed as anode catalysts in the hydrogen oxidation in the presence of CO.

3.2.1.2. AuNPs

AuNP is another kind of promising catalytic NP in fuel cell. AuNPs were proved to be catalytic for the reduction of oxygen/hydrogen peroxide to water directly in acid [163-168] or base solution [169,170] and in the oxidation of carbon monoxide or methanol. Reference [171] reported the catalytic activation of core-shell assembled AuNPs as catalyst for methanol electro-oxidation; they found the oxidized layer plays a very important role in this process.

3.2.1.3. Carbon NPs

Because of economic advantages and excellent properties of carbon materials [172] such as accessibility, low price, good electric conductivity, good thermal conductivity, low density, corrosion resistance, oxidation resistance and reduction resistance, carbon electrolytes were widely used as catalyst supporting materials [148]. The other forms of carbon (nano-horns, nano-tubes, nano-fibers, nano-sheets, powers, clothes or papers) nanostructures were also adopted as catalyst supporting materials [173,174] Moreover, carbon nanotubes are also used in the hydrogen storage [175-177].

3.2.2. Supercapacitor

A supercapacitor is an electrochemical capacitor that offers very high capacitance in a small package [178]. They are able to store a large amount of charge (energy) that can be released very quickly. This means they are superior in short term, high-energy applications, such as when an appliance is switched on or an electric car accelerates. The electrochemical supercapacitors (ES) are also called Electrochemical Capacitors, (EC), Supercapacitors or Ultracapacitors. ES can be classified as chat 2 shows.

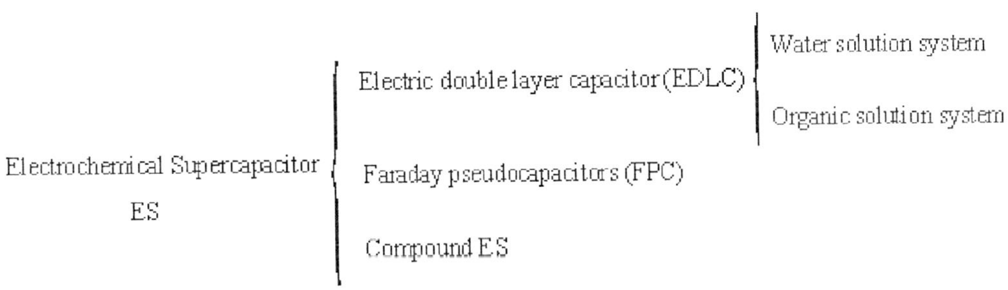

Chart 2. the classification of electrochemical supercapacitors.

The merits of ES including:

1. Quick recharge and discharge (seconds);
2. Super capacitance
   a. Gravimetric specific energy >2.5Wh/kg
   b. Gravimetric specific power>500W/kg;
3. Superior recyclability (over 500 000 recharge cycles);
4. No memory effect;
5. Free size and shape;
6. Low-toxicity

Applications of ES including:

1. Hybrid-electric vehicles
2. In notebook computers
3. Industrial actuators and controls
4. Uninterruptible power supplies and power quality devices
5. Automotive auxiliary systems and engine starting
6. Two-way pagers, cellular phones and other wireless communications devices
7. Digital cameras, power tools, etc
8. Motive and standby power.
9. Military affairs, to apply very high power instantly

High specific surface area materials such as CNTs, carbon powder, carbon black are commonly used electrode materials. FPC is based on the rapid reaction on solid surface. Metal oxides NPs are commonly used as electrode materials. This kind of ES system possesses good conductivity, high electric capacitor to weight, long life, but high price and toxicity. Compound ES is combining EDLC and FPC together through proper design for given purpose. Because the construction of compound ES is most an engineering issue, we don't discuss compound ES here.

The capacitance of double layers is typically dozens of $\mu F\ cm^{-2}$ grade. If high surface area electrodes are employed for capacitor devices, the capacitance could reach hundreds of Faradays. Electrochemical supercapacitor has revolutionized enormous energy storage if high specific-surface area materials were used as electrode materials [178]. High specific surface area of carbon powders [179] and hybrid materials are utilized in commercial electrochemical capacitors.

3.2.2.1.CNTs

CNTs were firstly utilized as electrode for electrochemical supercapacitors in 1997 [180]. They prepared multi-walled CNTs with specific surface of 430 $m^2\ g^{-1}$ and the maximum specific capacitance and energy density reached 113 F/g and 0.56Wh/kg correspondingly. Other researches also demonstrated the CNTs are very suitable for the supercapacitor devices. [181,182]

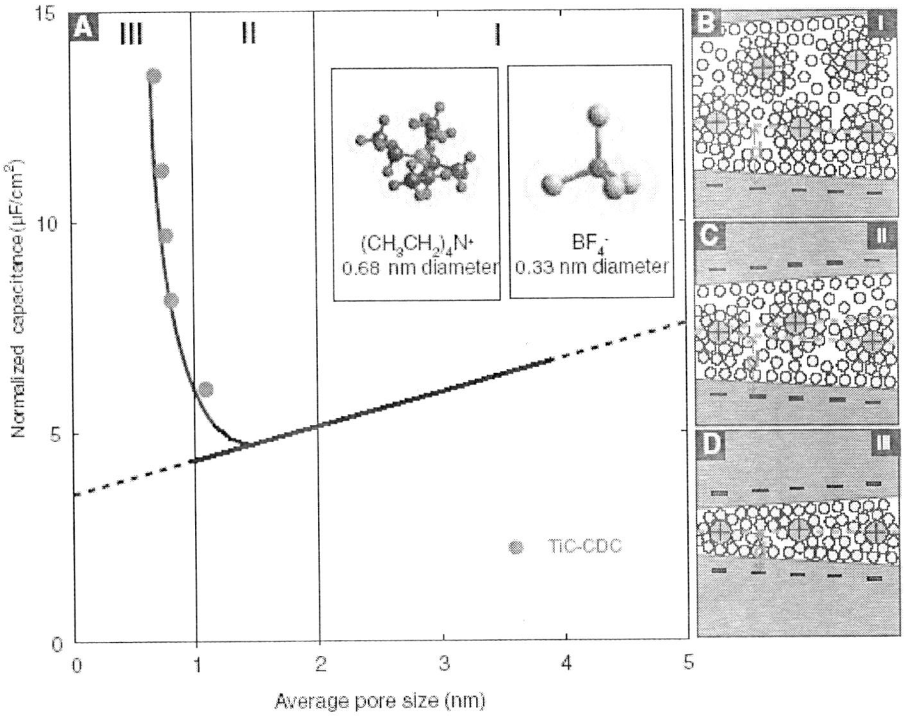

Figure 9. (A) Plot of specific capacitance normalized by BET SSA for nanoporous TiC-CDC with average diameter from ~ 0.6 nm to 1.25 nm (red dot) in contrast to traditional result (straight line) Drawings of solvated ions residing in pores with distance between adjacent pore walls (B) greater than 2 nm, (C) between 1 and 2 nm, and (D) less than 1 nm illustrate this behavior schematically. The capacitor can be calculated with equation $C=\varepsilon A/d$; ε is the electrolyte dielectric constant; A is the electrode area accessible to electrolyte ions, d is the electrode separation distance. (Reprinted from Ref [183] Copyright 2006 Science).

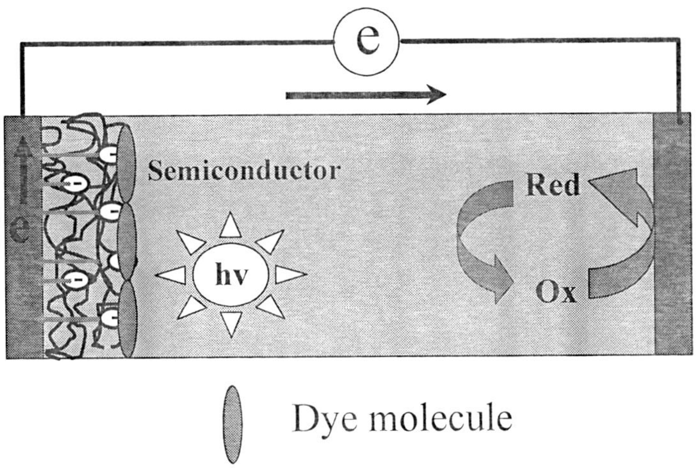

Figure 9. Photocurrent generation process in DSSCs.

3.2.2.2. Mesoporous Hybrid Nano-materials

C. Portet etc studied the supercapacitor with pore size smaller than 1nm porous material as electrode [183]. They synthesized the titanium carbide-derived carbon (TiC-CDC) with pore size range from 0.6nm to 2.25nm (with 0.05nm accuracy). They found when the nanopore size is smaller than the solvated ion, the specific capacitor increase dramatically because of the compact and squeezing structure of solvated ion in the pores (figure 9). The increase of capacitor arises from the dramatic decrease of virtual counter electrode distance (d). Their report firstly demonstrated 1) the charge storage in pores smaller than the size of solvated electrolyte ions could lead to specific capacitor increase 2) ions could squeeze into nanopores whose size is similar or even smaller than the ion size (this result beyond the exception). This amazing and unbelievable research makes super-capacitor become explicitly promising power device.

3.2.3. Solar Cell

The most commonly used NPs in nanocrystalline solar cells are semiconductive NPs, such as TiO_2, CdS, ZnS, CdSe, CNTs, C_{60}, etc [184]. Photosensitized semiconductor solar cells have different fundamental principles in compare to p-n junction solar cell. In reference to high-cost silicon solar, the nanocrystalline solar cells are more attractive because of low-cost and easy fabrication. Dye-sensitized nanocrystalline solar cells (DSSCs) system is one of the examples. The most widely studied semiconductive NPs in solar cell are titanium dioxide (TiO_2) NPs. Grätzel firstly reported the dye-sensitized TiO_2 NPs solar cell in 1991 and the efficiency can get about as high as 7.5%. [185]

Figure 9 shows the structure of typical DSSCs. DSSCs consist of anode, cathode and electro-active electrolyte. The irradiation light is absorbed by dye molecules and the excited electrons transfer to the conduction band of semiconductor nanocrystals. The positively charged dye molecules immediately react with electro-active electrolyte and get compensated. At the same time, the collected electrons on anode (semiconductor electrode) go through the out loop and generate the photocurrent (The electro-active electrolyte gets electrons from the cathode).

4. Conclusion

Similarly to that the bulk electrodes used so far, NPs modified electrode (including gold, platinum, copper and nickel) become deactivated through fouling [76-78,140,186]. The NPs surface fouling is a deadly and serious problem for the application of NPs in electrochemistry. The possible approach to solve this problem is to design NPs structure and surface functionalization. Reference [187] described the cleaning of platinum NPs without loss of crystalline surface structure. The NPs could be electrochemically refreshed in sulfuric acid solution after cyclic voltammograms. NPs doped carbon paste electrode surface could be renewed just by removing the top surface [12,19,187].

The other problem of NPs in electrochemistry is the reproducibility. One of the drawbacks of electrochemical methods is the bad reproducibility which arises from the difficult reproduction of electrode surface. It seems that the introduction of NPs in electrochemistry aggravates this situation but not making it better. That's because the bad reproducibility of NPs modification or doping as electrode element. To get better reproducibility, the precise fabrication of nanostructure and functional surface is necessary. The related techniques are still under development.

Another topic of NPs in electrochemistry is the safety, which is a comprehensive topic in NPs applications. Because of small size, NPs can penetrate cell membranes and integrate themselves into larger molecules. The fatalness of the invasion of NPs to bio-system is still under investigation and attracting more and more attentions.

REFERENCES

[1] Allen J. Bard, L. R. F. *John Wiley & Sons, Inc.* 2001.

[2] David Herna¬ndez-Santos, M. B. G. l.-G., Agustỉn Costa Garcỉa *Electroanalysis* 2002, *14*, 1225-1235.

[3] Compton, C. M. W. R. G. *Anal Bioanal Chem* 2006, *384*, 601-619.

[4] Pingang He, Y. X., and Yuzhi Fang *Microchim Acta* 2006, *152*, 175.

[5] Kuanping Gong, Y. Y., Meining Zhang, Lei Su, ShaoXiong , Lanqun Mao *Anal Sci* 2005, *21*, 1383-1393.

[6] Adam K. Wanekaya, W. C., Nosang V. Myung, Ashok Mulchandani *Electroanalysis* 2006, *18*, 533 - 550.

[7] Fabio Raimondi, G. n. G. S., R diger K tz, and Alexander Wokaun *Angew. Chem. Int. Ed.* 2005, *44*, 2190-2209.

[8] Shaojun Dong, E. W. *Anal. Chim. Acta* 2007 *598*, 181-192.

[9] Frens, G. *Nature Phys. Sci.* 1973, *241*, 20-22.

[10] Mathias Brust, M. W., Donald Bethell, David J. Schiffrin, Robin Whyman *J. Chem. Soc., Chem. Commun.* 1994, *7*, 801-802.

[11] Allen C. Templeton, W. P. W., and Royce W. Murray. *Acc.Chem.Res* 2000, *33*, 27-36.

[12] Liu S, J. H. *Electroanalysis* 2003, *15*, 1488.

[13] Song Y, J. V., McKinney C, Donkers R, Murray RW *Anal Chem* 2003, *75*, 5088.

[14] Maye M M, L. J., Lin Y, Engelhard MH, Hepel M, Zhong C-J *Langmuir* 2003, *19*, 125.

[15] Wilcoxon JP, M. J., Provencio P *Langmuir* 2000, *16*, 9912.

[16] Daniel, M.-C. A., D. *Chem. Rev.* 2004, *104*, 293.

[17] Sutherland, W. S. W., J. D. *J. Colloid. Intelface Sci.* 1992, *148*, 129-141.

[18] Kenneth R. Brown, D. G. W., and Michael J. Natan *Chem. Mater.* 2000, *12*, 306-313.

[19] Miscoria SA, B. G., Rivas GA *Electroanalysis* 2005, *17*, 1578.

[20] Becker JA, S. R., Festag R, Ruland W, Wendorff JH, Pebler J, Quaiser SA, Helbig W, Reetz MT *J Chem Phys* 1995, *103*, 2520.

[21] Khilar, R. P. B. a. K. C. *Langmuir* 2000, *16*, 905-910.

[22] Abhijit Manna, T. I., Masayasu Iida, and Naoko Hisamatsu *Langmuir* 2001, *17*, 6000-6004.

[23] Vidal-Iglesias FJ, S.-G. J., Rodriguez P, Herrero E, Montiel V, Feliu JM, Aldaz A *Electrochem Commun* 2004, *6*, 1080.

[24] Harriman A, M. G., Neta P, Richoux MC, Thomas JM *J Phys Chem* 1988, *92:*, 1286.

[25] Thomas Hirsch, M. Z. *Angew. Chem. Int. Ed.* 2005, *44*, 6775 -6778.

[26] Helbig, M. T. R. a. W. *J. Am. Chem. Soc.* 1994, *116*, 7401-7402.

[27] Yu-Ying Yu, S.-S. C., Chien-Liang Lee, and C. R. Chris Wang *J. Phys. Chem. B* 1997, *101*, 6661-6664.
[28] L. Rodrı´guez-Sa´nchez, M. C. B., and M. A. Lo´pez-Quintela *J. Phys. Chem. B* 2000, *104*, 9683-9688.
[29] Zoval JV, S. R., Biernacki PR, Penner RM *J Phys Chem* 1996, *100*, 837.
[30] Ueda M, D. H., Anders A, Kneppe H, Meixner A, Plieth W *Electrochim. Acta* 2002, *48*, 377.
[31] Adora S, S. J. P., Soldo-Olivier Y, Faure R, Chainet E, Durand R *ChemPhysChem* 2004, *5*, 1178.
[32] Liu S, T. Z., Wang E, Dong S *Electrochem Commun* 2000, *2*, 800.
[33] Lu G, Z. G. *J Phys Chem B* 2005, *109*, 7998.
[34] He Z, C. J., Liu D, Tang H, Deng W, Kuang Y *Mater Chem Phys Lett* 2004, *85*, 396.
[35] Maillard F, E. M., Cherstiouk OV, Schreier S, Savinova E, Stimming U *Faraday Discuss* 2004, *125*, 357.
[36] Sarkar DK, Z. X., Tannous A, Louie M, Leung KT *Solid State Commun* 2003, *125*, 365.
[37] Leger, J. M. *Electrochim. Acta* 2005, *50*, 3123.
[38] Przenioslo R, S. I., Rousse G, Hempelmann R: *Phys Rev B* 2002, *66*, 014404/1.
[39] Reetz MT, M. M. *Adv Mater* 1999, *11*, 773.
[40] Y. J. Yang, L. Y. H., H. Xiang *Russian Journal of Electrochemistry* 2006, *42*, 954-958.
[41] You T, N. O., Tomita M, Hirono S *Anal Chem* 2003, *75*, 2080.
[42] You T, N. O., Horiuchi T, Tomita M, Iwasaki Y, Ueno Y, Hirono S *Chem Mater* 2002, *14*, 4796.
[43] Tianyan You, O. N., Zilin Chen, Katsuyoshi Hayashi, Masato Tomita, and Shigeru Hirono *Anal Chem.* 2003, *75*, 5191-5196.
[44] Guo L, H. Q., Li X-y, Yang S *Phys Chem Chem Phys* 2001, *3*, 1661.
[45] Dubois E, C. J. *Langmuir* 2003, *19*, 10892.
[46] Grunes J, Z. J., Anderson EA, Somorjai GA *J Phys Chem B* 2002, *106*, 11463.
[47] Suslick KS, C. S., Cichowlas AA, Grinstaff MW *Nature* 1991, *353*, 414.
[48] Serp P, F. R., Kihn Y, Kalck P, Faria JL, Figueiredo JL *J Mater Chem* 2001, *11*, 1980.
[49] Heiz U, V. F., Sanchez A, Schneider WD *J. Am. Chem. Soc.* 1998, *120*, 9668.
[50] U, H. *Appl Phys A* 1998, *67*, 621.
[51] Liu Y-C, L. L.-H. *Electrochem Commun* 2004, *6*, 1163.
[52] Dimitrios Tasis, N. T., Alberto Bianco, and Maurizio Prato *Chem. Rev.* 2006, *106*, 1105-1136.
[53] P.M.Ajayan *Chem. Rev.* 1999, *99*, 1787.
[54] J.C.Charlier *Acc.Chem.Res* 2002, *35*, 1063.
[55] Iijima, S. *Nature* 1991, *354*, 56.
[56] S.Iijima, T. I. *Nature* 1993, *363*, 603.
[57] Rubinstein, E. S. a. I. *J. Phys. Chem.* 1987, *91*, 6663.
[58] Stulik, K., Amatore, C., Holub, K., Marecek, V., Kutner, W *Pure & Appl.Chem.* 2000, *72*, 1483.
[59] Alan M Cassell, Q., Brett A Cruden, Jun Li, Philippe C Sarrazin, Hou Tee Ng, JieHan, and M Meyyappan *Nanotechnology* 2004, *15*, 9.
[60] Baranski., A. S. *J. Electroanal. Chem.* 1991, *300*, 309
[61] W.M.Arrigan, D. *Analyst.* 2004, *129*, 1157.

[62] Faraday, M. *Philos. Trans. R. Soc. London* 1857, *147*, 145.
[63] Christophe Demaille, M. B., Michael Tsionsky, and Allen J. Bard *Anal Chem.* 1997, *69*, 2323-2328.
[64] Wenlong Cheng, S. D., and Erkang Wang, *Anal Chem.* 2002, *74*, 3599.
[65] Ru-hai Tian, T. N. R., Yasuaki Einaga, and Jin-fang Zhi *Chem. Mater.* 2006, *18*, 939-945.
[66] Ru-hai Tian, J.-f. Z. *Electrochem Commun* 2007, *9*, 1120-1126.
[67] Jun Li, J. E. K., Alan M. Cassell, Hua Chen, Hou Tee Ng, Qi Ye, Wendy Fan, Jie Han, M. Meyyappan *Electroanalysis* 2005, *17*, 15-27.
[68] Wang, J. *Electroanalysis* 2005, *17*, 7-14.
[69] Y. Tu, Y. L., Z. F. Ren *Nano Letters* 2003, *3*, 107-109.
[70] Yi Tu, Y. L., Wassana Yantasee, Zhifeng Ren *Electroanalysis* 2005, *17*, 79-84.
[71] Yuehe Lin, F. L., Yi Tu, and Zhifeng Ren *Nano Letters* 2004, *4*, 191-195.
[72] Yuehe Lin, W. Y., and Joseph Wang *Frontiers in Bioscience* 2005, *10*, 492-505.
[73] Jessica Koehne, J. L., Alan M. Cassell, Hua Chen, Qi Ye, Hou Tee Ng, Jie Han and M. Meyyappan *J. Mater. Chem* 2004, *14*, 676-684.
[74] Salata, O. *Journal of Nanobiotechnology* 2004, *2*, http://www.pubmedcentral.nih.gov/articlerender.fcgi?artid=419715.
[75] Raj CR, O. T., Ohsaka T *J. Electroanal. Chem* 2003, *543*, 127.
[76] Pihel K, W. Q., Wightman RM *Anal Chem* 1996, *68*, 2084.
[77] Malem F, M. D. *Anal Chem* 1993, *65*, 37.
[78] Montenegro MCBSM, S. M. *J Pharm Sci* 2000, *89*, 876.
[79] Liping Lu, S., Wang, Xiangqin Lin *Anal Sci* 2004, *20*, 1131-1135.
[80] Yu A, L. Z., Cho J, Caruso F *Nano Letters* 2003, *3*, 1203.
[81] Liu S, Y. J., Ju H *J Electroanal Chem* 2003, *540*, 61.
[82] Liu S, J. H. *Biosens. Bioelect* 2003, *19*, 177.
[83] Lei Zhang, X. J., Erkang Wang, Shaojun Dong *Biosens. Bioelect* 2005, *21*, 337-345.
[84] Xiaojun Han, W. C., Zheling Zhang, Shaojun Dong, Erkang Wang *Biochimica et Biophysica Acta* 2002, *1556*, 273- 277.
[85] S. Bharathi, M. N. *Analyst* 2001, *126*, 1919-1922.
[86] Jingdong Zhang, M. K., Munetaka Oyama *Electrochem Commun* 2004, *6*, 683-688.
[87] Liping Lu, X. L. *Anal Sci* 2004, *20*, 527-530.
[88] L. Agü'ı, J. M., P. Yáñez-Sedeño, J.M. Pingarrón *Talanta* 2004, *64*, 1041-1047.
[89] I. G. Casella, A. D., E. Desimoni *Analyst* 1996, *121*, 249.
[90] C.-Y. Chiou, T.-C. C. *Electroanalysis* 1996, *8*, 1179.
[91] A. L. Crumbliss, J. S., R. W. Henkens, J. Zhao, J. P. O'Daly *Biosens. Bioelect* 1993, *8*, 331.
[92] Xuan Dai, O. N., Michael E. Hyde, and Richard G. Compton *Anal Chem.* 2004, *76*, 5924-5929.
[93] M. Dequaire, C. D., B. Limoges *Anal Chem.* 2000, *72*, 5521-5528.
[94] Hong Cai, C. X., Pingang He, Yuzhi Fang *J. Electroanal. Chem.* 2001, *510*, 78-85.
[95] Joseph Wang, R. P., and Danke Xu *Langmuir* 2001 *17*, 5739-5741.
[96] Wang, J. *Chem. Eur. J.* 1999, *5*, 1681.
[97] J. Wang, D. X., A.-N. Kawde, R. Polsky *Anal Chem.* 2001, *73*, 5576.
[98] J. Wang, R. P., D. Xu *Langmuir* 2001, *17*, 5739.
[99] K. R. Brown, A. P. F., M. J. Natan *J. Am. Chem. Soc.* 1996, *118*, 1154.

[100] M. B. Gonzalez Garcìa, A. C.-G. *Bioelectrochem. Bioenerg* 1995, *38*, 389.
[101] M. B. Gonzalez Garcìa, C. F. n.-S. n., A. Costa-Garcìa *Biosens. Bioelect.* 2000, *15*, 315.
[102] Craig E. Banks, T. J. D., Gregory G. Wildgoose and Richard G. Compton *Chem Comm* 2005, *7*, 829-841.
[103] Craig E. Banks, R. R. M., Trevor J. Davies and Richard G. Compton *Chem Comm* 2004 1804-1805.
[104] Yazhen Wang, Q. L., Shengshui Hu *Bioelectrochemistry* 2005, *65*, 135.
[105] F.Wu, G. Z., X.Wei *Electrochem.Commun* 2002, *4* 690.
[106] M.Musameh, N. S. L., J.Wang *Electrochem Commun* 2005 *7* 14.
[107] J. Wang, M. M. *Anal Chem.* 2003, *75* 2075.
[108] F.Valentini, S. O., M.L.Terranova,A.Amine, G.Palleschi *Sens, Actuators, B* 2004, *100*, 117.
[109] M.D.Rubianes, G. A. R. *Electrochem.Commun* 2003, *5*, 689.
[110] K.Wu, J. F., S.Hu *Anal.Biochem* 2003, *318* 100.
[111] W.C.Poh, K. P. L., W.De Zhang, S.Triparthy, J.Ye, F.Sheu *Langmuir* 2004, *20*, 5484.
[112] H.Zhang, X. W., L.wang, Y.Liu, C.Bai *Electrochim. Acta* 2004, *49*, 715.
[113] W.Zhang, Y. X., S.Ai, F.Wan, J.Wang, L.Jin, J.Jin *J.Chromatogr. B* 2003, *791*, 217.
[114] J.Ye, Y. W., W.Zhang *Electrochem Commun* 2004, *6*, 66-70.
[115] W.C.Poh, K. P. L., W.De Zhang, S.Triparthy, J.Ye, F.Shey *Langmuir* 2004, *20*, 5484.
[116] Y.zhao, W. Z., H.Chen and Q.Luo *Sens, Actuators, B* 2003, *92*, 279.
[117] G.Chen, L. Z., J.Wang *Talanta* 2004, *64*, 1018.
[118] K.Gong, Y. D., S.Xiong, Y.Chen, L.Mao *Biosens. Bioelect* 2004 *20* 253.
[119] K.Gong, M. Z., Y.Yan, L.Su, L.Mao, S.Xiong, Y.Chen *Anal Chem.* 2004, *76* 6500.
[120] M.Zhang, Y. Y., K.Gong, L.Mao, Z.Guo, Y.Chen *Langmuir* 2004 *20* 8781.
[121] M.L.Pedano, G. A. R. *Electrochem.Commun* 2004, *6*, 10.
[122] J.Wang, M. M. *Anal.Chim.Acta* 2004, *511*, 33.
[123] Z.Xu, X. C., X.Qu, S.Dong *Electroanalysis* 2004, *16*, 684.
[124] D.Nematollahi, M. A., S.W.Husainc *Electroanalysis* 2004, *16*, 1359.
[125] He P G, D. L. M. *Chem Comm* 2004, *3*, 348.
[126] Taft B J, L. A. D., etc *J. Am. Chem. Soc.* 2004, *126*, 12750.
[127] Xin Lu, Z. C. *Chem. Rev.* 2005, *105*, 3643-3696.
[128] Y.zhao, W. Z., H.Chen, Q.Luo, S.F.Y.Li *Sens, Actuators, B* 2002, *87*, 168.
[129] J.J.Davis, K. S. C., B.R.Azamian, C.B.Bagshqw, M.L.H.Green *Chem.Eur.J.* 2004, *9*, 3732.
[130] Wang J, L. G. D., Jan M R *J. Am. Chem. Soc.* 2004, *126*, 3010.
[131] Chen G F, Z. J., Tu Y H etc *Anal.Chim.Acta* 2005, *533*, 11.
[132] Buchberger, W. W., K.; Breitwieser, C. Fresenius' Z *Anal Chem* 1983, *315*, 518-520.
[133] Zen J-M, S. Y.-S., Chung H-H, Hsu C-T, Kumar AS *Anal Chem* 2002, *74*, 6126.
[134] Zen J-M, C. H.-H., Kumar AS *Analyst* 2000, *125*, 1633.
[135] Zen J-M, C. H.-H., Kumar AS *Anal. Chem.* 2002, *74*, 1202.
[136] I. G. Casella, M. G., M. R. Guascito, T. R. I. Cataldi *Anal.Chim.Acta* 1997, *357*, 63.
[137] Welch CM, H. M., Banks CE, Compton RG *Anal Sci* 2005, *21*, 1421.
[138] Jyh-Myng Zen, C.-T. H., Annamalai Senthil Kumar, Hueih-Jing Lyuu and Ker-Yun Lin *Analyst* 2004, *129*, 841-845.
[139] Elliot JM, B. P., Bartlett PN, Attard GS *Langmuir* 1999, *15*, 7411.

[140] Hrapovic S, L. Y., Male KB, Luong JHT *Anal Chem* 2004, *76*, 1083.
[141] Gan X, L. T., Zhu X, Li G *Anal Sci* 2004, *20*, 1271.
[142] Fang B, W. G., Zhang W, Li M, Kan X *Electroanalysis* 2005, *17*, 744.
[143] Goto T, O. T., Hirai T *Scr Mater* 2001, *44*, 1187.
[144] Martin Winter, R. J. B. *Chem. Rev* 2004, *104*, 4245-4269.
[145] B. Hammer, J. K. N. *Adv. Catal* 2000, *45*, 71.
[146] M. Valden, X. L., D. W. Goodman *Science* 1998, *28*, 1647-1650.
[147] M. Baumer, H.-J. F. *Prog. Surf. Sci.* 1999, *61*, 127.
[148] D. A. Landsman, F. J. L. *Handbook of Fuel Cells, Wiley, New York* 2003, *4*.
[149] JM, L. *Electrochim. Acta* 2005, *50*, 3123.
[150] Shen Y, L. J., Wu A, Jiang J, Bi L, Liu B, Li Z, Dong S *Langmuir* 2003, *19*, 5397.
[151] Zhang L, C. B., Samulski ET *Chem Phys Lett* 2004, *398*, 505.
[152] Nagahara Y, H. M., Yoshimoto S, Inukai J, Yau S-L, Itaya K *J Phys Chem B* 2004, *108*, 3224.
[153] L. Genie¡ s, R. F., R. Durand *Electrochim. Acta* 1998, *44*, 1317.
[154] O. Antoine, Y. B., R. Durand *J. Electroanal. Chem* 2001, *499*, 85.
[155] A. Gamez, D. R., P. Gallezot, F. Gloaguen, R. Faure, R. Durand *Electrochim. Acta* 1996, *41*, 307.
[156] Stonehart, P. *Ber. Bunsenges. Phys. Chem* 1990, *94*, 913.
[157] U. A. Paulus, A. W., G. G. Scherer, T. J. Schmidt, V.Stamenkovic, N. M. Markovic, P. N. Ross *Electrochim. Acta* 2002, *47*, 3787.
[158] U. A. Paulus, G. G. S., A. Wokaun, T. J. Schmidt, V. Stamenkovic, V. Radmilovic, N. M. Markovic, P. N. Ross *J. Phys. Chem. B* 2002, *106*, 4181.
[159] Arenz M, M. K., Stamenkovic V, Bilzanac BB, Tomoyuki T, Ross PN, Markovic NM *J. Am. Chem. Soc.* 2004, *127*, 6819.
[160] Y. Lou, M. M. M., L. Han, J. Luo, C.-J. Zhong *Chem Comm* 2001, 473.
[161] F. Hajbolouri, B. A., G. G. Scherer, A. Wokaun *Fuel Cells* 2004, *4*, 160.
[162] C. Roth, N. M., F. Hahn, J.-M. Leger, C. Lamy, H. Fuess *J.Electrochem. Soc.* 2002, *149*, 433.
[163] El-Deab MS, O. T. *Electrochim. Acta* 2002, *47*, 4255.
[164] El-Deab MS, O. T., Ohsaka T *J Electrochem Soc* 2003, *150*, A851.
[165] Mohamed S. El- Deab 1, T. O. *Electrochem Commun* 2002, *4* 288-292.
[166] Yanrong Zhang, S. A., Sachio Yoshihara, Takashi Shirakashi *Electrochim. Acta* 2003, *48*, 741-747.
[167] Ave Sarapuu, K. T. e. *Electrochem Commun* 2001, *3* 446-450.
[168] Ichizo Yagi, T. I., Kohei Uosaki *Electrochem Commun* 2004, *6*, 773-779.
[169] Mohamed S. El-Deab, T. O. *J. Electroanal. Chem.* 2003, *553*, 107-115.
[170] Y. Zhang, V. S., Nakazawa, S. Yoshihara* , T. Shirakashi *Electrochim. Acta* 2004, *49*, 5235-5240.
[171] Jin Luoa, M. M. M., Yongbing Loua, Li Hana, Maria Hepel b, Chuan Jian Zhonga, *Catalysis Today* 2002, *77*, 127-138.
[172] Kinoshita, K. *Wiley, New York* 1988.
[173] T. Vad, F. H., H. G. Haubold,G. G. Scherer, A.Wokaun *J. Phys. Chem. B* 2004, *108*, 12442.

[174] T. Yoshitake, Y. S., S. Kuroshima, H. Kimura, T. Ichihashi, Y. Kubo, D. Kasuya, K. Takahashi, F. Kokai, M.Yudasaka, S. Iijima *Phys. B* 2002, *323*, 124.

[175] Froudakis, G. E. *Pev. Aav. Mater. Sci* 2003, *5*, 259-264.

[176] Dillon, A. C. a. M. J. H. *Applied Physics A: Mateials Science & Processing* 2001, *72*, 133-142.

[177] Dillon, A. C., et al *Nature* 1997, *386*, 377-379.

[178] Conway, B. E. *Kluwer Academic/Plenum Publishers, New York* 1999.

[179] R. Kotz, M. C. *Electrochim. Acta* 2000, *45*, 2483.

[180] C. Niu, E. K. S., R. Hoch, D. Moy, H. Tennent *Appl. Phys.Lett.* 1997, *70*, 1480.

[181] Chong-yang Liu, A. J. B., Fred Wudl, Iris Weitz and James R. Heath *Electrochemical and Solid-State Letters* 1999, *2*, 557

[182] L. Diederich, E. B., P. Piseri, A. Podesta, P. Milani, A. Schneuwly and R. Gallay *Appl. Phys.Lett.* 1999, *75*, 2662.

[183] J. Chmiola, G. Y., Y. Gogotsi, C. Portet, P. Simon, P. L. Taberna *Science* 2006 *313*, 1760.

[184] Sandeep Kumar, G. D. S. *Microchim Acta* 2007, in press.

[185] M.K. Nazeeruddin, A. K., I. Rodicio, R. Humphrybaker, E. Muller, P. Liska, N. Vlachopoulos, M. Gra¨ tzel *J. Am. Chem. Soc.* 1993, *115*, 6382.

[186] Zen J-M, H. C.-T., Kumar AS, Lyuu H-J, Lin K-Y *Analyst* 2004, *129*, 841.

[187] J. Solla-Gullon, V. M., A. Aldaz, J. Clavilier, J *Electroanal. Chem* 2000, *491*, 69.

In: Encyclopedia of Nanotechnology
Editor: Elwood D. Carlson

ISBN 978-1-60692-079-4

Chapter 46

POLYMERIC NANOPARTICLES FOR ORAL DELIVERY OF PROTEIN DRUGS

Ji-Shan Quan, Hu- Lin Jiang, Jia-Hui Yu, Ding-Ding Guo, Rohidas Arote, Yun-Jai Choi, Chong-Su Cho*

Department of Agricultural Biotechnology, Seoul National University, Seoul 151-921, South Korea

ABSTRACT

Proteins and peptides as therapeutic agents have been used for clinical applications. While due to their short *in vivo* half life, easily undergoing hydrolysis, low permeability to biological membrane barriers and poor stability in gastrointestinal tract, parenteral administration such as subcutaneous, intravenous, or intramuscular injections is the common way for these drugs delivery. Alternative injection, the oral route presents many advantages such as avoidance injection pain and discomfort, elimination of possible infections caused by the use of needles. Therefore, many strategies to improve their oral delivery efficiency have been studied. One of the most potential strategies is usage of micro- and nanoparticles of functional polymers as protein drug carriers. This chapter summarizes the nanoparticles based on different polymers as an oral administration system for protein and peptide drugs delivery. It primarily describes the pH-sensitive nanoparticles. Depending on the fact that pH of the human gastrointestinal tract increases progressively from the stomach to the intestine, nanoparticles remain insoluble in the stomach and disintegrate at the higher pH. Secondly, mucoadhesive polymer-coated nanoparticles are described, which can intensify the contact between the polymer and the mucus surface either by non covalent bonds or covalent disulfide bonds. Thirdly, bacteria degradable polymer-coated nanoparticles for colon delivery are briefly reported. Fourthly, a pH-sensitive and mucoadhesive polymer based on nano- and microparticles is included. Finally, some success of oral protein drug *in vivo* deliveries are enumerates. All of these nanoparticles can be expected to improve the proteins absorption.

Keywords: Nanoparticles; Protein drugs; Oral delivery; Mucoadhesive; pH-sensitive

---

* To whom should be addressed. Tel: +82 2 880 4636; Fax: +82 2 875 2494; E-mail:chocs@plaza.snu.ac.kr

1. INTRODUCTION

With the advances in biotechnology and genetic engineering, peptides and proteins are attracting increasing interest for applying to the therapeutic agents for treatment of numerous diseases. These proteins have great potential to cure diseases with negligible side effects and merely symptomatic treatment [1]. Generally these proteins possesses shorter half lives *in vivo*, easily undergoes hydrolysis, low permeability to biological membrane barriers and also poor stability in gastrointestinal tract. Owing to above fact, these protein drugs are administered parenterally such as subcutaneous, intravenous, or intramuscular injections. These administration routes also reveal some disadvantages such as discomfort due to repeated and prolonged dosage regimes and difficulty to self-administration, and may cause infections by injectable dosage forms which could not be ignored [2]. Oral route of administration is so far easiest route for these protein drugs which avoids injection pain and elimination of the possible infections caused by improper sterilization techniques. However, proteins and peptides show poor bioavailability after oral administration. To improve their bioavailability various strategies have been employed such as:

a. Chemical modification of these peptides and proteins for physicochemical properties: structure modifications of these macromolecules can provide several advantages such as improvement in the intestinal absorption [3], enhancement of the stability [4] and protection from degradation by proteases [5].
b. Addition of novel functionality to the proteins and peptides: introduction of novel functionality to peptides and proteins using transport-carrier molecules that are recognized by endogenous cellular-transport systems in the gastrointestinal (GI) tract to increase the intestinal absorption [1, 7].
c. Administration of the protein drugs with the penetration enhancers or protease inhibitors: penetration enhancers can modify the barrier properties of the mucosa; as a result the drug absorption is promoted [8]. The protease inhibitors suppress the enzyme activity, thereby avoiding the degradation in GI by enzymes [9].
d. Utilization of the delivery carriers: liposome, nano-/ microparticles, micelles, and etc. protection of the proteins from degradation by stomach acid or enzymes. And they also can improve the absorption of proteins and peptides [10].

These approaches are successfully used to increase the bioavailability of protein drugs in the laboratory although it still shows some safety concerns when used in clinics. For example, the use of enzyme inhibitors in long-term therapy remains questionable because of possible absorption of unwanted proteins, disturbance of the digestion of nutritive proteins and stimulation of protease secretion as a result of feedback regulation [1] and some surfactants which enhance the absorption of poor absorptive drugs can also cause the mucosal damage in the gastrointestinal tract [11].

Carriers for oral protein and peptide delivery were found attractive owing to their property to protect proteins from degradation by the stomach acid or by the proteolytic enzymes in the GI tract, and especially some nanoparticles can deliver antigens to the payers patch for oral immunization [12, 13, and 14]. Polymeric nanoparticles have been studied as carriers for oral drug delivery since the latter half of the 1980s [15]. They include solid particles ranging in size from 10 nm to 1000 nm (1 μm) and consist of macromolecular materials in which the active principle (drug or biologically active material) is dissolved, entrapped, or encapsulated, and/or to which the active material is adsorbed or attached [16]. Of the delivery systems, polymeric nanoparticles have received more attention because of the subcellular size which makes them to cross the epithelium than do microparticles, and have the stability in biologic fluids as well as during storage than the liposomes have [17]. Polymeric nanoparticles have been used as oral drug carriers due to several reasons [18]:

1. Improvement of the bioavailability of drugs with poor absorption characteristics
2. Prolongation of the residence time of drugs in the intestine
3. High dispersion at the molecular level and consequently increase of absorption
4. Delivery of vaccine antigens to gut-associated lymphoid tissue
5. Control of the release of the drugs
6. Targeting of therapeutic agents to a particular organ and thus reducing toxicity
7. Reduction of the GI mucosal irritation caused by drugs
8. Assurance of the stability of drugs in the GIT

In this chapter, we will focus on the nanoparticles based on different functional polymers as an oral administration system for protein and peptide drugs delivery.

2. pH-Sensitive Polymeric Nanoparticles for Protein and Peptide Drugs Delivery

The pH-dependent systems are based on the fact that pH of the human gastrointestinal tract increases progressively from the stomach (pH 1-2), small intestine (pH 5-7) at the site of digestion and it increases to 7-8 in the distal ileum. pH-dependent polymers, containing ionizable groups in their backbone that can accept or donate protons in response to the environmental change in pH. The degree of ionization in a polymer bearing weakly ionizable groups is dramatically altered at a specific pH. This rapid change in net charge of pendant groups causes an alternation of the hydrodynamic volume of the polymer chains. The transition from collapsed state to expanded state is explained by the osmotic pressure exerted by mobile counter ions neutralizing the network charges. Thus these polymer- coated particles swell at high pH, further release the active drug more.

Hydrogels have been extensively exploited for biomedical applications due to their high water content and excellent biocompatibility The pH-sensitive hydrogels containing pendant acidic or basic groups such as carboxylic acids, sulphonic acids, primary amines, or ammonium salts which change ionization in response to change in the pH have become the subject matter of major interest for use as carriers in oral drug delivery research [19].

Weak polyacids (or polybases), which undergo an ionization/deionization transition from pH 4-8, are utilized as pH-responsive polymers such as poly(acrylic acid) (PAAc) and poly(methacrylic acid) (PMAAc). Their carboxylic pendant groups accept protons at low pH, while releasing them at high pH. This gives a momentum along with the hydrophobic interaction to govern precipitation or solubilization of molecular chains, deswelling/swelling of hydrogels, or hydrophobic/hydrophilic characteristics of surfaces. Therefore, at low pH (1–

2) these materials shrink and the encapsulated drug can be protected from the acidic environment of stomach due to its limited release. As they enter the small intestine, the pH changes from acidic to neutral (6 to 7.4) lead the matrix to swell and release the encapsulated drug [20].

The pH-sensitive polymers containing sulfonamide groups (derivatives of *p*-aminobenzene sulfonamide) are the weak polyacids that shows various pKa values ranging from pH 3 to 11, resulting from different pendant substituents at the sulfoneamide group acting as electron withdrawing or donating groups [21, 22]. The hydrogen atom of the amide nitrogen can be readily ionized to form a weak polyacid.

When these polymers including poly (methacrylic acid-co-methylacrylate) copolymers, such as Eudragit® E100, Eudragit®. L100-55, Eudragit®. L100, Eudragit®. S100 are used as particulate carriers with incorporation of CyA using an adaptation of the quasi-emulsion solvent diffusion technique. These polymers can be dissolved rapidly upon deprotonation of carboxylic acid groups at specific pH values. Thereby the release profiles of these nanoparticles exhibit significant pH-sensitivity, making CyA mainly released at its specific absorption part of the gastrointestinal tract, decreasing the degradation by gastric acid and the first-pass metabolism by gastrointestinal enzymes and increase the oral bioavailability of CyA [23].

Insulin and bovine serum albumin (BSA) are poorly suited for the purpose of the oral delivery due to their susceptibility towards enzymatic degradation and extremely poor diffusion across intestinal mucosa. But when these two kinds of drugs were incorporated into the polymeric nanoparticles based on the pH-sensitive polymethacrylic acid–chitosan–polyethylene glycol (PCP), release characteristics of model proteins were largely dependent on pH of the medium: less than 15 % of loaded-BSA and around 15 % of loaded-insulin was released at acidic pH in the first 2 h of study. On the other hand, at pH 7.4, more than 90% of loaded-BSA was released in 5 h and the insulin from PCP nanoparticles was completely released in 3 h. Thus these pH-sensitive polymeric nanoparticles can release drug at a specific pH within the GIT, as close as possible to the absorption window of the drug. It shows that these polymeric nanoparticles can serve as good candidate for oral peptide delivery [24].

Biocompatible nanoparticles from vinylpyrrolidone and acrylic acid monomers crosslinked with N, N'-methylene bisacrylamide were prepared [19] and FITC-dextran was used as a marker compound which was entrapped in these nanoparticles. It was found that the nanoparticles exhibited an initial burst of about 60 and 40% release for the first day at pH 10 and pH 7.0 solutions, respectively. In contrast, the solution of pH 1.0 showed release of only 2–3% of the dextran for the first day. The swelling behaviors of the solid copolymer gels were studied by measuring the increase in weight of the polymer. Swelling was increased quickly with the time of treatment, which was faster in the initial steps, reaching a constant value after a time depending on the pH of the medium and the time to reach equilibrium in acidic solution was much longer than that required in neutral and alkaline solutions [19].

3. MUCOADHESIVE POLYMERIC NANOPARTICLES FOR PROTEIN DRUGS DELIVERY

Over the past few decades, there has been considerable interest in developing new routes, alternative to injection, for delivering the macromolecules such as proteins and peptides [25]. However, peptide and protein drugs get degraded before they reach the blood stream which in turn unable to cross the mucosal barriers [26]. The mucoadhesive polymer-coated nanoparticles as colloidal carriers are a promising approach to solve these problems. Especially the mucoadhesive nanaoparticulate system is one of the most attractive applications for the oral administration of peptide or protein drugs because peptide or protein drugs are intrinsically poorly absorbable owing to their high molecular weight and hydrophilicity and they are also susceptible to enzymatic degradation in the gastrointestinal tract whereas the entrapped drug can be protected from enzymatic degradation using muchoadhesive nanoparticles [27]. In this context, we will focus the basic mechanisms by which mucoadhesive can adhere to a mucous membrane and peptide or protein drug delivery with mucoadhesive nanoparticulate systems.

3.1. Mechanisms of Mucoadhesion

3.1.1. Mucous Membranes

Mucous membranes are the moist surfaces lining the walls of various body cavities such as the gastrointestinal and respiratory tracts. They consist of a connective tissue layer above which is an epithelial layer, the surface of which is made moist usually by the presence of a mucus layer [28]. The epithelia may be either single layered or multilayered/stratified. The former contains goblet cells which secrete mucus directly onto the epithelial surfaces, the latter contains, or is adjacent to tissues containing specialized glands such as salivary glands that secret mucus onto the epithelial surface. Mucus is present as either a gel layer adherent to the mucosal surface of as a luminal soluble or suspended form. The major components of all mucus gels are mucin glycoproteins, lipids, inorganic salts and water, the latter accounting for more than 95% of its weight, making it a highly hydrated system [29]. The mucin glycoproteins are the most important structure-forming component of the mucus gel, resulting in its characteristic gel-like, cohesive and adhesive properties. The thickness of this mucus layer varies on different mucosal surface, from 50 to 450 μm in the stomach [30, 31], to less than 1 μm in the oral cavity [32]. The major functions of mucus are that of protection and lubrication.

3.1.2. The Mucoadhesive/mucosa Interaction

For adhesion to occur, molecules must bond across the interface. These bonds arises in the following ways [28].

(1) Ionic bonds: where two oppositely charged ions attract each other via electrostatic interactions to form a strong bond.
(2) Covalent bonds: where electrons are shared, in pairs, between the bonded atoms in order to "fill" the orbitals in both. These are also strong bonds.

(3) Hydrogen bonds: here a hydrogen atom, when covalently bonded to electronegative atoms such as oxygen, fluorine or nitrogen, carries a slight positively charge and is therefore attracted to other electronegative atoms. The hydrogen can therefore be thought of as being shared, and the formed bond is generally weaker than ionic or covalent bonds.

(4) Van-der-Waals bonds: these are some of the weakest forms of interaction that arise from dipole-dipole and dipole-induced dipole attractions in polar molecules, and dispersion forces with non-polar substances.

(5) Hydrophobic bonds: more accurately described as the hydrophobic effect, these are indirect bonds that occur when non-polar groups are present in an aqueous solution. Water molecules adjacent to non-polar groups form hydrogen bonded structures, which lowers the system entropy. There is therefore an increase in the tendency of non-polar groups to associate with each other to minimize this effect.

3.2. The Mucoadhesive Materials

The most widely investigated group of mucoadhesives is hydrophilic macromolecules containing numerous hydrogen bond forming groups [33, 34, and35]. Their initial use as mucoadhesives was in denture fixative powders or pastes. The presence of hydroxyl, carboxyl or amine groups on the molecules favors adhesion. They are called "wet" adhesives in that they are activated by moistening and will adhere non-specifically to many surfaces [36]. Once activated, they will show stronger adhesion to dry inert surfaces than those covered with mucus. Unless water uptake is restricted, they may overhydrate to form slippery mucilage. Like typical hydrocolloid glues, once formed, the adhesive joint is allowed to dry which in turn forms very strong adhesive bond. Typical examples are carbomers [37], chitosan [38], sodium alginate [39] and the cellulose [40] derivatives.

3.3. Mucoadhesive Polymeric Nanoparticles for Peptide or Protein Drugs Delivery

3.3.1. Mucoadhesive Polymeric Nanoparticles

Nanoparticles of biodegradable and biocompatible polymers are good candidates for particulate carriers to deliver peptide or protein drugs. Such particles are expected to be adsorbed in an intact form in the gastrointestinal tract after oral administration [41].

Chitosan is as a widely available mucoadhesive polymer able to increase cellular permeability and improve the bioavailability of orally administered peptide or protein drugs [42]. The mucoadhesive properties of chitosan have been illustrated by its ability to adhere to porcine gastric mucosa *in vitro* [43]. In the interactions between chitosan and mucus, the primary mechanism of action at the molecular level was found to be electrostatic [44]. This mucoadhesive property of chitosan was even significantly further improved by the immobilization of thiol groups on the polymer. The immobilization of thiol bearing moieties on the chitosan backbone yields the thiolated chitosan. The derivatization of the primary amino groups, mainly at the 2-position of the glucosamine subunits, with coupling reagents bearing thiol groups leads to the formation of thiolated chitosans [45]. Mainly four type of thiolated chitosans have been synthesized so far. They are chitosan-thioglycolic acid conjugates [46, 47, and 48], chitosan-cysteine conjugates [49], chitosan-4-thio butyl-amidine (chitosan-TBA) conjugates [50] and chitosan-thioethylamidine (TEA) derivative [51]. The mucoadhesive and permeation enhancing properties of chitosan are improved by thiolation [47, 49, 50, and 52]. The improved mucoadhesive properties of thiolated chitosans are mainly due to the formation of covalent bonds between thiol groups of the polymer and cysteine-rich subdomains of glycoproteins in the mucus layer [53]. These covalent bonds are supposedly stronger than non-covalent bonds, such as ionic interactions of chitosan with anionic substructures of the mucus layer.

3.3.2. Surface Modification of Nanoparticles with Mucoadhesive Polymers

One of the most promising strategies for designing mucoadhesive particulate systems is surface modification, or coating, of the drug carrier particles with mucoadhesive polymers. In reviewing the surface modification of colloidal particles, a number of investigations of the polymer-coating of liposomes have been published. Sunamoto et al. demonstrated an improvement in the chemical and physical stability of polymer-coated liposomes prepared with polysaccharide derivatives such as mannan or amylopectin [54, 55]. So far several compounds such as poloxamer [56], polysorbate 80 [57], carboxymethyl chitin [58], carboxymethyl chitosan [59] and dextran derivatives [60] have been reported to be suitable for the surface-coatings of the liposomal system. Kawashima et al. also reported that an increase in residence time and intimate contact of the chitosan-coated nanoparticles with the wall of the small intestine significantly improved the degree of drug absorption although they reported an enhancement of nanoparticles taken up by gut tissue [61].

The mucoadhesive nanoparticulate system was found to be useful for peptide or protein drugs to improve oral mucosal delivery due to their prolonged retention in the gastro-intestinal tract and excellent penetration into the mucus layer.

4. Bacteria Degradable Polymeric Particles for Protien and Peptide Delivery

Colon as a target site for protein and peptide drugs received an increasing interest because it offers distinct advantages. The colon has relatively low activities of proteolytic enzymes. Thus, the peptides and proteins can be more effectively absorbed in the colon. And there is around 10^{11}-10^{12}CFU ml^{-1} of microflora in the colon, which is much more than the other region of GI tract [62]. These bacteria utilize the foods that have not been digested in the small intestine, for instance di, tri-polysaccharides, and mucopolysaccharides [63], as a source of carbon produce a wide range of reductive and hydrolytic enzymes. These include a-arabinosidase, azoreductase, β- glucuronidase, β-galactosidase, β-xylosidase, nitroreductase, deaminase, urea hydroxylase etc [64]. Therefore, proteins and peptides can be encapsulated in the polymers which are refractory in the stomach and small intestine yet specifically degraded by those enzymes in the colon [65].

4.1. Azo Polymer-coating Particles

Varied azo polymers [65, 66] have been synthesized and evaluated as coating materials for colon targeting. When the polymers, crosslinked by the azo aromatic group $R-C_6H_4-N=N-C_6H_4-R$, are used for coating, particles can protect the drugs from digestion in the stomach and small intestine. Once they reached at the colon, the microflora reduced the azo bonds, and broke the crosslinks to form a pair of aromatic amines, $R-C_6H_4-NH_2$ + $H_2N--C_6H_4-R$. The polymer is degraded and the drug will be released [67]. However, the use of azo polymers and other synthetic polymers need to concern about the safety of these polymers. And a detailed toxicological study should be performed before being used commercially [62].

4.2. Polysaccharides-coating Particles

Polysaccharide-coated particles can also be used as colon targeting carriers for protein and peptide drugs. As we known, the polysaccharides can be degraded by the polysaccharidase in the colon, which is the main reason for these natural compounds used as colon specific carriers. They also have some other advantages including low prices and wide availability, as well as they can be modified chemically and biochemically, highly stable, safe and etc. Polysaccharides such as pectin, chitosan, inulin, dextran etc have been studied for this purpose. Pectins are nonstarch, linear polysaccharides that consist of α-1,4-D-galacturonic acid and 1,2 D-rhamnose with D-galactose and D-arabinose side chains having average molecular weights between 50,000 to 150,000 [68]. One major problem with the pectin and also with other polysaccharides is the high water solubility that would lead to drug early release in the tracts of the upper GI tract [69]. Therefore, they must be made water-insoluble by crosslinking or hydrophobic dramatization [68]. Atyabi and coworkers [70] prepared bovine serum albumin-loaded pectinate beads by adding the mixture of polymer and drug to the calcium chloride solution via a nozzle. The spheres were formed by ionotropic reaction between the divalent calcium ions and the negatively charged carboxyl groups of the pectin molecules. It was observed when the enzyme existed in the dissolution medium, BSA content released from the calcium pectinate beads was much faster.

5. pH-Sensitive and Mucoadhesive Polymereric Particles for Protein and Peptide Delivery

pH-sensitive polymer and mucoadhesive polymer-coating particles have been described at the above. While, Eudragit-cysteine conjugate which have both pH-dependent and mucoadhesive character was synthesized by Quan et al. [71]. The amount of cysteine in the conjugate calculated by ^1H nuclear magnetic resonance (^1H NMR) was about 8 mol-percent. Furthermore, the polymer was used to coat chitosan microspheres. The release rate of BSA from BSA-loaded thiolated Eudragit-coated chitosan microspheres (TECMs) at pH 7.4 PBS solution was significantly higher than at pH 2.0 PBS solution. Gamma camera imaging of Tc-99m labeled microsphere distribution in rats after oral administration suggested that TECMs had comparatively stronger mucoadhesive characters than either Eudragit microspheres or chitosan microspheres. Therefore, thiolated Eudragit-coated chitosan microspheres will be a good carrier for oral protein drug delivery.

6. In Vivo Results of Functional Polymeric Nanoparticles

For oral delivery of the protein and peptide drugs, the functional polymeric nanoparticles have been demonstrated to possess excellent property *in vitro*. However, it should be testified *in vivo* to see the promising relationship between two of them, as the *in vivo* condition has lots of difference from the *in vitro*. Many papers reported that functional polymeric nanoparticles as an oral dosage form can significantly improve bioavailability of some protein or peptide drugs. Jaeghere et al. reported that pH-sensitive nanoparticles prepared by the poly (methacrylic acid-co-methacrylate) copolymer Eudragit® . L100-55 improved the oral bioavailability of HIV-1 protease inhibitor-CGP 70726 in dogs [72]. Similar result was obtained by Dai et al., when CyA-loaded pH-sensitive nanoparticles made by Eudragit copolymer was given to SD rats, in spite of some difference of relative bioavailability for each polymer [23].

Mucoadhesive nanoparticles for oral delivery of protein and peptide drugs also have been evaluated by *in vivo*. Pan et al. [73] and Ma et al. [74] used chitosan nanoparticles (CNS) to incorporate insulin. These formulations were orally administered to streptozotocin-induced diabetic rats. After orally administration, the decrease in plasma glucose levels was significantly different from that induced by the control insulin solution, and they sustained the serum glucose at pre-diabetic levels for 15h and 11h, respectively. The average pharmacological bioavailability relative to SC injection of insulin solution was up to 14.9% [73]. Confocal micrographs showed strong interaction between rat intestinal epithelium and chitosan nanoparticles after 3 h post-oral administration [74]. Sarmento et al. [75] also successfully delivered the insulin by alginate/chitosan nanoparticles by oral administration. The basal serum glucose levels lowered were more than 40% with 50 and 100 IU/kg doses and hypoglycemia was sustained for over 18 h.

7. Conclusion

The development of a dosage form that improves the oral absorption of peptide and protein drugs with an extremely low bioavailability because of instability in the gastro-intestinal tract and low permeability through the intestinal membrane is one of the greatest challenges in the pharmaceutical field. From the discussion mentioned above, functional polymer-coated nanoparticulate systems were found to be a promising strategy for oral peptide or protein drug carriers as these materials possesses excellent absorption enhancement/proteolytic enzyme inhibition capabilities, which avoid the proper drugs' potential risks of degradation by gastric acid and the first-pass metabolism by gastrointestinal enzymes. The same bioavailability improving effects can be obtained using mucoadhesive material, pH-dependent material or bacteria degradable polymers/copolymer as carriers due to their unique characterizations. Overall the polymeric nanopaticles have high potential in protein drug delivery. However, the success of these special vehicles is greatly dependent upon the availability of superior nontoxic materials with high biocompatibility. Also, methods for preparing these nanoparticles need to be developed persistently. Moreover, additional *in vivo* experiments must be performed to substantiate the success of polymeric drug delivery systems in the real life applications.

8. ACKNOWLEDGEMENTS

We would like to acknowledge Biogreen 21 (20050401034696) for providing funding. Ji-Shan Quan and Hu-Lin Jiang were supported by BK21 program.

REFERENCES

[1] Morishita, M. and Peppas, N.A., (2006). Is the oral route possible for peptide and protein drug delivery? *Drug Discovery Today,* 11, 905-910.

[2] Hamman, J.H., Enslin, G.M. and Kotzé, A.F., (2005).Oral delivery of peptide drugs: barriers and developments. *Bio Drugs,* 19, 165-177.

[3] Fujita, T., Fujita, T., Morikawa, K., Tanaka, H., Iemura, O., Yamamoto, A. and Muranishi, S., (1996). Improvement of intestinal absorption of human calcitonin by chemical modification with fatty acids: Synergistic effects of acylation and absorption enhancers. *Int. J. Pharm.*, 134, 47-57.

[4] Calceti, P., Salmaso, S., Walker, G. and Bernkop-Schnurch, A., (2004). Development and invivo evaluation of an oral insulin-PEG delivery system. *Eur. J. Pharm. Sci.,* 22 , 315-323.

[5] Bundgaard, H. and Møss, J., (1990). Prodrugs of peptides. 6. Bioreversible derivatives of thyrotropin-releasing hormone (TRH) with increased lipophilicity and resistance to cleavage by the TRH-specific serum enzyme. *Pharm Res.,* 7, 885-892.

[6] Lee, C.H., Woo, J.H., Cho, K.K., Kang, S.H., Kang,S.K. and Choi, Y.J., (2007). Expression and characterization of human growth hormone-Fc fusion proteins for transcytosis induction. *Biotechnol. Appl. Biochem.,* 46, 211-217.

[7] Eaimtrakarn, S., Rama Prasad, Y.V., Ohno, T., Konishi, T., Yoshikawa, Y., Shibata, N. and Takada, K., (2002). Absorption enhancing effect of labrasol on the intestinal absorption of insulin in rats. *J. Drug Target,* 10, 255-260.

[8] Fix, J.A., (1987). Absorption enhancing agents for the GI system. *J. Control. Release,* 6, 151-156.

[9] Lee, V.H.L., (1990). Protease ihibitors and penetration enhancers as approaches to modify peptide absorption. *J. Control. Release,* 13, 213-223.

[10] Sakuma, S., Hayashi, M. and Akashi, M., (2001). Design of nanoparticles composed of graft copolymers for oral peptide delivery. *Adv Drug Deliv Rev,* 47, 21-37.

[11] Oberle, R.L., Moore, T.J. and Krummel, D.A.P., (1995). Evaluation of mucosal damage of surfactants in rat jejunum and colon. *J. Pharmacol. Toxicol. Methods,* 33, 75-81.

[12] Jani, P.U., Florence, A.T., and McCarthy, D.E., (1992). Further histological evidence of the gastro-intestinal absorption of polystyrene nanospheres in the rat. *Int. J. Pharm.*, 84, 245-252.

[13] Florence, A.T., Hillery, A.M., Hussain, N. and Jani, P.U., (1995). Nanoparticles as carriers for oral peptide absorption: studies on particle uptake and fate. *J. Control. Release*, 36, 39-46.

[14] des Rieux A, Fievez, V., Garinot, M., Schneider, Y.J. and Préat, V., (2006). Nanoparticles as potential oral delivery systems of proteins and vaccines: A mechanistic approach. *J. Control. Release*, 116, 1-27.

[15] Sakuma, S., Hayashi, M. and Akashi, M., (2001). Design of nanoparticles composed of graft copolymers for oral peptide delivery. *Adv Drug Deliv Rev*, 47, 21-37.

[16] Kreuter, J., (1996). Nanoparticles and microparticles for drug and vaccine delivery. *J. Anat.*, 189, 503-505.

[17] Pinto, Reis C., Neufeld, R.J., Ribeiro, A.J. and Veiga, F., (2006). Nanoencapsulation II. Biomedical applications and current status of peptide and protein nanoparticulate delivery systems. *Nanomedicine: Nanotechnology, Biology, and Medicine*, 2, 53-65.

[18] Shah, R.B., Ahsan, F. and Khan, M.A., (2002). Oral delivery of proteins progress and prognostication. *Drit. Rev. ther. Drug carrier syst.*, 19, 135-169.

[19] Sahoo, S.K., De, T.K., Ghosh, P.K. and Maitra, A.J., (1998). pH- and thermo-sensitive hydrogel nanoparticles. *Colloid Interface Sci.*, 206, 361-368.

[20] Gil, Eun Seok. and Hudson, Samuel M., (2004). Stimuli-reponsive polymers and their bioconjugates. *Prog. Polym. Sci.*, 29, 1173–1222.

[21] Park, S.Y. and Bae, Y.H., (1999). Novel pH-sensitive polymers containing sulfonamide groups. *Macromol Rapid Commun*, 20, 269–273.

[22] Kang, S.I. and Bae, Y.H., (2003). A sulfonamide based glucose-responsive hydrogel with covalently immobilized glucose oxidase and catalase. *J. Control. Release*, 86, 115–121.

[23] Dai, J., Nagai, T., Wang, X., Zhang, T., Meng, M. and Zhang, Q., (2004). pH-sensitive nanoparticles for improving the oral bioavailability of cyclosporine A. *Int. J. Pharm.*, 280, 229-240.

[24] Sajeesh, S. and Sharma, C.P., (2006). Novel pH responsive polymethacrylic acid-chitosan-polyethylene glycol nanoparticles for oral peptide delivery. *J Biomed Mater Res B Appl Biomater*, 76, 298-305.

[25] Cui, F.Y., Qian, F. and Yin, C.H., (2006). Preparation and characterization of mucoadhesive polymer-coated nanoparticles. *Int. J. Pharm.*, 316, 154-161.

[26] Prego, C., Garci a, M., Torres, D., and Alonso, M.J., (2005). Transmucosal macromolecular drug delivery. *J. Control. Release*, 101, 151-162.

[27] Hirofumi, T., Hiromitsu, Y. and Yoshiaki, K., (2001). Mucoadhesive nanaoparticulate systems for peptide drug delivery. *Adv Drug Deliv Rev*, 47, 39-54.

[28] John, D. S., (2005). The basics and underlying mechanisms of mucoadhesion. *Adv Drug Deliv Rev*, 57, 1556-1568.

[29] Marriott, C. and Gregory, N.P., (1990). Mucus physiology and pathology, in: V. Lanaerts, R. Gurny (Eds.). *Bioadhesive Drug Delivery Systems*, CRC Press, Florida, 1-24.

[30] Allen, A., Cunliffe, W.J., Pearson, J.P., and Venables, C.W., (1990). The adherent gastric mucus gel barrier in man and changes n peptic ulceration. *J. Intern. Med,.* 228, 83-90.

[31] Kerss, S., Allen, A., and Garner, A., (1982). A simple method for measuring the thickness of the mucus gel layer adherent to rat, frog, and human gastric mucosa: influence of feeding, prostaglandin, N-acetylcysteine and other agents. *Clin. Sci.,* 63, 187-195.

[32] Sonju, T., Cristensen, T.B., Kornstad, L., and Rolla, G., (1974). Electron microscopy, carbohydrate analysis and biological activities of the proteins adsorbed in two hours to tooth surfaces in vivo. *Caries Res.,* 8, 113-122.

[33] Smart, J.D., Kellaway, I.W. and Worthington, H.E.C., (1984). An in vitro investigation of mucosa-adhesive materials for use in controlled drug delivery. *J. Pharm. Pharmacol.,* 36, 295-299.

[34] Smart, J.D., (1993). Drug delivery using buccal adhesive systems. *Adv Drug Deliv Rev,* 23, 253-270.

[35] Lee, J.W., Park, J.H. and Robinson, J.R., (2000). Bioadhesive-based dosage forms: the next generation. *J. Pharm. Sci.,* 89, 850-866.

[36] Ben Zion, O. and Nussinovitch, A., (1997). Physical properties of hydrocolloid wet glues. *Food Hydrocoll.,* 11, 429-442.

[37] Llabot, J.M., Manzo, R.H. and Allemandi, D.A., (2004). Drug release from carbomer:carbomer sodium salt matrices with potential use as mucoadhesive drug delivery system. *Int J Pharm.,* 276, 59-66.

[38] George, M. and Abraham, T.E., (2006). Polyionic hydrocolloids for the intestinal delivery of protein drugs: alginate and chitosan--a review. *J. Control. Release,* 114, 1-14.

[39] Ryu, J.M., Chung, S.J., Lee, M.H., Kim, C.K. and Shim, C.K., (1999). Increased bioavailability of propranolol in rats by retaining thermally gelling liquid suppositories in the rectum. *J. Control. Release,* 59, 163-172.

[40] Miyazaki, Y., Ogihara, K., Yakou, S., Nagai, T. and Takayama, K., (2003). Bioavailability of theophylline and thiamine disulfide incorporated into mucoadhesive microspheres consisting of dextran derivatives and cellulose acetate butyrate. *Biol Pharm Bull.* 26, 1744-1747.

[41] Florence, A.T., Hillery, A.M., Hussain, N. and Jani, P.Y., (1995). Nanoparticles as carriers for oral peptide absorption: studies on particle uptake and fate. *J. Control. Release,* 36, 39-46.

[42] Bowman, K. and Leong, K.W., (2006). Chitosan nanoparticles for oral drug and gene delivery. *Int. J. Nanomedicine,* 1, 117-128.

[43] Gåserød, O., Jolliffe, I.G., Hampson, F.C., Dettmar, P.W. and Skjåk-Bræk, G., (1998). The enhancement of the bioadhesive properties of calcium alginate gel beads by coating with chitosan. *Int. J. Pharm.,* 175, 237-246.

[44] Deacon, M.P., McGurk, S., Roberts, C.J., Williams, P.M., Tendler, S.J., Davies, M.C., Davis, S.S. and Harding, S.E., (2000). Atomic force microscopy of gastric mucin and chitosan mucoadhesive systems. *Biochem. J.,* 348, 557-563.

[45] Schnürch, A.B., Hornof, M. and Guggi, D., (2004). Thiolated chitosans. *Eur. J. Pharm. Biopharm.,* 57, 9-17.

[46] Schnürch, A.B. and Hopf, T.E., (2001). Synthesis and in vitro evaluation of chitosan–thioglycolic acid conjugates. *Sci. Pharm.,* 69, 109-118.

[47] Kast, C.E. and Schnürch, A.B., (2001). Thiolated polymers-thiomers: development and in vitro evaluation of chitosan–thioglycolic acid conjugates. *Biomaterials,* 22, 2345-2352.

[48] Hornof, M.D., Kast, C.E. and Schnürch, A.B., (2003). In vitro evaluation of the viscoelastic behavior of chitosan-thioglycolic acid conjugates. *Eur. J. Pharm. Biopharm.,* 55, 185-190.

[49] Schnürch, A.B., Brandt, U.M. and Clausen, A.E., (1999). Synthesis and in vitro evaluation of chitosancysteine conjugates. *Sci. Pharm.,* 67, 196-208.

[50] Schnürch, A.B., Hornof, M. and Zoidl, T., (2003). Thiolated polymers-thiomers: modification of chitosanwith 2-iminothiolane. *Int. J. Pharm.,* 260, 229-237.

[51] Kafedjiiski, K., Krauland, A.H., Hoffer, M.H. and Schnurch, A.B., (2005). Synthesis and in vitro evaluation of a novel thiolated chitosan. *Biomaterials,* 26, 819-826.

[52] Langoth, N., Guggi, D., Pinter, Y. and Schnürch, A.B., (2003).Thiolated chitosan: in vitro evaluation of its permeation enhancing. *Properties, 30th Annual Meeting and Exposition of the Controlled Release Society, Glasgow, UK,* 034.

[53] Leitner, V.M., Walker, G.F. and Schnürch, A.B., (2003). Thiolated polymers: evidence for the formation of disulphide bonds with mucus glycoproteins. *Eur. J. Pharm. Biopharm.,* 56, 207-214.

[54] Sunamoto, J., Iwamoto, K., Takada, M., Yuzuriha, T. and Katayama, K., (1983). Improved drug delivery to target specific organs using liposomes as coated with polysacchrides. *Polymer Sci. Technol.,* 23, 157-168.

[55] Sunamoto, J., Iwamoto, K., Takada, M., Yuzuriha, T. and Katayama, K., (1984). Polymer coated liposomes for drug delivery to target specific organs. In: J.M. Anderson and S.W. Kim, Editors, *Recent Advances in Drug Delivery Systems,* 153-162.

[56] Jamshaid, M., Farr, S.J., Kearney, P. and Kellaway, I.W., (1988). Poloxamer sorption on liposomes: comparison with polystyrene latex and influence on solute efflux. *Int. J. Pharm.,* 48, 125-131.

[57] Kronberg, B., Dahlman, A., Carlfors, J., Karlsson, J., and Artursson, P., (1990). Preparation and evaluation of sterically stabilized liposomes: colloidal stability, serum stability, macrophage uptake, and toxicity. *J. Pharm. Sci.,* 79, 667-671.

[58] Dong, C. and Rogers, J.A., (1991). Polymer-coated liposomes: stability and release of ASA from carboxymethyl chitin-coated liposomes. *J. Control. Release,* 17, 217-224.

[59] Alamelu, S. and Rao, K.P., (1991). Studies on the carboxymethyl chitosan-containing liposomes for their stability and controlled release of dapson. *J. Microencapsul.,* 8, 505-515.

[60] Elferink, W.G.L., de Wit, J.G., In't Veld, G., Reichert, A., Driessen, A.J.M., Ringsdorf, H. and Konings, W.N., (1992). The stability and functional properties of proteoliposomes mixed with dextran derivatives bearing hydrophobic anchor groups. *Biochim. Biophys. Acta.,* 1106, 23-30.

[61] Kawashima, Y., Yamamoto, H., Takeuchi, H. and Kuno, Y., (2000). Mucoadhesive DL-lactide/glycolide copolymer nanospheres coated with chitosan to improve oral delivery of elcatonin. *Pharm Dev Technol.,* 5, 77-85.

[62] Sinha, V.R. and Kumaria, R., (2003). Microbially triggered drug delivery to the colon. *Eur J Pharm Sci.,* 18, 3-18.

[63] Rubinstein, A., (1990). Microbially controlled drug delivery to the colon. *Biopharm. Drug Dispos.*, 11, 465-475.

[64] Kinget, R., Kalala, W., Vervoort, L. and Mooter, G.U., (1998). Colonic drug targeting. *J. Drug Target,* 6, 129-149.

[65] Tozaki, H., Nishioka, J., Komoike, J., Okada, N., Fujita, T., Muranishi, S., Kim, S.H., Terashima, H. and Yamamoto, A., (2001). Enhanced absorption of inslin and (Asu^{1,7}) Eel-calcitonin using novel azopolymer-coated pellets for colon-specific drug delivery. *J Pharm Sci.,* 90, 89-97.

[66] Kimura, Y., Makita, Y., Kumagai, T., Yamane, H. and Kitao, T., (1992). Degradation of azo-containing polyurethane by the action of intestinal flora: Its mechanism and application as a drug delivery system. *Polymer,* 33, 5294-5299.

[67] Saffran, M., Kumar, G.S., Savariar, C., Burnham, J.C., Williams, F. and Neckers, D.C., (1986). A new approach to the oral administration of insulin and other peptide drugs. *Science,* 233, 1081-1084.

[68] Chourasia, M.K. and Jain, S.K., (2004). Polysaccharides for colon targeted drug delivery. *Drug Delivery,* 129-148.

[69] Sinha, V.R. and Kumria, R., (2003). Microbially triggered drug delivery to the colon. *Eur J Pharm Sci.,* 18, 3-18.

[70] Atyabi, F., Inanloo, K. and Dinarvand, R., (2005). Bovine serum albumin-loaded pecinate beads as colonic peptide delivery system: preparation and in vitro characterization. *Drug Delivery,* 12, 367-375.

[71] Quan, J.S., Jiang, H.L., Choi, Y.J., Yoo, M.K. and Cho, C.S., (2007). Thiolated Eudragit-coated chitosan microspheres as an oral drug delivery system. *Key Engineering Materials, 342-343,* 445-448.

[72] De Jaeghere, F., Allémann, E., Kubel, F., Galli, B., Cozens, R., Doelker, E. and Gurny, R., (2000). Oral bioavailability of a poorly water soluble HIV-1 protease inhibitor incorporated into pH-sensitive particles: effect of the particle size and nutritional state. *J. Control. Release,* 68, 291-298.

[73] Pan, Y., Li,Y.J., Zhao, H.Y., Zheng ,J.M., Xu, H., Wei, G., Hao, J.S. and Cui, F.D., (2002). Bioadhesive polysaccharide in protein delivery system: chitosan nanoparticles improve the intestinal absorption of insulin in vivo. *Int J Pharm.,* 249, 139-147.

[74] Ma, Z.S., Lim, T. M. and Lima, L.Y., (2005). Pharmacological activity of peroral chitosan–insulin nanoparticles in diabetic rats. *Int J Pharm.,* 293, 271–280.

[75] Sarmento, B., Ribeiro, A., Veiga, F., Sampaio, P., Neufeld, R. and Ferreira, D., (2007). Alginate/chitosan nanoparticles are effective for oral insulin delivery. *Pharm Res. Dec.,* 24, 2198-2206.

In: Encyclopedia of Nanotechnology
Editor: Elwood D. Carlson

ISBN 978-1-60692-079-4

Chapter 47

FABRICATION OF MICRO/NANO-COMPOSITE STRUCTURES AND THEIR PHOTOVOLTAIC BEHAVIORS

Jin Zhai, Xiangliang Sheng, Yong Zhao and Liping Heng
Institute of Chemistry, Chinese Academy of Sciences,
Beijing 100080

ABSTRACT

Photovoltaic behavior in both organic thin films and organic/inorganic hybrid systems is an important subject not only in basic research but also in application fields such as organic solar cells, dyes sensitized solar cells etc. In order to improve the light collection efficiency, tuning film morphology is an efficient way to improve light harvest, then conversion efficiency. In this paper, some fabrication methods for different photoanodes with controlled morphology and their photovoltaic behaviors will be reviewed. While using electro-hydrodynamics (EHD) technique to prepare lotus leaf-like photoanodes will be discussed in detail to illustrate the effect of micro-/nano- composite structure on the enhancement of photo-electric conversion behavior in dye-sensitized solar cells (DSSCs). Further more, there will some discussion on the morphology control for the organic thin film system to increase photocurrent generation.

1. INTRODUCTION

In the past century, the consumption of non-renewable sources of energy has caused more environmental damage than any other human activity. Electricity generated from fossil fuels such as coal and crude oil has led to high concentrations of harmful gases in the atmosphere. This has in turn led to many problems being faced today such as ozone depletion and global warming. That do harm to the human life. Therefore, alternative and clean sources of energy have become very important and relevant to today's world. These sources, such as the sun and wind, can never be exhausted and therefore are called renewable. They cause less emission and are available locally. Their use can, to a large extent, reduce chemical, radioactive, and thermal pollution. Among all the renewable energy sources, photovoltaic technology utilizing solar energy is considered as the most promising one. Now, silicon-based and other inorganic semiconductor solar cells have made much progress not only in basic research but also in industrialization with the total conversion efficiencies between 15 and 20% [1]. But, the relatively high cost and complicated manufacturing have prevented their widespread application. Photovoltaic behavior in organic thin films and organic/inorganic hybrid systems is an important subject in organic solar cells and dye-sensitized solar cells (DSSCs). In order to improve the cells' conversion efficiencies, tuning film morphology, such as coupling the photonic crystal layer to electrode as dielectric mirror [2, 3], hollow sphere voids [4] and large size solid particles as scattering center [5-7], is an efficient way to improve light harvest. Therefore, porous structured photo-anodes with predominant penetration effect of viscous and solid electrolytes, fast charge transport and effective light collection efficiency are studious orientation for researchers.

Here some fabrication methods for different photo-anode with controlled morphology and their photovoltaic behaviors will be reviewed. At the same time, using electro-hydrodynamics (EHD) technique to prepare lotus leaf-like photo-anodes will be discussed in detail to illustrate the effect of micro-/nano- composite structure on the enhancement of photo-electric conversion behavior in DSSCs. Furthermore, the morphology control for the organic thin film system to increase photocurrent generation will be shown in this review.

2. Photo-Anodes with Different Morphologies in DSSCs

In 2002, A. P. Alivisatos et al [13] fabricated polymer solar cell by blending CdSe nanorods and P3HT (polythioene derivative). Their results illustrated that semiconductor nanorods showed better electron transport behavior than nanoparticles. Thereafter, P. D. Yang et al [14] successfully fabricated a dense array of oriented crystalline ZnO nanowires as photo-anode by mild aqueous chemistry and featured a surface area up to one-fifth as large as that of nanoparticle cell. The direct electrical pathways provided by the nanowires ensured the rapid collection of carriers generated throughout the device. As an ordered topology that increases the rate of electron transport, a nanowire electrode may provide a means to improve the quantum efficiency of DSSCs in the red region of the spectrum. Important differences in transport, internal electric field distribution and light scattering should make comparative studies of wire and particle devices fruitfully.

Due to the advantages of nanowires in electron transport, oriented one-dimensional nanomaterials were more and more attractive. In the same year, S. Uchida et al [15] synthesize TiO_2-based nanotubes through TiO_2 nanoparticles and fabricated photo-anodes. M. Adachi et al [16] formed single-crystal-like anatase TiO_2 nanowires in a network structure by surfactant-assisted self-assembling processes at low temperature. The crystal lattice planes of the nanowires and networks of such wires composed of many nanoparticles were almost perfectly aligned with each other due to the "oriented attachment" mechanism, resulting in the high rate of electron transfer through the TiO_2 nano-network with single-crystal-like anatase nanowires. A single-crystalline anatase TiO_2 exposing mainly the {101} plane has been prepared, which adsorbed ruthenium dye over 4 times higher than P25, and a high light-to-electricity conversion yield of 9.3% was achieved. C. A. Grimes et al [17] described the use of highly ordered transparent TiO_2 nanotube arrays in dye-sensitized solar cells (DSSCs) with

2.9% of overall conversion efficiency. Highly ordered nanotube arrays were grown perpendicular to a fluorine-doped tin oxide-coated glass substrate by anodic oxidation of a titanium thin film. Voltage-decay measurements indicated that the highly ordered TiO_2 nanotube arrays, in comparison to nanoparticulate systems, have superior electron lifetimes and provide excellent pathways for electron percolation. Some other ordered nanomaterials used in DSSCs include nanorods [18], nanosheet [19], nanotrees [20]. In addition, spinning TiO_2 fibers was also successful prepared and fabricated cells [21, 22]. All those results showed that one-dimensional nanomaterials have excellent electron transport and separation rate, which can boost the development of DSSCs.

Effective utilization of sunlight (especially above 600 nm light) can improve the performance of DSSCs, and multiple scattering structures are effective. A. Usami [23, 24] once pointed out that submicron multiple structures were in favor of light scattering and collecting through theory simulation. Theoretical calculation showed that the introduction of scattering layer or scattering center can effectively achieve useful photoelectric conversion efficiency [4-7, 25, 26]. T. E. Mallouk et al [2, 3] reported that the light harvesting efficiency of dye-sensitized photo-anodes was enhanced by coupling a TiO2 photonic crystal layer to a conventional film composed of TiO2 nanoparticles. In addition to act as a dielectric mirror, the inverse opal photonic crystal caused a significant change in dye absorbance that depended on the position of the stop band. Absorbance was suppressed at wavelengths shorter than the maximum stop band and enhanced at longer wavelengths. This effect arises from the slow group velocity of light in the vicinity of the stop band, and the consequent localization of light intensity in the voids (to the blue) or in the dye-sensitized TiO2 (to the red) portions of the photonic crystal. By coupling photonic crystals, the short circuit photocurrent efficiency across the visible spectrum (400-750 nm) could be increased by about 26%, comparing to an ordinary dye-sensitized nanocrystalline TiO2 photo-anode.

Spherical voids as light scattering centers in nanocrystalline TiO2 films were realized by S. Hore et al [4] with polystyrene particles of diameter 400 nm through sintering. As reported, spherical voids enhanced the scattering properties of the films and hence the light capture inside the device. Thus, it increased the overall conversion efficiency of the devices by 25%. Additionally, light scattering layers [25, 26] were also used to enhance light utilization and raised the performance of cells.

In all, there are great achievements of the photo-anode as the pivotal part in DSSCs during the past a few years. As a perfect photo-anode, it must have the following features such as fast electron injection and separation, slow electron recombination, excellent electron transport, high surface area, outstanding light collecting and so on. Composite structure, ordered one-dimensional nanomaterials and multiple structures provide the opportunities to further development in DSSCs.

3. Electro-Hydrodynamics (EHD) Technique in DSSCs

Since traditional preparation methods of photo-anodes such as doctor blade method, layer-by-layer deposition method and screening print method have the disadvantages of complex operation and processing, development of novel methods for fabricating photo-anodes in DSSCs with simple, efficient, large-scale traits is pursued by many researchers. At the same time, although DSSCs based on liquid electrolytes have reached efficiency as high as 10% under AM 1.5 (1000W/m2) [27], the devices sealing and long-term stability with organic liquid electrolytes (OLE) remain an important issue in view of the practical application. Consequently viscous molten-salt/polymer electrolytes without volatility, quasi-solid-state and solid-state electrolytes [28] have been introduced to substitute the organic liquid electrolytes. However superior penetration of those electrolytes into the nanocrystalline network structure and the quick charge transport in pure nano-structured films can't be realized easily. Therefore, porous structured photo-anodes with predominant penetration effect of viscous and solid electrolytes, fast charge transport and effective light collection efficiency are studious orientation for researchers.

The electro-hydrodynamic (EHD) technique has been confirmed to be effective and simple for fabricating micro to nano-scale fibers or particles from a variety of materials, such as organic [29–31], inorganic [32–34] and hybrid organic/inorganic [35, 36] compounds. It is widely applied in membrane technology [37, 38], drug delivery [39], tissue engineering [40], optical sensors [41] and biosensors [42]. Figure 1 shows the schematic diagram of the experimental setup used for EHD processing.

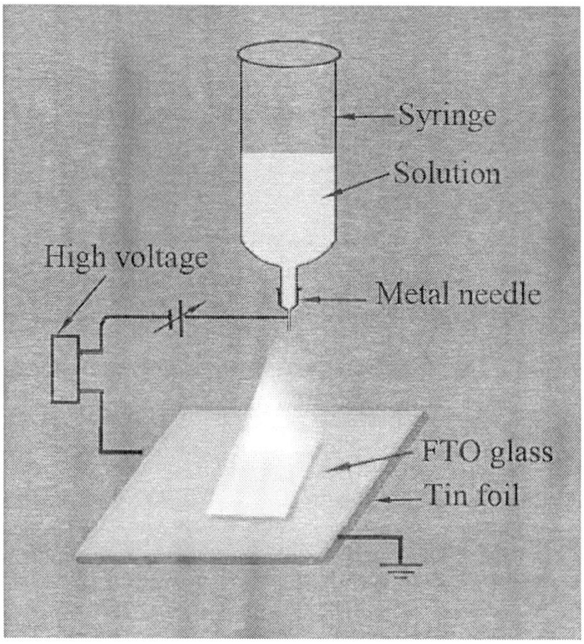

Figure 1. The schematic diagram of the experimental setup used for EHD processing.

In a typical process, the mixed solution containing the colloidal solution of nano-particles and polymer is sprayed from a nozzle under an applied electric field as strong as 1.5-3 kV/cm. The build-up of electrostatic charges on the surface of the liquid droplet induces the formation of jet, which is subsequently separated to form the mist. The solvent was removed from the solution under the influence of high voltage before depositing on the substrate. After the film was dried in the air, the calcinations were carried out to remove polymers. Hence, EHD technique can be used to realize not only large area but also continuous fabrication, which may be an alternative technique for future cell industrialization.

A one-dimensional TiO_2 nano-fibre structure prepared by electrospun and calcined was applied as an electrode in a DSSC and it facilitated the penetration of the viscous polymer gel electrolytes into the electrode. The photocurrent generation of the DSSC using the nano-fibre film as a photo-anode with a polymer gel electrolyte was over 90% of the value for an equivalent liquid DSSC. A η of 6.2% and an ff of 60% were obtained [21, 22]. As reported recently, TiO_2 photo-anode prepared by electro- spraying method had higher photocurrent than conventional photo-anode because of better ionic paths in the TiO_2 layer [43].

Very recently, EHD technique was used in our group and the photo-electric conversion behavior was studied in detail [44, 45].

Figure 2(a) and (b) show TiO2 composite electrodes prepared by EHD method, which has been confirmed to be effective and simple for fabricating micro to nano-scale fibers or particles from a variety of materials. From the SEM image, it can be seen clearly that the films was composed of the micro-scale and sub-micro-scale near spherical clusters with the diameter of 0.5-1 μm. Except for the spherical clusters, lots of nano-pores and micropores with the size of 2 nm -1 μm exist in this film. From the magnified image (Figure 2(b)), it can be seen that the macro-sphere was assembled by TiO2 nanoparticles and the size of nano-pores among the particles centralizes at ca. 10-20 nm. Micro-pores (micro-spheres) and nano-pores (nanoparticles) form a micro/nano composite structured film. Figure 2(c) illustrates the ideal sketch map of the composite electrode to show the structural specialties. Lots of submicro-pores are formed among the interspace of TiO2 submicro-spheres and hence the continuous micro-channels (red line). The inset is the magnified image. Micro-spheres formed by process of EHD method are composed of nanoparticles. Nano-pores exist among the nano-particles to form continuous nano-channels (blue line). The hierarchical branched inner channels composed of continuous micropores (red line) and nano-pores (blue line) in electrode, just like the blood circulation system of human body constructed by main vessel and capillary vessel, maybe hold nice capability of charge transport in viscous electrolytes.

Figure 2. (a) SEM images of film prepared by EHD method. (b) The magnified image of (a). (c) Simulative pore structure of electrodes with micro-channels and nano-channels.

Based on the composite hierarchical porous structure, we used the micro-nano electrode to assemble the DSSCs with viscous ionic liquid electrolytes, and also a nano-structured film was selected as a reference electrode to compare different performance of micro-nano structure and nano-structure. The analyzed data indicated that the photoelectrical parameters (fill factor and total conversion efficiency) were enhanced largely after incorporating composite hierarchical porous structure into the TiO_2 photo-anode, about increasing 10% for fill factor, 7% for open-circuit photovoltage, 6% for short-circuit photocurrent, and resulting enhancement of 20% for total conversion efficiency with viscous ionic liquid electrolytes.

The improved conversion efficiency was ascribed to enhancement of light collection efficiency and ionic diffusion rate in viscous electrolyte, which was induced by hierarchical porous channels in the micro-nano composite thin film.

The light-to-electric conversion efficiency of DSSCs is determined by light collection efficiency, capability of charge separation and charge transport. So the kinetics of ionic transportation and light scattering effect, which can be simultaneously tuned by morphology structure of the electrodes, are important parts in the system of DSSCs. They influence each other and there must be a balance for the best photovoltaic performance. If the pores are too large, despite of ions transporting easily for viscous electrolytes and possessing light-scattering effect, the thin films can't absorb enough dye molecules for light harvesting; if the pores are too small, the transportation kinetics of redox species limits the charge transfer rate and absence of light scattering effect though the amount of dye absorption is increased. Composite structure with hierarchical channels balances these two effects of ionic transportation and light scattering, and then obtains the optimal photoelectrical action. Except for TiO2 composite porous electrode with P25 and hydrothermal anatase nano-particles, we also extend the EHD technique to fabricate ZnO porous electrode, the improvement of photoelectrical performance was also observed compared to that of nano-structured ZnO electrode [46]. Based on the research of micro/nano electrodes, it can be deduced that the composite hierarchical porous structure is benefit for the high efficiency device with viscous electrolytes, such as ionic liquid and quasi-solid-state electrolyte. The idea of film structure modulation will provide a novel space for high efficiency devices.

4. Photocurrent Generation of Honeycomb Structures in Organic Thin Films

Many studies on organic solar cells have been conducted in the donor-acceptor linked molecules or hetero-structured films prepared by various methods. However, the surface structure effect has never been taken into account. Considering that light scattering can be achieved by adding pores or photonic crystals in the inorganic semiconductor films of TiO2 [2-4], which plays an important role in the efficiency of light harvesting in dye sensitized solar cells, fabrication of ordered honeycomb structure films with photo-electric functional polymers may benefit the photocurrent generation in organic thin film system accordingly. Breath figure formation is a simple and useful technique to prepare porous polymer films [47-51]. Under humid conditions, water vapors are condensed into water droplets at the solution surface during solvent evaporation, and then the water droplets are self-organized and closely packed hexagonally by the convectional flow or the capillary force generated at the solution front. After solvent and water evaporate completely, traces of water droplets remain in the polymer film to form honeycomb structures. The copolymer with the rod-coil block-poly (distyrylbenzene - co - triethylene glycol) (PDTG) was chosen and ordered honeycomb structures were introduced in the film by breath figure formation method [52].

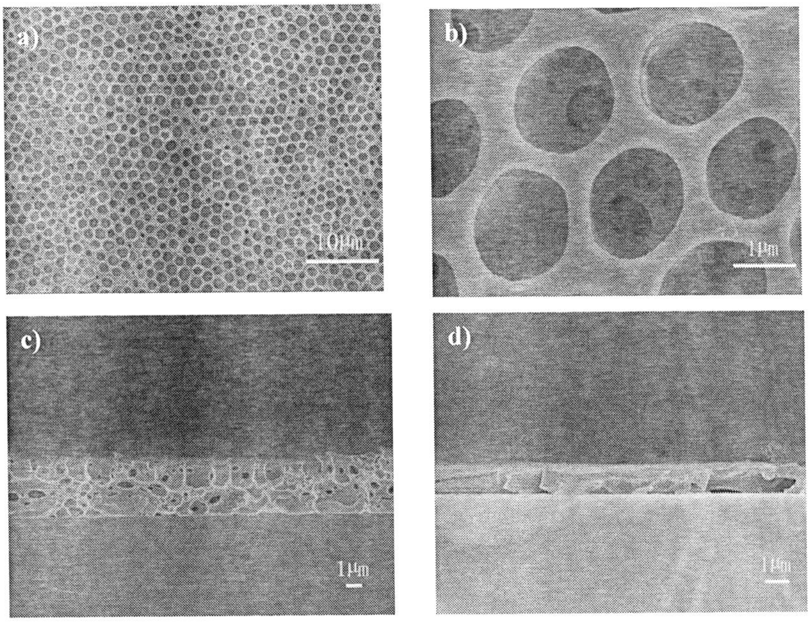

Figure 3. FE-SEM images of the as-prepared honeycomb film (a) low-magnification (b) high magnification. (c) cross-section and (d) cross-sectional FE-SEM image of the smooth honeycomb film.

Figure 3(a-c) shows the typical FE-SEM images of the as-prepared honeycomb structure films of PDTG under a humid environment (RH~70%) and (d) cross-sectional FE-SEM image of the smooth honeycomb film. It can be clearly seen that the porous films can be prepared over a large area, and uniform pores with hexagonal arrangement are formed with an average diameter of ca. 1.4 μm (Fig.3a). From the magnified image (Fig. 3b), the under layer can be observed through the pores of the top layer. The cross-sectional image (Fig. 3c) confirms that this porous film is made of double layers of empty spheres voids with the total thickness of 3.2 μm.

A steady and rapid cathodic photocurrent of 293 nA/cm2 was produced when the irradiation of the honeycomb structure film was switched on and off. The cathodic photocurrent response of its smooth film was 103 nA/cm2, which was about one third of the value obtained by the ordered porous film. It can be deduced that in the photo-electric conversion systems, the porous structures not only significantly improve light capture ability by enhance the light absorption, but also benefit the electron/hole separation and charge transfer by introducing PDTG/electrolyte interface. Therefore through the molecular design and the introducing of micro-/nano- structures on the photo-electric active surface, there will be a great prospect for improving the photo-electric conversion performance in the organic thin film system.

5. Conclusion

Photo-electric conversion is one of the most important issues in the field of energy conversion field. The era of nanotechnology, especially the bridging of the disparate fields, offers opportunities to challenge the conventional wisdom regarding the efficiency of alternative energy sources. According to the research achievements reviewed above, we are sure that by introducing the micro-/nano- structures on various photo-electric active surfaces, there will be a great prospect for improving the photo-electric conversion performance in the photovoltaic systems.

REFERENCES

[1] Green, M. A.; Emery, K.; King, D. L., Hishikawa, Y.; Warta, W. *Prog. Photovolt: Res. Appl. 2006*, 14, 455–461.

[2] Nishimura, S.; Abrams, N.; Lewis, B. A.; Halaoui, L. I.; Mallouk, T. E.; Benkstein, K. D.; Lagemaat, J.; Frank, A. J. *J. Am. Chem. Soc. 2003*, 125, 6306-6310.

[3] Halaoui, L. I.; Abrams, N. M.; Mallouk, T. E. *J. Phys. Chem. B 2005*, 109, 6334-6342.

[4] Hore, S.; Nitz, P.; Vetter, C.; Prahl, C.; Niggemannb, M.; Kern, R. *Chem. Commun. 2005*, 15, 2011-2013.

[5] Ferber, J.; Luther, J. Sol. Energy Mater. *Sol. Cells 1998*, 54, 265-275.

[6] Rothenberger, G.; Comte, P.; Grätzel, M. Sol. Energy Mater. *Sol. Cells 1999*, 58, 321-336.

[7] Wang, Z.-S.; Kawauchi, H.; Kashima, T.; Arakawa, H. *Coordinat. Chem. Rev. 2004*, 248, 1381-1389.

[8] Mobius, D. *Acc. Chem. Res. 1981*, 14, 63– 68.

[9] Souza, F. D.; Deviprasad, G. R.; Rahman, M. S.; Choi, J. P. *Inorg. Chem. 1999*, 38, 2157-2160.

[10] Tkachenko, N. V.; Rantala, L.; Tauber, A. Y.; Helaja, J. P.; Hynninen, H.; Lemmetyinen, H. *J. Am. Chem. Soc. 1999*, 121, 9378-9387.

[11] Ikeda, A.; Hatano, T.; Shinkai, S.; Akiyama, T.; Yamada, S. *J. Am. Chem. Soc. 2001*, 123, 4855-4856.

[12] Fushimi, T.; Oda, A.; Ohkita, H.; Ito, S. *Langmuir 2005*, 21, 1584-1589.

[13] Huynh, W. U.; Dittmer J. J.; Alivisatos, A. P. *Science 2002*, 295, 2425-2427.

[14] Law, M.; Greene, L. E.; Johnson, J. C.; Saykally, R.; Yang, P. D. *Nat. Mater. 2005*, 4, 455-459.

[15] Uchida, S.; Chiba, R.; Tomiha, M.; Masaki, N.; Shirai, M. *Electrochemistry 2002*, 70, 418-420.

[16] Adachi, M.; Murata, Y.; Okada, I.; Yoshikawa, S. *J. Electrochem.Soc. 2003*, 150, G488-G493.

[17] Mor, G. K.; Shankar, K.; Paulose, M.; Varghese, O. K.; Grimes, C. A. *Nano. Lett. 2006*, 6, 215-218.

[18] Guo, M.; Diao, P.; Wang, X.; D. Cai, S. M. *J. Solid State Chem. 2005*, 178, 3210 -3215.

[19] Hosono, E.; Fujihara, S.; Honma, I.; Zhou, H. S. *Adv. Mater. 2005*, 17, 2091 -2094.

[20] Baxter, J. B.; Aydil, E. S. *Appl. Phys. Lett. 2005*, 86, 053114

[21] Song, M. Y.; Kim, D. K.; Ihn, K. J.; Jo, S. M.; Kim, D. Y. *Nanotechnology 2004*, 15,1861-1865.

[22] Song, M. Y.; Ahn, Y. R.; Jo, S. M.; Kim, D. Y. *Appl. Phys. Lett. 2005*, 87, 113113.

[23] Usami, A. Sol. Energy Mater. *Sol. Cells 1999*, 59, 163-166.

[24] Usami, A. Sol. Energy Mater. *Sol. Cells 2000*, 62, 239-246.

[25] Nazeeruddin, M. K.; Kay, A; Rodicio, I; Humphry-Baker, R.; Müller, E.; Liska, P.; Vlachopoulos, N.; Grätzel, M. *J. Am. Chem. Soc. 1993*, 115, 6328-6390.

[26] Hore, S.; Vetter, C.; Kern, R.; Smit, H.; Hinsch, A. Sol. Energy Mater. *Sol. Cells 2006*, 90, 1176-1188.

[27] Grätzel, M. J. Photochem. *Photobiol. A-Chem. 2004*, 164, 3-14.

[28] Li, B.; Wang, L. D.; Kang, B. N.; Wang, P.; Qiu, Y. Sol. Energy Mater. *Sol. Cells 2006*, 90, 549-573.

[29] Madhugiri, S.; Dalton, A.; Gutierrez, J.; Ferraris, J. P.; Balkus, K. J. *J. Am. Chem. Soc. 2003*, 125, 14531-14538.

[30] Bognitzki, M.; Czado, W.; Frese, T.; Schaper, A.; Hellwig, M.; Steinhart, M.; Greiner, A.; Wendorff, J. H. *Adv. Mater. 2001*, 13, 70-72.

[31] Jiang, L.; Zhao, Y.; Zhai, J. *Angew. Chem. Int. Ed. 2004*, 43, 4338-4341.

[32] Ko, F.; Gogotsi, Y.; Ali, A.; Naguib, N.; Ye, H.; Yang, G.; Li, C.; Wills, P. *Adv. Mater. 2003*, 15, 1161-1165.

[33] Li, D.; Xia, Y. *Nano Lett. 2003*, 3, 555-560.

[34] Viswanathamurthi, P.; Bhattarai, N.; Kim, C. K.; Kim, H. Y.; Lee, D. R. *Inorg. Chem.* Commun. *2004*, 7, 679-682.

[35] Caruso, R. A.; Schattka, J. H.; Greiner, A. *Adv. Mater. 2001*, 13, 1577-1579.

[36] Demir, M. M.; Gulgun, M. A.; Menceloglu, Y. Z.; Erman, B.; Abramchuk, S. S.; Makhaeva, E. E.; Khokhlov, A. R.; Matveeva, V. G.; Sulman, M. G. *Macromolecules 2004*, 37, 1787-1792.

[37] Chen, C. H.; Emond, M. H. J.; Kelder, E. M.; Meester, B. Schoonman, J. *J. Aerosol Sci. 1999*, 30, 959-967.

[38] Benítez, R.; Soler, J.; Daza, L. J. *Power Sources 2005*, 151, 108-113.

[39] Kenawy, E. R.; Bowlin, G. L.; Mansfield, K.; Layman, J.; Simpson, D. G.; Sanders, E. H.; Wnek, G. E. *J. Control. Release. 2002*, 81, 57-64.

[40] Wnek, G. E.; Carr, M. E.; Simpson, D. G.; Bowlin, G. L. *Nano Lett. 2003*, 3, 213-216.

[41] Wang, X. Y.; Drew, C.; Lee, S. H.; Senecal, K. J.; Kumar, J.; Sarnuelson, L. A. *Nano Lett. 2002*, 2, 1273-1275.

[42] Wang, X. Y.; Drew, C.; Lee, S. H.; Senecal, K. J.; Kumar, J.; Sarnuelson, L. A. Nano Lett. *2004*, 4, 331-334.

[43] Fujimoto, M.; Kado, T.; Takashima, W.; Kaneto, K.; Hayase, S. *J. Electrochem. Soc. 2006*, 153, A826-A829.

[44] Zhao, Y.; Zhai, J.; Tan, S. X.; Wang, L.F.; Jiang, L.; Zhu, D. B. *Nanotechnology 2006*, 17, 2090-2097.

[45] Zhao, Y.; Sheng, X. L.; Zhai, J. Jiang, L. Yang, C. H.; Sun, Z. W.; Li, Y. F.; Zhu, D. B. *ChemPhysChem 2007*, 8, 856-861.

[46] Sheng, X. L.; Zhao, Y.; Zhai, J.; Jiang, L.; Zhu, D. B. *Appl. Phys.* A *2007*, 87(4), 715-719

[47] Widawski, G.; Rawiso, M.; Franois, B. *Nature 1994*, 369, 387 –389.

[48] Francois,B.; Pitois, O.; Francois, *J. Adv. Mater. 1995*, 7, 1041 –1044.

[49] Jenekhe, S.; Chen, X. *Science 1999*, 283, 372–375.

[50] Stenzel, M. H. *Aust. J. Chem. 2002*, 55, 239– 243.

[51] Karthaus, O.; Maruyama, N.; Cieren, X.; Shimomura, M.; Hasegawa, H.; Hashimoto, T. *Langmuir 2000*, 16, 6071 –6076.

[52] Heng, L.; Zhai, J.; Zhao, Y.; Xu, J.; Sheng, X.; Jiang, L. *ChemPhysChem 2006*, 7, 2520-2525.

In: Encyclopedia of Nanotechnology
Editor: Elwood D. Carlson

ISBN 978-1-60692-079-4

Chapter 48

Charge Separation and Transport in Organic/Inorganic Nanocomposites and Its Applications in Photovoltaic Devices

Chun-Wei Chen, Yun-Yue Lin, Yu-Ting Lin, Tsung-Wei Zeng, Chia-Hao Chang and Wei-Fang Su

Department of Materials Science and Engineering,
National Taiwan University, Taipei, 106, Taiwan

Abstract

All solution processed photovoltaic devices based on conjugated polymer/nanocrystal hybrid materials offer the potential toward fabricating low-cost, large area and mechanically flexible solar cells. In this article, the mechanisms of charge separation and transport in the solar cells based on polymer/nanocrystal (nanoparticles, nanorods) composites will be present. Carrier dynamics in the polymer/nanocrystal composites are investigated by temperature dependent and time-resolved photoluminescence (PL) spectroscopy and time of flight (TOF) technique. Rapid exciton dissociation at the polymer/nanocrystal interfaces leading to efficient charge separation, accompanying with a significant quench in the photoluminescence (PL) intensity and a decrease of PL lifetime with increasing nanocrystal concentrations. The separated carriers are transported through the device to the electrodes without recombining with oppositely charged carriers. The improvement in carrier transport can be further achieved by the inclusion of one-dimensional semiconductor nanorods which can offer direct pathways for electric conduction, supported by TOF mobility measurement. In addition, the efficiency of charge separation and transport in the polymer/nanocrystal composite can be further improved by surface treatment at polymer/nanocrystal interfaces.

1. Introduction

Composites of conjugated organic polymers and inorganic nanocrystals have attracted a great of interest due to their potential application in developing the low-cost, large-area, mechanically flexible photovoltaic devices [1,2]. Due to the low dielectric constant of organic materials, the dominant photogenerated species in most conjugated polymer is a neutral bound electron–hole pair (exciton). These neutral excitons can be dissociated from Coulomb attraction by offering an energetically favorable pathway for the electron from polymer (donor) to transfer onto an electron-accepting specie (acceptor). Subsequently, these dissociated carriers (electrons and holes) are transported to the electrodes before back recombination can occur and then generate electric current. The mechanism of charge separation and transport in polymer/nanocrystal composites is depicted schematically in Figure.1.

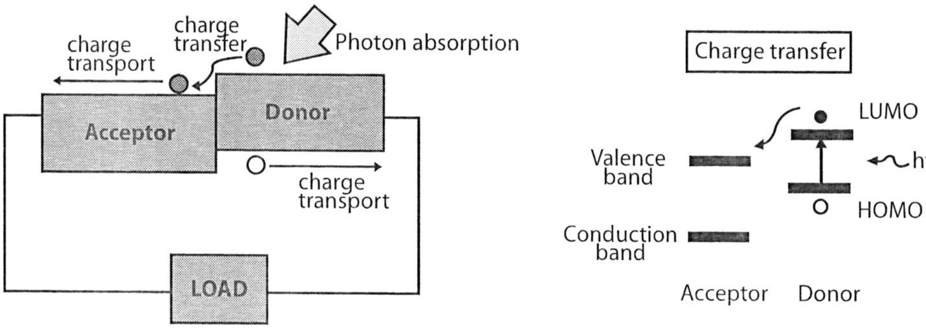

Figure 1. (a) Schematic representation of charge separation and transport in the donor-acceptor type photovoltaic device. (b) Charge separation at the interface between donor (polymer) and acceptor (nanocrystal).

It has been found that the inclusion of nanomaterials (such as C_{60} derivatives [3,4], CdSe[1,2]or PbS[5] nanocrystals) with a higher electron affinity within the conjugated polymer semiconductors facilitates the charge separation of photogenerated electron-hole pairs at an interface between the polymer and the electron-accepting specie [6,7]. The photocurrent in these composite materials is typically increased by several orders of magnitude with regard to polymer-only devices [1,2].

Although efficient charge separation in the conjugated polymer/nanocrystal composites can be achieved in the bulk heterojunction due to the large number of interfaces for charge transfer, the performance of photovoltaic applications is largely limited by the poor carrier transport in composites after charge separation. Electron transport is limited by inefficient hoping along poorly formed conduction paths. Thus, an enhanced charge transport route is desirable to achieve efficient electric conduction. One dimensional semiconductor nanorods are preferable for offering direct pathways for electric conduction [1, 9-11]. In this study, carrier dynamics with respect to charge separation and transport in conjugated polymer / nanocrystal composites are investigated by temperature dependent and time-resolved photoluminescence (PL) spectroscopy and time of flight (TOF) mobility measurement technique. Two kinds of nanocomposites consisting of conjugated polymer (MEH-PPV) /CdSe nanoparticles and MEH-PPV/TiO2 nanorods will be discussed in this chapter. Finally, photovoltaic devices based on these nanocomposites will be also presented.

2. Material Preparation and Experiment Setup

In this study, the MEH-PPV polymer was purchased from Aldrich Chemical company (U.S.A.) with molecular weight 58,000(Mn). The molecular structure of MEH-PPV is shown in Figure 2(a). The CdSe nanoparticles are grown according to the procedure described in the literature [12]. In brief, CdO 0.06 g, trioctylphosphine oxide (TOPO) 5 g, and tetradecylphosphonic acid (TDPA) 0.26 g are loaded in a three neck flask. At about 300 °C, reddish CdO powder is dissolved into colorless homogeneous solution. By cooling down to 270 °C, then with an injection of 0.9 ml Se solution (1.987 g Se dissolved in 50 ml tributylphosphine (TBP)), high-quality CdSe nanoparticles are obtained. The size of CdSe nanoparticles is evaluated to be 5.0 nm as shown in Figure 2(b). For the synthesis of TiO2 nanorods, high aspect ratio anatase titanium dioxide nanorods are made by the hydrolysis of titanium tetraisopropoxide according to the literature [13]. Typically, oleic acid (OLEA) (120 g, Aldrich, 90 %) is stirred vigorously at 120 °C for 1 hour in a three-neck flask under Ar flow, then allowed to cool to 90 °C and maintained at this temperature. Titanium isopropoxide (17 mmole, Aldrich, 99.999 %) is then added into the flask. After stirring for 5 min, trimethylamine-N-oxide dihydrate (34 mmole, ACROS, 98 %) in 17 ml water is rapidly injected. The trimethylamine-N-oxide dihydrate is used as a catalyst for polycondensation. The reaction is continued for several hours to have complete hydrolysis and crystallization. Finally, TiO_2 nanorods are obtained with a size about 4 nm in diameter and 20-40 nm in length are shown in Figure 2(c).

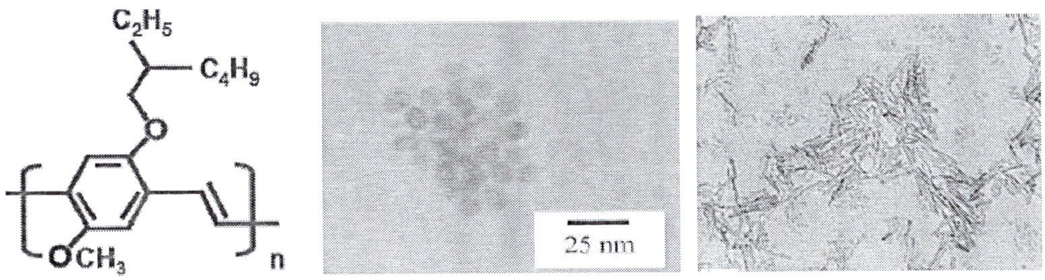

Figure 2. (a) Molecular structure of MEH-PPV (b) TEM image of CdSe nanoparticles with a size of 5.0 nm. (c) TEM image of TiO_2 nanorods with a size of 4-5 nm in diameter and 20-40 nm in length.

The polymer is dissolved in toluene with a concentration of 1 wt%. The hybrid materials are prepared by adding appropriate amount of nanoparticles into 1 wt% polymer solution to make various ratios of nanocrystal/polymer composite samples. The steady state PL spectra are taken by pumping the samples with a continuous wave (CW) He-Cd laser (325 nm) and the emission spectra are analyzed with a Jobin-Yvon TRIAX 0.55 m monochromator and detected by a photomultiplier tube and standard photon-counting electronics. Time-resolved photoluminescence spectroscopy is performed with a time-correlated single photon counting (TCSPC) spectrometer (Picoquant, Inc.). A pulse laser (375 nm) with an average power of 1 mW operating at 40 MHz with a duration of 70 ps is used for excitation. During the temperature-dependent PL measurements, the samples are maintained under vacuum inside a helium cryostat allowing temperature variation between 10K and ambient temperature. Photoluminescence excitation (PLE) spectroscopy is measured using the FluoroLog®-3 spectrofluorometer (Jobin-Yvon). The current-voltage characterizations (Keithley 2410 source meter) are performed under 10^{-3} torr vacuum and under illumination at defined beam size (Oriel Inc.).

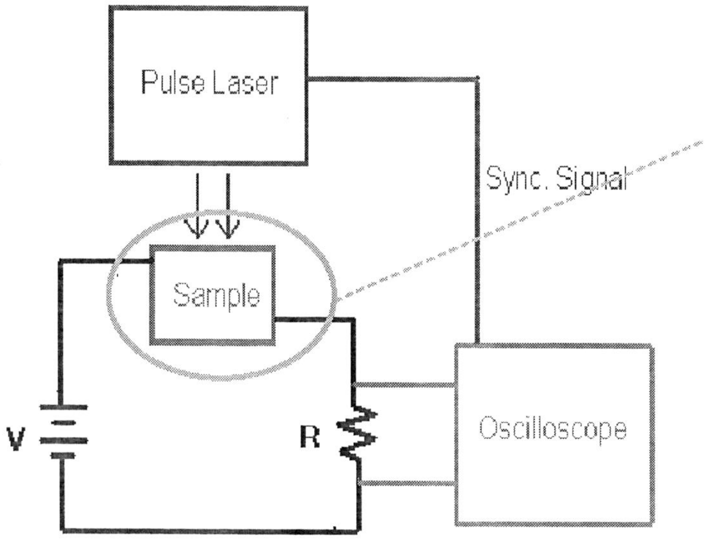

Figure 3. Experimental setup of time-of-flight mobility measurement.

The mobility of different types of carriers (electron or hole) is measured by the time of flight (TOF) technique as described schematically in Figure.3. Thick films for the TOF measurement are prepared by drop-casting the MEHPPV/CdSe nanoparticle blend solutions on the glass substrates precoated with semitransparent aluminum electrodes (35 nm) by thermal evaporation at a vacuum about 10^{-6} torr. The film thickness is measured by means of the Veeco M6 surface profiler to be ranged from 1.5 μm to 3.0 μm. The measured samples are then completed by thermal evaporation of thicker aluminum cathodes (~200nm) through a shadow mask. For the TOF transient photocurrent measurement, a thin layer of charge carriers are generated under illumination through the semi-transparent electrode using a frequency-tripled Nd:YAG pulsed laser ($\lambda = 355nm$). Under the applied electric field, these carriers drift toward the counter electrode, giving the transient photocurrent signal recorded by a digital oscilloscope (Tetronix TDS5052B). The carrier types (electron or hole) under study are selected by changing the polarity and magnitude of the applied electric field using a Kiethley 2410 source meter. The samples are mounted in a vacuum chamber under 10^{-3} torr during measurement. The mobility μ is then calculated according to

$$\mu = \frac{d}{Et_{tr}} \tag{1}$$

where d is the thickness of the film , E is the applied electric field and t_{tr} is the transit time for the arrival of carriers at the collecting electrode.

3. Charge Separation and Transport in MEH-PPV/CdSe Nanoparticle Composite

3.1. Exciton Dissociation and Migration in MEH-PPV/CdSe Nanoparticle Composite

The UV-VIS spectrum of the CdSe nanoparticles shows the sharp absorption peak that indicates the nanoparticles close to monodisperse. The excitonic absorption peak (~2.1 eV) as shown in Figure 4 exhibits a blue shift with respect to that of the bulk CdSe [14] due to the quantum confinement effect. For comparison, the representative absorption and PL spectra of the pristine MEH-PPV in this study are also shown. All spectra have been normalized for clarity of presentation.

Figure 5(a) shows the room temperature absorption and PL spectra of the various blends of MEH-PPV with 0, 20, 40, 60, 80 wt% of CdSe nanoparticles respectively. To avoid the thickness and geometry effects, the optical density of absorption and PL intensity were measured in solution in cuvette cells. As the concentration of the CdSe nanoparticles increases from 20 wt% to 80 wt%, the optical density of the absorption spectra in the composites increases with respect to the pristine MEH-PPV while the yield of the PL emission decreases substantially, suggesting the occurrence of PL quenching in the composites. The PL spectra of the composites are found to be similar to that of the pristine MEH-PPV, indicating the remaining luminescence predominantly results from excitons which radiatively recombine in polymer due to the much lower PL efficiency of the CdSe nanoparticles compared to that of MEH-PPV [2].

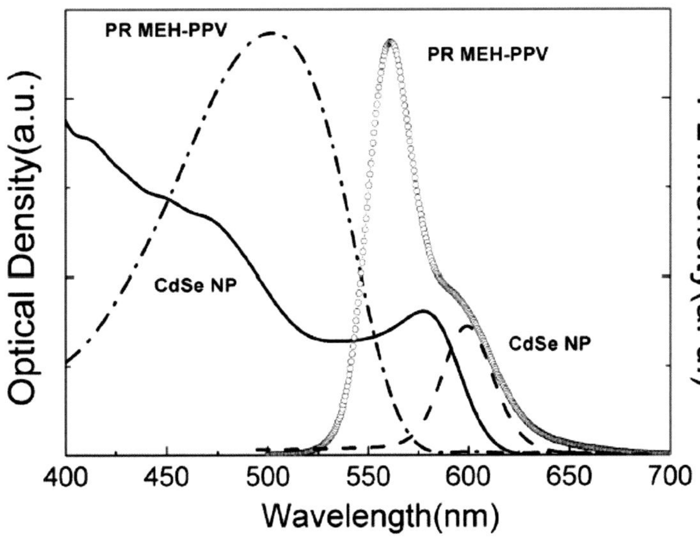

Figure 4. Absorption spectra of CdSe nanoparticles (solid line), pristine MEH-PPV (dash-dot line). The photoluminescence spectra of CdSe nanoparticles (dash line) and pristine MEH-PPV (circle) are also shown. Reprinted with permission from Lin, Y.-Y. et al., "Exciton dissociation and migration in enhanced order conjugated polymer/nanoparticle hybrid materials", *Nanotechnology*, 17, 1260, 2006, Copyright (2006) Institute of Physics.

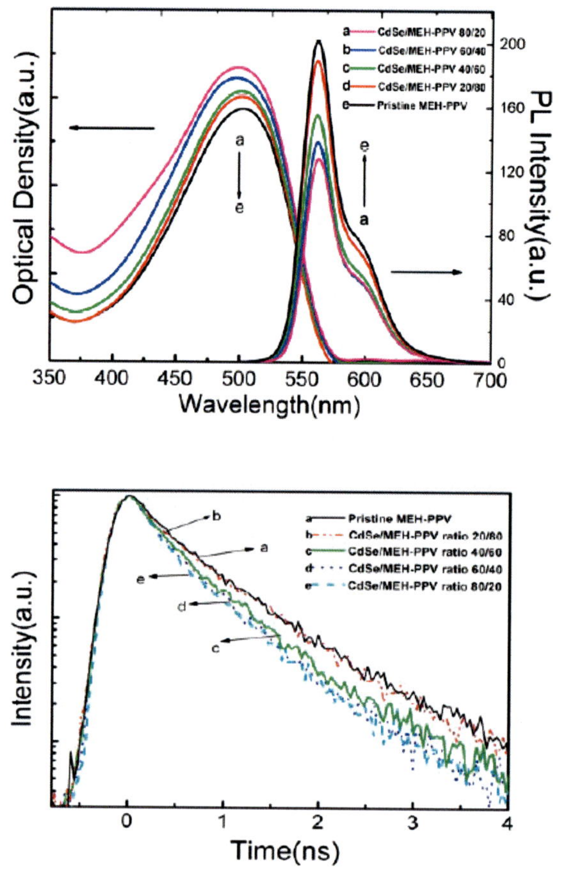

Figure 5. (a) Absorption and photoluminescence spectra of blends of MEH-PPV with CdSe nanoparticles for increasing concentrations of CdSe nanoparticles:0 wt%, 20 wt%, 40 wt%, 60 wt% , and 80 wt%. (b) Photoluminescence lifetime decay of the above samples at room temperature. Reprinted with permission from Lin, Y.-Y. et al., "Exciton dissociation and migration in enhanced order conjugated polymer/nanoparticle hybrid materials", *Nanotechnology*, 17, 1260, 2006, Copyright (2006) Institute of Physics.

The decrease in PL yield is attributed to quenching of the MEH-PPV PL emission by the CdSe nanoparticles acting as an electron accepting specie, where significant charge separation takes place. Figure 5 (b) presents the luminescence decays measured at 565 nm. The PL decay in the composite films demonstrates a monotonic decrease in lifetime by increasing nanoparticle content. The measured lifetime of the luminescence τ_{PL} can be related to the rate constants for radiative and nonradiative decays k_R and k_{NR} by

$$1/\tau_{PL} = k_R + k_{NR}, \tag{2}$$

The efficiency η is then given by

$$\eta = f\frac{k_R}{k_R + k_{NR}} = f\,{}^{\tau_{PL}}\!/_{\tau_R}, \tag{3}$$

where $\tau_R = 1/k_R$ is the natural radiative lifetime i.e. the lifetime that the luminescence would have in the absence of competing non-radiative decay process. f is the fraction of absorbed photons generating the emissive species. Analysis of the PL lifetime decay τ_{PL} reveals a bi-exponential emission, summarized in Table I.

As the nanoparticle concentration increases, a new relaxation process is present, which provides to the donor a further non-radiative process as the result of the separation of photogenerated electron-hole pairs. The enhancement of the non-radiative decay rate k_{NR} with increasing nanoparticle concentrations leads to shortening of the measured lifetime τ_{PL} and quenching of the PL efficiency η. By taking the PL efficiency of the pristine MEH-PPV as 1 and the natural radiative lifetime as τ_0, the relative natural radiative lifetime τ_R for the composite films, which can be estimated from Eq.(2) and (3), are about 1.04, 1.11, 1.18 and 1.21 τ_0 for the samples containing 20,40 60 and 80 wt % nanoparticles respectively.

Table I. Summary of the measured lifetime τ_{PL} and the relative nature radiative lifetime for the pristine MEH-PPV and the composites containing 20, 40, 60, 80 wt% CdSe nanoparticles respectively

CdSe/MEH-PPV Ratio	τ_{PL} (ps)	τ_R
Pristine MEH-PPV	450	τ_0
20:80 (wt %)	420	1.04 τ_0
40:60	350	1.11 τ_0
60:40	340	1.18 τ_0
80:20	330	1.21 τ_0

The increase in natural radiative lifetime τ_R for the composites with respect to the pristine MEH-PPV implies that the excitons, which contribute to the remaining luminescence by radiative recombination in polymer, will migrate along the polymer chains over a larger region before recombination.

Microscopic mechanisms for exciton migration in the polymer/nanoparticle composites can be further investigated by temperature dependent steady-state and time-resolved photoluminescence spectroscopy. Figure 6(a) exhibits the PL spectra of the composite films measured at 10 K. The composite films were made by spin coating from the solutions containing various ratios of nanoparticle to polymer. The film thicknesses are 240 nm, 256nm, 238nm, and 250nm for the samples containing 20 wt% , 40wt% , 60 wt% and 80 wt%

CdSe nanoparticles respectively. The PL spectra in Figure 6(a) consistently move toward a lower energy with respect to those in Figure 5(a) measured at room temperature.

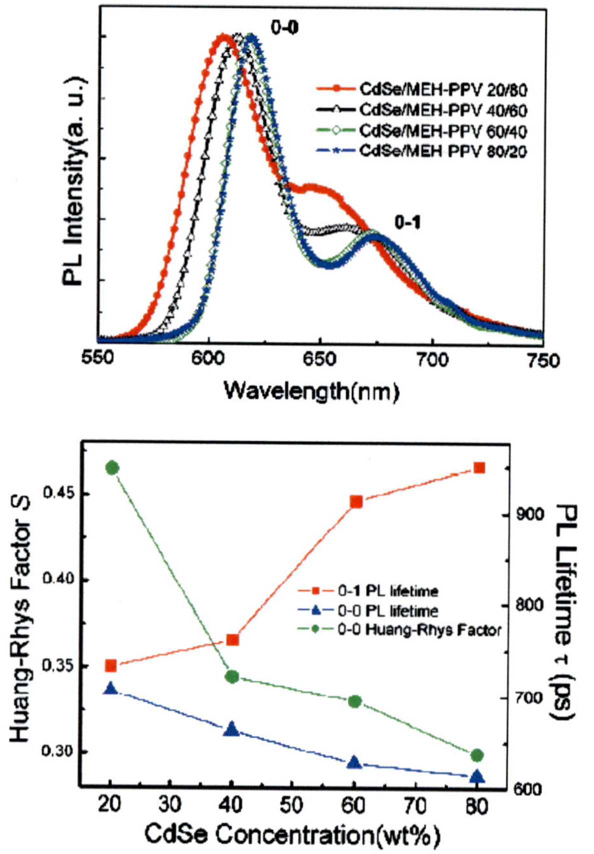

Figure 6. Normalized Photoluminescence spectra of CdSe nanoparticle/MEH-PPV hybrid materials measured at 10K. 0-0 and 0-1 peaks are ascribed to the zero-phonon line and the first vibronic band. (b) The Huang-Rhys factor *S,* 0-0 and 0-1 photoluminescence decay time of the hybrid materials as a function of CdSe concentrations. Reprinted with permission from Lin, Y.-Y. et al., "Exciton dissociation and migration in enhanced order conjugated polymer/nanoparticle hybrid materials", *Nanotechnology*, 17, 1260, 2006, Copyright (2006) Institute of Physics.

The observed redshift in PL spectra results from an increase in the effective conjugated length while the temperature decreases due to the reduction in thermal disorder in polymer. This enhances the delocalization of the π-electrons and lowers the band gap energy. The PL spectra were fitted with Gaussians in order to obtain the individual peak positions and linewidth broadening parameters (full width of half maximum, FWHM). As the energy separation of the PL spectrum peaks corresponds to a phonon mode, the observed zero-phonon line (ZPL) and the first and second vibronic bands are label as 0-0, 0-1 and 0-2 respectively. A progressively shift toward a lower energy as well as the reduction in the FWHM in PL spectra can be observed by increasing CdSe nanoparticle content in composite films. The dimensionless Huang-Rhys factor [10] *S*, described as $I_{0\to n} = \frac{e^{-S} S^n}{n!}$, where n is the number of phonons involved, corresponding to the average number of phonons involved during relaxation, is directly related to the effective conjugated length of the polymer system. Figure 6(b) shows that S monotonically decreases with increasing CdSe content as the result of enhanced conjugation in polymer by reducing disorder in conformation. The excited states are therefore delocalized over the more conjugated polymer chain, giving lower-energy emission. Smaller molecules usually exhibit a larger S factor due to their large normal coordinate displacements [15]. It is believed that the nanoparticles may either be in direct contact with the polymer, or alternatively, forms a thin barrier between the nanoparticle core and the polymer with a surfactant molecule, trioctylphosphineoxide (TOPO) coated on the surface of each nanoparticle. For the composite films containing host conjugated polymer MEH-PPV in blend with the high-molecular-weight CdSe nanoparticles, equivalently larger molecules are formed. Longer conjugated sequences can be formed in the more rigid structures owing to the inclusion of CdSe nanoparticles in the host polymer. Similar observation in the poly(3-octylthiophene) consisting of thiophene rings having large alkyl groups has also been reported [16]. Enhanced electronic delocalization in the composite films can be further evident from the time-resolved PL analysis. Figure 6(b) shows the PL lifetime decays of the 0-0 and 0-1 emission peaks measured at 10K for the composite film with 20,40,60 and 80 wt% CdSe nanoparticles respectively. The PL lifetime decays for the dominant 0-0 emission decrease while those for the 0-1 emission increase with increasing CdSe nanoparticle content. The decreasing PL lifetime in the 0-0 emission, as mentioned above, is attributed to efficient charge separation from the increase of non-radiative pathway created by the inclusion of nanoparticles. The *surviving* excitons which contribute to the remaining luminescence in polymer will migrate over more conjugated region before recombination as the result of the increasing PL decay lifetime for the 0-1 emission peaks with increasing CdSe nanoparticle concentrations. This result reveals is the characters of enhanced conjugation in the polymer/nanoparticle composite systems. Exciton dissociation and migration due to enhanced electronic delocalization in polymer after adding CdSe nanoparticles is depicted schematically in Figure 7.

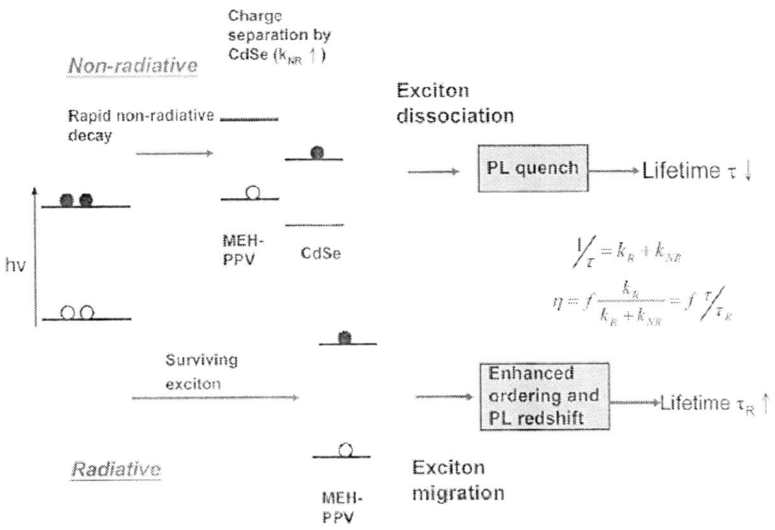

Figure 7. Exciton dissociation and migration due to enhanced electronic delocalization in polymer after adding CdSe nanoparticles.

3.2. Carrier Transport in MEH-PPV/CdSe Nanoparticle Composites by Time of Flight Mobility Measurement

Enhance electronic delocalization in MEH-PPV/CdSe nanoparticles composites is further investigated by the TOF mobility measurement. Figure 8(a) shows the typical transient hole photocurrent of the pristine MEH-PPV recorded at $E = 5.5\times10^{5} V/m$. The hole transient photocurrent for the pristine MEH-PPV is highly dispersive (inset) and the transit time t_{tr} can be estimated from the intersection point in the double-logarithmic plot. The hole mobility for the pristine MEH-PPV is deduced according to the Eq.(1) $\mu_h = 4\times10^{-6} cm^2/V.s$, which is very similar to the previous report in the TOF measurement on MEH-PPV[17]. Figure 8(b) show the evolution of transient hole photocurrent of the composite film with 80wt % CdSe nanoparticles under different applied electric fields. It is found that the corresponding hole photocurrent for the composite film consisting becomes less dispersive compared to that of the pristine MEH-PPV and the transient time t_{tr} becomes shortened with increasing electric field.

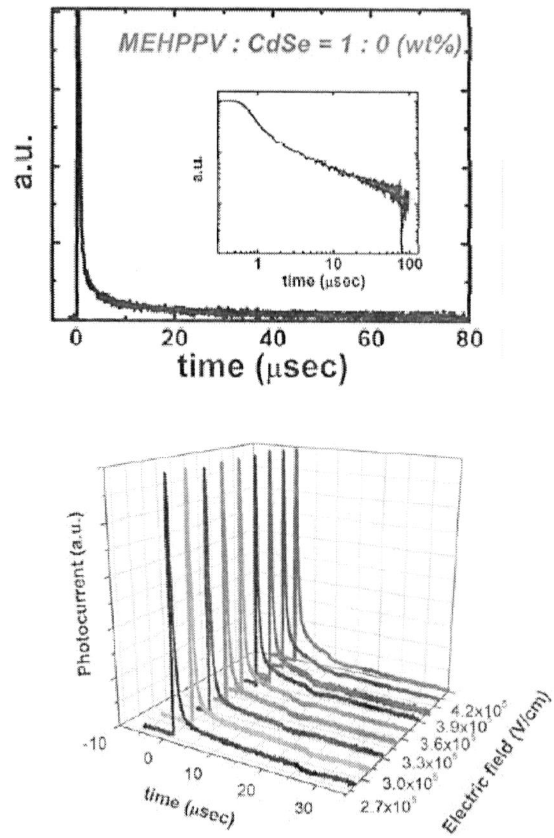

Figure.8. Typical TOF hole transient for (a) the pristine MEH-PPV film and (b) the evolution of transient hole photocurrent of the composite film with 80wt % CdSe nanoparticles under different applied electric fields. The insets show the double-logarithmic plots for the transient photocurrent.

In general, the photocurrent profile with a less dispersive (or non-dispersive) characteristic indicates that the carriers transport through a more ordered structure [18].The field dependence of the hole mobility for the composite films with different compositions is shown in Figure 9.

Figure 9(a), and 9(b) show the field dependence of the electron and hole mobilities for the composite with various CdSe nanoparticle concentrations. The curves can be fit according to the Poole Frenkel law [18] by plotting the mobility data as a function of the square root of electric field ($\sqrt{E}$). A linear relation can be found according to

$$\ln \mu = s\sqrt{E} \tag{4}$$

where s is a constant of proportionality. The square-root field dependence is often observed in the molecular materials [18,19] and other organic-inorganic hybrid systems[20,21]. A very weak dependence on the electric field on the mobilities of electrons and holes can be seen for all the compositions. As CdSe nanoparticles are added into MEH-PPV, both electron and hole mobilities increase compared to those in the pristine MEH-PPV.

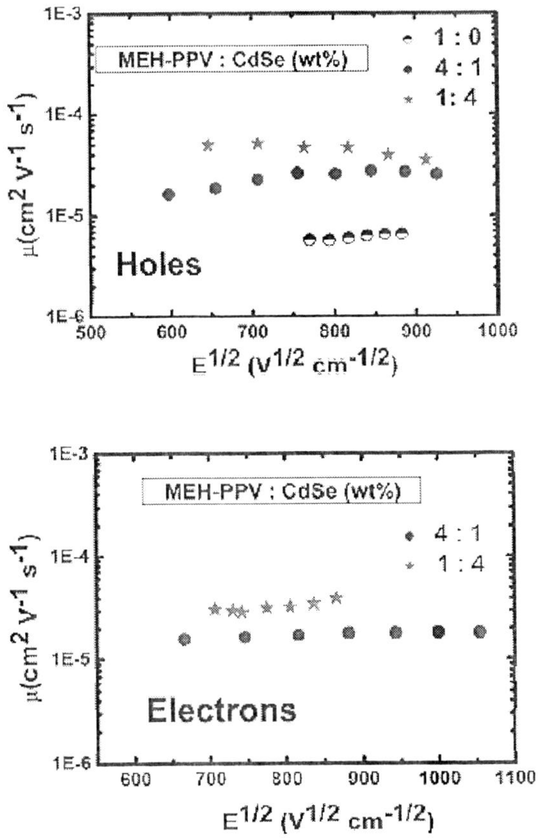

Figure 9. Field dependence of the (a) electron (b) hole mobilities for samples with different compositions. The transient electron photocurrent for the pristine MEH-PPV is highly dispersive and difficult to extract for applied electric field $E < 1.2\times10^{6} V/m$.

For the pristine MEH-PPV, the transport of electron is also highly dispersive and the electron mobility is very difficult to extract (for . $E < 1.2\times10^{6} V/m$). In composite films, due to the higher electron affinity of CdSe, electrons are transported through CdSe nanoparticles and holes are transported through the polymer matrix. When an adequate amount of CdSe nanoparticles are added for charge percolation, the conduction paths for electron transport are formed, leading to an increase in electron mobility in the composite films ($\mu_e = 3\times10^{-5} cm^2/V.s$, at $E = 5\times10^{5} V/m$ for the composite film with 80 wt% CdSe nanoparticles). However, electron transport in the polymer/nanoparticle blend is usually limited by poorly formed conduction path through hopping between nanoparticles. Thus, nanorods [1] and branched tetrapod nanocrystals [22] act to provide a direct transport pathway across the blend film, which can further improve the electron transport and device performance. In addition, the coating of insulating surfactant molecules of trioctylphosphine oxide (TOPO) on the nanoparticle surface from aggregation can further hinder carrier transport by forming a barrier between the nanocrystals and polymer. The dependence of carrier transport properties by removing the surfactant or replacing more conductive organic ligands on the nanocrystal surface will be discussed later.

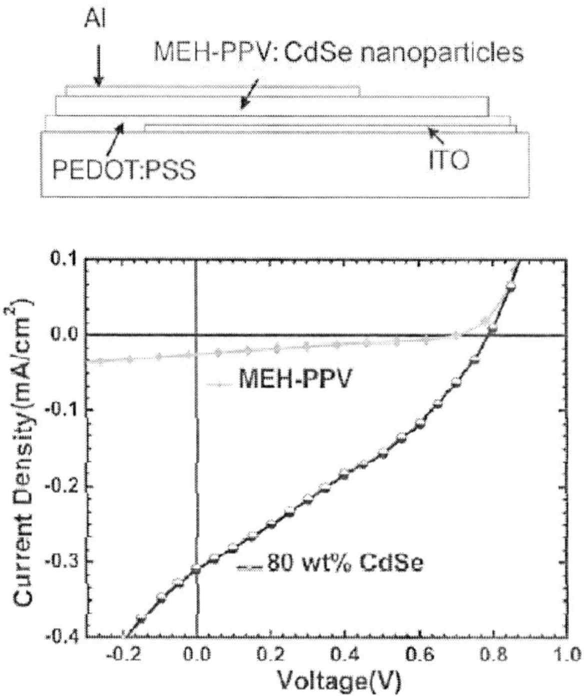

Figure 10. (a) The schematic structure of standard configuration MEH-PPV:CdSe nanoparticles hybrid photovoltaic device. (b) Current-voltage characteristics of the devices based on the pristine MEH-PPV and MEH-PPV:CdSe nanoparticle hybrid(80 wt% CdSe nanoparticles) materials.

The hole mobility for the blend films also shows a very weak field dependence and has increased by more than one order of magnitude compared to that in the pristine MEH-PPV, following with a non-dispersive transient photocurrent as mentioned above. The results can be attributed to the following reasons. Firstly, the disordered π-electron system might be qualitatively thought of as a distribution of finite oligmers with effective length equal to the mean conjugation length. In conjugated polymers such as PPV and its derivatives, the orientation of the phenylene rings can easily fluctuate [23]. The inclusion of CdSe nanoparticles in the host polymer enhances the rigidity in response to thermal induced disorder, as the result of the increasing effective conjugation length and a more-ordered structure for hole transport. In addition, the presence of CdSe nanoparticles in the host polymer matrix can also improve the interaction between neighboring polymer chains, which can also assist the hole transport. The result is consistent with the redshift in PL emission peaks after adding CdSe nanoparticles in polymer as shown above, indicating enhanced electronic delocalization.

3.3. Performance of Photovoltaic Devices

The photovoltaic devices based on MEH-PPV/CdSe nanoparticles with different compositions are fabricated in air with an active layer sandwiched between electrodes of indium-tin-oxide (ITO) and aluminum as shown schematically in Figure 10(a). Figure 10(b) shows the current-voltage characteristics of the devices under AM1.5 illumination. It is found that the short circuit current J_{sc} in the composite film (80 wt% CdSe) is enhanced about 20 times as a result of increasing charge separation and transport in composite films with respect to the pristine MEH-PPV. For the device consisting of 80 wt% CdSe nanoparticles, the open-circuit voltage is 0.78 V and the fill factor is approximately 0.4, giving a power conversion efficiency of 0.1 %. Although enhanced electronic delocalization in polymer can be achieved by adding CdSe nanoparticles as mentioned above, the performance of photovoltaic devices based on MEHP-PPV/CdSe nanoparticles is largely limited by two reasons. Firstly, the poorly formed conduction path through hopping between nanoparticles limits efficient electron transport. Secondly, the coating of insulating surfactant molecules TOPO on the nanoparticle surface from aggregation can further hinder charge separation and carrier transport by forming a barrier between the nanoparticles and polymer. In the next section, one-dimensional nanorods accompanying with the removal of insulating surfactant by pyridine treatment will be empolyed to enhance the efficiency of charge separation and transport in polymer/nanocrystal composites.

4. Efficient Charge Transfer and Transport in MEH-PPV/TiO2 Nanorods Composites

4.1. Efficient Charge Transfer in MEH-PPV/TiO2 Nanorod Nanocomposites

As mentioned above, electron transport in the polymer/nanoparticle hybrid is usually limited by poorly formed conduction path. Thus, one-dimensional semiconductor nanorods are preferable over nanoparticles for offering direct pathways for electric conduction. It has been demonstrated that the solar cell based on the CdSe nanorods/poly(3-hexylthiophene)(P3HT) hybrid material exhibits a better power conversion efficiency than its CdSe nanoparticle counterpart [1]. The environmental friendly and low-cost TiO_2 nanocrystal is another promising material in hybrid polymer/nanocrystal solar cell applications. In this section, the microscopic mechanisms of photo-induced charge transfer and transport in the MEH-PPV/TiO2-nanorod hybrid materials will be present. Finally, the current-voltage characteristics of photovoltaic devices based on the MEH-PPV/ TiO_2 nanorod hybrid materials with different surface treatments will be also reported.

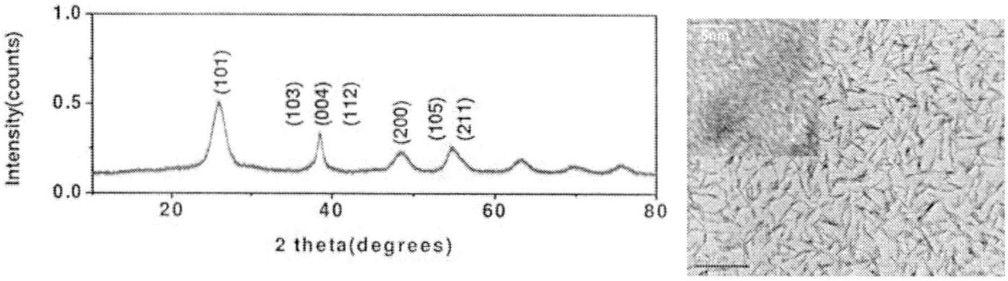

Figure 11. (a) Powder X-ray diffraction pattern of anatase TiO_2 nanorod. (b) Anatase TiO_2 nanorod structure images observed by TEM and HRTEM (inset). Reprinted with permission from Zeng, T.-Y. et al., "Large Interconnecting Network in Hybrid MEH-PPV/ TiO_2 Nanorods Photovoltaic Devices", *Nanotechnology*, Vol. 17, 5387, 2006. Copyright (2006) Institute of Physics.

Figure 11(a) shows the XRD pattern of TiO_2 nanorods, indicating the formation of the anatase phase. A sharp peak intensity and narrow width of (004) imply the formation of nanorods. The TEM image of TiO_2 nanorods reveals that the dimension of TiO_2 nanorods is 20-40 nm in length and 4-5 nm in diameter. The high-resolution TEM (HRTEM) image of TiO_2 nanorods is shown in the inset of Figure 11(b), indicating that the TiO_2 nanorods were growth in high crystallinity.

Figure 12 shows the room temperature UV-Visible absorption and PL spectra of the various blends of MEH-PPV with 0, 40, and 80 wt% of TiO_2 nanorods respectively.

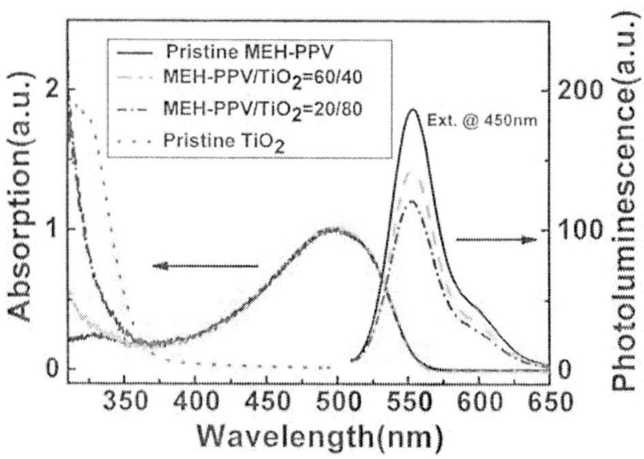

Figure 12. Absorption and photoluminescence spectra of blends of MEH-PPV with TiO2 nanorods for increasing concentrations of TiO2 nanorods: a: 0 wt%, b:40 wt%, c: 80 wt%. d: Absorption spectrum of TiO2 nanorods only. Reprinted with permission from Lin, Y-T et al, "Efficient photoinduced charge transfer in TiO_2 nanorods/conjugated polymer hybrid materials, *Nanotechnology*, Vol.17, 5781, 2006. Copyright (2006) Institute of Physics.

The pristine MEH-PPV exhibits a broad absorption spectrum peaked at about 490 nm and TiO_2 nanorods have a sharp absorption edge at about 350 nm. As the concentration of the TiO_2 nanorods increases, the optical density of the absorption spectra in the composites is simply the sum of the absorption spectra of constituent parts. No evidence of any additional absorption peaks in the measured spectral range is found. With excitation at 450 nm, only the polymer is excited and the PL spectra of the composites show similar emission features with respect to that of the pristine MEH-PPV, indicating the luminescence predominantly results from excitons which radiatively recombine in polymer. The yield of the PL emission in the composites decreases substantially with increasing TiO_2-nanorod content, indicating charge separation occurs at the interface between the two materials. This produces a charge-separated state with an electron on the TiO_2 nanorods and a hole on the polymer, which is responsible for the polymer emission quenching. Figure 13(a) presents the PL decay spectroscopy for the hybrid materials measured at 550 nm. The PL decay in the composite materials demonstrates a monotonic decrease in lifetime with increasing TiO_2-nanorod content.

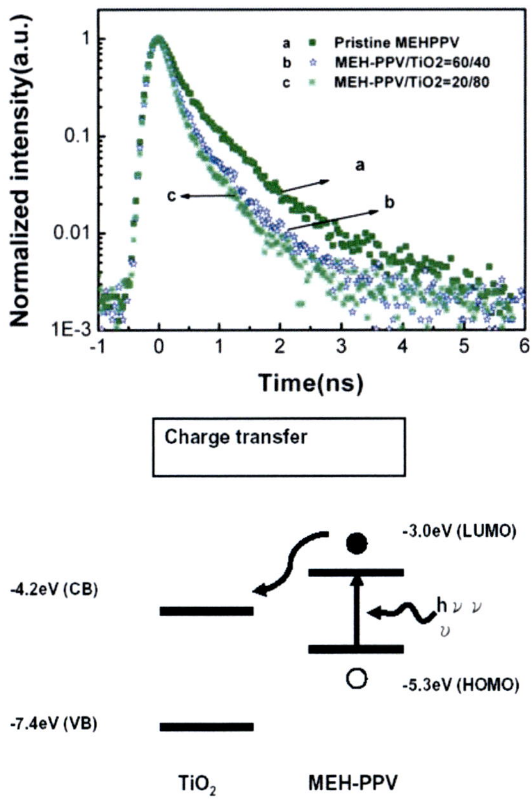

Figure 13. (a) Photoluminescence lifetime decay of theses samples at room temperature. The lifetime τ_{PL} and for the pristine MEH-PPV and the composites containing 40, and 80 wt% TiO_2 nanorods arc 738 ps, 520 ps and 497 ps respectively by a bi-exponential fitting. (b) Schematic representation of charge transfer at the MEH-PPV/TiO2-nanorod interfaces. Reprinted with permission from Lin, Y-T et al, “Efficient photoinduced charge transfer in TiO_2 nanorods/conjugated polymer hybrid materials, *Nanotechnology*, Vol.17, 5781, 2006. Copyright (2006) Institute of Physics.

Figure 13(b) exhibits the mechanism of charge transfer at the MEH-PPV/TiO2-nanorod interfaces schematically. As the TiO_2-nanorod concentration increases, a new relaxation process is present, which provides to the donor a further non-radiative process, leading to the enhancement of the non-radiative decay rate, leading to shortening of the measured lifetime τ_{PL} and quenching of the PL efficiency η.

Efficient charge transfer at MEH-PPV/TiO2 nanorod interfaces is further investigated by the PLE spectroscopy. The two-dimensional PL contour plots of the composite thin films with 0, 40, and 80 wt% of TiO_2-nanorod concentrations as a function of emission and excitation wavelengths are shown in Figure 14(a), (b) and (c) respectively. The excitation wavelengths were chosen to keep the condition that only polymer absorbs the light intensity. These samples were prepared by spin-coating the mixture of MEH-PPV and TiO_2 nanorods onto Si substrates. For the pristine MEH-PPV, the PL maxima (spots) on the map in Figure 14(a) are clearly seen at about 570-590 nm, consistent with the PL emission peak in Figure 12. As the concentrations of TiO_2 nanorods are increased up to 40 and 80 wt% as shown in Figure 14(b) and 14(c) respectively, the intensities of PLE signals quench at all emission wavelengths, indicating efficient charge separation occurs at the TiO_2 nanorods/MEH-PPV interfaces. Figure 15 shows the quenching efficiency Q(λ) as a function of excitation wavelengths from 400-550 nm for the composites with respect to the pristine MEH-PPV. The quenching rate is enhanced with increasing TiO_2 nanorod content in the composite films. For excitation with a longer wavelength which is closer to the PL emission peak, the quench rate is also increased. In addition, it is worth noting that the fluorescence quenching for the composite thin films is more efficient with respect to the mixed solutions in our case. Similar observation is also reported previously [24], where the possible formation of nanocrystal aggregation may take place in solutions or thicker films.

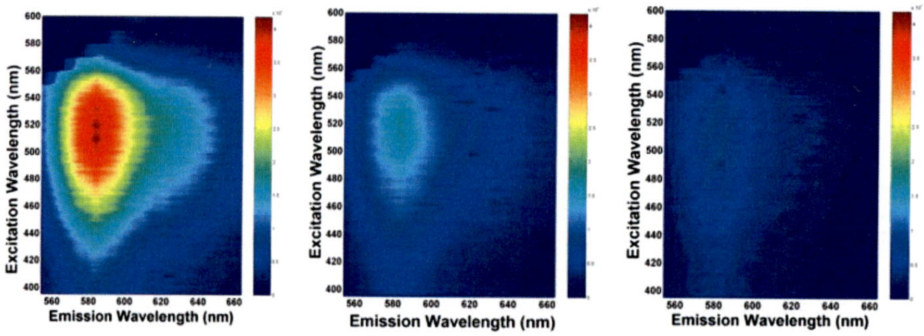

Figure 14. Two-dimensional contour plots of the photoluminescence mapping as a function of emission and excitation wavelengths for the composite films with (a) 0 wt%, (b) 40 wt%, and (c) 80 wt% of TiO_2-nanorod concentrations. To avoid the second harmonic effect, the detection of emission wavelength is limited by less than two times of the excitation wavelength. Reprinted with permission from Lin, Y-T et al, "Efficient photoinduced charge transfer in TiO_2 nanorods/conjugated polymer hybrid materials, *Nanotechnology*, Vol.17, 5781, 2006. Copyright (2006) Institute of Physics.

We have further characterized the optical properties of the composite films at low temperature. Figure 16 shows the absorption spectra measured at 20K of the pristine MEH-PPV and the composite film with a 80 wt% TiO_2-nanorod concentration respectively. As shown in Figure 12, a similar absorption peak at about 500 nm is observed for these two samples at room temperature.

The effective conjugation length in polymer chains increases with decreasing temperature, owing to the reduction in thermal disorder. Accordingly, enhancement in the delocalization of the π-electrons brings the absorption peaks to lower energies as temperature is lowered [25]. For the composite film, a larger redshift in the absorption peak at low temperature with respect to the pristine MEH-PPV indicates that the inclusion of TiO_2 nanorods in polymer can further reduce disorder in conformation and enhance conjugation in polymer chains at low temperature. A redshift in the normalized PL spectrum for the composite film with respect to that of the pristine MEH-PPV is also shown in inset, which is similar to the observation in MEHP-PPV/CdSe-nanoparticle composite films.

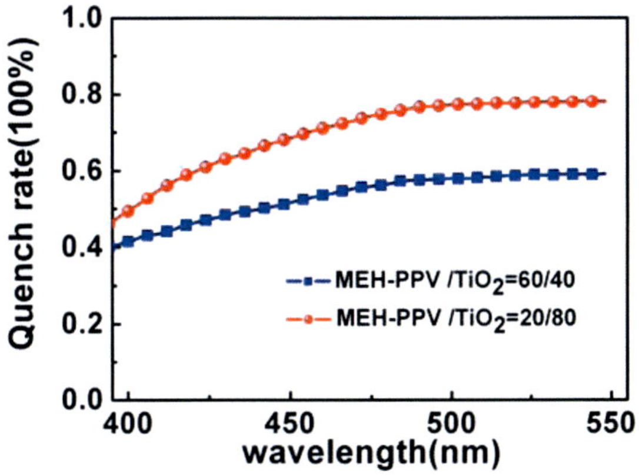

Figure 15. The quenching efficiency Q(λ) as a function of excitation wavelengths for the composites with respect to the pristine MEH-PPV. Reprinted with permission from Lin, Y-T et al, "Efficient photoinduced charge transfer in TiO_2 nanorods/conjugated polymer hybrid materials, *Nanotechnology*, Vol.17, 5781, 2006. Copyright (2006) Institute of Physics.

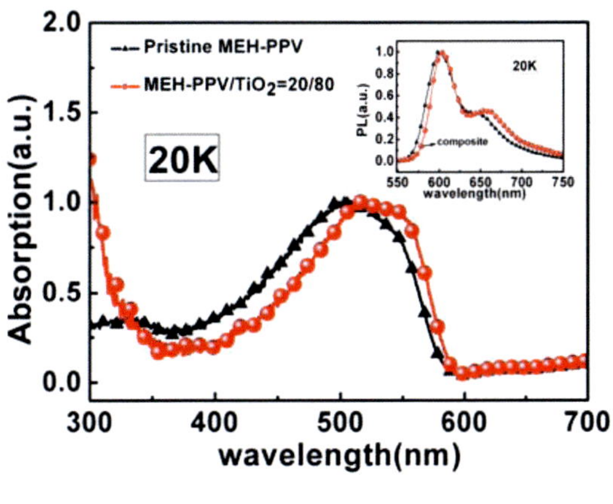

Figure 16. Normalized absorption spectra measured at 20K of the pristine MEH-PPV and the composite film with a 80 wt% of TiO_2 nanorods. The inset exhibits the normalized PL spectra for the composite and pristine MEH-PPV measured at 20K. Reprinted with permission from Lin, Y-T et al, "Efficient photoinduced charge transfer in TiO_2 nanorods/conjugated polymer hybrid materials, *Nanotechnology*, Vol.17, 5781, 2006. Copyright (2006) Institute of Physics.

4.2. Dependence of Surface Treatment on Carrier Transport

As the temperature is lowered to 20K, there is a significant redshift in the absorption spectrum for the composite film with respect to the pristine MEH-PPV.

The surface ligand is also another important factor in determining the morphology and physical properties of polymer/nanocrystal composite. The TiO2 nanorods can be either coated with a surfactant molecule, oleic acid (OLEA) which forms a thin barrier or in direct contact with polymer by removing the surfactant using pyridine treatment. The schematic diagrams of OA coated and non-coated TiO2 nanorods blending with MEH-PPV are shown in Figure 17. The addition of surfactant molecule OLEA prevents nanorods from aggregation, which increases the number of interfaces between nanorods and polymer. However, the thin barrier of the surfactant molecules can play a role to hinder charge transfer and transport. After removing surfactant molecules by pyridine treatment, the charge transfer efficiency between nanorods and polymer is enhanced but the possibility of aggregation of neighboring nanorods can lead to the reduction of the number of interfaces for charge transfer. In addition, the aggregation of neighboring nanorods is also likely to improve electron transport between nanorods, which usually occurs through thermally activated hopping [26]. The corresponding transport properties of the hybrid materials with different surface treatment are investigated by the time-of-flight mobility measurement are summarized in Table II. It is found that as the TiO2 nanorods are coated with OLEA molecules, both electron and hole show dispersive transport behavior. In contrast, the removal of surfactant molecules leads to a significant increase both in mobilities of electron and holes by an order of magnitude, accompanying with non-dispersive transport, indicating efficient charge transport is achieved after removing surfactant.

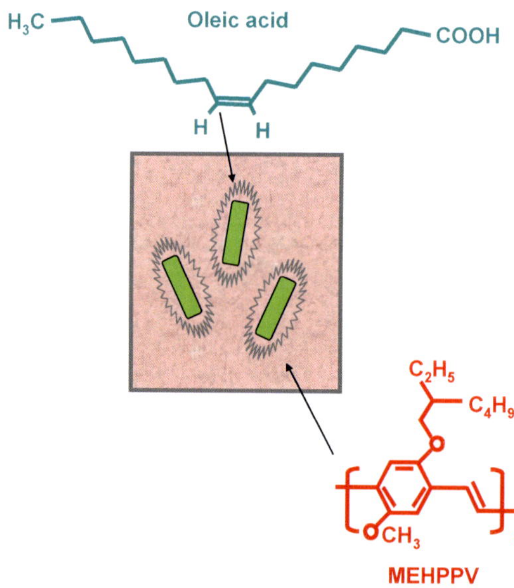

Figure 17. Schematic representation of MEH-PPV/TiO2 nanorod hybrids (a) with and (b) without surfactant molecules (OLEA).

Table II. Electron and hole transport properties of the hybrid materials with different surface treatment investigated by the time-of-flight mobility measurement

	Surface treatment	Mobility ($cm^2/V.s$)	transport
Electron			
Pristine MEH-PPV	None	$4x10^{-6}$	dispersive
MEH-PPV/TiO2 (1:1)	OLEA	$1.81x10^{-5}$	dispersive
MEH-PPV/TiO2 (1:1)	Pyridine	$3.29x10^{-4}$	Non-dispersive
Hole			
Pristine MEH-PPV	None	$8x10^{-6}$	dispersive
MEH-PPV/TiO2 (1:1)	OLEA	$2.51x10^{-5}$	dispersive
MEH-PPV/TiO2 (1:1)	Pyridine	$1.51x10^{-4}$	Non-dispersive

4.3. Performance of the Photovoltaic Devices

As mentioned above, efficient charge separation and transport can take place at MEH-PPV:TiO_2 nanorods hybrid materials due to the formation of the large interconnecting network by TiO2 nanorods within the polymer matrix. Here, we have fabricated MEH-PPV:TiO_2 nanorods heterojunction photovoltaic device. The TiO2 nanorods used for the device fabrication are capped with OLEA surfactant molecules. All the devices studied here are fabricated in air. As a starting point, a standard hybrid device structure similar to the previous device configuration in Figure 10(a) is fabricated, which consists of a transparent indium-tin-oxide (ITO) conducting electrode, poly (3,4-ethylenedioxythiophene)-poly(styrenesulfonate) (PEDOT:PSS), the MEH-PPV:TiO_2 nanorod hybrid film, and an aluminum (Al) electrode. We have further modified the device configuration by including an additional electron conducting layer of TiO_2 nanorods sandwiched between the active layer and the aluminum electrode to improve device performance as shown in Figure 18.

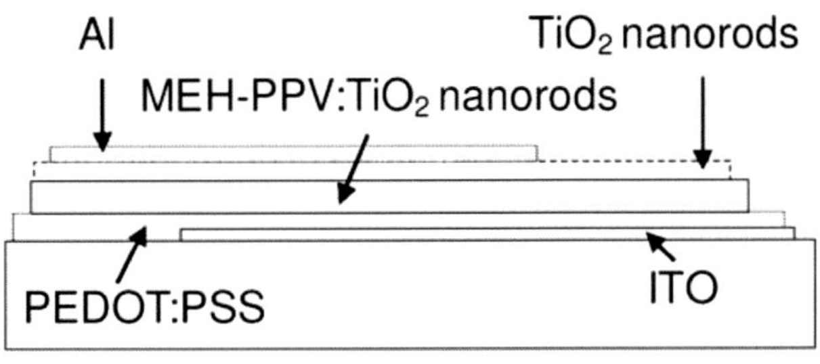

Figure 18. Schematic structure of MEH-PPV:TiO_2 nanorods hybrid photovoltaic device included a TiO_2 nanorods layer.

Figure 19(a) shows a smooth topography of MEH-PPV: TiO_2 nanorods hybrid with roughness 2nm. The TiO_2 nanorods were randomly distributed in the polymer matrix for the interconnecting work formation. Figure 19(b) shows the phase image of MEH-PPV: TiO_2 nanorods hybrid. Tapping-mode AFM can also give the information about the materials at the film surface via phase images. Because a hard material generally shows a positive phase shift with respect to a soft material due to the cantilever oscillation is related to the power dissipated in a nonelastic tip-sample interaction [27], therefore the bright areas in Figure 19(b) are interpreted as the harder material of TiO_2 nanorods and the darker areas as the soft material of polymer. Homogenous distribution of TiO_2 nanorods in polymer was observed in Figure 19(b). Figure 19(c) shows the surface topography of spin-cast TiO_2 nanorods layer on MEH-PPV: TiO_2 nanorods hybrid. Feature of aggregation of nanorod structure was found. The TiO_2 nanorods thin film exhibits a porous structure of relatively high film roughness. Figure 19(d), the phase image of TiO_2 nanorods thin film, shows a single phase of the bright areas consist of TiO_2 nanorods. The dark region in Figure 19(d) could be seen as deep pores of the TiO_2 nanorods thin film which is consistent as the surface topography observed in Figure 19(b).

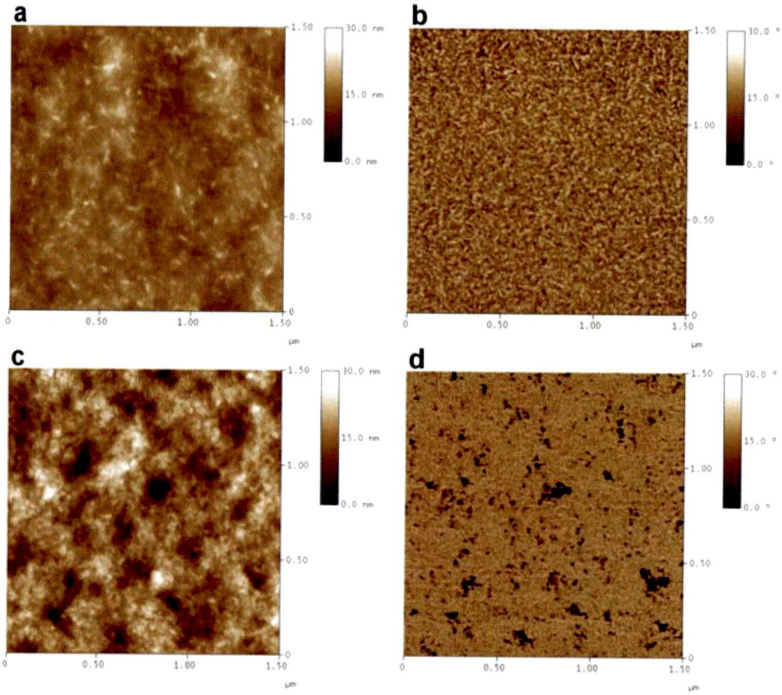

Figure 19. AFM images showing the surface morphology of MEH-PPV: TiO2 nanorods hybrid film and TiO2 nanorods film. (a) Height image of spin-cast film of MEH-PPV: TiO2 (52 wt%) nanorods hybrid. The image size is 1.5µm x 1.5 µm, and the vertical scale is 30 nm. (b) Phase image of MEH-PPV: TiO2 (52 wt%) nanorods hybrid film. The image size is 1.5µm x 1.5 µm, and the vertical scale is 30o. (c) Height image of spin-cast film of TiO2 nanorods on MEH-PPV: TiO2 nanorods hybrid film. The image size is 1.5µm x 1.5 µm, and the vertical scale is 30 nm. (d) Phase image of of TiO2 nanorods film on MEH-PPV: TiO2 nanorods hybrid film. The image size is 1.5µm x 1.5 µm, and the vertical scale is 30o. Reprinted with permission from Zeng, T.-Y. et al., "Large Interconnecting Network in Hybrid MEH-PPV/ TiO_2 Nanorods Photovoltaic Devices", *Nanotechnology*, Vol. 17, 5387, 2006. Copyright (2006) Institute of Physics.

From the results above, we have constructed an interconnecting network in MEH-PPV: TiO_2 nanorods photoactive hybrid and a thin film composed of mere TiO_2 nanorods sandwiched between hybrid layer and Al electrode through our process conditions.

Based upon considering the energy levels of respective materials in the device, a TiO_2 nanorod layer inserted between the active layer and the aluminum electrode is satisfied and appropriate for offering a better connectivity of electron transport path to the electrode. Functions of the TiO_2 nanorod layer can be explained by the band diagram in Figure 20. The energy level diagram demonstrates that the TiO_2 nanorod layer acts as hole-blocking electron-transporting layer in this device. As the electron-hole pairs are generated by incident light, an efficient charge separation occurs at the interface of the MEH-PPV:TiO_2 nanorod hybrid. Electrons move toward the aluminum electrode and holes move toward the ITO electrode. The addition of the continuous TiO_2 nanorod thin film allows for the current to be conducted effectively and also prevents electrons from back recombination with holes in the MEH-PPV. The TiO_2 nanorod layer acts as a hole blocking layer because of lower valence band value. In contrast, if inserting a thin MEH-PPV layer instead, it is energetically unfavorable for electron transport.

Figure 21 shows the current-voltage response of the devices with and without TiO_2 nanorod layer. The device containing TiO_2 nanorod layer increases short-circuit current density by a factor of 2.5 with respect to a device without the layer. For a comparison, the thin TiO_2 nanorod layer was replaced with a thin MEH-PPV layer and a ~3 order magnitude of decrease in the short-circuit current was found.

Apart form the hole-blocking electron-transporting function of TiO_2 nanorod layer mentioned above, the interfaces introduced, (MEH-PPV:TiO_2 /TiO_2 and TiO_2 /Al), seems are more beneficial to charge transport as compared to the MEH-PPV:TiO_2 /Al contact. The TiO_2 nanorod layer can be connected to the TiO_2 nanorods in the active hybrid.

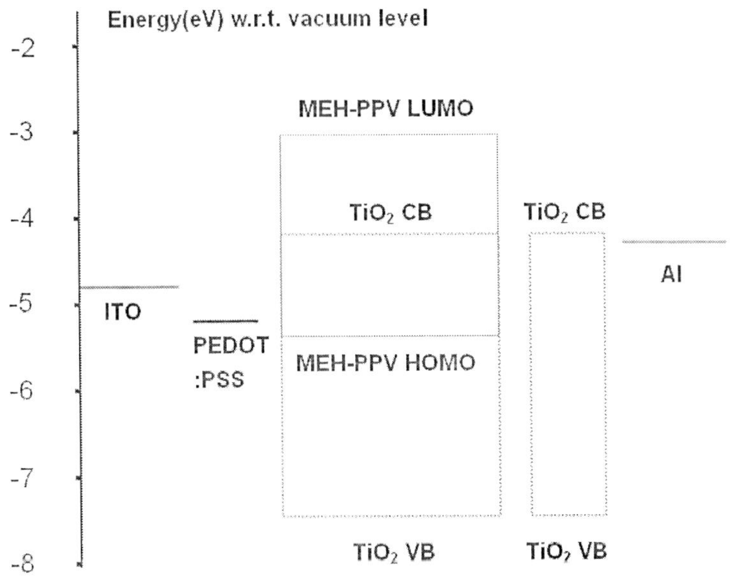

Figure 20. (a) Flat band energy-level diagram of ITO/PEDOT: PSS/ MEH-PPV: TiO2 nanorods / TiO2 nanorods /Al devices.

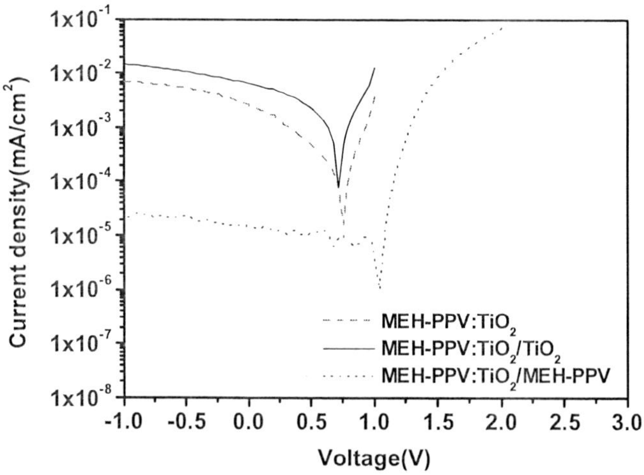

Figure 21. Plots of current density as the function of applied voltage for three different configuration devices under 0.09 mW/cm^2 illumination at 560 nm. (MEH-PPV:TiO_2 nanorods (52 wt%) (dashed line); MEH-PPV:TiO_2 nanorods (52 wt%)/ TiO_2 nanorods (solid line) and MEH-PPV: TiO_2 nanorods(52 wt%)/ MEH-PPV (dotted line)). Reprinted with permission from Zeng, T.-Y. et al., "Large Interconnecting Network in Hybrid MEH-PPV/ TiO_2 Nanorods Photovoltaic Devices", *Nanotechnology*, Vol. 17, 5387, 2006. Copyright (2006) Institute of Physics.

In addition, the rough surface of TiO_2 layer can lead to stronger contact and increased contact area to Al electrode. Besides, inserting this layer can create a second interfacial area for exciton dissociation that might increase the charge transfer rate. To introduce an additional titanium oxide thin film as hole-blocking electron-transporting layer through various approaches have been presented in producing higher efficiencies heterojunction organic solar cells [28-31]. Here we present a thin film of crystalline TiO_2 nanorods that made via fully solution process can lead to improvement in device performance. The TiO_2 nanorod thin film could be explored as a promising hole-blocking electron-transporting layer in the photovoltaic devices.

An equivalent circuit was frequently used to describe the electric behavior of a photovoltaic device. We further analyzed characteristics of the devices based upon this equivalent circuit, the current density versus voltage characteristics can be described by the following equation:

$$I = I_0 \times \left[\exp\left(e \frac{U - IR_S}{nkT} \right) - 1 \right] + \frac{U - IR_S}{R_{SH}} - I_{PH} \tag{5}$$

where I_0 is the saturation current, e is the magnitude of the electronic charge, U is the applied voltage, n is the ideality factor, k is Boltzmann's constant, T is the absolute temperature, R_S is the series resistance, R_{SH} is the shunt resistance and I_{PH} is the photocurrent. The current-voltage characteristics are largely dependent on the series and shunt resistance. A lower series resistance means that higher current will flow through the device. High shunt resistance corresponds to less shorts or leaks in the device. The ideal cell would have a series resistance approaching zero and shunt resistance approaching infinity. The series resistance can be estimated from the inverse slope at a positive voltage where the I-V curves become linear. The shunt resistance can be derived by taking the inverse slope of the I-V curves around 0V.

$$R_S = \lim_{V \to \infty}\left(\frac{dV}{dI}\right) \tag{6}$$

$$R_{SH} \approx \frac{dV}{dI}(V = 0) \quad R_S << R_{SH} \tag{7}$$

The R_S and R_{SH} were analyzed from the I-V curves of devices (Figure 22(b)); it is found that a significant, nearly 60%, R_S reduced as the TiO_2 nanorod layer introduced into the device. Slightly reducing of the shunt resistance was observed also. The series resistance can be expressed as the sum of the bulk and interfacial resistance. It is likely that two interfaces that have been introduced (MEH-PPV:TiO_2/ TiO_2 and TiO_2/Al) combined with the TiO_2 nanorod layer offer a much lower magnitude of series resistance as compared to the MEH-PPV:TiO_2 /Al contact. The introducing of TiO_2 layer decreases series resistance in the device and thereby increases the current.

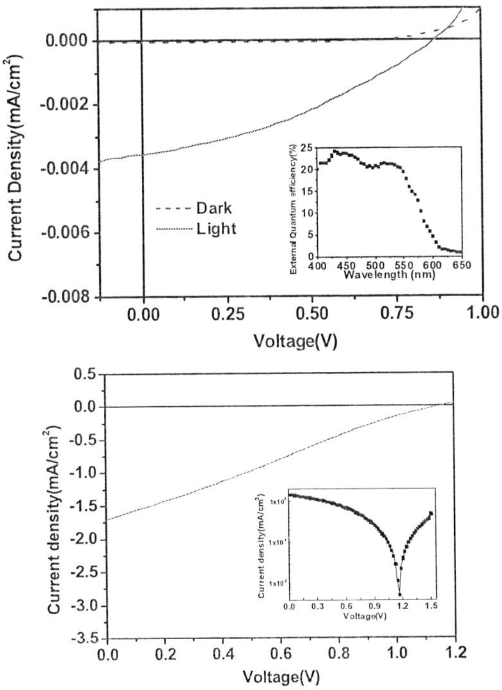

Figure 22. (a) Plot of current density versus voltage in the dark (dashed line); and under 0.05 mW/cm^2 illumination at 565 nm (solid line, V_{oc}=0.86 V, J_{sc}=-0.0035 mA/cm^2, FF =0.35 and η=2.2 %). The insert shows the external quantum efficiency versus wavelength of the device. The device structure is ITO/PEDOT: PSS/ MEH-PPV:TiO_2 nanorods(52 wt%)/ TiO_2 nanorods /Al. (b) The corresponding I-V curve of the ITO/PEDOT: PSS/ MEH-PPV:TiO_2 nanorods(52 wt%)/ TiO_2 nanorods /Al device at A.M. 1.5 illumination(100mW/cm^2). (V_{oc}=1.15 V, J_{sc}=-1.7 mA/cm^2, FF =0.25 and η=0.49 %). Logarithmic I-V characteristic of the device is shown in the inset. Reprinted with permission from Zeng, T.-Y. et al., "Large Interconnecting Network in Hybrid MEH-PPV/ TiO_2 Nanorods Photovoltaic Devices", *Nanotechnology*, Vol. 17, 5387, 2006. Copyright (2006) Institute of Physics.

The device performance with a structure of ITO/PEDOT: PSS/ MEH-PPV:TiO_2 nanorods(52 wt%)/ TiO_2 nanorods /Al is evaluated. The I-V characteristic of the device exhibits a short-circuit current density (J_{sc}) of -0.0035 mA/cm^2, an open circuit voltage (V_{oc}) of 0.86 V and a fill factor (FF) of 0.35. A power conversion efficiency (η) of 2.2 % is achieved under 0.05 mW/cm^2 illumination at 565 nm (Figure 22(a)). The insert shows the external quantum efficiency (EQE) of the device under illumination. A maximum EQE of 24 % under 0.07 mW/cm^2 at 430 nm is achieved. Figure 22(b) presents the characteristics of the device tested under A.M. 1.5 illuminations with an intensity of 100mW/cm^2. The J_{sc}, FF, and V_{oc} are -1.7 mA/cm^2, 0.25, and 1.15 V respectively for the device, yielding a power conversion efficiency of 0.5 %. Although efficient charge separation and transport have been found in MEH-PPV/TiO2 nanorod composites, the devices shown here are still far from optimization. It is worth mentioning that the devices studied here are fabricated in air, which may cause the degrading of device performance dramatically since polymer is highly sensitive to oxygen and moisture. The work to optimize the device efficiency is still undergoing to achieve better device performance such as heat and surface treatment, alignment of TiO2 nanorods, and improvement of carrier mobility etc.

5. Conclusions

In this chapter we have demonstrate the mechanism of charge separation and transport in organic polymer/inorganic nanocrystal (CdSe nanoparticles and TiO2 nanorods) nanocomposites. It is found enhanced electronic delocalization in MEHPPV/CdSe nanoparticle hybrid composite materials has been observed. The inclusion of CdSe nanoparticles in the host polymer enhances the rigidity in response to thermal induced torsion, leading to increase in the effective conjugation length and enhancement in electronic delocalization in polymer. By using one-dimensional TiO_2 nanorods as efficient electron acceptors and transport components, it is found that efficient charge separation and transport have been found in MEH-PPV/TiO2 nanorod composites due to the formation of large interconnecting network within polymer matrix, which significantly improves electric transport. The results suggest that one-dimensional TiO_2 nanorods are a promising material for hybrid polymer/nanocrystal bulk heterojunction solar cell applications. Further improvements in the device performance could be accomplished by controlling the nanorod sizes and by improving the polymer:TiO_2 nanorod interface.

References

[1] W. U. Huynh, J. J. Dittmer, and A. P. Alivisatos, *Science* 295 (2002) 2425.
[2] N.C.Greenham, X. Peng, and A. P. Alivisatos, *Phys. Rev.* B, 54 (1996) 17628.
[3] G.Yu, A.J.Heeger, *J. Appl. Phys*. 78, 4510, (2001).
[4] A.Haugeneder, M. Neges, C. Kallinger, W. Spirkl, U. Lemmer, J. Felmann, *Phys. Rev.* B, 59, 15346, (1999).

[5] A. Watt, T. Eichmann, H. Rubinsztein-Dunlop, and P. Meredith, *Appl. Phys. Lett.* 87, A.Haugeneder, M. Neges, C. Kallinger, W. Spirkl, U. Lemmer, J. Felmann, *Phys. Rev.* B, 59 (1999) 15346.

[6] J. H. Warner, A.R. Watt, E. Thomsen, N. Heckenberg, P. Meredith, H. R. Dunlop, *J. Phys. Chem.*, 109 (2005) 9001.

[7] Y.Y. Lin, C.W. Chen, J. Chang, T.Y. Lin, I.S. Liu and W.F. Su, *Nanotechnology*, 17, 1260, (2006).

[8] Huynh W U, Dittmer J J, Libby W C, Whiting G L and Alivisatos A P 2003 *Adv. Func. Mater.* 13 73.

[9] Sun B, Marx E and Greenham N C 2003 *Nano Lett.* 3 961.

[10] Jiu J, Isoda S, Wang F and Adachi M 2006 *J. Phys. Chem. B* 110 2087.

[11] C-M Chuang, W-B Lu, W-F Su, C-M Lin and Y-F Chen, 2005, *J. Appl. Phys.*, 97, 96104.

[12] Cozzoli P D, Kornowski A, and Weller H 2003. *J. Am. Chem. Soc.* 125 14539.

[13] E. Lifshitz, I. Dag, I. Litvin, G. Holdes, S. Gorer, R. Reisfeld, M. Zelner, H. Minti, 1998, *Chem. Phys. Lett.* ,288, 188.

[14] . W. Hagler, K. Pakbaz, K. F. Voss, and A. J. Heeger, *Phys. Rev.* B 44 (1991) 8652.

[15] A. Watanabe, T. Kodaira, and O. Ito, *Chem. Phys. Lett.* 273 (1997) 227.

[16] A. R. Inigo, C. H. Tan, Wunshain Fann, Y-S Huang, G-Y Perng and S-Y Chen, *Adv. Mater.,* 13, 504, (2001).

[17] H. Bässler, *Phys. Status Solidi* (b) 175, 15 (1993).

[18] P.W.M.Blom, *Phys. Rev.* B 55, R656 (1997).

[19] J. Huang, G. Li and Y.Yang, *Appl. Phys. Lett.*, 87, 112105, (2005).

[20] S. M. Tuladar, Dmitry Poplavskyy, S. A. Choulis, J.R.Durrant, Donal D.C.Bradley, *Adv. Fun. Mater.* , 15, 1171, (2005).

[21] L. Manna, D.J. Milliron, A. Meisel, E.C. Scher, A.P. Alivisatos, *Nature Mater.* 2, 382 (2003).

[22] Z. G. Yu, D. L. Smith, A. Saxena, R. L. Martin, and A. R. Bishop, *Phys. Rev. Lett.* 84, 721, (2000).

[23] A. Petrella, M.Tamborra, M.L.Curri, P.Cosma, M.Striccoli, P.D.Cozzoli, and A.Agostiano, 2005 *J. Phys. Chem., B*, 109,1554.

[24] T.W.Hagler, K.Pakbaz, K.F.Voss and A.J.Heeger, 1991 *Phys. Rev. B*, 44, 8652.

[25] D.S.Ginger and N.C. Greenham, *J. Appl. Phys.*, 2000, 87, 1361.

[26] Beek W J E, Wienk M M, Kemerink M, Yang X and Janssen R A J 2005 *J. Phys. Chem. B* 109 9505.

[27] Wang H, Oey C C, Djurišic A B, Man K K Y, Chan W K, Xie M H, Leung Y H, Chui P C, Pandey A and Nunzi J-M 2005 *Appl. Phys. Lett.* 87 023507.

[28] Oey C C, Djurišic A B, Wang H, Man K K Y, Chan W K, Xie M H, Leung Y H, Pandey A , Nunzi J-M and Chui P C *Nanotechnology* 2006 17 706.

[29] Thelakkat M, Schmitz C and Schmidt H-W 2002 *Adv. Mater.* 14 577.

[30] Kim J Y, Kim S H, Lee H-H, Lee K, Ma W, Gong X and Heeger A J 2006 *Adv. Mater.* 18 572.

In: Encyclopedia of Nanotechnology
Editor: Elwood D. Carlson

ISBN 978-1-60692-079-4

Chapter 49

POLYMERIC NANOCOMPOSITES, STABILIZED ORGANIC DERIVATIVES OF FIVE-VALENT PHOSPHORUS

A. Kh. Shaov*, Kh. Kh. Gurdaliev and A. M. Kharaev
Kabardino-Balkarian State University, Nalchik, KBR, Russia

ABSTRACT

On the basis of aromatic polyesters and the block-copolyesters, and also polyethylene of high density compositions c were prepared the contents of organic compounds of five-valent phosphorus with cyclohexyl a radical at atom of phosphorus. The sizes of molecules phosphor organic compounds make 8-10 Å3, that allows them to relate to so-called nanoparticles. It is established, that investigated phosphorus organic compounds appeared effective stabilizers thermal and mechanical properties of the investigated polymers, and also good antipyrens.

Keywords: aromatic polyesters; polyolefin's; phosphor organic compounds; stabilization; nanoparticles.

The basic directions of researches in the field of increase thermo stability and fire resistance of polymeric materials are replacement of antioxidizers and antipyrens from compounds of bromine (and in general halogens) and threehidrate ammonium on more effective and less carcinogenic substances. Growing cost of compounds of antimony also has resulted in necessity of their replacement on phosphorus containing antioxidizers and antipyrens which in a combination to compounds of bromine, provide synergetic effect, that is obviously expressed in oxygenic polymers.

*Kabardino-Balkarian State University, 360004, Nalchik, Chernyshevsky st. 173, KBR, Russia. Fax: (095) 337-99-55; E-mail: ah_shaov@mail.ru

The general tendency in the given area of a science are questions of compatibility of additives with polymers, their influence on coloring of materials, shock durability and adaptability to manufacture, and also development of target additives for concrete types of polymeric materials.

Nanocomposites a new class of the polymeric systems having extraordinary properties. The term "nanocomposites" means biphasic materials in which filled it is allocated in a polymeric matrix on nanolevel (10^{-9} m). Some more years back, before occurrence of the term "nanocomposites", the effects notice in polymeric compositions, additives of the most various origin containing small amounts, explained so-called «effect of small additives».

In the present work results of the researches received at studying of polymeric materials, containing some organic compounds are discussed on the basis of five-valent phosphorus in small concentration.

For an estimation of efficiency of stabilizers of polymeric materials use chemical, physical and chemical and physical-mechanical methods of research, but as most objective of them and important in the practical relation are considered physical-mechanical.

In this compounds, one axis stretching at speed of 40 mm/mines at 20^0 C us were investigated by a method mechanical properties of film samples of polymeric compositions. They are prepared by a method pour from a solution.

Polymeric compositions on a basis the block-copolyesters contain phosphor organic compounds (PhOC) in quantity of 0,5-30,0 % from weight of polymer, in case of a polycarbonate - in quantity of 0,1-30,0 %, and in PEHD - 0,1-0,5 %.

For an estimation of influence phosphor organic compounds on mechanical characteristics of polymers were used the following parameters: a breaking point (σ_b), the maximal relative lengthening (ε_b), the module of elasticity (E), and also impact strength (A_{st}) which are designed by known techniques. Influence phosphinic acids on mechanical characteristics the block-copolyesters can be looked after on change of values of explosive durability and the maximal relative lengthening depending on concentration phosphinic acids. It is established, that the greatest explosive durability among compositions from block-copolyester on a basis bisphenol A (BCP-7D) the structure containing 1,0 % phosphinic of an acid has. The further increase of concentration PhOC results in gradual decrease of breaking strength.

For block-copolyester on a basis phenolphthalein (BCP-7F) the maximal value of explosive durability falls to a composition with the contents of 10,0 % phosphinic to acids and is equal 139,0 MPa against 116,0 MPa for initial polymer. The greatest value of the maximal relative lengthening for compositions on basis BCP-7D also falls to structure with the contents of 1,0 % phosphinic to acids. For a composition on a basis phenolphthalein the block-copolyester this maximum is observed at the contents of 0,5 % of a researched acid.

The minimal value of the module of elasticity for compositions on the basis of both block-copolyesters falls to the contents phosphor organic to compounds in small quantities. It speaks that such structures are more elastic, than initial polymers.

With the purpose of finding - out of the mechanism of strengthening influence phosphinic acids at its use in rather small amounts on mechanical properties the block-copolyesters was carried out roentgen-structural the analysis of compositions on a basis diphenylolpropane polymer with the contents of various quantities phosphinic acids. Diffractogrames were removed on $Cu_{k\alpha}$ radiation (λ=1,5405 Å) on device DRON-3. It was established, that on diffractogrames both initial polymer, and the compositions containing various quantities phosphinic of an acid, it is not observed crystal areas.

Proceeding from experimental results, it is possible to assume, that small quantities phosphinic acids form the most thin monomolecular layers in interstructural areas macromolecules polymers which carry out a role so-called boundary greasing. In compositions there is "colloids" a system in which PhOC plays a role falling, i.e. phosphinic the acid represents itself as some softener of interstructural action. Besides since in a molecule phosphororganic compounds are available polar P=O and P-OH - groups it, carrying out a role falling, will participate in formation of intermolecular connections with polymeric groupings polymer macromolecules. The this interaction, the is stronger, in view of role PhOC as surface-activity compounds (SAC), probably, mobility over molecules is higher than structures meaning that orientation is facilitated at a stretching, that always promotes hardening of polymer. The further increase of concentration phosphinic acids, apparently, because of stereo the difficulties caused by presence in compositions of a plenty of rather voluminous molecules PhOC, results in easing intermolecular connections in polymers therefore the composite material collapses at smaller loadings. As to compositions on a basis phenolphthalein the block-copolyester and phosphinic acids when the maximal value of a breaking point falls to 10 % the contents phosphororganic to compounds in this case, apparently, for formation of a "continuous" most thin monomolecular layer, i.e. the boundary greasing facilitating mobility over molecules of structures, it is required a lot of surface-active substance (phosphinic acids) because of less dense packing polymer macromolecule on a basis phenolphthalein.

On curve dependences of a pressure on relative lengthening for compositions on a basis diphenylolpropane polymer also at the contents of small quantities phosphinic acids there is a displacement of curves in area of high values σ_b. At transition from phosphinic acids to her potassium salt character of influence phosphororganic compounds on mechanical properties the block-copolyester does not vary. Optimum concentration are small additives potassium phosphinate. But, as against the acid, at her use potassium salts of value of the module of elasticity in all a concentration interval are lower, and the maximal relative lengthening is higher, than in a case phosphinic acids. It is possible to explain greater polarity P-O-K - bond in a molecule phosphinate, in comparison with P-O-H - bond in a molecule of the acid.

With potassium salt in comparison with the structures containing phosphinic an acid it is possible to explain some decrease of durability of compositions larger in volume of an ion potassium, than atom of hydrogen in hydroxyl to group of an acid therefore arise additional stereo difficulties. It also it is possible to explain and some increase of values of the maximal relative lengthening for the compositions containing potassium phosphinate, i.e. because of increase of volume PhOC and, apparently, its polarity, force of intermolecular interaction between molecules potassium phosphinate as SAC and polar groupings macromolecules the block-copolyesters operate on larger distances in comparison with phosphinic an acid.

Use of nickel salt phosphinic acids as the additive to the block-copolyesters brings the contribution, though and not so distinguished from other compounds of five-valent phosphorus. It, first of all, is appreciable by consideration of influence nickel phosphinate on explosive durability of compositions. For diphenylol a block-copolymer of appreciable changes, as against the compositions containing phosphinic an acid or her potassium salt, it is not observed. For phenolphthalein the block-copolyester almost in all a concentration interval value of a breaking point is lower, than at initial polymer. It, most likely, is connected to the greater rigidity macromolecules itself phenolphthalein a block-copolymer in comparison with diphenylolpropane, and also stereo the difficulties brought by phosphinate molecules of nickel because of increase of number cycloalkyl and butoxyphenyl of groups as against phosphinic of an acid or potassium phosphinate.

As well as in a case potassium phosphinate, nickel salt phosphinic acids in small concentration raises value of the maximal relative lengthening, and the module of elasticity in all a concentration interval is lower, than size E of initial polymers.

Modifying influence nickel phosphinate on aromatic the block-copolyesters, as well as in a case with others PhOC, is more shown in relation to plastic deformation of polymers.

Phosphinate bivalent iron the block-copolyesters practically influences on mechanical properties the same as and nickel phosphinate, except for the module of elasticity - he starts to grow at the big concentration iron (II) phosphinate.

Such identity of character of influence nickel phosphinate and bivalent iron, apparently, is connected to identical assistants at cations both bivalent metals and approximately equal polarity of these molecules.

Character of influence phosphinate trivalent iron on mechanical properties the block-copolyesters practically reflects influence of other salts of bivalent metals: the tendency to reduction of explosive durability of compositions on a basis phenolphthalein a block-copolymer is evidently shown at increase of number phosphinate radicals in phosphor organically salts, that in turn depends on valency of metal. As to compositions on a basis bisphenol A the block-copolyester and iron (III) phosphinate here mechanical parameters remain on enough high level.

Research of influence phosphinic and its salts on mechanical properties of two poly-(arylatsulfonoxides) block-copolymers has shown acids, that for all phosphororganic compounds by the optimal concentration their small quantities, basically up to 3,0 % are.

In view of that processing polymers is improved, and at small quantities PhOC also are considerably improved mechanical characteristics of compositions on a basis the block-copolyesters and investigated derivative five-valent phosphorus, it is possible to assume, that offered compositions can find application as high-strength constructional polymeric materials.

After change of character of influence PhOC on mechanical properties the block-copolyesters was considered at transition from phosphinic acids to its salts, changing it the quantity phosphinate groups in a molecule of derivative five-valent phosphorus, was of interest research of influence on mechanical properties of polymers of replacement hydroxyl groups in a molecule phosphine acids on the second butoxyphenyl a grouping. Appeared, that at small concentration phosphinoxyde his action on mechanical properties the block-copolyesters comes nearer to those at use as modifiers phosphinic acids and her salts.

With increase of the contents phosphinoxyde in compositions on the basis of aromatic the block-copolyesters is shown the tendency to decrease of the investigated parameters. As well as others phosphororganic compounds, phosphinoxyde is more effective in relation to BCP-7D.

In a case phenolphthalein polymer, already since 1,0 % additives phosphinoxyde, plastic deformation starts to be reduced, but mechanical characteristics remain a little bit above, than with initial polymer. The further increase of concentration phosphinoxyde in compositions results in fragile destruction of samples.

In conclusion of consideration mechanical properties of compositions on the basis of aromatic block-copolyesters and phosphinic acids and her salts, and also phosphinoxyde, it is possible to notice the block-copolyesters, that all PhOC at their use in small concentration are the effective additives raising mechanical characteristics of investigated polymers. Such strengthening mechanical properties influence on polymers is shown more poorly in case of compositions on a basis phenolphthalein a block-copolymer. It first of all is connected to distinction in above molecules structures the block-copolyesters, caused by presence of the rests of molecules phenolphthalein in polymer on his basis.

In connection with that the stabilization considered as objects aromatic the block-copolyesters are only laboratory development was of interest to study, as the compounds of five-valent phosphorus received during researches will influence physical and chemical properties of industrial polyesters. As one of such polymers we choose a polycarbonate on a basis bisphenol A marks of the PC-3. It is established, that phosphinic the acid at her introduction in a polycarbonate does not render such appreciable strengthening influence on mechanical the characteristic of polymer as it took place in case of compositions on a basis the block-copolyesters.

On σ_b phosphinic the acid influences explosive durability of a polycarbonate practically a little though at her low concentration in compositions some increase of values of a breaking point is observed. Values of the maximal relative lengthening of compositions raise in comparison with initial polymer at use phosphinic acids up to 0,5 % a little. So, relative lengthening of a composition of the PC-3+0,5 % of phosphinic on 4,8 % is more than acid, than at an initial polycarbonate. The module of elasticity of compositions also is a little bit higher in comparison with the initial PC.

As the polycarbonate concerns to in part crystal polymers was of interest to find out influence PhOC on a degree crystallite a polycarbonate.

With this purpose in the same conditions, as for compositions on basis BCP-7D, was carried out X-ray structural the analysis polycarbonate compositions with the contents of various quantities phosphinic acids.

Compositions prepared as films a method pour from a solution in chloride methylene. It is established, that introduction phosphinic in small quantities results acids in some increase of a degree crystallite a polycarbonate. The increase of the contents phosphor organic up to 10,0 % appreciably reduces compounds crystallite polymer. The further increase of concentration practically does not render appreciable influence on a degree crystallite polymer. It is established, that on diffractogrames both a clean polycarbonate, and the compositions containing various quantities phosphinic of an acid, the crystal peak is on 17^0, that corresponds to the intermolecular distance equal 5,21 Å.

Proceeding from the received data it is possible to assume, that rather ineffective influence phosphinic acids on mechanical characteristics of a polycarbonate is connected to some infringement of orderliness of crystal packing macromolecules in a polycarbonate caused stereo by difficulties, voluminous molecules brought in compositions phosphororganic compounds.

It is known, that forces of an intermolecular attraction which bring the certain contribution in mechanical characteristics of polymers, operate on distance 3-4 Å. As in our case the intermolecular distance appeared little bit more than this parameter also the intermolecular attraction will be reduced a little, that, apparently, and results, along with stereo factors, to some downturn of strengthening influence phosphororganic compounds on a polycarbonate, in comparison with compositions on the basis of aromatic the block-copolyesters.

It is revealed, that at small concentration phosphinic acids plastic deformation and explosive durability of a polycarbonate remain on a high level, but the further increase of the contents phosphororganic results compounds in fragile destruction of samples.

Influence potassium phoshinate on mechanical properties of a polycarbonate, as well as in case of compositions on a basis the block-copolyesters, is more sharply expressed in comparison with influence of the acid.

The module of elasticity of compositions at the contents in them up to 3,0 % potassium phosphinate remains to higher in comparison with initial polymer, but further the module of elasticity of compositions is lower, than at a polycarbonate. Apparently, it is connected to increase of polarity phosphororganic compounds at transition from an acid to her potassium salts. Small additives potassium phosphinate raise plastic and mechanical characteristics of a polycarbonate a little.

Character of influence phosphinates bivalent metals (nickel and iron) on mechanical properties of a polycarbonate is practically identical: with increase of concentration of salts value of a breaking point tends to decrease, and the maximal relative lengthening raises at small quantities phosphinates a little; the module of elasticity of both systems gradually raises. Such identity of character of influence of these two salts on mechanical properties of a polycarbonate, probably, is connected to an identical structure of molecules phosphinates.

Phosphinate trivalent iron monotonously reduces explosive durability of a polycarbonate, and value of relative lengthening at break for the contents of salt in quantity of 5,0 % practically does not vary, but further is reduced. The module of elasticity of salt at small concentration varies also a little, and at increase of quantity phosphinate iron (III) this parameter grows.

Introduction phosphinoxyde in a polycarbonate reduces mechanical characteristics of polymer a little. It is possible to explain it to that given phosphororganic compounds less, apparently, as least polar of all investigated compounds of phosphorus, forms (in comparison with phosphinic an acid and her salts) intermolecular bonds with macromolecules a polycarbonate, but at the same time molecules phosphinoxyde have rather big volume and they can represent itself as loosening macromolecules some polymer of the agent therefore the intermolecular distance is a little increased, as results in some decrease of mechanical properties of compositions in comparison with initial polymer.

In connection with that in the industry the stabilized variant of a polycarbonate is issued, and as the stabilizer against destruction of thermo oxidation apply derivative trivalent phosphorus - three(p-nonylphenyl)phosphit (Polygard), was of interest to study, as influences Polygard on mechanical properties of a polycarbonate and to compare him to character of influence investigated by us phosphor organic compounds. With this purpose, and also for revealing an opportunity of display synergetic effect at simultaneous presence at compositions Polygard and synthesized phosphororganic compounds, we investigated mechanical properties of the industrial stabilized polycarbonate with the contents of 1,0 % Polygard, and the compositions containing derivative trivalent phosphorus and investigated compounds of five-valent phosphorus in quantity of 0,1-30,0 %. It is established, that the stabilized variant of a polycarbonate containing Polygard, has the worse parameters mechanical characteristics in comparison with not stabilized and modified offered PhOC (Table. 1).

Table 1. Mechanical properties of compositions on the basis of a polycarbonate and phosphor organic compounds

No	Composition	σ_b, MPa	ε_b, %	E, GPa
1	Polycarbonate (PC)	88	17,6	1,6
2	PC + 1,0 % Polygard	82	14,7	2,3
3	PC + 0,5 % phosphinic acids	108	34,4	2,0
4	PC + 1,0 % potassium phosphinate	101	62,1	1,5
5	PC + 3,0 % nickel phosphinate	104	56,4	1,5
6	PC + 10,0 % phosphinate Fe (II)	103	51,6	1,6
7	PC + 0,1 % phosphinate Fe (III)	155	91,4	1,8
8	PC + 1,0 % phosphinoxyde	94	14,7	2,1

Besides apparently from tab.1, small concentration phosphororganic compounds render appreciable improving effect at their sharing with Polygard. The highest parameter of mechanical properties from all investigated compositions with phosphororganic compounds has trivalent iron phosphinate at his use in quantity of 0,1 %.

Thus, comparison of compositions on the basis of the industrial polycarbonate, containing simultaneously offered by us phosphororganic compounds on the basis of five-valent phosphorus and industrial stabilizer Polygard, allows to approve, that composite materials with improved mechanical properties are received.

In connection with that the stabilization used as objects aromatic the block-copolyesters have high impact strength (in particular for BSP-7D it not less than 117-118 kJ/m^2), were of interest to determine character of influence received phosphororganic compounds on impact strength the block-copolyesters. With this purpose compositions on a basis bisphenol A a block-copolymer and phosphororganic compounds in quantity of 1,0 % from weight of polymer were prepared. The choice of such concentration phosphororganic compounds is caused by the optimal contents of these compound by consideration quasi static mechanical properties of compositions. It is established, that investigated phosphororganic acids reduce impact strength of polymer. The highest values of impact strength were received for compositions with the contents) of salts of iron and phosphinic acids. It is interesting to note, that among salts the most effective appeared derivative cyclohexyl phosphonic acids. So, the composition with the contents nickel hydrocyclohexylphosphonate has shown value A_{st}=107 kJ/m^2, with the contents bivalent iron hydrocyclohexylphosphonate - 115 kJ/m^2, and compositions with trivalent iron cyclohexylphosphonate had A_{st}=136 kJ/m^2.

To one of the major factors determining resistance of polymers to shock loadings, plasticity of polymeric materials and improvement of this characteristic of polymers concerns will promote increase of shock durability of samples. Told practically proves to be true the data received as a result of research of influence PhOC on impact strength the block-copolyester on a basis diphenylolpropane, and also industrial polyarilate DV- salts are more effective modifiers by way of improvement of shock durability and as it was shown above, plasticity of polymers.

The difference in the mechanism of influence phosphororganic compounds on mechanical behavior of polymers, most likely, is connected from them above molecule by structure. Block-copolyester BSP-7F it is worse "diphenylol" than a block-copolymer on mechanical to characteristics for this reason.

It is necessary to allocate the following factors influencing properties of a composition. Except for above molecule structures (AMS) of initial polymer, it is chemical structure PhOC, its volume, a degree of polarity and a way of distribution of the additives- modifiers, changes of defectiveness influencing a degree AMS of polymer. Besides it is necessary to take into account that circumstance, that the mentioned above factors can render various influence on mechanical behavior of a composition at different stages of deformation.

In case BSP-7D (to a lesser degree BSP-7F, meaning) introduction PhOC in small quantities results initial parameters to hardening a block-copolymer, is increased σ_b and ε_b. The module of elasticity simultaneously decreases. At an explanation of such character of change of mechanical properties it is necessary to mean, that these characteristics of polymers are determined by the total contribution as inside, and inter chain interaction. Their parity depends and is determined as actually by deformation (size ε), and a temperature-time mode mechanical loading. In conditions quasi static the tests used in the present work, speed of deformation is sufficient smaller ($\sim 10^{-2}$-10^{-4} c^{-1}) to consider, as inside, and inter chain interaction "have completely time" to realize the influence on mechanical behavior of polymer. The above efficiency of intermolecular interaction, the more its contribution to size of the module of elasticity and rigidity as a whole a composition at all stages of deformation. We shall specify, that the module of elasticity is determined on an initial site of the diagram "σ-ε" at very small values of lengthening. For this reason the assumption of role PhOC as SAC and «boundary greasing» is quite allowable. The second mechanism - strengthening of intermolecular interaction as each of considered phosphororganic compounds to some extent is active in this sense in parallel operates.

In view of small contents PhOC and, all the same, reduction of size E, it is logical to assume, that in this concentration interval and in the field of deformations where the module of elasticity is determined, the contribution to its formation of the first mechanism prevails. In process of development of deformation «boundary greasing» facilitates reorganization AMS of a block-copolymer resulting in its steadier to destruction to structure. At the deformations close to ε_b, on durability all mentioned above factors have to the full an effect. Influence of polarity PhOC in this case is evidently shown: the maximal values of a breaking point decrease in process of its growth. The same dependence σ_b and from volume of molecules phosphor organic compounds. These experimental facts confirm the assumption that at small contents PhOC the mechanism of interstructural plasticization the block-copolyesters prevails. Comparing BSP-7D and BSP-7F, influence and mechanisms of influence on mechanical behavior BSP-7F of small quantities PhOC more ambiguous is possible to come to a conclusion, that.

Decrease with increase of contents PhOC mechanical parameters the block-copolyesters speaks redistribution of the relative contribution of plasticization and strengthening of intermolecular interaction (IMI) both in durability, and in the module of elasticity. Thus it is important to note, that phosphinic an acid and phosphinoxyde at high contents render more effective complex influence on mechanical behavior of compositions: rather high mechanical parameters and growth of the module of elasticity. General, though and in a different degree expressed, result - reduction ε_b, and influence on parameters of plasticity of polymer.

Loosen structures of blocks-copolymers can occur at increase of contents PhOC on various mechanisms. However it is obvious, that efficiency of the additive now in the greater degree is determined, except for structural features, volumetric characteristics (steric the factor), activity phosphor organic compounds in relation to IMI. In this respect higher efficiency of some PhOC can be caused by an optimum parity of parameters of the additive and distribution of free volume in polymer. It is possible to assume, that the sizes appropriate of ~150-250 sm^3/mol – Van der Waals to volume (V_w) phosphinic acids and phosphinoxyde are such. Similar assumptions can be considered and concerning polarity phosphor organic compounds. That strengthening of efficiency IMI ambiguously influences on mechanical properties of compositions that is visible from comparison of some compositions on basis BSP-7D and BSP-7F proves to be true also.

In a polycarbonate the free volume is less, than in the block-copolyesters. On the other hand is in part crystal polymer with rather crystal high degree. In this case again it is necessary to find an optimum parity of properties of additive PhOC (for example, its volume) with distribution of free volume in polymer. Besides the polycarbonate is: more loosen packing and denser packing areas of an amorphous part, crystal and inter crystal areas. Therefore is unequivocal to determine the mechanism of influence PhOC on properties of a polycarbonate more difficulty. For example, the size dense packing areas of the PC is estimated by different authors, different experimental methods in limits 30÷100 Å besides, that its volume fraction of ~50 %. On the other hand molecules phosphor organic additives ($V_w \approx$ 8-10 $Å^3$) borrow, most likely, free volume in more loosen packing parts of amorphous area and, with smaller probability, inter crystal areas.

In view of told above, it was possible to expect less effective strengthening influence of all phosphor organic compounds on a polycarbonate, and laws at his studying can be others, is especial regarding influence of salts of metals phosphinic acids. Here the combination of small concentration and properties PhOC (for example,) has given some salt of trivalent iron the best result.

It is necessary to note, that shock tests differ that in conditions of high-speed deformation the relative contribution to durability and plasticity of effective intermolecular interaction sharply grows. Here it is necessary to take into account that impact strength A_{st} - the integrated power characteristic of durability of polymer or a composition, showing power expenses for destruction, σ_b and E - power.

Results of shock tests (A_{st}) in very big degree depend on plasticity of samples. On the other hand, in plasticity brings the contribution not all free volume, but only his "effective" part. Efficiency of influence of additives PhOC on shock durability depends, thus, on a parity of the sizes of a molecule of the additive with "effective" free volume of polymer and, thus, from as far as concentration PhOC is picked optimum up. In a case block-copolyester BSP-7D in conditions of shock tests are salts of iron and phosphinic acids; for the PC-trivalent iron phosphinate.

One of the basic operational characteristics of polymeric materials is shock durability in this connection structures on the basis of polyethylene of high density (PEHD) are investigated by a technique of shock tests on Sharpy.

Experimental installations with which help polymeric systems are investigated represents pendulum koper UT-1/4, supplied by the piezoelectric gauge of loading, the signal with which moved directly on remembering oscillograph of model from C 8-12. Character of influence PhOC on physical-mechanical properties PEHD in conditions shock loading estimated under such characteristics, as A_{st}, E, ε_b, σ_b, σ_{ce} and ε_{ce} which values are calculated with use of settlement formulas: $A_{st} = W_{st}/B(D\text{-}a)$; $E = (P\times L^3)/4\delta B\ (D\text{-}a)^3$; $\varepsilon = [6\delta(D\text{-}a)]/L2$; $\sigma_b = 3P_{st}L/2B(D\text{-}a)^2$; $\sigma_{ce} = 3PbL/2B(D\text{-}a)^2$, where A_{st}- impact strength; W_{st}- energy of destruction of a sample; B- width of a sample; D- thickness of a sample; and - size of a cut; P - loading on an initial linear site of the diagram; δ- a deflection of a sample in its middle; L- distance between support of a pendulum; ε_b- deformation of destruction; σ_{ce}- a limit of the compelled elasticity; ε_{ce}- deformation of the compelled elasticity; σ_b- a breaking point; P_{ce}- loading on diagram P-t (time) appropriate to macroscopically compelled elasticity; P_{st}- loading of destruction.

The received results testify that phosphinic an acid and her salts render plastically influence on polyethylene of high density. To draw such conclusion some increase of impact strength allows at simultaneous decrease of the module of elasticity and increase of relative deformation at destruction of samples.

All phosphor organic compounds appeared practically more effective additives to polyethylene in comparison with the industrial stabilizer polyolefin's Irganox-1010.

At the same time it is necessary to note, that phosphinic acid salts reduce a breaking point whereas the acid practically does not render essential influence on the given parameter a little.

Character of change A_{st} at introduction in polymer PhOC correlates with change of values of a pressure of the compelled elasticity. The given fact can speak about some increase of intermolecular interaction in polymer, that, apparently, is connected by that phosphor organic compounds, occupying "free" volumes in macro chain, the polar groupings strengthen intermolecular interaction a little.

Such behavior of modifiers should result in some improvement of orderliness in macro chain, that practically and occurs, if it to judge on increase of a degree of crystallinity (α_m) (Tables 2, 5). Thus the degree of crystallinity raises with transition from an acid to its salts, i.e. with increase of polarity phosphor organic the modifier.

By consideration of character of influence cyclohexyl phosphonic acids, her potassium salts, and also phosphine oxide on physical-mechanical properties PEHD in conditions shock loading it is noticed (Tables 3, 4), that the behavior of data PhOC practically does not differ from those phosphinic acids and its salts.

The results received at research of character of influence phosphor organic of compounds on physical-mechanical characteristics PEHD in conditions shock loading, allow to assert with the big share of reliability, that irrespective of molecular weight PhOC show plastic property, i.e. the mechanism of influence of modifiers, probably, does not vary. It is necessary to note, that salts phosphinic acids reduce value σ_b polyethylene whereas the acid practically does not render essential influence on the given parameter a little. Character of change of impact strength PEHD at introduction in him PhOC correlates with change of values of a pressure of the compelled elasticity that can speak about some increase of intermolecular interaction in polymer. It, apparently, is connected by that phosphor organic compounds occupying free volumes in macro chain, the polar groups strengthen intermolecular interaction a little. As confirmation of such assumption that circumstance can serve, that at Van der Waals volume (V_W) polyethylene in 20,6 sm^3/mol, found on a known technique the share of free volume (V_E) makes 7,6 sm^3/mol and shares V_W phosphor organic compounds, their falling investigated dosages (0,05-0,5 %), make only 0,005-0,088 sm^3/100 r polymer. The effect of small additives, probably, is connected by that at such dosages PhOC they in the optimum image "find room" in free volume of polymer, though own Van der Waalsoves volumes phosphor organic compounds are greater (58,1-423,0 sm^3/mol), than at polyethylene. For the industrial stabilizer polyolefin's Irganox-1010 value V_W even above also makes 711,1 sm^3/100 r polyethylene.

Table 2. Influence phosphor organic modifiers on density and degree of crystallinity PEHD

No	Composition	d, g/sm^3	α_m
1	PEHD	0,951	0,701
2	PEHD0,1 % Irganox-1010	0,958	0,746
3	PEHD0,05 % phosphine acids	0,952	0,707
4	- « - + 0,1 % - «-	0,957	0,739
5	- « - + 0,3 % - «-	0,951	0,701
6	- « - + 0,5 % - «-	0,957	0,739
7	PEHD0,1 % - « - + 0,1 % Irganox-1010	0,957	0,739
8	PEHD + 0,05 % nickel phosphinate	0,952	0,707
9	- « - + 0,1 % - «-	0,959	0,752
10	- « - + 0,3 % - «-	0,955	0,727
11	- « - + 0,5 % - «-	0,957	0,739
12	PEHD0,1 %-«-+ 0,1 % Irganox-1010	0,955	0,727
13	PEHD + 0,05 % iron (II) phosphinate	0,958	0,746
14	- « - + 0,1 % - «-	0,954	0,720
15	- « - + 0,3 % - «-	0,956	0,733
16	PEHD + 0,05 % iron (III) phosphinate	0,957	0,739
17	- « - + 0,1 % - «-	0,960	0,758
18	- « - + 0,3 % - «-	0,952	0,707
19	- « - + 0,5 % - «-	0,958	0,746

Note: the degree of crystallinity α_m is designed proceeding from density.

Table 3. Physical-mechanical properties of compositions on a basis PEHD and PhOC in conditions of shock test

No	Composition	A_{st}, kJ/m^2	E, GPa	σ_b, MPa	ε_b, %	σ_{ce}, MPa	ε_{ce}, %
1	PEHD	11,0	1,06	21,1	5,7	27,9	4,8
2	PEHD0,1 % Irganox-1010	12,8	1,55	18,6	6,4	30,0	3,6
3	PEHD0,05 % phosphinic acids	22,6	0,81	19,9	7,9	29,1	4,7
4	- « - + 0,1 % - «-	18,5	0,79	17,8	8,3	29,3	4,6
5	- « - + 0,3 % - «-	18,8	0,79	18,1	9,1	29,7	5,1
6	- « - + 0,5 % - «-	11,5	0,94	20,2	6,4	28,9	4,4
7	PEHD0,05 % nickel phosphinate	14,6	0,85	15,9	9,0	25,5	4,2
8	- « - + 0,1 % - «-	17,9	0,89	14,1	10,6	26,9	4,6
9	- « - + 0,3 % - «-	22,1	0,85	12,8	8,3	26,9	4,7
10	- « - + 0,5 % - «-	17,4	0,72	17,5	7,9	27,1	5,3
11	PEHD0,05 % iron (II) phosphinate	12,8	0,85	16,0	7,9	28,7	4,3
12	- « - + 0,1 % - «-	11,6	0,85	16,3	7,7	27,1	4,7
13	- « - + 0,3 % - «-	9,8	0,77	12,0	7,5	23,9	4,5
14	PEHD0,05 % iron (III) phosphinate	13,3	0,77	11,2	8,4	23,9	4,7
15	- « - + 0,1 % - «-	17,6	0,82	15,9	8,1	24,7	4,5
16	- « - + 0,3 % - «-	14,4	0,78	12,5	9,4	21,0	4,7
17	- « - + 0,5 % - «-	6,4	0,62	13,9	8,3	23,0	4,9

It is necessary to note, that polyethylene - more "simple" structure, than the block-copolyesters and a polycarbonate, with the big factor of packing, besides in part - crystal. His studying is complicated that amorphous his phase at room temperatures no glassing, and at shock tests is necessary to bring artificial defect - a cut to have an opportunity supervision of the full diagram of destruction in coordinates "force-deformation".

Table 4. Physical-mechanical properties of compositions on a basis PEHD and PhOC in conditions of shock test

No	Composition	A_{st},KJ/m^2	E, GPa	σ_b, MPa	ε_b, %	σ_{ce}, MPa	ε_{ce}, %
1	PEHD	11,0	1,06	21,1	5,7	27,9	4,8
2	PEHD + 0,1 % Irganox-1010	12,8	1,55	18,6	6,4	30,0	3,6
3	- « - +0,05 % phosphinic acid	19,5	0,82	14,1	9,5	28,1	5,0
4	- « - + 0,1 % - «-	16,3	0,90	18,6	7,6	28,5	4,6
5	- « - + 0,3 % - «-	18,8	0,81	17,1	7,6	28,5	4,7
6	- « - + 0,5 % - «-	20,0	0,87	16,5	8,7	28,5	4,8
7	- « - +0,1 %-«-+ 0,1% Irganox	22,6	0,85	16,2	8,3	28,7	4,8
8	PEHD0,05 % potassium monophosphonate	20,0	1,05	29,9	5,1	30,6	3,9
9	- « - + 0,1 % - «-	19,3	0,85	16,0	8,1	27,9	4,9
10	- « - + 0,3 % - «-	15,0	0,82	16,0	7,5	27,1	4,8
11	- « - + 0,5 % - «-	12,1	0,57	18,7	6,9	27,3	4,7
12	- « - +0,1 %-«-+ 0,1 % Irganox	23,1	1,03	15,2	9,8	30,4	5,3
13	PEHD0,05 % potassium phosphonate	20,5	0,92	16,0	8,6	25,4	4,1
14	- « - + 0,1 % - «-	19,0	0,85	18,0	7,6	26,7	4,6
15	- « + 0,3 % - «-	15,5	0,91	17,3	8,4	26,7	4,8
16	- « + 0,5 % - «-	15,4	0,82	18,3	8,5	27,9	4,6
17	- « + 0,1 %-«-+ 0,1 % Irganox	19,5	0,84	15,7	7,9	26,3	4,6
18	PEHD0,05 % phosphinoxyde	21,8	0,90	12,8	8,8	27,1	4,6
19	- « - + 0,1 % - «-	20,7	0,90	13,3	9,5	28,5	4,7
20	- « + 0,3 % - «-	18,2	0,80	16,0	7,9	26,9	5,2
21	- « + 0,5 % - «-	19,7	0,85	18,3	7,9	27,1	4,4
22	- « + 0,1 %-«-+ 0,1 % Irganox	17,4	0,94	16,5	7,7	27,7	4,4

Therefore it is necessary to take into account not only structure, volume, properties of additives PhOC, but also as far as the put cut (~5 mm) is close or far from structural defect PE. By influence on σ_b, the critical border between PhOC in sense of their influence on defectiveness of a matrix passes between phosphinic an acid and her salts. The integrated estimation of influence PhOC on mechanical properties of polyethylene at impact results in the same conclusions, as earlier, in case of consideration the block-copolyesters and a polycarbonate.

Table 5. Influence cyclohexyl phosphonic acids and her derivativeson density and degree of crystallinity PEHD

No	Composition	d, g/sm^3	α_m
1	PEHD	0,951	0,701
2	PEHD + 0,1 % Irganox-1010	0,958	0,746
3	- « - + 0,05 % phosphonic acid	0,954	0,720
4	- « - + 0,1 % - «-	0,946	0,669
5	- « - + 0,3 % - «-	0,956	0,733
6	- « - + 0,5 % - «-	0,951	0,701
7	- « - + 0,1 % - « - + 0,1 % Irganox-1010	0,954	0,720
8	PEHD + 0,05 % potassium monophosphonate	0,957	0,739
9	- « - + 0,1 % - «-	0,958	0,746
10	- « - + 0,3 % - «-	0,951	0,701
11	- « - + 0,5 % - «-	0,957	0,739
12	- « - + 0,1 % + 0,1 % Irganox-1010	0,954	0,720
13	ПЭВП + 0,05 % potassium phosphonate	0,954	0,720
14	- « - + 0,1 % - «-	0,949	0,688
15	- « - + 0,3 % - «-	0,951	0,701
16	- « - + 0,5 % - «-	0,954	0,720
17	- « - + 0,1 % - « - + 0,1 % Irganox-1010	0,955	0,727
18	PEHD + 0,05 % phosphinoxyde	0,962	0,771
19	- « - + 0,1 % - «-	0,950	0,695
20	- « - + 0,3 % - «-	0,951	0,701
21	- « - + 0,5 % - «-	0,954	0,720
22	- « - + 0,1 % - « - + 0,1 % Irganox-1010	0,946	0,669

In the conclusion it is possible to ascertain, that high efficiency synthesized phosphor organic compounds in quality inhibitors thermo oxidation destruction aromatic the block-copolyesters is established - the appreciable effect is observed at the contents of these compounds in compositions in quantity up to 1,0 %.

On increase of stabilizing effect phosphor organic compounds form a line: p-butoxyphenyl cyclohexyl phosphinic an acid, her potassium salt, di-(p-butoxyphenyl) cyclohexylphosphinoxyde; higher efficiency of the synthesized new compounds of five-valent phosphorus - di-(p-butoxyphenyl) cyclohexylphosphinoxyde - is found out as the stabilizer for a polycarbonate in comparison with the traditional stabilizer for this polymer - Polygard, being derivative of trivalent phosphorus - [tris-(p-nonylphenyl) phosphit], that is shown, in particular, in the best ability to preservation of durability and elasticity after long tests of polymer at 423 K; It is established, that synthesized phosphor organic the compounds containing in molecules atoms of five-valent phosphorus, influence on polyethylene of the low pressure, similar plastifying to effect that is shown in increase of impact strength and relative lengthening at simultaneous decrease of the module of elasticity in conditions of shock tests.

References

[1] Wilku, Ch. A. Polymer degradation and stabilization.-*Polym. News*. - 1995.-20.-No 10. P.316-317.

[2] Schetlick, K., Habicher, W.D. Organophosphores antioxidants action mechanismes and new trends: Pap. 16 th Int. Conf. Adv. and Degrad. Polym. Luzern.-June 14-17.-1994.- *Angew. makromol. Chem*. - 1995. - 232. - P. 239-246.

[3] Foigt, I. Stabilization of synthetic polymers against action of light and heat.-L.: *Chemistry*. - 1972.-544 p.

[4] Shrainer, R., Fyuzon, R., Kertin, D., Moryll T. *Identification of organic compounds. - M*. - 1983. - 704 p.

[5] Ozden, S., Shaov, A.H., Charayev, A.M., Bidanikov A.Y. Compositions Based in Aromatic Block Copolyester and p-Bytoxyphenyl Cyclohexyl Phosphinic Acid.-*Polym. and polym. Comp*.-1998.-Vol. 6.-No 2.-P.103-107.

[6] Ozden, S., Shaov, A.H., Charaev, A.M., Gurdaliyev X.X. The effects of p-butoxyphenyl-cyclohexylphosphinic acid on the properties of PC based on bisphenol A. - *J. Appl. Polym. Sci*.-2001.-Vol. 80.-P. 2113 - 2119.

[7] Ozden, S., Shaov, A.H., Charayev, A.M., Mikitaev A.K., Bedanokov A.Y. Aromatic block copolyesters stabilised with metallic salts of phosphinic acid.-*Polymers and Polymer Composites*.-2001.-Vol. 9.-No 3.-P. 213-219.

[8] A.Kh. Shaov, N.N. Amerkhanova, A.M. Kharaev. Phosphorous Organic Compounds as the stabilizers of the Aromatic Blockpolyethers. - *Aging of polymers, polymer blends and polymer composites*.-Nova Sci. Publ.-New York.-2002.-Vol. 2.- P. 161-166.

[9] Shaov, A.Kh., Sarieva, Z.I., Gurdaliev, Kh. Kh. The Research of the Character of the Influence of Phosphinates and Phosphinoxyde on Mechanical Properties of Polyethylene High Density. - Nova Sci. Publ. - *Aging of polymers, polymer blends and polymer composites*. - New York. - 2002.-Vol. 2.-P. 167-170.

[10] Shaov, A.Kh., Dubayeva, E.Kh., Mikitaev, A. K. Stabilization of Polycarbonate with some Organic Derivatives of the Five-Valent Phosphorus. - *Aging of polymers, polymer blends and polymer composites*. - Nova Sci. Publ. - New York. - 2002.-Vol. 2.-P. 171-174.

[11] Shaov, A.Kh., Kodzokova, E.Kh. Organic derivative five-valent phosphorus as stabilizers and modifiers of polymeric materials (Review).-P.1 – *Plasticheskie Massy*. - 2004. - No 12. – P. 21-34.

[12] Shaov, A.Kh., Kodzokova, E.Kh. Organic derivative five-valent phosphorus as stabilizers and modifiers of polymeric materials (Review).-P.2. - *Plasticheskie Massy*. - 2005. - No 3. – P. 33-41.

[13] Shaov, A.Kh., Borukaev, T.A., Begretov, M.M. Last achievements in the field of creation of fire-resistant polymeric materials with use phosphor organics compounds (Review.) - *Chemical physics of pyrolysis, combustion and oxidation*.-Nova Sci. Publ., New York.-2005.-P.19-31.

In: Encyclopedia of Nanotechnology
Editor: Elwood D. Carlson

ISBN 978-1-60692-079-4

Chapter 50

POLYMER/SILICATE NANOCOMPOSITES BASED ON ORGANOMODIFIED CLAYS

***A. K. Mikitaev*[1], *O. B. Lednev*[2*], *A. Yu. Bedanokov*[1] and *M. A. Mikitaev*[3]**

[1]A.N. Nesmeyanov Institute of Organoelement Compounds of RAS, 119991 Vaviliva st., 28, Moscow, Russia

[2]D.I. Mendeleev University of Chemical Technology of Russia, 125047 Miusskaya Square 9, Moscow, Russia

[3]State Scientific Institution "Composite Center", 125047 Miusskii Square 9, Moscow, Russia

ABSTRACT

It should be known that a lot of studies devoted to the preparation of polymer nanocomposite materials have been investigated at resent years. The amount of such works increases intensively. The possibility of preparation such materials was shown for practically all kinds of polymerized and polycondensed polymer materials. Investigators demonstrate particular interest to the organomodified montmorillonyte as an element of nanotechnology and bearer of nanostructure with great differ between its length and thickness. In this case the organomodification is carried out by using of ionic surfactants. The using of nonionic surfactants for hydrofobization of clay's surface also found reflection in some works. Common steady understanding is formed about investigation technique and structure of polymer nanocomposite materials and how thermo-mechanical properties depend on its structure. The increasing in amount of such investigations shows that this perspective technology will find reflection in industrial application.

Keywords: nanocomposite, organoclay, fire retardant polymer.

[*]Correspondence to: Oleg B. Lednev, D.I. Mendeleev University of Chemical technology of Russia, 125047 Miusskaya sq., 9, Moscow, Russia. mailto: lednev_oleg@mail.ru

During last years one of the most perspective fields of polymer science there is a preparation of polymer materials that have a lot of improved and new properties. Such properties can be attributed to the new kind of materials calling polymer nanocomposites that can be used in different branches of polymer applications.

To achieve improved properties in polymer composites have to use such additives as pigments, inhibitors, antioxidants, plasticizers and other compounds. Materials including the inorganic particles (oxides, nitrides, carbides, silicates etc.) are introduced to the polymer matrix in case of nanocomposites. Main our interest devoted to the polymer nanocomposite materials based on organomodified layered silicates [1]. Incompatibility of these inorganic and organic components – main problem has to be solved. There is method to overcome this problem. It's a modification of clays by organic ionic or nonionic compounds. Modified clay (organoclay) has some advantages in comparison with simple clay:

1) Organoclays can be well dispersed in polymer matrix [2].
2) Organoclays interacted with polymer's chain [3].

For preparation of such nanocomposites based on organoclays have to be used layered natural inorganic structures as montmorillonite [4, 5, 6], hectorite [3], vermiculite [7], saponin [8], kaolin, etc. Length of these layers about 220 nm, and thickness – 1nm [9, 10].

Their crystal structure consists of two fused silica tetrahedral sheets sandwiching an edge-shared octahedral sheet of either alumina or magnesia. Stacking of the layers leads to a regular Van der Waals gap between the layers, called the *interlayer* or *gallery*. Isomorphous substitution of Si4+ for Al3+ in the tetrahedral lattice, and of Al3+ for Mg2+ in the octahedral sheet generates and excess of negative charges that are normally counterbalanced by cations (Na+ or K+) residing in the interlayers (Figure1) [11].

The organically modified clays are prepared by the addition of long chain aliphatic quarternary ammonium or phosfonium cations to sheet mineral inorganic clays. An ion exchange process is utilized to displays the inorganic cations (e.g. sodium) with organic cations, thus improving the compatibility of the organosilicate with an organic environment [2].

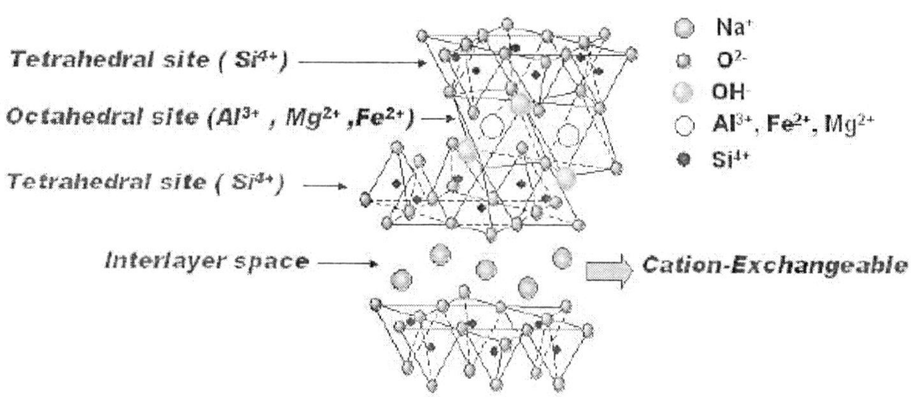

Figure 1. Structure of layered clays.

When mixed into the host polymer, exfoliation (breaking apart) of the nanophase organoclay can occur whereby the silicate sheets lose their attraction to each other. A very large increase in surface area occurs and, if the chemistry properly designed, the polymer chains can become attracted to the clay sheets. A hybrid inorganic-organic material is produced with altered properties that vary depending on the level of dispertion, the organic cation, the silicate, and the host polymer. Also can be used nonionic substances to modify clays that can be attracted to clay's surface mainly by hydrogen bonds. In some cases organoclays obtained by nonionic surfactants are more chemically stable than organoclays obtained using cation modifiers (Figure 2) [12].

Desorbability of surfactants from bentonite depends on the mechanism of adsorption. A desorption hysteresis is generally observed when cationic surfactant is adsorbed via ion exchange (Figure 2 II). The desorbability of surfactants from bentonite was compared by consecutively washing the organobentonite with deionized water. The nonionic surfactants in bentonite are relatively resistant tot desorption; >80% were still adsorbed after seven consecutive washes. However, for cationic surfactants, washing resulted in 25% desorption. Apparently, the cumulative effect of hydrogen bonding between individual ethylene oxide units and the bentonite surface makes organobentonite derived from nonionic surfactant chemically more stable than organobentonite derived cationic surfactant.

Their makeup is such that they can be transformed into new materials possessing the advantages of both organic materials, such as light-weight, flexibility, and good moldability, and inorganic materials, such as high strength, heat stability, and chemical resistance. The incorporation of organic/inorganic hybrids can result in materials possessing excellent of stiffness, strength and gas barrier properties with far less inorganic content than is used in conventionally filled polymer composites: the higher the degree of delamination in polymer/clay nanocomposites, the greater the enhancement of these properties [5].

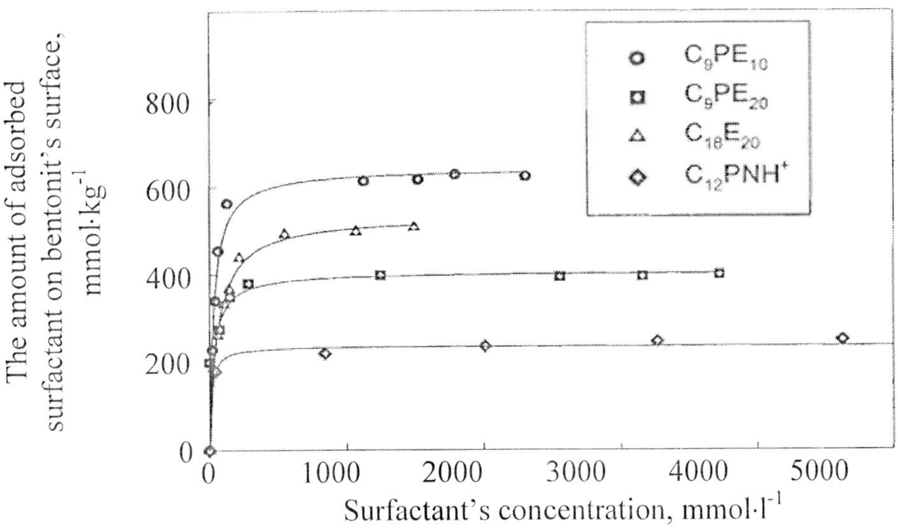

Figure 2. I – the adsorptions of different modifiers on clay's surface.

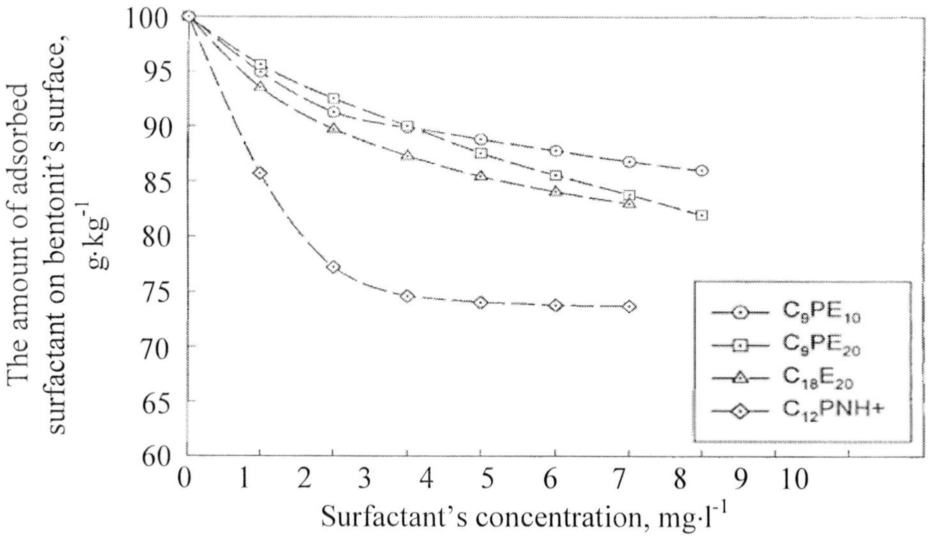

Figure 2. II – the desorptions of different modifiers on clay's surface, where: C9PE10 – C9H19C6H4(CH2 CH2O)10OH; C9PE20 – C9H19C6H4(CH2 CH2O)20OH; C18E20 – C18H37(CH2 CH2O)20OH; C12PNH+ – C12H25C6H4NH+Cl-.

At present time have been synthesized a lot of different polymer nanocomposite materials based on various kinds of polymer materials and natural inorganic fillers. The amount of such works increases very intensively (Table 1).

Table 1. Nanocomposites based on organoclay

Name of polymer	*Shorthand notation*	*Literature*
Polyacrilate	PAC_r	[13]
Polyamide	PA	[3, 14, 15]
Polybenzoksazole	PBO	[16]
Polybutylenetherephtalate	PBT	[2, 4, 18]
Polyimide	PI	[19]
Polycarbonate	PC	[20]
Polymethilmetacrilate	PMMA	[21]
Polypropylene	PP	[22, 23]
Polystyrene	PS	[21]
Polysulfone	PS_n	[24]
Polyurethane	PU	[25]
Polyethyleneterephtalate	PET	[26]
Polyethylene	PE	[27]
Epoxy	EP	[28]

Several methods have been used to obtain polymer nanocomposites by using organoclays [29-32], i.e. solution intercalation [33-39], melt intercalation [40, 41], and in situ interlayer intercalation [30, 42, 43]. Among them, in situ interlayer polymerization relies on swelling of the organoclay by the monomer, followed by in situ polymerization initiated thermally or by the addition of a suitable compound. The chain growth in the clay galleries accelerates clay exfoliation and nanocomposite formation.

This technique of in situ interlayer polymerization is also particularly attractive due to its versatility and compatibility with reactive monomers and is beginning to be used for commercial applications. However, there is ample evidence that nanocomposites can also be formed by melt processing in extruders. There are many reasons why melt processing may be more preferred method for producing nanocomposites for commercial use. Additionally, other approaches, such as the sol–gel process [44, 47] and monomer/polymer grafting to clay layers, have resulted in organic/inorganic polymer hybrids.

In the process of melt intercalation, the layered silicate is mixed with a molten polymer matrix. If the silicate surfaces are sufficiently compatible with the chosen polymer, then the polymer can enter the interlayer space and form an intercalated or an exfoliated nanocomposite. Otherwise, in situ intercalation polymerization is a method based on the use of one or more monomers that may be in situ linearly polymerized or cross linked and was the first method used to synthesize polymer-layered silicate nanocomposites based on polyamide 6. The in situ intercalation method relies on the swelling of the organoclay due to by the monomer, followed by in situ polymerization initiated thermally or by the addition of a suitable compound. The chain growth in the clay galleries triggers clay exfoliation and nanocomposite formation. Thus, an advantage of the in situ method is the preparation of polymer hybrids without physical or chemical interactions between the organic polymer and the inorganic material.

According to the early work of Giannelis [49], in general two types of hybrid structures can be obtained upon PCN preparation: *intercalated*, in which a single, extended polymer chain is intercalated between the silicate layers, resulting in a well-ordered multilayer with alternating polymer/inorganic host layers and a repeat distance of a few nanometers, and *disordered* or *delaminated*, in which the silicate layers (1 nm thick) are exfoliated and dispersed in a continuous polymer matrix (Fig 3). The best performances are commonly observed for the exfoliated nanocomposites; the two situations can, however, coexist in the same material. In any case, to make a successful nanocomposite it is very important to be able to disperse the inorganic material throughout the polymer. If a uniform dispersion is not achieved, agglomerates of inorganic materials are found within the host polymer matrix, thus limiting improvement.

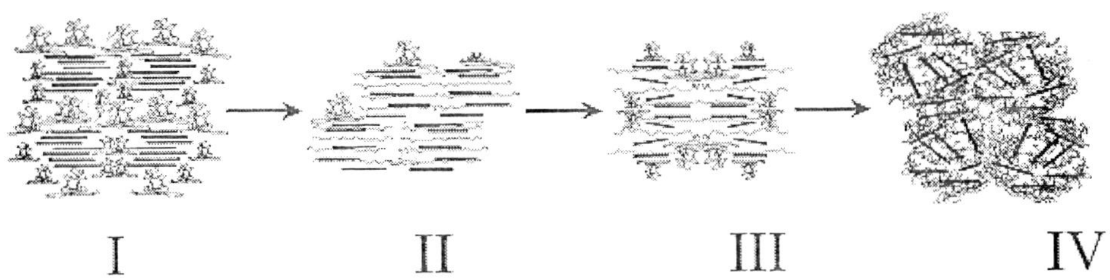

Figure 3. Schematic illustration of nanocomposite formation.

Unseparated MMT layers, after introduction into the polymer, are often referred to as tactoids. The term intercalated describes the case where a small amount of polymer moves into the gallery spacing between clay platelets, but causes less than 2-3 nm separation between the platelets. Exfoliation or delamination occurs when polymer further separates the clay platelets e.g. by 8-10 nm or more. A well-delaminated and dispersed nanocomposite consists of delaminated platelets distributed homogeneously in the polymer [4, 5, 49, and 50].

One of the methods to study the dispertion of organoclay in nanocomposite is wide-angle XRD-diffraction. Thus, on figure 4 are shown XRD-data obtained for Na^{+}MMT, modified clay and PBT-fibers with different loading of organoclay.

The characteristic peak for pristine clay, Na^{+}MMT, appears at $2\theta=8.56°$ (d=1.2 nm). For Na^{+}MMT reacted with alkyl amine, NCT-MMT, this peak is broadened and shifted to $2\theta=3.98°$ (d=2.6 nm), suggesting that the clay is swollen to the range of the d spacing. In general, a greater interlayer spacing should be advantageous in the intercalation of polymer chains. It should also lead to easier dissociation of the clay, which should result in hybrids with better dispersions of clay. In addition to the main diffraction peak, an additional small peak is observed at $2\theta=7.13°$ (d=1.43 nm). This secondary peak may be related to the XRD spectra of the organoclay itself. Figure3 also shows the X-ray diffraction (XRD curves of pure PBT and of PBT hybrid fibers with 2–5 wt% organoclay loadings. Pure PBT synthesized with an MMT interlayer exhibits its usual XRD peaks. However, in the cases of the 2 and the 3 wt% PBT hybrids, the curves show no characteristic organoclay peaks in the range of the $2\theta=2$–$8°$; that is, the peak corresponding to the basal spacing has disappeared. In the cases of the PBT hybrids with 4 and 5 wt% organoclay loadings, however, a small peak is observed at $2\theta=5.44°$ (d=1.88 nm). This indicates that agglomeration of a small part of the clay has occurred in the PBT matrix [4].

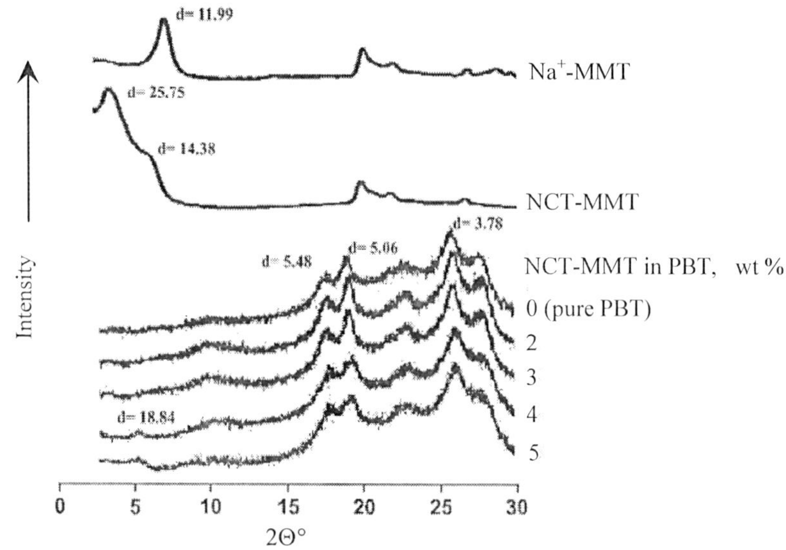

Figure 4. XRD patterns for clay, organoclay, and PBT hybrid fibers with various organoclay contents.

XRD-data also were obtained for other polymers (Figure 5) [4, 49, and 50]. XRD is most useful for the measurement of the d-spacing of ordered immiscible and ordered intercalated polymer nanocomposites with clay, but it may be insufficient for the measurement of disordered and exfoliated materials that give no peak. The organoclay dispersion in also has to be crosschecked further by using the SEM and TEM data [53, 54].

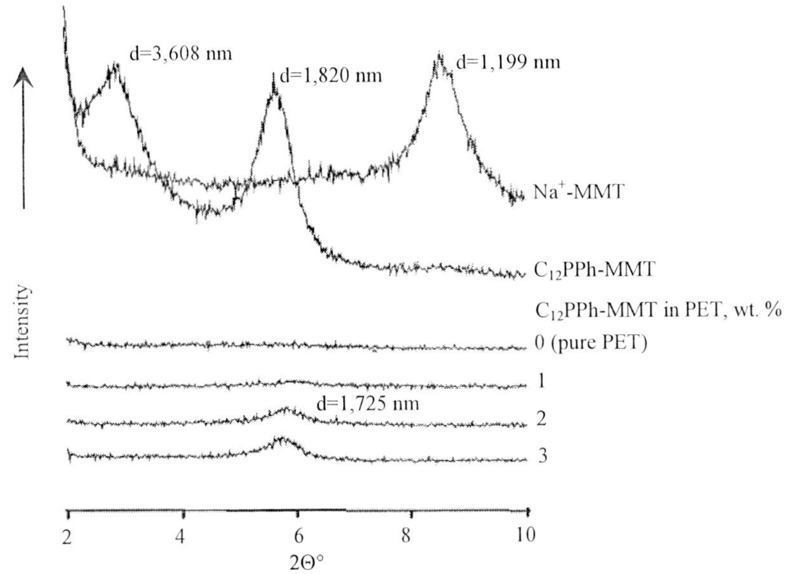

Figure 5. XRD-data for clay, organoclay and nanocomposite PET/organoclay.

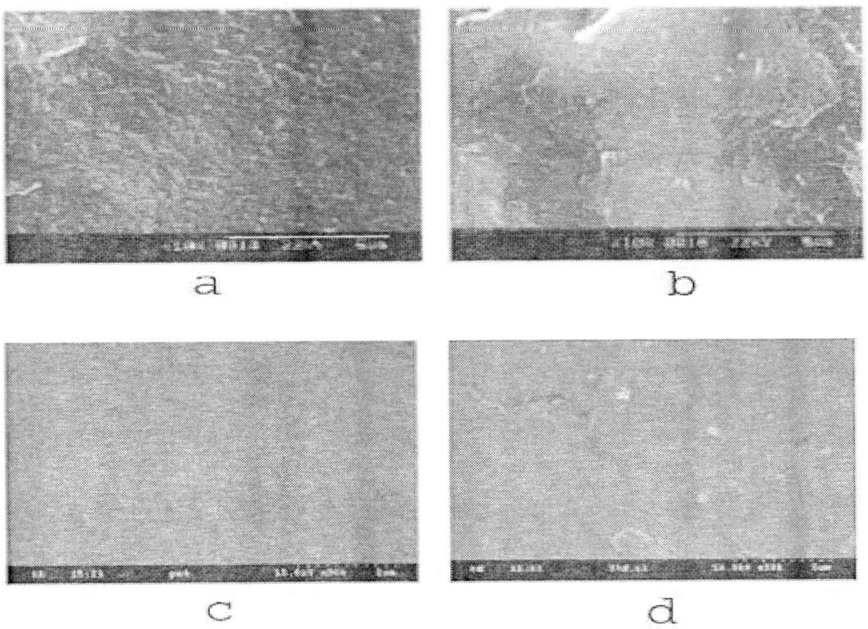

Figure 6. SEM photomicrographs of (a) 0% (pure PBT); (b) 3% organic-MMT in PBT hybrid fibers; (c) 0% (pure PET); (d) 3% organic-MMT in PET hybrid fibers.

SEM micrographs of the fractured surfaces of PBT hybrid fibers prepared with different clay contents are compared in Figure 6, 7. The micrographs of the pure PBT/PET (Figure 6(a, c)) and the PBT/PET hybrid fiber containing 3 wt% organoclay (Figure 6(b, d)) show smooth surfaces due to better dispersed clay particles.

Conversely, Figure 7(a, b, c, d) show voids and some deformed regions that may result from the coarseness of the fractured surface. However, the fractured surfaces were more deformed when higher contents of organoclay were used in the hybrids. This is probably a consequence of the agglomeration of clay particles [55, 56].

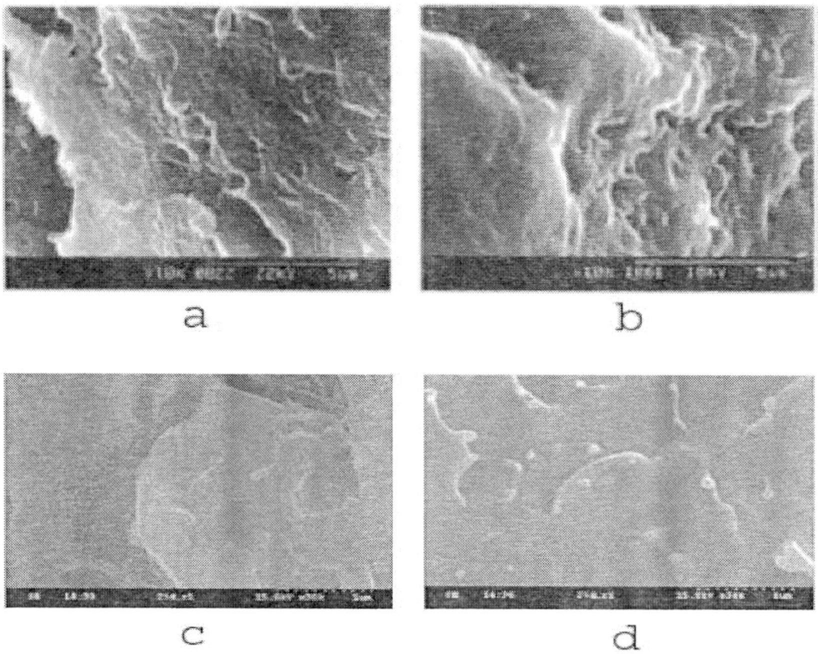

Figure 7. SEM photomicrographs of (a) 4% (pure PBT); (b) 5% organic-MMT in PBT hybrid fibers; (c) 4% (pure PET); (d) 5% organic-MMT in PET hybrid fibers.

More direct evidence for the formation of a true nano-scaled composite was provided by TEM analysis of an ultramicrotomed section. The TEM micrographs are presented in Figure 8, 9. The dark lines are the intersections of 1 nm-thick clay sheets, and the spaces between the dark lines are the interlayer spaces. Some of the clay layers of Figure 6 show individual dispersion of delaminated sheets in the matrix, as well as regions where the regular stacking arrangement is maintained with a layer of polymer between the sheets. Although a face-to-face layer morphology is retained, the layers are irregularly separated by, 4–10 nm of polymer.

For the 4 and the 5 wt% organoclay-loaded PBT/PET hybrid fibers (Figure 9) however, some of the clay is well dispersed in the PBT/PET matrix, and some of it is agglomerated to a size of approximately 4-8 nm. This is consistent with the XRD results shown in Figure 4, 5.

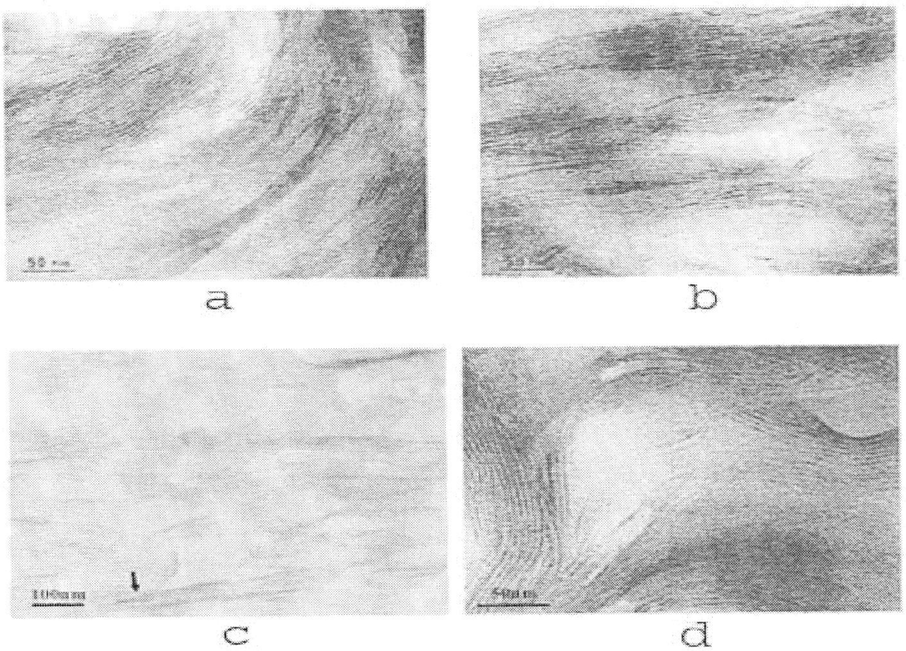

Figure 8. TEM photomicrographs of (a) 2% organoclay in PBT; (b) 3% organoclay in PBT; (c) 1% organoclay in PET; (d) 1% organoclay in PET.

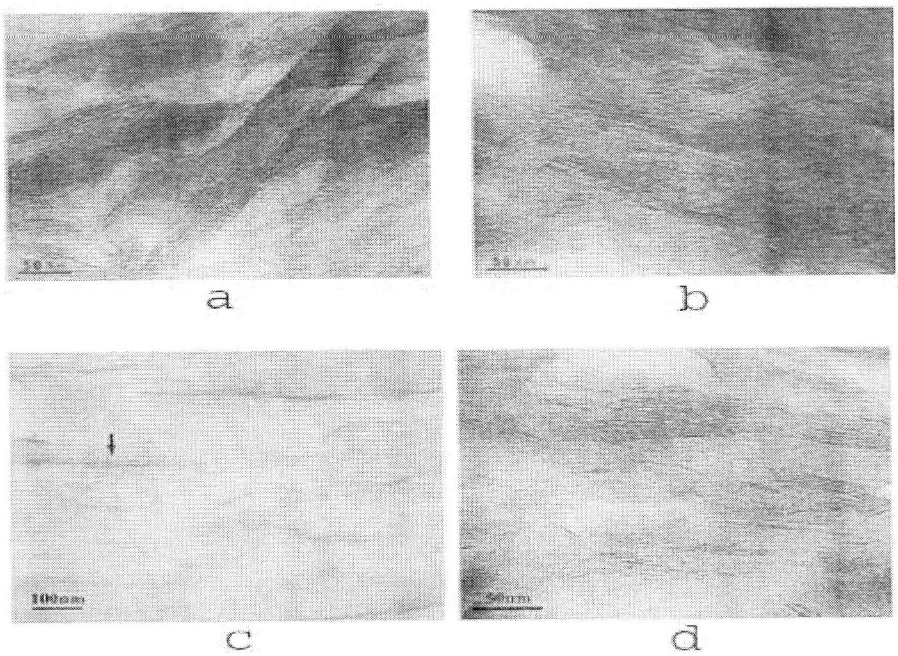

Figure 9. TEM photomicrographs of (a) 4% organoclay in PBT; (b) 5% organoclay in PBT; (c) 3% organoclay in PET; (d) 3% organoclay in PET.

From the results of XRD and electron micrographs, the morphology at a low organoclay content (<3 wt %) presents a mixture of intercalated and partially exfoliated features. The dispersion is better at a lower organoclay loading than at a high organoclay loading.

Using of organoclays as nanoparticles in polymer materials can change such properties as glass transition temperature (T_g), melting temperature (T_m) and decomposition temperature (T_D), ultimate tensile strength (Ult. Str.), initial modulus (Ini. Mod.), percent of elongation at the break (E.B.), gas-, moisture barrier properties.

The thermal and mechanical properties of PBT and PET (polyethylene terephthalate) hybrids with different contents of organoclay are listed in Table 2.

The glass transition temperatures (Tg) of the PBT hybrids increased from 27 to 33 °C with increasing clay loading from 0 to 2 wt% and then remained fairly constant up to 5 wt% organoclay. The increase in the Tg of these hybrids may be the result of two factors [57-60]. First, the effect of a small amount of dispersion of the clays on the free volume of PBT is significant and has an influence on the glass transition temperature of the PBT hybrids. Second, confinement of intercalated polymer chains within the clay galleries prevents segmental motions of the polymer chains. The endothermic peak of pure PBT appears at 222 °C and corresponds to the melting temperature (Tm). Similar to the result for Tg; the DSC thermograms show that the value of Tm increases from 222 to 230 °C with increasing organoclay content up to 2 wt%, and then remained constant for additional organoclay loading up to 5 wt% (Table 2). This increase in the thermal behavior of the hybrids may result from the heat insulation effect of the clay layer structure, as well as from the interaction between the organoclay and PBT molecular chains [59, 60].

Table 2. Main properties of polymer nanocomposites

Properties	Composite structure								
	Polybutylene terephthalate + NCT-montmorillionite					Polyethylene terephthalate + C_{12}PPh-montmorillonite			
	Organoclay quantity, %								
	0	2	3	4	5	0	1	2	3
Viscosity, Dl/g	0,84	1,16	0,77	0,88	0,86	1,02	1,26	0,98	1,23
T_g, °C	27	33	34	33	33	---	---	---	---
T_m, °C	222	230	230	229	231	245	247	245	246
T_d, °C	371	390	388	390	389	370	375	384	386
W_{tR}^{600c}, %	1	6	7	7	9	1	8	15	21
Ultimate strength MPa	41	50	60	53	49	46	58	68	71
Initial modulus GPa	1,37	1,66	1,76	1,80	1,86	2,21	2,88	3,31	4,10
Percent of elongation at break, %	5	7	6	7	7	3	3	3	3

Thermal stability of nanocomposites based on PBT/PET studied by thermo gravimetric analysis (TGA) is shown in table 2 and figures 10, 11 [4, 5].

The weight loss due to the decomposition of PBT and its hybrids was nearly the same until a temperature of 350°C (Figure 10). Above 350°C, $T°_D$ was influenced by the organoclay loading in the hybrids. The addition of clay enhanced the performance by acting as a superior insulator and as a mass-transport barrier to the volatile products generated during decomposition [63, 64, and 65]. This kind of improvement in thermal stability has also been observed in many systems of hybrids [66, 67]. The weight of the residue at 600°C increased with increasing clay loading from 0 to 5%, ranging from 1 to 9%. This enhancement of the char formation is ascribed to the high heat resistance due to the clay itself. Considering the above results, we find it consistently believable that the introduction of an inorganic clay component into an organic polymer can improve polymer's thermal properties due to the good thermal stability of the clay.

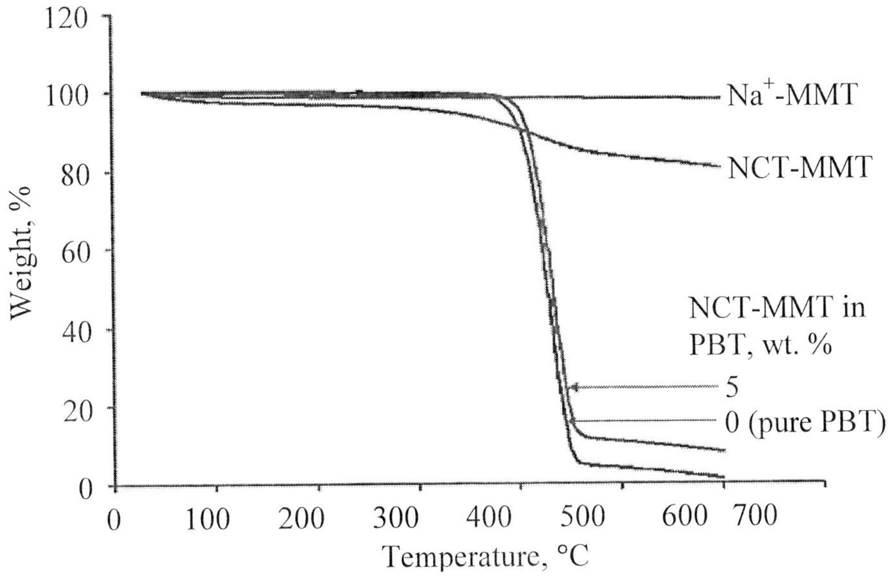

Figure 10. TGA curves of montmorillonites, PBT and nanocomposite PBT/organoclay.

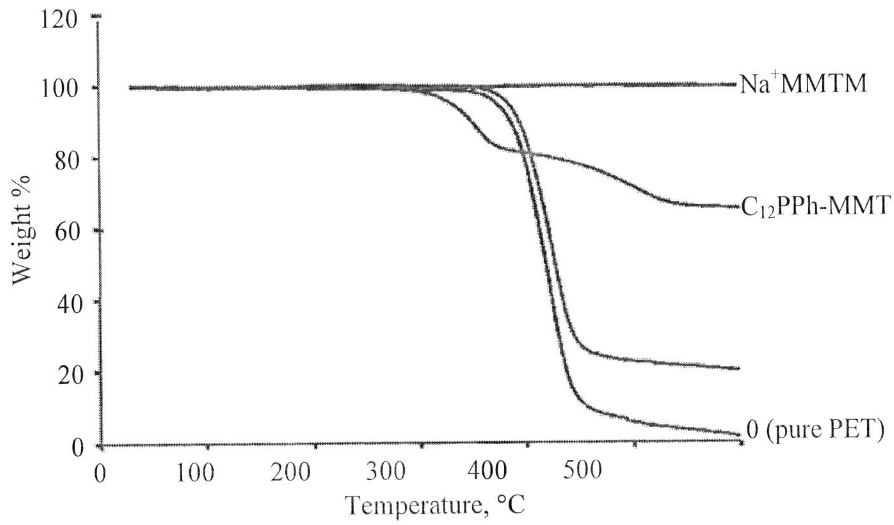

Figure 11. TGA curves of montmorillonites, PET and nanocomposite PET/organoclay.

As it shown in table 2 at DR (draw rate) = 1, the ultimate tensile strength of the PBT/NCTMMT hybrid fibers increases with the addition of clay up to a critical clay loading, and then decreases above that critical content. For example, the strength of 3 wt% PBT hybrid fibers is 60 MPa, which is about 50% higher than that of pure PBT (41 MPa). When the amount of organoclay in PBT reaches to 5 wt%, the strength has decreased again to 49 MPa. This suggests that the NCT-MMT domains are more agglomerated above 3 wt% organoclay content in the PBT matrix [61, 68, and 69].

However, the initial modulus monotonically increased with increasing organoclay content in the PBT (Table 2). The value of the initial modulus increased constantly from 1.37 to 1.86 GPa with increasing NCTMMT content up to 5 wt%. This enhancement modulus is ascribed to the high resistance exerted clay itself. Additionally, the stretching resistance and oriented backbone of the polymer chain contributes to the enhancement of the modulus. Percent elongations at the break of all hybrids were 7%. These values remained constant for organoclay loadings from 2 to 5 wt%.

As Table 2 shows, the values of the strength and the initial modulus for pure PBT fibers were enhanced with increasing DR. However, as expected for the case of a flexible coil polymer, the increases in the tensile strength and the modulus with increasing DR were insignificant for pure PBT. Table 3 shows that for pure PBT, the strength and the modulus values increased. On the other hand, the values of the strength and the initial modulus of the hybrid fibers decreased with increasing DR. For the hybrid with 3 wt% organoclay, for example, when the DR was increased from 1 to 6, the ultimate strength and the initial modulus decreased. These trends with increasing DR were observed in all systems containing 2–5 wt% organoclay. This declination in the tensile properties seems to be the result of debonding between the organoclay and the PBT and the presence of many micro-sized voids due to excess stretching of the fibers [71-73].

Table 3. Effect of the DR on the tensile properties of PBT nanocomposite fibers

Organoclay loading, %	Ultimate strength MPa			Initial modulus GPa		
	DR=1	DR=3	DR=6	DR=1	DR=3	DR=6
0 (pure PBT)	41	50	52	1,37	1,49	1,52
3	60	35	29	1,76	1,46	1,39

Finally it has to be noted that the amount of studies devoted to the polymer nanocomposite materials increases very intensively. It was shown the ability of using of practically all kinds of polymer materials. Common understanding about investigation methods and structure of nanocomposite polymer materials, about relationships between properties and specific structure of nanocomposites is formed. Increasing of studies in this field proves that it will become very perspective approach to prepare in industrial scales new polymer materials with a lot of improved properties.

REFERENCES

[1] Kaladzhyan, A.A., Lednev, O.B., Mikitaev, M.A. Nanocomposite polymer materials based on organoclay // *Thesis of report on International Conference* Dedicated to 50th Anniversary of A.N. Nesmeyanov Institute of Organoelement Compounds (INEOS) Russian Academy of Sciences. Moscow, 2004, P31.

[2] Delozier, D.M., Orwoll, R.A., Cahoon, J.F., Ladislaw, J.S., Smith, J.G., Connell, J.W.. *Polymer*, 2003;44:2231-2241.

[3] Delozier, D.M., Orwoll, R.A., Cahoon, J.F., Johnston, N.J., Smith, J.G., Connell, J.W.. *Polymer*, 2002;43:813-822.

[4] Chang, J.-H., An, Y.U., Kim, S.J., *S. Im. Polymer*, 2003;44:5655-5661.

[5] Chang, J.-H., Kim, S.J., Joo, Y.L., *S. Im.. Polymer*, 2004;45:919-926.

[6] Mikitaev, A.K., Kaladzhyan, A.A., Lednev, O.B., Mikitaev, M.A. Nanocomposite polymer materials based on organoclays.// *Electronic journal "Investigated in Russia"*, 83, p. 912-922, 2004. http://zhurnal.ape.relarn.ru/articles/2004/083.pdf

[7] Kelly, P., Akelah, A., Moet, A. *J. Mater. Sci.* 1994. V.29. P.2274—2280.

[8] Chang, J.-H., An, Y.U., Cho, D., Giannelis, E.P.. *Polymer*, 2003;44:3715-3720.

[9] Yano, K, Usuki, A, Okada, A. *J Polym Sci, Part A*: *Polym. Chem.* 1997;35:2289.

[10] Garcia-Martinez, JM, Laguna, O, Areso, S, Collar, EP. *J Polym. Sci, Part B: Polym Phys.* 2000;38:1564.

[11] Pinnavaia, TJ, Beall, GW, editors. *Polymer–Clay Nanocomposites*. New York: Wiley; 2000.

[12] Shen, Y.-H.. *Chemosphere*, 2001;44:989-995

[13] Chen, Z, Huang, C, Liu, S, Zhang, Y, Gong, K. *J. Apply Polym. Sci.* 2000;75:796-801.

[14] Okado, A, Kawasumi, M, Kojima, Y, Kurauchi, T, Kamigato, O. *Mater Res Soc Symp Proc* 1990;171:45.

[15] Leszek, A. Utracki, Jorgen Lyngaae-Jorgensen. *Rheologica Acta*, 2002;41: 394-407

[16] Hsu, S.L.-C., Chang, K.-C. *Polymer*, 2002;43:4097-4101.

[17] Wagener, R., Reisinger, T.J.G. *Polymer*, 2003;44:7513-7518.

[18] Li, X, Kang, T, Cho, WJ, Lee, JK, Ha, CS. *Macromol Rapid Commun.*

[19] Tyan, H.-L., Liu, Y.-C., Wei, K.-H. *Polymer*, 1999;40:4877-4886.

[20] Vaia, R, Huang, X, Lewis, S, Brittain, W. *Macromolecules* 2000;33:2000-4.

[21] Okamoto, M, Morita, S, Taguchi, H, Kim, Y, Kotaka, T, Tateyama, H. *Polymer* 2000;41:3887-90.

[22] Chow, W.S., Mohd Ishak, Z.A., Karger-Kocsis, J., Apostolov, A.A., Ishiaku, U.S. *Polymer*, 2003;44:7427-7440.

[23] Antipov, E.M., Huseva, M.A., Herasin, V.A., Korolyov, Yu.M., Rebrov, A.V., Fisher, H.R., Razumovskaya, I.V. Structure and deformation behavior of nanocomposites based on polypropylene and modified clays. // *Highmolecular compounds*. A. 2003. V.45. №11. P. 1885-1899.

[24] Sur, G, Sun, H, Lyu, S, Mark, J. *Polymer* 2001;42:9783-9.

[25] Wang, Z, Pinnavaia, T. *Chem. Mater* 1998;10:3769-71.

[26] Chang, J.-H., Kim, S.J., Joo, Y.L., *S. Im. Polymer*, 2004;45:919-926.

[27] Antipov, E.M., Huseva, M.A., Herasin, V.A., Korolyov, Yu.M., Rebrov, A.V., Fisher, H.R., Razumovskaya, I.V. Structure and deformation behavior of nanocomposites

based on polyethylene and modified clays. // *Highmolecular compounds*. A. 2003. V.45. №11. P. 1874-1884.
[28] Lan, T, Kaviartna, P, Pinnavaia, T. *Proceedings of the ACS PMSE* 1994;71:527-8.
[29] Pinnavaia, TJ. *Science* 1983;220:365.
[30] Messersmith, PB, Giannelis, EP. *Chem. Mater.* 1993;5:1064.
[31] Giannelis, EP. *Adv. Mater.* 1996;8:29.
[32] Gilman, JW. *Appl. Clay Sci.* 1999;15:31.
[33] Greenland, DG. *J. Colloid Sci.* 1963;18:647.
[34] Chang, JH, Park, KM. *Polym. Engng Sci.* 2001;41:2226.
[35] Greenland, DG. *J. Colloid Sci.* 1963;18:647.
[36] Chang, JH, Park, KM. *Polym. Eng. Sci.* 2001;41:2226.
[37] Vaia, RA, Ishii, H, Giannelis, EP. *Adv. Mater.* 1996;8:29.
[38] Vaia, RA, Jandt, KD, Kramer, EJ, Giannelis, EP. *Macromolecules* 1995;28:8080.
[39] Fukushima, Y, Okada, A, Kawasumi, M, Kurauchi, T, Kamigaito, O. *Clay Miner* 1988;23:27.
[40] Vaia, RA, Ishii, H, Giannelis, EP. *Adv. Mater.* 1996;8:29.
[41] Vaia, RA, Jandt, KD, Kramer, EJ, Giannelis, EP. *Macromolecules* 1995;28:8080.
[42] Fukushima, Y, Okada, A, Kawasumi, M, Kurauchi, T, Kamigaito, O. *Clay Miner* 1988;23:27.
[43] Akelah, A, Moet, A. *J. Mater Sci.* 1996;31:3589.
[44] Chvalun, S.N. *Nature* 2000, №7.
[45] Brinker, C.J., Scherer, G.W. *Sol-Gel Science.* Boston, 1990.
[46] Mascia, L, Tang, T. *Polymer* 1998;39:3045.
[47] Tamaki, R, Chujo, Y. *Chem. Mater.* 1999;11:1719.
[48] Dennis, H.R., Hunter, D.L., Chang, D., Kim, S., White, J.L., Cho, J.W., Paul, D.R. *Polymer,* 2001;42:9513-9522.
[49] Giannelis, E.P., *Adv. Mater.* 8 (1996) 29–35.
[50] Kornmann, X., Lindberg, H., Berglund, L.A., *Polymer* 42 (2001) 1303–1310.
[51] Voulgaris, D., Petridis, D.. *Polymer*, 2002;43:2213-2218.
[52] Tyan, H.-L., Liu, Y.-C., Wei, K.-H.. *Polymer*, 1999;40:4877-4886.
[53] Davis, CH, Mathias, LJ, Gilman, JW, Schiraldi, DA, Shields, JR, Trulove, P, Sutto, TE, Delong, HC. *J. Polym. Sci., Part B: Polym. Phys.* 2002;40:2661.
[54] Morgan, AB, Gilman, JW. *J. Appl. Polym. Sci.* 2003;87:1329.
[55] Chang, JH, An, YU, Sur, GS. *J. Polym. Sci. Part B: Polym Phys* 2003;41:94.
[56] Chang, JH, Park, DK, Ihn, KJ. *J. App.l Polym. Sci.* 2002;84:2294.
[57] Xu, H, Kuo, SW, Lee, JS, Chang, FC. *Macromolecules* 2002;35:8788.
[58] Haddad, TS, Lichtenhan, JD. *Macromolecules* 1996;29:7302.
[59] Mather, PT, Jeon, HG, Romo-Uribe, A, Haddad, TS, Lichtenhan, JD. *Macromolecules* 1996;29:7302.
[60] Hsu, SLC, Chang, KC. *Polymer* 2002;43:4097.
[61] Chang, JH, Seo, BS, Hwang, DH. *Polymer* 2002;43:2969.
[62] Fornes, TD, Yoon, PJ, Hunter, DL, Keskkula, H, Paul, DR. *Polymer* 2002;43:5915.
[63] Chang, JH, Seo, BS, Hwang, DH. *Polymer* 2002;43:2969.
[64] Fornes, TD, Yoon, PJ, Hunter, DL, Keskkula, H, Paul, DR. *Polymer* 2002;43:5915.
[65] Mikitaev, A.K., Kaladzhyan, A.A., Lednev, O.B., Mikitaev, M.A., Davydov, E.M. Nanocomposite polymer materials based on organoclays with enhanced fire resistance.

// *Electronic journal "Investigated in Russia",* 129, p. 1365-1370, 2004. http://zhurnal.ape.relarn.ru/articles /2004/129.pdf.

[66] Wen, J, Wikes, GL. *Chem. Mater.* 1996;8:1667.

[67] Zhu, ZK, Yang, Y, Yin, J, Wang, X, Ke, Y, Qi, Z. *J. Appl. Polym. Sci.* 1999;3:2063.

[68] Lan, T, Pinnavaia, TJ. *Chem. Mater.* 1994;6:2216.

[69] Masenelli-Varlot, K, Reynaud, E, Vigier, G, Varlet, J. *J. Polym. Sci. Part B: Polym. Phys.* 2002;40:272.

[70] Yano, K, Usuki, A, Okada, A. *J. Polym. Sci. Part A: Polym. Chem.* 1997;35:2289.

[71] Shia, D, Hui, Y, Burnside, SD, Giannelis, EP. *Polym. Engng Sci.* 1987;27:887.

[72] Curtin, WA. *J. Am. Ceram. Soc.* 1991;74:2837.

[73] Chawla, KK. *Composite materials science and engineering*. New York: Springer; 1987.

In: Encyclopedia of Nanotechnology
Editor: Elwood D. Carlson

ISBN 978-1-60692-079-4

Chapter 51

FUNDAMENTAL ASPECTS OF FILLING OF NANOCOMPOSITES WITH HIGH-ELASTICITY MATRIX: FRACTAL MODELS

Georgi V. Kozlov[1], Yurii G. Yanovskii[1]* and Gennadi E. Zaikov[2]
[1]Institute of Applied Mechanics of Russian Academy of Sciences, 119991, Leninskii pr. 32 A, Moscow, Russian Federation
[2]N.M. Emanuel Institute of Biochemical Physics of Russian Academy of Sciences, 119991 Kosygin St. 4, Moscow, Russian Federation

ABSTRACT

In the present paper are considered three fundamental factors controlling filling processes of nanocomposites with high-elasticity matrix. They are: filler particles aggregation, strain localization and change of structure type from Euclidean object up to fractal. The last postulate is the most physically fundamental, since it demonstrates noncorrectness of description of structure and properties of nanocomposites with high-elasticity matrix within the framework of classical theory of entropic high-elasticity.

Keywords: Nanocomposite; high-elasticity matrix; filler; aggregation; strain localization; fractal structure.

INTRODUCTION

The practical importance and complexity of structure of polymer composites with matrix, which is in high-elasticity state, are due to a large number of works, devoted to studies of structure and properties of the mentioned composites. Lipatov [1] summarized obtained

*Correspondence to: Yurii G. Yanovskii, Institute of Applied Mechanics of Russian Academy of Sciences, Moscow, Russian Federation, Leninskii pr. 32 A, 119991. mailto: i_dolbin@mail.ru

results and came to the conclusion that the main factors of reinforcement of the considered composites are:

(1) the features of polymer chemical constitution and strong intermolecular interaction of it with filler particles surface;
(2) the formation of second structures by filler particles;
(3) reorganization at deformation of filled material molecular structure without its failure;
(4) the failure of network structure – rupture of filler – polymer bonds or molecular network transverse bonds.

Among other factors, influenced on the reinforcement, Lipatov marks form and size of filler particles, character of their distribution and aggregation and also a number of common physical-chemical causes, typical for all polymer composites [1].

DISCUSSION

Not rejecting correctness of these principles, becoming already classical, we nevertheless distinguished three fundamental factors, the role of which happens obvious owing to the fractal analysis application for the description of polymer composites structure and properties [2-4].

The first of these factors is filler particles aggregation. The detailed description of this effect influence on structure and properties of composites, having glassy matrix, is given in review [3]. For the considered composites this effect is important first of all for determination of filler volume fraction φ_f, which is calculated by division of filler mass fraction into its density. However, filler particles in real composites exist in the form of fractal aggregates, which density can be more smaller than monolithic object of such filler. Below the example of value φ_f calculation for concrete composite will be given.

The second aspect of filler particles aggregation is the possibility of taking into consideration occlusive rubber in aggregates of these particles. As it is known [5], the relationship between fractal aggregate radius R_{ag}, and particles number N in it will look like:

$$R_{ag} \sim N^{1/d_f} \quad (1)$$

where d_f is fractal dimension of filler particles aggregate.

The rise of occlusive rubber fraction in aggregate at condition R_{ag}=const means decrease N and reduction d_f. This gives the possibility of quantitative calculation of occlusive rubber fraction in aggregate.

And at last, the third important aspect of aggregation process is its mechanism. As it is known [6], within the framework of irreversible aggregation models are possible two aggregation mechanisms: cluster-cluster and particle-cluster, the physical sense of which follows from its name. Within the framework of the first mechanism the value R_{ag} is determined by the relationship [3]:

$$R_{ag} \sim \left(\frac{4c_0 kT}{3\eta m_0} \right)^{1/d_f} t^{1/d_f} \tag{2}$$

where c_0 is initial concentration of filler particles, k is Boltzmann constant, T is temperature, η is medium viscosity, m_0 is initial particle mass, t is aggregation duration.

For particle-cluster mechanism the value R_{ag} is determined like this [6]:

$$R_{ag} \sim c_0^{-1/(d-d_f)} \tag{3}$$

where d is dimension of Euclidean space, in which a fractal (obviously, in our case d=3) is considered.

From the relationship (2) follows, that at cluster-cluster aggregation (merging of small aggregates in largerones) the value R_{ag} can be controlled by variation of parameters T, η, m_0, d_f and t. But the most interesting is c_0 role, i.e., filler content, in regulation of value R_{ag}: for cluster-cluster mechanism the c_0 rise results to increase R_{ag} and for particle-cluster mechanism – to its reduction. In this aspect the most interesting is the relationship (3), since from it follows, that at definite conditions the aggregation process can be suppressed by c_0. Therefore, the ability of aggregation mechanism change is the perspective instrument of control of structure of filler particles aggregates. Besides, the value d_f ($\approx$ 2.50) for particle-cluster aggregation mechanism is essentially more than corresponding dimension ($d_f \approx$1.80-2.10 [6]) for cluster-cluster mechanism, i.e., the first of the mentioned mechanisms allows to obtain more compact aggregates, practically not containing occlusive rubber.

The second important factor is the calculation of draw ratio λ for composites. This problem is due to heterogeneity of composite structure, consisting of two phases elasticity modulus of one of them is essentially (in some orders of value) more, than the second. Differently speaking, during composite tension process only one phase is extended, namely, rubber, owing to nominal value λ not already does describe adequately composite deformation process. This effect is well known and for its quantitative accounting for there are two models. The first of them uses the equation [7]:

$$\lambda_{mol} = \frac{\lambda}{1-\varphi_f} \tag{4}$$

where λ_{mol} is molecular draw ratio.

The equation (4) recommended itself well at the description of extrudates of componors based on the ultrahighmolecular polyethylene [2]. The second model uses more precise equation for calculation of local draw ratio of two-phase polymeric material [8]:

$$\lambda = \lambda_p \left(1-\varphi_f\right) + \lambda_f \varphi_f \tag{5}$$

where λ_p and λ_f are draw ratio for polymeric matrix and filler, accordingly.

Returning to the question of filler particles aggregation, it should be noted, that for compact aggregates of high-modulus filler λ_f=1 and then from comparison of the equations

(4) and (5) follows, that for large enough λ $\lambda_{mol}\approx\lambda_p$. However, this situation can be essentially changed for loose (having rather small dimension d_f) aggregates, containing of occlusiv rubber significant fraction. For them $\lambda_f>1$ can be expected, that reduces the value λ_p at λ=const, i.e., weakens reinforcing the action of the filler.

And at last, the third and the most fundamental factor is the change of nanocomposite structure at the introduction of particulate filler in high-elasticity polymeric matrix. As Balankin showed [9], classical theory of entropic high-elasticity has a number of principal deficiencies due to non-fulfilment for real rubbers of two main postulates of this theory, namely, essentially non-Gaussian statistics of real polymeric networks and lack of co-ordination of postulates about Gaussian statistics and incompressibility of elastic materials. Last postulate means, that Poisson's ratio ν of these materials must be equal to 0.5. As it is known [10], Gaussian statistics of macromolecular coil is correct only in case of its dimension D_f=2.0, i.e., for coil in θ-solvent. Since between value D_f and fractal dimension d_f^p of polymer's condensed state structure exists the relationship [11]:

$$d_f^p = 1.5 D_f \tag{6}$$

then for execution of chains Gaussian statistics condition in volume polymer the criterion d_f^p=3 realization is required, i.e., the polymer structure must be Euclidean object. If the last condition with definite provisos is carried out for nonfilled rubbers, then the introduction in rubber filler with significantly smaller Poisson's ratio ν_f will decrease the value ν of composite as system. Since the value of fractal dimension of composite structure d_f^c is determined like that [12]:

$$d_f^c = (d-1)(1+\nu) \tag{7}$$

then this means, that the classical entropic high-elasticity theory is applied only to Euclidean objects (ν=0.5, d_f^c=3.0) and, accordingly, inapplicable for description of rubbers, filled by particulate filler, of which $\nu<0.5$ and $d_f<d=3$, i.e., which are fractal objects. Therefore, for description of such objects behaviour the fractal theory of elasticity and entropic high-elasticity should be used [9].

In order to demonstrate the correctness of expounded above general postulates, let's make quantitative estimations of the dependence of stress σ on φ_f and T for rubber SKI-3 filled by technical carbon P-234 according to the data of paper [13]. For this let's make approximated, by plausible assumption, that the technical carbon particles with diameter d_{part}=50 mnm [13] form the aggregates from five particles. Then the radius of such aggregate can be estimated according to the equation [14]:

$$R_{ag} = \left(\frac{n_{part} S}{\pi \eta_p} \right)^{1/2} \tag{8}$$

where n_{part} is particles number in aggregate, S is cross-sectional area of particle, η_p is packing density, which is equal to 0.74 for monodisperse circles [15].

At n_{part}=5 and d_{part}=50 mnm R_{ag}≈65 mnm. The density of such aggregate ρ_{ag} is estimated as follows [16]:

$$\rho_{ag} = \rho_{dens} \left(\frac{R_{ag}}{a} \right)^{d_f - d} \tag{9}$$

where ρ_{dens} is aggregate substance density in dense packing state (ρ_{dens}=2700 kg/m^3 for carbon [3]), a is lower boundary of aggregate fractal behaviour accepted equal to 15 mnm [17]. For d_f=2.5, d=3 ρ_{ag}≈1300 kg/m^3 will be obtained, that is more than two times smaller than density of filler mass content into value ρ_{ag} and calculate molecular (real) draw ratio $\lambda_{mol}=\lambda_p$ of polymeric matrix according to the equation (5). Estimating shear modulus G according to the known equations of high-elasticity classical theory [18], let's calculate Poisson's ratio according to the equation [19]:

$$\nu = \nu_r \left(1 - \varphi_f\right) + \varphi_f \nu_f \tag{10}$$

where ν_r and ν_f are values of Poisson's ratio for rubber and filler, accepted equal to 0.50 and 0.25, accordingly.

Further the main equation of high-elasticity fractal theory for σ calculation in its simplified form will be used [9]:

$$\sigma = \frac{E}{1 + 2\nu + 4\nu^3} \left[\lambda_{mol}^{1+2\nu} - 2\nu \lambda_{mol}^{-1-2\nu(1+\nu)} - (1 - 2\nu) \lambda_{mol}^{-2\nu} \right] \tag{11}$$

where elasticity modulus is defined this way [20]:

$$E = 2(1 + \nu) G \tag{12}$$

As follows from the data of Figure 1, where the comparison of experimental σ^e is made and calculated according to the considered above method σ^T stress of SKI-3 and based on it composites with technical carbon content 20, 30, 40 and 60 mass. % at two testing temperatures (250 and 380 K), good enough correspondence of theory and experiment is obtained, despite arbitrary choice of important parameters number. Two interesting moments should be noted. Firstly, for nonfilled rubber SKI-3 the calculation is made at ν=0.5. Secondly, from the plot of Fig 1 follows, that at small φ_f the values σ^T are somewhat more σ^e and at large φ_f the relation is opposite. This observation can assume an aggregation amplification at filler's content increase, i.e., the R_{ag} rise or aggregates compactness reduction, i.e., d_f decrease.

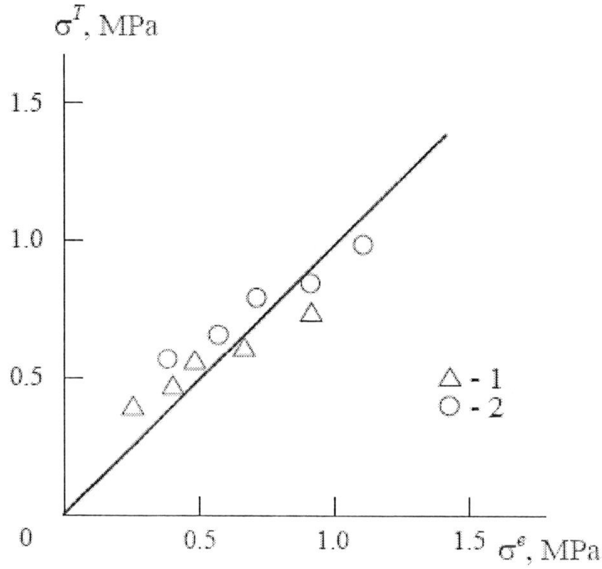

Figure 1. The comparison of experimental σ^e and calculated according to the equation (11) σ^T stress values at temperatures 250 (1) and 380 K (2) for rubber SKI-3, filled by technical carbon.

As it is known [18], within the framework of high-elasticity classical theory the value σ is described with the aid of the two following equations:

$$\sigma = G\left(\lambda^2 - \lambda^{-1}\right) \tag{13}$$

$$\sigma = A\left(\lambda - \lambda^{-1/2}\right) \tag{14}$$

where A is material constant.

In Figure 2 and 3 the dependences σ on generalized stress for studied rubbers, corresponding to the equations (13) and (14), are shown. As can be seen, in case of composites the linearity of these dependences is violated, i.e., at least, the filled rubbers behaviour does not corresponded to high-elasticity classical theory, that is assumed above. Differently speaking, filled rubbers are impossible to consider as ideal, for which internal energy change ΔU is equal to zero in deformation process.

Within the framework of high-elasticity theory the value G in the equation (13) is determined as follows [18]:

$$G = NkT \tag{15}$$

where N is active chains number on rubber volume unit.

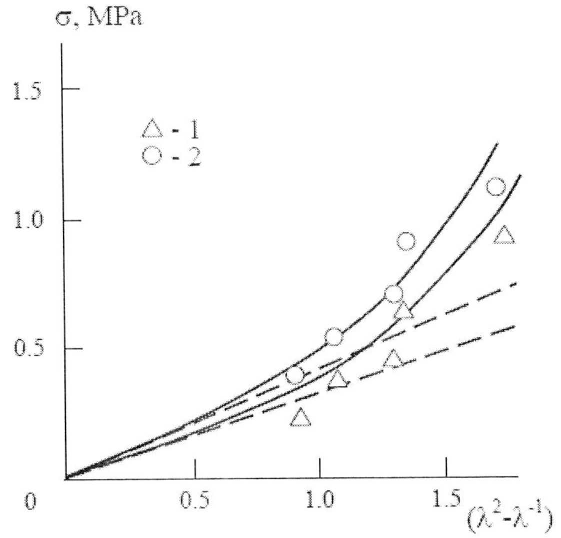

Figure 2. The dependence of stress σ on generalized strain $(\lambda^2-\lambda^{-1})$ at temperatures 250 (1) and 380 K (2) for rubber SKI-3, filled by technical carbon. The shaded lines are shown the dependencies assumed by high-elasticity classical theory.

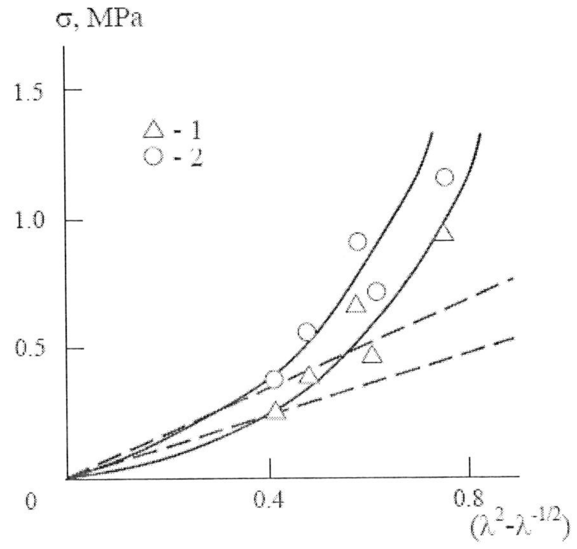

Figure 3. The dependence of stress σ on generalized strain (λ-λ-1/2) at temperatures 250 (1) and 380 K (2) for rubber SKI-3, filled by technical carbon. The shaded lines are shown the dependencies assumed by high-elasticity classical theory.

From combination of the equations (13) and (15) follows, that at T=0 K G=0 and σ=0. However, as it is shown in paper [13], it is true only for nonfilled SKI-3, but for filled samples the linear dependence $\sigma(T)$ at T=0 K is extrapolated to nonzero value $\sigma(\sigma_0)$, which is raised at φ_f increase. Besides, from the data of Figure 2 the deviation value of the dependence $\sigma(\lambda^2-\lambda^{-1})$ Δσ from its linear extrapolation (it is shown in Figure 2 by shaded line) can be estimated, which is supposed according to the high-elasticity classical theory (see the equation (13)). In a Figure 4 and 5 the dependences σ_0 and $\Delta\sigma$ on φ_f and $(d-d_f)$, accordingly are given (for convenience they were linearized by using quadratic form of dependence). As can be seen, in both cases linear correlation passing thourgh coordinates origin for $\varphi_f=0$ and $d_f=d$ is obtained, i.e., for nonfilled rubber, which is Euclidean object.

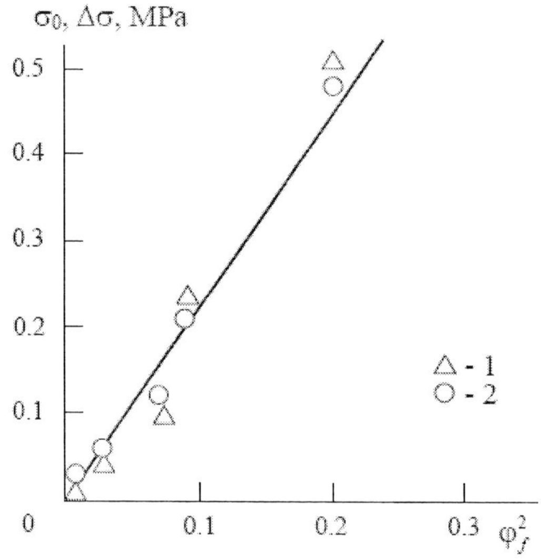

Figure 4. The dependence of deviations from high-elasticity classical theory σ_0 (1) and $\Delta\sigma$ (2) (the explanations give in the text) on filler volumetric degree φ_f for rubber SKI-3, filled by technical carbon.

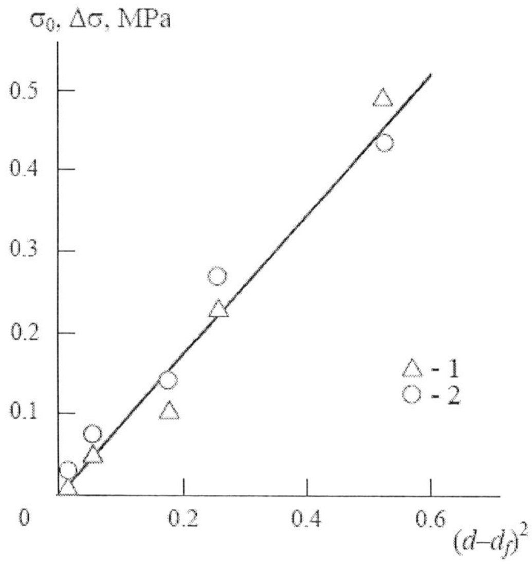

Figure 5. The dependence of deviations from high-elasticity classical theory σ_0 (1) and $\Delta\sigma$ (2) (the explanations give in the text) on Euclidean and fractal dimensions difference $(d–d_f)$ for rubber SKI-3, filled by technical carbon.

CONCLUSION

Therefore, in the present paper three fundamental factors controlling filling processes of nanocomposites with high-elasticity matrix are considered. They are: filler particles aggregation, strain localization and change of structure type from Euclidean object up to fractal. The last postulate is the most physically fundamental, since it demonstrates noncorrectness of description of structure and properties of nanocomposites with high-elasticity matrix within the framework of classical theory of entropic high-elasticity.

REFERENCES

[1] Lipatov, Yu.S. *in Kompozitsionnye Polimernye Materialy.* Kiev, Naukova Dumka, p. 75 (1975).

[2] Aloev, V.Z., Kozlov, G.V., *Physics of Orientational Phenomenas in Polymeric Materials*. Nal'chik, Polygraphservis and T, 2002, 288 p.

[3] Kozlov, G.V., Yanovskii, Yu.G., Lipatov, Yu.S., *Mekhanika Kompozitsionnych Materialov i Konstruktsii*, 9: 398 (2003).

[4] Kozlov, G.V., Lipatov, Yu.S., *Poverchnost'*, 8: 81 (2003).

[5] Feder, F., *Fractaly.* Moscow, Mir, 1991, 256 p.

[6] Shogenov, V.N., Kozlov, G.V., *Fraktal'nye Klastery v Fiziko-Khimii Polimerov.* Nal'chik, Polygraphservis and T, 2002, 268 p.

[7] Watts, M.P.C., Zachariades, A.E., Porter, P.S., *J. Mater. Sci.*, 15: 426 (1980).

[8] Castellani, L., Maestrini, C., *Polymer*, 31: 2278 (1990).

[9] Balankin, A.S., *Pis'ma v ZhTF*, 17: 68 (1991).

[10] Kozlov, G.V., Zaikov, G.E., *Structure of the Polymer Amorphous State*. Utrecht-Boston, Brill Academic Publisher, 2004, 465 p.

[11] Kozlov, G.V., Temiraev, K.B., Shustov, G.B., Mashukov, N.I., *J. Appl. Polymer Sci.*, 85: 1137 (2002).

[12] Balankin, A.S., *Synergetics of Deformable Body.* Moscow, Publisher of Ministry for Defence SSSR, 1991, 404 p.

[13] Yanovskii, Yu.G., Zgaevskii, V.E., *Fizicheskaya Mezomekhanika*, 4: 63 (2001).

[14] Kozlov, G.V., Shogenov, V.N., Mikitaev, A.K., *Inzhenerno-Fizicheskii Zhurnal*, 71: 1012 (1998).

[15] Bobryshev, A.N., Kozomasov, V.N., Babin, L.O., Solomatov, V.I., *Sinergetika Kompozitsionnych Materialov*. Lipetsk, NPO ORIUS, 1994, 154 p.

[16] Brady, R.M., Ball, R.C., *Nature*, 309: 225 (1984).

[17] Avnir, D., Farin, D., Pfeifer, P., *Nature*, 308: 261 (1984).

[18] Bartenev, G.M., Frenkel, S.Ya., *Fizika Polimerov*. Leningrad, Khimiya, 1990, 432 p.

[19] Shustov, G.B., Afaunova, Z.I., Kozlov, G.V., *Vestnik KBGU. Khimicheskie nauki*, 3: 45 (1990).

[20] Kozlov, G.V., Sanditov, D.S., *Anharmonic Effects and Physical mechanical Properties of polymers.* Novosibirsk, Nauka, 1994, 261 p.

In: Encyclopedia of Nanotechnology
Editor: Elwood D. Carlson

ISBN 978-1-60692-079-4

Chapter 52

STRUCTURE FORMATION OF POLYMER NANOCOMPOSITES BASED ON POLYPROPYLENE

Ahmed Kh. Malamatov[1]*, Georgi V. Kozlov[1] and Eugeni M. Antipov[2]

[1]Kabardino-Balkarian State University, 360004 Chernyshevski st. 173, Nal'chik, Russia
[2]Topchiev Institute of Petrochemical Synthesis of Russian Academy of Sciences, 119991 Leninskii pr. 29, Moscow, Russia

ABSTRACT

It is shown, that the structure formation of nanocomposites based on the polypropylene takes in Euclidean space with dimension d=3. Within the framework of the offered for these material structural model is shown identity of local order regions, interfacial areas and crystalline β-phase. This model allows quantitative description of nanocomposites properties.

Keywords: Nanocomposite; polypropylene; structure; Euclidean space; local order; crystalline β-phase.

INTRODUCTION

The introduction of disperce filler in polymeric matrix changes significantly properties of the last [1]. This fact is well known, but there are definite difficulties with its quantitative estimation. For particulate-filled composites polyhydroxiether-graphite (PHE-Gr) was found the structure fractal dimension d_f increase from 2.6 to 2.8 within the volume content of filler interval φ_f=0-0.09. This structure change is accompanied by corresponding increase of

*Correspondence to: Ahmed Kh. Malamatov, Kabardino-Balkarian State University, Nal'chik, Russia, Chernyshevski 173, 360004, mailto: deanchem@ns.kbsu.ru

composites mechanical characteristics (elasticity modulus, yielding stress, fracture stress and so on) [2]. The mentioned change of composites PHE-Gr structure is due to its formation in fractal space, which creates a filler network of particles (particle aggregates) [3,4]. As studies of mechanical properties of nanocomposites based on the polypropylene (PP), filled by modified natural clay (NC) Na^+-montmorillonite in interval φ_f=0.025-0.10, show that for them elasticity modulus raising is not observed and yielding stress and strength are even a little decreased [5]. Proceeded from these differences of mechanical properties behaviour of composites PHE-Gr and nanocomposites based on the PP, presents interest of the study of structural changes, due to introduction of layered filler and determines mechanical behaviour of nanocomposites, that is the purpose of present paper.

EXPERIMENTAL

The data of paper [5] for nanocomposites based on the isotactic industrial PP (Shell Co.), filled by Na^+-montmorillonite with last contents φ_f=0.025, 0.050 and 0.10 were used. As modificators: dioctadecyldimethyl ammonium bromide (DODAB) plus block-copolymer polyethyleneoxide-polyethylene (PEO-PE) (conventional sign of nanocomposite PP-NC-1); PEO-PE (PP-NC-2); DODAB plus PEO-PE with isobutylene (PP-NC-3); PEO-PE plus isobutylene (PP-NC-4) are used. The detailed description of specimens preparation methodics is cited in the paper [5].

The mechanical tests were made on film specimens with length 10 mm, width 3 mm and thickness 0.2 mm, preparated by pressing from melt at 473 K. The tests were made at temperature 293 K and strain rate 8.3×10^{-3} s^{-1} [5].

The experimental values d_f(d_f^e) were determined according to the equation [6]

$$d_f^e = (d-1)(1+\nu) \tag{1}$$

where d is dimension of Euclidean space, in which a fractal is considered (obviously, in our case d=3), ν is Poisson's ratio, the value of which can be estimated on mechanical testing results with the aid of the relationship [7]

$$\frac{\sigma_Y}{E} = \frac{1-2\nu}{6(1+\nu)} \tag{2}$$

where σ_Y is yielding stress, E is elasticity modulus.

RESULTS AND DISCUSSION

As it is mentioned above, the change d_f at increase φ_f for composites PHE-Gr is due to formation of polymeric matrix structure in fractal space with dimension D_{lat} [3,4]. The value can be determined with the aid of the following equation [8]

$$\nu_F = \frac{3}{2 + D_{lat}} \tag{3}$$

where ν_F is Flory exponent connected with dimension d_f in the linear polymers case by the following relationship [9]

$$\nu_F = \frac{1.5}{d_f} \tag{4}$$

The calculation of value d_f^e according to the equation (1) shows this dimension variation within the limits 2.678-2.739 with average value 2.709. The estimation D_{lat} according to the equation (3) give in this case the value ~ 3.42, that is does not correspond to condition $D_{lat} \leq 3$, which is common for all objects in three-dimensional Euclidean space [6]. Therefore, it is necessary to assume $D_{lat}=3$ and value d_f in this case is equal to 2.5 according to the equation (3). This value d_f is excellently coordinated with calculation according to the equation (1) for initial PP (d_f^e=2.518). Thus, the cited estimations allow to assume, that the formation of structure of initial PP and polymeric matrix of nanocomposites on its basis is realized in the three-dimensional Euclidean space, and increase d_f^e for nanocomposites in comparison with PP is due to introduction of the filler. It is necessary to mark, that at condition $D_{lat}<3$ (structure formation in fractal space) the decrease d_f in comparison with $D_{lat}=3$ for nanocomposites in comparison with initial PP must take place, that contradicts to experimental data.

Let's consider physical base of the mentioned increase d_f^e. As it is shown in paper [5], the introduction of layered filler (Na^+-montmorillonite) in PP results to the formation of crystalline β-phase, the crystallites of which have primary orientation inseparably connected with flat texture of siliceous layers, on which they are crystallized. This allows to consider mentioned crystallites both as local order regions, and as interfacial areas at the same time. Let's perform quantitative estimation of these regions contents at both of their definitions. The relative fraction of local order regions (clusters) φ_{cl} for semi-crystalline polymers can be calculated according to the following percolation relationship [9]

$$\varphi_{cl} = 0.03(1-K)(T_m - T)^{0.55} \tag{5}$$

where K is orthorhombic crystallinity degree, T_m and T are temperature of melting and testing accordingly.

For PP $T_m \approx 440$ K [10] and value K can be determined according to the equation [11]

$$K = 0.32 C_\infty^{1/3} \tag{6}$$

where C_∞ is characteristic ratio, which is an indicator of polymer chain statistical flexibility [12] and connected with dimension d_f by the relationship [9]

$$C_\infty = \frac{2d_f}{d(d-1)(d-d_f)} + \frac{4}{3} \quad (7)$$

The relative fraction of interfacial areas φ_{inf} was calculated with the aid of the equation [13]

$$\frac{E_c}{E_m} = 1 + 11(\varphi_f + \varphi_{inf}) \quad (8)$$

where E_c and E_m are elasticity modulus of composite and matrix polymer, accordingly.

The connection between dimension d_f and crystallinity degree is defined by the following empirical approximation [11]

$$d_f = 2 + K \quad (9)$$

It is obvious, that for the studied nanocomposites with appreciation of crystalline β-phase formation the following version of the equation (9) should be written

$$d_f = 2 + K + \varphi_{inf} \quad (10)$$

According to the equation (10) the formation of nanocomposites structure in fractal space, defining decrease d_f, must decrease value K, that also does not correspond to the experiment.

In Figure 1 the comparison of dimensions d_f^e and d_f^T calculated according to the equations (1) and (10), accordingly, is shown. As can be seen, between these values d_f the good correspondence is received (average discrepancy by fractional part of dimensions, which performs main information about structure, is equal to 4.6 %). This allows to assume, that the structure of nanocomposites based on the PP consists of the following structural regions: orthorhombic crystallites, amorphous phase, filler and crystalline β-phase, forming interfacial areas. Let's note, that the calculation φ_{inf} according to the equation (8) and nanocomposites crystallinity degree rise in comparison with initial PP [5] give close values: ~ 0.15-0.20.

Another method of calculation of theoretical value d_f(d_f^T) gives the following equation [9]

$$d_f^T = 3 - 6\left(\frac{\varphi_{cl}}{SC_\infty}\right)^{1/2} \quad (11)$$

where S is cross-section area of macromolecule, which is equal to 34.27 Å^2 for PP [14].

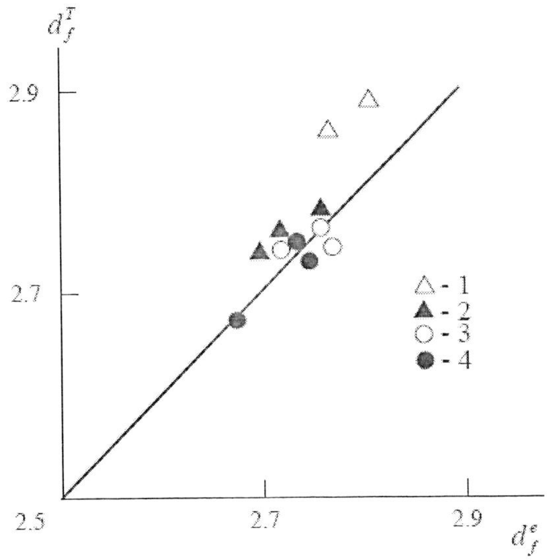

Figure 1. The comparison of experimental d_f^e and calculated according to the equation (10) d_f^T values of structure fractal dimension for nanocomposites PP-NC-1 (1), PP-NC-2 (2), PP-NC-3 (3) and PP-NC-4 (4).

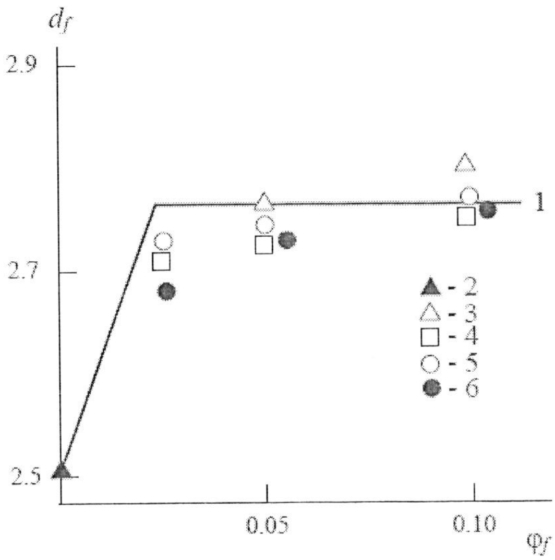

Figure 2. The comparison of calculated according to the equations (4) and (11) (1) and experimental (2-6) the dependences of fractal dimension d_f on filler contents φ_f for PP (2), PP-NC-1 (3), PP-NC-2 (4), PP-NC-3 (5) and PP-NC-4 (6).

In Figure 2 the dependence d_f^T on φ_f calculated according to the equation (4) for PP and according to the equation (11) for nanocomposites based on the PP at φ_{cl} estimated according to the equation (5) is shown. The comparison with experimental data for four groups of nanocomposites and initial PP shows correctness of this estimation.

CONCLUSION

Therefore, the results of the present paper showed, that the structure formation of nanocomposites based on the polypropylene takes place in Euclidean space with dimension d=3. Within the framework of offered for these materials structural model is shown identity of local order regions, interfacial areas and crystalline β-phase. This model allows quantitative description of nanocomposites properties.

REFERENCES

[1] Lipatov, Yu.S., *Physical-mechanical Principles of Polymer Filling*. Moscow, Khimiya, 1991, 259 p.

[2] Novikov, V.U., Kozlov, G.V., *Mekhanika Kompozitn. Mater.*, 35: 269 (1999).

[3] Kozlov, G.V., Yanovskii, Yu.G., Lipatov, Yu.S., *Mekhanika Kompozitsion. Mater. and Konstruk.*, 8: 467 (2002).

[4] Kozlov, G.V., Lipatov, Yu.S., *Mekhanika Kompozitn. Mater.*, 40: 827 (2004).

[5] Antipov, E.M., Barannikov, A.A., Gerasin, V.A., Shklyaruk, B.F., Tsamalashvili, L.A., Fisher, H.R., Razumovskaya, I.V., *Vysokomol. Soed* A, 45: 1885 (2003).

[6] Balankin, A.S., *Synergetics of Deformable Body*. Moscow, Publisher of Ministry for Defence SSSR, 1991, 404 p.

[7] Kozlov, G.V., Sanditov, D.S., *Anharmonic Effects and Physical-mechanical Properties of polymers*. Novosibirsk, Nauka, 1994, 261 p.

[8] Vannimenus, J., *Physica D*, 38: 351 (1989).

[9] Kozlov, G.V., Zaikov, G.E., *Structure of the Polymer Amorphous State*. Utrecht-Boston, Brill Academic Publisher, 2004, 465 p.

[10] Kalinchev, E.L., Sakovtseva, M.B., *Properties and Processing of Thermoplasts*. Leningrad, Khimiya, 1983, 288 p.

[11] Aloev, V.Z., Kozlov, G.V., *Physics of Orientational Phenomenas in Polymeric Materials*. Nal'chik, Polygraphservis and T, 2002, 288 p.

[12] Budtov, V.P., *Physical Chemistry of Polymer Solutions*. Sankt-Peterburg, Khimiya, 1992, 384 p.

[13] Kozlov, G.V., Burya, A.I., Aloev, V.Z., Grineva, L.G., in *Mechaniks and Management Processes,* Volume 1, p. 97, Ekaterinburg (2004).

[14] Aharoni, S.M., *Macromolecules*, 18: 2624 (1985).

In: Encyclopedia of Nanotechnology
Editor: Elwood D. Carlson

ISBN 978-1-60692-079-4

Chapter 53

THE INTERRELATION OF ELASTICITY MODULUS AND AMORPHOUS CHAIN'S TIGHTNESS FOR NANOCOMPOSITES BASED ON THE POLYPROPYLENE

Georgi V. Kozlov[1], Ahmed Kh. Malamatov[1]*, Eugeni M. Antipov[2] and Abdulah K. Mikitaev[3]

[1]Kabardino-Balkarian State University, Nal'chik, Russia
[2]Topchiev Institute of Petrochemical Synthesis of Russian Academy of Sciences, 119991 Leninskii pr. 29, Moscow, Russia
[3]State Scientific Institution "Compositecenter", 125047 Miusskii Square 9, Moscow, Russia

ABSTRACT

The interrelation of elasticity modulus and amorphous chain's tightness characterized by fractal dimension of chain part between its fixation points for nanocomposites based on the polypropylene is shown. This assumes the polymeric matrix structure change in comparison with initial polymer: the role of densely-packed regions for it is played by interphase areas. An offered fractal model allows estimation of elasticity modulus limiting values.

Keywords: Nanocomposite; polypropylene; elasticity modulus; amorphous chain tightness; fractal analysis.

*Correspondence to: Ahmed Kh. Malamatov, Kabardino-Balkarian State University, Nal'chik, Russia, Chernyshevski 173, 360004, mailto: deanchem@ns.kbsu.ru

INTRODUCTION

At present time it is assumed [1], that the elasticity modulus E of semi-crystalline polymers is controled by structure (topology) of noncrystalline areas. The authors [2] were offered the model, according which value E increases at amplification of chains tightness in amorphous areas (amorphous chains) of the mentioned polymers. This model was applied succesfully for description of elasticity modulus behaviour of compositions based on the high density polyethylene modified by highly dispersed mixture Fe/FeO [1]. However, the degree of amorphous chains tightness can also be experessed within the framework of fractal analysis using for this dimension D of chain part between its fixation point (physical entanglements, chemical cross-linking nodes and so on) [3]. The value D is varied in limits $1<D\leq 2$ and at $D=1$ the chain is fully streched between its fixation points (mobility of this chain part is freezed). At $D=2$ the chain has maximally possible mobility which is typical for rubber state of polymers [3]. This allows enough simple and adequate description of amorphous chains tightness degree with the aid of dimension D. The purpose of present paper is to give the development of fractal analoque of model [2] and with its aid to receive the description of behaviour of elasticity modulus of nanocomposites based on the polypropylene.

EXPERIMENTAL

The data of paper [4] for nanocomposites based on industrial isotactic polypropylene (PP) of manufacture Shell Co. is used. As a filler natural clay Na^+-montmorillonite is used modified by four modificators, with volume contents φ_f=0.025, 0.050 and 0.10. The more detailed description of nanocomposites in paper [4] is cited. Specimens for mechanical testing in the form of dog-bone with base lenght 10 mm and thichness 0.2 mm were produced by pressing from the melt at 473 K. The tension testing was made at temperatures 293 and 373 K and strain rate 8.3×10^{-3} s^{-1} [4].

RESULTS AND DISCUSSION

As it is known [4], Na^+-montmorillonite is layered silicate the plates of which have smaller thichness than their width and length. Therefore it is assumed, that the structure of nanocomposite having that filler besides actually Na^+-montmorillonite has crystalline regions, amorphous (rubber) phase and interfacial areas. In this case reinforcement degree E_c/E_m can be written in the following form [5]

$$\frac{E_c}{E_m} = \left[1 + 11\left(\varphi_f + \varphi_{\text{int}}\right)^{1.7}\right] \tag{1}$$

where E_c and E_m are elasticity modulus of nanocomposite and matrix polymer accordingly, φ_{int} – relative fraction of interfacial areas. The equation (1) allows to determine the φ_{int} value on the base of experimental values E_c, E_m and φ_f.

The calculation of dimension D was made according to following method. First ofall fractal (Hausdorff) dimension of nanocomposite structure d_f was determined according to the equation [3]

$$d_f = (d-1)(1+\nu) \tag{2}$$

where d is dimension of Euclidean space, in which the fractal is considered (obviously, in our case d=3), ν is Poisson's ratio estimated on the mechanical testing results with the aid of relationship [1]

$$\frac{\sigma_Y}{E} = \frac{1-2\nu}{6(1+\nu)} \tag{3}$$

where σ_Y is yielding stress.

The value of characteristic ratio C_∞, which is the indicator of polymeric chain statistical flexibility, is determined according to the following equation [3]

$$C_\infty = \frac{2d_f}{d(d-1)(d-d_f)} + \frac{4}{3} \tag{4}$$

Orthorhombic crystallinity degree K of polymer is connected with parameter C_∞ by the equation [6]

$$K = 0.32C_\infty^{1/3} \tag{5}$$

Now the value of fractal dimension D can be calculated with the aid of the following equation [7]

$$\frac{2(1-K)}{\varphi_{\text{int}}} = C_\infty^D \tag{6}$$

In a Fig. 1 the dependences of elasticity modulus E_c on dimension D at two mentioned above testing temperatures T for nanocomposites based on the PP are shown. Just as it was expected, linear increase of E_c at decrease D or amplification of amorphous chains tightness is observed. Two various straight lines for two testing temperatures are obtained, according which the maximally reached values $E_c^{\max}$ can be determined: at T=293 K $E_c^{\max}$=1.16 GPa, at T=373 K $E_c^{\max}$=0.53 GPa. It should be noted, that one of the advantages of fractal analysis is availability of precisely determined dimension limits (for example, $1<D\leq2$ [7]), that allows estimation of limiting values of any property. The data of Figure 1 allow to receive analytical relationship between E_c and D in the following form

$$E_c = E_c^{\max} - E_c^{\max}(D-1) \text{ or} \tag{7}$$

$$E_c = E_c^{\max}(2-D) \tag{8}$$

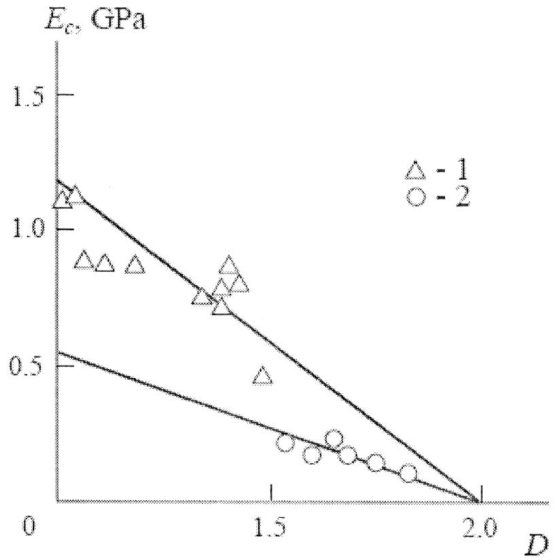

Figure 1. The elasticity modulus E_c vs. fractal dimension of chain part between interfacial areas D at testing temperatures 293 (1) and 373 K (2) for nanocomposites based on the polypropylene.

Conclusion

Therefore, the results of the present paper showed interrelation of elasticity modulus and amorphous chains tightness characterized by fractal dimension of chain part between its fixation points for nanocomposites based on the semi-crystalline polypropylene. This assumes the polymeric matrix structure change in comparison with initial polymer: the role of densely-packed regions for it is played by interfacial areas. An offered fractal model allows estimation of elasticity modulus limiting values.

References

[1] Kozlov, G.V., Sanditov, D.S., *Anharmonic Effects and Physical-mechanical Properties of polymers.* Novosibirsk, Nauka, 1994, 261 p.

[2] Krigbaum, W.R., Roe, R.-G., Smith, K.J., *Polymer*, 5: 533 (1964).

[3] Novikov, V.U., Kozlov, G.V., *Polymer's Fracture Analysis within the Framework of Fractals Conception.* Moscow, MSOU Publisher, 2001, 136 p.

[4] Antipov, E.M., Barannikov, A.A., Gerasin, V.A., Shklyaruk, B.F., Tsamalashvili, L.A., Fisher, H.R., Razumovskaya, I.V., *Vysokomolek. Soed.* A, 45: 1885 (2003).

[5] Kozlov, G.V., Burya, A.I., Aloev, V.Z., Grineva, L.G., in *Mechaniks and Management Processes*, Ekaterinburg, p. 97 (2004).

[6] Aloev, V.Z., Kozlov, G.V., *Physics of Orientational Phenomenas in Polymeric Materials*. Nal'chik, Polygraphservis and T, 2002, 288 p.

[7] Kozlov, G.V., Zaikov, G.E., *Structure of the Polymer Amorphous State*. Utrecht-Boston, Brill Academic Publisher, 2004, 465 p.

In: Encyclopedia of Nanotechnology
Editor: Elwood D. Carlson

ISBN 978-1-60692-079-4

Chapter 54

THERMAL DEGRADATION AND COMBUSTION OF POLYPROPYLENE NANOCOMPOSITE

S. M. Lomakin[1]*, I. L. Dubnikova[2], S. M. Berezina[2], G. E. Zaikov[1], R. Kozlowski[3], Gyeong-Man Kim[4] and G. H. Michler[4]

[1]N.M. Emanuel Institute of Biochemical Physics of Russian Academy of Sciences, 119991 Kosygin 4, Moscow, Russia

[2]N.N. Semenov Institute of Chemical Physics of Russian 119991 Kosygin 4,Moscow, Russia

[3]Institute of Natural Fibres, Poznan, ul. Wojska Polskiego 71 b, Poland

[4]Martin-Luther-Universität Halle-Wittenberg, Geusaer Straße, D-06217 Merseburg, Germany

ABSTRACT

Polypropylene (PP) has wide acceptance for use in many application areas. However, low thermal resistance complicates its general practice. The new approach in thermal stabilization of PP is based on the synthesis of PP nanocomposites. This paper discusses new advances in the study of the thermo-oxidative degradation of PP nanocomposite. The observed results are interpreted by a proposed kinetic model, and the predominant role of the one-dimensional diffusion type reaction. According to the kinetic analysis, PP nanocomposites had superior thermal and fireproof behavior compared with neat PP. Evidently, the mechanism of nanocomposite flame retardancy is based on shielding role of high-performance carbonaceous-silicate char which insulates the underlying polymeric material and slows down the mass loss rate of decomposition products.

Keywords: charring; diffusion; flammability; kinetics; nanocomposite; polypropylene; thermal degradation.

---

*Correspondence to: Sergei M. Lomakin, N.M. Emanuel Institute of Biochemical Physics ofRussian Academy of Sciences, Moscow, Russia, Kosygin 4, 119991, mailto:lomakin@sky.chph.ras.ru

Introduction

Polypropylene (PP) has wide acceptance for use in many application areas. However, low thermal and flame resistance complicates its general practice.

Much interest has been devoted to the study of PP thermal degradation [1-5]. Earlier studies on the thermal breakdown of polypropylene and its mechanism have been reviewed by Bockhorn et al [1]. The thermal depolymerization of polypropylene was considered as a radical process, which includes initiation, propagation and termination, which is analogous to the mechanism of polyethylene degradation [1]. After the bond scission into primary and secondary radicals, tertiary radicals are formed *via* rearrangements reactions. Subsequent β-scission leads to volatile alkenes and the chain carriers [1]. A β-scission to the other side leads to a short secondary radical and a polymer chain with a terminated double bond. This short secondary radical is saturated *via* intramolecular hydrogen transfer and results in an alkane. Due to the low alkane concentration, transfer reactions seem to play a minor role in polypropylene degradation. The apparent kinetic parameters for the overall thermal degradation of PP were determined under isothermal conditions using a gradient free reactor with on-line mass spectrometry [1]. For pure PP, different investigators have reported values of activation energy from 220–270 kJ/mol [1-3].

Chan and Balke applied thermogravimetric data for the pyrolysis of PP to provide a kinetic model of thermal degradation [4]. However, the wide temperature range used in this work (45–580°C) encompassed a change in the decomposition mechanism and this greatly limited the utility of the methods. To accommodate this mechanism change, the data were treated as a pseudo first-order reaction [4]. It was observed that the data conformed to a first-order fit at temperatures of less than 404 to 421°C (depending upon heating rate) with activation energy of 98.3 ± 3.1 kJ/mol [4]. At higher temperatures the data could again be fit as a pseudo first-order reaction, but with an activation energy of 327.9 ± 8.6 kJ/mol. The two regions were separated by a relatively narrow transition region. The lower activation energy occurring at lower degradation temperatures is attributed to scission of 'weak links' in the polymer. The higher activation energy was similar to the carbon-carbon bond dissociation energy and is associated with random scission throughout the polymer [4].

In contrast with many previous publications, Gao et al. showed that a first-order reaction model cannot be applied to describe thermal degradation of polypropylene [5]. The appropriate reaction order was determined to be 0.35 by using the degree of conversion at the maximum degradation rate measured under dynamic conditions. The validity of the reaction order thus determined was verified by similarity of activation energy between the single heating rate plot and the isoconversional plot. The reaction order was also supported by the consistency of the Arrhenius parameters of isothermal degradation with dynamic degradation [5].

The new approach in thermal stabilization of PP is based on polymer nanotechnology [6, 7]. Over the past decade, polymer nanocomposites have received considerable interest as an effective way for developing new composite materials, and they have been studied widely. Because of the larger surface area and surface energy of the additives when individual particles become smaller, it is not an easy task to obtain homogeneously dispersed organic / inorganic composites when the additives are down sized to nano-scale [8]. Melt intercalation has been successful in preparing polymer clay nanocomposites [9].

Recently the thermal degradation behavior of nanocomposites based upon PP-organoclay was studied by Zanetti, Camino *et. al.* using isothermal and dynamic thermogravimetry [10]. They suggested that the oxygen charring action and scavenging effect in the nanocomposite increases as the volatilization proceeds and that in the nanocomposite a catalytic role is played by the intimate polymer-silicate contact that may further favor the oxidative dehydrogenation-crosslinking-charring process.

This silicate morphology may act as an efficient barrier to oxygen diffusion towards the bulk of the polymer. Surface polymer molecules trapped within the silicate are thus brought to a close contact with oxygen to produce the thermally and oxidative stable charred material providing a new char-layered silicate nanocomposite acting as an effective surface shield [10]. In the present study a routine set of TGA analytical data was accomplished to provide the kinetic analysis of thermal degradation in air of PP nanocomposite based on layered organoclay (Cloisite 20A). The results obtained in this study gave an additional evidence of diffusion-controlled character of thermal degradation of PP nanocomposite caused by the catalytic-charring effect of nanosilicate clay.

Lately Qin at al. reported data on polypropylene/montmorillonite (PP/MMT) microcomposites thermal degradation and flammability [11]. They mentioned that PP microcomposites exhibit higher thermal stability and considerably reduced peak heat release rate due to physico-chemical adsorption of the volatile degradation products on the silicates [11]. On the other hand, the addition of MMT can catalyze the initial decomposition of PP matrix and accelerate its ignition in combustion. It has been observed that a ceramic-like char formed on the surface of the composites during the combustion test. The characterization of the char surface prior to ignition indicates that it is an inorganic-rich surface, leading to improvement of the thermal stability and reduction of flammability of the composites because of its better barrier properties [11].

EXPERIMENTAL

Synthesis

Polypropylene (PP) by Moscow petroleum refinery (MFI = 0.7) and maleic anhydride-modified oligomer (MAPP - Licomont AR 504 by Clariant co.) with M_n~2900; MA-content~4 wt.%. were blended/mixed using a laboratory Brabender mixing chamber for 2 min. at the first stage.

The concentration of MAPP was 20% by weight of original PP. Then the 7% wt. of organoclay (Cloisite 20A - Na^+ montmorillonite modified by dimethyl, dihydrogenatedtallow ammonium chloride by Southern Clay Co.) was added to the PP-MAPP - melt at a rotor speed of 60 rpm and set temperature of 190°C (PP-MAPP- Cloisite 20A). 10 min. mixing time was used in all the experiments.

Characterization

WAXS analysis of nanocomposite layered structure was carried out with a DRON-2 X-ray Diffractometer with Cu-Kα radiation. Diffraction patterns were collected in reflection-mode geometry from 2° to 10°2θ.

AFM studies were performed with commercial scanning probe microscope Nanoscope IIIA and IV MultiMode (Digital Instruments/Veeco Metrology Group, USA) in tapping mode, at ambient conditions. Conventional etched Si probes (stiffness ~40 N/m, resonant frequency 160-170 kHz) were used. The amplitude of the free-oscillating probe, A_0, was varied in the 10-20 nm range while the set-point amplitude, A_{sp}, ranged from 0.5 to 0.8 A_0. Imaging was conducted on the flat sample surfaces being prepared at –80°C with an ultramicrotome MS-01 (MicroStar Inc., USA) equipped with a diamond knife.

The thermal decomposition studies were performed over a temperature range of 20-600°C using a MOM Q 1500 thermogravimetric analysis (TGA) system under air environment at the scan rates of 3, 5 and 10K/min.

Kinetic analysis of PP compositions thermal degradation was carried out using Thermokinetics software by NETZSCH-Gerätebau GmbH.

Results and Discussion

Structure Characterization

It is always necessary to carefully characterize the polymer structure in order to ensure a sort of the dispersion for the nanoclays in polymers. XRD analysis and TEM would provide some information on the nanocomposite structural morphology. The diffraction patterns for nanoclay and nanocomposites are displayed in Figure 1 (a). The Cloisite 20A itself has a single peak at around 3.6° with d-space of 2.4 nm.

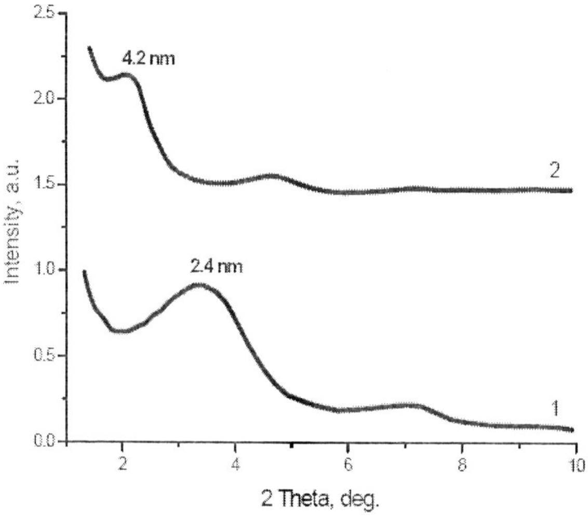

Figure 1. WAXS analysis for Cloisite 20A (1) and PP-MAPP-Cloisite 20A (2).

The shift of the clay basal spacing d_{001} from 3.6° to 2.2° in PP-MAPP-Cloisite 20A sample suggests the intercalated nanocomposite sample have higher d-space (4.2 nm) than that in the original clay (2.4 nm), it may have some exfoliated structures considering the smearing of peak in nanocomposite sample.

However, the XRD can only detect the periodically stacked montmorillonite layers (A); for all these nanocomposites there also exists a small number of exfoliated layers (B) as well, which can be directly observed by transmission electron microscopy (TEM).

In Figure 2 we present TEM images with different magnification which designate the presence of intercalated tactoids (A) and apparently exfoliated monolayers (B) coexisting in the nanocomposite structure. The intercalated structures are characterized by a parallel registry that gives rise to the XRD reflection of Figure 1.

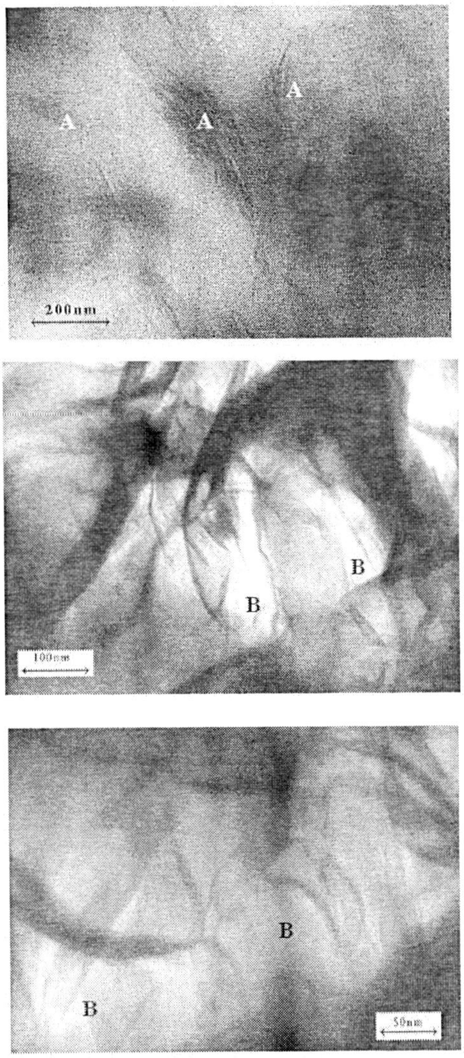

Figure 2.TEM photos with different magnification of PP-MAPP-Cloisite 20A where: A - stacks of layers (intercalated tactoids), B - exfoliated monolayers.

AFM studies demonstrate the similar morphology characteristics for PP-MAPP-Cloisite 20A (Figure 3). Height and phase images were simultaneously recorded on polymer surfaces. Height image presents surface topography, whereas phase images provide a sharp contrast of fine structural features and emphasize differences in sample components. We presume the availability of essentially intercalated structure with the complex multilayered tactoids (A) and small amount of exfoliated monolayer units (B) (Figure 3).

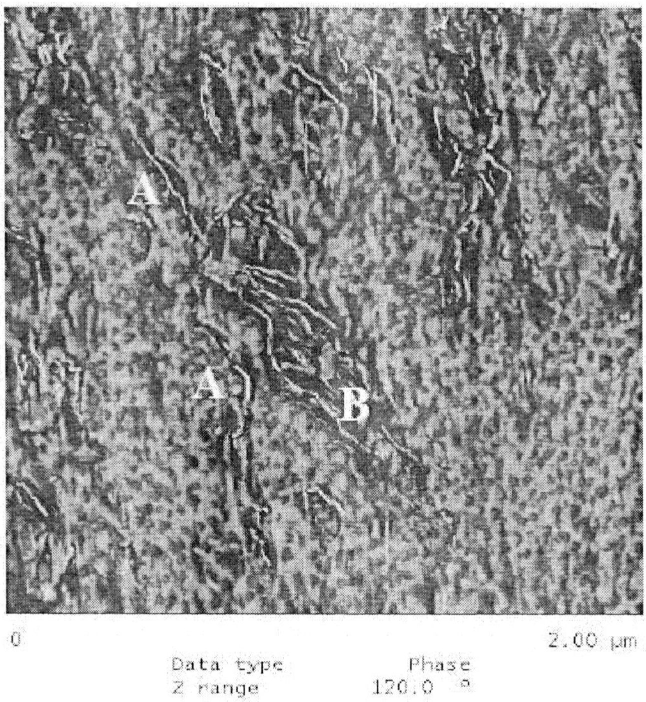

Figure 3. AFM image of PP-mPP with 7% Cloisite where: A - stacks of layers (intercalated tactoids), B-exfoliated monolayers.

Thermo-Oxidative Degradation Study

TG analysis of PP and PP nanocomposite (PP-MAPP-Cloisite 20A) show that at heating in air at 10°C/min, PP volatilizes completely, in two steps beginning at about 300°C with maximum rate at 400°C through a radical chain process propagated by carbon centered radicals originated by carbon-carbon bond scission (Figure4) [12]. Below 200°C, the hydroperoxidation on C-H bonds, in which oxygen addition occurs to the carbon radicals created within the polymer chain by H abstraction initiates radical-chain degradation of PP [13], whereas above 200-250°C, oxidative dehydrogenation of PP takes over [14]. Depolymerization and random scission by direct thermal clevege of carbon-carbon bonds becomes possible in air as in nitrogen, above 300°C.

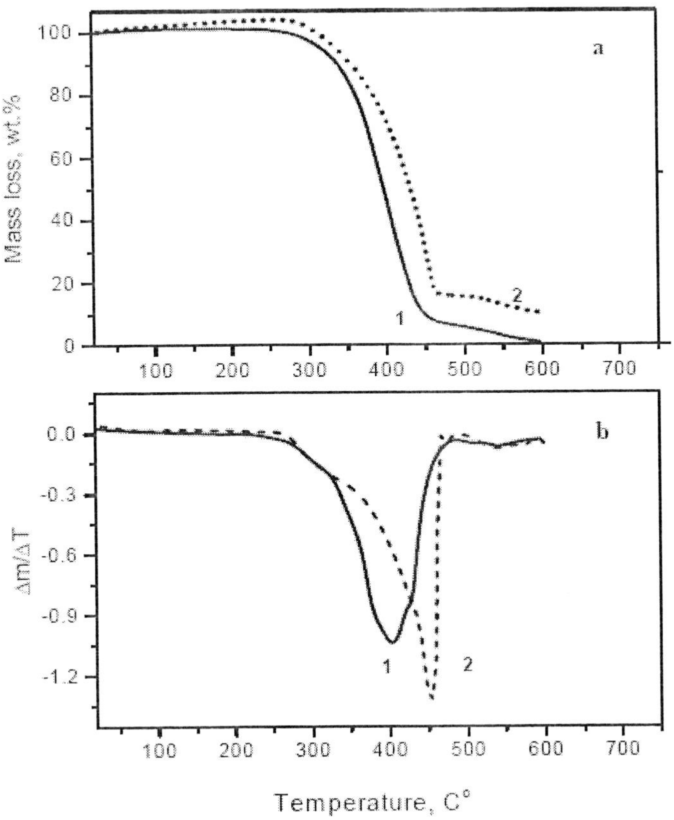

Figure 4. TG – (a) and DTG – (b) curves of PP (1) and PP-MAPP-Cloisite 20A (2) in air at the heating rate of 10K/min.

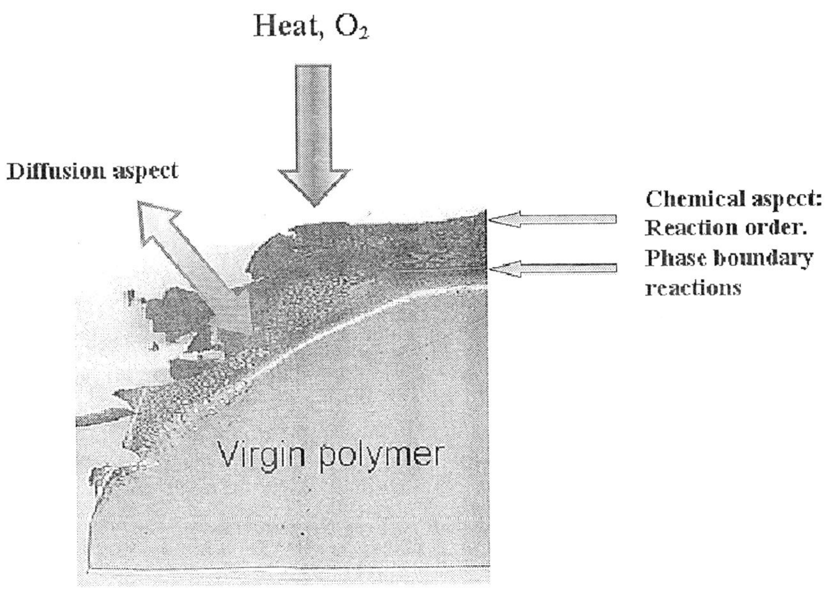

Figure 5. Scheme of the thermo-oxidative degradation of PP nanocomposite (PP-MAPP-Cloisite 20A).

Stabilizing effect of Δ50°C (Figure4 b) of PP-MAPP-Cloisite 20A over neat PP calculated with the maximum rate of mass loss can be explain by means of the barrier effect of the silicate nanolayers which operate in the nanocomposite level against oxygen diffusion, shielding the polymer from its action.

A char residue from neat PP is left at 450°C (5%), due to charring promoted by oxidative dehydrogenation. Then it slowly decomposes on heating up to 600°C in air (Figure4). On the other hand, thermal degradation of PP-MAPP-Cloisite 20A in air results to much more stable char form which doesn't oxidize even at 600°C (Figure 4 and 5).

Silicate nanostructure executes a role of efficient barrier to oxygen diffusion towards the native polymer. Surface polymer molecules trapped within the silicate are thus brought to a close contact with oxygen and catalytic - silicate layers to produce the thermally and oxidative steady carbonized structures (Scheme on Figure5).

Kinetic Analysis Using TGA Data

Kinetic studies of materials degradation have been carried out for many years using numerous techniques to analyze the data. Most often, TGA is the experimental method of choice and the only technique to be explored here. TGA involves placing a sample of polymer on a microbalance within a furnace and monitoring the weight of the sample during some temperature program. It is generally accepted that materials degradation obeys the basic equation (1) [15]

$$dc/dt = - F(t,T\ c_o\ c_f) \quad (1)$$

where: t - time, T - temperature, c_o - initial concentration of the reactant, and c_f - concentration of the final product. Equation $F(t,T,c_o,c_f)$ can be described by two separable functions, $k(T)$ and $f(c_o,c_f)$:

$$F(t,T,c_o,c_f) = k(T(t)\cdot f(c_o,c_f) \quad (2)$$

Arrhenius equation (4) will be assumed to be valid for the following:

$$k(T) = A\cdot \exp(-E/RT) \quad (3)$$

Therefore,

$$dc/dt= - A\cdot \exp(-E/RT)\cdot f(c_o,c_f) \quad (4)$$

A series of reactions types: classic homogeneous reactions and typical solid state reactions, is listed in Table 1 [15].

The analytical output must fit of measurements with different temperature profiles by means of a common kinetic model.

Kinetic analysis of PP compositions thermo-oxidative degradation at heating rates of 3, 5 and 10K/min. was carried out using NETZSCH Thermokinetics software in order to provide an extra evidence of the diffusion-stabilizing effect of nanoclay structure.

Table 1. Reaction types and corresponding reaction equations, $dc/dt = -A \cdot exp(-E/RT) \cdot f(c_o, c_f)$

Name	$f(c_o, c_f)$	Reaction type
F_1	c	first-order reaction
F_2	c^2	second-order reaction
F_n	c^n	nth-order reaction
R_2	$2 \cdot c^{1/2}$	two-dimensional phase boundary reaction
R_3	$3 \cdot c^{2/3}$	three-dimensional phase boundary reaction
D_1	$0.5/(1 - c)$	one-dimensional diffusion
D_2	$-1/ln(c)$	two-dimensional diffusion
D_3	$1.5 \cdot e^{1/3}(c^{-1/3} - 1)$	three-dimensional diffusion (Jander's type)
D_4	$1.5/(c^{-1/3} - 1)$	three-dimensional diffusion (Ginstling-Brounstein type)
B_1	$c_o \cdot c_f$	simple Prout-Tompkins equation
B_{na}	$c_o^n \cdot c_f^a$	expanded Prout-Tompkins equation (na)
$C_{1\text{-}X}$	$c \cdot (1+K_{cat} \cdot X)$	first-order reaction with autocatalysis through the reactants, X. $X = c_f$.
$C_{n\text{-}X}$	$c^n \cdot (1+K_{cat} \cdot X)$	nth-order reaction with autocatalysis through the reactants, X
A_2	$2 \cdot c \cdot (-ln(c))^{1/2}$	two-dimensional nucleation
A_3	$3 \cdot c \cdot (-ln(c))^{2/3}$	three-dimensional nucleation
A_n	$N \cdot c \cdot (-ln(c))^{(n-1)/n}$	n-dimensional nucleationnucleus growth according to Avrami/Erofeev

The curves show higher values at the beginning of the sintering process, i.e. at lower partial-length-change values, and considerably higher values at the end of the process (a) and particularly (b). This indicates the presence of a multiple-step process.

Model-free methods evaluations were chosen as the starting points in kinetic analysis of neat PP and PP-mPP with 7% Cloisite 20A for determining the activation energy in the development of the model. Figure 6 shows a corresponding Friedman analysis, where the activation energy is a function of partial mass loss change [16].

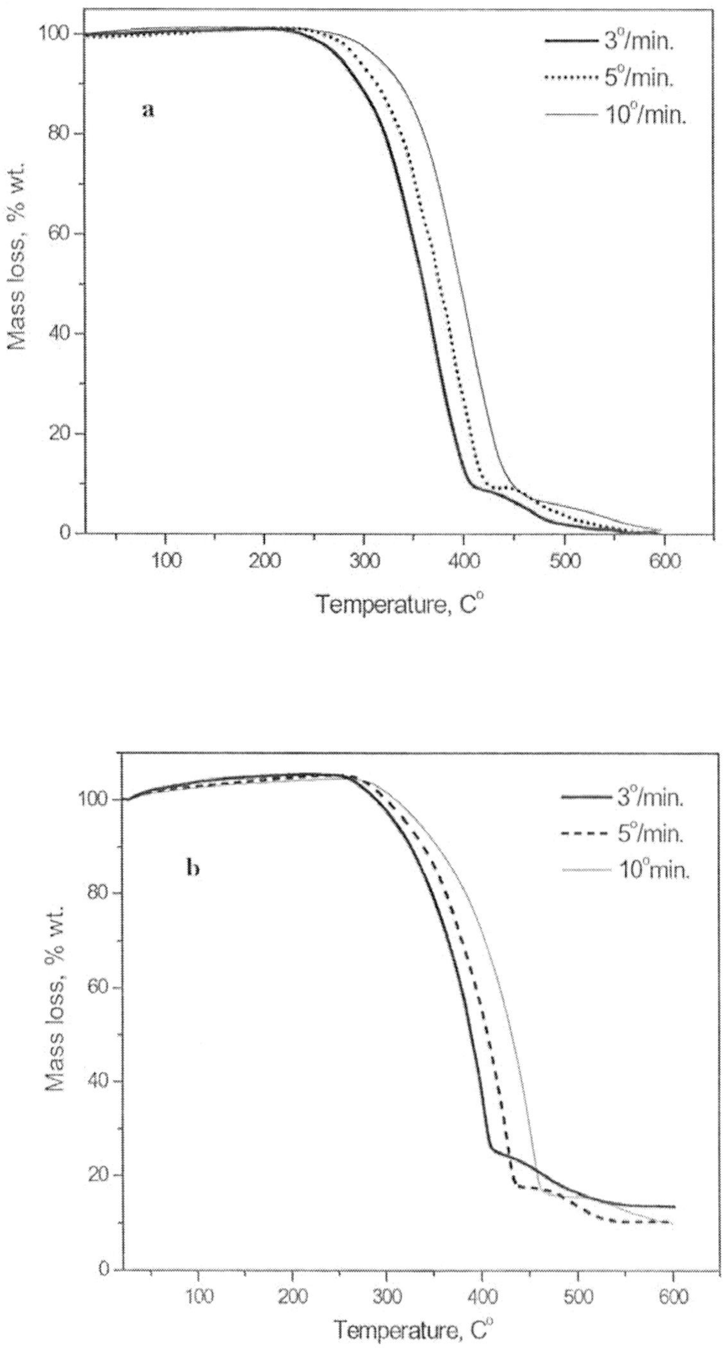

Figure 6. Friedman Analysis of neat PP – (a) and PP-MAPP-Cloisite 20A – (b).

First round analysis by Friedman method indicates a complexity of the scheme for neat PP and PP-MAPP-Cloisite 20A thermal degradation in air [16]. Nonlinear fitting procedure established the two - stage scheme for neat PP

$$A \rightarrow X_1 \rightarrow B \rightarrow X_2 \rightarrow C \tag{4}$$

and the triple - stage scheme for PP-mPP with 7% Cloisite 20A (Figure 4 a, b) [15,17].

$$A \rightarrow X_1 \rightarrow B \rightarrow X_2 \rightarrow C \rightarrow X_3 \rightarrow D \qquad (5)$$

Taking these findings into consideration for neat PP, a fit was attempted using nonlinear regression with model (4), where the nth-order (F_n) reaction type was used for all steps of the reaction (Figure 7, Table 2).

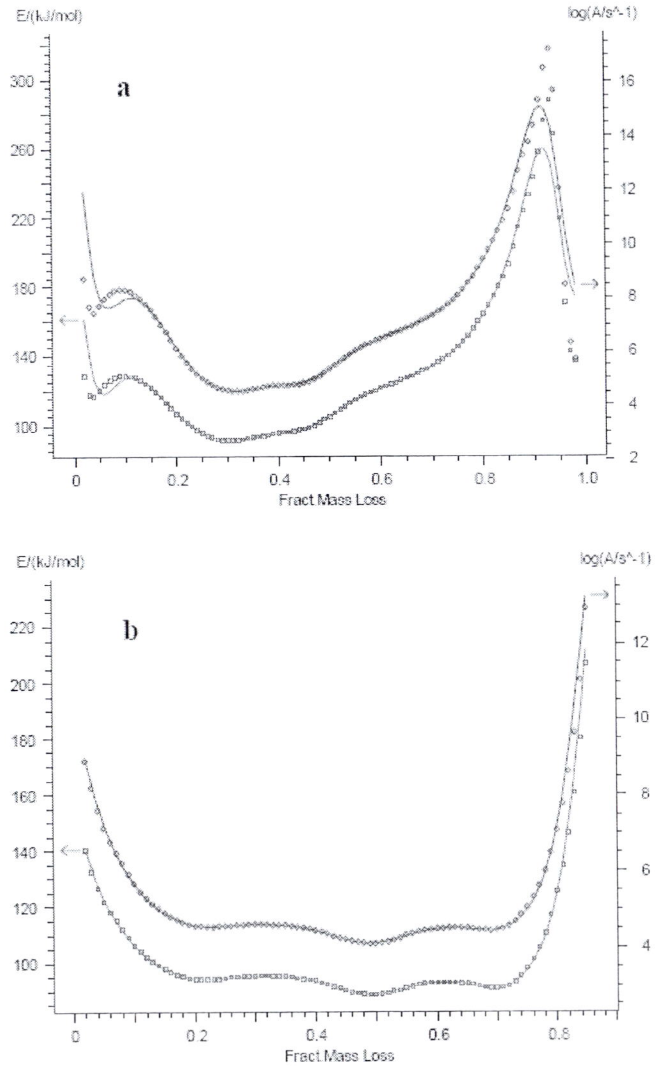

Figure 7. Nonlinear kinetic modelling for neat PP. Comparison between experimental TG data (dots) and model results (lines) at the heating rates of 3, 5 and 10K/min.

With this expanded model, an excellent fit is possible for all three measurements. The kinetic parameters are listed in Table 2.

Table 2. Kinetic parameters resulting from multiple-curve analyses (heating rates 3, 5 and 10 K/min) with reaction model (A→X_1→B →X_2→C) from TG measurement of neat PP

Reaction models (types, X*i*)	Parameter	Value	Corr. Coeff.
$F_n \rightarrow F_n$	$\log A_1$, s^{-1}	6.4	0.9994
	E_1, kJ/mol	110.3	
	n_1	1.13	
	$\log A_2$, s^{-1}	9.9	
	E_2, kJ/mol	151.6	
	n_2	2.59	

A more sophisticated model (5), based on different reaction types, was chosen for PP-MAPP-Cloisite 20A thermo-oxidative degradation. The parameters are listed in Table 3.

Table 3. Kinetic parameters resulting from multiple-curve analyses (heating rates 3, 5 and 10 K/min) with different reaction models (A→X_1→B →X_2→C→X_3→D) from TG measurement of PP-MAPP-Cloisite 20A

Reaction models (types, X*i*)	Parameter	Value	Corr. Coeff.
$F_n \rightarrow F_n \rightarrow F_n$	$\log A_1$, s^{-1}	6.3	0.9974
	E_1, kJ/mol	113.4	
	n_1	1.16	
	$\log A_2$, s^{-1}	8.8	
	E_2, kJ/mol	150.9	
	n_2	2.46	
	$\log A_3$, s^{-1}	11.5	
	E_3, kJ/mol	188.5	
	n_3	0.78	
$F_n \rightarrow D_1 \rightarrow F_n$	$\log A_1$, s^{-1}	6.9	0.9988
	E_1, kJ/mol	113.4	
	n_1	1.21	
	$\log A_2$, s^{-1}	4.7	
	E_2, kJ/mol	100.0	
	$\log A_3$, s^{-1}	12.0	
	E_3, kJ/mol	199.8	
	n_3	1.17	

Table 3. Continued

Reaction models (types, Xi)	Parameter	Value	Corr. Coeff.
$F_n \rightarrow D_2 \rightarrow F_n$	$\log A_1$, s^{-1}	6.4	0.9974
	E_1, kJ/mol	113.3	
	n_1	1.68	
	$\log A_2$, s^{-1}	6.2	
	E_2, kJ/mol	118.4	
	$\log A_3$, s^{-1}	11.7	
	E_3, kJ/mol	197.2	
	n_3	0.95	
$F_n \rightarrow D_3 \rightarrow F_n$	$\log A_1$, s^{-1}	6.5	0.9973
	E_1, kJ/mol	113.6	
	n_1	2.04	
	$\log A_2$, s^{-1}	8.3	
	E_2, kJ/mol	152.3	
	$\log A_3$, s^{-1}	11.8	
	E_3, kJ/mol	197.1	
	n_3	0.94	
$F_n \rightarrow D_4 \rightarrow F_n$	$\log A_1$, s^{-1}	6.6	0.9973
	E_1, kJ/mol	113.7	
	n_1	2.01	
	$\log A_2$, s^{-1}	7.2	
	E_2, kJ/mol	138.9	
	$\log A_3$, s^{-1}	11.9	
	E_3, kJ/mol	195.2	
	n_3	0.98	
$F_n \rightarrow A_2 \rightarrow F_n$	$\log A_1$, s^{-1}	6.6	0.9975
	E_1, kJ/mol	114.4	
	n_1	2.10	
	$\log A_2$, s^{-1}	5.6	
	E_2, kJ/mol	105.2	
	$\log A_3$, s^{-1}	12.0	
	E_3, kJ/mol	199.3	
	n_3	1.23	
$F_n \rightarrow A_3 \rightarrow F_n$	$\log A_1$, s^{-1}	6.5	0.9976
	E_1, kJ/mol	114.1	
	n_1	2.02	
	$\log A_2$, s^{-1}	4.8	
	E_2, kJ/mol	95.3	
	$\log A_3$, s^{-1}	12.4	
	E_3, kJ/mol	200.1	
	n_3	1.46	

Table 3. Continued

Reaction models (types, X*i*)	Parameter	Value	Corr. Coeff.
$F_n \rightarrow A_n \rightarrow F_n$	$\log A_1$, s^{-1}	6.7	0.9983
	E_1, kJ/mol	114.7	
	n_1	0.97	
	$\log A_2$, s^{-1}	4.7	
	E_2, kJ/mol	97.7	
	$\log A_3$, s^{-1}	11.9	
	E_3, kJ/mol	200.0	
	n_3	1.26	
$F_n \rightarrow R_2 \rightarrow F_n$	$\log A_1$, s^{-1}	7.1	0.9979
	E_1, kJ/mol	114.9	
	n_1	1.27	
	$\log A_2$, s^{-1}	5.1	
	E_2, kJ/mol	105.7	
	$\log A_3$, s^{-1}	12.1	
	E_3, kJ/mol	200.3	
	n_3	1.76	
$F_n \rightarrow R_3 \rightarrow F_n$	$\log A_1$, s^{-1}	7.2	0.9978
	E_1, kJ/mol	114.6	
	n_1	1.10	
	$\log A_2$, s^{-1}	5.2	
	E_2, kJ/mol	108.9	
	$\log A_3$, s^{-1}	11.6	
	E_3, kJ/mol	188.5	
	n_3	0.78	

Taking these fittings for PP-MAPP-Cloisite 20A, a best approximation was attempted using nonlinear regression with model (5), based on the best fit quality (correlation coefficient) (Figure 8), where the one-dimensional diffusion (D_1) reaction type was used for the second step of the reaction (Table 3), whereas the nth-order reaction models were chosen for the first and third steps respectively.

These results show that the second step in thermo-oxidative degradation of PP-MAPP-Cloisite 20A is described by one-dimensional diffusion (D_1) reaction type which is liable for the overall process of the carbonization in nanocomposite polypropylene structure.

Flame Resistant Properties

The recent interest in the reported char-promoting functionalized dispersed nanoclays to yield nanocomposite structures having enhanced fire and mechanical properties, when the clays are present only at levels of 2~10%, prompts their investigation as potential fire retardants. Because of its wholly aliphatic hydrocarbon structure, neat polypropylene by itself burns very rapidly with a relatively smoke-free flame and without leaving a char residue. It has a high self-ignition temperature (570°C), a rapid decomposition rates and hence has a high flammability.

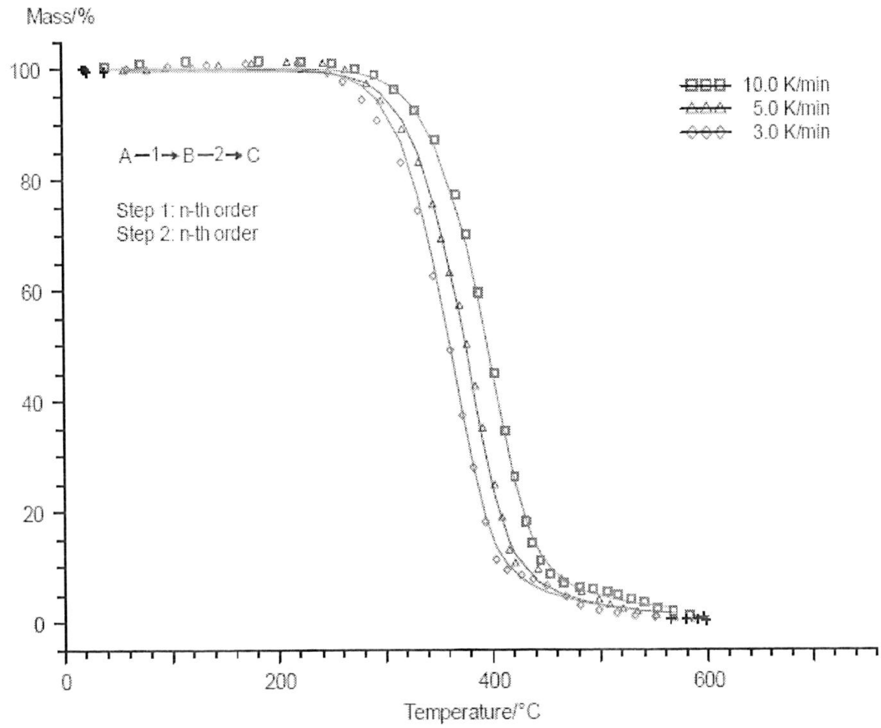

Figure 8. Nonlinear kinetic modelling for PP-MAPP-Cloisite 20A. Comparison between experimental TG data (dots) and model results (lines) at the heating rates of 3, 5 and 10K/min.

Polypropylene nanocomposites have attracted more and more interest in flame retardant area in recent years due to their improved fire properties [18-20]. It is suggested that the presence of clay can enhance the char formation providing a transient protective barrier and hence slowing down the degradation of the matrix [19, 20].

The kinetic results of the present study let us the starting point to predict the mass loss of material under isothermal pyrolysis conditions using the same thermokinetics software. Figure 9 shows the fractional mass loss curves as a function of time with temperature (400 – 600°C) as a parameter.

It is clearly seen that under conditions of polymer ignition and initial surface combustion, the mass loss for PP-MAPP-Cloisite 20A and its rate are noticeably lower then adequate values for the neat PP. An improvement in flame resistance of PP-MAPP-Cloisite 20A over the neat PP happens as a result of the char formation providing a transient protective barrier. In the present study this phenomena was interpreted in terms of isothermal kinetic analysis.

Apart from this information, the graphs of mass loss rates (*dm/dt*) vs. time for neat PP and PP-MAPP-Cloisite 20A indicate the depression of the degradation (fuel) products under the isothermal pyrolysis conditions at 600°C (Figure 10). It is well known that the temperature of 600°C corresponds to an incident heat flux of 35 kW/m^2; this is referred to the real scale fire scenario [21].

The cone calorimeter is one of the most effective bench-scale methods for studying the flammability properties of materials. Fire-relevant properties, measured by the cone calorimeter, such as heat release rate (RHR), maximum RHR, smoke and carbon monoxide yield, are vital to the evaluation of the fire safety of materials [22].

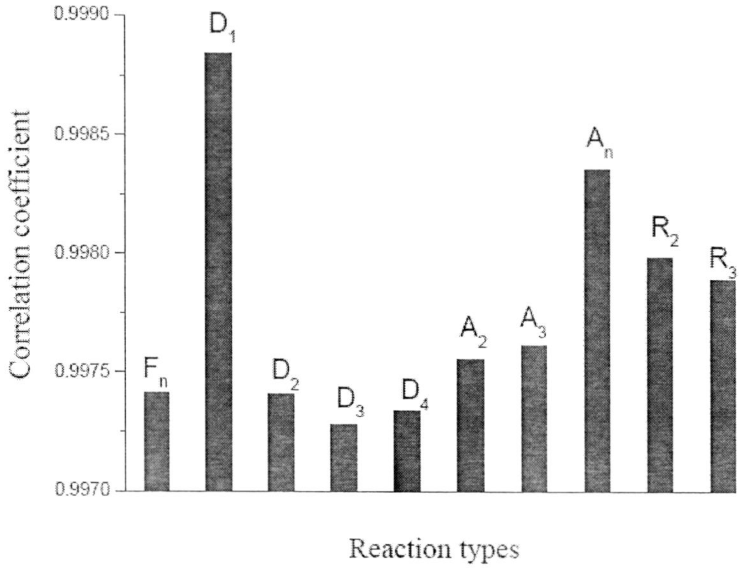

Figure 9. Fractional reaction vs. time for neat PP (a) and PP-MAPP-Cloisite 20A (b).

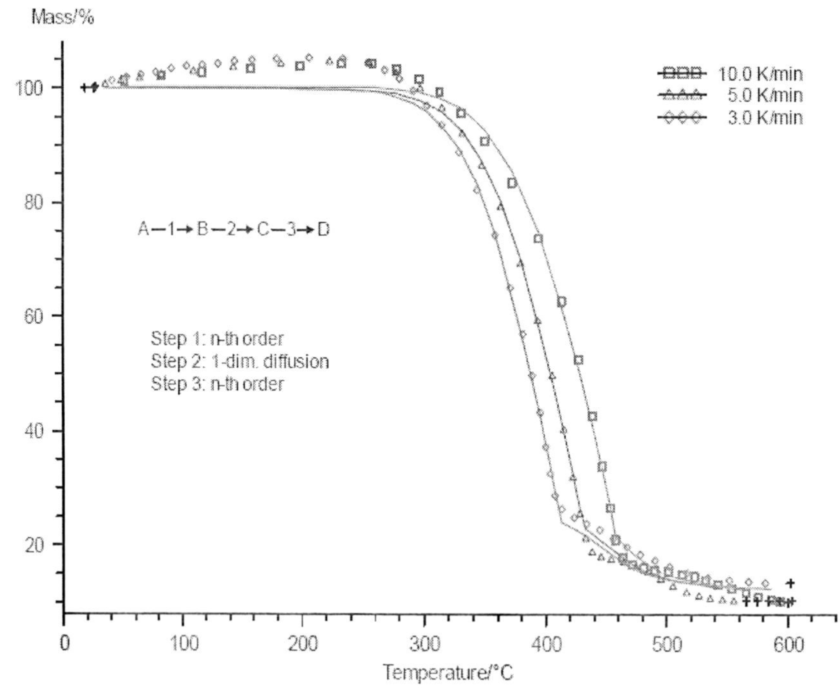

Figure 10. Mass loss rates vs. time for neat PP-MAPP-Cloisite 20A (2) under the isothermal heating condition of 600oC.

In the present study the combustibility of polypropylene nanocomposite was evaluated by a cone calorimeter. The tests were performed at an incident heat flux of 35 kW/m^2 using the cone heater [21]. Peak heat release rate (RHR), mass loss rate (MLR), specific extinction area (SEA) data, carbon monoxide and heat of combustion data measured at 35 kW/m2, are presented in Figs. 11 and 12.

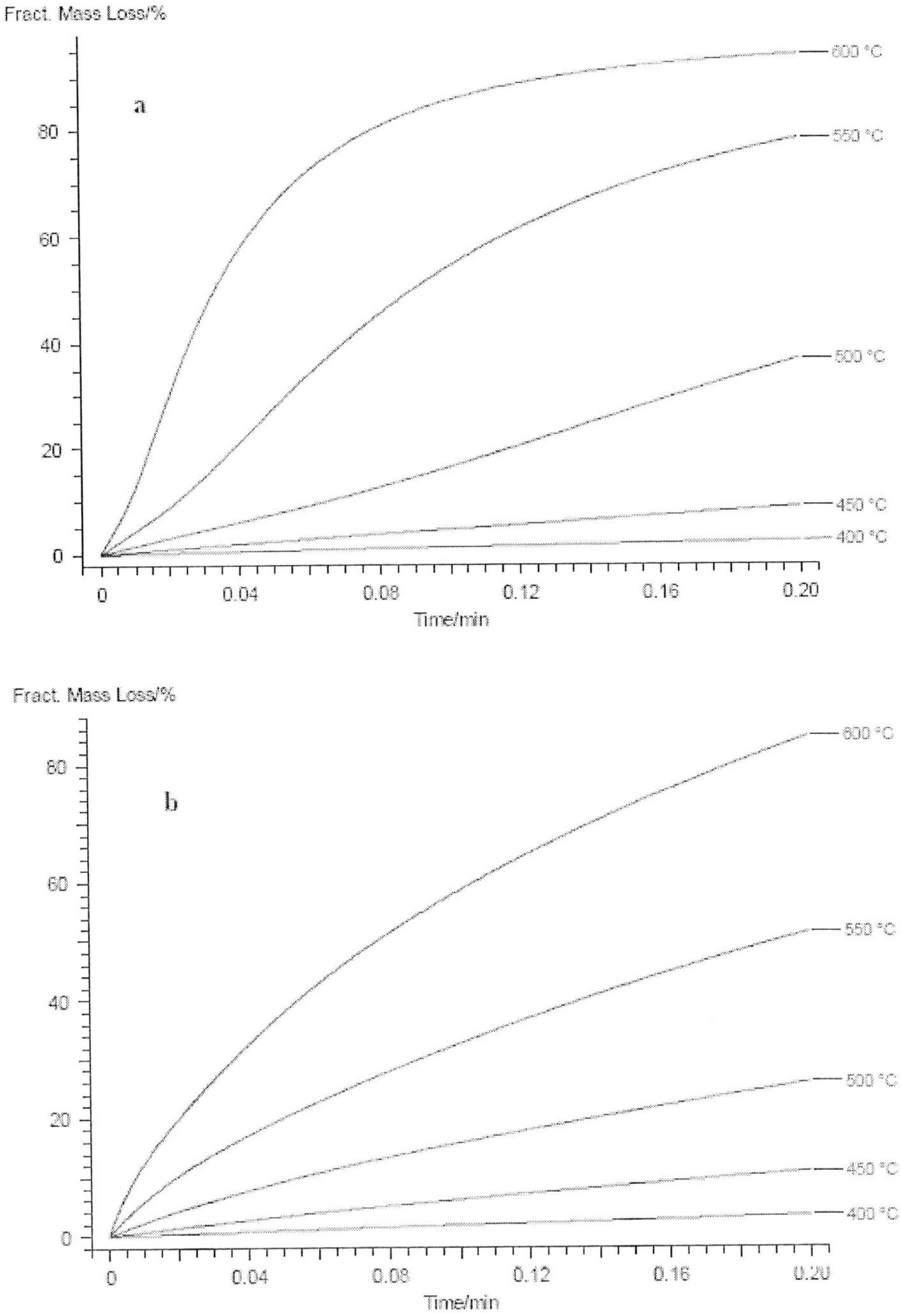

Figure 11. Rate of heat release vs. time for PP and PP-MAPP-Cloisite 20A.

The RHR plots for PP-MAPP-Cloisite 20A nanocomposite and PP at 35 kW/m^2 heat flux shown in Figure indicate a 60% - decrease of peak of RHR (Figure11). Comparison of the Cone calorimeter data PP and PP-MAPP- 7% Cloisite 20A reveals that the specific heat of combustion (H_c), specific extinction area (SEA), a measure of smoke yield, and carbon monoxide yields are practically unchanged; this suggests that the source of the improved flammability properties of these materials is due to differences in condensed-phase decomposition processes and not to a gas-phase effect. The primary parameter responsible for the lower RHR of the nanocomposites is the mass loss rate (MLR) during combustion, which is significantly reduced from the value observed for the pure PP (Figure12).

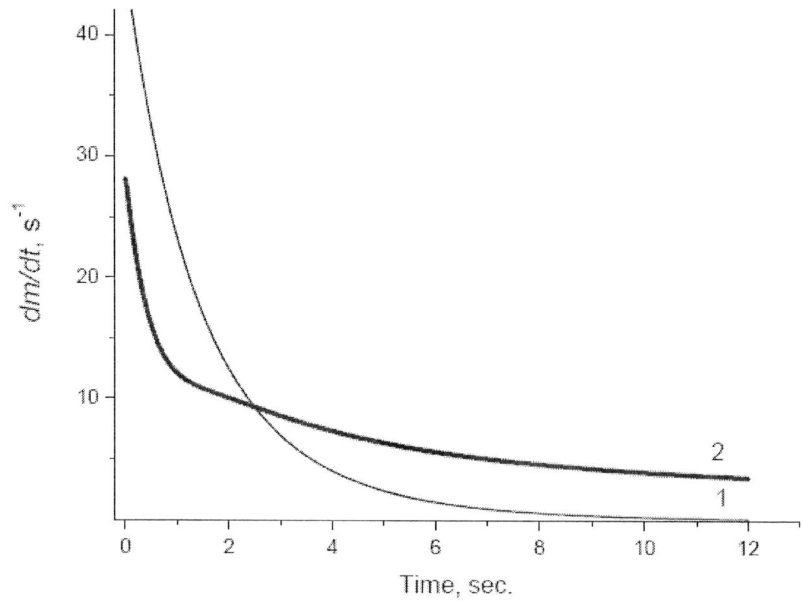

Figure 12. Mass loss rate vs. time for PP and PP-MAPP-Cloisite 20A.

It is supposed, that this effect is caused by ability to initiate the formation of char barrier on a surface of burning polymeric nanocomposites that drastically limits the heat and mass transfer in a burning zone.

CONCLUSION

The kinetic data obtained by dynamic TGA designate thermal stabilization effect of nanoclay structure into a polymer matrix, caused in one-dimensional diffusion process of catalytic-charring throughout the thermal degradation of PP nanocomposition. According to provided kinetic analysis, polypropylene nanocomposite demonstrated the transcendent thermal and fireproof behaviour in relation to neat polypropylene.

Based on modern concepts of the mechanisms of polymer nanocomposites flame retardancy and char formation, it is possible to assume that simulation of dynamic behavior of anisotropic particles with the different types of morphology and relationship of geometric dimensions presented in viscous polymeric melt will be one of the promising trends in this field of research.

On the other hand, the unique barrier properties of nano-dispersed polymeric composites are of interest of polymer combustion due to specific laminar morphology. This type of structure is especially effective in comparison with the other forms of fillers because of the "labyrinth effect". Researches in this area will allow defining an influence of diffusion of low-molecular products of pyrolysis on the process of micro-intumescence in a superficial layer of burning polymers.

ACKNOWLEDGEMENTS

This work was supported by the Russian Foundation for Basic Research, project N 04-03-32052.

The authors are pleased to acknowledge the Moscow division of NETZSCH-Gerätebau GmbH for Thermokinetics software.

REFERENCES

[1] Bockhorn, H, A. Hornung, A, Hornung, U, Schawaller, D, *Journal of Analytical and Applied.*

[2] Pyrolysis, 48(2):93 (1999).2. Ballice, L, Reimert, R, *Chemical Engineering and Processing,* 41:289 (2002).

[3] Murty, MVS, Rangarajan, P, Grulke, EA, Bhattacharyya, D, *Fuel Processing and Technology,* 49:75 (1996).

[4] Chan, JH, Balke, ST, *Polymer Degradation and Stability,* Volume 57:135 (1997).

[5] Gao, Z, Kaneko, T, Amasak,i I, Nakada, M, *Polymer Degradation and Stability,* 80:269 (2003).

[6] Giannelis, E, *Adv Mater*, 8:29 (1996).

[7] Gilman, JW., Kashiwagi, T, Nyden, MR, Brown, JET, Jackson, C L, Lomakin, SM, Giannelis, E P, Manias, E, in *Chemistry and Technology of Polymer Additives*, Chapter 14, ed by Ak-Malaika S, Golovoy A, Wilkie CA, p 249, Blackwell Science Inc., Malden MA (1999).

[8] Zanetti, M, Lomakin, S, Camino, G, *Macromol Mater Eng*, 279:1-9 (2000).

[9] Kojima, Y., Usuki, A., Kawasumi, M., Okada, A., Fukushima, Y., Kurauch,i T., Kamigaito O., *J. Mater. Res.*, 8:1185 (1993).

[10] Zanetti, M, Camino, G, Reichert, P, Mülhaupt, R, *Macromolecular Rapid Communications* 22:176-180 (2001).

[11] Qin, H, Zhang, S, Zhao, C, Feng, M, Yang, M, Shu, Z, Yang, S, *Polymer Degradation and Stability*, 85:807 (2004).

[12] Grassie, N, Scott, G, in *Polymer degradation and stabilization*, Cambridge University Press, Cambridge, p 275 (1985).

[13] March, J, in *Advanced Organic Chemistry*, McGraw-Hill Kogakusa Ltd, Tokyo, p.367 (1977).

[14] Benson, SW, Nogia, PS, *Account of Chemical Research*, 12:233 (1979).
[15] Opfermann, J, *J. Thermal Anal. Cal.*, 60: 641 (2000).
[16] Friedman, HL, *J. Polym. Sci.*, C6:175 (1965).
[17] Opfermann, J, Kaisersberger, E, *Thermochim Acta*, 11:167 (1992).
[18] Lomakin, SM, Zaikov, GE, in *Modern Polymer Flame Retardancy*, VSP Int. Sci. Publ. Utrecht, Boston, p. 272 (2003).
[19] Gilman, JW, *Applied Clay Sci*, 15:31 (1999).
[20] Gilman, GW, Jackson, C L., Morgan, A B, Harris, R H, Manias, E, Giannelis, E P, Wuthenow, M, Hilton, D, Phillips, S, *Chem. Mater*, 12:1866 (2000).
[21] Babrauskas, V, *Fire and Materials*, 19:243 (1995).
[22] Babrauskas, V., Peacock, R. D. *Fire Safety J.*, 18:255 (1992).

In: Encyclopedia of Nanotechnology
Editor: Elwood D. Carlson

ISBN 978-1-60692-079-4

Chapter 55

BIOMEDICAL APPLICATION OF NANOTECHNOLOGY IN CANCER EARLY DIAGNOSIS

Min Song and Xuemei Wang*
State Key Laboratory of Bioelectronics,
Department of Bioscience and Medical Engineering,
Southeast University, Nanjing 210096,
People's Republic of China

ABSTRACT

Nanomaterials have attracted much attention in the biomedical areas due to their unique advantages of application in the biomedical related fields. It is already known that nanoparticles could present a versatile nanoscale interface for respective biomolecular recognition, drug delivery, etc. The strategy of nanotechnology could result in an increased concentration of anticancer drugs in selected target tissues and a reduction of systemic side effects and toxicity. Furthermore, with the development of the advanced nanotechnology, nanocomposites functioned with different biomarkers or relevant functional groups could be used to adsorb different biomolecules for specifically targeting tumor cells, which may provide a new way for the cancer diagnosis and efficiently facilitate drug accumulation on target cancer cells and inhibit the multidrug resistance of the cancer chemotherapy.

In the meanwhile, nanotechnology is in a unique position to transform cancer diagnostics and produce a new generation of biosensors and medical imaging techniques with higher sensitivity and precision of recognition. Nanomaterial-based biomolecule delivery carriers show promising perspective in target biomarker and cancer therapy. Based on the recent development of nanotechnology, nanomaterials display promising and far-ranging prospect in the fields of biomedical engineering, especially in the diagnosis and treatment of cancers.

* Tel/Fax: +86-25-83792177; Email: xuewang@seu.edu.cn

1. INTRODUCTION

Nanomaterials have attracted much attention in the areas of biomedical and bioengineering due to their unique properties of the relative material in the nanoscaled level. Nanoparticles are chiefly classified into inorganic, polymeric and hydrogel particles in the biomedical fields, such as disease diagnostics and therapeutics [1]. Among inorganic materials, magnetite, quantum dots of CdS and ZnS, hydroxyl apatite, gold, silica, calcium phosphate and others have been explored widely for the drug delivery, photodynamic therapy, and vaccination, etc [2-7]. Besides, the respective biocompatibility and biodegradability of polymeric nanoparticles have received much attention in controlled drug delivery, tissue growth, and temporary therapeutic applications in pharmacology and surgery [8-11]. It is reported that some biodegradable nanoparticles are well-known carriers that could carry specific bioactive molecules to tissues of interest, due to the relative potential for control of target release of the drug, and to the well-established methods of the preparation [12-15].

In recent years, as the development of nanocomposites, many methods have been explored to prepare the new nanocomposites to enhance the functions of materials. For example, different metal nanoparticles have been coated or deposited (covalently or non-covalently) on the sidewall of carbon nanotubes [16, 17]. Furthermore, some researchers have explored the method of embedding nanoparticles into polymer matrices, which combines the advantages of both polymer and nanoparticles [18]. Ding et al [19] prepared the gold nanoparticle/PNIPAM composite in one-step by using Au nanoparticles functionalized with a vinyl group. As for their biomedical application, a variety of nanoparticulate designs have been proposed for cancer therapy and diagnosis. For example, nanoparticles can be conjugated to a biospecific ligand, which could direct them to the target tissue or organ. Moreover, nanoparticles have become an important research area in the field of drug delivery for the reason that they have the ability to deliver a wide range of drugs to varying areas of the body. Various factors such as particle size, surface charge, hydrophilicity / hydrophobicity, composition, administration route and others will directly affect the relative efficiency of the drug delivery, cellular and tissue biomarker [20, 22].

Cancer is one of the most serious and difficult diseases that threaten people's health and life, so it has attracted intensive interest in various areas of the biomedical fields [23-25]. Now it is already known that cancer is made up of a hodgepodge of cells containing different genetic mutations, each of which allows it to wreak a different brand of havoc. In order to reduce the morbidity and mortality of cancers, early diagnosis and the development of novel cancer prevention and systemic therapies for advanced disease are of paramount importance. Meanwhile, it is still a big challenge to develop the strategy to mark and detect the target cancer cells for the early diagnosis and target drug delivery of the respective cancers and thus follows the development and proliferation of cancer cells. One of the difficult problems in such disease diagnosis and treatment is the drug resistance. The emergence of drug resistance is a worldwide problem in the related diseases' therapies and treatments. Several different biophysical processes contribute to the related drug resistance, which includes the decrease in drug uptake, the increase in drug efflux, the increase in drug metabolism rate, or alterations in drug-target properties [26].

Although much effort has been done for the chemotherapy over the past years, yet there still exist many problems. The disadvantage of the chemotherapy is the modest sensitivity due to the occurrence of false negative results in malignant cases by strict diagnostic criteria. Another major disadvantage is non-specific. So, it is a key point for how to improve the sensitivity and reliability in diagnosis target treatment of cancers. The sensitive detection and efficiently targeted drug delivery for drug resistant cancer cells may afford a new strategy of the early diagnosis and efficient treatment of targeted cancers. And nanomaterials could be used to improve the sensitivity and precision of cancer diagnosis due to their unique properties like good biocompatibility, ease of surface chemistry modification, very high surface area, etc. Besides, associating antitumor drugs with nanomaterials could be utilized to overcome drug resistance and to increase the selectivity of drugs towards cancer cells while reducing their toxicity towards normal tissues. Recent reports have demonstrated that some drug coated polymer nanospheres and nanoparticles could increase the anticancer drug delivery and inhibit multidrug resistance [27, 28].

On the basis of these observations, the application of the different aspects of nanoparticles as biomarkers of cancers and on reversing multidrug resistance and facilitating the drug delivery has been reviewed in this paper. Since nanotechnology could provide very useful and promising platform for clinical diagnosis and therapy of some difficult-to-treat diseases, the function and advantage of nanotechnology in the relevant cancer diagnosis and therapy has also been explored in this review.

2. The Application of Nanotechnology for Cancer Early Diagnosis

Due to the unique properties of nanomaterials, nanotechnology has the potential to facilitate major advances in the cancer early diagnosis and treatment. In the past years, a variety of nanoparticles, including metal nanoparticles, oxide nanoparticles, semiconductor nanoparticles, and even composite nanoparticles, have been under development for cancer diagnosis and therapy. Nanoparticle-based drug delivery systems offer the potential to mark cancer cells /or tissues and facilitate drug delivery while reducing drug side effects [29, 30]. Many kinds of nanoscale platforms have been used for a wide variety of applications in cancer diagnostics and therapeutics.

2.1. Application of Nanomaterials for Targeted Biomarkers of Cancers

In view of some major drawbacks of conventional chemotherapy such as poor specificity, side effects, and reduced therapeutic efficacy, application of nanotechnology in cancer diagnosis and target drug delivery has received much attention in recent years. Nanotechnology can integrate molecular recognition and site-specific delivery of the therapeutic molecules to facilitate an effective drug therapy. It has been demonstrated that the nanoparticles could enhance efficient drug accumulation at the target site in the body where they can penetrate within even small capillaries and are taken up within cells for their small size [31-33]. Moreover, nanoparticles which are prepared for biodegradable materials allow for sustained drug release within the target site over a period of days or even weeks after injection [34]. Targeting the drug to specific site and overcoming the tissue barrier or cellular barrier in nanomaterial applications was their major purpose as relevant biomarker and for the controlled drug delivery [35]. In the past years, some relative nanoparticles have been used as drug carriers to enhance the ability of anticancer drug delivery [36-38]. Magnetic nanoparticles have been investigated most widely among inorganic materials in the area of cancer therapy and diagnosis [39-47]. Some reports have illustrated that the iron oxide magnetic particles may act as the good system for drug delivery and offer new perspectives for diagnostic and targeted therapeutic approaches, where those nanoparticles could be phagocytosed by cancer cells and interact with components inside the relative cells [48,49]. In addition, some researchers have immobilized poly (ethylene glycol) (PEG) and folic acid on the surfaces of superparamagnetic magnetite nanoparticles to further improve their intracellular uptake and ability to target specific cells [50]. Furthermore, polymeric drug carriers have attracted increasing attention in recent years in the field of drug delivery, tissue engineering and temporary therapeutic applications in pharmacology and surgery due to their biodegradability and biocompatibility [51].

Recently, carbon nanotubes are carbon cylinders composed of benzene rings, and their use in biologic applications has been widely illustrated in the literature. Nanotubes with different sizes have been shown to be internalized intracellularly through endocytosis [52]. Functionalized nanotubes have been investigated for biologic applications by adsorbing different molecules and antigens to their surface so that they may specifically target tumor cells. Kam et al. used folic acid (FA) and fluorescent tag-conjugated, single-walled carbon nanotubes (SWNTs), to specifically target HeLa cells [53]. Meanwhile, gold nanoparticles have been used as a platform for cancer diagnosis and therapy. It was demonstrated that systemically delivered gold nanoparticles conjugated to tumor necrosis factor (TNF) accumulated in tumors [54].

To improve the efficiency of drug uptake and minimize toxic side effects, many efforts have been made by the addition of surface modifications to nanoparticles. Wender et al. has demonstrated that single-walled nanotubes (SWNT) and SWNT-streptavidin conjugates can enter human promyelocytic leukemia (HL60) cells and human T cells via the endocytosis pathway and by themselves are not toxic [55]. Inert substances such as PEO segments have been grafted onto or within existing polymers such as polyurethanes in order to enhance biocompatibility and reducing protein adsorption [56, 57].

2.2. Application of Nanomaterials for Monitoring and Reversing Multidrug Resistance

It is already known that the development of tumor resistance to the chemotherapeutic agents and the occurrence of multidrug resistance (MDR) phenomenon is the major obstacle to the success of the chemotherapy of tumors. MDR is mainly due to the overexpression of glycoprotein (P-glycoprotein P-gp, a integral membrane protein) on drug resistance cell membrane, which is responsible for the ATP-dependent export of certain structurally unrelated compounds including many chemotherapeutic drugs [58-61]. Many cancers may express multidrug resistance gene and multidrug resistance proteins. It has been reported that several different biophysical processes contribute to the related drug resistance, which includes the decrease in drug uptake, the increase in drug efflux and drug metabolism rate, or alterations in drug-target properties, etc. [62, 63].

Nanotechnology has been utilized to solve relevant drug resistance problems. It is reported that polyisobutylcyanoacrylate (PIBCA) and polyisohexylcyanoacrylate (PIHCA) nanoparticles, loaded with doxorubicin, could efficiently overcome Pgp-mediated MDR [64]. The adsorption of nanoparticles to the cell surface and an increased diffusion of doxorubicin across the plasma membrane were observed due to the formation of an ion pair between the negatively charged cyanoacrylic acid and the positively charged doxorubicin [64]. Besides, it has also been observed that some P-glycoprotein inhibitors could increase oral absorption of drugs and reduces drug excretion [65].

In our studies, functionalized magnetic nanoparticles of Fe_3O_4, Fe_2O_3 and Ni have been utilized to facilitate the in vitro anticancer drug daunorubicin uptake of leukemia K562 cells. Our observations indicate that the presence of Fe_3O_4 nanoparticles could remarkably increase the intracellular drug accumulation of daunorubicin, while the relevant enhancement to drug resistant cells is much stronger than that of drug sensitive cells. It appears that these functionalized magnetic nanoparticles may act as the inhibitor of P-glycoprotein and could be used as a new strategy in respective cancer diagnosis and target therapy [66, 67].

It has been reported that various factors such as particle size, surface charge, hydrophilicity/hydrophobicity, composition, administration route and others, will directly affect the biodistribution of nanoparticles and lead to probable targeted drug release systems [68-70] (as shown in Figure 1). So nanoparticles with different size have different ability to enter target cancer cells. Our studies also illustrate that both the size and the unique properties of magnetic nanoparticles themselves could contribute to the synergistic enhanced effect of the drug uptake of target cancer cells, which may provide a new strategy to inhibit the multidrug resistance of the chemotherapy of cancer [66].

Meanwhile, nano titanium dioxide (TiO_2), as one of the most useful and well-studied photocatalytic materials with weak or no toxicity in vitro and in vivo, possess the specific ability to advance photochemical applications and can efficiently separate photogenerated charges that could facilitate some redox chemical reactions with attached biomolecules [71-77]. With UV illumination, TiO_2 photocatalyst in aqueous solution can generate reactive oxygen species (ROS) such as hydroxyl radical, hydrogen peroxide and superoxide, which makes it possible for the application of TiO_2 nanoparticles in the decomposition of some bioorganic compounds and apply in the photodynamic therapies of some kind of cancers [78-81].

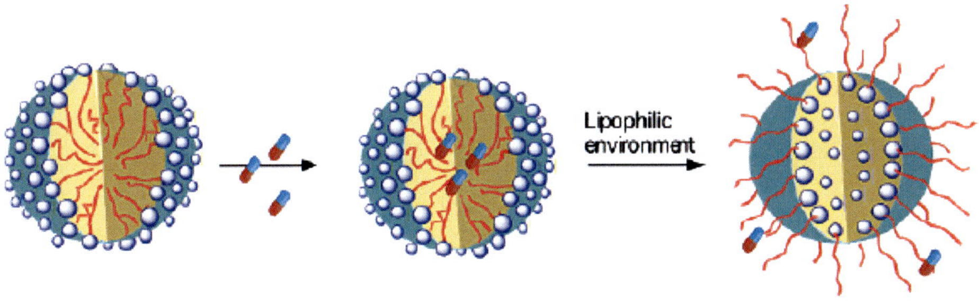

Figure 1. Enviroment-dependent conformational changes in biaryl-based amphilpilic dendrimers leading to probable targeted drug release systems. Reprinted with permission from ref. [70], Ambade, A. V.; Savariar, E. N.; Thayumanavan, S. (2005). Dendrimeric micelles for controlled drug release and targeted delivery. Molecular Pharmaceutics 2, 264-272., Copyright (2005) American Chemical Society.

Some researches illustrated that human colon carcinoma cells treated with photoexcited TiO_2 nanoparticles can be effectively affected and damaged [82]. We have also explored the combination of the high reactivity of TiO_2 nanoparticles and daunorubicin to inhibit multidrug resistance (MDR) of leukemia K562 cells. These observations indicate that TiO_2 nanoparticles could play an important role to facilitate the drug uptake of cancer cells and may also act as an effective anti-MDR agent to accompany with other anticancer drugs to inhibit the relative drug resistance and facilitate the relative chemotherapy efficiency. The rationale behind this strategy is to increase the intracellular concentration of the targeted drug by the synergistic enhancement effect of nano TiO_2 [83]. With the help of nanoparticles TiO_2 and UV irradiation, the anticancer drug daunorubicin could readily provoke some considerable changes of the cell membrane of the target leukemia cells, suggesting that TiO_2 nanoparticles could be fixed at the ailing tissue during treatment of the tumor followed by UV irradiation to operate photodynamic therapy (PDT) simultaneously with the relative efficient chemotherapy (as shown in Figure 2).

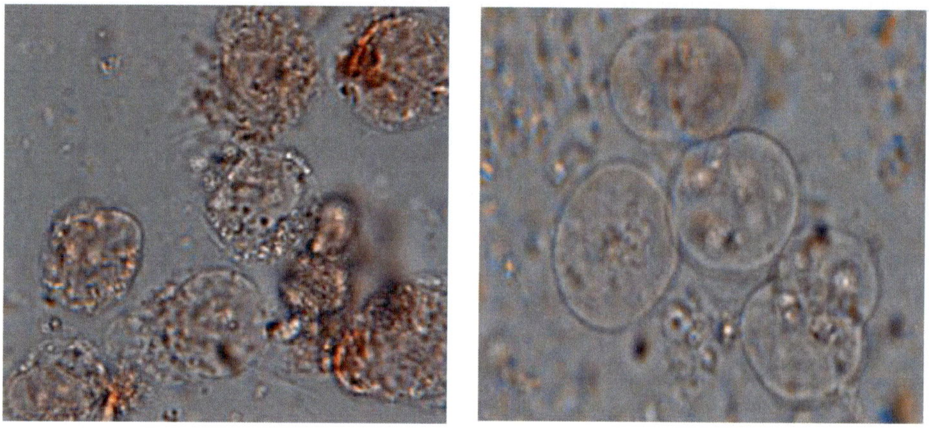

Figure 2. Microscopic image of drug resistant (left) and sensitive (right) leukemia K562 cells treated with daunorubicin in the presence of TiO2 nanoparticles after irradiation with UV for 100 seconds. All images were obtained after incubating the cells for 10 min. Scale Bar: 100×100μm. Reprinted with permission from ref. [83], Song, M.; Zhang, R. Y.; Dai, Y. Y.; Gao, F.; Chi, H. M.; Lv, G.; Chen, B. A.; Wang, X. M. (2006). The in Vitro Inhibition of Multidrug Resistance by Combined Nanoparticulate Titanium Dioxide and UV Irradition. Biomaterials 27, 4230-4238., Copyright (2006) Elsevier.

Based on the above observations, it is evident that the nanoparticles could not only increase the drug accumulation in target cancer cells but also act as an effective anti-MDR agent to inhibit the relative drug resistance, suggesting a promising perspective in relevant cancer diagnosis and therapy.

2.3. Functionalized Nanomaterials for Bio-Diagnostics

With the occurrence and development of nanotechnology, an increasing number of new nanomaterials with unique characteristics have been produced, some of which show promising and far-ranging prospect in the fields of medicine, especially in diagnosis and treatment of cancers.

2.3.1. Imaging of Nonfluorescent Nanoparticles

It is important to develop a technique to visualize the location of nanoparticles and assess their therapeutic efficacy in nano-oriented therapeutic interventions. Imaging techniques which show the ability to detect molecular signatures will not only target these abnormalities for therapy at the earliest possible stages but also will prove useful in further unraveling the molecular origins of cancer. Therefore, many efforts have been made to explore different imaging modalities [84-86].

Some particular nanoparticles with the unique properties have been potentially attractive vehicles for in vivo and in vitro imaging. To further enhance and widen diagnostic utility, contrast agents have been loaded onto nanoparticles for tumour diagnosis purposes. The physical and chemical features of the nanoparticles such as particle size, surface charge, surface coating and stability could affect the redirection and the concentration of the biomarker at the specific site of interest [87]. Smaller sized contrast agents less than 3 nm in diameter were easy to perfuse throughout the body. Contrast agents 3-6 nm in diameter that was quickly excreted through the kidney was suitable as functional renal contrast agents. Contrast agents 7-12 nm in diameter were retained in circulation and could be used as blood pool contrast agents. Due to the different hydrophobic / hydrophilic properties, it has been reported that hydrophobic variants of contrast agents could be quickly accumulated in the liver and used as liver contrast agents. Larger hydrophilic agents have suitable characteristics for lymphatic imaging.

Recently, nanosized contrast agents have been under development to improve the utility of MRI and computed tomography (CT) in imaging cancer. Labelled colloidal particles have already been used as radiodiagnostic agents. On the other hand, some non-labelled colloidal systems have been already in use and some are still being tested as contrast agents in related diagnosis procedures such as computed tomography and NMR imaging [87]. Moreover, NIR quantum dots have been used for in vivo lymphatic mapping in several animal models, while Peng et al. reported anti-ODNs carrier based on amino silica nanoparticles (NH_2SiNPs) and its preliminary applications in cancer cells [88-91] (as shown in Figure 3 and Figure 4). Meanwhile, superparamagnetic iron oxide contrast agents have been developed as MR contrast agents and used in clinical diagnosis. Furthermore, it has been demonstrated that nanoshells have been used in vivo as a contrast agent for imaging with optical coherence tomography [92] and photoacoustic tomography [93]. In addition, contrast agents conjugated with either monoclonal antibody could function as tumor-specific contrast agents and might also be used as either gadolinium neutron capture therapy or in conjunction with radioimmunotherapy [94]. SWNTs coupled to paramagnetic gadolinium particles are being developed for use as a high-performance MRI contrast agent [95]. Song et al. have developed a surface-modulated and highly biocompatible magnetic iron oxide (Fe_3O_4) nanocrystal probe that can be used for efficient intracellular labeling and their MRI applications [96].

2.3.2. Optical Diagnostics with Contrast Nanoagents

As we know, fluorescence labeling has been used as the most efficient tool to study living cells and targeted tissues [97-102]. However, conventional organic fluorescent dyes have some disadvantages such as photobleaching and a quick loss of staining due to diffusion of the fluorescent agent out of the cells that contribute to distort genuine cell properties [103-112].

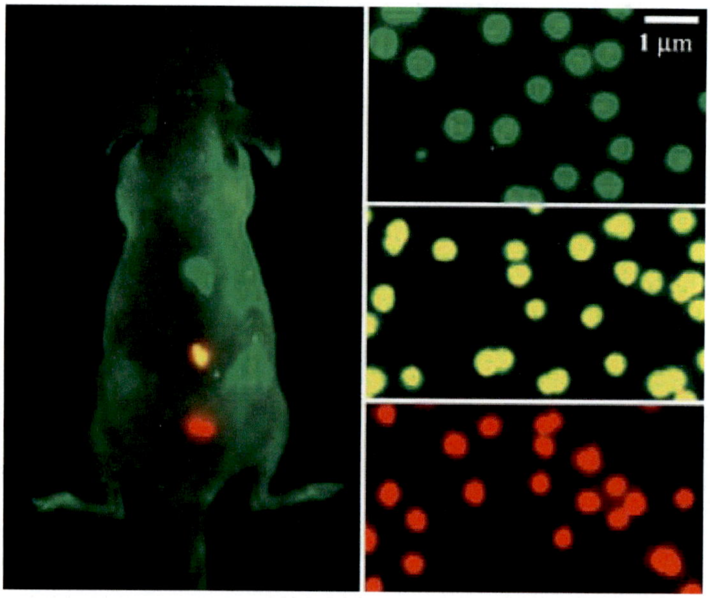

Figure 3. Multicolor quantum dot (QD) capability of QD imaging in live animals. Approximately 1 to 2 million in each color were injected subcutaneously at 3 adjacent locations on a host animal. Images were obtained with tungsten or mercury lamp excitation. Reprinted with permission from ref.[88], Gao, X.; Cui, Y.; Levenson, R. M.; Chung, L. W.; Nie, S. (2004). In vivo cancer targeting and imaging with semiconductor quantum dots. Nat Biotechnol. 22, 969-976., Copyright (2004) Nature Publishing Group.

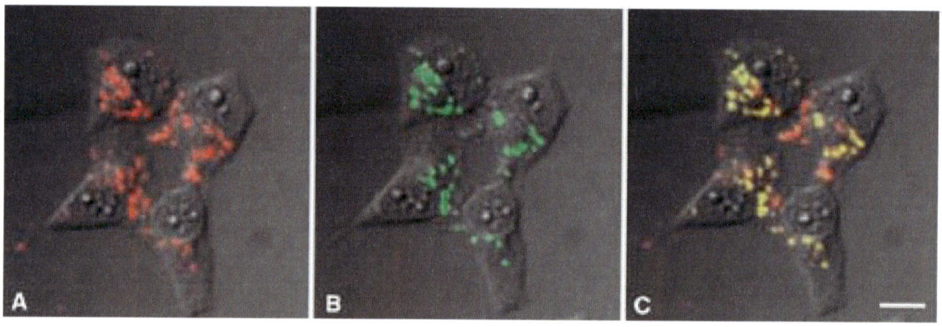

Figure 4. Localization of FNH2SiNPs-FODNs complexes internalized in Hela cells. A, RITC indicates the position of the NH2SiNPs; B, FITC indicates the position of the anti-ODNs; C, Merged image of A and B indicates that both fluorescent signals are superimposed and co-localized species appear as yellow. Scale bar in figure represented 10 Am. Reprinted with permission from ref. [89], Peng, J.; He, X.; Wang, K.; Tan, W.; Li, H.; Xing, X.; Wang, Y. (2006). An antisense oligonucleotide carrier based on amino silica nanoparticles for antisense inhibition of cancer cells. *Nanomedicine: Nanotechnology, Biology, and Medicine* 2, 113-120., Copyright (2006) Elsevier.

Recently, much attention has been paid on developing nanomaterials such as semiconductor nanocrystals as fluorescent contrast agents. It has been demonstrated that these fluorescent contrast nano-agents could efficiently resist bleaching longer than conventional fluorescent dyes [113-116]. In addition, metal nanoparticles could be readily utilized for optical labeling for the reason that they do not photobleach or blink (i.e., photostable), and can be easily conjugated with many proteins and antibodies [117-119]. Gold nanoshells show promise for many biomedical applications, in particular hyperthermia induction for the treatment of cancers [120]. Chen et al have illustrated that gold nanocages are promising as both a contrast agent for optical diagnostics of cancers and as a potential thermal therapeutic agent for cancer treatment [121]. Sokolov et al. successfully used gold nanoparticles conjugated to EGFR antibodies to label cervical biopsies for identification of precancerous lesions [122]. Lee et al. have explored novel antifouling poly (TMSMA-*r*-PEGMA) SPION and demonstrated the feasibility of the nanoprobe as an MR contrast agent for in vivo cancer imaging [123]. Furthermore, SWNTs were engineered to absorb NIR light and subsequently were used for photothermal ablation applications in vitro [53].

2.3.3. Nanomaterials Based Biosensors for Cancer Monitoring

The identification of molecular signatures of cancers will enhance our ability to identify and accurately diagnose disease. Our goal is to use this information to identify cancers or precancers at the earliest point and intervene before symptomatic disease becomes apparent. Therefore, it is necessary to seek technologies that will support the earliest detection of the molecular signatures of cancer.

Molecular biomarkers, which play a key role in facilitating advances in disease detection and prevention, have been used for some time in laboratories and clinics to monitor metabolic status and disease in tissues and cells. The sensitive and specific biomarker may afford a new strategy for the early diagnosis of targeted cancers to detect and prevent the progression of precancerous lesions. Nanometer-sized particles such as semiconductor quantum dots and iron oxide nanocrystals, have extended application in cancer diagnosis and therapy due to their novel optical, electronic, magnetic or structural properties. These nanoparticles that have been linked with tumor-targeting ligands, such as monoclonal antibodies [94], peptide fragments of tumor-specific proteins or small molecules [124,125], can be used to target tumor antigens (biomarkers) and tumor vasculatures with high affinity and specificity.

Recent advances in the semiconductor nanocrystals (quantum dots) open a promising field toward the development of a new generation of luminescent biomarkers. These luminescent dots with large surface areas and functional groups have been functionalized to couple biomolecules and therapeutic (e.g., anticancer) agents [126-128]. Moreover, some researchers have developed novel luminophore-doped silica nanoparticles as efficient and uniform biomarkers that combine the excellent biospecificity of designated biomolecules with optically stable luminophores [129]. In addition, nanotubes have been functionalized to adsorb different molecules and antigens to their surface for specifically targeting tumor cells [53]. Recently, Fuente and Berry reported that Au NPs functionalized with a Tat protein-derived peptide could pass through nucleus pores with the potential application of cell imaging and other biomedical detection [130].

In our study, we have explored the possibility of adopting the 3-mercaptopropionic acid (MPA) capped Au NPs as a sensitive and efficient way to facilitate the respective biomarking/and accumulating of anticancer drug daunorubicin in drug-resistant tumor cells of human leukemia K562 cell lines. Our observations demonstrate that the combination of the anticancer drug daunorubicin with the 3-mercaptopropionic acid (MPA) capped Au nanoparticles could be further explored as a new way to inhibit the multidrug resistance of the targeted tumor or cancer cells and as a sensitive biomarker for the early diagnosis of the respective cancers [131].

Biosensors that provide quantitative or semi-quantitative analytical information have enormous potential for the early diagnosis and classification of different forms of cancers [132,133]. The relevant biosensors must be simple to handle, small, cheap, and able to provide reliable information; especially, they need to be sensitive and selective for the target bio-analyte and suitable for in situ monitoring [134]. Thus, biosensors can be used to not only identify and quantify new biological molecules but also study biological processes involved in cancer development and progression. The possibility of using noninvasive analytical systems to monitor biochemical process within a single cell could provide great advances in understanding cellular function and may offer an enormous potential in relevant cell biology [135,136].

Nanotechnology is in a unique position to transform cancer diagnostics and to produce a new generation of biosensors and medical imaging techniques with higher sensitivity and precision of recognition. The combination of nanotechnology with biology has received much attention in recent years widely due to the desire to use biomolecular recognition to aid nanoscale assembly and to create sensitive biosensors based on nanoscale devices [137-142]. For example, colloidal gold nanoparticle matrix has been used as a mimetic interface for cell proliferation and assembly of human hepatocytes to construct a cellular biosensor [138]. Vo-Dinh et al. designed an antibody-based fiber optic nanosensor for in situ monitoring of benzo[a]pyrene (BaP) in a single MCF-7 human breast carcinoma cell line [139]. Schmidt et al. explored a biosensor system for measuring human telomerase activity [140]. Torabi et al. designed an amperometric reticulated vitreous carbon enzymatic sensor for coulometric detection of NADH/NAD^+ in normal and cancerous cells [142]. Many studies have focused on electric behavior and characters of living cells using electrochemical methods for the reason that a series of very important biochemical and physiological processes in a living cell may have very intimate relation with the electron behavior [143] (as shown in Figure 5).

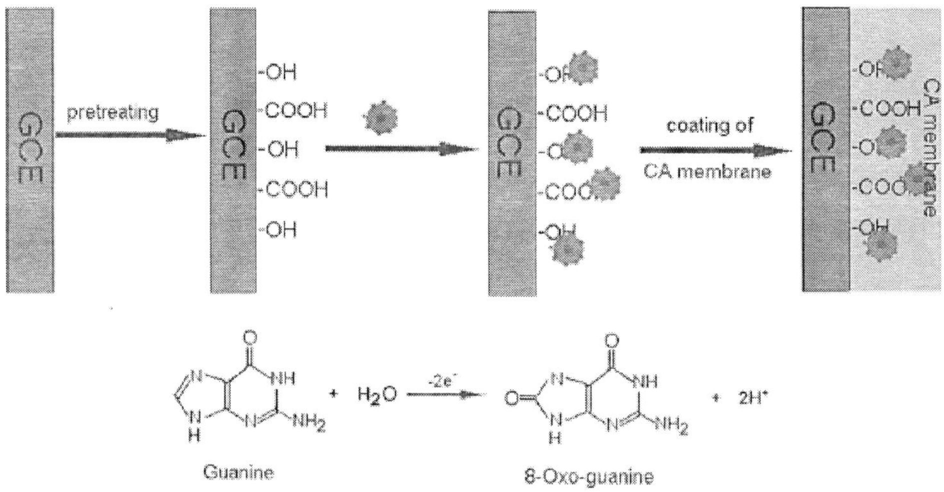

Figure 5. Preparation process of cellular biosensor (CA/leukemia K562 cells-Au/GCE) and the oxidization of guanine. are colloidal gold nanoparticles modified leukemia K562 cell. Reprinted with permission from ref. [143], Yan, F.; Chen, J.; Ju, H. (2007). Immobilization and electrochemical behavior of gold nanoparticles modified leukemia K562 cells and application in drug sensitivity test. Electrochemistry Communications 9, 293-298., Copyright (2007) Elsevier.

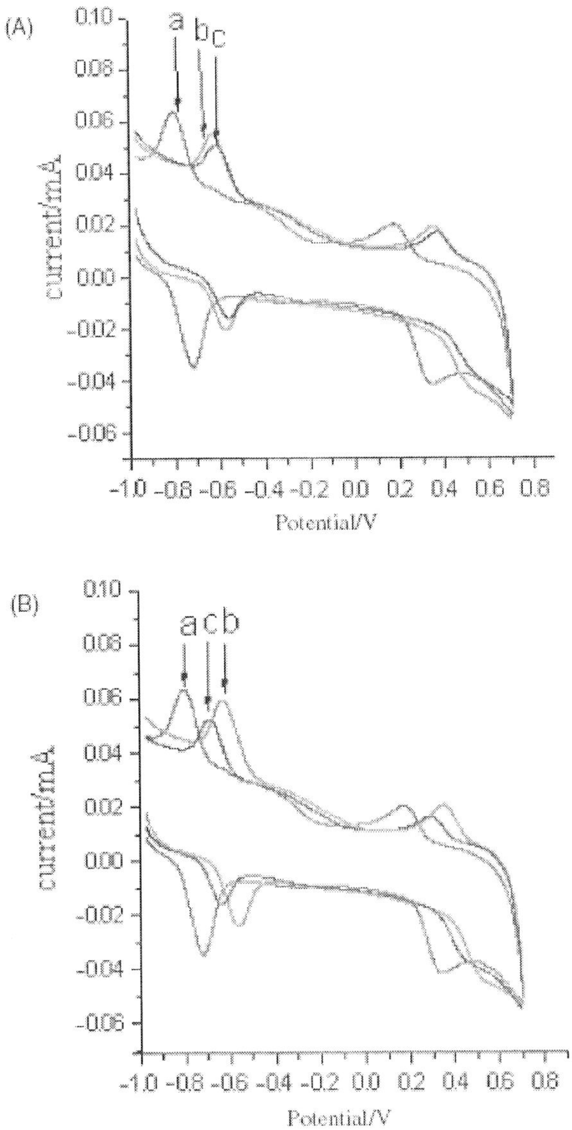

Figure 6. Cyclic voltammetry (CV) of doxorubicin (DOX) residue outside cells in the absence and presence of CdS NPs (2.0×10-6 M), A: (a) CV of DOX (1.0×10-4 M), (b) CV of DOX residue outside K562/B.W with free CdS NPs, (c) CV of DOX residue outside K562/B.W with CdS NPs. B: (a) CV of DOX (1.0×10-4 M), (b) CV of DOX residue outside K562/ADM with free CdS NPs, (c) CV of DOX residue outside K562/ADM with CdS NPs. Scan Rate: 0.1V/s. Reprinted with permission from ref. [148], Li, J.; Wu, C.; Dai, Y.; Zhang, R.; Wang, X.; Fu, D.; Chen, B. (2007). Doxorubicin-CdS nanoparticles: a potential anticancer agent for enhancing the drug uptake of cancer cells. *J Nanosci Nanotechno*. 7, 1-5., Copyright (2007) American Scientific Publishers.

The electrochemical biosensor was used to detect and monitor real changes in cell behavior using changes in the electrochemical potential at the cell/surface interface in the extracellular microenvironment. Intracellular redox activity of initiation of apoptosis and the sensitivity of tumor cells to antitumor drugs can be also explored by cyclic voltammetric behavior [144, 145]. With the help of electron transfer mediators, amperometry can be applied for real-time monitoring of intracellular redox enzyme activity [146]. A novel electrochemical technique was designed to study the effect of several anticancer drugs on the human ovarian carcinoma cell line A2780 and its adriamycin (A2780adr) and cisplatin (A2780cispt) resistant variant [147]. The electrochemical studies could also provide fresh evidence for the synergistic enhancement effect of nanoparticles on the drug uptake of anticancer drug in leukemia K562 cell lines [83, 148]. As shown in Figure 6, the corresponding electrochemical results suggest that much more doxorubicin molecules have been absorbed into leukemia K562 cells in the presence of CdS nanoparticles [148].

Meanwhile, scanning electrochemical microscopy (SECM) has been developed as novel technique for monitor living cells by measuring the intracellular redox activity [149]. Scanning electrochemical microscope has been used to probe the redox activity of individual purple bacteria [150] and detect the nitric oxide release from human umbilical vein endothelial cells with NO-sensors [151]. Electrochemical impedance methods have been utilized for rapid detection of Escherichia coli O157:H7 [152] and the study of the attachment and spreading of fibroblast cells [153].

3. Conclusion

In summary, nanotechnology plays a vital role in various biomedical applications. The strategy of nanomaterials based biomarker and targeting could result in an efficient drug delivery of anticancer drugs in selected target tissues and a reduction of systemic side effects and toxicity. Different kinds of nanomaterials could act as different /or specific biomarkers and drug carriers to effectively facilitate drug accumulation and reverse the drug resistance. Besides, nanomaterials could be functionalized with various kinds of target molecules for the potential application of cell imaging and other biomedical detection. Based on the recent development of nanotechnology, nanomaterials display promising and far-ranging prospect in the fields of biomedical engineering, especially in early diagnosis and treatment of cancers. However, many efforts are still needed to explore in clinical research. And it is desirable to develop the in vivo detection strategy as both new therapeutic and diagnostic tools.

Acknowledgements

The support from NCET (050462) and NSFC (20675014, 20535010 and 60121101) is greatly appreciated.

References

[1] Farokhzad, O. C.; Khademhosseini, A.; Jon, S.; Hermmann, A.; Cheng, J.; Chin, C.; Kiselyuk, A.; Teply, B.; Eng, G.; Langer, R. (2005). Microfluidic system for studying the interaction of nanoparticles and microparticles with cells . *Anal. Chem.* 77, 5453-5559.

[2] Akerman, M. E.; Chan, W. C.; Laakkonen, P.; Bhatia, S. N.; Ruoslahti, E. (2002). Nanocrystal targeting in vivo. *Proc. Natl. Acad. Sci. USA 99,* 12617-12621.

[3] Li, J.; Wu, C.; Gao, F.; Zhang, R.; Lv, G.; Fu, D.; Chen, B.; Wang, X. (2006). In vitro study of drug accumulation in cancer cells via specific association with CdS nanoparticles. *Bioorg. Med. Chem. Lett. 16,* 4808-4812.

[4] Liu, Z. S.; Tang, S. L.; Ai, Z. L. (2003). Effects of hydroxyapatite nanoparticles on proliferation and apoptosis of human hepatoma BEL-7402cells. *World J. Gastroenterol. 9,* 1968-1971.

[5] Tkachenko, A. G.; Xie, H.; Coleman, D.; Glomm, W.; Ryan, J.; Anderson, M. F.; Franzen, S.; Feldheim, D. L. (2003). Multifunctional gold nanoparticle-peptide complexes for nuclear targeting. *J. Am. Chem. Soc. 125,* 4700-4701.

[6] Santra, S.; Zhang, P.; Wang, K.; Tapec, R.; Tan, W. (2001). Conjugation of biomolecules with luminophore-doped silica nanoparticles for photostable biomarkers. *Anal. Chem. 73,* 4988-4993.

[7] He, Q.; Mitchell, A. R.; Johnson, S. L.; Wagner-Bartak, C.; Morcol, T.; Bell, S. J. (2000). Calcium phosphate nanoparticle adjuvant. *Clin. Diagn. Lab. Immunol. 7,* 899-903.

[8] Hans, M. L.; Lowman, A. M. (2002). Biodegradable nanoparticles for drug delivery and targeting. *Current Opinion in Solid State and Materials Science 6,* 319-327.

[9] Qiu, H.; Rieger, J.; Gilbert, B.; Jérôme, R.; Jérôme, C. (2004). PLA-coated gold nanoparticles for the labeling of PLA biocarriers. *Chem. Mater. 16,* 850-856.

[10] Nam, J. Y.; Ray, S. S.; Okamoto, M. (2003). Crystallization behavior and morphology of biodegradable polylactide/layered silicate nanocomposite. *Macromolecules. 36,* 7126-7131.

[11] Kim, S. H.; Jeong, J. H.; Chun, K. W.; Park, T. G. (2005). Target-specific cellular uptake of PLGA nanoparticles coated with poly (L-lysine)-poly (ethylene glycol)-folate conjugate. Langmuir 21, 8852-8857.

[12] Soppimath, K. S.; Aminabhavi, T. M.; Kulkarni, A. R.; Rudzinski, W. E. (2001). Biodegradable polymeric nanoparticles as drug delivery devices. *J. Controlled Release. 70,* 1-20.

[13] Sakhalkar, H. S.; Dalal, M. K.; Salem, A. K.; Ansari, R.; Fu, J.; Kiani, M. F.; Kurjiaka, D. T.; Hanes, J.; Shakesheff, K. M.; Goetz, D. J. (2003). Leukocyte-inspired biodegradable particles that selectively and avidly adhere to inflamed endothelium in vitro and in vivo. *Proc. Natl. Acad. Sci. U.S.A. 100,* 15895-15900.

[14] Peppas, N. A.; Bures, P.; Leobandung, W.; Ichikawa, H. (2000). Hydrogels in pharmaceutical formulations. *Eur. J. Pharm. Biopharm. 50,* 27-46.

[15] Kiserab, P. F.; Wilson, G.; Needham, D. (2000). Lipid-coated microgels. for triggered release of doxorubicin. *J. Control Release 68,* 9-22.

[16] Zhang, R. Y.; Wang, X. M. (2007). One step synthesis of multi-walled carbon nanotube /gold (MWCNT-Au) nanocomposites for enhancing electrochemical response, *Chem. Mater., 19,* 976-978.

[17] Ye, X. R.; Lin, Y. H.; Wang, C. M.; Wai, C. M. (2003). Supercritical fluid fabrication of metal nanowires and nanorods templated by multiwalled carbon nanotubes. *Adv. Mater. 15,* 316-319.

[18] Xu, S.; Zhang, J.; Paquet, C.; Lin, Y; Kumacheva, E. (2003). From hybrid microgels to photonic crystals. *Adv. Funct. Mater. 13,* 468–472.

[19] Zhao, X.; Ding, X.; Deng, Z.; Zheng, Z.; Peng, Y.; Long, X. (2005). Thermoswitchable Electronic Properties of a. Gold Nanoparticle/Hydrogel Composite. *Macromol. Rapid Commun. 26,* 1784-1787.

[20] Sadzuka, Y.; Nakai, S.; Miyagishima, A.; Nozawa, Y.; Hirota, S. (1997). Effects of administered route on tissue distribution and antitumor activity of polyethyleneglycol-coated liposomes containing adriamycin. *Cancer Lett. 111,* 77-86.

[21] Zou, Y.; Ling, Y.; Reddy, S.; Priebe, W.; Perez-soler, R. (1995). Effect of vesicle size and lipid composition on the in vivo tumor selectivity and toxicity of the non-cross-resistant anthracycline Anamycin incorporated in liposomes. *Int. J. Cancer 61,* 666-671.

[22] Klibanov, A. L.; Maruyama, K.; Beckerleg, A. M.; Torchilin, V. P.; Huang, L. (1991). Activity of amphipathic poly(ethylene glycol) 5000 to prolong the circulation time of liposomes depends on the liposome size and is unfavorable for immunoliposome binding to target. *Biochim.Biophys. Acta 1062,* 142-148.

[23] Ramachandra, M.; Ambudkar, S. V.; Chen, D.; Hrycyna, C. A.; Dey, S.; Gottesman, M. M.; Pastan, I. (1998). Human P-Glycoprotein Exhibits Reduced Affinity for Substrates during a Catalytic Transition State. *Biochemistry* 37, 5010-5019.

[24] Némati, F.; Dubernet, C.; Fessi, H.; Colin de Verdière, A.; Poupon, M. F.; Puisieux, F.; Couvreur, P. (1996). Reversion of multidrug resistance using nanoparticles in vitro: Influence of the nature of the polymer. *International Journal of Pharmaceutics 138,* 237-246.

[25] Kruh, G. D. (2003). Introduction to resistance to anticancer agents. *Oncogene. 22,* 7262-7264.

[26] Chen, Y.; Gratzl, M. (1998). Continuous in situ electrochemical monitoring of doxorubicin efflux from sensitive and drug-resistant cancer cells. *Biophys. J. 75,* 2255-2261.

[27] Lambert, G.; Fatta, E.; Pinto-Alphandary, H.; Gulik, A.; Couvreur, P. (2000). Polyisobutylcyanoacrylate nanocapsules containing an aqueous core as a novel colloidal carrier for the delivery of oligonucleotides. *Pharm. Res. 17,* 707-714.

[28] Astier, A.; Doat, B.; Ferrer, M. J.; Benoit, G.; Fleury, J.; Rolland, A.; Leverge, R. (1988). Enhancement of adriamycin antitumor activity by its binding with an intracellular sustained release form, polymethymethacrylate nanospheres, in U-937 cells. *Cancer Res.* 48, 1835-1841.

[29] Jain, K. K. (2005). Nanotechnology-based drug delivery for cancer. *Technol Cancer Res Treat. 4,* 407-416.

[30] Hu, Z.; Jin, S.; Scotto, K. W. (2000). Transcriptional activation of the MDR1 gene by UV irradiation: Role of NF-Y and SP1. *Journal of Biological Chemistry 275,* 2979-2985.

[31] Clement, O.; Rety, F.; Cuenod, C. A.; Siauve, N.; Carnot, F.; Bordat, C.; Siche, M.; Frija, G. (1998). MR lymphography: evidence of extravasation of superparamagnetic nanoparticles into the lymph. *Acad. Radiol. 5, Suppl 1.* S170-172.

[32] Monsky, W. L.; Fukumura, D.; Gohongi, T.; Ancukiewcz, M.; Weich, H. A.; Torchilin, V. P.; Yuan, F.; Jain, R. K. (1999). Augmentation of transvascular transport of macromolecules and nanoparticles in tumors using vascular endothelial growth factor. *Cancer Res. 59,* 4129-4135.

[33] Winet, H.; Hollinger, J. O.; Stevanovic, M. (1995). Incorporation of polylactide-polyglycolide in a cortical defect: neoangiogenesis and blood supply in a bone chamber. *J. Orthop. Res. 13,* 679-689.

[34] Guzman, L. A.; Labhasetwar, V.; Song, C.; Jang, Y.; Lincoff, A. M.; Levy, R.; Topol, E. J. (1996). Local intraluminal infusion of biodegradable polymeric nanoparticles. A novel approach for prolonged drug delivery after balloon angioplasty. *Circulation 94,* 1441-1448.

[35] Langer, R.; Peppas, N. A. (2003). Advances in Biomaterials, Drug Delivery, and Bionanotechnology. *AIChE J. 49,* 2990-3006.

[36] Vauthier, C.; Dubernet, C.; Chauvierre, C.; Brigger, I.; Couvreur, P. (2003). Drug delivery to resistant tumors: The potential of poly (alkyl cyanoacrylate) nanoparticles. *Journal of Controlled Release 93,* 151-160.

[37] Treupel, L.; Poupon, M. F.; Couvreur, P.; Puisieux, F. (1991). Vectorisation of doxorubicin in nanospheres and reversion of pleiotropic resistance of tumor cells. *C. R. Acad. Sci. III 313,* 171-174.

[38] Cuvier, C.; Roblot-Treupel, L.; Millot, J. M.; Lizard, G.; Chevillard, S.; Manfait, M.; Couvreur, P.; Poupon, M. F. (1992). Doxorubicin-loaded nanospheres bypass tumor cell multidrug resistance. *Biochem. Pharmacol. 44,* 509-517.

[39] Griffiths, J. R.; Glickson, J. D. (2000). Monitoring pharmacokinetics of anticancer drugs: non-invasive investigation using magnetic resonance spectroscopy. *Adv. Drug. Deliv. Rev. 41,* 75-90.

[40] Patolsky, F.; Weizmann, Y.; Katz, E.; Willner, I. (2004). Amplified telomerase analysis by using rotating magnetic particles: The rapid and sensitive detection of cancer cells. *ChemBioChem* 5, 943-948.

[41] Moffat, B. A.; Reddy, G. R.; McConville, P.; Hall, D. E.; Chenevert, T. L.; Kopelman, R. R.; Philbert, M.; Weissleder, R.; Rehemtulla, A.; Ross, B. D. (2003). A novel polyacrylamide magnetic nanoparticle contrast agent for molecular imaging using MRI. *Mol. Imaging 2,* 324-332.

[42] Bjornerud, A.; Johansson; L. O.; Ahlstrom, H. K. (2001). Pre-clinical results with Clariscan (NC100150 Injection); experience from different disease models. *MAGMA 12,* 99-103.

[43] Kircher, M. F.; Allport, J. R.; Graves, E. E.; Love, V.; Josephson, L.; Lichtman, A. H.; Weissleder, R. (2003). In vivo high resolution three-dimensional imaging of antigen-specific cytotoxic T-lymphocyte trafficking to tumors. *Cancer Res. 63,* 6838-6846.

[44] Josephson, L.; Tung, C. H.; Moore, A.; Weissleder, R. (1999). High-efficiency intercellular magnetic labeling with novel superparamagnetic-Tat peptide conjugates. *Bioconjugate Chem. 10,* 186-191.

[45] Zhao, M.; Kircher, M. F.; Josephson, L.; Weissleder, R. (2002). Differential conjugation of tat peptide to. superparamagnetic nanoparticles and its effect on cellular uptake. *Bioconjugate Chem. 13,* 840-844.

[46] Schellenberger, E. A.; Bogdanov, A.Jr.; Hogemann, D.; Tait, J.; Weissleder, R.; Josephson, L. (2002). A Nanopartical for Detection Apoptosis by MRI. *Mol. Imaging 1* 102-107.

[47] Högemann, D.; Josephson, L.; Weissleder, R.; Basilion, J. P. (2000). Improvement of MRI probes to allow efficient detection of gene expression. *Bioconjugate Chem. 11,* 941-946.

[48] Jain, T. K.; Morales, M. A.; Sahoo, S. K.; Leslie-Pelecky, D. L.; Labhasetwar, V. (2005). Iron-oxide nanoparticles for sustained delivery of anticancer agents. *Molecular Pharmaceutics 2,* 194-205.

[49] Wagner, K.; Kautz, A.; Röder, M.; Schwalbe, M.; Pachmann, K.; Clement, J. H.; Schnabelrauch, M. (2004). Synthesis of oligonucleotide-functionalized magnetic nanoparticles and study on their *in vitro* cell uptake. *Applied Organometallic Chemistry 18,* 514-519.

[50] Zhang, Y.; Kohler, N.; Zhang, M. Q. (2002). Surface modification ofsuperparamagnetic magnetite nanoparticles and their intracellular uptake. *Biomaterials 23,* 1553-1561.

[51] Lambert, G.; Fattal, E.; Couvreur, P. (2001). Nanoparticlate systems for the delivery of antisense oligonucleotides. *Adv. Drug. Deliv. Rev. 47,* 99-112.

[52] Kam, N. W.; Dai, H. (2005). Carbon nanotubes as intracellular protein transporters: generality and biological functionality. *J. Am. Chem. Soc. 127,* 6021-6026.

[53] Kam, N.W.; O'Connell, M.; Wisdom, J.A.; Dai, H. (2005). Carbon nanotubes as multifunctional biological transporters and nearinfrared agents for selective cancer cell destruction. *Proc Natl Acad Sci USA. 102,* 11600-111605.

[54] Paciotti, G.F.; Myer, L.; Weinreich, D.; Goia, D.; Pavel, N; McLaughlin, R. E.; Tamarkin, L. (2004). Colloidal gold: a novel nanoparticle vector for tumor directed drug delivery. *Drug Deliv. 11,* 169-183.

[55] Kam, N. W. S.; Jessop, T. C.; Wender, P. A.; Dai, H. (2004). Nanotube molecular transporters:Internalization of carbon nanotube-protein conjugates into mammalian cells. *J. Am. Chem. Soc. 126,* 6850-6851.

[56] Peppas, N. A.; Keys, K. B.; Torres-Lugo, M.; Lowman, A. M. (1999). Poly (ethylene glycol)-containing hydrogels in drug delivery. *J. Controlled Release 62,* 81-88.

[57] Zhang, H.; Mardyani, S.; Chan, W. C. W.; Kumacheva, E. (2006). Design of biocompatible chitosan microgels for targeted pH-mediated intracellular release of cancer therapeutics. *Biomacromolecules 7,* 1568-1572.

[58] Krishna, R.; Mayer, L. D. (2000). Multidrug resistance (MDR) in cancer: mechanisms,reversal using modulators of MDR and the role of MDR modulators in influencing the pharmacokinetics of anticancer drugs. *Eur. J. Cancer Sci. 11,* 265-283.

[59] Brigger, I.; Dubernet, C.; Couvreur, P. (2002). Nanoparticles in cancer therapy and diagnosis. Adv. *Drug Deliv. Rev. 54,* 631-651.

[60] Roninson, I.B. (1987). Molecular mechanism of multidrug resistance in tumor cells. *Clin Physiol Biochem 5,* 140-51.

[61] Larsen, A.K.; Escargueil, A. E.; Skladanowski, A. (2000). Resistance mechanisms associated with altered intracellular distribution of anticancer agents. *Pharmacol. Ther. 85,* 217-229.

[62] Goodrich, G. P.; Helfrich, M. R.; Overberg, J. J.; Keating, C. D. (2004). The effect of macromolecular crowding on DNA: Au nanoparticle bioconjugate assembly. *Langmuir 20,* 10246-10251.

[63] Tsuruo, T.; Iida, H.; Tsukagoshi, S.; Sakurai, Y. (1981). Overcoming of vincristine resistance in P388 leukemia in vivo and in vitro through enhanced cytotoxicity of vincristine and vinblastine by verapamil. *Cancer Res. 41,* 1967-1972.

[64] Sibley, L. D.; Andrews, N. W. (2000). Cell invasion by un-palatable parasites. *Traffic 1,* 100-106.

[65] Van Asperen, J.; vans Tellingen, O.; van der Valk, M.; Rozenhart, M.; Beijnen, J. (1998). Enhanced oral absorption and decreased elimination of paclitaxel in mice cotreated with cyclosporin A. *Clin. Cancer. Res. 4,* 2293-2297.

[66] Zhang, R.; Wang, X.; Wu, C.; Song, M.; Li, J.; Lv, G.; Zhou, J.; Chen, C.; Dai, Y.; Gao, F.; Fu, D.; Li, X. Guan, Z.; Chen, B. (2006). Synergistic enhancement effect of magnetic nanoparticles on anticancer drug accumulation in cancer cells. *Nanotechnology 17,* 3622-3626.

[67] Wang, X.; Zhang, R.; Wu, C.; Dai, Y.; Song, M.; Gutmann, S.; Gao, F.; Lv, G.; Li, J.; Li, X.; Guan, Z.; Fu, D.; Chen, B. (2007). The application of Fe_3O_4 nanoparticles in cancer research: A new strategy to inhibit drug resistance. *Journal of Biomedical Materials Research Part A. 80A,* 852-860.

[68] Sadzuka, Y.; Nakai, S.; Miyagishima, A.; Nozawa, Y.; Hirota, S. (1997). Effects of administered route on tissue distribution and antitumor activity of polyethyleneglycol-coated liposomes containing adriamycin. *Cancer Lett. 111,* 77-86.

[69] Zou, Y.; Ling, Y.; Reddy, S.; Priebe, W.; Perez-soler, R. (1995). Effect of vesicle size and lipid composition on the in vivo tumor selectivity and toxicity of the non-cross-resistant anthracycline anamycin incorporated in liposomes. *Int. J. Cancer 61,* 666-671.

[70] Ambade, A. V.; Savariar, E. N.; Thayumanavan, S. (2005). Dendrimeric micelles for controlled drug release and targeted delivery. *Molecular Pharmaceutics 2,* 264-272.

[71] Chowdhury, D.; Paul, A.; Chattopadhyay, A. (2005). Photocatalytic polypyrrole-TiO_2-nanoparticles composite thin film generated at the air-water interface. *Langmuir 21,* 4123-4128.

[72] Furuzono, T.; Iwasaki, M.; Yasuda, S.; Korematsu, A.; Yoshioka, T.; Ito, S.; Kishida, A. (2003). Photoreactivity and cell adhesiveness of amino-group-modified titanium dioxide nano-particles on silicone substrate coated by covalent linkage. *Journal of Materials Science Letters 22,* 1737-1740.

[73] Bernard, B. K.; Osheroff, M. R.; Hofman, A.; Mennear, J.H. (1990). Toxicology and carcinogenesis studies of dietary titanium dioxide-coated mica in male and female Fischer 344 rats. *J Toxicol Environ Health 29,* 417-429.

[74] Miller, B. M.; Pujads, E.; Gocke, E. (1995). Evaluation of the micronucleus test in vitro using Chinese hamster cells: results of four chemicals weakly positive in the in vivo micronucleus test. *Environ. Mol. Mutagen. 26,* 240-247.

[75] Tennant, R. W.; Margolin, B. H.; Shelby, M. D.; Zeiger, E.; Haseman, J. K.; Spalding, J.; Caspary, W.; Resnick, M.; Stasiewicz, S.; Anderson, B.; Minor, R. (1987). Prediction of chemical carcinogenicity in rodents from in vitro genetic toxicity assays. *Science 236,* 933-941.

[76] Barnard, A. S.; Zapol, P.; Curtiss, L. A. (2005). Modeling the Morphology and Phase Stability of TiO_2 Nanocrystals in Water. *J. Chem. Theory Comput. 1,* 107-116.

[77] Dimitrijevic, N. M.; Saponjic, Z. V.; Rabatic, B. M.; Rajh, T. (2005). Assembly and Charge Transfer in Hybrid TiO_2 Architectures Using Biotin-Avidin as a Connector. *J. Am. Chem. Soc.127,* 1344-1345.

[78] Lu, Z. X.; Zhou, L.; Zhang, Z. L.; Shi, W. L.; Xie, Z. X.; Xie, H. Y.; Pang, D. W.; Shen, P. (2003). Cell damage induced by photocatalysis of TiO_2 thin films. *Langmuir 19,* 8765-8768.

[79] Wang, K. H.; Hsieh, Y. H.; Ko, R. C.; Chang, C. Y. (1999). Photocatalytic degradation of wastewater from manufactured fiber by titanium dioxide suspensions in aqueous solution. *Environ. Int. 25,* 671-676.

[80] Muneer, M.; Singh, H. K.; Bahnemann, D. (2002). Semiconductor-mediated photocatalysed degradation of two selected priority organic pollutants, benzidine and 1, 2-diphenylhydrazine, in aqueous suspension. *Chemosphere 49,* 193-203.

[81] Cai, R.; Kubota, Y.; Shiumi, T.; Sakai, H.; Hashimoto, K.; Fujishima, A. (1992). Induction of cytotoxicity by photoexited titanium dioxide particles. *Cancer Res. 52,* 2346-2348.

[82] Zhang, A. P.; Sun, Y. P. (2004). Photocatalytic killing effect of TiO2nanoparticles on Ls-174-t human colon carcinoma cells. *World J Gastroenterol 10,* 3191-3193.

[83] Song, M.; Zhang, R. Y.; Dai, Y. Y.; Gao, F.; Chi, H. M.; Lv, G.; Chen, B. A.; Wang, X. M. (2006). The in Vitro Inhibition of Multidrug Resistance by Combined Nanoparticulate Titanium Dioxide and UV Irradition. *Biomaterials 27,* 4230-4238.

[84] Slayter, E. M. (1993). Light and electron microscopy. Cambridge7 Cambridge University Press.

[85] Engel, A.; Lyubchenko, Y.; Muller, D. (1999) Atomic force microscopy: a powerful tool to observe biomolecules at work. *Trends Cell Biol 9,* 77-80.

[86] Zharov, V. P.; Kim, J. W.; Curiel, D. T.; Everts, M. (2005). Self-assembling nanoclusters in living systems -application for integrated photothermal nanodiagnostics and therapy. *Nanomedicine: Nanotechnology, Biology, and Medicine 1,* 326-345.

[87] Ernest, H.; Shetty, R. (2005). Impact of nanotechnology on biomedical sciences: review of current concepts on convergence of nanotechnology with biology. *AZoNano-Online Journal of Nanotechnology 1,* 1-14.

[88] Gao, X.; Cui, Y.; Levenson, R. M.; Chung, L. W.; Nie, S. (2004). In vivo cancer targeting and imaging with semiconductor quantum dots. *Nat Biotechnol. 22,* 969-976.

[89] Peng, J.; He, X.; Wang, K.; Tan, W.; Li, H.; Xing, X.; Wang, Y. (2006). An antisense oligonucleotide carrier based on amino silica nanoparticles for antisense inhibition of cancer cells. *Nanomedicine: Nanotechnology, Biology, and Medicine 2,* 113-120.

[90] Chen, F. Y.; Colson, Y. L.; Cohn, L. H.; Bawendi, M. G.; Frangioni, J. V. (2005). Intraoperative identification of esophageal sentinel lymph nodes with nearinfrared fluorescence imaging. *J Thorac Cardiovasc Surg. 129,* 844-850.

[91] Parungo, C. P.; Ohnishi, S.; De Grand, A. M.; Laurence, R.G.; Soltesz, E.G.; Colson, Y.L.; Kang, P.M.; Mihaljevic, T.; Cohn, L. H.; Frangioni, J. V. (2004). In vivo optical imaging of pleural space drainage to lymph nodes of prognostic significance. *Ann Surg Oncol. 11,* 1085-1092.

[92] Barton, J.; Halas, N. J.; West, J.; Drezek, R. (2004). Nanoshells as an optical coherence tomography contrast agent. Tuchin, V.; Izatt, J.; Fujifoto, J. editors. *Proc SPIE. 5316,* 99-106.

[93] Wang, Y.; Xie, X.; Wang, X.; Ku, G.; Bill, K.L.; O'Neal, D.P.; Stoica, G.; Wang, L.V. (2004). Photoacoustic tomography of a nanoshell contrast agent in the in vivo rat brain. *Nano Lett. 4,* 1689-1692.

[94] Kobayashi, H.; Brechbiel, M. W. (2005). Nano-sized MRI contrast agents with dendrimer cores. *Adv Drug Deliv Rev. 57,* 2271-2286.

[95] Sitharaman, B.; Kissell, K. R.; Hartman, K. B.; Tran, L. A.; Baikalov, A.; Rusakova, I.; Sun, Y.; Khant, H. A.; Ludtke, S. J.; Chiu, W.; Laus, S.; Tóth, É.; Helm, L.; Merbach,

A. E.; Wilson, L. J. (2005). Superparamagnetic gadonanotubes are high-performance MRI contrast agents. *Chem Commun (Camb).* 3915-3917.

[96] Song, H. T.; Choi, J.; Huh, Y. M.; Kim, S.; Jun, Y.; Suh, J. S.; Cheon, J. (2005). Surface modulation of magnetic nanocrystals in the development of highly efficient magnetic resonance probes for intracellular labeling. *J. Am. Chem. Soc. 127,* 9992-9993.

[97] Konig, K. (2000). Multiphoton microscopy in life sciences. *J Microsc 200,* 83- 104.

[98] So, P. T.; Konig, K.; Berland, K.; Dong, C. Y.; French, T.; Buhler, C.; Ragan, T.; Gratton, E. (1998). New time-resolved techniques in two-photon microscopy. *Cell Mol Biol 44,* 771-793.

[99] Denk, W.; Strickler, J. H.; Webb, W. W. (1990). Two-photon laser scanning fluorescence microscopy. *Science 248,* 73-76.

[100] Hutchinson, C. L.; Lakowicz, J. R.; Sevick-Muraca, E. M. (1995). Fluorescence lifetime-based sensing in tissues: a computational study. *Biophys J. 68,* 1574-1582.

[101] Pawley, J. B. Handbook of biological confocal microscopy. 2nd ed. New York7 Plenum Press; 1995.

[102] Webb, R. H. (1996). Confocal optical microscopy. *Rep Prog Phys 59,* 427- 471.

[103] Lakowicz, J. R. Principles of fluorescence spectroscopy. New York7 Plenum Press; 1993.

[104] Martinez-Zaguilan, R.; Parnami, G.; Lynch, R. M. (1996). Selection of fluorescent ion indicators for simultaneous measurements of pH and Ca^{2+}. *Cell Calcium 19,* 337-349.

[105] Muschol, M.; Dasgupta, B. R.; Salzberg, B. M. (1999). Caffeine Interaction with Fluorescent Calcium Indicator Dyes. *Biophys J. 77,* 577-586.

[106] Jeannot, V.; Salmon, J. M.; Deumie. M.; Viallet, P. (1997). Intracellular accumulation of. rhodamine 110 in single living cells. *J Histochem Cytochem 45,* 403–412.

[107] Oh, D. J.; Lee, G. M.; Francis. K.; Palsson, B. O. (1999). Phototoxicity of the fluorescent membrane dyes PKH2 and PKH26 on the human hematopoietic KG 1a progenitor cell line. *Cytometry 36,* 312-318.

[108] Lattanzio, F. A.; Bartschat, D. K. (1991). The effect of pH, rate constants, ion selectivity and thermodynamic properties of fluorescent calcium and magnesium indicators. *Biochem Biophys Res Commun 177,* 184-191.

[109] Blackwood, A. M.; Sagnella, G. A.; Markandu, N. D.; MacGregor, G. A. (1997). Problems associated with using Fura-2 to measure free intracellular calcium concentrations in human red blood cells. *J Hum Hypertens 11,* 601-604.

[110] Lee, S. K.; Lee, J. Y.; Lee, M. Y.; Chung, S. M.; Chung, J. H. (1999). Advantages of calcium green-1 over other fluorescent dyes in measuring cytosolic calcium in platelets. *Anal Biochem 273,* 186-191.

[111] Johnson, P. L.; Smith, W.; Baynham, T. C.; Knisley, S. B. (1999). Errors caused by combination of di-4-ANEPPS and Fluo3/4 for simultaneous measurements of transmembrane potentials and intracellular calcium. *Ann Biomed Eng 27,* 563-571.

[112] Oliver, A. E.; Baker, G. A.; Fugate, R. D.; Tablin, F.; Crowe, J. H. (2000). Effects of temperature on calcium-sensitive fluorescent probes. *Biophys J. 78,* 2116-2126.

[113] Bruchez, M.; Moronne, M.; Gin, P.; Weiss, S.; Alivisatos, A. P. (1998). Semiconductor nanocrystals as fluorescent biological labels. *Science 281,* 2013-2016.

[114] Chen, X.; Nazzal, A. Y.; Xiao, M.; Peng, Z. A.; Peng, X. (2002). Photoluminescence from single CdSe quantum rods. *J Luminescence 97,* 205-211.

[115] Chan, W. C.; Maxwell, D. J.; Gao, X.; Bailey, R. E.; Han, M.; Nie, S. (2002). Luminescent quantum dots for multiplexed biological detection and imaging. *Curr Opin Biotechnol 13,* 40-46.

[116] Chan, W. C. W.; Nie, S. M. (1998). Quantum dot bioconjugates for ultra-sensitive nonisotopic detection. *Science 281,* 2016-2018.

[117] West, J. L.; Halas, N. J. (2003). Engineered nanomaterials for biophotonics applications: improving sensing, imaging, and therapeutics. *Ann Rev Biomed Eng 5,* 285-292.

[118] Jain, K. K. (2003). Nanodiagnostics: application of nanotechnology in molecular diagnostics. *Expert Rev Mol Diagn 3,* 153-161.

[119] West, J. L.; Halas, N. J. (2000). Applications of nanotechnology to biotechnology commentary. *Curr Opin Biotechnol 11,* 215-217.

[120] Hirsch, L. R.; Jackson, J. B.; Lee, A.; Halas, N. J.; West, J. L. (2003). A whole blood immunoassay using gold nanoshells. *Anal Chem 75,* 2377-2381.

[121] Chen, J.; Saeki, F.; Wiley, B. J.; Cang, H.; Cobb, M. J.; Li, Z. Y.; Au, L.; Zhang, H.; Kimmey, M. B.; Li, X.; Xia, Y. (2005). Gold nanocages: bioconjugation and their potential use as optical imaging contrast agents. *Nano Lett. 5,* 473-477.

[122] Sokolov, K.; Follen, M.; Aaron, J.; Pavlova, I.; Malpica, A.; Lotan, R.; Richards-Kortum, R. (2003). Real-time vital optical imaging of precancer using anti-epidermal growth factor receptor antibodies conjugated to gold nanoparticles. *Cancer Res. 63,* 1999-2004.

[123] Lee, H.; Lee, E.; Kim, D. K.; Jang, N. K.; Jeong, Y. Y.; Jon, S. (2006). Antibiofouling polymer-coated superparamagnetic iron oxide nanoparticles as potential magnetic resonance contrast agents for in vivo cancer imaging. *J. Am. Chem. Soc. 128,* 7383-7389.

[124] Schally, A. V.; Nagy, A. (1999). Cancer chemotherapy based on targeting of cytotoxic peptide conjugates to their receptors on tumors. *Eur. J. Endocrinol. 141*, 1-14.

[125] Schally, A. V.; Nagy, A. (2003). New approaches to treatment of various cancers based on cytotoxic analogs of LHRH, somatostatin and bombesin. *Life Sci. 72*, 2305-2320.

[126] Chaloin, L.; Bigey, P.; Loup, C.; Marin, M.;Galeotti, N.; Piechaczyk, M.; Heitz, F.; Meunier, B. (2001). Improvement of porphyrin cellular delivery and activity by conjugation to a carrier peptide . *Bioconjugate Chem.* 12, 691-700.

[127] Taylor, J. R.; Fang, M. M.; Nie, S. M. (2000). Probing specific sequences on single DNA molecules with bioconjugated fluorescent nanoparticles. *Anal. Chem. 72*, 1979-1986.

[128] Mitchell, G. P.; Mirkin, C. A.; Letsinger, R. L. (1999). Programmed assembly of DNA dunctionalized quantum dots. *J. Am. Chem. Soc.* 121, 8122-8123.

[129] Santra, S.; Zhang, P.; Wang, K.; Tapec, R.; Tan, W. (2001). Conjugation of biomolecules with luminophore-doped silica nanoparticles for photostable biomarkers. *Anal. Chem. 73,* 4988-4993.

[130] De la Fuente, J.; Berry, C. C. (2005). Tat Peptide as an efficient molecule to translocate gold nanoparticles into the cell nucleus. *Bioconjugate Chem.* 16, 1176-1180.

[131] Li, J.; Wang, X.; Wang, C.; Chen, B.; Dai, Y.; Zhang, R.; Song, M.; Lv, G.; Fu, D. (2007). The enhancement effect of gold nanoparticles in drug delivery and as biomarkers of drug-resistant cancer cells. *ChemMedChem* 2, 374-378.

[132] Thevénot, D. R.; Toth, K.; Durst, R. A.; Wilson. G. S. (2001). Electrochemical Biosensors: Recommended Definitions and Classification. *Biosens. Bioelectron. 16,* 121-131.

[133] Turner, A. P. F.; Karube, I.; Wilson, G. S. *Biosensors: Fundamentals and Applications.* In: Turner A. P. F., Editor. Oxford: Oxford University Press; 1987; 770.

[134] Andreescu, S.; Sadik, O. A. (2004). Trends and challenges in biochemical sensors for clinical and environmental monitoring. *Pure Appl. Chem. 76*, 861-878.

[135] Vo-Dinh, T.; Cullum, B. M.; Stokes, D. L. (2001). Nanosensors and biochips: frontiers in biomolecular diagnostics. *Sens. Actuators B 74,* 2-11.

[136] Van Dongen, G. A.; Visser, G. W.; Vrouenraets, M. B. (2004). Photosensitizer-antibody conjugates for detection and therapy of cancer. *Adv. Drug. Deliv. Rev. 56*, 31-52.

[137] Streifer, J. A.; Kim, H.; Nichols, B. M.; Hamers, R. J. (2005). Covalent functionalization and biomolecular recognition properties of DNA-modified silicon nanowires. *Nanotechnology 16,* 1868-1873.

[138] Gu, H. Y.; Chen, Z.; Sa, R. X.; Yuan, S. S.; Chen, H.Y.; Ding, Y. T.; Yu, A. M. (2004). The immobilization of hepatocytes on 24 nm-sized gold colloid for enhanced hepatocytes proliferation. *Biomaterials 25,* 3445-3451.

[139] Kasili, P. M.; Cullum, B. M.; Vo-Dinh, T. (2002). Nanosensor for in vivo measurement of the carcinogen benzo [a] pyrene in a single cell. *J. Nanosci. Nanotechnol. 2,* 653-658.

[140] Schmidt, P. M.; Lehmann, C.; Matthes, E.; Bier. F. F. (2002). Detection of activity of telomerase in tumor cells using fiber optical biosensors. *Biosens. Bioelectron. 17,* 1081–1087.

[141] Wang, J.; Kawde, A. N. (2001). Pencil-based renewable biosensor for label-free. Electrochemical detection of DNA hybridisation. *Anal. Chim. Acta 431,* 219-224.

[142] Torabi, F.; Ramanathan, K.; Larson, P. O.; Gorton, L.; Svanberg, K.; Okamoto, Y.; Danielsson, B.; Khayyami, M. (1999). Coulometric determination of NAD+ and NADH in normal and cancer cells using LDH, RVC and a polymer mediator. *Talanta 50,* 787-797.

[143] Yan, F.; Chen, J.; Ju, H. (2007). Immobilization and electrochemical behavior of gold nanoparticles modified leukemia K562 cells and application in drug sensitivity test. *Electrochemistry Communications 9,* 293-298.

[144] Li, H. N.; Ci, Y. X. (2000). Electrochemical method for analyzing intracellular redox activity changes of the etoposide-induced apoptosis in HL-60 cells. *Anal. Chim. Acta 416,* 221-226.

[145] Chen, J.; Du, D.; Yan, F.; Ju, H. X.; Lian, H.Z. (2005). Sensitivity test for leukemia K562 cells at a carbon-nanotube-modified electrode. *Chem. Eur. J. 11,* 1467-1472.

[146] Heiskanen, A.; Yakovleva, J.; Spégel, C.; Taboryski, R.; Koudelka-Hep, M.; Emnéus, J.; Ruzgas, T. (2004). Amperometric monitoring of redox activity in living yeast cells: comparison of menadione and menadione sodium bisulfite as electron transfer mediators. *Electrochem. Commun. 6,* 219-224.

[147] Woolley, D. E.; Tetlow, L. C.; Adlam, D. J.; Gearey, D.; Eden, R. D.; Ward, T. H.; Allen, T. D. (2002). Electrochemical monitoring of anticancer compounds on the human ovarian carcinoma cell line A2780 and its adriamycin- and Cisplatin-resistant variants. *Exp.Cell Res. 273,* 65-72.

[148] Li, J.; Wu, C.; Dai, Y.; Zhang, R.; Wang, X.; Fu, D.; Chen, B. (2007). Doxorubicin-CdS nanoparticles: a potential anticancer agent for enhancing the drug uptake of cancer cells. *J. Nanosci. Nanotechno. 7,* 1-5.

[149] Liu, B.; Rotenberg, S. A.; Mirkin, M. V. (2002). Scanning electrochemical microscopy of living cells. 4. Mechanistic study of charge transfer reactions in human rreast cells. *Anal. Chem. 74,* 6340-6348. (b) Liu, B.; Rotenberg, S. A.; Mirkin, M. V. (2000). Scanning electrochemical microscopy of living cells: Different redox activities of nonmetastatic and metastatic human breast cells. *Proc. Natl. Acad. Sci. 97,* 9855-9860.

[150] Cai, C.; Liu, B.; Mirkin, M. V.; Frank, H. A.; Rusling, J. F. (2002). Scanning Electrochemical Microscopy of Living Cells. 3. Rhodobacter sphaeroides. *Anal. Chem. 74,* 114-119.

[151] Pailleret, A.; Oni, J.; Reiter, S.; Isik, S.; Etienne, M.; Bedioui, F.; Schuhmann, W. (2003). In situ formation and scanning electrochemical microscopy assisted positioning of NO-sensors above human umbilical vein endothelial cells for the detection of nitric oxide release. *Electrochem. Commun. 5,* 847-852.

[152] Yang, L. J.; Li, Y. B.; Erf, G. F. (2004). Interdigitated array microelectrode-based electrochemical impedance immunosensor for detection of Escherichia coli O157: H7. *Anal. Chem. 76,* 1107-1103.

[153] Xiao, C. D.; Lachance, B.; Sunahara, G.; Luong, J. H. T. (2002). An in-depth analysis of electric cell-substrate impedance sensing to study the attachment and spreading of mammalian cells. *Anal. Chem.* 74, 1333-1339.

In: Encyclopedia of Nanotechnology
Editor: Elwood D. Carlson

ISBN 978-1-60692-079-4

Chapter 56

GOLD NANORODS: SYNTHESIS, OPTICAL PROPERTIES, ASSEMBLY AND BIOAPPLICATIONS

Zhanfang Ma*[*1] and Chungang Wang*[2]

[1]Department of Chemistry,
Capital Normal University, Beijing 100037, P. R. China
[2]Department of Chemistry, Northeast Normal University,
Changchun 130024, P. R. China

ABSTRACT

The considerable interest in designing anisotropic metal nanoparticles, especially gold nanorods, stems from their unique properties and potential applications in biosensing, catalysis, biomedical imaging, cancer diagnosis and therapy and drug delivery. This chapter describes recent advances in the synthesis, assembly and bioapplications of gold naorods. At the beginning, we outline briefly the most popular solution methods to synthesize gold nanorods, as well as their shape-dependent optical properties. In addition, gold nanorods as attractive building blocks to form hybrid nanomaterials such as the assembly of gold nanorods on inorganic materials or core-shell particles are elucidated. Finally, particular emphasis is placed on the rapid development of the intergration gold nanorods with biomolecules such as protein and DNA and the utilization of such bionanohybrid assemblies for bioanalytical applications.

INTRODUCTION

Nanoscale metal nanostructures with specific size and morphology have been the focus of intensive research owing to their potential applications in the fabrication of electronic, optical, optoelectronic, and magnetic devices. The intrinsic properties of a metal nanostructure can be tuned by finely controlling its size and shape [1-4]. Especially, gold nanostructures such as nanocages [5-9], nanowires [10-12], nanorings [13, 14], nanocubes

---

* Email: mazhanfang@yahoo.com (Dr. Z. F. Ma)

[15, 16], nanoplates [17, 18] and nanobelts [19] have attracted much attention due to their numerous potential applications in science and engineering. In particular, gold nanorods have been intensively studied because of their unique optical properties such as tunable surface plasmon resonance (SPR), as well as the synthesis of gold nanorods in high yield and monodispersity, compared with other shapes [20-23]. A variety of synthetic methods for the preparation of gold nanorods are available. Moreover, gold nanorods functionalized with various chem/biomolecular have also been assembled into one-, two-and three-dimensional (1D, 2D and 3D) hybrid systems on surfaces and in solution [24-27]. The high sensitive dependence of the unique SPR of gold nanorods on the dielectric environment constructs the platform for currently developed biomedical applications of gold nanorods as elementary optical biosensors based on transforming specific interactions between probe molecules attached to the sensor and target molecules into a detectable optical signal. The combination of gold nanorods with a variety of biomolecules paves the way to novel elements in bioelectronic and biosensing systems [28]. More importantly, gold nanorods are promising as contrast agents for biological imaging and photothermal cancer therapy [29].

It is the aim of this chapter to summarize recent advances in the synthesis, assembly and bioapplications of gold nanorods. Specifically, the use of biomolecules–gold nanorods hybrid systems in biomedicine and biosensors is highlighted.

1. Synthesis of Gold Nanorods

A variety of methods have been developed, such as the template-directed method pioneered by Martin and others [20, 21], the electrochemical method introduced by Wang, the photochemical reduction method recently reported by Yang [22], and the seed-mediated growth method demonstrated by Murphy [23], for preparing gold nanorods. Three of which will be further discussed, but particular emphasis is placed on the seed-mediated growth method, which has been extensively used.

1.1. Electrochemical Method

This synthesis can be conducted in a simple, two-electrode type electrochemical cell, with a gold plate serving as the anode and a platinum plate as the cathode, as shown in the schematic diagram in Figure 1. Both metal electrodes are immersed in an electrolyte solution containing a mixture of surfactants such as cetyltrimethylammoniumbromide (CTAB) and tetradodecylammonium bromide (TOAB). The CTAB serves not only as a supporting electrolyte but also as a capping agent to prevent the nanoparticles from lateral growth while the TOAB acts as a rod-inducing cosurfactant. The electrolytic cell is then placed in an ultrasonic bath at 36 °C. A small amount of acetone and cyclohexane are added into the electrolytic solution before electrolysis. Acetone is used for loosening the micellar structure, and cyclohexane is necessary for facilitating the formation of elongated, rod-like CTAB micelles [30]. Electrolysis is carried out under a constant current mode for a typical period of 20 min. To better control the aspect-ratio of the gold nanorods, an additional silver plate is added to the electrolytic solution, which is gradually immersed behind the Pt electrode. The redox reaction between gold ions generated from the anode and silver metal leads to the formation of silver ions as shown in Figure 1. It is found that the concentration of silver ions and their release rate can determine the length of the gold nanorods. However, this method is limited in applications such as the highly complex electrochemical process and various effects involved such as sonication and electric field, which makes it difficult to understand the role of the silver ions.

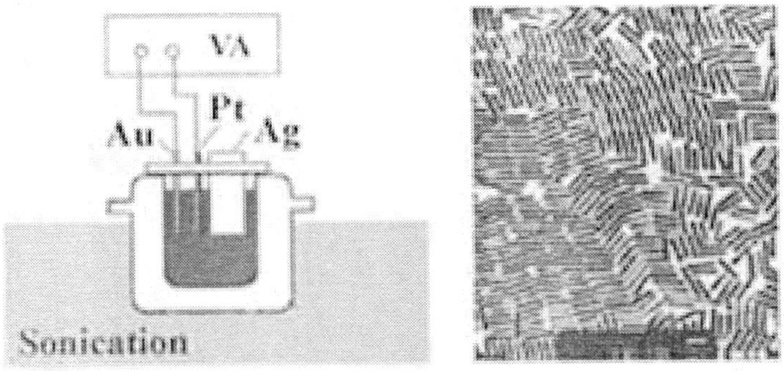

Figure 1. Schematic diagram of the set-up for preparation of gold nanorods via the electrochemical method. VA, power supply, (left). TEM micrographs of gold nanorods with average aspect ratio 6.1 (right). Scale bar represents 50 nm. Reproduced from [21] with permission.

1.2. Photochemical Method

Yang *et al.* have demonstrated photochemical routes to synthesize gold nanorods [22]. Briefly, the method is performed in a growth solution similar to that described for the electrochemical method, in the presence of different concentrations of silver ions and without chemical reducing agent. Gold nanorod formation is induced by 254 nm UV light irridation for 30 h (420 $\mu W/cm^2$), although it is not clear what is being oxidized in response.

1.3. Seed-Mediated Growth Method

Murphy *et al.* developed a seed-mediated synthesis of gold nanorods in the presence of surfactants [22, 31-44]. In a typical process, a solution of ca. 3.5 nm gold seed particles is first prepared by reducing $HAuCl_4$ with a strong reducing agent such as sodium borohydride ($NaBH_4$) in the presence of citrate sodium, which serves only as the capping agent since it cannot reduce gold salt at room temperature. A growth solution containing $HAuCl_4$ and CTAB is mixed with freshly prepared ascorbic acid solution (a milder reducing agent) and then added to a seed solution to generate gold nanorods. Although gold nanospheres are also formed in this synthesis, they can be readily removed via centrifugation. Recently, Mulvaney *et al.* investigated the factors affecting the formation of gold nanorods and modified the solution-phase synthesis to adjust the length of the rods from 25–170 nm (Figure2) [45, 46].

The above seed-mediated growth method of gold nanorod synthesis was later improved by El-Sayed group (Figure 3) [47]. The main modifications exist: (1) $AgNO_3$ is added in the growth stage with different amounts; and (2) replacement of the citrate with CTAB molecules in the seed formation step.

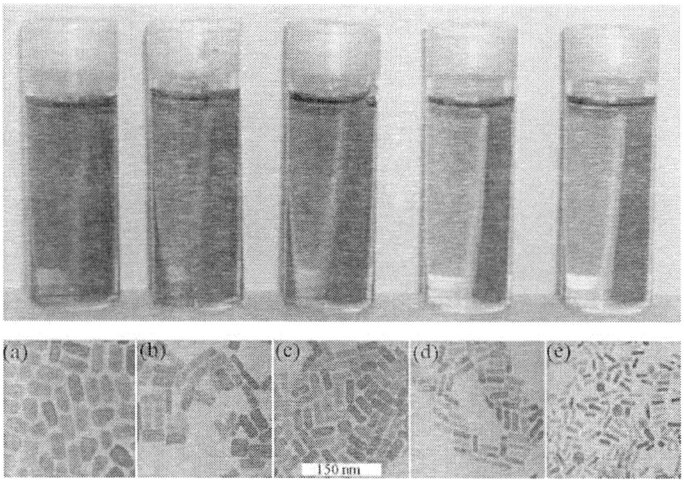

Figure 2. TEM images of the gold nanorods (down) and the corresponding photos (top) of such gold nanorods with different aspect-ratios. Reproduced from [45] with permission.

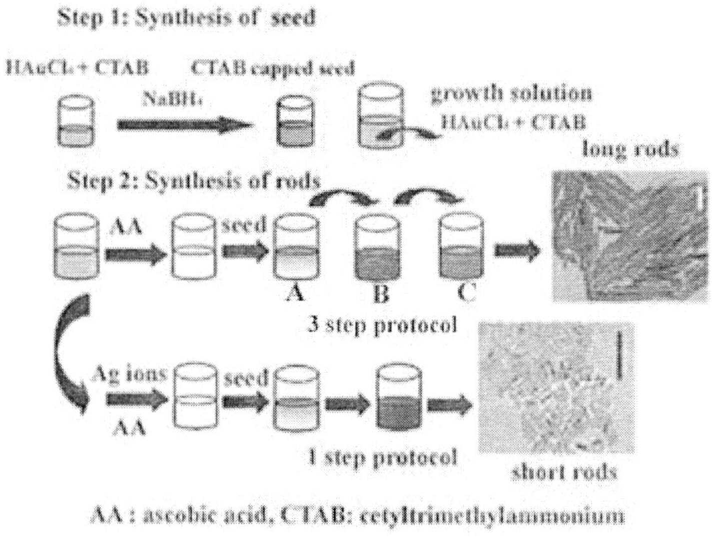

Figure 3. Schematic of the seed-mediated growth of gold nanorods with controlled aspect ratio. The seed is capped with citrate or CTAB. Subsequently, a growth solution of HAuCl4 and surfactant, CTAB, is prepared, to which is added a weak reducing agent (ascorbic acid), followed by the seed. Upon addition of ascorbic acid, the growth solution turns from orange to colorless, indicating a reduction of Au3+ to Au+. In the three-step seeding protocol that yields high aspect ratio gold nanorods, A-C are each growth solutions; the seed is added to A, and then after the color changes, an aliquot of solution A is added to B, etc. For short gold nanorods, silver nitrate and ascorbic acid are added to the growth solution, followed by the seed solution. TEM images of gold nanorod (right). Scale bars represent 200 nm. Reproduced from [24] with permission.

The improved method has profound effect on the yields, shape and crystallinity of the nanorods. Short gold nanorods in nearly quantitative yields (ca. 97%) can be obtained by the addition of $AgNO_3$ to the growth solution. $AgNO_3$ can also be used to fine-tune the aspect ratio of the nanorods for a given volume of seed solution. It was noted that the lower concentrations of Ag ions could first increase the aspect ratios but the trend was reversed at the higher concentrations. Similarly, the presence of $AgNO_3$ is considered to be helpful in achieving single crystal gold nanorods. It should be noted that the addition of silver ions to the reaction mixture to make gold nanorods significantly increases the yield of the gold nanorods but limits the achievable aspect ratios to ca.6.

The presence of silver nitrate significantly enhanced the yields of gold nanorods and fine-tuned the aspect ratio. However, according to the data of some research groups and our observations, the plasmon resonances of such gold nanorods are often unstable and blue shift with increasing storage time in aqueous environment [48]. The optical absorption instability of gold nanorod synthesized in the presence of $AgNO_3$ is an intrinsic problem, and cannot be suppressed by changing CTAB or seed concentration. So it is necessary to develop a process to prevent or retard the reshaping of the nanorod after synthesis to apply them as contrast agents and other bioapplications. This important problem has been solved by the addition of millimolar concentrations of Na_2S to the growth solution reported by Wei *et al.* [49].

On the contrary, high aspect ratio gold nanorods are usually prepared via a three-step seed mediated approach in the absence of $AgNO_3$. The yields of gold nanorods are lower due to the formation of non-rod shaped particles in large quantities [22]. This limitation can be overcome to some extent by increasing the pH of the growth solution [36]. More recently, Huang *et al.* report a new method for a modified seed-mediated synthesis of gold nanorods with high aspect ratio with the addition of an appropriate amount of nitric acid during nanorod growth [50, 51]. Various amounts of nitric acid were added during nanorod growth to significantly enhance the production of gold nanorods with high aspect ratio. The resulting nanorods with average aspect ratios 21-23 are uniform and monodispersed in large quantity. The role of $AgNO_3$ in gold nanorod synthesis is yet elusive, despite some possible growth mechanisms of gold nanorods have been proposed [52-55].

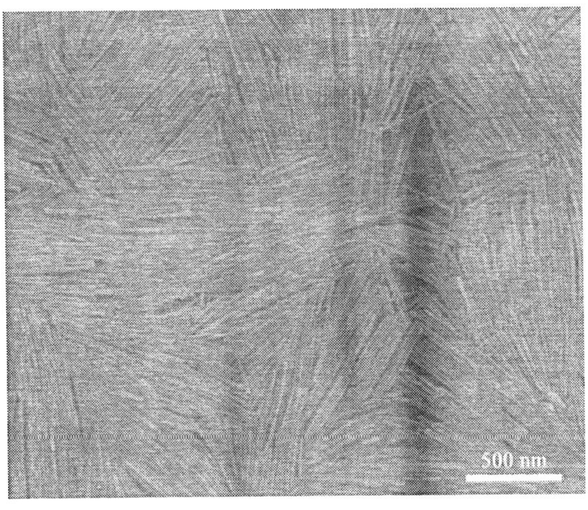

Figure 4. SEM image of high aspect ratio gold nanorods prepared with the addition of nitric acid. Reproduced from [51] with permission.

To date, gold nanorods with aspect ratios larger than 15 obtained by using a multistep seed-mediated growth process are multiply twinned. Wang *et al.* developed one-step synthesis of large-aspect-ratio single crystalline gold nanorods in high yields by using cetyltripropylammonium bromide (CTPAB) and cetyltributylammonium bromide (CTBAB) surfactants in the presence of silver nitrate [56]. The diameters of the nanorods range from 3 to 11 nm, their lengths are in the range of 15 to 350 nm, and their aspect ratios are in the range of 2 to 70. The diameters of the gold nanorods obtained from one growth batch in CTPAB solutions decrease as their lengths increase, and their volumes decrease as the aspect ratios increase. These gold nanorods are single crystalline and can be found to be oriented in either the {100} or {110} direction, irrespective of their sizes.

2. Optical Properties of Gold Nanorods

The free electrons in the metal (d electrons in silver and gold) are free to travel through the material. The mean free path in gold and silver is ca. 50 nm, therefore in particles smaller than this size, no scattering is expected from the bulk. Thus, all interactions are expected to take place on the surface. When the wavelength of light is much larger than the size of nanoparticle, it can set up standing resonance conditions. Light in resonance with the surface plasmon oscillation causes the free-electrons in the metal to oscillate [57]. As the wave front of the light passes, the electron density in the particle is polarized to one surface and oscillates in resonance with the light's frequency causing a standing oscillation as represented in Figure 5 (top). The resonance condition is determined from absorption and scattering spectroscopy and is found to depend on the shape, size, and dielectric constants of both the metal and the surrounding material. This is referred to the surface plasmon resonance, since it is located at the surface [58].

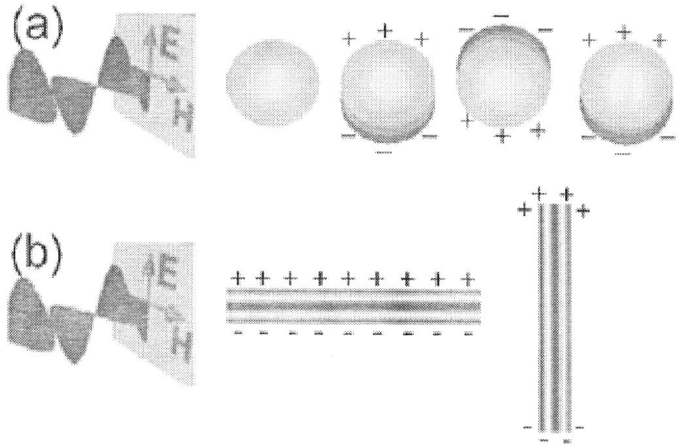

Figure 5 Schematic origin of surface plasmon resonance due to coherent interaction of the electrons in the conduction band with the incoming light (electromagnetic wave). (a) metal nanospheres; (b) transversal and longitudinal oscillation of electrons in a metal nanorod. Reproduced from [59] with permission.

As the shape or size of the nanoparticle changes, the surface geometry changes causing a shift in the electric field density on the surface. This causes a change in the oscillation frequency of the electrons, generating different cross-sections for the optical properties including absorption and scattering. The shape of a particle can drastically change the SPR properties. For gold nanorods, the resonance wavelength depends on the orientation of the electric field. Therefore, two oscillations, transverse and longitudinal are possible (Figure 5, bottom) [59].

Mie theory predicts the optical properties of metallic spherical particles [60]. For very small particles of radius R (where 2R<< λ, and λ is the wavelength of light), the extinction cross-section (C_{ext}) of the metallic spherical particles can be obtained as a series of multipole oscillations if the boundary conditions are specified. So C_{ext} can be expressed as follows:

$$C_{ext} = \frac{24\pi^2 R^3 \varepsilon_m^{3/2}}{\lambda} \frac{\varepsilon_2}{\left(\varepsilon_1 + 2\varepsilon_m\right)^2 + \varepsilon_2^2} \qquad (1)$$

where ε_m is the dielectric constant of the surrounding medium and $\varepsilon_m = \varepsilon_1 + i\varepsilon_2$ is the wavelength-dependent, complex dielectric constant of the nanoparticle material. From eqn (1), a resonance peak occurs whenever the condition of $\varepsilon_1 = -2\varepsilon_m$ is satisfied resulting from that the light field induces a resonant coherent dipolar oscillation of free electrons across the nanoparticles. For Au, Ag and Cu nanoparticles, the resonance condition is satisfied at visible wavelengths accounting for the brilliant colours of various metal nanoparticles.

For nanorods, the boundary conditions in Mie theory have to be modified to describe their plasmonic properties. Unfortunately, it is not possible to obtain analytic expressions for particles with an arbitrary shape. In 1912, Gans modified Mie theory and predicted that for rod-shaped particles the SPR band would split into two modes due to different orientations of the rod with respect to the electric field of incident light (Figure 5b and Figure 6) [61]. According to Gans, the extinction cross-section can be expressed as:

$$C_{ext} = \frac{2\pi V}{3\lambda} \varepsilon_m^{3/2} \sum_j \frac{\left(\frac{1}{P_j^2}\right)\varepsilon_2}{\left(\varepsilon_1 + \frac{1-P_j}{P_j}\varepsilon_m\right)^2 + \varepsilon_2^2}$$

where V is the volume of the rod, Pj (j = A, B, C; A > B = C, A = length, B = C = width) are the depolarization factors for the elongated particles (gold nanorod), which are given by:

$$P_A = \frac{1-e^2}{e^2}\left[\frac{1}{2e}\ln\left(\frac{1+e}{1-e}\right) - 1\right]$$

$$P_B = P_C = \frac{1-P_A}{2}$$

where e is referred to as the rod ellipticity given by $e^2 = 1-(A/B)^2$, and A / B is the aspect ratio of the gold nanorod. These equations can be used to calculate the extinction spectra of gold nanorods of any aspect-ratio, and the results were summarized by El-Sayed *et al.* [62].

Gold nanorods have two principle plasmon absorption peaks: one at shorter wavelength at round 520–530 nm that corresponds to the absorption and scattering of light along the short axis of the nanorod (the transverse plasmon band), and the other stronger band at longer wavelength that corresponds to the absorption and scattering of light along the long axis of the nanorod (the longitudinal plasmon band) (Figure 6). It has been well established that the longitudinal plasmon band can be shifted from the visible into the near–infrared (NIR) region by an increase in the nanorod aspect ratio (Figure 7).

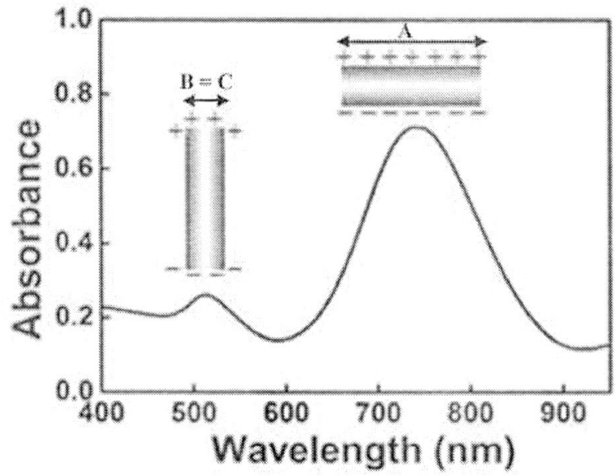

Figure 6. UV-vis spectra for gold nanorods with average aspect ratio 3. 0 (unpublished result).

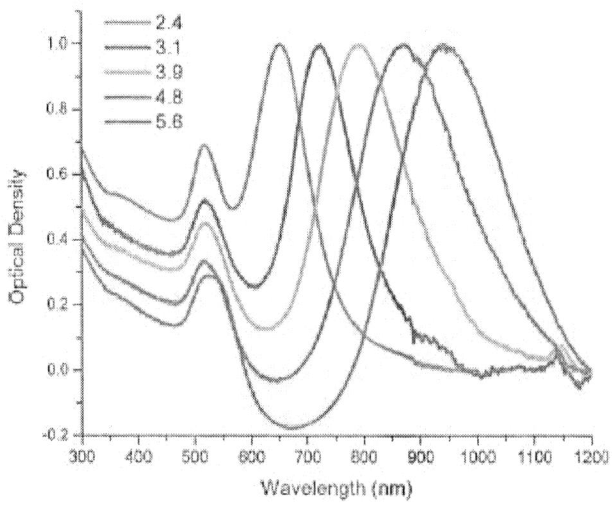

Figure 7. UV-vis spectra for gold nanorods with different aspect ratios. Reproduced from [103] with permission.

3. One-, Two- and Three-Dimensional Assembly of Gold Nanorods

3.1. End-to-End Assembly of Gold Nanorods

Fabricating 1D linear arrays, or end-to-end assemblies, of gold nanorods on surfaces and in solution have extremely interesting applications in nanoelectronic circuits [63], photonics [64] and sensing [65, 66]. Functionalizing the ends only of gold nanorods is possible, if the sides are physically protected from reaction, which could then result in the possibility of designing the end-to-end fashion. El-Sayed *et al.* have used thermogravimetric analysis to quantitate the amount of surfactant on the surface of gold nanorods [67, 68]. Coupled with Fourier transform infrared measurements, they established that the surfactant, CTAB, was associated with gold nanorods in the form of a bilayer. The CTAB binding is thought to place via a gold-bromide surfactant complex. The bromide ion is thought to be the bridge between the gold surface and the positively charged quaternary nitrogen of the surfactant. Moreover, CTAB molecules preferentially attach to their {110} and {100} side face of gold nanorods. The attachment of CTAB to the {111} faces (gold nanorod ends) is weaker and as such this face is more exposed and accessible to the chemical and biological linkers.

Biomaterials were utilized in the fabrication of gold nanorods assemblies by biological linkages including protein host–guest pairs such as biotin–streptavidin, antigen–antibody, and complementary oligonucleotide pairs. Murphy *et al.* have prepared end-to-end assemblies of aspect ratio ca. 20 gold nanorods driven by biotin and streptavidin interactions [69]. In that report, the specific binding of biotin disulfide to the gold nanorod edges was achieved due to the preferred binding of thiol molecules to the Au {111} surface as opposed to the gold nanorod side faces. The addition of a streptavidin linker to the biotin-modified Au nanorods produced an unexpectedly high proportion of end-to-end linked Au nanorods that were separated by 4–5 nm spacing, corresponding to the size of a streptavidin molecule (ca. 5 nm) as observes by TEM (Figure 8). This overwhelming preference for end-to-end nanorod assembly is attributed to greater accessibility of the gold surface at the ends of the nanorods by thiol modifiers, relative to the sides of the rods protected by a more densely packed bilayer of CTAB.

More recently, Tan *et al.* produced the same end-to-end assemblies of shorter gold nanorods using different biological linkages, IgG and anti-IgG. The gold nanorods could be assembled to nanorods chain, which was as long as 3 μm [70]. This highlights the biorecognition capabilities of guiding the assembly of the modified gold nanorods. In addition to biologically aligned nanorods into 1D structure, Kamat *et al.* has prepared aligned nanorods based on hydrogen bonding with end-to-end fashion using chemical linkages. Gold nanorods in acetonitrile/water solutions, with CTAB molecules that are presumably on the sides of the nanorods i.e. {110} and {100} faces, are modified with acid-terminated thiols such as 11-mercaptoundecanoic acid, which selectively attached to the ends of nanorods, {111} and formed chains through hydrogen bonding [71].

Another similar example is that gold nanorods can be linked to each other in an end-to-end way by using cysteine as a molecular bridge via two-point electrostatic interaction due to the selective binding of cysteine at the two ends of the gold nanorods, resulting in considerable inter-plasmon coupling [66, 72].

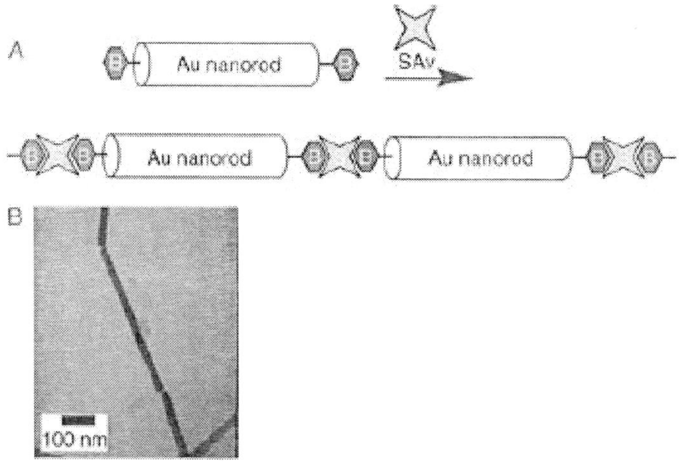

Figure 8. (A) End-to-end assembly of biotin-functionalized gold nanorods that are cross-linked by streptavidin. (B) Transmission electron micrograph of the end-to-end assembly of gold nanorods. Reproduced from [69] with permission.

The formation of the end-to-end assembly of gold nanorods is also achieved by the polyelectrolyte layer-by-layer (LBL) approach, which is mainly based on the use of polyelectrolytes as molecular glue between the substrate and nanoparticles or between successive nanoparticle layers [73,74]. Liz-marzán *et al.* demonstrated the uniform electrostatic assembly of gold nanorods on multiwall carbon nnaotubes (MWNT) in such a way that the rods form strings with end-to-end contacts based on LBL method (Figure 9) [75]. Given the universality of the LBL process, the procedure has the potential to be applied for the creation of a large variety of 1D anisotropic metal nanoparticles with end-to-end assembly.

In addition to the assembly of gold nanorods with end-to-end way in solution, Zamborini *et al.* demonstrated a novel and simple procedure for promoting the growth of highly aligned gold nanorods directly on Si (100) substrate [76]. A facile chemical amidation reaction on NH_2-functionalized Si (100) was used for the growth of highly aligned gold nanorods, which were observed over large areas. However, the alignment mechanism is unclear. Honda *et al.* demonstrated gold nanorods modified by phosphatidylcholineby adsorbed on polyanion-modified glass substrates [77]. The gold nanorods formed end-to-end assemblies on glass surfaces, also showing red-shifted and broaden longitudinal surface plasmon bands.

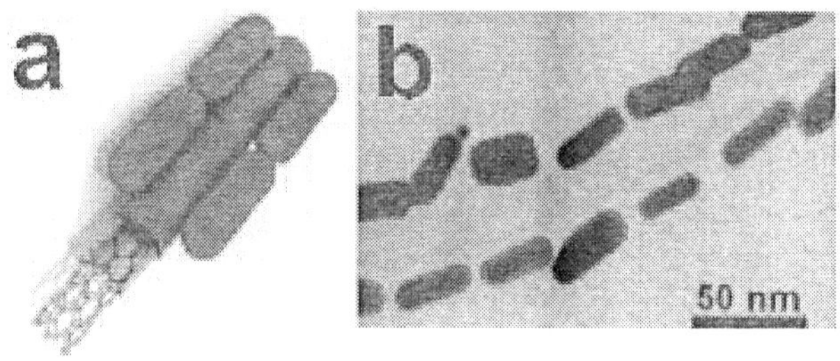

Figure 9. (a) Drawing and (b) TEM micrograph of a selected area of a gold nanorod-MWNT composite. Reproduced from [75] with permission.

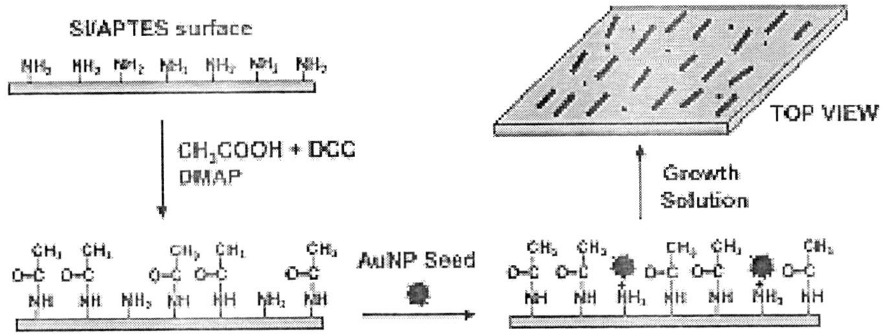

Figure 10. Cartoon showing the growth of aligned gold nanorods. Reproduced from [76] with permission.

More recently, ring-like assemblies (end-to-end) have been also found in gold nanorods. Zubarev *et al.* described spontaneous assembly of hybrid gold/polymer core–shell nanorods into ring-like arrays (Figure 11) [78]. First, gold nanorods were modified by polystyrene (PS) chains (inset, Figure 11). Then the nanorods were put in a solution of chloroform. When the chloroform evaporated, its surface temperature decreased enough to cause the condensation of water vapors from the air, forming dew-like droplets. Thousands of water microdroplets pressed the nanorods from the solution up against them, making the nanorods build rings around them. The polymer shell prevented the gold nanorods from being absorbed into the water droplets because of the insolubility of the PS shell. When the water droplets evaporated, the nanorings remained stable. When nanorods are assembled into a ring superstructure, their optical and electromagnetic properties undergo considerable changes. The new and facile method can fabricate thousands of well-defined rings in a matter of seconds in high yield. This method can be utilized for organizing nanocrystals of various shapes, sizes and chemical composition into circular arrays.

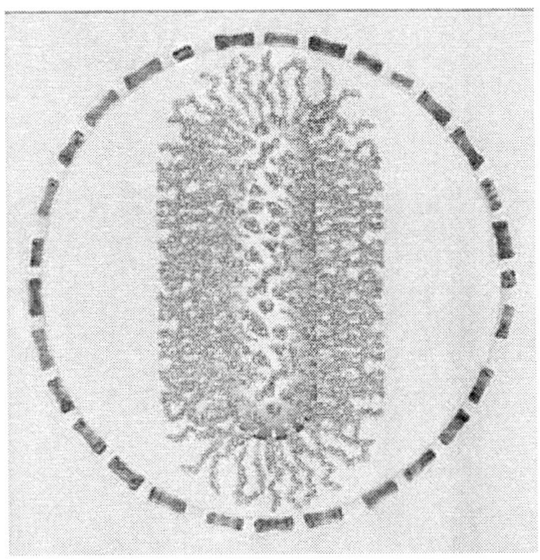

Figure 11. TEM image of rings formed by gold nanorods (PS)n from a solution in CH2Cl2. Inset shows the synthesis of gold nanorods with covalently attached polystyrene arms. Reproduced from [78] with permission.

An alternative strategy for 1D assembly of gold nanorods is the use of physical means such as electric fields [79] or stretching a polymer films containing gold nanorods to align them along the stretch direction. Recently, Liz-Marzán *et al.* reported gold nanorods were uniformly distributed in poly(vinyl alcohol) thin films and could be aligned by heating and stretching the films [80]. In addition, irradiation of the films with a nanosecond laser (1064 nm), using a TEM grid as mask, selectively reshapes the nanorods into nanospheres. The similar approach was reported by Murphy *et al.* [81].

An additional method for physically assembling gold nanorods would be to align them into grooves or channels on substrates. This has been demonstrated for depositing liquid crystals into grooves on a substrate created by rubbing of the substrate in one direction [82].

3.2. Side-To-Side Assembly of Gold Nanorods

The organization of metal nanoparticles into 2D and 3D functional structures is necessary for the assembly of chemical, optical, magnetic, and electronic devices. DNA molecules are very powerful and versatile linkers for the controlled assembly of nanoparticles. DNA chains are of a controllable length and enormous control over complementarity. To demonstrate the DNA-mediated assembling of anisotropic metal nanoparticles, Mann *et al.* reported that gold nanorods by DNA modified could be assembled into 3D structure in side-to-side way (Figure 12) [83]. Gold nanorods were cross-linked with two different DNA sequences, then a third complementary linking DNA strand was added that could hydrogen-bond to both nanorod batches. Thus, the spacing between gold nanorods can be governed by the length of the DNA linkers.

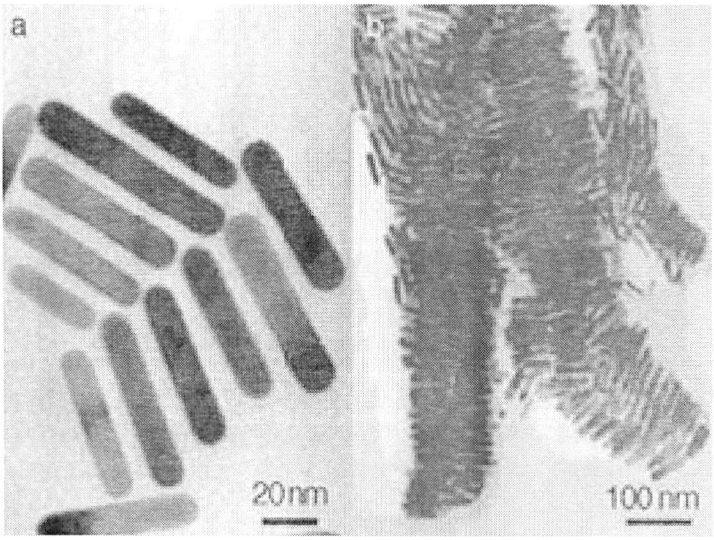

Figure 12. Transmission electron micrographs of gold nanorods that are organized by DNA hybridization. Reproduced from [83] with permission.

Another biomolecule-based recognition systems, biotin and streptavidin, can be used to induce gold nanorods into side-to-side assemblies. Murphy *et al.* demonstrated a simple procedure to biotinylate the entire gold nanorod surface and subsequently form a 3D assembly by addition of streptavidin [84]. The surface of gold nanorods was further modified by a layer of a weak polyelectrolyte, poly(acrylic acid), PAA. A biotin molecule which has an amine group at one end was anchored to the carboxylic acid group of the polyelectrolyte using the well-known carbodiimide chemistry. This process biotinylates the entire gold nanorod surface. Addition of streptavidin further leads to aggregation of gold nanorods, showing a preferential side-to-side assembly of gold nanorods.

More recently, Nakashima reported the control assembly of gold nanorods functionalized by thiol modified crown ether derivatives with side-to-side and end-to-end fashion induced by the selective recognition of cations (Na^+ and K^+) as shown in Figure13 [85]. Furthermore, in the aggregates preferential end-to-end or side-to-side assembly of gold nanorods was observed which was dependent on the concentrations of added crown ethers and cations.

The above reports covered the side-to-side assembly of gold nanorods based on the chemical linkages. Recently, Murphy *et al.* reported the side-to-side assemblies of shorter gold nanorods could be prepared in a pH-dependent manner [86]. The pH-dependent assembly was directed by electrostatic interaction between the positively charged nanorods and negatively charged, deprotonated adipic acid added in the gold nanorod solution.

An alternative approach to self-assembling gold nanorods into ordered 2D superstructures by solvent evaporation is fulfilled [87].

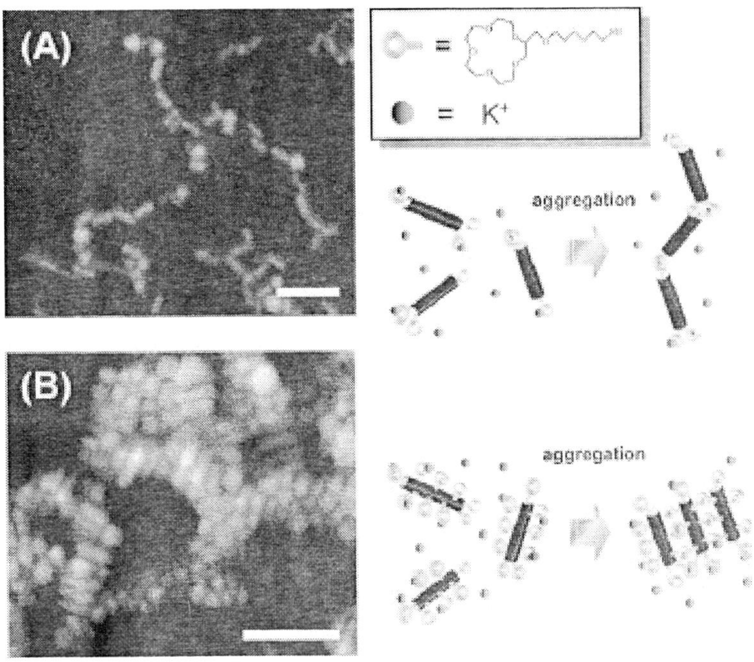

Figure 13. SEM images of the aggregates of the gold nanorod complex on silicon substrate: (A) end-to-end type by addition of 5 mL of 15-crown-5- SH in KCl solution; and (B) side-to-side type by addition of 50 mL of 15-crown-5-SH in KCl solution. Scale bar: 100 nm. Reproduced from [85] with permission.

3.3. Random Orientation Assembly of Gold Nanorods

Interaction of gold nanorods with biomolecules is of importance for a range of applications such as sensitive electronic devices. Berry *et al.* have shown that positively charged CTAB coating on gold nanorods could self-assemble randomly on the gram-positive bacteria, *Bacillus cereus*, with highly negatively charged teichoic acid brushes on the surfaces by the strong electrostatic interaction (Figure 14) [88]. This organization resulted in electronic coupling between nanorod monolayer with bacteria, which opened the possibility of novel hybrid devices utilizing the machinery of the biological system.

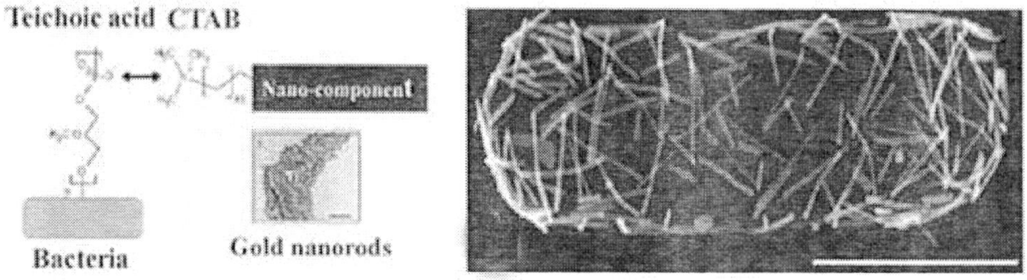

Figure 14. Electrostatic deposition of gold nanorods is accomplished by capping the nanorods with CTAB that electrostatically binds to the teichoic acid brush on the gram-positive bacterial surface. Insets show the TEM images of nanorods. Scale bar: 500 nm (left); TEM image of the deposition of gold nanorods (length: 400 nm, width: 25 nm) on bacteria (right). Scale bar is 1μm. Reproduced from [88] with permission.

Murphy *et al.* have demonstrated that assembly of gold nanorods onto acid-terminated self-assembled monolayers by electrostatic interactions between the CTAB bilayer around the nanorods and the negatively charged surface [89]. Additionally, the presence of a bilayer of CTAB on the gold nanorod surface imparts a net positive charged to the nanorods. Alternate adsorption of anionic and cationic polyelectrolytes on the gold nanorods can be achieved by electrostatic interaction. Depending on the surface functionality of the coated gold naorods, they can be selectively assembled onto either cationic or anionic surfaces by layer-by-layer (LBL) method [90].

A universal and facile approach for the self-assembly of surfactant–coated anisotropic metal nanoparticles on the amino-functionalized planar and spherical surfaces based on electrostatic attraction has also been explored [91].Recently, we have fabricated large-scale and different surface coverage of monolayer film and metallodielectric core-shell nanostructures of anisotropic metal nanoparticles, including gold nanorods (Figure15), $Au_{core}Ag_{shell}$ nanorods and Ag nanoprisms [92]. This method based on electrostatic interaction between slightly negatively charged poly(vinylpyrrolidone) (PVP) and the protonated amines shows several advantages compared with layer-by-layer (LBL) techniques using oppositely charged polyelectrolytes, such as it can effectively avoid the aggregation of the anisotropic metal nanoparticles and be accomplished in simpler experimental steps. Moreover, controlling the surface coverage of self-assembled anisotropic metal nanoparticles to amino-functionalized surfaces with adjustable optical response in the visible and the NIR by varying the experimental parameters such as the concentration of the anisotropic solution, the amount of silica particles and the immersion time of the substrates in the nanoparticle solution have been demonstrated.

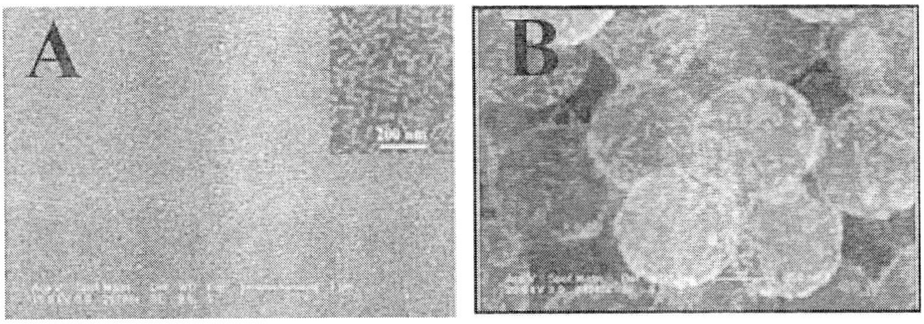

Figure 15. FE-SEM images of (A) the random assembly of gold nanorods with aspect ratio 3.0 ± 0.25. on the glass surface, and (b) the silica particles with 340 nm diameter coated with gold nanorods. (unpublished results).

Recently, we decribed that silica coated gold nanorods (Au@SiO_2), having the pure-silica surface, straightforwardly and rapidly assembled onto poly(4-vinylpyridine) (PVP) functionalized quartz substrates (Figure 16) [93]. The large-scale assemblies exhibit highly reversible optical changes that can be observed spectroscopically or with the naked eye when repeatedly cycled between dry and wet process. The optical properties of two-dimensional assemblies was affected by the aspect ratio and shape of gold nanoparticles and environmental refractive index.

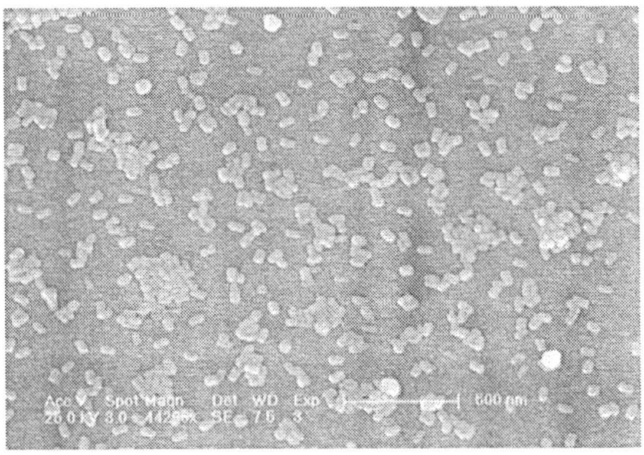

Figure 16. SEM image of silica coated gold nanorods (aspect ratio 2.25 ± 0.48) self-assembled on PVP functionalized quartz substrate. Reproduced from [93] with permission.

4. Gold Nanorods in Bio- and Biomedical Applications

The synthesis of gold nanorods with various aspect ratios, which enable tunable absorption wavelength in the NIR region, is quite simple in decent yields and well-established. In addition, the biological inertness of gold is well known. These properties suggest several advantages of gold nanorods for the applications in biological sensing, imaging and cancer therapy which should benefit from the nanorods geometry compared with similar sized spherical nanoparticles. It should be noted that rod-shaped gold nanoparticles have an inherently high sensitivity to the local dielectric environment including the substrate, solvents, and adsorbates, as well as the inter-particle spacing of the gold nanorods. However, the use of gold nanorods as biosensors for analytes has not been widely pursued. This is due to that the CTAB bilayers on the surfaces of gold nanorods are obstructive for displacing by the interest of biomolecules. On the other hand, the large amount of CTAB dispersed in the solution interferes with biological processes and displays significant cytotoxicity. The toxicity is not due to the CTAB layers statically bound to the nanoparticles in human cells, which have been discussed in detail by Wyatt *et al.* [94]. Additionally, the unbound CTAB was proved to be toxic to cultivated cells in their paper. For further biological applications, therefore, excess CTAB should be removed, which can be achieved by repeated centrifugation of the nanorod solution. After the centrifugation, although it is possible to partially modify gold nanorods and the CTAB bilayers on the nanorod surfaces remained, the gold naorods obtained are likely not steady, which resulted mainly from the desorption of CTAB molecules from the surfaces of gold nanorods. Moreover, further removal of the desorbed CTAB will lead to the aggregation of the nanorods because of the lack of a repulsive interaction between the gold nanorods. In this case, it is very difficult to eliminate all of the uncoordinated CTAB without aggregation of nanorods. Thus, gold nanorods are limited to applications in biological sensing, imaging and therapy. The controlled modification of the surface chemistry of the nanorods is of critical importance, as issues of cell-specific targeting and nonspecific uptake must be addressed prior to clinical evaluation.

Hafner *et al.* have demonstrated that CTAB bilayer surrounding gold nanorods was replaced by thiol terminated methoxypoly(ethylene glycol) and subsequently conjugated to IgG antibodies [95]. More recently, Yamada *et al.* have reported a method for replacing CTAB with phospholipids by extraction using a chloroform phase [96]. But bioconjugates have not been performed in the presence of phospholipids that are very difficult to displace by analyte of interest. It is thus necessary to develop the surface chemistry that will not only protect gold nanorods from aggregation but also render them amenable to assembly , surface functionalization and bioconjugation. We reported an efficient silica coating process using an improved Stöber method to prepare silica-coated gold nanorods ($Au_{rod}@SiO_2$) [97]. The resulting pure and clean silica surface of gold nanorods is important for straightforward self-assembling $Au_{rod}@SiO_2$ films onto poly(4-vinylpyridine) modified quartz substrates and offers versatile functional groups such as amino group via well-established silica surface modification for the immobilization of biomolecules. Subsequent covalent bioconjugation of the amino-functionalized $Au_{rod}@SiO_2$ films with goat anti-h-IgG is successfully employed for colorimetric detection of h-IgG as a model based on specific binding affinity between proteins. More importantly, the protein-recognition event on $Au_{rod}@SiO_2$ films is not only monitored via changes of light absorption but also observed by the naked eye.

More recently, Wang *et al.* demonstrated the feasibility of the utilizing the longitudinal plasmon band of gold nanorods (aspect ratio 5.2) in biomolecular sensing based on biotin and streptavidin biorecognition [98]. The biomolecule detection was performed by placing the probe glass slide, glass/gold nanorods-biotin, into the streptavidin aqueous solution for a certain period of time and measuring its absorption spectrum. The results revealed that the shifts of the longitudinal bands were linear to the concentrations of the streptavidin (0.025-20.0 μg/mL). The longitudinal plasmon band sensing mode of the gold nanorods exhibits much higher sensitivity (366 nm/RIU) compared with that of the gold spheres (76.4 nm/ RIU).

In addition to the construction of biosensing platform based on the gold nanorods on the glass substrate, we presented that the surface of gold nanorods with average aspect ratio 3.5 functionalized by poly(styrenesulfonate) for the attachment of goat anti-h-IgG yielded nanorod molecular probes, which exhibited the rapid and sensitive detection of h-IgG in solution by localized surface plasmon absorption (LSPR) as shown in Figure 17 [99]. Plasmonic sensors are highly rapid and sensitive and can be used to monitor both changes in refractive index caused by molecular interactions in the vicinity of the gold nanorods and the biorecognition-driven assembly of gold nanorod process. A very small amount (30 ng) quantity of h-IgG added into nanorod molecular probe solution (500 μL) leads to a marked red shift of LSPR in a very short time (ca. 100 s). The detection limit of the present method is 60 ng mL^{-1}, thereby rivaling the limits of fluorescent, radioactive, and enzyme colorimetric methods. This novel method using GNRs with different aspect ratios is simple, sensitive, and will provide a novel optical multiplex biosensors platform and has broad potential applications in immunoassay and disease diagnosis. A further detailed investigation is being carried out in the laboratory at the present time.

In addition, Luong *et al.* have found that high aspect ratio gold nanorods (aspect ratio > 13) exhibited enhanced fluorescence [100]. This unique optical property of gold nanorods was employed to the DNA hybridization biosensing based on monitoring the fluorescence intensity of gold nanorods. The results showed that the unique fluorescence properties of gold nanorods could potentially develop as a sensitive probe in fluorescence-based microassay and optical biosensor.

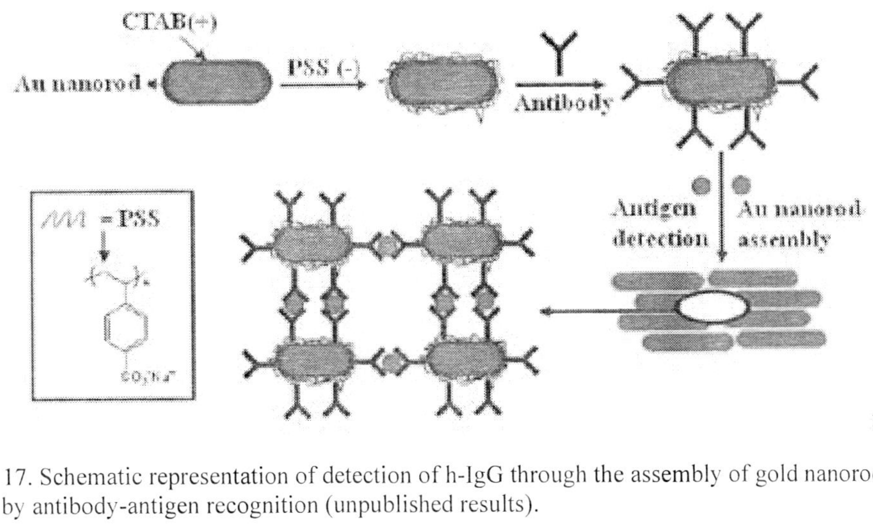

Figure 17. Schematic representation of detection of h-IgG through the assembly of gold nanorods driven by antibody-antigen recognition (unpublished results).

More importantly, gold nanorods exhibit strongly enhanced fluorescence allowing *in vivo* imaging using a two-photon luminescence method. Wei *et al.* demonstrated the application of the dumbbell-shaped gold nanorods with an average aspect ratio of ~3.1 as two-photon luminescence (TPL) imaging agents by *in vivo* imaging of single gold nanorod flowing through a mouse's ear in order to track blood [101]. As well as enabling 3D imaging in a far-field laser-scanning microscope with a Ti:sapphire laser beam with a wavelength of around

830 nm to generate TPL from the rods. A single gold nanorod provided the TPL signal that's 58 times brighter than the two-photon fluorescence signal from a single rhodamine molecule. The gold nanorods remained detectable for around half an hour after their injection into the mouse due to the mouse's kidneys eventually filtered the nanorods out of the blood. Using this technique with labelled gold nanorods may allow us to detect diseases at an early stage of development. The results show that the longitudinal plasmon resonance at near-infrared wavelengths of the gold nanorods boosts the TPL signal. Experiments also revealed that the TPL excitation spectrum overlapped with the longitudinal plasmon band of gold nanorods whose frequencies are ideal for biological imaging as water and biological molecules have relatively low absorption in the range.

Another highly important topic concerns the simultaneous molecular imaging and photothermal cancer therapy of gold nanorods with suitable aspect ratios. Many cancer cells have a protein, known as Epidermal Growth Factor Receptor (EFGR), all over their surface, while healthy cells typically do not express the protein as strongly, so the detection of EFGR at the early stage is a key problem. More recently, Durr *et al.* have also demonstrated the effective use of gold nanorods as bright contrast agents for two-photon luminescence (TPL) imaging of cancer cells (Figure18) [102]. By labeling cancer cells with anti-EGFR nanorod conjugates, molecular-specific imaging could be carried out with very high signal-to-noise ratios deep into a tissue phantom down to 75 μm deep. The TPL intensity from gold-nanorod-labeled cancer cells is 3 orders of magnitude brighter than the two-photon autofluorescence (TPAF) emission intensity from unlabeled cancer cells at 760 nm excitation light. Additionally, Durr *et al.* have also shown that the use of gold nanorods can expand the capabilities of TPL to allow noninvasive three dimensional imaging of a variety of new molecular signatures. The strong signal of gold nanorods, resistance to photobleaching, chemical stability, ease of synthesis, simplicity of conjugation chemistry, and biocompatibility make gold nanorods an attractive contrast agent for two-photon imaging of epithelial cancer.

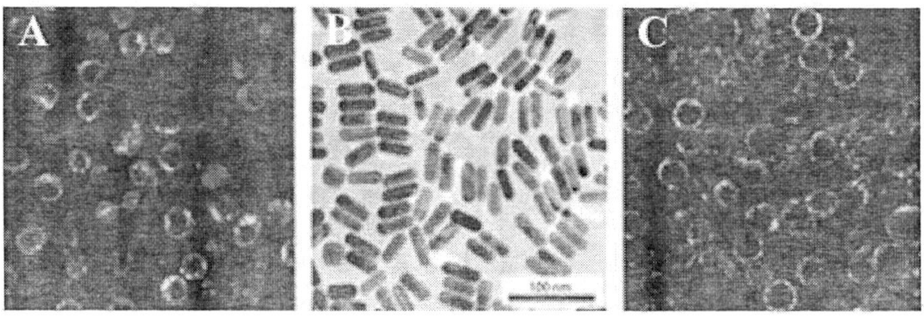

Figure 18. (A) The two-photon autofluorescence (TPAF) imaging of unlabeled cancer cells embedded in a collagen matrix at increasing depths, (B) TEM image of gold nanorods with aspect ratio 3.4 ± 0.6 (length: 48.1 ± 5.5 nm and width: 14.3 ± 2.2 nm), and (C) The two-photon luminescence (TPL) imaging of nanorod-labeled cancer cells embedded in a collagen matrix at increasing depths. Reproduced from [102] with permission.

Besides the enhanced fluorescence of gold nanorods, the strong scattering and absorption in the NIR region can also make gold nanorods as a novel class of optically active dual imaging/therapy agent. Huang *et al.* employed gold nanorods conjugated to anti-EGFR antibodies for NIR cancer cell imaging and selective photothermal therapy [103]. Gold nanorods with an average aspect ratio of ~3.9 were used which showed a strong longitudinal absorption band around 800 nm. As a result of intense red scattering from the gold nanorods, malignant HOC and HSC cells labeled with the gold nanorod bioconjugates were clearly identified in optical microscopy images under dark-field white light illumination, allowing their differentiation from nonmalignant HaCaT cells. Then the labeled cells were exposed to a continuous wave Ti:sapphire laser at 800 nm having a strong spectral overlap with the intense longitudinal SPR absorption band of the nanorods. The results show the labeled malignant cells (10 W/cm^2) only require nearly half the laser energy to kill the nonmalignant cells (20 W/cm^2). For photothermal therapy of most of the solid cancer deep under the skin, like brain, stomach and breast cancer, the use of gold nanorods treatment is a very good choice.

Multiplex analysis, in which two or more analysts are detected simultaneously, is a powerful method to study protein expression. Specifically, gold nanorods with different aspect ratios could be easily fabricated and bioconjugated to different antibodies, unique "multiplexing" advantage could be realized. To demonstrate multiplex detection capability, Wang *et al.* demonstrated the use of gold nanorods of different aspect ratios to achieve multiple targeting and simultaneous detection in photoacoustic imaging by using laser irradiation *in vitro* [104]. Antibodies such as anti-HER2 and anti-CXCR4 are conjugated to two kinds of nanorods of two different aspect ratios (3.7 and 5.9). Thus, the specific targeting for tumor cells with the antigens can be demonstrated with a wavelength tunable Ti-Sapphire laser irradiation at 800 and 940 nm that corresponds to the longitudinal plasmon band of the gold nanorods with aspect ratios 3.7 and 5.9, respectively. The results clearly show the potential of photoacoustic molecular imaging with multiple targeting in revealing different oncogene expression levels of the cancer cells.

Another important area concerns that irradiation with suitable laser can induce structural and morphological transitions of gold nanorods into spheres [105]. It is expected that the photo induced reshaping of gold nanorods into spherical nanoparticles will be a key reaction for achieving gene delivery and photothermal cancer therapy. Chen *et al.* achieved controlled release of plasmid DNA using gold nanorods, promoting enhancement of localized gene expression (Figure 19) [106]. Gold nanorods were attached to the gene of enhanced green fluorescence protein (EGFP) for the remote control of gene expression in living cells. Upon NIR irradiation, the gold nanorods of EGFP-GNR conjugates underwent shape transformation that resulted in the release of EGFP DNA (Figure 20). When EGFP-GNR conjugates were delivered to cultured HeLa cells, induced GFP expression was specifically observed in cells that were locally exposed to NIR irradiation.

Niidome et al. have also reported that pulsed laser irradiation onto the longitudinal plasmon bands induced the release of plasmid DNA from complexes of gold nanorods and DNA. The morphological changes of gold nanorods induced by laser irradiation at 1064nm could trigger the release of plasmid DNA from gold nanorods-DNA complexes [107]. Subsequently, Niidome et al. also demonstrated that photothermal conversion and reshaping of gold nanorods (aspect ratio 5.9) passivated by phosphatidylcholine (PC) with pulsed NIR irradiation resulted in cell death [108].

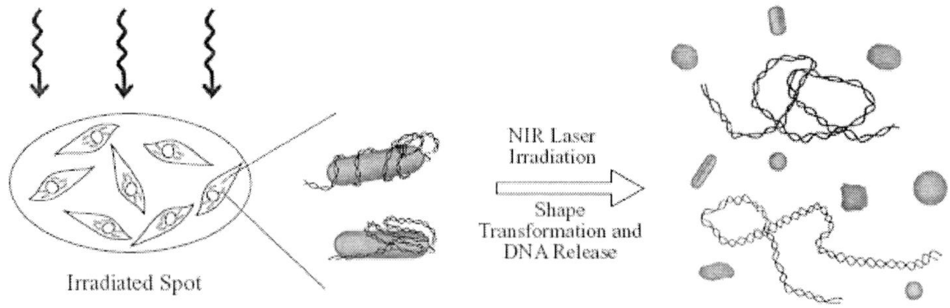

Figure 19. Schematic illustration to demonstrate that the cells containing EGFP-GNR conjugates within a spot (3.5mm in diameter) are irradiated by NIR laser (left). After laser irradiation, the gold nanorods of EGFP-GNR conjugates undergo shape transformation that results in the release of EGFP DNA (right). Reproduced from [106] with permission.

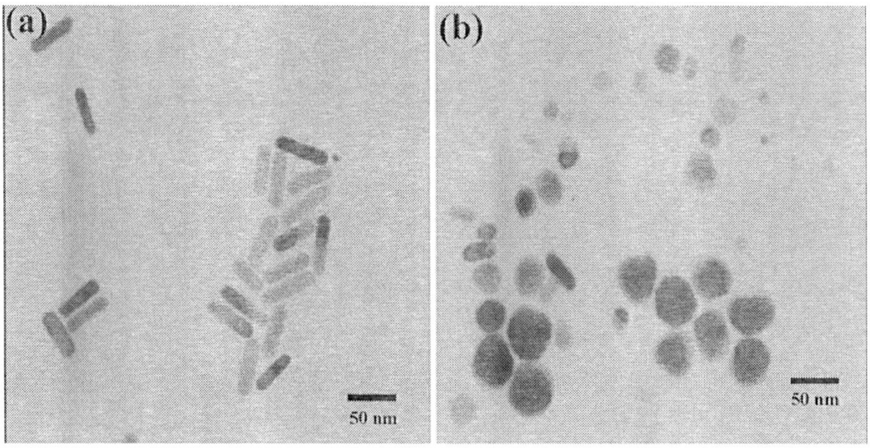

Figure 20. TEM images of EGFP-GNR conjugates (a) before and (b) after irradiation with laser beam. Reproduced from [106] with permission.

Photothermal reaction of PC-gold nanorods with pulsed near-IR laser irradiation induced cell death and simultaneously induced reshaping of PC-gold nanorods into spherical gold nanoparticles. For the laser irradiation experiments, HeLa cells were plated at a 96-well glass plate. Cells were irradiated using the fundamental light of a Q-switched Nd:YAG laser (Continuum Surelite 1064 nm). The results show that the unique photoreaction of gold nanorods with suitable aspect ratios has the potential applications in producing selective cell destruction and preventing unwanted cell damage.

CONCLUSION

This chapter addresses recent advances in the synthesis, assembly and bioapplications of gold naorods. Gold nanorods are of great interest for scientific research because of their attractive applications in photonics, electronics, optical sensing and imaging, and gene express. These interesting applications are built upon the fascinating SPR features of nanorods, which can be tuned over a broad range from the visible to the near infrared. The srongly enhanced SPR scattering and adsorption, biocompatibility and ease of surface modification of gold nanorods make them a novel and highly effective class of contrast agents for *in vivo* optical imaging-based cancer diagnostics and photothermal cancer treatment. Gold nanorods as attractive building blocks to form hybrid nanomaterials such as the assembly of gold nanorods on inorganic materials or core-shell particles have been addressed. Particular emphasis is placed on the rapid development of the integration gold nanorods with biomolecules such as protein and DNA and the utilization of such bionanohybrid assemblies for bioanalytical applications. Gold nanorods have been integrated with biomaterials to yield hybrid materials with unique functional properties. The bionanohybrid systems possess great promise for the development for novel biosensors, molecular imaging, biocatalytic systems and biocomputing elements. Bio-and biomedical applications of biofunctionalized anisotropic metal nanoparticles are developing and opening a new scientific area.

REFERENCES

[1] Jin, R.; Cao, Y.; Mirkin, C. A.; Kelly, K. L.; Schatz, G. C.; Zheng, J. G. *Science* 2001, *294*, 1901-1903.

[2] El-Sayed, M. A. *Acc. Chem. Res.* 2001, *34*, 257-264.

[3] Wang, Z. L.; Gao, R. P.; Pan, Z. W.; Dai, Z. R. *Adv. Eng. Mater.* 2001, *3*, 657-661.

[4] Burda, C.; Chen, X. B.; Narayanan, R.; El-Sayed, M. A. *Chem. Rev.* 2005, *4*, 1025-1102.

[5] Sun, Y.; Xia, Y. *Science* 2002, *298*, 2176-2179.

[6] Sun, Y.; Xia, Y. *J. Am. Chem. Soc.* 2004, *126*, 3892-3901.

[7] Chen, J.; Saeki, F.; Wiley, B. J.; Cang, H.; Cobb, M. J.; Li, Z. Y.; Au, L.; Zhang, H.; Kimmey, M. B.; Li, X.; Xia, Y. *Nano. Lett* .2005, *5*, 473-477.

[8] Cang, H.; Sun, T.; Li, Z. Y.; Chen, J.; Wiley, B. J.; Xia, Y.; Li, X. *Opt. Lett.* 2005, *30*, 3048-3050.

[9] Chen, J.; Wiley, B.; Li, Z. Y.; Campbell, D.; Saeki, F.; Cang, H.; Au, L.; Lee, J.; Li, X.; Xia, Y. *Adv. Mater.* 2005, *17*, 2255-2261.

[10] Tosatti, E.; Prestipino, S. *Science* 2000, *289*, 561-563.

[11] Sainsbury, T.; Fitzmaurice, D. *Chem. Mater.* 2004, *16*, 2174-2179.

[12] Hutchinson, T. O.; Liu, Y. P.; Kiely, C.; Kiely, C. J.; Brust, M. *Adv. Mater.* 2001, *13*, 1800-1803.

[13] Sun, Q.; Wang, Q.; Jena, P.; Note, R.; Yu, J. Z.; Kawazoe, Y. *Phys. Rev. B* 2004, *70*, 245411-1-245411-5.

[14] Sun, Y.; Xia, Y. *Adv. Mater* 2003, *15*, 695-699.

[15] Sun, Y.; Xia, Y. *Science* 2002, *298*, 2176-2179.

[16] Chen, J.; Wiley, B.; Li, Z. Y.; Campbell, D.; Saeki, F.; Cang, H.; Au, L.; Lee, J.; Li, X.; Xia, Y. *Adv. Mater.* 2005, *17*, 2255-2261.

[17] Sun, X.; Dong, S.; Wang, E. *Angew. Chem. Int. Ed.* 2004, *43*, 6360-6363.

[18] Brohm, D.; Metzger, S.; Bhargava, A.; MMller, O.; Lieb, F.; Waldmann, H. *Angew. Chem. Int. Ed.* 2002, *41*, 195-195.

[19] Zhang, J. L.; Du, J. M.; Han, B. X.; Liu, Z. M.; Jiang, T.; Zhang, Z. F. *Angew. Chem. Int. Ed.* 2006, *45*, 1116-1119.

[20] Martin, C. R. *Chem. Mater.* 1996, *8*, 1739-1746.

[21] Chang, S. S.; Shih, C. W.; Chen, C. D.; Lai, W. C.; Wang, C. R. C. *Langmuir* 1999, *15*, 701-709.

[22] Jana, N. R.; Gearheart, L.; Murphy, C. J. *J. Phys. Chem. B.* 2001, *105*, 4065-4067.

[23] Kim, F.; Song, J. H.; Yang, P. D. *J. Am. Chem. Soc.* 2002, *124*, 14316-14317.

[24] Murphy, C. J.; Gole, A. M.; Hunyadi, S. E.; Orendorff, C. J. *Inorg. Chem.* 2006, *45*, 7544-7554.

[25] Murphy, C. J.; Sau, T. K.; Gole, A. M.; Orendorff, C. J.; Gao, J. X.; Gou, L. F.; Hunyadi, S. E.; Li, T. *J. Phys. Chem. B* 2005, *109*, 13857-13870.

[26] Pérez-Juste, J.; Pastoriza-Santos, I.; Liz-Marzán, L.M.; Mulvaney, P. *Coord. Chem. Rev.* 2005, *249*, 1870-1901.

[27] Alekseeva, A. V.; Bogatyrev, V. A.; Khlebtsov, B. N.; Mel'nikov, A. G.; Dykman, L. A.; Khlebtsov, N. G. *Colloid. J.* 2006, *68,* 725-744.

[28] Katz, E.; Willner, I. *Angew. Chem. Int. Ed.* 2004, *43*, 6042-6108.

[29] Jain, P. K.; El-Sayed, I. H.; El-Sayed, M. A. *Nanotoday* 2007, *1*, 18-29.

[30] Toernblom, M.; Henriksson, U. *J. Phys. Chem. B.* 1997, *101*, 6028-6035.

[31] Jana, N. R.; Gearheart, L.; Murphy, C. J. *Adv. Mater.* 2001, *13*, 1389-1393.

[32] Jana, N. R.; Gearheart, L.; Murphy, C. J. *J. Phys. Chem. B* 2001, *105*, 4065-4067.

[33] Jana, N. R.; Gearheart, L.; Murphy, C. J. *Chem. Commun.* 2001, 617-618.

[34] Murphy, C. J.; Jana, N. R. *Adv. Mater.* 2002, *14*, 80-82.

[35] Johnson, C. J.; Dujardin, E.; Davis, S. A.; Murphy, C. J.; Mann, S. *J. Mater. Chem.* 2002, *12*, 1765-1770.

[36] Busbee, B. D.; Obare, S. O.; Murphy, C. J. *Adv. Mater.* 2003, *15*, 414-416.

[37] Gao, J.; Bender, C. M.; Murphy, C. J. *Langmuir* 2003, *19*, 9065-9070.

[38] Sau, T. K.; Murphy, C. J. *Mater Res. Soc. Symp. Proc* 2004, *789*, 203-218.

[39] Sau, T. K.; Murphy, C. J. *Langmuir* 2004, *20*, 6414-6420.

[40] Sau, T. K.; Murphy, C. J. *J. Am. Chem. Soc* 2004, *126*, 8648-8649.

[41] Gole, A.; Murphy, C. J. *Chem. Mater.* 2004, *16*, 3633-3640.

[42] Murphy, C. J.; Sau, T. K.; Gole, A.; Orendorff, C. J. *MRS Bull.* 2005, *30*, 349.

[43] Gou, L.; Murphy, C. J. *Chem. Mater.* 2005, *17*, 3668-3672.

[44] Caswell, K. K.; Bender, C. M.; Murphy, C. J. *Nano. Lett.* 2003, *3*, 667-669.

[45] Pérez-Juste, J.; Correa-Duarte, M. A.; Liz-Marzán, L. M. *Appl. Surf. Sci.* 2004, *226*,137-143.

[46] Hu, M.; Wang, X.; Hartland, G. V.; Mulvaney, P.; Pérez-Juste, J.; Sader, J. E. *J. Am. Chem. Soc.* 2003, *125*, 14925-14933.

[47] Nikoobakht, B.; El-Sayed, M. A. *Chem. Mater.* 2003, *15*, 1957-1962.

[48] Iqbal, M.; Tae, G. *J. Nanosci. Nanotech.* 2006, *6*, 3355-3359.

[49] Zweifel, D. A.; Wei, A. *Chem. Mater.* 2005, *17*, 4256-4261.

[50] Wu, H.Y.; Chu, H. C.; Kuo, T. J.; Kuo, C. L.; Huang, M. H. *Chem. Mater.* 2005, *17*, 6447-6451.

[51] Wu, H. Y.; Huang, W. L.; Huang, M. H. *Crystal. Growth and Design* 2007, *ASAP.*

[52] Orendorff, C. J.; Murphy, C. J. *J. Phys. Chem. B* 2006, *110,* 3990-3994.

[53] Liu, M. ; Guyot-Sionnest, P. *J. Phys. Chem. B* 2005, *109*, 22192-22200.

[54] Pérez-Juste, J.; Liz-Marzán, L. M.; Carnie, S.; Chan, D. Y. C. ; Mulvaney, P. *Adv. Funct. Mater.* 2004, *14*, 571-579.

[55] Rodriguez-Fernandez, J.; Pérez-Juste, J.; Mulvaney, P.; Liz-Marzán, L. M. *J. Phys. Chem. B* 2005, *109*, 14257-14261.

[56] Kou, X. S.; Zhang, S. Z.; Tsung, C. K.; Yang, Z.; Yeung, M. H.; Stucky, G. D.; Sun, L. D.; Wang, J. F.; Yan, C. H.; *Chem. Eur. J.* 2007, *13*, 2929-2936.

[57] Eustis, S.; El-Sayed, M. A. *Chem. Soc. Rev.* 2006, *35*, 1084-1094.

[58] Henglein, A. *J. Phys. Chem.* 1993, *97*, 5457-5471.

[59] Liz-Marzán, L. M. *Langmuir* 2006, *22*, 32-41.

[60] Bohren, C. F.; Huffman, D. F.; *Absorption and Scattering of Light by Small Particles*; Wiley: New York, 1983.

[61] Gans, R. *Ann Phys* 1915, *47*, 270.

[62] Link, S.; El-Sayed, M. A. *Int. Rev. Phys. Chem.* 2000, *19*, 409-453. and references therein.

[63] Park, S. J.; Taton, T. A.; Mirkin, C. A. *Science* 2002, *295*, 1503-1506.

[64] Panoiu, N. C.; Osgood Jr., R. M. *Nano. Lett.* 2004, *4*, 2427-2430.

[65] Sudeep, P. K.; Shibu, S. T.; Thomas, K. G. *J. Am. Chem. Soc.* 2005, *127*, 6516-6517.

[66] Pan, B. F.; Ao, L. M.; Gao, F.; Tian, H. Y.; He, R.; Cui, D. X. *Nanotechnology* 2005, *16*, 1776-1780.

[67] Nikoobakht, B.; El-Sayed, M. A. *Langmuir* 2001, *17*, 6368-6374.

[68] Nikoobakht, B.; Wang, J.; El-Sayed, M. A. *Chem. Phys. Lett.* 2002, *366*, 17-23.

[69] Caswell, K. K.; Wilson, J. N.; Bunz, U. H. F.; Murphy, C. J. *J. Am. Chem. Soc.* 2003, *125*, 13914-13915.

[70] Chang, J. Y.; Wu, H.; Chen, H.; Ling, Y. C.; Tan, W. H. *Chem. Commun.* 2005, 1092-1094.

[71] Thomas, K. G.; Barazzouk, S.; Ipe, B. I.; Joseph, S. T. S.; Kamat, P. V. *J. Phys. Chem. B.* 2004, *108*, 13066-13068.

[72] Hu, X. G.; Cheng, W. L.; Wang, T.; Wang, E. K.; Dong, S. J. *Nanotechnology* 2005, *16*, 2164-2169.

[73] Kotov, N. A.; Dekany, I.; Fendler, J. H. *J. Phys. Chem.* 1995, *99*, 13065-13069.

[74] Decher, G. *Science* 1997, *277*, 1232-1237.

[75] Correa-Duarte, M. A.; Pérez-Juste, J.; Sánchez-Iglesias, A.; Giersig, M.; Liz-Marzán, L. M. *Angew. Chem. Int. Ed.* 2005, *44*, 4375-4378.

[76] Mieszawska, A. J.; Slawinski, G. W.; Zamborini, F. P. *J. Am. Chem. Soc.* 2006, *128*, 5622-5623.

[77] Honda, K.; Niidome, Y., Nakashima, N., Kawazumi, H., Yamada, S. *Chem. Lett.* 2006, *35*, 854-855.

[78] Khanal, B. P.; Zubarev, E. R. *Angew. Chem. Int. Ed.* 2007, *46*, 2195-2198.

[79] van der Zande, B. M. I.; Koper, G. J. M.; Lekkerkerker, H. N. W. *J. Phys. Chem.* 1999, *103*, 5754-5760.

[80] Pérez-Juste, J.; Rodríguez-González, B.; Mulvaney, P.; Liz-Marzán, L. M. *Adv. Funct. Mater.* 2005, *15*, 1065-1071.

[81] Murphy, C. J.; Orendorff, C. J. *Adv. Mater.* 2005, *17*, 2173-2177.

[82] Varghese, S.; Narayanankutty, S.; Bastiaansen, C. W. M.; Crawford, G. P.; Broer, D. J. *Adv. Mater.* 2004, *16*, 1600-1605.

[83] E. Dujardin, L. B. Hsin, C. R. C. Wang, S. Mann, *Chem. Commun.* 2001, 1264-1265.

[84] Gole, A.; Murphy, C. J. *Langmuir* 2006, *21*, 10756-10762.

[85] Nakashima, H.; Furukawa, K.; Kashimura, Y.; Torimitsu, K. *Chem. Commun.* 2007, 1080-1082.

[86] Orendorff, C. J.; Hankins, P.; Murphy, C. J. *Langmuir* 2005, *21*, 2022-2026.

[87] Sau, T. K.; Murphy, C. J. *Langmuir* 2005, *21*, 2923-2929.

[88] Berry, V.; Gole, A.; Kundu, S.; Murphy, C. J.; Saraf, R. F. *J. Am. Chem. Soc.* 2005, *127*, 17600-17601.

[89] Gole, A.; Orendorff, C. J.; Murphy, C. J. *Langmuir* 2004, *20*, 7177-7182.

[90] Gole, A.; Murphy, C. J. *Chem. Mater.* 2005, *17*, 1325-1330.

[91] Pastoriza-Santos, I.; Gomez, D.; Pérez-Juste, J.; Liz-Marzán, L.M.; Mulvaney, P. *Phys. Chem. Chem. Phys.* 2004, 6, 5056-5060.

[92] Wang, C. G..; Ma, Z. F.; Su, Z. M. *unpublished.*

[93] Wang, C. G..; Ma, Z. F.; Su, Z. M. *Nanotechnology* 2006, *17*, 1819-1824.

[94] Connor, E. E.; Mwamuka, J.; Gole, A.; Murphy, C. J.; Wyatt, M. D. *Small* 2005, *1*, 325-327.

[95] Liao, H. W.; Hafner, J. H. *Chem. Mater.* 2005, *17*, 4636-4641.

[96] Takahashi, H.; Niidome, Y.; Niidome, T.; Kaneko, K.; Kawasaki, H.; Yamada, S. *Langmuir* 2006, *22*, 2-5.

[97] Wang, C. G..; Ma, Z. F.; Wang, T. T.; Su, Z. M. *Adv. Funct. Mater.* 2006, *16*, 1673-1678.

[98] Chen, C. D.; Cheng, S. F.; Chau, L. K.; Wang, C. R. C. *Biosens and Bioelectron* 2007, *22*, 926-932.

[99] Wang, C. G..; Ma, Z. F.; Wang, T. T. *unpublished.*

[100] Li, C. Z.; Male, K. B.; Hrapovic, S.; Luong, J. H. T. *Chem. Commun.* 2005, 3924-3926.

[101] Wang, H.; Huff, T. B.; Zweifel, D. A.; He, W.; Low, P. S.; Wei, A.; Cheng, J. X. *Proc Natl. Acad. Sci. USA* 2005, *102*, 15752-15756.

[102] Durr, N. J.; Larson, T.; Smith, D. K.; Korgel, B. A.; Sokolov, K.; Yakar, A. B. *Nano. Lett.* 2007, *ASAP.*

[103] Huang, X.; El-Sayed, I. H.; Qian, W.; El-Sayed, M. A. *J. Am. Chem. Soc.* 2006, *128*, 2115-2120.

[104] Li, P. C.; Wei, C.W.; Liao, C. K.; Chen, C. D.; Pao, K. C.; Wang, C. R. C.; Wu, Y. N.; Shieh, D. B. *Proc. SPIE* 2006, *6068*, 60860M-1-60860M-10.

[105] Petrova, H.; Pérez-Juste, J.; Pastoriza-Santos, I.; Hartland, G. V.; Liz-Marzán, L. M.; Mulvaney, P. *Phys. Chem. Chem. Phys.* 2006, *8*, 814-821.

[106] Chen, C. C.; Lin, Y. P.; Wang, C. W.; Tzeng, H. C.; Wu, C. H.; Chen, Y. C.; Chen, C. P.; Chen, L. C.; Wu, Y. C. *J Am Chem Soc* 2006, *128*, 3709-3715.

[107] Takahashi, H.; Niidome, Y.; Yamada, S. *Chem. Commun.* 2005, 2247-2249.

[108] Takahashi, H.; Niidome, T.; Nariai, A.; Niidome, Y.; Yamada, S. *Nanotechnology* 2006, *17*, 4431-4435.

In: Encyclopedia of Nanotechnology
Editor: Elwood D. Carlson
ISBN 978-1-60692-079-4

Chapter 57

NANOTECHNOLOGY BASED GAS SENSORS

John T.W. Yeow* and Yun Wang
Systems Design Engineering, University of Waterloo
200 University Ave West, Waterloo, Ontario, Canada, N2L 3G1

ABSTRACT

This chapter focuses on novel gas/chemical sensing systems that are based on nanotechnology. Recently, there has been an extensive interest in nanomaterials such as carbon nanotubes, nanowires, nanofibers, and nanoparticles. Their application on gas sensing systems has also been widely studied. Due to their unique micro/nano porous structures and ultra high surface-to-volume ratio, nanostructured materials have a high potential to enhance conventional gas sensors' performance, including the sensitivity, the operation temperature and the recovery time. In this chapter, various nanomaterials-based gas/chemical sensors are presented, including their structure, fabrication, and operation principles.

Gas sensing systems, or chemical sensors, are attracting tremendous interest because of their widespread applications in industry, environmental monitoring, biomedicine, pharmaceutics, and space exploration. Gas sensors with high sensitivity and selectivity are required for leakage detection of explosive gases such as hydrogen, and for real-time detection of toxic or pathogenic gases in industry. There is also a strong demand for the ability to monitor and control of our ambient environment, especially with the increasing concern of the globe warming. NASA researchers can also use similar high performance gas/chemical sensors for identification of atmospheric components of various planets. In addition, nerve agent sensing for homeland security is also at the centre of public concern [1].

Generally, there are several basic criteria for good and efficient gas sensing system: i) high sensitivity and selectivity; ii) fast response time and recovery time; iii) low analyte consumption; iv) low operating temperature and temperature independence; and v) stability in performances. Commonly used gas sensing systems include polymer gas sensors,

* E-mail: jyeow@engmail.uwaterloo.ca

semiconductor metal oxide gas sensors and surface acoustic wave sensors. Usually polymer gas sensors have limited sensitivity, while semiconductor metal oxide gas sensors require high temperatures (approximately 200-500oC). Since the most common gas sensing principle is the adsorption and desorption of gas molecules on sensing materials, it is quite understandable that by increasing the contact interfaces between the analyte and sensing material, the sensitivity can be significantly enhanced.

Recent development of nanotechnology has created huge potential to build highly sensitive, low cost, portable sensors with low power consumption. The extremely high surface-to-volume ratio and hollow structure of nanomaterials is ideal for gas/chemical molecules adsorption and storage. Particularly, the advent of carbon nanotubes (CNTs), which is, arguably, the most widely investigated nanomaterial, has fuelled the inventions of gas sensors that exploit CNTs' unique geometry, morphology and material properties. Upon exposure to certain gases, the change in CNTs' properties or in the sensing medium which is induced by the morphology of CNTs, can be detected by various methods. As a result, CNT-based gas sensing systems and the theoretical analyses of gas adsorption and collision effects on the nanotubes have been subjects of intense research. In addition, nanomaterials other than CNTs such as nanowires, nanofibers, and nanoparticles have also been investigated widely as potential materials for gas sensing applications. This chapter includes a broad but yet in-depth survey of nanomaterials that are used for gas sensing applications. The underlying sensing mechanisms and their corresponding advantages will also be discussed.

1. Carbon Nanotubes Based Gas Sensors

Since being discovered in 1991, CNTs have captured researchers' attention worldwidely. Considerable research have been conducted for new synthesis and preparation techniques, as well as to investigate the properties and potential applications of CNTs. In this section, a brief introduction of CNTs is presented. Illustration of the mechanism of gas molecules adsorption on nanotubes will be discussed and some CNT-based gas sensing systems that have been developed recently will be introduced.

1.1. Brief Introduction of CNTs

CNTs belong to the family of fullerene structures. There are two types of nanotubes: (i) single-walled carbon nanotubes (SWCNTs); and (ii) multi-walled carbon nanotubes (MWCNTs). A SWCNT can be considered as a one-atom-thick layer of graphite rolled up into a seamless cylinder with a diameter of several nanometers, and length on the order of 1-100 microns [2]. Similarly, MWCNTs consist of multiple layers of graphite wrapped up together to form a tube shape, sharing the same central axis. MWCNTs were first observed in 1991, by Iijima of NEC Laboratory in Japan when he saw fine threads in carbon-soot grown by arc-discharge [3]. In 1992, Ebbesen and Ajayan managed to grow MWCNTs at the gram level by the arc-discharge method [4]. In the following year, SWCNTs were first synthesized in small amount using the same method by Bethune et al. [5]. Since then, tremendous CNT related research, including both experimental and theoretical aspects have been conducted to better understand the growth of CNTs and their unique properties

1.1.1. Preparation of Carbon Nanotubes

There are three main methods to prepare CNTs: i) arc-discharge technique; ii) laser ablation technique; and iii) chemical vapor deposition (CVD) technique.

The carbon arc-discharge method involves two carbon electrodes as carbon source. When high DC voltage is applied between the carbon anode and cathode in a vacuum chamber, plasma of the insert gas (typically helium) is generated to evaporate the carbon atoms. The ejected carbon atoms are then deposited on the negative electrode to form CNTs. It is the principal method to produce high quality CNTs, both SWCNTs and MWCNTs, with nearly perfect structures.

The laser ablation technique requires a carbon target to be ablated by intense laser pulses in a furnace. In the presence of an inert gas and a catalyst, CNTs are formed and collected on a cold substrate. Both the arc-discharge and laser-ablation methods require high growth temperature, which is about 3000-4000oC for the evaporation of carbon atoms from solid carbon source [6].

In a CVD system, a gas hydrocarbon source (usually methane, acetylene or ethylene) flows into the reaction chamber and molecules are broken into reactive species at the temperature range of 550-1000oC. The reactive species react with the catalysts (metal nanoparticles) that are coated on the substrate, leading to the formation of CNTs.

Compared with the first two techniques, CNTs can be synthesized at relative low temperature using CVD method. Therefore, this technique is more efficient and allows for scale-up growth of SWCNTs. By modification and calculated control of the growth parameters, vertically aligned MWCNTs growth can be achieved by CVD technique. This may enhance CNTs electronic properties in different applications. High quality SWCNTs can also be obtained by optimization of the catalysts. However, one of the main disadvantages of CVD technique is the relative high defect densities in MWCNTs, which can be attributed to the lack of sufficient thermal energy [6].

In all of these three growth methods, CNTs come with a number of impurities, which may have negative effects on CNTs' inherent properties. In order to make CNT-based devices more efficient and consistent, purification is an important process to obtain high purity nanotubes. Generally, the most commonly observed impurities are carbonaceous impurities and metallic impurities. The carbonaceous components are byproducts of the growth process and the metallic impurities are residual catalysts. To eliminate the carbonaceous impurities, the main method is oxidation. Two approaches are developed: i) gas phase purification, which involves high temperature and has low yield production; and ii) liquid phase purification, where nanotubes are washed in acidic solutions. Nitric acid, sulphuric acid, and their mixture are all commonly used for liquid purification [7]. Metallic impurities can be removed by heating the samples up to the evaporation temperature of the impurity. With the development and modification of different purification procedure, so far CNTs samples with a purity of up to 99.6% can be obtained [8]. The quality of purification can be inspected by microscopic techniques such as scanning electron microscopy or spectroscopic techniques such as Raman spectroscopy [9].

1.1.2. Properties of Carbon Nanotubes

The structure of carbon nanotubes provides them with inherently unique electrical, physical and chemical properties. In terms of electrical properties, CNTs can be either metallic or semiconducting, depending upon the tube diameter and the chirality (the direction in which the graphite sheet is rolled to form the tube) [10]. The chirality is usually represented by an integer pair (n, m). Nanotubes with $n - m = 3j$ (j being a nonzero integer) are metallic while all the others are semiconducting. The dielectric property of nanotubes is highly anisotropic due to their nearly one dimensional structures, which may enable nanotubes to carry high currents with negligible heating effect[11].

For mechanical and thermal properties, CNTs are the strongest and stiffest fibers that are known currently, because the C-C bond in graphite is one of the strongest in nature. Due to the limitation of measure techniques and the nonuniformity between CNTs prepared by different method, it is very challenging to obtain the accurate value of the Young's modulus and tensile strength. Lourie and Wagner reported the Young's modulus of 2800–3600 GPa for SWCNTs and 1700–2400 GPa for MWCNTs [12]. Yu et al. found the tensile strength of SWCNT bundles in the range of 13 to 52 GPa, while the value for MWCNTs was in the range of 11 to 63 GPa [13]. CNTs have high thermal stability in both vacuum and air. The thermal conductivity is dependent on the chirality. Theoretically, they are expected to be good thermal conductors along the tube direction and good insulators laterally to the tube axis.

1.2. Gas Adsorption on Carbon Nanotubes

The extremely high surface-to-volume ratio and hollow structures of CNTs create extraordinarily large adsorption capacity for gas molecules. Therefore, CNTs are considered to be very good candidates for the application of gas storage and gas sensing. However, the potential application of CNTs for gas/chemical sensing did not gain serious attention until 2000. Kong et al. demonstrated that upon exposure to a very small concentration of certain gas molecules (NO2 and NH3), the electrical resistance of a single semiconducting SWCNT can be dramatically increased or decreased [14]. Inspired by this finding, the effect of O2, Ar, H2 and other gases on the field emission or electrical property of SWCNTs and MWCNTs are widely studied in the following years [15]. Simultaneously, there were also significant developments on theoretical and simulating studies on the interaction between gas molecules and CNTs [15-21].

In a bundle of SWCNTs, there are generally four distinct sites for gas molecules to be adsorbed onto: i) the external *surface* of the bundle; ii) the *groove* formed at the contact between adjacent tubes on the outside of the bundle; iii) the interior *pore* of individual tubes; and iv) interstitial *channel* formed between three adjacent tubes within the bundle [16]. The gas adsorption on these sites is decided by the binding energy of the gas molecule as well as the site availability. Some of these sites may not be available for certain gases because of the gas molecules' dimension and the site diameter. The interior *pore* is only accessible when the SWCNT is uncapped or has defects on the tube walls. Zhao et al. studied the adsorption of various gas molecules (NO2, O2, NH3, N2, CO2, CH4, H2O, H2, Ar) on single SWCNT and SWCNT bundles using first principles method. Calculations of the binding energy (E_B) for hydrogen adsorption have showed that E_B for channels and grooves are larger than that for pores and surfaces; however, pores and surface have much larger available adsorption area than the other two (as shown in Figure 1). Similar results were obtained for other gas molecules that they studied as well. The larger adsorption energy on groove and channels can be explained by the fact that there are more carbon molecules within a specific volume along these two sites [17]. It is also demonstrated by theoretical studies that gases such as NO2, O2 and NH3 have larger binding energies with SWCNT than other gases, which is coincident with experimental results that CNTs are more sensitive to these three gases.

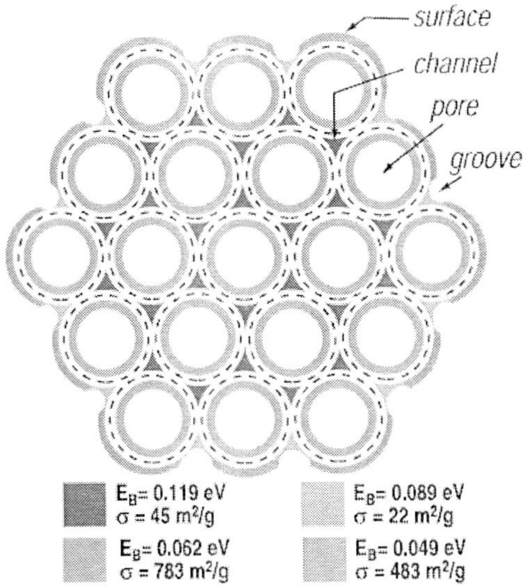

Figure 1. Schematic structure of a SWCNT bundle showing different sites for gas adsorption [16], Copyright Elsevier.

Gas molecule adsorption has effects on both thermoelectric power (S0) and conductance (or resistance) of CNTs. In general, the change in conductance is easier to detecte, so it is more commonly used as the sensing parameter. It is believed that changes in the electronic property of semiconducting SWCNTs are caused by the charge transfer between nanotubes and the adsorbates, which can act as dopants and consequently shift the Fermi energy of the nanotubes. As a result, the nanotubes can be either n-type or p-type doped, depending on whether the adsorbed gas is a charge-donator or a charge-acceptor. O2 and NO2, for example, are charge acceptors with negative charge transfer (-0.06 to -0.14 electron per molecule) and relatively large binding energy compared to other gases [9]. This is consistent with experimental results indicating that the conductance of semiconducting SWCNTs increases significantly upon exposure to O2 and NO2 [18]. Gas sensing properties of MWCNTs and MWCNT films are also reported but much less theoretical work have been done due to the complexity in their tube structures. However, the gas adsorption and charge transfer mechanism may also work for MWCNTs, but metallic nanotubes may play a more important role in MWCNT bundles and films.

1.3. Gas Sensing Devices Based on Carbon Nanotubes

1.3.1. Semiconducting SWCNT (S-SWCNT) FET for Gas Sensing

As addressed in former section, SWCNTs can be either semiconducting or metallic. If two metal contacts are connected to each end of an individual S-SWCNT, the metal/S-SWCNT/metal device exhibits *p*-type transistor characteristics. Based on this, Kong and his coworkers [14] have built an S-SWCNT transistor for chemical sensing. The device was fabricated by controlled CVD growth of individual SWCNTs on SiO2/Si substrates with patterned catalyst islands. Polysilicon in the back side acted as gate of the transistor. The schematic structure is shown in Figure 2. Figure 3 shows the conductance versus time curves of the transistor upon exposure to NO2 and NH3 at room temperature. A sharp increase in conductance of the S-SWCNT FET by about three orders of magnitude was observed after introducing 200 ppm of NO2 into the testing chamber (Figure3a). The response time was about 2-10 second, and the sensitivity (the ratio between resistance after and before gas exposure) was about 100 to 1000. When the same sample was exposed to a 1% NH3 flow after recovery, the conductance of the device decreased dramatically within 1 to 2 minute, and the sensitivity was about 10 to 100 (Figure 3b).

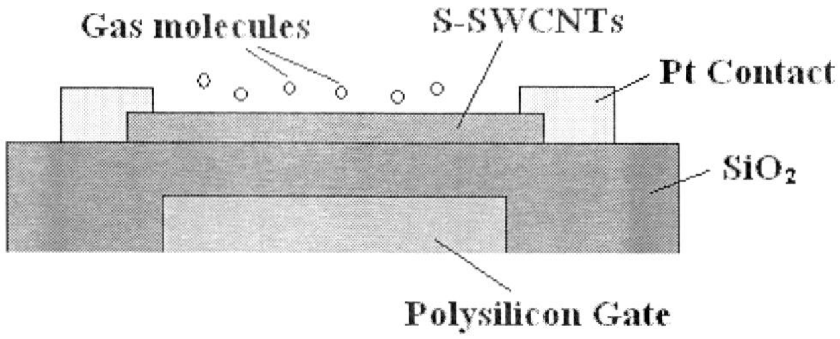

Figure 2. Schematic illustration of Single semiconducting SWCNT FET [14]. Reprinted with permission from AAAS.

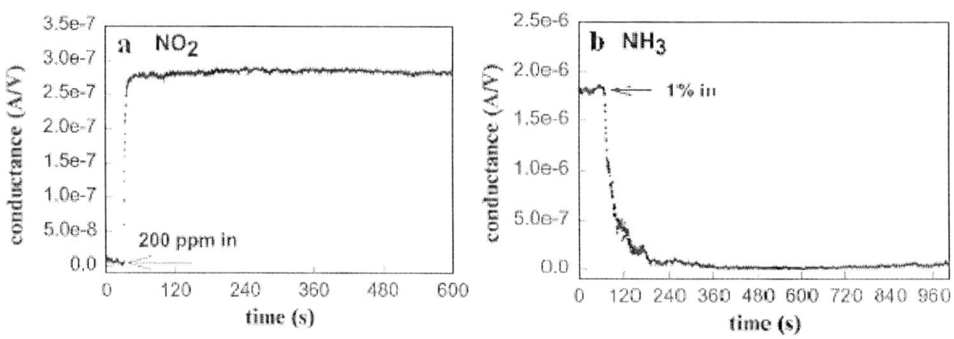

Figure 3. Change of sensor's conductance when exposed to (a) NO2 and (b) NH3 gases [14]. Reprinted with permission from AAAS.

Takao Someya et al. characterized the influence of alcohols adsorption on the resistance of SWCNTs using the same transistor structure [22]. The measurements were carried out under flow of dry nitrogen with and without addition of alcoholic vapor. Figure 4 shows the drain current versus time curves of the device upon exposure to different saturated alcoholic vapor at room temperature (except for tertiary-butanol) with Vg = -20V. For the vapor of methanol, tertiary-butanol, 1-propanol, 2-propanol, and ethanol, the response time were within 5-15 min and the current reduction were in the range of 50%-90%. Short recovery time (less than few seconds) and excellent repeatability can be obtained if the gate voltage is released briefly and then restored before the next cycle, and no degradation of the sensor performance was observed over 1 hour's operation. Moreover, the different response characteristic of the device to different alcoholic vapor also suggests its potential application on the identification of gas sources.

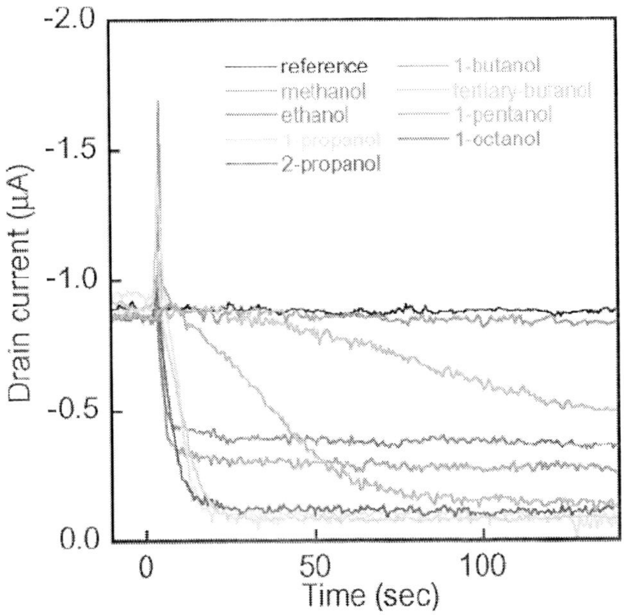

Figure 4. Drain currents a function of time upon exposure to saturated vaporof various kinds of alcohols: methanol, ethanol, 1-propanol, 2-propanol, 1-butanol, tertiary-butanol, 1-pentanol, and 1-octanol. Reprinted with permission from [22]. Copyright (2003) ACS

Although these individual semiconducting SWCNT transistors show pronounced response upon exposure to gas molecules at room temperature, it is generally very difficult to obtain pure and individual semiconducting SWCNTs. CNTs samples grown by all the three techniques mentioned earlier are mixture of both metallic and semiconducting nanotubes, and there is no efficient technique to realize the selective growth of semiconducting or metallic nanotubes up to now. This is one of the main issues related to S-SWCNT FET devices.

1.3.2. Resistance Gas Sensors Based on CNT Films

In contrast with individual nanotubes, preparation of CNTs mat or CNTs films are much less complex and more efficient. Therefore, they are able to provide much higher yield. Based on the change in electrical properties of CNTs by gas molecule adsorption, CNTs resistance gas sensors for different gases have been successfully developed by researchers with simple fabrication processes [23-26].

Jing Li et al. developed a resistance gas sensor by simply casting SWCNTs on an interdigitated electrode (IDE) [8], which provided a large sensing area and sufficient contact between the electrodes and coated CNTs. The electrodes were fabricated by photolithography and evaporation of Ti and Au (60 nm in thickness together) on silicon oxide (Figure 5(a)). As-grown SWCNTs were purified with acid first and then by air oxidation before integrated with IDE. As a result, the final SWCNTs had a relative high purity up to 99.6%, and the impact of impurities on the sensor's characteristic was minimized. The purified nanotubes were then dispersed in dimethylformamide (DMF) and drop-deposited onto the electrode area. A network of nanotubes subsequently formed after the evaporation of DMF, as shown in figure 4(b). Upon exposure to NO2, the conductance change of the sensor showed a linear response to the gas concentration with a detection limit of <4 ppb. The recovery time was about 10 hours but can be intensively shortened by UV light treatment. The sensor's response to organic vapor such as benzene, acetone, and nitrotoluene were also tested.

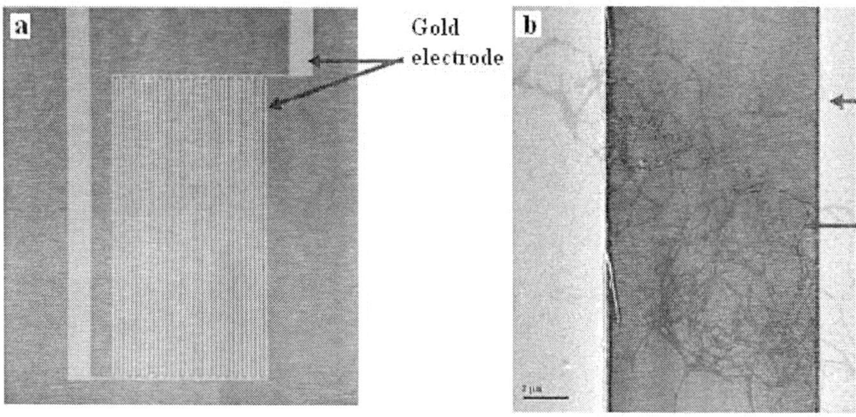

Figure 5. (a) interdigitated electrodes; (b) SEM image of SWCNTs across two gold electrodes. (Reprinted with permission from [8]. Copyright (2003) ACS)

Compared with SWCNTs, the mechanism of MWCNTs' response to gas adsorption is more complicated due to the multi-layer tube structure. However, they also show high sensitivity to specific gases experimentally. A MWCNTs resistance gas sensor was developed by laterally growth of nanotubes between two electrode arrays using CVD [23]. The growth of MWCNTs was guided by electrical field applied between the electrode arms to achieve the nano-wiring. Figure 6(a) shows the SEM image of aligned CNTs grown on silicon dioxide between electrodes, and the insert figure shows the image of the developed device. The resistance of the sensor increased when cycling the test chamber atmosphere from air to vacuum (shown in Figure 6b). This phenomenon is consistent with the fact that CNTs can be *p*-doped by oxygen in air because O2 acts as charge acceptor to the nanotubes. The sensor also showed good response to NH3 with almost linear relationship between the resistance and the gas concentration. By heating the sensor after exposure to ammonia, the recovery time can be intensively shortened from ~20h to 3h. This improvement in performance can be explained by the accelerated gas molecules desorption due to increased the ambient temperature.

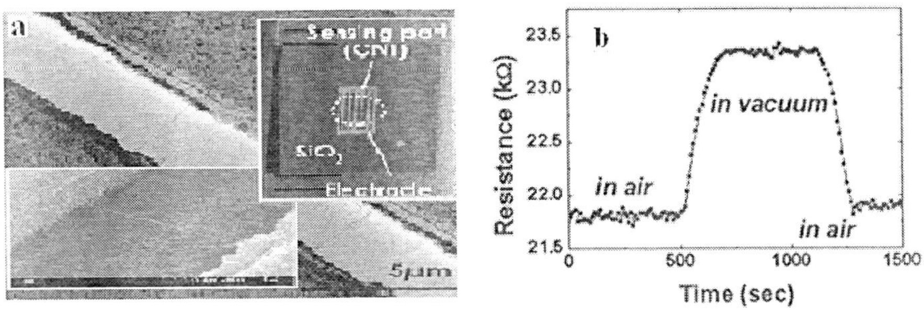

Figure 6. (a) SEM image of aligned CNTs resting on SiO_2 surface after directed growth; (b) electrical resistance of sensor by cycling chamber atmosphere from air to vacuum [23], Copyright Elsevier.

In general, slow recovery time is a common problem related to gas sensing based on adsorption and desorption of gas molecules. To overcome this issue, Woo-Sung Cho et al. integrated a micro heater under the MWCNTs sensing layer of a resistance gas sensor [24]. As shown in Figure 7(a), the sensor consists of a micro heater, an insulating layer, and a pair of electrodes with MWCNTs sensing film grown on top of them. The function of the heater was to accelerate the gas desorption rate during operation. Figure 7(b) shows the resistance change of the sensor in vacuum with and without heat treatment. The recovery time decreased significantly from approximate 6 hours to 10 minutes (the recovery time was defined as the time for 90% of total resistance change). When tested under cycles of NO2 exposure and evacuation (100 ppm), the sensor showed excellent reproducibility and stability with the resistance change being about 220% (Figure 7c).

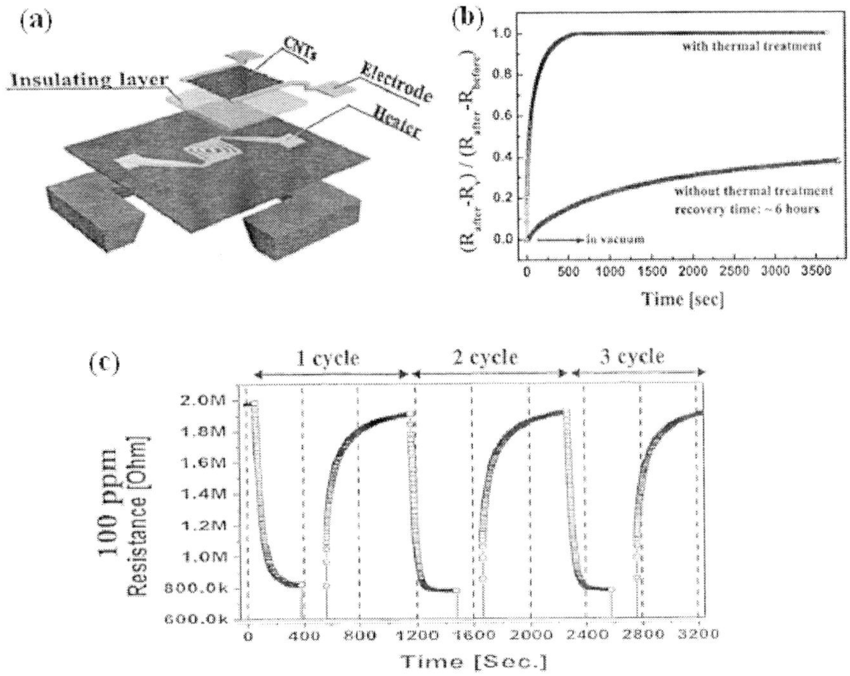

Figure 7. (a) Schematic of the sensor; (b) Recovery time of sensor with thermal treatment and without thermal treatment in vacuum; (c) Resistance changes of the sensor as a function of time when exposed to three cycles of a sequence of NO [24] (@ [2005] IEEE).

The main advantage of resistance gas sensors is their simple fabrication process. By using IDE or electrode arm arrays, good and efficient contact between CNTs and the electrodes are obtained to ensure large signals. All the sensors mentioned above operated at room temperature and fast recovery time can be achieved by UV light or thermal treatment.

1.3.3. CNTs/Polymer Gas Sensors

Organic polymers are one of the principal materials applied in gas sensing systems. Some conducting polymers can behave like semiconductors due to their heterocyclic compounds which display physicochemical characteristics. As a result, reversible changes in the sensing layer's conductivity can be detected upon polar chemicals' adsorption on the surfaces at room temperature [27]. This effect is believed to be caused by the charge transfer between gas molecules and the polymer or the polymer film's swelling [28]. This sensing response has intensively driven the motivation to develop high sensitive and selective gas/chemical sensors by tailoring the compounds of different organic polymers.

Non-conductive polymers can also be used as gas sensors. For some polymers, the bulk dissolution of gas into the film can cause changes of their physical properties [28]. Since most physical property changes are not readily detectable; researchers have tried to integrate polymers with other functional sensing elements to achieve easier detection process. For example, a recent experimental and theoretical study on the chemical sensing characteristic of polymer coated SAW and STW resonators was reported by Ivan D. Avramov et al. [29,30].

Another promising sensing material based on polymers is conducting particles–insulating matrix composition. The polymer acts as the insulating matrix while dispersed conducting particles provides the conducting path for sensing [31]. Due to adsorption of interested analyte, there are volumetric changes of the matrix polymer. This can lead to a distinct change in percolation-type conductivity around a critical composition of the material, which is known as "percolation threshold". Generally, the percolation threshold is dependent on the shape of the conducting particle. Composite consisted of particles with higher aspect ratio shows lower threshold and higher sensitivity [32].

CNTs, with almost one dimensional thread like structure and good conductivity, are ideal as the dispersed particles in this conducting particles–insulating matrix composition for gas sensing systems. Seong M. Cho et al. investigated the chemical sensing characteristic of SWCNT-Ethyl cellulose (EC) composite [31]. To get uniform dispersion of CNTs and stable composite, the material was prepared carefully with several steps. Poly (p-phenylenevinylene-co-2,5-dioctoxy-m-phenylenevinylene) (PmPV) was used as the dispersant of the nanotubes and chloroform was used as the media. The nanotubes sol was sonicated for 4 hours before mixed with EC, and then the mixture was sonicated for another 6 hours. The mixed sol were sprayed on interdigitated electrodes and annealed at 80oC for 24 hours to form the thin sensing film. The sensor's response to benzene and ethanol was measured separately as a function of gas concentration and SWCNTs' content in the composite. The change of resistance showed almost linear relationship with concentration. The maximum sensitivity was 9% for the film with 5wt.% SWCNT contents when exposured to 5,000ppm of benzene. The sensor's response to ethanol was less obvious and with some noise, however it showed good linearity within the concerned range with short response and recovery time. Based on the same idea, Biju Philip et al. developed composite thin film of polymethylmethacrylate (PMMA) with both MWCNTs and oxidation modified MWCNTs dispersed in it for gas sensing. The resistance change of the composites was evaluated upon exposure to different gases including dichloromethane, chloroform and acetone et al. A 102-103 order of magnitude increase in resistance of the surface-modified-CNT/PMMA composite was obtained at room temperature with fast response and recovery time [32].

In addition of CNT/polymer composite, the sensing characteristics of polymer-coated CNTs film to specific gases were also investigated. In general, CNTs do not have sensing response to all gases and vapours but only the ones with high adsorption energy or that can interact with them. Therefore, coating or doping of an enhance element on CNTs may broaden the application range. Based on IDE SWCNTs resistance gas sensor that they developed, Jing Li et al. studied the sensing character of the same structure but with polymer coatings on CNTs [33]. Chlorosulfonated polyethylene was dissolved in tetrahydrofuran (THF) solvent as a coating solution for Cl2 sensing, and Hydroxypropyl cellulose was dissolved in chloroform as coating to detect HCl. The resistance of polymer coated SWCNT film changed upon exposure to Cl2 and HCl while uncoated pure SWCNTs showed no sensing signal. The sensing response demonstrates a huge potential of using modified CNTs gas sensing materials for a broad range of gases and chemical vapors.

1.3.4. CNTs Enhanced Ionization Chamber for Gas Sensing

All gas sensing systems introduced in the previous sections are based on chemical or physical adsorption of analytes on the sensing layers. Therefore, they are limited by the low adsorption energy and poor charge transfer of the gas species with CNTs. For ionization gas sensors, the sensing principle is based on the ionization characteristics of different gases, which is independent of the chemical interactions between gases and the sensing materials. Generally, ionization of neutral gas is carried out by collisions of molecules with electrons that are accelerated by the electrical field between the anode and cathode. The issue related to conventional ionization gas sensors is their considerable high breakdown voltage (approximately thousands of volts). This is quite inefficient and risky in operation. CNTs, due to their extremely sharp tip structure, can induce a large field enhancement factor and therefore intensively increase the electric field around the tips to initiate corona discharge at low voltage [34].

A miniaturized gas ionization sensor with aligned MWNT film as anode was developed by Ashish Modi et al. in 2003 [35]. Figure 8 (a) shows a diagram of the sensor's structure. It consistes of MWNT film anode, Al plate cathode and a 150μm thick glass insulator between them. The vertical aligned MWNT film was prepared by CVD on SiO2 substrate with tip diameter in the range of 25~30nm and 30μm in length (Figure8 (c)). As shown in Figure8 (b), the current between anode and cathode was tested under increasing DC voltage. By using CNTs film as anode, the breakdown voltage of air decreased dramatically from about 1000V to 346V. Different gases including NH3, CO2, N2, O2, He, Ar and air were tested and showed distinct and precise breakdown voltage (Figure 9).

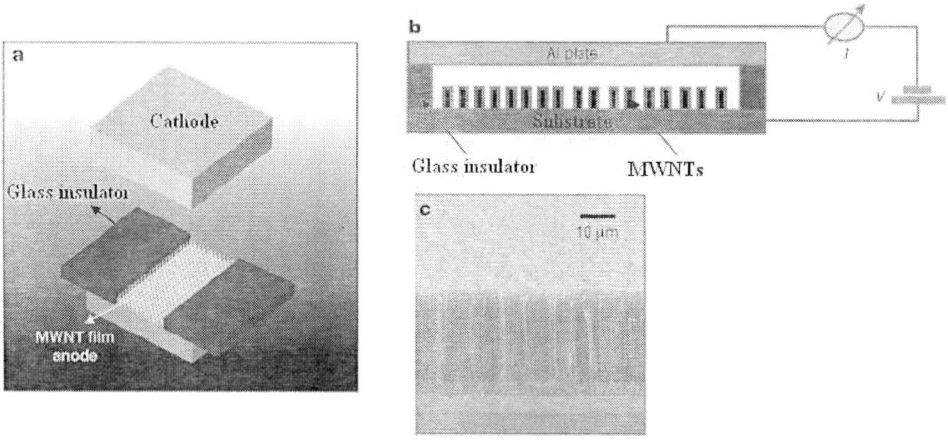

Figure 8. (a) Schematic diagram of the nanotube sensor device; (b) Diagram of actual test set-up; (c) SEM micrograph of vertically aligned MWCNT film used as the anode. Reprinted by permission from Macmillan Publishers Ltd: [23], copyright (2003).

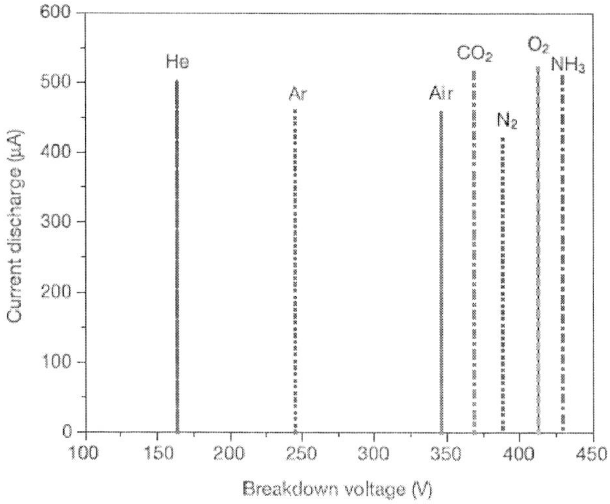

Figure 9. Current–voltage curves for electrical breakdown for NH3,CO2, N2, O2, He, Ar and air. Reprinted by permission from Macmillan Publishers Ltd: [23], copyright (2003).

The distinct breakdown voltage signatures indicated its potential for gas identification within a mixture of gases. The discharge current of each gas showed almost linear relationship with the gas concentration, but the breakdown voltage was rarely affected by the concentration. This can be explained by the fact that the discharge current is dominated by the number of molecules per unit volume and the breakdown voltage depends mainly on the intensity of electric field and the bonding energy of the gas molecules.

Nanotubes are good electron emitter due to their sharp tip curvature and low electron escaping work function. If CNT films are deposited as cathode of the device instead of anode, the breakdown process will be initiated by a negative corona. Field emission may occur under low pressure to enhance the ionization, and lower breakdown voltage can be obtained as a result [36]. Another important factor that dominates the ionization characteristic is the distance between anode and cathode. According to Paschen's curve, a certain gap distance is required for breakdown phenomenon to be induced, and the breakdown voltage decreases by bringing the electrodes closer. In the capacitor-architecture device, it is very difficult to control the electrode gap under 10μm to gain a safe operation voltage. Zhongyu Hou et al. [34] introduced a novel hollow slot electrode system with CNT sidewalls and electrode spacing of 6-12μm for gas ionization sensors, which dramatically decreased the threshold voltage to approximately 5V-40V.

2. Gas Sensors Based on Other Nanotechnologies

Besides carbon nanotubes, other nanomaterials such as nanowires, nanofibers, nanoparticles, and nano porous structures have also attracted significant interests due to their special morphology, property and the potential applications on various systems. With the development of synthesize methods, nano structures of metal, metal oxide and polymers have been obtained successfully and their properties have been intensively investigated.

In gas sensing systems that are based on the adsorption and desorption mechanism, the interface or interaction area between the sensing layer and analytes is a key factor to gain a large signal and to achieve a lowe detection limit. Compared to bulk or film materials, nanomaterials possess ultra high surface to volume ratio and porous structures. This can provide much larger sensing surface and easier diffusion of gas molecules through the sensing material. For sensing materials whose reaction with gas molecules mainly occurs on the grain boundaries, nanocrystalline structure can intensively increase their sensitivity. And in terms of functional modification of some gas sensing materials, the ultra small size of nanoparticals permits more efficient dispersion into other sensing elements. All of these advantages make nanomaterials excellent candidates in gas sensing systems. In this section, we will introuduce the application of nanomaterials other than CNTs on gas/chemical sensing systems.

2.1. Nanosized Semiconductor Metal Oxide Gas Sensors

Semiconductor metal oxide (SMO) gas sensors are one of the most commonly used chemical sensing systems due to their advantages of low cost, simple structure, fast response and recovery time, and insensitivity to moisture and temperature. Their sensing layer generally consists of polycrystalline metal oxide films, such as SnO2, ZnO, TiO2, and CeO2 et al. SMO gas sensing systems can be widely used for the detection of various gases such as O2, CO, NH3, NO, H2 et al. The detection principle of SMO gas sensors is based on the variations of energy barriers in presence of reducing or oxidizing gases, which is caused by the change of the depletion layer at the grain boundaries. The conductivity of the polycrystalline metal oxide, for example, increases when the magnitude of these barriers is reduced by adsorbed gases [37]. One of the main issues with SMO gas sensors is their high operation temperature (300~500°C) [38]. It is believed that by reducing the grain size of the metal-oxide semiconductor materials, the detection threshold can be lowered. As a result, a higher sensitivity and lower operation temperature can be achieved [40]. This is mainly due to the increase of surface area to volume ratio of the particles and the formation of grains with dimensions in the same order of the depletion layer depth [41]. Therefore, synthesize of nanosized metal oxide materials and their sensing characteristic has attracted extensive research interest.

Zinc oxide (ZnO) is a multifunctional semiconductor oxide material that exhibits piezoelectric and pyroelectric properties. It has wide direct band gap of 3.36 eV and large exciton binding energy (60 meV), which makes it a promising smart material for diverse optical and electronic devices. For SMO gas sensing devices, nanostructured ZnO has been synthesized and their sensing characteristics have been investigated recently. The properties of ZnO film depend strongly on the stoichiometry, the nature and quantity of impurities, and the growth parameters. By doing gowth modifications, ZnO with better sensing properties can be achieved. T.Shishiyanu et al. [41] developed Sn and Cu doped nanostructured ZnO thin film for the detecting of volatile organic compounds (VOC) gas (ethanol and methanol). Nanostructured thin films with various %Sn and %Cu were prepared on a glass substrate by chemical deposition, followed by rapid photothermal processing (RPP) at 350-550℃ for 20s to form homogeneous crystal. The synthesized particle size was in the range of 100~150nm. The sensors sensing characteristics were detected at different temperatures and gas concentrations. When the synthesized thin films were exposed to VOC gas, the conduction electrons were consumed. And the subsequent reactions resulted in the increase of the barrier height and decrease of the surface conductance. The highest sensitivity (defined as resistance change over original resistance) was approximately 28%/ppm when exposed to ethanol at room temperature. The Sn and Cu concentration of this film was 2% and 1% respectively. It was demonstrated that the Sn and Cu impurities improved the ZnO film's sensitivity, and shortened the response time as well. The temperature where the maximum sensitivity was shown was determined by the grain size of the nanostructure.

ZnO nanowires were also synthesized by researchers for gas sensing application based on their semiconducting property. Zhiyong Fan [39] built a field-effect transistor with single-crystalline ZnO nanowires that were grown via vapor trapping chemical vapor deposition method. The as-synthesized nanowires were suspended in alcohol and deposited onto Si substrates with a SiO2 insulating layer on the top. The metal contacts were fabricated using photolithography and ebeam evaporation. Figure 10(a) illustrates the AFM image of the device and the schematic of the circuit. The device's sensing characteristic to various gases including O2, NO2, NH3, and CO were studied. This nanowire FET sensor showed oxidizing property to O2, NO2, and NH3 with decrease conductance at room temperature. However, due to the shift in Fermi energy at elevated temperature, it showed reducing sensing property for NH3 at 500K. Figure 10(b) shows the sensor's response to different concentration of NO2. It was also demonstrated that by reducing the nanowires diameter, the material's sensitivity to O2 increased significantly.

Yin Hu and her coworkers prepared nanosized-SrTi1±xO3−δ series powders for oxygen sensing. The nanostructured powder was synthesized through high energy ball-milling technique. The powder were formed into pastes using organic binding vehicle and then screen printed onto Al2O3 substrates with precoated interdigital Au electrodes on the top. All the samples were annealed in air before tested. The sensors showed optimal operating temperature of 40℃. The deviating stoichiometry of the nanosized materials had obvious effects on their sensing properties [40].

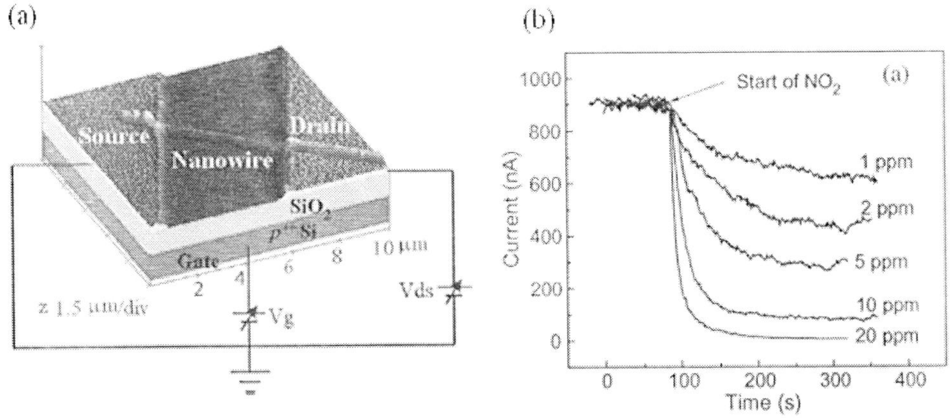

Figure 10. (a) AFM image of ZnO nanowire FET and the schematic of the measurement circuit; (b) ZnO nanowire sensing response to different concentration of NO_2 [39] (@ [2006] IEEE).

SnO2 is another commonly used metal oxide material in SMO sensors. Liqin Shi et al. [42] developed a highly sensitive SnO2 based gas sensor for indoor air quality monitoring. The nanostructured SnO2 was grown on top of Au electrodes using RF induction plasma deposition (IPD) method. It showed a good sensitivity to indoor air pollutants (formaldehyde and nitrogen oxide) at extremely low gas concentration (approximately 20 ppb). Modified SnO2 films also showed improved sensitivity for certain gases. Arijit Chowdhuri et al. [43] used CuO for H2S sensing and obtained improved response by dispersing CuO nanoparticles onto SnO2 film. Agrawal Rajnikant [44] developed nanoparticle layer that consists of indium oxide (In2O3)-doped tin oxide (SnO2) semiconductor nanocrystalline particles with platinum (Pt) nanoclusters for hydrogen sensing. The nanoparticle layer was deposited on interdigitated electrodes and the device showed resistance increase by five orders of magnitude upon exposure to H2 at room temperature.

2.2. Metallic Nanomaterials and Polyaniline Nanofiber for Gas Sensing

Palladium (Pd) metal film is widely utilized for hydrogen sensing due to its incremental change in resistance by exposure to hydrogen. The development of synthesis technology of nanomaterials motivated researchers interest to fabricate nanostructured Pd based sensors.

Pd nanowires or nanoporous Pd film have been fabricated and their sensing characteristic to hydrogen has been studied by different groups in past few years [45-47]. Figure 11 shows a bundle of vertical aligned Pd nanowires fabricated by electrodeposition into anodized aluminium oxide (AAO) films for hydrogen detecting [45]. The diameter of the nanowires was 80~90nm while the length of the bundle was about 1μm. Compared to Pd thick film hydrogen sensor, the resistance of the Pd nanowires decreased when H2 is adsorbed into the sensing layer. This is due to the expansion of Pd grains caused by the adsorption of H2. The swelling causes the gap reducing in grain boundaries. As a result, the intergranular resistance was lowered in a higher portion than the resistance increase factor due to the formation of Pd hydride [46].

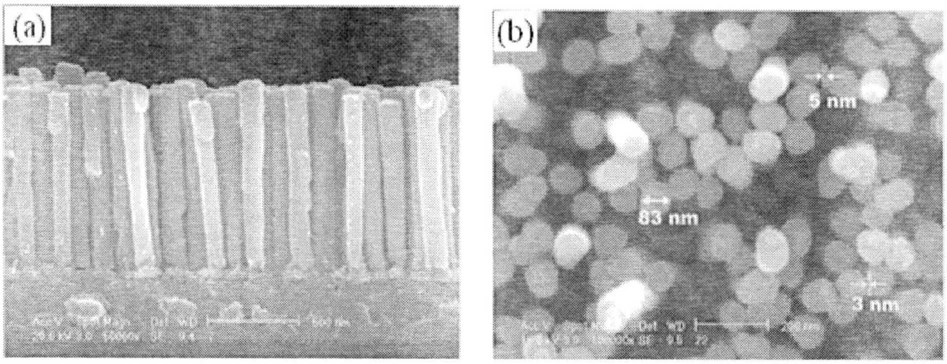

Figure 11. (a) SEM image of the freestanding Pd nanowires on Si substrate; (b) Top view image of Pd nanowires [45] (@ [2006] IEEE).

High sensitivity (approximate two orders decreasing in resistance), fast response time (700ms) and recovery time (2s), and room temperature operation were reported on this Pd nanowire sensors, which demonstrated significant improvement to conventional Pd thick film sensors.

As introduced in former section, conducting polymer is one of the commonly used materials in gas sensing systems. Polymer nanofibers, possessing all the advantages of nanostructures, is another promising material for gas sensing. Polyaniline nanofiber based SAW gas sensor, for example, was developed by Abu Z. Sadek et al [48]. The polyaniline nanofibers were synthesized using chemical oxidative polymerization of aniline by a template-free rapidly mixed reaction. The nanofibers were deposited onto layered ZnO/64o YX LiNbO3 surface-acoustic-wave transducers. The device showed enhanced sensitivity and linear response to hydrogen at room temperature. Fast response and recovery time with good repeatability were also obtained.

Summary

In this chapter, we introduced various gas/chemical sensing systems based on nanomaterials. The utilization of carbon nanotubes and other nanomaterials shows extensive enhancement on the performance of gas/chemical sensors. High sensitivity, room-temperature operation, and fast response and recovery time can be achieved simultaneously by nanotechnology based sensors. However, there are still some issues related with this technique. For example, the mechanisms of gas sensing with CNTs are not completely understood yet, especially for MWCNTs and nanotubes films. The effect of defects and impurities, the relative narrow sensing range of gases, and improvements on the stability and reversibility of the systems all demand further and in-depth research. The development of nanotechnology based gas sensing systems has a challenging but promising prospect.

REFERENCES

[1] M. Meyyappan and Harry Partridge, *Nano and Biotechnology R*esearch at NASA Ames, http://www.ipt.arc.nasa.gov.

[2] M. S. Dresselhaus, G. Dresselhaus, and P. C. Eklund, *Science of Fullerenes and Carbon Nanotubes*, Academic Press, New York (1996).

[3] S. Ijima, *Nature*, 354, 56 (1991).

[4] T.W .Ebbesen and P.M. Ajayan, *Nature* 358, 220 (1992).

[5] D. S. Bethune, C. H. Kiang, M. S.de Vries, G. Gorman, R. Sa voy, J.Vazquez, and R. Beyers, *Nature* 363, 305 (1993).

[6] Hongjie Dai, Nanotube Growth and Characterization. In: M.S. Dresselhaus, G. Dresselhaus, *Carbon Nanotubes: Synthesis, structure,properties and applications*. New York: Springer, 2000, 29-54.

[7] Thomas W. Ebbesen, Production and Purification of Carbon Nanotubes. In: Thomas W. Ebbesen, *Carbon Nanotubes: Preparation and Properties*. CRC Press, 1997, 139-162.

[8] Jing Li, Yijiang Lu, and Qi Ye et al, *Nano Letters*, Vol.3, No.7 (2003): 929-933.

[9] Jun Li, Hou Tee Ng, *Encyclopedia of Nanoscience and Nanotechnology*, Volume 1: Pages (591–601).

[10] R. Saito, M. Fujita, G. Dresselhaus, and M. S. Dresselhaus, *Appl. Phys. Lett.* 60, 2204 (1992)

[11] Niraj Sinha, Jiazhi Ma, John T. W. Yeow, Journal of Nanoscience and Nanotechnology, Vol.6, 573-590, 2006.

[12] O. Lourie and H. D. Wagner, *J. Mater. Res.* 13, 2418 (1998).

[13] M. F. Yu, O. Lourie, M. J. Dyer, et al., *Science* 287, 637 (2000).

[14] J. Kong, Nathan R. Franklin, Chongwu Zhou, et al., *Science* 287, 622 (2000).

[15] A. Wadhawan, R. E. Stallcup, K. F. Stephens, and J. M. Pereza, *Appl. Phys. Lett.* 79, 1867 (2001).

[16] C. K. W. Adu, G. U. Sumanasekera, and B. K. Pradhan, et al., *Chem. Phys. Lett.* 337, 31 (2001).

[17] J. Zhao, A. Buldum, J. Han, and J. P. Lu, *Nanotechnology* 13, 195(2002).

[18] Shu Peng, James O'Keeffe, Chengyu Wei, and K. Cho, the 3rd International Workshop on Structural Health Monitoring, 2001.

[19] Seung-Hoon Jhi, Steven G. Louie, and Marvin L. Cohen, *Phys. Rev. Lett.* 85, 1710 (2000);

[20] G. U. Sumanasekera, C. K.W. Adu, S. Fang,2 and P. C. Eklund, *Phys. Rev. Lett.* 85, 1096 (2000);

[21] H E Romero, G U Sumanasekera, SKishore1 and P C Eklund, *J. Phys.: Condens. Matter* 16 (2004) 1939–1949.

[22] Takao Someya, Joshua Small, Philip Kim, Colin Nuckolls, and James T. Yardley, *Nano Letters*, Vol.3, No.7 (2003): 877-881.

[23] Yoon-Taek Jang, Seung-Il Moona, and Jin-Ho Ahnb, et al., *Sensors and Actuators* B 99 (2004) 118–122.

[24] Woo-Sung Cho, Seung-Il Moon, Yang-Doo Lee, et al., *IEEE Electron Device Letters*, Vol.26, No.7, JULY 2005.

[25] O.K. Varghese, P.D. Kichambre, D. Gong, et al., *Sensors and Actuators* B 81 (2001) 32-41.

[26] L. Valentini, C. Cantalini, I. Armentano, et al., *Diamond and Related Materials* 13 (2004) 1301–1305.

[27] Krishna C Persaud, *IEE Colloquium on Molecular Electronics*, Vol.19, 4/1 - 4/3 (1990).

[28] M.C. Petty and R. Casalini, *Eng. Sci. and Edu. J.* June (2001) 99-105.

[29] Ivan D. Avramov, Shigeru Kurosawa, and Michael Rapp, *IEEE Trans. on Microwave Theory and Tech.*, Vol. 49, No. 4, April 2001.

[30] Ventzislav M. Yantchev, Vesseline L. Strashilov, Michael Rapp, *IEEE Sensors J.*, Vol.2, No.4, August 2002.

[31] Seong M. Cho, Young Jun Kim, Yong Shin Kim, Sensors, 2004. *Proceedings of IEEE*, Vol.2, 701 -704 (2004).

[32] Biju Philip, Jose K Abraham, Anupama Chandrasekhar and Vijay K Varadan, *Smart Mater. Struct.* 12 (2003) 935–939.

[33] Jing Li, Yijiang Lu, and M. Meyyappan, *IEEE Sensors J.*, Vol.6, No.5, 2006.

[34] Zhongyu Hou, Dong Xu, and Bingchu Cai, *Appl. Phys. Lett.* 89, 213502(2006).

[35] Ashish Modi, Nikhil Koratkar, Eric Lass, Bingqing Wei, and Pulickel M. Ajayan, *Nature*, Vol 424, 171(2003).

[36] S J Kim, *J. Phys. D: Appl. Phys.* 39 (2006) 3026–3029.

[37] G. Bläser, Th. Rühl, C. Diehl, M. Ulrich, D. Kohl, *Physica* A 266 (1999) 218-223.

[38] Ying Hu, Ooi Kiang Tan, and Weiguang Zhu, *IEEE Sensors J.*, Vol.5, No.5, 2006.

[39] Zhiyong Fan and Jia G. Lu, *IEEE Trans. on Nanotechnology*, Vol. 5, No.4, July,2006.

[40] Ying Hu, Ooi Kiang Tan, and Weiguang Zhu, *IEEE Sensors J.*, Vol.6, No.6, 2006.

[41] T. Shishiyanu, S. Shishiyanu, 0. Lupan, V. Sontea, A. Bragorenco, *International Semiconductor Conference*, Vol.1, 201-204(2006).

[42] Liqin Shil, Y. Hasegawa, T. Katsubel, *The 13th international Conference on Solid-State Sensors, Actuators and Microsystems*, 1203-1206(2005).

[43] Chowdhuri, A., Gupta, V., Kumar, R., et al., Sensors, 2003. *Proceedings of IEEE*, Vol 1, 201 – 205(2003).

[44] Rajnikant, A., Shukla, S., Ludwig, L., et al. Sensors, 2004. *Proceedings of IEEE*, Vol 1, 395 – 398(2004).

[45] Kyun Tae Kim, Sang Jun Sim, and Sung Min Cho, *IEEE Sensors J.*, Vol.6, No.3, 2006.

[46] Massood Z. Atashbar, Deep Banerji, and Srikanth Singamaneni, *IEEE Sensors J.*, Vol.5, No.5, 2006.

[47] Dongyan Ding and Zhi Chen, *6th IEEE Conference on Nanotechnology*, 2006, Vol.2, 777- 780(2006).

[48] Abu Z. Sadek, Christina O. Baker, David A. Powell, *IEEE Sensors J.*, Vol.7, No.2, 2007.

INDEX

A

absorption spectra, xvii, 172, 383, 483, 484, 488, 489, 553, 1405, 1415, 1416, 1417
absorption spectroscopy, xviii, 172, 175, 547, 553
accelerator, 465
access, 115, 156, 184, 475, 587, 692, 796
accessibility, 830, 1529
accounting, 234, 626, 1459, 1527
accumulation, 982, 983, 985, 986
accuracy, 126, 314, 359, 416, 478, 689, 692, 703, 707, 953
acetaminophen, 786
acetic acid, 372, 779, 915, 923
acetone, xi, 7, 11, 14, 15, 16, 17, 18, 19, 20, 31, 397, 420, 444, 549, 564, 649, 757, 923, 924, 925, 1522, 1552, 1555
acetonitrile, 118, 907, 1529
acetylcholinesterase, 183, 789
acetylene, xxi, 809, 810, 811, 813, 815, 817, 818, 819, 820, 821, 822, 825, 826, 827, 828, 829, 830, 831, 832, 834, 835, 837
acetylene gas, 870
achievement, 1, 2, 3, 211, 759, 790, 792
acidic, 760
acidity, 337, 338, 341, 342, 343, 351, 353, 597, 607, 760, 924
acrylate, 116, 130, 131, 751, 762, 785, 922
acrylic acid, 167, 1533
activated carbon, 2, 830, 842
activation, 40, 42, 45, 46, 49, 50, 51, 55, 57, 59, 60, 148, 175, 208, 223, 366, 383, 646, 756, 757, 829, 875, 1480, 1487, 1512
activation energy, 42, 46, 50, 59, 60, 208, 223, 366, 1480, 1487
activation entropy, 208
active oxygen, 19
active site, xxi, 796, 809, 823
actuators, 428
adamantane, 115, 528, 529, 530, 532, 533, 534, 535, 536, 537, 538, 539, 540
adaptability, 1428
additives, 503, 514, 564, 639, 758, 960, 979, 1428, 1429, 1430, 1431, 1432, 1434, 1435, 1436, 1438, 1442, 1480
adenine, 750, 751, 754, 765, 795, 800
adenocarcinoma, 149
ADH, 749, 787
adhesion, 38, 41, 43, 49, 51, 52, 53, 58, 163, 760, 763
adjustment, 760, 850, 873, 1159
adriamycin, 1510, 1512, 1515, 1519
adsorption isotherms, 834, 835
aerospace, xvi, 393
affect, 18, 20, 66, 851, 964, 972, 974, 981, 985
AFM, xviii, xix, 279, 280, 281, 282, 284, 286, 287, 289, 290, 291, 292, 293, 294, 295, 296, 297, 434, 475, 532, 535, 547, 548, 551, 552, 553, 558, 623, 624, 625, 627, 628, 629, 630, 631, 751, 784, 793, 817, 1420, 1482, 1484, 1558, 1559
age, 536, 742
ageing, 588
agent, 67, 73, 116, 120, 128, 133, 135, 137, 138, 143, 148, 151, 153, 166, 179, 383, 399, 400, 402, 403, 549, 760, 772, 790, 796, 915, 963, 964, 965, 966, 967, 969, 1164, 1432, 1504, 1505, 1507, 1509, 1520, 1522, 1523, 1524, 1538, 1545
aggregates, 3, 4, 16, 76, 135, 167, 313, 396, 399, 487, 555, 667, 679, 973, 976, 1162, 1458, 1459, 1460, 1462, 1468, 1533
aggregation, xxxi, 3, 70, 76, 124, 148, 164, 242, 308, 316, 351, 395, 396, 399, 400, 402, 403, 533, 908, 915, 940, 1412, 1413, 1416, 1418, 1420, 1457, 1458, 1459, 1461, 1465, 1533, 1534, 1536
aggregation process, 1458, 1459
aging, xxvi, 171, 312, 1157, 1179
agriculture, 909

air pollutants, 1559
air quality, 1559
alanine, 385, 387
alcohol, 3, 103, 119, 178, 385, 564, 583, 589, 593, 666, 667, 749, 787, 788, 789, 867, 907, 908, 909, 922, 923, 925, 933, 1532, 1558
alcohol use, 908
alcohols, 116, 121, 767, 1550, 1551
aldehydes, 175
algorithm, 688, 691, 695, 699, 700, 707, 714, 721, 743
alkalescent property, xix, 581, 596, 618
alkaline, 764, 774, 780, 789
alkaline earth metals, 583
alkaline media, 780
alkaline phosphatase, 764, 789
alkalinity, 596
alkane, 57, 1480
alkenes, 1480
allergic reaction, 538
alloys, xx, 309, 427, 462, 633, 636, 644, 646, 651, 659, 660, 666, 694, 739, 742, 747, 813, 839
allylamine, 185
ALS, 972, 975
alternative, xxiii, 18, 72, 176, 245, 274, 314, 324, 382, 764, 905, 909, 1391, 1394, 1398, 1532, 1533
alternative energy, 1398
alternatives, 314, 582
aluminium, 1559
aluminum, xvii, 174, 243, 264, 418, 459, 460, 467, 468, 469, 470, 471, 472, 474, 690, 703, 909, 1404, 1413, 1419, 1421
aluminum oxide, xvii, 459, 460, 469, 470, 474, 909
aluminum surface, 467
Alzheimer's disease, 164
ambient air, 850, 851, 852, 853, 854
ambivalent, 383
amide, 757, 765
amines, 116, 121, 171, 179, 384, 387, 765, 796, 1534
amino acid, xvi, xviii, 4, 127, 147, 151, 191, 329, 341, 351, 365, 366, 371, 375, 376, 378, 382, 383, 385, 387, 388, 390, 527, 534, 759, 777, 794
amino acids, xvi, 151, 329, 351, 371, 375, 378, 382, 387, 390, 534, 759, 777, 794
ammonia, 67, 76, 121, 311, 312, 313, 315, 316, 317, 319, 320, 321, 322, 323, 324, 593, 609, 876, 1552
ammonium, 400, 401, 404, 751, 763, 1427, 1442, 1468, 1481
ammonium chloride, 1481
amplitude, 39, 50, 639, 707, 708, 709, 714, 832, 1482
amyloid plaques, 164
anatase, xi, 7, 8, 10, 11, 12, 24, 27, 28, 29, 31, 336, 337, 341, 938, 961, 1392, 1396, 1403, 1414
aniline, 1560
animal models, 1505
animals, 778, 1506
anion, 156, 161, 172, 762, 768, 798
anisotropy, 69, 162, 168, 176, 726, 728, 731, 914
annealing, xix, 9, 23, 24, 25, 27, 29, 96, 361, 435, 446, 486, 487, 488, 489, 562, 564, 568, 570, 573, 581, 592, 596, 597, 609, 610, 611, 612, 613, 619, 644, 646, 659, 679, 721, 744, 755, 756
annihilation, 172
anomalous diffusion, 53
anoscope, 1482
antibody, 129, 137, 1508, 1519, 1529, 1537
anticancer drug, xxxii, 1499, 1501, 1502, 1503, 1504, 1507, 1510, 1513, 1514, 1515
antigen, 1513, 1529, 1537
antimony, 1427
antioxidants, 1440, 1442
antisense, 141, 142, 191, 1506, 1514, 1516
antisense oligonucleotides, 141, 191, 1514
antitumor, 1501, 1509, 1512, 1515
apoptosis, 137, 1509, 1511, 1519
applied research, 8
aqueous solutions, 44, 166, 383, 462, 759, 760, 763, 896, 906, 912, 915, 1158, 1165, 1166
aqueous suspension, 10, 422, 424, 759, 792, 795, 898, 1516
ARC, 756
arginine, 139, 375, 388, 777
argon, xi, 7, 11, 12, 432, 441, 464, 589, 597, 649, 654, 755, 813, 815, 859, 861, 862, 864
argument, 970
aromatic compounds, 182
aromatic rings, 757
aromatics, 9
arrest, 341, 717, 719, 743
Arrhenius equation, 1486
Arrhenius parameters, 1480
arsenic, 298
arteries, 133
arteriosclerosis, 786
ascorbic acid, 549, 749, 759, 911, 1523, 1524
ASI, 32
aspect ratio, 249, 253, 271, 283, 292, 301, 408, 412, 417, 418, 419, 421, 519, 810, 823, 909, 954, 955, 956, 957, 1403, 1523, 1524, 1525, 1526, 1528, 1529, 1535, 1536, 1537, 1538, 1539, 1540, 1554
assessment, 3, 405
assignment, 314
assumptions, 45, 214, 627, 1435
asymmetry, 314, 762, 849

atmospheric pressure, 420, 503, 856, 857, 981
atomic distances, 693
atomic force, xii, xviii, xix, xxi, 18, 37, 61, 118, 434, 532, 538, 547, 566, 623, 624, 631, 687, 700, 751, 793, 817, 868, 878, 962
atomic force microscope, 434, 532, 566, 868
atomic positions, 700, 850
ATP, xviii, 547, 548, 549, 551, 552, 553, 555, 556, 1502
attachment, 3, 4, 58, 131, 133, 141, 145, 159, 178, 222, 374, 538, 541, 548, 759, 763, 765, 797, 800, 938, 966, 1392, 1510, 1520, 1529, 1537
attacks, 676
attention, xi, xvii, 7, 8, 10, 55, 86, 88, 501, 538, 562, 666, 675, 680, 695, 707, 716, 753, 779, 790, 798, 824, 960
attribution, 49
Au nanoparticles, 87, 153, 180, 411, 1500, 1507
Au substrate, 179
Australia, 422, 566
autocatalysis, 376, 1487
availability, 108, 361, 538, 791, 792, 793, 830, 969, 1475, 1484, 1548
averaging, xvi, 459, 531, 704

B

backscattering, 405
bacteria, 4, 146, 149, 151, 371, 382, 914, 1510, 1534
balloon angioplasty, 1513
band gap, 443, 498, 502, 517, 535, 1160, 1408, 1558
bandgap, 104, 239, 263, 345, 428, 439, 446, 447, 517, 605
barium, 583, 589, 592, 593
barriers, xxii, 53, 54, 175, 278, 300, 301, 302, 303, 613, 725, 824, 845, 846, 866, 1557
basic research, xx, xxx, 665, 1391, 1392
basis set, 693
baths, 462, 465
batteries, 8, 69, 77, 823, 824, 841
BBB, 140
beams, 748
Beijing, 423, 547, 549, 581, 633, 660, 662, 687, 923, 941, 1391, 1521
Belgium, 459
beliefs, 948
bending, 86, 87, 100, 101, 349, 351, 369, 453, 456, 556, 628, 693
beneficial effect, 986
benign, 538
benzene, 117, 124, 130, 400, 811, 963, 1502, 1552, 1554
beverages, 795
bias, 9, 245, 264, 448, 450, 451, 452, 615, 618, 757, 760, 771, 846, 852, 963
bicarbonate, 966, 967, 970
bile, 173
binding, 30, 53, 58, 107, 121, 123, 129, 130, 131, 140, 141, 148, 151, 157, 163, 164, 165, 166, 169, 176, 179, 195, 314, 339, 340, 341, 346, 351, 375, 382, 502, 517, 549, 553, 562, 641, 704, 751, 798, 847, 850, 869, 911, 912, 914, 915, 1165, 1512, 1529, 1536, 1548, 1549, 1558
binding energies, 869, 1165, 1549
binding energy, 58, 107, 314, 346, 502, 517, 562, 641, 850, 869, 1548, 1549, 1558
bioanalytical, xxxii, 1521, 1541
bioavailability, 149, 192
biocatalysts, 770, 781
biochemistry, xxiii, 905
biocompatibility, xiii, xxi, 30, 114, 146, 191, 378, 428, 548, 752, 758, 761, 772, 780, 1500, 1501, 1502, 1538, 1541
biodegradability, 117, 1500, 1502
biodegradable, xiii, 114, 142, 143, 146, 191, 531, 1500, 1501, 1511, 1513
biodegradable materials, 1501
biodegradation, 144, 148
biofuel, 760, 794, 795
biological activity, 30
biological behavior, 146
biological processes, 1508, 1536
biological responses, 147
biological systems, 173, 330, 792, 915
biomacromolecules, xvi, 329, 330, 331, 332, 336, 339, 345, 349, 366, 371, 372, 380, 381, 390
biomarker, xxxii, 1499, 1500, 1502, 1505, 1507, 1510
biomarkers, xxxii, 772, 1499, 1501, 1507, 1510, 1511, 1518
biomaterials, xv, 140, 329, 331, 351, 390, 1541
biomedical, xxi, 502, 538, 752, 779
biomedical applications, 146, 191, 502, 538, 1507, 1510, 1522, 1541
biomolecules, xxxii, 181, 339, 353, 366, 381, 385, 535, 760, 761, 767, 774, 789, 800, 909, 923, 1179, 1499, 1503, 1507, 1511, 1516, 1518, 1521, 1522, 1534, 1536, 1541
bionanohybrid, xxxii, 1521, 1541
biopolymer, 55, 760
bioremediation, 181
biosensors, xxxii, 181, 182, 184, 479, 752, 758, 759, 760, 761, 770, 781, 782, 783, 785, 786, 787, 788, 789, 790, 792, 794, 795, 798, 1394, 1499, 1508, 1519, 1522, 1536, 1537, 1541
biosynthesis, 382

biotechnology, xv, xvi, xxiii, 329, 390, 905, 1518
biotin, 143, 1529, 1530, 1532, 1536
birefringence, 168
birth, 699
bisphenol, 465, 1428, 1430, 1431, 1433, 1440
biuret, 116
bladder, 133
bleaching, 1176, 1177, 1180, 1506
blend films, 1412
blends, 130, 1405, 1406, 1414, 1440
blocks, xviii, 90, 144, 161, 163, 184, 322, 332, 387, 466, 506, 515, 527, 531, 533, 534, 535, 540, 1434
blood, 133, 140, 147, 148, 539, 774, 781, 792, 1395, 1505, 1513, 1518, 1537
blood flow, 147
blood stream, 140
blood supply, 1513
blood vessels, 133
blueshift, xxvi, 1157, 1158
body, xi, 37, 61, 850, 854
body fluid, 795
Boltzmann constant, 208, 217, 222, 229, 641, 1459
Boltzmann distribution, 692
bonding, 51, 58, 66, 67, 76, 80, 99, 114, 168, 175, 314, 315, 400, 533, 536, 537, 548, 572, 598, 599, 600, 698, 765, 781, 908, 910, 947, 964, 969, 970, 971, 1165, 1443, 1529, 1556
bonds, xxiv, 43, 49, 74, 77, 314, 315, 339, 346, 351, 375, 377, 530, 536, 537, 539, 540, 549, 551, 628, 693, 698, 765, 914, 943, 953, 959, 964, 969, 970, 1432, 1443, 1458, 1484
borate, 796
boundary surface, 651
brain, 140, 786, 1516, 1539
branched polymers, xiii, 114, 146
branching, xiii, 113, 115, 121, 127, 139, 145, 151, 178, 945, 948, 949
breakdown, xxiv, 428, 449, 450, 813, 959, 960, 962, 963, 972, 977, 979, 980, 981, 982, 984, 986, 1480, 1555, 1556
breast cancer, 1539
breast carcinoma, 1508
breathing, 857, 860, 863, 870, 871, 872
broadband, 428
bromine, 121, 1427
Brownian motion, 50, 76, 395, 396, 414, 416, 424
Brownian particle, 50
buffer, 12, 279, 437, 512, 549, 751, 759, 769, 773, 777, 781, 785, 788, 792, 795, 796, 797
building blocks, xv, xviii, xxxii, 4, 76, 88, 90, 123, 126, 130, 155, 160, 163, 166, 178, 202, 307, 311, 313, 316, 320, 321, 347, 371, 380, 382, 506, 515, 527, 531, 533, 534, 535, 540, 548, 550, 846, 906, 1521, 1541
bulk materials, xx, 595, 633, 649, 652, 659, 711
bulk polymerization, 936
burn, 476
burning, 649, 759, 817, 867, 869, 870, 872, 873, 874, 875, 1496, 1497
butyl methacrylate, 922, 936
by-products, 761

C

C_{60}, 532, 534, 754, 755, 836
Ca^{2+}, 1517
cadmium, 71, 80, 171, 759, 772, 777
calcination temperature, 332, 335, 337, 339, 341, 342, 343, 347, 349, 352, 356, 361, 367, 390, 606
calcium, 961, 1500, 1517
calibration, 30, 535, 784, 788, 789, 833, 964, 1174
calorimetry, 164
Canada, 7, 32, 987, 989, 1545
cancer, xxxii, 138, 140, 141, 142, 147, 174, 192, 896, 1499, 1500, 1501, 1502, 1503, 1504, 1505, 1506, 1507, 1508, 1509, 1511, 1512, 1513, 1514, 1515, 1516, 1518, 1519, 1520, 1521, 1522, 1535, 1538, 1539, 1541
cancer cells, xxxii, 138, 141, 142, 1499, 1500, 1501, 1502, 1503, 1504, 1505, 1506, 1507, 1509, 1511, 1512, 1513, 1515, 1516, 1518, 1519, 1520, 1538, 1539
cancer treatment, 1507, 1541
cancerous cells, 1508
candidates, xxi, xxiii, 55, 90, 412, 465, 534, 752, 905, 906, 1158, 1548, 1557
capacitance, 583, 767, 778, 830, 831, 832
capillary, 203, 204, 213, 319, 342, 343, 370, 417, 430, 749, 767, 773, 774, 912, 915, 1395, 1396
carbides, 397, 421, 823, 840, 1442
carbohydrate, 774
carbohydrates, 759, 767, 774
carbon atoms, 57, 59, 60, 528, 529, 530, 532, 534, 537, 628, 629, 630, 631, 810, 811, 821, 851, 950, 967, 968, 1547
carbon dioxide, 589, 967
carbon film, 12
carbon materials, 746, 753, 762, 772, 824, 830, 833, 843
carbon molecule, 1549
carbon monoxide, 70, 780, 1494, 1495, 1496
carbonization, 1492
carbonyl groups, 339
carbosilanes, 121
carboxylic acids, 933

carboxylic groups, xxiii, 825, 921
carcinogen, 1519
carcinogenesis, 1515
carcinogenicity, 1515
carcinoma, 1504, 1510, 1516, 1519
cardiovascular disease, 772
carotene, 138
carrier, xiv, xxx, 2, 139, 140, 143, 145, 190, 191, 225, 237, 261, 269, 278, 279, 283, 292, 345, 433, 446, 452, 456, 498, 505, 657, 752, 754, 790, 792, 798, 800, 815, 848, 849, 854, 863, 879, 985, 1401, 1402, 1404, 1412, 1413, 1424, 1505, 1506, 1512, 1516, 1518
cartilage, 144, 145
cast, 695, 759, 785, 787, 789, 1420
casting, 164, 331, 432, 754, 757, 760, 761, 763, 786, 792, 794, 795, 798, 1404, 1551
catalase, 184, 750, 790
catalysis, xiii, xxxii, 3, 114, 151, 371, 382, 538, 548, 762, 841, 1521
catalytic activity, 151, 557, 558, 787, 790, 792, 794, 874
catalytic properties, 152, 382, 548, 754, 841
cathepsin B, 142
cathode materials, 794, 824
cation, 66, 67, 68, 79, 91, 169, 172, 762, 896, 1443
C-C, 57, 58, 59, 151, 314, 530, 536, 628, 953, 954, 1548
CCC, 556
cell death, 1539, 1540
cell division, 699
cell line, 137, 147, 149, 192, 1507, 1508, 1510, 1517, 1519
cell lines, 137, 147, 1507, 1510
cell surface, 1503
cellulose, 1554, 1555
ceramic, xix, 1, 11, 332, 366, 367, 371, 407, 410, 417, 534, 563, 568, 582, 583, 589, 592, 594, 602, 604, 606, 615, 616, 617, 619, 746, 749, 761, 1481
ceramics, xvi, xix, 309, 329, 332, 534, 581, 582, 583, 587, 592, 604, 605, 606, 607, 608, 609, 610, 615, 618, 619, 651, 836
cesium, 530
chain transfer, 116
chalcogenides, xii, xvi, xx, 65, 329, 371, 377, 390, 665, 666, 682
channels, xxiii, 95, 169, 321, 323, 336, 339, 342, 430, 465, 835, 905, 945, 948, 949, 950, 1172, 1395, 1396, 1532, 1548
charge density, 185, 366, 694
charge trapping, 851
charring, 1479, 1481, 1486, 1496
chemical bonds, 104, 968
chemical composition, xiii, 114, 146, 607, 1531
chemical deposition, 1558
chemical etching, 465, 468
chemical industry, 9
chemical interaction, 695, 1445, 1555
chemical properties, xi, xii, 3, 7, 55, 57, 85, 88, 96, 331, 582, 587, 595, 634, 675, 676, 810, 1431, 1548
chemical reactions, 49, 67, 71, 95, 308, 324, 345, 359, 360, 395, 599, 703, 1503
chemical reactivity, xix, 435, 633, 634
chemical stability, xxi, 8, 428, 752, 779, 801, 823, 1538
chemical vapor deposition, 10, 66, 221, 223, 308, 430, 431, 469, 477, 512, 562, 675, 810, 838, 839, 840, 846, 855, 858, 862, 866, 875, 1547, 1558
chemical vapour, xxi, 750, 756, 809, 836, 839
chemical vapour deposition, xxi, 750, 756, 809, 836, 839
chemokines, 144
chemotherapeutic agent, 1502
chemotherapeutic drugs, 137, 1502
chemotherapy, xxxii, 138, 1499, 1500, 1501, 1502, 1503, 1504, 1518
Chicago, 527
China, 65, 80, 113, 201, 232, 329, 393, 423, 547, 558, 581, 633, 660, 665, 741, 749, 801, 808, 905, 918, 921, 935, 941, 1157, 1180, 1499, 1521
Chinese, 137, 423, 660, 738, 740, 741, 742, 744, 803, 805, 936, 957, 1174, 1391, 1515
chirality, 531, 846, 866, 871, 872, 908, 952, 1548
chitosan, 749, 760
chloride, 117, 123, 156, 159, 160, 378, 438, 530, 751, 763, 1431, 1481
chloroform, 116, 923, 1531, 1536, 1554, 1555
chlorophyll, xxvii, 1157, 1171, 1172, 1173, 1175
cholesterol, 749, 761, 763, 768, 786, 911
chondrocyte, 144
chondrocytes, 144
chromatography, 750, 772, 776, 777, 923, 939
circular dichroism, 797
circulation, 38, 1395, 1505, 1512
cisplatin, 1510
cladding, 381
classes, 4, 176, 318, 319, 652
classification, 117, 342, 370, 722, 1508
cleaning, xi, 8, 443, 462, 464
cleavage, 41, 148, 741
clinical diagnosis, 775, 786, 1501, 1505
cluster-cluster, 1458, 1459
cluster-cluster mechanism, 1459
clustering, 202, 225, 409, 410, 411, 414, 415, 421, 422

clusters, xv, xvii, xx, 71, 92, 118, 141, 147, 154, 160, 161, 166, 176, 191, 202, 203, 204, 214, 218, 221, 225, 281, 283, 291, 295, 307, 308, 385, 388, 398, 409, 411, 414, 415, 417, 421, 483, 484, 486, 487, 488, 489, 507, 517, 533, 633, 651, 652, 659, 699, 715, 722, 725, 726, 728, 730, 743, 748, 1395, 1469
C-N, 314, 315
CO2, xiv, 44, 153, 168, 238, 269, 270, 271, 359, 360, 600, 672, 673, 876, 966, 1160, 1171, 1548, 1555, 1556
coagulation, 648
coal, 1391
coatings, xi, 5, 8, 309, 345, 382, 562, 675, 774, 787, 909, 1555
cobalt, 167, 182, 366, 367, 368, 750, 780, 785, 811, 821, 855
coding, 1172
coherence, xvi, 459, 553, 1505, 1516
cohesion, 55, 58, 60, 625, 627, 628
collaboration, xvi, 329
collagen, 145, 330, 339, 375, 377, 380, 1538
collateral, 339, 347
collisions, 396, 416, 746, 856, 1555
colloid particles, 341, 342
colloidal particles, 1505
colloids, 1, 336, 339, 349, 351, 353, 382, 555, 1429
colon, 149, 1504, 1516
combined effect, 53, 58, 94
combustibility, 1495
combustion, 9, 359, 438, 1440, 1481, 1493, 1495, 1496, 1497
communication, 427, 701, 792
community, 692, 703
comparative research, 199
compatibility, 461, 722, 858, 876, 923, 960, 973, 981, 986, 1428, 1442, 1445
compensation, 2, 492
competence, 330
competition, 164, 360, 861, 1164
complement, 532, 973
complementarity, 1532
complementary DNA, 141, 798
complexity, 4, 53, 308, 312, 323, 324, 531, 737, 976, 980, 1457, 1488, 1549
compliance, 625
complications, 40, 54, 146, 625
components, xii, 37, 39, 46, 50, 56, 61, 124, 133, 136, 148, 151, 314, 318, 322, 331, 339, 345, 356, 361, 371, 380, 385, 390, 485, 492, 533, 539, 540, 564, 571, 634, 639, 660, 691, 693, 696, 776, 778, 811, 823, 869, 871, 902, 912, 1424, 1442, 1484, 1502, 1545, 1547
composites, xvi, xxx, 1, 181, 329, 373, 378, 381, 384, 389, 417, 425, 531, 592, 602, 619, 651, 659, 689, 761, 763, 774, 785, 810, 823, 842, 936, 937, 942, 972, 973, 981, 1401, 1402, 1405, 1407, 1410, 1413, 1415, 1416, 1417, 1424, 1440, 1442, 1443, 1457, 1458, 1459, 1461, 1462, 1468, 1469, 1480, 1481, 1497, 1554
composition, xiii, xxii, 16, 22, 23, 38, 54, 72, 90, 98, 99, 104, 114, 151, 204, 213, 215, 226, 302, 313, 375, 447, 484, 506, 562, 587, 588, 589, 592, 593, 600, 602, 616, 623, 670, 673, 767, 828, 829, 830, 895, 896, 900, 902, 909, 1428, 1431, 1433, 1434, 1435, 1500, 1503, 1512, 1515, 1554
compounds, xxxi, 66, 71, 72, 96, 114, 138, 141, 152, 160, 176, 182, 194, 271, 308, 324, 350, 492, 528, 530, 536, 538, 549, 634, 651, 666, 674, 678, 680, 762, 772, 773, 778, 780, 790, 922, 1394, 1427, 1428, 1429, 1430, 1431, 1432, 1433, 1434, 1435, 1436, 1439, 1440, 1442, 1453, 1454, 1502, 1503, 1519, 1554
computation, 691, 694, 699, 702, 703, 704, 706, 714, 721
computational grid, 709
computed tomography, 133, 1505
computer simulation, 690, 691, 739, 747
computer simulations, xxi, 716, 737
computing, 691, 692, 697, 699
concrete, 1428, 1458
condensation, xx, 91, 114, 130, 165, 226, 331, 342, 343, 370, 375, 376, 377, 399, 401, 410, 512, 633, 647, 648, 649, 651, 652, 653, 659, 725, 835, 910, 964, 968, 969, 971, 1531
conduction, xxx, 8, 257, 264, 300, 302, 406, 414, 418, 424, 453, 476, 535, 851, 854, 865, 879, 1401, 1402, 1412, 1413, 1526, 1558
conductivity, xvi, xxi, 283, 293, 393, 394, 406, 407, 408, 409, 410, 411, 412, 413, 414, 415, 416, 417, 419, 421, 422, 423, 424, 425, 476, 478, 492, 494, 497, 519, 562, 654, 656, 657, 660, 752, 761, 762, 772, 780, 782, 792, 798, 823, 829, 981, 1548, 1554, 1557
conductor, 979
configuration, xx, 40, 49, 67, 76, 173, 322, 332, 338, 339, 341, 347, 372, 379, 403, 452, 630, 633, 634, 651, 659, 688, 698, 712, 713, 721, 722, 724, 725, 728, 731, 732, 739, 790, 797, 953, 1412, 1419, 1422
confinement, xiii, xiv, xv, 113, 202, 232, 237, 238, 261, 263, 269, 277, 278, 302, 309, 460, 635, 960, 1450
Congress, 662
conjugate gradient method, 953

conjugation, 127, 133, 138, 139, 140, 149, 172, 1179, 1409, 1413, 1417, 1424, 1513, 1518, 1538
connectivity, 343, 1421
consensus, 314
conservation, xxiii, 704, 705, 706, 707, 714, 746, 941, 942, 943, 944, 945, 946, 948, 953, 957
consolidation, 534, 639, 647, 649, 651, 659, 688, 724
constraints, 909
construction, 126, 139, 166, 171, 176, 330, 349, 353, 540, 721, 752, 758, 764, 771, 772, 781, 783, 787, 1537
consumption, xv, 238, 269, 271, 273, 274, 281, 1391, 1545, 1546
contact time, 962
contaminant, 763
contamination, 245, 314, 462, 464
contrast agent, 133, 134, 189, 382, 1505, 1506, 1513, 1516, 1517, 1518, 1522, 1525, 1538, 1541
conventional composite, 604
convergence, 704, 1516
conversion, xv, xxx, 20, 27, 38, 72, 89, 95, 119, 171, 307, 353, 371, 563, 564, 680, 798, 852, 853, 906, 930, 931, 1391, 1392, 1393, 1395, 1396, 1397, 1413, 1424, 1480, 1539
cooling, 279, 393, 403, 414, 420, 443, 492, 493, 564, 647, 666, 675, 879, 906, 907, 908, 915, 1159, 1403
cooling process, 1159
copolymers, 3, 115, 116, 132, 136, 166, 167, 460, 465, 466, 922, 1430, 1434
copper, xxi, 12, 72, 153, 176, 181, 242, 264, 312, 317, 394, 399, 402, 403, 410, 417, 418, 420, 422, 423, 425, 426, 476, 484, 486, 487, 488, 489, 649, 670, 673, 688, 698, 713, 720, 721, 737, 739, 740, 742, 744, 747, 759, 767, 774, 781, 859, 923, 939
copper oxide, 489
core-shell, xix, xxxii, 116, 166, 225, 311, 320, 321, 443, 581, 582, 595, 597, 598, 599, 602, 603, 604, 605, 607, 608, 609, 613, 614, 618, 619, 929, 1521, 1534, 1541
corona discharge, 1555
coronary heart disease, 786
correlation, 2, 3, 31, 52, 79, 173, 357, 359, 360, 361, 416, 424, 742, 789, 850, 971, 1464, 1492
correlation coefficient, 31, 789, 1492
correlations, 339
corrosion, 46, 634, 689
cost saving, 461
costimulatory molecules, 144
costs, 52, 456, 701, 731
Coulomb interaction, 494
couples, 184, 767
coupling, xxiii, 104, 121, 137, 150, 151, 381, 614, 704, 707, 738, 745, 757, 765, 871, 935, 938, 939, 959, 961, 962, 963, 964, 965, 966, 967, 969, 1392, 1393, 1529, 1534
covalent bond, 58, 530, 539, 693, 765, 781, 954, 964, 968, 969, 970
covalent bonding, 765, 781, 969, 970
coverage, 16, 140, 185, 296, 298, 303, 397, 793, 965, 969, 971, 1534
covering, xiii, 80, 113, 301, 312, 313, 353, 384, 711, 752
CPU, 704
crack, 47, 563, 689, 702, 703, 704, 711, 735, 741, 743, 745
creep, 688, 740
critical value, 42, 43, 214, 643, 645, 646
crude oil, 1391
crystal growth, 67, 76, 80, 90, 93, 96, 335, 567, 568, 676, 702, 744
crystal polymers, 1431
crystal structure, 368, 614, 635, 637, 638, 649, 652, 655, 897, 1442
crystalline, xxxi, 168, 207, 213, 222, 223, 239, 295, 296, 308, 311, 312, 313, 314, 315, 316, 318, 320, 324, 337, 348, 354, 364, 365, 366, 367, 373, 377, 378, 379, 384, 389, 394, 435, 492, 497, 505, 516, 530, 538, 584, 602, 609, 610, 698, 699, 718, 720, 722, 725, 727, 728, 729, 730, 741, 744, 763, 906, 910, 1392, 1422, 1467, 1469, 1470, 1472, 1474, 1476, 1526, 1558
crystalline solids, 394
crystallinity, 86, 364, 373, 565, 569, 583, 592, 678, 682, 908, 974, 982, 1414, 1436, 1437, 1439, 1469, 1470, 1475, 1525
crystallisation, 762
crystallites, 2, 314, 316, 318, 332, 336, 339, 341, 367, 373, 375, 376, 381, 385, 387, 389, 497, 567, 667, 714, 715, 972, 974, 1469, 1470
crystallization, 332, 336, 339, 341, 345, 351, 361, 365, 367, 371, 380, 567, 568, 579, 598, 599, 611, 640, 649, 1403
crystals, xi, 7, 8, 10, 14, 18, 31, 44, 49, 61, 66, 67, 70, 77, 88, 221, 262, 295, 308, 321, 322, 330, 363, 364, 365, 366, 444, 446, 494, 507, 510, 512, 517, 530, 535, 568, 672, 673, 675, 676, 677, 678, 681, 682, 690, 693, 703, 711, 713, 716, 731, 745, 746, 975
Cu cluster, 725, 731
culture, 1160, 1171, 1172
culture conditions, 1171
curiosity, 2, 5
current ratio, 258, 879

CVD, xxi, 66, 209, 432, 433, 442, 469, 473, 750, 756, 769, 770, 771, 774, 785, 801, 809, 810, 811, 815, 816, 818, 821, 835, 837, 838, 839, 853, 855, 858, 862, 864, 866, 870, 874, 875, 876, 880, 1547, 1550, 1552, 1555
cycles, 38, 56, 58, 454, 813, 1171, 1553
cycling, 831, 1552, 1553
cyclodextrins, 749
cysteine residues, 385
cytochrome, 165, 181, 750, 790
cytokines, 144
cytometry, 140
cytoplasm, 1172
cytotoxicity, 139, 1179, 1514, 1516, 1536

D

damage, xii, 37, 42, 43, 49, 56, 60, 876
damping, 52, 55, 553, 625, 709, 712, 714, 737
data analysis, 465, 979
data communication, 700
data distribution, 982
database, 531, 655
decay, 105, 172, 618, 1158, 1393, 1406, 1407, 1408, 1409, 1415, 1416
decay times, 1158
decomposition, xxxi, 27, 28, 281, 302, 345, 350, 443, 446, 462, 517, 564, 567, 568, 598, 599, 672, 673, 682, 696, 703, 762, 811, 813, 817, 819, 835, 856, 858, 912, 1450, 1479, 1480, 1481, 1482, 1493, 1496, 1503
decomposition temperature, 762, 1450
deduction, 207
defects, xiii, xvi, 41, 42, 43, 49, 55, 57, 58, 59, 91, 97, 100, 103, 104, 107, 113, 144, 295, 296, 317, 320, 360, 429, 456, 459, 465, 471, 472, 492, 494, 497, 498, 518, 519, 563, 564, 570, 575, 625, 626, 627, 629, 697, 699, 703, 716, 725, 730, 732, 744, 753, 754, 755, 757, 817, 820, 854, 857, 943, 982, 1163, 1164, 1548, 1560
deficiency, 730, 757, 782
definition, xi, 37, 38, 712
deformation, xx, xxi, 39, 42, 221, 350, 469, 599, 600, 625, 687, 688, 690, 693, 698, 709, 711, 716, 721, 722, 723, 724, 725, 728, 729, 730, 731, 732, 737, 740, 741, 742, 746, 747, 813, 864, 879, 1430, 1432, 1434, 1435, 1436, 1438, 1453, 1458, 1459, 1462
degenerate, 127, 582
degradation, xxxi, 8, 27, 28, 31, 117, 182, 419, 420, 476, 497, 854, 880, 960, 1440, 1479, 1480, 1481, 1482, 1484, 1485, 1486, 1488, 1490, 1492, 1493, 1496, 1497, 1516, 1551
degradation rate, 117, 1480
degree of crystallinity, 963, 972, 974, 976, 980, 981, 1436, 1437, 1439
degrees of freedom, 695, 711, 712, 737
degumming, 382, 385
dehydration, 76, 336
dehydrogenases, 787
delivery, xiii, xviii, xxxii, 114, 135, 136, 137, 138, 139, 140, 141, 142, 146, 147, 148, 150, 164, 165, 185, 190, 191, 221, 527, 539, 541, 812, 915, 1499, 1500, 1501, 1503, 1512, 1513, 1514, 1515, 1518, 1539
demand, 634, 810, 1545, 1560
denaturation, 797
dendrimers, xiii, 1, 113, 122, 126, 130, 138, 139, 140, 146, 148, 149, 151, 155, 159, 161, 163, 164, 166, 168, 171, 181, 184, 186, 187, 188, 190, 191, 192, 193, 194, 195, 196, 197, 198, 199
dendrite, 681
dendrites, 319, 796
dendritic cell, 144
density functional theory, 541, 693
deoxyribonucleic acid, 163
Department of Energy, 541
depolarization, 1527
depolymerization, 1480
deposition, xiv, xviii, xxi, xxiii, 1, 10, 91, 93, 95, 100, 179, 201, 208, 212, 217, 221, 224, 269, 271, 273, 281, 285, 287, 309, 318, 323, 432, 433, 440, 454, 460, 461, 462, 463, 471, 472, 473, 474, 501, 503, 506, 562, 563, 648, 688, 750, 756, 763, 764, 771, 782, 785, 786, 796, 809, 810, 816, 823, 836, 838, 839, 840, 856, 858, 905, 906, 909, 910, 936, 1393, 1534, 1559
deposition rate, 461, 471, 472
deposits, 3, 815, 816
depression, 1493
derivatives, 127, 140, 166, 191, 531, 540, 541, 548, 709, 938, 1402, 1413, 1533
desire, 1508
desorption, 130, 342, 344, 352, 360, 370, 436, 813, 817, 1443, 1536, 1546, 1552, 1553, 1557
destruction, 9, 1430, 1432, 1434, 1435, 1436, 1438, 1439, 1514, 1540
detachment, 49, 53, 59
detection, xviii, 31, 71, 129, 133, 172, 173, 178, 181, 182, 184, 185, 278, 359, 465, 547, 555, 558, 750, 760, 771, 772, 773, 774, 776, 777, 778, 780, 781, 782, 783, 784, 785, 786, 787, 789, 790, 792, 793, 796, 797, 798, 800, 1158, 1416, 1501, 1507, 1508, 1510, 1513, 1518, 1519, 1520, 1536, 1537, 1538, 1539, 1545, 1552, 1554, 1557

deviation, 58, 283, 284, 285, 360, 574, 640, 644, 712, 713, 714, 716, 725, 1463
dew, 279, 1531
DFT, 541, 851
diabetes, 781
diabetes mellitus, 781
diagnostic criteria, 1501
dialysis, 548
diamondoids, xviii, 527, 528, 529, 530, 531, 533, 534, 535, 536, 537, 538, 539, 540, 541
dielectric constant, 130, 485, 486, 553, 587, 592, 594, 615, 616, 617, 618, 619, 1402, 1526, 1527
dielectric strength, 960, 979
dielectrics, 130, 592
differential equations, 743
differential scanning, 597, 763, 867, 876
differential scanning calorimeter, 763
differential scanning calorimetry, 597, 876
differentiation, 3, 1539
diffraction, 15, 21, 97, 225, 313, 320, 332, 337, 345, 348, 361, 364, 367, 378, 384, 439, 492, 494, 496, 497, 503, 515, 528, 569, 585, 587, 597, 602, 609, 666, 817, 825, 963, 974, 1446, 1482
diffraction spectrum, 503
diffuse reflectance, 12, 27, 28
diffusion, xiii, xxxi, 18, 19, 31, 48, 49, 70, 76, 179, 201, 208, 211, 213, 239, 241, 242, 284, 296, 298, 320, 323, 388, 389, 395, 416, 436, 498, 675, 688, 723, 744, 771, 782, 793, 813, 821, 915, 1396, 1479, 1481, 1486, 1487, 1492, 1496, 1497, 1503, 1505, 1557
diffusion process, 18, 48, 208, 782, 1496
diffusion rates, 19
diffusion time, 208, 211
diffusivities, 127
diffusivity, 406, 407
digestion, 129, 188
dilute gas, 309
dimensionality, xv, xvi, 277, 427, 452
dimer, 172
dimethylformamide, 750, 757, 923, 1552
diodes, 108, 261, 446, 476, 823
dipeptides, 123, 167, 169, 170, 196
diphenylolpropane, 1428, 1429, 1430, 1433
dipole, 127, 502, 1167
dipole moment, 127, 502, 1167
direct adsorption, 800
direct observation, 403, 589
discontinuity, 102, 261
discretization, 706, 712
discrimination, 130
discs, 418, 600
disease model, 1513
dislocation, xxi, 225, 613, 688, 689, 690, 691, 703, 714, 716, 718, 721, 725, 730, 732, 733, 734, 735, 736, 737, 741, 742, 744, 745
disorder, 71, 72, 176, 355, 361, 415, 698, 699, 716, 722, 877, 976, 1408, 1413, 1417
dispersion, 2, 278, 282, 283, 288, 291, 294, 295, 338, 356, 395, 396, 397, 398, 399, 408, 410, 421, 423, 595, 596, 682, 758, 759, 760, 761, 767, 769, 782, 789, 794, 825, 826, 842, 855, 866, 909, 910, 936, 937, 960, 963, 972, 973, 981, 982, 1445, 1447, 1448, 1450, 1482, 1554, 1557
displacement, 41, 46, 57, 689, 690, 704, 707, 710, 716, 739, 740, 744, 1429
dissociation, xxx, 68, 79, 165, 342, 353, 439, 442, 741, 821, 964, 982, 1168, 1401, 1405, 1406, 1408, 1409, 1422, 1446, 1480
dissociation temperature, 439
dissolved oxygen, 464, 782
distillation, 923, 962
distilled water, 11, 332, 345, 361, 378, 385, 386, 401, 444, 607, 666, 667, 785, 792, 896, 906
distortions, 353
distribution function, 415, 699
divergence, 502, 824, 948
division, 699, 1458, 1497
DMF, 750, 757, 765, 766, 923, 924, 925, 1552
DNA, xviii, xxxii, 136, 141, 143, 148, 163, 164, 165, 179, 185, 189, 190, 191, 194, 195, 198, 199, 330, 371, 527, 535, 540, 750, 751, 763, 764, 767, 772, 793, 798, 800, 909, 911, 1158, 1172, 1514, 1518, 1519, 1521, 1532, 1537, 1539, 1540, 1541
DNA damage, 764, 800
donors, 79, 171, 175, 176, 185, 1167
dopamine, 750, 759, 840
dopants, 105, 108, 447, 605, 769, 854, 1549
doping, xvi, 9, 427, 429, 446, 448, 456, 484, 502, 519, 562, 594, 605, 755, 761, 792, 850, 851, 854, 1555
dosage, 352
double bonds, 121, 123, 178
drainage, 1516
drinking water, 778
drug carriers, 149, 190, 1502, 1510
drug delivery, xviii, xxxii, 1, 138, 139, 140, 142, 145, 146, 148, 168, 191, 527, 539, 1394, 1499, 1500, 1501, 1510, 1511, 1512, 1513, 1514, 1518, 1521
drug delivery systems, 1, 139, 142, 148, 1501
drug efflux, 149, 1500, 1502
drug metabolism, 1500, 1502
drug release, 1501, 1503, 1515
drug resistance, 1500, 1501, 1502, 1503, 1504, 1510, 1515

drug side effects, 1501
drug therapy, 1501
drugs, xiii, 114, 136, 139, 141, 146, 147, 148, 168, 190, 760, 779, 780, 1500, 1501, 1503, 1509
drying, 308, 312, 318, 319, 320, 323, 583, 584, 592, 924, 925, 963, 964, 966, 968, 970
DSC, 589, 596, 597, 598, 599, 639, 654, 763, 867, 876, 963, 974, 975, 987, 1450
DTA curve, 345, 346, 361
ductility, 534, 688
durability, 1428, 1429, 1430, 1431, 1432, 1433, 1434, 1435, 1439
duration, xi, 7, 10, 18, 31, 295, 309, 312, 323, 435, 1403, 1459
DWNT, 861
dyes, xxx, 382, 1160, 1391, 1505, 1506, 1517
dynamic viscosity, 417

E

earth, xix, 104, 105, 539, 588, 589, 633, 634, 635, 637, 638, 639, 643, 647, 648, 649, 650, 651, 652, 653, 654, 655, 656, 657, 658, 659, 660
economics, 9
education, 935
efflux transporters, 149, 192
egg, 330
EIA, 592
elaboration, 782
elastic deformation, 698, 711
elasticity, xxxi, 52, 654, 657, 658, 660, 688, 690, 693, 703, 707, 711, 737, 1428, 1429, 1430, 1431, 1432, 1434, 1436, 1439, 1457, 1459, 1460, 1461, 1462, 1463, 1464, 1465, 1468, 1470, 1473, 1474, 1475, 1476
elasticity modulus, xxxi, 1459, 1461, 1468, 1470, 1473, 1474, 1475, 1476
electric charge, 397
electric current, 1402
electric field, xiv, 237, 238, 239, 250, 254, 255, 256, 261, 274, 428, 449, 450, 452, 453, 467, 492, 520, 550, 797, 851, 979, 981, 984, 1392, 1394, 1404, 1410, 1411, 1523, 1527, 1532, 1555, 1556
electrical breakdown, 1556
electrical conductivity, 360, 519, 634, 651, 829
electrical power, xiv, 238, 269
electrical properties, xxii, 2, 238, 360, 498, 562, 582, 603, 605, 634, 845, 878, 960, 1548, 1551
electrical resistance, 354, 1548, 1553
electricity, 1392
electroanalysis, 772, 780, 801, 802, 803, 804, 805, 806, 807, 808
electrocatalysis, xviii, 184, 199, 547, 549, 557, 558, 773
electrocatalytic, xxi, 181, 199, 752, 754, 759, 764, 767, 768, 769, 770, 772, 773, 774, 777, 780, 782, 783, 786, 787, 788, 789, 792, 793, 795, 801, 825, 828, 842
electrochemical, xxi, 750, 752, 753, 754, 755, 756, 757, 759, 760, 761, 762, 763, 764, 767, 768, 769, 770, 771, 772, 773, 774, 775, 776, 777, 778, 780, 781, 782, 783, 785, 786, 787, 788, 789, 792, 793, 795, 796, 797, 798, 800, 801, 824, 830, 841, 842
electrochemical deposition, 3, 179, 461, 477, 479, 771, 782
electrochemical detection, 179, 180, 750, 774, 776, 781, 786, 798
electrochemical impedance, 750, 793, 798, 800, 1520
electrochemical reaction, 253, 258
electrochemistry, xviii, 66, 184, 199, 238, 547, 558, 752, 753, 757, 759, 762, 765, 767, 768, 793
electrodeposition, 160, 460, 462, 463, 464, 465, 469, 473, 474, 477, 480, 503, 649, 759, 760, 771, 781, 785, 786, 1559
electro-hydrodynamics (EHD), xxx, 1391, 1392
electroless deposition, 914
electroluminescence, xii, xiv, 85, 86, 117, 171, 237, 239, 273, 447, 562
electrolysis, 761, 1522
electrolyte, 12, 29, 30, 239, 242, 244, 245, 254, 256, 268, 460, 462, 467, 477, 775, 776, 778, 779, 788, 824, 829, 830, 906, 1395, 1396, 1397, 1522
electromagnetic, 73, 401, 555, 557, 914, 1526, 1531
electromagnetic fields, 73, 557
electromagnetic wave, 401, 1526
electron beam lithography, 852
electron density, 155, 536, 694, 709, 1526
electron density distribution, 536
electron diffraction, 100, 348, 389, 439, 503, 515, 587, 590, 673
electron microscopy, xviii, xxi, 100, 382, 501, 503, 666, 687, 750, 751, 792, 817, 923, 1483, 1516
electron pairs, 66, 537
electronic circuits, 470
electronic materials, 169, 170, 196, 738
electronic structure, xxii, 8, 536, 692, 693, 754, 845, 847, 879
electrons, 8, 72, 97, 99, 103, 245, 360, 361, 447, 452, 486, 492, 494, 553, 634, 692, 755, 757, 793, 797, 829, 848, 850, 851, 854, 871, 981, 983, 1402, 1408, 1411, 1412, 1417, 1421, 1526, 1527, 1555, 1558
electrophoresis, 749, 773, 774
electroreduction, 774

electrospinning, 2, 4
electrostatic interactions, 166, 185, 381, 512, 533, 765, 915, 1534
elongation, 688, 1450
embedded atom method, 742
emergence, 38, 74
emitters, 8, 108, 519, 823, 840
empirical interatomic potentials, 692
empirical potential, 713, 746
employment, 27, 166
emulsion polymerization, 2
encapsulation, 4, 114, 116, 145, 759, 811
endocytosis, 140, 142, 149, 1175, 1502
endothelial cells, 1510, 1520
endothelium, 1511
endothermic, 589, 598, 599, 1450
energy density, xix, 209, 218, 582, 594, 615, 619
energy emission, 1409
energy supply, 393
energy transfer, xiii, 114, 117, 156, 157, 172, 174, 175, 176, 1171, 1180
enlargement, 58, 351, 468, 711
entanglement, 706
entanglements, 1474
entrapment, 769, 781, 785
entropy, 639, 641, 643, 644, 688, 739
environment, xii, xxvii, 3, 5, 37, 41, 44, 52, 53, 61, 71, 76, 77, 97, 119, 182, 244, 245, 388, 395, 428, 442, 553, 634, 682, 698, 711, 762, 777, 779, 823, 833, 853, 854, 909, 912, 966, 1157, 1165, 1397, 1442, 1482, 1522, 1525, 1536, 1545
environmental change, 375
environmental conditions, 907
enzyme, 135, 142, 148, 182, 183, 198, 353, 371, 761, 767, 769, 774, 780, 781, 783, 784, 786, 790, 792, 793, 794, 795, 796, 797, 914, 1510, 1537
enzyme immobilization, 198, 790
enzyme sensor, 183
enzymes, 4, 150, 752, 760, 761, 766, 770, 774, 780, 782, 783, 789, 790, 791, 792, 794, 797
epidermal growth factor, 142, 1518
epinephrine, 750, 754
epitaxial growth, 47, 429, 614
epoxy, xxiii, 767, 771, 773, 778, 921, 922, 923, 924, 925, 931, 933, 935
epoxy groups, xxiii, 921, 922, 923, 924, 925, 931, 933, 935
EPR, 140
equality, 42, 710
equilibrium, 50, 51, 53, 176, 204, 209, 215, 222, 226, 376, 395, 418, 467, 531, 630, 637, 639, 640, 641, 642, 647, 691, 692, 707, 710, 725, 730, 848, 946, 947, 948, 953, 955
equipment, 264, 269, 270, 308, 585, 613, 651, 652, 856, 909, 987
erosion, 48, 242, 394
Escherichia coli, 130, 151, 751, 798, 1510, 1520
ester, 117, 118, 142, 152, 167, 193
esters, 765
estimating, 184
etching, xiv, 29, 237, 238, 239, 242, 244, 245, 246, 248, 249, 250, 251, 252, 253, 254, 256, 257, 258, 261, 263, 264, 268, 274, 431, 465, 468, 475, 477, 493, 538, 875, 1158
ethanol, xi, 7, 11, 12, 14, 16, 19, 20, 27, 31, 56, 95, 101, 354, 355, 356, 358, 359, 360, 361, 401, 549, 584, 588, 593, 597, 599, 600, 757, 765, 789, 796, 906, 908, 910, 1159, 1551, 1554, 1558
ethers, 121, 1533
ethylene, 72, 115, 133, 135, 137, 138, 139, 141, 144, 147, 148, 149, 155, 164, 167, 189, 190, 192, 193, 393, 394, 397, 399, 402, 403, 407, 410, 422, 669, 670, 672, 678, 682, 811, 924, 1443, 1502, 1511, 1512, 1514, 1536, 1547
ethylene glycol, 72, 133, 135, 139, 141, 144, 147, 148, 149, 155, 190, 192, 193, 393, 394, 397, 399, 402, 403, 407, 410, 422, 669, 670, 672, 678, 682, 1502, 1511, 1512, 1514, 1536
ethylene oxide, 115, 138, 167, 1443
Euclidean object, xxxi, 1457, 1460, 1464, 1465
Euclidean space, xxxi, 1459, 1467, 1468, 1469, 1472, 1475
evacuation, 476, 478, 1553
evaporation, 10, 90, 91, 92, 105, 318, 319, 322, 323, 324, 399, 401, 403, 410, 433, 434, 461, 472, 493, 502, 503, 504, 505, 506, 507, 508, 512, 598, 647, 811, 969, 1396, 1404, 1533, 1547, 1552, 1558
evidence, 77, 100, 688, 856, 1445, 1448, 1481, 1486
evolution, xiv, 56, 61, 67, 90, 117, 201, 202, 203, 226, 232, 280, 354, 356, 362, 363, 385, 488, 645, 688, 691, 702, 703, 714, 715, 718, 724, 725, 726, 728, 731, 744, 900, 1164, 1165, 1166, 1167, 1168, 1169, 1170, 1410
excitability, 535
excitation, 105, 162, 172, 174, 238, 249, 253, 258, 269, 374, 377, 383, 447, 448, 494, 502, 535, 555, 599, 666, 863, 870, 871, 872, 981, 986, 1158, 1160, 1161, 1162, 1168, 1174, 1175, 1403, 1415, 1416, 1417, 1506, 1538
exciton, xxx, 107, 178, 345, 497, 498, 517, 562, 1160, 1401, 1402, 1407, 1422, 1558
exclusion, 148, 776, 781
excretion, 147, 1503
execution, 380, 1460
exothermic, 348, 590, 599
expectation, 46, 964

experimental condition, xxiv, 94, 96, 97, 317, 338, 347, 502, 856, 908, 959, 970
exposure, 146, 314, 817, 852, 854, 867, 915, 1546, 1548, 1549, 1550, 1551, 1552, 1553, 1554, 1555, 1559
expression, 46, 53, 54, 578
extinction, 174, 1495, 1496, 1527, 1528
extracellular matrix, 4, 330
extraction, 154, 168, 271, 492, 493, 824, 923, 1174, 1536
extrapolation, 576, 1463
extravasation, 1512
extrusion, 909

F

FAD, 750, 765, 795, 797
failure, xx, 46, 450, 687, 741, 742, 960, 964, 979, 981, 1458
false negative, 1501
family, xvi, xx, 121, 124, 126, 132, 137, 157, 175, 329, 531, 633, 635, 637, 643, 652, 654, 659, 676, 698, 1546
fatigue, xii, 37, 38, 39, 689
FCC, 698, 699, 710, 711, 716, 720, 721, 722, 725, 727, 728, 730
feedback, 447
FEM, 704, 953
fermentation, 795
Fermi level, 494, 848, 851, 854, 855
ferrite, xxii, 895, 896, 899, 900, 902
ferritin, 130, 915
ferrocenyl, 121, 161, 750, 781
ferrofluids, 896
FFT, 44, 341, 591, 615
fibers, 14, 331, 332, 333, 335, 336, 337, 339, 342, 344, 347, 353, 354, 362, 367, 368, 372, 375, 376, 377, 378, 379, 381, 382, 384, 385, 387, 389, 390, 811, 836, 837, 1393, 1394, 1395, 1446, 1447, 1448, 1452, 1548
fibroblasts, 4
fidelity, 939
field emission scanning electron microscopy, 597
filament, 822
filled polymers, 972
filler particles, xxxi, 982, 985, 1457, 1458, 1459, 1465
fillers, 4, 960, 986, 1444, 1497
film thickness, 563, 577, 781, 1404, 1407
filters, 810, 823, 837, 1172
filtration, 430, 588, 856
financial support, 835, 918
finite element method, xxiii, 704, 745, 941, 942, 953, 957
fire resistance, 1427, 1454
fire retardants, 1492
fireproof, xxxi, 1479, 1496
first generation, 184
fission, 465
fixation, xxxi, 759, 1473, 1474, 1476
flame, xxxi, 1479, 1480, 1492, 1493, 1496
flammability, 1479, 1481, 1493, 1494, 1496
flatness, 566
flexibility, 72, 162, 172, 460, 628, 823, 1443, 1470, 1475
flight, xxx, 130, 963, 973, 1401, 1402, 1404, 1418, 1419
flocculation, 397
flood, 963
fluctuations, 50, 53, 713
fluid, xiv, xxiii, 142, 238, 269, 394, 395, 396, 397, 402, 407, 408, 412, 414, 415, 416, 417, 418, 419, 420, 421, 423, 424, 425, 702, 823, 895, 896, 1511
fluorescence, xxvii, 117, 119, 121, 124, 127, 135, 137, 157, 158, 162, 165, 171, 172, 174, 176, 178, 666, 915, 923, 1157, 1160, 1161, 1162, 1171, 1172, 1174, 1175, 1416, 1505, 1516, 1517, 1537, 1538, 1539
fluorescence decay, 171
fluorine, 1393
fluorophores, 1158, 1175
focusing, xxii, 66, 308, 587, 845
foils, 468, 469, 472
folate, 138, 139, 140, 141, 143, 169, 190
folic acid, 137, 138, 141, 143, 1502
food, 774
food industry, 774
forests, 766
formaldehyde, 358, 359, 360, 1559
fossil, 1391
fossil fuels, 1391
fouling, 767, 774, 789
Fourier, 751, 763
Fourier transform infrared spectroscopy, 597
four-wave mixing, 127
fractal analysis, 1458, 1473, 1474, 1475
fractal dimension, xxxi, 1458, 1460, 1464, 1467, 1471, 1473, 1475, 1476
fractal objects, 1460
fractal space, 1468, 1469, 1470
fractal structure, 942, 1457
fractal theory, 415, 1460, 1461
fracture stress, 1468
fragmentation, 116, 148, 166, 185
France, 190, 480

free energy, xiv, 58, 76, 91, 96, 164, 201, 203, 204, 206, 207, 210, 214, 215, 217, 218, 219, 223, 226, 227, 229, 230, 231, 232, 624, 636, 638, 641, 643, 644, 645, 646, 647, 659, 693
free radicals, 57, 794
free volume, 1435, 1436, 1450
freedom, 695, 696, 711, 712, 737
freezing, 715, 741, 743, 746, 910
friction, xi, xix, xxi, 37, 38, 39, 40, 41, 43, 44, 47, 52, 53, 55, 56, 59, 418, 539, 623, 624, 626, 628, 630, 687, 739
fructose, 750, 774, 775, 787, 789
FTIR, xxiii, 339, 340, 349, 350, 351, 368, 369, 374, 385, 386, 597, 598, 599, 600, 751, 763, 794, 921, 923, 927, 928, 931, 932, 933, 934, 963, 964, 965, 969, 970, 971
fuel, 5, 825, 826, 829, 841, 1493
fuel cell, 825, 826, 829, 841
fulfillment, 666
fullerene, 118, 127, 175, 176, 534, 743, 837, 1546
functionalization, 2, 27, 121, 380, 912, 921, 922, 923, 936, 937, 1519, 1536
furan, 118
fusion, 906, 907
futures, 468

G

gadolinium, 1505
gallium, 449, 451, 452, 512, 562, 567
gas phase, 203, 222, 445, 446, 492, 647, 812, 833, 1547
gas sensors, xxxii, 239, 345, 1545, 1546, 1551, 1554, 1555, 1557
gases, xiv, 221, 238, 278, 309, 354, 357, 359, 360, 441, 504, 813, 1391, 1545, 1546, 1548, 1549, 1550, 1551, 1552, 1555, 1556, 1557, 1558, 1559, 1560
gasoline, 354, 355, 356, 358, 359, 360, 361
gastrointestinal tract, 157
Gaussian, 312, 313, 314, 315, 357, 542, 610, 612, 713, 714, 725, 1460
GCE, 750, 759, 763, 765, 768, 769, 770, 773, 774, 775, 776, 777, 778, 779, 780, 781, 782, 783, 784, 786, 787, 789, 790, 791, 793, 794, 797, 798, 800, 1508
gel, 2, 117, 176, 331, 332, 333, 336, 339, 340, 345, 349, 351, 563, 567, 568, 579, 582, 621, 761, 763, 768, 783, 786, 789, 794, 906, 910, 1395, 1445
gel permeation chromatography, 117, 176
gelation, 4, 331, 336, 349
gene, xiii, 114, 135, 136, 137, 139, 141, 143, 145, 148, 164, 165, 189, 190, 192, 1502, 1512, 1513, 1539, 1540
gene expression, 148, 192, 1513, 1539
gene silencing, 141
gene therapy, 136
gene transfer, 136, 143, 189
generation, xiii, xxx, xxxii, 113, 114, 117, 118, 121, 123, 124, 126, 127, 129, 132, 137, 138, 140, 141, 143, 147, 148, 149, 151, 152, 153, 155, 156, 159, 164, 166, 167, 168, 171, 173, 176, 178, 180, 182, 183, 184, 185, 242, 258, 308, 506, 515, 701, 714, 763, 786, 909, 954, 986, 1168, 1391, 1392, 1395, 1396, 1499, 1507, 1508
genetic disease, 798
genetic mutations, 1500
geometrical parameters, 479
Georgia, 491, 498
germanium, 742
Germany, 32, 537, 918, 987, 1479
Gibbs energy, 203, 636, 637
Gibbs free, xiv, 201, 203, 204, 206, 207, 210, 214, 215, 217, 218, 223, 226, 230, 231, 232, 636, 638
glass, xvii, xviii, 470, 471, 483, 484, 486, 487, 488, 489, 515, 547, 548, 549, 555, 558, 597, 651, 659, 839, 1160, 1172, 1179, 1393, 1404, 1450, 1530, 1535, 1536, 1537, 1540, 1555, 1558
glass transition, 651, 1450
glass transition temperature, 651, 1450
glasses, xvii, 460, 483, 484, 486, 488, 489
glucose, 72, 121, 140, 181, 182, 184, 198, 750, 759, 760, 763, 767, 768, 769, 770, 771, 774, 775, 781, 782, 783, 785, 786, 787, 788, 789, 795
glucose oxidase, 181, 182, 750, 759, 781
glutamic acid, 143, 165, 191
glutathione, 154, 750, 768, 772
glycerol, 144, 666, 908, 1172
glycine, 387
glycol, 4, 141, 143, 148, 666, 667, 669, 670, 672, 678, 911, 1396
glycoprotein, 339, 340, 341, 351, 353, 365, 366, 371, 372, 374, 375, 376, 377, 380, 381, 390, 1502, 1503
glycoside, 166
gold nanoparticles, xviii, 180, 181, 182, 198, 515, 547, 548, 549, 550, 551, 552, 553, 554, 555, 556, 557, 558, 911, 914, 915, 938, 1502, 1507, 1508, 1511, 1518, 1519, 1535, 1536, 1540
GPC, 927, 930
gracilis, 1171
grain boundaries, 18, 19, 31, 360, 608, 634, 639, 640, 643, 644, 645, 646, 654, 657, 688, 690, 691,

714, 716, 717, 718, 719, 721, 722, 723, 747, 748, 1557, 1559
grain boundary structure, 640
grain refinement, 723
grains, xi, xix, 7, 13, 18, 20, 21, 31, 330, 361, 385, 386, 506, 561, 562, 567, 573, 574, 578, 584, 602, 608, 619, 634, 644, 645, 646, 649, 652, 653, 688, 689, 691, 714, 715, 716, 721, 722, 723, 724, 725, 1557, 1559
gram-positive bacteria, 1534
granules, 331, 333
graph, 405, 435, 716, 717, 724, 726, 727, 728, 859
graphene sheet, 753, 861, 873
graphite, 92, 311, 312, 313, 314, 320, 324, 398, 400, 404, 405, 412, 413, 419, 422, 534, 624, 651, 652, 666, 743, 749, 750, 751, 753, 754, 755, 766, 767, 768, 772, 773, 774, 785, 787, 790, 794, 797, 819, 820, 821, 857, 860, 862, 871, 877, 942, 947, 949, 950, 953, 954, 955, 1467, 1546, 1548
grass, 318, 322
gravimetric analysis, 923, 1450
gravity, 395, 405
grids, 317, 667, 705, 706, 707, 708, 709, 726
Grignard reagents, 120, 158
grouping, 1430
growth factor, 137, 1512
growth mechanism, xx, 10, 18, 88, 91, 92, 100, 227, 228, 309, 323, 433, 435, 443, 444, 445, 512, 565, 665, 676, 716, 810, 822, 823, 858, 1525
growth rate, xiv, 19, 45, 46, 91, 97, 201, 202, 203, 223, 224, 227, 228, 230, 279, 286, 289, 290, 291, 292, 293, 294, 295, 296, 297, 375, 435, 474, 646, 659
growth temperature, 91, 222, 229, 281, 283, 284, 287, 293, 296, 437, 442, 493, 811, 846, 855, 858, 859, 860, 861, 862, 873, 1547
growth time, 230, 279, 280, 281, 285, 286, 289, 291, 292, 811, 864
Guangdong, 232
Guangzhou, 201
guanine, 799, 800, 1508
guidance, 303, 543, 747
guidelines, 563
Guinea, 662

H

Hall-effect, xiv, 237, 238, 245, 246, 247, 248, 249, 250, 264, 265, 266, 267, 268, 269, 274
halogen, 71, 859
halogens, 160, 1427
Hamiltonian, 696, 704
hands, 520
hardener, 767
hardness, xx, 39, 52, 428, 534, 538, 654, 657, 658, 660, 687, 689, 690, 695, 740
harm, 1391
harmful effects, 39
harvesting, 174, 175, 176, 178, 198, 1393, 1396
hate, 572
H-bonding, 339
healing, 742
health, 3, 4, 146, 1500
health problems, 3
heart disease, 786
heat, xvi, xvii, xviii, 8, 331, 332, 341, 346, 349, 361, 363, 364, 368, 393, 394, 400, 406, 407, 414, 416, 417, 418, 419, 420, 421, 422, 423, 424, 425, 426, 476, 478, 483, 489, 561, 563, 564, 568, 569, 570, 571, 572, 573, 574, 575, 576, 577, 578, 583, 602, 624, 634, 643, 651, 654, 659, 737, 753, 817, 847, 852, 963, 1424, 1440, 1443, 1450, 1451, 1481, 1493, 1494, 1495, 1496, 1553
heat capacity, 643, 654
heat conductivity, 651
heat release, 1481, 1494, 1495
heat transfer, xvi, 393, 394, 400, 407, 414, 416, 417, 418, 419, 420, 421, 422, 425, 426
heating, 11, 74, 96, 332, 333, 335, 336, 339, 343, 361, 363, 364, 365, 393, 436, 440, 442, 503, 506, 507, 515, 567, 568, 573, 579, 597, 599, 647, 651, 652, 666, 680, 682, 756, 811, 817, 876, 877, 923, 963, 968, 1480, 1484, 1485, 1486, 1489, 1490, 1493, 1494, 1532, 1547, 1548, 1552
heating rate, 11, 332, 333, 335, 336, 339, 343, 567, 568, 597, 651, 652, 876, 923, 963, 1480, 1485, 1486, 1489, 1490, 1493
height, 20, 48, 52, 53, 57, 226, 227, 229, 230, 264, 280, 282, 283, 284, 285, 288, 290, 291, 292, 299, 301, 302, 446, 452, 510, 565, 566, 574, 585, 613, 626, 1558
helium, 654, 1403, 1547
heme, 184, 791, 792, 793, 794
hemoglobin, 750, 759, 790, 792
hepatocytes, 143, 1508, 1519
hepatoma, 1511
heptane, 910
heterogeneity, 1459
heterogeneous catalysis, 151, 841
heterojunctions, 840, 938
hexagonal lattice, 570, 847
hexane, 924
high density polyethylene, 1474
histamine, 796
histidine, 151, 777, 1166, 1167, 1168
histogram, 285, 290, 291, 292, 865

HIV, 147, 191
HIV infection, 147
homeland security, 1545
homocysteine, 750, 759
homogeneity, 132, 331, 857, 915
Honda, 27, 32, 33, 1530, 1543
Hong Kong, 581
hormone, 760
Horseradish peroxidase, 793
hospitals, 634
host, xvii, xviii, 105, 130, 163, 459, 460, 473, 477, 478, 527, 694, 915, 1409, 1413, 1424, 1443, 1445, 1506, 1529
hot spots, 414
House, 661
HPLC, 750, 772, 774, 939, 940
HRTEM, xix, 12, 15, 16, 22, 24, 25, 87, 89, 90, 91, 92, 96, 100, 101, 225, 317, 336, 337, 341, 343, 349, 365, 367, 370, 373, 379, 380, 381, 441, 512, 513, 516, 517, 582, 613, 614, 619, 649, 650, 653, 819, 821, 862, 864, 865, 880, 1414
human activity, 1391
hybrid, xxiii, xxx, xxxii, 4, 137, 161, 167, 190, 372, 378, 380, 381, 390, 582, 594, 628, 702, 703, 745, 761, 762, 786, 789, 921, 922, 1391, 1392, 1394, 1401, 1403, 1405, 1406, 1408, 1411, 1412, 1413, 1414, 1415, 1416, 1417, 1418, 1419, 1420, 1421, 1424, 1443, 1445, 1446, 1447, 1448, 1452, 1511, 1521, 1522, 1531, 1534, 1541
hybridization, 772, 798, 1532, 1537
hydrazine, 403, 768
hydride, xxi, 809, 810, 811, 813, 814, 815, 816, 817, 818, 819, 820, 821, 822, 823, 825, 826, 827, 828, 829, 830, 831, 832, 834, 835, 837, 839, 842
hydrides, 537, 813, 814, 823
hydrocarbons, 528, 534, 535, 537, 540, 811
hydrochloric acid, xxiii, 921, 924
hydrofobization, xxxi, 1441
hydrogels, 4, 1514
hydrogen atoms, 529, 532, 533, 540
hydrogen bonds, 4, 70, 74, 156, 340, 351, 530, 968, 1443
hydrogen fluoride, xiv, 237, 238
hydrogen gas, 279, 283, 813, 853, 870, 876, 909
hydrogen peroxide, 182, 183, 184, 199, 759, 780, 782, 783, 784, 787, 790, 793, 794, 795, 797, 1503
hydrogenation, 151, 813, 841
hydrolysis, 66, 68, 130, 131, 156, 178, 331, 341, 356, 790, 910, 966, 968, 969, 1403
hydroperoxides, 152
hydrophilic groups, 371, 759
hydrophilicity, 1500, 1503
hydrophility, 21
hydrophobic, 533, 540, 753, 758, 759, 763, 764, 780, 782, 797, 800
hydrophobic groups, 2
hydrophobic interactions, 168, 387, 758, 760, 764, 782
hydrophobic properties, 148
hydrophobicity, 4, 20, 21, 168, 756, 973, 1500, 1503
hydroquinone, 129, 183
hydrosilylation, 123
hydrothermal process, 86, 595
hydroxide, xix, 11, 341, 564, 581, 582, 588, 592, 593, 597, 599, 619, 777, 869
hydroxides, 351, 589
hydroxyapatite, 937, 1511
hydroxyl, xix, xxiii, 117, 133, 171, 346, 369, 385, 390, 530, 551, 581, 589, 595, 597, 618, 760, 921, 923, 924, 925, 933, 935, 964, 965, 969, 1429, 1430, 1500, 1503
hydroxyl groups, xxiii, 117, 171, 369, 530, 551, 589, 595, 597, 618, 921, 924, 925, 933, 964, 965, 1430
hyperbranched polymers, xiii, 114, 139, 937
hyperthermia, xxiii, 895, 896, 1507
hypothesis, 53, 140, 970, 982, 1171
hysteresis, xix, 342, 343, 370, 420, 623, 625, 626, 627, 628, 631, 824, 1443
hysteresis loop, 342, 343, 370, 625, 626, 628, 631

I

ibuprofen, 136, 140
idealization, 943
identification, 97, 100, 172, 314, 538, 1507, 1516, 1545, 1551, 1556
identity, xxxi, 98, 1430, 1432, 1467, 1472
III-nitrides, 427, 428, 429, 456
IL-6, 144
IL-8, 144
illumination, 9, 244, 245, 246, 249, 250, 251, 252, 253, 254, 256, 258, 259, 260, 261, 263, 264, 274, 1165, 1166, 1404, 1413, 1422, 1423, 1424, 1503, 1539
IMA, 542
imaging, xiii, xxxii, 114, 126, 133, 135, 138, 139, 140, 141, 145, 189, 427, 535, 924, 1175, 1180, 1499, 1505, 1506, 1507, 1508, 1510, 1513, 1516, 1518, 1521, 1522, 1535, 1537, 1538, 1539, 1540
imaging modalities, 135, 1505
imaging techniques, xxxii, 1499, 1508
immersion, 55, 350, 1535
immobilization, xxi, 4, 10, 30, 182, 183, 185, 371, 540, 757, 765, 767, 768, 769, 770, 772, 780, 781, 782, 785, 789, 790, 792, 793, 794, 795, 796, 798, 800, 801, 1519, 1536

immunogenicity, 146, 149
impact strength, 1428, 1433, 1435, 1436, 1439
impregnation, 940
imprinting, 130, 131, 189
impurities, xvii, xxii, 91, 104, 348, 361, 491, 492, 498, 519, 669, 753, 760, 767, 811, 817, 845, 846, 855, 862, 863, 864, 874, 875, 876, 877, 879, 880, 970, 1547, 1552, 1558, 1560
in situ, xxiii, 144, 153, 154, 175, 179, 363, 375, 376, 377, 380, 381, 382, 383, 385, 388, 390, 797, 838, 911, 921, 922, 923, 924, 925, 935, 936, 937, 1444, 1445, 1508, 1512
in vitro, 136, 137, 138, 140, 142, 146, 147, 150, 366, 1503, 1505, 1507, 1511, 1512, 1514, 1515, 1539
in vivo, xiii, 3, 114, 135, 136, 137, 138, 140, 142, 143, 144, 146, 174, 191, 777, 1503, 1505, 1507, 1510, 1511, 1512, 1514, 1515, 1516, 1518, 1519, 1537, 1541
InAs/GaAs, 281
incidence, 43, 53, 54
inclusion, xxx, 315, 605, 1401, 1402, 1409, 1413, 1417, 1424
incompatibility, 973, 976
incubation time, 912
independence, 59, 1545
indexing, 649, 650, 653
India, 85, 483, 809, 835
indication, 92, 101, 339, 367, 404, 405, 532
indicators, 1517
indices, 698
indium, xv, 180, 182, 277, 278, 298, 300, 301, 302, 303, 431, 442, 443, 514, 519, 548, 562, 1413, 1419, 1559
indomethacin, 139
induction, 168, 1507, 1559
industrial application, xxxi, 278, 394, 1441
industrial production, 592
industrial sectors, 393, 394
industry, xxii, 69, 533, 666, 774, 845, 1432, 1545
inelastic, 553, 711, 748, 819
infancy, 5, 146, 324
infarction, 786
infinite, 74, 77, 79, 330, 712
influence, 11, 20, 27, 38, 565, 670, 675, 852, 964, 971, 977, 980
information technology, xx, 687
infrared spectroscopy, 96
inhibition, 44, 67, 68, 74, 80, 148, 164, 790, 1506, 1516
inhibitor, 149, 1503
inhibitory effect, 183
initial state, 696, 697
initiation, 914, 1480, 1509
innovation, xiii, 113
inorganic fillers, 1444
input, 107
insertion, 689, 824, 841
insight, 4, 18, 159, 164, 324, 563, 911, 1166
inspiration, 330, 949
instability, 9, 214, 342, 425, 429, 563, 646, 695, 1171, 1525
instruments, xii, 1, 37, 40, 61, 535, 584, 634, 647
insulation, 406, 971, 1450
insulators, 1548
integrated circuits, 130
integration, xvii, xxii, 1, 51, 254, 390, 433, 460, 470, 474, 479, 660, 692, 695, 697, 707, 831, 845, 846, 854, 855, 866, 876, 879, 935, 1541
integrity, 184, 615
intercalation, 824, 841, 1444, 1445, 1446, 1480
interest, xii, xx, 27, 50, 65, 85, 88, 89, 563, 665, 666, 847, 960, 972
interface, xxxi, 4, 18, 19, 31, 51, 76, 91, 204, 205, 206, 207, 209, 210, 214, 223, 224, 225, 226, 256, 296, 309, 366, 388, 414, 417, 421, 424, 432, 452, 516, 517, 548, 549, 553, 614, 634, 635, 651, 689, 703, 709, 710, 731, 740, 741, 748, 968, 969, 970, 972, 981, 982, 984, 985, 986, 1163, 1397, 1402, 1415, 1421, 1424, 1499, 1508, 1509, 1515, 1557
interface energy, 206, 214, 223
interfacial bonding, 968
interfacial layer, 414, 415, 424
interfacial properties, 168
interference, 181, 576, 776, 783, 789, 796, 976, 1161
intermetallic compounds, 680
intermolecular interactions, xviii, 117, 127, 154, 527, 531, 532
internalization, 136, 137, 141, 143, 1175
interphase, xxxi, 1473
interpretation, 494, 716
interval, 332, 361, 387, 859, 1429, 1430, 1434, 1467
intervention, 308
intestine, 157
intrinsic value, 455
intrinsic viscosity, 167
intuition, 562
inventions, 1546
inversion, 245
ion channels, 169
ion implantation, 744
ionic conduction, 985
ionic liquids, 151, 751, 762, 794
ionization, 4, 130, 383, 386, 397, 823, 836, 1555, 1556
ions, xv, 9, 24, 29, 31, 41, 46, 47, 67, 71, 77, 96, 104, 105, 107, 108, 153, 155, 156, 157, 179, 193,

307, 308, 339, 351, 356, 366, 371, 376, 380, 381, 382, 383, 385, 386, 387, 388, 441, 461, 465, 484, 492, 502, 507, 517, 535, 594, 597, 598, 599, 600, 611, 672, 676, 760, 769, 778, 798, 830, 841, 864, 896, 910, 915, 922, 933, 963, 970, 985, 1159, 1396, 1523, 1525
IR spectra, 73, 79, 755, 756, 794, 965
iridium, 127, 158, 781
iron, 68, 157, 234, 309, 689, 694, 714, 740, 742, 744, 746, 750, 782, 792, 794, 821, 838, 840, 855, 867, 869, 870, 873, 1430, 1432, 1433, 1435, 1437, 1502, 1505, 1507, 1518
irradiation, xxvii, 9, 12, 27, 28, 29, 30, 67, 121, 136, 153, 168, 174, 185, 319, 398, 399, 401, 402, 423, 465, 487, 666, 680, 682, 856, 912, 1157, 1162, 1165, 1166, 1168, 1170, 1171, 1175, 1176, 1177, 1178, 1180, 1397, 1504, 1532, 1539, 1540
irreversible aggregation, 1458
irreversible aggregation models, 1458
Islam, 190
isobutylene, 1468
isolation, 117, 124
isomers, 529
isophthalic acid, 114
isotactic polypropylene, 1474
isothermal, 644, 646, 659, 1480, 1481, 1493, 1494
isothermal heating, 1494
isotherms, 168, 342, 343, 344, 352, 370, 834, 835
Italy, 1, 37, 623
I-V curves, 1423

J

Japan, 423, 501, 563, 564, 665, 666, 674, 743, 905, 918, 988, 989, 1546
Jordan, 837, 903
judgment, 405, 947

K

K^+, 1442, 1533
KBr, 47, 48, 963
kidney, 140, 143, 146, 147, 1171, 1505
kidneys, 1538
killing, 1516
kinetic energy, 702
kinetic model, xxxi, 203, 221, 1479, 1480, 1486, 1489, 1493
kinetic parameters, 172, 225, 1480, 1489
kinetics, xiii, 128, 165, 172, 179, 201, 209, 211, 213, 214, 216, 218, 220, 221, 224, 226, 283, 309, 359, 639, 640, 646, 688, 691, 702, 753, 754, 765, 829, 930, 1163, 1164, 1166, 1396, 1479
kinks, 46
knowledge, xii, 37, 55, 61, 876
KOH, 30
Korea, 422, 501, 561, 845

L

labeling, 666, 939, 1158, 1172, 1179, 1505, 1506, 1511, 1513, 1517, 1538
lactose, 121, 166, 775
laminar, 418, 419, 422, 425, 1497
land, xx, 5, 687
Langmuir-Blodgett, 631
lanthanum, 843
laser, 503, 512, 811, 837
laser ablation, xv, 87, 212, 307, 308, 309, 318, 324, 433, 435, 512, 811, 837, 855, 856, 866, 875, 1547
lasers, xii, xvii, 85, 86, 90, 105, 108, 278, 303, 371, 427, 448, 501, 537, 675, 811
lasing threshold, 448, 449, 450
lattice parameters, 97, 585, 589, 600, 678
lattices, 44, 58, 78, 79, 123, 178, 317, 369, 376, 534, 589
laws, xxiii, 39, 704, 941, 942, 943, 944, 945, 946, 948, 953, 957, 1435
layered architecture, xiii, 113, 146
layering, 414, 417, 421, 424
leaching, 783
lead, 9, 52, 57, 73, 77, 79, 108, 519, 682, 702, 723, 737, 755, 772, 777, 833, 981, 1446
leakage, 429, 778, 846, 855, 880, 1545
leaks, 833, 1422
LED, 108, 243
lens, 312, 1162
leukemia, 137, 1502, 1503, 1504, 1507, 1508, 1510, 1514, 1519
leukemia cells, 1504
life sciences, 1517
lifetime, xxx, 46, 103, 373, 862, 874, 1170, 1401, 1406, 1407, 1409, 1415, 1416, 1517
ligand, xiii, 113, 119, 156, 157, 158, 160, 161, 171, 176, 345, 1158, 1418, 1500
ligands, 66, 67, 68, 76, 77, 79, 80, 121, 123, 141, 151, 156, 157, 159, 160, 161, 171, 193, 375, 551, 583, 676, 1158, 1168, 1169, 1412, 1507
light emitting diode, 2, 239, 243, 427, 675
light scattering, 72, 118, 405, 541, 972, 975, 1392, 1393, 1396
light transmission, 73
light-emitting diodes, 86

limitation, 173, 244, 249, 251, 253, 254, 258, 274, 783, 1165, 1525, 1548
linear dependence, 412, 414, 576, 577, 1463
linear function, 359
linear polymers, 1469
linkage, 321, 704, 908, 922, 1515
links, 691, 1480
lipase, 4
lipids, 147, 164, 907, 908, 911
liposomes, 147, 192, 1512, 1515
liquid chromatography, 750, 772, 776
liquid crystals, 1532
liquid phase, xv, 307, 435, 636, 638, 647, 867, 876, 1547
liquid-air interface, 548
liquids, xiv, xvi, 238, 309, 310, 393, 394, 403, 407, 423, 537, 699, 751, 762, 794
literature, 519, 520, 533, 825
lithium, 69, 824, 841
lithography, xiv, 237, 238, 254, 274, 431, 433, 460, 467, 475, 477, 478, 515, 799
liver, 133, 140, 143, 147, 148, 192, 1160, 1505
liver cancer, 1160
living radical polymerization, 939
local order, xxxi, 747, 1467, 1469, 1472
localization, xxxi, 126, 178, 198, 477, 781, 915, 1393, 1457, 1465
location, 50, 155, 318, 346, 372, 388, 390, 438, 690, 697, 716, 732, 908, 945, 1505
London, 748, 840
low temperatures, 436, 472, 498, 579, 583
LPG, 354, 355, 356, 358, 359, 360, 361
lubricants, 38, 55
luminescence, 103, 104, 105, 117, 156, 160, 171, 175, 239, 249, 257, 262, 264, 268, 274, 377, 494, 562, 934, 1158, 1160, 1162, 1164, 1168, 1172, 1173, 1174, 1177, 1405, 1406, 1407, 1409, 1415, 1537, 1538
luminescence efficiency, 1168
lung, 143, 148
lying, 51, 54, 88, 533, 725, 871
lymph, 135, 1512, 1516
lymph node, 135, 1516
lysine, 133, 139, 140, 147, 148, 151, 164, 190, 192, 375, 388, 777, 1511
lysis, 914
lysozyme, 130, 181, 339, 342, 353

M

mAb, 137, 190
machinery, 540, 1534
Mackintosh, 919
macromolecular coil, 1460
macromolecules, xiii, 1, 113, 114, 130, 144, 150, 159, 181, 183, 330, 342, 351, 353, 365, 366, 376, 380, 381, 385, 387, 388, 390, 532, 743, 764, 780, 1429, 1430, 1431, 1432, 1512
macrophages, 144
macropores, 253, 343, 370
macroradicals, 58
magnesium, 446, 595, 597, 602, 607, 1517
magnet, 896
magnetic field, 168, 245, 246, 250, 258, 264, 274, 476, 493, 896, 899
magnetic fluids, 896
magnetic moment, 634, 654, 897, 899, 900, 901, 902
magnetic particles, 896, 914, 938, 1502, 1513
magnetic properties, xxii, 86, 153, 634, 654, 895, 896, 902
magnetic resonance, 135, 634, 1513, 1517, 1518
magnetic resonance imaging, 634
magnetic resonance spectroscopy, 1513
magnetism, 160
magnetite, 798, 900, 901
magnetization, 69, 476, 634
magnetoresistance, 476
magnets, 634
malignant cells, 1539
malignant tumors, 896
maltose, 775
mammal, 1175
mammalian cells, 1514, 1520
manganese, 77, 78, 105
manifolds, 943, 949
manipulation, 121, 346, 515, 533, 538, 763, 951
mannitol, 142
mantle, 339, 340, 341, 342, 343, 351, 353, 372, 374, 375, 376, 377, 378, 380
manufacturer, 1172
manufacturing, xiv, xvii, 237, 238, 250, 251, 254, 258, 274, 460, 538, 539, 540, 951, 1392
mapping, 1416, 1505
market, 1, 146
Marx, 1425
Maryland, 543
MAS, 350, 385, 386
masking, xiv, 237, 238, 254, 274, 477, 478
mass, xi, 37, 49, 50, 51, 61, 68, 866, 876, 963, 965, 987
mass loss, xxxi, 599, 965, 1479, 1486, 1487, 1493, 1495, 1496
mass spectrometry, 130, 876, 1480
mastery, 1
material sciences, xx, 687, 942
materials science, xv, xxiii, 329, 390, 538, 905, 1455

mathematical methods, 612
meanings, 711
measurement, xxx, 107, 182, 273, 279, 281, 298, 403, 405, 406, 413, 423, 494, 497, 520, 577, 646, 689, 773, 796, 798, 800, 850, 854, 962, 1160, 1164, 1171, 1179, 1401, 1402, 1404, 1410, 1418, 1419, 1446, 1490, 1519, 1559
measures, 395, 577, 690
mechanical behavior, xxi, 690, 737, 1433, 1434
mechanical properties, xx, xxiii, xxxi, 8, 144, 532, 654, 657, 660, 687, 688, 689, 702, 747, 810, 941, 942, 953, 954, 957, 960, 1427, 1428, 1429, 1430, 1431, 1432, 1433, 1434, 1435, 1436, 1437, 1438, 1441, 1450, 1468, 1492
mechanical stress, 909
mechanical testing, 1468, 1474, 1475
media, xiii, 2, 72, 113, 148, 151, 153, 174, 332, 337, 339, 340, 341, 342, 343, 345, 353, 382, 390, 403, 671, 682, 754, 768, 780, 906, 1554
mediation, 548
medicine, xv, xvi, xviii, xxiii, 3, 5, 307, 393, 527, 777, 905, 909, 1504
melamine, 142, 191
melt, xxiii, 484, 530, 762, 910, 959, 962, 973, 981, 982, 1444, 1445, 1468, 1474, 1481, 1497
melting, 18, 442, 469, 604, 619, 636, 637, 638, 639, 647, 648, 649, 651, 652, 659, 723, 743, 813, 823, 876, 906, 963, 974, 975, 1450, 1469
melting temperature, 636, 637, 638, 639, 647, 652, 659, 723, 823, 975, 1450
melts, 164, 466, 638
membranes, 151, 169, 181, 185, 221, 239, 330, 346, 460, 465, 468, 469, 470, 548, 915
memory, xxii, 1, 3, 261, 453, 454, 476, 587, 673, 704, 845, 846
menadione, 1519
Mendeleev, 1441
mercury, 91, 778, 923, 1506
messages, 701
metabolism, 148
metabolites, 772
metal extraction, 493
metal ions, 484, 760, 778, 841
metal nanoparticles, xxxii, 151, 153, 154, 381, 382, 385, 389, 410, 548, 636, 647, 659, 811, 840, 841, 910, 938, 1500, 1501, 1506, 1521, 1527, 1530, 1532, 1534, 1541, 1547
metal organic chemical vapor deposition, 512
metal oxide, 309, 331, 367, 369, 370, 371, 824, 910, 1546, 1557, 1559
metal oxides, 309, 331, 367, 369, 370, 371, 824, 910
metal salts, 331, 667, 668, 670, 672, 673, 676, 680, 681, 909
metallocenes, 811
metallurgy, 639, 649, 659
metal-oxide-semiconductor, 450
metals, xii, xix, 9, 37, 39, 54, 65, 66, 159, 160, 212, 221, 309, 382, 397, 421, 461, 462, 487, 492, 534, 563, 633, 634, 635, 636, 637, 644, 647, 648, 651, 652, 653, 654, 655, 656, 657, 658, 659, 660, 666, 675, 688, 694, 698, 702, 716, 725, 739, 740, 742, 744, 746, 747, 748, 781, 813, 821, 823, 866, 922, 1430, 1432, 1435
metal-semiconductor, 256
methacrylates, 144
methane, 126, 811, 838, 1547
methanol, 70, 167, 766, 842, 908, 923, 924, 925, 1551, 1558
methionine, 388, 777
methyl group, 625, 626, 627, 628, 629, 630
methyl groups, 58, 59, 60, 625, 626, 627, 628
methylene, 124, 751, 764, 1431
MFI, 1481
Mg^{2+}, 1442
mice, 133, 135, 140, 146, 147, 190, 1515
micelle formation, 388
micelles, 116, 166, 167, 387, 759, 1503, 1515, 1522
microcavity, 239
microcutting, 39
microdialysates, 773
microelectronics, 239, 393, 535, 583
microemulsion, 548
microenvironment, 184, 1509
microgels, 1511, 1514
micrometer, xvi, 4, 9, 14, 89, 394, 407, 440, 459, 461, 649, 976
micronucleus, 1515
micropatterns, 70
microscope, xviii, xx, 171, 184, 238, 271, 403, 431, 438, 532, 535, 561, 563, 590, 665, 666, 771, 856, 923, 963, 973, 1161, 1162, 1165, 1172, 1174, 1175, 1482, 1510, 1537
microscopy, xii, xviii, xix, xxi, 3, 18, 37, 55, 61, 141, 176, 503, 532, 538, 540, 547, 554, 623, 624, 631, 687, 751, 793, 810, 817, 840, 923, 962, 1510, 1516, 1517, 1520
microspheres, 2, 548
microstructure, xxiv, 2, 24, 359, 565, 590, 592, 635, 657, 659, 672, 716, 739, 908, 960, 980, 982, 985
microstructures, xx, 312, 324, 466, 534, 634, 652, 653, 654, 660, 907, 908, 909
microtubes, xviii, 501, 503, 505, 506
microtubules, 836, 837, 908, 909, 935
microwave, xix, 261, 398, 401, 423, 476, 582, 592, 594, 615, 616, 618, 619, 785, 812, 839, 909
microwaves, 682

migration, 159, 160, 162, 278, 640, 657, 724, 982, 1405, 1406, 1407, 1408, 1409
military, 278
minerals, 538
miniaturization, xii, 85, 394, 583, 635, 752
mining, 382
Ministry of Education, 113, 232, 633, 660, 935
minority, 851
MIP, 144
misfit dislocations, 281, 741
Missouri, 527
mixing, xxiii, 9, 96, 123, 301, 398, 399, 400, 669, 761, 762, 767, 773, 793, 907, 922, 959, 963, 973, 981, 982, 1481
mobility, xiv, xvii, xxx, 2, 91, 237, 261, 269, 395, 427, 446, 454, 455, 456, 491, 497, 498, 723, 966, 983, 985, 1401, 1402, 1404, 1410, 1411, 1412, 1418, 1419, 1424, 1429, 1474
MOCVD, 289, 430, 431, 432, 438, 448, 449, 562, 675
mode, 21, 67, 74, 104, 566, 852, 857, 859, 860, 863, 871, 877, 962, 963, 966
model system, 88, 792
modeling, 3, 702, 703, 737, 744, 745, 748, 880
models, xii, xiii, xviii, xx, 20, 38, 61, 130, 151, 201, 203, 204, 205, 208, 239, 414, 415, 417, 418, 421, 446, 501, 514, 535, 633, 635, 640, 646, 660, 691, 694, 702, 703, 704, 737, 742, 907, 1163, 1458, 1459, 1490, 1491, 1492
modulus, xx, xxxi, 46, 59, 229, 641, 642, 654, 657, 658, 660, 687, 740, 954, 955, 956, 957, 1450, 1452, 1459, 1461, 1468, 1470, 1473, 1474, 1475, 1476, 1548
moieties, 118, 121, 127, 133, 138, 139, 140, 141, 143, 157, 164, 171, 185, 773, 777
moisture, 331, 593, 603, 968, 1424, 1450, 1557
mold, 913, 962
moldings, 962
mole, 204, 215, 217, 218, 223, 226, 227, 234, 636, 638, 926, 927
molecular beam, 461, 675
molecular beam epitaxy, 461, 675
molecular dynamics, 117, 164, 639, 692, 699, 704, 721, 742, 743, 744, 745, 746, 747, 748, 953
molecular mass, 133, 151, 192
molecular oxygen, 9, 360, 492, 792
molecular structure, 3, 66, 148, 162, 532, 1403, 1458
molecular weight, xiii, 66, 114, 116, 120, 121, 137, 139, 146, 147, 151, 397, 467, 679, 792, 793, 794, 795, 926, 927, 928, 930, 931, 961, 970, 980, 1403, 1436
molecular weight distribution, 120
molybdenum, 160, 161
MOM, 1482
momentum, 693, 696, 705, 706, 707, 709
monitoring, 44
monoclonal antibody, 137, 1505
monolayer, xxiv, 45, 57, 129, 154, 179, 183, 343, 370, 423, 548, 566, 765, 793, 959, 969, 970, 971, 1484, 1534
monomer, xiii, 2, 113, 116, 137, 139, 762, 769, 781, 925, 926, 927, 930, 931, 939, 1444, 1445
monomers, xiii, 113, 115, 728, 769, 770, 781, 922, 1445
Monte Carlo, 691
morbidity, 1500
morphine, 768
mortality, 1500
mosaic, 909
Moscow, 895, 1441, 1453, 1457, 1465, 1467, 1472, 1473, 1476, 1479, 1481, 1497
motion, xi, 37, 38, 39, 49, 50, 51, 57, 60, 61, 76, 330, 414, 416, 625, 626, 630, 657, 689, 691, 695, 696, 697, 702, 703, 704, 717, 737, 741, 972
motivation, 1554
movement, 405, 414, 417, 421, 492, 604, 607, 626, 689, 716
MOVPE, 277, 278, 279, 283, 285, 289, 291, 292, 302, 303
MRI, 133, 135, 634, 1505, 1513, 1516, 1517
mRNA, 142
MTS, 695, 731
multilayer films, 179, 763, 764, 780, 788, 937
multilayered structure, 460
multiplication, 691
multiwalled carbon nanotubes, xxi, xxiii, 422, 809, 841, 842, 921, 923, 925, 935, 936, 938, 1511
mutation, 337
myocardial infarction, 786
myoglobin, 184, 751, 790, 791

N

Na^+, 153, 271, 484, 1442, 1446, 1468, 1469, 1474, 1481, 1533
Na2SO4, 484, 676
NaCl, 46, 605, 762, 775, 784, 796
NAD, 751, 787, 788, 789, 1508, 1519
NADH, 751, 754, 764, 767, 770, 787, 788, 789, 1508, 1519
nanobelts, xii, 85, 86, 89, 91, 94, 95, 99, 100, 104, 105, 108, 225, 226, 227, 229, 230, 353, 502, 504, 506, 507, 508, 512, 517, 1522
nanocomb, 512
nanocomposites, xxxi, xxxii, 4, 38, 127, 330, 371, 372, 377, 378, 379, 380, 381, 390, 836, 915, 922,

923, 933, 935, 936, 937, 960, 972, 975, 976, 977, 978, 980, 982, 986, 1402, 1424, 1427, 1428, 1441, 1442, 1443, 1444, 1445, 1446, 1450, 1452, 1453, 1457, 1465, 1467, 1468, 1469, 1470, 1471, 1472, 1473, 1474, 1475, 1476, 1479, 1480, 1481, 1482, 1483, 1493, 1496, 1499, 1500, 1511
nanocrystal, xxx, 548, 914, 938, 1401, 1402, 1403, 1412, 1413, 1416, 1418, 1424, 1505
nanocrystalline metals, 634, 640, 643, 644, 645, 647, 688, 725, 739, 742, 747
nanocrystals, xii, xiv, xxi, xxiii, 8, 65, 67, 86, 105, 107, 153, 237, 309, 311, 312, 318, 321, 388, 534, 582, 679, 688, 691, 715, 716, 721, 737, 910, 913, 914, 921, 925, 935, 1401, 1402, 1412, 1506, 1507, 1517, 1531
nano-crystals, 238
nanocubes, 373, 375, 376, 378, 1521
nanodevices, xxiii, 1, 140, 190, 371, 909, 917, 921, 922, 923
nanodots, xvii, 16, 17, 19, 25, 501, 915
nanoelectronics, 823
nanofabrication, 330, 715
nanofibers, xi, xxxii, 2, 3, 4, 7, 10, 14, 15, 16, 860, 1545, 1546, 1557, 1560
nanofilm, xviii, 547, 551, 552, 555
nanofluids, xvi, 393, 394, 395, 397, 398, 399, 400, 401, 402, 403, 404, 405, 406, 407, 408, 409, 410, 411, 412, 413, 414, 415, 416, 417, 418, 419, 420, 421, 422, 423, 424, 425, 426
nanoimprint, 460
nanoindentation, 467, 474, 689, 690, 716, 717, 718, 720, 730, 740, 743, 746
nanolayers, 465, 1486
nanolithography, xvii, 1, 429, 460, 474, 478, 548
nanomaterials, xi, xv, xvi, xvii, xx, xxi, xxiii, xxxii, 7, 9, 10, 12, 14, 24, 27, 28, 31, 66, 79, 100, 307, 308, 311, 324, 329, 330, 349, 353, 361, 367, 378, 380, 382, 390, 459, 515, 538, 634, 635, 659, 660, 666, 752, 780, 801, 905, 906, 917, 921, 922, 923, 960, 1392, 1393, 1402, 1499, 1501, 1504, 1506, 1510, 1518, 1521, 1541, 1545, 1546, 1557, 1559, 1560
nanomechanics, xxi, 687, 738, 741
nanometer, xiii, xx, 113, 202, 213, 226, 345, 370, 381, 394, 430, 460, 462, 594, 634, 639, 687, 689, 691, 739, 746, 761, 823, 912
nanometer scale, xiii, xx, 113, 202, 213, 226, 370, 594, 639, 687, 739
nanometers, 2, 203, 415, 504, 512, 533, 602, 645, 688, 691, 912, 1445, 1546
nanonails, xviii, 501, 508, 511, 512, 513, 514, 520, 521
nanoparticle synthesis, 403, 423
nanoribbons, xii, xviii, 69, 85, 86, 89, 91, 92, 94, 95, 96, 97, 100, 101, 108, 501
nanoscale materials, xvii, 501, 515, 634, 635, 792
nanoscale structures, 38, 55, 960
nanostructure, 71, 72, 74, 88, 91
nanostructured materials, xi, xv, xxxii, 7, 28, 31, 86, 166, 307, 635, 647, 652, 688, 780, 1545
nanosystems, 515, 532, 539, 744
Nanowires, 503
naphthalene, 124
National Science Foundation, 232
NATO, 32
NCS, 75, 76
Nd, 310, 311, 312, 435, 637, 638, 649, 650, 652, 653, 654, 655, 656, 657, 658, 1404, 1540
needs, 18, 90, 854
neoangiogenesis, 1513
nerve, 178, 790, 1545
Netherlands, 32, 621, 989
network, xi, 7, 19, 26, 27, 31, 77, 80, 147, 156, 175, 179, 192, 239, 314, 315, 334, 347, 379, 528, 530, 672, 679, 762, 774, 782, 796, 864, 865, 866, 949, 950, 951, 969, 1392, 1394, 1419, 1421, 1424, 1458, 1468, 1552
neurodegenerative disorders, 164
neurotransmitters, 775
new media, 789
New York, 234, 421, 489, 541, 542, 543, 544, 545, 622, 660, 738, 743, 748, 801, 1440, 1453, 1455, 1517, 1543, 1561
Newton, 737
next generation, xvi, xxii, 393, 845, 846
nickel, 153, 155, 156, 403, 506, 669, 723, 739, 759, 777, 811, 821, 840, 843, 855, 1429, 1430, 1432, 1433, 1437
nicotinamide, 751, 754
NIR, 484, 563, 566, 1505, 1507, 1528, 1534, 1535, 1538, 1539, 1540
nitrate, xix, 157, 271, 345, 382, 389, 581, 595, 597, 599, 601, 602, 607, 609, 619, 1524, 1525, 1526
nitrates, 563, 599, 601, 602, 603
nitric acid, 412, 444, 754, 755, 1525
nitric oxide, 185, 764, 792, 800, 1510, 1520
nitrides, 86, 428, 1442
nitrogen, xxvii, 9, 77, 159, 279, 283, 314, 320, 343, 370, 437, 442, 445, 446, 472, 551, 589, 792, 910, 923, 938, 962, 963, 1157, 1162, 1164, 1165, 1168, 1169, 1179, 1484, 1529, 1550, 1559
nitrogen gas, 279, 283, 1164, 1165, 1168, 1169, 1179
NMR, xxiii, xxiv, 117, 129, 164, 189, 350, 351, 385, 386, 537, 921, 923, 930, 959, 962, 964, 966, 967, 968, 969, 970, 971, 972, 1505

noble metals, xvi, 155, 329, 382, 781
nodes, 706, 708, 1474
noise, 53, 624, 625, 709, 751, 771, 772, 773, 778, 783, 1554
nonionic surfactants, xxxi, 1441, 1443
nonlinear optics, 582
norepinephrine, 751, 754
novel materials, xv, 307, 308, 309, 666, 810
novelty, 1
nuclear magnetic resonance, 923, 962
nucleation, xiii, xvi, xx, xxiii, 42, 45, 46, 49, 55, 57, 58, 76, 87, 90, 91, 92, 160, 201, 202, 203, 204, 205, 206, 207, 208, 209, 211, 212, 213, 214, 216, 219, 220, 221, 226, 227, 232, 283, 293, 294, 309, 317, 318, 319, 321, 324, 329, 330, 366, 376, 377, 379, 387, 388, 436, 444, 445, 446, 506, 627, 633, 651, 652, 659, 682, 716, 717, 741, 742, 747, 748, 856, 905, 906, 909, 910, 912, 1487
nuclei, xiii, 201, 202, 204, 207, 209, 210, 213, 214, 215, 221, 222, 226, 227, 230, 231, 283, 286, 288, 291, 294, 296, 298, 341, 351, 365, 366, 373, 376, 377, 388, 389, 390, 445, 446, 692, 912, 915
nucleic acid, 798, 800
nucleus, 204, 206, 207, 210, 221, 222, 223, 224, 226, 227, 321, 375, 376, 540, 1507, 1518

O

observations, 15, 16, 20, 92, 93, 94, 159, 164, 227, 231, 314, 335, 399, 403, 506, 512, 564, 594, 647, 723, 725, 850, 854, 856, 929, 933, 973, 975, 983, 1501, 1503, 1504, 1507, 1525
ODN, 142
oil, 393, 394, 397, 398, 399, 400, 403, 408, 410, 411, 412, 415, 421, 548, 564, 767, 789
olefins, 8, 9
oligomers, 155, 169
one dimension, 74, 960, 1548, 1554
operator, 696, 697
optical coatings, 439
optical density, 1405, 1415
optical microscopy, 973, 976, 1517, 1539
optical properties, xxi, xxxii, 10, 27, 160, 181, 238, 239, 382, 484, 492, 498, 503, 521, 562, 752, 1158, 1160, 1416, 1521, 1522, 1527, 1535
optics, xv, 5, 307, 332, 535, 548
optimization, 2, 562, 744, 986, 1424, 1547
optoelectronic, xvii, xxi, 501, 502, 515, 738, 752, 770
optoelectronic devices, xv, 274, 277, 278, 439, 515, 770, 1158
optoelectronic properties, 738
optoelectronics, xii, 65, 80, 239, 502
orbit, 104, 346
ores, 718
organ, 66, 161, 194, 675, 762, 1158, 1441, 1442, 1443, 1444, 1445, 1446, 1447, 1448, 1449, 1450, 1451, 1452, 1453, 1481, 1500
organelles, 1174, 1179
organic compounds, xxx, 184, 1427, 1428, 1432, 1433, 1434, 1435, 1436, 1439, 1440, 1558
organic polymers, 314, 1401, 1554
organic solvent, xvii, 130, 132, 153, 154, 155, 459, 464, 598, 600, 907, 915
organic solvents, xvii, 130, 132, 154, 155, 459, 598, 600, 867, 907
organic thin films, xxx, 1391, 1392
organization, xvii, 56, 170, 183, 196, 371, 387, 459, 506, 533, 715, 746, 911, 912, 964, 972, 976, 1532, 1534
organometallic, 161, 194, 762, 1158
orientation, xxi, 57, 58, 68, 71, 72, 73, 76, 80, 93, 308, 317, 364, 366, 431, 497, 516, 556, 563, 565, 567, 568, 573, 575, 579, 614, 625, 628, 724, 731, 765, 793, 801, 810, 908, 976, 1392, 1394, 1413, 1429, 1469, 1527
oscillation, 52, 406, 413, 1420, 1526, 1527
oscillograph, 1435
overload, 157
oviduct, 330
oxalate, 77, 582, 583
oxidation rate, 19
oxide, xvii, 501, 502, 505, 506, 815, 838, 1436
oxide clusters, 351
oxide nanoparticles, 407, 421, 1501, 1514, 1518
oxides, xii, xvi, 31, 39, 65, 77, 86, 151, 213, 309, 329, 330, 331, 366, 371, 397, 421, 462, 562, 582, 594, 604, 634, 666, 817, 824, 866, 1167, 1442
oxygen plasma, 477, 497
oxygen sensors, 146
oyster, 330
ozone, 1391

P

PAA, 922, 1533
packaging, xx, 58, 629, 687, 966
paclitaxel, 138, 1515
paints, 8
palisade layer, 330
palladium, 159, 366, 367, 368, 378, 379, 380, 381, 723, 781, 786
pancreas, 147
parallel algorithm, 699, 743, 744
parallel cluster, 699
parallelization, 744

parameter, 27, 74, 102, 164, 358, 359, 407, 417, 466, 510, 565, 569, 570, 574, 585, 586, 588, 591, 600, 609, 611, 612, 613, 615, 619, 640, 642, 690, 697, 698, 702, 714, 858, 897, 979, 982, 1431, 1432, 1433, 1436, 1475, 1493, 1496, 1549
parasites, 149, 1514
particle collisions, 76
particle mass, 1459
particle-cluster, 1458, 1459
particulate-filled composites, 1467
passivation, 452, 456, 689, 778, 852, 1163, 1164
passive, 140, 484, 847, 866
patents, 538, 689
pathogenic, 798
pathogens, 151
pathways, xxx, 162, 176, 436, 866, 1392, 1393, 1401, 1402, 1413
patterning, xvii, 254, 460, 474, 477, 478
PCP, 937
PEHD, 1428, 1435, 1436, 1437, 1438, 1439
penicillin, 148, 1160, 1171
peptide chain, 339, 351
peptides, 130, 136, 146, 149, 150, 151, 165, 192, 366, 381, 382, 910, 914, 940
percolation, 343, 865, 1393, 1412, 1469, 1554
percolation theory, 343
performance, xxi, xxxi, 712, 750, 752, 772, 780, 782, 783, 786, 789, 825, 826, 828, 829, 830, 841, 1450, 1479
periodicity, 41, 44, 624, 626, 628, 629
permeability, 140, 146, 760, 786
permit, 460, 472, 760
permittivity, 615, 961, 962, 963, 977
perovskite, xix, 581, 582, 583, 588, 592, 593, 594, 595, 596, 597, 598, 603, 604, 609, 610, 615, 618, 619
perovskite oxide, 582, 592
peroxide, 183, 759, 780, 782, 783, 784, 787, 790, 793, 794, 795, 797
perylene, 118
pesticide, 183, 790
pesticides, 182
PET, 1444, 1447, 1448, 1449, 1450, 1451
pH, 30, 31, 43, 44, 72, 164, 165, 168, 171, 179, 183, 195, 198, 332, 333, 336, 337, 338, 340, 341, 342, 343, 344, 345, 349, 353, 366, 378, 381, 383, 387, 390, 398, 400, 404, 405, 408, 467, 549, 557, 596, 607, 760, 769, 770, 773, 775, 776, 777, 778, 779, 781, 783, 787, 788, 789, 791, 792, 793, 794, 795, 796, 800, 896, 906, 909, 924, 936, 937, 1159, 1168, 1171, 1514, 1517, 1525, 1533
pharmaceutical, xxi, 752
pharmaceuticals, 538
pharmacokinetics, xiii, 114, 139, 142, 146, 148, 190, 1513, 1514
pharmacology, 777, 779, 1500, 1502
phase decomposition, 583, 1496
phase diagram, 204, 207, 209, 212, 215, 218, 220, 435, 466, 605, 636
phase transformation, 106, 203, 214, 226, 332, 341, 469, 598, 640, 655
phase transitions, 49, 587, 598, 635
PHE-Gr, 1467, 1469
phenol, 777
phenolphthalein, 1428, 1429, 1430, 1431
phenylalanine, 777
phonons, 104, 377, 498, 709, 711, 737, 871, 1408
phosphate, 750, 751, 759, 762, 764, 769, 773, 788, 792, 795, 796
phosphatidylcholine, 1539
phospholipids, 1536
phosphorus, xxx, 124, 125, 147, 1427, 1428, 1429, 1430, 1431, 1432, 1433, 1439, 1440
photobleaching, xxvii, 126, 915, 1157, 1175, 1176, 1177, 1178, 1180, 1505, 1538
photocatalysis, xvi, 8, 9, 10, 86, 329, 332, 1515
photocatalysts, 8, 9, 151
photodetectors, 239, 428
photoelectron spectroscopy, 752, 817
photoemission, 869
photographs, 571, 572
photolithography, 533, 812, 853, 868, 1551, 1558
photoluminescence, xii, xiv, xv, xviii, xx, xxx, 3, 85, 86, 96, 105, 127, 171, 237, 238, 254, 264, 277, 278, 279, 300, 373, 374, 442, 443, 448, 492, 494, 495, 496, 501, 517, 518, 519, 562, 665, 1158, 1401, 1402, 1403, 1405, 1406, 1407, 1408, 1414, 1416
photoluminescence spectra, 300, 374, 1405, 1406, 1414
photomicrographs, 1447, 1448, 1449
photonic crystals, 2, 3, 239, 1393, 1396, 1511
photonic devices, xvii, 501
photonics, 378, 456, 535, 1529, 1540
photons, 97, 173, 178, 198, 447, 1407
photooxidation, xxvi, 1157, 1162, 1164, 1165, 1166, 1168, 1169, 1171, 1178, 1180
photovoltaic cells, 8, 10
photovoltaic devices, xxx, 264, 345, 1401, 1402, 1413, 1414, 1422
physical and mechanical properties, xx, 633, 635, 647, 654, 658, 660
physical mechanisms, 232
physical properties, xvi, 1, 4, 86, 159, 309, 329, 332, 530, 635, 654, 660, 725, 812, 972, 1418, 1554
physical treatments, 754

physical-mechanical properties, 1435, 1436
physicochemical properties, xviii, 147, 149, 547, 558
physics, 1, 50, 85, 178, 202, 213, 232, 838, 942, 1440
piezoelectric, 502, 751, 800, 1435
piezoelectricity, 502, 753
pigments, 634, 1442
pitch, 431, 907
PL spectrum, xvii, 103, 278, 294, 295, 303, 491, 494, 497, 1160, 1408, 1417
planets, 1545
plants, 330, 382, 778
plasma, 73, 146, 148, 437, 447, 475, 477, 492, 497, 562, 648, 651, 680, 771, 772, 785, 812, 839, 846, 855, 858, 859, 860, 861, 862, 863, 864, 865, 866, 1172, 1503, 1547, 1559
plasma membrane, 1172, 1503
plasmid, 165, 1172, 1539
plastic deformation, 649, 688, 698, 711, 722, 746, 747, 813, 1430, 1432
plasticity, 688, 690, 716, 722, 730, 737, 741, 746, 1433, 1434, 1435
plasticization, 1434
plastics, 634
platelets, 363, 436, 437, 1445, 1517
platinum, 12, 160, 182, 242, 264, 406, 471, 493, 550, 759, 781, 824, 841, 842, 963, 1522, 1559
plurality, 480
PMMA, 467, 477, 867, 922, 1444, 1554
point defects, 104, 492, 498, 562
poison, 874
Poland, 1479
polar groups, 757, 981, 984, 985, 986, 1436
polarity, 172, 256, 675, 762, 852, 982, 983, 1404, 1429, 1430, 1432, 1434, 1435, 1436
polarization, 415, 417, 502, 577, 692, 826, 829, 962, 986
polarized light, 168
polarized light microscopy, 975
pollen, 330
pollutants, 27, 1516
pollution, 254, 1392
poly(2-hydroxyethyl methacrylate), 922
poly(methyl methacrylate), 922, 936
poly(vinylpyrrolidone), 1534
polyacrylamide, 1513
polyamide, 1445
polycarbonate, 465, 858, 909, 1428, 1431, 1432, 1433, 1435, 1438, 1439
polycondensation, 341, 1403
polydispersity, 931
polyelectrolytes, 763
polyesters, xxx, 1427, 1431
polyether, 138, 165
polyethylene, xxx, 1427, 1435, 1436, 1438, 1439, 1450, 1454, 1459, 1468, 1474, 1480
polyimide, 922
polymer blends, 1440
polymer chains, 3, 76, 115, 397, 922, 929, 933, 1407, 1413, 1417, 1443, 1446, 1450
polymer combustion, 1497
polymer composites, 2, 982, 1440, 1442, 1443, 1457, 1458
polymer electrolytes, 1394
polymer film, 770, 783, 1396, 1532, 1554
polymer films, 770, 783, 1396, 1532
polymer materials, xxxi, 982, 1441, 1442, 1444, 1450, 1452, 1453, 1454
polymer matrix, 973, 974, 976, 984, 1412, 1413, 1419, 1424, 1442, 1445, 1496
polymer media, 1519
polymer molecule, 396, 397, 1481, 1486
polymer nanocomposites, xxiii, 171, 836, 959, 986, 1442, 1444, 1446, 1450, 1480, 1496
polymer structure, 55, 1460, 1482
polymeric composites, 1497
polymeric films, 127
polymeric materials, 54, 782, 1427, 1428, 1430, 1433, 1435, 1440
polymeric melt, 1497
polymerization, xxiii, 3, 115, 116, 120, 137, 142, 769, 921, 922, 924, 925, 927, 930, 931, 936, 937, 939, 1444, 1445, 1560
polymerization kinetics, xxiii, 921
polymerization process, 930, 931
polymerization processes, 930
polymerization time, 930
polymers, xii, xviii, xxxi, 1, 2, 3, 4, 5, 37, 66, 74, 77, 79, 80, 114, 116, 120, 127, 137, 142, 146, 151, 159, 164, 165, 167, 186, 387, 397, 465, 466, 527, 535, 538, 551, 758, 769, 780, 782, 922, 923, 925, 927, 929, 933, 935, 939, 940, 980, 982, 985, 1394, 1396, 1413, 1427, 1428, 1429, 1430, 1431, 1433, 1434, 1440, 1446, 1465, 1469, 1472, 1474, 1476, 1482, 1497, 1502, 1554, 1557
polymethylmethacrylate, 2, 467, 477, 867, 1554
polyolefin, 1427, 1436
polyp, 155, 159
polypeptide, 164, 341, 351, 387, 792, 794, 914
polypeptides, 339, 341, 351
polypropylene, xxxi, 1453, 1467, 1468, 1472, 1473, 1474, 1476, 1479, 1480, 1481, 1492, 1495, 1496
polysaccharide, 760
polystyrene, 2, 116, 118, 467, 867, 922, 935, 1393, 1531
polyurethanes, 116, 1502

polyvinylpyrrolidone, 400
poor, 9, 72, 239, 244, 254, 264, 273, 274, 419, 420, 461, 693, 761, 775, 782, 783, 847, 878, 981, 1402, 1501, 1555
population, 360
porosity, xv, 238, 249, 250, 269, 575
porphyrins, 172
portability, 798
potassium, 961, 1429, 1430, 1432, 1433, 1436, 1438, 1439
potential energy, 396, 709, 732, 947
power, xv, 107, 108, 119, 127, 238, 255, 269, 271, 273, 274, 308, 311, 315, 317, 319, 321, 322, 407, 418, 427, 448, 449, 476, 538, 565, 615, 624, 659, 675, 688, 692, 716, 738, 760, 824, 842, 850, 858, 860, 861, 862, 864, 865, 961, 1162, 1165, 1166, 1168, 1175, 1176, 1177, 1178, 1180, 1403, 1413, 1420, 1424, 1435, 1523, 1546, 1549
precancerous lesions, 1507
precipitation, xxii, 121, 208, 342, 353, 400, 582, 584, 588, 593, 823, 895, 896, 897, 908, 1158
prediction, 209, 212, 214, 312, 416, 536, 638, 639, 652, 810
preference, 158, 1529
preparation, xxiv, 9, 56, 67, 71, 562, 563, 564, 666, 680, 682, 856, 875, 959, 971, 973
pressure, xiv, xv, 10, 18, 42, 44, 46, 50, 96, 203, 204, 208, 211, 212, 214, 215, 217, 218, 219, 223, 224, 225, 226, 230, 238, 271, 273, 274, 279, 307, 308, 309, 311, 343, 370, 417, 430, 431, 433, 435, 436, 438, 442, 444, 464, 503, 505, 531, 562, 583, 584, 604, 641, 642, 643, 647, 648, 651, 652, 690, 746, 762, 813, 833, 834, 854, 856, 859, 863, 875, 879, 925, 962, 1429, 1436, 1439, 1556
prevention, 1500, 1507
principle, xii, 65, 66, 80, 95, 99
prior knowledge, 359, 738
probability, 39, 49, 50, 55, 58, 211, 293, 360, 494, 657, 979, 1170, 1435
probe, xii, xviii, xix, 1, 3, 12, 27, 37, 38, 40, 41, 44, 46, 48, 49, 52, 53, 54, 55, 56, 57, 58, 59, 60, 124, 135, 165, 181, 195, 532, 535, 537, 540, 561, 563, 623, 624, 625, 626, 627, 628, 629, 630, 631, 689, 716, 753, 798, 800, 810, 819, 823, 837, 840, 856, 961, 1482, 1505, 1510, 1522, 1536, 1537
prodrugs, 148, 149
production, 8, 9, 10, 41, 43, 49, 52, 59, 60, 68, 76, 146, 147, 271, 324, 363, 393, 498, 538, 541, 592, 648, 649, 682, 744, 755, 770, 795, 810, 811, 830, 837, 840, 841, 866, 876, 912, 1525, 1547
production technology, 830
productivity, 838
program, 167, 465, 744, 1486
pro-inflammatory, 144
prokaryotes, 382
proliferation, 4, 38, 151, 1500, 1508, 1511, 1519
promote, 538, 761, 767, 793, 1433
promoter, 121, 799
propagation, 702, 703, 705, 724, 727, 729, 731, 743, 745, 747, 787, 1480
propagators, 743
proportionality, 1411
propranolol, 149, 192
propylene, 132, 133, 168, 171, 189
protective coating, 319, 763, 786
protein, xxvii, xxxii, 4, 129, 135, 139, 140, 163, 165, 181, 183, 184, 199, 331, 333, 339, 341, 342, 343, 345, 353, 382, 384, 385, 387, 389, 751, 762, 764, 767, 789, 791, 792, 793, 794, 798, 909, 914, 937, 1157, 1158, 1172, 1502, 1507, 1514, 1521, 1529, 1536, 1538, 1539, 1541
protein blocks, 389
protein conformations, 342, 353
protein films, 184, 199
protein folding, 49
protein immobilization, 937
protein sequence, 4
protein structure, 345
proteinase, 914
proteins, xvi, xxi, 3, 129, 150, 163, 166, 179, 181, 184, 188, 198, 329, 330, 332, 338, 339, 342, 349, 353, 366, 371, 375, 381, 382, 385, 387, 389, 743, 752, 757, 766, 790, 791, 792, 797, 801, 940, 1172, 1502, 1507, 1536
proteoglycans, 145, 342, 353
protocol, 583, 764, 1172, 1524
proton exchange membrane, 841, 842
protons, 824
prototype, 453, 456
PSD, 565, 566, 567, 574, 575
Pseudomonas aeruginosa, 4
pulse, 309, 310, 311, 312, 313, 315, 316, 317, 320, 323, 406, 750, 770, 962, 964, 1403
pulses, 811, 837, 1547
pure water, 11, 418, 419, 420, 464, 862
purification, xxii, 8, 31, 152, 308, 309, 312, 666, 753, 757, 767, 817, 819, 833, 838, 839, 845, 856, 862, 876, 878, 879, 880, 936, 1547
PVA, 922
PVP, 73, 398, 399, 400, 402, 761, 1534, 1535
PVS, 180, 181, 182
pyrene, 176, 1508, 1519
pyrimidine, 151
pyrolysis, xviii, xxi, 69, 561, 568, 571, 579, 647, 809, 810, 811, 815, 818, 819, 820, 821, 822, 825,

826, 827, 828, 829, 830, 831, 832, 834, 835, 837, 838, 843, 1440, 1480, 1493, 1497
pyrolytic graphite, 184, 749, 750, 751, 753, 754, 755, 766, 772

Q

quality control, 146
quantitative estimation, 1460, 1467, 1469
quantum confinement, xxvi, 87, 242, 262, 269, 301, 656, 1157, 1158, 1160, 1163, 1405
quantum dot, xv, 1, 3, 171, 197, 277, 278, 279, 377, 738, 922, 938, 1158, 1500, 1505, 1506, 1507, 1516, 1518
quantum dots, xv, 3, 171, 197, 277, 278, 377, 738, 922, 938, 1158, 1500, 1505, 1506, 1507, 1516, 1518
quantum mechanics, 691, 692, 703, 704
quantum well, xv, 277, 278, 302
quantum yields, 9, 121, 156, 172
quantum-chemical calculations, 178
quartz, 91, 184, 312, 433, 440, 441, 442, 444, 503, 751, 785, 800, 811, 815, 816, 817, 859, 863, 1535, 1536
quasi-equilibrium, 50, 730
quinone, 128, 385, 756

R

radial distribution, 321, 699
radiation, 12, 67, 389, 428, 465, 597, 666, 672, 748, 814, 897, 1428, 1482
radiation damage, 465
radical polymerization, 115, 119, 923, 924, 925, 926, 935, 936, 939
radio, 839, 858, 859, 863
radius, 42, 46, 53, 58, 59, 130, 204, 206, 207, 208, 209, 212, 215, 217, 218, 219, 220, 223, 224, 226, 227, 229, 345, 395, 407, 416, 486, 487, 519, 636, 641, 690, 695, 699, 945, 954, 1158, 1458, 1460, 1527
rain, 723
Raman, 503, 755, 763, 817, 819, 820, 822
Raman scattering measurements, 377
Raman spectra, xiii, 17, 85, 104, 354, 357, 359, 361, 369, 377, 755, 763, 820, 822, 859, 860, 861, 863, 864, 870, 871, 872, 874, 877, 879
Raman spectroscopy, 96, 341, 359, 377, 817, 819, 857, 871, 873, 876, 878, 1547
random walk, 242
raw materials, 583, 639
reactant, 49, 373, 390, 1486
reactants, 213, 385, 430, 535, 1487
reaction mechanism, xii, 65, 536, 968, 970
reaction order, 1480
reaction rate, 1166
reaction temperature, 16, 17, 18, 27, 31, 441, 503, 506, 508, 514, 582, 584, 669, 678, 811, 1158, 1159
reaction time, 2, 19, 27, 312, 313, 324, 376, 378, 379, 438, 515, 678, 816, 912, 930, 931
reactive groups, 760
reactive ion, 477
reactive oxygen, 1168, 1503
reactivity, xxi, 114, 117, 354, 488, 536, 634, 752, 756, 761, 809, 812, 823, 829, 1504
reading, 405
reagents, 66, 81, 141, 384, 388, 549, 583, 666, 758, 765, 783, 909, 923
reality, 51
receptors, 138, 143, 150, 163, 166, 1518
reciprocity, 378
recognition, xiii, xxxii, 113, 114, 165, 178, 798, 906, 1499, 1501, 1508, 1519, 1532, 1533, 1536, 1537
recombination, 8, 172, 262, 283, 429, 517, 519, 1393, 1402, 1407, 1409, 1421
reconstruction, 44
recovery, xxxii, 151, 1545, 1550, 1551, 1552, 1553, 1554, 1557, 1560
recrystallization, 469
rectification, 261
recycling, 10
red blood cell, 539, 1517
red blood cells, 1517
red shift, 301, 339, 349, 351, 375, 447, 1537
redistribution, 1434
redox-active, 176, 380, 381, 382, 386, 389, 781, 797
redshift, 302, 448, 1408, 1413, 1417, 1418
reference system, 695, 743
reflectance spectra, 29
reflection, xxxi, 27, 337, 372, 378, 384, 564, 569, 711, 712, 725, 737, 933, 963, 973, 981, 1441, 1482, 1483
reflectivity, 599, 600
refractive index, 8, 485, 541, 577, 578, 1535, 1537
regeneration, 4
regression, 31, 979, 1489, 1492
regression equation, 31
regulation, 4, 902, 1459
regulations, 942, 943, 946, 948, 957
reinforcement, 1458, 1474
relationship, 3, 4, 9, 41, 112, 129, 162, 209, 210, 211, 217, 219, 220, 223, 256, 315, 317, 360, 406, 407, 408, 411, 412, 428, 498, 516, 531, 563, 582, 591, 614, 640, 642, 652, 670, 978, 1177, 1458,

1459, 1460, 1468, 1469, 1470, 1475, 1496, 1552, 1554, 1556
relationships, 230, 354, 359, 535, 647, 1452
relaxation, 50, 51, 55, 127, 579, 625, 639, 692, 714, 721, 962, 1407, 1409, 1416
relaxation process, 1407, 1416
relaxation rate, 50
relevance, 38
reliability, 603, 692, 960, 1436, 1501
remote sensing, 278
renewable energy, 1392
repair, xiii, 3, 114, 144, 145, 146
reparation, xxi, 484, 809, 812, 823
replacement, 77
replication, 331, 333, 335, 337, 347, 390, 910
reputation, 555
Reserpine, 779
residues, 165, 332, 339, 340, 342, 345, 348, 351, 353, 371, 372, 375, 380, 381, 382, 385, 386, 387, 388, 475, 602, 755, 757, 800, 870, 874
resistance, xiv, xxxi, xxxii, 9, 60, 185, 238, 268, 269, 273, 274, 354, 360, 394, 406, 417, 446, 455, 476, 478, 536, 615, 634, 657, 753, 774, 798, 879, 960, 1422, 1423, 1427, 1433, 1443, 1451, 1452, 1454, 1479, 1480, 1493, 1499, 1500, 1501, 1502, 1503, 1504, 1507, 1512, 1513, 1514, 1538, 1549, 1550, 1551, 1552, 1553, 1554, 1555, 1558, 1559, 1560
resolution, xviii, xxi, 3, 12, 38, 42, 100, 225, 271, 317, 374, 441, 501, 535, 591, 649, 687, 819, 856, 868, 869, 873, 912, 914, 929, 963, 1165, 1166, 1414, 1513
respiratory, 772, 793
response time, 782, 786, 1545, 1550, 1551, 1558, 1560
restructuring, 43
retention, 30, 140, 453, 454
rhenium, 160
rhodium, 159
rigidity, 164, 534, 538, 540, 1413, 1424, 1430, 1434
rings, 130, 308, 312, 337, 348, 364, 379, 389, 528, 531, 539, 757, 1409, 1413, 1502, 1531
risk, 772
RNA, 742, 772, 911
rodents, 1515
rods, 3, 19, 48, 66, 67, 72, 74, 202, 232, 320, 335, 540, 1517, 1523, 1529, 1530, 1538
roentgen, 1428
rolling, 510, 688, 950
rotations, 716
roughness, xiv, 20, 238, 269, 377, 419, 472, 565, 572, 573, 574, 575, 576, 868, 1419
Royal Society, 74, 75, 319, 748, 755, 768, 799
rubber, 1458, 1459, 1460, 1461, 1462, 1463, 1464, 1474
rubbers, 1460, 1462
Russia, 544, 1427, 1441, 1453, 1455, 1467, 1473, 1479
ruthenium, 159, 160, 176, 1392
rutile, xi, 7, 8, 10, 11, 12, 15, 16, 17, 18, 19, 20, 24, 27, 28, 29, 31, 332, 337, 341, 345, 348, 354, 356

S

SAD, 225
safety, 146, 245, 279, 1494
salt, xix, 151, 153, 308, 345, 372, 373, 398, 400, 484, 581, 597, 599, 619, 749, 769, 825, 915, 1394, 1429, 1430, 1432, 1435, 1439, 1523
salts, xix, 157, 465, 564, 581, 597, 619, 672, 673, 760, 762, 1429, 1430, 1431, 1432, 1433, 1435, 1436, 1438, 1440
sampling, 383, 738, 748
Samsung, 845
SAP, 563
saponin, 1442
sapphire, 429, 431, 432, 437, 438, 439, 471, 515, 1537, 1539
satellite, 428
saturated hydrocarbons, 528, 537
saturation, 45, 46, 93, 94, 208, 209, 211, 212, 216, 217, 223, 234, 428, 450, 823, 1422
scalability, 699
scaling, 127, 176, 641, 716, 847
scanning calorimetry, 963
scanning electron microscopy, xviii, xx, 12, 96, 225, 503, 547, 584, 590, 665, 750, 751, 817, 1547
scanning tunnelling microscopy, 40, 817
scar tissue, 144, 191
scatter, 565, 690, 712, 979, 982
scattering, xii, xvi, xviii, 65, 104, 130, 178, 198, 225, 261, 369, 403, 429, 446, 456, 459, 498, 541, 547, 550, 553, 555, 657, 702, 722, 748, 819, 974, 981, 986, 1392, 1393, 1396, 1526, 1527, 1528, 1538, 1541
schema, 701
Schottky barriers, 846, 847, 849
science, xx, 531, 538, 687, 692, 702, 740, 762, 810, 841, 1428, 1442, 1455
scientific community, 1
SEA, 1495, 1496
search, xx, 687, 692, 698
searching, 314
second generation, 117, 166
security, 833
sediment, 404

sedimentation, 395, 400, 404, 405, 420
sediments, 312
seed, xviii, 90, 312, 318, 324, 341, 547, 549, 550, 558, 1522, 1523, 1524, 1525, 1526
seeding, 312, 1524
segregation, 301, 466, 467, 534, 637, 646
selected area electron diffraction, 312, 368, 389, 503, 515, 649, 667, 677
selecting, 77, 382
selectivity, 9, 148, 151, 152, 156, 157, 179, 354, 465, 769, 778, 780, 876, 923, 1501, 1512, 1515, 1517, 1545
selenium, 676
self, xi, 8, 30, 55, 57, 61, 66, 68, 69, 104, 107, 667, 866
self-assembly, xiii, xv, xviii, xxiii, 3, 55, 114, 124, 141, 164, 166, 167, 169, 181, 202, 225, 232, 277, 279, 286, 303, 309, 321, 323, 324, 330, 371, 506, 527, 540, 548, 553, 555, 738, 764, 765, 866, 905, 907, 910, 912, 922, 931, 936, 1534
self-ordering, xvii, 459, 467
self-organization, 169, 715, 746, 942, 948
self-similarity, 950
SEM micrographs, 460, 468, 472, 478, 565, 973, 1448
semiconductor, xv, xvi, xxx, 3, 69, 86, 88, 102, 103, 105, 107, 202, 213, 245, 274, 277, 278, 309, 319, 329, 345, 354, 377, 428, 429, 439, 446, 491, 492, 517, 531, 533, 576, 674, 676, 680, 846, 847, 912, 914, 1158, 1160, 1161, 1167, 1171, 1392, 1396, 1401, 1402, 1413, 1501, 1506, 1507, 1516, 1546, 1557, 1558, 1559
semiconductor lasers, xv, 277, 278
semiconductors, 2, 221, 309, 345, 427, 428, 429, 456, 462, 492, 534, 535, 576, 666, 678, 690, 715, 866, 1402, 1554
semi-crystalline polymers, 1469, 1474
semimetals, 462
sensing, xii, xv, xvi, xxxii, 65, 73, 174, 179, 277, 329, 345, 354, 359, 360, 414, 427, 553, 689, 740, 764, 770, 773, 774, 775, 789, 1517, 1518, 1520, 1529, 1535, 1536, 1540, 1545, 1546, 1548, 1549, 1550, 1551, 1553, 1554, 1555, 1557, 1558, 1559, 1560
sensitivity, xxi, xxxii, 164, 178, 181, 183, 354, 355, 356, 360, 361, 553, 555, 690, 702, 752, 769, 772, 774, 780, 782, 783, 785, 786, 789, 798, 801, 974, 1499, 1500, 1508, 1509, 1519, 1536, 1537, 1545, 1550, 1552, 1554, 1557, 1558, 1559, 1560
sensors, xi, xii, xiii, xvi, xviii, xxi, xxxii, 2, 3, 5, 8, 31, 85, 114, 183, 199, 239, 345, 354, 361, 382, 427, 428, 439, 477, 479, 502, 547, 548, 558, 562, 634, 752, 755, 757, 759, 760, 761, 763, 764, 767, 769, 770, 772, 773, 774, 775, 777, 778, 780, 781, 783, 787, 795, 798, 800, 801, 810, 823, 836, 911, 937, 1394, 1510, 1519, 1520, 1537, 1545, 1546, 1554, 1555, 1557, 1558, 1559, 1560
separation, xxx, 3, 50, 53, 172, 176, 430, 464, 515, 640, 750, 753, 760, 773, 794, 795, 796, 875, 940, 1393, 1396, 1397, 1401, 1402, 1406, 1407, 1408, 1413, 1415, 1416, 1419, 1421, 1424, 1445
series, xv, 43, 44, 47, 116, 118, 119, 120, 121, 147, 151, 155, 156, 159, 164, 166, 168, 169, 172, 175, 176, 178, 202, 221, 232, 329, 331, 340, 351, 354, 359, 367, 371, 378, 388, 390, 508, 529, 565, 582, 646, 648, 652, 700, 760, 915, 925, 928, 935, 1422, 1423, 1486, 1508, 1527, 1558
serine, 387
serotonin, 751, 768
serum, 130, 142, 164, 182, 1160, 1171
serum albumin, 130, 182
SFT, 269, 270
shaping, 3
shares, 1436
sharing, 79, 322, 1433, 1546
shear, 38, 44, 46, 52, 53, 58, 398, 400, 417, 419, 693, 731, 732, 973, 1461
shear deformation, 731
shock, 705, 730, 1428, 1433, 1435, 1436, 1437, 1438, 1439
shock waves, 730
shoulders, 969, 970
Si3N4, 224
sialic acid, 147
side effects, xxxii, 1499, 1501, 1510
sign, 570, 612, 1468
signals, 315, 339, 351, 427, 475, 535, 773, 775, 783, 1172, 1416, 1506, 1554
signal-to-noise ratio, 126, 751, 771, 772, 778, 783, 1538
silane, xxiii, 749, 795, 959, 961, 962, 963, 964, 965, 966, 967, 968, 969, 970, 971, 972, 973, 977, 981, 986
silica, xviii, 39, 95, 330, 425, 474, 484, 548, 561, 563, 564, 567, 568, 572, 573, 575, 579, 761, 786, 811, 856, 857, 910, 911, 1442, 1500, 1505, 1506, 1507, 1511, 1516, 1518, 1535, 1536
silicon, xiv, xvii, 42, 47, 209, 213, 218, 234, 237, 238, 239, 318, 322, 430, 431, 459, 470, 504, 505, 506, 510, 512, 514, 538, 565, 566, 680, 690, 691, 704, 713, 739, 741, 742, 746, 747, 764, 785, 838, 858, 863, 866, 924, 940, 966, 967, 968, 969, 970, 1165, 1392, 1519, 1533, 1552
silicon dioxide, 1552
silk, xv, 329, 330, 382, 383, 384, 385, 386, 387, 389, 390

silkworm, 332, 382
silver, xiv, xxiii, 68, 73, 121, 166, 174, 238, 271, 382, 383, 385, 386, 387, 388, 389, 399, 403, 423, 484, 486, 487, 488, 746, 770, 914, 921, 938, 1522, 1523, 1524, 1525, 1526
similarity, 536, 950, 1480
simulation, 127, 416, 425, 532, 630, 688, 690, 691, 692, 693, 695, 699, 701, 702, 703, 704, 709, 711, 712, 713, 714, 716, 717, 720, 721, 722, 723, 724, 725, 731, 732, 737, 738, 739, 740, 741, 742, 743, 744, 745, 746, 747, 748, 850, 1393, 1496
sine wave, 566, 574
Singapore, 277, 303, 989
single crystals, 8, 23, 97, 104, 354, 690, 716
single walled carbon nanotubes, 810
sintering, xix, 581, 583, 592, 600, 603, 604, 605, 607, 608, 609, 613, 619, 639, 651, 652, 659, 1393, 1487
SiO2, 4, 167, 213, 324, 331, 397, 420, 421, 423, 448, 460, 471, 477, 484, 605, 771, 799, 812, 838, 852, 853, 863, 868, 869, 870, 871, 872, 873, 875, 911, 913, 1535, 1536, 1550, 1553, 1555, 1558
SiO2 films, 1536
SiO2 surface, 1553
sites, xxi, 91, 92, 104, 105, 107, 506, 538, 541, 692, 698, 702, 707, 709, 711, 716, 718, 754, 755, 785, 797, 809, 812, 823, 825, 970, 981, 986
skeleton, 528
skin, 357, 384, 385, 1539
Slovakia, 457
smoke, 1492, 1494, 1496
smoothing, 39
smoothness, 709, 712
society, 4
sodium, 11, 129, 166, 325, 400, 413, 484, 549, 749, 751, 759, 769, 774, 915, 922, 936, 1159, 1166, 1442, 1519, 1523
sodium dodecyl sulfate (SDS), 400
sodium hydroxide, 11
soft lithography, 855, 866
softener, 1429
software, 357, 465, 566, 1482, 1486, 1493, 1497
SOI, 239
soil, 8
solar cell, xxx, 5, 119, 239, 345, 428, 439, 1391, 1392, 1396, 1401, 1413, 1422, 1424
solar cells, xxx, 5, 8, 27, 119, 239, 345, 428, 439, 562, 675, 680, 1391, 1392, 1396, 1401, 1422
solar energy, 1392
sol-gel, 10, 95, 334, 340, 341, 342, 344, 345, 353, 390, 592, 594, 647, 761, 768, 780, 782, 783, 786, 789, 910
solid phase, 90, 435
solid solutions, xix, 581, 583, 593, 618, 838
solid state, 103, 117, 127, 160, 194, 530, 692, 745, 1486
solidification, 366, 506
solubility, 2, 66, 76, 96, 127, 132, 138, 142, 149, 155, 168, 181, 271, 446, 587, 588, 672, 675, 757, 758, 760, 922
solvent, 2, 120, 124, 153, 172, 175, 242, 245, 254, 264, 269, 271, 433, 464, 593, 598, 757, 909, 925, 1394, 1396, 1460, 1533, 1555
solvents, 8, 73, 74, 154, 155, 181, 564, 567, 568, 579, 599, 602, 670, 675, 679, 745, 755, 757, 758, 761, 762, 908, 923, 1536
sorption, 833, 835
sounds, 515
South Korea, 427, 565
space charge distribution, xxiv, 959, 960, 964, 972, 977, 982, 983, 984, 985, 986, 987
space charge measurements, 962
spatial frequency, 565, 566, 574
spatial location, 49
specialization, 635, 659
species, xv, xvii, 9, 18, 19, 90, 93, 100, 104, 156, 158, 160, 181, 183, 222, 307, 308, 330, 340, 341, 351, 358, 359, 388, 389, 430, 461, 462, 483, 488, 489, 752, 754, 758, 760, 762, 765, 767, 770, 773, 774, 775, 776, 777, 778, 780, 782, 786, 790, 796, 811, 864, 1167, 1168, 1177, 1396, 1402, 1407, 1503, 1506, 1547, 1555
specific heat, 417, 637, 642, 643, 654, 655, 660, 1496
specific surface, 66, 357, 358, 359, 361, 366, 394, 534, 753, 830
specificity, 137, 1501, 1507
spectroscopy, xviii, xxvi, xxx, 96, 97, 99, 117, 129, 164, 165, 278, 346, 537, 547, 553, 600, 639, 750, 752, 753, 793, 800, 817, 819, 873, 963, 987, 1157, 1158, 1160, 1165, 1401, 1402, 1403, 1407, 1415, 1416, 1517, 1526
spectrum, xv, xvii, 17, 23, 24, 27, 69, 72, 98, 99, 100, 101, 102, 105, 124, 129, 277, 301, 315, 339, 340, 349, 354, 355, 356, 369, 373, 377, 383, 385, 386, 447, 448, 450, 491, 494, 497, 503, 554, 555, 564, 577, 671, 672, 674, 691, 733, 820, 825, 897, 933, 963, 966, 967, 969, 971, 1160, 1162, 1392, 1393, 1405, 1414, 1415, 1418, 1536, 1538
speed, 42, 96, 319, 335, 376, 400, 563, 625, 675, 703, 709, 730, 747, 962, 1428, 1434, 1435, 1481
spherulite, 976, 980, 981, 982
spin, xviii, 10, 68, 104, 130, 346, 461, 475, 476, 561, 563, 564, 579, 778, 848, 850, 851, 852, 867, 1407, 1416, 1420
spindle, 70

spine, 896
spleen, 133, 140, 143
sputtering, 433, 440, 442, 472, 725, 728
St. Petersburg, 544
stability, xvi, xxi, 4, 31, 114, 148, 150, 153, 154, 156, 167, 181, 183, 192, 203, 214, 366, 393, 394, 395, 396, 397, 398, 399, 400, 401, 404, 405, 478, 519, 534, 536, 538, 562, 564, 639, 640, 645, 659, 672, 704, 716, 741, 752, 757, 758, 761, 762, 772, 774, 776, 779, 783, 786, 787, 789, 790, 801, 823, 825, 830, 960, 1394, 1427, 1443, 1450, 1451, 1481, 1505, 1545, 1553, 1560
stabilization, xvi, xxxi, 197, 353, 393, 395, 396, 397, 400, 403, 405, 423, 859, 1427, 1431, 1433, 1440, 1479, 1480, 1496, 1497
stabilizers, xiii, xxxi, 113, 151, 180, 402, 1427, 1428, 1440
stages, xiv, 94, 201, 232, 308, 435, 446, 646, 690, 724, 866, 1434, 1505
standard deviation, 281, 282, 283, 284, 285, 288, 293, 294, 295, 296, 566, 713, 714, 725
standards, 566
star polymers, 115, 116, 939
starch, 760
stars, 115, 161, 194
statistical analysis, 312
statistics, 225, 565, 1460
steel, 312, 689, 740, 748, 963, 1159
stem cells, 3
stimulus, 44
STM, 40, 176, 817
stock, 811, 815, 1166
stoichiometry, 96, 98, 158, 346, 360, 600, 666, 1558
stomach, 1539
storage, xii, xx, 2, 3, 4, 65, 80, 185, 202, 232, 308, 371, 427, 587, 624, 634, 643, 659, 687, 786, 791, 792, 810, 813, 823, 824, 833, 835, 837, 838, 841, 842, 843, 966, 968, 971, 1525, 1546, 1548
storage media, 810, 823, 835
strain, xxxi, 45, 48, 227, 229, 281, 283, 292, 294, 296, 301, 530, 531, 570, 612, 688, 690, 693, 706, 707, 721, 722, 724, 738, 741, 747, 753, 954, 955, 956, 975, 1457, 1463, 1465, 1468, 1474
strategies, xii, xv, 3, 65, 66, 117, 168, 312, 329, 330, 548
strength, xxiv, 2, 72, 183, 274, 530, 534, 536, 538, 539, 540, 565, 572, 605, 634, 652, 689, 716, 810, 823, 911, 959, 960, 963, 972, 977, 979, 980, 981, 982, 984, 986, 1428, 1430, 1433, 1435, 1436, 1439, 1443, 1450, 1452, 1468, 1548
stress, xx, 38, 42, 44, 45, 46, 49, 54, 57, 58, 59, 60, 97, 149, 294, 570, 571, 611, 612, 613, 630, 687, 688, 689, 690, 693, 705, 706, 707, 708, 709, 710, 712, 713, 716, 721, 722, 732, 735, 955, 956, 981, 1460, 1461, 1462, 1463, 1468, 1475
stress fields, 690
stress-strain curves, 956
stretching, xxi, 339, 351, 374, 385, 556, 557, 599, 600, 720, 721, 722, 724, 737, 927, 965, 966, 1428, 1429, 1452, 1532
striatum, 774
strong interaction, 172, 489
structural changes, 106, 179, 1468
structural characteristics, 66, 68, 531, 532
structural defects, 104, 337, 349, 367
structural relaxation, 567
structure formation, xxxi, 1467, 1469, 1472
structuring, 515
styrene, 116, 152, 167, 548
substitutes, 540
substitution, 41, 117, 151, 161, 592, 902, 1442
substrates, xvi, xviii, xxi, 10, 25, 26, 31, 55, 66, 94, 180, 203, 207, 210, 224, 252, 277, 278, 303, 329, 371, 383, 385, 387, 389, 390, 429, 431, 433, 434, 437, 438, 439, 442, 444, 456, 470, 473, 484, 515, 555, 561, 562, 563, 564, 567, 568, 579, 754, 757, 763, 770, 785, 789, 801, 810, 812, 846, 855, 859, 861, 863, 867, 871, 875, 939, 940, 1404, 1416, 1530, 1532, 1535, 1536, 1550, 1558
subtraction, 585, 613
sucrose, 775
Sudan, 843
sugar, 341, 351, 939
sulfate, 143, 325, 339, 402, 774
sulfonamides, 121
sulfur, 70, 104, 517, 549, 551, 624, 876, 877, 878, 879
sulfuric acid, 181, 473, 754, 765
supercapacitor, 810, 823, 830, 831, 835
supercritical carbon dioxide, 153, 269
superimposition, 251, 258, 261, 263, 274
superiority, 140, 551, 823
superlattice, 548, 628
superoxide, 768, 1503
superplasticity, 739, 740
supervision, 1438
supply, xxiii, 53, 227, 229, 503, 709, 760, 905, 1523
suppression, 273, 846, 849, 851, 854, 879, 985
surface area, 9, 20, 21, 27, 55, 76, 91, 230, 239, 351, 352, 357, 359, 360, 361, 389, 416, 473, 492, 534, 558, 624, 634, 635, 639, 790, 823, 825, 830, 850, 960, 961, 965, 1392, 1393, 1443, 1480, 1501, 1507, 1557
surface chemistry, 421, 1501, 1536
surface diffusion, 208, 278, 437

surface energy, 2, 55, 204, 206, 207, 209, 218, 223, 227, 229, 230, 366, 375, 395, 502, 534, 624, 636, 1480
surface layer, 244, 245, 251, 253, 268, 269, 497, 689, 740
surface modification, xxiv, 309, 361, 400, 937, 959, 960, 966, 975, 978, 979, 981, 982, 985, 986, 1158, 1179, 1502, 1536, 1541
surface properties, xii, 37, 61, 636, 753
surface reactions, 221, 462, 464
surface region, 753
surface structure, xxiv, 10, 21, 26, 31, 573, 575, 959, 970, 1396
surface tension, xiv, 76, 201, 318, 636
surface treatment, xxx, 452, 976, 1158, 1179, 1401, 1414, 1418, 1419, 1424
surface-initiated atom transfer radical polymerization, xxiii, 921
surfactant, 55, 72, 73, 298, 351, 353, 372, 399, 402, 420, 548, 549, 550, 551, 552, 555, 558, 679, 1392, 1409, 1412, 1413, 1418, 1419, 1443, 1524, 1529, 1534
surfactants, xxxi, 2, 151, 168, 196, 308, 376, 378, 390, 758, 1441, 1443, 1522, 1523, 1526
surprise, 948
susceptibility, 127, 760
suspensions, 312, 394, 407, 414, 422, 423, 424, 425, 596, 757, 758, 759, 761, 1516
swelling, 144, 1444, 1445, 1554, 1559
switching, 261, 371, 461, 474, 673, 675, 688, 696, 697, 876, 878, 879, 922
Switzerland, 40
SWNTs, xxii, 120, 750, 751, 753, 755, 756, 757, 759, 760, 761, 762, 763, 764, 765, 766, 767, 774, 775, 776, 777, 778, 779, 781, 783, 784, 788, 789, 793, 794, 795, 797, 800, 810, 811, 819, 824, 833, 845, 855, 856, 857, 858, 859, 860, 861, 862, 863, 864, 865, 866, 867, 870, 871, 872, 873, 874, 875, 1502, 1505, 1507
symbols, 45, 455, 860, 861, 872
symmetry, 91, 104, 316, 318, 322, 355, 364, 369, 377, 431, 510, 532, 588, 589, 595, 605, 619, 628, 629, 715, 836, 948, 949, 950
synthetic polymers, xiii, 2, 3, 113, 146, 1440

T

T cell, 1502
Taiwan, 237, 1401
tanks, 462, 464, 824
tantalum, 770, 785, 910
targets, 137, 138, 309
TBP, 279, 1403
T_C, xxii, 748, 895, 896, 897, 901, 902
technical carbon, 1460, 1461, 1462, 1463, 1464
technology, xiii, xv, xvi, xvii, xx, xxii, xxiii, xxxi, 3, 8, 10, 69, 112, 113, 136, 238, 243, 250, 308, 329, 382, 408, 423, 429, 446, 456, 491, 493, 531, 538, 553, 633, 651, 659, 666, 687, 688, 762, 767, 810, 830, 841, 845, 846, 854, 858, 905, 950, 960, 1392, 1394, 1441, 1559
teeth, 90, 507, 508, 515, 517
TEG, 679
telecommunications, 594
tellurium, 676, 1159
temperature annealing, 603
temperature dependence, 69, 169, 416, 594, 610, 615, 616, 617
temperature gradient, 415
tensile strength, 1450, 1452, 1548
tension, 636, 710, 732, 736, 740, 954, 1459, 1474
TEOS, 910
terminals, 175, 256
ternary complex, 136
terraces, 566
tetrad, 915
tetrahydrofuran, 153, 161, 1555
tetrapod, 1412
Texas, 544
textiles, 8
TGA, xxiii, 332, 345, 346, 348, 361, 597, 598, 599, 751, 760, 817, 921, 923, 926, 927, 928, 930, 931, 933, 963, 964, 965, 969, 970, 971, 987, 1159, 1450, 1451, 1481, 1482, 1486, 1496
theory, xxxi, 3, 55, 93, 104, 156, 172, 202, 203, 208, 212, 213, 214, 221, 269, 294, 405, 414, 415, 416, 417, 421, 484, 485, 521, 531, 540, 541, 556, 577, 690, 693, 695, 701, 716, 737, 742, 745, 907, 927, 942, 948, 1393, 1457, 1460, 1461, 1462, 1463, 1464, 1465, 1527
therapeutic agents, 146
therapeutic approaches, 1502
therapeutic interventions, 1505
therapeutics, 141, 146, 1500, 1501, 1514, 1518
therapy, xxxii, 174, 896, 1499, 1500, 1501, 1502, 1503, 1504, 1505, 1507, 1514, 1516, 1519, 1521, 1522, 1535, 1538, 1539
thermal activation, 49, 645
thermal decomposition, 130, 332, 399, 403, 568, 595, 672, 682, 1482
thermal degradation, 1479, 1480, 1481, 1482, 1486, 1488, 1496
thermal energy, 54, 394, 626, 646, 1547
thermal evaporation, 88, 89, 90, 91, 105, 108, 502, 503, 504, 505, 506, 507, 508, 512, 853, 1404
thermal expansion, 655, 705

thermal oxidation, 11, 517
thermal properties, xvi, 393, 395, 423, 639, 640, 654, 660, 1451, 1548
thermal resistance, xxxi, 417, 420, 421, 1479
thermal stability, xx, 167, 375, 592, 633, 635, 639, 640, 643, 645, 647, 659, 761, 762, 1450, 1481, 1548
thermal treatment, 469, 472, 473, 589, 596, 1553, 1554
thermodynamic calculations, 643
thermodynamic equilibrium, 636
thermodynamic function, 635, 639, 640, 643, 659
thermodynamic method, 209
thermodynamic parameters, 216, 218, 223, 224, 640, 643
thermodynamic properties, 635, 640, 643, 647, 659, 691, 742, 793, 1517
thermodynamics, xiii, 201, 202, 203, 207, 209, 213, 214, 216, 219, 221, 226, 309, 635, 640, 659, 691
thermograms, 964, 965, 974, 975, 1450
thermogravimetric, 597, 751, 760, 817, 1480, 1482, 1529
thermogravimetric analysis, 597, 751, 760, 817, 1482, 1529
thermogravimetry, 876, 1481
thin films, xviii, 2, 9, 10, 18, 38, 121, 309, 460, 497, 514, 520, 540, 561, 562, 563, 565, 567, 568, 569, 571, 573, 574, 575, 579, 680, 689, 690, 866, 962, 963, 1396, 1416, 1515, 1532, 1558
threonine, 130
threshold, 42, 44, 58, 108, 208, 209, 212, 278, 447, 448, 449, 450, 566, 688, 865, 878, 879, 1554, 1557
thrombosis, 786
thymine, 763
thymus, 763, 800
tics, 701
time periods, 405
tin, 11, 120, 155, 159, 180, 182, 345, 350, 351, 359, 548, 562, 1393, 1413, 1419, 1559
tin oxide, 159, 180, 182, 345, 350, 351, 359, 562, 1393, 1559
tissue, xiii, 3, 4, 114, 133, 140, 144, 145, 146, 174, 253, 258, 261, 1171, 1394, 1500, 1501, 1504, 1512, 1515, 1538
titania, 332, 333, 334, 335, 336, 337, 338, 340, 341, 342, 343
titanium, xi, 7, 8, 10, 11, 14, 15, 16, 18, 21, 24, 29, 31, 332, 583, 588, 592, 593, 856, 910, 1393, 1403, 1422, 1503, 1515, 1516
titanium isopropoxide, 583
TNF, 144, 1502
TNF-α, 144
tobacco, 909
Tokyo, 841, 1497
toluene, 118, 130, 410, 411, 416, 563, 564, 568, 757, 961, 962, 968, 969, 986, 1403
topology, 117, 178, 952, 953, 954, 1392, 1474
total energy, 628, 694
toxic effect, 778, 794
toxic side effect, 1502
toxicity, xxxii, 30, 142, 146, 148, 149, 151, 191, 1499, 1501, 1503, 1510, 1512, 1515, 1536
toxicology, 146
TPA, 172, 173
trade, 158
trade-off, 158
tradition, 382
traffic, 427
traits, 1393
trajectory, 628, 697
transducer, xviii, 40, 547, 780
transfection, 136, 139, 145, 164, 1172
transfer performance, 394, 419
transformation, xiv, 202, 226, 232, 298, 339, 380, 386, 640, 647, 655, 906, 1539, 1540
transgene, 136
transistor, xxii, 427, 836, 845, 846, 847, 850, 862, 878, 879, 1550, 1558
transistors, 531, 810, 823, 836
transition, xiv, xx, 9, 21, 44, 50, 51, 52, 53, 55, 58, 66, 67, 72, 92, 101, 102, 103, 105, 107, 124, 157, 160, 201, 203, 221, 222, 223, 226, 230, 231, 232, 301, 302, 309, 505, 519, 576, 583, 635, 648, 651, 654, 665, 667, 680, 682, 691, 694, 703, 738, 742, 748, 756, 811, 813, 823, 824, 860, 863, 866, 871, 872, 879, 906, 910, 915, 938, 943, 979, 1167, 1429, 1430, 1432, 1436, 1450, 1480
transition elements, 160
transition metal, xx, 66, 67, 72, 157, 583, 665, 667, 680, 682, 694, 742, 811, 813, 823, 824, 866, 910, 915
transition metal ions, 157
transition temperature, 92, 124, 648, 654, 906, 1450
transitions, 54, 102, 230, 373, 519, 626, 871, 952, 1539
translation, 339
transmission, xviii, xx, 12, 96, 101, 153, 225, 312, 403, 405, 438, 454, 455, 501, 503, 547, 554, 575, 584, 587, 590, 597, 649, 665, 666, 700, 701, 711, 713, 714, 751, 755, 792, 817, 847, 848, 856, 877, 961, 963, 1161, 1172, 1483
transmission electron microscopy, xviii, 12, 96, 153, 225, 312, 403, 454, 455, 501, 503, 584, 590, 597, 649, 751, 792, 817, 961, 1483
Transmission Electron Microscopy (TEM), 897

transparency, 535
transport, xii, xiii, xxii, xxx, 2, 3, 65, 80, 114, 136, 137, 140, 149, 181, 190, 201, 202, 213, 221, 226, 232, 308, 345, 414, 416, 424, 425, 430, 444, 445, 446, 452, 456, 461, 475, 656, 738, 760, 791, 792, 797, 829, 845, 846, 847, 848, 850, 852, 853, 854, 879, 880, 985, 1392, 1393, 1394, 1395, 1396, 1401, 1402, 1411, 1412, 1413, 1418, 1419, 1421, 1424, 1450, 1512
transportation, xvi, 393, 731, 824, 833, 1396
trees, 797
trend, 86, 123, 283, 315, 316, 343, 357, 359, 394, 566, 574, 583, 593, 690, 720, 722, 847, 1525
tribology, xi, 37, 61
triggers, 1445
trimer, 173
trimethylamine, 1403
trust, 739
tryptophan, 777
tumor, xxxii, 133, 137, 140, 143, 190, 1499, 1502, 1504, 1505, 1507, 1509, 1512, 1513, 1514, 1515, 1519, 1539
tumor cells, xxxii, 143, 1499, 1502, 1507, 1509, 1513, 1514, 1519, 1539
tumor necrosis factor, 1502
tumor resistance, 1502
tumors, 137, 146, 896, 1502, 1512, 1513, 1518
tungsten, 160, 624, 690, 1506
tunneling, 450, 452, 656, 797, 817, 848, 849
turbulence, 702
twinning, 912
twins, 337, 349
tyrosine, 382, 385, 387, 777

U

ultrasound, 68, 856
uncertainty, xxii, 845, 846, 866
underlying mechanisms, xxiv, 959
uniaxial tension, 955
unit plan, 217
United Kingdom (UK), 307, 424
United Nations, 543
United States, 544
universality, 1530
updating, 707
uranium, 465
urea, 789
urease, 789
uric acid, 752, 759, 796
urine, 772, 789
UV, xvii, xviii, 8, 9, 10, 12, 20, 27, 28, 29, 69, 72, 103, 108, 117, 121, 124, 153, 158, 168, 250, 254, 256, 258, 383, 384, 427, 447, 465, 484, 501, 517, 519, 547, 553, 554, 555, 558, 562, 563, 566, 797, 799, 923, 1160, 1405, 1414, 1503, 1504, 1512, 1516, 1523, 1528, 1552, 1554
UV irradiation, 20, 28, 153, 1504, 1512
UV light, 8, 10, 27, 29, 121, 168, 256, 1523, 1552, 1554

V

vacancies, xxiv, 360, 446, 492, 494, 519, 562, 959, 970
vacancy defects, 595, 600
vacuum, 11, 40, 51, 91, 308, 309, 403, 464, 493, 498, 667, 755, 756, 811, 813, 817, 852, 853, 854, 858, 863, 876, 879, 923, 924, 925, 1165, 1403, 1404, 1547, 1548, 1552, 1553
valence, 8, 104, 494, 530, 535, 536, 537, 628, 850, 851, 1421
validation, 711
validity, 946, 1480
Van der Waals, 1435, 1436, 1442
Van der Waals forces, 533, 763
vanadium, 505, 910
vapor, xii, xiii, xv, xviii, 10, 19, 65, 66, 80, 85, 86, 90, 91, 92, 94, 95, 201, 202, 204, 206, 213, 214, 215, 216, 217, 221, 223, 224, 226, 242, 269, 277, 278, 399, 401, 402, 429, 430, 431, 432, 433, 440, 442, 443, 444, 445, 446, 471, 501, 502, 503, 504, 508, 512, 515, 647, 648, 675, 810, 837, 838, 839, 840, 858, 874, 915, 1550, 1551, 1552, 1558
vapor-liquid-solid, xii, xiii, 19, 66, 85, 90, 201, 202, 213, 429, 431, 508, 512
vapor-solid (VS), 90, 91, 430
variable, 53, 896
variables, 414, 704, 706, 979
variation, 249, 257, 264, 266, 267, 268, 278, 302, 315, 343, 360, 416, 576, 577, 578, 612, 613, 616, 619, 628, 690, 721, 854, 861, 877, 908, 975, 985, 1403, 1459, 1469
vdW, 628
vector, xiii, 113, 143, 148, 164, 693, 695, 847, 943, 945, 1172, 1514
vehicles, xiii, 114, 135, 138, 142, 345, 1505
vein, 1510, 1520
velocity, 45, 47, 48, 50, 52, 217, 222, 223, 224, 395, 416, 418, 419, 428, 486, 624, 626, 627, 629, 689, 691, 692, 695, 705, 706, 709, 725, 731, 737, 1393
venue, 781
verapamil, 1514
versatility, 465, 774, 1445
vertebrates, 792
vesicle, 908, 1512, 1515

vibration, 51, 217, 218, 222, 223, 351, 354, 356, 361, 369, 374, 385, 399, 400, 413, 555, 598, 599, 600, 601, 625, 639, 692, 704, 737
vincristine, 1514
viruses, 146, 149, 330, 382
viscoelastic materials, 624
viscosity, 52, 53, 283, 293, 395, 417, 419, 421, 423, 425, 563, 669, 670, 762, 772, 962, 1459
vision, 1, 51
visualization, 135, 176, 699, 725, 939
VLS, xiii, xxi, 19, 66, 80, 81, 86, 87, 90, 93, 94, 95, 97, 98, 100, 101, 104, 201, 202, 203, 209, 212, 213, 216, 217, 226, 227, 232, 429, 431, 433, 434, 435, 437, 439, 508, 512, 809, 823, 835
volatility, 876, 1394
volatilization, 563, 570, 1481
volumetric changes, 1554

W

war, 298
warrants, 142
wastewater, 31, 182, 1516
water, xi, 7, 8, 10, 11, 13, 14, 16, 19, 20, 27, 31, 46, 70, 72, 73, 76, 79, 96, 564, 670, 673, 679, 851, 862, 863, 864, 865, 866, 874, 876, 961, 963, 964, 965, 966, 968, 969, 970, 971
water absorption, 46
water evaporation, 319, 322
water permeability, 760
water vapor, xi, 7, 14, 16, 31, 314, 433, 851, 863, 874, 876, 1396, 1531
water-soluble polymers, 758
wave number, 857
wave propagation, 747
wavelengths, 174, 278, 302, 447, 565, 566, 577, 863, 1393, 1416, 1417, 1527, 1538
WAXS, 1482
weak interaction, 488, 758
wear, xi, xii, xix, xxi, 37, 38, 39, 40, 41, 43, 44, 45, 46, 47, 48, 49, 50, 51, 52, 53, 54, 55, 56, 58, 60, 61, 309, 623, 624, 626, 627, 628, 630, 687
weight gain, 815, 817
weight loss, 345, 361, 589, 598, 817, 867, 923, 970, 1450
weight ratio, 508, 510, 530, 763, 926
welding, 945
wells, 49, 50, 52
wettability, 20, 754
wetting, xi, 7, 10, 20, 21, 31, 280, 281, 282, 283, 284, 291, 296, 446, 754
wide band gap, 86, 105, 498, 502, 562
wind, 48, 1391
windows, 86, 879
wireless networks, 428
wires, xxi, 3, 66, 202, 227, 232, 269, 476, 738, 792, 797, 801, 823, 836, 1392
withdrawal, 563
WKB approximation, 848
wood, 39, 330
wool, 441
work, xi, 10, 20, 32, 37, 39, 66, 80, 82, 87, 94, 563, 566, 567, 568, 570, 571, 573, 666, 848, 849, 854, 860, 878, 880, 960, 972, 979, 981, 983, 984, 986, 987
workers, 10, 29, 45, 47, 309, 314, 502, 515, 553, 562, 570, 777, 778, 785, 789, 800, 908
working conditions, 395
workstation, 550

X

X-axis, 566
xenon, 1160
x-ray, 535, 537
X-ray analysis, 74
X-ray photoelectron spectroscopy (XPS), xx, 96, 99, 313, 314, 315, 327, 346, 562, 665, 752, 754, 763, 817, 825, 869, 1165, 1166
96, 313, 1165

Y

yang, 307
Y-axis, 566
yeast, 371, 1519
yield, xiv, xxii, xxiv, 8, 9, 16, 46, 121, 127, 130, 135, 153, 156, 163, 166, 171, 185, 214, 237, 238, 271, 308, 379, 534, 566, 678, 728, 741, 809, 810, 811, 815, 817, 823, 835, 837, 838, 839, 910, 923, 959, 1158, 1159, 1167, 1392, 1405, 1406, 1415, 1492, 1494, 1496, 1522, 1525, 1531, 1541, 1547, 1551

Z

zeolites, 811, 858
zeta potential, 405
zinc, xvii, xviii, 80, 86, 87, 104, 105, 107, 162, 165, 172, 174, 176, 323, 361, 365, 366, 367, 368, 501, 502, 503, 504, 506, 512, 514, 517, 518, 519, 561, 562, 563, 564, 567, 568, 569, 579, 607, 675, 798
zinc oxide, 323, 365, 366, 367, 368, 562, 567, 568
zirconium, 454, 592, 593
ZnO, xvii, xviii, 73, 74, 76, 77, 89, 95, 107, 311, 324, 331, 361, 362, 363, 364, 365, 366, 367, 368,

369, 370, 371, 432, 491, 492, 494, 495, 496, 497, 498, 501, 502, 503, 504, 505, 506, 507, 508, 509, 510, 511, 512, 513, 514, 515, 516, 517, 518, 519, 520, 521, 561, 562, 563, 564, 567, 568, 569, 570, 571, 572, 573, 574, 575, 576, 577, 578, 579, 605, 606, 607, 608, 615, 1392, 1396, 1557, 1558, 1559, 1560

ZnO nanorods, 364, 512, 515, 520

ZnO nanostructures, xvii, 77, 501, 502, 503, 506, 508, 512, 514, 518, 519, 520, 521